## NOBEL PRIZES AWARDED FOR RESEARCH IN GENETICS OR GENETICS-RELATED AREAS

| Year | Recipients | Nobel Prize | Research Topic |
|---|---|---|---|
| 2007 | M. R. Capecchi<br>M. J. Evans<br>O. Smithies | Medicine or Physiology | Development of gene-targeting technology essential to the creation of knockout mice serving as animal models of human disease |
| 2006 | R. Kornberg | Chemistry | Molecular basis of eukaryotic transcription |
| 2006 | A. Z. Fire<br>C. C. Mello | Medicine or Physiology | RNA interference (RNAi) |
| 2002 | S. Brenner<br>H. R. Horvitz<br>J. E. Sulston | Medicine or Physiology | Genetic regulation of organ development and programmed cell death |
| 2001 | L. Hartwell<br>T. Hunt<br>P. Nurse | Medicine or Physiology | Genes and regulatory molecules controlling the cell cycle |
| 1999 | G. Blobel | Medicine or Physiology | Genetically-encoded amino acid sequences in proteins that guide their cellular transport |
| 1997 | S. Prusiner | Medicine or Physiology | Prions—a new biological principle of infection |
| 1995 | E. B. Lewis<br>C. Nusslein-Volhard<br>E. Wieschaus | Medicine or Physiology | Genetic control of early development in *Drosophila* |
| 1993 | R. Roberts<br>P. Sharp | Medicine or Physiology | RNA processing of split genes |
| | K. Mullis<br>M. Smith | Chemistry | Development of polymerase chain reaction (PCR) and site-directed mutagenesis (SDM) |
| 1989 | J. M. Bishop<br>H. E. Varmus | Medicine or Physiology | Role of retroviruses and oncogenes in cancer |
| | T. R. Cech<br>S. Altman | Chemistry | Ribozyme function during RNA splicing |
| 1987 | S. Tonegawa | Medicine or Physiology | Genetic basis of antibody diversity |
| 1985 | M. S. Brown<br>J. L. Goldstein | Medicine or Physiology | Genetic regulation of cholesterol metabolism |
| 1983 | B. McClintock | Medicine or Physiology | Mobile genetic elements in maize |
| 1982 | A. Klug | Chemistry | Crystalline structure analysis of significant complexes, including tRNA and nucleosomes |
| 1980 | P. Berg<br>W. Gilbert<br>F. Sanger | Chemistry | Development of recombinant DNA and DNA sequencing technology |
| 1978 | W. Arber<br>D. Nathans<br>H. O. Smith | Medicine or Physiology | Recombinant DNA technology using restriction endonuclease technology |

| Year | Recipients | Nobel Prize | Research Topic |
|---|---|---|---|
| 1976 | B. S. Blumberg<br>D. C. Gajdusek | Medicine or Physiology | Elucidation of the human prior-based diseases, kuru and Creutzfeldt-Jakob disease |
| 1975 | D. Baltimore<br>R. Dulbecco<br>H. Temin | Medicine or Physiology | Molecular genetics of tumor viruses |
| 1972 | G. M. Edelman<br>R. R. Porter | Medicine or Physiology | Chemical structure of immunoglobulins |
| | C. Anfinsen | Chemistry | Relationship between primary and tertiary structure of proteins |
| 1970 | N. Borlaug | Peace Prize | Genetic improvement of Mexican wheat |
| 1969 | M. Delbrück<br>A. D. Hershey<br>S. E. Luria | Medicine or Physiology | Replication mechanisms and genetic structure of bacteriophages |
| 1968 | H. G. Khorana<br>M. W. Nirenberg | Medicine or Physiology | Deciphering the genetic code |
| | R. W. Holley | Medicine or Physiology | Structure and nucleotide sequence of transfer RNA |
| 1966 | P. F. Rous | Medicine or Physiology | Viral induction of cancer in chickens |
| 1965 | F. Jacob<br>A. M. Lwoff<br>J. L. Monod | Medicine or Physiology | Genetic regulation of enzyme synthesis in bacteria |
| 1962 | F. H. C. Crick<br>J. D. Watson<br>M. H. F. Wilkins | Medicine or Physiology | Double helical model of DNA |
| | J. C. Kendrew<br>M. F. Perutz | Chemistry | Three-dimensional structure of globular proteins |
| 1959 | A. Kornberg<br>S. Ochoa | Medicine or Physiology | Biological synthesis of DNA and RNA |
| 1958 | G. W. Beadle<br>E. L. Tatum | Medicine or Physiology | Genetic control of biochemical processes |
| | J. Lederberg | Medicine or Physiology | Genetic recombination in bacteria |
| | F. Sanger | Chemistry | Primary structure of proteins |
| 1954 | L. Pauling | Chemistry | Alpha helical structure of proteins |
| 1946 | H. J. Muller | Medicine or Physiology | X-ray induction of mutations in *Drosophila* |
| 1933 | T. H. Morgan | Medicine or Physiology | Chromosomal theory of genetics |
| 1930 | K. Landsteiner | Medicine or Physiology | Discovery of human blood groups |

国外生命科学优秀教材

# 遗传学概要（影印版）

# Concepts of Genetics

NINTH EDITION


William S. Klug
THE COLLEGE OF NEW JERSEY

Michael R. Cummings
ILLINOIS INSTITUTE OF TECHNOLOGY

Charlotte A. Spencer
UNIVERSITY OF ALBERTA

Michael A. Palladino
MONMOUTH UNIVERSITY

WITH CONTRIBUTIONS BY
SARAH M. WARD,
COLORADO STATE UNIVERSITY



科学出版社

北京

图字：01-2011-3863

## 内 容 简 介

本书保持了前八版的编写特色：注重基础知识，概念清晰准确，探讨解决专业问题，师生易学易教。第九版内容拓展到基因组、生物信息、蛋白质组及相关前沿领域。该书在亚马逊专业教材销售排行榜长期名列前茅，被许多北美、欧洲高校教学选用。

第九版内容主要包括：遗传学概况，有丝分裂和减数分裂，孟德尔遗传学，孟德尔比率，真核生物染色体图谱，细菌和噬菌体的遗传分析，性别决定和性染色体，染色体突变：染色体数量和分布，核外遗传，DNA结构与分析，DNA复制与重组，染色体DNA的组织，重组DNA技术和基因克隆，遗传密码和转录，翻译，基因突变和DNA修复，原核生物基因表达，真核生物基因表达调控，模式生物的发育遗传学，癌症和细胞周期调控，基因组学，蛋白质组学，生物信息学，基因组动力学：转座子，免疫遗传学，真核病毒，基因组分析——基因功能，基因工程的应用及生物伦理剖析，数量遗传学和多因子性状，行为遗传学，群体遗传学，进化遗传学，保护遗传学。

本书适合生命科学相关专业教学选用，也可供从业人员参考使用。



**图书在版编目(CIP)数据**

遗传学概要＝Concepts of Genetics：第9版：英文/(美)克卢格(Klug, W. S.)等著. —影印本. —北京：科学出版社，2011.6

国外生命科学优秀教材

ISBN 978-7-03-031760-5

Ⅰ.①生… Ⅱ.①克… Ⅲ. ①遗传学-教材-英文 Ⅳ. ①Q3

中国版本图书馆CIP数据核字(2011)第119703号

责任编辑：王玉时

责任印制：张 伟 / 封面设计：陈 敬

科学出版社出版

北京东黄城根北街16号

邮政编码：100717

http://www.sciencep.com

北京虎彩文化传播有限公司印刷

科学出版社发行 各地新华书店经销

*

2011年7月第 一 版 开本：A4(890×1240)

2020年5月第二次印刷 印张：56

字数：1 500 000

**定价：198.00元**

（如有印装质量问题，我社负责调换）

## DEDICATION

In his 1973 article, the pioneering evolutionary geneticist Theodosius Dobzhansky published a thesis that is now widely quoted: "Nothing in biology makes sense except in the light of evolution." In the first decade of the twenty-first century, as we continue to move further into the era of genomics, it is equally relevant to state that "Our understanding of all things biological will remain incomplete until the genetic basis of every living process is made clear." It is with a great sense of excitement and wonderment that we acknowledge our rapid movement toward realizing this goal.

How lucky are all of us who study genetics and become enlightened by the findings, both past and present, that make up this discipline. Not only is each discovery a continued source of interest, feeding our hunger for knowledge, but collectively these findings are permeated with a display of analytical thinking and discovery, which are the cornerstones of science. Not a week passes without something of great genetic significance being reported both in the scientific literature and by the media at large.

We thus dedicate this book to all those who have come to appreciate genetics as the "core discipline" in biology. In particular, we single out the students just beginning their studies, who will soon join these ranks. We hope that this text provides valuable insights and inspiration as they expand their scientific horizons.

W. S. Klug
M. R. Cummings
C. A. Spencer
M. A. Palladino

# About the Authors

*William S. Klug* is Professor of Biology at The College of New Jersey (formerly Trenton State College) in Ewing, New Jersey. He served as Chair of the Biology Department for 17 years. He received his B.A. degree in Biology from Wabash College in Crawfordsville, Indiana, and his Ph.D. from Northwestern University in Evanston, Illinois. Prior to coming to The College of New Jersey, he was on the faculty of Wabash College as an Assistant Professor, where he first taught genetics, as well as general biology and electron microscopy. His research interests have involved ultrastructural and molecular genetic studies of development, utilizing oogenesis in *Drosophila* as a model system. He has taught the genetics course as well as the senior capstone seminar course in human and molecular genetics to undergraduate biology majors for many years. He was the recent recipient of the first annual teaching award given at The College of New Jersey granted to the faculty member who "most challenges students to achieve high standards." He also received the 2004 Outstanding Professor Award from Sigma Pi International, and in the same year, he was nominated as the Educator of the Year, an award given by the Research and Development Council of New Jersey.

*Michael R. Cummings* is Research Professor in the Department of Biological, Chemical, and Physical Sciences at Illinois Institute of Technology, Chicago, Illinois. For more than 25 years, he was a faculty member in the Department of Biological Sciences and in the Department of Molecular Genetics at the University of Illinois at Chicago. He has also served on the faculties of Northwestern University and Florida State University. He received his B.A. from St. Mary's College in Winona, Minnesota, and his M.S. and Ph.D. from Northwestern University in Evanston, Illinois. In addition to this text and its companion volumes, he has also written textbooks in human genetics and general biology for nonmajors. His research interests center on the molecular organization and physical mapping of the heterochromatic regions of human acrocentric chromosomes. At the undergraduate level, he teaches courses in Mendelian and molecular genetics, human genetics, and general biology, and has received numerous awards for teaching excellence given by university faculty, student organizations, and graduating seniors.

*Charlotte A. Spencer* is currently Associate Professor in the Department of Oncology at the University of Alberta in Edmonton, Alberta, Canada. She has also served as a faculty member in the Department of Biochemistry at the University of Alberta. She received her B.Sc. in Microbiology from the University of British Columbia and her Ph.D. in Genetics from the University of Alberta, followed by postdoctoral training at the Fred Hutchinson Cancer Research Center in Seattle, Washington. Her research interests involve the regulation of RNA polymerase II transcription in cancer cells, cells infected with DNA viruses, and cells traversing the mitotic phase of the cell cycle. She has taught courses in Biochemistry, Genetics, Molecular Biology, and Oncology, at both undergraduate and graduate levels. She has contributed Genetics, Technology, and Society essays for several editions of *Concepts of Genetics* and *Essentials of Genetics*. In addition, she has written booklets in the Prentice Hall Exploring Biology series, which are aimed at the undergraduate nonmajor level.

*Michael A. Palladino* is Associate Professor in the Department of Biology at Monmouth University in West Long Branch, New Jersey. He received his B.S. degree in Biology from Trenton State College (now known as The College of New Jersey) and his Ph.D. in Anatomy and Cell Biology from the University of Virginia. He directs an active laboratory of undergraduate student researchers studying molecular mechanisms involved in innate immunity of mammalian male reproductive organs and genes involved in oxygen homeostasis and ischemic injury of the testis. He has taught a wide range of courses for both majors and nonmajors and currently teaches genetics, biotechnology, endocrinology, and laboratory in cell and molecular biology. He has received several awards for research and teaching, including the New Investigator Award of the American Society of Andrology, the 2005 Distinguished Teacher Award from Monmouth University, and the 2005 Caring Heart Award from the New Jersey Association for Biomedical Research. He is co-author of the undergraduate textbook *Introduction to Biotechnology,* Series Editor for the Benjamin Cummings *Special Topics in Biology* booklet series, and author of the first booklet in the series, *Understanding the Human Genome Project.*

# Brief Contents

# Contents

## 5 Chromosome Mapping in Eukaryotes *105*

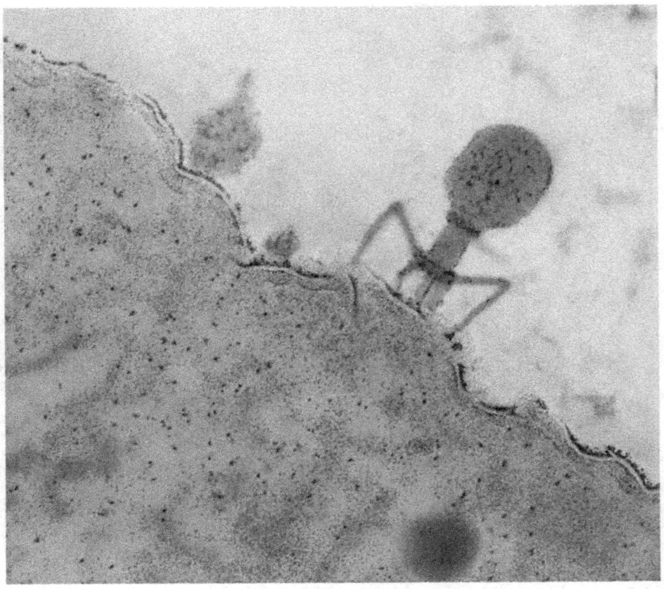

# 6 Genetic Analysis and Mapping in Bacteria and Bacteriophages 143



# 7 Sex Determination and Sex Chromosomes 173

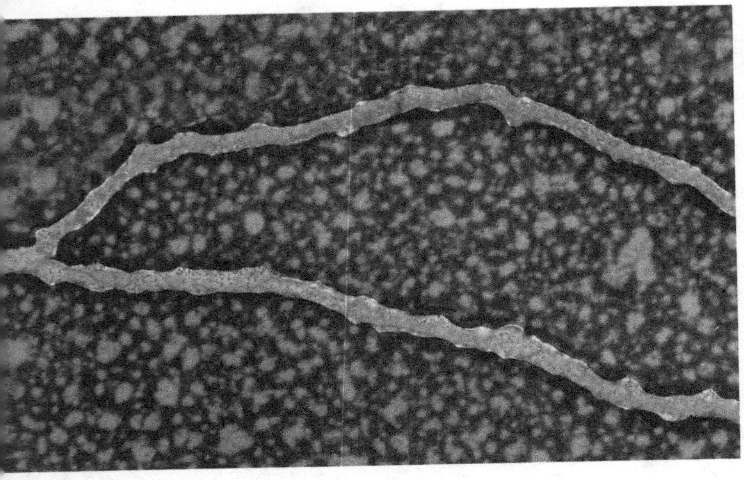

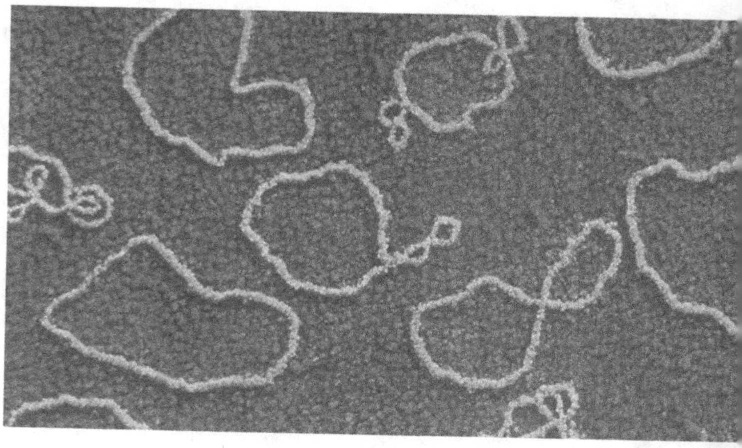

# 13 Recombinant DNA Technology and Gene Cloning 322



## PART THREE GENE EXPRESSION, REGULATION, AND DEVELOPMENT

# 14 The Genetic Code and Transcription 352

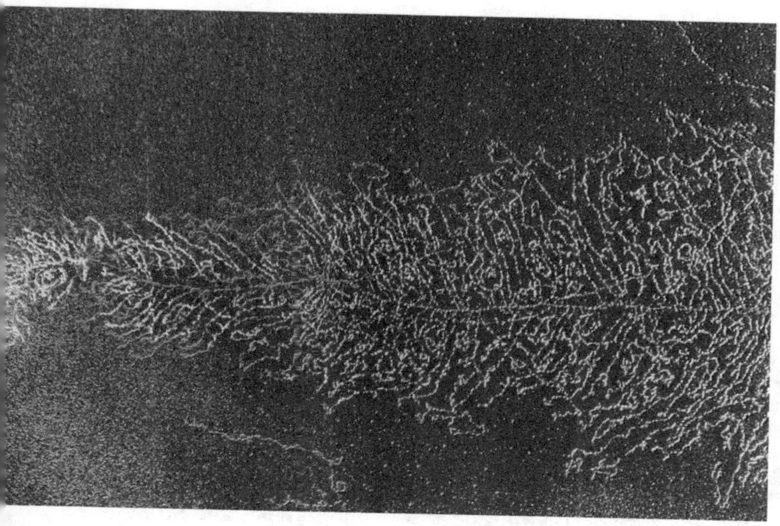

# 16 Gene Mutation and DNA Repair 410



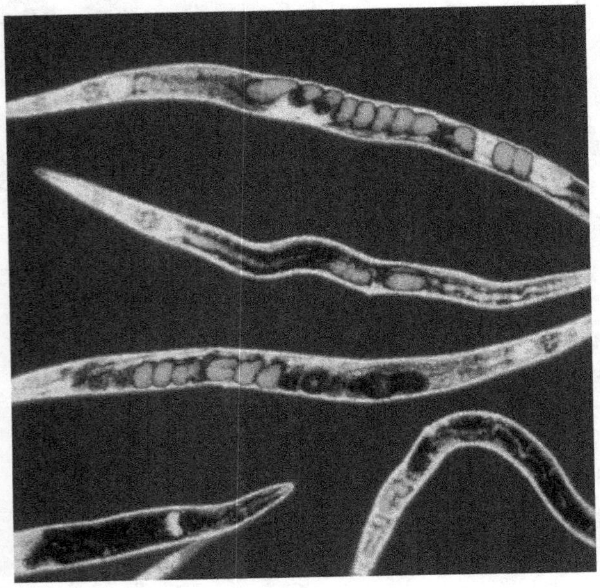



# 17 Regulation of Gene Expression in Prokaryotes 435

## PART FIVE
## GENETICS OF ORGANISMS AND POPULATION

# Preface

It is essential that textbook authors step back and look with fresh eyes as each edition of their work goes into planning and preparation. In doing so, they always need to pose two main questions: (1) How has the body of information in their field—in this case genetics—shifted since the last edition? (2) What pedagogic innovations might they add that will unquestionably enhance students' learning? The preparation of the 9th edition of *Concepts of Genetics*, now well into its third decade of providing support for students studying in this field, occasioned such a fresh look. And what we clearly saw is that in the past three years, the rapid expansion of the study of genomics, and the impact of that information at all levels in the field of genetics, represent the major advances in the field. In keeping with these observations, we have placed particular emphasis on genomics as we carefully revised and updated the entire text. This was accomplished not only by adding a new chapter related to genomics, but also by devising a new pedagogic feature that brings genomic information into each and every chapter in the text. Called *Exploring Genomics*, this innovation provides the information necessary for students to explore on the Web one or more databases closely related to the chapter topics being studied. We will discuss the details of our added coverage of genomic information as well as *Exploring Genomics* later in this preface.

The field of genetics has grown tremendously since our book was first published, both in what we know and what we want beginning students to comprehend. In creating this edition, we sought not only to continue to familiarize students with the most important discoveries of the past 150 years, but also to help them relate this information to the underlying genetic mechanisms that explain cellular processes, biological diversity, and evolution. We have also emphasized connections that link transmission genetics, molecular genetics, genomics, and proteomics.

In the first decade of this new millennium, discoveries in genetics continue to be numerous and profound. As students of genetics, the thrill of being part of this era must be balanced by a strong sense of responsibility and careful attention to the many scientific, social, and ethical issues that have already arisen, and others that will undoubtedly arise in the future. Policy makers, legislators, and an informed public will increasingly depend on knowledge of the details of genetics in order to address these issues. As a result, there has never been a greater need for a genetics textbook that clearly explains the principles of genetics.

## Goals

In the 9th edition of *Concepts of Genetics*, as in all past editions, we had six major goals. Specifically, we sought to:

- Emphasize the basic concepts of genetics.
- Write clearly and directly to students in order to provide understandable explanations of complex, analytical topics.
- Establish a careful organization within and between chapters.
- Maintain constant emphasis on science as a way of illustrating how we know what we know.
- Propagate the rich history of genetics, which so beautifully illustrates how information is acquired during scientific investigation.
- Create inviting, engaging, and pedagogically useful full-color figures enhanced by equally helpful photographs to support concept development.

These goals collectively serve as the cornerstone of *Concepts of Genetics*. This pedagogic foundation allows the book to be used in courses with many different approaches and lecture formats. Although the chapters are presented in a coherent order that represents one approach to offering a course in genetics, they are nevertheless written to be independent of one another, allowing instructors to utilize them in various sequences. We believe that the varied approaches embodied in these goals together provide students with optimal support for their study of genetics.

Writing a textbook that achieves these goals and having the opportunity to continually improve on each new edition has been a labor of love for us. The creation of each of the nine editions is a reflection not only of our passion for teaching genetics, but also of the constructive feedback and encouragement provided by adopters, reviewers, and our students over the past three decades.

## Major Innovations and Strengths of This Edition

- **Organization**—A revised organization, both within and between chapters, better illustrates how genetics is taught in the era of genomics. The introductory chapter provides an essential overview of molecular biology as a way to connect the early transmission genetics chapters to the molecular topics that follow. Enhanced coverage of model organisms is woven throughout many chapters but is especially prominent in the Introduction to Genetics (Chapter 1), as well as the chapters that consider Developmental Genetics of Model Organisms (Chapter 19) and the Genomic Analysis: Dissection of Gene Function (Chapter 23).

  The table of contents marks a number of changes from the 8th edition. The chapter introducing Recombinant DNA and Gene Cloning, the foundation on which genomic information is initially obtained, is more suitably located in Part Two of the text (Chapter 13), which focuses specifically on DNA technology.

The chapters that address Developmental Genetics and Cancer and Cell-Cycle Regulation have been relocated so that they now follow one another (Chapters 19 and 20) and are placed just after the chapters on Gene Regulation (Chapters 17 and 18). This reorganization recognizes the common links between these topics and integrates them into more cohesive coverage.

- **Pedagogy**—For this edition we have created an exciting new feature that appears in every chapter: *Exploring Genomics*. The presence and execution of this feature confirm for students that genomics impacts every aspect of genetics. Introduced in each entry are one or more genomics-related Web sites that collectively are among the best publicly available resources and databases that scientists around the world rely on for current information in genomics. The student is led through a series of interactive exercises that ensure their familiarity with the type of genomic or proteomic information available through the site and with applications of this information. The exercises instruct students on how to explore specific topics and how to access significant data. Questions are provided to guide student exploration, and the student is challenged to further explore the sites on their own. Their participation in these entries ensures that students become knowledgeable about "cutting-edge" genetic topics in genomics, proteomics, bioinformatics, and related areas as well as introducing them to the impact and application of the field of genomics to every aspect of genetics. Most importantly, the *Exploring Genomics* feature integrates genomics throughout the text, and each exercise is connected to chapter content in order to expand or reinforce genomics-related topics from the chapter.

  Another valued pedagogic feature, first introduced in the 8th edition, continues to appear in each chapter: *How Do We Know?* Previously appearing in the text of each chapter, entries have been consolidated and moved to the *Problems and Discussion Questions* section found at the end of each chapter. The *How Do We Know?* logo identifies this feature among the other problems. Each entry asks the student to identify and examine the experimental basis underlying important concepts and conclusions presented in the chapter. Addressing these questions will aid the student in more fully understanding, rather than only memorizing, the end-point of each body of research. This feature is an extension of the learning approach in biology often referred to as "Science as a Way of Knowing."

  Finally, a third feature, *Now Solve This,* has been maintained and is integrated within the text of each chapter. Each entry directs the student to a problem found at the end of the chapter that is closely related to the current text discussion. In each case, a pedagogic hint is provided to aid in solving the problem. This feature more closely links the text discussions to the problems.

  All three of these features, which appear throughout each chapter, seek to challenge students to think more deeply about, and thus understand more comprehensively, the information he or she has just finished studying.

- **New Chapters**—In keeping with our intent to offer information that represents the "cutting edge" of genetics and, as well, to increase the coverage of genomics and of model organisms utilized in genetic study, we have created two new chapters that ensure that we meet these goals. The first new chapter, Genome Dynamics—Transposons, Immunogenetics, and Eukaryotic Viruses (Chapter 22), provides modern coverage of three important topics essential to the study of modern genetics and also establishes that the genome is not a static entity. The second new chapter, Genetics and Behavior (Chapter 26), reflects our growing knowledge of the way genes impact many aspects of an organism's existence within the environment in which it finds itself. This topic is as interesting as any in the text and is important because the findings surrounding it intersect our knowledge of our own species.

  In addition, a third chapter, that relates to genomics, has received an important update and shift in emphasis: Genomic Analysis—Dissection of Gene Function (Chapter 23). This chapter establishes the important concept that genomic analysis allows us to explore more deeply the nature of the gene and how it functions. Relying on mutational studies of genomes, this topic represents one of the most important applications of genomic study.

  We have also given particular attention to quantitative genetics, population genetics, and evolutionary genetics (Chapters 25, 27, and 28). The coverage in these chapters has been extensively reviewed, and their revision is the product of the best thinking of many colleagues specialized in these fields.

- **Modernization of Topics**—Although we have updated each chapter in the text so that we report the most current and significant findings in genetics, we have especially focused on modernizing the discussions found in the chapters entitled Cancer and the Regulation of the Cell Cycle (Chapter 20), Genomics, Bioinformatics, and Proteomics (Chapter 21), and Applications and Ethics of Genetic Engineering and Biotechnology (Chapter 24). An in-depth consideration of Conservation Genetics (Chapter 29) continues to be another hallmark of our modern genetic coverage. This field, which attempts to assess and maintain genetic diversity in endangered species, remains at the forefront of genetic studies.

- **New/Revised *Genetics, Technology, and Society* Essays**—We have added several new essays that relate genetics to popular culture topics, and we have revised many that embody recent findings in genetics and their impact on society. The four new essays consider Gene Silencing (Chapter 14), Targeted Cancer Therapies (Chapter 20), The Quest for the $1000 Genome (Chapter 21), and Genetics of Sexual Orientation (Chapter 26). Those essays that have been updated and refined include Tay-Sachs Disease (Chapter 3), Purebred Dogs (Chapter 4), Fragile Sites and Cancer (Chapter 8), Telomerase and Aging (Chapter 11), Prions and Mad Cow Disease (Chapter 15), Gene Regulation and Human Disorders (Chapter 18), and Stem Cell Wars (Chapter 19).

These new or revised essays supplement those that discuss edible vaccines, human sex selection, genetically modified foods, gene therapy, and endangered species such as the Florida panther, among other topics.

- **New Illustrations**—The 9th edition includes many new figures and refines many of the existing figures in order to enhance their pedagogic value and artistic quality. Many figures feature "flow diagrams" that visually guide a student through experimental protocols and techniques.

- **Section Numbers**—All major sections of each chapter are numbered, making it easier to assign and locate topics within chapters.

- **Instructor and Student Media Address Real Needs**—Support for lecture presentations and other teaching responsibilities has been increased, including electronic access to more text photos and tables and a greater variety of PowerPoint offerings on the book's Instructor Resource Center on CD/DVD. Media found on the revamped Companion Web Site reflect the growing awareness that today's students must use their limited study time as wisely as possible.

## Emphasis on Concepts

*Concepts of Genetics,* as its title implies, emphasizes the conceptual framework of genetics. Our experience with this book, reinforced by the many adopters with whom we have been in contact over the years, demonstrates quite conclusively that students whose primary focus is on concepts more easily comprehend and take with them to succeeding courses the most important ideas in genetics as well as an analytic view of biological problem solving.

To aid students in identifying the conceptual aspects of a major topic, each chapter begins with a section called *Chapter Concepts,* which outlines the most important ideas about to be presented. In the *Problems and Discussion Questions* section, the *How Do We Know?* feature asks the student to connect concepts to experiments. In addition, the *Now Solve This* feature asks students to link conceptual understanding to problem solving in a more immediate way. Each chapter ends with a *Chapter Summary,* which enumerates the five to ten key points that have been discussed. Collectively, these features help to ensure that students easily become aware of and understand the major conceptual issues as they confront the extensive vocabulary and the many important details of genetics. Carefully designed figures support this approach throughout the book.

## Problem Solving and Insights and Solutions

To optimize the opportunities for student growth in the important areas of problem solving and analytical thinking, each chapter ends with an extensive collection of *Problems and Discussion Questions.* These include several levels of difficulty, with the most challenging (*Extra-Spicy Problems*) located at the end of each section. Brief answers to approximately half the problems are presented in Appendix B. The *Student Handbook and Solutions Manual* answers every problem and is available to students when faculty decide that it is appropriate. As the reader familiar with previous editions will see, about 75 new problems appear throughout the text.

As an aid to the student in learning to solve problems, the *Problems and Discussion Questions* section of each chapter is preceded by what has become an extremely popular and successful section called *Insights and Solutions.* This expanded section poses problems or questions and provides detailed solutions or answers. The questions and their solutions are designed to stress problem solving, quantitative analysis, analytical thinking, and experimental rationale. Collectively, these constitute the cornerstone of scientific inquiry and discovery. These feature primes students for moving on to the *Problems and Discussion Questions.*

The Genetics MediaLab section is available on the Companion Web Site. Each MediaLab contains several Web-linked problems designed to enhance and extend the topics presented in the chapter. To complete these problems, students must actively participate in the exercises and virtual experiments. For reference, the estimated time required to solve the problem is noted at the beginning of the exercise.

## Acknowledgments

### Contributors

We begin with special acknowledgments to those who have made direct contributions to this text. We particularly thank Sarah Ward at Colorado State University for creating Chapter 29 on Conservation Genetics and also for providing revised drafts of the chapters involving Quantitative and Population Genetics. We also thank David Kass of Eastern Michigan University, Chaoyang Zeng of the University of Wisconsin at Milwaukee, and Virginia McDonough of Hope College for their most useful input into both text-related topics and the revision of the Companion Web Site. In addition, Amanda Norvell revised several sections emphasizing eukaryotic molecular genetics, and Janet Morrison revised numerous aspects of our coverage of evolutionary genetics. Amanda and Janet are colleagues from The College of New Jersey. Katherine Uyhazi, now at Yale University Medical School, wrote the *Genetics, Technology, and Society* essay on Quorum Sensing in Bacteria (Chapter 17) and helped revise several other essays. David Kass also contributed the essay on Gene Regulation and Human Disorders (Chapter 18). Tamara Mans, currently teaching at North Hennepin Community College, wrote the essay on the Genetics of Sexual Orientation (Chapter 26). She also helped revise numerous essays. Mark Shotwell at Slippery Rock University contributed several essays. As with previous editions, Elliott Goldstein from Arizona State University was always readily available to consult with us concerning the most modern findings in molecular genetics. We also express special thanks to Harry Nickla, recently retired from Creighton University. In his role as author of the *Student Handbook and Solutions Manual* and the *Instructor's Resource Manual with Tests*, he has reviewed and edited the problems at the end of each chapter, and has written many of the new entries as well. He also provided the brief answers to selected problems that appear in Appendix B.

We are grateful to all of these contributors not only for sharing their genetic expertise, but for their dedication to this project as well as the pleasant interactions they provided.

### Proofreaders and Accuracy Checking

Proofreading the manuscript of an 800+ page text deserves more thanks than words can offer. Our utmost appreciation is extended to the three individuals who confronted this task with patience, diligence, and good humor:

Tamara Horton Mans, *North Hennepin Community College*
Sudhir Nayak, *The College of New Jersey*
Michael Rossa, *Proofreader*

### Reviewers

All comprehensive texts are dependent on the valuable input provided by many reviewers. While we take full responsibility for any errors in this book, we gratefully acknowledge the help provided by those individuals who reviewed the content and pedagogy of this and the previous edition:

Robert A. Angus, *University of Alabama, Birmingham*
Peta Bonham-Smith, *University of Saskatchewan*
Alan H. Christensen, *George Mason University*
Bert Ely, *University of South Carolina*
Elliott S. Goldstein, *Arizona State University*
Edward M. Golenberg, *Wayne State University*
Ashley Hagler, *University of North Carolina, Charlotte*
Jocelyn Krebs, *University of Alaska, Fairbanks*
Traci Lee, *University of Wisconsin, Parkside*
Paul F. Lurquin, *Washington State University*
Virginia McDonough, *Hope College*
Kim McKim, *Rutgers University*
Clint Magill, *Texas A&M University*
Harry Nickla, *Creighton University*
Mohamed Noor, *Duke University*
John C. Osterman, *University of Nebraska–Lincoln*
Gloria Regisford, *Prairie View A&M University*
Rodney Scott, *Wheaton College*
Barkur Shastry, *Oakland University*
Fang-sheng Wu, *Virginia Commonwealth University*
Chaoyang Zeng, *University of Wisconsin, Milwaukee*

Special thanks go to Mike Guidry of LightCone Interactive and Karen Hughes of the University of Tennessee for their original contributions to the media program.

As these acknowledgments make clear, a text such as this is a collective enterprise. All of the above individuals deserve to share in any success this text enjoys. We want them to know that our gratitude is equaled only by the extreme dedication evident in their efforts. Many, many thanks to them all.

### Editorial and Production Input

At Benjamin Cummings, we express appreciation and high praise for the editorial guidance and seminal input of Gary Carlson, whose ideas and efforts have helped to shape and refine the features of this and the previous editions of the text. In addition, our editorial team—Deborah Gale, Executive Director of Development, Leata Holloway, Project Editor, and our Media Producer, Laura Tomassi—has provided valuable input into the current edition. They have worked tirelessly to ensure that the pedagogy and design of the book and media package are at the cutting edge of a rapidly changing discipline. We were most fortunate to benefit from superb developmental editing provided by Moira Nelson, who proved to us that you are never too old to learn how to write more clearly, and outstanding copyediting performed by Betty Pessagno, for which we are most grateful. We also appreciate the production efforts of Lori Newman and those at Preparé Inc., whose quest for perfection is reflected throughout the text. In particular, Rosaria Cassinese provided an essential measure of sanity to the otherwise chaotic process of production. Without their work ethic and dedication, the text would never have come to fruition. Lauren Harp has professionally and enthusiastically managed the marketing of the text. Kaci Smith, Editorial Assistant, has worked efficiently to provide assistance to the editorial staff. Finally, the beauty and consistent presentation of the art work is the product of Imagineering of Toronto. We particularly thank Victor Ayers for his efforts.

## For the Student

### Companion Web Site—www.geneticsplace.com

Respect for the students' increasingly valuable study time is evident in the features of the Companion Web Site, which have been designed to enable users of the 9th edition to focus on those chapter sections and topics where they need review or further explanation. The Online Study Guide provides students with a focused, section-by-section review of topic coverage that features concise summary points accompanied by key illustrations and probing review questions that offer hints and feedback. The Web Tutorials offer today's learners the opportunity to quickly and conveniently visualize complex topics and dynamic processes—or to simply refamiliarize themselves with concepts they may have learned earlier but are encountering for the first time in the context of a genetics course. The media's strict adherence to both the principles and specific lessons of the textbook means that students and instructors can be assured that study time is not being squandered on media that confuse students and emphasize extraneous topics. The media tab on the outside margin of this page appears throughout

WEB TUTORIAL

the book to indicate when there is a Web Tutorial on a topic related to the coverage in the book.

All *Exploring Genomics* exercises also are available at the Companion Web Site, along with answers for each exercise. Although we have presented high-quality Web resources for *Exploring Genomics*, on occasion site addresses and navigation details may change that will affect instructions for an exercise. We encourage you to refer to *Exploring Genomics* at the Companion Web Site for the most up-to-date versions of these exercises. Another advantage of this approach is that for many exercises you can cut and paste nucleotide or amino acid sequence data to be analyzed rather than type long sequences from the exercise in the text into a Web site.

In addition, a Media Lab for each chapter is offered on the Companion Web Site for those who want to explore genetics beyond the boundaries of a book through the vast array of genetics-related resources available through the Web.

### Student Handbook and Solutions Manual

Authored by Harry Nickla, Creighton University (Emeritus) (0321544609)

This valuable handbook provides a detailed step-by-step solution or lengthy discussion for every problem in the text. The handbook also features additional study aids, including extra study problems, chapter outlines, vocabulary exercises, and an overview of how to study genetics.

## For the Instructor

### Instructor Resource Center on CD/DVD

(0321544633)

The Instructor Resource Center on CD/DVD for the 9th edition offers adopters of the text convenient access to the most comprehensive and innovative set of lecture presentation and teaching tools offered by any genetics textbook. Developed to meet the needs of veteran and newer instructors alike, these resources include:

- The JPEG files of all text line drawings with labels individually enhanced for optimal projection results (as well as unlabeled versions) and all text tables.
- Most of the text photos, including all photos with pedagogical significance, as JPEG files.
- The JPEG files of line drawings, photos, and tables preloaded into comprehensive PowerPoint® presentations for each chapter.
- A second set of PowerPoint® presentations consisting of a thorough lecture outline for each chapter augmented by key text illustrations.
- An impressive series of concise instructor animations adding depth and visual clarity to the most important topics and dynamic processes described in the text.
- The instructor animations preloaded into PowerPoint® presentation files for each chapter.
- PowerPoint® presentations containing a comprehensive set of in-class Classroom Response System (CRS) questions for each chapter.
- In Word files, a complete set of the assessment materials and study questions and answers from the testbank, the text's in-chapter text questions, and the student media practice questions, as well as files containing the entire *Instructor's Manual and Solutions Manual.*
- Finally, to help instructors keep track of all that is available in this media package, a printable Media Integration Guide in PDF format that lists each chapter's media offerings.

### Instructor's Resource Manual with Tests

(0321548485)

This manual and testbank contains over 1000 questions and problems for use in preparing exams. The manual also provides optional course sequences, a guide to audiovisual supplements, and a section on searching the Web. The testbank portion of the manual is also available in electronic format.

### TestGen EQ Computerized Testing Software

(0321550447)

In addition to the printed volume, the test questions are also available as part of the TestGen EQ Testing Software, a text-specific testing program that is networkable for administering tests. It also allows instructors to view and edit questions, export the questions as tests, and print them out in a variety of formats.

### Transparencies

(0321544617)

The transparency package includes 275 figures from the text: 225 four-color trasparencies from the text plus 50 transparency masters. The font size of the labels has been increased and boldfaced for easy viewing from the back of the classroom.

Human metaphase chromosomes, each composed of two sister chromatids joined at a common centromere. Metaphase is the stage of cell division when the members of each pair of chromatids are about to separate from one another and be distributed between two new cells.

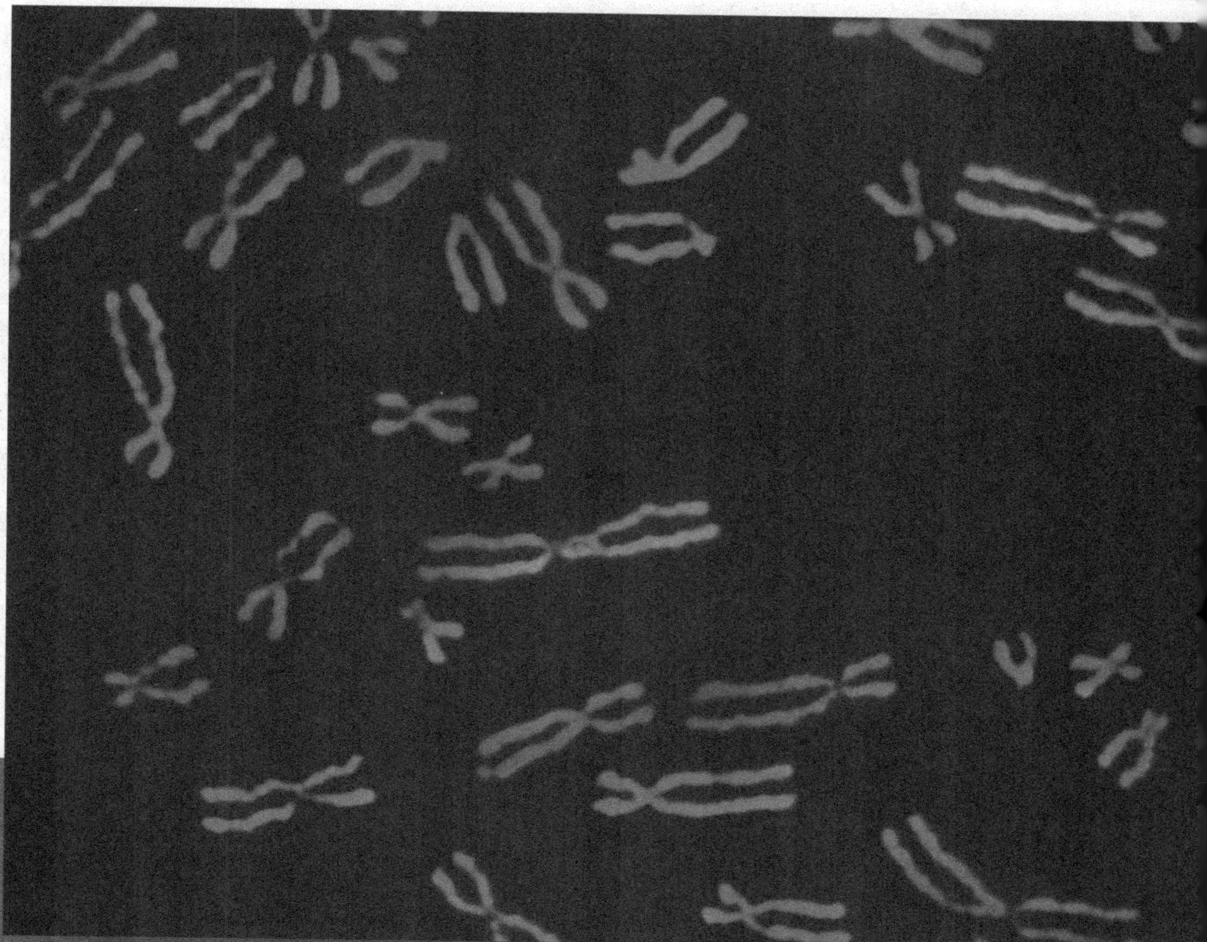

1

# Introduction to Genetics

## CHAPTER CONCEPTS

- Transmission genetics is the general process by which traits controlled by factors (genes) are transmitted through gametes from generation to generation. Its fundamental principles were first put forward by Gregor Mendel in the mid-nineteenth century. Later work by others showed that genes are on chromosomes and that mutant strains can be used to map genes on chromosomes.
- The recognition that DNA encodes genetic information, the discovery of DNA's structure, and elucidation of the mechanism of gene expression form the foundation of molecular genetics.
- Recombinant DNA technology, which allows scientists to prepare large quantities of specific DNA sequences, has revolutionized genetics, laying the foundation for new fields—and for endeavors such as the Human Genome Project—that combine genetics with information technology.
- Biotechnology includes the use of genetically modified organisms and their products in a wide range of activities involving agriculture, medicine, and industry.
- The model organisms employed in genetics research since the early part of the twentieth century are now used in combination with recombinant DNA technology and genomics to study human diseases.
- Genetic technology is developing faster than the policies, laws, and conventions that govern its use.

In December 1998, following months of heated debate, the Icelandic Parliament passed a law granting deCODE Genetics, a biotechnology company with headquarters in Iceland, a license to create and operate a database containing detailed information drawn from medical records of all of Iceland's 270,000 residents. The records in this Icelandic Health Sector (or HSD) database were encoded to ensure anonymity. The new law also allowed deCODE Genetics to cross-reference medical information from the HSD with a comprehensive genealogical database from the National Archives. In addition, deCODE Genetics would be able to correlate information in these two databases with results of deoxyribonucleic acid (DNA) profiles collected from Icelandic donors. This combination of medical, genealogical, and genetic information would be a powerful resource available exclusively to deCODE Genetics for marketing to researchers and companies for a period of 12 years, beginning in 2000.

This is not a science fiction scenario from a movie such as *Gattaca* but a real example of the increasingly complex interaction of genetics and society at the beginning of the twenty-first century. The development and use of these databases in Iceland has generated similar projects in other countries as well. The largest is the "UK Biobank" effort launched in Great Britain in 2003. There, a huge database containing the genetic information of 500,000 Britons will be compiled from an initial group of 1.2 million residents. The database will be used to search for susceptibility genes that control complex traits. Other projects have since been announced in Estonia, Latvia, Sweden, Singapore, and the Kingdom of Tonga, while in the United States, smaller-scale programs, involving tens of thousands of individuals, are underway at the Marshfield Clinic in Marshfield, Wisconsin; Northwestern University in Chicago, Illinois; and Howard University in Washington, D.C.

deCODE Genetics selected Iceland for this unprecedented project because the people of Iceland have a level of genetic uniformity seldom seen or accessible to scientific investigation. This high degree of genetic relatedness derives from the founding of Iceland about 1000 years ago by a small population drawn mainly from Scandinavian and Celtic sources. Subsequent periodic population reductions by disease and natural disasters further reduced genetic diversity there, and until the last few decades, few immigrants arrived to bring new genes into the population. Moreover, because Iceland's health-care system is state-supported, medical records for all residents go back as far as the early 1900s. Genealogical information is available in the National Archives and church records for almost every resident and for more than 500,000 of the estimated 750,000 individuals who have ever lived in Iceland. For all these reasons, the Icelandic data are a tremendous asset for geneticists in search of genes that control complex disorders. The project already has a number of successes to its credit. Scientists at deCODE Genetics have isolated 15 genes with 12 common diseases including asthma, heart disease, stroke, and osteoporosis.

On the flip side of these successes are questions of privacy, consent, and commercialization—issues at the heart of many controversies arising from the applications of genetic technology. Scientists and nonscientists alike are debating the fate and control of genetic information and the role of law, the individual, and society in decisions about how and when genetic technology is used. For example, how will knowledge of the complete nucleotide sequence of the human genome be used? Will disclosure of genetic information about individuals lead to discrimination in jobs or insurance? Should genetic technology such as prenatal diagnosis or gene therapy be available to all, regardless of ability to pay? More than at any other time in the history of science, addressing the ethical questions surrounding an emerging technology is as important as the information gained from that technology.

This introductory chapter provides an overview of genetics in which we survey some of the high points of its history and give preliminary descriptions of its central principles and emerging developments. All the topics discussed in this chapter will be explored in far greater detail elsewhere in the book. Later chapters will also revisit the controversies alluded to above and discuss many other issues that are current sources of debate. There has never been a more exciting time to be part of the science of inherited traits, but never has the need for caution and awareness of social consequences been more apparent. This text will enable you to achieve a thorough understanding of modern-day genetics and its underlying principles. Along the way, enjoy your studies, but take your responsibilities as a novice geneticist very seriously.

## 1.1 Genetics Progressed from Mendel to DNA in Less Than a Century

Because genetic processes are fundamental to life itself, the science of genetics unifies biology and serves as its core. Thus, it is not surprising that genetics has a long, rich history. Its starting point was a monastery garden in central Europe in the 1860s.

### Mendel's Work on Transmission of Traits

In this garden (Figure 1–1) Gregor Mendel, an Augustinian monk, conducted a decade-long series of experiments using pea plants. Mendel's work showed that traits of living things are passed from parents to offspring in predictable ways. He concluded that traits in pea plants, such as height and flower color, are controlled by discrete units of inheritance we now call **genes**. He further concluded that each trait in the plant is controlled by a pair of genes and that members of a gene pair separate from each other during gamete formation (the formation of egg cells and sperm). His work was published in 1866 but was largely unknown until it was partially duplicated and cited in papers by Carl Correns and others around 1900. Having been confirmed by others, Mendel's findings became recognized as

**FIGURE 1–1** The monastery garden where Gregor Mendel conducted his experiments with garden peas. In 1866, Mendel put forward the major postulates of transmission genetics.

explaining the transmission of traits in pea plants and all other higher organisms. His work forms the foundation for **genetics**, which is defined as the branch of biology concerned with the study of heredity and variation. The story of Gregor Mendel and the beginning of genetics is told in an engaging book, *The Monk in the Garden: The Lost and Found Genius of Gregor Mendel, the Father of Genetics*, by Robin M. Henig. Mendelian genetics will be discussed in Chapters 3 and 4.

## The Chromosome Theory of Inheritance: Uniting Mendel and Meiosis

Mendel did his experiments before the structure and role of chromosomes was known. About 20 years after his work was published, advances in microscopy allowed researchers to identify chromosomes (Figure 1–2) and establish that, in most eukaryotes, members of each species have a characteristic number of chromosomes called the **diploid number** (**2*n***) in most of its cells. For example, humans have a diploid number of 46 (Figure 1–3). Chromosomes in diploid cells exist in pairs, called **homologous chromosomes**. Members of a pair are identical in size and location of the centromere, a structure to which spindle fibers attach during cell division.

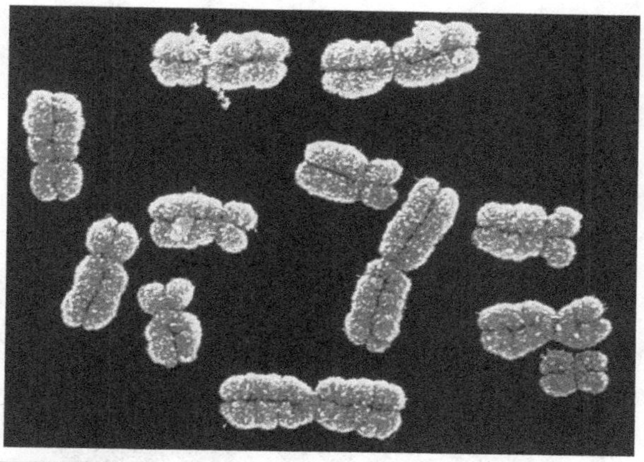

**FIGURE 1–2** Colorized image of human chromosomes that have duplicated in preparation for cell division, as visualized under the scanning electron microscope.

Researchers in the last decades of the nineteenth century also described the behavior of chromosomes during two forms of cell division, **mitosis** and **meiosis**. In mitosis (Figure 1–4), chromosomes are copied and distributed so that each daughter cell receives a diploid set of chromosomes. Meiosis is associated with gamete formation. Cells produced by meiosis receive only one chromosome from each chromosome pair, in which case the resulting number of chromosomes is called the **haploid** (***n***) **number**. This reduction in chromosome number is essential if the offspring arising from the union of two parental gametes are to maintain, over the generations, a constant number of chromosomes characteristic of their parents and other members of their species.

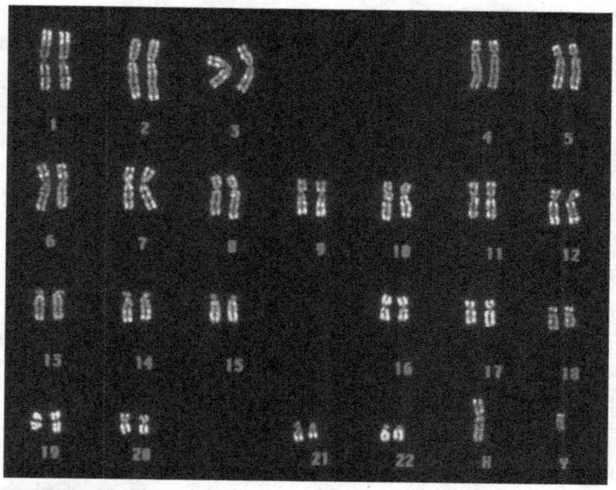

**FIGURE 1–3** A colorized image of the human male chromosome set. Arranged in this way, the set is called a karyotype.

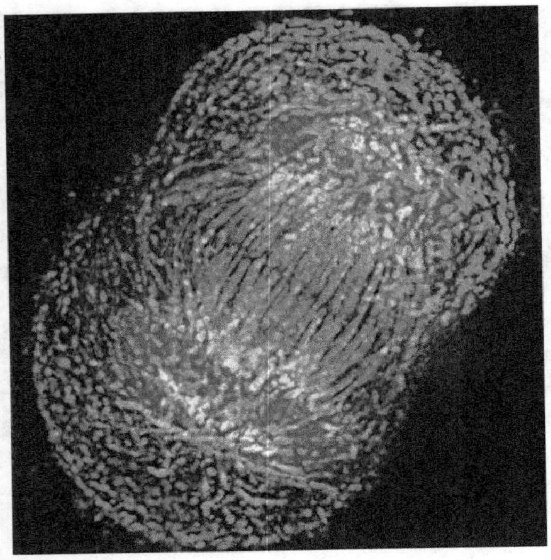

FIGURE 1–4 A stage in mitosis when the chromosomes (stained blue) move apart.

Early in the twentieth century, Walter Sutton and Theodore Boveri independently noted that genes, as hypothesized by Mendel, and chromosomes, as observed under the microscope, have several properties in common and that the behavior of chromosomes during meiosis is identical to the presumed behavior of genes during gamete formation. For example, genes and chromosomes exist in pairs, and members of a gene pair and members of a chromosome pair separate from each other during gamete formation. Based on these parallels, Sutton and Boveri each proposed that genes are carried on chromosomes (Figure 1–5). This proposal is the basis of the **chromosome theory of inheritance**, which states that inherited traits are controlled by genes residing on chromosomes faithfully transmitted through gametes, maintaining genetic continuity from generation to generation.

Geneticists encountered many different examples of inherited traits between 1910 and about 1940, allowing them to test the theory over and over. Patterns of inheritance sometimes varied from the simple examples described by Mendel, but the chromosome theory of inheritance could always be applied. It continues to explain how traits are passed from generation to generation in a variety of organisms, including humans.

## Genetic Variation

At about the same time as the chromosome theory of inheritance was proposed, scientists began studying the inheritance of traits in the fruit fly, *Drosophila melanogaster*. A white-eyed fly (Figure 1–6) was discovered in a bottle containing normal (wild-type) red-eyed flies. This variation was produced by a mutation in one of the genes controlling eye color. **Mutations** are defined as any heritable change and are the source of all genetic variation.

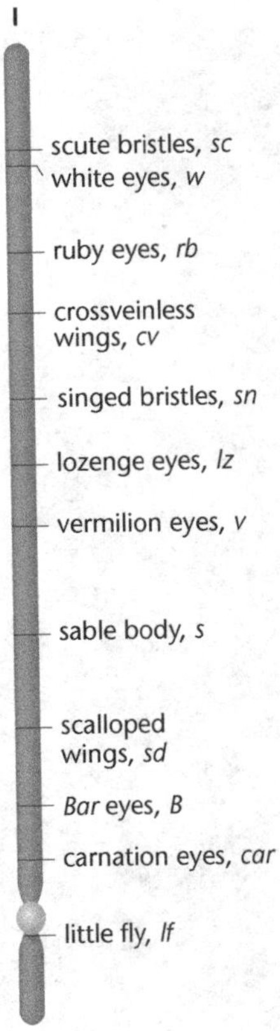


FIGURE 1–5 A drawing of chromosome I (the X chromosome, meaning one of the sex-determining chromosomes) of *D. melanogaster*, showing the locations of various genes. Chromosomes can contain hundreds of genes.

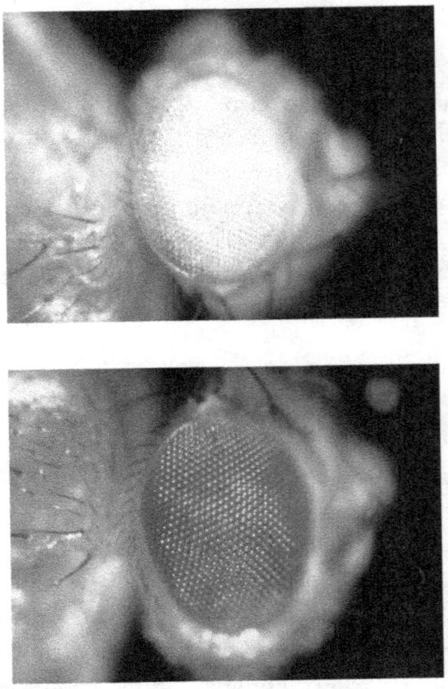

FIGURE 1–6 The normal red eye color in *D. melanogaster* (bottom) and the white-eyed mutant (top).

The variant eye color gene discovered in *Drosophila* is an **allele** of a gene controlling eye color. Alleles are defined as alternative forms of a gene. Different alleles may produce differences in the observable features, or **phenotype,** of an organism. The set of alleles for a given trait carried by an organism is called the **genotype.** Using mutant genes as markers, geneticists were able to map the location of genes on chromosomes.

### The Search for the Chemical Nature of Genes: DNA or Protein?

Work on white-eyed *Drosophila* showed that the mutant trait could be traced to a single chromosome, confirming the idea that genes are carried on chromosomes. Once this relationship was established, investigators turned their attention to identifying which chemical component of chromosomes carried genetic information. By the 1920s, scientists were aware that proteins and DNA were the major chemical components of chromosomes. Proteins are the most abundant component in cells. There are a large number of different proteins, and because of their universal distribution in the nucleus and cytoplasm, many researchers thought proteins would be shown to be the carriers of genetic information.

In 1944, Oswald Avery, Colin MacLeod, and Maclyn McCarty, three researchers at the Rockefeller Institute in New York, published experiments showing that DNA was the carrier of genetic information in bacteria. This evidence, though clear-cut, failed to convince many influential scientists. Additional evidence for the role of DNA as a carrier of genetic information came from other researchers who worked with viruses that infect and kill cells of the bacterium *Escherichia coli* (Figure 1–7). Viruses that attack bacteria are called **bacteriophages,** or **phages** for short, and like all viruses, consist of a protein coat surrounding a DNA core. Experiments showed that during infection the protein coat of the virus remains outside the bacterial cell, while the viral DNA enters the cell and directs the synthesis and assembly of more phage. This evidence that DNA carries genetic information, along with other research over the next few years, provided solid proof that DNA, not protein, is the genetic material, setting the stage for work to establish the structure of DNA.

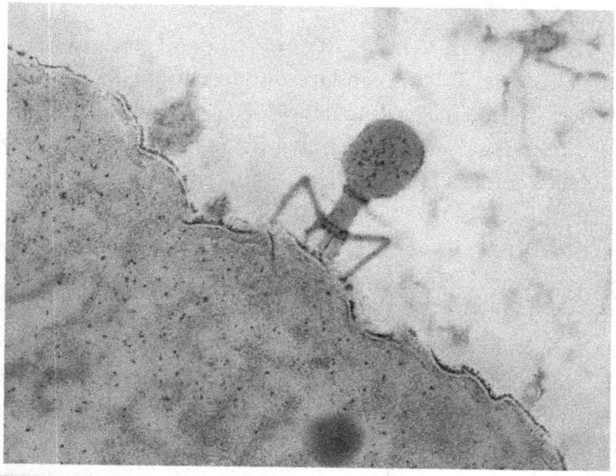

FIGURE 1–7 An electron micrograph showing T phage infecting a cell of the bacterium *E. coli.*

1.2

## Discovery of the Double Helix Launched the Era of Molecular Genetics

Once it was accepted that DNA carries genetic information, efforts were focused on deciphering the structure of the DNA molecule and the mechanism by which information stored in it is expressed to produce an observable trait, called the phenotype. In the years after this was accomplished, researchers learned how to isolate and make copies of specific regions of DNA molecules, opening the way for the era of recombinant DNA technology.

### The Structure of DNA and RNA

DNA is a long, ladder-like macromolecule that twists to form a double helix (Figure 1–8). Each strand of the helix is a linear polymer made up of subunits called **nucleotides.** In DNA, there are four different nucleotides. Each DNA nucleotide contains one of four nitrogenous bases, abbreviated A (adenine), G (guanine), T (thymine), or C (cytosine). These four bases, in various sequence combinations, ultimately specify the amino acid sequences of proteins. One of the great discoveries of the twentieth century was made in 1953 by James Watson and Francis Crick, who established that the two strands of DNA are exact complements of one another, so that the rungs of the ladder in the double helix always consist of $A = T$ and $G \equiv C$ base pairs. Along with Maurice Wilkins, Watson and Crick were awarded a Nobel Prize in 1962 for their work on the structure of DNA. A first-hand account of the race to discover the structure of DNA is told in the book *The Double Helix*, by James Watson. We will discuss the structure of DNA in Chapter 10.

As we shall see in later chapters, this **complementary relationship** between adenine and thymine and between guanine and cytosine is critical to genetic function. It serves as the basis for both the replication of DNA (Chapter 11) and for gene expression (Chapters 14 and 15). During both processes, DNA strands serve as templates for the synthesis of complementary molecules. Two depictions of the structure and components of DNA are shown in Figure 1–8.

RNA, another nucleic acid, is chemically similar to DNA but contains a different sugar (ribose rather than deoxyribose) in its nucleotides and contains the nitrogenous base uracil in place of thymine. In addition, in contrast to the double helix structure of DNA, RNA is generally single stranded. Importantly, RNA can form complementary structures with a strand of DNA.

### Gene Expression: From DNA to Phenotype

As noted earlier, nucleotide complementarity is the basis for gene expression, the chain of events that causes a gene to produce a phenotype. This process begins in the nucleus with **transcription,** in

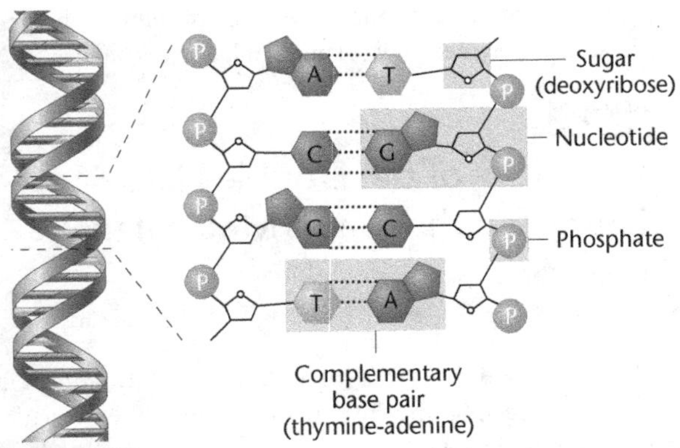


FIGURE 1-8 Summary of the structure of DNA, illustrating the arrangement of the double helix (on the left) and the chemical components making up each strand (on the right).

which the nucleotide sequence in one strand of DNA is used to construct a complementary RNA sequence (top part of Figure 1–9). Once an RNA molecule is produced, it moves to the cytoplasm. In protein synthesis, the RNA—called **messenger RNA**, or **mRNA** for short—binds to a **ribosome**. The synthesis of proteins under the direction of mRNA is called **translation** (bottom part of Figure 1–9). Proteins, the end product of many genes, are polymers made up of amino acid monomers. There are 20 different amino acids commonly found in proteins.

How can information contained in mRNA direct the addition of specific amino acids onto protein chains as they are synthesized? The information encoded in mRNA and called the **genetic code** consists of linear series of nucleotide triplets. Each triplet, called a **codon**, is complementary to the information stored in DNA and specifies the insertion of a specific amino acid into a protein. Protein assembly is accomplished with the aid of adapter molecules called **transfer RNA (tRNA)**. Within the ribosome, tRNAs recognize the information encoded in the mRNA codons and carry the proper amino acids for construction of the protein during translation.

As the preceding discussion shows, DNA makes RNA, which most often makes protein. This sequence of events, known as the **central dogma** of genetics, occurs with great specificity. Using an alphabet of only four letters (A, T, C, and G), genes direct the synthesis of highly specific proteins that collectively serve as the basis for all biological function.

## Proteins and Biological Function

As we have mentioned, proteins are the end products of gene expression. These molecules are responsible for imparting the properties of living systems. The diversity of proteins and of the biological functions they can perform—the diversity of life itself—arises from the fact that proteins are made from combinations of 20 different amino acids. Consider that a protein chain containing 100 amino acids can have at each position any one of 20 amino acids; the number of

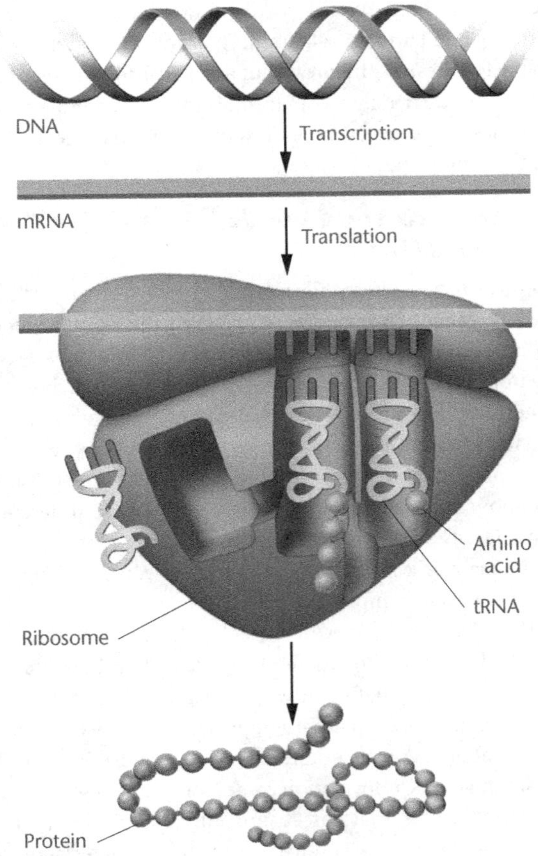


FIGURE 1-9 Gene expression consists of transcription of DNA into mRNA (top) and the translation (center) of mRNA (with the help of a ribosome) into a protein (bottom).

possible different 100 amino acid proteins, each with a unique sequence, is therefore equal to

$$20^{100}$$

Because $20^{10}$ exceeds $5 \times 10^{12}$, or 5 trillion, imagine how large a number $20^{100}$ is! The tremendous number of possible amino acid sequences in proteins leads to enormous variation in their possible three-dimensional conformations. Obviously, evolution has seized on a class of molecules with the potential for enormous structural diversity to serve as the mainstay of biological systems.

The largest category of proteins is the **enzymes** (Figure 1–10). These molecules serve as biological catalysts, essentially causing biochemical reactions to proceed at the rates that are necessary for sustaining life. By lowering the energy of activation in reactions, enzymes enable cellular metabolism to proceed at body temperatures, when otherwise those reactions would require intense heat or pressure in order to occur.

Countless proteins other than enzymes are critical components of cells and organisms. These include hemoglobin, the oxygen-binding pigment in red blood cells; insulin, the pancreatic hormone; collagen,

FIGURE 1–10 A three-dimensional conformation of a protein. The protein shown here is an enzyme.

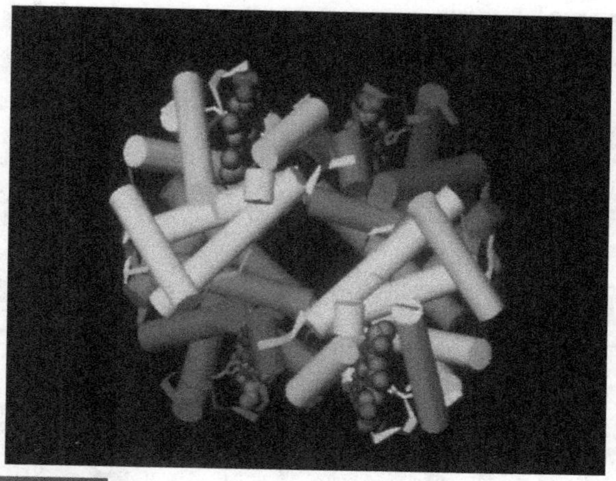

FIGURE 1–11 The hemoglobin molecule, showing the two alpha chains and the two beta chains. A mutation in the gene for the beta chain produces abnormal hemoglobin molecules and sickle-cell anemia.

the connective tissue molecule; keratin, the structural molecule in hair; histones, proteins integral to chromosome structure in eukaryotes (that is, organisms whose cells have nuclei); actin and myosin, the contractile muscle proteins; and immunoglobulins, the antibody molecules of the immune system. A protein's shape and chemical behavior are determined by its linear sequence of amino acids, which is dictated by the stored information in the DNA of a gene that is transferred to RNA, which then directs the protein's synthesis. To repeat, DNA makes RNA, which then makes protein.

## Linking Genotype to Phenotype: Sickle-Cell Anemia

Once a protein is constructed, its biochemical or structural behavior in a cell plays a role in producing a phenotype. When mutation alters a gene, it may modify or even eliminate the encoded protein's usual function and cause an altered phenotype. To trace the chain of events leading from the synthesis of a given protein to the presence of a certain phenotype, we will examine sickle-cell anemia, a human genetic disorder.

Sickle-cell anemia is caused by a mutant form of hemoglobin, the protein that transports oxygen from the lungs to cells in the body (Figure 1–11). Hemoglobin is a composite molecule made up of two different proteins, $\alpha$-globin and $\beta$-globin, each encoded by a different gene. Each functional hemoglobin molecule contains two $\alpha$-globin and two $\beta$-globin proteins. In sickle-cell anemia, a mutation in the gene encoding $\beta$-globin causes an amino acid substitution in 1 of the 146 amino acids in the protein. Figure 1–12 shows part of the DNA sequence, and the corresponding mRNA codons and amino acid sequence, for the normal and mutant forms of $\beta$-globin. Notice that the mutation in sickle-cell anemia consists of a change in one DNA nucleotide, which leads to a change in codon 6 in mRNA from GAG to GUG, which in turn changes amino acid number 6 in $\beta$-globin from glutamic acid to valine. The other 145 amino acids in the protein are not changed by this mutation.

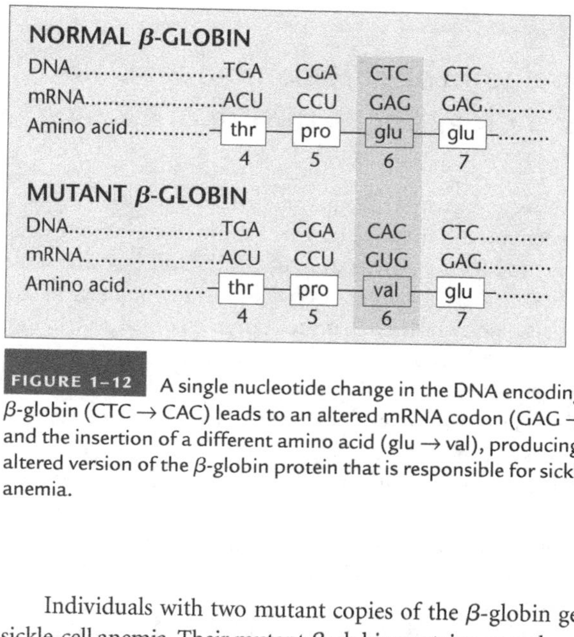


FIGURE 1–12 A single nucleotide change in the DNA encoding $\beta$-globin (CTC $\rightarrow$ CAC) leads to an altered mRNA codon (GAG $\rightarrow$ GUG) and the insertion of a different amino acid (glu $\rightarrow$ val), producing the altered version of the $\beta$-globin protein that is responsible for sickle-cell anemia.

Individuals with two mutant copies of the $\beta$-globin gene have sickle-cell anemia. Their mutant $\beta$-globin proteins cause hemoglobin molecules in red blood cells to polymerize when the blood's oxygen concentration is low, forming long chains of hemoglobin that distort the shape of red blood cells (Figure 1–13). The deformed cells are fragile and break easily, so that the number of red blood cells in circulation is reduced (anemia is an insufficiency of red blood cells). Moreover, when blood cells are sickle shaped, they block blood flow in capillaries and small blood vessels, causing severe pain and damage to the heart, brain, muscles, and kidneys. Sickle-cell anemia can cause heart attacks and stroke, and can be fatal if left untreated. All the symptoms of this disorder are caused by a change in a single nucleotide in a gene that changes one amino acid out of 146 in the $\beta$-globin molecule, demonstrating the close relationship between genotype and phenotype.

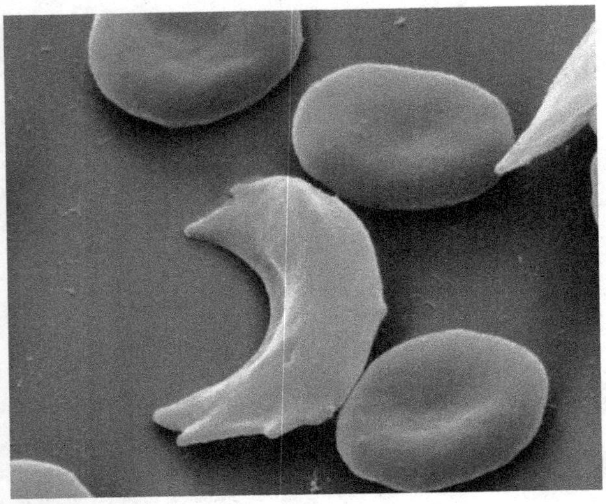

FIGURE 1–13 Normal red blood cells (round) and sickled red blood cells. The sickled cells block capillaries and small blood vessels.

1.3

## Development of Recombinant DNA Technology Began the Era of Cloning

The era of recombinant DNA began in the early 1970s, when researchers discovered that bacteria protect themselves from viral infection by producing enzymes that cut viral DNA at specific sites. When cut, the viral DNA cannot direct the synthesis of phage particles. Scientists quickly realized that such enzymes, called **restriction enzymes**, could be used to cut any organism's DNA at specific nucleotide sequences, producing a reproducible set of fragments. This set the stage for the development of DNA cloning, or making large numbers of copies of DNA sequences.

Soon after researchers discovered that restriction enzymes produce specific DNA fragments, methods were developed to insert these fragments into carrier DNA molecules called vectors to make **recombinant DNA** molecules and transfer them into bacterial cells. As the bacterial cells reproduce, thousands of copies, or **clones**, of the combined vector and DNA fragments are produced (Figure 1–14). These cloned copies can be recovered from the bacterial cells, and large amounts of the cloned DNA fragment can be isolated. Once large quantities of specific DNA fragments became available by cloning, they were used in many different ways: to isolate genes, to study their organization and expression, and to study their nucleotide sequence and evolution.

As techniques became more refined, it became possible to clone larger and larger DNA fragments, paving the way to establish collections of clones that represented an organism's **genome**, which is the complete haploid content of DNA specific to that organism. Collections of clones that contain an entire genome are called genomic libraries. Genomic libraries are now available for hundreds of organisms.

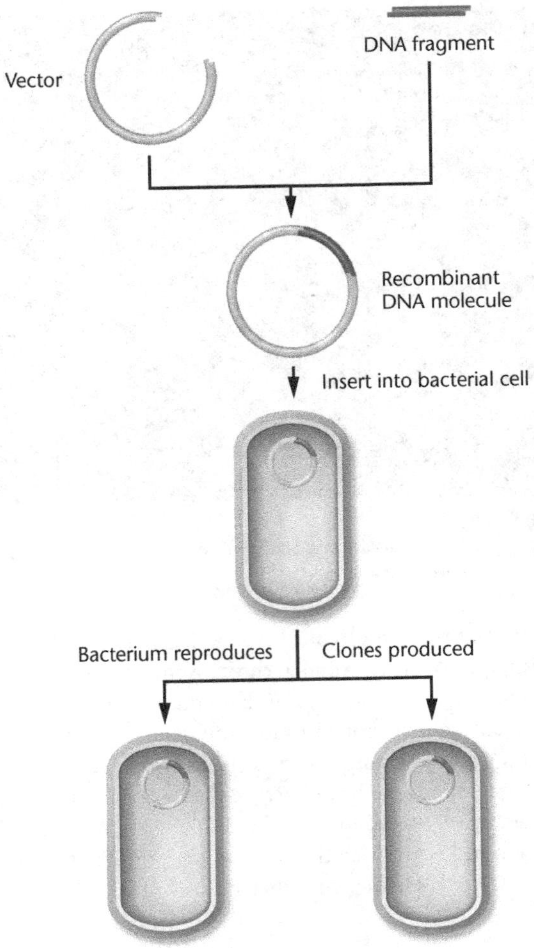


FIGURE 1–14 In cloning, a vector and a DNA fragment produced by cutting with a restriction enzyme are joined to produce a recombinant DNA molecule. The recombinant DNA is transferred into a bacterial cell, where it is cloned into many copies by replication of the recombinant molecule and by division of the bacterial cell.

Recombinant DNA technology has not only greatly accelerated the pace of research but has also given rise to the biotechnology industry, which has grown over the last 25 years to become a major contributor to the U.S. economy.

1.4

## The Impact of Biotechnology Is Continually Expanding

Quietly and without arousing much notice in the United States, biotechnology has revolutionized many aspects of everyday life. Humans have used microorganisms, plants, and animals for thousands of years, but the development of recombinant DNA technology and associated techniques allows us to genetically modify organisms in new ways and use them or their products to enhance our lives. **Biotechnology** is the use of these modified organisms or their products. It is now in evidence at the supermarket; in doctors' offices; at drug stores, department stores, hospitals, and clinics; on farms and in orchards; in law enforcement and court-ordered child support; and even in industrial chemicals. There is a detailed discussion of biotechnology in Chapter 24, but for now, let's look at biotechnology's impact on just a small sampling of everyday examples.

## Plants, Animals, and the Food Supply

The genetic modification of crop plants is one of the most rapidly expanding areas of biotechnology. Efforts have been focused on traits such as resistance to herbicides, insects, and viruses; enhancement of oil content; and delay of ripening (Table 1.1). Currently, over a dozen genetically modified crop plants have been approved for commercial use in the United States, with over 75 more being tested in field trials. Herbicide-resistant corn and soybeans were first planted in the mid-1990s, and now about 45 percent of the U.S. corn crop and 85 percent of the U.S. soybean crop is genetically modified. In addition, more than 50 percent of the canola crop and 75 percent of the cotton crop are grown from genetically modified strains. It is estimated that more than 60 percent of the processed food in the United States contains ingredients from genetically modified crop plants.

This agricultural transformation is a source of controversy. Critics are concerned that the use of herbicide-resistant crop plants will lead to dependence on chemical weed management and may eventually result in the emergence of herbicide-resistant weeds. They also worry that traits in genetically engineered crops could be transferred to wild plants in a way that leads to irreversible changes in the ecosystem.

Biotechnology is also being used to enhance the nutritional value of crop plants. More than one-third of the world's population uses rice as a dietary staple, but most varieties of rice contain little or no vitamin A. Vitamin A deficiency causes more than 500,000 cases of blindness in children each year. A genetically engineered strain, called golden rice, has high levels of two compounds that the body converts to vitamin A. Golden rice should be available for planting in the near future, with the aim of reducing this burden of disease. Other crops, including wheat, corn, beans, and cassava, are also being modified to enhance nutritional value by increasing their vitamin and mineral content.

Livestock such as sheep and cattle have been commercially cloned for more than 25 years, mainly by a method called embryo splitting. In 1996, Dolly the sheep (Figure 1–15) was cloned by nuclear transfer, a method in which the nucleus of a differentiated adult cell (meaning a cell recognizable as belonging to some type of tissue) is transferred into an egg that has had its nucleus removed. This nuclear transfer method makes it possible to produce dozens or hundreds of offspring with desirable traits. Cloning by nuclear transfer has many applications in agriculture, sports, and medicine. Some desirable traits, such as high milk production in cows, or speed in race horses, do not appear until adulthood; rather than mating two adults and waiting to see if their offspring inherit the desired characteristics, animals that are known to have these traits can now be produced by cloning differentiated cells from an adult with a desirable trait. For medical applications, researchers have transferred human genes into animals—so-called transgenic animals—so that as adults, they produce human proteins in their milk. By selecting and cloning animals with high levels of human protein production, biopharmaceutical companies can produce a herd with uniformly high rates of protein production. Human proteins from transgenic animals are now being tested as drug treatments for diseases such as emphysema. If successful, these proteins will soon be commercially available.

## Who Owns Transgenic Organisms?

Once produced, can a transgenic plant or animal be patented? The answer is yes. In 1980 the United States Supreme Court ruled that

**TABLE 1.1**

**Some Genetically Altered Traits in Crop Plants**

| |
|---|
| **Herbicide Resistance** |
| Corn, soybeans, rice, cotton, sugarbeets, canola |
| **Insect Resistance** |
| Corn, cotton, potato |
| **Virus Resistance** |
| Potato, yellow squash, papaya |
| **Nutritional Enhancement** |
| Golden rice |
| **Altered Oil Content** |
| Soybeans, canola |
| **Delayed Ripening** |
| Tomato |

FIGURE 1–15 Dolly, a Finn Dorset sheep cloned from the genetic material of an adult mammary cell, shown next to her first-born lamb, Bonnie.

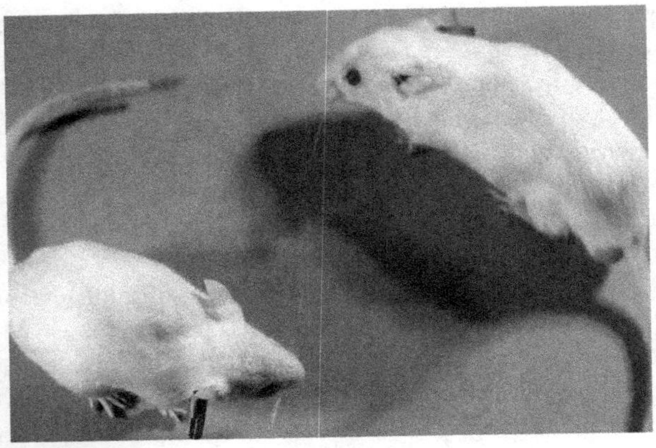

FIGURE 1–16 The first genetically altered organism to be patented, the *onc* strain of mouse, genetically engineered to be susceptible to many forms of cancer. These mice were designed for studying cancer development and the design of new anticancer drugs.

living organisms and individual genes can be patented, and in 1988 an organism modified by recombinant DNA technology was patented for the first time (Figure 1–16). Since then, dozens of plants and animals have been patented. The ethics of patenting living organisms is a contentious issue. Supporters of patenting argue that without the ability to patent the products of research to recover their costs, biotechnology companies will not invest in large-scale research and development. They further argue that patents represent an incentive to develop new products because companies will reap the benefits of taking risks to bring new products to market. Critics argue that patents for organisms such as crop plants will concentrate ownership of food production in the hands of a small number of biotechnology companies, making farmers economically dependent on seeds and pesticides produced by these companies, and reducing the genetic diversity of crop plants as farmers discard local crops that might harbor important genes for resistance to pests and disease. Resolution of these and other issues raised by biotechnology and its uses will require public awareness and education, enlightened social policy, and carefully written legislation.

### Biotechnology in Genetics and Medicine

Biotechnology in the form of genetic testing and gene therapy, already an important part of medicine, will be a leading force deciding the nature of medical practice in the twenty-first century. More than 10 million children or adults in the United States suffer from some form of genetic disorder, and every childbearing couple stands an approximately 3 percent risk of having a child with some form of genetic anomaly. The molecular basis for hundreds of genetic disorders is now known (Figure 1–17). Genes for sickle-cell anemia, cystic fibrosis, hemophilia, muscular dystrophy, phenylketonuria, and many other metabolic disorders have been cloned and are used for the prenatal detection of affected fetuses. In addition, tests are now available to inform parents of their status as "carriers" of a large number of inherited disorders. The combination of genetic testing and genetic counseling gives couples objective information on which they can base decisions about childbearing. At present, genetic testing is available for several hundred inherited disorders, and this number will grow as more genes are identified, isolated, and cloned. The use of genetic testing and other technologies, including gene therapy, raises ethical concerns that have yet to be resolved.

Instead of testing one gene at a time to discover whether someone carries a mutant gene that can produce a disorder in his or her offspring, a new technology is being developed that will allow screening of an entire genome to determine an individual's risk of developing a genetic disorder or of having a child with a genetic disorder. This technology uses devices called **DNA microarrays**, or **DNA chips** (Figure 1–18). Each microarray can carry thousands of genes. In fact, microarrays carrying the entire human genome are now commercially available and are being used to test for gene expression in cancer cells as a step in developing therapies tailored to specific forms of cancer. As the technology develops further, it will be possible to scan an individual's genome in one step to identify risks for genetic and environmental factors that may trigger disease.

In **gene therapy**, clinicians transfer normal genes into individuals affected with genetic disorders. Unfortunately, although many attempts at gene therapy appeared initially to be successful, therapeutic failures and patient deaths have slowed the development of this technology. New methods of gene transfer are expected to reduce these risks, however, so it seems certain that gene therapy will become an important tool in treating inherited disorders and that, as more is learned about the molecular basis of human diseases, more such therapies will be developed.

## 1.5 Genomics, Proteomics, and Bioinformatics Are New and Expanding Fields

Once genomic libraries became available, scientists began to consider ways to sequence all the clones in such a library so as to spell out the nucleotide sequence of an organism's genome. Laboratories around the world initiated projects to sequence and analyze the genomes of different organisms, including those that cause human diseases. To date, the genomes of over 550 organisms have been sequenced, and over a thousand additional genome projects are underway.

The Human Genome Project began in 1990 as an international, government-sponsored effort to sequence the human genome and the genomes of five of the model organisms used in genetics research (the importance of model organisms is discussed below). At about the same time, various industry-sponsored genome projects also got underway. The first sequenced genome from a free-living organism, a bacterium (Figure 1–19), was reported in 1995 by scientists at a biotechnology company.

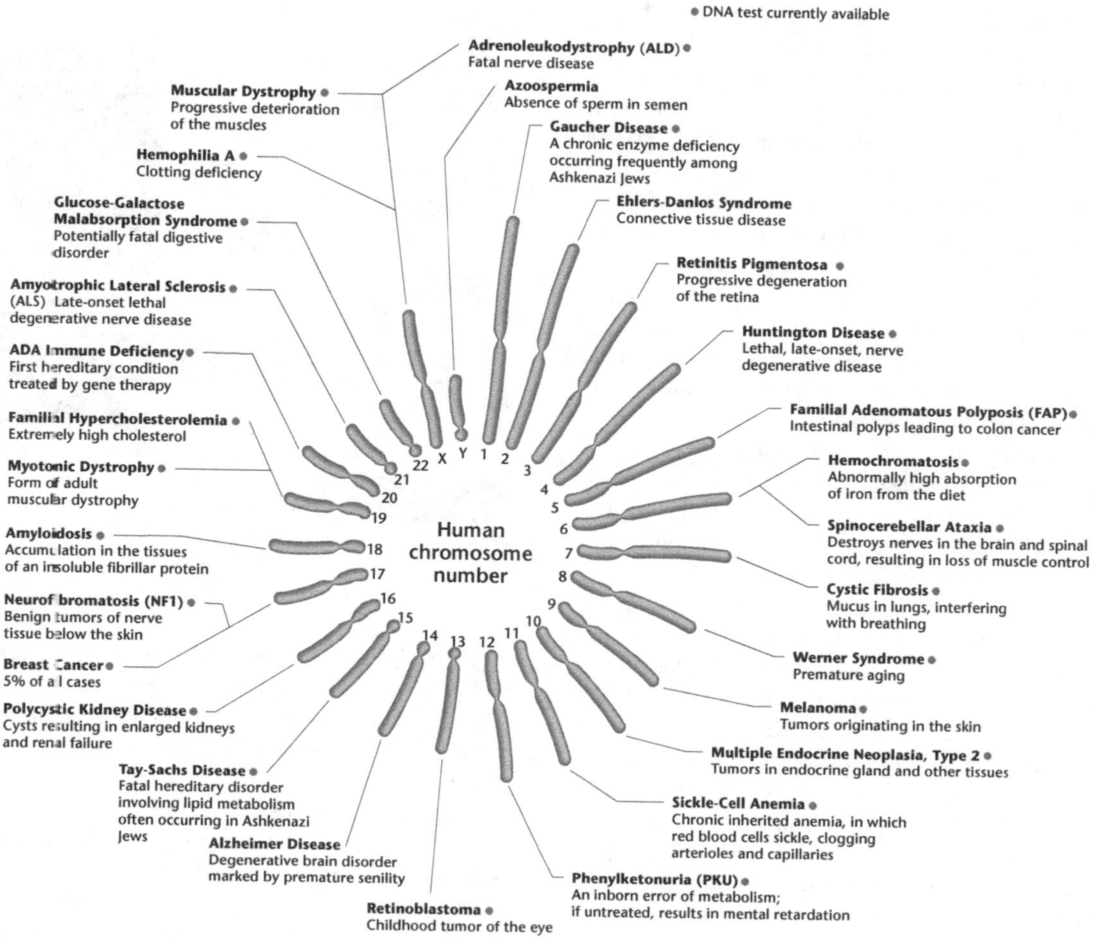


**FIGURE 1–17** Diagram of the human chromosome set, showing the location of some genes whose mutant forms cause hereditary diseases. Conditions that can be diagnosed using DNA analysis are indicated by a red dot.

**FIGURE 1–18** A portion of a DNA microarray. These arrays contain thousands of fields (the circles) to which DNA molecules are attached. Mounted on a microarray, DNA from an individual can be tested to detect mutant copies of genes.

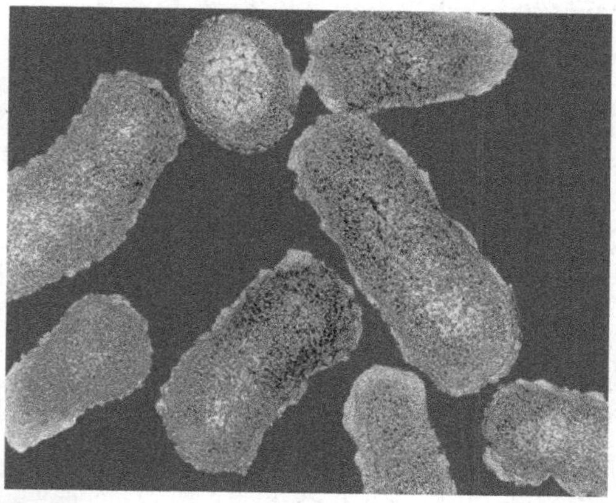

**FIGURE 1–19** A colorized electron micrograph of *Haemophilus influenzae*, a bacterium that was the first free-living organism to have its genome sequenced. This bacterium causes respiratory infections and bacterial meningitis in humans.

In 2001, the publicly funded Human Genome Project and a private genome project sponsored by Celera Corporation reported the first draft of the human genome sequence, covering about 96 percent of the gene-containing portion of the genome. In 2003, the remaining portion of the gene-coding sequence was completed and published. Efforts are now focused on sequencing the remaining noncoding regions of the genome. The five model organisms whose genomes were also sequenced by the Human Genome Project are *Escherichia coli* (a bacterium), *Saccharomyces cerevisiae* (a yeast), *Caenorhabditis elegans* (a roundworm), *Drosophila melanogaster* (the fruit fly), and *Mus musculus* (the mouse).

As genome projects multiplied and more and more genome sequences were acquired, several new biological disciplines arose. One, called **genomics** (the study of genomes), sequences genomes and studies the structure, function, and evolution of genes and genomes. A second field, **proteomics**, is an outgrowth of genomics. Proteomics identifies the set of proteins present in a cell under a given set of conditions and additionally studies the post-translational modification of these proteins, their location within cells, and the protein–protein interactions occurring in the cell. To store, retrieve, and analyze the massive amount of data generated by genomics and proteomics, a specialized subfield of information technology called **bioinformatics** was created to develop hardware and software for processing nucleotide and protein data. Consider that the human genome contains over 3 billion nucleotides, representing some 25,000 genes encoding tens of thousands of proteins, and you can appreciate the need for databases to store this information.

These new fields are drastically changing biology from a laboratory-based science to one that combines lab experiments with information technology. Geneticists and other biologists now use information in databases containing nucleic acid sequences, protein sequences, and gene interaction networks to answer experimental questions in a matter of minutes instead of months and years. A feature called **Exploring Genomics**, located at the end of all chapters in this textbook, gives you the opportunity to explore these databases for yourself while completing an interactive genetics exercise.

1.6

## Genetic Studies Rely on the Use of Model Organisms

After the rediscovery of Mendel's work in 1900, genetic research on a wide range of organisms confirmed that the principles of inheritance he described were of universal significance among plants and animals. Although work continued on the genetics of many different organisms, geneticists gradually came to focus particular attention on a small number of organisms, including the fruit fly (*Drosophila melanogaster*) and the mouse (*Mus musculus*) (Figure 1–20). This trend developed for two main reasons: first, it was clear that genetic mechanisms were the same in most organisms, and second, the preferred species had several characteristics that made them especially suitable for genetic research. They were easy to grow, had relatively short life cycles, produced many offspring, and their genetic analysis was fairly straightforward. Over time, researchers created a large catalog of mutant strains for the preferred species, and the mutations were carefully studied, characterized, and mapped. Because of their well-characterized genetics, these species became **model organisms**, defined as organisms used for the study of basic biological processes. Although originally developed to study genetic mechanisms, model organisms are now being used to study cellular events in general, as well as the origin and mechanisms of many human diseases (genetic or otherwise) and to develop new and innovative therapies to treat them. In later chapters, we will see how discoveries in model organisms are shedding light on many aspects of biology, including aging, cancer, the immune system, and behavior.

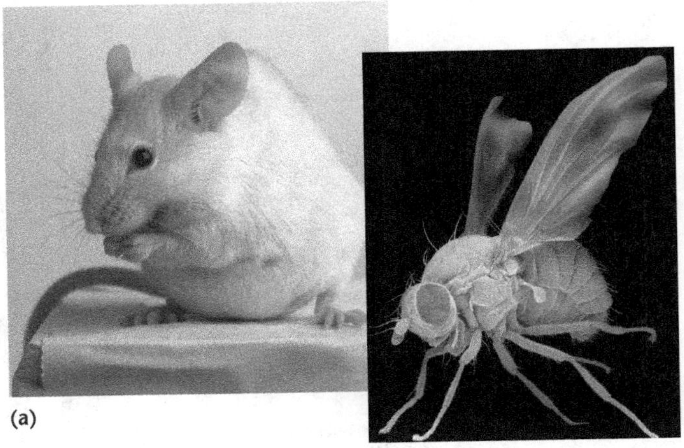

(a) (b)

FIGURE 1–20 The first generation of model organisms in genetic analysis included (a) the mouse and (b) the fruit fly.

### The Modern Set of Genetic Model Organisms

Gradually, geneticists added other species to their collection of model organisms, including viruses (such as the T phages and lambda phage) and microorganisms (the bacterium *Escherichia coli* and the yeast *Saccharomyces cerevisiae*) (Figure 1–21). Some of these were chosen for the reasons outlined above, while others were selected because of other characteristics that allowed certain aspects of genetics to be studied more easily.

More recently, three additional species have been developed as model organisms. Each was chosen to study some aspect of embryonic development. To study the nervous system and its role in behavior, the nematode *Caenorhabditis elegans* [Figure 1–22(a)] was chosen as a model system. It is small, it is easy to grow, and it has a nervous system with only a few hundred cells. *Arabidopsis thaliana* [Figure 1–22(b)] is a small plant with a short life cycle that can be grown in the laboratory. It was first used to study flower development but has

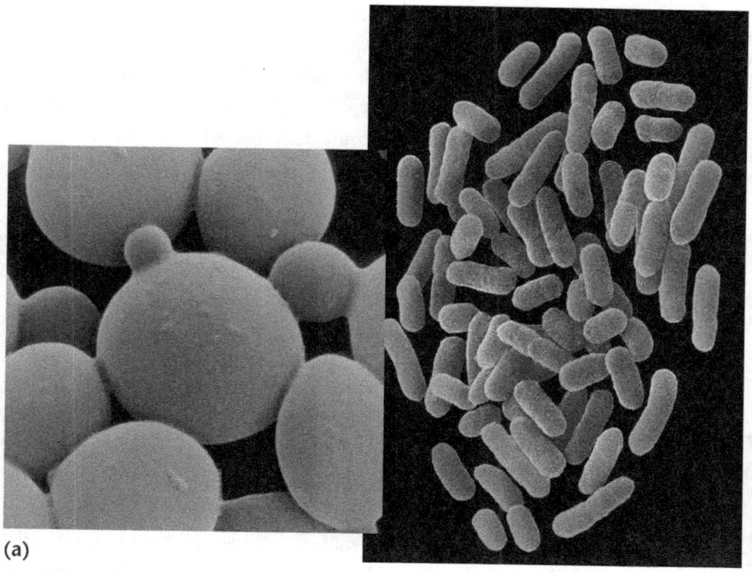

(a) (b)

**FIGURE 1-21** Microbes that have become model organisms for genetic studies include (a) the yeast *S. cerevisiae* and (b) the bacterium *E. coli*.

become a model organism for the study of many other aspects of plant biology. The zebrafish, *Danio rerio* [Figure 1–22(c)], has several advantages for the study of vertebrate development: it is small, it reproduces rapidly, and its egg, embryo, and larvae are all transparent.

## Model Organisms and Human Diseases

The development of recombinant DNA technology and the results of genome sequencing have confirmed that all life has a common origin. Because of this common origin, genes with similar functions in different organisms tend to be similar or identical in structure and nucleotide sequence. Much of what scientists learn by studying the genetics of other species can therefore be applied to humans and serve as the basis for understanding and treating human diseases. In addition, the ability to transfer genes between species has enabled scientists to develop models of human diseases in organisms ranging from bacteria to fungi, plants, and animals (Table 1.2).

The idea of studying a human disease such as colon cancer by using *E. coli* may strike you as strange, but the basic steps of DNA repair (a process that is defective in some forms of colon cancer) are the same in both organisms, and the gene involved (*mutL* in *E. coli* and *MLH1* in humans) is found in both organisms. More importantly, *E. coli* has the advantage of being easier to grow (the cells divide every 20 minutes), so that researchers can easily create and study new mutations in the bacterial *mutL* gene in order to figure out how it works. This knowledge may eventually lead to the development of drugs and other therapies to treat colon cancer in humans.

The fruit fly, *D. melanogaster*, is also being used to study specific human diseases. Mutant genes have been identified in

(a)

(b)

(c)

**FIGURE 1-22** Newer model organisms in genetics include (a) the roundworm *C. elegans*, (b) the plant *A. thaliana*, and (c) the zebrafish, *D. rerio*.

*D. melanogaster* that produce phenotypes with abnormalities of the nervous system, including abnormalities of brain structure, adult-onset degeneration of the nervous system, and visual defects such as retinal degeneration. The information from genome sequencing

TABLE 1.2

**Model Organisms Used to Study Human Diseases**

| Organism | Human Diseases |
|---|---|
| *E. coli* | Colon cancer and other cancers |
| *S. cerevisiae* | Cancer, Werner syndrome |
| *D. melanogaster* | Disorders of the nervous system, Cancer |
| *C. elegans* | Diabetes |
| *D. rerio* | Cardiovascular disease |
| *M. musculus* | Lesch–Nyhan disease, cystic fibrosis, fragile-X syndrome, and many other diseases |

projects indicates that almost all these genes have human counterparts. As an example, genes involved in a complex human disease of the retina called retinitis pigmentosa are identical to *Drosophila* genes involved in retinal degeneration. Study of these mutations in *Drosophila* is helping to dissect this complex disease and identify the function of the genes involved.

Another approach to using *Drosophila* for studying diseases of the human nervous system is to transfer human disease genes into the flies by means of recombinant DNA technology. The transgenic flies are then used for studying the mutant human genes themselves, the genes affecting the expression of the human disease genes, and the effects of therapeutic drugs on the action of those genes, all studies that are difficult or impossible to perform in humans. This gene transfer approach is being used to study almost a dozen human neurodegenerative disorders, including Huntington disease, Machado-Joseph disease, myotonic dystrophy, and Alzheimer disease.

As you read this textbook, you will encounter these model organisms again and again. Remember that, each time you meet them they not only have a rich history in basic genetics research but are also at the forefront in the study of human genetic disorders and infectious diseases.

The use of model organisms for understanding human health and disease is one of many ways genetics and biotechnology are rapidly changing everyday life. As discussed in the next section, however, we have yet to reach a consensus on how and when this technology is determined to be safe and ethically acceptable.

## 1.7 We Live in the Age of Genetics

Mendel described his decade-long project on inheritance in pea plants in an 1865 paper presented at a meeting of the Natural History Society of Brünn in Moravia. Just 100 years later, the 1965 Nobel Prize was awarded to François Jacob, André Lwoff, and Jacques Monod for their work on the molecular basis of gene regulation in bacteria. This time span encompassed the years leading up to the acceptance of Mendel's work, the discovery that genes are on chromosomes, the experiments that proved DNA encodes genetic information, and the elucidation of the molecular basis for DNA replication. The rapid development of genetics from Mendel's monastery garden to the Human Genome Project and beyond is summarized in a timeline in Figure 1–23.

### The Nobel Prize and Genetics

Although other scientific disciplines have also expanded in recent years, none has paralleled the explosion of information and excitement generated by the discoveries in genetics. Nowhere is this impact more apparent than in the list of Nobel Prizes related to genetics, beginning with those awarded in the early and mid-twentieth century and continuing into the present (see inside front cover). Nobel Prizes in the categories of Medicine or Physiology and Chemistry have been consistently awarded for work in genetics and associated fields. The first Nobel Prize awarded for such work was given to Thomas Morgan in 1933 for his research on the

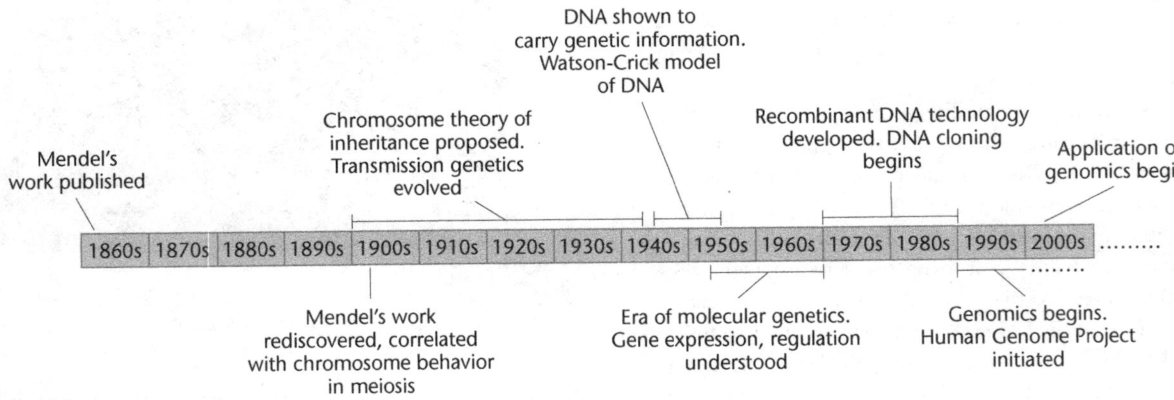


FIGURE 1–23 A timeline showing the development of genetics from Gregor Mendel's work on pea plants to the current era of genomics and its many applications in research, medicine, and society. Having a sense of the history of discovery in genetics should provide you with a useful framework as you proceed through this textbook.

chromosome theory of inheritance. That award was followed by many others, including prizes for the discovery of genetic recombination, the relationship between genes and proteins, the structure of DNA, and the genetic code. In this century, geneticists continue to be recognized for their impact on biology in the current millennium. The 2002 Prize for Medicine or Physiology was awarded to Sydney Brenner, H. Robert Horvitz, and John E. Sulston for their work on the genetic regulation of organ development and programmed cell death. In 2006, prizes went to Andrew Fire and Craig Mello for their discovery that RNA molecules play an important role in regulating gene expression and to Roger Kornberg for his work on the molecular basis of eukaryotic transcription. The 2007 Nobel Prize went to M.R. Capecchi, O. Smithies, and M.J. Evans for the development of gene-targeting technology essential to the creation of knockout mice serving as animal models of human disease.

## Genetics and Society

Just as there has never been a more exciting time to study genetics, the impact of this discipline on society has never been more profound. Genetics and its applications in biotechnology are developing much faster than the social conventions, public policies, and laws required to regulate their use (see this chapter's essay on "Genetics, Technology, and Society"). As a society, we are grappling with a host of sensitive genetics-related issues, including concerns about prenatal testing, insurance coverage, genetic discrimination, ownership of genes, access to and safety of gene therapy, and genetic privacy. By the time you finish this course, you will have seen more than enough evidence to convince you that the present is the Age of Genetics, and you will understand the need to think about and become a participant in the dialogue concerning genetic science and its use.

GENETICS, TECHNOLOGY, AND SOCIETY

### Genetics and Society: The Application and Impact of Science and Technology

One of the special features of this text is the series of essays on **Genetics, Technology, and Society** that you will find at the conclusion of most chapters. These essays explore genetics-related topics that have an impact on the lives of each of us and thus on society in general. Today, genetics touches all aspects of modern life, bringing rapid changes in medicine, agriculture, law, the pharmaceutical industry, and biotechnology. Physicians now use hundreds of genetic tests to diagnose and predict the course of disease and to detect genetic defects *in utero*. DNA-based methods allow scientists to trace the path of evolution taken by many species, including our own. Farmers grow disease-resistant and drought-resistant crops, and raise more productive farm animals, created by gene transfer techniques. DNA profiling methods are applied to paternity testing and murder investigations. Biotechnologies resulting from genomics research have had dramatic effects on industry in general. Meanwhile, the biotechnology industry itself generates over 700,000 jobs and \$50 billion in revenue each year and doubles in size every decade.

Along with these rapidly changing gene-based technologies come a challenging array of ethical dilemmas. Who owns and controls genetic information? Are gene-enhanced agricultural plants and animals safe for humans and the environment? Do we have the right to patent organisms and profit from their commercialization? How can we ensure that genomic technologies will be available to all and not just to the wealthy? What are the likely social consequences of the new reproductive technologies? It is a time when everyone needs to understand genetics in order to make complex personal and societal decisions.

The **Genetics, Technology, and Society** essays explore the interface of society and genetic technology. It is our hope that these essays will act as entry points for your exploration of the myriad applications of modern genetics and their social implications. Below, we list the topics that serve as the basis of many of these essays, followed by the number of the chapter in which each is found. Even should your genetics course not cover certain chapters, we hope that you will find the essays in those chapters to be of interest. Good reading!

## EXPLORING GENOMICS

# Internet Resources for Learning about the Genomes of Model Organisms

Genomics is one of the most rapidly changing disciplines of genetics. New information in this field is accumulating at an astounding rate. Keeping up with current developments in genomics, proteomics, bioinformatics, and other examples of the "omics" era of modern genetics is a challenging task indeed! As a result, geneticists, molecular biologists, and other scientists are relying on online databases to share and compare new information.

The purpose of the "Exploring Genomics" feature, which appears at the end of each chapter, is to introduce you to a range of Internet databases that scientists around the world depend on for sharing, analyzing, organizing, comparing, and storing data from studies in genomics, proteomics, and related fields. We will explore this incredible pool of new information—comprising some of the best publicly available resources in the world—and show you how to use bioinformatics approaches to analyze the sequence and structural data to be found there. Each set of Exploring Genomics exercises will provide a basic introduction to one or more especially relevant or useful databases or programs and then guide you through exercises that use the databases to expand on or reinforce important concepts discussed in the chapter. The exercises are designed to help you learn to navigate the databases, but your explorations need not be limited to these experiences. Part of the fun of learning about genomics is exploring these outstanding databases on your own, so that you can get the latest information on any topic that interests you. Enjoy your explorations!

In this chapter, we discussed the importance of model organisms to both classic and modern experimental approaches in genetics. In our first set of Exploring Genomics exercises, we introduce you to a number of Internet sites that are excellent resources for finding up-to-date information on a wide range of completed and ongoing genomics projects involving model organisms.

### ■ Exercise I – Genome News Network

Since 1995, when scientists unveiled the genome for *Haemophilus influenzae,* making this bacterium the first organism to have its genome sequenced, the sequences for more than 500 organisms have been completed. **Genome News Network** is a site that provides access to basic information about recently completed genome sequences.

1. Visit the Genome News Network at www.genomenewsnetwork.org.
2. Click on the "Quick Guide to Sequenced Genomes" link. Scroll down the page; click on the appropriate links to find information about the genomes for *Anopheles gambiae, Lactococcus lactis,* and *Pan troglodytes;* and answer the following questions for each organism:
   a. Who sequenced this organism's genome, and in what year was it completed?
   b. What is the size of each organism's genome in base pairs?
   c. Approximately how many genes are in each genome?
   d. Briefly describe why geneticists are interested in studying this organism's genome.

### ■ Exercise II – Exploring the Genomes of Model Organisms

A tremendous amount of information is available about the genomes of the many model organisms that have played invaluable roles in advancing our understanding of genetics. Following are links to several sites that are excellent resources for you as you study genetics. Visit the site for your favorite model organism to learn more about its genome!

- Ensembl Genome Browser: www.ensembl.org/index.html. Outstanding site for genome information on many model organisms.
- FlyBase: flybase.bio.indiana.edu. Great database on *Drosophila* genes and genomes.
- Gold™ Genomes OnLine Database: www.genomesonline.org/gold.cgi. Comprehensive access to completed and ongoing genome projects worldwide.
- Model Organisms for Biomedical Research: www.nih.gov/science/models/. National Institutes of Health site with a wealth of resources on model organisms.
- Mouse Genome Informatics: www. informatics.jax.org/. Genetics and genomics of lab mice.
- Rat Genome Project: www.hgsc.bcm.tmc.edu/projects/rat/. Baylor College of Medicine site on the rat genome.
- *Saccharomyces* Genome Database: www.yeastgenome.org/. Database for genetics of *Saccharomyces cerevisiae,* commonly known as baker's yeast.
- Science Functional Genomics: www.sciencemag.org/feature/plus/sfg/. Hosted by the journal *Science,* this is a good resource for information on model organism genomes and other current areas of genomics.
- The *Arabidopsis* Information Resource: www.arabidopsis.org/. Genetic database for the model plant *Arabidopsis thaliana.*
- WormBase: www.wormbase.org. Genome database for the nematode roundworm *Caenorhabditis elegans.*

## Chapter Summary

1. Mendel's work on pea plants established the principles of gene transmission from parents to offspring that are the foundation for the science of genetics.
2. Genes and chromosomes are the fundamental units in the chromosomal theory of inheritance. This theory explains the transmission of genetic information controlling phenotypic traits.
3. Molecular genetics—based on the central dogma that DNA is a template for making RNA, which encodes the linear structure of proteins—explains the phenomena described by Mendelian genetics, also referred to as transmission genetics.
4. Recombinant DNA technology, a far-reaching methodology used in molecular genetics, allows genes from one organism to be spliced into vectors and cloned.
5. Genomics, proteomics, and bioinformatics are new fields derived from recombinant DNA technology. These new fields combine genetics with information technology and allow scientists to explore genome sequences, the structure and function of genes, the protein set within cells, and the evolution of genomes. The Human Genome Project is one example of genomics.
6. Biotechnology has revolutionized agriculture, the pharmaceutical industry, and medicine. It has made possible the mass production of medically important gene products. Genetic testing allows detection of individuals with genetic disorders and those at risk of having affected children, and gene therapy offers hope for the treatment of serious genetic disorders.
7. The study of model organisms in genetics has advanced the understanding of genetic mechanisms and, coupled with recombinant DNA technology, has produced models of human genetic diseases.
8. The effects of genetic technology on society are profound, and the development of policy and legislation is lagging behind the resulting innovations.

## Problems and Discussion Questions

1. Describe Mendel's conclusions about how traits are passed from generation to generation.
2. What is the chromosome theory of inheritance, and how is it related to Mendel's findings?
3. Define genotype and phenotype, and describe how they are related.
4. What are alleles? Is it possible for more than two alleles of a gene to exist?
5. Given the state of knowledge at the time of the Avery, MacLeod, and McCarty experiment, why was it difficult for some scientists to accept that DNA is the carrier of genetic information?
6. Contrast chromosomes and genes.
7. How is genetic information encoded in a DNA molecule?
8. Describe the central dogma of molecular genetics and how it serves as the basis of modern genetics.
9. How many different proteins, each with a unique amino acid sequence, can be constructed with a length of five amino acids?
10. Outline the roles played by restriction enzymes and vectors in cloning DNA.
11. What are some of the impacts of biotechnology on crop plants in the United States?
12. Summarize the arguments for and against patenting genetically modified organisms.
13. We all carry 25,000 to 30,000 genes in our genome. So far, patents have been issued for more than 6000 of these genes. Do you think that companies or individuals should be able to patent human genes? Why or why not?
14. How has the use of model organisms advanced our knowledge of the genes that control human diseases?
15. If you knew that a devastating late-onset inherited disease runs in your family (in other words, a disease that does not appear until later in life) and you could be tested for it at the age of 20, would you want to know whether you are a carrier? Would your answer be likely to change when you reach age 40?
16. The "Age of Genetics" has been brought on by remarkable advances in the applications of biotechnology to manipulate plant and animal genomes. Given that the world population has topped 6 billion and is expected to double in the next 50 years, some scientists have proposed that only the world-wide introduction of genetically modified (GM) foods will make it possible for future nutritional demands to be met. Pest resistance, herbicide, cold, drought, and salinity tolerance, along with increased nutrition are seen as positive attributes of GM foods. However, some caution that unintended harm to other organisms, reduced effectiveness to pesticides, gene transfer to non-target species, allergenicity, and as yet, unknown effects to human health are potential concerns regarding GM foods. If you were in a position to control the introduction of a GM primary food product (rice, for example), what criteria would you establish before allowing such introduction?
17. The BIO (Biotechnology Industry Organization) meeting held in Philadelphia (June, 2005) brought together world-wide leaders from the biotechnology and pharmaceutical industries. Concurrently, BioDemocracy 2005, a group composed of people seeking to highlight hazards from widespread applications of biotechnology, met in Philadelphia. The benefits of biotechnology are outlined in your text. Predict some of the risks that were no doubt discussed at the BioDemocracy meeting.

Chromosomes in the prometaphase stage of mitosis, derived from a cell in the flower of *Haemanthus*.

2

# Mitosis and Meiosis

CHAPTER CONCEPTS

- Genetic continuity between generations of cells and between generations of sexually reproducing organisms is maintained through the processes of mitosis and meiosis, respectively.
- Diploid eukaryotic cells contain their genetic information in pairs of homologous chromosomes, with one member of each pair being derived from the maternal parent and one from the paternal parent.
- Mitosis provides a mechanism by which chromosomes, having been duplicated, are distributed into progeny cells during cell reproduction.
- Mitosis converts a diploid cell into two diploid daughter cells.
- The process of meiosis distributes one member of each homologous pair of chromosomes into each gamete or spore, thus reducing the diploid chromosome number to the haploid chromosome number.
- Meiosis generates genetic variability by distributing various combinations of maternal and paternal members of each homologous pair of chromosomes into gametes or spores.
- During the stages of mitosis and meiosis, the genetic material is condensed into discrete structures called chromosomes.

Every living thing contains a substance described as the genetic material. Except in certain viruses, this material is composed of the nucleic acid, DNA. DNA has an underlying linear structure possessing segments called genes, the products of which direct the metabolic activities of cells. An organism's DNA, with its arrays of genes, is organized into structures called **chromosomes**, which serve as vehicles for transmitting genetic information. The manner in which chromosomes are transmitted from one generation of cells to the next and from organisms to their descendants must be exceedingly precise. In this chapter we consider exactly how genetic continuity is maintained between cells and organisms.

Two major processes are involved in the genetic continuity of nucleated cells: **mitosis** and **meiosis**. Although the mechanisms of the two processes are similar in many ways, the outcomes are quite different. Mitosis leads to the production of two cells, each with the same number of chromosomes as the parent cell. In contrast, meiosis reduces the genetic content and the number of chromosomes by precisely half. This reduction is essential if sexual reproduction is to occur without doubling the amount of genetic material in each new generation. Strictly speaking, mitosis is that portion of the cell cycle during which the hereditary components are equally partitioned into daughter cells. Meiosis is part of a special type of cell division that leads to the production of sex cells: **gametes** or **spores**. This process is an essential step in the transmission of genetic information from an organism to its offspring.

Normally, chromosomes are visible only during mitosis and meiosis. When cells are not undergoing division, the genetic material making up chromosomes unfolds and uncoils into a diffuse network within the nucleus, generally referred to as **chromatin**. Before describing mitosis and meiosis, we will briefly review the structure of cells, emphasizing components that are of particular significance to genetic function. We will also compare the structural differences between the prokaryotic (nonnucleated) cells of bacteria and the eukaryotic cells of higher organisms. We then devote the remainder of the chapter to the behavior of chromosomes during cell division.

## 2.1 Cell Structure Is Closely Tied to Genetic Function

Before 1940, our knowledge of cell structure was limited to what we could see with the light microscope. Around 1940, the transmission electron microscope was in its early stages of development, and by 1950, many details of cell ultrastructure had emerged. Under the electron microscope, cells were seen as highly varied, highly organized structures whose form and function are dependent on specific genetic expression by each cell type. A new world of whorled membranes, organelles, microtubules, granules, and filaments was revealed. These discoveries revolutionized thinking in the entire field of biology. Many cell components, such as the nucleolus, ribosome, and centriole, are involved directly or indirectly with genetic processes. Other components—the mitochondria and chloroplasts—contain their own unique genetic information. Here, we will focus primarily on those aspects of cell structure that relate to genetic study. The generalized animal cell shown in Figure 2–1 illustrates most of the structures we will discuss.

All cells are surrounded by a **plasma membrane**, an outer covering that defines the cell boundary and delimits the cell from its immediate external environment. This membrane is not passive but instead actively controls the movement of materials into and out of the cell. In addition to this membrane, plant cells have an outer covering called the **cell wall** whose major component is a polysaccharide called **cellulose**.

Many, if not most, animal cells have a covering over the plasma membrane, referred to as the **cell coat**. Consisting of glycoproteins and polysaccharides, the cell coat has a chemical composition that differs from comparable structures in either plants or bacteria. The cell coat, among other functions, provides biochemical identity at the surface of cells, and the components of the coat that establish cellular identity are under genetic control. For example, various cell-identity markers that you may have heard of—the **AB**, **Rh**, and **MN antigens**—are found on the surface of red blood cells. On other cell surfaces, **histocompatibility antigens**, which elicit an immune response during tissue and organ transplants, are present. Various **receptor molecules** are also found on the surfaces of cells. These molecules act as recognition sites that transfer specific chemical signals across the cell membrane into the cell.

Living organisms are categorized into two major groups depending on whether or not their cells contain a nucleus. The presence of a nucleus and other membranous organelles is the defining characteristic of **eukaryotic organisms**. The **nucleus** in eukaryotic cells is a membrane-bound structure that houses the genetic material, DNA, which is complexed with an array of acidic and basic proteins into thin fibers. During nondivisional phases of the cell cycle, the fibers are uncoiled and dispersed into chromatin (as mentioned above). During mitosis and meiosis, chromatin fibers coil and condense into chromosomes. Also present in the nucleus is the **nucleolus**, an amorphous component where ribosomal RNA (rRNA) is synthesized and where the initial stages of ribosomal assembly occur. The portions of DNA that encode rRNA are collectively referred to as the **nucleolus organizer region**, or the **NOR**.

**Prokaryotic organisms** lack a nuclear envelope and membranous organelles. Most prokaryotes are bacteria. For example, in *Escherichia coli*, the genetic material is present as a long, circular DNA molecule that is compacted into an unenclosed region called the **nucleoid.** Part of the DNA may be attached to the cell membrane, but in general the nucleoid extends through a large part of

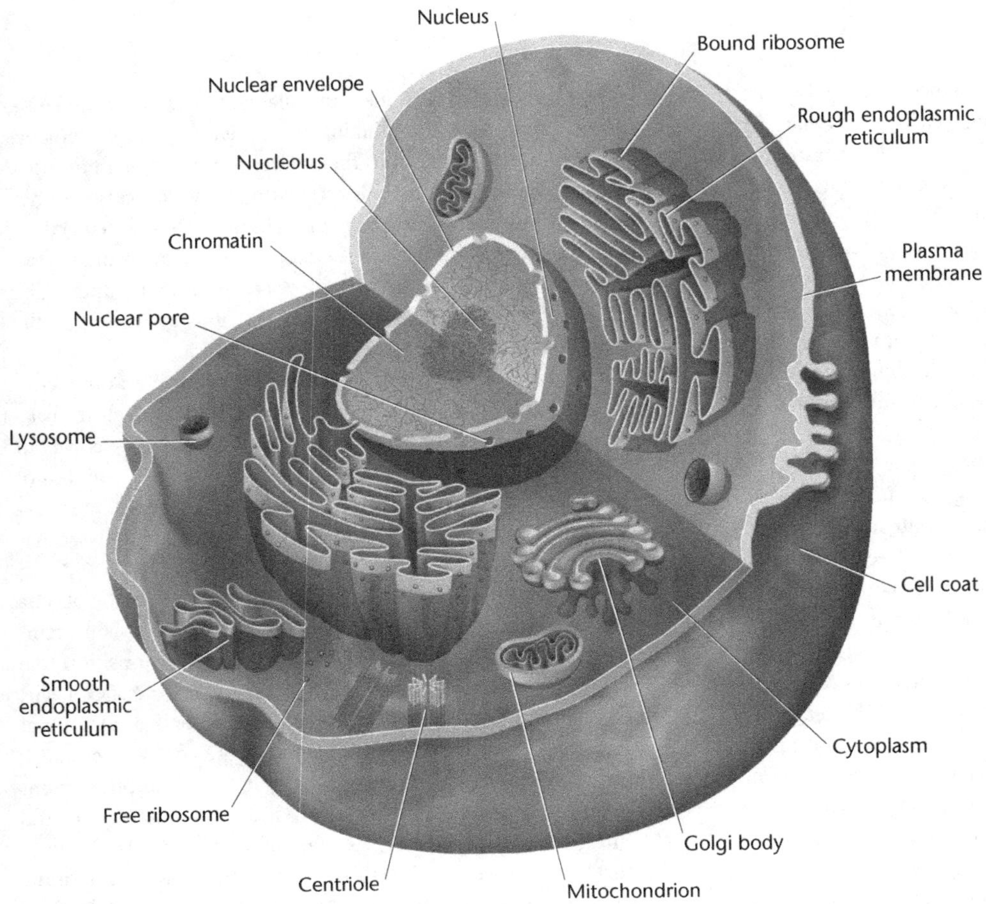


FIGURE 2–1 A generalized animal cell. The cellular components discussed in the text are emphasized here.

the cell. Although the DNA is compacted, it does not undergo the extensive coiling characteristic of the stages of mitosis, during which the chromosomes of eukaryotes become visible. Nor is the DNA in prokaryotes associated as extensively with proteins as is eukaryotic DNA. Figure 2–2, which shows two bacteria forming by cell division, illustrates the nucleoid regions where the bacterial chromosomes collect. Prokaryotic cells do not have a distinct nucleolus but do contain genes that specify rRNA molecules.

The remainder of the eukaryotic cell within the plasma membrane, excluding the nucleus, is referred to as **cytoplasm**, and includes a variety of extranuclear cellular organelles. In the cytoplasm, a nonparticulate, colloidal material referred to as the **cytosol** surrounds and encompasses the cellular organelles. The cytoplasm also includes an extensive system of tubules and filaments, comprising the cytoskeleton, which provides a lattice of support structures within the cell. Consisting primarily of **microtubules** made of the protein **tubulin** and **microfilaments** made of the protein **actin**, this structural framework maintains cell shape, facilitates cell mobility, and anchors the various organelles.

One organelle, the membranous **endoplasmic reticulum (ER)**, compartmentalizes the cytoplasm, greatly increasing the surface area available for biochemical synthesis. The ER appears smooth in places where it serves as the site for synthesizing fatty acids and phospholipids; in other places, it appears rough because it is studded with ribosomes. **Ribosomes** serve as sites where genetic information contained in messenger RNA (mRNA) is translated into proteins.

Three other cytoplasmic structures are very important in the eukaryotic cell's activities: mitochondria, chloroplasts, and centrioles. **Mitochondria** are found in most eukaryotes, including both animal and plant cells and are the sites of the oxidative phases of cell respiration. These chemical reactions generate large amounts of the energy-rich molecule adenosine triphosphate (ATP). **Chloroplasts**, which are found in plants, algae, and some protozoans, are associated with photosynthesis, the major energy-trapping process on Earth. Both mitochondria and chloroplasts contain DNA in a form distinct from that found in the nucleus. They are able to duplicate themselves and transcribe and translate their own genetic information. It is interesting to note that the genetic machinery of mitochondria and chloroplasts closely resembles that of prokaryotic cells. This and other observations have led to the proposal that these organelles were once primitive freeliving organisms that established symbiotic relationships with primitive eukaryotic cells. This theory concerning the evolutionary origin of these organelles is called the **endosymbiont hypothesis**.

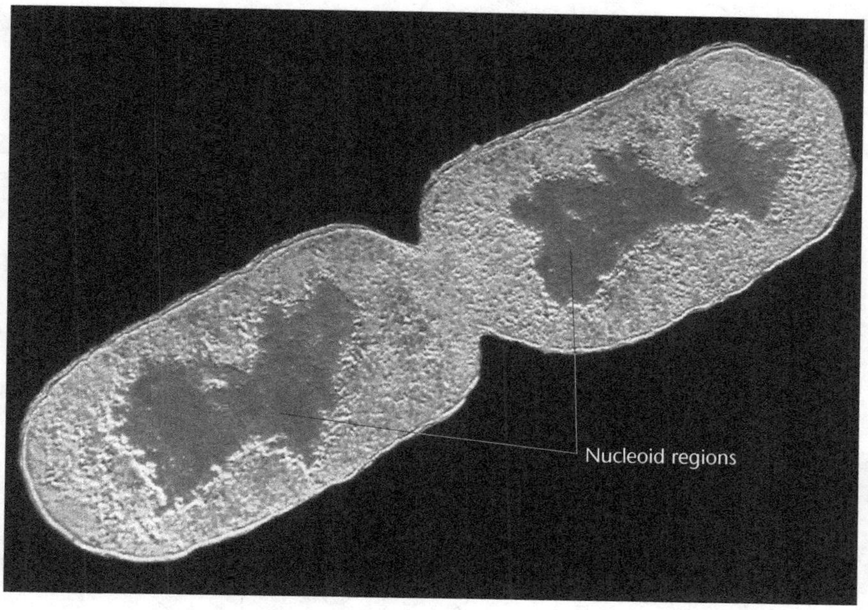


FIGURE 2–2 Color-enhanced electron micrograph of *E. coli* undergoing cell division. Particularly prominent are the two chromosomal areas (shown in red), called nucleoids, that have been partitioned into the daughter cells.

Animal cells and some plant cells also contain a pair of complex structures called **centrioles**. These cytoplasmic bodies, located in a specialized region called the centrosome, are associated with the organization of spindle fibers that function in mitosis and meiosis. In some organisms, the centriole is derived from another structure, the basal body, which is associated with the formation of cilia and flagella (hair-like and whip-like structures for propelling cells or moving materials). Over the years, many reports have suggested that centrioles and basal bodies contain DNA, which could be involved in the replication of these structures. This idea is still being investigated.

The organization of **spindle fibers** by the centrioles occurs during the early phases of mitosis and meiosis. These fibers play an important role in the movement of chromosomes as they separate during cell division. They are composed of arrays of microtubules consisting of polymers of the protein tubulin.

## 2.2 Chromosomes Exist in Homologous Pairs in Diploid Organisms

As we discuss the processes of mitosis and meiosis, it is important that you understand the concept of homologous chromosomes. Such an understanding will also be of critical importance in our future discussions of Mendelian genetics. Chromosomes are most easily visualized during mitosis. When they are examined carefully, distinctive lengths and shapes are apparent. Each chromosome contains a constricted region called the **centromere,** whose location establishes the general appearance of each chromosome. Figure 2–3 shows chromosomes with centromere placements at different distances along their length. Extending from either side of the centromere are the arms of the chromosome. Depending on the position of the centromere, different arm ratios are produced. As Figure 2–3 illustrates, chromosomes are classified as **metacentric**, **submetacentric**, **acrocentric**, or **telocentric** on the basis of the centromere location. The shorter arm, by convention, is shown above the centromere and is called the **p arm** (p, for "petite"). The longer arm is shown below the centromere and is called the **q arm** (because q is the next letter in the alphabet).

In the study of mitosis, several other observations are of particular relevance. First, all somatic cells derived from members of the same species contain an identical number of chromosomes. In most cases, this represents the **diploid number (2*n*)**, whose meaning will become clearer below. When the lengths and centromere placements of all such chromosomes are examined, a second general feature is apparent. Nearly all chromosomes exist in pairs with regard to these two properties, and the members of each pair are called **homologous chromosomes.** So, for each chromosome exhibiting a specific length and centromere placement, another exists with identical features.

There are exceptions to this rule. Many bacteria and viruses have but one chromosome, and organisms such as yeasts and molds, and certain plants such as bryophytes (mosses), spend the predominant phase of their life cycle in the haploid stage. That is, they contain only one member of each homologous pair of chromosomes during most of their lives.

Figure 2–4 illustrates the physical appearance of different pairs of homologous chromosomes. There, the human mitotic chromosomes have been photographed, cut out of the print, and matched up, creating a display called a **karyotype.** As you can see, humans have a 2*n* number of 46 chromosomes, which on close examination exhibit a diversity of sizes and centromere placements. Note also that each of the 46 chromosomes in this karyotype is clearly a double structure consisting of two parallel **sister chromatids** connected by a common

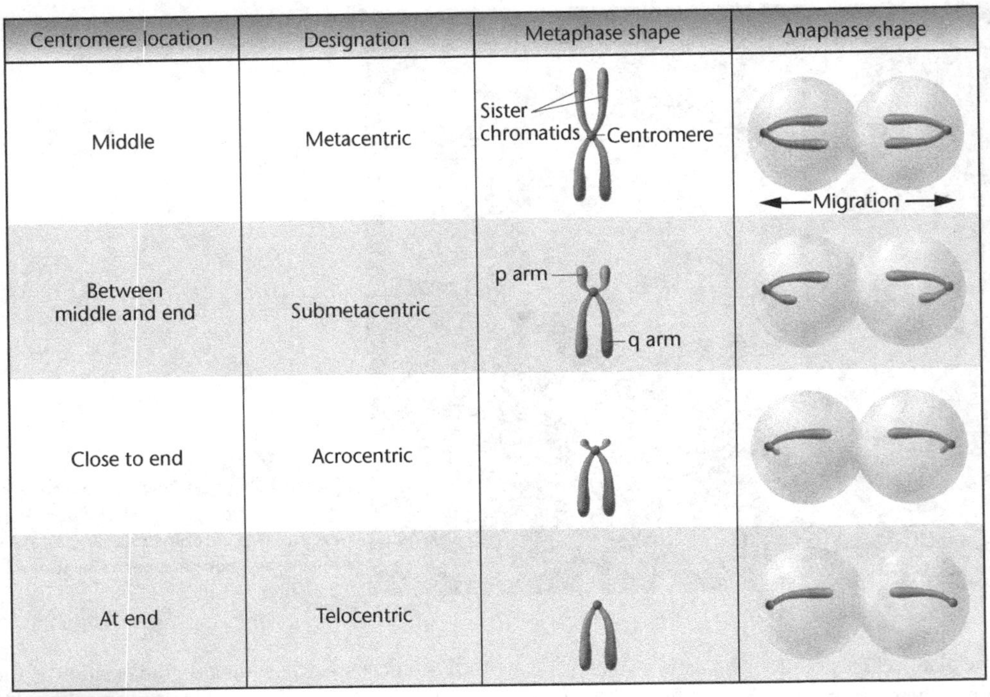


**FIGURE 2-3** Centromere locations and the chromosome designations that are based on them. Note that the shape of the chromosome during anaphase is determined by the position of the centromere during metaphase.

centromere. Had these chromosomes been allowed to continue dividing, the sister chromatids, which are replicas of one another, would have separated into the two new cells as division continued.

The haploid number ($n$) of chromosomes is equal to one-half the diploid number. Collectively, the genetic information contained in a haploid set of chromosomes constitutes the **genome** of the species. This, of course, includes copies of all genes as well as a large amount of noncoding DNA. The examples listed in Table 2.1 demonstrate the wide range of $n$ values found in plants and animals.

Homologous chromosomes have important genetic similarities. They contain identical gene sites along their lengths, each site called a **locus** (pl. loci). Thus, they are identical in the traits that they

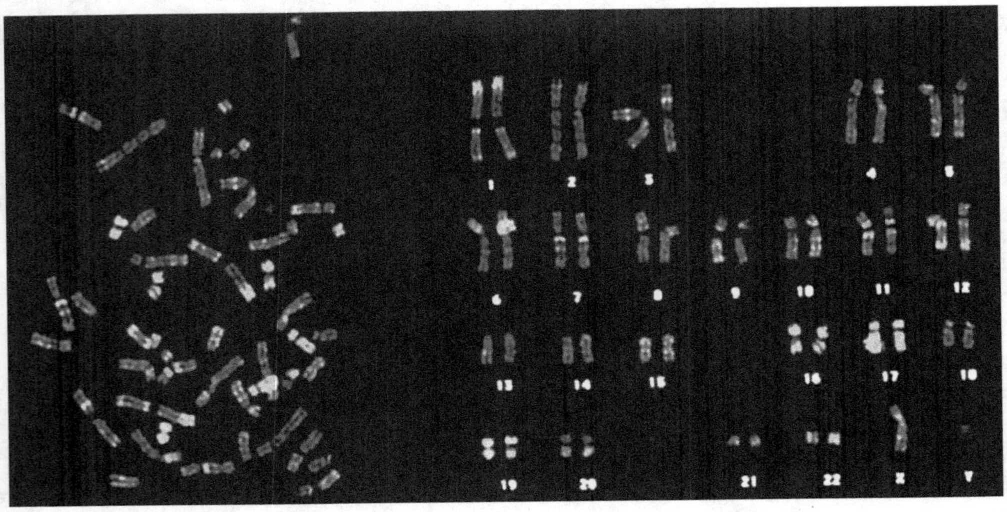

**FIGURE 2-4** A metaphase preparation of chromosomes derived from a dividing cell of a human male (left), and the karyotype derived from the metaphase preparation (right). All but the X and Y chromosomes are present in homologous pairs. Each chromosome is clearly a double structure consisting of a pair of sister chromatids joined by a common centromere.

TABLE 2.1

The Haploid Number of Chromosomes for a Variety of Organisms

| Common Name | Scientific Name | Haploid Number |
|---|---|---|
| Black bread mold | *Aspergillus nidulans* | 8 |
| Broad bean | *Vicia faba* | 6 |
| Cat | *Felis domesticus* | 19 |
| Cattle | *Bos Taurus* | 30 |
| Chicken | *Gallus domesticus* | 39 |
| Chimpanzee | *Pan troglodytes* | 24 |
| Corn | *Zea mays* | 10 |
| Cotton | *Gossypium hirsutum* | 26 |
| Dog | *Canis familiaris* | 39 |
| Evening primrose | *Oenothera biennis* | 7 |
| Frog | *Rana pipiens* | 13 |
| Fruit fly | *Drosophila melanogaster* | 4 |
| Garden onion | *Allium cepa* | 8 |
| Garden pea | *Pisum sativum* | 7 |
| Grasshopper | *Melanoplus differentialis* | 12 |
| Green alga | *Chlamydomonas reinhardii* | 18 |
| Horse | *Equus caballus* | 32 |
| House fly | *Musca domestica* | 6 |
| House mouse | *Mus musculus* | 20 |
| Human | *Homo sapiens* | 23 |
| Jimson weed | *Datura stramonium* | 12 |
| Mosquito | *Culex pipiens* | 3 |
| Mustard plant | *Arabidopsis thaliana* | 5 |
| Pink bread mold | *Neurospora crassa* | 7 |
| Potato | *Solanum tuberosum* | 24 |
| Rhesus monkey | *Macaca mulatta* | 21 |
| Roundworm | *Caenorhabditis elegans* | 6 |
| Silkworm | *Bombyx mori* | 28 |
| Slime mold | *Dictyostelium discoideum* | 7 |
| Snapdragon | *Antirrhinum majus* | 8 |
| Tobacco | *Nicotiana tabacum* | 24 |
| Tomato | *Lycopersicon esculentum* | 12 |
| Water fly | *Nymphaea alba* | 80 |
| Wheat | *Triticum aestivum* | 21 |
| Yeast | *Saccharomyces cerevisiae* | 16 |
| Zebrafish | *Danio rerio* | 25 |

influence and their genetic potential. In sexually reproducing organisms, one member of each pair is derived from the maternal parent (through the ovum) and one is derived from the paternal parent (through the sperm). Therefore, each diploid organism contains two copies of each gene as a consequence of **biparental inheritance**, inheritance from two parents. As we shall see in the chapters on transmission genetics, the members of each pair of genes, while influencing the same characteristic or trait, need not be identical. In a population of members of the same species, many different alternative forms of the same gene, called **alleles**, can exist.

The concepts of haploid number, diploid number, and homologous chromosomes are important for understanding the process of meiosis. During the formation of gametes or spores, meiosis converts the diploid number of chromosomes to the haploid number. As a result, haploid gametes or spores contain precisely one member of each homologous pair of chromosomes—that is, one complete haploid set. Following fusion of two gametes at fertilization, the diploid number is reestablished; that is, the zygote contains two complete haploid sets of chromosomes. The constancy of genetic material is thus maintained from generation to generation.

There is one important exception to the concept of homologous pairs of chromosomes. In many species, one pair, consisting of the **sex-determining chromosomes**, is often not homologous in size, centromere placement, arm ratio, or genetic content. For example, in humans, while females carry two homologous X chromosomes, males carry one Y chromosome in addition to one X chromosome (Figure 2–4). These X and Y chromosomes are not strictly homologous. The Y is considerably smaller and lacks most of the gene sites contained on the X. Nevertheless, they contain homologous regions and behave as homologs in meiosis so that gametes produced by males receive either one X or one Y chromosome.

## 2.3 Mitosis Partitions Chromosomes into Dividing Cells

The process of mitosis is critical to all eukaryotic organisms. In some single-celled organisms, such as protozoans and some fungi and algae, mitosis (as a part of cell division) provides the basis for asexual reproduction. Multicellular diploid organisms begin life as single-celled fertilized eggs called **zygotes**. The mitotic activity of the zygote and the subsequent daughter cells is the foundation for the development and growth of the organism. In adult organisms, mitotic activity is the basis for wound healing and other forms of cell replacement in certain tissues. For example, the epidermal cells of the skin and the intestinal lining of humans are continuously sloughed off and replaced. Cell division also results in the continuous production of reticulocytes that eventually shed their nuclei and replenish the supply of red blood cells in vertebrates. In abnormal situations, somatic cells may lose control of cell division, and form a tumor.

The genetic material is partitioned into daughter cells during nuclear division, or **karyokinesis**. This process is quite complex and requires great precision. The chromosomes must first be exactly replicated and then accurately partitioned. The end result is the production of two daughter nuclei, each with a chromosome composition identical to that of the parent cell.

Karyokinesis is followed by cytoplasmic division, or **cytokinesis**. This less complex process requires a mechanism that partitions the volume into two parts, then encloses each new cell in a distinct plasma membrane. As the cytoplasm is reconstituted, organelles either replicate themselves, arise from existing membrane structures, or are synthesized *de novo* (anew) in each cell.

Following cell division, the initial size of each new daughter cell is approximately one-half the size of the parent cell. However, the nucleus of each new cell is not appreciably smaller than the nucleus of the original cell. Quantitative measurements of DNA confirm that there is an amount of genetic material in the daughter nuclei equivalent to that in the parent cell.

## Interphase and the Cell Cycle

Many cells undergo a continuous alternation between division and nondivision. The events that occur from the completion of one division until the completion of the next division constitute the **cell cycle** (Figure 2–5). We will consider **interphase**, the initial stage of the cell cycle, as the interval between divisions. It was once thought that the biochemical activity during interphase was devoted solely to the cell's growth and its normal function. However, we now know that another biochemical step critical to the ensuing mitosis occurs during interphase: *the replication of the DNA of each chromosome.* This period, during which DNA is synthesized, occurs before the cell enters mitosis and is called the **S phase**. The initiation and completion of synthesis can be detected by monitoring the incorporation of radioactive precursors into DNA.

Investigations of this nature demonstrate two periods during interphase when no DNA synthesis occurs, one before and one after the S phase. These are designated **G1 (gap I)** and **G2 (gap II)**, respectively. During both of these intervals, as well as during S, intensive metabolic activity, cell growth, and cell differentiation are evident. By the end of G2, the volume of the cell has roughly doubled, DNA has been replicated, and mitosis (M) is initiated. Following mitosis, continuously dividing cells then repeat this cycle (G1, S, G2, M) over and over, as shown in Figure 2–5.

Much is known about the cell cycle based on *in vitro* (literally, "in glass") studies. When grown in culture, many cell types in different organisms traverse the complete cycle in about 16 hours. The actual process of mitosis occupies only a small part of the overall cycle, often less than an hour. The lengths of the S and G2 phases of interphase are fairly consistent in different cell types. Most variation is seen in the length of time spent in the G1 stage. Figure 2–6 shows the relative length of these intervals in a human cell in culture.

G1 is of great interest in the study of cell proliferation and its control. At a point late in G1, all cells follow one of two paths. They either withdraw from the cycle, become quiescent, and enter the **G0 stage** (see Figure 2–5), or they become committed to initiating DNA synthesis and completing the cycle. Cells that enter G0 remain viable and metabolically active but are not proliferative. Cancer cells apparently avoid entering G0 or pass through it very quickly. Other cells enter G0 and never reenter the cell cycle. Still other cells in G0 can be stimulated to return to G1 and thereby reenter the cell cycle.

Cytologically, interphase is characterized by the absence of visible chromosomes. Instead, the nucleus is filled with chromatin fibers that are formed as the chromosomes uncoil and disperse after the previous mitosis [Figure 2–7(a)]. Once G1, S, and G2 are completed, mitosis is initiated. Mitosis is a dynamic period of vigorous and continual activity. For discussion purposes, the entire process is subdivided into discrete stages, and specific events are assigned to each one. These stages, in order of occurrence, are prophase, prometaphase, metaphase, anaphase, and telophase. They are diagrammed wit corresponding photomicrographs in Figure 2–7.

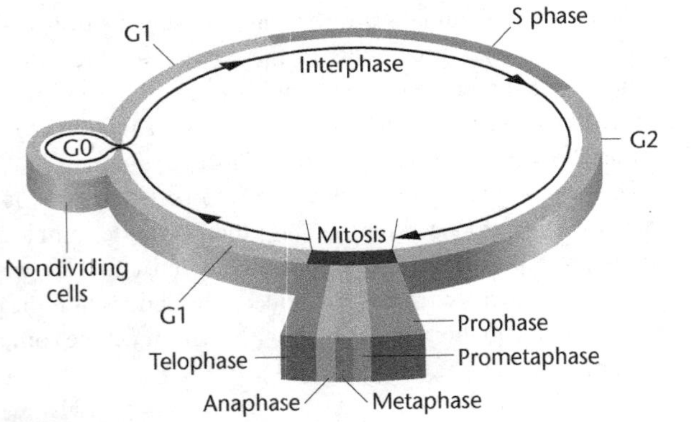


FIGURE 2–5 The stages comprising an arbitrary cell cycle. Following mitosis, cells enter the G1 stage of interphase, initiating a new cycle. Cells may become nondividing (G0) or continue through G1, where they become committed to begin DNA synthesis (S) and complete the cycle (G2 and mitosis). Following mitosis, two daughter cells are produced, and the cycle begins anew for both of them.

## Prophase

Often, over half of mitosis is spent in prophase [Figure 2–7(b)], a stage characterized by several significant occurrences. One of the early events in prophase of all animal cells is the migration of two pairs of centrioles to opposite ends of the cell. These structures are found just outside the nuclear envelope in an area of differentiated cytoplasm called the centrosome (introduced in Section 2.1). It is believed that each pair of centrioles consists of one mature unit and a smaller, newly formed centriole.

The centrioles migrate to establish poles at opposite ends of the cell. After migrating, the centrioles are responsible for organizing cytoplasmic microtubules into the spindle fibers that run between these poles, creating an axis along which chromosomal separation occurs. Interestingly, the cells of most plants (there are a few exceptions),

| Interphase | | | Mitosis |
|---|---|---|---|
| G1 | S | G2 | M |
| 5 | 7 | 3 | 1 |

Hours

| Pro | Met | Ana | Tel |
|---|---|---|---|
| 36 | 3 | 3 | 18 |

Minutes

FIGURE 2–6 The time spent in each interval of one complete cell cycle of a human cell in culture. Times vary according to cell types and conditions.

fungi, and certain algae seem to lack centrioles. Spindle fibers are nevertheless apparent during mitosis. Therefore, centrioles are not universally responsible for the organization of spindle fibers.

As the centrioles migrate, the nuclear envelope begins to break down and gradually disappears. In a similar fashion, the nucleolus disintegrates within the nucleus. While these events are taking place, the diffuse chromatin fibers begin to condense, until distinct thread-like structures, the chromosomes, become visible. It becomes apparent near the end of prophase that each chromosome is actually a double structure split longitudinally except at a single point of constriction, the centromere. The two parts of each chromosome are called chromatids. Because the DNA contained in each pair of chromatids represents the duplication of a single chromosome, the two chromatids are genetically identical. This is why they are called sister chromatids. Sister chromatids are held together by a protein called **cohesin.** It is originally produced during the S phase of the cell cycle when the DNA of each chromosome is replicated. Thus, even though we cannot see chromatids in interphase because the chromatin is uncoiled and dispersed in the nucleus, the chromosomes are already double structures (although it doesn't become apparent until late prophase). In humans, with a diploid number of 46, a cytological preparation of late prophase reveals 46 chromosomes randomly distributed in the area formerly occupied by the nucleus.

## Prometaphase and Metaphase

The distinguishing event of the two ensuing stages is the migration of every chromosome, led by its centromeric region, to the equatorial plane. The equatorial plane, also referred to as the metaphase plate, is the midline region of the cell, a plane that lies perpendicular to the axis established by the spindle fibers. In some descriptions, the term **prometaphase** refers to the period of chromosome movement [Figure 2-7(c)], and **metaphase** is applied strictly to the chromosome configuration following migration.

Migration is made possible by the binding of spindle fibers to the chromosome's **kinetochore,** an assembly of multilayered plates of proteins associated with the centromere. This structure forms on opposite sides of each centromere, in intimate association with the two sister chromatids. Once attached to the spindle fibers, the sister chromatids are ready to be pulled to opposite poles during the ensuing anaphase stage.

We know a great deal about spindle fibers. They consist of microtubules, which themselves consist of molecular subunits of the protein tubulin (we noted earlier that tubulin-derived microtubules also make up part of the cytoskeleton). Microtubules seem to originate and "grow" out of the two centrosome regions (which contain the centrioles) at opposite poles of the cell. They are dynamic structures that lengthen and shorten as a result of the addition or loss of polarized tubulin subunits. The microtubules most directly responsible for chromosome migration make contact with, and adhere to, kinetochores as they grow from the centrosome region. They are referred to as **kinetochore microtubules** and have one end near the centrosome region (at one of the poles of the cell) and the other anchored to the kinetochore. The number of microtubules that bind to the kinetochore varies greatly between organisms. Yeast (*Saccharomyces*) have only a single microtubule bound to each platelike structure of the kinetochore. Mitotic cells of mammals, at the other extreme, reveal 30 to 40 microtubules bound to each portion of the kinetochore.

At the completion of metaphase, each centromere is aligned at the metaphase plate with the chromosome arms extending outward in a random array. This configuration is shown in Figure 2–7(d).

## Anaphase

Events critical to chromosome distribution during mitosis occur during **anaphase**, the shortest stage of mitosis. During this phase, sister chromatids of each chromosome *disjoin* (separate) from each other—an event described as **disjunction**—and migrate to opposite ends of the cell. For complete disjunction to occur, each centromeric region must split in two. This splitting signals the initiation of anaphase. Once it occurs, each chromatid is referred to as a **daughter chromosome.**

Movement of daughter chromosomes to the opposite poles of the cell is dependent on the centromere–spindle fiber attachment. Recent investigations reveal that chromosome migration results from the activity of a series of specific molecules called motor proteins found at several locations within the dividing cell. These proteins, described as **molecular motors,** use the energy generated by the hydrolysis of ATP. Their effect on the activity of microtubules serves ultimately to shorten the spindle fibers, drawing the chromosomes to opposite ends of the cell. The centromeres of each chromosome *appear* to lead the way during migration, with the chromosome arms trailing behind. Several models have been proposed to account for the shortening of spindle fibers. They share in common the selective removal of tubulin subunits at the ends of the spindle fibers. The removal process is accomplished by the molecular motor proteins described above.

The location of the centromere determines the shape of the chromosome during separation, as you saw in Figure 2–3. The steps that occur during anaphase are critical in providing each subsequent daughter cell with an identical set of chromosomes. In human cells, there would now be 46 chromosomes at each pole, one from each original sister pair. Figure 2–7(e) shows anaphase prior to its completion.

**NOW SOLVE THIS**

With the initial appearance of the feature we call "Now Solve This," a short introduction is in order. The feature occurs several times in this and all ensuing chapters, each time introducing a problem from the "Problems and Discussion Questions" at the end of the chapter. In every case, the problem is related to the discussion just presented. A comment is made about the nature of the problem, and then a Hint is offered that may help you decide how to solve the problem. Here is the first one.

Problem 5 on page 40 involves an understanding of what happens to each pair of homologous chromosomes during mitosis.

■ HINT: *The key to solving this problem is to understand that throughout mitosis, the members of each homologous pair do not pair up, but instead behave independently.*

MITOSIS AND THE CELL CYCLE
WEB TUTORIAL 2.1

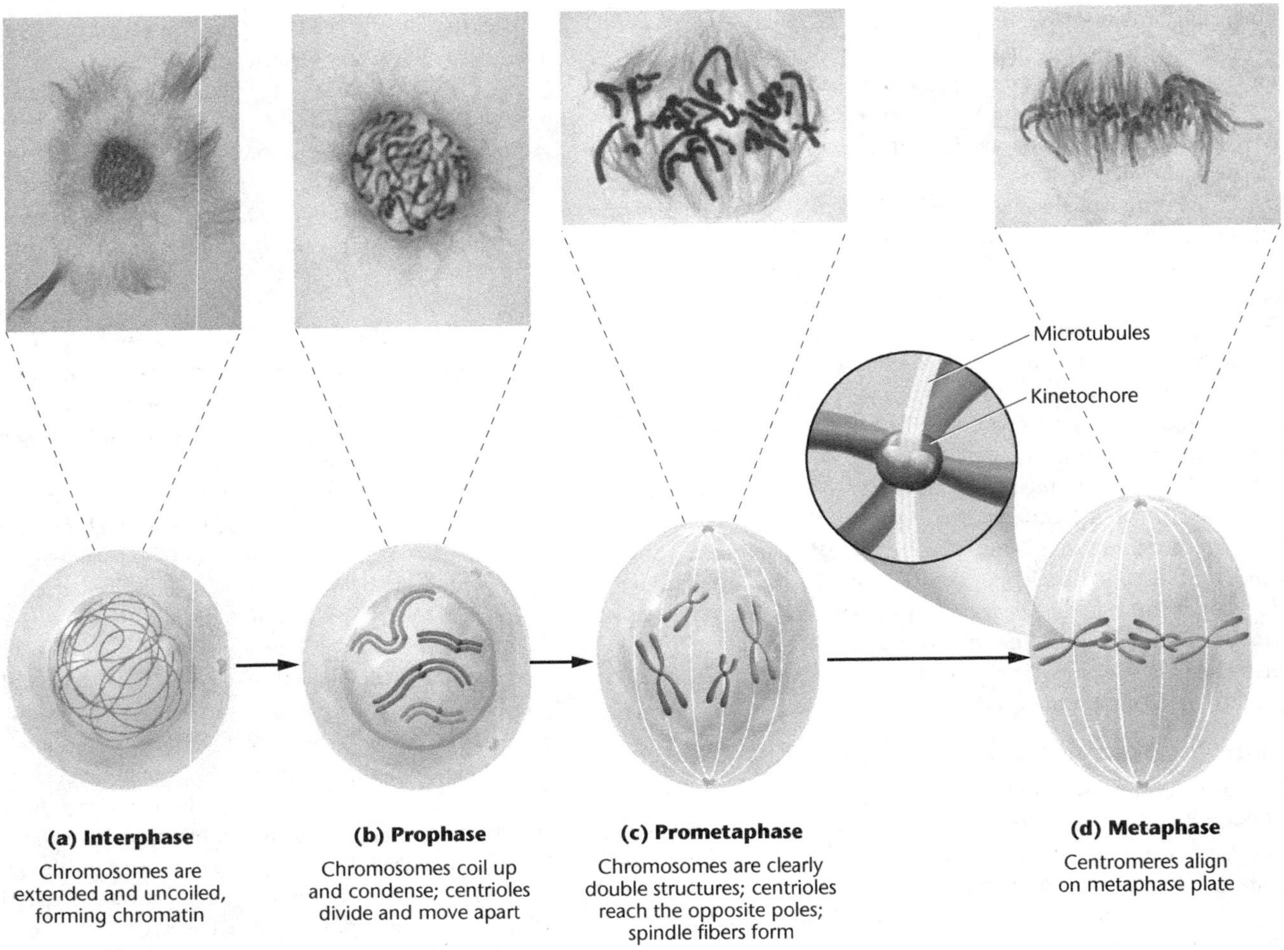


FIGURE 2–7 Drawings depicting mitosis in an animal cell with a diploid number of 4. The events occurring in each stage are described in the text. Of the two homologous pairs of chromosomes, one pair consists of longer, metacentric members and the other of shorter, submetacentric members. The maternal chromosome and the paternal chromosome of each pair are shown in different colors. In (f), a drawing of late telophase in a plant cell shows the formation of the cell plate and lack of centrioles. The cells shown in the light micrographs came from the flower of *Haemanthus*, a plant that has a diploid number of 8.

## Telophase

**Telophase** is the final stage of mitosis and is depicted in Figure 2–7(f). At its beginning, two complete sets of chromosomes are present, one set at each pole. The most significant event of this stage is cytokinesis, the division or partitioning of the cytoplasm. Cytokinesis is essential if two new cells are to be produced from one cell. The mechanism of cytokinesis differs greatly in plant and animal cells, but the end result is the same: two new cells are produced. In plant cells, a **cell plate** is synthesized and laid down across the region of the metaphase plate. Animal cells, however, undergo a constriction of the cytoplasm, in much the way that a loop of string might be tightened around the middle of a balloon.

It is not surprising that the process of cytokinesis varies in different organisms. Plant cells, which are more regularly shaped and structurally rigid, require a mechanism for depositing new cell wall material around the plasma membrane. The cell plate laid down during telophase becomes a structure called the **middle lamella.** Subsequently, the primary and secondary layers of the cell wall are deposited between the cell membrane and middle lamella in each of the resulting daughter cells. In animals, complete constriction of the cell membrane produces the **cell furrow** characteristic of newly divided cells.

Other events necessary for the transition from mitosis to interphase are initiated during late telophase. They generally constitute a reversal of events that occurred during prophase. In each new cell, the chromosomes begin to uncoil and become diffuse chromatin once again, while the nuclear envelope reforms around them, the spindle fibers disappear, and the nucleolus gradually reforms and becomes visible in the nucleus during early interphase. At the completion of telophase, the cell enters interphase.

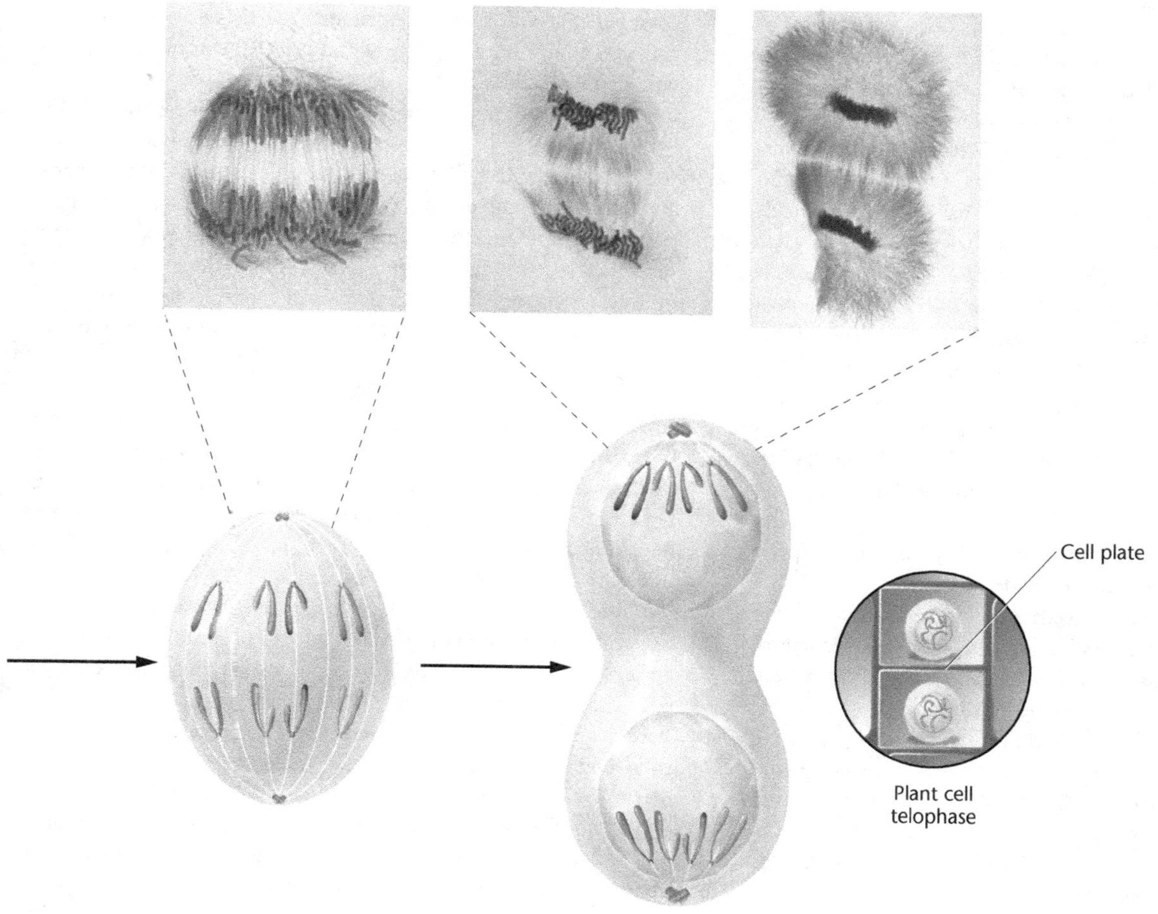


**(e) Anaphase**
Centromeres split and daughter chromosomes migrate to opposite poles

**(f) Telophase**
Daughter chromosomes arrive at the poles; cytokinesis commences

FIGURE 2-7 *(Continued)*

## Cell-Cycle Regulation and Checkpoints

The cell cycle, culminating in mitosis, is fundamentally the same in all eukaryotic organisms. This similarity in many diverse organisms suggests that the cell cycle is governed by a genetically regulated program that has been conserved throughout evolution. Because disruption of this regulation may underlie the uncontrolled cell division characterizing malignancy, interest in how genes regulate the cell cycle is particularly strong.

A mammoth research effort over the past 15 years has paid high dividends, and we now have knowledge of many genes involved in the control of the cell cycle. This work was recognized by the awarding of the 2001 Nobel Prize in Medicine or Physiology to Lee Hartwell, Paul Nurse, and Tim Hunt. As with other studies of genetic control over essential biological processes, investigation has focussed on the discovery of mutations that interrupt the cell cycle and on the effects of those mutations. As we shall return to this subject in Chapter 20, during our consideration of cancer, what follows is a very brief overview.

Many mutations are now known that exert an effect at one or another stage of the cell cycle. First discovered in yeast, but now evident in all organisms, including humans, such mutations were originally designated as ***cell division cycle (cdc)* mutations.** The normal products of many of the mutated genes are enzymes called **kinases** that can add phosphates to other proteins. They serve as "master control" molecules functioning in conjunction with proteins called **cyclins.** Cyclins bind to these kinases, activating them at appropriate times during the cell cycle. Activated kinases then phosphorylate other target proteins that regulate the progress of the cell cycle. The study of *cdc* mutations has established that the cell cycle contains at least three major **checkpoints**, when the processes culminating in normal mitosis are monitored, or "checked," by these master control molecules before the next stage of the cycle commences.

The first of the three checkpoints, the **G1/S checkpoint**, monitors the size the cell has achieved since its previous mitosis and also evaluates the condition of the DNA. If the cell has not reached an adequate size or if the DNA has been damaged, further progress through the cycle is arrested until these conditions are "corrected." If both conditions are "normal" at G1/S, then the cell is allowed to proceed to the S phase of the cycle. The second important checkpoint is the **G2/M checkpoint**, where DNA is monitored prior to the start of mitosis. If DNA replication is incomplete or any DNA damage is detected and has not been repaired, the cell cycle is arrested. The final checkpoint occurs during mitosis and is called the **M checkpoint.** Here, both the successful formation of the spindle fiber system and the attachment of spindle fibers to the kinetochores associated with the centromeres are monitored. If spindle fibers are not properly formed or attachment is inadequate, mitosis is arrested.

The importance of cell-cycle control and these checkpoints can be demonstrated by considering what happens when this regulatory system is impaired. Let's assume, for example, that the DNA of a cell has incurred damage leading to one or more mutations impairing cell-cycle control. If allowed to proceed through the cell cycle as one of the population of dividing cells, this genetically altered cell would divide uncontrollably—precisely the definition of a cancerous cell. If instead the cell cycle is arrested at one of the checkpoints, the cell may effectively be removed from the population of dividing cells, preventing its potential malignancy.

2.4

## Meiosis Reduces the Chromosome Number from Diploid to Haploid in Germ Cells and Spores

The process of meiosis, unlike mitosis, reduces the amount of genetic material by one-half. Whereas in diploids mitosis produces daughter cells with a full diploid complement, meiosis produces gametes or spores with only one haploid set of chromosomes. During sexual reproduction, gametes then combine through fertilization to reconstitute the diploid complement found in parental cells. Figure 2–8 compares the two processes by following two pairs of homologous chromosomes.

The events of meiosis must be highly specific since by definition, haploid gametes or spores contain precisely one member of each homologous pair of chromosomes. If successfully completed, meiosis ensures genetic continuity from generation to generation.

The process of sexual reproduction also ensures genetic variety among members of a species. As you study meiosis, you will see that this process results in gametes that each contain unique combinations of maternally and paternally derived chromosomes in their haploid complement. With such a tremendous genetic variation among the gametes, a huge number of maternal-paternal chromosome combinations are possible at fertilization. Furthermore, you will see that the meiotic event referred to as **crossing over** results in genetic exchange between members of each homologous pair of chromosomes. This process creates intact chromosomes that are mosaics of the maternal and paternal homologs from which they arise, further enhancing the potential genetic variation in gametes and the offspring derived from them. Sexual reproduction therefore reshuffles the genetic material, producing offspring that often differ greatly from either parent. Thus meiosis is the major source of genetic recombination within species.

### An Overview of Meiosis

In the preceding discussion, we established what might be considered the goal of meiosis: the reduction to the haploid complement of chromosomes. Before we consider the phases of this process systematically, we will briefly summarize how diploid cells give rise to haploid gametes or spores. You should refer to the right-hand side of Figure 2–8 during the following discussion.

You have seen that in mitosis each paternally and maternally derived member of any given homologous pair of chromosomes behaves autonomously during division. By contrast, early in meiosis, homologous chromosomes form pairs; that is, they **synapse.** Each synapsed structure, initially called a **bivalent**, eventually gives rise to a **tetrad** consisting of four chromatids. The presence of four chromatids demonstrates that both homologs (making up the bivalent) have, in fact, duplicated. Therefore, to achieve haploidy, two divisions are necessary. The first division occurs in meiosis I and is described as a **reductional division** (because the number of centromeres, each representing one chromosome, is *reduced* by one-half). Components of each tetrad—representing the two homologs—separate, yielding two **dyads.** Each dyad is composed of two sister chromatids joined at a common centromere. The second division occurs during meiosis II and is described as an **equational division** (because the number of centromeres remains *equal*). Here each dyad splits into two **monads** of one chromosome each. Thus, the two divisions potentially produce four haploid cells.

### The First Meiotic Division: Prophase I

We turn now to a detailed account of meiosis. Like mitosis, meiosis is a continuous process. We assign names to its stages and substages only to facilitate discussion. From a genetic standpoint, three events characterize the initial stage, **prophase I** (Figure 2–9). First, as in mitosis, chromatin present in interphase thickens and coils into visible chromosomes. Second, unlike mitosis, members of each homologous pair of chromosomes undergo synapsis. Third, crossing over occurs between synapsed homologs. Because of the complexity of these genetic events, this stage of meiosis is divided into five substages: leptonema, zygonema, pachynema, diplonema,* and diakinesis. As we discuss these substages, be aware that, even though it is

*These are the noun forms of these substages. The adjective forms (leptotene, zygotene, pachytene, and diplotene) are also used in the text.

OVERVIEW OF MEIOSIS WEB TUTORIAL 2.2

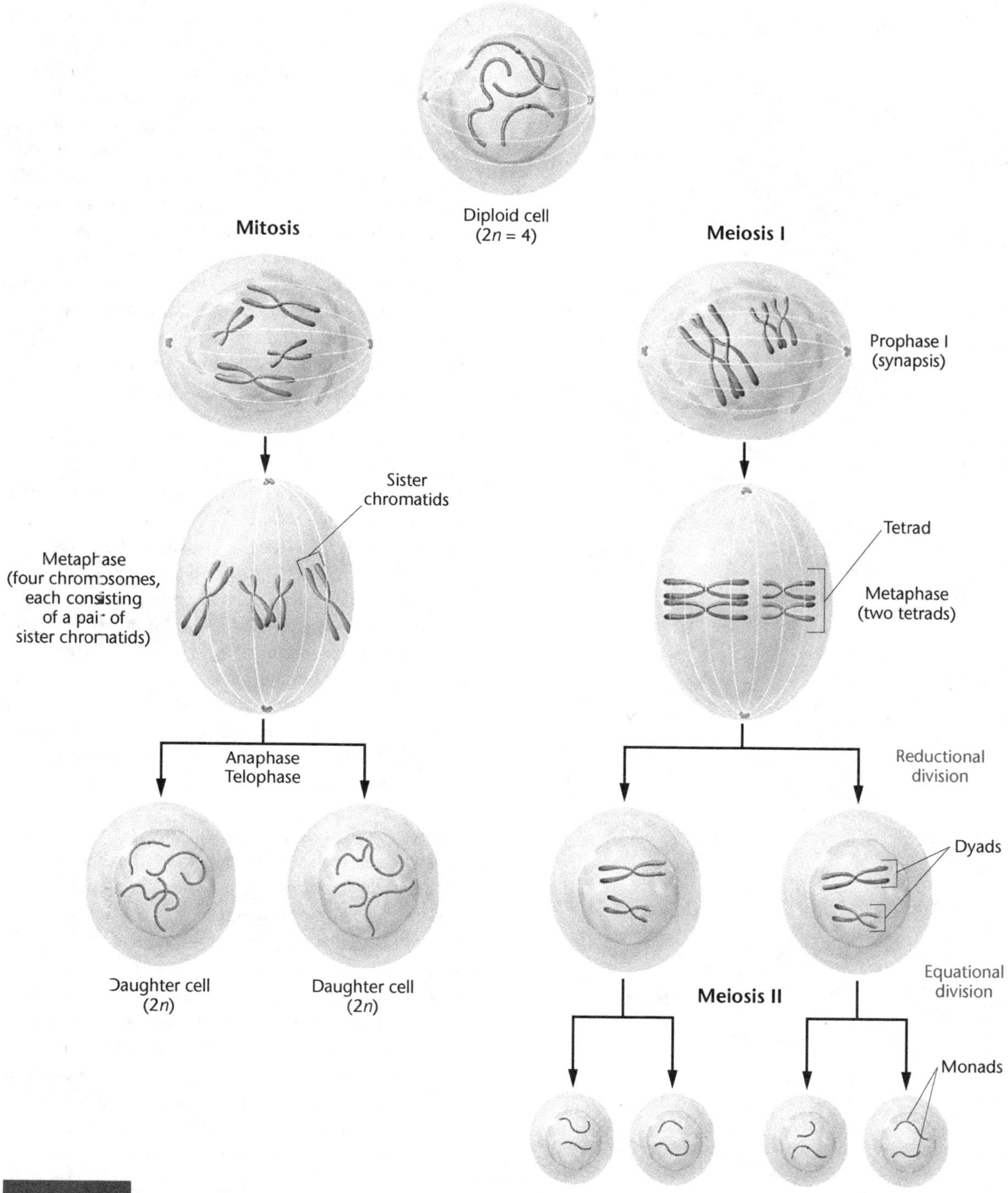


FIGURE 2-8 Overview of the major events and outcomes of mitosis and meiosis. As in Figure 2-7, two pairs of homologous chromosomes are followed.

not immediately apparent in the earliest phases of meiosis, the DNA of chromosomes has been replicated during the prior interphase.

**Leptonema** During the **leptotene stage**, the interphase chromatin material begins to condense, and the chromosomes, although still extended, become visible. Along each chromosome are **chromomeres**, localized condensations that resemble beads on a string. Recent evidence suggests that a process called **homology search**, which precedes and is essential to the initial pairing of homologs, begins during leptonema.

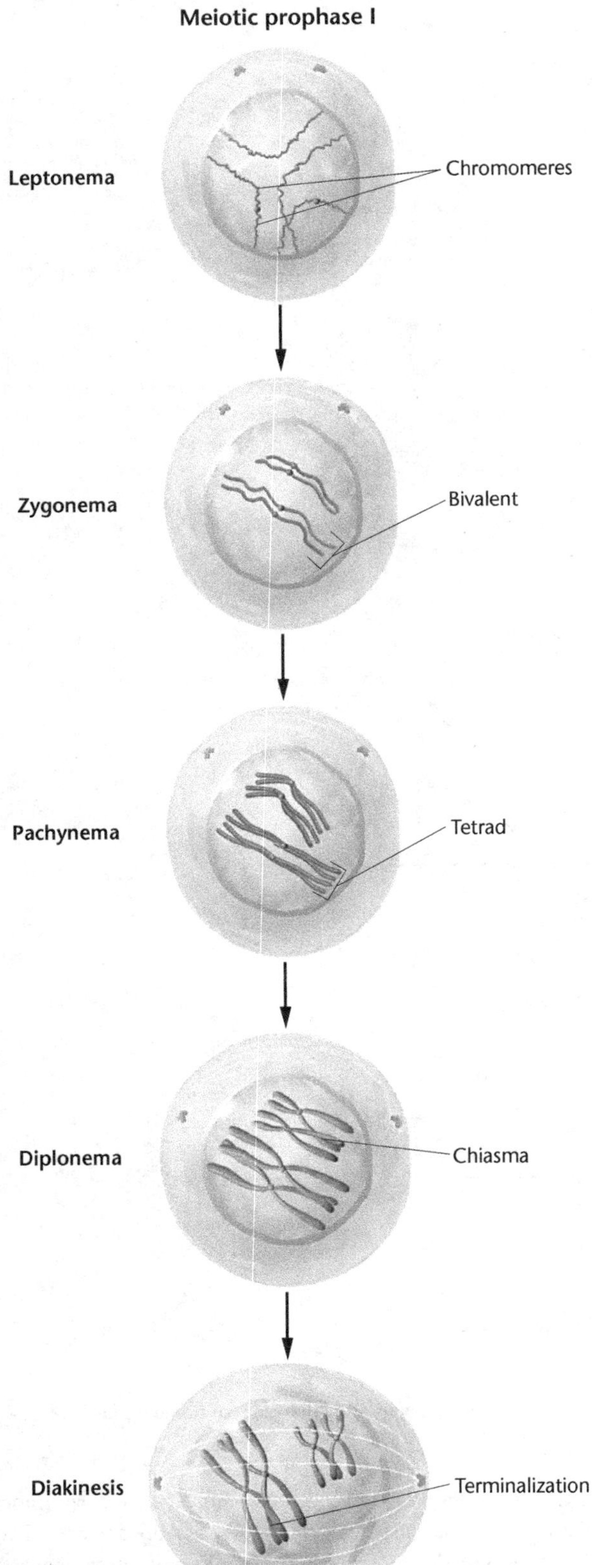


FIGURE 2-9 The substages of meiotic prophase I for the chromosomes depicted in Figure 2-8.

**Zygonema** The chromosomes continue to shorten and thicken during the **zygotene stage**. During the process of homology search, homologous chromosomes undergo initial alignment with one another. This so-called *rough pairing* is complete by the end of zygonema. In yeast, homologs are separated by about 300 nm, and near the end of zygonema, structures called lateral elements are visible between paired homologs. As meiosis proceeds, the overall length of the lateral elements along the chromosome increases, and a more extensive ultrastructural component called the **synaptonemal complex** begins to form between the homologs.

It is at the completion of zygonema that the paired homologs are referred to as bivalents. Although both members of each bivalent have already replicated their DNA, it is not yet visually apparent that each member is a double structure. The number of bivalents in each species is equal to the haploid ($n$) number.

**Pachynema** In the transition from the zygotene to the **pachytene stage**, the chromosomes continue to coil and shorten, and further development of the synaptonemal complex occurs between the two members of each bivalent. This leads to synapsis, a more intimate pairing. Compared to the rough-pairing characteristic of zygonema, homologs are now separated by only 100 nm.

During pachynema, each homolog is now evident as a double structure, providing visual evidence of the earlier replication of the DNA of each chromosome. Thus, each bivalent contains four member chromatids. As in mitosis, replicates are called sister chromatids, whereas chromatids from maternal and paternal members of a homologous pair are called nonsister chromatids. The four-membered structure, also referred to as a tetrad, contains two pairs of sister chromatids.

**Diplonema** During the ensuing **diplotene stage**, it is even more apparent that each tetrad consists of two pairs of sister chromatids. Within each tetrad, each pair of sister chromatids begins to separate. However, one or more areas remain in contact where chromatids are intertwined. Each such area, called a **chiasma** (pl. **chiasmata**), is thought to represent a point where nonsister chromatids have undergone genetic exchange through the process referred to above as crossing over. Although the physical exchange between chromosome areas occurred during the previous pachytene stage, the result of crossing over is visible only when the duplicated chromosomes begin to separate. Crossing over is an important source of genetic variability, and as indicated earlier, new combinations of genetic material are formed during this process.

**Diakinesis** The final stage of prophase I is **diakinesis**. The chromosomes pull farther apart, but nonsister chromatids remain loosely associated at the chiasmata. As separation proceeds, the chiasmata move toward the ends of the tetrad. This process of **terminalization** begins in late diplonema and is completed during diakinesis. During this final substage, the nucleolus and nuclear envelope break down, and the two centromeres of each tetrad attach to the recently formed spindle fibers. By the completion of prophase I, the centromeres of each tetrad structure are present on the metaphase plate of the cell.

### Metaphase, Anaphase, and Telophase I

The remainder of the meiotic process is depicted in Figure 2–10. After meiotic prophase I, steps similar to those of mitosis occur. In the first division, **metaphase I,** the chromosomes have maximally shortened and thickened. The terminal chiasmata of each tetrad are visible and appear to be the only factor holding the nonsister chromatids together. Each tetrad interacts with spindle fibers, facilitating its movement to the metaphase plate. The alignment of each tetrad prior to the first anaphase is random: half of the tetrad will be pulled to one or the other pole, and the other half moves to the opposite pole.

During the stages of meiosis I, a single centromere holds each pair of sister chromatids together. It does *not* divide. At **anaphase I,** one-half of each tetrad (a dyad) is pulled toward each pole of the dividing cell. This separation process is the physical basis of disjunction, the separation of homologous chromosomes from one another. Occasionally, errors in meiosis occur and separation is not achieved. The term **nondisjunction** describes such an error. At the completion of the normal anaphase I, a series of dyads equal to the haploid number is present at each pole.

If crossing over had not occurred in the first meiotic prophase, each dyad at each pole would consist solely of either paternal or maternal chromatids. However, the exchanges produced by crossing over create mosaic chromatids of paternal and maternal origin.

In many organisms, **telophase I** reveals a nuclear membrane forming around the dyads. In this case, the nucleus next enters into a short interphase period. If interphase occurs, the chromosomes do not replicate because they already consist of two chromatids. In other organisms, the cells go directly from anaphase I to meiosis II. In general, meiotic telophase is much shorter than the corresponding stage in mitosis.

### The Second Meiotic Division

A second division, referred to as **meiosis II,** is essential if each gamete or spore is to receive only one chromatid from each original tetrad. The stages characterizing meiosis II are shown on the right side of Figure 2–10. During **prophase II,** each dyad is composed of one pair of sister chromatids attached by a common centromere. During **metaphase II,** the centromeres are positioned on the equatorial plate. When they divide, **anaphase II** is initiated, and the sister chromatids of each dyad are pulled to opposite poles. Because the number of dyads is equal to the haploid number, **telophase II** reveals one member of each pair of homologous chromosomes present at each pole. Each chromosome is now a monad. Following cytokinesis in telophase II, four haploid gametes may result from a single meiotic event. At the conclusion of meiosis II, not only has the haploid state been achieved, but if crossing over has occurred, each monad is a combination of maternal and paternal genetic information. As a result, the offspring produced by any gamete will receive a mixture of genetic information originally present in his or her grandparents. Meiosis thus significantly increases the level of genetic variation in each ensuing generation.

**NOW SOLVE THIS**

Problem 14 on page 40 involves an understanding of what happens to the maternal and paternal members of each pair of homologous chromosomes during meiosis.

■ HINT: *The key to solving this problem is to understand that maternal and paternal homologs synapse during meiosis. Once each chromatid has duplicated, creating a tetrad in the early phases of meiosis, each original pair behaves as a unit and leads to two dyads during anaphase I.*

## 2.5 The Development of Gametes Varies in Spermatogenesis Compared to Oogenesis

Although events that occur during the meiotic divisions are similar in all cells participating in gametogenesis in most animal species, there are certain differences between the production of a male gamete (spermatogenesis) and a female gamete (oogenesis). Figure 2–11 summarizes these processes.

**Spermatogenesis** takes place in the testes, the male reproductive organs. The process begins with the enlargement of an undifferentiated diploid germ cell called a **spermatogonium.** This cell grows to become a **primary spermatocyte,** which undergoes the first meiotic division. The products of this division, called **secondary spermatocytes,** contain a haploid number of dyads. The secondary spermatocytes then undergo meiosis II, and each of these cells produces two haploid **spermatids.** Spermatids go through a series of developmental changes, **spermiogenesis,** to become highly specialized, motile **spermatozoa,** or **sperm.** All sperm cells produced during spermatogenesis contain the haploid number of chromosomes and equal amounts of cytoplasm.

Spermatogenesis may be continuous or may occur periodically in mature male animals; its onset is determined by the species' reproductive cycles. Animals that reproduce year-round produce sperm continuously, whereas those whose breeding period is confined to a particular season produce sperm only during that time.

In animal **oogenesis,** the formation of **ova** (sing. **ovum**), or eggs, occurs in the ovaries, the female reproductive organs. The daughter cells resulting from the two meiotic divisions of this process receive equal amounts of genetic material, but they do *not* receive equal amounts of cytoplasm. Instead, during each division, almost all the cytoplasm of the **primary oocyte,** itself derived from the **oogonium,** is concentrated in one of the two daughter cells. The concentration of cytoplasm is necessary because a major function of the mature ovum is to nourish the developing embryo following fertilization.

MEIOSIS WEB TUTORIAL 2.3

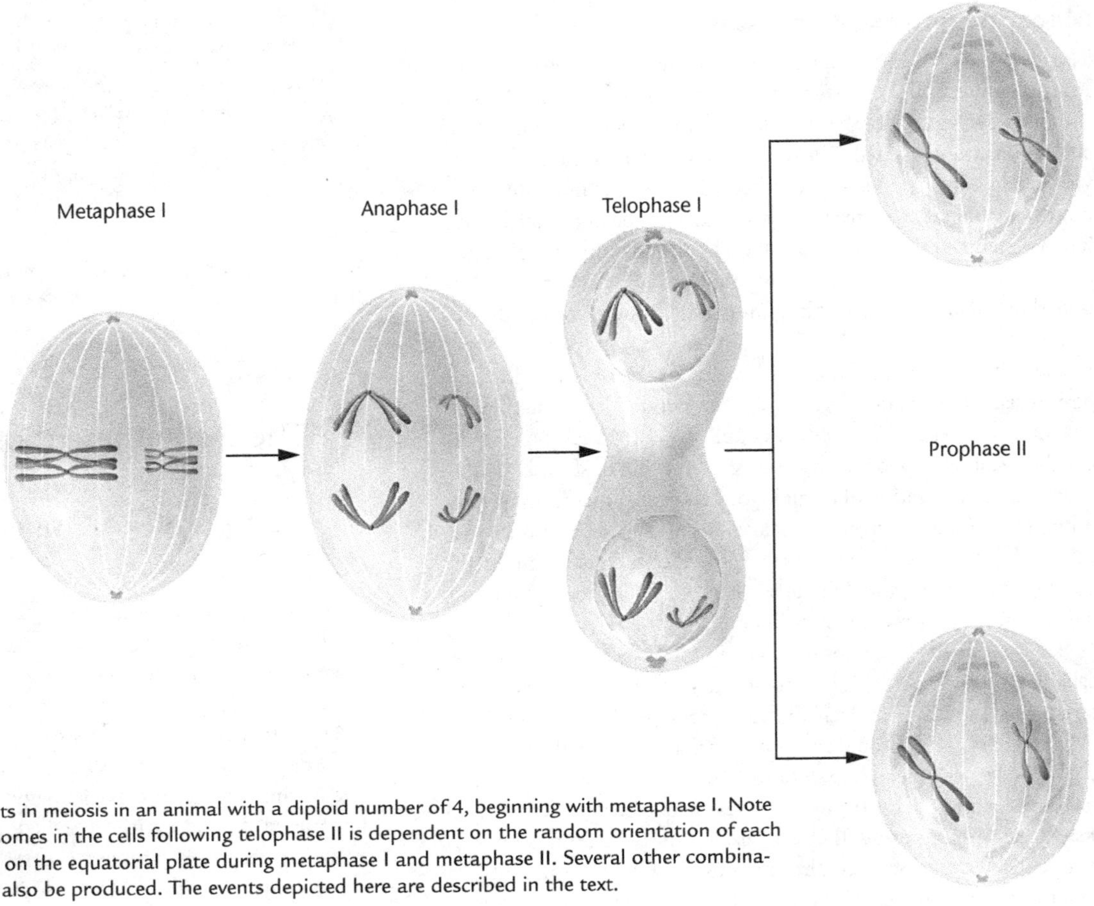


FIGURE 2–10 The major events in meiosis in an animal with a diploid number of 4, beginning with metaphase I. Note that the combination of chromosomes in the cells following telophase II is dependent on the random orientation of each tetrad and dyad when they align on the equatorial plate during metaphase I and metaphase II. Several other combinations, which are not shown, can also be produced. The events depicted here are described in the text.

During the anaphase I in oogenesis, the tetrads of the primary oocyte separate, and the dyads move toward opposite poles. During telophase I, the dyads at one pole are pinched off with very little surrounding cytoplasm to form the **first polar body**. The first polar body may or may not divide again to produce two small haploid cells. The other daughter cell produced by this first meiotic division contains most of the cytoplasm and is called the **secondary oocyte.** The mature ovum will be produced from the secondary oocyte during the second meiotic division. During this division, the cytoplasm of the secondary oocyte again divides unequally, producing an **ootid** and a **second polar body**. The ootid then differentiates into the mature ovum.

Unlike the divisions of spermatogenesis, the two meiotic divisions of oogenesis may not be continuous. In some animal species, the second division may directly follow the first. In others, including humans, the first division of all oocytes begins in the embryonic ovary but arrests in prophase I. Many years later, meiosis resumes in each oocyte just prior to its ovulation. The second division is completed only after fertilization.

**NOW SOLVE THIS**

Problem 9 on page 40 involves an understanding of meiosis during oogenesis.

■ HINT: *To answer this question, you must take into account that crossing over occurred between each pair of homologs during meiosis I.*

## 2.6 Meiosis Is Critical to the Successful Sexual Reproduction of All Diploid Organisms

The process of meiosis is critical to the successful sexual reproduction of all diploid organisms. It is the mechanism by which the diploid amount of genetic information is reduced to the haploid amount. In animals, meiosis leads to the formation of gametes,

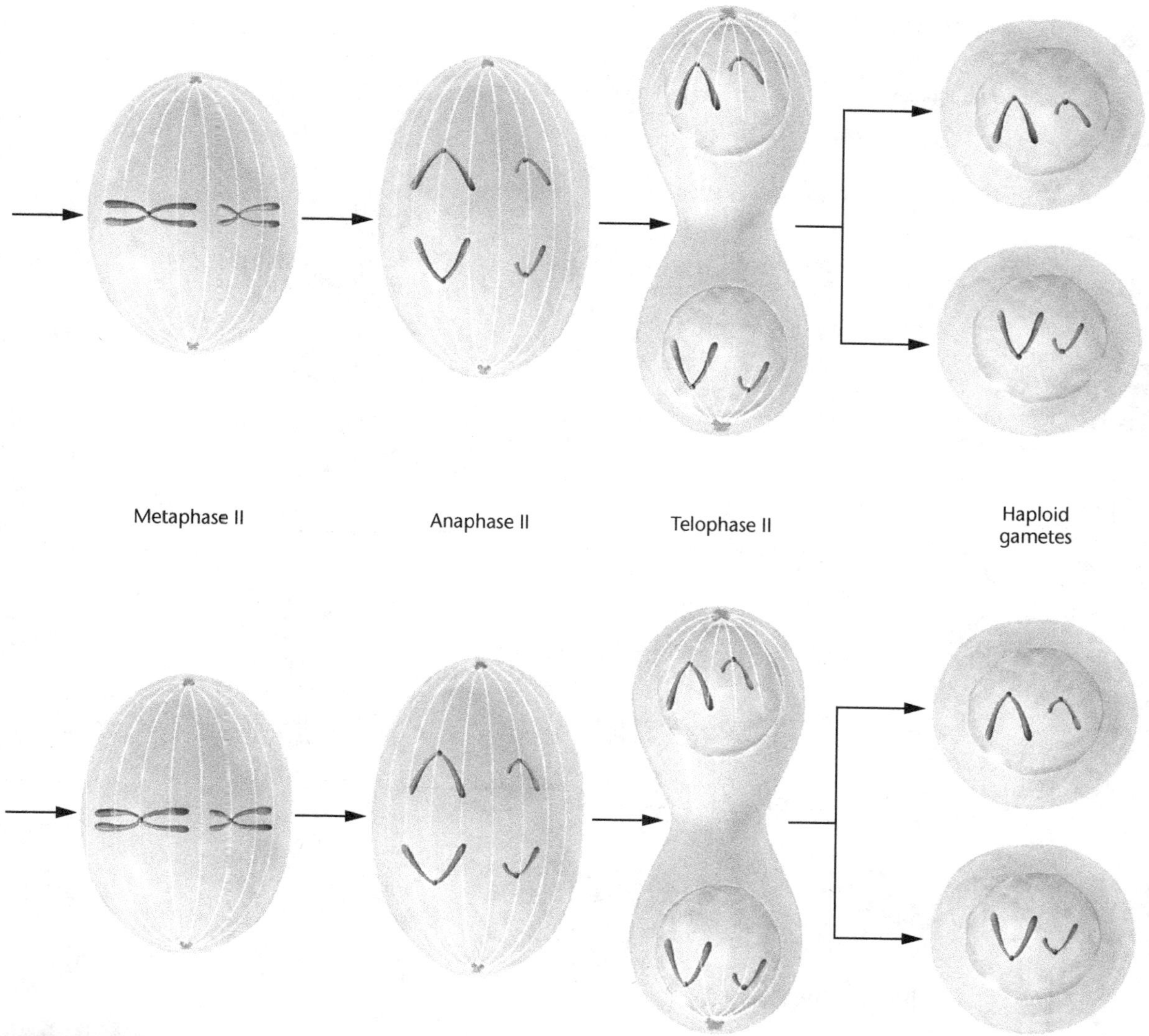


**FIGURE 2–10** (*Continued*)

whereas in plants haploid spores are produced, which in turn lead to the formation of haploid gametes.

Each diploid organism stores its genetic information in the form of homologous pairs of chromosomes. Each pair consists of one member derived from the maternal parent and one from the paternal parent. Following meiosis, haploid cells potentially contain either the paternal or the maternal representative of every homologous pair of chromosomes. However, the process of crossing over, which occurs in the first meiotic prophase, further reshuffles the alleles between the maternal and paternal members of each homologous pair, which then segregate and assort independently into gametes. These events result in the great amounts of genetic variation in gametes.

It is important to touch briefly on the significant role that meiosis plays in the life cycles of fungi and plants. In many fungi, the predominant stage of the life cycle consists of haploid vegetative cells. They arise through meiosis and proliferate by mitotic cell division. In multicellular plants, the life cycle alternates between the diploid **sporophyte stage** and the haploid **gametophyte stage** (Figure 2–12). While one or the other predominates in different plant groups during this "alternation of generations," the processes of meiosis and fertilization constitute the "bridges" between the sporophyte and gametophyte stages. Therefore, meiosis is an essential component of the life cycle of plants.

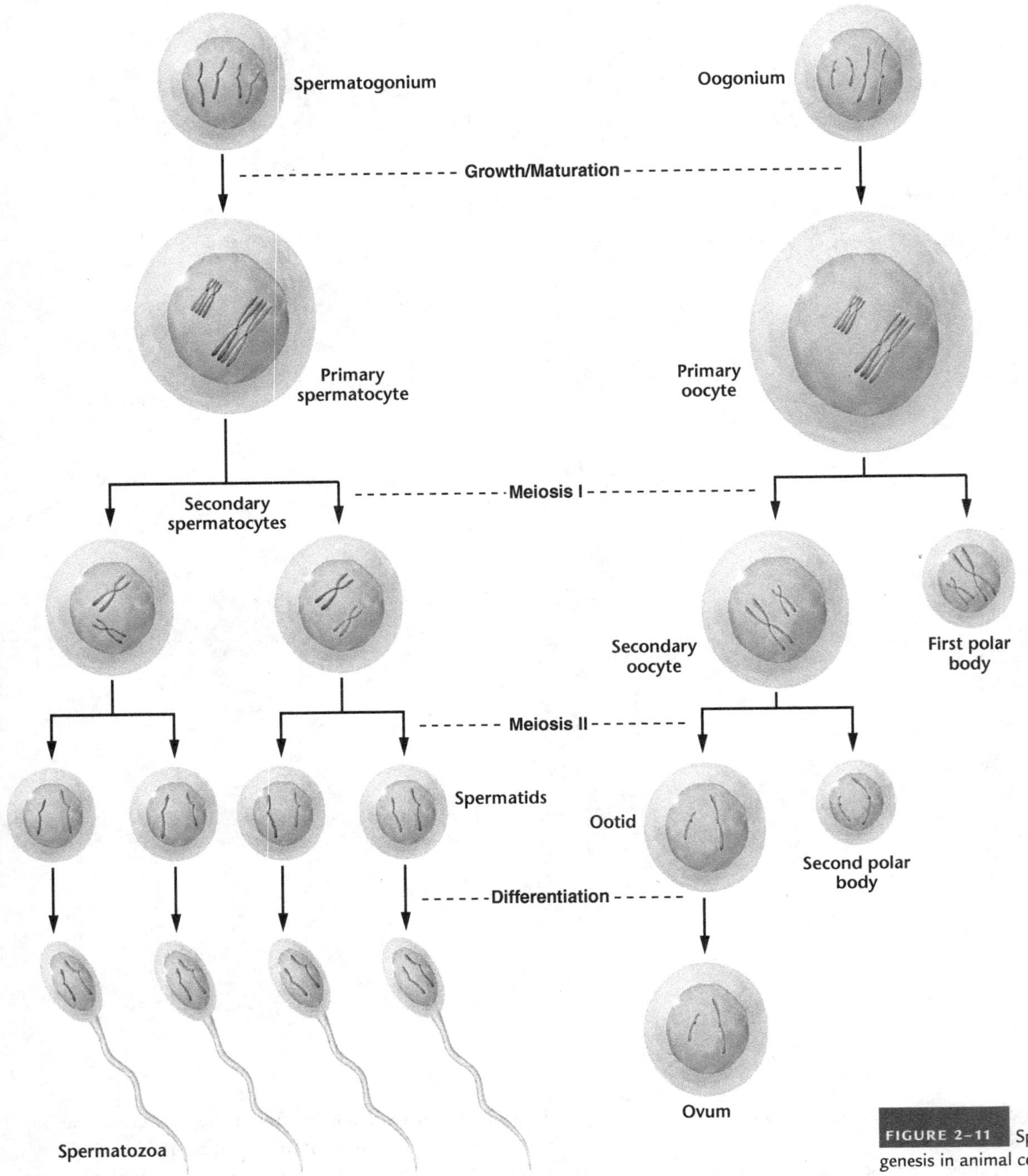


FIGURE 2–11 Spermatogenesis and oogenesis in animal cells.

2.7

## Electron Microscopy Has Revealed the Physical Structure of Mitotic and Meiotic Chromosomes

Thus far in this chapter, we have focused on mitotic and meiotic chromosomes, emphasizing their behavior during cell division and gamete formation. An interesting question is why chromosomes are invisible during interphase but visible during the various stages of mitosis and meiosis. Studies using electron microscopy clearly show why this is the case.

Recall that, during interphase, only dispersed chromatin fibers are present in the nucleus [Figure 2–13(a)]. Once mitosis begins, however, the fibers coil and fold, condensing into typical mitotic chromosomes [Figure 2–13(b)]. If the fibers comprising a mitotic chromosome are loosened, the areas of greatest spreading

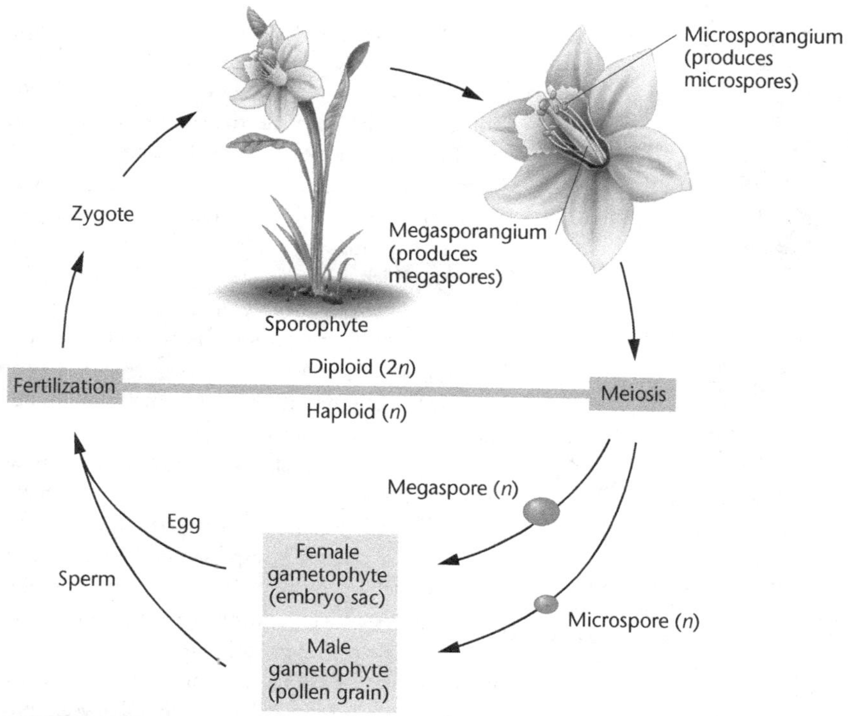


FIGURE 2–12 Alternation of generations between the diploid sporophyte ($2n$) and the haploid gametophyte ($n$) in a multicellular plant. The processes of meiosis and fertilization bridge the two phases of the life cycle. In angiosperms (flowering plants), like the one shown here, the sporophyte stage is the predominant phase.

reveal individual fibers similar to those seen in interphase chromatin [Figure 2–13(c)]. Very few fiber ends seem to be present, and in some cases, none can be seen. Instead, individual fibers always seem to loop back into the interior. Such fibers are obviously twisted and coiled around one another, forming the regular pattern of folding in the mitotic chromosome. Starting in late telophase of mitosis and continuing during G1 of interphase, chromosomes unwind to form the long fibers characteristic of chromatin, which consist of DNA and associated proteins, particularly proteins called histones. It is in this physical arrangement that DNA can most efficiently function during transcription and replication.

Electron microscopic observations of metaphase chromosomes in varying degrees of coiling led Ernest DuPraw to postulate the **folded-fiber model,** shown in Figure 2–13(d). During metaphase, each chromosome consists of two sister chromatids joined at the centromeric region. Each arm of the chromatid appears to be a single fiber wound much like a skein of yarn. The fiber is composed of tightly coiled double-stranded DNA and protein. An orderly coiling–twisting–condensing process appears to effect the transition of the interphase chromatin into the more condensed mitotic chromosomes. Geneticists believe that during the transition from interphase to

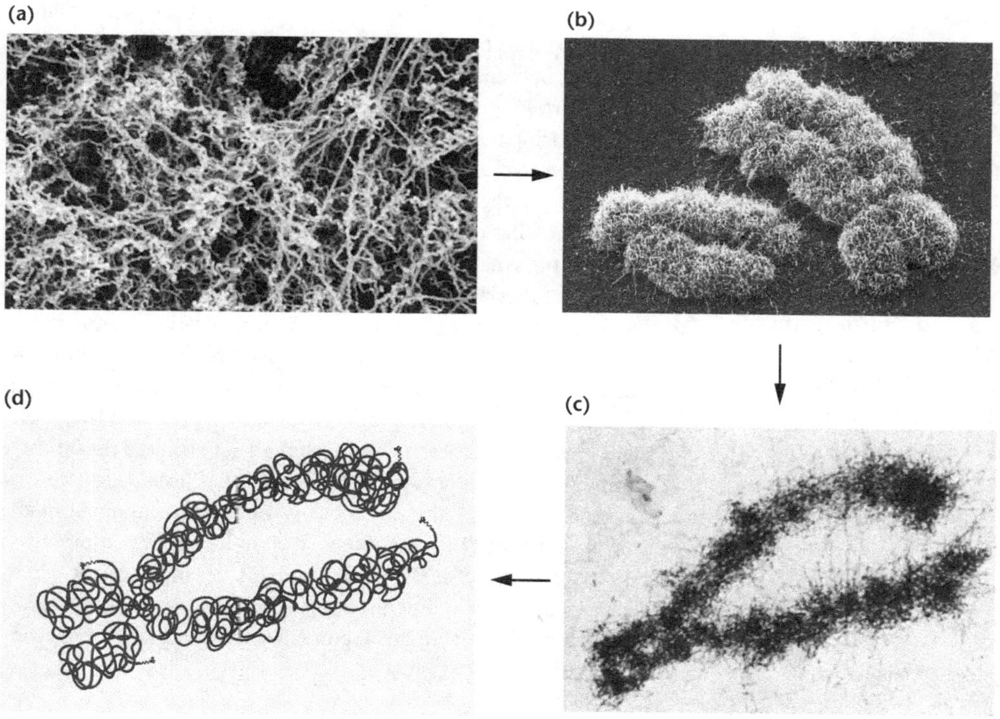


FIGURE 2–13 Comparison of (a) the chromatin fibers characteristic of the interphase nucleus with (b) and (c) metaphase chromosomes that are derived from chromatin during mitosis. Part (d) diagrams the mitotic chromosome and its various components, showing how chromatin is condensed to produce it. Parts (a) and (c) are transmission electron micrographs, while part (b) is a scanning electron micrograph.

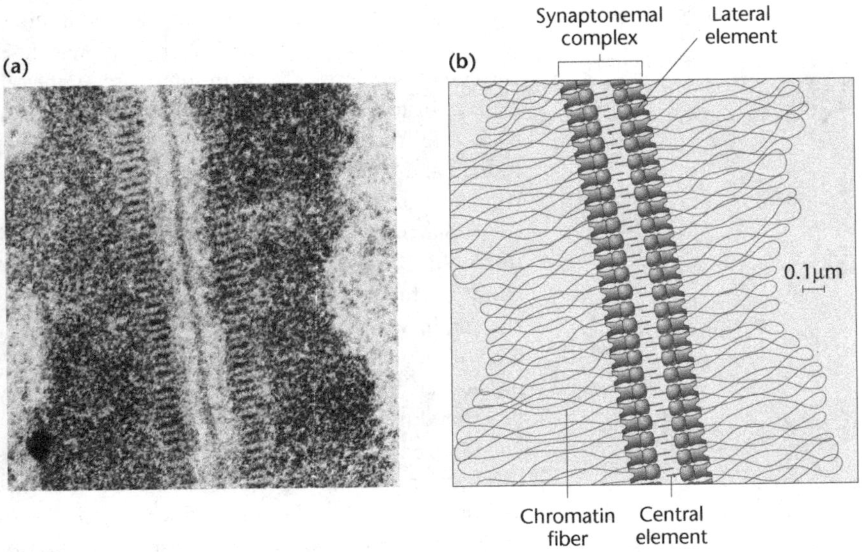


FIGURE 2–14 (a) Electron micrograph of a portion of a synaptonemal complex found between synapsed bivalents of the fungus *Neotiella rutilans*. (b) Schematic interpretation of the components making up the synaptonemal complex. *D. von Wettstein. Annual Reviews, Inc. With permission, from* Annual Review of Genetics, *Volume 6, 1972 by Annual Reviews, Inc. www.annualreviews.org.*

prophase, a 5000-fold compaction occurs in the length of DNA within the chromatin fiber! This process must be extremely precise given the highly ordered and consistent appearance of mitotic chromosomes in all eukaryotes. Note particularly in the micrographs the clear distinction between the sister chromatids constituting each chromosome. They are joined only by the common centromere that they share prior to anaphase.

## The Synaptonemal Complex

The electron microscope has also been used to visualize another structural component of the chromosome found only in cells undergoing meiosis. This structure, first introduced during our earlier discussion of the first meiotic prophase stage, is seen connecting synapsed homologs and is called the synaptonemal complex.* In 1956, Montrose Moses observed this complex in spermatocytes of crayfish, and Don Fawcett saw it in pigeon and human spermatocytes. Because there was not yet any satisfactory explanation of the mechanism of synapsis or of crossing over and chiasma formation, many researchers became interested in this structure. With few exceptions, the ensuing studies revealed the synaptonemal complex to be present in most plant and animal cells visualized during meiosis.

As you can see in the electron micrograph in Figure 2–14(a), the synaptonemal complex is a tripartite structure. Its central element is usually less dense and thinner (100–150 Å) than the two outer elements (500 Å), which are identical to one another. The outer structures, the lateral elements, are intimately associated with the synapsed homologs on either side. Selective staining has revealed that these lateral elements consist primarily of DNA and protein, suggesting that chromatin is an essential part of them. Some DNA fibrils traverse the lateral elements, making connections with the central element, which is composed primarily of protein. Figure 2–14(b) provides a diagrammatic interpretation of the electron micrograph consistent with the foregoing description.

The formation of the synaptonemal complex begins prior to the pachytene stage. As early as leptonema of the first meiotic prophase, lateral elements are seen in association with sister chromatids. Homologs have yet to associate with one another and are randomly dispersed in the nucleus. As we saw earlier, by the next stage, zygonema, homologous chromosomes begin to align with one another in what is called rough pairing, but they remain distinctly apart by some 300 nm. Then, during pachynema, the intimate association between homologs, characteristic of synapsis, occurs as formation of the complex is completed. In some diploid organisms, this occurs in a zipperlike fashion, beginning at the ends of the chromosomes, which may be attached to the nuclear envelope.

The synaptonemal complex is the vehicle for the pairing of homologs and also for their subsequent segregation during meiosis. However, some degree of synapsis can occur in certain cases where no synaptonemal complexes are formed. Thus, it is possible that the function of this structure may go beyond its involvement in the formation of bivalents.

In certain instances where no synaptonemal complexes are formed during meiosis, synapsis is not complete and crossing over is reduced or eliminated. For example, in male *Drosophila melanogaster*, where synaptonemal complexes are not usually seen, meiotic crossing over rarely, if ever, occurs. This observation suggests that the synaptonemal complex may be important in order for chiasmata to form and crossing over to occur.

The study of *zip1*, a mutation in the yeast *Saccharomyces cerevisiae*, has provided further insights into chromosome pairing. Cells bearing this mutation can undergo the initial alignment stage (rough pairing) and full-length central and lateral element formation, but their chromosomes fail to achieve the intimate pairing that is characteristic of synapsis. It has been suggested that the gene product of the *zip1* locus is a protein component of the central element of the synaptonemal complex, since that protein is absent in mutant cells. This observation further suggests that a complete and intact synaptonemal complex is essential during the transition from the initial rough alignment stage to the intimate pairing of synapsis.

*An alternative spelling of this term is synaptinemal complex.

GENETICS, TECHNOLOGY, AND SOCIETY

# Breast Cancer: The Double-Edged Sword of Genetic Testing

These are exhilarating times for genetics and biotechnology. The completion of the Human Genome Project has brought a rush of optimism about future applications of the resulting data. Scientists and the media predict that gene technologies will soon diagnose and cure diseases as diverse as diabetes, asthma, heart disease, and Parkinson disease.

The prospect of using genetics to prevent and cure a wide range of diseases is exciting. However, in our enthusiasm, we often forget that these new technologies still have significant limitations and profound ethical complexities. The story of genetic testing for breast cancer illustrates how we must temper our high expectations with respect for uncertainty.

Breast cancer is the most common cancer among women and the third leading cause of cancer deaths (after lung and colon cancer). Each year, more than 190,000 new cases are diagnosed in the United States. Breast cancer is not limited to women; about 1400 men are also diagnosed with the disease each year. A woman's lifetime risk of developing breast cancer is about 12 percent, and the risk increases with age.

Approximately 5 to 10 percent of breast cancers are familial, a category defined by the appearance of several cases of breast or ovarian cancer among near blood relatives and the early onset of these diseases. In 1994, two genes were identified that show linkage to familial breast cancers: *BRCA1* and *BRCA2*. Germline mutations (that is, inheritable mutations) in these genes are associated with the majority of familial breast cancers. The molecular functions of *BRCA1* and *BRCA2* are still uncertain, although they appear to be involved in repairing damaged DNA. Mutations in them are autosomal dominant with variable penetrance. Women with mutations in *BRCA1* or *BRCA2* have a 36 to 85 percent lifetime risk of developing breast cancer and a 16 to 60 percent risk of developing ovarian cancer. Men with germline mutations in *BRCA2* have a 6 percent lifetime breast cancer risk—a hundred-fold increase over the general male population.

*BRCA1* and *BRCA2* genetic tests detect any of the over 2000 different mutations that are known to occur within the coding regions of these genes, but the tests have limitations. They do not detect mutations in regulatory regions outside the coding region—mutations that could cause aberrant expression of these genes. Also, little is known about the cancer risk connected with any particular mutation, or about how the effects of each mutation may be modified by environmental factors or by interactions with other genes that confer susceptibility to cancer.

Many patients at risk for familial breast cancer opt to undergo genetic testing. These patients feel that test results could motivate them to take steps to prevent breast or ovarian cancers, guide them in childbearing decisions, and provide information concerning the risk of close relatives. But all these potential benefits are fraught with uncertainties.

A woman whose *BRCA* test results are negative may feel relieved and assume that she is not subject to familial breast cancer. However, her risk of developing breast cancer is still 12 percent (the population risk), and she should continue to monitor herself for the disease. Also, a negative *BRCA* genetic test does not eliminate the possibility that she carries an inherited mutation in another gene that increases breast cancer risk or that *BRCA1* or *BRCA2* gene mutations exist in regions of the genes that are inaccessible to current genetic tests.

A woman whose test results are positive faces difficult choices. Her treatment options are poor, consisting of close monitoring, prophylactic mastectomy or oophorectomy (removal of breasts and ovaries, respectively), and taking prophylactic drugs such as tamoxifen. Prophylactic surgery reduces her risks but does not eliminate them, as cancers can still occur in tissues that remain after surgery. Drugs such as tamoxifen reduce her risks but have serious side effects. Genetic tests not only affect the patient but also affect the patient's entire family. People often experience fear, anxiety, and guilt on learning that they are carriers of a genetic disease. Studies show that people who refuse genetic test results often suffer even more anxiety than those who opt to be informed of the results. Confidentiality is also a major concern. Patients fear that their genetic test results may be leaked to insurance companies or employers, jeopardizing their prospects for jobs or affordable health and life insurance. One study shows that a quarter of eligible patients refuse *BRCA* gene testing because of concerns about cost, confidentiality, and potential discrimination.

Genetic testing is such a new development that its use by the health system has lagged behind the science. For example, genetic testing should always be accompanied by genetic counseling to help patients and their families deal with both the psychological and the medical ambiguities of test results. However, there are insufficient numbers of genetic counselors with experience in genetic testing, and even the most qualified of these find the issues to be complex and difficult. Physicians often have limited knowledge of human clinical genetics and feel inadequate to advise their patients. The federal government and the insurance industries have yet to develop comprehensive policies concerning genetic tests and genetic information. Given the ambiguities of *BRCA* genetic tests, the relatively ineffective treatment options, and the potential for psychological and social side effects, it is not surprising that only about 60 percent of familial breast cancer patients and their families decide to undergo the genetic tests.

The unanswered questions about *BRCA1* and *BRCA2* genetic testing are many and important. What cancer risks are associated with which mutations? Should all people have access to *BRCA* tests, or only those at high risk? How can we ensure that the high costs of genetic tests and counseling do not limit this new technology to only a portion of the population? As we develop genetic tests for more and more diseases over the next few decades, our struggle with these kinds of issues will continue.

## References

Surbone, A. 2001. Ethical implications of genetic testing for breast cancer susceptibility. *Crit. Rev. in Onc./Hem.* 40: 149–157.

## Web Sites

Genetic Testing for BRCA1 and BRCA2: It's Your Choice [online]. National Institutes of Health. *http://cis.nci.nih.gov/fact/3_62.htm*

EXPLORING GENOMICS

## PubMed: Exploring and Retrieving Biomedical Literature

In this era of rapidly expanding information on genomics and the biomedical sciences, scientists must be conversant in the use of multiple online databases. These resources provide access to DNA and protein sequences, genomic data, chromosome maps, microarray gene expression networks, and molecular structures, as well as to the bioinformatics tools necessary for data manipulation. Perhaps the most central database resource is **PubMed**, an online tool for conducting literature searches and accessing biomedical publications.

PubMed is an Internet-based search system developed by the National Center of Biotechnology Information (NCBI) at the National Library of Medicine. Using PubMed, one can access over 15 million articles in over 4600 biomedical journals. The full text of many of the journals can be obtained electronically through college or university libraries, and some journals (such as *Proceedings of the National Academy of Sciences USA; Genome Biology;* and *Science*) provide free public access to articles within certain time frames.

In this exercise, we will explore PubMed to answer questions about relationships between tubulin, human cancers, and cancer therapies, as well as the genetics of spermatogenesis.

### ■ Exercise I – Tubulin, Cancer, and Mitosis

In this chapter we were introduced to tubulin and the dynamic behavior of microtubules during the cell cycle. Cancer cells are characterized by continuous and uncontrolled mitotic divisions.

Is it possible that tubulin and microtubules contribute to the development of cancer? Could these important structures be targets for cancer therapies?

1. To begin your search for the answers, access the PubMed site at www.ncbi.nlm.nih.gov/entrez/query.fcgi?DB=pubmed.
2. In the SEARCH box, type "tubulin cancer" and then select the "Go" button to perform the search.
3. Select several research papers and read the abstracts.

To answer the question about tubulin's association with cancer, you may want to limit your search to fewer papers, perhaps those that are review articles. To do this:

1. Select the "Limits" tab near the top of the page.
2. Scroll down the page and select "Review" in the "Type of Article" list.
3. Select "Go" to perform the search.

Explore some of the articles, as abstracts or as full text, if access is available through your library, by personal subscription, or by free public access. Prepare a brief report or verbally share your experiences with your class. Describe two of the most important things you learned during your exploration and identify the information sources you encountered during the search.

### ■ Exercise II – Human Disorders of Spermatogenesis

Using the methods described in Exercise I, identify some human disorders associated with defective spermatogenesis. Which human genes are involved in spermatogenesis? How do defects in these genes result in fertility disorders? Prepare a brief written or verbal report on what you have learned and what sources you used to acquire your information.

## Chapter Summary

1. The structure of cells is elaborate and complex. Many components of cells are involved directly or indirectly with genetic processes.
2. In diploid organisms, chromosomes exist in homologous pairs. Each homologous pair shares the same size, centromere placement, and gene sites. One member of each pair is derived from the maternal parent, and one is derived from the paternal parent.
3. Mitosis and meiosis are mechanisms by which cells distribute the genetic information contained in their chromosomes to progeny cells in a precise, orderly fashion.
4. Mitosis is but one part of the cell cycle, which is characteristic of all eukaryotes. The cell cycle also consists of the stages G1, S, and G2, which precede mitosis.
5. Mitosis, or nuclear division, is the basis of cellular reproduction. Daughter cells are produced that are genetically identical to their progenitor cell.
6. Mitosis may be subdivided into discrete stages: prophase, prometaphase, metaphase, anaphase, and telophase. Condensation of chromatin into chromosome structures occurs during prophase. During prometaphase, chromosomes appear as double structures, each composed of a pair of sister chromatids. In metaphase, chromosomes line up on the equatorial plane of the cell. During anaphase, sister chromatids of each chromosome are pulled apart and directed toward opposite poles. Daughter cell formation is completed at telophase and is characterized by cytokinesis, the division of the cytoplasm.

7. Meiosis converts a diploid cell into a haploid gamete or spore, making sexual reproduction possible. As a result of chromosome duplication and two subsequent meiotic divisions, each haploid cell receives one member of each homologous pair of chromosomes.
8. There is a major difference between meiosis in males and in females. On the one hand, spermatogenesis partitions the cytoplasmic volume equally and produces four haploid sperm cells. Oogenesis, on the other hand, collects the bulk of cytoplasm in one egg cell and reduces the other haploid products to polar bodies. The extra cytoplasm in the egg contributes to zygote development following fertilization.
9. Meiosis results in extensive genetic variation by virtue of the exchange during crossing over between maternal and paternal chromatids and their random segregation into gametes. In addition, meiosis plays an important role in the life cycles of fungi and plants, serving as the bridge between alternating generations.
10. Mitotic chromosomes are produced as a result of the coiling and condensation of chromatin fibers characteristic of interphase.

# INSIGHTS AND SOLUTIONS

*This initial appearance of "Insights and Solutions" begins a feature that will have great value to you as a student. From this point on, "Insights and Solutions" precedes the "Problems and Discussion Questions" at each chapter's end to provide sample problems and solutions that demonstrate approaches you will find useful in genetic analysis. The insights you gain by working through the sample problems will improve your ability to solve the ensuing problems in each chapter.*

1. In an organism with a diploid number of $2n = 6$, how many individual chromosomal structures will align on the metaphase plate during (a) mitosis, (b) meiosis I, and (c) meiosis II? Describe each configuration.

   **Solution:**
   (a) Remember that in mitosis, homologous chromosomes do not synapse, so there will be six double structures, each consisting of a pair of sister chromatids. In other words, the number of structures is equivalent to the diploid number.

   (b) In meiosis I, the homologs have synapsed, reducing the number of structures to three. Each is called a tetrad and consists of two pairs of sister chromatids.

   (c) In meiosis II, the same number of structures exist (three), but in this case they are called dyads. Each dyad is a pair of sister chromatids. When crossing over has occurred, each chromatid may contain parts of one of its nonsister chromatids, obtained during exchange in prophase I.

2. Disregarding crossing over, draw all possible alignment configurations that can occur during metaphase for the chromosomes shown in Figure 2–10.

   **Solution:** As shown in the following diagram, four configurations are possible when $n = 2$.

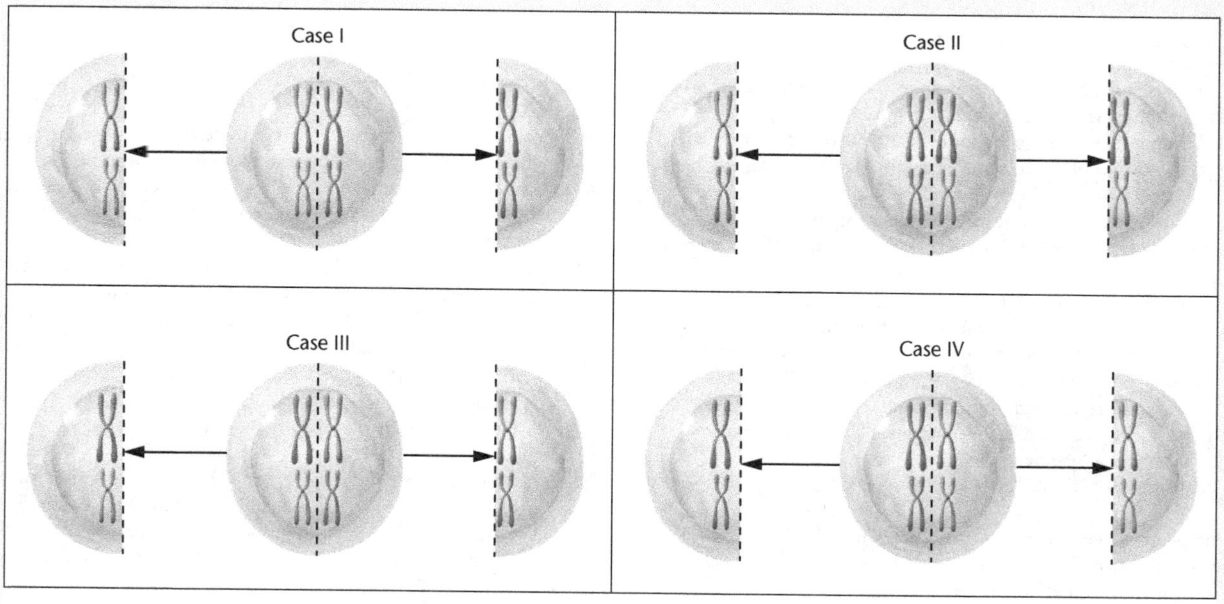


Solution for #2

*Continued on next page*

*Insights and Solutions, continued*

3. For the chromosomes in the previous problem, assume that each of the larger chromosomes has a different allele for a given gene: *A* and *a*, as shown. Also assume that each of the smaller chromosomes has a different allele for a second gene: *B* and *b*. Calculate the probability of generating each possible combination of these alleles (*AB*, *Ab*, *aB*, *ab*) following meiosis I.

**Solution:** As shown in the accompanying diagram:

| | |
|---|---|
| Case I | *AB* and *ab* |
| Case II | *Ab* and *aB* |
| Case III | aB and Ab |
| Case IV | *ab* and *AB* |

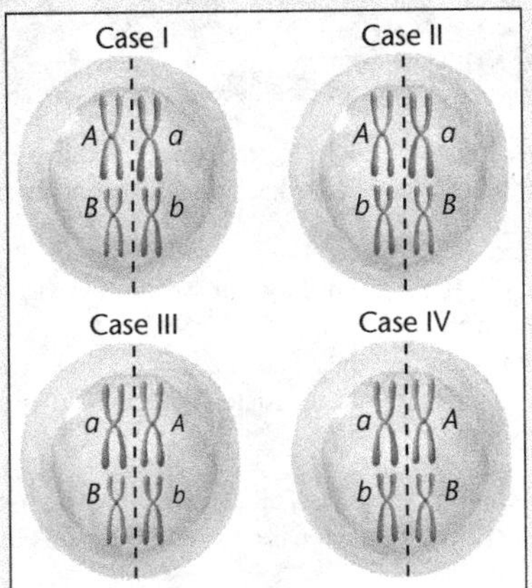

Solution for #3

| | | |
|---|---|---|
| Total: | $AB = 2$ | $(p = 1/4)$ |
| | $Ab = 2$ | $(p = 1/4)$ |
| | $aB = 2$ | $(p = 1/4)$ |
| | $ab = 2$ | $(p = 1/4)$ |

4. How many different chromosome configurations can occur following meiosis I if three different pairs of chromosomes are present ($n = 3$)?

**Solution:** If $n = 3$, then eight different configurations would be possible. The formula $2^n$, where $n$ equals the haploid number, represents the number of potential alignment patterns. As we will see in the next chapter, these patterns are produced according to the Mendelian postulate of *segregation*, and they serve as the physical basis of another Mendelian postulate called *independent assortment.*

5. Describe the composition of a meiotic tetrad during prophase I, assuming no crossover event has occurred. What impact would a single crossover event have on this structure?

**Solution:** Such a tetrad contains four chromatids, existing as two pairs. Members of each pair are sister chromatids. They are held together by a common centromere. Members of one pair are maternally derived, whereas members of the other are paternally derived. Maternal and paternal members are called nonsister chromatids. A single crossover event has the effect of exchanging a portion of a maternal and a paternal chromatid, leading to a chiasma, where the two involved chromatids overlap physically in the tetrad. The process of exchange is referred to as crossing over.

## Problems and Discussion Questions

1. What role do the following cellular components play in the storage, expression, or transmission of genetic information: (a) chromatin, (b) nucleolus, (c) ribosome, (d) mitochondrion, (e) centriole, (f) centromere?
2. Discuss the concepts of homologous chromosomes, diploidy, and haploidy. What characteristics do two homologous chromosomes share?
3. If two chromosomes of a species are the same length and have similar centromere placements and yet are not homologous, what is different about them?
4. Describe the events that characterize each stage of mitosis.
5. If an organism has a diploid number of 16, how many chromatids are visible at the end of mitotic prophase? How many chromosomes are moving to each pole during anaphase of mitosis?
6. What designations are assigned to chromosomes on the basis of their centromere placement, and where is the centromere located in each case?
7. Contrast telophase in plant and animal mitosis.
8. Describe the phases of the cell cycle and the events that characterize each phase.
9. Examine Figure 2–11, which shows oogenesis in animal cells. Will the genotype of the second polar body (derived from meiosis II) always be identical to that of the ootid? Why or why not?
10. Contrast the end results of meiosis with those of mitosis.
11. Define and discuss these terms: (a) synapsis, (b) bivalents, (c) chiasmata, (d) crossing over, (e) chromomeres, (f) sister chromatids, (g) tetrads, (h) dyads, (i) monads.
12. Contrast the genetic content and the origin of sister versus nonsister chromatids during their earliest appearance in prophase I of meiosis. How might the genetic content of these change by the time tetrads have aligned at the equatorial plate during metaphase I?
13. Given the end results of the two types of division, why is it necessary for homologs to pair during meiosis and not desirable for them to pair during mitosis?
14. An organism has a diploid number of 16 in a primary oocyte. (a) How many tetrads are present in the first meiotic prophase? (b) How many dyads are present in the second meiotic prophase? (c) How many monads migrate to each pole during the second meiotic anaphase?
15. Contrast spermatogenesis and oogenesis. What is the significance of the formation of polar bodies?
16. Explain why meiosis leads to significant genetic variation while mitosis does not.
17. A diploid cell contains three pairs of homologous chromosomes designated C1 and C2, M1 and M2, and S1 and S2. No crossing over occurs. What combinations of chromosomes are possible in (a) daughter cells following mitosis? (b) cells undergoing the first meiotic metaphase? (c) haploid cells following both divisions of meiosis?
18. Considering the preceding problem, predict the number of different haploid cells that could be produced by meiosis if a fourth chromosome pair (W1 and W2) were added.

19. During oogenesis in an animal species with a haploid number of 6, one dyad undergoes nondisjunction during meiosis II. Following the second meiotic division, this dyad ends up intact in the ovum. How many chromosomes are present in (a) the mature ovum and (b) the second polar body? (c) Following fertilization by a normal sperm, what chromosome condition is created?
20. What is the probability that, in an organism with a haploid number of 10, a sperm will be formed that contains all 10 chromosomes whose centromeres were derived from maternal homologs?
21. During the first meiotic prophase, (a) when does crossing over occur; (b) when does synapsis occur; (c) during which stage are the chromosomes least condensed; and (d) when are chiasmata first visible?
22. Describe the role of meiosis in the life cycle of a vascular plant.
23. Contrast the chromatin fiber with the mitotic chromosome. How are the two structures related?
24. Describe the "folded-fiber" model of the mitotic chromosome.
25. You are given a metaphase chromosome preparation (a slide) from an unknown organism that contains 12 chromosomes. Two that are clearly smaller than the rest appear identical in length and centromere placement. Describe all that you can about these chromosomes.

**HOW DO WE KNOW?**

26. In this chapter, we focused on how chromosomes are distributed during cell division, both in dividing somatic cells (mitosis) and in gamete- and spore-forming cells (meiosis). At the same time, we found many opportunities to consider the methods and reasoning by which much of this information was acquired. From the explanations given in the chapter, what answers would you propose to the following fundamental questions:
    (a) How do we know that chromosomes exist in homologous pairs?
    (b) How do we know that DNA replication occurs during interphase, not early in mitosis?
    (c) How do we know that mitotic chromosomes are derived from chromatin?

## Extra-Spicy Problems

*As part of the "Problems and Discussion Questions" section in each chapter, we shall present a number of "Extra-Spicy" genetics problems. We have chosen to set these apart in order to identify problems that are particularly challenging. You may be asked to examine and assess actual data, to design genetics experiments, or to engage in cooperative learning. Like genetic varieties of peppers, some of these experiences are just spicy and some are very hot. Hopefully, all of them will leave an aftertaste that is pleasing to those willing to give them a try.*

For Questions 27–32, consider a diploid cell that contains three pairs of chromosomes designated AA, BB, and CC. Each pair contains a maternal and a paternal member (e.g., $A^m$ and $A^p$). Using these designations, demonstrate your understanding of mitosis and meiosis by drawing chromatid combinations as requested. Be sure to indicate when chromatids are paired as a result of replication and/or synapsis. You may wish to use a large piece of brown manila wrapping paper or a cut-up paper grocery bag for this project and to work in partnership with another student. We recommend cooperative learning as an efficacious way to develop the skills you will need for solving the problems presented throughout this text.

27. In mitosis, what chromatid combination(s) will be present during metaphase? What combination(s) will be present at each pole at the completion of anaphase?
28. During meiosis I, assuming no crossing over, what chromatid combination(s) will be present at the completion of prophase? Draw all possible alignments of chromatids as migration begins during early anaphase.
29. Are there any possible combinations present during prophase of meiosis II other than those that you drew in Problem 28? If so, draw them.
30. Draw all possible combinations of chromatids during the early phases of anaphase in meiosis II.
31. Assume that during meiosis I none of the *C* chromosomes disjoin at metaphase, but they separate into dyads (instead of monads) during meiosis II. How would this change the alignments that you constructed during the anaphase stages in meiosis I and II? Draw them.
32. Assume that each gamete resulting from Problem 31 fuses, in fertilization, with a normal haploid gamete. What combinations will result? What percentage of zygotes will be diploid, containing one paternal and one maternal member of each chromosome pair?
33. A species of cereal rye (*Secale cereale*) has a chromosome number of 14, while a species of Canadian wild rye (*Elymus canadensis*) has a chromosome number of 28. Sterile hybrids can be produced by crossing *Secale* with *Elymus*.
    (a) What would be the expected chromosome number in the somatic cells of the hybrids?
    (b) Assume that the G1 nuclear DNA content of *Elymus* is 25.5 picograms and that the G1 nuclear DNA content of *Secale* is 16.8 picograms. What would be the expected DNA content in a metaphase somatic cell of the hybrid?
    (c) Given that none of the chromosomes pair at meiosis I in the sterile hybrid (Hang and Franckowlak, 1984), speculate on the anaphase I separation patterns of these chromosomes.
34. An interesting procedure has been applied for assessing the chromosomal balance of potential secondary oocytes for use in human *in vitro* fertilization. Using fluorescence *in situ* hybridization (FISH), Kuliev and Verlinsky (2004) were able to identify individual chromosomes in first polar bodies and thereby infer the chromosomal makeup of "sister" oocytes.
    (a) Assume that when examining a first polar body you saw that it had one copy (dyad) of each chromosome but two dyads of chromosome 21. What would you expect to be the chromosomal 21 complement in the secondary oocyte? What consequences are likely in the resulting zygote, if the secondary oocyte was fertilized?
    (b) Assume that you were examining a first polar body and noted that it had one copy (dyad) of each chromosome except chromosome 21. Chromosome 21 was completely absent. What would you expect to be the chromosome 21 complement (only with respect to chromosome 21) in the secondary oocyte? What consequences are likely in the resulting zygote if the secondary oocyte was fertilized?
    (c) The authors state that there was a relatively high number of separation errors at meiosis I. In these cases the centromere underwent a premature division, occurring at meiosis I rather than meiosis II. Regarding chromosome 21, what would you expect to be the chromosome 21 complement in the secondary oocyte in which you saw a single chromatid (monad) for chromosome 21 in the first polar body? If this secondary oocyte was involved in fertilization, what would be the expected consequences?

Gregor Johann Mendel, who in 1866 put forward the major postulates of transmission genetics as a result of experiments with the garden pea.

3

# Mendelian Genetics

## CHAPTER CONCEPTS

- Inheritance is governed by information stored in discrete factors called genes.
- Genes are transmitted from generation to generation on vehicles called chromosomes.
- Chromosomes, which exist in pairs in diploid organisms, provide the basis of biparental inheritance.
- During gamete formation, chromosomes are distributed according to postulates first described by Gregor Mendel, based on his nineteenth-century research with the garden pea.
- Mendelian postulates prescribe that homologous chromosomes segregate from one another and assort independently with other segregating homologs during gamete formation.
- Genetic ratios, expressed as probabilities, are subject to chance deviation and may be evaluated statistically.
- The analysis of pedigrees allows predictions concerning the genetic nature of human traits.

Although inheritance of biological traits has been recognized for thousands of years, the first significant insights into how it takes place only occurred about 140 years ago. In 1866, Gregor Johann Mendel published the results of a series of experiments that would lay the foundation for the formal discipline of genetics. Mendel's work went largely unnoticed until the turn of the century, but eventually, the concept of the gene as a distinct hereditary unit was established. Since then, the ways in which genes, as segments of chromosomes, are transmitted to offspring and control traits have been clarified. Research has continued unabated throughout the twentieth century and into the present—indeed, studies in genetics, most recently at the molecular level, have remained at the forefront of biological research since the early 1900s.

When Mendel began his studies of inheritance using *Pisum sativum*, the garden pea, chromosomes and the role and mechanism of meiosis were totally unknown. Nevertheless, he determined that discrete *units of inheritance* exist and predicted their behavior in the formation of gametes. Subsequent investigators, with access to cytological data, saw a relationship between their own observations of chromosome behavior during meiosis and Mendel's principles of inheritance. Once this correlation was recognized, Mendel's postulates were accepted as the basis for the study of what is known as **transmission genetics**, how genes are transmitted from parents to offspring. These principles were derived directly from Mendel's experimentation. Even today, they serve as the cornerstone of the study of inheritance. In this chapter, we focus on the development of Mendel's principles.

3.1

## Mendel Used a Model Experimental Approach to Study Patterns of Inheritance

Johann Mendel was born in 1822 to a peasant family in the Central European village of Heinzendorf. An excellent student in high school, he studied philosophy for several years afterward and in 1843, taking the name Gregor, was admitted to the Augustinian Monastery of St. Thomas in Brno, now part of the Czech Republic. In 1849, he was relieved of pastoral duties and received a teaching appointment that lasted a number of years. From 1851 to 1853, he attended the University of Vienna, where he studied physics and botany. He returned to Brno in 1854, where he taught physics and natural science for the next 16 years. Mendel received support from the monastery for his studies and research throughout his life.

In 1856, Mendel performed his first set of hybridization experiments with the garden pea, launching the research phase of his career. His experiments continued until 1868, when he was elected abbot of the monastery. Although he retained his interest in genetics, his new responsibilities demanded most of his time. In 1884, Mendel died of a kidney disorder. The local newspaper paid him the following tribute:

> "His death deprives the poor of a benefactor, and mankind at large of a man of the noblest character, one who was a warm friend, a promoter of the natural sciences, and an exemplary priest."

Mendel first reported the results of some simple genetic crosses between certain strains of the garden pea in 1865. Although his was not the first attempt to provide experimental evidence pertaining to inheritance, Mendel's success where others had failed can be attributed, at least in part, to his elegant experimental design and analysis.

Mendel showed remarkable insight into the methodology necessary for good experimental biology. First, he chose an organism that was easy to grow and to hybridize artificially. The pea plant is self-fertilizing in nature, but it is easy to cross-breed experimentally. It reproduces well and grows to maturity in a single season. Mendel followed seven visible features (we refer to them as characters, or characteristics), each represented by two contrasting properties, or **traits** (Figure 3–1). For the character stem height, for example, he experimented with the traits *tall* and *dwarf*. He selected six other visibly contrasting pairs of traits involving seed shape and color, pod shape and color, and flower color and position. From local seed merchants, Mendel obtained true-breeding strains, those in which each trait appeared unchanged generation after generation in self-fertilizing plants.

There were several other reasons for Mendel's success. In addition to his choice of a suitable organism, he restricted his examination to one or very few pairs of contrasting traits in each experiment. He also kept accurate quantitative records, a necessity in genetic experiments. From the analysis of his data, Mendel derived certain postulates that have become the principles of transmission genetics.

The results of Mendel's experiments went unappreciated until the turn of the century, well after his death. However, once Mendel's publications were rediscovered by geneticists investigating the function and behavior of chromosomes, the implications of his postulates were immediately apparent. He had discovered the basis for the transmission of hereditary traits!

3.2

## The Monohybrid Cross Reveals How One Trait Is Transmitted from Generation to Generation

Mendel's simplest crosses involved only one pair of contrasting traits. Each such experiment is called a **monohybrid cross**. A monohybrid cross is made by mating true-breeding individuals from two parent strains, each exhibiting one of the two contrasting forms of

| Character | Contrasting traits | $F_1$ results | $F_2$ results | $F_2$ ratio |
|---|---|---|---|---|
| Seed shape | round/wrinkled | all round | 5474 round<br>1850 wrinkled | 2.96:1 |
| Seed color | yellow/green | all yellow | 6022 yellow<br>2001 green | 3.01:1 |
| Pod shape | full/constricted | all full | 882 full<br>299 constricted | 2.95:1 |
| Pod color | green/yellow | all green | 428 green<br>152 yellow | 2.82:1 |
| Flower color | violet/white | all violet | 705 violet<br>224 white | 3.15:1 |
| Flower position | axial/terminal | all axial | 651 axial<br>207 terminal | 3.14:1 |
| Stem height | tall/dwarf | all tall | 787 tall<br>277 dwarf | 2.84:1 |

FIGURE 3–1 Seven pairs of contrasting traits and the results of Mendel's seven monohybrid crosses of the garden pea (*Pisum sativum*). In each case, pollen derived from plants exhibiting one trait was used to fertilize the ova of plants exhibiting the other trait. In the $F_1$ generation, one of the two traits was exhibited by all plants. The contrasting trait reappeared in approximately 1/4 of the $F_2$ plants.

the character under study. Initially, we examine the first generation of offspring of such a cross, and then we consider the offspring of **selfing**, that is, of self-fertilization of individuals from this first generation. The original parents constitute the $\mathbf{P_1}$, or **parental generation**; their offspring are the $\mathbf{F_1}$, or **first filial generation**; the individuals resulting from the selfed $F_1$ generation are the $\mathbf{F_2}$, or **second filial generation**; and so on.

The cross between true-breeding pea plants with tall stems and dwarf stems is representative of Mendel's monohybrid crosses. *Tall* and *dwarf* are contrasting traits of the character of stem height. Unless tall or dwarf plants are crossed together or with another strain, they will undergo self-fertilization and breed true, producing their respective traits generation after generation. However, when Mendel crossed tall plants with dwarf plants, the resulting $F_1$ generation consisted of only tall plants. When members of the $F_1$ generation were selfed, Mendel observed that 787 of 1064 $F_2$ plants were tall, while 277 of 1064 were dwarf. Note that in this cross (Figure 3–1), the dwarf trait disappeared in the $F_1$ generation, only to reappear in the $F_2$ generation. Mendel made similar crosses between pea plants exhibiting each of the other pairs of contrasting traits. Results of these crosses are also shown in Figure 3–1. In every case, the outcome was similar to the tall/dwarf cross.

Genetic data are usually expressed and analyzed as ratios. In this particular example, many identical $P_1$ crosses were made and many $F_1$ plants—all tall—were produced. Of the 1064 $F_2$ offspring, 787 were tall and 277 were dwarf—a ratio of approximately 2.8:1.0, or about 3:1.

Mendel made similar crosses between pea plants exhibiting each of the other pairs of contrasting traits; the results of these crosses are shown in Figure 3–1. In every case, the outcome was similar to the tall/dwarf cross just described. For the character of interest, all $F_1$ offspring had the same trait exhibited by one of the parents, but in the $F_2$ offspring, an approximate ratio of 3:1 was obtained. That is, three-fourths looked like the $F_1$ plants, while one-fourth exhibited the contrasting trait, which had disappeared in the $F_1$ generation.

We should note one further aspect of Mendel's monohybrid crosses. In each cross, the $F_1$ and $F_2$ patterns of inheritance were similar regardless of which $P_1$ plant served as the source of pollen (sperm) and which served as the source of the ovum (egg). The crosses could be made either way—pollination of dwarf plants by tall plants, or vice versa. Crosses made in both these ways are called **reciprocal crosses**. Therefore, the results of Mendel's monohybrid crosses were not sex-dependent.

To explain these results, Mendel proposed the existence of *particulate unit factors* for each trait. He suggested that these factors serve as the basic units of heredity and are passed unchanged from generation to generation, determining various traits expressed by each individual plant. Using these general ideas, Mendel proceeded to hypothesize precisely how such factors could account for the results of the monohybrid crosses.

## Mendel's First Three Postulates

Using the consistent pattern of results in the monohybrid crosses, Mendel derived the following three postulates, or principles, of inheritance.

1. UNIT FACTORS IN PAIRS

*Genetic characters are controlled by unit factors existing in pairs in individual organisms.*

In the monohybrid cross involving tall and dwarf stems, a specific unit factor exists for each trait. Each diploid individual receives one factor from each parent. Because the factors occur in pairs, three combinations are possible: two factors for tall stems, two factors for dwarf stems, or one of each factor. Every individual possesses one of these three combinations, which determines stem height.

2. DOMINANCE/RECESSIVENESS

*When two unlike unit factors responsible for a single character are present in a single individual, one unit factor is dominant to the other, which is said to be recessive.*

In each monohybrid cross, the trait expressed in the $F_1$ generation is controlled by the dominant unit factor. The trait not expressed is controlled by the recessive unit factor. The terms *dominant* and *recessive* are also used to designate traits. In this case, tall stems are said to be dominant over recessive dwarf stems.

3. SEGREGATION

*During the formation of gametes, the paired unit factors separate, or segregate, randomly so that each gamete receives one or the other with equal likelihood.*

If an individual contains a pair of like unit factors (e.g., both specific for tall), then all its gametes receive one of that same kind of unit factor (in this case, tall). If an individual contains unlike unit factors (e.g., one for tall and one for dwarf), then each gamete has a 50 percent probability of receiving either kind of unit factor (either the tall or the dwarf).

These postulates provide a suitable explanation for the results of the monohybrid crosses. Let's use the tall/dwarf cross to illustrate. Mendel reasoned that $P_1$ tall plants contained identical paired unit factors, as did the $P_1$ dwarf plants. The gametes of tall plants all receive one tall unit factor as a result of **segregation**. Similarly, the gametes of dwarf plants all receive one dwarf unit factor. Following fertilization, all $F_1$ plants receive one unit factor from each parent—a tall factor from one and a dwarf factor from the other—reestablishing the paired relationship, but because tall is dominant to dwarf, all $F_1$ plants are tall.

When $F_1$ plants form gametes, the postulate of segregation demands that each gamete randomly receives either the tall *or* dwarf unit factor. Following random fertilization events during $F_1$ selfing, four $F_2$ combinations will result with equal frequency:

1. tall/tall
2. tall/dwarf
3. dwarf/tall
4. dwarf/dwarf

Combinations (1) and (4) will clearly result in tall and dwarf plants, respectively. According to the postulate of dominance/recessiveness, combinations (2) and (3) will both yield tall plants. Therefore, the $F_2$ is predicted to consist of 3/4 tall and 1/4 dwarf, or a ratio of 3:1. This is approximately what Mendel observed in his cross between tall and dwarf plants. A similar pattern was observed in each of the other monohybrid crosses (Figure 3–1).

## Modern Genetic Terminology

To analyze the monohybrid cross and Mendel's first three postulates, we must first introduce several new terms as well as a symbol convention for the unit factors.

Traits such as tall or dwarf are physical expressions of the information contained in unit factors. The physical expression of a trait is the **phenotype** of the individual. Mendel's unit factors represent units of inheritance called **genes** by modern geneticists. For any given character, such as plant height, the phenotype is determined by alternative forms of a single gene, called **alleles**. For example, the unit factors representing tall and dwarf are alleles determining the height of the pea plant.

Geneticists have several different systems for using symbols to represent genes. In Chapter 4, we will review a number of these conventions, but for now, we will adopt one to use consistently throughout this chapter. According to this convention, the first letter of the recessive trait symbolizes the character in question; in lowercase italic, it designates the allele for the recessive trait, and in uppercase italic, it designates the allele for the dominant trait. Thus for Mendel's pea plants, we use *d* for the *d*warf allele and *D* for the tall allele. When alleles are written in pairs to represent the two unit factors present in any individual (*DD*, *Dd*, or *dd*), the resulting symbol is called the **genotype**. The genotype designates the genetic makeup of an individual for the trait or traits it describes, whether the individual is haploid or diploid. By reading the genotype, we know the phenotype of the individual: *DD* and *Dd* are tall, and *dd* is dwarf. When both alleles are the same (*DD* or *dd*), the individual is **homozygous** for the trait, or a **homozygote**; when the alleles are different (*Dd*), we use the terms **heterozygous** and **heterozygote**. These symbols and terms are used in Figure 3–2 to describe the monohybrid cross.

## Mendel's Analytical Approach

What led Mendel to deduce that unit factors exist in pairs? Because there were two contrasting traits for each of the characters he chose, it seemed logical that two distinct factors must exist. However, why does one of the two traits or phenotypes disappear in the $F_1$ generation? Observation of the $F_2$ generation helps to answer this question. The recessive trait and its unit factor do not actually disappear in the $F_1$; they are merely hidden or masked, only to reappear in one-fourth of the $F_2$ offspring. Therefore, Mendel concluded that one unit factor for tall and one for dwarf were transmitted to each $F_1$ individual, but that because the tall factor or allele is dominant to the dwarf

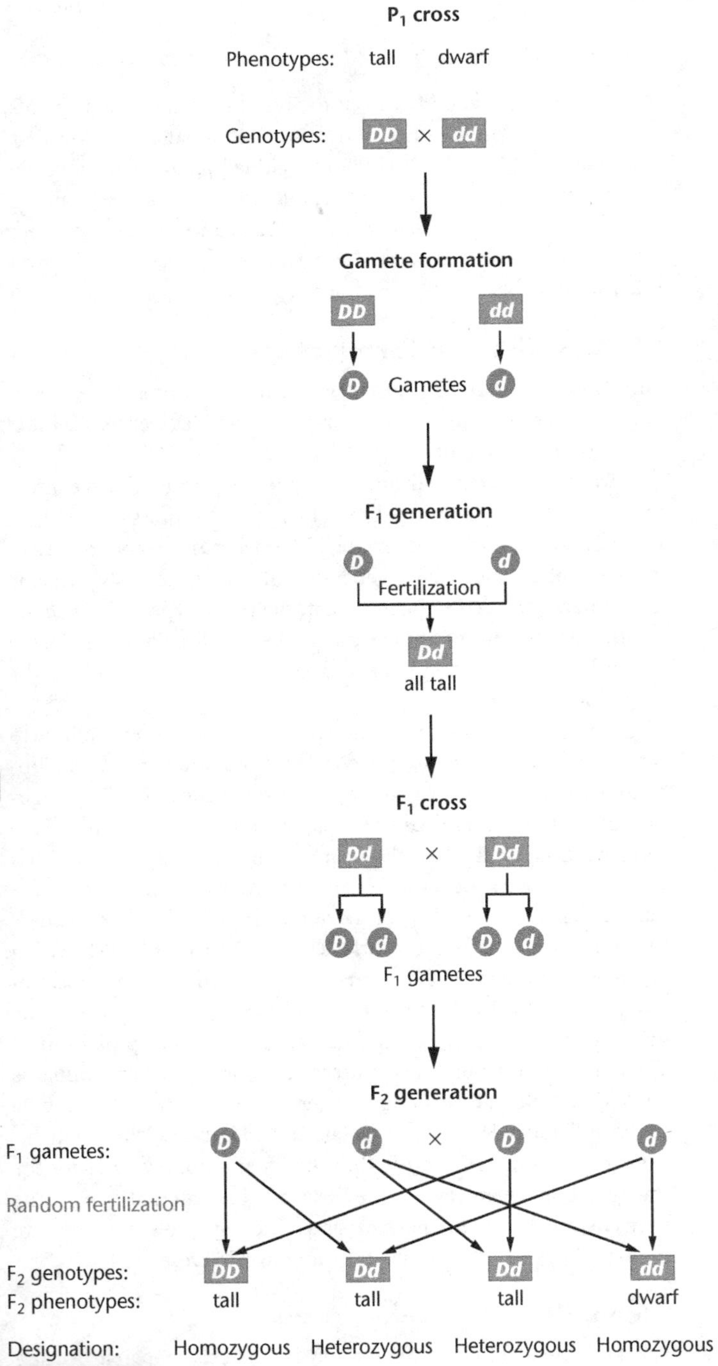


FIGURE 3-2 The monohybrid cross between tall (*D*) and dwarf (*d*) pea plants. Individuals are shown in rectangles, and gametes are shown in circles.

MONOHYBRID CROSS AND PUNNETT SQUARES WEB TUTORIAL 3.1

factor or allele, all $F_1$ plants are tall. Given this information, we can ask how Mendel explained the 3:1 $F_2$ ratio. As shown in Figure 3–2, Mendel deduced that the tall and dwarf alleles of the $F_1$ heterozygote segregate randomly into gametes. If fertilization is random, this ratio is predicted. If a large population of offspring is generated, the outcome of such a cross should reflect the 3:1 ratio.

Because he operated without the hindsight that modern geneticists enjoy, Mendel's analytical reasoning must be considered a truly outstanding scientific achievement. On the basis of rather simple, but precisely executed breeding experiments, he not only proposed that discrete particulate units of heredity exist, but he also explained how they are transmitted from one generation to the next.

NOW SOLVE THIS

Problem 6 on page 66 describes a set of crosses in pigeons and asks you to determine the mode of inheritance and the genotypes of the parents and offspring in a number of instances.

■ HINT: *The first step is to determine whether there is more than one gene pair involved. To do so, convert the data to ratios that are characteristic of Mendelian crosses. In the case of this problem, ask first whether any of the $F_2$ ratios match Mendel's 3:1 monohybrid ratio.*

## Punnett Squares

The genotypes and phenotypes resulting from combining gametes during fertilization can be easily visualized by constructing a diagram called a **Punnett square**, named after the person who first devised this approach, Reginald C. Punnett. Figure 3–3 illustrates this method of analysis for our $F_1 \times F_1$ monohybrid cross. Each of the possible gametes is assigned a column or a row; the vertical columns represent those of the female parent, and the horizontal rows represent those of the male parent. After assigning the gametes to the rows and columns, we predict the new generation by entering the male and female gametic information into each box and thus producing every possible resulting genotype. By filling out the Punnett square, we are listing all possible random fertilization events. The genotypes and phenotypes of all potential offspring are ascertained by reading the combinations in the boxes.

The Punnett square method is particularly useful when you are first learning about genetics and how to solve genetics problems. Note the ease with which the 3:1 phenotypic ratio and the 1:2:1 genotypic ratio may be derived for the $F_2$ generation in Figure 3–3.

## The Testcross: One Character

Tall plants produced in the $F_2$ generation are predicted to have either the *DD* or the *Dd* genotype. You might ask if there is a way to distinguish the genotype. Mendel devised a rather simple method that is still used today to discover the genotype of plants and animals: the **testcross.** The organism expressing the dominant phenotype but having an unknown genotype is crossed with a known *homozygous recessive individual.* For example, as shown in Figure 3–4(a), if a tall plant of genotype *DD* is testcrossed with a dwarf plant, which must have the *dd* genotype, all offspring will be tall phenotypically and *Dd* genotypically. However, as shown in Figure 3–4(b), if a tall plant is *Dd* and is crossed with a dwarf plant (*dd*), then

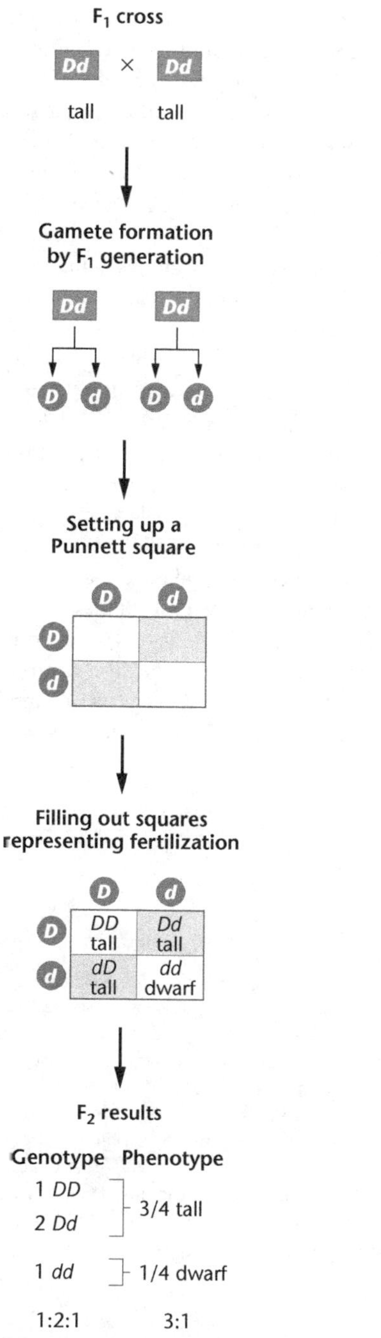


FIGURE 3-3 A Punnett square generating the $F_2$ ratio of the $F_1 \times F_1$ cross shown in Figure 3-2.

one-half of the offspring will be tall (*Dd*) and the other half will be dwarf (*dd*). Therefore, a 1:1 tall/dwarf ratio demonstrates the heterozygous nature of the tall plant of unknown genotype. The results of the testcross reinforced Mendel's conclusion that separate unit factors control traits.

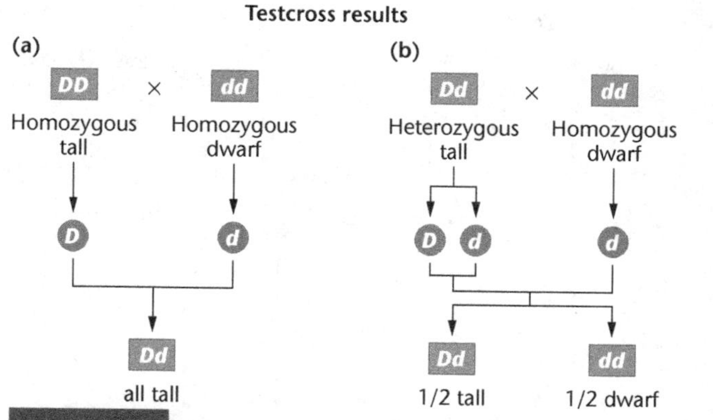


FIGURE 3-4 Testcross of a single character. In (a), the tall parent is homozygous, but in (b), the tall parent is heterozygous. The genotype of each tall $P_1$ plant can be determined by examining the offspring when each is crossed with the homozygous recessive dwarf plant.

3.3

## Mendel's Dihybrid Cross Generated a Unique $F_2$ Ratio

WEB TUTORIAL 3.2 INDEPENDENT ASSORTMENT

As a natural extension of the monohybrid cross, Mendel also designed experiments in which he examined two characters simultaneously. Such a cross, involving two pairs of contrasting traits, is a **dihybrid cross**, or a *two-factor cross*. For example, if pea plants having yellow seeds that are round were bred with those having green seeds that are wrinkled, the results shown in Figure 3–5 would occur: the $F_1$ offspring would all be yellow and round. It is therefore apparent that yellow is dominant to green and that round is dominant to wrinkled. When the $F_1$ individuals are selfed, approximately 9/16 of the $F_2$ plants express the yellow and round traits, 3/16 express yellow and wrinkled, 3/16 express green and round, and 1/16 express green and wrinkled.

A variation of this cross is also shown in Figure 3–5. Instead of crossing one $P_1$ parent with both dominant traits (yellow, round) to one with both recessive traits (green, wrinkled), plants with yellow, wrinkled seeds are crossed with those with green, round seeds. In spite of the change in the $P_1$ phenotypes, both the $F_1$ and $F_2$ results remain unchanged. It will become clear in the next section why this is so.

### Mendel's Fourth Postulate: Independent Assortment

We can most easily understand the results of a dihybrid cross if we consider it theoretically as consisting of two monohybrid crosses conducted separately. Think of the two sets of traits as being inherited independently of each other; that is, the chance of any plant having yellow or green seeds is not at all influenced by the chance that this plant will have round or wrinkled seeds. Thus, because yellow is dominant to green, all $F_1$ plants in the first theoretical cross would have yellow seeds. In the second theoretical cross, all $F_1$ plants would have round seeds because round is dominant to wrinkled. When Mendel

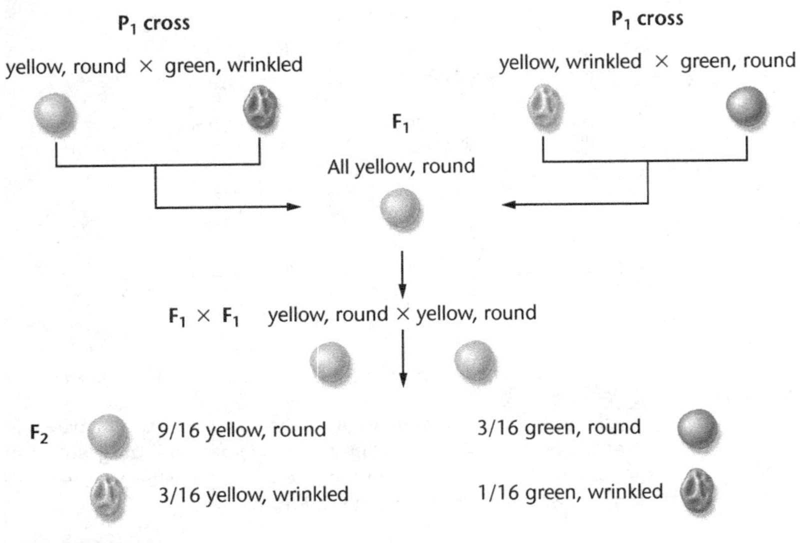


**FIGURE 3–5** $F_1$ and $F_2$ results of Mendel's dihybrid crosses in which the plants on the top left with yellow, round seeds are crossed with plants having green, wrinkled seeds, and the plants on the top right with yellow, wrinkled seeds are crossed with plants having green, round seeds.

examined the $F_1$ plants of the dihybrid cross, all were yellow and round, as our theoretical cross just predicted.

The predicted $F_2$ results of the first cross are 3/4 yellow and 1/4 green. Similarly, the second cross would yield 3/4 round and 1/4 wrinkled. Figure 3–5 shows that in the dihybrid cross, 12/16 $F_2$ plants are yellow, while 4/16 are green, exhibiting the expected 3:1 (3/4:1/4) ratio. Similarly, 12/16 of all $F_2$ plants have round seeds, while 4/16 have wrinkled seeds, again revealing the 3:1 ratio.

These numbers show that the two pairs of contrasting traits are inherited independently, so we can predict the frequencies of all possible $F_2$ phenotypes by applying the **product law** of probabilities: *When two independent events occur simultaneously, the probability of the two outcomes occurring in combination is equal to the product of their individual probabilities of occurrence.* For example, the probability of an $F_2$ plant having yellow and round seeds is (3/4)(3/4), or 9/16, because 3/4 of all $F_2$ plants should be yellow and 3/4 of all $F_2$ plants should be round.

In a like manner, the probabilities of the other three $F_2$ phenotypes can be calculated: yellow (3/4) and wrinkled (1/4) are predicted to be present together 3/16 of the time; green (1/4) and round (3/4) are predicted 3/16 of the time; and green (1/4) and wrinkled (1/4) are predicted 1/16 of the time. These calculations are shown in Figure 3–6.

## How Mendel's Peas Become Wrinkled: A Molecular Explanation

Only recently, well over a hundred years after Mendel used wrinkled peas in his groundbreaking hybridization experiments, have we come to find out how the *wrinkled* gene makes peas wrinkled. The wild-type allele of the gene encodes a protein called *starch-branching enzyme (SBEI)*. This enzyme catalyzes the formation of highly branched starch molecules as the seed matures.

Wrinkled peas, which result from the homozygous presence of the mutant form of the gene, lack the activity of this enzyme. As a consequence, the production of branch points is inhibited during the synthesis of starch within the seed, which in turn leads to the accumulation of more sucrose and a higher water content while the seed develops. Osmotic pressure inside the seed rises, causing the seed to lose water, ultimately resulting in a wrinkled appearance at maturity. In contrast, developing seeds that bear at least one copy of the normal gene (being either homozygous or heterozygous for the dominant allele) synthesize starch and achieve an osmotic balance that minimizes the loss of water. The end result for them is a smooth-textured outer coat.

Cloning and analyis of the *SBEI* gene has provided new insight into the relationships between genotypes and phenotypes. Interestingly, the mutant gene contains a foreign sequence of some 800 base pairs that disrupts the normal coding sequence. This foreign segment closely resembles sequences called *transposable elements* that have been discovered to have the ability to move from place to place in the genome of certain organisms. Transposable elements have been found in maize (corn), parsley, snapdragons, and fruit flies, among many other organisms.

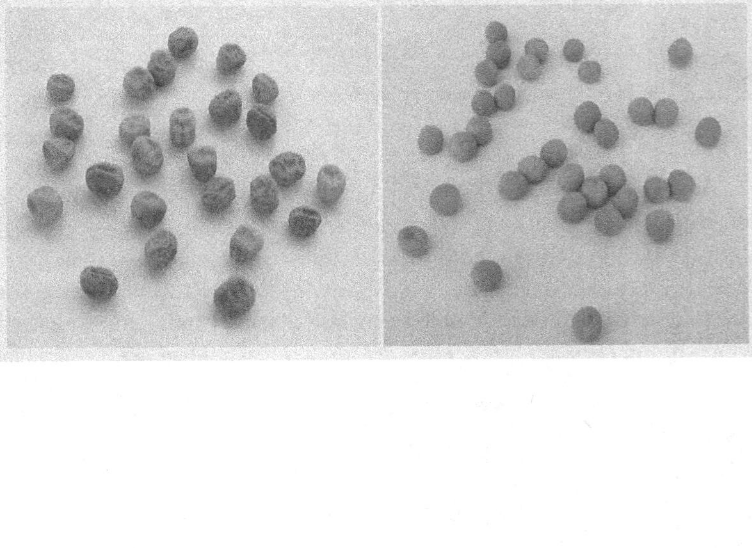

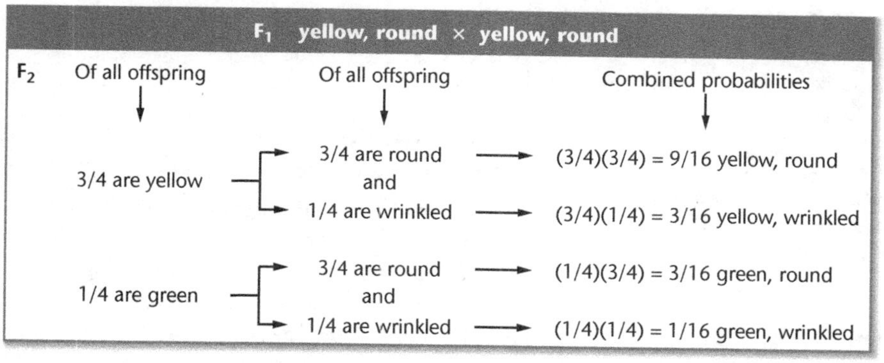


**FIGURE 3–6** Computation of the combined probabilities of each $F_2$ phenotype for two independently inherited characters. The probability of each plant being yellow or green is independent of the probability of it bearing round or wrinkled seeds.

It is now apparent why the $F_1$ and $F_2$ results are identical whether the initial cross is yellow round plants bred with green, wrinkled plants, or whether yellow, wrinkled plants are bred with green, round plants. In both crosses, the $F_1$ genotype of all offspring is identical. As a result, the $F_2$ generation is also identical in both crosses.

On the basis of similar results in numerous dihybrid crosses, Mendel proposed a fourth postulate:

4. INDEPENDENT ASSORTMENT
   *During gamete formation, segregating pairs of unit factors assort independently of each other.*
   This postulate stipulates that segregation of any pair of unit factors occurs independently of all others. As a result of random segregation, each gamete receives one member of every pair of unit factors. For one pair, whichever unit factor is received does not influence the outcome of segregation of any other pair. Thus, according to the postulate of **independent assortment**, all possible combinations of gametes should be formed in equal frequency.

The Punnett square in Figure 3–7 shows how independent assortment works in the formation of the $F_2$ generation. Examine the formation of gametes by the $F_1$ plants; segregation prescribes that every gamete receives either a *G* or *g* allele and a *W* or *w* allele. Independent assortment stipulates that all four combinations (*GW*, *Gw*, *gW*, and *gw*) will be formed with equal probabilities.

In every $F_1 \times F_1$ fertilization event, each zygote has an equal probability of receiving one of the four combinations from each parent. If many offspring are produced, 9/16 have yellow, round seeds, 3/16 have yellow, wrinkled seeds, 3/16 have green, round seeds, and 1/16 have green, wrinkled seeds, yielding what is designated as **Mendel's 9:3:3:1 dihybrid ratio.** This is an ideal ratio based on probability events involving segregation, independent assortment, and random fertilization. Because of deviation due strictly to chance, particularly if small numbers of offspring are produced, actual results are highly unlikely to match the ideal ratio.

### The Testcross: Two Characters

The testcross may also be applied to individuals that express two dominant traits but whose genotypes are unknown. For example, the expression of the yellow, round seed phenotype in the $F_2$ generation just described may result from the *GGWW*, *GGWw*, *GgWW*, or *GgWw* genotypes. If an $F_2$ yellow, round plant is crossed with the homozygous recessive green, wrinkled plant (*ggww*), analysis of the offspring will indicate the exact genotype of that yellow, round plant. Each of the above genotypes results in a different set of gametes and, in a testcross, a different set of phenotypes in the resulting offspring. You may wish to work out the results of each of these four crosses before examining the predicted outcomes shown in Figure 3–8, where three cases are illustrated.

**NOW SOLVE THIS**

Problem 9 on page 66 presents a series of Mendelian dihybrid crosses and asks you to determine the genotypes of the parents.

■ HINT: *In each case, write down everything that you know for certain. This reduces the problem to its bare essentials and clarifies what remains to be figured out. For example, the wrinkled, yellow plant in case (b) must be homozygous for the recessive wrinkled alleles and bear at least one dominant allele for the yellow trait. Having established this, you need only determine the remaining allele for cotyledon color.*

## 3.4 The Trihybrid Cross Demonstrates That Mendel's Principles Apply to Inheritance of Multiple Traits

Thus far, we have considered inheritance of up to two pairs of contrasting traits. Mendel demonstrated that the processes of segregation and independent assortment also apply to three pairs of contrasting traits, in what is called a **trihybrid cross**, or *three-factor cross.*

Although a trihybrid cross is somewhat more complex than a dihybrid cross, its results are easily calculated if the principles of segregation and independent assortment are followed. For example, consider the cross shown in Figure 3–9 where the gene pairs of theoretical contrasting traits are represented by the symbols *A*, *a*, *B*, *b*, *C*, and *c*. In the cross between *AABBCC* and *aabbcc* individuals, all $F_1$ individuals are heterozygous for all three gene pairs. Their genotype, *AaBbCc*, results in the phenotypic expression of the dominant *A*, *B*, and *C* traits. When $F_1$ individuals serve as parents, each produces eight different gametes in equal frequencies. At this point, we could construct a Punnett square with 64 separate boxes and read out the phenotypes—but such a

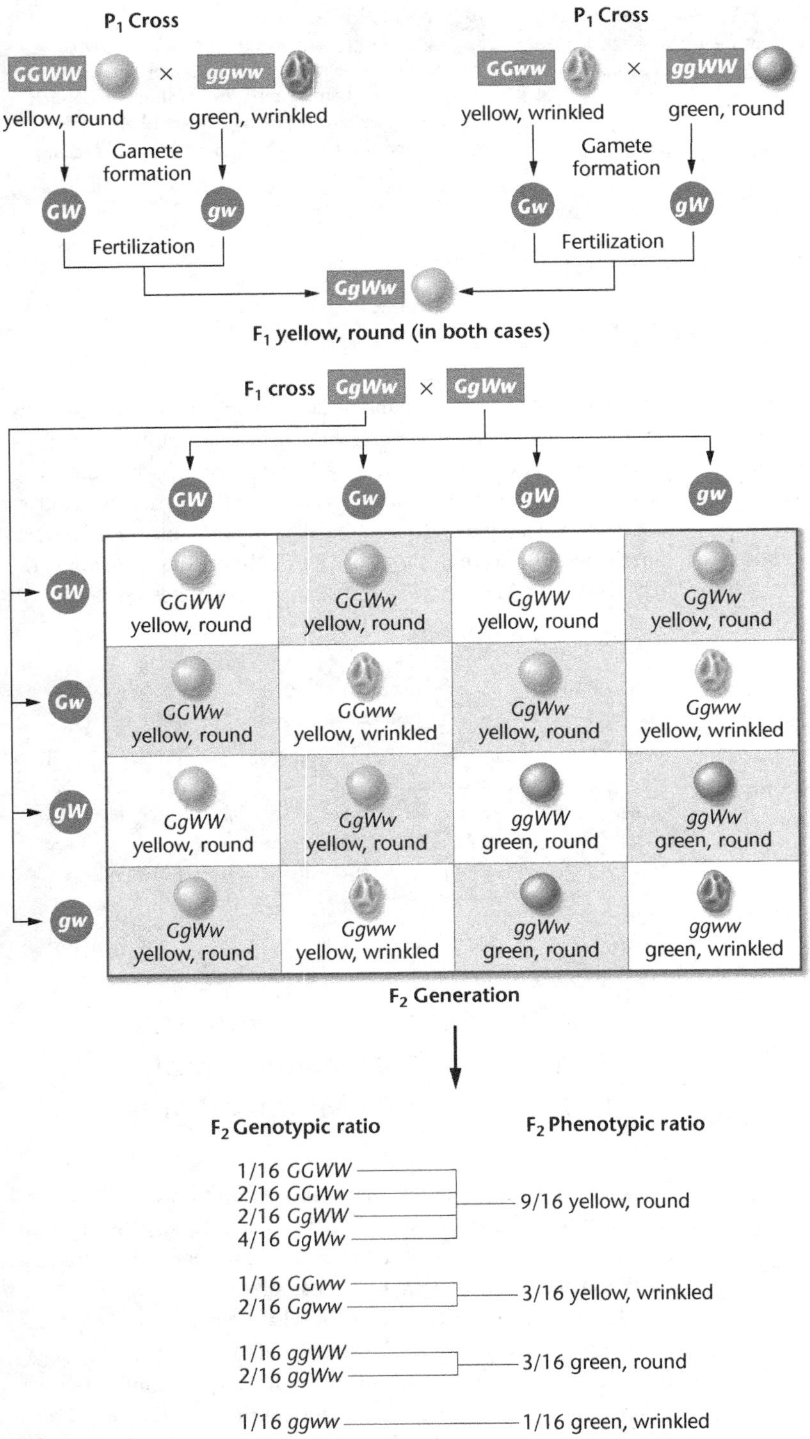


FIGURE 3–7 Analysis of the dihybrid crosses shown in Figure 3–5. The $F_1$ heterozygous plants are self-fertilized to produce an $F_2$ generation, which is computed using a Punnett square. Both the phenotypic and genotypic $F_2$ ratios are shown.

method is cumbersome in a cross involving so many factors. Therefore, another method has been devised to calculate the predicted ratio.

## The Forked-Line Method, or Branch Diagram

It is much less difficult to consider each contrasting pair of traits separately and then to combine these results by using the **forked-line method**, first shown in Figure 3–6. This method, also called a **branch diagram**, relies on the simple application of the laws of probability established for the dihybrid cross. Each gene pair is assumed to behave independently during gamete formation.

When the monohybrid cross $AA \times aa$ is made, we know that:

1. All $F_1$ individuals have the genotype *Aa* and express the phenotype represented by the *A* allele, which is called the *A* phenotype in the discussion that follows.

2. The $F_2$ generation consists of individuals with either the *A* phenotype or the *a* phenotype in the ratio of 3:1.

The same generalizations can be made for the $BB \times bb$ and $CC \times cc$ crosses. Thus, in the $F_2$ generation, 3/4 of all organisms will express phenotype *A*, 3/4 will express *B*, and 3/4 express *C*. Similarly, 1/4 of all organisms will express *a*, 1/4 will express *b*, and 1/4 will express *c*. The proportions of organisms that express each phenotypic combination can be predicted by assuming that fertilization, following the independent assortment of these three gene pairs during gamete formation, is a random process. We apply the product law of probabilities once again. Figure 3–10 uses the forked-line method to calculate the phenotypic proportions of the $F_2$ generation. They fall into the trihybrid ratio of 27:9:9:9:3:3:3:1. The same method can be used to solve crosses involving any number of gene pairs, *provided that all gene pairs assort independently from each other*. We shall see later that gene pairs do not always assort with complete independence. However, it appeared to be true for all of Mendel's characters.

Note that in Figure 3–10, only phenotypic ratios of the $F_2$ generation have been derived. It is possible to generate genotypic ratios as well. To do so, we again consider the *A/a*, *B/b*, and *C/c* gene pairs separately. For example, for the *A/a* pair, the $F_1$ cross is $Aa \times Aa$. Phenotypically, an $F_2$ ratio of 3/4 *A*:1/4 *a* is produced. Genotypically, however, the $F_2$ ratio is different—1/4 *AA*:1/2 *Aa*:1/4 *aa* will result. Using Figure 3–10 as a model, we would enter these genotypic frequencies in the leftmost column of the diagram. Each would be connected by three lines to 1/4 *BB*, 1/2 *Bb*, and 1/4 *bb*, respectively. From each of these nine designations, three more lines would extend to the 1/4 *CC*, 1/2 *Cc*, and 1/4 *cc* genotypes. On the right side of the completed diagram, 27 genotypes and their frequencies of occurrence would appear.

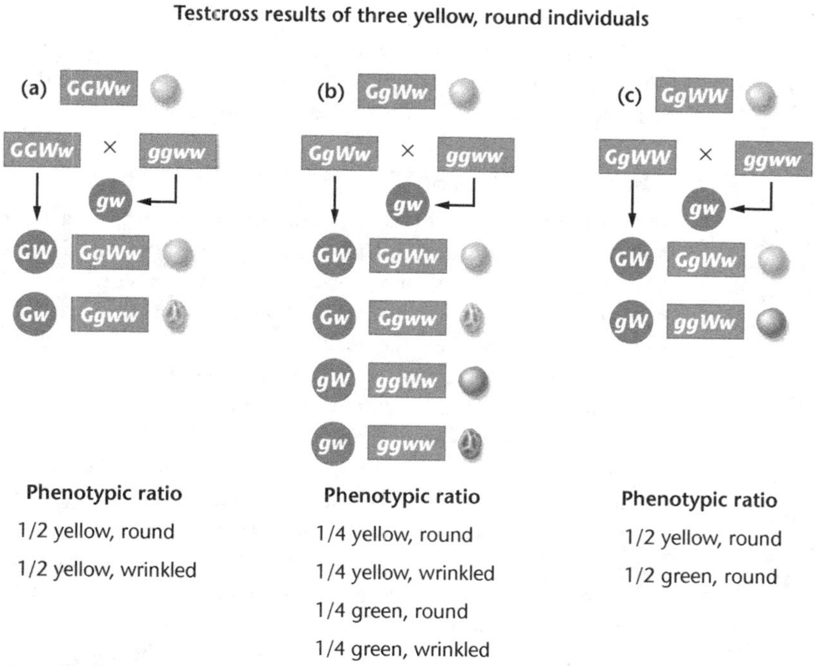


FIGURE 3-8 The testcross illustrated with two independent characters.

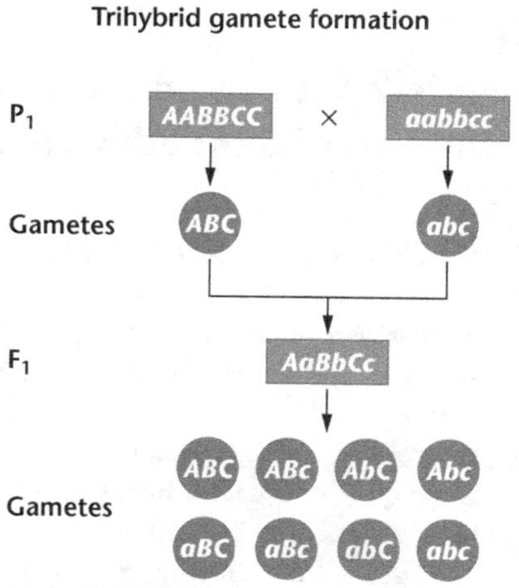


FIGURE 3-9 Formation of $P_1$ and $F_1$ gametes in a trihybrid cross.

**Generation of $F_2$ trihybrid phenotypes**

| A or a | B or b | C or c | Combined proportion |
|---|---|---|---|
| 3/4 A | 3/4 B | 3/4 C | → (3/4)(3/4)(3/4) *ABC* = 27/64 *ABC* |
| | | 1/4 c | → (3/4)(3/4)(1/4) *ABc* = 9/64 *ABc* |
| | 1/4 b | 3/4 C | → (3/4)(1/4)(3/4) *AbC* = 9/64 *AbC* |
| | | 1/4 c | → (3/4)(1/4)(1/4) *Abc* = 3/64 *Abc* |
| 1/4 a | 3/4 B | 3/4 C | → (1/4)(3/4)(3/4) *aBC* = 9/64 *aBC* |
| | | 1/4 c | → (1/4)(3/4)(1/4) *aBc* = 3/64 *aBc* |
| | 1/4 b | 3/4 C | → (1/4)(1/4)(3/4) *abC* = 3/64 *abC* |
| | | 1/4 c | → (1/4)(1/4)(1/4) *abc* = 1/64 *abc* |

FIGURE 3-10 Generation of the $F_2$ trihybrid phenotypic ratio using the forked-line method. This method is based on the expected probability of occurrence of each phenotype.

**NOW SOLVE THIS**

Problem 17 on page 67 asks you to use the forked-line method to determine the outcome of a number of trihybrid crosses.

HINT: *In using the forked-line method, consider each gene pair separately. For example, for each cross, first predict the outcome of the A/a genes; for each of those outcomes, write predictions for the B/b genes; and finally, for each of those outcomes, write predictions for the C/c genes. At that point, you will be ready to multiply across to determine the proportionate numbers of all the different possible combinations.*

In crosses involving two or more gene pairs, the calculation of gametes and genotypic and phenotypic results is quite complex. Several simple mathematical rules will enable you to check the accuracy of various steps required in working these problems. First, you must determine the number of different *heterozygous* gene pairs ($n$) involved in the cross. For example, where *AaBb* × *AaBb*represents the cross, $n = 2$; for *AaBbCc* × *AaBcCc*, $n = 3$; for *AaBBCcDd* × *AaBBCcDd*, $n = 3$ (because the *B* genes are not heterozygous). Once $n$ is determined, $2^n$ is the number of different gametes that can be formed by each parent; $3^n$ is the number of different genotypes that result following fertilization; and $2^n$ is the number of different phenotypes that are produced from these genotypes. Table 3.1 summarizes these rules, which may be applied to crosses involving any number of genes, *provided that they assort independently from one another.*

TABLE 3.1

**Simple Mathematical Rules Useful in Working Genetics Problems**

| Crosses between Organisms Heterozygous for Genes Exhibiting Independent Assortment | | | |
|---|---|---|---|
| Number of Heterozygous Gene Pairs | Number of Different Types of Gametes Formed | Number of Different Genotypes Produced | Number of Different Phenotypes Produced* |
| $n$ | $2^n$ | $3^n$ | $2^n$ |
| 1 | 2 | 3 | 2 |
| 2 | 4 | 9 | 4 |
| 3 | 8 | 27 | 8 |
| 4 | 16 | 81 | 16 |

*The fourth column assumes a simple dominant-recessive relationship in each gene pair.

3.5

## Mendel's Work Was Rediscovered in the Early Twentieth Century

Mendel initiated his work in 1856, presented it to the Brünn Society of Natural Science in 1865, and published it the following year. While his findings were often cited and discussed, their significance went unappreciated for about 35 years. Many explanations have been proposed for this delay.

First, Mendel's adherence to mathematical analysis of probability events was quite unusual for biological studies in those days. Perhaps it seemed foreign to his contemporaries. More important, his conclusions did not fit well with existing hypotheses concerning the cause of variation among organisms. The topic of natural variation intrigued students of evolutionary theory. This group, stimulated by the proposal developed by Charles Darwin and Alfred Russel Wallace, ascribed to the theory of **continuous variation**, which held that offspring were a blend of their parents' phenotypes. As we mentioned earlier, Mendel theorized that variation was due to a dominance–recessive relationship between discrete or particulate units, resulting in **discontinuous variation**. For example, note that the $F_2$ flowers in Figure 3–1 are either white or violet, never something intermediate. Mendel proposed that the $F_2$ offspring of a dihybrid cross are expressing traits produced by new combinations of previously existing unit factors. As a result, Mendel's hypotheses did not fit well with the evolutionists' preconceptions about causes of variation.

It is also likely that Mendel's contemporaries failed to realize that Mendel's postulates explained *how* variation was transmitted to offspring. Instead, they may have attempted to interpret his work in a way that addressed the issue of *why* certain phenotypes survive preferentially. It was this latter question that had been addressed in the theory of natural selection, but it was not addressed by Mendel. The collective vision of Mendel's scientific colleagues may have been obscured by the impact of Darwin's extraordinary theory of organic evolution.

3.6

## The Correlation of Mendel's Postulates with the Behavior of Chromosomes Provided the Foundation of Modern Transmission Genetics

In the latter part of the nineteenth century, a remarkable observation set the scene for the recognition of Mendel's work: Walter Flemming's discovery of chromosomes in the nuclei of salamander cells. In 1879, Flemming described the behavior of these threadlike structures during cell division. As a result of his findings and the work of many other cytologists, the presence of discrete units within the nucleus soon became an integral part of scientists' ideas about inheritance. It was this mind-set that prompted a reexamination of Mendel's findings.

### The Chromosomal Theory of Inheritance

In the early twentieth century, hybridization experiments similar to Mendel's were performed independently by three botanists, Hugo de Vries, Karl Correns, and Erich Tschermak. De Vries's work demonstrated the principle of segregation in several plant species. Apparently, he searched the existing literature and found that Mendel's work had anticipated his own conclusions! Correns and Tschermak also reached conclusions similar to those of Mendel.

In 1902, two cytologists, Walter Sutton and Theodor Boveri, independently published papers linking their discoveries of the behavior of chromosomes during meiosis to the Mendelian principles of segregation and independent assortment. They pointed out that the separation of chromosomes during meiosis could serve as the cytological basis of these two postulates. Although they thought that Mendel's unit factors were probably chromosomes rather than genes on chromosomes, their findings reestablished the importance of Mendel's work and led to many ensuing genetic investigations. Sutton and Boveri are credited with initiating the **chromosomal theory of inheritance**, the idea that the genetic material in living organisms is contained in chromosomes, which was developed during the next two decades. As we will see in subsequent chapters, work by Thomas H. Morgan, Alfred H. Sturtevant, Calvin Bridges, and others established beyond a reasonable doubt that Sutton's and Boveri's hypothesis was correct.

### Unit Factors, Genes, and Homologous Chromosomes

Because the correlation between Sutton's and Boveri's observations and Mendelian principles serves as the foundation for the modern description of transmission genetics, we will examine this correlation in some depth before moving on to other topics.

As we know, each species possesses a specific number of chromosomes in each somatic cell nucleus. For diploid organisms, this number is called the **diploid number ($2n$)** and is characteristic of that species. During the formation of gametes (meiosis), the number is precisely halved ($n$), and when two gametes combine during fertilization, the diploid number is reestablished. During meiosis, however, the chromosome number is not reduced in a random manner. It was apparent to early cytologists that the diploid number of chromosomes is composed of homologous pairs identifiable by their morphological appearance and behavior. The gametes contain one member of each pair—thus the chromosome complement of a gamete is quite specific, and the number of chromosomes in each gamete is equal to the haploid number.

With this basic information, we can see the correlation between the behavior of unit factors and chromosomes and genes. Figure 3–11 shows three of Mendel's postulates and the chromosomal

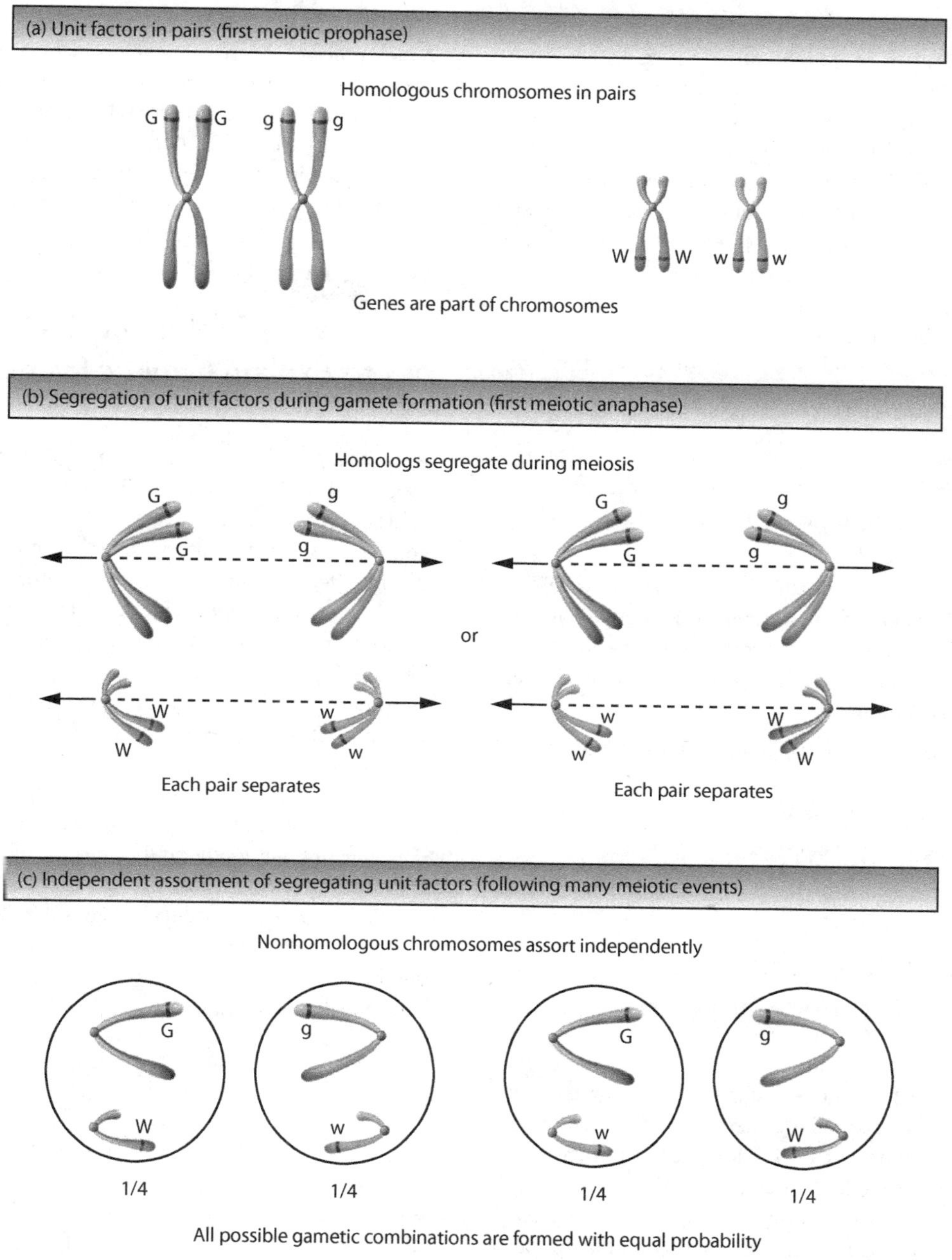


**FIGURE 3–11** Illustrated correlation between the Mendelian postulates of (a) unit factors in pairs, (b) segregation, and (c) independent assortment, showing the presence of genes located on homologous chromosomes and their behavior during meiosis.

explanation of each. Unit factors are really genes located on homologous pairs of chromosomes [Figure 3–11(a)]. Members of each pair of homologs separate, or segregate, during gamete formation [Figure 3–11(b)]. In the figure, two different alignments are possible, both of which are shown.

To illustrate the principle of independent assortment, it is important to distinguish between members of any given homologous pair of chromosomes. One member of each pair is derived from the **maternal parent**, whereas the other comes from the **paternal parent**. (We represent the different parental origins with different colors.) As shown in Figure 3–11(c), following independent segregation of each pair of homologs, each gamete receives one member from each pair of chromosomes. All possible combinations are formed with equal probability. If we add the symbols used in Mendel's dihybrid cross (*G*, *g* and *W*, *w*) to the diagram, we can see why equal numbers of the four types of gametes are formed. The independent behavior of

Mendel's pairs of unit factors (*G* and *W* in this example) is due to their presence on separate pairs of homologous chromosomes.

Observations of the phenotypic diversity of living organisms make it logical to assume that there are many more genes than chromosomes. Therefore, each homolog must carry genetic information for more than one trait. The currently accepted concept is that a chromosome is composed of a large number of linearly ordered, information-containing genes. Mendel's paired unit factors (which determine tall or dwarf stems, for example) actually constitute a pair of genes located on one pair of homologous chromosomes. The location on a given chromosome where any particular gene occurs is called its **locus** (pl. loci). The different alleles of a given gene (for example, *G* and *g*) contain slightly different genetic information (green or yellow) that determines the same character (seed color in this case). Although we have examined only genes with two alternative alleles, most genes have more than two allelic forms. We conclude this section by reviewing the criteria necessary to classify two chromosomes as a homologous pair:

1. During mitosis and meiosis, when chromosomes are visible in their characteristic shapes, both members of a homologous pair are the same size and exhibit identical centromere locations. The sex chromosomes (e.g., the X and the Y chromosomes in mammals) are an exception.
2. During early stages of meiosis, homologous chromosomes form pairs, or synapse.
3. Although it is not generally visible under the microscope, homologs contain the identical linear order of gene loci.

## 3.7 Independent Assortment Leads to Extensive Genetic Variation

One consequence of independent assortment is the production by an individual of genetically dissimilar gametes. Genetic variation results because the two members of any homologous pair of chromosomes are rarely, if ever, genetically identical. As the maternal and paternal members of all pairs are distributed to gametes through independent assortment, all possible chromosome combinations are produced, leading to extensive genetic diversity.

We have seen that the number of possible gametes, each with different chromosome compositions, is $2^n$, where *n* equals the haploid number. Thus, if a species has a haploid number of 4, then $2^4$, or 16, different gamete combinations can be formed as a result of independent assortment. Although this number is not high, consider the human species, where $n = 23$. When $2^{23}$ is calculated, we find that in excess of $8 \times 10^6$, or over 8 million, different types of gametes are possible through independent assortment. Because fertilization represents an event involving only one of approximately $8 \times 10^6$ possible gametes from each of two parents, each offspring represents only one of $(8 \times 10^6)^2$ or one of only $64 \times 10^{12}$ potential genetic combinations. Given that this probability is less than one in one trillion, it is no wonder that, except for identical twins, each member of the human species exhibits a distinctive set of traits—this number of combinations of chromosomes is far greater than the number of humans who have ever lived on Earth! Genetic variation resulting from independent assortment has been extremely important to the process of evolution in all sexually reproducing organisms.

## 3.8 Laws of Probability Help to Explain Genetic Events

Recall that genetic ratios—for example, 3/4 tall:1/4 dwarf—are most properly thought of as probabilities. These values predict the outcome of each fertilization event, such that the probability of each zygote having the genetic potential for becoming tall is 3/4, whereas the potential for its being a dwarf is 1/4. Probabilities range from 0.0, where an event *is certain not to occur*, to 1.0, where an event *is certain to occur*. In this section, we consider the relation of probability to genetics. When two or more events with known probabilities occur independently but at the same time, we can calculate the probability of their possible outcomes occurring together. This is accomplished by applying the *product law*, which says that the probability of two or more events occurring simultaneously is equal to the product of their individual probabilities (see Section 3.3). Two or more events are independent of one another if the outcome of each one does not affect the outcome of any of the others under consideration.

To illustrate the product law, consider the possible results if you toss a penny (*P*) and a nickel (*N*) at the same time and examine all combinations of heads (*H*) and tails (*T*) that can occur. There are four possible outcomes:

$$(P_H{:}N_H) = (1/2)(1/2) = 1/4$$
$$(P_T{:}N_H) = (1/2)(1/2) = 1/4$$
$$(P_H{:}N_T) = (1/2)(1/2) = 1/4$$
$$(P_T{:}N_T) = (1/2)(1/2) = 1/4$$

The probability of obtaining a head or a tail in the toss of either coin is 1/2 and is unrelated to the outcome for the other coin. Thus, all four possible combinations are predicted to occur with equal probability.

If we want to calculate the probability when the possible outcomes of two events are independent of one another but can be accomplished in more than one way, we can apply the **sum law**. For example, what is the probability of tossing our penny and nickel and obtaining one head and one tail? In such a case, we do not care whether it is the penny or the nickel that comes up heads, provided that the other coin has the alternative outcome. As we saw above, there are two ways in which the desired outcome can be accomplished, each with a probability of 1/4. The sum law states that the probability of obtaining any single outcome, where that outcome can be achieved by two or more events, is equal to the sum of the individual probabilities of all such events. Thus, according to the sum law, the overall probability in our example is equal to

$$(1/4) + (1/4) = 1/2$$

One-half of all two-coin tosses are predicted to yield the desired outcome.

These simple probability laws will be useful throughout our discussions of transmission genetics and for solving genetics problems. In fact, we already applied the product law when we used the forked-line method to calculate the phenotypic results of Mendel's dihybrid and trihybrid crosses. When we wish to know the results of a cross, we need only calculate the probability of each possible outcome. The results of this calculation then allow us to predict the proportion of offspring expressing each phenotype or each genotype.

An important point to remember when you deal with probability is that predictions of possible outcomes are based on large sample sizes. If we predict that 9/16 of the offspring of a dihybrid cross will express both dominant traits, it is very unlikely that, in a small sample, exactly 9 of every 16 will express this phenotype. Instead, our prediction is that, of a large number of offspring, approximately 9/16 will do so. The deviation from the predicted ratio in smaller sample sizes is attributed to chance, a subject we examine in our discussion of statistics in the next section. As you shall see, the impact of deviation due strictly to chance diminishes as the sample size increases.

## Conditional Probability

Sometimes we may wish to calculate the probability of an outcome that is dependent on a specific condition related to that outcome. For example, in the $F_2$ of Mendel's monohybrid cross involving tall and dwarf plants, what is the probability that a tall plant is heterozygous (and not homozygous)? The condition we have set is to consider only tall $F_2$ offspring since we know that all dwarf plants are homozygous.

Because the outcome and specific condition are not independent, we cannot apply the product law of probability. The likelihood of the outcome in such a case is referred to as a **conditional probability**. In its simplest terms we are asking what is the probability that one outcome will occur, given the specific condition upon which this outcome is dependent. Let us call this probability $p_c$.

To solve for $p_c$ we must consider both the probability of the outcome of interest and that of the specific condition that produces the outcome. These are (a) the probability of an $F_2$ plant being heterozygous as a result of receiving both a dominant and a recessive allele ($p_a$) and (b) the probability of the condition under which the event is being assessed, that is, being tall ($p_b$).

*Probability of outcome:*

$p_a$ = plant inheriting one dominant and one recessive allele (i.e., being a heterozygote)
= 1/2

*Probability of condition*

$p_b$ = probability of an $F_2$ plant of a monohybrid cross being tall
= 3/4

To calculate the conditional probability ($p_c$), we divide $p_a$ by $p_b$:

$$\begin{aligned} p_c &= p_a/p_b \\ &= (1/2)/(3/4) \\ &= (1/2)(4/3) \\ &= 4/6 \\ p_c &= 2/3 \end{aligned}$$

The conditional probability of any tall plant being heterozygous is two-thirds (2/3). On the average, two-thirds of the $F_2$ tall plants will be heterozygous. We can confirm this calculation by reexamining Figure 3–3.

Conditional probability has many applications in genetics. In genetic counseling, for example, it is possible to calculate the probability $p_c$ that an unaffected sibling of a brother or sister expressing a recessive disorder is a carrier of the disease-causing allele (i.e., a heterozygote). Assuming that both parents are unaffected (and are therefore carriers), the calculation of $p_c$ is identical to the preceding example. The value of $p_c = 2/3$

WEB TUTORIAL 3.3 PROBABILITY

## The Binomial Theorem

Finally, probability can be used to analyze cases where one of two alternative outcomes is possible during each of a number of trials. By applying the **binomial theorem**, we can rather quickly calculate the probability of any specific combination of the outcomes for any given number of potential events. For example, for families of any size, we can calculate the probability of any combination of male and female children. In a family with four children, then, we can calculate the probability that two will be male and two will be female.

The expression of the binomial theorem is

$$(a + b)^n = 1$$

| *n* | Binomial | Expanded Binomial |
|---|---|---|
| 1 | $(a+b)^1$ | $a + b$ |
| 2 | $(a+b)^2$ | $a^2 + 2ab + b^2$ |
| 3 | $(a+b)^3$ | $a^3 + 3a^2b + 3ab^2 + b^3$ |
| 4 | $(a+b)^4$ | $a^4 + 4a^3b + 6a^2b^2 + 4ab^3 + b^4$ |
| 5 | $(a+b)^5$ | $a^5 + 5a^4b + 10a^3b^2 + 10a^2b^3 + 5ab^4 + b^5$ |
| | etc. | etc. |

where $a$ and $b$ are the respective probabilities of the two alternative outcomes and $n$ equals the number of trials.

As the value of $n$ increases and the expanded binomial becomes more complex, Pascal's triangle, shown in Table 3.2, is useful in determining the numerical coefficient of each term in the expanded equation. Starting with the third line from the top of this triangle, each number is the sum of the two numbers immediately above it.

To expand any binomial, the various exponents of $a$ and $b$ (e.g., $a^3b^2$) are determined by using the pattern

$$(a + b)^n = a^n, a^{n-1}b, a^{n-2}b^2, a^{n-3}b^3, \ldots, b^n$$

TABLE 3.2

**Pascal's Triangle**

| *n* | Numerical Coefficients |
|---|---|
| | 1 |
| 1 | 1 1 |
| 2 | 1 2 1 |
| 3 | 1 3 3 1 |
| 4 | 1 4 6 4 1 |
| 5 | 1 5 10 10 5 1 |
| 6 | 1 6 15 20 15 6 1 |
| 7 | 1 7 21 35 35 21 7 1 |
| etc. | etc. |

*Notice that all numbers other than the 1's are equal to the sum of the two numbers directly above them.

Using these methods for setting up the expression, we find that the expansion of $(a + b)^7$ is

$$a^7 + 7a^6b + 21a^5b^2 + 35a^4b^3 + \cdots + b^7$$

Let's now return to our original question: *What is the probability that in a family with four children, two are male and two are female?*

First, assign initial probabilities to each outcome:

$$a = \text{male} = 1/2$$
$$b = \text{female} = 1/2$$

Then write out the expanded binomial for the value of $n = 4$,

$$(a + b)^4 = a^4 + 4a^3b + 6a^2b^2 + 4ab^3 + b^4$$

Each term represents a possible outcome, with the exponent of $a$ representing the number of males and the exponent of $b$ representing the number of females. Therefore, the term describing the outcome of two males and two females—the expression of the probability ($p$) we are looking for—is

$$\begin{aligned} p &= 6a^2b^2 \\ &= 6(1/2)^2(1/2)^2 \\ &= 6(1/2)^4 \\ &= 6(1/16) \\ &= 6/16 \\ p &= 3/8 \end{aligned}$$

Thus, the probability of families of four children having two boys and two girls is 3/8. Of all families with four children, 3 out of 8 are predicted to have two boys and two girls.

Before examining one other example, we should note that a single formula can be used to determine the numerical coefficient for any set of exponents,

$$n!/(s!t!)$$

where

$n$ = the total number of events

$s$ = the number of times outcome *a occurs*

$v$ = the number of times outcome $b$ occurs

Therefore, $n = s + t$

The symbol ! denotes a factorial, which is the product of all the positive integers from 1 through some positive integer. For example,

$$5! = (5)(4)(3)(2)(1) = 120.$$

Note that in factorials, $0! = 1$.

Using the formula, let's determine the probability that in a family with seven children, five will be males and two females. In this case, $n = 7$, $s = 5$, and $v = 2$. We begin by setting up our equation to find the term for five events having outcome $a$ and two events having outcome $b$:

$$\begin{aligned} p &= \frac{n!}{s!t!}a^sb^t \\ &= \frac{7!}{5!2!}(1/2)^5(1/2)^2 \\ &= \frac{(7)\cdot(6)\cdot(5)\cdot(4)\cdot(3)\cdot(2)\cdot(1)}{(5)\cdot(4)\cdot(3)\cdot(2)\cdot(1)\cdot(2)\cdot(1)}(1/2)^7 \\ &= \frac{(7)\cdot(6)}{(2)\cdot(1)}(1/2)^7 \\ &= \frac{42}{2}(1/2)^7 \\ &= 21(1/2)^7 \\ &= 21(1/128) \\ p &= 21/128 \end{aligned}$$

Of families with seven children, on the average, 21/128 are predicted to have five males and two females.

Calculations using the binomial theorem have various applications in genetics, including the analysis of polygenic traits (Chapter 25) and studies of population equilibrium (Chapter 27).

## 3.9 Chi-Square Analysis Evaluates the Influence of Chance on Genetic Data

Mendel's 3:1 monohybrid and 9:3:3:1 dihybrid ratios are hypothetical predictions based on the following assumptions: (1) each allele is dominant or recessive, (2) segregation is unimpeded, (3) independent assortment occurs, and (4) fertilization is random. The final two assumptions are influenced by chance events and therefore are subject to random fluctuation. This concept of **chance deviation** is most easily illustrated by tossing a single coin numerous times and recording the number of heads and tails observed. In each toss, there is a probability of 1/2 that a head will occur and a probability of 1/2 that a tail will occur. Therefore, the expected ratio of many tosses is 1/2:1/2, or 1:1. If a coin is tossed 1000 times, usually *about* 500 heads and 500 tails will be observed. Any reasonable fluctuation from this hypothetical ratio (e.g., 486 heads and 514 tails) is attributed to chance.

As the total number of tosses is reduced, the impact of chance deviation increases. For example, if a coin is tossed only four times, you would not be too surprised if all four tosses resulted in only heads or only tails. But, for 1000 tosses, 1000 heads or 1000 tails would be most unexpected. In fact, you might believe that such a result would be impossible. Actually, all heads or all tails in 1000 tosses can be predicted to occur with a probability of $(1/2)^{1000}$. Since $(1/2)^{20}$ is less than one in a million times, an event occurring with a probability as small as $(1/2)^{1000}$ is virtually impossible. Two major points to keep in mind when predicting or analyzing genetic outcomes are:

1. The outcomes of independent assortment and fertilization, like coin tossing, are subject to random fluctuations from their predicted occurrences as a result of chance deviation.
2. As the sample size increases, the average deviation from the expected results decreases. Therefore, a larger sample size diminishes the impact of chance deviation on the final outcome.

## Chi-Square Calculations and the Null Hypothesis

In genetics, being able to evaluate observed deviation is a crucial skill. When we assume that data will fit a given ratio such as 1:1, 3:1, or 9:3:3:1, we establish what is called the **null hypothesis ($H_0$)**. It is so named because the hypothesis assumes that there is *no real difference* between the *measured values* (or ratio) and the *predicted values* (or ratio). Any apparent difference can be attributed purely to chance. The validity of the null hypothesis for a given set of data is measured using statistical analysis. Depending on the results of this analysis, the null hypothesis may either (1) be rejected or (2) fail to be rejected. If it is rejected, the observed deviation from the expected result is judged not to be attributable to chance alone. In this case, the null hypothesis and the underlying assumptions leading to it must be reexamined. If the null hypothesis fails to be rejected, any observed deviations are attributed to chance.

One of the simplest statistical tests for assessing the goodness of fit of the null hypothesis is **chi-square ($\chi^2$) analysis.** This test takes into account the observed deviation in each component of a ratio (from what was expected) as well as the sample size and reduces them to a single numerical value. The value for $\chi^2$ is then used to estimate how frequently the observed deviation can be expected to occur strictly as a result of chance. The formula used in chi-square analysis is

$$\chi^2 = \Sigma \frac{(o - e)^2}{e}$$

where $o$ is the observed value for a given category, $e$ is the expected value for that category, and $\Sigma$ (the Greek letter sigma) represents the sum of the calculated values for each category in the ratio. Because $(o - e)$ is the deviation $(d)$ in each case, the equation reduces to

$$\chi^2 = \Sigma \frac{d^2}{e}$$

Table 3.3(a) shows the steps in the $\chi^2$ calculation for the $F_2$ results of a hypothetical monohybrid cross. To analyze the data obtained from this cross, work from left to right across the table, verifying the calculations as appropriate. Note that regardless of whether the deviation $d$ is positive or negative, $d^2$ always becomes positive after the number is squared. In Table 3.3(b) $F_2$ results of a hypothetical dihybrid cross are analyzed. Make sure that you understand how each number was calculated in this example.

The final step in chi-square analysis is to interpret the $\chi^2$ value. To do so, you must initially determine a value called the **degrees of freedom (df)**, which is equal to $n - 1$, where $n$ is the number of different categories into which the data are divided, in other words, the number of possible outcomes. For the 3:1 ratio, $n = 2$, so $df = 1$. For the 9:3:3:1 ratio, $n = 4$ and $df = 3$. Degrees of freedom must be taken into account because the greater the number of categories, the more deviation is expected as a result of chance.

**TABLE 3.3**

**Chi-Square Analysis**

**(a) Monohybrid Cross**

| Expected Ratio | Observed ($o$) | Expected ($e$) | Deviation ($o - e$) | Deviation ($d^2$) | $d^2/e$ |
|---|---|---|---|---|---|
| 3/4 | 740 | 3/4(1000) = 750 | 740 − 750 = −10 | $(-10^2) = 100$ | 100/750 = 0.13 |
| 1/4 | 260 | 1/4(1000) = 250 | 260 − 250 = +10 | $(+10)^2 = 100$ | 100/250 = 0.40 |
| | Total = 1000 | | | | $\chi^2 = 0.53$ |
| | | | | | $p = 0.48$ |

**(b) Dihybrid Cross**

| Expected Ratio | $o$ | $e$ | $(o - e)$ | $d^2$ | $d^2/e$ |
|---|---|---|---|---|---|
| 9/16 | 587 | 567 | +20 | 400 | 0.71 |
| 3/16 | 197 | 189 | +8 | 64 | 0.34 |
| 3/16 | 168 | 189 | −21 | 441 | 2.33 |
| 1/16 | 56 | 63 | −7 | 49 | 0.78 |
| | Total = 1008 | | | | $\chi^2 = 4.16$ |
| | | | | | $p = 0.26$ |

Once you have determined the degrees of freedom, you can interpret the $\chi^2$ value in terms of a corresponding **probability value (*p*)**. Since this calculation is complex, we usually take the *p* value from a standard table or graph. Figure 3–12 shows a wide range of $\chi^2$ values and the corresponding *p* values for various degrees of freedom in both a graph and a table. Let's use the graph to explain how to determine the *p* value. The caption for Figure 3–12(b) explains how to use the table.

To determine *p* using the graph, execute the following steps:

1. Locate the $\chi^2$ value on the abscissa (the horizontal axis, or *x*-axis).
2. Draw a vertical line from this point up to the line on the graph representing the appropriate *df*.
3. From there, extend a horizontal line to the left until it intersects the ordinate (the vertical axis, or *y*-axis).
4. Estimate, by interpolation, the corresponding *p* value.

We used these steps for the monohybrid cross in Table 3.3(a) to estimate the *p* value of 0.48, as shown in Figure 3–12(a). Now try this method to see if you can determine the *p* value for the dihybrid cross [Table 3.3(b)]. Since the $\chi^2$ value is 4.16 and $df = 3$, an approximate *p* value is 0.26. Checking this result in the table confirms that *p* values for both the monohybrid and dihybrid crosses are between 0.20 and 0.50.

## Interpreting Probability Values

So far, we have been concerned with calculating $\chi^2$ values and determining the corresponding *p* values. These steps bring us to the most important aspect of chi-square analysis: understanding the meaning of the *p* value. It is simplest to think of the *p* value as a percentage. Let's use the example of the dihybrid cross in Table 3.1(b) where $p = 0.26$, which can be thought of as 26 percent. In our example, the *p* value indicates that if we repeat the same experiment many times, 26 percent of the trials would be expected to exhibit chance deviation as great as or greater than that seen in the initial trial. Conversely, 74 percent of the repeats would show less deviation than initially observed as a result of chance. Thus, the *p* value reveals that a null hypothesis (concerning the 9:3:3:1 ratio, in this case) is never proved or disproved absolutely. Instead, a relative standard is set that we use to either *reject* or *fail to reject* the null hypothesis. This standard is most often a *p* value of 0.05. When applied to chi-square analysis, a *p* value less than 0.05 means that the observed deviation in the set of results will be obtained by chance alone less than 5 percent of the time. Such a *p* value indicates that the difference between the observed and predicted results is substantial and requires us to reject the null hypothesis.

(a)

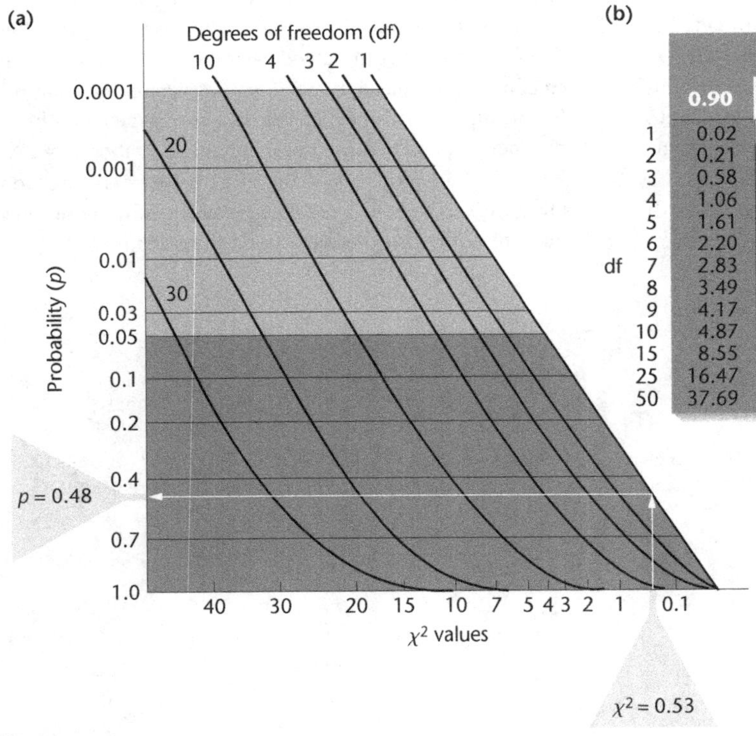


(b)

| df | Probability (*p*) 0.90 | 0.50 | 0.20 | 0.05 | 0.01 | 0.001 |
|---|---|---|---|---|---|---|
| 1 | 0.02 | 0.46 | 1.64 | 3.84 | 6.64 | 10.83 |
| 2 | 0.21 | 1.39 | 3.22 | 5.99 | 9.21 | 13.82 |
| 3 | 0.58 | 2.37 | 4.64 | 7.82 | 11.35 | 16.27 |
| 4 | 1.06 | 3.36 | 5.99 | 9.49 | 13.28 | 18.47 |
| 5 | 1.61 | 4.35 | 7.29 | 11.07 | 15.09 | 20.52 |
| 6 | 2.20 | 5.35 | 8.56 | 12.59 | 16.81 | 22.46 |
| 7 | 2.83 | 6.35 | 9.80 | 14.07 | 18.48 | 24.32 |
| 8 | 3.49 | 7.34 | 11.03 | 15.51 | 20.09 | 26.13 |
| 9 | 4.17 | 8.34 | 12.24 | 16.92 | 21.67 | 27.88 |
| 10 | 4.87 | 9.34 | 13.44 | 18.31 | 23.21 | 29.59 |
| 15 | 8.55 | 14.34 | 19.31 | 25.00 | 30.58 | 37.30 |
| 25 | 16.47 | 24.34 | 30.68 | 37.65 | 44.31 | 52.62 |
| 50 | 37.69 | 49.34 | 58.16 | 67.51 | 76.15 | 86.60 |

$\chi^2$ values

Fails to reject the null hypothesis
Rejects the null hypothesis

FIGURE 3-12 (a) Graph for converting $\chi^2$ values to *p* values. (b) Table of $\chi^2$ values for selected values of *df* and *p*. $\chi^2$ values that lead to a *p* value of 0.05 or greater (darker blue areas) justify failure to reject the null hypothesis. Values leading to a *p* value of less than 0.05 (lighter blue areas) justify rejecting the null hypothesis. For example, the table in part (b) shows that for $\chi^2 = 0.53$ with 1 degree of freedom, the corresponding *p* value is between 0.20 and 0.50. The graph in (a) gives a more precise *p* value of 0.48 by interpolation. Thus, we fail to reject the null hypothesis.

On the other hand, $p$ values of 0.05 or greater (0.05 to 1.0) indicate that the observed deviation will be obtained by chance alone 5 percent or more of the time. This conclusion allows us not to reject the null hypothesis (when we are using $p = 0.05$ as our standard). Thus, with its $p$ value of 0.26, the null hypothesis that independent assortment accounts for the results fails to be rejected. Therefore, the observed deviation can be reasonably attributed to chance.

A final note is relevant here concerning the case where the null hypothesis is rejected, that is, where $p < 0.05$. Suppose we had tested a data set to assess a possible 9:3:3:1 ratio, as in Table 3.3(b), but we rejected the null hypothesis based on our $\chi^2$ calculation. What are alternative interpretations of the data? Researchers will reassess the assumptions that underlie the null hypothesis. In our dyhibrid cross, we assumed that segregation operates faithfully for both gene pairs. We also assumed that fertilization is random and that the viability of all gametes is equal regardless of genotype—that is, all gametes are equally likely to participate in fertilization. Finally, we assumed that, following fertilization, all preadult stages and adult offspring are equally viable, regardless of their genotype. If any of these assumptions is incorrect, then the original hypothesis is not necessarily invalid.

For example, suppose our null hypothesis is that a dihybrid cross between fruit flies will result in 3/16 mutant wingless flies. However, perhaps fewer of the mutant embryos are able to survive their preadult development or young adulthood compared to flies whose genotype gives rise to wings. As a result, when the data are gathered, there will be fewer than 3/16 wingless flies. Rejection of the null hypothesis is not in itself cause for us to reject the validity of the postulates of segregation and independent assortment, because other factors we are unaware of may also be affecting the outcome.

The point of the foregoing discussion is that statistical information must be assessed carefully on a case-by-case basis. When we reject a null hypothesis, we must examine all underlying assumptions. If there is no concern about their validity, then we must consider alternative hypotheses to explain the results.

**NOW SOLVE THIS**

Problem 23 on page 67 asks you to apply $\chi^2$ analysis to a set of data and determine whether the data fit certain ratios.

■ HINT: *In calculating $\chi^2$, first determine the expected outcomes using the predicted ratios. Then, following a stepwise approach, determine the deviation in each case, and calculate $d^2/e$ for each category.*

## 3.10 Pedigrees Reveal Patterns of Inheritance of Human Traits

We now explore how to determine the mode of inheritance of phenotypes in humans, where experimental matings are not made and where relatively few offspring are available for study. The traditional way to study inheritance has been to construct a family tree, indicating the presence or absence of the trait in question for each member of each generation. Such a family tree is called a **pedigree.** By analyzing a pedigree, we may be able to predict how the trait under study is inherited—for example, is it due to a dominant or recessive allele? When many pedigrees for the same trait are studied, we can often ascertain the mode of inheritance.

### Pedigree Conventions

Figure 3–13 illustrates some of the conventions geneticists follow in constructing pedigrees. Circles represent females and squares designate males. If the sex of an individual is unknown, a diamond is used. Parents are generally connected to each other by a single horizontal line, and vertical lines lead to their offspring. If the parents are related—that is, **consanguineous**—such as first cousins, they are connected by a double line. Offspring are called **sibs** (short for **siblings**) and are connected by a horizontal **sibship line.** Sibs are placed in birth order from left to right and are labeled with Arabic numerals. Parents also receive an Arabic number designation. Each generation is indicated by a Roman numeral. When a pedigree traces only a single trait, the circles, squares, and diamonds are shaded if the phenotype being considered is expressed and unshaded if not. In some pedigrees, those individuals that fail to ex-

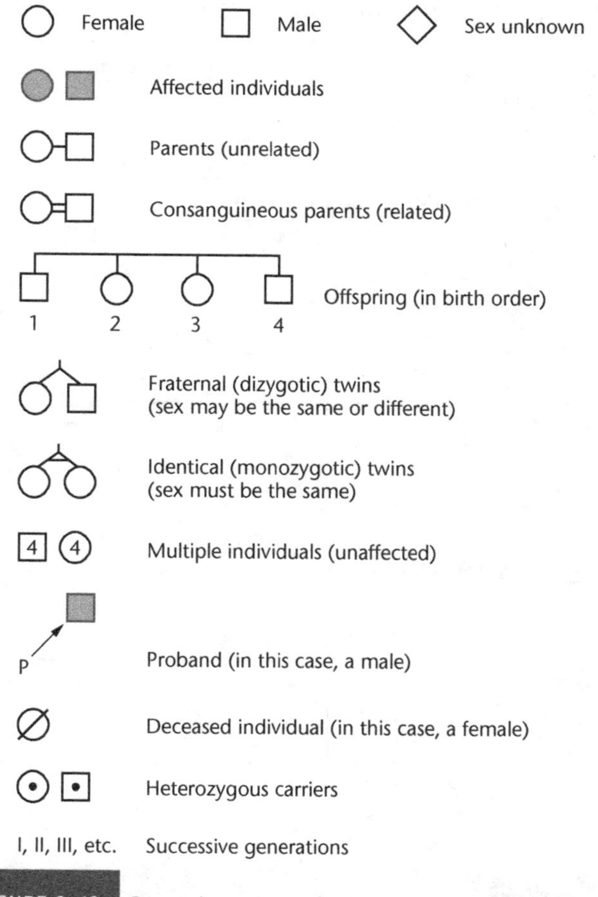


FIGURE 3–13 Conventions commonly encountered in human pedigrees.

press a recessive trait but are known with certainty to be heterozygous carriers have a shaded dot within their unshaded circle or square. If an individual is deceased and the phenotype is unknown, a diagonal line is placed over the circle or square.

Twins are indicated by diagonal lines stemming from a vertical line connected to the sibship line. For identical, or **monozygotic**, twins, the diagonal lines are linked by a horizontal line. Fraternal, or **dizygotic**, twins lack this connecting line. A number within one of the symbols represents that number of sibs of the same sex and of the same or unknown phenotypes. The individual whose phenotype first brought attention to the family is called the **proband** and is indicated by an arrow connected to the designation **p**. This term applies to either a male or a female.

## Pedigree Analysis

In Figure 3–14, two pedigrees are shown. The first is a representative pedigree for a trait that demonstrates autosomal recessive inheritance, such as **albinism**, where synthesis of the pigment melanin in obstructed. The male parent of the first generation (I-1) is affected. Characteristic of a situation in which a parent has a rare recessive trait, the trait "disappears" in the offspring of the next generation. Assuming recessiveness, we might predict that the unaffected female parent (I-2) is a homozygous normal individual because none of the offspring show the disorder. Had she been heterozygous, one-half of the offspring would be expected to exhibit albinism, but none do. However, such a small sample (three offspring) prevents our knowing for certain.

Further evidence supports the prediction of a recessive trait. If albinism were inherited as a dominant trait, individual II-3 would have to express the disorder in order to pass it to his offspring (III-3 and III-4), but he does not. Inspection of the offspring constituting the third generation (row III) provides still further support for the hypothesis that albinism is a recessive trait. If it is, parents II-3 and II-4 are both heterozygous, and approximately one-fourth of their offspring should be affected. Two of the six offspring do show albinism. This deviation from the expected ratio is not unexpected in crosses with few offspring. Once we are confident that albinism is inherited as an autosomal recessive trait, we could portray the II-3 and II-4 individuals with a shaded dot within their larger square and circle. Finally, we can note that, characteristic of pedigrees for autosomal traits, both males and females are affected with equal probability. In Chapter 4, we will examine a pedigree representing a gene located on the sex-determining X chromosome. We will see certain patterns characteristic of the transmission of X-linked traits, such as that these traits are more prevalent in male offspring and are never passed from affected fathers to their sons.

The second pedigree illustrates the pattern of inheritance for a trait such as Huntington disease, which is caused by an autosomal dominant allele. The key to identifying a pedigree that reflects a dominant trait is that all affected offspring will have a parent that also expresses the trait. It is also possible, by chance, that none of the offspring will inherit the dominant allele. If so, the trait will cease to exist in future generations. Like recessive traits, provided that the gene is autosomal, both males and females are equally affected.

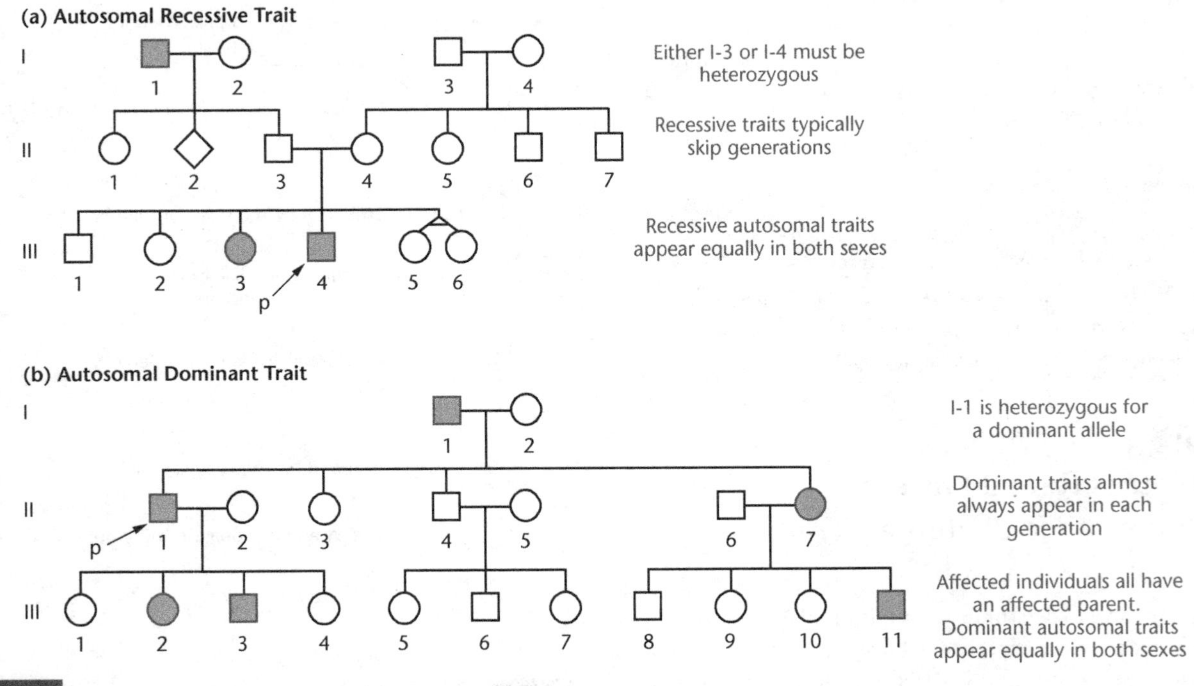


ANALYSIS OF HUMAN PEDIGREES
WEB TUTORIAL 3.4

FIGURE 3–14 Representative pedigrees for two characteristics, each followed through three generations.

When a given autosomal dominant disease is rare within the population, and most are, then it is highly unlikely that affected individuals will inherit a copy of the mutant gene from both parents. Therefore, in most cases, affected individuals are heterozygous for the dominant allele. As a result, approximately one-half of the offspring inherit it. This is borne out in the second pedigree in Figure 3–14. Furthermore, if a mutation is dominant, and a single copy is sufficient to produce a mutant phenotype, homozygotes are likely to be even more severely affected, perhaps even failing to survive. An illustration of this is the dominant gene for **familial hypercholesterolemia.** Heterozygotes display a defect in their receptors for low-density lipoproteins, the so-called LDLs (known popularly as "bad cholesterol"). As a result, too little cholesterol is taken up by cells from the blood, and elevated plasma levels of LDLs result. Such heterozygous individuals almost always have heart attacks during the fourth decade of their life, or before. While heterozygotes have LDL levels about double that of a normal individual, rare homozygotes have been detected. They lack LDL receptors altogether, and their LDL levels are nearly ten times above the normal range. They are likely to have a heart attack very early in life, even before age 5, and almost inevitably before they reach the age of 20.

Pedigree analysis of many traits has historically been an extremely valuable research technique in human genetic studies. However, the approach does not usually provide the certainty of the conclusions obtained through experimental crosses yielding large numbers of offspring. Nevertheless, when many independent pedigrees of the same trait or disorder are analyzed, consistent conclusions can often be drawn. Table 3.4 lists numerous human traits and classifies them according to their recessive or dominant expression.

**TABLE 3.4**

**Representative Recessive and Dominant Human Traits**

| Recessive Traits | Dominant Traits |
|---|---|
| Albinism | Achondroplasia |
| Alkaptonuria | Brachydactyly |
| Ataxia telangiectasia | Congenital stationary night blindness |
| Color blindness | Ehler–Danlos syndrome |
| Cystic fibrosis | Hypotrichosis |
| Duchenne muscular dystrophy | Huntington disease |
| Galactosemia | Hypercholesterolemia |
| Hemophilia | Marfan syndrome |
| Lesch–Nyhan syndrome | Neurofibromatosis |
| Phenylketonuria | Phenylthiocarbamide tasting |
| Sickle-cell anemia | Porphyria (some forms) |
| Tay–Sachs disease | Widow's peak |

**NOW SOLVE THIS**

Problem 27 on page 67 asks you to examine a pedigree for myopia and predict whether the trait is dominant or recessive.

■ HINT: *One of the first steps in analyzing a pedigree is to look for individuals who express the trait of interest but neither of whose parents also express the trait. Such an observation makes it highly unlikely that the trait is dominant.*

## GENETICS, TECHNOLOGY, AND SOCIETY

### Tay–Sachs Disease: The Molecular Basis of a Recessive Disorder in Humans

Tay–Sachs disease (TSD) is an inherited disorder that causes unalterable destruction of the central nervous system. This condition is particularly tragic because infants with TSD are unaffected at birth and appear to develop normally until they are about six months old. Parents, having believed that their child was normal, then must witness the progressive deterioration of mental and physical abilities. The disorder is severe, and afflicted infants eventually become blind, deaf, mentally retarded, and paralyzed, often within only a year or two. Most do not live beyond age 5. Named for Warren Tay and Bernard Sachs who first described the symptoms and associated them with the disorder in the late 1800s, Tay–Sachs disease clearly demonstrates the classical Mendelian pattern of autosomal recessive inheritance. Two unaffected heterozygous parents, who most often have no immediate family history of the disorder, have a probability of one in four of having a Tay–Sachs child.

The protein product of the affected gene has been identified, and we now have a clear understanding of the underlying molecular basis of the disorder. TSD results from the loss of activity of the enzyme hexosaminidase A (Hex-A). This enzyme is normally found in lysosomes, organelles that break down large molecules for recycling by the cell. Hex-A is needed to degrade the ganglioside $G_{M2}$, a lipid component of nerve cell membranes. Without functional Hex-A, gangliosides accumulate within neurons in the brain and cause deterioration of the nervous system. Heterozygous carriers of TSD, with one normal copy of the gene, produce only about 50 percent of the normal amount of Hex-A, but they show no symptoms of the disorder. The observation that the activity of only one gene (one wild-type allele) is sufficient for the normal development and function of the nervous system explains and illustrates the molecular basis of recessive mutations. Only when both genes are disrupted by mutation is the mutant phenotype evident.

The gene responsible for Tay–Sachs disease (*HEXA*) has now been localized on chromosome 15 and codes for the alpha subunit of the Hex-A enzyme. (Hex-A, like many enzymes, displays a quaternary protein structure, which means it consists of multiple polypeptide sub-

*Continued on next page*

*Genetics, Technology, and Society, continued*

units.) Since the gene was isolated in 1985, more than 50 different mutations have been identified within it that lead to TSD. Although the most common form of the disease is the infantile form, where no functional Hex-A is produced, there is also a rare late-onset form that occurs in patients with greatly reduced Hex-A activity. Late-onset TSD is not detectable until patients are in their twenties or thirties, and it is generally much less severe than the infantile form. Symptoms include hand tremors, speech impediments, muscle weakness, and loss of balance.

Tay-Sachs disease is almost a hundred times more common in Ashkenazi Jews—Jews of central or eastern European descent—than in the general population, and it also has a higher incidence in French Canadians and in members of the Cajun population in Louisiana. In the United States, approximately one in every 27 Ashkenazi Jews is a heterozygous carrier of TSD. By contrast, the carrier rate in the general population and in Jews of Sephardic (Spanish or Portuguese) origin is approximately one in every 250.

Although there are currently no effective treatments for TSD, recent advances in carrier screening have helped to reduce the prevalence of the disorder in high-risk populations. Carriers can be identified by tests that measure Hex-A activity or by DNA-based tests that detect specific gene mutations. When both parents are carriers, prenatal diagnosis can be utilized with each pregnancy in order to detect affected fetuses. In addition, ganglioside synthesis inhibitors and Hex-A enzyme replacement therapy are currently being investigated as potential treatments for TSD newborns.

The most encouraging news is that genetic counseling and educational programs, particularly directed toward Ashkenazi Jews in the United States, have now all but eliminated TSD in this country. It is also heartening to know that two carrier parents do not need to forego procreating. They may conceive normally and undergo prenatal testing as mentioned above. However, the most modern approach open to potential carriers desiring unaffected offspring involves DNA-based testing of a single cell of an early embryo created using *in vitro* fertilization. Prospective parents undergo *in vitro* fertilization procedures, and at a very early post-fertilization stage, such as the eight-cell stage, one of the cells is removed from the dividing cell mass and tested for the presence of the mutant alleles. If the DNA testing reveals homozygosity for a mutant allele, the pre-embryo from which it was derived is discarded. If at least one normal allele is detected, the embryo is implanted and carried to term. In spite of the loss of one cell at this early stage, development may proceed normally, leading to the birth of an unaffected baby.

The approach described here is but one example of the way our knowledge of genetics has led to the development and subsequent application of technology designed to solve some problem encountered by individuals in our society today. One can only imagine the advances in genetic technology that will be commonplace just a few decades from now!

### ■ References

Fernandes, F., and Shapiro, B. 2004. Tay-Sachs disease. *Arch. Neurol.* 61: 1466–1468.

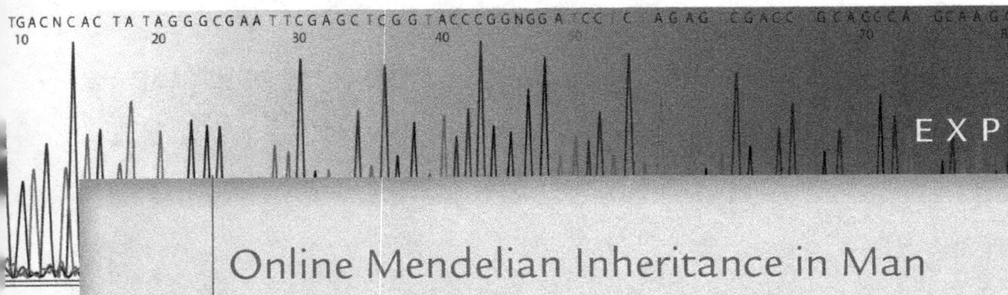

EXPLORING GENOMICS

## Online Mendelian Inheritance in Man

The **Online Mendelian Inheritance in Man (OMIM)** database is a catalog of human genes and human genetic disorders that are inherited in a Mendelian manner. Genetic disorders that arise from major chromosomal aberrations, such as monosomy or trisomy (the loss of a chromosome or the presence of a superfluous chromosome, respectively), are not included. The OMIM database is a daily-updated version of the book *Mendelian Inheritance in Man*, edited by Dr. Victor McKusick of Johns Hopkins University. Scientists use OMIM as an important information source to accompany the sequence data generated by the **Human Genome Project**.

The OMIM entries will give you links to a wealth of information, including DNA and protein sequences, chromosomal maps, disease descriptions, and relevant scientific publications. In this exercise, you will explore OMIM to answer questions about the recessive human disease sickle-cell anemia and other Mendelian inherited disorders.

### ■ Exercise I –Sickle-cell Anemia

In this chapter, you were introduced to sickle-cell anemia as an example of a single-gene recessive disease. You will now discover more about sickle-cell anemia by exploring the OMIM database.

1. To begin the search, access the OMIM site at: www.ncbi.nlm.nih.gov/entrez/query.fcgi?db=OMIM&itool=toolbar.
2. In the "SEARCH" box, type "sickle-cell anemia" and click on the "Go" button to perform the search.
3. Select the first entry (#603903).
4. Examine the list of subject headings in the left-hand column and read some of the information about sickle-cell anemia.
5. Select one or two references at the bottom of the page and follow them to their abstracts in PubMed.

6. Using the information in this entry, answer the following questions:

a. Which gene is mutated in individuals with sickle-cell anemia?

b. What are the major symptoms of this disorder?

c. What was the first published scientific description of sickle-cell anemia?

d. Describe two other features of this disorder that you learned from the OMIM database and state where in the database you found this information.

**■ Exercise II – Other Recessive or Dominant Disorders**

Select another human disorder that is inherited as either a dominant or recessive trait and investigate its features, following the general procedure presented above. Follow links from OMIM to other databases if you choose.

Describe several interesting pieces of information you acquired during your exploration and cite the information sources you encountered during the search.

## Chapter Summary

1. Over a century ago, Mendel studied inheritance patterns in the garden pea and established the principles of transmission genetics.
2. Mendel's postulates help describe the basis for the inheritance of phenotypic expression. He showed that unit factors, later called alleles of individual genes, exist in pairs and exhibit a dominant–recessive relationship in determining the expression of traits.
3. Mendel postulated that unit factors must segregate during gamete formation such that each gamete receives only one of the two factors with equal probability.
4. Mendel's postulate of independent assortment states that each pair of unit factors segregates independently of other such pairs. As a result, all possible combinations of gametes will be formed with equal probability.
5. The discovery of chromosomes in the late 1800s, along with subsequent studies of their behavior during meiosis, led to the rediscovery of Mendel's work, linking the behavior of his unit factors to that of chromosomes during meiosis.
6. The Punnett square and the forked-line methods are used to predict the probabilities of phenotypes and genotypes from crosses involving two or more gene pairs.
7. Genetic ratios are expressed as probabilities. Thus, deriving outcomes of genetic crosses requires an understanding of the laws of probability.
8. Statistical analysis is used to test the validity of experimental outcomes. In genetics, variations from the expected ratios due to chance deviations can be anticipated.
9. Chi-square analysis allows us to assess the null hypothesis, namely, that there is no real difference between the expected and observed values. Specifically, it tests the probability of whether observed variations can be attributed to chance deviation.
10. Pedigree analysis is a method for studying the inheritance patterns of human traits over several generations. It frequently provides the basis for determining the mode of inheritance of human characteristics and disorders.

## INSIGHTS AND SOLUTIONS

*As a genetics student, you will be asked to demonstrate your knowledge of transmission genetics by solving various problems. Success at this task requires not only comprehension of theory but also its application to more practical genetic situations. Most students find problem solving in genetics to be both challenging and rewarding. This section is designed to provide basic insights into the reasoning essential to this process.*

Genetics problems are in many ways similar to word problems in algebra. The approach to solving them is identical: (1) analyze the problem carefully; (2) translate words into symbols and define each symbol precisely; and (3) choose and apply a specific technique to solve the problem. The first two steps are the most critical. The third step is largely mechanical.

The simplest problems state all necessary information about a $P_1$ generation and ask you to find the expected ratios of the $F_1$ and $F_2$ genotypes and/or phenotypes. Always follow these steps when you encounter this type of problem:

(a) Determine insofar as possible the genotypes of the individuals in the $P_1$ generation.

(b) Determine what gametes may be formed by the $P_1$ parents.

(c) Recombine the gametes by the Punnett square or the forked-line methods, or if the situation is very simple, by inspection. From the genotypes of the $F_1$ generation, determine the phenotypes. Read the $F_1$ phenotypes.

(d) Repeat the process to obtain information about the $F_2$ generation.

Determining the genotypes from the given information requires that you understand the basic theory of transmission genetics. Consider

*Continued on next page*

*Insights and Solutions, continued*

this problem: *A recessive mutant allele, black, causes a very dark body in* Drosophila *when homozygous. The normal wild-type color is described as gray. What $F_1$ phenotypic ratio is predicted when a black female is crossed to a gray male whose father was black?*

To work out this problem, you must understand dominance and recessiveness, as well as the principle of segregation. Furthermore, you must use the information about the male parent's father. Here is one way to solve this problem:

1. The female parent is black, so she must be homozygous for the mutant allele (*bb*).
2. The male parent is gray; therefore, he must have at least one dominant allele (*B*). His father was black (*bb*), and he received one of the chromosomes bearing these alleles, so the male parent must be heterozygous (*Bb*).

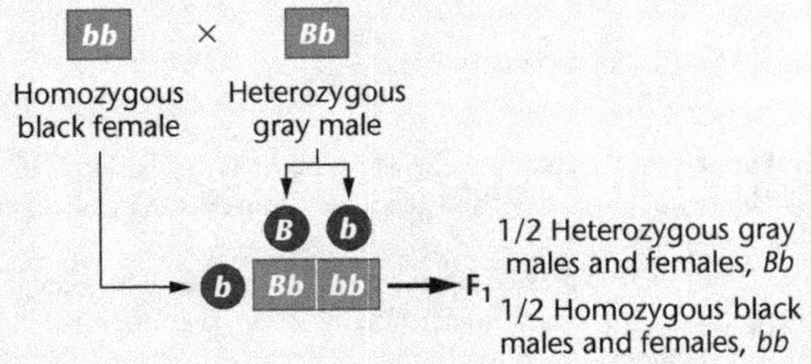


From this point, solving the problem is simple:

Apply the approach we just studied to the following problems.

1. Mendel found that full pea pods are dominant over constricted pods, while round seeds are dominant over wrinkled seeds. One of his crosses was between full, round plants and constricted, wrinkled plants. From this cross, he obtained an $F_1$ generation that was all full and round. In the $F_2$ generation, Mendel obtained his classic 9:3:3:1 ratio. Using this information, determine the expected $F_1$ and $F_2$ results of a cross between homozygous constricted, round and full, wrinkled plants.

Solution: First, assign gene symbols to each pair of contrasting traits. Use the lowercase first letter of each recessive trait to designate that trait, and use the same letter in uppercase to designate the dominant trait. Thus, *C* and *c* indicate full and constricted pods, respectively, and *W* and *w* indicate the round and wrinkled phenotypes, respectively.

Determine the genotypes of the $P_1$ generation, form the gametes, combine them in the $F_1$ generation, and read off the phenotype(s):

| | | | |
|---|---|---|---|
| $P_1$ | *ccWW* constricted, round | × | *CCww* full, wrinkled |
| | ↓ | | ↓ |
| Gametes | *cW* | | *Cw* |
| $F_1$ | | *CcWw* full, round | |

You can immediately see that the $F_1$ generation expresses both dominant phenotypes and is heterozygous for both gene pairs. Thus, you expect that the $F_2$ generation will yield the classic Mendelian ratio of 9:3:3:1. Let's work it out anyway, just to confirm this expectation, using the forked-line method. Both gene pairs are heterozygous and can be expected to assort independently, so we can predict the $F_2$ outcomes from each gene pair separately and then proceed with the forked-line method.

The $F_2$ offspring should exhibit the individual traits in the following proportions:

| *Cc* × *Cc* | | *Ww* × *Ww* | |
|---|---|---|---|
| ↓ | | ↓ | |
| *CC*, *Cc*, *cC* | full | *WW*, *Ww*, *wW* | round |
| *cc* | constricted | *ww* | wrinkled |

1. Using these proportions to complete a forked-line diagram confirms the 9:3:3:1 phenotypic ratio. (Remember that this ratio represents proportions of 9/16:3/16:3/16:1/16.) Note that we are applying the product law as we compute the final probabilities:

3/4 full
- 3/4 round —(3/4) (3/4)→ 9/16 full, round
- 1/4 wrinkled —(3/4) (1/4)→ 3/16 full, wrinkled

1/4 constricted
- 3/4 round —(1/4) (3/4)→ 3/16 constricted, round
- 1/4 wrinkled —(1/4) (1/4)→ 1/16 constricted, wrinkled

2. Determine the probability that a plant of genotype *CcWw* will be produced from parental plants of the genotypes *CcWw* and *Ccww*.

Solution: The two gene pairs demonstrate straightforward dominance and recessiveness and assort independently during gamete formation. We need only calculate the individual probabilities of the two separate events (*Cc* and *Ww*) and apply the product law to calculate the final probability:

$$Cc \times Cc \rightarrow 1/4\ CC{:}1/2Cc{:}1/4cc$$

$$Ww \times ww \rightarrow 1/2\ Ww{:}1/2\ ww$$

$$p = (1/2\ Cc)(1/2\ Ww) = 1/4\ CcWw$$

3. In another cross, involving parent plants of unknown genotype and phenotype, the following offspring were obtained.

| | |
|---|---|
| 3/8 | full, round |
| 3/8 | full, wrinkled |
| 1/8 | constricted, round |
| 1/8 | constricted, wrinkled |

Determine the genotypes and phenotypes of the parents.

Solution: This problem is more difficult and requires keener insight because you must work backward to arrive at the answer. The best approach is to consider the outcomes of pod shape separately from those of seed texture.

Of all the plants, 3/8 + 3/8 = 3/4 are full and 1/8 + 1/8 = 1/4 are constricted. Of the various genotypic combinations that can serve as parents, which will give rise to a ratio of 3/4:1/4? This ratio is identical to Mendel's monohybrid $F_2$ results, and we can propose that both unknown parents share the same genetic characteristic as the monohybrid $F_1$ parents: they must both be heterozygous for the genes controlling pod shape and thus are *Cc*.

Before we accept this hypothesis, let's consider the possible genotypic combinations that control seed texture. If we consider this characteristic alone, we can see that the traits are expressed in a ratio of 3/8 + 1/8 = 1/2 round: 3/8 + 1/8 = 1/2 wrinkled. To generate such a ratio, the parents cannot both be heterozygous or their offspring would yield a 3/4:1/4 phenotypic ratio. They cannot both be homozygous or all offspring would express a single phenotype. Thus, we are left with testing the hypothesis that one parent is homozygous and one is heterozygous for the alleles controlling texture. The potential case of $WW \times Ww$ does not work, because it would also yield only a single phenotype. This leaves us with the potential case of $ww \times Ww$. Offspring in such a mating will yield 1/2 *Ww* (round): 1/2 *ww* (wrinkled), exactly the outcome we are seeking.

Now, let's combine our hypotheses and predict the outcome of the cross. In our solution, we use a dash (–) to indicate that the second allele may be dominant or recessive, since we are only predicting phenotypes.

3/4 *C–*
- 1/2 *Ww* → 3/8 *C–Ww* full, round
- 1/2 *ww* → 3/8 *C–ww* full, wrinkled

1/4 *cc*
- 1/2 *Ww* → 1/8 *ccWw* constricted, round
- 1/2 *ww* → 1/8 *ccww* constricted, wrinkled

As you can see, this cross produces offspring in proportions that match our initial information, and we have solved the problem. Note that, in the solution, we have used genotypes in the forked-line method, in contrast to the use of phenotypes in Solution 1.

4. In the laboratory, a genetics student crossed flies with normal long wings with flies expressing the *dumpy* mutation (truncated wings), which she believed was a recessive trait. In the $F_1$ generation, all flies had long wings. The following results were obtained in the $F_2$ generation:

792 long-winged flies
208 dumpy-winged flies

The student tested the hypothesis that the dumpy wing is inherited as a recessive trait using $\chi^2$ analysis of the $F_2$ data.

(a) What ratio was hypothesized?

(b) Did the analysis support the hypothesis?

(c) What do the data suggest about the *dumpy* mutation?

Solution:

(a) The student hypothesized that the $F_2$ data (792:208) fit Mendel's 3:1 monohybrid ratio for recessive genes.

(b) The initial step in $\chi^2$ analysis is to calculate the expected results ($e$) for a ratio of 3:1. Then we can compute deviation $o - e$ ($d$) and the remaining numbers.

| Ratio | $o$ | $e$ | $d$ | $d^2$ | $d^2/e$ |
|---|---|---|---|---|---|
| 3/4 | 792 | 750 | 42 | 1764 | 2.35 |
| 1/4 | 208 | 250 | −42 | 1764 | 7.06 |
| Total = 1000 | | | | | |

$$\chi^2 = \Sigma \frac{d^2}{e}$$
$$= 2.35 + 7.06$$
$$= 9.41$$

We consult Figure 3–12 to determine the probability ($p$) and to decide whether the deviations can be attributed to chance. There are two possible outcomes ($n = 2$), so the degrees of freedom ($df$) $= n - 1$, or 1. The table in Figure 3–12(b) shows that $p$ is a value between 0.01 and 0.001; the graph in Figure 3–12(a) gives an estimate of about 0.001. Since $p < 0.05$, we reject the null hypothesis. The data do not fit a 3:1 ratio.

(c) When the student hypothesized that Mendel's 3:1 ratio was a valid expression of the monohybrid cross, she was tacitly making numerous assumptions. Examining these underlying assumptions may explain why the null hypothesis was rejected. For one thing, she assumed that all the genotypes resulting from the cross were equally viable—that genotypes yielding long wings are equally likely to survive from fertilization through adulthood as the genotype yielding dumpy wings. Further study would reveal that dumpy-winged flies are somewhat less viable than normal flies. As a result, we would expect *less* than 1/4 of the total offspring to express dumpy wings. This observation is borne out in the data, although we have not proven that this is true.

5. If two parents, both heterozygous carriers of the autosomal recessive gene causing cystic fibrosis, have five children, what is the probability that exactly three will be normal?

**Solution:**

This is an opportunity to use the binomial theorem. To do so requires two facts you already possess: the probability of having a normal child during each pregnancy is

$$p_a = \text{normal} = 3/4$$

and the probability of having an afflicted child is

$$p_b = \text{afflicted} = 1/4$$

Insert these into the formula

$$\frac{n!}{s!t!}a^s b^t$$

where $n = 5$, $s = 3$, and $t = 2$

$$p = \frac{(5)\cdot(4)\cdot(3)\cdot(2)\cdot(1)}{(3)\cdot(2)\cdot(1)\cdot(2)\cdot(1)}(3/4)^3(1/4)^2$$
$$= \frac{(5)\cdot(4)}{(2)\cdot(1)}(3/4)^3(1/4)^2$$
$$= 10(27/64)\cdot(1/16)$$
$$= 10(27/1024)$$
$$= 270/1024$$
$$p = \sim 0.26$$

# Problems and Discussion Questions

*When working genetics problems in this and succeeding chapters, always assume that members of the* $P_1$ *generation are homozygous, unless the information or data you are given require you to do otherwise.*

1. In a cross between a black and a white guinea pig, all members of the $F_1$ generation are black. The $F_2$ generation is made up of approximately 3/4 black and 1/4 white guinea pigs.
   (a) Diagram this cross, showing the genotypes and phenotypes.
   (b) What will the offspring be like if two $F_2$ white guinea pigs are mated?
   (c) Two different matings were made between black members of the $F_2$ generation, with the following results.

| Cross | Offspring |
| --- | --- |
| Cross 1 | All black |
| Cross 2 | 3/4 black, 1/4 white |

   Diagram each of the crosses.

2. Albinism in humans is inherited as a simple recessive trait. For the following families, determine the genotypes of the parents and offspring. (When two alternative genotypes are possible, list both.)
   (a) Two normal parents have five children, four normal and one albino.
   (b) A normal male and an albino female have six children, all normal.
   (c) A normal male and an albino female have six children, three normal and three albino.
   (d) Construct a pedigree of the families in (b) and (c). Assume that one of the normal children in (b) and one of the albino children in (c) become the parents of eight children. Add these children to the pedigree, predicting their phenotypes (normal or albino).
3. Which of Mendel's postulates are illustrated by the pedigree in Problem 2? List and define these postulates.
4. Discuss how Mendel's monohybrid results served as the basis for all but one of his postulates. Which postulate was not based on these results? Why?
5. What advantages were provided by Mendel's choice of the garden pea in his experiments?
6. Pigeons may exhibit a checkered or plain color pattern. In a series of controlled matings, the following data were obtained.

| $P_1$ Cross | $F_1$ Progeny Checkered | $F_1$ Progeny Plain |
| --- | --- | --- |
| (a) checkered × checkered | 36 | 0 |
| (b) checkered × plain | 38 | 0 |
| (c) plain × plain | 0 | 35 |

   Then $F_1$ offspring were selectively mated with the following results. (The $P_1$ cross giving rise to each $F_1$ pigeon is indicated in parentheses.)

| $F_1 \times F_1$ Crosses | $F_2$ Progeny Checkered | $F_2$ Progeny Plain |
| --- | --- | --- |
| (d) checkered (a) × plain (c) | 34 | 0 |
| (e) checkered (b) × plain (c) | 17 | 14 |
| (f) checkered (b) × checkered (b) | 28 | 9 |
| (g) checkered (a) × checkered (b) | 39 | 0 |

   How are the checkered and plain patterns inherited? Select and assign symbols for the genes involved, and determine the genotypes of the parents and offspring in each cross.

7. Mendel crossed peas having round seeds and yellow cotyledons (seed leaves) with peas having wrinkled seeds and green cotyledons. All the $F_1$ plants had round seeds with yellow cotyledons. Diagram this cross through the $F_2$ generation, using both the Punnett square and forked-line, or branch diagram, methods.
8. Based on the preceding cross, what is the probability that an organism in the $F_2$ generation will have round seeds and green cotyledons *and* be true breeding?
9. Based on the same characters and traits as in Problem 7, determine the genotypes of the parental plants involved in the crosses shown here by analyzing the phenotypes of their offspring.

| Parental Plants | Offspring |
| --- | --- |
| (a) round, yellow × round, yellow | 3/4 round, yellow<br>1/4 wrinkled, yellow |
| (b) wrinkled, yellow × round, yellow | 6/16 wrinkled, yellow<br>2/16 wrinkled, green<br>6/16 round, yellow<br>2/16 round, green |
| (c) round, yellow × round, yellow | 9/16 round, yellow<br>3/16 round, green<br>3/16 wrinkled, yellow<br>1/16 wrinkled, green |
| (d) round, yellow × wrinkled, green | 1/4 round, yellow<br>1/4 round, green<br>1/4 wrinkled, yellow<br>1/4 wrinkled, green |

10. Are any of the crosses in Problem 9 testcrosses? If so, which one(s)?
11. Which of Mendel's postulates can only be demonstrated in crosses involving at least two pairs of traits? State the postulate.

12. Correlate Mendel's four postulates with what is now known about homologous chromosomes, genes, alleles, and the process of meiosis.
13. What is the basis for homology among chromosomes?
14. Distinguish between homozygosity and heterozygosity.
15. In *Drosophila*, *gray* body color is dominant to *ebony* body color, while *long* wings are dominant to *vestigial* wings. Assuming that the $P_1$ individuals are homozygous, work the following crosses through the $F_2$ generation, and determine the genotypic and phenotypic ratios for each generation.

    (a) gray, long × ebony, vestigial

    (b) gray, vestigial × ebony, long

    (c) gray, long × gray, vestigial
16. How many different types of gametes can be formed by individuals of the following genotypes: (a) *AaBb*, (b) *AaBB*, (c) *AaBbCc*, (d) *AaBBcc*, (e) *AaBbcc*, and (f) *AaBbCcDdEe*? What are the gametes in each case?
17. Using the forked-line, or branch diagram, method, determine the genotypic and phenotypic ratios of these trihybrid crosses: (a) *AaBbCc* × *AaBBCC*, (b) *AaBBCc* × *aaBBCc*, and (c) *AaBbCc* × *AaBbCc*.
18. Mendel crossed peas having green seeds with peas having yellow seeds. The $F_1$ generation produced only yellow seeds. In the $F_2$, the progeny consisted of 6022 plants with yellow seeds and 2001 plants with green seeds. Of the $F_2$ yellow-seeded plants, 519 were self-fertilized with the following results: 166 bred true for yellow and 353 produced an $F_3$ ratio of 3/4 yellow: 1/4 green. Explain these results by diagramming the crosses.
19. In a study of black guinea pigs and white guinea pigs, 100 black animals were crossed with 100 white animals, and each cross was carried to an $F_2$ generation. In 94 of the crosses, all the $F_1$ offspring were black and an $F_2$ ratio of 3 black:1 white was obtained. In the other 6 cases, half of the $F_1$ animals were black and the other half were white. Why? Predict the results of crossing the black and white $F_1$ guinea pigs from the 6 exceptional cases.
20. Mendel crossed peas having round green seeds with peas having wrinkled yellow seeds. All $F_1$ plants had seeds that were round and yellow. Predict the results of testcrossing these $F_1$ plants.
21. Thalassemia is an inherited anemic disorder in humans. Affected individuals exhibit either a minor anemia or a major anemia. Assuming that only a single gene pair and two alleles are involved in the inheritance of these conditions, is thalassemia a dominant or recessive disorder?
22. The following are $F_2$ results of two of Mendel's monohybrid crosses.

| | | |
|---|---|---|
| (a) | full pods | 882 |
| | constricted pods | 299 |
| (b) | violet flowers | 705 |
| | white flowers | 224 |

    For each cross, state a null hypothesis to be tested using $\chi^2$ analysis. Calculate the $\chi^2$ value and determine the $p$ value for both. Interpret the $p$ values. Can the deviation in each case be attributed to chance or not? Which of the two crosses shows a greater amount of deviation?
23. In one of Mendel's dihybrid crosses, he observed 315 round yellow, 108 round green, 101 wrinkled yellow, and 32 wrinkled green $F_2$ plants. Analyze these data using the $\chi^2$ test to see if

    (a) they fit a 9:3:3:1 ratio.

    (b) the round:wrinkled data fit a 3:1 ratio.

    (c) the yellow:green data fit a 3:1 ratio.
24. In assessing data that fell into two phenotypic classes, a geneticist observed values of 250:150. She decided to perform a $\chi^2$ analysis by using the following two different null hypotheses: (a) the data fit a 3:1 ratio, and (b) the data fit a 1:1 ratio. Calculate the $\chi^2$ values for each hypothesis. What can be concluded about each hypothesis?
25. The basis for rejecting any null hypothesis is arbitrary. The researcher can set more or less stringent standards by deciding to raise or lower the $p$ value used to reject or not reject the hypothesis. In the case of the chi-square analysis of genetic crosses, would the use of a standard of $p = 0.10$ be more or less stringent about not rejecting the null hypothesis? Explain.
26. Consider the following pedigree.

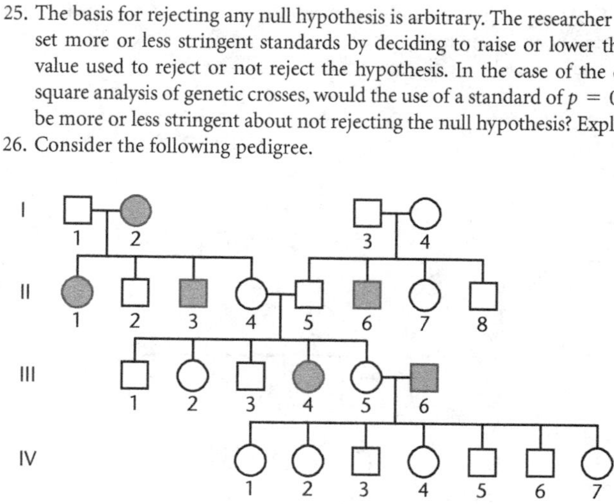


    Predict the mode of inheritance of the trait of interest and the most probable genotype of each individual. Assume that the alleles *A* and *a* control the expression.
27. The following pedigree is for myopia (nearsightedness) in humans.

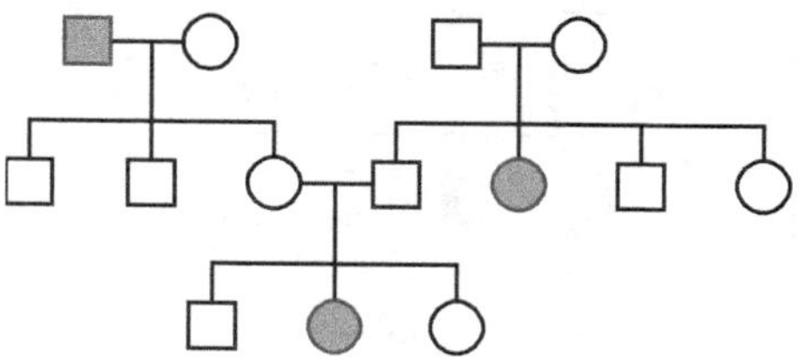

    Predict whether the disorder is inherited as the result of a dominant or recessive trait. Determine the most probable genotype for each individual based on your prediction.
28. Consider three independently assorting gene pairs, *A/a*, *B/b*, and *C/c*. What is the probability of obtaining an offspring that is *AABbCc* from parents that are *AaBbCC* and *AABbCc*?
29. What is the probability of obtaining a triply recessive individual from the parents shown in Problem 28?
30. Of all offspring of the parents in Problem 28, what proportion will express all three dominant traits?
31. When a die (one of a pair of dice) is rolled, it has an equal probability of landing on any of its six sides.

    (a) What is the probability of rolling a 3 with a single throw?

    (b) When a die is rolled twice, what is the probability that the first throw will be a 3 and the second will be a 6?

    (c) When a die is rolled twice, what is the probability that one throw will result in a 3 and the other throw will result in a 6?

    (d) If two dice are rolled together, what is the combined probability that one will be a 3 and the other will be a 6?

    (e) If one die is rolled and it comes up as an odd number, what is the probability that it is a 5?
32. Consider the $F_2$ offspring of Mendel's dihybrid cross. Determine the conditional probability that $F_2$ plants expressing both dominant traits are heterozygous at both loci.

33. Draw all possible conclusions concerning the mode of inheritance of the trait portrayed in each of the following limited pedigrees. (Each of the 4 cases is based on a different trait.)

(a)

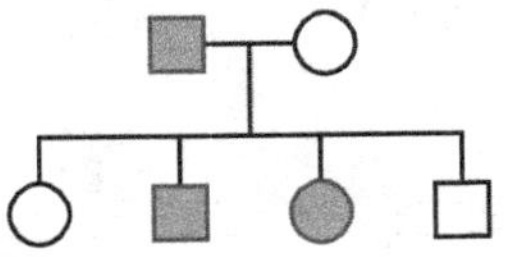

(b)

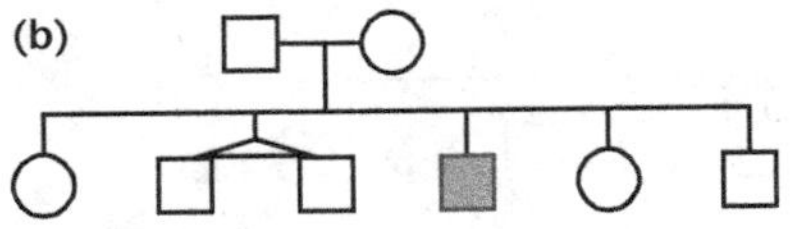

(c)

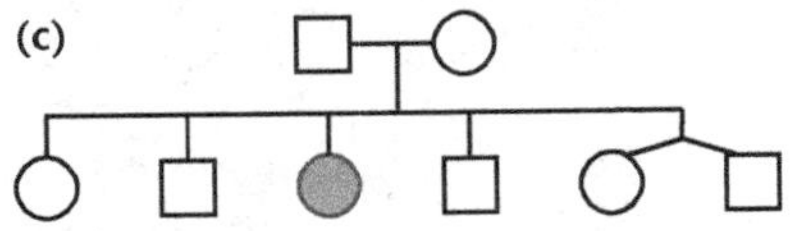

(d)

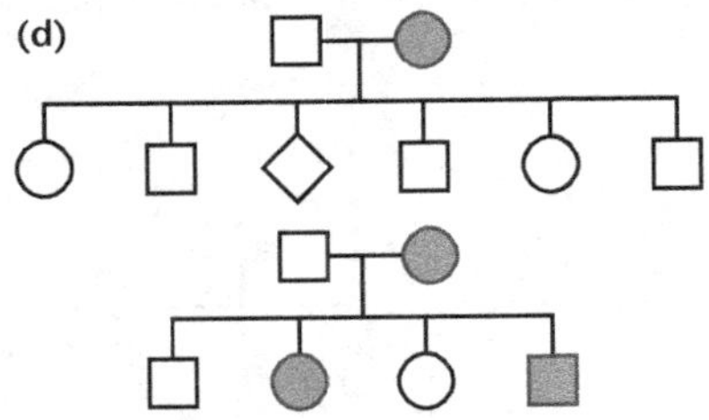

34. Cystic fibrosis is an autosomal recessive disorder. A male whose brother has the disease has children with a female whose sister has the disease. It is not known if either the male or the female is a carrier. If the male and female have one child, what is the probability that the child will have cystic fibrosis?
35. In a family of five children, what is the probability that
    (a) all are males?
    (b) three are males and two are females?
    (c) two are males and three are females?
    (d) all are the same sex?
    Assume that the probability of a male child is equal to the probability of a female child ($p = 1/2$).
36. In a family of eight children, where both parents are heterozygous for albinism, what mathematical expression predicts the probability that six are normal and two are albinos?

HOW DO WE KNOW?

37. In this chapter, we focused on the Mendelian postulates, probability, and pedigree analysis. We also considered some of the methods and reasoning by which these ideas, concepts, and techniques were developed. On the basis of these discussions, what answers would you propose to the following questions:
    (a) How was Mendel able to derive postulates concerning the behavior of "unit factors" during gamete formation, when he could not directly observe them?
    (b) How can we know whether an organism expressing a dominant trait is homozygous or heterozygous?
    (c) In analyzing genetic data, how do we know whether deviation from the expected ratio is due to chance rather than to another, independent factor?
    (d) Since experimental crosses are not performed in humans, how do we know how traits are inherited?

## Extra-Spicy Problems

38. Two true-breeding pea plants were crossed. One parent is round, terminal, violet, constricted, while the other expresses the respective contrasting phenotypes of wrinkled, axial, white, full. The four pairs of contrasting traits are controlled by four genes, each located on a separate chromosome. In the $F_1$ only round, axial, violet, and full were expressed. In the $F_2$, all possible combinations of these traits were expressed in ratios consistent with Mendelian inheritance.
    (a) What conclusion about the inheritance of the traits can be drawn based on the $F_1$ results?
    (b) In the $F_2$ results, which phenotype appeared most frequently? Write a mathematical expression that predicts the probability of occurrence of this phenotype.
    (c) Which $F_2$ phenotype is expected to occur least frequently? Write a mathematical expression that predicts this probability.
    (d) In the $F_2$ generation, how often is either of the $P_1$ phenotypes likely to occur?
    (e) If the $F_1$ plants were testcrossed, how many different phenotypes would be produced? How does this number compare with the number of different phenotypes in the $F_2$ generation just discussed?
39. Tay–Sachs disease (TSD) is an inborn error of metabolism that results in death, often by the age of 2. You are a genetic counselor interviewing a phenotypically normal couple who tell you the male had a female first cousin (on his father's side) who died from TSD and the female had a maternal uncle with TSD. There are no other known cases in either of the families, and none of the matings have been between related individuals. Assume that this trait is very rare.
    (a) Draw a pedigree of the families of this couple, showing the relevant individuals.
    (b) Calculate the probability that both the male and female are carriers for TSD.
    (c) What is the probability that neither of them is a carrier?
    (d) What is the probability that one of them is a carrier and the other is not? [*Hint:* The *p* values in (b), (c), and (d) should equal 1.]
40. *Datura stramonium* (the Jimsonweed) expresses flower colors of purple and white and pod textures of smooth and spiny. The results of two crosses in which the parents were not necessarily true-breeding were observed to be

    white spiny × white spiny → 3/4 white spiny : 1/4 white smooth
    purple smooth × purple smooth → 3/4 purple smooth : 1/4 white smooth

    (a) Based on these results, put forward a hypothesis for the inheritance of the purple/white and smooth/spiny traits.
    (b) Assuming that true-breeding strains of all combinations of traits are available, what single cross could you execute and carry to an F2 generation that will prove or disprove your hypothesis? Assuming your hypothesis is correct, what results of this cross will support it?

41. The wild-type (normal) fruit fly, *Drosophila melanogaster*, has straight wings and long bristles. Mutant strains have been isolated that have either curled wings or short bristles. The genes representing these two mutant traits are located on separate autosomes. Carefully examine the data from the following five crosses shown below (running across both columns).
    (a) Identify each mutation as either dominant or recessive. In each case, indicate which crosses support your answer.
    (b) Assign gene symbols and, for each cross, determine the genotypes of the parents.

| Cross | Progeny: straight wings, long bristles | straight wings, short bristles | curled wings, long bristles | curled wings, short bristles |
|---|---|---|---|---|
| 1. straight, short × straight, short | 30 | 90 | 10 | 30 |
| 2. straight, long × straight, long | 120 | 0 | 40 | 0 |
| 3. curled, long × straight, short | 40 | 40 | 40 | 40 |
| 4. straight, short × straight, short | 40 | 120 | 0 | 0 |
| 5. curled, short × straight, short | 20 | 60 | 20 | 60 |

42. An alternative to using the expanded binomial equation and Pascal's triangle in determining probabilities of phenotypes in a subsequent generation when the parents' genotypes are known is to use the following equation:

$$\frac{n!}{s!t!}a^s b^t$$

where $n$ is the total number of offspring, $s$ is the number of offspring in one phenotypic category, $t$ is the number of offspring in the other phenotypic category, $a$ is the probability of occurrence of the first phenotype, and $b$ is the probability of the second phenotype. Using this equation, determine the probability of a family of 5 offspring having exactly 2 children afflicted with sickle-cell anemia (an autosomal recessive disease) when both parents are heterozygous for the sickle-cell allele.

43. Considering the information in Problem 42, to what do you suppose the following mathematical expression applies?

$$\frac{n!}{s!t!u!}a^s b^t c^u$$

Can you think of a genetic example where it might have application?

44. To assess Mendel's law of segregation using tomatoes, a true-breeding tall variety (*SS*) is crossed with a true-breeding short variety (*ss*). The heterozygous $F_1$ tall plants (*Ss*) were crossed to produce two sets of $F_2$ data, as follows.

| Set I | Set II |
|---|---|
| 30 tall | 300 tall |
| 5 short | 50 short |

    (a) Using the $\chi^2$ test, analyze the results for both data sets. Calculate $\chi^2$ values and estimate the $p$ values in both cases.
    (b) From the above analysis, what can you conclude about the importance of generating large data sets in experimental conditions?

45. When examining Sutton's drawings of chromosomes of the grasshopper, *Brachystola magna*, Eleanor Carothers (1913) noted a pair of unlike chromosomes—one large dyad and one small dyad—making up a tetrad in each of 300 primary spermatocytes. In addition, an accessory chromosome (unpaired and later called the X chromosome) was identified in females, such that males had 23 chromosomes and females had 24 chromosomes. Carothers found that the larger dyad in each unlike pair went to the same pole as the accessory chromosome in 154 anaphases, while the smaller dyad went with the accessory chromosome in the remaining 146 anaphases. (a) How do these findings relate to Mendel's postulates, and (b) how do they support the chromosome theory of heredity?

46. *Dentinogenesis imperfecta* is a tooth disorder involving the production of dentin sialophosphoprotein, a bone-like component of the protective middle layer of teeth. The trait is inherited as an autosomal dominant allele located on chromosome 4 in humans and occurs in about 1 in 6000 to 8000 people. Assume that a man with *dentinogenesis imperfecta*, whose father had the disease but whose mother had normal teeth, married a woman with normal teeth. They have six children. What is the probability that their first child will be a male with *dentinogenesis imperfecta*? What is the probability that three of their six children will have the disease?

Labrador retrievers expressing brown (chocolate), golden (yellow), and black coat colors, traits controlled by two gene pairs.

# 4

# Extensions of Mendelian Genetics

## CHAPTER CONCEPTS

- While alleles are transmitted from parent to offspring according to Mendelian principles, they often do not display the clear-cut dominant/recessive relationship observed by Mendel.
- In many cases, in a departure from Mendelian genetics, two or more genes are known to influence the phenotype of a single characteristic.
- Still another exception to Mendelian inheritance occurs when genes are located on the X chromosome, because one of the sexes receives only one copy of that chromosome, eliminating the possibility of heterozygosity.
- Phenotypes are often the combined result of genetics and the environment within which genes are expressed.
- The result of the various exceptions to Mendelian principles is the occurrence of phenotypic ratios that differ from those produced by standard monohybrid, dihybrid, and trihybrid crosses.

In Chapter 3, we discussed the fundamental principles of transmission genetics. We saw that genes are present on homologous chromosomes and that these chromosomes segregate from each other and assort independently from other segregating chromosomes during gamete formation. These two postulates are the basic principles of gene transmission from parent to offspring. Once an offspring has received the total set of genes, it is the expression of genes that determines the organism's phenotype. When gene expression does not adhere to a simple dominant/recessive mode, or when more than one pair of genes influences the expression of a single character, the classic 3:1 and 9:3:3:1 $F_2$ ratios are usually modified. In this and the next several chapters, we consider more complex modes of inheritance. In spite of the greater complexity of these situations, the fundamental principles set down by Mendel still hold.

In this chapter, we restrict our initial discussion to the inheritance of traits controlled by only one set of genes. In diploid organisms, which have homologous pairs of chromosomes, two copies of each gene influence such traits. The copies need not be identical since alternative forms of genes, **alleles**, occur within populations. How alleles influence phenotypes will be our primary focus. We will then consider **gene interaction**, a situation in which a single phenotype is affected by more than one set of genes. Numerous examples will be presented to illustrate a variety of heritable patterns observed in such situations.

Thus far, we have restricted our discussion to chromosomes other than the X and Y pair. By examining cases where genes are present on the X chromosome, illustrating **X-linkage**, we will see yet another modification of Mendelian ratios. Our discussion of modified ratios also includes the consideration of sex-limited and sex-influenced inheritance, cases where the sex of the individual, but not necessarily the genes on the X chromosome, influences the phenotype. We conclude the chapter by showing how a given phenotype often varies depending on the overall environment in which a gene, a cell, or an organism finds itself. This discussion points out that phenotypic expression depends on more than just the genotype of an organism.

## 4.1 Alleles Alter Phenotypes in Different Ways

Following the rediscovery of Mendel's work in the early 1900s, research focused on the many ways in which genes influence an individual's phenotype. This course of investigation, stemming from Mendel's findings, is called neo-Mendelian genetics (*neo* from the Greek word meaning *since* or *new*).

Each type of inheritance described in this chapter was investigated when observations of genetic data did not conform precisely to the expected Mendelian ratios. Hypotheses that modified and extended the Mendelian principles were proposed and tested with specifically designed crosses. The explanations proffered to account for these observations were constructed in accordance with the principle that a phenotype is under the influence of one or more genes located at specific loci on one or more pairs of homologous chromosomes.

To understand the various modes of inheritance, we must first consider the potential function of an allele. An allele is an alternative form of a gene. The allele that occurs most frequently in a population, the one that we arbitrarily designate as normal, is called the **wild-type allele**. This is often, but not always, dominant. Wild-type alleles are responsible for the corresponding wild-type phenotype and are the standards against which all other mutations occurring at a particular locus are compared.

A mutant allele contains modified genetic information and often specifies an altered gene product. For example, in human populations, there are many known alleles of the gene encoding the $\beta$ chain of human hemoglobin. All such alleles store information necessary for the synthesis of the $\beta$ chain polypeptide, but each allele specifies a slightly different form of the same molecule. Once the allele's product has been manufactured, the product's function may or may not be altered.

The process of mutation is the source of alleles. For a new allele to be recognized by observation of an organism, the allele must cause a change in the phenotype. A new phenotype results from a change in functional activity of the cellular product specified by that gene. Often, the mutation causes the diminution or the loss of the specific wild-type function. For example, if a gene is responsible for the synthesis of a specific enzyme, a mutation in that gene may ultimately change the conformation of this enzyme and reduce or eliminate its affinity for the substrate. Such a mutation is designated as a **loss-of-function mutation**. If the loss is complete, the mutation has resulted in what is called a **null allele**.

Conversely, other mutations may enhance the function of the wild-type product. Most often when this occurs, it is the result of increasing the quantity of the gene product. For example, the mutation may be affecting the regulation of transcription of the gene under consideration. Such mutations, designated **gain-of-function mutations**, generally result in dominant alleles, since one copy of the mutation in a diploid organism is sufficient to alter the normal phenotype. Examples of gain-of-function mutations include the genetic conversion of proto-oncogenes, which regulate the cell cycle, to oncogenes, where regulation is overridden by excess gene product. The result is the creation of a cancerous cell.

Having introduced the concepts of gain- and loss-of-function mutations, we should note the possibility that a mutation will create an allele that produces no detectable change in function. In this case, the mutation would not be immediately apparent since no phenotypic variation would be evident. However, such a mutation could be detected if the DNA sequence of the gene was examined directly. These are sometimes referred to as **neutral mutations** since the gene product presents no change to either the phenotype or to the evolutionary fitness of the organism.

Finally, we note that while a phenotypic trait may be affected by a single mutation in one gene, traits are often influenced by many gene products. For example, enzymatic reactions are most often part of complex metabolic pathways leading to the synthesis of an end product, such as an amino acid. Mutations in any of a pathway's reactions can have a common effect—the failure to synthesize the end product. Therefore, phenotypic traits related to the end product are often influenced by more than one gene. Such is the case in *Drosophila* eye color mutations. Eye color results from the synthesis and deposition of a brown and a bright red pigment in the facets of the compound eye. This causes the wild-type eye color to appear brick red. There are a series of recessive loss-of-function mutations that interrupt the multistep pathway leading to the synthesis of the brown pigment. While these mutations represent genes located on different chromosomes, they all result in the same phenotype: a bright red eye whose color is due to the absence of the brown pigment. Examples are the mutations *vermilion, cinnabar*, and *scarlet*, which are indistinguishable phenotypically.

In each of the many crosses discussed in the next few chapters, only one or a few gene pairs are involved. Keep in mind that in each cross, all genes that are not under consideration are assumed to have no effect on the inheritance patterns described.

4.2

## Geneticists Use a Variety of Symbols for Alleles

In Chapter 3, we learned a standard convention used to symbolize alleles for very simple Mendelian traits. The initial letter of the name of a recessive trait, lowercased and italicized, denotes the recessive allele, and the same letter in uppercase refers to the dominant allele. Thus, in the case of *tall* and *dwarf*, where *dwarf* is recessive, *D* and *d* represent the alleles responsible for these respective traits. Mendel used upper- and lowercase letters such as these to symbolize his unit factors.

Another useful system was developed in genetic studies of the fruit fly *Drosophila melanogaster* to discriminate between wild-type and mutant traits. This system uses the initial letter, or a combination of two or three letters, from the name of the mutant trait. If the trait is recessive, lowercase is used; if it is dominant, uppercase is used. The contrasting wild-type trait is denoted by the same letters, but with a superscript +. For example, *ebony* is a recessive body color mutation in *Drosophila*. The normal wild-type body color is gray. Using this system, we denote *ebony* by the symbol *e*, while gray is denoted by $e^+$. The responsible locus may be occupied by either the wild-type allele ($e^+$) or the mutant allele (*e*). A diploid fly may thus exhibit one of three possible genotypes (the two phenotypes are indicated parenthetically):

| | |
|---|---|
| $e^+/e^+$ | gray homozygote (wild type) |
| $e^+/e$ | gray heterozygote (wild type) |
| $e/e$ | ebony homozygote (mutant) |

The slash between the letters indicates that the two allele designations represent the same locus on two homologous chromosomes. If we instead consider a mutant allele that is dominant to the normal wild-type allele, such as *Wrinkled* wing in *Drosophila*, the three possible genotypes are $Wr/Wr$, $Wr/Wr^+$, and $Wr^+/Wr^+$. The initial two genotypes express the mutant wrinkled-wing phenotype.

One advantage of this system is that further abbreviation can be used when convenient: The wild-type allele may simply be denoted by the + symbol. With *ebony* as an example, the designations of the three possible genotypes become

| | |
|---|---|
| $+/+$ | gray homozygote (wild type) |
| $+/e$ | gray heterozygote (wild type) |
| $e/e$ | ebony homozygote (mutant) |

Another variation is utilized when no dominance exists between alleles (a situation we will explore in Section 4.3). We simply use uppercase letters and superscripts to denote alternative alleles (e.g., $R^1$ and $R^2$, $L^M$ and $L^N$, and $I^A$ and $I^B$).

Many diverse systems of genetic nomenclature are used to identify genes in various organisms. Usually, the symbol selected reflects the function of the gene or even a disorder caused by a mutant gene. For example, in yeast, *cdk* is the abbreviation for the *c*yclin-*d*ependent *k*inase gene, whose product is involved in the cell-cycle regulation mechanism discussed in Chapter 2. In bacteria, $leu^-$ refers to a mutation that interrupts the biosynthesis of the amino acid leucine, and the wild-type gene is designated $leu^+$. The symbol *dnaA* represents a bacterial gene involved in DNA replication (and DnaA, without italics, is the protein made by that gene). In humans, italicized capital letters are used to name genes: *BRCA1* represents one of the genes associated with susceptibility to *br*east *ca*ncer. Although these different systems may seem complex, they are useful ways to symbolize genes.

4.3

## Neither Allele Is Dominant in Incomplete, or Partial, Dominance

Unlike the Mendelian crosses reported in Chapter 3, a cross between parents with contrasting traits may sometimes generate offspring with an intermediate phenotype. For example, if a four-o'clock or a snapdragon plant with red flowers is crossed with a white-flowered plant, the offspring have pink flowers. Because some red pigment is produced in the $F_1$ intermediate-colored plant, neither the red nor white flower color is dominant. Such a situation is known as **incomplete**, or **partial, dominance.**

If the phenotype is under the control of a single gene and two alleles, where neither is dominant, the results of the $F_1$ (pink) × $F_1$ (pink) cross can be predicted. The resulting $F_2$ generation shown

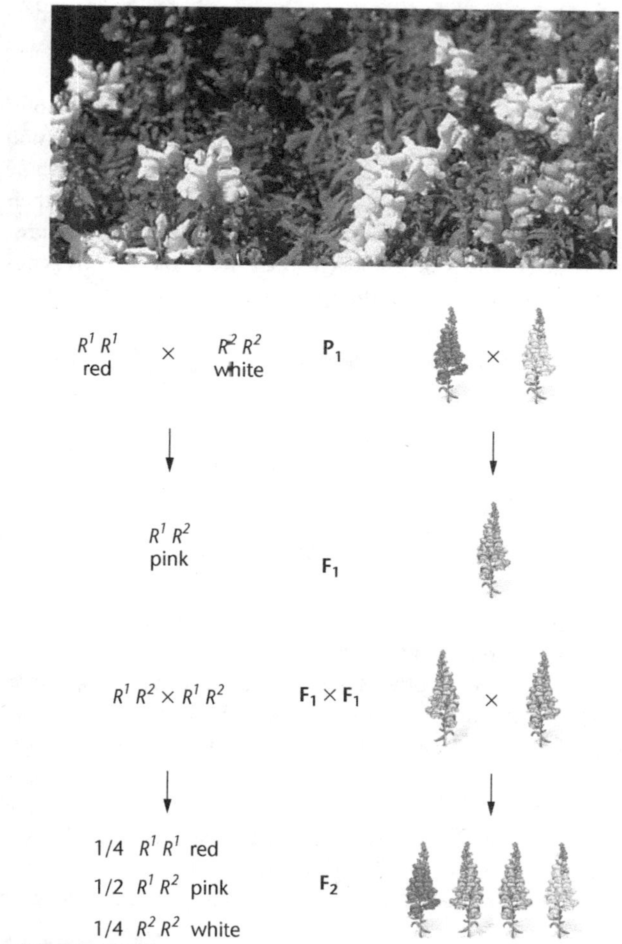


FIGURE 4–1 Incomplete dominance shown in the flower color of snapdragons.

in Figure 4–1 confirms the hypothesis that only one pair of alleles determines these phenotypes. The genotypic ratio (1:2:1) of the $F_2$ generation is identical to that of Mendel's monohybrid cross. However, because neither allele is dominant, the phenotypic ratio is identical to the genotypic ratio (in contrast to the 3:1 phenotypic ratio of a Mendelian monohybrid cross). Note that because neither allele is recessive, we have chosen not to use upper- and lowercase letters as symbols. Instead, we denote the red and white alleles as $R^1$ and $R^2$. We could have chosen $W^1$ and $W^2$ or still other designations such as $C^W$ and $C^R$, where $C$ indicates "color" and the $W$ and $R$ superscripts indicate white and red, respectively.

How are we to interpret lack of dominance whereby an intermediate phenotype characterizes heterozygotes? The most accurate way is to consider gene expression in a quantitative way. In the case of flower color above, the mutation causing white flowers is most likely one where complete "loss of function" occurs. In this case, it is likely that the gene product of the wild-type allele ($R^1$) is an enzyme that participates in a reaction leading to the synthesis of a red pigment. The mutant allele ($R^2$) produces an enzyme that cannot catalyze the reaction leading to pigment. The end result is that the heterozygote produces only about half the pigment of the red-flowered plant and the phenotype is pink.

Clear-cut cases of incomplete dominance are relatively rare. However, even when one allele seems to have complete dominance over the other, careful examination of the gene product, rather than the phenotype, often reveals an intermediate level of gene expression. An example is the human biochemical disorder **Tay–Sachs disease**, in which homozygous recessive individuals are severely affected with a fatal lipid-storage disorder and neonates die during their first one to three years of life. (Recall the extensive discussion of this human malady in the Genetics, Technology, and Society essay at the end of Chapter 3.) In afflicted individuals, there is almost no activity of the enzyme **hexosaminidase A**, an enzyme normally involved in lipid metabolism. Heterozygotes, with only a single copy of the mutant gene, are phenotypically normal, but with only about 50 percent of the enzyme activity found in homozygous normal individuals. Fortunately, this level of enzyme activity is adequate to achieve normal biochemical function. This situation is not uncommon in enzyme disorders and illustrates the concept of the **threshold effect**, whereby normal phenotypic expression occurs anytime a certain level of gene product is attained. Most often, and in particular in Tay–Sachs disease, the threshold is less than 50 percent.

WEB TUTORIAL 4.1 EXTENSIONS OF MENDELIAN GENETICS

## 4.4 In Codominance, the Influence of Both Alleles in a Heterozygote Is Clearly Evident

If two alleles of a single gene are responsible for producing two distinct, detectable gene products, a situation different from incomplete dominance or dominance/recessiveness arises. In this case, the joint expression of both alleles in a heterozygote is called **codominance.** The **MN blood group** in humans illustrates this phenomenon. Karl Landsteiner and Philip Levin discovered a glycoprotein molecule found on the surface of red blood cells that acts as a native antigen, providing biochemical and immunological identity to individuals. In the human population, two forms of this glycoprotein exist, designated M and N; an individual may exhibit either one or both of them.

The MN system is under the control of a locus found on chromosome 4, with two alleles designated $L^M$ and $L^N$. Because humans are diploid, three combinations are possible, each resulting in a distinct blood type:

| Genotype | Phenotype |
|---|---|
| $L^M L^M$ | M |
| $L^M L^N$ | MN |
| $L^N L^N$ | N |

As predicted, a mating between two heterozygous MN parents may produce children of all three blood types, as follows:

$$L^M L^N \times L^M L^N$$
$$\downarrow$$
$$1/4\ L^M L^M$$
$$1/2\ L^M L^N$$
$$1/4\ L^N L^N$$

Once again, the genotypic ratio 1:2:1 is upheld.

Codominant inheritance is characterized by *distinct expression of the gene products of both alleles.* This characteristic distinguishes codominance from incomplete dominance, where heterozygotes express an intermediate, blended, phenotype. For codominance to be studied, both products must be phenotypically detectable. We shall see another example of codominance when we examine the ABO blood-type system.

## 4.5 Multiple Alleles of a Gene May Exist in a Population

The information stored in any gene is extensive, and mutations can modify this information in many ways. Each change produces a different allele. Therefore, for any gene, the number of alleles within members of a population need not be restricted to two. When three or more alleles of the same gene—which we designate as **multiple alleles**—are present in a population, the resulting mode of inheritance may be unique. It is important to realize that *multiple alleles can be studied only in populations.* Any individual diploid organism has, at most, two homologous gene loci that may be occupied by different alleles of the same gene. However, among members of a species, numerous alternative forms of the same gene can exist.

### The ABO Blood Groups

The simplest case of multiple alleles occurs when three alternative alleles of one gene exist. This situation is illustrated in the inheritance of the **ABO blood groups** in humans, discovered by Karl Landsteiner in the early 1900s. The ABO system, like the MN blood types, is characterized by the presence of antigens on the surface of red blood cells. The A and B antigens are distinct from the MN antigens and are under the control of a different gene, located on chromosome 9. As in the MN system, one combination of alleles in the ABO system exhibits a codominant mode of inheritance.

The ABO phenotype of any individual is ascertained by mixing a blood sample with an antiserum containing type A or type B antibodies. If an antigen is present on the surface of the person's red blood cells, it will react with the corresponding antibody and cause clumping, or agglutination, of the red blood cells. When an individual is tested in this way, one of four phenotypes may be revealed. Each individual has either the A antigen (A phenotype), the B antigen (B phenotype), the A and B antigens (AB phenotype), or neither antigen (O phenotype).

In 1924, it was hypothesized that these phenotypes were inherited as the result of three alleles of a single gene. This hypothesis was based on studies of the blood types of many different families. Although different designations can be used, we will use the symbols $I^A$, $I^B$, and $I^O$ to distinguish these three alleles. The $I$ designation stands for **isoagglutinogen**, another term for antigen. If we assume that the $I^A$ and $I^B$ alleles are responsible for the production of their respective A and B antigens and that $I^O$ is an allele that does not produce any detectable A or B antigens, we can list the various genotypic possibilities and assign the appropriate phenotype to each:

| Genotype | Antigen | Phenotype |
|---|---|---|
| $I^A I^A$ | A | A |
| $I^A I^O$ | A | A |
| $I^B I^B$ | B | B |
| $I^B I^O$ | B | B |
| $I^A I^B$ | A, B | AB |
| $I^O I^O$ | Neither | O |

In these assignments, the $I^A$ and $I^B$ alleles are dominant to the $I^O$ allele, but codominant to each other.

We can test the hypothesis that three alleles control ABO blood groups by examining potential offspring from the various combinations of matings, as shown in Table 4.1. If we assume heterozygosity wherever possible, we can predict which phenotypes can occur. These theoretical predictions have been upheld in numerous studies examining the blood types of children of parents with all possible phenotypic combinations. The hypothesis that three alleles control ABO blood types in the human population is now universally accepted.

**TABLE 4.1**

**Potential Phenotypes in the Offspring of Parents with All Possible ABO Blood Group Combinations, Assuming Heterozygosity Whenever Possible**

| Parents | | Potential Offspring | | | |
|---|---|---|---|---|---|
| Phenotypes | Genotypes | A | B | AB | O |
| A × A | $I^A I^O \times I^A I^O$ | 3/4 | — | — | 1/4 |
| B × B | $I^B I^O \times I^B I^O$ | — | 3/4 | — | 1/4 |
| O × O | $I^O I^O \times I^O I^O$ | — | — | — | all |
| A × B | $I^A I^O \times I^B I^O$ | 1/4 | 1/4 | 1/4 | 1/4 |
| A × AB | $I^A I^O \times I^A I^B$ | 1/2 | 1/4 | 1/4 | — |
| A × O | $I^A I^O \times I^O I^O$ | 1/2 | — | — | 1/2 |
| B × AB | $I^B I^O \times I^A I^B$ | 1/4 | 1/2 | 1/4 | — |
| B × O | $I^B I^O \times I^O I^O$ | — | 1/2 | — | 1/2 |
| AB × O | $I^A I^B \times I^O I^O$ | 1/2 | 1/2 | — | — |
| AB × AB | $I^A I^B \times I^A I^B$ | 1/4 | 1/4 | 1/2 | — |

Our knowledge of human blood types has several practical applications. One of the most important is testing the compatibility of blood transfusions. Another application involves cases of disputed parentage, where newborns are inadvertently mixed up in the hospital, or when it is uncertain whether a specific male is the father of a child. An examination of the ABO blood groups as well as other inherited antigens of the possible parents and the child may help to resolve the situation. For example, of all the matings shown in Table 4.1, the only one that can result in offspring with all four phenotypes is that between two heterozygous individuals, one showing the A phenotype and the other showing the B phenotype. On genetic grounds alone, a male or female may be unequivocally ruled out as the parent of a certain child. However, this type of genetic evidence never proves parenthood.

## The A and B Antigens

The biochemical basis of the ABO blood type system has now been carefully worked out. The A and B antigens are actually carbohydrate groups (sugars) that are bound to lipid molecules (fatty acids) protruding from the membrane of the red blood cell. The specificity of the A and B antigens is based on the terminal sugar of the carbohydrate group.

Almost all individuals possess what is called the **H substance**, to which one or two terminal sugars are added. As shown in Figure 4–2,

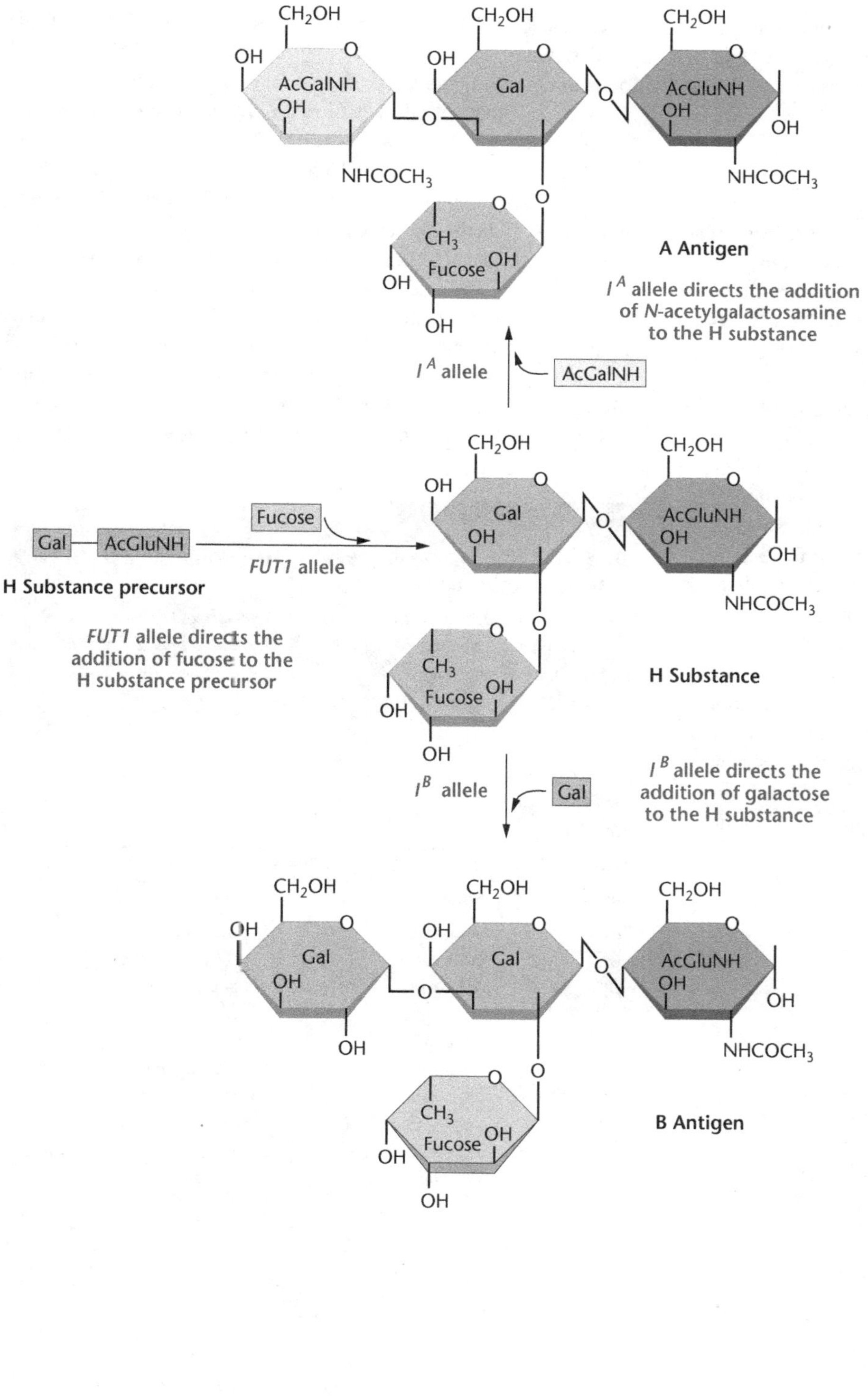


FIGURE 4–2 The biochemical basis of the ABO blood groups. The wild-type *FUT1* allele, present in almost all humans, directs the conversion of a precursor molecule to the H substance by adding a molecule of fucose to it. The $I^A$ and $I^B$ alleles are then able to direct the addition of terminal sugar residues to the H substance. The $I^O$ allele is unable to direct either of these terminal additions. Failure to produce the H substance results in the Bombay phenotype, in which individuals are type O regardless of the presence of an $I^A$ or $I^B$ allele. Gal: galactose; AcGluNH: *N*-acetylglucosamine; AcGalNH: *N*-acetylgalactosamine.

the H substance itself contains three sugar molecules—galactose (Gal), *N*-acetylglucosamine (AcGluNH), and fucose—chemically linked together. The $I^A$ allele is responsible for an enzyme that can add the terminal sugar *N*-acetylgalactosamine (AcGalNH) to the H substance. The $I^B$ allele is responsible for a modified enzyme that cannot add *N*-acetylgalactosamine, but instead can add a terminal galactose. Heterozygotes ($I^AI^B$) add either one or the other sugar at the many sites (substrates) available on the surface of the red blood cell, illustrating the biochemical basis of codominance in individuals of the AB blood type. Finally, persons of type O ($I^OI^O$) cannot add either terminal sugar; these persons have only the H substance protruding from the surface of their red blood cells.

The molecular genetic basis of the mutations leading to the $I^A$, $I^B$, and $I^O$ alleles has also been clarified. We will describe it in Chapter 16 when we discuss mutation and mutagenesis.

## The Bombay Phenotype

In 1952, a very unusual situation provided information concerning the genetic basis of the H substance. A woman in Bombay displayed a unique genetic history inconsistent with her blood type. In need of a transfusion, she was found to lack both the A and B antigens and was thus typed as O. However, as shown in the partial pedigree in Figure 4–3, one of her parents was type AB, and she herself was the obvious donor of an $I^B$ allele to two of her offspring. Thus, she was genetically type B but functionally type O!

This woman was subsequently shown to be homozygous for a rare recessive mutation in a gene designated *FUT1* (encoding an enzyme, fucosyl transferase), which prevented her from synthesizing the complete H substance. In this mutation, the terminal portion of the carbohydrate chain protruding from the red cell membrane lacks fucose, normally added by the enzyme. In the absence of fucose, the enzymes specified by the $I^A$ and $I^B$ alleles apparently are unable to recognize the incomplete H substance as a proper substrate. Thus, neither the terminal galactose nor *N*-acetylgalactosamine can be added, even though the appropriate enzymes capable of doing so are present and functional. As a result, the ABO system genotype cannot be expressed in individuals homozygous for the mutant form of the *FUT1* gene; even though they may have the $I^A$ and/or the $I^B$ alleles, neither antigen is added to the cell surface, and they are functionally type O. To distinguish them from the rest of the population, they are

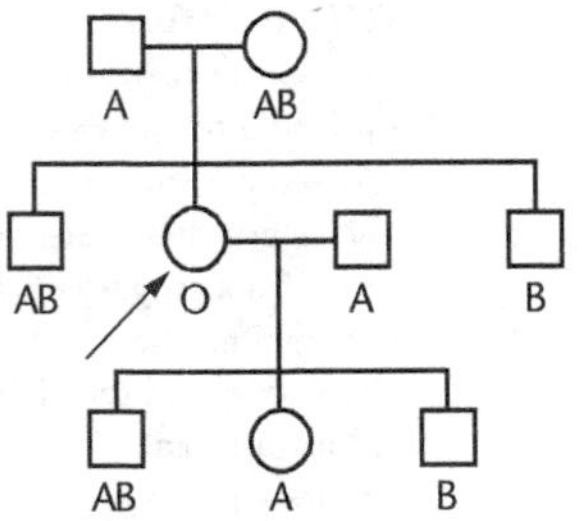


FIGURE 4–3 A partial pedigree of a woman with the Bombay phenotype. Functionally, her ABO blood group behaves as type O. Genetically, she is type B.

said to demonstrate the **Bombay phenotype**. The frequency of the mutant *FUT1* allele is exceedingly low. Hence, the vast majority of the human population can synthesize the H substance.

## The *white* Locus in *Drosophila*

Many other phenotypes in plants and animals are influenced by multiple allelic inheritance. In *Drosophila*, many alleles are present at practically every locus. The recessive mutation that causes white eyes, discovered by Thomas H. Morgan and Calvin Bridges in 1912, is one of over 100 alleles that can occupy this locus. In this allelic series, eye colors range from complete absence of pigment in the *white* allele to deep ruby in the *white-satsuma* allele, orange in the *white-apricot* allele, and a buff color in the *white-buff* allele. These alleles are designated $w$, $w^{sat}$, $w^a$, and $w^{bf}$, respectively. In each case, the total amount of pigment in these mutant eyes is reduced to less than 20 percent of that found in the brick-red wild-type eye. Table 4.2 lists these and other *white* alleles and their color phenotypes.

It is interesting to note the biological basis of the original *white* mutation in *Drosophila*. Given what we know about eye color in this organism, it might be logical to presume that the mutant allele somehow interrupts the biochemical synthesis of pigments making up the brick red eye of the wild-type fly. However, it is now clear that the product of the *white* locus is a protein that is involved in transporting pigments into the ommatidia (the individual units) comprising the compound eye. While flies expressing the *white* mutation can synthesize eye pigments normally, they cannot transport them into these structural units of the eye, thus rendering the white phenotype.

NOW SOLVE THIS

Problem 10 on page 99 involves a series of multiple alleles controlling coat color in rabbits.

■ HINT: *Note particularly the hierarchy of dominance of the various alleles. Remember also that even though there can be more than two alleles in a population, an individual can have at most two of these. Thus, the allelic distribution into gametes adheres to the principle of segregation.*

TABLE 4.2

**Some of the Alleles Present at the *White* Locus of *Drosophila***

| Allele | Name | Eye Color |
|---|---|---|
| $w$ | *white* | pure white |
| $w^a$ | *white-apricot* | yellowish orange |
| $w^{bf}$ | *white-buff* | light buff |
| $w^{bl}$ | *white-blood* | yellowish ruby |
| $w^{cf}$ | *white-coffee* | deep ruby |
| $w^e$ | *white-eosin* | yellowish pink |
| $w^{mo}$ | *white-mottled orange* | light mottled orange |
| $w^{sat}$ | *white-satsuma* | deep ruby |
| $w^{sp}$ | *white-spotted* | fine grain, yellow mottling |
| $w^t$ | *white-tinged* | light pink |

4.6

## Lethal Alleles Represent Essential Genes

Many gene products are essential to an organism's normal development and survival. When such genes mutate, the premature death of an organism may be the result. As we will see below, in some cases, the complete absence of the gene product is the cause of lethality. If both alleles in a diploid organism must be mutated to cause death, the mutation is behaving as a *recessive lethal*. In other cases, less well understood, just a single copy of a mutant allele is sufficient to cause lethality. Such a mutation creates a *dominant lethal* condition. We will examine both kinds of lethal mutations.

### Recessive Lethal Mutations

When the complete absence of a gene product is lethal, the genetic change is most often a loss-of-function mutation that creates a nonfunctional product. Such a mutation can often be tolerated in the heterozygous state where one wild-type allele may produce a sufficient quantity of the product to allow normal development. However, such a mutation behaves as a **recessive lethal allele**, and individuals who are homozygous for the recessive allele will not survive. The time of death will depend on when the product is essential. In mammals, for example, this might occur during development, early childhood, or even during adulthood.

In some cases, the allele responsible for a lethal effect when it is homozygous may also result in a distinctive mutant phenotype when it is present heterozygously. *Such an allele is behaving as a recessive lethal but is dominant with respect to the phenotype.* This obviously creates a very interesting genetic situation. For example, a mutation that causes yellow coat color in mice was discovered in the early part of the twentieth century. The yellow coat differs from the normal agouti coat phenotype, as shown in Figure 4–4. Crosses between the various combinations of the two strains yield unusual results:

| Crosses | | | | |
|---|---|---|---|---|
| (A) agouti | × | agouti | ⟶ | all agouti |
| (B) yellow | × | yellow | ⟶ | 2/3 yellow: 1/3 agouti |
| (C) agouti | × | yellow | ⟶ | 1/2 yellow: 1/2 agouti |

These results are explained on the basis of a single pair of alleles. With regard to coat color, the mutant yellow allele ($A^Y$) is dominant to the wild-type agouti allele ($A$), so heterozygous mice will have yellow coats. However, the yellow allele also behaves as a homozygous recessive lethal. Mice of the genotype $A^Y A^Y$ die before birth. Thus, no homozygous yellow mice are ever recovered. The genetic basis for these three crosses is provided in Figure 4–4.

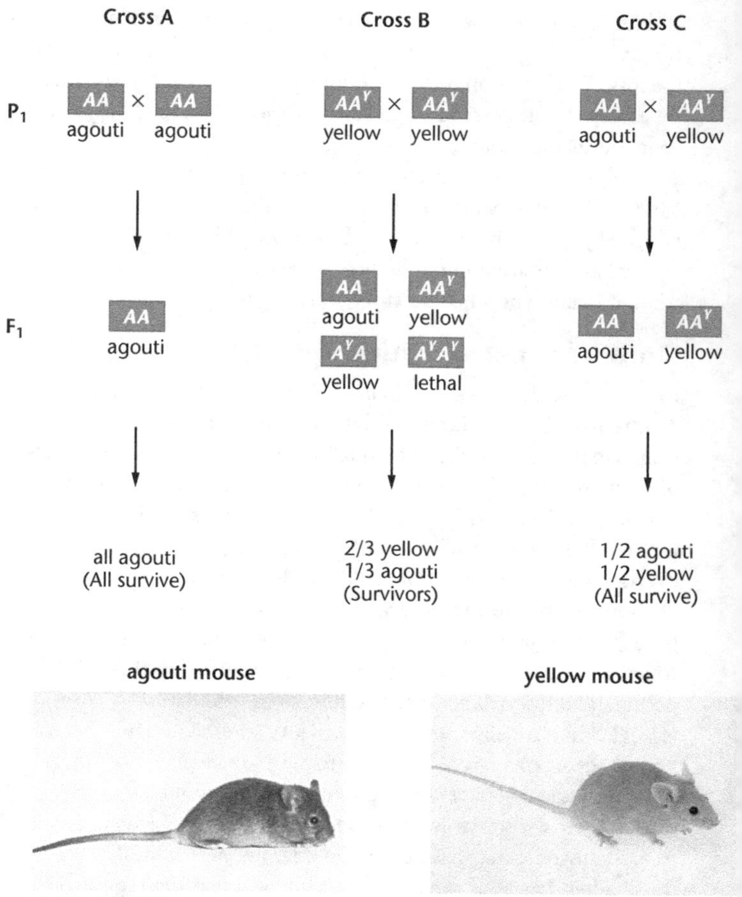


FIGURE 4–4 Inheritance patterns in three crosses involving the normal wild-type *agouti* allele ($A$) and the mutant *yellow* allele ($A^Y$) in the mouse. Note that the mutant allele behaves dominantly to the normal allele in controlling coat color, but it also behaves as a homozygous recessive lethal allele. The genotype $A^Y A^Y$ does not survive.

Molecular analysis of the $A$ gene in both normal agouti and mutant yellow mice has provided insight into how a mutation can be both dominant for one phenotypic effect (hair color) and recessive for another (embryonic development). The $A^Y$ allele is a classic example of a gain-of-function mutation. Animals homozygous for the wild-type $A$ allele have yellow pigment deposited as a band on the otherwise black hair shaft, resulting in the agouti phenotype (see Figure 4–4). Heterozygotes deposit yellow pigment along the entire length of hair shafts as a result of the deletion of the regulatory region preceding the DNA coding region of the $A^Y$ allele. Without any means to regulate its expression, one copy of the $A^Y$ allele is always turned on in heterozygotes, resulting in the gain of function leading to the dominant effect.

The homozygous lethal effect has also been explained by molecular analysis of the mutant gene. The extensive deletion of genetic material that produced the $A^Y$ allele actually extends into the coding region of an adjacent gene (*Merc*), rendering it nonfunctional. It is this gene that is critical to embryonic development, and the loss of its function in $A^Y/A^Y$ homozygotes is what causes lethality. Heterozygotes exceed the threshold level of the wild-type *Merc* gene product and thus survive.

Many genes are known to exhibit similar properties in other organisms. In *Drosophila*, *Curly* wing (*Cy*), *Plum* eye (*Pm*), *Dichaete* wing (*D*), *Stubble* bristle (*Sb*), and *Lyra* wing (*Ly*) behave as recessive lethals but are dominant with respect to the expression of the mutant phenotype when heterozygous.

## Dominant Lethal Mutations

In other cases, lethal alleles behave dominantly to their wild-type counterpart. In such **dominant lethal alleles**, one copy of the allele results in the death of the individual. When this occurs, the presence of only one normal allele encoding the gene product may be insufficient to achieve a critical threshold level of an essential gene product. Or the presence of the mutant gene product may somehow override the normal function of the wild-type product.

One of the most tragic examples of a dominant lethal gene is that responsible for **Huntington disease** in humans (once referred to as Huntington's chorea). Caused by the dominant autosomal allele *H*, the disease in heterozygotes (*Hh*) does not usually appear until well into adulthood. The typical age of onset is about 40. Affected individuals then undergo gradual nervous and motor degeneration until they die. This lethal disorder is particularly tragic because an affected individual may have produced a family, and each of the children has a 50 percent probability of inheriting the lethal allele and developing the disease. The American folk singer and composer Woody Guthrie, father of modern-day folk singer Arlo Guthrie, died from this disease at age 39.

Dominant lethal alleles are rarely observed. For these alleles to persist in a population, the affected individuals must reproduce before the lethal allele is expressed, as can occur in Huntington disease. If all affected individuals die before reaching the reproductive age, the mutant allele will not be passed to future generations and will disappear from the population unless it arises again as a result of a new mutation.

## 4.7 Combinations of Two Gene Pairs with Two Modes of Inheritance Modify the 9:3:3:1 Ratio

Each example discussed so far modifies Mendel's 3:1 $F_2$ monohybrid ratio. Therefore, combining any two of these modes of inheritance in a dihybrid cross will also modify the classical 9:3:3:1 dihybrid ratio. Having established the foundation of the modes of inheritance of incomplete dominance, codominance, multiple alleles, and lethal alleles, we can now deal with the situation of two modes of inheritance occurring simultaneously. Mendel's principle of independent assortment applies to these situations, provided that the genes controlling each character are not located on the same chromosome—in other words, that they do not demonstrate what is called *genetic linkage*.

Consider, for example, a mating between two humans who are both heterozygous for the autosomal recessive gene that causes albinism and who are both of blood type AB. What is the probability of a particular phenotypic combination occurring in each of their children? Albinism is inherited in the simple Mendelian fashion, and the blood types are determined by the series of three multiple alleles, $I^A$, $I^B$, and $I^O$. The solution to this problem is diagrammed in Figure 4–5, using the forked-line method. This dihybrid cross does

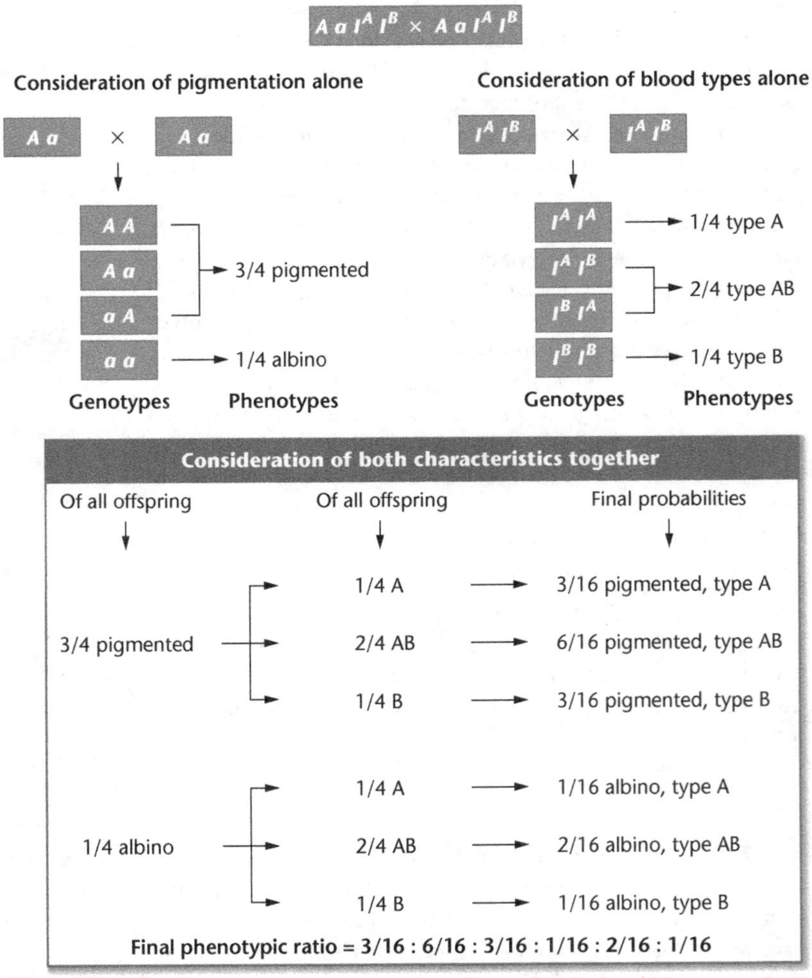


**FIGURE 4–5** Calculation of the probabilities in a mating involving the ABO blood type and albinism in humans, using the forked-line method.

not yield four phenotypes in the classical 9:3:3:1 ratio. Instead, six phenotypes occur in a 3:6:3:1:2:1 ratio, establishing the expected probability for each phenotype.

This example is just one of many variants of modified ratios possible when different modes of inheritance are combined. We can predict the outcome in a similar way for any combination of two modes of inheritance. You will be asked to determine the phenotypes and their expected probabilities for many of these combinations in the problems at the end of the chapter. In each case, the final phenotypic ratio is a modification of the 9:3:3:1 dihybrid ratio.

## 4.8 Phenotypes Are Often Affected by More Than One Gene

Soon after Mendel's work was rediscovered, experimentation revealed that in many cases a given phenotype is affected by more than one gene. This was a significant discovery because it revealed that genetic influence on the phenotype is often much more complex than the situations Mendel encountered in his crosses with the garden pea. Instead of single genes controlling the development of individual parts of a plant or animal body, it soon became clear that phenotypic characters such as eye color, hair color, or fruit shape can be influenced by many different genes and their products.

The term **gene interaction** is often used to express the idea that several genes influence a particular characteristic. This does not mean, however, that two or more genes or their products necessarily interact directly with one another to influence a particular phenotype. Rather, the term means that the cellular function of numerous gene products contributes to the development of a common phenotype. For example, the development of an organ such as the eye of an insect is exceedingly complex and leads to a structure with multiple phenotypic manifestations, for example, to an eye having a specific size, shape, texture, and color. The development of the eye is a complex cascade of developmental events leading to that organ's formation. This process illustrates the developmental concept of **epigenesis**, whereby each step of development increases the complexity of the organ or feature of interest and is under the control and influence of many genes.

An enlightening example of epigenesis and multiple gene interaction involves the formation of the inner ear in mammals, allowing organisms to detect and interpret sound. The structure and function of the inner ear is exceedingly complex. Its formation includes not only distinctive anatomical features to capture, funnel and transmit external sound toward and through the middle ear, but also to convert sound waves into nerve impulses within the inner ear. Thus, the ear forms as a result of a cascade of intricate developmental events influenced by many genes. Mutations that interrupt many of the steps of ear development lead to a common phenotype: **hereditary deafness**. In a sense, these many genes "interact" to produce a common phenotype. In such situations, the mutant phenotype is described as a **heterogeneous trait**, reflecting the many genes involved. In humans, while a few common alleles are responsible for the vast majority of cases of hereditary deafness, over 50 genes are involved in the development of the ability to discern sound.

### Epistasis

We turn now to consideration of specific inheritance patterns produced when more than one gene affects the same characteristic. Some of the best examples of gene interaction are those showing the phenomenon of **epistasis** (Greek for *stoppage*). In epistasis, the effect of one gene or gene pair masks or modifies the effect of another gene or gene pair. Sometimes the genes involved influence the same general phenotypic characteristic in an antagonistic manner, which leads to masking. In other cases, however, the genes involved exert their influence on one another in a complementary, or cooperative, fashion.

For example, the homozygous presence of a recessive allele may prevent or override the expression of other alleles at a second locus (or several other loci). In this case, the alleles at the first locus are said to be *epistatic* to those at the second locus, and the alleles at the second locus are *hypostatic* to those at the first locus. As we will see, there are several variations on this theme. In another example, a single dominant allele at the first locus may be epistatic to the expression of the alleles at a second gene locus. In a third example, two gene pairs may **complement** one another such that at least one dominant allele in each pair is required to express a particular phenotype.

The Bombay phenotype discussed earlier is an example of the homozygous recessive condition at one locus masking the expression of a second locus. There we established that the homozygous presence of the mutant form of the *FUT1* gene masks the expression of the $I^A$ and $I^B$ alleles. Only individuals containing at least one wild-type *FUT1* allele can form the A or B antigen. As a result, individuals whose genotypes include the $I^A$ or $I^B$ allele and who have no wild-type *FUT1* allele are of the type O phenotype, regardless of their potential to make either antigen. An example of the outcome of matings between individuals heterozygous at both loci is illustrated in Figure 4–6. If many such individuals have children, the phenotypic ratio of 3 A: 6 AB: 3 B: 4 O is expected in their offspring.

It is important to note two things when examining this cross and the predicted phenotypic ratio:

1. A key distinction exists between this cross and the modified dihybrid cross shown in Figure 4–5: *only one characteristic—blood type—is being followed.* In the modified dihybrid cross in Figure 4–5, blood type *and* skin pigmentation are followed as separate phenotypic characteristics.
2. Even though only a single character was followed, the phenotypic ratio comes out in sixteenths. If we knew nothing about the H substance and the gene controlling it, we could still be confident (because the proportions are in sixteenths) that a second gene pair, other than that controlling the A and B antigens, was involved in the phenotypic expression. *When a single character is being studied, a ratio that is expressed in 16 parts*

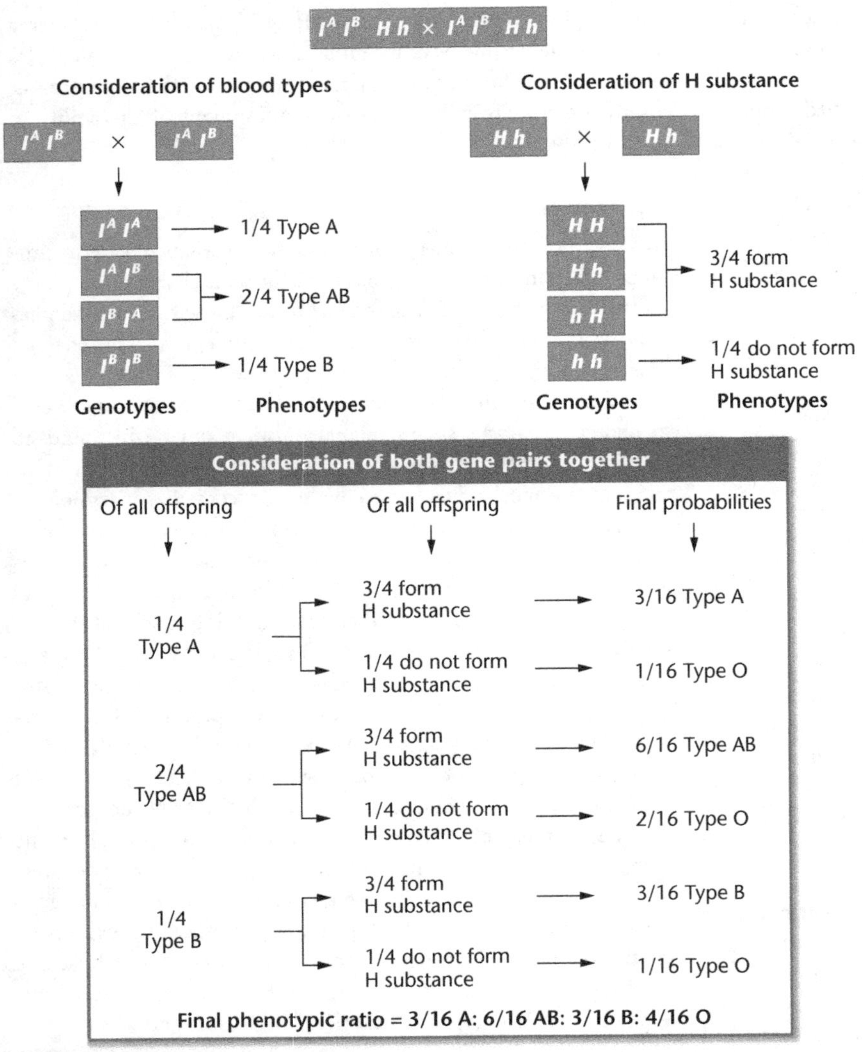


**FIGURE 4–6** The outcome of a mating between individuals heterozygous at two genes determining their ABO blood type. Final phenotypes are calculated by considering each gene separately and then combining the results using the forked-line method.

*(e.g., 3:6:3:4) suggests that two gene pairs are "interacting" in the expression of the phenotype under consideration.*

The study of gene interaction reveals a number of inheritance patterns that are modifications of the Mendelian dihybrid $F_2$ ratio (9:3:3:1). In several of the subsequent examples, epistasis has the effect of combining one or more of the four phenotypic categories in various ways. The generation of these four groups is reviewed in Figure 4–7, along with several modified ratios.

As we discuss these and other examples (see Figure 4–8), we will make several assumptions and adopt certain conventions:

1. In each case, distinct phenotypic classes are produced, each clearly discernible from all others. Such traits illustrate discontinuous variation, where phenotypic categories are discrete and qualitatively different from one another.
2. The genes considered in each cross are on different chromosomes and therefore assort independently of one another during gamete formation. To allow you to easily compare the results of different crosses, we designated alleles as *A*, *a* and *B*, *b* in each case.
3. When we assume that complete dominance exists within a gene pair, such that *AA* and *Aa* or *BB* and *Bb* are equivalent in their genetic effects, we use the designations *A*– or *B*– for both combinations, where the dash (–) indicates that either allele may be present without consequence to the phenotype.
4. All $P_1$ crosses involve homozygous individuals (e.g., *AABB* × *aabb*, *AAbb* × *aaBB* or *aaBB* × *AAbb*). Therefore, each F1 generation consists of only heterozygotes of genotype *AaBb*.
5. In each example, the $F_2$ generation produced from these heterozygous parents is our main focus of analysis. When two genes are involved (Figure 4–7), the $F_2$ genotypes fall into four categories: 9/16 *A–B–*, 3/16 *A–bb*, 3/16 *aaB–*, and 1/16 *aabb*. Because of dominance, all genotypes in each category are equivalent in their effect on the phenotype.

Case 1 is the inheritance of coat color in mice (Figure 4–8). Normal wild-type coat color is agouti, a grayish pattern formed by alternating bands of pigment on each hair (see Figure 4–4). Agouti is dominant to black (nonagouti) hair, which results from the homozygous expression of a recessive mutation that we designate *a*. Thus, *A*– results in agouti, whereas *aa* yields black coat color. When a recessive mutation, *b*, at a separate locus is homozygous, it eliminates pigmentation altogether, yielding albino mice (*bb*), regardless of the genotype at the *a* locus. The presence of at least one *B* allele allows pigmentation to occur in much the same way that the *FUT1* allele in humans allows the expression of the ABO blood types. In a cross between agouti (*AABB*) and albino (*aabb*) parents, members of the $F_1$ are all *AaBb* and have agouti coat color. In the $F_2$ progeny of a cross between two $F_1$ double heterozygotes, the following genotypes and phenotypes are observed:

$$F_1: AaBb \times AaBb$$
$$\downarrow$$

| $F_2$ Ratio | Genotype | Phenotype | Final Phenotypic Ratio |
|---|---|---|---|
| 9/16 | *A–B–* | agouti | 9/16 agouti |
| 3/16 | *A–bb* | albino | 4/16 albino |
| 3/16 | *aa B–* | black | 3/16 black |
| 1/16 | *aa bb* | albino | |

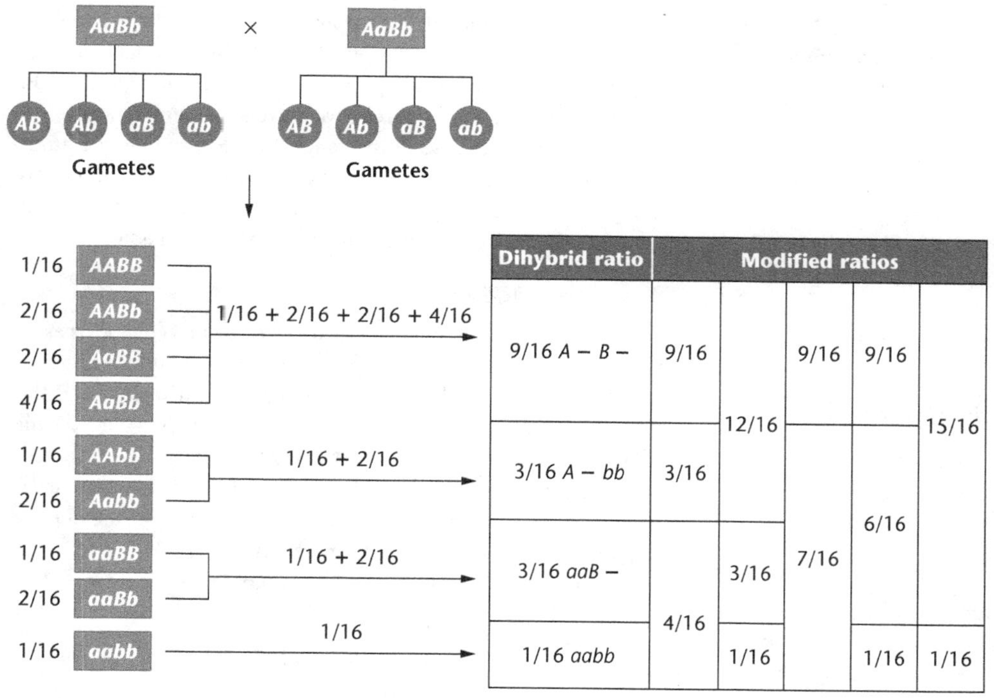


**FIGURE 4–7** Generation of various modified dihybrid ratios from the nine unique genotypes produced in a cross between individuals heterozygous at two genes.

We can envision gene interaction yielding the observed 9:3:4 $F_2$ ratio as a two-step process:

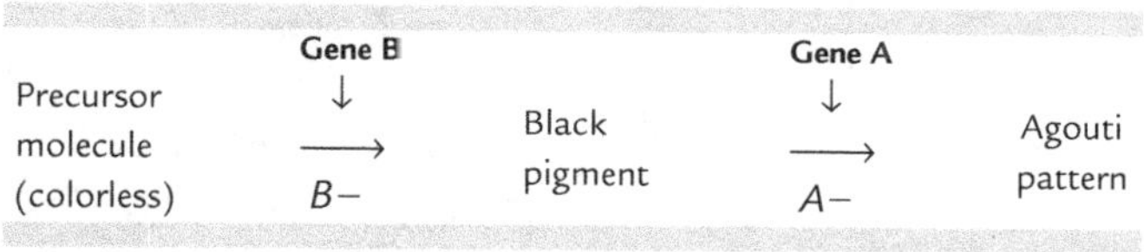


In the presence of a *B* allele, black pigment can be made from a colorless substance. In the presence of an *A* allele, the black pigment is deposited during the development of hair in a pattern that produces the agouti phenotype. If the *aa* genotype occurs, all of the hair remains black. If the *bb* genotype occurs, no black pigment is produced, regardless of the presence of the *A* or *a* alleles, and the mouse is albino. Therefore, the *bb* genotype masks or suppresses the expression of the *A* allele. As a result, this is referred to as *recessive epistasis.*

<table>
<tr><th rowspan="2">Case</th><th rowspan="2">Organism</th><th rowspan="2">Character</th><th colspan="4">$F_2$ Phenotypes</th><th rowspan="2">Modified ratio</th></tr>
<tr><th>9/16</th><th>3/16</th><th>3/16</th><th>1/16</th></tr>
<tr><td>1</td><td>Mouse</td><td>Coat color</td><td>agouti</td><td>albino</td><td>black</td><td>albino</td><td>9:3:4</td></tr>
<tr><td>2</td><td>Squash</td><td>Color</td><td colspan="2">white</td><td>yellow</td><td>green</td><td>12:3:1</td></tr>
<tr><td>3</td><td>Pea</td><td>Flower color</td><td>purple</td><td colspan="3">white</td><td>9:7</td></tr>
<tr><td>4</td><td>Squash</td><td>Fruit shape</td><td>disc</td><td colspan="2">sphere</td><td>long</td><td>9:6:1</td></tr>
<tr><td>5</td><td>Chicken</td><td>Color</td><td colspan="2">white</td><td>colored</td><td>white</td><td>13:3</td></tr>
<tr><td>6</td><td>Mouse</td><td>Color</td><td>white-spotted</td><td>white</td><td>colored</td><td>white-spotted</td><td>10:3:3</td></tr>
<tr><td>7</td><td>Shepherd's purse</td><td>Seed capsule</td><td colspan="3">triangular</td><td>ovoid</td><td>15:1</td></tr>
<tr><td>8</td><td>Flour beetle</td><td>Color</td><td>6/16 sooty and 3/16 red</td><td>black</td><td>jet</td><td>black</td><td>6:3:3:4</td></tr>
</table>

**FIGURE 4–8** The basis of modified dihybrid $F_2$ phenotypic ratios resulting from crosses between doubly heterozygous $F_1$ individuals. The four groupings of the $F_2$ genotypes shown in Figure 4–7 and across the top of this figure are combined in various ways to produce these ratios.

A second type of epistasis, called *dominant epistasis*, occurs when a dominant allele at one genetic locus masks the expression of the alleles of a second locus. For instance, Case 2 of Figure 4–8 deals with the inheritance of fruit color in summer squash. Here, the dominant allele *A* results in white fruit color regardless of the genotype at a second locus, *B*. In the absence of a dominant *A* allele (the *aa* genotype), *BB* or *Bb* results in yellow color, while *bb* results in green color. Therefore, if two white-colored double heterozygotes (*AaBb*) are crossed, this type of epistasis generates an interesting phenotypic ratio:

$$F_1: AaBb \times AaBb$$

$$\downarrow$$

| $F_2$ Ratio | Genotype | Phenotype | Final Phenotypic Ratio |
|---|---|---|---|
| 9/16 | $A- B-$ | White | 12/16 white |
| 3/16 | $A- bb$ | White | 3/16 yellow |
| 3/16 | $aa B-$ | Yellow | 1/16 green |
| 1/16 | $aa bb$ | Green | |

Of the offspring, 9/16 are *A–B–* and are thus white. The 3/16 bearing the genotypes *A–bb* are also white. Of the remaining squash, 3/16 are yellow (*aaB–*), while 1/16 are green (*aabb*). Thus, the modified phenotypic ratio of 12:3:1 occurs.

Our third example (Case 3 of Figure 4–8), first discovered by William Bateson and Reginald Punnett (of Punnett square fame), is demonstrated in a cross between two true-breeding strains of white-flowered sweet peas. Unexpectedly, the results of this cross yield all purple $F_1$ plants, and the $F_2$ plants occur in a ratio of 9/16 purple to 7/16 white. The proposed explanation suggests that the presence of at least one dominant allele of each of two gene pairs is essential in order for flowers to be purple. Thus, this cross represents a case of *complementary gene interaction*. All other genotype combinations yield white flowers because the homozygous condition of either recessive allele masks the expression of the dominant allele at the other locus.

The cross is shown as follows:

$$P_1: AAbb \times aaBB$$
white white

$$\downarrow$$

$F_1$: All *AaBb* (purple)

| $F_2$ Ratio | Genotype | Phenotype | Final Phenotypic Ratio |
|---|---|---|---|
| 9/16 | $A- B-$ | purple | |
| 3/16 | $A- bb$ | white | 9/16 purple |
| 3/16 | $aa B-$ | white | 7/16 white |
| 1/16 | $aa bb$ | white | |

We can now envision how two gene pairs might yield such results:

| | Gene A | | Gene B | |
|---|---|---|---|---|
| Precursor substance (colorless) | ↓ ⟶ $A-$ | Intermediate product (colorless) | ↓ ⟶ $B-$ | Final product (purple) |

At least one dominant allele from each pair of genes is necessary to ensure both biochemical conversions to the final product, yielding purple flowers. In the preceding cross, this will occur in 9/16 of the $F_2$ offspring. All other plants (7/16) have flowers that remain white.

These three examples illustrate in a simple way how the products of two genes interact to influence the development of a common phenotype. In other instances, more than two genes and their products are involved in controlling phenotypic expression.

## Novel Phenotypes

Other cases of gene interaction yield novel, or new, phenotypes in the $F_2$ generation, in addition to producing modified dihybrid ratios. Case 4 in Figure 4–8 depicts the inheritance of fruit shape in the summer squash *Cucurbita pepo*. When plants with disc-shaped fruit (*AABB*) are crossed with plants with long fruit (*aabb*), the $F_1$ generation all have disc fruit. However, in the $F_2$ progeny, fruit with a novel shape—sphere—appear, as well as fruit exhibiting the parental phenotypes. A variety of fruit shapes are shown in Figure 4–9.

The $F_2$ generation, with a modified 9:6:1 ratio, is generated as follows:

$$F_1: AaBb \times AaBb$$
disc disc

$$\downarrow$$

| $F_2$ Ratio | Genotype | Phenotype | Final Phenotypic Ratio |
|---|---|---|---|
| 9/16 | $A- B-$ | disc | 9/16 disc |
| 3/16 | $A- bb$ | sphere | 6/16 sphere |
| 3/16 | $aa B-$ | sphere | 1/16 long |
| 1/16 | $aa bb$ | long | |

In this example of gene interaction, both gene pairs influence fruit shape equally. A dominant allele at either locus ensures a sphere-shaped fruit. In the absence of dominant alleles, the fruit is long. However, if both dominant alleles (*A* and *B*) are present, the fruit displays a flattened, disc shape.

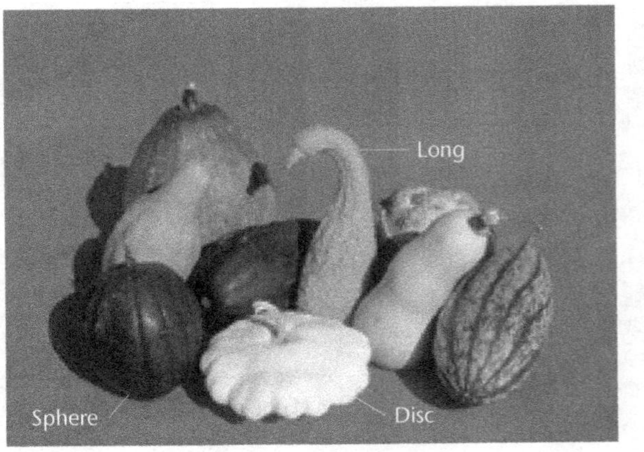


FIGURE 4–9 Summer squash exhibiting various fruit-shape phenotypes disc (white), long (orange gooseneck), and sphere (bottom left).

**NOW SOLVE THIS**

Problem 17 on page 100 describes a plant in which flower color, a single characteristic, can take on one of three variations. You are asked to determine how many genes are involved in the inheritance of this characteristic and what genotypes are responsible for what phenotypes.

■ HINT: *The most important information is the data provided. You must analyze the raw data and convert the numbers to a meaningful ratio. This will guide you in determining how many gene pairs are involved. Then you can group the genotypes in a way that corresponds to the phenotypic ratio.*

Another interesting example of an unexpected phenotype arising in the $F_2$ generation is the inheritance of eye color in *Drosophila melanogaster*. As mentioned earlier, the wild-type eye color is brick red. When two autosomal recessive mutants, *brown* and *scarlet*, are crossed, the $F_1$ generation consists of flies with wild-type eye color. In the $F_2$ generation, wild, scarlet, brown, and white-eyed flies are found in a 9:3:3:1 ratio. While this ratio is numerically the same as Mendel's dihybrid ratio, the *Drosophila* cross involves only one character: eye color. This is an important distinction to make when modified dihybrid ratios resulting from gene interaction are studied.

The *Drosophila* cross is an excellent example of gene interaction because the biochemical basis of eye color in this organism has been determined (Figure 4–10). *Drosophila*, as a typical arthropod, has compound eyes made up of hundreds of individual visual units called ommatidia. The wild-type eye color is due to the deposition and mixing of two separate pigment groups in each ommatidium—the bright-red **drosopterins** and the brown **xanthommatins.** Each type of pigment is produced by a separate biosynthetic pathway. Each step of each pathway is catalyzed by a separate enzyme and is thus under the control of a separate gene. As shown in Figure 4–10, the *brown* mutation, when homozygous, interrupts the pathway leading to the synthesis of the bright-red pigments. Because only xanthommatin pigments are present, the eye is brown. The *scarlet* mutation, affecting a gene located on a separate autosome, interrupts the pathway leading to the synthesis of the brown xanthommatins and renders the eye color bright red in homozygous mutant flies. Each mutation apparently causes the production of a nonfunctional enzyme. Flies that are double mutants and thus homozygous for both *brown* and *scarlet* lack both functional enzymes and can make

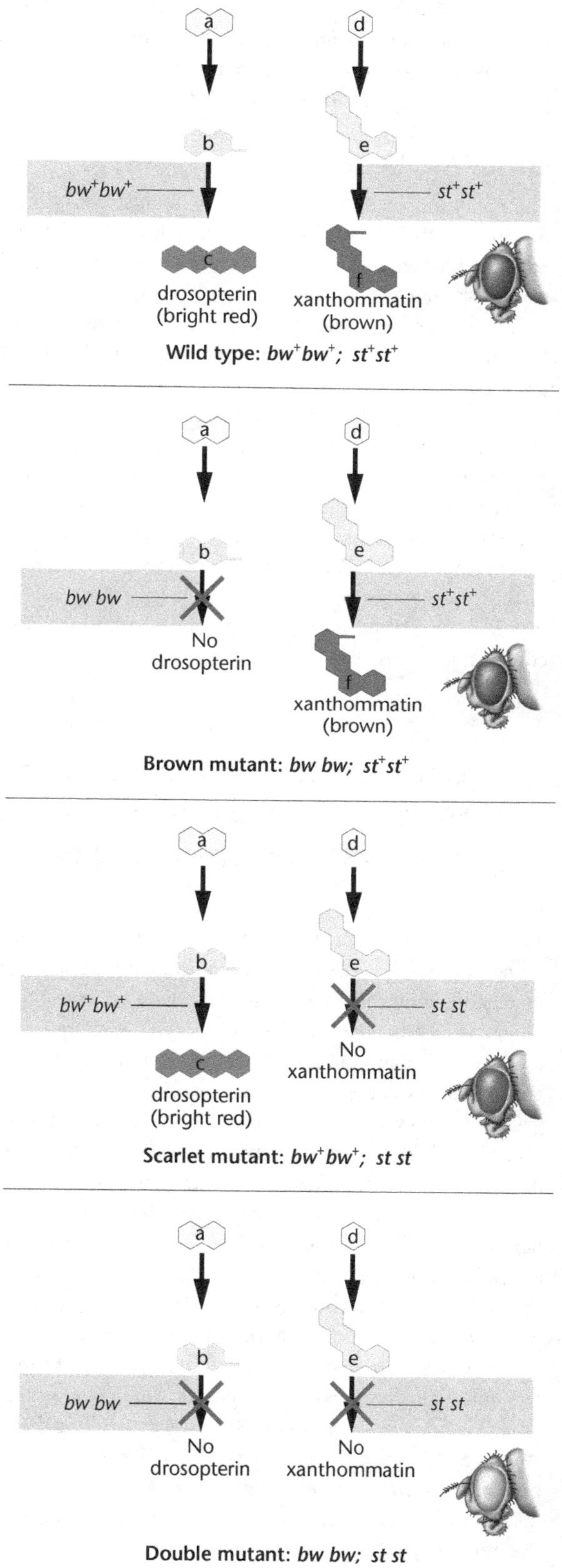


**FIGURE 4–10** A theoretical explanation of the biochemical basis of the four eye color phenotypes produced in a cross between *Drosophila* with brown eyes and scarlet eyes. In the presence of at least one wild-type $bw^+$ allele, an enzyme is produced that converts substance b to c, and the pigment drosopterin is synthesized. In the presence of at least one wild-type $st^+$ allele, substance e is converted to f, and the pigment xanthommatin is synthesized. The homozygous presence of the recessive *st* or *bw* mutant allele blocks the synthesis of the respective pigment molecule. Either one, both, or neither of these pathways can be blocked, depending on the genotype.

neither of the pigments; they represent the novel white-eyed flies appearing in 1/16 of the $F_2$ generation. Note that the absence of pigment in these flies is not due to the X-linked *white* mutation, in which pigments can be synthesized but the necessary precursors cannot be transported into the cells making up the ommatidia.

## Other Modified Dihybrid Ratios

The remaining cases (5–8) in Figure 4–8 illustrate additional modifications of the dihybrid ratio and provide still other examples of gene interactions. As you will note, ratios of 13:3, 10:3:3; 15:1, and 6:3:3:4 are illustrated. These cases, like the four preceding them, have two things in common. First, we need not violate the principles of segregation and independent assortment to explain the inheritance pattern of each case. Therefore, the added complexity of inheritance in these examples does not detract from the validity of Mendel's conclusions. Second, the $F_2$ phenotypic ratio in each example has been expressed in sixteenths. When sixteenths are seen in the ratios of crosses where the inheritance pattern is unknown, they suggest to geneticists that two gene pairs are controlling the observed phenotypes. You should make the same inference in your analysis of genetics problems. Other insights into solving genetics problems are provided in the "Insights and Solutions" section at the conclusion of this chapter.

4.9

## Complementation Analysis Can Determine If Two Mutations Causing a Similar Phenotype Are Alleles

An interesting situation arises when two mutations that both produce a similar phenotype are isolated independently. Suppose that two investigators independently isolate and establish a true-breeding strain of wingless *Drosophila* and demonstrate that each mutant phenotype is due to a recessive mutation. We might assume that both strains contain mutations in the same gene. However, since we know that many genes are involved in the formation of wings, we must consider the possibility that mutations in any one of them might inhibit wing formation during development. This is the case with any *heterogeneous trait*, a concept introduced earlier in this chapter in our discussion of hereditary deafness. An analytical procedure called **complementation analysis** allows us to determine whether two independently isolated mutations are in the same gene—that is, whether they are alleles—or whether they represent mutations in separate genes.

To repeat, our analysis seeks to answer this simple question: *Are two mutations that yield similar phenotypes present in the same gene or in two different genes?* To find the answer, we cross the two mutant strains and analyze the $F_1$ generation. The two possible alternative outcomes and their interpretations are shown in Figure 4–11. To discuss these possibilities (Case 1 and Case 2), we designate one of the mutations $m^a$ and the other $m^b$.

**Case 1.** *All offspring develop normal wings.*

**Interpretation:** The two recessive mutations are in separate genes and are not alleles of one another. Following the cross, all $F_1$ flies are heterozygous for both genes. Since each mutation is in a separate gene and each $F_1$ fly is heterozygous at both loci, the normal products of both genes are produced (by the one normal copy of each gene), and wings develop. Under such circumstances, the genes complement one another in restoration of the wild-type phenotype, and complementation is said to occur because the two mutations are in different genes.

**Case 2.** *All offspring fail to develop wings.*

**Interpretation:** The two mutations affect the same gene and are alleles of one another. Complementation does not occur. Since the two mutations affect the same gene, the $F_1$ flies are homozygous for the two mutant alleles (the $m^a$ allele and the $m^b$ allele). No normal product of the gene is produced, and in the absence of this essential product, wings do not form.

Complementation analysis, as originally devised by the *Drosophila* geneticist Edward B. Lewis, may be used to screen any number of individual mutations that result in the same phenotype. Such an analysis may reveal that only a single gene is involved or that two or more genes are involved. All mutations determined to be present in any single gene are said to fall into the same **complementation group**, and they will complement mutations in all other groups. When large numbers of mutations affecting the same trait are available and studied using complementation analysis, it is possible to predict the total number of genes involved in the determination of that trait.

4.10

## Expression of a Single Gene May Have Multiple Effects

While the previous sections have focused on the effects of two or more genes on a single characteristic, the converse situation, where expression of a single gene has multiple phenotypic effects, is also quite common. This phenomenon, which often becomes apparent when phenotypes are examined carefully, is referred to as **pleiotropy**. Many excellent examples can be drawn from human disorders, and we will review two such cases to illustrate this point.

The first disorder is **Marfan syndrome**, a human malady resulting from an autosomal dominant mutation in the gene encoding the connective tissue protein fibrillin. Because this protein is widespread in many tissues in the body, one would expect multiple effects of such a defect. In fact, fibrillin is important to the structural integrity of the lens of the eye, to the lining of vessels such as the aorta, and to bones, among other tissues. As a result, the phenotype associated with Marfan syndrome includes lens dislocation, in-

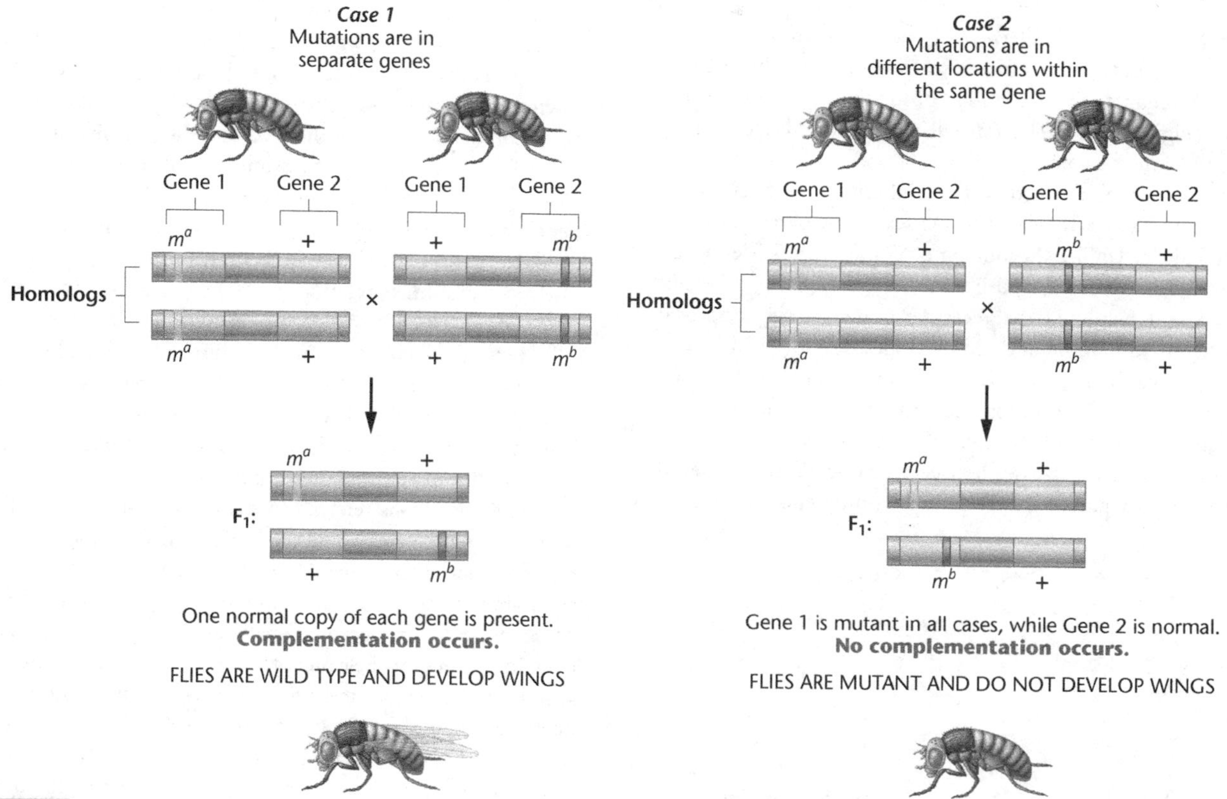


**FIGURE 4–11** Complementation analysis of alternative outcomes of two wingless mutations in *Drosophila* ($m^a$ and $m^b$). In Case 1, the mutations are not alleles of the same gene, while in Case 2, the mutations are alleles of the same gene.

creased risk of aortic aneurysm, and lengthened long bones in limbs. This disorder is of historical interest in that speculation abounds that Abraham Lincoln was afflicted.

A second example involves another human autosomal dominant disorder, **porphyria variegata.** Afflicted individuals cannot adequately metabolize the porphyrin component of hemoglobin when this respiratory pigment is broken down as red blood cells are replaced. The accumulation of excess porphyrins is immediately evident in the urine, which takes on a deep red color. However, this phenotypic characteristic is merely diagnostic. The severe features of the disorder are due to the toxicity of the buildup of porphyrins in the body, particularly in the brain. Complete phenotypic characterization includes abdominal pain, muscular weakness, fever, a racing pulse, insomnia, headaches, vision problems (that can lead to blindness), delirium, and ultimately convulsions. As you can see, deciding which phenotypic trait best characterizes the disorder is impossible.

Like Marfan syndrome, porphyria variegata is also of historical significance. George III, king of England during the American Revolution, is believed to have suffered from episodes involving all of the above symptoms. He ultimately became blind and senile prior to his death.

We could cite many other examples to illustrate pleiotropy, but suffice it to say that if one looks carefully, most mutations display more than a single manifestation when expressed.

## 4.11 X-Linkage Describes Genes on the X Chromosome

In many animals and some plant species, one of the sexes contains a pair of unlike chromosomes that are involved in sex determination. In many cases, these are designated as X and Y. For example, in both *Drosophila* and humans, males contain an X and a Y chromosome, whereas females contain two X chromosomes. The Y chromosome must contain a region of pairing homology with the X chromosome if the two are to synapse and segregate during meiosis, but a major portion of the Y chromosome in humans as well as other species is considered to be relatively inert genetically. While we now recognize a number of male-specific genes on the human Y chromosome, it lacks copies of most genes present on the X chromosome. As a result, genes present on the X chromosome exhibit patterns of inheritance that are very different from those seen with autosomal genes. The term **X-linkage** is used to describe these situations.

In the following discussion, we will focus on inheritance patterns resulting from genes present on the X but absent from the Y chromosome. This situation results in a modification of Mendelian ratios, the central theme of this chapter.

## X-Linkage in *Drosophila*

One of the first cases of X-linkage was documented in 1910 by Thomas H. Morgan during his studies of the *white* eye mutation in *Drosophila* (Figure 4–12). The normal wild-type red eye color is dominant to white eye color.

Morgan's work established that the inheritance pattern of the white-eye trait was clearly related to the sex of the parent carrying the mutant allele. Unlike the outcome of the typical Mendelian monohybrid cross where $F_1$ and $F_2$ data were similar regardless of which $P_1$ parent exhibited the recessive mutant trait, reciprocal crosses between white-eyed and red-eyed flies did not yield identical results. Morgan's analysis led to the conclusion that the *white* locus is present on the X chromosome rather than on one of the autosomes. Both the gene and the trait are said to be X-linked.

Results of reciprocal crosses between white-eyed and red-eyed flies are shown in Figure 4–12. The obvious differences in phenotypic ratios in both the $F_1$ and $F_2$ generations are dependent on whether or not the $P_1$ white-eyed parent was male or female.

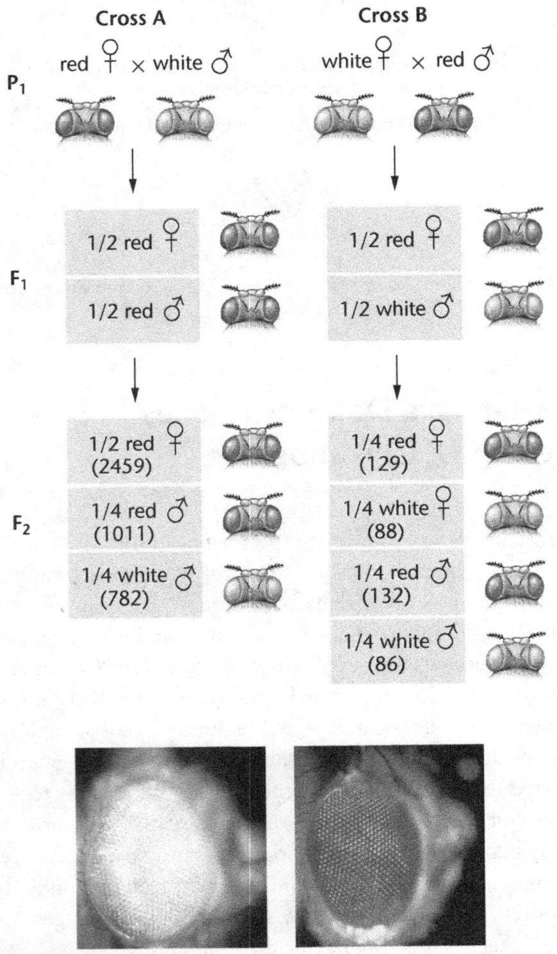


**FIGURE 4–12** The $F_1$ and $F_2$ results of T. H. Morgan's reciprocal crosses involving the X-linked *white* mutation in *Drosophila melanogaster*. The actual data are shown in parentheses. The photographs show white eye and the brick red wild-type eye color.

Morgan was able to correlate these observations with the difference found in the sex-chromosome composition of male and female *Drosophila*. He hypothesized that the recessive allele for white eye is found on the X chromosome, but its corresponding locus is absent from the Y chromosome. Females thus have two available gene loci, one on each X chromosome, whereas males have only one available locus, on their single X chromosome.

Morgan's interpretation of X-linked inheritance, shown in Figure 4–13, provides a suitable theoretical explanation for his results. Since the Y chromosome lacks homology with almost all genes on the X chromosome, these alleles present on the X chromosome of the males will be directly expressed in the phenotype. Males cannot be either homozygous or heterozygous for X-linked genes; instead, their condition—possession of only one copy of a gene in an otherwise diploid cell—is referred to as **hemizygosity**. The individual is said to be **hemizygous.** One result of X-linkage is the **crisscross pattern of inheritance**, in which phenotypic traits controlled by recessive X-linked genes are passed from homozygous mothers to all sons. This pattern occurs because females exhibiting a recessive trait must contain the mutant allele on both X chromosomes. Because male offspring receive one of their mother's two X chromosomes and are hemizygous for all alleles present on that X, all sons will express the same recessive X-linked traits as their mother.

Morgan's work has taken on great historical significance. By 1910, the correlation between Mendel's work and the behavior of chromosomes during meiosis had provided the basis for the **chromosome theory of inheritance**, as postulated by Sutton and Boveri (see Chapter 3). Morgan's work, and subsequently that of his student, Calvin Bridges, around 1920, provided direct evidence that genes are transmitted on specific chromosomes, and is considered the first solid experimental evidence in support of this theory. In the ensuing two decades, the outcome of research inspired by these findings provided indisputable evidence in support of this theory.

## X-Linkage in Humans

In humans, many genes and the traits they control are recognized as being linked to the X chromosome (see Table 4.3). These X-linked traits can be easily identified in a pedigree, because of the crisscross pattern of inheritance. A pedigree for one form of human color blindness is shown in Figure 4–14. The mother in generation I passes the trait on to all her sons but to none of her daughters. If the offspring in generation II have children by normal individuals, the color-blind sons will produce all normal male and female offspring (III-1, -2, and -3); the normal-vision daughters will produce normal-vision female offspring (III-4, -6, and -7), as well as color-blind (III-8) and normal-vision (III-5) male offspring.

Many X-linked human genes have now been identified, as shown in Table 4.3. For example, the genes controlling two forms of hemophilia and two forms of muscular dystrophy are located on the

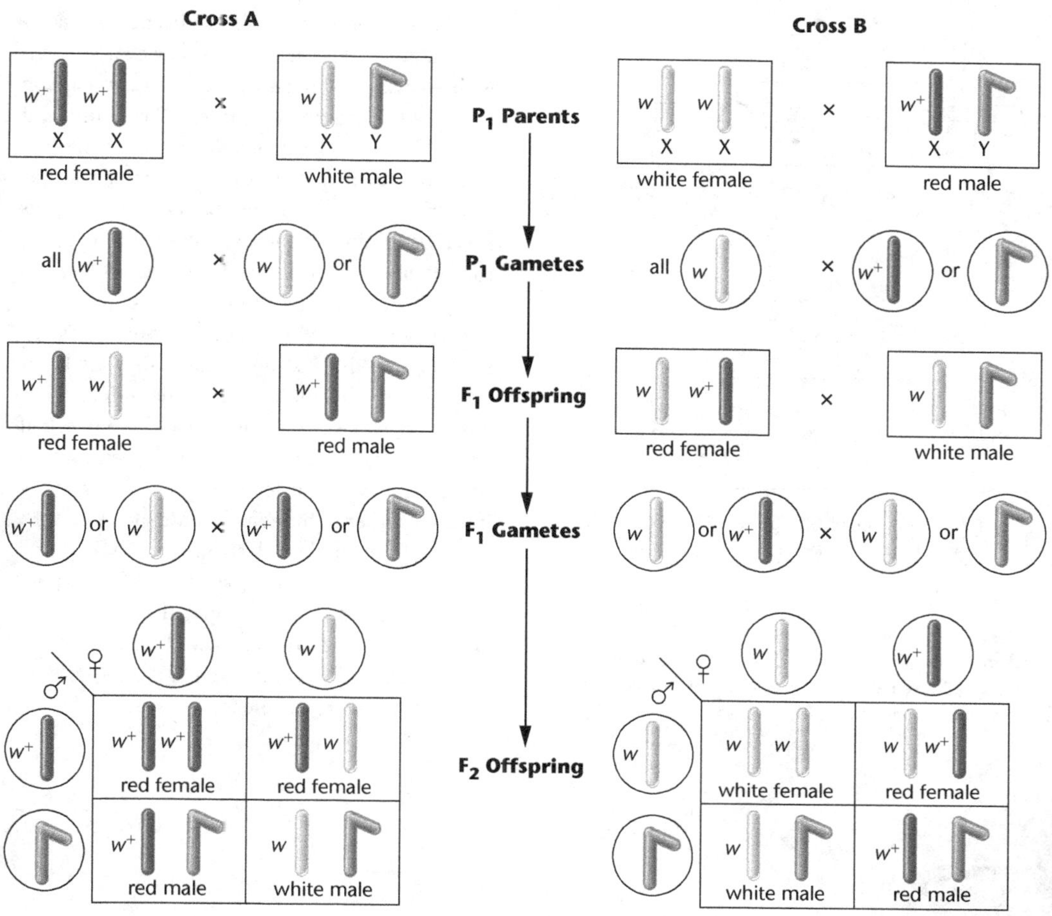


FIGURE 4-13 The chromosomal explanation of the results of the X-linked crosses shown in Figure 4-12.

X chromosome. In addition, numerous genes whose expression yields well-studied enzymes are X-linked. Glucose-6-phosphate dehydrogenase and hypoxanthine-guanine-phosphoribosyl transferase are two examples. In the latter case, the severe Lesch–Nyhan syndrome (discussed later in this chapter) results from the mutant form of the X-linked gene product.

TABLE 4.3

**Human X-Linked Traits**

| Condition | Characteristics |
|---|---|
| Color blindness, deutan type | Insensitivity to green light |
| Color blindness, protan type | Insensitivity to red light |
| Fabry's disease | Deficiency of galactosidase A; heart and kidney defects, early death |
| G-6-PD deficiency | Deficiency of glucose-6-phosphate dehydrogenase; severe anemic reaction following intake of primaquines in drugs and certain foods, including fava beans |
| Hemophilia A | Classic form of clotting deficiency; deficiency of clotting factor VIII |
| Hemophilia B | Christmas disease; deficiency of clotting factor IX |
| Hunter syndrome | Mucopolysaccharide storage disease resulting from iduronate sulfatase enzyme deficiency; short stature, clawlike fingers, coarse facial features, slow mental deterioration, and deafness |
| Ichthyosis | Deficiency of steroid sulfatase enzyme; scaly dry skin, particularly on extremities |
| Lesch–Nyhan syndrome | Deficiency of hypoxanthine-guanine phosphoribosyltransferase enzyme (HPRT) leading to motor and mental retardation, self-mutilation, and early death |
| Muscular dystrophy | Progressive, life-shortening disorder characterized by muscle degeneration and weakness; (Duchenne type) sometimes associated with mental retardation; deficiency of the protein dystrophin |

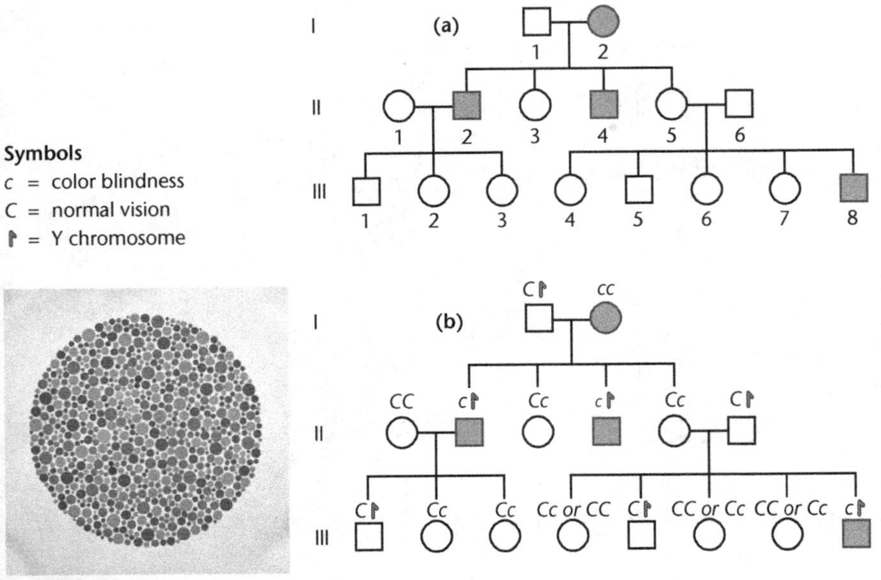


**FIGURE 4–14** (a) A human pedigree of the X-linked color blindness trait. (b) The most probable genotypes of each individual in the pedigree. The photograph is of an Ishihara color blindness chart, which tests for red-green color blindness. Red-green color-blind individuals see a 3 rather than the 8 visualized by those with normal color vision.

Because of the way X-linked genes are transmitted, unusual circumstances may be associated with recessive X-linked disorders in comparison to recessive autosomal disorders. For example, if anX-linked disorder debilitates or is lethal to the affected individual prior to reproductive maturation, the disorder occurs exclusively in males. This is because almost the only sources of the lethal allele in the population are heterozygous females who are "carriers" and do not express the disorder. They pass the allele to one-half of their sons, who develop the disorder because they are hemizygous but who rarely, if ever, reproduce. Heterozygous females also pass the allele to one-half of their daughters, who become carriers but do not develop the disorder. Examples of such an X-linked disorder in humans include the Duchenne form of muscular dystrophy (DMD) and Lesch–Nyhan syndrome. (See the box below for a description of the molecular basis of Lesch–Nyhan syndrome.) DMD has an onset prior to age 6 and is often lethal around age 20. Affected males are unable to reproduce.

## Lesch–Nyhan Syndrome: The Molecular Basis of a Rare X-Linked Recessive Disorder

*Lesch–Nyhan syndrome (LNS)* is a devastating disease that is first apparent in infants at age 3 to 6 months, when orange particles (sometimes referred to as orange sand) appear in the urine and discolor the affected infant's diaper. These urinary stones consist of urate crystals, and they are a harbinger of many future difficulties that will ultimately lead to premature death. LNS occurs only in males and is the result of the complete or nearly complete loss of activity of a critical enzyme, *hypoxanthine-guanine phosphoribosyltransferase (HPRT)*. This enzyme imparts the ability to metabolically recycle purines, one of the two major types of nitrogenous bases that make up nucleotides in DNA. While the purines adenine and guanine can be synthesized from basic chemical components, mammals have evolved the ability to extract them from DNA that is being degraded, recovering them in the form of the purine hypoxanthine. Under the direction of HPRT, hypoxanthine can be converted back to adenine and guanine-containing nucleotides. When this mechanism fails as a result of mutation, the excess hypoxanthine is converted to uric acid, which accumulates well beyond the body's ability to excrete it.

This so-called metabolic or biochemical disorder has numerous effects, the most severe being mental retardation, seizures, and aggressive, uncontrolled spastic movements (resembling cerebral palsy) that include self-mutilation of the fingers and lips. Patients require 24-hour care throughout their lives and almost always die prior to age 30, usually as a result of kidney failure. In February 2007, one of the oldest living LNS patients, Philip Barker, celebrated his thirty-sixth birthday in Bayview, New York.

The gene involved in LNS is located on the long arm of the X chromosome and consists of 44,000 base pairs (44 Kb).However, the HPRT gene product is only 218 amino acids long, thus requiring only 654 base pairs to encode it. Analysis of the cloned version of the gene reveals it to contain 9 exons and 8 introns (the latter are segments that are not translated to produce a protein, as explained in Chapter 15). Mice have a nearly identical gene that is 95 percent homologous to its humancounterpart.

In normal individuals the enzyme is ubiquitous in tissues throughout the body but is present in greatest concentration in the basal ganglia of the brain. No doubt this somehow relates to the behavioral phenotype characterizing LNS patients, who lack enzyme activity in the brain and elsewhere in their bodies. In spite of extensive research efforts, there is no known cure. Because the responsible gene is recessive and X-linked, and since affected males never reproduce, females, while they can be carriers of the mutant gene, never become homozygous and never develop LNS.

**NOW SOLVE THIS**

Problem 32 on page 101 asks you to determine if each of three pedigrees is consistent with X-linkage.

■ HINT: *In X-linkage, because of hemizygosity, the genotype of males is immediately evident. Therefore, the key to solving this type of problem is to consider the possible genotypes of females that do not express the trait.*

4.12

## In Sex-Limited and Sex-Influenced Inheritance, an Individual's Sex Influences the Phenotype

In contrast to X-linked inheritance, patterns of gene expression may be affected by the sex of an individual even when the genes are not on the X chromosome. In numerous examples in different organisms, the sex of the individual plays a determining role in the expression of a phenotype. In some cases, the expression of a specific phenotype is absolutely limited to one sex; in others, the sex of an individual influences the expression of a phenotype that is not limited to one sex or the other. This distinction differentiates **sex-limited inheritance** from **sex-influenced inheritance.**

In both types of inheritance, autosomal genes are responsible for the existence of contrasting phenotypes, but the expression of these genes is dependent on the hormone constitution of the individual. Thus, the heterozygous genotype may exhibit one phenotype in males and the contrasting one in females. In domestic fowl, for example, tail and neck plumage is often distinctly different in males and females (Figure 4–15), demonstrating *sex-limited inheritance.* Cock feathering is longer, more curved, and pointed, whereas hen feathering is shorter and less curved. Inheritance of these feather phenotypes is controlled by a single pair of autosomal alleles whose expression is modified by the individual's sex hormones. As shown in the following chart, hen feathering is due to a dominant allele, *H*, but regardless of the homozygous presence of the recessive *h* allele, all females remain hen-feathered. Only in males does the *hh* genotype result in cock feathering.

| Genotype | Phenotype | |
|---|---|---|
| | ♀ | ♂ |
| *HH* | Hen-feathered | Hen-feathered |
| *Hh* | Hen-feathered | Hen-feathered |
| *hh* | Hen-feathered | Cock-feathered |

In certain breeds of fowl, the hen feathering or cock feathering allele has become fixed in the population. In the Leghorn breed, all individuals are of the *hh* genotype; as a result, males always differ from females in their plumage. Seabright bantams are all *HH*, showing no sexual distinction in feathering phenotypes.

FIGURE 4–15 Hen feathering (left) and cock feathering (right) in domestic fowl. The hen's feathers are shorter and less curved.

Another example of sex-limited inheritance involves the autosomal genes responsible for milk yield in dairy cattle. Regardless of the overall genotype that influences the quantity of milk production, those genes are obviously expressed only in females.

Cases of *sex-influenced inheritance* include pattern baldness in humans, horn formation in certain breeds of sheep (e.g., Dorsett Horn sheep), and certain coat patterns in cattle. In such cases, autosomal genes are responsible for the contrasting phenotypes, and while the trait may be displayed by both males and females, the expression of these genes is dependent on the hormone constitution of the individual. Thus, the heterozygous genotype exhibits one phenotype in one sex and the contrasting one in the other. For example, pattern baldness in humans, where the hair is very thin or

**NOW SOLVE THIS**

Problem 33 on page 102 involves the inheritance of the beard in goats and asks you to analyze the $F_1$ and $F_2$ ratios to determine the mode of inheritance.

■ HINT: *Note particularly that the data are differentiated into male and female offspring and that the ratios in the $F_2$ vary according to sex (i.e., 3/8 of the males are bearded while only 1/8 of the females are bearded, etc.). This should immediately alert you to consider the possible influences of sex differences on the outcome of crosses. In this case, you should consider whether X-linkage, sex-limited inheritance, or sex-influenced inheritance might be involved.*

FIGURE 4–16 Pattern baldness, a sex-influenced autosomal trait in humans.

absent on the top of the head (Figure 4–16), is inherited in the following way:

| Genotype | Phenotype ♀ | ♂ |
|---|---|---|
| *BB* | Bald | Bald |
| *Bb* | Not bald | Bald |
| *bb* | Not bald | Not bald |

Females can display pattern baldness, but this phenotype is much more prevalent in males. When females do inherit the *BB* genotype, the phenotype is less pronounced than in males and is expressed later in life.

## 4.13 Genetic Background and the Environment May Alter Phenotypic Expression

We conclude this chapter with reconsideration of *phenotypic expression*. In Chapters 2 and 3, we assumed that the genotype of an organism is always directly expressed in its phenotype. For example, pea plants homozygous for the recessive *d* allele (*dd*) will always be dwarf. There we discussed gene expression as though the genes operate in a closed system in which the presence or absence of functional products directly determines the collective phenotype of an individual. The situation is actually much more complex. Most gene products function within the internal milieu of the cell, and cells interact with one another in various ways. Furthermore, the organism exists under diverse environmental influences. Thus, gene expression and the resultant phenotype are often modified through the interaction between an individual's particular genotype and the external environment. In this final section of this chapter, we will deal with some of the variables that are known to modify gene expression.

### Penetrance and Expressivity

Some mutant genotypes are always expressed as a distinct phenotype, whereas others produce a proportion of individuals whose phenotypes cannot be distinguished from normal (wild type). The degree of expression of a particular trait can be studied quantitatively by determining the *penetrance* and *expressivity* of the genotype under investigation.

The percentage of individuals that show at least some degree of expression of a mutant genotype defines the **penetrance** of the mutation. For example, the phenotypic expression of many of the mutant alleles found in *Drosophila* can overlap with wild-type expression. If 15 percent of flies with a given mutant genotype show the wild-type appearance, the mutant gene is said to have a penetrance of 85 percent.

By contrast, **expressivity** reflects the *range of expression* of the mutant genotype. Flies homozygous for the recessive mutant gene *eyeless* exhibit phenotypes that range from the presence of normal eyes to a partial reduction in size to the complete absence of one or both eyes (Figure 4–17). Although the average reduction of eye size is one-fourth to one-half, expressivity ranges from complete loss of both eyes to completely normal eyes.

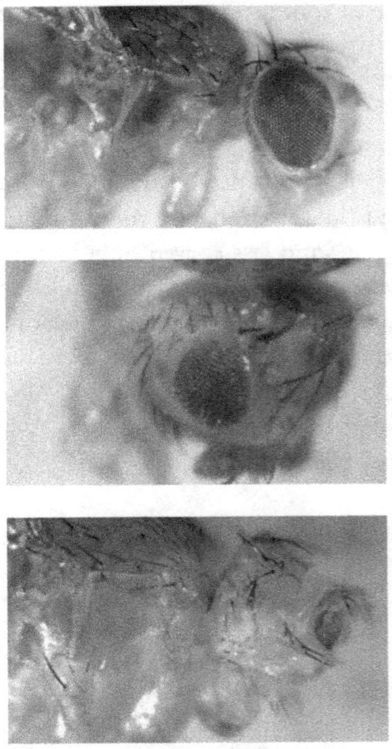

FIGURE 4–17 Variable expressivity as shown in flies homozygous for the *eyeless* mutation in *Drosophila*. Gradations in phenotype range from wild type to partial reduction to eyeless.

Examples such as the expression of the *eyeless* gene have provided the basis for experiments to determine the causes of phenotypic variation. If the laboratory environment is held constant and extensive variation is still observed, other genes may be influencing or modifying the phenotype. On the other hand, if the genetic background is not the cause of the phenotypic variation, environmental factors such as temperature, humidity, and nutrition may be involved. In the case of the *eyeless* phenotype, experiments have shown that both genetic background and environmental factors influence its expression.

## Genetic Background: Suppression and Position Effects

It is difficult to assess the specific effect of the rest of the genome—that is, the **genetic background**—on the expression of a gene responsible for determining a potential phenotype. Nevertheless, two effects of genetic background have been well characterized.

One of these effects is the phenomenon of **genetic suppression**, in which the effect of one mutant gene is counteracted by the effect of a second mutant gene. Mutant genes such as *suppressor of sable (su-s), suppressor of forked (su-f),* and *suppressor of Hairy-wing (su-Hw)* in *Drosophila* completely or partially restore the normal phenotype in an organism that is homozygous (or hemizygous) for the *sable, forked,* and *Hairy-wing* mutations, respectively. For example, flies hemizygous for both *forked* (a bristle mutation) and *su-f* have normal bristles. In each case, the suppressor gene causes the complete reversal of the expected phenotypic expression of the original mutation. Suppressor genes are excellent examples of the genetic background modifying primary gene effects. In addition, in combination with the genes that they suppress, they represent examples of *epistasis*, discussed earlier in this chapter.

Second, the physical location of a gene in relation to other genetic material may influence its expression. Such a situation is called a **position effect**. For example, if a region of a chromosome is relocated or rearranged (called a translocation or inversion event), normal expression of genes in that chromosomal region may be modified. This is particularly true if the gene is relocated to or near certain areas of the chromosome that are prematurely condensed and genetically inert, referred to as **heterochromatin**.

An example of a position effect involves female *Drosophila* heterozygous for the X-linked recessive eye color mutant *white* ($w$). The $w^+/w$ genotype normally results in a wild-type brick red eye color. However, if the region of the X chromosome containing the wild-type $w^+$ allele is translocated so that it is close to a heterochromatic region, expression of the $w^+$ allele is modified. Instead of having a red color, the eyes are variegated, or mottled with red and white patches (Figure 4–18). Therefore, following translocation, the dominant effect of the normal $w^+$ allele is intermittent. A similar position effect is produced if a heterochromatic region is relocated next to the *white* locus on the X chromosome. Apparently, heterochromatic regions inhibit the expression of adjacent genes. Loci in many other organisms also exhibit position effects, providing proof that alteration of the normal arrangement of genetic information can modify its expression.

(a)

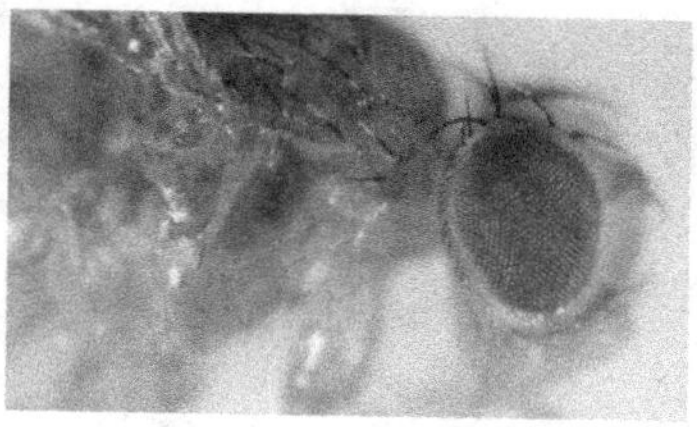

(b)

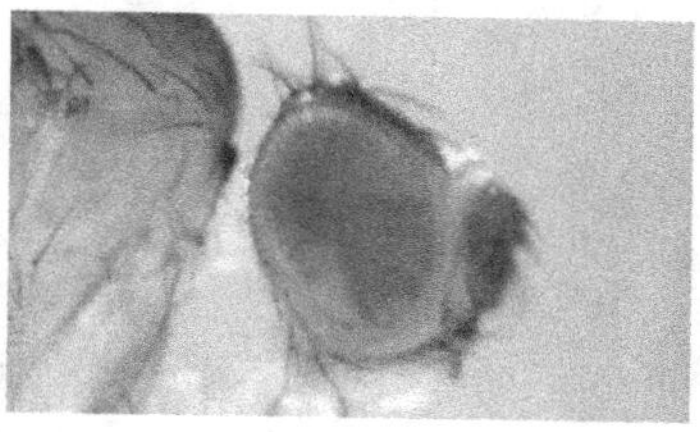

FIGURE 4–18 Position effect, as illustrated in the eye phenotype in two female *Drosophila* heterozygous for the gene *white*. (a) Normal dominant phenotype showing brick red eye color. (b) Variegated color of an eye caused by translocation of the *white* gene to another location in the genome.

## Temperature Effects—An Introduction to Conditional Mutations

Chemical activity depends on the kinetic energy of the reacting substances, which in turn depends on the surrounding temperature. We can thus expect temperature to influence phenotypes. An example is seen in the evening primrose, which produces red flowers when grown at 23°C and white flowers when grown at 18°C. An even more striking example is seen in Siamese cats and Himalayan rabbits, which exhibit dark fur in certain regions where their body temperature is slightly cooler, particularly the nose, ears, and paws (Figure 4–19). In these cases, it appears that the enzyme normally responsible for pigment production is functional only at the lower temperatures present in the extremities, but it loses its catalytic function at the slightly higher temperatures found throughout the rest of the body.

Mutations whose expression is affected by temperature, called **temperature-sensitive mutations**, are examples of **conditional mutations**, whereby phenotypic expression is determined by environmental conditions. Examples of temperature-sensitive mutations are known in viruses and a variety of organisms, including bacteria, fungi, and *Drosophila*. In extreme cases, an organism carrying a mutant allele may express a mutant phenotype when grown at one temperature but express the wild-type phenotype when reared at another temperature. This type of temperature effect is useful in studying mutations that interrupt essential processes during development and are thus normally detrimental or lethal. For example, if bacterial viruses are cultured under *permissive conditions* of 25°C, the mutant gene product is functional, infection proceeds normally, and new viruses are produced and can be studied. However, if bacterial viruses carrying temperature-sensitive mutations infect bacteria cultured at 42°C—the *restrictive condition*—infection

(a)

(b)

FIGURE 4–19 (a) A Himalayan rabbit. (b) A Siamese cat. Both show dark fur color on the snout, ears, and paws. These patches are due to the effect of a temperature-sensitive allele responsible for pigment production.

progresses up to the point where the essential gene product is required (e.g., for viral assembly) and then arrests. Temperature-sensitive mutations are easily induced and isolated in viruses, and have added immensely to the study of viral genetics.

Another temperature effect involves genes that are activated only when the organism finds itself under the stress of elevated environmental temperatures. First discovered in *Drosophila*, these are called **heat-shock genes**, and are responsible for producing a group of proteins believed to provide protection from heat stress. The coordinated activation of these genes in eukaryotes is attributed to shared promotional elements involved in their transcriptional regulation.

## Nutritional Effects

Another category of phenotypes that are not always a direct reflection of the organism's genotype consists of **nutritional mutations.** In microorganisms, mutations that prevent synthesis of nutrient molecules are quite common, such as when an enzyme essential to a biosynthetic pathway becomes inactive. A microorganism bearing such a mutation is called an **auxotroph.** If the end product of a biochemical pathway can no longer be synthesized, and if that molecule is essential to normal growth and development, the mutation prevents growth and may be lethal. For example, if the bread mold *Neurospora* can no longer synthesize the amino acid leucine, proteins cannot be synthesized. If leucine is present in the growth medium, the detrimental effect is overcome. Nutritional mutants have been crucial to genetic studies in bacteria and also served as the basis for George Beadle and Edward Tatum's proposal, in the early 1940s, that one gene functions to produce one enzyme. (See Chapter 15.)

A slightly different set of circumstances exists in humans. The ingestion of certain dietary substances that normal individuals may consume without harm can adversely affect individuals with abnormal genetic constitutions. Often, a mutation may prevent an individual from metabolizing some substance commonly found in normal diets. For example, those afflicted with the genetic disorder **phenylketonuria** cannot metabolize the amino acid phenylalanine. Those with **galactosemia** cannot metabolize galactose. However, if the dietary intake of the molecule is drastically reduced or eliminated, the associated phenotype may be ameliorated.

The fairly common case of **lactose intolerance,** in which individuals are intolerant of the milk sugar lactose, illustrates the general principles involved. Lactose is a disaccharide consisting of a molecule of glucose linked to a molecule of galactose, and makes up 7 percent of human milk and 4 percent of cow's milk. To metabolize lactose, humans require the enzyme **lactase,** which cleaves the disaccharide. Adequate amounts of lactase are produced during the first few years after birth. However, in many people, the level of this enzyme soon drops drastically. As adults, these individuals become intolerant of milk. The major phenotypic effects include severe intestinal diarrhea, flatulence, and abdominal cramps. This condition is particularly prevalent in (though not limited to) people of Eskimo, African, or Asian heritage. In some of these cultures, milk is converted to cheese, butter, and yogurt, which significantly reduces the amount of lactose, thus lessening the adverse effects. In the United States, milk low in lactose is commercially available, and ingestible lactase preparations are available commercially to aid in the digestion of other lactose-containing foods.

## Onset of Genetic Expression

Not all genetic traits become apparent at the same time during an organism's life span. In most cases, the age at which a mutant gene exerts a noticeable phenotype depends on events during the normal sequence of growth and development. In humans, the prenatal, infant, preadult, and adult phases require different genetic information. As a result, many severe inherited disorders are not manifested until after birth. For example, as we saw in Chapter 3, **Tay–Sachs disease,** inherited as an autosomal recessive, is a lethal lipid-metabolism disease involving an abnormal enzyme, hexosaminidase A. Newborns appear to be phenotypically normal for the first few months. Then, developmental retardation, paralysis, and blindness ensue, and most affected children die around the age of 3.

The **Lesch–Nyhan syndrome** (see page 88), inherited as an X-linked recessive disease, is characterized by abnormal nucleic acid metabolism (inability to salvage nitrogenous purine bases), leading to the accumulation of uric acid in blood and tissues, mental retardation, palsy, and self-mutilation of the lips and fingers. The disorder is due to a mutation in the gene encoding hypoxanthine-guanine phosphoribosyl transferase (HPRT). Newborns are normal for six to eight months prior to the onset of the first symptoms.

Still another example is **Duchenne muscular dystrophy (DMD),** an X-linked recessive disorder associated with progressive muscular wasting. It is not usually diagnosed until a child is 3 to 5 years old. Even with modern medical intervention, the disease is often fatal in the early twenties.

Perhaps the most variable age of onset for an inherited human disorder is seen in **Huntington disease.** Inherited as an autosomal dominant disorder, Huntington disease affects the frontal lobes of the cerebral cortex, where progressive cell death occurs over a period of more than a decade. Brain deterioration is accompanied by spastic uncontrolled movements, intellectual and emotional deterioration, and ultimately death. While onset has been reported at all ages, it most frequently occurs between ages 30 and 50, with a mean onset age of 38 years.

These examples support the concept that gene products may play more essential roles at certain times during the life cycle of an organism. One may be able to tolerate the impact of a mutant gene for a considerable period of time without noticeable effect. At some point, however, a mutant phenotype is manifested. Perhaps this is the result of the internal physiological environment of an organism changing during development and with age.

## Genetic Anticipation

Interest in studying the genetic onset of phenotypic expression has intensified with the discovery of heritable disorders that *exhibit a progressively earlier age of onset and an increased severity of the disorder in each successive generation*. This phenomenon is referred to as **genetic anticipation**.

**Myotonic dystrophy** (**DM**), the most common type of adult muscular dystrophy, clearly illustrates genetic anticipation. Individuals afflicted with this autosomal dominant disorder exhibit extreme variation in the severity of symptoms. Mildly affected individuals develop cataracts as adults, but have little or no muscular weakness. Severely affected individuals demonstrate more extensive weakness, as well as myotonia (muscle hyperexcitability) and in some cases mental retardation. In its most extreme form, the disease is fatal just after birth. A great deal of excitement was generated in 1989, when C. J. Howeler and colleagues confirmed the correlation of increased severity and earlier onset with successive generations of inheritance. The researchers studied 61 parent–child pairs, and in 60 of the cases, age of onset was earlier and more severe in the child than in his or her affected parent.

In 1992, an explanation was put forward to explain both the molecular cause of the mutation responsible for DM and the basis of genetic anticipation in the disorder. As we will see in Chapter 16, a short (3-nucleotide) DNA sequence of the DM gene is repeated a variable number of times and is unstable. Normal individuals have about five copies of this region; minimally affected individuals have about 50 copies; and severely affected individuals possess over 1000 copies. The most remarkable observation was that, in successive generations of DM individuals, the size of the repeated segment increases. Although it is not yet clear exactly how the expansion in size affects onset and phenotypic expression, the correlation is extremely strong. Several other inherited human disorders, including the fragile-X syndrome, Kennedy disease, and Huntington disease, also reveal an association between the size of specific regions of the responsible gene and disease severity. We will return to this general topic and discuss the molecular explanation in detail in Chapter 16.

## Genomic (Parental) Imprinting

Our final example of modification of the laws of Mendelian inheritance involves the variation of phenotypic expression that results during early development after one or the other member of a gene pair has been silenced, depending on the parental origin of the chromosome on which a particular allele is located. This phenomenon is called **genomic**, or **parental**, **imprinting**. In some species, certain chromosomal regions and the genes contained within them are somehow "imprinted," depending on their parental origin, in a way that determines whether specific genes will be expressed or remain genetically silent. Such "silencing," for example, leads to the direct phenotypic expression of the allele that is not being silenced.

The imprinting step, the critical issue in understanding this phenomenon, is thought to occur before or during gamete formation, leading to differentially marked genes (or chromosome regions) in sperm-forming versus egg-forming tissues. The process is different from mutation because the imprint is eventually erased and can be reversed in succeeding generations as genes pass from a parent of one sex to an offspring of the other, and so on.

The first example of genomic imprinting was discovered in 1991, in three specific mouse genes. One is the gene encoding insulin-like growth factor II (*Igf2*). A mouse that carries two nonmutant alleles of this gene is normal in size, whereas a mouse that carries two mutant alleles lacks a growth factor and is a dwarf. The size of a heterozygous mouse—one allele normal and one mutant (Figure 4–20)—depends on the parental origin of the wild-type allele. The mouse is

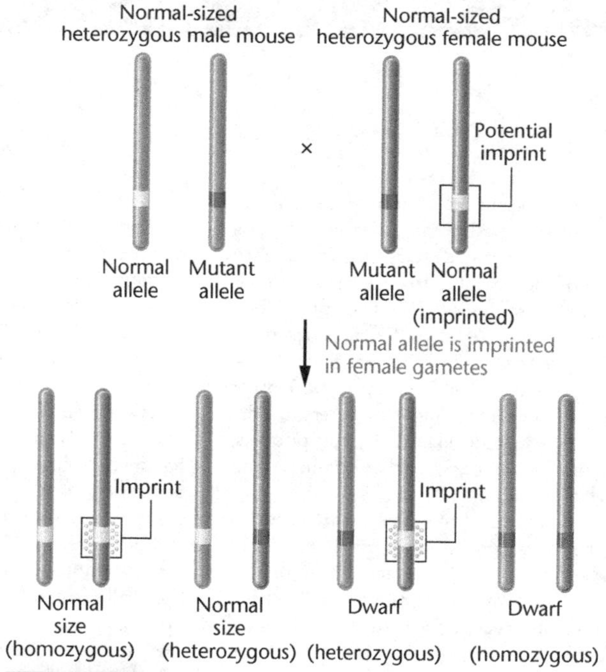


**FIGURE 4–20** The effect of imprinting on the mouse *Igf2* gene, which produces dwarf mice in the homozygous condition. Heterozygous offspring that receive the normal allele from their father are normal in size. Heterozygotes that receive an imprinted normal allele from their mother are dwarf.

normal in size if the normal allele came from the father, but it is dwarf if the normal allele came from the mother. From this, we can deduce that the normal *Igf2* gene is imprinted during egg production in a way that causes it to be silenced, but it functions normally when it has passed through sperm-producing tissue in males.

Imprinting in the next generation depends on whether the gene now passes through sperm-producing or egg-forming tissue. For example, a heterozygous normal-sized male will, on average, donate to half his offspring a normal-functioning wild-type allele that will counteract a mutant allele received from the mother. In humans, two distinct genetic disorders are thought to be caused by differential imprinting of the same region of chromosome 15 (15q1). In both cases, the disorders *appear* to be due to an identical deletion of this region in one member of the chromosome 15 pair. The first disorder, **Prader–Willi syndrome (PWS)**, results when only an undeleted maternal chromosome remains. If only an undeleted paternal chromosome remains, an entirely different disorder, **Angelman syndrome (AS)**, results.

These two conditions exhibit different phenotypes. PWS entails mental retardation, a severe eating disorder marked by an uncontrollable appetite, obesity, diabetes, and growth retardation. Angelman syndrome also involves mental retardation, but involuntary muscle contractions (chorea) and seizures accompany the disorder. We can conclude that the involved region of chromosome 15 is imprinted differently in male and female gametes and that both an undeleted maternal and paternal region are required for normal development.

Although numerous questions remain unanswered regarding genomic imprinting, it is now clear that many genes are subject to this process. More than 50 have been identified in mammals thus far. It appears that regions of chromosomes rather than specific genes are imprinted. The molecular mechanism of imprinting is still a matter for conjecture, but it seems certain that **DNA methylation** is involved. In vertebrates, methyl groups can be added to the carbon atom at position 5 in cytosine (see Chapter 10) as a result of the activity of the enzyme DNA methyltransferase. Methyl groups are added when the dinucleotide CpG or groups of CpG units (called CpG islands) are present along a DNA chain.

DNA methylation is a reasonable mechanism for establishing a molecular imprint, since there is evidence that a high level of methylation can inhibit gene activity and that active genes (or their regulatory sequences) are often undermethylated. Whatever the cause of this phenomenon, it is a fascinating topic and one that clearly establishes the requirement for epigenetic asymmetry between the maternal and paternal genome following fertilization. No doubt, the general phenomenon of genomic imprinting will remain an active area of research in the future.

## GENETICS, TECHNOLOGY, AND SOCIETY

### Improving the Genetic Fate of Purebred Dogs

For dog lovers, nothing is quite so heartbreaking as watching a dog slowly go blind, struggling to adapt to a life of perpetual darkness. That's what happens in progressive retinal atrophy (PRA), a group of inherited disorders first described in Gordon setters in 1909. Since then, PRA has been detected in more than 100 other breeds of dogs, including Irish setters, border collies, Norwegian elkhounds, toy poodles, miniature schnauzers, cocker spaniels, and Siberian huskies.

The products of many genes are required for the development and maintenance of healthy retinas, and a defect in any one of these genes may cause retinal dysfunction. Decades of research have led to the identification of five such genes (*PDE6A, PDE6B, PRCD, rhodopsin,* and *PRGR*), and more may be discovered. Different mutant alleles are found in different breeds, and each allele is associated with a different form of PRA that varies slightly in its clinical symptoms and rate of progression. Mutations of *PDE6A, PDE6B,* and *PRCD* genes are inherited in a recessive pattern, mutations of the *rhodopsin* gene (found in Mastiffs) are dominant, and *PRGR* mutations (in Siberian huskies and Samoyeds) are X-linked.

PRA is almost ten times more common in certain purebred dogs than in mixed breeds. The development of distinct breeds of dogs has involved intensive selection for desirable attributes, such as a particular size, shape, color, or behavior. Many desired characteristics are determined by recessive alleles. The fastest way to increase the homozygosity of these alleles is to mate close relatives, which are likely to carry the same alleles. For example, dogs may be mated to a cousin or a grandparent. Some breeders, in an attempt to profit from impressive pedigrees, also produce hundreds of offspring from individual dogs that have won major prizes at dog shows. This "popular sire effect," as it has been termed, further increases the homozygosity of alleles in purebred dogs.

Unfortunately, the generations of inbreeding that have established favorable characteristics in purebreeds have also increased the homozygosity of certain harmful recessive alleles, resulting in a high incidence of inherited diseases. Many breeds, such as German shepherds, are plagued with inherited hip dysplasia. Deafness and kidney disorders are common genetic maladies in dalmatians. More than 300 genetic diseases have been characterized in purebred dogs, and many breeds have a predisposition to more than 20 of them. According to researchers at Cornell University, purebred dogs suffer the highest incidence of inherited disease of any animal: 25 percent of the 20 million purebred dogs in America are affected with one genetic ailment or another.

Fortunately, advances in canine genetics are beginning to provide new tools to increase the health of purebred dogs. As of 2007, genetic tests are available to detect 30 different retinal diseases in dogs. Most of these tests are specific for varieties of PRA that affect a particular dog breed. In contrast, the test for *PRCD* mutations, responsible for progressive rod-cone degeneration (the most common form of PRA), is useful for at least 18 different breeds that bear mutations in the *PRCD* gene. The *PRCD* test is now being used to identify heterozygous carriers of *PRCD* mutations—dogs that show no symptoms of PRA but, if mated with other carriers, pass the trait on to about 25 percent of their offspring. Eliminating PRA carriers from breeding programs has almost eradicated this condition from Portuguese Water Dogs and has greatly reduced its prevalence in other breeds. OptiGen, a company devoted solely to testing and preventing inherited diseases in purebred dogs, offers blood-based tests for all known forms of PRA and is developing tests for similar inherited disorders. The increased availability of genetic tests will help ensure that breeding animals are free from harmful recessive alleles and, with hope, will offer a solution to the problems caused by inbreeding.

The identification of genes underlying canine inherited disease will be faster, thanks to the completion of the Dog Genome Project in 2005. The dog genome is smaller than the human genome but contains more than 18,000 homologs to human genes. Much to the delight of researchers and dog owners alike, information provided by the Dog Genome Project has already helped combat retinal degenerative disease. Researchers at the University of Pennsylvania used gene therapy to treat Leber's Congenital Amaurosis (LCA), a disease similar to PRA. By injecting copies of normal genes into the retina of a congenitally blind puppy, researchers were able to partially restore the dog's vision. Since this trial, more than a dozen dogs have been treated with similar gene therapy, with restored vision lasting at least three years.

The Dog Genome Project may have benefits for humans beyond the reduction of disease in their canine companions. Eighty-five percent of the genes in the dog genome have equivalents in humans, and over 300 diseases affecting dogs also affect humans. In fact, the ten most common disorders of purebred dogs are all major human health concerns, including heart disease, epilepsy, allergies, and cancer. The identification of a disease-causing dog gene can be a shortcut to the isolation of the corresponding gene in humans. For example, the *PRCD* gene isolated in dogs has a human version, affected in some cases of human retinitis pigmentosa (RP), which afflicts about 1.5 million people worldwide. In fact, an identical single base-substitution mutation in *PRCD* was found in affected dogs and humans. Despite much research, human RP remains poorly understood, and current treatments only slow its progress. Understanding the genetic basis of PRA will lead to breakthroughs in the diagnosis and treatment of RP, potentially saving the sight of thousands of people every year. By contributing to the cure of human diseases, dogs may prove to be "man's best friend" in an entirely new way.

### References

Kirkness, E.F., et al. 2003. The dog genome: survey sequencing and comparative analysis. *Science* 301: 1898–1903.

Zangerl, B. 2006. Identical mutation in a novel retinal gene causes progressive rod-cone degeneration in dogs and retinitis pigmentosa in humans. *Genomics* 88: 551–563.

# EXPLORING GENOMICS

## The Human Epigenome Project

A consortium of scientists from the Wellcome Trust Sanger Institute in the United Kingdom, Epigenomics AG in Germany, and the Centre National de Génotypage in France are working on the **Human Epigenome Project (HEP)**. A major goal of the HEP is to catalog methylation sites in the human genome in different tissues. So far chromosomes 6, 20, and 22 have been cataloged in about a dozen different tissues. Because DNA methylation, which usually occurs on CpG islands, affects the expression of genes, often by silencing them, HEP scientists are using methylation profiling to understand how methylation controls and influences gene expression patterns in different cell types, particularly those genes involved in cancer. Epigenomics scientists eventually hope to customize drug treatments based on a patient's epigenetic (methylation) profile.

In this exercise, we will explore the Human Epigenome Project Web site to learn about methylation patterns on chromosome 6, one of the first targets of the HEP because many human cancer genes are located on this chromosome.

### Exercise I – Methylation Variable Positions and Chromosome 6

1. Access the HEP site at www.sanger.ac.uk/PostGenomics/epigenome/.

2. At the top right of the screen is a box with chromosome images. To learn how methylation data is presented at this site, click the link below the chromosome box, "For instructions click here." This will take you to a brief tutorial on navigating the

*Continued on next page*

*Exploring Genomics, continued*

site. It explains that methylation patterns are shown as blue squares indicating methylation variable positions (MVP) in the genome. MVPs represent differences in methylation patterns of DNA in different tissues.

3. Return to the HEP home page and click on chromosome 6 (the next page may load slowly, so be patient!). At the top of the screen, you will see boxes indicating the base-pair length for the region of chromosome 6 with an MVP. What part of chromosome 6 is shown?

4. In the center of the page you will see a vertical gray box containing blue squares indicating MVPs at this location on chromosome 6. Left click on each of the thin lines to the left of the gray box, and text will appear telling you which tissues were analyzed for MVPs. Which cells and tissues have been analyzed so far?

5. Use the right and left arrow buttons to move 1 or 2 Mb (megabases) along chromosome 6 to explore other methylation sites on this chromosome, or type in a range of base pairs to explore using the boxes at the top of the screen.

**Exercise II – Functions of the *HUS1B* Gene**

At the bottom of the "MVP Viewer" screen for chromosome 6 that you opened in step 3 above, the Ensembl trans line displays locations of RNA transcripts for genes in this region of chromosome 6. Notice that one of the genes located in an MVP is called *HUS1B*. Click on the gene name and a small box will appear. Click the "Gene Info" link, then follow the instructions given below, and answer the corresponding questions:

1. Use the link for "Genomic Location" or click on the "*HUS1B*" link. Is the locus for *HUS1B* on the p- or q-arm of chromosome 6? What is the specific band where *HUS1B* is located?

2. Go to the bottom of the screen. Potential orthologs (similar genes in different species that are thought to have evolved from a common ancestor) for *HUS1B* have been found in several other organisms. Review the list to see which organisms may have a *HUS1B* ortholog.

3. Under the "Gene" category at the top of the Ensembl report page, click the *HUS1B* link. This link will take you to a gene name and symbol database from the Human Genome Nomenclature Committee (HGNC). As indicated by the approved name, in what organism was this gene first identified? While at the HGNC site, click the PubMed ID (PMID) link and read the abstract from the paper describing the identification of *HUS1B*. What is the function of this gene?

4. Based on the cell types that you identified in Exercise I, in which *HUS1B* is known to be methylated (click on the SwissProt link from the HGNC site for more information about expression in different cell types), and on the function of *HUS1B*, explain why you think methylation of this gene may play a role in certain forms of leukemia.

## Chapter Summary

1. Since Mendel's work was rediscovered, the study of transmission genetics has expanded to include many alternative modes of inheritance involving various numbers of genes.
2. Incomplete, or partial, dominance is exhibited when intermediate phenotypic expression of a trait occurs in an organism that is heterozygous for two alleles.
3. Codominance is exhibited when both alleles in a heterozygous organism are expressed.
4. The concept of multiple alleles applies to populations, since a diploid organism may host only two alleles for any given locus. However, within a population, many alternative alleles of the same gene can occur.
5. Lethal mutations usually result in the inactivation or the lack of synthesis of gene products that are essential during an organism's development. Such mutations may be recessive or dominant. Some lethal genes, such as the gene causing Huntington disease, are not expressed until adulthood.
6. Mendel's classic $F_2$ ratio is often modified in instances where gene interaction controls phenotypic variation.
7. Epistasis may occur when two or more genes influence a single characteristic. Usually, the expression of one of the genes masks the expression of the other gene or genes.
8. Genes located on the X chromosome result in a characteristic mode of inheritance referred to as X-linkage. Hemizygous individuals (those with an X and a Y chromosome) express all alleles present on their X chromosome.
9. Sex-limited and sex-influenced inheritance occurs when the sex of the organism affects the phenotype controlled by a gene located on an autosome.
10. Phenotypic expression is not always the direct reflection of the genotype. Penetrance measures the percentage of organisms in a given population exhibiting evidence of the corresponding mutant phenotype. Expressivity, on the other hand, measures the range of phenotypic expression of a given genotype.
11. Phenotypic expression can be modified by genetic background, temperature, and nutrition. Position effects illustrate the genetic background affecting phenotypic expression.
12. Genetic anticipation refers to the phenomenon where the onset of phenotypic expression occurs earlier and becomes more severe in each ensuing generation.
13. Genomic imprinting is a process whereby a region of either the paternal or maternal chromosome is modified (marked or imprinted), thereby affecting phenotypic expression. Expression therefore depends on which parent contributes a mutant allele.

# INSIGHTS AND SOLUTIONS

*Genetic problems take on added complexity if they involve two independent characters and multiple alleles, incomplete dominance, or epistasis. The most difficult types of problems are those that pioneering geneticists faced during laboratory or field studies. They had to determine the mode of inheritance by working backward from the observations of offspring to parents of unknown genotype.*

1. Consider the problem of comb-shape inheritance in chickens, where walnut, rose, pea, and single are observed as distinct phenotypes. These variations are shown in the accompanying photographs. Considering the following data, determine *how comb shape is inherited and what genotypes are present in the $P_1$ generation of each cross.*

| | | | |
|---|---|---|---|
| Cross 1: | single × single | ⟶ | all single |
| Cross 2: | walnut × walnut | ⟶ | all walnut |
| Cross 3: | rose × pea | ⟶ | all walnut |
| Cross 4: | $F_1 \times F_1$ of Cross 3 | | |
| | walnut × walnut | ⟶ | 93 walnut |
| | | | 28 rose |
| | | | 32 pea |
| | | | 10 single |

**Solution:** At first glance, this problem appears quite difficult. However, working systematically and breaking the analysis into steps simplifies it. To start, look at the data carefully for any useful information. Once you identify something that is clearly helpful, follow an empirical approach; that is, formulate a hypothesis and test it against the given data. Look for a pattern of inheritance that is consistent with all cases.

This problem gives two immediately useful facts. First, in cross 1, $P_1$ singles breed true. Second, while $P_1$ walnut breeds true in cross 2, a walnut phenotype is also produced in cross 3 between rose and pea. When these $F_1$ walnuts are mated in cross 4, all four comb shapes are produced in a ratio that approximates 9:3:3:1. This observation immediately suggests a cross involving two gene pairs, because the resulting data display the same ratio as in Mendel's dihybrid crosses. Since only one character is involved (comb shape), epistasis may be occurring. This could serve as your working hypothesis, and you must now propose how the two gene pairs "interact" to produce each phenotype.

If you call the allele pairs *A*, *a* and *B*, *b*, you might predict that because walnut represents 9/16 of the offspring in cross 4, *A–B–* will produce walnut. (Recall that *A–* and *B–* mean *AA* or *Aa* and *BB* or *Bb*, respectively.) You might also hypothesize that in cross 2, the genotypes are *AABB* × *AABB* where walnut bred true.

The phenotype representing 1/16 of the offspring of cross 4 is single; therefore you could predict that the single phenotype is the result of the *aabb* genotype. This is consistent with cross 1.

Now you have only to determine the genotypes for rose and pea. The most logical prediction is that at least one dominant *A* or *B* allele combined with the double recessive condition of the other allele pair accounts for these phenotypes. For example,

| | | |
|---|---|---|
| *A–bb* | ⟶ | rose |
| *aaB–* | ⟶ | pea |

If *AAbb* (rose) is crossed with *aaBB* (pea) in cross 3, all offspring would be *AaBb* (walnut). This is consistent with the data, and you need now look at only cross 4. We predict these walnut genotypes to be *AaBb* (as above), and from the cross

*AaBb* (walnut) × *AaBb* (walnut)

we expect

| | |
|---|---|
| 9/16 | *A–B–* (walnut) |
| 3/16 | *A–bb* (rose) |
| 3/16 | *aaB–* (pea) |
| 1/16 | *aabb* (single) |

Our prediction is consistent with the data given. The initial hypothesis of the interaction of two gene pairs proves consistent throughout, and the problem is solved.

This problem demonstrates the usefulness of a basic theoretical knowledge of transmission genetics. With such knowledge, you can search for clues that will enable you to proceed in a stepwise fashion toward a solution. Mastering problem-solving requires practice, but can give you a great deal of satisfaction. Apply the same general approach to the following problems.

2. In radishes, flower color may be red, purple, or white. The edible portion of the radish may be long or oval. When only flower color is studied, no dominance is evident, and red × white crosses yield all purple. If these $F_1$ purples are interbred, the $F_2$ generation consists of 1/4 red: 1/2 purple: 1/4 white. Regarding radish shape, long is dominant to oval in a normal Mendelian fashion.

(a) Determine the $F_1$ and $F_2$ phenotypes from a cross between a true-breeding red, long radish and one that is white and oval. Be sure to define all gene symbols at the start.

(b) A red oval plant was crossed with a plant of unknown genotype and phenotype, yielding the following offspring:

103 red long: 101 red oval
98 purple long: 100 purple oval

Determine the genotype and phenotype of the unknown plant.

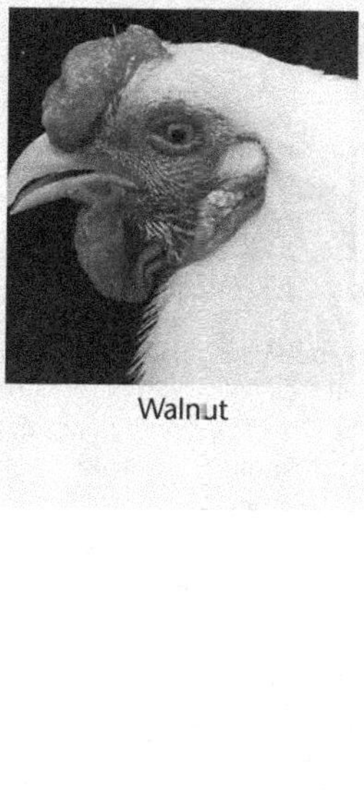
Walnut

Pea

Rose

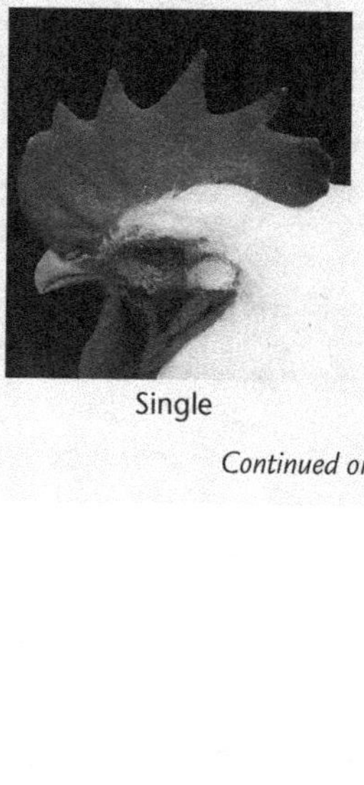
Single

*Continued on next page*

*Insights and Solutions, continued*

**Solution:** First, establish gene symbols:

$RR$ = red  $O-$ = long
$Rr$ = purple  $oo$ = oval
$rr$ = white

(a) This is a modified dihybrid cross where the gene pair controlling color exhibits incomplete dominance. Shape is controlled conventionally.

$P_1$: *RROO* × *rroo*
(red long) (white oval)

$F_1$: all *RrOo* (purple long)

$F_1 \times F_1$: *RrOo* × *RrOo*

| | | | | |
|---|---|---|---|---|
| $F_2$: | 1/4 *RR* | 3/4 *O*– | 3/16 *RR O*– | red long |
| | | 1/4 *oo* | 1/16 *RR oo* | red oval |
| | 2/4 *Rr* | 3/4 *O*– | 6/16 *Rr O*– | purple long |
| | | 1/4 *oo* | 2/16 *Rr oo* | purple oval |
| | 1/4 *rr* | 3/4 *O*– | 3/16 *rr O*– | white long |
| | | 1/4 *oo* | 1/16 *rr oo* | white oval |

Note that to generate the $F_2$ results, we have used the forked-line method. First, we consider the outcome of crossing $F_1$ parents for the color genes (*Rr* × *Rr*). Then the outcome of shape is considered (*Oo* × *Oo*).

(b) The two characters appear to be inherited independently, so consider them separately. The data indicate a 1/4: 1/4: 1/4: 1/4 proportion. First, consider color:

$P_1$: red × ??? (unknown)

$F_1$: 204 red (1/2)
198 purple (1/2)

Because the red parent must be *RR*, the unknown must have a genotype of *Rr* to produce these results. Thus it is purple. Now, consider shape:

$P_1$: oval × ??? (unknown)

$F_1$: 201 long (1/2)
201 oval (1/2)

Since the oval plant must be *oo*, the unknown plant must have a genotype of *Oo* to produce these results. Thus it is long. The unknown plant is

*RrOo* purple long

3. In humans, red–green color-blindness is inherited as an X-linked recessive trait. A woman with normal vision whose father is color blind marries a male who has normal vision. Predict the color vision of their male and female offspring.

**Solution:** The female is heterozygous, since she inherited an X chromosome with the mutant allele from her father. Her husband is normal. Therefore, the parental genotypes are

*Cc* × *C*↾ (↾ represents the Y chromosome)

All female offspring are normal (*CC* or *Cc*). One-half of the male children will be color-blind (*c*↾), and the other half will have normal vision (*C*↾).

4. Consider the two very limited unrelated pedigrees shown here. Of the four combinations of X-linked recessive, X-linked dominant, autosomal recessive, and autosomal dominant, which modes of inheritance can be absolutely ruled out in each case?

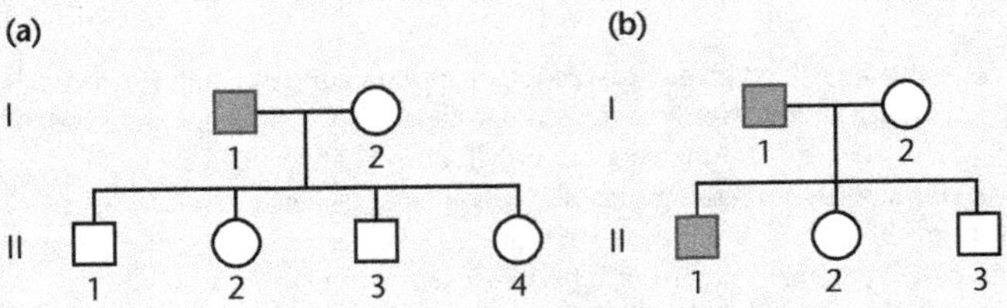


**Solution:** For both pedigrees, X-linked recessive and autosomal recessive remain possible, provided that the maternal parent is heterozygous in pedigree (b). Autosomal dominance seems at first glance unlikely in pedigree (a), since at least half of the offspring should express a dominant trait expressed by one of their parents. However, while it is true that if the affected parent carries an autosomal dominant gene heterozygously, each offspring has a 50 percent chance of inheriting and expressing the mutant gene, the sample size of four offspring is too small to rule this possibility out. In Pedigree (b), autosomal dominance is clearly possible. In both cases, one can rule out X-linked dominance because the female offspring would inherit and express the dominant allele, and they do not express the trait in either pedigree.

## Problems and Discussion Questions

1. In shorthorn cattle, coat color may be red, white, or roan. Roan is an intermediate phenotype expressed as a mixture of red and white hairs. The following data were obtained from various crosses:

| | | | | |
|---|---|---|---|---|
| red | × | red | → | all red |
| white | × | white | → | all white |
| red | × | white | → | all roan |
| roan | × | roan | → | 1/4 red:1/2 roan:1/4 white |

How is coat color inherited? What are the genotypes of parents and offspring for each cross?

2. Contrast incomplete dominance and codominance. Define the phenomenon of epistasis in the context of the concept of gene interaction.

3. In foxes, two alleles of a single gene, *P* and *p*, may result in lethality (*PP*), platinum coat (*Pp*), or silver coat (*pp*). What ratio is obtained when platinum foxes are interbred? Is the *P* allele behaving dominantly or recessively in causing (a) lethality; (b) platinum coat color?

4. In mice, a short-tailed mutant was discovered. When it was crossed to a normal long-tailed mouse, 4 offspring were short-tailed and 3 were long-tailed. Two short-tailed mice from the $F_1$ generation were selected and crossed. They produced 6 short-tailed and 3 long-tailed mice. These genetic experiments were repeated three times with approximately the same results. What genetic ratios are illustrated? Hypothesize the mode of inheritance and diagram the crosses.
5. List all possible genotypes for the A, B, AB, and O phenotypes. Is the mode of inheritance of the ABO blood types representative of dominance? of recessiveness? of codominance?
6. With regard to the ABO blood types in humans, determine the genotype of the male parent and female parent shown here:

   Male parent: Blood type B; mother type O
   Female parent: Blood type A; father type B

   Predict the blood types of the offspring that this couple may have and the expected proportion of each.
7. In a disputed parentage case, the child is blood type O, while the mother is blood type A. What blood type would exclude a male from being the father? Would the other blood types prove that a particular male was the father?
8. The A and B antigens in humans may be found in water-soluble form in secretions, including saliva, of some individuals (*Se*/*Se* and *Se*/*se*) but not in others (*se*/*se*). The population thus contains "secretors" and "nonsecretors."
   (a) Determine the proportion of various phenotypes (blood type and ability to secrete) in matings between individuals that are blood type AB and type O, both of whom are *Se*/*se*.
   (b) How will the results of such matings change if both parents are heterozygous for the gene controlling the synthesis of the H substance (*Hh*)?
9. In chickens, a condition referred to as "creeper" exists whereby the bird has very short legs and wings, and appears to be creeping when it walks. If creepers are bred to normal chickens, one-half of the offspring are normal and one-half are creepers. Creepers never breed true. If bred together, they yield two-thirds creepers and one-third normal. Propose an explanation for the inheritance of this condition.
10. In rabbits, a series of multiple alleles controls coat color in the following way: *C* is dominant to all other alleles and causes full color. The chinchilla phenotype is due to the $c^{ch}$ allele, which is dominant to all alleles other than *C*. The $c^h$ allele, dominant only to $c^a$ (albino), results in the Himalayan coat color. Thus, the order of dominance is $C > c^{ch} > c^h > c^a$. For each of the following three cases, the phenotypes of the $P_1$ generations of two crosses are shown, as well as the phenotype of *one member* of the $F_1$ generation.

| | $P_1$ Phenotypes | | $F_1$ Phenotypes |
|---|---|---|---|
| (a) | Himalayan × Himalayan | ⟶ | albino |
| | | | × ⟶ ?? |
| | full color × albino | ⟶ | chinchilla |
| (b) | albino × chinchilla | ⟶ | albino |
| | | | × ⟶ ?? |
| | full color × albino | ⟶ | full color |
| (c) | chinchilla × albino | ⟶ | Himalayan |
| | | | × ⟶ ?? |
| | full color × albino | ⟶ | Himalayan |

For each case, determine the genotypes of the $P_1$ generation and the $F_1$ offspring, and predict the results of making each indicated cross between $F_1$ individuals.

11. In the guinea pig, one locus involved in the control of coat color may be occupied by any of four alleles: *C* (full color), $c^k$ (sepia), $c^d$ (cream), or $c^a$ (albino). Like coat color in rabbits (Problem 10), an order of dominance exists: $C > c^k > c^d > c^a$. In the following crosses, write the parental genotypes and predict the phenotypic ratios that would result:
    (a) sepia × cream, where both guinea pigs had an albino parent
    (b) sepia × cream, where the sepia guinea pig had an albino parent and the cream guinea pig had two sepia parents
    (c) sepia × cream, where the sepia guinea pig had two full-color parents and the cream guinea pig had two sepia parents
    (d) sepia × cream, where the sepia guinea pig had a full-color parent and an albino parent and the cream guinea pig had two full-color parents
12. Three gene pairs located on separate autosomes determine flower color and shape as well as plant height. The first pair exhibits incomplete dominance, where the color can be red, pink (the heterozygote), or white. The second pair leads to personate (dominant) or peloric (recessive) flower shape, while the third gene pair produces either the dominant tall trait or the recessive dwarf trait. Homozygous plants that are red, personate, and tall are crossed to those that are white, peloric, and dwarf. Determine the $F_1$ genotype(s) and phenotype(s). If the $F_1$ plants are interbred, what proportion of the offspring will exhibit the same phenotype as the $F_1$ plants?

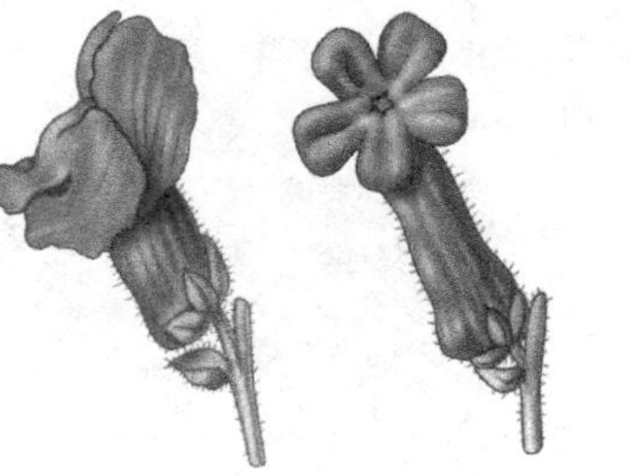


13. As in Problem 12, flower color may be red, white, or pink, and flower shape may be personate or peloric. For the following crosses, determine the $P_1$ and $F_1$ genotypes:
    (a) red, peloric × white, personate
    ↓
    $F_1$: all pink, personate
    (b) red, personate × white, peloric
    ↓
    $F_1$: all pink, personate

(c) pink, personate × red, peloric → $F_1$ { 1/4 red, personate; 1/4 red, peloric; 1/4 pink, peloric; 1/4 pink, personate }

(d) pink, personate × white, peloric → $F_1$ { 1/4 white, personate; 1/4 white, peloric; 1/4 pink, personate; 1/4 pink, peloric }

(e) What phenotypic ratios would result from crossing the $F_1$ of (a) to the $F_1$ of (b)?

14. Horses can be cremello (a light cream color), chestnut (a brownish color), or palomino (a golden color with white in the horse's tail and mane). Of these phenotypes, only palominos never breed true.

| | | |
|---|---|---|
| cremello × palomino | ⟶ | 1/2 cremello<br>1/2 palomino |
| chestnut × palomino | ⟶ | 1/2 chestnut<br>1/2 palomino |
| palomino × palomino | ⟶ | 1/4 chestnut<br>1/2 palomino<br>1/4 cremello |

(a) From the results given above, determine the mode of inheritance by assigning gene symbols and indicating which genotypes yield which phenotypes.

(b) Predict the $F_1$ and $F_2$ results of many initial matings between cremello and chestnut horses.

Chestnut

Palomino

Cremello

15. With reference to the eye color phenotypes produced by the recessive, autosomal, unlinked *brown* and *scarlet* loci in *Drosophila* (see Figure 4–10), predict the $F_1$ and $F_2$ results of the following $P_1$ crosses. (Recall that when both the *brown* and *scarlet* alleles are homozygous, no pigment is produced, and the eyes are white.)
(a) wild type × white
(b) wild type × scarlet
(c) brown × white

16. Pigment in mouse fur is only produced when the *C* allele is present. Individuals of the *cc* genotype are white. If color is present, it may be determined by the *A*, *a* alleles. *AA* or *Aa* results in agouti color, while *aa* results in black coats.
(a) What $F_1$ and $F_2$ genotypic and phenotypic ratios are obtained from a cross between *AACC* and *aacc* mice?
(b) In three crosses between agouti females whose genotypes were unknown and males of the *aacc* genotype, the following phenotypic ratios were obtained:

| (1) | (2) | (3) |
|---|---|---|
| 8 agouti | 9 agouti | 4 agouti |
| 8 white | 10 black | 5 black |
| | | 10 white |

What are the genotypes of these female parents?

17. In some plants a red pigment, cyanidin, is synthesized from a colorless precursor. The addition of a hydroxyl group ($OH^-$) to the cyanidin molecule causes it to become purple. In a cross between two randomly selected purple plants, the following results were obtained:

94 purple
31 red
43 white

How many genes are involved in the determination of these flower colors? Which genotypic combinations produce which phenotypes? Diagram the purple × purple cross.

18. In rats, the following genotypes of two independently assorting autosomal genes determine coat color:

| | |
|---|---|
| *A–B–* | (gray) |
| *A–bb* | (yellow) |
| *aaB–* | (black) |
| *aabb* | (cream) |

A third gene pair on a separate autosome determines whether or not any color will be produced. The *CC* and *Cc* genotypes allow color according to the expression of the *A* and *B* alleles. However, the *cc* genotype results in albino rats regardless of the *A* and *B* alleles present. Determine the $F_1$ phenotypic ratio of the following crosses:
(a) *AAbbCC* × *aaBBcc*
(b) *AaBBCC* × *AABbcc*
(c) *AaBbCc* × *AaBbcc*
(d) *AaBBCc* × *AaBBCc*
(e) *AABbCc* × *AABbcc*

19. Given the inheritance pattern of coat color in rats described in Problem 18, predict the genotype and phenotype of the parents who produced the following offspring:
(a) 9/16 gray: 3/16 yellow: 3/16 black: 1/16 cream
(b) 9/16 gray: 3/16 yellow: 4/16 albino
(c) 27/64 gray: 16/64 albino: 9/64 yellow: 9/64 black: 3/64 cream
(d) 3/8 black: 3/8 cream: 2/8 albino
(e) 3/8 black: 4/8 albino: 1/8 cream

20. In a species of the cat family, eye color can be gray, blue, green, or brown, and each trait is true breeding. In separate crosses involving homozygous parents, the following data were obtained:

| Cross | $P_1$ | $F_1$ | $F_2$ |
|---|---|---|---|
| A | green × gray | all green | 3/4 green: 1/4 gray |
| B | green × brown | all green | 3/4 green: 1/4 brown |
| C | gray × brown | all green | 9/16 green: 3/16 brown<br>3/16 gray: 1/16 blue |

(a) Analyze the data. How many genes are involved? Define gene symbols and indicate which genotypes yield each phenotype.
(b) In a cross between a gray-eyed cat and one of unknown genotype and phenotype, the $F_1$ generation was not observed. However, the $F_2$ resulted in the same $F_2$ ratio as in cross C. Determine the genotypes and phenotypes of the unknown $P_1$ and $F_1$ cats.

21. In a plant, a tall variety was crossed with a dwarf variety. All $F_1$ plants were tall. When $F_1 \times F_1$ plants were interbred, 9/16 of the $F_2$ were tall and 7/16 were dwarf.
(a) Explain the inheritance of height by indicating the number of gene pairs involved and by designating which genotypes yield tall and which yield dwarf. (Use dashes where appropriate.)
(b) What proportion of the $F_2$ plants will be true breeding if self-fertilized? List these genotypes.

22. In a unique species of plants, flowers may be yellow, blue, red, or mauve. All colors may be true breeding. If plants with blue flowers are crossed to red-flowered plants, all $F_1$ plants have yellow flowers. When these produced an $F_2$ generation, the following ratio was observed:

9/16 yellow: 3/16 blue: 3/16 red: 1/16 mauve

In still another cross using true-breeding parents, yellow-flowered plants are crossed with mauve-flowered plants. Again, all $F_1$ plants had yellow flowers and the $F_2$ showed a 9:3:3:1 ratio, as just shown.
(a) Describe the inheritance of flower color by defining gene symbols and designating which genotypes give rise to each of the four phenotypes.
(b) Determine the $F_1$ and $F_2$ results of a cross between true-breeding red and true-breeding mauve-flowered plants.

23. Five human matings (1–5), identified by both maternal and paternal phenotypes for ABO and MN blood-group antigen status, are shown on the left side of the following table:

| Parental Phenotypes | Offspring |
|---|---|
| (1) A, M × A, N | (a) A, N |
| (2) B, M × B, M | (b) O, N |
| (3) O, N × B, N | (c) O, MN |
| (4) AB, M × O, N | (d) B, M |
| (5) AB, MN × AB, MN | (e) B, MN |

Each mating resulted in one of the five offspring shown in the right-hand column (a–e). Match each offspring with one correct set of parents, using each parental set only once. Is there more than one set of correct answers?

24. A husband and wife have normal vision, although both of their fathers are red–green color-blind, an inherited X-linked recessive condition. What is the probability that their first child will be (a) a normal son? (b) a normal daughter? (c) a color-blind son? (d) a color-blind daughter?

25. In humans, the ABO blood type is under the control of autosomal multiple alleles. Color blindness is a recessive X-linked trait. If two parents who are both type A and have normal vision produce a son who is color-blind and is type O, what is the probability that their next child will be a female who has normal vision and is type O?

26. In *Drosophila*, an X-linked recessive mutation, *scalloped (sd)*, causes irregular wing margins. Diagram the $F_1$ and $F_2$ results if (a) a scalloped female is crossed with a normal male; (b) a scalloped male is crossed with a normal female. Compare these results with those that would be obtained if the *scalloped* gene were autosomal.

27. Another recessive mutation in *Drosophila*, *ebony (e)*, is on an autosome (chromosome 3) and causes darkening of the body compared with wild-type flies. What phenotypic $F_1$ and $F_2$ male and female ratios will result if a scalloped-winged female with normal body color is crossed with a normal-winged ebony male? Work this problem by both the Punnett square method and the forked-line method.

28. In *Drosophila*, the X-linked recessive mutation *vermilion (v)* causes bright red eyes, in contrast to the brick-red eyes of wild type. A separate autosomal recessive mutation, *suppressor of vermilion (su-v)*, causes flies homozygous or hemizygous for *v* to have wild-type eyes. In the absence of *vermilion* alleles, *su-v* has no effect on eye color. Determine the $F_1$ and $F_2$ phenotypic ratios from a cross between a female with wild-type alleles at the *vermilion* locus, but who is homozygous for *su-v*, with a *vermilion* male who has wild-type alleles at the *su-v* locus.

29. While *vermilion* is X-linked in *Drosophila* and causes the eye color to be bright red, *brown* is an autosomal recessive mutation that causes the eye to be brown. Flies carrying both mutations lose all pigmentation and are white-eyed. Predict the $F_1$ and $F_2$ results of the following crosses:
(a) *vermilion* females × *brown* males
(b) *brown* females × *vermilion* males
(c) *white* females × wild-type males

30. In a cross in *Drosophila* involving the X-linked recessive eye mutation *white* and the autosomally linked recessive eye mutation *sepia* (resulting in a dark eye), predict the $F_1$ and $F_2$ results of crossing true-breeding parents of the following phenotypes:
(a) white females × sepia males
(b) sepia females × white males
Note that white is epistatic to the expression of sepia.

31. Consider the following three pedigrees, all involving a single human trait:

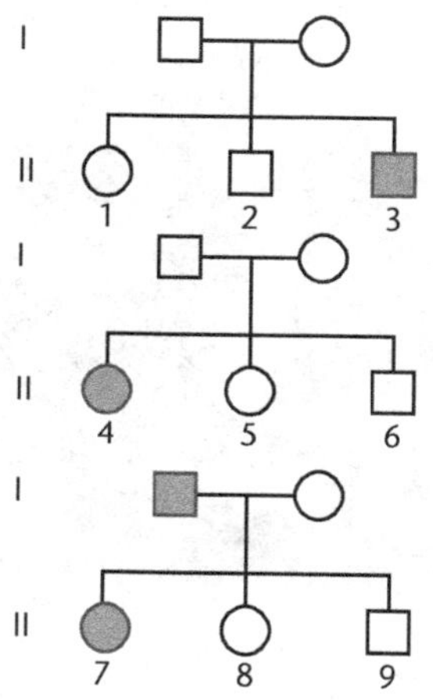


(a) Which conditions, if any, can be excluded?
dominant and X-linked
dominant and autosomal
recessive and X-linked
recessive and autosomal
(b) For each condition that you excluded, indicate the single individual in generation II (e.g., II-1, II-2) that was most instrumental in your decision to exclude that condition. If none were excluded, answer "none apply."
(c) Given your conclusions in part (a), indicate the genotype of the following individuals:
II-1, II-6, II-9
If more than one possibility applies, list all possibilities. Use the symbols *A* and *a* for the genotypes.

32. Following are three pedigrees. For each, consider whether it could or could not be consistent with an X-linked recessive trait. In a sentence or two, indicate why or why not.

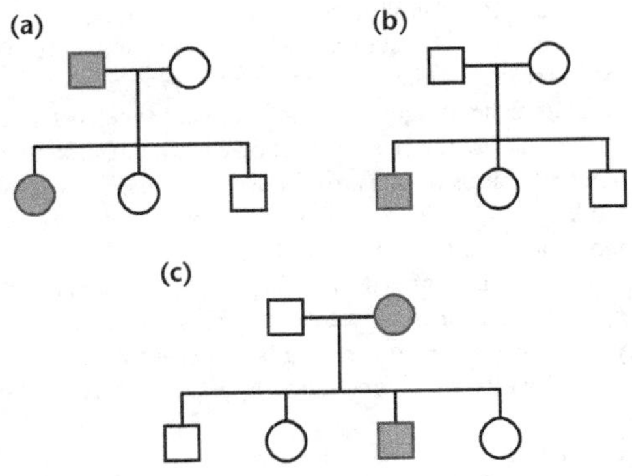


33. In goats, the development of the beard is due to a recessive gene. The following cross involving true-breeding goats was made and carried to the $F_2$ generation:

$P_1$: bearded female × beardless male

↓

$F_1$: all bearded males and beardless females

$F_1 \times F_1 \rightarrow$
- 1/8 beardless males
- 3/8 bearded males
- 3/8 beardless females
- 1/8 bearded females

Offer an explanation for the inheritance and expression of this trait, diagramming the cross. Propose one or more crosses to test your hypothesis.

34. Predict the $F_1$ and $F_2$ results of crossing a male fowl that is cock feathered with a true-breeding hen-feathered female fowl. Recall that these traits are sex limited.
35. Two mothers give birth to sons at the same time at a busy urban hospital. The son of mother 1 is afflicted with hemophilia, a disease caused by an X-linked recessive allele. Neither parent has the disease. Mother 2 has a normal son, despite the fact that the father has hemophilia. Several years later, couple 1 sues the hospital, claiming that these two newborns were swapped in the nursery following their birth. As a genetic counselor, you are called to testify. What information can you provide the jury concerning the allegation?
36. Discuss the topic of phenotypic expression and the many factors that impinge on it.
37. Contrast penetrance and expressivity as the terms relate to phenotypic expression.
38. Contrast the phenomena of genetic anticipation and genomic imprinting.

**HOW DO WE KNOW?**

39. In this chapter, we focused on many extensions and modifications of Mendelian principles and ratios. In the process, we found many opportunities to consider how this information was acquired. From the explanations given in the chapter, what answers would you propose to the following questions:
    (a) How were early geneticists able to ascertain inheritance patterns that did not fit typical Mendelian ratios?
    (b) How did geneticists determine that inheritance of some phenotypic characteristics involves the interactions of two or more gene pairs? How were they able to determine how many gene pairs were involved?
    (c) How do we know that specific genes are located on the sex-determining chromosomes rather than on autosomes?
    (d) For genes whose expression seems to be tied to the sex of individuals, how do we know whether a gene is X-linked in contrast to exhibiting sex-limited or sex-influenced inheritance?

## Extra-Spicy Problems

40. Labrador retrievers may be black, brown (chocolate), or golden (yellow) in color. While each color may breed true, many different outcomes are seen when numerous litters are examined from a variety of matings where the parents are not necessarily true breeding. Shown below are just some of the many possibilities.

| | | | | | |
|---|---|---|---|---|---|
| (a) | black | × | brown | → | all black |
| (b) | black | × | brown | → | 1/2 black<br>1/2 brown |
| (c) | black | × | brown | → | 3/4 black<br>1/4 golden |
| (d) | black | × | golden | → | all black |
| (e) | black | × | golden | → | 4/8 golden<br>3/8 black<br>1/8 brown |
| (f) | black | × | golden | → | 2/4 golden<br>1/4 black<br>1/4 brown |
| (g) | brown | × | brown | → | 3/4 brown<br>1/4 golden |
| (h) | black | × | black | → | 9/16 black<br>4/16 golden<br>3/16 brown |

Propose a mode of inheritance that is consistent with these data, and indicate the corresponding genotypes of the parents in each mating. Indicate as well the genotypes of dogs that breed true for each color.

41. A true-breeding purple-leafed plant isolated from one side of El Yunque, the rain forest in Puerto Rico, was crossed to a true-breeding white variety found on the other side. The $F_1$ offspring were all purple. A large number of $F_1 \times F_1$ crosses produced the following results:

purple: 4219 white: 5781 (Total = 10,000)

Propose an explanation for the inheritance of leaf color. As a geneticist, how might you go about testing your hypothesis? Describe the genetic experiments that you would conduct.

42. In Dexter and Kerry cattle, animals may be polled (hornless) or horned. The Dexter animals have short legs, whereas the Kerry animals have long legs. When many offspring were obtained from matings between polled Kerrys and horned Dexters, half were found to be polled Dexters and half polled Kerrys. When these two types of $F_1$ cattle were mated to one another, the following $F_2$ data were obtained:

3/8 polled Dexters
3/8 polled Kerrys
1/8 horned Dexters
1/8 horned Kerrys

Kerry cow

Dexter bull

A geneticist was puzzled by these data and interviewed farmers who had bred these cattle for decades. She learned that Kerrys were true breeding. Dexters, on the other hand, were not true breeding and never produced as many offspring as Kerrys. Provide a genetic explanation for these observations.

43. A geneticist from an alien planet that prohibits genetic research brought with him to Earth two pure-breeding lines of frogs. One line croaks by *uttering* "rib-it rib-it" and has purple eyes. The other line croaks more softly by *muttering* "knee-deep knee-deep" and has green eyes. With a newfound freedom of inquiry, the geneticist mated the two types of frogs, producing $F_1$ frogs that were all utterers and had blue eyes. A large $F_2$ generation then yielded the following ratios:

| | |
|---|---|
| 27/64 | blue-eyed, rib-it utterer |
| 12/64 | green-eyed, rib-it utterer |
| 9/64 | blue-eyed, knee-deep mutterer |
| 9/64 | purple-eyed, rib-it utterer |
| 4/64 | green-eyed, knee-deep mutterer |
| 3/64 | purple-eyed, knee-deep mutterer |

(a) How many total gene pairs are involved in the inheritance of both traits? Support your answer.
(b) Of these, how many are controlling eye color? How can you tell? How many are controlling croaking?
(c) Assign gene symbols for all phenotypes and indicate the genotypes of the $P_1$ and $F_1$ frogs.
(d) Indicate the genotypes of the six $F_2$ phenotypes.
(e) After years of experiments, the geneticist isolated pure-breeding strains of all six $F_2$ phenotypes. Indicate the $F_1$ and $F_2$ phenotypic ratios of the following cross using these pure-breeding strains:
blue-eyed, "knee-deep" mutterer × purple-eyed, "rib-it" utterer
(f) One set of crosses with his true-breeding lines initially caused the geneticist some confusion. When he crossed true-breeding purple-eyed, "knee-deep" mutterers with true-breeding green-eyed, "knee-deep" mutterers, he often got different results. In some matings, all offspring were blue-eyed, "knee-deep" mutterers, but in other matings all offspring were purple-eyed, "knee-deep" mutterers. In still a third mating, 1/2 blue-eyed, "knee-deep" mutterers and 1/2 purple-eyed, "knee-deep" mutterers were observed. Explain why the results differed.
(g) In another experiment, the geneticist crossed two purple-eyed, "rib-it" *utterers* together with the results shown here:

| | |
|---|---|
| 9/16 | purple-eyed, "rib-it" utterer |
| 3/16 | purple-eyed, "knee-deep" mutterer |
| 3/16 | green-eyed, "rib-it" utterer |
| 1/16 | green-eyed, "knee-deep" mutterer |

What were the genotypes of the two parents?

44. The following pedigree is characteristic of an inherited condition known as male precocious puberty, where affected males show signs of puberty by age 4. Propose a genetic explanation of this phenotype.

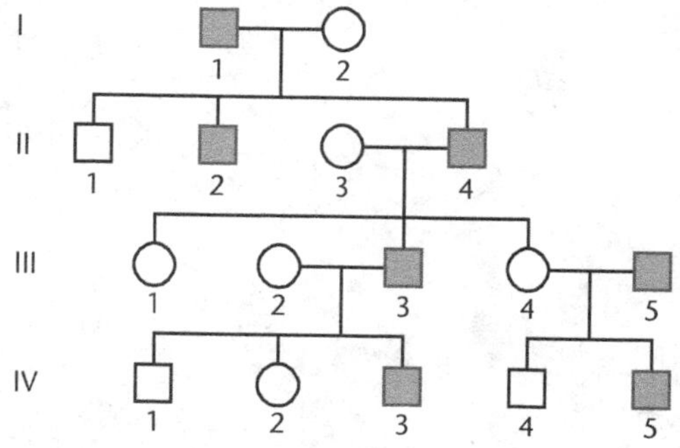


45. In birds, the male contains two identical sex chromosomes designated ZZ. The female contains one Z as well as a nearly blank chromosome, designated W (females are ZW). In budgerigars, two genes control feather color. The presence of the dominant *Y* allele at the first gene locus results in the production of a yellow pigment. The dominant *B* allele at the second controls melanin production. When both genes are ac-

tive, a green pigment results. If only the *Y* gene is active, the feathers are yellow. If only the *B* gene is active, a blue color is exhibited. If neither gene is active, the birds are albinos. Therefore, with our conventional designations, phenotypes are produced as follows:

| | |
|---|---|
| *Y–B–* | green |
| *Y–bb* | yellow |
| *yyB–* | blue |
| *yybb* | albino |

(a) A series of crosses established that one of the genes is autosomal and one is Z-linked. Based on the results of the following cross shown here, where both parents are true breeding, determine which gene is Z-linked.

| | |
|---|---|
| $P_1$: | green male × albino female |
| $F_1$: | 1/2 green males: 1/2 green females |
| $F_2$: | 6/16 green males |
| | 2/16 yellow males |
| | 3/16 green females |
| | 1/16 yellow females |
| | 3/16 blue females |
| | 1/16 albino females |

Support your answer by establishing the genotypes of the $P_1$ parents and working the cross through the $F_2$ generation.

(b) In a cross where the parental genotypes and whether the parents were true breeding were unknown, the offspring from repeated matings were recorded, as shown here:

| | |
|---|---|
| 13 | green males |
| 3 | yellow males |
| 11 | blue females |
| 5 | albino females |

Based on the results, determine the phenotypes and genotypes of the parents.

46. Students taking a genetics exam were expected to answer the following question by converting data to a "meaningful ratio" and then solving the problem. The instructor assumed that the final ratio would reflect two gene pairs, and most correct answers did. Here is the exam question: "Flowers may be white, orange, or brown. When plants with white flowers are crossed with plants with brown flowers, all the $F_1$ flowers are white. For $F_2$ flowers, the following data were obtained:

| | |
|---|---|
| 48 | white |
| 12 | orange |
| 4 | brown |

Convert the $F_2$ data to a meaningful ratio that allows you to explain the inheritance of color. Determine the number of genes involved and the genotypes that yield each phenotype."

(a) Solve the problem for two gene pairs. What is the final $F_2$ ratio?

(b) A number of students failed to reduce the ratio for two gene pairs as described above and solved the problem using three gene pairs. When examined carefully, their solution was deemed a valid response by the instructor. Solve the problem using three gene pairs.

(c) We now have a dilemma. The data are consistent with two alternative mechanisms of inheritance. Propose an experiment that executes crosses involving the original parents that would distinguish between the two solutions proposed by the students. Explain how this experiment would resolve the dilemma.

47. In four o'clock plants, many flower colors are observed. In a cross involving two true-breeding strains, one crimson and the other white, all of the $F_1$ generation were rose color. In the $F_2$, four new phenotypes appeared along with the $P_1$ and $F_1$ parental colors. The following ratio was obtained:

| | |
|---|---|
| 1/16 crimson | 4/16 rose |
| 2/16 orange | 2/16 pale yellow |
| 1/16 yellow | 4/16 white |
| 2/16 magenta | |

Propose an explanation for the inheritance of these flower colors.

48. Proto-oncogenes stimulate cells to progress through the cell cycle and begin mitosis. In cells that stop dividing, transcription of proto-oncogenes is inhibited by regulatory molecules. As is typical of all genes, proto-oncogenes contain a regulatory DNA region followed by a coding DNA region that specifies the amino acid sequence of the gene product. Consider two types of mutation in a proto-oncogene, one in the regulatory region that eliminates transcriptional control and the other in the coding region that renders the gene product inactive. Characterize both of these mutant alleles as either gain-of-function or loss-of-function mutations and indicate whether each would be dominant or recessive.

49. In the human disorder sickle-cell anemia, many phenotypic traits are evident besides the sickling and clumping of red blood cells under low oxygen tension, which reduces the cells' half-life, thus causing anemia. The overall phenotype includes episodes of severe abdominal pain; weakness and fatigue; lengthened long bones; heart enlargement; kidney failure; and respiratory difficulties, including pneumonia. (a) What term describes cases of multiple phenotypic manifestations of a single gene? (b) Relate each of these phenotypic responses to either the sickling phenomenon or the anemia.

Chiasmata present between synapsed homologs during the first meiotic prophase.

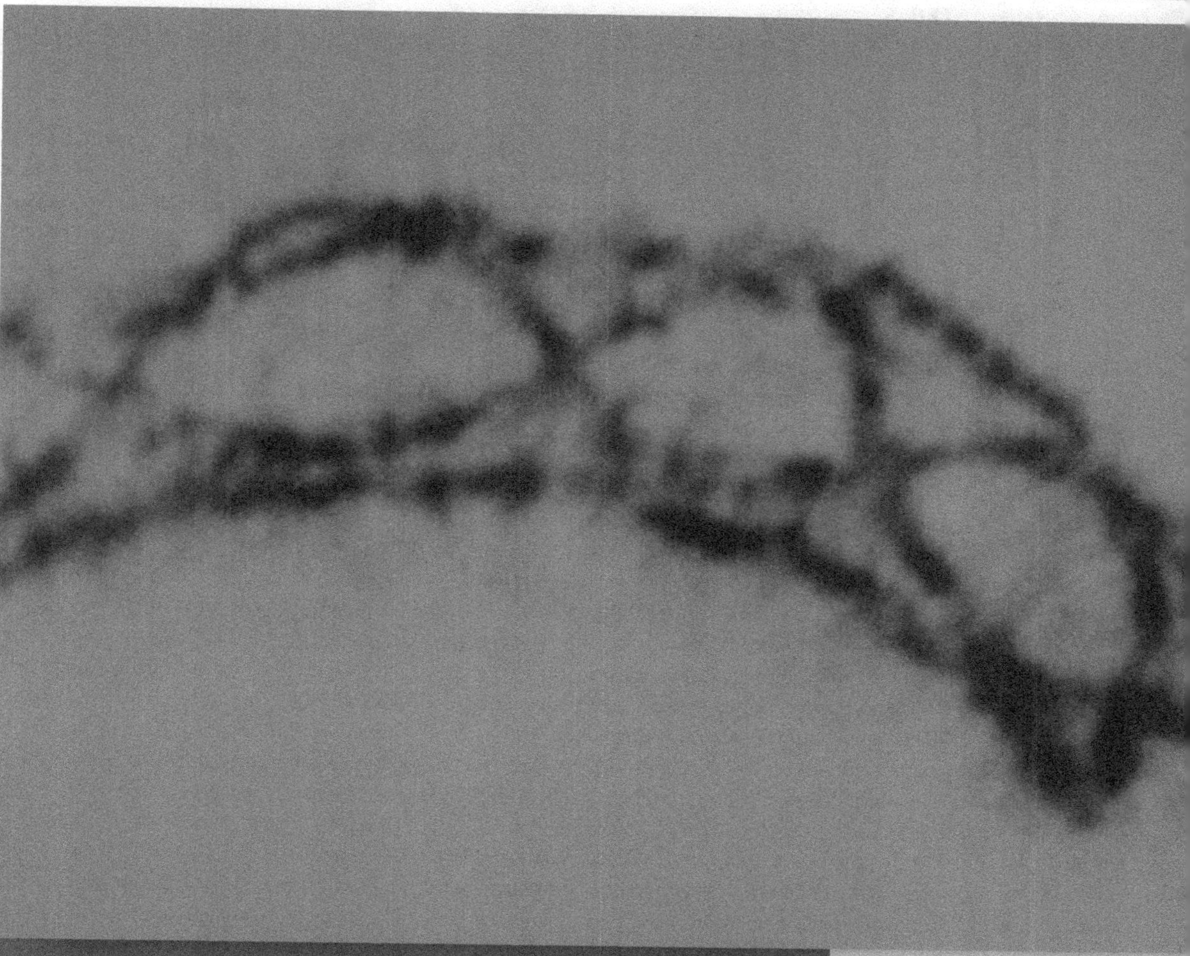

5

# Chromosome Mapping in Eukaryotes

## CHAPTER CONCEPTS

- Chromosomes in eukaryotes contain large numbers of genes, whose locations are fixed along the length of the chromosomes.
- Unless separated by crossing over, alleles on the same chromosome segregate as a unit during gamete formation.
- Crossing over between homologs during meiosis creates recombinant gametes with different combinations of alleles that enhance genetic variation.
- Crossing over between homologs serves as the basis for the construction of chromosome maps. The greater the distance between two genes on a chromosome, the higher the frequency of crossing over between them.
- Recombination also occurs between mitotic chromosomes and between sister chromatids.
- Linkage analysis and mapping can be performed for haploid organisms as well as diploid organisms.

Walter Sutton, along with Theodor Boveri, was instrumental in uniting the fields of cytology and genetics. As early as 1903, Sutton pointed out the likelihood that there must be many more "unit factors" than chromosomes in most organisms. Soon thereafter, genetic studies with several organisms revealed that certain genes segregate as if they were somehow joined or linked together. Further investigations showed that such genes are part of the same chromosome, and they may indeed be transmitted as a single unit. We now know that most chromosomes contain a very large number of genes. Those that are part of the same chromosome are said to be *linked* and to demonstrate **linkage** in genetic crosses.

Because the chromosome, not the gene, is the unit of transmission during meiosis, linked genes are not free to undergo independent assortment. Instead, the alleles at all loci of one chromosome should, in theory, be transmitted as a unit during gamete formation. However, in many instances this does not occur. As we saw in Chapter 2, during the first meiotic prophase, when homologs are paired, or synapsed, a reciprocal exchange of chromosome segments may take place. This **crossing over** results in the reshuffling, or **recombination**, of the alleles between homologs and always occurs during the tetrad stage.

Crossing over is currently viewed as an actual physical breaking and rejoining process that occurs during meiosis. You can see an example in the micrograph that opens this chapter. The exchange of chromosome segments provides an enormous potential for genetic variation in the gametes formed by any individual. This type of variation, in combination with that resulting from independent assortment, ensures that all offspring will contain a diverse mixture of maternal and paternal alleles.

The frequency of crossing over between any two loci on a single chromosome is proportional to the distance between them, known as the **interlocus distance.** Thus, depending on which loci are being considered, the percentage of recombinant gametes varies. This correlation allows us to construct **chromosome maps**, which indicate the relative locations of genes on the chromosomes.

In this chapter, we will discuss linkage, crossing over, and chromosome mapping in more detail. We will also consider a variety of other topics involving the exchange of genetic information, concluding the chapter with the rather intriguing question of why Mendel, who studied seven genes in an organism with seven chromosomes, did not encounter linkage. Or did he?

## 5.1 Genes Linked on the Same Chromosome Segregate Together

A simplified overview of the major theme of this chapter is given in Figure 5–1, which contrasts the meiotic consequences of (a) independent assortment, (b) linkage *without* crossing over, and (c) linkage *with* crossing over. In Figure 5–1(a) we see the results of independent assortment of two pairs of chromosomes, each containing one heterozygous gene pair. No linkage is exhibited. When these same two chromosomes are observed in a large number of meiotic events, they are seen to form four genetically different gametes in equal proportions, each containing a different combination of alleles of the two genes.

Now let's compare these results with what occurs if the same genes are linked on the same chromosome. If no crossing over occurs between the two genes [Figure 5–1(b)], only two genetically different kinds of gametes are formed. Each gamete receives the alleles present on one homolog or the other, which is transmitted intact as the result of segregation. This case illustrates *complete linkage*, which produces only **parental,** or **noncrossover, gametes.** The two parental gametes are formed in equal proportions. Though complete linkage between two genes seldom occurs, it is useful to consider the theoretical consequences of this concept.

Figure 5–1(c) shows the results of crossing over between two linked genes. As you can see, this crossover involves only two nonsister chromatids of the four chromatids present in the tetrad. This exchange generates two new allele combinations, called **recombinant,** or **crossover, gametes.** The two chromatids not involved in the exchange result in noncrossover gametes, like those in Figure 5–1(b). The frequency with which crossing over occurs between any two linked genes is generally proportional to the distance separating the respective loci along the chromosome. In theory, two randomly selected genes can be so close to each other that crossover events are too infrequent to be easily detected. As shown in Figure 5–1(b), this complete linkage produces only parental gametes. On the other hand, if a small, but distinct, distance separates two genes, few recombinant and many parental gametes will be formed. As the distance between the two genes increases, the proportion of recombinant gametes increases and that of the parental gametes decreases.

As we will discuss again later in this chapter, when the loci of two linked genes are far apart, the number of recombinant gametes approaches, but does not exceed, 50 percent. If 50 percent recombinants occur, the result is a 1:1:1:1 ratio of the four types (two parental and two recombinant gametes). In this case, transmission of two linked genes is indistinguishable from that of two unlinked, independently assorting genes. That is, the proportion of the four possible genotypes would be identical, as shown in Figure 5–1(a) and 5–1(c).

NOW SOLVE THIS

Problem 9 on page 137 asks you to contrast the results of a testcross when two genes are unlinked versus linked, and when they are linked, if they are very far apart or relatively close together.

HINT: *The results are indistinguishable when two genes are unlinked compared to the case where they are linked but so far apart that crossing over always intervenes between them during meiosis.*

WEB TUTORIAL 5.1

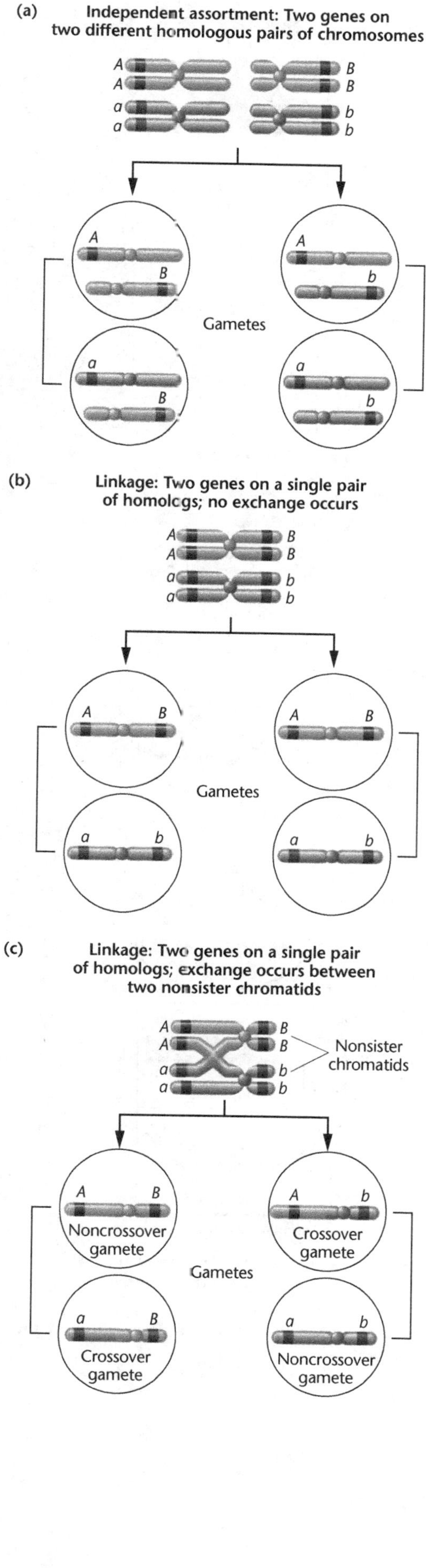


## The Linkage Ratio

If complete linkage exists between two genes because of their close proximity, and organisms heterozygous at both loci are mated, a unique $F_2$ phenotypic ratio results, which we designate the **linkage ratio**. To illustrate this ratio, let's consider a cross involving the closely linked, recessive, mutant genes *heavy* wing vein (*hv*) and *brown* eye (*bw*) in *Drosophila melanogaster* (Figure 5–2). The normal, wild-type alleles $hv^+$ and $bw^+$ are both dominant and result in thin wing veins and red eyes, respectively.

In this cross, flies with normal thin wing veins and mutant brown eyes are mated to flies with mutant heavy wing veins and normal red eyes. In more concise terms, heavy-veined flies are crossed with brown-eyed flies. Linked genes are represented by placing their allele designations (the genetic symbols established in Chapter 4) above and below a single or double horizontal line. Those above the line are located at loci on one homolog, and those below the line are located at the homologous loci on the other homolog. Thus, we represent the $P_1$ generation as follows:

$$P_1: \frac{hv^+\ bw}{hv^+\ bw} \times \frac{hv\ bw^+}{hv\ bw^+}$$

thin, brown    heavy, red

Because these genes are located on an autosome, no designation of male or female is necessary.

In the $F_1$ generation, each fly receives one chromosome of each pair from each parent. All flies are heterozygous for both gene pairs and exhibit the dominant traits of thin veins and red eyes:

$$F_1: \frac{hv^+\ bw}{hv\ bw^+}$$

thin, red

As shown in Figure 5–2(a), when the $F_1$ generation is interbred, each $F_1$ individual forms only parental gametes because of complete linkage. Following fertilization, the $F_2$ generation is produced in a 1:2:1 phenotypic and genotypic ratio. One-fourth of this generation shows thin wing veins and brown eyes; one-half shows both wild-type traits, namely, thin veins and red eyes; and one-fourth will show heavy wing veins and red eyes. Therefore, the ratio is 1 heavy: 2 wild: 1 brown. Such a 1:2:1 ratio is characteristic of complete linkage. Complete linkage is usually observed only when genes are very close together and the number of progeny is relatively small.

Figure 5–2(b) demonstrates the results of a testcross with the $F_1$ flies. Such a cross produces a 1:1 ratio of thin, brown and heavy, red flies. Had the genes controlling these traits been incompletely linked or located on separate autosomes, the testcross would have produced four phenotypes rather than two.

**FIGURE 5–1** Results of gamete formation when two heterozygous genes are (a) on two different pairs of chromosomes; (b) on the same pair of homologs, but with no exchange occurring between them; and (c) on the same pair of homologs, but with an exchange occurring between two nonsister chromatids. Note that in this and the following figures that members of homologous pairs of chromosomes are shown in two different colors. This convention was established in Chapter 2 (see, for example, Figure 2–7 and Figure 2–10).

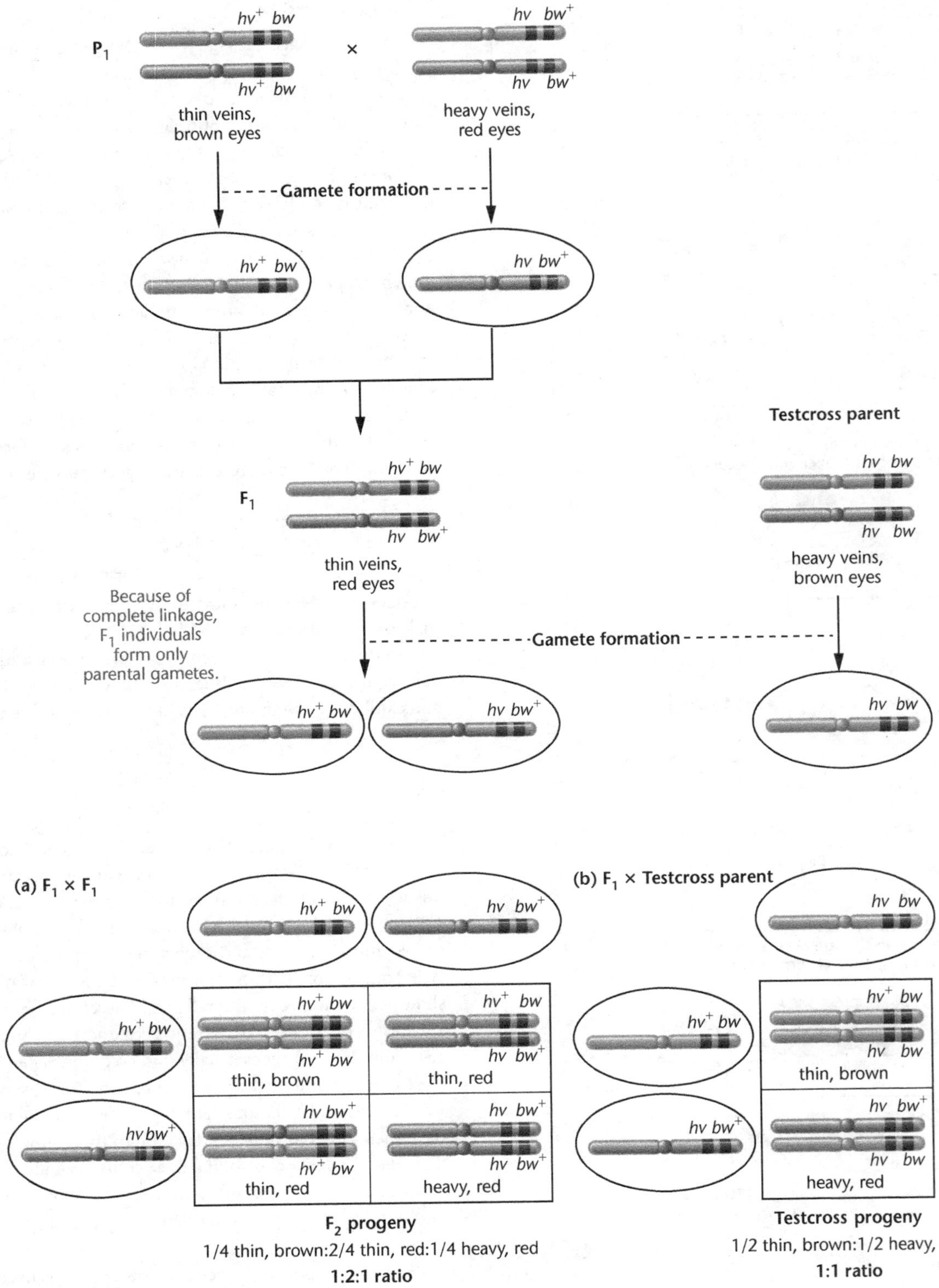


FIGURE 5-2 Results of a cross involving two genes located on the same chromosome and demonstrating complete linkage. (a) The $F_2$ results of the cross. (b) The results of a testcross involving the $F_1$ progeny.

LINKED GENES AND MAPPING WEB TUTORIAL 5.2

When large numbers of mutant genes in any given species are investigated, genes located on the same chromosome show evidence of linkage to one another. As a result, **linkage groups** can be identified, one for each chromosome. In theory, the number of linkage groups should correspond to the haploid number of chromosomes. In diploid organisms in which large numbers of mutant genes are available for genetic study, this correlation has been confirmed.

5.2

## Crossing Over Serves as the Basis for Determining the Distance between Genes in Chromosome Mapping

It is highly improbable that two randomly selected genes linked on the same chromosome will be so close to one another along the chromosome that they demonstrate complete linkage. Instead, crosses involving two such genes will almost always produce a percentage of offspring resulting from recombinant gametes. The percentage will vary depending on the distance between the two genes along the chromosome. This phenomenon was first explained in 1911 by two *Drosophila* geneticists, Thomas H. Morgan and his undergraduate student, Alfred H. Sturtevant.

### Morgan and Crossing Over

As you may recall from our discussion in Chapter 4, Morgan was the first to discover the phenomenon of X-linkage. In his studies, he investigated numerous *Drosophila* mutations located on the X chromosome. His original analysis, based on crosses involving only one gene on the X chromosome, led to the discovery of X-linked inheritance. However, when he made crosses involving two X-linked genes, his results were initially puzzling. For example, female flies expressing the mutant *yellow* body ($y$) and *white* eyes ($w$) alleles were crossed with wild-type males (gray body and red eyes). The $F_1$ females were wild type, while the $F_1$ males expressed both mutant traits. In the $F_2$ the vast majority of the total offspring showed the expected parental phenotypes—yellow-bodied, white-eyed flies and wild-type flies (gray-bodied, red-eyed). The remaining flies, less than 1.0 percent, were either yellow-bodied with red eyes or gray-bodied with white eyes. It was as if the two mutant alleles had somehow separated from each other on the homolog during gamete formation in the $F_1$ female flies. This cross is illustrated in cross A of Figure 5–3, using data later compiled by Sturtevant.

When Morgan studied other X-linked genes, the same basic pattern was observed, but the proportion of $F_2$ phenotypes differed. For example, when he crossed *white*-eye, *miniature*-wing mutants with wild-type flies, only 65.5 percent of all the $F_2$ flies showed the parental phenotypes, while 34.5 percent of the offspring appeared as if the mutant genes had been separated during gamete formation. This is illustrated in cross B of Figure 5–3, again using data subsequently compiled by Sturtevant.

Morgan was faced with two questions: (1) What was the source of gene separation and (2) why did the frequency of the apparent separation vary depending on the genes being studied? The answer Morgan proposed for the first question was based on his knowledge of earlier cytological observations made by F. A. Janssens and others. Janssens had observed that synapsed homologous chromosomes in meiosis wrapped around each other, creating **chiasmata** (sing. *chiasma*), X-shaped intersections where points of overlap are evident (see the photo on p. 105). Morgan proposed that these chiasmata could represent points of genetic exchange.

Regarding the crosses shown in Figure 5–3, Morgan postulated that if an exchange of chromosome material occurs during gamete formation, at a chiasma between the mutant genes on the two X chromosomes of the $F_1$ females, the unique phenotypes will occur. He suggested that such exchanges led to recombinant gametes in both the *yellow–white* cross and the *white–miniature* cross, as compared to the parental gametes that underwent no exchange. On the basis of this and other experimentation, Morgan concluded that linked genes are arranged in a linear sequence along the chromosome and that a variable frequency of exchange occurs between any two genes during gamete formation.

In answer to the second question, Morgan proposed that two genes located relatively close to each other along a chromosome are less likely to have a chiasma form between them than if the two genes are farther apart on the chromosome. Therefore, the closer two genes are, the less likely that a genetic exchange will occur between them. Morgan was the first to propose the term *crossing over* to describe the physical exchange leading to recombination.

### Sturtevant and Mapping

Morgan's student, Alfred H. Sturtevant, was the first to realize that his mentor's proposal could be used to map the sequence of linked genes. According to Sturtevant,

> "In a conversation with Morgan . . . I suddenly realized that the variations in strength of linkage, already attributed by Morgan to differences in the spatial separation of the genes, offered the possibility of determining sequences in the linear dimension of a chromosome. I went home and spent most of the night (to the neglect of my undergraduate homework) in producing the first chromosomal map."

Sturtevant, in a paper published in 1913, compiled data from numerous crosses made by Morgan and other geneticists involving recombination between the genes represented by the *yellow, white*, and *miniature* mutants. A subset of these data are shown in Figure 5–3. The frequencies of recombination between each pair of these three genes are as follows:

| | |
|---|---|
| (1) *yellow, white* | 0.5% |
| (2) *white, miniature* | 34.5% |
| (3) *yellow, miniature* | 35.4% |

Because the sum of (1) and (2) approximately equals (3), Sturtevant suggested that the recombination frequencies between linked genes are additive. On this basis, he predicted that the order of the genes on the X chromosome is *yellow–white–miniature*. In arriving at this

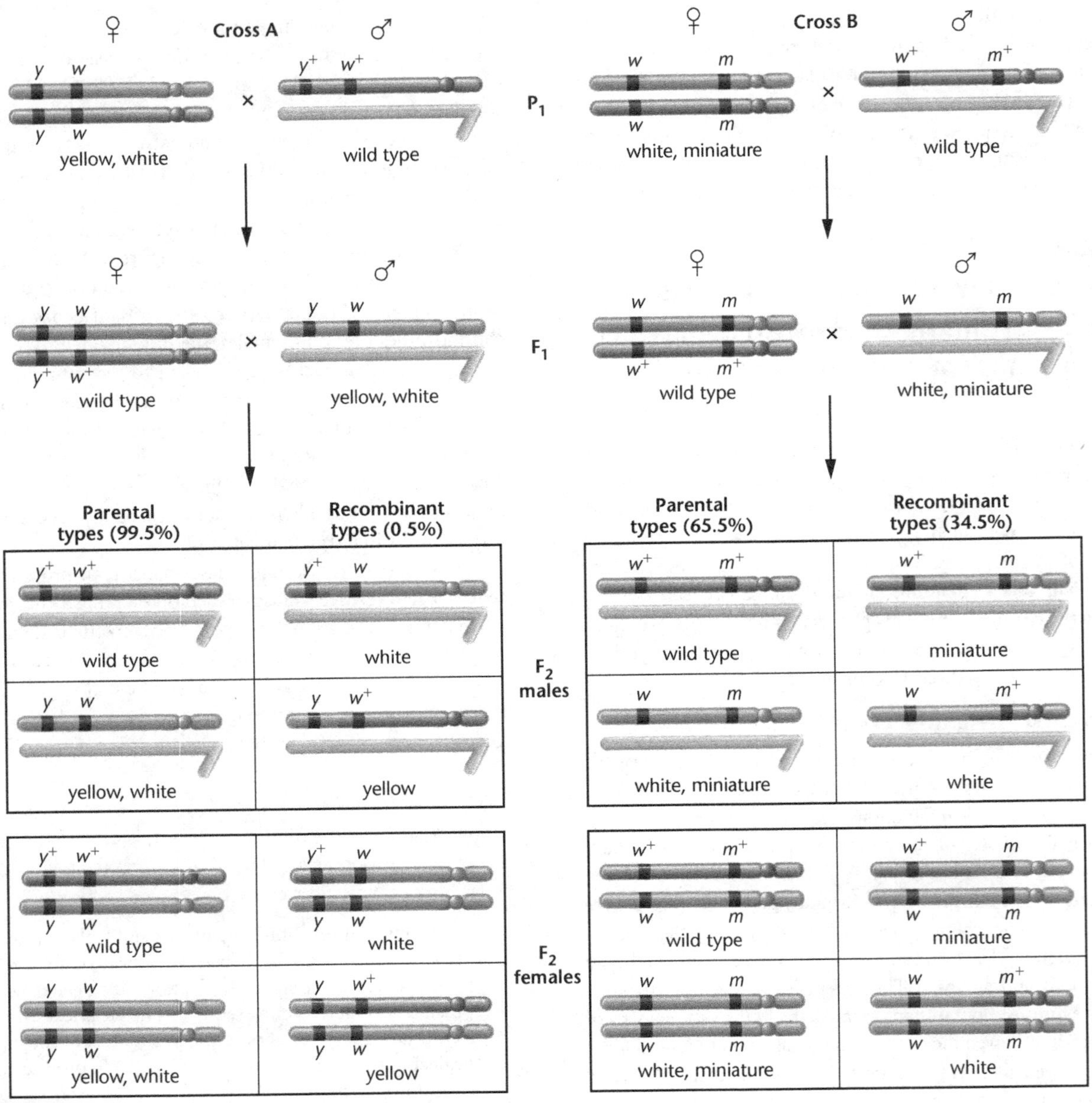


FIGURE 5-3 The $F_1$ and $F_2$ results of crosses involving the *yellow* (*y*), *white* (*w*) mutations (cross A), and the *white*, *miniature* (*m*) mutations (cross B), as compiled by Sturtevant. In cross A, 0.5 percent of the $F_2$ flies (males and females) demonstrate recombinant phenotypes, which express either *white* or *yellow*. In cross B, 34.5 percent of the $F_2$ flies (males and females) demonstrate recombinant phenotypes, which are either *miniature* or *white* mutants.

conclusion, he reasoned as follows: The *yellow* and *white* genes are apparently close to each other because the recombination frequency is low. However, both of these genes are quite far from the *miniature* gene, because the *white–miniature* and *yellow–miniature* combinations show larger recombination frequencies. Because *miniature* shows more recombination with *yellow* than with *white* (35.4 percent vs. 34.5 percent), it follows that *white* is located between the other two genes, not outside of them.

Sturtevant knew from Morgan's work that the frequency of exchange could be used as an estimate of the distance between two genes or loci along the chromosome. He constructed a **chromosome map** of the three genes on the X chromosome, setting one map unit (mu) equal to 1 percent recombination between two genes.* The distance between *yellow* and *white* is thus 0.5 mu, and the distance between *yellow* and *miniature* is 35.4 mu. It follows that the distance between *white* and *miniature* should be 35.4 − 0.5 = 34.9 mu. This estimate is close to the actual frequency of recombination be-

*In honor of Morgan's work, map units are often referred to as centiMorgans (cM).

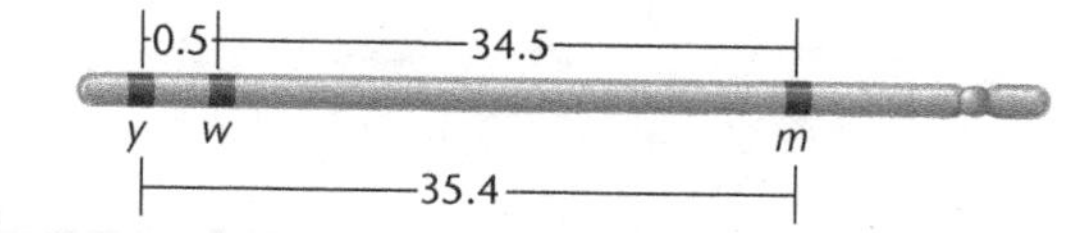


FIGURE 5-4 A map of the *yellow* (*y*), *white* (*w*), and *miniature* (*m*) genes on the X chromosome of *Drosophila melanogaster*. Each number represents the percentage of recombinant offspring produced in one of three crosses, each involving two different genes.

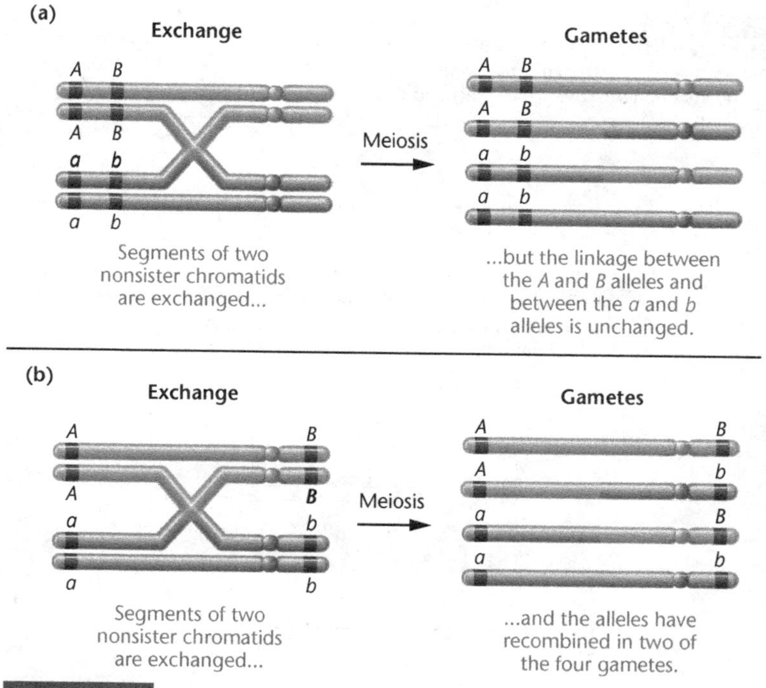


FIGURE 5-5 Two examples of a single crossover between two nonsister chromatids and the gametes subsequently produced. In (a) the exchange does not alter the linkage arrangement between the alleles of the two genes, only parental gametes are formed, and the exchange goes undetected. In (b) the exchange separates the alleles, resulting in recombinant gametes, which are detectable.

tween *white* and *miniature* (34.5 mu). The map for these three genes is shown in Figure 5–4. The fact that these numbers do not add up perfectly is due to normal variation that one would expect between crosses, leading to the minor imprecisions encountered in independently conducted mapping experiments.

In addition to these three genes, Sturtevant considered crosses involving two other genes on the X chromosome and produced a more extensive map that included all five genes. He and a colleague, Calvin Bridges, soon began a search for autosomal linkage in *Drosophila*. By 1923, they had clearly shown that linkage and crossing over are not restricted to X-linked genes but could also be demonstrated with autosomes. During this work, they made another interesting observation. In *Drosophila*, crossing over was shown to occur only in females. The fact that no crossing over occurs in males made genetic mapping much less complex to analyze in *Drosophila*. While crossing over does occur in both sexes in most other organisms, crossing over in males is often observed to occur less frequently than in females. For example, in humans, such recombination occurs only about 60 percent as often in males compared to females.

Although many refinements have been added to chromosome mapping since Sturtevant's initial work, his basic principles are accepted as correct. These principles are used to produce detailed chromosome maps of organisms for which large numbers of linked mutant genes are known. Sturtevant's findings are also historically significant to the broader field of genetics. In 1910, the **chromosomal theory of inheritance** was still widely disputed—even Morgan was skeptical of this theory before he conducted his experiments. Research has now firmly established that chromosomes contain genes in a linear order and that these genes are the equivalent of Mendel's unit factors.

## Single Crossovers

Why should the relative distance between two loci influence the amount of crossing over and recombination observed between them? During meiosis, a limited number of crossover events occur in each tetrad. These recombinant events occur randomly along the length of the tetrad. Therefore, the closer that two loci reside along the axis of the chromosome, the less likely that any **single crossover** event will occur between them. The same reasoning suggests that the farther apart two linked loci, the more likely a random crossover event will occur in between them.

In Figure 5–5(a), a single crossover occurs between two nonsister chromatids, but not in between the two loci being studied; therefore, the crossover is undetected because no recombinant gametes are produced for the two traits of interest. In Figure 5–5(b), where the two loci under study are quite far apart, the crossover does occur between them, yielding gametes in which the traits of interest are recombined.

When a single crossover occurs between two nonsister chromatids, the other two chromatids of the tetrad are not involved in the exchange and enter the gamete unchanged. Even if a single crossover occurs 100 percent of the time between two linked genes, recombination is subsequently observed in only 50 percent of the potential gametes formed. This concept is diagrammed in Figure 5–6. Theoretically, if we assume only single exchanges between a given pair of loci and observe 20 percent recombinant gametes, we will conclude that crossing over actually occurs between these two loci in 40 percent of the tetrads. The general rule is that, under these conditions, the percentage of tetrads involved in an exchange between two genes is twice as great as the percentage of recombinant gametes produced. Therefore, the theoretical limit of observed recombination due to crossing over is 50 percent.

When two linked genes are more than 50 map units apart, a crossover can theoretically be expected to occur between them in 100 percent of the tetrads. If this prediction were achieved, each tetrad would yield equal proportions of the four gametes shown in Figure 5–6, just as if the genes were on different chromosomes and assorting independently. For a variety of reasons, this theoretical limit is seldom achieved.

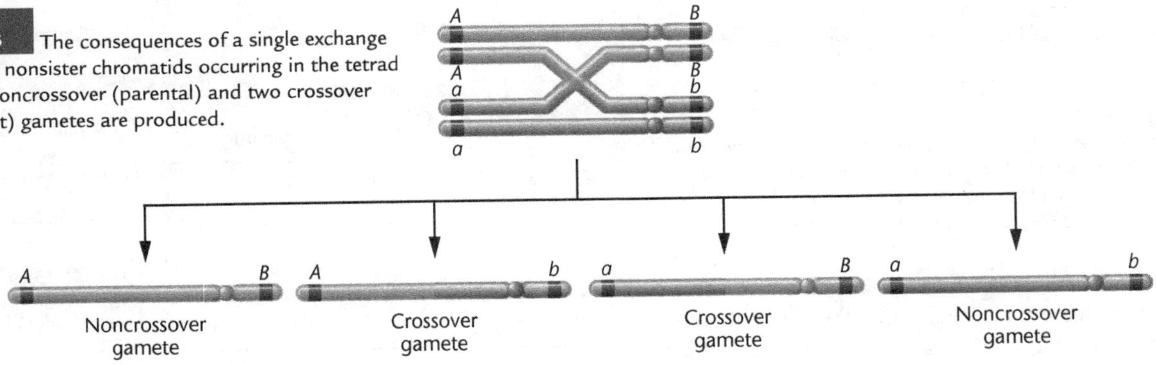


FIGURE 5-6 The consequences of a single exchange between two nonsister chromatids occurring in the tetrad stage. Two noncrossover (parental) and two crossover (recombinant) gametes are produced.

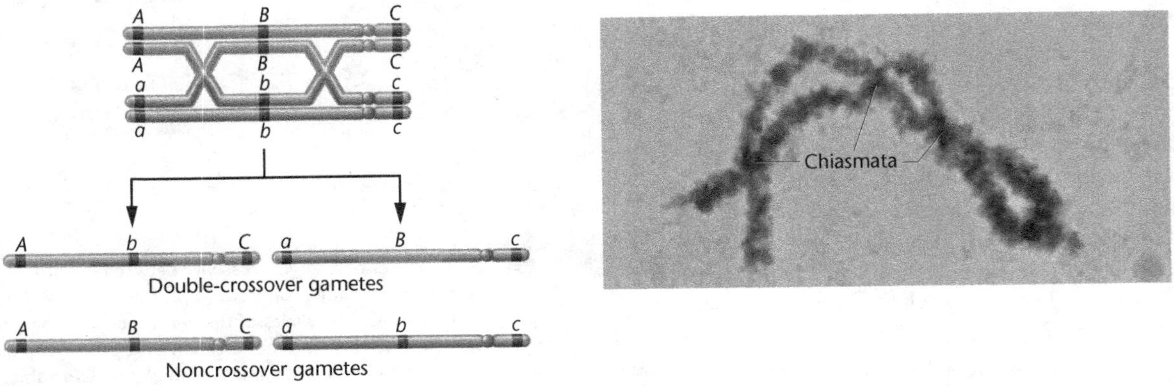


FIGURE 5-7 Consequences of a double exchange occurring between two nonsister chromatids. Because the exchanges involve only two chromatids, two noncrossover gametes and two double-crossover gametes are produced. The photograph illustrates several chiasmata found in a tetrad isolated during the first meiotic prophase stage. See also the Chapter Opening photograph on p. 105.

5.3

## Determining the Gene Sequence during Mapping Requires the Analysis of Multiple Crossovers

The study of single crossovers between two linked genes provides a basis for determining the *distance* between them. However, when many linked genes are studied, their *sequence* along the chromosome is more difficult to determine. Fortunately, the discovery that multiple crossovers occur between the chromatids of a tetrad has facilitated the process of producing more extensive chromosome maps. As we shall see next, when three or more linked genes are investigated simultaneously, it is possible to determine first the sequence of and then the distances between genes.

### Multiple Exchanges

It is possible that in a single tetrad, two, three, or more exchanges will occur between nonsister chromatids as a result of several crossing over events. Double exchanges of genetic material result from **double crossovers (DCOs)**, as shown in Figure 5–7. To study a double exchange, three gene pairs must be investigated, each heterozygous for two alleles. Before we determine the frequency of recombination among all three loci, let's review some simple probability calculations.

As we have seen, the probability of a single exchange occurring in between the *A* and *B* or the *B* and *C* genes is related directly to the distance between the respective loci. The closer *A* is to *B* and *B* is to *C*, the less likely it is that a single exchange will occur in between either of the two sets of loci. In the case of a double crossover, two separate and independent events or exchanges must occur simultaneously. The mathematical probability of two independent events occurring simultaneously is equal to the product of the individual probabilities. This is the *product law* introduced in Chapter 3.

Suppose that crossover gametes resulting from single exchanges are recovered 20 percent of the time ($p = 0.20$) between *A* and *B*, and 30 percent of the time ($p = 0.30$) between *B* and *C*. The probability of recovering a double-crossover gamete arising from two exchanges (between *A* and *B* and between *B* and *C*) is predicted to be $(0.20)(0.30) = 0.06$, or 6 percent. It is apparent from this calculation that the expected frequency of double-crossover gametes is always expected to be much lower than that of either single-crossover class of gametes.

If three genes are relatively close together along one chromosome, the expected frequency of double-crossover gametes is extremely low. For example, suppose that the *A–B* distance in Figure 5–7 is 3 mu and the *B–C* distance is 2 mu. The expected double-crossover frequency is (0.03) (0.02) = 0.0006, or 0.06 percent. This translates to only 6 events in 10,000. Thus in a mapping experiment where closely linked genes are involved, very large numbers of offspring are required to detect double-crossover events. In this example, it is unlikely that a double crossover will be observed even if 1000 offspring are examined. Thus, it is evident that if four or five genes are being mapped, even fewer triple and quadruple crossovers can be expected to occur.

**NOW SOLVE THIS**

Problem 14 on page 138 asks you to contrast the results of crossing over when the arrangement of alleles along the homologs differs in two organisms heterozygous for three genes.

HINT: *Homologs enter noncrossover gametes unchanged, and all crossover results must be derived from the noncrossover sequence of alleles.*

## Three-Point Mapping in *Drosophila*

The information presented in the previous section enables us to map three or more linked genes in a single cross. To illustrate the mapping process in its entirety, we examine two situations involving three linked genes in two quite different organisms.

To execute a successful mapping cross, three criteria must be met:

1. The genotype of the organism producing the crossover gametes must be heterozygous at all loci under consideration. If homozygosity occurred at any locus, all gametes produced would contain the same allele, precluding mapping analysis.
2. The cross must be constructed so that the genotypes of all gametes can be accurately determined by observing the phenotypes of the resulting offspring. This is necessary because the gametes and their genotypes can never be observed directly. To overcome this problem, each phenotypic class must reflect the genotype of the gametes of the parents producing it.
3. A sufficient number of offspring must be produced in the mapping experiment to recover a representative sample of all crossover classes.

These criteria are met in the three-point mapping cross of *Drosophila melanogaster* shown in Figure 5–8. In this cross three X-linked recessive mutant genes—*yellow* body color, *white* eye color, and *echinus* eye shape—are considered. To diagram the cross, *we must assume some theoretical sequence, even though we do not yet know if it is correct.* In Figure 5–8, we initially assume the sequence of the three genes to be *y–w–ec*. If this is incorrect, our analysis shall demonstrate it and reveal the correct sequence.

In the $P_1$ generation males hemizygous for all three wild-type alleles are crossed to females that are homozygous for all three recessive mutant alleles. Therefore, the $P_1$ males are wild type with respect to body color, eye color, and eye shape. They are said to have a *wild-type phenotype*. The females, on the other hand, exhibit the three mutant traits: yellow body color, white eyes, and echinus eye shape.

This cross produces an $F_1$ generation consisting of females that are heterozygous at all three loci and males that, because of the Y chromosome, are hemizygous for the three mutant alleles. Phenotypically, all $F_1$ females are wild type, while all $F_1$ males are yellow, white, and echinus. The genotype of the $F_1$ females fulfills the first criterion for constructing a map of the three linked genes; that is, it is heterozygous at the three loci and may serve as the source of recombinant gametes generated by crossing over. Note that, because of the genotypes of the $P_1$ parents, all three of the mutant alleles are on one homolog and all three wild-type alleles are on the other homolog. With other parents, *other arrangements would be possible that could produce a heterozygous genotype.* For example, a heterozygous female could have the *y* and *ec* mutant alleles on one homolog and the *w* allele on the other. This would occur if one of her parents was *yellow, echinus* and the other parent was *white.*

In our cross, the second criterion is met as a result of the gametes formed by the $F_1$ males. Every gamete contains either an X chromosome bearing the three mutant alleles or a Y chromosome, which does not contain any of the three loci being considered. Whichever type participates in fertilization, the genotype of the gamete produced by the $F_1$ female will be expressed phenotypically in the $F_2$ female and male offspring derived from it. As a result, all noncrossover and crossover gametes produced by the $F_1$ female parent can be determined by observing the $F_2$ phenotypes.

With these two criteria met, we can construct a chromosome map from the crosses illustrated in Figure 5–8. First, we must determine which $F_2$ phenotypes correspond to the various noncrossover and crossover categories. To determine the **noncrossover** $F_2$ phenotypes, we must identify individuals derived from the parental gametes formed by the $F_1$ female. Each such gamete contains *an X chromosome unaffected by crossing over.* As a result of segregation, approximately equal proportions of the two types of gametes, and subsequently their $F_2$ phenotypes, are produced. Because they derive from a heterozygote, the genotypes of the two parental gametes and the $F_2$ phenotypes complement one another. For example, if one is wild type, the other is mutant for all three genes. This is the case in the cross being considered. In other situations, if one chromosome shows one mutant allele, the second chromosome shows the other two mutant alleles, and so on. These are therefore called **reciprocal classes** of gametes and phenotypes.

The two noncrossover phenotypes are most easily recognized because *they occur in the greatest proportion of offspring.* Figure 5–8 shows that gametes (1) and (2) are present in the greatest numbers. Therefore, flies that are yellow, white, and echinus and those that are normal, or wild type, for all three characters constitute the noncrossover category and represent 94.44 percent of the $F_2$ offspring.

The second category that can be easily detected is represented by the double-crossover phenotypes. Because of their low probability of occurrence, *they must be present in the least numbers.* Remember that

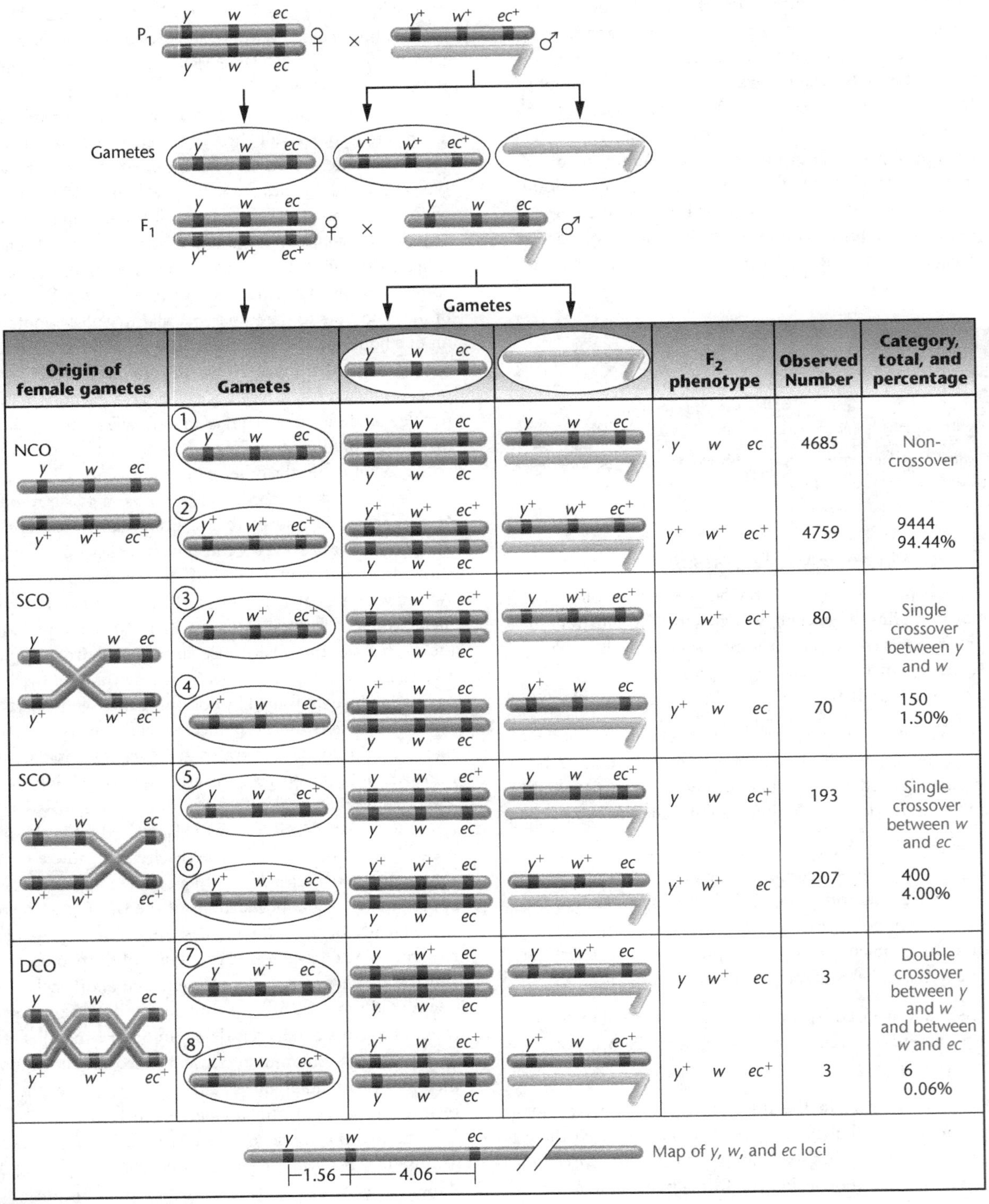


**FIGURE 5-8** A three-point mapping cross involving the *yellow* ($y$ or $y^+$), *white* ($w$ or $w^+$), and *echinus* ($ec$ or $ec^+$) genes in *Drosophila melanogaster*. NCO, SCO, and DCO refer to noncrossover, single-crossover, and double-crossover groups, respectively. Centromeres are not drawn on the chromosomes, and only two nonsister chromatids are initially shown in the left-hand column.

THREE-POINT MAPPING

WEB TUTORIAL 5.3

this group represents two independent but simultaneous single-crossover events. Two reciprocal phenotypes can be identified: gamete 7, which shows the mutant traits yellow and echinus, but normal eye color; and gamete 8, which shows the mutant trait white, but normal body color and eye shape. Together these double-crossover phenotypes constitute only 0.06 percent of the $F_2$ offspring.

The remaining four phenotypic classes fall into two categories resulting from single crossovers. Gametes 3 and 4, reciprocal phenotypes produced by single-crossover events occurring between the *yellow* and *white* loci, are equal to 1.50 percent of the $F_2$ offspring. Gametes 5 and 6, constituting 4.00 percent of the $F_2$ offspring, represent the reciprocal phenotypes resulting from single-crossover events occurring between the *white* and *echinus* loci.

We can now calculate the map distances between the three loci. The distance between *y* and *w*, or between *w* and *ec*, is equal to the percentage of all detectable exchanges occurring between them. For any two genes under consideration, this includes all related single crossovers as well as all double crossovers. *The latter are included because they represent two simultaneous single crossovers.* For the *y* and *w* genes, this includes gametes 3, 4, 7, and 8, totaling 1.50% + 0.06%, or 1.56 mu. Similarly, the distance between *w* and *ec* is equal to the percentage of offspring resulting from an exchange between these two loci: gametes 5, 6, 7, and 8, totaling 4.00% + 0.06%, or 4.06 mu. The map of these three loci on the X chromosome is shown at the bottom of Figure 5–8.

## Determining the Gene Sequence

In the preceding example, we assumed that the sequence (or order) of the three genes along the chromosome was *y–w–ec*. Our analysis established that the sequence is consistent with the data. However, in most mapping experiments, the gene sequence is not known, and this constitutes another variable in the analysis. In our example, had the gene order been unknown, we could have used one of two methods (which we will study next) to determine it. In your own work, you should select one of these methods and use it consistently.

**Method I** This method is based on the fact that there are only three possible arrangements, each containing a different one of the three genes between the other two:

(I) *w–y–ec* (*y* is in the middle)
(II) *y–ec–w* (*ec* is in the middle)
(III) *y–w–ec* (*w* is in the middle)

Use the following steps during your analysis to determine the gene order:

1. Assuming any of the three orders, first determine the *arrangement of alleles* along each homolog of the heterozygous parent giving rise to noncrossover and crossover gametes (the $F_1$ female in our example).

2. Determine whether a double-crossover event occurring within that arrangement will produce the *observed double-crossover phenotypes.* Remember that these phenotypes occur least frequently and are easily identified.

3. If this order does not produce the correct phenotypes, try each of the other two orders. One must work!

These steps are shown in Figure 5–9, using our *y–w–ec* cross. The three possible arrangements are labeled I, II, and III, as shown above.

1. Assuming that *y* is between *w* and *ec* (arrangement I), the distribution of alleles between the homologs of the $F_1$ heterozygote is:

$$\frac{w \quad y \quad ec}{w^+ \quad y^+ \quad ec^+}$$

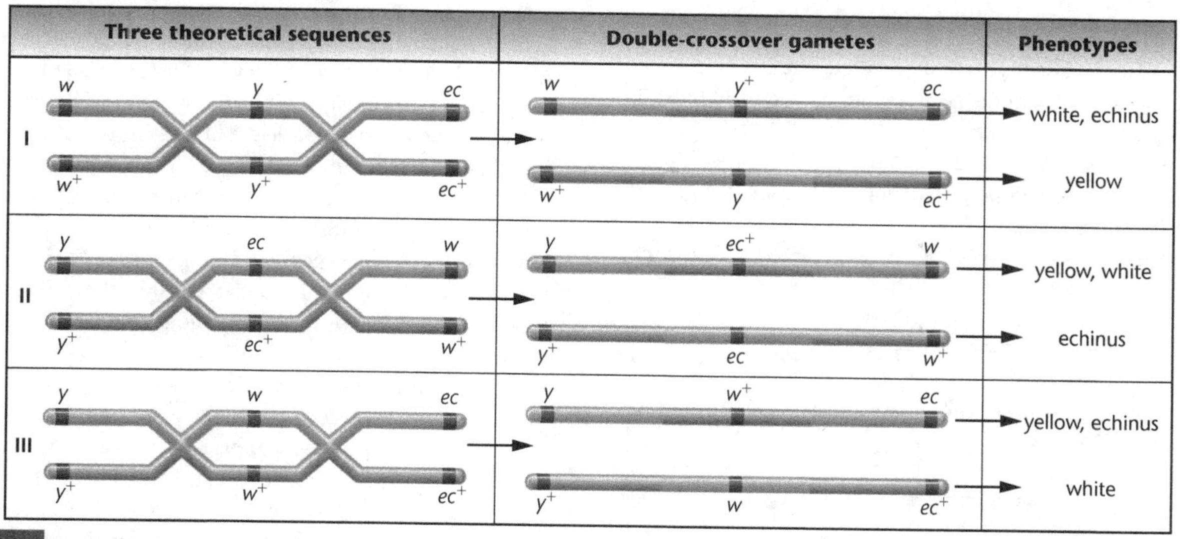


FIGURE 5-9 The three possible sequences of the *white*, *yellow*, and *echinus* genes, the results of a double crossover in each case, and the resulting phenotypes produced in a testcross. For simplicity, the two noncrossover chromatids of each tetrad are omitted.

We know this because of the way in which the $P_1$ generation was crossed: The $P_1$ female contributes an X chromosome bearing the *w*, *y*, and *ec* alleles, while the $P_1$ male contributed an X chromosome bearing the $w^+$, $y^+$, and $ec^+$ alleles.

2. A double crossover within that arrangement yields the following gametes:

$$\underline{w\ y^+\ ec} \text{ and } \underline{w^+\ y\ ec^+}$$

Following fertilization, if *y* is in the middle, the $F_2$ double-crossover phenotypes will correspond to these gametic genotypes, yielding offspring that express the white, echinus phenotype and offspring that express the yellow phenotype. Instead, determination of the actual double crossovers reveals them to be yellow, echinus flies and white flies. *Therefore, our assumed order is incorrect.*

3. If we consider arrangement II, with the $ec/ec^+$ alleles in the middle, or arrangement III, with the $w/w^+$ alleles in the middle:

$$\text{(II)}\ \frac{y\quad ec\quad w}{y^+\quad ec^+\quad w^+} \quad \text{or} \quad \text{(III)}\ \frac{y\quad w\quad ec}{y^+\quad w^+\quad ec^+}$$

we see that arrangement II again provides *predicted* double-crossover phenotypes that *do not* correspond to the *actual* (observed) double-crossover phenotypes. The predicted phenotypes are yellow, white flies and echinus flies in the $F_2$ generation. *Therefore, this order is also incorrect.* However, arrangement III produces the observed phenotypes—yellow, echinus flies and white flies. *Therefore, this arrangement, with the w gene in the middle, is correct.*

To summarize Method I: First, determine the arrangement of alleles on the homologs of the heterozygote yielding the crossover gametes by identifying the reciprocal noncrossover phenotypes. Then, test each of the three possible orders to determine which one yields the observed double-crossover phenotypes—*the one that does so represents the correct order.* This method is summarized in Figure 5–9.

**Method II** Method II also begins by determining the arrangement of alleles along each homolog of the heterozygous parent. In addition, it requires one further assumption:

*Following a double-crossover event, the allele in the middle position will fall between the outside, or flanking, alleles that were present on the opposite parental homolog.*

To illustrate, assume order I, *w–y–ec*, in the following arrangement:

$$\frac{w\quad y\quad ec}{w^+\quad y^+\quad ec^+}$$

Following a double-crossover event, the *y* and $y^+$ alleles would be switched to this arrangement:

$$\frac{w\quad y^+\quad ec}{w^+\quad y\quad ec^+}$$

After segregation, two gametes would be formed:

$$\underline{w\ y^+\ ec} \text{ and } \underline{w^+\ y\ ec^+}$$

Because the genotype of the gamete will be expressed directly in the phenotype following fertilization, the double-crossover phenotypes will be:

white, echinus flies and yellow flies

Note that the *yellow* allele, assumed to be in the middle, is now associated with the two outside markers of the other homolog, $w^+$ and $ec^+$. However, these predicted phenotypes do not coincide with the observed double-crossover phenotypes. Therefore, the *yellow* gene is not in the middle.

This same reasoning can be applied to the assumption that the *echinus* gene or the *white* gene is in the middle. In the former case, we will reach a negative conclusion. If we assume that the *white* gene is in the middle, the *predicted* and *actual* double crossovers coincide. Therefore, we conclude that the *white* gene is located between the *yellow* and *echinus* genes.

To summarize Method II, determine the arrangement of alleles on the homologs of the heterozygote yielding crossover gametes. Then examine the actual double-crossover phenotypes and identify the single allele that has been switched so that it is now no longer associated with its original neighboring alleles. That allele will be the one located between the other two in the sequence.

In our example *y*, *ec*, and *w* are on one homolog in the $F_1$ heterozygote, and $y^+$, $ec^+$, and $w^+$ are on the other. In the $F_2$ double-crossover classes, it is *w* and $w^+$ that have been switched. The *w* allele is now associated with $y^+$ and $ec^+$, while the $w^+$ allele is now associated with the *y* and *ec* alleles. Therefore, the *white* gene is in the middle, and the *yellow* and *echinus* genes are the flanking markers.

## A Mapping Problem in Maize

Having established the basic principles of chromosome mapping, we will now consider a related problem in maize (corn). This analysis differs from the preceding example in two ways. First, the previous mapping cross involved X-linked genes. Here, we consider autosomal genes. Second, in the discussion of this cross, we will change our use of symbols, as first suggested in Chapter 4. Instead of using the gene symbols and superscripts (e.g., $bm^+$, $v^+$, and $pr^+$), we simply use + to denote each wild-type allele. This system is easier to manipulate but requires a better understanding of mapping procedures.

When we look at three autosomally linked genes in maize, our experimental cross must still meet the same three criteria we established for the X-linked genes in *Drosophila:* (1) One parent must be heterozygous for all traits under consideration; (2) the gametic genotypes produced by the heterozygote must be apparent from observing the phenotypes of the offspring; and (3) a sufficient sample size must be available for complete analysis.

In maize, the recessive mutant genes *bm* (*brown* midrib), *v* (*virescent* seedling), and *pr* (*purple* aleurone) are linked on chromosome 5. Assume that a female plant is known to be heterozygous for all three traits, but we do not know (1) the arrangement of the mutant alleles on the maternal and paternal homologs of this heterozygote, (2) the sequence of genes, or (3) the map distances between the genes. What genotype must the male plant have to allow successful

**(a) Some possible allele arrangements and gene sequences in a heterozygous female**

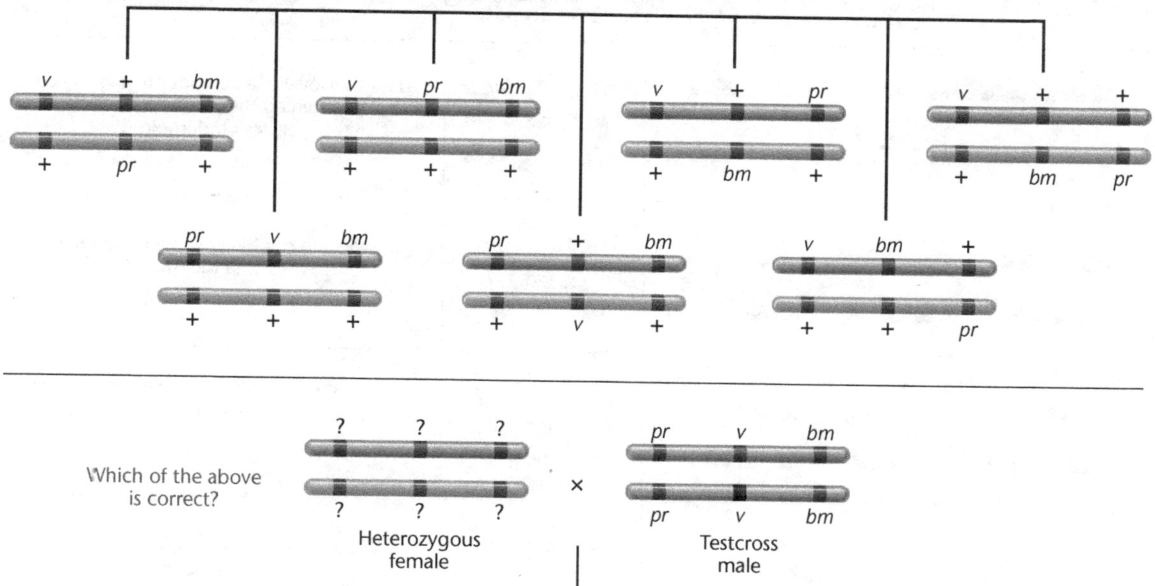


**(b) Actual results of mapping cross***

| Phenotypes of offspring | Number | Total and percentage | Exchange classification |
|---|---|---|---|
| + *v* *bm* | 230 | 467<br>42.1% | Noncrossover (NCO) |
| *pr* + + | 237 | | |
| + + *bm* | 82 | 161<br>14.5% | Single crossover (SCO) |
| *pr* *v* + | 79 | | |
| + *v* + | 200 | 395<br>35.6% | Single crossover (SCO) |
| *pr* + *bm* | 195 | | |
| *pr* *v* *bm* | 44 | 86<br>7.8% | Double crossover (DCO) |
| + + + | 42 | | |

* The sequence *pr* – *v* – *bm* may or may not be correct.

**FIGURE 5–10** (a) Some possible allele arrangements and gene sequences in a heterozygous female. The data from a three-point mapping cross, depicted in (b), where the female is testcrossed, provide the basis for determining which combination of arrangement and sequence is correct. [See Figure 5–11(d).]

mapping? To meet the second criterion, the male must be homozygous for all three recessive mutant alleles. Otherwise, offspring of this cross showing a given phenotype might represent more than one genotype, making accurate mapping impossible. Note that this is equivalent to performing a testcross.

Figure 5–10 diagrams this cross. As shown, we know neither the arrangement of alleles nor the sequence of loci in the heterozygous female. Several possibilities are shown, but we have yet to determine which is correct. We don't know the sequence in the testcross male parent either, so we must designate it randomly. Note that we initially placed *v* in the middle. *This may or may not be correct.*

The offspring have been arranged in groups of two, representing each pair of reciprocal phenotypic classes. The four reciprocal classes are derived from no crossing over (NCO), each of two possible single-crossover events (SCO), and a double-crossover event (DCO).

To solve this problem, refer to Figures 5–10 and 5–11 as you consider the following questions:

1. *What is the correct heterozygous arrangement of alleles in the female parent?*
   Determine the two noncrossover classes, those that occur with the highest frequency. In this case, they are $\underline{+\ v\ bm}$ and $\underline{pr\ +\ +}$.

VIRTUAL CROSSOVER LABORATORY

WEB TUTORIAL 5.4

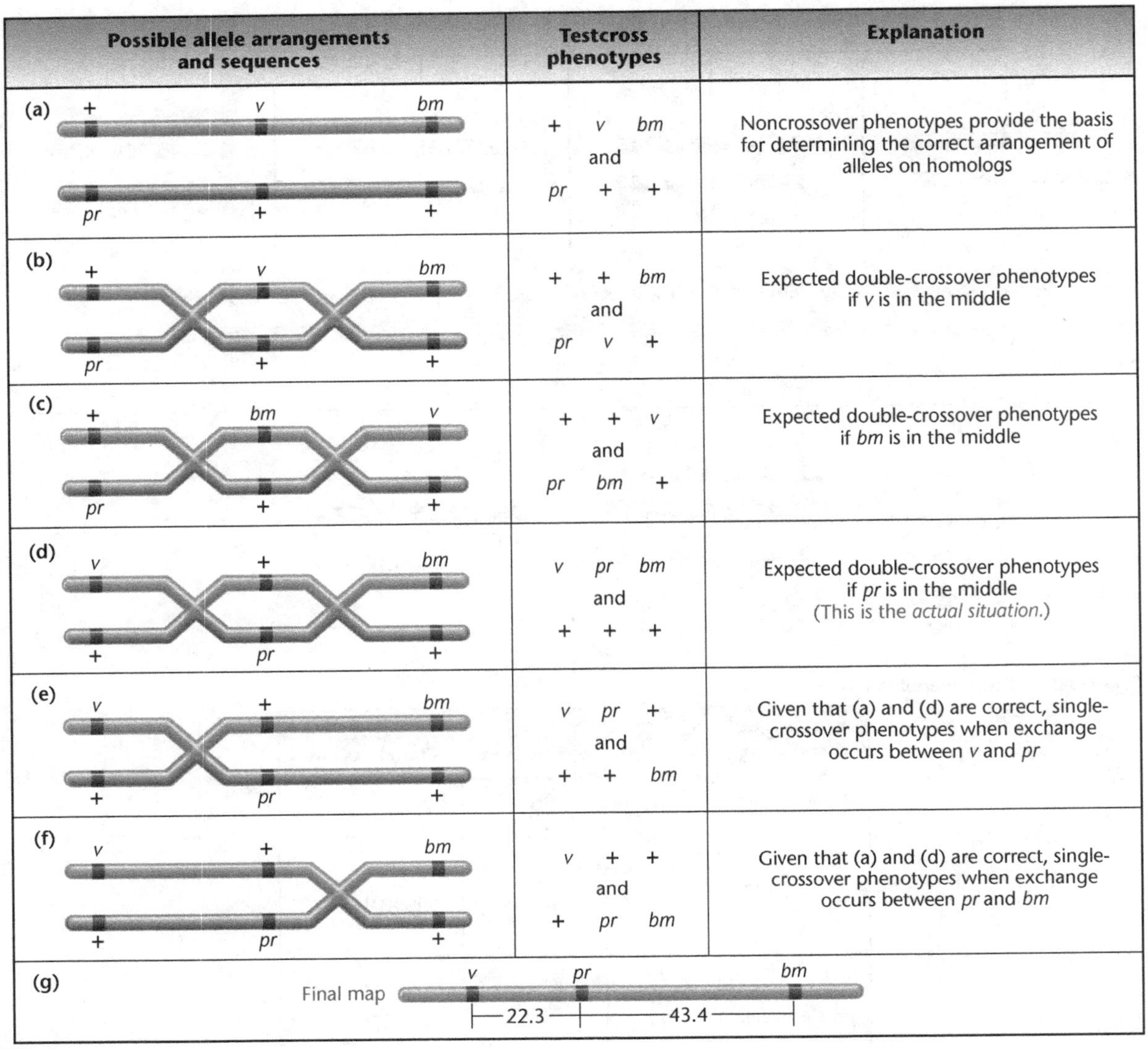


**FIGURE 5–11** Producing a map of the three genes in the cross in Figure 5–10, where neither the arrangement of alleles nor the sequence of genes in the heterozygous female parent is known.

Therefore, the alleles on the homologs of the female parent must be distributed as shown in Figure 5–11(a). These homologs segregate into gametes, unaffected by any recombination event. Any other arrangement of alleles will not yield the observed noncrossover classes. (Remember that $\underline{+\ v\ bm}$ is equivalent to $\underline{pr^+\ v\ bm}$ and that $\underline{pr\ +\ +}$ is equivalent to $\underline{pr\ v^+\ bm^+}$.)

2. *What is the correct sequence of genes?*
To answer this question, we will first use the approach described in Method I. We know, based on the answer to question 1, that the correct arrangement of alleles is

$$\frac{+\quad v\quad bm}{pr\quad +\quad +}$$

But is the gene sequence correct? That is, will a double-crossover event yield the observed double-crossover phenotypes following fertilization? *Observation shows that it will not* [Figure 5–11(b)]. Now try the other two orders [Figure 5–11(c) and 5–11(d)], *keeping the same allelic arrangement:*

$$\frac{+\quad bm\quad v}{pr\quad +\quad +}\quad \text{or}\quad \frac{v\quad +\quad bm}{+\quad pr\quad +}$$

*Only the order on the right yields the observed double-crossover gametes* [Figure 5–11(d)]. Therefore, the *pr* gene is in the middle.

The same conclusion is reached if we used Method II to analyze the problem. In this case, no assumption of gene sequence is necessary. The arrangement of alleles along homologs in the heterozygous parent is

$$\frac{+ \quad v \quad bm}{pr \quad + \quad +}$$

The double-crossover gametes are also known:

$$\underline{pr\ v\ bm} \quad \text{and} \quad \underline{+ \quad + \quad +}$$

We can see that it is the *pr* allele that has shifted relative to its noncrossover arrangement, so as to be associated with *v* and *bm* following a double crossover. The latter two alleles (*v* and *bm*) were present together on one homolog, and they stayed together. Therefore, *pr* is the odd gene, so to speak, and is located in the middle. Thus, we arrive at the same arrangement and sequence as we did with Method I:

$$\frac{v \quad + \quad bm}{+ \quad pr \quad +}$$

3. *What is the distance between each pair of genes?*
Having established the correct sequence of loci as *v–pr–bm*, we can now determine the distance between *v* and *pr* and between *pr* and *bm*. Remember that the map distance between two genes is calculated on the basis of all detectable recombinational events occurring between them. This includes both the single- and double-crossover events.
Figure 5–11(e) shows that the phenotypes $\overline{v\ pr\ +}$ and $\overline{+\ +\ bm}$ result from single crossovers between *v* and *pr*, and Figure 5–10 shows that those single crossovers account for 14.5 percent of the offspring. By adding the percentage of double crossovers (7.8 percent) to the number obtained for those single crossovers, we calculate the total distance between *v* and *pr* to be 22.3 mu.
Figure 5–11(f) shows that the phenotypes $\overline{v + +}$ and $\overline{+\ pr\ bm}$ result from single crossovers between the *pr* and *bm* loci, totaling 35.6 percent, according to Figure 5–10. Adding the double-crossover classes (7.8 percent), we compute the distance between *pr* and *bm* as 43.4 mu. The final map for all three genes in this example is shown in Figure 5–11(g).

5.4

## Interference Affects the Recovery of Multiple Exchanges

As the review of the product law in Section 5.3 would indicate, the expected frequency of multiple exchanges, such as double crossovers, can be predicted once the distance between genes is established. For example, in the maize cross of the previous section, the distance between *v* and *pr* is 22.3 mu, and the distance between *pr* and *bm* is 43.4 mu. If the two single crossovers that make up a double crossover occur independently of one another, we can calculate the expected frequency of double crossovers ($DCO_{exp}$) as follows:

$$DCO_{exp} = (0.223) \times (0.434) = 0.097 = 9.7\%$$

Often in mapping experiments, the observed DCO frequency is less than the expected number of DCOs. In the maize cross, for example, only 7.8 percent DCOs are observed when 9.7 percent are expected. **Interference (*I*)**, the inhibition of further crossover events by a crossover event in a nearby region of the chromosome, causes this reduction.

To quantify the disparities that result from interference, we calculate the **coefficient of coincidence (*C*)**:

$$C = \frac{\text{Observed DCO}}{\text{Expected DCO}}$$

In the maize cross, we have

$$C = \frac{0.078}{0.097} = 0.804$$

Once we have found *C*, we can quantify interference (*I*) by using this simple equation

$$I = 1 - C$$

In the maize cross, we have

$$I = 1.000 - 0.804 = 0.196$$

If interference is complete and no double crossovers occur, then $I = 1.0$. If fewer DCOs than expected occur, *I* is a positive number and **positive interference** has occurred. If more DCOs than expected occur, *I* is a negative number and **negative interference** has occurred. In this example, *I* is a positive number (0.196), indicating that 19.6 percent fewer double crossovers occurred than expected.

Positive interference is most often observed in eukaryotic systems. In general, the closer genes are to one another along the chromosome, the more positive interference occurs. In fact, interference in *Drosophila* is often complete within a distance of 10 map units, and no multiple crossovers are recovered. This observation suggests that physical constraints preventing the formation of closely spaced chiasmata contribute to interference. The interpretation is consistent with the finding that interference decreases as the genes in question are located farther apart. In the maize cross illustrated in Figures 5–10 and 5–11, the three genes are relatively far apart, and 80 percent of the expected double crossovers are observed.

NOW SOLVE THIS

Problem 20 on page 139 asks you to solve a three-point mapping problem in which only six phenotypic categories are observed even though eight categories are typical of such a cross.

■ HINT: *If the distances between the loci are relatively small, the sample size may be too small for the predicted number of double crossovers to be recovered, even though reciprocal pairs of single crossovers are seen. You should write the missing gametes down as double crossovers and record zeros for their frequency of appearance.*

5.5

## As the Distance between Two Genes Increases, the Results of Mapping Experiments Become Less Accurate

In theory, the frequency of crossing over between any two genes in a mapping experiment should be directly proportional to the actual distance between the genes. However, in most cases, the experimentally derived mapping distance between two genes is an underestimate, and the farther apart the two genes are, the greater the inaccuracy. The discrepancy is due primarily to multiple exchanges that are predicted to occur between the two genes, but that are not detected during experimental mapping. As we will explain next, this inaccuracy is the result of probability events that can be described using the **Poisson distribution.**

First, let us examine a mapping experiment that involves two exchanges between two genes that are far apart on a chromosome. As shown in Figure 5–12, there are three possible ways that two exchanges (equivalent to a double-crossover event) can occur between nonsister chromatids within a tetrad. A **two-strand double exchange** yields no recombinant chromatids (as far as the two genes of interest are concerned), a **three-strand double exchange** yields 50 percent recombinant chromatids, and a **four-strand double exchange** yields 100 percent recombinant chromatids. In the aggregate, therefore, these uncommon multiple events "even out," so that two genes far apart on the chromosome theoretically yield the maximum of 50 percent recombination essential for *accurate* gene mapping.

In such a mapping experiment, double exchanges (and for that matter, all multiple exchanges) occurring between two genes are relatively infrequent in comparison to the total number of single crossovers. As a result, the *actual* occurrence of the infrequent events is subject to probability considerations based on the Poisson distribution. In our case, this distribution allows us to predict the mathematical frequency of samples that will *actually* undergo double exchanges. It is the failure of such exchanges to occur that leads to the underestimation of mapping distance.

The Poisson distribution is a mathematical function that depicts the probability of observing various numbers of a specific event in a sample. To illustrate the Poisson distribution, let's consider the analogy of an Easter egg hunt where 1000 children randomly search a large area for 1000 randomly hidden eggs. In one hour, all eggs are recovered. If all children are equally adept in the search, we can safely predict not only that many children will have one egg, but also that many will have either no eggs or more than one egg. The Poisson distribution allows us to predict the frequency (probability) of each outcome, that is, the frequency of children within the sample that recovered 0, 1, 2, 3, 4, . . . eggs. The Poisson distribution applies when the average number of events is small (most of the children find an egg once), while the total number of times the event that can occur within the sample is relatively large (1000 eggs can be found).

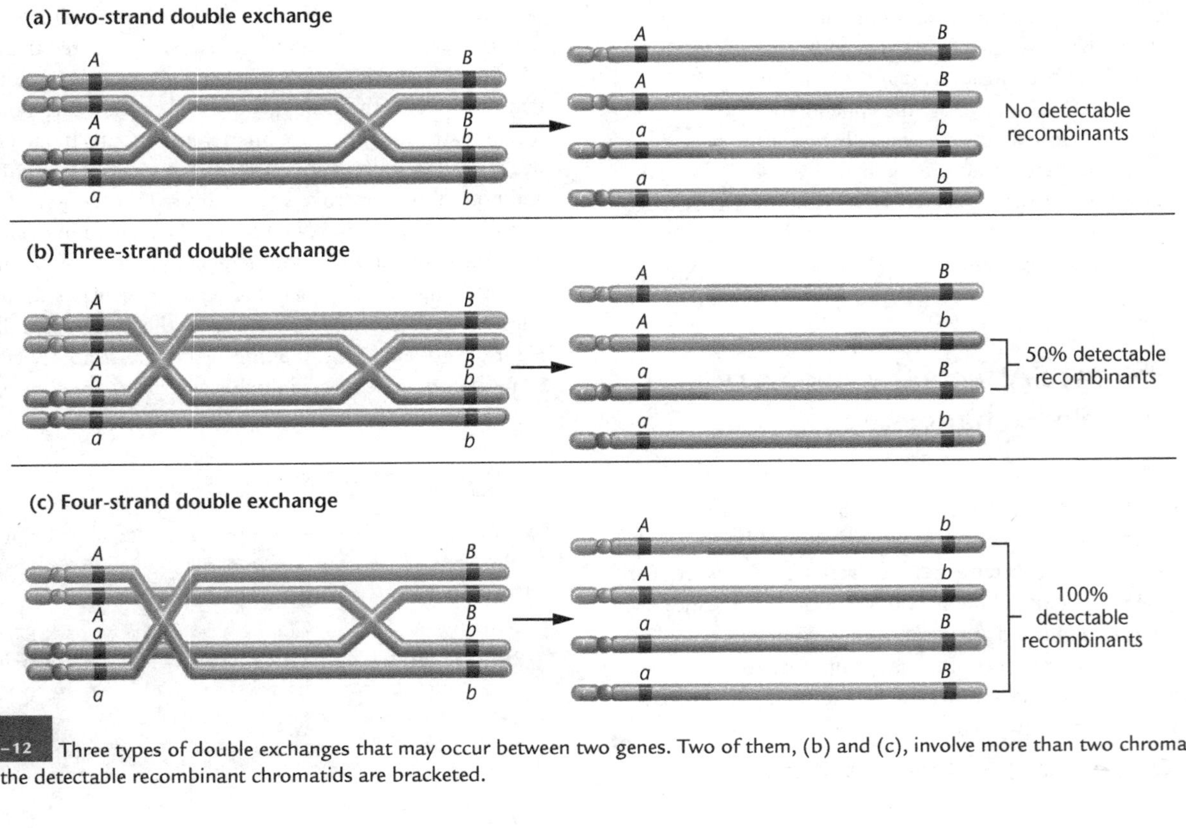


FIGURE 5-12 Three types of double exchanges that may occur between two genes. Two of them, (b) and (c), involve more than two chromatids. In each case, the detectable recombinant chromatids are bracketed.

The Poisson terms used to calculate predicted distributions of events are

| Distribution of Events | Probability |
|---|---|
| 0 | $e^{-m}$ |
| 1 | $me^{-m}$ |
| 2 | $(m^2/2)\ (e^{-m})$ |
| 3 | $(m^3/6)\ (e^{-m})$ |
| etc. | |

where the mean number of independently occurring events is $m$ and $e$ represents the base of natural logarithms ($e$ = about 2.7). For the Easter egg analogy, the calculation will reveal that over 300 children will fail to find an egg. Had we attempted to estimate the total number of youngsters in the hunt by assuming it must be close to the number who found at least one egg, we would have seriously underestimated the number of participants in the hunt.

In chromosome mapping, we must take into account the number of cases in which double exchanges had the potential to occur between two genes but did not in actuality occur, as predicted by Poisson distribution. Such an analysis creates what is called a **mapping function** that relates recombination (crossover) frequency (RF) to map distance.

To apply the Poisson distribution, we must assume that no interference occurs. Any class where $m$ is 1 or more (one or more random crossovers) will yield, on average, 50 percent recombinant chromatids. Thus, we are interested in the zero term, which effectively reduces the number of recombinant chromatids. The proportion of meioses with one or more crossovers is equal to 1 minus the fraction of zero crossovers ($1 - e^{-m}$), whereby 50 percent recombinant chromatids will occur. Therefore, percent observed recombination frequency (RF) $= 0.5(1 - e^{-m}) \times 100$.

Solving this equation generates the curve (mapping function) labeled "Actual" in Figure 5–13. This contrasts markedly with the "Theoretical" case of Figure 5–13, where recombination is presumed to be directly proportional to mapping distance—that is, where interference is complete and no multiple exchanges occur.

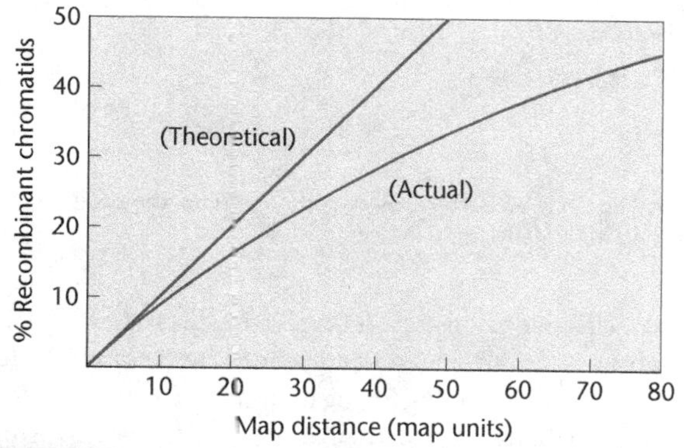


FIGURE 5–13 A comparison of the theoretical vs. the actual percentage of recombinant chromatids produced as map distance increases. In theory, a direct relationship is assumed between the frequency of recombination and map distance. However, once distance exceeds 7 mu, this relationship declines, as established in studies of *Drosophila*, *Neurospora*, and *Zea mays*.

Careful examination of the graph reveals two important features. When the actual map distance is low (i.e., 0–7 mu), the two lines coincide. Thus, *when two genes are close together, the accuracy of a mapping experiment is very high! However, as the distance between two genes increases, the accuracy of the experiment diminishes.* As predicted by the Poisson distribution, the absence of multiple exchanges has a very significant impact. For example, when 25 percent recombinant chromatids are detected, actual map distance is almost 35 mu! When just over 30 percent recombinants are detected, the true distance, discounting any interference, approaches 50 mu! Such inaccuracy has been well documented in a number of studies involving various organisms, including *Zea mays*, *Drosophila*, and *Neurospora*.

## 5.6 *Drosophila* Genes Have Been Extensively Mapped

In organisms such as fruit flies, maize, and the mouse, where large numbers of mutants have been discovered and where mapping crosses are possible, extensive maps of each chromosome have been constructed. Figure 5–14 presents partial maps of the four chromosomes of *Drosophila melanogaster*. Virtually every morphological feature of the fruit fly has been subjected to mutation. Each locus affected by mutation is first localized to one of the four chromosomes, or linkage groups, and then mapped in relation to other genes present on that chromosome. As you can see, the genetic map of the X chromosome is somewhat less extensive than that of autosome II or III. In comparison to these three, autosome IV is miniscule. Cytological evidence has shown that the relative lengths of the genetic maps correlate roughly with the relative physical lengths of these chromosomes.

## 5.7 Lod Score Analysis and Somatic Cell Hybridization Were Historically Important in Creating Human Chromosome Maps

In humans, genetic experiments involving carefully planned crosses and large numbers of offspring are neither ethical nor feasible, so the earliest linkage studies were based on pedigree analysis. These studies attempted to establish whether certain traits were X-linked or autosomal. As we showed in Chapter 4, traits determined by genes located on the X chromosome result in characteristic pedigrees; thus, such genes were easier to identify. For autosomal traits, geneticists tried to distinguish clearly whether pairs of traits demonstrated linkage or independent assortment. When extensive pedigrees are available, it is possible to conclude that two genes under consideration are closely

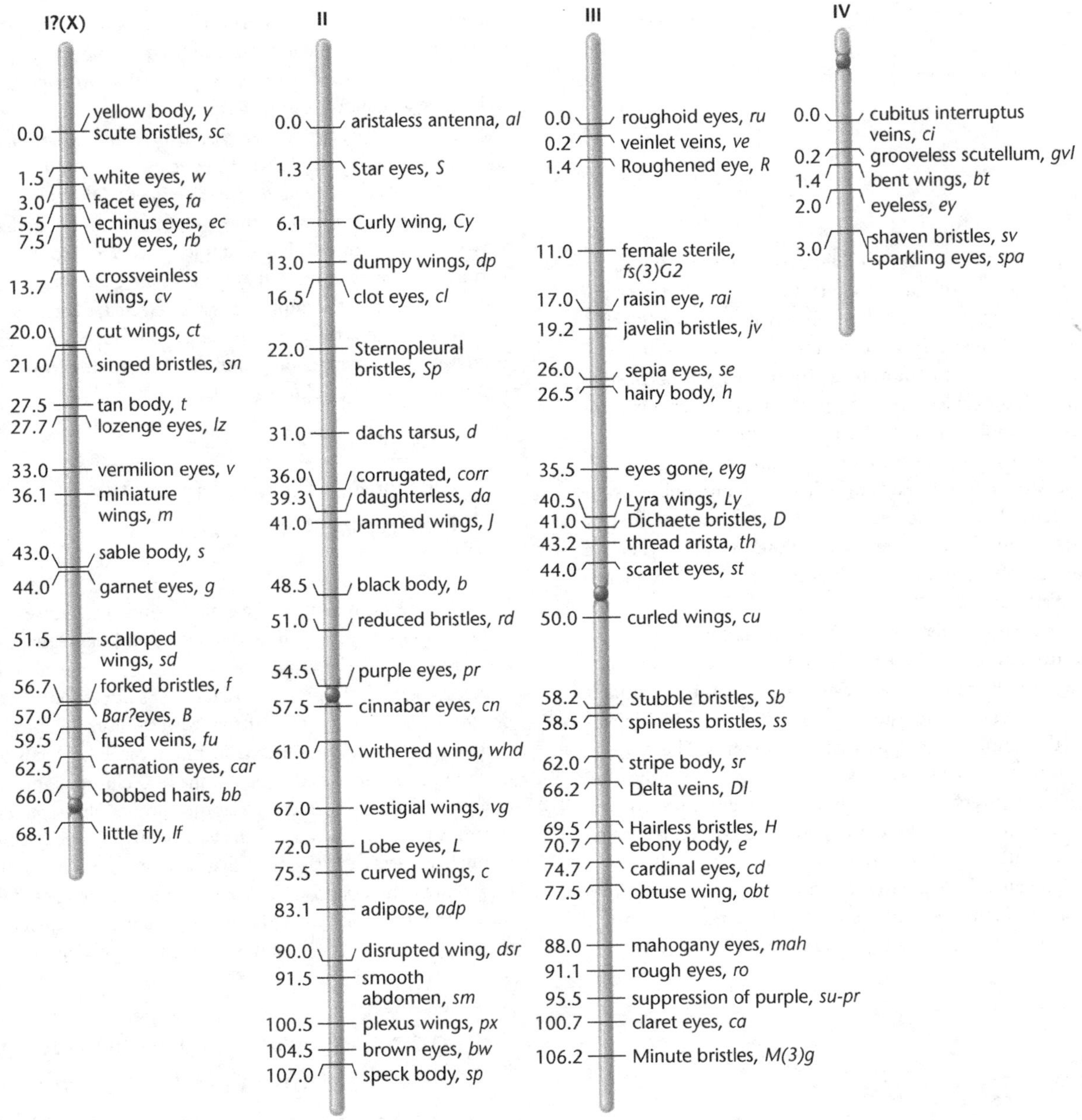


**FIGURE 5–14** A partial genetic map of the four chromosomes of *Drosophila melanogaster*. The circle on each chromosome represents the position of the centromere. Chromosome I is the X chromosome. Chromosome IV is not drawn to scale, i.e., it is smaller than indicated.

linked (i.e., rarely separated by crossing over) from the fact that the two traits segregate together. This approach established linkage between the genes encoding the **Rh antigens** and the gene responsible for the phenotype referred to as **elliptocytosis,** where the shape of erythrocytes is oval. It was hoped that from these kinds of observations a human gene map could be created.

A difficulty arises, however, when two genes of interest are separated on a chromosome to the degree that recombinant gametes are formed, obscuring linkage in a pedigree. In these cases, an approach relying on probability calculations, called the **lod score method**, helps to demonstrate linkage. First devised by J. B. S. Haldane and C. A. Smith in 1947 and refined by Newton Morton in 1955, the lod score (standing for *l*og of the *od*ds favoring linkage) assesses the probability that a particular pedigree (or several pedigrees for the same traits of interest) involving two traits reflects genetic linkage between them. First, the probability is calculated that the family (pedigree) data concerning two traits conform to transmission without linkage—that is, the traits appear to be independently assorting. Then the probability is calculated that the identical family data for these same traits result from linkage with a specified recombination frequency. These probability calculations factor in the statistical significance at the $p = 0.05$ level. The ratio of these probability values is then calculated and converted to the logarithm of this value, which reflects the "odds" for, and against, linkage. Traditionally, a value of 3.0 or higher strongly indicates linkage, whereas a value of $<2.0$ argues strongly against linkage.

The lod score method represented an important advance in assigning human genes to specific chromosomes and in constructing preliminary human chromosome maps. However, its accuracy is limited by the extent of the pedigree, and the initial results were discouraging—both because of this limitation and because of the relatively high haploid number of human chromosomes (23). By 1960, very little autosomal linkage information had become available. Today, however, in contrast to its restricted impact when originally developed, this elegant technique has become important in human linkage analysis, owing to the discovery of countless molecular *DNA markers* along every human chromosome. The discovery of these markers, which behave just as genes do in occupying a particular locus along a chromosome, was a result of recombinant DNA techniques (Chapter 13) and genomic analysis (Chapter 21). Any human trait may now be tested for linkage with such markers. We will return to a consideration of DNA markers in Section 5.8.

In the 1960s, a new technique, **somatic cell hybridization**, proved to be an immense aid in assigning human genes to their respective chromosomes. This technique, first discovered by Georges Barsky, relies on the fact that two cells in culture can be induced to fuse into a single hybrid cell. Barsky used two mouse-cell lines, but it soon became evident that cells from different organisms could also be fused. When fusion occurs, an initial cell type called a **heterokaryon** is produced. The hybrid cell contains two nuclei in a common cytoplasm. Using the proper techniques, we can fuse human and mouse cells, for example, and isolate the hybrids from the parental cells.

As the heterokaryons are cultured *in vitro*, two interesting changes occur. Eventually, the nuclei fuse together, creating a **synkaryon**. Then, as culturing is continued for many generations, chromosomes from one of the two parental species are gradually lost. In the case of the human–mouse hybrid, human chromosomes are lost randomly until eventually the synkaryon has a full complement of mouse chromosomes and only a few human chromosomes. It is the preferential loss of human chromosomes (rather than mouse chromosomes) that makes possible the assignment of human genes to the chromosomes on which they reside.

The experimental rationale is straightforward. If a specific human gene product is synthesized in a synkaryon containing three human chromosomes, then the gene responsible for that product must reside on one of the three human chromosomes remaining in the hybrid cell. On the other hand, if the human gene product is not synthesized in the synkaryon, the responsible gene cannot be present on any of the remaining three human chromosomes. Ideally, one would have a panel of 23 hybrid cell lines, each with a different human chromosome, allowing the immediate assignment to a particular chromosome of any human gene for which the product could be characterized.

In practice, a panel of cell lines that each contain several remaining human chromosomes is most often used. The correlation of the presence or absence of each chromosome with the presence or absence of each gene product is called **synteny testing**. Consider, for example, the hypothetical data provided in Figure 5–15, where four gene products (A, B, C, and D) are tested in relationship to eight human chromosomes. Let us carefully analyze the results to locate the gene that produces product A.

1. Product A is not produced by cell line 23, but chromosomes 1, 2, 3, and 4 are present in cell line 23. Therefore, we can rule out the presence of gene *A* on those four chromosomes and conclude that it must be on chromosome 5, 6, 7, or 8.
2. Product A is produced by cell line 34, which contains chromosomes 5 and 6, but not 7 and 8. Therefore, gene *A* is on chromosome 5 or 6, but cannot be on 7 or 8 because they are absent, even though product A is produced.
3. Product A is also produced by cell line 41, which contains chromosome 5, but not chromosome 6. Therefore, gene *A* is on chromosome 5, according to this analysis.

Using a similar approach, we can assign gene *B* to chromosome 3. Perform the analysis for yourself to demonstrate that this is correct.

Gene *C* presents a unique situation. The data indicate that it is not present on chromosomes 1–7. While it might be on chromosome 8, no direct evidence supports this conclusion. Other panels are needed. We leave gene *D* for you to analyze. Upon what chromosome does it reside?

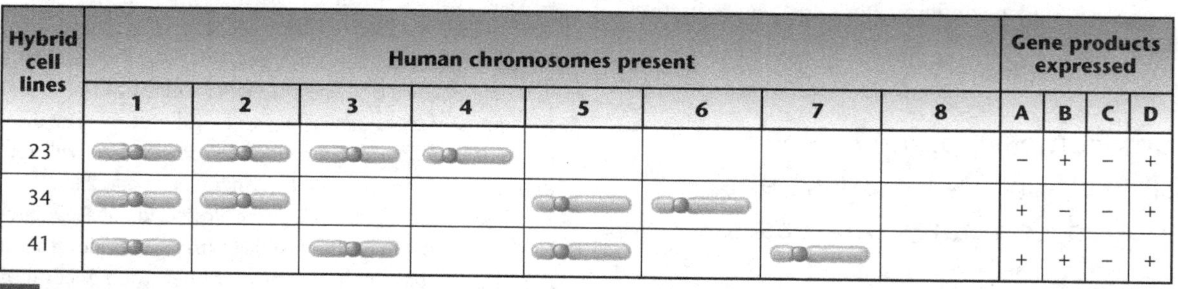


**FIGURE 5–15** A hypothetical grid of data used in synteny testing to assign genes to their appropriate human chromosomes. Three somatic hybrid cell lines, designated 23, 34, and 41, have each been scored for the presence, or absence, of human chromosomes 1 through 8, as well as for their ability to produce the hypothetical human gene products A, B, C, and D.

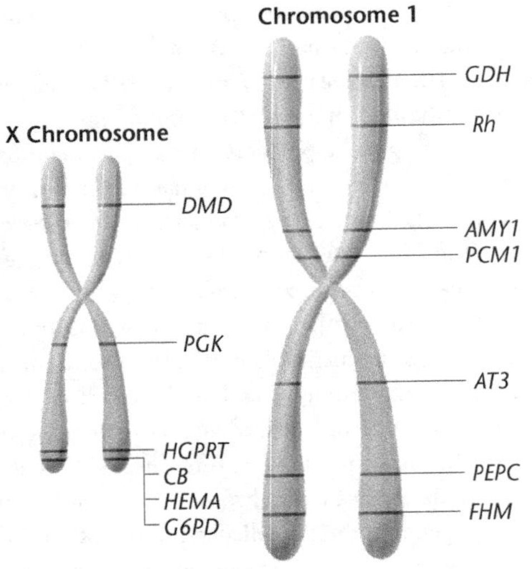


| Key | | |
|---|---|---|
| | *AMY* | Amylase (salivary and pancreatic) |
| | *AT3* | Antithrombin (clotting factor IV) |
| | *CB* | Color blindness |
| | *DMD* | Duchenne muscular dystrophy |
| | *FHM* | Fumarate hydratase (mitochondrial) |
| | *GDH* | Glucose dehydrogenase |
| | *G6PD* | Glucose-6-phosphate dehydrogenase |
| | *HEMA* | Hemophilia A (classic) |
| | *HGPRT* | Hypoxanthine-Guanine-Phosphoribosyl Transferase (Lesch–Nyhan syndrome) |
| | *PEPC* | Peptidase C |
| | *PGK* | Phosphoglycerate kinase |
| | *PGM* | Phosphoglucomutase |
| | *Rh* | Rhesus blood group (erythroblastosis fetalis) |

**FIGURE 5–16** Representative regional gene assignments for human chromosome 1 and the X chromosome. Many assignments were initially derived using somatic cell hybridization techniques.

By using the approach just described, researchers have assigned literally hundreds of human genes to one chromosome or another. Figure 5–16 illustrates some gene locations on two human chromosomes: X and 1. The gene assignments shown were either derived or confirmed with the use of somatic cell hybridization techniques. To map genes for which the products have yet to be discovered, researchers have had to rely on other approaches. For example, by combining recombinant DNA technology with pedigree analysis, it has been possible to assign the genes responsible for Huntington disease, cystic fibrosis, and neurofibromatosis to their respective chromosomes 4, 7, and 17.

## 5.8 Chromosome Mapping Is Now Possible Using DNA Markers and Annotated Computer Databases

While traditional methods based on recombination analysis have produced detailed chromosomal maps in several organisms, such maps in other organisms (including humans) that do not lend themselves to such studies are greatly limited. Fortunately, the development of technology allowing direct analysis of DNA has greatly enhanced mapping in those organisms. We will address this topic using humans as an example.

Progress has initially relied on the discovery of **DNA markers** (mentioned earlier) that have been identified during recombinant DNA and genomic studies. These markers are short segments of DNA whose sequence and location are known, making them useful *landmarks* for mapping purposes. The analysis of human genes in relation to these markers has extended our knowledge of the location within the genome of countless genes, which is the ultimate goal of mapping.

The earliest examples are the DNA markers referred to as **restriction fragment length polymorphisms (RFLPs)** (see Chapter 24) and **microsatellites** (see Chapter 12). RFLPS are polymorphic sites generated when specific DNA sequences are recognized and cut by restriction enzymes. Microsatellites are short repetitive sequences that are found throughout the genome and they vary in the number of repeats at any given site. For example, the two-nucleotide sequence CA is repeated 5–50 times per site [$(CA)_n$] and appears throughout the genome approximately every 10,000 bases, on average. Microsatellites may be identified not only by the number of repeats but by the DNA sequences that flank them. More recently, variation in single nucleotides (called **single nucleotide polymorphisms** or **SNPs**) has been utilized. Found throughout the genome, up to several million of these variations may be screened for an association with a disease or trait of interest, thus providing geneticists with a means to identify and locate related genes.

**Cystic fibrosis** offers an early example of a gene located by using DNA markers. It is a life-shortening autosomal recessive exocrine disorder resulting in excessive, thick mucus that impedes the function of organs such as the lung and pancreas. After scientists established that the gene causing this disorder is located on chromosome 7, they were then able to pinpoint its exact location on the long arm (the q arm) of that chromosome.

Very recently (June, 2007) using SNPs as DNA markers, associations between 24 genomic locations have been established with seven common human diseases: Type 1 (insulin dependent) and Type 2 diabetes, Crohn's disease (inflammatory bowel disease), hypertension, coronary artery disease, bipolar (manic-depressive) disorder, and rheumatoid arthritis. In each case, an inherited susceptibility effect has been mapped to a specific location on a specific chromosome within the genome. In some cases, this has either confirmed or led to the identification of a specific gene involved in the cause of the disease. In other cases, new genes will no doubt soon be discovered as a result of the identification of their location. We will return to this topic in much greater detail in Chapters 23 and 24.

The many Human Genome Project databases that have been completed now make it possible to map genes along a human chromosome in base-pair distances rather than recombination frequency. This distinguishes what is referred to as a **physical map** of the genome from the genetic maps described above. Distances can then be determined relative to other genes and to features such as the DNA markers discussed above. The power of this approach is that geneticists will soon be able to construct chromosome maps for individuals that designate specific allele combinations at each gene site.

5.9

## Crossing Over Involves a Physical Exchange between Chromatids

Once genetic mapping techniques had been developed, they were used to study the relationship between the chiasmata observed in meiotic prophase I and crossing over. For example, are chiasmata visible manifestations of crossover events? If so, then crossing over in higher organisms appears to be the result of an actual physical exchange between homologous chromosomes. That this is the case was demonstrated independently in the 1930s by Harriet Creighton and Barbara McClintock in *Zea mays* (maize) and by Curt Stern in *Drosophila.*

Because the experiments are similar, we will consider only one of them, the work with maize. Creighton and McClintock studied two linked genes on chromosome 9 of the maize plant. At one locus, the alleles *colorless* (*c*) and *colored* (*C*) control endosperm coloration (the endosperm is the nutritive tissue inside the corn kernel). At the other locus, the alleles *starchy* (*Wx*) and *waxy* (*wx*) control the carbohydrate characteristics of the endosperm. The maize plant studied was heterozygous at both loci. The key to this experiment is that one of the homologs contained two unique cytological markers. The markers consisted of a densely stained knob at one end of the chromosome and a translocated piece of another chromosome (8) at the other end. The arrangements of these alleles and markers could be detected cytologically and are shown in Figure 5–17.

Creighton and McClintock crossed this plant to one homozygous for the *colorless* allele (*c*) and heterozygous for the *waxy/starchy* alleles. They obtained a variety of different phenotypes in the offspring, but they were most interested in one that occurred as a result of a crossover involving the chromosome with the unique cytological markers. They examined the chromosomes of this plant, having a colorless, waxy phenotype (case I in Figure 5–17), for the presence of the cytological markers. If genetic crossing over was accompanied by a physical exchange between homologs, the translocated chromosome would still be present, but the knob would not. This was the case! In a second plant (case II), the phenotype colored, starchy should result from either nonrecombinant gametes or crossing over. Some of the cases then ought to contain chromosomes with the dense knob but not the translocated chromosome. This condition was also found, and the conclusion that a physical exchange had taken place was again supported. Along with Curt Stern's findings in *Drosophila,* this work clearly established that crossing over has a cytological basis.

Once we have introduced the chemical structure and replication of DNA (Chapters 10 and 11), we will return to the topic of crossing over and look at how breakage and reunion occur between the strands of DNA making up chromatids in Chapter 11. This discussion will provide a better understanding of genetic recombination.

5.10

## Recombination Occurs between Mitotic Chromosomes

In 1936, Curt Stern studied the question of whether exchanges similar to crossing over occur during mitosis. He was able to demonstrate that it indeed is the case in *Drosophila.* This finding, the first demonstration of **mitotic recombination**, was considered unusual because homologs do not normally pair up during mitosis in most organisms. However, such synapsis appears to be the rule in *Drosophila.* Since Stern's discovery, genetic exchange during mitosis has also been shown to be a general event in certain fungi.

Stern observed small patches of mutant tissue in female *Drosophila* heterozygous for the X-linked recessive mutations *yellow* body and *singed* bristles. Under normal circumstances, a heterozygous female is completely wild type (gray-bodied with straight, long bristles). He explained the appearance of the mutant patches by postulating that, during mitosis in certain cells during development, homologous exchanges could occur between the loci for *yellow* (*y*)

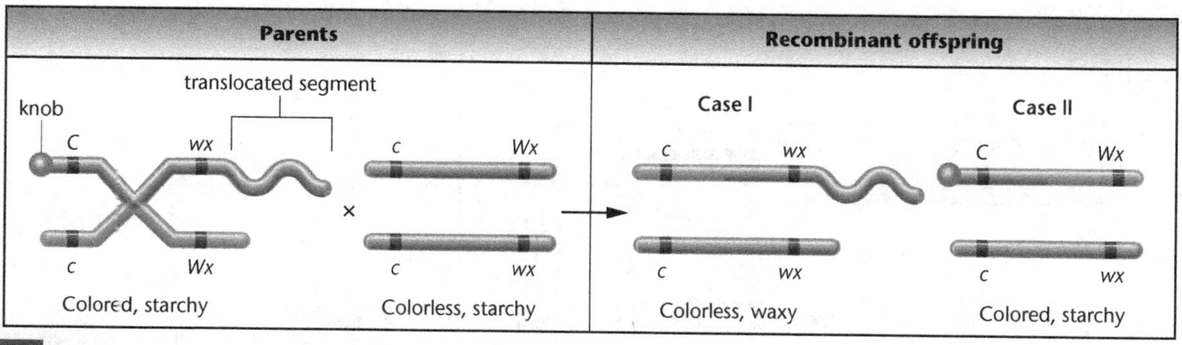


FIGURE 5–17 The phenotypes and chromosome compositions of parents and recombinant offspring in Creighton and McClintock's experiment in maize. The knob and translocated segment served as cytological markers, which established that crossing over involves an actual exchange of chromosome arms.

and *singed* (*sn*) or between *singed* and the centromere. Figure 5–18 diagrams the nature of the proposed exchanges, in contrast to the case where no exchange occurs. The mutant patches that result from the different exchanges are also depicted. When no exchange occurs, all tissue is wild type (gray body and gray, straight bristles). After an exchange, tissue is produced either with a *yellow* patch *or* with side-by-side *yellow* and *singed* patches (called a twin spot).

In 1958, George Pontecorvo and others described a similar phenomenon in the fungus *Aspergillus.* Although the vegetative stage is normally haploid, some cells fuse. The resultant diploid cells then divide mitotically. As in *Drosophila,* crossing over occasionally occurs between linked genes during mitosis in this diploid stage, resulting in recombinant cells. Pontecorvo referred to these events that produce genetic variability as the **parasexual cycle.** On the basis of such exchanges, genes can be mapped by estimating the frequency of recombinant classes.

As a rule, if mitotic recombination occurs at all in an organism, it does so at a much lower frequency than meiotic crossing over. We assume that there is always at least one exchange per meiotic tetrad. By contrast, in organisms demonstrating mitotic exchange, it occurs in 1 percent or fewer of mitotic divisions.

## 5.11 Exchanges Also Occur between Sister Chromatids

Considering that crossing over occurs between synapsed homologs in meiosis, we might ask whether such a physical exchange occurs between sister chromatids that are aligned together during mitosis. Each individual chromosome in prophase and metaphase of mito-

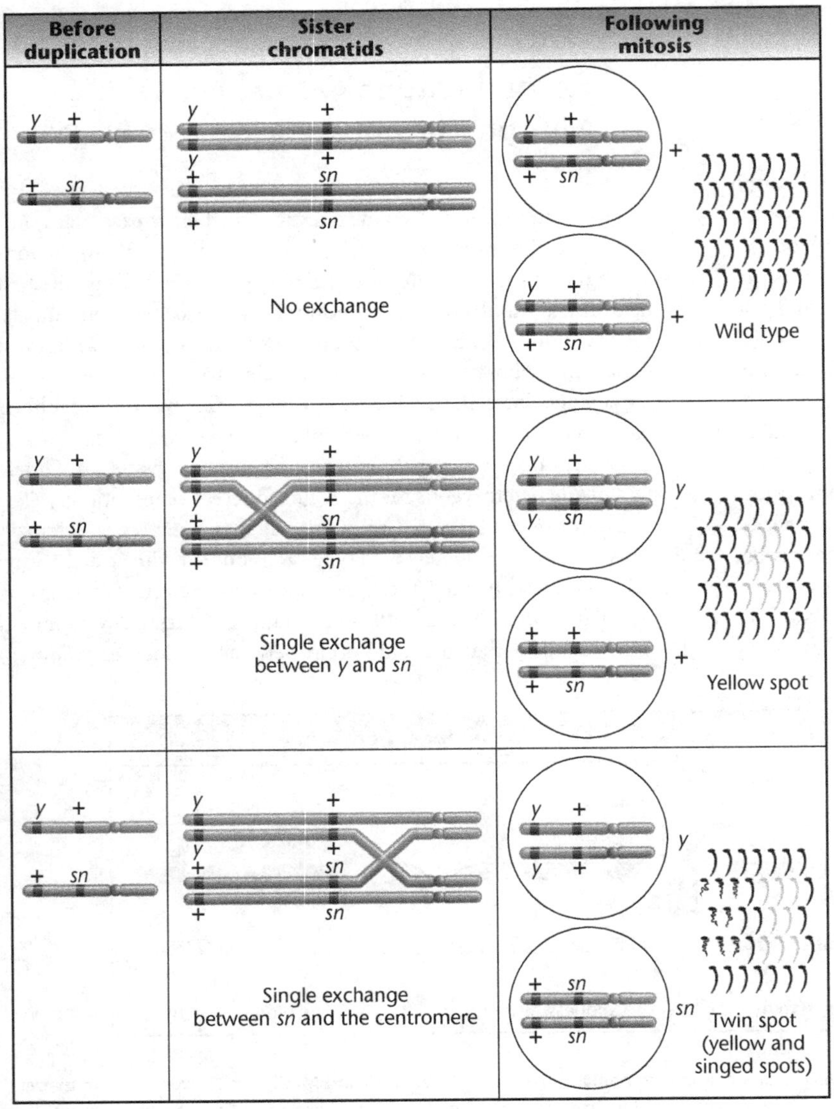


**FIGURE 5–18** The production of mutant tissue as a result of mitotic recombination in a female *Drosophila* heterozygous for the recessive *yellow* (*y*) and *singed* (*sn*) alleles.

sis consists of two identical sister chromatids, joined at a common centromere. A number of experiments have demonstrated that reciprocal exchanges similar to crossing over do occur between sister chromatids. While these **sister chromatid exchanges (SCEs)** do not produce new allelic combinations, evidence is accumulating that attaches significance to these events.

Identification and study of SCEs are facilitated by several modern staining techniques. In one approach, cells are allowed to replicate for two generations in the presence of the thymidine analog bromodeoxyuridine (BrdU). Following two rounds of replication, each pair of sister chromatids has one member with one strand of DNA "labeled" with BrdU, and the other member with both strands labeled with BrdU. Using a differential stain, chromatids with the analog in both strands stain *less* brightly than chromatids with BrdU in only one strand. As a result, any SCEs are readily detectable. In Figure 5–19, numerous instances of SCE events are clearly evident. Because of their patterns of alternating patches, these sister chromatids are sometimes referred to as **harlequin chromosomes.**

The significance of SCEs is still uncertain, but several observations have led to great interest in this phenomenon. We know, for example, that agents that induce chromosome damage (e.g., viruses, X rays, ultraviolet light, and certain chemical mutagens) also increase the frequency of SCEs. Further, the frequency of SCEs is elevated in **Bloom syndrome,** a human disorder caused by a mutation in the *BLM* gene on chromosome 15. This rare, recessively inherited disease is characterized by prenatal and postnatal retardation of growth, a great sensitivity of the facial skin to the sun, immune deficiency, a predisposition to malignant and benign tumors, and abnormal behavior patterns. The chromosomes from cultured leukocytes, bone marrow cells, and fibroblasts derived from homozygotes are very fragile and unstable when compared with those derived from homozygous and heterozygous normal individuals. Increased breaks and rearrangements between nonhomologous chromosomes are observed in addition to excessive amounts of sister chromatid exchanges. Work by James German and colleagues suggests that the *BLM* gene encodes an enzyme called **DNA helicase,** which is best known for its role in DNA replication (see Chapter 11).

The mechanisms of exchange between nonhomologous chromosomes and between sister chromatids may prove to share common features, because the frequency of both events increases substantially in individuals with certain genetic disorders. These findings suggest that further study of sister chromatid exchange may contribute to the understanding of recombination mechanisms and to the relative stability of normal and genetically abnormal chromosomes. We shall encounter still another demonstration of SCEs in Chapter 11 when we consider replication of DNA (see Figure 11–5).

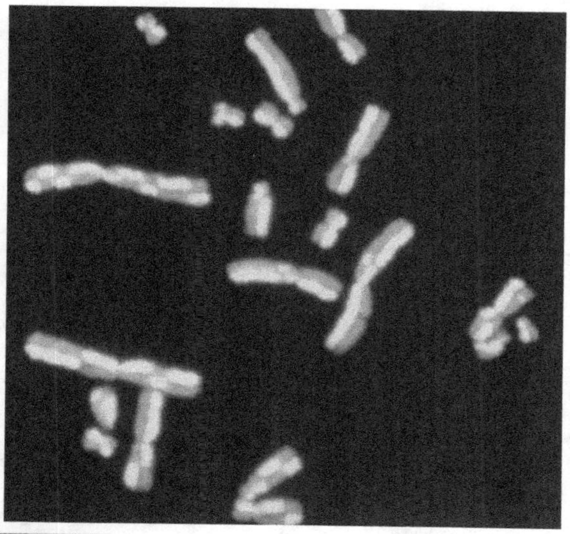

**FIGURE 5–19** Demonstration of sister chromatid exchanges (SCEs) in mitotic chromosomes. Sometimes called harlequin chromosomes because of the alternating patterns they exhibit, sister chromatids containing the thymidine analog BrdU are seen to fluoresce *less* brightly where they contain the analog in both DNA strands than where they contain the analog in only one strand. These chromosomes were stained with 33258-Hoechst reagent and acridine orange and then viewed under fluorescence microscopy.

## 5.12 Linkage and Mapping Studies Can Be Performed in Haploid Organisms

We now turn to yet another extension of transmission genetics: linkage analysis and chromosome mapping in *haploid* eukaryotes. As we shall see, even though analysis of the location of genes relative to one another in the genome of haploid organisms may *seem* a bit more complex than in diploid organisms, the underlying principles are the same. In fact, many basic principles of inheritance were established through the study of haploid fungi.

Many of the single-celled eukaryotes are haploid during the vegetative stages of their life cycle. The alga *Chlamydomonas* and the mold *Neurospora* demonstrate this genetic condition. These organisms do form reproductive cells that fuse during fertilization, forming a diploid zygote; however, the zygote soon undergoes meiosis and reestablishes haploidy. The haploid meiotic products are the progenitors of the subsequent members of the vegetative phase of the life cycle. Figure 5–20 illustrates this type of cycle in the green alga *Chlamydomonas.* Even though the haploid cells that fuse during fertilization *look* identical, and are thus called **isogametes,** a chemical identity that distinguishes two distinct types exists on their surface. As a result, all strains are either "+" or "–," and fertilization occurs only between unlike cells.

To perform genetic experiments with haploid organisms, researchers isolate genetic strains of different genotypes and cross them with one another. Following fertilization and meiosis, the meiotic products remain close together and can be analyzed. Such is the case in *Chlamydomonas* as well as in the fungus *Neurospora,* which we shall use as an example in the ensuing discussion. Following

**FIGURE 5-20** The life cycle of *Chlamydomonas*. The diploid zygote (in the center) undergoes meiosis, producing "+" or "−" haploid cells that undergo mitosis, yielding vegetative colonies. Unfavorable conditions stimulate them to form isogametes, which fuse in fertilization, producing a zygote that repeats the cycle. Vegetative colonies are illustrated photographically.

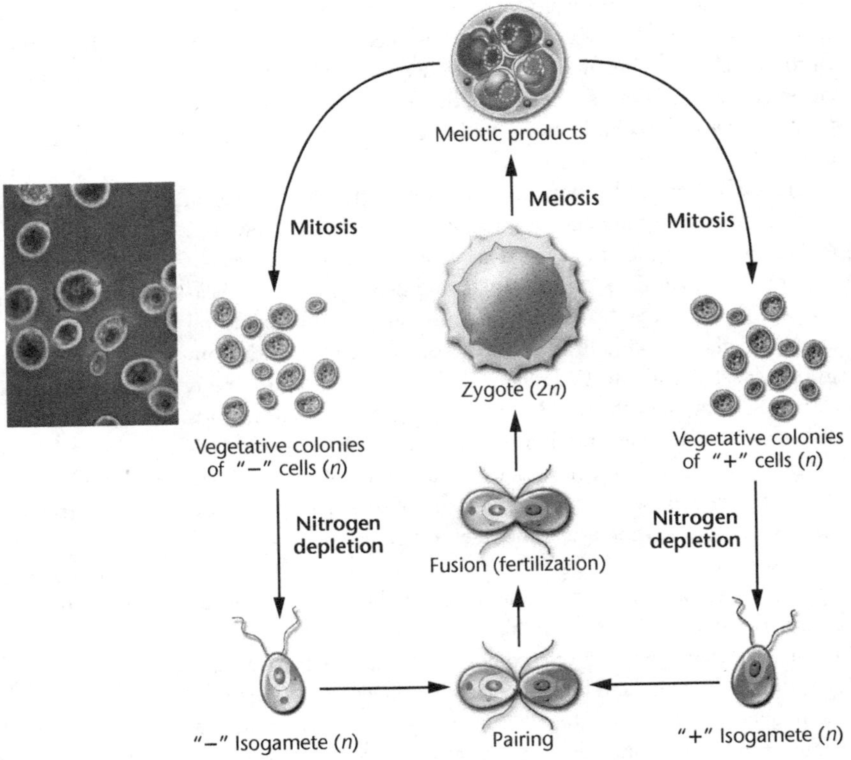


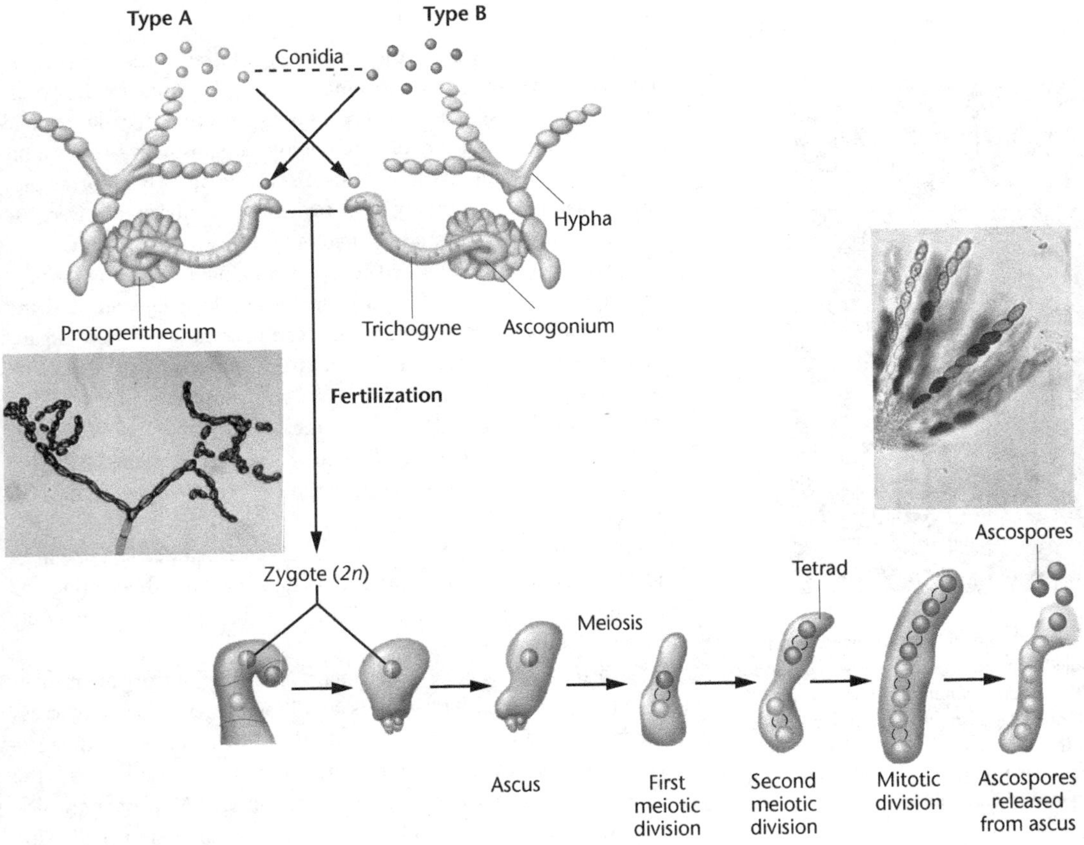


**FIGURE 5-21** Sexual reproduction during the life cycle of *Neurospora* is initiated following fusion of conidia (asexual spores) of opposite mating types. After fertilization, each diploid zygote becomes enclosed in an ascus where meiosis occurs, leading to four haploid cells, two of each mating type. A mitotic division then occurs, and the eight haploid ascospores are later released. Upon germination, the cycle may be repeated. The photographs show the vegetative stage of the organism and several asci that may form in a single structure, even though we have illustrated the events occurring in only one ascus.

fertilization in *Neurospora* (Figure 5–21), meiosis occurs in a saclike structure called the **ascus** (pl. asci), within which the initial set of haploid products, called a **tetrad**, are retained. *The term* tetrad *has a different meaning here than when it is used to describe the four-stranded chromosome configuration characteristic of meiotic prophase I in diploids.*

Following meiosis in *Neurospora*, each cell in the ascus divides mitotically, producing eight haploid **ascospores**. These can be dissected and examined morphologically or tested to determine their genotypes and phenotypes. Because the arrangement of the eight cells reflects the *sequence* of their formation following meiosis, the tetrad is "ordered" and we can do *ordered tetrad analysis*. This process is critical to our subsequent discussion.

## Gene-to-Centromere Mapping

When the ascospore pattern for a single pair of alleles ($a/+$) is analyzed in *Neurospora*, as diagrammed in Figure 5–22, the data can be used to calculate the map distance between that gene locus and the centromere. This process is sometimes referred to as **mapping the centromere**. It is accomplished by experimentally determining the frequency of recombination using tetrad data. Note in Figure 5–21 that once the four meiotic products of the tetrad are formed, a mitotic division occurs, resulting in eight ordered products (ascospores). If no crossover event occurs between the gene under study and the centromere, the pattern of ascospores (contained within an ascus) appears as shown in Figure 5–22(a) ($aaaa$++++).*

This pattern represents **first-division segregation**, because the two alleles are separated during the first meiotic division. However, crossover events will alter the pattern, as shown in Figure 5–22(b)

* The pattern (++++$aaaa$) can also be formed, but it is indistinguishable from ($aaaa$++++).

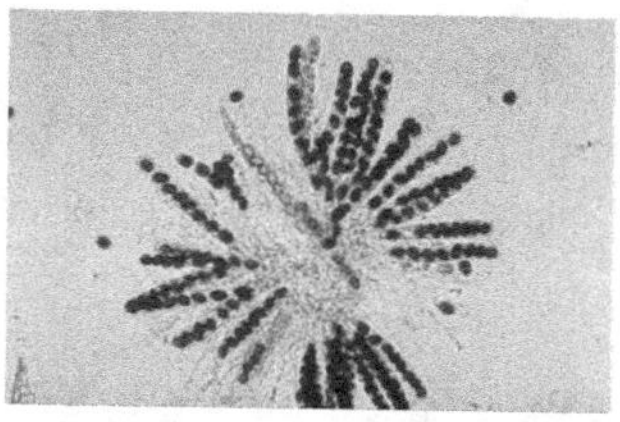

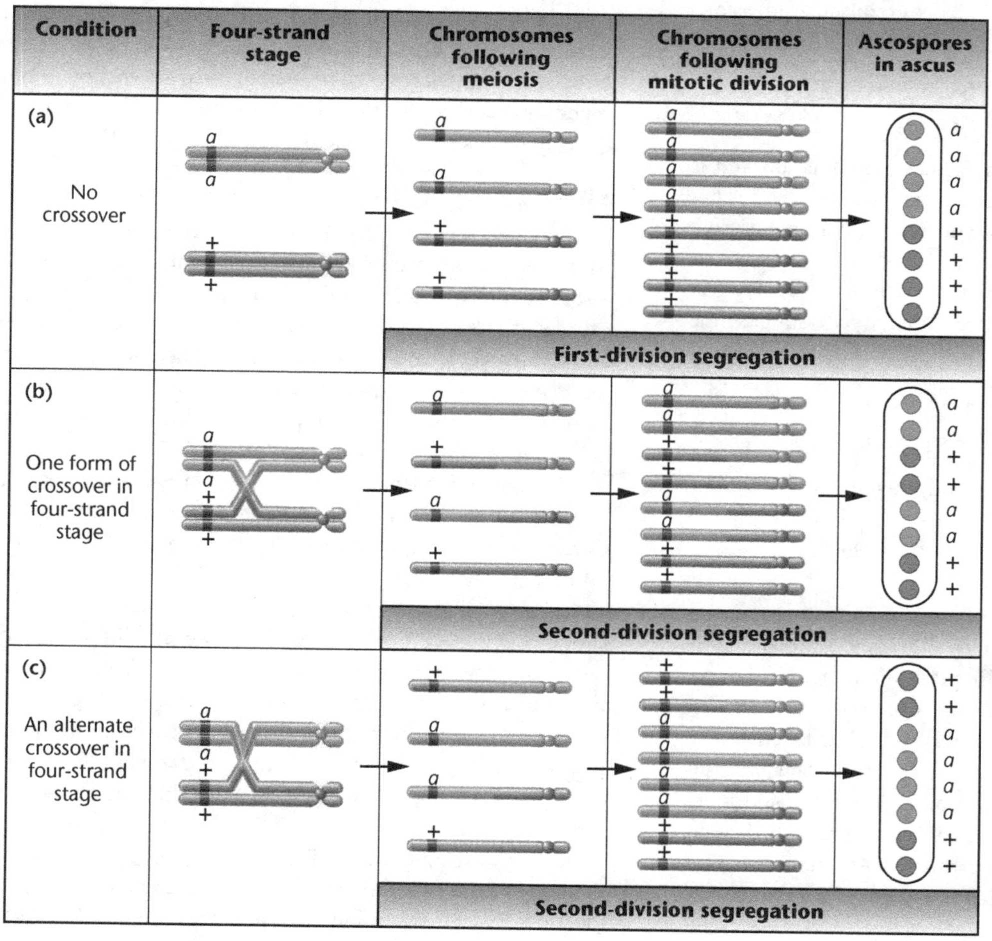


FIGURE 5–22 Three ways in which different ascospore patterns can be generated in *Neurospora*. Analysis of these patterns can serve as the basis of gene-to-centromere mapping. The photograph shows a variety of ascospore arrangements within *Neurospora* asci.

($aa{+}{+}aa{+}{+}$) and 5–22(c) ($++aaaa++$). Two other recombinant patterns also occur, depending on the chromatid orientation during the second meiotic division: ($++aa++aa$) and ($aa++++aa$). All four patterns, resulting from a crossover event between the *a* gene and the centromere, reflect **second-division segregation**, because the two alleles are not separated until the second meiotic division. Since the mitotic division simply replicates the patterns (increasing the 4 ascospores to 8), ordered tetrad data are usually condensed to reflect the genotypes of the four ascospore pairs, and six unique combinations are possible.

| | | | | |
|---|---|---|---|---|
| | **First-Division Segregation** | | | |
| (1) | *a* | *a* | + | + |
| (2) | + | + | *a* | *a* |
| | **Second-Division Segregation** | | | |
| (3) | *a* | + | *a* | + |
| (4) | + | *a* | + | *a* |
| (5) | + | *a* | *a* | + |
| (6) | *a* | + | + | *a* |

To calculate the distance between the gene and the centromere, data must be tabulated from a large number of asci resulting from a controlled cross. We then use these data to calculate the distance ($d$):

$$d = \frac{1/2\ (\text{second division segregant asci})}{\text{total asci scored}} \times 100$$

The distance ($d$) reflects the percentage of recombination and is only half the number of second-division segregant asci. This is because crossing over occurs in only two of the four chromatids during meiosis.

To illustrate, we use *a* for albino and + for wild type in *Neurospora*. In crosses between the two genetic types, suppose the following data are observed:

65 first-division segregants

70 second-division segregants

Thus, the distance between *a* and the centromere is

$$d = \frac{(1/2)(70)}{135} = 0.259 \times 100 = 25.9$$

or about 26 mu.

As the distance increases to 50 units, all asci should theoretically reflect second-division segregation. However, numerous factors prevent it in actuality. As in diploid organisms, accuracy is greatest when the gene and the centromere are relatively close together.

As we will discuss in the next section, we can also analyze haploid organisms in order to distinguish between linkage and independent assortment of two genes. Once linkage is established, mapping distances between gene loci are calculated. As a result, detailed maps of organisms such as *Saccharomyces, Neurospora,* and *Chlamydomonas* are now available.

## Ordered versus Unordered Tetrad Analysis

In our previous discussion, we assumed that the genotype of each ascospore and its position in the tetrad can be determined. To perform such an **ordered tetrad analysis**, individual asci must be dissected, and each ascospore must be tracked as it germinates. This is a tedious process, but it is essential for two types of analysis:

1. To distinguish between first-division segregation and second-division segregation of alleles in meiosis.
2. To determine whether or not recombination events are reciprocal. Such information is essential for "mapping the centromere," as we have just discussed. Thus, ordered tetrad analysis must be performed in order to map the distance between a gene and the centromere.

Ordered tetrad analysis has revealed that recombination events are not always reciprocal, particularly when the genes under study are closely linked. This observation has led to the investigation of the phenomenon called *gene conversion.* Because its discussion requires a background in DNA structure and analysis, we will return to this topic in Chapter 11.

Much less tedious than ordered tetrad analysis is to isolate individual asci, allow them to mature, and then determine the genotypes of each ascospore in no particular order. This approach is referred to as **unordered tetrad analysis.** As we shall see in the next section, such an analysis can be used to discover whether or not two genes are linked on the same chromosome and, if so, to determine the map distance between them.

## Linkage and Mapping

To show how analysis of genetic data derived from haploid organisms can be used to distinguish between linkage and independent assortment of two genes, and then allows mapping distances to be calculated between gene loci, we shall consider tetrad analysis in the alga *Chlamydomonas.* Except that the four meiotic products are not ordered and *do not* undergo a mitotic division following the completion of meiosis, the general principles discussed for *Neurospora* also apply to *Chlamydomonas.*

To compare independent assortment and linkage, imagine two mutant alleles, *a* and *b*, representing two distinct loci in *Chlamydomonas.* Suppose that 100 tetrads derived from the cross $ab \times ++$ yield the tetrad data shown in Table 5.1. As you can see, all tetrads produce one of three patterns. For example, all tetrads in category I produce two + + cells and two *ab* cells and are designated as **parental ditypes (P)**. Category II tetrads produce two $a+$ cells and two $+b$ cells and are called **nonparental ditypes (NP)**. Category III tetrads produce four cells that each have one of the four possible genotypes and are thus termed **tetratypes (T)**.

These data support the hypothesis that the genes represented by the *a* and *b* alleles are located on separate chromosomes. To understand why, you should refer to Figure 5–23. In parts (a) and (b) of that figure, the origin of parental (P) and nonparental (NP) ditypes is demonstrated for two unlinked genes. According to the Mendelian principle of independent assortment of unlinked genes, approximately equal proportions of these tetrad types are predicted. Thus, when the parental ditypes are equal to the nonparental ditypes, the two genes are not linked. The data in Table 5.1 confirm this prediction. Because independent assortment has occurred, it can be concluded that the two genes are located on separate chromosomes.

The origin of category III, the tetratypes, is diagrammed in Figure 5–23(c,d). The genotypes of tetrads in this category can be generated in two possible ways. Both involve a crossover event between one of the genes and the centromere. In Figure 5–23(c), the exchange involves one of the two chromosomes and occurs between gene *a* and the centromere; in Figure 5–23(d), the other chromosome is involved, and the exchange occurs between gene *b* and the centromere.

**TABLE 5.1**

**Tetrad Analysis in *Chlamydomonas***

| Category | I | II | III |
|---|---|---|---|
| Tetrad type | Parental (P) | Nonparental (NP) | Tetratypes (T) |
| | + + | *a* + | + + |
| Genotypes | + + | *a* + | *a* + |
| present | *a b* | + *b* | + *b* |
| | *a b* | + *b* | *a b* |
| Number of tetrads | 43 | 43 | 14 |

Production of tetratype tetrads does not alter the final ratio of the four genotypes present in all meiotic products. If the genotypes from 100 tetrads (which yield 400 cells) are computed, 100 of each genotype are found. This 1:1:1:1 ratio is predicted according to independent assortment.

Now consider the case where the genes *a* and *b* are linked (Figure 5–24). The same categories of tetrads will be produced. However, parental and nonparental ditypes will not necessarily occur in equal proportions; nor will the four genotypic combinations be found in equal numbers. For example, the following data might be encountered:

| Category I | Category II | Category III |
|---|---|---|
| P | NP | T |
| 64 | 6 | 30 |

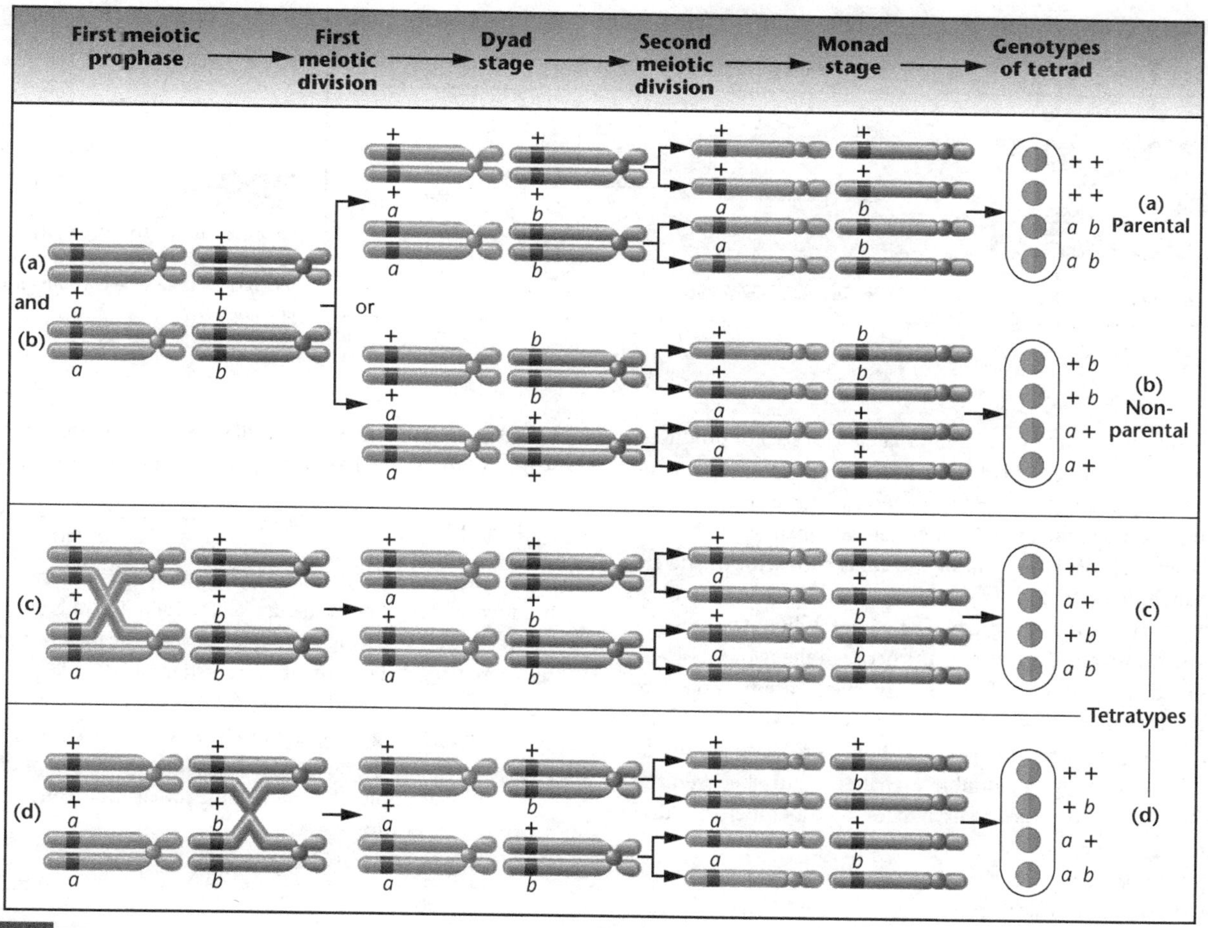


**FIGURE 5-23** The origin of various genotypes found in tetrads in *Chlamydomonas* when two genes located on separate chromosomes are considered.

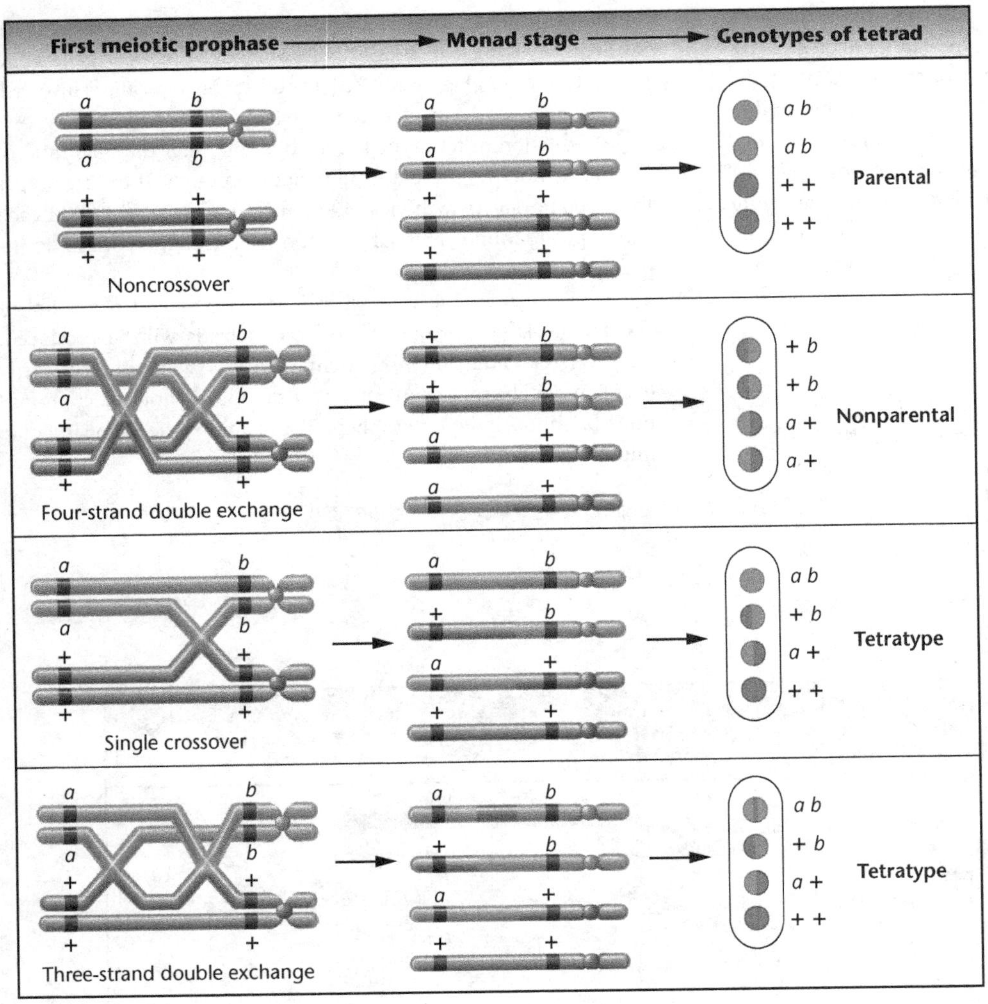


FIGURE 5–24 The various types of exchanges leading to the genotypes found in tetrads in *Chlamydomonas* when two genes located on the same chromosome are considered.

Since the parental and nonparental categories are not produced in equal proportions, we can conclude that independent assortment is not in operation and that the two genes are linked. We can then proceed to determine the map distance between them.

In the analysis of these data, we are concerned with the determination of which tetrad types represent genetic exchanges within the interval between the two genes. The parental ditype tetrads (P) arise only when no crossing over occurs between the two genes. The nonparental ditype tetrads (NP) arise only when a double exchange involving all four chromatids occurs between two genes. The tetratype tetrads (T) arise when either a single crossover occurs or when an alternative type of double exchange occurs between the two genes. The various types of exchanges described here are diagrammed in Figure 5–24.

When the proportion of the three tetrad types has been determined, it is possible to calculate the map distance between the two linked genes. The following formula computes the exchange frequency, which is proportional to the map distance between the two genes:

$$\text{exchange frequency (\%)} = \frac{\text{NP} + 1/2(\text{T})}{\text{total number of tetrads}} \times 100$$

In this formula, NP represents the nonparental tetrads; all meiotic products represent an exchange. The tetratype tetrads are represented by T; assuming only single exchanges, half of the meiotic products represent exchanges. The sum of the scored tetrads that fall into these categories is then divided by the total number of tetrads examined (P+NP+T). Multiplying that result by 100 converts it to a percentage, which is directly equivalent to the map distance between the genes.

In our example, the calculation reveals that genes $a$ and $b$ are separated by 21 mu:

$$\frac{6 + 1/2(30)}{100} = \frac{6 + 15}{100} = \frac{21}{100} = 0.21 \times 100 = 21\%$$

Although we have considered linkage analysis and mapping of only two genes at a time, such studies often involve three or more genes. In these cases, both gene sequence and map distances can be determined.

## 5.13 Did Mendel Encounter Linkage?

We conclude this chapter by examining a modern-day interpretation of the experiments that form the cornerstone of transmission genetics—Mendel's crosses with garden peas.

Some observers believe that Mendel had extremely good fortune in his classic experiments with the garden pea. In their view, he did not encounter any linkage relationships between the seven mutant characters in his crosses. Had Mendel obtained highly variable data characteristic of linkage and crossing over, these observers say, he might not have succeeded in recognizing the basic patterns of inheritance and interpreting them correctly.

The article by Stig Blixt, reprinted in its entirety in the box that follows, demonstrates the inadequacy of this hypothesis. As we shall see, some of Mendel's genes were indeed linked. We shall leave it to Stig Blixt to enlighten you as to why Mendel did not detect linkage.

### Why Didn't Gregor Mendel Find Linkage?

It is quite often said that Mendel was very fortunate not to run into the complication of linkage during his experiments. He used seven genes, and the pea has only seven chromosomes. Some have said that had he taken just one more, he would have had problems. This, however, is a gross oversimplification. The actual situation, most probably, is that Mendel worked with three genes in chromosome 4, two genes in chromosome 1, and one gene in each of chromosomes 5 and 7. (See Table 1.) It seems at first glance that, out of the 21 dihybrid combinations Mendel theoretically could have studied, no less than four (that is, *a–i, v–fa, v–le, fa–le*) ought to have resulted in linkages. As found, however, in hundreds of crosses and shown by the genetic map of the pea, *a* and *i* in chromosome 1 are so distantly located on the chromosome that no linkage is normally detected. The same is true for *v* and *le* on the one hand, and *fa* on the other, in chromosome 4. This leaves *v–le*, which ought to have shown linkage.

Mendel, however, does not seem to have published this particular combination and thus, presumably, never made the appropriate cross to obtain both genes segregating simultaneously. It is therefore not so astonishing that Mendel did not run into the complication of linkage, although he did not avoid it by choosing one gene from each chromosome.

**Stig Blixt**

*Weibullsholm Plant Breeding Institute, Landskrona, Sweden, and Centro de Energia Nuclear na Agricultura, Piracicaba, SP, Brazil.*

*Source:* Reprinted by permission from Stig Blixt. Why didn't Gregor Mendel find linkage? *Nature*, Vol. 256, p. 206.


**TABLE 1**

**Relationship between Modern Genetic Terminology and Character Pairs Used by Mendel**

| Character Pair Used by Mendel | Alleles in Modern Terminology | Located in Chromosome |
|---|---|---|
| Seed color, yellow-green | *I–i* | 1 |
| Seed coat and flowers, colored-white | *A–a* | 1 |
| Mature pods, smooth expanded-wrinkled indented | *V–v* | 4 |
| Inflorescences, from leaf axis-umbellate in top of plant | *Fa–fa* | 4 |
| Plant height, 0.5–1 m | *Le–le* | 4 |
| Unripe pods, green yellow | *Gp–gp* | 5 |
| Mature seeds, smooth wrinkled | *R–r* | 7 |

EXPLORING GENOMICS

# Human Chromosome Maps on the Internet

In Chapter 5 we discussed how recombination data can be analyzed to develop chromosome maps based on linkage. Although linkage analysis and chromosome mapping continue to be important approaches in genetics, chromosome maps are increasingly being developed for many species using genomics techniques to sequence entire chromosomes. As a result of the Human Genome Project, maps of human chromosomes are now freely available on the Internet. With the click of a mouse you can have immediate access to an incredible wealth of information. In this exercise we will explore the **National Center for Biotechnology Information (NCBI) Genes and Disease** Web site to learn more about human chromosome maps.

## ■ Exercise I – NCBI Genes and Disease

The NCBI Web site is an outstanding resource for genome data, and we will use it for several exercises in Exploring Genomics. Here we explore the Genes and Disease site, which presents human chromosome maps that show the locations of specific disease genes.

1. Access the Genes and Disease site at www.ncbi.nlm.nih.gov/books/bv.fcgi?rid=gnd&ref=sidebar.
2. Click on any of the chromosomes at the top of the page to see a map showing selected genes on that chromosome that are associated with genetic diseases. For example, click on chromosome 7; and then click on the OB gene label to learn about the role of this gene in obesity. Notice that the top of each chromosome page displays the number of genes on the chromosome and the number of base pairs the chromosome contains.
3. Look again at chromosome 7. At first you might think there are only five disease genes on this chromosome because the initial view shows only selected disease genes. However, if you click the "MapViewer" link for the chromosome (just above the drawing), you will see detailed information about the chromosome, including a complete "Master Map" of the genes it contains. Map Viewer information includes the following:

   Ideogram: This shows the G-banding staining pattern of the chromosome (explained in Chapter 12), used for cytogenetic maps. Click on a band in the ideogram to zoom in to that region of the chromosome and see the gene locations corresponding to that part of the cytogenetic map.

   Gene Symbols: Clicking on the gene symbols takes you to the **NCBI Entrez Gene database**, a searchable tool for information on genes in the NCBI database.

   Links: The items in the "Links" column provide access to OMIM (Online Mendelian Inheritance in Man, discussed in the Exploring Genomics feature for Chapter 3) data for a particular gene, as well as to protein information (*pn*) and lists of homologous genes (*hm;* these are other genes that have similar sequences).

4. Click on the links in MapViewer to learn more about a gene of interest.
5. Scan the chromosome maps in MapViewer until you see one of the genes that is listed as a "hypothetical gene or protein."
   a. What does it mean if a gene or protein is referred to as hypothetical?
   b. What information do you think genome scientists use to assign a gene locus for a gene encoding a hypothetical protein?

Visit the **NCBI Map Viewer** home page (www.ncbi.nlm.nih.gov/mapview/) for an excellent database containing chromosome maps for a wide variety of different organisms. Search this database to learn more about chromosome maps for an organism you are interested in.

## ■ Exercise II – Exploring Chromosome 2

One of the many valuable features of chromosome maps is their ability to reveal groups of related genes, or gene clusters, that may play a role in a particular pathway such as a human disease condition.

1. For an example, return to the Genes and Disease site, click on chromosome 2, and then click on the "Map Viewer" link. There, next to "Query" (near the top of the page) click the "[clear]" link to obtain access to all loci on this chromosome. On the Ideogram, click on band 2p24.3 and select "zoom in ×4." Look for the genes identified as *MSH2* and *MSH6* (you may need to scroll up and down the ideogram to locate each one) and then answer the following questions about MSH genes:
   a. What is the locus for *MSH2*? for *MSH6*?
   b. What are the functions of the proteins encoded by these genes?
   c. What human disease condition are these genes involved in?
   d. What functions of *MSH* genes support their role in the human disease you identified as the answer to part c above?
   e. *E. coli* obviously do not have a colon. Would you expect them to have an MSH homolog? Find out.
2. Use the Genes and Disease Web site to view a chromosome you are interested in, and explore the MapViewer links to learn more about genes of interest on the chromosome you selected.

## Chapter Summary

1. Genes located on the same chromosome are said to be linked. Alleles located close together on the same homolog are usually transmitted together during gamete formation. However, the mechanism of crossing over between homologs during meiosis results in the reshuffling of alleles, thereby contributing to genetic variability within gametes.
2. Early in the twentieth century, geneticists realized that crossover frequency is proportional to the distance between genes. This relationship provides an experimental basis for mapping the location of linked genes relative to one another along the chromosome.
3. Interference is a phenomenon describing the extent to which a crossover in one region of a chromosome influences the occurrence of a crossover in an adjacent region of the chromosome. The coefficient of coincidence ($C$) is a quantitative estimate of interference calculated by dividing the observed double crossovers by the expected double crossovers.
4. Due to statistical considerations described by the Poisson distribution, as the actual distance between two genes increases, experimentally determined mapping distances become more and more inaccurate (underestimated).
5. Extensive genetic maps have been created for organisms such as maize, mice, and *Drosophila*.
6. Cytological investigations of both maize and *Drosophila* reveal that crossing over involves a physical exchange of segments between nonsister chromatids.
7. Human linkage studies, initially relying on lod score analysis and somatic cell hybridization techniques, are now enhanced by the use of newly discovered molecular DNA markers.
8. In a few organisms, including *Drosophila* and *Aspergillus*, homologs pair during mitosis and undergo crossing over, though at a frequency far lower than during meiosis.
9. An exchange of genetic material between sister chromatids can occur during mitosis as well. These events are referred to as sister chromatid exchanges (SCEs). An elevated frequency of such events is seen in the human disorder, Bloom syndrome.
10. Linkage analysis and chromosome mapping are possible in haploid eukaryotes. Our discussion has described haploid gene-to-centromere and gene-to-gene mapping, as well as how to distinguish between linkage and independent assortment.
11. Evidence now suggests that several of the genes studied by Mendel are, in fact, linked. However, in such cases, the genes are sufficiently far apart to prevent the detection of linkage.

## INSIGHTS AND SOLUTIONS

1. In a series of two-point mapping crosses involving three genes linked on chromosome III in *Drosophila*, the following distances were calculated:

*cd*–*sr* 13 mu
*cd*–*ro* 16 mu

(a) Why can't we determine the sequence and construct a map of these three genes?

(b) What mapping data will resolve the issue?

(c) Can we tell which of the sequences shown here is correct?

*ro* —16— *cd* —13— *sr*

or

*sr* —13— *cd* —16— *ro*

**Solution:**

(a) It is impossible to do so because there are two possibilities based on these limited data:

Case 1: *cd* —13— *sr* —3— *ro*

or

Case 2: *ro* —16— *cd* —13— *sr*

(b) The map distance determined by crossing over between *ro* and *sr*. If case 1 is correct, it should be 3 mu, and if case 2 is correct, it should be 29 mu. In fact, this distance is 29 mu, demonstrating that case 2 is correct.

(c) No; based on the mapping data, they are equivalent.

2. In *Drosophila*, *Lyra (Ly)* and *Stubble (Sb)* are dominant mutations located at loci 40 and 58, respectively, on chromosome III. A recessive mutation with bright red eyes was discovered and shown also to be on chromosome III. A map is obtained by crossing a female who is heterozygous for all three mutations to a male homozygous for the bright red mutation (which we refer to here as *br*). The data in the table are generated. Determine the location of the *br* mutation on chromosome III. By referring to Figure 5–14, predict what mutation has been discovered. How could you be sure?

| Phenotype | Number |
|---|---|
| (1) *Ly Sb br* | 404 |
| (2) + + + | 422 |
| (3) *Ly* + + | 18 |
| (4) + *Sb br* | 16 |
| (5) *Ly* + *br* | 75 |
| (6) + *Sb* + | 59 |
| (7) *Ly Sb* + | 4 |
| (8) + + *br* | 2 |
| Total | = 1000 |

**Solution:** First determine the distribution of the alleles between the homologs of the heterozygous crossover parent (the female in this case). To do this, locate the most frequent reciprocal phenotypes, which arise from the noncrossover gametes. These are phenotypes 1 and 2. Each phenotype represents the alleles on one of the homologs. Therefore, the distribution of alleles is

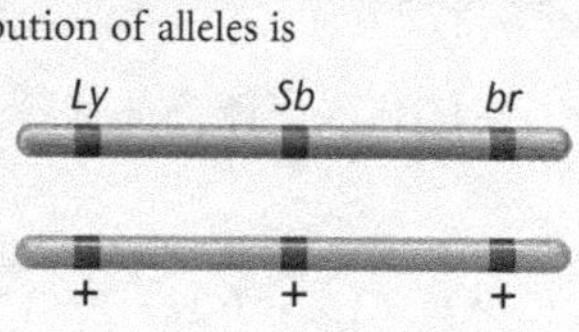


*Continued on next page*

*Insights and Solutions, continued*

Second, determine the correct *sequence* of the three loci along the chromosome. This is done by determining which sequence yields the observed double-crossover phenotypes that are the least frequent reciprocal phenotypes (7 and 8). If the sequence is correct as written, then a double crossover, depicted here,

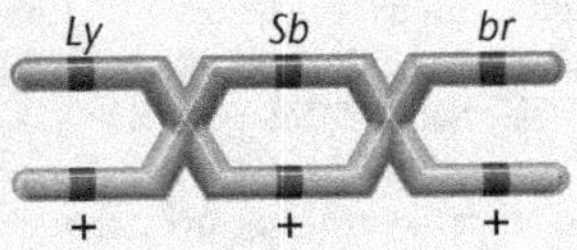


would yield *Ly* + *br* and + *Sb* + as phenotypes. Inspection shows that these categories (5 and 6) are actually single crossovers, not double crossovers. Therefore, the sequence, as written, is incorrect. There are only two other possible sequences. Either the *Ly* gene (Case A below) or the *br* gene (Case B below) is in the middle between the other two genes.

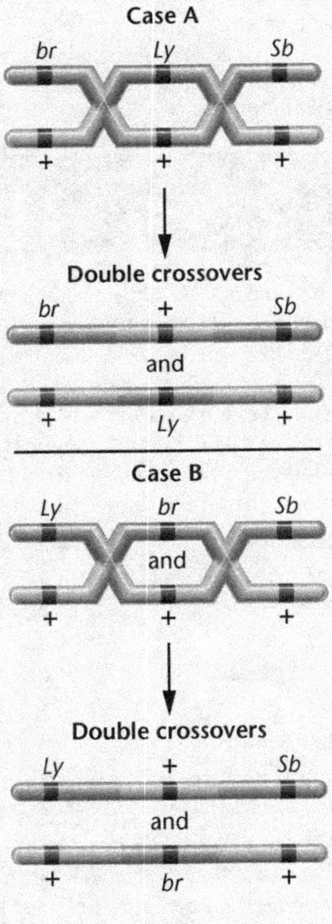


Comparison with the actual data shows that case B is correct. The double-crossover gametes 7 and 8 yield flies that express *Ly* and *Sb* but not *br*, or express *br* but not *Ly* and *Sb*. Therefore, the correct arrangement and sequence are as follows:

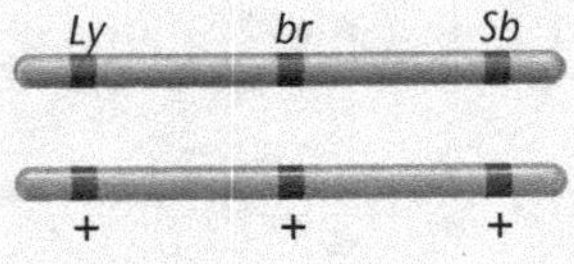


Once the sequence is found, determine the location of *br* relative to *Ly* and *Sb*. A single crossover between *Ly* and *br*, as shown here,

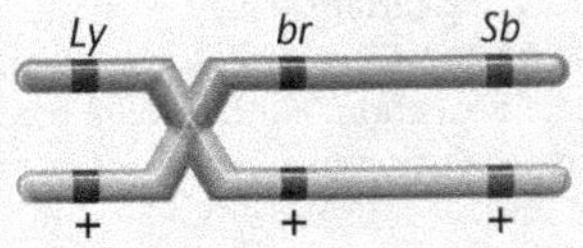


yields flies that are *Ly* + + and + *br Sb* (phenotypes 3 and 4). Therefore, the distance between the *Ly* and *br* loci is equal to

$$\frac{18 + 16 + 4 + 2}{1000} = \frac{40}{1000} = 0.04 = 4 \text{ mu}$$

Remember that, because we need to know the frequency of all crossovers between *Ly* and *br*, we must add in the double crossovers, since they represent two single crossovers occurring simultaneously. Similarly, the distance between the *br* and *Sb* loci is derived mainly from single crossovers between them.

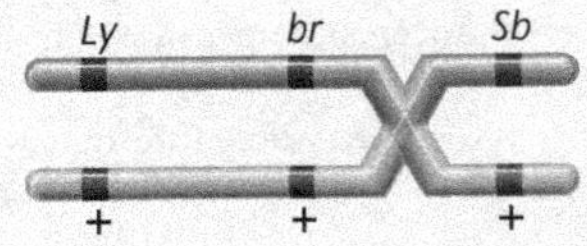


This event yields *Ly br* + and + + *Sb* phenotypes (phenotypes 5 and 6). Therefore, the distance equals

$$\frac{75 + 59 + 4 + 2}{1000} = \frac{140}{1000} = 0.14 = 14 \text{ mu}$$

The final map shows that *br* is located at locus 44, since *Lyra* and *Stubble* are known:

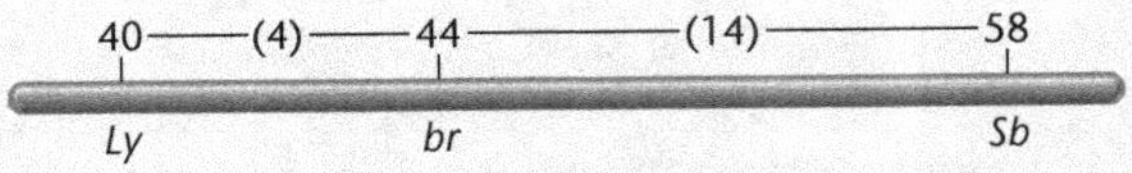


Inspection of Figure 5–14 reveals that the mutation *scarlet*, which produces bright red eyes, is known to sit at locus 44, so it is reasonable to hypothesize that the bright red eye mutation is an allele of *scarlet*. To test this hypothesis, we could cross females of our bright-red mutant with known *scarlet* males. If the two mutations are alleles, no complementation will occur, and all progeny will reveal a bright-red mutant eye phenotype. If complementation occurs, all progeny will show normal brick red (wild-type) eyes, since the bright-red mutation and *scarlet* are at different loci. (They are probably very close together.) In such a case, all progeny will be heterozygous at both the bright eye and the *scarlet* loci and will not express either mutation because they are both recessive. This cross represents what is called an *allelism test*.

3. In rabbits, black color (*B*) is dominant to brown (*b*), while full color (*C*) is dominant to chinchilla ($c^{ch}$). The genes controlling these traits are linked. Rabbits that are heterozygous for both traits and express black, full color were crossed with rabbits that express brown, chinchilla, with the following results:

31 brown, chinchilla

34 black, full color

16 brown, full color

19 black, chinchilla

Determine the arrangement of alleles in the heterozygous parents and the map distance between the two genes.

**Solution:** This is a two-point mapping problem. The two most prevalent reciprocal phenotypes are the noncrossovers, and the less frequent reciprocal phenotypes arise from a single crossover. The distribution of alleles is derived from the noncrossover phenotypes, because they enter gametes intact.

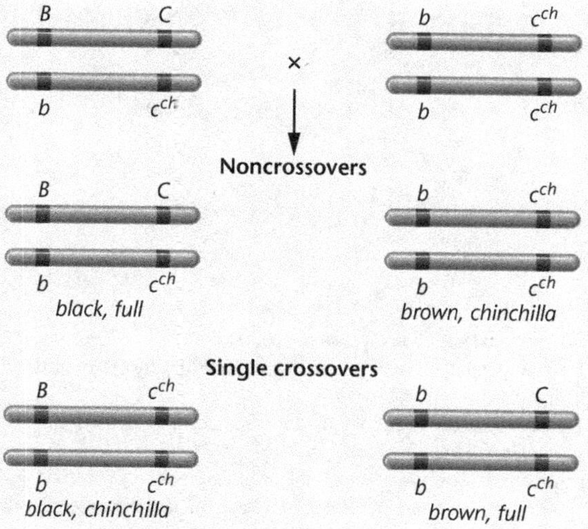


The single crossovers give rise to 35/100 offspring (35 percent). Therefore, the distance between the two genes is 35 mu.

4. In a cross in *Neurospora* where one parent expresses the mutant allele *a* and the other expresses a wild-type phenotype (+), the following data were obtained in the analysis of ascospores:

| Sequence of ascospores in ascus | Ascus Types 1 | 2 | 3 | 4 | 5 | 6 | |
|---|---|---|---|---|---|---|---|
| | + | *a* | *a* | + | *a* | + | |
| | + | *a* | *a* | + | *a* | + | |
| | + | *a* | + | *a* | + | *a* | |
| | + | *a* | + | *a* | + | *a* | |
| | *a* | + | *a* | + | + | *a* | |
| | *a* | + | *a* | + | + | *a* | |
| | *a* | + | + | *a* | *a* | + | |
| | *a* | + | + | *a* | *a* | + | |
| | 39 | 33 | 5 | 4 | 9 | 10 | Total = 100 |

Calculate the gene-to-centromere distance.

**Solution:** Ascus types 1 and 2 represent first-division segregation (fds), where no crossing over occurred between the *a* locus and the centromere. All others (3–6) represent second-division segregation (sds). By applying the formula

$$\text{distance} = \frac{1/2 \text{ sds}}{\text{total asci}}$$

we obtain the following result:

$$\begin{aligned} d &= 1/2(5 + 4 + 9 + 10)/100 \\ &= 1/2(28)/100 \\ &= 0.14 \\ &= 14 \text{ mu} \end{aligned}$$

## Problems and Discussion Questions

1. What is the significance of crossing over (which leads to genetic recombination) to the process of evolution?
2. Describe the cytological observation that suggests that crossing over occurs during the first meiotic prophase.
3. Why does more crossing over occur between two distantly linked genes than between two genes that are very close together on the same chromosome?
4. Why is a 50 percent recovery of single-crossover products the upper limit, even when crossing over *always* occurs between two linked genes?
5. Why are double-crossover events expected less frequently than single-crossover events?
6. What is the proposed basis for positive interference?
7. What two essential criteria must be met in order to execute a successful mapping cross?
8. The genes *dumpy* (*dp*), *clot* (*cl*), and *apterous* (*ap*) are linked on chromosome II of *Drosophila*. In a series of two-point mapping crosses, the following genetic distances were determined. What is the sequence of the three genes?

| | |
|---|---|
| *dp* – *ap* | 42 |
| *dp* – *cl* | 3 |
| *ap* – *cl* | 39 |

9. Consider two hypothetical recessive autosomal genes *a* and *b*, where a heterozygote is testcrossed to a double-homozygous mutant. Predict the phenotypic ratios under the following conditions:
(a) *a* and *b* are located on separate autosomes.
(b) *a* and *b* are linked on the same autosome, but are so far apart that a crossover always occurs between them.
(c) *a* and *b* are linked on the same autosome, but are so close together that a crossover almost never occurs.
(d) *a* and *b* are linked on the same autosome about 10 mu apart.
10. Colored aleurone in the kernels of corn is due to the dominant allele *R*. The recessive allele *r*, when homozygous, produces colorless aleurone. The plant color (not the kernel color) is controlled by another gene with two alleles, *Y* and *y*. The dominant *Y* allele results in green color, whereas the homozygous presence of the recessive *y* allele causes the plant to appear yellow. In a testcross between a plant of unknown genotype and phenotype and a plant that is homozygous recessive for both traits, the following progeny were obtained:

| | |
|---|---|
| Colored, green | 88 |
| Colored, yellow | 12 |
| Colorless, green | 8 |
| Colorless, yellow | 92 |

Explain how these results were obtained by determining the exact genotype and phenotype of the unknown plant, including the precise arrangement of the allleles on the homologs.

11. In the cross shown here, involving two linked genes, *ebony* (*e*) and *claret* (*ca*), in *Drosophila*, where crossing over does not occur in males, offspring were produced in a 2 + :1 *ca*:1 *e* phenotypic ratio:

| ♀ | | ♂ |
|---|---|---|
| $\dfrac{e \quad ca^+}{e^+ \quad ca}$ | × | $\dfrac{e \quad ca^+}{e^+ \quad ca}$ |

These genes are 30 units apart on chromosome III. What did crossing over in the female contribute to these phenotypes?

12. With two pairs of genes involved (*P*/*p* and *Z*/*z*), a testcross (*ppzz*) with an organism of unknown genotype indicated that the gametes produced were in the following proportions:

*PZ*, 42.4%; *Pz*, 6.9%; *pZ*, 7.1%; *pz*, 43.6%

Draw all possible conclusions from these data.

13. In a series of two-point mapping crosses involving five genes located on chromosome II in *Drosophila*, the following recombinant (single-crossover) frequencies were observed:

| | |
|---|---|
| *pr–adp* | 29% |
| *pr–vg* | 13 |
| *pr–c* | 21 |
| *pr–b* | 6 |
| *adp–b* | 35 |
| *adp–c* | 8 |
| *adp–vg* | 16 |
| *vg–b* | 19 |
| *vg–c* | 8 |
| *c–b* | 27 |

(a) Given that the *adp* gene is near the end of chromosome II (locus 83), construct a map of these genes.

(b) In another set of experiments, a sixth gene, *d*, was tested against *b* and *pr*:

| | |
|---|---|
| *d–b* | 17% |
| *d–pr* | 23% |

Predict the results of two-point mapping between *d* and *c*, *d* and *vg*, and *d* and *adp*.

14. Two different female *Drosophila* were isolated, each heterozygous for the autosomally linked genes *b* (*black body*), *d* (*dachs tarsus*), and *c* (*curved wings*). These genes are in the order *d–b–c*, with *b* being closer to *d* than to *c*. Shown here is the genotypic arrangement for each female along with the various gametes formed by both:

| Female A | | Gamete formation | Female B | |
|---|---|---|---|---|
| $\dfrac{d\ b\ +}{+\ +\ c}$ ↓ | | | $\dfrac{d\ +\ +}{+\ b\ c}$ ↓ | |
| (1) *d b c* | (5) *d* + + | | (1) *d b* + | (5) *d b c* |
| (2) + + + | (6) + *b c* | | (2) + + *c* | (6) + + + |
| (3) + + *c* | (7) *d* + *c* | | (3) *d* + *c* | (7) *d* + + |
| (4) *d b* + | (8) + *b* + | | (4) + *b* + | (8) + *b c* |

Identify which categories are noncrossovers (NCO), single crossovers (SCO), and double crossovers (DCO) in each case. Then, indicate the relative frequency in which each will be produced.

15. In *Drosophila*, a cross was made between females—all expressing the three X-linked recessive traits *scute* bristles (*sc*), *sable* body (*s*), and *vermilion* eyes (*v*)—and wild type males. In the $F_1$, all females were wild type, while all males expressed all three mutant traits. The cross was carried to the $F_2$ generation, and 1000 offspring were counted, with the results shown in the table.

| Phenotype | | | Offspring |
|---|---|---|---|
| *sc* | *s* | *v* | 314 |
| + | + | + | 280 |
| + | *s* | *v* | 150 |
| *sc* | + | + | 156 |
| *sc* | + | *v* | 46 |
| + | *s* | + | 30 |
| *sc* | *s* | + | 10 |
| + | + | *v* | 14 |

No determination of sex was made in the data.

(a) Using proper nomenclature, determine the genotypes of the $P_1$ and $F_1$ parents.

(b) Determine the sequence of the three genes and the map distances between them.

(c) Are there more or fewer double crossovers than expected?

(d) Calculate the coefficient of coincidence. Does it represent positive or negative interference?

16. Another cross in *Drosophila* involved the recessive, X-linked genes *yellow* (*y*), *white* (*w*), and *cut* (*ct*). A yellow-bodied, white-eyed female with normal wings was crossed to a male whose eyes and body were normal but whose wings were cut. The $F_1$ females were wild type for all three traits, while the $F_1$ males expressed the yellow-body and white-eye traits. The cross was carried to an $F_2$ progeny, and only male offspring were tallied. On the basis of the data shown here, a genetic map was constructed.

| Phenotype | | | Male Offspring |
|---|---|---|---|
| *y* | + | *ct* | 9 |
| + | *w* | + | 6 |
| *y* | *w* | *ct* | 90 |
| + | + | + | 95 |
| + | + | *ct* | 424 |
| *y* | *w* | + | 376 |
| *y* | + | + | 0 |
| + | *w* | *ct* | 0 |

(a) Diagram the genotypes of the $F_1$ parents.

(b) Construct a map, assuming that *white* is at locus 1.5 on the X chromosome.

(c) Were any double-crossover offspring expected?

(d) Could the $F_2$ female offspring be used to construct the map? Why or why not?

17. In *Drosophila, Dichaete* (*D*) is a mutation on chromosome III with a dominant effect on wing shape. It is lethal when homozygous. The genes *ebony* body (*e*) and *pink* eye (*p*) are recessive mutations on chromosome III. Flies from a *Dichaete* stock were crossed to homozygous ebony, pink flies, and the $F_1$ progeny, with a Dichaete phenotype, were backcrossed to the ebony, pink homozygotes. Using the results of this backcross shown in the table,

(a) Diagram this cross, showing the genotypes of the parents and offspring of both crosses.

(b) What is the sequence and interlocus distance between these three genes?

| Phenotype | Number |
|---|---|
| Dichaete | 401 |
| ebony, pink | 389 |
| Dichaete, ebony | 84 |
| pink | 96 |
| Dichaete, pink | 2 |
| ebony | 3 |
| Dichaete, ebony, pink | 12 |
| wild type | 13 |

18. *Drosophila* females homozygous for the third chromosomal genes *pink* and *ebony* (the same genes from Problem 17) were crossed with males homozygous for the second chromosomal gene *dumpy*. Because these genes are recessive, all offspring were wild type (normal). $F_1$ females were testcrossed to triply recessive males. If we assume that the two linked genes, *pink* and *ebony*, are 20 mu apart, predict the results of this cross. If the reciprocal cross were made ($F_1$ males—where no crossing over occurs—with triply recessive females), how would the results vary, if at all?
19. In *Drosophila*, two mutations, *Stubble* (*Sb*) and *curled* (*cu*), are linked on chromosome III. *Stubble* is a dominant gene that is lethal in a homozygous state, and *curled* is a recessive gene. If a female of the genotype

$$\frac{Sb \quad cu}{+ \quad +}$$

is to be mated to detect recombinants among her offspring, what male genotype would you choose as a mate?
20. In *Drosophila*, a heterozygous female for the X-linked recessive traits *a*, *b*, and *c* was crossed to a male that phenotypically expressed *a*, *b*, and *c*. The offspring occurred in the following phenotypic ratios.

| | | | |
|---|---|---|---|
| + | *b* | *c* | 460 |
| *a* | + | + | 450 |
| *a* | *b* | *c* | 32 |
| + | + | + | 38 |
| *a* | + | *c* | 11 |
| + | *b* | + | 9 |

No other phenotypes were observed.
(a) What is the genotypic arrangement of the alleles of these genes on the X chromosome of the female?
(b) Determine the correct sequence and construct a map of these genes on the X chromosome.
(c) What progeny phenotypes are missing? Why?
21. Why did Stern observe more "twin spots" than *singed* spots in his study of somatic crossing over? If he had been studying *tan* body color (locus 27.5) and *forked* bristles (locus 56.7) on the X chromosome of heterozygous females, what relative frequencies of tan spots, forked spots, and "twin spots" would you predict he might have found?
22. Are mitotic recombinations and sister chromatid exchanges effective in producing genetic variability in an individual? in the offspring of individuals?
23. What possible conclusions can be drawn from the observations that in male *Drosophila*, no crossing over occurs, and that during meiosis, synaptonemal complexes are not seen in males but are observed in females where crossing over occurs?
24. An organism of the genotype *AaBbCc* was testcrossed to a triply recessive organism (*aabbcc*). The genotypes of the progeny are in the following table.

| | | | |
|---|---|---|---|
| 20 | *AaBbCc* | 20 | *AaBbcc* |
| 20 | *aabbCc* | 20 | *aabbcc* |
| 5 | *AabbCc* | 5 | *Aabbcc* |
| 5 | *aaBbCc* | 5 | *aaBbcc* |

(a) If these three genes were all assorting independently, how many genotypic and phenotypic classes would result in the offspring, and in what proportion, assuming simple dominance and recessiveness in each gene pair?
(b) Answer part (a) again, assuming the three genes are so tightly linked on a single chromosome that no crossover gametes were recovered in the sample of offspring.
(c) What can you conclude from the *actual* data about the location of the three genes in relation to one another?
25. Based on our discussion of the potential inaccuracy of mapping (see Figure 5–13), would you revise your answer to Problem 24? If so, how?
26. In a certain plant, fruit is either red or yellow, and fruit shape is either oval or long. Red and oval are the dominant traits. Two plants, both heterozygous for these traits, were testcrossed, with the following results.

| Phenotype | Progeny Plant A | Progeny Plant B |
|---|---|---|
| red, long | 46 | 4 |
| yellow, oval | 44 | 6 |
| red, oval | 5 | 43 |
| yellow, long | 5 | 47 |
| | 100 | 100 |

Determine the location of the genes relative to one another and the genotypes of the two parental plants.
27. Two plants in a cross were each heterozygous for two gene pairs (*Ab*/*aB*) whose loci are linked and 25 mu apart. Assuming that crossing over occurs during the formation of both male and female gametes and that the *A* and *B* alleles are dominant, determine the phenotypic ratio of their offspring.
28. In a cross in *Neurospora* involving two alleles, *B* and *b*, the following tetrad patterns were observed. Calculate the distance between the gene and the centromere.

| Tetrad Pattern | Number |
|---|---|
| *BBbb* | 36 |
| *bbBB* | 44 |
| *BbBb* | 4 |
| *bBbB* | 6 |
| *BbbB* | 3 |
| *bBBb* | 7 |

29. In *Neurospora*, the cross *a*+ × +*b* yielded only two types of ordered tetrads in approximately equal numbers:
What can be concluded?

| | Spore Pair 1–2 | 3–4 | 5–6 | 7–8 |
|---|---|---|---|---|
| Tetrad Type 1 | *a* + | *a* + | + *b* | + *b* |
| Tetrad Type 2 | + + | + + | *a b* | *a b* |

30. Here are two sets of data derived from crosses in *Chlamydomonas* involving three genes represented by the mutant alleles *a*, *b*, and *c*:

| Genes | Cross | P | NP | T |
|---|---|---|---|---|
| 1 | *a* and *b* | 36 | 36 | 28 |
| 2 | *b* and *c* | 79 | 3 | 18 |
| 3 | *a* and *c* | ? | ? | ? |

Determine as much as you can concerning the arrangement of these three genes relative to one another. Assuming that *a* and *c* are linked and are 38 mu apart and that 100 tetrads are produced, describe the expected results of cross 3.

31. In *Chlamydomonas*, a cross *ab* × ++ yielded the following unordered tetrad data where *a* and *b* are linked:

| | | | | | | | |
|---|---|---|---|---|---|---|---|
| (1) | + +<br>+ +<br>*a b*<br>*a b* | 38 | (3) | *a* +<br>*a* +<br>+ *b*<br>+ *b* | 6 | (5) | *a b*<br>+ +<br>+ *b*<br>*a* + | 2 |
| (2) | + +<br>*a b*<br>+ +<br>*a b* | 5 | (4) | *a b*<br>*a* +<br>+ *b*<br>+ + | 17 | (6) | *a b*<br>+ *b*<br>*a* +<br>+ + | 3 |

(a) Identify the tetrads representing parental ditypes (P), nonparental ditypes (NP), and tetratypes (T).
(b) Explain the origin of tetrad (2).
(c) Determine the map distance between *a* and *b*.

32. The following results are ordered tetrad pairs from a cross between strain *cd* and strain + +:

| Tetrad Class | | | | | | |
|---|---|---|---|---|---|---|
| 1 | 2 | 3 | 4 | 5 | 6 | 7 |
| *c* + | *c* + | *c d* | + *d* | *c* + | *c d* | *c* + |
| *c* + | *c d* | *c d* | *c* + | + + | + + | + *d* |
| + *d* | + + | + + | *c* + | *c d* | *c d* | *c d* |
| + *d* | + *d* | + + | + *d* | + *d* | + + | + + |
| 1 | 17 | 41 | 1 | 5 | 3 | 1 |

They are summarized by tetrad classes.
(a) Name the ascus type of each class from 1 to 7 (P, NP, or T).
(b) The data support the conclusion that the *c* and *d* loci are linked. State the evidence in support of this conclusion.
(c) Calculate the gene–centromere distance for each locus.
(d) Calculate the distance between the two linked loci.
(e) Draw a chromosome map, including the centromere, and explain the discrepancy between the distances determined by the two different methods in parts (c) and (d).
(f) Describe the arrangement of crossovers needed to produce the ascus class 6.

33. In a cross in *Chlamydomonas*, *AB* × *ab*, 211 unordered asci were recovered:

| | | | | |
|---|---|---|---|---|
| 10 | *AB* | *Ab* | *aB* | *ab* |
| 102 | *Ab* | *aB* | *Ab* | *aB* |
| 99 | *AB* | *AB* | *ab* | *ab* |

(a) Correlate each of the three tetrad types in the problem with their appropriate tetrad designations (names).
(b) Are genes *A* and *B* linked?
(c) If they are linked, determine the map distance between the two genes. If they are unlinked, provide the maximum information you can about why you drew this conclusion.

**HOW DO WE KNOW?**

34. In this chapter, we focused on linkage, chromosomal mapping, and many associated phenomena. In the process, we found many opportunities to consider the methods and reasoning by which much of this information was acquired. From the explanations given in the chapter, what answers would you propose to the following fundamental questions?
(a) How was it established experimentally that the frequency of recombination (crossing over) between two genes is related to the distance between them along the chromosome?
(b) How do we know that specific genes are linked on a single chromosome, in contrast to being located on separate chromosomes?
(c) How do we know that crossing over results from a physical exchange between chromatids?
(d) How do we know that sister chromatids undergo recombination during mitosis?
(e) When designed matings cannot be conducted in an organism (for example, in humans), how do we learn that genes are linked, and how do we map them?

## Extra-Spicy Problems

35. A number of human–mouse somatic cell hybrid clones were examined for the expression of specific human genes and the presence of human chromosomes. The results are summarized in the following table. Assign each gene to the chromosome upon which it is located.

| | Hybrid Cell Clone | | | | | |
|---|---|---|---|---|---|---|
| | A | B | C | D | E | F |
| **Genes expressed** | | | | | | |
| *ENO1* (*enolase-1*) | − | + | − | + | + | − |
| *MDH1* (*malate dehydrogenase-1*) | + | + | − | + | − | + |
| *PEPS* (*peptidase S*) | + | − | + | − | − | − |
| *PGM1* (*phosphoglucomutase-1*) | − | + | − | + | + | − |
| **Chromosomes (present or absent)** | | | | | | |
| 1 | − | + | − | + | + | − |
| 2 | + | + | − | + | − | + |
| 3 | + | + | − | − | + | − |
| 4 | + | − | + | − | − | − |
| 5 | − | + | + | + | + | + |

36. A female of genotype

$$\frac{a \quad b \quad c}{+ \quad + \quad +}$$

produces 100 meiotic tetrads. Of these, 68 show no crossover events. Of the remaining 32, 20 show a crossover between *a* and *b*, 10 show a crossover between *b* and *c*, and 2 show a double crossover between *a* and *b* and between *b* and *c*. Of the 400 gametes produced, how many of each of the 8 different genotypes will be produced? Assuming the order *a–b–c* and the allele arrangement previously shown, what is the map distance between these loci?

37. In laboratory class, a genetics student was assigned to study an unknown mutation in *Drosophila* that had a whitish eye. He crossed females from his true-breeding mutant stock to wild-type (brick red–eyed) males, recovering all wild-type $F_1$ flies. In the $F_2$ generation, the following offspring were recovered in the following proportions:

| | |
|---|---|
| wild type | 5/8 |
| bright red | 1/8 |
| brown eye | 1/8 |
| white eye | 1/8 |

The student was stumped until the instructor suggested that perhaps the whitish eye in the original stock was the result of homozygosity for a mutation causing brown eyes *and* a mutation causing bright red eyes, illustrating gene interaction (see Chapter 4). After much thought, the student was able to analyze the data, explain the results, and learn several things about the location of the two genes relative to one another. One key to his understanding was that crossing over occurs in *Drosophila* females but not in males. Based on his analysis, what did the student learn about the two genes?

38. *Drosophila melanogaster* has one pair of sex chromosomes (XX or XY) and three pairs of autosomes, referred to as chromosomes II, III, and IV. A genetics student discovered a male fly with very short (*sh*) legs. Using this male, the student was able to establish a pure breeding stock of this mutant and found that it was recessive. She then incorporated the mutant into a stock containing the recessive gene *black* (*b*, body color located on chromosome II) and the recessive gene *pink* (*p*, eye color located on chromosome III). A female from the homozygous black, pink, short stock was then mated to a wild-type male. The $F_1$ males of this cross were all wild type and were then backcrossed to the homozygous *b, p, sh* females. The $F_2$ results appeared as shown in the table that follows. No other phenotypes were observed.

| | Wild | Pink* | Black, Short* | Black, Pink, Short |
|---|---|---|---|---|
| Females | 63 | 58 | 55 | 69 |
| Males | 59 | 65 | 51 | 60 |

*Other trait or traits are wild type.

(a) Based on these results, the student was able to assign *short* to a linkage group (a chromosome). Which one was it? Include your step-by-step reasoning.

(b) The student repeated the experiment, making the reciprocal cross, $F_1$ females backcrossed to homozygous *b, p, sh* males. She observed that 85 percent of the offspring fell into the given classes, but that 15 percent of the offspring were equally divided among *b* + *p*, *b* + +, + *sh p*, and + *sh* + phenotypic males and females. How can these results be explained, and what information can be derived from the data?

39. In *Drosophila*, a female fly is heterozygous for three mutations, *Bar* eyes (*B*), *miniature* wings (*m*), and *ebony* body (*e*). Note that *Bar* is a dominant mutation. The fly is crossed to a male with normal eyes, miniature wings, and ebony body. The results of the cross are as follows.

| | |
|---|---|
| 111 miniature | 101 Bar, ebony |
| 29 wild type | 31 Bar, miniature, ebony |
| 117 Bar | 35 ebony |
| 26 Bar, miniature | 115 miniature, ebony |

Interpret the results of this cross. If you conclude that linkage is involved between any of the genes, determine the map distance(s) between them.

40. The gene controlling the Xg blood group alleles ($Xg^+$ and $Xg^-$) and the gene controlling a newly described form of inherited recessive muscle weakness called *episodic muscle weakness* (*EMWX*) (Ryan et al., 1999) are closely linked on the X chromosome in humans at position Xp22.3 (the tip of the short arm). A male with EMWX who is $Xg^-$ marries a woman who is $Xg^+$, and they have eight daughters and one son all of whom are normal for muscle function, the male being $Xg^+$ and all the daughters being heterozygous at both the *EMWX* and *Xg* loci. Following is a table that lists three of the daughters with the phenotypes of their husbands

and children. (a) Create a pedigree that represents all data stated above and in the table below. (b) For each of the offspring, indicate whether or not a crossover was required to produce the phenotypes that are given.

| | Husband's Phenotype | Offspring's Sex | Offspring's Phenotype |
|---|---|---|---|
| *Daughter 1:* | $Xg^+$ | female | $Xg^+$ |
| | | male | EMWX, $Xg^+$ |
| *Daughter 2:* | $Xg^-$ | male | $Xg^-$ |
| | | female | $Xg^+$ |
| | | male | EMWX, $Xg^-$ |
| *Daughter 3:* | $Xg^-$ | male | EMWX, $Xg^-$ |
| | | male | $Xg^+$ |
| | | male | $Xg^-$ |
| | | male | EMWX, $Xg^+$ |
| | | male | $Xg^-$ |
| | | male | EMWX, $Xg^-$ |
| | | female | $Xg^+$ |
| | | female | $Xg^-$ |
| | | female | $Xg^+$ |

41. Because of the relatively high frequency of meiotic errors that lead to developmental abnormalities in humans, many research efforts have focused on identifying correlations between error frequency and chromosome morphology and behavior. Tease et al. (2002) studied human fetal oocytes of chromosomes 21, 18, and 13 using an immunocytological approach that allowed a direct estimate of the frequency and position of meiotic recombination. Below is a summary of information (modified from Tease et al., 2002) that compares recombination frequency with the frequency of trisomy for chromosomes 21, 18, and 13. (*Note:* You may want to read appropriate portions of Chapter 8 for descriptions of these trisomic conditions.)

| Trisomic | Mean Recombination Frequency | Live-born Frequency |
|---|---|---|
| *Chromosome 21* | 1.23 | 1/700 |
| *Chromosome 18* | 2.36 | 1/3000–1/8000 |
| *Chromosome 13* | 2.50 | 1/5000–1/19000 |

(a) What conclusions can be drawn from these data in terms of recombination and nondisjunction frequencies? How might recombination frequencies influence trisomic frequencies?

(b) Other studies indicate that the number of crossovers per oocyte is somewhat constant, and it has been suggested that positive chromosomal interference acts to spread out a limited number of crossovers among as many chromosomes as possible. Considering information in part (a), speculate on the selective advantage positive chromosomal interference might confer.

Transmission electron micrograph of conjugating *E. coli*

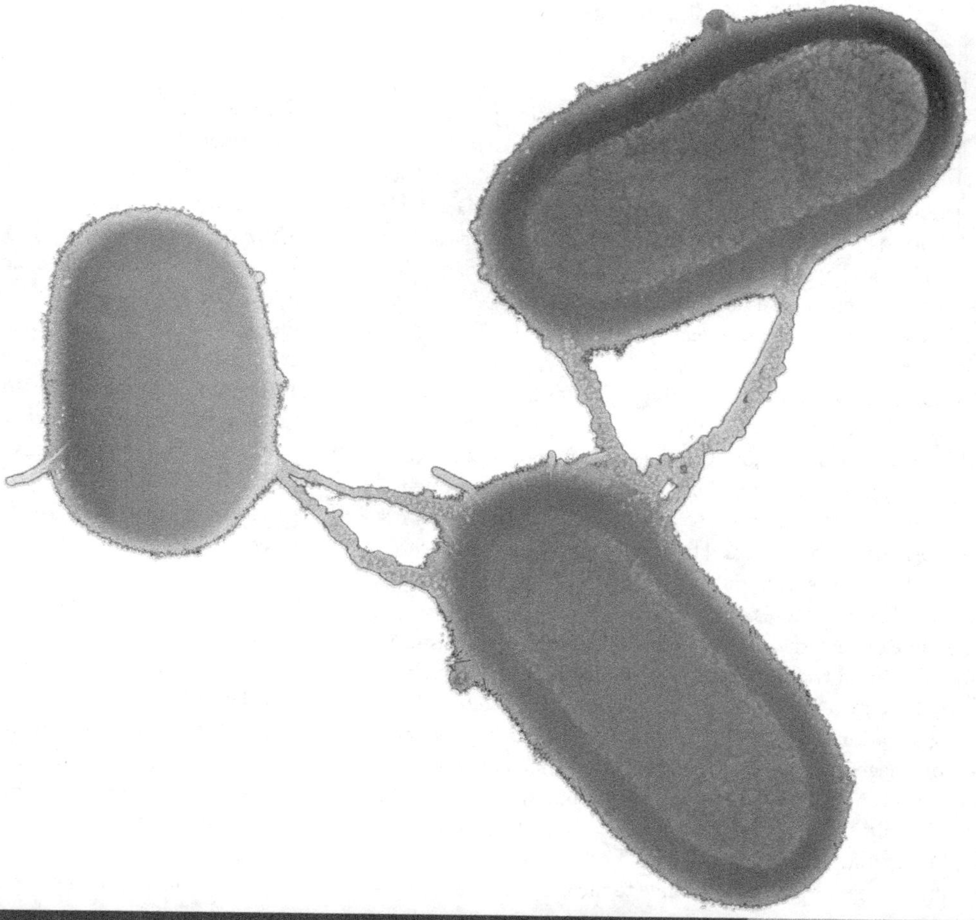

# 6

# Genetic Analysis and Mapping in Bacteria and Bacteriophages

## CHAPTER CONCEPTS

- Bacterial genomes are most often contained in a single circular chromosome.
- Bacteria have developed numerous ways of exchanging and recombining genetic information between individual cells, including conjugation, transformation, and transduction.
- The ability to undergo conjugation and to transfer the bacterial chromosome from one cell to another is governed by genetic information contained in the DNA of a "fertility," or F, factor.
- The F factor can exist autonomously in the bacterial cytoplasm as a plasmid, or it can integrate into the bacterial chromosome, where it facilitates the transfer of the host chromosome to the recipient cell, leading to genetic recombination.
- Genetic recombination during conjugation provides a means of mapping bacterial genes.
- Bacteriophages are viruses that have bacteria as their hosts.
- During infection of the bacterial host, bacteriophage DNA is injected into the host cell, where it is replicated and directs the reproduction of the bacteriophage.
- During bacteriophage infection, replication of the phage DNA may be followed by its recombination, which may serve as the basis for intergenic mapping.

In this chapter, we shift from the consideration of transmission genetics and mapping in eukaryotes to a discussion of the analysis of genetic recombination and mapping in bacteria and bacteriophages, viruses that have bacteria as their host. As we focus on these topics, it will become clear that complex processes have evolved in bacteria and bacteriophages that transfer genetic information between individual cells within populations. These processes provide geneticists with the basis for chromosome mapping.

The study of bacteria and bacteriophages has been essential to the accumulation of knowledge in many areas of genetic study. For example, much of what we know about the expression and regulation of genetic information was initially derived from experimental work with them. Furthermore, as we shall see in Chapter 13, our extensive knowledge of bacteria and their resident plasmids has served as the basis for their widespread use in DNA cloning and other recombinant DNA procedures.

The value of bacteria and their viruses as research organisms in genetics is based on two important characteristics that they display. First, they have extremely short reproductive cycles. Literally hundreds of generations, amounting to billions of genetically identical bacteria or phages, can be produced in short periods of time. Second, they can also be studied in pure culture. That is, a single species or mutant strain of bacteria or one type of virus can with ease be isolated and investigated independently of other similar organisms. As a result, they have been indispensable to the progress made in genetics over the past half century.

## 6.1 Bacteria Mutate Spontaneously and Grow at an Exponential Rate

Genetic studies using bacteria depend on our ability to study mutations in these organisms. It has long been known that genetically homogeneous cultures of bacteria occasionally give rise to cells exhibiting heritable variation, particularly with respect to growth under unique environmental conditions. Prior to 1943, the source of this variation was hotly debated. The majority of bacteriologists believed that environmental factors induced changes in certain bacteria, leading to their survival or adaptation to the new conditions. For example, strains of *E. coli* are known to be sensitive to infection by the bacteriophage T1. Infection by the bacteriophage T1 leads to reproduction of the virus at the expense of the bacterial host, from which new phages are released as the host cell is disrupted, or lysed. If a plate of *E. coli* is uniformly sprayed with T1, almost all cells are lysed. Rare *E. coli* cells, however, survive infection and are not lysed. If these cells are isolated and established in pure culture, all their descendants are resistant to T1 infection. The **adaptation hypothesis**, put forth to explain this type of observation, implies that the interaction of the phage and bacterium is essential to the acquisition of immunity. In other words, exposure to the phage "induces" resistance in the bacteria.

BACTERIAL GENETICS WEB TUTORIAL 6.1

On the other hand, the occurrence of **spontaneous mutations**, which occur regardless of the presence or absence of bacteriophage T1, suggested an alternative model to explain the origin of resistance in *E. coli*. In 1943, Salvador Luria and Max Delbruck presented the first convincing evidence that bacteria, like eukaryotic organisms, are capable of spontaneous mutation. Their experiment, referred to as the **fluctuation test**, marks the initiation of modern bacterial genetic study. We will explore this discovery in Chapter 16. Spontaneous mutation is now considered the primary source of genetic variation in bacteria.

Mutant cells that arise spontaneously in otherwise pure cultures can be isolated and established independently from the parent strain by the use of selection techniques. *Selection* refers to growing the organism under conditions where only the desired mutant grows well, while the wild type does not grow. With carefully designed selection, mutations for almost any desired characteristic can now be isolated. Because bacteria and viruses usually contain only single chromosomes and are therefore haploid, all mutations are expressed directly in the descendants of mutant cells, adding to the ease with which these microorganisms can be studied.

Bacteria are grown either in a liquid culture medium or in a petri dish on a semisolid agar surface. If the nutrient components of the growth medium are very simple and consist only of an organic carbon source (such as glucose or lactose) and various inorganic ions, including $Na^+$, $K^+$, $Mg^{++}$, $Ca^{++}$, and $NH_4^+$ present as inorganic salts, it is called **minimal medium**. To grow on such a medium, a bacterium must be able to synthesize all essential organic compounds (e.g., amino acids, purines, pyrimidines, sugars, vitamins, and fatty acids). A bacterium that can accomplish this remarkable biosynthetic feat—one that the human body cannot duplicate—is a **prototroph**. It is said to be wild type for all growth requirements. On the other hand, if a bacterium loses, through mutation, the ability to synthesize one or more organic components, it is an **auxotroph**. For example, a bacterium that loses the ability to make histidine is designated as a $his^-$ auxotroph, in contrast to its prototrophic $his^+$ counterpart. For the $his^-$ bacterium to grow, this amino acid must be added as a supplement to the minimal medium. Medium that has been extensively supplemented is called *complete medium*.

To study mutant bacteria quantitatively, an inoculum of bacteria—a small amount of a bacteria-containing solution, for example, 0.1 or 1.0 mL—is placed in liquid culture medium. A graph of the characteristic growth pattern for a bacteria culture is shown in Figure 6–1. Initially, during the **lag phase**, growth is slow. Then, a period of rapid growth, called the **logarithmic (log) phase**, ensues. During this phase, cells divide continually with a fixed time interval between cell divisions, resulting in exponential growth. When a cell density of about $10^9$ cells/mL of culture medium is reached, nutrients become limiting and cells cease dividing; at this point, the cells

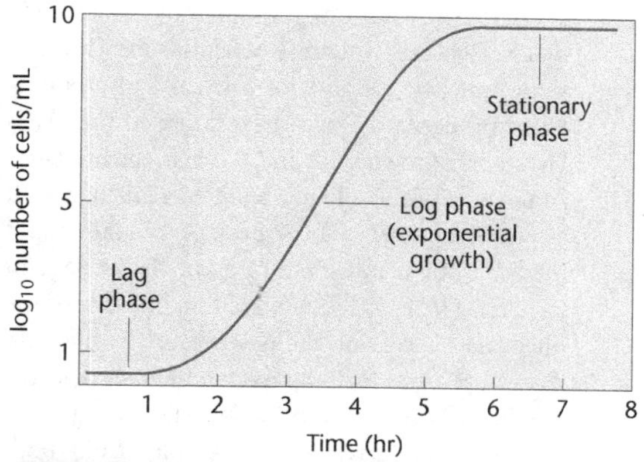


**FIGURE 6–1** Typical bacterial population growth curve showing the initial lag phase, the subsequent log phase where exponential growth occurs, and the stationary phase that occurs when nutrients are exhausted. Eventually, all cells will die.

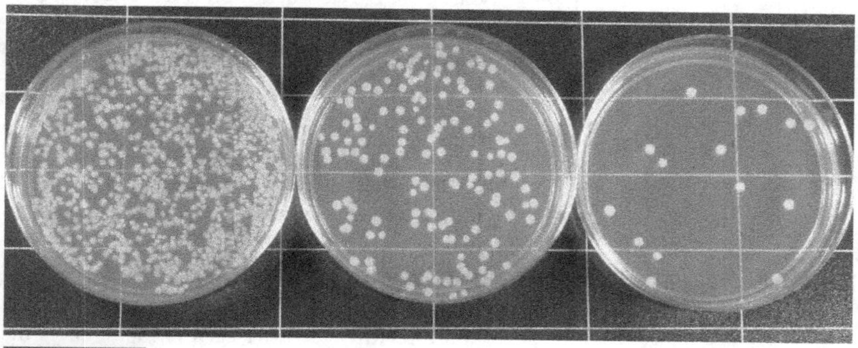

**FIGURE 6–2** Results of the serial dilution technique and subsequent culture of bacteria. Each dilution varies by a factor of 10. Each colony was derived from a single bacterial cell.

enter the **stationary phase**. The doubling time during the log phase can be as short as 20 minutes. Thus, an initial inoculum of a few thousand cells added to the culture easily achieves maximum cell density during an overnight incubation.

Cells grown in liquid medium can be quantified by transferring them to the semisolid medium of a petri dish. Following incubation and many divisions, each cell gives rise to a colony visible on the surface of the medium. By counting colonies, it is possible to estimate the number of bacteria present in the original culture. If the number of colonies is too great to count, then successive dilutions (in a technique called serial dilution) of the original liquid culture are made and plated, until the colony number is reduced to the point where it can be counted (Figure 6–2). This technique allows the number of bacteria present in the original culture to be calculated.

For example, let's assume that the three petri dishes in Figure 6–2 represent dilutions of the liquid culture by $10^{-3}$, $10^{-4}$, and $10^{-5}$ (from left to right).* We need only select the dish in which the number of colonies can be counted accurately. Because each colony presumably arose from a single bacterium, the number of colonies times the dilution factor represents the number of bacteria in each milliliter (mL) of the initial inoculum before it was diluted. In Figure 6–2, the rightmost dish contains 15 colonies. The dilution factor for a $10^{-5}$ dilution is $10^5$. Therefore, the initial number of bacteria was $15 \times 10^5$ per mL.

*$10^{-5}$ represents a 1:100,000 dilution.

6.2

## Conjugation Is One Means of Genetic Recombination in Bacteria

Development of techniques that allowed the identification and study of bacterial mutations led to detailed investigations of the arrangement of genes on the bacterial chromosome. Such studies began in 1946, when Joshua Lederberg and Edward Tatum showed that bacteria undergo **conjugation**, a process by which genetic information from one bacterium is transferred to and recombined with that of another bacterium (see the chapter opening photograph). Like meiotic crossing over in eukaryotes, this process of **genetic recombination** in bacteria provided the basis for the development of chromosome mapping methodology. Note that the term *genetic recombination*, as applied to bacteria and bacteriophages, refers to the *replacement* of one or more genes present in one strain with those from a genetically distinct strain. While this is somewhat different from our use of the term in eukaryotes—where it describes crossing over resulting in reciprocal exchange events—the overall effect is the same: Genetic information is transferred from one chromosome to another, resulting in an altered genotype. Two other phenomena that result in the transfer of genetic information from one bacterium to another, *transformation* and *transduction*, have also helped us determine the arrangement of genes on the bacterial chromosome. We shall discuss these processes later in this chapter.

Lederberg and Tatum's initial experiments were performed with two multiple-auxotroph strains (nutritional mutants) of *E. coli* strain K12. As shown in Figure 6–3, strain A required methionine (met) and biotin (bio) in order to grow, whereas strain B required threonine (thr), leucine (leu), and thiamine (thi). Neither strain would grow on minimal medium. The two strains were first grown separately in supplemented media, and then cells from both were mixed and grown together for several more generations. They were then plated on minimal medium. Any cells that grew on minimal medium were prototrophs. It is highly improbable that any of the cells containing two or three mutant genes would undergo spontaneous mutation simultaneously at two or three independent locations to become wild-type cells. Therefore, the researchers assumed that any prototrophs recovered must have arisen as a result of some form of genetic exchange and recombination between the two mutant strains.

In this experiment, prototrophs were recovered at a rate of $1/10^7$ (or $10^{-7}$) cells plated. The controls for this experiment

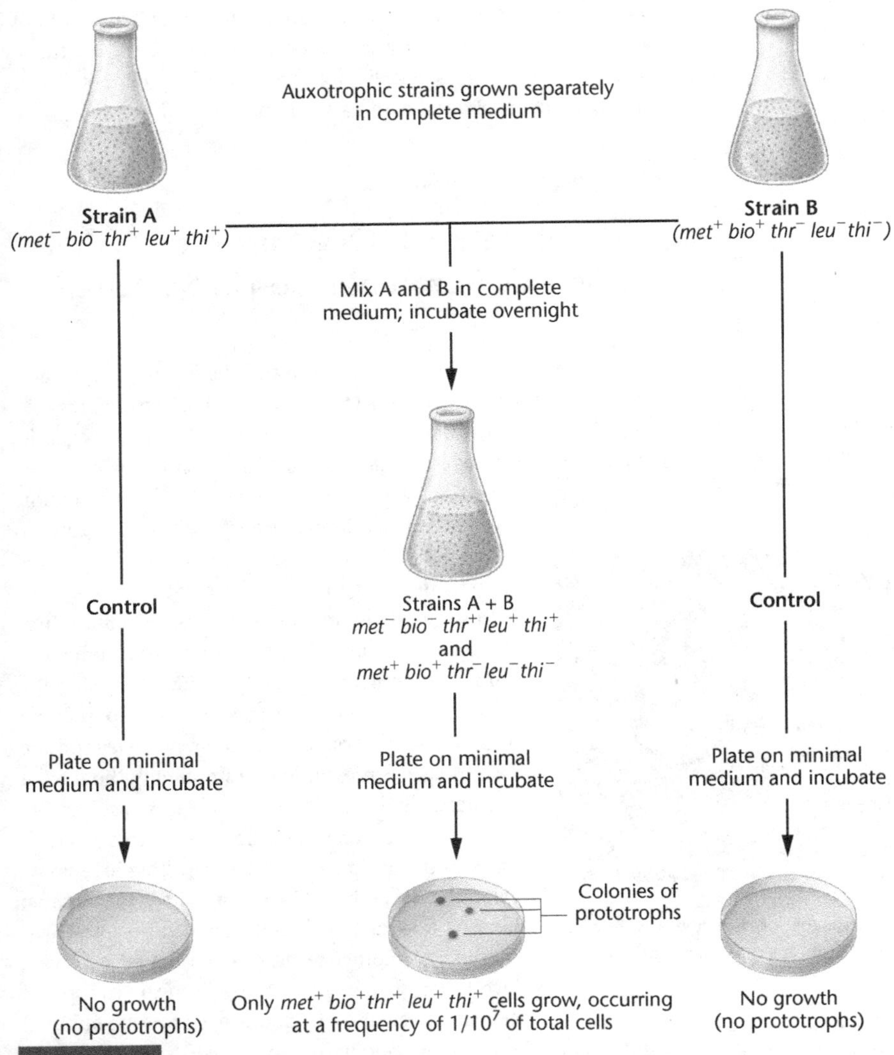


**FIGURE 6–3** Production of prototrophs as a result of genetic recombination between two auxotrophic strains. Neither auxotrophic strain will grow on minimal medium, but prototrophs do, suggesting that genetic recombination has occurred.

consisted of separate plating of cells from strains A and B on minimal medium. No prototrophs were recovered. On the basis of these observations, Lederberg and Tatum proposed that, while the events were indeed rare, genetic recombination had occurred.

## $F^+$ and $F^-$ Bacteria

Lederberg and Tatum's findings were soon followed by numerous experiments that elucidated the physical nature and the genetic basis of conjugation. It quickly became evident that different strains of bacteria were capable of effecting a unidirectional transfer of genetic material. When cells serve as donors of parts of their chromosomes, they are designated as **$F^+$ cells** (F for "fertility"). Recipient bacteria, designated as **$F^-$ cells**, receive the donor chromosome material (now known to be DNA), and recombine it with part of their own chromosome.

Experimentation subsequently established that cell-to-cell contact is essential for chromosome transfer. Support for this concept was provided by Bernard Davis, who designed the Davis U-tube for growing $F^+$ and $F^-$ cells (Figure 6–4). At the base of the tube is a sintered glass filter with a pore size that allows passage of the liquid medium but is too small to allow the passage of bacteria. The $F^+$ cells are placed on one side of the filter and $F^-$ cells on the other side. The medium passes back and forth across the filter so that it is shared by both sets of bacterial cells during incubation. When Davis plated samples from both sides of the tube on minimal medium, no prototrophs were found, and he logically concluded that *physical contact between cells of the two strains is essential to genetic recombination*. We now know that this physical interaction is the initial stage of the process of conjugation and is mediated by a structure called the **F pilus** (or **sex pilus**; pl. pili), a microscopic tubular extension of the cell. Bacteria often have many pili of different types performing different cellular functions, but all pili are involved in some way with adhesion (the binding together of cells). After contact has been initiated between mating pairs through F pili (Figure 6–5), chromosome transfer is possible.

Later evidence established that $F^+$ cells contain a **fertility factor (F factor)** that confers the ability to donate part of their chromosome during conjugation. Experiments by Joshua and Esther Lederberg and by William Hayes and Luca Cavalli-Sforza showed that certain environmental conditions eliminate the F factor from otherwise fertile cells. However, if these "infertile" cells are then grown with fertile donor cells, the F factor is regained. These findings led to the hypothesis that the F factor is a mobile element, a conclusion further supported by the observation that, after conjugation, recipient $F^-$ cells always become $F^+$. Thus, in addition to the rare cases of gene transfer (genetic recombination) that result from conjugation, the F factor itself is passed to *all* recipient cells. Accordingly, the initial cross of Lederberg and Tatum (Figure 6–3) can be described as follows:

| Strain A | | Strain B |
|---|---|---|
| $F^+$ | $\times$ | $F^-$ |
| (DONOR) | | (RECIPIENT) |

Characterization of the F factor confirmed these conclusions. Like the bacterial chromosome, though distinct from it, the F factor has been shown to consist of a circular, double-stranded DNA

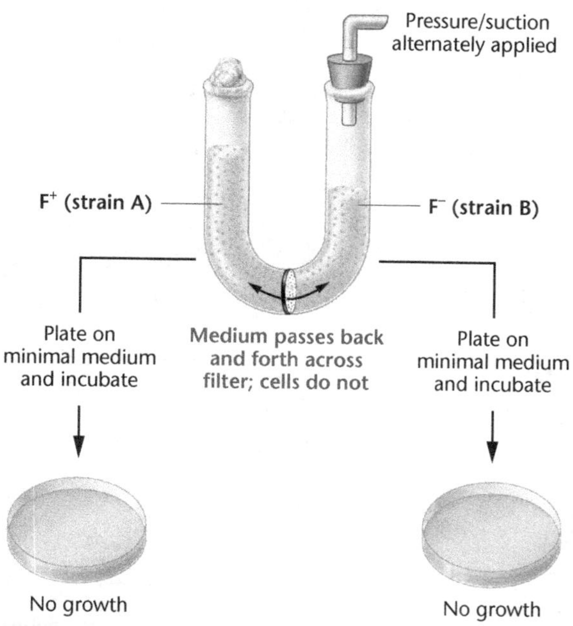


FIGURE 6–4 When strain A and strain B auxotrophs are grown in a common medium but separated by a filter, as in this Davis U-tube apparatus, no genetic recombination occurs and no prototrophs are produced.

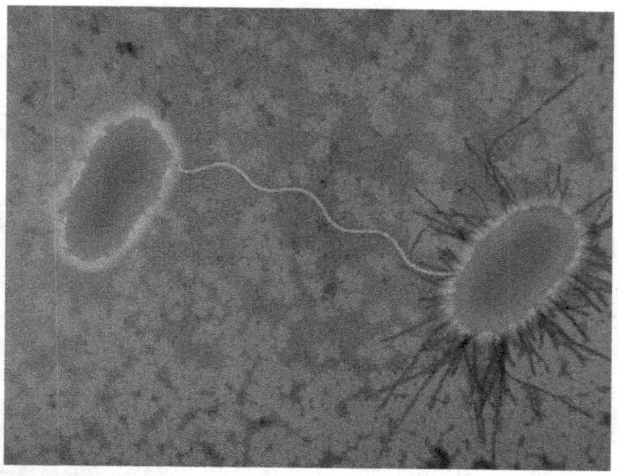

FIGURE 6–5 An electron micrograph of conjugation between an $F^+$ and an $F^-$ *E. coli* cell. The sex pilus linking them is clearly visible.

molecule, equivalent in size to about 2 percent of the bacterial chromosome (about 100,000 nucleotide pairs). There are as many as 40 genes contained within the F factor. Many are *tra* genes, whose products are involved in the transfer of genetic information, including the genes essential to the formation of the sex pilus.

Geneticists believe that transfer of the F factor during conjugation involves separation of the two strands of its double helix and movement of one of the two strands into the recipient cell. The other strand remains in the donor cell. Both strands, one moving across the conjugation tube and one remaining in the donor cell, are replicated. The result is that both the donor and the recipient cells become $F^+$. This process is diagrammed in Figure 6–6.

To summarize, an *E. coli* cell may or may not contain the F factor. When it is present, the cell is able to form a sex pilus and potentially serve as a donor of genetic information. During conjugation, a copy of the F factor is almost always transferred from the $F^+$ cell to the $F^-$ recipient, converting the recipient to the $F^+$ state. The question remained as to exactly why such a low proportion of these matings ($10^{-7}$) also results in genetic recombination. Also, it was unclear what the transfer of the F factor had to do with the transfer and recombination of particular genes. The answers to these questions awaited further experimentation.

As you soon shall see, the F factor is in reality an autonomous genetic unit referred to as a *plasmid*. However, in covering the history of its discovery, in this chapter we will continue to refer to it as a "factor."

## Hfr Bacteria and Chromosome Mapping

Subsequent discoveries not only clarified how genetic recombination occurs but also defined a mechanism by which the *E. coli* chromosome could be mapped. Let's address chromosome mapping first.

In 1950, Cavalli-Sforza treated an $F^+$ strain of *E. coli* K12 with nitrogen mustard, a potent chemical known to induce mutations. From these treated cells, he recovered a genetically altered strain of donor bacteria that underwent recombination at a rate of $1/10^4$ (or $10^{-4}$), 1000 times more frequently than the original $F^+$ strains. In 1953, William Hayes isolated another strain that demonstrated a similarly elevated frequency of recombination. Both strains were designated **Hfr**, for **high-frequency recombination.** Hfr cells constitute a special class of $F^+$ cells.

In addition to the higher frequency of recombination, another important difference was noted between Hfr strains and the original $F^+$ strains. If a donor cell is from an Hfr strain, recipient cells, though sometimes displaying genetic recombination, never become Hfr; in fact, they remain $F^-$. In comparison, then,

$$F^+ \times F^- \longrightarrow \text{recipient becomes } F^+ \text{ (low rate of recombination)}$$
$$\text{Hfr} \times F^- \longrightarrow \text{recipient remains } F^- \text{ (high rate of recombination)}$$

Perhaps the most significant characteristic of Hfr strains is the *specific nature of recombination* in each case. In a given Hfr strain, certain genes are more frequently recombined than others, and some do not recombine at all. This nonrandom pattern of gene transfer was shown to vary among Hfr strains. Although these results were puzzling, Hayes interpreted them to mean that some physiological alteration of the F factor had occurred to produce Hfr strains of *E. coli.*

In the mid-1950s, experimentation by Ellie Wollman and François Jacob explained the differences between Hfr cells and $F^+$ cells and showed how Hfr strains would allow genetic mapping of the *E. coli* chromosome. In Wollman's and Jacob's experiments, Hfr and antibiotic-resistant $F^-$ strains with suitable marker genes were mixed, and recombination of these genes was assayed at different

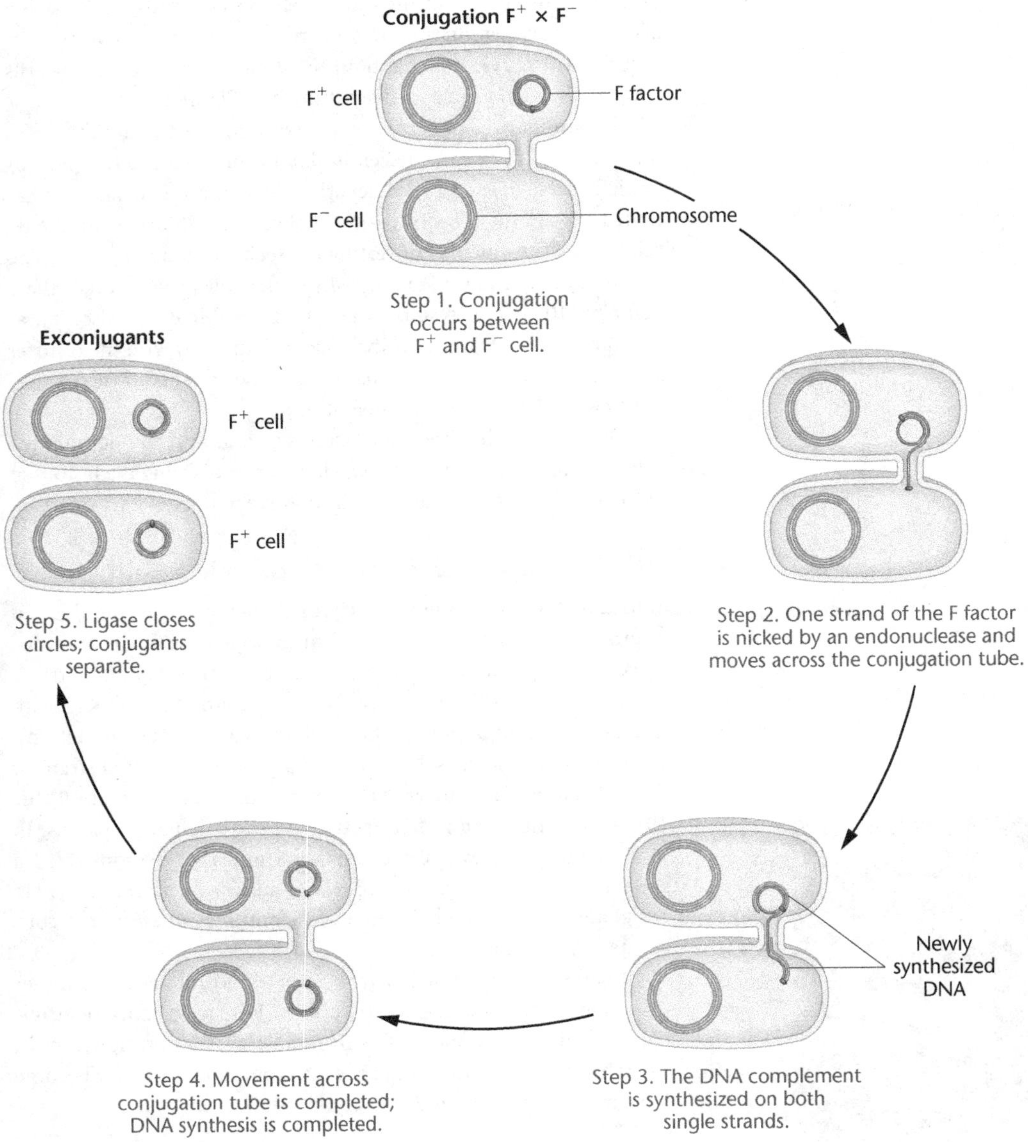


**FIGURE 6–6** An F$^+$ × F$^-$ mating, demonstrating how the recipient F$^-$ cell is converted to F$^+$. During conjugation, the DNA of the F factor is replicated, with one new copy entering the recipient cell, converting it to F$^+$. The bars drawn on the F factors indicate their clockwise rotation during replication. Newly replicated DNA is depicted by a lighter shade of blue as the F factor is transferred.

times. Specifically, a culture containing a mixture of an Hfr and an F$^-$ strain was incubated, and samples were removed at intervals and placed in a blender. The shear forces created in the blender separated conjugating bacteria so that the transfer of the chromosome was terminated. Then the sampled cells were grown on medium containing the antibiotic, so that only recipient cells would be recovered. These cells were subsequently tested for the transfer of specific genes.

This process, called the **interrupted mating technique**, demonstrated that, depending on the specific Hfr strain, certain genes are transferred and recombined sooner than others. The graph in Figure 6–7 illustrates this point. During the first eight minutes after the two strains were mixed, no genetic recombination was detected. At about 10 minutes, recombination of the *azi*$^R$ gene could be detected, but no transfer of the *ton*$^S$, *lac*$^+$, or *gal*$^+$ genes was noted. By 15 minutes, 50 percent of the recombinants were *azi*$^R$ and 15 percent were also *ton*$^S$; but none was *lac*$^+$ or *gal*$^+$. Within 20 minutes, the *lac*$^+$ gene was found among the recombinants; and within 25 minutes, *gal*$^+$ was also beginning to be transferred. Wollman and Jacob had demonstrated an *ordered transfer of genes* that correlated with the length of time conjugation proceeded.

It appeared that the chromosome of the Hfr bacterium was transferred linearly, so that the gene order and distance between genes, as measured in minutes, could be predicted from experiments such as Wollman and Jacob's (Figure 6–8). This information, sometimes referred to as **time mapping**, served as the basis for the first genetic map of the *E. coli* chromosome. Minutes in bacterial mapping provide a measure similar to map units in eukaryotes.

Wollman and Jacob repeated the same type of experiment with other Hfr strains, obtaining similar results but with one important difference. Although genes were always transferred linearly with time, as in their original experiment, the order in which genes entered the recipient seemed to vary from Hfr strain to Hfr strain [Figure 6–9(a)]. Nevertheless, when the researchers reexamined the entry rate of genes, and thus the different genetic maps for each strain, a definite pattern emerged. The major difference between each strain was simply the point of the origin (*O*)—the first part of the donor chromosome to enter the recipient—and the direction in which entry proceeded from that point [Figure 6–9(b)].

To explain these results, Wollman and Jacob postulated that the *E. coli* chromosome is circular (a closed circle, with no free ends). If the point of origin (*O*) varies from strain to strain, a different sequence of genes will be transferred in each case. But what determines *O*? They proposed that, in various Hfr strains, the F factor integrates into the chromosome at different points, and its position determines the *O* site. One such case of integration is shown in step 1 of Figure 6–10. During conjugation between an Hfr and an $F^-$ cell, the position of the F factor determines the initial point of transfer (steps 2 and 3). Those genes adjacent to *O* are transferred first, and the F factor becomes the last part that can be transferred (step 4). However, conjugation rarely, if ever, lasts long enough to allow the entire chromosome to pass across the conjugation tube (step 5). *This proposal explains why recipient cells, when mated with Hfr cells, remain* $F^-$.

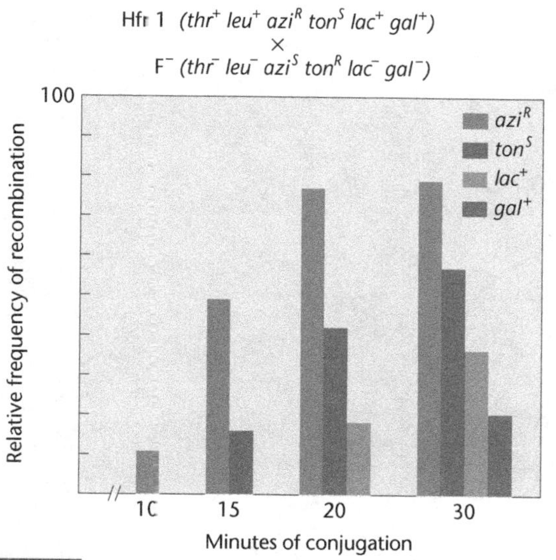


FIGURE 6–7 The progressive transfer during conjugation of various genes from a specific Hfr strain of *E. coli* to an $F^-$ strain. Certain genes (*azi* and *ton*) transfer sooner than others and recombine more frequently. Others (*lac* and *gal*) transfer later, and recombinants are found at a lower frequency. Still others (*thr* and *leu*) are always transferred and were used in the initial screen for recombinants, but are not shown here.

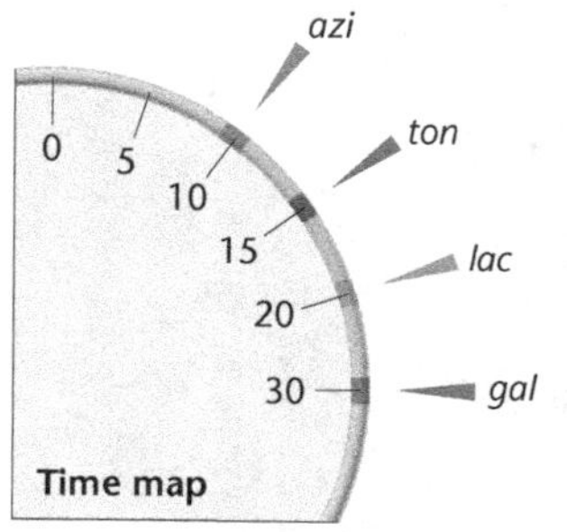


FIGURE 6–8 A time map of the genes studied in the experiment depicted in Figure 6–7.

(a)

| Hfr strain | (latest) | | | Order of transfer | | | | (earliest) |
|---|---|---|---|---|---|---|---|---|
| H | *thr* | *leu* | *azi* | *ton* | *pro* | *lac* | *gal* | *thi* |
| 1 | *leu* | *thr* | *thi* | *gal* | *lac* | *pro* | *ton* | *azi* |
| 2 | *pro* | *ton* | *azi* | *leu* | *thr* | *thi* | *gal* | *lac* |
| 7 | *ton* | *azi* | *leu* | *thr* | *thi* | *gal* | *lac* | *pro* |

(b)

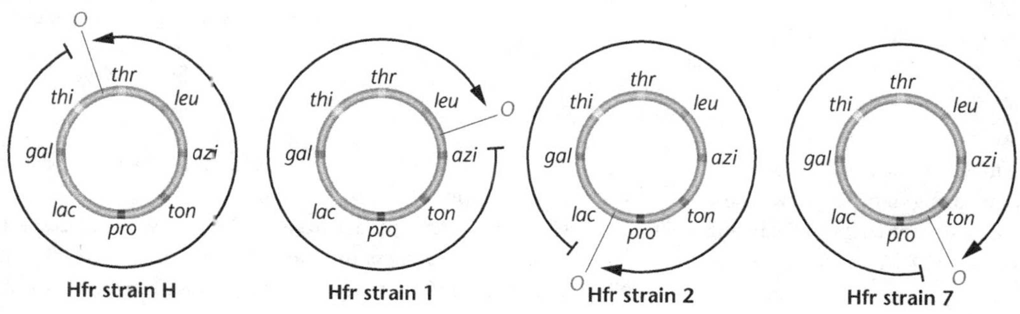


FIGURE 6–9 (a) The order of gene transfer in four Hfr strains, suggesting that the *E. coli* chromosome is circular. (b) The point where transfer originates (*O*) is identified in each strain. The origin is the point of integration of the F factor into the chromosome; the direction of transfer is determined by the orientation of the F factor as it integrates. The arrowheads indicate the points of initial transfer.

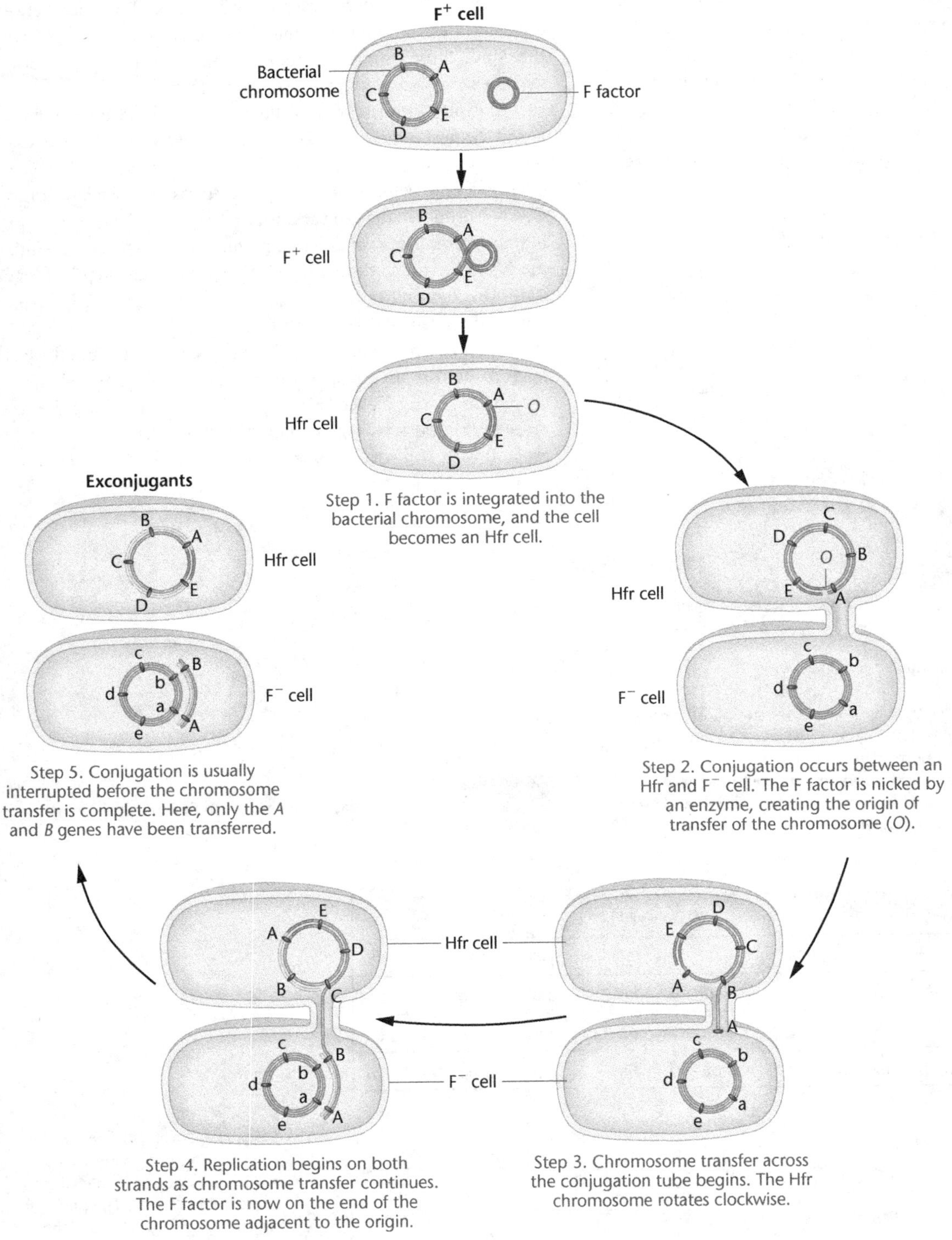


**FIGURE 6–10** Conversion of $F^+$ to an Hfr state occurs by integration of the F factor into the bacterial chromosome. The point of integration determines the origin (*O*) of transfer. During conjugation, an enzyme nicks the F factor, now integrated into the host chromosome, initiating the transfer of the chromosome at that point. Conjugation is usually interrupted prior to complete transfer. Here, only the *A* and *B* genes are transferred to the $F^-$ cell; they may recombine with the host chromosome. Newly replicated DNA of the chromosome is depicted by a lighter shade of orange.

Figure 6–10 also depicts the way in which the two strands making up a donor's DNA molecule behave during transfer, allowing for the entry of one strand of DNA into the recipient (step 3). Following its replication in the recipient, the entering DNA has the potential to recombine with the region homologous to it on the host chromosome. The DNA strand that remains in the donor also undergoes replication.

Use of the interrupted mating technique with different Hfr strains allowed researchers to map the entire *E. coli* chromosome. Mapped in time units, strain K12 (or *E. coli* K12) was shown to be

100 minutes long. While modern genome analysis of the *E. coli* chromosome has now established the presence of just over 4000 protein-coding sequences, this original mapping procedure established the location of approximately 1000 genes.

NOW SOLVE THIS

Problem 22 on page 170 involves an understanding of how the bacterial chromosome is transferred during conjugation, leading to recombination and providing data for mapping. You are asked to interpret data and draw a chromosome map.

■ HINT: *Chromosome transfer is strain-specific and depends on where in the chromosome the F factor has integrated.*

## Recombination in $F^+ \times F^-$ Matings: A Reexamination

The preceding model helped geneticists better understand how genetic recombination occurs during the $F^+ \times F^-$ matings. Recall that recombination occurs much less frequently in them than in Hfr $\times F^-$ matings and that random gene transfer is involved. The current belief is that when $F^+$ and $F^-$ cells are mixed, conjugation occurs readily, and each $F^-$ cell involved in conjugation with an $F^+$ cell receives a copy of the F factor, *but no genetic recombination occurs*. However, at an extremely low frequency in a population of $F^+$ cells, the F factor integrates spontaneously into a random point in the bacterial chromosome, converting the $F^+$ cell to the Hfr state as we saw in Figure 6–10. Therefore, in $F^+ \times F^-$ matings, the extremely low frequency of genetic recombination ($10^{-7}$) is attributed to the rare, newly formed Hfr cells, which then undergo conjugation with $F^-$ cells. Because the point of integration of the F factor is random, the genes transferred by any newly formed Hfr donor *will also appear to be random within the larger $F^+/F^-$ population*. The recipient bacterium will appear as a recombinant but will, in fact, remain $F^-$. If it subsequently undergoes conjugation with an $F^+$ cell, it will be converted to $F^+$.

## The F′ State and Merozygotes

In 1959, during experiments with Hfr strains of *E. coli*, Edward Adelberg discovered that the F factor could lose its integrated status, causing the cell to revert to the $F^+$ state (Figure 6–11, step 1). When this occurs, the F factor frequently carries several adjacent bacterial genes along with it (step 2). Adelberg designated this condition **F′** to distinguish it from $F^+$ and Hfr. F′, like Hfr, is thus another special case of $F^+$.

The presence of bacterial genes within a cytoplasmic F factor creates an interesting situation. An F′ bacterium behaves like an $F^+$ cell by initiating conjugation with $F^-$ cells (Figure 6–11, step 3). When this occurs, the F factor, containing chromosomal genes, is transferred to the $F^-$ cell (step 4). As a result, whatever chromosomal genes are part of the F factor are now present as duplicates in the recipient cell (step 5), because the recipient still has a complete chromosome. This creates a partially diploid cell called a **merozygote**. Pure cultures of F′ merozygotes can be established. They have been extremely useful in the study of genetic regulation in bacteria, as we will discuss in Chapter 17.

# 6.3 Rec Proteins Are Essential to Bacterial Recombination

Once researchers established that a unidirectional transfer of DNA occurs between bacteria, they became interested in determining how the actual recombination event occurs in the recipient cell. Just how does the donor DNA replace the homologous region in the recipient chromosome? As with many systems, the biochemical mechanism by which recombination occurs was deciphered through genetic studies. Major insights were gained as a result of the isolation of a group of mutations that impaired the process of recombination and led to the discovery of *rec* (for recombination) genes.

The first relevant observation in this case involved a series of mutant genes labeled *recA*, *recB*, *recC*, and *recD*. The first mutant gene, *recA*, diminished genetic recombination in bacteria 1000-fold, nearly eliminating it altogether; the other *rec* mutations each reduced recombination by about 100 times. Clearly, the normal wild-type products of these genes play some essential role in the process of recombination.

Researchers looked for, and subsequently isolated, several functional gene products present in normal cells but missing in *rec* mutant cells and showed that they played a role in genetic recombination. The first product is called the **RecA protein.*** This protein plays an important role in recombination involving either a single-stranded DNA molecule or the linear end of a double-stranded DNA molecule that has unwound. As it turns out, **single-strand displacement** is a common form of recombination in many bacterial species. When double-stranded DNA enters a recipient cell, one strand is often degraded, leaving the complementary strand as the only source of recombination. This strand must find its homologous region along the host chromosome, and once it does, RecA facilitates recombination.

The second related gene product is a more complex protein called the **RecBCD protein**, an enzyme consisting of polypeptide subunits encoded by three other *rec* genes. This protein is important when double-stranded DNA serves as the source of genetic recombination. RecBCD unwinds the helix, facilitating recombination that involves RecA. These discoveries have extended our knowledge of the process of recombination considerably and underscore the value of isolating mutations, establishing their phenotypes, and determining the biological role of the normal, wild-type genes. The

*Note that the names of bacterial genes use lowercase letters and are italicized, while the names of the corresponding gene products begin with capital letters and are not italicized. For example, the *recA* gene encodes the RecA protein.

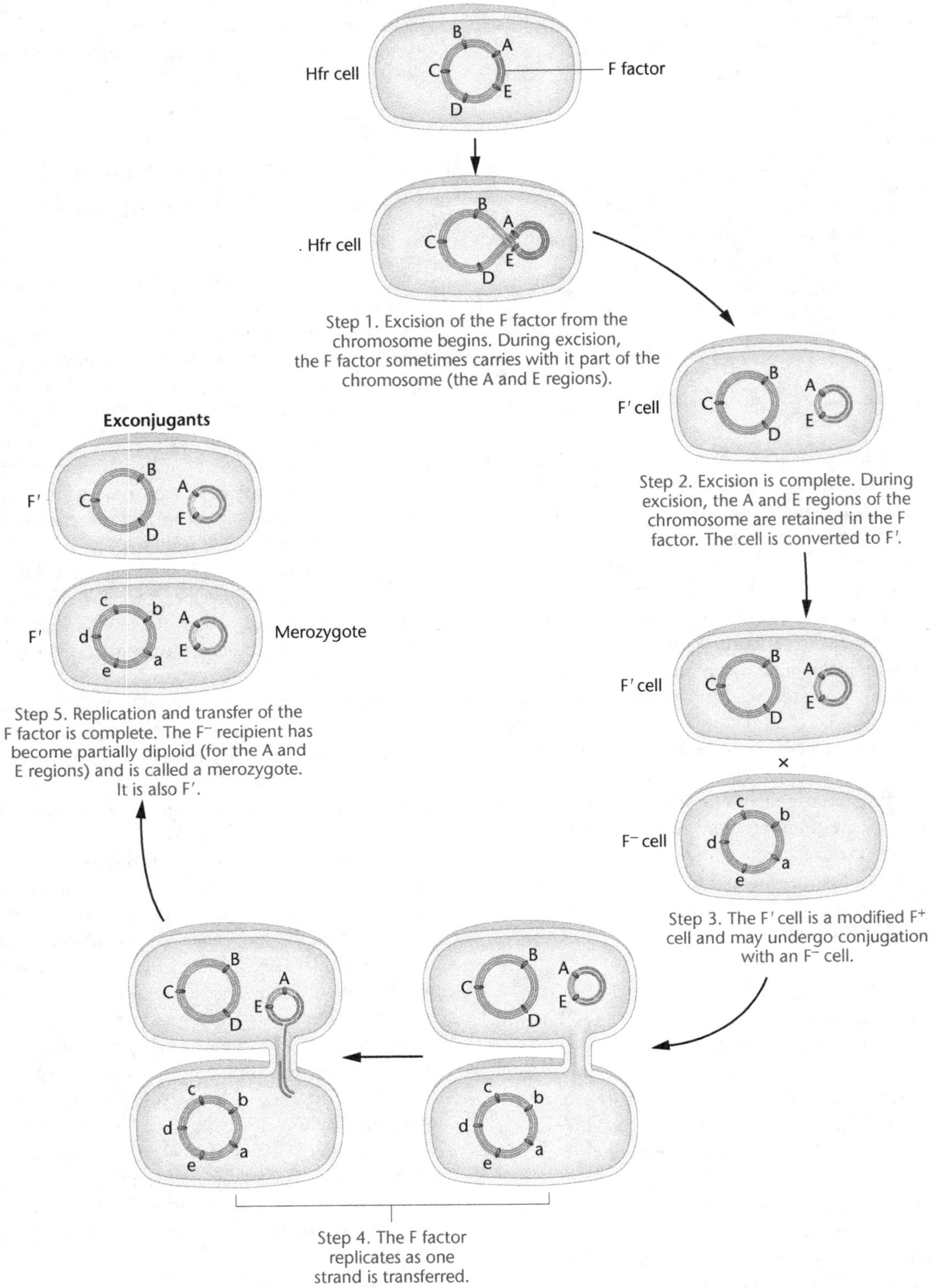


**FIGURE 6–11** Conversion of an Hfr bacterium to F′ and its subsequent mating with an $F^-$ cell. The conversion occurs when the F factor loses its integrated status. During excision from the chromosome, the F factor may carry with it one or more chromosomal genes (in this case, *A* and *E*). Following conjugation, the recipient cell becomes partially diploid and is called a merozygote. It also behaves as an $F^+$ donor cell.

model of recombination based on the *rec* discoveries also applies to eukaryotes: eukaryotic proteins similar to RecA have been isolated and studied. We will return to this topic in Chapter 11 and there discuss two models of DNA recombination.

6.4

## The F Factor Is an Example of a Plasmid

The preceding sections introduced the extrachromosomal heredity unit called the F factor that bacteria require for conjugation. When it exists autonomously in the bacterial cytoplasm, it is composed of a double-stranded closed circle of DNA. These characteristics place the F factor in the more general category of genetic structures called **plasmids** [Figure 6–12(a)]. Plasmids may contain one or more genes, and often quite a few. Their replication depends on the same enzymes that replicate the chromosome of the host cell (many of which will be described in Chapter 11), and they are distributed to daughter cells along with the host chromosome during cell division. Most often, the cell has multiple copies of each type of plasmid it possesses.

Plasmids can be classified according to the genetic information specified by their DNA. The F factor plasmid confers fertility and contains genes essential for sex pilus formation, on which conjugation and subsequent genetic recombination depend. Other examples of plasmids include the R and the Col plasmids.

Most **R plasmids** consist of two components: the **resistance transfer factor (RTF)** and one or more **r-determinants** [Figure 6–12(b)]. The RTF encodes genetic information essential to transferring the plasmid between bacteria, and the r-determinants are genes conferring resistance to antibiotics or heavy metals such as mercury. While RTFs are quite similar in a variety of plasmids from different bacterial species, there is wide variation in r-determinants, each of which is specific for resistance to one class of antibiotic. Determinants with resistance to tetracycline, streptomycin, ampicillin, sulfanilamide, kanamycin, or chloramphenicol are the most frequently encountered. Sometimes plasmids contain many r-determinants, conferring resistance to several antibiotics [Figure 6–12(b)]. Bacteria bearing such plasmids are of great medical significance, not only because of their multiple resistance but also because of the ease with which the plasmids may be transferred to other pathogenic bacteria, rendering those bacteria resistant to a wide range of antibiotics.

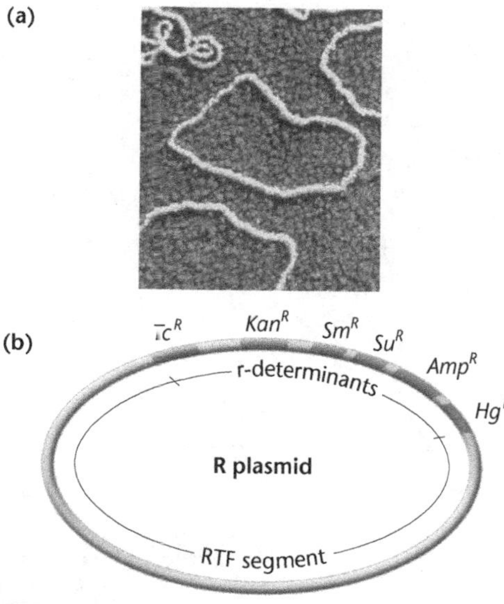


**FIGURE 6–12** (a) Electron micrograph of plasmids isolated from *E. coli*. (b) An R plasmid containing a resistance transfer factor (RTF) and multiple r-determinants (Tc, tetracycline; Kan, kanamycin; Sm, streptomycin; Su, sulfonamide; Amp, ampicillin; and Hg, mercury).

The first known case of such a plasmid occurred in Japan in the 1950s in the bacterium *Shigella*, which causes dysentery. In hospitals, bacteria were isolated that were resistant to as many as five of the above antibiotics. Obviously, this phenomenon represents a major health threat. Fortunately, a bacterial cell sometimes contains r-determinant plasmids but no RTF. Although such a cell is resistant, it cannot transfer the genetic information for resistance to recipient cells. The most commonly studied plasmids, however, contain the RTF as well as one or more r-determinants.

The **Col plasmid**, ColE1 (derived from *E. coli*), is clearly distinct from R plasmids. It encodes one or more proteins that are highly toxic to bacterial strains that do not harbor the same plasmid. These proteins, called **colicins**, can kill neighboring bacteria, and bacteria that carry the plasmid are said to be *colicinogenic*. Present in 10 to 20 copies per cell, the Col plasmid also contains a gene encoding an immunity protein that protects the host cell from the toxin. Unlike an R plasmid, the Col plasmid is not usually transmissible to other cells.

Interest in plasmids has increased dramatically because of their role in recombinant DNA research. As we will see in Chapter 13, specific genes from any source can be inserted into a plasmid, which may then be inserted into a bacterial cell. As the altered cell replicates its DNA and undergoes division, the foreign gene is also replicated, thus being cloned.

6.5

## Transformation Is Another Process Leading to Genetic Recombination in Bacteria

**Transformation** provides another mechanism for recombining genetic information in some bacteria. Small pieces of extracellular (exogenous) DNA are taken up by a living bacterium, potentially leading to a stable genetic change in the recipient cell. We discuss transformation in this chapter because in those bacterial species in which it occurs, the process can be used to map bacterial genes, though in a more limited way than conjugation. As we will see in Chapter 10, the process of transformation was also instrumental in proving that DNA is the genetic material.

## The Transformation Process

The process of transformation (Figure 6–13) consists of numerous steps that achieve two basic outcomes: (1) entry of foreign DNA into a recipient cell; and (2) recombination between the foreign DNA and its homologous region in the recipient chromosome. While completion of both outcomes is required for genetic recombination, the first step of transformation can occur without the second step, resulting in the addition of foreign DNA to the bacterial cytoplasm but not to its chromosome.

In a population of bacterial cells, only those in a particular physiological state of **competence** take up DNA. Entry is thought to occur at a limited number of receptor sites on the surface of the bacterial cell (Figure 6–13, step 1). Passage into the cell is thought to be an active process that requires energy and specific transport molecules. This model is supported by the fact that substances that inhibit energy production or protein synthesis also inhibit transformation.

Soon after entry, one of the two strands of the double helix is digested by nucleases, leaving only a single strand to participate in

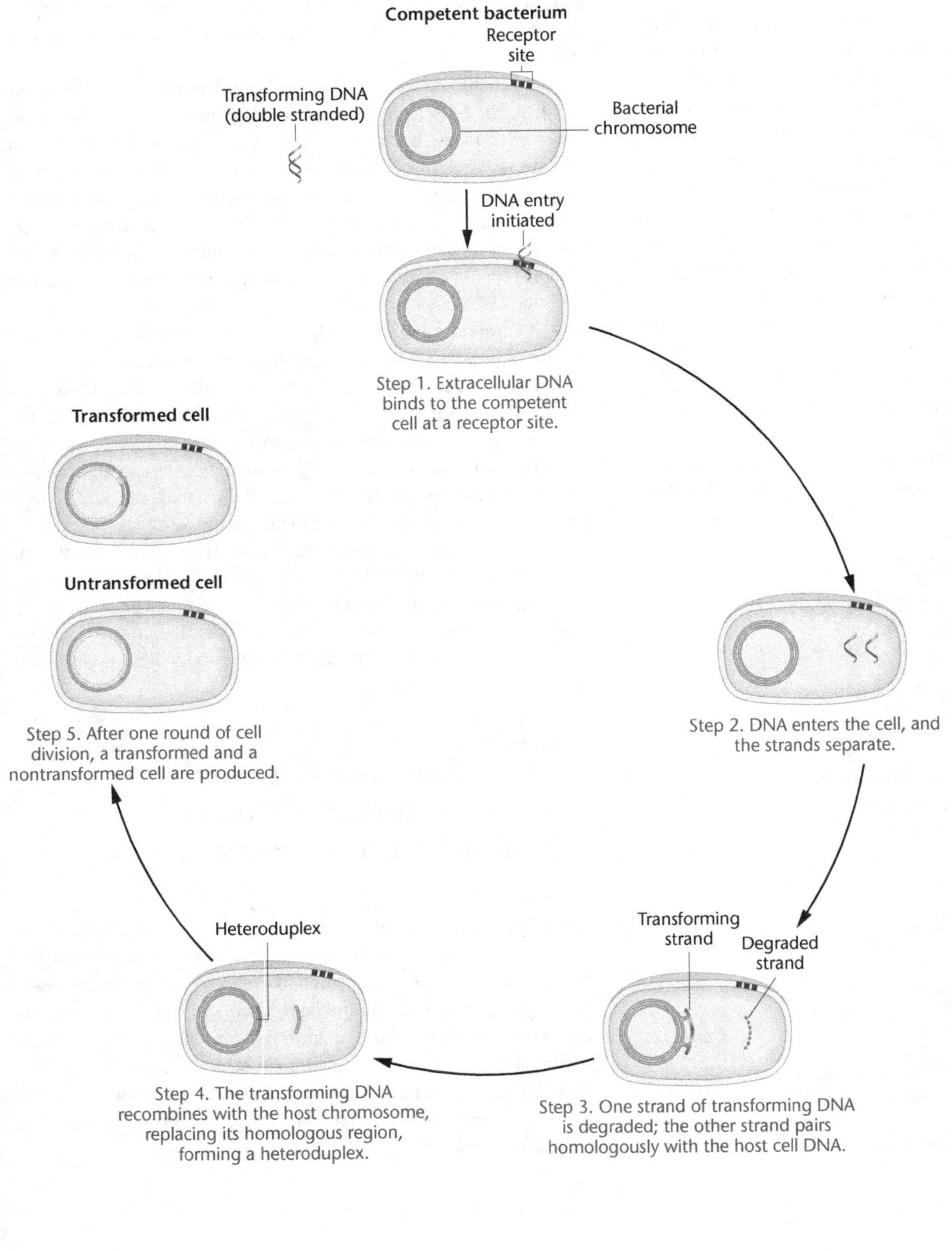


**FIGURE 6–13** Proposed steps for transformation of a bacterial cell by exogenous DNA. Only one of the two strands of the entering DNA is involved in the transformation event, which is completed following cell division.

transformation (Figure 6–13, steps 2 and 3). The surviving DNA strand aligns with the complementary region of the bacterial chromosome. In a process involving several enzymes, this segment of DNA replaces its counterpart in the chromosome (step 4), which is excised and degraded.

For recombination to be detected, the transforming DNA must be derived from a different strain of bacteria that bears some distinguishing genetic variation, such as a mutation. Once this is integrated into the chromosome, the recombinant region contains one host strand (present originally) and one mutant strand. Because these strands are from different sources, the region is referred to as a **heteroduplex**. Following one round of DNA replication, one chromosome is restored to its original DNA sequence, identical to that of the original recipient cell, and the other contains the mutant gene. Following cell division, one nontransformed cell (nonmutant) and one transformed cell (mutant) are produced (step 5).

### Transformation and Linked Genes

Ideally, for transformation to occur, the exogenous DNA includes between 10,000 and 20,000 nucleotide pairs, a length equal to about 1/200 of the *E. coli* chromosome. This size is sufficient to encode several genes. Genes adjacent to or very close to one another on the bacterial chromosome can be carried on a single segment of this size. Consequently, a single transfer event can result in the **cotransformation** of several genes simultaneously. Genes that are close enough to each other to be cotransformed are *linked*. In contrast to *linkage groups* in eukaryotes, which consist of all genes on a single chromosome, note that here *linkage* refers to the proximity of genes that permits cotransformation (i.e., the genes are next to, or close to, one another).

If two genes are not linked, simultaneous transformation occurs only as a result of two independent events involving two distinct segments of DNA. As with double crossovers in eukaryotes, the probability of two independent events occurring simultaneously is equal to the product of the individual probabilities. Thus, the frequency of two unlinked genes being transformed simultaneously is much lower than if they are linked.

Studies have shown that various kinds of bacteria readily undergo transformation (e.g., *Haemophilus influenzae, Bacillus subtilis, Shigella paradysenteriae, Diplococcus pneumoniae,* and *E. coli*). Under certain conditions, relative distances between linked genes can be determined from transformation data in a manner analogous to chromosome mapping in eukaryotes, although somewhat more complex.

**NOW SOLVE THIS**

Problem 8 on page 170 involves an understanding of how transformation can be used to determine if bacterial genes are closely "linked" to one another. You are asked to predict the location of two genes relative to one another.

■ HINT: *Cotransformation occurs according to the laws of probability. Two "unlinked" genes are transformed as a result of two separate events. In such a case, the probability of that occurrence is equal to the product of the individual probabilities.*

6.6

## Bacteriophages Are Bacterial Viruses

WEB TUTORIAL 6.2 PHAGE GENETICS

**Bacteriophages**, or **phages** as they are commonly known, are viruses that have bacteria as their hosts. The reproduction of phages can lead to still another mode of bacterial genetic recombination, called transduction. To understand this process, we first must consider the genetics of bacteriophages, which themselves can undergo recombination.

A great deal of genetic research has been done using bacteriophages as a model system. In this section, we will first examine the structure and life cycle of one type of bacteriophage. We then discuss how these phages are studied during their infection of bacteria. Finally, we contrast two possible modes of behavior once initial phage infection occurs. This information is background for our discussion of *transduction* and *bacteriophage recombination*.

### Phage T4: Structure and Life Cycle

Bacteriophage T4 is one of a group of related bacterial viruses referred to as T-even phages. It exhibits the intricate structure shown in Figure 6–14. Its genetic material, DNA, is contained within an icosahedral (referring to a polyhedron with 20 faces) protein coat, making up the head of the virus. The DNA is sufficient in quantity to encode more than 150 average-sized genes. The head is connected to a tail that contains a collar and a contractile sheath surrounding a central core. Tail fibers, which protrude from the tail, contain

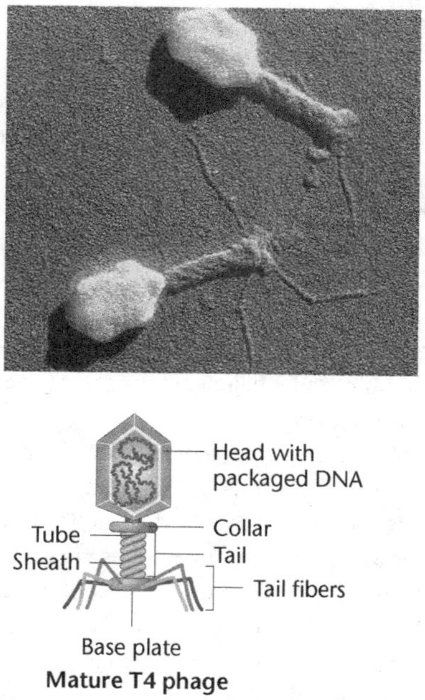


FIGURE 6–14 The structure of bacteriophage T4 which includes an icosahedral head filled with DNA; a tail consisting of a collar, tube, and sheath; and a base plate with tail fibers. During assembly, the tail components are added to the head and then tail fibers are added.

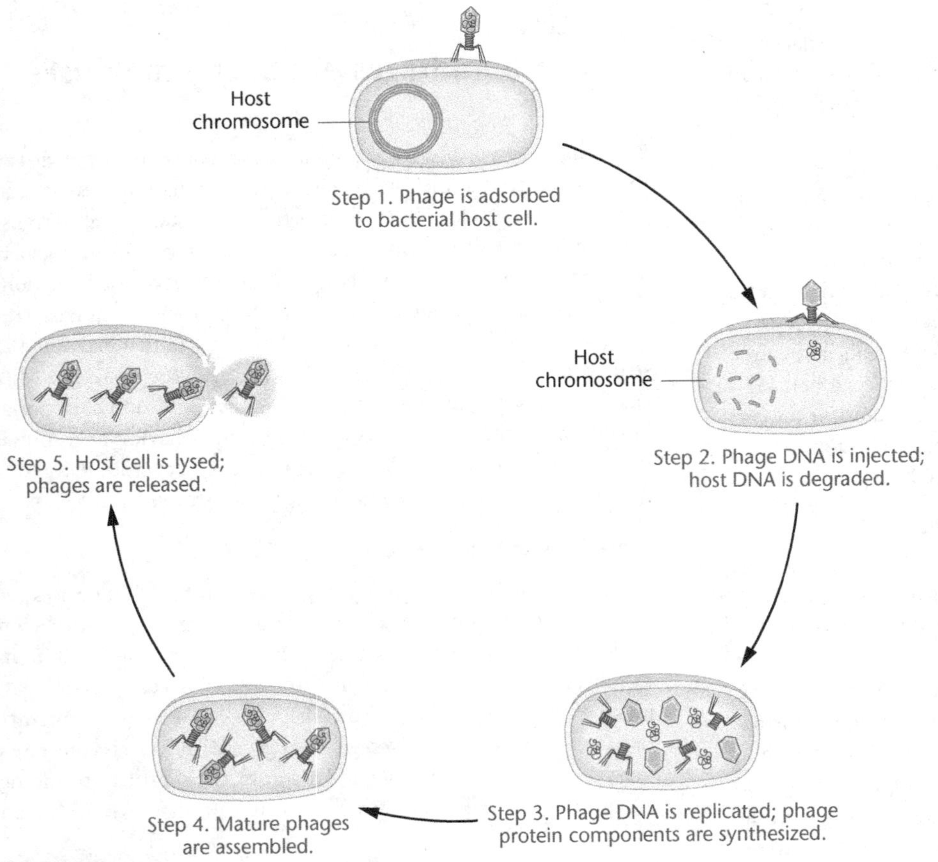


FIGURE 6-15 Life cycle of bacteriophage T4.

binding sites in their tips that specifically recognize unique areas of the outer surface of the cell wall of the bacterial host, *E. coli*.

The life cycle of phage T4 (Figure 6–15) is initiated when the virus binds by adsorption to the bacterial host cell. Then, an ATP-driven contraction of the tail sheath causes the central core to penetrate the cell wall. The DNA in the head is extruded, and it moves across the cell membrane into the bacterial cytoplasm. Within minutes, all bacterial DNA, RNA, and protein synthesis is inhibited, and synthesis of viral molecules begins. At the same time, degradation of the host DNA is initiated.

A period of intensive viral gene activity characterizes infection. Initially, phage DNA replication occurs, leading to a pool of viral DNA molecules. Then, the components of the head, tail, and tail fibers are synthesized. The assembly of mature viruses is a complex process that has been well studied by William Wood, Robert Edgar, and others. Three sequential pathways take part: (1) DNA packaging as the viral heads are assembled, (2) tail assembly, and (3) tail-fiber assembly. Once DNA is packaged into the head, that structure combines with the tail components, to which tail fibers are added. Total construction is a combination of self-assembly and enzyme-directed processes.

When approximately 200 new viruses are constructed, the bacterial cell is ruptured by the action of lysozyme (a phage gene product), and the mature phages are released from the host cell. This step during infection is referred to as **lysis**, and it completes what is referred to as the **lytic cycle**. The 200 new phages infect other available bacterial cells, and the process repeats itself over and over again.

## The Plaque Assay

Bacteriophages and other viruses have played a critical role in our understanding of molecular genetics. During infection of bacteria, enormous quantities of bacteriophages may be obtained for investigation. Often, more than $10^{10}$ viruses are produced per milliliter of culture medium. Many genetic studies have relied on our ability to determine the number of phages produced following infection under specific culture conditions. The **plaque assay**, routinely used for such determinations, is invaluable in quantitative analysis during mutational and recombinational studies of bacteriophages.

This assay is illustrated in Figure 6–16, where actual plaque morphology is also shown. A serial dilution of the original virally infected bacterial culture is performed. Then, a 0.1-mL sample (an aliquot, meaning a fractional portion) from a dilution is added to a small volume of melted nutrient agar (about 3 mL) into which a few drops of a healthy bacterial culture have been added. The solution is then poured evenly over a base of solid nutrient agar in a Petri dish and allowed to solidify before incubation. A clear area called a **plaque** occurs wherever a single virus initially infected one bacterium in the culture (the lawn) that has grown up during incubation. The plaque represents clones of the single infecting bacteriophage, created as reproduction cycles are repeated. If the dilution factor is too low, the plaques will be plentiful, and they may fuse, lysing the entire lawn of bacteria. This has occurred in the $10^{-3}$ dilution in Figure 6–16. However, if the dilution factor is increased appropriately, plaques can be counted, and the density of viruses in the initial culture can be estimated,

$$\text{initial phage density} = (\text{plaque number/mL}) \times (\text{dilution factor})$$

Figure 6–16 shows that 23 phage plaques were derived from the 0.1-mL aliquot of the $10^{-5}$ dilution. Therefore, we estimate a density of 230 phages/mL *at this dilution* (since the initial aliquot was 0.1 mL). The initial phage density in the undiluted sample, given that 23 plaques were observed from 0.1 mL of the $10^{-5}$ dilution, is calculated as

$$\text{initial phage density} = (230/\text{mL}) \times (10^5) = (230 \times 10^5)/\text{mL}$$

Because this figure is derived from the $10^{-5}$ dilution, we can also estimate that there would be only 0.23 phage/0.1 mL in the $10^{-7}$ dilution. Thus, if 0.1 mL from this tube were assayed, we would pre-

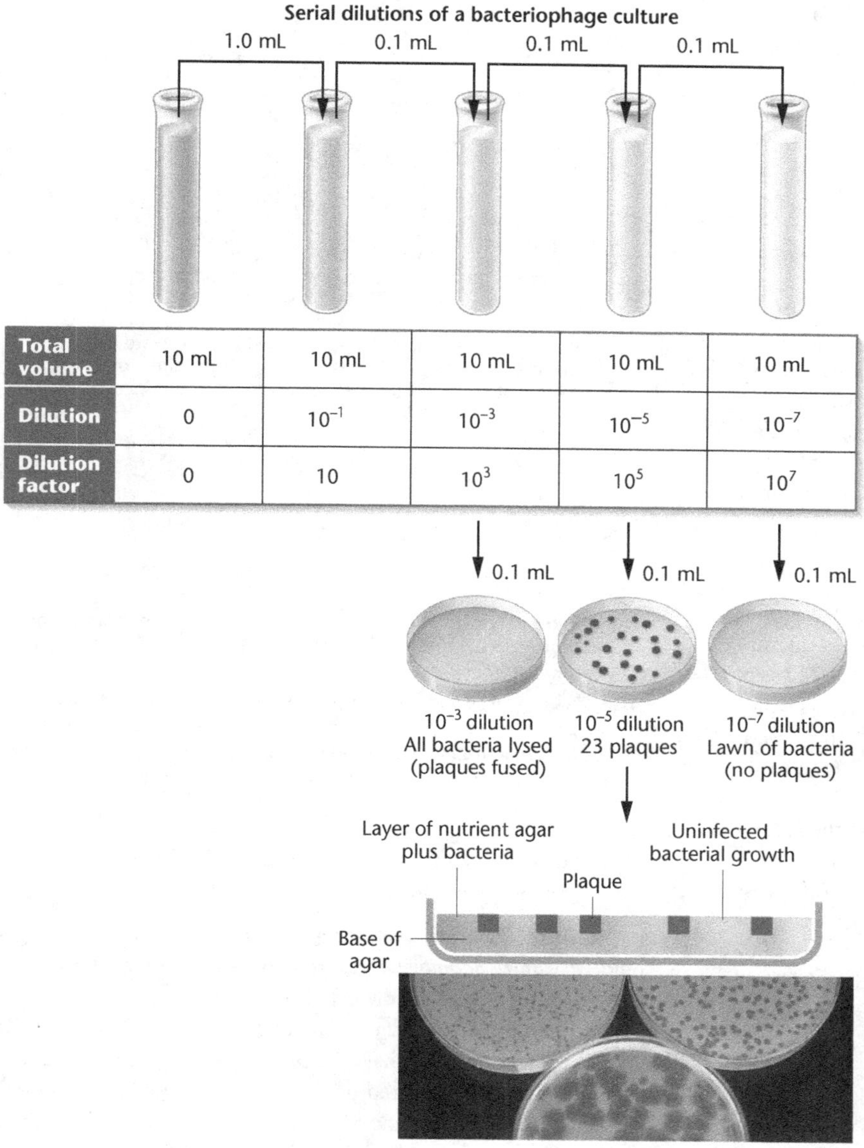


**FIGURE 6–16** A plaque assay for bacteriophage analysis. First, serial dilutions are made of a bacterial culture infected with bacteriophages. Then, three of the dilutions ($10^{-3}$, $10^{-5}$, and $10^{-7}$) are analyzed using the plaque assay technique. Each plaque represents the initial infection of one bacterial cell by one bacteriophage. In the $10^{-3}$ dilution, so many phages are present that all bacteria are lysed. In the $10^{-5}$ dilution, 23 plaques are produced. In the $10^{-7}$ dilution, the dilution factor is so great that no phages are present in the 0.1-mL sample, and thus no plaques form. From the 0.1-mL sample of the $10^{-5}$ dilution, the original bacteriophage density is calculated to be $23 \times 10 \times 10^5$ phages/mL ($230 \times 10^5$, or $23 \times 10^6$). The photograph shows phage T2 plaques on lawns of *E. coli*.

dict that no phage particles would be present. This prediction is borne out in Figure 6–16, where an intact lawn of bacteria lacking any plaques is depicted. The dilution factor is simply too great.

The use of the plaque assay has been invaluable in mutational and recombinational studies of bacteriophages. We will apply this technique more directly later in this chapter when we discuss Seymour Benzer's elegant genetic analysis of a single gene in phage T4.

## Lysogeny

Infection of a bacterium by a virus does not always result in viral reproduction and lysis. As early as the 1920s, it was known that a virus can enter a bacterial cell and coexist with it. The precise molecular basis of this relationship is now well understood. Upon entry, the viral DNA is integrated into the bacterial chromosome instead of replicating in the bacterial cytoplasm; this integration characterizes the developmental stage referred to as **lysogeny**. Subsequently, each time the bacterial chromosome is replicated, the viral DNA is also replicated and passed to daughter bacterial cells following division. No new viruses are produced, and no lysis of the bacterial cell occurs. However, under certain stimuli, such as chemical or ultraviolet-light treatment, the viral DNA loses its integrated status and initiates replication, phage reproduction, and lysis of the bacterium.

Several terms are used in describing this relationship. The viral DNA integrated into the bacterial chromosome is called a **prophage**. Viruses that can either lyse the cell or behave as a prophage are called **temperate phages**. Those that can only lyse the cell are referred to as **virulent phages**. A bacterium harboring a prophage has been **lysogenized** and is said to be **lysogenic**; that is, it is capable of being lysed as a result of induced viral reproduction. The viral DNA is classified as an **episome**, meaning a genetic particle that can replicate either extrachromosomally or as part of the chromosome.

## 6.7 Transduction Is Virus-Mediated Bacterial DNA Transfer

In 1952, Norton Zinder and Joshua Lederberg were investigating possible recombination in the bacterium *Salmonella typhimurium*. Although they recovered prototrophs from mixed cultures of two different auxotrophic strains, subsequent investigations showed that recombination was not due to the presence of an F factor and conjugation, as in *E. coli*. What they discovered was a process of bacterial recombination mediated by bacteriophages and now called **transduction**.

### The Lederberg–Zinder Experiment

Lederberg and Zinder mixed the *Salmonella* auxotrophic strains LA-22 and LA-2 together, and when the mixture was plated on minimal medium, they recovered prototrophic cells. The LA-22 strain was unable to synthesize the amino acids phenylalanine and tryptophan ($phe^-$ $trp^-$), and LA-2 could not synthesize the amino acids methionine and histidine ($met^-$ $his^-$). Prototrophs ($phe^+$ $trp^+$ $met^+$ $his^+$) were recovered at a rate of about $1/10^5$ (or $10^{-5}$) cells.

Although these observations at first suggested that the recombination was the type observed earlier in conjugative strains of *E. coli*, experiments using the Davis U-tube soon showed otherwise (Figure 6–17). The two auxotrophic strains were separated by a sintered glass filter, thus preventing contact between the strains while allowing them to grow in a common medium. Surprisingly, when samples were removed from both sides of the filter and plated independently on minimal medium, prototrophs *were* recovered, but only from the side of the tube containing LA-22 bacteria. Recall that if conjugation were responsible, the Davis U-tube should have *prevented* recombination altogether (see Figure 6–4).

Since LA-2 cells appeared to be the source of the new genetic information ($phe^+$ and $trp^+$), how that information crossed the filter from the LA-2 cells to the LA-22 cells, allowing recombination to occur, was a mystery. The unknown source was designated simply as a **filterable agent (FA)**.

Three observations were used to identify the FA:

1. The FA was produced by the LA-2 cells only when they were grown in association with LA-22 cells. If LA-2 cells were grown independently in a culture medium that was later added to LA-22 cells, recombination did not occur. Therefore, the LA-22 cells played some role in the production of FA by LA-2 cells but did so only when the two strains were sharing a common growth medium.
2. The addition of DNase, which enzymatically digests DNA, did not render the FA ineffective. Therefore, the FA is not DNA, ruling out transformation.

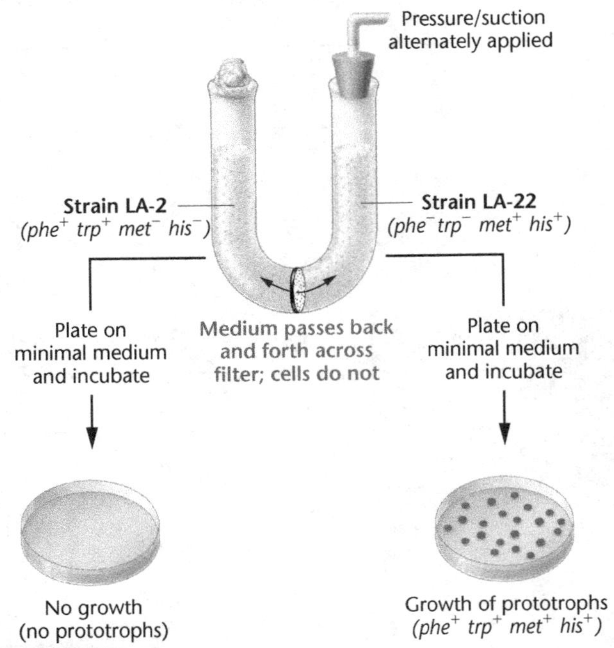


FIGURE 6–17 The Lederberg–Zinder experiment using *Salmonella*. After placing two auxotrophic strains on opposite sides of a Davis U-tube, Lederberg and Zinder recovered prototrophs from the side with the LA-22 strain but not from the side containing the LA-2 strain.

3. The FA could not pass across the filter of the Davis U-tube when the pore size was reduced below the size of bacteriophages.

Aided by these observations and aware that temperate phages could lysogenize *Salmonella*, researchers proposed that the genetic recombination event was mediated by bacteriophage P22, present initially as a prophage in the chromosome of the LA-22 *Salmonella* cells. They hypothesized that P22 prophages sometimes enter the vegetative, or lytic, phase, reproduce, and are released by the LA-22 cells. Such P22 phages, being much smaller than a bacterium, then cross the filter of the U-tube and subsequently infect and lyse some of the LA-2 cells. In the process of lysis of LA-2, the P22 phages occasionally package a region of the LA-2 chromosome in their heads. If this region contains the $phe^+$ and $trp^+$ genes, and if the phages subsequently pass back across the filter and infect LA-22 cells, these newly lysogenized cells will behave as prototrophs. This process of transduction, whereby bacterial recombination is mediated by bacteriophage P22, is diagrammed in Figure 6–18.

### The Nature of Transduction

Further studies revealed the existence of **transducing phages** in other species of bacteria. For example, *E. coli* can be transduced by phages P1 and λ, and *Bacillus subtilis* and *Pseudomonas aeruginosa* can be transduced by the phages SPO1 and F116, respectively. The details of several different modes of transduction have also been established. Even though the initial discovery of transduction

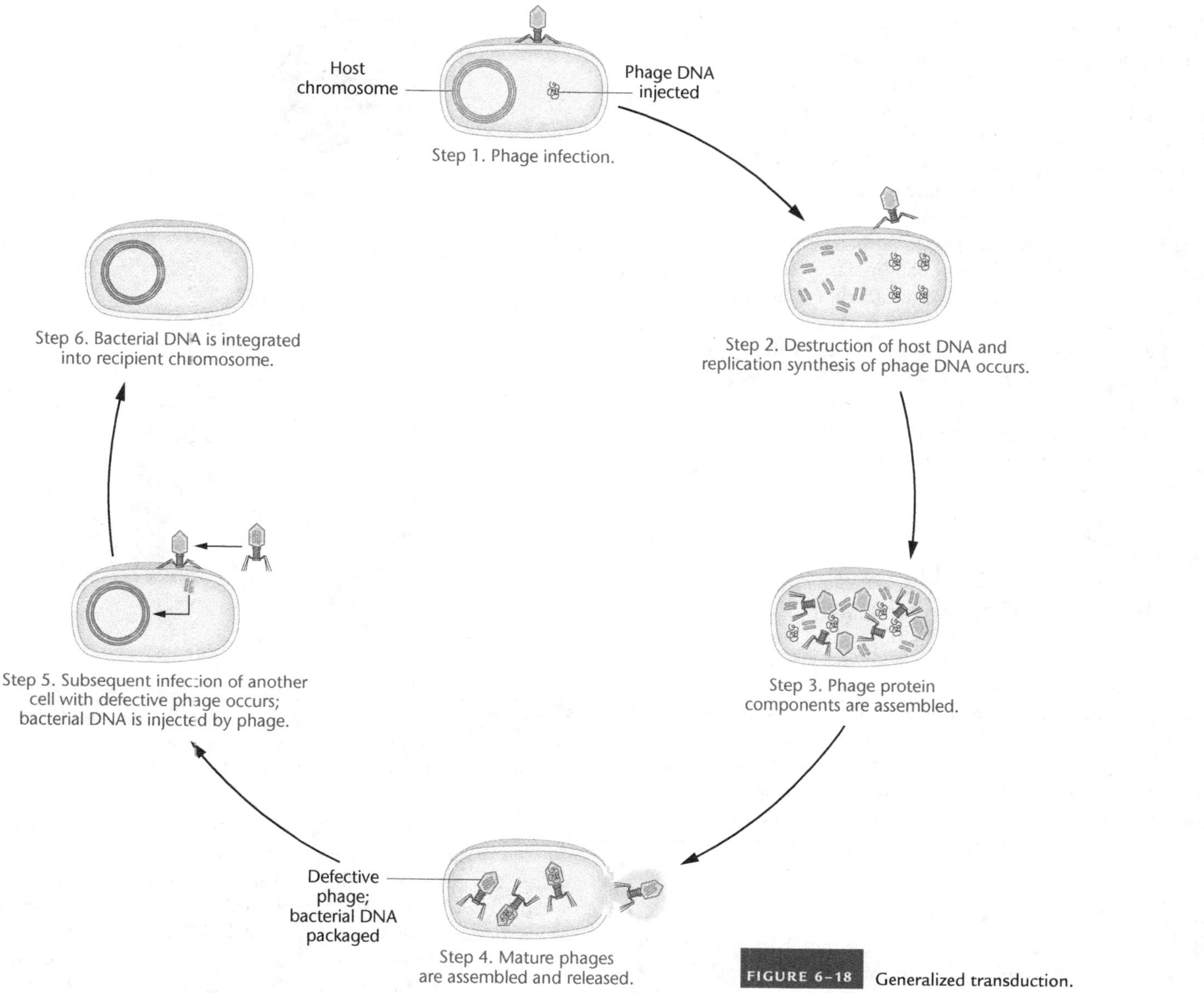


**FIGURE 6–18** Generalized transduction.

involved a temperate phage and a lysogenized bacterium, the same process can occur during the normal lytic cycle. Sometimes, during what is called **specialized transduction**, a small piece of bacterial DNA is packaged *along with* the viral chromosome. In such cases, only a few specific bacterial genes are present in the transducing phage. In other cases, referred to as **generalized transduction**, the phage DNA is excluded completely and *only* bacterial DNA is packaged. Regions as large as 1 percent of the bacterial chromosome may become enclosed in the viral head. In both types of transduction, the ability to infect host cells is unaffected by the presence of foreign DNA, making transduction possible.

During generalized transduction, when a defective phage injects bacterial rather than viral DNA into a bacterium, the DNA either remains in the bacterial cytoplasm or recombines with the homologous region of the bacterial chromosome. If the bacterial DNA remains in the cytoplasm, it does not replicate but is transmitted to one progeny cell following each division. When this happens, only a single cell, partially diploid for the transduced genes, is produced—a phenomenon called **abortive transduction.** If the bacterial DNA recombines with its homologous region of the bacterial chromosome, **complete transduction** occurs, where the transduced genes are replicated as part of the chromosome and passed to all daughter cells.

Both abortive and complete transduction are characterized by randomness in the DNA fragments and genes transduced. Each fragment of the bacterial chromosome has a finite but small chance of being packaged in the phage head. Most cases of generalized transduction are of the abortive type; some data suggest that complete transduction occurs 10 to 20 times less frequently.

## Transduction and Mapping

Like transformation, generalized transduction has been used in linkage and mapping studies of the bacterial chromosome. The fragment of bacterial DNA involved in a transduction event may be large enough to include numerous genes. As a result, two genes that are close to one another along the bacterial chromosome (i.e., are linked) can be transduced simultaneously, a process called **cotransduction.** If two genes are not close enough to one another along the chromosome to be included on a single DNA fragment, two independent transduction events must occur to carry them into a single cell. Since this occurs with a much lower probability than cotransduction, linkage can be determined by comparing the frequency of specific simultaneous recombinations.

By concentrating on two or three linked genes, transduction studies can also determine the precise order of these genes. The closer linked genes are to each other, the greater the frequency of cotransduction. Mapping studies can be done on three closely aligned genes, predicated on the same rationale that underlies other mapping techniques.

6.8

# Bacteriophages Undergo Intergenic Recombination

Around 1947, several research teams demonstrated that genetic recombination can be detected in bacteriophages. This led to the discovery that gene mapping can be performed in these viruses. Such studies relied on finding numerous phage mutations that could be visualized or assayed. As in bacteria and eukaryotes, these mutations allow genes to be identified and followed in mapping experiments. Before considering recombination and mapping in these bacterial viruses, we briefly introduce several of the mutations that were studied.

## Bacteriophage Mutations

Phage mutations often affect the morphology of the plaques formed following lysis of bacterial cells. For example, in 1946, Alfred Hershey observed unusual T2 plaques on plates of *E. coli* strain B. Normal T2 plaques are small and have a clear center surrounded by a diffuse (nearly invisible) halo. In contrast, the unusual plaques were larger and possessed a distinctive outer perimeter (compare the lighter plaques in Figure 6–19). When the viruses were isolated from these plaques and replated on *E. coli* B cells, the resulting plaque appearance was identical. Thus, the plaque phenotype was an inherited trait resulting from the reproduction of mutant phages. Hershey named the mutant *rapid lysis* (*r*) because the plaques' larger size was thought to be due to a more rapid or more efficient life cycle of the phage. We now know that, in wild-type phages, reproduction is inhibited once a particular-sized plaque has been formed. The *r* mutant T2 phages overcome this inhibition, producing larger plaques.

Salvador Luria discovered another bacteriophage mutation, *host range* (*h*). This mutation extends the range of bacterial hosts that the phage can infect. Although wild-type T2 phages can infect *E. coli* B

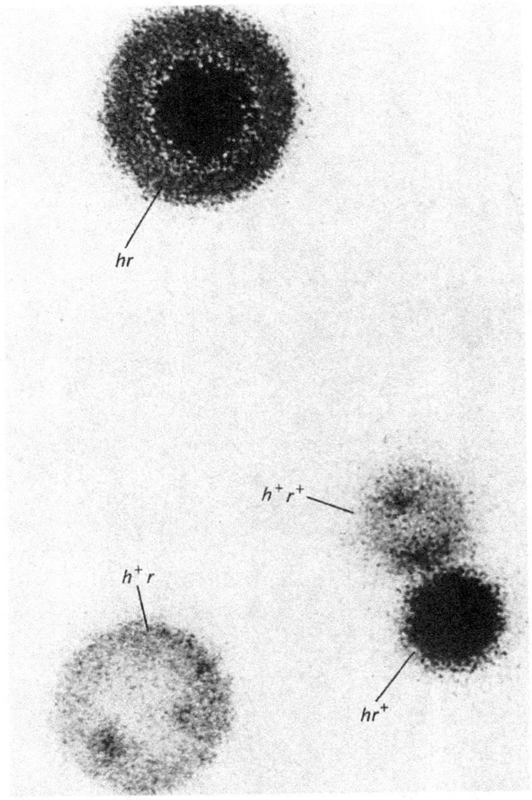


FIGURE 6–19 Plaque morphology phenotypes observed following simultaneous infection of *E. coli* by two strains of phage *T2*, $h^+r$ and $hr^+$. In addition to the parental genotypes, recombinant plaques *hr* and $h^+r^+$ are shown.

(a unique strain), they normally cannot attach or be adsorbed to the surface of *E. coli* B-2 (a different strain). The *h* mutation, however, confers the ability to adsorb to and subsequently infect *E. coli* B-2. When grown on a mixture of *E. coli* B and B-2, the *h* plaque has a center that appears much darker than that of the $h^+$ plaque (Figure 6–19).

Table 6.1 lists other types of mutations that have been isolated and studied in the T-even series of bacteriophages (e.g., T2, T4, T6).

TABLE 6.1

Some Mutant Types of T-Even Phages

| Name | Description |
|---|---|
| *minute* | Small plaques |
| *turbid* | Turbid plaques on *E. coli* B |
| *star* | Irregular plaques |
| *UV-sensitive* | Alters UV sensitivity |
| *acriflavin-resistant* | Forms plaques on acriflavin agar |
| *osmotic shock* | Withstands rapid dilution into distilled water |
| *lysozyme* | Does not produce lysozyme |
| *amber* | Grows in *E. coli* K12 but not B |
| *temperature-sensitive* | Grows at 25°C but not at 42°C |

These mutations are important to the study of genetic phenomena in bacteriophages.

## Mapping in Bacteriophages

Genetic recombination in bacteriophages was discovered during **mixed infection experiments**, in which two distinct mutant strains were allowed to *simultaneously* infect the same bacterial culture. These studies were designed so that the number of viral particles sufficiently exceeded the number of bacterial cells to ensure simultaneous infection of most cells by both viral strains. If two loci are involved, recombination is referred to as **intergenic**.

For example, in one study using the T2/*E. coli* system, the parental viruses were of either the $h^+r$ (wild-type host range, rapid lysis) or the $hr^+$ (extended host range, normal lysis) genotype. If no recombination occurred, these two parental genotypes would be the only expected phage progeny. However, the recombinants $h^+r^+$ and $hr$ were detected in addition to the parental genotypes (see Figure 6–19). As with eukaryotes, the percentage of recombinant plaques divided by the total number of plaques reflects the relative distance between the genes:

$$\text{recombinational frequency} = (h^+r^+ + hr)/\text{total plaques} \times 100$$

Sample data for the *h* and *r* loci are shown in Table 6.2.

TABLE 6.2

**Results of a Cross Involving the *h* and *r* Genes in Phage T2 ($hr^+ \times h^+r$)**

| Genotype | Plaques | Designation |
|---|---|---|
| $h\ r^+$ | 42 | Parental progeny 76% |
| $h^+r$ | 34 | |
| $h^+r^+$ | 12 | Recombinants 24% |
| $h\ r$ | 12 | |

*Source:* Data derived from Hershey and Rotman (1949).

Similar recombinational studies have been conducted with numerous mutant genes in a variety of bacteriophages. Data are analyzed in much the same way as in eukaryotic mapping experiments. Two- and three-point mapping crosses are possible, and the percentage of recombinants in the total number of phage progeny is calculated. This value is proportional to the relative distance between two genes along the DNA molecule constituting the chromosome.

Investigations into phage recombination support a model similar to that of eukaryotic crossing over—a breakage and reunion process between the viral chromosomes. A fairly clear picture of the dynamics of viral recombination has emerged. Following the early phase of infection, the chromosomes of the phages begin replication. As this stage progresses, a pool of chromosomes accumulates in the bacterial cytoplasm. If double infection by phages of two genotypes has occurred, then the pool of chromosomes initially consists of the two parental types. Genetic exchange between these two types will occur before, during, and after replication, producing recombinant chromosomes.

In the case of the $h^+r$ and $hr^+$ example discussed here, recombinant $h^+r^+$ and $hr$ chromosomes are produced. Each of these chromosomes can undergo replication, with new replicates undergoing exchange with each other and with parental chromosomes. Furthermore, recombination is not restricted to exchange between two chromosomes—three or more may be involved simultaneously. As phage development progresses, chromosomes are randomly removed from the pool and packed into the phage head, forming mature phage particles. Thus, a variety of parental and recombinant genotypes are represented in progeny phages.

As we will see in the next section, powerful selection systems have made it possible to detect *intragenic* recombination in viruses, where exchanges occur at points within a single gene, as opposed to intergenic recombination, where exchanges occur at points located between genes. Such studies have led to what has been called the fine-structure analysis of the gene.

NOW SOLVE THIS

Problem 13 on page 170 involves an understanding of how intergenic mapping is performed in bacteriophages.

■ HINT: *Mapping is based on recombination occurring during simultaneous infection of a bacterium by two or more strain-specific bacteriophages. The frequency of recombination varies directly with the distance between two genes along the bacteriophage chromosome. Mapping theory in phages is therefore similar to mapping in eukaryotes.*

## 6.9 Intragenic Recombination Occurs in Phage T4

We conclude this chapter with an account of an ingenious example of genetic analysis. In the early 1950s, Seymour Benzer undertook a detailed examination of a single locus, *rII*, in phage T4. Benzer successfully designed experiments to recover the extremely rare genetic recombinants arising as a result of intragenic exchange. Such recombination is equivalent to eukaryotic crossing over, but in this case, within a gene rather than at a point between two genes. Benzer demonstrated that such recombination occurs between the DNA of individual bacteriophages during simultaneous infection of the host bacterium *E. coli*.

The end result of Benzer's work was the production of a detailed map of the *rII* locus. Because of the extremely detailed information provided from his analysis, and because these experiments occurred decades before DNA-sequencing techniques were developed, the insights concerning the internal structure of the gene were particularly noteworthy.

## The *rII* Locus of Phage T4

The primary requirement in genetic analysis is the isolation of a large number of mutations in the gene being investigated. Mutants at the *rII* locus produce distinctive plaques when plated on *E. coli* strain B, allowing their easy identification. Figure 6–19 illustrates mutant *r* plaques compared to their wild-type $r^+$ counterparts in the related T2 phage. Benzer's approach was to isolate many independent *rII* mutants—he eventually obtained about 20,000—and to perform recombinational studies so as to produce a genetic map of this locus. Benzer assumed that most of these mutations, because they were randomly isolated, would represent different locations within the *rII* locus and would thus provide an ample basis for mapping studies.

The key to Benzer's analysis was that *rII* mutant phages, though capable of infecting and lysing *E. coli* B, could not successfully lyse a second related strain, *E. coli* K12(λ).* Wild-type phages, by contrast, could lyse both the B and the K12 strains. Benzer reasoned that these conditions provided the potential for a highly sensitive screening system. If phages from any two different mutant strains were allowed to simultaneously infect *E. coli* B, exchanges between the two mutant sites within the locus would produce rare wild-type recombinants (Figure 6–20). If the progeny phage population, which contained more than 99.9 percent *rII* phages and less than 0.1 percent wild-type phages, were then allowed to infect strain K12, the wild-type recombinants would successfully reproduce and produce wild-type plaques. This is the critical step in recovering and quantifying rare recombinants.

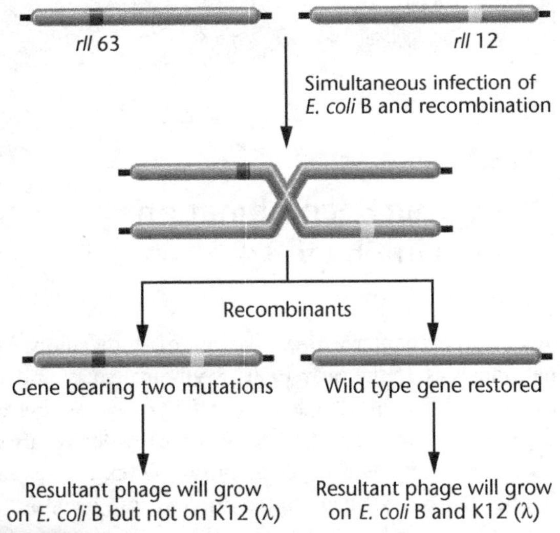


**FIGURE 6–20** Illustration of intragenic recombination between two mutations in the *rII* locus of phage T4. The result is the production of a wild-type phage, which will grow on both *E. coli* B and K12, and of a phage that has incorporated both mutations into the *rII* locus. The latter will grow on *E. coli* B but not on *E. coli* K12.

*The inclusion of "(λ)" in the designation of K12 indicates that this bacterial strain is lysogenized by phage λ. This, in fact, is the reason that *rII* mutants cannot lyse such bacteria. In future discussions, this strain will simply be abbreviated as *E. coli* K12.

By using serial dilution techniques, Benzer was able to determine the total number of mutant *rII* phages produced on *E. coli* B and the total number of recombinant wild-type phages that would lyse *E. coli* K12. These data provided the basis for calculating the frequency of recombination, a value proportional to the distance within the gene between the two mutations being studied. As we will see, this experimental design was extraordinarily sensitive. Remarkably, it was possible for Benzer to detect as few as one recombinant wild-type phage among 100 million mutant phages. When information from many such experiments is combined, a detailed map of the locus is possible.

Before we discuss this mapping, we need to describe an important discovery Benzer made during the early development of his screen—a discovery that led to the development of a technique used widely in genetics labs today, the **complementation assay** you learned about in Chapter 4.

## Complementation by *rII* Mutations

Before Benzer was able to initiate these intragenic recombination studies, he had to resolve a problem encountered during the early stages of his experimentation. While doing a control study in which K12 bacteria were simultaneously infected with pairs of different *rII* mutant strains, Benzer sometimes found that certain pairs of the *rII* mutant strains lysed the K12 bacteria. This was initially quite puzzling, since only the wild-type *rII* was supposed to be capable of lysing K12 bacteria. How could two mutant strains of *rII*, each of which was thought to contain a defect in the same gene, show a wild-type function?

Benzer reasoned that, during simultaneous infection, each mutant strain provided something that the other lacked, thus restoring wild-type function. This phenomenon, which he called **complementation**, is illustrated in Figure 6–21(a). When many pairs of mutations were tested, each mutation fell into one of two possible **complementation groups**, A or B. Those that failed to complement one another were placed in the same complementation group, while those that did complement one another were each assigned to a different complementation group. Benzer coined the term **cistron**, which he defined as the smallest functional genetic unit, to describe a complementation group. In modern terminology, we know that a cistron represents a gene.

We now know that Benzer's A and B cistrons represent two separate genes in what we originally referred to as the *rII* locus (because of the initial assumption that it was a single gene). Complementation occurs when K12 bacteria are infected with two *rII* mutants, one with a mutation in the A gene and one with a mutation in the B gene. Therefore, there is a source of both wild-type gene products, since the A mutant provides wild-type B and the B mutant provides wild-type A. We can also explain why two strains that fail to complement, say two A-cistron mutants, are actually mutations in the same gene. In this case, if two A-cistron mutants are combined, there will be an immediate source of the wild-type B product, but no immediate source of the wild-type A product [Figure 6–21(b)].

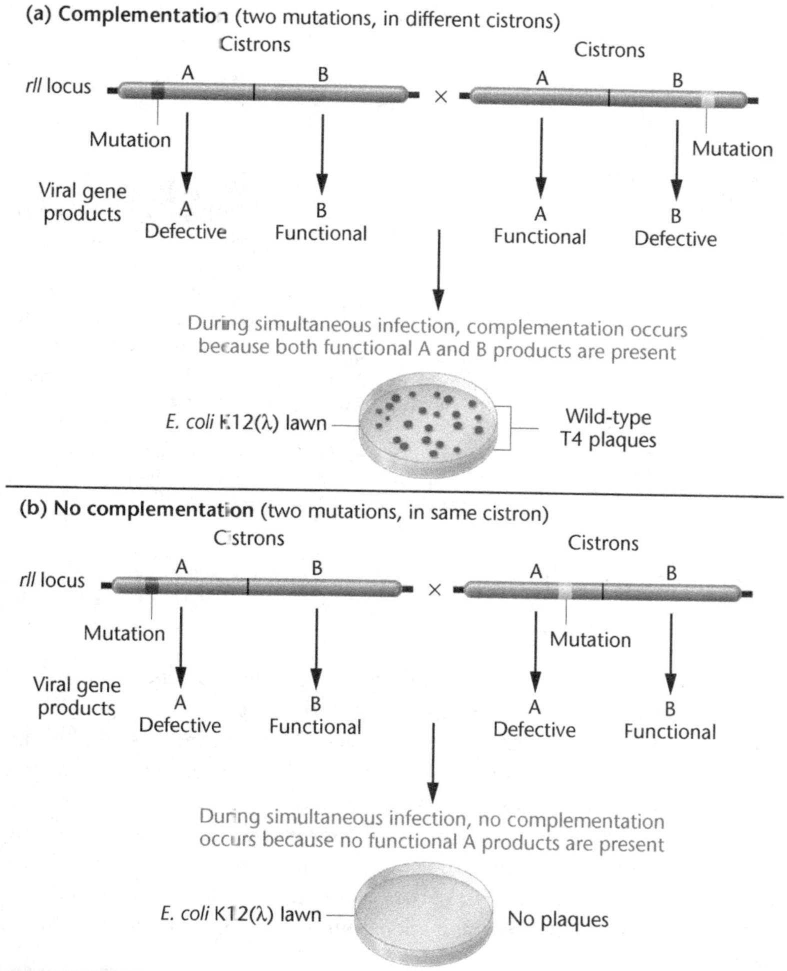


**FIGURE 6–21** Comparison of two pairs of *rII* mutations. (a) In one case, they complement one another. (b) In the other case, they do not complement one another. Complementation occurs when each mutation is in a separate cistron. Failure to complement occurs when the two mutations are in the same cistron.

Once Benzer was able to place all *rII* mutations in either the A or the B cistron, he was set to return to his intragenic recombination studies, testing mutations in the A cistron against each other and testing mutations in the B cistron against each other.

NOW SOLVE THIS

Problem 18 on page 170 involves an understanding of why complementation occurs during simultaneous infection of a bacterial cell by two bacteriophage strains, each with a different mutation within the *rII* locus. You are asked to examine data of a complementation experiment and predict the results.

■ HINT: *If each mutation alters a different genetic product, then each strain will provide the product that the other is missing, thus leading to complementation.*

## Recombinational Analysis

Of the approximately 20,000 *rII* mutations, roughly half fell into each cistron. Benzer set about mapping the mutations within each one. For example, if two *rII* A mutants (i.e., two phage strains with different mutations in the A cistron) were first allowed to infect *E. coli* B in a liquid culture, and if a recombination event occurred between the mutational sites in the A cistron, then wild-type progeny viruses would be produced at low frequency. If samples of the progeny viruses from such an experiment were then plated on *E. coli* K12, only the wild-type recombinants would lyse the bacteria and produce plaques. The total number of nonrecombinant progeny viruses would be determined by plating samples on *E. coli* B.

This experimental protocol is illustrated in Figure 6–22. The percentage of recombinants can be determined by counting the plaques at the appropriate dilution in each case. As in eukaryotic mapping experiments, the frequency of recombination is an estimate of the distance between the two mutations within the cistron. For example, if the number of recombinants is equal to $4 \times 10^3$/mL, and the total number of progeny is $8 \times 10^9$/mL, then the frequency of recombination between the two mutants is

$$2\left(\frac{4 \times 10^3}{8 \times 10^9}\right) = 2(0.5 \times 10^{-6})$$
$$= 10^{-6}$$
$$= 0.000001$$

Multiplying by 2 is necessary because each recombinant event yields two reciprocal products, only one of which—the wild type—is detected.

NOW SOLVE THIS

Problem 20 on page xxx involves intragenic mapping within each cistron of the *rII* locus. You are asked to examine experimental data and calculate the recombination frequency.

■ HINT: *Recombination occurs within genes in the same way that it occurs between genes, but at a much lower frequency.*

## Deletion Testing of the *rII* Locus

Although the system for assessing recombination frequencies described earlier allowed for mapping mutations within each cistron, testing 1000 mutants two at a time in all combinations would have required millions of experiments. Fortunately, Benzer was able to overcome this obstacle when he devised an analytical approach referred to as **deletion testing**. He discovered that some of the *rII* mutations were, in reality, deletions of small parts of both cistrons. That is, the genetic changes giving rise to the *rII* properties were not a characteristic of point mutations. Most importantly, when a deletion mutation

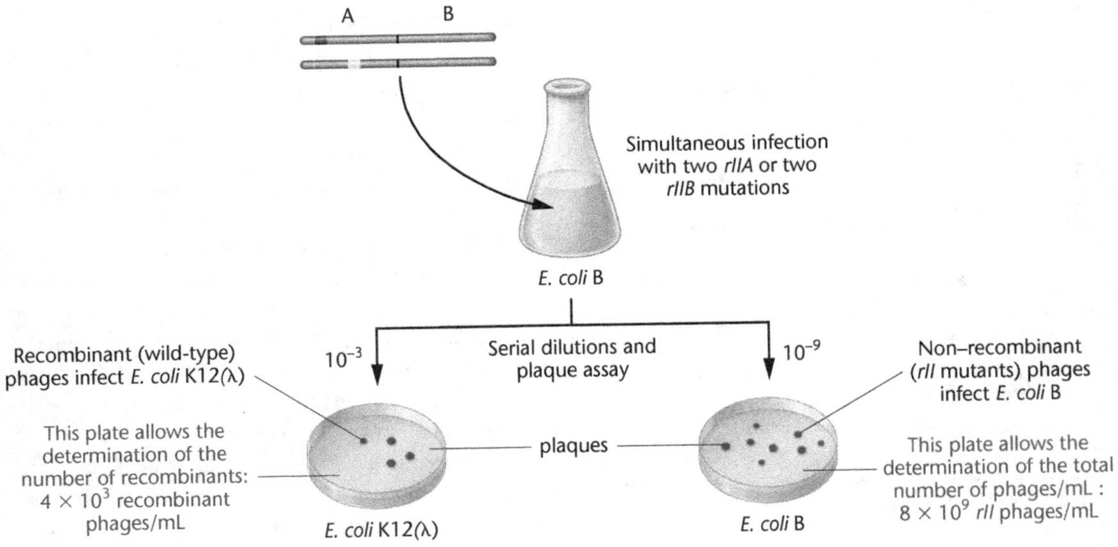


**FIGURE 6–22** The experimental protocol for recombination studies between pairs of mutations in the same cistron. In this figure, all phage infecting *E. coli* B (in the flask) contain one of two mutations in the A cistron, as shown in the depiction of their chromosomes to the left of the flask.

was tested using simultaneous infection by two phage strains, one having the deletion mutation and the other having a point mutation located in the deleted part of the same cistron, the test never yielded wild-type recombinants. The reason is illustrated in Figure 6–23. Because the deleted area is lacking the area of DNA containing the point mutation, no recombination is possible. Thus, a method was available that could roughly, but quickly, localize any mutation, provided it was contained within a region covered by a deletion.

Deletion testing could thus provide data for the initial localization of each mutation. For example, as shown in Figure 6–24, seven overlapping deletions spanning various regions of the A cistron were used for the initial screening of point mutations in that cistron. Depending on whether the viral chromosome bearing a point mutation does or does not undergo recombination with the chromosome bearing a deletion, each point mutation can be assigned to a specific area of the cistron. Further deletions within each of the seven areas can be used to localize, or map, each *rII* point mutation more precisely. Remember that, in each case, a point mutation is localized in the area of a deletion when it fails to give rise to any wild-type recombinants.

## The *rII* Gene Map

After several years of work, Benzer produced a genetic map of the two cistrons composing the *rII* locus of phage T4 (Figure 6–25). From the 20,000 mutations analyzed, 307 distinct sites within this locus were mapped in relation to one another. Areas containing many mutations, designated as **hot spots**, were apparently more susceptible to mutation than were areas in which only one or a few mutations were found. In addition, Benzer discovered areas within the cistrons in which no mutations were localized. He estimated that as many as 200 recombinational units had not been localized by his studies.

The significance of Benzer's work is his application of genetic analysis to what had previously been considered an abstract unit—the gene. Benzer had demonstrated in 1955 that a gene is not an indivisible particle, but instead consists of mutational and recombinational units that are arranged in a specific order. Today, we know these are nucleotides composing DNA. His analysis, performed prior to the detailed molecular studies of the gene in the 1960s, is considered a classic example of genetic experimentation.

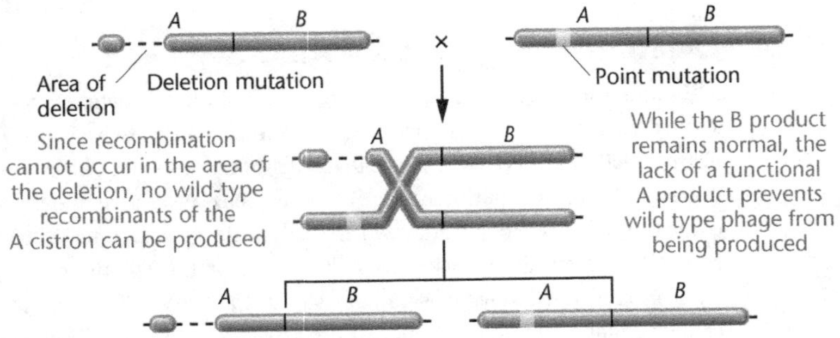


**FIGURE 6–23** Demonstration that recombination between a phage chromosome with a deletion in the A cistron and another phage with a point mutation overlapped by that deletion cannot yield a chromosome with wild-type A and B cistrons.

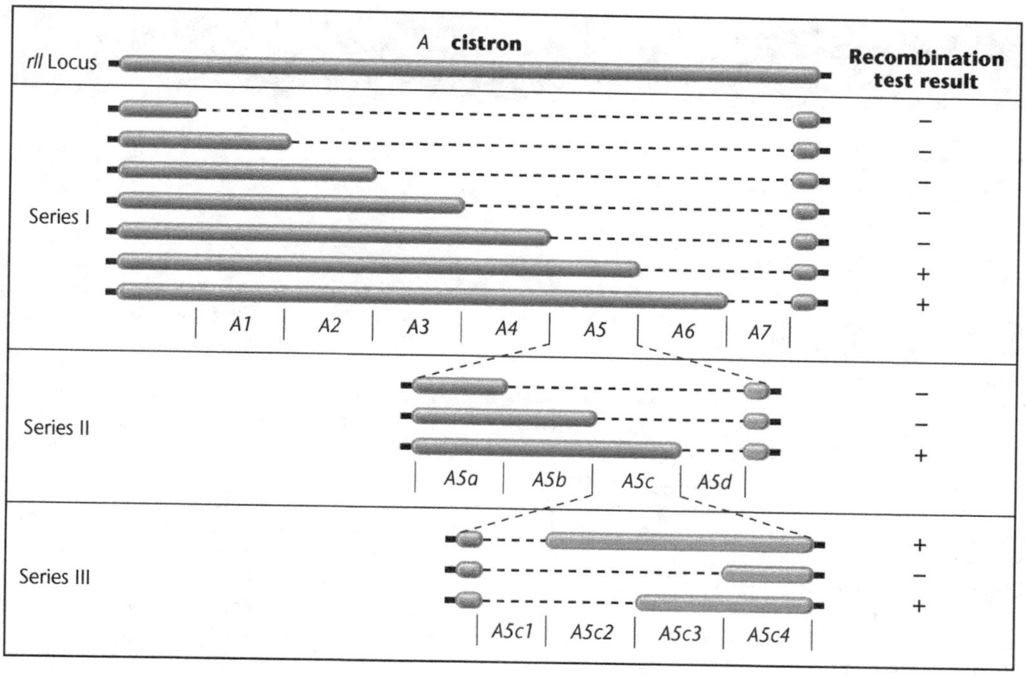


FIGURE 6–24 Three series of overlapping deletions in the *A* cistron of the *rII* locus used to localize the position of an unknown *rII* mutation. For example, if a mutant strain tested against each deletion (dashed areas) in Series I for the production of recombinant wild-type progeny shows the results at the right (+ or −), the mutation must be in segment *A5*. In Series II, the mutation is further narrowed to segment *A5c*, and in Series III to segment *A5c3*.

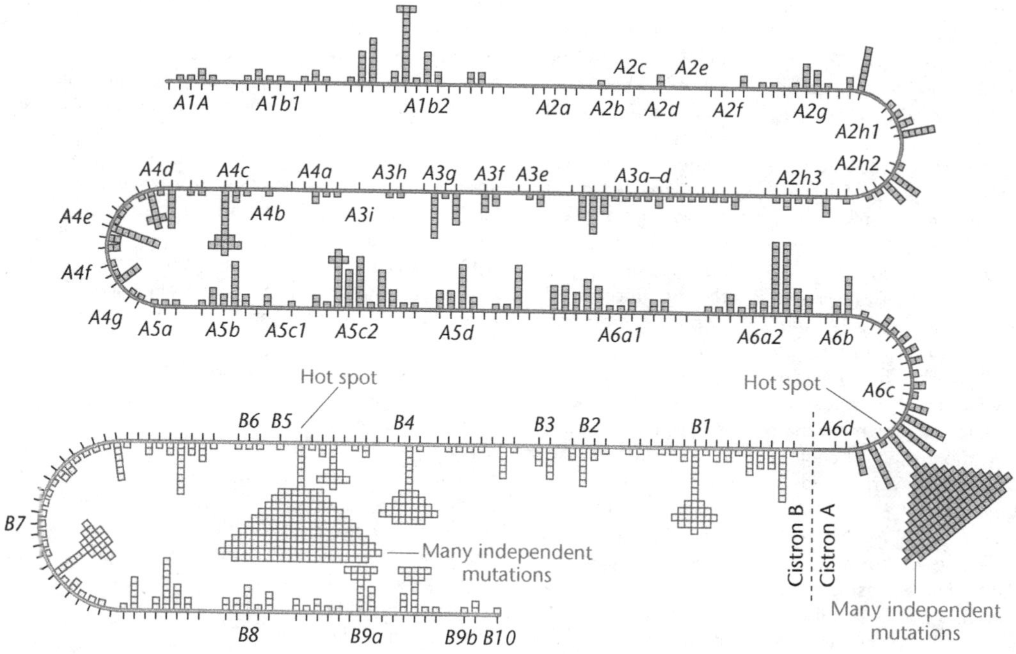


FIGURE 6–25 A partial map of mutations in the *A* and *B* cistrons of the *rII* locus of phage *T4*. Each square represents an independently isolated mutation. Note the two areas in which the largest number of mutations are present, referred to as "hot spots" (*A6cd* and *B5*).

## GENETICS, TECHNOLOGY, AND SOCIETY

# Bacterial Genes and Disease: From Gene Expression to Edible Vaccines

Using an expanding toolbox of molecular genetic tools, scientists are tackling some of the most serious bacterial diseases affecting our species. As an example, a new understanding of bacterial genes is leading directly to exciting new treatments based on edible vaccines. The story of vaccines against cholera and hepatitis B are models for this research. We will focus here on cholera.

The causative agent of cholera is *Vibrio cholerae*, a curved, rod-shaped bacterium found mostly in rivers and oceans. Most genetic strains of *V. cholerae* are harmless; only a few are pathogenic. Infection occurs when a person drinks water or eats food contaminated with pathogenic *V. cholerae*. Once in the digestive system, these bacteria colonize the small intestine and produce proteins called enterotoxins that invade the mucosal cells lining the intestine. This triggers a massive secretion of water and dissolved salts resulting in violent diarrhea, severe dehydration, muscle cramps, lethargy, and often death. The enterotoxin consists of two polypeptides, called the A and B subunits, encoded by two separate genes.

Cholera remains a leading cause of death of infants and children throughout the Third World, where basic sanitation is lacking and water supplies are often contaminated. For example, in July 1994, 70,000 cases of cholera leading to 12,000 fatalities were reported among the Rwandans crowded into refugee camps in Goma, Zaire. And after an absence of over 100 years, cholera reappeared in Latin America in 1991, spreading from Peru to Mexico and claiming more than 10,000 lives. In 2001, more than 40 cholera outbreaks in 28 countries were reported to the World Health Organization.

A new genetic technology is emerging to attack cholera. This technology centers on genetically engineered plants that act as vaccines. When a foreign gene—a gene for a disease antigen—is inserted into the plant genome, transcribed, and translated, a transgenic plant is produced that contains the foreign gene product. Immunity is then acquired by eating these plants; the foreign gene product acts as an antigen, stimulating the production of antibodies to protect against bacterial infection. Since it is the B subunit of the cholera enterotoxin that binds to intestinal cells, attention has focused on using this polypeptide as the antigen, with the expectation that antibodies against it will prevent toxin binding and render the bacteria harmless.

Leading the efforts to develop an edible vaccine are Charles Arntzen and associates at Cornell University. To test the system, Arntzen is using the B subunit of an *E. coli* enterotoxin, which is similar in structure and immunological properties to the cholera protein. The first step was to obtain the DNA clone of the gene encoding the B subunit and to attach it to a promoter that would induce transcription in all tissues of the plant. Second, the hybrid gene was introduced into potato plants by means of *Agrobacterium*-mediated transformation. Arntzen and his colleagues chose the potato because its large edible tubers would be convenient for testing the antigen's effectiveness. Analysis showed that the engineered plants expressed their new gene and produced the enterotoxin B subunit. The third step was to feed mice a few grams of the genetically engineered tubers. Arntzen found that the mice produced specific antibodies against the B subunit and secreted them into the small intestine. Most critically, mice that were later fed purified enterotoxin were protected from its effects and did not develop the symptoms of cholera. In clinical trials conducted using humans in 1998, almost all of the volunteers developed an immune response, and none experienced adverse side effects.

The Arntzen group is also producing edible vaccines in bananas and tomatoes. Bananas have several advantages over potatoes. Bananas can be grown almost anywhere throughout the tropical or subtropical developing countries of the world. Unlike potatoes, bananas are usually eaten raw, avoiding the potential inactivation of the antigenic proteins by cooking. Finally, bananas are well liked by infants and children, making this approach to immunization a more feasible one. If all goes as planned, it may someday be possible to immunize all Third World children against cholera and other intestinal diseases.

Arntzen's experimentation has served as a model for other research efforts involving a variety of human diseases. A group from the John P. Robarts Institute in Ontario, Canada, has shown that potato-produced vaccines can prevent juvenile diabetes in mice. Meristem Therapeutics, based in France, is in clinical trials using corn engineered to alleviate the effects of cystic fibrosis. An Australian research team has successfully produced tobacco plants that contain a protein found in the measles virus. After demonstrating the induction of immunity by feeding mice extracts of the tobacco leaves, researchers have begun to test this measles vaccine on primates. Tobacco plants are also being used to produce the antiviral protein interleukin 10 to treat Crohn's disease. Similar efforts are underway to create vaccines against rabies, anthrax, tetanus, and AIDS.

Although edible vaccines appear to have a promising future, many issues remain to be addressed. One problem has been dosage. Plants may vary in the concentration of the antigen. The amount of the potato or the banana ingested may also vary, altering the dose consumed. One solution might be to prepare capsules containing plant extracts in order to standardize doses of the antigen. There are also concerns over potential environmental hazards associated with growing transgenic crops. In particular, measures must be taken to prevent transgenes from being introduced into native plant populations. In addition, citizens of some countries are morally opposed to genetically engineered foods. If these obstacles can be overcome, edible vaccines hold great promise for the amelioration of many human diseases.

### References

Haq, T.A., Mason, H.S., Clements, J.D., and Arntzen, C.J. 1995. Oral immunization with a recombinant bacterial antigen produced in transgenic plants. *Science* 268: 714–776.

Mason, H.S., Warzecha, H., Mor, T., and Arntzen, C.J. 2002. Edible plant vaccines: Applications for prophylactic and therapeutic molecular medicine. *Trends Mol. Med.* 8: 324–329.

Webster, D.E., et al. 2002. Appetizing solutions: An edible vaccine for measles. *Med. J. Australia* 176: 434–437.

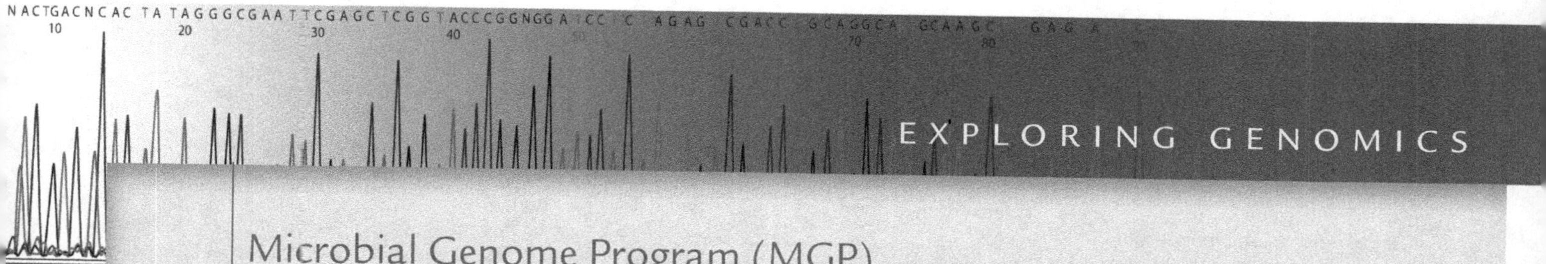

EXPLORING GENOMICS

# Microbial Genome Program (MGP)

Bacteria are the most abundant life forms on Earth, and they can live and thrive in some of the harshest environments on the planet. From polar ice caps to deserts, as well as under extreme conditions of radiation, pressure, darkness, and pH, microbes are supremely adapted to their environments. Because, in part, of their tremendous genetic diversity and remarkable adaptations, microbes have received a great deal of attention from genomics researchers. A large number of microbial genomes have already been completely sequenced, providing fascinating insight into microbial genetics. In this set of exercises, we will explore the **Microbial Genome Program (MGP)** Web site to learn about completed and ongoing microbial genome projects.

## ■ Exercise I – The Microbial Genome Program

In 1994, as an extension of the Human Genome Project, the U.S. Department of Energy launched the Microbial Genome Program, a comprehensive effort to sequence microbial genomes.

1. Visit the MGP Web site at http://microbialgenomics.energy.gov/index.shtml, and use the information presented at the home page to answer the following:
    a. What are some of the primary reasons for sequencing microbial genomes? What do scientists expect to learn from studying them?
    b. Provide examples of applications that may result from understanding microbial genomes.
2. Click on the "Research by Microbe" link and review the list of microbes that appears. Notice that the MGP is dedicated not only to completing bacterial genomes but also to sequencing the genomes of other categories of microorganisms. Under the "Bacteria" category, locate *Deinococcus*. Visit the link to "*Deinococcus radiodurans* Genome Database from TIGR" and the link to "Microbial Genomes from NCBI" (you will then need to scroll down and click on "*D. radiodurans*"), and answer the following questions:
    a. What is a unique ability of *D. radiodurans?*
    b. What genetic features of the *D. radiodurans* genome explain this microbe's ability to survive under extreme conditions?
    c. Propose a biotechnology application for *D. radiodurans*.

## ■ Exercise II – Metagenomics and a Global Expedition to Sequence Microbial Genomes: The *Sorcerer II* Project

Another excellent database for microbial genomes can be found at **The Institute for Genomic Research (TIGR) Microbial Sequencing Center** (http://msc.tigr.org/projects.shtml). In 1995, TIGR scientists, led by Dr. J. Craig Venter, sequenced the genome for the bacterium *Haemophilus influenzae*. (In Chapter 21 you will read more about this pioneering work.) The novel shotgun sequencing methods developed by TIGR played a major role in completing the Human Genome Project, and without question these techniques led to the genomics revolution. Currently, Dr. Venter is the director of the J. Craig Venter Institute. One of its main initiatives is the *Sorcerer II* project, a global sailing expedition that is being carried out on a research yacht in order to sequence the genomes for marine and terrestrial microorganisms around the world. Studying genomes of organisms collected from the environment is known as *metagenomics*, or *environmental genomics*.

The *Sorcerer II* project is sequencing microbial genomes at an astounding rate. It has already discovered more than 7 million previously unidentified sequences from a total of over 6 billion bp collected from over 400 uncharacterized microbial species!

1. Track the journey of *Sorcerer II* at www.sorcerer2expedition.org/version1/HTML/main.htm. After entering the site, click on the "Sampling Methods" tab for a basic overview of the techniques involved in sampling microbes from water samples, DNA sequencing, and database assembly and annotation.
2. Recently, *Sorcerer II* scientists published two papers in the free, online journal *Public Library of Science (PLoS) Biology*. Visit the *PLoS Biology* site at www.plos.org/journals/index.html and search the journal for Sorcerer II, or access the collection of articles on the project at http://collections.plos.org/plosbiology/gos-2007.php. Find and read the essay by Jonathan A. Eisen, "Environmental Shotgun Sequencing: Its Potential and Challenges for Studying the Hidden World of Microbes," *PLoS Biol*. 5(3) 0384-0388, for an excellent overview of metagenomics.
3. Search the PLoS site to find the following papers from the project: Rusch, D. B., et al., The *Sorcerer II* Global Ocean Sampling Expedition: Northwest Atlantic through Eastern Tropical Pacific, *PLoS Biol.* 5(3): 0398-0431, and Yooseph, S., et al., The *Sorcerer II* Global Ocean Sampling Expedition: Expanding the Universe of Protein Families, *PLoS Biol.* 5(3): 0432-0466. Review each paper to learn about some of the most significant findings of this project.

## Chapter Summary

1. Inherited phenotypic variation in bacteria results from spontaneous mutation.
2. Genetic recombination in bacteria takes place in three ways: conjugation, transformation, and transduction.
3. Conjugation is initiated by a bacterium housing a plasmid called the F factor. If the F factor is in the cytoplasm of a donor ($F^+$) cell, the recipient ($F^-$) cell receives a copy of the F factor and is converted to the $F^+$ status.
4. If the F factor is integrated into the donor cell chromosome (making that cell Hfr), the donor chromosome moves unidirectionally into the recipient, initiating recombination. Time mapping of the bacterial chromosome is based on the location and orientation of the F factor in the donor chromosome.
5. The products of a group of genes designated *rec* are directly involved in the process of recombination between the invading DNA and the recipient bacterial chromosome.
6. Plasmids, such as the F factor, are autonomously replicating DNA molecules found in the bacterial cytoplasm. Some plasmids contain unique genes conferring antibiotic resistance, as well as the genes necessary for plasmid transfer during conjugation.
7. In the phenomenon of transformation, which does not require cell-to-cell contact, exogenous DNA enters a recipient bacterium and recombines with the host's chromosome. Linkage mapping of closely aligned genes may be performed using this process.
8. Bacteriophages (viruses that infect bacteria) demonstrate a well-defined life cycle during which they reproduce within the host cell. They are studied using the plaque assay.
9. Bacteriophages can be lytic, meaning they infect the host cell, reproduce, and then lyse the host cell; or, in contrast, they can lysogenize the host cell, which means they infect it and integrate their DNA into the host chromosome but without reproducing.
10. Transduction is virus-mediated bacterial DNA recombination. When a lysogenized bacterium subsequently reenters the lytic cycle, the new bacteriophages serve as vehicles for the transfer of host (bacterial) DNA. In the process of generalized transduction, a random part of the bacterial chromosome is transferred. In specialized transduction, only specific genes adjacent to the point of insertion of the prophage are transferred.
11. Transduction is also used for bacterial linkage and mapping studies.
12. Various mutant phenotypes, including mutations in plaque morphology and host range, have been studied in bacteriophages. These have served as the basis for investigating genetic exchange and mapping in these viruses.
13. Genetic analysis of the *rII* locus in bacteriophage T4 allowed Seymour Benzer to study intragenic recombination. By isolating *rII* mutants and performing complementation analysis, recombinational studies, and deletion mapping, Benzer was able to locate and map more than 300 distinct sites within the two cistrons of the *rII* locus.

## INSIGHTS AND SOLUTIONS

1. Time mapping is performed in a cross involving the genes *his*, *leu*, *mal*, and *xyl*. The recipient cells were auxotrophic for all four genes. After 25 minutes, mating was interrupted with the following results in recipient cells. Diagram the positions of these genes relative to the origin (*O*) of the F factor and to one another.

(a) 90% were $xyl^+$

(b) 80% were $mal^+$

(c) 20% were $his^+$

(d) none were $leu^+$

**Solution:** The *xyl* gene was transferred most frequently, which shows it is closest to *O* (very close). The *mal* gene is next closest and reasonably near *xyl*, followed by the more distant *his* gene. The *leu* gene is far beyond these three, since no recombinants are recovered that include it. The diagram shows these relative locations along a piece of the circular chromosome.

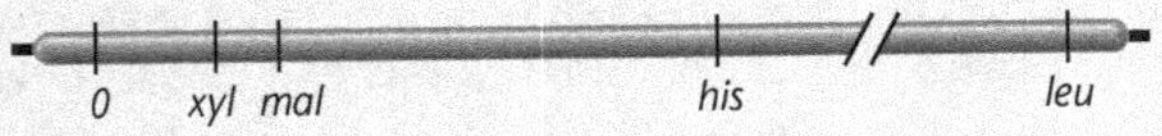


2. Three strains of bacteria, each bearing a separate mutation, $a^-$, $b^-$, or $c^-$, are the sources of donor DNA in a transformation experiment. Recipient cells are wild type for those genes but express the mutation $d^-$.

(a) Based on the following data, and assuming that the location of the *d* gene precedes the *a*, *b*, and *c* genes, propose a linkage map for the four genes.

| DNA Donor | Recipient | Transformants | Frequency of Transformants |
|---|---|---|---|
| $a^-\ d^+$ | $a^+\ d^-$ | $a^+\ d^+$ | 0.21 |
| $b^-\ d^+$ | $b^+\ d^-$ | $b^+\ d^+$ | 0.18 |
| $c^-\ d^+$ | $c^+\ d^-$ | $c^+\ d^+$ | 0.63 |

(b) If the donor DNA were wild type and the recipient cells were either $a^-\ b^-$, $a^-\ c^-$, or $b^-\ c^-$, which of the crosses would be expected to produce the greatest number of wild-type transformants?

**Solution:**

(a) These data reflect the relative distances between the *a*, *b*, and *c* genes, individually, and the *d* gene. The *a* and *b* genes are about the same distance from the *d* gene and are thus tightly linked to one another. The *c* gene is more distant. Assuming that the *d* gene precedes the others, the map looks like this:

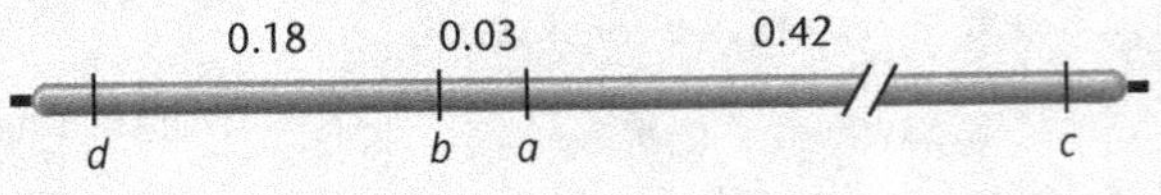


(b) Because the *a* and *b* genes are closely linked, they most likely cotransform in a single event. Thus, recipient cells $a^-\ b^-$ are most likely to convert to wild type.

3. In four Hfr strains of bacteria, all derived from an original $F^+$ culture grown over several months, a group of hypothetical genes was studied and shown to be transferred in the order shown in the following table.

| Hfr Strain | Order of Transfer | | | | | |
|---|---|---|---|---|---|---|
| 1 | *E* | *R* | *I* | *U* | *M* | *B* |
| 2 | *U* | *M* | *B* | *A* | *C* | *T* |
| 3 | *C* | *T* | *E* | *R* | *I* | *U* |
| 4 | *R* | *E* | *T* | *C* | *A* | *B* |

(a) Assuming that *B* is the first gene along the chromosome, determine the sequence of all genes shown.

(b) One strain creates an apparent dilemma. Which one is it? Explain why the dilemma is only apparent, not real.

**Solution:**

(a) The solution is found by overlapping the genes in each strain.

| | | | | | | | | | | | | |
|---|---|---|---|---|---|---|---|---|---|---|---|---|
| | 2 | *U* | *M* | *B* | *A* | *C* | *T* | | | | | |
| Strain: | 3 | | | | | *C* | *T* | *E* | *R* | *I* | *U* | |
| | 1 | | | | | | | *E* | *R* | *I* | *U* | *M* *B* |

Starting with *B*, the genes sequence is *BACTERIUM*.

(b) Strain 4 creates an apparent dilemma in that the genes initially appear to be out of order. The dilemma is resolved when we realize that the F factor is integrated in the opposite orientation. Thus, the genes enter in the opposite sequence, starting with gene *R*:

RETCAB
⟶

4. For his fine-structure analysis of the *rII* locus in phage T4, Benzer was able to perform complementation testing of any pair of mutations once it was clear that the locus contained two cistrons. Complementation was assayed by simultaneously infecting *E. coli* K12 with two phage strains, each with an independent mutation, neither of which could alone lyse K12. From the data that follow, determine which mutations are in which cistron, assuming that mutation 1 (M-1) is in the A cistron and mutation 2 (M-2) is in the B cistron. Are there any cases where the mutation cannot be properly assigned?

| Test Pair | Results* |
|---|---|
| 1, 2 | + |
| 1, 3 | − |
| 1, 4 | − |
| 1, 5 | + |
| 2, 3 | − |
| 2, 4 | + |
| 2, 5 | − |

*+ or − indicates complementation or the failure of complementation, respectively.

**Solution:** M-1 and M-5 complement one another and, therefore, are not in the same cistron. Thus, M-5 must be in the B cistron. M-2 and M-4 complement one another. By the same reasoning, M-4 is not with M-2 and, therefore, is in the A cistron. M-3 fails to complement either M-1 or M-2, and so it would seem to be in both cistrons. One explanation is that the physical cause of M-3 somehow overlaps both the A and the B cistrons. It might be a double mutation with one sequence change in each cistron. It might also be a deletion that overlaps both cistrons and thus could not complement either M-1 or M-2.

5. Another mutation, M-6, was tested with the results shown here:

| Test Pair | Results |
|---|---|
| 1, 6 | + |
| 2, 6 | − |
| 3, 6 | − |
| 4, 6 | + |
| 5, 6 | − |

Draw all possible conclusions about M-6.

**Solution:** These results are consistent with assigning M-6 to the B cistron.

6. Recombination testing was then performed for M-2, M-5, and M-6 so as to map the B cistron. Recombination analysis using both *E. coli* B and K12 showed that recombination occurred between M-2 and M-5 and between M-5 and M-6, but not between M-2 and M-6. Why not?

**Solution:** Either M-2 and M-6 represent identical mutations, or one of them may be a deletion that overlaps the other but does not overlap M-5. Furthermore, the data cannot rule out the possibility that both are deletions.

7. In recombination studies of *rII* locus in phage T4, what is the significance of the value determined by calculating phage growth in the K12 versus the B strains of *E. coli* following simultaneous infection in *E. coli* B? Which value is always greater?

**Solution:** When plaque analysis is performed on *E. coli* B, in which the wild-type and mutant phages are both lytic, the total number of phages per milliliter can be determined. Because almost all cells are *rII* mutants of one type or another, this value is much larger than the value obtained with K12. To avoid total lysis of the plate, extensive dilution is necessary. In K12, *rII* mutations will not grow, but wild-type phages will. Because wild-type phages are the rare recombinants, there are relatively few of them and extensive dilution is not required.

## Problems and Discussion Questions

1. Distinguish among the three modes of recombination in bacteria.
2. With respect to $F^+$ and $F^-$ bacterial matings, answer the following questions:
   (a) How was it established that physical contact between cells was necessary?
   (b) How was it established that chromosome transfer was unidirectional?
   (c) What is the genetic basis for a bacterium's being $F^+$?
3. List all major differences between (a) the $F^+ \times F^-$ and the Hfr $\times F^-$ bacterial crosses; and (b) the $F^+$, $F^-$, Hfr, and F′ bacteria.
4. Describe the basis for chromosome mapping in the Hfr $\times F^-$ crosses.
5. Why are the recombinants produced from an Hfr $\times F^-$ cross rarely, if ever, $F^+$?
6. Describe the origin of F′ bacteria and merozygotes.

7. Describe what is known about the mechanism of transformation.
8. In a transformation experiment involving a recipient bacterial strain of genotype $a^- b^-$, the following results were obtained. What can you conclude about the location of the *a* and *b* genes relative to each other?

| Transforming DNA | Transformants (%) | | |
|---|---|---|---|
| | $a^+ b^-$ | $a^- b^+$ | $a^+ b^+$ |
| $a^+ b^+$ | 3.1 | 1.2 | 0.04 |
| $a^+ b^-$ and $a^- b^+$ | 2.4 | 1.4 | 0.03 |

9. In a transformation experiment, donor DNA was obtained from a prototroph bacterial strain ($a^+ b^+ c^+$), and the recipient was a triple auxotroph ($a^- b^- c^-$). What general conclusions can you draw about the linkage relationships among the three genes from the following transformant classes that were recovered?

| | |
|---|---|
| $a^+ b^- c^-$ | 180 |
| $a^- b^+ c^-$ | 150 |
| $a^+ b^+ c^-$ | 210 |
| $a^- b^- c^+$ | 179 |
| $a^+ b^- c^+$ | 2 |
| $a^- b^+ c^+$ | 1 |
| $a^+ b^+ c^+$ | 3 |

10. Explain the observations that led Zinder and Lederberg to conclude that the prototrophs recovered in their transduction experiments were not the result of $F^+$ mediated conjugation.
11. Define plaque, lysogeny, and prophage.
12. Differentiate between generalized and specialized transduction.
13. Two theoretical genetic strains of a virus ($a^- b^- c^-$ and $a^+ b^+ c^+$) were used to simultaneously infect a culture of host bacteria. Of 10,000 plaques scored, the following genotypes were observed. Determine the genetic map of these three genes on the viral chromosome. Decide whether interference was positive or negative.

| | | | |
|---|---|---|---|
| $a^+ b^+ c^+$ | 4100 | $a^- b^+ c^-$ | 160 |
| $a^- b^- c^-$ | 3990 | $a^+ b^- c^+$ | 140 |
| $a^+ b^- c^-$ | 740 | $a^- b^- c^+$ | 90 |
| $a^- b^+ c^+$ | 670 | $a^+ b^+ c^-$ | 110 |

14. Describe the conditions under which genetic recombination may occur in bacteriophages.
15. The bacteriophage genome consists primarily of genes encoding proteins that make up the head, collar, tail, and tail fibers. When these genes are transcribed following phage infection, how are these proteins synthesized, since the phage genome lacks genes essential to ribosome structure?
16. If a single bacteriophage infects one *E. coli* cell present on a lawn of bacteria and, upon lysis, yields 200 viable viruses, how many phages will exist in a single plaque if only three more lytic cycles occur?
17. A phage-infected bacterial culture was subjected to a series of dilutions, and a plaque assay was performed in each case, with the results shown in the following table. What conclusion can be drawn in the case of each dilution, assuming that 0.1 mL was used in each plaque assay?

| | Dilution Factor | Assay Results |
|---|---|---|
| (a) | $10^4$ | All bacteria lysed |
| (b) | $10^5$ | 14 plaques |
| (c) | $10^6$ | 0 plaques |

18. In complementation studies of the *rII* locus of phage T4, three groups of three different mutations were tested. For each group, only two combinations were tested. On the basis of each set of data (shown here), predict the results of the third experiment for each group.

| Group A | Group B | Group C |
|---|---|---|
| $d \times e$—lysis | $g \times b$—no lysis | $j \times k$—lysis |
| $d \times f$—no lysis | $g \times i$—no lysis | $j \times l$—lysis |
| $e \times f$—? | $b \times i$—? | $k \times l$—? |

19. In an analysis of other *rII* mutants, complementation testing yielded the following results:

| Mutants | Results (+/− lysis) |
|---|---|
| 1, 2 | + |
| 1, 3 | + |
| 1, 4 | − |
| 1, 5 | − |

(a) Predict the results of testing 2 and 3, 2 and 4, and 3 and 4 together.
(b) If further testing yielded the following results, what would you conclude about mutant 5?

| Mutants | Results |
|---|---|
| 2, 5 | − |
| 3, 5 | − |
| 4, 5 | − |

20. Using mutants 2 and 3 from the previous problem, following mixed infection on *E. coli* B, progeny viruses were plated in a series of dilutions on both *E. coli* B and K12 with the following results. What is the recombination frequency between the two mutants?

| Strain Plated | Dilution | Plaques |
|---|---|---|
| *E. coli* B | $10^{-5}$ | 2 |
| *E. coli* K12 | $10^{-1}$ | 5 |

21. Another mutation, 6, was tested in relation to mutations 1 through 5 from the previous problem. In initial testing, mutant 6 complemented mutants 2 and 3. In recombination testing with 1, 4, and 5, mutant 6 yielded recombinants with 1 and 5, but not with 4. What can you conclude about mutation 6?
22. When the interrupted mating technique was used with five different strains of Hfr bacteria, the following orders of gene entry and recombination were observed. On the basis of these data, draw a map of the bacterial chromosome. Do the data support the concept of circularity?

| Hfr Strain | Order | | | | |
|---|---|---|---|---|---|
| 1 | *T* | *C* | *H* | *R* | *O* |
| 2 | *H* | *R* | *O* | *M* | *B* |
| 3 | *M* | *O* | *R* | *H* | *C* |
| 4 | *M* | *B* | *A* | *K* | *T* |
| 5 | *C* | *T* | *K* | *A* | *B* |

HOW DO WE KNOW?

23. In this chapter, we have focused on genetic systems present in bacteria and on the viruses that use bacteria as hosts (bacteriophages). In particular, we discussed mechanisms by which bacteria and their phages undergo genetic recombination, which allows geneticists to map bacterial and bacteriophage chromosomes. In the process, we found many opportunities to consider how this information was acquired. From the expla-

nations given in the chapter, what answers would you propose to the following questions?

(a) How do we know that genes exist in bacteria and bacteriophages?

(b) How do we know that bacteria undergo genetic recombination, allowing the transfer of genes from one organism to another?

(c) How do we know that genetic recombination between bacteria involves cell-to-cell contact and that such contact precedes the transfer of genes from one bacterium to another?

(d) How do we know that bacteriophages recombine genetic material through transduction and that cell-to-cell contact is not essential for transduction to occur?

(e) How do we know that intergenic exchange occurs in bacteriophages?

(f) How do we know that in bacteriophage T4 the *rII* locus is subdivided into two regions, or cistrons?

## Extra-Spicy Problems

24. During the analysis of seven *rII* mutations in phage T4, mutants 1, 2, and 6 were in cistron A, while mutants 3, 4, and 5 were in cistron B. Of these, mutant 4 was a deletion overlapping mutant 5. The remainder were point mutations. Nothing was known about mutant 7. Predict the results of complementation (+ or −) between 1 and 2; 1 and 3; 2 and 4; and 4 and 5.

25. In studies of recombination between mutants 1 and 2 from the previous problem, the results shown below were obtained.

| Strain | Dilution | Plaques | Phenotypes |
|---|---|---|---|
| *E. coli* B | $10^{-7}$ | 4 | *r* |
| *E. coli* K12 | $10^{-2}$ | 8 | + |

(a) Calculate the recombination frequency.

(b) When mutant 6 was tested for recombination with mutant 1, the data were the same as shown above for strain B, but not for K12. The researcher lost the K12 data, but remembered that recombination was 10 times more frequent than when mutants 1 and 2 were tested. What were the lost values (dilution and colony numbers)?

(c) Mutant 7 failed to complement any of the other mutants (1–6). Define the nature of mutant 7.

26. In *Bacillus subtilis*, linkage analysis of two mutant genes affecting the synthesis of two amino acids, tryptophan ($trp_2^-$) and tyrosine ($tyr_1^-$), was performed using transformation. Examine the following data and draw all possible conclusions regarding linkage. What is the purpose of Part B of the experiment? [Reference: E. Nester, M. Schafer, and J. Lederberg (1963).]

| Donor DNA | Recipient Cell | Transformants | No. |
|---|---|---|---|
| A. $trp_2^+\ tyr_1^+$ | $trp_2^-\ tyr_1^-$ | $trp^+\ tyr^-$ | 196 |
| | | $trp^-\ tyr^+$ | 328 |
| | | $trp^+\ tyr^+$ | 367 |
| B. $trp_2^+\ tyr_1^-$ and $trp_2^-\ tyr_1^+$ | $trp_2^-\ tyr_1^-$ | $trp^+\ tyr^-$ | 190 |
| | | $trp^-\ tyr^+$ | 256 |
| | | $trp^+\ tyr^+$ | 2 |

27. An Hfr strain is used to map three genes in an interrupted mating experiment. The cross is $Hfr/a^+\ b^+\ c^+\ rif \times F^-/a^-\ b^-\ c^-\ rif^r$. (No map order is implied in the listing of the alleles; $rif^r$ is resistance to the antibiotic rifampicin.) The $a^+$ gene is required for the biosynthesis of nutrient A, the $b^+$ gene for nutrient B, and $c^+$ for nutrient C. The minus alleles are auxotrophs for these nutrients. The cross is initiated at time = 0, and at various times, the mating mixture is plated on three types of medium. Each plate contains minimal medium (MM) plus rifampicin plus specific supplements that are indicated in the following table. (The results for each time interval are shown as the number of colonies growing on each plate.)

| | Time of Interruption | | | |
|---|---|---|---|---|
| Supplements Added to MM | 5 min | 10 min | 15 min | 20 min |
| Nutrients A and B | 0 | 0 | 4 | 21 |
| Nutrients B and C | 0 | 5 | 23 | 40 |
| Nutrients A and C | 4 | 25 | 60 | 82 |

(a) What is the purpose of rifampicin in the experiment?

(b) Based on these data, determine the approximate location on the chromosome of the *a*, *b*, and *c* genes relative to one another and to the F factor.

(c) Can the location of the *rif* gene be determined in this experiment? If not, design an experiment to determine the location of *rif* relative to the F factor and to gene *b*.

28. A plaque assay is performed beginning with 1 mL of a solution containing bacteriophages. This solution is serially diluted three times by combining 0.1 mL of each sequential dilution with 9.9 mL of liquid medium. Then 0.1 mL of the final dilution is plated in the plaque assay and yields 17 plaques. What is the initial density of bacteriophages in the original 1 mL?

29. In a cotransformation experiment, using various combinations of genes two at a time, the following data were produced. Determine which genes are "linked" to which others.

| Successful Cotransformation | Unsuccessful Cotransformation |
|---|---|
| *a* and *d*; *b* and *c*; | *a* and *b*; *a* and *c*; *a* and *f*; |
| *b* and *f* | *d* and *b*; *d* and *c*; *d* and *f* |
| | *a* and *e*; *b* and *e*; *c* and *e* |
| | *d* and *e*; *f* and *e* |

30. For the experiment in Problem 29, another gene, *g*, was studied. It demonstrated positive cotransformation when tested with gene *f*. Predict the results of testing gene *g* with genes *a*, *b*, *c*, *d*, and *e*.

31. Bacterial conjugation, mediated mainly by conjugative plasmids such as F, represents a potential health threat through the sharing of genes for pathogenicity or antibiotic resistance. Given that more than 400 different species of bacteria coinhabit a healthy human gut and more than 200 coinhabit human skin, Francisco Dionisio [*Genetics* (2002) 162:1525–1532] investigated the ability of plasmids to undergo between-species conjugal transfer. The following data are presented for various species of the enterobacterial genus *Escherichia*. The data are presented as "log base 10" values; for example, −2.0 would be equivalent to $10^{-2}$ as a rate of transfer. Assume that all differences between values presented are statistically significant.

(a) What general conclusion(s) can be drawn from these data?

(b) In what species is within-species transfer most likely? In what species pair is between-species transfer most likely?

(c) What is the significance of these findings in terms of human health?

| Recipient | Donor | | | |
|---|---|---|---|---|
| | *E. chrysanthemi* | *E. blattae* | *E. fergusonii* | *E. coli* |
| *E. chrysanthemi* | −2.4 | −4.7 | −5.8 | −3.7 |
| *E. blattae* | −2.0 | −3.4 | −5.2 | −3.4 |
| *E. fergusonii* | −3.4 | −5.0 | −5.8 | −4.2 |
| *E. coli* | −1.7 | −3.7 | −5.3 | −3.5 |

32. A study was conducted in an attempt to determine which functional regions of a particular conjugative transfer gene (*tra1*) are involved in the transfer of plasmid R27 in *Salmonella enterica*. The R27 plasmid is of significant clinical interest because it is capable of encoding multiple-antibiotic resistance to typhoid fever. To identify functional regions responsible for conjugal transfer, an analysis by Lawley et al. (2002. *J. Bacteriol.* 184:2173–2180) was conducted in which particular regions of the *tra1* gene were mutated and tested for their impact on conjugation. Shown here is a map of the regions tested and believed to be involved in conjugative transfer of the plasmid. Similar coloring indicates related function. Numbers correspond to each functional region subjected to mutation analysis.

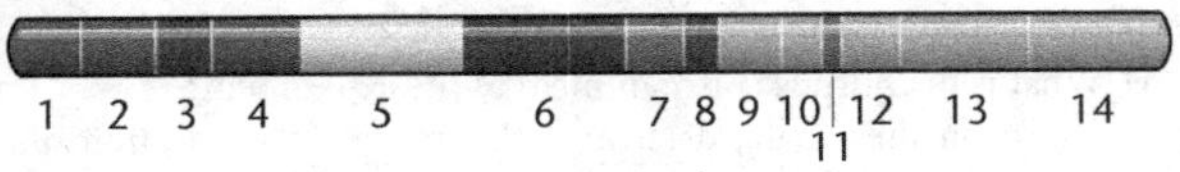


Accompanying the map is a table showing the effects of these mutations on R27 conjugation.

**Effects of Mutations in Functional Regions of Transfer Region 1(*tra1*) on R27 Conjugation**

| R27 Mutation in Region | Conjugative Transfer | Relative Conjugation Frequency (%)* |
|---|---|---|
| 1 | + | 100 |
| 2 | + | 100 |
| 3 | − | 0 |
| 4 | + | 100 |
| 5 | − | 0 |
| 6 | − | 0 |
| 7 | + | 12 |
| 8 | − | 0 |
| 9 | − | 0 |
| 10 | − | 0 |
| 11 | + | 13 |
| 12 | − | 0 |
| 13 | − | 0 |
| 14 | − | 0 |

(a) Given the data, do all functional regions appear to influence conjugative transfer?
(b) Which regions appear to have the most impact on conjugation?
(c) Which regions appear to have a limited impact on conjugation?
(d) What general conclusions might one draw from these data?

33. Influenza (the flu) is responsible for approximately 250,000 to 500,000 deaths annually, but periodically its toll has been much higher. For example, the 1918 flu pandemic killed approximately 30 million people worldwide and is considered the worst spread of a deadly illness in recorded history. With highly virulent flu strains emerging periodically, it is little wonder that the scientific community is actively studying influenza biology. In 2007, the National Institute of Allergy and Infectious Diseases completed sequencing of 2035 human and avian influenza virus strains. Influenza strains undergo recombination as described in this chapter, and they have a high mutation rate owing to the error-prone replication of their genome (which consists of RNA rather than DNA). In addition, they are capable of chromosome reassortment in which various combinations of their eight chromosomes (or portions thereof) can be packaged into progeny viruses when two or more strains infect the same cell. The end result is that we can make vaccines, but they must change annually, and even then, we can only guess at what specific viral strains will be prevalent in any given year. Based on the above information, consider the following questions:
(a) Of what evolutionary value to influenza viruses are high mutation and recombination rates coupled with chromosome reassortment?
(b) Why can't humans combat influenza just as they do mumps, measles, or chicken pox?
(c) Why are vaccines available for many viral diseases but not influenza?

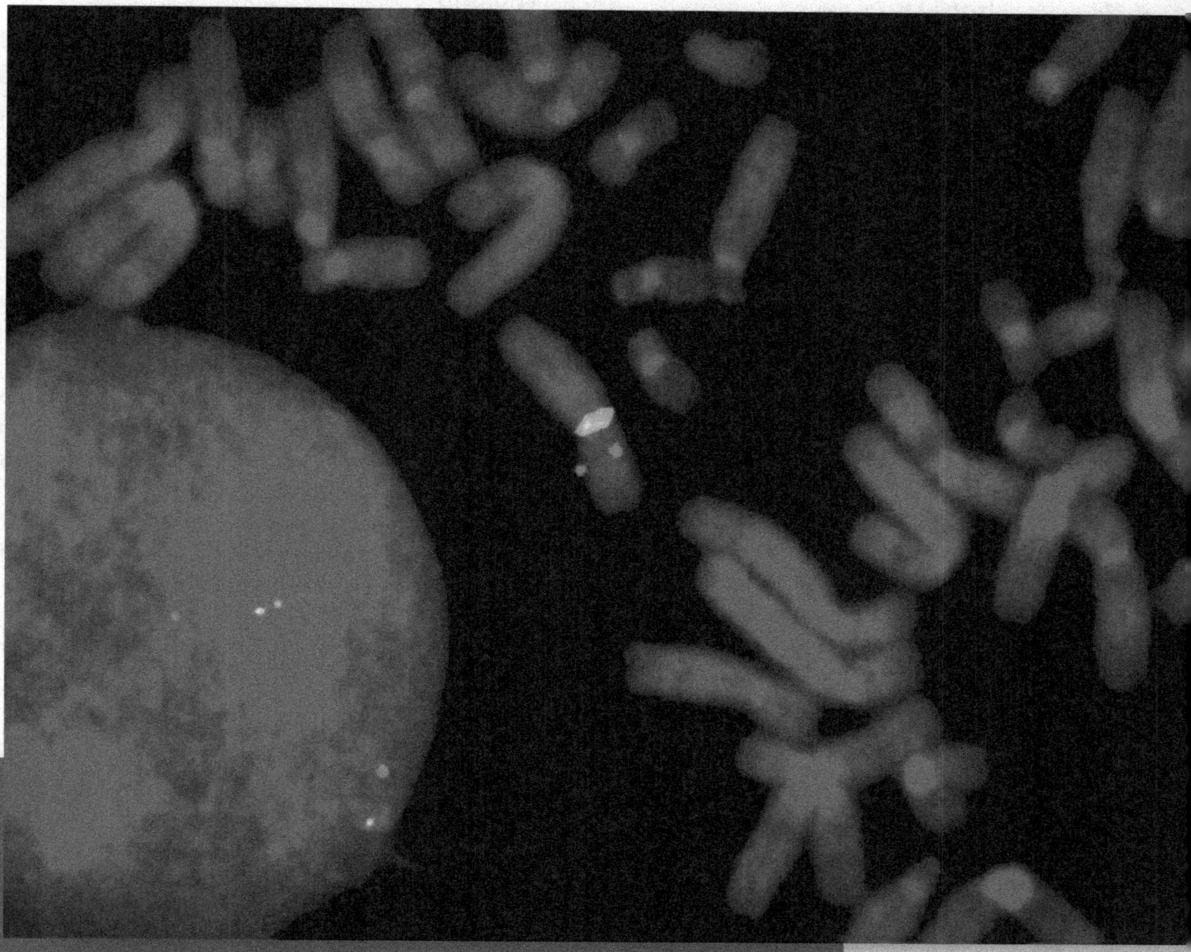

Human X chromosomes highlighted using fluorescence *in situ* hybridization (FISH), a method in which specific probes bind to specific sequences of DNA. The probe producing green fluoresence binds to the DNA of the X chromosome centromeres. The probe producing red fluorescence binds to the DNA sequence of the Duchenne muscular dystrophy (DMD) gene, an X-linked gene.

7

# Sex Determination and Sex Chromosomes

## CHAPTER CONCEPTS

- Sexual reproduction, which greatly enhances genetic variation within species, requires mechanisms that result in sexual differentiation.
- A wide variety of genetic mechanisms lead to sexual dimorphism.
- Often, specific genes, usually on a single chromosome, cause maleness or femaleness during development.
- In humans, the presence of extra X or Y chromosomes beyond the diploid number may be tolerated but often leads to syndromes demonstrating distinctive phenotypes.
- While segregation of sex-determining chromosomes should theoretically lead to a one-to-one sex ratio of males to females, in humans the actual ratio greatly favors males at conception.
- In mammals, females inherit two X chromosomes compared to one in males, but the extra genetic information in females is compensated for by random inactivation of one of the X chromosomes early in development.
- In some reptilian species, temperature during incubation of eggs determines the sex of offspring.

In the biological world, a wide range of reproductive modes and life cycles are observed. Some organisms are entirely asexual, displaying no evidence of sexual reproduction. Some organisms alternate between short periods of sexual reproduction and prolonged periods of asexual reproduction. In most diploid eukaryotes, however, sexual reproduction is the only natural mechanism for producing new members of the species. Orderly transmission of genetic material from parents to offspring, and the resultant phenotypic variability, relies on the processes of segregation and independent assortment that occur during meiosis. Meiosis produces haploid gametes so that, following fertilization, the resulting offspring maintain the diploid number of chromosomes characteristic of their kind. Thus, meiosis ensures genetic constancy within members of the same species.

These events, seen in the perpetuation of all sexually reproducing organisms, depend ultimately on an efficient union of gametes during fertilization. In turn, successful fertilization depends on some form of sexual differentiation in the reproductive organisms. Even though it is not overtly evident, this differentiation occurs in organisms as low on the evolutionary scale as bacteria and single-celled eukaryotic algae. In more complex forms of life, the differentiation of the sexes is more evident as phenotypic dimorphism of males and females. The ancient symbol for iron and for Mars, depicting a shield and spear (♂), and the ancient symbol for copper and for Venus, depicting a mirror (♀), have also come to symbolize maleness and femaleness, respectively.

Dissimilar, or **heteromorphic, chromosomes**, such as the XY pair in mammals, characterize one sex or the other in a wide range of species, resulting in their label as **sex chromosomes**. Nevertheless, it is genes, rather than chromosomes, that ultimately serve as the underlying basis of **sex determination**. As we will see, some of these genes are present on sex chromosomes, but others are autosomal. Extensive investigation has revealed a wide variation in sex-chromosome systems—even in closely related organisms—suggesting that mechanisms controlling sex determination have undergone rapid evolution many times in the history of life.

In this chapter, we review some representative modes of sexual differentiation by examining the life cycles of three model organisms often studied in genetics: the green alga *Chlamydomonas*; the maize plant, *Zea mays*; and the nematode (roundworm), *Caenorhabditis elegans*. These organisms contrast the different roles that sexual differentiation plays in the lives of diverse organisms. Then, we delve more deeply into what is known about the genetic basis for the determination of sexual differences, with a particular emphasis on two organisms: our own species, representative of mammals; and *Drosophila*, on which pioneering sex-determining studies were performed.

## 7.1 Life Cycles Depend on Sexual Differentiation

In describing sexual dimorphism (differences between males and females) in multicellular animals, biologists distinguish between **primary sexual differentiation**, which involves only the gonads, where gametes are produced, and **secondary sexual differentiation**, which involves the overall appearance of the organism, including clear differences in such organs as mammary glands and external genitalia as well as in nonreproductive organs. In plants and animals, the terms **unisexual**, **dioecious**, and **gonochoric** are equivalent; they all refer to an individual containing only male *or* only female reproductive organs. Conversely, the terms **bisexual**, **monoecious**, and **hermaphroditic** refer to individuals containing both male *and* female reproductive organs, a common occurrence in both the plant and animal kingdoms. These organisms can produce both eggs and sperm. The term **intersex** is usually reserved for individuals of an intermediate sexual condition; these are most often sterile.

### *Chlamydomonas*

The life cycle of the green alga *Chlamydomonas*, shown in Figure 7–1, is representative of organisms exhibiting only infrequent periods of sexual reproduction. Such organisms spend most of their life cycle in a haploid phase, asexually producing daughter cells by mitotic divisions. However, under unfavorable nutrient conditions, such as nitrogen depletion, certain daughter cells function as gametes, joining together in fertilization. Following fertilization, a diploid zygote, which may withstand the unfavorable environment, is formed. When conditions change for the better, meiosis ensues and haploid vegetative cells are again produced.

In such species, there is little visible difference between the haploid vegetative cells that reproduce asexually and the haploid gametes that are involved in sexual reproduction. Moreover, the two gametes that fuse during mating are not usually morphologically distinguishable from one another, which is why they are called **isogametes** (*iso-* means equal, or uniform). Species producing them are said to be **isogamous**.

In 1954, Ruth Sager and Sam Granik demonstrated that gametes in *Chlamydomonas* could be subdivided into two **mating types**. Working with clones derived from single haploid cells, they showed that cells from a given clone mate with cells from some but not all other clones. When they tested the mating abilities of large numbers of clones, all could be placed into one of two mating categories, either $mt^+$ or $mt^-$ cells. "Plus" cells mate only with "minus" cells, and vice versa, as represented in Figure 7–2. Following fertilization and meiosis, the four haploid cells, or **zoospores**, produced (see the top of Figure 7–1) were found to consist of two plus types and two minus types.

Further experimentation established that plus and minus cells differ chemically. When extracts are prepared from cloned *Chlamydomonas* cells (or their flagella) of one type and then added

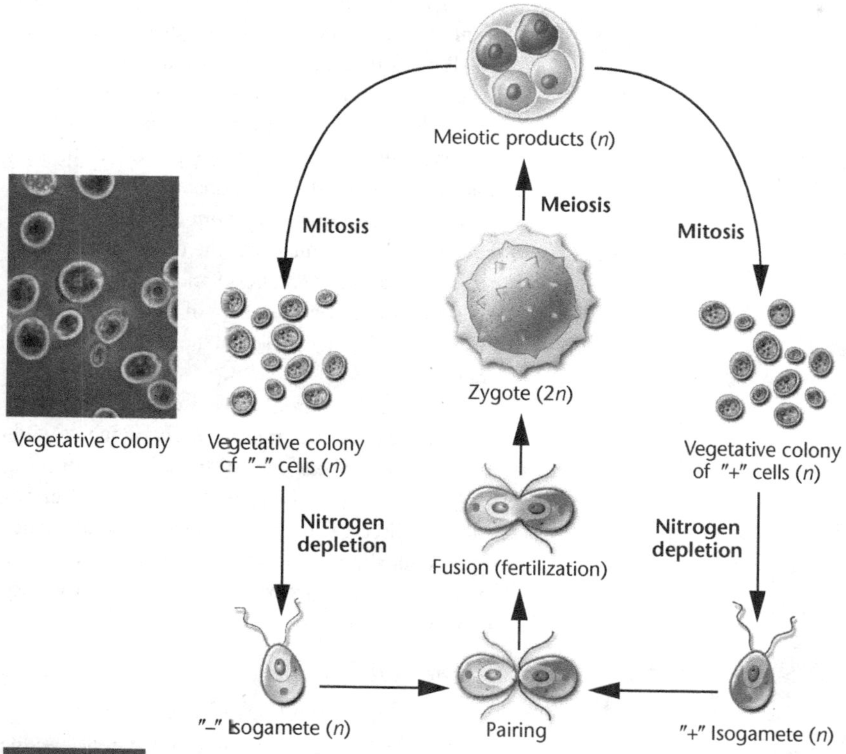


**FIGURE 7–1** The life cycle of *Chlamydomonas*. Unfavorable conditions stimulate the formation of isogametes of opposite mating types that may fuse in fertilization. The resulting zygote undergoes meiosis, producing two haploid cells of each mating type. The photograph shows vegetative cells of this green alga.

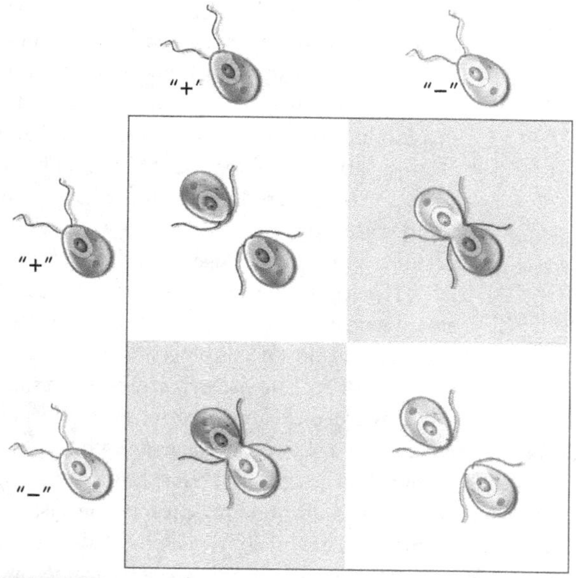


**FIGURE 7–2** Illustration of mating types during fertilization in *Chlamydomonas*. Mating will occur only when plus (+) and minus (–) cells meet.

to cells of the opposite mating type, clumping, or agglutination, occurs. No such agglutination occurs if the extracts are added to cells of the mating type from which they were derived. These observations suggest that despite the morphological similarities between isogametes, they are differentiated chemically. Therefore, in this alga, a primitive means of sex differentiation exists, even though there is no morphological indication that such differentiation has occurred. Further research has pinpointed the *mt* locus to *Chlamydomonas* chromosome VI and has identified the gene that mediates the expression of the $mt^-$ mating type, which is essential for cell fusion in response to nitrogen depletion.

## *Zea mays*

The life cycles of many plants alternate between the haploid gametophyte stage and the diploid sporophyte stage (see Figure 2–12). The processes of meiosis and fertilization link the two phases during the life cycle. The relative amount of time spent in the two phases varies between the major plant groups. In some nonseed plants, such as mosses, the haploid gametophyte phase and the morphological structures representing this stage predominate. The reverse is true in seed plants.

Maize (*Zea mays*), familiar to you as corn, exemplifies a monoecious seed plant, meaning a plant in which the sporophyte phase and the morphological structures representing that phase predominate during the life cycle. Both male and female structures are present on the adult plant. Thus, sex determination occurs differently in different tissues of the same organism, as shown in the life cycle of this plant (Figure 7–3). The stamens, which collectively constitute the tassel, produce diploid microspore mother cells, each of which undergoes meiosis and gives rise to four haploid microspores. Each haploid microspore in turn develops into a mature male microgametophyte—the pollen grain—which contains two haploid sperm nuclei.

Equivalent female diploid cells, known as megaspore mother cells, are produced in the pistil of the sporophyte. Following meiosis, only one of the four haploid megaspores survives. It usually divides mitotically three times, producing a total of eight haploid nuclei enclosed in the embryo sac. Two of these nuclei unite near the center of the embryo sac, becoming the endosperm nuclei. At the micropyle end of the sac, where the sperm enters, sit three other nuclei: the oocyte nucleus and two synergids. The remaining three, antipodal nuclei, cluster at the opposite end of the embryo sac.

Pollination occurs when pollen grains make contact with the silks (or stigma) of the pistil and develop long pollen tubes that grow toward the embryo sac. When contact is made at the micropyle, the two sperm nuclei enter the embryo sac. One sperm nucleus unites with the haploid oocyte nucleus, and the other sperm nucleus unites with two endosperm nuclei. This process, known as *double fertilization*, results in the diploid zygote nucleus and the triploid endosperm

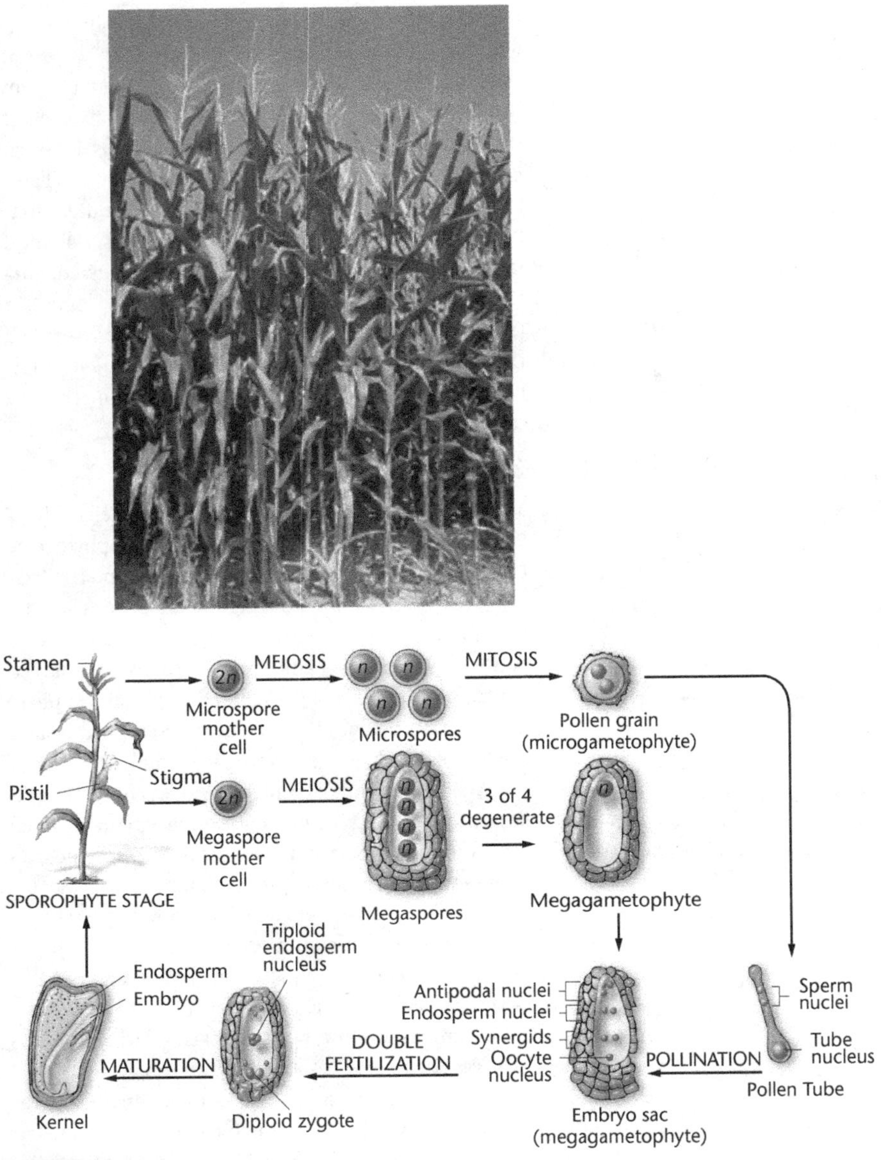


**FIGURE 7–3** The life cycle of maize (*Zea mays*). The diploid sporophyte bears stamens and pistils that give rise to haploid microspores and megaspores, which develop into the pollen grain and the embryo sac that ultimately house the sperm and oocyte, respectively. Following fertilization, the embryo develops within the kernel and is nourished by the endosperm. Germination of the kernel gives rise to a new sporophyte (the mature corn plant), and the cycle repeats itself.

nucleus, respectively. Each ear of corn can contain as many as 1000 of these structures, each of which develops into a single kernel. Each kernel, if allowed to germinate, gives rise to a new plant, the *sporophyte*.

The mechanism of sex determination and differentiation in a monoecious plant such as *Zea mays*, where the tissues that form the male and female gametes have the same genetic constitution, was difficult to unravel at first. However, the discovery of a large number of mutant genes that disrupt normal tassel and pistil formation supports the concept that normal products of these genes play an important role in sex determination by affecting the differentiation of male or female tissue in several ways.

For example, mutant genes that cause sex reversal provide valuable information. When homozygous, all mutations classified as *tassel seed (ts)* interfere with tassel production and induce the formation of female structures instead. Thus, a single gene can cause a normally monoecious plant to become exclusively female. On the other hand, the recessive mutations *silkless (sk)* and *barren stalk (ba)* interfere with the development of the pistil, resulting in plants with only male-functioning reproductive organs.

Data gathered from studies of these and other mutants suggest that the products of many wild-type alleles of these genes interact in controlling sex determination. During development, certain cells are "directed" to become male or female structures. Following sexual differentiation into either male or female structures, male or female gametes are produced.

## *Caenorhabditis elegans*

The nematode worm *Caenorhabditis elegans* [*C. elegans*, for short; Figure 7–4(a)] has become a popular organism in genetic studies, particularly for investigating the genetic control of development. Its usefulness is based on the fact that adults consist of approximately 1000 cells, the precise lineage of which can be traced back to specific embryonic origins. Among many interesting mutant phenotypes that have been studied, those representing behavioral modifications are a favorite subject of inquiry.

There are two sexual phenotypes in these worms: males, which have only testes, and hermaphrodites, which contain both testes and ovaries. During larval development of hermaphrodites, testes form that produce sperm, which is stored. Ovaries are also produced, but oogenesis does not occur until the adult stage is reached several days later. The eggs that are produced are fertilized by the stored sperm in a process of self-fertilization.

The outcome of this process is quite interesting [Figure 7–4(b)]. The vast majority of organisms that result are hermaphrodites, like the parental worm; less than 1 percent of the offspring are males. As adults, males can mate with hermaphrodites, producing about half male and half hermaphrodite offspring.

The genetic signal that determines maleness in contrast to hermaphroditic development is provided by genes located on both the X chromosome and autosomes. *C. elegans* lacks a Y chromosome altogether—hermaphrodites have two X chromosomes, while males have only one X chromosome. It is believed that the ratio of X chromosomes to the number of sets of autosomes ultimately determines

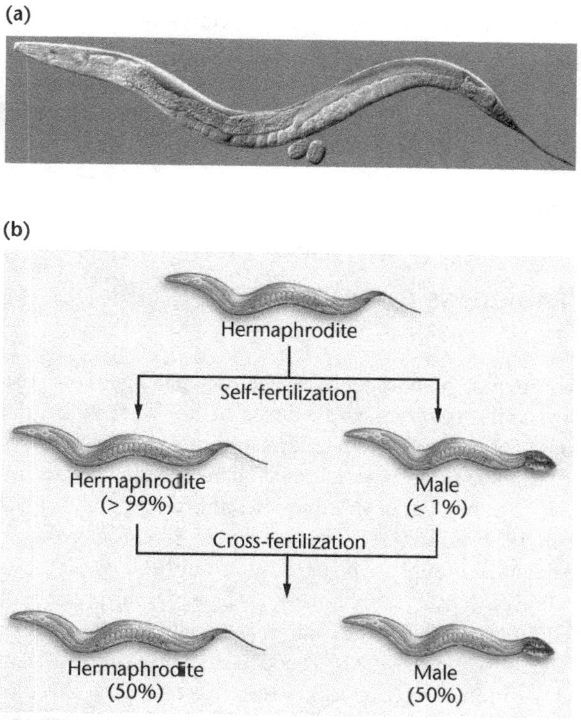


FIGURE 7–4 (a) Photomicrograph of a hermaphroditic nematode, *C. elegans*; (b) the outcomes of self-fertilization in a hermaphrodite, and a mating of a hermaphrodite and a male worm.

the sex of these worms. A ratio of 1.0 (two X chromosomes and two copies of each autosome) results in hermaphrodites, and a ratio of 0.5 results in males. The absence of a heteromorphic Y chromosome is not uncommon in organisms.

**NOW SOLVE THIS**

Problem 24 on page 195 asks you to devise an experimental approach to elucidate the original findings regarding sex determination in a marine worm.

■ HINT: *An obvious approach would be to attempt to isolate the unknown factor affecting sex determination and devise experiments making use of it. An alternative approach, more in line with genetic analysis, could involve the study of mutations that alter the normal outcomes.*

## 7.2 X and Y Chromosomes Were First Linked to Sex Determination Early in the Twentieth Century

How sex is determined has long intrigued geneticists. In 1891, H. Henking identified a nuclear structure in the sperm of certain insects, which he labeled the X-body. Several years later, Clarance McClung showed that some of the sperm in grasshoppers contain an unusual genetic structure, which he called a *heterochromosome*, but the remainder of the sperm lack such a structure. He mistakenly associated the presence of the heterochromosome with the production of male progeny. In 1906, Edmund B. Wilson clarified Henking and McClung's findings when he demonstrated that female somatic cells in the butterfly *Protenor* contain 14 chromosomes, including two X chromosomes. During oogenesis, an even reduction occurs, producing gametes with seven chromosomes, including one X chromosome. Male somatic cells, on the other hand, contain only 13 chromosomes, including one X chromosome. During spermatogenesis, gametes are produced containing either six chromosomes, without an X, or seven chromosomes, one of which is an X. Fertilization by X-bearing sperm results in female offspring, and fertilization by X-deficient sperm results in male offspring [Figure 7–5(a)].

The presence or absence of the X chromosome in male gametes provides an efficient mechanism for sex determination in this species and also produces a 1:1 sex ratio in the resulting offspring. This mechanism, now called the **XX/XO**, or ***Protenor*, mode of sex determination,** depends on the random distribution of the X chromosome into one-half of the male gametes during segregation. As we saw earlier, *C. elegans* exhibits this system of sex determination.

Wilson also experimented with the milkweed bug *Lygaeus turcicus*, in which both sexes have 14 chromosomes. Twelve of these are autosomes (A). In addition, the females have two X chromosomes, while the males have only a single X and a smaller heterochromosome labeled the **Y chromosome.** Females in this species produce only gametes of the (6A + X) constitution, but males produce two types of gametes in equal proportions, (6A + X) and (6A + Y). Therefore, following random fertilization, equal numbers of male and female progeny will be produced with distinct chromosome complements. This mode of sex determination is called the ***Lygaeus*,** or **XX/XY, mode of sex determination** [Figure 7–5(b)].

In *Protenor* and *Lygaeus* insects, males produce unlike gametes. As a result, they are described as the **heterogametic sex**, and in effect, their gametes ultimately determine the sex of the progeny in those species. In such cases, the female, who has like sex chromosomes, is the **homogametic sex**, producing uniform gametes with regard to chromosome numbers and types.

The male is not always the heterogametic sex. In some organisms, the female produces unlike gametes, exhibiting either the *Protenor* XX/XO or *Lygaeus* XX/XY mode of sex determination. Examples include certain moths and butterflies; most birds; some fish; reptiles; amphibians; and at least one species of plants (*Fragaria orientalis*). To immediately distinguish situations in which the female is the heterogametic sex, some geneticists use the notation **ZZ/ZW**, where ZW is the heterogamous female, instead of the XX/XY notation.

Geneticists' experience with fowl (chickens) illustrates the difficulty of establishing which sex is heterogametic and whether the *Protenor* or *Lygaeus* mode is operable. While genetic evidence supported the hypothesis that the female is the heterogametic sex, the cytological identification of the sex chromosome was not accomplished until 1961, because of the large number of chromosomes (78)

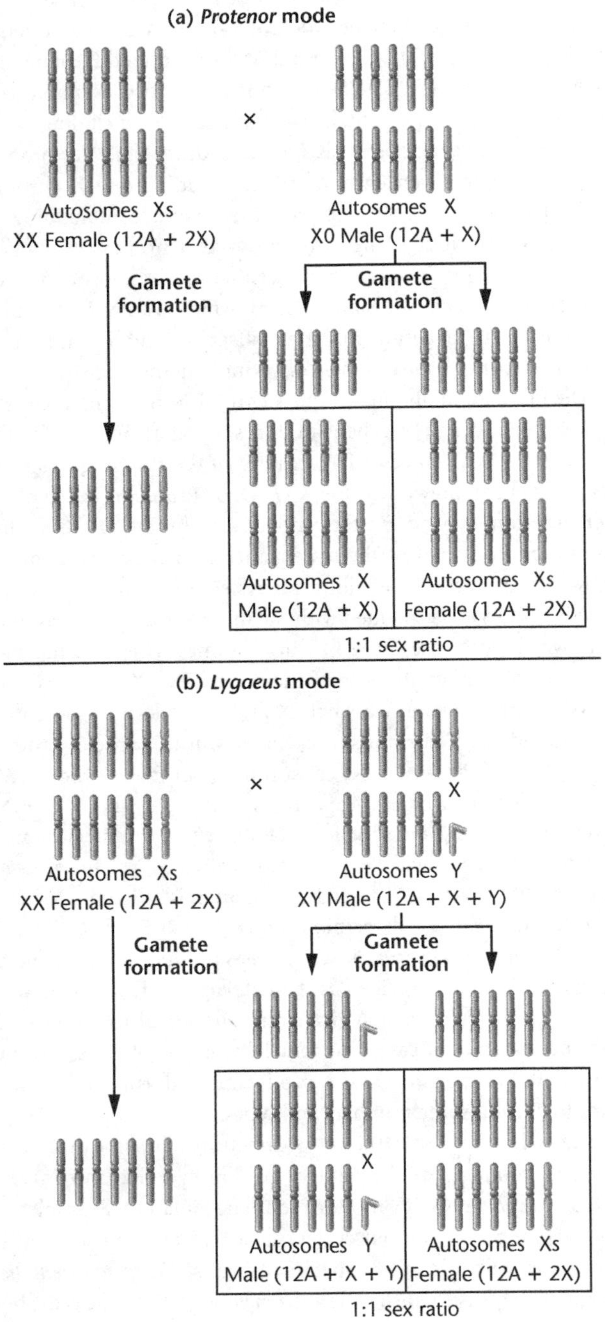


FIGURE 7-5 (a) The *Protenor* mode of sex determination where the heterogametic sex (the male in this example) is X0 and produces gametes with or without the X chromosome; (b) the *Lygaeus* mode of sex determination, where the heterogametic sex (again, the male in this example) is XY and produces gametes with either an X or a Y chromosome. In both cases, the chromosome composition of the offspring determines its sex.

characteristic of chickens. When the sex chromosomes were finally identified, the female was shown to contain an unlike chromosome pair including a heteromorphic chromosome (the W chromosome). Thus, in fowl, the female is indeed heterogametic and is characterized by the *Lygaeus* type of sex determination.

## 7.3 The Y Chromosome Determines Maleness in Humans

The first attempt to understand sex determination in our own species occurred almost 100 years ago and involved the visual examination of chromosomes in dividing cells. Efforts were made to accurately determine the diploid chromosome number of humans, but because of the relatively large number of chromosomes, this proved to be quite difficult. In 1912, H. von Winiwarter counted 47 chromosomes in a spermatogonial metaphase preparation. It was believed that the sex-determining mechanism in humans was based on the presence of an extra chromosome in females, who were thought to have 48 chromosomes. However, in the 1920s, Theophilus Painter counted between 45 and 48 chromosomes in cells of testicular tissue and also discovered the small Y chromosome, which is now known to occur only in males. In his original paper, Painter favored 46 as the diploid number in humans, but he later concluded incorrectly that 48 was the chromosome number in both males and females.

For 30 years, this number was accepted. Then, in 1956, Joe Hin Tjio and Albert Levan discovered a better way to prepare chromosomes for viewing. This improved technique led to a strikingly clear demonstration of metaphase stages showing that 46 was indeed the human diploid number. Later that same year, C. E. Ford and John L. Hamerton, also working with testicular tissue, confirmed this finding. The familiar karyotypes of humans (Figure 7–6) are prepared using Tjio and Levan's technique.

Of the normal 23 pairs of human chromosomes, one pair was shown to vary in configuration in males and females. These two chromosomes were designated the X and Y sex chromosomes. The human female has two X chromosomes, and the human male has one X and one Y chromosome.

We might believe that this observation is sufficient to conclude that the Y chromosome determines maleness. However, several other interpretations are possible. The Y could play no role in sex determination; the presence of two X chromosomes could cause femaleness; or maleness could result from the lack of a second X chromosome. The evidence that clarified which explanation was correct came from study of the effects of human sex chromosome variations, described below. As such investigations revealed, the Y chromosome does indeed determine maleness in humans.

### Klinefelter and Turner Syndromes

About 1940, scientists identified two human abnormalities characterized by aberrant sexual development, **Klinefelter syndrome**

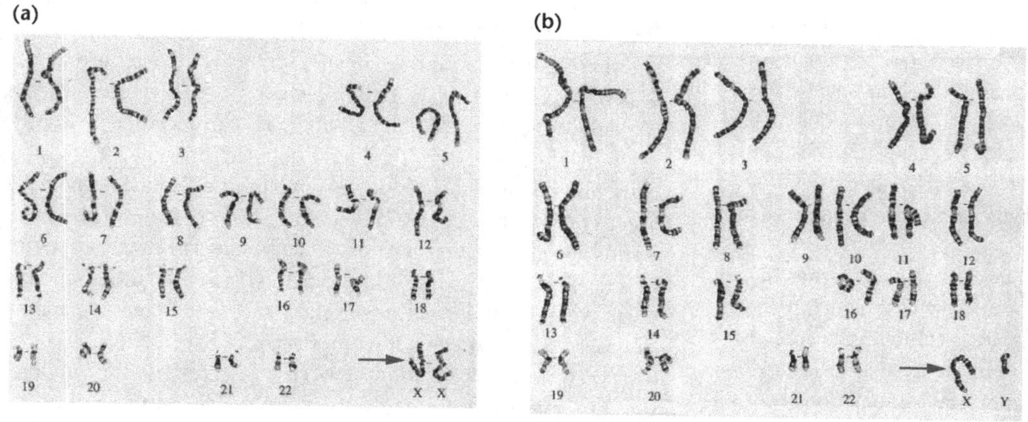


FIGURE 7–6 The traditional human karyotypes derived from a normal female and a normal male. Each contains 22 pairs of autosomes and two sex chromosomes. The female (a) contains two X chromosomes, while the male (b) contains one X and one Y chromosome.

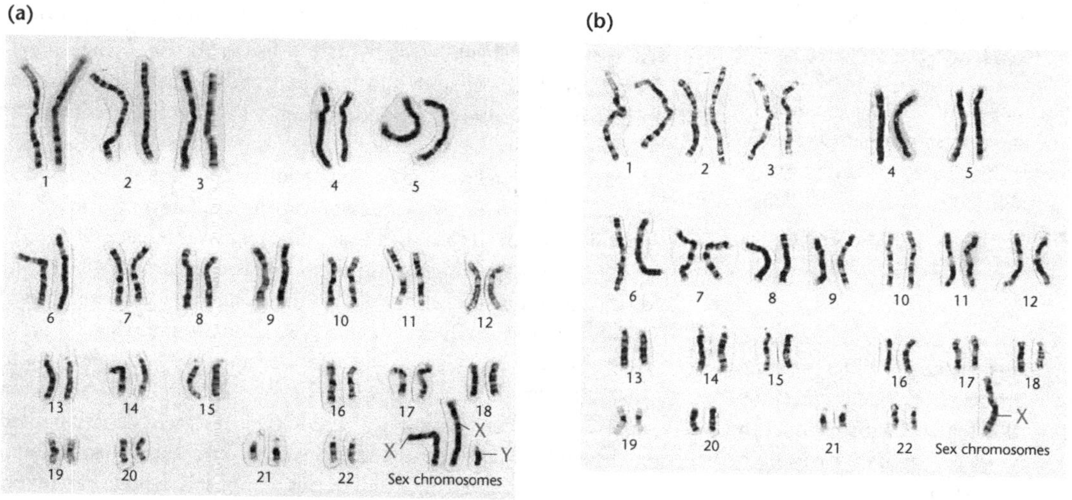


FIGURE 7–7 The karyotypes of individuals with (a) Klinefelter syndrome (47,XXY) and (b) Turner syndrome (45,X).

**(47,XXY)** and **Turner syndrome (45,X)**.* Individuals with Klinefelter syndrome are generally tall and have long arms and legs and large hands and feet. They usually have genitalia and internal ducts that are male, but their testes are rudimentary and fail to produce sperm. At the same time, feminine sexual development is not entirely suppressed. Slight enlargement of the breasts (gynecomastia) is common, and the hips are often rounded. This ambiguous sexual development, referred to as intersexuality, can lead to abnormal social development. Intelligence is often below the normal range as well.

In Turner syndrome, the affected individual has female external genitalia and internal ducts, but the ovaries are rudimentary. Other characteristic abnormalities include short stature (usually under 5 feet), skin flaps on the back of the neck, and underdeveloped breasts. A broad, shieldlike chest is sometimes noted. Intelligence is usually normal.

In 1959, the karyotypes of individuals with these syndromes were determined to be abnormal with respect to the sex chromosomes. Individuals with Klinefelter syndrome have more than one X chromosome. Most often they have an XXY complement in addition to 44 autosomes [Figure 7–7(a)], which is why people with this karyotype are designated 47,XXY. Individuals with Turner syndrome most often have only 45 chromosomes, including just a single X chromosome; thus, they are designated 45,X [Figure 7(b)]. Note the convention used in designating these chromosome compositions: the number states the total number of chromosomes present, and the symbols after the comma indicate the deviation from the normal diploid content. Both conditions result from **nondisjunction**, the failure of the X chromosomes to segregate properly during meiosis (nondisjunction is described in Chapter 8 and illustrated in Figure 8–1).

These Klinefelter and Turner karyotypes and their corresponding sexual phenotypes led scientists to conclude that the Y chromosome determines maleness in humans. In its absence, the person's sex is female, even if only a single X chromosome is present. The presence of the Y chromosome in the individual with Klinefelter syndrome is sufficient to determine maleness, even though male development is not complete. Similarly, in the absence of a Y chromosome, as in the case of individuals with Turner syndrome, no masculinization occurs. Note that we cannot conclude anything

*Although the possessive form of the names of eponymous syndromes is sometimes used (e.g., Klinefelter's syndrome), the current preference is to use the nonpossessive form.

regarding sex determination under circumstances where a Y chromosome is present without an X because Y-containing human embryos lacking an X chromosome (designated 45,Y) do not survive.

Klinefelter syndrome occurs in about 1 of every 660 male births. The karyotypes **48,XXXY, 48,XXYY, 49,XXXXY,** and **49,XXXYY** are similar phenotypically to 47,XXY, but manifestations are often more severe in individuals with a greater number of X chromosomes.

Turner syndrome can also result from karyotypes other than 45,X, including individuals called **mosaics,** whose somatic cells display two different genetic cell lines, each exhibiting a different karyotype. Such cell lines result from a mitotic error during early development, the most common chromosome combinations being **45,X/46,XY** and **45,X/46,XX.** Thus, an embryo that began life with a normal karyotype can give rise to an individual whose cells show a mixture of karyotypes and who exhibits varying aspects of this syndrome.

Turner syndrome is observed in about 1 in 2000 female births, a frequency much lower than that for Klinefelter syndrome. One explanation for this difference is the observation that a substantial majority of 45,X fetuses die *in utero* and are aborted spontaneously. Thus, a similar frequency of the two syndromes may occur at conception.

## 47,XXX Syndrome

The abnormal presence of three X chromosomes along with a normal set of autosomes (**47,XXX**) results in female differentiation. The highly variable syndrome that accompanies this genotype, often called **triplo-X,** occurs in about 1 of 1000 female births. Frequently, 47,XXX women are perfectly normal and may remain unaware of their abnormality in chromosome number unless a karyotype is done. In other cases, underdeveloped secondary sex characteristics, sterility, delayed development of language and motor skills, and mental retardation may occur. In rare instances, **48,XXXX** (tetra-X) and **49,XXXXX** (penta-X) karyotypes have been reported. The syndromes associated with these karyotypes are similar to but more pronounced than the 47,XXX syndrome. Thus, in many cases, the presence of additional X chromosomes appears to disrupt the delicate balance of genetic information essential to normal female development.

## 47,XYY Condition

Another human condition involving the sex chromosomes is **47,XYY.** Studies of this condition, where the only deviation from diploidy is the presence of an additional Y chromosome in an otherwise normal male karyotype, were initiated in 1965 by Patricia Jacobs. She discovered that 9 of 315 males in a Scottish maximum security prison had the 47,XYY karyotype. These males were significantly above average in height and had been incarcerated as a result of antisocial (nonviolent) criminal acts. Of the nine males studied, seven were of subnormal intelligence, and all suffered personality disorders. Several other studies produced similar findings.

The possible correlation between this chromosome composition and criminal behavior piqued considerable interest, and extensive investigation of the phenotype and frequency of the 47,XYY condition in both criminal and noncriminal populations ensued. Above-average height (usually over 6 feet) and subnormal intelligence have been generally substantiated, and the frequency of males displaying this karyotype is indeed higher in penal and mental institutions compared with unincarcerated populations (see Table 7.1). A particularly relevant question involves the characteristics displayed by XYY males who are not incarcerated. The only nearly constant association is that such individuals are over 6 feet tall.

A study addressing this issue was initiated to identify 47,XYY individuals at birth and to follow their behavioral patterns during preadult and adult development. By 1974, the two investigators, Stanley Walzer and Park Gerald, had identified about 20 XYY newborns in 15,000 births at Boston Hospital for Women. However, they soon came under great pressure to abandon their research. Those opposed to the study argued that the investigation could not be justified and might cause great harm to individuals who displayed this karyotype. The opponents argued that (1) no association between the additional Y chromosome and abnormal behavior had been previously established in the population at large, and (2) "labeling" the individuals in the study might create a self-fulfilling prophecy. That is, as a result of participation in the study, parents, relatives, and friends might treat individuals identified as 47,XYY differently, ultimately producing the expected antisocial behavior. Despite the support of a government funding agency and the faculty at Harvard Medical School, Walzer and Gerald abandoned the investigation in 1975.

TABLE 7.1

Frequency of XYY Individuals in Various Settings

| Setting | Restriction | Number Studied | Number XYY | Frequency XYY |
|---|---|---|---|---|
| Control population | Newborns | 28,366 | 29 | 0.10% |
| Mental-penal | No height restriction | 4,239 | 82 | 1.93 |
| Penal | No height restriction | 5,805 | 26 | 0.44 |
| Mental | No height restriction | 2,562 | 8 | 0.31 |
| Mental-penal | Height restriction | 1,048 | 48 | 4.61 |
| Penal | Height restriction | 1,683 | 31 | 1.84 |
| Mental | Height restriction | 649 | 9 | 1.38 |

*Source:* Compiled from data presented in Hook, 1973, Tables 1-8. Copyright 1973 by the American Association for the Advancement of Science.

Since Walzer and Gerald's work, it has become clear that many XYY males are present in the population who do not exhibit antisocial behavior and who lead normal lives. Therefore, we must conclude that there is a high, but not constant, correlation between the extra Y chromosome and the predisposition of these males to exhibit behavioral problems.

## Sexual Differentiation in Humans

Once researchers had established that, in humans, it is the Y chromosome that houses genetic information necessary for maleness, they attempted to pinpoint a specific gene or genes capable of providing the "signal" responsible for sex determination. Before we delve into this topic, it is useful to consider how sexual differentiation occurs in order to better comprehend how humans develop into sexually dimorphic males and females. During early development, every human embryo undergoes a period when it is potentially hermaphroditic. By the fifth week of gestation, gonadal primordia (the tissues that will form the gonad) arise as a pair of **gonadal (genital) ridges** associated with each embryonic kidney (Figure 7–8). The embryo is potentially hermaphroditic because at this stage its gonadal phenotype is sexually indifferent—male or female reproductive structures cannot be distinguished, and the gonadal ridge tissue can develop to form male or female gonads. As development progresses, primordial germ cells migrate to these ridges, where an outer cortex and inner medulla form (*cortex* and *medulla* are the outer and inner tissues of an organ, respectively). The cortex is capable of developing into an ovary, while the medulla may develop into a testis. In addition, two sets of undifferentiated ducts called the Wolffian and Müllerian ducts exist in each embryo. Wolffian ducts differentiate into other organs of the male reproductive tract, while Müllerian ducts differentiate into structures of the female reproductive tract (Figure 7–8).

Because gonadal ridges can form either ovaries or testes, they are commonly referred to as **bipotential gonads.** What is the switch that triggers gonadal ridge development into testes or ovaries? The presence or absence of a Y chromosome is the key. If cells of the ridge have an XY constitution, development of the medulla into a testis is initiated around the seventh week. However, in the absence of the Y chromosome, no male development occurs, the cortex of the ridge subsequently forms ovarian tissue, and the Müllerian duct forms oviducts (Fallopian tubes), uterus, cervix, and portions of the vagina. Depending on which pathway is initiated, parallel development of the appropriate male or female duct system then occurs, and the other duct system degenerates. If testes differentiation is initiated, the embryonic testicular tissue secretes hormones that are essential for continued male sexual differentiation. As we will discuss in the next section, presence of a Y chromosome and development of the testes also inhibits formation of female reproductive organs.

In females, as the 12th week of fetal development approaches, oogonia within the ovaries begin meiosis, and primary oocytes can be

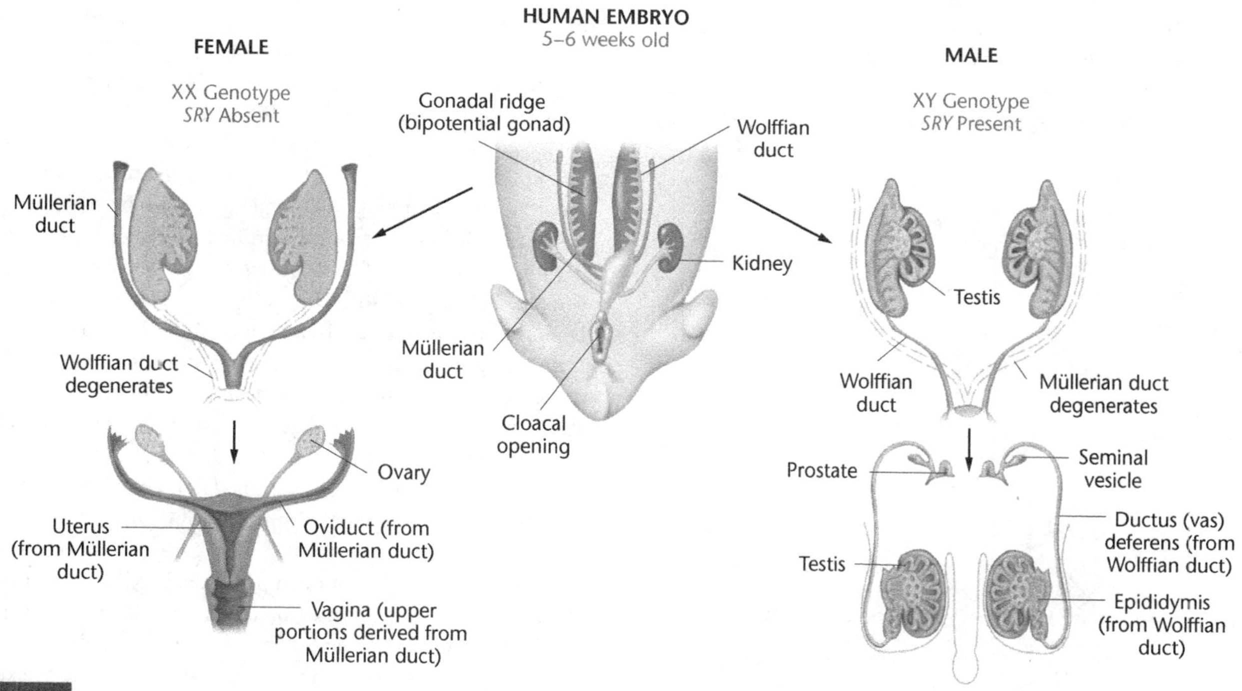


FIGURE 7–8 The presence or absence of the Y chromosome and *SRY* determines sexual differentiation in humans. Early human embryos are sexually indifferent. Bipotential gonads can form either ovaries or testes depending on the presence or absence of the *SRY* genotype. With its presence in XY embryos, bipotential gonads form testes and, subsequently, male reproductive organs and ducts. In the absence of *SRY*, in XX embryos, the female reproductive organs and ducts are formed.

detected. By the 25th week of gestation, all oocytes become arrested in meiosis and remain dormant until puberty is reached some 10 to 15 years later. In males, on the other hand, primary spermatocytes are not produced until puberty is reached (refer to Figure 2–11).

## The Y Chromosome and Male Development

The human Y chromosome, unlike the X, was long thought to be mostly blank genetically. It is now known that this is not true, even though the Y chromosome contains far fewer genes than does the X. Data from the Human Genome Project indicate that the Y chromosome has at least 75 genes, compared to 900–1400 genes on the X. Current analysis of these genes and regions with potential genetic function reveals that some have homologous counterparts on the X chromosome and others do not. For example, present on both ends of the Y chromosome are so-called **pseudoautosomal regions (PARs)** that share homology with regions on the X chromosome and synapse and recombine with it during meiosis. The presence of such a pairing region is critical to segregation of the X and Y chromosomes during male gametogenesis. The remainder of the chromosome, about 95 percent of it, does not synapse or recombine with the X chromosome. As a result, it was originally referred to as the *nonrecombining region of the Y (NRY)*. More recently, researchers have designated this region as the **male-specific region of the Y (MSY)**. As you will see, some portions of the MSY share homology with genes on the X chromosome, and others do not.

The human Y chromosome is diagrammed in Figure 7–9. The MSY is divided about equally between *euchromatic* regions, containing functional genes, and *heterochromatic* regions, lacking genes. Within euchromatin, adjacent to the PAR of the short arm of the Y chromosome, is a critical gene that controls male sexual development, called the ***sex-determining region Y (SRY)***. In humans, the absence of a Y chromosome almost always leads to female development; thus, this gene is absent from the X chromosome. At 6 to 8 weeks of development, the *SRY* gene becomes active in XY embryos. *SRY* encodes a protein that causes the undifferentiated gonadal tissue of the embryo to form testes. This protein is called the **testis-determining factor (TDF)**. *SRY* (or a closely related version) is present in all mammals thus far examined, indicative of its essential function throughout this diverse group of animals.

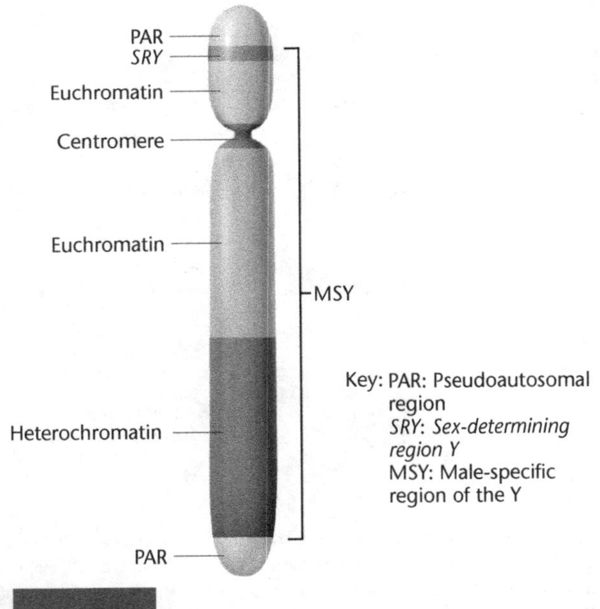


FIGURE 7–9 The regions of the human Y chromosome.

Our ability to identify the presence or absence of DNA sequences in rare individuals whose sex chromosome composition does not correspond to their sexual phenotype has provided evidence that *SRY* is the gene responsible for male sex determination. For example, there are human males who have two X and no Y chromosomes. Often, attached to one of their X chromosomes is the region of the Y that contains *SRY*. There are also females who have one X and one Y chromosome. Their Y is almost always missing the *SRY* gene. These observations argue strongly in favor of the role of *SRY* in providing the primary signal for male development.

Further support of this conclusion involves an experiment using **transgenic mice**. These animals are produced from fertilized eggs injected with foreign DNA that is subsequently incorporated into the genetic composition of the developing embryo. In normal mice, a chromosome region designated *Sry* has been identified that is comparable to *SRY* in humans. When mouse DNA containing *Sry* is injected into normal XX mouse eggs, most of the offspring develop into males.

The question of how the product of this gene triggers development of embryonic gonadal tissue into testes rather than ovaries has been under investigation for a number of years. TDF is now believed to function as a *transcription factor*, a DNA-binding protein that interacts directly with regulatory sequences of other genes to stimulate their expression. Thus, while TDF behaves as a master switch that controls other genes downstream in the process of sexual differentiation, identifying TDF target genes has been elusive. To date no targets for TDF have been identified. However, one potential target for activation by TDF that has been extensively studied is the gene for **Müllerian inhibiting substance (MIS**, also called Müllerian inhibiting hormone, MIH, or anti-Müllerian hormone). Cells of the developing testes secrete MIS. As its name suggests, MIS protein causes regression (atrophy) of cells in the Müllerian duct. Degeneration of the duct, shown in Figure 7–8, prevents formation of the female reproductive tract.

Other autosomal genes, as studied in humans, are believed to be part of a cascade of genetic expression initiated by *SRY*. Examples include the *SOX9* gene and the *WT1* gene (on chromosome 11), originally identified as an oncogene associated with Wilms tumor, which affects the kidney and gonads. Another gene, *SF1*, is involved in the regulation of enzymes affecting steroid metabolism. In mice, this gene is initially active in both the male and female bisexual genital ridge, persisting until the point in development when testis formation is apparent. At that time, its expression persists in males but is extinguished in females. Establishment of the link between these various genes and sex determination has brought us closer to a complete understanding of how males and females arise in humans, but much work remains to be done.

Findings by David Page and his many colleagues have now provided a reasonably complete picture of the MSY region of the human Y chromosome. This work, completed in 2003, is based on information gained through the Human Genome Project, in which the DNA of all chromosomes of a representative human genome has been sequenced. Page has spearheaded the detailed study of the Y chromosome for the past several decades.

The MSY consists of about 23 million base pairs (23 Mb) and can be divided into three regions. The first region is the *X-transposed region*. It comprises about 15 percent of the MSY and was originally derived from the X chromosome in the course of human evolution (about 3 to 4 million years ago). The X-transposed region is 99 percent identical to region Xq21 of the modern human X chromosome. Two genes, both with X chromosome homologs, are present in this region.

The second area is designated the *X-degenerative region*. Comprising about 20 percent of the MSY, this region contains DNA sequences that are even more distantly related to those present on the X chromosome. The X-degenerative region contains 27 single-copy genes and a number of *pseudogenes* (genes whose sequences have degenerated sufficiently during evolution to render them nonfunctional). As with the genes present in the X-transposed region, all share some homology with counterparts on the X chromosome. These 27 genetic units include 14 that are capable of being transcribed, and each is present as a single copy. One of these is the *SRY* gene, discussed earlier. Other X-degenerative genes that encode protein products are expressed ubiquitously in all tissues in the body, but *SRY* is expressed only in the testes.

The third area, the *ampliconic region*, contains about 30 percent of the MSY, including most of the genes closely associated with testes development. These genes lack counterparts on the X chromosome, and their expression is limited to the testes. There are 60 transcription units (genes that yield a product) divided among 9 gene families in this region, most represented by multiple copies. Members of each family have nearly identical (>98 percent) DNA sequences. Each repeat unit is an **amplicon** and is contained within seven segments scattered across the euchromatic regions of both the short and long arms of the Y chromosome. Genes in the ampliconic region encode proteins specific to the development and function of the testes, and the products of many of these genes are directly related to fertility in males. It is currently believed that a great deal of male sterility in our population can be linked to mutations in these genes. Research by David Page and others has also revealed that sequences called **palindromes**—sequences of base pairs that read the same but in the opposite direction on complementary strands—are present throughout the MSY. Recombination between palindromes on sister chromatids of the Y during replication is a mechanism used to repair mutations in the Y. This discovery has fascinating implications concerning how the Y chromosome may maintain its size and structure, given that homologous recombination between the X and Y occurs primarily in PARs.

This recent work has greatly expanded our picture of the genetic information carried by this unique chromosome. It clearly refutes the so-called "wasteland" theory, prevalent only 20 years ago, that depicted the human Y chromosome as almost devoid of genetic information other than a few genes that cause maleness. The knowledge we have gained provides the basis for a much clearer picture of how maleness is determined. In addition, it provides important clues to the origin of the Y chromosome during human evolution.

**NOW SOLVE THIS**

Problem 31 on page 196 concerns the autosomal gene *SOX9*, which when mutated appears to inhibit normal human male development. You are asked to analyze a sequence of observations involving individuals with campomelic dysplasia (CMD1) and to draw appropriate conclusions.

■ HINT: *Some genes are activated and produce their normal product as a result of expression of products of other genes found on different chromosomes—in this case, perhaps one that is on the Y chromosome.*

## 7.4 The Ratio of Males to Females in Humans Is Not 1.0

The presence of heteromorphic sex chromosomes in one sex of a species but not the other provides a potential mechanism for producing equal proportions of male and female offspring. This potential depends on the segregation of the X and Y (or Z and W) chromosomes during meiosis, such that half of the gametes of the heterogametic sex receive one of the chromosomes and half receive the other one. As we learned in the previous section, small pseudoautosomal regions of pairing homology do exist at both ends of the human X and Y chromosomes, suggesting that the X and Y chromosomes do synapse and then segregate into different gametes. Provided that both types of gametes are equally successful in fertilization and that the two sexes are equally viable during development, a 1:1 ratio of male and female offspring should result.

The actual proportion of male to female offspring, referred to as the **sex ratio**, has been assessed in two ways. The **primary sex ratio** reflects the proportion of males to females conceived in a population. The **secondary sex ratio** reflects the proportion of each sex that is born. The secondary sex ratio is much easier to determine but has the disadvantage of not accounting for any disproportionate embryonic or fetal mortality.

When the secondary sex ratio in the human population was determined in 1969 by using worldwide census data, it was found not to equal 1.0. For example, in the Caucasian population in the United States, the secondary ratio was a little less than 1.06, indicating that about 106 males were born for each 100 females. (In 1995, this ratio dropped to slightly less than 1.05.) In the African-American population in the United States, the ratio was 1.025. In other countries the excess of male births is even greater than is reflected in these values. For example, in Korea, the secondary sex ratio was 1.15.

Despite these ratios, it is possible that the *primary sex ratio* is 1.0 and is later altered between conception and birth. For the secondary ratio to exceed 1.0, then, prenatal female mortality would have to be greater than prenatal male mortality. However, this hypothesis has been examined and shown to be false. In fact, just the opposite occurs. In a Carnegie Institute study, reported in 1948, the sex of approximately 6000 embryos and fetuses recovered from miscarriages and abortions was determined, and fetal mortality was actually higher in males. On the basis of the data derived from that study, the primary sex ratio in U.S. Caucasians was estimated to be 1.079. It is now believed that the figure is much higher—between 1.20 and 1.60, suggesting that many more males than females are conceived in the human population.

It is not clear why such a radical departure from the expected primary sex ratio of 1.0 occurs. To come up with a suitable explanation, researchers must examine the assumptions on which the theoretical ratio is based:

1. Because of segregation, males produce equal numbers of X- and Y-bearing sperm.
2. Each type of sperm has equivalent viability and motility in the female reproductive tract.
3. The egg surface is equally receptive to both X- and Y-bearing sperm.

No direct experimental evidence contradicts any of these assumptions; however, the human Y chromosome is smaller than the X chromosome and therefore of less mass. Thus, it has been speculated that Y-bearing sperm are more motile than X-bearing sperm. If this is true, then the probability of a fertilization event leading to a male zygote is increased, providing one possible explanation for the observed primary ratio.

7.5

## Dosage Compensation Prevents Excessive Expression of X-Linked Genes in Humans and Other Mammals

The presence of two X chromosomes in normal human females and only one X in normal human males is unique compared with the equal numbers of autosomes present in the cells of both sexes. On theoretical grounds alone, it is possible to speculate that this disparity should create a "genetic dosage" difference between males and females, with attendant problems, for all X-linked genes. There is the potential for females to produce twice as much of each product of all X-linked genes. The additional X chromosomes in both males and females exhibiting the various syndromes discussed earlier in this chapter are thought to compound this dosage problem. Embryonic development depends on proper timing and precisely regulated levels of gene expression. Otherwise, disease phenotypes or embryonic lethality can occur. In this section, we will describe research findings regarding X-linked gene expression that demonstrate a genetic mechanism of **dosage compensation** that balances the dose of X chromosome gene expression in females and males.

### Barr Bodies

Murray L. Barr and Ewart G. Bertram's experiments with female cats, as well as Keith Moore and Barr's subsequent study in humans, demonstrate a genetic mechanism in mammals that compensates for X chromosome dosage disparities. Barr and Bertram observed a darkly staining body in interphase nerve cells of female cats that was absent in similar cells of males. In humans, this body can be easily demonstrated in female cells derived from the buccal mucosa (cheek cells) or in fibroblasts (undifferentiated connective tissue cells), but not in similar male cells (Figure 7–10). This highly condensed structure, about in 1 $\mu$m in diameter, lies against the nuclear envelope of interphase cells. It stains positively in the Feulgen reaction, a cytochemical test for DNA.

Current experimental evidence demonstrates that this body, called a **sex chromatin body,** or simply a **Barr body,** is an

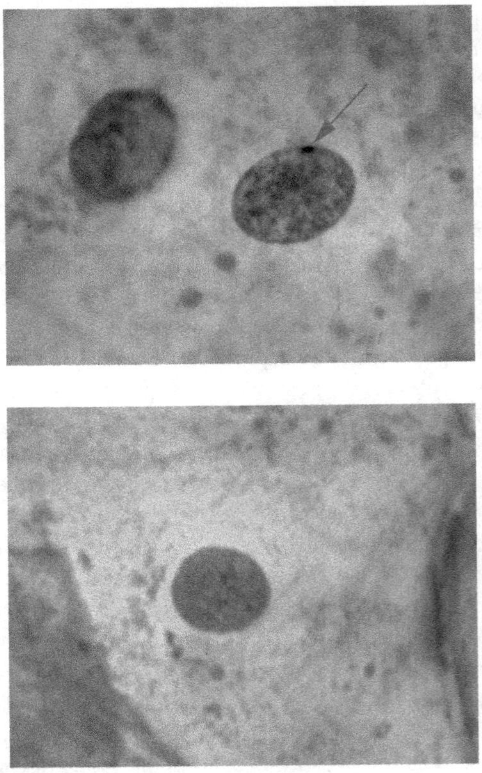

FIGURE 7–10 Photomicrographs comparing cheek epithelial cell nuclei from a male that fails to reveal Barr bodies (bottom) with a nucleus from a female that demonstrates a Barr body (indicated by the arrow in the top image). This structure, also called a sex chromatin body, represents an inactivated X chromosome.

inactivated X chromosome. Susumo Ohno was the first to suggest that the Barr body arises from one of the two X chromosomes. This hypothesis is attractive because it provides a possible mechanism for dosage compensation. If one of the two X chromosomes is inactive in the cells of females, the dosage of genetic information that can be expressed in males and females will be equivalent. Convincing, though indirect, evidence for this hypothesis comes from the study of the sex-chromosome syndromes described earlier in this chapter. Regardless of how many X chromosomes a somatic cell possesses, all but one of them appear to be inactivated and can be seen as Barr bodies. For example no Barr body is seen in the somatic cells of Turner 45,X females; one is seen in Klinefelter 47,XXY males; two in 47,XXX females; three in 48,XXXX females; and so on (Figure 7–11). Therefore, the number of Barr bodies follows an $N - 1$ rule, where $N$ is the total number of X chromosomes present.

Although this apparent inactivation of all but one X chromosome increases our understanding of dosage compensation, it further complicates our perception of other matters. For example, because one of the two X chromosomes is inactivated in normal human females, why then is the Turner 45,X individual not entirely normal? Why aren't females with the triplo-X and tetra-X karyotypes (47,XXX and 48,XXXX) completely unaffected by the additional X chromosome? Furthermore, in Klinefelter syndrome (47,XXY), X chromosome inactivation effectively renders the person 46,XY. Why aren't these males unaffected by the extra X chromosome in their nuclei?

One possible explanation is that chromosome inactivation does not normally occur in the very early stages of development of those cells destined to form gonadal tissues. Another possible explanation is that not all of each X chromosome forming a Barr body is inactivated. Recent studies have indeed demonstrated that as many as 15 percent of the human X-chromosomal genes actually escape inactivation. Clearly, then, not every gene on the X requires inactivation. In either case, excessive expression of certain X-linked genes might still occur at critical times during development despite apparent inactivation of superfluous X chromosomes.

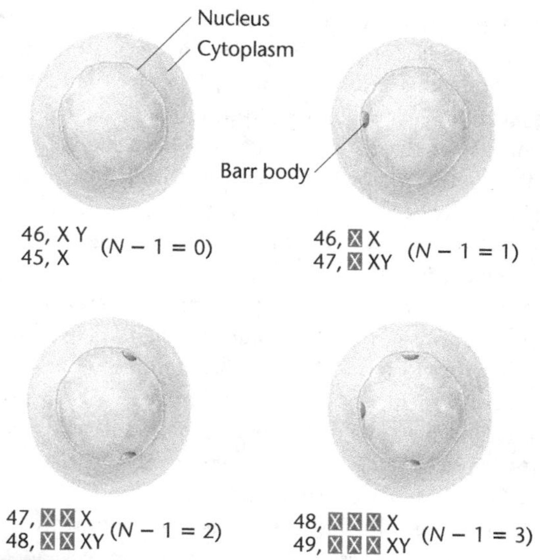


FIGURE 7–11 Occurrence of Barr bodies in various human karyotypes, where all X chromosomes except one ($N - 1$) are inactivated.

## The Lyon Hypothesis

In mammalian females, one X chromosome is of maternal origin, and the other is of paternal origin. Which one is inactivated? Is the inactivation random? Is the same chromosome inactive in all somatic cells? In 1961, Mary Lyon and Liane Russell independently proposed a hypothesis that answers these questions. They postulated that the inactivation of X chromosomes occurs randomly in somatic cells at a point early in embryonic development, most likely sometime during the blastocyst stage of development. Once inactivation has occurred, all descendant cells have the same X chromosome inactivated as their initial progenitor cell.

This explanation, which has come to be called the **Lyon hypothesis,** was initially based on observations of female mice heterozygous for X-linked coat color genes. The pigmentation of these heterozygous females was mottled, with large patches expressing the color allele on one X and other patches expressing the allele on the other X. This is the phenotypic pattern that would be expected if different X chromosomes were inactive in adjacent patches of cells. Similar mosaic patterns occur in the black and yellow-orange patches of female tortoiseshell and calico cats (Figure 7–12). Such X-linked coat color patterns do not occur in male cats because all their cells contain the single maternal X chromosome and are therefore hemizygous for only one X-linked coat color allele.

FIGURE 7–12 (a) The random distribution of orange and black patches in a calico cat illustrates the Lyon hypothesis. The white patches are due to another gene, distinguishing calico cats from tortoiseshell cats (b), which lack the white patches.

The most direct evidence in support of the Lyon hypothesis comes from studies of gene expression in clones of human fibroblast cells. Individual cells are isolated following biopsy and cultured *in vitro*. A culture of cells derived from a single cell is called a **clone**. The synthesis of the enzyme glucose-6-phosphate dehydrogenase (G6PD) is controlled by an X-linked gene. Numerous mutant alleles of this gene have been detected, and their gene products can be differentiated from the wild-type enzyme by their migration pattern in an electrophoretic field.

Fibroblasts have been taken from females heterozygous for different allelic forms of *G6PD* and studied. The Lyon hypothesis predicts that if inactivation of an X chromosome occurs randomly early in development, and thereafter all progeny cells have the same X chromosome inactivated as their progenitor, such a female should show two types of clones, each containing only one electrophoretic form of *G6PD*, in approximately equal proportions.

In 1963, Ronald Davidson and colleagues performed an experiment involving 14 clones from a single heterozygous female. Seven showed only one form of the enzyme, and 7 showed only the other form. Most important was the finding that none of the 14 clones showed both forms of the enzyme. Studies of *G6PD* mutants thus provide strong support for the random permanent inactivation of either the maternal or paternal X chromosome.

The Lyon hypothesis is generally accepted as valid; in fact, the inactivation of an X chromosome into a Barr body is sometimes referred to as **lyonization**. One extension of the hypothesis is that mammalian females are mosaics for all heterozygous X-linked alleles—some areas of the body express only the maternally derived alleles, and others express only the paternally derived alleles. Two especially interesting examples involve **red-green color blindness** and **anhidrotic ectodermal dysplasia**, both X-linked recessive disorders. In the former case, hemizygous males are fully colorblind in all retinal cells. However, heterozygous females display mosaic retinas, with patches of defective color perception and surrounding areas with normal color perception. Males hemizygous for anhidrotic ectodermal dysplasia show absence of teeth, sparse hair growth, and lack of sweat glands. The skin of females heterozygous for this disorder reveals random patterns of tissue with and without sweat glands (Figure 7–13). In both examples, random inactivation of one or the other X chromosome early in the development of heterozygous females has led to these occurrences.

NOW SOLVE THIS

Problem 35 on page 196 describes Carbon Copy, the first cat created by cloning, who was derived from a somatic nucleus of a calico cat. You are asked to comment on the likelihood that CC will appear identical to her genetic donor.

■ HINT: *The donor nucleus was from a differentiated ovarian cell of an adult female cat, which itself had inactivated one of its X chromosomes.*

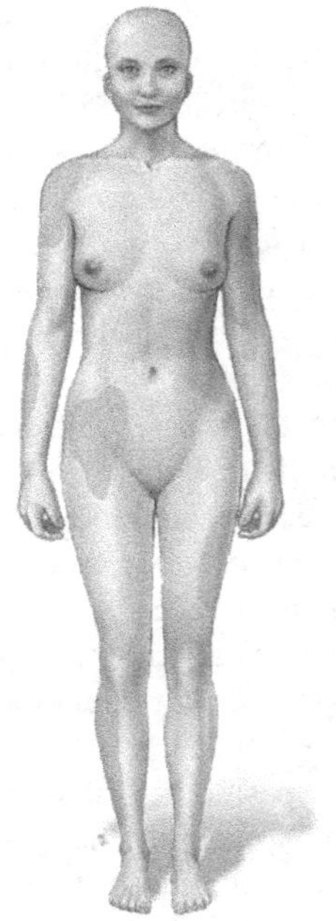

FIGURE 7–13 Depiction of the absence of sweat glands (shaded regions) in a female heterozygous for the X-linked condition anhidrotic ectodermal dysplasia. The locations vary from female to female, based on the random pattern of X chromosome inactivation during early development, resulting in unique mosaic distributions of sweat glands in heterozygotes.

## The Mechanism of Inactivation

The least understood aspect of the Lyon hypothesis is the mechanism of chromosome inactivation in mammals. Somehow, either the DNA or the attached histone proteins, or both, of one (or more) of the mammalian X chromosomes of females is modified, silencing most genes that are part of that chromosome. Whatever the modification, a memory is created that keeps the same chromosome inactivated in the following chromosome replications and cell divisions. Such a process, whereby genetic expression of one homolog, but not the other, is affected, is referred to as **imprinting**. This term also applies to a number of similar processes in which genetic information is modified epigenetically.

Recent investigations are beginning to clarify this issue. A region of the mammalian X chromosome is the major control unit. This region, located on the proximal end of the p arm in humans (the end toward the centromere), is called the **X inactivation center (*Xic*)**, and its genetic expression occurs only on the X chromosome that is inactivated. The *Xic* is about 1 Mb ($10^6$ base pairs) in length and is known to contain several putative regulatory units and four genes. One of these, ***X-inactive specific transcript (XIST)***, is now known to be a critical gene for X-inactivation.

Some interesting observations have been made regarding the RNA that is transcribed from the *XIST* gene, many coming from experiments that used the equivalent gene in the mouse (*Xist*). First, the RNA product is quite large and lacks what is called an extended **open reading frame (ORF)**. An ORF is comprised of the information necessary for translation of the RNA product into a protein. Thus, in this case, the RNA is transcribed but is not translated. It appears to serve a structural role in the nucleus, presumably in the mechanism of chromosome inactivation. This finding has led to the belief that the RNA products of *Xist* spread over and coat the X chromosome bearing the gene that produced them, creating some sort of molecular "cage" that entraps and inactivates the chromosome. Inactivation is therefore said to be *cis*-acting. Two other noncoding genes at the *Xic* locus, *Tsix* (an antisense partner of *Xist*) and *Xite* are also believed to play important roles in X-inactivation.

Second, transcription of *Xist* initially occurs at low levels on all X chromosomes. As the inactivation process begins, however, transcription continues, and is enhanced, only on the X chromosome that becomes inactivated. In 1996, a research group led by Graeme Penny provided convincing evidence that transcription of *Xist* is the critical event in chromosome inactivation. These researchers were able to introduce a targeted deletion (7 kb) into this gene. As a result, the chromosome bearing the mutation lost its ability to become inactivated.

Several interesting questions remain regarding imprinting and inactivation. In cells with more than two X chromosomes, what sort of "counting" mechanism exists that designates all but one X chromosome to be inactivated? Recent studies by Jeannie T. Lee and colleagues suggest that maternal and paternal X chromosomes must first pair briefly and align at their *Xic* loci as a mechanism for counting the number of X chromosomes prior to X-inactivation (Figure 7–14). Using mouse embryonic stem cells, Lee's group deleted the *Tsix* gene contained in the *Xic* locus. This deletion blocked X–X pairing and resulted in chaotic inactivation of 0, 1, or 2 X chromosomes. Lee and colleagues provided further evidence for the role of *Xic* locus in chromosome counting by adding copies of genetically engineered non-X chromosomes containing multiple copies of *Tsix* or *Xite* (these are referred to as **transgenes** because they are artificially introduced into the organism). This experimental procedure effectively blocked X–X pairing and prevented X-chromosome inactivation (Figure 7–14).

What "blocks" the *Xic* locus of the active chromosome, preventing further transcription of *Xist*? How does imprinting impart a memory such that inactivation of the same X chromosome or chromosomes is subsequently maintained in progeny cells, as the Lyon hypothesis calls for? The inactivation signal must somehow remain stable as cells proceed through chromosome replication. Methylation patterns of DNA, changes in acetylation of histones, and other chromosomal modifications may play roles in designating the inactivated X. Whatever the answers to these questions, scientists have taken exciting steps toward understanding how dosage compensation is accomplished in mammals.

7.6

## The Ratio of X Chromosomes to Sets of Autosomes Determines Sex in *Drosophila*

Because males and females in *Drosophila melanogaster* (and other *Drosophila* species) have the same general sex chromosome composition as humans (males are XY and females are XX), we might assume that the Y chromosome also causes maleness in these flies. However, the elegant work of Calvin Bridges in 1916 showed this not to be true. His studies of flies with quite varied chromosome compositions led him to the conclusion that the Y chromosome is not involved in sex determination in this organism. Instead, Bridges proposed that the X chromosomes and autosomes together play a critical role in sex determination. Recall that in the nematode *C. elegans*, which lacks a Y chromosome, the sex chromosomes and autosomes are also both critical to sex determination.

Bridges' work can be divided into two phases: (1) A study of offspring resulting from nondisjunction of the X chromosomes during meiosis in females and (2) subsequent work with progeny of females containing three copies of each chromosome, called triploid ($3n$) females. As we have seen previously in this chapter (and as you will see in Figure 8–1), nondisjunction is the failure of paired chromosomes to segregate or separate during the anaphase stage of the first or second meiotic divisions. The result is the production of two types of abnormal gametes, one of which contains an extra chromosome ($n + 1$)

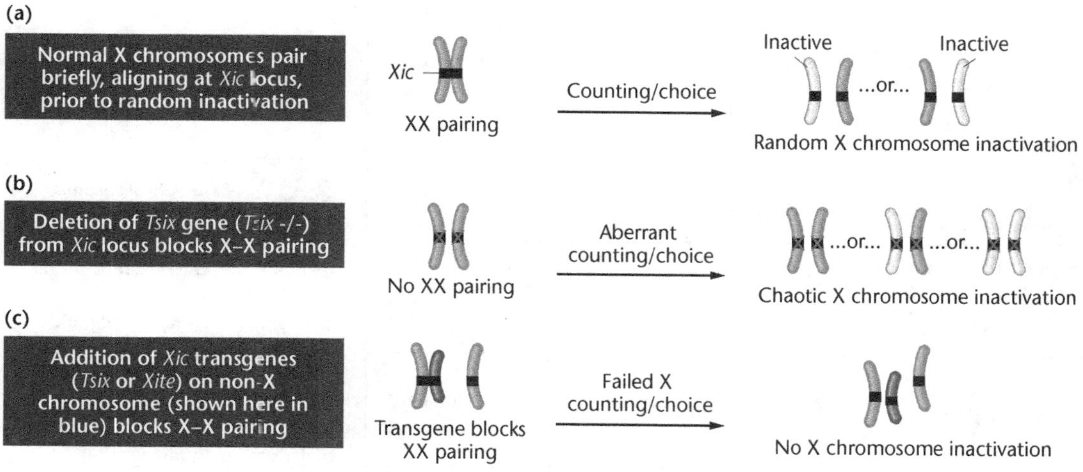


**FIGURE 7–14** (a) Transient pairing of X chromosomes may be required for initiating X-inactivation. (b) Deleting the *Tsix* gene of the *Xic* locus prevents X–X pairing and leads to chaotic X-inactivation. (c) Blocking X–X pairing by addition of *Xic*-containing transgenes blocks X–X pairing and prevents X-inactivation.

and the other of which lacks a chromosome ($n - 1$). Fertilization of such gametes with a haploid gamete produces $2n + 1$ or $2n - 1$ zygotes. As in humans, if nondisjunction involves the X chromosome, in addition to the normal complement of autosomes, both an XXY and an X0 sex-chromosome composition may result. (The "0" signifies that neither a second X nor a Y chromosome is present, as occurs in X0 genotypes of individuals with Turner syndrome.)

Contrary to what was later discovered in humans, Bridges found that the XXY flies were normal females and the X0 flies were sterile males. The presence of the Y chromosome in the XXY flies did not cause maleness, and its absence in the X0 flies did not produce femaleness. From these data, he concluded that the Y chromosome in *Drosophila* lacks male-determining factors, but since the X0 males were sterile, it does contain genetic information essential to male fertility.

Bridges was able to clarify the mode of sex determination in *Drosophila* by studying the progeny of triploid females ($3n$), which have three copies each of the haploid complement of chromosomes. *Drosophila* has a haploid number of 4, thereby possessing three pairs of autosomes in addition to its pair of sex chromosomes. Triploid females apparently originate from rare diploid eggs fertilized by normal haploid sperm. Triploid females have heavy-set bodies, coarse bristles, and coarse eyes, and they may be fertile. Because of the odd number of each chromosome (3), during meiosis, a variety of different chromosome complements are distributed into gametes that give rise to offspring with a variety of abnormal chromosome constitutions. Correlations between the sexual morphology and chromosome composition, along with Bridges' interpretation, are shown in Figure 7–15.

Bridges realized that the critical factor in determining sex is the ratio of X chromosomes to the number of haploid sets of autosomes (A) present. Normal (2X:2A) and triploid (3X:3A) females each have a ratio equal to 1.0, and both are fertile. As the ratio exceeds unity (3X:2A, or 1.5, for example), what was once called a *superfemale* is produced. Because such females are most often inviable, they are now more appropriately called **metafemales.**

Normal (XY:2A) and sterile (X0:2A) males each have a ratio of 1:2, or 0.5. When the ratio decreases to 1:3, or 0.33, as in the case of an XY:3A male, infertile **metamales** result. Other flies recovered by Bridges in these studies had an X:A ratio intermediate between 0.5 and 1.0. These flies were generally larger, and they exhibited a variety of morphological abnormalities and rudimentary bisexual gonads

**Normal diploid male**

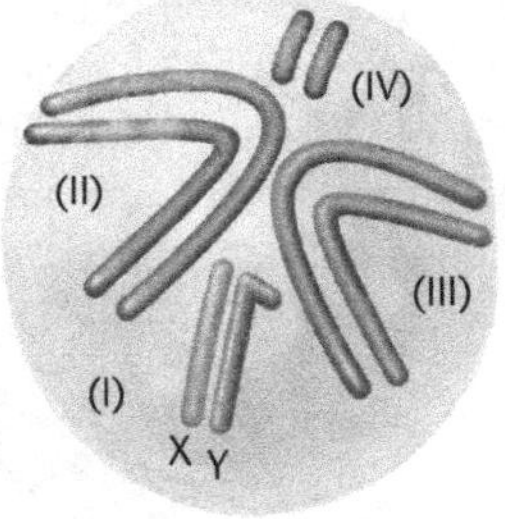


2 sets of autosomes
+
X Y

| Chromosome composition | Chromosome formulation | Ratio of X chromosomes to autosome sets | Sexual morphology |
|---|---|---|---|
| 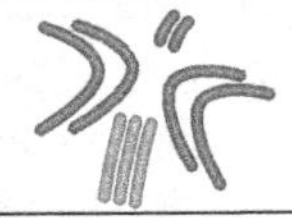 | 3X/2A | 1.5 | Metafemale |
| 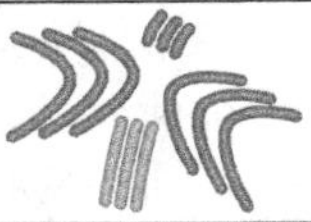 | 3X/3A | 1.0 | Female |
| 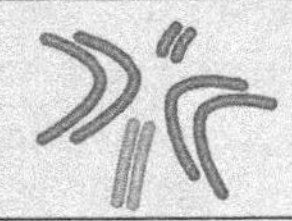 | 2X/2A | 1.0 | Female |
| 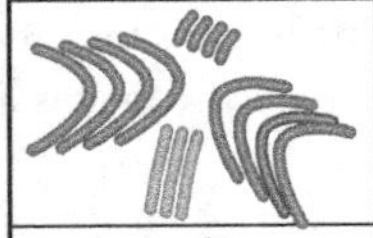 | 3X/4A | 0.75 | Intersex |
| 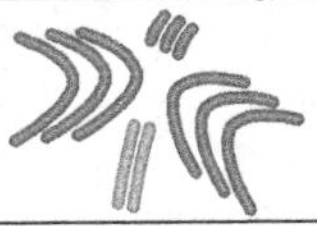 | 2X/3A | 0.67 | Intersex |
| 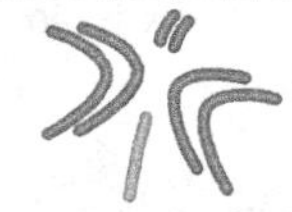 | X/2A | 0.50 | Male |
| 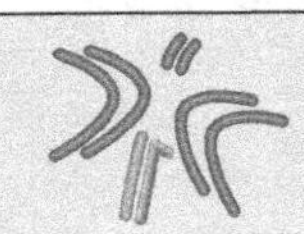 | XY/2A | 0.50 | Male |
| 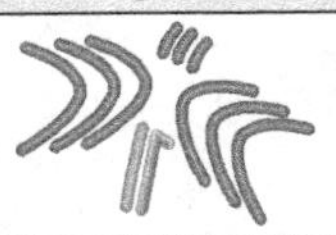 | XY/3A | 0.33 | Metamale |

FIGURE 7–15 Chromosome compositions, the corresponding ratios of X chromosomes to sets of autosomes, and the resultant sexual morphology seen in *Drosophila melanogaster.* The normal diploid male chromosome composition is shown as a reference on the left (XY/2A). The rows representing normally-occuring females and males are lightly shaded.

and genitalia. They were invariably sterile and expressed both male and female morphology, thus being designated as **intersexes.**

Bridges' results indicate that in *Drosophila*, factors that cause a fly to develop into a male are not located on the sex chromosomes but are instead found on the autosomes. Some female-determining factors, however, are located on the X chromosomes. Thus, with respect to primary sex determination, male gametes containing one of each autosome plus a Y chromosome result in male offspring not because of the presence of the Y but because they fail to contribute an X chromosome. This mode of sex determination is explained by the **genic balance theory**. Bridges proposed that a threshold for maleness is reached when the X:A ratio is 1:2 (X:2A), but that the presence of an additional X (XX:2A) alters the balance and results in female differentiation.

Numerous mutant genes have been identified that are involved in sex determination in *Drosophila.* The recessive autosomal gene *transformer* (*tra*), discovered over 50 years ago by Alfred H. Sturtevant, clearly demonstrated that a single autosomal gene could have a profound impact on sex determination. Females homozygous for *tra* are transformed into sterile males, but homozygous males are unaffected.

More recently, another gene, *Sex-lethal* (*Sxl*), has been shown to play a critical role, serving as a "master switch" in sex determination. Activation of the X-linked *Sxl* gene, which relies on a ratio of X chromosomes to sets of autosomes that equals 1.0, is essential to female development. In the absence of activation—as when, for example, the X:A ratio is 0.5—male development occurs. It is interesting to note that mutations that inactivate the *Sxl* gene, as originally studied in 1960 by Hermann J. Muller, kill female embryos but have no effect on male embryos, consistent with the role of the gene. Although it is not yet exactly clear how this ratio influences the *Sxl* locus, we do have some insights into the question. The *Sxl* locus is part of a hierarchy of gene expression and exerts control over other genes, including *tra* (discussed in the previous paragraph) and *dsx* (*doublesex*). The wild-type allele of *tra* is activated by the product of *Sxl* only in females and in turn influences the expression of *dsx*. Depending on how the initial RNA transcript of *dsx* is processed (spliced, as explained below), the resultant dsx protein activates either male- or female-specific genes required for sexual differentiation. Each step in this regulatory cascade requires a form of processing called **RNA splicing**, in which portions of the RNA are removed and the remaining fragments are "spliced" back together prior to translation into a protein. In the case of the *Sxl* gene, the RNA transcript may be spliced in different ways, a phenomenon called **alternative splicing**. A different RNA transcript is produced in females than in males. In potential females, the transcript is active and initiates a cascade of regulatory gene expression, ultimately leading to female differentiation. In potential males, the transcript is inactive, leading to a different pattern of gene activity, whereby male differentiation occurs. We will return to this topic in Chapter 18, where alternative splicing is again addressed as one of the mechanisms involved in the regulation of genetic expression in eukaryotes.

## Dosage Compensation in *Drosophila*

Since *Drosophila* females contain two copies of X-linked genes, whereas males contain only one copy, a dosage problem exists as in mammals such as humans and mice. However, the mechanism of dosage compensation in *Drosophila* differs considerably from that in mammals, since X chromosome inactivation is not observed. Instead, male X-linked genes are transcribed at twice the level of the comparable genes in females. Interestingly, if groups of X-linked genes are moved (translocated) to autosomes, dosage compensation still affects them, even though they are no longer part of the X chromosome.

As in mammals, considerable gains have been made recently in understanding the process of dosage compensation in *Drosophila.* At least four autosomal genes are known to be involved, under the same master-switch gene, *Sxl*, that induces female differentiation during sex determination. Mutations in any of these genes severely reduce the increased expression of X-linked genes in males, causing lethality.

Evidence supporting a mechanism of increased genetic activity in males is now available. The well-accepted model proposes that one of the autosomal genes, *mle* (*maleless*), encodes a protein that binds to numerous sites along the X chromosome, causing enhancement of genetic expression. The products of three other autosomal genes also participate in and are required for *mle* binding. In addition, proteins called male-specific lethals (MSLs) have been shown to bind to gene-rich regions of the X to increase gene expression in male flies. Collectively, this cluster of gene-activating proteins is called the **dosage compensation complex (DCC)**. Figure 7–16

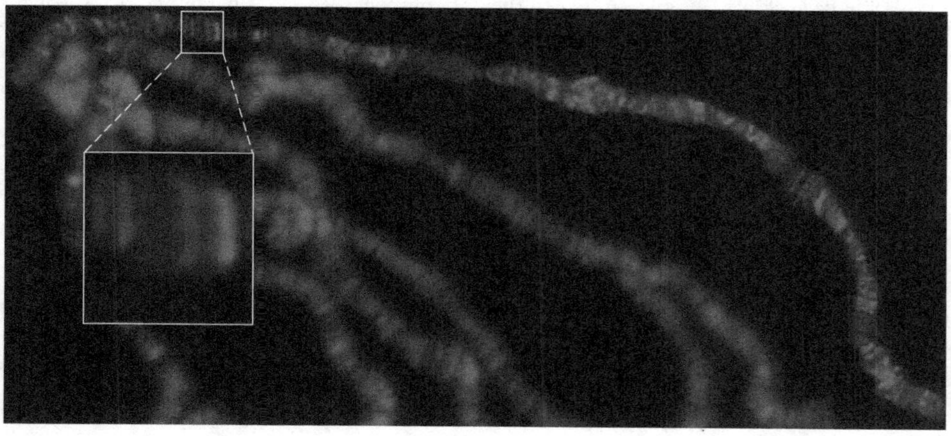

**FIGURE 7–16** Fluorescent antibodies against proteins in the dosage compensation complex (DCC) bind only to the X chromosome in *Drosophila* polytene chromosome preparations, providing evidence concerning the role of the DCC in increasing the expression of X-linked genes.

illustrates the presence of these proteins bound to the X chromosome in *Drosophila* in contrast to their failure to bind to the autosomes. The location of DCC proteins may be identified when fluorescent antibodies against these proteins are added to preparations of the large polytene chromosomes characteristic of salivary gland cells in fly larvae (see Chapter 12).

This model predicts that the master-switch *Sxl* gene plays an important role during dosage compensation. In XY flies, *Sxl* is inactive; therefore, the autosomal genes are activated, causing enhanced X chromosome activity. On the other hand, *Sxl* is active in XX females and functions to inactivate one or more of the male-specific autosomal genes, perhaps *mle*. By dampening the activity of these autosomal genes, it ensures that they will not serve to double the expression of X-linked genes in females, which would further compound the dosage problem.

Tom Cline has proposed that, before the aforementioned dosage compensation mechanism is activated, *Sxl* acts as a sensor for the expression of several other X-linked genes. In a way, *Sxl* counts X chromosomes. When it registers the dose of their expression as being high—for example, as the result of the presence of two X chromosomes—the *Sxl* gene product is modified to dampen the expression of the autosomal genes. Although this model may yet be revised or refined, it is useful for guiding future research.

Clearly, an entirely different mechanism of dosage compensation exists in *Drosophila* (and probably many related organisms) than that seen in mammals. The development of elaborate mechanisms to equalize the expression of X-linked genes demonstrates the critical importance of level of gene expression. A delicate balance of gene products is necessary to maintain normal development of both males and females.

### *Drosophila* Mosaics

Our knowledge of sex determination and of X-linkage in *Drosophila* (Chapter 4) helps us to understand the unusual appearance of a unique fruit fly, shown in Figure 7–17. This fly was recovered from a stock in which all other females were heterozygous for the X-linked genes *white* eye ($w$) and *miniature* wing ($m$). It is a **bilateral gynandromorph,** which means that one-half of its body (the left half) has developed as a male and the other half (the right half) as a female.

We can account for the presence of both sexes in a single fly in the following way. If a female zygote (heterozygous for *white* eye and *miniature* wing) were to lose one of its X chromosomes during the first mitotic division, the two cells would be of the XX and X0 constitution, respectively. Thus, one cell would be female and the other would be male. Each of these cells is responsible for producing all progeny cells that make up either the right half or the left half of the body during embryogenesis.

In the case of the bilateral gynandromorph, the original cell of X0 constitution apparently produced only identical progeny cells and gave rise to the left half of the fly, which, because of its chromosomal constitution, was male. Since the male half demonstrated the

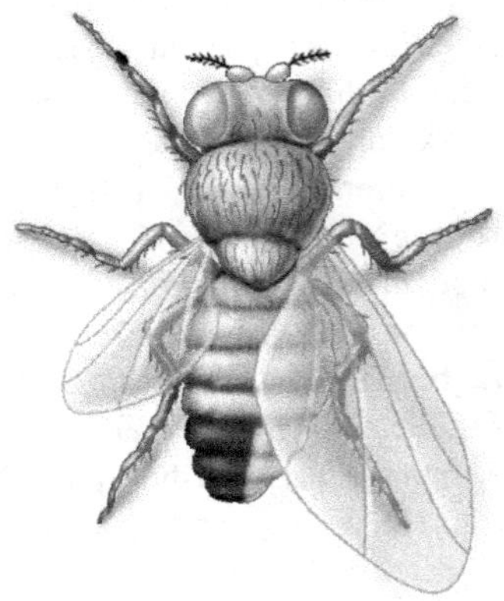

FIGURE 7–17 A bilateral gynandromorph of *Drosophila melanogaster* formed following the loss of one X chromosome in one of the two cells during the first mitotic division. The left side of the fly, composed of male cells containing a single X, expresses the mutant *white*-eye and *miniature*-wing alleles. The right side is composed of female cells containing two X chromosomes heterozygous for the two recessive alleles.

*white, miniature* phenotype, the X chromosome bearing the $w^+$, $m^+$ alleles was lost, while the $w$, $m$-bearing homolog was retained. All cells on the right side of the body were derived from the original XX cell, leading to female development. These cells, which remained heterozygous for both mutant genes, expressed the wild-type eye–wing phenotypes.

Depending on the orientation of the spindle during the first mitotic division, gynandromorphs can be produced that have the "line" demarcating male versus female development along or across any axis of the fly's body.

## 7.7 Temperature Variation Controls Sex Determination in Reptiles

We conclude this chapter by discussing several cases involving reptiles, in which the environment—specifically temperature—has a profound influence on sex determination. In contrast to **chromosomal**, or **genotypic, sex determination** (**CSD** or **GSD**), in which sex is determined genetically (as is true of all examples thus far presented in the chapter), the cases that we will now discuss are categorized as **temperature-dependent sex determination (TSD)**. As we shall see, the investigations leading to this information may well have come closer to revealing the true nature of the underlying basis of sex determination than any findings previously discussed.

In many species of reptiles, sex is predetermined at conception by sex-chromosome composition, as is the case in many organisms already considered in this chapter. For example, in many snakes, including vipers, a ZZ/ZW mode is in effect, in which the female is the

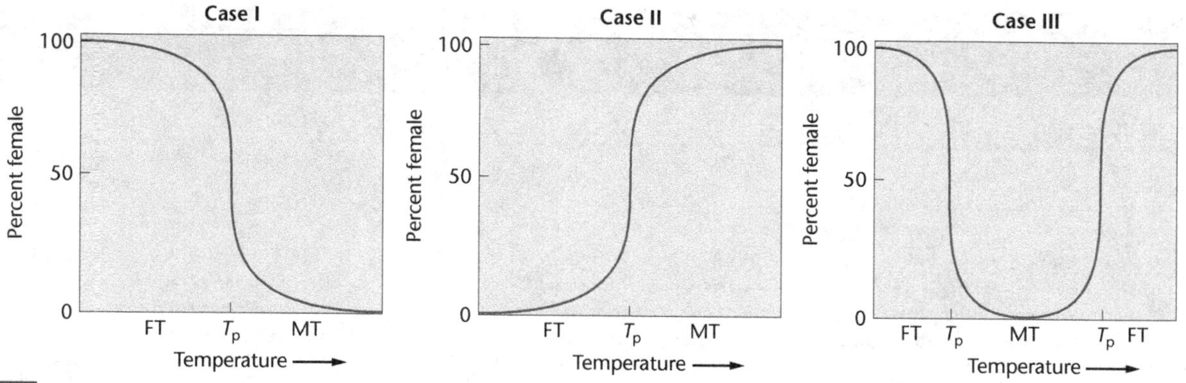


**FIGURE 7-18** Three different patterns of temperature-dependent sex determination (TSD) in reptiles, as described in the text. The relative pivotal temperature $T_P$ is crucial to sex determination during a critical point during embryonic development. FT = female-determining temperature; MT = male-determining temperature.

heterogamous sex (ZW). However, in boas and pythons, it is impossible to distinguish one sex chromosome from the other in either sex. In many lizards, both the XX/XY and ZZ/ZW systems are found, depending on the species. In still other reptilian species, including all crocodiles, most turtles, and some lizards, sex determination is achieved according to the incubation temperature of eggs during a critical period of embryonic development.

Three distinct patterns of TSD emerge (cases I–III in Figure 7–18). In case I, low temperatures yield 100 percent females, and high temperatures yield 100 percent males. Just the opposite occurs in case II. In case III, low *and* high temperatures yield 100 percent females, while intermediate temperatures yield various proportions of males. The third pattern is seen in various species of crocodiles, turtles, and lizards, although other members of these groups are known to exhibit the other patterns.

Two observations are noteworthy. First, in all three patterns, certain temperatures result in both male and female offspring; second, this pivotal temperature ($T_P$) range is fairly narrow, usually spanning less than 5°C, and sometimes only 1°C. The central question raised by these observations is this: What are the metabolic or physiological parameters affected by temperature that lead to the differentiation of one sex or the other?

The answer is thought to involve steroids (mainly estrogens) and the enzymes involved in their synthesis. Studies clearly demonstrate that the effects of temperature on estrogens, androgens, and inhibitors of the enzymes controlling their synthesis are involved in the sexual differentiation of ovaries and testes. One enzyme in particular, **aromatase**, converts androgens (male hormones such as testosterone) to estrogens (female hormones such as estradiol). The activity of this enzyme is correlated with the pathway of reactions that occurs during gonadal differentiation activity and is high in developing ovaries and low in developing testes. Researchers in this field, including Claude Pieau and colleagues, have proposed that a thermosensitive factor mediates the transcription of the reptilian aromatase gene, leading to temperature-dependent sex determination. Several other genes are likely to be involved in this mediation.

The involvement of sex steroids in gonadal differentiation has also been documented in birds, fishes, and amphibians. Thus, sex-determining mechanisms involving estrogens seem to be characteristic of nonmammalian vertebrates. The regulation of such systems, while temperature dependent in many reptiles, appears to be controlled by sex chromosomes (XX/XY or ZZ/ZW) in many of these other organisms. A final intriguing thought on this matter is that the product of *SRY*, a key component in mammalian sex determination, has been shown to bind *in vitro* to a regulatory portion of the aromatase gene, suggesting a mechanism whereby it could act as a repressor of ovarian development.

GENETICS, TECHNOLOGY, AND SOCIETY

## A Question of Gender: Sex Selection in Humans

Whether or not they admit it, and whether or not they agree with their partner, many prospective parents awaiting the birth of a baby have a preference for a child of a certain sex. Throughout history, people have resorted to varied and sometimes bizarre methods of influencing the gender of their offspring. In medieval Europe, prospective parents would place a hammer under the bed to help them conceive a boy, or a pair of scissors to conceive a girl. Other practices were based on the ancient belief that semen from the right testicle created male offspring and that from the left testicle created females. As late as the eighteenth century, European men might tie off or remove their left testicle to increase the chances of getting a male heir.

In some cultures, efforts to control the sex of offspring has had a darker outcome—female infanticide. In ancient Greece, the murder of female infants was so common that the male:female ratio in some areas approached 4:1. Some societies, even in present times, still kill female infants. In some parts of rural India, hundreds of families admitted to this practice as late as the 1990s. In 1997, the World Health Organization reported population data showing that about 50 million women were "missing" in China, likely because of selective abortion of female fetuses and institutionalized neglect of female children. Such behavior arises from poverty and age-old traditions. In some Indian cultures, sons work to provide income and security, whereas daughters not only contribute no income but require large dowries when they marry.

In recent times, sex-specific abortion has replaced much of the traditional female infanticide. Amniocentesis and ultrasound techniques have become lucrative businesses that provide prenatal sex determination. Studies in India estimate that hundreds of thousands of fetuses are aborted each year because they are female. As a result of sex-selective abortion, the female:male ratio in India was 927:1000 in 1991. In some northern states, the ratio is as low as 600:1000. Although sex determination and selective abortion of female fetuses was outlawed in India and China in the mid-1990s, the practice is thought to continue.

In Western industrial countries, advances in genetics and reproductive technology offer parents ways to select their children's gender prior to implantation of the embryo in the uterus—called *preimplantation gender selection (PGS)*. Following *in vitro* fertilization, embryos can be biopsied and assessed for gender. Only sex-selected embryos are then implanted. The simplest method involves separating X and Y chromosome-bearing spermatozoa. Sperm are sorted based on their DNA content. Because of the difference in size of the X and Y chromosomes, X-bearing sperm contain 2.8–3.0 percent more DNA than Y-bearing sperm. Sperm samples are treated with a fluorescent DNA stain, then passed through a laser beam in a Fluorescence-Activated Cell Sorter (FACS) machine that separates the sperm into two fractions based on the intensity of their DNA-fluorescence. Human sperm can be separated into X and Y chromosome fractions, with enrichments of about 85 percent and 75 percent, respectively. The sorted sperm are then used for standard intrauterine insemination. The Genetics and IVF Institute (Fairfax, Virginia) is presently using this PGS technique in an FDA-approved clinical trial. As of January 2006, over 1000 human pregnancies have resulted. The company reports an approximately 80 percent success rate in producing the desired gender.

The emerging PGS methods raise a number of legal and ethical issues. Some people feel that prospective parents have the legal right to use sex-selection techniques as part of their fundamental procreative liberty. Others believe that this liberty does not extend to custom designing a child to the parents' specifications. Proponents state that the benefits far outweigh any dangers to offspring or society. For example, people at risk for transmitting X-linked diseases such as hemophilia or Duchenne muscular dystrophy can now enhance their chance of conceiving a female child, who will not express the disease. As there are more than 500 known X-linked diseases and they are expressed in about 1 in 1000 live births, PGS could greatly reduce suffering for many families.

The majority of people who undertake PGS, however, do so for nonmedical reasons—to "balance" their families. A possible argument in favor of this use is that the ability to intentionally select the sex of an offspring may reduce overpopulation and economic burdens for families who would repeatedly reproduce to get the desired gender. By the same token, PGS may reduce the number of abortions. It is also possible that PGS may increase the happiness of both parents and children, as the children would be more "wanted."

On the other hand, some argue that PGS serves neither the individual nor the common good. It is argued that PGS is inherently sexist, having its basis in the idea that one sex is superior to the other, and leads to an increase in linking a child's worth to gender. Other critics of PGS argue that this technology—if it is available only to those who can afford it—may contribute to social and economic inequality. Other critics fear that social approval of PGS will open the door to people's accepting other genetic manipulations of children's characteristics. It is difficult to predict the full effects that PGS will bring to the world. But the gender- selection genie is now out of the bottle and is unwilling to return.

### References

Sills, E.S., Kirman, I., Thatcher, S.S., III, and Palermo, G.D. 1998. Sex-selection of human spermatozoa: Evolution of current techniques and applications. *Arch. Gynecol. Obstet.* 261: 109–115.

Robertson, J.A. 2001. Preconception Gender Selection. *Am. J. Bioethics* 1: 2–9.

### Web Sites

Microsort technique, Genetics & IVF Institute, Fairfax, Virginia. http://www.microsort.net

Female Infanticide, Gendercide Watch. http://www.gendercide.org/case_infanticide.html

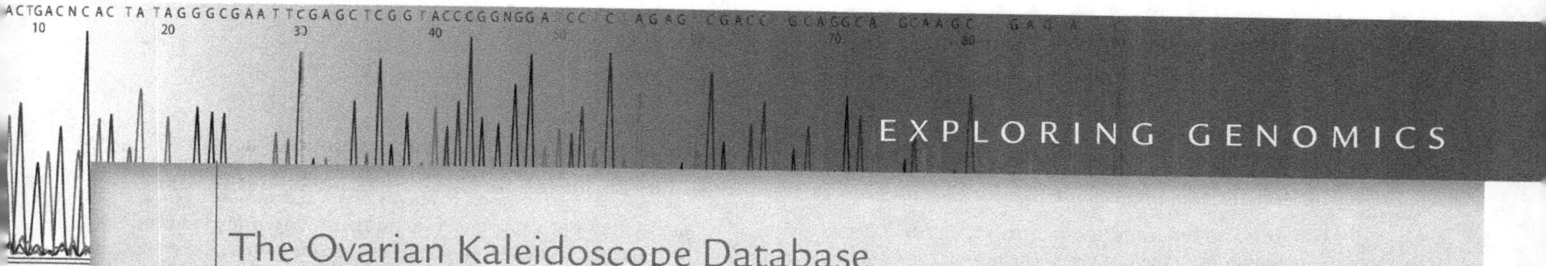


# The Ovarian Kaleidoscope Database (OKDB)

In this chapter we discussed mechanisms of sex determination and aspects of sex chromosomes that contribute to sexual differentiation. The genetics of sexual differentiation is a very active area of research. Many studies are currently focusing on identifying genes that are expressed in the testis and ovary and are involved in differentiation of reproductive organs or that are important for gamete formation and development. In this exercise we will explore the **Ovarian Kaleidoscope Database (OKDB)** to learn more about the *STRA8* gene involved in development of the ovary.

## ■ Exercise I – Ovarian Kaleidoscope Database (OKDB)

The OKDB was developed and is maintained by Stanford University scientists as a resource for reproductive biologists, developmental biologists, and other scientists interested in gene expression in the ovary. It is an excellent resource for new information on gene expression in the ovary.

1. Access the OKDB at http://ovary.stanford.edu/4_home.html.
2. Recently, the gene *STRA8* has been identified as an early marker of ovarian development. Y chromosome researcher David Page, whom you learned about in this chapter, and colleagues have obtained very interesting findings about the function of *STRA8*. Use the "Gene Name" search feature to find information on *STRA8*, and then answer the following questions:
   a. In humans, on what chromosome is *STRA8* located?
   b. How does *STRA8* expression differ in the developing testis compared to the developing ovary?
   c. What vitamin-derived hormone is important for stimulating *STRA8* expression and for initiating meiosis in the developing ovary?
3. Use the "Recent Publications" link at the bottom of the page to find PubMed-indexed recent publications on *STRA8* by David Page and other researchers.
4. Currently, there is no equivalent database for Y-specific genes, but the NCBI Genes and Disease Site (www.ncbi.nlm.nih.gov/books/bv.fcgi?rid=gnd&ref=sidebar) that you were introduced to in Exploring Genomics for Chapter 5 has excellent maps of the human Y chromosome. Visit this site to explore Y-linked genes. Choose a Y chromosome gene for further exploration and write a short report about the function of that gene.

## ■ Exercise II – *STRA8* Orthologs and Exploring *Stra8* in Mice

Orthologs, or orthologous genes, are genes with similar sequences from different species. Orthologs arise from the same gene in a common ancestor.

1. Humans and mice are not the only species expressing *STRA8*. Click on the "Orthologous Genes" link at the top of the *STRA8* page in OKDB to see a table of *STRA8* orthologs. Are you surprised by the results of this search?
2. Let's learn more about *Stra8* in mice (*Mus musculus*). Much of what we know about the role of retinoic acid and *Stra8* in development of the ovary and testis has been learned from studies in mice. At the top of the ortholog table, click on the "HomoloGene:49197" link and then the "*M. musculus Stra8*" link to enter the NCBI Entrez Gene database (a searchable tool for genes in the NCBI database). Feel free to browse Entrez Gene.
3. In Entrez Gene, under the "Summary" category, find the primary source link "MGI: 107917." Use this link to access the **Mouse Genome Information Site (MGI)**. MGI is a database of mouse genes maintained by the Jackson Laboratory, in Bar Harbor, Maine. Explore links on the MGI site to answer the following questions about the mouse *Stra8* gene.
   a. On what chromosome is *Stra8* located in mice?
   b. How many amino acids are in the mouse Stra8 protein?
   c. What structural feature or domain of the Stra8 protein may suggests its function?
   d. Is Stra8 expressed in other mouse tissues in addition to the ovary and testis?
   e. An understanding of the role of retinoic acid and *Stra8* was developed in part through gene knockout experiments in mice (a technique we will discuss in detail in Chapter 19). What phenotypes are observed in *Stra8-/-* knockout mice?
4. From the MGI page for the mouse *Stra8* gene, go to the "Sequence Map" category of information and explore the links to "Ensembl," "UCSC" (University of California Santa Cruz), and "NCBI" (National Center for Biotechnology Information). Each of these browsers provides different representations of detailed information on *Stra8*. These sites are excellent resources for accessing genomic data.

## Chapter Summary

1. Sexual reproduction ultimately relies on some form of sexual differentiation, which is achieved by a variety of sex-determining mechanisms.
2. The genetic basis of sexual differentiation is usually related to different chromosome compositions in the two sexes. The heterogametic sex either lacks one chromosome or contains a unique heteromorphic chromosome, usually referred to as the Y or W chromosome.
3. In humans, the study of individuals with altered sex-chromosome compositions has established that the Y chromosome is responsible for male differentiation. The absence of the Y leads to female differentiation. Similar studies in *Drosophila* have excluded the Y in such a role, instead demonstrating that a balance between the number of X chromosomes and sets of autosomes is the critical factor.
4. The primary sex ratio in humans substantially favors males at conception. During embryonic and fetal development, male mortality is higher than that of females. The secondary sex ratio at birth still favors males by a small margin.
5. In mammals and fruit flies, dosage compensation mechanisms exist to equilibrate the expression of X-linked genes, given that females have two X chromosomes and males have only one X chromosome. Mammalian females inactivate all but one X chromosome early in development. Fruit fly males double the expression of genes on their single X chromosome in comparison to expression on the female X chromosomes.
6. The Lyon hypothesis states that, early in development, inactivation is random between the maternal and paternal X chromosomes. All subsequent progeny cells inactivate the same X as their progenitor cell. Mammalian females thus develop as mosaics with respect to their expression of heterozygous X-linked alleles.
7. In many reptiles, the incubation temperature at a critical time during embryogenesis is responsible for sex determination. Temperature influences the activity of enzymes involved in the metabolism of steroids related to sexual differentiation.

## INSIGHTS AND SOLUTIONS

1. In *Drosophila*, the X chromosomes may become attached to one another ($\widehat{XX}$) such that they always segregate together. Some flies thus contain a set of attached X chromosomes plus a Y chromosome.

(a) What sex would such a fly be? Explain why this is so.

(b) Given the answer to part (a), predict the sex of the offspring that would occur in a cross between this fly and a normal one of the opposite sex.

(c) If the offspring described in part (b) are allowed to interbreed, what will be the outcome?

**Solution:**

(a) The fly will be a female. The ratio of X chromosomes to sets of autosomes—which determines sex in *Drosophila*—will be 1.0, leading to normal female development. The Y chromosome has no influence on sex determination in *Drosophila*.

(b) All progeny flies will have two sets of autosomes along with one of the following sex-chromosome compositions:

(1) $\widehat{XX}X$ → a metafemale with 3 X's (called a trisomic)

(2) $\widehat{XX}Y$ → a female like her mother

(3) XY → a normal male

(4) YY → no development occurs

(c) A stock will be created that maintains attached-X females generation after generation.

2. The Xg cell-surface antigen is coded for by a gene located on the X chromosome. No equivalent gene exists on the Y chromosome. Two codominant alleles of this gene have been identified: *Xg1* and *Xg2*. A woman of genotype *Xg2/Xg2* bears children with a man of genotype *Xg1/Y*, and they produce a son with Klinefelter syndrome of genotype *Xg1/Xg2Y* Using proper genetic terminology, briefly explain how this individual was generated. In which parent and in which meiotic division did the mistake occur?

**Solution:** Because the son with Klinefelter syndrome is *Xg1/Xg2Y*, he must have received both the *Xg1* allele and the Y chromosome from his father. Therefore, nondisjunction must have occurred during meiosis I in the father.

## Problems and Discussion Questions

1. As related to sex determination, what is meant by
   (a) homomorphic and heteromorphic chromosomes and
   (b) isogamous and heterogamous organisms?
2. Contrast the life cycle of a plant such as *Zea mays* with an animal such as *C. elegans*.
3. Discuss the role of sexual differentiation in the life cycles of *Chlamydomonas*, *Zea mays*, and *C. elegans*.
4. Distinguish between the concepts of sexual differentiation and sex determination.
5. Contrast the *Protenor* and *Lygaeus* modes of sex determination.
6. Describe the major difference between sex determination in *Drosophila* and in humans.
7. How do mammals, including humans, solve the "dosage problem" caused by the presence of an X and Y chromosome in one sex and two X chromosomes in the other sex?

8. The phenotype of an early-stage human embryo is considered sexually indifferent. Explain why this is so even though the embryo's genotypic sex is already fixed.
9. What specific observations (evidence) support the conclusions about sex determination in *Drosophila* and humans?
10. Describe how nondisjunction in human female gametes can give rise to Klinefelter and Turner syndrome offspring following fertilization by a normal male gamete.
11. An insect species is discovered in which the heterogametic sex is unknown. An X-linked recessive mutation for *reduced wing* (*rw*) is discovered. Contrast the $F_1$ and $F_2$ generations from a cross between a female with reduced wings and a male with normal-sized wings when
    (a) the female is the heterogametic sex and
    (b) the male is the heterogametic sex.
12. Given your answers to Problem 11, is it possible to distinguish between the *Protenor* and *Lygaeus* mode of sex determination based on the outcome of these crosses?
13. When cows have twin calves of unlike sex (fraternal twins), the female twin is usually sterile and has masculinized reproductive organs. This calf is referred to as a freemartin. In cows, twins may share a common placenta and thus fetal circulation. Predict why a freemartin develops.
14. An attached-X female fly, $\widehat{XX}Y$ (see the "Insights and Solutions" box), expresses the recessive X-linked *white*-eye phenotype. It is crossed to a male fly that expresses the X-linked recessive miniature wing phenotype. Determine the outcome of this cross in terms of sex, eye color, and wing size of the offspring.
15. Assume that on rare occasions the attached X chromosomes in female gametes become unattached. Based on the parental phenotypes in Problem 14, what outcomes in the $F_1$ generation would indicate that this has occurred during female meiosis?
16. It has been suggested that any male-determining genes contained on the Y chromosome in humans cannot be located in the limited region that synapses with the X chromosome during meiosis. What might be the outcome if such genes were located in this region?
17. What is a Barr body, and where is it found in a cell?
18. Indicate the expected number of Barr bodies in interphase cells of individuals with Klinefelter syndrome; Turner syndrome; and karyotypes 47,XYY, 47,XXX, and 48,XXXX.
19. Define the Lyon hypothesis.
20. Can the Lyon hypothesis be tested in a human female who is homozygous for one allele of the X-linked *G6PD* gene? Why, or why not?
21. Predict the potential effect of the Lyon hypothesis on the retina of a human female heterozygous for the X-linked red-green color-blindness trait.
22. Cat breeders are aware that kittens expressing the X-linked calico coat pattern and tortoiseshell pattern are almost invariably females. Why?
23. What does the apparent need for dosage compensation mechanisms suggest about the expression of genetic information in normal diploid individuals?
24. The marine echiurid worm *Bonellia viridis* is an extreme example of environmental influence on sex determination. Undifferentiated larvae either remain free-swimming and differentiate into females or they settle on the proboscis of an adult female and become males. If larvae that have been on a female proboscis for a short period are removed and placed in seawater, they develop as intersexes. If larvae are forced to develop in an aquarium where pieces of proboscises have been placed, they develop into males. Contrast this mode of sexual differentiation with that of mammals. Suggest further experimentation to elucidate the mechanism of sex determination in *B. viridis*.
25. What type of evidence supports the conclusion that the primary sex ratio in humans is as high as 1.20 to 1.60?
26. Devise as many hypotheses as you can that might explain why so many more human male conceptions than human female conceptions occur.
27. In mice, the *Sry* gene (see Section 7.3) is located on the Y chromosome very close to one of the pseudoautosomal regions that pairs with the X chromosome during male meiosis. Given this information, propose a model to explain the generation of unusual males who have two X chromosomes (with an *Sry*-containing piece of the Y chromosome attached to one X chromosome).
28. The genes encoding the red- and green-color-detecting proteins of the human eye are located next to one another on the X chromosome and probably evolved from a common ancestral pigment gene. The two proteins demonstrate 76 percent homology in their amino acid sequences. A normal-visioned woman with both genes on each of her two X chromosomes has a red-color-blind son who was shown to have one copy of the green-detecting gene and no copies of the red-detecting gene. Devise an explanation for these observations at the chromosomal level (involving meiosis).

**HOW DO WE KNOW?**

29. In this chapter, we focused on sex differentiation, sex chromosomes, and genetic mechanisms involved in sex determination. At the same time, we found many opportunities to consider the methods and reasoning by which much of this information was acquired. From the explanations given in the chapter, what answers would you propose to the following fundamental questions?
    (a) How do we know that specific genes in maize play a role in sexual differentiation?
    (b) How do we know whether or not a heteromorphic chromosome such as the Y chromosome plays a crucial role in the determination of sex?
    (c) How do we know that in humans the X chromosomes play no role in human sex determination, while the Y chromosome causes maleness and its absence causes femaleness?
    (d) How did we learn that, although the sex ratio at birth in humans favors males slightly, the sex ratio at conception favors them much more?
    (e) How do we know that X chromosomal inactivation of either the paternal or maternal homolog is a random event during early development in mammalian females?
    (f) How do we know that *Drosophila* utilizes a different sex-determination mechanism than mammals, even though it has the same sex-chromosome compositions in males and females?

## Extra-Spicy Problems

30. In mice, the X-linked dominant mutation *Testicular feminization* (*Tfm*) eliminates the normal response to the testicular hormone testosterone during sexual differentiation. An XY mouse bearing the *Tfm* allele on the X chromosome develops testes, but no further male differentiation occurs—the external genitalia of such an animal are female. From this information, what might you conclude about the role of the *Tfm* gene product and the X and Y chromosomes in sex determination and sexual differentiation in mammals? Can you devise an experiment, assuming you can "genetically engineer" the chromosomes of mice, to test and confirm your explanation?

31. Campomelic dysplasia (CMD1) is a congenital human syndrome featuring malformation of bone and cartilage. It is caused by an autosomal dominant mutation of a gene located on chromosome 17. Consider the following observations in sequence, and in each case, draw whatever appropriate conclusions are warranted.
   (a) Of those with the syndrome who are karyotypically 46,XY, approximately 75 percent are sex reversed, exhibiting a wide range of female characteristics.
   (b) The nonmutant form of the gene, called *SOX9*, is expressed in the developing gonad of the XY male, but not the XX female.
   (c) The *SOX9* gene shares 71 percent amino acid coding sequence homology with the Y-linked *SRY* gene.
   (d) CMD1 patients who exhibit a 46,XX karyotype develop as females, with no gonadal abnormalities.
32. In the wasp *Bracon hebetor*, a form of parthenogenesis (the development of unfertilized eggs into progeny) resulting in haploid organisms is not uncommon. All haploids are males. When offspring arise from fertilization, females almost invariably result. P. W. Whiting has shown that an X-linked gene with nine multiple alleles ($X_a$, $X_b$, etc.) controls sex determination. Any homozygous or hemizygous condition results in males, and any heterozygous condition results in females. If an $X_a/X_b$ female mates with an $X_a$ male and lays 50 percent fertilized and 50 percent unfertilized eggs, what proportion of male and female offspring will result?
33. Shown below are two graphs that plot the percentage of fertilized eggs containing males against the atmospheric temperature during early development in (a) snapping turtles and (b) most lizards. Interpret these data as they relate to the effect of temperature on sex determination.

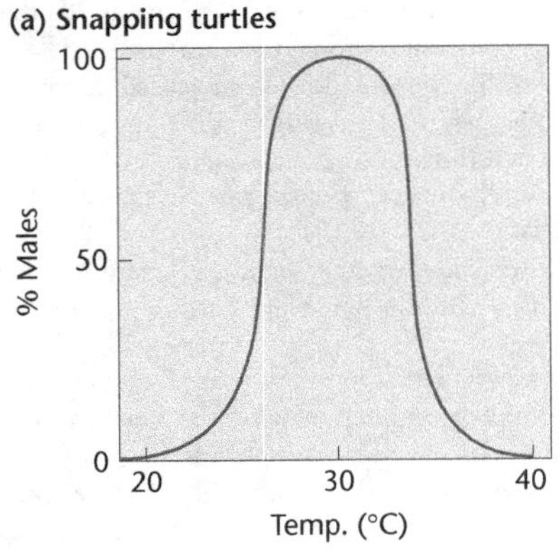


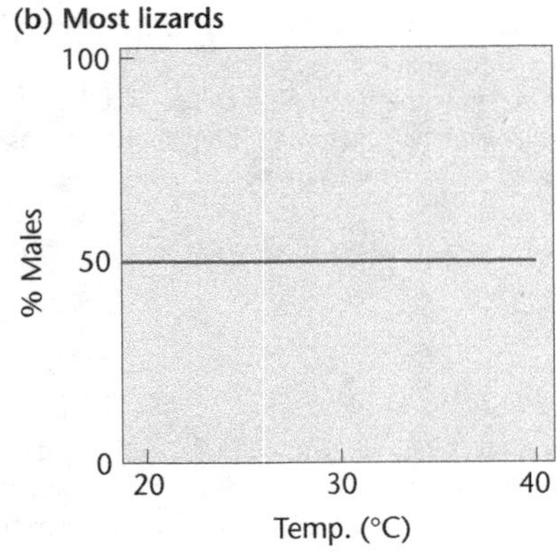


34. CC (Carbon Copy), the first cat produced from a clone, was created from an ovarian cell taken from her genetic donor, Rainbow. The diploid nucleus from the cell was extracted and then injected into an enucleated egg. The resulting zygote was then allowed to develop in a petri dish, and the cloned embryo was implanted in the uterus of a surrogate mother cat, who gave birth to CC. Rainbow is a calico cat. CC's surrogate mother is a tabby. Geneticists were very interested in the outcome of cloning a calico cat, because they were not certain if the cat would have patches of orange and black, just orange, or just black. Taking into account the Lyon hypothesis, explain the basis of the uncertainty.

Carbon Copy with her surrogate mother.

35. Let's assume hypothetically that Carbon Copy (see Problem 34) is indeed a calico with black and orange patches, along with the patches of white characterizing a calico cat. Would you expect CC to appear identical to Rainbow? Explain why or why not.
36. When Carbon Copy was born (see Problem 34), she had black patches and white patches, but completely lacked any orange patches. The knowledgeable students of genetics were not surprised at this outcome. Starting with the somatic ovarian cell used as the source of the nucleus in the cloning process, explain how this outcome occurred.
37. Clearly, a number of sex-determination schemes have evolved in different taxa, ranging from temperature-dependent sex determination (TSD) to chromosomal or genotypic sex determination (CSD or GSD, respectively). Below is a list of various organisms that we have discussed in this chapter.

   *Bonellia viridis*
   *Bracon herbetor*
   *Homo sapiens*
   *Drosophila melanogaster*
   *Caenorhabditis elegans*
   *Protenor*
   *Lygaeus*
   Birds, moths, butterflies
   Lizards
   Crocodiles
   Turtles

   (a) For each species or group listed, provide a brief description of the sex-determining mechanism, and then arrange the listings into groups (TSD, CSD, GSD, or OTHER).
   (b) While little information is available about sex determination in amphibians, Eggert reports that in addition to parthenogenesis, a variety of CSD and GSD mechanisms have been documented. However, sex differentiation can be overridden by variations in temperature. Based on this information, which of the categories specified in part (a) is(are) most closely related to amphibians?

(c) Among all the listings in part (a), *Homo sapiens* is the only one that maintains a constant body temperature (is endothermic), and some scientists argue that hidden within the sex-determination mechanisms of mammals is a form of temperature-dependent sex determination. Would you consider this to be a reasonable suggestion?

38. In a number of organisms, including *Drosophila* and butterflies, genes that alter the sex-ratio have been described. In the pest species *Musca domesticus* (the house fly), *Aedes aegypti* (the mosquito that is the vector for yellow fever), and *Culex pipiens* (the mosquito vector for filariasis and some viral diseases), scientists are especially interested in such genes. Sex in *Culex* is determined by a single gene pair, *Mm* being male and *mm* being female. Males homozygous for the recessive gene *dd* never produce many female offspring. The *dd* combination in males causes fragmentation of the *m*-bearing dyad during the first meiotic division, hence its failure to complete spermatogenesis.

(a) Account for this sex-ratio distortion by drawing labeled chromosome arrangements in primary and secondary spermatocytes for each of the following genotypes: *Mm Dd* and *Mm dd*. How do meiotic products differ between *Dd* and *dd* genotypes? Note that the diploid chromosome number is 6 in *Culex pipiens* and both *D* and *M* loci are linked on the same chromosome.

(b) How might a sex-ratio distorter such as *dd* be used to control pest population numbers?

Spectral karyotyping of human chromosomes, utilizing differentially labeled "painting" probes.

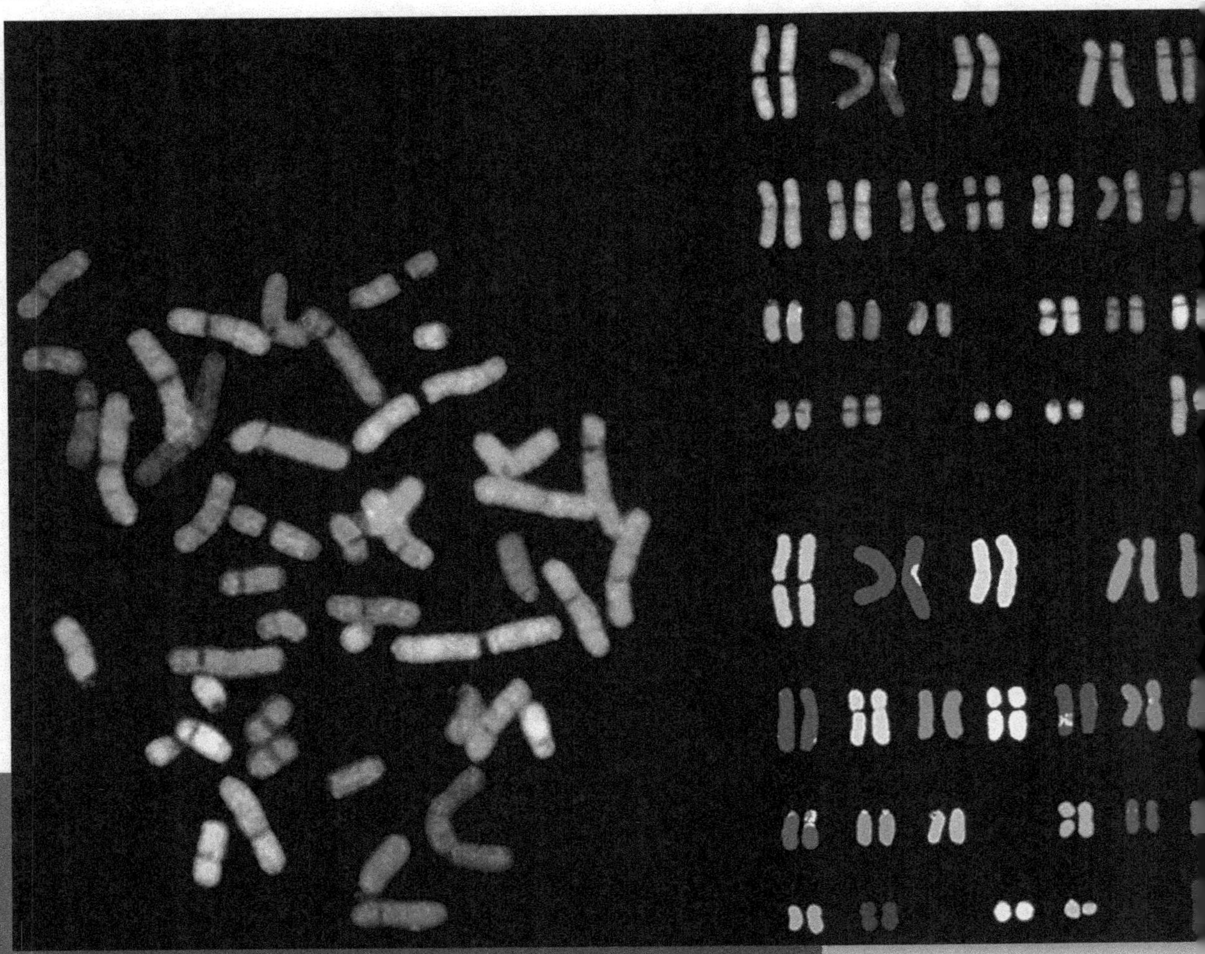

8

# Chromosome Mutations: Variation in Chromosome Number and Arrangement

## CHAPTER CONCEPTS

- The failure of chromosomes to properly separate during meiosis results in variation in the chromosome content of gametes and subsequently in offspring arising from such gametes.
- Plants often tolerate an abnormal genetic content, but, as a result, they often manifest unique phenotypes. Such genetic variation has been an important factor in the evolution of plants.
- In animals, genetic information is in a delicate equilibrium whereby the gain or loss of a chromosome, or part of a chromosome, in an otherwise diploid organism often leads to lethality or to an abnormal phenotype.
- The rearrangement of genetic information within the genome of a diploid organism may be tolerated by that organism but may affect the viability of gametes and the phenotypes of organisms arising from those gametes.
- Chromosomes in humans contain fragile sites—regions susceptible to breakage, which leads to abnormal phenotypes.

Thus far, we have emphasized how mutations and the resulting alleles affect an organism's phenotype and how traits are passed from parents to offspring according to Mendelian principles. In this chapter, we shall look at phenotypic variation that results from changes that are more substantial than alterations of individual genes—modifications at the level of the chromosome.

Although most members of diploid species normally contain precisely two haploid chromosome sets, many known cases vary from this pattern. Modifications have occurred through a change in the total number of chromosomes, the deletion or duplication of genes or segments of a chromosome, or rearrangements of the genetic material either within or among chromosomes. Collectively, such changes are called **chromosome mutations** or **chromosome aberrations**, to distinguish them from gene mutations. Because, according to Mendelian laws, the chromosome is the unit of genetic transmission, chromosome aberrations are passed on to offspring in a predictable manner, resulting in many unique genetic outcomes.

Inasmuch as the genetic component of an organism is delicately balanced, even minor alterations of either the content or location of genetic information within the genome can result in some form of phenotypic variation. More substantial changes may be lethal, particularly in animals. In this chapter, we consider many well-studied types of chromosomal aberrations, the phenotypic consequences for the organism that harbors an aberration, and the impact of the aberration on offspring of the affected individual. We will also discuss the role of chromosome aberrations in the evolutionary process.

8.1

## Specific Terminology Describes Variations in Chromosome Number

Variation in chromosome number ranges from the addition or loss of one or more chromosomes to the addition of one or more haploid sets of chromosomes. Before embarking on your study of these variations, you should learn some of the terminology that geneticists use to describe such changes. For example, in the general condition known as **aneuploidy**, an organism has gained or lost one or more chromosomes but not a complete set. The absence of a single chromosome from an otherwise diploid genome is called *monosomy*. The gain of one extra chromosome results in *trisomy*. Such changes are contrasted with the condition of **euploidy**, where all chromosomes belong to complete haploid sets. If more than two sets are present, the term **polyploidy** applies. Organisms with three sets are specifically *triploid*; those with four sets are *tetraploid*; and so on. Table 8.1 provides an organizational framework for you to follow as we discuss each of these categories of aneuploid and euploid variation and the subsets within them.

TABLE 8.1

**Terminology for Variation in Chromosome Numbers**

| Term | Explanation |
|---|---|
| Aneuploidy | $2n \pm x$ chromosomes |
| Monosomy | $2n - 1$ |
| Disomy | $2n$ |
| Trisomy | $2n + 1$ |
| Tetrasomy, pentasomy, etc. | $2n + 2$, $2n + 3$, etc. |
| Euploidy | Multiples of $n$ |
| Diploidy | $2n$ |
| Polyploidy | $3n$, $4n$, $5n$, . . . |
| Triploidy | $3n$ |
| Tetraploidy, pentaploidy, etc. | $4n$, $5n$, etc. |
| Autopolyploidy | Multiples of the same genome |
| Allopolyploidy (Amphidiploidy) | Multiples of closely related genomes |

### Variation in the Number of Chromosomes Results from Nondisjunction

As we consider cases that result from the gain or loss of chromosomes, it is useful to examine how such aberrations originate. For instance, how do the syndromes described in Chapter 7 arise, in which the number of sex-determining chromosomes in humans is altered? The gain (47,XXY) or the loss (45,X) of a sex-determining chromosome from an otherwise diploid genome alters the normal phenotype (in spite of a mechanism in somatic cells that inactivates all X chromosomes in excess of one), resulting in **Klinefelter syndrome** or **Turner syndrome**, respectively (see Figure 7–7). Human females may have extra X chromosomes (e.g., 47,XXX and 48,XXXX), and some males have an extra Y chromosome (47,XYY).

Such chromosomal variation originates as a random error during the production of gametes, a phenomenon referred to as **nondisjunction**, whereby paired homologs fail to disjoin during segregation. This process disrupts the normal distribution of chromosomes into gametes. The results of nondisjunction during meiosis I and meiosis II for a single chromosome of a diploid organism are shown in Figure 8–1. As you can see, abnormal gametes can form containing either two members of the affected chromosome or none at all. Fertilizing these with a normal haploid gamete

NOW SOLVE THIS

Problem 13 on page 224 considers a female with Turner syndrome who exhibits hemophilia, as did her father. You are asked which of her parents was responsible for the nondisjunction event leading to her syndrome.

■ HINT: *The parent who contributed a gamete with an X chromosome underwent normal meiosis.*

CHROMOSOME SEPARATION IN MEIOSIS: DISJUNCTION
WEB TUTORIAL 8.1

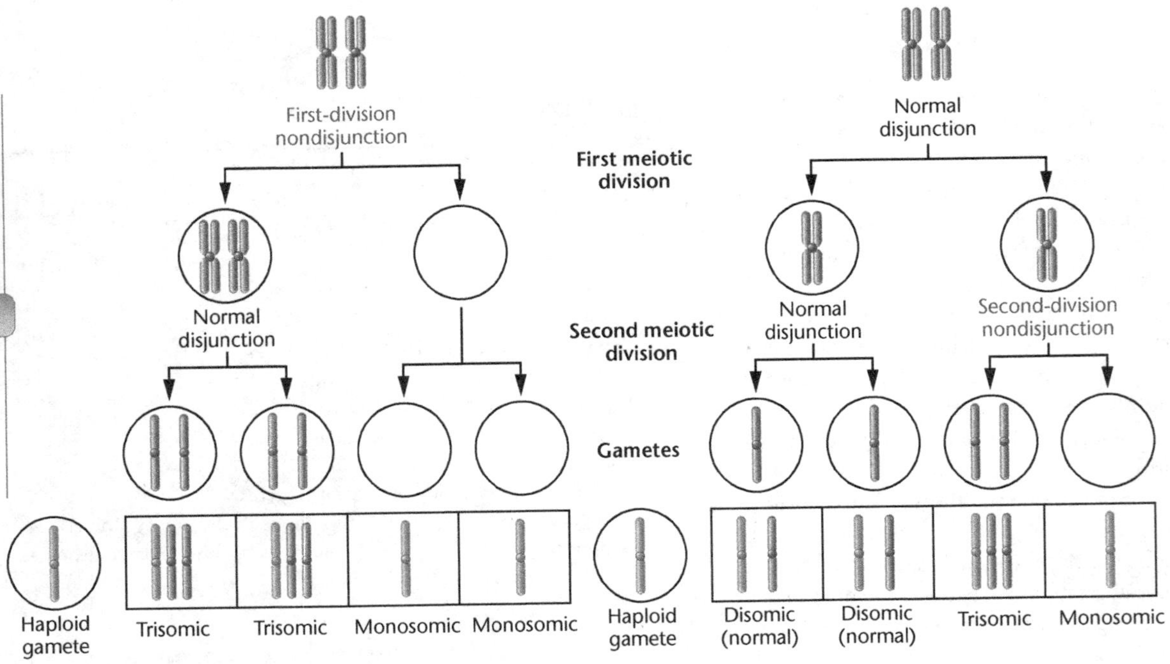


**FIGURE 8-1** Nondisjunction during the first and second meiotic divisions. In both cases, some of the gametes formed either contain two members of a specific chromosome or lack that chromosome. After fertilization by a gamete with a normal haploid content, monosomic, disomic (normal), or trisomic zygotes are produced.

produces a zygote with either three members (trisomy) or only one member (monosomy) of this chromosome. Nondisjunction leads to a variety of aneuploid conditions in humans and other organisms.

## 8.2 Monosomy, the Loss of a Single Chromosome, May Have Severe Phenotypic Effects

We turn now to consideration of variations in the number of autosomes and the genetic consequences of such changes. The most common examples of aneuploidy, where an organism has a chromosome number other than an exact multiple of the haploid set, are cases in which a single chromosome is either added to or lost from a normal diploid set. The loss of one chromosome produces a $2n - 1$ complement called **monosomy**.

Although monosomy for the X chromosome occurs in humans, as we have seen in 45,X Turner syndrome, monosomy for any of the autosomes is not usually tolerated in humans or other animals. In *Drosophila*, flies monosomic for the very small chromosome IV—a condition referred to as **Haplo-IV**—survive, but they develop more slowly, exhibit reduced body size, and have impaired viability. Chromosome IV contains no more than 5 percent of the genome of *Drosophila*. Monosomy for the larger chromosomes II and III is lethal.

The failure of monosomic individuals to survive in many animal species is at first quite puzzling, since at least a single copy of every gene is present on the remaining homolog. However, as with expression of genes on the X chromosome in animals, a balance of expression must be achieved by genes on the autosomes as well. Having a single copy of each gene in a monosomic organism does not produce an acceptable "dosage" of gene expression. With autosomes, there is no suitable compensatory mechanism that has evolved. Note, too, that if just one of the genes present on the single remaining chromosome is a lethal allele, it will be hemizygous, and the condition will lead to the death of the organism. Thus, monosomy has the potential to unmask recessive lethals that are tolerated in heterozygotes carrying the corresponding wild-type alleles.

While animals do not often survive monosomy, it is often tolerated in the plant kingdom, where the dosage requirement does not appear to be as stringent. Monosomy for autosomal chromosomes has been observed in maize, tobacco, the evening primrose (*Oenothera*), and the Jimson weed (*Datura*), among other plants. Nevertheless, monosomic plants of these species are often less viable than their diploid relatives. Haploid pollen grains, which undergo extensive development before participating in fertilization, are particularly sensitive to the lack of one chromosome and are usually inviable.

## 8.3 Trisomy Involves the Addition of a Chromosome to a Diploid Genome

In general, the effects of **trisomy** ($2n + 1$) parallel those of monosomy. However, the addition of an extra chromosome produces somewhat more viable individuals in both animal and plant species than does the loss of a chromosome. In animals, this is often true, provided that the chromosome involved is relatively small.

As in monosomy, sex-chromosome variation of the trisomic type has a less dramatic effect on the phenotype than does autosomal variation. Recall from our previous discussion that *Drosophila* females with three X chromosomes and a normal complement of two sets of autosomes (3X:2A) survive and reproduce, but are less viable than normal 2X:2A females. In humans, the addition of an extra X or Y chromosome to an otherwise normal male or female chromosome constitution (47,XXY, 47,XYY, and 47,XXX) leads to viable individuals exhibiting various syndromes. However, the addition of a large autosome to the diploid complement in both *Drosophila* and humans has severe effects and is usually lethal during development.

In plants, trisomic individuals are usually viable, but their phenotype may be altered. A classic example involves the Jimson weed *Datura*, a plant long known for its narcotic effect, whose diploid number is 24. Twelve primary trisomic conditions are possible, and examples of each one have been recovered. Each trisomy alters the phenotype of the plant's capsule sufficiently to produce a unique phenotype (Figure 8–2). These capsule phenotypes were first thought to be caused by mutations in one or more genes.

Still another example is seen in the rice plant (*Oryza sativa*), which has a haploid number of 12. Trisomic strains for each chromosome have been isolated and studied. The plants of 11 of the strains can be distinguished from one another and from wild type. Trisomics for the longer chromosomes are the most distinctive and grow more slowly than the rest. This is in keeping with the belief that gains or losses of larger chromosomes cause greater genetic imbalance than gains or losses of smaller ones. Leaf structure, foliage, stems, grain morphology, and plant height also vary among the different trisomies.

In plants and animals, trisomy may be detected during cytological observations of meiotic divisions. Since three copies of one of the chromosomes are present, pairing configurations are usually irregular. Only two of the three homologs can synapse at any given region along the chromosome length, but different pairs within the trio may be synapsed at different places. When three copies of a chromosome are synapsed, the configuration is called a **trivalent**, and it may be arranged on the spindle so that during anaphase, one member moves to one pole and two go to the opposite pole (Figure 8–3). When one **bivalent** and one **univalent** (an unpaired chromosome) are present instead of a trivalent prior to the first meiotic division, meiosis produces gametes with a chromosome composition of $(n + 1)$, which can perpetuate the trisomic condition.

## Down Syndrome

**Down syndrome**, the only human autosomal trisomy in which a significant number of individuals survive longer than a year past birth, was first reported in 1865 by John Langdon Down. A contemporary of Charles Darwin, Down had a grandson with this condition. It was not until the late 1950s that it was shown to result from an extra copy

FIGURE 8–2 Drawings of capsule phenotypes of the fruits of the Jimson weed *Datura stramonium*. In comparison with wild type, each phenotype is the result of trisomy of 1 of the 12 chromosomes characteristic of the haploid genome. The photograph illustrates the wild type fruit.

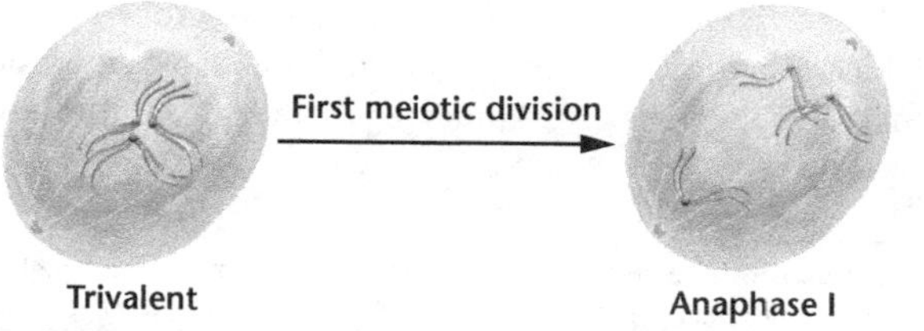


FIGURE 8–3 Diagrammatic representation of one possible pairing arrangement during meiosis I of three copies of a single chromosome, forming a trivalent configuration. During anaphase I, two chromosomes move toward one pole, and one chromosome moves toward the other pole.

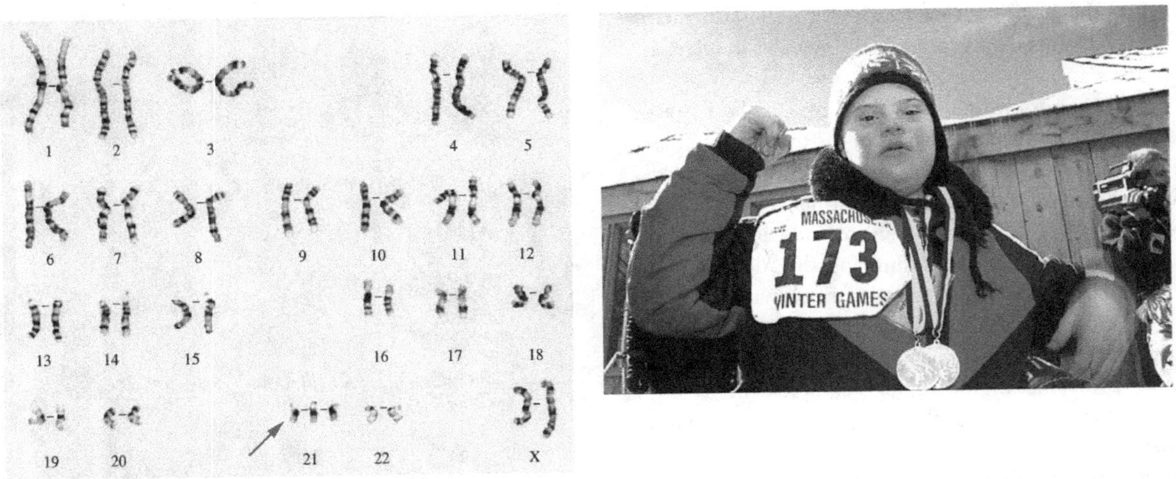


FIGURE 8–4 The karyotype and a photograph of a child with Down syndrome. In the karyotype, three members of the G group chromosome 21 are present, creating the 47,21+ condition.

of chromosome 21, one of the G group* (Figure 8–4). Also sometimes referred to as **trisomy 21** (designated **47,21 +**), this syndrome is found in approximately one infant in every 800 live births. While this might seem to be a rare, improbable event, there are approximately 5500 such births annually in the United States, and there are currently 350,000 individuals with Down syndrome.

Typical of other conditions classified as syndromes, many phenotypic characteristics *may* be present in trisomy 21, but any single affected individual usually exhibits only a subset of these. In the case of Down syndrome, there are 12 to 14 such characteristics, with each individual, on average, expressing 6 to 8 of them. Nevertheless, the outward appearance of these individuals is very similar, and they bear a striking resemblance to one another. This is, for the most part, due to a prominent epicanthic fold in each eye** and the typically flat face and round head. People with Down syndrome are also characteristically short and may have a protruding, furrowed tongue (which causes the mouth to remain partially open) and short, broad hands with characteristic palm and fingerprint patterns. Physical, psychomotor, and mental development are retarded, and poor muscle tone is characteristic. While life expectancy is shortened to an average of about 50 years, individuals are known to survive into their sixties.

Children afflicted with Down syndrome are prone to respiratory disease and heart malformations, and they show an incidence of leukemia approximately 20 times higher than that of the normal population. However, careful medical scrutiny and treatment throughout their lives can extend their survival significantly. A striking observation is that death in older Down syndrome adults is frequently due to Alzheimer's disease. The onset of this disease occurs at a much earlier age than in the normal population.

Because Down syndrome is common in our population, a comprehensive understanding of the underlying genetic basis has long been a research goal. Investigations have given rise to the idea that a critical region of chromosome 21 contains the genes that are dosage sensitive in this trisomy and responsible for the many phenotypes associated with the syndrome. This hypothetical portion of the chromosome has been called the **Down syndrome critical region (DSCR)**. A mouse model was created in 2004 that is trisomic for the DSCR, but some mice did not exhibit the characteristics of the syndrome. Nevertheless, this remains an important investigative approach.

Most frequently, this trisomic condition occurs through nondisjunction of chromosome 21 during meiosis. Failure of paired homologs to disjoin during either anaphase I or II may lead to gametes with the $n + 1$ chromosome composition. About 75 percent of these errors leading to Down syndrome are attributed to nondisjunction during meiosis I. Subsequent fertilization with a normal gamete creates the trisomic condition.

Chromosome analysis has shown that, while the additional chromosome may be derived from either the mother or father, the ovum is the source in about 95 percent of 47,21+ trisomy cases. Before the development of techniques using polymorphic markers to distinguish paternal from maternal homologs, this conclusion was supported by the more indirect evidence derived from studies of the age of mothers giving birth to infants afflicted with Down syndrome. Figure 8–5 shows the relationship between the incidence of Down syndrome births and maternal age, illustrating the dramatic increase as the age of the mother increases. While the frequency is about 1 in 1000 at maternal age 30, a tenfold increase to a frequency of 1 in 100 is noted at age 40. The frequency increases still further to about 1 in

*On the basis of size and centromere placement, human autosomal chromosomes are divided into seven groups: A (1–3), B (4–5), C (6–12), D (13–15), E (16–18), F (19–20), and G (21–22).

**The epicanthic fold, or epicanthus, is a skin fold of the upper eyelid, extending from the nose to the inner side of the eyebrow. It covers and appears to lower the inner corner of the eye, giving the eye a slanted, or almond-shaped, appearance. The epicanthus is a prominent normal component of the eyes in many Asian groups.

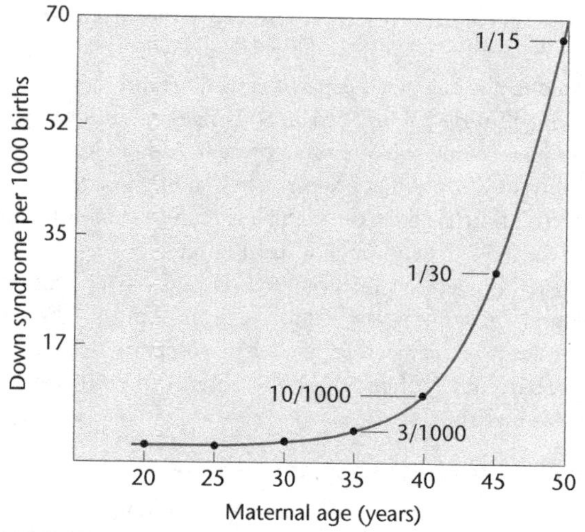


FIGURE 8–5 Incidence of Down syndrome births related to maternal age.

30 at age 45. A very alarming statistic is that as the age of childbearing women exceeds 45, the probability of a Down syndrome birth continues to increase substantially. In spite of this high probability, more than half of Down syndrome births occur to women younger than 35 years, because the overwhelming proportion of pregnancies in the general population involve women under 35.

Although the nondisjunctional event that produces Down syndrome seems more likely to occur during oogenesis in women over the age of 35, we do not know with certainty why this is so. However, one observation may be relevant. Meiosis is initiated in all the eggs of a human female when she is still a fetus, until the point where the homologs synapse and recombination has begun. Then oocyte development is arrested in meiosis I. Thus, all primary oocytes have been formed by birth. When ovulation begins, at puberty, meiosis is reinitiated in one egg during each ovulatory cycle and continues into meiosis II. The process is once again arrested after ovulation and is not completed unless fertilization occurs.

The end result of this progression is that each ovum that is released has been arrested in meiosis I for about a month longer than the one released during the preceding cycle. As a consequence, women 30 or 40 years old produce ova that are significantly older and that have been arrested longer than those they ovulated 10 or 20 years previously. However, no direct evidence proves that ovum age is the cause of the increased incidence of nondisjunction leading to Down syndrome.

These statistics obviously pose a serious problem for the woman who becomes pregnant late in her reproductive years. Genetic counseling early in such pregnancies is highly recommended. Counseling informs prospective parents about the probability that their child will be affected and educates them about Down syndrome. Although some individuals with Down syndrome must be institutionalized, others benefit greatly from special education programs and may be cared for at home. (Down syndrome children in general are noted for their affectionate, loving nature.) A genetic counselor may also recommend a prenatal diagnostic technique in which fetal cells are isolated and cultured. In **amniocentesis** and **chorionic villus sampling (CVS)**, the two most familiar approaches, fetal cells are obtained from the amniotic fluid or the chorion of the placenta, respectively. In a newer approach, fetal cells are derived directly from the maternal circulation. Once perfected, this approach will be preferable, because it is noninvasive, posing no risk to the fetus. After fetal cells are obtained, the karyotype can be determined by cytogenetic analysis. If the fetus is diagnosed as having Down syndrome, a therapeutic abortion is one option currently available to parents. Obviously, this is a difficult decision involving a number of religious and ethical issues.

Since Down syndrome is caused by a random error—nondisjunction of chromosome 21 during maternal or paternal meiosis—the occurrence of the disorder is *not* expected to be inherited. Nevertheless, Down syndrome occasionally runs in families. These instances, referred to as *familial Down syndrome*, involve a translocation of chromosome 21, another type of chromosomal aberration, which we will discuss later in the chapter.

## Patau Syndrome

In 1960, Klaus Patau and his associates observed an infant with severe developmental malformations and a karyotype of 47 chromosomes (Figure 8–6). The additional chromosome was medium-sized,

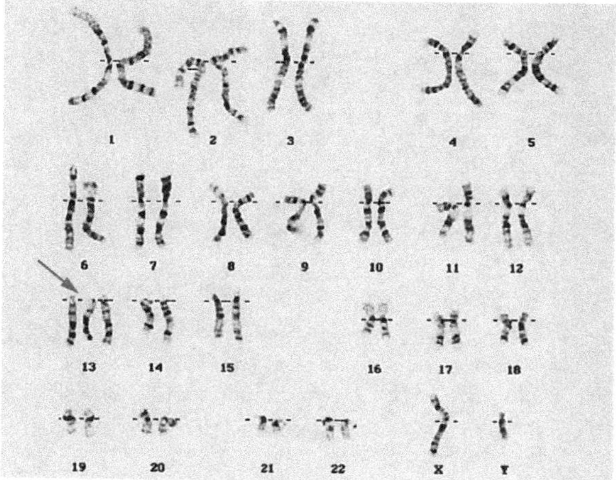


| | |
|---|---|
| Mental retardation | Microcephaly |
| Growth failure | Cleft lip and palate |
| Low-set, deformed ears | Polydactyly |
| Deafness | Deformed finger nails |
| Atrial septal defect | Kidney cysts |
| Ventricular septal defect | Double ureter |
| Abnormal polymorphonuclear granulocytes | Umbilical hernia |
| | Developmental uterine abnormalities |
| | Cryptorchidism |

FIGURE 8–6 The karyotype and potential phenotypic description of an infant with Patau syndrome, where three members of the D group chromosome 13 are present, creating the 47,13+ condition.

one of the acrocentric D group. It is now designated as chromosome 13. This **trisomy 13** condition has since been described in many newborns and is called **Patau syndrome (47,13 + )**. Affected infants are not mentally alert, are thought to be deaf, and characteristically have a harelip, cleft palate, and polydactyly. Autopsies have revealed congenital malformation of most organ systems, a condition indicative of abnormal developmental events occurring as early as five to six weeks of gestation. The average survival of these infants is about three months.

The average maternal and paternal ages of parents of Patau infants are higher than the ages of parents of normal children, but they are not as high as the average maternal age in cases of Down syndrome. Both male and female parents average about 32 years of age when the affected child is born. Because the condition is so rare, occurring as infrequently as 1 in 19,000 live births, it is not known whether the origin of the extra chromosome is more often maternal, more often paternal, or arises equally from either parent.

## Edwards Syndrome

In 1960, John H. Edwards and his colleagues reported on an infant trisomic for a chromosome in the E group, now known to be chromosome 18 (Figure 8–7). Referred to as **trisomy 18 (47,18+)**, this aberration is also named **Edwards syndrome** after its discoverer. The phenotype of this child, like that of individuals with Down and Patau syndromes, illustrates that the presence of an extra autosome produces congenital malformations and reduced life expectancy. These infants are smaller than the average newborn. Their skulls are elongated in the anterior-posterior direction, and their ears are set low and malformed. A webbed neck, congenital dislocation of the hips, and a receding chin are often present. Although the frequency of trisomy 18 is somewhat greater than that of trisomy 13, the average survival time is about the same, less than four months. Death is usually caused by pneumonia or heart failure.

Again, the average maternal age is high—34.7 years by one calculation. In contrast to Patau syndrome, most Edwards syndrome infants are females. In one set of observations based on 143 cases, 80 percent were female. Overall, about 1 in 8000 live births exhibits this malady.

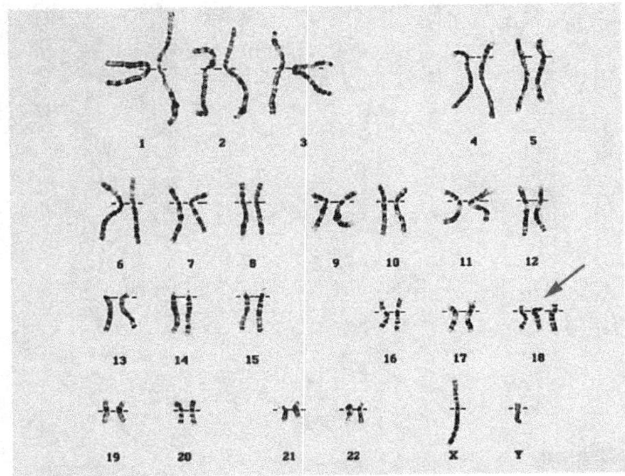


Growth failure
Mental retardation
Open skull sutures at birth
High, arched eyebrows
Low-set, deformed ears
Short sternum
Ventricular septal defect
Flexion deformities of fingers

Abnormal kidneys
Persistent ductus arteriosus
Deformity of hips
Prominent external genitalia
Muscular hypertonus
Prominent heel
Dorsal flexion of big toes

FIGURE 8–7 The karyotype and potential phenotypic description of an infant with Edwards syndrome. Three members of the E group chromosome 18 are present, creating the 47,18+ condition.

## Viability in Human Aneuploidy

The reduced viability of individuals with recognized monosomic and trisomic conditions suggests that many other aneuploid conditions may arise, but the affected fetuses do not survive to term. This hypothesis has been confirmed by karyotypic analysis of spontaneously aborted fetuses. In an extensive review of this subject in 1971 by David H. Carr, it was shown that a significant percentage of abortuses are trisomic for one or another of the autosomal chromosomes. Trisomies for every human chromosome were recovered. Monosomies, however, were seldom found in the Carr study, even though nondisjunction should produce $n - 1$ gametes with a frequency equal to $n + 1$ gametes. This finding leads us to believe that gametes lacking a single chromosome are either so functionally impaired that they never participate in fertilization or that the monosomic embryo dies so early in its development that recovery occurs infrequently. Various forms of polyploidy and other miscellaneous chromosomal anomalies were also found in Carr's study.

Carr's study and similar investigations have produced some striking statistics. We now estimate that about 25 percent of all human conceptions are terminated by spontaneous abortion and that about 50 percent of all spontaneously aborted fetuses demonstrate some form of chromosomal anomaly. A calculation using these figures ($0.25 \times 0.50 = 0.125$) predicts that up to 12.5 percent of all pregnancies originate with an abnormal number of chromosomes. Approximately 90 percent of chromosomal anomalies are spontaneously terminated prior to birth. More recently derived figures are even more dramatic. It is now estimated that as many as 10 to 30 percent of all fertilized eggs in humans contain some error in chromosome number.

The largest percentage of chromosomal abnormalities are aneuploidies. Surprisingly, an aneuploidy with one of the highest incidence rates among abortuses is the 45,X condition, which produces an infant with Turner syndrome if the fetus survives to term. About 70 to 80 percent of aborted and live-born 45,X conditions exhibit the maternal X chromosome. Thus, the meiotic error leading to this syndrome occurs during spermatogenesis.

Collectively, these observations support the hypothesis that normal embryonic development requires a precise diploid complement of chromosomes to maintain the delicate equilibrium required for the correct expression of genetic information. The prenatal mortality of most aneuploids provides a barrier against the introduction of these anomalies into the human population.

8.4

## Polyploidy, in Which More Than Two Haploid Sets of Chromosomes Are Present, Is Prevalent in Plants

The term *polyploidy* describes instances in which more than two haploid chromosome sets are found. The naming of polyploids is based on the number of sets of chromosomes in the nucleus: A **triploid** has $3n$ chromosomes; a **tetraploid** has $4n$; a **pentaploid**, $5n$; and so forth. Several general statements may be made about polyploidy. This condition is relatively infrequent in many animal species but is well known in lizards, amphibians, and fish. It is much more common in plant species. Usually, odd numbers of chromosome sets are not maintained reliably from generation to generation, because a polyploid organism with an uneven number of homologs does not produce genetically balanced gametes. For this reason, triploids, pentaploids, and so on are not usually found in species that depend solely on sexual reproduction for propagation.

Polyploidy can originate in two ways: (1) The addition of one or more extra sets of chromosomes, identical to the normal haploid complement of the same species, results in **autopolyploidy**; and (2) the combination of chromosome sets from different species may occur as a consequence of hybridization, resulting in **allopolyploidy** (from the Greek word *allo*, meaning other or different). The distinction between auto- and allopolyploidy lies in the genetic origin of the extra chromosome sets, as illustrated in Figure 8–8.

In our discussion of polyploidy, we use the following symbol convention to clarify the origin of additional chromosome sets. If $A$ represents the haploid set of chromosomes of any organism, then

$$A = a_1 + a_2 + a_3 + a_4 + \cdots + a_n$$

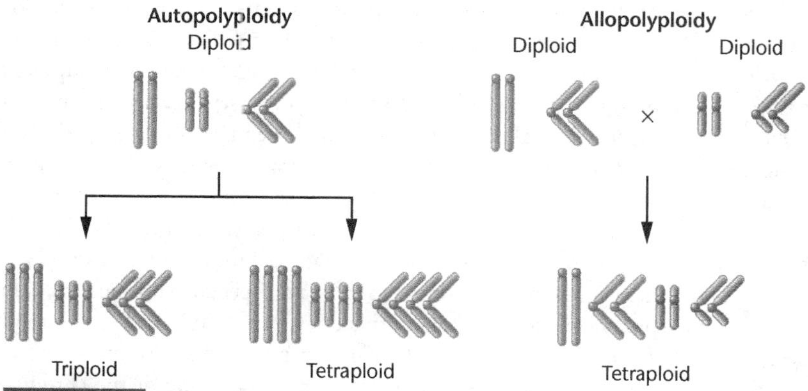


FIGURE 8–8 Contrasting chromosome origins of an autopolyploid with an allopolyploid karyotype.

where $a_1$, $a_2$, and so on, are individual chromosomes, and where $n$ is the haploid number. A normal diploid organism is represented simply as $AA$.

### Autopolyploidy

In autopolyploidy, each additional set of chromosomes is one of the kinds produced by the parent species. Therefore, triploids are represented as $AAA$, tetraploids are $AAAA$, and so forth.

**Autotriploids** arise in several ways. A failure of all chromosomes to segregate during meiotic divisions (first-division or second-division nondisjunction) can produce a diploid gamete. If such a gamete is fertilized by a haploid gamete, a zygote with three sets of chromosomes is produced. Or, rarely, two sperm may fertilize an ovum, resulting in a triploid zygote. Triploids are produced under experimental conditions by crossing diploids with tetraploids. Diploid organisms produce gametes with $n$ chromosomes, whereas tetraploids produce $2n$ gametes. Upon fertilization, the desired triploid is produced.

Because they have an even number of chromosomes, **autotetraploids** ($4n$) are theoretically more likely to be found in nature than are autotriploids. Unlike triploids, which often produce genetically unbalanced gametes with various numbers of chromosomes, tetraploids are more likely to produce balanced gametes for participation in sexual reproduction.

How polyploidy arises in nature is of great interest. In theory, if chromosomes have replicated but the parent cell never divides before reentering interphase, the chromosome number will be doubled. That this occurs is supported by the observation that tetraploid cells can be produced experimentally from diploid cells by applying cold or heat shock to meiotic cells, or by applying colchicine to somatic cells undergoing mitosis. **Colchicine**, an alkaloid derived from the autumn crocus, inhibits micotubule polymerization during spindle formation, preventing replicated chromosomes from migrating to the poles. When colchicine is removed, the cell can reenter interphase. When the paired sister chromatids separate and uncoil, the nucleus will contain twice the diploid number of chromosomes and is therefore $4n$. This process is illustrated in Figure 8–9.

In general, autopolyploid organisms are larger than their diploid relatives. Such an increase seems to be due to a larger cell size rather than a greater number of cells. Although autopolyploids do not contain new or unique information compared with their diploid relatives, the flower and fruit of autopolyploid plants are often increased in size, making such varieties of greater horticultural or commercial value. Economically important triploid plants include several potato species of the genus *Solanum*, Winesap apples, commercial bananas, seedless watermelons, and the cultivated tiger lily *Lilium tigrinum*. These plants are propagated asexually. Diploid bananas contain hard seeds, but the commercial, triploid, "seedless" variety has edible seeds. Tetraploid alfalfa, coffee, peanuts, and McIntosh apples are also of economic value, because they are either larger or grow more vigorously

WEB TUTORIAL 8.2 POLYPLOIDY

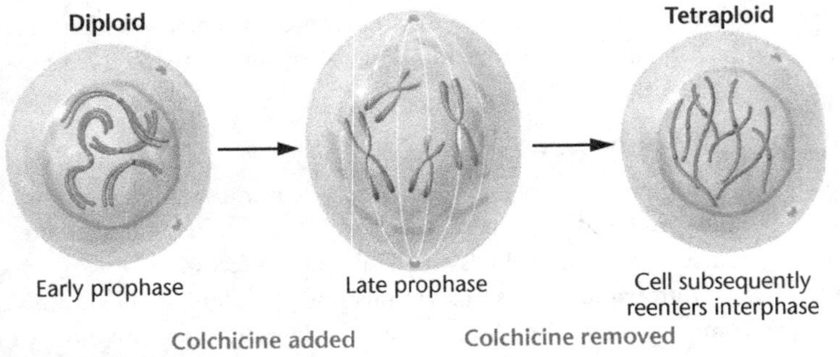


FIGURE 8–9 The presumed involvement of colchicine in doubling the chromosome number. Two pairs of homologous chromosomes are shown. While each chromosome had replicated its DNA earlier during interphase, the chromosomes do not appear as double structures until late prophase. When anaphase fails to occur normally (owing to the presence of colchicine), the chromosome number doubles if the cell reenters interphase.

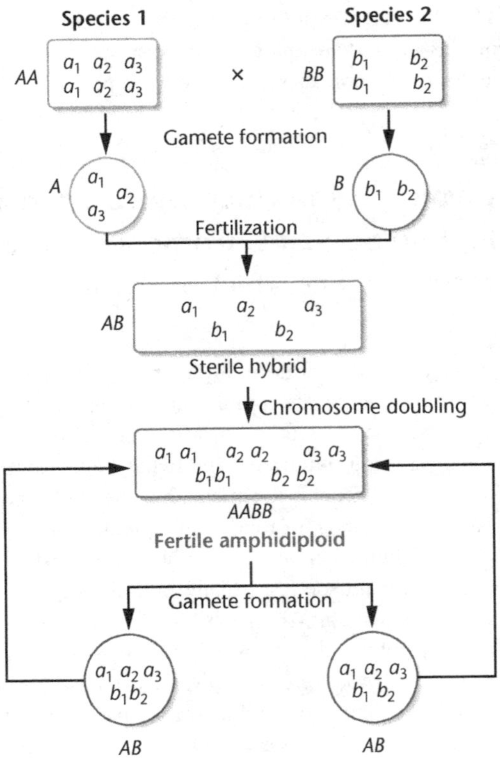


FIGURE 8–10 The origin and propagation of an amphidiploid. Species 1 contains genome *A* consisting of three distinct chromosomes, $a_1$, $a_2$, and $a_3$. Species 2 contains genome *B* consisting of two distinct chromosomes, $b_1$ and $b_2$. Following fertilization between members of the two species and chromosome doubling, a fertile amphidiploid containing two complete diploid genomes (*AABB*) is formed.

than their diploid or triploid counterparts. The commercial strawberry is an octoploid.

We have long been curious about how cells with increased ploidy values express different phenotypes than their diploid counterparts even though they have no new genes. Our improved ability to examine gene expression using modern biotechnology has provided some interesting insights. For example, Gerald Fink and his colleagues were able to create strains of the yeast *Saccharomyces cerevisiae* with one, two, three, or four copies of the genome. Thus, each strain contains identical genes (they are said to be *isogenic*) but different ploidy values. The researchers then proceeded to examine expression levels of all genes at each stage in the cell cycle of the organism. Using the rather stringent standards of a tenfold increase or decrease of gene expression, Fink and colleagues identified ten cases where, as ploidy increased, gene expression was increased at least tenfold, and seven cases where it was reduced by a similar level.

One of these genes provides insights into how polyploid cells become larger than their haploid or diploid counterparts. In polyploid yeast, two **G1 cyclins**, Cln1 and Pcl1, are repressed as ploidy increases, while the size of the yeast cells increases. This is explained by the observation that G1 cyclins facilitate the cell's movement through G1, which is delayed when expression of their genes is repressed. The polyploid cell stays in G1 longer and, on average, grows to a larger size before it moves beyond the G1 stage of the cell cycle. Yeast cells also show different morphology as ploidy increases. Several other genes, repressed as ploidy increases, have been linked to cytoskeletal dynamics that account for the morphological changes. Note that we first introduced cyclins in Chapter 2 during our discussion of cell-cycle regulation and that more expanded coverage is upcoming in Chapter 20 in our discussion of cancer genetics.

## Allopolyploidy

Polyploidy can also result from hybridizing two closely related species. If a haploid ovum from a species with chromosome sets *AA* is fertilized by a haploid sperm from a species with sets *BB*, the resulting hybrid is *AB*, where $A = a_1, a_2, a_3, \ldots a_n$ and $B = b_1, b_2, b_3, \ldots b_n$. The hybrid plant may be sterile because of its inability to produce viable gametes. Most often, this occurs when some or all of the *a* and *b* chromosomes are not homologous and therefore cannot synapse in meiosis. As a result, unbalanced genetic conditions result. If, however, the new *AB* genetic combination undergoes a natural or induced chromosomal doubling, two copies of all *a* chromosomes and two copies of all *b* chromosomes will be present, and they will pair during meiosis. As a result, a fertile *AABB* tetraploid is produced. These events are illustrated in Figure 8–10. Since this polyploid contains the equivalent of four haploid genomes derived from two separate species, such an organism is called an **allotetraploid.** When both original species are known, an equivalent term, **amphidiploid,** is preferred for describing the allotetraploid.

Amphidiploid plants are often found in nature. Their reproductive success is based on their potential for forming balanced gametes. Since two homologs of each specific chromosome are present, meiosis occurs normally (Figure 8–10), and fertilization successfully propagates the plant sexually. This description, however, assumes the simplest situation, where none of the chromosomes in set *A* are homologous to any in set *B*. In amphidiploids formed from closely related species, some homology between *a* and

FIGURE 8–11 The pods of the amphidiploid form of *Gossypium*, the cultivated cotton plant.

*b* chromosomes is likely. In that situation, meiotic pairing is more complex. During synapsis, multivalents will be formed, resulting in the production of unbalanced gametes. In such cases, aneuploid varieties of amphidiploids may arise. Allopolyploids are rare among animals because mating behavior is most often species specific; thus, the initial step in hybridization is unlikely to occur.

A classic example of amphidiploidy in plants is the cultivated species of American cotton, *Gossypium* (Figure 8–11). This species has 26 pairs of chromosomes: 13 are large and 13 are much smaller. When it was discovered that Old World cotton had only 13 pairs of large chromosomes, allopolyploidy was suspected. After an examination of wild American cotton revealed 13 pairs of small chromosomes, this speculation was strengthened. J. O. Beasley reconstructed the origin of cultivated cotton experimentally by crossing the Old World strain with the wild American strain, then treating the hybrid with colchicine to double the chromosome number. The result of these treatments was a fertile amphidiploid variety of cotton. It contained 26 pairs of chromosomes and characteristics similar to the cultivated variety.

Amphidiploids often exhibit traits of both parental species. An interesting example, but one with no practical economic importance, is that of the hybrid formed between the radish *Raphanus sativus* and the cabbage *Brassica oleracea*. Both species have a haploid number $n = 9$. The initial hybrid consists of 9 *Raphanus* and 9 *Brassica* chromosomes (9R + 9B). Although hybrids are almost always sterile, some fertile amphidiploids (18R + 18B) have been produced. Unfortunately, the root of this plant is more like the cabbage and its shoot is more like the radish. Had the converse occurred, the hybrid might have become a useful crop of economic importance.

A much more successful commercial hybridization involves the grasses wheat and rye. Wheat (genus *Triticum*) has a basic haploid genome of 7 chromosomes. In addition to normal diploids ($2n = 14$), cultivated allopolyploids exist, including tetraploid ($4n = 28$) and hexaploid ($6n = 42$) species. Rye (genus *Secale*) also has a genome consisting of 7 chromosomes. The only cultivated species is the diploid plant ($2n = 14$).

Using the technique outlined in Figure 8–10, geneticists have produced various hybrids. When tetraploid wheat is crossed with diploid rye and the $F_1$ is treated with colchicine, a hexaploid variety ($6n = 42$) is derived. This hybrid, designated *Triticale*, represents a new genus. Fertile hybrid varieties derived from various wheat and rye species can be crossed together or backcrossed. Such crosses have created many variations of the genus *Triticale*. For example, a useful octaploid (an allooctaploid) variety with 56 chromosomes has been produced from a diploid rye and a hexaploid wheat plant.

The hybrid plants demonstrate characteristics of both wheat and rye. For example, certain hybrids combine the high protein content of wheat with the high content of the amino acid lysine in rye. (The lysine content is low in wheat and thus is a limiting nutritional factor.) Wheat is considered a high-yielding grain, whereas rye is noted for its versatility of growth in unfavorable environments. *Triticale* species, combining both traits, have the potential of significantly increasing grain production. Programs designed to improve crops through hybridization have long been underway in several developing countries, as discussed in the "Genetics, Technology, and Society" essay in Chapter 25.

Recall that in Chapter 5, we discussed the use of **somatic cell hybridization** to map human genes. This technique has also been applied to the production of amphidiploid plants (Figure 8–12). Cells from the developing leaves of plants can be treated to remove their cell wall. These altered cells, called **protoplasts,** can be maintained in culture and stimulated to fuse with other protoplasts, producing somatic cell hybrids. If cells from different plant species are fused in this way, hybrid amphidiploid cells can be produced. Because protoplasts can be induced to divide and differentiate into stems that develop leaves, the potential for producing allopolyploids is available in the research laboratory. In some cases, entire plants can be derived from cultured protoplasts. If only stems and leaves are produced, these can be grafted onto the stem of another plant. If flowers are formed, fertilization may yield mature seeds, which, upon germination, yield an allopolyploid plant.

Many examples of allopolyploids have been created commercially by using the aforementioned approach. Although most of them are not true amphidiploids (one or more chromosomes are missing), full allotetraploids are sometimes produced.

NOW SOLVE THIS

In Problem 5 on page 224 you are asked to consider a hybrid plant derived from two different species that is more ornate than either of its parents but is sterile.

■ HINT: *Allopolyploid plants are often sterile when they contain an odd number of each chromosome, resulting in unbalanced gametes during meiosis.*

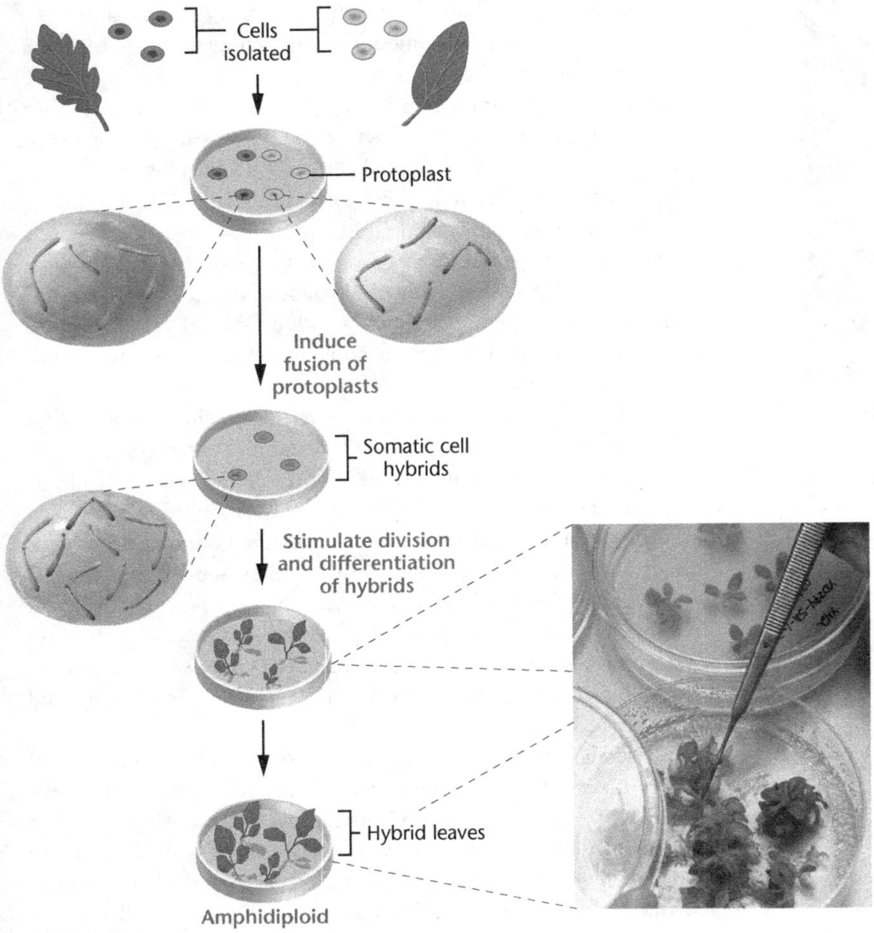


FIGURE 8–12 Application of the somatic cell hybridization technique in the production of an amphidiploid. Cells from the leaves of two species of plants are removed and cultured. The cell walls are digested away and the resultant protoplasts are induced to undergo cell fusion. The hybrid cell is selected and stimulated to divide and differentiate, as illustrated in the photograph. An amphidiploid has a complete set of chromosomes from each parental cell type and displays phenotypic characteristics of each. Two pairs of chromosomes from each species are depicted.

## Endopolyploidy

**Endopolyploidy** is the condition in which only certain cells in an otherwise diploid organism are polyploid. In such cells, replication and separation of chromosomes occur without nuclear division. Numerous examples of naturally occurring endopolyploidy have been observed. For example, vertebrate liver cell nuclei, including human ones, often contain $4n$, $8n$, or $16n$ chromosome sets. The stem and parenchymal tissue of apical regions of flowering plants are also often endopolyploid. Cells lining the gut of mosquito larvae attain a $16n$ ploidy, but during the pupal stages, such cells undergo very quick reduction divisions, giving rise to smaller diploid cells. In the water strider *Gerris*, wide variations in chromosome numbers are found in different tissues, with as many as 1024 to 2048 copies of each chromosome in the salivary gland cells. Since the diploid number in this organism is 22, the nuclei of these cells may contain more than 40,000 chromosomes.

Although the role of endopolyploidy is not clear, the proliferation of chromosome copies often occurs in cells where high levels of certain gene products are required. In fact, it is well established that certain genes whose product is in high demand in *every* cell exist naturally in multiple copies in the genome. Ribosomal and transfer RNA genes are examples of multiple-copy genes. In certain cells of organisms, where even this condition may not allow for a sufficient amount of a particular gene product, it may be necessary to replicate the entire genome, allowing an even greater rate of expression of that gene.

## 8.5 Variation Occurs in the Internal Composition and Arrangement of Chromosomes

The second general class of chromosome aberrations includes structural changes that delete, add, or rearrange substantial portions of one or more chromosomes. Included in this broad category are deletions and duplications of genes or part of a chromosome and rearrangements of genetic material in which a chromosome segment is inverted, exchanged with a segment of a nonhomologous chromosome, or merely transferred to another chromosome. These exchanges and transfers, in which the location of a gene is altered within the genome, are called *translocations*. These types of chromosome alterations are illustrated in Figure 8–13.

In most instances, these structural changes are due to one or more breaks along the axis of a chromosome, followed by either the loss or rearrangement of genetic material. Chromosomes can break spontaneously, but the rate of breakage may increase in cells exposed to chemicals or radiation. Although the actual ends of chromosomes, known as *telomeres*, do not readily fuse with newly created ends of "broken" chromosomes or with other telomeres, the ends produced at points of breakage are "sticky" and can rejoin other broken ends. If breakage and rejoining do not reestablish the original relationship, and if the alteration occurs in germ plasm, the structural rearrangement will persist in the gametes and be heritable.

If the aberration is found in one homolog but not the other, the individual is said to be *heterozygous for the aberration*. In such cases, as described in the following sections, unusual but characteristic pairing configurations are formed during meiotic synapsis. These patterns are useful in identifying the type of change that has

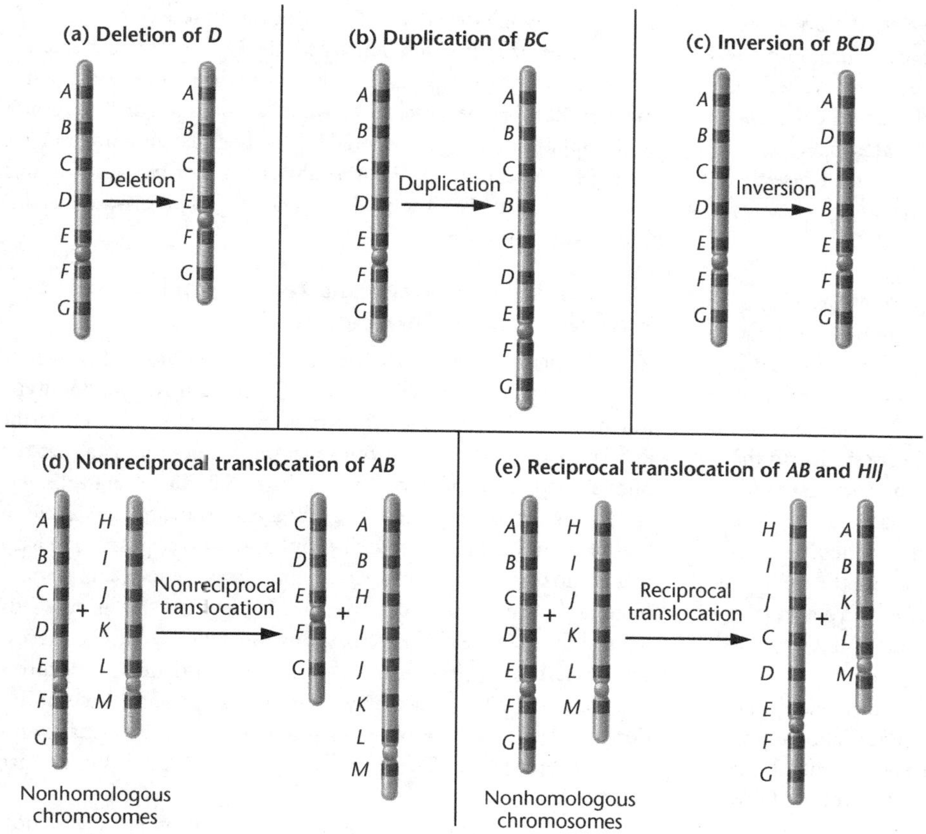


**FIGURE 8–13** An overview of the five different types of gain, loss, or rearrangement of chromosome segments.

WEB TUTORIAL 8.3
ABERRATIONS

occurred. If there has been no loss or gain of genetic material, individuals bearing the aberration "heterozygously" are likely to be unaffected phenotypically. However, the unusual pairing arrangements often lead to gametes that are duplicated or deficient for some chromosomal regions. When this occurs, the offspring of "carriers" of certain aberrations have an increased probability of demonstrating phenotypic manifestations.

**FIGURE 8–14** Origins of (a) a terminal and (b) an intercalary deletion. Part (c) shows that pairing can occur between a normal chromosome and one with an intercalary deletion if the undeleted portion buckles out to form a deletion (or a compensation) loop.

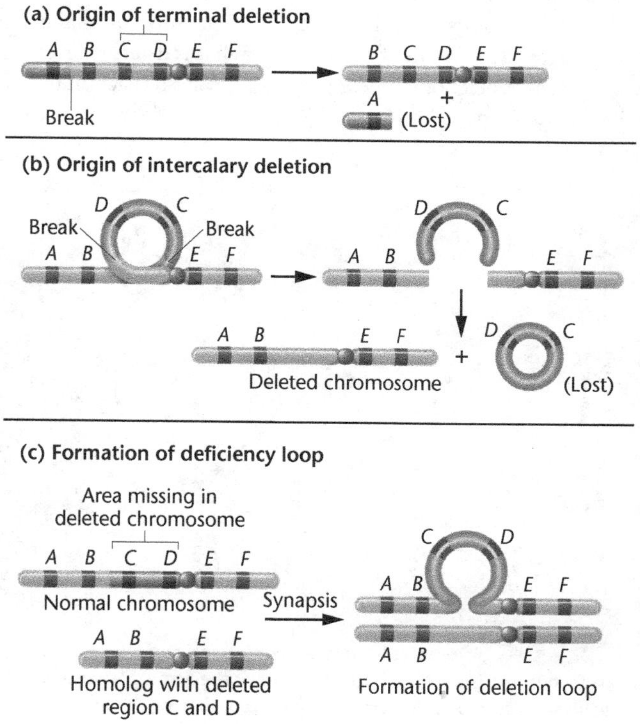


8.6

## A Deletion Is a Missing Region of a Chromosome

When a chromosome breaks in one or more places and a portion of it is lost, the missing piece is referred to as a **deletion** (or a **deficiency**). The deletion can occur either near one end or from the interior of the chromosome. These are called **terminal** and **intercalary deletions**, respectively [Figure 8–14(a) and (b)]. The portion of the chromosome that retains the centromere region is usually maintained when the cell divides, whereas the segment without the centromere is eventually lost in progeny cells following mitosis or meiosis. For synapsis to occur between a chromosome with a large intercalary deficiency and a normal complete homolog, the unpaired region of the normal homolog must "buckle out" into a **deletion** or **compensation loop** [Figure 8–14(c)].

If only a small part of a chromosome is deleted, the organism might survive. However, a deletion of a portion of a chromosome need not be very great before the effects become severe. We see an example of this in the following discussion of the cri du chat syndrome in humans. If even more genetic information is lost as a result of a deletion, the aberration is often lethal, in which case the chromosome mutation never becomes available for study.

## Cri du Chat Syndrome in Humans

In humans, the **cri du chat syndrome** results from the deletion of a small terminal portion of chromosome 5. It might be considered a case of *partial monosomy*, but since the region that is missing is so small, it is better referred to as a **segmental deletion**. This syndrome was first reported by Jérôme LeJeune in 1963, when he described the clinical symptoms, including an eerie cry similar to the meowing of a cat, after which the syndrome is named. This syndrome is associated with the loss of a small, variable part of the short arm of chromosome 5 (Figure 8–15). Thus, the genetic constitution may be designated as ***46,5p−***, meaning that the individual has all 46 chromosomes but that some or all of the p arm (the petite, or short, arm) of one member of the chromosome 5 pair is missing.

Infants with this syndrome may exhibit anatomic malformations, including gastrointestinal and cardiac complications, and they are often mentally retarded. Abnormal development of the glottis and larynx (leading to the characteristic cry) is typical of this syndrome.

Since 1963, hundreds of cases of cri-du-chat syndrome have been reported worldwide. An incidence of 1 in 25,000–50,000 live births has been estimated. Most often, the condition is not inherited but instead results from the sporadic loss of chromosomal material in gametes. The length of the short arm that is deleted varies somewhat; longer deletions appear to have a greater impact on the physical, psychomotor, and mental skill levels of those children who survive. Although the effects of the syndrome are severe, most individuals achieve motor and language skills and may be home-cared. In 2004, it was reported that the portion of the chromosome that is missing contains the *TERT* gene, which encodes telomerase reverse transcriptase, an enzyme essential for the maintenance of telomeres during DNA replication. Whether the absence of this gene on one homolog is related to the multiple phenotypes of cri-du-chat infants is still unknown.

## *Drosophila* Heterozygous for Deficiencies May Exhibit Pseudodominance

A final consequence of deletions can be noted in organisms heterozygous for a deficiency. Consider the mutant *Notch* phenotype in *Drosophila*. In these flies, the wings are notched on the posterior and lateral margins. Data from breeding studies indicate that the phenotype is controlled by an X-linked dominant mutation: heterozygous females have notched wings and transmit this allele to half of their female progeny. The mutation also appears to behave as a homozygous and hemizygous lethal, because such females and males are never recovered. It has also been noted that if notched-winged females are also heterozygous for the closely linked recessive mutations *white*-eye, *facet*-eye, or *split*-bristle, they express these mutant phenotypes as well as *Notch*. Because these mutations are recessive, heterozygotes should express the normal, wild-type phenotypes. These genotypes and phenotypes are summarized in Table 8.2.

These observations have been explained through a cytological examination of the particularly large chromosomes characterizing certain larval cells of many insects. Called **polytene chromosomes** (see Chapter 12), they exhibit banding patterns specific to each region of each chromosome. In such cells of heterozygous *Notch* females, a deficiency loop was found along the X chromosome, from the band designated 3C2 through band 3C11, as shown in Figure 8–16. These bands had previously been shown to include the loci

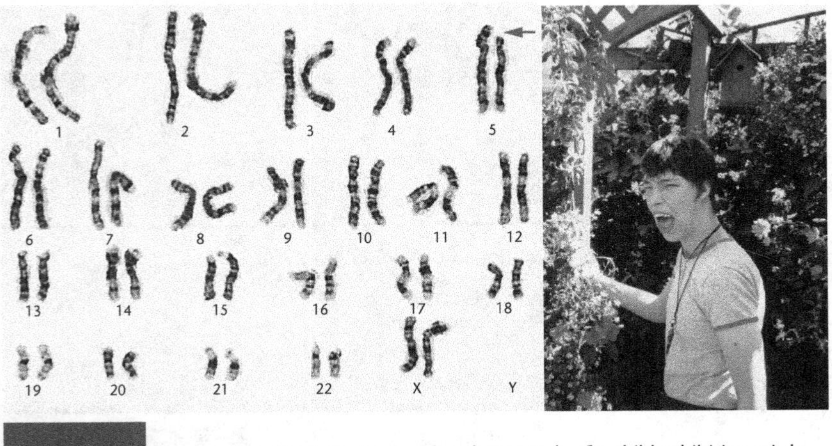


**FIGURE 8–15** A representative karyotype and a photograph of a child exhibiting cri-du-chat syndrome (46,5p−). In the karyotype, the arrow identifies the absence of a small part of the short arm of one member of the chromosome 5 homologs.

**TABLE 8.2**

**Notch Genotypes and Phenotypes**

| Genotype | Phenotype |
|---|---|
| $\dfrac{N^+}{N}$ | *Notch* female |
| $\dfrac{N}{N}$ | Lethal |
| $\dfrac{N}{\rightharpoondown}$ | Lethal |
| $\dfrac{N^+ w}{N(w^+)^*}$ | *Notch*, *white* female |
| $\dfrac{N^+ \; fa \; spl}{N(fa^+ spl^+)^*}$ | *Notch*, *facet*, *split* female |

*(genes deleted)

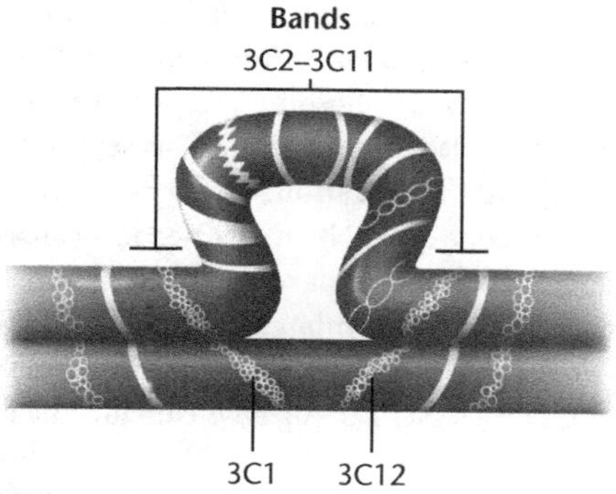


**FIGURE 8–16** Deficiency loop formed in polytene chromosomes from the salivary glands of *Drosophila melanogaster* where the fly is heterozygous for a deletion. The deletion encompasses bands 3C2 through 3C11, corresponding to a portion of the region associated with the Notch phenotype.

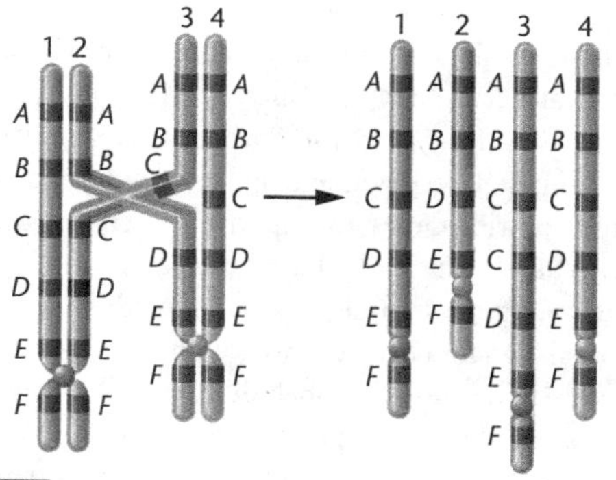

**FIGURE 8–17** The origin of duplicated and deficient regions of chromosomes as a result of unequal crossing over. The tetrad at the left is mispaired during synapsis. A single crossover between chromatids 2 and 3 results in deficient (chromosome 2) and duplicated (chromosome 3) chromosomal regions. The two chromosomes uninvolved in the crossover event remain normal in their gene sequence and content.

for the *white, facet,* and *split* genes, among others. This region's deficiency in one of the two homologous X chromosomes has two distinct effects. First, it results in the *Notch* phenotype. Second, because the deleted segment includes the loci for genes whose mutant alleles are present on the other X chromosome, the deficiency creates a partially hemizygous condition whereby the recessive *white, facet,* or *split* alleles are expressed whenever present. This type of phenotypic expression of recessive genes in association with a deletion is an example of the phenomenon called **pseudodominance.**

Many independently arising *Notch* phenotypes have been investigated. The common deficient band for all *Notch* phenotypes has now been designated as 3C7. In every case in which *white* was also expressed pseudodominantly, the band 3C2 was also missing.

## 8.7 A Duplication Is a Repeated Segment of the Genetic Material

The presence of any part of the genetic material—a single locus or a large piece of a chromosome—more than once in the genome is called a **duplication.** As with deletions, the pairing of an affected and a nonaffected homolog will produce a compensation loop. Duplications can arise as the result of an error during genetic exchanges (referred to as **unequal crossing over**) between synapsed chromosomes during meiosis, or through a replication error prior to meiosis. In the former case, misalignment of the homologs occurs, and following crossing over, both a duplication and a deletion are produced (Figure 8–17).

We consider three interesting aspects of duplications. First, they may result in gene redundancy. Second, as with deletions, duplications may produce phenotypic variation. Third, according to one convincing theory, duplications have also been an important source of genetic variability during evolution.

### Gene Redundancy and Amplification: Ribosomal RNA Genes

Although many gene products are not needed in every cell of an organism, other gene products are known to be essential components of all cells. For example, ribosomal RNA must be present in abundance to support protein synthesis. The more metabolically active a cell is, the higher the demand for this molecule, which becomes a part of each ribosome. We might hypothesize that a single copy of a gene encoding rRNA would be insufficient for many cells. Studies using the technique of molecular hybridization, which allows determination of the percentage of the genome coding for specific RNA sequences, show that our hypothesis is correct and that cells contain multiple copies of the genes coding for rRNA. Collectively such DNA is called **rDNA,** and the general phenomenon is called *gene redundancy.* For example, in the common intestinal bacterium *Escherichia coli* (*E. coli*), about 0.7 percent of the haploid genome consists of rDNA—this is equivalent to 7 copies of the gene. In *Drosophila melanogaster,* 0.3 percent of the haploid genome, equivalent to 130 gene copies, consists of rDNA. Although the presence of multiple copies of the same gene is not restricted to genes coding for rRNA, we focus on those genes in this section.

Studies of *Drosophila* have documented the critical importance of the extensive amounts of rRNA and ribosomes that result from the presence of multiple copies of these genes. In *Drosophila,* the X-linked mutation *bobbed* has, in fact, been shown to be due to a deletion of a *variable number* of the 130 genes coding for rRNA.

When the number of these genes is reduced, and inadequate rRNA is produced, the result is mutant flies that have low viability, are underdeveloped, and have bristles reduced in size. Both general development and bristle formation, which occurs very rapidly during normal pupal development, apparently depend on having a great number of ribosomes to support protein synthesis. Many *bobbed* alleles have been studied, and each has been shown to involve a deletion, often of a unique size. The reduction of both viability and bristle length correlates well with the relative number of rRNA genes deleted. We may conclude that the normal rDNA redundancy observed in wild-type *Drosophila* is near the minimum required for adequate ribosome production during normal development.

In some cells, particularly oocytes, even the normal redundancy of rDNA may be insufficient to provide adequate amounts of rRNA and ribosomes. Oocytes store abundant nutrients, including huge quantitites of ribosomes, in the ooplasm for use by the embryo during early development. In fact, more ribosomes are contained in oocytes than in any other cell type. By considering how the amphibian *Xenopus laevis* (the South African clawed frog) acquires this abundance of ribosomes, we will see a second way in which the amount of rRNA is increased. This second phenomenon is referred to as **gene amplification.**

The genes that code for rRNA are located in an area of the chromosome known as the **nucleolar organizer region (NOR)**. The NOR is intimately associated with the nucleolus, which is a processing center for ribosome production. Molecular hybridization analysis has shown that each NOR in *Xenopus* contains the equivalent of 400 redundant gene copies coding for rRNA. Even this number of genes is apparently inadequate to synthesize the vast amount of ribosomes that must accumulate in the amphibian oocyte to support development following fertilization.

To further amplify the number of rRNA genes, the rDNA is selectively replicated, and each new set of genes is released from its template. Because each new copy is equivalent to one NOR, multiple small nucleoli are formed around each NOR in the oocyte. As many as 1500 of these "micronucleoli" have been observed in a single oocyte. If we multiply the number of micronucleoli (1500) by the number of gene copies in each NOR (400), we see that amplification in *Xenopus* oocytes can result in more than half a million gene copies! If each copy is transcribed only 20 times during the maturation of a single oocyte, in theory, sufficient copies of rRNA are produced to result in more than 12 million ribosomes.

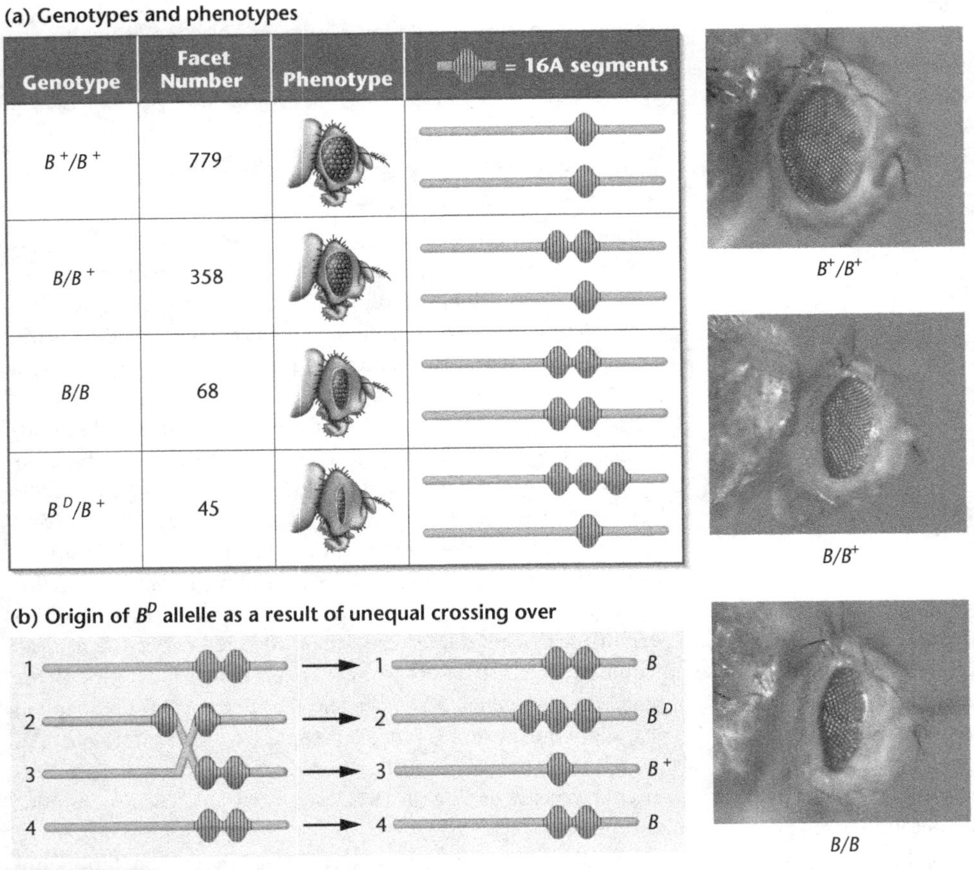


FIGURE 8–18 (a) Duplication genotypes and resultant *Bar*-eye phenotypes in *Drosophila*. Photographs show two *Bar* eye phenotypes and the wild type ($B^+/B^+$). (b) The origin of the $B^D$ (*double Bar*) allele as a result of unequal crossing over.

## The *Bar* Mutation in *Drosophila*

Duplications can cause phenotypic variations that might at first appear to be caused by a simple gene mutation. The *Bar*-eye phenotype (Figure 8–18) in *Drosophila* is a classic example. Instead of the normal oval eye shape, *Bar*-eyed flies have narrow, slitlike eyes. This phenotype is inherited in the same way as a dominant X-linked mutation. However, while both heterozygous females and hemizygous males exhibit the trait, homozygous females show a more pronounced phenotype than either of these cases.

In the early 1920s, Alfred H. Sturtevant and Thomas H. Morgan discovered and investigated this mutation. As illustrated in Figure 8–18(a), normal wild-type females ($B^+/B^+$) have about 800 facets in each eye. Heterozygous females ($B/B^+$) have about 350 facets, while females homozygous for $B$ ($B/B$) average only about 70 facets. Females are occasionally recovered with even fewer facets and are designated as *double Bar* ($B^D/B^+$).

About 10 years later, Calvin Bridges and Herman J. Muller compared the polytene X-chromosome-banding pattern of the *Bar* fly with that of the wild-type fly. The specific banding patterns of each region of these chromosomes have been well characterized. As shown in Figure 8–18, their studies revealed that one copy of region 16A is present on both X chromosomes of wild-type flies, but that this region was duplicated in *Bar* flies and triplicated in *double Bar* flies. These observations provided evidence that the *Bar* phenotype is not the result of a simple chemical change in the gene but is instead a duplication. The *double Bar* condition originates as a result of unequal crossing over, which produces the triplicated 16A region [Figure 8–18(b)].

Figure 8–18 also illustrates what is referred to as a **position effect**. You may recall from our introduction of this term in Chapter 4 that a position effect refers to altered gene expression resulting from new positioning of a gene within the genome. When the eye facet phenotypes of $B/B$ and $B^D/B^+$ flies are compared, an average of 68 and 45 facets are found, respectively. In both cases, there are two extra 16A regions. However, when the two extra segments are located on the same homolog instead of being positioned one on each homolog, the phenotype is more pronounced. Thus, the same amount of genetic information produces two distinct phenotypes, depending on the position of the genes.

## The Role of Gene Duplication in Evolution

As part of the study of evolution, it is intriguing to speculate on the possible mechanisms of genetic variation. The origin of unique gene products present in more recently evolved organisms but absent in ancestral forms is a topic of particular interest. In other words, how do "new" genes arise?

In 1970, Susumo Ohno published a provocative monograph, *Evolution by Gene Duplication*, in which he suggested that gene duplication is essential to the origin of new genes during evolution. While he was not the first to suggest that duplications might provide a "reservoir" from which new genes might arise, Ohno provided a detailed description of how the process might work. Ohno's thesis is based on the supposition that the products of many genes present as only a single copy in the genome are indispensable to the survival of members of any species during evolution. Therefore, unique genes are not free to accumulate mutations that alter their primary function and potentially give rise to new genes.

However, if an essential gene were to become duplicated in the germ line, major mutational changes in this extra copy would be tolerated in future generations because the original gene provides the genetic information for its essential function. The duplicated copy would be free to acquire many mutational changes over extended periods of time. Over short intervals, the new genetic information may be of no practical advantage. However, over long evolutionary periods, the duplicated gene may change sufficiently that its product assumes a new role in the cell. The new function may impart an "adaptive" advantage to organisms, enhancing their fitness. Ohno has thus outlined a mechanism through which sustained genetic variability may have originated.

Ohno's thesis is supported by the discovery of groups of genes in which substantial stretches of the DNA sequence are identical but whose gene products are distinct. The vertebrate digestive enzymes trypsin and chymotrypsin fit this description, as do the respiratory proteins myoglobin and hemoglobin. The DNA sequence homology is so similar in each case that we may conclude that members of each gene pair arose from a common ancestral gene through duplication. During evolution, the related genes diverged sufficiently that their products became unique.

Other support includes the presence of **gene families**, groups of closely linked genes whose products perform the same general function. Again, members of a family show sufficient DNA sequence homology for geneticists to conclude that they share a common origin. One example is the various types of human globin chains. While the various globin chains function at different times during development, they are all part of hemoglobin, whose role is to transport oxygen. Other examples include the immunologically important T-cell receptors and antigens encoded by a set of genes called the *major histocompatibility complex (MHC)*.

Many recent findings derived from our ability to sequence entire genomes have further supported the idea that gene duplication has been a common feature of evolutionary progression. For example, Jurg Spring compared a large number of genes in *Drosophila* with their counterparts in humans. For 50 genes studied, the fruit fly has only one copy of each, while there are multiple copies present in the human genome. Many other investigations have provided similar findings. For example, among the approximately 5000 genes in yeast are 55 duplicated regions containing a total of 376 pairs of duplicated genes. In the mustard plant *Arabidopsis thaliana*, approximately 70 percent of the genome is duplicated. In humans, there are 1077 duplicated blocks of genes, with 781 of them containing five or more copies. Chromosomes 18 and 20 contain large duplicated areas accounting for almost half of each chromosome.

A new debate has begun concerning a second aspect of Ohno's thesis—that major evolutionary jumps, such as the transition from invertebrates to vertebrates, may have involved the duplication of entire genomes. Ohno suggested that this might have occurred several times during the course of evolution. There is not yet any compelling evidence to convince evolutionary biologists that this was clearly the case. However, from the standpoint of gene expression, duplicating all genes proportionately seems more likely to be tolerated than duplicating just a portion of them. Whatever the causes of such jumps may have been, it is exciting when evidence obtained from new technology sheds light on a hypothesis that was put forward more than 40 years ago, long before that technology was available.

## Copy Number Variants (CNVs)—Duplications and Deletions of Specific DNA Sequences

While many examples of duplications and deletions have been discovered as a result of investigations of classical mutant phenotypes (e.g., *Bar* eye, *Notch* wing, and *bobbed* bristles in *Drosophila*), genomic investigations that focus on the DNA sequences in humans are providing new insights into these aberrations as well as into significant variation that exists between individuals in the human population. Importantly, as we shall see in this essay, these variations, often involving thousands, or even millions of base pairs, may prove to play crucial roles in many of our individual attributes, such as sensitivity to drugs and susceptibility to disease. These differences, because they represent duplications and deletions of large DNA sequences, are termed copy number variants (CNVs), and are found in both coding and noncoding regions of the genome. Mutant CNVs show an increase or a decrease in copy number in comparison to a reference genome from a normal individual.

In 2004, two research groups independently described the presence of CNVs in the genomes of healthy individuals with no known genetic disorders. CNVs were defined as regions of DNA at least 1kb in length (1000 base pairs) that display at least 90 percent sequence identity. This initial study revealed 50 loci consisting of CNVs, but likely underestimated their prevalence owing to small sample size and limitations of the screening technology. In 2005, several other groups began sifting through the genome in search of CNVs and found almost 300 additional sites. The current number of CNV sites has now risen to 1447, covering about 12 percent of the human genome. It is still unclear exactly how variable CNVs are among individuals, how closely the variations are associated with phenotypic differences or diseases, and how they are inherited. Questions such as these will likely be answered as we continue to develop better methods to identify and categorize large-scale variations.

Thanks to investigators such as Nigel Carter, whose research team developed a "CNV-finder" algorithm used during genomic analysis, a great deal of progress has already been made. Data from Carter and other groups have been compiled into a comprehensive map of CNVs in the human genome. This atlas will expand as more discoveries are made and will include correlations of CNVs to overt phenotypes, disease susceptibility, drug sensitivities, and interactions with epigenetic factors.

The focus of CNV studies most recently has been on finding associations with human diseases. CNVs appear to have both positive and negative associations with many diseases in which the genetics are not yet fully understood. For example, in 2007 Jonathan Sebat and colleagues reported an association between CNVs and autism, a neurodevelopmental disorder that impairs communication, behavior, and social interaction in affected individuals. Interestingly, a mutant CNV site has been found to appear *de novo* in 10 percent of so-called sporadic cases of autism, in which an affected child has unaffected parents lacking the mutation. This is in contrast to only 2 percent of affected individuals where the disease appears to be familial (run in the family). Similarly, Gonzalez and colleagues reported differences in the copy number of the gene *CCL3L1*, which imparts an HIV-suppressive effect during viral infection. Individuals with a higher than average copy number of the gene designated *CCL3L1* exhibit a reduction in susceptibility to HIV infection progressing to AIDS. Another research group has associated specific mutant CNV sites with certain subset populations of individuals with lung cancer—the greater number of copies of the *EGFR* (*Epidermal Growth Factor Receptor*) gene, the more responsive are patients with non-small cell lung cancer to treatment. Finally, the greater the reduction in copy number of the gene designated *DEFB*, the greater the risk of developing Crohn's disease, a condition affecting the colon. These results are still preliminary but are extremely encouraging for the future of CNV research. And, relevant to this chapter, they reveal that duplications and deletions are no longer restricted to textbook examples of these chromosomal mutations.

8.8

## Inversions Rearrange the Linear Gene Sequence

The **inversion**, another class of structural variation, is a type of chromosomal aberration in which a segment of a chromosome is turned around 180 degrees within a chromosome. An inversion does not involve a loss of genetic information, but simply rearranges the linear gene sequence. An inversion requires two breaks along the length of the chromosome and subsequent reinsertion of the inverted segment. Figure 8–19 illustrates how an inversion might arise. Formation of a chromosomal loop prior to breakage brings the newly created "sticky" ends close together so that they rejoin.

The inverted segment may be short or quite long and may or may not include the centromere. If the centromere *is* included in the inverted segment (as it is in Figure 8–19), the term **pericentric** is used to describe the inversion. If the centromere is *not* part of the rearranged chromosome segment, the inversion is said to be **paracentric**. Although the gene sequence has been reversed in the paracentric inversion, the ratio of arm lengths extending from the centromere is unchanged (Figure 8–20). In contrast, some pericentric inversions create chromosomes with arms of different lengths from those of the noninverted chromosome, thereby changing the arm ratio. The change in arm length may sometimes be visualized during the metaphase stage of mitotic or meiotic divisions.

Although inversions may seem to have minimal impact on the individuals in whom they occur, their consequences are of great interest to geneticists. Organisms heterozygous for inversions may

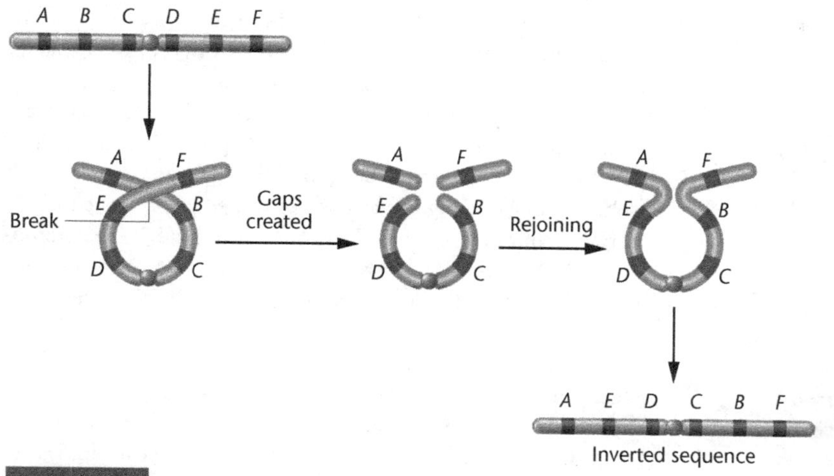


FIGURE 8–19 One possible origin of a pericentric inversion.

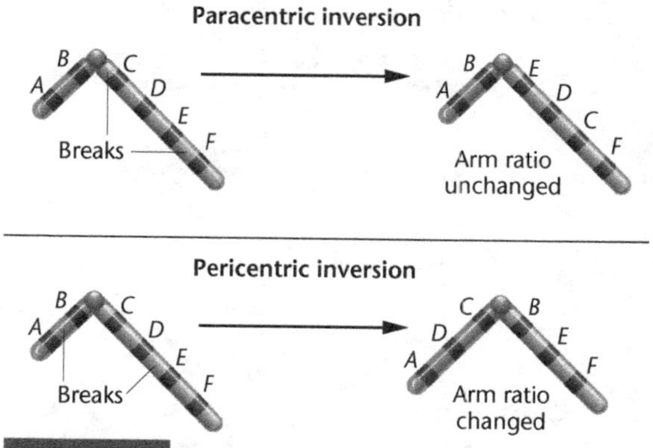


FIGURE 8–20 A comparison of the arm ratios of a submetacentric chromosome before and after the occurrence of a paracentric and a pericentric inversion. Only the pericentric inversion results in an alteration of the original ratio.

produce aberrant gametes that have a major impact on their offspring. Inversions may also result in position effects and play an important role in the evolutionary process.

## Consequences of Inversions during Gamete Formation

If only one member of a homologous pair of chromosomes has an inverted segment, normal linear synapsis during meiosis is not possible. Organisms with one inverted chromosome and one noninverted homolog are called **inversion heterozygotes**. Pairing between two such chromosomes in meiosis may be accomplished only if they form an **inversion loop** (Figure 8–21).

If crossing over does not occur within the inverted segment of the inversion heterozygote, the homologs will segregate and result in two normal and two inverted chromatids that are distributed into gametes, and the inversion will be passed on to half of the offspring. However, if crossing over occurs within the inversion loop, abnormal chromatids are produced. For example, the effect of a single crossover event (SCO) within a paracentric inversion is diagrammed in Figure 8–22(a).

In any meiotic tetrad, a single crossover between nonsister chromatids produces two parental chromatids and two recombinant chromatids. When the crossover occurs within a paracentric inversion, however, one recombinant **dicentric chromatid**—having two centromeres—and one recombinant **acentric chromatid**—lacking a centromere—are produced. Both contain duplications and deletions of chromosome segments as well. During anaphase, an acentric chromatid moves randomly to one pole or the other, or it may be lost, whereas a dicentric chromatid is pulled in two directions. This tug-of-war produces a **dicentric bridge** that is recognizable under the microscope. A dicentric chromatid usually breaks at some point so that part of the chromatid goes into one gamete and part into another gamete during the reduction divisions. Therefore, gametes containing either recombinant chromatid are deficient in

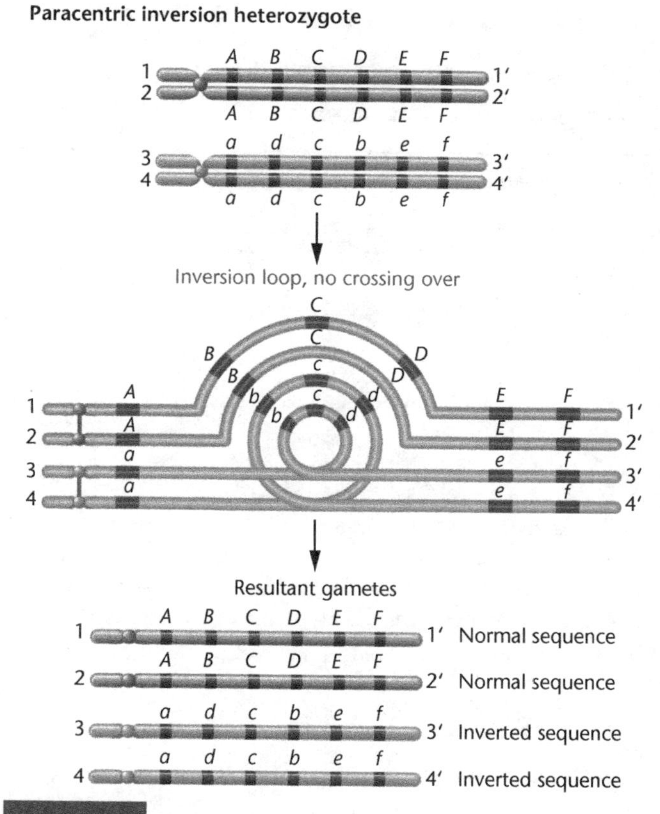


FIGURE 8–21 Illustration of how synapsis occurs in a paracentric inversion heterozygote by virtue of the formation of an inversion loop.

genetic material. When such a gamete participates in fertilization, the zygote most often develops abnormally, if at all.

A similar chromosomal imbalance is produced as a result of a crossover event between a chromatid bearing a pericentric inversion

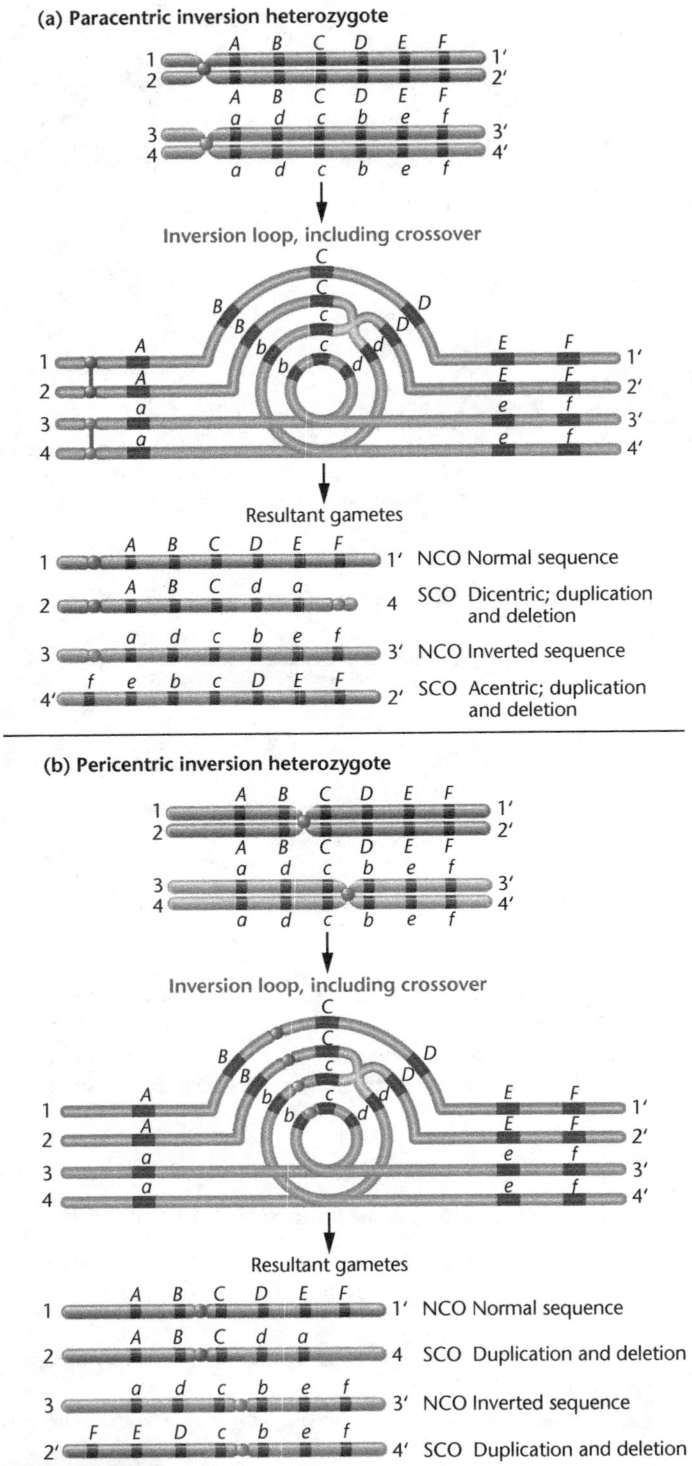


FIGURE 8-22 (a) The effects of a single crossover (SCO) within an inversion loop in a paracentric inversion heterozygote, where two altered chromosomes are produced, one acentric and one dicentric. Both chromosomes also contain duplicated and deficient regions. (b) The effects of a crossover in a pericentric inversion heterozygote, where two altered chromosomes are produced, both with duplicated and deficient regions.

and its noninverted homolog, as illustrated in Figure 8–22(b). The recombinant chromatids that are directly involved in the exchange have duplications and deletions. However, no acentric or dicentric chromatids are produced. In plants, gametes receiving such aberrant chromatids fail to develop normally, leading to aborted pollen or ovules. Thus, lethality occurs prior to fertilization, and inviable seeds result. In animals, the gametes have developed prior to the meiotic error, so fertilization is more likely to occur in spite of the chromosome error. However, the end result is the production of inviable embryos following fertilization. In both cases, viability is reduced.

Because offspring bearing crossover gametes are inviable and not recovered, it *appears* as if the inversion suppresses crossing over. Actually, in inversion heterozygotes, the inversion has the effect of *suppressing the recovery of crossover products* when chromosome exchange occurs within the inverted region. If crossing over always occurred within a paracentric or pericentric inversion, 50 percent of the gametes would be ineffective. The viability of the resulting zygotes is therefore greatly diminished. Nevertheless, up to half of the viable gametes have the inverted chromosome, and the inversion will be perpetuated within the species. The cycle will be repeated continuously during meiosis in future generations.

**NOW SOLVE THIS**

Problem 21 on page 224 considers a prospective male parent with a family history of stillbirths and malformed babies. He was shown to have an inversion covering 70 percent of one homolog of chromosome 1. You are asked to explain the cause of the stillbirths.

■ HINT: *Human chromosome 1 is metacentric; therefore, an inversion covering 70 percent of the chromosome is no doubt a pericentric inversion.*

## Position Effects of Inversions

Some consequences of inversions result from the new positioning of genes relative to other genes and particularly to areas of the chromosome that do not contain genes, such as the centromere. If the expression of the gene is altered as a result of its relocation, a change in phenotype may result. Such a change is an example of a position effect, as introduced in our earlier discussion of the *Bar* duplication.

We also introduced the general topic of position effect in Chapter 4, during our discussion of phenotypic expression. As we saw there, in *Drosophila* females heterozygous for the X-linked recessive mutation *white* eye ($w^+/w$), the X chromosome bearing the wild-type allele ($w^+$) may be inverted such that the *white* locus is moved to a point adjacent to the centromere. If the inversion is not present, a heterozygous female has wild-type red eyes, since the *white* allele is recessive. Females with the X chromosome inversion just described have eyes that are mottled or variegated, that is, with red and white patches (see Figure 4–17). Placement of the $w^+$ allele next to the centromere apparently inhibits wild-type gene expression, resulting in the loss of complete dominance over the *w* allele. Other genes, also located on the X chromosome, behave in the same manner when they are similarly relocated. Reversion to wild-type expression has sometimes been noted. When this has occurred, cytological examination has shown that the inversion has reestablished the normal sequence along the chromosome.

## Evolutionary Advantages of Inversions

One major effect of an inversion is the preservation of a set of specific alleles at a series of adjacent loci, provided that they are contained within the inversion. Because the recovery of crossover products is suppressed in inversion heterozygotes, a particular combination of alleles is preserved intact in the viable gametes. If the alleles of the involved genes confer a survival advantage on organisms maintaining them, the inversion is beneficial to the evolutionary survival of the species.

For example, say that the set of alleles *ABcDef* is more adaptive than the sets *AbCdeF* or *abcdEF*. Effective gametes will contain this favorable set of genes, undisrupted by crossing over, if it is located within a heterozygous inversion. As we will see in Chapter 28, there are documented examples where inversions are adaptive in this way. Specifically, Theodosius Dobzhansky has shown that the maintenance of different inversions on chromosome III of *Drosophila pseudoobscura* through many generations has been highly adaptive to this species. Certain inversions seem to be associated with enhanced survival under specific environmental conditions.

8.9

# Translocations Alter the Location of Chromosomal Segments in the Genome

**Translocation**, as the name implies, is the movement of a chromosomal segment to a new location in the genome. A **reciprocal translocation**, for example, involves the exchange of segments between two nonhomologous chromosomes. The least complex way for this event to occur is for two nonhomologous chromosome arms to come close to each other so that an exchange is facilitated. Figure 8–23(a) illustrates a simple reciprocal translocation in which only two breaks are required. If the exchange includes internal chromosome segments, four breaks are required, two on each chromosome.

The genetic consequences of reciprocal translocations are in some ways similar to those of inversions. For example, genetic information is not lost or gained. Rather, there is only a rearrangement of genetic material. The presence of a translocation does not therefore directly alter the viability of individuals bearing it. Like an inversion, a translocation may also produce a position effect, because it may realign certain genes in relation to other genes. This exchange may create a new genetic linkage relationship that can be detected experimentally.

Homologs that are heterozygous for a reciprocal translocation undergo unorthodox synapsis during meiosis. As shown in Figure 8–23(b), pairing results in a cruciform, or crosslike, configuration. As with inversions, some of the gametes produced are genetically unbalanced as a result of this unusual alignment during meiosis. In the case of translocations, however, aberrant gametes are not necessarily the

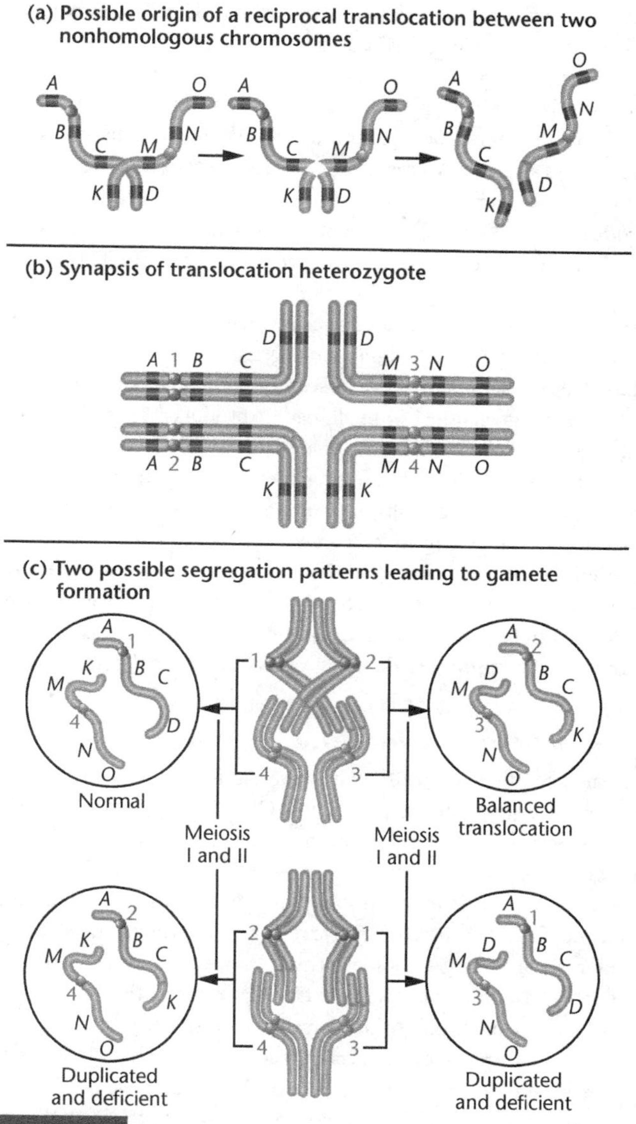


**FIGURE 8–23** (a) Possible origin of a reciprocal translocation. (b) Synaptic configuration formed during meiosis in an individual that is heterozygous for the translocation. (c) Two possible segregation patterns, one of which leads to a normal and a balanced gamete (called alternate segregation) and one that leads to gametes containing duplications and deficiencies (called adjacent segregation).

result of crossing over. To see how unbalanced gametes are produced, focus on the homologous centromeres in Figure 8–23(b) and (c). According to the principle of independent assortment, the chromosome containing centromere 1 migrates randomly toward one pole of the spindle during the first meiotic anaphase; it travels along with *either* the chromosome having centromere 3 *or* the chromosome having centromere 4. The chromosome with centromere 2 moves to the other pole, along with *either* the chromosome containing centromere 3 *or* centromere 4. This results in four potential meiotic products. The 1,4 combination contains chromosomes that are not involved in the translocation and are normal. The 2,3 combination, however, contains the translocated chromosomes, but these contain a complete complement of genetic information and are balanced. When this result is achieved [the top configuration in Figure 8–23(c)], the segregation pattern at the first meiotic division is referred to as **alternate segregation.** A second pattern [the bottom configuration in Figure 8–23(c)] produces the other two potential products, the 1,3 and 2,4 combinations, which contain chromosomes displaying duplicated and deleted (deficient) segments. This pattern is called **adjacent segregation-I.** For simplicity, a third type of arrangement, where homologous centromeres segregate to the same pole during meiosis (called *adjacent segregation-II*), has not been included in this figure. This type of segregation has an outcome similar to adjacent segregation-I, with all meiotic products containing duplicated and deleted chromosomal material.

When genetically unbalanced gametes participate in fertilization in animals, the resultant offspring do not usually survive. Fewer than 50 percent of the progeny of parents heterozygous for a reciprocal translocation survive. This condition in a parent is called **semisterility,** and its impact on the reproductive fitness of organisms plays a role in evolution. In humans, such an unbalanced condition results in partial monosomy or trisomy, leading to a variety of birth defects.

### Translocations in Humans: Familial Down Syndrome

Research conducted since 1959 has revealed numerous translocations in members of the human population. One common type of translocation involves breaks at the extreme ends of the short arms of two nonhomologous acrocentric chromosomes (13, 14, 15, 21, or 22). These small acentric fragments are lost, and the larger chromosomal segments fuse at their centromeric region. This type of translocation produces a new, larger submetacentric or metacentric chromosome (Figure 8–24), often called a **Robertsonian translocation.** It is a fairly common type of chromosome rearrangement in humans.

One such translocation accounts for cases in which Down syndrome is *familial* (inherited). Earlier in this chapter, we pointed out that most instances of Down syndrome are due to trisomy 21. This chromosome composition results from nondisjunction during meiosis in one parent. Trisomy accounts for over 95 percent of all cases of Down syndrome. In such instances, the chance of the same parents producing a second afflicted child is extremely low. However, in the remaining families with a Down child, the syndrome occurs at a much higher frequency over several generations.

Cytogenetic studies of the parents and their offspring from these unusual cases explain the cause of **familial Down syndrome.** Analysis reveals that one of the parents contains a **14/21, D/G translocation** (Figure 8–25). That is, one parent has the majority of the G-group chromosome 21 translocated to one end of the D-group chromosome 14. This individual is phenotypically normal, even though he or she has only 45 chromosomes. After meiosis, with homologous centromeres segregating to opposite poles, one-fourth of the individual's gametes will have two copies of chromosome 21: a normal chromosome and a second copy translocated to chromosome 14. When such a gamete is fertilized by a standard haploid gamete, the resulting zygote has

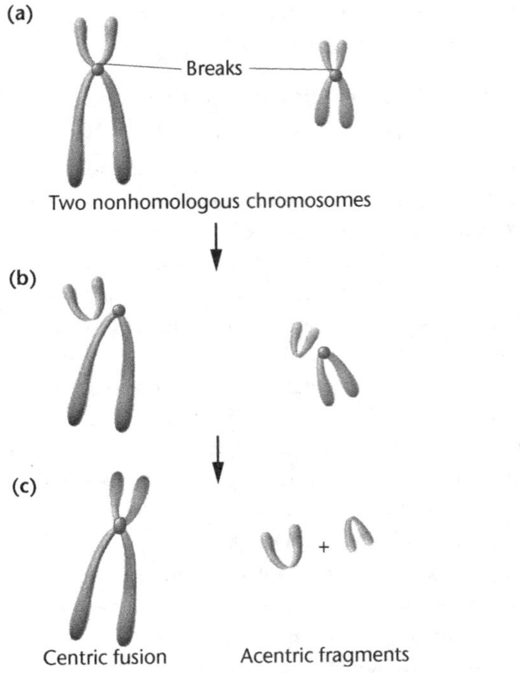


**FIGURE 8–24** The possible origin of a Robertsonian translocation. Two independent breaks occur within the centromeric region on two nonhomologous chromosomes. Centric fusion of the long arms of the two acrocentric chromosomes creates the unique chromosome. Two acentric fragments remain.

46 chromosomes but three copies of chromosome 21. These individuals exhibit Down syndrome. Other potential surviving offspring contain either the standard diploid genome (without a translocation) or the balanced translocation like the parent. Both cases result in normal individuals. Knowledge of translocations has allowed geneticists to resolve the seeming paradox of an inherited trisomic phenotype in an individual with an apparently diploid number of chromosomes.

It is interesting to note that the "carrier," who has 45 chromosomes and exhibits a normal phenotype, does not contain the *complete* diploid amount of genetic material. A small region is lost from both chromosomes 14 and 21 during the translocation event. This occurs because the ends of both chromosomes have broken off prior to their fusion. These specific regions are known to be two of many chromosomal locations housing multiple copies of the genes encoding rRNA, the major component of ribosomes. Despite the loss of up to 20 percent of these genes, the carrier is unaffected.

## 8.10 Fragile Sites in Humans Are Susceptible to Chromosome Breakage

We conclude this chapter with a brief discussion of the results of an interesting discovery made around 1970 during observations of metaphase chromosomes prepared following human cell culture.

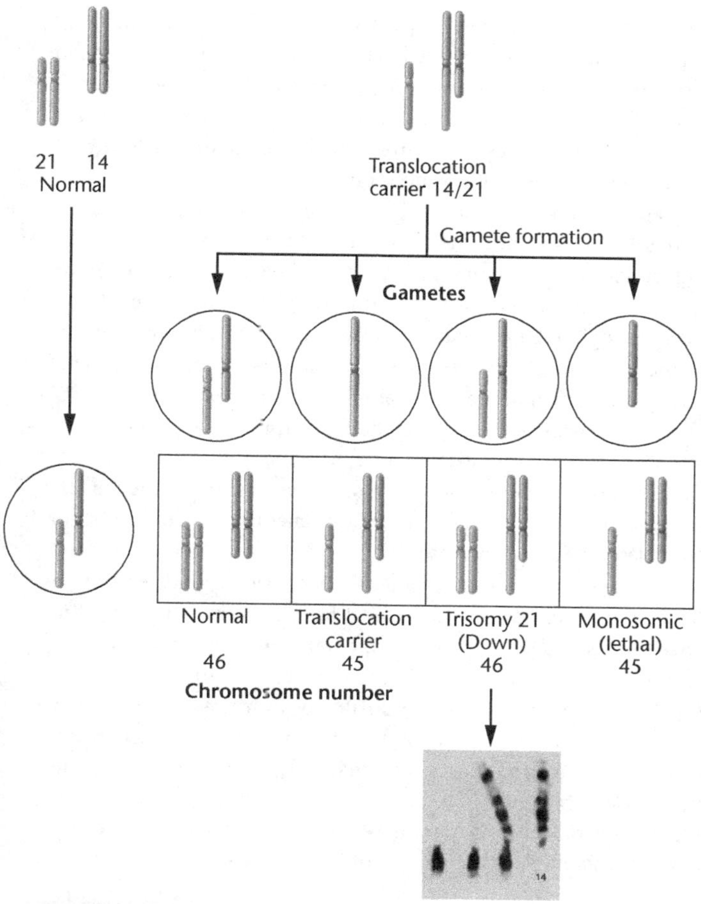


FIGURE 8-25 Chromosomal involvement in familial Down syndrome. The photograph illustrates the relevant chromosomes from a trisomy 21 offspring produced by a translocation carrier parent.

In cells derived from certain individuals, a specific area along one of the chromosomes failed to stain, giving the appearance of a gap. In other individuals whose chromosomes displayed such morphology, the gaps appeared at other places within the chromosome set. Such areas eventually became known as **fragile sites**, since they appeared to be susceptible to chromosome breakage when cultured in the absence of certain chemicals, such as folic acid, which is normally present in the culture medium. Fragile sites were at first considered curiosities, until a strong association was subsequently shown to exist between one of the sites and a form of mental retardation.

The cause of the fragility at these sites (and others) is unknown. Because they represent points along the chromosome that are susceptible to breakage, these sites may indicate regions where chromatin is not tightly coiled or compacted. Note that even though almost all studies of fragile sites have been carried out *in vitro*, using mitotically dividing cells, clear associations have been established between several of these sites and the corresponding altered phenotypes, including mental retardation and cancer. We will focus our subsequent discussion on one particular site on the human X chromosome that leads to the Martin–Bell syndrome. Then, in the "Genetics, Technology, and Society" essay, we will consider the link between fragile sites and cancer.

## Fragile X Syndrome (Martin–Bell Syndrome)

Many fragile sites do not appear to be associated with any clinical syndrome. However, individuals bearing a *folate-sensitive site* on the X chromosome (Figure 8–26) exhibit **fragile X syndrome** (or **Martin–Bell syndrome**), the most common form of inherited mental retardation. This syndrome affects about 1 in 4000 males and 1 in 8000 females. Because it is a dominant trait, females carrying only one fragile X chromosome can be mentally retarded. Fortunately, the trait is not fully expressed, as only about 30 percent of fragile X females are retarded, whereas about 80 percent of fragile X males are mentally retarded. In addition to mental retardation, affected males have characteristic long, narrow faces with protruding chins and enlarged ears, and increased testicular size.

A gene that spans the fragile site may be responsible for this syndrome. This gene, known as *FMR-1*, is one of a growing number of genes that have been discovered in which a sequence of three nucleotides is repeated many times, expanding the size of the gene. This phenomenon, called **trinucleotide repeats**, has been recognized in other human disorders, including Huntington disease. In *FMR-1*, the trinucleotide sequence CGG is repeated in an untranslated area prior to and adjacent to the coding sequence of the gene (called the "5′-upstream" region). The number of repeats varies immensely within the human population, and a high number correlates directly with expression of fragile X syndrome. Normal individuals have between 6 and 54 repeats, whereas those with 55 to 230 repeats are considered to be "carriers" of the disorder. A level above 230 repeats leads to expression of the syndrome.

It is thought that when the number of repeats reaches this level, the CGG regions of the gene become chemically modified so that the bases within and around the repeat are methylated, causing inactivation of the gene. The normal product of the gene (FMRP—fragile X mental retardation protein) is an RNA-binding protein known to be expressed in the brain. Evidence is now accumulating

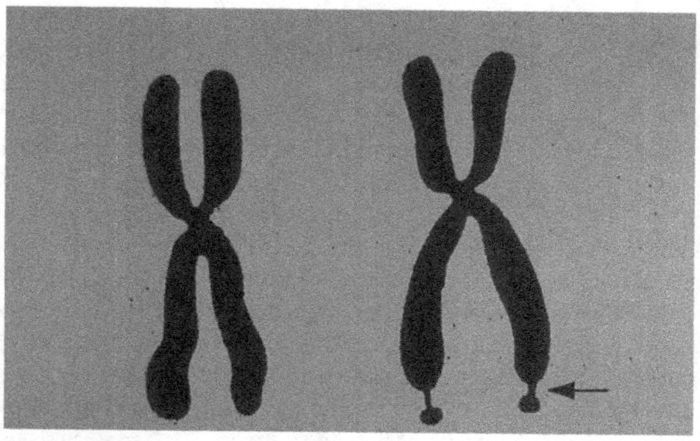

FIGURE 8-26 A normal human X chromosome (left) contrasted with a fragile X chromosome (right). The "gap" region (near the bottom of the chromosome) identifies the fragile X site and is associated with the syndrome.

that directly links the absence of the protein with the cognitive defects associated with the syndrome.

The protein is prominent in cells of the developing brain and normally shuttles between the nucleus and cytoplasm, transporting mRNAs to ribosomal complexes. Its absence in the brain's dendritic cells seems to prevent the translation of other critical proteins during brain development, presumably those produced from mRNAs that are transported by FMRP. Ultimately, the result is impairment of synaptic activity related to learning and memory. Molecular analysis has shown that FMRP is very selective, binding specifically to mRNAs that have a high guanine content. These RNAs form unique secondary structures called **G-quartets**, for which the protein has a strong binding affinity. That possession of such quartets is important to recognition by FMRP is evident from the study of cases where the formation of these stacked quartets is inhibited (by $Li^+$). In such cases, FMRP binding to these RNAs is abolished.

The *Drosophila* homolog (*dfrx*) of the human *FMR-1* gene has now been identified and used as a model for experimental investigation. The normal protein product of the fruit fly displays chemical properties that parallel its human counterpart. Mutations have been generated and studied that either overexpress the gene or eliminate its expression altogether. In the latter type, designated a null mutation, the gene has been "knocked out" using a molecular procedure now common in DNA biotechnology. The results of these studies show that the normal gene functions during the formation of synapses and that abnormal synaptic function accompanies the loss of *dfrx* expression. **Knockout mice** (mice that have had one or more genes experimentally excised from their genome; see Chapter 13) lacking this gene also display a similar phenotype. Ultimately, then, a closely related version of this gene is conserved in humans, *Drosophila*, and mice. In all three species, it has been shown to be essential to synaptic maturation and the subsequent establishment of proper cell communication in the brain. In humans, the neurological features of the syndrome are linked to the absence of its genetic expression.

From a genetic standpoint, one of the most interesting aspects of fragile X syndrome is the instability of the CGG repeats. An individual with 6 to 54 repeats transmits a gene containing the same number to his or her offspring. However, those with 55 to 230 repeats, though not at risk of developing the syndrome, may transmit to their offspring a gene containing more repeats than they themselves originally inherited. The number of repeats continues to increase from generation to generation, illustrating the phenomenon known as **genetic anticipation**, first introduced in Chapter 4. Once the threshold of 230 repeats is exceeded, expression of the malady becomes more severe in each successive generation as the number of trinucleotide repeats increases.

While the mechanism that leads to the trinucleotide expansion has not yet been established, several factors are known that influence the instability. Most significant is the observation that expansion from the carrier status (55 to 230 repeats) to the syndrome status (over 230 repeats) occurs during the transmission of the gene by the maternal parent, but not by the paternal parent. Furthermore, several reports suggest that male offspring are more likely to receive the increased repeat size leading to the syndrome than are female offspring. Obviously, we have more to learn about the genetic basis of instability and expansion of DNA sequences.

## GENETICS, TECHNOLOGY, AND SOCIETY

### The Link between Fragile Sites and Cancer

While the study of the fragile X syndrome first brought unstable chromosome regions to the attention of geneticists, a second link between a fragile site and a human disorder was reported in 1996 by Carlo Croce, Kay Huebner, and their colleagues. They demonstrated an association between an autosomal fragile site and lung cancer. Many years of investigation led this research group to postulate that the defect associated with this fragile site contributes to the formation of a variety of different tumor types.

Croce and Huebner first showed that the *FHIT* gene (standing for *f*ragile *hi*stidine *t*riad), located within the well-defined fragile site designated as *FRA3B* (on the p arm of chromosome 3), is often altered or missing in cells taken from tumors of individuals with lung cancer. Molecular analysis of numerous mutations showed that the DNA had been broken and incorrectly fused back together, resulting in deletions within the *FHIT* gene. In most cases, these mutations resulted in the loss of the normal protein product of this gene. More extensive studies have now revealed that this protein is absent in cells of many other cancers, including those of the esophagus, breast, cervix, liver, kidney, pancreas, colon, and stomach. Genes such as *FHIT* that are located within fragile regions undoubtedly have an increased susceptibility to mutations and deletions.

We now have a better understanding of fragile sites, with more than 80 such regions identified in the Human Genome Database. Fragile sites are said to be "recombinogenic," based on the observations that within these chromosomal regions, alterations such as translocations, sister-chromatid exchanges (SCEs), and other rearrangements between chromosomes frequently occur. It is now believed that breaks and gaps associated with fragile sites are induced by agents that inhibit DNA replication. The gaps and breaks are thought to be the result of DNA that has been incompletely replicated. Synthesis of DNA within these regions during the S phase of the cell cycle is delayed as compared to nonfragile chromosomal regions. This hypothesis has been strengthened in the case of the *FRA3B* region, now shown to be characterized by delayed DNA replication. As a result, the DNA in this region is not replicated by the time the cell enters the $G_2$ phase and subsequently initiates mitosis. The resulting metaphase chromosomes reveal distinct gaps and breaks within the fragile region. If these breaks are incorrectly repaired during cell-cycle checkpoints, cancer-specific mutations may occur if the region contains genes involved in cell-cycle control.

*FHIT*, a good example of a mutant gene within a fragile site, has now been designated a tumor-suppressor gene. The product of the normal gene is believed to recognize genetic damage in cells and to induce apoptosis, in which a potentially malignant cell is targeted to undergo programmed cell death and is effectively eliminated. The failure to remove such cells as a result of mutations in the *FHIT* gene causes individuals to be particularly sensitive to carcinogen-induced damage, creating a susceptibility to cancer. For example, cigarette smokers who develop lung cancer demonstrate an increased expression of the *FRA3B* fragile region compared to the normal population. Still in question, however, is whether the alteration in the fragile site, leading to inactivation of the gene, causes cancer, or whether the behavior of malignant cells somehow induces breaks at fragile sites, subsequently inactivating or deleting this gene. Another important question under investigation is the extent of genetic polymorphism at the fragile site within the human population, causing some individuals to be more susceptible to the effects of carcinogens than others.

More recently, Muller Fabbri and Kay Huebner, working with others in Croce's lab, have identified and studied another fragile site, with most interesting results. Found within the *FRA16D* site on chromosome 16 is the *WWOX* gene. Like the *FHIT* gene, it has been implicated in a range of human cancers. In particular, like *FHIT*, it has been found to be either lost or genetically silenced in the large majority of lung tumors, as well as in cancer tissue of the breast, ovary, prostate, bladder, esophagus, and pancreas. When the gene is present but silent, its DNA is thought to be heavily methylated, rendering it inactive. Furthermore, the gene is also thought to behave as a tumor suppressor, providing a surveillance function by recognizing cancer cells and inducing apoptosis, effectively eliminating them before malignant tumors can be initiated.

Fabbri, Huebner, and colleagues have attempted to demonstrate the tumor suppressor role of the normal *WWOX* gene by delivering it into cancer cells (through recombinant DNA technology) and monitoring for apoptotic action. Using a number of cancer cell lines in culture, they were able to successfully introduce the gene into a viral vector and infect the cells. The majority of those with working copies of the gene underwent apoptosis, while those that lacked working copies of the gene did not. Encouraged by these results, Fabbri et al. then turned to *in vivo* experiments, transplanting treated and untreated cancer cells under the skin of "nude" mice, which lack a functional immune system and are favorites in this type of experiment. The results were striking. Control mice that were infected with cells containing the virus that lacked the *WWOX* gene developed skin tumors, while about 70 percent of the mice that had received cells with an active *WWOX* gene remained tumor-free. These findings reveal that restoration of gene function in cells that had previously lost it will not only inhibit further reproduction of cancer cells but, in fact, destroy them.

This is an important example of pioneering work in gene therapy, representing significant and exciting studies that may prove valuable in cancer therapy. While applications of the research may not prevent the initiation of cancer, approaches using these genes found in fragile sites may someday play an important role, combined with other cancer therapies, in treating a variety of malignancies.

### References


Fabbri, M., et al. 2005. *WWOX* gene restoration prevents lung cancer growth *in vitro* and *in vivo*. *Proc. Nat. Acad. Sci.* 102: 15611–15616.

Huebner, K., and Croce, C. 2003. Cancer and the *FRA3B/FHIT* fragile locus: it's a HIT. *British J. Cancer* 88: 1501–1506.


## EXPLORING GENOMICS

## Atlas of Genetics and Cytogenetics in Oncology and Haematology

In this chapter, we discussed how variations in chromosome number and alterations in chromosome structure can affect the chromosome content of gametes to create genetic alterations in offspring and abnormal phenotypes. The **Atlas of Genetics and Cytogenetics in Oncology and Haematology** is a peer-reviewed, online journal and database of cytogenetics that specializes in cataloging chromosome abnormalities and genes involved in different cancers. Hematology is the study of blood. Much of the information presented in the atlas has been provided from clinical studies of patients with blood cancers, such as different forms of leukemia (cancer of white blood cells), that have revealed variations in chromosome number and chromosome structural defects (duplications, deletions, and translocations). In this exercise we explore the Atlas of Genetics site to learn more about human chromosome abnormalities.

*Continued on next page*

*Exploring Genomics, continued*

### ■ Exercise I – Exploring Chromosome 9

1. Access the Atlas of Genetics and Cytogenetics in Oncology and Haematology at http://atlasgeneticsoncology.org/.
2. Notice that the home page lists database entries in several ways, including an alphabetical listing of cancer genes, a listing by chromosome, and a catalog of case reports. We will explore each of these features here.
3. Under "Entities by Chromosome" click on chromosome 9. Explore the many links to abnormalities of chromosome 9 involved in different leukemias. Notice that many of these are translocations. For example, t9;12 (q34;p13) symbolizes a translocation between chromosome 9, band 34 of the q-arm, and chromosome 12, band 13 of the p-arm.
4. Find the link to "Familial melanoma," review the information on this condition, and then address the following items:
   a. Describe this disease condition.
   b. What locus on chromosome 9 is implicated in this disease? What gene is found at this locus? What is the function of this gene? (Use the disease gene links from the familial melanoma page or search the "Genes" feature of this database to learn about the gene's function.)
5. Notice that under the heading "External links" each chromosome page displays links to complete chromosome maps such as the Genes and Disease maps we visited in the Exploring Genomics exercises for Chapter 5.

### ■ Exercise II – Case Reports: Rare Examples of Chromosome Alterations

The "Case reports" section of the atlas (see the link at the top of the page) provides reports on examples of rare conditions caused by chromosome abnormalities that have been observed in different patients. Many of these case reports contain excellent examples of data generated using the cytogenetic techniques discussed in this chapter and elsewhere in the book.

1. Find the link to the case report by Shambhu Roy, Sonal Bakshi, Shailesh Patel, and colleagues. Explore the information presented in their report and then answer the following questions:
   a. What chromosome abnormality is reported by this group?
   b. What techniques were used to diagnose this condition, and what tissue samples were used for the diagnosis?
   c. What is the disease condition associated with this patient?
   d. Why is this disease referred to as "atypical CML"? (Refer to Chapter 20 and Figure 20-5 to find the answer to this question.)
   e. Name another genetic condition associated with extra copies of chromosome 21.
2. Explore more case reports presented in the atlas to learn about other rare examples of chromosome abnormalities that have been observed by scientists and physicians around the world.

## Chapter Summary

1. Investigations into the uniqueness of each organism's chromosomal constitution have further enhanced our understanding of genetic variation. Alterations of the precise diploid content of chromosomes are referred to as chromosomal aberrations or chromosomal mutations.
2. Deviations from the expected chromosomal number, or mutations in the structure of the chromosome, are inherited in predictable Mendelian fashion; they often result in inviable organisms or substantial changes in the phenotype.
3. Aneuploidy is the gain or loss of one or more chromosomes from the diploid complement, resulting in conditions of monosomy, trisomy, tetrasomy, and so on. Studies of monosomic and trisomic disorders are increasing our understanding of the delicate genetic balance that enables normal development.
4. When complete sets of chromosomes are added to the diploid genome, polyploidy occurs. These sets can have identical or diverse genetic origin, creating either autopolyploidy or allopolyploidy, respectively.
5. Large segments of a chromosome can be modified by deletions or duplications. Deletions can produce serious conditions such as the cri-du-chat syndrome in humans, whereas duplications can be particularly important as a source of redundant or new genes.
6. Inversions and translocations, while altering the gene order along chromosomes, initially cause little or no loss of genetic information or deleterious effects. However, heterozygous combinations may result in genetically abnormal gametes following meiosis, with lethality or inviability often ensuing.
7. Fragile sites in human mitotic chromosomes have sparked research interest because one such site on the X chromosome is associated with the most common form of inherited mental retardation.

# INSIGHTS AND SOLUTIONS

1. In a cross involving three linked genes, *a*, *b*, and *c*, in maize, a heterozygote (*abc*/+++) is testcrossed to *abc*/*abc*. Even though the three genes are separated along the chromosome, so that crossover gametes and the resultant phenotypes would be expected, only two phenotypes are recovered: *abc* and +++. In addition, the cross produced significantly fewer viable plants than expected. Can you propose why no other phenotypes were recovered and why the viability was reduced?

   **Solution:** One of the two chromosomes may contain an inversion that overlaps all three genes, effectively precluding the recovery of any "crossover" offspring. If this is a paracentric inversion and the genes are clearly separated (ensuring that a significant number of crossovers occurs between them), then numerous acentric and dicentric chromosomes will form, resulting in the observed reduction in viability.

2. A male *Drosophila* from a wild-type stock is discovered to have only seven chromosomes rather than the normal $2n = 8$. Close examination reveals that one member of chromosome IV (the smallest chromosome) is attached to (translocated to) the distal end of chromosome II and is missing its centromere, thus accounting for the reduction in chromosome number.

   (a) Diagram all homologs of chromosomes II and IV during synapsis in meiosis I.

   **Solution:**

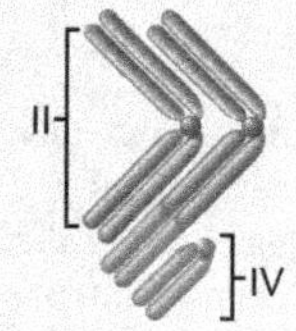


   (b) If this male mates with a female with a normal chromosome composition who is homozygous for the recessive chromosome IV mutation *eyeless* (*ey*), what chromosome II and IV compositions will occur in the offspring?

   **Solution:**

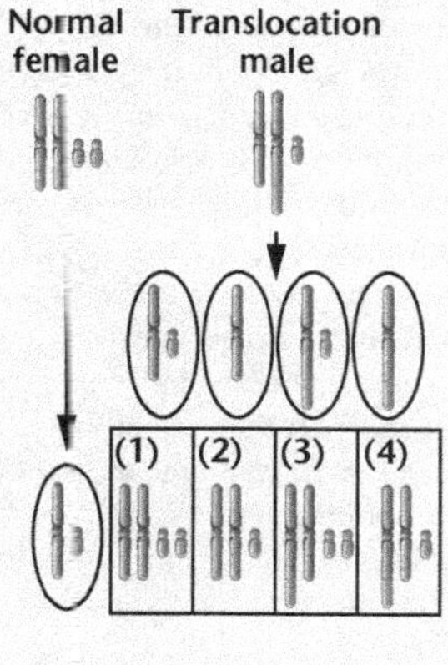


   (c) Referring to the diagram in the solution to part (b), predict what phenotypic ratio of the eyeless and normal gene will result, assuming all chromosome compositions survive.

   **Solution:**

   (1) normal (heterozygous)

   (2) eyeless (monosomic; contains chromosome IV from mother)

   (3) normal (heterozygous; trisomic and may die)

   (4) normal (heterozygous; balanced translocation)

   The final ratio is 3/4 normal : 1/4 eyeless

3. If a Haplo-IV female *Drosophila* (containing only one chromosome IV but an otherwise normal set of chromosomes) that has white eyes (an X-linked trait) and normal bristles is crossed with a male with a diploid set of chromosomes and normal red eye color but homozygous for the recessive chromosome IV bristle mutation *shaven* (*sv*), what $F_1$ phenotypic ratio might be expected?

   **Solution:** Let us first consider only the eye-color phenotypes. This is a straightforward X-linked cross. Offspring will appear as 1/2 red females:1/2 white males as shown here:

| | | | |
|---|---|---|---|
| $P_1$: | $w/w$<br>white female | × | $w^+/Y$<br>red male |
| $F_1$: | 1/2 $w/w^+$<br>red female | : | 1/2 $w/Y$<br>white male |

   The bristle phenotypes will be governed by the fact that of the gametes produced by the normal-bristle $P_1$ female, half contain a chromosome IV ($sv^+$) and half have no chromosome IV ("–" designates no chromosome). Following fertilization by sperm from the *shaven* male, half of the offspring will receive two members of chromosome IV and be heterozygous for the bristle type, expressing normal bristles. The other half will have only one copy of chromosome IV. Because its origin is the male parent, where the chromosome bears the *sv* allele, these flies will express *shaven*, there being no wild-type allele present to mask this recessive allele:

| | | | |
|---|---|---|---|
| $P_1$: | $sv^+/-$<br>normal female | × | $sv/sv$<br>shaven male |

$F_1$: 1/2 $sv^+/sv$ males and females = 1/2 wild type
1/2 $sv/-$ males and females = 1/2 shaven

   Using the forked-line method, we can consider both eye color and bristle phenotypes together:

| | | |
|---|---|---|
| 1/2 red-eyed females | → 1/2 normal bristles | → 1/4 red-eyed, normal-bristled females |
| | → 1/2 shaven bristles | → 1/4 red-eyed, shaven-bristled females |
| 1/2 white-eyed males | → 1/2 normal bristles | → 1/4 white-eyed, normal-bristled males |
| | → 1/2 shaven bristles | → 1/4 white-eyed, shaven-bristled males |

## Problems and Discussion Questions

1. For a species with a diploid number of 18, indicate how many chromosomes will be present in the somatic nuclei of individuals who are haploid, triploid, tetraploid, trisomic, and monosomic.
2. Define the following pairs of terms and distinguish between them.
   aneuploidy/euploidy
   monosomy/trisomy
   Patau syndrome/Edwards syndrome
   autopolyploidy/allopolyploidy
   autotetraploid/amphidiploid
   paracentric inversion/pericentric inversion
3. Contrast the relative survival times of individuals with Down syndrome, Patau syndrome, and Edwards syndrome. Speculate as to why such differences exist.
4. Contrast the fertility of an allotetraploid with an autotriploid and an autotetraploid.
5. When two plants belonging to the same genus but different species are crossed, the $F_1$ hybrid is more viable and has more ornate flowers. Unfortunately, this hybrid is sterile and can only be propagated by vegetative cuttings. Explain the sterility of the hybrid. How might a horticulturist attempt to reverse its sterility?
6. Describe the origin of cultivated American cotton.
7. Predict how the synaptic configurations of homologous pairs of chromosomes might appear when one member is normal and the other member has sustained a deletion or duplication.
8. Inversions are said to "suppress crossing over." Is this terminology technically correct? If not, restate the description accurately.
9. Contrast the genetic composition of gametes derived from tetrads of inversion heterozygotes where crossing over occurs within a paracentric versus a pericentric inversion.
10. Contrast the *Notch* locus versus the *Bar* locus in *Drosophila*. What phenotypic ratios would be produced in a cross between *Notch* females and *Bar* males?
11. Discuss Ohno's hypothesis on the role of gene duplication in the process of evolution and the evidence in support of the hypothesis.
12. What roles have inversions and translocations played in the evolutionary process?
13. A human female with Turner syndrome (45,X) also expresses the X-linked trait hemophilia, as did her father. Which of her parents underwent nondisjunction during meiosis, giving rise to the gamete responsible for the syndrome?
14. The primrose, *Primula kewensis*, has 36 chromosomes that are similar in appearance to the chromosomes in two related species, *Primula floribunda* ($2n = 18$) and *Primula verticillata* ($2n = 18$). How could *P. kewensis* arise from these species? How would you describe *P. kewensis* in genetic terms?
15. Certain varieties of chrysanthemums contain 18, 36, 54, 72, and 90 chromosomes; all are multiples of a basic set of 9 chromosomes. How would you describe these varieties genetically? What feature is shared by the karyotypes of each variety? A variety with 27 chromosomes was discovered, but it was sterile. Why?
16. What is the effect of a rare double crossover within a pericentric inversion present heterozygously? What is the effect within a paracentric inversion present heterozygously?
17. *Drosophila* may be monosomic for chromosome IV and remain fertile. Contrast the $F_1$ and $F_2$ results of the following crosses involving the recessive chromosome IV trait *bent* bristles:
   (a) monosomic IV, bent bristles × diploid, normal bristles
   (b) monosomic IV, normal bristles × diploid, bent bristles
18. *Drosophila* may also be trisomic for chromosome IV and remain fertile. Predict the $F_1$ and $F_2$ results of the following cross: trisomic, bent bristles $b/b/b$ × diploid, normal bristles ($b^+/b^+$).
19. Mendelian ratios are modified in crosses involving autotetraploids. Assume that one plant expresses the dominant trait green (seeds) and is homozygous (*WWWW*). This plant is crossed to one with white seeds that is also homozygous (*wwww*). If only one dominant allele is needed to produce green seeds, predict the $F_1$ and $F_2$ results of such a cross. Assume that synapsis between chromosome pairs is random during meiosis.
20. Having correctly established the $F_2$ ratio in Problem 19, predict the $F_2$ ratio of a "dihybrid" cross involving two independently assorting characteristics (e.g., $P_1$ = *WWWWAAAA* × *wwwwaaaa*).
21. A couple planning their family are aware that through the past three generations on the husband's side a substantial number of stillbirths have occurred and several malformed babies were born who died early in childhood. The wife has studied genetics and urges her husband to visit a genetic counseling clinic, where a complete karyotype-banding analysis is performed. Although the karyotype shows that he has a normal complement of 46 chromosomes, banding analysis reveals that one member of the chromosome 1 pair (in Group A) contains an inversion covering 70 percent of its length. All other chromosomes, including the homolog of chromosome 1, show the normal banding sequences.
   (a) How would you explain the high incidence of past stillbirths?
   (b) What would you predict about the probability of abnormality/normality of the couple's future children?
   (c) Would you advise the woman that she will have to bring each pregnancy to term to determine whether the fetus is normal? If not, what else can you suggest?

**HOW DO WE KNOW?**

22. In this chapter, we focused on chromosome mutations caused by alterations of an otherwise diploid genome. Either the chromosome number is changed or the arrangement of the genetic material is altered. At the same time, we found many opportunities to consider the methods and reasoning by which much of this information was acquired. From the explanations given in the chapter, what answers would you propose to the following fundamental questions?
   (a) How do we know that the extra chromosome causing Down syndrome is usually maternal in origin?
   (b) How do we know that human aneuploidy for each of the 22 autosomes occurs at conception, even though most often human aneuploids do not survive embryonic or fetal development and thus are never observed at birth?
   (c) How do we know that changes in chromosome number or structure result in specific mutant phenotypes?
   (d) How do we know that the mutant *Bar*-eye phenotype in *Drosophila* is due to a duplicated gene region rather than a change in the nucleotide sequence of a gene?
   (e) How do we know that gene duplication is a source of new genes during evolution?

# Extra-Spicy Problems

23. In a cross in *Drosophila*, a female heterozygous for the autosomally linked genes *a*, *b*, *c*, *d*, and *e* (*abcde*/+++++) was testcrossed with a male homozygous for all recessive alleles. Even though the distance between each of the loci was at least 3 map units, only four phenotypes were recovered, yielding the following data:

| Phenotype | No. of Flies |
|---|---|
| + + + + + | 440 |
| *a b c d e* | 460 |
| + + + + *e* | 48 |
| *a b c d* + | 52 |
| Total = | 1000 |

Why are many expected crossover phenotypes missing? Can any of these loci be mapped from the data given here? If so, determine map distances.

24. A woman who sought genetic counseling is found to be heterozygous for a chromosomal rearrangement between the second and third chromosomes. Her chromosomes are diagrammed here:

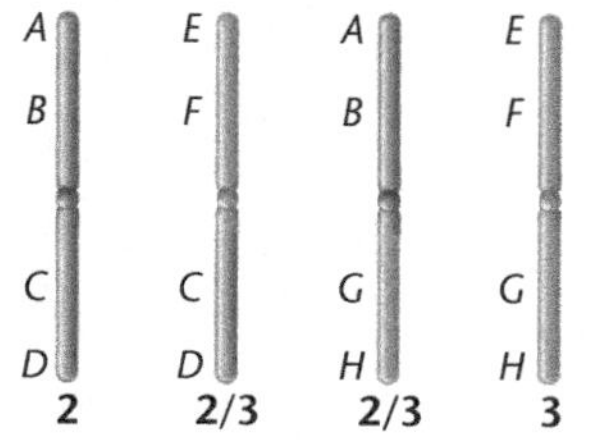


(a) What kind of chromosomal aberration is shown?
(b) Using a drawing, demonstrate how these chromosomes would pair during meiosis. Be sure to label the different segments of the chromosomes.
(c) This woman is phenotypically normal. Does that surprise you? Why or why not? Under what circumstances might you expect a phenotypic effect from such a rearrangement?

25. The woman in Problem 24 has had two miscarriages. She has come to you, an established genetic counselor, for advice. She raises the following questions: Is there a genetic explanation for her frequent miscarriages? Should she abandon her attempts to have a child of her own? If not, what is the chance that she could have a normal child? Form a small discussion group and after exploring these questions, provide an informed response to each of her concerns.

26. A boy with Klinefelter syndrome (47,XXY) is born to a mother who is phenotypically normal and a father who has the X-linked skin condition called anhidrotic ectodermal dysplasia. The mother's skin is completely normal, with no signs of the skin condition. In contrast, her son has patches of normal skin and patches of abnormal skin.
(a) Which parent contributed the abnormal gamete?
(b) Using the appropriate genetic terminology, describe the meiotic mistake that occurred. Be sure to indicate which division the mistake occurred in.
(c) Using the appropriate genetic terminology, explain the son's skin phenotype.

27. To investigate the origin of nondisjunction, 200 human oocytes that had failed to be fertilized during *in vitro* fertilization procedures were examined (Angel, R. 1997. *Am. J. Hum. Genet.* 61:23–32). These oocytes had completed meiosis I and were arrested in metaphase II (MII). The majority (67 percent) had a normal MII metaphase complement, showing 23 chromosomes, each consisting of two sister chromatids joined at a common centromere. The remaining oocytes all had abnormal chromosome compositions. Surprisingly, when trisomy was considered, none of the abnormal oocytes had 24 chromosomes.
(a) Interpret these results in regard to the origin of trisomy, as it relates to nondisjunction and when it occurs. Why are the results surprising? A large number of the abnormal oocytes contained 22 1/2 chromosomes—that is, 22 chromosomes plus a single chromatid representing the 1/2 chromosome.
(b) What chromosome compositions will result in the zygote if such oocytes proceed through meiosis and are fertilized by normal sperm?
(c) How could the complement of 22 1/2 chromosomes arise? Answer this question with a drawing that includes several pairs of MII chromosomes.
(d) Do your answers support or refute the generally accepted theory regarding nondisjunction and trisomy, as outlined in Figure 8–1?

28. In a human genetic study, a family with five phenotypically normal children was investigated. Two were "homozygous" for a Robertsonian translocation between chromosomes 19 and 20 (they contained two identical copies of the fused chromosome). These children have only 44 chromosomes but a complete genetic complement. Three of the children were "heterozygous" for the translocation and had 45 chromosomes, including one translocated chromosome plus a normal copy of both chromosome 19 and 20. Two other pregnancies resulted in stillbirths. It was discovered that the parents were first cousins. Based on the above information, determine the chromosome compositions of the parents. What led to the stillbirths? Why was the discovery that the parents were first cousins a key piece of information in understanding the genetics of this family?

29. A 3-year-old child exhibited some early indication of Turner syndrome, which results from a 45,X chromosome composition. Karyotypic analysis demonstrated two cell types: 46,XX (normal) and 45,X. Propose a mechanism for the origin of this mosaicism.

30. A normal female is discovered with 45 chromosomes, one of which exhibits a Robertsonian translocation containing most of chromosomes 18 and 21. Discuss the possible outcomes in her offspring when her husband contains a normal karyotype.

31. In humans, cystic fibrosis (CF) is a recessive disorder characterized by abnormal mucous secretions that affect the lungs. Because it is a rare disorder, most affected children have two parents that are heterozygous carriers. However, an occasional CF child has only one parent that is a carrier. Propose an explanation for this phenomenon.

32. While gene duplication may play a role in the origin of new genes, and endopolyploidy may provide certain gene products in abundance in certain cell lines, diploidization is often seen as a mechanism for the rapid origin of new species. Approximately 50 to 70 percent of all plant species were probably polyploid at one time in their ancestry. Mammals probably result from diploidization of genomes of amphibian or fish ancestors and as such contain polyploid amounts of genetic material but diploid levels of gene expression and chromosomal structure and behavior. Many species of polyploid *Arabidopsis*, *Brassica*, and *Arabis* (all related to cabbages in the family Brassicaceae) have a chromosome number that is a multiple of 8 ($n = 8$) and a likely nonpolyploid ancestor of $n = 4$. However, even though its closest relatives are $n = 8$, the model organism *Arabidopsis thaliana* (see Chapter 19) has $n = 5$, with only 80 percent of

the genome duplicated, the extensive duplication created by polyploidy having been reduced by diploidization.

(a) What evolutionary advantage might be conferred by polyploidization?

(b) How might selection favor diploidization?

(c) By what genetic and chromosomal processes might diploidization occur?

33. Prader-Willi and Angelman syndromes (see Chapter 4) are human conditions that result in part through the process of genomic imprinting, where a particular gene is "marked" during gamete formation in the parent of origin. In the case of these two syndromes, a portion of the long arm of chromosome 15 (15q11 to 15q13) is imprinted. Another condition, Beckwith-Wiedemann syndrome (accelerated growth and increased risk of cancer), is associated with abnormalities of imprinted genes on the short arm of chromosome 11. All three of the above syndromes occur when imprinted sequences become abnormally exposed by some genetic or chromosomal event. One such event is uniparental disomy (UPD), in which a person receives two homologs of one chromosome (or part of a chromosome) from one parent and no homolog from the other. In many cases, UPD is without consequence; however if chromosomes 11 or 15 are involved, then, coupled with genomic imprinting, complications can arise. Listed below are possible origins of UPD. Provide a diagram and explanation for each.

1. *Trisomic rescue:* loss of a chromosome in a trisomic zygote
2. *Monosomic rescue:* gain of a chromosome in a monosomic zygote
3. *Gamete complementation:* fertilization of a gamete with two copies of a chromosome by a gamete with no copies of that chromosome
4. *Isochromosome formation:* a chromosome that contains two copies of an arm (15q:15q, for example) rather than one

Variegated leaves like those of the shrub *Acanthopanax* led geneticists to discover extranuclear inheritance.

9

# Extranuclear Inheritance

## CHAPTER CONCEPTS

- Many traits exhibit a pattern of inheritance that is not biparental, demonstrating what is called extranuclear inheritance.
- Extranuclear inheritance is often due to the expression of genetic information contained in mitochondria or chloroplasts.
- Extranuclear traits determined by mitochondrial DNA are most often transmitted through the maternal gamete.
- Extranuclear traits determined by chloroplast DNA may be transmitted uniparentally or biparentally.
- Expression of the maternal genotype during gametogenesis and during early development may have a strong influence on the phenotype of an organism.

Throughout the history of genetics, occasional reports have challenged the basic tenet of Mendelian transmission genetics—that the phenotype is transmitted by nuclear genes located on the chromosomes of both parents. Observations have revealed inheritance patterns that fail to reflect Mendelian principles, and some indicate an apparent extranuclear influence on the phenotype. Before the role of DNA in genetics was firmly established, such observations were commonly regarded with skepticism. However, with the increasing knowledge of molecular genetics and the discovery of DNA in mitochondria and chloroplasts, extranuclear inheritance is now recognized as an important aspect of genetics.

There are several varieties of extranuclear inheritance. One major type, referred to above, is also described as **organelle heredity**. In this type of inheritance, DNA contained in mitochondria or chloroplasts determines certain phenotypic characteristics of the offspring. Examples are often recognized on the basis of the uniparental transmission of these organelles, usually from the female parent through the egg to progeny. A second type, called **infectious heredity**, results from a symbiotic or parasitic association with a microorganism. In such cases, an inherited phenotype is affected by the presence of the microorganism in the cytoplasm of the host cells. A third variety involves the **maternal effect** on the phenotype, whereby nuclear gene products are stored in the egg and then transmitted through the ooplasm to offspring. These gene products are distributed to cells of the developing embryo and influence its phenotype.

The common element in all of these examples is the transmission of genetic information to offspring through the cytoplasm rather than through the nucleus, most often from only one of the parents. This shall constitute our definition of **extranuclear inheritance**.

9.1

## Organelle Heredity Involves DNA in Chloroplasts and Mitochondria

In this section and the next, we will examine examples of inheritance patterns arising from chloroplast and mitochondrial function. Such patterns are now appropriately called organelle heredity. Before DNA was discovered in these organelles, the exact mechanism of transmission of the traits we are about to discuss was not clear, except that their inheritance appeared to be linked to something in the cytoplasm rather than to genes in the nucleus. Most often (but not in all cases), the traits appeared to be transmitted from the maternal parent through the ooplasm, causing the results of reciprocal crosses to vary.

Analysis of the inheritance patterns resulting from mutant alleles in chloroplasts and mitochondria has been difficult for two major reasons. First, the function of these organelles is dependent on gene products from both nuclear and organelle DNA, making the discovery of the genetic origin of mutations affecting organelle function difficult. Second, large numbers of these organelles are contributed to each progeny cell following cell division. If only one or a few of the organelles acquire a new mutation or contain an existing one in a cell with a population of mostly normal organelles, the corresponding mutant phenotype may not be revealed, since the organelles lacking the mutation perform the wild-type function for the cell. Such variation in the genetic content of organelles is called **heteroplasmy**. Analysis is thus much more complex than for Mendelian characters controlled by nuclear genes.

We will begin our discussion with several of the classical examples that ultimately called attention to organelle heredity. After that, we will discuss information concerning the DNA and the resultant genetic function in each organelle.

### Chloroplasts: Variegation in Four O'Clock Plants

In 1908, Carl Correns (one of the rediscoverers of Mendel's work) provided the earliest example of inheritance linked to chloroplast transmission. Correns discovered a variant of the four o'clock plant, *Mirabilis jalapa*, in which some branches had white leaves, some had green, and some had variegated leaves. The completely white leaves and the white areas in variegated leaves lack chlorophyll that provides the green color to normal leaves. Chlorophyll is the light-absorbing pigment made within chloroplasts.

Correns was curious about how inheritance of this phenotypic trait occurred. As shown in Figure 9–1, inheritance in all possible combinations of crosses is strictly determined by the phenotype of the ovule source. For example, if the seeds (representing the progeny) were derived from ovules on branches with green leaves, all progeny plants bore only green leaves, regardless of the phenotype of the source of pollen. Correns concluded that inheritance was transmitted through the cytoplasm of the maternal parent because the pollen, which contributes little or no cytoplasm to the zygote, had no apparent influence on the progeny phenotypes.

Since leaf coloration is a function of the chloroplast, genetic information either contained in that organelle or somehow present in the cytoplasm and influencing the chloroplast must be responsible for the inheritance pattern. It now seems certain that the genetic "defect" that eliminates the green chlorophyll in the white patches on leaves is a mutation in the DNA housed in the chloroplast.

### Chloroplast Mutations in *Chlamydomonas*

The unicellular green alga *Chlamydomonas reinhardtii* has provided an excellent system for the investigation of plastid inheritance. This haploid eukaryotic organism (Figure 5–20) has a single large chloroplast containing about 75 copies of a circular double-stranded DNA molecule. Matings that reestablish diploidy are immediately followed by meiosis, and the various stages of the life cycle are easily studied in culture in the laboratory. The first known cytoplasmic mutation, streptomycin resistance ($str^R$) in *Chlamydomonas*, was reported in 1954 by Ruth Sager. Although *Chlamydomonas*'s two mating types,

| Source of Pollen | Location of Ovule | | |
|---|---|---|---|
| | White branch | Green branch | Variegated branch |
| White branch | White | Green | White, green, or variegated |
| Green branch | White | Green | White, green, or variegated |
| Variegated branch | White | Green | White, green, or variegated |

FIGURE 9–1 Offspring of crosses involving flowers from various branches of variegated four o'clock plants. The photograph illustrates variegation in the flowers of the four o'clock plant.

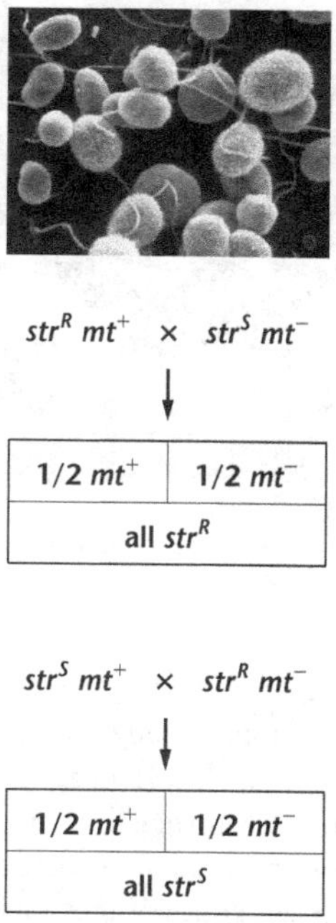


FIGURE 9–2 The results of reciprocal crosses between streptomycin-resistant ($str^R$) and streptomycin-sensitive ($str^S$) strains in the green alga *Chlamydomonas* (shown in the photograph).

$mt^+$ and $mt^-$, appear to make equal cytoplasmic contributions to the zygote, Sager determined that the $str^R$ phenotype is transmitted only through the $mt^+$ parent (Figure 9–2). Reciprocal crosses between sensitive and resistant strains yield different results depending on the genotype of the $mt^+$ parent, which is expressed in all offspring. As shown, one-half of the offspring are $mt^+$ and one-half of them are $mt^-$, indicating that mating type is controlled by a nuclear gene that segregates in a Mendelian fashion.

Since Sager's discovery, a number of other *Chlamydomonas* mutations (including resistance to, or dependence on, a variety of bacterial antibiotics) that show a similar uniparental inheritance pattern have been discovered. These mutations have all been linked to the transmission of the chloroplast, and their study has extended our knowledge of chloroplast inheritance.

Following fertilization, which involves the fusion of two cells of opposite mating type, the single chloroplasts of the two mating types fuse. After the resulting zygote has undergone meiosis and haploid cells are produced, it is apparent that the genetic information in the chloroplast of progeny cells is derived only from the $mt^+$ parent. The genetic information originally present within the $mt^-$ chloroplast has degenerated.

The inheritance of phenotypes influenced by mitochondria is also uniparental in *Chlamydomonas*. However, studies of the transmission of several cases of antibiotic resistance governed by mitochondria have shown that it is the $mt^-$ parent that transmits the mitochondrial genetic information to progeny cells. This is just the opposite of what occurs with chloroplast-derived phenotypes, such as $str^R$. The significance of inheriting one organelle from one parent and the other organelle from the other parent is not yet established.

NOW SOLVE THIS

In Problem 8(b) on page 242 you are asked to interpret the genetic outcome of crosses involving a mutation in *Chlamydomonas* and to propose how inheritance of the trait occurs.

■ HINT: *Consider the results you would expect from two possibilities: that inheritance of the trait is uniparental or that it is biparental.*

## Mitochondrial Mutations: The Case of *poky* in *Neurospora*

As alluded to earlier, mutations affecting mitochondrial function have also been discovered and studied, revealing that mitochondria, too, contain a distinctive genetic system. As with chloroplasts, mitochondrial mutations are transmitted through the cytoplasm. In the ensuing discussion, we will emphasize the link between mitochondrial mutations and the resultant extranuclear inheritance patterns.

In 1952, Mary B. Mitchell and Hershel K. Mitchell studied the pink bread mold *Neurospora crassa* (Figure 9–3). They discovered a slow-growing mutant strain and named it *poky*. (It is also designated *mi-1*, for *maternal inheritance*.) Slow growth is associated with impaired mitochondrial function, specifically caused by the absence of several cytochrome proteins essential for electron transport. In the absence of cytochromes, aerobic respiration leading to ATP synthesis is curtailed.

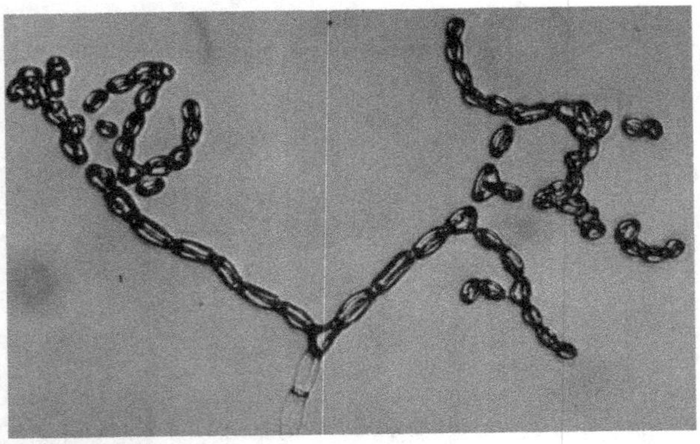

FIGURE 9–3 Micrograph illustrating the growth of the bread mold *Neurospora crassa*.

Results of genetic crosses between wild-type and *poky* strains suggest that the trait is maternally inherited. If one mating type is *poky* and the other is wild type, all progeny colonies are *poky*, yet the reciprocal cross produces normal wild-type colonies.

Studies with *poky* mutants show that occasionally hyphae from separate mycelia fuse with one another, giving rise to structures containing two or more nuclei in a common cytoplasm. If the hyphae contain nuclei of different genotypes, the structure is called a **heterokaryon**. The cytoplasm will contain mitochondria derived from both initial mycelia. A heterokaryon may give rise to haploid spores, or conidia, that produce new mycelia.

Heterokaryons produced by the fusion of *poky* and wild-type hyphae initially show normal rates of growth and respiration. However, mycelia produced through conidia formation become progressively more abnormal until they show the *poky* phenotype. This occurs despite the presumed presence of both wild-type and *poky* mitochondria in the cytoplasm of the hyphae.

To explain the initial growth and respiration pattern, we assume that the wild-type mitochondria support the respiratory needs of the hyphae. The subsequent expression of the *poky* phenotype suggests that the presence of the *poky* mitochondria may somehow prevent or depress the function of these wild-type mitochondria. Perhaps the *poky* mitochondria replicate more rapidly and "wash out" or dilute wild-type mitochondria numerically. Another possibility is that *poky* mitochondria produce a substance that inactivates the wild-type organelle or interferes with the replication of its DNA (mtDNA). This type of interaction makes *poky* an example of a broader category referred to as **suppressive mutations** (sometimes called *dominant-negative mutations*), which may also include many other mitochondrial mutations of *Neurospora* and yeast.

## Petites in *Saccharomyces*

Another extensive study of mitochondrial mutations has been performed with the yeast *Saccharomyces cerevisiae*. The first such mutation, described by Boris Ephrussi and his coworkers in 1956, was named *petite* because of the small size of the yeast colonies (Figure 9–4). Many independent petite mutations have since been discovered and studied, and all have a common characteristic: a deficiency in cellular respiration involving abnormal electron transport. This organism is a *facultative anaerobe* (an organism that can function without as well as with the presence of oxygen) and in the absence of oxygen can grow by fermenting glucose through glycolysis; thus, it may survive the loss of mitochondrial function by generating energy anaerobically.

The complex genetics of petite mutations is diagrammed in Figure 9–5. A small proportion of these mutants are the result of nuclear mutations. They exhibit Mendelian inheritance and are thus called ***segregational petites***. The remainder demonstrate cytoplasmic transmission, indicating mutations in the DNA of the mitochondria. They produce one of two effects in matings. ***Neutral petites***, when crossed to wild type, produce meiotic products (called ascospores) that give rise only to wild-type, or normal, colonies. The same pattern continues if progeny of such crosses are backcrossed to neutral petites. The majority of neutrals lack mtDNA completely or have lost a substantial portion of it, so for their offspring to be normal, the neutrals must also be inheriting normal mitochondria capable of aerobic respiration following reproduction. Thus, in yeast, mitochondria are inherited from both parental cells. The functional mitochondria are replicated in offspring and support normal mitochondrial function.

The third mutational type, the ***suppressive petite***, behaves like *poky* in *Neurospora*. Crosses between mutant and wild type give rise to mutant diploid zygotes, which after meiosis immediately yield haploid cells that are all mutant. Assuming that the offspring have received mitochondria from both parents, the *petite* mutation behaves "dominantly" and seems to suppress the function of the wild-type mitochondria. *Suppressive petites* also have deletions of mtDNA, but they are not nearly as extensive as deletions in the *neutral petites*.

Two major hypotheses concerning the organellar DNA have been advanced to explain this suppressiveness. One explanation suggests that the mutant (or deleted) DNA in the mitochondria (mtDNA) replicates more rapidly, resulting in the mutant mito-

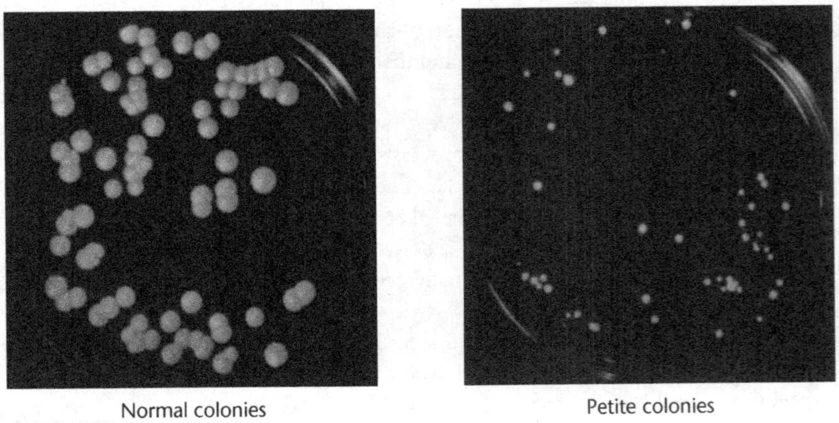

Normal colonies

Petite colonies

FIGURE 9–4 A comparison of normal versus petite colonies in the yeast *Saccharomyces cerevisiae*.

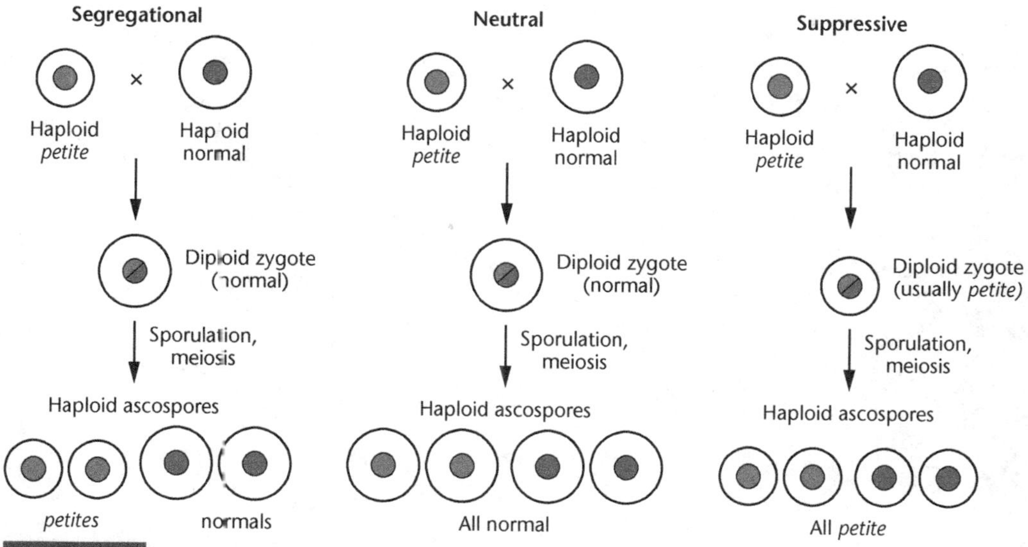


FIGURE 9–5 The outcome of crosses involving the three types of *petite* mutations affecting mitochondrial function in the yeast *Saccharomyces cerevisiae*.

chondria "taking over" or dominating the phenotype by numbers alone. The second explanation suggests that recombination occurs between the mutant and wild-type mtDNA, introducing errors into or disrupting the normal mtDNA. It is not yet clear which one, if either, of these explanations is correct.

**NOW SOLVE THIS**

In Problem 4 on page 242, involving a newly isolated *petite* mutation crossed to a normal yeast strain, you are asked to draw conclusions about the *petite* mutation.

■ HINT: *Remember that in yeast, inheritance of mitochondria is biparental.*

## 9.2 Knowledge of Mitochondrial and Chloroplast DNA Helps Explain Organelle Heredity

That both mitochondria and chloroplasts contain their own DNA and a system for expressing genetic information was first suggested by the discovery of mutations and the resultant inheritance patterns in plants, yeast, and other fungi, as already discussed. Because both mitochondria and chloroplasts are inherited through the maternal cytoplasm in most organisms, and because each of the above-mentioned examples of mutations could be linked hypothetically to the altered function of either chloroplasts or mitochondria, geneticists set out to look for more direct evidence of DNA in these organelles. Not only was unique DNA found to be a normal component of both mitochondria and chloroplasts, but careful examination of the nature of this genetic information was to provide essential clues as to the evolutionary origin of these organelles.

### Organelle DNA and the Endosymbiotic Theory

Electron microscopists not only documented the presence of DNA in mitochondria and chloroplasts, but they also saw that it exists there in a form quite unlike the form seen in the nucleus of the eukaryotic cells that house these organelles (Figures 9–6 and 9–7). The DNA in chloroplasts and mitochondria looks remarkably similar to the DNA seen in bacteria! This similarity, along with the observation of the presence of a unique genetic system capable of organelle-specific transcription and translation, led Lynn Margulis and others to the postulate known as the **endosymbiotic theory**. Basically, the theory states that mitochondria and chloroplasts arose independently about 2 billion years ago from free-living protobacteria (primitive bacteria). Progenitors possessed the abilities now attributed to these organelles—aerobic respiration and photosynthesis, respectively. This idea proposes that these ancient bacteria-like cells were engulfed by larger primitive eukaryotic cells, which originally lacked the ability to respire aerobically or to capture energy from sunlight. A beneficial, symbiotic relationship subsequently developed, whereby the bacteria eventually lost their ability to function autonomously, while the eukaryotic host cells gained the ability to perform either oxidative respiration or photosynthesis, as the case may be. Although some questions remain unanswered, evidence continues to accumulate in support of this theory, and its basic tenets are now widely accepted.

A brief examination of the modern-day mitochondria will help us better understand endosymbiotic theory. During the course of evolution subsequent to the invasion event, distinct branches of diverse eukaryotic organisms arose. As the evolution of the host cells progressed, the companion bacteria also underwent their own independent changes. The primary alteration was the transfer of many of the genes from the invading bacterium to the nucleus of the host. The *products* of these genes, while still functioning in the organelle, are nevertheless now encoded and transcribed in the nucleus and translated in the cytoplasm prior to their transport into the organelle. The amount of DNA remaining today in the typical mitochondrial genome is miniscule compared with that in the free-living bacteria from which it was derived. The most gene-rich organelles now have fewer than 10 percent of the genes present in the smallest bacterium known.

Similar changes have characterized the evolution of chloroplasts. In the subsequent sections, we will explore in some detail what is known about modern-day chloroplasts and mitochondria.

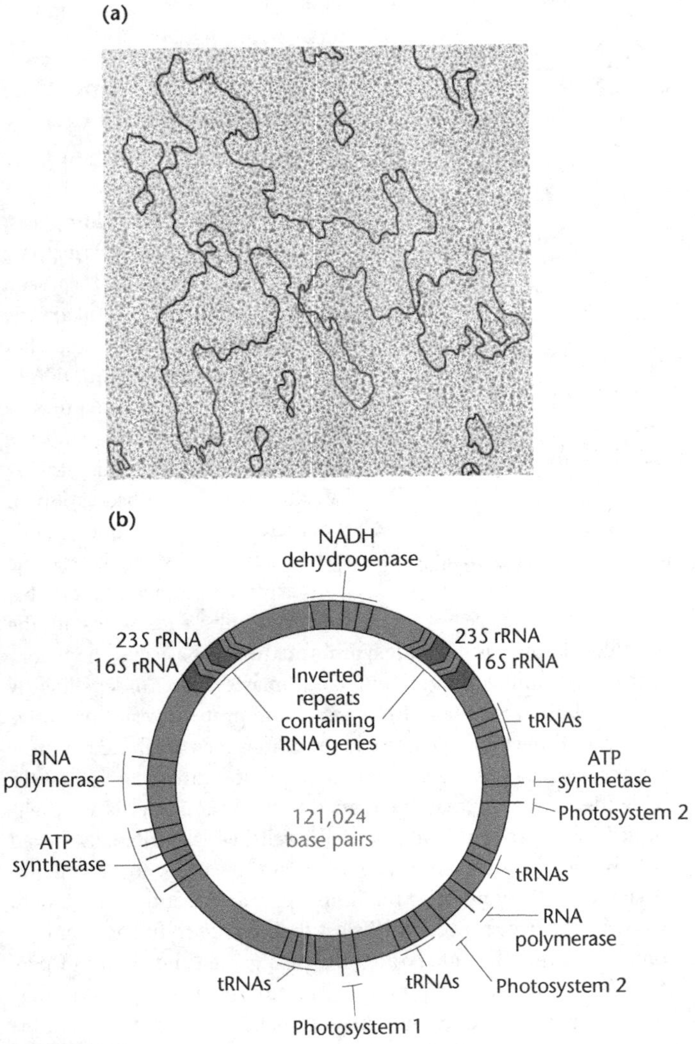


FIGURE 9–6 Examples of chloroplast DNA. (a) Electron micrograph of chloroplast DNA derived from lettuce. (b) Diagram illustrating the arrangement of many of the genes encoded by cpDNA of the moss, *Marchantia polymorpha*. Photosystems 1 and 2 are groups of genes with photosynthetic functions.

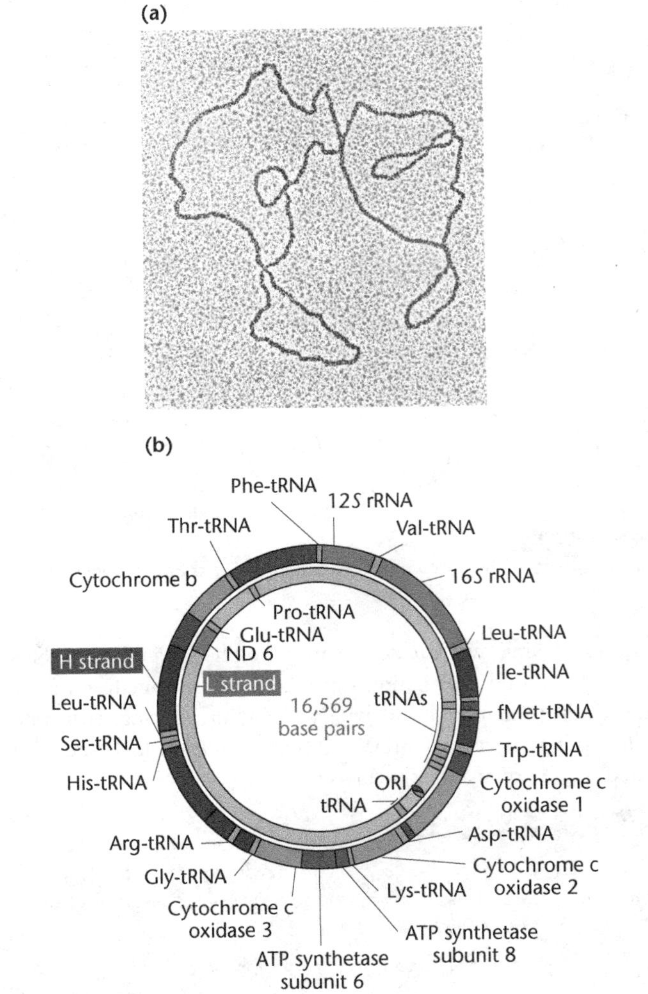


FIGURE 9–7 Examples of mitochondrial DNA. (a) Electron micrograph of mitochondrial DNA derived from the frog *Xenopus laevis*. (b) Diagram illustrating the arrangement of many of the genes encoded by human mtDNA.

## Molecular Organization and Gene Products of Chloroplast DNA

The details of the autonomous genetic system of chloroplasts have now been worked out, providing further support of the endosymbiotic theory. The chloroplast, responsible for photosynthesis, contains both DNA (as a source of genetic information) and a complete protein-synthesizing apparatus. The molecular components of the chloroplast's translation apparatus are derived from both nuclear and organelle genetic information.

Chloroplast DNA (cpDNA), shown in Figure 9–6(a), is fairly uniform in size among different organisms, ranging between 100 and 225 kb in length. It shares many similarities to DNA found in prokaryotic cells. It is circular and double stranded, and is free of the associated proteins characteristic of eukaryotic DNA. Compared with nuclear DNA from the same organism, it invariably shows a different buoyant density and base composition.

The size of cpDNA is much larger than that of mtDNA. To some extent, this can be accounted for by a larger number of genes. However, most of the difference appears to be due to the presence in cpDNA of many long noncoding nucleotide sequences both between and within genes, the latter being introns (noncoding DNA sequences, also characteristic of nuclear DNA, as you'll see in Chapter 14). Duplications of many DNA sequences are also present. Since such noncoding sequences vary in different plants, they are indicative that independent evolution occurred in chloroplasts following their initial invasion of a primitive eukaryotic-like cell.

In the green alga *Chlamydomonas*, there are about 75 copies of the chloroplast DNA molecule per organelle. Each copy consists of a length of DNA that contains 195,000 base pairs, or bp (195 kb). In higher plants, such as the sweet pea, multiple copies of the DNA molecule are also present in each organelle, but the molecule (134 kb) is considerably smaller than that in *Chlamydomonas*. Interestingly, genetic recombination between the multiple copies of DNA within chloroplasts has been documented in some organisms.

Numerous gene products encoded by chloroplast DNA function during translation within the organelle. Figure 9–6(b) illustrates genes that are present on cpDNA of the moss, which is representative of a variety of higher plants. Two sets each of the genes coding for the ribosomal RNAs—16*S*, and 23*S* rRNA—are present (*S* refers to the Svedberg coefficient, which is described in Chapter 10 and is related to the molecule's size and shape). In addition, cpDNA encodes numerous transfer RNAs (tRNAs), as well as many ribosomal proteins specific to the chloroplast ribosomes. For example, in the liverwort, whose cpDNA was the first to be sequenced, there are genes encoding 30 tRNAs, RNA polymerase, multiple rRNAs, and numerous ribosomal proteins. The variations in the gene products encoded in the cpDNA of different plants again attest to the independent evolution that occurred within chloroplasts.

Chloroplast ribosomes have a Svedberg coefficient slightly less than 70*S*, similar, but not identical, to bacterial ribosomes. Even though some chloroplast ribosomal proteins are encoded by chloroplast DNA and some by nuclear DNA, most, if not all, such proteins are distinct from their counterparts present in cytoplasmic ribosomes. Both observations provide direct support for the endosymbiotic theory.

Still other chloroplast genes specific to the photosynthetic function have been identified, as illustrated in Figure 9–6(b). For example, in the moss, there are 92 chloroplast genes encoding proteins that are part of the thylakoid membrane, a cellular component integral to the light-dependent reactions of photosynthesis. Mutations in these genes may inactivate photosynthesis. A typical distribution of genes between the nucleus and the chloroplast is illustrated by one of the major photosynthetic enzymes, ribulose-1-5-bisphosphate carboxylase (known as Rubisco). This enzyme has its small subunit encoded by a nuclear gene, whereas the large subunit is encoded by cpDNA.

## Molecular Organization and Gene Products of Mitochondrial DNA

Extensive information is also available concerning the structure and gene products of mitochondrial DNA (mtDNA). In most eukaryotes, mtDNA exists as a double-stranded, closed circle [Figure 9–7(a)] that, like cpDNA, is free of the chromosomal proteins characteristic of eukaryotic chromosomal DNA. An exception is found in some ciliated protozoans, in which the DNA is linear.

In size, mtDNA is much smaller than cpDNA and varies greatly among organisms, as demonstrated in Table 9.1. In a variety of animals, including humans, mtDNA consists of about 16,000 to 18,000 bp (16 to 18 kb). However, yeast (*Saccharomyces*) mtDNA consists of 75 kb. Plants typically exceed this amount—367 kb is present in mitochondria in the mustard plant, *Arabidopsis*. Vertebrates have 5 to 10 such DNA molecules per organelle, while plants have 20 to 40 copies per organelle.

**TABLE 9.1**

**The Size of mtDNA in Different Organisms**

| Organisms | Size (kb) |
|---|---|
| *Homo sapiens* (human) | 16.6 |
| *Mus musculus* (mouse) | 16.2 |
| *Xenopus laevis* (frog) | 18.4 |
| *Drosophila melanogaster* (fruit fly) | 18.4 |
| *Saccharomyces cerevisiae* (yeast) | 75.0 |
| *Pisum sativum* (pea) | 110.0 |
| *Arabidopsis thaliana* (mustard plant) | 367.0 |

There are several other noteworthy aspects of mtDNA. With only rare exceptions, introns are absent from mitochondrial genes, and gene repetitions are seldom present. Nor is there usually much in the way of intergenic spacer DNA. This is particularly true in species whose mtDNA is fairly small in size, such as humans. In *Saccharomyces*, with a much larger mtDNA molecule, much of the excess DNA is accounted for by introns and intergenic spacer DNA. As will be discussed in Chapter 14, the expression of mitochondrial genes uses several modifications of the otherwise standard genetic code. Also of interest is the fact that replication in mitochondria is dependent on enzymes encoded by nuclear DNA.

Human mtDNA [Figure 9–7(b)] encodes two ribosomal RNAs (rRNAs), 22 transfer RNAs (tRNAs), and also 13 polypeptides essential to the oxidative respiratory functions of the organelle. For instance, mitochondrial-encoded gene products are present in all of the protein complexes of the electron transport chain found in the inner membrane of mitochondria. In most cases, these polypeptides are part of multichain proteins, many of which also contain subunits that were encoded in the nucleus, synthesized in the cytoplasm, and then transported into the organelle. Thus, the protein-synthesizing apparatus and the molecular components for cellular respiration are jointly derived from nuclear and mitochondrial genes.

Another interesting observation is that in vertebrate mtDNA, the two strands vary in density, as revealed by centrifugation. This provides researchers with a way to isolate the strands for study, designating one heavy (H) and the other light (L). While most of the mitochondrial genes are encoded by the H strand, several are encoded by the complementary L strand.

As might be predicted by the endosymbiotic theory, ribosomes found in the organelle differ from those present in the neighboring cytoplasm. Table 9.2 shows that mitochondrial ribosomes of different species vary considerably in their Svedberg coefficients, ranging from 55*S* to 80*S*, while cytoplasmic ribosomes are uniformly 80*S*.

The majority of proteins that function in mitochondria are encoded by nuclear genes. In fact, over 1000 nuclear-coded gene products are essential to biological activity in the organelle. They include, for example, DNA and RNA polymerases, initiation and elongation factors essential for translation, ribosomal proteins,

TABLE 9.2

**Sedimentation Coefficients of Mitochondrial Ribosomes**

| Kingdom | Examples | Svedberg Coefficient (*S*) |
|---|---|---|
| Animals | Vertebrates | 55–60 |
| | Insects | 60–71 |
| Protists | *Euglena* | 71 |
| | *Tetrahymena* | 80 |
| Fungi | *Neurospora* | 73–80 |
| | *Saccharomyces* | 72–80 |
| Plants | Maize | 77 |

aminoacyl tRNA synthetases, and several tRNA species. These imported components are distinct from their cytoplasmic counterparts, even though both sets are coded by nuclear genes. For example, the synthetase enzymes essential for charging mitochondrial tRNA molecules (a process essential to translation) show a distinct affinity for the mitochondrial tRNA species as compared with the cytoplasmic tRNAs. Similar affinity has been shown for the initiation and elongation factors. Furthermore, while bacterial and nuclear RNA polymerases are known to be composed of numerous subunits, the mitochondrial variety consists of only one polypeptide chain. This polymerase is generally sensitive to antibiotics that inhibit bacterial RNA synthesis, but not to eukaryotic inhibitors. The various contributions of nuclear and mitochondrial gene products are contrasted in Figure 9–8.

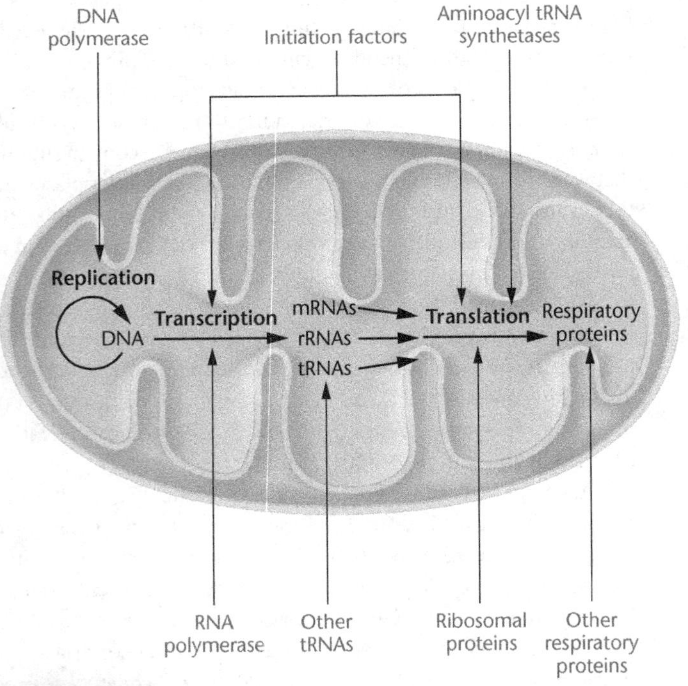


FIGURE 9–8 Gene products that are essential to mitochondrial function. Those shown entering the organelle are derived from the cytoplasm and encoded by nuclear genes.

NOW SOLVE THIS

In Problem 13 on page 243 you are asked to explain the number of different tRNA molecules encoded by human mitochondrial DNA in contrast to the actual number required for translation of proteins within the organelle.

■ HINT: *Mutations in certain nuclear genes have an impact on mitochondrial function.*

## 9.3 Mutations in Mitochondrial DNA Cause Human Disorders

The DNA found in human mitochondria has been completely sequenced and contains 16,569 base pairs. As mentioned earlier, mtDNA gene products include 13 proteins required for aerobic cellular respiration. Because a cell's energy supply is largely dependent on aerobic cellular respiration to generate ATP, disruption of any mitochondrial gene by mutation may potentially have a severe impact on that organism. We have seen this in our previous discussion of the *petite* mutations in yeast, which would be lethal were it not for this organism's ability to respire anaerobically. In fact, mtDNA is particularly vulnerable to mutations, for two possible reasons. First, in contrast to nuclear DNA, mtDNA does not have the structural protection from mutations provided by histone proteins in nuclear DNA. Second, the concentration of highly mutagenic free radicals generated by cell respiration that accumulate in such a confined space very likely raises the mutation rate in mtDNA.

The number of copies of mtDNA in human cells can range from several hundred in somatic cells to approximately 100,000 copies of mtDNA in an oocyte. Fortunately, a zygote receives a large number of organelles through the egg, so if only one organelle or a few of them contain a mutation, its impact is greatly diluted by the many mitochondria that lack the mutation and function normally. During early development, cell division disperses the initial population of mitochondria present in the zygote, and in the newly formed cells, these organelles reproduce autonomously. Therefore, if a deleterious mutation arises or is present in the initial population of organelles, adults will have cells with a variable mixture of both normal and abnormal organelles. This variation in the genetic content of organelles, as indicated earlier in the chapter, is called *heteroplasmy*.

In order for a human disorder to be attributable to genetically altered mitochondria, several criteria must be met:

1. Inheritance must exhibit a maternal rather than a Mendelian pattern.
2. The disorder must reflect a deficiency in the bioenergetic function of the organelle.
3. There must be a mutation in one or more of the mitochondrial genes.

Thus far, several disorders in humans are known to demonstrate these characteristics. For example, **myoclonic epilepsy and ragged red fiber disease (MERRF)** demonstrates a pattern of inheritance consistent with maternal transmission. Only offspring of affected mothers inherit the disorder; the offspring of affected fathers are normal. Individuals with this rare disorder express ataxia (lack of muscular coordination), deafness, dementia, and epileptic seizures. The disease is so named because of the presence of "ragged-red" skeletal muscle fibers that exhibit blotchy red patches resulting from the proliferation of aberrant mitochondria (Figure 9–9). Brain function, which has a high energy demand, is affected in this disorder, leading to the neurological symptoms described.

Analysis of mtDNA from patients with MERFF has revealed a mutation in one of the 22 mitochondrial genes encoding a transfer RNA. Specifically, the gene encoding $tRNA^{lys}$ (the tRNA that delivers lysine during translation) contains an A-to-G transition within its sequence. This genetic alteration apparently interferes with the capacity for translation within the organelle, which in turn leads to the various manifestations of the disorder.

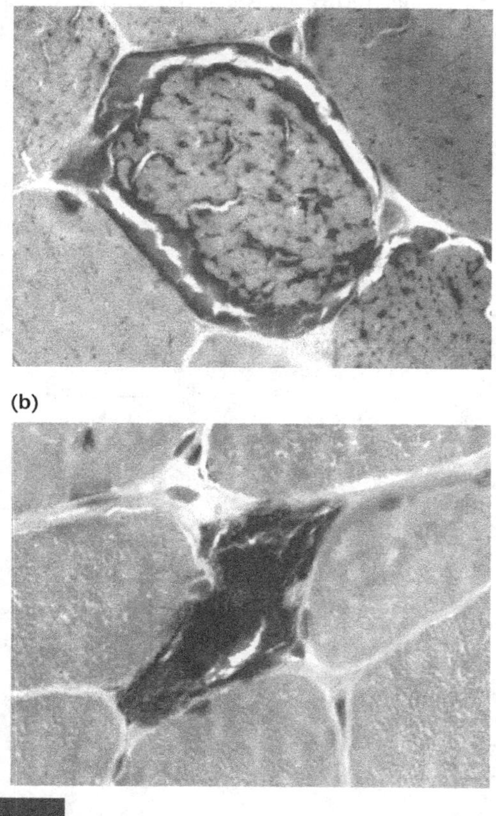

FIGURE 9–9 Ragged red fibers in skeletal muscle cells from patients with the mitochondrial disease MERRF. (a) The muscle fiber has mild proliferation of mitochondria. (See red rim and speckled cytoplasm.) (b) Marked proliferation in which mitochondria have replaced most cellular structures.

The cells of affected individuals exhibit heteroplasmy, containing a mixture of normal and abnormal mitochondria. Different patients contain different proportions of the two, and even different cells from the same patient exhibit various levels of abnormal mitochondria. Were it not for heteroplasmy, the mutation would very likely be lethal, testifying to the essential nature of mitochondrial function and its reliance on the genes encoded by mtDNA within the organelle.

A second disorder, **Leber's hereditary optic neuropathy (LHON)**, also exhibits maternal inheritance as well as mtDNA lesions. The disorder is characterized by sudden bilateral blindness. The average age of vision loss is 27, but onset is quite variable. Four mutations have been identified, all of which disrupt normal oxidative phosphorylation, the final pathway of respiration in cells. More than 50 percent of cases are due to a mutation at a specific position in the mitochondrial gene encoding a subunit of NADH dehydrogenase. This mutation is transmitted maternally through the mitochondria to all offspring. Noteworthy is the observation that in many instances of LHON, there is no family history; a significant number of cases are "sporadic," resulting from newly arisen mutations.

Individuals severely affected by a third disorder, **Kearns–Sayre syndrome (KSS)**, lose their vision, experience hearing loss, and display heart conditions. The genetic basis of KSS involves deletions at various positions within mtDNA. Many KSS patients are symptom-free as children but display progressive symptoms as adults. The proportion of mtDNAs that reveal deletion mutations increases as the severity of symptoms increases.

The study of hereditary mitochondrial-based disorders provides insights into the critical importance of this organelle during normal development. In fact, mitochondrial dysfunction seems to be implicated in most all major disease conditions, including Type II (late-onset) diabetes; atherosclerosis; neurodegenerative diseases such as Parkinson disease, Alzheimer disease, and Huntington disease; schizophrenia and bipolar disorders; and several cancers. It is becoming evident, for example, that mutations in mtDNA are present in such human malignancies as skin, colorectal, liver, breast, pancreatic, lung, prostate, and bladder cancers. Genetic tests for detecting mutations in the mtDNA genome that may serve as early-stage disease markers have been developed. For example, mtDNA mutations in skin cells have been detected as a biomarker of cumulative exposure of ultraviolet light and development of skin cancer. It is still unclear whether mtDNA mutations are causative effects contributing to development of malignant tumors or consequences of tumor formation. Nonetheless, there is an interesting link between mtDNA mutations and cancer. Studying interactions between nuclear and mtDNA gene products will be important for advancing our understanding of the possible roles of mitochondria in tumor formation. Moreover, the study of hereditary mitochondrial-based disorders has even suggested a hypothesis for aging, based on the progressive accumulation of mtDNA mutations and the accompanying loss of mitochondrial function.

9.4

## Infectious Heredity Is Based on a Symbiotic Relationship between Host Organism and Invader

Examples abound in eukaryotes of phenotypes being transmitted cytoplasmically by an invading microorganism or particle. The foreign invader coexists with its eukaryotic host in a symbiotic or a parasitic relationship, is usually passed through the maternal ooplasm to progeny cells or organisms, and confers a specific phenotype. We shall consider several examples of this phenomenon.

### Kappa in *Paramecium*

First described by Tracy Sonneborn, certain strains of *Paramecium aurelia* are called **Killers** because they release a cytoplasmic substance called **paramecin** that is toxic and sometimes lethal to sensitive strains. This substance is produced by so-called kappa particles that replicate in the Killer cytoplasm; one cell may contain 100 to 200 such particles. The kappa particles contain DNA and protein, and depend on a dominant nuclear gene, *K*, for their maintenance. To understand how the *K* gene and kappa particles are transmitted, we must look at the way paramecia reproduce.

Paramecia are diploid protozoans that can undergo sexual exchange of genetic information through the process of conjugation, shown in Figure 9–10. Each paramecium contains one macronucleus and two diploid micronuclei. Early in conjugation, both micronuclei in each mating pair undergo meiosis, resulting in eight haploid micronuclei. Seven of these degenerate (as does the macronucleus), and the remaining one undergoes a single mitotic division. Each cell then donates one of the two haploid micronuclei to the other, re-creating the diploid condition in both cells. As a result, diploid exconjugates typically have identical genotypes. In some instances, conjugation may be accompanied by cytoplasmic exchange.

Paramecia may also undergo **autogamy**, a similar process except that it involves only a single cell. Following meiosis of both micronuclei, seven products degenerate and one survives. This nucleus divides, and the resulting nuclei fuse to re-create the diploid condition. If the original cell was heterozygous, autogamy results in homozygosity, because the newly formed diploid nucleus is derived solely from a single haploid meiotic product. In a population of cells originally heterozygous, half of the new cells express one allele and half express the other allele.

Figure 9–11 illustrates the results of crosses between *KK* and *kk* cells, with and without cytoplasmic exchange. Sometimes, no cytoplasmic exchange occurs, even though the resultant cells may be *Kk* (or *KK*, following autogamy). When none occurs, no kappa particles are transmitted and cells remain sensitive. When exchange occurs, the cells become Killers, provided the kappa particles are supported by at least one dominant *K* allele.

Kappa particles are bacteria-like and may contain temperate bacteriophages. One theory holds that these viruses of kappa may enter the vegetative state and proceed to reproduce. During this multiplication, they produce the toxic products that are released and kill sensitive strains.

### Infective Particles in *Drosophila*

Two examples of infectious heredity are known in *Drosophila*, **$CO_2$ sensitivity** and **sex ratio**. In the former, flies that would normally

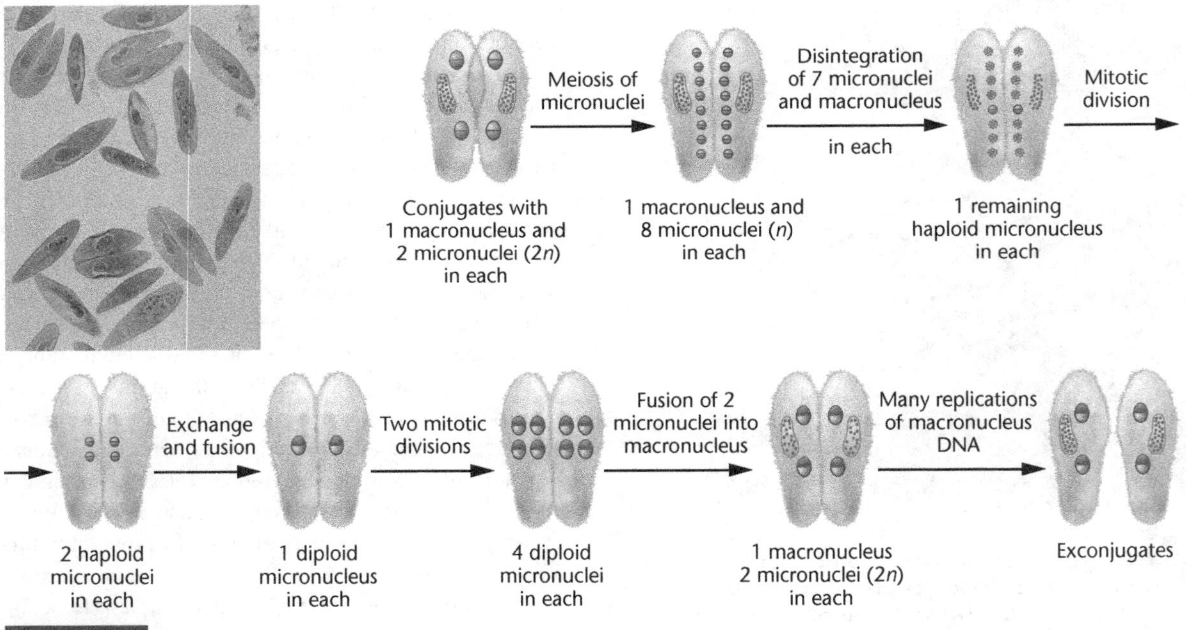


FIGURE 9–10 Genetic events occurring during conjugation in *Paramecium*. The photographic inset shows several pairs of organisms undergoing conjugation.

recover from carbon dioxide anesthetization instead become permanently paralyzed and are killed by $CO_2$. Sensitive mothers pass this trait to all offspring. Furthermore, extracts of sensitive flies induce the trait when injected into resistant flies. Phillip L'Heritier has postulated that sensitivity is due to the presence of a virus called **sigma**. The particle has been visualized under the electron microscope and is smaller than kappa. Attempts to transfer the virus to other insects have been unsuccessful, demonstrating that specific genes support the presence of sigma in *Drosophila*.

Our second example of infective particles comes from the study of *Drosophila bifasciata*. A small number of these flies produce predominantly female offspring if reared at 21°C or lower. This condition, designated sex ratio, is transmitted to daughters but not to the low percentage of males produced. Such a phenomenon was subsequently investigated in *Drosophila willistoni*. In these flies, the injection of ooplasm from sex-ratio females into the abdomens of normal females induced the condition, suggesting that an extrachromosomal element is responsible for the sex-ratio phenotype. The agent has now been isolated and shown to be a protozoan. While the protozoan has been found in both males and females, it is lethal primarily to developing male larvae. There is now some evidence that a virus harbored by the protozoan may be responsible for producing a male-lethal toxin.

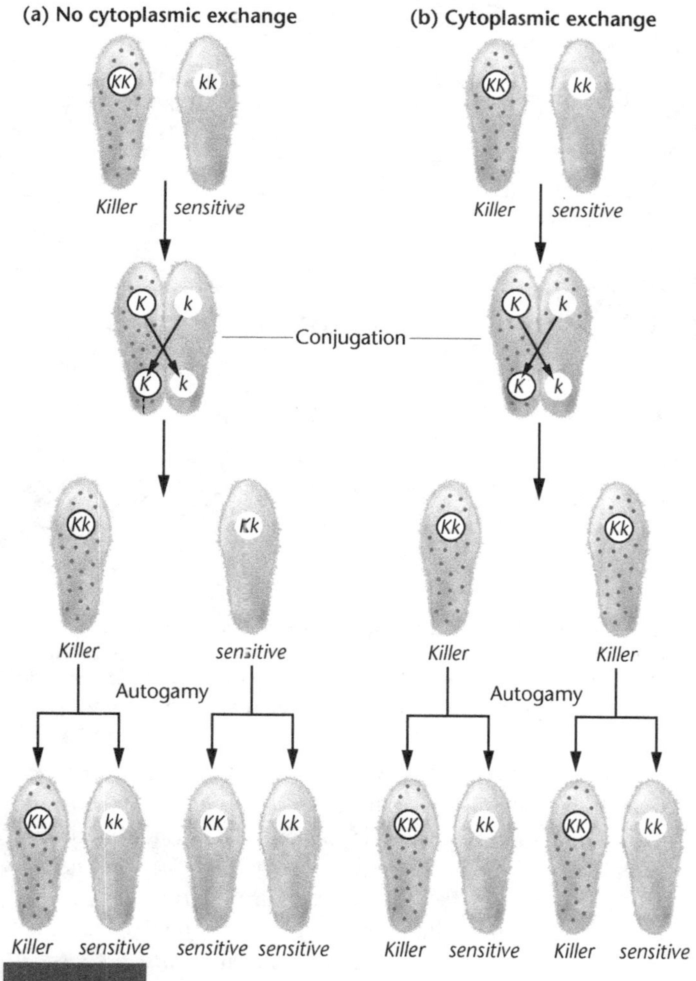


FIGURE 9–11 Results of crosses between Killer (*KK*) and sensitive (*kk*) strains of *Paramecium*, with and without cytoplasmic exchange during conjugation. The kappa particles (dots) are maintained only when a *K* allele is present.

## 9.5 In Maternal Effect, the Maternal Genotype Has a Strong Influence during Early Development

In **maternal effect**, also referred to as maternal influence, an offspring's phenotype for a particular trait is under the control of nuclear gene products present in the egg. This is in contrast to biparental inheritance, where both parents transmit information on genes in the nucleus that determines the offspring's phenotype. In cases of maternal effect, the nuclear genes of the female gamete are transcribed, and the genetic products (either proteins or yet untranslated mRNAs) accumulate in the egg cytoplasm. After fertilization, these products are distributed among newly formed cells and influence the patterns or traits established during early development. Three examples will illustrate such an influence of the maternal genome on particular traits.

### *Ephestia* Pigmentation

A very straightforward illustration of a maternal effect is seen in the Mediterranean meal moth, *Ephestia kuehniella*. The wild-type larva of this moth has a pigmented skin and brown eyes as a result of the dominant gene *A*. The pigment is derived from a precursor molecule, kynurenine, which is in turn a derivative of the amino acid tryptophan. A mutation, *a*, interrupts the synthesis of kynurenine and, when homozygous, may result in red eyes and little pigmentation in larvae. However, as illustrated in Figure 9–12, results of the cross

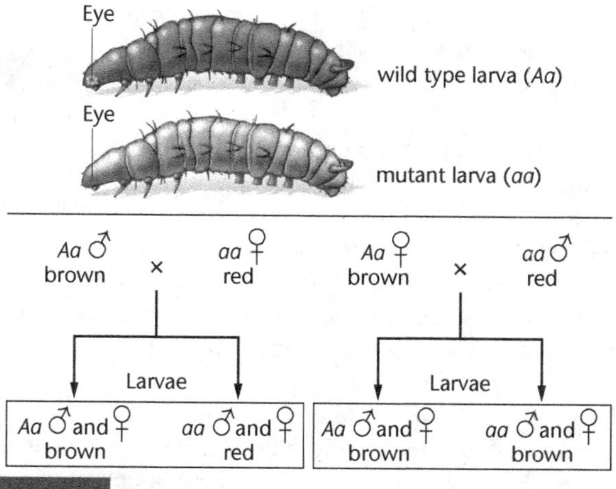


FIGURE 9–12 Maternal influence in the inheritance of eye pigment in the meal moth *Ephestia kuehniella*. Multiple light receptor structures (eyes) are present on each side of the anterior portion of larvae.

*Aa* × *aa* depend on which parent carries the dominant gene. When the male is the heterozygous parent, a 1:1 brown- to red-eyed ratio is observed in larvae, as predicted by Mendelian segregation. When the female is heterozygous for the *A* gene, however, all larvae are pigmented and have brown eyes, in spite of half of them being *aa*. As these larvae develop into adults, one-half of them gradually develop red eyes, reestablishing the 1:1 ratio.

One explanation for these results is that the *Aa* oocytes synthesize kynurenine or an enzyme necessary for its synthesis and accumulate it in the ooplasm prior to the completion of meiosis. Even in *aa* progeny, if the mothers were *Aa*, this pigment is distributed in the cytoplasm of the cells of the developing larvae; thus, they develop pigmentation and brown eyes. In these progeny, however, the pigment is eventually diluted among many cells and depleted, resulting in the conversion to red eyes as adults. The *Ephestia* example demonstrates the maternal effect in which a cytoplasmically stored nuclear gene product influences the larval phenotype and, at least temporarily, overrides the genotype of the progeny.

## *Limnaea* Coiling

Shell coiling in the snail *Limnaea peregra* is an excellent example of maternal effect on a permanent rather than a transitory phenotype. Some strains of this snail have left-handed, or sinistrally, coiled shells (*dd*), while others have right-handed, or dextrally, coiled shells (*DD* or *Dd*). These snails are hermaphroditic and may undergo either cross- or self-fertilization, providing a variety of types of matings.

Figure 9–13 illustrates the results of reciprocal crosses between true-breeding snails. As you can see, these crosses yield different outcomes, even though both are between sinistral and dextral organisms and produce all heterozygous offspring. Examination of the progeny reveals that their phenotypes depend on the genotypes of the female parents. If we adopt that conclusion as a working hypothesis, we can test it by examining the offspring in subsequent generations of self-fertilization events. In each case, the hypothesis is upheld. Ovum donors that are *DD* or *Dd* produce only dextrally coiled progeny. Maternal parents that are *dd* produce only sinistrally coiled progeny. The coiling pattern of the progeny snails is determined by the genotype of the parent producing the egg, *regardless of the phenotype of that parent.*

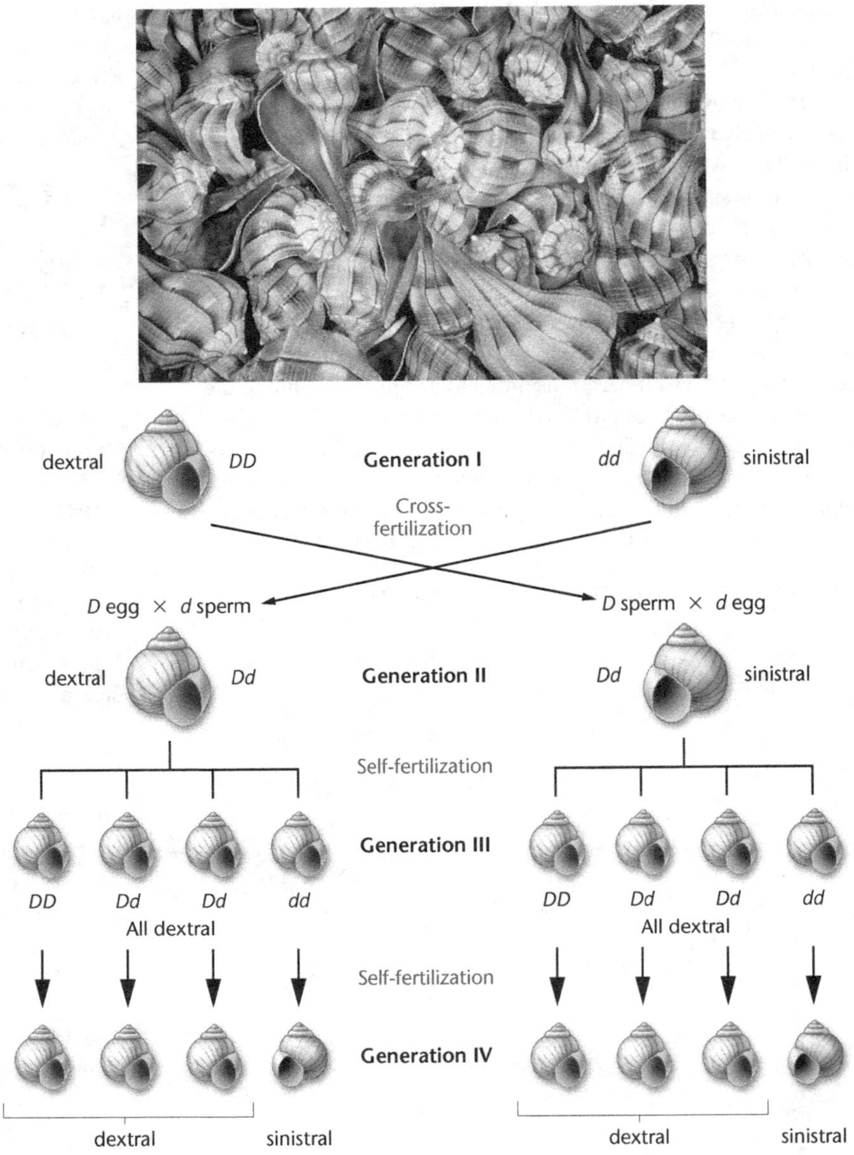


FIGURE 9–13 Inheritance of coiling in the snail *Limnaea peregra*. Coiling is either dextral (right handed) or sinistral (left handed). A maternal effect is evident in generations II and III, where the genotype of the maternal parent, rather than the offspring's own genotype, controls the phenotype of the offspring. The photograph illustrates a mixture of right- vs. left-handed coiled snails.

Investigation of the developmental events in *Limnaea* reveals that the orientation of the spindle in the first cleavage division after fertilization determines the direction of coiling. Spindle orientation appears to be controlled by maternal genes acting on the developing eggs in the ovary. The orientation of the spindle, in turn, influences cell divisions following fertilization and establishes the permanent adult coiling pattern. The dextral allele (*D*) produces an active gene product that causes right-handed coiling. If ooplasm from dextral eggs is injected into uncleaved sinistral eggs, they cleave in a dextral pattern. However, in the converse experiment, sinistral ooplasm has no effect when injected into dextral eggs. Apparently, the sinistral allele is the result of a classic recessive mutation that encodes an inactive gene product.

We can conclude, then, that females that are either *DD* or *Dd* produce oocytes that synthesize the *D* gene product, which is stored in the ooplasm. Even if the oocyte contains only the *d* allele following meiosis and is fertilized by a *d*-bearing sperm, the resulting *dd* snail will be dextrally coiled (right handed).

## Embryonic Development in *Drosophila*

A more recently documented example of maternal effect involves various genes that control embryonic development in *Drosophila melanogaster*. The genetic control of embryonic development in *Drosophila*, discussed in greater detail in Chapter 19, is a fascinating story. The protein products of the maternal-effect genes function to activate other genes, which may in turn activate still other genes. This cascade of gene activity leads to a normal embryo whose subsequent development yields a normal adult fly. The extensive work by Edward B. Lewis, Christiane Nüsslein-Volhard, and Eric Wieschaus (who shared the 1995 Nobel Prize for Physiology or Medicine for their findings) has clarified how these and other genes function. Genes that illustrate maternal effect have products that are synthesized by the developing egg and stored in the oocyte prior to fertilization. Following fertilization, these products create molecular gradients that determine spatial organization as development proceeds.

For example, the gene *bicoid* (*bcd*) plays an important role in specifying the development of the anterior portion of the fly. Embryos derived from mothers who are homozygous for this mutation ($bcd^-/bcd^-$) fail to develop anterior areas that normally give rise to the head and thorax of the adult fly. Embryos whose mothers contain at least one wild-type allele ($bcd^+$) develop normally, even if the genotype of the embryo is homozygous for the mutation. Consistent with the concept of *maternal effect*, the *genotype of the female parent*, not the genotype of the embryo, determines the phenotype of the offspring.

When we return to our discussion of this general topic in Chapter 19, we will see examples of other genes illustrating maternal effect, as well as many "zygotic" genes whose expression occurs during early development and that behave genetically in the conventional Mendelian fashion.

## GENETICS, TECHNOLOGY, AND SOCIETY

### Mitochondrial DNA and the Mystery of the Romanovs

By most accounts, Nicholas II, the last Tsar of Russia, was a substandard monarch. He was accused of bungling during the Russo-Japanese War of 1904–1905, and his regime was plagued by corruption and incompetence. Even so, he probably didn't deserve the fate that befell him and his family one summer night in 1918. As we shall see, a full understanding of that event has relied on, of all things, mitochondrial DNA.

After being forced to abdicate in 1917, ending 300 years of Romanov rule, Tsar Nicholas and the imperial family were banished to Ekaterinburg in western Siberia. There, it was believed, they would be out of reach of the fiercely anti-imperialist Bolsheviks, who were then fighting to gain control of the country. But the Bolsheviks eventually caught up with the Romanovs. During a July night in 1918, Tsar Nicholas, Tsarina Alexandra (granddaughter of Queen Victoria), their five children (Olga, 22; Tatiana, 21; Marie, 19; Anastasia, 17; and Alexei, 13), their family doctor, and three of their servants were awakened and brought to a downstairs room of the house where they were held prisoner. There they were made to form a double row against the wall, presumably so that a photograph could be taken. Instead, 11 men with revolvers burst into the room and opened fire. After exhausting their ammunition, they proceeded to bayonet the bodies and smash their faces in with rifle butts. The corpses were then hauled away and flung down a mineshaft, only to be pulled out two days later and dumped into a shallow grave, doused with sulfuric acid, and covered over. There they rested for more than 60 years.

An air of mystery soon developed around the demise of the Romanovs. Did all of the children of Nicholas and Alexandra die with their parents that bloody night, or did the youngest daughter, Anastasia, get away? Over the years, the possibility that Anastasia miraculously escaped execution has inspired countless books, a classic Hollywood movie, a ballet, a Broadway play, and, most recently, an animated movie. In all of these retellings, Anastasia reemerges to claim her birthright as the only surviving member of the Romanovs. Adding to the puzzle was one Anna Anderson. Two years after being dragged from a Berlin canal after a suicide attempt in 1920, she began to claim to be the Grand Duchess Anastasia. Despite a history of mental instability and a curious inability to speak Russian, she managed to convince many people.

The unraveling of the mystery began in 1979, when a Siberian geologist and a Moscow filmmaker discovered four skulls they believed to belong to the Tsar's family. It wasn't until the summer of 1991, after the establishment of *glasnost* in the former Soviet Union, that exhumation began. Altogether, almost 1000 bone fragments were recovered, which were reassembled into nine skeletons—five females and four males. Based on measurements of the bones and computer-assisted superimposition of the skulls onto photographs, the remains were tentatively identified as belonging to the murdered Romanovs. But there were still two missing bodies, one of the daughters (believed to be Anastasia) and the boy Alexei.

The next step in authenticating the remains involved studies of DNA that were conducted by Pavel Ivanov, the leading Russian forensic DNA analyst, in collaboration with Peter Gill of the British Forensic Science Service. Their goals were to establish family relationships among the remains, and then to determine, by comparisons with living relatives, whether the family group was in fact the Romanovs. They froze bone fragments from the nine skeletons in liquid nitrogen, ground them into a fine powder, and extracted small amounts of DNA, which included both nuclear DNA and mitochondrial DNA (mtDNA). Genomic DNA typing of each skeleton confirmed the familial relationships

*Continued on next page*

*Genetics, Technology, and Society, continued*

and showed that, indeed, one of the princesses and Alexei were missing. Now final proof that the bones belonged to the Romanovs awaited the analysis of mtDNA.

Mitochondrial DNA is ideal for forensic studies for several reasons. First, since all the mitochondria in a human cell are descended from the mitochondria present in the egg, mtDNA is transmitted strictly from mother to offspring, never from father to offspring. Therefore, mtDNA sequences can be used to trace maternal lineages without the complicating effects of meiotic crossing over, which recombines maternal and paternal nuclear genes every generation. In addition, mtDNA is small (~16,600 base pairs) and present in 500 to 1000 copies per cell, so it is much easier to recover intact than nuclear DNA.

Ivanov's group amplified two highly variable regions of the mtDNA isolated from all nine bone samples and determined the nucleotide sequences of these regions. By comparing these sequences with living relatives of the Romanovs, they hoped to establish the identity of the Ekaterinburg remains once and for all. The sequences from Tsarina Alexandra were an exact match with those from Prince Philip of England, who is her grandnephew, verifying her identity. However, authentication was more complicated for the Tsar.

The Tsar's sequences were compared with those from the only two living relatives who could be persuaded to participate in the study, Countess Xenia Cheremeteff-Sfiri (his great-grandniece) and James George Alexander Bannerman Carnegie, third Duke of Fife (a more distant relative, descended from a line of women stretching back to the Tsar's grandmother). These comparisons produced a surprise. At position 16,169 of the mtDNA, the Tsar seemed to have either one or another base, a C or a T. The Countess and the Duke, in contrast, both had only T. The conclusion was that Tsar Nicholas had two different populations of mitochondria in his cells, each with a different base at position 16,169 of its DNA. This condition, called heteroplasmy, is now believed to occur in 10 to 20 percent of humans.

This difference between the mtDNA of the presumed Tsar and that of his two living maternal relatives cast doubt on the identification of the remains. Fortunately, the Russian government granted a request to analyze the remains of the Tsar's younger brother, Grand Duke Georgij Romanov, who died in 1899 of tuberculosis. The Grand Duke's mtDNA was found to have the same heteroplasmic variant at position 16,169, either a C or a T. It was concluded that the Ekaterinburg bones were the doomed imperial family. With years of controversy finally resolved, the now-authenticated remains of Tsar Nicholas II, Tsarina Alexandra, and three of their daughters were buried in the St. Peter and Paul Cathedral in St. Petersburg on July 17, 1998, 80 years to the day after they were murdered.

This DNA analysis did not solve the mystery of the fate of Anastasia, however. Did she die with her parents, sisters, and brother in 1918? Or is it possible that Anna Anderson was telling the truth, that she was the escaped duchess? In a separate study, Anna Anderson's nuclear and mtDNA were recovered from intestinal tissue preserved from an operation performed five years before her death in 1984. Analysis of this DNA proved that she was not Anastasia, but rather a Polish peasant named Franziska Schanzkowska.

If Anastasia's remains were not among those of her parents and sisters, and if Anna Anderson was an imposter, what did happen to Anastasia? Most evidence suggests that Anastasia and her brother Alexei were not found with the others in the mass grave because their bodies were burned over the grave site two days after the killings and the ashes scattered, never to be found again. Not exactly a Hollywood ending.

### References

Gibbons, A. 1998. Calibrating the mitochondrial clock. *Science* 279: 28–29.

Ivanov, P., et al. 1996. Mitochondrial DNA sequence heteroplasmy in the Grand Duke of Russia Georgij Romanov establishes the authenticity of the remains of Tsar Nicholas II. *Nat. Genet.* 12: 417–420.

Masse, R. 1996. *The Romanovs: The final chapter.* New York: Ballantine Books.

Stoneking, M., et al. 1995. Establishing the identity of Anna Anderson Manahan. *Nat. Genet.* 9: 9–10.

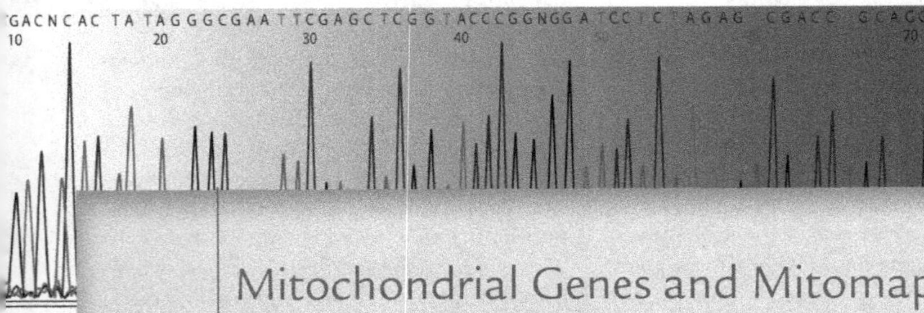

## EXPLORING GENOMICS

# Mitochondrial Genes and Mitomap

As you learned in this chapter, sequences in the human mitochondrial DNA (mtDNA) genome can vary from individual to individual and show variation within individuals. When abnormal copies of mtDNA predominate in a cell, disorders may be observed, particularly in cells with a high dependency on ATP, such as neurons, muscle cells, kidney cells, and cells of the pancreas. **Mitomap** is a database of mtDNA sequences from different species. In this exercise, we explore Mitomap to learn more about the human mitochondrial genome and disorders associated with mitochondrial genes.

**Exercise I – Mutations of the Human Mitochondrial Locus *MT-CO1* and the *COI* Gene**

1. Access Mitomap at www.mitomap.org.
2. Scroll to the bottom of the Mitomap home page to see a map of human mtDNA. The labels point to mitochondrial genes involved in human disease. Locate the part of the map labeled "DEAF."
3. To learn more about the gene involved in this condition, return to the top of the home page and in the "Search Mitomap for information on" box, type in "DEAF." Click on "Perform search." On the next page that appears, under the "Basic Search" column, the button labeled "All of Mitomap" will be indicated as the default option. Leave this button checked, then click on "search."
4. Scroll down the page to the tables of DEAF mutations. Review the first table and answer these questions (following additional instructions as required):
   a. What is the locus for this gene in the mtDNA genome?
   b. Click on the locus link to see a page of mtDNA loci and gene names. What is the gene symbol (abbreviation) and name of this gene?
   c. Refer to the "Map Position" column, which shows gene positions according to nucleotides in the mtDNA genome. What are the first and last nucleotides for this gene?
   d. Refer back to the table on the previous page. Which nucleotides in the coding region of this gene are most commonly mutated by base-pair substitutions?
   e. What are the three most common nucleotide changes in the coding region of this gene?
   f. What is the resulting amino acid change created by each mutation? (*Note:* Mitomap shows amino acid changes according to the single-letter code for an amino acid. Refer to Figure 15–16 for a table of amino acids listed by their single-letter code. "Ter" refers to a termination or stop codon. Also see Figure 14–7).

**Exercise II – Functions of *COI* in Human Disease**

1. To learn more about *COI*, click on the disease link "DEAF" in the top row of the table. This will take you to a page of Online Mendelian Inheritance in Man (OMIM) links for mitochondrial genes.
2. Scroll to the DEAF links and click on the "aminoglycoside-induced DEAFness" link. When the OMIM page appears, click on the link for "Deafness, Aminoglycoside-Induced" ("#580000") and answer the following:
   a. This condition is a main reason that the antibiotics kanamycin, neomycin, and streptomycin are generally not prescribed for treating ear infections in young children. Explain why this is so.
   b. Name two other genes thought to be involved in this condition.
3. Return to the first OMIM page to appear when you left Mitomap and use the links there to learn about the *TRMU* gene at locus 22q13 and its role in aminoglycoside-induced deafness.
4. Refer back to the table you viewed in Exercise I. Is this genetic disorder due to heteroplasmy (the presence of organelles with a normal and abnormal genotype for a particular allele) or homoplasmy (all organelles have the same genotype)?

# Chapter Summary

1. Patterns of inheritance sometimes vary from those expected from the biparental transmission of nuclear genes. In such instances, phenotypes most often appear to result from genetic information transmitted through the egg.
2. Organelle heredity is based on the genotypes of chloroplast and mitochondrial DNA, as these organelles are transmitted to offspring. Chloroplast mutations affect the photosynthetic capabilities of plants, whereas mitochondrial mutations affect cells highly dependent on energy generated through cellular respiration. The resulting mutants display phenotypes related to the loss of function of these organelles.
3. Both chloroplasts and mitochondria first appeared in eukaryotic cells some 2 billion years ago, originating as invading protobacteria. As evolution proceeded, many of the genes of the bacteria were transferred to the nucleus of the cell, and a symbiotic relationship developed in which these organelles enhanced the energetic capacity of the cell. Evidence in support of this endosymbiotic theory is extensive and centers around many observations involving the DNA and genetic machinery present in modern-day chloroplasts and mitochondria.
4. Another form of extranuclear inheritance is attributable to the transmission of infectious microorganisms. Kappa particles, $CO_2$-sensitivity, and sex-ratio determinants are examples.
5. Maternal-effect patterns result when nuclear gene products controlled by the maternal genotype of the egg influence early development. *Ephestia* pigmentation, coiling in snails, and gene expression during early development in *Drosophila* are examples.

# INSIGHTS AND SOLUTIONS

1. Analyze the following hypothetical pedigree, determine the most consistent interpretation of how the trait is inherited, and point out any inconsistencies:

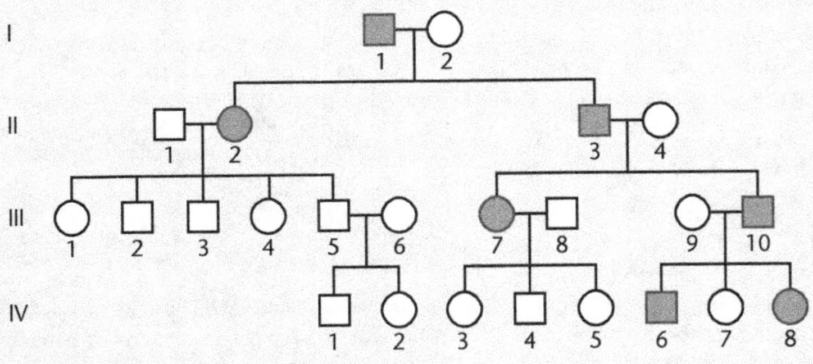


**Solution:** The trait is passed from all affected male parents to all but one offspring, but it is *never* passed maternally. Individual IV-7 (a female) is the only exception.

2. Can the explanation in Solution 1 be attributed to a gene on the Y chromosome? Defend your answer.

**Solution:** No, because male parents pass the trait to their daughters as well as to their sons.

3. Is the above pedigree an example of a paternal effect or of paternal inheritance?

**Solution:** It has all the characteristics of paternal inheritance because males pass the trait to almost all of their offspring. To assess whether the trait is due to a paternal effect (resulting from a nuclear gene in the male gamete), analysis of further matings would be needed.

## Problems and Discussion Questions

1. What genetic criteria distinguish a case of extranuclear inheritance from a case of Mendelian autosomal inheritance? from a case of X-linked inheritance?
2. Streptomycin resistance in *Chlamydomonas* may result from a mutation in either a chloroplast gene or a nuclear gene. What phenotypic results would occur in a cross between a member of an $mt^+$ strain resistant in both genes and a member of a strain sensitive to the antibiotic? What results would occur in the reciprocal cross?
3. A plant may have green, white, or green-and-white (variegated) leaves on its branches, owing to a mutation in the chloroplast that prevents color from developing. Predict the results of the following crosses:

| Ovule Source | | Pollen Source |
|---|---|---|
| (a) Green branch | × | White branch |
| (b) White branch | × | Green branch |
| (c) Variegated branch | × | Green branch |
| (d) Green branch | × | Variegated branch |

4. In aerobically cultured yeast, a *petite* mutant is isolated. To determine the type of mutation causing this phenotype, the *petite* and wild-type strains are crossed. Shown here are three potential outcomes of such a cross. For each set of results, what conclusion about the type of *petite* mutation is justified?
   (a) all wild type
   (b) some *petite*: some wild type
   (c) all *petite*
5. In diploid yeast strains, sporulation and subsequent meiosis can produce haploid ascospores, which may fuse to reestablish diploid cells. When ascospores from a *segregational petite* strain fuse with those of a normal wild-type strain, the diploid zygotes are all normal. Following meiosis, ascospores are *petite* and normal. Is the *segregational petite* phenotype inherited as a dominant or a recessive trait?
6. Predict the results of a cross between ascospores from a *segregational petite* strain and a *neutral petite* strain. Indicate the phenotype of the zygote and the ascospores it may subsequently produce.
7. Described here are the results of three crosses between strains of *Paramecium*:

| | | |
|---|---|---|
| (a) Killer × sensitive | → | 1/2 Killer: 1/2 sensitive |
| (b) Killer × sensitive | → | all Killer |
| (c) Killer × sensitive | → | 3/4 Killer: 1/4 sensitive |

   Determine the genotypes of the parental strains.
8. *Chlamydomonas*, a eukaryotic green alga, is sensitive to the antibiotic erythromycin, which inhibits protein synthesis in prokaryotes.
   (a) Explain why.
   (b) There are two mating types in this alga, $mt^+$ and $mt^-$. If an $mt^+$ cell sensitive to the antibiotic is crossed with an $mt^-$ cell that is resistant, all progeny cells are sensitive. The reciprocal cross ($mt^+$ resistant and $mt^-$ sensitive) yields all resistant progeny cells. Assuming that the mutation for resistance is in the chloroplast DNA, what can you conclude from the results of these crosses?
9. In *Limnaea*, what results would you expect in a cross between a *Dd* dextrally coiled and a *Dd* sinistrally coiled snail, assuming cross-fertilization occurs as shown in Figure 9–13? What results would occur if the *Dd* dextral produced only eggs and the *Dd* sinistral produced only sperm?

10. In a cross of *Limnaea*, the snail contributing the eggs was dextral but of unknown genotype. Both the genotype and the phenotype of the other snail are unknown. All $F_1$ offspring exhibited dextral coiling. Ten of the $F_1$ snails were allowed to undergo self-fertilization. One-half produced only dextrally coiled offspring, whereas the other half produced only sinistrally coiled offspring. What were the genotypes of the original parents?
11. In *Drosophila subobscura*, the presence of a recessive gene called *grandchildless* (*gs*) causes the offspring of homozygous females, but not those of homozygous males, to be sterile. Can you offer an explanation as to why females and not males are affected by the mutant gene?
12. A male mouse from a true-breeding strain of hyperactive animals is crossed with a female mouse from a true-breeding strain of lethargic animals. (These are both hypothetical strains.) All the progeny are lethargic. In the $F_2$ generation, all offspring are lethargic. What is the best genetic explanation for these observations? Propose a cross to test your explanation.
13. DNA in human mitochondria encode 22 different transfer RNA (tRNA) molecules. However, 32 different tRNA molecules are required for translation of proteins within mitochondria. How is this possible?
14. Consider the case where a mutation occurs that disrupts translation in a single human mitochondrion found in the oocyte participating in fertilization. What is the likely impact of this mutation on the offspring arising from this oocyte?
15. What is the endosymbiotic theory, and why is this theory relevant to the study of extranuclear DNA in eukaryotic organelles?
16. In the Problems and Discussion Questions in Chapter 7, questions 34–36, we described CC, the cat created by nuclear transfer cloning, whereby a diploid nucleus from one cell is injected into an enucleated egg cell to create an embryo. Cattle, sheep, rats, dogs, and several other species have been cloned using nuclei from somatic cells. Embryos and adults produced by this approach often show a number of different mitochondrial defects. Explain possible reasons for the prevalence of mitochondrial defects in embryos created by nuclear transfer cloning.
17. Carbon dioxide sensitivity in *Drosophila* was first discovered in 1937 and shown to be caused by an inherited Rhabdovirus named sigma. While its transmission is primarily maternal, males too can pass the virus, though at a lower frequency. Some *Drosophila* strains carry resistance genes to sigma. $CO_2$ sensitivity has also been described in some strains of the mosquito *Culex*. Assume that a series of reciprocal crosses between two *Culex* strains, A and B, produce the following results:

| Female Parent | Male Parent | Number of crosses | Percent offspring sensitive* |
|---|---|---|---|
| A | B | 20 | 84% |
| B | A | 20 | 42% |

*These percentages represent the number of crosses having more than 10% of the offspring sensitive to $CO_2$.

Compare these results with what you would expect to see for transmission and resistance of the sigma virus in *Drosophila*.

**HOW DO WE KNOW?**

18. In this chapter, we focused on extranuclear inheritance and how traits can be determined by genetic information contained in mitochondria and chloroplasts, and we discussed how expression of maternal genotypes can affect the phenotype of an organism. At the same time, we found many opportunities to consider the methods and reasoning by which much of this information was acquired. From the explanations given in the chapter, what answers would you propose to the following fundamental questions?
    (a) How was it established that particular phenotypes are inherited as a result of genetic information present in the chloroplast rather than in the nucleus?
    (b) How did the discovery of three categories of *petite* mutations in yeast lead researchers to postulate extranuclear inheritance of colony size?
    (c) What experimental observations support the endosymbiotic theory?
    (d) What key observations in crosses between dextrally and sinistrally coiled snails support the explanation that this phenotype is the result of maternal-effect inheritance?

## Extra-Spicy Problems

19. The specification of the anterior–posterior axis in *Drosophila* embryos is initially controlled by various gene products that are synthesized and stored in the mature egg following oogenesis. Mutations in these genes result in abnormalities of the axis during embryogenesis. These mutations illustrate *maternal effect*. How do such mutations vary from those produced by organelle heredity? Devise a set of parallel crosses and expected outcomes involving mutant genes that contrast maternal effect and organelle heredity.
20. The maternal-effect mutation *bicoid* (*bcd*) is recessive. In the absence of the bicoid protein product, embryogenesis is not completed. Consider a cross between a female heterozygous for the *bicoid* alleles ($bcd^+/bcd^-$) and a male homozygous for the mutation ($bcd^-/bcd^-$).
    (a) How is it possible for a male homozygous for the mutation to exist?
    (b) Predict the outcome (normal vs. failed embryogenesis) in the $F_1$ and $F_2$ generations of the cross described.
21. Shown here is a pedigree for a hypothetical human disorder:

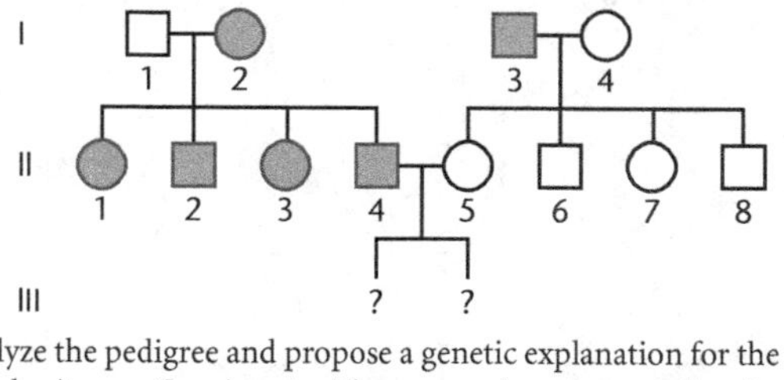


Analyze the pedigree and propose a genetic explanation for the nature of its inheritance. Consistent with your explanation, predict the outcome of a mating between individuals II-4 and II-5.

22. In the late 1950s, Yuichiro Hiraizumi, a postdoctoral fellow at the University of Wisconsin, crossed a laboratory strain of *Drosophila melanogaster* that was homozygous for the recessive second chromosomal

mutations *cinnabar* (*cn*) and *brown* (*bw*) with a wild-type strain collected in Madison, Wisconsin (Hiraizumi, Y. 1959. *Genetics* 44: 232–250). The lab strain had white eyes, while the wild strain had red eyes and was subsequently designated *SD*. The resulting progeny, being heterozygous for *cn* and *bw*, had red eyes. When Hiraizumi backcrossed $F_1$ females with *cn bw*/*cn bw* males, 50 percent of the offspring had white eyes, as expected. When the reciprocal backcross was made ($F_1$ males with *cn bw*/*cn bw* females), less than 2 percent of the flies had white eyes.

(a) Propose an explanation that is consistent with these results.
(b) Design a genetic experiment to test your hypothesis.
(c) *SD* stands for "segregation distortion." What is the significance of this description?

23. Extrachromosomally inherited traits are widespread among arthropods. In the two-spotted ladybird beetle, *Adalia bipunctata*, a male-killing trait has been discovered in which certain strains of females display a distorted sex ratio that favors female offspring (Werren, J., et al. 1994. *J. Bacteriol.* 176: 388–394). Unaffected strains show a normal one-to-one sex ratio. Two key observations are that affected strains can be cured by antibiotics, and that in addition to their normal 18*S* and 28*S* rRNA, 16*S* rDNA can be detected by PCR (polymerase chain reaction) analysis. Of the modes of extranuclear inheritance described in the text (organelle heredity, infectious heredity, and maternal effect), which is most likely to be causing this altered sex ratio in *Adalia*?

24. The *abnormal oocyte* (*abo*) gene in *Drosophila melanogaster* causes a recessive maternal effect that reduces the viability of offspring. Lethality generally occurs during late embryogenesis; however some larvae do hatch. Interestingly, genetic manipulation of histone gene clusters in the mother's genome alters the severity of the *abo* maternal effect, as shown in the following table (Berloco, M., et al. 2001. *Proc. Natl. Acad. Sci.* 98: 12126–12131).

**Influence of Deficiencies of Histone Gene Clusters on the *abo* Maternal Effect**

| Maternal genotype | | No. eggs | Progeny | Survival (adults/eggs) | Relative survival (E/C) |
|---|---|---|---|---|---|
| *Df(2L)DS5**, *abo*/*abo* | (E)** | 2905 | 969 | 0.33 | |
| *Df(2L)DS5*, *abo*/+ | (C)** | 2895 | 2400 | 0.83 | 0.40 |
| *Df(2L)DS6*, *abo*/*abo* | (E) | 1988 | 626 | 0.32 | |
| *Df(2L)DS6*, *abo*/+ | (C) | 2000 | 1565 | 0.78 | 0.41 |
| *Df(2L)DS9*, *abo*/*abo* | (E) | 3450 | 570 | 0.17 | |
| *Df(2L)DS9*, *abo*/+ | (C) | 3300 | 2970 | 0.90 | 0.19 |

**Df(2L)DS5* and *Df(2L)DS6* are histone deficiencies; *Df(2L)DS9* does not affect histone genes.
**(E) = experimental; (C) = control

Analyze the information in the table, and draw all possible conclusions regarding the *abo* maternal effect and the role of the wild type *abo* gene.

25. Researchers examined a family with an interesting distribution of Leigh syndrome symptoms. In this disorder, individuals may show a progressive loss of motor function (ataxia, A) with peripheral neuropathy (PN, meaning impairment of the peripheral nerves). A mitochondrial DNA (mtDNA) mutation that reduces ATPase activity was identified in various tissues of affected individuals. The accompanying table summarizes the presence of symptoms in an extended family.

| Person | Condition | Percent Mitochondria with Mutation |
|---|---|---|
| Proband | A and PN | >90% |
| Brother | A and PN | >90% |
| Brother | Asymptomatic | 17% |
| Mother | PN | 86% |
| Maternal uncle | PN | 85% |
| Maternal cousin | A and PN | 90% |
| Maternal cousin | A and PN | 91% |
| Maternal grandmother | Asymptomatic | 56% |

(a) Develop a pedigree that summarizes the information presented in the table.
(b) Provide an explanation for the pattern of inheritance of the disease. What term describes this pattern?
(c) How can some individuals in the same family show such variation in symptoms? What term, as related to organelle heredity, describes such variation?
(d) In what way does a condition caused by mtDNA differ in expression and transmission from a mutation that causes albinism?

Bronze sculpture of the Watson–Crick model for double-helical DNA.

10

# DNA Structure and Analysis

## CHAPTER CONCEPTS

- Except in some viruses, DNA serves as the genetic material in all living organisms on Earth.
- According to the Watson–Crick model, DNA exists in the form of a right-handed double helix.
- The strands of the double helix are antiparallel and held together by hydrogen bonding between complementary nitrogenous bases.
- The structure of DNA provides the means of storing and expressing genetic information.
- RNA has many similarities to DNA but exists mostly as a single-stranded molecule.
- In some viruses, RNA serves as the genetic material.

Up to this point in the text, we have described chromosomes as containing genes that control phenotypic traits that are transmitted through gametes to future offspring. Logically, genes must contain some sort of information that, when passed to a new generation, influences the form and characteristics of each individual. We refer to that information as the **genetic material**. Logic also suggests that this same information in some way directs the many complex processes that lead to an organism's adult form.

Until 1944, it was not clear what chemical component of the chromosome makes up genes and constitutes the genetic material. Because chromosomes were known to have both a nucleic acid and a protein component, both were candidates. In 1944, however, direct experimental evidence emerged showing that the nucleic acid DNA serves as the informational basis for the process of heredity.

Once the importance of DNA to genetic processes was realized, work was intensified with the hope of discerning not only the structure of this molecule but also the relationship of its structure to its function. Between 1944 and 1953, many scientists sought information that might answer the most significant and intriguing question in the history of biology: How does DNA serve as the genetic basis for living processes? Researchers believed the answer must depend strongly on the chemical structure of the DNA molecule, given the complex but orderly functions ascribed to it.

These efforts were rewarded in 1953, when James Watson and Francis Crick put forth their hypothesis for the double-helical nature of DNA. The assumption that the molecule's functions would be easier to clarify once its general structure was determined proved to be correct. In this chapter, we first review the evidence that DNA is the genetic material and then discuss the elucidation of its structure. We conclude the chapter with a discussion of various analytical techniques useful during the investigation of the nucleic acids, DNA and RNA.

## 10.1 The Genetic Material Must Exhibit Four Characteristics

For a molecule to serve as the genetic material, it must exhibit four crucial characteristics: **replication, storage of information, expression of information**, and **variation by mutation**. *Replication* of the genetic material is one facet of the cell cycle, and as such is a fundamental property of all living organisms. Once the genetic material of cells replicates and is doubled in amount, it must then be partitioned equally—through mitosis—into daughter cells. The genetic material is also replicated during the formation of gametes, but is partitioned so that each cell gets only one-half of the original amount of genetic material—the process of *meiosis*, discussed in Chapter 2. Although the products of mitosis and meiosis are different, these processes are both part of the more general phenomenon of cellular reproduction.

*Storage of information* requires the molecule to act as a repository of genetic information that may or may not be expressed by the cell in which it resides. It is clear that while most cells contain a complete copy of the organism's genome, at any point in time they express only a part of this genetic potential. For example, in bacteria many genes "turn on" in response to specific environmental conditions and "turn off" when conditions change. In vertebrates, skin cells may display active melanin genes but never activate their hemoglobin genes; in contrast, digestive cells activate many genes specific to their function but do not activate their melanin genes.

Inherent in the concept of storage is the need for the genetic material to be able to encode the vast variety of gene products found among the countless forms of life on our planet. The chemical language of the genetic material must have the capability of storing such diverse information and transmitting it to progeny cells and organisms.

*Expression* of the stored genetic information is a complex process that is the underlying basis for the concept of **information flow** within the cell (Figure 10–1). The initial event in this flow of information is the **transcription** of DNA, in which three main types of RNA molecules are synthesized: messenger RNA (mRNA), transfer RNA (tRNA), and ribosomal RNA (rRNA). Of these, mRNAs are translated into proteins, by means of a process mediated by the tRNA and rRNA. Each mRNA is the product of a specific gene and leads to the synthesis of a different protein. In **translation**, the chemical information in mRNA directs the construction of a chain of amino acids, called a polypeptide, which then folds into a protein. Collectively, these processes serve as the foundation for the **central dogma of molecular genetics:** "DNA makes RNA, which makes proteins."

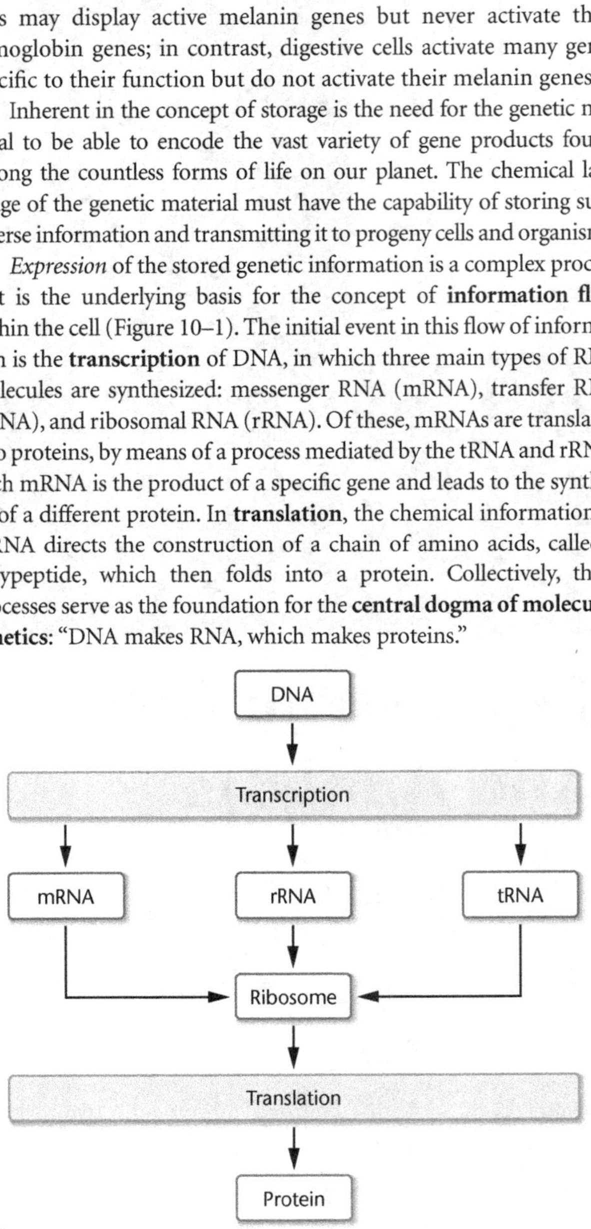


FIGURE 10–1 Simplified diagram of information flow (the central dogma) from DNA to RNA to produce the proteins within cells.

The genetic material is also the source of *variability* among organisms, through the process of mutation. If a mutation—a change in the chemical composition of DNA—occurs, the alteration is reflected during transcription and translation, affecting the specific protein. If a mutation is present in a gamete, it may be passed to future generations and, with time, become distributed in the population. Genetic variation, which also includes alterations of chromosome number and rearrangements within and between chromosomes (as discussed in Chapter 8), provides the raw material for the process of evolution.

10.2

## Until 1944, Observations Favored Protein as the Genetic Material

The idea that genetic material is physically transmitted from parent to offspring has been accepted for as long as the concept of inheritance has existed. Beginning in the late nineteenth century, research into the structure of biomolecules progressed considerably, setting the stage for describing the genetic material in chemical terms. Although proteins and nucleic acid were both considered major candidates for the role of genetic material, until the 1940s many geneticists favored proteins. This is not surprising, since proteins were known to be both diverse and abundant in cells, and much more was known about protein than about nucleic acid chemistry.

DNA was first studied in 1868 by a Swiss chemist, Friedrich Miescher. He isolated cell nuclei and derived an acidic substance, now known to contain DNA, that he called **nuclein**. As investigations of DNA progressed, however, showing it to be present in chromosomes, the substance seemed to lack the chemical diversity necessary to store extensive genetic information. This conclusion was based largely on Phoebus A. Levene's observations in 1910 that DNA contained approximately equal amounts of four similar molecular building blocks called *nucleotides*. Levene postulated, incorrectly, that identical groups consisting of these four components were repeated over and over, a supposition that formed the basis of his **tetranucleotide hypothesis** for DNA structure (Figure 10–2). Levene based his proposal on studies of the composition of the four types of nucleotides. Although his actual data revealed considerable variation in the relative amounts of the four nucleotides, he assumed they were present in a 1:1:1:1 ratio. He ascribed the discrepancy to faults in the analytical technique.

Because a single covalently bonded tetranucleotide was a relatively simple structure, geneticists believed that nucleic acids could not provide the large amount of chemical variation expected for the genetic material. Proteins, on the other hand, contain 20 different amino acids, thus providing a basis for substantial variation. As a result, attention was directed away from nucleic acids and in favor of the speculation that proteins served as the genetic material.

It was not until the 1940s that Erwin Chargaff proved Levene's proposal incorrect by demonstrating that most organisms do not contain precisely equal proportions of the four nucleotides. We shall see later that the structure of DNA accounts for Chargaff's observations.

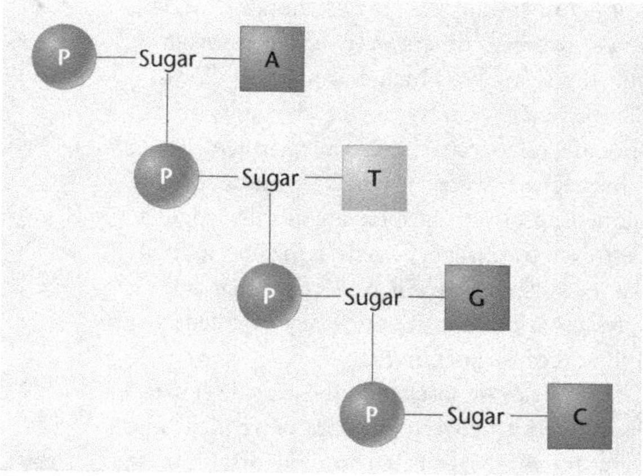


FIGURE 10–2 Diagrammatic depiction of Levene's proposed tetranucleotide, containing one molecule each of the four nitrogenous bases: adenine (A), thymine (T), guanine (G), and cytosine (C). The four blocks, each representing a different nitrogenous base, could have been shown in other potential sequences.

10.3

## Evidence Favoring DNA as the Genetic Material Was First Obtained during the Study of Bacteria and Bacteriophages

The 1944 publication by Oswald Avery, Colin MacLeod, and Maclyn McCarty concerning the chemical nature of a "transforming principle" in bacteria was the initial event leading to the acceptance of DNA as the genetic material. Their work, along with subsequent findings of other research teams, constituted the first direct experimental proof that DNA, and not protein, is the biomolecule responsible for heredity. It marked the beginning of the *era of molecular genetics*, a period of discovery in biology that made biotechnology feasible and has moved us closer to an understanding of the basis of life. The impact of their initial findings on future research and thinking paralleled that of the publication of Darwin's theory of evolution and the subsequent rediscovery of Mendel's postulates of transmission genetics. Together, these events constitute three great revolutions in biology.

### Transformation: Early Studies

The research that provided the foundation for Avery, MacLeod, and McCarty's work was initiated in 1927 by Frederick Griffith, a medical officer in the British Ministry of Health. He performed experiments with several different strains of the bacterium *Diplococcus pneumoniae*.* Some were *virulent*, that is, infectious, strains that cause

*This organism is now named *Streptococcus pneumoniae*.

pneumonia in certain vertebrates (notably humans and mice), whereas others were *avirulent*, meaning noninfectious, strains, which do not cause illness.

The difference in virulence depends on the presence of a polysaccharide capsule; virulent strains have this capsule, whereas avirulent strains do not. The nonencapsulated bacteria are readily engulfed and destroyed by phagocytic cells in the host animal's circulatory system. Virulent bacteria, which possess the polysaccharide coat, are not easily engulfed; they multiply and cause pneumonia.

The presence or absence of the capsule causes a visible difference between colonies of virulent and avirulent strains. Encapsulated bacteria form smooth, shiny-surfaced colonies (*S*) when grown on an agar culture plate; nonencapsulated strains produce rough colonies (*R*) (Figure 10–3). Thus, virulent and avirulent strains are easily distinguished by standard microbiological culture techniques.

Each strain of *Diplococcus* may be one of dozens of different types called **serotypes** that differ in the precise chemical structure of the polysaccharide constituent of the thick, slimy capsule. Serotypes are identified by immunological techniques and are usually designated by Roman numerals. In the United States, types I and II are the most common in causing pneumonia. Griffith used types II*R* and III*S* in his critical experiments that led to new concepts about the genetic material. Table 10.1 summarizes the characteristics of Griffith's two strains.

Griffith knew from the work of others that only living virulent cells would produce pneumonia in mice. If heat-killed virulent bacteria are injected into mice, no pneumonia results, just as living avirulent bacteria fail to produce the disease. Griffith's critical experiment (Figure 10–3) involved an injection into mice of living II*R* (avirulent) cells combined with heat-killed III*S* (virulent) cells. Since neither cell type caused death in mice when injected alone, Griffith expected that the double injection would not kill the mice. But, after five days, all of the mice that received both types of cells were dead. Paradoxically, analysis of their blood revealed a large number of living type III*S* (virulent) bacteria.

As far as could be determined, these III*S* bacteria were identical to the III*S* strain from which the heat-killed cell preparation had been made. The control mice, injected only with living avirulent II*R* bacteria for this set of experiments, did not develop pneumonia and remained healthy. This ruled out the possibility that the avirulent II*R* cells simply changed (or mutated) to virulent III*S* cells in the absence of the heat-killed III*S* bacteria. Instead, some type of interaction had taken place between living II*R* and heat-killed III*S* cells.

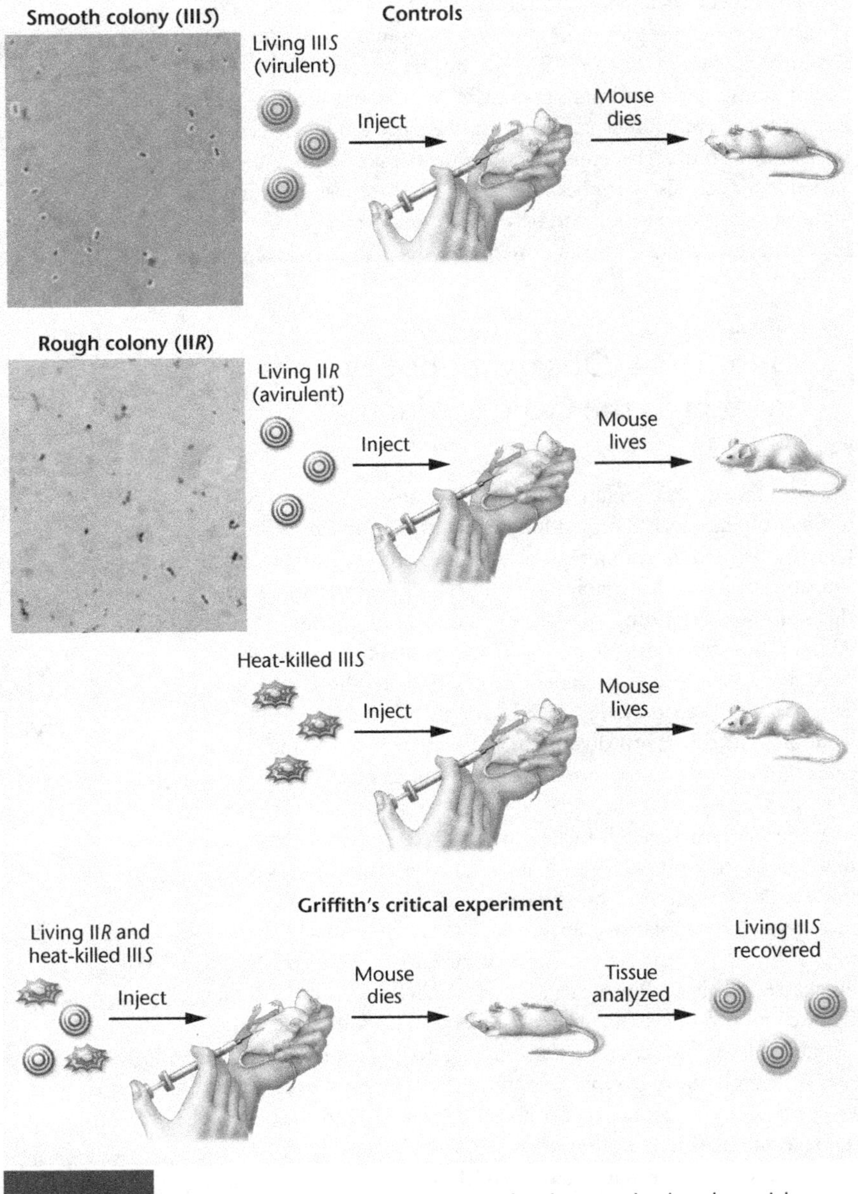


FIGURE 10–3 Griffith's transformation experiment. The photographs show bacterial colonies containing cells with capsules (type III*S*) and without capsules (type II*R*).

TABLE 10.1

**Strains of *Diplococcus pneumoniae* Used by Frederick Griffith in His Original Transformation Experiments**

| Serotype | Colony Morphology | Capsule | Virulence |
|---|---|---|---|
| II*R* | Rough | Absent | Avirulent |
| III*S* | Smooth | Present | Virulent |

Griffith concluded that the heat-killed III*S* bacteria somehow converted live avirulent II*R* cells into virulent III*S* cells. Calling the phenomenon **transformation**, he suggested that the **transforming principle** might be some part of the polysaccharide capsule or a compound required for capsule synthesis, although the capsule alone did not cause pneumonia. To use Griffith's term, the transforming principle from the dead III*S* cells served as a "pabulum"—that is, a nutrient source—for the II*R* cells.

Griffith's work led other physicians and bacteriologists to research the phenomenon of transformation. By 1931, Henry Dawson at the Rockefeller Institute had confirmed Griffith's observations and extended his work one step further. Dawson and his coworkers showed that transformation could occur *in vitro* (in a test tube). When heat-killed III*S* cells were incubated with living II*R* cells, living III*S* cells were recovered. Therefore, injection into mice was not necessary for transformation to occur. By 1933, Lionel J. Alloway had refined the *in vitro* experiments by using crude extracts of III*S* cells and living II*R* cells. The soluble filtrate from the heat-killed III*S* cells was as effective in inducing transformation as were the intact cells. Alloway and others did not view transformation as a genetic event, but rather as a physiological modification of some sort. Nevertheless, the experimental evidence that a chemical substance was responsible for transformation was quite convincing.

## Transformation: The Avery, MacLeod, and McCarty Experiment

The critical question, of course, was what molecule serves as the transforming principle? In 1944, after 10 years of work, Avery, MacLeod, and McCarty published their results in what is now regarded as a classic paper in the field of molecular genetics. They reported that they had obtained the transforming principle in a purified state, and that beyond reasonable doubt it was DNA.

The details of their work, sometimes called the Avery, MacLeod, and McCarty experiment, are outlined in Figure 10–4. These

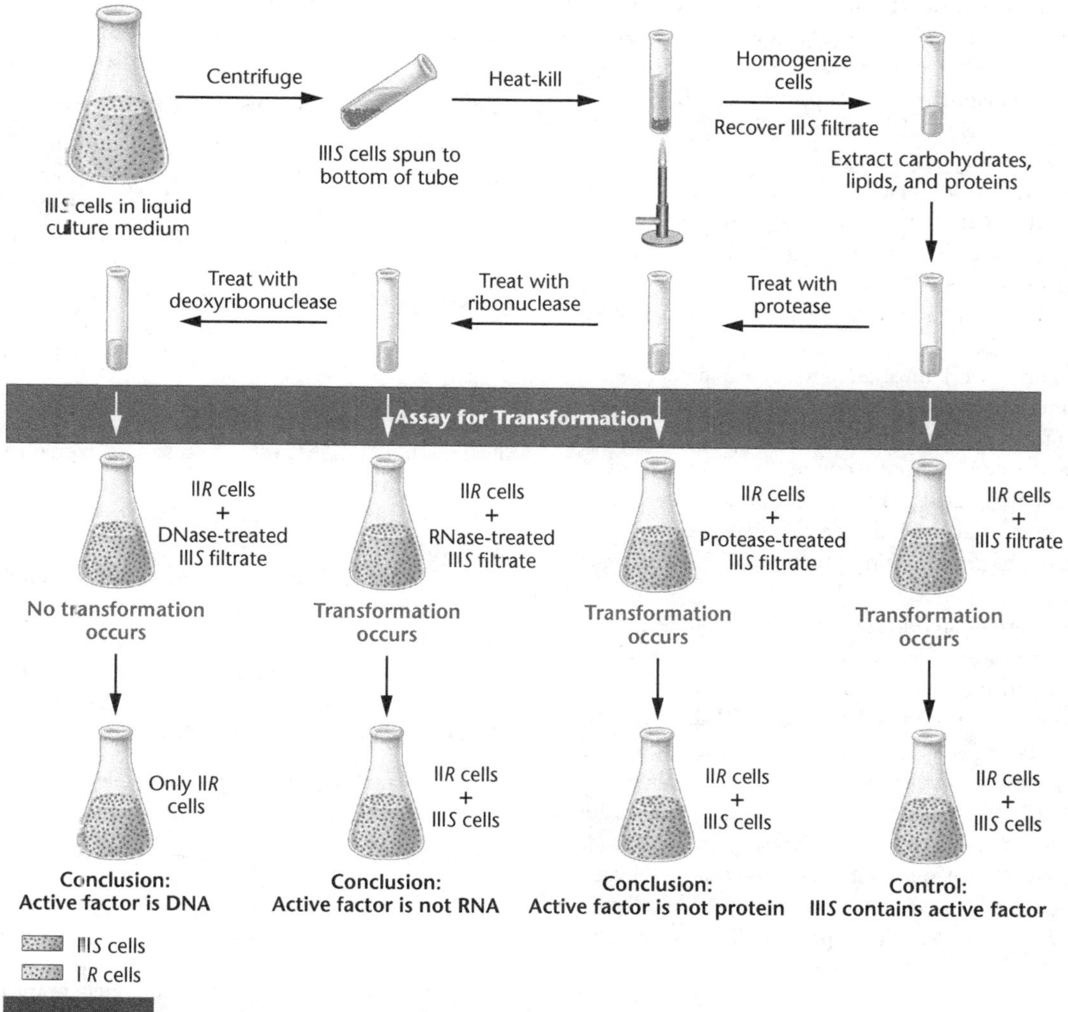


**FIGURE 10–4** Summary of Avery, MacLeod, and McCarty's experiment demonstrating that DNA is the transforming principle.

researchers began their isolation procedure with large quantities (50–75 liters) of liquid cultures of type III*S* virulent cells. The cells were centrifuged, collected, and heat killed. Following homogenization and several extractions with the detergent deoxycholate (DOC), the researchers obtained a soluble filtrate that retained the ability to induce transformation of type II*R* avirulent cells. Protein was removed from the active filtrate by several chloroform extractions, and polysaccharides were enzymatically digested and removed. Finally, precipitation with ethanol yielded a fibrous mass that still retained the ability to induce transformation of type II*R* avirulent cells. From the original 75-liter sample, the procedure yielded 10 to 25 mg of this "active factor."

Further testing clearly established that the transforming principle was DNA. The fibrous mass was first analyzed for its nitrogen–phosphorus ratio, which was shown to coincide with the ratio of "sodium desoxyribonucleate," the chemical name then used to describe DNA. To solidify their findings, Avery, MacLeod, and McCarty sought to eliminate, to the greatest extent possible, all probable contaminants from their final product. Thus, it was treated with the proteolytic enzymes trypsin and chymotrypsin and then with an RNA-digesting enzyme, called **ribonuclease (RNase)**. Such treatments destroyed any remaining activity of proteins and RNA. Nevertheless, transforming activity still remained. Chemical testing of the final product gave strong positive reactions for DNA. The final confirmation came with experiments using crude samples of the DNA-digesting enzyme **deoxyribonuclease (DNase)**, which was isolated from dog and rabbit sera. Digestion with this enzyme destroyed the transforming activity of the filtrate—thus Avery and his coworkers were certain that the active transforming principle in these experiments was DNA.

The great amount of work involved in this research, the confirmation and reconfirmation of the conclusions drawn, and the unambiguous logic of the experimental design are truly impressive. Avery, MacLeod, and McCarty's conclusion in the 1944 publication was, however, very simply stated: "The evidence presented supports the belief that a nucleic acid of the desoxyribose* type is the fundamental unit of the transforming principle of *Pneumococcus* Type III."

Avery and his colleagues recognized the genetic and biochemical implications of their work. They observed that "nucleic acids of this type must be regarded not merely as structurally important but as functionally active in determining the biochemical activities and specific characteristics of pneumococcal cells." This suggested that the transforming principle interacts with the II*R* cell and gives rise to a coordinated series of enzymatic reactions culminating in the synthesis of the type III*S* capsular polysaccharide. Avery, MacLeod, and McCarty emphasized that, once transformation occurs, the capsular polysaccharide is produced in successive generations. Transformation is therefore heritable, and the process affects the genetic material.

*Desoxyribose is now spelled deoxyribose.

Immediately after the publication of the report, several investigators turned to, or intensified, their studies of transformation in order to clarify the role of DNA in genetic mechanisms. In particular, the work of Rollin Hotchkiss was instrumental in confirming that the critical factor in transformation was DNA and not protein. In 1949, in a separate study, Harriet Taylor isolated an **extremely rough (*ER*)** mutant strain from a rough (*R*) strain. This *ER* strain produced colonies that were more irregular than the *R* strain. The DNA from *R* accomplished the transformation of *ER* to *R*. Thus, the *R* strain, which served as the recipient in the Avery experiments, was shown also to be able to serve as the DNA donor in transformation.

Transformation has now been shown to occur in *Haemophilus influenzae, Bacillus subtilis, Shigella paradysenteriae*, and *Escherichia coli*, among many other microorganisms. Transformation of numerous genetic traits other than colony morphology has also been demonstrated, including traits involving resistance to antibiotics. These observations further strengthened the belief that transformation by DNA is primarily a genetic event rather than simply a physiological change. We will pursue this idea again in the "Insights and Solutions" section at the end of this chapter.

## The Hershey–Chase Experiment

The second major piece of evidence supporting DNA as the genetic material was provided during the study of the bacterium *Escherichia coli* and one of its infecting viruses, **bacteriophage T2**. Often referred to simply as a **phage**, the virus consists of a protein coat surrounding a core of DNA. Electron micrographs reveal that the phage's external structure is composed of a hexagonal head plus a tail. Figure 10–5 shows as much of the life cycle as was known in 1952 for a T-even bacteriophage such as T2. Briefly, the phage adsorbs to the bacterial cell, and some genetic component of the phage enters the bacterial cell. Following infection, the viral component "commandeers" the cellular machinery of the host and causes viral reproduction. In a reasonably short time, many new phages are constructed and the bacterial cell is lysed, releasing the progeny viruses. This process is referred to as the **lytic cycle**.

In 1952, Alfred Hershey and Martha Chase published the results of experiments designed to clarify the events leading to phage reproduction. Several of the experiments clearly established the independent functions of phage protein and nucleic acid in the reproduction process associated with the bacterial cell. Hershey and Chase knew from existing data that:

1. T2 phages consist of approximately 50 percent protein and 50 percent DNA.
2. Infection is initiated by adsorption of the phage by its tail fibers to the bacterial cell.
3. The production of new viruses occurs within the bacterial cell.

It appeared that some molecular component of the phage—DNA or protein (or both)—entered the bacterial cell and directed viral reproduction. Which was it?

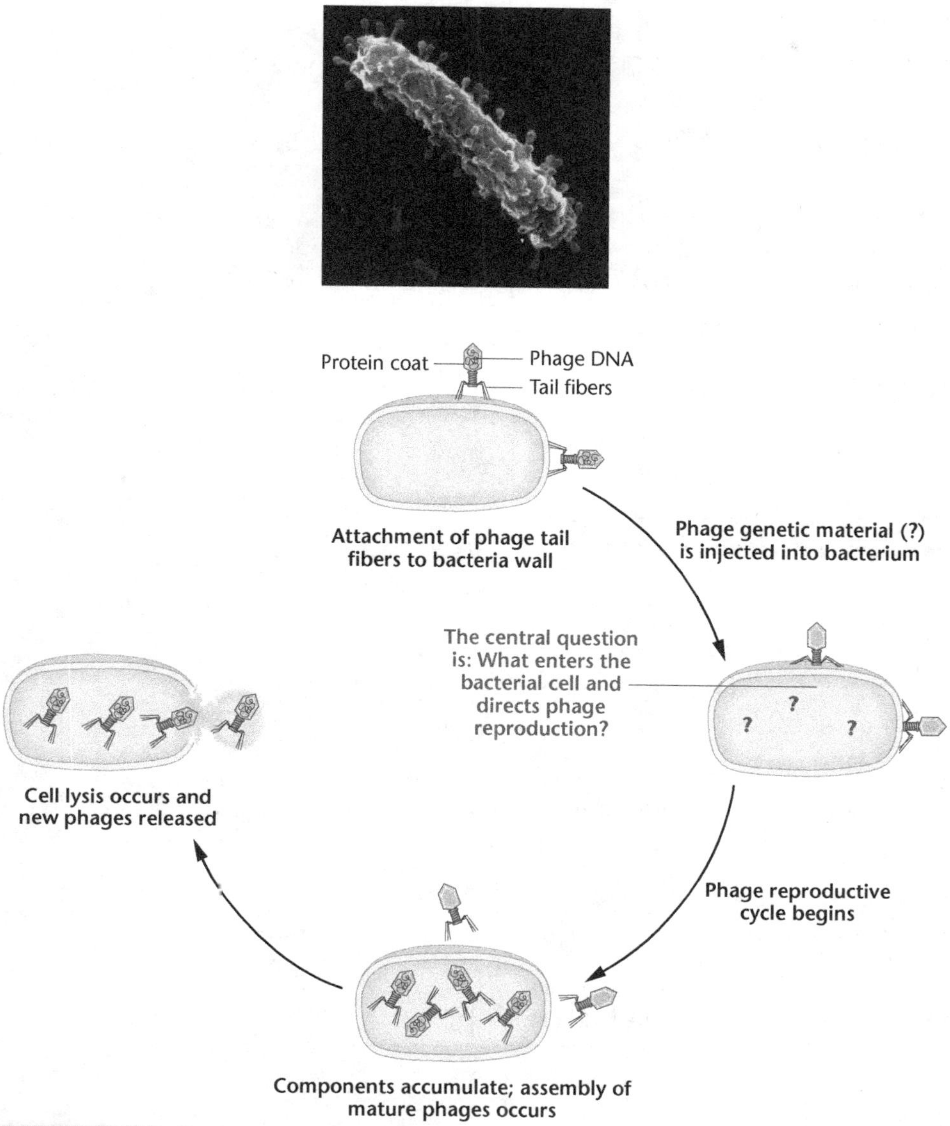


**FIGURE 10–5** Life cycle of a T-even bacteriophage, as known in 1952. The electron micrograph shows an *E. coli* cell during infection by numerous T2 phages (shown in blue).

Hershey and Chase used the radioisotopes $^{32}P$ and $^{35}S$ to follow the molecular components of phages during infection. Because DNA contains phosphorus (P) but not sulfur, $^{32}P$ effectively labels DNA; because proteins contain sulfur (S) but not phosphorus, $^{35}S$ labels protein. *This is a key feature of the experiment.* If *E. coli* cells are first grown in the presence of $^{32}P$ or $^{35}S$ and then infected with T2 viruses, the progeny phages will have *either* a radioactively labeled DNA core *or* a radioactively labeled protein coat, respectively. These labeled phages can be isolated and used to infect unlabeled bacteria (Figure 10–6).

When labeled phages and unlabeled bacteria were mixed, an adsorption complex was formed as the phages attached their tail fibers to the bacterial wall. These complexes were isolated and subjected to a high shear force in a blender. The force stripped off the attached phages so that the phages and bacteria could be analyzed separately. Centrifugation separated the lighter phage particles from the heavier bacterial cells (Figure 10–6). By tracing the radioisotopes, Hershey and Chase were able to demonstrate that most of the $^{32}P$-labeled DNA had been transferred into the bacterial cell following adsorption; on the other hand, almost all of the $^{35}S$-labeled protein remained outside the bacterial cell and was recovered in the phage "ghosts" (empty phage coats) after the blender treatment. Following this separation, the bacterial cells, which now contained viral DNA, were eventually lysed as new phages were produced. These progeny phages contained $^{32}P$, but not $^{35}S$.

Hershey and Chase interpreted these results as indicating that the protein of the phage coat remains outside the host cell and is not involved in directing the production of new phages. On the other hand, and most important, phage DNA enters the host cell and directs phage reproduction. Hershey and Chase had demonstrated that the genetic material in phage T2 is DNA, not protein.

These experiments, along with those of Avery and his colleagues, provided convincing evidence that DNA was the molecule responsible for heredity. This conclusion has since served as the cornerstone of the field of molecular genetics.

## Transfection Experiments

During the eight years following publication of the Hershey–Chase experiment, additional research using bacterial viruses provided even more solid proof that DNA is the genetic material. In 1957, several reports demonstrated that if *E. coli* is treated with the enzyme lysozyme, the

**NOW SOLVE THIS**

Problem 7 on page 275 asks about applying the protocol of the Hershey–Chase experiment to the investigation of transformation. You are asked to comment about the possible success of attempting to do so.

■ **HINT:** *As you attempt to apply the Hershey–Chase protocol, remember that in transformation, exogenous DNA enters the soon-to-be transformed cell and no cell-to-cell contact is involved in the process.*

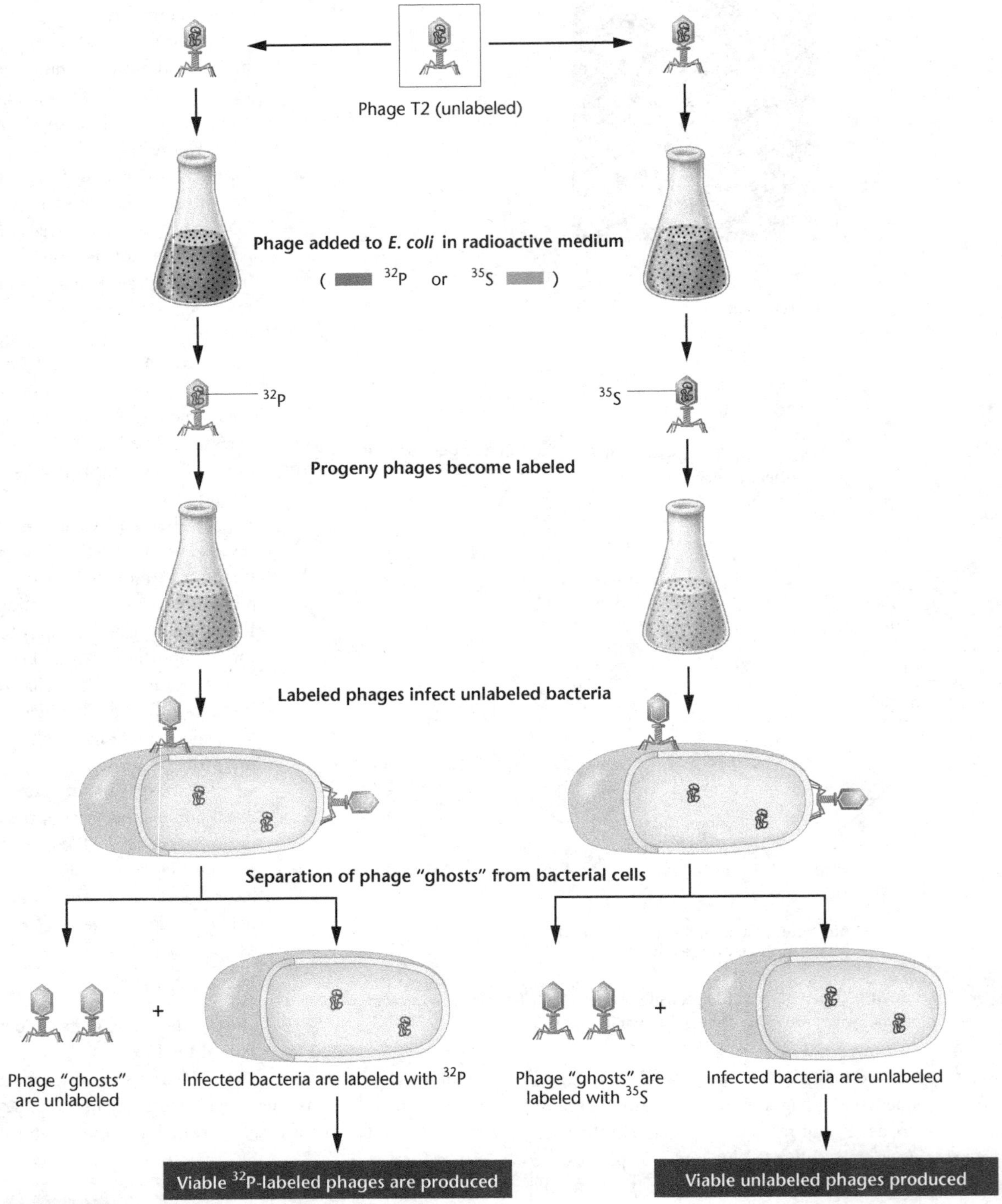


**FIGURE 10-6** Summary of the Hershey-Chase experiment demonstrating that DNA, and not protein, is responsible for directing the reproduction of phage T2 during the infection of *E. coli*.

outer wall of the cell can be removed without destroying the bacterium. Enzymatically treated cells are naked, so to speak, and contain only the cell membrane as their outer boundary. Such structures are called **protoplasts** (or **spheroplasts**). John Spizizen and Dean Fraser independently reported that by using protoplasts, they were able to initiate phage reproduction with disrupted T2 particles. That is, provided protoplasts were used, a virus did not have to be intact for infection to occur. Thus, the outer protein coat structure may be essential to the movement of DNA through the intact cell wall, but it is not essential for infection when protoplasts are used.

Similar, but more refined, experiments were reported in 1960 by George Guthrie and Robert Sinsheimer. DNA was purified from bacteriophage $\phi$X174, a small phage that contains a single-stranded circular DNA molecule of some 5386 nucleotides. When added to *E. coli* protoplasts, the purified DNA resulted in the production of complete $\phi$X174 bacteriophages. This process of infection by only the viral nucleic acid, called **transfection**, proves conclusively that $\phi$X174 DNA alone contains all the necessary information for production of mature viruses. Thus, the evidence that DNA serves as the genetic material was further strengthened, even though all direct evidence to that point had been obtained from bacterial and viral studies.

10.4

## Indirect and Direct Evidence Supports the Concept that DNA Is the Genetic Material in Eukaryotes

In 1950, eukaryotic organisms were not amenable to the types of experiments that used bacteria and viruses to demonstrate that DNA is the genetic material. Nevertheless, it was generally assumed that the genetic material would be a universal substance serving the same role in eukaryotes. Initially, support for this assumption relied on several circumstantial observations that, taken together, indicated that DNA does serve as the genetic material in eukaryotes. Subsequently, direct evidence established unequivocally the central role of DNA in genetic processes.

### Indirect Evidence: Distribution of DNA

The genetic material should be found where it functions—in the nucleus as part of chromosomes. Both DNA and protein fit this criterion. However, protein is also abundant in the cytoplasm, whereas DNA is not. Both mitochondria and chloroplasts are known to perform genetic functions, and DNA is also present in these organelles. Thus, DNA is found only where primary genetic functions occur. Protein, on the other hand, is found everywhere in the cell. These observations are consistent with the interpretation favoring DNA over proteins as the genetic material.

Because it had earlier been established that chromosomes within the nucleus contain the genetic material, a correlation was expected to exist between the ploidy ($n$, $2n$, etc.) of a cell and the quantity of the substance that functions as the genetic material. Meaningful comparisons can be made between gametes (sperm and eggs) and somatic or body cells. The latter are recognized as being diploid ($2n$) and containing twice the number of chromosomes as gametes, which are haploid ($n$).

Table 10.2 compares, for a variety of organisms, the amount of DNA found in haploid sperm to the amount found in diploid nucleated precursors of red blood cells. The amount of DNA and the number of sets of chromosomes is closely correlated. No such consistent correlation can be observed between gametes and diploid cells for proteins. These data thus provide further circumstantial evidence favoring DNA over proteins as the genetic material of eukaryotes.

TABLE 10.2

**DNA Content of Haploid versus Diploid Cells of Various Species***

| Organism | $n$ (pg) | $2n$ (pg) |
|---|---|---|
| Human | 3.25 | 7.30 |
| Chicken | 1.26 | 2.49 |
| Trout | 2.67 | 5.79 |
| Carp | 1.65 | 3.49 |
| Shad | 0.91 | 1.97 |

*Sperm ($n$) and nucleated precursors to red blood cells ($2n$) were used to contrast ploidy levels.

### Indirect Evidence: Mutagenesis

**Ultraviolet (UV) light** is one of a number of agents capable of inducing mutations in the genetic material. Simple organisms such as yeast and other fungi can be irradiated with various wavelengths of ultraviolet light and the effectiveness of each wavelength measured by the number of mutations it induces. When the data are plotted, an **action spectrum** of UV light as a mutagenic agent is obtained. This action spectrum can then be compared with the **absorption spectrum** of any molecule suspected to be the genetic material (Figure 10–7). *The molecule serving as the genetic material is expected to absorb at the wavelength(s) found to be mutagenic.*

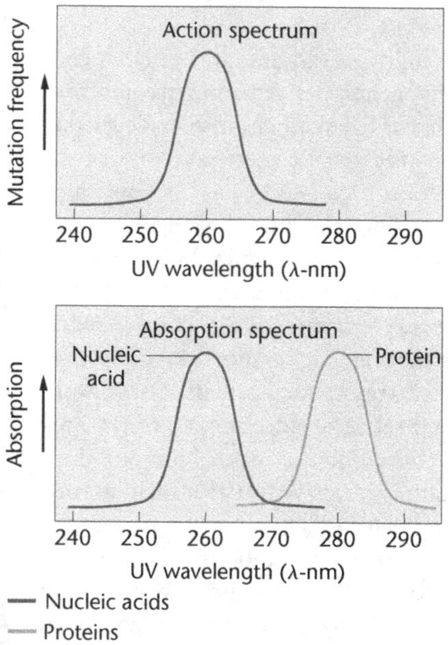


FIGURE 10–7 Comparison of the action spectrum (which determines the most effective mutagenic UV wavelength) and the absorption spectrum (which shows the range of wavelength where nucleic acids and proteins absorb UV light).

UV light is most mutagenic at the wavelength ($\lambda$) of 260 nanometers (nm), and both DNA and RNA absorb UV light most strongly at 260 nm. On the other hand, protein absorbs most strongly at 280 nm, yet no significant mutagenic effects are observed at that wavelength. This indirect evidence supports the idea that a nucleic acid, rather than protein, is the genetic material.

### Direct Evidence: Recombinant DNA Studies

Although the circumstantial evidence just described does not constitute direct proof that DNA is the genetic material in eukaryotes, those observations spurred researchers to forge ahead using this supposition as the underlying hypothesis. Today, there is no doubt of its validity; DNA *is* the genetic material in all eukaryotes. The strongest evidence is provided by molecular analysis utilizing **recombinant DNA technology**. In this procedure, segments of eukaryotic DNA corresponding to specific genes are isolated and spliced into bacterial DNA. The resulting complex can be inserted into a bacterial cell, and then its genetic expression is monitored. If a eukaryotic gene is introduced, the subsequent production of the corresponding eukaryotic protein product demonstrates directly that the eukaryotic DNA is now present and functional in the bacterial cell. This has been shown to be the case in countless instances. For example, the products of the human genes specifying insulin and interferon are produced by bacteria after the human genes that encode these proteins are inserted. As the bacterium divides, the eukaryotic DNA replicates along with the bacterial DNA and is distributed to the daughter cells, which also express the human genes by creating the corresponding proteins.

The availability of vast amounts of DNA coding for specific genes, derived from recombinant DNA research, has led to other direct evidence that DNA serves as the genetic material. Work in the laboratory of Beatrice Mintz has demonstrated that DNA encoding the human *β-globin* gene, when microinjected into a fertilized mouse egg, is later found to be present and expressed in adult mouse tissue and transmitted to and expressed in that mouse's progeny. These mice are examples of what are called **transgenic animals**. Other work has introduced rat DNA encoding a growth hormone into fertilized mouse eggs. About one-third of the resultant mice grew to twice their normal size, indicating that foreign DNA was present and functional. Subsequent generations of mice inherited this genetic information and also grew to a large size. This clearly demonstrates that DNA meets the requirement of expression of genetic information in eukaryotes. Later, we will see exactly how DNA is stored, replicated, expressed, and mutated.

## 10.5 RNA Serves as the Genetic Material in Some Viruses

Some viruses contain an RNA core rather than a DNA core. In these viruses, it appears that RNA serves as the genetic material—an exception to the general rule that DNA performs this function. In 1956, it was demonstrated that when purified RNA from **tobacco mosaic virus (TMV)** is spread on tobacco leaves, the characteristic lesions caused by this virus subsequently appear on the leaves. Thus, it was concluded that RNA is the genetic material of this virus.

Soon afterward, another type of experiment with TMV was reported by Heinz Fraenkel-Conrat and B. Singer, as illustrated in Figure 10–8. These scientists discovered that the RNA core and the protein coat from wild-type TMV and other viral strains could be isolated separately. In their work, RNA and coat proteins were separated and isolated from TMV and a second viral strain, **Holmes ribgrass (HR)**. Then, mixed viruses were reconstituted from the RNA of one strain and the protein of the other. When this "hybrid" virus was spread on tobacco leaves, the lesions that developed corresponded to the type of RNA in the reconstituted virus—that is, viruses with wild-type TMV RNA and HR protein coats produced TMV lesions, and vice versa. Again, it was concluded that RNA serves as the genetic material in these viruses.

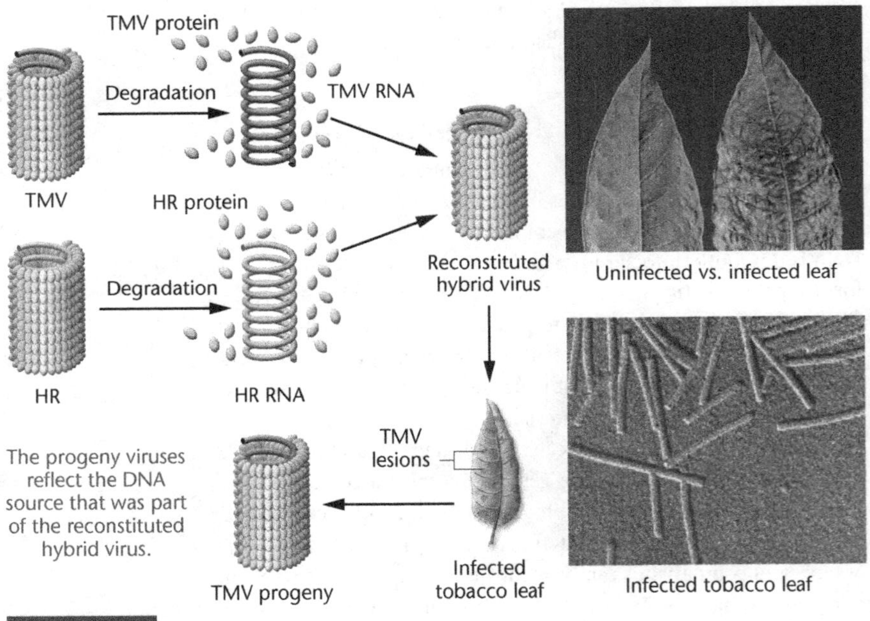


**FIGURE 10–8** Reconstitution of hybrid tobacco mosaic viruses. In the hybrid, the RNA is derived from the wild-type TMV virus, while the protein subunits are derived from the HR strain. Following infection, viruses are produced with protein subunits characteristic of the wild-type TMV strain and not those of the HR strain. The top photograph shows TMV lesions on a tobacco leaf compared with an uninfected leaf. At the bottom is an electron micrograph of mature viruses.

In 1965 and 1966, Norman R. Pace and Sol Spiegelman further demonstrated that RNA from the phage Q$\beta$ could be isolated and replicated *in vitro*. The replication depends on an enzyme, **RNA replicase**, which is isolated from host *E. coli* cells following normal infection. When the RNA replicated *in vitro* is added to *E. coli* protoplasts, infection and viral multiplication occur. Thus, RNA synthesized in a test tube serves as the genetic material in these phages by directing the production of all the components necessary for viral reproduction.

Finally, one other group of RNA-containing viruses bears mention. These are the **retroviruses**, which replicate in an unusual way. Their RNA serves as a template for the synthesis of the complementary DNA molecule. The process, **reverse transcription**, occurs under the direction of an RNA-dependent DNA polymerase enzyme called **reverse transcriptase**. This DNA intermediate can be incorporated into the genome of the host cell, and when the host DNA is transcribed, copies of the original retroviral RNA chromosomes are produced. Retroviruses include the human immunodeficiency virus (HIV), which causes AIDS, as well as the RNA tumor viruses.

WEB TUTORIAL 10.1 DNA STRUCTURE

## 10.6 Knowledge of Nucleic Acid Chemistry Is Essential to the Understanding of DNA Structure

Having established the critical importance of DNA and RNA in genetic processes, we will now take a brief look at the chemical structures of these molecules. As we shall see, the structural components of DNA and RNA are very similar. This chemical similarity is important in the coordinated functions played by these molecules during gene expression. Like the other major groups of organic biomolecules (proteins, carbohydrates, and lipids), nucleic acid chemistry is based on a variety of similar building blocks that are polymerized into chains of varying lengths.

### Nucleotides: Building Blocks of Nucleic Acids

DNA is a nucleic acid, and **nucleotides** are the building blocks of all nucleic acid molecules. Sometimes called mononucleotides, these structural units consist of three essential components: a **nitrogenous base**, a **pentose sugar** (a 5-carbon sugar), and a **phosphate group**. There are two kinds of nitrogenous bases: the nine-member double-ring **purines** and the six-member single-ring **pyrimidines**.

Two types of purines and three types of pyrimidines are commonly found in nucleic acids. The two purines are **adenine** and **guanine**, abbreviated **A** and **G**. The three pyrimidines are **cytosine, thymine**, and **uracil**, abbreviated **C, T**, and **U**. The chemical structures of A, G, C, T, and U are shown in Figure 10–9(a). Both DNA and RNA contain A, C, and G, but only DNA contains the base T and only RNA contains the base U. Each nitrogen or carbon atom of the ring structures of purines and pyrimidines is designated by an unprimed number. Note that corresponding atoms in the two rings are numbered differently in most cases.

FIGURE 10–9 (a) Chemical structures of the pyrimidines and purines that serve as the nitrogenous bases in RNA and DNA. The convention for numbering carbon and nitrogen atoms making up the two categories of bases is shown within the structures that appear on the left. (b) Chemical ring structures of ribose and 2-deoxyribose, which serve as the pentose sugars in RNA and DNA, respectively.

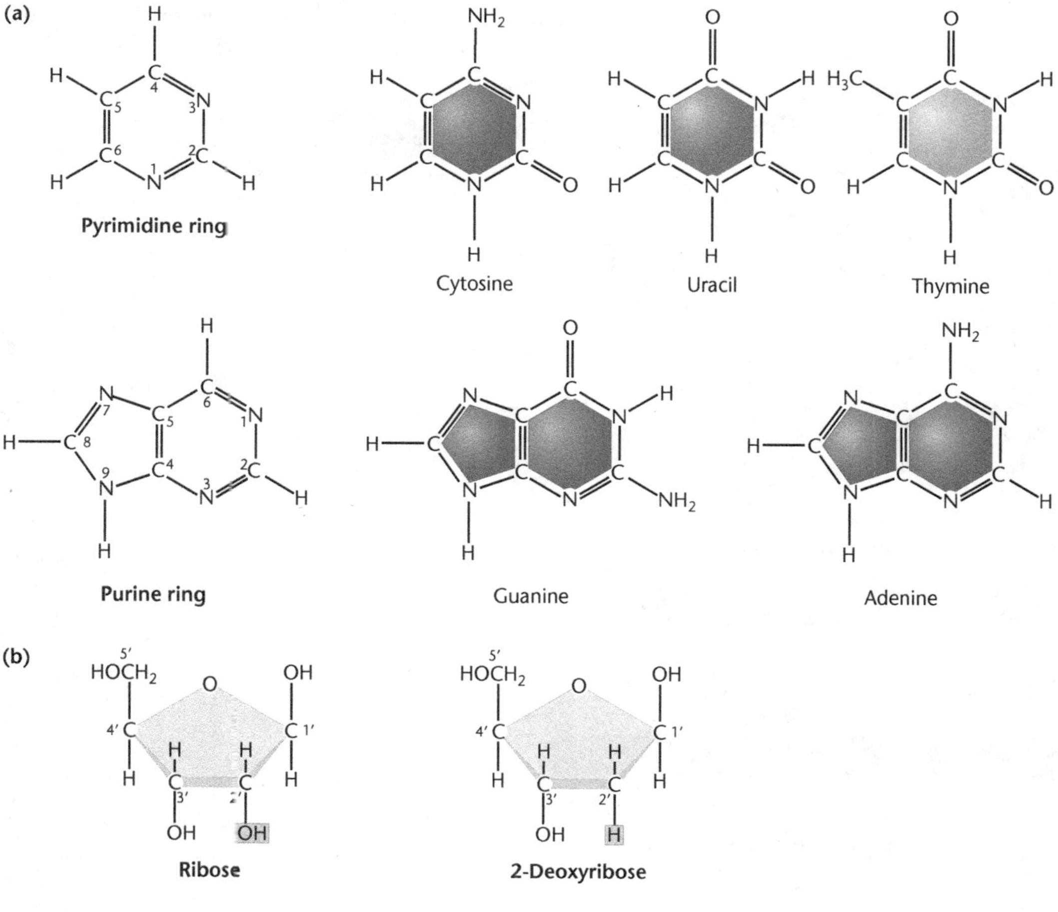

The pentose sugars found in nucleic acids give them their names. Ribonucleic acids (RNA) contain **ribose**, while deoxyribonucleic acids (DNA) contain **deoxyribose.** Figure 10–9(b) shows the ring structures for these two pentose sugars. Each carbon atom is distinguished by a number with a prime sign (e.g., C-1′, C-2′). Compared with ribose, deoxyribose has a hydrogen atom rather than a hydroxyl group at the C-2′ position. The absence of a hydroxyl group at the C-2′ position thus distinguishes DNA from RNA. In the absence of the C-2′ hydroxyl group, the sugar is more specifically named **2-deoxyribose.**

If a molecule is composed of a purine or pyrimidine base and a ribose or deoxyribose sugar, the chemical unit is called a **nucleoside.** If a phosphate group is added to the nucleoside, the molecule is now called a **nucleotide.** Nucleosides and nucleotides are named according to the specific nitrogenous base (A, T, G, C, or U) that is part of the molecule. The structures of a nucleoside and a nucleotide and the nomenclature used in naming nucleosides and nucleotides are given in Figure 10–10.

The bonding between components of a nucleotide is highly specific. The C-1′ atom of the sugar is involved in the chemical linkage to the nitrogenous base. If the base is a purine, the N-9 atom is covalently bonded to the sugar; if the base is a pyrimidine, the N-1 atom bonds to the sugar. In deoxyribonucleotides, the phosphate group may be bonded to the C-2′, C-3′, or C-5′ atom of the sugar. The C-5′ phosphate configuration is shown in Figure 10–10. It is by far the prevalent form in biological systems and the one found in DNA and RNA.

## Nucleoside Diphosphates and Triphosphates

Nucleotides are also described by the term **nucleoside monophosphate (NMP)**. The addition of one or two phosphate groups results in **nucleoside diphosphates (NDPs)** and **triphosphates (NTPs)**, respectively, as shown in Figure 10–11. The triphosphate form is significant because it serves as the precursor molecule during nucleic acid synthesis within the cell (see Chapter 11). In addition, **adenosine triphosphate (ATP)** and **guanosine triphosphate (GTP)** are important in cell bioenergetics because of the large amount of energy involved in adding or removing the terminal phosphate group. The hydrolysis of ATP or GTP to ADP or GDP and inorganic phosphate ($P_i$) is accompanied by the release of a large amount of energy in the cell. When these chemical conversions are coupled to other reactions, the energy produced is used to drive the reactions. As a result, both ATP and GTP are involved in many cellular activities, including numerous genetic events.

## Polynucleotides

The linkage between two mononucleotides consists of a phosphate group linked to two sugars. It is called a **phosphodiester bond** because phosphoric acid has been joined to two alcohols (the hydroxyl groups on the two sugars) by an ester linkage on both sides. Figure 10–12(a) shows the phosphodiester bond in DNA. The same bond is found in RNA. Each structure has a **C-5′ end** and a **C-3′ end.** Two joined nucleotides form a **dinucleotide**; three nucleotides, a **trinucleotide**; and so forth. Short chains consisting of up

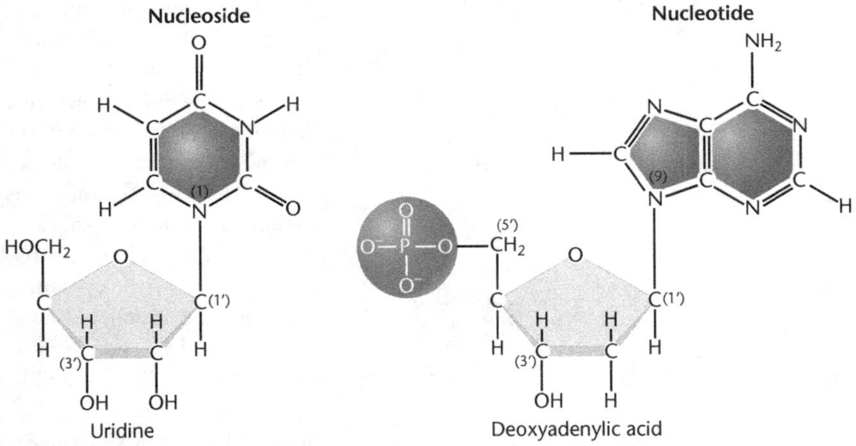


| Ribonucleosides | Ribonucleotides |
|---|---|
| Adenosine<br>Cytidine<br>Guanosine<br>Uridine | Adenylic acid<br>Cytidylic acid<br>Guanylic acid<br>Uridylic acid |
| **Deoxyribonucleosides** | **Deoxyribonucleotides** |
| Deoxyadenosine<br>Deoxycytidine<br>Deoxyguanosine<br>Deoxythymidine | Deoxyadenylic acid<br>Deoxycytidylic acid<br>Deoxyguanylic acid<br>Deoxythymidylic acid |

FIGURE 10–10 Structures and names of the nucleosides and nucleotides of RNA and DNA.

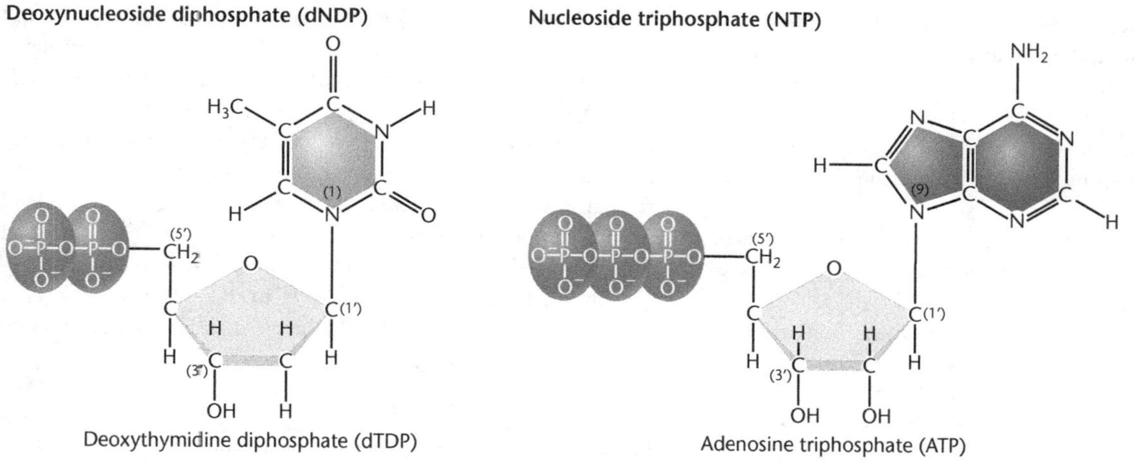


FIGURE 10–11 Structures of nucleoside diphosphates and triphosphates. Deoxythymidine diphosphate and adenosine triphosphate are diagrammed here.

to approximately 20 nucleotides linked together are called **oligonucleotides**; longer chains are called **polynucleotides**.

Because drawing polynucleotide structures, as shown in Figure 10–12(a), is time consuming and complex, a schematic shorthand method has been devised [Figure 10–12(b)]. The nearly vertical lines represent the pentose sugar; the nitrogenous base is attached at the top, in the C-1′ position. A diagonal line with the P in the middle of it is attached to the C-3′ position of one sugar and the C-5′ position of the neighboring sugar; it represents the phosphodiester bond. Several modifications of this shorthand method are in use, and they can be understood in terms of these guidelines.

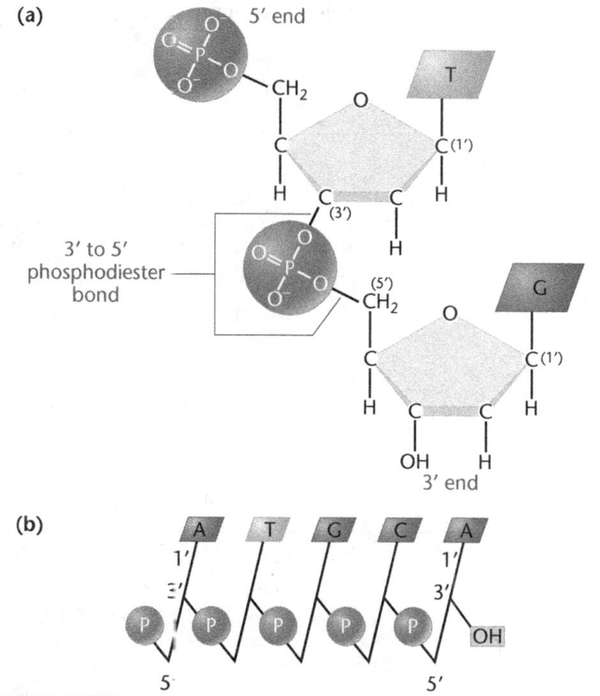


FIGURE 10–12 (a) Linkage of two nucleotides by the formation of a C-3′–C-5′ (3′–5′) phosphodiester bond, producing a dinucleotide. (b) Shorthand notation for a polynucleotide chain.

Although Levene's tetranucleotide hypothesis (described earlier in this chapter) was generally accepted before 1940, research in subsequent decades revealed it to be incorrect. It was shown that DNA does not necessarily contain equimolar quantities of the four bases. In addition, the molecular weight of DNA molecules was determined to be in the range of $10^6$ to $10^9$ daltons, far in excess of that of a tetranucleotide. The current view of DNA is that it consists of exceedingly long polynucleotide chains.

Long polynucleotide chains account for the large molecular weight of DNA and explain its most important property—storage of vast quantities of genetic information. If each nucleotide position in this long chain can be occupied by any one of four nucleotides, extraordinary variation is possible. For example, a polynucleotide only 1000 nucleotides in length can be arranged $4^{1000}$ different ways, each one different from all other possible sequences. This potential variation in molecular structure is essential if DNA is to store the vast amounts of chemical information necessary to direct cellular activities.

## 10.7 The Structure of DNA Holds the Key to Understanding Its Function

The previous sections in this chapter have established that DNA is the genetic material in all organisms (with certain viruses being the exception) and have provided details as to the basic chemical components making up nucleic acids. What remained to be deciphered was the precise structure of DNA. That is, how are polynucleotide chains organized into DNA, which serves as the genetic material? Is DNA composed of a single chain or more than one? If the latter is the case, how do the chains relate chemically to one another? Do the chains branch? And more important, how does the structure of this molecule relate to the various genetic functions served by DNA (i.e., storage, expression, replication, and mutation)?

From 1940 to 1953, many scientists were interested in solving the structure of DNA. Among others, Erwin Chargaff, Maurice

Wilkins, Rosalind Franklin, Linus Pauling, Francis Crick, and James Watson sought information that might answer what many consider to be the most significant and intriguing question in the history of biology: *How does DNA serve as the genetic basis for life?* The answer was believed to depend strongly on the chemical structure and organization of the DNA molecule, given the complex but orderly functions ascribed to it.

In 1953, James Watson and Francis Crick proposed that the structure of DNA is in the form of a double helix. Their model was described in a short paper published in the journal *Nature*. (The article is reprinted in its entirety on page 263.) In a sense, this publication was the finish of a highly competitive scientific race. Watson's book *The Double Helix* recounts the human side of the scientific drama that eventually led to the elucidation of DNA structure.

The data available to Watson and Crick, crucial to the development of their proposal, came primarily from two sources: (1) base composition analysis of hydrolyzed samples of DNA and (2) X-ray diffraction studies of DNA. Watson and Crick's analytical success can be attributed to their focus on building a model that conformed to the existing data. If the correct solution to the structure of DNA is viewed as a puzzle, Watson and Crick, working at the Cavendish Laboratory in Cambridge, England, were the first to fit the pieces together successfully.

Before learning the details of this far-reaching discovery, you may find some of the background on James Watson and Francis Crick to be of interest. Watson began his undergraduate studies at the University of Chicago at age 15, and was originally interested in ornithology. He then pursued his Ph.D. at Indiana University, where he studied viruses. He was only 24 years old in 1953, when he and Crick proposed the double-helix theory. Crick, now considered one of the great theoretical biologists of our time, had studied undergraduate physics at University College, London, and went on to perform military research during World War II. At the time of his collaboration with Watson, he was 35 years old and was performing X-ray diffraction studies of polypeptides and proteins as a graduate student. Immediately after they made their major discovery, Crick is reputed to have walked into the Eagle Pub in Cambridge, where the two frequently lunched, and announced for all to hear, "We have discovered the secret of life." It turns out that more than 50 years later, many scientists would quite agree!

## Base-Composition Studies

Between 1949 and 1953, Erwin Chargaff and his colleagues used chromatographic methods to separate the four bases in DNA samples from various organisms. Quantitative methods were then used to determine the amounts of the four bases from each source. Table 10.3(a) lists some of Chargaff's original data. Parts (b) and (c) of the table show more recently derived base-composition information that reinforces Chargaff's findings. As we shall see, Chargaff's data were critical to the creation of the successful model of DNA put forward by Watson and Crick. On the basis of these data, the following conclusions may be drawn:

1. As shown in Table 10.3(b), the amount of adenine residues is proportional to the amount of thymine residues in DNA (columns 1, 2, and 5). Also, the amount of guanine residues is proportional to the amount of cytosine residues (columns 3, 4, and 6).

TABLE 10.3

**DNA Base-Composition Data**

**(a) Chargaff's data***

| | Molar proportions[a] | | | |
|---|---|---|---|---|
| | 1 | 2 | 3 | 4 |
| Organism/Source | A | T | G | C |
| Ox thymus | 26 | 25 | 21 | 16 |
| Ox spleen | 25 | 24 | 20 | 15 |
| Yeast | 24 | 25 | 14 | 13 |
| Avian tubercle bacilli | 12 | 11 | 28 | 26 |
| Human sperm | 29 | 31 | 18 | 18 |

**(c) G + C content in several organisms**

| Organism | % G + C |
|---|---|
| Phage T2 | 36.0 |
| *Drosophila* | 45.0 |
| Maize | 49.1 |
| *Euglena* | 53.5 |
| *Neurospora* | 53.7 |

**(b) Base compositions of DNAs from various sources**

| | Base composition | | | | Base ratio | | Combined Base Ratios | |
|---|---|---|---|---|---|---|---|---|
| | 1 | 2 | 3 | 4 | 5 | 6 | 7 | 8 |
| Organism | A | T | G | C | A/T | G/C | (A + G)/(C + T) | (A + T)/(C + G) |
| Human | 30.9 | 29.4 | 19.9 | 19.8 | 1.05 | 1.00 | 1.04 | 1.52 |
| Sea urchin | 32.8 | 32.1 | 17.7 | 17.3 | 1.02 | 1.02 | 1.02 | 1.58 |
| *E. coli* | 24.7 | 23.6 | 26.0 | 25.7 | 1.04 | 1.01 | 1.03 | 0.93 |
| *Sarcina lutea* | 13.4 | 12.4 | 37.1 | 37.1 | 1.08 | 1.00 | 1.04 | 0.35 |
| T7 bacteriophage | 26.0 | 26.0 | 24.0 | 24.0 | 1.00 | 1.00 | 1.00 | 1.08 |

**Source*: From Chargaff, 1950.

[a]Moles of nitrogenous constituent per mole of P. (Often, the recovery was less than 100 percent.)

2. Based on this proportionality, the sum of the purines (A+G) equals the sum of the pyrimidines (C+T) as shown in column 7.

3. The percentage of (G+C) does not necessarily equal the percentage of (A+T). As you can see, this ratio varies greatly among organisms, as shown in column 8 and in Table 10.3(c).

These conclusions indicate a definite pattern of base composition in DNA molecules. The data provided the initial clue to the DNA puzzle. In addition, they directly refute Levene's tetranucleotide hypothesis, which stated that all four bases are present in equal amounts.

## X-Ray Diffraction Analysis

When fibers of a DNA molecule are subjected to X-ray bombardment, the X rays scatter (diffract) in a pattern that depends on the molecule's atomic structure. The pattern of diffraction can be captured as spots on photographic film and analyzed for clues to the overall shape of and regularities within the molecule. This process, **X-ray diffraction analysis**, was applied successfully to the study of protein structure by Linus Pauling and other chemists. The technique had been attempted on DNA as early as 1938 by William Astbury. By 1947, he had detected a periodicity of 3.4 angstroms (3.4-Å) repetitions* within the structure of the molecule, which suggested to him that the bases were stacked like coins on top of one another.

Between 1950 and 1953, Rosalind Franklin, working in the laboratory of Maurice Wilkins, obtained improved X-ray data from more purified samples of DNA (Figure 10–13). Her work confirmed the 3.4 Å periodicity seen by Astbury and suggested that the structure of DNA was some sort of helix. However, she did not propose a definitive model. Pauling had analyzed the work of Astbury and others and incorrectly proposed that DNA was a triple helix.

## The Watson–Crick Model

Watson and Crick published their analysis of DNA structure in 1953. By building models based on the above-mentioned parameters, they arrived at the double-helical form of DNA shown in Figure 10–14(a). This model has the following major features:

1. Two long polynucleotide chains are coiled around a central axis, forming a right-handed double helix.

2. The two chains are **antiparallel**; that is, their C-5′-to-C-3′ orientations run in opposite directions.

3. The bases of both chains are flat structures lying perpendicular to the axis; they are "stacked" on one another, 3.4 Å (0.34 nm) apart, on the inside of the double helix.

4. The nitrogenous bases of opposite chains are *paired* as the result of the formation of hydrogen bonds; in DNA, only A = T and G ≡ C pairs occur.

5. Each complete turn of the helix is 34 Å (3.4 nm) long; thus, each turn of the helix is the length of a series of 10 base pairs.

*Today, measurement in nanometers (nm) is favored (1 nm = 10 Å).

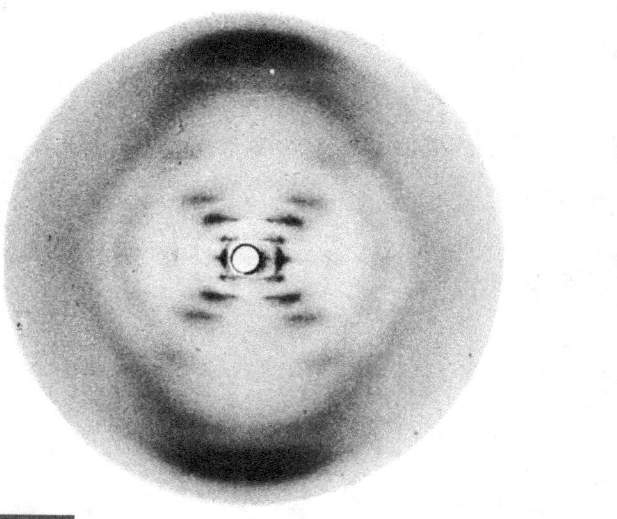

**FIGURE 10–13** X-ray diffraction photograph by Rosalind Franklin using the B form of purified DNA fibers. The strong arcs on the periphery represent closely spaced aspects of the molecule, allowing scientists to estimate the periodicity of the nitrogenous bases, which are 3.4 Å apart. The inner cross pattern of spots reveals the grosser aspects of the molecule, indicating its helical nature.

6. A larger **major groove** alternating with a smaller **minor groove** winds along the length of the molecule.

7. The double helix has a diameter of 20 Å (2.0 nm).

The nature of the base pairing (point 4 above) is the model's most significant feature in terms of explaining its genetic functions. Before we discuss it, however, several other important features warrant emphasis. First, the antiparallel arrangement of the two chains is a key part of the double-helix model. While one chain runs in the 5′-to-3′ orientation (what seems right side up to us), the other chain goes in the 3′-to- 5′ orientation (and thus appears upside down). This is indicated in Figure 10–14(b) and (c). Given the bond angles in the structures of the various nucleotide components, the double helix could not be constructed easily if both chains ran parallel to one another.

Second, the right-handed nature of the helix modeled by Watson and Crick is best appreciated by comparison with its left-handed counterpart, which is a mirror image, as shown in Figure 10–15. The conformation in space of the right-handed helix is most consistent with the data that were available to Watson and Crick, although an alternative form of DNA (Z-DNA) does exist as a left-handed helix, as we discuss below.

The key to the model proposed by Watson and Crick is the specificity of base pairing. Chargaff's data suggested that A was equal in amount to T and that G was equal in amount to C. Watson and Crick realized that pairing A with T and C with G would account for these proportions, and that such pairing could occur as a result of hydrogen bonds between base pairs [Figure 10–14(b)], which would also provide the chemical stability necessary to hold the two chains together. Arrangement of the components in this way produces the

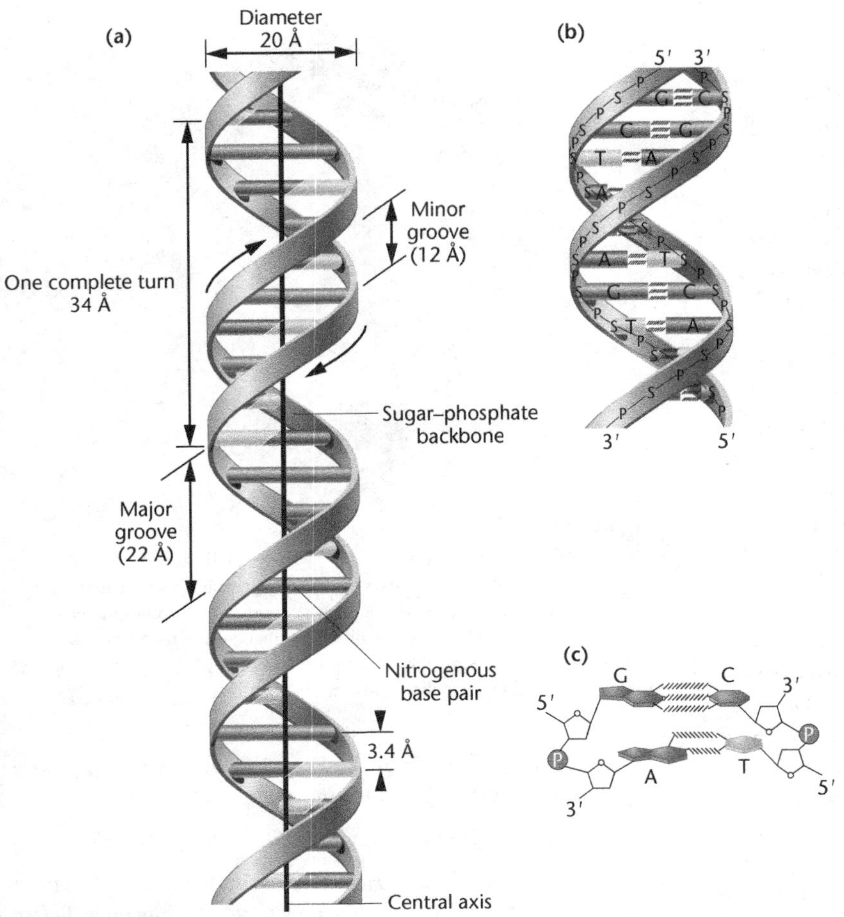


FIGURE 10–14 (a) The DNA double helix as proposed by Watson and Crick. The ribbon-like strands represent the sugar-phosphate backbones, and the horizontal rungs depict the nitrogenous base pairs, of which there are 10 per complete turn. The major and minor grooves are apparent. A solid vertical line shows the central axis. (b) A detailed view depicting the bases, sugars, phosphates, and hydrogen bonds of the helix. (c) A demonstration of the antiparallel arrangement of the chains and the horizontal stacking of the bases.

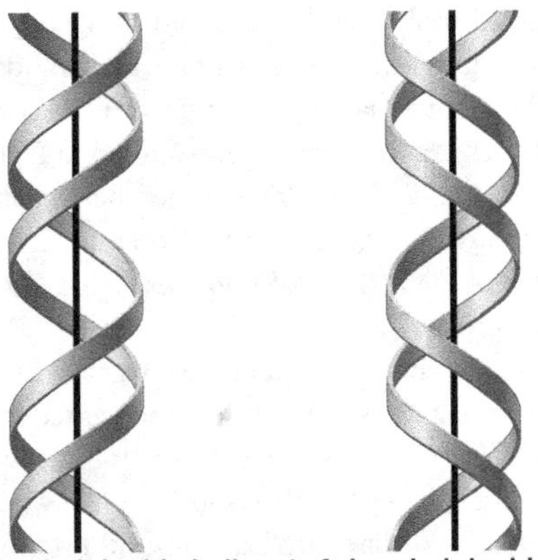


FIGURE 10–15 The right- and left-handed helical forms of DNA. Note that they are mirror images of one another.

major and minor grooves along the molecule's length. Furthermore, a purine (A or G) opposite a pyrimidine (T or C) on each "rung of the spiral staircase" in the proposed helix accounts for the 20 Å (2-nm) diameter suggested by X-ray diffraction studies.

The specific A $=$ T and G $\equiv$ C base pairing is described as **complementarity** and results from the chemical affinity that produces the hydrogen bonds in each pair of bases. As we will see, complementarity is very important in the processes of DNA replication and gene expression.

Two questions are particularly worthy of discussion. First, why aren't other base pairs possible? Watson and Crick discounted the pairing of A with G or of C with T because these would represent purine–purine and pyrimidine–pyrimidine pairings, respectively. Such pairings would lead to aberrant diameters of, in one case, more than and, in the other case, less than 20 Å because of the respective sizes of the purine and pyrimidine rings. In addition, the three-dimensional configurations that would be formed by such pairings would not produce an alignment that allows sufficient hydrogen-bond formation. It is for this reason that A $=$ C and G $\equiv$ T pairings were also discounted, even though those pairs would each consist of one purine and one pyrimidine.

The second question concerns the properties of hydrogen bonds. Just what is the nature of such a bond, and is it strong enough to stabilize the helix? A **hydrogen bond** is a very weak electrostatic attraction between a covalently bonded hydrogen atom and an atom with an unshared electron pair. The hydrogen atom assumes a partial positive charge, while the unshared electron pair—characteristic of covalently bonded oxygen and nitrogen atoms—assumes a partial negative charge. These opposite charges are responsible for the weak chemical attraction that is the basis of the hydrogen bond. As oriented in the double helix, adenine forms two hydrogen bonds with thymine, and guanine forms three hydrogen bonds with cytosine (Figure 10–16). Although two or three hydrogen bonds taken alone are energetically very weak, 2000 to 3000 bonds in tandem (typical of the number in two long polynucleotide chains) provide great stability to the helix.

Another stabilizing factor is the arrangement of sugars and bases along the axis. In the Watson–Crick model, the hydrophobic ("water-fearing") nitrogenous bases are stacked almost horizontally on the interior of the axis and are thus shielded from the watery environment that surrounds the molecule within the cell. The hydrophilic ("water-loving") sugar-phosphate backbones are on the outside of the axis, where both components may interact with water. These molecular arrangements provide significant chemical stabilization to the helix.

A more recent and accurate analysis of the form of DNA that served as the basis for the Watson–Crick model has revealed a minor structural difference between the substance and the model. A precise

## Molecular Structure of Nucleic Acids: A Structure for Deoxyribose Nucleic Acid

We wish to suggest a structure for the salt of deoxyribose nucleic acid (DNA). This structure has novel features which are of considerable biological interest. A structure for nucleic acid has already been proposed by Pauling and Corey.[1] They kindly made their manuscript available to us in advance of publication. Their model consists of three intertwined chains, with the phosphates near the fibre axis, and the bases on the outside. In our opinion, this structure is unsatisfactory for two reasons: (1) We believe that the material which gives the X-ray diagrams is the salt, not the free acid. Without the acidic hydrogen atoms it is not clear what forces would hold the structure together, especially as the negatively charged phosphates near the axis will repel each other. (2) Some of the van der Waals distances appear to be too small.

Another three-chain structure has also been suggested by Fraser (in the press). In his model the phosphates are on the outside and the bases on the inside, linked together by hydrogen bonds. This structure as described is rather ill-defined, and for this reason we shall not comment on it.

We wish to put forward a radically different structure for the salt of deoxyribose nucleic acid. This structure has two helical chains each coiled round the same axis. We have made the usual chemical assumptions, namely, that each chain consists of phosphate diester groups joining α-D-deoxyribofuranose residues with 3′, 5′ linkages. The two chains (but not their bases) are related by a dyad perpendicular to the fibre axis. Both chains follow right-handed helices, but owing to the dyad the sequences of the atoms in the two chains run in opposite directions. Each chain loosely resembles Furberg's[2] model No. 1; that is, the bases are on the inside of the helix and the phosphates on the outside. The configuration of the sugar and the atoms near it is close to Furberg's "standard configuration," the sugar being roughly perpendicular to the attached base. There is a residue on each chain every 3.4 Å in the $z$-direction. We have assumed an angle of 36° between adjacent residues in the same chain, so that the structure repeats after 10 residues on each chain, that is, after 34 Å the distance of a phosphorus atom from the fibre axis is 10 Å. As the phosphates are on the outside, cations have easy access to them.

The structure is an open one, and its water content is rather high. At lower water contents we would expect the bases to tilt so that the structure could become more compact.

The novel feature of the structure is the manner in which the two chains are held together by the purine and pyrimidine bases. The planes of the bases are perpendicular to the fibre axis. They are joined together in pairs, a single base from one chain being hydrogen-bonded to a single base from the other chain, so that the two lie side by side with identical $z$-coordinates. One of the pair must be a purine and the other a pyrimidine for bonding to occur. The hydrogen bonds are made as follows: purine position 1 to pyrimidine position 1; purine position 6 to pyrimidine position 6.

If it is assumed that the bases only occur in the structure in the most plausible tautomeric forms (that is, with the keto rather than the enol configurations) it is found that only specific pairs of bases can bond together. These pairs are: adenine (purine) with thymine (pyrimidine), and guanine (purine) with cytosine (pyrimidine).

In other words, if an adenine forms one member of a pair, on either chain, then on these assumptions the other member must be thymine; similarly for guanine and cytosine. The sequence of bases on a single chain does not appear to be restricted in any way. However, if only specific pairs of bases can be formed, it follows that if the sequence of bases on one chain is given, then the sequence on the other chain is automatically determined.

It has been found experimentally[3,4] that the ratio of the amounts of adenine to thymine, and the ratio of guanine to cytosine, are always very close to unity for deoxyribose nucleic acid.

It is probably impossible to build this structure with a ribose sugar in place of the deoxyribose, as the extra oxygen atom would make too close a van der Waals contact.

The previously published X-ray data[5,6] on deoxyribose nucleic acid are insufficient for a rigorous test of our structure. So far as we can tell, it is roughly compatible with the experimental data, but it must be regarded as unproved until it has been checked against more exact results. Some of these are given in the following communications. We were not aware of the details of the results presented there when we devised our structure, which rests mainly though not entirely on published experimental data and stereochemical arguments.

It has not escaped our notice that the specific pairing we have postulated immediately suggests a possible copying mechanism for the genetic material. Full details of the structure, including the conditions assumed in building it, together with a set of co-ordinates for the atoms, will be published elsewhere.

We are much indebted to Dr. Jerry Donohue for constant advice and criticism, especially on interatomic distances. We have also been stimulated by a knowledge of the general nature of the unpublished experimental results and ideas of Dr. M. H. F. Wilkins, Dr. R. E. Franklin and their co-workers at King's College, London. One of us (J. D. W.) has been aided by a fellowship from the National Foundation for Infantile Paralysis.

—J.D. Watson
—F.H.C. Crick

Medical Research Council Unit for the Study of the Molecular Structure of Biological Systems, Cavendish Laboratory, Cambridge, England.

[1] Pauling, L., and Corey, R. B., *Nature*, 171, 346 (1953); *Proc. U.S. Nat. Acad. Sci.*, 39, 84 (1953).
[2] Furberg, S., *Acta Chem. Scand.*, 6, 634 (1952).
[3] Chargaff, E. For references see Zamenhof, S., Brawerman, G., and Chargaff, E., *Biochim. et Biophys. Acta*, 9, 402 (1952).
[4] Wyatt, G. R., *J. Gen. Physiol.*, 36, 201 (1952).
[5] Astbury, W. T., *Symp. Soc. Exp. Biol. 1, Nucleic Acid*, 66 (Camb. Univ. Press, 1947).
[6] Wilkins, M.H.F., and Randall, J. T., *Biochim. et Biophys. Acta*, 10, 192 (1953).

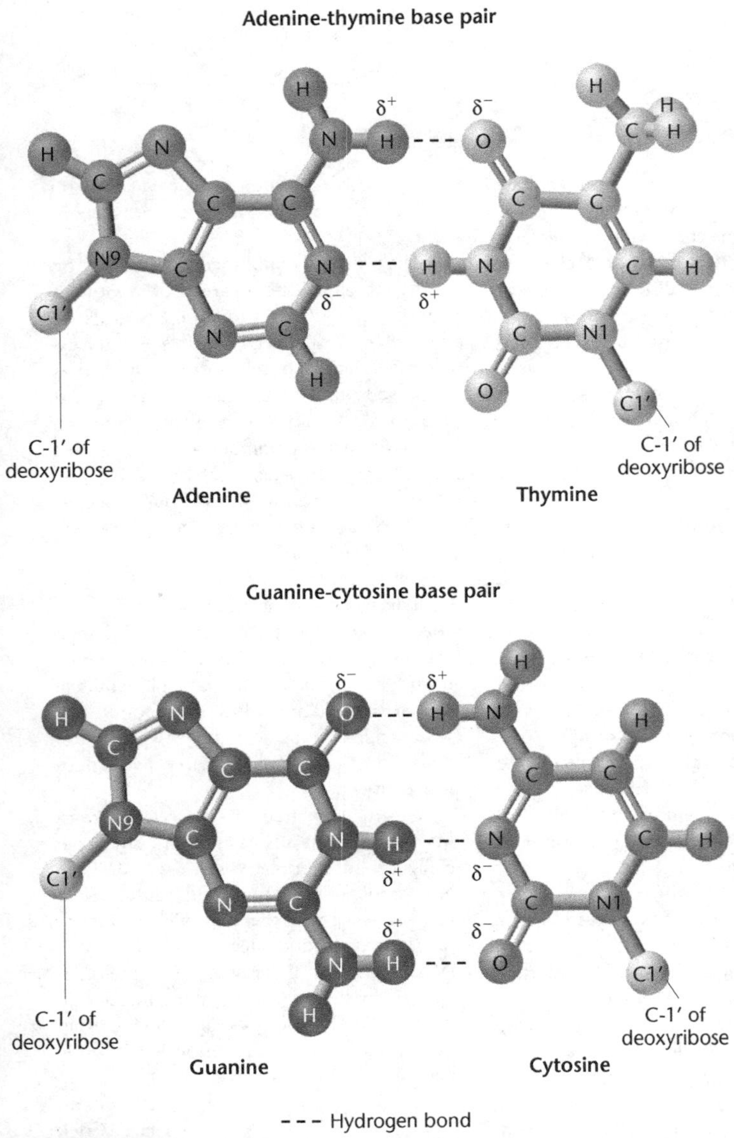


**FIGURE 10-16** Ball-and-stick models of A ═ T and G ≡ C base pairs. The dashes (- - -) represent the hydrogen bonds that form between bases.

measurement of the number of base pairs per turn has demonstrated a value of 10.4, rather than the 10.0 predicted by Watson and Crick. In the classic model, each base pair is rotated 36° around the helical axis relative to the adjacent base pair, but the new finding requires a rotation of 34.6°. This results in slightly more than 10 base pairs per 360° turn.

The Watson–Crick model had an instant effect on the emerging discipline of molecular biology. Even in their initial 1953 article in *Nature*, the authors observed, "It has not escaped our notice that the specific pairing we have postulated immediately suggests a possible copying mechanism for the genetic material." Two months later, Watson and Crick pursued this idea in a second article in *Nature*, suggesting a specific mechanism of replication of *DNA—the semiconservative mode of replication* (described in Chapter 11). The second article also alluded to two new concepts: (1) the storage of genetic information in the sequence of the bases and (2) the mutations or genetic changes that would result from an alteration of bases. These ideas have received vast amounts of experimental support since 1953 and are now universally accepted.

Watson and Crick's synthesis of ideas was highly significant with regard to subsequent studies of genetics and biology. The nature of the gene and its role in genetic mechanisms could now be viewed and studied in biochemical terms. Recognition of their work, along with that of Wilkins, led to all three receiving the Nobel Prize in Physiology and Medicine in 1962. Unfortunately, Rosalind Franklin had died in 1958 at the age of 37, making her contributions ineligible for consideration, since the award is not given posthumously. The Nobel Prize was to be one of many such awards bestowed for work in the field of molecular genetics.

## 10.8 Alternative Forms of DNA Exist

Under different conditions of isolation, different conformations of DNA are seen. At the time when Watson and Crick performed their analysis, two forms—**A-DNA** and **B-DNA**—were known. Watson and Crick's analysis was based on Rosalind Franklin's X-ray studies of the B form, which is seen under aqueous, low-salt conditions and is believed to be the biologically significant conformation.

While DNA studies around 1950 relied on the use of X-ray diffraction, more recent investigations have been performed using **single-crystal X-ray analysis.** The earlier studies achieved resolution of about 5 Å, but single crystals diffract X rays at about 1-Å intervals, near atomic resolution. As a result, every atom is "visible," and much greater structural detail is available.

With this modern technique, A-DNA, which is prevalent under high-salt or dehydration conditions, has now been scrutinized. In comparison to B-DNA, A-DNA is slightly more compact, with 9 base pairs in each complete turn of the helix, which is 23 Å (2.3 nm) in diameter (Figure 10–17). While it is also a right-handed helix, the orientation of the bases is somewhat different—they are tilted and displaced laterally in relation to the axis of the helix. As a result, the appearance of the major and minor grooves is modified. It seems doubtful that A-DNA occurs *in vivo* (under physiological conditions).

Still other forms of DNA right-handed helices have been discovered when investigated under various laboratory conditions. These have been designated C-, D-, E-, and more recently P-DNA. **C-DNA** is found under even greater dehydration conditions than those observed during the isolation of A- and B-DNA. It has only 9.3 base pairs per turn and is, thus, less compact. Its helical diameter is 19 Å. Like A-DNA, C-DNA does not have its base pairs lying flat; rather, they are tilted relative to the axis of the helix. Two other forms, **D-DNA** and **E-DNA**, occur in helices lacking guanine in their base composition. They have even fewer base pairs per turn: 8 and 7, respectively.

Recent research by Jean-François Allemand and colleagues has shown that if DNA is artificially stretched, still another conformation

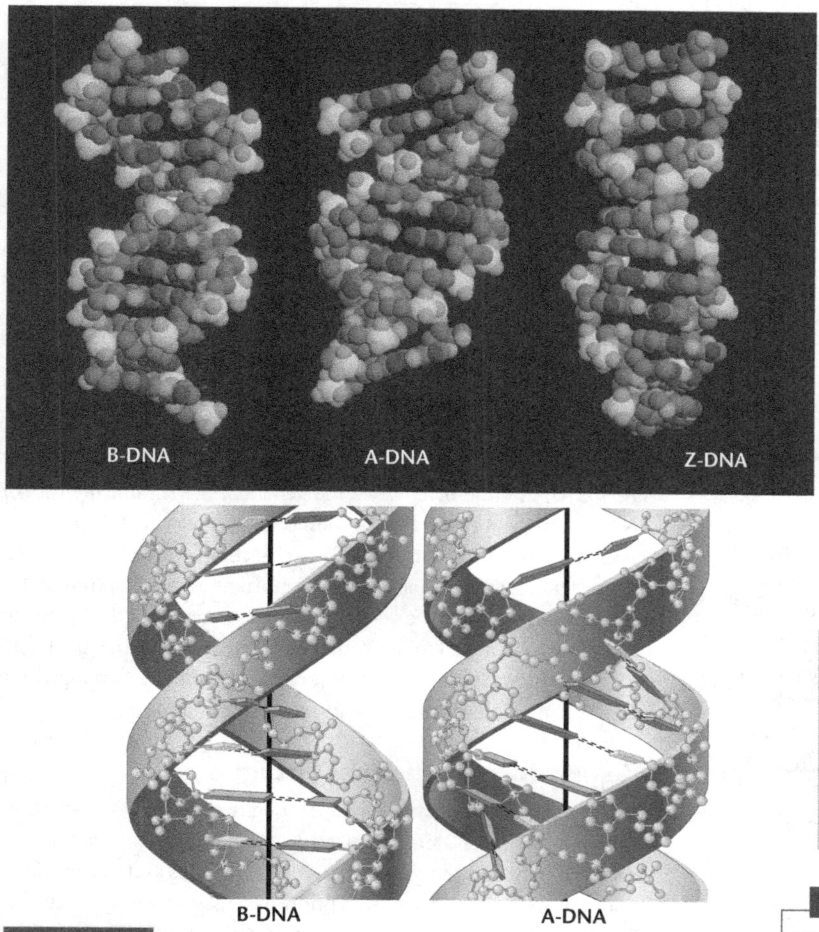


FIGURE 10–17 The top half of the figure shows computer-generated space-filling models of B-DNA, A-DNA, and Z-DNA. Below the photograph is an artist's depiction illustrating the orientation of the base pairs of B-DNA and A-DNA. Note that in B-DNA the base pairs are perpendicular to the helix, while they are tilted and pulled away from the helix in A-DNA.

is assumed, called **P-DNA** (named for Linus Pauling). Contrasted with the B form of DNA, P-DNA is quite interesting, since it is longer and narrower, and the phosphate groups, found on the outside of B-DNA, are located inside the molecule. The nitrogenous bases, present inside the helix in B-DNA, are found closer to the external surface in P-DNA, and there are fewer hydrogen bonds formed as a result. There are 2.62 bases per turn, in contrast to the 10.4 per turn in B-DNA.

Still another form of DNA, called **Z-DNA**, was discovered by Andrew Wang, Alexander Rich, and their colleagues in 1979, when they examined a small synthetic DNA oligonucleotide containing only $G \equiv C$ base pairs. Z-DNA takes on the rather remarkable configuration of a *left-handed double helix* (Figure 10–17). Like A- and B-DNA, Z-DNA consists of two antiparallel chains held together by Watson–Crick base pairs. Beyond these characteristics, Z-DNA is quite different. The left-handed helix is 18 Å (1.8 nm) in diameter, contains 12 base pairs per turn, and has a zigzag conformation (hence its name). The major groove present in B-DNA is nearly eliminated in Z-DNA. Speculation abounds over the possibility that regions of Z-DNA exist in the chromosomes of living organisms. The unique helical arrangement has the potential to provide an important recognition site for interaction with DNA-binding molecules. However, the extent to which Z-DNA occurs *in vivo* is still not clear.

The interest in alternative forms of DNA, such as Z and P, stems from the belief that DNA might have to assume a structure other than the B-form for some of its genetic functions. During both replication and transcription (when its RNA complement is synthesized during gene expression), the strands of the helix must separate and become accessible to large enzymes, as well as to a variety of other proteins involved in these processes. It is possible that changes in the shape of the DNA facilitate these functions. However, verification of the biological significance of alternative forms awaits further study.

NOW SOLVE THIS

Problem 36 on page 276 asks you to construct a protocol in order to analyze DNA from a newly discovered source.

■ HINT: *Knowing the nature and relative proportions of the nitrogenous bases of the unknown DNA will provide particularly important experimental information at the outset.*

10.9

## The Structure of RNA Is Chemically Similar to DNA, but Single Stranded

The structure of RNA molecules resembles DNA, with several important exceptions. Although RNA also has nucleotides linked into polynucleotide chains, the sugar ribose replaces deoxyribose, and the nitrogenous base uracil replaces thymine. Another important difference is that most RNA is single stranded, although there are two important exceptions. First, RNA molecules sometimes fold back on themselves to form double-stranded regions of complementary base pairs. Second, some animal viruses that have RNA as their genetic material contain double-stranded helices.

As established earlier (see Figure 10–1), three major classes of cellular RNA molecules function during the expression of genetic information: **ribosomal RNA (rRNA), messenger RNA (mRNA)**, and **transfer RNA (tRNA)**. These molecules all originate as complementary copies of one of the two strands of DNA segments during the process of transcription. That is, their nucleotide sequence is complementary to the deoxyribonucleotide sequence of DNA that served as the template for their synthesis. Because uracil replaces thymine in RNA, uracil is complementary to adenine during transcription and during RNA base pairing.

TABLE 10.4

**Principal Classes of RNA**

| Class | % of Total RNA* | Svedberg Coefficient | Eukaryotic (E) or Prokaryotic (P) | Number of Nucleotides |
|---|---|---|---|---|
| Ribosomal (rRNA) | 80 | 5 | P and E | 120 |
| | | 5.8 | E | 160 |
| | | 16 | P | 1542 |
| | | 18 | E | 1874 |
| | | 23 | P | 2904 |
| | | 28 | E | 4718 |
| Transfer (tRNA) | 15 | 4 | P and E | 75–90 |
| Messenger (mRNA) | 5 | varies | P and E | 100–10,000 |

*In *E. coli*

Table 10.4 characterizes these major forms of RNA as found in prokaryotic and eukaryotic cells. Different RNAs are distinguished according to their sedimentation behavior in a centrifugal field and by their size (the number of nucleotides each contains). Sedimentation behavior depends on a molecule's density, mass, and shape, and its measure is called the **Svedberg coefficient** (*S*). While higher *S* values almost always designate molecules of greater molecular weight, the correlation is not direct; that is, a twofold increase in molecular weight does not lead to a twofold increase in *S*. This is because, in addition to a molecule's mass, the size and shape of the molecule also affect its rate of sedimentation (*S*). As you can see in Table 10.4, RNA molecules come in a wide range of sizes.

Ribosomal RNA is generally the largest of these molecules (as is generally reflected in its *S* values) and usually constitutes about 80 percent of all RNA in an *E. coli* cell. Ribosomal RNAs are important structural components of **ribosomes**, which function as nonspecific workbenches where proteins are synthesized during translation. The various forms of rRNA found in prokaryotes and eukaryotes differ distinctly in size.

Messenger RNA molecules carry genetic information from the DNA of the gene to the ribosome. The mRNA molecules vary considerably in size, reflecting the range in the sizes of the proteins encoded by the mRNA as well as the different sizes of the genes serving as the templates for transcription of mRNA. While Table 10.4 shows that about 5 percent of RNA is mRNA in *E. coli*, this percentage varies from cell to cell and even at different times in the life of the same cell.

Transfer RNA, accounting for up to 15 percent of the RNA in a typical cell, is the smallest class (in terms of average molecule size) of these RNA molecules and carries amino acids to the ribosome during translation. Because more than one tRNA molecule interacts simultaneously with the ribosome, the molecule's smaller size facilitates these interactions.

We will discuss the functions of these three classes of RNA in much greater detail in Chapters 14 and 15. These RNAs represent the major forms of the molecule involved in genetic expression, but other unique RNAs exist that perform various roles, especially in eukaryotes. For example, **telomerase RNA** is involved in DNA replication at the ends of chromosomes (Chapter 11), **small nuclear RNA (snRNA)** participates in processing mRNAs (Chapter 13), and **antisense RNA, microRNA (miRNA),** and **short interfering RNA (siRNA)** are involved in gene regulation (Chapter 18). DNA stores genetic information, while RNA most often functions in the expression of that information.

## 10.10 Many Analytical Techniques Have Been Useful during the Investigation of DNA and RNA

Since 1953, the role of DNA as the genetic material and the role of RNA in transcription and translation have been clarified through detailed analysis of nucleic acids. In this section, we consider several methods that have been particularly important for accomplishing this analysis. Many of them make use of the unique nature of the hydrogen bond that is so integral to the structure of nucleic acids. In Chapter 13 we will consider still other techniques that have been developed for manipulating DNA in various ways, including cloning.

### Absorption of Ultraviolet Light

Nucleic acids absorb ultraviolet (UV) light most strongly at wavelengths of 254 to 260 nm (Figure 10–7), due to the interaction between UV light and the ring systems of the purines and pyrimidines. In aqueous solution, peak absorption by DNA and RNA occurs at 260 nm. Thus, UV light can be used in the localization, isolation, and characterization of molecules that contain nitrogenous bases (i.e., nucleosides, nucleotides, and polynucleotides).

Ultraviolet analysis is used in conjunction with many standard procedures that separate and identify molecules. As we shall see in the next section, the use of UV absorption is critical to the isolation of nucleic acids following their separation.

### Sedimentation Behavior

Nucleic acid mixtures can be separated into different components by several possible centrifugation procedures (Figure 10–18). For example, the mixture can be loaded on top of a solution in which a concentration gradient has been formed from top to bottom. Then the tube holding the two is spun at high speeds in an ultracentrifuge so that the molecules of the mixture migrate downward through the concentration gradient, with each kind of molecule moving at a different rate. After centrifugation is stopped, successive fractions are eluted from the tube and measured spectrophotometrically for absorption at 260 nm. In this way, the position of a nucleic acid fraction along the gradient can be determined and the fraction isolated and studied further.

These gradient centrifugations rely on the sedimentation behavior of molecules in solution. Two major types of gradient

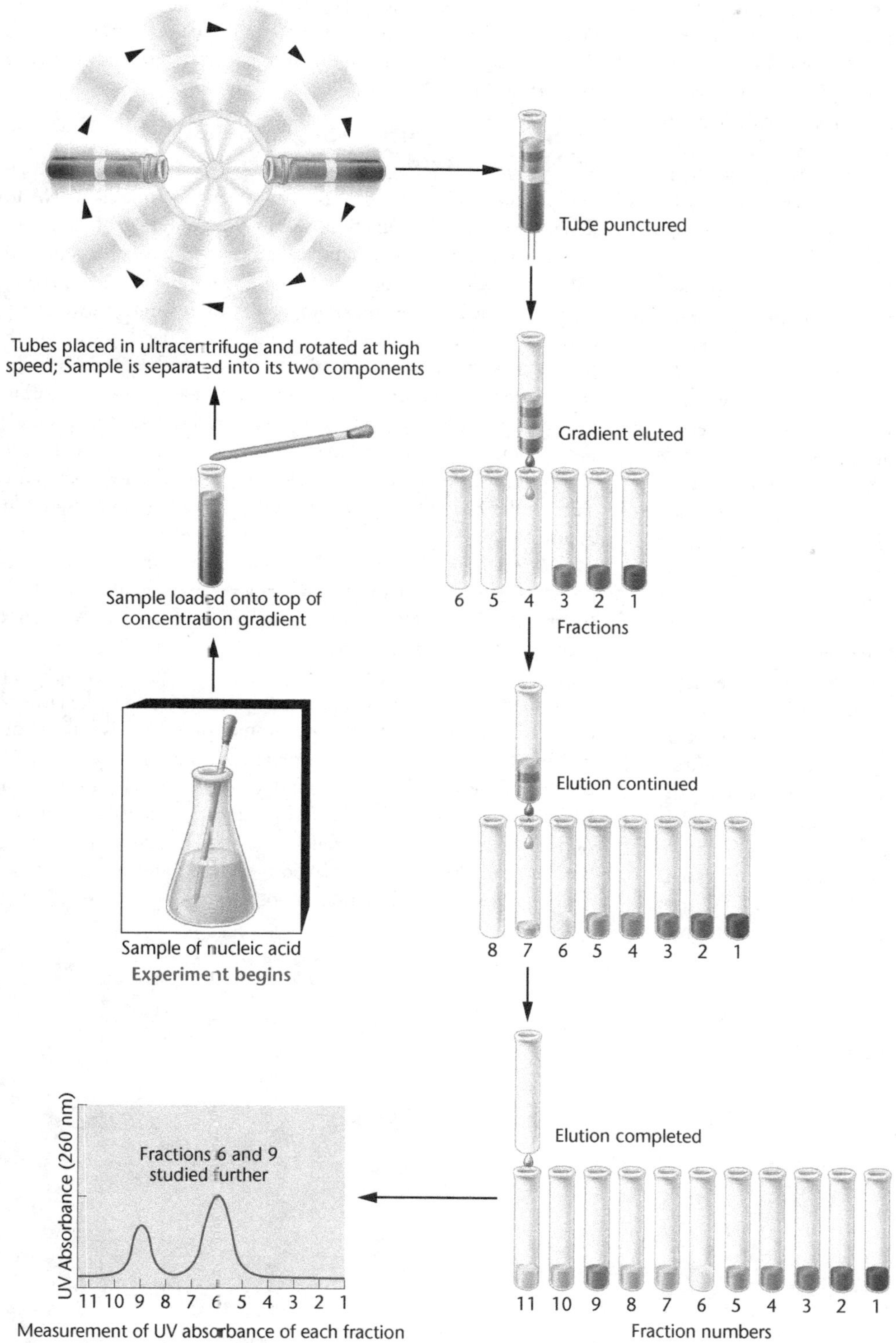


**FIGURE 10-18** Separation of a mixture of two types of nucleic acid by gradient centrifugation. To fractionate the gradient, successive samples are eluted from the bottom of the tube. Each is measured for absorbance of ultraviolet light at 260 nm, producing a profile of the sample in graphic form.

centrifugation techniques are employed in the analysis of nucleic acids: *sedimentation equilibrium* and *sedimentation velocity*. Both require the use of high-speed centrifugation to create large centrifugal forces upon molecules in a gradient solution.

In **sedimentation equilibrium centrifugation** (sometimes called density gradient centrifugation), a density gradient is created that overlaps the densities of the individual components of a mixture of molecules. Usually, the gradient is made of a solution of a heavy metal salt, such as cesium chloride (CsCl). During centrifugation, the molecules migrate until they reach a point of neutral buoyant density. At this point, the centrifugal force on them is equal and opposite to the upward diffusion force, and no further migration occurs. If DNAs of different densities are present, they will separate as the molecules of each density reach equilibrium with the corresponding density of CsCl. The gradient may then be fractionated and the components isolated (Figure 10–18). When properly executed, this technique provides high resolution in separating mixtures of molecules varying only slightly in density.

Sedimentation equilibrium centrifugation studies can also be used to generate data on the base composition of double-stranded DNA. The G≡C base pairs are more compact and dense than the A═T pairs. As shown in Figure 10–19, the percentage of G≡C pairs in DNA is directly proportional to the molecule's buoyant density. Using this relationship, scientists can make useful molecular characterizations of DNA from different sources.

The second technique, **sedimentation velocity centrifugation**, employs an analytical centrifuge that uses ultraviolet absorption optics to monitor the migration of the molecules during centrifugation and determine the "velocity of sedimentation." As mentioned earlier, this velocity has been standardized in units called Svedberg coefficients ($S$).

In this technique, the molecules are loaded on top of the gradient, and the gravitational forces created by centrifugation drive them toward the bottom of the tube. Two forces work against this downward movement: (1) the viscosity of the solution creates a frictional resistance, and (2) part of the force of diffusion is directed upward. Under these conditions, the key variables are the mass and shape of the molecules being examined. In general, the greater the mass, the greater the sedimentation velocity. However, the molecule's shape affects the frictional resistance. Therefore, two molecules of equal mass but different in shape will sediment at different rates.

One use of the sedimentation velocity technique is the determination of **molecular weight (MW)**. If certain physical-chemical properties of a molecule under study are also known, the MW can be calculated based on the sedimentation velocity. As noted earlier, the $S$ values increase with molecular weight, but they are not directly proportional to it.

## Denaturation and Renaturation of Nucleic Acids

Heat or other stresses can cause a complex molecule like DNA to **denature**, or lose its function due to the unfolding of its three-dimensional structure. In the denaturation of double-stranded DNA, the hydrogen bonds of the duplex (double-stranded) structure break, the helix unwinds, and the strands separate. However, no covalent bonds break. The viscosity of the DNA decreases, and both the UV absorption and the buoyant density increase. In laboratory studies, denaturation may be caused intentionally either by heating or chemical treatment. (Denaturation as a result of heating is sometimes referred to as *melting*.) In such studies, the increase in UV absorption of heated DNA in solution, called the **hyperchromic shift**, is the easiest change to measure. This effect is illustrated in Figure 10–20.

Because G≡C base pairs have one more hydrogen bond than do A═T pairs (see Figure 10–16), they are more stable under heat treatment. Thus, DNA with a greater proportion of G≡C pairs than A═T pairs requires higher temperatures to denature completely. When absorption at 260 nm is monitored and plotted against temperature during heating, a **melting profile** of the DNA is obtained. The midpoint of this profile, or curve, is called the **melting temperature** ($T_m$) and represents the point at which 50 percent of the strands are unwound, or denatured (Figure 10–20). When the curve plateaus at its maximum optical density, denaturation is complete, and only single strands exist. Analysis of melting profiles provides a characterization of DNA and an alternative method of estimating its base composition.

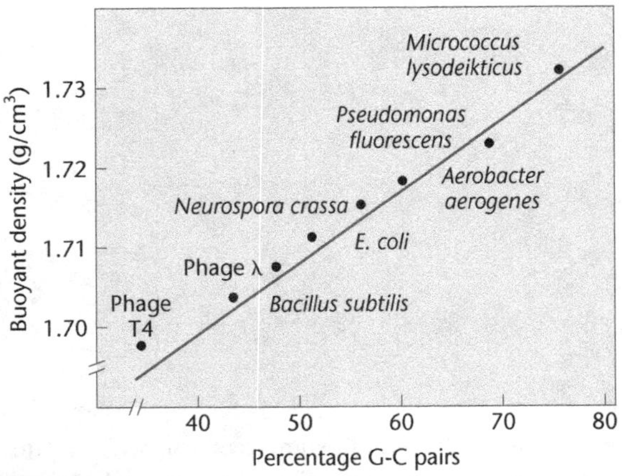


**FIGURE 10–19** Percentage of guanine-cytosine (G≡C) base pairs in the DNA of different microorganisms, plotted against buoyant density.

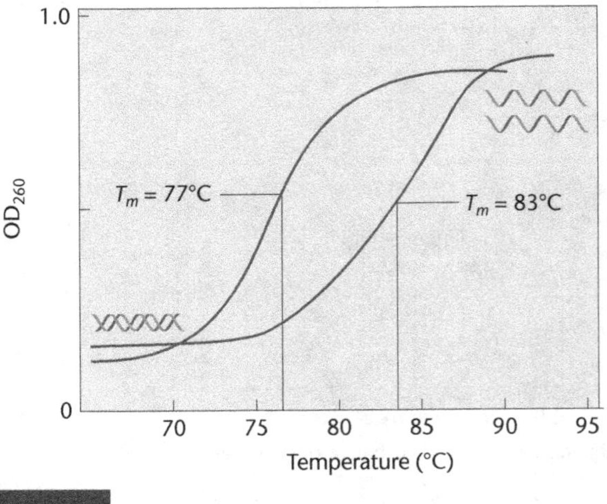


**FIGURE 10–20** A melting profile shows the increase in UV absorption versus temperature (the hyperchromic effect) for two DNA molecules with different G≡C contents. The molecule with a melting point ($T_m$) of 83°C has a greater G≡C content than the molecule with a $T_m$ of 77°C.

One might ask whether the denaturation process can be reversed; that is, can single strands of nucleic acids re-form a double helix, provided that each strand's complement is present? Not only is the answer yes, but such reassociation provides the basis for several important analytical techniques that have contributed much valuable information during genetic experimentation.

If DNA that has been denatured thermally is cooled slowly, random collisions between complementary strands will result in their reassociation. At the proper temperature, hydrogen bonds will re-form, securing pairs of strands into duplex structures. With time during cooling, more and more duplexes will form. Depending on the conditions, a complete match (that is, perfect complementarity) is not essential for duplex formation, provided there are stretches of base pairing on the two reassociating strands.

## Molecular Hybridization

The denaturation and renaturation of nucleic acids is the basis for one of the most powerful and useful techniques in molecular genetics—**molecular hybridization.** This technique derives its name from the fact that single strands need not originate from the same nucleic acid source in order to combine to form duplex structures. For example, if DNA strands are isolated from two distinct organisms and a reasonable degree of base complementarity exists between them, under the proper temperature conditions, double-stranded molecular hybrids will form during renaturation. Such hybridization may also occur between single strands of DNA and RNA. A case in point is when RNA and the DNA from which it has been transcribed are mixed together (Figure 10–21). The RNA will find and attach to its single-stranded DNA complement. In this example, DNA molecules

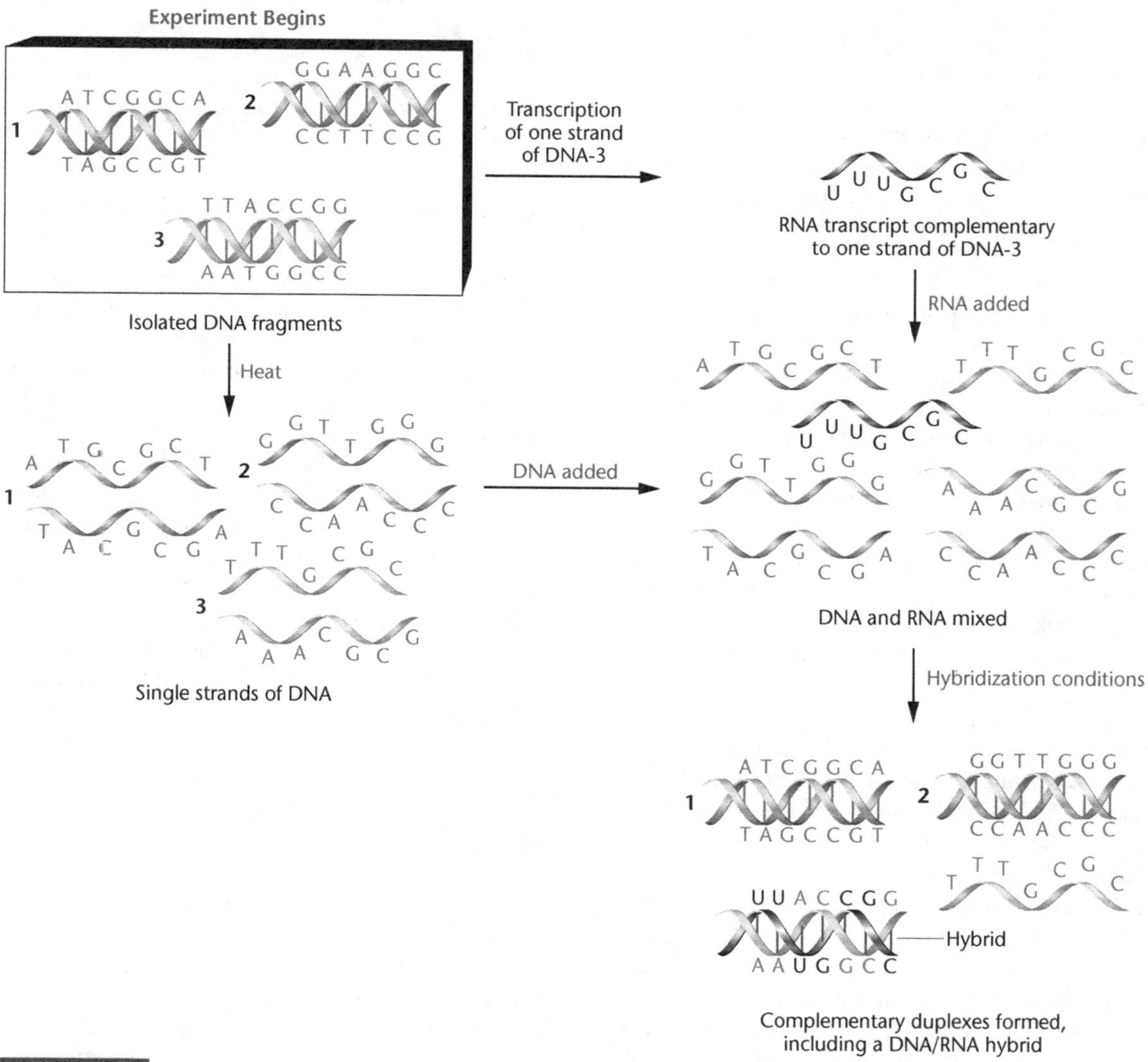


**FIGURE 10–21** Diagrammatic representation of the process of molecular hybridization between DNA fragments and RNA that has been transcribed on one of the single-stranded fragments.

are heated, causing strand separation, and then the strands are slowly cooled in the presence of single-stranded RNA. If the RNA has been transcribed from the DNA used in the experiment, and is therefore complementary to it, molecular hybridization will occur, creating a DNA:RNA duplex. Several methods are available for monitoring the amount of double-stranded molecules produced following strand separation. In early studies, radioisotopes were utilized to "tag" one of the strands and monitor its presence in hybrid duplexes that formed.

In the 1960s, molecular hybridization techniques contributed to our increased understanding of transcriptional events occurring at the gene level. Refinements of this process have occurred continually, helping to advance the study of both molecular evolution and the organization of DNA in chromosomes. Hybridization can occur in solution or when DNA is bound either to a gel or to a specialized binding filter. Such filters are used in a variety of **DNA blotting procedures**, whereby hybridization serves as a way to "probe" for complementary nucleic acid sequences. Blotting is used routinely in modern genomic analysis. In addition, hybridization will occur even when DNA is present in tissue affixed to a slide, as in the FISH procedure (discussed in the next section), or when affixed to a glass chip, the basis of **DNA microarray analysis** (discussed in Chapter 24). Microarray analysis allows mass screening for a specific DNA sequence from among thousands of cloned genes in a single assay.

## Fluorescent *in situ* Hybridization (FISH)

A refinement in the molecular hybridization technique has led to the use of DNA present in cytological preparations as the "target" for hybrid formation. When this approach is combined with the use of fluorescent probes to monitor hybridization, the technique is called **fluorescent *in situ* hybridization**, or simply by the acronym **FISH**. In this procedure, mitotic or interphase cells are fixed to slides and subjected to hybridization conditions. Single-stranded DNA or RNA is added, and hybridization is monitored. The added nucleic acid serves as a "probe," since it will hybridize only with the specific chromosomal areas for which it is sufficiently complementary. Before the use of fluorescent probes was refined, radioactive probes were used in these *in situ* procedures to allow detection on the slide by autoradiography.

Fluorescent probes are prepared in a unique way. When DNA is used, it is first coupled to the small organic molecule biotin (creating biotinylated DNA). Another molecule (avidin or streptavidin) that has a high binding affinity for biotin is linked with a fluorescent molecule such as fluorescein, and this complex is reacted with the cytological preparation after the *in situ* hybridization has occurred. This procedure represents an extremely sensitive method for localizing the hybridized DNA.

Figure 10–22 illustrates the results of using FISH to identify the DNA specific to the centromeres of human chromosomes. The resolution of FISH is great enough to detect just a single gene within an entire set of chromosomes. The use of this technique to identify chromosomal locations of specific genetic information has been a valuable addition to the repertoire of experimental geneticists.

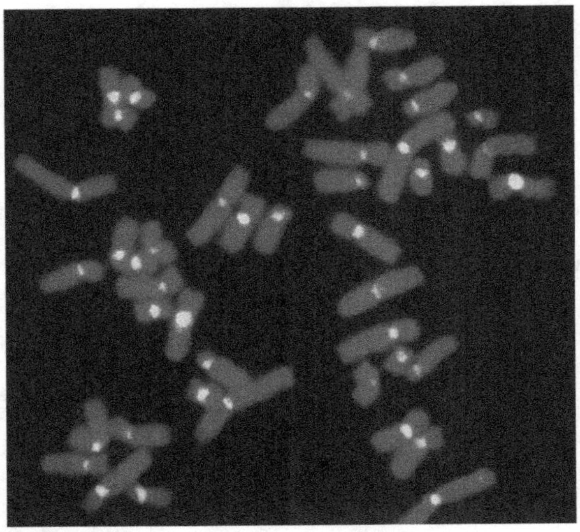

FIGURE 10–22 Fluorescent *in situ* hybridization (FISH) of human metaphase chromosomes. The probe, specific to centromeric DNA, produces a yellow fluorescence signal indicating hybridization. The red fluorescence is produced by propidium iodide counterstaining of chromosomal DNA.

## Reassociation Kinetics and Repetitive DNA

In one extension of molecular hybridization procedures, the *rate of reassociation* of complementary single DNA strands is analyzed. This technique, called **reassociation kinetics**, was first refined and studied by Roy Britten and David Kohne.

The DNA used in such studies is first fragmented into small pieces by shearing forces introduced during its isolation. The resultant DNA fragments have an average size of several hundred base pairs. These fragments of DNA are then dissociated into single strands by heating (denatured), and when the temperature is lowered, their reassociation is monitored. During reassociation, pieces of single-stranded DNA collide randomly. If they are complementary, a stable double strand is formed; if not, they separate and are free to encounter other DNA fragments. The process continues until all matches are made.

The results of one such experiment are presented in Figure 10–23. The percentage of reassociated DNA fragments is plotted against a logarithmic scale of normalized time, a function referred to as $C_0t$, where $C_0$ is the initial concentration of DNA single strands in moles per liter of nucleotides, and $t$ is time, usually measured in seconds. The process of renaturation follows second-order rate kinetics according to the equation

$$\frac{C}{C_0} = \frac{1}{1 + kC_0t}$$

where $C$ is the single-stranded DNA concentration remaining after time $t$ has elapsed and $k$ is the second-order rate constant. Initially, $C$ equals $C_0$ and the fraction remaining single stranded is 100 percent.

The initial shape of the curve reflects the fact that in a mixture of unique-sequence fragments, each with one complement, the fragments are slow at first to find other fragments whose sequences

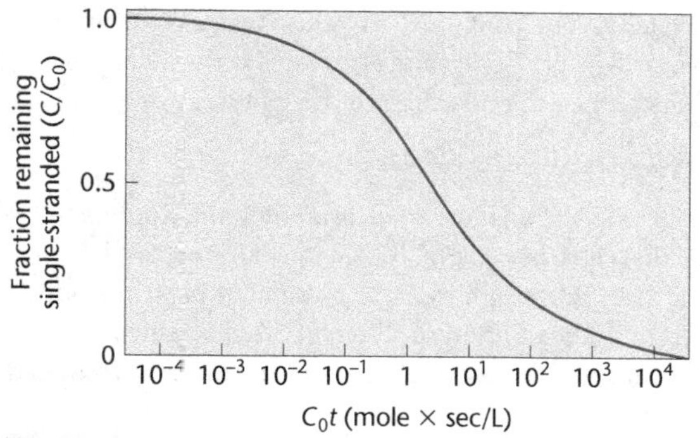


**FIGURE 10–23** The ideal time course for reassociation of DNA ($C/C_0$) when, at time zero, all DNA consists of unique fragments of single-stranded complements. Note that the abscissa ($C_0t$) is plotted logarithmically.

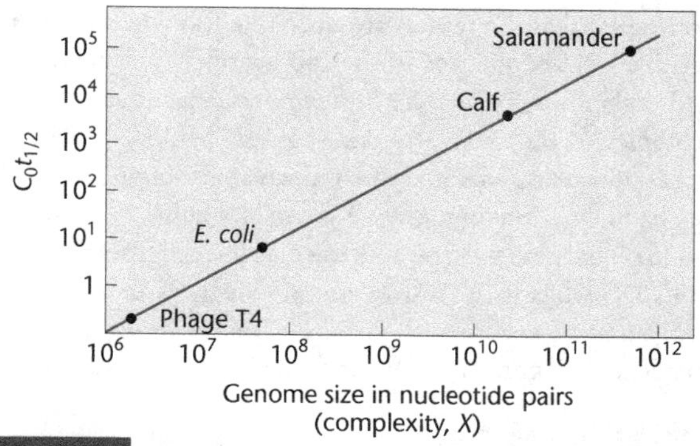


**FIGURE 10–25** Comparison of $C_0t_{1/2}$ and genome size for phage T4, *E. coli*, calf, and salamander.

are complementary. As time passes, and some of the single strands find their complements and form duplexes, matches start to take place more quickly, producing a steepening in the "slope" of the curve. Near the end of the reaction, however, the few remaining single strands require relatively more time to make the final matches.

A great deal of information can be obtained from studies comparing the reassociation profiles of DNA of different organisms. For example, we may compare the point in the reaction when one-half of the DNA is present as double-stranded fragments. This point is called the $C_0t_{1/2}$, or *half-reaction time.* Provided that all pairs of single-stranded DNA complements contain unique nucleotide sequences and all are about the same size, $C_0t_{1/2}$ varies directly with the complexity of the DNA. Designated as *X*, complexity represents the total length in nucleotide pairs of all unique DNA fragments laid end to end. If the DNA used in an experiment represents the entire genome, and if all DNA sequences are different from one another, then *X* is equal to the size of the haploid genome.

Figure 10–24 compares DNAs from two bacteriophages and one bacterial source, each with a different genome size. We see that as genome size increases, the curves obtained are shifted farther and farther to the right, indicative of an extended reassociation time.

As shown in Figure 10–25, $C_0t_{1/2}$ is directly proportional to the size of the genome. Reassociation occurs at a reduced rate in larger genomes because it takes longer for initial matches to occur if there are greater numbers of unique DNA fragments. This is because collisions are random; the more sequences there are, the greater the number of mismatches before all correct matchings occur. The reassociation method has been useful in assessing genome size in viruses and bacteria.

When reassociation kinetics in eukaryotic organisms with much larger genome sizes was first studied, a surprising observation was made. The data revealed that, rather than exhibiting a reduced rate of reassociation, *some* DNA segments from larger genomes reassociate even more rapidly than those derived from *E. coli*. The remaining DNA behaved as expected, taking longer to reassociate because of its greater complexity.

For example, Britten and Kohne examined DNA derived from calf thymus tissue (Figure 10–26). Based on these observations, they

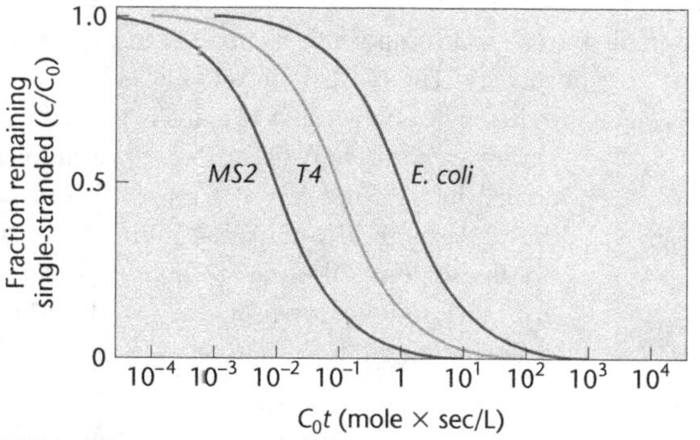


**FIGURE 10–24** The reassociation rates ($C/C_0$) of DNA derived from phage MS2, phage T4, and *E. coli*. The genome of T4 is larger than MS2 and that of *E. coli* is larger than T4.

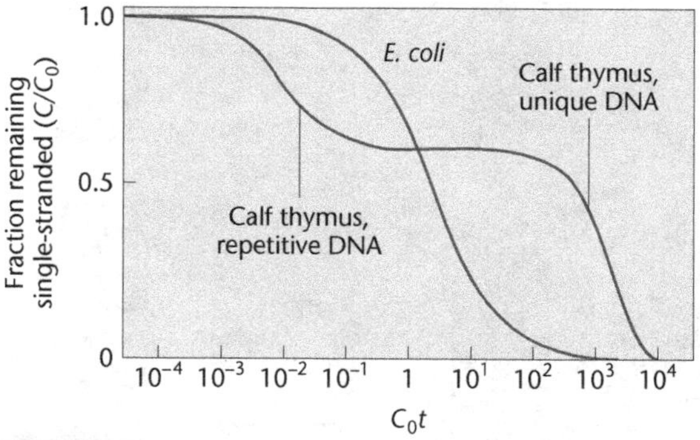


**FIGURE 10–26** The $C_0t$ curve of calf thymus DNA compared with *E. coli*. The repetitive fraction of calf DNA reassociates more quickly than *E. coli* DNA, while the more complex, unique parts of the calf DNA take longer to reassociate than *E. coli* DNA.

hypothesized that the rapidly reassociating fraction must represent **repetitive DNA sequences** in the calf genome. This interpretation would explain why these DNA segments reassociate so rapidly. Multiple copies of the same sequence are much more likely to make matches, thus reassociating more quickly than single copies. On the other hand, the remaining DNA segments consist of unique nucleotide sequences (present only once in the genome). Because calf thymus DNA has many more unique sequences than *E. coli*, their reassociation takes longer than *E. coli* DNA. The *E. coli* curve has been added to Figure 10–26 for the sake of comparison.

**NOW SOLVE THIS**

Problem 30 on page 276 asks you to extrapolate the overall size of the DNA molecule from information about $C_0t$ analysis of DNA.

■ HINT: *Absolute $C_0t$ values are directly proportional to the number of base pairs making up a DNA molecule.*

It is now clear that repetitive DNA sequences are prevalent in the genome of eukaryotes and are key to our understanding of how genetic information is organized in chromosomes. Careful study has shown that various levels of repetition exist. In some cases, short DNA sequences are repeated over a million times. In other cases, longer sequences are repeated only a few times, or intermediate levels of sequence redundancy are present. We will return to this topic in Chapter 12, where we will discuss the organization of DNA in genes and chromosomes. For now, we will conclude our discussion of repetitive DNA sequences by pointing out that the discovery of repetitive DNA was one of the first clues that much of the DNA in eukaryotes is not contained in genes that encode proteins. This concept will be developed and elaborated on as we proceed with our coverage of the molecular basis of heredity.

## Electrophoresis of Nucleic Acids

We conclude the chapter by considering **electrophoresis,** a technique that has made essential contributions to the analysis of nucleic acids. Electrophoresis, also useful in protein studies, can be applied to the separation of different-sized fragments of DNA and RNA chains and is invaluable in current research investigations in molecular genetics.

In general, electrophoresis separates, or *resolves*, molecules in a mixture by causing them to migrate under the influence of an electric field. A sample to be analyzed is placed on a porous substance (a piece of filter paper or a semisolid gel) that in turn is placed in a solution that conducts electricity. If two molecules have approximately the same shape and mass, the one with the greatest net charge will migrate more rapidly toward the electrode of opposite polarity.

As electrophoretic technology developed from its initial application (which was protein separation), researchers discovered that using gels of varying pore sizes significantly improved the method's resolution. This advance is particularly useful for mixtures of molecules with a similar charge-mass ratio but different sizes. For example, two polynucleotide chains of different lengths (say, 10 vs. 20 nucleotides) are both negatively charged because of the phosphate groups of the nucleotides. Thus, both chains move to the positively charged pole (the anode), and because they have the same charge-mass ratio, the electric field moves them at similar speeds. Consequently, the separation between the two chains is minimal. However, using a porous medium such as a **polyacrylamide gel** or an **agarose gel,** which can be prepared with various pore sizes, enables us to separate the two molecules. *Smaller molecules migrate at a faster rate through the gel than larger molecules* (Figure 10–27). The key to separation is the porous gel matrix, which restricts migration of larger molecules more than it restricts smaller molecules. The resolving power of this method is so great that polynucleotides that vary by even one nucleotide in length may be separated. Once electrophoresis is complete, bands representing the variously sized molecules are identified either by autoradiography (if a component of the molecule is radioactive) or by the use of a fluorescent dye that binds to nucleic acids.

Electrophoretic separation of nucleic acids is at the heart of a variety of commonly used research techniques discussed later in the text (Chapters 13 and 24). Of particular note, as you will see in those discussions, are the various "blotting" techniques (e.g., Southern blots and Northern blots), as well as DNA sequencing methods.

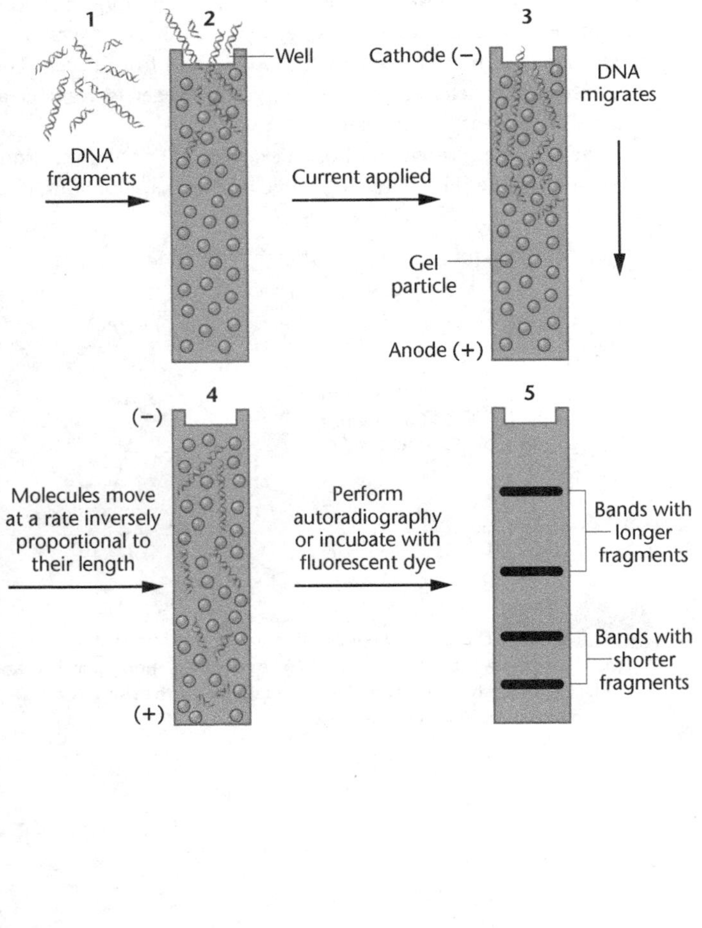


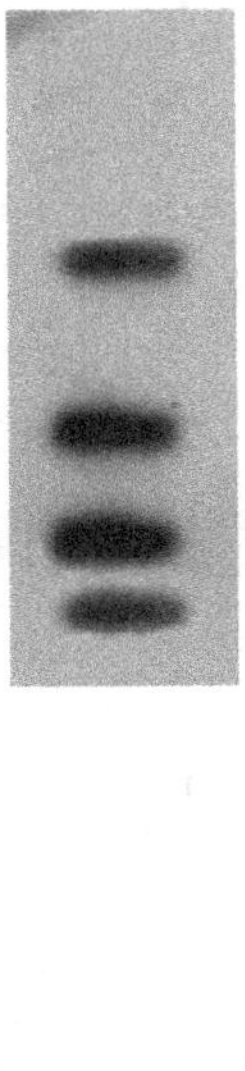

**FIGURE 10–27** Electrophoretic separation of a mixture of DNA fragments that vary in length. The photograph shows an agarose gel with DNA bands.

GENETICS, TECHNOLOGY, AND SOCIETY

# The Twists and Turns of the Helical Revolution

Western civilization is frequently transformed by new scientific ideas that overturn our self-concepts and permanently alter our relationships with each other and the rest of the animate world. For over 50 years, we have been in the midst of such a revolution—one as significant as those triggered by Darwin's theory of evolution and by the Copernican rejection of Ptolemy's Earth-centered universe.

The revolution began in April 1953 with Watson and Crick's discovery of the molecular structure of DNA. Their discovery that the DNA molecule consists of a twisted double helix, held together by weak bonds between specific pairs of bases, suddenly provided elegant solutions to age-old questions about the mechanisms of heredity, mutation, and evolution. Some of the greatest mysteries of life could be explained by the beauty and simplicity of a helix that replicates and shuffles the code of life.

After 1953, the double helix rapidly became the focus of modern science. Aware of DNA's helical structure, molecular biologists quickly devised methods to purify, mutate, cut, and paste DNA in the test tube. They spliced DNA molecules from one organism into those of another, and then introduced these chimeric molecules into bacteria or cells in culture. They read the nucleotide sequences of genes and modified the traits of bacteria, fungi, fruit flies, and mice by removing and mutating their genes, or by introducing genes from other organisms. On the 50th anniversary of Watson and Crick's double-helical DNA model, the Human Genome Project announced the completion of the largest DNA project so far—sequencing the entire human genome.

In a mere 50 years, the helical revolution has touched the lives of millions of people. We can now test for simple genetic diseases, such as Tay-Sachs, cystic fibrosis, and sickle-cell anemia. We can manufacture large quantities of medically important proteins, such as insulin and growth hormone, using DNA technologies. DNA forensic tests help convict criminals, exonerate the innocent, and establish paternity. By following a trail of DNA sequences, anthropologists can now trace human origins back in time and place.

The helical revolution has profoundly altered our view of the living world. Although scientists dismiss the idea that humans are simply the products of their genes, popular culture endows DNA with almost magical powers, expecting genes to explain personality, career choice, criminality, intelligence—even fashion preferences and political attitudes. Advertisements hijack the language of genetics in order to grant inanimate objects a "genealogy" or "genetic advantage." DNA is regarded as an immortal force—indeed, the essence of life—with the ability to affect morality and shape our future. Simple genetic explanations for our behavior appear to have more resonance for us than explanations involving social influences, economic factors, or free will. The beauty, symmetry, and biological significance of the double helix has insinuated itself into art, movies, advertising, and music. Paintings, sculpture, films—even video games and perfumes—use the language and imagery of genetics to confer upon the DNA molecule all the power and fears of modern technology.

But what of the future? Can we predict how the double helix and genetics will shape our world over the next 50 years? Although prophecy is a risky business, some scientific developments seem assured. With the completion of the Human Genome Project, we will undoubtedly identify more and more of the genes that control normal and abnormal processes. In turn, this will enhance our ability to diagnose and predict genetic diseases. Over the next 50 years, we can look forward to increasingly sophisticated gene therapies, prenatal diagnoses, and screening programs for susceptibilities to diseases as complicated as cancer and heart disease. We will continue to expand the applications of genetic engineering to agriculture as we manipulate plant and animal genes for enhanced productivity, disease resistance, and flavor.

The helical revolution will also continue to transform our concepts of ourselves and other creatures. As we compare the human genome to the genomes of other animals, it will become increasingly evident that we are closely related genetically to the rest of the animate world. The nucleotide sequence of the human genome differs only about 1 percent from that of chimpanzees, and some of our genes are virtually identical to homologous genes in plants, animals, and bacteria. As we realize the extent of our genetic kinship, stretching back over billions of years to the first forms of life on Earth, it is possible that this knowledge will alter our relationships with animals and with each other. When more genes are identified that contribute to phenotypic traits as simple as eye color and as complicated as intelligence or sexual orientation, it is possible that we will define ourselves even more as genetic beings and even less as creatures of free will or as the products of our environment.

During the first half of the century, we will inevitably be faced with the practical and philosophical consequences of the DNA revolution. Will society harness DNA for everyone's benefit, or will this new genetic knowledge be used as a vehicle for discrimination? At the same time that modern genetics grants us more dominion over life, will it paradoxically increase our feelings of powerlessness? Will our new DNA-centered self-concept increase our compassion for all life forms, or will it somehow separate us further from the natural world? We will make our choices, and human history will proceed.

## ■ References

Dennis, C., and Campbell, P. 2003. The eternal molecule. (Introduction to a series of feature articles commemorating the 50th anniversary of the discovery of DNA structure.) *Nature* 421: 396.

## ■ Web Site

A Revolution at 50. [*New York Times* article on the 50th anniversary of the discovery of DNA structure.]

*http://www.nytimes.com/indexes/2003/02/25/health/genetics/index.html*

EXPLORING GENOMICS

# Introduction to Bioinformatics: BLAST

In this chapter, we focused on the structural details of DNA, the genetic material for living organisms. In Chapter 13, you will learn how scientists can clone and sequence DNA—a routine technique in molecular biology and genetics laboratories. The explosion of DNA and protein sequence data that has occurred in the last 15 years has launched the field of *bioinformatics*, an interdisciplinary science that applies mathematics and computing technology to develop hardware and software for storing, sharing, comparing, and analyzing nucleic acid and protein sequence data.

A large number of sequence databases that make use of bioinformatics have been developed. An example is **GenBank** (www.ncbi.nlm.nih.gov/Genbank/index.html), which is the National Institutes of Health sequence database. This global resource, with access to databases in Europe and Japan, currently contains more than 65 billion base pairs of sequence data!

In the Exploring Genomics exercises for Chapter 5, you were introduced to the National Center for Biotechnology Information (NCBI) Genes and Disease site. Now we will use an NCBI application called **BLAST, Basic Local Alignment Search Tool**. BLAST is an invaluable program for searching through GenBank and other databases to find DNA- and protein-sequence similarities between cloned substances. It has many additional functions that we will explore in other exercises.

## Exercise I – Introduction to BLAST

1. Access BLAST from the NCBI Web site at www.ncbi.nlm.nih.gov/BLAST/.

2. Click on "nucleotide blast." This feature allows you to search DNA databases to look for a similarity between a sequence you enter and other sequences in the database. You will do a nucleotide search with the following sequence:

   CCAGAGTCCAGCTGCTGCTCATACTACT
   GATACTGCTGGG

3. Imagine that this sequence is a short part of a gene you cloned in your laboratory. You want to know if anyone else has cloned this gene and if other genes with similar or identical sequences have been discovered. Enter this sequence into the "Enter Query Sequence" text box at the top of the page. Near the bottom of the page, under the "Program Selection" category, choose "blastn"; then click on the "BLAST" button at the bottom of the page to run the search. It may take several minutes for results to be available because BLAST is using powerful algorithms to scroll through billions of bases of sequence data! A new page will appear with the results of your search.

4. Near the top of this page you will see a table showing significant matches to the sequence you searched with (called the query sequence). BLAST determines significant matches based on statistical measures that consider the length of the query sequence, the number of matches with sequences in the database (called subject sequences), and other factors. We will not discuss these measures in detail except to say that scores such as the E value (expect value) are calculated based on the number of matches in aligned sequences that would be expected by chance. Significant *alignments*, regions of significant similarity in the query and subject sequences, typically have E values less than 1.0.

5. The top part of the table lists matches to transcripts (mRNA sequences), and the lower part lists matches to genomic DNA sequences, in order of highest to lowest number of matches. Use the "Links" column to the far right of this table to explore gene and chromosome databases relevant to the matched sequences.

6. Alignments are indicated by horizontal lines. BLAST adjusts for gaps in the sequences, that is, for areas that may not align precisely because of missing bases in otherwise similar sequences. Scroll below the table to see the aligned sequences from this search, and then answer the following questions:

   a. What were the top three matches to your query sequence?

   b. For each alignment, BLAST also indicates the percent *identity* (a measure of similarity) and the number of gaps in the match between the query and subject sequences, in the column under "Max ident." Identity is determined by the sum of identical matches between aligned sequences divided by the total number of bases aligned (gaps are usually ignored in similarity scores). What was the percent identity for the top three matches? What percentage of each aligned sequence showed gaps indicating sequence differences?

   c. Click on the links for the first matched sequence (far-right column). These will take you to a wealth of information, including the size of the sequence; the species it was derived from; a PubMed-linked chronology of research publications pertaining to this sequence; the complete sequence; and if the sequence encodes a polypeptide, the predicted amino acid sequence coded by the gene. Skim through the information presented for this gene. What is the gene's function?

7. A BLAST search can also be done by entering the *accession number* for a sequence, which is a unique identifying number assigned to a sequence before it can be put into a database. For example, search with the accession number NM_007305. What did you find?

## Exercise II – BLAST: Searching DNA and Protein Databases for Sequence Matches

1. Run a BLAST search using the sequences or accession numbers listed below. In each case, after entering the accession number or sequence in the "Enter Query

Sequence" box, go to the "Choose Search Set" box and click on the "Others" button for database. Then go to the "Program Selection" box and click "megablast" before running your search. These features will allow you to align the query sequence with similar genes from a number of other species. When each search is completed, explore the information BLAST provides so that you can identify and learn about the gene encoded by the sequence.

a. NM_001006650. What is the top sequence that aligns with the query sequence of this accession number and shows 100 percent sequence identity?

b. *Orthologs* are genes of similar sequence and function in different organisms. Orthologs will show high similarity scores, indicating that the sequences may have been conserved during, and may be related by, evolutionary descent. Scroll down the alignments until you see the first alignment between the query sequence and a human gene sequence that is an ortholog of the query sequence. What human gene aligns with the query sequence?

c. How similar is this gene to the *polycystin 1* gene in *Bos taurus* (cattle), another ortholog of the query sequence?

d. DQ991619. What gene is encoded by this sequence?

e. NC_007596. What living animal has a sequence similar to this one?

f. TCTGCAATTGCTTAGGATGTTTTTCA TGAA. One of the alignments for this sequence shows you which chromosome this gene is located on. What is the gene, and where is it located?

g. ACCTGTAGCACAAGAGGCCTTAAAA AAGGAACTTGAAACTCTAACCACCAACT ACCAGTG. What gene does this sequence represent?

2. Type in a random sequence of bases for your search. What did you find? Did your "sequence" align with any sequences in the database?

3. As genome data for different organisms have been added to GenBank, users of BLAST have been able to limit their searches to organism-specific databases. In Chapter 1 and many other parts of the text, we have discussed how *Drosophila melanogaster* is being used to study human genetic conditions. Run a BLAST search with the accession number NM_001014741, which is specific for a *Drosophila* gene. Enter the accession number in the "Enter Query Sequence" box; then, in the "Choose Search Set" box, click on the "Human genomic + transcript" database; and in the "Program Selection" box, click on "blastn." This will enable you to search the human genome database for sequences similar to the *Drosophila* gene.

a. What *Drosophila* gene is represented by the accession number you used for the query?

b. What is the human ortholog identified as the top sequence alignment in the human genome?

c. How identical are the human and *Drosophila* orthologs for *rdgB*?

d. What disease condition is the *rdgB* gene involved in?

4. BLAST can also be used to search protein sequence databases. Return to the BLAST home page and use the "protein blast" feature to search protein sequence databases for similarities with the amino acid sequence MDLSALRVEEVQN. (This sequence uses the single-letter code presented in Figure 15–16 to represent amino acids). What did you find?

## Chapter Summary

1. The existence of a genetic material capable of replication, storage, expression, and mutation is deducible from the observed patterns of inheritance in organisms.
2. Both proteins and nucleic acids were initially considered as possible candidates for genetic material. Proteins are more diverse than nucleic acids (diversity being a requirement for the genetic material) and were favored owing to the advances being made in protein chemistry at the time. In addition, Levene's tetranucleotide hypothesis had underestimated the magnitude of chemical diversity inherent in nucleic acids.
3. By 1952, transformation studies and experiments using bacteria infected with bacteriophages strongly suggested that DNA is the genetic material in bacteria and most viruses.
4. Initially, only indirect observations supported the hypothesis that DNA controls inheritance in eukaryotes. These included DNA distribution in the cell, quantitative analysis of DNA, and UV-induced mutagenesis. More recent recombinant DNA techniques, as well as experiments with transgenic mice, have provided direct experimental evidence that the eukaryotic genetic material is DNA.
5. RNA serves as the genetic material in some viruses, including bacteriophages and some plant and animal viruses.
6. Establishment of DNA as the genetic material paved the way for the expansion of molecular genetics research and has served as the cornerstone for further important studies for nearly half a century.
7. During the late 1940s and early 1950s, considerable effort was made to integrate accumulated information about the chemical structure of nucleic acids into a model of the molecular structure of DNA. The X-ray diffraction data of Franklin and Wilkins suggested that DNA was some sort of helix. In 1953, Watson and Crick were able to assemble a model

of the double-helical DNA structure based on these X-ray diffraction studies as well as on Chargaff's analysis of base composition of DNA.

8. The DNA molecule exhibits antiparallel orientation of the two polynucleotide chains, along with adenine–thymine and guanine–cytosine base-pairing complementarity. This structure suggests a straightforward mechanism for DNA replication. The helical conformation appears to be a function of the nucleotide sequence and its chemical environment. Several alternative forms of the DNA helical structure exist. Watson and Crick described a B configuration, one of several right-handed helices. Wang and Rich discovered the left-handed Z-DNA currently being investigated for possible physiological and genetic significance.
9. The second category of nucleic acids important in genetic function is RNA, which is similar to DNA with the exceptions that it is usually single stranded, the sugar ribose replaces the deoxyribose, and the pyrimidine uracil replaces thymine. Classes of RNA—ribosomal, transfer, and messenger—facilitate the flow of information from DNA to RNA to proteins, which are the end products of most genes. Other forms of RNA are known to be involved in genetic regulation.
10. The structure of DNA lends itself to various methods of analysis, which have in turn led to studies of the functional aspects of the genetic machinery. Absorption of UV light, sedimentation properties, denaturation–reassociation, and electrophoresis procedures are among the important tools for the study of nucleic acids. Reassociation kinetics analysis enabled geneticists to postulate the existence of repetitive DNA ineukaryotes, where certain nucleotide sequences are present many times in the genome.

# INSIGHTS AND SOLUTIONS

*The current chapter, in contrast to preceding chapters, does not emphasize genetic problem solving. Instead, it recounts some of the initial experimental analyses that launched the era of molecular genetics. Accordingly, our "Insights and Solutions" section shifts its emphasis to experimental rationale and analytical thinking, an approach that will continue to be used in later chapters whenever appropriate.*

1. (a) Based strictly on your scrutiny of the transformation data of Avery, MacLeod, and McCarty, what objection might be made to the conclusion that DNA is the genetic material? What other conclusion might be considered? (b) What observations, including later ones, argue against this objection?

**Solution:** (a) Based solely on their results, it may be concluded that DNA is essential for transformation. However, DNA might have been a substance that caused capsular formation by *directly* converting nonencapsulated cells to ones with a capsule. That is, DNA may simply have played a catalytic role in capsular synthesis, leading to cells displaying smooth type III colonies.

(b) First, transformed cells pass the trait onto their progeny cells, thus supporting the conclusion that DNA is responsible for heredity, not for the direct production of polysaccharide coats. Second, subsequent transformation studies over a period of five years showed that other traits, such as antibiotic resistance, could be transformed. Therefore, the transforming factor has a broad general effect, not one specific to polysaccharide synthesis. This observation is more in keeping with the conclusion that DNA is the genetic material.

2. If RNA were the universal genetic material, how would it have affected the Avery experiment and the Hershey–Chase experiment?

**Solution:** In the Avery experiment, digestion of the soluble filtrate with RNase, rather than DNase, would have eliminated transformation. Had this occurred, Avery and his colleagues would have concluded that RNA was the transforming factor. Hershey and Chase would have obtained identical results, since $^{32}P$ would also label RNA but not protein. Had they been using a bacteriophage with RNA as its nucleic acid, and had they known this, they would have concluded that RNA was responsible for directing the reproduction of their bacteriophage.

3. In sea urchin DNA, which is double stranded, 17.5 percent of the bases were shown to be cytosine (C). What percentages of the other three bases are expected to be present in this DNA?

**Solution:** The amount of C equals G, so guanine is also present as 17.5 percent. The remaining bases, A and T, are present in equal amounts, and together they represent the rest of the bases (100 – 35). Therefore, A = T = 65/2 = 32.5 percent.

4. A quest to isolate an important disease-causing organism was successful, and molecular biologists were hard at work analyzing the results. The organism contained as its genetic material a remarkable nucleic acid with a base composition of A = 21 percent, C = 29 percent, G = 29 percent, U = 21 percent. When heated, it showed a major hyperchromic shift, and when the reassociation kinetics were studied, the nucleic acid of this organism provided the $C_0t$ curve shown below in contrast to that of phage T4 and *E. coli*. T4 contains $10^5$ nucleotide pairs.

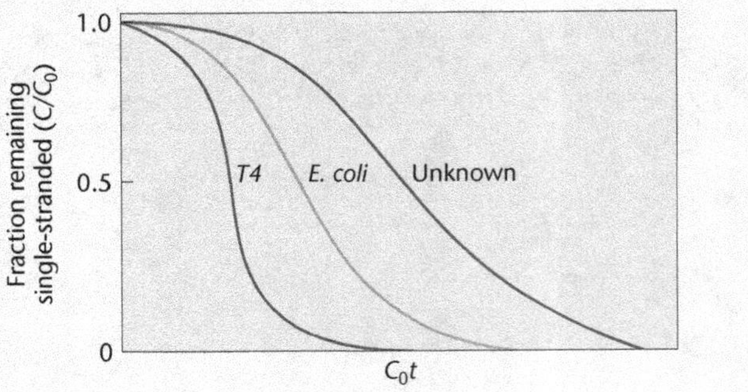


Analyze this information carefully, and draw *all* possible conclusions about the genetic material of this organism, based strictly on the preceding observations. As a test of your model, make one prediction that if upheld would strengthen your hypothesis about the nature of this molecule.

**Solution:** First of all, because of the presence of uracil (U), the molecule appears to be RNA. In contrast to normal RNA, this one has base ratios of A/U = G/C = 1, suggesting that the molecule may be a double helix. The hyperchromic shift and reassociation kinetics support this hypothesis. The kinetic study, based on the shape of the $C_0t$ curve, reveals that there is no repetitive sequence DNA. Furthermore, the complexity (*X*), or total length of unique-sequence DNA, is greater than that of either phage T4 ($10^5$ nucleotide pairs) or *E. coli*. There *may* be a greater number of genes present compared to T4 or *E. coli*, but the excessive unique-sequence DNA may serve some other role, or simply play no genetic role. A prediction might be made concerning the sugars. Our model suggests that ribose rather than deoxyribose should be present. If so, this observation would support the hypothesis that RNA is the genetic material in this organism.

## Problems and Discussion Questions

1. The functions ascribed to the genetic material are replication, expression, storage, and mutation. What does each of these terms mean in the context of genetics?
2. Discuss the reasons proteins were generally favored over DNA as the genetic material before 1940. What was the role of the tetranucleotide hypothesis in this controversy?
3. Contrast the various contributions made to an understanding of transformation by Griffith, by Avery and his colleagues, and by Taylor.
4. When Avery and his colleagues had obtained what was concluded to be the transforming factor from the IIIS virulent cells, they treated the fraction with proteases, RNase, and DNase, followed in each case by the assay for retention or loss of transforming ability. What were the purpose and results of these experiments? What conclusions were drawn?
5. Why were $^{32}P$ and $^{35}S$ chosen for use in the Hershey–Chase experiment? Discuss the rationale and conclusions of this experiment.
6. Does the design of the Hershey–Chase experiment distinguish between DNA and RNA as the molecule serving as the genetic material? Why or why not?
7. Would an experiment similar to that performed by Hershey and Chase work if the basic design were applied to the phenomenon of transformation? Explain why or why not.
8. What observations are consistent with the conclusion that DNA serves as the genetic material in eukaryotes? List and discuss them.
9. What are the exceptions to the general rule that DNA is the genetic material in all organisms? What evidence supports these exceptions?
10. Draw the chemical structure of the three components of a nucleotide, and then link the three together. What atoms are removed from the structures when the linkages are formed?
11. How are the carbon and nitrogen atoms of the sugars, purines, and pyrimidines numbered?
12. Adenine may also be named 6-amino purine. How would you name the other four nitrogenous bases, using this alternative system? (O is indicated by "oxy-," and $CH_3$ by "methyl.")
13. Draw the chemical structure of a dinucleotide composed of A and G. Opposite this structure, draw the dinucleotide composed of T and C in an antiparallel (or upside-down) fashion. Form the possible hydrogen bonds.
14. Describe the various characteristics of the Watson–Crick double-helix model for DNA.
15. What evidence did Watson and Crick have at their disposal in 1953? What was their approach in arriving at the structure of DNA?
16. What might Watson and Crick have concluded had Chargaff's data from a single source indicated the following?

| | A | T | G | C |
|---|---|---|---|---|
| % | 29 | 19 | 21 | 31 |

Why would this conclusion be contradictory to Wilkins' and Franklin's data?

17. How do covalent bonds differ from hydrogen bonds? Define base complementarity.
18. List three main differences between DNA and RNA.
19. What are the three major types of RNA molecules? How is each related to the concept of information flow?
20. What component of the nucleotide is responsible for the absorption of ultraviolet light? How is this technique important in the analysis of nucleic acids?
21. Distinguish between sedimentation velocity and sedimentation equilibrium centrifugation (density gradient centrifugation).
22. What is the basis for determining base composition using density gradient centrifugation?
23. What is the physical state of DNA following denaturation?
24. Compare the following curves representing reassociation kinetics.

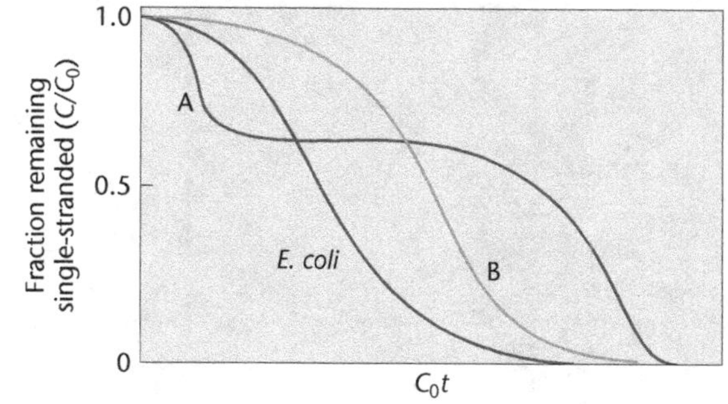


What can be said about the DNAs represented by each set of data compared with *E. coli?*

25. What is the hyperchromic effect? How is it measured? What does $T_m$ imply?
26. Why is $T_m$ related to base composition?
27. What is the chemical basis of molecular hybridization?
28. What did the Watson–Crick model suggest about the replication of DNA?
29. A genetics student was asked to draw the chemical structure of an adenine- and thymine-containing dinucleotide derived from DNA. His answer is shown here:

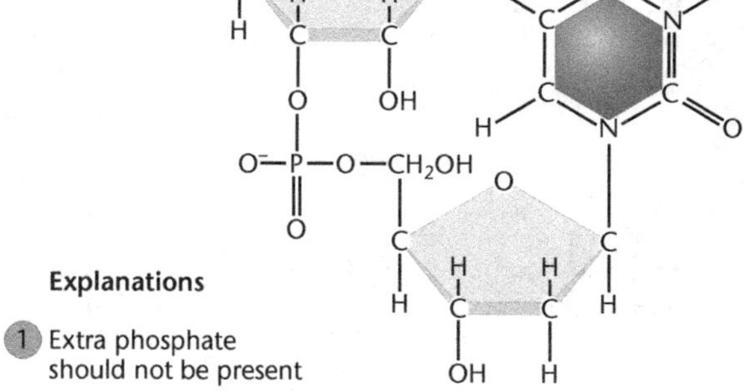


The student made more than six major errors. One of them is circled, numbered 1, and explained. Find five others. Circle them, number them 2 through 6, and briefly explain each in the manner of the example given.

30. The DNA of the bacterial virus T4 produces a $C_0 t_{1/2}$ of about 0.5 and contains $10^5$ nucleotide pairs in its genome.
How many nucleotide pairs are present in the genome of the virus *MS2* and the bacterium *E. coli*, whose respective DNAs produce $C_0 t_{1/2}$ values of 0.001 and 10.0?
31. Considering the information in this chapter on B- and Z-DNA and right- and left-handed helices, carefully analyze structures (a) and (b) below and draw conclusions about their helical nature. Which is right handed and which is left handed?

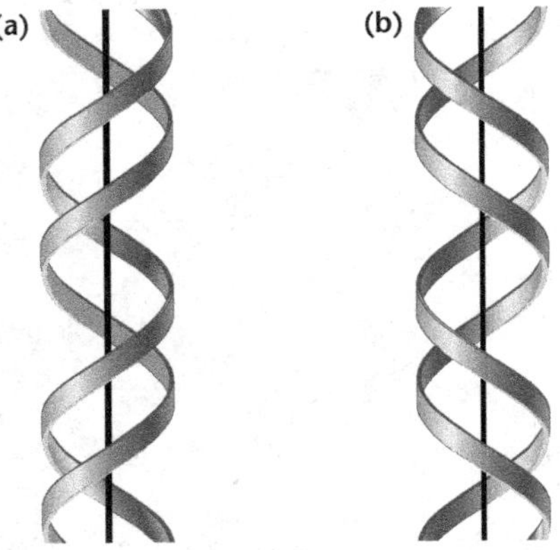


32. One of the most common spontaneous lesions that occurs in DNA under physiological conditions is the hydrolysis of the amino group of cytosine, converting the cytosine to uracil. What would be the effect on DNA structure of a uracil group replacing cytosine?
33. In some organisms, cytosine is methylated at carbon 5 of the pyrimidine ring after it is incorporated into DNA. If a 5-methyl cytosine is then hydrolyzed, as described in Problem 32, what base will be generated?
34. Because of its rapid turnaround time, fluorescent *in situ* hybridization (FISH) is commonly used in hospitals and laboratories as an aneuploid screen of cells retrieved from amniocentesis and chorionic villus sampling (CVS). Chromosomes 13, 18, 21, X, and Y (see Chapter 8) are typically screened for aneuploidy in this way. Explain how FISH might be accomplished using amniotic or CVS samples and why the above chromosomes have been chosen for screening.

**HOW DO WE KNOW?**

35. In this chapter, we first focused on the information that showed DNA to be the genetic material and then discussed the structure of DNA as proposed by Watson and Crick. We concluded the chapter by describing various techniques developed to study DNA. Along the way, we found many opportunities to consider the methods and reasoning by which much of this information was acquired. From the explanations given in the chapter, what answers would you propose to the following fundamental questions:
(a) How were scientists able to determine that DNA, and not some other molecule, serves as the genetic material in bacteria and bacteriophages?
(b) How do we know that DNA also serves as the genetic material in eukaryotes such as humans?
(c) How was it determined that the structure of DNA is a double helix with the two strands held together by hydrogen bonds formed between complementary nitrogenous bases?
(d) How do we know that G pairs with C and that A pairs with T as complementary base pairs are formed?
(e) How do we know that repetitive DNA sequences exist in eukaryotes?

## Extra-Spicy Problems

36. A primitive eukaryote was discovered that displayed a unique nucleic acid as its genetic material. Analysis provided the following information:
(a) The general X-ray diffraction pattern is similar to that of DNA, but with somewhat different dimensions and more irregularity.
(b) A major hyperchromic shift is evident upon heating and monitoring UV absorption at 260 nm.
(c) Base-composition analysis reveals four bases in the following proportions:

| | | |
|---|---|---|
| Adenine | = | 8% |
| Guanine | = | 37% |
| Xanthine | = | 37% |
| Hypoxanthine | = | 18% |

(d) About 75 percent of the sugars are deoxyribose, while 25 percent are ribose.
Postulate a model for the structure of this molecule that is consistent with the foregoing observations.
37. *Newsdate: March 1, 2015.* A unique creature has been discovered during exploration of outer space. Recently, its genetic material has been isolated and analyzed. This material is similar in some ways to DNA in its chemical makeup. It contains in abundance the 4-carbon sugar erythrose and a molar equivalent of phosphate groups. In addition, it contains six nitrogenous bases: adenine (A), guanine (G), thymine (T), cytosine (C), hypoxanthine (H), and xanthine (X). These bases exist in the following relative proportions:

$$A = T = H \text{ and } C = G = X$$

X-ray diffraction studies have established a regularity in the molecule and a constant diameter of about 30 Å.

Together, these data have suggested a model for the structure of this molecule.

(a) Propose a general model of this molecule. Describe it briefly.

(b) What base-pairing properties must exist for H and for X in the model?

(c) Given the constant diameter of 30 Å, do you think that *either* (i) both H and X are purines or both pyrimidines, *or* (ii) one is a purine and one is a pyrimidine?

38. You are provided with DNA samples from two newly discovered bacterial viruses. Based on the various analytical techniques discussed in this chapter, construct a research protocol that would be useful in characterizing and contrasting the DNA of both viruses. For each technique that you include in the protocol, indicate the type of information you hope to obtain.

39. During gel electrophoresis, DNA molecules can easily be separated according to size because all DNA molecules have the same charge-to-mass ratio and the same shape (long rod). Would you expect RNA molecules to behave in the same manner as DNA during gel electrophoresis? Why or why not?

40. Electrophoresis is an extremely useful procedure when applied to analysis of nucleic acids as it can resolve molecules of different sizes with relative ease and accuracy. Large molecules migrate more slowly than small molecules in agarose and polyacrylamide gels. However, the fact that nucleic acids of the same length may exist in a variety of conformations can often complicate the interpretation of electrophoretic separations. For instance, when a single species of a bacterial plasmid is isolated from cells, the individual plasmids may exist in three forms (depending on the genotype of their host and conditions of isolation): superhelical/supercoiled (form I), nicked/open circle (form II), and linear (form III). Form I is compact and very tightly coiled, with both DNA strands continuous. Form II exists as a loose circle because one of the two DNA strands has been broken, thus releasing the supercoil. All three have the same mass, but each will migrate at a different rate through a gel. Based on your understanding of gel composition and DNA migration, predict the relative rates of migration of the various DNA structures mentioned above.

41. Following is a table (modified from Kropinski, 1973) that presents the $T_m$ and chemical composition (%G≡C) of DNA from certain bacteriophages. From these data develop a graph that presents %G≡C (ordinate) and $T_m$ (abscissa). What is the relationship between $T_m$ and %G≡C for these samples? What might be the molecular basis of this relationship?

| Phage | $T_m$ | %G≡C |
|---|---|---|
| $\alpha$ | 86.5 | 44.0 |
| $\kappa$ | 91.5 | 53.8 |
| $\lambda$ | 89 | 49.2 |
| $\phi$ 80 | 90.5 | 53.0 |
| $\chi$ | 92.1 | 57.4 |
| Mu-1 | 88 | 51.4 |
| T1 | 89 | 48.0 |
| T3 | 90 | 49.6 |
| T7 | 89.5 | 48.0 |

Transmission electron micrograph of human DNA from a HeLa cell, illustrating a replication fork characteristic of active DNA replication. HeLa cells are a line of "immortal" human cells widely used in research.

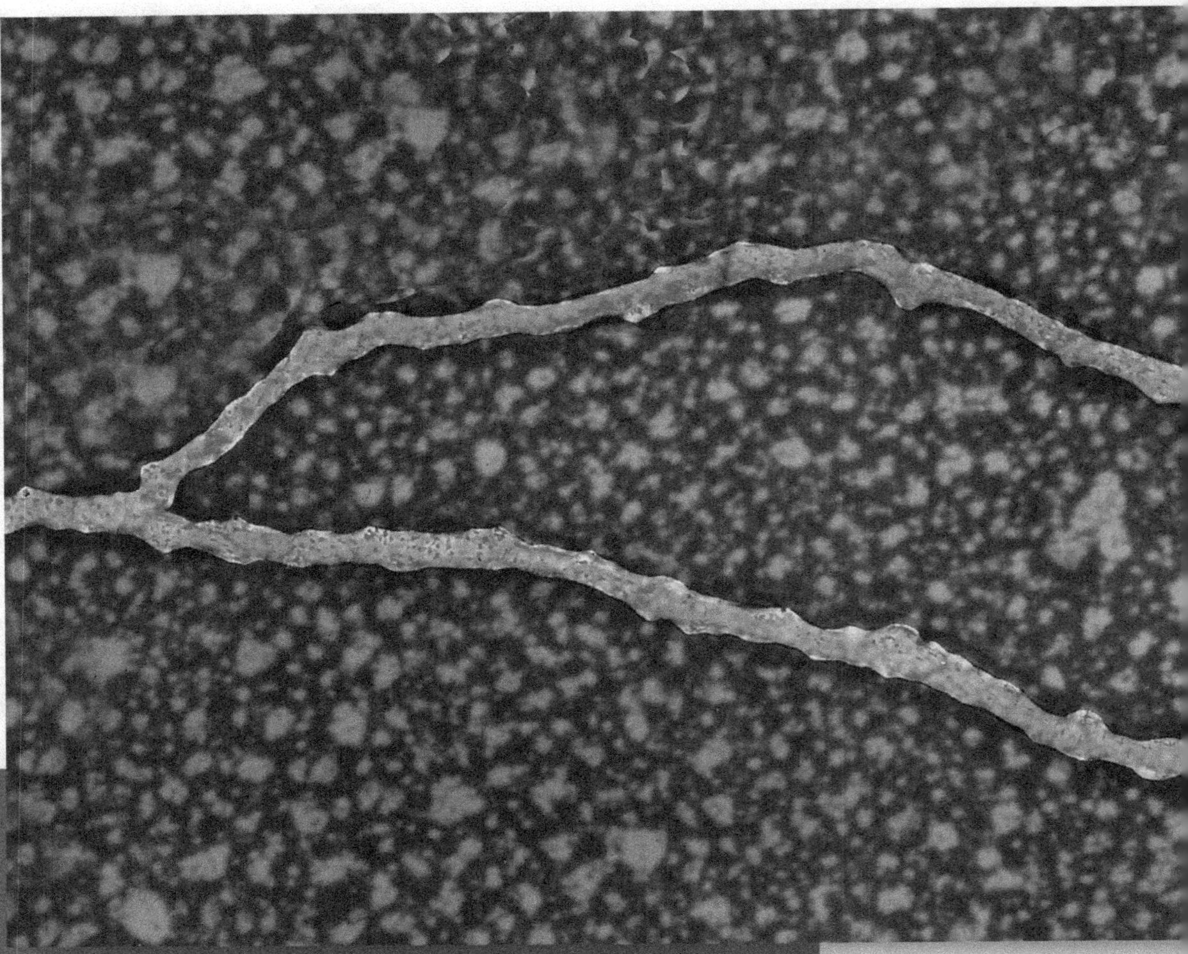

11

# DNA Replication and Recombination

## CHAPTER CONCEPTS

- Genetic continuity between parental and progeny cells is maintained by semiconservative replication of DNA, as predicted by the Watson–Crick model.
- Semiconservative replication uses each strand of the parent double helix as a template, and each newly replicated double helix includes one "old" and one "new" strand of DNA.
- DNA synthesis is a complex but orderly process, occurring under the direction of a myriad of enzymes and other proteins.
- DNA synthesis involves the polymerization of nucleotides into polynucleotide chains.
- DNA synthesis is similar in prokaryotes and eukaryotes, but more complex in eukaryotes.
- In eukaryotes, DNA synthesis at the ends of chromosomes poses a special problem, solved by end regions called telomeres and a unique RNA-containing enzyme called telomerase.
- Genetic recombination, an important process leading to the exchange of segments between DNA molecules, occurs under the direction of a group of enzymes.

In the wake of Watson and Crick's proposal for the structure of DNA, scientists focused their attention on how the DNA molecule is replicated. Replication is not only an essential function of the genetic material, one that must be executed precisely if genetic continuity between cells is to be maintained following cell division; it is also a huge and enormously complex task. Consider for a moment that the human genome's 24 chromosomes (22 autosomes, plus one X and one Y chromosome) contain more than $3 \times 10^9$ (3 billion) base pairs. To duplicate faithfully the DNA of just one of these chromosomes requires an extremely accurate mechanism. Even an error rate as low as $10^{-6}$ (one in a million) will still create 3000 errors during each replication cycle of the genome. Although the process is not error free—much of evolution would not have occurred if it were—an extremely reliable system of DNA replication has evolved in all organisms.

As Watson and Crick noted in the concluding paragraph of their 1953 paper (reprinted on page 261 in Chapter 10), their proposed model of the double helix provided the initial insight into the general process by which replication might occur. Called *semiconservative replication*, this mode of DNA duplication was soon to receive strong support from numerous studies of viruses, prokaryotes, and eukaryotes. Once the general mode of DNA replication was clarified, research to determine the precise details of DNA synthesis intensified. What has since been discovered is that numerous enzymes and other proteins are needed to copy a DNA helix. Because of the complexity of the chemical events during synthesis, this subject remains an extremely active area of research.

In this chapter, we will discuss the general mode of replication, as well as many specific details of DNA synthesis. The research leading to this knowledge comprises another link in our understanding of life processes at the molecular level.

## 11.1 DNA Is Reproduced by Semiconservative Replication

Watson and Crick recognized that, because of the arrangement and chemical properties of the nitrogenous bases, each strand of a DNA double helix could serve as a template for the synthesis of its complement (Figure 11–1). They proposed that, if the helix were unwound, each nucleotide along the two parent strands would have an affinity for its complementary nucleotide. As we learned in Chapter 10, the complementarity is due to the hydrogen bonds that can be formed: the nucleotide deoxythymidylic acid (T) would "attract" the nucleotide deoxyadenylic acid (A); similarly, deoxyguanylic acid (G) would attract the nucleotide deoxycytidylic acid (C); likewise, A would attract T, and C would attract G. If these nucleotides that had been attracted to the two parent strands were then covalently linked into polynucleotide chains along those templates, the result would be the production of two identical double strands of DNA. Each replicated DNA molecule would consist of one "old" and one "new" strand, hence the name **semiconservative replication.**

Two other theoretical modes of replication are possible that also rely on using the parental strands as templates (Figure 11–2). In **conservative replication**, complementary polynucleotide chains would be synthesized as described for the semiconservative mode above. Following synthesis, however, the two newly created strands would come together, and the parental strands would reassociate. The original helix would thus be "conserved."

In the second alternative mode, called **dispersive replication**, the parental strands would disperse into both strands of the two new double helices following replication. Hence, each strand would

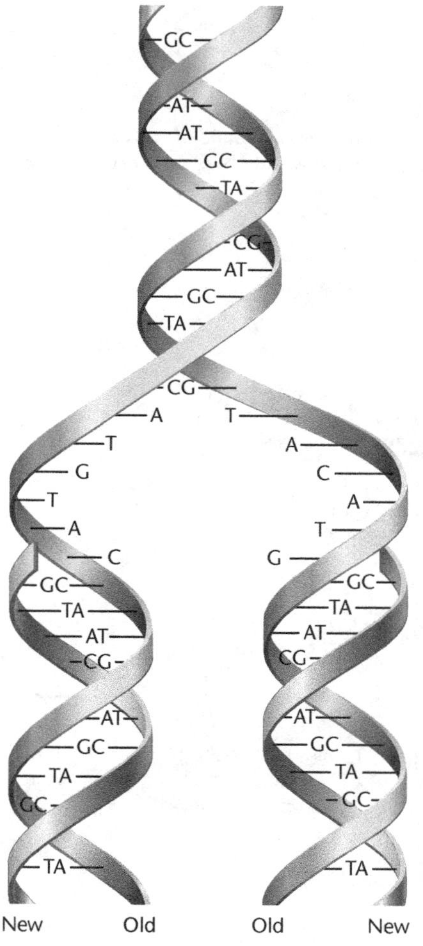


FIGURE 11–1 Generalized model of semiconservative replication of DNA. New synthesis is shown in blue.

DNA REPLICATION WEB TUTORIAL 11.1

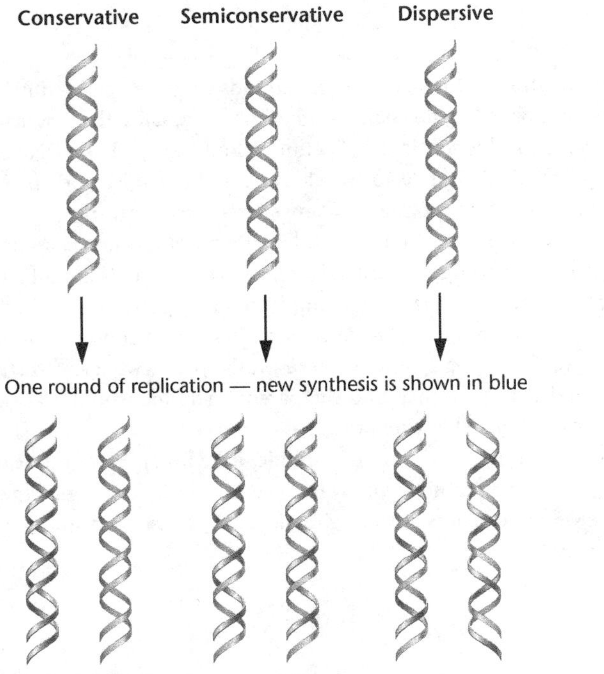


FIGURE 11-2 Results of one round of replication of DNA for each of the three possible modes by which replication could be accomplished.

consist of both old and new DNA. This mode would involve cleavage of the parental strands during replication. It is the most complex of the three possibilities and is therefore considered to be least likely to occur. It could not, however, be ruled out initially as an experimental model. Figure 11–2 shows the theoretical results of a single round of replication by each of the three different modes.

## The Meselson–Stahl Experiment

In 1958, Matthew Meselson and Franklin Stahl published the results of an experiment providing strong evidence that semiconservative replication is the mode used by bacterial cells to produce new DNA molecules. They grew *E. coli* cells for many generations in a medium that had $^{15}NH_4Cl$ (ammonium chloride) as the only nitrogen source. A "heavy" isotope of nitrogen, $^{15}N$ contains one more neutron than the naturally occurring $^{14}N$ isotope; thus, molecules containing $^{15}N$ are more dense than those containing $^{14}N$. Unlike radioactive isotopes, $^{15}N$ is stable. After many generations in this medium, almost all nitrogen-containing molecules in the *E. coli* cells, including the nitrogenous bases of DNA, contained the heavier isotope.

Critical to the success of this experiment was the researchers' ability to distinguish DNA containing $^{15}N$ from DNA containing $^{14}N$. They used a technique referred to as **sedimentation equilibrium centrifugation**, also called *density gradient centrifugation* (discussed in Chapter 10), in which centrifugation forces samples through a density gradient of a heavy metal salt, such as cesium chloride, until the sample substances reach a point of equilibrium and stop moving through the gradient. Molecules of DNA will reach equilibrium when their density equals the density of the gradient medium. $^{15}N$-DNA will reach this point at a position closer to the bottom of the tube than will $^{14}N$-DNA.

In the Meselson–Stahl experiment (Figure 11–3), uniformly labeled $^{15}N$ cells were transferred to a medium containing only $^{14}NH_4Cl$. Thus, all "new" synthesis of DNA during replication would contain only the "lighter" isotope of nitrogen. The time of transfer to the new medium was designated time zero ($t = 0$). The

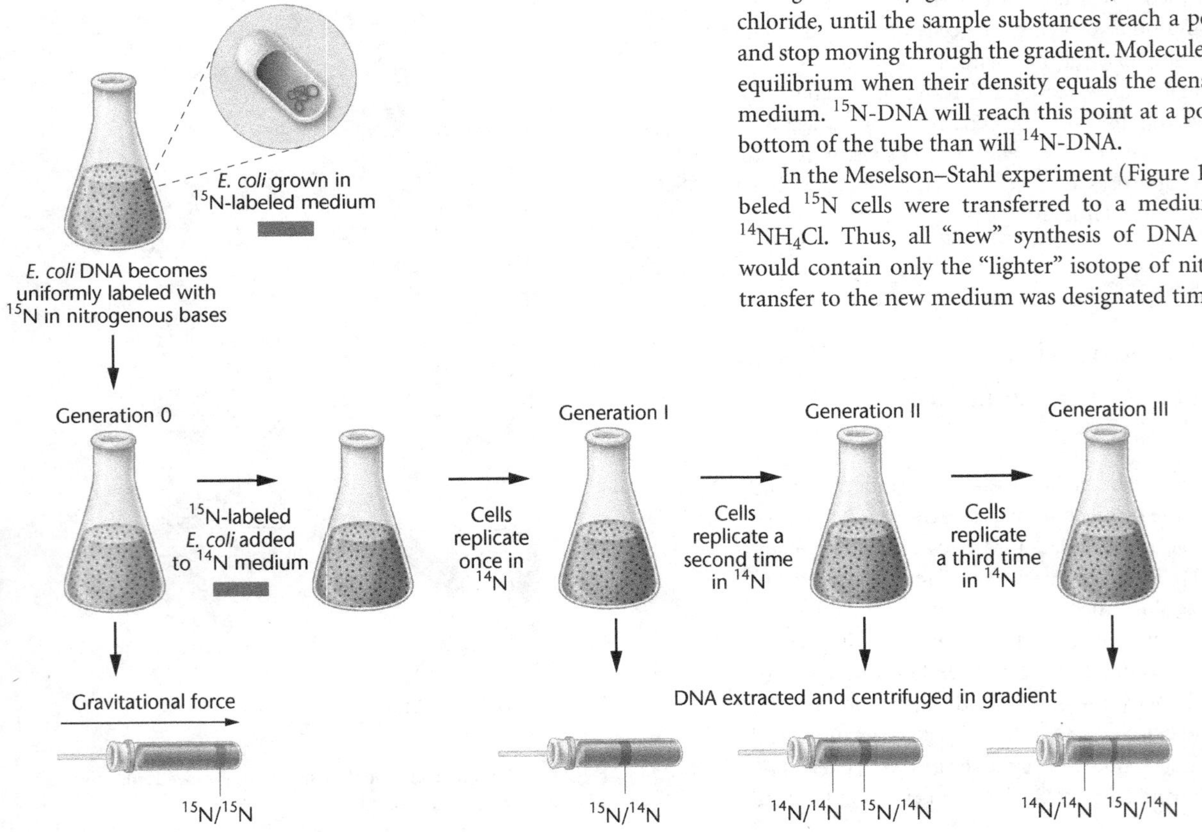


FIGURE 11-3 The Meselson–Stahl experiment.

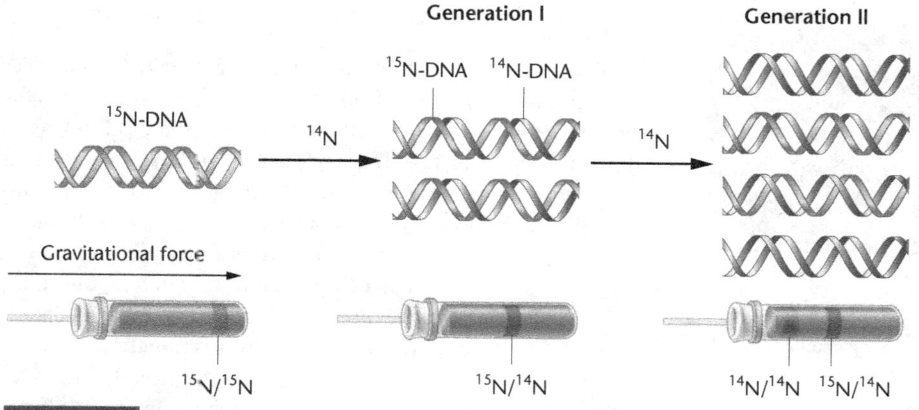


FIGURE 11-4 The expected results of two generations of semiconservative replication in the Meselson–Stahl experiment.

*E. coli* cells were then allowed to replicate over several generations, with cell samples removed after each replication cycle. DNA was isolated from each sample and subjected to sedimentation equilibrium centrifugation.

After one generation, the isolated DNA, when centrifuged, formed a single band, of intermediate density—the expected result for semiconservative replication in which each replicated molecule was composed of one new $^{14}N$-strand and one old $^{15}N$-strand (Figure 11–4). This result was not consistent with the prediction of conservative replication, in which two distinct bands would occur, and thus the conservative mode could be rejected.

After two cell divisions, DNA samples showed two density bands—one intermediate band and one lighter band, the latter corresponding to the $^{14}N$ position in the gradient. Similar results occurred after a third generation, except that the proportionate size of the $^{14}N$ band increased. This was again consistent with the interpretation that replication is semiconservative.

You may have realized that a molecule exhibiting intermediate density is also consistent with dispersive replication. However, Meselson and Stahl ruled out this mode of replication on the basis of two observations. First, after the first generation of replication in a $^{14}N$-containing medium, they isolated the hybrid molecule and heat-denatured it. Recall from Chapter 10 that heating will separate a duplex into single strands. When the densities of the single strands of the hybrid were determined, they exhibited *either* a $^{15}N$ profile *or* a $^{14}N$ profile, but *not* an intermediate density. This observation is consistent with the semiconservative mode but inconsistent with the dispersive mode.

Furthermore, if replication were dispersive, *all* generations after $t = 0$ would demonstrate DNA of an intermediate density. In each generation after the first, the ratio of $^{15}N/^{14}N$ would decrease, and the hybrid band would become lighter and lighter, eventually approaching the density of the $^{14}N$ band. This result was not observed. The Meselson–Stahl experiment provided conclusive support for semiconservative replication in bacteria and tended to rule out both the conservative and dispersive modes.

## Semiconservative Replication in Eukaryotes

In 1957, the year before the work of Meselson and his colleagues was published, J. Herbert Taylor, Philip Woods, and Walter Hughes presented evidence that semiconservative replication also occurs in eukaryotic organisms. They experimented with root tips of the broad bean *Vicia faba*, which are an excellent source of dividing cells. These researchers were able to monitor the process of replication by labeling DNA with $^{3}H$-thymidine, a radioactive precursor of DNA, and performing autoradiography.

**Autoradiography** is a laboratory technique with many uses, including pinpointing the location of a radioisotope in a cell. In this procedure, a photographic emulsion is placed over a histological preparation containing cellular material (root tips, in this experiment), and the preparation is stored in the dark. The slide is then developed, much as photographic film is processed. Because the radioisotope emits energy, the processed slide shows dark spots, or "grains," at the approximate points of radioactive emission, identifying the location of newly synthesized DNA within the cell.

Taylor and his colleagues grew root tips for approximately one generation in the presence of radioactively labeled thymidine and then placed them in unlabeled medium in which cell division continued. At the conclusion of each generation, they arrested the cultures at metaphase by adding colchicine (a chemical derived from the crocus plant that poisons the spindle fibers) and then examined the chromosomes by autoradiography. After the first round of replication, they found radioactive thymidine only in association with chromatids that contained newly synthesized DNA. Figure 11–5 illustrates the replication of a single chromosome over two division cycles in this experiment, including the distribution of grains.

These results are compatible with the semiconservative mode of replication. After the first replication cycle in the presence of the isotope, both sister chromatids show radioactivity, indicating that each chromatid contains one new radioactive DNA strand and one old unlabeled strand. After the second replication cycle, *which takes place in unlabeled medium*, only one of the two sister chromatids of each chromosome should be radioactive, because half of the parent strands are unlabeled. With only the minor exceptions of *sister chromatid exchanges* (discussed in Chapter 6), this result was observed.

Together, the Meselson–Stahl experiment and the experiment by Taylor, Woods, and Hughes soon led to the general acceptance of the semiconservative mode of replication. Later studies with other organisms reached the same conclusion and also strongly supported Watson and Crick's proposal for the double-helix model of DNA.

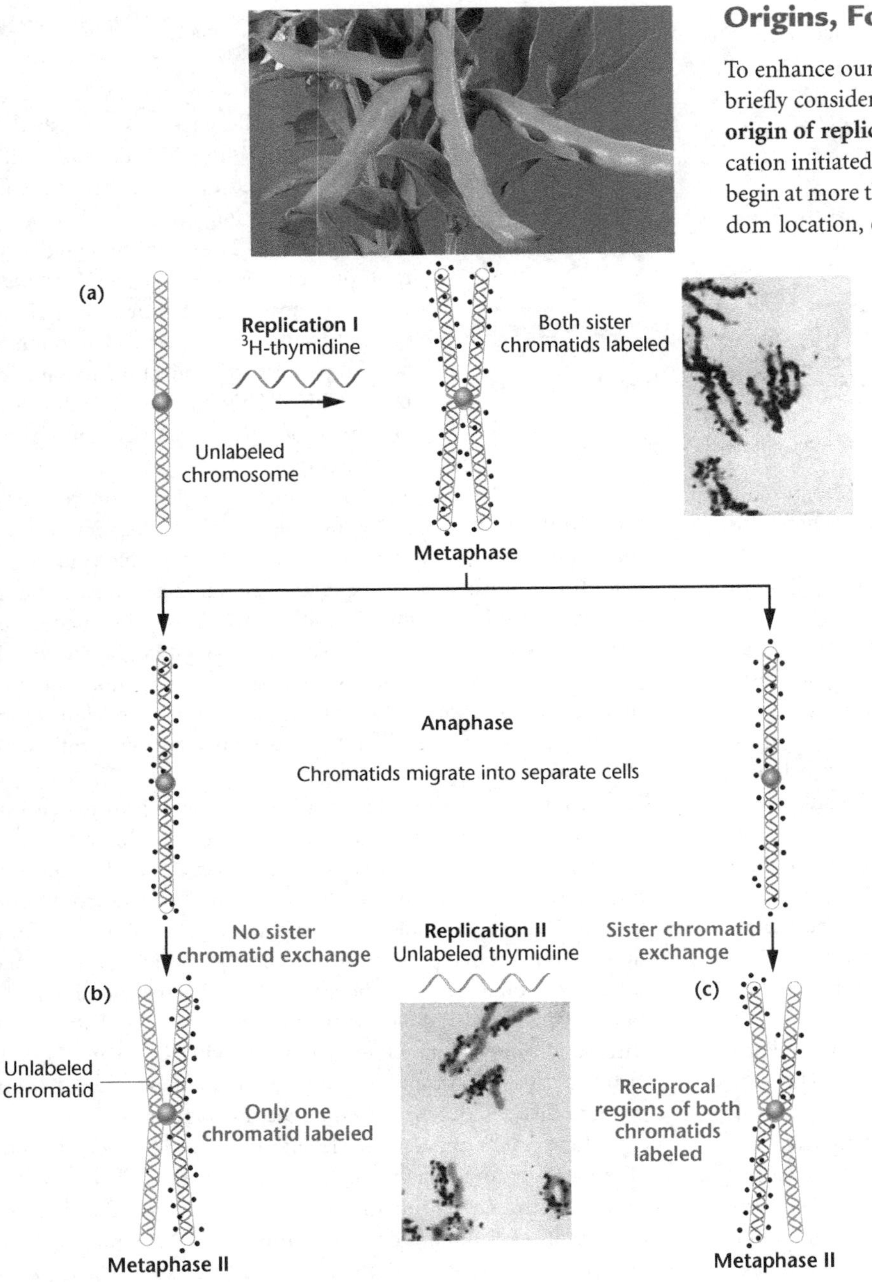


**FIGURE 11–5** The Taylor–Woods–Hughes experiment, demonstrating the semiconservative mode of replication of DNA in root tips of *Vicia faba* (known as the fava bean). A portion of the plant is shown in the top photograph. (a) An unlabeled chromosome proceeds through the cell cycle in the presence of $^3$H-thymidine. As it enters mitosis, both sister chromatids of the chromosome are labeled, as shown by autoradiography. (b) After a second round of replication, this time in the absence of $^3$H-thymidine, only one chromatid of each chromosome is expected to produce an image surrounded by grains. Except where a reciprocal exchange occurred between sister chromatids (c), the expectation was upheld. The micrographs are of the actual autoradiograms obtained in the experiment.

## Origins, Forks, and Units of Replication

To enhance our understanding of semiconservative replication, let's briefly consider a number of relevant issues. The first concerns the **origin of replication.** Where along the chromosome is DNA replication initiated? Is there only a single origin, or does DNA synthesis begin at more than one point? Does a point of origin occur in a random location, or does it occur at a specific region along the chromosome? Second, once replication begins, does it proceed in a single direction or in both directions away from the origin? In other words, is replication **unidirectional** or **bidirectional?**

To address these issues, we need to introduce two terms. First, at any point along the chromosome where replication is occurring, the strands of the helix are unwound, creating what is called a **replication fork.** Such a fork will initially appear at the point of origin of synthesis and then move along the DNA duplex as replication proceeds. If replication should prove to be bidirectional, two such forks will be present, migrating in opposite directions away from the origin. The second term pertains to the segment of DNA that is replicated following one initiation event at a single origin. This segment is referred to as the **replicon.**

The evidence is clear regarding the number of origins and direction of replication in prokaryotes. John Cairns tracked replication in *E. coli*, using radioactive precursors of DNA and autoradiography. He was able to demonstrate that in *E. coli* there is only a single origin, in a single region called *oriC*, where replication is initiated. The presence of only a single origin is characteristic of bacteria, which have only one circular chromosome. Since DNA synthesis in bacteriophages and bacteria originates at a single point, the entire chromosome constitutes one replicon. In *E. coli*, the replicon consists of the entire genome of 4.2 Mb (4.2 million base pairs) of DNA.

Figure 11–6 illustrates Cairns's interpretation of DNA replication in *E. coli*. This interpretation (and the accompanying micrograph) does not answer the question of unidirectional versus bidirectional synthesis. However, other results, derived from studies of bacteriophage lambda, demonstrated that replication is bidirectional, moving away from *oriC* in both directions. Figure 11–6 therefore interprets Cairns's work with that understanding. Bidirectional replication creates two replication forks that migrate farther and farther apart as replication proceeds. These forks eventually merge, as semiconservative replication of the entire chromosome is completed, at a termination region, called *ter*.

Later in this chapter we will see that in eukaryotes, replication is also bidirectional but that each eukaryotic chromosome contains multiple points of origin.

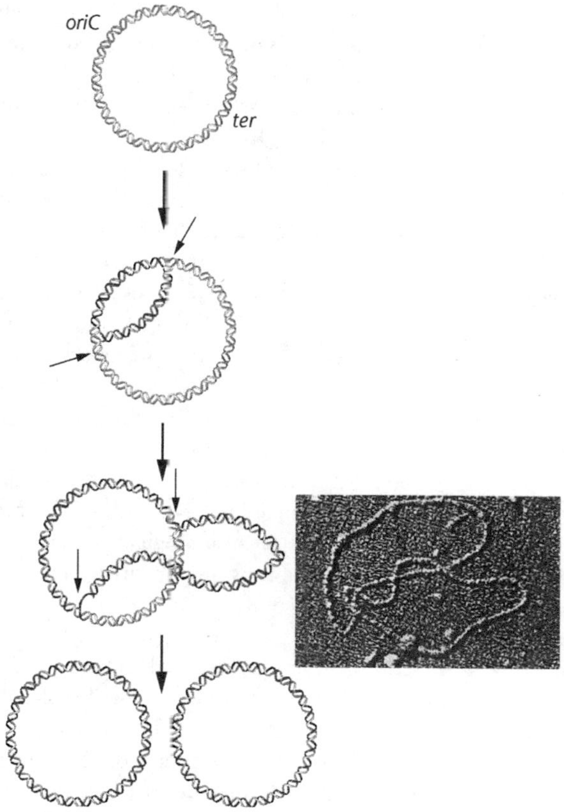


FIGURE 11–6 Bidirectional replication of the *E. coli* chromosome. The thin black arrows identify the advancing replication forks. The micrograph is of a bacterial chromosome in the stage of replication comparable to the figure next to it.

FIGURE 11–7 The chemical reaction catalyzed by DNA polymerase I. During each step, a single nucleotide is added to the growing complement of the DNA template (consisting of dinucleotide monophosphates, dNMPs), using a nucleoside triphosphate (dNTP) as the substrate. The release of inorganic pyrophosphate (P—P) drives the reaction energetically.

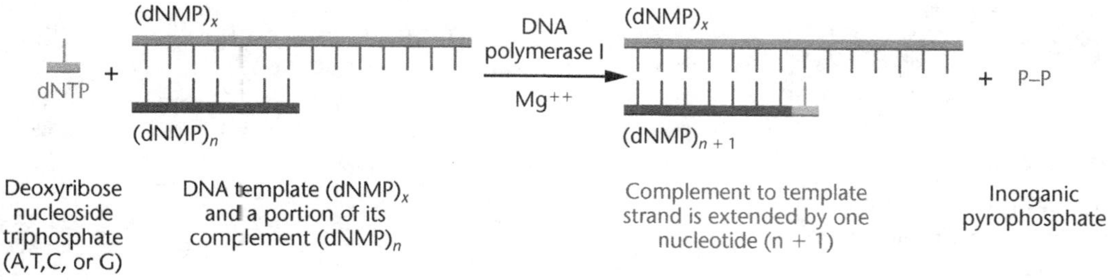


11.2

# DNA Synthesis in Bacteria Involves Five Polymerases, as Well as Other Enzymes

To say that replication is semiconservative and bidirectional describes the overall *pattern* of DNA duplication and the association of finished strands with one another once synthesis is completed. However, it says little about the more complex issue of how the actual *synthesis* of long complementary polynucleotide chains occurs on a DNA template. Like most questions in molecular biology, this one was first studied using microorganisms. Research on DNA synthesis began about the same time as the Meselson–Stahl work, and the topic is still an active area of investigation. What is most apparent in this research is the tremendous complexity of the biological synthesis of DNA.

## DNA Polymerase I

Studies of the enzymology of DNA replication were first reported by Arthur Kornberg and colleagues in 1957. These researchers isolated an enzyme from *E. coli* that was able to direct DNA synthesis in a cell-free (*in vitro*) system. The enzyme is called **DNA polymerase I**, as it was the first of several similar enzymes to be isolated.

Kornberg determined that there were two major requirements for *in vitro* DNA synthesis under the direction of DNA polymerase I: (1) the presence of all four deoxyribonucleoside triphosphates (dNTPs) and (2) template DNA. If any one of the four deoxyribonucleoside triphosphates was omitted from the reaction, no measurable synthesis occurred. If derivatives of these precursor molecules other than the nucleoside triphosphate were used (nucleotides or nucleoside diphosphates), synthesis also did not occur. If no template DNA was added, synthesis of DNA occurred but was reduced greatly.

Most of the synthesis directed by Kornberg's enzyme appeared to be exactly the type required for semiconservative replication. The catalyzed reaction is summarized in Figure 11–7, which depicts the addition of a single nucleotide. The enzyme has since been shown to consist of a single polypeptide containing 928 amino acids.

The way in which each nucleotide is added to the growing chain is a function of the specificity of DNA polymerase I. As shown in Figure 11–8, the precursor dNTP contains the three phosphate groups attached to the 5′-carbon of deoxyribose. As the two terminal phosphates are removed during synthesis, the remaining phosphate attached to the 5′-carbon is covalently linked to the 3′-OH group of the deoxyribose to which it is added. Thus, *chain elongation occurs in the 5′ to 3′ direction* by the addition of

FIGURE 11-8 Demonstration of 5′ to 3′ synthesis of DNA.

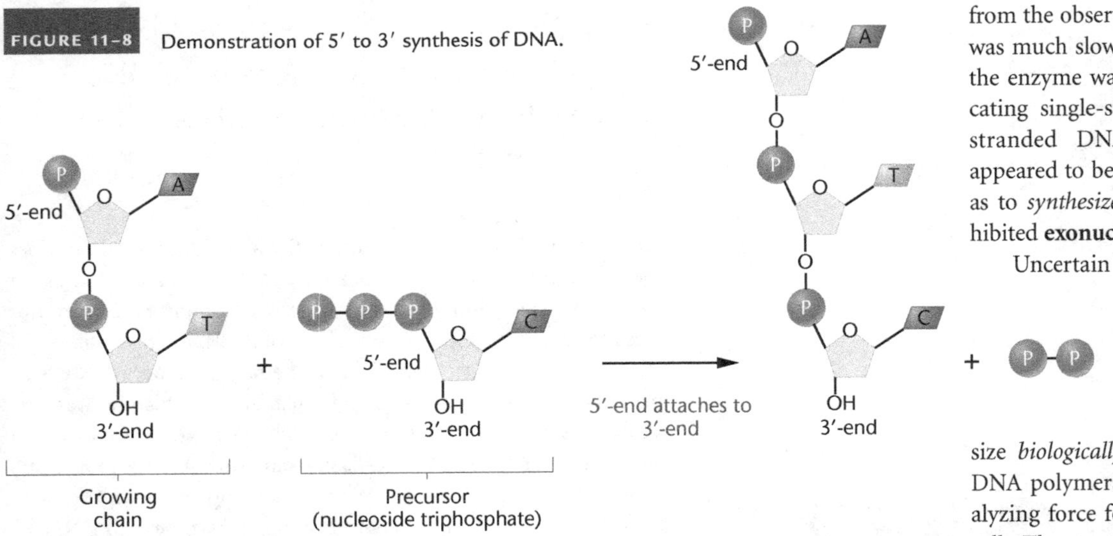


one nucleotide at a time to the growing 3′ end. Each step provides a newly exposed 3′-OH group that can participate in the next addition of a nucleotide as DNA synthesis proceeds.

Having isolated DNA polymerase I and demonstrated its catalytic activity, Kornberg next sought to demonstrate the accuracy, or fidelity, with which the enzyme replicated the DNA template. Because technology for ascertaining the nucleotide sequences of the template and newly synthesized strand was not yet available in 1957, he initially had to rely on several indirect methods.

One of Kornberg's approaches was to compare the nitrogenous base compositions of the DNA template with those of the recovered DNA product. Table 11.1 shows Kornberg's base-composition analysis of three DNA templates. Within experimental error, the base composition of each product agreed with the template DNAs used. These data, along with other types of comparisons of template and product, suggested that the templates were replicated faithfully.

NOW SOLVE THIS

Problem 40, on page 301, describes a procedure called nearest-neighbor analysis, an experimental approach that Kornberg used to test the fidelity of copying by DNA polymerase I.

■ HINT: *As you analyze this procedure, remember that during synthesis of DNA, DNA polymerase I adds 5′-nucleotides to the 3′-OH group of the existing polynucleotide chain—that is, each nucleotide added has a phosphate on the C-5′ of deoxyribose. However, the phosphodiesterase enzyme, which is used in this experiment, cleaves between the phosphate and the C-5′ atom, thereby producing 3′-nucleotides. As a result, the phosphate group is transferred to its "nearest neighbor," which is the key to your analysis.*

## Synthesis of Biologically Active DNA

Despite Kornberg's extensive work, not all researchers were convinced that DNA polymerase I was the enzyme that replicates DNA within bacteria cells (that is, *in vivo*). Their reservations stemmed from the observations that *in vitro* synthesis was much slower than the *in vivo* rate, that the enzyme was much more effective replicating single-stranded DNA than double-stranded DNA, and that the enzyme appeared to be able to *degrade* DNA as well as to *synthesize* it—that is, the enzyme exhibited **exonuclease activity**.

Uncertain of the true cellular function of DNA polymerase I, Kornberg pursued a new line of reasoning. He posited that if the enzyme could be used to synthesize *biologically active DNA in vitro*, then DNA polymerase I must be the major catalyzing force for DNA synthesis within the cell. The term **biological activity** means that the DNA synthesized is capable of supporting metabolic activities and directs the reproduction of the organism from which it was originally duplicated.

In 1967, Kornberg, Mehran Goulian, and Robert Sinsheimer experimented with the small bacteriophage $\phi$X174. This phage is an ideal experimental system, because its genetic material is a very small (5386 nucleotides), circular, single-stranded DNA molecule. In the normal course of $\phi$X174 infection, the circular single-stranded DNA, referred to as the (+) **strand**, enters an *E. coli* cell and serves as a template for the synthesis of the complementary (−) **strand**. The two strands (+ and −) remain together in a circular double helix, or duplex, called the **replicative form (RF)**. The RF serves as the template for its own replication, during which only (+) strands are produced. These strands are then packaged into viral coat proteins to form mature virus particles.

The experiment was carefully designed so that each newly synthesized strand could be distinguished and isolated from the template strand. The protocol of the experiment ensured that these (+) strands must have been synthesized *in vitro* under the direction of Kornberg's enzyme.

TABLE 11.1

**Base Composition of the DNA Template and the Product of Replication in Kornberg's Early Work**

| Organism | Template or Product | %A | %T | %G | %C |
|---|---|---|---|---|---|
| T2 | Template | 32.7 | 33.0 | 16.8 | 17.5 |
| | Product | 33.2 | 32.1 | 17.2 | 17.5 |
| *E. coli* | Template | 25.0 | 24.3 | 24.5 | 26.2 |
| | Product | 26.1 | 25.1 | 24.3 | 24.5 |
| Calf | Template | 28.9 | 26.7 | 22.8 | 21.6 |
| | Product | 28.7 | 27.7 | 21.8 | 21.8 |

*Source:* Kornberg (1960).

The critical test of biological activity was the DNA's success at transfection (Chapter 10), in which newly synthesized (+) strands were added to bacterial protoplasts (bacterial cells minus their cell wall). Following infection by the synthetic DNA, mature phages were produced, indicating that the synthetic DNA had successfully directed reproduction.

This demonstration of biological activity represented convincing proof of faithful copying. If even a single error had occurred to alter the base sequence of any of the 5386 nucleotides constituting the $\phi$X174 chromosome, the change might easily have caused a mutation that would prohibit the production of viable phages.

## DNA Polymerases II, III, IV, and V

Although DNA synthesized under the direction of polymerase I demonstrated biological activity, a more serious reservation about the enzyme's true biological role was raised in 1969. Paula DeLucia and John Cairns discovered a mutant strain of *E. coli* that was deficient in polymerase I activity. The mutation was designated *polA1*. In the absence of the functional enzyme, this mutant strain of *E. coli* still duplicated its DNA and successfully reproduced. However, the cells were impaired in their ability to repair DNA. For example, the mutant strain is highly sensitive to ultraviolet light (UV) and ionizing radiation, both of which damage DNA and are mutagenic. Nonmutant bacteria are able to repair a great deal of UV-induced damage.

These observations led to two conclusions:

1. At least one other enzyme that is responsible for replicating DNA *in vivo* is present in *E. coli* cells.
2. DNA polymerase I serves a secondary function *in vivo*. Kornberg and others now believe that this function is critical to the *fidelity* of DNA synthesis.

To date, four other unique DNA polymerases have been isolated from bacteria cells lacking polymerase I activity and from normal cells that contain polymerase I. Table 11.2 compares several characteristics of DNA polymerase I with those of **DNA polymerases II** and **III**. While none of the three can *initiate* DNA synthesis on a template, all three can *elongate* an existing DNA strand, called a **primer**.

The DNA polymerase enzymes are all large proteins exhibiting a molecular weight in excess of 100,000 Daltons (Da). All three possess 3′ to 5′ exonuclease activity, which means that they have the potential to polymerize in one direction and then pause, reverse their direction, and excise nucleotides just added. As we will discuss later in the chapter, this activity provides a capacity to proofread newly synthesized DNA and to remove and replace incorrect nucleotides.

DNA polymerase I also demonstrates 5′ to 3′ exonuclease activity. This activity allows the enzyme to excise nucleotides starting at the end at which synthesis begins and proceeding in the same direction as synthesis. Two final observations probably explain why Kornberg isolated polymerase I and not polymerase III: polymerase I is present in greater amounts than is polymerase III, and it is also much more stable.

What then are the roles of the polymerases *in vivo*? Polymerase III is the enzyme responsible for the 5′ to 3′ polymerization essential to *in vivo* replication. Its 3′ to 5′ exonuclease activity also provides a proofreading function that is activated when the enzyme inserts an incorrect nucleotide. When such a mistake occurs, synthesis stalls and the polymerase "reverses course," excising the incorrect nucleotide. Then, it resumes polymerization in the 5′ to 3′ direction, synthesizing the complement of the template strand. Polymerase I is believed to be responsible for removing the primer, as well as for the synthesis that fills gaps produced by this removal. Its exonuclease activities also allow for its participation in DNA repair. Polymerase II, as well as **polymerases IV** and **V**, is involved in various aspects of repair of DNA that has been damaged by external forces, such as ultraviolet light. Polymerase II is encoded by a gene activated by disruption of DNA synthesis at the replication fork.

We end this section by emphasizing the complexity of the DNA polymerase III molecule. In contrast to DNA polymerase I, which is but a single polypeptide, the active form of DNA polymerase III is a **holoenzyme**—a complex enzyme made up of multiple subunits. Polymerase III consists of 10 kinds of polypeptide subunits (Table 11.3) and has a molecular weight of 900,000 Da. The largest subunit, $\alpha$, has molecular weight of 140,000 Da and, along with subunits $\epsilon$ and $\theta$, constitutes a **core enzyme** responsible for the polymerization activity. The $\alpha$ subunit is responsible for nucleotide polymerization on the template strands, whereas the $\epsilon$ subunit of the core enzyme possesses the 3′ to 5′ exonuclease activity. A single

TABLE 11.2

**Properties Of Bacterial DNA Polymerases I, II, and III**

| Properties | I | II | III |
|---|---|---|---|
| Initiation of chain synthesis | − | − | − |
| 5′−3′ polymerization | + | + | + |
| 3′−5′exonuclease activity | + | + | + |
| 5′−3′exonuclease activity | + | − | − |
| Molecules of polymerase per cell | 400 | ? | 15 |

TABLE 11.3

**Subunits of the DNA Polymerase III Holoenzyme**

| Subunit | Function | Groupings |
|---|---|---|
| $\alpha$ | 5′−3′ polymerization | Core enzyme: elongates polynucleotide chain and proofreads |
| $\epsilon$ | 3′−5′ exonuclease | |
| $\theta$ | Core assembly | |
| $\gamma$ | Loads enzyme on template (serves as clamp loader) | $\gamma$ complex |
| $\delta$ | | |
| $\delta'$ | | |
| $\chi$ | | |
| $\psi$ | | |
| $\beta$ | Sliding clamp structure (processivity factor) | |
| $\tau$ | Dimerizes core complex | |

DNA polymerase III holoenzyme contains, along with other components, two core enzymes combined into a dimer.

A second group of five subunits ($\gamma$, $\delta$, $\delta'$, $\chi$, and $\psi$) forms what is called the $\gamma$ complex, which is involved in "loading" the enzyme onto the template at the replication fork. This enzymatic function requires energy and is dependent on the hydrolysis of ATP. The $\beta$ subunit serves as a "clamp" and prevents the core enzyme from falling off the template during polymerization. Finally, the $\tau$ subunit functions to dimerize two core polymerases, facilitating simultaneous synthesis of both strands of the helix at the replication fork. The holoenzyme and several other proteins at the replication fork together form a huge complex (nearly as large as a ribosome) known as the **replisome**. We consider the function of DNA polymerase III in more detail later in this chapter.

## 11.3 Many Complex Tasks Must Be Performed during DNA Replication

We have thus far established that in bacteria and viruses, replication is semiconservative and bidirectional along a single replicon. We also know that synthesis is catalyzed by DNA polymerase III and occurs in the 5′ to 3′ direction. Bidirectional synthesis creates two replication forks that move in opposite directions away from the origin of synthesis. As we can see from the following list, the tasks that must be executed as part of this overall process are many and diverse:

1. The helix must undergo localized unwinding, and the resulting "open" configuration must be stabilized so that synthesis may proceed along both strands.
2. As unwinding and subsequent DNA synthesis proceed, increased coiling creates tension further down the helix, and this tension must be reduced to allow the unwinding process to continue.
3. A primer of some sort must be synthesized so that polymerization can commence under the direction of DNA polymerase III. Surprisingly, RNA, not DNA, serves as the primer.
4. Once the RNA primers have been synthesized, DNA polymerase III begins to synthesize the DNA complement of both strands of the parent molecule. Because the two strands are antiparallel to one another, continuous synthesis in the direction that the replication fork moves is possible along only one of the two strands. On the other strand, synthesis must be discontinuous and thus involve a somewhat different process.
5. The RNA primers must be removed prior to completion of replication. The gaps that are temporarily created must be filled with DNA complementary to the template at each location.
6. The newly synthesized DNA strand that fills each temporary gap must be joined to the adjacent strand of DNA.
7. While DNA polymerases insert complementary bases along the template with a high degree of accuracy, no process is perfect, and, occasionally, incorrect bases are added to the growing strand. A proofreading mechanism that also corrects errors is therefore integral to the process of DNA synthesis.

As we consider these points, examine Figures 11–9, 11–10, 11–11, and 11–12 to see how each of these needs is met. Figure 11–13 summarizes the overall process of DNA synthesis.

### Unwinding the DNA Helix

As discussed earlier, there is a single point of origin along the circular chromosome of most bacteria and viruses at which DNA

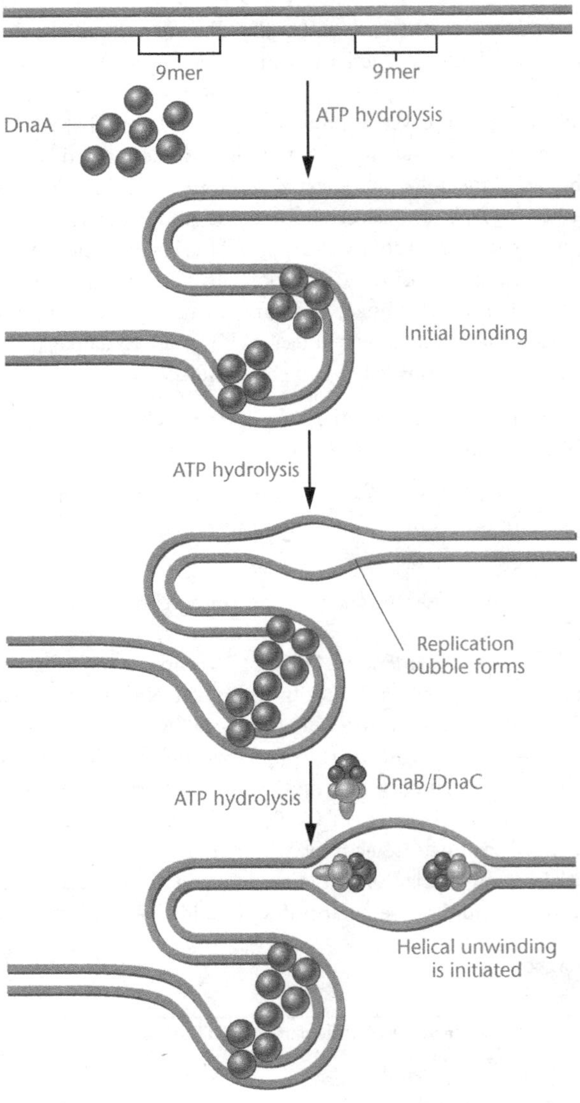


FIGURE 11–9 Helical unwinding of DNA during replication as accomplished by DnaA, DnaB, and DnaC proteins. Initial binding of many monomers of DnaA occurs at DNA sites containing repeating sequences of 9 nucleotides, called 9mers. Not illustrated are 13mers, which are also involved.

synthesis is initiated. This region of the *E. coli* chromosome has been particularly well studied. Called ***oriC***, it consists of 245 base pairs and is characterized by repeating sequences of 9 and 13 bases, called **9mers** and **13mers**. As shown in Figure 11–9, one particular protein, called **DnaA** (because it is encoded by the gene called *dnaA*), is responsible for the initial step in unwinding the helix. A number of subunits of the DnaA protein bind to each of several 9mers. This step facilitates the subsequent binding of **DnaB** and **DnaC** proteins that further open and destabilize the helix. Proteins such as these, which require the energy supplied by the hydrolysis of ATP in order to break hydrogen bonds and denature the double helix, are called **helicases.** Other proteins, called **single-stranded binding proteins (SSBPs)**, stabilize this open conformation.

As unwinding proceeds, a coiling tension is created ahead of the replication fork, often producing **supercoiling,** additional twisting that contorts nearby stretches of unwound DNA. In circular molecules, the added twists and turns look much like the coils you can create in a rubber band by stretching it out and then twisting one end. Supercoiling can be relaxed by **DNA gyrase**, a member of a larger group of enzymes referred to as **DNA topoisomerases**. The gyrase makes either single- or double-stranded "cuts" and also catalyzes localized movements that have the effect of "undoing" the twists and knots of supercoiling. The strands are then resealed. These various reactions are driven by the energy released during ATP hydrolysis.

Together, the DNA, the polymerase complex, and associated enzymes make up an array of molecules that participate in DNA synthesis and are part of what we have previously called the *replisome.*

## Initiation of DNA Synthesis with an RNA Primer

Once a small portion of the helix is unwound, what else is needed to initiate synthesis? As we have seen, DNA polymerase III requires a primer with a free 3′-hydroxyl group in order to construct (by elongation) a polynucleotide chain. The obvious absence of such a primer (a 3′end) in a closed circular chromosome prompted researchers to investigate how the first nucleotide could be added. It is now clear that RNA serves as the primer that initiates DNA synthesis.

A short segment of RNA (about 10 to 12 nucleotides long), complementary to DNA, is first synthesized on the DNA template. Synthesis of the RNA is directed by a form of RNA polymerase called **primase**, which does not require a free 3′ end to initiate synthesis. It is to this short segment of RNA that DNA polymerase III begins to add deoxyribonucleotides, initiating DNA synthesis. A conceptual diagram of initiation on a DNA template is shown in Figure 11–10. Later, the RNA primer is clipped out and replaced with DNA. This is thought to occur under the direction of DNA polymerase I. Recognized in viruses, bacteria, and several eukaryotic organisms, RNA priming is a universal feature of the initiation of DNA synthesis.

## Continuous and Discontinuous DNA Synthesis of Antiparallel Strands

We must now revisit the fact that the two strands of a double helix are **antiparallel** to each other—that is, one runs in the 5′–3′ direction, while the other has the opposite, 3′–5′, polarity. Because DNA polymerase III synthesizes DNA in only the 5′–3′ direction, synthesis along an advancing replication fork occurs in one direction on one strand and in the opposite direction on the other.

As a result, as the strands unwind and the replication fork progresses down the helix (Figure 11–11), only one strand can

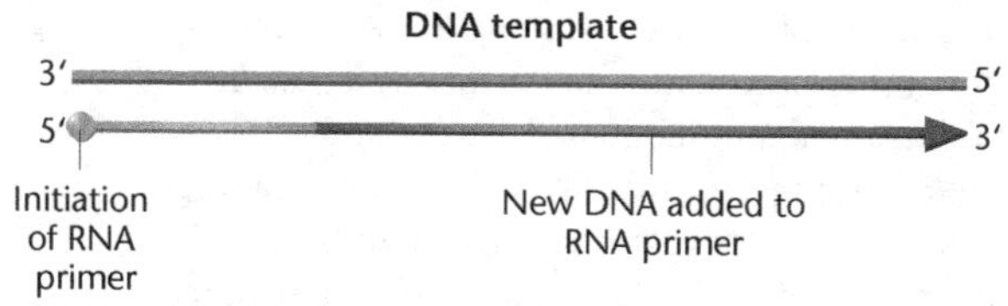


FIGURE 11–10 The initiation of DNA synthesis. A complementary RNA primer is first synthesized, to which DNA is added. All synthesis is in the 5′ to 3′ direction. Eventually, the RNA primer is replaced with DNA under the direction of DNA polymerase I.

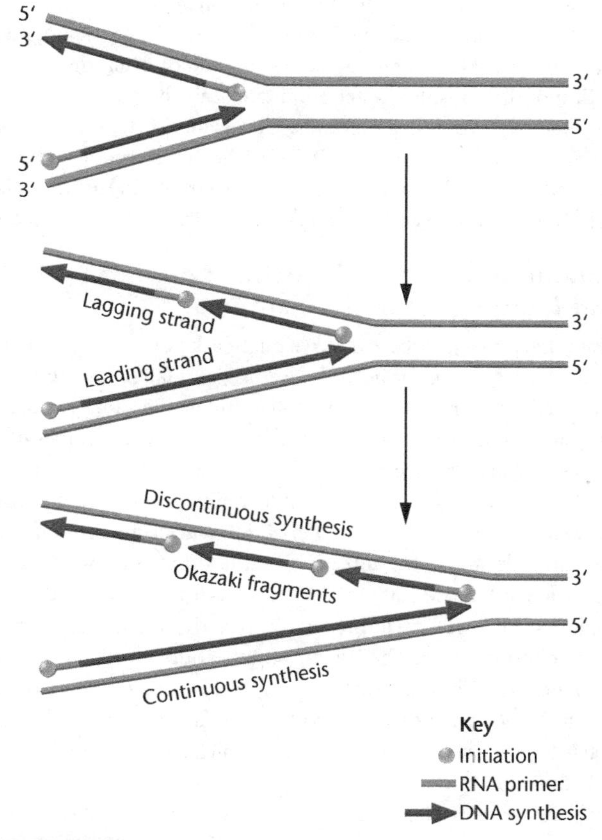


FIGURE 11–11 Opposite progression of DNA synthesis along the two strands, necessary because the two strands of DNA run antiparallel to one another and DNA polymerase III synthesizes only in one direction (5′ to 3′). On the lagging strand, synthesis must be discontinuous, resulting in the production of Okazaki fragments. On the leading strand, synthesis is continuous. RNA primers are used to initiate synthesis on both strands.

serve as a template for **continuous DNA synthesis**, synthesis in which construction of the replicon is not interrupted and reinitiated. The strand of DNA being synthesized in this way is called the **leading strand.** As the fork progresses, many points of initiation are necessary on the opposite DNA template, resulting in **discontinuous DNA synthesis** of the **lagging strand.***

Evidence supporting the occurrence of discontinuous DNA synthesis was first provided by Reiji and Tuneko Okazaki. They discovered that when bacteriophage DNA is replicated in *E. coli*, some of the newly formed DNA that is hydrogen bonded to the template strand is present as small fragments containing 1000–2000 nucleotides. RNA primers are part of each such fragment. These pieces, now called **Okazaki fragments,** are converted into longer and longer DNA strands of higher molecular weight as synthesis proceeds.

Discontinuous synthesis of DNA requires enzymes that remove the RNA primers and unite the Okazaki fragments into a continuous strand. As we have noted, DNA polymerase I removes the primers and replaces the missing nucleotides. Joining the fragments is the work of **DNA ligase,** which is capable of catalyzing the formation of the phosphodiester bond that seals the nick between discontinuously synthesized strands. The evidence that DNA ligase performs this function during DNA synthesis is strengthened by the observation of a ligase-deficient mutant strain (*lig*) of *E. coli*, in which a large number of unjoined Okazaki fragments accumulate.

## Concurrent Synthesis on the Leading and Lagging Strands

Given the model just discussed, we might ask how DNA polymerase III synthesizes DNA on both the leading and lagging strands. Can both strands be replicated simultaneously at the same replication fork, or are the events distinct, involving two separate copies of the enzyme? Evidence suggests that both strands can be replicated simultaneously. As Figure 11–12 illustrates, if the lagging strand forms a loop, nucleotide polymerization can occur on both template strands under the direction of the dimer of core enzymes. After the synthesis of 1000 to 2000 nucleotides, the monomer on the lagging strand will encounter a completed Okazaki fragment, at which point it releases the lagging strand. It then forms the lagging template strand into a new loop, and the process repeats. Looping inverts the physical orientation of the template but not the direction of synthesis viewed in chemical terms, which always occurs in the 5′ to 3′ direction.

Another important feature of the holoenzyme that facilitates synthesis at the replication fork is a dimer of the $\beta$ subunit that forms

* Because DNA synthesis is continuous on one strand and discontinuous on the other, the term *semidiscontinuous synthesis* is sometimes used to describe the replication of both strands.

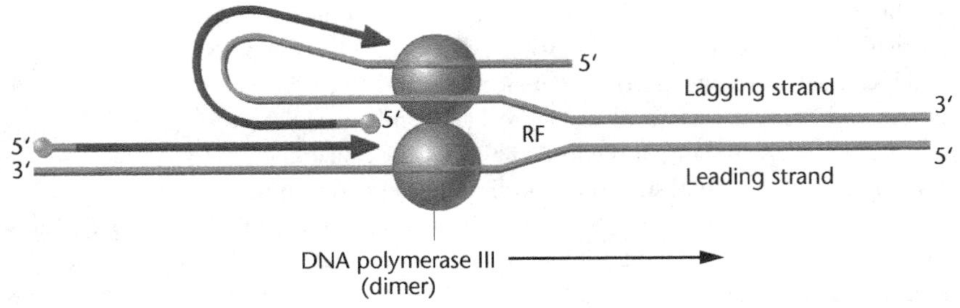


FIGURE 11–12 Illustration of how concurrent DNA synthesis may be achieved on both the leading and lagging strands at a single replication fork (RF). The lagging template strand is "looped" in order to invert the physical direction of synthesis but not the biochemical direction. The holoenzyme contains a dimer consisting of two core enzymes, each of which conducts synthesis on a different strand.

a clamplike structure around the newly formed DNA duplex. This $\beta$ subunit clamp prevents the core enzyme (the $\alpha$, $\epsilon$, and $\theta$ subunits that are responsible for catalysis of nucleotide addition) from falling off the template as polymerization proceeds. Because the entire holoenzyme moves along the parent duplex, advancing the replication fork, the $\beta$-subunit dimer is often referred to as a *sliding clamp*.

NOW SOLVE THIS

In Problem 33 on page 300, a hypothetical organism is observed in which no Okazaki fragments are produced. You are asked to suggest a DNA model consistent with this information.

■ HINT: *The lack of Okazaki fragments suggests that DNA synthesis is continuous on both strands.*

## Integrated Proofreading and Error Correction

The immediate purpose of DNA replication is the synthesis of a new strand that is precisely complementary to the template strand at each nucleotide position. Although the action of DNA polymerases is very accurate, synthesis is not perfect and a noncomplementary nucleotide is occasionally inserted erroneously. To compensate for such inaccuracies, the DNA polymerases all possess 3′ to 5′ exonuclease activity. This property allows them to detect and excise a mismatched nucleotide (in the 3′–5′ direction). Once the mismatched nucleotide is removed, 5′ to 3′ synthesis can again proceed. This process, called **proofreading,** increases the fidelity of synthesis by a factor of about 100. In the DNA polymerase III holoenzyme, the epsilon ($\epsilon$) subunit is directly involved in the proofreading step. In strains of *E. coli* with a mutation that renders the $\epsilon$ subunit nonfunctional, the error rate (the mutation rate) during DNA synthesis is increased substantially.

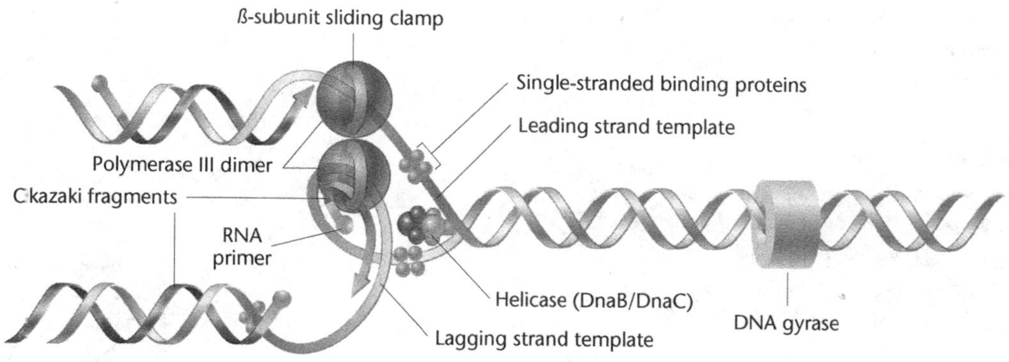


FIGURE 11–13 Summary of DNA synthesis at a single replication fork. Various enzymes and proteins essential to the process are shown.

11.4

## A Summary of DNA Replication in Prokaryotes

We can now combine the various aspects of DNA replication occurring at a single replication fork into a coherent model, as shown in Figure 11–13. At the advancing fork, a helicase is unwinding the double helix. Once a part of the helix has unwound, single-stranded binding proteins associate with the strands, preventing helix reformation. In advance of the replication fork, DNA gyrase functions to diminish the tension created as the helix supercoils. Each of the core enzymes in the dimer is bound to one of the template strands by a $\beta$-subunit sliding clamp. Continuous synthesis occurs on the leading strand, while the lagging strand must loop around in order for simultaneous (concurrent) synthesis to occur on both strands. Not shown in the figure, but essential to replication on the lagging strand, is the action of DNA polymerase I and DNA ligase, which replace the RNA primers with DNA and join the Okazaki fragments, respectively.

Because the investigation of DNA synthesis is still an extremely active area of research, this model will no doubt be extended in the future. In the meantime, it provides a summary of DNA synthesis against which genetic phenomena can be interpreted.

11.5

## Replication in Prokaryotes Is Controlled by a Variety of Genes

Much of what we know about DNA replication in viruses and bacteria is based on genetic analysis of the process. For example, we have already discussed studies involving the *polA1* mutation, which revealed that DNA polymerase I is not the major enzyme responsible for replication. Many other mutations interrupt or seriously impair some aspect of replication, such as the ligase-deficient and the proofreading-deficient mutations mentioned previously. Because mutations such as these tend to be lethal, genetic analysis frequently relies on **conditional mutations**, which are expressed under certain conditions but not others. For example, a **temperature-sensitive mutation** may not be expressed at a particular, *permissive* temperature, yet when the mutant organism is grown at a *restrictive* temperature, the mutant phenotype is expressed and can be studied. By pinpointing the loss of function brought about by a conditional mutation, researchers can acquire insight into the product and the associated function of the normal, nonmutant gene.

As shown in Table 11.4, a variety of genes in *E. coli* specify the subunits of the DNA polymerases and encode products involved in specification of the origin of synthesis, helix unwinding and stabilization, initiation and priming, relaxation of supercoiling, repair, and ligation. The existence of such a large group of genes attests to the complexity of the process of replication, even in the relatively simple prokaryote. Given the enormous quantity of DNA that must be unerringly replicated in a very brief time, this

TABLE 11.4

**Some of the *E. coli* Genes Having a Role in Replication**

| Gene | Product or Role |
|---|---|
| *polA* | DNA polymerase I |
| *polB* | DNA polymerase II |
| *dnaE, N, Q, X, Z* | DNA polymerase III subunits |
| *dnaG* | Primase |
| *dnaA, I, P* | Initiation |
| *dnaB, C* | Helicase at *oriC* |
| *gyrA, B* | Gyrase subunits |
| *lig* | DNA ligase |
| *rep* | DNA helicase |
| *ssb* | Single-stranded binding proteins |
| *rpoB* | RNA polymerase subunit |

level of complexity is not unexpected. As we will see in the next section, the process is even more involved, and therefore more difficult to investigate, in eukaryotes.

**NOW SOLVE THIS**

Problem 25 on page 299 involves several temperature-sensitive mutations in *E. coli*, asking you to interpret the action of the genes that have mutated based on the phenotypes that result.

■ HINT: *Each mutation has disrupted one of the many steps essential to DNA synthesis. In each case, the mutant phenotype provides the clue as to which enzyme or function is affected.*

11.6

## Eukaryotic DNA Synthesis Is Similar to Synthesis in Prokaryotes, but More Complex

Research has shown that eukaryotic DNA is replicated in a manner similar to that of bacteria. In both systems, double-stranded DNA is unwound at replication origins, replication forks are formed, and bidirectional DNA synthesis creates leading and lagging strands from templates under the direction of DNA polymerase. Eukaryotic polymerases have the same fundamental requirements for DNA synthesis as do bacterial systems: four deoxyribonucleoside triphosphates, a template, and a primer. However, because eukaryotic cells contain much more DNA per cell, because eukaryotic chromosomes are linear rather than circular, and because this DNA is complexed with proteins, DNA synthesis is more complicated in eukaryotes and more difficult to study. However, a great deal is now known about the process.

### Multiple Replication Origins

The most obvious difference between eukaryotic and prokaryotic DNA replication is that eukaryotic chromosomes contain multiple replication origins, in contrast to the single site that is part of the *E. coli* chromosome. Multiple origins, visible under the electron microscope as "replication bubbles" that form as the helix opens up (Figure 11–14), each provide two potential replication forks. Multiple origins are essential if replication of the entire genome of a typical eukaryote is to be completed in a reasonable time. Recall that (1) eukaryotes have much greater amounts of DNA than bacteria do—for example, yeast has three times as much DNA, and *Drosophila* has 40 times as much, as *E. coli*—and (2) the rate of synthesis by eukaryotic DNA polymerase is much slower—only about 2000 nucleotides per minute, a rate 25 times less than the comparable bacterial enzyme. Under these conditions, replication from a single origin of a typical eukaryotic chromosome might take days to

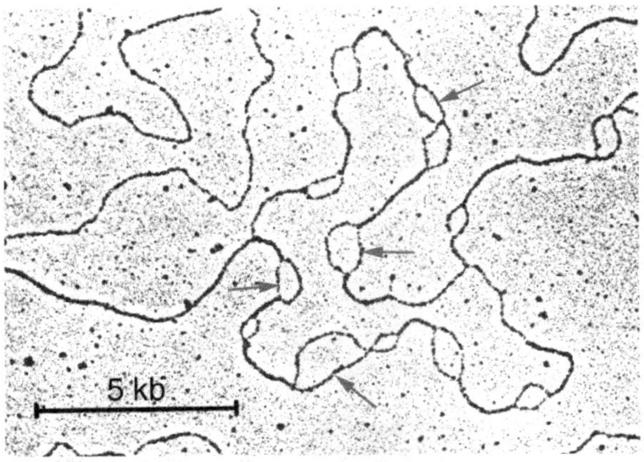


FIGURE 11–14 Micrograph showing multiple origins of replication along a eukaryotic chromosome. Each origin is apparent as a replication bubble along the axis of the chromosome. Arrows identify some of these replication bubbles.

complete! However, replication of entire eukaryotic genomes is usually accomplished in a matter of hours.

Information is now available concerning the molecular structure of the multiple origins and the initiation of DNA synthesis at these sites. Most of this information was originally derived from the study of yeast (e.g., *Saccharomyces cerevisiae*), which has between 250 and 400 replicons per genome; subsequent studies have used mammalian cells, which have as many as 25,000 replicons. The origins in yeast have been isolated and are called **autonomously replicating sequences (ARSs)**. They consist of a unit containing a consensus sequence (meaning a sequence that is the same, or nearly the same, in all yeast ARSs) of 11 base pairs, flanked by other short sequences involved in efficient initiation. As we know from Chapter 2, DNA synthesis is restricted to the S phase of the eukaryotic cell cycle. Research has shown that the many origins are not all activated at once; instead, clusters of 20–80 adjacent replicons are activated sequentially throughout the S phase until all DNA is replicated.

For the polymerase to find the ARSs among so much DNA is a prodigious feat of recognition. The mechanism that performs this feat is initiated prior to the S phase. During the G1 phase of the cell cycle, all ARSs are initially bound by a group of specific proteins (six in yeast), forming what is called an **origin recognition complex (ORC)**. Mutations either in the ARSs or in any of the genes encoding these proteins of the ORC abolish or reduce initiation of DNA synthesis. Since these recognition complexes are formed in G1 but synthesis is not initiated at these sites until S, yet other proteins must be involved in the actual initiation signal. The most important of these proteins are specific kinases, key enzymes involved in phosphorylation, which are integral parts of cell-cycle control. When these kinases are bound along with ORCs, a prereplication complex is formed that is accessible to DNA polymerase. After the kinases are activated, they serve to complete the initiation complex, directing localized unwinding and triggering DNA synthesis. Activation also inhibits reformation of the prereplication complexes once DNA synthesis has been completed at each replicon. This is an important mechanism, since it distinguishes segments of DNA that have completed replication from segments of unreplicated DNA, thus maintaining orderly and efficient replication. It ensures that replication only occurs once along each stretch of DNA during each cell cycle.

TABLE 11.5

**Properties of Eukaryotic DNA Polymerases**

| | Polymerase $\alpha$ | Polymerase $\beta$ | Polymerase $\delta$ | Polymerase $\epsilon$ | Polymerase $\gamma$ | Polymerase $\zeta$ |
|---|---|---|---|---|---|---|
| Location | Nucleus | Nucleus | Nucleus | Nucleus | Mitochondrion | Nucleus |
| $3'-5'$ Exonuclease Activity | No | No | Yes | Yes | Yes | No |
| Essential to Nuclear Replication | Yes | No | Yes | Yes | No | No |

## Eukaryotic DNA Polymerases

The most complex aspect of eukaryotic replication is the array of polymerases involved in directing DNA synthesis. As we will see, many different forms of polymerase have been isolated and studied. However, only four are actually involved in replication of DNA, while the remainder are involved in repair processes. In order for the polymerases to have access to DNA, the topology of the helix must first be modified. At each origin site, synthesis initiation begins with the double strands being opened up within an AT-rich region, allowing the entry of a helicase enzyme that proceeds to further unwind the double-stranded DNA. Before polymerases can begin synthesis, histone proteins complexed to the DNA (which form the characteristic nucleosomes of chromatin, as discussed in Chapter 12) also must be stripped away or otherwise modified. As DNA synthesis then proceeds, histones reassociate with the newly formed duplexes, reestablishing the characteristic nucleosome pattern (Figure 11–15). In eukaryotes, the synthesis of new histone proteins is tightly coupled to DNA synthesis during the S phase of the cell cycle.

The nomenclature and characteristics of six different polymerases are summarized in Table 11.5. Of these, three (Pol $\alpha$, $\delta$, and $\epsilon$) are essential to nuclear DNA replication in eukaryotic cells. Two (Pol $\beta$ and $\zeta$) are thought to be involved in DNA repair (and still other repair forms have been discovered). The sixth form (Pol $\gamma$) is involved in the synthesis of mitochondrial DNA. Presumably, its replication function is limited to that organelle, even though it is encoded by a nuclear gene. All but one of the six forms of the enzyme (the $\beta$ form) consist of multiple subunits. Different subunits perform different functions during replication.

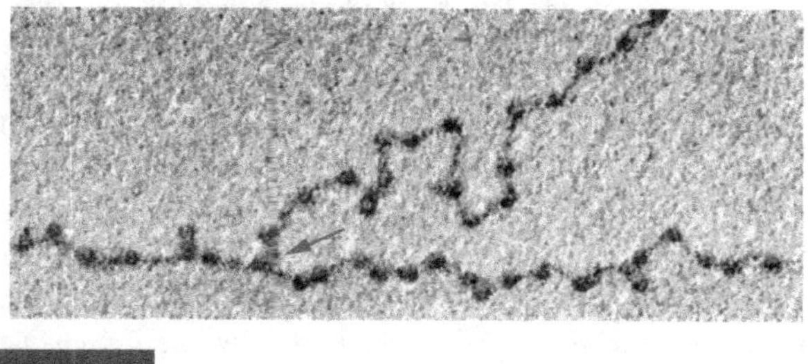

FIGURE 11–15 An electron micrograph of a eukaryotic replicating fork, demonstrating the presence of histone-protein-containing nucleosomes on both branches.

Pol $\alpha$ and $\delta$ are the major forms of the enzyme involved in initiation and elongation during nuclear DNA synthesis, so we concentrate our discussion on these. Two of the four subunits of the Pol $\alpha$ enzyme function in the synthesis of the RNA primers during the initiation of synthesis of both the leading and lagging strands. After the RNA primer reaches a length of about 10 ribonucleotides, another subunit functions to elongate it further by adding 20–30 complementary deoxyribonucleotides. Pol $\alpha$ is said to possess low **processivity**, a term that essentially reflects the length of DNA that is synthesized by an enzyme before it dissociates from the template. Once the primer is in place, an event known as **polymerase switching** occurs, whereby Pol $\alpha$ dissociates from the template and is replaced by Pol $\delta$. This form of the enzyme possesses *high processivity* and elongates the leading and lagging strands. It also possesses $3'$ to $5'$ exonuclease activity and thus has the potential to proofread. Pol $\epsilon$, the third essential form, possesses the same general characteristics as Pol $\delta$, and is also believed to be involved in elongation once the primer has been extended. In yeast, mutations that render Pol $\epsilon$ inactive are lethal, attesting to the importance of its function during replication.

The process described applies to both leading- and lagging-strand synthesis. On both strands, RNA primers must be replaced with DNA. On the lagging strand, the Okazaki fragments, which are about 10 times smaller (100–150 nucleotides) in eukaryotes than in prokaryotes, must be ligated.

To accommodate the increased number of replicons, eukaryotic cells contain many more DNA polymerase molecules than do bacteria. While *E. coli* has about 15 copies of DNA polymerase III per cell, there may be up to 50,000 copies of the $\alpha$ form of DNA polymerase in animal cells. As has been pointed out, the presence of greater numbers of smaller replicons in eukaryotes compared with bacteria compensates for the slower rate of DNA synthesis in eukaryotes. *E. coli* requires 20 to 40 minutes to replicate its chromosome, while *Drosophila*, with 40 times more DNA, is known during embryonic cell divisions to accomplish the same task in only 3 minutes.

11.7

## Telomeres Provide Structural Integrity at Chromosome Ends but Are Problematic to Replicate

A final difference between prokaryotic and eukaryotic DNA synthesis stems from the structural differences in their chromosomes. Unlike the closed, circular DNA of bacteria and most bacteriophages, eukaryotic chromosomes are linear. During replication, a special problem arises at the "ends" of these linear molecules.

Eukaryotic chromosomes end in distinctive sequences called **telomeres** that help preserve the integrity and stability of the chromosome. Telomeres are necessary because the double-stranded "ends" of DNA molecules at the termini of linear chromosomes potentially resemble the **double-stranded breaks** (**DSBs**) that can occur when a chromosome becomes fragmented internally. In such cases, the double-stranded loose ends can fuse to other such ends; if they don't fuse, they are vulnerable to degradation by nucleases. Either outcome eventuality can lead to problems. Telomeres are believed to create inert chromosome ends, protecting intact eukaryotic chromosomes from improper fusion or degradation.

### Telomere Structure

We could speculate that there must be something unique about the DNA sequence or the proteins that bind to it that confers this protective property to telomeres. Indeed, this has been shown to be the case. First discovered by Elizabeth Blackburn and Joe Gall in their study of micronuclei—the smaller of two nuclei in the ciliated protozoan *Tetrahymena*—the DNA at the protozoan's chromosome ends consists of the short tandem repeating sequence TTGGGG, present many times on one of the two helical strands making up each telomere. This strand is referred to as the G-rich strand, in contrast to its complementary strand, displaying the repeated sequence AACCCC, called the C-rich strand. In a similar way, all vertebrates contain the sequence TTAGGG at the ends of G-rich strands, repeated several thousand times in somatic cells. Since each linear chromosome ends with two helical DNA strands running antiparallel to one another, one strand has a 3′ ending and the other has a 5′-ending. It is the 3′-strand that is the G-rich one. As we will see later in this chapter, this has special significance during telomere replication.

But first, let's describe how this tandemly repeated DNA confers inertness to the chromosome ends. One model is based on the discovery that the 3′-ending G-rich strand extends as an overhang, lacking a complement, and thus forming a single-stranded tail at the terminus of each telomere. In *Tetrahymena*, this tail is only 12–16 nucleotides long. However, in vertebrates, it may be several hundred nucleotides long. The final conformation of these tails has been correlated with chromosome inertness. Though not considered complementary in the same way as A-T and G-C base pairs are, G-containing nucleotides are nevertheless capable of base pairing with one another when several are aligned opposite another G-rich sequence. Thus, the G-rich single-stranded tails are capable of looping back on themselves, forming multiple G-G hydrogen bonds to create what are referred to as **G-quartets**. Such quartets have been observed during *in vitro* crystal structure analysis of telomeric DNA. The resulting loops at the chromosome ends (called **t-loops** since they are part of the telomere) are much like those created when you tie your shoelaces into a bow. It is believed that these structures, in combination with specific proteins that bind to them, essentially close off the ends of chromosomes and make them inert. In another more recent model, t-loops are formed when the 3′ repeating terminus loops over and displaces the same DNA sequence in an upstream telomeric region. Whichever model is correct, there is currently agreement that the t-loop contributes significantly to the inertness of the chromosome end.

### Replication at the Telomere

Now let's consider the problem that semiconservative replication poses at the end of a double-stranded DNA molecule. While 5′ to 3′ synthesis on the leading-strand template may proceed to the end, a difficulty arises on the lagging strand once the final RNA primer is removed (Figure 11–16). Normally, the newly created gap would be filled in starting with the addition of a nucleotide to the adjacent 3′-OH group [the group to the right of gap (a) in Figure 11–16]. However, since the final gap [gap (b) in Figure 11–16] is at the end of the strand being synthesized, there is no Okazaki fragment present to provide the needed 3′-OH group. Thus, in the situation depicted in Figure 11–16, a gap remains on the lagging strand produced in each successive round of synthesis, shortening the double-stranded end of the chromosome by the length of the RNA primer. With each round of replication, the shortening becomes more severe in each daughter cell, eventually extending beyond the telomere to potentially delete gene-coding regions. Because of this significant problem, we can expect a molecular solution to have been developed early in evolution and subsequently conserved and shared by almost all eukaryotes. Indeed, DNA analysis shows this was the case.

A unique eukaryotic enzyme called **telomerase**, first discovered by Elizabeth Blackburn and Carol Grieder in subsequent studies of *Tetrahymena*, has helped us understand the solution to the problem of telomere shortening. As noted above, telomeric DNA in eukaryotes is always found to consist of many short, repeated nucleotide sequences, with the G-rich strand overhanging in the form of a single-stranded tail. In *Tetrahymena* the tail contains several repeats of the sequence 5′-TTGGGG-3′. As we will see, telomerase is capable of adding several more repeats of this six-nucleotide sequence to the 3′ end of the G-rich strand (using 5′–3′ synthesis). Detailed investigation by Blackburn and Greider of how the *Tetrahymena* telomerase enzyme accomplishes this synthesis yielded an extraordinary finding. The enzyme is highly unusual in that it is a *ribonucleoprotein*, containing within its molecular structure a short piece of RNA that is essential to its catalytic activity. The RNA

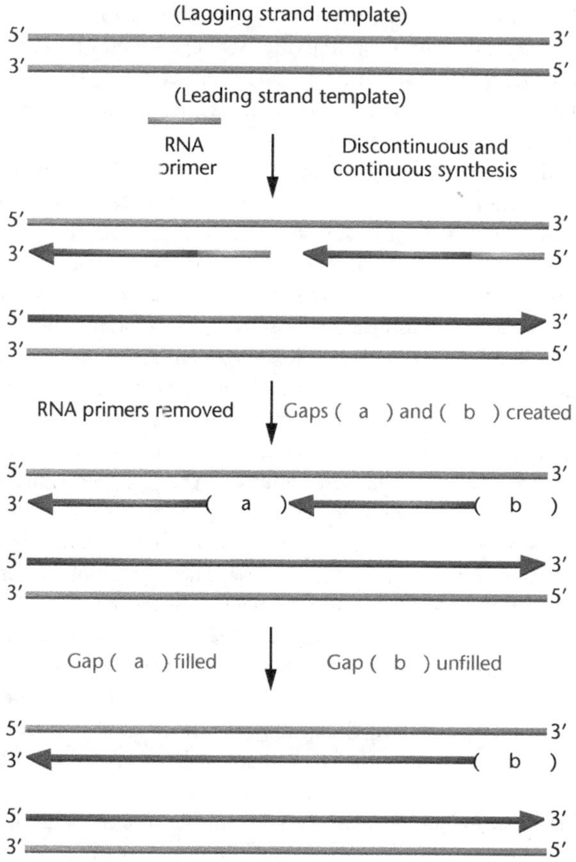


FIGURE 11-16 Diagram illustrating the problem presented by replication at the ends of linear chromosomes. A gap ( b ) is left after synthesis on the lagging strand.

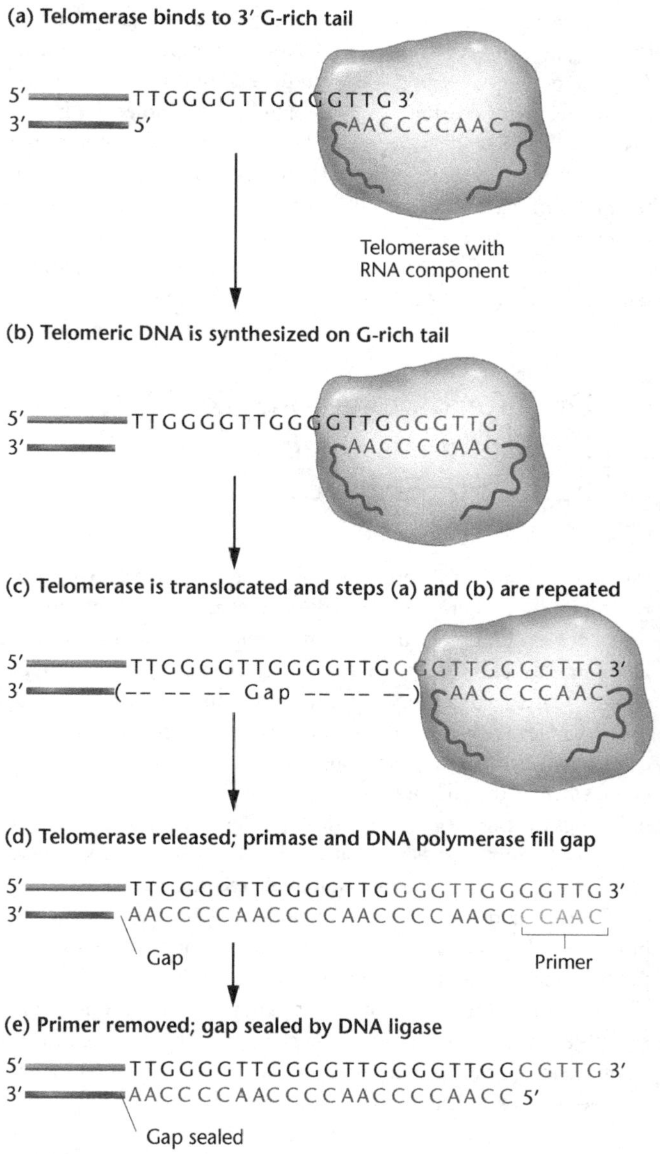


FIGURE 11-17 The predicted solution to the problem posed in Figure 11-16. The enzyme telomerase is a ribonucleoprotein containing an RNA sequence (shown in red) that base pairs with a small part of the single-stranded, 3′ overhanging DNA. There the enzyme behaves as a reverse transcriptase by synthesizing DNA and extending the 3′, overhanging, single-stranded G-rich tail. The enzyme is then translocated and the process repeated. Subsequently, the gap on the complementary C-rich strand can be filled by DNA primase and polymerase.

component serves as both a "guide" (to proper attachment of the enzyme to the telomere) and a template for the synthesis of its DNA complement, the latter being a process called **reverse transcription.** In *Tetrahymena*, the RNA contains the sequence AACCCCAAC, within which is found the complement of the repeating telomeric DNA sequence that must be synthesized (TTGGGG).

Figure 11–17 shows one model of how researchers envisioned the enzyme working. Part of the RNA sequence of the enzyme (shown in red) basepairs with the ending sequence of the single-stranded overhanging DNA, while the remainder of the RNA extends beyond the overhang. Next, reverse transcription of this extending RNA sequence—synthesizing DNA on an RNA template—extends the length of the G-rich lagging strand. It is believed that the enzyme is then translocated toward the (newly formed) end of the strand and the same events are repeated, continuing the extension process. In yeast, where the process has been investigated in detail, specific proteins are responsible for terminating this synthesis, and they determine the number of repeating units added. Similar proteins no doubt function in the same way in other eukaryotes. Once extension of the overhanging strand is completed, DNA primase and polymerase can synthesize the complement on the new C-rich strand, filling the gap that was created initially.

Similar enzyme function has now been found in almost all eukaryotes studied. As mentioned earlier, in humans, the telomeric DNA sequence that creates the lagging strand and is repeated is 5′-TTAGGG-3′, differing from *Tetrahymena* by only one nucleotide.

As we shall see in Chapter 12, telomeric DNA sequences have been highly conserved throughout evolution, reflecting the critical function of telomeres. In the essay at the end of this chapter, we will see that telomere shortening has been linked to a molecular mechanism involved in cellular aging. In fact, in most eukaryotic somatic cells, telomerase is not active, and thus, with each cell division, the telomeres of each chromosome shorten. After many divisions, the telomere is seriously eroded, and the cell loses the capacity for further division. Malignant cells, on the other hand, maintain telomerase activity and in this way are immortalized.

## 11.8 DNA Recombination, Like DNA Replication, Is Directed by Specific Enzymes

We now turn to a topic previously discussed in Chapter 5—genetic recombination. There, we pointed out that the process of crossing over between homologs depends on breakage and rejoining of the DNA strands. Now that we have discussed the chemistry and replication of DNA, we can consider how recombination can occur at the molecular level. In general, our discussion pertains to genetic exchange between any two homologous double-stranded DNA molecules, whether they be viral or bacterial chromosomes or eukaryotic homologs during meiosis. Genetic exchange at equivalent positions along two chromosomes with substantial DNA sequence homology is referred to as **general,** or **homologous, recombination**.

Several models attempt to explain homologous recombination, but they all have certain features in common. First, all are based on proposals put forth independently by Robin Holliday and Harold L. K. Whitehouse in 1964. Second, they all depend on the complementarity between DNA strands to explain the precision of the exchange. Finally, each model relies on a series of enzymatic processes in order to accomplish genetic recombination.

One such model is shown in Figure 11–18. It begins with two paired DNA duplexes, or homologs [Step (a)], in each of which an endonuclease introduces a single-stranded nick at an identical position [Step (b)]. The internal strand endings produced by these cuts are then displaced and subsequently pair with their complements on the opposite duplex [Step (c)]. Next, a ligase seals the loose ends [Step (d)], creating hybrid duplexes called **heteroduplex DNA molecules**, held together by a cross-bridge structure. The position of this cross bridge can then move down the chromosome by a process referred to as branch migration [Step (e)], which occurs as a result of a zipperlike action as hydrogen bonds are broken and then re-formed between complementary bases of the displaced strands of each duplex. This migration yields an increased length of heteroduplex DNA on both homologs.

DNA RECOMBINATION ◀ WEB TUTORIAL 11.2

If the duplexes now move in opposite directions [Step (f)] and the bottom portion shown in the figure rotates 180° [Step (g)], an intermediate planar structure called a $\chi$ form—the characteristic **Holliday structure**—is created. If the two strands on opposite homologs previously uninvolved in the exchange are now nicked by an endonuclease [Step (h)] and ligation occurs as in Step (i), recombinant duplexes are created. Note that the arrangement of alleles is altered as a result of this recombination.

Evidence supporting this model includes the electron microscopic visualization of $\chi$-form planar DNA molecules from bacteria, showing four duplex arms joined at a single point of exchange [Figure 11–18 (Step g)]. Further important evidence comes from the discovery of the **RecA protein** in *E. coli*. This molecule promotes the exchange of reciprocal single-stranded DNA molecules as occurs in Step (c) of the model. RecA also enhances the hydrogen-bond formation during strand displacement, thus initiating heteroduplex formation. Finally, many other enzymes essential to the nicking and ligation process have also been discovered and investigated. For example, the products of the *recB, recC,* and *recD* genes are thought to be involved in the nicking and unwinding of DNA. Numerous mutations that prevent genetic recombination have been found in viruses and bacteria. These mutations represent genes whose products play an essential role in this process.

Many investigators believe that the model above, involving *single-stranded breaks*, is but one mechanism by which DNA recombination occurs. Studies in yeast have led to a model, proposed by Franklin Stahl and Jack Szostak in the early 1980s, involving a *double-stranded break* in the DNA double helix of one of the two homologs, which leads to genetic exchange. Gaps are created as a result of this break and are then enlarged. Meanwhile, the two broken ends of one of the strands invade the intact double helix of the other homolog, leading to exchanges with one of its strands. DNA repair synthesis then fills all gaps, and exchange is completed. Two Holliday junctions are formed, and endonucleases are involved in finalizing the exchange. The end result is the same as our original model: reciprocal exchange as occurs during crossing over in meiotic tetrads.

This same general scheme is believed to be responsible for the repair of double-stranded breaks that occur in somatic cells, particularly during mitosis. Such damage to DNA can result from numerous causes, including the energy of ionizing radiation. The repair activity is referred to as **DNA double-stranded break repair (DSB repair)**. We will return to this topic again when we discuss DNA repair mechanisms in Chapter 16. Figure 16–16, depicting the *double-stranded break repair mechanism*, is thus also relevant to the Stahl-Szostak proposal for recombination. You may wish to correlate the above discussion with the depiction in that figure.

## 11.9 Gene Conversion Is a Consequence of DNA Recombination

A modification of the model in Figure 11–18 has helped us to better understand a unique genetic phenomenon known as **gene conversion**. Initially found in yeast by Carl Lindegren and in

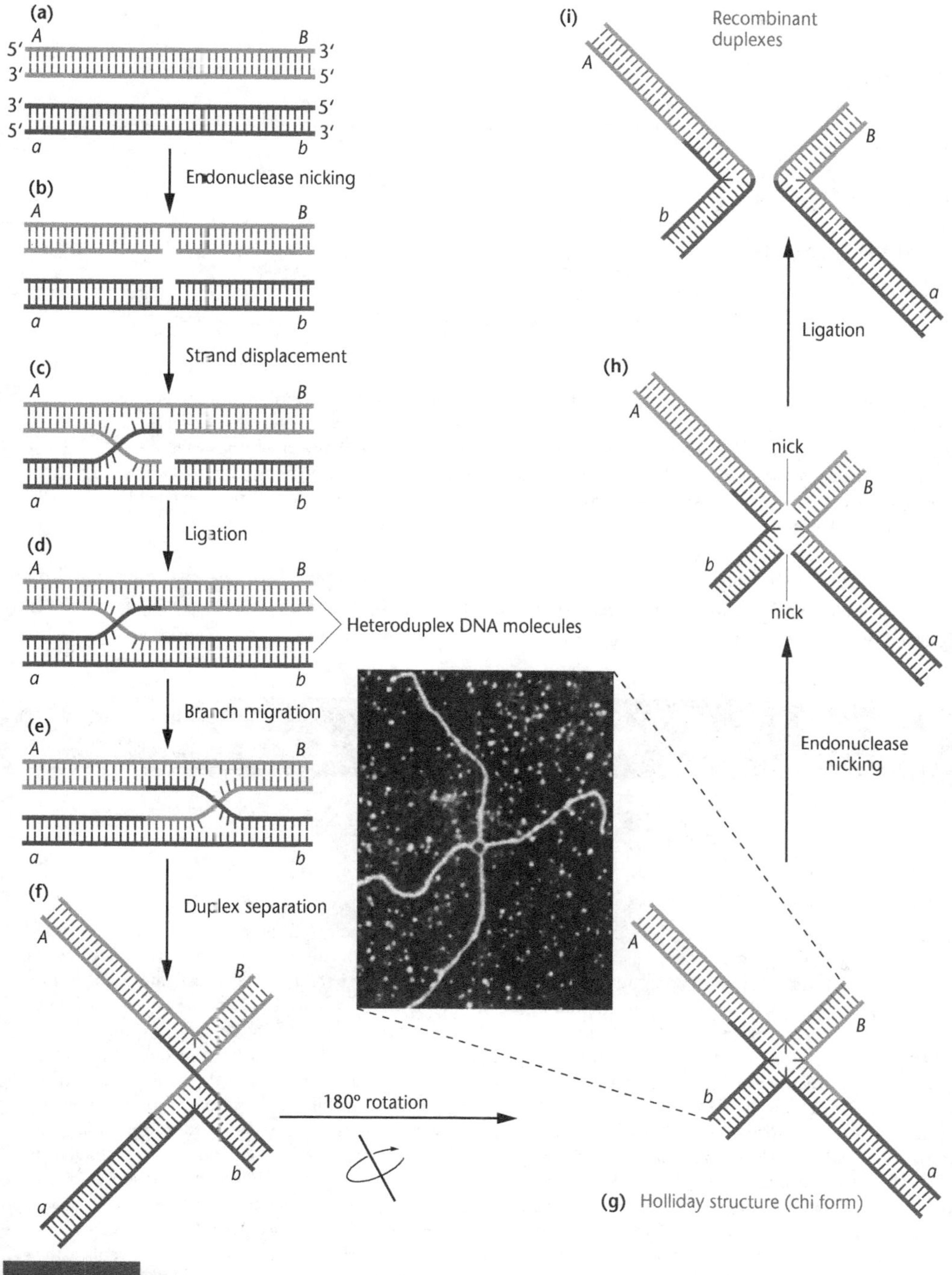


FIGURE 11–18 Model depicting how genetic recombination can occur as a result of the single-strand breaks and subsequent rejoining of heterologous DNA strands. Each stage is described in the text. The electron micrograph shows DNA in a $\chi$-form structure similar to the diagram in (g); the DNA is an extended Holliday structure derived from the *ColE1* plasmid of *E. coli*. *David Dressler, Oxford University, England*

*Neurospora* by Mary Mitchell, gene conversion is characterized by a *nonreciprocal* genetic exchange between two closely linked genes. For example, if we were to cross two *Neurospora* strains, each bearing a separate mutation ($a+ \times +b$), a *reciprocal* recombination between the genes would yield spore pairs of the ++ and the *ab* genotypes. However, a nonreciprocal exhange yields one pair without the other. Working with pyridoxine mutants, Mitchell observed several asci-containing spore pairs with the ++ genotype but not the reciprocal product (*ab*). Because the frequency of these events was higher than the predicted mutation rate and consequently could not be accounted for by mutation, they were called "gene conversions." They were so named because it appeared that one allele had somehow been "converted" to another in which genetic exchange had also occurred. Similar findings come from studies of other fungi as well.

Gene conversion is now considered to be a consequence of the process of DNA recombination. One possible explanation interprets conversion as resulting from a mismatch between bases paired during heteroduplex formation, as shown in Figure 11–19. Mismatched regions of hybrid strands can be repaired by the excision of one of the strands and the synthesis of the complement by using the remaining strand as a template. Excision may occur in either one of the strands, yielding two possible "corrections." One repairs the mismatched base pair by restoring the original sequence. The other also corrects the mismatch, but does so by copying the mutant strand, creating a base-pair substitution. Conversion may have the effect of creating identical alleles on the two homologs that were different initially.

In our example in Figure 11–19, suppose that the G≡C pair on one of the two homologs was responsible for the mutant allele, while the A=T pair was part of the wild-type gene sequence on the other homolog. Conversion of the G≡C pair to A=T would have the effect of changing the mutant allele to wild type, just as Mitchell originally observed.

Gene-conversion events have helped to explain other puzzling genetic phenomena in fungi. For example, when mutant and

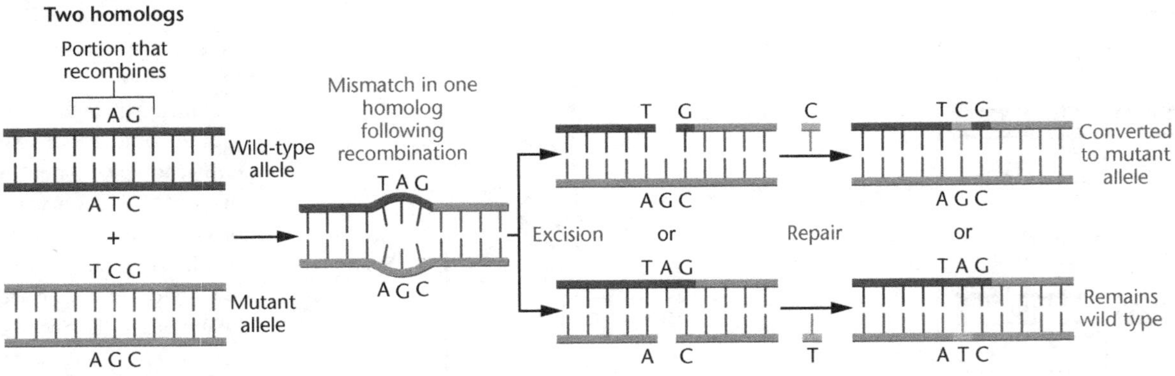


**FIGURE 11–19** A proposed mechanism that accounts for the phenomenon of gene conversion during recombination in meiosis. A base-pair mismatch occurs in a recombining homolog (owing to the presence of a mutant allele) during heteroduplex formation. During excision repair, one of the two mismatched bases is removed and the complement is synthesized. In one case, the mutant base pair is preserved. When it is subsequently included in a recombinant spore, the mutant genotype will be maintained. In the other case, the mutant base pair is converted to the wild-type sequence. When included in a recombinant spore, the wild-type genotype will be expressed, leading to a nonreciprocal exchange ratio.

wild-type alleles of a single gene are crossed, asci should yield equal numbers of mutant and wild-type spores. However, exceptional asci with 3:1 or 1:3 ratios are sometimes observed. These ratios can be understood in terms of gene conversion. The phenomenon also has been detected during mitotic events in fungi, as well as in the study of unique compound chromosomes in *Drosophila*.

## GENETICS, TECHNOLOGY, AND SOCIETY

### Telomeres: Defining the End of the Line?

Humans, like all multicellular organisms, grow old and die. As we age, our immune systems become less efficient, wound healing is impaired, and tissues and organs lose resilience. It has always been a mystery why we go through these age-related declines and why each species has a characteristic finite life span. Why do we grow old? Can we reverse this march to mortality? Some recent discoveries suggest that the answers to these questions may lie at the ends of our chromosomes.

The study of human aging begins with a study of human cells growing in culture dishes. Like the organisms from which the cells are taken, cells in culture have a finite life span. This "*replicative senescence*" was noted as early as the 1960s by Leonard Hayflick. He reported that average human fibroblasts lose their ability to grow and divide after about 50 cell divisions. These senescent cells remain metabolically active but can no longer proliferate. Eventually, they die. Although we don't know whether cellular senescence directly causes organismal aging, the evidence is suggestive. For example, cells from young people go through more divisions in culture than do cells from older people; human fetal cells divide 60–80 times before exhibiting senescence, whereas cells from older adults divide only 10–20 times. In addition, cells from species with short life spans stop growing after fewer divisions than do cells from species with longer life spans; mouse cells divide 11–15 times in culture, but tortoise cells undergo over 100 divisions. Moreover, cells from patients with genetic premature aging syndromes (such as Werner syndrome) undergo fewer divisions in culture than do cells from normal patients.

Another characteristic of aging cells is that their telomeres become shorter. This chapter described telomeres as the tips of linear chromosomes, consisting of several thousand repeats of a short DNA sequence (TTAGGG in humans). These DNA sequences interact with a complex of proteins, the *telosome,* to create a loop structure that helps protect chromosome ends from degrading or from fusing to other chromosomes.

In most mammalian somatic cells, telomeres become shorter with each DNA replication because DNA polymerase cannot synthesize new DNA at the 3′ ends of each parent strand. In certain populations of human cells that undergo extensive proliferation, like embryonic cells, germ cells, and tumors, telomere length is maintained by *telomerase*—a remarkable RNA-containing enzyme that adds telomeric DNA sequences onto the ends of linear chromosomes. However, most somatic cells contain little, if any, active telomerase. Their progressively shortening telomeres eventually trigger replication-related senescence in these cells.

Could we gain perpetual youth and vitality by increasing our telomere lengths? Studies suggest that it may be possible to reverse senescence by artificially increasing the amount of telomerase in our cells. When investigators introduce cloned telomerase genes into some types of normal human cells in culture, telomeres lengthen by thousands of base pairs, and the cells continue to grow long past their typical senescence point.

These observations suggest that telomere integrity acts as a cellular clock. In addition, they suggest that some of the atrophy of tissues that accompanies old age may someday be reversed by activating telomerase genes. However, before we rush out to buy telomerase pills, we must consider a possible corollary of cellular immortality: cancer.

Although normal cells undergo senescence after a specific number of cell divisions, cancer cells do not. It is thought that cancers arise after several genetic mutations accumulate in a cell. These mutations disrupt the normal checks and balances that control cell growth, division, and senescence. For example, cells with mutations in *p53* or *Rb* gene pathways (tumor suppressors) do not senesce. As these cells divide beyond their normal senescence point, their telomeres dwindle, and a state called "crisis" is reached. Most cells in crisis undergo cell death; however, a small proportion ($1/10^7$) may overcome the crisis state, to achieve immortality. These immortal cells have gained mechanisms for stably maintaining their telomeres. More than 80 percent of immortal human tumor cells contain telomerase activity, and the others use a less well understood mechanism known as ALT (for "alternative lengthening of telomeres").

The correlation between uncontrolled tumor cell growth and the presence of telomerase activity has led to the use of telomerase assays as diagnostic markers for cancer. Another offshoot of this correlation is the attempt to target telomerase in anticancer treatments. Agents that inhibit telomerase might destroy cancer cells by allowing telomeres to shorten, thereby forcing the cells into senescence. Because most normal human cells do not express telomerase, such a therapy might be relatively specific for tumor cells and hence less toxic than most current anticancer drugs. Many such anti-telomerase anticancer treatments are currently under development. Although it is too soon to determine whether this approach can be successful in animals or humans, it appears to work in cultured tumor cells.

Several uncertainties concerning these anti-telomerase treatments remain. Telomerase is thought to be required by some normal human cells such as lymphocytes and germ cells. If not carefully delivered, anti-telomerase drugs may be unacceptably toxic. Even if we inhibit telomerase activity in tumor cells, they may undergo so many divisions before reaching senescence that the tumor may still damage the host. Some cancer cells may compensate for the loss of telomerase by using other telomere-lengthening mechanisms, such as ALT. A newer line of attack hinges on the importance of telomere loop structures. Studies show that agents that interfere with the formation of normal telomere loop structures cause the death of cultured cancer cells but not normal cells.

Will a deeper understanding of telomeres allow us to both arrest cancers *and* reverse the descent into old age? Time will tell.

### ■ References


Boukamp, P., and Mirancea, N. 2006. Telomeres rather than telomerase a key target for anti-cancer therapy? *Experimental Dermatology* 16: 71–79.

Stewart, S. A., and Weinberg, R. A. 2006. Telomeres: Cancer to human aging. *Annu. Rev. Cell Dev. Biol.* 22: 531–557.


EXPLORING GENOMICS

# Entrez: A Gateway to Genome Resources

One of the challenges involved in trying to stay abreast of new developments in genomics is keeping track of the overwhelming number of useful sites and databases for genome information. In this Exploring Genomics, we turn again to the National Center for Biotechnology Information (NCBI) as a key resource. This time we look at **Entrez, The Life Sciences Search Engine**, a gateway maintained by NCBI for accessing an incredible wealth of genome-related databases. In many ways, Entrez is a "one-stop shop" for connecting to genome resources on the Internet. If you can't find something about a particular genome-related topic using Entrez, the information you are looking for probably isn't available on the Internet! Entrez accesses many of the databases we have visited in previous exercises—PubMed, OMIM, GenBank, and others.

This set of exercises, which explores Entrez to learn more about telomerase and DNA polymerase, will introduce you to Entrez as a great resource for genetic information.

### ■ Exercise I – Telomerase and Telomerase-Associated Proteins

1. Access Entrez at http://www.ncbi.nlm.nih.gov/gquery/gquery.fcgi.
2. Use the "Search across databases" feature to search for information about "telomerase." Visit some of the databases listed below:
   a. See PubMed for links to abstracts of recent publications involving telomerase.
   b. Visit Online Mendelian Inheritance in Man (OMIM) for links to human genetic conditions associated with telomerase gene mutations and alterations in telomerase enzymatic activity.
   c. Use the "Nucleotide" and "Protein" links, which will take you to information on telomerase sequences in different species studied to date.
   d. Explore HomoloGene, which provides a wealth of information on homologous genes for telomerase in different species.

*Continued on next page*

*Exploring Genomics, continued*

3. Search across the database for information about the human protein POT1. What information did you find about this protein?

■ **Exercise II – DNA Polymerase**

1. Search Entrez for "DNA polymerase" and you will find an even larger set of information than was available for telomerase. Explore the site to visit links that interest you.

2. Be sure to visit the "PubChem Compound" link. Explore some of the information on the compounds presented in this database. Why do you think a search for DNA polymerase leads to these compounds? For what purposes are many of them used?

3. As we discussed in this chapter, DNA polymerases in both prokaryotes and eukaryotes are complex, multi-subunit enzymes. Search Entrez resources to see if you can find the amino acid sequence for the human DNA polymerase lambda ($\lambda$) subunit. How many amino acids are in the $\lambda$ subunit? How does the size of the $\lambda$ subunit in humans compare with the size of the $\lambda$ subunit in chimpanzees (*Pan troglodytes*)?

## Chapter Summary

1. Researchers had to discover which of three potential modes of DNA replication—semiconservative, conservative, or dispersive—is the process by which DNA replication actually occurs. Although all three modes depend on the principle of base complementarity, semiconservative replication is the most straightforward and was predicted by Watson and Crick.
2. In 1958, Meselson and Stahl resolved this question in favor of semiconservative replication in *E. coli*, showing that newly synthesized DNA consists of one old strand and one new strand. Taylor, Woods, and Hughes used the root tips of the broad bean to demonstrate semiconservative replication in eukaryotes.
3. During the same period, Kornberg isolated the enzyme DNA polymerase I from *E. coli* and showed that it is capable of directing *in vitro* DNA synthesis, provided that a template and precursor nucleoside triphosphates are supplied.
4. The subsequent discovery of the *polA1* mutant strain of *E. coli*, capable of DNA replication despite its lack of polymerase I activity, cast doubt on the enzyme's hypothesized *in vivo* replicative function. DNA polymerases II and III were then isolated. Polymerase III has been identified as the enzyme responsible for DNA replication *in vivo*.
5. During the process of DNA synthesis, the double helix unwinds, forming a replication fork at which synthesis begins. Proteins stabilize the unwound helix and assist in relaxing the coiling tension created ahead of the replication fork.
6. Synthesis is initiated at specific sites along each template strand by the enzyme primase, resulting in short segments of RNA that provide suitable 3′ ends upon which DNA polymerase III can begin polymerization.
7. Because the strands are antiparallel in the double helix, polymerase III synthesizes DNA continuously on the leading strand in a 5′–3′ direction. On the opposite strand, called the lagging strand, synthesis results in short Okazaki fragments that are later joined by DNA ligase.
8. DNA polymerase I removes the RNA primer and replaces it with DNA, which is joined to the adjacent polynucleotide by DNA ligase.
9. The isolation of numerous phage and bacterial mutant genes affecting many of the molecules involved in the replication of DNA has helped to define the complex genetic control of the entire process.
10. DNA replication in eukaryotes is similar to, but more complex than, replication in prokaryotes. Multiple replication origins exist, and multiple forms of DNA polymerase direct DNA synthesis.
11. Replication at the ends of linear chromosomes in eukaryotes poses a special problem that can be solved by the presence there of telomeres and by a unique RNA-containing enzyme called telomerase.
12. Homologous recombination between DNA molecules relies on a series of enzymes that can cut, realign, and reseal DNA strands. The phenomenon of gene conversion may be best explained in terms of mismatch repair synthesis following these exchanges.

## INSIGHTS AND SOLUTIONS

1. Predict the theoretical results of conservative and dispersive replication of DNA under the conditions of the Meselson–Stahl experiment. Follow the results through two generations of replication after cells have been shifted to a $^{14}N$-containing medium, using the following sedimentation pattern.

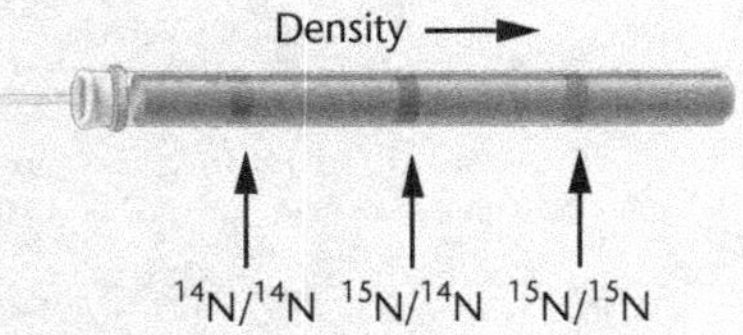


**Solution:**

Conservative replication

Generation I Generation II

Dispersive replication

Generation I Generation II

2. Mutations in the *dnaA* gene of *E. coli* are lethal and can only be studied following the isolation of conditional, temperature-sensitive mutations. Such mutant strains grow nicely and replicate their DNA at the permissive temperature of 18°C, but they do not grow or replicate their DNA at the restrictive temperature of 37°C. Two observations were useful in determining the function of the DnaA protein product. First, *in vitro* studies using DNA templates that have unwound do not require the DnaA protein. Second, if intact cells are grown at 18°C and are then shifted to a 37°C environment, DNA synthesis continues at the new temperature until one round of replication is completed, and then it stops. What do these observations suggest about the role of the *dnaA* gene product?

**Solution:** At 18°C (the permissive temperature), the mutation is not expressed and DNA synthesis begins. Following the shift to the restrictive temperature, the already-initiated DNA synthesis continues, but no new synthesis can begin. If the DnaA protein is not required for synthesis of unwound DNA and is not required after synthesis has been initiated, then the DnaA protein's essential role in DNA synthesis *in vivo* must be an interaction of some kind with the intact helix to facilitate the localized denaturation necessary for synthesis to proceed.

## Problems and Discussion Questions

1. Compare conservative, semiconservative, and dispersive modes of DNA replication.
2. Describe the role of $^{15}$N in the Meselson–Stahl experiment.
3. In the Meselson–Stahl experiment, which of the three modes of replication could be ruled out after one round of replication? after two rounds?
4. Predict the results of the experiment by Taylor, Woods, and Hughes if replication were (a) conservative and (b) dispersive.
5. Reconsider Problem 36 in Chapter 10. In the model you proposed, could the molecule be replicated semiconservatively? Why? Would other modes of replication work?
6. What are the requirements for *in vitro* synthesis of DNA under the direction of DNA polymerase I?
7. Of Kornberg's initial experiments, conducted at Washington University in St. Louis, it was rumored (inaccurately, as it turned out) that he grew *E. coli* in large Anheuser-Busch beer vats. Why do you think the use of beer vats might have been helpful to his research effort involving polymerase I?
8. How did Kornberg test the fidelity of copying DNA by polymerase I?
9. Which of Kornberg's tests is the more stringent assay? Why?
10. Which characteristics of DNA polymerase I raised doubts that its *in vivo* function is the synthesis of DNA leading to complete replication?
11. Explain the theory of nearest-neighbor frequency.
12. What is meant by "biologically active" DNA?
13. Why was the phage $\phi$X174 chosen for the experiment demonstrating biological activity?
14. Outline the experimental design of Kornberg's biological activity demonstration.
15. What was the significance of the *polA1* mutation?
16. Summarize and compare the properties of DNA polymerase I, II, and III.
17. List and describe the function of the 10 subunits constituting DNA polymerase III. Distinguish between the holoenzyme and the core enzyme.
18. Distinguish between (a) unidirectional and bidirectional synthesis and (b) continuous and discontinuous synthesis of DNA.
19. List the proteins that unwind DNA during *in vivo* DNA synthesis. How do they function?
20. Define and indicate the significance of (a) Okazaki fragments, (b) DNA ligase, and (c) primer RNA during DNA replication.
21. Outline the current model for DNA synthesis.
22. Why is DNA synthesis expected to be more complex in eukaryotes than in bacteria? How is DNA synthesis similar in the two types of organisms?
23. If an analysis of DNA from two different microorganisms demonstrates very similar base compositions, are the DNA sequences of the two organisms also nearly identical?
24. Suppose that *E. coli* synthesizes DNA at a rate of 100,000 nucleotides per minute and takes 40 minutes to replicate its chromosome.
    (a) How many base pairs are present in the entire *E. coli* chromosome?
    (b) What is the physical length of the chromosome in its helical configuration—that is, what is the circumference of the chromosome if it were opened into a circle?
25. Each of several different temperature-sensitive mutant strains of *E. coli* displays a different problem in synthesizing DNA. Predict what enzyme or function is being affected by each mutation.
    (a) Newly synthesized DNA contains many mismatched base pairs.
    (b) Okazaki fragments accumulate, and DNA synthesis is never completed.
    (c) No initiation occurs.
    (d) Synthesis is very slow.
    (e) Strands remain in a supercoiled state following initiation of replication, which is never completed.
26. Define gene conversion and describe how this phenomenon is related to genetic recombination.
27. Many of the gene products involved in DNA synthesis were initially defined by studying mutant *E. coli* strains that could not synthesize DNA.
    (a) The *dnaE* gene encodes the $\alpha$ subunit of DNA polymerase III. What effect is expected from a mutation in this gene? How could the mutant strain be maintained?
    (b) The *dnaQ* gene encodes the $\epsilon$ subunit of DNA polymerase. What effect is expected from a mutation in this gene?
28. In 1994, telomerase activity was discovered in human cancer cells in culture (*in vitro*). Although telomerase is not active in human somatic tissue, this discovery indicated that humans do contain the genes for telomerase proteins and telomerase RNA. Since inappropriate activation of telomerase can cause cancer, why do you think the genes coding for this enzyme have been maintained in the human genome throughout evolution? Are there any types of human body cells where telomerase activation would be advantageous or even necessary? Explain.

**HOW DO WE KNOW?**

29. In this chapter, we focused on DNA replication and the synthetic mechanisms that achieve it. We have also considered DNA recombination. Along the way, we found many opportunities to consider the methods and reasoning by which much of this information was acquired. From

the explanations given in the chapter, what answers would you propose to the following fundamental questions:

(a) What is the experimental basis for concluding that DNA replicates semiconservatively in both prokaryotes and eukaryotes?

(b) How was it determined that *in vivo* DNA synthesis in bacteria occurs under the direction of DNA polymerase III and not DNA polymerase I?

(c) How do we know that *in vivo* DNA synthesis occurs in the 5′ to 3′ direction?

(d) How was it discovered that DNA synthesis is discontinuous on one of the two template strands?

(e) What observations reveal the "telomere problem" posed by eukaryotic DNA replication, and how did we learn of the solution to this problem?

## Extra-Spicy Problems

30. The genome of the fruit fly *Drosophila melanogaster* consists of approximately $1.7 \times 10^8$ base pairs. DNA synthesis occurs at a rate of 30 base pairs per second. In the early embryo, the entire genome is replicated in five minutes. How many bidirectional origins of synthesis are required to accomplish this feat?

31. Assume a hypothetical organism in which DNA replication is conservative. Design an experiment similar to that of Taylor, Woods, and Hughes that will unequivocally establish this fact. Using the format established in Figure 11–5, draw sister chromatids and illustrate the expected results confirming this mode of replication.

32. DNA polymerases in all organisms only add 5′ nucleotides to the 3′ end of a growing DNA strand, never to the 5′ end. One possible reason for this is the fact that most DNA polymerases have a proofreading function that would not be *energetically* possible if DNA synthesis occurred in the 3′ to 5′ direction.

(a) Draw chemical structures showing the reaction that DNA polymerase would have to catalyze if DNA synthesis occurred in the 3′ to 5′ direction.

(b) Considering the information present in your drawing, speculate as to why proofreading would be problematic, while it isn't if synthesis occurs in the 5′ to 3′ direction.

33. Scientists investigated an alien organism that displayed the characteristics of eukaryotes. When DNA replication was examined, two unique features were apparent: (1) no Okazaki fragments were observed; and (2) there was a telomere problem; that is, telomeres shortened, but only on one end of the chromosome. Put forward a model of DNA that is consistent with both of these observations.

34. Assume that the sequence of bases given below is present on one nucleotide chain of a DNA duplex and that the chain has opened up at a replication fork. Synthesis of an RNA primer occurs on this template starting at the base that is underlined. (a) If the RNA primer consists of eight nucleotides, what is its base sequence? (b) In the intact RNA primer, which nucleotide has a free 3′-OH terminus?

**3′. . . .GGCTACCTGGATTCA. . . .5′**

35. Consider the following diagram. Assume that the phase G1 chromosome on the left underwent one round of replication in the presence of $^3$H-thymidine and that the metaphase chromosome on the right had both chromatids labeled. Which of the replicative models (conservative, dispersive, semiconservative) could be eliminated by this observation?

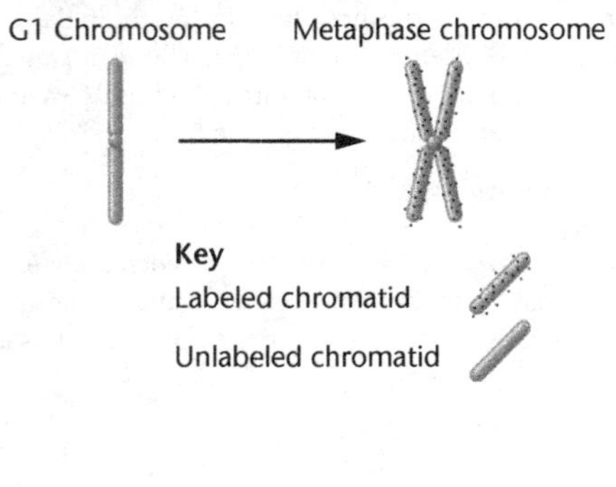


36. Consider the drawing of a dinucleotide below. (a) Is it DNA or RNA? (b) Is the arrow closest to the 5′ or the 3′ end? (c) Suppose that the molecule was cleaved with the enzyme spleen diesterase, which breaks the covalent bond connecting the phosphate to C-5′. After cleavage, to which nucleoside is the phosphate now attached (A or T)?

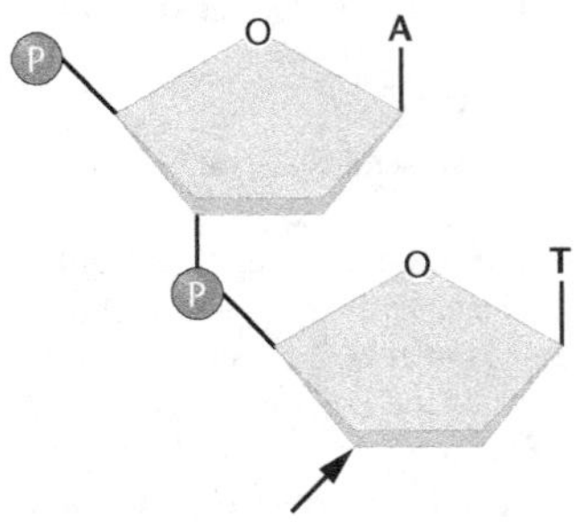


37. DNA is allowed to replicate in moderately radioactive $^3$H-thymidine for several minutes and is then switched to a highly radioactive medium for several more minutes. Synthesis is stopped, and the DNA is subjected to autoradiography and electron microscopy. On the basis of the drawing of the electron micrograph presented here, interpret as much as you can regarding DNA replication.

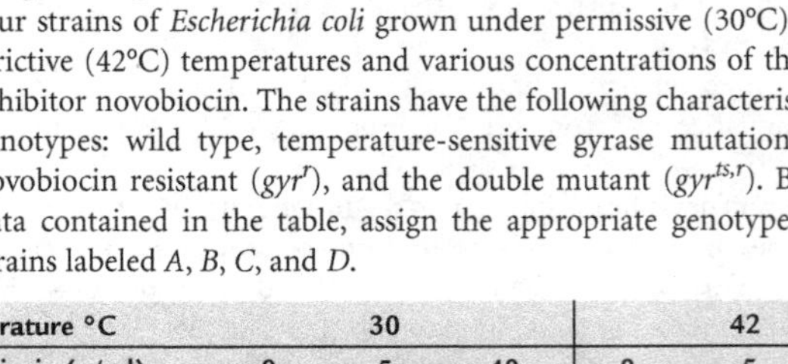

38. The following table (data adapted from Khodursky et al., 2000) presents the percentage of DNA synthesis after 15 minutes from initiation for four strains of *Escherichia coli* grown under permissive (30°C) and restrictive (42°C) temperatures and various concentrations of the gyrase inhibitor novobiocin. The strains have the following characteristics and genotypes: wild type, temperature-sensitive gyrase mutation ($gyr^{ts}$), novobiocin resistant ($gyr^r$), and the double mutant ($gyr^{ts,r}$). Based on data contained in the table, assign the appropriate genotypes to the strains labeled *A*, *B*, *C*, and *D*.

| Temperature °C | 30 | | | 42 | | |
|---|---|---|---|---|---|---|
| Novobiocin (g/ml) | 0 | 5 | 40 | 0 | 5 | 40 |
| Bacterial Strain | | | | | | |
| A | 100 | 56 | 9 | 40 | 26 | 1 |
| B | 100 | 100 | 100 | 41 | 43 | 40 |
| C | 100 | 58 | 4 | 100 | 56 | 3 |
| D | 100 | 100 | 100 | 100 | 100 | 100 |

39. Numerous comparisons have been made between the DNA replicative processes of prokaryotes and eukaryotes, noting differences as well as similarities. A third domain of living organisms, the Archaea, have genomic replicative processes that are an interesting blend of both. Matsunaga et al. (2003) report that the hyperthermophilic archaean *Pyrococcus abyssi* replicates its circular chromosome from a single origin, using RNA-primed replication intermediates. Many of its replicative proteins are eukaryotic-like, while the rate of synthesis is prokaryotic-like. The Okazaki fragments of *P. abyssi* are approximately 100 nucleotides in length with an RNA primer of 10–30 bases. Given this information, present a table that depicts this species as being more like either eukaryotes or prokaryotes, or equally similar to both.

40. To gauge the fidelity of DNA synthesis, Arthur Kornberg and colleagues devised the nearest-neighbor frequency test, which determines the frequency with which any two bases occur adjacent to each other along the polynucleotide chain (*J. Biol. Chem.* 236: 864–875). This test relies on the enzyme spleen phosphodiesterase. As we saw in Figure 11–8, DNA is synthesized by polymerization of 5′-nucleotides—that is, each nucleotide is added with the phosphate on the C-5′ of deoxyribose. However, as shown in the accompanying figure, the phosphodiesterase enzyme cleaves DNA between the phosphate and the C-5′ atom, thereby producing 3′-nucleotides. In this test the phosphates on only one of the four nucleotide precursors of DNA (cytidylic acid, for example) are made radioactive with $^{32}P$, and DNA is synthesized. Then the DNA is subjected to enzymatic cleavage, in which the radioactive phosphate is transferred to the base that is the "nearest neighbor" on the 5′ side of all cytidylic acid nucleotides.

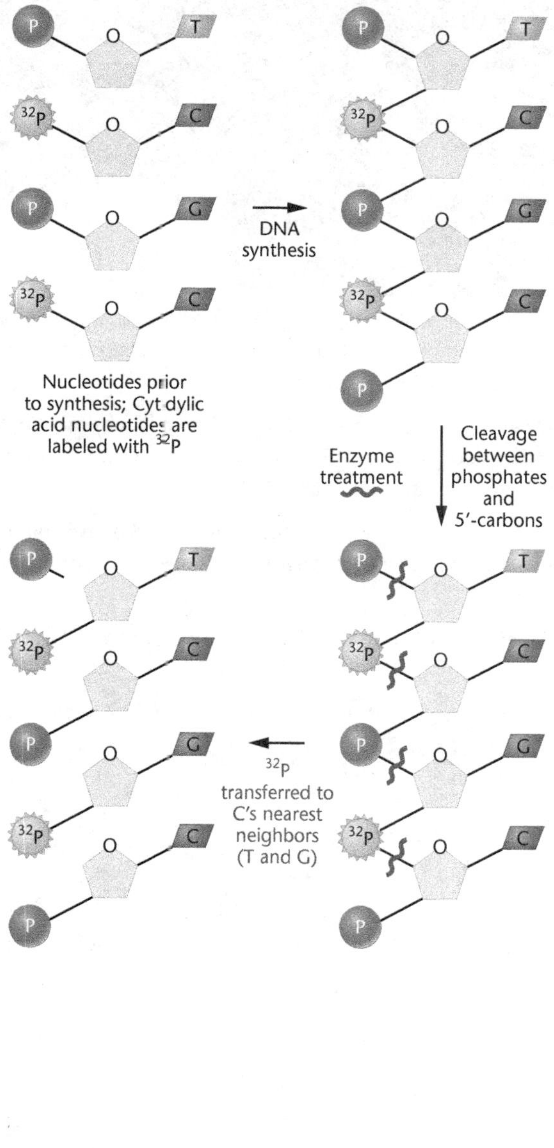


Following four separate experiments, in each of which a different one of the four nucleotide types is radioactive, the frequency of all 16 possible nearest neighbors can be calculated. When Kornberg applied the nearest-neighbor frequency test to the DNA template and resultant product from a variety of experiments, he found general agreement between the nearest-neighbor frequencies of the two.

Analysis of nearest-neighbor data led Kornberg to conclude that the two strands of the double helix are in opposite polarity to one another. Demonstrate this approach by determining the outcome of such an analysis if the strands of DNA shown here are (a) antiparallel versus (b) parallel:

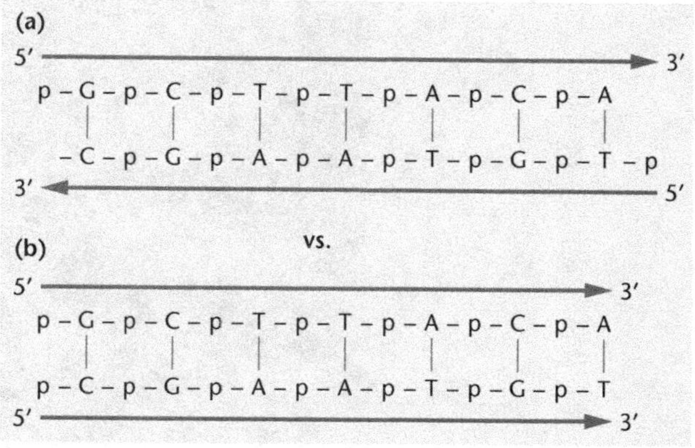

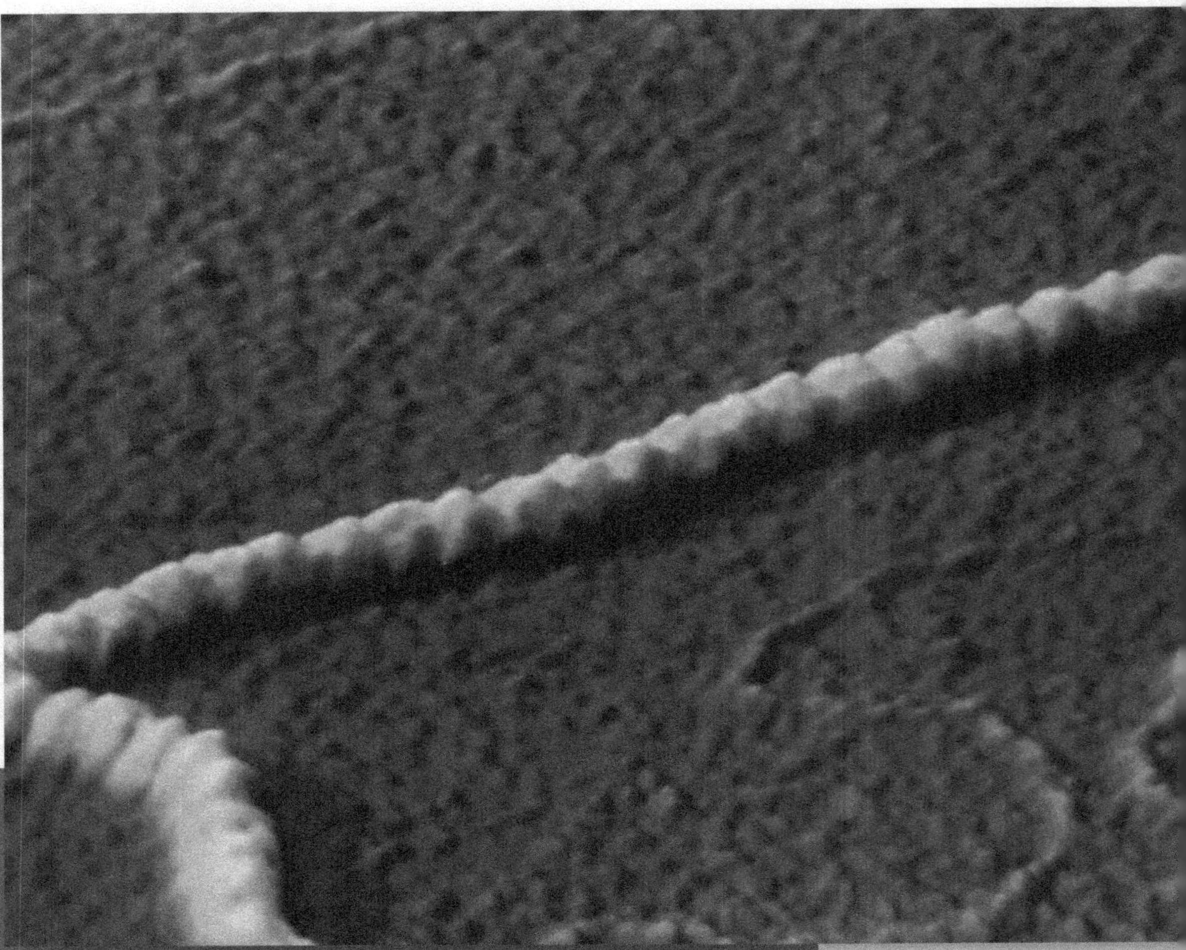

A chromatin fiber viewed using a scanning transmission electron microscope (STEM).

12

# DNA Organization in Chromosomes

## CHAPTER CONCEPTS

- Genetic information in viruses, bacteria, mitochondria, and chloroplasts is most often contained in a short, circular DNA molecule relatively free of associated proteins.
- Eukaryotic cells, in contrast to viruses and bacteria, contain large amounts of DNA organized into nucleosomes and present during most of the cell cycle as chromatin fibers.
- The uncoiled chromatin fibers characteristic of interphase coil up and condense into chromosomes during the stages of eukaryotic cell division.
- Eukaryotic genomes contain both unique and repetitive DNA sequences.
- Eukaryotic genomes consist mostly of noncoding DNA sequences.

Once geneticists understood that DNA houses genetic information, they focused their energies on discovering how DNA is organized into genes and how these basic units of genetic function are organized into chromosomes. In short, the next major questions they tackled had to do with how the genetic material is organized within the genome. These issues have a bearing on many areas of genetic inquiry. For example, the ways in which genomic organization varies in different organisms—from viruses to bacteria to eukaryotes—will undoubtedly provide a better understanding of the evolution of organisms on Earth.

In this chapter, we focus on the various ways DNA is organized into chromosomes. These structures have been studied using numerous techniques, instruments, and approaches, including analysis by light microscopy and electron microscopy. More recently, molecular analysis has provided significant insights into chromosome organization. In the first half of the chapter, after surveying what we know about chromosomes in viruses and bacteria, we examine the large specialized eukaryotic structures called polytene and lampbrush chromosomes. Then, in the second half, we discuss how eukaryotic chromosomes are organized at the molecular level—for example, how DNA is complexed with proteins to form chromatin, and how the chromatin fibers characteristic of interphase are condensed into chromosome structures visible during mitosis and meiosis. We conclude the chapter by examining certain aspects of DNA sequence organization in eukaryotic genomes.

## 12.1 Viral and Bacterial Chromosomes Are Relatively Simple DNA Molecules

The chromosomes of viruses and bacteria are much less complicated than those in eukaryotes. They usually consist of a single nucleic acid molecule quite different from the multiple chromosomes constituting the genome of higher forms. Prokaryotic chromosomes are largely devoid of associated proteins and contain relatively less genetic information. These characteristics have greatly simplified genetic analysis in prokaryotic organisms, and we now have a fairly comprehensive view of the structure of their chromosomes.

The chromosomes of viruses consist of a nucleic acid molecule—either DNA or RNA—that can be either single or double stranded. They can exist as circular structures (covalently closed circles), or they can take the form of linear molecules. The single-stranded DNA of the **$\phi$X174 bacteriophage** and the double-stranded DNA of the polyoma virus are closed circles housed within the protein coat of the mature viruses. The **bacteriophage lambda ($\lambda$)**, on the other hand, possesses a linear double-stranded DNA molecule prior to infection, but it closes to form a ring upon infection of the host cell. Still other viruses, such as the T-even series of bacteriophages, have linear double-stranded chromosomes of DNA that do not form circles inside the bacterial host. Thus, circularity is not an absolute requirement for replication in viruses.

Viral nucleic acid molecules have been visualized with the electron microscope. Figure 12–1 shows a mature bacteriophage $\lambda$ and its double-stranded DNA molecule in the circular configuration. One constant feature shared by viruses, bacteria, and eukaryotic cells is the ability to package an exceedingly long DNA molecule into a relatively small volume. In $\lambda$, the DNA is 17 $\mu$m long and must fit into the phage head, which is less than 0.1 $\mu$m on any side. Table 12.1 compares the length of the chromosomes of several viruses with the size of their head structure. In each case, a similar packaging feat must be accomplished. Compare the dimensions given for phage T2 with the micrograph of both the DNA and the viral particle shown in Figure 12–2. Seldom does the space available in the head of a virus exceed the chromosome volume by more than a factor of two. In many cases, almost all space is filled, indicating

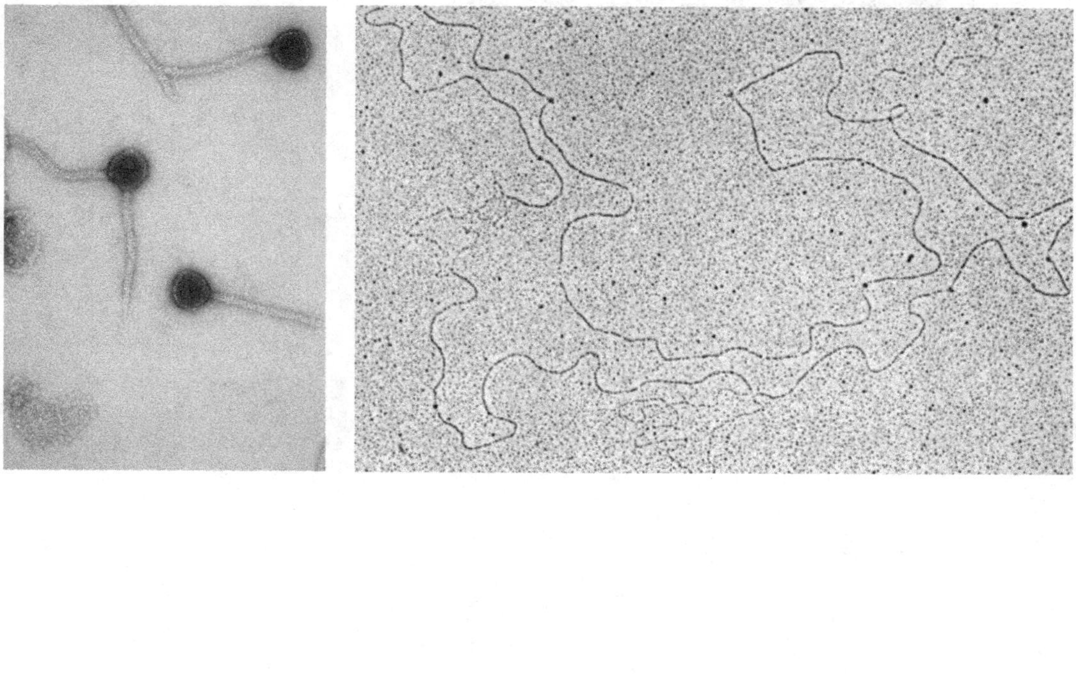


FIGURE 12–1 Electron micrographs of (a) phage $\lambda$ and (b) the DNA isolated from it. The chromosome is 17 $\mu$m long. The phages are magnified about five times more than the DNA.

TABLE 12.1

**The Genetic Material of Representative Viruses and Bacteria**

| Source | | Nucleic Acid Type | SS or DS* | Length ($\mu m$) | Overall Size of Viral Head or Bacterial Cell ($\mu m$) |
|---|---|---|---|---|---|
| Viruses | $\phi$X174 | DNA | SS | 2.0 | 0.025 × 0.025 |
| | Tobacco mosaic virus | RNA | SS | 3.3 | 0.30 × 0.02 |
| | Lambda phage | DNA | DS | 17.0 | 0.07 × 0.07 |
| | T2 phage | DNA | DS | 52.0 | 0.07 × 0.10 |
| Bacteria | *Haemophilus influenzae* | DNA | DS | 832.0 | 1.00 × 0.30 |
| | *Escherichia coli* | DNA | DS | 1200.0 | 2.00 × 0.50 |

*SS = single-stranded; DS = double-stranded

nearly perfect packing. Once packed within the head, the virus's genetic material is functionally inert until it is released into a host cell.

Bacterial chromosomes are also relatively simple in form. They always consist of a double-stranded DNA molecule, compacted into a structure sometimes referred to as the **nucleoid**. *Escherichia coli*, the most extensively studied bacterium, has a large, circular chromosome measuring approximately 1200 $\mu m$ (1.2 mm) in length. When the cell is gently lysed and the chromosome released, the chromosome can be visualized under the electron microscope (Figure 12–3).

DNA in bacterial chromosomes is found to be associated with several types of **DNA-binding proteins**. Two, called **HU** and **H1 proteins**, are small but abundant in the cell and contain a high percentage of positively charged amino acids that can bond ionically to the negative charges of the phosphate groups in DNA. Although these proteins are structurally similar to molecules called *histones*

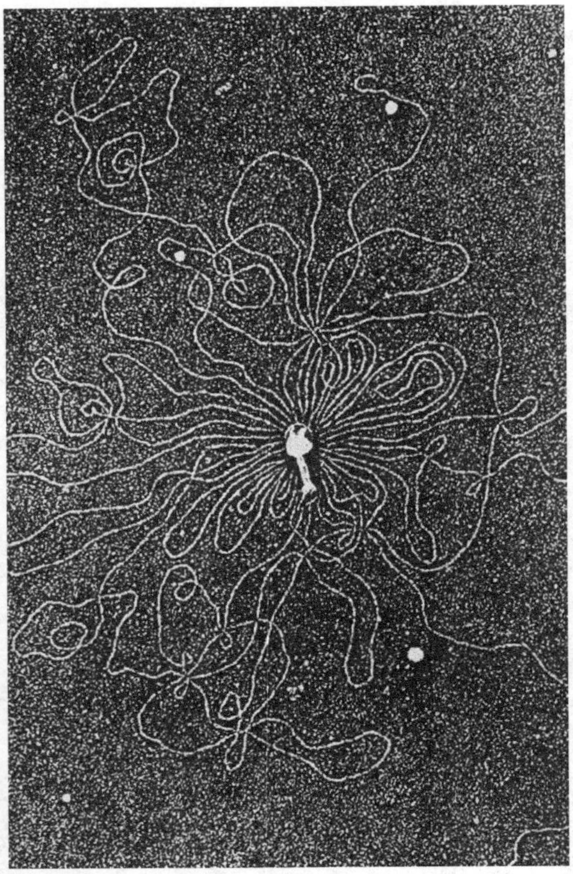

FIGURE 12–2 Electron micrograph of bacteriophage T2, which has had its DNA released by osmotic shock. The chromosome is 52 $\mu m$ long.

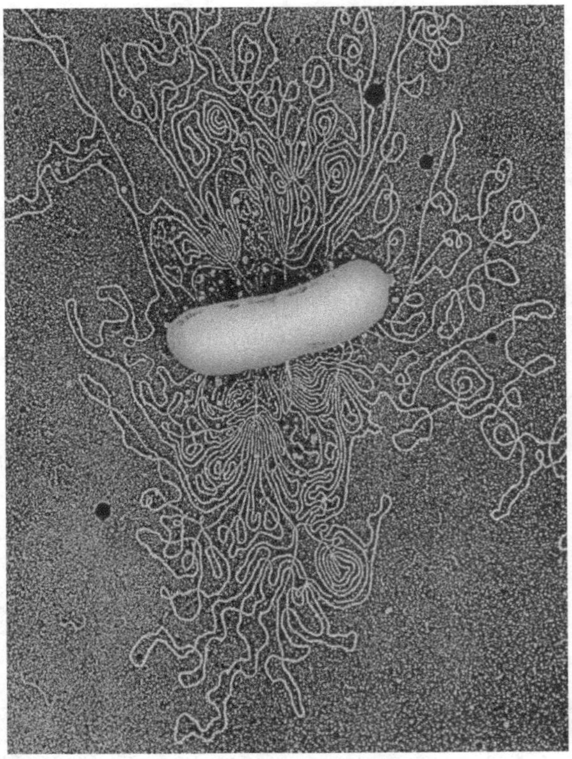

FIGURE 12–3 Electron micrograph of the bacterium *Escherichia coli*, which has had its DNA released by osmotic shock. The chromosome is 1200 $\mu m$ long.

that are associated with eukaryotic DNA (as described in Section 12.4), they are not involved in compacting DNA in a similar way. Unlike the tightly packed chromosome present in the head of a virus, the bacterial chromosome is *not* functionally inert, and can be readily replicated and transcribed.

**NOW SOLVE THIS**

Problem 2 on page 319 involves viral chromosomes that are linear in the bacteriophage but circularize after they enter the bacterial host cell. You are asked to consider the advantages of circular DNA molecules versus linear molecules during replication.

■ HINT: *Recall from Chapter 11 that the enzyme telomerase, absent in bacteria and viruses, is involved in replication in eukaryotes.*

12.2

## Supercoiling Facilitates Compaction of the DNA of Viral and Bacterial Chromosomes

One major insight into the way DNA is organized and packaged in viral and bacterial chromosomes has come from the discovery of **supercoiled DNA**, which is particularly characteristic of closed circular molecules. Supercoiled DNA was first proposed as a result of a study of double-stranded DNA molecules derived from the polyoma virus, which causes tumors in mice. In 1963, it was observed that when the polyoma DNA was subjected to high-speed centrifugation, it was resolved into three distinct components, each of different density and compactness. The one that was least compact, and thus least dense, demonstrated a decreased sedimentation velocity; the other two fractions each showed greater velocities owing to their greater compaction and density. All three were of identical molecular weight.

In 1965, Jerome Vinograd proposed an explanation for these observations. He postulated that the two fractions of greatest sedimentation velocity both consisted of circular DNA molecules, whereas the fraction of lower sedimentation velocity contained linear DNA molecules. Closed circular molecules are more compact and sediment more rapidly than do the same molecules in linear form.

Vinograd proposed further that the denser of the two fractions of circular molecules consisted of covalently closed DNA helices that are slightly *underwound* in comparison to the less dense circular molecules. Energetic forces stabilizing the double helix resist this underwinding, causing the molecule as a whole to **supercoil**, that is, to contort in a certain way, in order to retain normal base pairing. Vinograd proposed that it is the supercoiled shape that causes tighter packing and thus the increase in sedimentation velocity.

The transitions just described are illustrated in Figure 12–4. Consider a double-stranded linear molecule existing in the normal Watson–Crick right-handed helix [Figure 12–4(a)]. This helix contains 20 complete turns, which means the **linking number (*L*)** of this molecule is 20, or $L = 20$. Suppose we were to change this linear molecule into a closed circle by bringing the opposite ends together and joining them [Figure 12–4(b)]. If the closed circle still has a linking number of 20 (if we haven't introduced or eliminated any turns when joining the ends), we define the molecule as being *energetically relaxed*. Now suppose that the circle were subsequently cut open, underwound by two full turns, and then resealed [Figure 12–4(c)]. Such a structure, in which $L$ has been changed to 18, would be *energetically strained* and, as a result, would change its form to relieve the strain.

In order to assume a more energetically favorable conformation, an underwound molecule will form supercoils in the direction opposite to that of the underwinding. In our case [Figure 12–4(d)], two negative supercoils are introduced spontaneously, reestablishing, in total, the original number of turns in the helix. Use of the term *negative* refers to the fact that, by definition, the supercoils are left-handed

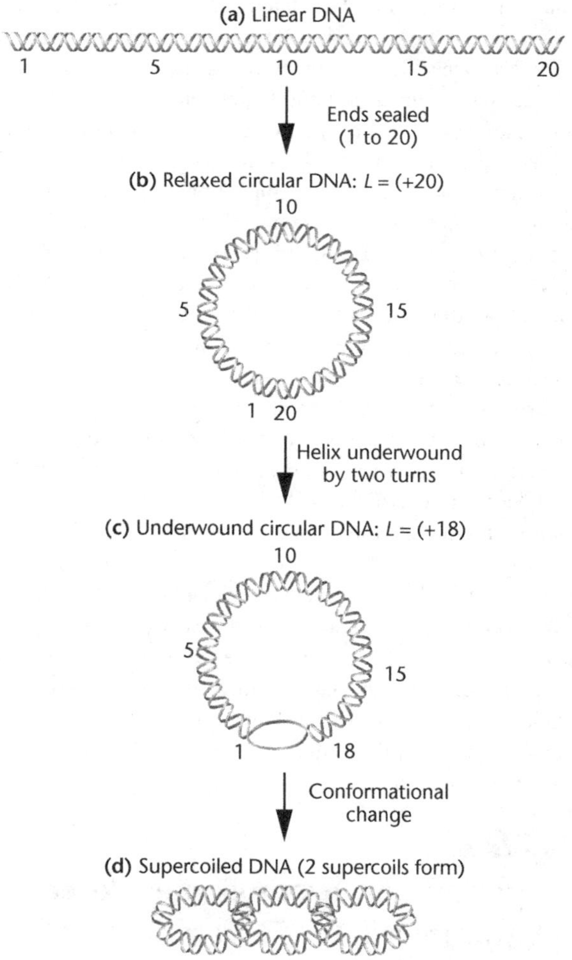


FIGURE 12-4 Depictions of the transformations leading to the supercoiling of circular DNA. $L$ signifies linking number.

(whereas the helix is right-handed). The end result is the formation of a more compact structure with enhanced physical stability.

In most closed circular DNA molecules in bacteria and their phages, the DNA helix is slightly underwound [as in Figure 12–4(c)]. For example, the virus SV40 contains 5200 base pairs. In energetically relaxed DNA, 10.4 base pairs occupy each complete turn of the helix, and the linking number can be calculated as

$$L = \frac{5200}{10.4} = 500$$

However, analysis of circular SV40 DNA reveals that it is underwound by 25 turns, so $L$ is equal to only 475. Predictably, 25 negative supercoils are observed in the molecule. In *E. coli*, an even larger number of supercoils is observed, greatly facilitating chromosome condensation in the nucleoid region.

Two otherwise identical molecules that differ only in their linking number are said to be **topoisomers** of one another. But how can a molecule convert from one topoisomer to the other if there are no free ends, as is the case in closed circles of DNA? Biologically, this may be accomplished by any one of a group of enzymes that cut one or both of the strands and wind or unwind the helix before resealing the ends.

Appropriately, these enzymes are called **topoisomerases.** First discovered by Martin Gellert and James Wang, these catalytic molecules are known as either type I or type II, depending on whether they cleave one or both strands in the helix, respectively. In *E. coli*, topoisomerase I serves to reduce the number of negative supercoils in a closed-circular DNA molecule. Topoisomerase II introduces negative supercoils into DNA. This latter enzyme is thought to bind to DNA, twist it, cleave both strands, and then pass them through the loop that it has created. Once the phosphodiester bonds are reformed, the linking number is decreased and one or more supercoils form spontaneously.

Supercoiled DNA and topoisomerases are also found in eukaryotes. While the chromosomes in these organisms are not usually circular, supercoils can occur when areas of DNA are embedded in a lattice of proteins associated with the chromatin fibers. This association creates "anchored" ends, providing the stability for the maintenance of supercoils once they are introduced by topoisomerases. In both prokaryotes and eukaryotes, DNA replication and transcription creates supercoils downstream as the double helix unwinds and becomes accessible to the appropriate enzyme.

Topoisomerases may play still other genetic roles involving eukaryotic DNA conformational changes. Interestingly, these enzymes are involved in separating (decatenating) the DNA of sister chromatids following replication.

## 12.3 Specialized Chromosomes Reveal Variations in the Organization of DNA

We move next to the consideration of chromosomes found in eukaryotes. Before discussing these at the molecular level, we first introduce two cases of highly specialized chromosomes. Both types, *polytene chromosomes* and *lampbrush chromosomes*, are so large that their organization was discerned using light microscopy long before we understood how mitotic chromosomes form from interphase chromatin. The study of these chromosomes provided many of our initial insights into the arrangement and function of the genetic information.

### Polytene Chromosomes

Giant **polytene chromosomes** are found in various tissues (salivary, midgut, rectal, and malpighian excretory tubules) in the larvae of some flies, as well as in several species of protozoans and plants. They were first observed by E. G. Balbiani in 1881. The large amount of information obtained from studies of these genetic structures provided a model system for subsequent investigations of chromosomes. What is particularly intriguing about polytene chromosomes is that they can be seen in the nuclei of interphase cells.

Each polytene chromosome is 200 to 600 $\mu$m long, and when they are observed under the light microscope, they exhibit a linear series of alternating bands and interbands (Figure 12–5 and 12–6). The banding pattern is distinctive for each chromosome in any given species. Individual bands are sometimes called **chromomeres,** a more generalized term describing lateral condensations of material along the axis of a chromosome.

Extensive study using electron microscopy and radioactive tracers led to an explanation for the unusual appearance of these chromosomes. First, polytene chromosomes represent paired homologs. This in itself is highly unusual, since they are present in somatic cells, where, in most organisms, chromosomal material is normally dispersed as chromatin and homologs are not paired. Second, their large size and distinctive appearance result from their being composed of large numbers of identical DNA strands. The DNA of these paired homologs undergoes many rounds of replication, *but without strand separation or cytoplasmic division.* As replication proceeds, chromosomes are created having 1000 to 5000 DNA strands

FIGURE 12–5 Polytene chromosomes derived from larval salivary gland cells of *Drosophila*.

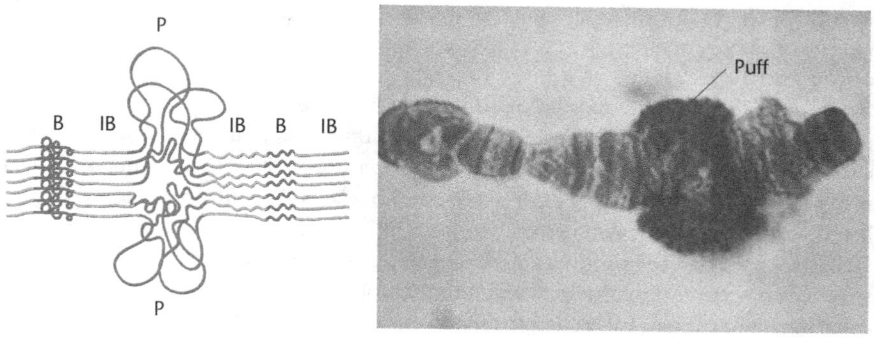


FIGURE 12-6 Photograph of a puff within a polytene chromosome. The diagram depicts the uncoiling of strands within a band region (B) to produce a puff (P) in a polytene chromosome. Each band (B) represents a chromomere. Interband regions (IB) are also labeled.

that remain in precise parallel alignment with one another. Apparently, the parallel register of so many DNA strands gives rise to the distinctive band pattern along the axis of the chromosome.

The presence of bands on polytene chromosomes was initially interpreted as the visible manifestation of individual genes. The discovery that the strands present in bands undergo localized uncoiling during genetic activity further strengthened this view. Each such uncoiling event results in a bulge called a **puff**, so labeled because of its appearance under the microscope (Figure 12–6). That puffs are visible manifestations of a high level of gene activity (transcription that produces RNA) is evidenced by their high rate of incorporation of radioactively labeled RNA precursors, as assayed by autoradiography. Bands that are not extended into puffs incorporate fewer radioactive precursors or none at all.

The study of bands during development in insects such as *Drosophila* and the midge fly *Chironomus* reveals differential gene activity. A characteristic pattern of band formation that is equated with gene activation is observed as development proceeds. Despite attempts to resolve the issue, it is not yet clear how many genes are contained in each band. However, we do know that a band may contain up to $10^7$ base pairs of DNA, certainly enough DNA to encode 50 to 100 average-size genes.

NOW SOLVE THIS

Problem 4 on page 319 involves polytene chromosomes that are cultured in $^3$H-thymidine and subjected to autoradiography. You are asked to predict the pattern of grains that will result.

■ HINT: *$^3$H-thymidine will only be incorporated during the synthesis of DNA.*

## Lampbrush Chromosomes

Another specialized chromosome that has given us insight into chromosomal structure is the **lampbrush chromosome**, so named because it resembles the brushes used to clean kerosene lamp chimneys in the nineteenth century. Lampbrush chromosomes were first discovered in 1892 in the oocytes of sharks and are now known to be characteristic of most vertebrate oocytes, as well as the spermatocytes of some insects. Therefore, they are meiotic chromosomes. Most of the experimental work on them has been done with material taken from amphibian oocytes.

These unique chromosomes are easily isolated from oocytes in the diplotene stage of the first prophase of meiosis, where they are active in directing the metabolic activities of the developing cell. The homologs are seen as synapsed pairs held together by chiasmata. However, instead of condensing, as most meiotic chromosomes do, lampbrush chromosomes are often extended to lengths of 500 to 800 $\mu$m. Later, in meiosis, they revert to their normal length of 15 to 20 $\mu$m. Based on these observations, lampbrush chromosomes are interpreted as being extended, uncoiled versions of the normal meiotic chromosomes.

The two views of lampbrush chromosomes in Figure 12–7 provide significant insights into their morphology. Part (a) shows the

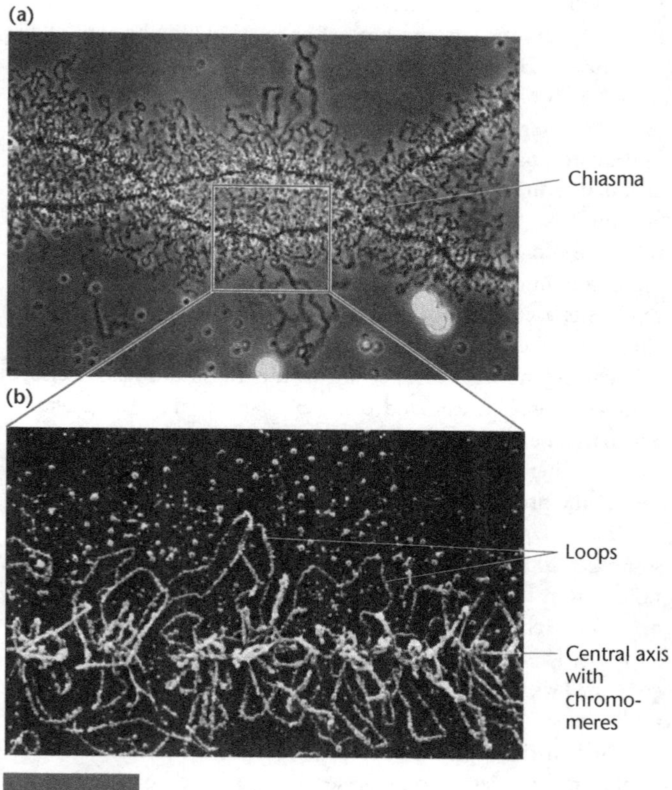


FIGURE 12-7 Lampbrush chromosomes derived from amphibian oocytes. (a) A photomicrograph. (b) A scanning electron micrograph.

meiotic configuration under the light microscope. The linear axis of each horizontal structure seen in the figure contains a large number of condensed areas, which, as with polytene chromosomes, are referred to as *chromomeres*. Emanating from each chromomere is a pair of lateral loops, giving the chromosome its distinctive appearance. In part (b), the scanning electron micrograph (SEM) shows many adjacent pairs of loops in detail along one of the axes. As with bands in polytene chromosomes, much more DNA is present in each loop than is needed to encode a single gene. Such an SEM provides a clear view of the chromomeres and the chromosomal fibers emanating from them. Each chromosomal loop is thought to be composed of one DNA double helix, while the central axis is made up of two DNA helices. This hypothesis is consistent with the belief that each meiotic chromosome is composed of a pair of sister chromatids. Studies using radioactive RNA precursors reveal that the loops are active in the synthesis of RNA. The lampbrush loops, in a manner similar to puffs in polytene chromosomes, represent DNA that has been reeled out from the central chromomere axis during transcription.

12.4

## DNA Is Organized into Chromatin in Eukaryotes

In the following more general discussion of the way DNA is organized in eukaryotic chromosomes, our focus will be on typical eukaryotic cells, in which chromosomes are visible only during mitosis. After chromosome separation and cell division, cells enter the interphase stage of the cell cycle, during which time the components of the chromosome uncoil and are present in the form referred to as **chromatin**. In interphase, the chromatin is dispersed in the nucleus, and the DNA of each chromosome undergoes replication. As the cell cycle progresses, most cells reenter mitosis, whereupon chromatin coils into visible chromosomes once again. This condensation represents a contraction in length of some 10,000 times for each chromatin fiber.

The organization of DNA during the transitions just described is much more intricate and complex than in viruses or bacteria, which never exhibit a complex process similar to mitosis. This is due to the greater amount of DNA per chromosome in eukaryotes, as well as the presence of a large number of proteins associated with eukaryotic DNA. For example, while DNA in the *E. coli* chromosome is 1200 $\mu$m long, the DNA in each human chromosome ranges from 19,000 to 73,000 $\mu$m in length. In a single human nucleus, all 46 chromosomes contain sufficient DNA to extend almost 2 meters. This genetic material, along with its associated proteins, is contained within a nucleus that usually measures about 5 to 10 $\mu$m in diameter.

Such intricacy parallels the structural and biochemical diversity of the many types of cells present in a multicellular eukaryotic organism. Different cells assume specific functions based on highly specific biochemical activity. Although all cells carry a full genetic complement, different cells activate different sets of genes. Clearly, then, a highly ordered regulatory system must exist to govern the use of the genetic information. Such a system must in some way be imposed on or related to the molecular structure of the genetic material.

Because of the limitations of light microscopy, early studies of the structure of eukaryotic genetic material concentrated on intact chromosomes, preferably large ones, such as the polytene and lampbrush chromosomes. Since those days, however, new techniques for biochemical analysis, as well as the ability of the electron microscope to provide images of relatively intact eukaryotic chromatin and mitotic chromosomes, have greatly enhanced our understanding of chromosome structure.

### Chromatin Structure and Nucleosomes

As we have seen, the genetic material of viruses and bacteria consists of strands of DNA or RNA nearly devoid of proteins. In contrast, eukaryotic chromatin has a substantial amount of protein associated with the chromosomal DNA in all phases of the cell cycle. The associated proteins can be categorized as either positively charged **histones** or less positively charged **nonhistones**. Of these two groups of proteins, the histones play the most essential structural role. Histones contain large amounts of the positively charged amino acids lysine and arginine, making it possible for them to bond electrostatically to the negatively charged phosphate groups of nucleotides. Recall that a similar interaction has been proposed for several bacterial proteins. The five main types of histones are shown in Table 12.2.

The general model for chromatin structure is based on the assumption that chromatin fibers, composed of DNA and protein, undergo extensive coiling and folding as they are condensed within the cell nucleus. Moreover, X-ray diffraction studies confirm that histones play an important role in chromatin structure. Chromatin produces regularly spaced diffraction rings, suggesting that repeating structural units occur along the chromatin axis. If the histone molecules are chemically removed from chromatin, the regularity of this diffraction pattern is disrupted.

TABLE 12.2

**Categories and Properties of Histone Proteins**

| Histone Type | Lysine-Arginine Content | Molecular Weight (Da) |
|---|---|---|
| H1 | Lysine-rich | 23,000 |
| H2A | Slightly lysine-rich | 14,000 |
| H2B | Slightly lysine-rich | 13,800 |
| H3 | Arginine-rich | 15,300 |
| H4 | Arginine-rich | 11,300 |

A basic model for chromatin structure was worked out in the mid-1970s. The following observations were particularly relevant to the development of this model:

1. Digestion of chromatin by certain endonucleases, such as micrococcal nuclease, yields DNA fragments that are approximately 200 base pairs in length or multiples thereof. This enzymatic digestion is not random, for if it were, we would expect a wide range of fragment sizes. Thus, chromatin consists of some type of repeating unit, each of which protects the DNA from enzymatic cleavage except where any two units are joined. It is the area between units that is attacked and cleaved by the endonuclease.
2. Electron microscopic observations of chromatin have revealed that chromatin fibers are composed of linear arrays of spherical particles (Figure 12–8). Discovered by Ada and Donald Olins, the particles occur regularly along the axis of a chromatin strand and resemble beads on a string. These particles, initially referred to as $\nu$-bodies ($\nu$ is the Greek letter nu), are now called nucleosomes. These findings conform to the above observation that suggests the existence of repeating units.
3. Studies of the chemical association between histone molecules and DNA in the nucleosomes of chromatin show that histones H2A, H2B, H3, and H4 occur as two types of tetramers, $(H2A)_2 \cdot (H2B)_2$ and $(H3)_2 \cdot (H4)_2$. Roger Kornberg predicted that each repeating nucleosome unit consists of one of each tetramer (creating an octomer) in association with about 200 base pairs of DNA. Such a structure is consistent with previous observations and provides the basis for a model that explains the interaction of histones and DNA in chromatin.
4. When nuclease digestion time is extended, some of the 200 base pairs of DNA are removed from the nucleosome, creating what is called a **nucleosome core particle** consisting of 147 base pairs. The DNA lost in the prolonged digestion is responsible for linking nucleosomes together. This **linker DNA** is associated with the fifth histone, H1.

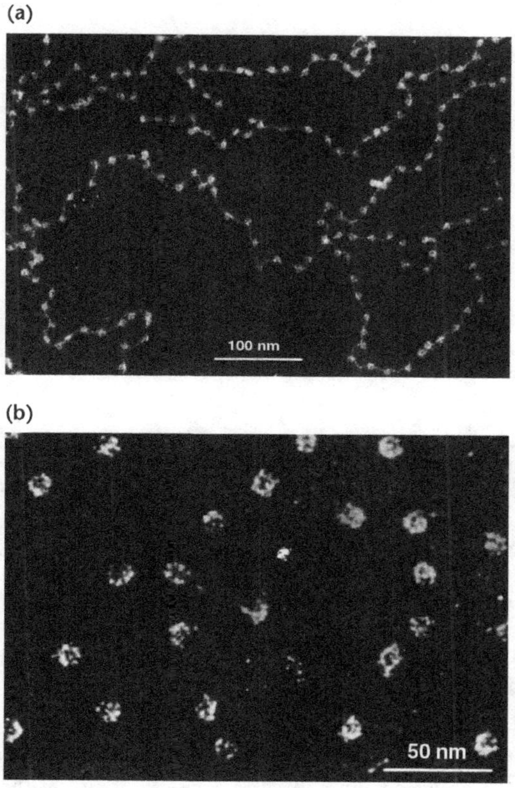


FIGURE 12–8 (a) Dark-field electron micrograph of nucleosomes present in chromatin derived from a chicken erythrocyte nucleus. (b) Dark-field electron micrograph of nucleosomes produced by micrococcal nuclease digestion.

On the basis of this information, as well as on X-ray and neutron-scattering analyses of crystallized core particles by John T. Finch, Aaron Klug, and others, a detailed model of the nucleosome was put forward in 1984, providing a basis for predicting chromatin structure and its condensation into chromosomes. In this model, illustrated in Figure 12–9, a 147-bp length of the 2-nm-diameter DNA molecule coils around an octamer of histones in a left-handed superhelix that completes about 1.7 turns per nucleosome. Each nucleosome, ellipsoidal in shape, measures about 11 nm at its longest point [Figure 12–9(a)]. Significantly, the formation of the nucleosome represents the first level of packing, whereby the DNA helix is reduced to about one-third of its original length by winding around the histones.

In the nucleus, the chromatin fiber seldom, if ever, exists in the extended form described in the previous paragraph (that is, as an extended chain of nucleosomes). Instead, the 11-nm-diameter fiber is further packed into a thicker, 30-nm-diameter structure that was initially called a **solenoid** [Figure 12–9(b)]. This thicker structure consists of numerous nucleosomes coiled closely together, creating the second level of packing. The exact details of the solenoid are not completely clear, but it represents a six-fold increase in compaction of the DNA and is characteristically seen under the electron microscope.

In the transition to the mitotic chromosome, still another level of packing occurs. The 30-nm structure forms a series of looped domains that further condense into the chromatin fiber, which is 300 nm in diameter [Figure 12–9(c)]. The condensed fibers are then coiled into the chromosome arms that constitute a chromatid, one of the longitudinal subunits of the metaphase chromosome [Figure 12–9(d)]. While Figure 12–9 shows the chromatid arms to be 700 nm in diameter, this value undoubtedly varies among different organisms. At a value of 700 nm, a pair of sister chromatids comprising a chromosome measures about 1400 nm.

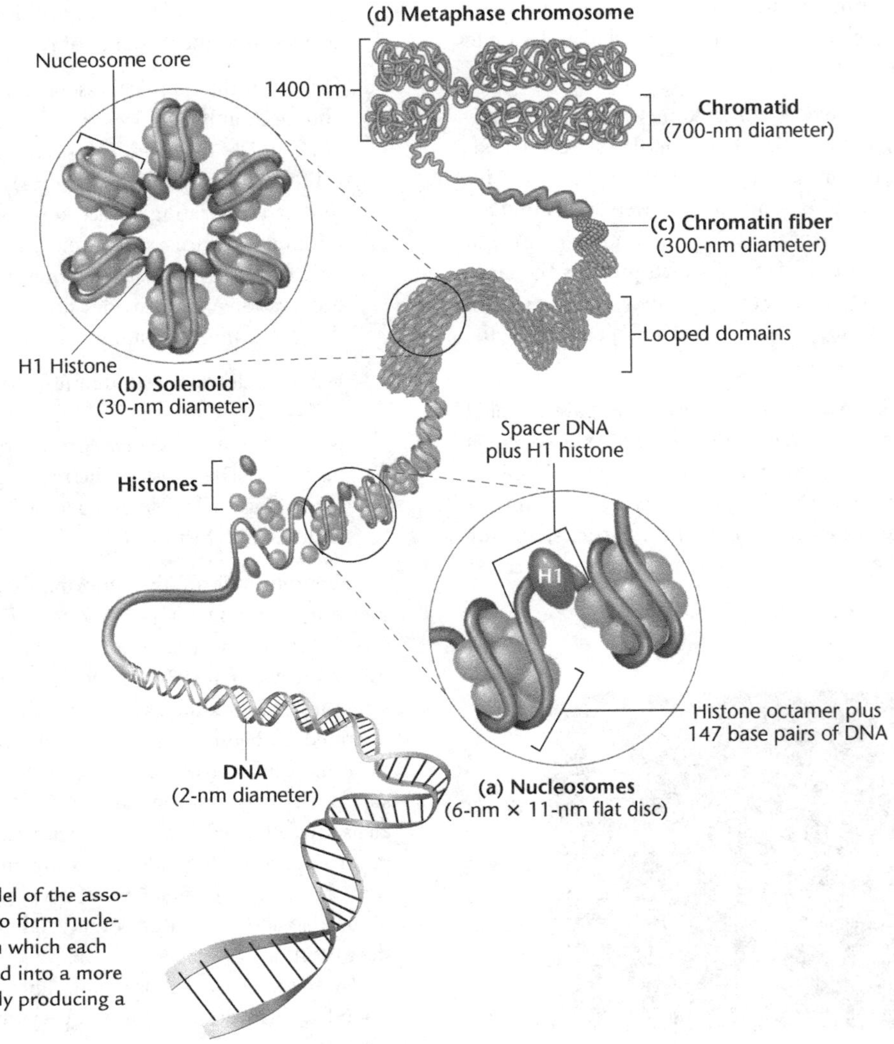


CHROMOSOME STRUCTURE
WEB TUTORIAL 12.1

FIGURE 12–9 General model of the association of histones and DNA to form nucleosomes, illustrating the way in which each thickness of fiber may be coiled into a more condensed structure, ultimately producing a metaphase chromosome.

The importance of the organization of DNA into chromatin and of chromatin into mitotic chromosomes can be illustrated by considering that a human cell stores its genetic material in a nucleus about 5 to 10 $\mu$m in diameter. The haploid genome contains more than 3 billion base pairs of DNA distributed among 23 chromosomes. The diploid cell contains twice that amount. At 0.34 nm per base pair, this amounts to an enormous length of DNA (as stated earlier, almost 2 meters)! One estimate is that the DNA inside a typical human nucleus is complexed with roughly $25 \times 10^6$ nucleosomes.

In the overall transition from a fully extended DNA helix to the extremely condensed status of the mitotic chromosome, a packing ratio (the ratio of DNA length to the length of the structure containing it) of about 500 to 1 must be achieved. In fact, our model accounts for a ratio of only about 50 to 1. Obviously, the larger fiber can be further bent, coiled, and packed to achieve even greater condensation during the formation of a mitotic chromosome.

NOW SOLVE THIS

Problem 16 on page 319 involves the extent to which the nucleus is filled by all of the diploid content of human DNA.

■ HINT: *Assuming the nucleus is a perfect sphere, start with the formula* $V = (4/3)\pi r^3$.

## High-Resolution Studies of the Nucleosome Core

As with many significant endeavors in genetics, the study of nucleosomes has answered some important questions but at the same time has opened up new ones. For example, in the preceding discussion, we established that histone proteins play an important structural role in packaging DNA into the nucleosomes that make up chromatin. While this discovery helped solve the structural problem of how the huge amount of DNA is organized within the eukaryotic nucleus, it brought another problem to the fore: *In chromatin fiber, complexed with histones and folded into various levels of compaction, the DNA is inaccessible to interaction with important nonhistone proteins.* For example, the various proteins that function in enzymatic and regulatory roles during the processes of replication and gene expression must interact directly with DNA. To accommodate these protein–DNA interactions, chromatin must be induced to change its structure, a process now referred to as **chromatin remodeling**. To allow replication and gene expression, chromatin must relax its compact structure and expose regions of DNA to regulatory proteins, but there must also be a mechanism for reversing the process during periods of inactivity.

Insights into how different states of chromatin structure might be achieved began to emerge in 1997, when Timothy Richmond and members of his research team were able to significantly improve the level of resolution in X-ray diffraction studies of nucleosome crystals, from 7 Å in the 1984 studies to 2.8 Å in the 1997 studies. One model based on their work is shown in Figure 12–10. At this resolution, most atoms are visible, thus revealing the subtle twists and turns of the superhelix of DNA encircling the histones. Recall that the double-helical ribbon in the figure represents 147 bp of DNA surrounding four pairs of histone proteins. This configuration is essentially repeated over and over in the chromatin fiber and is the principal packaging unit of DNA in the eukaryotic nucleus.

More recently, the work of Richmond and colleagues—having achieved a resolution of 1.9 Å in 2003—has revealed the details of the location of each histone entity within the nucleosome. Of particular relevance to the discussion of chromatin remodeling is the observation that there are unstructured **histone tails** that are *not* packed into the folded histone domains within the core of the nucleosomes but instead protrude from it. For example, tails devoid of any secondary structure extending from histones H3 and H2B protrude through the minor-groove channels of the DNA helix. You should look carefully at Figure 12–10 and locate examples of such tails. Other tails of histone H4 appear to make a connection with adjacent nucleosomes. The significance of histone tails is that they provide potential targets along the chromatin fiber for a variety of chemical modifications that may be linked to genetic functions, including chromatin remodeling and the possible regulation of gene expression.

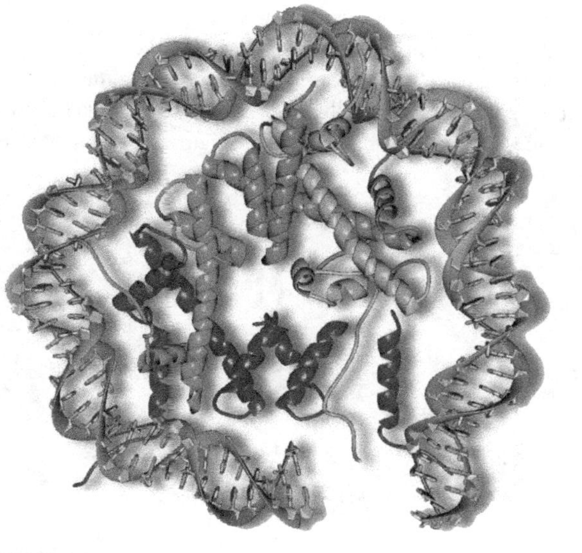

FIGURE 12–10 The nucleosome core particle derived from X-ray crystal analysis at 2.8 Å resolution. The double-helical DNA surrounds four pairs of histones.

Several of these potential chemical modifications are now recognized as important to genetic function. One of the most well-studied histone modifications involves **acetylation** by the action of the enzyme *histone acetyltransferase (HAT)*. This addition of an acetyl group to the positively charged amino group present on the side chain of the amino acid lysine effectively changes the net charge of the protein by neutralizing the positive charge. Lysine is in abundance in histones, and geneticists have known for some time that acetylation is linked to gene activation. It appears that high levels of acetylation open up (remodel) the chromatin fiber, an effect that increases in regions of active genes and decreases in inactive regions. In a well-studied example, the inactivation of the X chromosome in mammals, forming a Barr body (Chapter 7), histone H4 is known to be greatly underacetylated.

Two other important chemical modifications are the **methylation** and **phosphorylation** of amino acids that are part of histones. These chemical processes result from the action of enzymes called *methyltransferases* and *kinases*, respectively. Methyl groups can be added to both arginine and lysine residues in histones, and this change has been correlated with the activation of genes. Phosphate groups can be added to the hydroxyl groups of the amino acids serine and histidine, introducing a negative charge on the protein. During the cell cycle, increased phosphorylation, particularly of histone H3, is known to occur at characteristic times. Such chemical modification is believed to be related to the cycle of chromatin unfolding and condensation that occurs during and after DNA replication.

Interestingly, while methylation of histones in nucleosomes is positively correlated with gene activity in eukaryotes, methylation of the nitrogenous base cytosine within polynucleotide chains of DNA, forming **5-methyl cytosine**, is negatively correlated with gene activity. Methylation of cytosine occurs most often when the nucleotide cytidylic acid is next to the nucleotide guanylic acid, forming what is called a **CpG island**.

The research described above has extended our knowledge of nucleosomes and chromatin organization and serves here as a general introduction to the concept of chromatin remodeling. A great deal more work must be done, however, to elucidate the specific involvement of chromatin remodeling during genetic processes. In particular, the way in which the modifications are influenced by regulatory molecules within cells will provide important insights into the mechanisms of gene expression. What is clear is that the dynamic forms in which chromatin exists are vitally important to the way that all genetic processes directly involving DNA are executed. We will return to a more detailed discussion of the role of chromatin remodeling when we consider the regulation of eukaryotic gene expression in Chapter 18.

## Heterochromatin

Although we know that the DNA of the eukaryotic chromosome consists of one continuous double-helical fiber along its entire length, we also know that the whole chromosome is not structurally uniform from end to end. In the early part of the twentieth century, it was observed that some parts of the chromosome remain condensed and stain deeply during interphase, while most parts are uncoiled and do not stain. In 1928, the terms **euchromatin** and **heterochromatin** were coined to describe the parts of chromosomes that are uncoiled and those that remain condensed, respectively.

Subsequent investigation revealed a number of characteristics that distinguish heterochromatin from euchromatin. Heterochromatic areas are genetically inactive, because they either lack genes or contain genes that are repressed. Also, heterochromatin replicates later during the S phase of the cell cycle than does euchromatin. The discovery of heterochromatin provided the first clues that parts of eukaryotic chromosomes do not always encode proteins. For example, one particular heterochromatic region of the chromosome, the *telomere*, is thought to be involved in maintenance of the chromosome's structural integrity, and another region, the *centromere*, is involved in chromosome movement during cell division.

The presence of heterochromatin is unique to and characteristic of the genetic material of eukaryotes. In some cases, whole chromosomes are heterochromatic. A case in point is the mammalian Y chromosome, much of which is genetically inert. And, as we discussed in Chapter 7, the inactivated X chromosome in mammalian females is condensed into an inert heterochromatic Barr body. In some species, such as mealy bugs, all chromosomes of one entire haploid set are heterochromatic.

When certain heterochromatic areas from one chromosome are translocated to a new site on the same or another nonhomologous chromosome, genetically active areas sometimes become genetically inert if they now lie adjacent to the translocated heterochromatin. As we saw in Chapter 4, this influence on existing euchromatin is one example of what is more generally referred to as a **position effect**. That is, the position of a gene or group of genes relative to all other genetic material may affect their expression.

## 12.5 Chromosome Banding Differentiates Regions along the Mitotic Chromosome

Until about 1970, mitotic chromosomes viewed under the light microscope could be distinguished only by their relative sizes and the positions of their centromeres. Unfortunately, even in organisms with a low haploid number, two or more chromosomes are often visually indistinguishable from one another. Since that time, however, new cytological procedures made possible differential staining along the longitudinal axis of mitotic chromosomes. Such methods are now referred to as **chromosome-banding techniques**, because the staining patterns resemble the bands of polytene chromosomes.

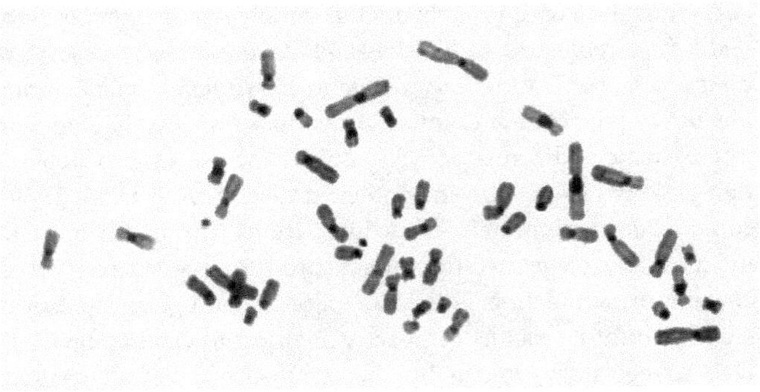

FIGURE 12–11 A human mitotic chromosome preparation processed to demonstrate C-banding. Only the centromeres stain.

One of the first chromosome-banding techniques was devised by Mary Lou Pardue and Joe Gall. They found that if chromosome preparations from mice were heat denatured and then treated with Giemsa stain, a unique staining pattern emerged: only the centromeric regions of mitotic chromosomes took up the stain! The staining pattern was thus referred to as **C-banding**. Relevant to our immediate discussion, this cytological technique identifies a specific area of the chromosome composed of heterochromatin. A micrograph of the human karyotype treated in this way is shown in Figure 12–11. Mouse chromosomes are all telocentric, thus localizing the stain at the end of each chromosome.

Other chromosome-banding techniques were developed about the same time. The most useful of these techniques produces a staining pattern differentially along the length of each chromosome. This method, producing **G-bands** (Figure 12–12), involves the

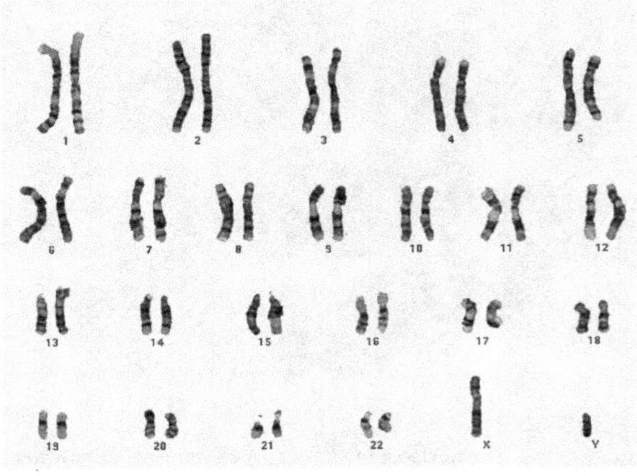


FIGURE 12–12 G-banded karyotype of a normal human male. Chromosomes were derived from cells in metaphase.

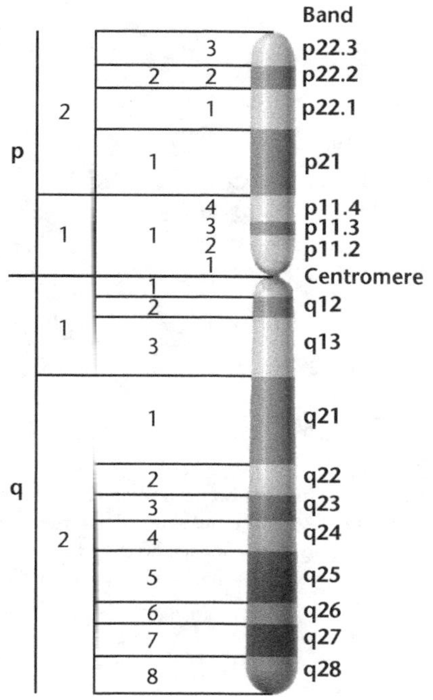


**FIGURE 12–13** The regions of the human X chromosome distinguished by its banding pattern. The designations on the right identify specific bands.

digestion of the mitotic chromosomes with the proteolytic enzyme trypsin, followed by Giemsa staining. The differential staining reactions reflect the heterogeneity and complexity of the chromosome along its length.

In 1971 a uniform nomenclature for human chromosome-banding patterns was established based on G-banding. Figure 12–13 illustrates the application of this nomenclature to the X chromosome. On the left of the chromosome are the various organizational levels of banding of the p and q arms that can be identified; the resulting designation for each of the specific regions is shown on the right side.

Although the molecular mechanisms involved in producing the various banding patterns are not well understood, the bands have played an important role in cytogenetic analysis, particularly in humans. The pattern of banding on each chromosome is unique, allowing a distinction to be made even between those chromosomes that are identical in size and centromere placement (e.g., human chromosomes 4 and 5 and 21 and 22). So precise is the banding pattern of each chromosome that homologs can be distinguished from one another, and when a segment of one chromosome has been translocated to another chromosome, its origin can be determined with great precision.

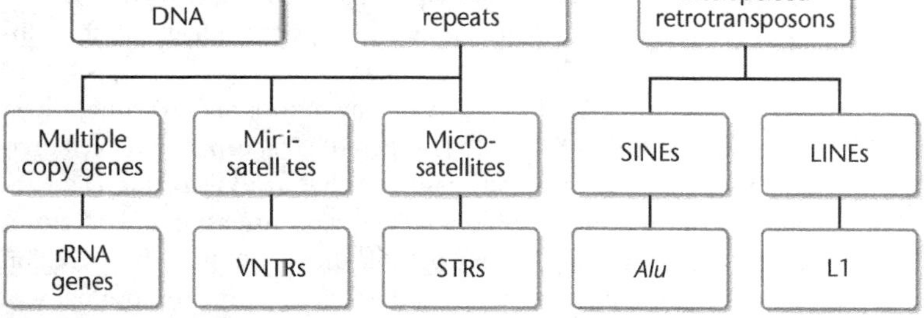


**FIGURE 12–14** An overview of the various categories of repetitive DNA.

## 12.6 Eukaryotic Chromosomes Demonstrate Complex Sequence Organization Characterized by Repetitive DNA

Thus far, we have examined the general structure of chromosomes in bacteriophages, bacteria, and eukaryotes. We now begin an examination of what we know about the organization of DNA sequences within the chromosomes making up an organism's genome, placing our emphasis on eukaryotes. Once we establish certain general aspects of this organization, we will focus on how genes themselves are organized within chromosomes (see Chapter 21).

We learned in Chapter 10 that, in addition to single copies of unique DNA sequences that make up genes, a great deal of the DNA sequencing within eukaryotic chromosomes is repetitive in nature and that various levels of repetition occur within the genomes of organisms. Many studies have now provided insights into **repetitive DNA**, demonstrating various classes of sequences and organization. Figure 12–14 schematizes these categories. Some functional genes are present in more than one copy (they are referred to as **multiple copy genes**) and so are repetitive in nature. However, the majority of repetitive sequences are nongenic and, most serve no known function. We will explore three main categories of repetitive sequences: (1) heterochromatin found to be associated with centromeres and making up telomeres; (2) tandem repeats of both short and long DNA sequences; and (3) transposable sequences that are interspersed throughout the genome of eukaryotes.

### Satellite DNA

The nucleotide composition (e.g., the percentage of G≡C versus A=T pairs) of the DNA of a particular species is reflected in the DNA's density, which can be measured with sedimentation equilibrium centrifugation (introduced in Chapter 10). When eukaryotic DNA is analyzed in this way, a graph describes its composition as a single main peak, representing a single main band, of fairly uniform density. However, one or more additional peaks indicate the presence of DNA that differs slightly in density. This component, called **satellite DNA**, makes up a variable proportion of the total DNA, depending on the species. For example, a profile of main-band and satellite DNA from the mouse is shown in Figure 12–15. By contrast, prokaryotes do not contain satellite DNA.

The significance of satellite DNA remained an enigma until the mid-1960s, when Roy Britten and David Kohne developed the technique for measuring the reassociation kinetics of DNA that had previously been dissociated into single strands (Chapter 10). The researchers demonstrated that certain portions of DNA reannealed more rapidly than others, and concluded that rapid reannealing was characteristic of multiple DNA fragments composed of identical or nearly identical nucleotide sequences—the basis for the descriptive term *repetitive DNA*. Recall that, in contrast, prokaryotic DNA is nearly devoid of anything other than unique, single-copy sequences.

When satellite DNA was subjected to analysis by reassociation kinetics (see Chapter 10), it fell into the category of *highly repetitive DNA*, consisting of short sequences repeated a large number of times. Further evidence suggested that these sequences are present as tandem (meaning adjacent) repeats clustered in very specific chromosomal areas known to be heterochromatic—the regions flanking centromeres. This was discovered in 1969 when several researchers, including Mary Lou Pardue and Joe Gall, applied the technique of ***in situ* molecular hybridization** to the study of satellite DNA. This technique (see Appendix A) involves molecular hybridization between an isolated fraction of radioactively labeled DNA or RNA probes and the DNA contained in the chromosomes of a cytological preparation. Following the hybridization procedure, autoradiography is performed to locate the chromosome areas complementary to the fraction of DNA or RNA.

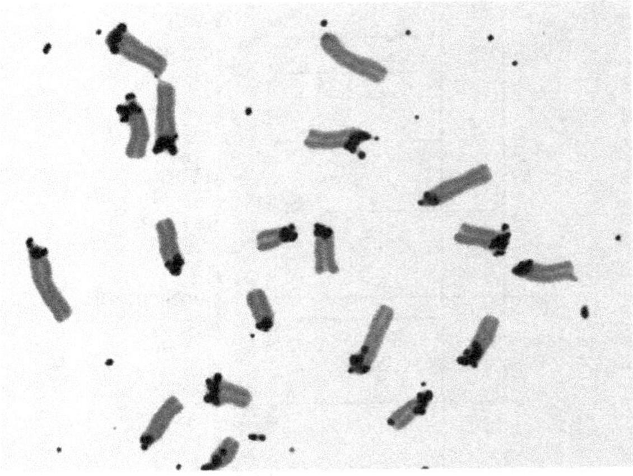

FIGURE 12–16 *In situ* hybridization between a radioactive probe representing mouse satellite DNA and mouse mitotic chromosomes. The grains in the autoradiograph are concentrated in the chromosome regions (the centromeres), revealing them to be the location of satellite DNA sequences.

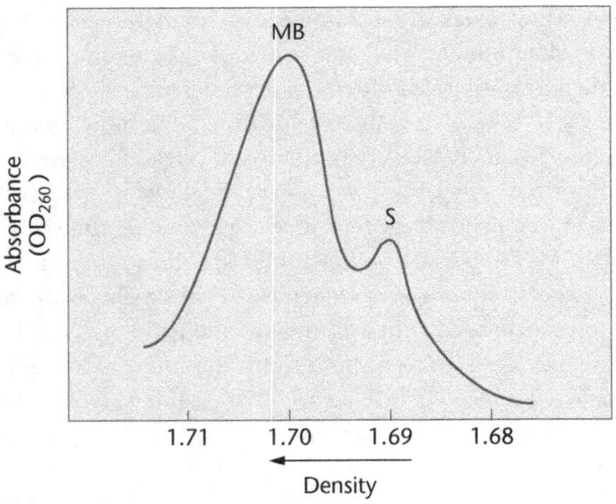


FIGURE 12–15 Separation of main-band (MB) and satellite (S) DNA from the mouse by using ultracentrifugation in a CsCl gradient.

Pardue and Gall demonstrated that radioactive molecular probes made from mouse satellite DNA hybridize with DNA of centromeric regions of mouse mitotic chromosomes (Figure 12–16). Several conclusions were drawn: Satellite DNA differs from main-band DNA in its molecular composition, as established by buoyant density studies. It is composed of short repetitive sequences. Finally, satellite DNA is found in the heterochromatic centromeric regions of chromosomes.

## Centromeric DNA Sequences

Centromeres, described cytologically in the late nineteenth century as the *primary constrictions* along eukaryotic chromosomes, play several crucial roles during mitosis and meiosis. First, they are responsible for the maintenance of sister chromatid cohesion prior to the anaphase stage. It is at the point of the centromere along the chromosome that sister chromatids remain paired together during the early stages of mitosis and meiosis. Second, centromeres are the site of formation of the kinetochore, the proteinaceous platform that is organized around the centromere and that attaches to the microtubules of spindle fibers. Hence, centromeres mediate chromosome migration during the anaphase stage. This process is essential to the separation of chromatids and thus the fidelity of chromosome distribution during cell division.

Most estimates of infidelity during mitosis are exceedingly low: $1 \times 10^{-5}$ to $1 \times 10^{-6}$, or 1 error per 100,000 to 1 million cell divisions. It has been generally assumed that analysis of the DNA sequence of centromeric regions will provide insights into the features responsible for this rather remarkable property. The minimal region of the centromere that supports the function of chromosomal segregation has been designated the **CEN region**. Within this heterochromatic region of the chromosome, the DNA binds a platform of proteins forming the centromere, which in multicellular organisms includes the *kinetochore* that binds to the spindle fiber during division.

The CEN regions of the yeast *Saccharomyces cerevisiae* were the first to be studied. Because each centromere serves an identical function, it is not surprising that all CENs were found to be remarkably similar in their organization. The region critical to centromere function for all 16 chromosomes in yeast consists of about 100 bp, divided into three contiguous regions (Figure 12–17). The first and third regions (I and III) are relatively short and highly conserved, consisting of only 9 and 11 bp, respectively. Region II, which is larger (~80 bp) and extremely A=T rich (up to 95 percent), varies in sequence in different chromosomes.

Mutational analysis suggests that regions I and II are less critical to centromere function than region III. Mutations in the first two regions often reduce centromere function (segregational activity), but do not inactivate it. However, a mutation in the central CCG triplet within region III inactivates centromere function. It has also been revealed that each of the three regions provides a specific DNA binding function to proteins critical to chromosome segregation. Thus, while the CEN appears to be essential to the binding of the chromosome to the spindle fiber (yeast do not have kinetochores), the DNA sequences are not unique to specific chromosomes. They can be experimentally exchanged between chromosomes without altering centromere function.

Based on the studies in yeast, it was assumed that similar findings would be obtained from multicellular eukaryotes. However, to the surprise of researchers, the CEN sequence in organisms such as mammals, including humans, varies considerably and has not been highly conserved, leaving in question the role of the CEN in these organisms. The amount of DNA associated with the centromeric region in multicellular organisms is much more extensive than in yeast. Recall from our prior discussion that highly repetitive satellite DNA is localized in the centromere regions of mice. Such sequences, absent from yeast, but characteristic of most multicellular organisms, vary considerably in size. For example, in *Drosophila*, the CEN region is made up of the 10-bp sequence AATAACATAG tandemly repeated within about 200 to 600 kb of DNA in the centromeres of all four chromosomes. In humans, one of the most recognized satellite DNA sequences is the **alphoid family**. Found mainly in the centromere regions, a 171-bp motif of alphoid DNA is present in tandem head-to-tail repeating arrays totaling up to 1 million base pairs. Embedded somewhere in this repetitive DNA are more specific sequences that are critical to centromere function. While such a motif is present in other closely related primates, neither the sequence nor the number of repeats of the human 171-bp sequence is conserved. While the precise role in centromere function of this highly repetitive DNA remains unclear, it is known that the sequences are not transcribed.

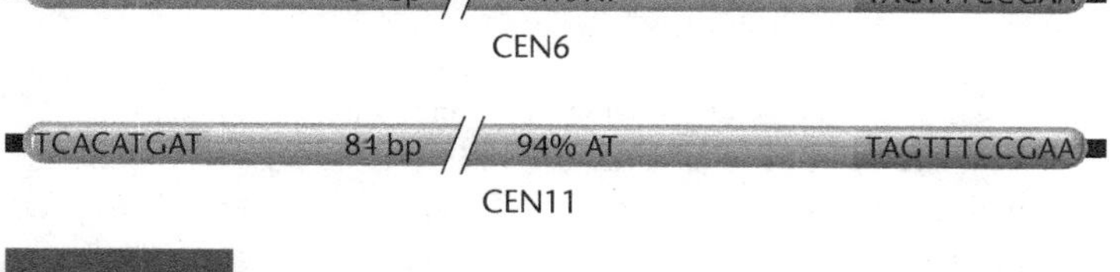


**FIGURE 12–17** Nucleotide sequence information derived from DNA of the three major centromere regions of yeast chromosomes 3, 4, 6, and 11.

## Telomeric DNA Sequences

We now return to a topic that was introduced in Chapter 11, the **telomere**—the structure that "caps" the ends of linear eukaryotic chromosomes. Our earlier discussion focused on replication of telomeric DNA as well as on the model describing how the DNA sequences in these structures function to maintain the stability of chromosomes. The cap structure renders chromosome ends inert in interactions with other chromosome ends and with enzymes that use double-stranded DNA ends as substrates (such as repair enzymes). As with centromeres, the analysis of telomeres was first approached by investigating the smaller chromosomes of simple eukaryotes, such as protozoans and yeast. The idea that all telomeres of all chromosomes in related organisms might share a common nucleotide sequence has now been borne out.

**Telomeric DNA sequences** consist of short tandem repeats. It is this group of repetitive sequences that contributes to the stability and integrity of the chromosome. In the ciliate *Tetrahymena*, more than 50 tandem repeats of the hexanucleotide sequence 5′-TTGGGG-3′ occur. In all vertebrates, including humans, the sequence 5′-TTAGGGG-3′ is repeated many times. The number of copies making up a telomere varies in different organisms, and there may be as many as 1000 repeats in some species. As discussed in Chapter 11, replication of the telomere requires a unique RNA-containing enzyme, telomerase. In its absence, the DNA at the ends of chromosomes becomes shorter at each replication. Because single-celled eukaryotes are immortalized cells, telomerase is critical for the survival of such species. In multicellular organisms, such as humans, telomerase is active in germ-line cells but is inactive in somatic cells. Chromosome shortening is considered part of the natural process of cell aging, serving as an internal clock. In human cancer cells, which have become immortalized, the transition to malignancy appears to require the activation of telomerase in order to overcome the normal senescence associated with chromosome shortening.

## Middle Repetitive Sequences: VNTRs and STRs

A brief look at still another prominent category of repetitive DNA sheds additional light on the organization of the eukaryotic genome. In addition to highly repetitive DNA, which constitutes about 5 percent of the human genome (and 10 percent of the mouse genome), a second category, **middle** (or **moderately**) **repetitive DNA**, recognized by $C_0t$ analysis, is fairly well characterized. Because we now know a great deal about the human genome, we will use our own species to illustrate this category of DNA in genome organization.

Although middle repetitive DNA does include some duplicated genes (such as those encoding ribosomal RNA), most prominent in this category are either noncoding tandemly repeated sequences or noncoding interspersed sequences. No function has been ascribed to these components of the genome. An example is DNA described as **variable number tandem repeats (VNTRs)**. These repeating DNA sequences may be 15 to 100 bp long and are found within and between genes. Many such clusters are dispersed throughout the genome, and they are often referred to as **minisatellites**.

The number of tandem copies of each specific sequence at each location varies from one individual to the next, creating localized regions of 1,000 to 20,000 bp (1–20 kb) in length. As we will see in Chapter 22, the variation in size (length) of these regions between individual humans was originally the basis for the forensic technique referred to as **DNA fingerprinting**.

Another group of tandemly repeated sequences consists of di-, tri-, tetra-, and pentanucleotides, also referred to as **microsatellites** or **short tandem repeats (STRs)**. Like VNTRs, they are dispersed throughout the genome and vary among individuals in the number of repeats present at any site. For example, in humans, the most common microsatellite is the dinucleotide $(CA)_n$, where $n$ equals the number of repeats. Most commonly, $n$ is between 5 and 50. These clusters have served as useful molecular markers for genome analysis.

## Repetitive Transposed Sequences: SINEs and LINEs

Still another category of repetitive DNA consists of sequences that are interspersed individually throughout the genome, rather than being tandemly repeated. They can be either short or long, and many have the added distinction of being **transposable sequences**, which are mobile and can potentially move to different locations within the genome. A large portion of the human genome is composed of such sequences.

For example, **short interspersed elements**, called **SINEs**, are less than 500 base pairs long and may be present 500,000 times or more in the human genome. The best characterized human SINE is a set of closely related sequences called the ***Alu* family** (the name is based on the presence of DNA sequences recognized by the restriction endonuclease *Alu*I). Members of this DNA family, also found in other mammals, are 200 to 300 base pairs long and are dispersed rather uniformly throughout the genome, both between and within genes. In humans, this family encompasses more than 5 percent of the entire genome.

*Alu* sequences are particularly interesting, although their function, if any, is yet undefined. Members of the *Alu* family are sometimes transcribed. The role of this RNA is not certain, but it is thought to be related to its mobility in the genome. *Alu* sequences are thought to have arisen from an RNA element whose DNA complement was dispersed throughout the genome as a result of the activity of reverse transcriptase (an enzyme that synthesizes DNA on an RNA template).

The group of **long interspersed elements (LINEs)** represents yet another category of repetitive transposable DNA sequences. LINEs are usually about 6 kb in length and in the human genome are present 850,000 times. The most prominent example in humans is the **L1 family**. Members of this sequence family are about 6400 base pairs long and are present up to 100,000 times. Their 5′-end is highly variable, and their role within the genome has yet to be defined.

The general mechanism for transposition of L1 elements is now clear. The L1 DNA sequence is first transcribed into an RNA molecule. The RNA then serves as the template for the synthesis of the DNA complement using the enzyme reverse transcriptase. This enzyme is encoded by a portion of the L1 sequence. The new L1 copy then integrates into the DNA of the chromosome at a new site. Because of the similarity of this transposition mechanism to that used by retroviruses, LINEs are referred to as **retrotransposons**.

SINEs and LINEs represent a significant portion of human DNA. SINEs constitute about 13 percent of the human genome, whereas LINEs constitute up to 21 percent. Within both types of elements, repeating sequences of DNA are present in combination with unique sequences.

## Middle Repetitive Multiple-Copy Genes

In some cases, middle repetitive DNA includes functional genes present tandemly in multiple copies. For example, many copies exist of the genes encoding ribosomal RNA. *Drosophila* has 120 copies per haploid genome. Single genetic units encode a large precursor molecule that is processed into the 5.8*S*, 18*S*, and 28*S* rRNA components. In humans, multiple copies of this gene are clustered on the p arm of the acrocentric chromosomes 13, 14, 15, 21, and 22. Multiple copies of the genes encoding 5*S* rRNA are transcribed separately from multiple clusters found together on the terminal portion of the p arm of chromosome 1.

## 12.7 The Vast Majority of a Eukaryotic Genome Does Not Encode Functional Genes

Given the preceding information concerning various forms of repetitive DNA in eukaryotes, it is of interest to pose an important question: *What proportion of the eukaryotic genome actually encodes functional genes?*

We have seen that, taken together, the various forms of highly repetitive and moderately repetitive DNA comprise a substantial portion of the human genome. In addition to repetitive DNA, a large amount of the DNA consists of single-copy sequences as defined by $C_0t$ analysis that appear to be noncoding. Included are many instances of what we call **pseudogenes**. These are DNA sequences representing evolutionary vestiges of duplicated copies of genes that have undergone significant mutational alteration. As a result, although they show some homology to their parent gene, they are usually not transcribed because of insertions and deletions throughout their structure.

While the proportion of the genome consisting of repetitive DNA varies among organisms, one feature seems to be shared: *Only a very small part of the genome actually codes for proteins.* For example, the 20,000 to 30,000 genes encoding proteins in sea urchin occupy less than 10 percent of the genome. In *Drosophila*, only 5 to 10 percent of the genome is occupied by genes coding for proteins. In humans, it appears that the estimated 20,000 to 25,000 functional genes occupy only about 2 percent of the total DNA sequence making up the genome.

Study of the various forms of repetitive DNA has significantly enhanced our understanding of genome organization. In the next chapter, we will explore the organization of genes within chromosomes.

# EXPLORING GENOMICS

## UniGene Transcript Maps

In this chapter, as well as in Chapter 5, we discussed how chromosome banding patterns revealed by stains such as the Giemsa stain provided the first maps of chromosome structure. Although they must be considered relatively low-resolution maps, chromosome banding patterns do reveal areas of heterochromatin and euchromatin. We now have access to much higher-resolution maps of chromosome structure—the sequence maps we have examined in several Exploring Genomics exercises. In Exploring Genomics for Chapter 5, for example, we explored some detailed maps of human chromosomes available on the Internet.

In this exercise, we explore **UniGene**, another of the databases provided by the National Center for Biotechnology Information. **UniGene** displays human chromosome maps of expressed mRNA sequences (transcript maps). We will explore these maps to see if regions of gene expression on a chromosome can be revealed by chromosome banding patterns.

### ■ Exercise I – UniGene: Comparing Clusters of Expressed Genes with Chromosome Banding Patterns

1. Begin this exercise by accessing the NCBI Human Genome Resources site http://www.ncbi.nlm.nih.gov/genome/guide/human/. Click on the X chromosome (under "Browse your Genome") and you will be taken to the Map Viewer feature of NCBI. You have seen the Map Viewer feature before, in Chapter 5.

2. Recall that the diagram of the X chromosome G-banding pattern you see at the far left of the screen is referred to here as an **ideogram**. Immediately to the right of the ideogram is a **UniGene cluster** map. The UniGene cluster map shows levels of mRNA expression characteristic of different regions of the ideogram, as well as expression data for ESTs (expressed sequence tags)—short, incomplete sequences of expressed genes (you will learn more about ESTs in Chapter 21). Specifically, the cluster map is a histogram showing the density of expressed gene clusters in each region of the chromosome. Names for expressed mRNAs are indicated in blue type, beginning with "Hs" (for *Homo sapiens*) followed by a number.

3. Let's explore a region of heterochromatin on the X chromosome. Refer to Figure 12–13 in your text for an illustration of G-banding patterns of the human X chromosome. Notice that band q21.1 is a prominent segment of heterochromatin.

4. Find band q21.1 on the ideogram. Click on the corresponding region of Hs Unigene and zoom in 8X and for a closer look. (Note that the ideogram, Unigene map, and the Genes sequence map do not always align exactly.) Answer the following questions:

   a. Based on the UniGene cluster map, does it appear that many genes are expressed in this region?

   b. Compare band q21.1 to band q13.3. Describe the relative numbers of genes. Is the number of expressed genes at band q13.3 consistent with the chromosome banding pattern for q13.3?

   c. Click on the ideogram band and zoom in again for another 8X enlargement. What do you find now?

5. Click on the names for any expressed mRNAs (transcripts) to learn more about the corresponding genes.

### ■ Exercise II – Exploring Centromeres Using UniGene

1. Based on what you know about centromeric DNA sequences, would you expect to find many expressed genes at the centromere of a chromosome? Use UniGene to explore the centromeric region of the Y chromosome and at least three other chromosomes to help you answer this question. Choose chromosomes whose centromeres are in different positions (for example, metacentric and telocentric chromosomes). Begin your examination with the UniGene histograms of expressed gene density, and then click on the centromere to zoom in for more detail. What did you discover?

## Chapter Summary

1. Understanding the organization of the molecular components that form chromosomes is essential to understanding the function of the genetic material. Largely devoid of associated proteins, the chromosomes of bacteriophages and bacteria contain DNA molecules in a form equivalent to the Watson–Crick model.
2. Polytene and lampbrush chromosomes are examples of specialized structures that have extended our knowledge of genetic organization and function.
3. Eukaryotic chromatin is a nucleoprotein organized into repeating units called nucleosomes. Composed of 200 base pairs of DNA, an octamer of four types of histones, plus one linker histone, the nucleosome is important for condensing the extensive chromatin fiber within the interphase nucleus into the highly condensed chromosome seen in mitosis.
4. The structural heterogeneity of the chromosome axis has been established as a result of both biochemical and cytological investigation. Heterochromatin, prematurely condensed in interphase, is for the most part genetically inert. The centromeric and telomeric regions, the Y chromosome, and the Barr body are examples.
5. DNA analysis has revealed unique nucleotide sequences in both the centromere and telomere regions of chromosomes which no doubt impart the heterochromatic characteristics of those regions and play a role in their respective functions.
6. Eukaryotic genomes demonstrate complex sequence organization characterized by numerous categories of repetitive DNA.
7. Repetitive DNA consists of either tandem repeats clustered in various regions of the genome or single sequences repeatedly interspersed at random throughout the genome. In the former type of repeat, the size of each cluster varies among individuals, providing one form of genetic identity. The latter type are transposable elements and may be short, like *Alu* sequences, or long, like L1 sequences.
8. The vast majority of the DNA in most eukaryotic genomes does not encode functional genes. In humans, for example, only about 2 percent of the genome is used to encode the 20,000 to 25,000 genes found there.

## INSIGHTS AND SOLUTIONS

*The following questions pertain to a previously undiscovered single-celled organism found living at a great depth on the ocean floor. Its nucleus contains only a single, linear chromosome consisting of $7 \times 10^6$ nucleotide pairs of DNA coalesced with three types of histonelike proteins.*

1. A short micrococcal nuclease digestion yielded DNA fractions consisting of 700, 1400, and 2100 base pairs. Predict what these fractions represent. What conclusions can be drawn?

**Solution:** The chromatin fiber may consist of a nucleosome variant containing 700 base pairs of DNA. The 1400- and 2100-bp fractions represent two and three of these nucleosomes, respectively, linked together. Enzymatic digestion may have been incomplete, leading to the latter two fractions.

2. The analysis of individual nucleosomes revealed that each unit contained one copy of each protein and that the short linker DNA had no protein bound to it. If the entire chromosome consists of nucleosomes (discounting any linker DNA), how many are there, and how many total proteins are needed to form them?

**Solution:** Since the chromosome contains $7 \times 10^6$ base pairs of DNA, the number of nucleosomes, each containing $7 \times 10^2$ base pairs, is equal to

$$7 \times 10^6/7 \times 10^2 = 10^4 \text{ nucleosomes}$$

The chromosome thus contains copies of each of the three proteins, for a total of $3 \times 10^4$ proteins.

3. Analysis then revealed the organism's DNA to be a double helix similar to the Watson–Crick model, but containing 20 base pairs per complete turn of the right-handed helix. The physical size of the nucleosome was exactly double the volume occupied by the nucleosome found in any other known eukaryote, and the nucleosome's axis length was greater by a factor of two. Compare the degree of compaction (the number of turns per nucleosome) of this organism's nucleosome with that found in other eukaryotes.

**Solution:** The unique organism compacts a length of DNA consisting of 35 complete turns of the helix (700 base pairs per nucleosome/20 base pairs per turn) into each nucleosome. The normal eukaryote compacts a length of DNA consisting of 20 complete turns of the helix (200 base pairs per nucleosome/10 base pairs per turn) into a nucleosome half the volume of that in the unique organism. The degree of compaction is therefore less in the unique organism.

4. No further coiling or compaction of this unique chromosome occurs in the newly discovered organism. Compare this situation with that of a eukaryotic chromosome. Do you think an interphase human chromosome $7 \times 10^6$ base pairs in length would be a shorter or longer chromatin fiber?

**Solution:** The eukaryotic chromosome contains still another level of condensation in the form of solenoids, which are coils consisting of nucleosomes connected with linker DNA. Solenoids condense the eukaryotic fiber by still another factor of five. The length of the unique chromosome is compacted into $10^4$ nucleosomes, each containing an axis length twice that of the eukaryotic fiber. The eukaryotic fiber consists of $7 \times 10^6/2 \times 10^2 = 3.5 \times 10^4$ nucleosomes, 3.5 times more than the unique organism. However, they are compacted by the factor of five in each solenoid. Therefore, the chromosome of the unique organism is a longer chromatin fiber.

## Problems and Discussion Questions

1. Contrast the size of the single chromosome in bacteriophage $\lambda$ and T2 with that of *E. coli*. How does this relate to the relative size and complexity of phages and bacteria?
2. In bacteriophages and bacteria, the DNA is almost always organized into circular (closed loops) chromosomes. Phage $\lambda$ is an exception, maintaining its DNA in a linear chromosome within the viral particle. However, as soon as this DNA is injected into a host cell, it circularizes before replication begins. Taking into account information in Chapter 11, what advantage exists in replicating circular DNA molecules compared to linear molecules?
3. Describe the structure of giant polytene chromosomes and how they arise.
4. After salivary gland cells from *Drosophila* are isolated and cultured in the presence of radioactive thymidylic acid, autoradiography is performed, revealing polytene chromosomes. Predict the distribution of the grains along the chromosomes.
5. What genetic process is occurring in a puff of a polytene chromosome? How do we know this experimentally?
6. Describe the structure of LINE sequences. Why are LINEs referred to as retrotransposons?
7. During what genetic process are lampbrush chromosomes present in vertebrates?
8. Why might we predict that the organization of eukaryotic genetic material will be more complex than that of viruses or bacteria?
9. Describe the sequence of research findings that led to the development of the model of chromatin structure.
10. Describe the molecular composition and arrangement of the components in the nucleosome.
11. Describe the transitions that occur as nucleosomes are coiled and folded, ultimately forming a chromatid.
12. Provide a comprehensive definition of heterochromatin and list as many examples as you can.
13. Mammals contain a diploid genome consisting of at least $10^9$ bp. If this amount of DNA is present as chromatin fibers, where each group of 200 bp of DNA is combined with 9 histones into a nucleosome and each group of 6 nucleosomes is combined into a solenoid, achieving a final packing ratio of 50, determine (a) the total number of nucleosomes in all fibers, (b) the total number of histone molecules combined with DNA in the diploid genome, and (c) the combined length of all fibers.
14. Assume that a viral DNA molecule is a 50-$\mu$m-long circular strand with a uniform 20-Å diameter. If this molecule is contained in a viral head that is a 0.08-$\mu$m-diameter sphere, will the DNA molecule fit into the viral head, assuming complete flexibility of the molecule? Justify your answer mathematically.
15. How many base pairs are in a molecule of phage T2 DNA 52 $\mu$m long?
16. If a human nucleus is 10 $\mu$m in diameter, and it must hold as much as 2 m of DNA, which is complexed into nucleosomes that during full extension are 11 nm in diameter, what percentage of the volume of the nucleus is occupied by the genetic material?
17. Examples of histone modifications are acetylation (by histone acetyltransferase, or HAT), which is often linked to gene activation, and deacetylation (by histone deacetylases, or HDACs), which often leads to gene silencing typical of heterochromatin. Such heterochromatinization is initiated from a nucleation site and spreads bidirectionally until encountering boundaries that delimit the silenced areas. Recall from Chapter 4 the brief discussion of position effect, where repositioning of the $w^+$ allele in *Drosophila* by translocation or inversion near heterochromatin produces intermittent $w^+$ activity. In the heterozygous state ($w^+/w$), a variegated eye is produced, with white and red patches. How might one explain position-effect variegation in terms of histone acetylation and/or deacetylation?
18. In light of indications that there are at least 19,000 pseudogenes in the human genome and perhaps as few as 21,000 protein-coding genes, some researchers suggest that as assays become more refined, the number of pseudogenes may outnumber protein-coding genes (Gerstein and Zheng, 2006).
    (a) Why would it take more time to locate pseudogenes than protein-coding genes?
    (b) Most pseudogenes are damaged remains of previously functional genes and are considered to be nonfunctional. However, it has been suggested that some pseudogenes do function. In light of the various ways in which DNA is organized in chromosomes, how might pseudogenes function?
19. Variable number tandem repeats (VNTRs) are repeating DNA sequences of about 15 to 100 bp in length, found both within and between genes. Why are they commonly used in forensics?

#### HOW DO WE KNOW?

20. In this chapter, we focused on how DNA is organized at the chromosomal level. Along the way, we found many opportunities to consider the methods and reasoning by which much of this information was acquired. From the explanations given in the chapter, what answers would you propose to the following fundamental questions:
    (a) What is the experimental basis for concluding that puffs in polytene chromosomes and loops in lampbrush chromosomes are areas of intense transcription of RNA?
    (b) How do we know about the organization of DNA within chromosomes, given that DNA has a diameter of only 2 nm?
    (c) What experimental evidence supports the idea that eukaryotic chromatin exists in the form of repeating nucleosomes, each consisting of about 200 base pairs and an octamer of histones?
    (d) How do we know that satellite DNA consists of repetitive sequences and has been derived from regions of the centromere?

## Extra-Spicy Problems

21. In a study of *Drosophila*, two normally active genes, $w^+$ (wild-type allele of the *white*-eye gene) and *hsp26* (a heat-shock gene), were introduced (using a plasmid vector) into euchromatic and heterochromatic chromosomal regions, and the relative activity of each gene was assessed (Sun et al., 2002). An approximation of the resulting data is shown in the following table. Which characteristic or characteristics of heterochromatin are supported by the experimental data?

| Gene | Activity (relative percentage) | |
|---|---|---|
| | Euchromatin | Heterochromatin |
| *hsp26* | 100% | 31% |
| $w^+$ | 100% | 8% |

22. Using molecular methods to "label" chromosomes with fluorescent dyes, researchers observed a precise nuclear positioning of chromosomes during an early stage of mitosis (prometaphase) in human fibroblast cells (Nagele et al., 1995). Below is a sketch modified from this research that describes the relative positions of chromosomes 7, 8, 16, and X. Homologous chromosomes share the same color. Assuming that this pattern is consistent in human cells, what conclusions can be drawn regarding the nuclear positions of the chromosomes during interphase and during the initial phases of mitosis? How could chromosomal localization influence gene function during interphase?

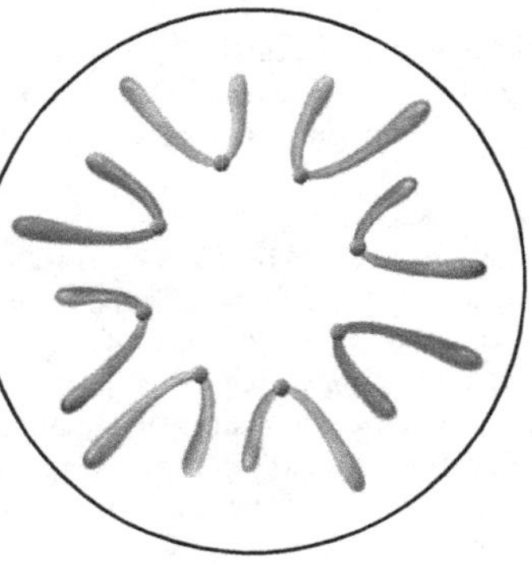

23. While much remains to be learned about the role of nucleosomes and chromatin structure and function, recent research indicates that *in vivo* chemical modification of histones is associated with changes in gene activity. One study determined that acetylation of H3 and H4 is associated with 21.1 percent and 13.8 percent increases in yeast gene activity, respectively, and that yeast heterochromatin is hypomethylated relative to the genome average (Bernstein et al., 2000). Speculate on the significance of these findings in terms of nucleosome–DNA interactions and gene activity.

24. An article entitled "Nucleosome Positioning at the Replication Fork" states, "both the 'old' randomly segregated nucleosomes as well as the 'new' assembled histone octamers rapidly position themselves (within seconds) on the newly replicated DNA strands" (Lucchini et al., 2002). Given this statement, how would one compare the distribution of nucleosomes and DNA in newly replicated chromatin? How could one experimentally test the distribution of nucleosomes on newly replicated chromosomes?

25. The human genome contains approximately $10^6$ copies of an *Alu* sequence, one of the best-studied classes of short interspersed elements (SINEs), per haploid genome. Individual *Alus* share a 282-nucleotide consensus sequence followed by a 3′-adenine-rich tail region (Schmid, 1998). Given that there are approximately $3 \times 10^9$ base pairs per human haploid genome, about how many base pairs are spaced between each *Alu* sequence?

26. Following is a diagram of the general structure of the bacteriophage λ chromosome. Speculate on the mechanism by which it forms a closed ring upon infection of the host cell.

5′GGGCGGCGACCT—double-stranded region—3′
3′—double-stranded region—CCCGCCGCTGGA5′

27. Tandemly repeated DNA sequences with a repeat sequence of one to six base pairs—for example, $(GACA)_n$—are called microsatellites and are common in eukaryotes. A particular subset of such sequences, the trinucleotide repeat, is of great interest because of the role such repeats play in human neurodegenerative disorders (Huntington disease, myotonic dystrophy, spinal-bulbar muscular atrophy, spinocerebellar ataxia, and fragile X syndrome). Following are data (modified from Toth et al., 2000) regarding the location of microsatellites within and between genes. What general conclusions can be drawn from these data?

**Percentage of Microsatellite DNA Sequences within Genes and between Genes**

| Taxonomic Group | Within Genes | Between Genes |
|---|---|---|
| Primates | 7.4 | 92.6 |
| Rodents | 33.7 | 66.3 |
| Arthropods | 46.7 | 53.3 |
| Yeasts | 77.0 | 23.0 |
| Other fungi | 66.7 | 33.3 |

28. More information from the research effort in Problem 27 produced data regarding the pattern of the length of such repeats within genes. Each value in the following table represents the number of times a microsatellite of a particular sequence length, one to six bases long, is found within genes. For instance, in primates, a dinucleotide sequence (GC, for example) is found 10 times, while a trinucleotide is found 1126 times. In fungi, a repeat motif composed of 6 nucleotides (GACACC, for example) is found 219 times, whereas a tetranucleotide repeat (GACA, for example) is found only 2 times. Analyze and interpret these data by indicating what general pattern is apparent for the distribution of various microsatellite lengths within genes. Of what significance might this general pattern be?

**Distribution of Microsatellites by Unit Length within Genes**

| Taxonomic Group | Length of Repeated Motif (bp) | | | | | |
|---|---|---|---|---|---|---|
| | 1 | 2 | 3 | 4 | 5 | 6 |
| Primates | 49 | 10 | 1126 | 29 | 57 | 244 |
| Rodents | 62 | 70 | 1557 | 63 | 116 | 620 |
| Arthropods | 12 | 34 | 1566 | 0 | 21 | 591 |
| Yeasts | 36 | 19 | 706 | 7 | 52 | 330 |
| Other fungi | 9 | 4 | 381 | 2 | 35 | 219 |

29. In spite of the considerable medical and biological significance of repetitive DNA sequences, the factors that determine their genesis and genomic distribution remain uncertain. Misalignment of repetitive DNA strands and DNA polymerase slippage have been described as mechanisms causing variation in the number of *existing* repeats. Until recently, there has been little information relating to the *initial* creation of a microsatellite genomic region. In 2001, researchers sequenced DNA surrounding numerous tetranucleotide microsatellite regions in several strains within two species of *Drosophila* and observed the sequences shown below (Wilder & Hollocher, 2001). (a) Identify the microsatellite tetranucleotide motif. Is it a perfect motif? Using "Pu" to represent a purine and "Py" to represent a pyrimidine, symbolize the tetranucleotide repeat [in a form similar to $(CPuGPy)_n$]. (b) What is the sequence of the nonmicrosatellite region? Is it a perfectly conserved region among all the species and strains listed?

| Species (strain) | Base Sequence |
|---|---|
| *D. nigrodunni*-1 | 5′-TCGATATAGCCATGTCCGTCTGTCCGTCTGT |
| *D. nigrodunni*-2 | 5′-TCGATATAGCCATGTCCGTCTGTCCGTCTGT |
| *D. nigrodunni*-3 | 5′-TCGATATAGCAATGTCCGTCTGTCCGTCTGT |
| *D. dunni*-1 | 5′-TCGATATAGCAATGTCCGTCTGTCCGTCTGT |
| *D. dunni*-2 | 5′-TCGATATAGCCATGTCCGTCTGTCCGTCTGT |

30. Regarding the findings and your analysis of data in Problem 29, what significance might there be to a highly conserved nonmicrosatellite region flanking a specific microsatellite type?

31. Microsatellites are currently exploited as markers for paternity testing. A sample paternity test is shown in the following table in which ten microsatellite markers were used to test samples from a mother, her child, and an alleged father. The name of the microsatellite locus is given in the left-hand column, and the genotype of each individual is recorded as the number of repeats he or she carries at that locus. For example, at locus D9S302, the mother carries 30 repeats on one of her chromosomes and 31 on the other. In cases where an individual carries the same number of repeats on both chromosomes, only a single number is recorded. (Some of the numbers are followed by a decimal point, for example, 20.2, to indicate a partial repeat in addition to the complete repeats.) Assuming that these markers are inherited in a simple Mendelian fashion, can the alleged father be excluded as the source of the sperm that produced the child? Why or why not? Explain.

| Microsatellite Locus-Chromosome Location | Mother | Child | Alleged Father |
|---|---|---|---|
| D9S302-9q31-q33 | 30<br>31 | 31<br>32 | 32<br>33 |
| D22S883-22pter-22qter | 17<br>22 | 20.2<br>22 | 20.2 |
| D18S535-18q12.2-q12.3 | 12<br>14 | 13<br>14 | 11<br>13 |
| D7SI 804-7pter-7qter | 27<br>30 | 26<br>30 | 26<br>27 |
| D3S2387-3p24.2.3pter | 23<br>25.2 | 24<br>25.2 | 20.2<br>24 |
| D4S2386-4pter-qter | 12 | 12 | 12<br>16 |
| D5S1719-5pter-5qter | 11<br>11.3 | 10.3<br>11 | 10<br>10.3 |
| CSF1PO-5q33.3.q34 | 11 | 11<br>12 | 10<br>12 |
| FESFPS-15q25-15qter | 11<br>12 | 12<br>13 | 10<br>13 |
| TH01-11p15.5 | 7 | 7 | 7<br>8 |

32. Recall from Chapter 10 (Figure 10–20) that when double-stranded DNA is heated, the optical density rises as the helix denatures and the strands separate. How might this technique be used to characterize different components and sequences of the eukaryotic genome?

33. If DNA from *Drosophila melanogaster* is analyzed using density gradient centrifugation and is found to contain a "satellite" peak distinct from main-band DNA, describe a set of experiments to determine as much as you can about the nature of this DNA in contrast to main-band DNA.

34. At the end of the short arm of human chromosome 16 (16p), several genes associated with disease are present, including thalassemia and polycystic kidney disease. When that region of chromosome 16 was sequenced, gene-coding regions were found to be very close to the telomere-associated sequences. Could there be a possible link between the location of these genes and the presence of the telomere-associated sequences? What further information concerning the disease genes would be useful in your analysis?

A Petri dish revealing the growth of host cells after uptake of recombinant plasmids.

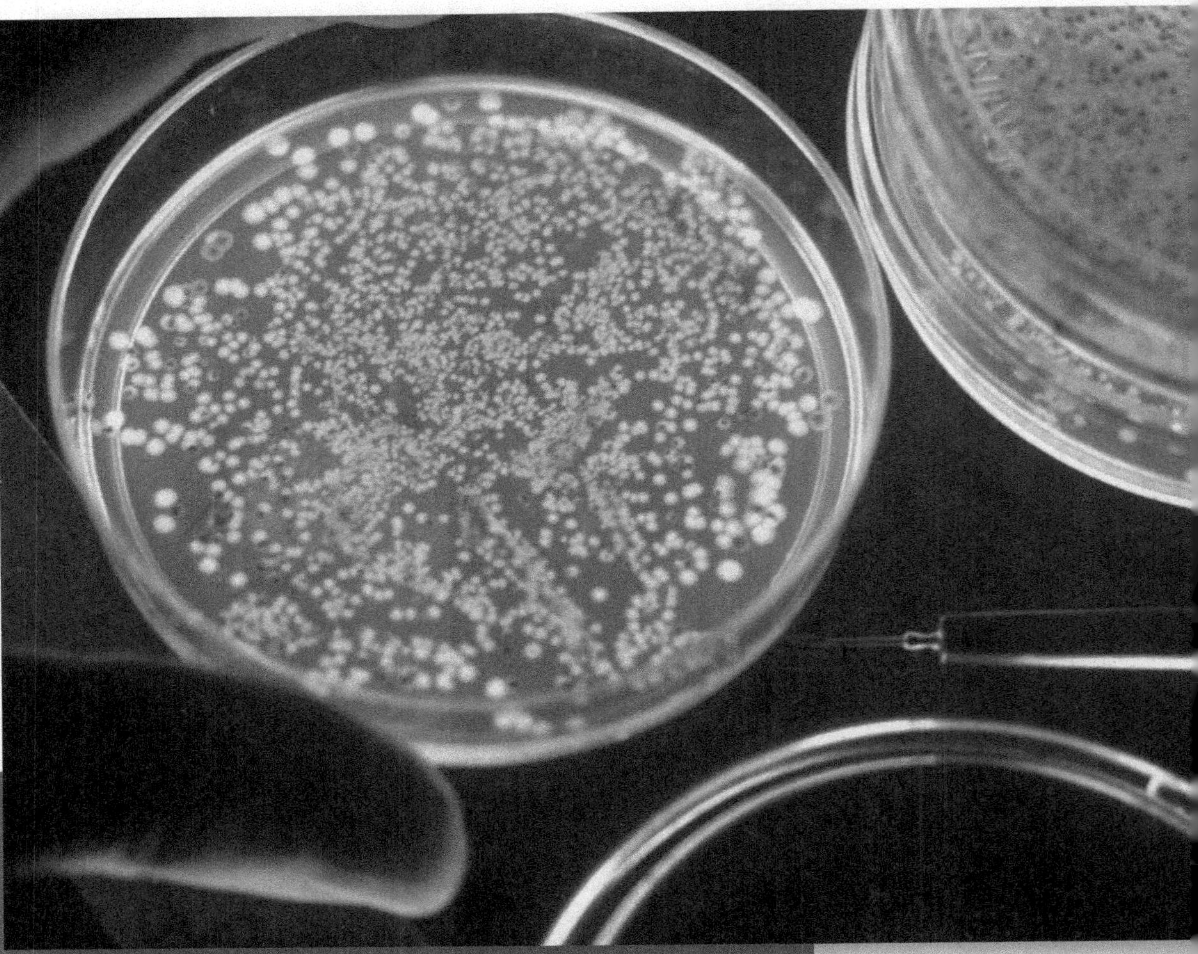

# 13

# Recombinant DNA Technology and Gene Cloning

## CHAPTER CONCEPTS

- Recombinant DNA technology creates artificial combinations of DNA molecules, usually from two different sources, most often from different species.
- Recombinant DNA technology depends in part on the ability to cleave and rejoin DNA segments at specific base sequences.
- The most useful application of recombinant DNA technology is to clone a DNA segment of interest.
- For cloning, specific DNA segments are inserted into vectors (such as plasmids), which are transferred into host cells (such as bacterial cells), where the recombinant molecules replicate as the host cells divide.
- Recombinant molecules can be cloned in prokaryotic or eukaryotic host cells.
- DNA segments can be quickly and efficiently amplified millions of times using the polymerase chain reaction (PCR).
- Cloned DNA segments are analyzed in several ways, the most specific being DNA sequencing.
- Recombinant DNA technology has revolutionized our ability to investigate the genomes of diverse species.

In 1971, a paper published by Kathleen Danna and Daniel Nathans marked the beginning of the recombinant DNA era. The paper described the isolation of an enzyme from a bacterial strain and the use of the enzyme to cleave viral DNA at specific nucleotide sequences. It contained the first published photograph of DNA cut with such an enzyme, now called a restriction enzyme. Using restriction enzymes and a number of other resources, researchers of the mid- to late 1970s developed various techniques to create, replicate, and analyze recombinant DNA molecules. This set of methods, called recombinant DNA technology, was a major advance in research, allowing scientists to isolate and study specific DNA sequences. For their contributions to the development of this technology, Nathans, Hamilton Smith, and Werner Arber were awarded the 1978 Nobel Prize for Physiology or Medicine.

The term **recombinant DNA** has two meanings in genetics. The more specific of the two is a DNA molecule formed in the laboratory by joining together DNA sequences from different biological sources. Such **recombinant DNA molecules** are artificial laboratory creations and are not found in nature. The term *recombinant DNA* is also used more loosely to refer to the technology that is utilized to create and study these hybrid molecules. The power of recombinant DNA technology is astonishing, enabling geneticists to identify and isolate a single gene or DNA segment of interest from the thousands or tens of thousands present in a genome. (The human genome, for example, contains more than 3 billion nucleotides and 20,000 to 25,000 genes.) Subsequently, through cloning, huge quantities of identical copies of this specific DNA molecule can be produced. These identical copies, or **clones,** can then be manipulated for numerous purposes, including research into the structure and organization of the DNA, or for the commercial production of its encoded protein. In this chapter, we review the basic methods of recombinant DNA technology used to isolate, replicate, and analyze genes. In Chapter 24, we will discuss some applications of this technology to research, medicine, the legal system, agriculture, and industry.

## 13.1 Recombinant DNA Technology Combines Several Laboratory Techniques

Although natural genetic processes such as crossing over produce recombined DNA molecules, the term *recombinant DNA* is generally reserved for molecules produced by artificially joining DNA obtained from different biological sources. The methods used to create these molecules are largely derived from nucleic acid biochemistry, coupled with genetic techniques developed for the study of bacteria and viruses. The basic method involves the following steps:

1. DNA to be cloned is purified from cells or tissues.
2. Proteins called **restriction enzymes** are used to generate specific DNA fragments. These enzymes recognize and cut DNA molecules at specific nucleotide sequences.
3. The fragments produced by restriction enzymes are joined to other DNA molecules that serve as vectors, or carrier molecules. A vector joined to a DNA fragment is a recombinant DNA molecule.
4. The recombinant DNA molecule is transferred to a host cell. Within the host cell, the recombinant molecule replicates, producing dozens of identical copies, or clones, of the recombinant molecule.
5. As host cells replicate, the recombinant DNA molecules within them are passed on to all their progeny, creating a population of host cells, each of which carries copies of the cloned DNA sequence.
6. The cloned DNA can be recovered from host cells, purified, and analyzed.
7. The cloned DNA can then be transcribed, its mRNA translated, and the encoded gene product isolated and used for research or sold commercially.

We will begin our discussion of this technology by considering two important components used to construct recombinant DNA molecules: restriction enzymes and vectors.

## 13.2 Restriction Enzymes Cut DNA at Specific Recognition Sequences

Restriction enzymes are produced by bacteria as a defense mechanism against infection by viruses. They restrict or prevent viral infection by degrading the DNA of invading viruses. More than 3500 restriction enzymes have been identified, and about 150 of these are commonly used by researchers. A restriction enzyme binds to DNA (Figure 13–1) and recognizes a specific nucleotide sequence called a **recognition sequence**. The enzyme then cuts both strands of the DNA within that sequence. The usefulness of restriction enzymes in cloning derives from their ability to accurately and reproducibly cut genomic DNA into fragments called **restriction fragments.** The size of restriction fragments is determined by the number of times a given restriction enzyme cuts the DNA. Enzymes with a four-base recognition sequence—such as the enzyme *Alu*I, which recognizes the sequence AGCT—will cut, on average, every 256 base pairs ($4^n = 4^4 = 256$) if all four nucleotides are present in equal proportions, producing many small fragments. Enzymes such as *Not*I have an eight-base recognition sequence (GCGGCCGC) and cut the

RECOMBINANT DNA TECHNOLOGY: RESTRICTION
WEB TUTORIAL 13.1

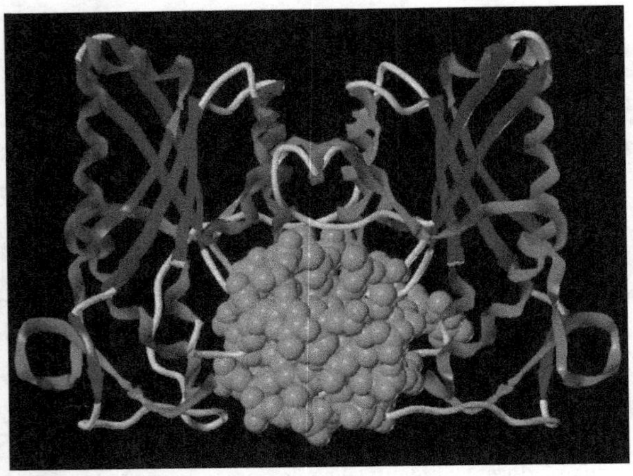

FIGURE 13–1 The restriction enzyme *Bam*H1 bound to a DNA molecule (green). Restriction enzymes cut DNA at specific sites.

DNA on average every 65,500 base pairs ($4^n = 4^8$), producing fewer, larger fragments. The actual fragment sizes produced by cutting DNA with a given restriction enzyme vary because the number and location of recognition sequences are not always distributed randomly in DNA.

Most recognition sequences exhibit a form of symmetry described as a **palindrome**: the nucleotide sequence reads the same on both strands of the DNA when read in the 5′ to 3′ direction. Each restriction enzyme recognizes its particular recognition sequence and cuts the DNA in a characteristic cleavage pattern. The most common recognition sequences are four or six nucleotides long, but some contain eight or more nucleotides. Enzymes such as *Eco*RI and *Hin*dIII make offset cuts in the DNA strands, thus producing fragments with single-stranded tails, while others such as *Alu*I and *Bal*I cut both strands at the same nucleotide pair, producing blunt ends. Some restriction enzymes and their recognition sequences are shown in Figure 13–2.

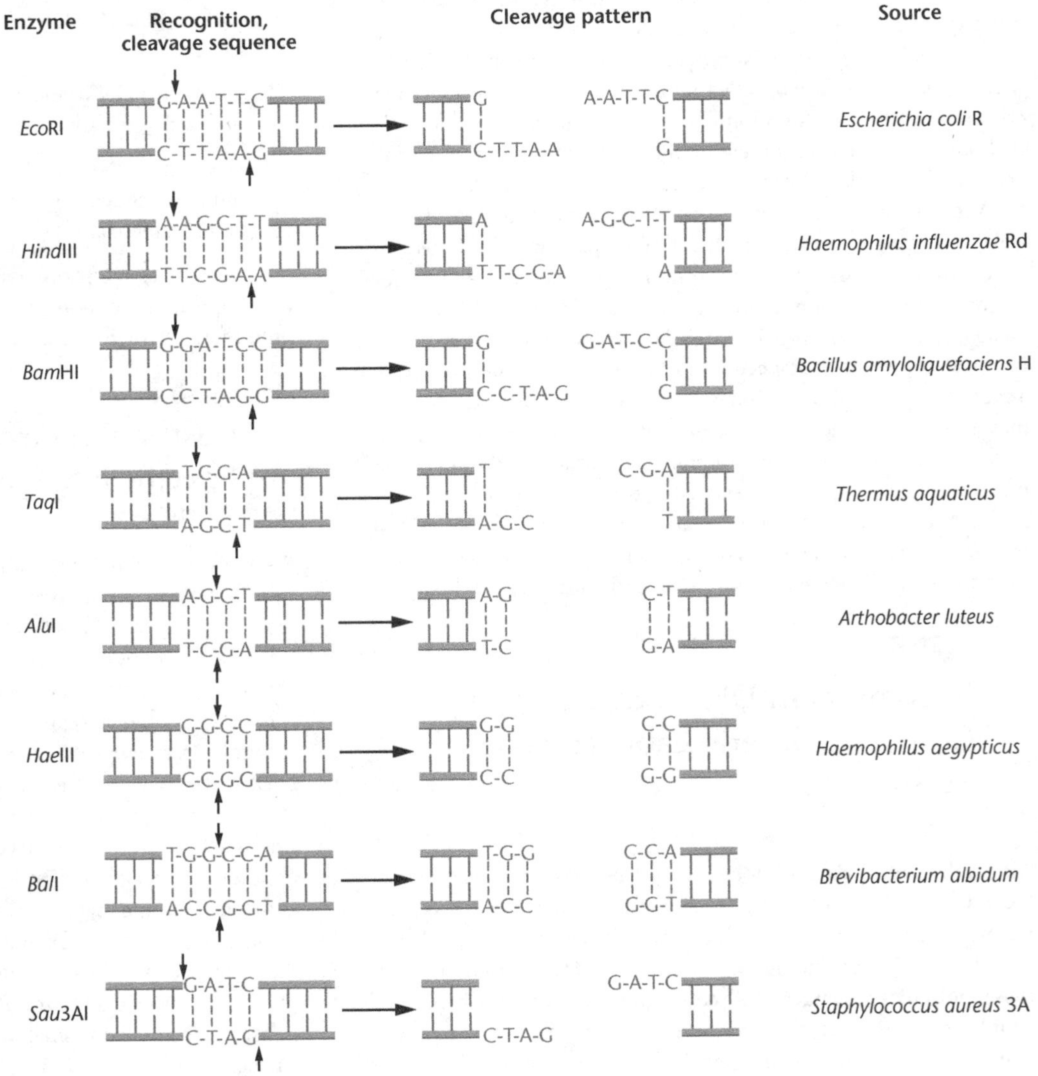


FIGURE 13–2 Some common restriction enzymes, with their recognition sequences, cleavage patterns, and sources. The arrows indicate the cutting sites for each enzyme.

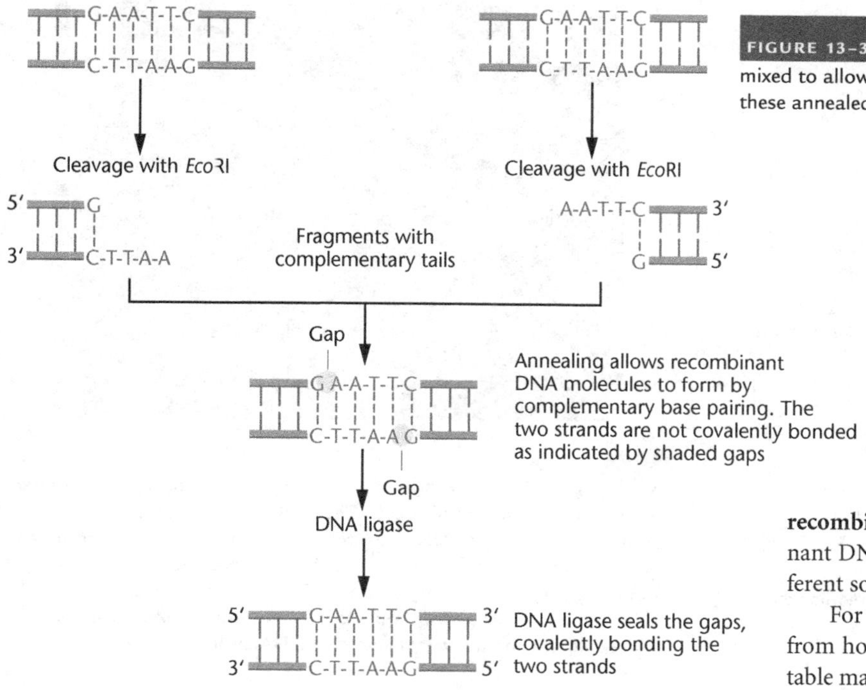


**FIGURE 13-3** DNA from different sources is cleaved with *Eco*RI and mixed to allow annealing. The enzyme DNA ligase then chemically bonds these annealed fragments into an intact recombinant DNA molecule.

One of the first restriction enzymes to be identified was isolated from *Escherichia coli* strain R and is designated *Eco*RI (pronounced echo-r-one or eeko-r-one). DNA fragments produced by *Eco*RI digestion (Figure 13–2) have overhanging single-stranded tails ("sticky ends") that can base-pair with complementary single-stranded tails on other DNA fragments. When mixed together in the correct proportions, complementary single-stranded ends of DNA fragments from different sources can **anneal**, or stick together, by base pairing of their single-stranded ends. Adding the enzyme **DNA ligase** to the solution causes covalent linkage of the fragments to form recombinant DNA molecules (Figure 13–3).

## 13.3 Vectors Carry DNA Molecules to Be Cloned

The fragments of DNA produced by restriction enzyme digestion are often destined to be copied (cloned) inside bacteria or some other host cell, but they cannot directly enter the bacterial cells for cloning without first being joined to a vector. **Vectors** are, in essence, carrier DNA molecules that transfer and help replicate inserted DNA fragments. Many different vectors are available for cloning; they differ in terms of the host cells they are able to enter, in the size of inserts they can carry, and also in other properties, such as the number of copies that can be produced, the number of recognition sequences available for cloning, and the number and type of marker genes they contain.

To serve as a vector, a DNA molecule must be able to independently replicate itself and any DNA fragment it carries once it is inside a host cell. The vector should also contain several restriction-enzyme cleavage sites that allow insertion of the DNA fragments to be cloned. For insertion of a DNA fragment, the vector is cut with a restriction enzyme and mixed with a collection of DNA fragments produced by cutting with the same enzyme. Vectors carrying an inserted fragment are called **recombinant vectors**, and each is an example of a recombinant DNA molecule, produced by joining DNA from two different sources.

For distinguishing host cells that have taken up vectors from host cells that have not, the vector should carry a selectable marker gene to identify cells that contain vectors (usually an antibiotic resistance gene or the gene for an enzyme absent from the host cell). Finally, the vector and its inserted DNA fragment should be easy to recover from the host cell.

### Plasmid Vectors

Genetically modified plasmids were the first vectors developed and are still widely used for cloning. These plasmid vectors were derived from naturally occurring plasmids, the extrachromosomal, double-stranded DNA molecules that replicate autonomously within bacterial cells (Figure 13–4). The genetics of plasmids and their host bacterial cells were introduced in Chapter 6 and will also be discussed in Chapter 17. In this section, we emphasize the use of plasmids as vectors for cloning DNA.

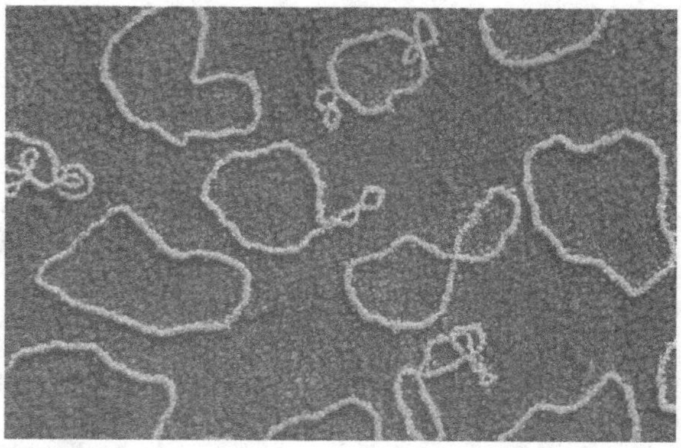

**FIGURE 13-4** A color-enhanced electron micrograph of circular plasmid molecules isolated from *E. coli*. Genetically engineered plasmids are used as vectors for cloning DNA.

RECOMBINANT DNA TECHNOLOGY: VECTORS
WEB TUTORIAL 13.2

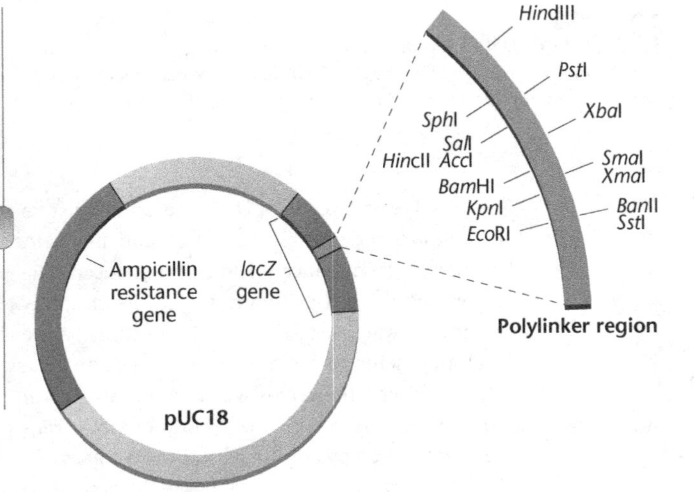


**FIGURE 13-5** A diagram of the plasmid pUC18 showing the polylinker region located within a *lacZ* gene. DNA inserted into the polylinker region disrupts the *lacZ* gene, resulting in white colonies that allow direct identification of bacterial colonies carrying cloned DNA inserts.

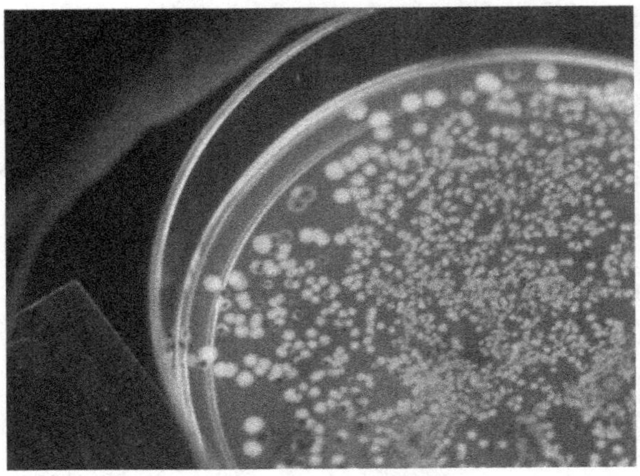

**FIGURE 13-6** A Petri dish showing the growth of bacterial cells after uptake of recombinant plasmids. The medium on the plate contains a compound called Xgal. DNA inserted into the pUC18 vector disrupts the gene responsible for the formation of blue colonies. As a result, it is easy to distinguish colonies carrying cloned DNA inserts. Cells in blue colonies contain vectors without cloned DNA inserts, whereas cells in white colonies contain vectors carrying DNA inserts.

Plasmids have been extensively modified by genetic engineering to serve as cloning vectors. Many are now available with a range of useful features. For example, although only a single plasmid generally enters a bacterial host cell, once inside, some plasmids can increase their number so that several hundred copies of the original plasmid are present. As vectors, these plasmids greatly enhance the number of DNA clones that can be produced. For added convenience, these vectors have also been genetically engineered to contain a number of convenient restriction-enzyme recognition sequences, as well as marker genes that reveal their presence in host cells.

One such plasmid is **pUC18** (Figure 13–5), which has several useful features as a vector.

1. It is small (2686 base pairs), so it can carry relatively large DNA inserts.

2. It has an origin of replication, necessary for initiating DNA synthesis, and it can produce up to 500 copies of inserted DNA fragments per cell.

3. A large number of restriction-enzyme recognition sequences have been engineered into pUC18, conveniently clustered in one region called a **polylinker site.**

4. Recombinant pUC18 plasmids are easily identified. For example, pUC18 carries a fragment of the bacterial *lacZ* gene as a selectable marker, and the polylinker site is inserted into this fragment. Expression of *lacZ* causes bacterial host cells carrying pUC18 to produce blue colonies when grown on medium containing a compound known as Xgal. If a DNA fragment is inserted anywhere in the polylinker site, the *lacZ* gene is disrupted and becomes inactive. Thus, a bacterial cell carrying pUC18 with an inserted DNA fragment forms white colonies on Xgal medium, whereas colonies carrying pUC18 plasmids without inserted DNA fragments form blue colonies (Figure 13–6).

The steps for generating a recombinant DNA molecule using pUC18 are illustrated in Figure 13–7.

## Lambda ($\lambda$) Phage Vectors

Plasmid vectors generally carry up to 25 kilobases (kb) of inserted DNA, but for many experiments, larger pieces of DNA are necessary. For these purposes, genetically modified strains of **$\lambda$ phage** are used as vectors. The genome of $\lambda$ phage, a virus that infects *E. coli*, has been completely mapped and sequenced. The central third of its chromosome can be replaced with foreign DNA without affecting the phage's ability to infect cells and form plaques (Figure 13–8).

To clone DNA using this vector, the phage DNA is isolated and cut with a restriction enzyme such as *Eco*RI, producing three chromosomal fragments: the left arm, the right arm, and the dispensable central region. The arms are then isolated and mixed with DNA from another source that also has been cut with *Eco*RI. Ligation with DNA ligase produces recombinant $\lambda$ vectors that are subsequently packaged into phage protein heads *in vitro* and introduced into bacterial host cells growing on Petri plates. Inside the bacteria, the vectors replicate and form many copies of infective phage, each of which carries a DNA insert. As they reproduce, they lyse their bacterial host cells, forming the clear spots known as plaques (described in Chapter 6), from which the cloned DNA can be recovered.

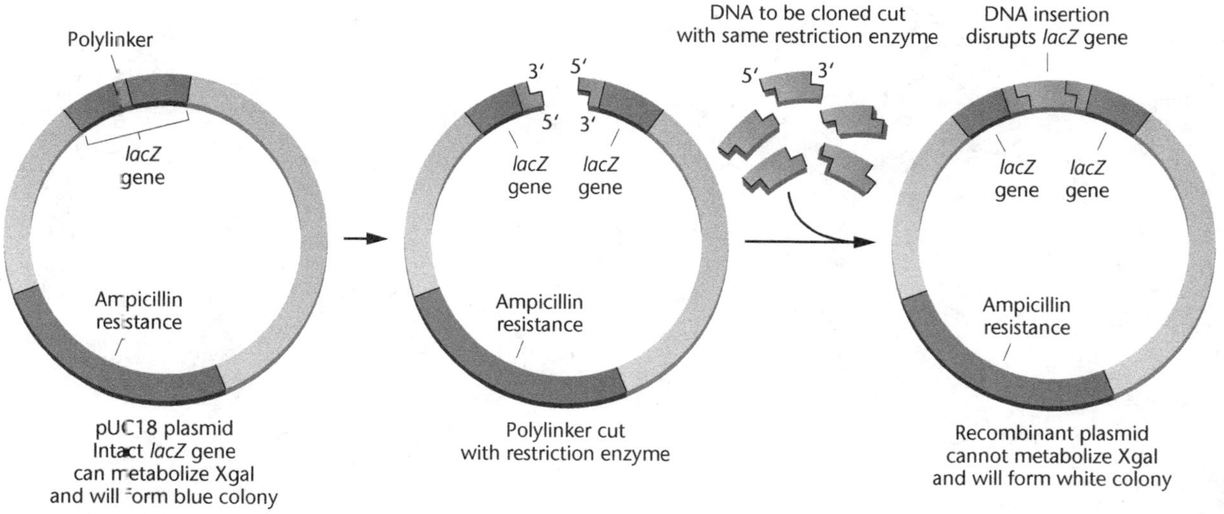


**FIGURE 13–7** The plasmid vector pUC18 carries unique restriction cleavage sites in a polylinker region within the *lacZ* gene. Cleavage of the plasmid and DNA to be cloned with a restriction enzyme, followed by insertion of a DNA fragment into the polylinker region, disrupts the *lacZ* gene, so that plasmids carrying inserts are unable to metabolize Xgal. These plasmids therefore form white colonies on nutrient plates containing Xgal.

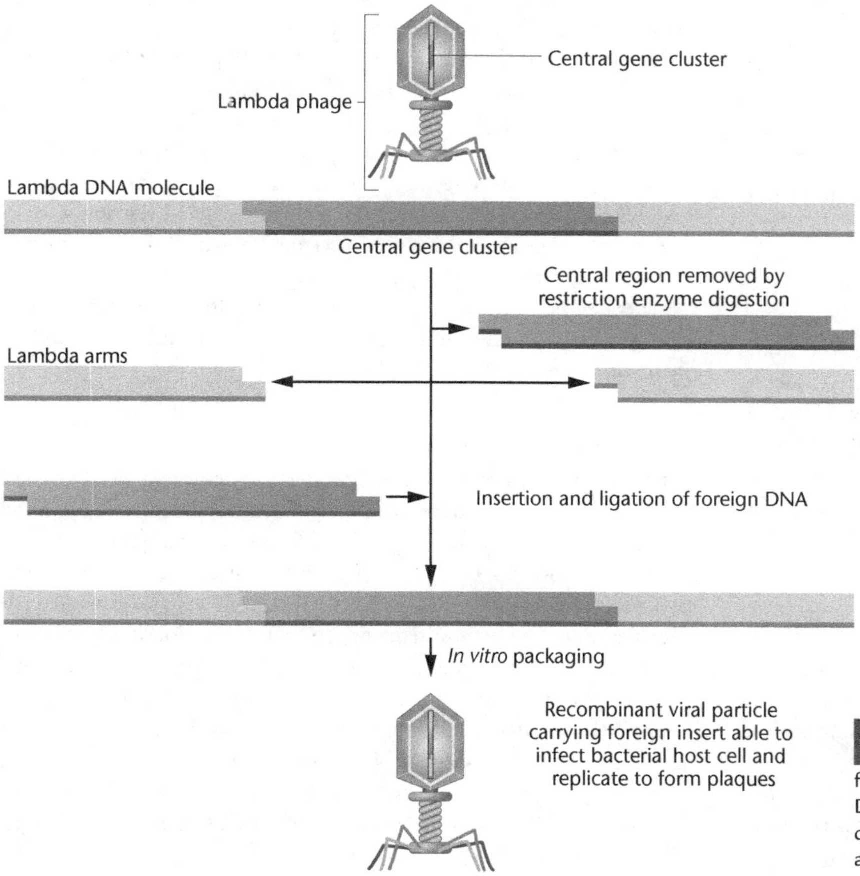


**FIGURE 13–8** λ phage as a vector. DNA is extracted from the phage, the central gene cluster is removed, and the DNA to be cloned is ligated into the arms of the λ chromosome. The recombinant chromosome is then packaged into phage proteins to form a recombinant virus.

Some phage vectors can carry inserts up to 45 kb, more than twice as long as DNA inserts in plasmid vectors. This is an important advantage for cloning large genes or small genomes. Other phage vectors will not accept inserts under a minimum size and thus do not end up carrying relatively useless small inserts only a few dozen or a few hundred nucleotides in length.

## Cosmid Vectors

**Cosmids** are hybrid vectors created by combining parts of the lambda chromosome with parts of plasmids. Cosmids contain the *cos* sequence of phage lambda, necessary for packing phage DNA into phage protein coats, and have plasmid sequences necessary for replication. They also contain a plasmid-derived antibiotic resistance gene (Figure 13–9) as a selectable marker to help identify host cells carrying recombinant cosmids. After insertion of DNA fragments, recombinant cosmids

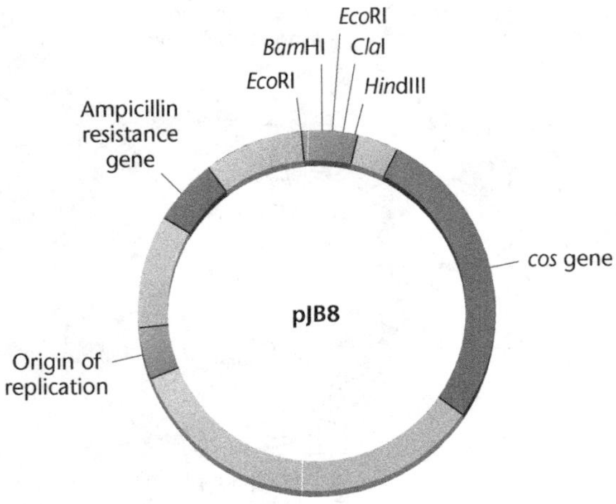


FIGURE 13–9 The cosmid pJB8 contains a bacterial origin of replication (*ori*), a single cos sequence (*cos*), an ampicillin resistance gene (*amp*, for selection of colonies that have taken up the cosmid), and a region containing four insertion sites for cloning (*Bam*HI, *Eco*RI, *Cla*I, and *Hind*III). Because the vector is small (5.4 kb long), it can accept foreign DNA segments between 33 and 46 kb in length. The *cos* sequence allows cosmids carrying large inserts to be packaged into lambda viral coat proteins as though they were viral chromosomes. The viral coats carrying the cosmid can be used to infect a suitable bacterial host, and the vector, carrying a DNA insert, will be transferred into the host cell. Once inside, the *ori* sequence allows the cosmid to replicate as a bacterial plasmid.

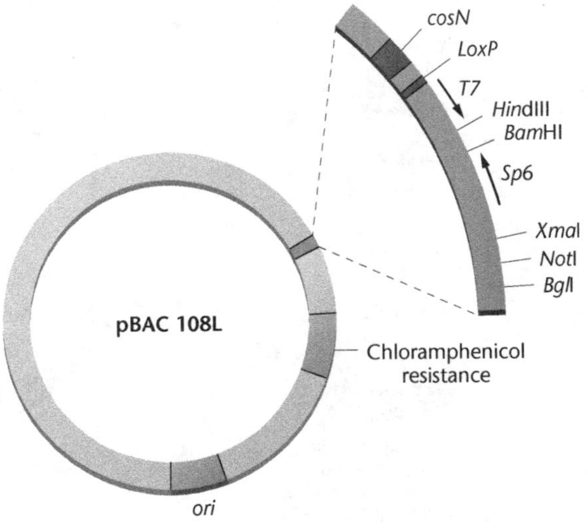


FIGURE 13–10 A bacterial artificial chromosome (BAC). The polylinker carries a number of unique sites for the insertion of foreign DNA. The arrows labeled T7 and Sp6 include regulatory sequences that allow expression of genes inserted between the two regions.

are packaged into lambda protein heads, forming infective phage particles. Once inside a bacterial host cell, the cosmid replicates as a plasmid. Because only a very small part of the lambda genome has been retained, cosmids can carry DNA inserts that are much larger than those carried by lambda vectors. Cosmids can carry almost 50 kb of inserted DNA, whereas phage vectors can accommodate DNA inserts 10–15 kb in length.

Other hybrid vectors with origins of replication derived from different sources (e.g., animal viruses such as SV40) can replicate in more than one type of host cell (both prokaryotic and eukaryotic) and are called *shuttle vectors*. These vectors usually contain genetic markers that are selectable in both types of host cells and are used to shuttle DNA inserts between *E. coli* and another type of host cell, such as yeast. Often such vectors are employed in studying gene expression.

## Bacterial Artificial Chromosomes

The mapping and analysis of large complex eukaryotic genomes require cloning vectors that can carry very large DNA fragments. Some human genes are 1000 kb to more than 2000 kb long, so the vectors that carry them must have large cloning capacities.

One vector with a large cloning capacity is based on the fertility plasmid (F factor) of bacteria and is called a **bacterial artificial chromosome (BAC)**. Recall from Chapter 6 that F factors are independently replicating plasmids that transfer genetic information during bacterial conjugation. Because F factors can carry fragments of the bacterial chromosome up to 1 Mb (1000 kb) in length, they have been engineered to act as vectors for eukaryotic DNA and can carry inserts of about 300 kb (Figure 13–10). BAC vectors carry F factor genes for replication and copy number, and have at least one antibiotic resistance gene as a selectable marker as well as a polylinker region containing a number of clustered restriction enzyme recognition sequences for inserting foreign DNA. In addition, the polylinker is flanked by promoter sequences that can be used to generate RNA molecules for the expression of the cloned gene, serve as probes in chromosome walking (Chapter 21), and sequence the cloned insert.

## Expression Vectors

**Expression vectors** are vectors designed to activate a cloned gene and produce many copies of the gene's encoded protein in a host cell. Expression vectors are available for both prokaryotic and eukaryotic host cells. One expression vector for use in an *E. coli* host cell is pET (Figure 13–11). In this vector, the gene to be expressed is cloned into a restriction sequence so that its location is adjacent to a viral regulatory sequence (called the T7 promoter) and a second regulatory gene (the bacterial *lac* operator gene) (Chapter 17). These sequences control expression of the cloned gene when a specific component is added to the growth medium. Expression is induced by adding the lactose analog IPTG to the medium. Host cells growing on IPTG activate expression of the cloned gene, producing large quantities of the encoded protein.

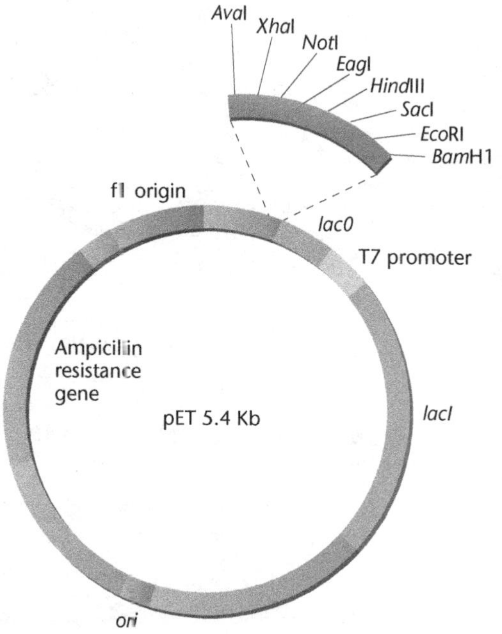


**FIGURE 13–11** A pET expression vector. This system uses a genetically engineered host cell. The host cell carries the viral T7 RNA polymerase gene under the control of two different regulatory sequences: (1) a *lac* promoter, which controls expression of the inserted gene, and (2) an operator, which expresses the gene only when the lactose analog IPTG is present in the growth medium.

## 13.4

## DNA Was First Cloned in Prokaryotic Host Cells

As discussed earlier, scientists use recombinant DNA technology to construct *and* replicate recombinant DNA molecules in order to clone specific DNA sequences. Replication takes place after transfer of recombinant molecules into host cells. This cloning method, known as cell-based cloning, was the first method developed and was carried out using first prokaryotic host cells and later eukaryotic host cells. It is still widely used to make cloned DNA.

After being exposed to the recombinant vectors, the prokaryotic host cells are plated on nutrient medium, where they form colonies. These are screened to identify colonies that have taken up the recombinant plasmids. Because the cells in each colony are derived from a single ancestral cell, all the cells in the colony, as well as the plasmids they contain, are genetically identical clones.

A variety of prokaryotic cells can serve as host for recombinant vector replication. One of the most commonly used hosts is a laboratory strain of the bacterium *E. coli* known as K12. *E. coli* strains such as K12 are genetically well characterized and can host a wide range of vectors. The general steps for cloning with a plasmid vector are summarized in Figure 13–12. Similarly, phages containing foreign DNA are used to infect *E. coli* host cells, and when the cells are plated on solid medium, each resulting plaque represents a cloned descendant of a single ancestral bacteriophage.

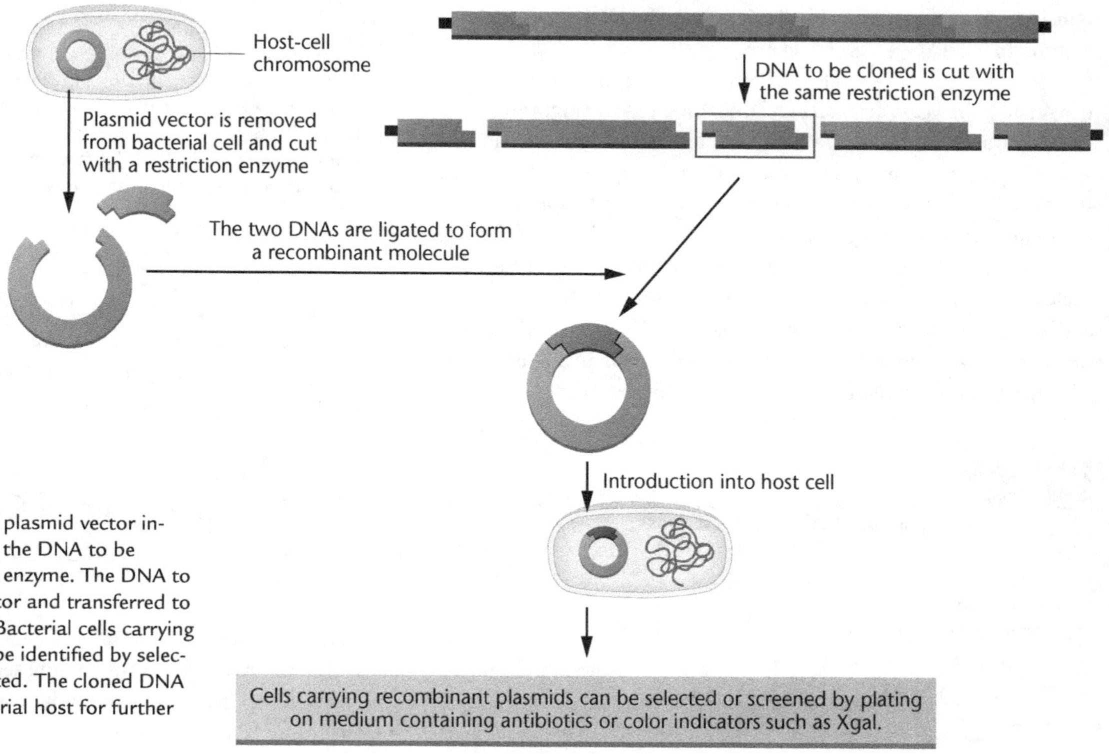


**FIGURE 13–12** Cloning with a plasmid vector involves cutting both plasmid and the DNA to be cloned with the same restriction enzyme. The DNA to be cloned is spliced into the vector and transferred to a bacterial host for replication. Bacterial cells carrying plasmids with DNA inserts can be identified by selection or screening and then isolated. The cloned DNA is then recovered from the bacterial host for further analysis.

**NOW SOLVE THIS**

Question 12 on page 348 relates to screening for successful insertions in a vector carrying two antibiotic resistance genes.

■ HINT: *Inserting foreign DNA into the vector disrupts one of the resistance genes in the plasmid. Bacteria taking up plasmids with inserts will be able to grow on medium containing only one of the antibiotics.*

13.5

## Yeast Cells Are Used as Eukaryotic Hosts for Cloning

Whereas *E. coli* is widely used as a prokaryotic host cell, the yeast *Saccharomyces cerevisiae* is extensively used as a host cell for the cloning and expression of eukaryotic genes. There are several reasons: (1) Although yeast is a eukaryotic organism, it can be grown and manipulated in much the same way as bacterial cells. (2) The genetics of yeast has been intensively studied, providing a large catalog of mutants and a highly developed genetic map. (3) The entire yeast genome has been sequenced, and most genes in the organism have been identified. (4) To study the function of some eukaryotic proteins, it is necessary to use a host cell that can modify the protein after it has been synthesized, to convert it to a functional form (bacteria cannot carry out some of these modifications). (5) Yeast has been used for centuries in the baking and brewing industries and is considered to be a safe organism for producing proteins for vaccines and therapeutic agents. Table 13.1 lists some of the products of cloning in yeast.

Several types of yeast cloning vectors have been developed, one of which is the **yeast artificial chromosome (YAC)** (Figure 13–13). Like natural chromosomes, a YAC has telomeres at each end, an origin of replication, and a centromere. These components are joined to selectable marker genes (*TRP1* and *URA3*) and to a cluster of restriction-enzyme recognition sequences for insertion of foreign DNA. Yeast chromosomes range in size from 230 kb to over 1900 kb, making it possible to clone DNA inserts from 100 to 1000 kb in YACs. The ability to clone large pieces of DNA in these vectors makes them an important tool in genome sequencing projects, including the Human Genome Project (see Chapter 21).

**TABLE 13.1**

| Recombinant Proteins Synthesized in Yeast Cells |
|---|
| Hepatitis B virus surface protein |
| Malaria parasite protein |
| Epidermal growth factor |
| Platelet-derived growth factor |
| $\alpha_1$-antitrypsin |
| Clotting factor XIIIA |

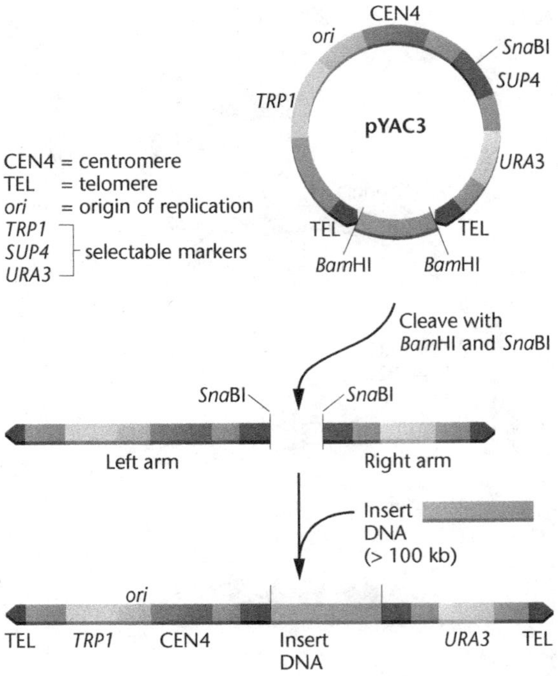


**FIGURE 13–13** The yeast artificial chromosome pYAC3 contains telomere sequences (TEL), a centromere (CEN4) derived from yeast chromosome 4, and an origin of replication (*ori*). These elements give the cloning vector the properties of a chromosome. *TRP1* and *URA3* are yeast genes that are selectable markers for the left and right arms of the chromosome. Within the *SUP4* gene is a restriction enzyme recognition sequence for the enzyme *Sna*B1. Two *Bam*H1 recognition sequences flank a spacer segment. Cleavage with *Sna*B1 and *Bam*H1 breaks the artificial chromosome into two arms. The DNA to be cloned is treated with *Sna*B1, producing a collection of fragments. The arms and fragments are ligated together, and the artificial chromosome is inserted into yeast host cells. Because naturally occurring yeast chromosomes range in size from 230 kb to over 1900 kb, it is possible to insert large DNA fragments in creating YACs. Insert sizes can be in the million base-pair range.

Yeast vectors and host cells are currently the most advanced eukaryotic systems used for cloning, but other systems are being developed, including human artificial chromosome vectors with mammalian cells as hosts, which we discuss in the following section.

13.6

## Plant and Animal Cells Can Be Used as Host Cells for Cloning

Like yeast, other eukaryotic cells, including both plant and animal cells, can serve as hosts for cloning segments of DNA that have been inserted into vectors. Moreover, several different types of vectors, including YACs, can be used to transfer DNA into eukaryotic cells.

When the vector is a plasmid, DNA transfer is referred to as **transformation.** When the vector is a virus, the term **transfection** is used to describe uptake.

## Plant Cell Hosts

In one widely used system, gene transfer into higher plants is accomplished by bacterial plasmid vectors. The bacterium *Agrobacterium tumifaciens* infects plant cells and produces tumors (called plant galls) in many species of plants. Tumor formation is associated with the presence of a tumor-inducing (Ti) plasmid carried in the bacteria (Figure 13–14). When Ti plasmid-carrying bacteria infect plant cells, a segment of the Ti plasmid, known as *T-DNA*, is transferred into the genome of the host plant cell. Genes in the T-DNA segment control tumor formation and the synthesis of compounds required for growth of the infecting bacteria. Foreign genes can be inserted into one of the multiple restriction sites and the recombinant plasmid can be transferred into plant cells by infection with *A. tumifaciens.* Once inside the cell, the foreign DNA is inserted into the plant genome when the T-DNA integrates into a host-cell chromosome. Plant cells carrying a recombinant Ti plasmid can be grown in tissue culture to form a cell mass called a *callus*. The presence of certain compounds in the culture medium can induce the callus to form roots and shoots, and eventually a mature plant carrying a foreign gene.

Plants (or animals) carrying a foreign gene are called **transgenic** organisms. In Chapter 24, we will see how gene transfer has been used to alter crop plants.

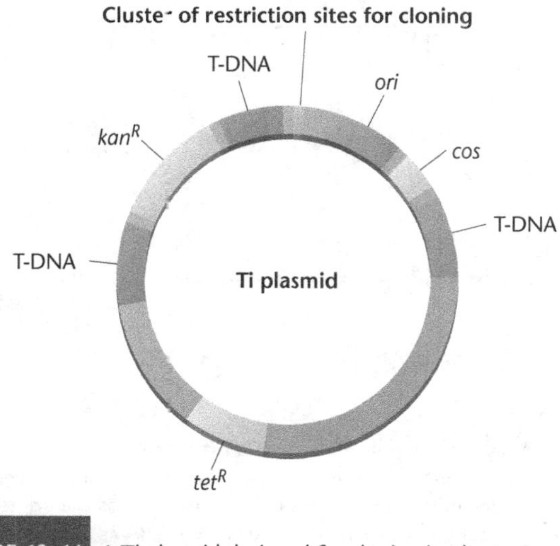


FIGURE 13–14 A Ti plasmid designed for cloning in plants. Segments of T-DNA, including those necessary for integration, are combined with bacterial segments that incorporate cloning sites and antibiotic resistance genes ($kan^R$ and $tet^R$). The vector also contains an origin of replication (*ori*), as well as a lambda *cos* sequence that permits recovery of cloned inserts from the host plant cell.

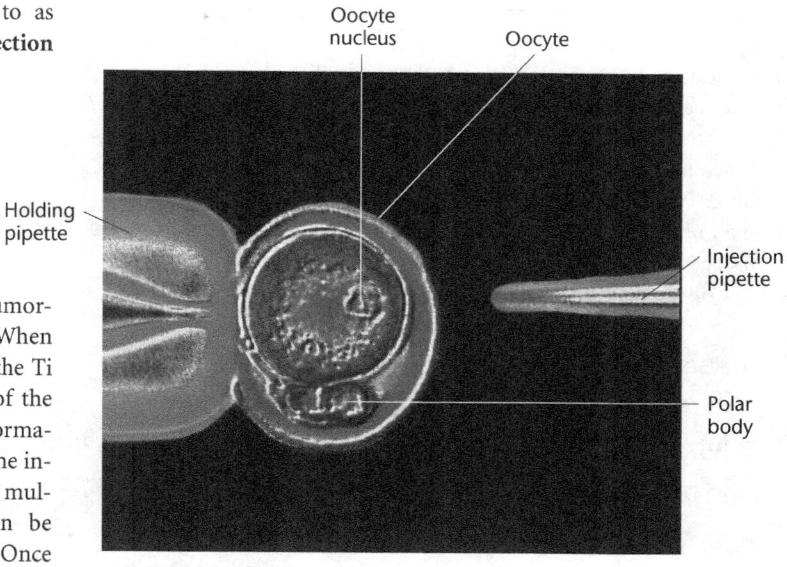


FIGURE 13–15 Cloned DNA can be transferred into mammals by direct injection into the oocytes.

## Mammalian Cell Hosts

DNA can be transferred into mammalian cells by several methods, including endocytosis, or encapsulation of DNA into artificial membranes (liposomes) that then fuse with cell membranes. DNA can also be transferred using YACs and vectors based on retroviruses. DNA introduced into a mammalian cell by any of these methods is usually integrated into the host genome. Genes may be transferred into fertilized eggs to produce transgenic animals. These same methods are also used to replicate cloned genes using mammalian cells as hosts.

YACs are used as vectors for several purposes, one being to increase the efficiency of gene transfer into the germ line of mice. The first step in producing transgenic mice involves transferring a recombinant YAC into the nucleus of an appropriate mouse cell, such as a fertilized egg or an undifferentiated embryonic cell (a stem cell), followed by the integration of the DNA into a chromosome.

YACs are transferred to mice in several ways. One method uses microinjection of purified YAC DNA into the nucleus of a mouse oocyte (Figure 13–15). Transgenic zygotes are then implanted in foster mothers for development. Another method transfers YACs into mouse embryonic stem (ES) cells by fusing a yeast cell carrying a YAC with a mouse stem cell, which transfers the inserted genes and all or most of the yeast genome into the stem cell. These transgenic ES cells are injected into early-stage mouse embryos, where they participate in the formation of adult tissues, including those that form germ cells. The ability to transfer large DNA segments into mice has applications in many areas of research. Some of these are described in Chapter 24.

Other vectors for mammalian cells are based on genetically engineered avian and mouse retroviruses. These retroviruses have single-stranded RNA molecules as their genomes. After infection of the host cell, the RNA is transcribed by reverse transcriptase into a double-stranded DNA (dsDNA) molecule. The dsDNA integrates into the host genome and is passed on to daughter cells during cell division. The retroviral genome can be engineered so that it lacks a number of viral genes, the better to accept foreign DNA, including human genes. These vectors are used in gene therapies for treating genetic disorders, a topic that is also discussed in Chapter 24.

POLYMERASE CHAIN REACTION
WEB TUTORIAL 13.3

## 13.7 The Polymerase Chain Reaction Makes DNA Copies Without Host Cells

The recombinant DNA techniques developed in the early 1970s also gave birth to the booming biotechnology industry. However, cloning DNA using vectors and host cells can be labor intensive and time-consuming. In 1986, another technique, called the **polymerase chain reaction (PCR)**, was developed. This advance again revolutionized recombinant DNA methodology and further accelerated the pace of biological research. The significance of this method was underscored by the awarding of the 1993 Nobel Prize in Chemistry to Kary Mullis, who developed the technique.

PCR is a rapid method of DNA cloning that extends the power of recombinant DNA research and in many cases eliminates the need to use host cells for cloning. Although cell-based cloning is still widely used, PCR is the method of choice for many applications, whether in molecular biology, human genetics, evolution, development, conservation, or forensics.

By copying a specific DNA sequence through a series of *in vitro* reactions, PCR can amplify target DNA sequences that are initially present in very small quantities in a population of other DNA molecules. As a prerequisite for PCR, some information about the nucleotide sequence of the target DNA is required. This sequence information is used to synthesize two oligonucleotide primers: one complementary to the 5′ end and one complementary to the 3′ end of the target DNA. When added to a sample of DNA that has been converted into single strands, the primers bind to complementary nucleotides flanking the sequence to be cloned. A heat-stable DNA polymerase is added after this hybridization, and it synthesizes a second strand of the target DNA (Figure 13–16). Repetition of the process produces large numbers of copies of the DNA very quickly.

In practice, the PCR reaction involves three steps. The amount of amplified DNA produced is theoretically limited only by the number of times these steps are repeated.

1. The DNA to be cloned is *denatured* into single strands. The DNA can come from many sources, including genomic DNA, mummified remains, fossils, or forensic samples such as dried

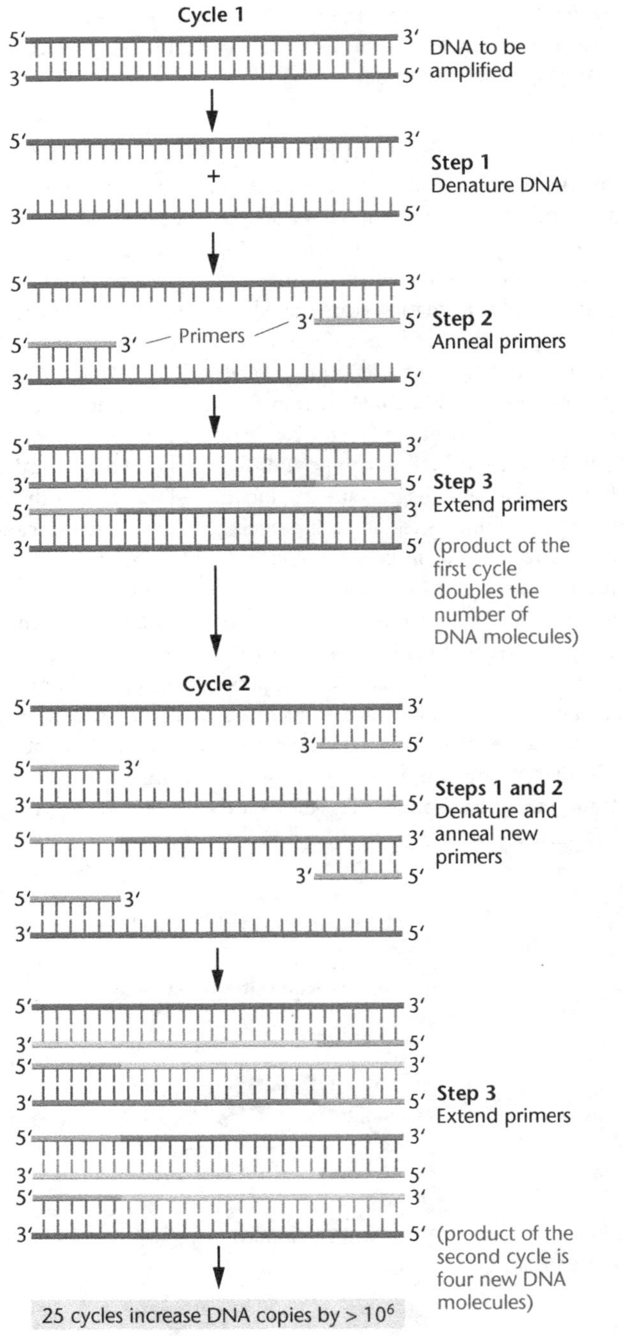


**FIGURE 13–16** In the polymerase chain reaction (PCR), the target DNA is denatured into single strands; each strand is then annealed to a short, complementary primer. DNA polymerase extends the primers in the 5′ to 3′ direction, using the single-stranded DNA as a template. The result is two newly synthesized double-stranded DNA molecules, with the primers incorporated into them. Repeated cycles of PCR can quickly amplify the original DNA sequence more than a millionfold.

blood or semen, single hairs, or dried samples from medical records. Heating to 90–95°C denatures the double-stranded DNA, which dissociates into single strands (usually in about 1 minute).

2. The temperature of the reaction is lowered to an *annealing temperature* between 50°C and 70°C, which causes the primers to bind to the denatured, single-stranded DNA. As described earlier, the primers are synthetic oligonucleotides (15–30 nucleotides long) complementary to sequences flanking the target DNA. The primers serve as starting points for synthesizing new DNA strands complementary to the target DNA.

3. A heat-stable form of DNA polymerase (such as *Taq* polymerase) is added to the reaction mixture, and DNA synthesis is carried out at temperatures between 70°C and 75°C (the optimal temperature for this form of polymerase). The *Taq* polymerase *extends* the primers by adding nucleotides in the 5′-to-3′ direction making a double-stranded copy of the target DNA.

Each set of three steps—*denaturation* of the double-stranded DNA, *primer annealing*, and *extension* by polymerase—is a cycle. PCR is a chain reaction because the number of new DNA strands is doubled in each cycle, and the new strands, along with the old strands, serve as templates in the next cycle. Each cycle takes 2 to 5 minutes and can be repeated immediately, so that in less than 3 hours, 25 to 30 cycles result in over a millionfold increase in the amount of DNA (Figure 13–16). This process is automated by machines called *thermocyclers* that can be programmed to carry out a predetermined number of cycles, yielding large amounts of a specific DNA sequence that can be used for many purposes, including cloning into plasmid vectors, DNA sequencing, clinical diagnosis, and genetic screening.

PCR-based DNA cloning has several advantages over cell-based cloning. PCR is rapid and can be carried out in a few hours, rather than the days required for cell-based cloning. In addition, the design of PCR primers is done automatically with computer software, and the commercial synthesis of the oligonucleotides is also fast and economical. If desired, the products of PCR can be cloned into plasmid vectors for further use.

PCR is also very sensitive and amplifies specific DNA sequences from vanishingly small DNA samples, including the DNA in a single cell. This feature of PCR is invaluable in several kinds of applications, including genetic testing, forensics, and molecular paleontology. With carefully designed primers, DNA samples that have been partially degraded, contaminated with other materials, or embedded in a matrix (such as amber) can be recovered and amplified, when conventional cloning would be difficult or impossible.

### Limitations of PCR

Although PCR is a valuable technique, it does have limitations: some information about the nucleotide sequence of the target DNA must be known, and even minor contamination of the sample with DNA from other sources can cause problems. For example, cells shed from the skin of a researcher can contaminate samples gathered from a crime scene or taken from fossils, making it difficult to obtain accurate results. PCR reactions must always be performed in parallel with carefully designed and appropriate controls.

### Other Applications of PCR

PCR DNA cloning is one of the most widely used techniques in genetics and molecular biology. PCR and its variations have many other applications as well. They quickly identify restriction-enzyme recognition sequence variants as well as variations in tandemly repeated DNA sequences that can be used as genetic markers in gene-mapping studies and forensic identification. Gene-specific primers provide a way of screening for mutations in genetic disorders, allowing the location and nature of mutation to be determined quickly. Primers can be designed to distinguish between target sequences that differ by only a single nucleotide. (This makes it possible to synthesize allele-specific probes for genetic testing.) Random primers indiscriminately amplify DNA and are particularly advantageous when studying samples from single cells, fossils, or a crime scene, where a single hair or even a saliva-moistened postage stamp is the source of the DNA. Using PCR, researchers can also explore uncharacterized DNA regions adjacent to known regions and even sequence DNA. PCR has been used to enforce the worldwide ban on the sale of certain whale products and to settle arguments about the pedigree background of purebred dogs. In short, PCR is one of the most versatile techniques in modern genetics.

## 13.8 Recombinant Libraries Are Collections of Cloned Sequences

Only relatively small DNA segments—representing only a single gene or even a portion of a gene—are produced through cloning in conventional plasmid vectors. As a result, a large collection of clones is needed for exploring even a small fraction of an organism's genome. A set of DNA clones derived from a single individual or a single population is called a cloned *library*. These libraries can represent an entire genome, a single chromosome, or a set of genes that are expressed in a single cell type.

### Genomic Libraries

Ideally, a **genomic library** contains at least one copy of every sequence in an organism's genome. Genomic libraries are constructed using host-cell cloning methods, since PCR-cloned DNA fragments are relatively small. In making a genomic library, DNA is extracted from cells or tissues and cut with restriction enzymes, and the resulting fragments are inserted into vectors. Since some vectors (such as plasmids) carry only a few thousand base pairs of inserted DNA, selecting the vector so that the library contains the whole genome in the smallest number of clones is an important consideration.

How big does a genomic library have to be to have a 95 or 99 percent chance of containing all the sequences in a genome? The number of clones required to contain a genome depends on several factors, including the average size of the cloned inserts, the size of the genome to be cloned, and the level of probability desired. The number of clones in a library can be calculated as

$$N = \frac{\ln(1 - P)}{\ln(1 - f)}$$

where $N$ is the number of required clones, $P$ is the probability of recovering a given sequence, and $f$ is the fraction of the genome in each clone.

Suppose we wish to prepare a human genome library large enough to have a 99 percent chance of containing all the sequences in the genome. Because the human genome is so large, the choice of vector is a primary consideration in making this library. If we construct the library using a plasmid vector with an average insert size of 5 kb, then more than 2.4 million clones would be required for a 99 percent probability of recovering any given sequence from the genome. Because of its size, this library would be difficult to use efficiently. If a phage vector with an average insert size of 17 kb is used to construct the library, then about 800,000 clones would be required for a 99 percent probability of finding any given human sequence. While it is much smaller than a plasmid library, screening a phage library of this size would still be a labor-intensive chore. However, if the library was constructed in a YAC vector with an average insert size of 1 Mb, then the library would only need to contain about 14,000 YACs, making it relatively easy to use. Vectors with large cloning capacities such as YACs are essential tools for the Human Genome Project.

## Chromosome-Specific Libraries

A library made from a subgenomic amount of DNA, such as that from a single chromosome, can be of great value in the cloning of specific genes and in the study of chromosome organization. In one case, *Drosophila* DNA from a small segment of the X chromosome about 50 polytene bands long was isolated by manually dissecting this chromosome region. The DNA in this chromosomal fragment was purified, cut with a restriction endonuclease, inserted into a lambda vector, and cloned to produce a library of the DNA sequences contained in a specific chromosome region. This X-chromosomal region contains the genes *white*, *zeste*, and *Notch*, as well as an insertion site for a transposable element that can translocate a chromosomal segment to more than 100 other loci scattered throughout the genome (transposable elements will be discussed in Chapter 16). This technically challenging procedure produced a library that contains only the genes of interest and their adjacent sequences, saving the time and effort that would otherwise have been needed to sort through a genomic library to recover all the clones from this region of the X chromosome.

Cloned libraries prepared from individual human chromosomes are made using a technique known as flow cytometry. To isolate individual chromosomes, mitotic cells are collected, and the metaphase chromosomes are stained with two fluorescent dyes, one that binds to AT pairs, the other to GC pairs. The stained chromosomes flow past a laser beam that stimulates them to fluoresce, and a photometer sorts and fractionates the chromosomes by differences in dye binding and light scattering (Figure 13–17). Once the chromosomes are isolated, DNA is extracted and cut with a restriction enzyme, and the DNA fragments are inserted into a vector and cloned. Libraries for each human chromosome are available; these libraries played an important role in the Human Genome Project (to be discussed in Chapter 21).

Individual chromosomes have been isolated for library construction in other ways. A version of gel electrophoresis known as **pulsed-field gel electrophoresis** that separates very large DNA molecules was used to isolate yeast chromosomes to be used in the con-

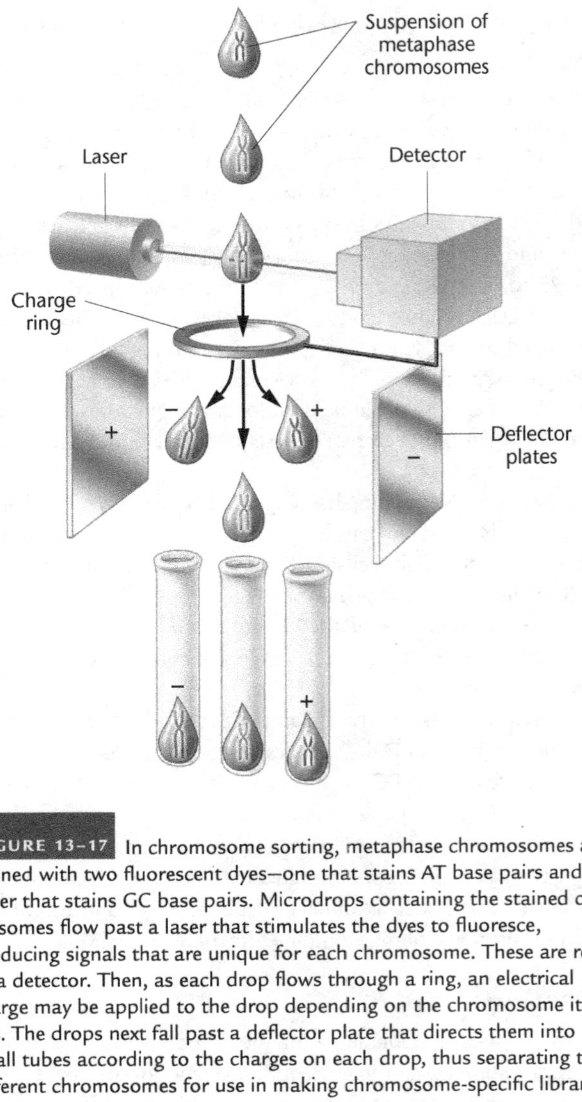


FIGURE 13–17 In chromosome sorting, metaphase chromosomes are stained with two fluorescent dyes—one that stains AT base pairs and another that stains GC base pairs. Microdrops containing the stained chromosomes flow past a laser that stimulates the dyes to fluoresce, producing signals that are unique for each chromosome. These are read by a detector. Then, as each drop flows through a ring, an electrical charge may be applied to the drop depending on the chromosome it carries. The drops next fall past a deflector plate that directs them into small tubes according to the charges on each drop, thus separating the different chromosomes for use in making chromosome-specific libraries.

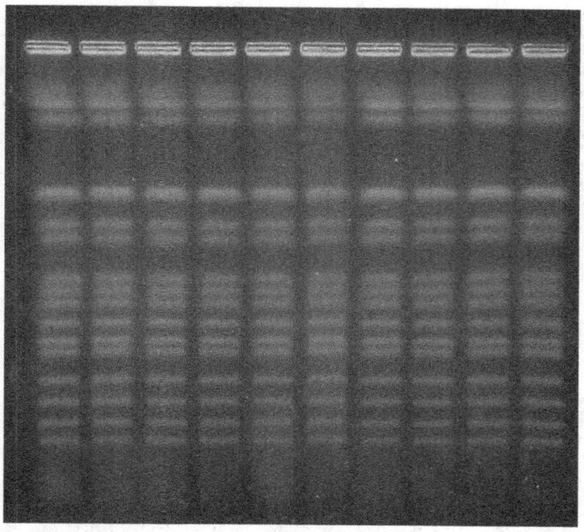

FIGURE 13–18 Intact yeast chromosomes separated using a method employing pulsed-field gel electrophoresis. In each lane, 15 of the 16 yeast chromosomes are visible, separated by size, with the largest chromosomes at the top.

struction of chromosome-specific libraries (Figure 13–18; electrophoresis was introduced in Chapter 10). A cloned library of yeast chromosome III (315 kb) produced in this way was the starting point for the Yeast Genome Project, in which a consortium of laboratories sequenced this chromosome and the rest of the yeast genome.

One of the unexpected results of sequencing chromosome III was the discovery that about half of all the genes on this chromosome were previously unknown. It was difficult for many geneticists to accept the fact that the time-tested methods of mutagenesis and gene mapping used for decades were so inefficient. However, this finding was confirmed and extended when the sequence of yeast chromosome XI (664 kb) was published and when the sequencing of the entire yeast genome was completed in 1996.

Single-chromosome libraries are valuable for gaining access to genetic loci when conventional methods such as mutagenesis and genetic analysis have been unsuccessful and when other probes, such as mRNA or gene products, are unavailable or unknown. In addition, chromosome-specific libraries provide a means for studying the molecular organization and even the nucleotide sequence in a defined region of the genome.

## cDNA Libraries

To study specific events in development, cell death, cancer, and other biological processes, a library of the subset of the genome that is expressed in a given cell type at a given time can be a valuable tool. Genomic libraries and chromosome libraries contain all the genes in a genome or on a chromosome, but these collections cannot be directly used to find genes that are actively expressed in a cell.

A **cDNA library** contains DNA copies made from the messenger RNA (mRNA) molecules present in a cell population at a given time, and therefore represents the genes being expressed in the cells at the time the library was made. (Recall from Chapter 1 that genes are transcribed to produce single-stranded messenger RNA molecules.) It is called a cDNA library because the DNA it contains—known as **complementary DNA**, or **cDNA**—is complementary to the nucleotide sequence of the mRNA.

Clones in a cDNA library are not the same as those in a genomic library. Eukaryotic mRNA transcripts are processed to remove some sequences (see Chapter 14), and a poly-A tail is added to the end of the mature mRNA molecule. Moreover, an mRNA molecule does not include the sequences adjacent to the gene that regulate its activity.

A cDNA library is prepared by isolating mRNA from a population of cells. This is possible because almost all eukaryotic mRNA molecules contain a poly-A tail at the 3′ end. mRNAs with poly-A tails are isolated and used as templates for the synthesis of cDNA molecules. The cDNA molecules are subsequently inserted into vectors and cloned to produce a library that is a snapshot of genes that were transcriptionally active at a given time.

The first step in making a cDNA library is to mix mRNAs having poly-A tails with oligo-dT primers, which anneal to the poly-A, forming a partially double-stranded product (Figure 13–19). The enzyme **reverse transcriptase** extends the primer and synthesizes a complementary DNA copy of the mRNA sequence. The product of this reaction is an mRNA–DNA double-stranded hybrid molecule. Action of the enzyme **RNAse H** introduces nicks in the RNA strand by partially digesting the RNA. The remaining RNA fragments serve as primers for the enzyme DNA polymerase I. (This situation is similar to synthesis of the lagging strand of DNA in prokaryotes.) DNA polymerase I synthesizes a second DNA strand and removes the RNA primers, producing double-stranded cDNA.

The cDNA can be inserted into a plasmid or phage vector by attaching linker sequences to the ends of the cDNA. Linker sequences are short double-stranded oligonucleotides containing a restriction enzyme recognition sequence (e.g., *Eco*RI). After attachment to the cDNAs, the linkers are cut with *Eco*RI and ligated to vectors treated with the same enzyme. Transfer of vectors carrying cDNA molecules to host cells and cloning is used to make a cDNA library. Many different cDNA libraries are available from cells and tissues in specific stages of development, or different organs such as brain, muscle, and kidney. These libraries provide an instant catalog of all the genes active in a cell at a specific time.

A cDNA library can also be prepared using a variation of PCR called **reverse transcriptase PCR (RT-PCR)**. In this procedure, reverse transcriptase is used to generate single-stranded cDNA copies of mRNA molecules as described earlier. This reaction is followed by PCR to copy the single-stranded DNA into double-stranded molecules and then amplify these into many copies. Taq polymerase and random DNA primers (instead of primers specific for a given gene) are added to the single-stranded cDNA. After primer binding,

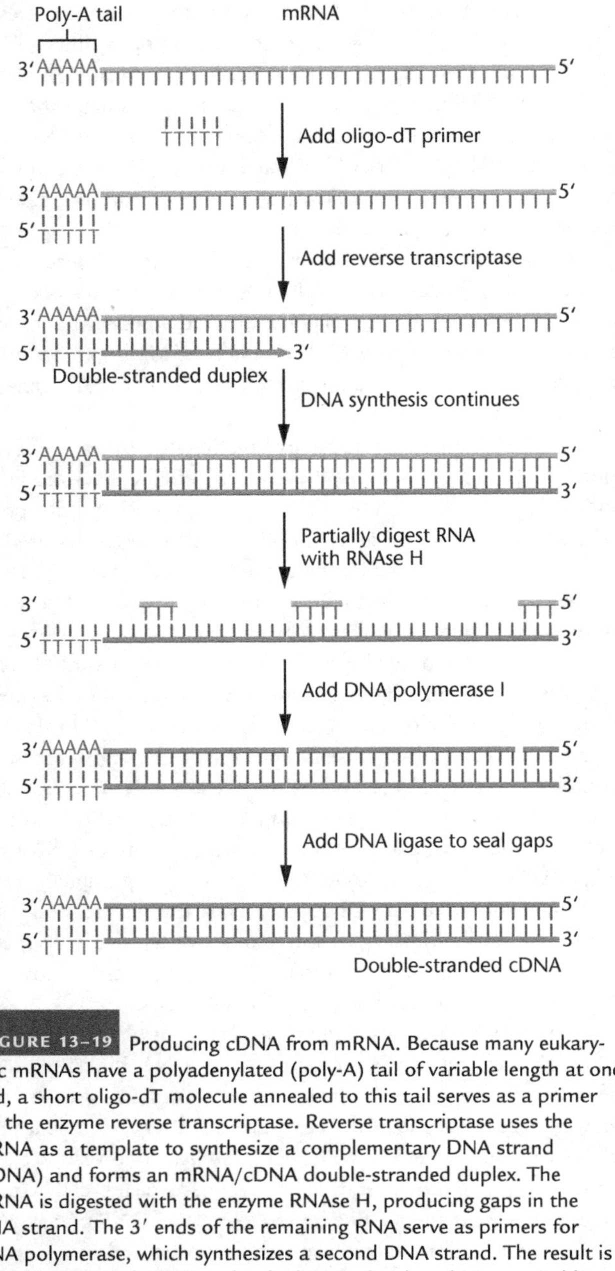


**FIGURE 13-19** Producing cDNA from mRNA. Because many eukaryotic mRNAs have a polyadenylated (poly-A) tail of variable length at one end, a short oligo-dT molecule annealed to this tail serves as a primer for the enzyme reverse transcriptase. Reverse transcriptase uses the mRNA as a template to synthesize a complementary DNA strand (cDNA) and forms an mRNA/cDNA double-stranded duplex. The mRNA is digested with the enzyme RNAse H, producing gaps in the RNA strand. The 3′ ends of the remaining RNA serve as primers for DNA polymerase, which synthesizes a second DNA strand. The result is a double-stranded cDNA molecule that can be cloned into a suitable vector or used directly as a probe for library screening.

the Taq polymerase extends the primers, making double-stranded cDNA. Additional cycles of PCR make many copies of the cDNA. The amplified cDNA can be inserted into plasmid vectors, which are replicated to produce a cDNA library. RT-PCR is more sensitive than conventional cDNA preparation and is a powerful tool for identifying mRNAs that may be present in only one or two copies per cell.

**NOW SOLVE THIS**

Question 13 on page 348 involves calculating how many different clones it takes to make a *Drosophila* genomic library using a plasmid vector.

- HINT: *Remember that there are three variables in this calculation: the size of the genome, the average size of the cloned inserts, and the probability of having a gene included in the library.*

## 13.9 Specific Clones Can Be Recovered from a Library

A genomic library often consists of several hundred thousand different clones. To find a specific gene, we need to identify and isolate only the clone or clones containing that gene. We must also determine whether a given clone contains all or only part of the gene we are studying. Several methods allow us to sort through a library—called *screening* the library—to recover clones of interest. The choice of method often depends on the circumstances and available information about the gene being sought.

### Probes Identify Specific Clones

Probes are used to screen a library to recover clones of a specific gene. A **probe** is any DNA or RNA sequence that has been labeled in some way and is complementary to some part of a cloned sequence present in the library.

When used in a hybridization reaction, the probe binds to any complementary DNA sequences present in one or more clones. Probes can be labeled with radioactive isotopes, or with compounds that undergo chemical or color reactions to indicate the location of a specific clone in a library.

Probes are derived from a variety of sources—even related genes isolated from other species can be used if enough of the DNA sequence is conserved. For example, extrachromosomal copies of the ribosomal RNA genes of the African clawed frog *Xenopus laevis* can be isolated by centrifugation, inserted into plasmid vectors, and cloned. Because ribosomal gene sequences are highly conserved (that is, are similar in different organisms), clones carrying human ribosomal genes can be recovered from a genomic library using cloned fragments of *Xenopus* ribosomal DNA as probes.

If the gene to be selected from a genomic library is expressed in certain cell types, a cDNA probe can be used. This technique is particularly helpful when purified or enriched mRNA for a gene product can be obtained. For example, $\beta$-globin mRNA is present in high concentrations in certain stages of red blood cell development. The mRNA purified from these cells can be copied by reverse transcriptase into a cDNA molecule for use as a probe. In fact, a cDNA probe was originally used to recover the structural gene for human $\beta$-globin from a cloned genomic library.

## Screening a Library

To screen a library constructed using a plasmid vector (this can be a genomic library, a chromosome-specific library, or a cDNA library) clones from the library are grown on nutrient agar plates, where they form hundreds or thousands of colonies (Figure 13–20). A replica of the colonies on each plate is made by gently pressing a nylon or nitrocellulose filter onto the plate's surface; this transfers the pattern of bacterial colonies from the plate to the filter. The filter is processed to lyse the bacterial cells, denature the double-stranded DNA released from the cells into single strands, and bind these strands to the filter.

The DNA on the filter is screened by incubation with a labeled nucleic acid probe. First, the probe is heated and quickly cooled to form single-stranded molecules, and then it is added to a solution containing the filter. If the nucleotide sequence of any of the DNA on the filter is complementary to the probe, a double-stranded DNA–DNA hybrid molecule will form (one strand from the probe and the other from the cloned DNA on the filter). For example, if a cDNA probe of $\beta$-globin is used, it will bind to cloned DNA sequences encoding the $\beta$-globin gene. After incubation of the probe and the filter, unbound probe molecules are washed away, and the filter is assayed to detect the hybrid molecules that remain. If a radioactive probe has been used, the filter is overlaid with a piece of X-ray film. Radioactive decay in the probe molecules bound to DNA on the filter will expose the film, producing dark spots on the film. These spots represent colonies on the plate containing the cloned gene of interest (Figure 13–20). The positions of spots on the film are used as a guide to identify and recover the corresponding colonies on the plate. The cloned DNA they contain can be used in further experiments. With some nonradioactive probes, a chemical reaction emits photons of light (chemiluminescence) to expose the photographic film and reveal the location of colonies carrying the gene of interest.

To screen a phage library, a slightly different method, called **plaque hybridization**, is used. A solution of phage carrying DNA inserts is spread over a lawn of bacteria growing on a plate. The phages infect the bacterial cells and form plaques as they replicate. Each plaque, which appears as a clear spot on the plate, represents the progeny of a single phage and is a clone. The plaques are transferred to a nylon membrane. The phages are

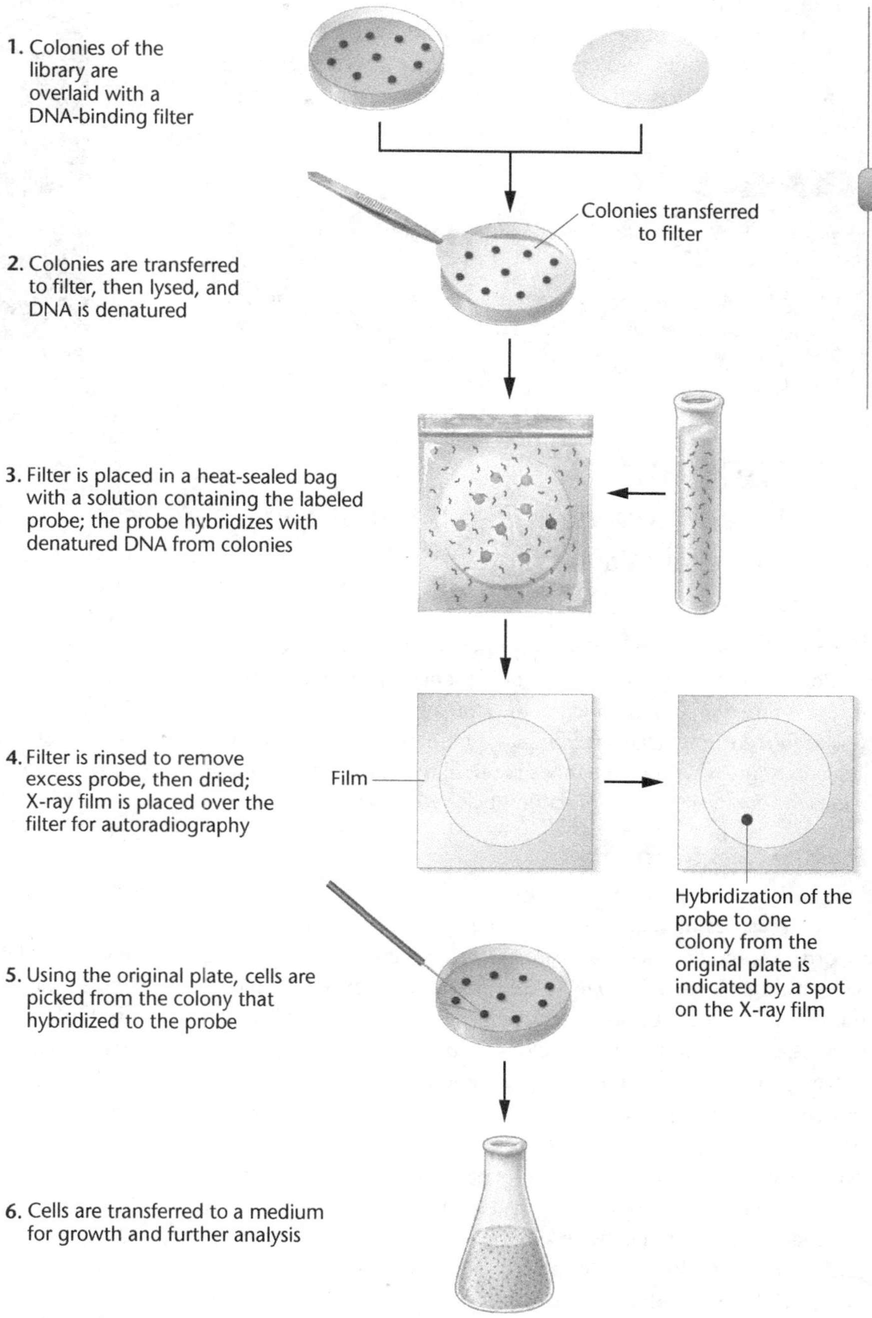


**FIGURE 13–20** Screening a library constructed using a plasmid vector to recover a specific gene. The library, present in bacteria on Petri plates, is overlaid with a DNA-binding filter, and colonies are transferred to the filter. Colonies on the filter are lysed, and the DNA is denatured to single strands. The filter is placed in a hybridization bag along with buffer and a labeled single-stranded DNA probe. During incubation, the probe forms a double-stranded hybrid with any complementary sequences on the filter. The filter is removed from the bag and washed to remove excess probe. Hybrids are detected by placing a piece of X-ray film over the filter and exposing it for a short time. The film is developed, and hybridization events are visualized as spots on the film. Colonies containing the insert that hybridized to the probe are identified from the orientation of the spots. Cells are picked from this colony for growth and further analysis.

disrupted, and the DNA on the filter is denatured into single strands and screened with a labeled probe. Phage plaques are much smaller than plasmid colonies, and many plaques can be screened on a single filter, making this method more efficient for screening large genomic libraries.

NOW SOLVE THIS

Question 18 on page 348 involves selecting a cloned gene from a cDNA library.

■ HINT: *cDNA clones do not have all the sequences of a genomic library but do have the coding sequences, and they can be selected with the proper probe.*

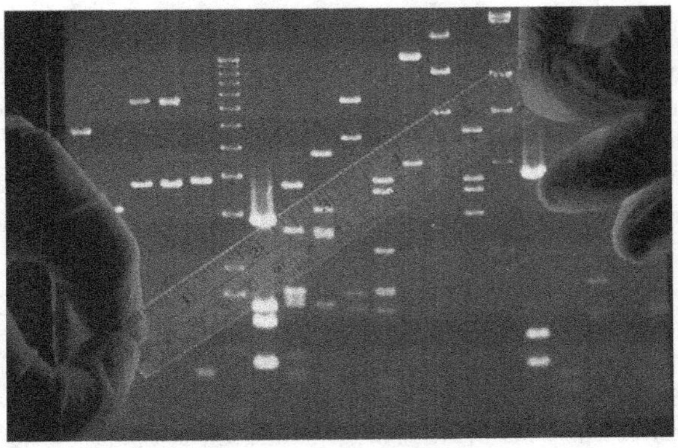

FIGURE 13–21 An agarose gel containing separated DNA fragments stained with a dye (ethidium bromide) and visualized under ultraviolet light. Smaller fragments migrate faster and farther than do larger fragments, resulting in the distribution shown.

13.10

## Cloned Sequences Can Be Analyzed in Several Ways

The recovery and identification of genes and other DNA sequences by cloning or by PCR is a powerful tool for analyzing genomic structure and function. In fact, much of the Human Genome Project data was acquired through such techniques. In the following sections, we consider some ways these methods are used to provide information about the organization and function of cloned sequences.

### Restriction Mapping

One of the first steps in characterizing a DNA clone is the construction of a **restriction map**. A restriction map establishes the number of, order of, and distances between restriction-enzyme cleavage sites along a cloned segment of DNA, thus providing information about the length of the cloned insert and the location of restriction-enzyme cleavage sites within the clone. The map units are expressed in base pairs (bp) or, for longer lengths, kilobase (kb) pairs. Restriction maps for different cloned DNAs are usually different enough to serve as identity tags for those clones. The data the maps provide can be used to reclone fragments of a gene or compare its internal organization with that of other cloned sequences.

Fragments generated by cutting DNA with restriction enzymes can be separated by gel electrophoresis, a method that separates fragments by size, with the smallest pieces moving farthest through the gel (see Chapter 10). The fragments form a series of bands that can be visualized by staining the DNA with ethidium bromide and illuminating it with ultraviolet light (Figure 13–21).

Figure 13–22 shows the construction of a restriction map from a cloned DNA segment. For the sake of our demonstration, let's say that the cloned DNA segment is 7.0 kb in length. Three samples of the cloned DNA are digested with restriction enzymes—one with *Hin*dIII, one with *Sal*I, and one with both *Hin*dIII and *Sal*I. The fragments are separated by gel electrophoresis and stained with ethidium bromide, producing a series of bands on the gel. The resulting bands are photographed or scanned for analysis. The molecular weights of the fragments are measured by comparing their location on the gel to a set of molecular weight standards run in an adjacent lane. The restriction map is constructed by analyzing the number and length of the fragments. When the DNA is cut with *Hin*dIII, two fragments (0.8 and 6.2 kb) are produced, confirming that the cloned insert is 7.0 kb in length and showing that it contains only one recognition sequence for this enzyme (located 0.8 kb from one end). When the DNA is cut with *Sal*I, two fragments (1.2 and 5.8 kb) result, indicating that the insert also has only one recognition sequence for this enzyme, located 1.2 kb from one end of the cloned DNA segment.

These results show that each enzyme has one restriction site in the cloned DNA, but the relative positions of the two restriction-enzyme cleavage sites are unknown. From this newly obtained data, two different maps are possible. One map, model 1 in Figure 13–22, shows the *Hin*dIII recognition sequence located 0.8 kb from one end and the *Sal*I recognition sequence 1.2 kb from the same end. The alternative map, model 2, locates the *Hin*dIII recognition sequence 0.8 kb from one end and the *Sal*I sequence 1.2 kb from the other end.

The correct model is determined by analyzing the results from the sample digested with *Hin*dIII *and Sal*I. Model 1 predicts that digestion with both enzymes will generate three fragments: 0.4, 0.8, and 5.8 kb in length; model 2 predicts that there will be three fragments of 0.8, 5.0, and 1.2 kb. The pattern and molecular weights seen on the gel after digestion with both enzymes indicate that model 1 is correct (see Figure 13–22).

Restriction maps are an important way of characterizing cloned DNA and can be constructed in the absence of any other information about the DNA, including whether or not it encodes a gene or has other functions. In conjunction with other techniques, restriction mapping can define the boundaries of a gene, dissect the internal organization of a gene and its flanking regions, and locate mutations within genes.

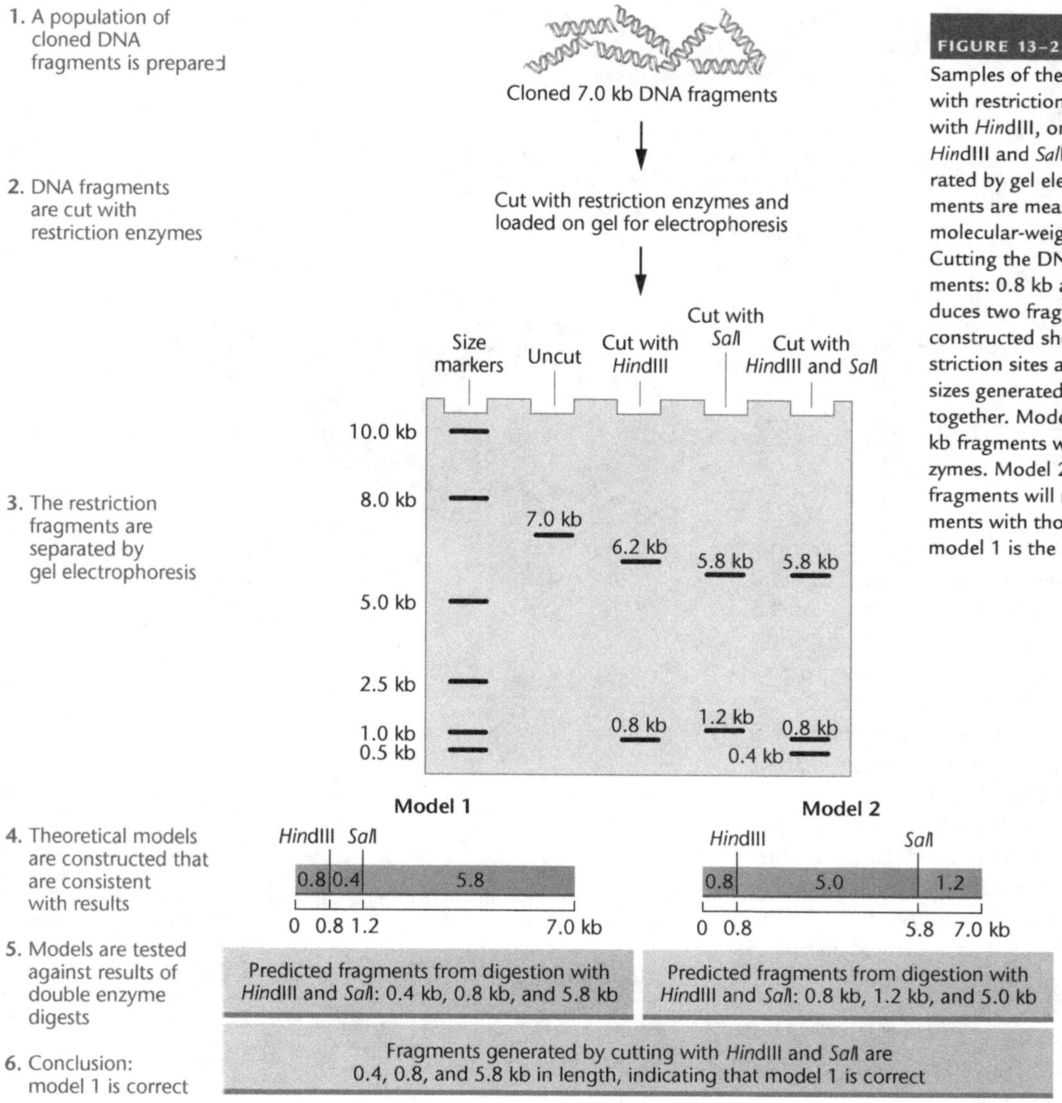


**FIGURE 13-22** Constructing a restriction map. Samples of the 7.0-kb DNA fragments are digested with restriction enzymes: one sample is digested with *Hind*III, one with *Sal*I, and one with both *Hind*III and *Sal*I. The resulting fragments are separated by gel electrophoresis. The separated fragments are measured by comparing them with molecular-weight standards in an adjacent lane. Cutting the DNA with *Hind*III generates two fragments: 0.8 kb and 6.2 kb. Cutting with *Sal*I produces two fragments: 1.2 kb and 5.8 kb. Models are constructed showing the possible locations of restriction sites and are used to predict the fragment sizes generated by cutting with both *Hind*III and *Sal*I together. Model 1 predicts that 0.4-, 0.8-, and 5.8-kb fragments will result from cutting with both enzymes. Model 2 predicts that 0.8-, 1.2-, and 5.0-kb fragments will result. Comparing the predicted fragments with those observed on the gel indicates that model 1 is the correct restriction map.

WEB TUTORIAL 13.5 RESTRICTION MAPPING

Restriction digestion of clones (that is, cutting the clones with restriction enzymes) plays an important role in mapping genes to specific human chromosomes and to defined regions of individual chromosomes. In addition, if a restriction site maps close to a mutant allele, this site can be used as a marker in genetic testing to identify carriers of recessively inherited disorders, or to prenatally diagnose a fetal genotype. This topic will be discussed in Chapter 24.

## Nucleic Acid Blotting

Many of the techniques described in this chapter rely on hybridization between complementary nucleic acid (DNA or RNA) molecules. One of the most widely used methods for detecting such hybrids is called Southern blotting (after Edwin Southern, who devised it). The **Southern blot** method can be used to identify which clones in a library contain a given DNA sequence (such as ribosomal DNA, a $\beta$-globin gene, etc.) and to characterize the size of the fragments, thus permitting a restriction map to be made. Southern blots can also be used to identify fragments carrying specific genes in genomic DNA digested with a restriction enzyme. Fragments of genomic clones isolated by Southern blots can in turn be isolated and recloned, providing a way to isolate parts of a gene.

Southern blotting has two components: separation of DNA fragments by gel electrophoresis and hybridization of the fragments using labeled probes. Gel electrophoresis can be used, as shown above, to characterize the number of fragments produced by restriction digestion and to estimate their molecular weights. However, restriction-enzyme digestion of large genomes—such as the human genome, with more than 3 billion nucleotides—will produce so many different fragments that they will run together on a gel to produce a continuous smear. The identification of specific fragments in these cases is accomplished in the next step:

hybridization characterizes the DNA sequences present in the fragments. The DNA to be characterized by Southern blot hybridization can come from several sources, including clones selected from a library or genomic DNA. Our discussion will use examples from both sources to show how Southern blots are used to characterize the number, size, organization, and base sequence of DNA fragments.

To make a Southern blot, DNA is cut into fragments with one or more restriction enzymes, and the fragments are separated by gel electrophoresis (Figure 13–23). In preparation for hybridization, the DNA in the gel is denatured with alkaline treatment to form single-stranded fragments. The gel is then overlaid with a DNA-binding membrane, usually nitrocellulose or a nylon derivative. Transfer of the DNA fragments to the membrane is accomplished by placing the membrane and gel on a wick (often a sponge) sitting in a buffer solution. Layers of paper towels or blotting paper are placed on top of the filter and held in place with a weight. Capillary action draws buffer up through the gel, transferring the DNA fragments from the gel to the membrane.

The filter is placed in a heat-sealed bag with a labeled, single-stranded DNA probe for hybridization. DNA fragments on the filter that are complementary to the probe's nucleotide sequence bind to the probe to form double-stranded hybrids. Excess probe is then washed away, and the hybridized fragments are visualized on a piece of film (Figures 13–23 and 13–24).

To produce Figure 13–24, researchers cut samples of genomic DNA with several restriction enzymes. The pattern of fragments obtained for each restriction enzyme is shown in Figure 13–24(a). A Southern blot of this gel is shown in Figure 13–24(b). The probe hybridized to complementary sequences, identifying fragments of interest.

In addition to characterizing cloned DNAs, Southern blots can be used to create restriction maps of segments within and near a gene and to identify DNA fragments carrying all or parts of a single gene in a mixture of fragments. By comparing the pattern of bands in normal cells to those from patients with genetic disorders or cancer, Southern blots also detect rearrangements, deletions, and duplications in genes associated with human genetic disorders and cancers.

To determine whether a gene is actively being expressed in a given cell or tissue type, a related blotting technique probes for the presence of mRNA complementary to a cloned gene. To do this, mRNA is extracted from a specific cell or tissue type and separated by gel electrophoresis. The resulting pattern of RNA bands is transferred to a membrane, as in Southern blotting. The membrane is then exposed to a labeled single-stranded DNA probe derived from

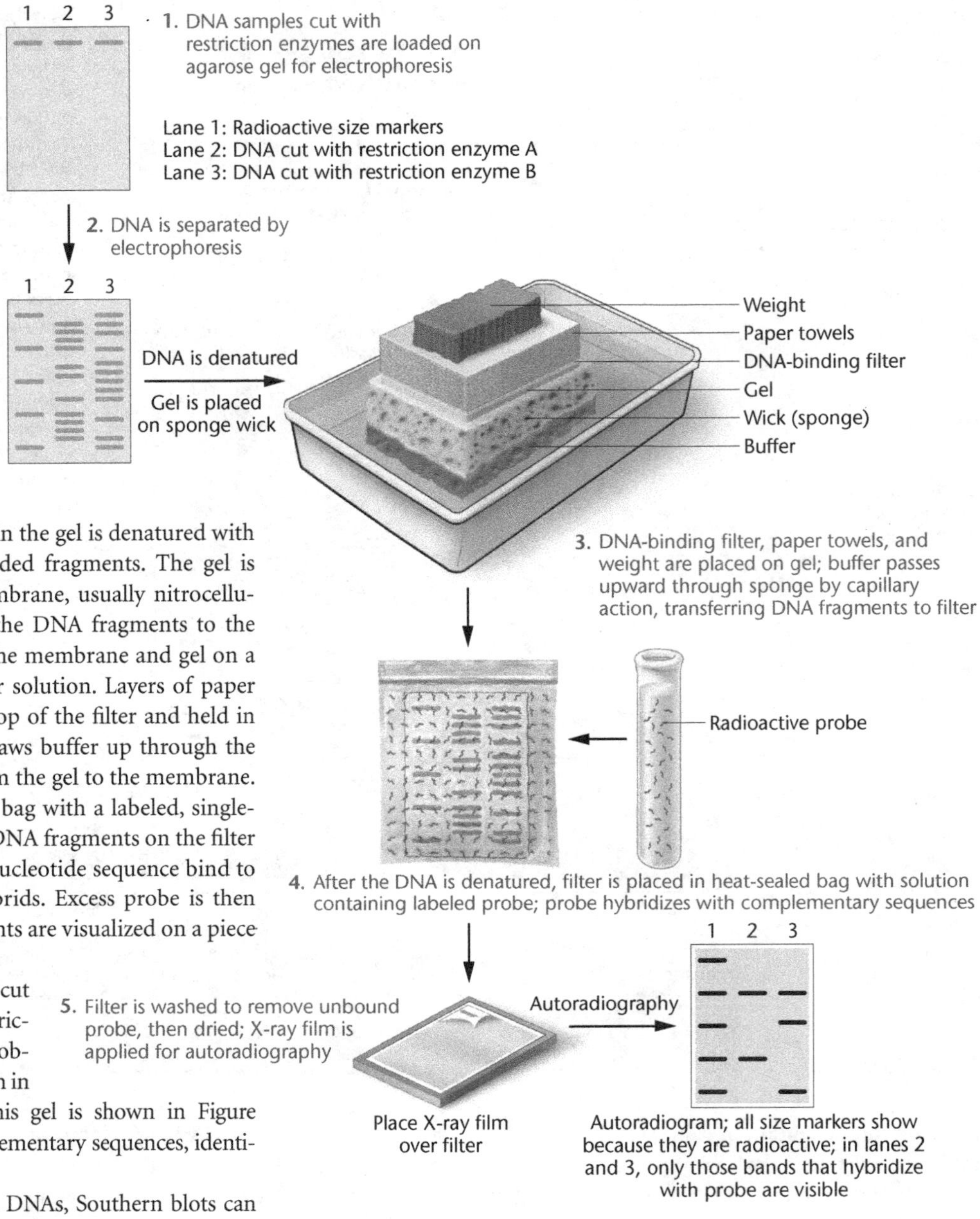


**FIGURE 13–23** In the Southern blotting technique, samples of the DNA to be probed are cut with restriction enzymes and the fragments are separated by gel electrophoresis. The pattern of fragments is visualized and photographed under ultraviolet illumination. Then the gel is placed on a sponge wick that is in contact with a buffer solution and covered with a DNA-binding filter. Layers of paper towels or blotting paper are placed on top of the filter and held in place with a weight. Capillary action draws the buffer through the gel, transferring the pattern of DNA fragments from the gel to the filter. The DNA fragments on the filter are then denatured into single strands and hybridized with a labeled DNA probe. The filter is washed to remove excess probe and overlaid with a piece of X-ray film for autoradiography. The hybridized fragments show up as bands on the X-ray film.

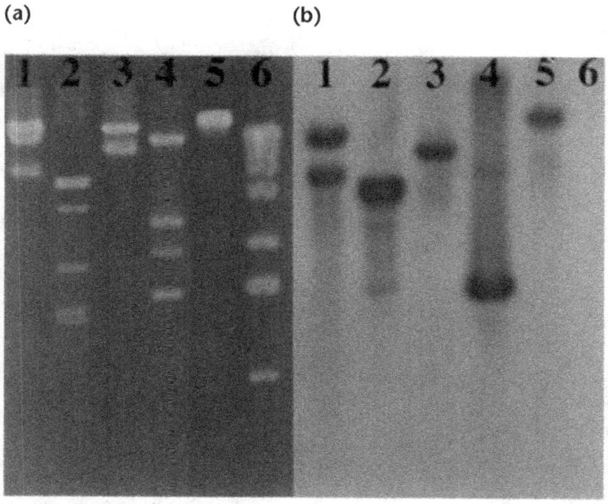


FIGURE 13-24 (a) Agarose gel stained with ethidium bromide to show DNA fragments. (b) Exposed X-ray film of a Southern blot prepared from the gel in part (a). Only those bands containing DNA sequences complementary to the probe show hybridization.

a cloned copy of the gene. If mRNA complementary to the DNA probe is present, the complementary sequences will hybridize and be detected as a band on the film. Because the original procedure (DNA bound to a filter) is known as a Southern blot, this variant procedure (RNA bound to a filter) is called a **northern blot**. (Following this somewhat perverse logic, another procedure involving proteins bound to a filter is known as a **western blot**.)

Northern blots provide information about the expression of specific genes and are used to study patterns of gene expression in embryonic tissues, cancer, and genetic disorders. Northern blots also detect alternatively spliced mRNAs (multiple types of transcripts derived from a single gene) and are used to derive other information about transcribed mRNAs. If marker RNAs of known size are run as controls, northern blots can be used to measure the size of a gene's mRNA transcripts. Measuring band density gives an estimate of the relative transcriptional activity of the gene. Thus, northern blots characterize and quantify the transcriptional activity of genes in different cells, tissues, and organisms.

## 13.11 DNA Sequencing Is the Ultimate Way to Characterize a Clone

In a sense, a cloned DNA molecule or any DNA, from a clone to a genome, is completely characterized only when its nucleotide sequence is known. The ability to sequence DNA has greatly enhanced our understanding of genome organization and increased our knowledge of gene structure, function, and mechanisms of regulation.

The most commonly used method of DNA sequencing was developed by Fred Sanger and his colleagues. In this procedure, a DNA molecule whose sequence is to be determined is converted to single strands that are used as a template for synthesizing a series of complementary strands. Each of these strands randomly terminates at a different, specific nucleotide. The resulting series of DNA fragments are separated by electrophoresis and analyzed to reveal the sequence of the DNA. In the first step of the procedure, DNA is heated to denature it and form single strands. The single-stranded DNA is mixed with primers that anneal to the 3′ end of the DNA. Samples of the primer-bound single-stranded DNA are distributed into four tubes. In the next step, DNA polymerase and the four deoxyribonucleotide triphosphates (dATP, dCTP, dGTP, and dTTP) are added to each tube. In addition, each tube receives a small amount of one modified deoxyribonucleotide (Figure 13–25), called a dideoxynucleotide (e.g., ddATP, ddCTP, ddGTP, and ddTTP). Dideoxynucleotides have a 3′ hydrogen instead of a 3′ hydroxyl group. One of the deoxyribonucleotides or the primer is labeled with radioactivity for later analysis of the sequence. DNA polymerase is added to each tube, and the primer is elongated in the 5′ to 3′ direction, forming a complementary strand on the template.

As DNA synthesis takes place, the polymerase occasionally inserts a dideoxynucleotide instead of a deoxyribonucleotide into a growing DNA strand. Since the dideoxynucleotide has no 3′-OH group, it cannot form a 3′ bond with another nucleotide, and DNA synthesis terminates. For example, in the tube with added ddATP, the polymerase inserts ddATP instead of dATP, causing termination of chain elongation (Figure 13–26). As the reaction proceeds, the tube with ddATP will accumulate DNA molecules that terminate at all positions containing A. In the other tubes, reactions terminate at G, C, and T, respectively. The DNA fragments from each reaction

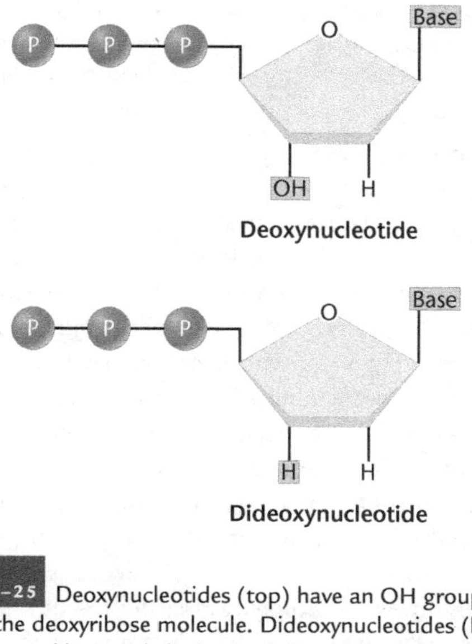


FIGURE 13-25 Deoxynucleotides (top) have an OH group at the 3′ position in the deoxyribose molecule. Dideoxynucleotides (bottom) lack an OH group and have only hydrogen (H) at this position. Dideoxynucleotides can be incorporated into a growing DNA strand, but the lack of a 3′-OH group prevents formation of a phosphodiester bond with another nucleotide, terminating further elongation of the template strand.

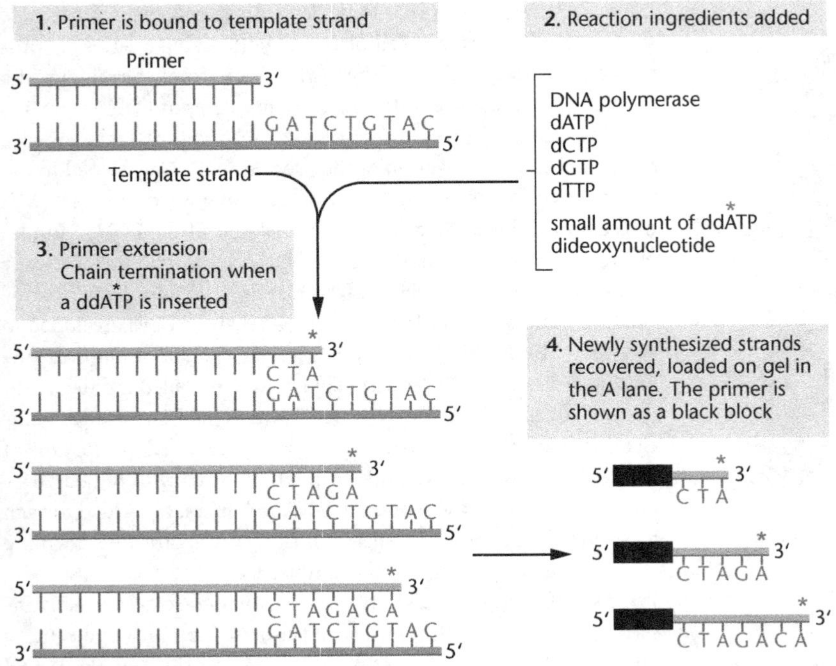


FIGURE 13-26 DNA sequencing using the chain-termination method. (1) A primer is annealed to a sequence adjacent to the DNA being sequenced (usually at the insertion site of a cloning vector). (2) A reaction mixture is added to the primer–template combination. This includes DNA polymerase, the four dNTPs (one of which is radioactively labeled), and a small amount of one dideoxynucleotide. Four tubes are used, each containing a different dideoxynucleotide (ddATP, ddCTP, etc.). (3) During primer extension, the polymerase occasionally inserts a ddNTP instead of a dNTP, terminating the synthesis of the chain, because the ddNTP does not have the OH group needed to attach the next nucleotide. In the figure, ddATP and the A inserted from this dideoxynucleotide are indicated with an asterisk. Over the course of the reaction, all possible termination sites will have a ddNTP inserted. (4) The newly synthesized strands are removed from the template, and the mixture is placed on a gel. DNA fragments from the reaction tube containing ddATP and terminating with A are loaded in the A lane, those ending in C are loaded in the C lane, and so forth.

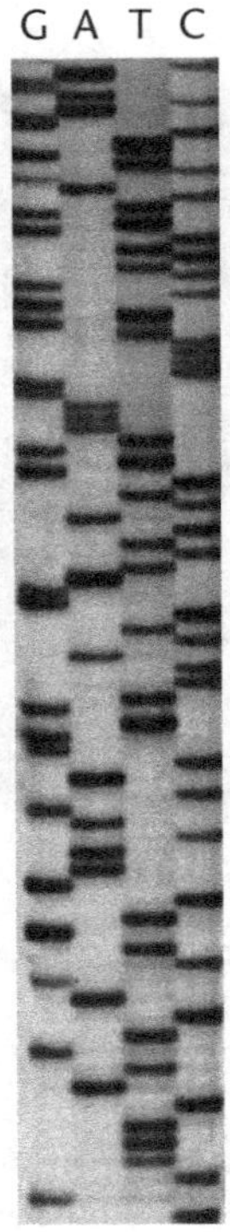


FIGURE 13-27 DNA sequencing gel showing the separation of newly synthesized fragments in the four sequencing reactions (one per lane). To obtain the base sequence of the DNA fragment, the gel is read from the bottom, beginning with the lowest band in any lane, then the next lowest, then the next, and so on. For example, the sequence of the DNA on this gel begins with 5′-CG at the very bottom of the gel, and proceeds upwards as 5′-CGCTTTCATGTCA, and so forth.

tube (one for each dideoxynucleotide) are separated in adjacent lanes by gel electrophoresis. The result is a series of bands forming a ladderlike pattern that is visualized by developing film exposed to the gel (Figure 13–27). The nucleotide sequence revealed by the bands is read directly from bottom to top, corresponding to the sequence of the DNA strand complementary to the template.

DNA SEQUENCING
WEB TUTORIAL 13.6

DNA sequencing is now largely automated and uses machines that can sequence several hundred thousand nucleotides per day. In the automated procedure, each of the four dideoxynucleotide analogs is labeled with a different colored fluorescent dye (Figure 13–28) so that chains terminating in adenosine are labeled with one color, those ending in cytosine with another color, and so forth. All four labeled dideoxynucleotides are added to a single tube, and after primer extension by DNA polymerase, the reaction products are loaded into one lane on a gel. The gel is scanned with a laser, causing each band to fluoresce a different color. A detector in the sequencing machine reads the color of each band and determines whether it represents an A, T, C, or G. The data are represented as a series of colored peaks, each corresponding to one nucleotide in the sequence (Figure 13–29).

## Recombinant DNA Technology and Genomics

DNA sequencing is one of the technologies that makes genome projects possible. Through a combination of recombinant DNA techniques and nucleotide sequencing, the genomes of more than 500 species have been sequenced, and almost 2000 additional projects are underway. The Human Genome Project sponsored by the U.S. Department of Energy and the National Institutes of Health and a private project sponsored by the biotechnology company Celera used a combination of genomic and chromosome-specific libraries to finish sequencing the coding portion of the human genome in 2003. Newer methods of DNA sequencing now under development will increase the speed, capacity, and accuracy of sequencing at a lower cost than present technology. Once these new methods of sequencing are available, it will be feasible to sequence the genomes of many more species.

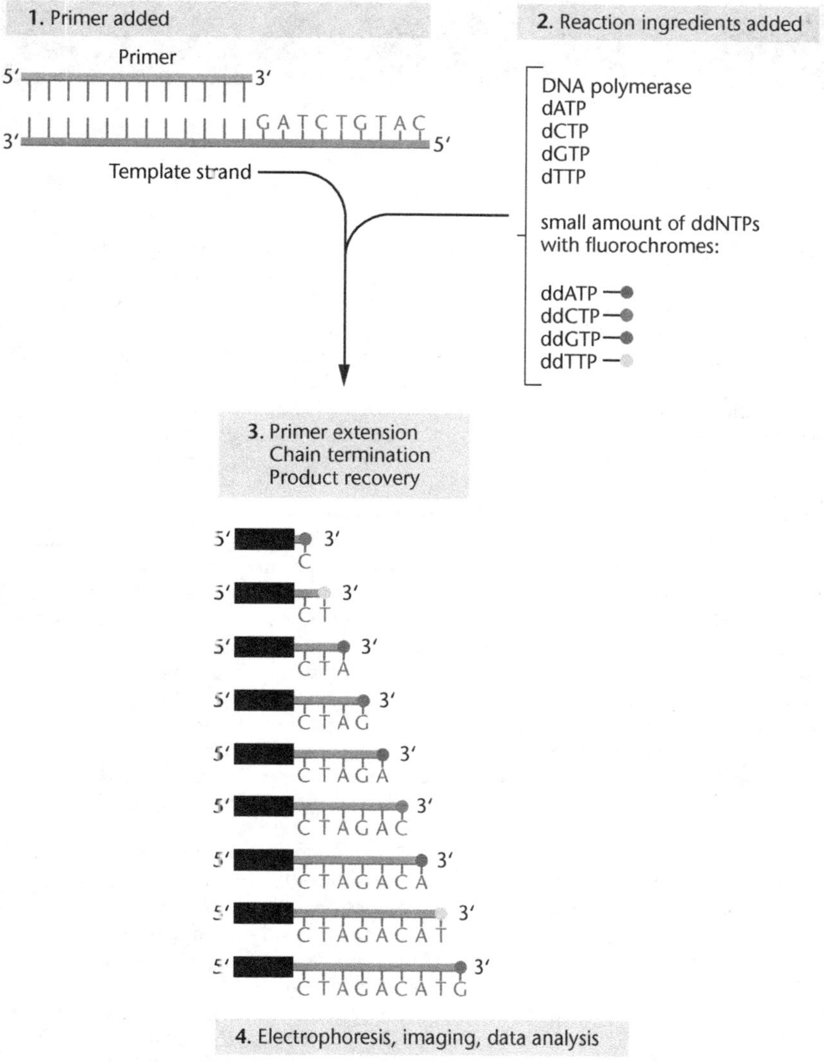


FIGURE 13-28 In DNA sequencing using dideoxynucleotides labeled with fluorescent dyes, all four ddNTPs are added to the same tube, and during primer extension, all possible lengths of chains are produced. The products of the reaction are added to a single lane on a gel, and the bands are read by a detector and imaging system. This process is now automated, and robotic machines, such as those used in the Human Genome Project, sequence several hundred thousand nucleotides in a 24-hour period and then store and analyze the data automatically. The sequence is obtained by extension of the primer and is read from the newly synthesized strand, not the template strand. Thus, the sequence obtained begins with 5′-CTAGACATG.

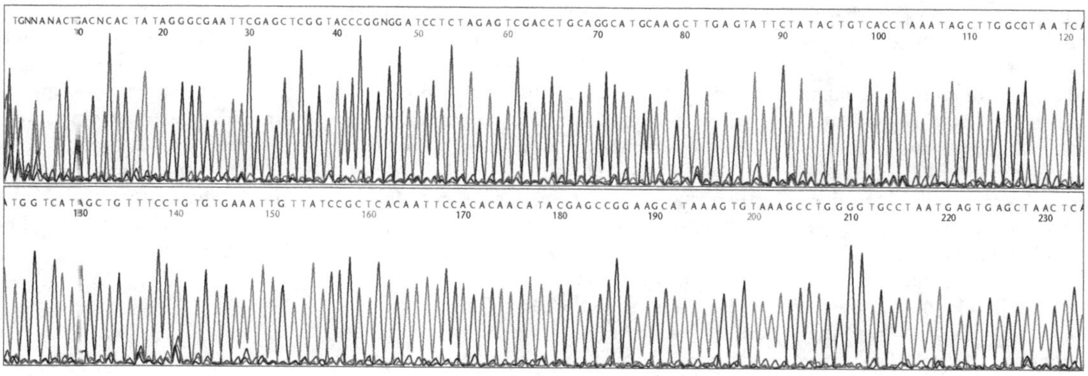

FIGURE 13-29 Automated DNA sequencing using fluorescent dyes, one for each base. Each peak represents the correct nucleotide in the sequence. The sequence extending from the primer (which is not shown here) starts at the upper left of the diagram and extends rightward. The bases labeled as N are ambiguous and cannot be identified with certainty. These ambiguous base readings are more likely to occur near the primer because the quality of sequence determination deteriorates the closer the sequence is to the primer. The separated bases are read in order along the axis from left to right. Thus, this sequence begins as 5′-TGNNANACTGACNCAC. Numbers below the bases indicate length of the sequence in base pairs.

GENETICS, TECHNOLOGY, AND SOCIETY

## Beyond Dolly: The Cloning of Humans

The death of Dolly the sheep, in February 2003, marked a poignant end to the beginning of the human cloning debate.

Six years earlier, her birth took the world by surprise. Before Dolly, the idea that an animal could be cloned from the cells of an adult animal was science fiction—something from *Brave New World* or *The Boys from Brazil*. For decades, scientists believed that it would be impossible to clone mammals, as DNA from adult cells could not be reprogrammed to code for the development of a new, complete organism. But then Dolly appeared.

Dolly, who was cloned from a frozen udder cell of a long-dead sheep, was brought into being by a group of embryologists led by Ian Wilmut and Keith Campbell at the Roslin Institute in Scotland. Their goal was to clone transgenic farm animals that secrete pharmaceutical products, such as blood-clotting factors or insulin, into their milk. In this way, herds of identical animals might be used as bioreactors to synthesize large quantities of medically important proteins.

The cloning method that Wilmut and Campbell used to create Dolly—a procedure called *nuclear transfer*—was first suggested by embryologist Hans Spemann in 1938. The method is simply to replace the nucleus of an egg with the nucleus from an adult cell, thereby creating a hybrid zygote. In theory, the genetic information in the donor nucleus should direct all further embryonic development, and the new organism should be a genetic replica of the adult that donated the nucleus. Although the procedure sounds simple in theory, it proved to be extremely difficult in practice because adult nuclei express only a small subset of the genes required for embryonic development. Wilmut and Campbell overcame this limitation by reprogramming the adult nuclei before transferring them into recipient eggs. They did this by starving the donor cells so they became quiescent. In addition, they passed an electric current through the recipient egg. For unknown reasons, these procedures turned on the silent genes within the differentiated cell nucleus. To create Dolly, over 200 udder-cell nuclei were transferred into eggs. Of these nuclear transfers, only 29 developed into embryos; and 13 of them were implanted into surrogate mother ewes. One pregnancy resulted, which culminated in the birth of Dolly. Although Dolly was the first mammal to be cloned from adult cells, the method has since been used to clone mice, pigs, cattle, rabbits, goats, an ox, a mule, a horse, and a cat.

Dolly's birth not only shattered scientists' views of mammalian cloning, but it triggered an avalanche of controversy. The idea that humans might be cloned was denounced as immoral, repugnant, and ethically wrong. Within days of the announcement of Dolly's birth, bills were introduced into the United States Congress to prohibit research into human cloning, and worldwide bans were called for. Frightening scenarios were proposed—rich and powerful people cloning themselves for reasons of vanity, people with serious illnesses cloning replicas to act as organ donors, and legions of human clones suffering loss of autonomy, individuality, and kinship ties.

Is it really possible to clone humans? And if so, *should* we create human clones? The answer to the first question is simple: The same technology used to create Dolly could be used to clone a human. Most of the technical procedures are already used for human in vitro fertilizations, and it seems likely that adult human nuclei could be reprogrammed similarly to the adult sheep nuclei that created Dolly. However, the cloning process is extremely inefficient. Nuclear transfers into hundreds of human eggs would be required to yield one full-term pregnancy. As yet, there have been no verified successful attempts to clone a human, despite a few high-profile claims by several doctors and a religious sect.

The answer to the second question is not as simple. To begin with, it is necessary to understand the differences between two kinds of cloning—reproductive and therapeutic. Reproductive cloning results in the creation of an animal, such as Dolly. In contrast, therapeutic cloning creates early-stage embryos for the purpose of harvesting stem cells—cells that have the potential to treat various diseases. As legislators and the public tend to equate these two forms of cloning, research into human therapeutic cloning has been affected by all-encompassing bans that target reproductive cloning.

Although most scientists believe that research into therapeutic cloning should continue, they conclude that we should refrain from human reproductive cloning for both scientific and ethical reasons. The most serious concerns are that many cloned animals appear to suffer developmental defects. About 12% of cloned mice and 38% of cloned goats show congenital abnormalities such as oversized internal organs, and respiratory, circulatory, and immunological defects. The ethical arguments against human reproductive cloning involve concerns about potential abuses of the technology and threats to the dignity of human procreation. Some people worry that human clones might be discriminated against or cannibalized for spare parts and even question whether they would have the same personhood as nonclones. In contrast, proponents of human cloning suggest that the technology should be considered merely another fertility treatment, like *in vitro* fertilization—enabling infertile couples to have children that are genetically related to them or who do not carry a genetic defect of one parent.

As the debate over human cloning continues, Dolly will take her place among the important players in the history of modern science. Her remains have been preserved and are now on display in the National Museum of Scotland in Edinburgh. She is survived by several of her six lambs and the sadness of those who knew her and cared for her.

### References

Jaenisch, R., and I. Wilmut, 2001. Don't clone humans! *Science* 291: 2552.

EXPLORING GENOMICS

# Manipulating Recombinant DNA: Restriction Mapping and Designing PCR Primers

As you learned in this chapter, restriction enzymes are sophisticated "scissors" that molecular biologists use to cut DNA, and they are routinely used in genetics and molecular biology laboratories for recombinant DNA experiments. Yet another advantage of the genomics revolution has been the development of a wide variety of online tools to assist scientists working with restriction enzymes and manipulating recombinant DNA for different applications, such as restriction mapping and designing primers for PCR experiments. Here we explore **Webcutter** and **Primer3**, two sites that make recombinant DNA experiments much easier.

### ■ Exercise I – Creating a Restriction Map in Webcutter

Suppose you had cloned and sequenced a gene and you wanted to design a probe approximately 600 bp long that could be used to analyze expression of this gene in different human tissues by Northern blot analysis. Not too long ago, you had primarily two ways to approach this task. You could digest the cloned DNA with whatever restriction enzymes were in your freezer, then run agarose gels and develop restriction maps in the hope of identifying cutting sites that would give you the size fragment you wanted. Or you could scan the sequence with your eyes, looking for restriction sites of interest—a very time-consuming and eye-straining effort! Internet sites such as **Webcutter** take the guesswork out of developing restriction maps and make it relatively easy to design experiments for manipulating recombinant DNA. In this exercise, you will use Webcutter to create a restriction map of human DNA with the enzymes *Eco*RI, *Bam*HI, and *Pst*I.

1. Access **Webcutter** at http://rna.lundberg.gu.se/cutter2. Copy the sequence of cloned human DNA shown below and paste it into the text box in Webcutter. (*Hint*: Access this sequence from the Companion Web site so you can copy and paste the sequence into Webcutter).

**Human DNA sequence**

CCCCAGGAGACCTGGTTGTGGAATTCTG
TGTGTGAGTGGTTGACCTTCCTCCATCC
CCTGGTCCTTCCCTTCCCTTCCCGAGGC
ACAGAGAGACAGGGCAGGATCCACGTG
CCCATTGTGGAGGCAGAGAAAAGAGAAA
GTGTTTTATATACGGTACTTATTTAATATC
CCTTTTTAATTAGAAATTAAAACAGTTAAT
TTAATTAAAGAGTAGGGTTTTTTTTCAGTA
TTCTTGGTTAATATTTAATTTCAACTATTTA
TGAGATGTATCTTTTGCTCTCTCTTGCTC
TCTTATTTGTACCGGTTTTTGTATATAAAA
TTCATGTTTCCAATCTCTCTCTCCCTGAT
CGGTGACAGTCACTAGCTTATCTTGAAC
AGATATTTAATTTTGCTAACACTCAGCTCT
GCCCTCCCCGATCCCCTGGCTCCCCAGC
ACACATTCCTTTGAAATAAGGTTTCAATA
TACATCTACATACTATATATATATATTTGGCA
ACTTGTATTTGTGTGTATATATATATATATA
TGTTTATGTATATATGTGATTCTGATAAAA
TAGACATTGCTATTCTGTTTTTTATATGTA
AAAACAAAACAAGAAAAAATAGAGAATTT
ACATACTAAATCTCTCTCCTTTTTTAATTT
TAATATTTGTTATCATTTATTTATTGGTGC
TACTGTTTATCCGTAATAATTGTGGGGAA
AAGATATTAACATCACGTCTTTGTCTCTA
GTGCAGTTTTTCGAGATATTCCGTAGTAC
ATATTTATTTTTAAACAACGACAAAGAAAT
ACAGATATATCTTAAAAAAAAAAAAGCAT
TTTGTATTAAAGAATTTAATTCTGATCTGC
AGCTCAAAAAAA AAAAAA

2. Scroll down to "Please indicate which enzymes to include in the analysis." Click the button indicating "Use only the following enzymes." Select the restriction enzymes *Eco*RI, *Bam*HI, and *Pst*I from the list provided, then click "Analyze sequence." (*Note*: Use the command, control, or shift key to select multiple restriction enzymes.)

3. After examining the results provided by Webcutter, create a table showing the number of cutting sites for each enzyme and the fragment sizes that would be generated by digesting with each enzyme. Draw a restriction map indicating cutting sites for each enzyme with distances between each site and the total size of this piece of human DNA.

### ■ Exercise II – Designing a Recombinant DNA Experiment

Now that you have created a restriction map of your piece of human DNA, you need to ligate the DNA into a plasmid DNA vector that you can use to make your probe (molecular biologists often refer to this as subcloning). To do this you will need to determine which restriction enzymes would best be suited for cutting both the plasmid and the human DNA.

1. Below is a plasmid DNA sequence. Copy this sequence into the text box in Webcutter and identify cutting sites for the same enzymes you used in Exercise I. Then answer the following questions:
   a. What is the total size of the plasmid DNA analyzed in Webcutter?
   b. Which enzyme(s) could be used in a recombinant DNA experiment to ligate the plasmid to the *largest* DNA fragment from the human gene? Briefly explain your answer.
   c. What size recombinant DNA molecule will be created by ligating these fragments?
   d. Draw a simple diagram showing the cloned DNA inserted into the plasmid and indicate the restriction-enzyme cutting site(s) used to create this recombinant plasmid.

**Plasmid DNA sequence**

TATAAATATAGAATAATGAATCATATAAAAC
ATATCATTATTCATTTATTTACATTTAAAATT
ATTGTTTCAGTATCTTTAATTTATTATGTAT
ATATAAAAATAACTTACAATTTTATTAATAA
ACAATATATGTTTATTAATTCATGTTTTGTA
ATTTATGGGATAGCGATTTTTTTTACTGTC
TGTATTTTTCTTTTTTAATTATGTTTTAATT
GTATTTTATTTTTATTATTGTTCTTTTTATAG
TATTATTTTAAAACAAAATGTATTTTCTAAG
AACTTATAATAATAATAAATATAAATTTTAA
TAAAAATTATATTTATCTTTTACAATATGAA
CATAAAGTACAACATTAATATATATAGCTTTTA
ATATTTTTATTCCTAATCATGTAAATCTTAA
ATTTTTCTTTTTAAACATATGTTAAATATTT
ATTTCTCATTATATATATAAGAACATATTTATT

*Continued on next page*

*Exploring Genomics, continued*

AAATCTAGAATTCTATAGTGAGTCGTATTA
CAATTCACTGGCCGTCGTTTTACAACGTC
GTGACTGGGAAAACCCTGGCGTTACCCA
ACTTAATCGCCTTGCAGCACATCCCCCTT
TCGCCAGCTGGCGTAATAGCGAAGAGGC
CCGCACCGATCGCCCTTCCCAACAGTTG
CGCAGCCTGAATGGCGAATGGCGCCTGA
TGCGGTATTTTCTCCTTACGCATCTGTGC
GGTATTTCACACCGCATATGGTGCACTCT
CAGTACAATCTGCTCTGATGCCGCATAGT
TAAGCCAGCCCCGACACCCGCCAACACC
CGCTGACGCGCCCTGACGGGCTTGTCTG
CTCCCGGCATCCGCTTACAGACAAGCTG
TGACCGTCTCCGGGAGCTGCATGTGTCA
GAGGTTTTCACCGTCATCACCGAAACGC
GCGAGACGAAAGGGCCTCGTGATACGCC
TATTTTTATAGGTTAATGTCATGATAATAAT
GGTTTCTTAGACGTCAGGTGGCACTTTTC
GGGGAAATGTGCGCGGAACCCCTATTTG
TTTATTTTTCTAAATACATTCAAATATGTAT
CCGCTCATGAGACAATAACCCTGATAAAT
GCTTCAATAATATTGAAAAAGGAAGAGTA
TGAGTATTCAACATTTCCGTGTCGCCCTT
ATTCCCTTTTTTGCGGCATTTTGCCTTCCT
GTTTTTGCTCACCCAGAAACGCTGGTGAA
AGTAAAAGATGCTGAAGATCAGTTGGGTG
CACGAGTGGGTTACATCGAACTGGATCTC
AACAGCGGTAAGATCCTTGAGAGTTTTCG
CCCCGAAGAACGTTTTCCAATGATGAGCA
CTTTTAAAGTTCTGCTATGTGGCGCGGTA
TTATCCCGTATTGACGCCGGGCAAGAGC
AACTCGGTCGCCGCATACACTATTCTCAG
AATGACTTGGTTGAGTACTCACCAGTCAC
AGAAAAGCATCTTACGGATGGCATGACAG
TAAGAGAATTATGCAGTGCTGCCATAACC
ATGAGTGATAACACTGCGGCCAACTTACT
TCTGACAACGATCGGAGGACCGAAGGAG
CTAACCGCTTTTTTGCACAACATGGGGGA
TCATGTAACTCGCCTTGATCGTTGGGAAC
CGGAGCTGAATGAAGCCATACCAAACGA
CGAGCGTGACACCACGATGCCTGTAGCA
ATGCCAACAACGTTGCGCAAACTATTAAC
TGGCGAACTACTTACTCTAGCTTCCCGGC
AACAATTAATAGACTGGATGGAGGCGGAT
AAAGTTGCAGGACCACTTCTGCGCTCGG
CCCTTCCGGCTGGCTGGTTTATTGCTGAT
AAATCTGGAGCCGGTGAGCGTGGGTCTC
GCGGTATCATTGCAGCACTGGGGCCAGA
TGGTAAGCCCTCCCGTATCGTAGTTATCT
ACACGACGGGGAGTCAGGCAACTATGGA
TGAACGAAATAGACAGATCGCTGAGATAG
GTGCCTCACTGATTAAGCATTGGTAACTG
TCAGACCAAGTTTACTCATATATACTTTAG
ATTGATTTAAAACTTCATTTTTAATTTAAAA
GGATCTAGGTGAAGATCCTTTTTGATAA
TCTCATGACCAAAATCCCTTAACGTGAGT
TTTCGTTCCACTGAGCGTCAGACCCCGTA
GAAAAGATCAAAGGATCTTCTTGAGATCC
TTTTTTTCTGCGCGTAATCTGCTGCTTGCA
AACAAAAAAACCACCGCTACCAGCGGTG
GTTTGTTTGCCGGATCAAGAGCTAC

2. As you prepare to carry out this subcloning experiment, you find that the expiration dates on most of your restriction enzymes have long since passed. Rather than run an experiment with old enzymes, you decide to purchase new enzymes. Fortunately, a site called **REBASE®: The Restriction Enzyme Database** can help you. Over 300 restriction enzymes are commercially available rather inexpensively, but scientists are always looking for ways to stretch their research budgets as far as possible. REBASE is excellent for locating enzyme suppliers and enzyme specifics, particularly if you needed to work with an enzyme that you are unfamiliar with. Visit **REBASE®** at http://rebase.neb.com/rebase/rebase.html to identify companies that sell the restriction enzyme(s) you need for this experiment.

### ■ Exercise III – Designing PCR Primers

Giving this experiment more thought, you decide to try reverse transcriptase PCR (RT-PCR) first instead of Northern blotting because RT-PCR is a faster and more sensitive way to detect gene expression. Picking correct primers for a PCR experiment is not a trivial process. You have to be sure the primers are complementary to the gene of interest, and you need to avoid primer self-annealing—or having primers bind to each other—among many other considerations. Fortunately, primer design is another task made much easier by the Internet. In this exercise, you will use **Primer3**, a PCR primer design site from the Whitehead Institute for Biomedical Research.

1. Access **Primer3** at http://frodo.wi.mit.edu/cgi-bin/primer3/primer3_www.cgi. Copy the human DNA sequence from Exercise I into the text box then click "Pick Primers."
2. On the next page, the sequences for the best recommended primers will appear at the top of the screen. Answer the following:
   a. What is the length, in base pairs, of the left (forward) primer and right (reverse) primer? Where does each of these primers bind in the gene sequence?
   b. The hybridization temperature for a PCR reaction is often set around 5 degrees below the melting temperature, or $T_m$ (refer to Chapter 10 for a discussion of melting temperature). Based on the $T_m$ for these primers, what might be the optimal hybridization temperature for this experiment?
   c. What size PCR product would you expect these primers to generate if you ran them on an agarose gel?

## Chapter Summary

1. The development of recombinant DNA technology was originally made possible by the discovery of a class of enzymes called restriction enzymes, which cut DNA at specific recognition sites. The fragments thus produced are joined with DNA vectors to form recombinant DNA molecules.
2. Vectors replicate autonomously in host cells and facilitate the manipulation of the newly created recombinant DNA molecules. Vector components come from many sources, including bacterial plasmids and phages.
3. Recombinant DNA molecules are transferred into a host cell, and cloned copies are produced during host-cell replication. Many kinds of host cells may be used for replication, including bacteria, yeast, and mammalian cells. Cloned copies of foreign DNA sequences are recovered, purified, and analyzed.
4. The polymerase chain reaction (PCR) is a method for amplifying a specific DNA sequence that is present in a collection of DNA sequences, such as genomic DNA. PCR allows DNA to be amplified without host cells and is a rapid, sensitive method with wide-ranging applications.
5. Once cloned, DNA sequences are analyzed through a variety of methods, including restriction mapping and DNA sequencing. Other methods, such as Southern blotting, use hybridization to identify genes and flanking regulatory regions within the cloned sequences.

## INSIGHTS AND SOLUTIONS

1. The recognition sequence for the restriction enzyme *Sau*3AI is GATC (see Figure 13–2); in the recognition sequence for the enzyme *Bam*HI—GGATCC—the four internal bases are identical to the *Sau*3A sequence. The single-stranded ends produced by the two enzymes are identical. Suppose you have a cloning vector that contains a *Bam*HI recognition sequence and you also have foreign DNA that was cut with *Sau*3AI. (a) Can this DNA be ligated into the *Bam*HI site of the vector, and if so, why? (b) Can the DNA segment cloned into this sequence be cut from the vector with *Sau*3AI? With *Bam*HI? What potential problems do you see with the use of *Bam*HI?

**Solution:** (a) DNA cut with *Sau*3AI can be ligated into the vector's *Bam*HI cutting site because the single-stranded ends generated by the two enzymes are identical. (b) The DNA can be cut from the vector with *Sau*3AI because the recognition sequence for this enzyme (GATC) is maintained on each side of the insert. Recovering the cloned insert with *Bam*HI is more problematic. In the ligated vector, the conserved sequences are GGATC (left) and GATCC (right). The correct base for recognition by *Bam*HI will *follow* the conserved sequence (to produce GGATCC on the left) only about 25 percent of the time, and the correct base will *precede* the conserved sequence (and produce GGATCC on the right) about 25 percent of the time as well. Thus, *Bam*HI will be able to cut the insert from the vector ($0.25 \times 0.25 = 0.0625$), or only about 6 percent, of the time.

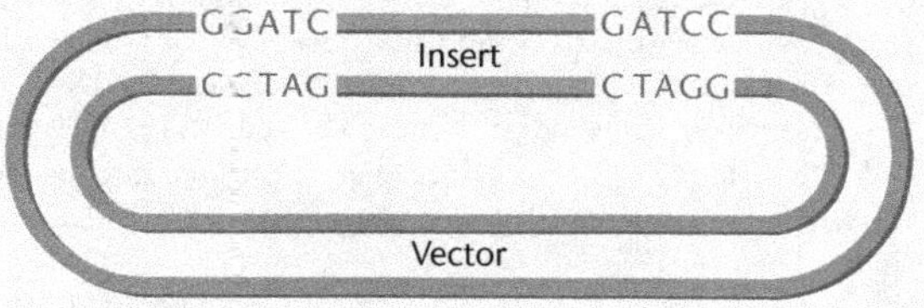


2. In setting up a PCR reaction, which set of primers (out of the three sets listed below) would you choose to amplify the sequence shown as a series of asterisks?

5′-TTAAGATCCGTTACGTATGC * * * * * * AACCCGTTCCTACGAACCTT-3′
3′-AATTCTAGGCAATGCATACG * * * * * * TTGGGCAAGGATGCTTGGAA-5′

Primers:

| | |
|---|---|
| Set 1: 5′-TTAAGATCCGTT-3′ | 5′-CGTTCCTACGAA-3′ |
| Set 2: 5′-GATCCGTTACGT-3′ | 5′-TTCGTAGGAACG-3′ |
| Set 3: 5′-CGTATGCATTGC-3′ | 5′-TTCCAAGCATCC-3′ |

**Solution:** A PCR reaction requires two primers, one for each strand. The primers are complementary to sequences near the end of each single-stranded DNA. In the first step of PCR, the DNA to be copied is heated to separate the molecule into single strands. Then the primers are annealed to the single strands, and elongated in the 5′ to 3′ direction by the addition of Taq polymerase. The only set of primers that will anneal properly is Set 2. One primer has a complementary sequence 4 base pairs in from the end of the lower strand of DNA, and the other primer has a complementary sequence 4 base pairs in from the end of the upper strand.

## Problems and Discussion Questions

1. What roles do restriction enzymes, vectors, and host cells play in recombinant DNA studies?
2. Why is oligo-dT an effective primer for reverse transcriptase?
3. The human insulin gene contains a number of sequences that are removed in the processing of the mRNA transcript. In spite of the fact that bacterial cells cannot excise these sequences from mRNA transcripts, explain how a gene like this can be cloned into a bacterial cell and produce insulin.
4. Restriction enzymes recognize palindromic sequences in intact DNA molecules and cleave the double-stranded helix at these sequences. Inasmuch as the bases are internal in a DNA double helix, how is this recognition accomplished?
5. Although the potential benefits of cloning in higher plants are obvious, the development of this field has lagged behind cloning in bacteria, yeast, and mammalian cells. Can you think of a reason for this?
6. Using DNA sequencing on a cloned DNA segment, you recover the nucleotide sequence shown below. Does this segment contain a palindromic recognition sequence for a restriction enzyme? If so, what is the double-stranded sequence of the palindrome, and what enzyme would cut at this sequence? (Consult Figure 13–2 for a list of restriction-enzyme recognition sequences.)

   CAGTATCCTAGGCAT

7. Restriction-enzyme sequences are palindromic; that is, they read the same in the 5′ to 3′ direction on each strand of DNA. What is the advantage of having restriction recognition sites organized in this way?
8. List the advantages and disadvantages of using plasmids and YACs as cloning vectors.

9. Listed here are several restriction enzymes with four- and six-base recognition sequences. For each of these enzymes, if we assume random distribution and equal amounts of each nucleotide in the DNA to be cut, what is the average distance between each of these recognition sequences?

| | | | |
|---|---|---|---|
| *Taq*I | TCGA | *Hind*III | AAGCTT |
| *Alu*I | AGCT | *Bal*I | TGGCCA |
| *Hae*III | GGCC | *Bam*HI | GGATCC |

10. Some restriction enzymes have recognition sites that are specific and unambiguous, whereas other recognition sites are ambiguous, meaning that any purine, pyrimidine, or nucleotide can occupy certain positions in the recognition sequence. In the following examples, *Not*I has an unambiguous recognition sequence, *Hin*fI has an ambiguous one (N = any nucleotide), and *Xho*II also has an ambiguous recognition sequence (Pu = any purine; Py = any pyrimidine). Assuming random distribution and equal amounts of each nucleotide in the DNA to be cut, what is the average distance between each of these restriction recognition sequences?

| | |
|---|---|
| *Not*I | GCGGCCGC |
| *Hin*fI | GANTC |
| *Xho*II | PuGATCPy |

11. What are the advantages of using a restriction enzyme whose recognition site is relatively rare? When would you use such enzymes?
12. An ampicillin-resistant, tetracycline-resistant plasmid, pBR322, is cleaved with *Pst*I, which cleaves within the ampicillin resistance gene. The cut plasmid is ligated with *Pst*I-digested *Drosophila* DNA to prepare a genomic library, and the mixture is used to transform *E. coli* K12.
 (a) Which antibiotic should be added to the medium to select cells that have incorporated a plasmid?
 (b) What growth pattern should be selected to obtain plasmids containing *Drosophila* inserts?
 (c) How can you explain the presence of colonies that are resistant to both antibiotics?

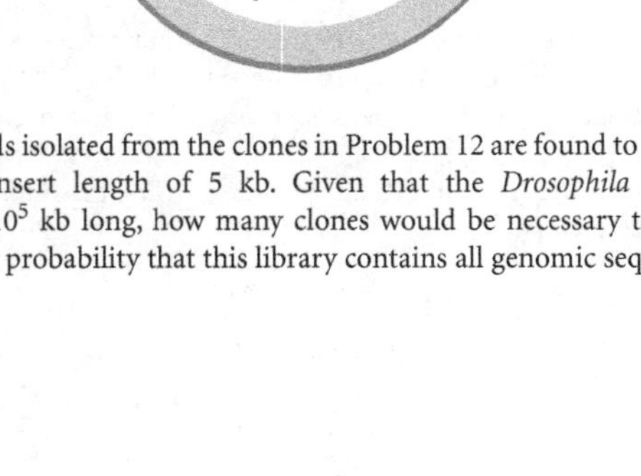


13. Plasmids isolated from the clones in Problem 12 are found to have an average insert length of 5 kb. Given that the *Drosophila* genome is $1.5 \times 10^5$ kb long, how many clones would be necessary to give a 99 percent probability that this library contains all genomic sequences?
14. In a control experiment, a plasmid containing a *Hin*dIII recognition sequence within a kanamycin resistance gene is cut with *Hin*dIII, religated, and used to transform *E. coli* K12 cells. Kanamycin-resistant colonies are selected, and plasmid DNA from these colonies is subjected to electrophoresis. Most of the colonies contain plasmids that produce single bands that migrate at the same rate as the original intact plasmid. A few colonies, however, produce two bands, one of original size and one that migrates much higher in the gel. Diagram the origin of this slow band as a product of ligation.
15. You have just created the world's first genomic library from the African okapi, a relative of the giraffe. No genes from this genome have been previously isolated or described. You wish to isolate the gene encoding the oxygen-transporting protein $\beta$-globin from the okapi library. This gene has been isolated from humans, and its nucleotide sequence and amino acid sequence are available in databases. Using the information available about the human $\beta$-globin gene, what two strategies can you use to isolate this gene from the okapi library?
16. In the production of cDNA, the single-stranded DNA produced by reverse transcriptase can be made double stranded by treatment with DNA polymerase I. However, no primer is required with the DNA polymerase. Why is this?
17. What should you consider in deciding which vector to use in constructing a genomic library of eukaryotic DNA?
18. You are given a cDNA library of human genes prepared in a bacterial plasmid vector. You are also given the cloned yeast gene that encodes EF-1a, a protein that is highly conserved among eukaryotes. Outline how you would use these resources to identify the human cDNA clone encoding EF-1a.
19. Once you have isolated the human cDNA clone for EF-1a in Problem 18, you sequence the clone and find that it is 1384 nucleotide pairs long. Using this cDNA clone as a probe, you isolate the DNA encoding EF-1a from a human genomic library. The genomic clone is sequenced and found to be 5282 nucleotides long. What accounts for the difference in length observed between the cDNA clone and the genomic clone?
20. You have recovered a cloned DNA segment from a vector and determine that the insert is 1300 bp in length. To characterize this cloned segment, you isolate the insert and decide to construct a restriction map. Using enzyme I and enzyme II, followed by gel electrophoresis, you determine the number and size of the fragments produced by enzymes I and II alone and in combination, as recorded in the following table. Construct a restriction map from these data, showing the positions of the restriction-enzyme cutting sites relative to one another and the distance between them in units of base pairs.
21. To create a cDNA library, cDNA can be inserted into vectors and cloned. In the analysis of cDNA clones, it is often difficult to find clones that are full length—that is, many clones are shorter than the mature mRNA molecules from which they are derived. Why is this so?

| Enzymes | Restriction Fragment Sizes (bp) |
|---|---|
| I | 350, 950 |
| II | 200, 1100 |
| I and II | 150, 200, 950 |

22. Although the capture and trading of great apes has been banned in 112 countries since 1973, it is estimated that about 1000 chimpanzees are removed annually from Africa and smuggled into Europe, the United States, and Japan. This illegal trade is often disguised by simulating births in captivity. Until recently, genetic identity tests to uncover these illegal activities were not used because of the lack of highly polymorphic

markers (markers that vary from one individual to the next) and the difficulties of obtaining chimpanzee blood samples. Recently, a study was reported in which DNA samples were extracted from freshly plucked chimpanzee hair roots and used as templates for PCR. The primers used in these studies flank highly polymorphic sites in human DNA that result from variable numbers of tandem nucleotide repeats. Several offspring and their putative parents were tested to determine whether the offspring were "legitimate" or the product of illegal trading. The data are shown in the following Southern blot.

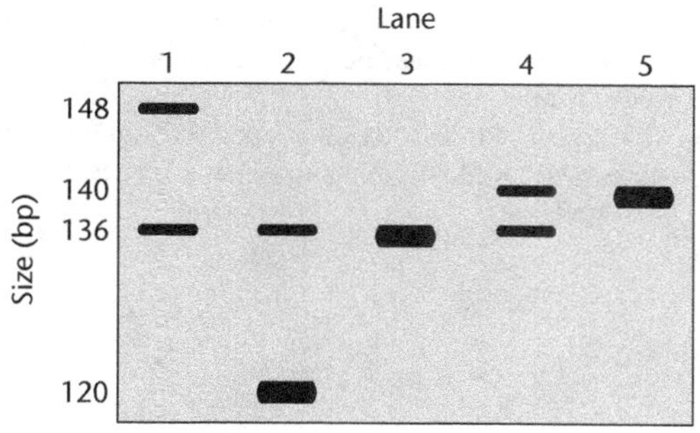


Lane 1: father chimpanzee
Lane 2: mother chimpanzee
Lanes 3–5: putative offspring A, B, C

Examine the data carefully, and choose the best conclusion.

(a) None of the offspring is legitimate.
(b) Offspring B and C are not the products of these parents and were probably purchased on the illegal market. The data are consistent with offspring A being legitimate.
(c) Offspring A and B are products of the parents shown, but C is not and was therefore probably purchased on the illegal market.
(d) There are not enough data to draw any conclusions. Additional polymorphic sites should be examined.
(e) No conclusion can be drawn because "human" primers were used.

23. The following partial restriction map shows a recombinant plasmid, pBIO220, formed by cloning a piece of *Drosophila* DNA (striped segment), including the gene *rosy*, into the vector pBR322, which also contains the penicillin resistance gene, *pen*. The vector part of the plasmid contains only the two E recognition sequences shown and no A or B sequences. The gel (in the next column) shows several restriction digests of pBIO220.

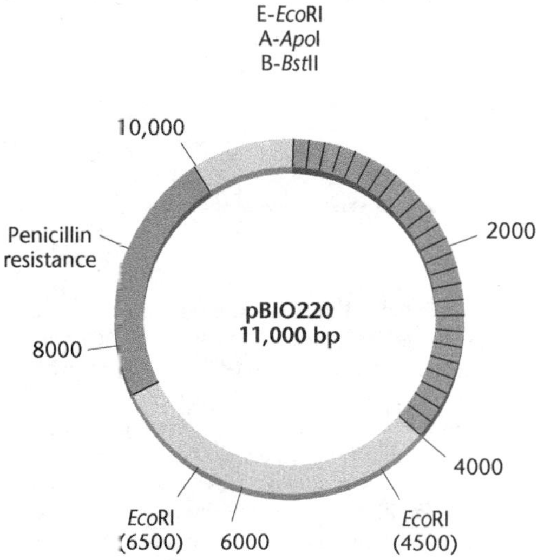


(a) Use the stained gel pattern to deduce where restriction-enzyme recognition sequences are located in the cloned fragment.
(b) A PCR-amplified copy of the entire 2000-bp *rosy* gene was used to probe a Southern blot of the same gel. Use the Southern blot results to deduce the location of *rosy* in the cloned fragment. Redraw the map showing the location of the *rosy* gene.

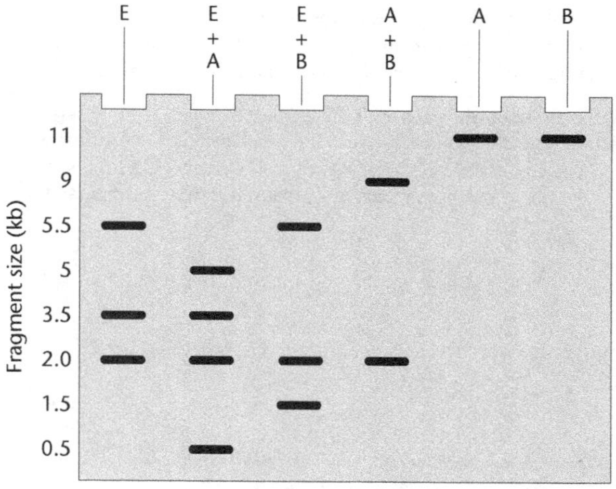


**Stained gel**

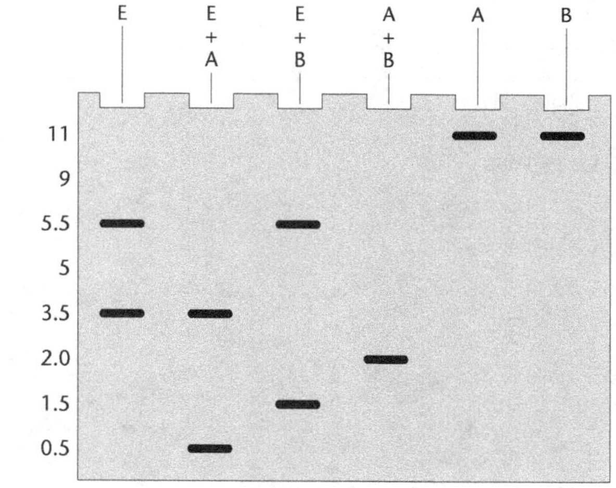


**Southern blot**

24. List the steps involved in screening a genomic library. What must be known before starting such a procedure? What are the potential problems with such a procedure, and how can they be overcome or minimized?
25. To estimate the number of cleavage sites in a particular piece of DNA with a known size, you can apply the formula $N/4^n$, where $N$ is the number of base pairs in the target DNA and $n$ is the number of bases in the recognition sequence of the restriction enzyme. If the recognition sequence for *Bam*HI is GGATCC and the λ phage DNA contains approximately 48,500 bp, how many cleavage sites would you expect?
26. In a typical PCR reaction, what phenomena are occurring at temperature ranges (a) 90–95°C, (b) 50–70°C, and (c) 70–75°C?

27. We usually think of enzymes as being most active at around 37°C, yet in PCR the DNA polymerase is subjected to multiple exposures of relatively high temperatures and seems to function appropriately at 70–75°C. What is special about the DNA polymerizing enzymes typically used in PCR?
28. How are dideoxynucleotides (ddNTPs) used in the chain-termination method of DNA sequencing?
29. Assume you have conducted a standard DNA a sequencing reaction using the chain-termination method. You performed all the steps correctly and electrophoresed the resulting DNA fragments correctly, but when you looked at the sequencing gel, many of the bands were duplicated (in terms of length) in other lanes. What might have happened?

**HOW DO WE KNOW?**

30. In this chapter we focused on how specific DNA sequences can be copied, identified, characterized, and sequenced. At the same time, we found many opportunities to consider the methods and reasoning underlying these techniques. From the explanations given in the chapter, what answers would you propose to the following fundamental questions?
    (a) How do we know whether a restriction enzyme will cut DNA frequently or infrequently when all bases are equally represented?
    (b) How can we experimentally determine the identity, location, and distance between restriction-enzyme recognition sites in a DNA segment?
    (c) What combination of techniques is used to establish that cells from different tissues are expressing different sets of genes?
    (d) What properties must a DNA segment possess in order to function as an artificial chromosome vector for DNA cloning? How are such vectors constructed?
    (e) What steps make PCR a chain reaction that can produce millions of copies of a specific DNA molecule in a matter of hours without using host cells?

## Extra-Spicy Problems

31. The gel presented here shows the pattern of bands of fragments produced with several restriction enzymes. The enzymes used are identified above and below the gel.

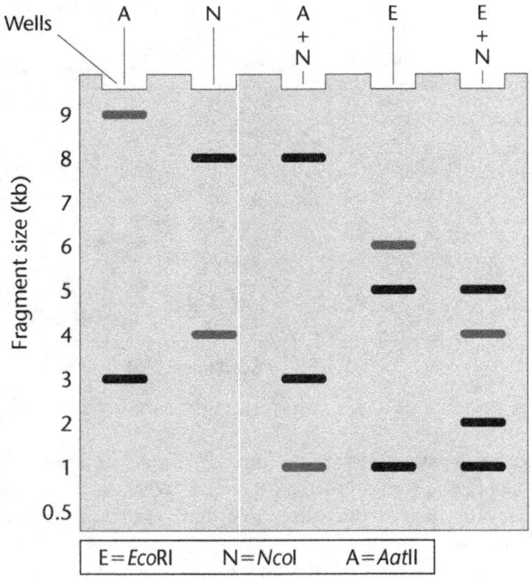


One of the six restriction maps shown in the next column is consistent with the pattern of bands shown in the gel.
(a) From your analysis of the pattern of bands on the gel, select the correct map and explain your reasoning.
(b) In a Southern blot prepared from this gel, the highlighted bands (pink) hybridized with the gene *pep*. Where is the *pep* gene located?

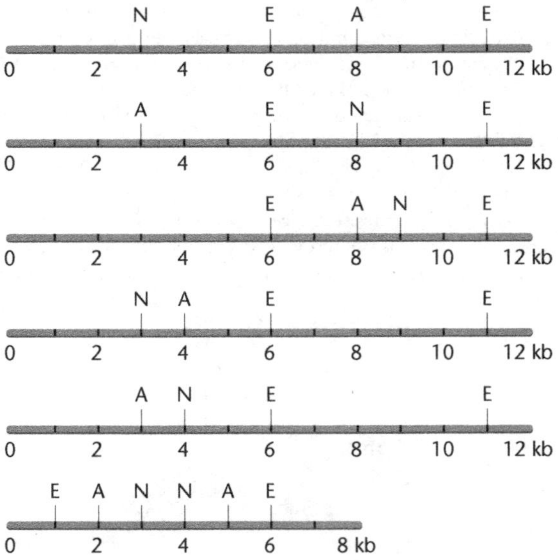


32. A widely used method for calculating the annealing temperature for a primer used in PCR is 5 degrees below the $T_m$ (°C), which is computed by the equation $81.5 + 0.41\ (\%\ GC) - (675/N)$, where %GC is the percentage of GC nucleotides in the oligonucleotide and $N$ is the length of the oligonucleotide. Notice from the formula that both the GC content *and* the length of the oligonucleotide are variables. Assuming you have the following oligonucleotide as a primer, compute the annealing temperature for PCR. What is the relationship between $T_m$ (°C) and %GC? Why? (*Note:* In reality, this computation provides only a starting

point for empirical determination of the most useful annealing temperature.)

5′-TTGAAAATATTTCCCATTGCC-3′

33. The diagram below represents a simplified genetic map of the bacteriophage λ chromosome.

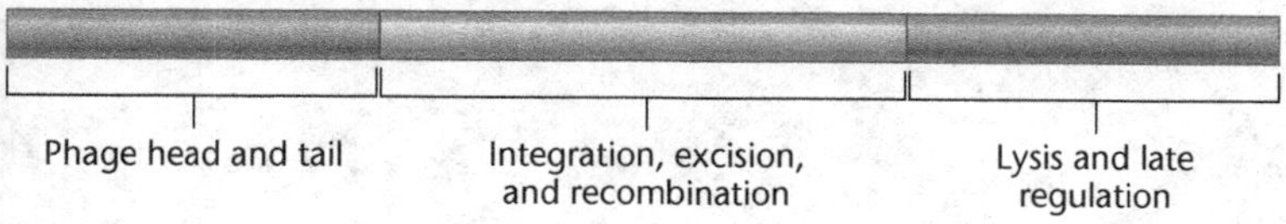


As stated in the text, when λ phage is used as a cloning vector, the central gene cluster is removed and the DNA to be cloned essentially replaces the removed section. A recombinant λ therefore contains the two ends with the cloned DNA sandwiched between. Why is the central region, rather than either or both ends, removed and replaced?

34. Most of the techniques described in this chapter (blotting, cloning, PCR, etc.) are dependent on intermolecular attractions (annealing) between different populations of nucleic acids. Length of the strands, temperature, and percentage of GC nucleotides weigh considerably on intermolecular associations. Two other components commonly used in hybridization protocols are monovalent ions and formamide. A formula that takes monovalent ion ($Na^+$) and formamide concentrations into consideration to compute a $T_m$ (temperature of melting) is given below:

$$T_m = 81.5 + 16.6(\log M[Na^+]) + 0.41\%\ GC - 0.72(\%\ formamide)$$

(a) For the following concentrations of $Na^+$ and formamide, calculate the $T_m$. Assume 45% GC content.

| $Na^+$ | % Formamide |
|---|---|
| 0.825 | 20 |
| 0.825 | 40 |
| 0.165 | 20 |
| 0.165 | 40 |

(b) Given that formamide competes for hydrogen bond locations of nucleic acid bases and monovalent cations are attracted to the negative charges of nucleic acids explain why the $T_m$ varies as described in part (a).

35. There are a variety of circumstances under which rapid results using multiple markers in PCR amplifications is highly desired, such as in forensics, pathogen analysis, or detection of genetically modified organisms. In multiplex PCR, multiple sets of primers are used, often with less success than when applied to PCR as individual sets. Numerous studies have been conducted to optimize procedures, but each has described the process as time consuming and often unsuccessful. Considering the information given in question 34, why should multiplex PCR be any different than single primer set PCR in terms of dependability and ease of optimization?

36. The U.S. Department of Justice has recently sponsored research to establish a database that catalogs PCR amplification products from short tandem repeats of the Y (Y-STRs) chromosome in humans. The database contains polymorphisms of five U.S. ethnic groups (African Americans, European Americans, Hispanics, Native Americans, and Asian Americans) as well as worldwide population.

(a) Given that STRs are repeats of varying lengths, for example $(TCTG)_{9-17}$ or $(TAT)_{6-14}$, explain how PCR could reveal differences (polymorphisms) among individuals. How could the Department of Justice make use of those differences?

(b) Y-STRs from the nonrecombining region of the Y chromosome (NRY) have special relevance for forensic purposes. Why?

(c) What would be the value of knowing the ethnic population differences for Y-STR polymorphisms?

(d) For forensic applications, the probability of a "match" for a crime scene DNA sample and a suspect's DNA often culminates in a guilty or innocent verdict. How is a "match" determined, and what are the uses and limitations of such probabilities?

Electron micrograph of a segment of DNA undergoing transcription.

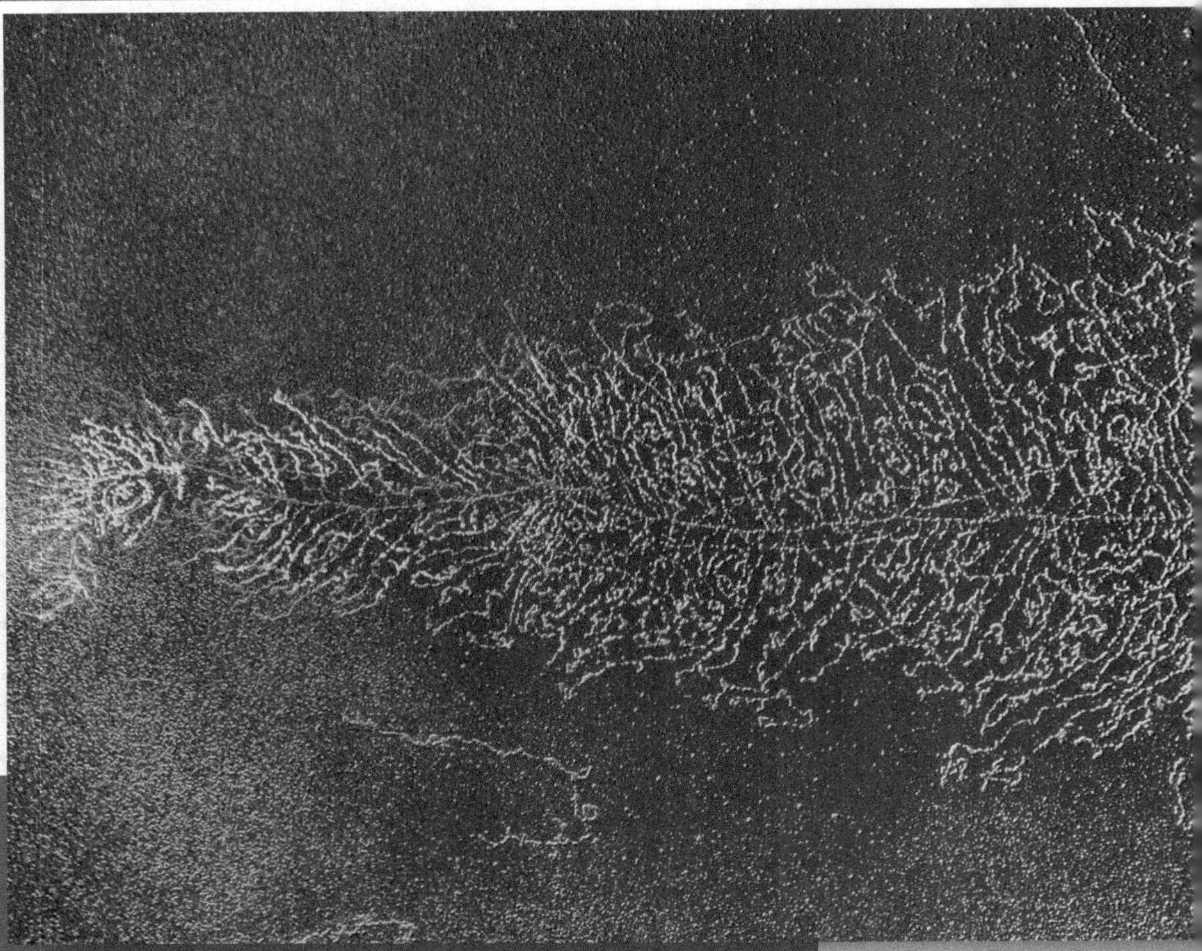

14

# The Genetic Code and Transcription

## CHAPTER CONCEPTS

- Genetic information is stored in DNA by means of a triplet code that is nearly universal to all living things on Earth.
- The genetic code is initially transferred from DNA to RNA, in the process of transcription.
- Once transferred to RNA, the genetic code exists as triplet codons, which are sets of three nucleotides in which each nucleotide is one of the four kinds of ribonucleotides composing RNA.
- RNA's four ribonucleotides, analogous to an alphabet of four "letters," can be arranged into 64 different three-letter sequences. Most of the triplets in RNA encode one of the 20 amino acids present in proteins, which are the end products of most genes.
- Several codons act as signals that initiate or terminate protein synthesis.
- In eukaryotes, the process of transcription is similar to, but more complex than, that in prokaryotes and in the bacteriophages that infect them.

As we saw in Chapter 10, the structure of DNA consists of a linear sequence of deoxyribonucleotides. This sequence ultimately dictates the components of proteins, the end products of most genes. A central issue is how information stored as a nucleic acid can be decoded into a protein. Figure 14–1 provides a simple overview of how this transfer of information, resulting ultimately in gene expression, occurs. In the first step, information present on one of the two strands of DNA (the template strand) is transferred into an RNA complement through the process of transcription. Once synthesized, this RNA acts as a "messenger" molecule, transporting the coded information out of the nucleus—hence its name, **messenger RNA (mRNA)**. The mRNAs then associate with ribosomes, where decoding into proteins occurs.

In this chapter, we will focus on the initial phases of gene expression by addressing two major questions. First, how is genetic information encoded? Second, how does the transfer from DNA to RNA occur, thus explaining the process of transcription? As you shall see, ingenious analytical research has established that the genetic code is written in units of three letters—triplets of ribonucleotides in mRNA that reflect the stored information in genes. Most of the triplet code words direct the incorporation of a specific amino acid into a protein as it is synthesized. As we can predict based on the complexity of the replication of DNA (discussed in Chapter 11), transcription is also a complex process dependent on a major polymerase enzyme and a cast of supporting proteins. We will explore what is known about transcription in bacteria and then contrast this prokaryotic model with transcription in eukaryotes.

In Chapter 15, we will continue our discussion of gene expression by addressing how translation occurs and then describing the structure and function of proteins. Together, these chapters provide a comprehensive picture of molecular genetics, which serves as the most basic foundation for the understanding of living organisms.

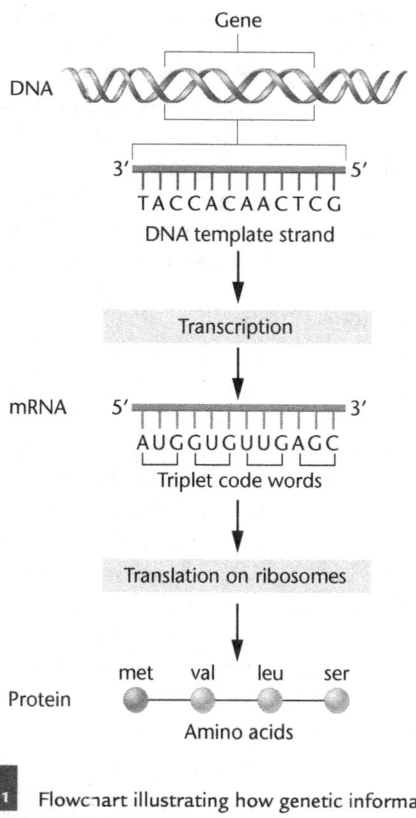


**FIGURE 14–1** Flowchart illustrating how genetic information encoded in DNA produces protein.

## 14.1 The Genetic Code Uses Ribonucleotide Bases as "Letters"

Before we consider the various analytical approaches that led to our current understanding of the genetic code, let's summarize the general features that characterize it:

1. The genetic code is written in linear form, using as "letters" the ribonucleotide bases that compose mRNA molecules. The ribonucleotide sequence is derived from the complementary nucleotide bases in DNA.
2. Each "word" within the mRNA consists of three ribonucleotide letters, thus referred to as a **triplet code**. With only three exceptions, each group of *three* ribonucleotides, called a **codon**, specifies *one* amino acid.
3. The code is **unambiguous**—each triplet specifies only a single amino acid.
4. The code is **degenerate**, meaning that a given amino acid can be specified by more than one triplet codon. This is the case for 18 of the 20 amino acids.
5. The code contains one "start" and three "stop" signals, triplets that **initiate** and **terminate** translation, respectively.
6. No internal punctuation (analogous, for example, to a comma) is used in the code. Thus, the code is said to be **commaless**. Once translation of mRNA begins, the codons are read one after the other with no breaks between them (until a stop signal is reached).
7. The code is **nonoverlapping**. After translation commences, any single ribonucleotide within the mRNA is part of only one triplet.
8. The code is nearly **universal**. With only minor exceptions, a single coding dictionary is used by almost all viruses, prokaryotes, archaea, and eukaryotes.

WEB TUTORIAL 14.1 TRANSCRIPTION

14.2

# Early Studies Established the Basic Operational Patterns of the Code

In the late 1950s, before it became clear that mRNA serves as an intermediate in the transfer of genetic information from DNA to proteins, researchers thought that DNA might encode protein synthesis more directly. Because ribosomes had already been identified, the initial thinking was that information in DNA was transferred in the nucleus to the RNA of the ribosome, which served as the template for protein synthesis in the cytoplasm. This concept soon became untenable, as accumulating evidence indicated that there was an unstable intermediate template. The RNA of ribosomes, on the other hand, was extremely stable. In 1961, François Jacob and Jacques Monod postulated the existence of messenger RNA (mRNA). After mRNA was discovered, it soon became clear that, even though genetic information is stored in DNA, the code that is translated into proteins resides in RNA. The central question then was, how could only four letters—the four nucleotides—specify 20 words, the amino acids? Scientists considered the possibility that DNA codes could overlap in such a way that each nucleotide would be part of more than one code word, but accumulating evidence was inconsistent with the constraints such a code would place on the resultant amino acid sequences. We will address some of this evidence below.

## The Triplet Nature of the Code

In the early 1960s, Sidney Brenner argued on theoretical grounds that the code must be a triplet, since three-letter words represent the minimal length needed for using four letters to specify 20 amino acids. A code of four nucleotides combined into two-letter words, for example, provides only 16 unique code words ($4^2$). A triplet code yields 64 words ($4^3$) and therefore is sufficient for the 20 needed but far less cumbersome than a four-letter code ($4^4$), which would specify 256 words.

The ingenious experimental work of Francis Crick, Leslie Barnett, Brenner, and R. J. Watts-Tobin provided the first solid evidence for a triplet code. These researchers induced insertion and deletion mutations in the *rII* locus of phage T4 (see Chapter 6). Mutations in this locus cause rapid lysis and distinctive plaques. Such mutants will successfully infect strain B of *E. coli* but cannot reproduce in a different strain of *E. coli* designated K12. Crick and his colleagues used the acridine dye proflavin to induce the mutations. This mutagenic agent intercalates within the double helix of DNA, causing the insertion or deletion of one or more nucleotides during replication. As shown in Figure 14–2(a), an insertion of a single nucleotide causes the **reading frame** (the contiguous sequence of nucleotides encoding a polypeptide) to shift, changing all subsequent downstream codons (codons to the right of the insertion in the figure). Such mutations are referred to as **frameshift mutations.** Upon translation, the amino acid sequence of the encoded protein will be altered radically starting at the point where the mutation occurred.

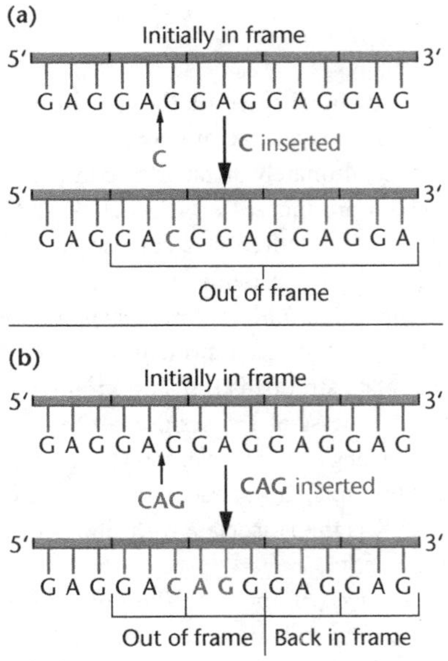


FIGURE 14–2 The effect of frameshift mutations on a DNA sequence repeating the triplet sequence GAG. (a) The insertion of a single nucleotide shifts all subsequent triplets out of the reading frame. (b) The insertion of three nucleotides changes only two triplets, after which the original reading frame is reestablished.

When these mutations are present at the *rII* locus, T4 will not reproduce on *E. coli* K12.

Crick and his colleagues reasoned that if phages with these induced frameshift mutations were treated again with proflavin, still other insertions or deletions would occur. A second change might result in a revertant phage, which would display wild-type behavior and successfully infect *E. coli* K12. For example, if the original mutant contained an insertion (+), a second event causing a deletion (−) close to the insertion would restore the original reading frame. In the same way, an event resulting in an insertion (+) might correct an original deletion (−).

By studying many mutations of this type, these researchers were able to compare combinations of various frameshift mutations within the same DNA sequence. They found that various combinations of one plus (+) and one minus (−) caused reversion to wild-type behavior. Still other observations shed light on the number of nucleotides constituting the genetic code. When two pluses or two minuses occurred in the same sequence, the correct reading frame *was not* reestablished. This argued against a doublet (two-letter) code. However, when three pluses [Figure 14–2(b)] or three minuses were present together, the original frame *was* reestablished. These observations strongly supported the triplet nature of the code.

## The Nonoverlapping Nature of the Code

Work by Sydney Brenner and others soon established that the code could not be overlapping for any one transcript. For example, beginning by assuming a triplet code, Brenner deduced the restrictions that might be expected if the code were overlapping. He imagined theoretical nucleotide sequences encoding a protein consisting of three amino acids. If the nucleotide sequence were, say, GTACA, and parts of the central codon, TAC, were shared by the outer codons, GTA and ACA, then only certain amino acids (those with codons ending in TA or beginning with AC) should be found adjacent to the one encoded by the central codon. This consideration led Brenner to conclude that if the code were overlapping, the compositions of tripeptide sequences within proteins should be somewhat limited.

For example, when any particular central amino acid is considered, only 16 combinations ($2^4$) of three amino acids (tripeptide sequences) would theoretically be possible. Looking at the amino acid sequences of proteins that were known at that time, Brenner failed to find such restrictions in tripeptide sequences. For any central amino acid, he found many more than 16 different tripeptides. This observation led him to conclude that the code is not overlapping.

A second major argument against an overlapping code concerned the effect of a single nucleotide change. With an overlapping code, two adjacent amino acids would be affected by such a point mutation. However, mutations in the genes coding for the protein coat of tobacco mosaic virus (TMV), human hemoglobin, and the bacterial enzyme tryptophan synthetase invariably revealed only single amino acid changes.

The third argument against an overlapping code was presented by Francis Crick in 1957, when he predicted that DNA would not serve as a direct template for the formation of proteins. Crick reasoned that any affinity between nucleotides and an amino acid would require hydrogen bonding. Chemically, however, such specific affinities seemed unlikely. Instead, Crick proposed that there must be an "adaptor molecule" that could covalently bind to the amino acid, yet also be capable of hydrogen bonding to a nucleotide sequence. With an overlapping code, various adaptors would somehow have to overlap one another at nucleotide sites during translation, making the translation process overly complex, in Crick's opinion, and possibly inefficient. As we will see later in this chapter, Crick's prediction was correct—transfer RNA (tRNA) serves as the adaptor in protein synthesis, and the ribosome accommodates two tRNA molecules at a time.

Crick's and Brenner's arguments, taken together, strongly suggested that, during translation, the genetic code is nonoverlapping. Without exception, this concept has been upheld.

### The Commaless and Degenerate Nature of the Code

Between 1958 and 1960, information relating to the genetic code continued to accumulate. In addition to his adaptor proposal, Crick hypothesized, on the basis of genetic evidence, that the code would be commaless—that is, he believed no internal punctuation would occur along the reading frame. Crick also speculated that only 20 of the 64 possible codons would specify an amino acid and that the remaining 44 would carry no coding assignment.

Was Crick wrong with respect to the 44 "blank" codes? Or is the code degenerate, meaning that more than one codon specifies the same amino acid? Crick's frameshift studies (discussed earlier in this chapter) suggested that, in fact, contrary to his earlier proposal, the code is degenerate. For the cases in which wild-type function is restored—that is, (+) with (−); (++) with (− −); and (+++) with (− − −)—the original frame of reading is also restored. However, in between the insertion and deletion, there may be numerous codons that would still be out of frame. If 44 of the 64 possible codons were blank, referred to as **nonsense codons**, and did not specify an amino acid, at least one blank codon would very likely occur in the string of nucleotides still out of frame. If such a nonsense codon was encountered during protein synthesis, the process would probably stop or be terminated at that point. If so, the product of the *rII* locus would not be made, and restoration would not occur. Because the various mutant combinations were able to reproduce on *E. coli* K12, Crick and his colleagues concluded that, in all likelihood, most, if not all, of the remaining 44 triplets were not blank. It followed that the genetic code was degenerate. As we shall see, this reasoning proved to be correct.

## 14.3 Studies by Nirenberg, Matthaei, and Others Led to Deciphering of the Code

In 1961, Marshall Nirenberg and J. Heinrich Matthaei became the first to characterize specific coding sequences, laying a cornerstone for the complete analysis of the genetic code. Their success, as well as that of others who made important contributions in deciphering the code, was dependent on the use of two experimental tools, an *in vitro (cell-free) protein-synthesizing system* and the enzyme **polynucleotide phosphorylase**, which allowed the production of synthetic mRNAs. These mRNAs served as templates for polypeptide synthesis in the cell-free system.

### Synthesizing Polypeptides in a Cell-Free System

In the cell-free protein-synthesizing system, amino acids are incorporated into polypeptide chains. The process begins with an *in vitro* mixture containing all the essential factors for protein synthesis in the cell: ribosomes, tRNAs, amino acids, and other molecules essential to translation (see Chapter 15). To allow scientists to follow (or trace) the progress of protein synthesis, one or more of the amino acids must be radioactive. Finally, an mRNA must be added, to serve as the template to be translated.

In 1961, mRNA had yet to be isolated. However, use of the enzyme polynucleotide phosphorylase allowed the artificial synthesis of RNA templates, which could be added to the cell-free system. This enzyme, isolated from bacteria, catalyzes the reaction shown in Figure 14–3. Discovered in 1955 by Marianne Grunberg-Manago and Severo Ochoa, the enzyme functions metabolically in bacterial cells to degrade RNA. However, *in vitro*, in the presence of high concentrations of ribonucleoside diphosphates, the reaction can be

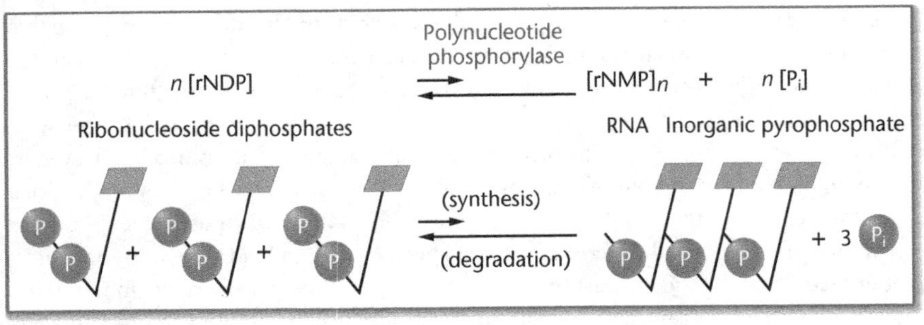


FIGURE 14-3 The reaction catalyzed by the enzyme polynucleotide phosphorylase. Note that the equilibrium of the reaction favors the degradation of RNA but that the reaction can be "forced" in the direction favoring synthesis.

"forced" in the opposite direction, to synthesize RNA, as illustrated in the figure.

In contrast to RNA polymerase (which constructs mRNA *in vivo*), polynucleotide phosphorylase does not require a DNA template. As a result, the order in which ribonucleotides are added is random, depending on the relative concentration of the four ribonucleoside diphosphates present in the reaction mixture. The probability of the insertion of a specific ribonucleotide is proportional to the availability of that molecule relative to other available ribonucleotides. *This point is absolutely critical to understanding the work of Nirenberg and others in the ensuing discussion.*

Together, the cell-free system for protein synthesis and the availability of synthetic mRNAs provided a means of deciphering the ribonucleotide composition of various codons encoding specific amino acids.

## Homopolymer Codes

For their initial experiments, Nirenberg and Matthaei synthesized **RNA homopolymers**, RNA molecules containing only one type of ribonucleotide, and used them for synthesizing polypeptides *in vitro*. In other words, the mRNA they used in their cell-free protein-synthesizing system was either UUUUUU. . ., AAAAAA. . ., CCCCCC. . ., or GGGGGG. . . . They tested each of these types of mRNA to see which, if any, amino acids were consequently incorporated into newly synthesized proteins. The method they used was to conduct many experimental syntheses with each homopolymer. They always made all 20 amino acids available, but for each experiment they attached a radioactive label to a different amino acid and thus could tell when that amino acid had been incorporated into the resulting polypeptide.

For example, in experiments using $^{14}$C-phenylalanine (Table 14.1), Nirenberg and Matthaei concluded that the RNA homopolymer UUUUU. . . (polyuridylic acid) directed the incorporation of only phenylalanine into the peptide homopolymer polyphenylalanine. Assuming the validity of a triplet code, they made the first specific codon assignment: UUU codes for phenylalanine. Using similar experiments, they quickly found that AAA codes for lysine and CCC codes for proline. Poly G was not a functional template, probably because the molecule folds back on itself. Thus, the assignment for GGG had to await other approaches.

Note that the specific triplet codon assignments were possible only because homopolymers were used. In this method, only the general nucleotide composition of the template is known, not the specific order of the nucleotides in each triplet, but since three identical letters can have only one possible sequence (e.g., UUU), three of the actual codons for phenylalanine, lysine, and proline could be identified.

## Mixed Copolymers

With the initial success of these techniques, Nirenberg and Matthaei, and Ochoa and coworkers turned to the use of **RNA heteropolymers**. In their next experiments, two or more different ribonucleoside diphosphates were used in combination to form the artificial message. The researchers reasoned that if they knew the relative proportion of each type of ribonucleoside diphosphate in their synthetic mRNA, they could predict the frequency of each of the possible triplet codons it contained. If they then added the mRNA to the cell-free system and ascertained the percentage of each amino acid present in the resulting polypeptide, they could analyze the results and predict the *composition* of the triplets that had specified those particular amino acids.

This approach is illustrated in Figure 14–4. Suppose that only A and C are used for synthesizing the mRNA, in a ratio of 1A:5C. The insertion of a ribonucleotide at any position along the RNA molecule during its synthesis is determined by the ratio of A:C. Therefore, there is a 1/6 possibility for an A and a 5/6 chance for a C to occupy each position. On this basis, we can calculate the frequency of any given triplet appearing in the message.

For AAA, the frequency is $(1/6)^3$ or about 0.4 percent. For AAC, ACA, and CAA, the frequencies are identical—that is, $(1/6)^3$ $(5/6)$ or about 2.3 percent for each triplet. Together, all three 2A:1C triplets account for 6.9 percent of the total three-letter sequences. In the same way, each of three 1A:2C triplets accounts for $(1/6)$ $(5/6)^2$ or 11.6 percent (or a total of 34.8 percent); CCC is represented by $(5/6)^3$, or 57.9 percent of the triplets.

By examining the percentages of the different amino acids incorporated into the polypeptide synthesized under the direction of this message, we can propose probable base compositions for each

TABLE 14.1

**Incorporation of $^{14}$C-Phenylalanine into Protein**

| Artificial mRNA | Radioactivity (counts/min) |
|---|---|
| None | 44 |
| Poly U | 39,800 |
| Poly A | 50 |
| Poly C | 38 |

SOURCE: After Nirenberg and Matthaei (1961).

| Possible compositions | Possible triplets | Probability of occurrence of any triplet | Final % |
|---|---|---|---|
| 3A | AAA | $(1/6)^3 = 1/216 = 0.4\%$ | 0.4 |
| 1C:2A | AAC ACA CAA | $(5/6)(1/6)^2 = 5/216 = 2.3\%$ | 3 x 2.3 = 6.9 |
| 2C:1A | ACC CAC CCA | $(5/6)^2(1/6) = 25/216 = 11.6\%$ | 3 x 11.6 = 34.8 |
| 3C | CCC | $(5/6)^3 = 125/216 = 57.9\%$ | 57.9 |
| | | | 100.0 |

Chemical synthesis of message ↓

RNA
C C C C C C C C A C C C C C C A A C C A C C C C C A C C C C C A C C C A A

Translation of message ↓

| Percentage of amino acids in protein | | Probable base-composition assignments |
|---|---|---|
| Lysine | <1 | AAA |
| Glutamine | 2 | 1C:2A |
| Asparagine | 2 | 1C:2A |
| Threonine | 12 | 2C:1A |
| Histidine | 14 | 2C:1A, 1C:2A |
| Proline | 69 | CCC, 2C:1A |

FIGURE 14–4 Results and interpretation of a mixed copolymer experiment in which a ratio of 1A:5C is used (1/6A:5/6C).

of those amino acids (Figure 14–4). Because proline appears 69 percent of the time, we could propose that proline is encoded by CCC (57.9 percent) and also by one of the codons consisting of 2C:1A (11.6 percent). Histidine, at 14 percent, is probably coded by one 2C:1A codon (11.6 percent) and one 1C:2A codon (2.3 percent). Threonine, at 12 percent, is likely coded by only one 2C:1A codon. Asparagine and glutamine each appear to be coded by one of the 1C:2A codons, and lysine appears to be coded by AAA.

Using as many as all four ribonucleotides to construct the mRNA, the researchers conducted many similar experiments. Although the determination by this means of the *composition* of triplet code words corresponding to all 20 amino acids represented a very significant breakthrough, the *specific sequences* of triplets were still unknown—other approaches were still needed.

**NOW SOLVE THIS**

Problem 28 on page 379 asks you to analyze a reciprocal pair of mixed copolymer experiments and to predict codon compositions for the amino acids the copolymers encode.

HINT: *The reciprocity of these two experiments is essential to solving the problem, because analysis of the data set from only one of them leads to more than one possible answer. However, there is only one answer consistent with both sets of data.*

## The Triplet Binding Assay

It was not long before more advanced techniques for discovering codons were developed. In 1964, Nirenberg and Philip Leder developed the **triplet binding assay**, leading to specific assignments of triplet codons. This technique took advantage of the observation that ribosomes, when presented *in vitro* with an RNA sequence as short as three ribonucleotides, will bind to it and form a complex similar to what is found *in vivo*. The triplet RNA sequence acts like a codon in mRNA, attracting a tRNA molecule containing a complementary sequence (Figure 14–5). Such a triplet sequence in

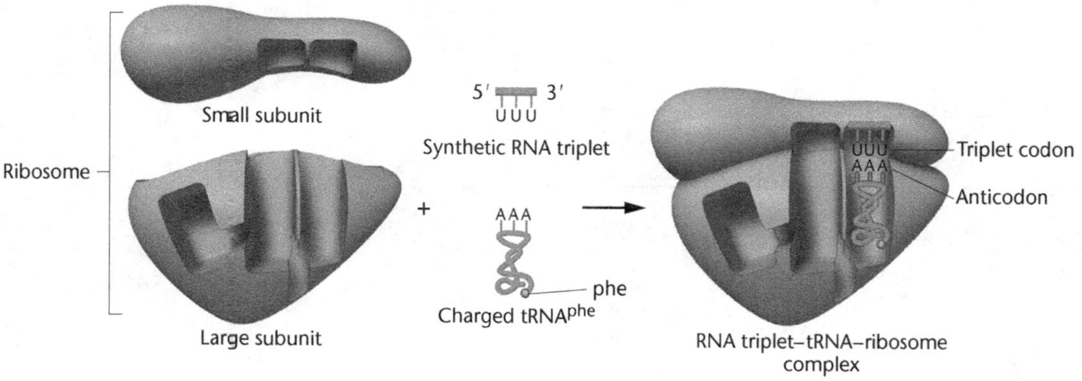


FIGURE 14–5 The behavior of the components during the triplet-binding assay. When the UUU triplet is positioned in the ribosome, it acts as a codon, attracting the complementary AAA anticodon of the charged $tRNA^{phe}$.

TABLE 14.2

**Amino Acid Assignments to Specific Trinucleotides Derived from the Triplet-Binding Assay**

| Trinucleotides | Amino Acid |
|---|---|
| AAA AAG | Lysine |
| AUG | Methionine |
| AUU AUG AUA | Isoleucine |
| CCU CCC CCG CCA | Proline |
| CUC CUA CUG CUU | Leucine |
| GAA GAG | Glutamic acid |
| UCU UCC UCA UCG | Serine |
| UGU UGC | Cysteine |
| UUA UUG | Leucine |
| UUU UUC | Phenylalanine |

tRNA, that is, complementary to a codon of mRNA, is known as an **anticodon.**

Although it was not yet feasible to chemically synthesize long stretches of RNA, specific triplet sequences could be synthesized in the laboratory to serve as templates. All that was needed was a method to determine which tRNA–amino acid was bound to the triplet RNA-ribosome complex. The test system Nirenberg and Leder devised was quite simple. The amino acid to be tested was made radioactive and combined with its cognate tRNA, creating a "charged" tRNA. Because codon compositions (though not exact sequences) were known, it was possible to narrow the decision as to which amino acids should be tested for each specific triplet.

The radioactively charged tRNA, the RNA triplet, and ribosomes were incubated together on a nitrocellulose filter, which retains ribosomes (because of their larger size) but not the other, smaller components, such as charged tRNA. If radioactivity is not retained on the filter, an incorrect amino acid has been tested. If radioactivity remains on the filter, it does so because the charged tRNA has bound to the RNA triplet associated with the ribosome, which itself remains on the filter. In such a case, a specific codon assignment can be made.

Work proceeded in several laboratories, and in many cases clear-cut, unambiguous results were obtained. For example, Table 14.2 shows 26 triplet codons assigned to nine different amino acids. However, in some cases the triplet binding was inefficient, and assignments were not possible. Eventually, about 50 of the 64 triplets were assigned. These specific assignments led to two major conclusions. First, the genetic code is degenerate—that is, one amino acid may be specified by more than one triplet. Second, the code is also *unambiguous*—that is, a single codon specifies only one amino acid. As we shall see later in this chapter, these conclusions have been upheld with only minor exceptions. The triplet binding technique was a major innovation in the effort to decipher the genetic code.

## Repeating Copolymers

Yet another innovative technique for deciphering the genetic code was developed in the early 1960s by Gobind Khorana, who was able to chemically synthesize long RNA molecules consisting of short sequences repeated many times. First, he created the individual short sequences (e.g., di-, tri-, and tetranucleotides); then he replicated them many times and finally joined them enzymatically to form the long polynucleotides. In Figure 14–6, a dinucleotide made in this way is converted to an mRNA with two repeating triplet codons. A trinucleotide is converted into an mRNA that can be read as containing one of three potential repeating triplets, depending on the point at which initiation occurs. Finally, a tetranucleotide creates a message with four repeating triplet sequences.

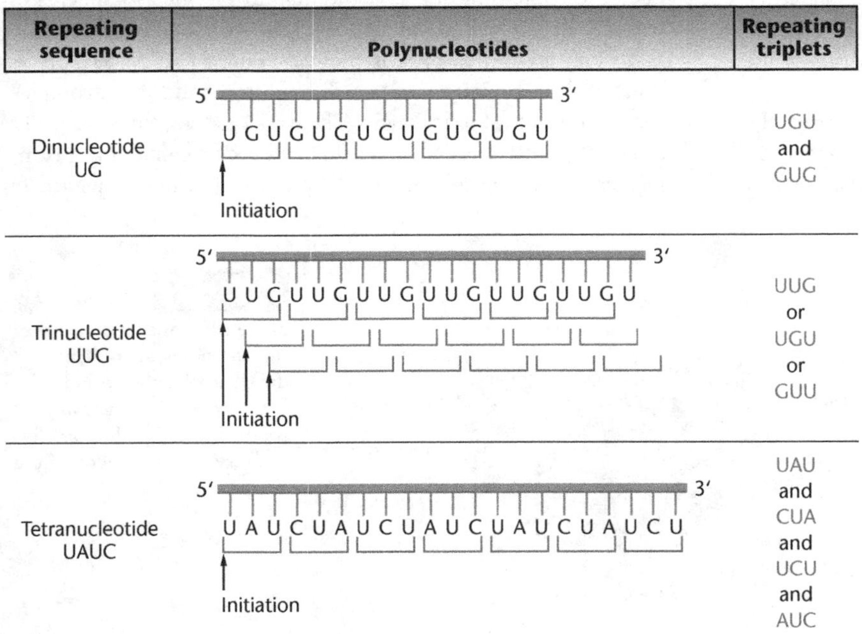


FIGURE 14–6 The conversion of di-, tri-, and tetranucleotides into repeating copolymers. The triplet codons produced in each case are shown.

When these synthetic mRNAs were added to a cell-free system, the predicted proportions of amino acids were found to be incorporated in the resulting polypeptides. When these data were combined with data drawn from mixed copolymer and triplet binding experiments, specific assignments were possible.

One example of specific assignments made in this way will illustrate the value of Khorana's approach. Consider the following three experiments in concert with one another: (1) The repeating *trinucleotide sequence* UUCUUCUUC... can be read as three possible repeating triplets—UUC, UCU, and CUU—depending on the initiation point. When placed in a cell-free translation system, three different polypeptide homopolymers—containing either phenylalanine (phe), serine (ser), or leucine (leu)—are produced. Thus, we know that each of the three triplets encodes one of the three amino acids, but we do not know which codes which. (2) On the other hand, the *repeating dinucleotide sequence* UCUCUCUC... produces the triplets UCU and CUC and, when used in an experiment, leads to the incorporation of leucine and serine into a polypeptide. Thus, the triplets UCU and CUC specify leucine and serine, but we still do not know which triplet specifies which amino acid. However, when considering both sets of results in concert, we can conclude that UCU, which is common to both experiments, must encode either leucine or serine but not phenylalanine. Thus, either CUU *or* UUC encodes leucine *or* serine, while the other encodes phenylalanine. (3) To derive more specific information, we can examine the results of using the repeating tetranucleotide sequence UUAC, which produces the triplets UUA, UAC, ACU, and CUU. The CUU triplet is one of the two in which we are interested. Three amino acids are incorporated by this experiment: leucine, threonine, and tyrosine. Because CUU must specify only serine or leucine, and because, of these two, only leucine appears in the resulting polypeptide, we may conclude that CUU specifies leucine.

Once this assignment is established, we can logically determine all others. Of the two triplet pairs remaining (UUC and UCU from the first experiment *and* UCU and CUC from the second experiment), whichever triplet is common to both must encode serine. This is UCU. By elimination, UUC is determined to encode phenylalanine and CUC is determined to encode leucine. Thus, through painstaking logical analysis, four specific triplets encoding three different amino acids have been assigned from these experiments.

From these interpretations, Khorana reaffirmed the identity of triplets that had already been deciphered and filled in gaps left from other approaches. A number of examples are shown in Table 14.3. For example, the use of two tetranucleotide sequences, GAUA and GUAA, suggested that at least two triplets were *termination codons*. Khorana reached this conclusion because neither of these repeating sequences directed the incorporation of more than a few amino acids into a polypeptide, too few to detect. There are no triplets common to both messages, and both seemed to contain at least one triplet that terminates protein synthesis. Of the possible triplets in the poly-(GAUA) sequence, shown in Table 14.3, UAG was later shown to be a termination codon.

TABLE 14.3

**Amino Acids Incorporated Using Repeated Synthetic Copolymers of RNA**

| Repeating Copolymer | Codons Produced | Amino Acids in Resulting Polypeptides |
|---|---|---|
| UG | UGU | Cysteine |
| | GUG | Valine |
| AC | ACA | Threonine |
| | CAC | Histidine |
| UUC | UUC | Phenylalanine |
| | UCU | Serine |
| | CUU | Leucine |
| AUC | AUC | Isoleucine |
| | UCA | Serine |
| | CAU | Histidine |
| UAUC | UAU | Tyrosine |
| | CUA | Leucine |
| | UCU | Serine |
| | AUC | Isoleucine |
| GAUA | GAU | None |
| | AGA | None |
| | UAG | None |
| | AUA | None |

NOW SOLVE THIS

Problem 4 on page 377 asks you to consider different outcomes of repeating copolymer experiments.

■ HINT: *On a repeating copolymer of RNA, translation can be initiated at different ribonucleotides. Determining the number of triplet codons produced by initiation at each of the different ribonucleotides is the key to solving this problem.*

## 14.4 The Coding Dictionary Reveals Several Interesting Patterns among the 64 Codons

The various techniques applied to decipher the genetic code have yielded a dictionary of 61 triplet codons assigned to amino acids. The remaining three codons are termination signals, not specifying any amino acid.

### Degeneracy and the Wobble Hypothesis

A general pattern of triplet codon assignments becomes apparent when we look at the genetic coding dictionary. Figure 14–7 displays the assignments in a particularly revealing form first suggested by Francis Crick.

Most evident is that the code is degenerate, as the early researchers predicted. That is, almost all amino acids are specified by

| First position (5′-end) | Second position: U | C | A | G | Third position (3′-end) |
|---|---|---|---|---|---|
| U | UUU *phe* | UCU *ser* | UAU *tyr* | UGU *cys* | U |
| | UUC *phe* | UCC *ser* | UAC *tyr* | UGC *cys* | C |
| | UUA *leu* | UCA *ser* | UAA *Stop* | UGA *Stop* | A |
| | UUG *leu* | UCG *ser* | UAG *Stop* | UGG *trp* | G |
| C | CUU *leu* | CCU *pro* | CAU *his* | CGU *arg* | U |
| | CUC *leu* | CCC *pro* | CAC *his* | CGC *arg* | C |
| | CUA *leu* | CCA *pro* | CAA *gln* | CGA *arg* | A |
| | CUG *leu* | CCG *pro* | CAG *gln* | CGG *arg* | G |
| A | AUU *ile* | ACU *thr* | AAU *asn* | AGU *ser* | U |
| | AUC *ile* | ACC *thr* | AAC *asn* | AGC *ser* | C |
| | AUA *ile* | ACA *thr* | AAA *lys* | AGA *arg* | A |
| | AUG *met* | ACG *thr* | AAG *lys* | AGG *arg* | G |
| G | GUU *val* | GCU *ala* | GAU *asp* | GGU *gly* | U |
| | GUC *val* | GCC *ala* | GAC *asp* | GGC *gly* | C |
| | GUA *val* | GCA *ala* | GAA *glu* | GGA *gly* | A |
| | GUG *val* | GCG *ala* | GAG *glu* | GGG *gly* | G |

Initiation Termination

FIGURE 14-7 The coding dictionary. AUG encodes methionine, which initiates most polypeptide chains. All other amino acids except tryptophan, which is encoded only by UGG, are represented by two to six triplets. The triplets UAA, UAG, and UGA are termination signals and do not encode any amino acids.

two, three, or four different codons. Three amino acids (serine, arginine, and leucine) are each encoded by six different codons. Only tryptophan and methionine are encoded by single codons.

Also evident is the *pattern* of degeneracy. Most often in a set of codons specifying the same amino acid, the first two letters are the same, with only the third differing. For example, as you can see in Figure 14–7, the codons for phenylalanine (UUU and UUC in the top left corner of the coding table) differ only by their third letter. Either U or C in the third position specifies phenylalanine. Four codons specify valine (GUU, GUC, GUA, and GUG, in the bottom left corner), and they differ only by their third letter. In this case, all four letters in the third position specify valine.

Crick observed this pattern in the degeneracy throughout the code, and in 1966, he postulated the **wobble hypothesis.** Crick's hypothesis predicted that the initial two ribonucleotides of triplet codes are often more critical than the third member in attracting the correct tRNA. He postulated that hydrogen bonding at the third position of the codon–anticodon interaction would be *less* constrained and need not adhere as strictly to the established base-pairing rules. The wobble hypothesis proposes a more flexible set of base-pairing rules at the third position of the codon (Table 14.4).

This relaxed base-pairing requirement, or "wobble," allows the anticodon of a single form of tRNA to pair with more than one triplet in mRNA. Consistent with the wobble hypothesis and the degeneracy of the code, U at the first position (the 5′ end) of the tRNA anticodon may pair with A or G at the third position (the 3′ end) of the mRNA codon, and G may likewise pair with U or C. Inosine (I), one of the modified bases found in tRNA (described in Chapter 15), may pair with C, U, or A. As a result of these wobble rules, only about 30 different tRNA species are necessary to accommodate the 61 codons specifying an amino acid. If nothing else, wobble can be considered an economy measure, assuming that the fidelity of translation is not compromised. Current estimates are that 30 to 40 tRNA species are present in bacteria and up to 50 tRNA species in animal and plant cells.

TABLE 14.4

**Anticodon–Codon Base-Pairing Rules**

| *Base at first position (5′ end) of tRNA* | *Base at third position (3′ end) of mRNA* |
|---|---|
| A | U |
| C | G |
| G | C or U |
| U | A or G |
| I | A, U, or C |

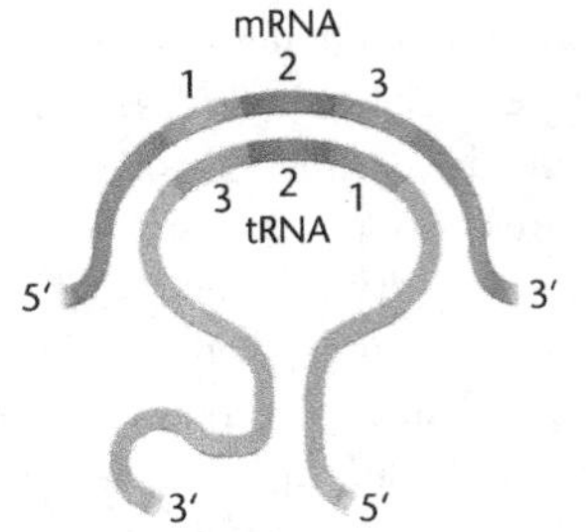


## The Ordered Nature of the Code

Still another observation has been made concerning the pattern of codon sequences and their corresponding amino acids, leading to the description of the genetic code as **ordered.** Chemically similar amino acids often share one or two "middle" bases in the different triplets encoding them. For example, either U or C is often present in the second position of triplets that specify hydrophobic amino acids, including valine and alanine, among others. Two codons (AAA and AAG) specify the positively charged amino acid lysine. If only the middle letter of these codons is changed from A to G (AGA and AGG), the positively charged amino acid arginine is specified. Hydrophilic amino acids, such as serine or threonine are specified by triplet codons with G or C in the second position.

The chemical properties of amino acids will be discussed in more detail in Chapter 15. The end result of an "ordered" code is that it buffers the potential effect of mutation on protein function. While many mutations of the second base of triplet codons result in a change of one amino acid to another, the change is often to an amino acid with similar chemical properties. In such cases, protein function may not be noticeably altered.

### Initiation, Termination, and Suppression

In contrast to the *in vitro* experiments discussed earlier, initiation of protein synthesis *in vivo* is a highly specific process. In bacteria, the initial amino acid inserted into all polypeptide chains is a modified form of methionine—***N*-formylmethionine (fmet)**. Only one codon, AUG, codes for methionine, and it is sometimes called the **initiator codon**. However, when AUG appears internally in mRNA, rather than at an initiating position, unformylated methionine is inserted into the polypeptide chain. Rarely, another codon, GUG, specifies methionine during initiation, though it is not clear why this happens, since GUG normally encodes valine.

In bacteria, either the formyl group is removed from the initial methionine upon the completion of synthesis of a protein, or the entire formylmethionine residue is removed. In eukaryotes, unformylated methionine is the initial amino acid of polypeptide synthesis. As in bacteria, this initial methionine residue may be cleared from the polypeptide.

As mentioned in the preceding section, three other codons (UAG, UAA, and UGA) serve as **termination codons**, punctuation signals that do not code for any amino acid.* They are not recognized by a tRNA molecule, and translation terminates when they are encountered. Mutations that produce any of the three codons internally in a gene also result in termination. In that case, only a partial polypeptide is synthesized, since it is prematurely released from the ribosome. When such a change occurs in the DNA, it is called a **nonsense mutation.**

Interestingly, a distinct mutation in a second gene may trigger suppression of premature termination resulting from a nonsense mutation; the second mutation is referred to as a **suppressor mutation**. These mutations cause the chain-termination signal to be read as a "sense" codon. The "correction" usually inserts an amino acid other than that found in the wild-type protein. But if the protein's structure is not altered drastically, it may function almost normally. Therefore, this second mutation has "suppressed" the mutant character resulting from the initial change to the termination codon.

Nonsense-suppressor mutations occur in genes specifying tRNAs. If the mutation results in a change in the anticodon such that it becomes complementary to a termination codon, there is the potential for insertion of an amino acid and suppression.

## 14.5 The Genetic Code Has Been Confirmed in Studies of Phage MS2

The various aspects of the genetic code discussed so far yield a fairly complete picture. The code is triplet in nature, degenerate, unambiguous, and commaless, although it contains punctuation in the form of start and stop signals. These individual principles have been confirmed by the detailed analysis of the RNA-containing bacteriophage MS2 by Walter Fiers and his coworkers.

**MS2** is a bacteriophage that infects *E. coli*. Its nucleic acid (RNA) contains only about 3500 ribonucleotides, making up only three genes. These genes specify a coat protein, an RNA-directed replicase, and a maturation protein (the A protein). This simple system of a small genome and few gene products allowed Fiers and his colleagues to sequence the genes and their products. The amino acid sequence of the coat protein was completed in 1970, and the nucleotide sequence of the gene and a number of nucleotides on each end of it were reported in 1972.

When the chemical constitutions of this gene and its encoded protein are compared, they are found to exhibit colinearity. That is, the linear sequence of triplet codons formed by the nucleotides corresponds precisely with the linear sequence of amino acids in the protein. Furthermore, the codon for the first amino acid is AUG, the common initiator codon; the codon for the last amino acid is followed by two consecutive termination codons, UAA and UAG.

By 1976, the other two genes of MS2 and their protein products were sequenced. The analysis clearly showed that the genetic code in this virus was identical to that established in bacterial systems. Other evidence suggests that the code is also identical in eukaryotes, thus providing confirmation of what seemed to be a universal code.

## 14.6 The Genetic Code Is Nearly Universal

Between 1960 and 1978, it was generally assumed that the genetic code would be found to be universal, applying equally to viruses, bacteria, archaea, and eukaryotes. Certainly, the nature of mRNA and the translation machinery seemed to be very similar in these organisms. For example, cell-free systems derived from bacteria could translate eukaryotic mRNAs. Poly U stimulates synthesis of polyphenylalanine in cell-free systems when the components are derived from eukaryotes. Many recent studies involving recombinant DNA technology (Chapter 13) reveal that eukaryotic genes can be inserted into bacterial cells, which are then transcribed and translated. Within eukaryotes, mRNAs from mice and rabbits have been injected into amphibian eggs and efficiently translated. For the many eukaryotic genes that have been sequenced, notably those for hemoglobin molecules, the amino acid sequence of the encoded proteins adheres to the coding dictionary established from bacterial studies.

However, several 1979 reports on the coding properties of DNA derived from mitochondria (mtDNA) of yeast and humans undermined the hypothesis of the universality of the genetic language. Since then, mtDNA has been examined in many other organisms.

Cloned mtDNA fragments were sequenced and compared with the amino acid sequences of various mitochondrial proteins,

* Historically, the terms *amber* (UAG), *ochre* (UAA), and *opal* (UGA) were used to distinguish mutations producing any of the three termination codons.

TABLE 14.5

**Exceptions to the Universal Code**

| Codon | Normal Code Word | Altered Code Word | Source |
|---|---|---|---|
| UGA | Termination | trp | Human and yeast mitochondria; *Mycoplasma* |
| CUA | leu | thr | Yeast mitochondria |
| AUA | ile | met | Human mitochondria |
| AGA | arg | Termination | Human mitochondria |
| AGG | arg | Termination | Human mitochondria |
| UAA | Termination | gln | *Paramecium, Tetrahymena,* and *Stylonychia* |
| UAG | Termination | gln | *Paramecium* |

revealing several exceptions to the coding dictionary (Table 14.5). Most surprising was that the codon UGA, normally specifying termination, specifies the insertion of tryptophan during translation in yeast and human mitochondria. In human mitochondria, AUA, which normally specifies isoleucine, directs the internal insertion of methionine. In yeast mitochondria, threonine is inserted instead of leucine when CUA is encountered in mRNA.

In 1985, several other exceptions to the standard coding dictionary were discovered in the bacterium *Mycoplasma capricolum*, and in the nuclear genes of the protozoan ciliates *Paramecium, Tetrahymena*, and *Stylonychia*. For example, as shown in Table 14.5, one alteration converts the termination codon (UGA) to tryptophan. Several other code alterations convert a termination codon to glutamine (gln). These changes are significant because they are seen in both a prokaryote and several eukaryotes, that is, in distinct species that have evolved separately over a long period of time.

Note the pattern apparent in several of the altered codon assignments: the change in coding capacity involves only a shift in recognition of the third, or wobble, position. For example, AUA specifies isoleucine in the cytoplasm and methionine in the mitochondrion, but in the cytoplasm, methionine is specified by AUG. In a similar example, UGA calls for termination in the cytoplasm, but calls for tryptophan in the mitochondrion; in the cytoplasm, tryptophan is specified by UGG. It has been suggested that such changes in codon recognition may represent an evolutionary trend toward reducing the number of tRNAs needed in mitochondria; only 22 tRNA species are encoded in human mitochondria, for example. However, until more examples are discovered, the differences must be considered to be exceptions to the previously established general coding rules.

## 14.7 Different Initiation Points Create Overlapping Genes

Earlier we stated that the genetic code is nonoverlapping, meaning that each ribonucleotide in the code for a given polypeptide is part of only one codon. However, this characteristic of the code does not rule out the possibility that a single mRNA may have multiple initiation points for translation. If so, these points could theoretically create several different reading frames within the same mRNA, thus specifying more than one polypeptide. This concept of **overlapping genes** is illustrated in Figure 14–8(a).

That this might actually occur in some viruses was suspected when phage $\phi$X174 was carefully investigated. The circular DNA chromosome consists of 5386 nucleotides, which should encode a maximum of 1795 amino acids, sufficient for five or six proteins. However, this small virus in fact synthesizes 11 proteins consisting of more than 2300 amino acids. A comparison of the nucleotide sequence of the DNA and the amino acid sequences of the polypeptides synthesized has clarified the apparent paradox. At least four instances of multiple initiation have been discovered, creating overlapping genes [Figure 14–8(b)].

The sequences specifying the K and B polypeptides are initiated in separate reading frames within the sequence specifying the A polypeptide. The *K* gene sequence overlaps into the adjacent sequence specifying the C polypeptide. The E sequence is out of frame with, but initiated within, that of the D polypeptide. Finally, the A′ sequence, while in frame with the A sequence, is initiated in the middle of the A sequence. They both terminate at the identical point. In all, seven different polypeptides are created from a DNA sequence that might otherwise have specified only three (A, C, and D).

A similar situation has been observed in other viruses, including phage G4 and the animal virus SV40. Like $\phi$X174, phage G4 contains a circular single-stranded DNA molecule. The employment of overlapping reading frames optimizes the limited amount of DNA present in these small viruses. However, such an approach

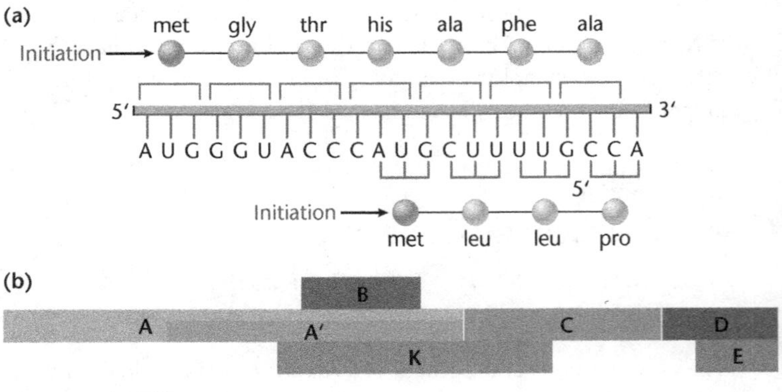


FIGURE 14–8 Illustration of the concept of overlapping genes. (a) Translation initiated at two different AUG positions out of frame with one another will give rise to two distinct amino acid sequences. (b) The relative positions of the sequences encoding seven polypeptides of the phage $\phi$X174.

to storing information has a distinct disadvantage in that a single mutation may affect more than one protein and thus increase the chances that the change will be deleterious or lethal. In the case we just discussed [Figure 14–8(b)], a single mutation in the middle of the *B* gene could potentially affect three other proteins (the A, A′, and K proteins). It may be for this reason that overlapping genes are not common in other organisms.

## 14.8 Transcription Synthesizes RNA on a DNA Template

Even while the genetic code was being studied, it was quite clear that proteins were the end products of many genes. Hence, while some geneticists were attempting to elucidate the code, other research efforts were directed toward the nature of genetic expression. The central question was how DNA, a nucleic acid, is able to specify a protein composed of amino acids.

The complex, multistep process begins with the transfer of genetic information stored in DNA to RNA. The process by which RNA molecules are synthesized on a DNA template is called **transcription.** It results in an mRNA molecule complementary to the gene sequence of one of the two strands of the double helix. Each triplet codon in the mRNA is, in turn, complementary to the anticodon region of its corresponding tRNA, which inserts the correct amino acid into the polypeptide chain during translation. The significance of transcription is enormous, for it is the initial step in the process of *information flow* within the cell. The idea that RNA is involved as an intermediate molecule in the process of information flow between DNA and protein is suggested by the following observations:

1. DNA is, for the most part, associated with chromosomes in the nucleus of the eukaryotic cell. However, protein synthesis occurs in association with ribosomes located outside the nucleus, in the cytoplasm. Therefore, DNA does not appear to participate directly in protein synthesis.
2. RNA is synthesized in the nucleus of eukaryotic cells, in which DNA is found, and is chemically similar to DNA.
3. Following its synthesis, most RNA migrates to the cytoplasm, in which protein synthesis (translation) occurs.
4. The amount of RNA is generally proportional to the amount of protein in a cell.

Collectively, these observations suggested that genetic information, stored in DNA, is transferred to an RNA intermediate, which directs the synthesis of the proteins. As with most new ideas in molecular genetics, the initial supporting experimental evidence for an RNA intermediate was based on studies of bacteria and their phages.

## 14.9 Studies with Bacteria and Phages Provided Evidence for the Existence of mRNA

In two papers published in 1956 and 1958, Elliot Volkin and his colleagues reported their analysis of RNA produced immediately after bacteriophage infection of *E. coli*. Using the isotope $^{32}P$ to follow newly synthesized RNA, they found that its base composition closely resembled that of the phage DNA, but was different from that of bacterial RNA (Table 14.6). This newly synthesized RNA was unstable (short lived); however, its production was shown to precede the synthesis of new phage proteins. Thus, Volkin and his coworkers considered the possibility that synthesis of RNA is a preliminary step in the process of protein synthesis.

Although ribosomes were known to participate in protein synthesis, their role in this process was not clear. As we noted earlier, one possibility was that each ribosome is specific for the protein synthesized in association with it. That is, perhaps genetic information in DNA is transferred to the RNA of a ribosome (rRNA) during the latter's synthesis so that each ribosome is restricted to the production of a particular protein. The alternative hypothesis was that ribosomes are nonspecific "workbenches" for protein synthesis and that specific genetic information rests with a messenger RNA.

In an elegant experiment using the *E. coli*–phage system, the results of which were reported in 1961, Sidney Brenner, François

TABLE 14.6

**Base Compositions (in mole percents) of RNA Produced Immediately Following Infection of *E. coli* by the Bacteriophages T2 and T7 in Contrast to the Composition of RNA of Uninfected *E. coli***

| Source | Adenine | Thymine | Uracil | Cytosine | Guanine |
|---|---|---|---|---|---|
| Post-infection RNA in T2-infected cells | 33 | — | 32 | 18 | 18 |
| T2 DNA | 32 | 32 | — | 17* | 18 |
| Post-infection RNA in T7-infected cells | 27 | — | 28 | 24 | 22 |
| T7 DNA | 26 | 26 | — | 24 | 22 |
| *E. coli* RNA | 23 | — | 22 | 18 | 17 |

*5-hydroxymethyl cytosine.

SOURCE: From Volkin and Astrachan (1956); and Volkin, Astrachan, and Countryman (1958).

Jacob, and Matthew Meselson clarified this question. They labeled uninfected *E. coli* ribosomes with heavy isotopes and then allowed phage infection to occur in the presence of radioactive RNA precursors. By following these components during translation, the researchers demonstrated that the synthesis of phage proteins (under the direction of newly synthesized RNA) occurred on bacterial ribosomes that were present prior to infection. The ribosomes appeared to be nonspecific, strengthening the case that another type of RNA serves as an intermediary in the process of protein synthesis.

That same year, Sol Spiegelman and his colleagues reached the same conclusion when they isolated $^{32}$P-labeled RNA following the infection of bacteria and used it in molecular hybridization studies. They tried hybridizing this RNA to the DNA of both phages and bacteria in separate experiments. The RNA hybridized only with the phage DNA, showing that it was complementary in base sequence to the viral genetic information.

The results of these experiments agree with the concept of a messenger RNA (mRNA) being made on a DNA template and then directing the synthesis of specific proteins in association with ribosomes. This concept was formally proposed by François Jacob and Jacques Monod in 1961 as part of a model for gene regulation in bacteria. Since then, mRNA has been isolated and thoroughly studied. There is no longer any question about its role in genetic processes.

## 14.10 RNA Polymerase Directs RNA Synthesis

To prove that RNA can be synthesized on a DNA template, it was necessary to demonstrate that there is an enzyme capable of directing this synthesis. By 1959, several investigators, including Samuel Weiss, had independently discovered such a molecule in rat liver. Called **RNA polymerase**, it has the same general substrate requirements as does DNA polymerase, the major exception being that the substrate nucleotides contain the ribose rather than the deoxyribose form of the sugar. Unlike DNA polymerase, no primer is required to initiate synthesis. The overall reaction summarizing the synthesis of RNA on a DNA template can be expressed as

$$n(\text{NTP}) \xrightarrow[\text{enzyme}]{\text{DNA}} (\text{NMP})_n + n(\text{PP}_i)$$

As the equation reveals, nucleoside triphosphates (NTPs) serve as substrates for the enzyme, which catalyzes the polymerization of nucleoside monophosphates (NMPs), or nucleotides, into a polynucleotide chain $(\text{NMP})_n$. Nucleotides are linked during synthesis by 5′ to 3′ phosphodiester bonds (see Figure 10–12). The energy created by cleaving the triphosphate precursor into the monophosphate form drives the reaction, and inorganic phosphates ($\text{PP}_i$) are produced.

A second equation summarizes the sequential addition of each ribonucleotide as the process of transcription progresses:

$$(\text{NMP})_n + \text{NTP} \xrightarrow[\text{enzyme}]{\text{DNA}} (\text{NMP})_{n+1} + \text{PP}_i$$

As this equation shows, each step of transcription involves the addition of one ribonucleotide (NMP) to the growing polyribonucleotide chain $(\text{NMP})_{n+1}$, using a nucleoside triphosphate (NTP) as the precursor.

RNA polymerase from *E. coli* has been extensively characterized and shown to consist of subunits designated $\alpha$, $\beta$, $\beta'$, $\omega$, and $\sigma$. The complex, active form of the enzyme, the **holoenzyme**, contains the subunits $\alpha_2\beta\beta'\sigma$ and has a molecular weight of almost 500,000 Da. Of these subunits, it is the $\beta$ and $\beta'$ polypeptides that provide the catalytic mechanism and active site for transcription. As we will see, the **$\sigma$ (sigma) factor** [Figure 14–9(a)] plays a regulatory function in the initiation of RNA transcription.

While there is but a single form of the enzyme in *E. coli*, there are several different $\sigma$ factors, creating variations of the polymerase holoenzyme. On the other hand, eukaryotes display three distinct forms of RNA polymerase, each consisting of a greater number of polypeptide subunits than in bacteria. We shall return to a discussion of the eukaryotic form of the enzyme later in this chapter.

**NOW SOLVE THIS**

Problem 23 on page 378 asks you to consider the outcome of the transfer of complementary information from DNA to RNA and the amino acids encoded by this information.

■ HINT: *In RNA, uracil is complementary to adenine, and while DNA stores genetic information in the cell, the code that is translated is contained in the RNA complementary to the template strand of DNA making up a gene.*

### Promoters, Template Binding, and the $\sigma$ Subunit

Transcription results in the synthesis of a single-stranded RNA molecule complementary to a region along only one of the two strands of the DNA double helix. When discussing transcription, scientists call the DNA strand that is transcribed the *template strand* and its complement the *partner strand*.

The initial step in transcription is referred to as **template binding** [Figure 14–9(b)]. In bacteria, the site of this initial binding is established when the RNA polymerase $\sigma$ subunit recognizes specific DNA sequences called **promoters**. These sequences are located in the 5′ region, upstream from the point of initial transcription of a gene. It is believed that the enzyme "explores" a length of DNA until it encounters a promoter region and binds there to about 60 nucleotide pairs along the helix, 40 of which are upstream from the point of initial transcription. Once this occurs, the helix is denatured, or unwound, locally, making the template strand of the DNA accessible to the action of the enzyme. The point at which transcription actually begins is called the **transcription start site.**

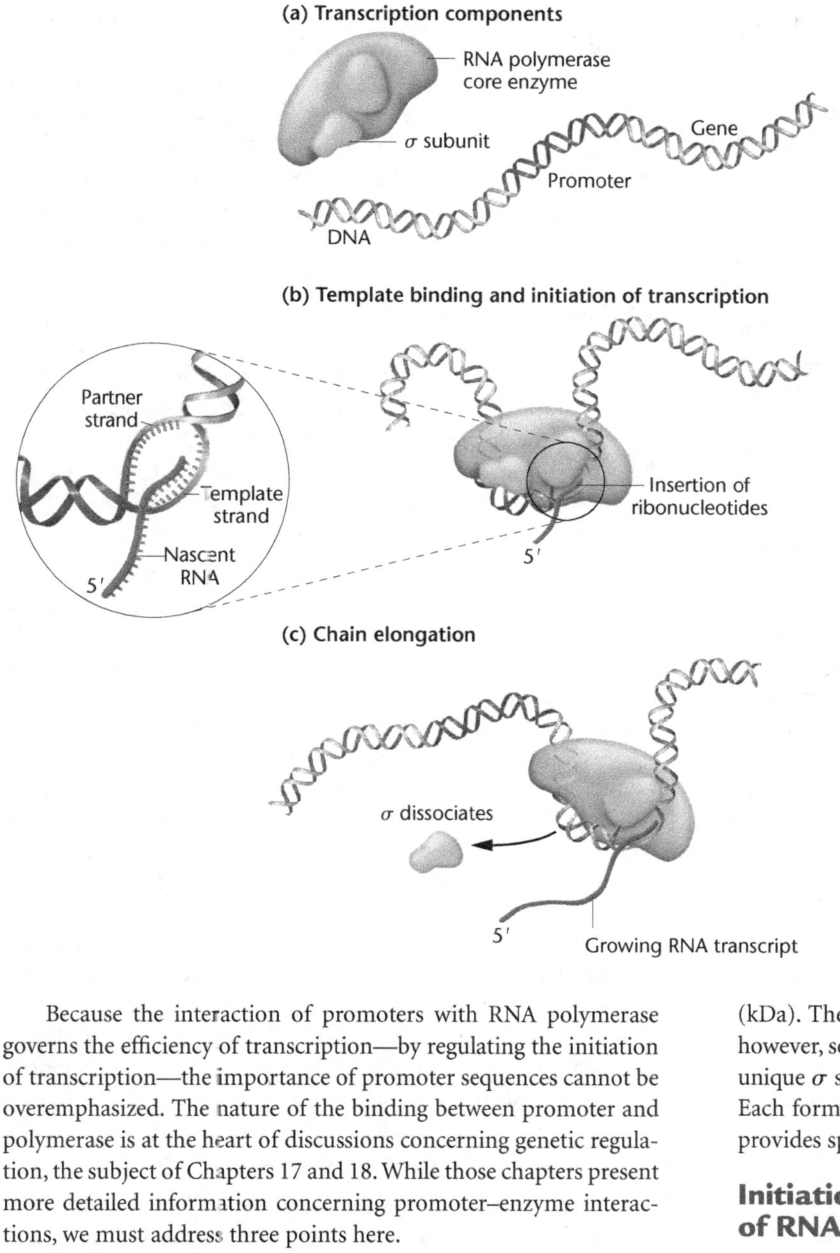


**FIGURE 14–9** The early stages of transcription in prokaryotes, showing (a) the components of the process; (b) template binding at the −10 site involving the sigma subunit of RNA polymerase and subsequent initiation of RNA synthesis; and (c) chain elongation, after the $\sigma$ subunit has dissociated from the transcription complex and the enzyme moves along the DNA template.

Because the interaction of promoters with RNA polymerase governs the efficiency of transcription—by regulating the initiation of transcription—the importance of promoter sequences cannot be overemphasized. The nature of the binding between promoter and polymerase is at the heart of discussions concerning genetic regulation, the subject of Chapters 17 and 18. While those chapters present more detailed information concerning promoter–enzyme interactions, we must address three points here.

The first point is the concept of **consensus sequences** of DNA, sequences that are similar (homologous) in different genes of the same organism or in one or more genes of related organisms. Their conservation during evolution attests to the critical nature of their role in biological processes. Two consensus sequences have been found in bacterial promoters. One, TATAAT, is located 10 nucleotides upstream from the site of initial transcription (the −10 region, or **Pribnow box**). The other, TTGACA, is located 35 nucleotides upstream (the −35 region). Mutations in either region diminish transcription, often severely.

Sequences such as these, in regions adjacent to the gene itself, are said to be ***cis*-acting elements**. The term *cis*, drawn from organic chemistry nomenclature, means "next to" or on the same side as other functional groups, in contrast to being *trans* to or "across from," them. In molecular genetics, then, *cis*-elements are adjacent parts of the same DNA molecule. In contrast, ***trans*-acting factors** are molecules that bind to these DNA elements. As we will soon see, in most eukaryotic genes studied, a consensus sequence comparable to that in the −10 region has been recognized. Because it is rich in adenine and thymine residues, it is called the *TATA box*.

The second point is that the degree of RNA polymerase binding to different promoters varies greatly, causing variable gene expression. Currently, this is attributed to sequence variation in the promoters. In bacteria, both strong promoters and weak promoters have been discovered, causing a variation in time of initiation from once every 1 to 2 seconds to as little as once every 10 to 20 minutes. Mutations in promoter sequences may severely reduce the initiation of gene expression.

A final general point to be made involves the $\sigma$ subunit in bacteria. The major form is designated $\sigma^{70}$ based on its molecular weight of 70 kilodaltons (kDa). The promoters of most bacterial genes recognize this form; however, several alternative forms of RNA polymerase in *E. coli* have unique $\sigma$ subunits associated with them (e.g., $\sigma^{32}$, $\sigma^{54}$, $\sigma^{S}$, and $\sigma^{E}$). Each form recognizes different promoter sequences, which in turn provides specificity to the initiation of transcription.

## Initiation, Elongation, and Termination of RNA Synthesis

Once it has recognized and bound to the promoter, RNA polymerase catalyzes **initiation**, the insertion of the first 5′-ribonucleoside triphosphate, which is complementary to the first nucleotide at the start site of the DNA template strand. As we noted, no primer is required. Subsequent ribonucleotide complements are inserted and linked together by phosphodiester bonds as RNA polymerization proceeds. This process continues in a 5′ to 3′ direction (in terms of the nascent RNA), creating a temporary 8-bp DNA/RNA duplex whose chains run antiparallel to one another [Figure 14–9(b)].

After these ribonucleotides have been added to the growing RNA chain, the $\sigma$ subunit dissociates from the holoenzyme, and **chain elongation** proceeds under the direction of the core enzyme

[Figure 14–9(c)]. In *E. coli*, this process proceeds at the rate of about 50 nucleotides/second at 37°C.

The enzyme traverses the entire gene until eventually it encounters a specific nucleotide sequence that acts as a termination signal. Such termination sequences, about 40 base pairs in length, are extremely important in prokaryotes because of the close proximity of the end of one gene to the upstream sequences of the adjacent gene. An interesting aspect of termination in bacteria is that the termination sequence alluded to above is actually transcribed into RNA. The unique sequence of nucleotides in this termination region causes the newly formed transcript to fold back on itself, forming what is called a **hairpin secondary structure**, held together by hydrogen bonds. The hairpin is important to termination. In some cases, the termination of synthesis is also dependent upon the **termination factor, rho** ($\rho$). Rho is a large hexameric protein that physically interacts with the growing RNA transcript, facilitating termination of transcription.

When termination is achieved, the transcribed RNA molecule is released from the DNA template, and the core polymerase enzyme dissociates. The synthesized RNA molecule is precisely complementary to the DNA sequence of the template strand of the gene. Wherever an A, T, C, or G residue was encountered there, a corresponding U, A, G, or C residue, respectively, was incorporated into the RNA molecule. These RNA molecules ultimately provide the information leading to the synthesis of all proteins in the cell.

In bacteria, groups of genes whose protein products are involved in the same metabolic pathway are often clustered together along the chromosome. In many such cases, the genes are contiguous and all but the last gene lack the encoded signals for termination of transcription. The result is that during transcription, a large mRNA is produced, encoding more than one protein. Since genes in bacteria are sometimes called *cistrons*, the RNA is called a **polycistronic mRNA**. The products of genes transcribed in this fashion are usually all needed by the cell at the same time, so this is an efficient way to transcribe and subsequently translate the needed genetic information. In eukaryotes, **monocistronic** mRNAs are the rule, although an increasing number of exceptions are being reported.

## 14.11 Transcription in Eukaryotes Differs from Prokaryotic Transcription in Several Ways

Much of our knowledge of transcription has been derived from studies of prokaryotes. Most of the general aspects of the mechanics of these processes are similar in eukaryotes, but there are several notable differences:

1. Transcription in eukaryotes occurs within the nucleus under the direction of three separate forms of RNA polymerase. Unlike the prokaryotic process, in eukaryotes the RNA transcript is not free to associate with ribosomes prior to the completion of transcription. For the mRNA to be translated, it must move out of the nucleus into the cytoplasm.
2. Initiation of transcription of eukaryotic genes requires the compact chromatin fiber, characterized by nucleosome coiling (Chapter 12), to be uncoiled and the DNA to be made accessible to RNA polymerase and other regulatory proteins. This transition, referred to as **chromatin remodeling**, reflects the dynamics involved in the conformational change that occurs as the DNA helix is opened.
3. Initiation and regulation of transcription entail a more extensive interaction between *cis*-acting upstream DNA sequences and *trans*-acting protein factors involved in stimulating and initiating transcription. In addition to promoters, other control units, called **enhancers**, may be located in the 5′ regulatory region upstream from the initiation point, but they have also been found within the gene or even in the 3′ downstream region, beyond the coding sequence.
4. Alteration of the primary RNA transcript to produce mature eukaryotic mRNA involves many complex stages referred to generally as "processing." An initial processing step involves the addition of a 5′ cap and a 3′ tail to most transcripts destined to become mRNAs. Other extensive modifications occur to the internal nucleotide sequence of eukaryotic RNA transcripts that eventually serve as mRNAs. The initial (or primary) transcripts are most often much larger than those that are eventually translated into protein. Sometimes called **pre-mRNAs**, these primary transcripts belong to a group of molecules found only in the nucleus and referred to collectively as **heterogeneous nuclear RNA (hnRNA)**. Such RNA molecules are of variable but large size and are complexed with proteins, forming **heterogeneous nuclear ribonucleoprotein particles (hnRNPs)**. Only about 25 percent of hnRNA molecules are converted to mRNA. From those that are converted, substantial amounts of the ribonucleotide sequence are excised, and the remaining segments are spliced back together prior to nuclear export and translation. This phenomenon has given rise to the concepts of *split genes* and *splicing* in eukaryotes (discussed in Section 14.12).

In the remainder of this chapter we will look at the basic details of transcription in eukaryotic cells. The process of transcription is highly regulated, determining which DNA sequences are copied into RNA and when and how frequently they are transcribed. We will return to topics directly related to regulation of eukaryotic transcription in Chapter 18.

### Initiation of Transcription in Eukaryotes

The recognition of certain highly specific DNA regions by RNA polymerase is the basis of orderly genetic function in all cells. Eukaryotic RNA polymerase exists in three distinct forms. While the three forms of the enzyme share certain polypeptide subunits, each nevertheless transcribes different types of genes, as indicated in Table 14.7. Each enzyme is larger and more complex than the single

TABLE 14.7

**RNA Polymerases in Eukaryotes**

| Form | Product | Location |
|---|---|---|
| I | rRNA | Nucleolus |
| II | mRNA, snRNA | Nucleoplasm |
| III | 5S rRNA, tRNA | Nucleoplasm |

prokaryotic polymerase. For example, in yeast, the holoenzyme consists of two large subunits and 10 smaller subunits.

In regard to the initial template-binding step and promoter regions, most is known about **RNA polymerase II (RNP II)**, which is responsible for the production of all mRNAs in eukaryotes. The activity of RNP II is dependent upon both *cis*-acting elements in the gene itself and a number of *trans*-acting transciption factors that bind to these DNA elements (we will return to the topic of transcription factors below). At least three *cis*-acting DNA elements regulate the initiation of transcription by RNP II. The first of these sequences, called a **core-promoter element**, determines where RNP II binds to the DNA and where it begins copying the DNA into RNA. The other two types of regulatory DNA sequences, called **promoter** and **enhancer elements**, influence the efficiency or the rate of transcription by RNP II as the process proceeds from the core promoter element. Recall that in prokaryotes, the DNA sequence recognized by RNA polymerase is also called the promoter. In eukaryotes, however, transcriptional initiation is controlled by this group of *cis*-acting DNA elements, and they are *collectively* referred to as the promoter. Thus, in eukaryotes the term *promoter* refers to both the core-promoter sequence where RNP II binds and to the promoter and enhancer elements in the DNA that influence RNP II activity.

In most, if not all, genes studied, the *cis*-acting core-promoter element is the **Goldberg–Hogness**, or **TATA, box** present in almost all eukaryotic genes. Located about 35 nucleotide pairs upstream (−35) from the start point of transcription, the TATA boxes share a consensus sequence that is a heptanucleotide consisting solely of A and T residues (TATAAAA). The sequence and function are analogous to that found in the −10 promoter region of prokaryotic genes. Because this region is common to most eukaryotic genes, the TATA box is thought to be nonspecific and to be responsible only for fixing the site of transcription initiation by facilitating denaturation of the helix. Such a conclusion is supported by the fact that A = T base pairs are less stable than G ≡ C pairs.

Another *cis*-acting DNA sequence that is part of the eukaryotic promoter is the **CAAT box.** This specific DNA sequence, which contains the consensus sequence GGCCAATCT, is typically located upstream (in the 5′ region) of the gene at about 80 nucleotides from the start of transcription (−80). Still other upstream regulatory regions have been found, and most genes contain one or more of them. They influence the efficiency of the promoter, along with the TATA box and CAAT box. The locations of these elements have been identified through studies of deletions of particular regions of promoters that reduce the efficiency of transcription.

DNA regions called enhancers represent another *cis*-acting element. Although their locations can vary, enhancers are often found farther upstream than the regions already mentioned, or even downstream or within the gene. Thus they can modulate transcription from a distance. Although they may not participate directly in RNP II binding to the core promoter, they are essential to highly efficient initiation of transcription.

Complementing the *cis*-acting regulatory sequences are various *trans*-acting factors that facilitate RNP II binding and, therefore, the initiation of transcription. These are proteins referred to as **transcription factors.** There are two broad categories of transcription factors: the **general transcription factors** that are absolutely required for all RNP II–mediated transcription, and the **specific transcription factors** that influence the efficiency or the rate of RNP II transcription. We will consider the second category of transcription factors, the specific transcription factors, in more detail in Chapter 18. The general transcription factors, however, are essential because RNA polymerase II cannot bind directly to eukaryotic promoter sites and initiate transcription without their presence. The general transcription factors involved with human RNP II binding are well characterized and designated **TFIIA, TFIIB,** and so on. One of these, **TFIID,** binds directly to the TATA-box sequence. TFIID consists of about 10 polypeptide subunits, one of which is sometimes called the **TATA-binding protein (TBP).** Once initial binding to DNA occurs, at least seven other general transcription factors bind sequentially to TFIID, forming an extensive pre-initiation complex, which is then bound by RNA polymerase II.

Transcription factors with similar activity have been discovered in a variety of eukaryotes, including *Drosophila* and yeast. These factors appear to supplant the role of the $\sigma$ factor seen in the prokaryotic enzyme and play an important part in eukaryotic gene regulation. In Chapter 18, we will consider the role of transcription factors in eukaryotic gene regulation, as well as the various DNA-binding domains that characterize them.

## Recent Discoveries Concerning RNA Polymerase Function

One approach to extending our understanding of the process of transcription has been to study the details of the structure and function of RNA. The ability to crystallize large nucleic acid–protein structures and perform X-ray diffraction analysis at a resolution below 5 Å has led to some remarkable observations. The work of Roger Kornberg and colleagues, using the enzyme isolated from yeast, has been particularly informative. It is useful to note here that achieving a resolution below 2.8 Å allows the visualization of each amino acid of every protein in the complex!

What Kornberg has discovered provides a highly detailed account of the most critical processes of transcription. RNA polymerase II in yeast contains two large subunits and ten smaller ones, forming a huge three-dimensional complex with a molecular weight of about 500 kDa. The promoter region of the DNA duplex that is to be transcribed enters a positively charged cleft between the two large subunits of the enzyme. The subunit assemblage in fact resembles a pair of jaws that can open and partially close. Prior to association with DNA, the cleft is open; once associated with DNA, the cleft partially closes, securing the duplex during the initiation of transcription. The part of the enzyme that is critical for this transition is about 50 kDa in size and is called the *clamp*.

Once secured by the clamp, the strands of a small duplex region of DNA separate at a position within the enzyme referred to as the *active center*, and complementary RNA synthesis is initiated on the DNA template strand. However, the entire complex remains unstable, and transcription usually terminates following the incorporation of only a few ribonucleotides. It is not clear why, but this so-called *abortive transcription* is repeated a number of times before a stable DNA:RNA hybrid containing a transcript of 11 ribonucleotides is formed. Once this occurs, abortive transcription is overcome, a stable complex is achieved, and elongation of the RNA transcript proceeds in earnest. Transcription at this point is said to have achieved a level of highly processive RNA polymerization.

As transcription proceeds, the enzyme moves along the DNA, and at any given time, about 40 base pairs of DNA and 18 residues of the growing RNA chain are part of the enzyme complex. The RNA synthesized earliest runs through a groove in the enzyme and exits under a structure at the top and back designated as the *lid*. Another area, called the *pore*, has been identified at the bottom of the enzyme. It serves as the point through which RNA precursors gain entry into the complex.

Eventually, as transcription proceeds, the portion of the DNA that signals termination is encountered, and the complex once again becomes unstable, much as it was during the earlier state of abortive transcription. The clamp opens, and both DNA and RNA are released from the enzyme as transcription is terminated. This completes the cycle that constitutes transcription: an unstable complex is formed during the initiation of transcription, stability is established once elongation manages to create a duplex of sufficient size, elongation proceeds, and then instability again characterizes termination of transcription.

Clearly, Kornberg's findings extended our knowledge of transcription considerably. For this work, he was awarded the Nobel Prize in Chemistry in 2006. As you think back on this cycle, try to visualize the process mentally, from the time the DNA first associates with the enzyme until the transcript is released from the large molecular complex. If these images are clear to you, you no doubt have acquired a firm understanding of transcription in eukaryotes, which is more complex than in prokaryotes.

## Heterogeneous Nuclear RNA and Its Processing: Caps and Tails

Our discussion continues from the point at which an initial transcript has been produced. The genetic code that is now contained in the ribonucleotide sequence of this newly synthesized chain of RNA originated in the template strand of a DNA molecule, where complementary sequences of deoxyribonucleotides are stored. In bacteria, the relationship between DNA and mRNA appears to be quite direct. The DNA base sequence is transcribed into an mRNA sequence, which is then immediately translated into an amino acid sequence according to the genetic code. In eukaryotes, by contrast, the RNA must undergo significant processing before being transported to the cytoplasm as mRNA to participate in translation.

By 1970, accumulating evidence showed that eukaryotic mRNA is transcribed initially as a precursor molecule much larger than that which is translated into protein. This notion was based on the observation by James Darnell and his coworkers of the large heterogeneous nuclear RNA (hnRNA) in mammalian nuclei that contained nucleotide sequences common to the smaller mRNA molecules present in the cytoplasm. They proposed that the initial transcript of a gene results in a large RNA molecule that must first be processed in the nucleus before it appears in the cytoplasm as a mature mRNA molecule. The various processing steps, discussed in the sections that follow, are summarized in Figure 14–10.

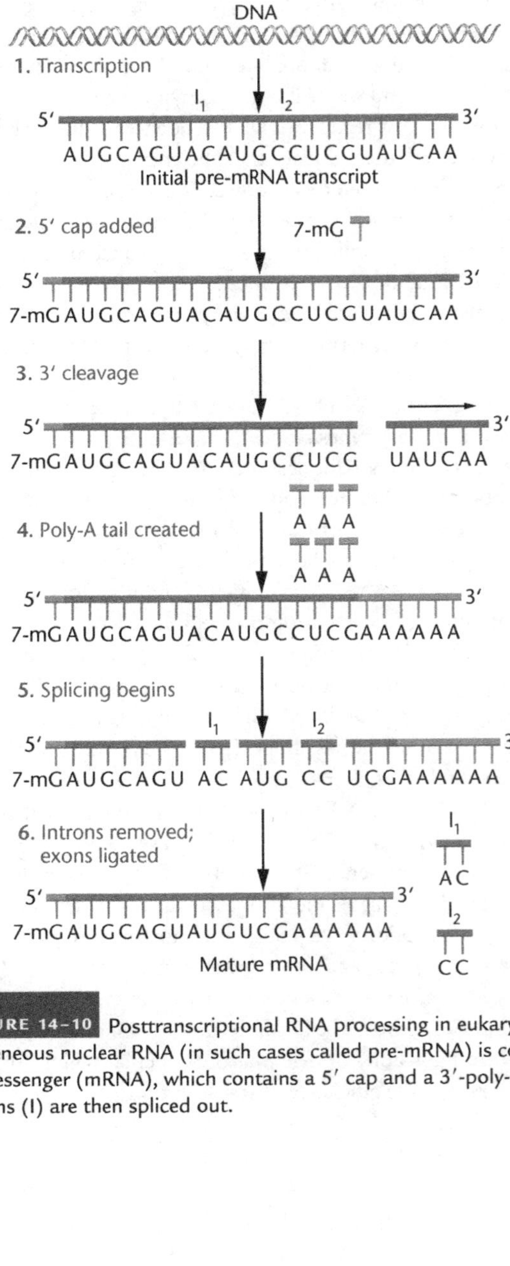


FIGURE 14–10 Posttranscriptional RNA processing in eukaryotes. Heterogeneous nuclear RNA (in such cases called pre-mRNA) is converted to messenger (mRNA), which contains a 5′ cap and a 3′-poly-A tail. The introns (I) are then spliced out.

The initial **posttranscriptional modification** of eukaryotic RNA transcripts destined to become mRNAs occurs at the 5′-end of these molecules, where a **7-methylguanosine (7-mG) cap** is added (Figure 14–10, Step 2). The cap is added even before the initial transcript is complete and appears to be important to the subsequent processing within the nucleus. Perhaps it protects the 5′-end of the molecule from nuclease attack. Subsequently, it may be involved in the transport of mature mRNAs across the nuclear membrane into the cytoplasm and in the initiation of translation of the mRNA into protein. The cap is fairly complex and is distinguished by the unique 5′-5′ bonding that connects it to the initial ribonucleotide of the RNA. Some eukaryotes also acquire a methyl group ($CH_3$) at the 2′-carbon of the ribose sugars of the first two ribonucleotides of the RNA.

Further insights into the processing of RNA transcripts during the maturation of mRNA came from the discovery that both pre-RNAs and mRNAs contain at their 3′ end a stretch of as many as 250 adenylic acid residues. This **poly-A sequence** is added after the 7-mG cap has been attached. First, the 3′ end of the initial transcript is cleaved enzymatically at a point some 10 to 35 ribonucleotides from a highly conserved AAUAAA sequence (Step 3). Then, polyadenylation occurs by the sequential addition of adenylic acid residues (Step 4). Poly A has now been found at the 3′ end of almost all mRNAs studied in a variety of eukaryotic organisms. In fact, poly-A tails have also been detected in some prokaryotic mRNAs. The exceptions in eukaryotes seem to be the RNAs that encode the histone proteins.

While the AAUAAA sequence is not found on all eukaryotic transcripts, it appears to be essential to those that have it. If the sequence is changed as a result of a mutation, those transcripts that would normally have it cannot add the poly-A tail. In the absence of this tail, these RNA transcripts are rapidly degraded. Therefore, both the 5′ cap and the 3′ poly-A tail are critical if an mRNA transcript is to be further processed and transported to the cytoplasm.

14.12

## The Coding Regions of Eukaryotic Genes Are Interrupted by Intervening Sequences

One of the most exciting breakthroughs in the history of molecular genetics occurred in 1977, when Susan Berget, Philip Sharp, and Richard Roberts presented direct evidence that the genes of animal viruses contain *internal* nucleotide sequences that are not expressed in the amino acid sequence of the proteins they encode. These internal DNA sequences are represented in initial RNA transcripts, but they are removed before the mature mRNA is translated (Figure 14–10, Steps 5 and 6). Such nucleotide segments are called **intervening sequences,** and the genes that contain them are **split genes.** DNA sequences that are not represented in the final mRNA product are also called **introns** ("int" for intervening), and those retained and expressed are called **exons** ("ex" for expressed). Splicing involves the removal of the corresponding ribonucleotide sequences representing introns as a result of an excision process and the rejoining of the regions representing exons.

Similar discoveries were soon made in many other eukaryotic genes. Two approaches have been most fruitful for this purpose. The first involves the molecular hybridization of purified, functionally mature mRNAs with DNA containing the genes from which the RNA was originally transcribed. Hybridization between nucleic acids that are not perfectly complementary results in **heteroduplexes,** in which introns present in the DNA but absent in the mRNA loop out and remain unpaired. Such structures can be visualized with the electron microscope, as shown in Figure 14–11. The chicken ovalbumin shown in the figure is a heteroduplex with seven loops (A through G), representing seven introns whose sequences are present in the DNA but not in the final mRNA.

The second approach provides more specific information. It involves a direct comparison of nucleotide sequences of DNA with those of mRNA and their correlation with amino acid sequences. Such an approach allows the precise identification of all intervening sequences.

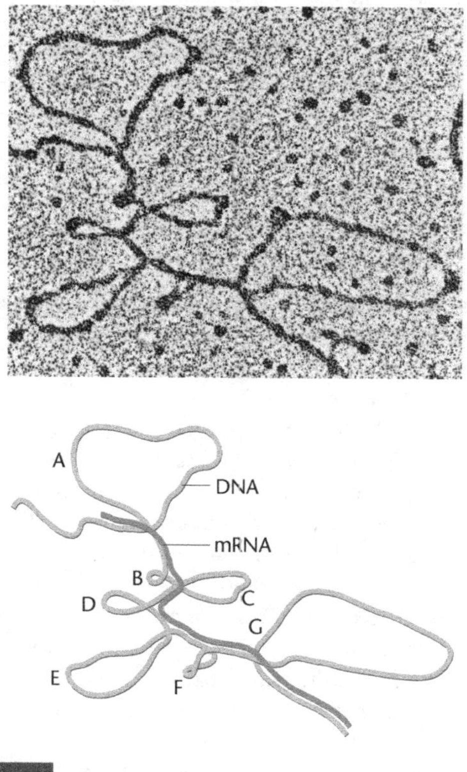


FIGURE 14–11 An electron micrograph and an interpretive drawing of the hybrid molecule (heteroduplex) formed between the template DNA strand of the chicken ovalbumin gene and the mature ovalbumin mRNA. Seven DNA introns, A–G, produce unpaired loops.

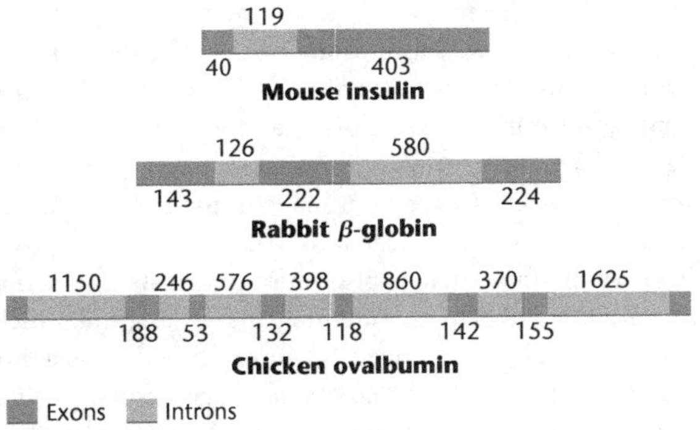


FIGURE 14–12 Intervening sequences in various eukaryotic genes. The numbers indicate the number of nucleotides present in various intron and exon regions.

Thus far, most eukaryotic genes have been shown to contain introns (Figure 14–12). One of the first so identified was the **$\beta$-globin gene** in mice and rabbits, studied independently by Philip Leder and Richard Flavell. The mouse gene contains an intron 550 nucleotides long, beginning immediately after the codon specifying the 104th amino acid. In the rabbit, there is an intron of 580 base pairs near the codon for the 110th amino acid. In addition, another intron of about 120 nucleotides exists earlier in both genes. Similar introns have been found in the $\beta$-globin gene in all mammals examined.

The **ovalbumin gene** of chickens has been extensively characterized by Bert O'Malley in the United States and Pierre Chambon in France. As shown in Figure 14–12, the gene contains seven introns. In fact, the majority of the gene's DNA sequence is composed of introns and is thus "silent." The initial RNA transcript is nearly three times the length of the mature mRNA. Compare the ovalbumin gene in Figures 14–11 and 14–12. Can you match the unpaired loops in Figure 14–11 with the order of introns specified in Figure 14–12?

The list of genes containing intervening sequences is long. In fact, few eukaryotic genes seem to lack introns. An extreme example of the number of introns in a single gene is provided by the gene coding for one of the subunits of collagen, the major connective tissue protein in vertebrates. The *pro-$\alpha$-2(1) collagen* gene contains 50 introns. The precision of cutting and splicing that occurs must be extraordinary if errors are not to be introduced into the mature mRNA. Equally noteworthy is the difference between the size of a typical gene with the size of the final mRNA transcribed from it once introns are removed. As shown in Table 14.8, only about 15 percent of the collagen gene consists of exons that finally appear in mRNA. For other proteins, an even more extreme picture emerges. Only about 8 percent of the albumin gene remains to be translated, and in the largest human gene known, dystrophin (which is the protein product absent in Duchenne muscular dystrophy), less than 1 percent of the gene sequence is retained in the mRNA. Two other human genes are also contrasted in Table 14.8.

Although the vast majority of eukaryotic genes examined thus far contain introns, there are several exceptions. Notably, the genes coding for histones and for interferon appear to contain no introns. It is not clear why or how the genes encoding these molecules have been maintained throughout evolution without acquiring the extraneous information characteristic of almost all other genes.

TABLE 14.8

**Contrasting Human Gene Size, mRNA Size, and Number of Introns**

| Gene | Gene Size (kb) | mRNA Size (kb) | Number of Introns |
|---|---|---|---|
| Insulin | 1.7 | 0.4 | 2 |
| Collagen [*pro-$\alpha$-2(1)*] | 38.0 | 5.0 | 50 |
| Albumin | 25.0 | 2.1 | 14 |
| Phenylalanine hydroxylase | 90.0 | 2.4 | 12 |
| Dystrophin | 2000.0 | 17.0 | 50 |

## Splicing Mechanisms: Autocatalytic RNAs

The discovery of split genes led to intensive attempts to elucidate the mechanism by which introns of RNA are excised and exons are spliced back together. A great deal of progress has already been made, relying heavily on *in vitro* studies. Interestingly, it appears that somewhat different mechanisms exist for different types of RNA, as well as for RNAs produced in mitochondria and chloroplasts. Note that splicing has also been found during the maturation of tRNAs in prokaryotes. However, we will focus here on eukaryotic splicing.

We might envision the simplest possible mechanism for removing an intron to be as illustrated in Steps 5 and 6 of Figure 14–10: after an endonucleolytic "cut" is made at each end of an intron, the intron is removed, and the terminal ends of the adjacent exons are ligated by an enzyme (in short, the intron is snipped out and the exon ends are rejoined). This is apparently what happens to the introns present in transfer RNAs (tRNAs) in yeast. A specific endonuclease recognizes the intron termini and excises the intervening sequences. Then DNA ligase seals the exon ends to complete each splicing event. However, in the studies of all other RNAs—tRNA in higher eukaryotes and rRNAs and pre-mRNAs in all eukaryotes—precise excision of introns is much more complex and a much more interesting story.

Introns can be categorized into several groups based on their splicing mechanisms. Group I, represented by introns that are part of the primary transcript of rRNAs, require no additional components for intron excision; the intron itself is the source of the enzymatic activity necessary for removal. This amazing discovery was made in 1982 by Thomas Cech and his colleagues during a study of the ciliate protozoan *Tetrahymena*. RNAs that are capable of catalytic activity are referred to as **ribozymes**. The autocatalytic tRNA capable of self-splicing and thus functioning like an enzyme is an excellent example. As we will see in Chapter 15, the RNAs present in ribosomes (rRNA) are another.

The **self-excision process** is shown in Figure 14–13. Chemically, two nucleophilic reactions take place—that is, reactions caused by the presence of electron-rich chemical species (in this

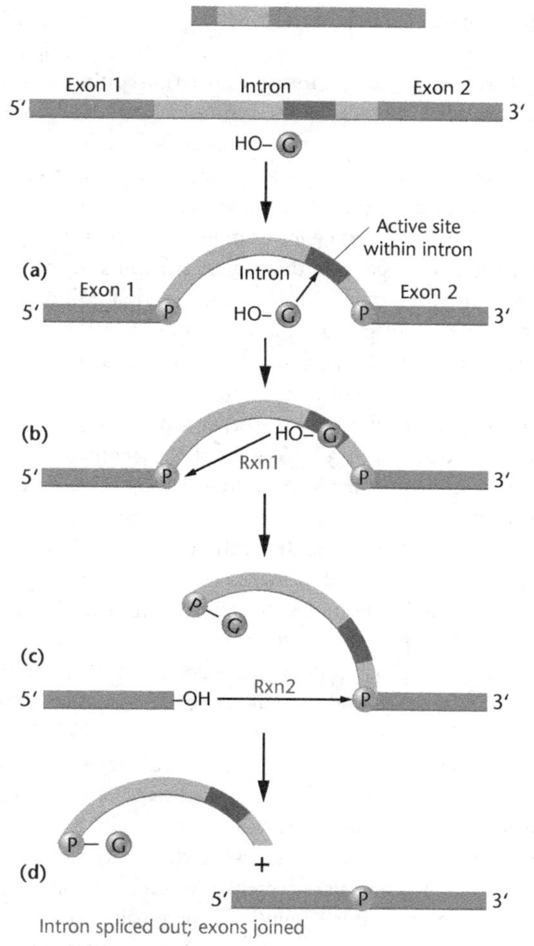


**FIGURE 14–13** Splicing mechanism for removal of group I introns from the initial rRNA transcript (pre-rRNA). The process is one of self-excision involving two transesterification reactions.

case, they are transesterification reactions). The first is an interaction between guanosine, which acts as a cofactor in the reaction, and the primary transcript [Figure 14–13(a)]. The 3′-OH group of guanosine is transferred to the nucleotide adjacent to the 5′-end of the intron [Figures 14–13(b) and 14–13(c)]. The second reaction involves the interaction of the newly acquired 3′-OH group on the left-hand exon and the phosphate on the 3′-end of the intron [Figure 14–13(c)]. The intron is spliced out and the two exon regions are ligated, leading to the mature RNA [Figure 14–13(d)].

Self-excision of group I introns, as described above, is now known to apply to pre-rRNAs from other protozoans besides *Tetrahymena*. Self-excision also seems to govern the removal of introns from the primary mRNA and tRNA transcripts produced in mitochondria and chloroplasts. These are referred to as group II introns. As in group I molecules, splicing here involves two autocatalytic reactions leading to the excision of introns. However, guanosine is not involved as a cofactor with group II introns.

## Splicing Mechanisms: The Spliceosome

Introns are a major component of nuclear-derived pre-mRNA transcripts. Compared to the group I and group II introns discussed above, introns in nuclear-derived mRNA can be much larger—up to 20,000 nucleotides—and they are more plentiful. Their removal appears to require a much more complex mechanism that has been difficult to discover.

Nevertheless, many clues are emerging, as illustrated by the model diagrammed in Figure 14–14, showing the removal of one

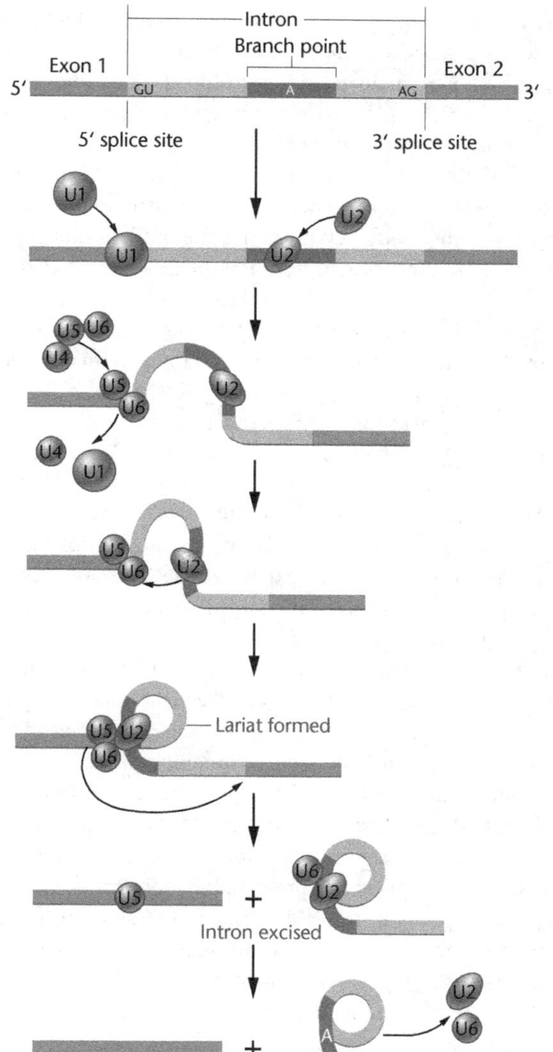


**FIGURE 14–14** A model of the splicing mechanism for removal of an intron from nuclear-derived pre-mRNA. Excision is dependent on various snRNAs (U1, U2, . . ., U6) that combine with proteins to form snRNPs (snurps), which function as part of a large structure referred to as the spliceosome. The lariat structure in the intermediate stage is characteristic of this mechanism.

intron. The nucleotide sequences near the ends of this type of intron are all identical. They begin at the 5′ end with a GU dinucleotide sequence, called the *donor sequence*, and terminate at the 3′ end with an AG dinucleotide, called the *acceptor sequence*. These, as well as other consensus sequences shared by introns, attract specific molecules that form a molecular complex, a **spliceosome**, essential to this form of splicing. Spliceosomes have been identified in extracts of yeast as well as in mammalian cells, and are very large, 40*S* in yeast and 60*S* in mammals. Perhaps the most essential component of spliceosomes is the unique set of **small nuclear RNAs (snRNAs)** they contain. These RNAs are usually 100 to 200 nucleotides long or less and are often complexed with proteins to form **small nuclear ribonucleoproteins (snRNPs)**. These are found only in the nucleus. Because they are rich in uridine residues, the snRNAs have been arbitrarily designated U1, U2, . . ., U6.

The snRNA U1 bears a nucleotide sequence that is homologous to the 5′-splice donor sequence end of the intron. Base pairing resulting from this homology promotes the binding that represents the initial step in the formation of the spliceosome. After the other snRNPs (U2, U4, U5, and U6) are added, splicing commences. As with group I splicing, two *transesterification reactions* occur. The first involves the interaction of the 3′-OH group from an adenine (A) residue present within the **branch point** region of the intron. The A residue attacks the 5′-splice site, cutting the RNA chain. In a subsequent step involving several other "snurps," an intermediate structure is formed and the second reaction ensues, linking the cut 5′ end of the intron to the A. This results in the formation of a characteristic loop structure called a lariat, which contains the excised intron. The exons are then ligated and the snRNPs are released.

The processing involved in splicing, which occurs within the nucleus, represents a potential regulatory step in gene expression in eukaryotes. For instance, several cases are known wherein introns present in pre-mRNAs *derived from the same gene* are spliced *in more than one way*, thereby yielding different collections of exons in the mature mRNA. Such **alternative splicing** yields a group of similar but nonidentical mRNAs that, upon translation, result in a series of related proteins called **isoforms**. Many examples have been encountered in organisms ranging from viruses to *Drosophila* to humans. Alternative splicing of pre-mRNAs represents a way of producing related proteins from a single gene, increasing the number of gene products that can be derived from an organism's genome. We shall return to this topic in our Chapter 18 discussion of the regulation of gene expression in eukaryotes.

## RNA Editing Modifies the Final Transcript

In the late 1980s, still another unexpected form of posttranscriptional RNA processing was discovered in several organisms. In this form, referred to as **RNA editing**, the nucleotide sequence of a pre-mRNA is actually changed prior to translation. As a result, the ribonucleotide sequence of the mature RNA differs from the sequence encoded in the exons of the DNA from which the RNA was transcribed.

Although other variations exist, there are two main types of RNA editing: **substitution editing**, in which the identities of individual nucleotide bases are altered; and **insertion/deletion editing**, in which nucleotides are added to or subtracted from the total number of bases. Substitution editing is used in some nuclear-derived eukaryotic RNAs and is prevalent in mitochondrial and chloroplast RNAs transcribed in plants. *Physarum polycephalum*, a slime mold, uses both substitution and insertion/deletion editing for its mitochondrial mRNAs.

*Trypanosoma*, a parasite that causes African sleeping sickness, and its relatives use extensive insertion/deletion editing in mitochondrial RNAs. The uridines added to an individual transcript can make up more than 60 percent of the coding sequence, usually forming the initiation codon and bringing the rest of the sequence into the proper reading frame. Insertion/deletion editing in trypanosomes is directed by **gRNA (guide RNA)** templates, which are also transcribed from the mitochondrial genome. These small RNAs are complementary to the edited region of the final, edited mRNAs. They base-pair with the preedited mRNAs to direct the editing machinery to make the correct changes.

The best-studied examples of substitutional editing occur in mammalian nuclear-encoded mRNA transcripts. One such example is the protein apolipoprotein B (apo B), which exists in both a long and a short form that are encoded by the same gene. In human intestinal cells, apo B mRNA is edited by a single C-to-U change, which converts a CAA glutamine codon into a UAA stop codon and terminates the polypeptide at approximately half its genomically encoded length. The editing is performed by a complex of proteins that bind to a "mooring sequence" on the mRNA transcript just downstream of the editing site. A second example occurs in the synthesis of subunits constituting the glutamate receptor channels (GluR) in mammalian brain tissue. In this case, adenosine (A) to inosine (I) editing occurs in pre-mRNAs prior to their translation, during which I is read as guanosine (G). A family of three ADAR (*a*denosine *d*eaminase *a*cting on *R*NA) enzymes is believed to be responsible for the editing of various sites within the glutamate channel subunits. The double-stranded RNAs required for editing by the ADAR enzymes are provided by intron/exon pairing of the GluR mRNA transcripts. The editing changes alter the physiological parameters (solute permeability and desensitization response time) of the receptors containing the subunits.

The importance of RNA editing resulting from the action of ADARs is most apparent in situations where these enzymes have lost their functional capacity as a result of mutation. In several investigations, the loss of function was shown to have a lethal impact in mice. In one study, embryos heterozygous for a defective *ADAR1* gene died during embryonic development as a result of a defective hematopoietic system. In another study, mice with two defective copies of *ADAR2* progressed through development normally but were prone to epileptic seizures and died while still in the weaning stage. Their tissues contained the unedited version of one of the GluR products. The defect leading to death was believed to be in the brain. Heterozygotes for the mutation were normal.

Findings such as these in mammals have established that RNA editing provides still another important mechanism of posttranscrip-

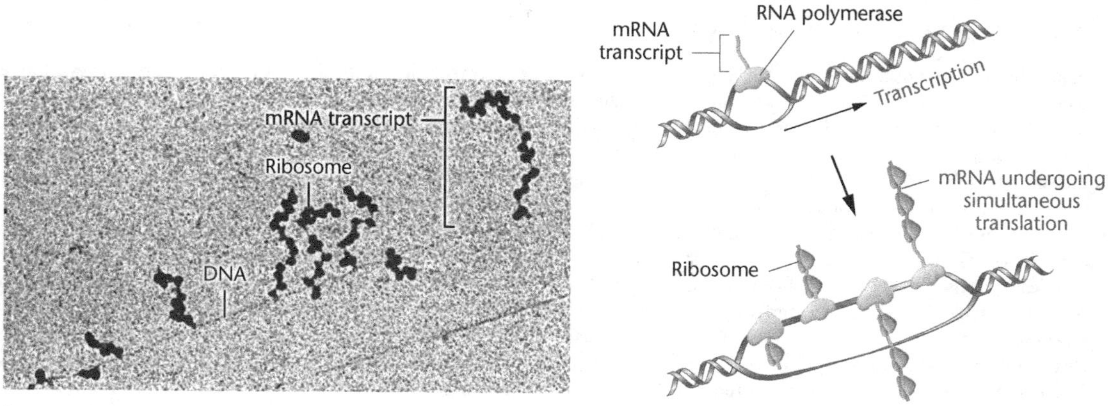


**FIGURE 14–15** Electron micrograph and interpretive drawing of simultaneous transcription of genes in *E. coli*. As each transcript is forming, ribosomes attach, initiating simultaneous translation along each strand. *O.L. Miller, Jr. Barbara A. Hamkalo, C.A. Thomas, Jr.* Science *169: 392–395, 1970 by the American Association for the Advancement of Science. F:2.*

tional modification, and that this process is not restricted to small or asexually reproducing genomes, such as those in mitochondria. Several new examples of RNA editing have been found each year since its discovery, and the trend is likely to continue. These discoveries, too, have important implications for the regulation of genetic expression.

## 14.13 Transcription Has Been Visualized by Electron Microscopy

We conclude this chapter by presenting a striking visual demonstration of the transcription process based on the electron microscope studies of Oscar Miller, Jr., Barbara Hamkalo, and Charles Thomas. Their combined work has captured the transcription process in both prokaryotes and eukaryotes. Figure 14–15 shows a micrograph and interpretive drawings of transcription in *E. coli*. In the micrograph, multiple strands of RNA are seen to emanate from different points along a central DNA template. Many RNA strands result because numerous transcription events are occurring simultaneously along each gene. Progressively longer RNA strands are found farther downstream from the point of initiation of transcription along a given gene, whereas the shortest strands are closest to the point of initiation.

An interesting picture emerges from the study of the *E. coli* micrograph. Because prokaryotes lack nuclei, cytoplasmic ribosomes are not separated physically from the chromosome. As a result, ribosomes are free to attach to *partially* transcribed mRNA molecules and initiate translation. The longer the RNA strand, the greater the number of ribosomes attached to it. As we will see in Chapter 15, these structures are called **polyribosomes**. Visualization of transcription confirms many of the predictions scientists had made from the biochemical analysis of this process.

## GENETICS, TECHNOLOGY, AND SOCIETY

### Nucleic Acid-Based Gene Silencing: Attacking the Messenger

Standard chemotherapies for diseases such as cancer and AIDS are often accompanied by toxic side effects. Conventional therapeutic drugs target both normal and diseased cells, with diseased or infected cells being only slightly more susceptible than the patient's normal cells. Scientists have long wished for a magic bullet that could seek out and destroy the virus or cancer cells, leaving normal cells alive and healthy. Over the last decade, a group of promising candidates for magic-bullet status has emerged, collectively described as *nucleic acid-based gene silencing*.

The two chief nucleic acid-based approaches currently being investigated as potential therapies are the use of antisense oligonucleotides and the use of RNAi. Both have arisen through an understanding of the molecular biology of gene expression. Gene expression, we have seen, is a two-step process. First, a single-stranded messenger RNA (mRNA) is copied from the template strand of the duplex DNA molecule. Second, the mRNA is complexed with ribosomes, and its coded information is translated into the amino acid sequence of a polypeptide.

Normally, a gene is transcribed into RNA from only one strand of the DNA duplex. The resulting RNA is known as *sense RNA*. However, it is possible for the other DNA strand to be copied into RNA, and this RNA, produced by the transcription of the "wrong" strand of DNA, is called *antisense RNA*. Complementary strands of RNA can form double-stranded molecules, and the formation of such duplex structures between a sense and an antisense RNA may affect the sense RNA by physically blocking ribosome binding or elongation,

*Continued on next page*

*Genetics, Technology, and Society, continued*

hence inhibiting translation. Also, binding antisense RNA to sense RNA may trigger the degradation of the sense (and antisense) RNA, because double-stranded RNA molecules are attacked by intracellular ribonucleases. By either method, gene expression is blocked.

The antisense approach excited researchers because of its potential specificity. Scientists can design antisense RNA (or DNA) molecules of known nucleotide sequence and can then synthesize large amounts of these antisense nucleic acids *in vitro*. Usually, oligonucleotides of up to 20 nucleotides are used. It is theoretically possible to treat cells with these synthetic antisense oligonucleotides, so that they enter the cell, bind to precise target mRNAs, and in this way turn off the synthesis of specific proteins. If the specific protein is necessary for virus reproduction or cancer cell growth (but is not necessary in normal cells), the antisense oligonucleotide should have only therapeutic effects.

A major challenge in the use of antisense oligonucleotides in the treatment of disease is their low activity *in vivo*. Initially, one reason for this weakness was that the antisense oligonucleotides administered were being destroyed along with the sense mRNAs by the very mechanism of degradation that was being exploited: The intracellular ribonucleases were attacking the double-stranded molecule, and thus, large amounts of antisense oligonucleotides were required for an effective response. Since then, a potential fix for this problem has emerged: modification of the phosphate backbone to give it a chemical linkage that resists the degradation. When these newer, "stabilized" oligonucleotides are used, the antisense strand silences the gene by physically blocking translation (although now the duplex structure is not subsequently broken down). These stabilized antisense oligonucleotides were shown to work effectively in more reasonable doses in model systems.

More recently, another gene-silencing technology, analogous to the antisense principle, has emerged. This technology, known as RNA interference (RNAi), uses one of two types of short double-stranded RNA molecules (~20-25 nucleotides long) to target specific mRNAs within cells. These are *short interfering RNAs* (siRNA) and *microRNAs* (miRNA).

siRNAs, first discovered as natural antiviral defenses in plants and worms, can be artificially introduced into mammalian cells. One strand of the siRNA binds to its complementary sequence within the target mRNA, aided by an *RNA-induced silencing complex* (RISC). RISC is found naturally in all mammalian cells. When RISC and the siRNA are bound to the target mRNA, the RISC degrades the mRNA. miRNAs, in contrast, were found to exist naturally in mammalian cells. They, too, join up with RISC and pair with complementary mRNA sequences. However, unlike siRNAs, miRNAs that are bound to their "targets" will often contain several mismatched bases. Furthermore, when miRNAs and RISC pair with the targets, the mRNAs are not destroyed but are merely prevented from being translated.

In some direct comparisons, it has been estimated that RNAi could be 10 or even 100 times more effective than antisense therapy at similar doses. This is because the siRNAs and miRNAs seem to be protected from degradation by association with RISC. Thus, these short double-stranded RNAs are "recycled" and have a longer lasting effect. In addition, RNAi seems to be universally effective with all genes, while antisense gene silencing has thus far been successful only with particular genes and triggers more nonspecific effects. RNAi is so promising that clinical trials are being conducted to study its use in combating the eye disease macular degeneration, and positive results are already being trumpeted. Other areas of high interest for RNAi-based treatments are cancers, diseases of the nervous system, and viral infections, such as hepatitis B and C and HIV-1. Most researchers mention a few challenges, such as developing better methods of delivery of the treatments, increasing stability of the siRNAs, and reducing side effects. However, there is much optimism that these obstacles will soon be overcome, and we may then indeed have acquired a molecular magic bullet to use in the battle against a variety of diseases.

## References

Dykxhoorn, D. M., et al. 2006. The silent treatment: siRNAs as small molecule drugs. *Gene Therapy*. 13:541–552.

Dykxhoorn, D. M., and Lieberman, J. 2006. Running interference: Prospects and obstacles to using small interfering RNAs as small molecule drugs. *Annu. Rev. Biomed. Eng.* 8:377–402.

Wilson, J. A., and Richardson, C. D. 2006. Future promise of siRNA and other nucleic acid based therapeutics for the treatment of chronic HCV. *Infectious Disorders—Drug Targets*. 6:34–56.

EXPLORING GENOMICS

# Transcriptome Databases and Noncoding RNA Databases

In this chapter, we discussed the process of transcription. One of the newly emerging areas of genomics is the study of *transcriptomes*, a genomics term for all of the genes expressed in a cell or tissue. Until relatively recently, geneticists studying transcription relied on recombinant DNA techniques that you learned about in Chapter 13, such as Northern blot analysis, to study the expression of one or a few genes at a time. Transcriptome analysis now provides geneticists with the ability to study expression *patterns* for thousands of genes simultaneously in order to provide a profile of gene-expression changes in a tissue in response to different conditions. For example, transcriptome analysis of normal and cancer cells is providing valuable information about gene-expression changes in cancer cells.

Transcriptome analysis often relies on cloning expressed genes from complementary DNA (cDNA) libraries and on gene-expression studies based on DNA microarray, or gene chip, data. You will learn more about DNA microarrays in the Exploring Genomics exercises in Chapter 17 and Chapter 21. Microarrays allow scientists to study the expression of thousands of genes simultaneously. In this exercise, we return to the **UniGene** site from the **National Center for Biotechnology Information (NCBI)** to view transcriptome databases, and then we will explore two databases of non-protein coding genes.

## ■ Exercise I – Transcriptome Databases

In Exploring Genomics for Chapter 12, we explored the **UniGene** database to view human chromosome maps of expressed mRNA sequences. In this exercise, we will use **Unigene: An Organized View of the Transcriptome**. This database provides collections of expressed genes from a number of different species. Most of the sequences in UniGene are full-length, complete gene sequences derived from cDNA libraries. These cDNAs represent all of the expressed mRNAs in a tissue, and the abundance of a particular cDNA reflects the amount of mRNA the tissue contains for that gene. Many other sequences in the database are *expressed sequence tags (ESTs)*—short pieces of cDNA that do not completely span the entire length of a gene. But overlapping EST fragments that were derived from the same mRNA can be pieced together and used to measure expression of a particular gene or different parts of a gene, such as particular exons that may be alternatively spliced to produce multiple mRNAs from the same gene.

1. Access UniGene at http://www.ncbi.nlm.nih.gov/entrez/query.fcgi?db=unigene. Notice that transcriptomes are listed according to different species in the database.
2. Click on the *Homo sapiens* link and explore the information presented on the human transcriptome.
3. Under "UniGene Links," click on "Library Browser" to see a list of cDNA libraries from different tissues and organs that were used to organize transcriptome data.
4. Find the set of libraries from spleen and click the link for the library SPLEN2 (Lib. 18474). Notice that EST data in these libraries are arranged from the most highly expressed transcripts to the least abundant transcripts expressed. Review the transcriptome data presented, then answer the following questions:
   a. What is the most abundant cDNA or EST expressed in this library?
   b. What are the least abundant cDNAs or ESTs expressed in this library?
   c. Click the link next to any of these genes to learn more about them. Information presented through these links includes the identification of the same gene in other species (based on protein-sequence similarity), expression of each gene in other tissues, and the chromosomal locus for the gene.
5. Return to the Library Browser page and pick another library of interest from the tissues used for the human transcriptome. Look at the top 10 most highly expressed genes in this library, and compare these to the top 10 genes expressed in the SPLEN2 library. Are any of the most highly expressed genes the same in these two libraries? Explain what you observed.
6. Return to the UniGene home page and explore other transcriptomes of interest.
7. Another trend in transcriptome research is the development of databases for genes expressed in different disease states of a tissue. For example, in the Exploring Genomics exercises for Chapters 15 and 20, you will learn about cancer gene transcriptome databases that have been established as resources for geneticists and physicians studying the gene-expression patterns in different cancers. Alzgene, http://www.alzforum.org/res/com/gen/alzgene/default.asp is a transcriptome database of candidate genes involved in Alzheimer disease. Alzgene connects to UniGene, GenBank, and other databases that we have used in Exploring Genomics exercises and that provide excellent resources for scientists, physicians, and the public. Explore the Alzgene databases to learn more about genes implicated in Alzheimer disease.

## ■ Exercise II – Noncoding RNA Databases

As we discussed in Chapter 14, transcription also produces non-protein coding RNAs such as tRNA, rRNA, and snRNA. Non-protein coding RNAs are gaining increased attention, in part due to the recent discovery of *short interfering RNAs (siRNAs)* and *microRNAs (miRNAs)*. These RNA molecules are short sequences that can bind to mRNAs and target them for degradation or block translation of the bound mRNA. Both of these mechanisms interfere with or silence gene expression. The relatively recent discoveries of siRNAs and miRNAs have inspired a new area of research in gene expression regulation. You will learn more about siRNAs, miRNAs, and gene silencing in Chapter 17. In this exercise, we explore the **Noncoding RNA Database** and the **microRNA (miRNA) Database** to learn more about non-protein coding RNAs.

1. Access the **Noncoding RNA Database** at http://biobases.ibch.poznan.pl/ncRNA/. Use the "Browse Information pages" link to see a list of noncoding RNAs. Explore information provided on the genes below to learn more about each noncoding RNA.
   a. *Tsix*
   b. *hsrω*
   c. *SZ-1*
2. Access the **microRNA (miRNA) Database** at http://microrna.sanger.ac.uk/sequences/. Use the "Browse" tab at the top of the page to see a list of miRNAs identified in different species. Links to each miRNA provide sequence data and loci for each miRNA.
3. Two RNAs, *lin-4* and *let-7*, were among the first miRNAs discovered. Use the "Search by miRNA name or keyword" feature of the database to learn more about each of these miRNAs. What organism expresses these miRNAs? On what chromosome is each located? What are their function?

## Chapter Summary

1. The genetic code stored in DNA is transcribed into RNA, which then directs the synthesis of polypeptide chains. The code is degenerate, unambiguous, nonoverlapping, and commaless.
2. The complete coding dictionary, determined using various experimental approaches, reveals that of the 64 possible codons, 61 encode the 20 amino acids found in proteins, while three triplets terminate translation. One of these 61 is the initiation codon and specifies methionine.
3. The observed pattern of degeneracy often involves only the third letter of a triplet series. It led Francis Crick to propose the wobble hypothesis.
4. Confirmation for the coding dictionary, including codons for initiation and termination, was obtained by comparing the complete nucleotide sequences of phage MS2 with the amino acid sequence of the corresponding proteins. Other findings support the belief that, with only minor exceptions, the code is universal for all organisms.
5. Some bacteriophages possess multiple initiation points for the transcription of RNA, resulting in multiple reading frames and overlapping genes.
6. Transcription—the initial step in gene expression—is the synthesis, under the direction of RNA polymerase, of a strand of RNA complementary to a DNA template.
7. Like DNA replication, the processes of transcription can be subdivided into the stages of initiation, elongation, and termination. Also like DNA replication, transcription relies on base-pairing affinities between complementary nucleotides.
8. Initiation of transcription is dependent on an upstream (5′) DNA region, called the promoter, that represents the initial binding site for RNA polymerase. Promoters contain specific DNA sequences, such as the TATA box, that are essential to polymerase binding.
9. Transcription is more complex in eukaryotes than in prokaryotes. The primary transcript is a pre-mRNA that must be modified in various ways before it can be efficiently translated. Processing, which produces a mature mRNA, includes the addition of a 7-mG cap and a poly-A tail, and the removal, through splicing, of intervening sequences, or introns. RNA editing of pre-mRNA prior to its translation also occurs in some systems.

## INSIGHTS AND SOLUTIONS

1. Calculate how many triplet codons would be possible had evolution seized on six bases (three complementary base pairs) rather than four bases with which to construct DNA. Would six bases accommodate a two-letter code, assuming 20 amino acids and start-and-stop codons?

**Solution:** Six bases taken three at a time would produce $(6)^3$ or 216, triplet codes. If the code was a doublet, there would be $(6)^2$ or 36, two-letter codes, more than enough to accommodate 20 amino acids and start and stop signals.

2. In a heteropolymer experiment using 1/2C : 1/4A : 1/4G, how many different triplets will occur in the synthetic RNA molecule? How often will the most frequent triplet occur?

**Solution:** There will be $(3)^3$ or 27, triplets produced. The most frequent will be CCC, present $(1/2)^3$ or 1/8 of the time.

3. In a regular copolymer experiment, in which UUAC is repeated over and over, how many different triplets will occur in the synthetic RNA, and how many amino acids will occur in the polypeptide when this RNA is translated? (Consult Figure 14–7.)

**Solution:** The synthetic RNA will repeat four triplets—UUA, CUU, ACU, and UAC—over and over. Because both UUA and CUU encode leucine, while ACU and UAC encode threonine and tyrosine, respectively, the polypeptides synthesized under the directions of such an RNA contain three amino acids in the repeating sequence leu-leu-thr-tyr.

4. Actinomycin D inhibits DNA-dependent RNA synthesis. This antibiotic is added to a bacterial culture in which a specific protein is being monitored. Compared to a control culture, into which no antibiotic is added, translation of the protein declines over a period of 20 minutes, until no further protein is made. Explain these results.

**Solution:** The mRNA, which is the basis for the translation of the protein, has a lifetime of about 20 minutes. When actinomycin D is added, transcription is inhibited, and no new mRNAs are made. Those already present support the translation of the protein for up to 20 minutes.

5. DNA and RNA base compositions were analyzed from a hypothetical bacterial species with the following results:

| | (A + G)/(T + C) | (A + T)/(C + G) | (A + G)/(U + C) | (A + U)/(C + G) |
|---|---|---|---|---|
| DNA | 1.0 | 1.2 | | |
| RNA | | | 1.3 | 1.2 |

On the basis of these data, what can we conclude about the DNA and RNA of the organism? Are the data consistent with the Watson–Crick model of DNA? Is the RNA single stranded or double stranded, or can't we tell? If we assume that the entire length of DNA has been transcribed, do the data suggest that RNA has been derived from the transcription of one or both DNA strands, or can't we tell from these data?

**Solution:** This problem is a theoretical exercise designed to get you to look at the consequences of base complementarity as it affects the base composition of DNA and RNA. The base composition of DNA is consistent with the Watson–Crick double helix. In a double helix, we expect A + G to equal T + C. (The number of purines should equal the number of pyrimidines.) In this case, there is a preponderance of A═T base pairs (120 A═T pairs to every 100 G≡C pairs).

Given what we know about RNA, there is no reason to expect the RNA to be double stranded, but if it were double stranded, then we would expect that A═U and C≡G. If so, then (A + G)/(U + C) = 1. Since it doesn't equal unity, we can conclude that the RNA is not double stranded.

If all the DNA from either one or both strands is transcribed, the ratio of (A + U)/(C + G) in RNA should be 1.2, which it is. Note that this ratio will not change, regardless of whether only one or both of the strands are transcribed. This is the case because, for example, for every A═T pair in DNA, transcription of RNA will yield one A *and* one U if both strands are transcribed, whereas if just one strand is transcribed, transcription will yield one A *or* one U. In either case, the (A + U)/(C + G) ratio in RNA will reflect the (A + T)/(C + G) ratio in the DNA from which it was transcribed. To prove this to yourself, draw out a DNA molecule with 12 A═T pairs and 10 C≡G pairs and transcribe *both* strands. Then, transcribe *either* strand. Count the bases in the RNAs produced in both cases and calculate the ratios. Thus, we cannot determine whether just one or both strands are transcribed from the (A + U)/(C + G) ratio.

However, if both strands are transcribed, then the ratio of (A + G)/(U + C) should equal 1.0, and it doesn't. It equals 1.3. To verify this conclusion, examine the theoretical data you drew out on paper. One explanation for the observed ratio of 1.3 is that only one of the two strands is transcribed. If this is the case, then the (A + G)/(U + C) will reflect the proportion of A═T pairs that are A and the proportion of the G≡C pairs that are G *on the DNA strand that is transcribed*. Another explanation is that transcription occurs on one strand for some genes and on the other strand for other genes.

## Problems and Discussion Questions

1. Early proposals regarding the genetic code considered the possibility that DNA served directly as the template for polypeptide synthesis. (See Gamow, 1954, in Selected Readings.) In eukaryotes, what difficulties would such a system pose? What observations and theoretical considerations argue against such a proposal?
2. In their studies of frameshift mutations, Crick, Barnett, Brenner, and Watts-Tobin found that either three "pluses" or three "minuses" restored the correct reading frame. (a) Assuming the code is a triplet, what effect would the addition or loss of six nucleotides have on the reading frame? (b) If the code were a sextuplet (consisting of six nucleotides), would the reading frame be restored by the addition or loss of three, six, or nine nucleotides?
3. In a mixed copolymer experiment using polynucleotide phosphorylase, 3/4G:1/4C was used to form the synthetic message. The amino acid composition of the resulting protein was determined to be:

| | | |
|---|---|---|
| Glycine | 36/64 | (56 percent) |
| Alanine | 12/64 | (19 percent) |
| Arginine | 12/64 | (19 percent) |
| Proline | 4/64 | ( 6 percent) |

From this information,
(a) indicate the percentage (or fraction) of the time each possible codon will occur in the message.
(b) determine one consistent base-composition assignment for the amino acids present.
(c) in view of the wobble hypothesis, predict as many specific codon assignments as possible.

4. When repeating copolymers are used to form synthetic mRNAs, dinucleotides produce a single type of polypeptide that contains only two different amino acids. On the other hand, using a trinucleotide sequence produces three different polypeptides, each consisting of only a single amino acid. Why? What will be produced when a repeating tetranucleotide is used?
5. The mRNA formed from the repeating tetranucleotide UUAC incorporates only three amino acids, but the use of UAUC incorporates four amino acids. Why?
6. In studies using repeating copolymers, AC . . . incorporates threonine and histidine, and CAACAA . . . incorporates glutamine, asparagine, and threonine. What triplet code can definitely be assigned to threonine?
7. In a coding experiment using repeating copolymers (as demonstrated in Table 14.3), the following data were obtained:

| Copolymer | Codons Produced | Amino Acids in Polypeptide |
|---|---|---|
| AG | AGA, GAG | Arg, Glu |
| AAG | AGA, AAG, GAA | Lys, Arg, Glu |

AGG is known to code for arginine. Taking into account the wobble hypothesis, assign each of the four codons produced in the experiment to its correct amino acid.

8. In the triplet-binding technique, radioactivity remains on the filter when the amino acid corresponding to the codon is labeled. Explain the rationale for this technique.
9. When the amino acid sequences of insulin isolated from different organisms were determined, some differences were noted. For example, alanine was substituted for threonine, serine was substituted for glycine, and valine was substituted for isoleucine at corresponding positions in the protein. List the single-base changes that could occur in codons of the genetic code to produce these amino acid changes.
10. In studies of the amino acid sequence of wild-type and mutant forms of tryptophan synthetase in *E. coli*, the following changes have been observed:

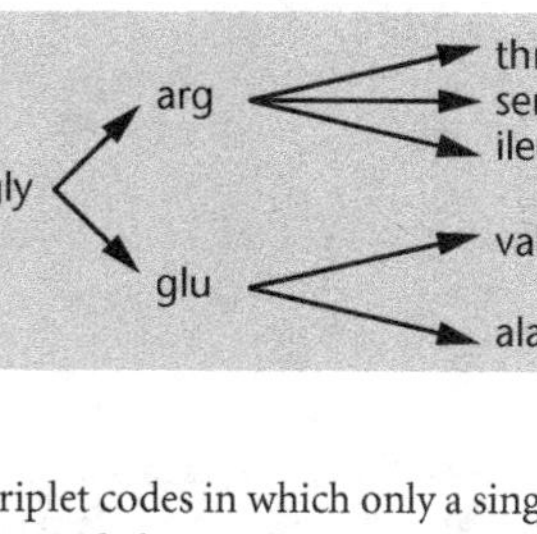


Determine a set of triplet codes in which only a single nucleotide change produces each amino acid change.

11. Why doesn't polynucleotide phosphorylase (Ochoa's enzyme) synthesize RNA *in vivo*?
12. Refer to Table 14.1. Can you hypothesize why a mixture of Poly U + Poly A would not stimulate incorporation of $^{14}$C-phenylalanine into protein?

13. Predict the amino acid sequence produced during translation by the following short hypothetical mRNA sequences (note that the second sequence was formed from the first by a deletion of only one nucleotide):

    Sequence 1: 5′-AUGCCGGAUUAUAGUUGA-3′

    Sequence 2: 5′-AUGCCGGAUUAAGUUGA-3′

    What type of mutation gave rise to Sequence 2?
14. A short RNA molecule was isolated that demonstrated a hyperchromic shift, indicating secondary structure. Its sequence was determined to be

    5′-AGGCGCCGACUCUACU-3′

    (a) Propose a two-dimensional model for this molecule.
    (b) What DNA sequence would give rise to this RNA molecule through transcription?
    (c) If the molecule were a tRNA fragment containing a CGA anticodon, what would the corresponding codon be?
    (d) If the molecule were an internal part of a message, what amino acid sequence would result from it following translation? (Refer to the code chart in Figure 14–7.)
15. A glycine residue is in position 210 of the tryptophan synthetase enzyme of wild-type *E. coli*. If the codon specifying glycine is GGA, how many single-base substitutions will result in an amino acid substitution at position 210? What are they? How many will result if the wild-type codon is GGU?
16. Refer to Figure 14–7 to respond to the following:
    (a) Shown here is a hypothetical viral mRNA sequence:

    5′-AUGCAUACCUAUGAGACCCUUGGA-3′

    Assuming that it could arise from overlapping genes, how many different polypeptide sequences can be produced? What are the sequences?
    (b) A base-substitution mutation that altered the sequence in (a) eliminated the synthesis of all but one polypeptide. The altered sequence is shown here:

    5′-AUGCAUACCUAUGUGACCCUUGGA-3′

    Determine why.
17. Most proteins have more leucine than histidine residues, but more histidine than tryptophan residues. Correlate the number of codons for these three amino acids with this information.
18. Define the process of transcription. Where does this process fit into the central dogma of molecular genetics (DNA makes RNA makes protein)?
19. What was the initial evidence for the existence of mRNA?
20. Describe the structure of RNA polymerase in bacteria. What is the core enzyme? What is the role of the $\sigma$ subunit?
21. Write a paragraph describing the abbreviated chemical reactions that summarize RNA polymerase-directed transcription.
22. Messenger RNA molecules are very difficult to isolate in prokaryotes because they are rather quickly degraded in the cell. Can you suggest a reason why this occurs? Eukaryotic mRNAs are more stable and exist longer in the cell than do prokaryotic mRNAs. Is this an advantage or disadvantage for a pancreatic cell making large quantities of insulin?
23. The following represent deoxyribonucleotide sequences in the template strand of DNA:

| | |
|---|---|
| Sequence 1: | 5′-CTTTTTTGCCAT-3′ |
| Sequence 2: | 5′-ACATCAATAACT-3′ |
| Sequence 3: | 5′-TACAAGGGTTCT-3′ |

    (a) For each strand, determine the mRNA sequence that would be derived from transcription.
    (b) Using Figure 14–7, determine the amino acid sequence that is encoded by these mRNAs.
    (c) For Sequence 1, what is the sequence of the partner DNA strand?
24. Present an overview of various forms of posttranscriptional processing in eukaryotes. For each, provide an example.
25. Describe the role of two forms of RNA editing that lead to changes in the size and sequence of pre-mRNAs. Briefly describe several examples of each form of editing, including their impact on respective protein products.

**HOW DO WE KNOW?**

26. In this chapter, we focused on the genetic code and the transcription of stored genetic information into RNA molecules. Along the way, we found many opportunities to consider the methods and reasoning by which much of this information was acquired. From the explanations given in the chapter, what answers would you propose to the following fundamental questions:
    (a) Why did geneticists believe, even before direct experimental evidence was obtained, that the genetic code would turn out to be composed of triplet sequences and nonoverlapping? How were these suppositions shown to be correct?
    (b) What experimental evidence provided the initial insights into the *compositions* of codons encoding specific amino acids?
    (c) How were the specific sequences of triplet codes determined experimentally?
    (d) How were the experimentally derived triplet codon assignments verified in studies using bacteriophage MS2?
    (e) What evidence do we have that the expression of the information encoded in DNA involves an RNA intermediate?
    (f) How do we know that the initial transcript of a eukaryotic gene contains noncoding sequences that must be removed before accurate translation into proteins can occur?

## Extra-Spicy Problems

27. It has been suggested that the present-day triplet genetic code evolved from a doublet code when there were fewer amino acids available for primitive protein synthesis.
    (a) Can you find any support for the doublet code notion in the existing coding dictionary?
    (b) The amino acids Ala, Val, Gly, Asp, and Glu are all early members of biosynthetic pathways (Taylor and Coates, 1989) and are more evolutionarily conserved than other amino acids (Brooks and Fresco, 2003). They therefore probably represent "early" amino acids. Of what significance is this information in terms of the evolution of the genetic code? Also, which base, of the first two, would likely have been the more significant in originally specifying these amino acids?
    (c) As determined by comparisons of ancient and recently evolved proteins, cysteine, tyrosine, and phenylalanine appear to be late-arriving

amino acids. In addition, they are considered to have been absent in the abiotic earth (Miller, 1987). All three of these amino acids have only two codons each, while many others, earlier in origin, have more. Is this mere coincidence, or might there be some underlying explanation?

28. In a mixed copolymer experiment, messages were created with either 4/5C:1/5A or 4/5A:1/5C. These messages yielded proteins with the following amino acid compositions.

| 4/5C:1/5A | | 4/5A:1/5C | |
|---|---|---|---|
| Proline | 63.0 percent | Proline | 3.5 percent |
| Histidine | 13.0 percent | Histidine | 3.0 percent |
| Threonine | 16.0 percent | Threonine | 16.6 percent |
| Glutamine | 3.0 percent | Glutamine | 13.0 percent |
| Asparagine | 3.0 percent | Asparagine | 13.0 percent |
| Lysine | 0.5 percent | Lysine | 50.0 percent |
| | 98.5 percent | | 99.1 percent |

Using these data, predict the most specific coding composition for each amino acid.

29. Shown here are the amino acid sequences of the wild-type and three mutant forms of a short protein. Use this information to answer the following questions:

| | |
|---|---|
| Wild type: | mer-trp-tyr-arg-gly-ser-pro-thr |
| Mutant 1: | met-trp |
| Mutant 2: | met-trp-his-arg-gly-ser-pro-thr |
| Mutant 3: | met-cys-ile-val-val-val-gln-his |

(a) Using Figure 14–7, predict the type of mutation that led to each altered protein.
(b) For each mutant protein, determine the specific ribonucleotide change that led to its synthesis.
(c) The wild-type RNA consists of nine triplets. What is the role of the ninth triplet?
(d) Of the first eight wild-type triplets, which, if any, can you determine specifically from an analysis of the mutant proteins? In each case, explain why or why not.
(e) Another mutation (Mutant 4) is isolated. Its amino acid sequence is unchanged, but the mutant cells produce abnormally low amounts of the wild-type proteins. As specifically as you can, predict where this mutation exists in the gene.

30. The genetic code is degenerate. Amino acids are encoded by either 1, 2, 3, 4, or 6 triplet codons. (See Figure 14–7.) An interesting question is whether the number of triplet codes for a given amino acid is in any way correlated with the frequency with which that amino acid appears in proteins. That is, is the genetic code optimized for its intended use? Some approximations of the frequency of appearance of nine amino acids in proteins in *E. coli* are

| Amino Acid | Percentage |
|---|---|
| Met | 2 |
| Cys | 2 |
| Gln | 5 |
| Pro | 5 |
| Arg | 5 |
| Ile | 6 |
| Glu | 7 |
| Ala | 8 |
| Leu | 10 |

(a) Determine how many triplets encode each amino acid.
(b) Devise a way to graphically compare the two sets of information (data).
(c) Analyze your data to determine what, if any, correlations can be drawn between the relative frequency of amino acids making up proteins and the number of codons for each. Write a paragraph that states your specific and general conclusions.
(d) How would you proceed with your analysis if you wanted to pursue this problem further?

31. As described in Chapter 12, *Alu* elements proliferate in the human genome by a process called retrotransposition, in which the *Alu* DNA sequence is transcribed into RNA, copied into double-stranded DNA, and then inserted back into the genome at a site distant from that of its "parent" *Alu* gene.

Clearly, this has been an extremely efficient process, since *Alu* genes have proliferated to about $10^6$ copies in the human genome! This efficiency is largely due to the fact that *Alu* elements, like many small structural RNAs, carry their promoter sequences *within the transcribed region of the gene*, rather than 5′ to the transcription start site.

If *Alu* elements carried promoters upstream of the transcription site, as do protein-coding genes, what would happen once they were retrotransposed? Would a retrotransposed *Alu* gene be able to proliferate? Explain.

32. M. Klemke and others (2001. *EMBO J.* 20: 3849–3860) discovered an interesting coding phenomenon in which an exon within a neurologic hormone receptor gene in mammals appears to produce two different protein entities (XL$\alpha$s ALEX). Below is the DNA sequence of the exon's 5′ end derived from a rat. The lowercase letters represent the initial coding portion for the XL$\alpha$s protein, and the uppercase letters indicate the portion where the ALEX entity is initiated. (For simplicity, and to correspond with the RNA coding dictionary, it is customary to represent the noncoding, nontemplate strand of the DNA segment.)

5′-gtcccaaccatgcccaccgatcttccgcctgcttctgaagATGCGGGCCCAG

(a) Convert the noncoding DNA sequence to the coding RNA sequence.
(b) Locate the initiator codon within the XL$\alpha$s segment.
(c) Locate the initiator codon within the ALEX segment. Are the two initiator codons in frame?
(d) Provide the amino acid sequence for each coding sequence. In the region of overlap, are the two amino acid sequences the same?
(e) Are there any evolutionary advantages to having the same DNA sequence code for two protein products? Are there any disadvantages?

33. The concept of consensus sequences of DNA was defined in this chapter as sequences that are similar (homologous) in different genes of the same organism or in genes of different organisms. Examples were the Pribnow box and the –35 region in prokaryotes and the TATA-box region in eukaryotes. One study found that among 73 isolates from the virus HIV-Type 1C (a major contributor to the AIDS epidemic), a GGGNNNNNCC consensus sequence exists (where N equals any nitrogenous base) in the promoter–enhancer region of the NF-$\kappa$B transcription factor, a *cis*-acting element that is critical for initiating HIV transcription in human macrophages (Novitsky et al., 2002. *J. Virol.* 76: 5435–5451). The authors contend that finding this and other conserved sequences may be of value in designing an AIDS vaccine. What advantages would knowing these consensus sequences confer? Are there disadvantages as a vaccine is designed?

34. Theoretically, antisense oligodeoxynucleotides (relatively short, 7- to 20-nucleotide, single-stranded DNAs) can selectively block disease-causing genes. They do so by base-pairing with complementary regions of the RNA transcripts and inhibiting their function. The RNA/DNA hybrid is recognized by intracellular RNase H, and the RNA portion of the hybrid is degraded. Cancer genes have often been chosen as potential targets for

antisense drugs. Data shown in the following table (from Cho et al., 2001. *Proc. Natl. Acad. Sci. [USA]* 98: 9819–9823) established that in response to a single population of antisense oligodeoxynucleotides that target a specific kinase mRNA in a prostate cancer cell line, there was as much as a twentyfold response differential (including both increases and decreases) in unrelated gene expression (as measured by mRNA production).

| Gene Assayed | Change in Expression |
|---|---|
| Myosin light chain | +14.4x |
| G protein receptor | +3.7x |
| Collagen | −4.0x |
| Catalase | −6.7x |

(a) Diagram your concept of the intracellular action of an antisense oligodeoxynucleotide.
(b) Diagram your concept of the action of RNase H.
(c) What might cause nonrelated gene expression to be decreased, as in the case of collagen and catalase?
(d) What might cause nonrelated gene expression to be increased, as in the case of myosin and G protein?
(e) Given the data presented here, what drawbacks might you expect in the use of antisense therapy for genetic diseases?

35. Recent observations indicate that alternative splicing is a common way for eukaryotes to expand their repertoire of gene functions. Studies indicate that approximately 50 percent of human genes exhibit alternative splicing and approximately 15 percent of disease-causing mutations involve aberrant alternative splicing. Different tissues show remarkably different frequencies of alternative splicing, with the brain accounting for approximately 18 percent of such events (Xu et al., 2002. *Nuc. Acids Res.* 30: 3754–3766).
(a) Define alternative splicing and speculate on the evolutionary strategy alternative splicing offers to organisms.
(b) Why might some tissues engage in more alternative splicing than others?

Crystal structure of a *Thermus thermophilus* 70S ribosome containing three bound transfer RNAs.

*Reprinted from the front cover of* **Science**, *Vol. 292, May 4, 2001. Crystal structure of a Thermus thermophilus 70S ribosome containing three bound transfer RNAs (top) and exploded views showing its different molecular components (middle and bottom). Image provided by Dr. Albion Baucom (baucom@biology.ucsc.edu). Copyright American Association for the Advancement of Science.*

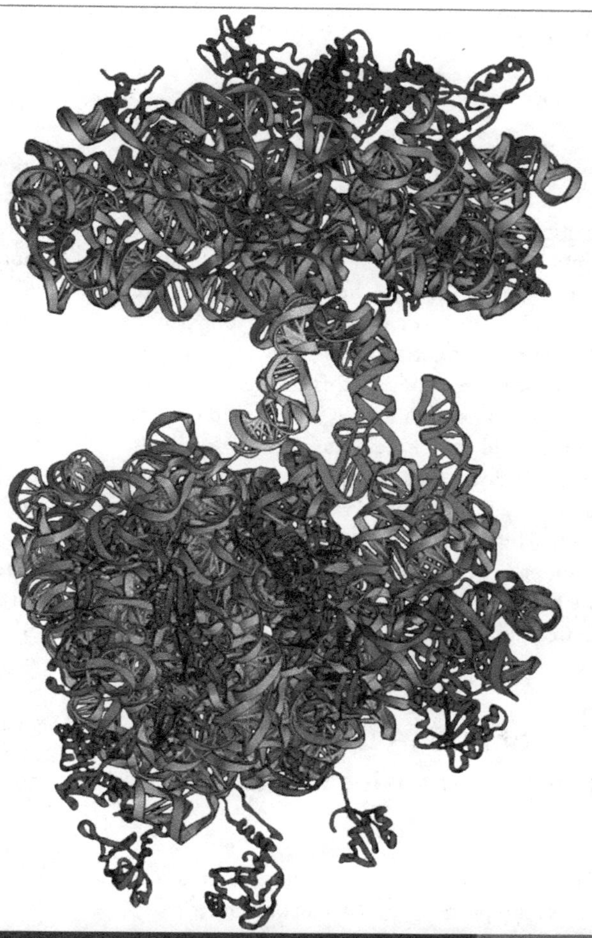

15

# Translation and Proteins

## CHAPTER CONCEPTS

- The ribonucleotide sequence of messenger RNA (mRNA) reflects genetic information stored in the DNA of genes and corresponds to the amino acid sequences in proteins encoded by those genes.
- The process of translation decodes the information in mRNA, leading to the synthesis of polypeptide chains.
- Translation involves the interactions of mRNA, tRNA, ribosomes, and a variety of translation factors essential to the initiation, elongation, and termination of the polypeptide chain.
- Proteins, the final product of most genes, achieve a three-dimensional conformation that arises from the primary amino acid sequences of the polypeptide chains making up each protein.
- The function of any protein is closely tied to its three-dimensional structure, which can be disrupted by mutation.

In Chapter 14, we established that a genetic code stores information in the form of triplet codons in DNA and that this information is initially expressed, through the process of transcription, as a messenger RNA complementary to the template strand of the DNA helix. However, the final product of gene expression, in most instances, is a polypeptide chain consisting of a linear series of amino acids whose sequence has been prescribed by the genetic code. In this chapter, we will examine how the information present in mRNA is translated to create polypeptides, which then fold into protein molecules. We will also review the evidence that confirms that proteins are the end products of genes and discuss briefly the various levels of protein structure, diversity, and function. This information extends our understanding of gene expression and provides an important foundation for interpreting how the mutations that arise in DNA can result in the diverse phenotypic effects observed in organisms.

## 15.1 Translation of mRNA Depends on Ribosomes and Transfer RNAs

**Translation** of mRNA is the biological polymerization of amino acids into polypeptide chains. This process, alluded to in our earlier discussion of the genetic code, occurs only in association with **ribosomes**, which serve as nonspecific workbenches. The central question in translation is how triplet codons of mRNA direct specific amino acids into their correct position in the polypeptide. That question was answered once **transfer RNA (tRNA)** was discovered. This class of molecules adapts genetic information present as specific triplet codons in mRNA to their corresponding amino acids. The requirement for some sort of "adaptor" was postulated by Francis Crick in 1957.

In association with a ribosome, mRNA presents a triplet codon that calls for a specific amino acid. A specific tRNA molecule contains within its nucleotide sequence three consecutive ribonucleotides complementary to the codon, called the **anticodon**, which can base-pair with the codon. Another region of this tRNA is covalently bonded to the codon's corresponding amino acid.

Inside the ribosome, hydrogen bonding of tRNAs to mRNA holds amino acids in proximity to each other so that a peptide bond can be formed between them. The process occurs over and over as mRNA runs through the ribosome, and amino acids are polymerized into a polypeptide. Before looking more closely at this process, we will first consider the structures of the ribosome and transfer RNA.

### Ribosomal Structure

Because of its essential role in the expression of genetic information, the ribosome has been extensively analyzed. One bacterial cell contains about 10,000 ribosomes, and a eukaryotic cell contains many times more. Electron microscopy has revealed that the bacterial ribosome is about 25 $\mu$m at its largest diameter and consists of two subunits, one large and one small. Both subunits consist of one or more molecules of rRNA and an array of **ribosomal proteins**. When the two subunits are associated with each other in a single ribosome, the structure is sometimes called a **monosome**.

The main differences between prokaryotic and eukaryotic ribosomes are summarized in Figure 15–1. The subunit and rRNA components are most easily isolated and characterized on the basis of their sedimentation behavior in sucrose gradients (their rate of migration when centrifuged, abbreviated *S* for Svedberg, as introduced in Chapter 10). In prokaryotes, the monosome is a 70*S* particle, and in eukaryotes it is approximately 80*S*. Sedimentation coefficients, which reflect differences in the rate of migration of different-sized particles and molecules, are not additive. For example, the prokaryotic 70*S* monosome consists of a 50*S* and a 30*S* subunit, and the eukaryotic 80*S* monosome consists of a 60*S* and a 40*S* subunit.

The larger subunit in prokaryotes consists of a 23*S* rRNA molecule, a 5*S* rRNA molecule, and 31 ribosomal proteins. In the eukaryotic equivalent, a 28*S* rRNA molecule is accompanied by a 5.8*S* and a 5*S* rRNA molecule and 49 proteins. The smaller prokaryotic subunits consist of a 16*S* rRNA component and 21 proteins. In the eukaryotic equivalent, an 18*S* rRNA component and about 33 proteins are found. The approximate molecular weights (in daltons, or Da) and number of nucleotides of these components are also shown in Figure 15–1.

It is now clear that the RNA components of the ribosome perform all-important catalytic functions associated with translation. The many ribosomal proteins, whose functions were long a mystery, are thought to promote the binding of the various molecules involved in translation and, in general, to fine-tune the process. This conclusion is based on the observation that some of the catalytic functions in ribosomes still occur in experiments involving "ribosomal protein-depleted" ribosomes.

Molecular hybridization studies have established the degree of redundancy of the genes coding for the rRNA components. The *E. coli* genome contains seven copies of a single sequence that encodes all three components—23*S*, 16*S*, and 5*S*. The initial transcript of each set of these genes produces a 30*S* RNA molecule that is enzymatically cleaved into these smaller components. The coupling of the genetic information encoding these three rRNA components ensures that, following multiple transcription events, equal quantities of all three will be present as ribosomes are assembled.

In eukaryotes, many more copies of a sequence encoding the 28*S*, 18*S*, and 5.8*S* components are present. In *Drosophila*, approximately 120 copies per haploid genome are each transcribed into a molecule of about 34*S*. This is processed into the 28*S*, 18*S*, and 5.8*S* rRNA species. These species are homologous to the three rRNA components of *E. coli*. In *Xenopus laevis*, more than 500 copies of the 34*S* component are present per haploid genome. In mammalian cells, the initial transcript is 45*S*. The rRNA genes, called **rDNA**, are part of the

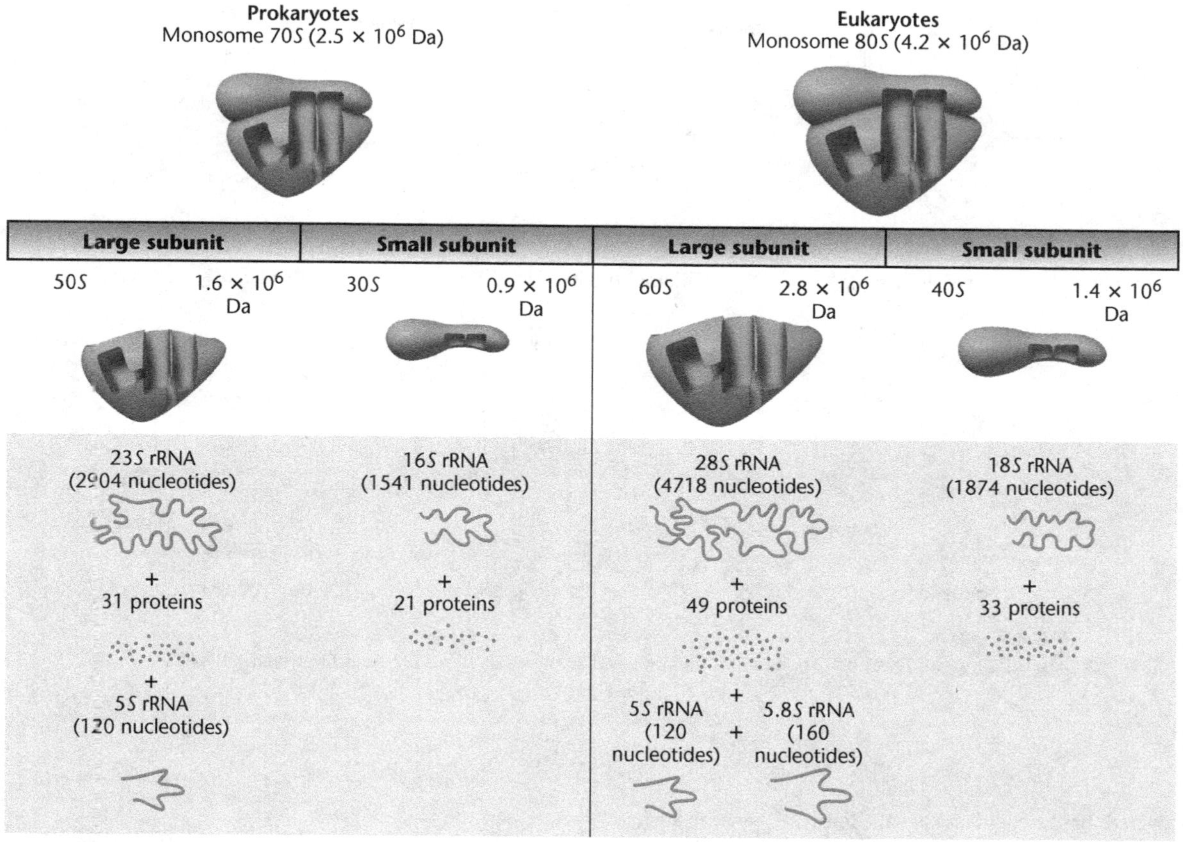


FIGURE 15–1 A comparison of the components in prokaryotic and eukaryotic ribosomes.

moderately repetitive DNA fraction and are present in clusters at various chromosomal sites.

Each cluster in eukaryotes consists of **tandem repeats**, with each unit separated by a noncoding **spacer DNA** sequence. In humans, these gene clusters have been localized near the ends of chromosomes 13, 14, 15, 21, and 22. The unique 5*S* rRNA component of eukaryotes is not part of this larger transcript. Instead, genes coding for the 5*S* ribosomal component are distinct and located separately. In humans, a gene cluster encoding 5*S* rRNA has been located on chromosome 1.

Despite their detailed knowledge of the structure and genetic origin of the ribosomal components, a complete understanding of the function of these components has eluded geneticists. This is not surprising; the ribosome is the largest and perhaps the most intricate of all cellular structures. For example, the bacterial monosome has a combined molecular weight of 2.5 million Da!

## tRNA Structure

Because of their small size and stability in the cell, transfer RNAs (tRNAs) have been investigated extensively and are the best characterized RNA molecules. They are composed of only 75 to 90 nucleotides, displaying a nearly identical structure in bacteria and eukaryotes. In both types of organisms, tRNAs are transcribed from DNA as larger precursors, which are cleaved into mature 4*S* tRNA molecules. In *E. coli*, for example, $tRNA^{tyr}$ (the superscript identifies the specific tRNA by the amino acid that binds to it, called its *cognate amino acid*) is composed of 77 nucleotides, yet its precursor contains 126 nucleotides.

In 1965, Robert Holley and his colleagues reported the complete sequence of $tRNA^{ala}$ isolated from yeast. Of great interest was their finding that a number of nucleotides are unique to tRNA. As illustrated in Figure 15–2, each of these nucleotides is a modification of one of the four nitrogenous bases expected in RNA (G, C, A, and U). Shown are inosinic acid (which contains the purine hypoxanthine), ribothymidylic acid, pseudouridine, and several others. These modified structures, variously referred to as *unusual*, *rare*, or *odd bases*, are created *after* transcription, illustrating the more general concept of **posttranscriptional modification**. In this case, the unmodified base is inserted during transcription of tRNA, and subsequently, enzymatic reactions catalyze the chemical modifications to the base.

Holley's sequence analysis led him to propose the two-dimensional **cloverleaf model of tRNA**. It had been known that tRNA has a characteristic secondary structure created by base pairing.

WEB TUTORIAL 15.1 TRANSFER RNAS

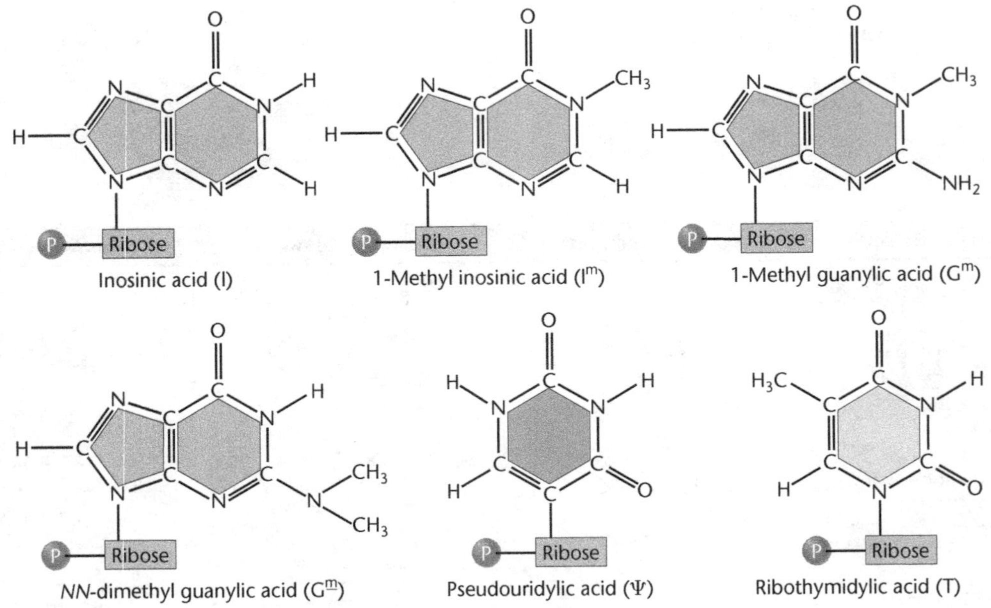


FIGURE 15–2 Ribonucleotides containing unusual nitrogenous bases are found in transfer RNA.

Holley discovered that he could arrange the linear sequence in such a way that several stretches of base pairing would result. His arrangement, with its series of paired stems and unpaired loops, resembled the shape of a cloverleaf. The loops consistently contained modified bases that did not generally form base pairs. Holley's model is shown in Figure 15–3.

The triplets GCU, GCC, and GCA specify alanine; therefore, Holley looked for an anticodon sequence complementary to one of these codons in his $tRNA^{ala}$ molecule. He found it in the form of CGI (the 3′ to 5′-direction) in one loop of the cloverleaf. The nitrogenous base I (inosinic acid) can form hydrogen bonds with U, C, or A, the third members of the alanine triplets. Thus, the **anticodon loop** was established.

Studies of other tRNA species revealed many constant features. First, at the 3′-end, all tRNAs contain the sequence (. . . pCpCpA-3′). At this end of the molecule, the amino acid is covalently joined to the terminal adenosine residue. All tRNAs contain the nucleotide (5′-pG. . .) at the other end of the molecule. In addition, the lengths of the analogous stems and loops in tRNA molecules are very similar. Each tRNA examined also contains an anticodon complementary to the known codon for the tRNA's cognate amino acid, and all anticodon loops are present in the same position of the cloverleaf.

Because the cloverleaf model was predicted strictly on the basis of nucleotide sequence, there was great interest in the three-dimensional structure that would be revealed by X-ray crystallographic examination of tRNA. By 1974, Alexander Rich and his colleagues in the United States, and J. Roberts, B. Clark, Aaron Klug, and their colleagues in

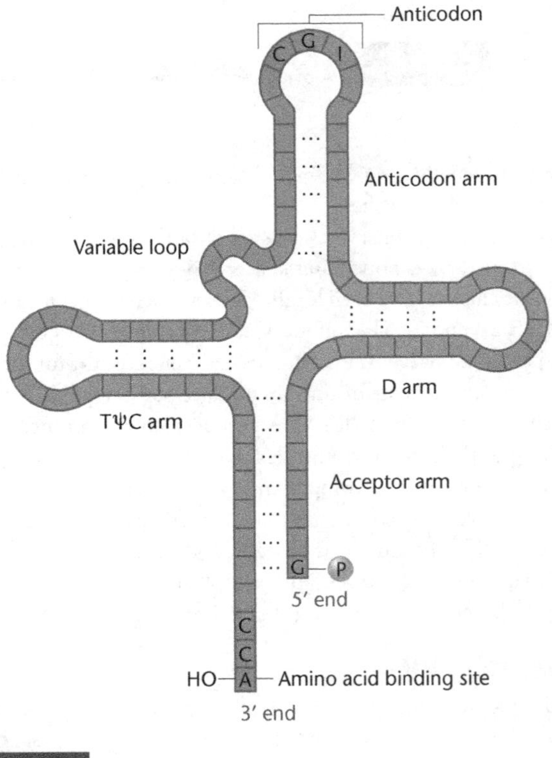


FIGURE 15–3 Holley's two-dimensional cloverleaf model of transfer RNA.

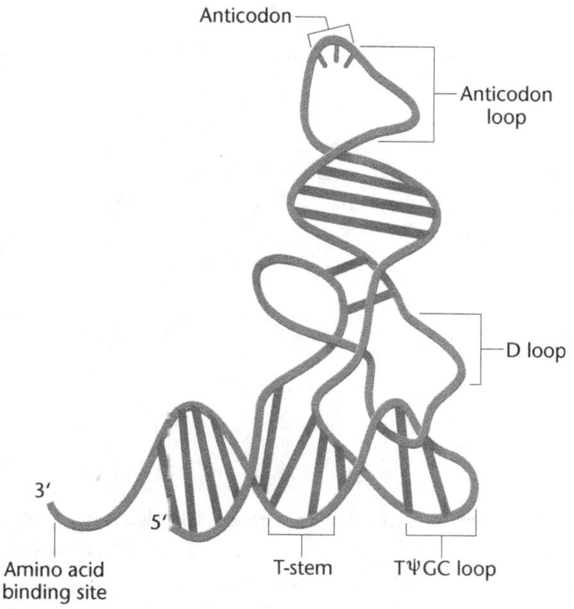


FIGURE 15–4 A three-dimensional model of transfer RNA.

England had succeeded in crystallizing tRNA and performing X-ray crystallography at a resolution of 3 Å. At such resolution, the pattern formed by individual nucleotides is discernible.

As a result of these studies, a complete three-dimensional model of tRNA is now available (Figure 15–4). Both the anticodon loop and the 3′-acceptor region (to which the amino acid is covalently linked) have been located. Geneticists speculate that the shapes of the intervening loops are recognized by the enzymes responsible for attaching amino acids to tRNAs—a subject to which we now turn our attention.

## Charging tRNA

Before translation can proceed, the tRNA molecules must be chemically linked to their respective amino acids. This activation process, called **charging**, or *aminoacylation*, occurs under the direction of enzymes called **aminoacyl tRNA synthetases**. Because there are 20 different amino acids, there must be at least 20 different tRNA molecules and as many different enzymes. In theory, because there are 61 triplets that encode amino acids, there could be 61 specific tRNAs and enzymes. However, because of the ability of the third member of a triplet code to "wobble," it is now thought that there are but 31 different tRNAs. It is also believed that there are only 20 synthetases, one for each amino acid, regardless of the greater number of corresponding tRNAs.

The charging process is outlined in Figure 15–5. In the initial step, the amino acid is converted to an activated form, reacting with ATP to create an **aminoacyladenylic acid**. A covalent linkage is formed between the 5 -phosphate group of ATP and the carboxyl end of the amino acid. This reaction occurs in association with the synthetase enzyme, forming a complex that then reacts with a specific tRNA molecule. During the next step, the amino acid is transferred to the appropriate tRNA and bonded covalently to the adenine residue at the 3′ end. The charged tRNA may then participate directly in protein synthesis. Aminoacyl tRNA synthetases are highly specific enzymes because they recognize only one amino acid and only the tRNAs corresponding to that amino acid, called **isoaccepting tRNAs**. This is a crucial point if fidelity of translation is to be maintained.

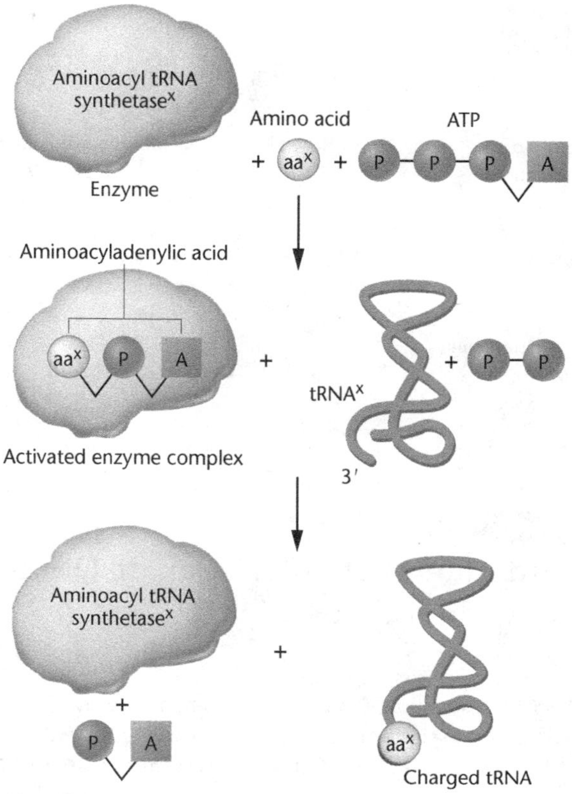


FIGURE 15–5 Steps involved in charging tRNA. The superscript x denotes that only the corresponding specific tRNA and specific aminoacyl tRNA synthetase enzyme are involved in the charging process for each amino acid.

**NOW SOLVE THIS**

Problem 32 on page 407 is concerned with establishing whether tRNA or the amino acid added to the tRNA during charging is responsible for the response of charged tRNA to mRNA during translation.

■ HINT: *In this experiment, when the triplet codon in mRNA calls for cysteine, alanine is inserted during translation, even though it is the "incorrect" amino acid.*

TABLE 15.1

**Various Protein Factors Involved during Translation in *E. coli***

| Process | Factor | Role |
|---|---|---|
| Initiation of translation | IF1 | Stabilizes 30*S* subunit |
| | IF2 | Binds f-met-tRNA to 30*S*-mRNA complex; binds to GTP and stimulates hydrolysis |
| | IF3 | Binds 30*S* subunit to mRNA; dissociates monosomes into subunits following termination |
| Elongation of polypeptide | EF-Tu | Binds GTP; brings aminoacyl-tRNA to the A site of the ribosome |
| | EF-Ts | Generates active EF-Tu |
| | EF-G | Stimulates translocation; GTP-dependent |
| Termination of translation and release of polypeptide | RF1 | Catalyzes release of the polypeptide chain from tRNA and dissociation of the translocation complex; specific for UAA and UAG termination codons |
| | RF2 | Behaves like RF1; specific for UGA and UAA codons |
| | RF3 | Stimulates RF1 and RF2 |

15.2

## Translation of mRNA Can Be Divided into Three Steps

Much like transcription, the process of translation can best be described by breaking it into discrete phases. We will consider three such phases, each with its own set of illustrations (Figures 15–6, 15–7, and 15–8), but keep in mind that translation is a dynamic, continuous process. As you read the following discussion, keep track of the step-by-step events depicted in the figures. Many of the protein factors and their roles in translation are presented in Table 15.1.

### Initiation

Initiation of prokaryotic translation is depicted in Figure 15–6. Recall that the ribosome serves as a nonspecific workbench for the translation process. Ribosomes, when they are not involved in

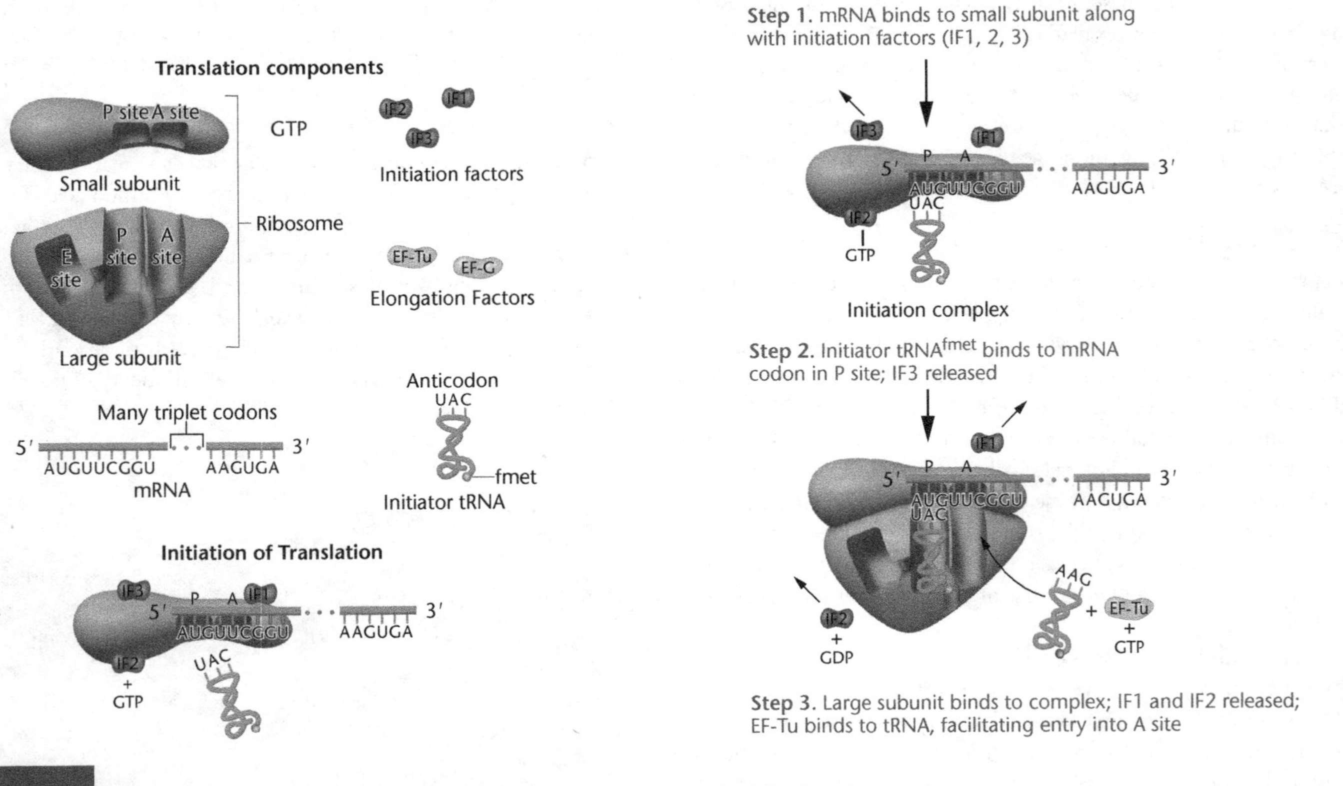


PROTEIN TRANSLATION | WEB TUTORIAL 15.2

FIGURE 15-6 Initiation of translation. The separate components are depicted at the left of the figure.

translation, are dissociated into their large and small subunits. Initiation of translation in *E. coli* involves the small ribosomal subunit, an mRNA molecule, a specific charged initiator tRNA, GTP, $Mg^{2+}$, and at least three proteinaceous **initiation factors (IFs)** that enhance the binding affinity of the various translational components.

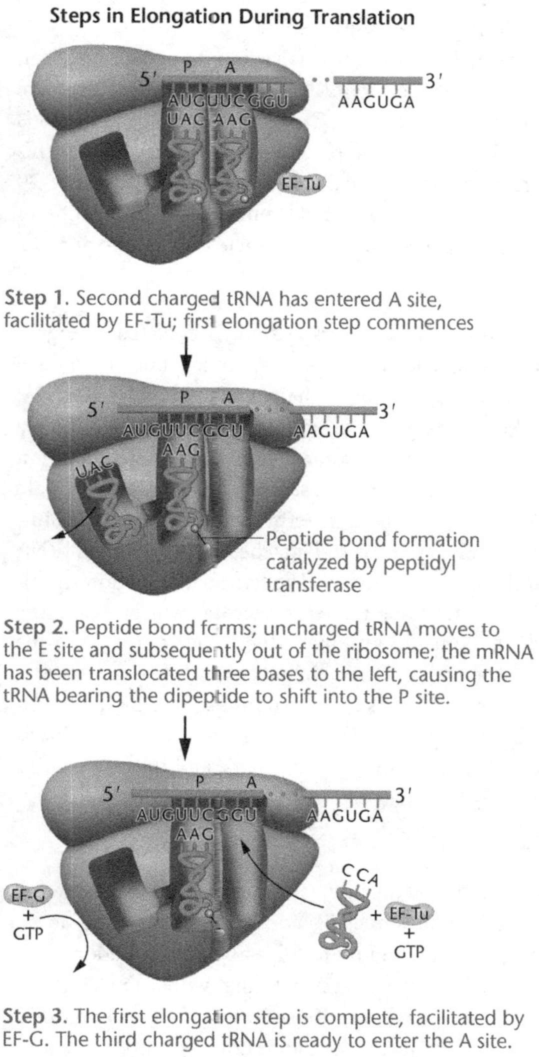

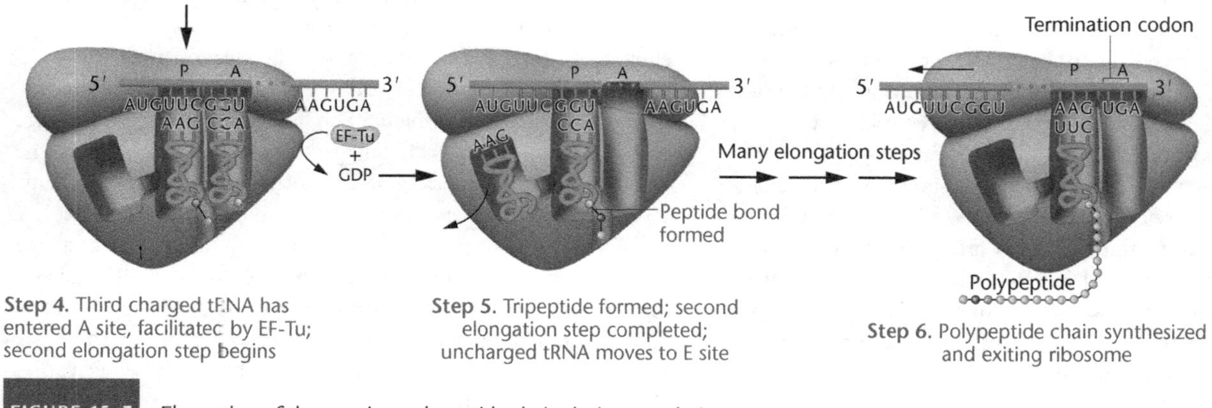


**FIGURE 15-7** Elongation of the growing polypeptide chain during translation.

Unlike ribosomal proteins, IFs are released from the ribosome once initiation is completed. In prokaryotes, the initiation codon of mRNA—AUG—calls for the modified amino acid ***N*-formylmethionine (f-met)**.

The small ribosomal subunit binds to several initiation factors, and this complex in turn binds to mRNA (Step 1). In bacteria, this binding involves a sequence of up to six ribonucleotides (AGGAGG, not shown in Figure 15–6) that *precedes* the initial AUG start codon of mRNA. This sequence—containing only purines and called the **Shine–Dalgarno sequence**—base-pairs with a region of the 16*S* rRNA of the small ribosomal subunit, facilitating initiation.

Another initiation protein then enhances the binding of charged formylmethionyl tRNA to the small subunit in response to the AUG triplet (Step 2). This step "sets" the reading frame so that all subsequent groups of three ribonucleotides are translated accurately. The aggregate represents the **initiation complex**, which then combines with the large ribosomal subunit. In this process, a molecule of GTP is hydrolyzed, providing the required energy, and the initiation factors are released (Step 3).

## Elongation

The second phase of translation, elongation, is depicted in Figure 15–7. Once both subunits of the ribosome are assembled with the mRNA, binding sites for two charged tRNA molecules are formed. These are the **P (peptidyl) site** and the **A (aminoacyl) site**. The charged initiator tRNA binds to the P site, provided that the AUG codon of mRNA is in the corresponding position of the small subunit.

The lengthening of the growing polypeptide chain by one amino acid is called **elongation**. The sequence of the second triplet in mRNA dictates which charged tRNA molecule will become positioned at the A site (Step 1). Once it is present, an enzymatic reaction occurs within the large subunit of the ribosome, catalyzing the formation of the peptide bond that links the two amino acids together. At the same time, the covalent bond between the tRNA occupying the P site and its amino acid is hydrolyzed (broken). The dipeptide remains attached to the end of the tRNA still residing in the A site (Step 2). These reactions were initially

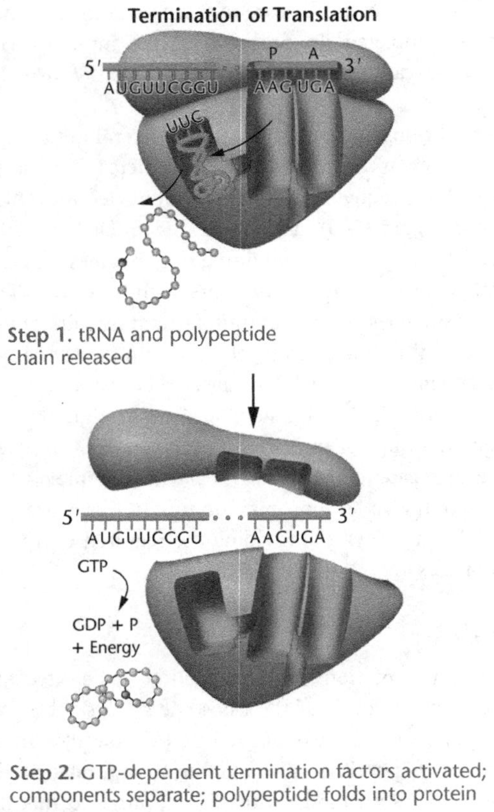


FIGURE 15–8 Termination of the process of translation.

believed to be catalyzed by an enzyme called **peptidyl transferase**, embedded in but never isolated from the large subunit of the ribosome. However, it is now clear that this catalytic activity is a function of the 23*S* rRNA of the large subunit, perhaps in conjunction with one or more of the ribosomal proteins. In such a case, as we saw with splicing of pre-mRNAs in Chapter 14, we refer to the complex as a **ribozyme**, recognizing the catalytic role that RNA plays in the process.

Before elongation can be repeated, the tRNA attached to the P site, which is now uncharged, must be released from the large subunit. The uncharged tRNA moves briefly into a third site on the ribosome called the **E (exit) site**. The entire **mRNA–tRNA–$aa_2$–$aa_1$** complex then shifts in the direction of the P site by a distance of three nucleotides (Step 3). This event requires several protein elongation factors (EFs) as well as the energy derived from hydrolysis of GTP. The result is that the third codon of mRNA is now in a position to accept its specific charged tRNA into the A site (Step 4). One simple way to distinguish the two sites in your mind is to remember that, *following the shift*, the P site (*P* for peptide) contains a tRNA attached to a peptide chain, whereas the A site (*A* for amino acid) contains a tRNA with an amino acid attached.

The sequence of elongation is repeated over and over (Steps 4 and 5). An additional amino acid is added to the growing polypeptide chain each time the mRNA advances through the ribosome. Once a polypeptide chain of reasonable size is assembled (about 30 amino acids), it begins to emerge from the base of the large subunit, as illustrated in Step 6. The large subunit contains a tunnel through which the elongating polypeptide emerges.

As we have seen, the role of the small subunit during elongation is to "decode" the codons in the mRNA, while the role of the large subunit is peptide-bond synthesis. The efficiency of the process is remarkably high: The observed error rate is only about $10^{-4}$. At this rate, an incorrect amino acid will occur only once in every 20 polypeptides of an average length of 500 amino acids! Elongation in *E. coli* proceeds at a rate of about 15 amino acids per second at 37°C.

## Termination

Termination, the third phase of translation, is depicted in Figure 15–8. The process is signaled by the presence of any one of the three possible triplet codons appearing in the A site: UAG, UAA, or UGA. These codons do not specify an amino acid, nor do they call for a tRNA in the A site. They are called **stop codons, termination codons**, or **nonsense codons**. Often, several consecutive codons are part of an mRNA. When one such termination codon is encountered, the finished polypeptide is still attached to the terminal tRNA at the P site, and the A site is empty. The termination codon signals the action of **GTP-dependent release factors**, which cleave the polypeptide chain from the terminal tRNA, releasing it from the translation complex (Step 1). Then, the tRNA is released from the ribosome, which then dissociates into its subunits (Step 2). If a termination codon should appear in the middle of an mRNA molecule as a result of mutation, the same process occurs, and the polypeptide chain is prematurely terminated.

## Polyribosomes

As elongation proceeds and the initial portion of an mRNA molecule has passed through the ribosome, this portion of mRNA is free to associate with another small subunit to form a second initiation complex. The process can be repeated several times with a single mRNA and results in what are called **polyribosomes**, or just **polysomes**.

After cells are gently lysed in the laboratory, polyribosomes can be isolated from them and analyzed. The photos in Figure 15–9 show these complexes as seen under the electron microscope. In Figure 15–9(a), you can see the thin lines of mRNA between the individual ribosomes. The micrograph in Figure 15–9(b) is even more remarkable, for it shows the polypeptide chains emerging from the ribosomes during translation. The formation of polysome complexes represents an efficient use of the components available for protein synthesis during a unit of time. Using the analogy of a tape and a tape recorder, in polysome complexes, we would thread and play one tape (mRNA) simultaneously through several recorders (the ribosomes). At any given moment, each recorder would be playing a different part of the song (the polypeptide being synthesized in each ribosome).

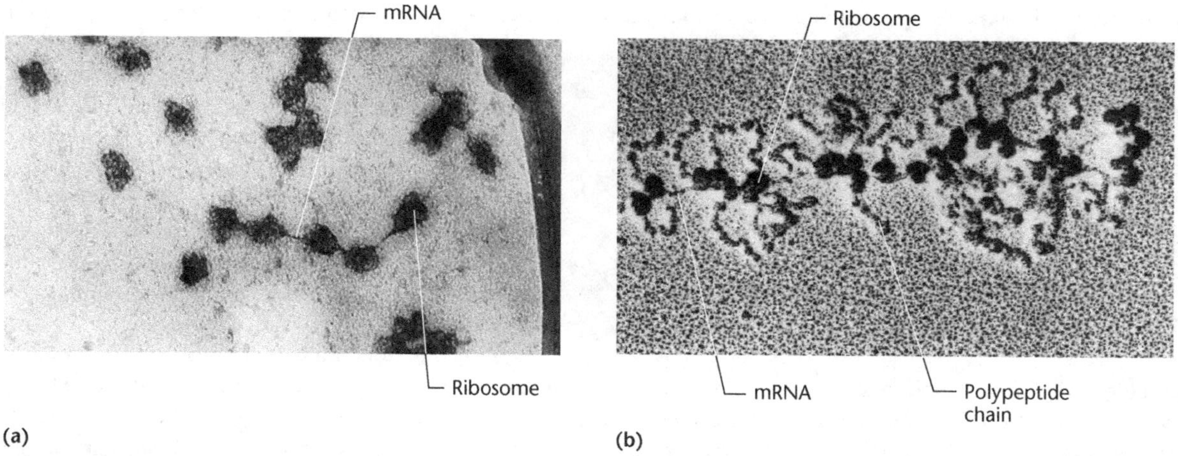


FIGURE 15-9 Polyribosomes as seen under the electron microscope. (a) Examples derived from rabbit reticulocytes engaged in the translation of hemoglobin mRNA. (b) Examples taken from salivary gland cells of the midgefly, *Chironomus thummi*. Note that the nascent polypeptide chains are apparent as they emerge from each ribosome. Their length increases as translation proceeds from left to right along the mRNA.

15.3

## Crystallographic Analysis Has Revealed Many Details about the Functional Prokaryotic Ribosome

Our knowledge of the process of translation and the structure of the ribosome, as described in the previous sections, is based primarily on biochemical and genetic observations, in addition to the visualization of ribosomes under the electron microscope. Because of the tremendous size and complexity of the functional ribosome during active translation, it was extremely difficult to obtain the crystals needed to perform X-ray diffraction studies. Nevertheless, great strides have been made in the past several years. First, the individual ribosomal subunits were crystallized and examined in several laboratories, most prominently that of V. Ramakrishnan. Then, in 2001, the crystal structure of the intact 70*S* ribosome, complete with associated mRNA and tRNAs, was examined by Harry Noller and colleagues. In essence, the entire translational complex was seen at the atomic level. Both Ramakrishnan and Noller derived the ribosomes from the bacterium *Thermus thermophilus*.

Many noteworthy observations have been made from these investigations. For example, the sizes and shapes of the subunits, measured at atomic dimensions, are in agreement with earlier estimates based on high-resolution electron microscopy. Furthermore, the shape of the ribosome changes during different functional states, attesting to the dynamic nature of the process of translation. A great deal has also been learned about the prominence and location of the RNA components of the subunits. For example, about one-third of the 16*S* RNA is responsible for producing a flat projection, referred to as the platform, within the smaller 30*S* subunit, and it modulates movement of the mRNA–tRNA complex during translocation. One of the models based on Noller's findings is shown in the opening photograph of this chapter (p. 381).

Crystallographic analysis also supports the concept that RNA is the real "player" in the ribosome during translation. The interface between the two subunits, considered to be the location in the ribosome where polymerization of amino acids occurs, is composed almost exclusively of RNA. In contrast, the numerous ribosomal proteins are found mostly on the periphery of the ribosome. These observations confirm what has been predicted on genetic grounds—the catalytic steps that join amino acids during translation occur under the direction of RNA, not proteins.

Another interesting finding involves the actual location of the three sites predicted to house tRNAs during translation. All three sites (A, P, and E), have been identified in X-ray diffraction studies, and in each case, the RNA of the ribosome makes direct contact with the various loops and domains of the tRNA molecule. This observation supports the hypotheses that had been developed concerning the roles of the different regions of tRNA and helps us understand why the distinctive three-dimensional conformation that is characteristic of all tRNA molecules has been preserved throughout evolution.

Still another noteworthy observation is that the intervals between the A, P, and E sites are at least 20 Å, and perhaps as much as 50 Å, wide, thus defining the atomic distance that the tRNA molecules must shift during each translocation event. This is considered a fairly large distance relative to the size of the tRNAs themselves. Further analysis has led to the identification of molecular (RNA–protein) bridges existing between the three sites and apparently involved in the translocation events. Other such bridges are present at other key locations and have been related to ribosome function. These observations provide us with a much more complete picture of the dynamic changes that must occur within the ribosome during translation. A final observation takes us back almost 50 years, to when Francis Crick proposed the **wobble hypothesis**, as introduced in Chapter 14. The Ramakrishnan group has identified the precise location along the 16*S* rRNA of the 30*S* subunit involved in the decoding step that connects mRNA to the proper tRNA. At this location, two particular nucleotides of the 16*S* rRNA actually flip out and probe the codon:anticodon region, and are believed to check for accuracy of base pairing during this interaction. According to the wobble hypothesis, the stringency of this step is high for the first two base pairs but less stringent for the third (or wobble) base pair.

Numerous questions about ribosome structure and function still remain. In particular, the role of the many ribosomal proteins is yet to be clarified. Nevertheless, the models that are emerging based on the work of Noller, Ramakrishnan, and their many colleagues provide us with a much better understanding of the mechanism of translation.

## 15.4 Translation Is More Complex in Eukaryotes

The general features of the model we just discussed were initially derived from investigations of the translation process in bacteria. As we saw, one main difference between translation in prokaryotes and eukaryotes is that in the latter, translation occurs on larger ribosomes whose rRNA and protein components are more complex than those of prokaryotes (see Figure 15–1). Another significant distinction is that whereas transcription and translation are coupled in prokaryotes, in eukaryotes these two processes are separated both spatially and temporally. In eukaryotic cells transcription occurs in the nucleus and translation in the cytoplasm. This separation provides multiple opportunities for regulation of genetic expression in eukaryotic cells.

Several other differences are also important. Eukaryotic mRNAs are much longer lived than are their prokaryotic counterparts. Most exist for hours rather than minutes prior to degradation by nucleases in the cell; thus they remain available much longer to orchestrate protein synthesis.

Several aspects of the initiation of translation are different in eukaryotes. First, as we discussed in our consideration of mRNA in eukaryotes, the 5′ end of mRNA is capped with a 7-methylguanosine (7-mG) residue at maturation. The presence of the 7-mG cap, absent in prokaryotes, is essential for efficient translation, because RNAs lacking the cap are translated poorly. In addition, most eukaryotic mRNAs contain a short recognition sequence that surrounds the initiating AUG codon—5′-ACCAUGG. Named after its discoverer Marilyn Kozak, this **Kozak sequence** functions during initiation by creating the proper context at the initiating AUG codon. If the Kozak sequence is missing, initiator tRNA does not select the AUG codon and continues scanning the mRNA until it finds another AUG that is accompanied by the Kozak sequence. Thus, the Kozak sequence is in a similar position to that of the Shine–Dalgarno sequence in prokaryotic mRNA. Although they work differently, both greatly facilitate the initial binding of mRNA to the small subunit of the ribosome.

Another difference is that the amino acid formylmethionine is not required for the initiation of eukaryotic translation. However, as in prokaryotes, the AUG triplet, which encodes methionine, is essential to the formation of the translational complex, and a unique transfer RNA ($tRNA_i^{met}$) is used during initiation.

Protein factors similar to those in prokaryotes guide the initiation, elongation, and termination of translation in eukaryotes. Many of these eukaryotic factors are clearly homologous to their counterparts in prokaryotes. However, a greater number of factors are usually required during each step, and some are more complex than in prokaryotes.

Finally, recall that in eukaryotes a great many of the cell's ribosomes are found in association with the membranes that make up the endoplasmic reticulum (forming the rough ER). Such membranes are absent from the cytoplasm of prokaryotic cells. This association in eukaryotes facilitates the secretion of newly synthesized proteins from the ribosomes directly into the channels of the endoplasmic reticulum. Recent studies using cryoelectron microscopy have established how this occurs. A tunnel in the large subunit of the ribosome begins near the point where the two subunits interface and exits near the back of the large subunit. The location of the tunnel within the large subunit is the basis for the belief that it provides the conduit for the movement of the newly synthesized polypeptide chain out of the ribosome. In studies in yeast, newly synthesized polypeptides enter the ER through a membrane channel formed by a specific protein, Sec61. This channel is perfectly aligned with the exit point of the ribosomal tunnel. In prokaryotes, the polypeptides are released by the ribosome directly into the cytoplasm.

## 15.5 The Initial Insight That Proteins Are Important in Heredity Was Provided by the Study of Inborn Errors of Metabolism

Now, let's consider how we know that proteins are the end products of genetic expression. The first insight into the role of proteins in genetic processes was provided by observations made by Sir Archibald Garrod and William Bateson early in the twentieth century. Garrod was born into an English family of medical scientists. His father was a physician with a strong interest in the chemical basis of rheumatoid arthritis, and his eldest brother was a leading zoologist in London. As a practicing physician, Garrod himself became interested in several human disorders that seemed to be inherited. Although he

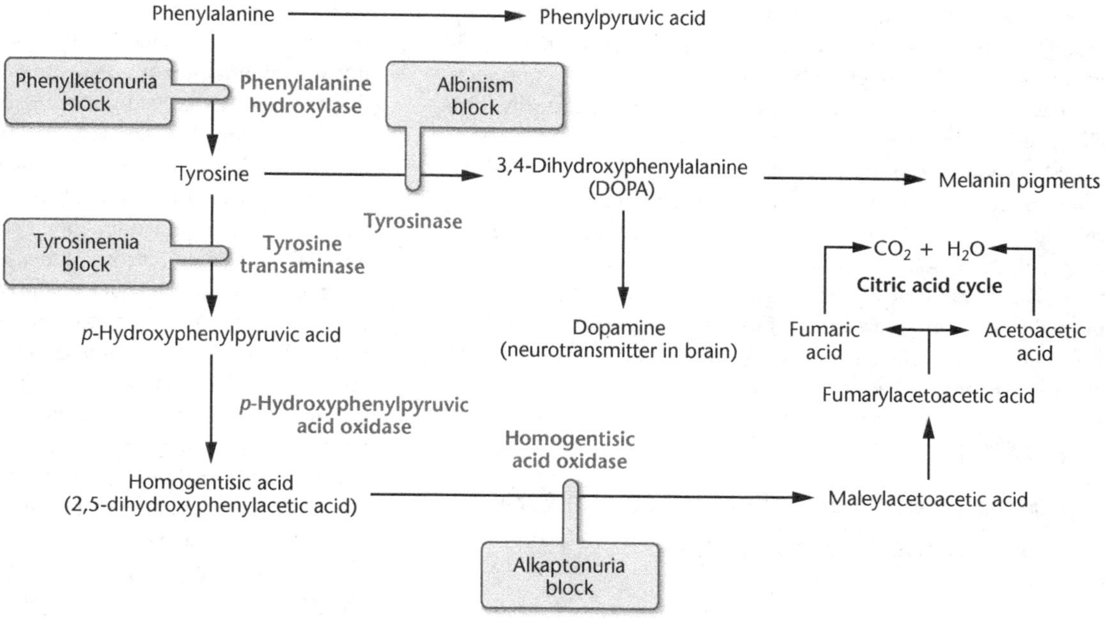


**FIGURE 15–10** Metabolic pathway involving phenylalanine and tyrosine. Various metabolic blocks resulting from mutations lead to the disorders phenylketonuria, alkaptonuria, albinism, and tyrosinemia.

also studied albinism and cystinuria, we will describe his investigation of the disorder **alkaptonuria**. Individuals afflicted with this disorder have a disruption in an important metabolic pathway (Figure 15–10). As a result, they cannot metabolize the alkapton 2,5-dihydroxyphenylacetic acid, also known as homogentisic acid. Homogentisic acid thus accumulates in their cells and tissues and is excreted in the urine. The molecule's oxidation products are black and easily detectable in the diapers of newborns. The unmetabolized products tend to accumulate in cartilaginous areas, causing a darkening of the ears and nose. In joints, this deposition leads to a benign arthritic condition. Alkaptonuria is a rare but not serious disease that persists throughout an individual's life.

Garrod studied alkaptonuria by looking for patterns of inheritance of this benign trait. Eventually he concluded that it was genetic in nature. Of 32 known cases, he ascertained that 19 were confined to 7 families, with one family having four affected siblings. In several instances, the parents were unaffected but known to be related as first cousins, and therefore **consanguine**, a term describing relatives having a common recent ancestor. Parents who are so related have a higher probability than unrelated parents of producing offspring that express recessive traits, because such parents are both more likely to be heterozygous for some of the same recessive traits (see Chapter 27). Garrod concluded that this inherited condition was the result of an alternative mode of metabolism, thus implying that hereditary information controls chemical reactions in the body. While *genes* and *enzymes* were not familiar terms during Garrod's time, he used the corresponding concepts of *unit factors* and *ferments*. Garrod published his initial observations in 1902.

Only a few geneticists, including Bateson, were familiar with and made reference to Garrod's work. Garrod's ideas fit nicely with Bateson's belief that inherited conditions were caused by the lack of some critical substance. In 1909, Bateson published *Mendel's Principles of Heredity*, in which he linked ferments with heredity. However, for almost 30 years, most geneticists failed to see the relationship between genes and enzymes. Garrod and Bateson, like Mendel, were ahead of their time.

## Phenylketonuria

The inherited human metabolic disorder **phenylketonuria (PKU)** results when another reaction in the pathway shown in Figure 15–10 is blocked. Described first in 1934, this disorder can result in mental retardation and is transmitted as an autosomal recessive disease. Afflicted individuals are unable to convert the amino acid phenylalanine to the amino acid tyrosine. These molecules differ by only a single hydroxyl group (OH), present in tyrosine but absent in phenylalanine. The reaction is catalyzed by the enzyme **phenylalanine hydroxylase**, which is inactive in affected individuals and active at a level of about 30 percent in heterozygotes. The enzyme functions in the liver. While the normal blood level of phenylalanine is about 1 mg/100 ml, people with phenylketonuria show a level as high as 50 mg/100 mL.

As phenylalanine accumulates, it may be converted to phenylpyruvic acid and, subsequently, to other derivatives. These are less efficiently resorbed by the kidney and tend to spill into the urine more quickly than phenylalanine. Both phenylalanine and its derivatives enter the cerebrospinal fluid, resulting in elevated levels in the brain. The presence of these substances during early development is thought to cause mental retardation.

Phenylketonuria occurs in approximately 1 in 11,000 births, and newborns are routinely screened for PKU throughout the

United States. When the condition is detected in the analysis of an infant's blood, a strict dietary regimen is instituted in time to prevent retardation. A low-phenylalanine diet can reduce by-products such as phenylpyruvic acid, and the development of abnormalities characterizing the disease can be diminished.

Our knowledge of inherited metabolic disorders such as alkaptonuria and phenylketonuria has caused a revolution in medical thinking and practice. Human disease, once thought to be solely attributable to the action of invading microorganisms, viruses, or parasites, clearly can have a genetic basis. We know now that hundreds of medical conditions are caused by errors in metabolism resulting from mutant genes. These human biochemical disorders include all classes of organic biomolecules.

## 15.6 Studies of *Neurospora* Led to the One-Gene:One-Enzyme Hypothesis

In two separate investigations beginning in 1933, George Beadle provided the first convincing experimental evidence that genes are directly responsible for the synthesis of enzymes. The first investigation, conducted in collaboration with Boris Ephrussi, involved *Drosophila* eye pigments. Together, Beadle and Ephrussi confirmed that mutant genes that altered the eye color of fruit flies could be linked to biochemical errors that, in all likelihood, involved the loss of enzyme function. Encouraged by these findings, Beadle then joined with Edward Tatum to investigate nutritional mutations in the pink bread mold *Neurospora crassa*. This investigation led to the **one-gene:one-enzyme hypothesis.**

### Analysis of *Neurospora* Mutants by Beadle and Tatum

In the early 1940s, Beadle and Tatum chose to work with *Neurospora* because much was known about its biochemistry, and mutations could be induced and isolated with relative ease. By inducing mutations, they produced strains that had genetic blocks of reactions essential to the growth of the organism.

Beadle and Tatum knew that this mold could manufacture nearly every biomolecule necessary for normal development. For example, using rudimentary carbon and nitrogen sources, the organism can synthesize nine water-soluble vitamins, 20 amino acids, numerous carotenoid pigments, and all essential purines and pyrimidines. Beadle and Tatum irradiated asexual conidia (spores) with X rays to increase the frequency of mutations and then grew the spores on "complete" medium containing all the necessary growth factors (e.g., vitamins and amino acids). Under such growth conditions, a mutant strain unable to grow on minimal medium would be able to grow by ingesting the supplements present in the enriched, complete medium. All the cultures were then transferred to minimal medium. Any organisms capable of growing on the minimal medium must be able to synthesize all the necessary growth factors themselves, and the researchers could conclude that the cultures from which those organisms came did not contain a nutritional mutation. If no growth occurred, then it was concluded that the culture that had not been able to grow contained a nutritional mutation. The next task was to determine the type of nutritional mutation. The results are shown in Figure 15–11(a).

Many thousands of individual spores derived by this procedure were isolated and grown on complete medium. In subsequent tests on minimal medium, many cultures failed to grow, indicating that a nutritional mutation had been induced. To identify the mutant type, the mutant strains were tested on a series of different incomplete media [Figure 15–11(b) and 15–11(c)], each containing different groups of supplements, and subsequently on media containing single vitamins, amino acids, purines, or pyrimidines as supplements, until one specific supplement that permitted growth was found. Beadle and Tatum reasoned that the supplement that restored growth was the molecule that the mutant strain could not synthesize.

The first mutant strain isolated required vitamin $B_6$ (pyridoxine) in the medium, and the second one required vitamin $B_1$ (thiamine). Using the same procedure, Beadle and Tatum eventually isolated and studied hundreds of mutants deficient in the ability to synthesize other vitamins, amino acids, nucleotides, or other substances.

The findings derived from testing more than 80,000 spores convinced Beadle and Tatum that genetics and biochemistry have much in common. It seemed likely that each nutritional mutation caused the loss of the enzymatic activity that facilitates an essential reaction in wild-type organisms. It also appeared that a mutation could be found for nearly any enzymatically controlled reaction. Beadle and Tatum had thus provided sound experimental evidence for the hypothesis that one gene specifies one enzyme, an idea alluded to more than 30 years earlier by Garrod and Bateson. With modifications, this concept was to become another major principle of genetics.

### Genes and Enzymes: Analysis of Biochemical Pathways

The one-gene:one-enzyme concept and its attendant research methods have been used over the years to work out many details of metabolism in *Neurospora, Escherichia coli*, and a number of other microorganisms. One of the first metabolic pathways to be investigated in detail was that leading to the synthesis of the amino acid arginine in *Neurospora*. By studying seven mutant strains, each requiring arginine for growth ($arg^-$), Adrian Srb and Norman Horowitz ascertained a partial biochemical pathway that leads to the synthesis of the amino acid. Their work demonstrates how genetic analysis can be used to establish biochemical information.

Srb and Horowitz tested each mutant strain's ability to reestablish growth if either citrulline or ornithine, two compounds with close chemical similarity to arginine, was used as a supplement to minimal medium. If either was able to substitute for arginine, they reasoned that it must be involved in the biosynthetic pathway of arginine. The researchers found that both molecules could be substituted in one or more strains.

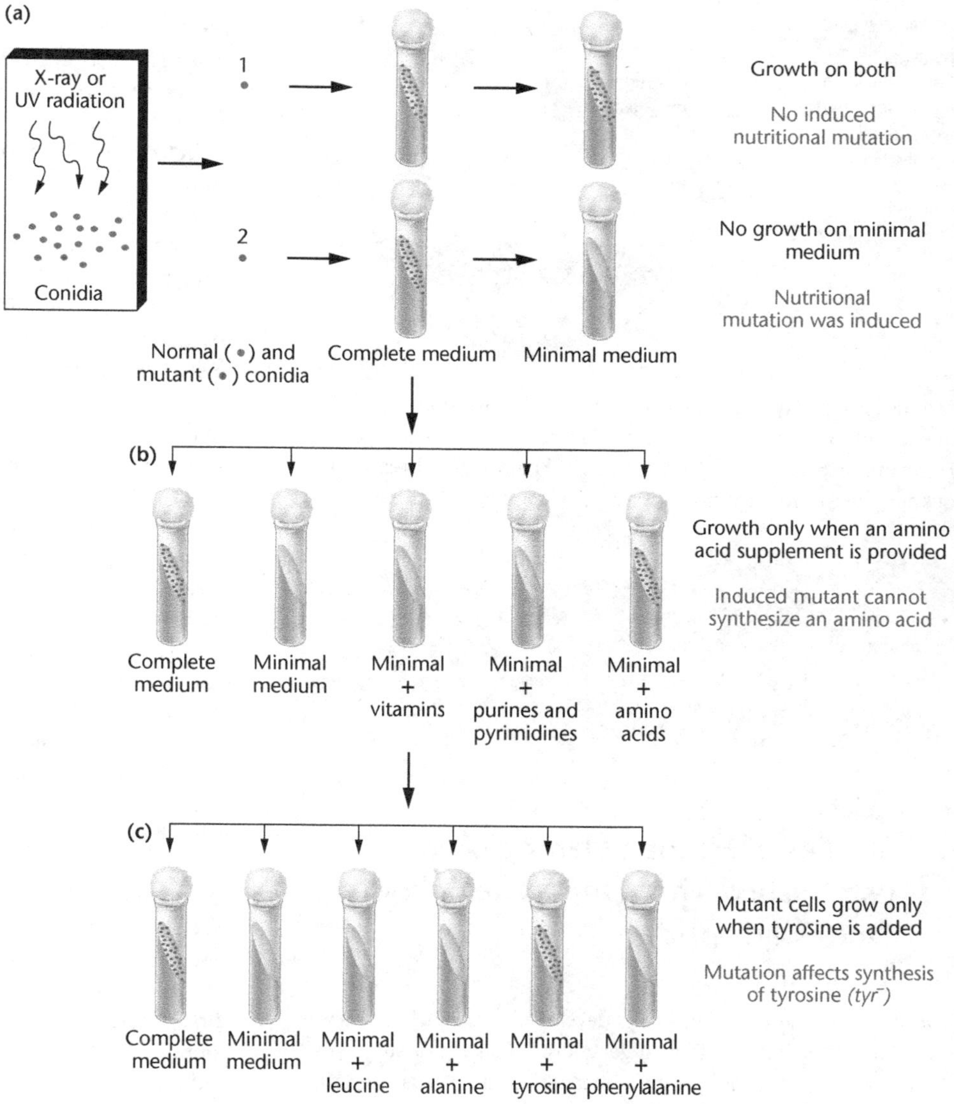


**FIGURE 15–11** Induction, isolation, and characterization of a nutritional auxotrophic mutation in *Neurospora*. (a) Most conidia are not affected, but one conidium (shown in red) contains a mutation. (b and c) The precise nature of the mutation is established and found to involve the biosynthesis of tyrosine.

Of the seven mutant strains, four of them (*arg 4* through *arg 7*) grew if supplied with either citrulline, ornithine, or arginine. Two of them (*arg 2* and *arg 3*) grew if supplied with citrulline or arginine. One strain (*arg 1*) would grow only if arginine were supplied; neither citrulline nor ornithine could substitute for it. From these experimental observations, the following pathway and metabolic blocks for each mutation were deduced:

$$\text{precursor} \xrightarrow[\text{Enzyme A}]{arg\ 4-7} \text{Ornithine} \xrightarrow[\text{Enzyme B}]{arg\ 2-3} \text{Citrulline} \xrightarrow[\text{Enzyme C}]{arg\ 1} \text{Arginine}$$

The logic supporting these conclusions is as follows: If mutants *arg 4* through *arg 7* can grow regardless of which of the three molecules is supplied as a supplement to minimal medium, the mutations preventing growth must cause a metabolic block that occurs *prior* to the involvement of ornithine, citrulline, or arginine in the pathway. When any one of these three molecules is added, *its presence bypasses the block*. As a result, both citrulline and ornithine appear to be involved in the biosynthesis of arginine. However, the sequence of their participation in the pathway cannot be determined on the basis of these data.

On the other hand, both the *arg 2* and the *arg 3* mutations grow if supplied citrulline, but not if they are supplied with only ornithine. Therefore, ornithine must be synthesized in the pathway *prior to the block*. Its presence will not overcome the block. Citrulline, however, does *overcome the block*, so it must be synthesized beyond the point of blockage. Therefore, the conversion of ornithine to citrulline represents the correct sequence in the pathway.

Finally, we can conclude that *arg 1* represents a mutation preventing the conversion of citrulline to arginine. Neither ornithine nor citrulline can overcome the metabolic block in this mutation because both participate earlier in the pathway.

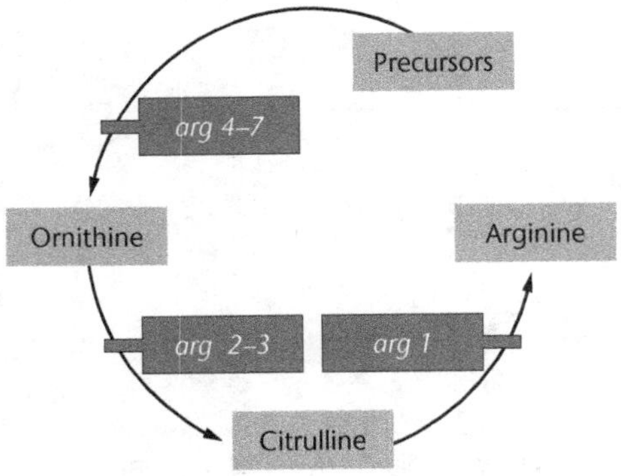


**FIGURE 15–12** Abbreviated pathway describing the biosynthesis of arginine in *Neurospora*.

Taken together, the analysis as described above supports the sequence of biosynthesis shown in Figure 15–12. Since Srb and Horowitz's experiments in 1944, the detailed pathway has been worked out and the enzymes controlling each step have been characterized.

**NOW SOLVE THIS**

Problem 14 on page 406 asks you to analyze data to establish a biochemical pathway in the bacterium *Salmonella*.

■ HINT: *Apply the same principles and approach that were used to decipher biochemical pathways in* Neurospora.

## 15.7 Studies of Human Hemoglobin Established That One Gene Encodes One Polypeptide

The one-gene:one-enzyme concept developed in the early 1940s was not immediately accepted by all geneticists. This is not surprising, since it was not yet clear how mutant enzymes could cause variation in the many different kinds of phenotypic traits. For example, *Drosophila* mutants demonstrated altered eye size, wing shape, wing vein pattern, and so on. Plants exhibited mutant varieties of seed texture, height, and fruit size. How an inactive mutant enzyme could result in such phenotypes was puzzling to many geneticists.

Two factors soon modified the one-gene:one-enzyme hypothesis. First, while *nearly all enzymes are proteins, not all proteins are enzymes.* As the study of biochemical genetics proceeded, it became clear that all proteins are specified by the information stored in genes, leading to the more accurate phraseology **one-gene:one-protein hypothesis.** Second, proteins often have a subunit structure consisting of two or more polypeptide chains. This is the basis of the quaternary structure of proteins, which we will discuss later in the chapter. Because each distinct polypeptide chain is encoded by a separate gene, a more modern statement of Beadle and Tatum's basic principle is **one-gene:one-polypeptide chain hypothesis.** The need for these modifications of the original hypothesis became apparent during the analysis of hemoglobin structure in individuals afflicted with sickle-cell anemia.

### Sickle-Cell Anemia

The first direct evidence that genes specify proteins other than enzymes came from the work on mutant hemoglobin molecules derived from humans afflicted with the disorder **sickle-cell anemia.** Affected individuals have erythrocytes that, under low oxygen tension, become elongated and curved because of the polymerization of hemoglobin. The sickle shape of these erythrocytes is in contrast to the biconcave disc shape characteristic in unaffected individuals (Figure 15–13). Those with the disease suffer attacks when red blood cells aggregate in the venous side of capillary systems, where oxygen tension is very low. As a result, a variety of tissues are deprived of oxygen and suffer severe damage. When this occurs, an individual is said to experience a sickle-cell crisis. If untreated, a crisis may be fatal. The kidneys, muscles, joints, brain, gastrointestinal tract, and lungs can be affected.

In addition to suffering crises, these individuals are anemic because their erythrocytes are destroyed more rapidly than normal red blood cells. Compensatory physiological mechanisms include increased red-cell production by bone marrow and accentuated heart action. These mechanisms lead to abnormal bone size and shape as well as dilation of the heart.

In 1949, James Neel and E. A. Beet demonstrated that the disease is inherited as a Mendelian trait. Pedigree analysis revealed three genotypes and phenotypes controlled by a single pair of alleles, $Hb^A$ and $Hb^S$. Unaffected and affected individuals result from the homozygous genotypes $Hb^A/Hb^A$ and $Hb^S/Hb^S$, respectively. The red blood cells of heterozygotes, who exhibit the **sickle-cell trait** but not the disease, undergo much less sickling because more than half of their hemoglobin is normal. Though largely unaffected, such heterozygotes are "carriers" of the defective gene, which is transmitted on average to 50 percent of their offspring.

In the same year, Linus Pauling and his coworkers provided the first insight into the molecular basis of sickle-cell anemia. They showed that hemoglobins isolated from diseased and normal individuals differ in their rates of electrophoretic migration. In electrophoresis (described in Chapter 10 and Appendix A), charged molecules migrate in an electric field. If the net charge of two molecules is different, their rates of migration will be different. Hence, Pauling and his colleagues concluded that a chemical difference exists between normal and sickle-cell hemoglobin. The two molecules are now designated **HbA** and **HbS**, respectively.

Figure 15–14(a) illustrates the migration pattern of hemoglobin derived from individuals of all three possible genotypes when subjected to **starch gel electrophoresis.** The gel provides the supporting medium for the molecules during migration. In this experiment, samples were placed at a point of origin between the cathode (−) and the anode (+), and an electric current was applied. The migration pattern revealed that all molecules moved toward the anode, indicating a net negative charge. However, HbA migrated farther than HbS, suggesting that its net negative charge was greater. The electrophoretic pattern of hemoglobin derived from individuals who were carriers revealed the presence of both HbA and HbS, and confirmed their heterozygous genotype.

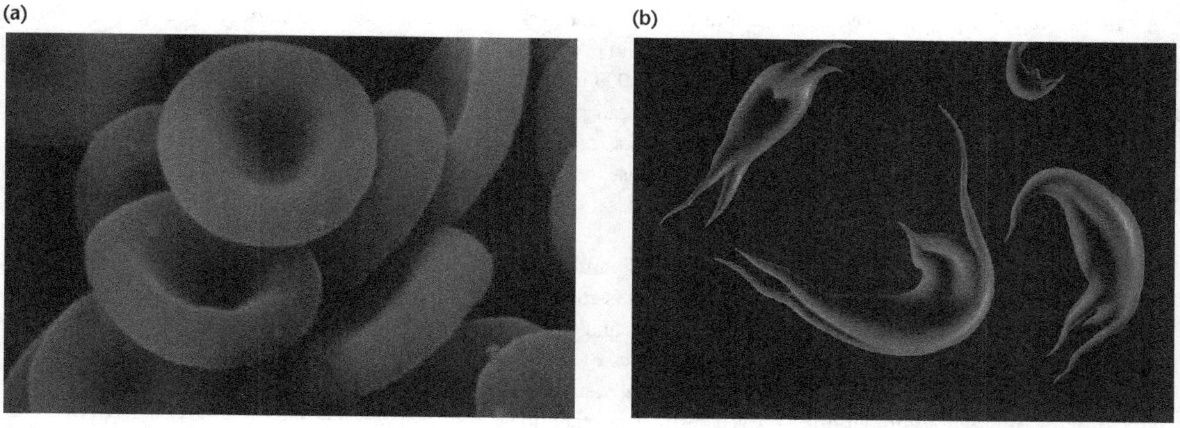


**FIGURE 15–13** A comparison of erythrocytes from (a) healthy individuals and (b) those with sickle-cell anemia.

Pauling's findings suggested two possibilities. It was known that hemoglobin consists of four nonproteinaceous, iron-containing *heme groups* and a *globin portion* that contains four polypeptide chains. The alteration in net charge in HbS had to be due, theoretically, to a chemical change in one of these components.

Work carried out between 1954 and 1957 by Vernon Ingram resolved this question. He demonstrated that the chemical change occurs in the primary structure of the globin portion of the hemoglobin molecule. Using the **fingerprinting technique** shown in Figure 15–14(b), Ingram showed that HbS differs in amino acid composition compared to HbA. Human adult hemoglobin contains two identical $\alpha$ chains of 141 amino acids and two identical $\beta$ chains of 146 amino acids in its quaternary structure.

The fingerprinting technique involves the enzymatic digestion of the protein into peptide fragments. The mixture is then placed on absorbent paper and exposed to an electric field, where migration occurs according to net charge. The paper is then turned at a right angle to its first exposure and placed in a solvent, in which chromatographic action causes the migration of the peptides in the second direction. The end result is a two-dimensional separation of the peptide fragments into a distinctive pattern of spots, or a "fingerprint." Ingram's work revealed that HbS and HbA differed by only a single peptide fragment [Figure 15–14(b)]. Further analysis then revealed a single amino acid change: valine was substituted for glutamic acid at the sixth position of the $\beta$ chain, accounting for the peptide difference [Figure 15–14(c)].

The significance of this discovery has been multifaceted. It clearly establishes that a single gene provides the genetic information for a single polypeptide chain. Studies of HbS also demonstrate that a mutation can affect the phenotype by directing a single amino acid substitution. Also, by providing the explanation for sickle-cell anemia, the concept of inherited **molecular disease** was firmly established. Finally, this work led to a thorough study of human hemoglobins, which has provided valuable genetic insights.

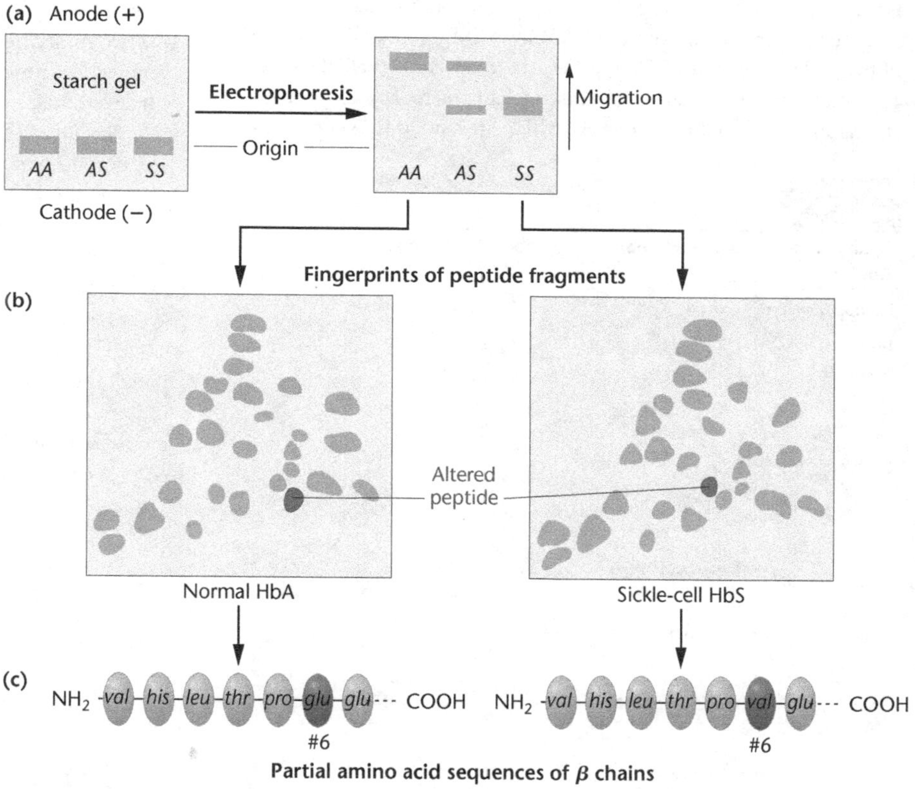


**FIGURE 15–14** Investigation of hemoglobin derived from $Hb^AHb^A$, $Hb^AHb^S$, and $Hb^SHb^S$ individuals using electrophoresis, fingerprinting, and amino acid analysis. Hemoglobin from individuals with sickle-cell anemia ($Hb^SHb^S$) (a) migrates differently in an electrophoretic field, (b) shows an altered peptide in fingerprint analysis, and (c) shows an altered amino acid, valine, at the sixth position in the $\beta$ chain. During electrophoresis, heterozygotes ($Hb^AHb^S$) are shown to have both forms of hemoglobin.

In the United States, sickle-cell anemia is found almost exclusively in the African-American population. It affects about one in every 625 African-American infants. Currently, about 50,000 to 75,000 individuals are afflicted. In about 1 of every 145 African-American married couples, both partners are heterozygous carriers. In these cases, each of their children has a 25 percent chance of having the disease.

## Human Hemoglobins

Having introduced human hemoglobins in a historical context, we now end the discussion by providing an update of what is currently known about these molecules in our species. Molecular analysis reveals that an individual human produces different types of hemoglobin molecules at different stages of the life cycle. All are tetramers consisting of different combinations of seven distinct polypeptide chains, each encoded by a separate gene. The expression of these various genes is developmentally regulated.

Almost all adult hemoglobin consists of **HbA**, which contains two **$\alpha$ and two $\beta$ chains.** Recall that the mutation in sickle-cell anemia involves the $\beta$ chain. HbA represents about 98 percent of all hemoglobin found in an adult's erythrocytes after the age of six months. The remaining 2 percent consists of **$HbA_2$**, a minor adult component. This molecule contains two $\alpha$ chains and two **delta ($\delta$) chains.** The $\delta$ chain is very similar to the $\beta$ chain, consisting of 146 amino acids.

During embryonic and fetal development, quite a different set of hemoglobins is found. The earliest set to develop is called **Gower 1**, containing two **zeta ($\zeta$) chains**, which are most similar to $\alpha$ chains, and two **epsilon ($\epsilon$) chains**, which are similar to $\beta$ chains. By eight weeks' gestation, the embryonic form is gradually replaced by still another hemoglobin molecule with still different chains. This molecule is called **HbF**, or **fetal hemoglobin**, and consists of two $\alpha$ chains and two **gamma ($\gamma$) chains.** There are two types of $\gamma$ chains, designated $^{G}\gamma$ and $^{A}\gamma$. Both are similar to $\beta$ chains and differ from each other by only a single amino acid. These persist until birth, after which HbF is, again gradually, replaced with HbA and $HbA_2$. The nomenclature and sequence of appearance of the five tetramers we have described are summarized in Table 15.2.

TABLE 15.2

**Chain Compositions of Human Hemoglobins from Conception to Adulthood**

| Hemoglobin Type | Chain Composition |
|---|---|
| Embryonic-Gower 1 | $\zeta_2 \epsilon_2$ |
| Fetal-HbF | $\alpha_2\ {}^{G}\gamma_2$ |
| | $\alpha_2\ {}^{A}\gamma_2$ |
| Adult-HbA | $\alpha_2 \beta_2$ |
| Minor adult-$HbA_2$ | $\alpha_2 \delta_2$ |

15.8

## The Nucleotide Sequence of a Gene and the Amino Acid Sequence of the Corresponding Protein Exhibit Colinearity

Once it was established that genes specify the synthesis of polypeptide chains, the next logical question was, how can the genetic information contained in the nucleotide sequence of a gene be transferred to the amino acid sequence of a polypeptide chain? It seemed most likely that a colinear relationship would exist between the two molecules. That is, the order of nucleotides in the DNA of a gene would correlate directly with the order of amino acids in the corresponding polypeptide—the concept of **colinearity.**

The initial experimental evidence in support of this concept was derived from studies by Charles Yanofsky of the *trpA* gene that encodes the A subunit of the enzyme **tryptophan synthetase** in *E. coli.* Yanofsky isolated many independent mutants that had lost the activity of the enzyme. He then mapped the various mutations, establishing their location with respect to one another within the gene. He also determined where the amino acid substitution had occurred in each mutant protein. When the two sets of data were compared, the colinear relationship was apparent. The location of each mutation in the *trpA* gene correlated with the position of the altered amino acid in the A polypeptide of tryptophan synthetase. This comparison is illustrated in Figure 15–15.

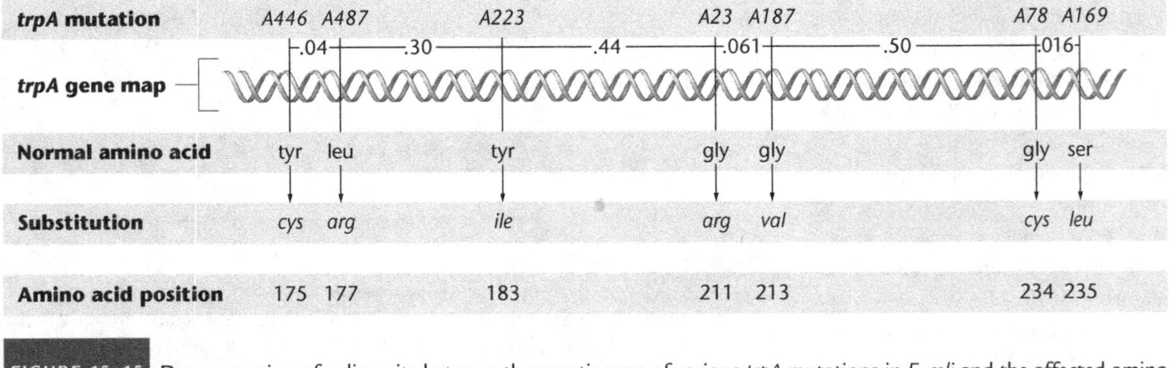


FIGURE 15–15 Demonstration of colinearity between the genetic map of various *trpA* mutations in *E. coli* and the affected amino acids in the protein product. The numbers shown between mutations represent linkage distances.

15.9

## Variation in Protein Structure Provides the Basis of Biological Diversity

In contrast to nucleic acids, which store and express genetic information, proteins, as end products of genetic expression, are more closely aligned with biological function. It is the variation in biological function that provides the basis of diversity between cell types and between organisms. What is it about proteins that enables them to perform or control enormous numbers of complex and important cellular activities in an organism? As we will see, the secret of the complexity of protein function lies in the incredible structural diversity of proteins.

At the outset of our discussion, we should differentiate between **polypeptides** and **proteins.** Polypeptides, most simply, are precursors of proteins. Thus, as the amino acid polymer is assembled on and then released from the ribosome during translation, it is called a *polypeptide.* Once a polypeptide subsequently folds up and assumes a functional three-dimensional conformation, it is called a *protein.* In most cases, several polypeptides combine during this process to produce an even higher order of protein structure. It is its three-dimensional conformation in space that is essential to a protein's specific function and that distinguishes it from other proteins.

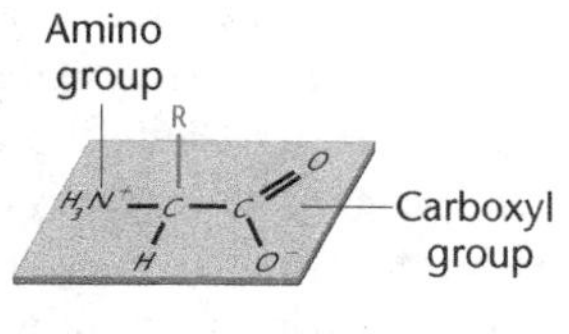

**Amino acid structure**

Like nucleic acids, the polypeptide chains comprising proteins are linear, nonbranched polymers. There are 20 amino acids that serve as the subunits (the building blocks) of proteins. Each amino acid has a **carboxyl group**, an **amino group**, and a **radical (R) group** (a side chain that determines the type of amino acid) bound covalently to a **central carbon (C) atom.** Figure 15–16 shows the 20 R groups that define the 20 amino acids in proteins. The R groups are varied in structure and can be divided into four main classes: (1) **nonpolar (hydrophobic)**, (2) **polar (hydrophilic)**, (3) **positively charged**, and (4) **negatively charged.** Because polypeptides are often long polymers, and because each unit in the polymer may be any 1 of 20 amino acids, each with unique chemical properties, enormous variation in the molecule's final conformation and chemical activity is possible. For

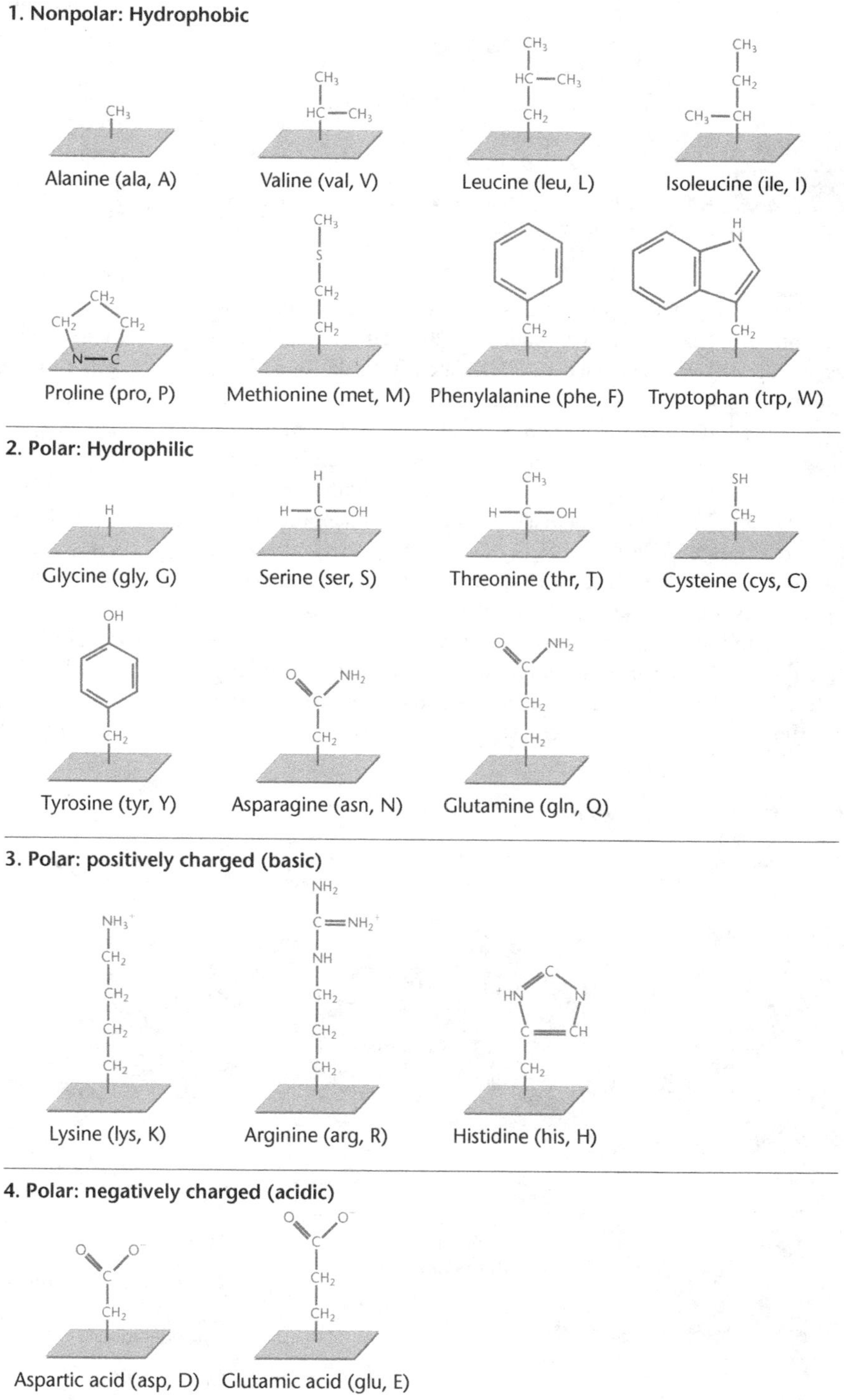


**FIGURE 15–16** Chemical structures and designations of the 20 amino acids encoded by living organisms, divided into four major categories. Each amino acid has two abbreviations in universal use; for example, alanine is designated either ala or A.

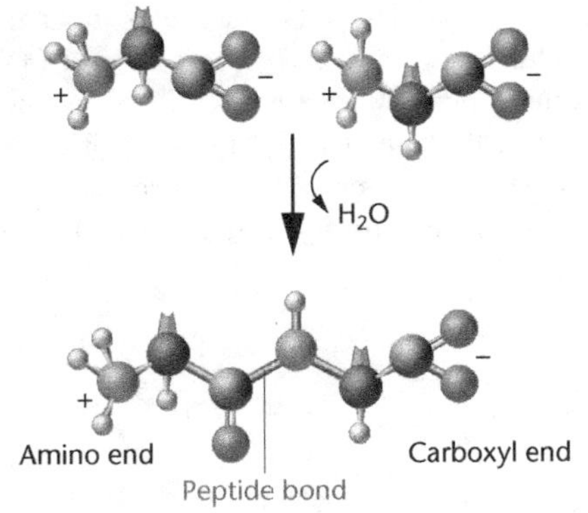


FIGURE 15–17 Peptide bond formation between two amino acids, resulting from a dehydration reaction.

example, if an average polypeptide is composed of 200 amino acids (a molecular weight of about 20,000 Da), $20^{200}$ different molecules, each with a unique sequence, can be created from the 20 different building blocks.

Around 1900, German chemist Emil Fischer determined the manner in which the amino acids are bonded together. He showed that the amino group of one amino acid can react with the carboxyl group of another amino acid in a dehydration (condensation) reaction, releasing a molecule of $H_2O$. The resulting covalent bond is called a **peptide bond** (Figure 15–17). Two amino acids linked together constitute a *dipeptide*, three a *tripeptide*, and so on. Once 10 or more amino acids are linked by peptide bonds, the chain is referred to as a *polypeptide*. Generally, no matter how long a polypeptide is, it will have an amino group at one end (the N-terminus) and a carboxyl group at the other end (the C-terminus).

Four levels of protein structure are recognized: primary, secondary, tertiary, and quaternary. The sequence of amino acids in the linear backbone of the polypeptide constitutes its **primary structure.** This sequence is specified by the sequence of deoxyribonucleotides in DNA through an mRNA intermediate. The primary structure of a polypeptide helps determine the specific characteristics of the higher orders of organization as a protein is formed.

**Secondary structures** are certain regular or repeating configurations in space assumed by amino acids lying close to one another in the polypeptide chain. In 1951, Linus Pauling and Robert Corey predicted, on theoretical grounds, an ***α* helix** as one type of secondary structure. The α-helix model [Figure 15–18(a)] has since been confirmed by X-ray crystallographic studies. It is rodlike and has the greatest possible theoretical stability. The helix is composed of a spiral chain of amino acids stabilized by hydrogen bonds.

The side chains (the R groups) of amino acids extend outward from the helix, and each amino acid residue occupies a distance of 1.5 Å in the length of the helix. There are 3.6 residues per turn. Although left-handed helices are theoretically possible, all α helices seen in proteins are right-handed.

Also in 1951, Pauling and Corey proposed another secondary structure, the ***β*-pleated sheet**. In this model, a single polypeptide chain folds back on itself, or several chains run in either parallel or antiparallel fashion next to one another. Each such structure is stabilized by hydrogen bonds formed between certain atoms on adjacent chains [Figure 15–18(b)]. In the zigzagging plane formation that results, amino acids in adjacent rows are 3.5 Å apart.

As a general rule, most proteins exhibit a mixture of α-helix and *β*-pleated sheet structures. Globular proteins, most of which are round in shape and water soluble, usually contain a *β*-pleated sheet structure at their core, as well as many areas with *α* helices. The more rigid structural proteins, many of which are water insoluble,

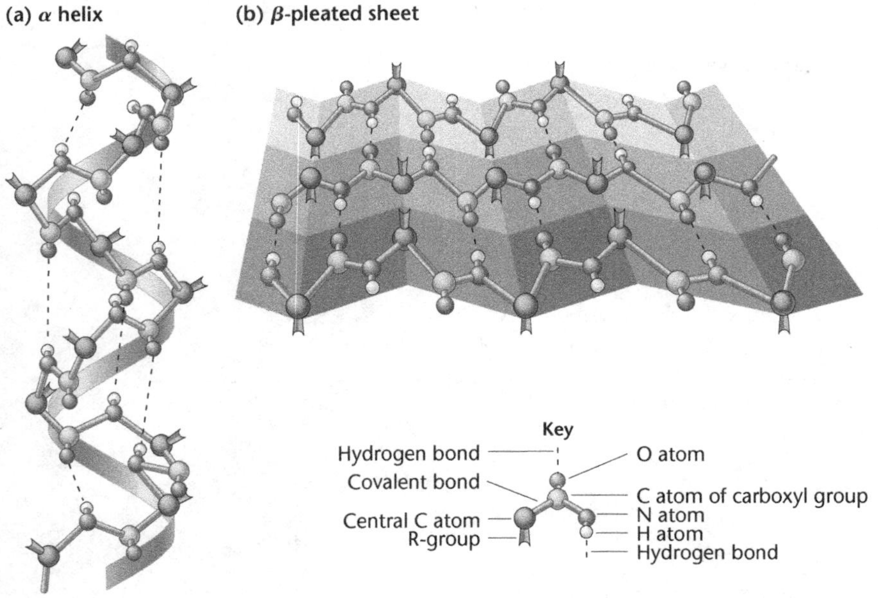


FIGURE 15–18 (a) The right-handed *α* helix, which represents one form of secondary structure of a polypeptide chain. (b) The *β*-pleated sheet, an alternative form of secondary structure of polypeptide chains. To maintain clarity, not all atoms are shown.

rely on more extensive $\beta$-pleated sheet regions for their rigidity. For example, **fibroin**, the protein made by the silk moth, depends extensively on this form of secondary structure.

The secondary structure describes the folding and interactions of amino acids in certain parts of a polypeptide chain, but the **tertiary structure** defines the three-dimensional spatial conformation of the chain as a whole. Each polypeptide twists and turns and loops around itself in a very specific fashion, characteristic of the particular protein. A model of a tertiary structure is shown in Figure 15–19. At this level of structure, three factors are most important in determining the conformation and in stabilizing the molecule:

1. Covalent disulfide bonds form between closely aligned cysteine residues to form the unique amino acid cystine.
2. Usually, the polar hydrophilic R groups are located on the surface of the configuration, where they can interact with water.
3. The nonpolar hydrophobic R groups are usually located on the inside of the molecule, where they interact with one another, avoiding interaction with water.

It is important to emphasize that the three-dimensional conformation achieved by any protein is a product of the *primary structure* of the polypeptide Thus, the genetic code need only specify the sequence of amino acids in order ultimately to produce the final configuration of proteins. The effects of the three stabilizing factors depend on the location of each amino acid relative to all others in the chain. As folding occurs, the most thermodynamically stable conformation possible results. This level of organization is extremely important because the specific function of any protein is directly related to its tertiary structure.

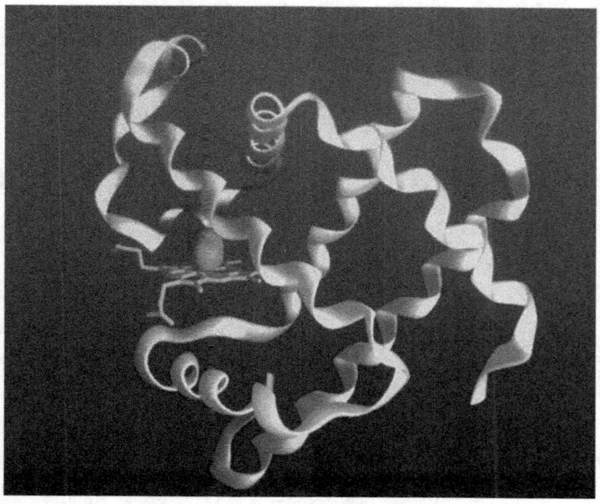

FIGURE 15–19 The tertiary level of protein structure for the respiratory pigment myoglobin. The bound oxygen atom is shown in red.

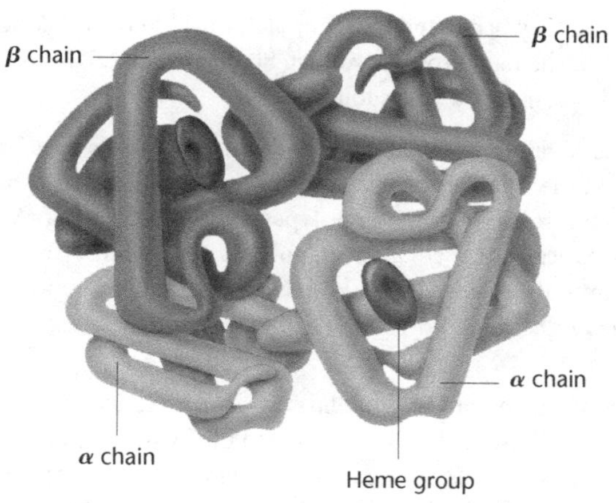


FIGURE 15–20 The quaternary level of protein structure as seen in hemoglobin. Four chains (two $\alpha$ and two $\beta$) interact with four heme groups to form the functional molecule.

The concept of **quaternary structure** of proteins applies only to those composed of more than one polypeptide chain and refers to the position of the various chains in relation to one another. This type of protein is *oligomeric*, and each chain in it is a *protomer*, or, less formally, a *subunit*. Protomers have conformations that facilitate their fitting together in a specific complementary fashion. Hemoglobin, an oligomeric protein consisting of four protomers (two $\alpha$ and two $\beta$ chains), has been studied in great detail. Its quaternary structure is shown in Figure 15–20. Most enzymes, including DNA and RNA polymerase, demonstrate quaternary structure.

NOW SOLVE THIS

Problem 35 on page 408 asks you to consider the potential impact of several amino acid substitutions that result from mutations in one of the genes encoding one of the chains making up human hemoglobin.

■ HINT: *When considering the three amino acids (glutamic acid, lysine, and valine), consider the net charge of each R group.*

## 15.10 Posttranslational Modification Alters the Final Protein Product

Before we turn to a discussion of protein function, it is important to point out that polypeptide chains, like RNA transcripts, are often modified once they have been synthesized. This additional processing is broadly described as **posttranslational modification.** Although many of these alterations are detailed biochemical transformations and beyond the scope of our discussion, you should be aware that they occur and that they are critical to the functional capability of the final protein product. Several examples of posttranslational modification are as follows:

1. *The N-terminus amino acid is usually removed or modified.* For example, either the formyl group or the entire formylmethionine residue in bacterial polypeptides is usually removed enzymatically. In eukaryotic polypeptide chains, the amino group of the initial methionine residue is often removed, and the amino group of the N-terminal residue may be modified (acetylated).
2. *Individual amino acid residues are sometimes modified.* For example, phosphates may be added to the hydroxyl groups of certain amino acids, such as tyrosine. Modifications such as these create negatively charged residues that may form an ionic bond with other molecules. The process of phosphorylation is extremely important in regulating many cellular activities and is a result of the action of enzymes called **kinases.** At other amino acid residues, methyl groups or acetyl groups may be added enzymatically, which can affect the function of the modified polypeptide chain.
3. *Carbohydrate side chains are sometimes attached.* These are added covalently, producing **glycoproteins**, an important category of cell-surface molecules, such as those specifying the antigens in the ABO blood-type system in humans.
4. *Polypeptide chains may be trimmed.* For example, the insulin gene is first translated into a longer molecule that is enzymatically trimmed to insulin's final 51-amino-acid form.
5. *Signal sequences are removed.* At the N-terminal end of some proteins is a sequence of up to 30 amino acids that plays an important role in directing the protein to the location in the cell in which it functions. This is called a **signal sequence**, and it determines the final destination of a protein in the cell. The process is called **protein targeting**. For example, proteins whose fate involves secretion or proteins that are to become part of the plasma membrane are dependent on specific sequences for their initial transport into the lumen of the endoplasmic reticulum. While the signal sequence of various proteins with a common destination might differ in their primary amino acid sequence, they share many chemical properties. For example, those destined for secretion all contain a string of up to 15 hydrophobic amino acids preceded by a positively charged amino acid at the N-terminus of the signal sequence. Once the polypeptides are transported, but before they achieve their functional status as proteins, the signal sequence is enzymatically removed from them.
6. *Polypeptide chains are often complexed with metals.* The tertiary and quaternary levels of protein structure often include and are dependent on metal atoms. The functional protein is thus a molecular complex that includes both polypeptide chains and metal atoms. Hemoglobin, which contains four iron atoms along with its four polypeptide chains, is a good example.

These types of posttranslational modifications are obviously important in achieving the functional status specific to any given protein. Because the final three-dimensional structure of the molecule is directly responsible for its specific function, how polypeptide chains ultimately fold into their final conformations is also an important topic. For many years, it was thought that protein folding was a spontaneous process whereby the molecule achieved maximum thermodynamic stability automatically as a result of the combined chemical properties inherent in the amino acid sequence of its polypeptide chain(s). However, numerous studies have shown that, for many proteins, folding is dependent on members of a family of still other, ubiquitous proteins called **chaperones.** Chaperone proteins (sometimes called *molecular chaperones* or *chaperonins*) function to facilitate the folding of other proteins. The mechanism by which chaperones function is not yet clear, but like enzymes, they do not become part of the final product. Initially discovered in *Drosophila*, in which they are called **heat-shock proteins**, chaperones have now been discovered in various organisms, including bacteria, animals, and plants. Ultimately, protein folding is a critically important process, not only because misfolded proteins may be nonfunctional, but also because improperly folded proteins can be dangerous. It is becoming clear that some disorders, such as the spongiform encephalopathies—**mad cow disease** in cattle and **Creutzfeldt–Jakob disease** in humans (see the Genetics, Technology, and Society essay later in this chapter)—are caused by the presence of incorrectly folded neural proteins. Currently, many laboratories are focused on trying to understand how protein folding occurs normally, as well as how the presence of misfolded polypeptides "poisons" the folding of normal polypeptides and ultimately causes cell death and disease.

## 15.11 Proteins Function in Many Diverse Roles

The essence of life on Earth resides at the level of diversity of cellular function. While DNA and RNA serve as vehicles for storing and expressing genetic information, proteins are the *means* of cellular function. And it is the capability of cells to assume diverse structures and functions that distinguishes most eukaryotes from less evolutionarily advanced organisms such as bacteria. Therefore, an introductory understanding of protein function is critical to a complete view of genetic processes.

Proteins are the most abundant macromolecules found in cells. As the end products of genes, they play many diverse roles. For example, the respiratory pigments **hemoglobin** and **myoglobin** transport oxygen, which is essential for cellular metabolism. **Collagen** and **keratin** are structural proteins associated with the skin, connective tissue, and hair of organisms. **Actin** and

**myosin** are contractile proteins, found in abundance in muscle tissue, while **tubulin** is the basis of the function of microtubules in mitotic and meiotic spindles. Still other examples are the **immunoglobulins**, which function in the immune system of vertebrates; **transport proteins**, involved in the movement of molecules across membranes; some of the **hormones** and their **receptors**, which regulate various types of chemical activity; **histones**, which bind to DNA in eukaryotic organisms; and **transcription factors** that regulate gene expression.

Nevertheless, the most diverse and extensive group of proteins (in terms of function) are the **enzymes**, to which we have referred throughout this chapter. Enzymes specialize in catalyzing chemical reactions within living cells. Like all catalysts, they increase the rate at which a chemical reaction reaches equilibrium, but they do not alter the end point of the chemical equilibrium. Their remarkable, highly specific catalytic properties largely determine the metabolic capacity of any cell type and provide the underlying basis of what we refer to as **biochemistry**. The specific functions of many enzymes involved in the genetic and cellular processes of cells are described throughout the text.

**Biological catalysis** is a process whereby enzymes lower the **energy of activation** for a given reaction (Figure 15–21). The energy of activation is the increased kinetic energy state that molecules must usually reach before they react with one another. This state can be attained as a result of elevated temperatures, but enzymes allow biological reactions to occur at lower, physiological temperatures. Thus, enzymes make life as we know it possible.

The catalytic properties of an enzyme are determined by the chemical configuration of the molecule's **active site**. This site is associated with a crevice, a cleft, or a pit on the surface of the enzyme that binds the reactants, or substrates, and facilitates their interaction.

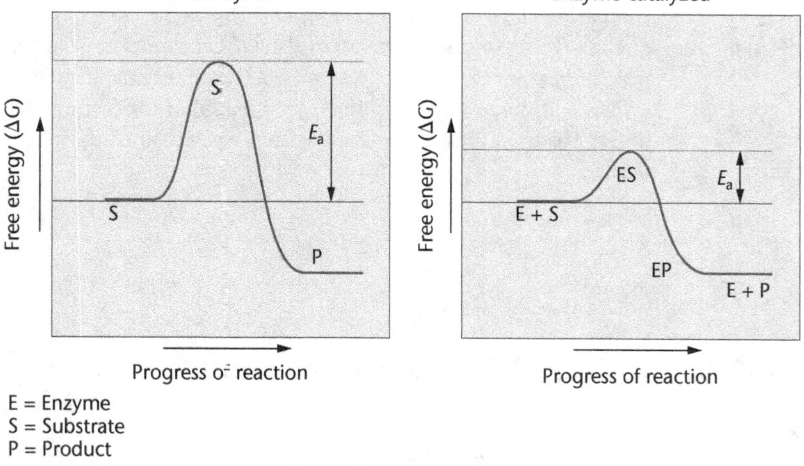


**FIGURE 15–21** Energy requirements of an uncatalyzed versus an enzymatically catalyzed chemical reaction. The energy of activation ($E_a$) necessary to initiate the reaction is substantially lower as a result of catalysis.

Enzymatically catalyzed reactions control metabolic activities in the cell. Each reaction is either *catabolic* or *anabolic*. **Catabolism** is the degradation of large molecules into smaller, simpler ones with the release of chemical energy. **Anabolism** is the synthetic phase of metabolism, building the various components that make up nucleic acids, proteins, lipids, and carbohydrates.

## 15.12 Proteins Are Made Up of One or More Functional Domains

We conclude this chapter by briefly discussing the important finding that regions made up of specific amino acid sequences are associated with specific functions in protein molecules. Such sequences, usually between 50 and 300 amino acids, constitute **protein domains** and represent modular portions of the protein that fold into stable, unique conformations independently of the rest of the molecule. Different domains impart different functional capabilities. Some proteins contain only a single domain, while others contain two or more.

The significance of domains resides in the tertiary structures of proteins. Each domain can contain a mixture of secondary structures, including $\alpha$ helices and $\beta$-pleated sheets. The unique conformation of a given domain imparts a specific function to the protein. For example, a domain may serve as the catalytic site of an enzyme, or it may impart an ability to bind to a specific ligand. Thus, discussions of proteins may mention *catalytic domains, DNA-binding domains*, and so on. In short, a protein must be seen as being composed of a series of structural and functional modules. Obviously, the presence of multiple domains in a single protein increases the versatility of each molecule and adds to its functional complexity.

### Exon Shuffling

An interesting hypothesis concerning the genetic origin of protein domains was put forward by Walter Gilbert in 1977. Gilbert suggested that the functional regions of genes in higher organisms consist of collections of exons originally present in ancestral genes and brought together through recombination during the course of evolution. Referring to the process as **exon shuffling**, Gilbert proposed that exons, like protein domains, are also modular, and that during evolution, exons may have been reshuffled between genes in eukaryotes, with the result that different genes have similar domains.

Several observations lend support to this proposal. Most exons are fairly small, averaging about 150 base pairs and encoding about 50 amino acids, consistent with the sizes of many functional domains in proteins. Second, recombinational events that lead to exon shuffling would be expected to occur within areas of genes represented by introns. Because introns are free to accumulate mutations without harm to the organism, recombinational events would tend to further randomize their nucleotide sequences. Over extended evolutionary periods, sequence diversity would increase. This is, in fact, what is observed: introns range from 50 to 20,000 bases in length and exhibit fairly random base sequences.

Since 1977, a serious research effort has been aimed at analyzing gene structure. In 1985, more direct evidence in favor of Gilbert's proposal of exon modules was presented. For example, the human gene encoding the membrane receptor for low-density lipoproteins (LDLs) was isolated and sequenced. The LDL receptor protein is essential to the transport of plasma cholesterol into the cell. It mediates endocytosis and is expected to have numerous functional domains. These include domains capable of binding specifically to the LDL substrates and interacting with other proteins located at different depths within the membrane as the LDL is being transported across it. In addition, the receptor molecule is modified posttranslationally by the addition of a carbohydrate; a domain must exist that links to this carbohydrate. Given these functional constraints, one would predict that the LDL receptor polypeptide would contain several distinct domains.

Detailed analysis of the gene encoding the LDL receptor supports the concept of exon modules and their shuffling during evolution. The gene is quite large—45,000 base pairs—and contains 18 exons, which in turn contain only slightly less than 2600 nucleotides. These exons code the various functional domains of the protein *and* appear to have been recruited from other genes during evolution.

Figure 15–22 shows these relationships. The first exon encodes a signal sequence that is removed from the protein before the LDL receptor becomes part of the membrane. The next five exons, collectively, represent the domain specifying the binding site for cholesterol. This domain is made up of a 40-amino acid sequence repeated seven times. The next domain, encoded by eight exons, consists of a sequence of 400 amino acids bearing a striking homology to the peptide-hormone epidermal growth factor (EGF) in mice. (A similar sequence is also found in three blood-clotting proteins). This region contains three repetitive sequences of 40 amino acids. The fifteenth exon specifies the domain for the posttranslational addition of the carbohydrate, while the next two specify regions of the protein that are integrated into the membrane, anchoring the receptor to specific sites called *coated pits* on the cell surface.

These observations concerning the LDL exons are fairly compelling in support of the theory of exon shuffling during evolution. Certainly, there is no disagreement concerning the concept of protein domains being responsible for specific molecular interactions.

## The Origin of Protein Domains

What remains controversial and evocative in the exon-shuffling theory is the question of when introns first appeared on the evolutionary scene. In 1978, W. Ford Doolittle proposed that these intervening sequences were part of the genome of the most primitive ancestors of modern-day eukaryotes. In support of this "intron-early" idea, Gilbert has argued that if similarities in intron DNA sequences are found in identical positions within genes shared by distantly related eukaryotes (such as humans, chickens, and corn), they must have also been present in primitive ancestral genomes.

If Doolittle's proposal is correct, why are introns absent in most prokaryotes and infrequent in yeast? Gilbert argues that they were present at one point during evolution, but as the genome of these primitive organisms evolved, they were lost. The loss resulted from strong selection pressure to streamline chromosomes so as to minimize the energy expenditure on replication and gene expression. Furthermore, streamlining led to fewer errors in mRNA production. However, supporters of the opposing "intron-late" school, including Jeffrey Palmer, argue that introns' first appearance came much later in evolution, when they were acquired by a single group of eukaryotes that are ancestral to modern-day eukaryotes but not to prokaryotes.

The advent of large-scale genome sequencing has served to increase the controversy. We now have the ability to decipher the complete nucleotide sequence of all DNA in specific organisms. By comparing the amino acid coding and the noncoding DNA sequences from evolutionarily spaced species, geneticists hope to gain insight into how these sequences evolved. Currently, no indisputable evidence in support of either the "intron-early" or the "intron-late" theory has been found, and the question has remained difficult to resolve.

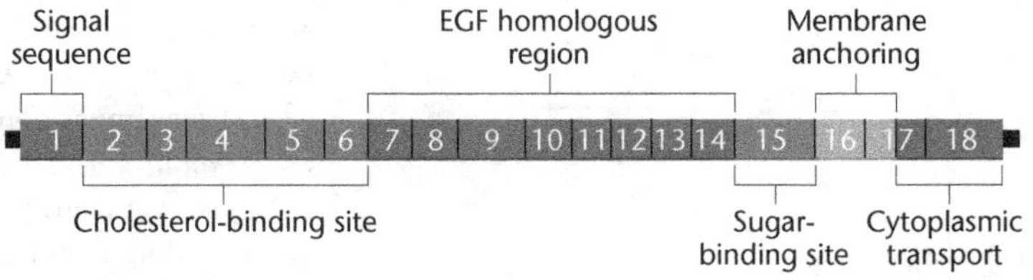


FIGURE 15-22 The 18 exons making up the gene encoding the LDL receptor protein are organized into five functional domains and one signal sequence.

## GENETICS, TECHNOLOGY, AND SOCIETY

# Mad Cow Disease: The Prion Story

In March 1996 the British government announced that a new brain disease had killed 10 young Britons and that the victims might have acquired the disease by eating affected beef. Bovine spongiform encephalopathy (BSE), popularly known as mad cow disease, slowly destroys brain cells and is always fatal. Recent studies confirm that BSE and the human disease known as new variant Creutzfeldt-Jakob disease (nvCJD) are so similar at the molecular and pathological levels that they are most likely the same. Mad cow disease triggered political turmoil in Europe, a worldwide ban on British beef, and the near-collapse of the $8.9 billion British beef industry. The European Union demanded the slaughter and incineration of 4.7 million British cattle, a campaign that cost the government more than $12 billion in milk imports, compensation to farmers, and the purchase of stock for new herds. Most European countries, as well as the Middle East, Japan, the United States, and Canada, have discovered BSE in at least one cow.

Although most nvCJD cases have occurred in Britain, cases have also appeared in France, Italy, Ireland, South America, Canada, and the United States. Nearly 200 people have now died of nvCJD, and epidemiologists predict that many more cases will appear in the next few decades. The long incubation time (10 years or more) of nvCJD makes calculations of its current and future frequency difficult.

BSE, nvCJD, and Creutzfeldt-Jakob disease (CJD) belong to a group of neurological diseases known as transmissible spongiform encephalopathies, which affect animals and humans. In this group of diseases, brain tissue eventually resembles a sponge (hence, spongiform) and is riddled with proteinaceous deposits. Victims lose motor function, become demented, and ultimately die. CJD and nvCJD differ symptomatically in that victims of nvCJD display psychological symptoms prior to developing neurological symptoms and are usually young (16–40 years old). CJD normally affects people over 55. CJD cases arise spontaneously and randomly, at a rate of one per million per year worldwide, but CJD can also be inherited as an autosomal dominant condition.

CJD can be transmitted through tissue grafts, or by injection of growth hormones derived from infected human pituitary glands. Blood transfusion may also be responsible for some cases of nvCJD. Kuru, a CJD-like disease of the Fore people of New Guinea, was transmitted through ritualistic cannibalism. Transmissible spongiform encephalopathies in animals include scrapie (sheep and goats), chronic wasting disease (deer and elk), and BSE. Like kuru, BSE is passed from animal to animal by ingestion of diseased animal remains, particularly neural tissue. The epidemic of BSE in Britain occurred because diseased cows and sheep were processed and fed to cattle as a protein supplement.

For many years, spongiform encephalopathies defied analysis. The diseases are difficult to study for a number of reasons. First, brain material has to be injected into the brains of experimental animals to induce the diseases, which then take months or years to develop. In addition, the infectious agent is apparently not a virus or bacterium, and affected animals do not develop antibodies against these mysterious agents.

The infectious material is unaffected by radiation or nucleases that damage nucleic acids; however, it is destroyed by reagents that hydrolyze or modify proteins. In the early 1980s, American scientist Stanley Prusiner purified the infectious agent and concluded that it consists of only protein. He proposed that scrapie is spread by an infectious protein particle that he called a *prion*. His hypothesis was initially dismissed by most scientists, as the idea of an infectious agent with no DNA or RNA as genetic material was heretical. However, Prusiner and others presented evidence supporting the prion hypothesis, and the notion that the disease can be transmitted by an infectious particle that contains no genetic material has gained acceptance.

If prions are composed of protein only, how do they cause disease? The answer may be as strange as the disease itself. The so-called PrP protein that makes up a prion is a version of a normal protein that is synthesized in neurons and found in the brains of all adult animals. The difference between normal PrP and prion PrP lies in their secondary protein structures. Normal, noninfectious PrP folds into $\alpha$ helices, whereas infectious prion PrP folds into $\beta$-pleated sheets. When a normal PrP molecule contacts a prion PrP molecule, the normal protein refolds into the abnormal PrP conformation. The newly abnormal PrP molecules then spread their lethal conformations to neighboring normal PrP molecules. The process of refolding takes off in a chain reaction, with potentially devastating results. Hence, spongiform encephalopathies can be considered diseases of secondary protein structure.

The outbreaks of BSE have led to major changes in agricultural practices. In 1998, the British government banned the use of cows and sheep as feed for other cows and sheep, and the epidemic subsided. The European Union has also banned the use of feeds containing animal products for all livestock. In contrast, the United States and Canada still allow nonruminant animals to consume feed containing ruminants, and ruminant animals to consume feed containing some nonruminants, as long as certain high-risk materials such as brain and nervous tissue are removed. Although changes in feeding methods have reduced the occurrence of BSE in cattle, other control methods are being developed. One proposal is to create genetically modified cattle that lack PrP. Although the functions of normal PrP are not certain, removal of the entire *PRNP* gene that encodes PrP apparently has no detectable ill effects. Cattle lacking the *PRNP* gene may be bred as BSE-resistant stock.

The nature of prion diseases makes them challenging to treat. Though several drugs appear to limit production of prion PrP in cultured cells, none have yet been found that are safe and also pass through the blood-brain barrier. Successful antiprion treatments will likely require a better understanding of protein folding. What a wrinkled tale prions have told thus far!

### ■ References

Aguzzi, A. 2006. Prion diseases of humans and farm animals: Epidemiology, genetics, and pathogenesis. *J Neurochem*. 97:1726–1739.

### ■ Web Sites

U.S. Food and Drug Administration collection of reports and information about BSE, http://www.fda.gov/oc/opacom/hottopics/bse.html

EXPLORING GENOMICS

# Translation Tools, Swiss-Prot, and Protein–Protein Interaction Databases

Many of the databases and bioinformatics programs we have used for Exploring Genomics exercises have focused on manipulating and analyzing DNA and RNA sequences. However, scientists working on various aspects of translation and protein structure and function also have a wide range of bioinformatics tools and databases at their disposal via the Internet. Many of these sites were developed as a consequence of the newly emerging discipline of *proteomics*, the study of all the proteins expressed in a cell or tissue. We will discuss proteomics in more detail in Chapter 21.

In this Exploring Genomics, we will use a program from **ExPASy (Expert Protein Analysis System)** to translate a segment of a gene into a possible polypeptide. We will then explore databases for studying protein–protein interaction networks, and use them to learn more about this polypeptide, including other polypeptides it associates with.

## ■ Exercise I – Translating a Nucleotide Sequence and Analyzing a Polypeptide

**ExPASy (Expert Protein Analysis System),** hosted by the Swiss Institute of Bioinformatics, provides a wealth of resources for studies in proteomics. In this exercise we will use a program from ExPASy called **Translate Tool** to translate a nucleotide sequence to a polypeptide sequence. Although many other programs are available on the Web for this purpose, ExPASy is one of the more student-friendly tools. Translate Tool allows you to make a predicted polypeptide sequence from a cloned gene and then look for open reading frames and variations in possible polypeptides.

1. Below is a partial sequence for a human gene based on a complementary DNA (cDNA) sequence. Recall from Chapter 13 that cDNA sequences are DNA copies of mRNA molecules expressed in a cell. Before you translate this sequence in ExPASy, run a nucleotide-nucleotide BLAST search from the NCBI Web site (http://www.ncbi.nlm.nih.gov/BLAST/) to identify the gene corresponding to this sequence. Refer to the Exploring Genomics exercise in Chapter 10 if you need help with BLAST searches.

   ACATTTGCTTCTGACACAATTGTGTTCAC
   TAGCAACCTCAAACAGACACCATGGTGC
   ATCTGACTCCTGAGGAGAAGTCTGCCGT
   TACTGCCCTGTGGGGCAAGGTGAACGT
   GGATGAAGTTGGTGGTGAGGCCCTGGG
   CAG

2. Access the ExPASy Translate Tool program at http://us.expasy.org/tools/dna.html. Copy and paste the cDNA sequence into Translate Tool and click "Translate Sequence" to generate possible polypeptide sequences encoded by this cDNA.

3. Review the translation results and then answer the following questions:

   a. Did Translate Tool provide one or multiple possible polypeptide sequences?

   b. If the translation results showed multiple polypeptide sequences, what does this mean? Explain.

   c. Refer to Figure 15–14 in the chapter. Based on this figure, which reading frame generated by Translate Tool appears to be correct?

4. ExPASy also provides access to a wealth of information about this polypeptide by connecting to a large number of different databases such as **UniProt KB/Swiss-Prot,** a protein sequence database maintained by the Swiss Institute for Bioinformatics (SIB) and the European Bioinformatics Institute (EBI), and a database called the Protein Data Bank. UniProt KB/Swiss-Prot is widely used by scientists around the world. Visit UniProt KB/Swiss-Prot (http://ca.expasy.org/sprot/hpi/) for a wealth of information on the human genome and proteomics.

5. To learn more about the features of this polypeptide, it is best to work with a complete sequence. To obtain a complete sequence, click on the reading frame you believe is correct.

6. To retrieve the amino acid sequence for the entire polypeptide, click methionine (m) as the first amino acid in the polypeptide and then use the "BLAST" link near the bottom of the next page to run a BLAST search. From the BLAST results page, locate UniProtKB/Swiss-Prot entry P68871; this is an accession number for the protein sequence, similar to the accession numbers assigned to DNA and RNA sequences, and it is the correct match for this sequence. Click on this link to reveal a comprehensive report about the protein. Be sure the identity of this protein agrees with what you discovered in step 1.

7. Scroll past the references in the Swiss-Prot report; then explore the following features:

   a. 3D Structure Databases—presents 3D modeling representations showing polypeptide folding arrangements.

   b. Under 2D Gel Databases, use the Swiss-2D PAGE link to view 2D gels of this polypeptide from different tissue samples. Refer to Figure 21–23 for a representation of 2D gels. When viewing a 2D gel image with this feature, click on links under "Map Locations" to identify specific spots on a gel that correspond to this polypeptide.

   c. The Family and Domain databases links will take you to a wealth of information about this polypeptide and related polypeptides and proteins.

   d. Explore the "Other" category and visit the DrugBank link that provides information on drugs that bind to and affect this polypeptide.

   e. At the very bottom of the page, under "Sequence analysis tools," use the "ProtParam" feature to learn more about predicted secondary structures formed by this polypeptide.

**Exercise II – Studying Protein-Protein "Interactomes"**

Another rapidly emerging area in the "omics" revolution is the study of protein-protein interaction networks, or so-called *interactomes*. As you will learn in Chapter 21, *systems biology* approaches to understanding cell functions and disease are based on the concept that cell physiology depends on complex interactions between gene products through protein interaction networks. Such networks might consist, for example, of proteins that interact in an enzymatic pathway or signaling pathway. Protein-protein interaction networks in diseased tissues often differ in comparison to those in normal tissues; therefore, scientists anticipate that identifying and understanding protein-protein interactions in cells and tissues will help us better understand the complexities of cell physiology and disease states. A number of protein interactome databases are developing. These gather information from around the world, cataloging protein interactions based on research demonstrating interactions between different proteins.

1. ExPASy provides access to a protein interactome database called **IntAct**. Look for the "IntAct" link provided in the Swiss Prot report you generated for step 6, exercise I.

2. Several other protein interactome databases are available on the Internet, including:

   **Pathguide**, http://www.pathguide.org. This database provides a collection of protein-protein interaction databases.

   **Human Interactome Map (HiMap)**, http://www.himap.org/main/index.jsp. This is an excellent browser for analyzing human protein-protein interactions. *Note:* To access HiMap, you must set up a free registration account.

3. Using either the "IntAct" link from ExPASy, any of the databases available in Pathguide, or the HiMap database, identify some of the polypeptides that are known to interact with the polypeptide you examined in Exercise I. Prior to searching these databases, can you think of at least one other polypeptide that you know interacts with this one, based on what you already know about this polypeptide?

## Chapter Summary

1. Translation is the synthesis of polypeptide chains under the direction of mRNA in association with ribosomes. This process ultimately converts the information stored in the genetic code of the DNA that makes up a gene into the corresponding sequence of amino acids making up the polypeptide.
2. Translation is a complex, energy-requiring process that also depends on charged tRNA molecules and numerous protein factors. Transfer RNA (tRNA) serves as the adaptor molecule between an mRNA codon and the appropriate amino acid.
3. The processes of translation, like transcription, can be subdivided into the stages of initiation, elongation, and termination. The process relies on base-pairing affinities between complementary nucleotides and is more complex in eukaryotes than in prokaryotes.
4. The first suggestion that proteins are the end products of genes was made by Garrod through his study of inherited metabolic disorders—especially cystinuria, albinism, and alkaptonuria—in humans early in the twentieth century.
5. The investigation of nutritional requirements in *Neurospora* by Beadle and colleagues made it clear that mutations cause the loss of enzyme activity. Their work led to the concept of the one-gene:one-enzyme hypothesis.
6. The one-gene:one-enzyme hypothesis was later revised. Pauling and Ingram's investigations of hemoglobins from patients with sickle-cell anemia led to the discovery that one gene directs the synthesis of only one polypeptide chain.
7. Thorough investigations have revealed the existence of several major types of human hemoglobin molecules found in the embryo, fetus, and adult. Specific genes control each polypeptide chain constituting these various hemoglobin molecules.
8. The proposal that a gene's nucleotide sequence specifies in a colinear way the sequence of amino acids in a polypeptide chain was confirmed by experiments involving mutations in the tryptophan synthetase gene in *E. coli*.
9. Variation in protein function is the basis for biological diversity. Structural variation of proteins is incredibly diverse and based on certain fundamental chemical principles.
10. Proteins, the end products of genes, demonstrate four levels of structural organization that together constitute their three-dimensional conformation, which is the basis of each molecule's function. The first level—the amino acid sequence—is the direct chemical basis for the other three.
11. Of the myriad functions performed by proteins, the most influential role belongs to enzymes. These highly specific biological catalysts play a central role in the production of all classes of molecules in living systems.
12. Proteins consist of one or more functional domains, many of which are found in many different molecules. The origin of these domains may have been the shuffling of exons during evolution.

## INSIGHTS AND SOLUTIONS

1. The growth responses in the following chart were obtained by growing four mutant strains of *Neurospora* on different media, each containing one of four related compounds, A, B, C, and D. None of the mutations grows on minimal medium. Draw all possible conclusions from this data.

| | Growth Product | | | |
|---|---|---|---|---|
| Mutation | A | B | C | D |
| 1 | – | – | – | – |
| 2 | + | + | – | + |
| 3 | + | + | – | – |
| 4 | – | + | – | – |

**Solution:** Nothing can be concluded about mutation *1* except that it is lacking some essential factor, perhaps even unrelated to any biochemical pathway in which A, B, C, and D participate. Nor can anything be concluded about compound C. If it is involved in a pathway with the other compounds, it is a product synthesized prior to the synthesis of A, B, and D. We must now analyze these three compounds and the control of their synthesis by the enzymes encoded by genes *2*, *3*, and *4*. Because product B allows growth in all three cases, it may be considered the "end product"—it bypasses the block in all three instances. Similar reasoning suggests that product A precedes B in the pathway, since A bypasses the block in two of the three steps. Product D precedes B, yielding a partial solution:

$$C(?) \longrightarrow D \longrightarrow A \longrightarrow B$$

Now let's determine which mutations control which steps. Since mutation *2* can be alleviated by products D, B, and A, it must control a step prior to all three products, perhaps the direct conversion to D, although we cannot be certain. Mutation *3* is alleviated by B and A, so its effect must precede theirs in the pathway. Thus, we will assign it a role controlling the conversion of D to A. Likewise, we can provisionally assign mutation *4* to the conversion of A to B, leading to the more complete solution

$$C(?) \xrightarrow{2(?)} D \xrightarrow{3} A \xrightarrow{4} B$$

## Problems and Discussion Questions

1. List and describe the role of all of the molecular constituents of a functional polyribosome.
2. Contrast the roles of tRNA and mRNA during translation and list all enzymes that participate in the transcription and translation process.
3. Francis Crick proposed the "adaptor hypothesis" for the function of tRNA. Why did he choose that description?
4. During translation, what molecule bears the codon? the anticodon?
5. The $\alpha$ chain of eukaryotic hemoglobin is composed of 141 amino acids. What is the minimum number of nucleotides in an mRNA coding for this polypeptide chain?
6. Assuming that each nucleotide is 0.34 nm long in the mRNA, how many triplet codes can occupy at one time the space in a ribosome that is 20 nm in diameter?
7. Summarize the steps involved in charging tRNAs with their appropriate amino acids.
8. To carry out its role, each transfer RNA requires at least four specific recognition sites that must be inherent in its tertiary structure. What are they?
9. Discuss the potential difficulties of designing a diet to alleviate the symptoms of phenylketonuria.
10. Phenylketonurics cannot convert phenylalanine to tyrosine. Why don't these individuals exhibit a deficiency of tyrosine?
11. Phenylketonurics are often more lightly pigmented than are normal individuals. Can you suggest a reason why this is so?
12. Early detection and adherence to a strict dietary regime has prevented much of the mental retardation that used to occur in those afflicted with phenylketonuria (PKU). Affected individuals now often lead normal lives and have families. For various reasons, such individuals adhere less rigorously to their diet as they get older. Predict the effect that mothers with PKU who neglect their diets might have on newborns.
13. The synthesis of flower pigments is known to be dependent on enzymatically controlled biosynthetic pathways. For the crosses shown here, postulate the role of mutant genes and their products in producing the observed phenotypes:
    (a) $P_1$: white strain A × white strain B
    $F_1$: all purple
    $F_2$: 9/16 purple: 7/16 white
    (b) $P_1$: white × pink
    $F_1$: all purple
    $F_2$: 9/16 purple: 3/16 pink: 4/16 white
14. A series of mutations in the bacterium *Salmonella typhimurium* results in the requirement of either tryptophan or some related molecule in order for growth to occur. From the data shown here, suggest a biosynthetic pathway for tryptophan:

| | Growth Supplement | | | | |
|---|---|---|---|---|---|
| Mutation | Minimal Medium | Anthranilic Acid | Indole Glycerol Phosphate | Indole | Tryptophan |
| *trp*-8 | – | + | + | + | + |
| *trp*-2 | – | – | + | + | + |
| *trp*-3 | – | – | – | + | + |
| *trp*-1 | – | – | – | – | + |

15. The study of biochemical mutants in organisms such as *Neurospora* has demonstrated that some pathways are branched. The data shown here illustrate the branched nature of the pathway resulting in the synthesis of thiamine:

| | Growth Supplement | | | |
|---|---|---|---|---|
| **Mutation** | **Minimal Medium** | **Pyrimidine** | **Thiazole** | **Thiamine** |
| *thi*-1 | – | – | + | + |
| *thi*-2 | – | + | – | + |
| *thi*-3 | – | – | – | + |

Why don't the data support a linear pathway? Can you postulate a pathway for the synthesis of thiamine in *Neurospora?*

16. Explain why the one-gene:one-enzyme concept is not considered totally accurate today.
17. Why is an alteration of electrophoretic mobility interpreted as a change in the primary structure of the protein under study?
18. Contrast the polypeptide-chain components of each of the hemoglobin molecules found in humans.
19. Using sickle-cell anemia as an example, describe what is meant by a molecular or genetic disease. What are the similarities and dissimilarities between this type of a disorder and a disease caused by an invading microorganism?
20. Contrast the contributions of Pauling and Ingram to our understanding of the genetic basis for sickle-cell anemia.
21. Hemoglobins from two individuals are compared by electrophoresis and by fingerprinting. Electrophoresis reveals no difference in migration, but fingerprinting shows an amino acid difference. How is this possible?
22. HbS results in anemia and resistance to malaria, whereas in those with HbA, the parasite *Plasmodium falciparum* invades red blood cells and causes the disease. Predict whether those with HbC are likely to be anemic and whether they would be resistant to malaria.
23. Shown here are several amino acid substitutions in the $\alpha$ and $\beta$ chains of human hemoglobin:

| Hb Type | Normal Amino Acid | Substituted Amino Acid |
|---|---|---|
| Hb Toronto | ala | asp ($\alpha$-5) |
| HbJ Oxford | gly | asp ($\alpha$-15) |
| Hb Mexico | gln | glu ($\alpha$-54) |
| Hb Bethesda | tyr | his ($\beta$-145) |
| Hb Sydney | val | ala ($\beta$-67) |
| HbM Saskatoon | his | tyr ($\beta$-63) |

Using the code table (Figure 14–7), determine how many of them can occur as a result of a single nucleotide change.

24. Certain mutations called *amber* in bacteria and viruses result in premature termination of polypeptide chains during translation. Many *amber* mutations have been detected at different points along the gene coding for a head protein in phage T4. How might this system be further investigated to demonstrate and support the concept of colinearity?
25. Describe what colinearity means. Of what significance is the concept of colinearity in the study of genetics?
26. Does Yanofsky's work with the *trpA* locus in *E. coli* (discussed in this chapter) constitute more or less direct evidence in support of colinearity than Fiers's work with phage MS2 (discussed in Chapter 14)? Explain.
27. Define and compare the four levels of protein organization.
28. List as many different categories of protein functions as you can. Wherever possible, give an example of each category.
29. How does an enzyme function? Why are enzymes essential for living organisms on Earth?
30. Exon shuffling is a proposal that relates exons in DNA to the repositioning of functional domains in proteins. What evidence exists in support of exon shuffling? Two schools of thought have emerged concerning the origin of exons, "intron-early" and "intron-late." Briefly describe both of them and present support for each.

**HOW DO WE KNOW?**

31. In this chapter, we focused on the translation of mRNA into proteins as well as on protein structure and function. Along the way, we found many opportunities to consider the methods and reasoning by which much of this information was acquired. From the explanations given in the chapter, what answers would you propose to the following fundamental questions:
    (a) What experimentally derived information led to Holley's proposal of the two-dimensional cloverleaf model of tRNA?
    (b) What experimental information verifies that certain codons in mRNA specify chain termination during translation?
    (c) How do we know, based on studies of *Neurospora* nutritional mutations, that one gene specifies one enzyme?
    (d) On what basis have we concluded that proteins are the end products of genetic expression?
    (e) What experimental information directly confirms that the genetic code, as shown in Figure 14–7, is correct?
    (f) How do we know that the structure of a protein is intimately related to the function of that protein?

## Extra-Spicy Problems

32. In 1962, F. Chapeville and others reported an experiment in which they isolated radioactive $^{14}$C-cysteinyl-tRNA$^{cys}$ (charged tRNA$^{cys}$ + cysteine). They then removed the sulfur group from the cysteine, creating alanyl-tRNA$^{cys}$ (charged tRNA$^{cys}$ + alanine). When alanyl-tRNA$^{cys}$ was added to a synthetic mRNA calling for cysteine, but not alanine, a polypeptide chain was synthesized containing alanine (Chapeville et al., 1962. *Proc. Natl. Acad. Sci.* [*USA*] 48:1086–1093). What can you conclude from this experiment?
33. Three independently assorting genes are known to control the following biochemical pathway that provides the basis for flower color in a hypothetical plant:

$$\text{Colorless} \xrightarrow{A-} \text{yellow} \xrightarrow{B-} \text{green} \xrightarrow{C-} \text{speckled}$$

Three homozygous recessive mutations are also known, each of which interrupts a different one of these steps. Determine the phenotypic

results in the $F_1$ and $F_2$ generations resulting from the $P_1$ crosses of true-breeding plants listed below:
(a) speckled (*AABBCC*) × yellow (*AabbCC*)
(b) yellow (*AAbbCC*) × green (*AABBcc*)
(c) colorless (*aaBBCC*) × green (*AABBcc*)

34. How would the results vary in cross (a) of Problem 33 if genes *A* and *B* were linked with no crossing over between them? How would the results of cross (a) vary if genes *A* and *B* were linked and 20 μm apart?

35. HbS results from the substitution of valine for glutamic acid at the number 6 position in the B chain of human hemoglobin. HbC is the result of a change at the same position in the B chain, but in this case lysine replaces glutamic acid. Return to the genetic code table (Figure 14–7) and determine whether single nucleotide changes can account for these mutations. Then view Figure 15–16 and examine the R groups in the amino acids glutamic acid, valine, and lysine. Describe the chemical differences between the three amino acids. Predict how the changes might alter the structure of the molecule and lead to altered hemoglobin function.

36. Most wood lily plants have orange flowers and are true breeding. A genetics student discovered both red and yellow true-breeding variants and proceeded to hybridize the three strains. Being knowledgeable about transmission genetics, the student surmised that two gene pairs were involved. On this basis, she proposed that these genes account for the various colors as shown here:

| $P_1$ | $F_1$ | $F_2$ |
|---|---|---|
| (1) orange × red | all orange | 3/4 orange<br>1/4 red |
| (2) orange × yellow | all orange | 3/4 orange<br>1/4 yellow |
| (3) red × yellow | all orange | 9/16 orange<br>4/16 yellow<br>3/16 red |

(a) What outcome(s) of the crosses suggested that two gene pairs were at work?
(b) Assuming that she was correct, and using the mutant-gene symbols *y* for yellow and *r* for red, propose which genotypes give rise to which phenotypes.
(c) The student then proposed several possible biochemical pathways for the production of these pigments, as well as which steps are enzymatically controlled by which genes. These are listed here.

I. white precursor I $\xrightarrow{y\text{ gene}}$ red pigment
white precursor II $\xrightarrow{r\text{ gene}}$ yellow pigment

II. white precursor I $\xrightarrow{r\text{ gene}}$ red pigment
white precursor II $\xrightarrow{y\text{ gene}}$ yellow pigment

III. white precursor $\longrightarrow$ red $\xrightarrow{y\text{ gene}}$ yellow $\xrightarrow{r\text{ gene}}$ orange

IV. white precursor $\longrightarrow$ red $\xrightarrow{r\text{ gene}}$ yellow $\xrightarrow{y\text{ gene}}$ orange

V. white precursor $\longrightarrow$ yellow $\xrightarrow{y\text{ gene}}$ red $\xrightarrow{r\text{ gene}}$ orange

VI. white precursor $\longrightarrow$ yellow $\xrightarrow{r\text{ gene}}$ red $\xrightarrow{y\text{ gene}}$ orange

Which one or more of these would be consistent with your proposal in (b), which was based on the original data? Assume that a mixture of red and yellow pigment results in orange flowers. Defend your answer.

37. Deep in a previously unexplored South American rain forest, a species of plants was discovered with true-breeding varieties whose flowers were either pink, rose, orange, or purple. A very astute plant geneticist made a single cross, carried to the $F_2$ generation, as shown:

| $P_1$: | purple × pink |
|---|---|
| $F_1$: | all purple |
| $F_2$: | 27/64 purple<br>16/64 pink<br>12/64 rose<br>9/64 orange |

Based solely on these data, he was able to propose both a mode of inheritance for flower pigmentation and a biochemical pathway for the synthesis of these pigments.

Carefully study the data. Create a hypothesis of your own to explain the mode of inheritance. Then propose a biochemical pathway consistent with your hypothesis. How could you test the hypothesis by making other crosses?

38. The emergence of antibiotic-resistant strains of *Enterococci* and transfer of resistant genes to other bacterial pathogens have highlighted the need for new generations of antibiotics to combat serious infections. To grasp the range of potential sites for the action of existing antibiotics, sketch the components of the translation machinery (e.g., see Step 3 of Figure 15–6), and using a series of numbered pointers, indicate the specific location for the action of the antibiotics shown in the following table.

| Antibiotic | Action |
|---|---|
| 1. Streptomycin | Binds to 30*S* ribosomal subunit |
| 2. Chloramphenicol | Inhibits peptidyl transferase of 70*S* ribosome |
| 3. Tetracycline | Inhibits binding of charged tRNA to ribosome |
| 4. Erythromycin | Binds to free 50*S* particle and prevents formation of 70*S* ribosome |
| 5. Kasugamycin | Inhibits binding of $tRNA^{fmet}$ |
| 6. Thiostrepton | Prevents translocation by inhibiting EF-G |

39. The development of antibiotic resistance by pathogenic bacteria represents a major health concern. One potential new antibiotic is evernimicin, which was isolated from *Micromonospora carbonaceae*. Evernimicin is an oligosaccharide with antibiotic activity against a broad range of gram-positive pathogenic bacteria. To determine the mode of action of this drug, researchers have analyzed 23*S* ribosomal DNA mutants that showed reduced sensitivity to evernimicin (e.g., Adrian et al., 2000. *Antimicrob. Ag. and Chemo.* 44: 3101–3106). They discovered two classes of mutants that conferred resistance: in one class, the mutation occurs in 23*S* rRNA nucleotides 2475–2483; in the other class, it occurs in ribosomal protein L16. This suggests that these two ribosomal components are structurally and functionally linked. It turns out that the tRNA anticodon stem-loop appears to bind to the A site of the ribosome at rRNA bases 2465–2485. This finding conforms to the proposed function of L16, which appears to be involved in attracting the aminoacyl stem of the tRNA to the ribosome at its A site. Using your sketch of the translation machinery from Problem 38 along with this information, designate where the proposed antibacterial action of evernimicin is likely to occur.

40. The flow of genetic information from DNA to protein is mediated by messenger RNA. If you introduce short DNA strands (called antisense oligonucleotides) that are complementary to mRNAs, hydrogen bonding may occur and "label" the DNA/RNA hybrid for ribonuclease-H degradation of the RNA. One study compared the effect of different-

length antisense oligonucleotides upon ribonuclease-H–mediated degradation of tumor necrosis factor (*TNFα*) mRNA. *TNFα* exhibits antitumor and proinflammatory activities (Lloyd et al., 2001. *Nuc. Acids Res.* 29: 3664–3673). The following graph indicates the efficacy of various-sized antisense oligonucleotides in causing ribonuclease-H cleavage.

(a) Describe how antisense oligonucleotides interrupt the flow of genetic information in a cell.

(b) What general conclusion can be drawn from the graph?

(c) What factors other than oligonucleotide length are likely to influence antisense efficacy *in vivo*?

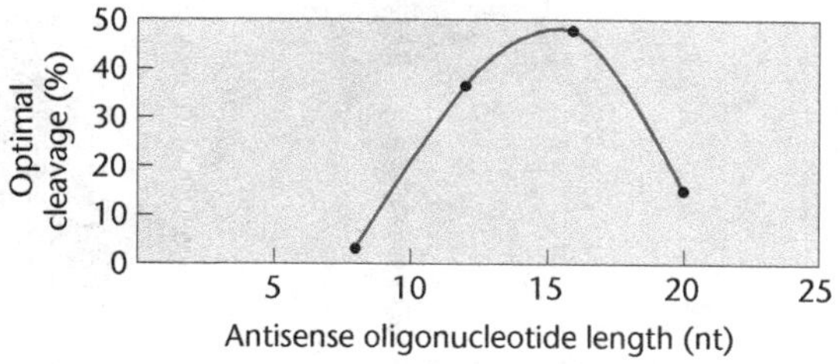


41. The fidelity of translation is dependent on the reliable action of aminoacyl tRNA synthetases that ensure the association of only one type of amino acid with a specific tRNA. Two relatively rare human conditions, Charcot-Marie-Tooth disease 2D (CMT2D) and distal spinal muscular atrophy type V (dSMA-V), are neuropathologies of the peripheral axons that map to a well-defined region of the short arm of chromosome 7, as does the gene for glycyl tRNA synthetase. Families with CMT2D or dSMA-V have been identified with missense mutations in the gene for glycyl tRNA synthetase; and the mutations present in CMT2D and dSMA-V have been found to result in a loss of activity of glycyl tRNA synthetase (Antonellis et al., 2003). A sample pedigree (from Antonellis et al., 2003) is presented here (to maintain anonymity, sexes are not provided).

(a) Considering the pedigree below and the function of aminoacyl tRNA synthetases in general, would you conclude that the genes causing these diseases are dominant or recessive in their action?

(b) Considering the vital role that synthetases play in protein synthesis, speculate as to how individuals might survive with such a defect in translational efficiency.

(c) Why might some tissues (neural) be more affected than others?

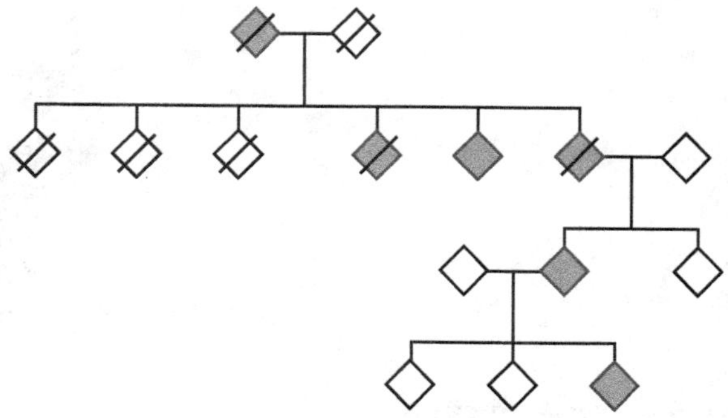

Mutant erythrocytes derived from an individual with sickle-cell anemia.

16

# Gene Mutation and DNA Repair

## CHAPTER CONCEPTS

- Mutation is a source of genetic variation and the basis for natural selection. It is also the source of genetic damage that contributes to cell death, genetic diseases, and cancer.
- Mutations have a wide range of effects on organisms depending on the type and location of the nucleotide change within the gene and the genome.
- Mutations can occur spontaneously as a result of natural biological and chemical processes, or they can be induced by external factors, such as chemicals or radiation.
- The rates of spontaneous mutation vary between organisms and between genes within an organism.
- Organisms rely on a number of DNA repair mechanisms to counteract mutations. These mechanisms range from proofreading of replication errors to base excision and homologous recombination repair.
- Mutations in genes whose products control DNA repair lead to genome hypermutability and human DNA repair diseases, such as xeroderma pigmentosum.
- Geneticists induce gene mutations as the first step in classical genetic analysis.

The ability of DNA molecules to store, replicate, transmit, and decode information is the basis of genetic function. But equally important is the capacity of DNA to make mistakes. Without the variation that arises from changes in DNA sequences, there would be no phenotypic variability, no adaptation to environmental changes, and no evolution. Gene mutations are the source of most new alleles and are the origin of genetic variation within populations. On the downside, they are also the source of genetic changes that can lead to cell death, genetic diseases, and cancer.

Mutations also provide the basis for genetic analysis. The phenotypic variability resulting from mutations allows geneticists to identify and study the genes responsible for the modified trait. In genetic investigations, mutations act as identifying "markers" for genes so that they can be followed during their transmission from parents to offspring. Without phenotypic variability, classical genetic analysis would be impossible. For example, if all pea plants displayed a uniform phenotype, Mendel would have had no foundation for his research.

In Chapter 8, we examined mutations in large regions of chromosomes—chromosomal mutations. In contrast, the mutations we will now explore are those occurring primarily in the base-pair sequence of DNA within individual genes—**gene mutations.** We will also describe how the cell defends itself from such mutations using various mechanisms of DNA repair. The chapter also describes how geneticists use mutations to identify genes and analyze gene functions, in humans and other organisms.

## 16.1 Gene Mutations Are Classified in Various Ways

A gene mutation can be defined as an alteration in DNA sequence. Any base-pair change in any part of a DNA molecule can be considered a mutation. A mutation may comprise a single base-pair substitution, a deletion or insertion of one or more base pairs, or a major alteration in the structure of a chromosome.

Mutations may occur within regions of a gene that code for protein or within noncoding regions of a gene such as introns and regulatory sequences. Mutations may or may not bring about a detectable change in phenotype. The extent to which a mutation changes the characteristics of an organism depends on where the mutation occurs and the degree to which the mutation alters the function of the gene product.

Mutations can occur in somatic cells or within germ cells. Those that occur in germ cells are heritable and are the basis for the transmission of genetic diversity and evolution, as well as genetic diseases. Those that occur in somatic cells are not transmitted to the next generation, but may lead to altered cellular function or tumors.

Because of the wide range of types and effects of mutations, geneticists classify mutations according to several different schemes. These organizational schemes are not mutually exclusive. In this section, we outline some of the ways in which gene mutations are classified.

### Spontaneous and Induced Mutations

Mutations can be classified as either spontaneous or induced, although these two categories overlap to some degree. **Spontaneous mutations** are changes in the nucleotide sequence of genes that appear to have no known cause. No specific agents are associated with their occurrence, and they are generally assumed to be accidental. Many of these mutations arise as a result of normal biological or chemical processes in the organism that alter the structure of nitrogenous bases. Often, spontaneous mutations occur during the enzymatic process of DNA replication, as we discuss later in this chapter.

In contrast to spontaneous mutations, mutations that result from the influence of extraneous factors are considered to be **induced mutations.** Induced mutations may be the result of either natural or artificial agents. For example, radiation from cosmic and mineral sources and ultraviolet radiation from the sun are energy sources to which most organisms are exposed and, as such, may be factors that cause induced mutations. The earliest demonstration of the artificial induction of mutations occurred in 1927, when Hermann J. Muller reported that X rays could cause mutations in *Drosophila*. In 1928, Lewis J. Stadler reported that X rays had the same effect on barley. In addition to various forms of radiation, numerous natural and synthetic chemical agents are also mutagenic.

Several generalizations can be made regarding spontaneous mutation rates in organisms (Table 16.1). First, the rate of spontaneous mutation is exceedingly low for all organisms. Second, the rate varies considerably between different organisms. Third, even within the same species, the spontaneous mutation rate varies from gene to gene.

Viral and bacterial genes undergo spontaneous mutation at an average of about 1 in 100 million ($10^{-8}$) cell divisions. *Neurospora* exhibits a similar rate, but maize, *Drosophila*, and humans demonstrate rates several orders of magnitude higher. The genes studied in these groups average between 1/1,000,000 and 1/100,000 ($10^{-6}$ and $10^{-5}$) mutations per gamete formed. Mouse genes are another order of magnitude higher in their spontaneous mutation rate, 1/100,000 to 1/10,000 ($10^{-5}$ to $10^{-4}$). It is not clear why such a large variation occurs in mutation rates. The variation between organisms may reflect the relative efficiencies of their DNA proofreading and repair systems. We will discuss these systems later in the chapter.

### The Luria-Delbruck Fluctuation Test: Are Mutations Spontaneous or Adaptive?

A concept described as **adaptive mutation** revolves around the controversial idea that organisms may in some way "select" or "direct" the mutation of their genes in order to adapt to a particular environmental pressure. In 1943, Salvador Luria and Max Delbrück presented the first direct evidence that mutations do not occur as

TABLE 16.1

**Rates of Spontaneous Mutations at Various Loci in Different Organisms**

| Organism | Character | Gene | Rate | Units |
|---|---|---|---|---|
| Bacteriophage T2 | Lysis inhibition | $r \rightarrow r^+$ | $1 \times 10^{-8}$ | Per gene replication |
| | Host range | $h^+ \rightarrow h$ | $3 \times 10^{-9}$ | |
| *E. coli* | Lactose fermentation | $lac^- \rightarrow lac^+$ | $2 \times 10^{-7}$ | Per cell division |
| | Lactose fermentation | $lac^+ \rightarrow lac^-$ | $2 \times 10^{-6}$ | |
| | Phage T1 resistance | $T1^s \rightarrow T1^r$ | $2 \times 10^{-8}$ | |
| | Histidine requirement | $his^+ \rightarrow his^-$ | $2 \times 10^{-6}$ | |
| | Streptomycin dependence | $str^s \rightarrow str^d$ | $1 \times 10^{-9}$ | |
| | Streptomycin sensitivity | $str^d \rightarrow str^s$ | $1 \times 10^{-8}$ | |
| | Radiation resistance | $rad^s \rightarrow rad^r$ | $1 \times 10^{-5}$ | |
| | Leucine independence | $leu^- \rightarrow leu^+$ | $7 \times 10^{-10}$ | |
| | Arginine independence | $arg^- \rightarrow arg^+$ | $4 \times 10^{-9}$ | |
| *Zea mays* | Shrunken seeds | $sh^+ \rightarrow sh^-$ | $1 \times 10^{-6}$ | Per gamete per generation |
| | Purple | $pr^+ \rightarrow pr^-$ | $1 \times 10^{-5}$ | |
| | Colorless | $c^+ \rightarrow c^-$ | $2 \times 10^{-6}$ | |
| | Sugary | $su^+ \rightarrow su^-$ | $2 \times 10^{-6}$ | |
| *Drosophila melanogaster* | Yellow body | $y^+ \rightarrow y$ | $1.2 \times 10^{-6}$ | Per gamete per generation |
| | White eye | $w^+ \rightarrow w$ | $4 \times 10^{-5}$ | |
| | Brown eye | $bw^+ \rightarrow bw$ | $3 \times 10^{-5}$ | |
| | Ebony body | $e^+ \rightarrow e$ | $2 \times 10^{-5}$ | |
| | Eyeless | $ey^+ \rightarrow ey$ | $6 \times 10^{-5}$ | |
| *Mus musculus* | Piebald coat | $s^+ \rightarrow s$ | $3 \times 10^{-5}$ | Per gamete per generation |
| | Dilute coat color | $d^+ \rightarrow d$ | $3 \times 10^{-5}$ | |
| | Brown coat | $b^+ \rightarrow b$ | $8.5 \times 10^{-4}$ | |
| | Pink eye | $p^+ \rightarrow p$ | $8.5 \times 10^{-4}$ | |
| *Homo sapiens* | Hemophilia | $h^+ \rightarrow h$ | $2 \times 10^{-5}$ | Per gamete per generation |
| | Huntington disease | $Hu^+ \rightarrow Hu$ | $5 \times 10^{-6}$ | |
| | Retinoblastoma | $R^+ \rightarrow R$ | $2 \times 10^{-5}$ | |
| | Anirida | $An^+ \rightarrow An$ | $5 \times 10^{-6}$ | |
| | Achondroplasia | $A^+ \rightarrow A$ | $5 \times 10^{-5}$ | |

part of an adaptive mechanism, but instead occur spontaneously. This experiment marked the beginning of modern bacterial genetic study. Their experiment, known as the **Luria–Delbrück fluctuation test**, is an example of exquisite analytical and theoretical work.

Luria and Delbrück carried out their experiments with the *E. coli*–T1 system. The T1 bacteriophage is a bacterial virus that infects *E. coli* cells and lyses the infected bacteria. A similar bacteriophage, T4, is described in Chapter 6. Luria and Delbrück grew many individual liquid cultures of phage-sensitive *E. coli* by inoculating the cultures with a small number of bacteria and allowing each of the cultures to grow to higher densities. At this point, they added numerous aliquots of each culture to petri dishes of agar medium containing T1 bacteriophages. To obtain precise quantitative data, they also determined the total number of bacteria added to each plate prior to incubation. Following incubation, each plate was scored for the number of phage-resistant bacterial colonies that grew in the presence of the bacteriophage. This was easy to ascertain because only mutant cells were not lysed and thus survived to be counted.

The experimental rationale for distinguishing between the two hypotheses (mutations occur spontaneously versus mutations occur as a result of adaptation) was as follows.

**Hypothesis 1: Adaptive Mutation.** In this scenario, bacteria are mutating in response to their incubation with the bacteriophage. Therefore, every bacterium in the petri dish has a small but constant probability of acquiring T1 resistance, and so the number of resistant cells will depend only on the number of bacteria and phages added to each plate. The final results should be independent of all other experimental conditions. Consequently, the adaptation hypothesis predicts that if a constant number of bacteria and phages is present on each plate, and if the incubation time is constant, there should be little fluctuation in the number of resistant cells from plate to plate and from experiment to experiment.

**Hypothesis 2: Spontaneous Mutation.** On the other hand, if resistance is acquired as a result of mutations that occur randomly—having nothing to do with any stimulus from being incubated with bacteriophage—resistance mutations will occur at a low rate during the

TABLE 16.2

**The Luria–Delbrück Experiment Demonstrating That Mutations Are Spontaneous**

| | Number of T1-Resistant Bacteria | |
|---|---|---|
| Sample No. | Same Culture (Control) | Different Cultures |
| 1 | 14 | 6 |
| 2 | 15 | 5 |
| 3 | 13 | 10 |
| 4 | 21 | 8 |
| 5 | 15 | 24 |
| 6 | 14 | 13 |
| 7 | 26 | 165 |
| 8 | 16 | 15 |
| 9 | 20 | 6 |
| 10 | 13 | 10 |
| Mean | 16.7 | 26.2 |
| Variance | 15.0 | 2178.0 |

SOURCE: After Luria and Delbrück (1943).

incubation in liquid medium *prior to plating*—that is, before any contact with the phage occurs. When mutations occur *early* during incubation, the subsequent growth and division of the mutant bacteria will produce a relatively large number of resistant cells. When mutations occur *later* during incubation, far fewer resistant cells will be produced. The random mutation hypothesis therefore predicts that the number of resistant cells will fluctuate significantly from experiment to experiment, and from tube to tube, reflecting the varying times at which most of the resistance mutations occurred spontaneously in liquid culture.

Table 16.2 shows a representative set of data from the Luria–Delbrück experiments. The middle column shows the number of mutants recovered from a series of aliquots derived from one large individual liquid culture. In the large culture, the mutants are evenly distributed because the culture is constantly mixed. As a result, these data serve as a control, because the number in each aliquot should be nearly identical. As predicted, little fluctuation is observed in these samples. In contrast, the right-hand column shows the number of resistant mutants recovered from each of 10 independently incubated liquid cultures. The amount of fluctuation in the data will support only one of the two alternative hypotheses. For this reason, the experiment has been designated the **fluctuation test**. Fluctuation is measured by the amount of variance, a statistical calculation.

As seen in the right-hand column, a great deal of fluctuation *was* observed between independently incubated cultures in the Luria–Delbrück experiment, thus supporting the hypothesis that mutations arise randomly, even in the absence of selective pressure, and are inherited in a stable fashion.

Although the concept of spontaneous mutation in viruses, bacteria, and higher organisms has been accepted for some time, the possibility that organisms might also be capable of inducing a specific set of mutations as a result of environmental pressures has long intrigued geneticists. Some recent and controversial research has suggested that under some stressful nutritional conditions such as starvation, bacteria may be capable of activating mechanisms that create a hypermutable state in genes that would, when mutated, enhance survival. The conclusions from these studies are still a source of debate, but they keep alive the interest in the possibility of adaptive mutation.

## Classification Based on Location of Mutation

Mutations may be classified according to the cell type or chromosomal locations in which they occur. **Somatic mutations** are those occurring in any cell in the body except germ cells. **Germ-line mutations** are those occurring in gametes. **Autosomal mutations** are mutations within genes located on the autosomes, whereas **X-linked mutations** are those within genes located on the X chromosome.

Mutations arising in somatic cells are not transmitted to future generations. When a recessive autosomal mutation occurs in a somatic cell of a diploid organism, it is unlikely to result in a detectable phenotype. The expression of most such mutations is likely to be masked by the wild-type allele within that cell. Somatic mutations will have a greater impact if they are dominant or, in males, if they are X-linked, since such mutations are most likely to be immediately expressed. Similarly, the impact of dominant or X-linked somatic mutations will be more noticeable if they occur early in development, when a small number of undifferentiated cells replicate to give rise to several differentiated tissues or organs. Dominant mutations that occur in cells of adult tissues are often masked by the thousands upon thousands of nonmutant cells in the same tissue that perform the normal function.

Mutations in gametes are of greater significance because they are transmitted to offspring as part of the germ line. They have the potential of being expressed in all cells of an offspring. Inherited dominant autosomal mutations will be expressed phenotypically in the first generation. X-linked recessive mutations arising in the gametes of a homogametic female may be expressed in hemizygous male offspring. This will occur provided that the male offspring receives the affected X chromosome. Because of heterozygosity, the occurrence of an autosomal recessive mutation in the gametes of either males or females (even one resulting in a lethal allele) may go unnoticed for many generations, until the resultant allele has become widespread in the population. Usually, the new allele will become evident only when a chance mating brings two copies of it together into the homozygous condition.

## Classification Based on Type of Molecular Change

Geneticists often classify gene mutations in terms of the nucleotide changes that constitute the mutation. A change of one base pair to another in a DNA molecule is known as a **point mutation**, or **base substitution** (Figure 16–1). A change of one nucleotide of a triplet within a protein-coding portion of a gene may result in the creation of a new triplet that codes for a different amino acid in the protein product. If this occurs, the mutation is known as a **missense mutation.** A second possible outcome is that the triplet will be changed into a stop codon, resulting in the termination of translation of the protein. This is known as a **nonsense mutation.** If the point

MUTATIONS AT THE DNA LEVEL
WEB TUTORIAL 16.1

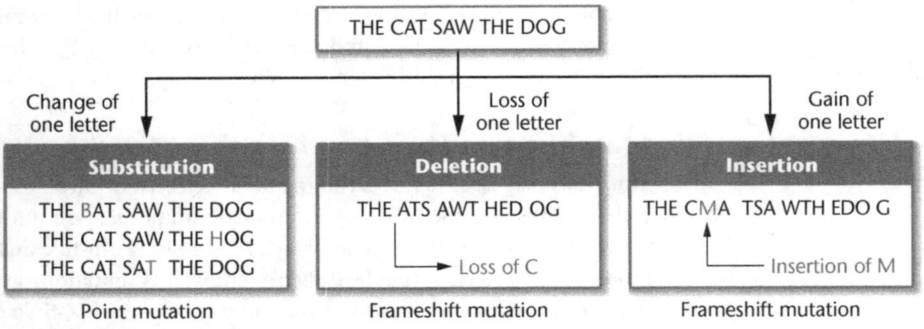


FIGURE 16–1 Analogy showing the effects of substitution, deletion, and insertion of one letter in a sentence composed of three-letter words to demonstrate point and frameshift mutations.

mutation alters a codon but does not result in a change in the amino acid at that position in the protein (due to degeneracy of the genetic code), it can be considered a **silent mutation**.

You will often see two other terms used to describe base substitutions. If a pyrimidine replaces a pyrimidine or a purine replaces a purine, a **transition** has occurred. If a purine replaces a pyrimidine, or vice versa, a **transversion** has occurred.

Another type of change is the insertion or deletion of one or more nucleotides at any point within the gene. As illustrated in Figure 16–1, the loss or addition of a single nucleotide causes all of the subsequent three-letter codons to be changed. These are called **frameshift mutations** because the frame of triplet reading during translation is altered. A frameshift mutation will occur when any number of bases are added or deleted, except multiples of three, which would reestablish the initial frame of reading. It is possible that one of the many altered triplets will be UAA, UAG, or UGA, the translation termination codons. When one of these triplets is encountered during translation, polypeptide synthesis is terminated at that point. Obviously, the results of frameshift mutations can be very severe, especially if they occur early in the coding sequence.

## Classification Based on Phenotypic Effects

Depending on their type and location, mutations can have a wide range of phenotypic effects, from none to severe.

As discussed in Chapter 4, a **loss-of-function mutation** is one that reduces or eliminates the function of the gene product. Any type of mutation, from a point mutation to deletion of the entire gene, may lead to a loss of function. Mutations that result in complete loss of function are known as **null mutations.** It is possible for a loss-of-function mutation to be either dominant or recessive. A dominant loss-of-function mutation may result from the presence of a defective protein product that binds to, or inhibits the action of, the normal gene product, which is also present in the same organism. A **gain-of-function** mutation results in a gene product with enhanced or new functions. This may be due to a change in the amino acid sequence of the protein that confers a new activity, or it may result from a mutation in a regulatory region of the gene, leading to expression of the gene at higher levels, or the synthesis of the gene product at abnormal times, or places. Most gain-of-function mutations are dominant. Because eukaryotic genomes consist mainly of noncoding regions, the vast majority of mutations are likely to occur in the large portions of the genome that do not contain genes. These are considered to be **neutral mutations,** if they do not affect gene products or gene expression.

The most easily observed mutations are those affecting a morphological trait. These mutations are known as **visible mutations** and are recognized by their ability to alter a normal or wild-type visible phenotype. For example, all of Mendel's pea characteristics and many genetic variations encountered in *Drosophila* fit this designation, since they cause obvious changes to the morphology of the organism.

Some mutations exhibit nutritional or biochemical effects. In bacteria and fungi, a typical **nutritional mutation** results in a loss of ability to synthesize an amino acid or vitamin. In humans, sickle-cell anemia and hemophilia are examples of diseases resulting from **biochemical mutations.** Although such mutations do not always affect morphological characters, they can have an effect on the well-being and survival of the affected individual.

Still another category consists of mutations that affect the behavior patterns of an organism. For example, the mating behavior or circadian rhythms of animals can be altered. The primary effect of **behavioral mutations** is often difficult to analyze. For example, the mating behavior of a fruit fly may be impaired if it cannot beat its wings. However, the defect may be in the flight muscles, the nerves leading to them, or the brain, where the nerve impulses that initiate wing movements originate.

Another group of mutations may affect the regulation of gene expression. For example, as we will see with the *lac* operon discussed in Chapter 17, a regulatory gene can produce a product that controls the transcription of other genes. A mutation in a regulatory gene or a gene control region can disrupt normal regulatory processes and inappropriately activate or inactivate expression of a gene. Our knowledge of genetic regulation has been dependent on the study of such **regulatory mutations.**

It is also possible that a mutation may interrupt a process that is essential to the survival of the organism. In this case, it is referred to as a **lethal mutation.** For example, a mutant bacterium that has lost the ability to synthesize an essential amino acid will cease to grow and eventually will die when placed in a medium lacking that amino acid. Various inherited human biochemical disorders are also examples of lethal mutations. For example, Tay–Sachs disease and Huntington disease are caused by mutations that result in lethality, but at different points in the life cycle of humans.

Another interesting class of mutations are those whose expression depends on the environment in which the organism finds itself. Such mutations are called **conditional mutations,** because the mutation is present in the genome of an organism but can be detected only under certain conditions. Among the best examples of conditional mutations are **temperature-sensitive mutations**. At a "permissive" temperature, the mutant gene product functions normally, but it loses its function at a different, "restrictive" temperature. Therefore, when the organism is shifted from the permissive to the restrictive temperature, the impact of the mutation becomes apparent.

As discussed previously, most mutations occur in parts of the genome that do not contain genes or gene regulatory regions. These are usually considered to be neutral mutations, if they do not affect gene products. While the rate at which neutral mutations occur is of great interest to geneticists, they are even more interested in the rate at which deleterious mutations occur, particularly in our own species. Recent molecular techniques, including DNA sequencing, have allowed geneticists to calculate the rates of deleterious mutations. The rate of deleterious mutations in humans is surprisingly high—at least 1.6 deleterious genetic changes per individual per generation.

NOW SOLVE THIS

Problem 17 on page 432 provides data concerning a bacterial mutation that interrupts the biosynthesis of the amino acid leucine. You are asked to calculate the spontaneous mutation rate.

■ HINT: *You must first determine the number of bacteria per milliliter in the original cultures grown under the two conditions by converting the "dilution" data back to the "undiluted" data. For example, if six colonies were observed in 1 mL after a 1000-fold dilution, then the original cell concentration was* $6 \times 10^3$ *bacteria/mL.*

16.2

## Spontaneous Mutations Arise from Replication Errors and Base Modifications

In this section, we will outline some of the processes that lead to spontaneous mutations. It is useful to keep in mind, however, that many of the base modifications that occur during spontaneous mutagenesis also occur, at a higher rate, during induced mutagenesis.

### DNA Replication Errors

As we learned in Chapter 11, the process of DNA replication is imperfect. Occasionally, DNA polymerases insert incorrect nucleotides during replication of a strand of DNA. Although DNA polymerases can correct most of these replication errors using their inherent 3′ to 5′ exonuclease proofreading capacity, misincorporated nucleotides may persist after replication. If these errors are not detected and corrected by DNA repair mechanisms, they may lead to mutations. Replication errors due to mispairing predominantly lead to point mutations. The fact that bases can take several forms, known as **tautomers**, also increases the chance of mispairing during DNA replication, as we explain below.

### Replication Slippage

In addition to point mutations, DNA replication can lead to the introduction of small insertions or deletions. These mutations can occur when one strand of the DNA template loops out and becomes displaced during replication, or when DNA polymerase slips or stutters during replication. If a loop occurs in the template strand during replication, DNA polymerase may miss the looped-out nucleotides, and a small deletion in the new strand will be introduced. If DNA polymerase repeatedly introduces nucleotides that are not present in the template strand, an insertion of one or more nucleotides will occur, creating an unpaired loop on the newly synthesized strand. Insertions and deletions may lead to frameshift mutations, amino acid additions, or deletions in the gene product.

**Replication slippage** can occur anywhere in the DNA but seems distinctly more common in regions containing repeated sequences. Repeat sequences are hot spots for DNA mutation and in some cases contribute to hereditary diseases, as discussed in Section 16.4. The hypermutability of repeat sequences in noncoding regions of the genome is the basis for current methods of forensic DNA analysis.

### Tautomeric Shifts

Purines and pyrimidines can exist in tautomeric forms—that is, in alternate chemical forms that differ by only a single proton shift in the molecule. The biologically important tautomers are the keto–enol forms of thymine and guanine and the amino–imino forms of cytosine and adenine. These shifts change the bonding structure of the molecule, allowing hydrogen bonding with noncomplementary bases. Hence, **tautomeric shifts** may lead to permanent base-pair changes and mutations. Figure 16–2 compares normal base-pairing relationships with rare unorthodox pairings. Anomalous T ≡ G and C = A pairs, among others, may be formed.

A mutation occurs during DNA replication when a transiently formed tautomer in the template strand pairs with a noncomplementary base. In the next round of replication, the "mismatched" members of the base pair are separated, and each becomes the template for its normal complementary base. The end result is a point mutation (Figure 16–3).

### Depurination and Deamination

Some of the most common causes of spontaneous mutations are two forms of DNA base damage: depurination and deamination. **Depurination** is the loss of one of the nitrogenous bases in an intact double-helical DNA molecule. Most frequently, the base is a purine—either guanine or adenine. These bases may be lost if the glycosidic bond linking the 1′-C of the deoxyribose and the number 9 position of the purine ring is broken, leaving an **apurinic site** on one strand of the DNA. Geneticists estimate that thousands of such spontaneous lesions are formed daily in the DNA of mammalian cells in culture. If apurinic sites are not repaired, there will be no base at that position to act as a template during DNA replication. As a result, DNA polymerase may introduce a nucleotide at random at that site.

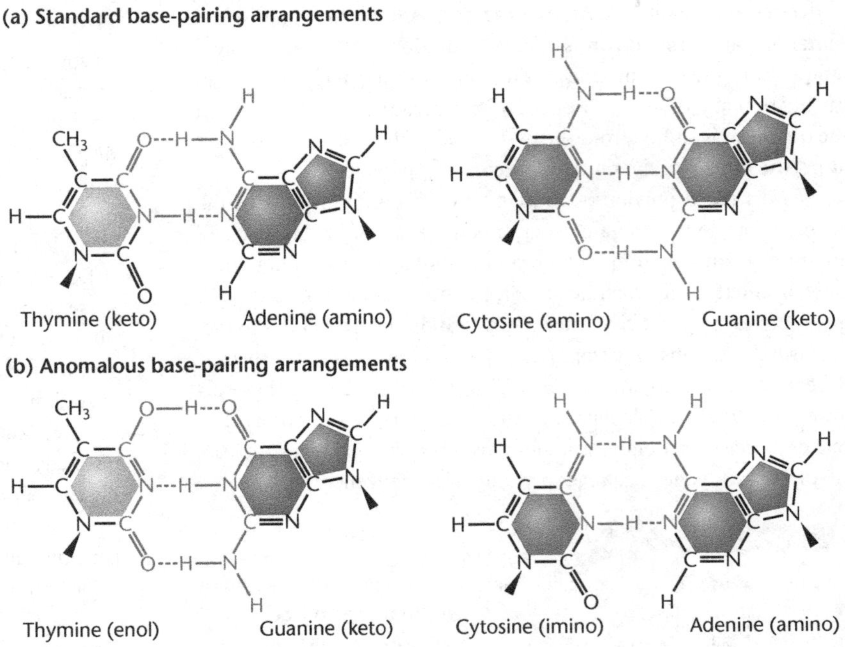


FIGURE 16-2 Standard base-pairing relationships (a) compared with examples of the anomalous base-pairing that occurs as a result of tautomeric shifts (b). The long triangle indicates the point at which the base bonds to the pentose sugar.

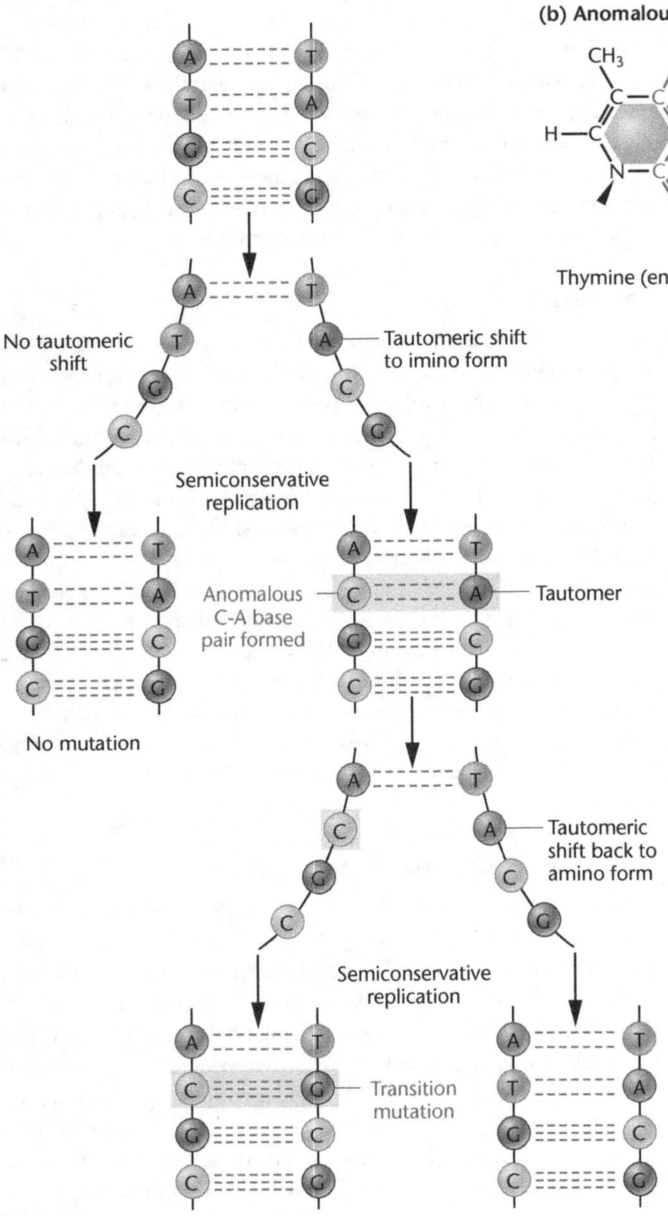


FIGURE 16-3 Formation of an A = T to G≡C transition mutation as a result of a tautomeric shift in adenine.

In **deamination**, an amino group in cytosine or adenine is converted to a keto group (Figure 16–4). In these cases, cytosine is converted to uracil, and adenine is changed to hypoxanthine. The major effect of these changes is an alteration in the base-pairing specificities of these two bases during DNA replication. For example, cytosine normally pairs with guanine. Following its conversion to uracil, which pairs with adenine, the original G ≡ C pair is converted to an A = U pair and then, in the next replication, is converted to an

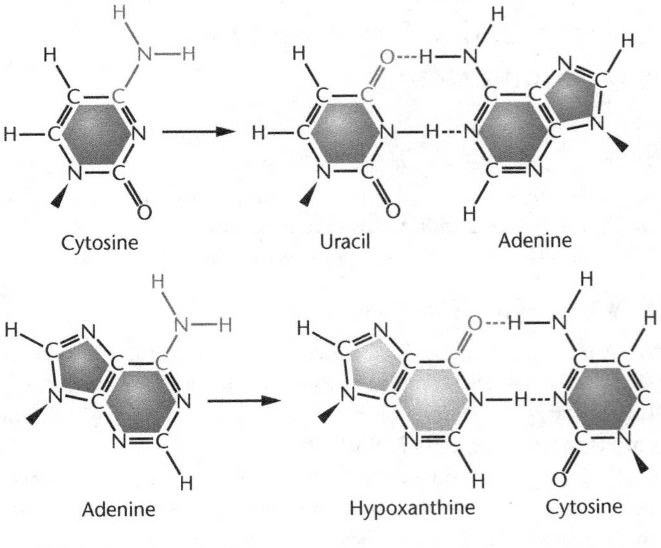


FIGURE 16-4 Deamination of cytosine and adenine, leading to new base pairing and mutation. Cytosine is converted to uracil, which base-pairs with adenine. Adenine is converted to hypoxanthine, which base-pairs with cytosine.

A = T pair. When adenine is deaminated, the original A = T pair is converted to a G ≡ C pair because hypoxanthine pairs naturally with cytosine. Deamination may occur spontaneously or as a result of treatment with chemical mutagens such as nitrous acid ($HNO_2$).

### Oxidative Damage

DNA may also suffer damage from the by-products of normal cellular processes. These by-products include reactive oxygen species (electrophilic oxidants) that are generated during normal aerobic respiration. For example, superoxides ($O_2^{\cdot -}$), hydroxyl radicals (·OH), and hydrogen peroxide ($H_2O_2$) are created during cellular metabolism and are constant threats to the integrity of DNA. Such **reactive oxidants**, also generated by exposure to high-energy radiation, can produce more than 100 different types of chemical modifications in DNA, including modifications to bases that lead to mispairing during replication.

### Transposons

Transposable genetic elements, or **transposons**, are DNA elements that can move within, or between, genomes. These elements are present in the genomes of all organisms, from bacteria to humans, and often comprise large portions of these genomes. Transposons can act as naturally occurring mutagens. If in moving to a new location they insert themselves into the coding region of a gene, they can alter the reading frame or introduce stop codons. If they insert into the regulatory region of a gene, they can disrupt proper expression of the gene. Transposons can also create chromosomal damage, including double-stranded breaks, inversions, and translocations. Transposable genetic elements are described in detail in Chapter 22.

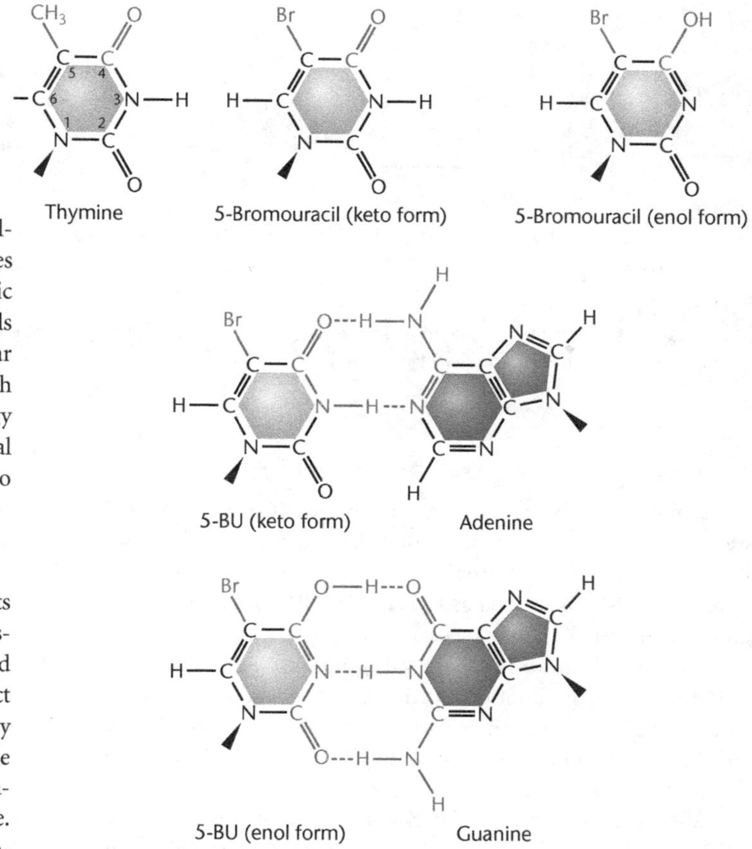


**FIGURE 16–5** Similarity of 5-bromouracil (5-BU) structure to thymine structure. In the common keto form, 5-BU base-pairs normally with adenine, behaving as a thymine analog. In the rare enol form, it pairs anomalously with guanine.

## 16.3 Induced Mutations Arise from DNA Damage Caused by Chemicals and Radiation

All cells on Earth are exposed to a plethora of agents called **mutagens**, which have the potential to damage DNA and cause mutations. Some of these agents, such as some fungal toxins, cosmic rays, and ultraviolet light, are natural components of our environment. Others, including some industrial pollutants, medical X rays, and chemicals within tobacco smoke, can be considered as unnatural or human-made additions to our modern world. On the positive side, geneticists harness some mutagens for use in analyzing genes and gene functions, as discussed in Chapter 23. The mechanisms by which some of these natural and unnatural agents lead to mutations are outlined in this section.

### Base Analogs

One category of mutagenic chemicals are **base analogs**, compounds that can substitute for purines or pyrimidines during nucleic acid biosynthesis. For example, **5-bromouracil (5-BU)**, a derivative of uracil, behaves as a thymine analog but is halogenated at the number 5 position of the pyrimidine ring. If 5-BU is chemically linked to deoxyribose, the nucleoside analog **bromodeoxyuridine (BrdU)** is formed. Figure 16–5 compares the structure of this analog with that of thymine. The presence of the bromine atom in place of the methyl group increases the probability that a tautomeric shift will occur. If 5-BU is incorporated into DNA in place of thymine and a tautomeric shift to the enol form occurs, 5-BU base-pairs with guanine. After one round of replication, an A = T to G ≡ C transition results. Furthermore, the presence of 5-BU within DNA increases the sensitivity of the molecule to ultraviolet (UV) light, which itself is mutagenic.

There are other base analogs that are mutagenic. For example, **2-amino purine (2-AP)** can act as an analog of adenine. In addition to its base-pairing affinity with thymine, 2-AP can also base-pair with cytosine, leading to possible transitions from A = T to G ≡ C following replication.

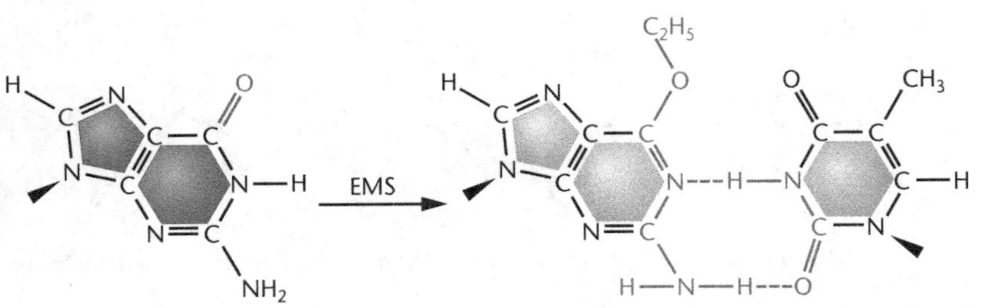


FIGURE 16-6 Conversion of guanine to 6-ethylguanine by the alkylating agent ethylmethane sulfonate (EMS). The 6-ethylguanine base-pairs with thymine.

## Alkylating Agents and Acridine Dyes

The sulfur-containing mustard gases, discovered during World War I, were some of the first chemical mutagens identified in chemical warfare studies. Mustard gases are **alkylating agents**—that is, they donate an alkyl group, such as $CH_3$ or $CH_3CH_2$, to amino or keto groups in nucleotides. Ethylmethane sulfonate (EMS), for example, alkylates the keto groups in the number 6 position of guanine and in the number 4 position of thymine. As with base analogs, base-pairing affinities are altered, and transition mutations result. For example, 6-ethylguanine acts as an analog of adenine and pairs with thymine (Figure 16–6).

Chemical mutagens called **acridine dyes** cause frameshift mutations. As illustrated in Figure 16–1, frameshift mutations result from the addition or removal of one or more base pairs in the polynucleotide sequence of the gene. Acridine dyes such as proflavin and acridine orange are about the same dimensions as nitrogenous base pairs and **intercalate**, or wedge, between the purines and pyrimidines of intact DNA. Intercalation introduces contortions in the DNA helix that may cause deletions and insertions during DNA replication or repair, leading to frameshift mutations.

## Ultraviolet Light

All electromagnetic radiation consists of energetic waves that we define by their different wavelengths (Figure 16–7). The full range of wavelengths is referred to as the **electromagnetic spectrum**, and the energy of any radiation in the spectrum varies inversely with its wavelength. Waves in the range of visible light and longer are benign when they interact with most organic molecules. However, waves of shorter length than visible light, being inherently more energetic, have the potential to disrupt organic molecules. As we know, purines and pyrimidines absorb **ultraviolet (UV) radiation** most intensely at a wavelength of about 260 nm. One major effect of UV radiation on DNA is the creation of **pyrimidine dimers**—chemical species consisting of two identical pyrimidines—particularly ones consisting of two thymine residues (Figure 16–8). The dimers distort the DNA conformation and inhibit normal replication. As a result, errors can be introduced in the base sequence of DNA during replication. When UV-induced dimerization is extensive, it is responsible (at least in part) for the killing effects of UV radiation on cells.

## Ionizing Radiation

As noted above, the energy of radiation varies inversely with wavelength. Therefore, **X rays**, **gamma rays**, and **cosmic rays** are more energetic than UV radiation (Figure 16–7). As a result, they penetrate deeply into tissues, causing ionization of the molecules encountered along the way.

As X rays penetrate cells, electrons are ejected from the atoms of molecules encountered by the radiation. Thus, stable molecules and atoms are transformed into **free radicals**—chemical species containing one or more unpaired electrons. Free radicals can directly or indirectly affect the genetic material, altering purines and pyrimidines in DNA and resulting in point mutations. **Ionizing radiation**, radiation energetic enough to produce ions and other high-energy particles, is also capable of breaking phosphodiester bonds, disrupting the integrity of chromosomes, and producing a variety of chromosomal aberrations, such as deletions, translocations, and chromosomal fragmentation.

Figure 16–9 shows a graph of the percentage of induced X-linked recessive lethal mutations

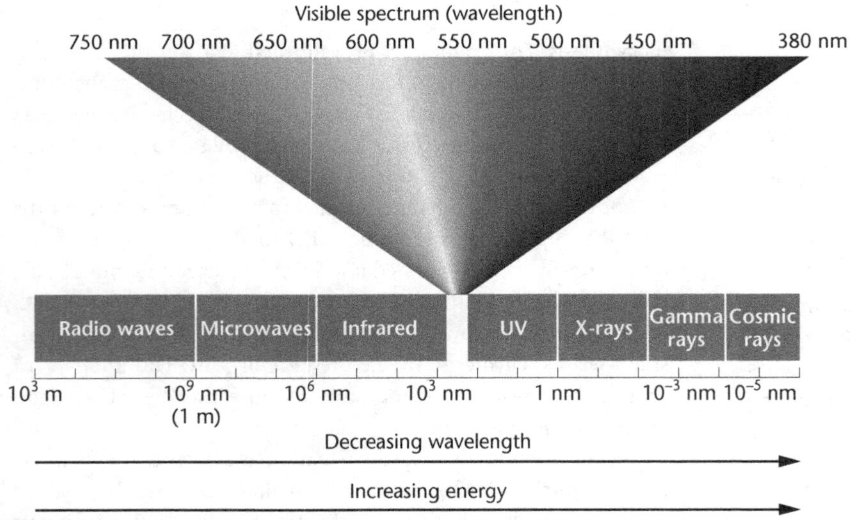


FIGURE 16-7 The regions of the electromagnetic spectrum and their associated wavelengths.

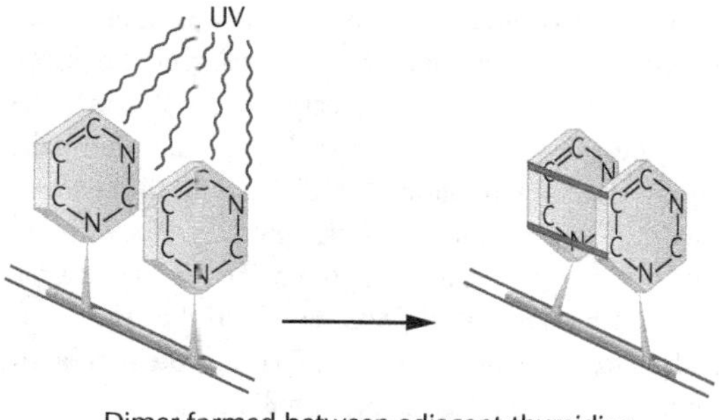


**FIGURE 16–8** Induction of a thymine dimer by UV radiation, leading to distortion of the DNA. The covalent crosslinks occur between the atoms of the pyrimidine ring.

versus the dose of X rays administered. There is a linear relationship between X ray dose and the induction of mutation; for each doubling of the dose, twice as many mutations are induced. Because the line intersects near the zero axis, this graph suggests that even very small doses of radiation are mutagenic.

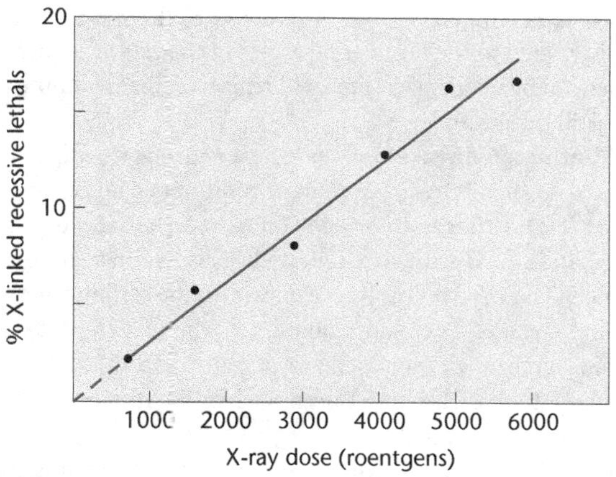


**FIGURE 16–9** Plot of the percentage of X-linked recessive mutations induced in *Drosophila* by increasing doses of X rays. If extrapolated, the graph intersects the zero axis as shown by the dashed line.

**NOW SOLVE THIS**

Problem 19 on page 432 asks you to describe ways in which the chemotherapy drug melphalan kills cancer cells and how these cells might repair the DNA damage caused by this drug.

■ HINT: *To answer this question, consider the effect of the alkylation of guanine on base pairing during DNA replication. In Section 16.7, you will learn about the ways in which cells repair the types of mutations introduced by alkylating agents.*

16.4

# Genomics and Gene Sequencing Have Enhanced Our Understanding of Mutations in Humans

Historically, scientists learned about mutations by analyzing the amino acid sequences of proteins within populations that show substantial diversity. This diversity, which arises during evolution, is a reflection of changes in the triplet codons following substitution, insertion, or deletion of one or more nucleotides in the DNA sequences of protein-coding genes.

As our ability to analyze DNA more directly increases, we are able to examine the actual nucleotide sequence of genes and surrounding DNA regions, and to gain greater insights into the nature of mutations. Modern techniques for rapid sequencing of DNA have greatly extended this knowledge. In this section, we describe several human genetic traits and what we have learned about the mutations responsible for them.

## ABO Blood Groups

The **ABO blood group system** is based on a series of antigenic determinants found on erythrocytes and other cells, particularly epithelial cells. As we discussed in Chapter 4, three alleles of the *I* gene encode the **glycosyltransferase** enzyme. The H substance is modified to either the A or B antigen as a result of the activity of this enzyme, depending on whether the enzyme was encoded by the $I^A$ or $I^B$ allele, respectively. Failure to modify the H substance results from the presence of the null, $I^O$, allele.

When the DNA sequences of the $I^A$ and $I^B$ alleles are compared, four consistent single-nucleotide substitutions are found. The resulting changes in the amino acid sequence of the glycosyltransferase gene product lead to altered gene product function and, consequently, to different modifications of the H substance.

The $I^O$ allele situation is unique and interesting. Individuals homozygous for this allele have type O blood, lack glycosyltransferase activity, and fail to modify the H substance. Analysis of the DNA of the $I^O$ allele shows one consistent change that is unique compared with the sequences of the other alleles—the deletion of a single nucleotide early in the coding sequence, causing a frameshift mutation. A complete messenger RNA is transcribed, but at translation, the reading frame shifts at the point of the deletion and continues out of frame for about 100 nucleotides before a stop codon is encountered. At this point, the glycosyltransferase polypeptide chain terminates prematurely, resulting in a nonfunctional product.

These findings provide a direct molecular explanation of the ABO allele system and the basis for the biosynthesis of the corresponding antigens. The molecular basis for the antigenic phenotypes is clearly a matter of base substitutions and frameshift mutations within the gene encoding the glycosyltransferase enzyme.

## Muscular Dystrophy

Muscular dystrophies are genetic diseases characterized by progressive muscle weakness and degeneration. There are many types of muscular dystrophy, and they differ in their severities, onsets, and genetic causes. Two related forms of muscular dystrophy—**Duchenne muscular dystrophy (DMD)** and **Becker muscular dystrophy (BMD)**—are recessive, X-linked conditions. DMD is the more severe of the two diseases, with a rapid progression of muscle degeneration and involvement of the heart and lungs. Males with DMD usually lose the ability to walk by the age of 12 and may die in their early 20s. The incidence of 1 in 5000 live male births makes DMD one of the most common life-shortening hereditary diseases. In contrast, BMD does not involve the heart or lungs and progresses slowly, from adolescence to the age of 50 or more.

The gene responsible for DMD and BMD—the *dystrophin* gene—is unusually large, consisting of about 2.5 million base pairs. In normal (unaffected) individuals, transcription and subsequent processing of the *dystrophin* initial transcript results in a messenger RNA containing only about 14,000 bases (14 kb). It is translated into the protein **dystrophin**, which consists of 3685 amino acids.

In recent years, geneticists have determined the molecular basis of mutations leading to BMD and DMD. With few exceptions, DMD mutations change the reading frame of the *dystrophin* gene, whereas BMD mutations usually do not. Studies have shown that about two-thirds of mutations in the *dystrophin* gene that lead to DMD and BMD are deletions and insertions. Only one-third are point mutations. The majority of DMD mutations lead to premature termination of translation. This, in turn, leads to degradation of the improperly translated *dystrophin* transcript and the nearly complete absence of dystrophin protein. In contrast, the majority of BMD gene mutations alter the internal sequence of the *dystrophin* transcript and protein, but do not alter the translation reading frame. As a result, a modified but somewhat functional dystrophin protein is produced, preventing the severe consequences of DMD.

## Fragile X Syndrome, Myotonic Dystrophy, and Huntington Disease

Beginning in about 1990, molecular analysis of the genes responsible for a number of inherited human disorders provided a remarkable set of observations. Researchers discovered that some mutant genes contain expansions of **trinucleotide repeat sequences**, specific trinucleotide DNA sequences repeated many times. Normal individuals usually have fewer than 15 repetitions of these sequences; but about 20 human disorders, including a number of developmental and degenerative diseases, appear to be caused by the presence of abnormally large numbers of repeat sequences within specific genes. For example, although trinucleotide repeat sequences are also present in the nonmutant (normal) alleles, the mutant alleles responsible for fragile X syndrome, myotonic dystrophy, and Huntington disease contain significant increases in the number of times the trinucleotide is repeated.

In **fragile X syndrome**, described in detail in Chapter 8, the responsible gene, *FMR-1*, may contain several to several thousand copies of the trinucleotide sequence CGG located in the 5′-untranslated region of the gene. Individuals with up to 55 copies are normal and do not display the mental retardation associated with the syndrome. Individuals with 55 to 230 copies are considered carriers. Although they are normal, their offspring may bear even more copies and express the syndrome. The large regions of CGG repeats in the gene's regulatory region result in loss of expression of the FMRP protein, thought to be an RNA-binding protein affecting brain cell function.

**Myotonic dystrophy** is the most common form of adult muscular dystrophy, affecting 1 in 8000 individuals. The dominantly inherited disorder is not as severe as DMD, and it is highly variable in both symptoms and age of onset. Mild myotonia (atrophy and weakness) of the musculature of the face and extremities is most common. Cataracts, reduced cognitive ability, and cutaneous and intestinal tumors are also part of the syndrome.

The affected gene, *MDPK*, encodes a serine–threonine protein kinase, MDPK. This protein is the product of 15 exons of the gene, the last of which encodes the 3′-untranslated sequence of the mRNA. It is this sequence that houses multiple copies of the trinucleotide CTG. Individuals with 5 to 35 copies of the CTG repeat are normal, and the number of copies is stable from generation to generation. Individuals with more than about 150 copies exhibit symptoms ranging from mild to severe, with the severity and onset directly related to the number of copies of the repeat sequence. The mechanism by which these repeated regions cause myotonic dystrophy is still uncertain.

**Huntington disease (HD)** is a fatal neurodegenerative disease, inherited as an autosomal dominant condition. The gene responsible for HD is located on chromosome 4 and contains the trinucleotide CAG sequence, repeated 11 to 34 times in normal individuals. The CAG repeat is located within the coding region of the gene and encodes a polyglutamine tract. The CAG repeat sequence exists in significantly increased numbers (up to 120) in diseased individuals. Much earlier onset occurs when the number of copies is closer to the upper range.

The role of repeated sequences in normal and mutant genes remains a mystery. Their locations within the gene vary in each of the disorders presented here. In Huntington disease, the repeats lie within the coding portion of the gene. This causes the mutant **huntingtin** protein to contain an excess of glutamine residues. In the gene responsible for fragile X syndrome, the repeat sequence is upstream (beyond the 5′ end) of the gene's coding region, in an area that is most often involved in regulating gene expression. In the case of myotonic dystrophy, the repeat is downstream of the coding region (beyond the 3′ end).

The mechanism by which the repeated sequence expands from generation to generation is of great interest. It is thought that expansion may result from either errors during replication or errors during DNA damage repair. Whatever the cause may be, the presence of these short and unstable repeat sequences seems to be prevalent in humans and in many other organisms.

**NOW SOLVE THIS**

Problem 20 on page 432 asks you to speculate on why the *dystrophin* gene appears to suffer a large number of mutations.

■ HINT: *In answering this question, you may want to think about the size of the gene and the organization of its introns and exons.*

## 16.5 The Ames Test Is Used to Assess the Mutagenicity of Compounds

There is great concern about the possible mutagenic properties of any chemical that enters the human body, whether through the skin, the digestive system, or the respiratory tract. Examples are chemicals that may be found in air and water pollution, food preservatives, artificial sweeteners, herbicides, pesticides, and pharmaceutical products. Mutagenicity may be tested in various organisms, including fungi, plants, and cultured mammalian cells; however, the most common test, which we describe here, uses bacteria.

The **Ames test** uses any of a dozen strains of the bacterium *Salmonella typhimurium* that have been selected for their ability to reveal the presence of specific types of mutations. For example, some strains are used to detect base-pair substitutions, and other strains detect various frameshift mutations. Each strain contains a mutation in one of the genes of the histidine operon. The mutant strains are unable to synthesize histidine ($his^-$ strains) and therefore require histidine for growth (Figure 16–10). The assay measures the frequency of reverse mutation within the mutant gene, which yields wild-type bacteria ($his^+$ revertants). These *Salmonella* strains also have an increased sensitivity to mutagens due to the presence of mutations in genes involved in DNA damage repair and synthesis of the lipopolysaccharide barrier that coats bacteria and protects them from external substances.

Many substances entering the human body are relatively innocuous until activated metabolically, usually in the liver, to a more chemically reactive product. Thus, the Ames test includes a step in which the test compound is incubated *in vitro* in the presence of a mammalian liver extract. Alternatively, test compounds may be injected into a mouse in which they are modified by liver enzymes and then recovered for use in the Ames test.

In the initial use of Ames testing in the 1970s, a large number of known **carcinogens**, or cancer-causing agents, were examined, and more than 80 percent of these were shown to be strong mutagens. This is not surprising, as the transformation of cells to the malignant state occurs as a result of mutations. Although a positive response in the Ames test does not prove that a compound is carcinogenic, the Ames test is useful as a preliminary screening device. The Ames test is used extensively during the development of industrial and pharmaceutical chemical compounds.

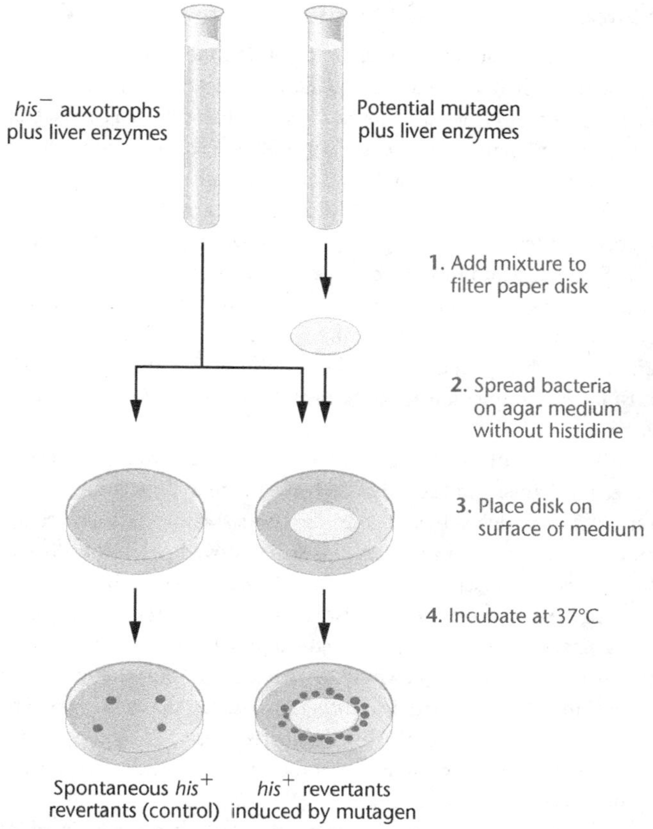


**FIGURE 16–10** The Ames test, which screens compounds for potential mutagenicity.

WEB TUTORIAL 16.2 CHEMICAL MUTAGENESIS: THE AMES TEST

## 16.6 Organisms Use DNA Repair Systems to Counteract Mutations

Living systems have evolved a variety of elaborate repair systems that counteract both spontaneous and induced DNA damage. These **DNA repair** systems are absolutely essential to the maintenance of the genetic integrity of organisms and, as such, to the survival of organisms on Earth. The balance between mutation and repair results in the observed mutation rates of individual genes and organisms. Of foremost interest in humans is the ability of these systems to counteract genetic damage that would otherwise result in genetic diseases and cancer. The link between defective DNA repair and cancer susceptibility is described in Chapter 20.

We now embark on a review of some systems of DNA repair. Since the field is expanding rapidly, our goal here is merely to survey the major approaches that organisms use to counteract genetic damage.

## Proofreading and Mismatch Repair

Some of the most common types of mutations arise during DNA replication when an incorrect nucleotide is inserted by DNA polymerase. The enzyme in bacteria (DNA polymerase III) makes an error approximately once every 100,000 insertions, leading to an error rate of $10^{-5}$. Fortunately, the enzyme polices its own synthesis by proofreading each step, catching 99 percent of those errors. If an incorrect nucleotide is inserted during polymerization, the enzyme has the potential to recognize the error and "reverse" its direction. It then behaves as a 3′ to 5′ exonuclease, cutting out the incorrect nucleotide and replacing it with the correct one. This improves the efficiency of replication one hundredfold, creating only $1/10^7$ mismatches immediately following DNA replication, for a final error rate of $10^{-7}$.

DNA REPAIR: PROOFREADING, MISMATCH, AND SOS
WEB TUTORIAL 16.3

To cope with those errors that remain after proofreading, another mechanism, called **mismatch repair**, may be activated. As in repair of other DNA lesions, the alterations, or mismatches, are detected, the incorrect nucleotides are removed, and the correct nucleotides are inserted in their place. But a special problem is confronted in the correction of a mismatch. How does the repair system recognize which strand is correct (the template) and which contains the incorrect base (the newly synthesized strand)? If the mismatch is recognized but no such discrimination occurs, the excision will be random, and the strand bearing the correct base will be clipped out 50 percent of the time. Hence, strand discrimination by a repair enzyme is a critical step.

The process of strand discrimination has been elucidated, at least in some bacteria, including *E. coli*, and is based on **DNA methylation**. These bacteria contain an enzyme, adenine methylase, which recognizes the DNA sequence

$$\begin{array}{c} 5' - \text{GATC} - 3' \\ 3' - \text{CTAG} - 5' \end{array}$$

as a substrate during DNA replication, adding a methyl group to each of the adenine residues.

Following a round of replication, the newly synthesized strand remains temporarily unmethylated, as the methylase lags behind the DNA polymerase. At this point, the repair enzyme recognizes the mismatch and binds to the unmethylated (newly synthesized) DNA strand. A nick is made by an endonuclease enzyme, either 5′ or 3′ to the mismatch on the unmethylated strand. The nicked DNA strand is then unwound and degraded by an exonuclease, until the region of the mismatch is reached. Finally, DNA polymerase fills in the gap created by the exonuclease, using the correct DNA strand as a template. DNA ligase then seals the gap.

A series of *E. coli* gene products, Mut H, L, and S, are involved in mismatch repair. Mutations in any of these genes result in bacterial strains deficient in mismatch repair. While the preceding mechanism is based on studies of *E. coli*, similar mechanisms involving homologous proteins exist in yeast and in mammals.

## Postreplication Repair and the SOS Repair System

Many other types of repair have been discovered, illustrating the diversity of mechanisms that have evolved to overcome DNA damage. One system, called **postreplication repair,** responds *after* damaged DNA has escaped repair and failed to be completely replicated (thus its name). As illustrated in Figure 16–11, when DNA bearing a lesion of some sort (such as a pyrimidine dimer) is being replicated, DNA polymerase may stall at the lesion and then skip over it, leaving a gap on the newly synthesized strand. To correct the gap, the RecA protein directs a recombinational exchange with the corresponding region on the undamaged parental strand

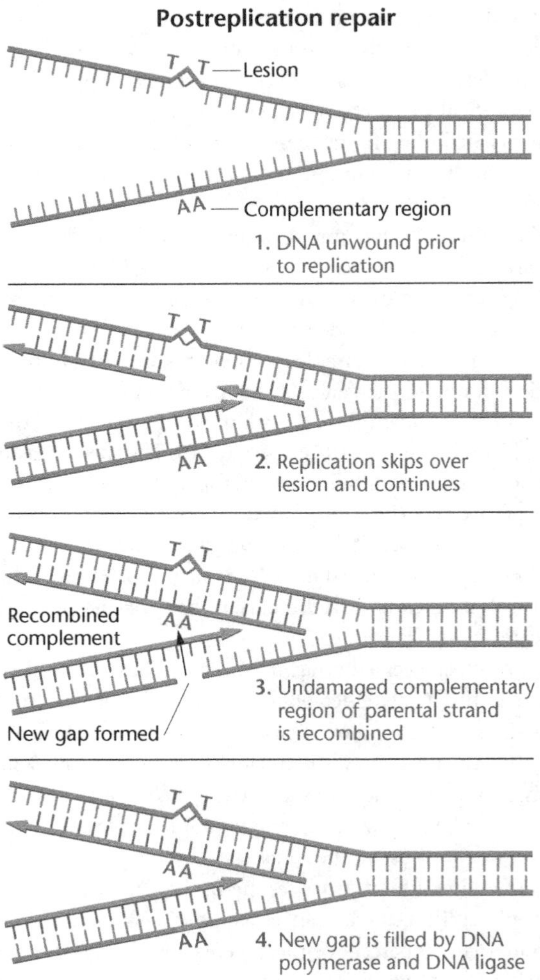


**FIGURE 16–11** Postreplication repair occurs if DNA replication has skipped over a lesion such as a thymine dimer. Through the process of recombination, the correct complementary sequence is recruited from the parental strand and inserted into the gap opposite the lesion. The new gap is filled by DNA polymerase and DNA ligase.

of the same polarity (the "donor" strand). When the undamaged segment of DNA replaces the gapped segment, this transfers the gap to the donor strand. The gap can be filled by repair synthesis as replication proceeds. Because a recombinational event is involved in this type of DNA repair, it is considered to be a form of homologous recombination repair.

Still another repair pathway in *E. coli*, the **SOS repair system**, also responds to damaged DNA, but in a different way. In the presence of a large number of DNA mismatches and gaps during replication, bacteria can induce the expression of about 20 genes (including *lexA*, *recA*, and *uvr*) whose products allow DNA replication to occur even in the presence of these types of lesions. This type of repair is a last resort to minimize DNA damage, hence its name. During SOS repair, DNA synthesis becomes error-prone, inserting random and possibly incorrect nucleotides in places that would normally stall DNA replication. As a result, SOS repair itself becomes mutagenic—although it may allow the cell to survive DNA damage that would otherwise kill it.

## Photoreactivation Repair: Reversal of UV Damage

As illustrated in Figure 16–8, UV light is mutagenic as a result of the creation of pyrimidine dimers. UV-induced damage to *E. coli* DNA can be partially reversed if, following irradiation, the cells are exposed briefly to light in the blue range of the visible spectrum. The process is dependent on the activity of a protein called **photoreactivation enzyme** (**PRE**). The enzyme's mode of action is to cleave the bonds between thymine dimers, thus directly reversing the effect of UV radiation on DNA (Figure 16–12). Although the enzyme will associate with a dimer in the dark, it must absorb a photon of light to cleave the dimer. In spite of its ability to reduce the number of UV-induced mutations, **photoreactivation repair** is not absolutely essential in *E. coli*; we know this because a mutation creating a null allele in the gene coding for PRE is not lethal. Nonetheless, the enzyme is detectable in many organisms, including bacteria, fungi, plants, and some vertebrates—although not in humans. Humans and other organisms that lack photoreactivation repair must rely on other repair mechanisms to reverse the effects of UV radiation.

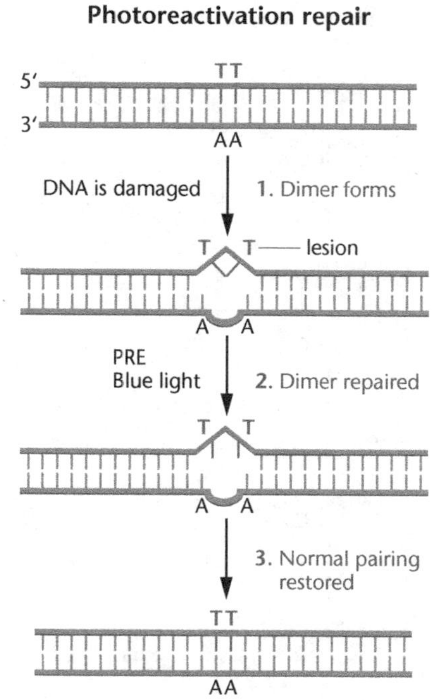


**FIGURE 16–12** Damaged DNA repaired by photoreactivation repair. The bond creating the thymine dimer is cleaved by the photoreactivation enzyme (PRE), which must be activated by blue light in the visible spectrum.



## Base and Nucleotide Excision Repair

In addition to PRE, light-independent repair systems exist in all prokaryotes and eukaryotes. The basic mechanisms involved in these types of repair—referred to as **excision repair**—consist of the following three steps and can be described generally as "cut-and-paste" systems.

1. The distortion or error present on one of the two strands of the DNA helix is recognized and enzymatically clipped out by an endonuclease. Excisions in the phosphodiester backbone usually include a number of nucleotides adjacent to the error as well, leaving a gap on one strand of the helix.
2. A DNA polymerase fills in the gap by inserting deoxyribonucleotides complementary to those on the intact strand, which it uses as a replicative template. The enzyme adds these bases to the free 3′-OH end of the clipped DNA. In *E. coli*, this step is usually performed by DNA polymerase I.
3. DNA ligase seals the final "nick" that remains at the 3′-OH end of the last base inserted, closing the gap.

There are two types of excision repair: base excision repair and nucleotide excision repair. **Base excision repair (BER)** corrects damage to nitrogenous bases created by spontaneous hydrolysis or by agents that chemically alter them. The first step in the BER pathway in *E. coli* involves the recognition of the chemically altered base by **DNA glycosylases**, which are specific to different types of DNA damage (Figure 16–13). For example, the enzyme uracil DNA glycosylase recognizes the presence of uracil in DNA. The enzyme first cuts the glycosidic bond between the base and the sugar, creating an apyrimidinic site. Such a sugar with a missing base is then recognized by an enzyme called **AP endonuclease**. The endonuclease makes a cut in the phosphodiester backbone at the apyrimidinic site. This creates a distortion in the DNA helix that is recognized by the excision-repair system, which is then activated, leading to the correction of the error.

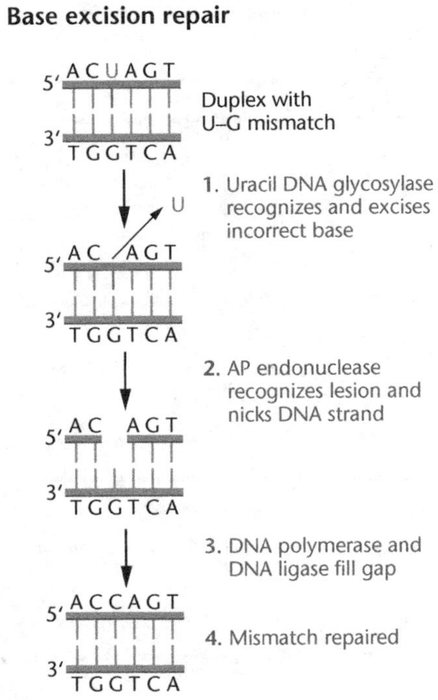


FIGURE 16–13 Base excision repair (BER) accomplished by uracil DNA glycosylase, AP endonuclease, DNA polymerase, and DNA ligase. Uracil is recognized as a noncomplementary base, excised, and replaced with the complementary base (C).

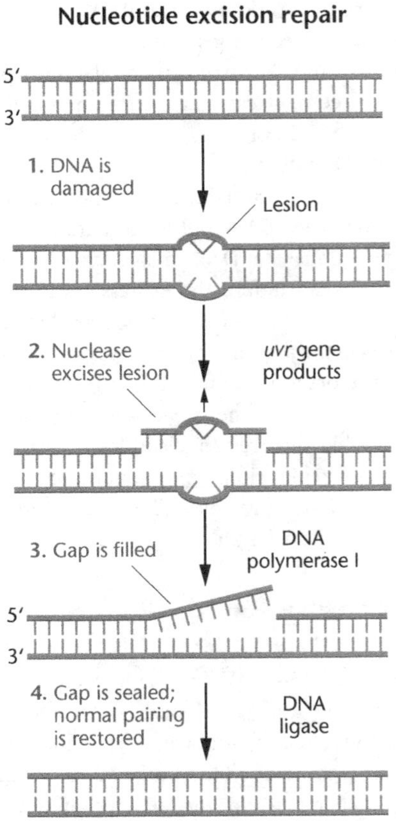


FIGURE 16–14 Nucleotide excision repair (NER) of a UV-induced thymine dimer. During repair, 13 bases are excised in prokaryotes, and 28 bases are excised in eukaryotes.

Although much has been learned about DNA glycosylases in *E. coli*, BER repair systems also operate in eukaryotes from yeast to humans. Experimental evidence shows that both mouse and human cells that are defective in BER activity are hypersensitive to the killing effects of gamma rays and oxidizing agents.

**Nucleotide excision repair (NER)** pathways repair "bulky" lesions in DNA that alter or distort the double helix, such as the UV-induced pyrimidine dimers discussed previously. The NER pathway (Figure 16–14) was first discovered in *E. coli* by Paul Howard-Flanders and coworkers, who isolated several independent mutants that are sensitive to UV radiation. One group of genes was designated *uvr* (ultraviolet repair) and included the *uvrA, uvrB,* and *uvrC* mutations. In the NER pathway, the *uvr* gene products are involved in recognizing and clipping out lesions in the DNA. Usually, a very specific number of nucleotides are clipped out around both sides of the lesion. In *E. coli*, usually a total of 13 nucleotides are removed, which include the lesion. The repair is then completed by DNA polymerase I and DNA ligase, in a manner similar to that occurring in BER. The undamaged strand opposite the lesion is used as a template for the replication, resulting in repair.

## Nucleotide Excision Repair and Xeroderma Pigmentosum in Humans

The mechanism of nucleotide excision repair (NER) in eukaryotes is much more complicated than that in prokaryotes and involves many more proteins, encoded by about 30 genes. Much of what is known about the system in humans has come from detailed studies of individuals with **xeroderma pigmentosum (XP)**, a rare recessive genetic disorder that predisposes individuals to severe skin abnormalities and cancers. These individuals have lost their ability to undergo NER; as a result, individuals suffering from XP who are exposed to the UV radiation in sunlight exhibit reactions that range from initial freckling and skin ulceration to the development of skin cancer.

The condition is severe and may be lethal, although early detection and protection from sunlight can arrest it (Figure 16–15). The repair of UV-induced lesions was investigated *in vitro*, using human fibroblast cell cultures derived from normal individuals and those with XP. (Fibroblasts are undifferentiated connective tissue cells.) The results of these studies suggested that the XP phenotype could be caused by more than one mutant gene.

In 1968, James Cleaver showed that cells from XP patients were deficient in **unscheduled DNA synthesis** (DNA synthesis other than that occurring during chromosome replication), which is elicited in normal cells by UV radiation. Because this type of synthesis is thought to represent the activity of DNA polymerization during excision repair, the lack of unscheduled DNA synthesis in XP patients suggested that XP may be a deficiency in excision repair.

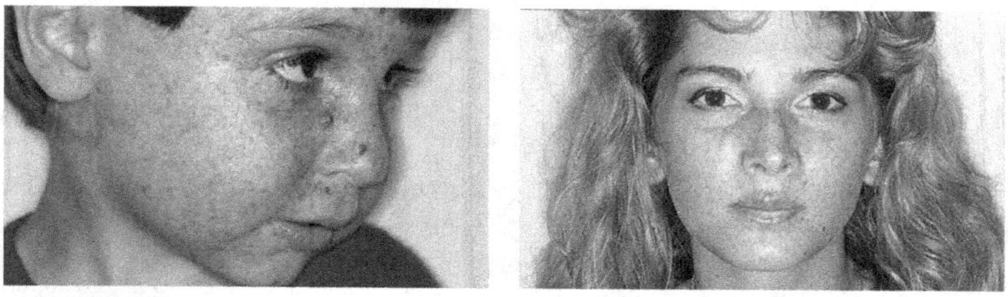

**FIGURE 16–15** Two individuals with xeroderma pigmentosum. The 4-year-old boy on the left shows marked skin lesions induced by sunlight. Mottled redness (erythema) and irregular pigment changes in response to cellular injury are apparent. Two nodular cancers are present on his nose. The 18-year-old girl on the right has been carefully protected from sunlight since her diagnosis of xeroderma pigmentosum in infancy. Several cancers have been removed, and she has worked as a successful model.

The link between XP and excision repair was further strengthened by studies using **somatic cell hybridization**. Fibroblast cells from any two unrelated XP patients, when grown together in tissue culture, can fuse together, forming a heterokaryon. A **heterokaryon** is a single cell with two nuclei from different organisms but a common cytoplasm. Excision repair in the heterokaryon can be measured by the level of unscheduled DNA synthesis. If the mutation in each of the two XP cells occurs in the same gene, the heterokaryon, like the cells that fused to form it, will still be unable to undergo excision repair. This is because there is no normal copy of the relevant gene present in the heterokaryon. However, if excision repair does occur in the heterokaryon, the mutations in the two XP cells must have been present in two different genes. Hence, the two mutants are said to demonstrate **complementation**, a concept also discussed in Chapters 4 and 6. Complementation occurs because the heterokaryon has at least one normal copy of each gene in the fused cell. By fusing XP cells from a large number of XP patients, researchers were able to determine how many genes contribute to the XP phenotype.

Based on these and other studies, XP patients have been divided into seven complementation groups, indicating that at least seven different genes are involved in excision repair in humans. A gene representing each of these complementation groups, *XPA* to *XPG* (*X*eroderma *P*igmentosum gene *A* to *G*), have now been identified, and a homologous gene for each has been identified in yeast. Approximately 20 percent of XP patients do not fall into any of the seven complementation groups. These patients manifest similar symptoms, but their fibroblasts do not demonstrate defective excision repair. There is some evidence that these XP cells are less efficient in normal DNA replication.

As a result of the study of defective genes in xeroderma pigmentosum, a great deal is now known about how NER counteracts DNA damage in normal cells. The first step in humans is the recognition of the damaged DNA by proteins encoded by the *XPC*, *XPE*, and *XPA* genes. These proteins then recruit the remainder of the repair proteins to the site of DNA damage. The *XPB* and *XPD* genes encode helicases, and the *XPF* and *XPG* genes encode nucleases. The excision repair complex containing these and other factors excises a 28-nucleotide-long fragment from the DNA strand that contains the lesion.

Another defect in the human NER pathway is **Cockayne syndrome (CS)**, which is characterized by impairment of both physical and neurological development. While CS patients, like XP patients, are highly photosensitive, they do not show a disposition to tumor formation. Five of the genes involved in CS are also involved in XP (*XPB*, *XPA*, *XPD*, *XPF*, and *XPG*). Products of the CS genes participate in **transcription-coupled repair**, a type of NER in which only the transcribed strands of actively transcribed genes are repaired.

**NOW SOLVE THIS**

Problem 24 on page 432 examines the results of heterokaryon analysis of seven cell lines derived from xeroderma pigmentosum patients. You are asked to determine the number of complementation groups represented by these data.

■ HINT: *Complementation (+) results only when the two mutations being examined are in different complementation groups (genes).*

## Double-Strand Break Repair in Eukaryotes

Thus far, we have discussed repair pathways that deal with damage or errors within one strand of DNA. We conclude our discussion of DNA repair by considering what happens when both strands of the DNA helix are cleaved—as a result of exposure to ionizing radiation, for example. These types of damage are extremely dangerous to cells, leading to chromosome rearrangements, disease syndromes, or cell death. In this section, we will discuss double-strand breaks in eukaryotic cells.

Specialized forms of DNA repair, the **DNA double-strand break repair (DSB repair)** pathways, are activated and are responsible for reattaching two broken DNA strands. Recently, interest in DSB repair has grown because defects in these pathways are associated with X-ray hypersensitivity and immune deficiency. Such defects may also underlie familial disposition to breast and ovarian cancer. Several human disease syndromes, such as Fanconi's anemia and ataxia telangiectasia, result from defects in DSB repair.

One pathway involved in double-strand break repair is **homologous recombination repair**. The first step in this process involves the activity of an enzyme that recognizes the double-strand break, then digests back the 5′ ends of the broken DNA helix, leaving overhanging 3′ ends (Figure 16–16). One 3′ overhanging end searches for a region of sequence complementarity on the sister chromatid and then invades the homologous DNA duplex, aligning the complementary sequences. Once aligned, DNA synthesis proceeds from the 3′ overhanging end, using the undamaged DNA strand as a template. The interaction of two sister chromatids is necessary because, when both strands of one helix are broken, there is no undamaged parental DNA strand available to use as a source of the complementary template DNA sequence during repair. After DNA repair synthesis, the resulting heteroduplex molecule is resolved and the two chromatids separate, as previously discussed in Section 11.8.

The process usually occurs during the late S or early G2 phase of the cell cycle, after DNA replication, a time when sister chromatids are available to be used as repair templates. Because an undamaged template is used during repair synthesis, homologous recombination repair is an accurate process.

A second pathway, called **nonhomologous end joining**, also repairs double-strand breaks. However, as the name implies, the

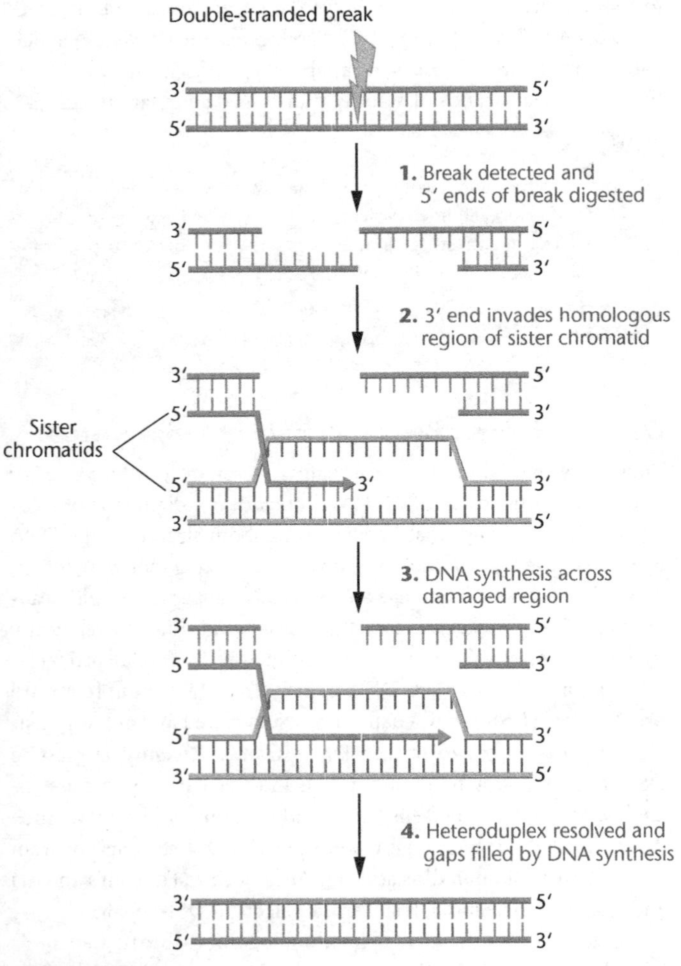


FIGURE 16–16 Steps in homologous recombination repair of double-stranded breaks.

mechanism does not recruit a homologous region of DNA during repair. This system is activated in G1, prior to DNA replication. End joining involves a complex of three proteins including DNA-dependent protein kinase. These proteins bind to the free ends of the broken DNA, trim the ends, and ligate them back together. Because some nucleotide sequences are lost in the process of end joining, it is an error-prone repair system. In addition, if more than one chromosome suffers a double-strand break, the wrong ends could be joined together, leading to abnormal chromosome structures, such as those discussed in Chapter 8.

## 16.7 Geneticists Use Mutations to Identify Genes and Study Gene Function

We conclude this chapter by describing how mutations contribute to genetic analysis. In order to identify the genes and processes that regulate biological functions, geneticists often induce mutations in model organisms. As described in Chapter 23, a good model organism is easy to grow, has a short generation time, produces abundant progeny, and is easy to mutagenize and to cross. The most extensively used model organisms are bacteria (*E. coli*), budding yeasts (*Saccharomyces cerevisiae*), fruit flies (*Drosophila melanogaster*), nematodes (*Caenorhabditis elegans*), mustard plants (*Arabidopsis thaliana*), and mice (*Mus musculus*).

Geneticists sometimes begin their analyses of genes by examining spontaneous, naturally occurring mutations. This is particularly true for genes involved in human genetic diseases. However, as mutations are generally rare in nature, researchers need to induce mutations in model organisms in order to increase the chances of detecting a relevant mutant. The goal of mutagenesis is to create one mutation at random in the genome of each individual in the experimental population, so that one gene product is disrupted in each individual, leaving the rest of the genome wild-type.

During genetic analyses, researchers use a wide range of different mutagens, depending on the type of mutation desired. For example, ionizing radiation can be used to create chromosome breaks, deletions, and translocations. Though useful for some types of studies, such mutations often have severe effects on the phenotype, making further genetic analysis difficult. In contrast, chemicals such as ethyl methane sulfonate (EMS) and nitrosoguanidine cause single base-pair changes and small deletions and insertions. With these mutagens, a range of mild to severe mutant phenotypes can be generated. It is more likely that single base-pair mutations will result in the creation of conditional mutations, such as temperature-sensitive mutations, which are particularly useful for the study of essential gene functions. Geneticists also use transposons to create mutations. If a transposon, such as a *Drosophila P* element, inserts into a gene's coding or regulatory regions, it can disrupt the gene's function.

Because humans are obviously not suitable experimental organisms (for both practical and ethical reasons), techniques developed for the detection of mutations in organisms such as bacteria and fruit flies are not available for human genetics research. To determine the mutational basis for any human characteristic or disorder, geneticists may first analyze a **pedigree** that traces the family history as far back as possible. If a trait is shown to be inherited, the pedigree may also be useful in determining whether the mutant allele is behaving as a dominant or a recessive mutation and whether it is X-linked or autosomal. An example of a human disease pedigree is described in the box below.

More recently, genomics and reverse genetic techniques have expanded the methods available for studying human mutations. Once a mutant gene has been identified and cloned, sequence analysis of mutant and normal genes from affected and unaffected individuals may reveal the molecular basis of the disease or mutant phenotype. In addition, knowledge of the mutant gene sequence may open the way to developing specific genetic tests and gene-based therapeutics. These new genomics-based methods are described in Chapters 21, 23, and 24.

NOW SOLVE THIS

Problem 22 on page 432 asks you to speculate about how X-linked hemophilia could have arisen in the British royal family.

■ HINT: *To answer this question, you might consider the ways in which mutations occur, the types of cells in which they can occur, and how they are inherited.*

## Hemophilia in the Royal Family

The most famous case of an X-linked recessive mutation in humans is that of **hemophilia** found in the descendants of Britain's Queen Victoria. Inspection of the pedigree, shown here, leaves little doubt that Victoria was heterozygous for the trait. Her father was not affected, and there is no evidence to show that her mother was a carrier, as was Victoria. The recessive mutation for hemophilia has occurred many times in human populations, but the political consequences of the mutation that occurred in the British royal family were sweeping. Two of Queen Victoria's daughters were carriers, and they passed hemophilia into the Russian and Spanish royal families. Many historians have posited that Tsarina Alexandra's attempts to deal with her son's hemophilia resulted in Rasputin's presence in the royal household and may have contributed to the start of the Russian Revolution.

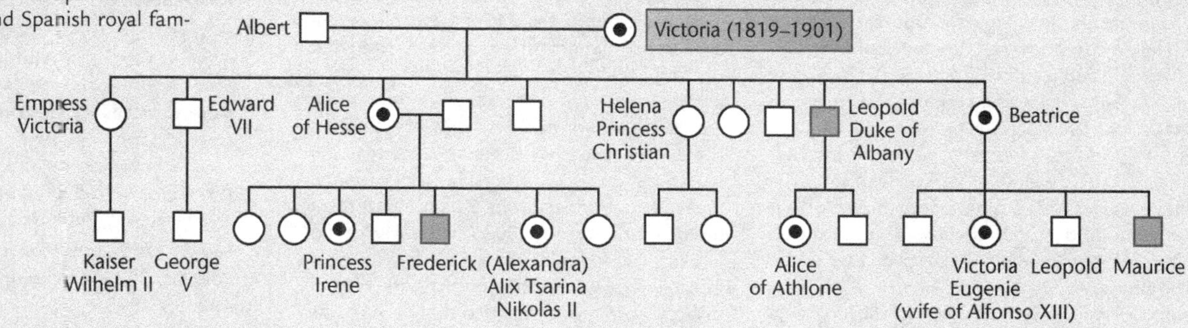


A partial pedigree of hemophilia in the British royal family descended from Queen Victoria. The pedigree is typical of the transmission of X-linked recessive traits. Circles with a dot in them indicate presumed female carriers heterozygous for the trait. The photograph shows Queen Victoria (seated at the table, center) and some of her immediate family.

## GENETICS, TECHNOLOGY, AND SOCIETY

# In the Shadow of Chernobyl

On April 26, 1986, Reactor 4 of the Chernobyl Nuclear Power Station exploded and ejected 6.7 tons of radioactive material into the surrounding countryside and throughout the Northern Hemisphere. The accident immediately killed 30 emergency workers and caused acute radiation sickness in more than 200 others. In the nine days following the initial explosion, as the reactor's temperature approached meltdown, radioactive fission products, including radioactive iodine, xenon, strontium, and cesium, were released into the atmosphere. Fallout traveled through central Europe, reaching Finland and Sweden three days after the initial explosion and the United Kingdom and North America within a week. Millions of people were exposed to measurable amounts of radioactivity. People living within 30 km (about 18 miles) of Chernobyl were exposed to high levels of radioactivity prior to their evacuation 36 hours after the accident. Equally high were the exposures of some of the 200,000 military and civilian workers who were sent to Chernobyl to decontaminate the area and encase the shattered reactor in a sarcophagus. A total of 62 deaths have so far been attributed to the Chernobyl accident.

The Chernobyl incident was the world's largest accidental release of radioactive material. The question that remains is whether Chernobyl's radioactive pollution directly threatens the long-term health of millions of people.

More is known about the effects of ionizing radiation on human health than any other toxic agent (except perhaps cigarette smoke). Ionizing radiation (such as X rays and gamma rays) is any radiation that possesses sufficient energy to eject electrons from atoms. Ionizing radiation can damage cellular components, alter nucleotides, and induce double-strand breaks in DNA. These DNA lesions can lead to mutations or chromosomal translocations.

High doses of ionizing radiation increase the risk of developing certain cancers. The survivors of the atomic-bomb blasts of Hiroshima and Nagasaki showed an increased incidence of leukemia within two years of the bombing. Breast cancers increased 10 years after exposure, as did cancers of the lung, thyroid, colon, ovary, stomach, and nervous system. Since ionizing radiation induces DNA damage, the offspring of the survivors were expected to show increases in birth defects, yet these increases were not detected.

The problem in extrapolating from Hiroshima to Chernobyl is the difference in dose. In atomic-bomb survivors, the cancer rate increased among people exposed to at least 200 mSv (mSv = millisievert, a unit dose of absorbed radiation). It is estimated that people living in the most contaminated areas close to Chernobyl may have received about 50 mSv, and some cleanup workers were exposed to approximately 250 mSv. Outside the Chernobyl area, radiation doses were estimated at 0.4 to 0.9 mSv in Germany and Finland, 0.01 mSv in the United Kingdom, and 0.0006 mSv in the United States.

To put these numbers in perspective, the average dose for medical diagnostic procedures (such as chest and dental X rays) is 0.39 mSv per year. A person's radiation exposure from natural background sources (cosmic rays, rocks, and radon gas) is about 2 to 3 mSv per year. Smokers expose themselves to an average dose of about 2.8 mSv per year from the intake of naturally occurring radioactive materials in tobacco smoke.

It appears that mutation rates among plants and animals exposed to Chernobyl's radioactive waste may be two- to tenfold higher than normal. Also, Chernobyl cleanup workers exhibit a 25 percent higher mutation frequency at the *HPRT* locus. Mutation rates at minisatellites among children born in polluted areas are twofold higher than those among controls in the United Kingdom. But do these increased mutation rates translate into health effects? So far, we cannot answer that question.

Estimates of 26 leukemia cases in addition to the 25 to 30 that would occur spontaneously were projected in the 115,000 people evacuated from Chernobyl. It was also estimated that up to 17,000 additional cancers might occur among Europeans, beyond the 123 million that would occur normally. At present, however, there have been no detectable increases in leukemias or solid tumors in contaminated areas of the former Soviet Union, Finland, or Sweden, or in the 200,000 Chernobyl cleanup workers. In addition, there have been no detectable fertility problems or increases in birth defects. Due to uncertainties about the effects of low doses of radiation, estimates of future cancer deaths due to Chernobyl vary from 4000 to more than 50,000.

Despite the lack of detectable increases in leukemias and solid tumors, one type of cancer has increased in the Chernobyl area. The rate of childhood thyroid cancer has reached more than 100 cases per million children per year, whereas normal rates are expected to be between 0.5 and 3 cases per million children per year. About 4000 cases of thyroid cancer have developed in people who were children at the time of the accident. These cases likely developed as a result of the children drinking milk contaminated with radioactive iodine. Luckily, the survival rate from thyroid cancer is about 99%, and only 15 of the 4000 have died from the disease.

The most profound effects of Chernobyl have been psychological. Chernobyl cleanup workers demonstrated a 50 percent increase in suicides and a detectable increase in smoking- and alcohol-related disease. This mirrors other studies showing that 45 percent of people living within 300 km of Chernobyl believe that they have a radiation-induced illness. Health effects such as depression, sleep disturbance, hypertension, and altered perception have been documented. Posttraumatic stress may be a greater threat to health than the actual radiation exposure from the accident. People feel that they live in constant danger and are simply awaiting the results of a cancer "lottery." Even if cancer rates and genetic defects do not increase dramatically, the indirect health effects from the Chernobyl Nuclear Power Station explosion have been and continue to be immense.

### ■ References

The Chernobyl Forum: 2003–2005. Chernobyl's Legacy: Health, Environmental, and Socio-economic Impacts. Report from eight UN Agencies including International Atomic Energy Agency, World Health Organization and the UN Development Program. September, 2005. **http://www.iaea.org/Publications/Booklets/Chernobyl/chernobyl.pdf**

Peplow, M. 2006. Counting the dead. *Nature* 440: 982–983.

Williams, D., and Baverstock, K. 2006. Too soon for a final diagnosis. *Nature* 440: 993–994.

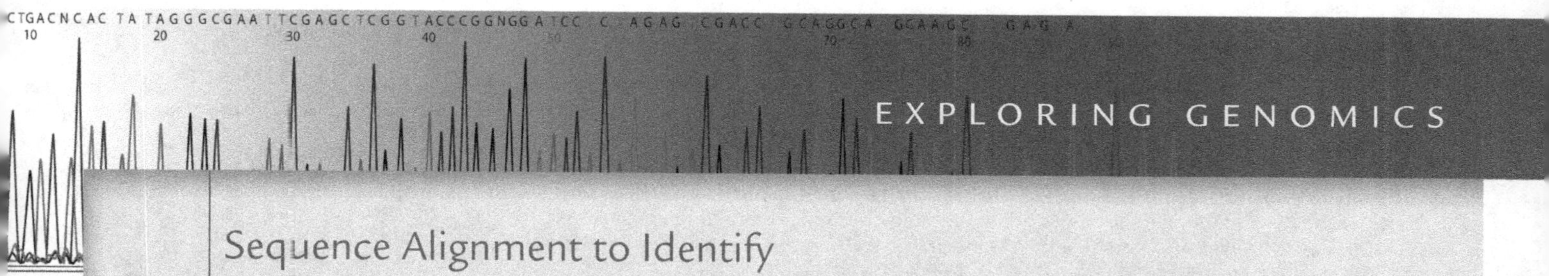


# Sequence Alignment to Identify a Mutation

In this chapter, we examined the causes of different types of mutations and how mutations affect phenotype by altering the structure and function of proteins. The emergence of genomics, bioinformatics, and proteomics as key disciplines in modern genetics has provided geneticists with an unprecedented set of tools for identifying and analyzing mutations in gene and protein sequences.

In this exercise we return to the **ExPASy (Expert Protein Analysis System)** you were introduced to in Chapter 15. Here we will use a program from ExPASy called SIM (for "similarity" in sequence) to compare two polypeptide sequences so as to pinpoint a mutation. Once the mutation has been identified, you will use ExPASy and its links to learn more about the gene encoding these polypeptides and about a human disease condition associated with the gene.

## ■ Exercise I – Identifying a Missense Mutation Affecting a Protein in Humans

1. Begin this exercise by accessing the ExPASy site at http://www.expasy.ch/tools/sim-prot.html. The SIM feature is an algorithm-based software program that allows us to compare multiple polypeptide sequences by looking for amino acid similarity in the sequences.

2. Below are amino acid sequences for polypeptides expressed in two different people.

**Person A**

MGAPACALALCVAVAIVAGASSESLGTEQ
RVVGRAAEVPGPEPGQQEQLVFGSGDAVE
LSCPPPGGGPMGPTVWVKDGTGLVPSERV
LVGPQRLQVLNASHEDSGAYSCRQRLTQR
VLCHFSVRVTDAPSSGDDEDGEDEAEDTG
VDTGAPYWTRPERMDKKLLAVPAANTVRF
RCPAAGNPTPSISWLKNGREFRGEHRIGGI
KLRHQQWSLVMESVVPSDRGNYTCVVENK
FGSIRQTYTLDVLERSPHRPILQAGLPANQT
AVLGSDVEFHCKVYSDAQPHIQWLKHVEV
NGSKVGPDGTPYVTVLKTAGANTTDKELEV
LSLHNVTFEDAGEYTCLAGNSIGFSHHSAW
LVVLPAEEELVEADEAGSVYAGILSYGVGFFLF
ILVVAAVTLCRLRSPPKKGLGSPTVHKISRFP
LKRQVSLESNASMSSNTPLVRIARLSSGEGP
TLANVSELELPADPKWELSRARLTLGKPLGE
GCFGQVVMAEAIGIDKDRAAKPVTVAVKM
LKDDATDKDLSDLVSEMEMMKMIGKHKN
IINLLGACTQGGPLYVLVEYAAKGNLREFLRA
RRPPGLDYSFDTCKPPEEQLTFKDLVSCAYQ
VARGMEYLASQKCIHRDLAARNVLVTEDN
VMKIADFGLARDVHNLDYYKKTTNGRLPVK
WMAPEALFDRVYTHQSDVWSFGVLLWEIF
TLGGSPYPGIPVEELFKLLKEGHRMDKPAN
CTHDLYMIMRECWHAAPSQRPTFKQLVED
LDRVLTVTSTDEYLDLSAPFEQYSPGGQDTPS
SSSGDDSVFAHDLLPPAPPSSGGSRT

**Person B**

MGAPACALALCVAVAIVAGASSESLGTEQR
VVGRAAEVPGPEPGQQEQLVFGSGDAVELS
CPPPGGGPMGPTVWVKDGTGLVPSERVLV
GPQRLQVLNASHEDSGAYSCRQRLTQRVL
CHFSVRVTDAPSSGDDEDGEDEAEDTGVD
TGAPYWTRPERMDKKLLAVPAANTVRFRC
PAAGNPTPSISWLKNGREFRGEHRIGGIKLR
HQQWSLVMESVVPSDRGNYTCVVENKFGSI
RQTYTLDVLERSPHRPILQAGLPANQTAVLGS
DVEFHCKVYSDAQPHIQWLKHVEVNGSKV
GPDGTPYVTVLKTAGANTTDKELEVLSLHN
VTFEDAGEYTCLAGNSIGFSHHSAWLVVLP
AEEELVEADEAGSVYAGILSYRVGFFLFILVVA
AVTLCRLRSPPKKGLGSPTVHKISRFPLKRQ
VSLESNASMSSNTPLVRIARLSSGEGPTLAN
VSELELPADPKWELSRARLTLGKPLGEGCFG
QVVMAEAIGIDKDRAAKPVTVAVKMLKDD
ATDKDLSDLVSEMEMMKMIGKHKNIINLL
GACTQGGPLYVLVEYAAKGNLREFLRARRP
PGLDYSFDTCKPPEEQLTFKDLVSCAYQVA
RGMEYLASQKCIHRDLAARNVLVTEDNV
MKIADFGLARDVHNLDYYKKTTNGRLPVK
WMAPEALFDRVYTHQSDVWSFGVLLWEIF
TLGGSPYPGIPVEELFKLLKEGHRMDKPAN
CTHDLYMIMRECWHAAPSQRPTFKQLVED
LDRVLTVTSTDEYLDLSAPFEQYSPGGQDTP
SSSSSGDDSVFAHDLLPPAPPSSGGSRT

3. Copy and paste each sequence into the "SEQUENCE" text boxes in SIM. (*Hint*: Access these sequences from the Companion Web site so you can copy and paste the sequence into SIM.) Use the Person A sequence for sequence 1 and the Person B sequence for sequence 2. Click the "User-entered sequence" button for each. Name the sequences "Person A" and "Person B" as appropriate. Submit the sequences for comparison and then answer the following questions:

   a. How many amino acids are in each sequence analyzed?

   b. Look carefully at the alignment results. Can you find any differences in amino acid sequence when comparing these two polypeptides? What did you find?

## ■ Exercise II – Identifying the Genetic Basis for a Human Genetic Disease Condition

1. Go to the ExPASy homepage and find the BLAST link. Run a protein BLAST (blastp) search to identify which polypeptide you have been studying. Explore the BLAST reports for the top three protein sequences that aligned with your query sequence by clicking on the link for each sequence. Pay particular attention to the "Comment" section of each report to help you answer the questions in the following part.

2. Now that you know what gene you are working with, go to PubMed (http://www.ncbi.nlm.nih.gov/entrez/query.fcgi?db=PubMed) and search for a review article from the authors Vajo, Z., Francomano, C. A., and Wilkin, D. J.

3. Answer the following questions:

   a. What gene codes for the polypeptides have you been studying?

   b. What is the function of this protein?

*Continued on next page*

*Exploring Genomics, continued*

c. Based on what you learned from the alignment results you analyzed in Exercise I, the BLAST reports, and your PubMed search, what human disease is caused by the mutation you identified in Exercise I? Explain your answer and briefly describe phenotypes associated with the disease.

4. The TMHMM Server (http://www.cbs.dtu.dk/services/TMHMM/) hosted by the Center for Biological Sequence Analysis at the Technical University of Denmark DTU is a useful resource for predicting the location of transmembrane helices in polypeptides. Visit the TMHMM server and run separate searches with the amino acid sequences from Exercise I. Run each search as "Extensive, with graphics." View the results page and then answer the following:

a. How many amino acids are predicted to be in the transmembrane helical region (TMH) for the protein from Person A? Person B?

b. Examine the graph on the results page to identify amino acids in each polypeptide that are predicted to reside outside the cell, within the plasma membrane, and inside the cell. Locate the approximate position of the amino acid difference you identified in Exercise I, part 3.

## Chapter Summary

1. Mutations provide the foundation for genetic variation, evolution, and genetic analysis.
2. A mutation is defined as an alteration in DNA sequence. A mutation may or may not create a detectable phenotype.
3. Mutations may be either spontaneous or induced, somatic or germ-line, autosomal or sex-linked.
4. Mutations can have many different effects on gene function depending on the type of nucleotide changes that comprise the mutation. Point mutations within coding regions of a gene may lead to missense mutations or nonsense mutations. Insertions or deletions within coding regions may lead to frameshift mutations and premature termination of translation.
5. Mutations have many different kinds of effects, ranging from loss of gene function to gain of a new function.
6. Spontaneous mutation rates vary considerably between organisms and between different genes in the same organism.
7. Spontaneous mutations occur in many ways, ranging from errors during DNA replication to damage caused to DNA bases as a result of normal cellular metabolism.
8. Mutations can be induced by many types of chemicals and radiation. These agents can damage both bases and the sugar-phosphate backbone of DNA molecules.
9. The human ABO blood groups owe their variation to point mutations or frameshift mutations within the gene encoding the enzyme that modifies the H substance.
10. The most severe forms of muscular dystrophy are caused by duplication or deletion mutations that create changes in the translation reading frame of the *dystrophin* gene.
11. Mutations in the numbers of trinucleotide repeats in some genes lead to human diseases such as the fragile X syndrome, myotonic dystrophy, and Huntington disease.
12. The Ames test allows scientists to estimate the mutagenicity and cancer-causing potential of chemical agents by following the rate of mutation in specific strains of bacteria.
13. Organisms counteract both spontaneous and induced mutations using DNA repair systems. Errors in DNA synthesis can be repaired by proofreading during replication, mismatch repair, and postreplication repair.
14. Other types of DNA damage can be repaired by photoreactivation repair, SOS repair, base excision repair, nucleotide excision repair, and double-strand break repair.
15. Mutations in genes controlling nucleotide excision repair in humans can lead to diseases such as xeroderma pigmentosum.
16. A range of genetic and biochemical techniques are used to induce and analyze mutations in model organisms. In contrast, the study of human mutations often begins with pedigree analysis and *in vitro* techniques.

# INSIGHTS AND SOLUTIONS

1. The base analog 2-amino purine (2-AP) substitutes for adenine during DNA replication, but it may base-pair with cytosine. The base analog 5-bromouracil (5-BU) substitutes for thymidine, but it may base-pair with guanine. Follow the double-stranded trinucleotide sequence shown here through three rounds of replication, assuming that, in the first round, both analogs are present and become incorporated wherever possible. Before the second and third round of replication, any unincorporated base analogs are removed. What final sequences occur?

**Solution:**

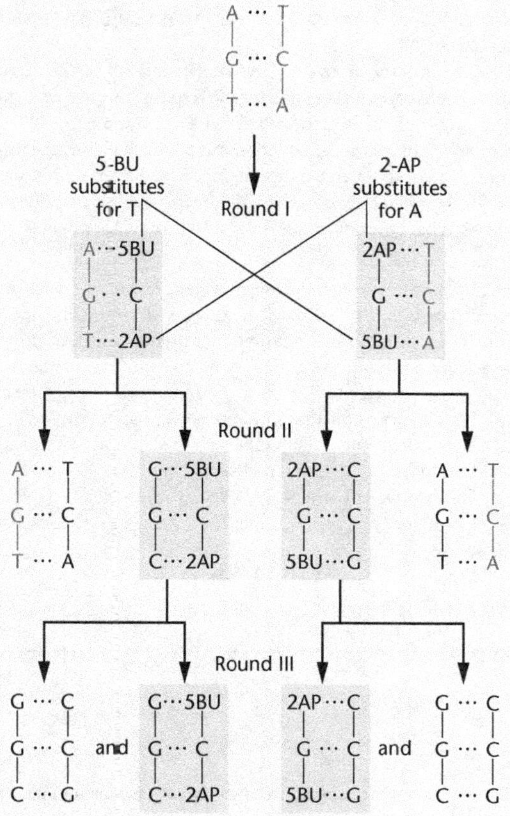


2. A rare dominant mutation expressed at birth was studied in humans. Records showed that six cases were discovered in 40,000 live births. Family histories revealed that in two cases, the mutation was already present in one of the parents. Calculate the spontaneous mutation rate for this mutation. What are some underlying assumptions that may affect our conclusions?

**Solution:** Only four cases represent a new mutation. Because each live birth represents two gametes, the sample size is from 80,000 meiotic events. The rate is equal to

$$\frac{4}{80{,}000} = \frac{1}{20{,}000} = 5 \times 10^{-5}$$

We have assumed that the mutant gene is fully penetrant and is expressed in each individual bearing it. If it is not fully penetrant, our calculation may be an underestimate because one or more mutations may have gone undetected. We have also assumed that the screening was 100 percent accurate. One or more mutant individuals may have been "missed," again leading to an underestimate. Finally, we assumed that the viability of the mutant and nonmutant individuals is equivalent and that they survive equally *in utero*. Therefore, our assumption is that the number of mutant individuals at birth is equal to the number at conception. If this were not true, our calculation would again be an underestimate.

3. Consider the following estimates:
   (a) There are $5.5 \times 10^9$ humans living on this planet.
   (b) Each individual has about 30,000 ($0.3 \times 10^5$) genes.
   (c) The average mutation rate at each locus is $10^{-5}$.

   How many spontaneous mutations are currently present in the human population? Assuming that these mutations are equally distributed among all genes, how many new mutations have arisen in each gene in the human population?

**Solution:** First, since each individual is diploid, there are two copies of each gene per person, each arising from a separate gamete. Therefore, the total number of spontaneous mutations is

$$(2 \times 0.3 \times 10^5 \text{ genes/individual})$$
$$\times (5.5 \times 10^9) \text{ individuals} \times (10^{-5} \text{ mutations/gene})$$
$$= (0.6 \times 10^5) \times (5.5 \times 10^9) \times (10^{-5}) \text{ mutations}$$
$$= 3.3 \times 10^9 \text{ mutations in the population}$$

$$3.3 \times 10^9 \text{ mutations}/0.3 \times 10^5 \text{ genes}$$
$$= 11 \times 10^4 \text{ mutations per gene in the population}$$

## Problems and Discussion Questions

1. Discuss the importance of mutations in genetic studies.
2. Why would a mutation in a somatic cell of a multicellular organism escape detection?
3. Most mutations are thought to be deleterious. Why, then, is it reasonable to state that mutations are essential to the evolutionary process?
4. Why is a random mutation more likely to be deleterious than beneficial?
5. Most mutations in a diploid organism are recessive. Why?
6. What is meant by a conditional mutation?
7. Describe a tautomeric shift and how it may lead to a mutation.

8. Contrast and compare the mutagenic effects of deaminating agents, alkylating agents, and base analogs.
9. Acridine dyes induce frameshift mutations. Why are frameshift mutations likely to be more detrimental than point mutations, in which a single pyrimidine or purine has been substituted?
10. Why are X rays more potent mutagens than UV radiation?
11. DNA damage brought on by a variety of natural and artificial agents elicits a wide variety of cellular responses involving numerous signaling pathways. In addition to the activation of DNA repair mechanisms, there can be activation of pathways leading to apoptosis (programmed cell death) and cell-cycle arrest. Why would apoptosis and cell-cycle arrest often be part of a cellular response to DNA damage?
12. Contrast the various types of DNA repair mechanisms known to counteract the effects of UV radiation. What is the role of visible light in repairing UV-induced mutations?
13. Mammography is an accurate screening technique for the early detection of breast cancer in humans. Because this technique uses X rays diagnostically, it has been highly controversial. Can you explain why? What reasons justify the use of X rays for such a medical screening technique?
14. Explain the molecular basis of fragile X syndrome, myotonic dystrophy, and Huntington disease and show how the types of mutations involved in each relate to the severity of the disorders.
15. Describe how the Ames test screens for potential environmental mutagens. Why is it thought that a compound that tests positively in the Ames test may also be carcinogenic?
16. What genetic defects result in the disorder xeroderma pigmentosum (XP) in humans? How do these defects create the phenotypes associated with the disorder?
17. In a bacterial culture in which all cells are unable to synthesize leucine (*leu*$^-$), a potent mutagen is added, and the cells are allowed to undergo one round of replication. At that point, samples are taken, a series of dilutions is made, and the cells are plated on either minimal medium or minimal medium containing leucine. The first culture condition (minimal medium) allows the growth of only *leu*$^+$ cells, while the second culture condition (minimum medium with leucine added) allows the growth of all cells. The results of the experiment are as follows:

| Culture Condition | Dilution | Colonies |
|---|---|---|
| minimal medium | $10^{-1}$ | 18 |
| minimal + leucine | $10^{-7}$ | 6 |

What is the rate of mutation at the locus associated with leucine biosynthesis?

18. Speculate on how improved living conditions and medical care in the developed nations might affect human mutation rates, both neutral and deleterious.
19. The cancer drug melphalan is an alkylating agent of the mustard gas family. It acts in two ways: by causing alkylation of guanine bases and by crosslinking DNA strands together. Describe two ways in which melphalan might kill cancer cells. What are two ways in which cancer cells could repair the DNA-damaging effects of melphalan?
20. Muscular dystrophies are some of the most common inherited diseases in humans, resulting from a large number of different mutations in the *dystrophin* gene. Speculate on why this gene appears to suffer so many mutations.
21. Describe how the process of DNA replication can lead to expansion of trinucleotide repeat regions in the gene responsible for Huntington disease.
22. The origin of the mutation that led to hemophilia in Queen Victoria's family is controversial. Her father did not have X-linked hemophilia, and there is no evidence that any members of her mother's family had the condition. What are some possible explanations of how the mutation arose? What types of mutations could lead to the disease?

**HOW DO WE KNOW?**

23. In this chapter, we focused on how gene mutations arise and how cells repair DNA damage. At the same time, we found many opportunities to consider the methods and reasoning by which much of this information was acquired. From the explanations given in the chapter, what answers would you propose to the following fundamental questions:
    (a) How do we know that mutations occur spontaneously?
    (b) How do we know that certain chemicals and wavelengths of radiation induce mutations in DNA?
    (c) How do we know that DNA repair mechanisms detect and correct the majority of spontaneous and induced mutations?

## Extra-Spicy Problems

24. Presented here are hypothetical findings from studies of heterokaryons formed from seven human xeroderma pigmentosum cell strains:

| | *XP1* | *XP2* | *XP3* | *XP4* | *XP5* | *XP6* | *XP7* |
|---|---|---|---|---|---|---|---|
| *XP1* | − | | | | | | |
| *XP2* | − | − | | | | | |
| *XP3* | − | − | − | | | | |
| *XP4* | + | + | + | − | | | |
| *XP5* | + | + | + | + | − | | |
| *XP6* | + | + | + | + | − | − | |
| *XP7* | + | + | + | + | − | − | − |

NOTE: + = complementing; − = noncomplementing

These data are measurements of the occurrence or nonoccurrence of unscheduled DNA synthesis in the fused heterokaryon. None of the strains alone shows any unscheduled DNA synthesis. What does unscheduled DNA synthesis represent? Which strains fall into the same complementation groups? How many different groups are revealed based on these data? What can we conclude about the genetic basis of XP from these data?

25. A variety of neural and muscular disorders are associated with expansions of trinucleotide repeat sequences. The following table lists several disorders, the repeat motifs, their locations, the normal number of repeats and the number of repeats in the full mutations.

| Disorder | Repeat | Location | Normal Number | Full Mutation |
|---|---|---|---|---|
| Fragile X | CCG | 5′ untranslated region | 6–230 | >230 |
| Huntington | CAG | Exon | 6–35 | 36–120 |
| Myotonic Dystrophy | CTG | 3′ untranslated region | 5–37 | 37–1500 |
| Oculopharyngeal Muscular Dystrophy | GCG | Exon | 6 | 8–13 |
| Friedreich Ataxia | AAG | Intron | 20 | 200–900 |

(a) Most disorders attributable to trinucleotide repeats result from expansion of the repeats. Two mechanisms are often proposed to explain repeat expansion: (1) unequal synapsis and crossing over and (2) errors in DNA replication where single-stranded, base-paired loops are formed that conflict with linear replication. Present a simple sketch of each mechanism.

(b) Notice that some of the repeats occur in areas of the gene that are not translated. How can a mutation occur if the alteration is not reflected in an altered amino acid sequence?

(c) In the two cases where the repeat expansions occur in exons, the extent of expansion is considerably less than when the expansion occurs outside exons. Present an explanation for this observation.

26. Cystic fibrosis (CF) is a severe autosomal recessive disorder in humans that results from a chloride ion–channel defect in epithelial cells. More than 500 mutations have been identified in the 24 exons of the responsible gene (*CFTR*, or cystic fibrosis transmembrane regulator), including dozens of different missense mutations and frameshift mutations, as well as numerous splice-site defects. Although all affected CF individuals demonstrate chronic obstructive lung disease, there is variation in whether or not they exhibit pancreatic enzyme insufficiency (PI). Speculate as to which types of mutations are likely to give rise to less severe symptoms of CF, including only minor PI. Some of the 300 sequence alterations that have been detected within the exon regions of the *CFTR* gene do not give rise to cystic fibrosis. Taking into account your accumulated knowledge of the genetic code, gene expression, protein function, and mutation, describe why this might be so, as if you were explaining it to a freshman biology major.

27. Electrophilic oxidants are known to create the modified base named 7,8-dihydro-8-oxoguanine (oxoG) in DNA. Whereas guanine base-pairs with cytosine, oxoG base-pairs with either cytosine or adenine.

(a) What are the sources of reactive oxidants within cells that cause this type of base alteration?

(b) Drawing on your knowledge of nucleotide chemistry, draw the structure of oxoG, and, below it, draw guanine. Opposite guanine, draw cytosine, including the hydrogen bonds that allow these two molecules to base-pair. Does the structure of oxoG, in contrast to guanine, provide any hint as to why it base-pairs with adenine?

(c) Assume that an unrepaired oxoG lesion is present in the helix of DNA opposite cytosine. Predict the type of mutation that will occur following several rounds of replication.

(d) Which DNA repair mechanisms might work to counteract an oxoG lesion? Which of these is likely to be most effective?

28. Mutations in the *IL2RG* gene cause approximately 30 percent of severe combined immunodeficiency disorder (SCID) cases. These mutations result in alterations to a protein component of cytokine receptors that are essential for proper development of the immune system. The *IL2RG* gene is composed of eight exons and contains upstream and downstream sequences that are necessary for proper transcription and translation. Below are some of the mutations observed. For each, explain its likely influence on the *IL2RG* gene product (assume its length to be 375 amino acids).

Nonsense mutation in coding regions
Insertion in Exon 1, causing frameshift
Insertion in Exon 7, causing frameshift
Missense mutation
Deletion in Exon 2, causing frameshift
Deletion in Exon 2, in frame
Large deletion covering Exons 2 and 3

29. Skin cancer carries a lifetime risk nearly equal to that of all other cancers combined. Following is a graph (modified from Kraemer. 1997. *Proc. Natl. Acad. Sci. (USA)* 94: 11–14) depicting the age of onset of skin cancers in patients with or without XP, where cumulative percentage of skin cancer is plotted against age. The non-XP curve is based on 29,757 cancers surveyed by the National Cancer Institute, and the curve representing those with XP is based on 63 skin cancers from the Xeroderma Pigmentosum Registry.

(a) Provide an overview of the information contained in the graph.

(b) Explain why individuals with XP show such an early age of onset.

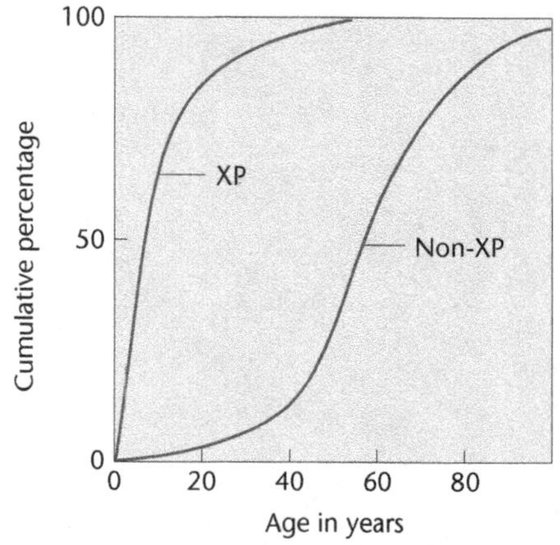


30. You have decided to use the Ames test to analyze a newly synthesized chemical (Compound Z) for its potential carcinogenicity. To do the test, you use a strain of *Salmonella typhimurium* that is *his*$^-$ and is designed to detect base substitution mutations that revert the bacteria to *his*$^+$. For your first test, you spread an equal number of tester bacteria onto the surface of five agar medium plates that contain no histidine. In the center of each plate, you place a disk of filter paper that has been soaked in buffer, several dilutions of Compound Z, or a high concentration of Compound Z that has been preincubated with rat

liver extract. After 24 hours of incubation, the plates appear as shown in the following figure:

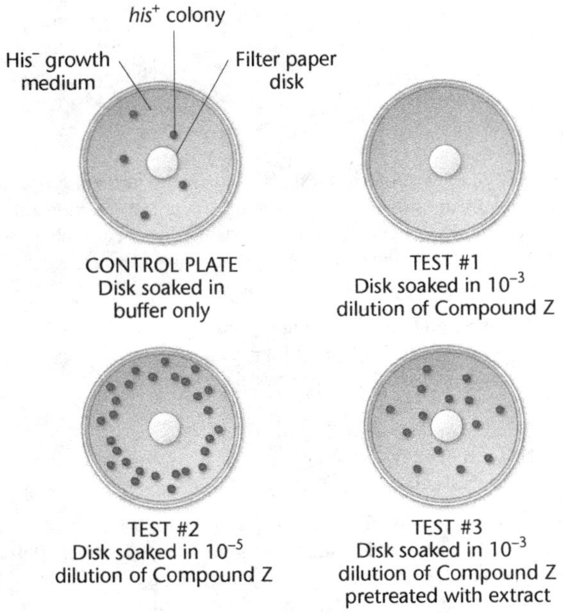


Based on the data shown in the figure, answer the following questions:

(a) Would you consider that Compound Z is carcinogenic? Explain your conclusion.

(b) How would you interpret the pattern of bacterial growth on each of the four plates containing Compound Z filter disks?

(c) Why is there a clear space with no bacterial growth surrounding the $10^{-5}$ dilution disk?

(d) Why is there no clear space surrounding the disk containing the extract-treated Compound Z?

(e) What do these data reveal about the limitations of the Ames test?

(f) Still using the Ames test, how would you further test Compound Z to determine its potential ability to cause cancer?

Model showing how the *lac* repressor (red) and catabolite-activating protein (dark blue in center of DNA loop) bind to the *lac* operon promoter, creating a 93-base-pair repression loop in the *lac* regulatory DNA.

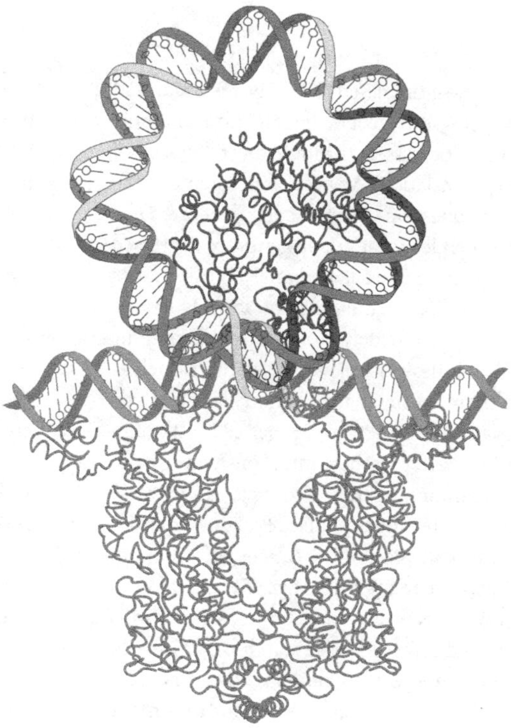

# 17

# Regulation of Gene Expression in Prokaryotes

## CHAPTER CONCEPTS

- In bacteria, regulation of gene expression is often linked to the metabolic needs of the cell.
- Efficient expression of genetic information in bacteria is dependent on intricate regulatory mechanisms that exert control over transcription.
- Mechanisms that regulate transcription are categorized as exerting either positive or negative control of gene expression.
- Prokaryotic genes that encode proteins with related functions tend to be organized in clusters and are often under coordinated control. Such clusters, including their adjacent regulatory sequences, are called operons.
- Transcription of genes within operons is either inducible or repressible.
- Often, the metabolic end product of a biosynthetic pathway induces or represses gene expression in that pathway.

Previous chapters have discussed how DNA is organized into genes, how genes store genetic information, and how this information is expressed through the processes of transcription and translation. We now consider one of the most fundamental questions in molecular genetics: *How is genetic expression regulated?* It is clear that not all genes are expressed at all times in all situations. For example, detailed analysis of proteins in *E. coli* shows that concentrations of the 4000 or so polypeptide chains encoded by the genome vary widely. Some proteins may be present in as few as 5 to 10 molecules per cell, whereas others, such as ribosomal proteins and the many proteins involved in the glycolytic pathway, are present in as many as 100,000 copies per cell. Although most prokaryotic gene products are present continuously at a basal level (a few copies), the concentration of these products can increase dramatically when required. Clearly, fundamental regulatory mechanisms must exist to control the expression of the genetic information.

In this chapter, we will explore regulation of gene expression in prokaryotes specifically. As we have seen in a number of previous chapters, these organisms served as excellent research organisms in many seminal investigations in molecular genetics. Bacteria have been especially useful research organisms in genetics for a number of reasons. For one thing, they have extremely short reproductive cycles. Literally hundreds of generations, giving rise to billions of genetically identical bacteria or phages, can be produced in overnight cultures. In addition, they can be studied in "pure culture," allowing mutant strains of genetically unique bacteria to be isolated and investigated separately.

Relevant to our current topic, bacteria also serve as an excellent model system for studies involving the induction of genetic transcription in response to changes in environmental conditions. Our focus will be on regulation at the level of the gene. Keep in mind that posttranscriptional regulation also occurs in bacteria. However, we will defer discussion of this level of regulation to Chapter 18, in which we consider eukaryotic regulation as well.

## 17.1 Prokaryotes Regulate Gene Expression in Response to Environmental Conditions

Regulation of gene expression has been extensively studied in prokaryotes, particularly in *E. coli*. Geneticists have learned that highly efficient genetic mechanisms have evolved in these organisms to turn transcription of specific genes on and off, depending on the cell's metabolic need for the respective gene products. Not only do bacteria respond to changes in their environment, but they also regulate gene activity associated with a variety of nonenvironmentally regulated cellular activities (including the replication, recombination, and repair of their DNA) and with cell division.

The idea that microorganisms regulate the synthesis of their gene products is not a new one. As early as 1900, it was shown that when lactose (a galactose and glucose-containing disaccharide) is present in the growth medium of yeast, the organisms synthesize enzymes required for lactose metabolism. When lactose is absent, the enzymes are not manufactured. Soon thereafter, investigators were able to generalize that bacteria also adapt to their environment, producing certain enzymes only when specific chemical substrates are present. These were thus referred to as **adaptive enzymes** (sometimes also called facultative). In contrast, enzymes that are produced continuously, regardless of the chemical makeup of the environment, were called **constitutive enzymes**. Since then, the term *adaptive* has been replaced with the more accurate term **inducible**, reflecting the role of the substrate, which serves as the **inducer** in enzyme production.

More recent investigation has revealed a contrasting system, whereby the presence of a specific molecule inhibits gene expression. Such molecules are usually end products of anabolic biosynthetic pathways. For example, the amino acid tryptophan can be synthesized by bacterial cells. If a sufficient supply of tryptophan is present in the environment or culture medium, then there is no reason for the organism to expend energy in synthesizing the enzymes necessary for tryptophan production. A mechanism has therefore evolved whereby tryptophan plays a role in repressing the transcription of mRNA needed for producing tryptophan-synthesizing enzymes. In contrast to the inducible system controlling lactose metabolism, the system governing tryptophan expression is said to be **repressible.**

Regulation, whether of the inducible or repressible type, may be under either **negative** or **positive control**. Under negative control, genetic expression occurs *unless it is shut off by some form of a regulator molecule.* In contrast, under positive control, transcription occurs *only if a regulator molecule directly stimulates RNA production.* In theory, either type of control or a combination of the two can govern inducible or repressible systems. Our discussion in the ensuing sections of this chapter will help clarify these contrasting systems of regulation. The enzymes involved in lactose digestion and tryptophan synthesis are under negative control.

## 17.2 Lactose Metabolism in *E. coli* Is Regulated by an Inducible System

Beginning in 1946 with the studies of Jacques Monod and continuing through the next decade with significant contributions by Joshua Lederberg, François Jacob, and Andre Lwoff, genetic and biochemical evidence concerning lactose metabolism was amassed. Research provided insights into the way in which the gene activity is repressed when lactose is absent but induced when it is available. In the presence of lactose, the concentration of the enzymes responsible

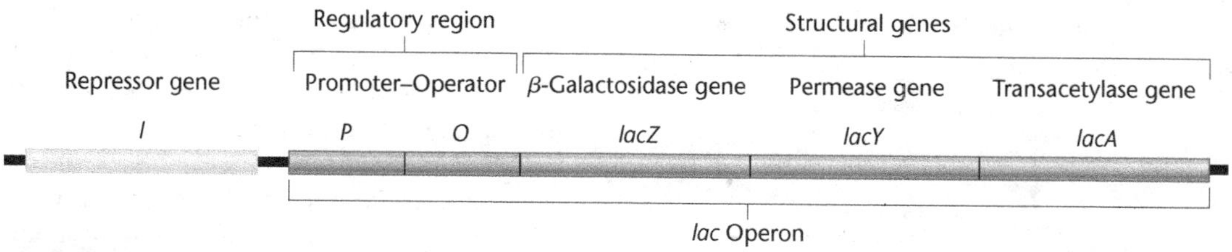


FIGURE 17–1 A simplified overview of the genes and regulatory units involved in the control of lactose metabolism. (The regions within this stretch of DNA are not drawn to scale.) A more detailed model will be developed later in this chapter. (See Figure 17–10.)

for its metabolism increases rapidly from a few molecules to thousands per cell. The enzymes responsible for lactose metabolism are thus inducible, and lactose serves as the inducer.

In prokaryotes, genes that code for enzymes with related functions (for example, the set of genes involved with lactose metabolism) tend to be organized in clusters on the bacterial chromosome, and transcription of these genes is often under the coordinated control of a single regulatory region. The location of this regulatory region is almost always upstream (5') of the gene cluster it controls. Because the regulatory region is on the same strand as those genes, we refer to it as a ***cis*-acting site**. *Cis*-acting regulatory regions bind molecules that control transcription of the gene cluster. Such molecules, as described in our Chapter 14 discussion of transcription, are called ***trans*-acting elements**. Events at the regulatory site determine whether the genes are transcribed into mRNA and thus whether the corresponding enzymes or other protein products may be synthesized from the genetic information in the mRNA. Binding of a *trans*-acting element at a *cis*-acting site can regulate the gene cluster either negatively (by turning off transcription) or positively (by turning on transcription of genes in the cluster). In this section, we discuss how transcription of such bacterial gene clusters is coordinately regulated.

The discovery of a regulatory gene and a regulatory site that are part of the gene cluster was paramount to the understanding of how gene expression is controlled in the system. Neither of these regulatory elements encodes enzymes necessary for lactose metabolism—the function of the three genes in the cluster. As illustrated in Figure 17–1, the three structural genes and the adjacent regulatory site constitute the **lactose,** or ***lac*, operon**. Together, the entire gene cluster functions in an integrated fashion to provide a rapid response to the presence or absence of lactose.

## Structural Genes

Genes coding for the primary structure of an enzyme are called **structural genes.** There are three structural genes in the *lac* operon. The *lacZ* gene encodes **β-galactosidase**, an enzyme whose primary role is to convert the disaccharide lactose to the monosaccharides glucose and galactose (Figure 17–2). This conversion is essential if lactose is to serve as the primary energy source in glycolysis. The second gene, *lacY*, specifies the primary structure of **permease**, an enzyme that facilitates the entry of lactose into the bacterial cell. The third gene, *lacA*, codes for the enzyme **transacetylase.** While its physiological role is still not completely clear, it may be involved in the removal of toxic by-products of lactose digestion from the cell.

To study the genes coding for these three enzymes, researchers isolated numerous mutations that lacked the function of one or the other enzyme. Such $lac^-$ mutants were first isolated and studied by Joshua Lederberg. Mutant cells that fail to produce active β-galactosidase ($lacZ^-$) or permease ($lacY^-$) are unable to use lactose as an energy source. Mutations were also found in the transacetylase gene. Mapping studies by Lederberg established that all three genes are closely linked or contiguous to one another on the bacterial chromosome, in the order *Z–Y–A* (see Figure 17–1).

Knowledge of their close linkage led to another discovery relevant to what later became known about the regulation of structural genes: All three genes are transcribed as a single unit, resulting in

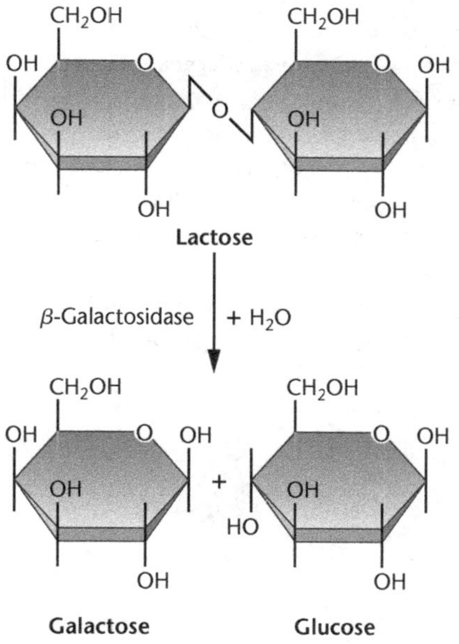


FIGURE 17–2 The catabolic conversion of the disaccharide lactose into its monosaccharide units, galactose and glucose.

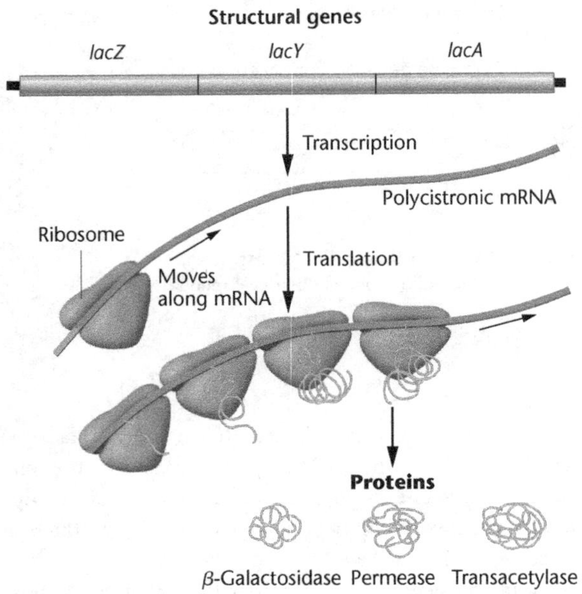


FIGURE 17-3 The structural genes of the *lac* operon are transcribed into a single polycistronic mRNA, which is translated simultaneously by several ribosomes into the three enzymes encoded by the operon.

aso-called *polycistronic mRNA* (Figure 17–3; recall that *cistron* refers to the part of a nucleotide sequence coding for a single gene). This results in the coordinate regulation of all three genes, since a single-message RNA is simultaneously translated into all three gene products.

## The Discovery of Regulatory Mutations

How does lactose stimulate transcription of the *lac* operon and induce the synthesis of the enzymes for which it codes? A partial answer came from studies using **gratuitous inducers**, chemical analogs of lactose such as the sulfur-containing analog **isopropylthiogalactoside (IPTG)**, shown in Figure 17–4. Gratuitous inducers behave like natural inducers, but they do not serve as substrates for the enzymes that are subsequently synthesized. Their discovery provides strong evidence that the primary induction event does *not* depend on the interaction between the inducer and the enzyme.

What, then, is the role of lactose in induction? The answer to this question required the study of another class of mutations

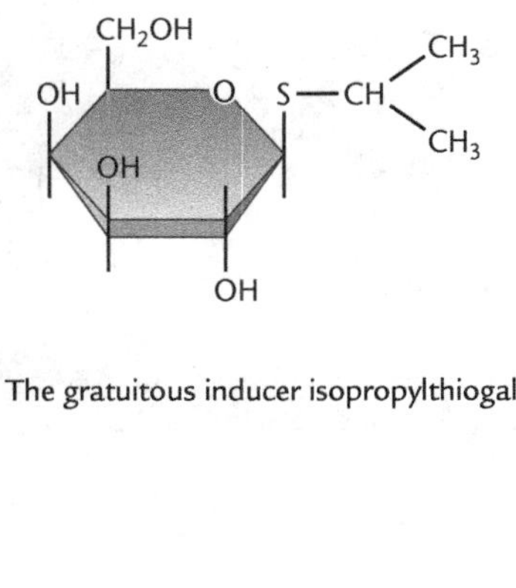


FIGURE 17-4 The gratuitous inducer isopropylthiogalactoside (IPTG).

described as **constitutive mutations.** In cells bearing these types of mutations, enzymes are produced regardless of the presence or absence of lactose. Studies of the constitutive mutation *lacI*$^-$ mapped the mutation to a site on the bacterial chromosome close to, but distinct from, the structural genes *lacZ, lacY,* and *lacA*. This mutation led researchers to discover the *lacI* gene, which is appropriately called a **repressor gene.** A second set of constitutive mutations producing effects identical to those of *lacI*$^-$ is present in a region immediately adjacent to the structural genes. This class of mutations, designated *lacO*$^C$, is located in the **operator region** of the operon. In both types of constitutive mutants, the enzymes are produced continually, inducibility is eliminated, and gene regulation has been lost.

## The Operon Model: Negative Control

Around 1960, Jacob and Monod proposed a hypothetical mechanism involving negative control that they called the **operon model,** in which a group of genes is regulated and expressed together as a unit. As we saw in Figure 17–1, the *lac* operon they proposed consists of the *Z, Y,* and *A* structural genes, as well as the adjacent sequences of DNA referred to as the *operator region.* They argued that the *lacI* gene regulates the transcription of the structural genes by producing a **repressor molecule,** and that the repressor is **allosteric,** meaning that the molecule reversibly interacts with another molecule, undergoing both a conformational change in three-dimensional shape and a change in chemical activity. Figure 17–5 illustrates the components of the *lac* operon as well as the action of the *lac* repressor in the presence and absence of lactose.

Jacob and Monod suggested that the repressor normally binds to the DNA sequence of the operator region. When it does so, it inhibits the action of RNA polymerase, effectively repressing the transcription of the structural genes [Figure 17–5(b)]. However, when lactose is present, this sugar binds to the repressor and causes an allosteric (conformational) change. The change alters the binding site of the repressor, rendering it incapable of interacting with operator DNA [Figure 17–5(c)]. In the absence of the repressor–operator interaction, RNA polymerase transcribes the structural genes, and the enzymes necessary for lactose metabolism are produced. Because transcription occurs only when the repressor *fails* to bind to the operator region, regulation is said to be under *negative control.*

To summarize, the operon model invokes a series of molecular interactions between proteins, inducers, and DNA to explain the efficient regulation of structural gene expression. In the absence of lactose, the enzymes encoded by the genes are not needed, and expression of genes encoding these enzymes is repressed. When lactose is present, it indirectly induces the activation of the genes by binding with the repressor. * If all lactose is metabolized, none is available to bind to the repressor, which is again free to bind to operator DNA and to repress transcription.

* Technically, the inducer is allolactose, an isomer of lactose. When lactose enters the bacterial cell, some of it is converted to allolactose by the β-galactosidase enzyme.

WEB TUTORIAL 17.1 REGULATION OF GENE EXPRESSION: PROKARYOTES

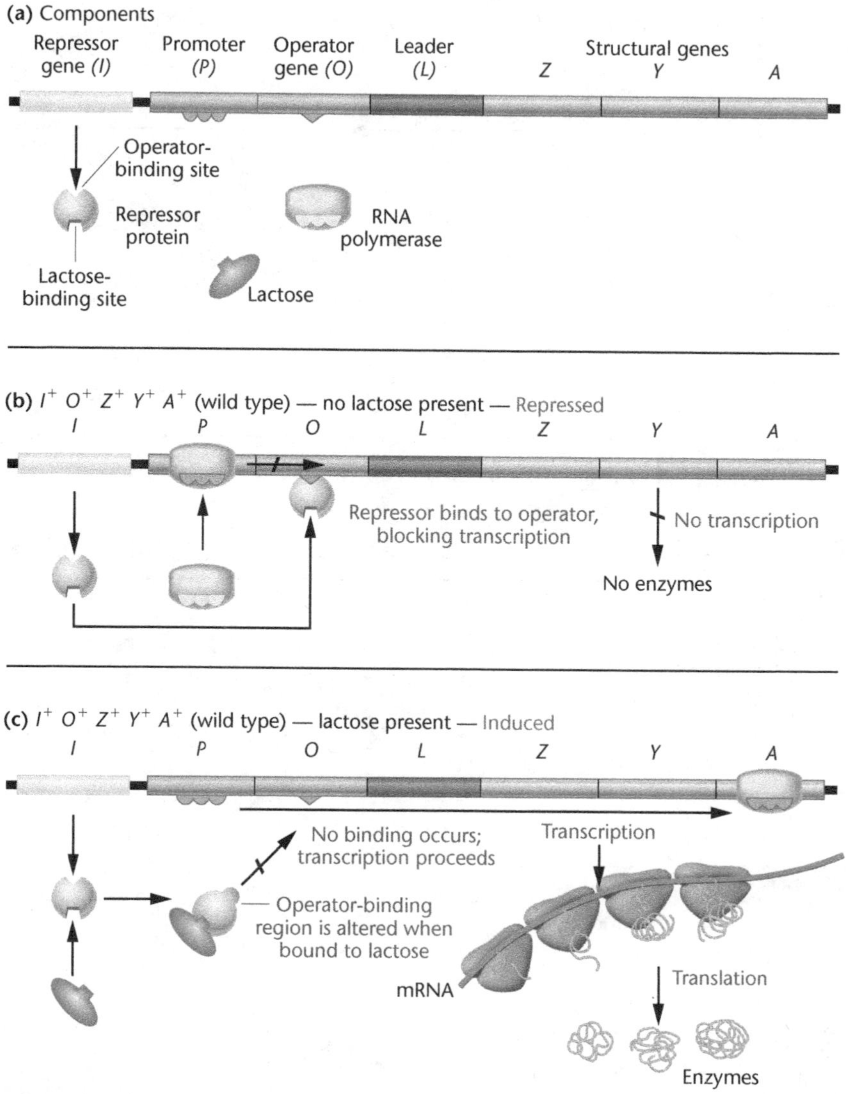


**FIGURE 17–5** The components of the wild-type *lac* operon and the response in the absence and the presence of lactose.

Both the $I^-$ and $O^C$ constitutive mutations interfere with these molecular interactions, allowing continuous transcription of the structural genes. In the case of the $I^-$ mutant, seen in Figure 17–6(a), the repressor protein is altered or absent and cannot bind to the operator region, so the structural genes are always turned on. In the case of the $O^C$ mutant [Figure 17–6(b)], the nucleotide sequence of the operator DNA is altered and will not bind with a normal repressor molecule. The result is the same: the structural genes are always transcribed.

## Genetic Proof of the Operon Model

The operon model is a good one because it leads to three major predictions that can be tested to determine its validity. The major predictions to be tested are that (1) the *I* gene produces a diffusible product (that is, a *trans*-acting product); (2) the *O* region is involved in regulation but does not produce a product (it is *cis*-acting); and (3) the *O* region must be adjacent to the structural genes in order to regulate transcription.

The creation of partially diploid bacteria allows us to assess these assumptions, particularly those that predict the presence of *trans*-acting regulatory elements. For example, as introduced in Chapter 6, the F plasmid may contain chromosomal genes, in which case it is designated F′. When an $F^-$ cell acquires such a plasmid, it contains its own chromosome plus one or more additional genes present in the plasmid. This host cell is thus a **merozygote**, a cell that is diploid for certain added genes (but not for the rest of the

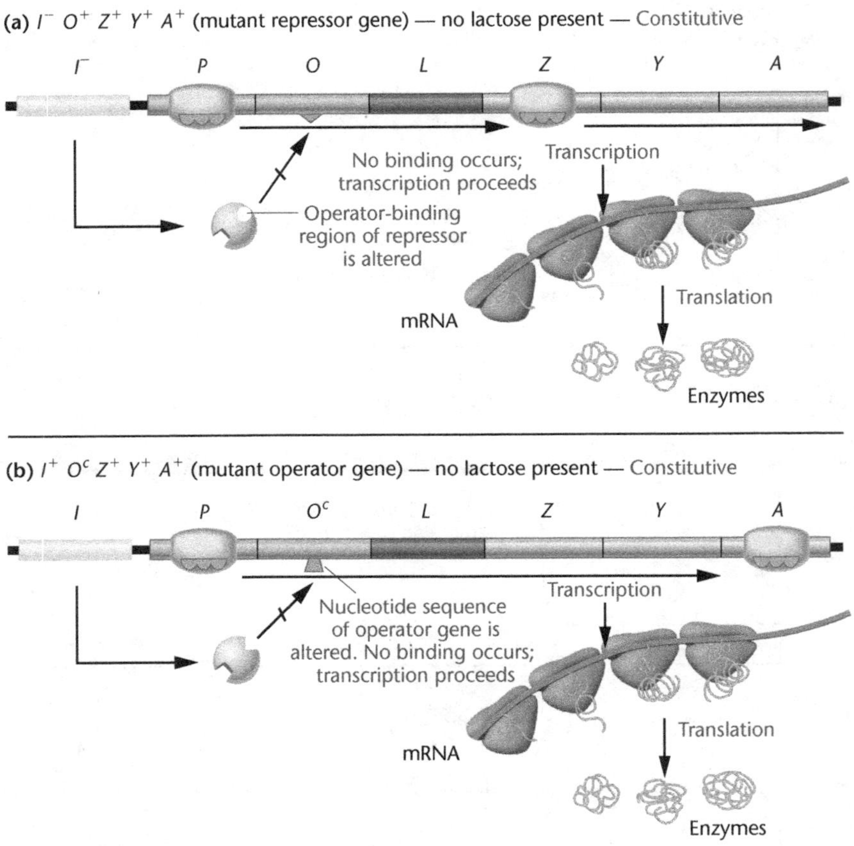


**FIGURE 17–6** The response of the *lac* operon in the absence of lactose when a cell bears either the $I^-$ or the $O^C$ mutation.

chromosome). The use of such a plasmid makes it possible, for example, to introduce an $I^+$ gene into a host cell whose genotype is $I^-$, or to introduce an $O^+$ region into a host cell of genotype $O^C$. The Jacob–Monod operon model predicts how regulation should be affected in such cells. Adding an $I^+$ gene to an $I^-$ cell should restore inducibility, because the normal wild-type repressor, which is a *trans*-acting factor, would be produced by the inserted $I^+$ gene. In contrast, adding an $O^+$ region to an $O^C$ cell should have no effect on constitutive enzyme production, since regulation depends on an $O^+$ region being located immediately adjacent to the structural genes—that is, $O^+$ a *cis*-acting regulator.

Results of these experiments are shown in Table 17.1, where $Z$ represents the structural genes (and the inserted genes are listed after the designation F′). In both cases described above, the Jacob–Monod model is upheld (part B of Table 17.1). Part C of the table shows the reverse experiments, where either an $I^-$ gene or an $O^C$ region is added to cells of normal inducible genotypes. As the model predicts, inducibility is maintained in these partial diploids.

Another prediction of the operon model is that certain mutations in the $I$ gene should have the opposite effect of $I^-$. That is,

**TABLE 17.1**

**A Comparison of Gene Activity (+ or −) in the Presence or Absence of Lactose for Various *E. coli* Genotypes**

| Genotype | Presence of β-Galactosidase Activity: Lactose Present | Lactose Absent |
|---|---|---|
| $I^+O^+Z^+$ | + | − |
| A. $I^+O^+Z^-$ | − | − |
| $I^-O^+Z^+$ | + | + |
| $I^+O^CZ^+$ | + | + |
| B. $I^-O^+Z^+/F'I^+$ | + | − |
| $I^+O^CZ^+/F'O^+$ | + | + |
| C. $I^+O^+Z^+/F'I^-$ | + | − |
| $I^+O^+Z^+/F'O^C$ | + | − |
| D. $I^SO^+Z^+$ | − | − |
| $I^SO^+Z^+/F'I^+$ | − | − |

NOTE: In parts B to D, most genotypes are partially diploid, containing an F factor plus attached genes (F′).

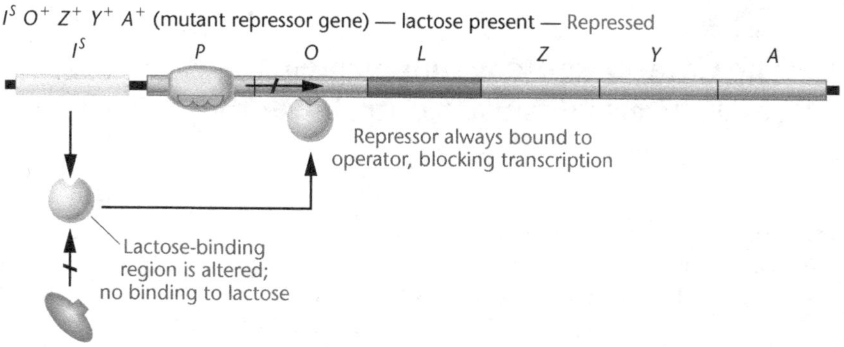


**FIGURE 17–7** The response of the *lac* operon in the presence of lactose in a cell bearing the $I^S$ mutation.

instead of being constitutive because the repressor can't bind the operator, mutant repressor molecules should be produced that cannot interact with the inducer, lactose. As a result, these repressors would always bind to the operator sequence, and the structural genes would be permanently repressed. In cases like this, the presence of an additional $I^+$ gene would have little or no effect on repression.

In fact, such a mutation, $I^S$, was discovered wherein the operon, as predicted, is "super-repressed," as shown in part D of Table 17.1 (and depicted in Figure 17–7). An additional $I^+$ gene does not effectively relieve repression of gene activity. These observations are consistent with the idea that the repressor contains separate DNA-binding domains and inducer-binding domains.

**NOW SOLVE THIS**

Problem 6 on page 454 asks you to predict the outcome of gene expression in varying genotypes and cellular conditions.

■ HINT: *Determine initially whether the repressor is active or inactive based on whether the gene encoding it is wild type or mutant. Then, consider the impact of the presence or absence of lactose.*

## Isolation of the Repressor

Although Jacob and Monod's operon theory succeeded in explaining many aspects of genetic regulation in prokaryotes, the nature of the repressor molecule was not known when their landmark paper was published in 1961. While they had assumed that the allosteric repressor was a protein, RNA was also a candidate because the activity of the molecule required the ability to bind to DNA. Despite many attempts to isolate and characterize the hypothetical repressor molecule, no direct chemical evidence was forthcoming. A single *E. coli* cell contains no more than 10 or so copies of the *lac* repressor; and direct chemical identification of 10 molecules in a population of millions of proteins and RNAs in a single cell presented a tremendous challenge.

In 1966, Walter Gilbert and Benno Müller-Hill reported the isolation of the *lac* repressor in partially purified form. To achieve the isolation, they used a *regulator quantity* ($I^q$) mutant strain that contains about 10 times as much repressor as do wild-type *E. coli* cells. Also instrumental in their success was the use of the gratuitous inducer IPTG, which binds to the repressor, and the technique of **equilibrium dialysis.** In this technique, extracts of $I^q$ cells were placed in a dialysis bag and allowed to attain equilibrium with an external solution of radioactive IPTG, a molecule small enough to diffuse freely in and out of the bag. At equilibrium, the concentration of radioactive IPTG was higher inside the bag than in the external solution, indicating that an IPTG-binding material was present in the cell extract and was too large to diffuse across the wall of the bag.

Ultimately, the IPTG-binding material was purified and shown to have various characteristics of a protein. In contrast, extracts of $I^-$ constitutive cells having no *lac* repressor *activity* did not exhibit IPTG binding, strongly suggesting that the isolated protein was the repressor molecule.

To confirm this thinking, Gilbert and Müller-Hill grew *E. coli* cells in a medium containing radioactive sulfur and then isolated the IPTG-binding protein, which was labeled in its sulfur-containing amino acids. Next, this protein was mixed with DNA from a strain of phage lambda ($\lambda$) carrying the $lacO^+$ gene. When the two substances are not mixed together, the DNA sediments at 40*S*, while the IPTG-binding protein sediments at 7*S*. However, when the DNA and protein were mixed and sedimented in a gradient, using ultracentrifugation, the radioactive protein sedimented at the same rate as did DNA, indicating that the protein binds to the DNA. Further experiments showed that the IPTG-binding, or repressor, protein binds only to DNA containing the *lac* region and does not bind to *lac* DNA containing an operator-constitutive $O^C$ mutation.

**NOW SOLVE THIS**

Problem 8 on page 454 asks you to describe how the *lac* repressor was isolated.

■ HINT: *You must first understand that equilibrium dialysis allows small molecules such as lactose or the gratuitous inducer IPTG to move freely back and forth across the membrane, while large molecules such as the repressor cannot cross freely but must remain on the same side of the membrane throughout the experiment.*

17.3

## The Catabolite-Activating Protein (CAP) Exerts Positive Control over the *lac* Operon

As described in the preceding discussion of the *lac* operon, the role of $\beta$-galactosidase is to cleave lactose into its components, glucose and galactose. Then, in order to be used by the cell, the galactose, too, must be converted to glucose. What if the cell found itself in an environment that contained ample amounts of both lactose *and* glucose? Given that glucose is the preferred carbon source for *E. coli*, it would not be energetically efficient for a cell to induce transcription of the *lac* operon, since what it really needs—glucose—is already present. As we shall see next, still another molecular component, called the **catabolite-activating protein (CAP)**, is involved in effectively repressing the expression of the *lac* operon when glucose is present. This inhibition is called **catabolite repression**.

To understand CAP and its role in regulation, let's backtrack for a moment to review the system depicted in Figure 17–5. When the *lac* repressor is bound to the inducer (lactose), the *lac* operon is activated, and RNA polymerase transcribes the structural genes. As stated in Chapter 14, transcription is initiated as a result of the binding that occurs between RNA polymerase and the nucleotide sequence of the **promoter region**, found upstream (5′) from the initial coding sequences. Within the *lac* operon, the promoter is found between the *I* gene and the operator region (*O*) (see Figure 17–1). Careful examination has revealed that polymerase binding is never very efficient unless CAP is also present to facilitate the process.

The mechanism is summarized in Figure 17–8. In the absence of glucose and under inducible conditions, CAP exerts positive control by binding to the CAP site, facilitating RNA-polymerase binding at the promoter, and thus transcription. Therefore, for maximal transcription of the structural genes, the repressor must be bound by lactose (so as not to repress operon expression), *and* CAP must be bound to the CAP-binding site.

This leads to the central question about CAP: how does the presence of glucose inhibit CAP binding? The answer involves still another molecule, **cyclic adenosine monophosphate (cAMP)**, upon which CAP binding is dependent. *In order to bind to the promoter, CAP must be bound to cAMP.* The level of cAMP is itself dependent on an enzyme, **adenyl cyclase**, which catalyzes the conversion of

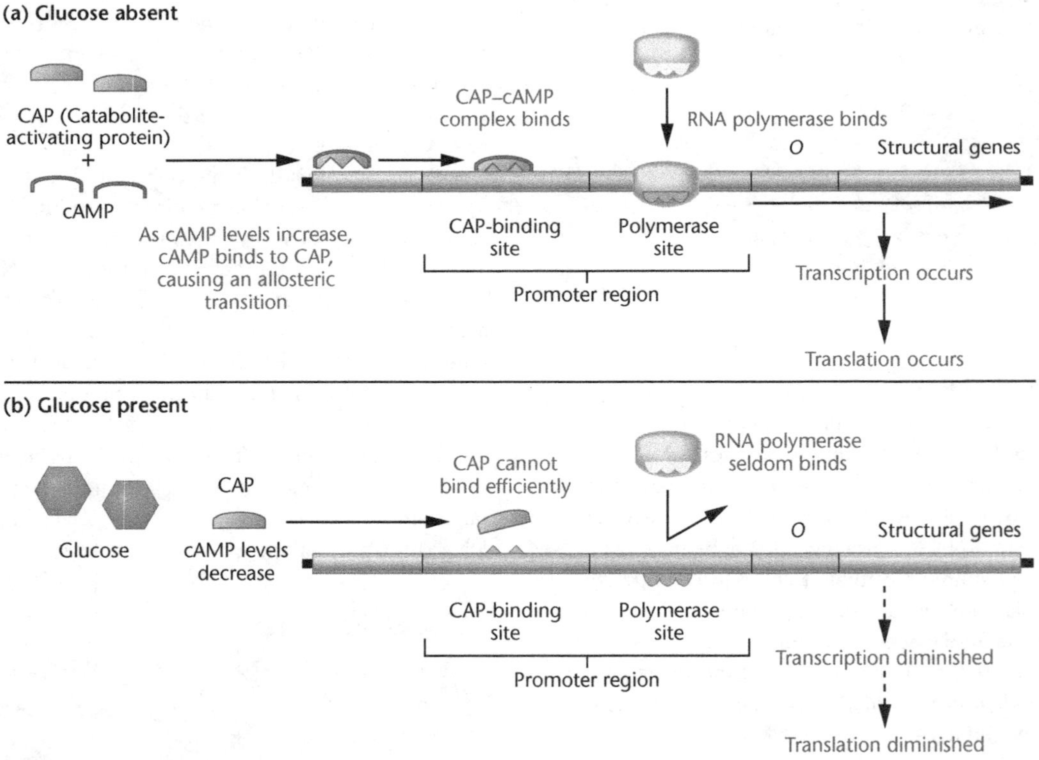


FIGURE 17–8 Catabolite repression. (a) In the absence of glucose, cAMP levels increase, resulting in the formation of a CAP–cAMP complex, which binds to the CAP site of the promoter, stimulating transcription. (b) In the presence of glucose, cAMP levels decrease, CAP–cAMP complexes are not formed, and transcription is not stimulated.

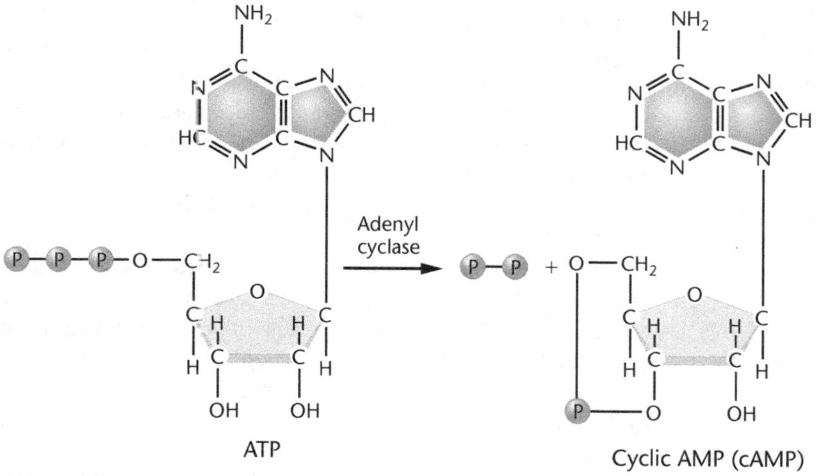


FIGURE 17–9 The formation of cAMP from ATP, catalyzed by adenyl cyclase.

ATP to cAMP (see Figure 17–9).* The role of glucose in catabolite repression is now clear. It inhibits the activity of adenyl cyclase, causing a decline in the level of cAMP in the cell. Under this condition, CAP cannot form the CAP–cAMP complex essential to the positive control of transcription of the *lac* operon.

Like the *lac* repressor, CAP and cAMP–CAP have been examined by X-ray crystallography. CAP is a dimer that inserts into adjacent regions of a specific nucleotide sequence of the DNA making up the *lac* promoter. The cAMP–CAP complex, when bound to DNA, bends it, causing it to assume a new conformation.

Binding studies in solution further clarify the mechanism of gene activation. Alone, neither cAMP–CAP nor RNA polymerase has a strong tendency to bind to *lac* promoter DNA, nor does either molecule have a strong affinity for the other. However, when both are together in the presence of the *lac* promoter DNA, a tightly bound complex is formed, an example of what is called **cooperative binding**. In the case of cAMP–CAP and the *lac* operon, this phenomenon illustrates the high degree of specificity that is involved in the genetic regulation of just one small group of genes.

Regulation of the *lac* operon by catabolite repression results in efficient energy use, because the presence of glucose will override the need for the metabolism of lactose, should the lactose also be available to the cell. In contrast to the negative regulation conferred by the *lac* repressor, the action of cAMP–CAP constitutes positive regulation. Thus, a combination of positive and negative regulatory mechanisms determine transcription levels of the *lac* operon. Catabolite repression involving CAP has also been observed for other inducible operons, including those controlling the metabolism of galactose and arabinose.

NOW SOLVE THIS

Problem 10 on page 454 asks you to assess the level of gene activity in the *lac* operon for various combinations of the presence or absence of lactose and glucose.

HINT: *You must keep in mind that regulation involving lactose is a negative control system, while regulation involving glucose is a positive control system.*

17.4

## Crystal Structure Analysis of Repressor Complexes Has Confirmed the Operon Model

We now have thorough knowledge of the biochemical nature of the regulatory region of the *lac* operon, including the precise locations of its various components relative to one another (Figure 17–10). In 1996, Mitchell Lewis, Ponzy Lu, and their colleagues succeeded in determining the crystal structure of the *lac* repressor, as well as the structure of the repressor bound to the inducer and to operator DNA. As a result, previous information that was based on genetic and biochemical data has now been complemented with the missing structural interpretation. Together, these contributions provide a nearly complete picture of the regulation of the operon.

The repressor, as the gene product of the *I* gene, is a monomer consisting of 360 amino acids. Within this monomer, the region of

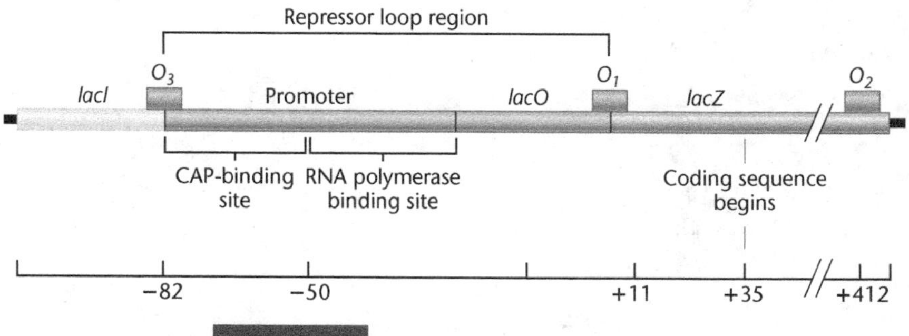


FIGURE 17–10 The various regulatory regions involved in the control of genetic expression of the *lac* operon, as described in the text. The numbers on the bottom scale represent nucleotide sites upstream and downstream from the initiation of transcription.

* Because of its involvement with cAMP, CAP is also called *cyclic AMP receptor protein (CRP)*, and the gene encoding the protein is named *crp*. Since the protein was first named CAP, we will adhere to the initial nomenclature.

inducer binding has been identified [Figure 17–11(a)]. While dimers [Figure 17–11(b)] can also bind the inducer, the functional repressor is a homotetramer (that is, it contains four copies of the monomer). The tetramer can be cleaved with a protease under controlled conditions to yield five fragments. Four are derived from the N-terminal ends of the tetramer subunits, and they bind to operator DNA. The fifth fragment is the remaining core of the tetramer, derived from the COOH-terminus ends; it binds to lactose and gratuitous inducers such as IPTG. Analysis has revealed that each tetramer can bind to two symmetrical operator DNA helices at a time.

The operator DNA that was previously defined by mutational studies (*lacO*$^{C}$) and confirmed by DNA-sequencing analysis is located just upstream from the beginning of the actual coding sequence of the *lacZ* gene. Crystallographic studies show that the actual region of repressor binding of this primary operator, $O_1$, consists of 21 base pairs. Two other auxiliary operator regions have been identified, as shown in Figure 17–10. One, $O_2$, is 401 base pairs downstream from the primary operator, within the *lacZ* gene. The other, $O_3$ , is 93 base pairs upstream from $O_1$, just beyond the CAP site. *In vivo*, all three operators must be bound for maximum repression.

Binding by the repressor simultaneously at two operator sites distorts the conformation of DNA, causing it to bend away from the repressor. When a model is created to demonstrate dual binding of operators $O_1$ and $O_3$ [Figure 17–11(c)], the 93 base pairs of DNA that intervene must jut out, forming what is called a **repression loop**. This model positions the promoter region that binds RNA polymerase on the inside of the loop, which prevents access by the polymerase during repression. In addition, the repression loop positions the CAP-binding site in a way that facilitates CAP interaction with RNA polymerase upon subsequent induction. The DNA looping caused by repression in this model is similar to configuration changes that are predicted to occur in eukaryotic systems (see Chapter 18).

Studies have also defined the three-dimensional conformational changes that accompany the interactions with the inducer molecules. Taken together, the crystallographic studies have brought a new level of understanding of the regulatory process occurring within the *lac* operon, confirming the findings and predictions of Jacob and Monod in their model set forth over 40 years ago, which was based strictly on genetic observations.

## 17.5 The Tryptophan (*trp*) Operon in *E. coli* Is a Repressible Gene System

Although the process of induction had been known for some time, it was not until 1953 that Monod and colleagues discovered a repressible operon. Wild-type *E. coli* are capable of producing the enzymes necessary for the biosynthesis of amino acids as well as other

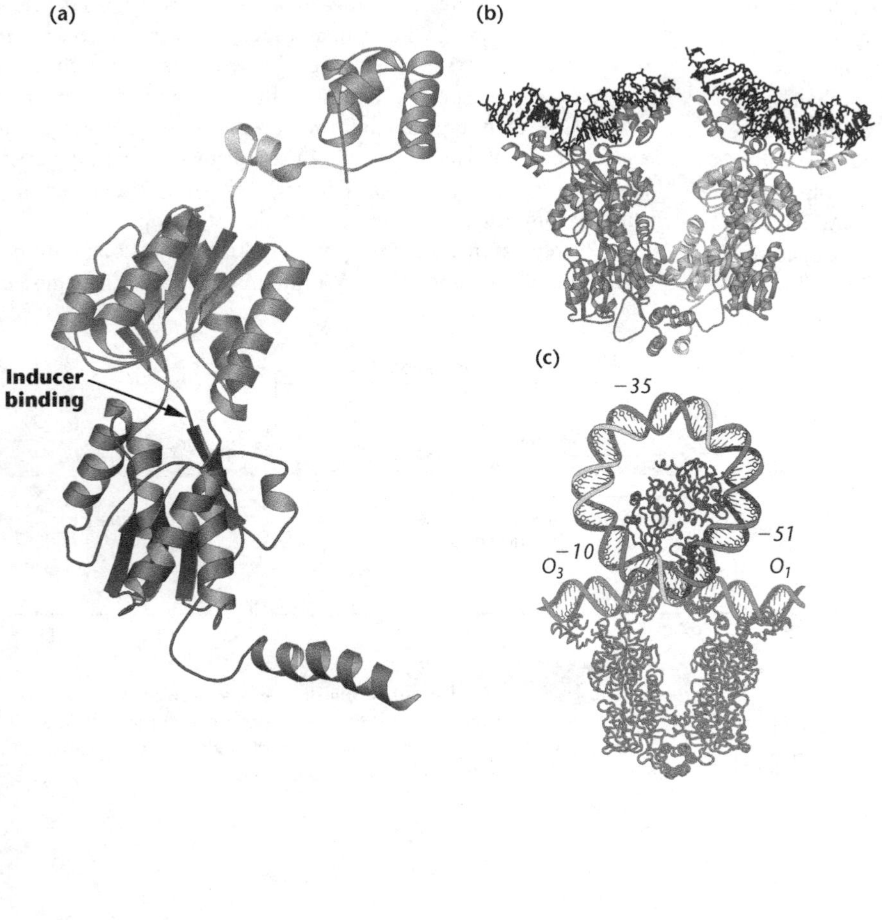


FIGURE 17–11 Models of the *lac* repressor and its binding to operator sites in DNA, as generated from crystal structure analysis. (a) The repressor monomer, showing the inducer-binding site. The DNA-binding region is shown in red. (b) The repressor dimer bound to two 21-base-pair segments of operator DNA (shown in dark blue). (c) The repressor (shown in pink) and CAP (shown in dark blue) bound to the *lac* DNA. Binding to operator regions $O_1$ and $O_3$ creates a 93-base-pair repression loop of promoter DNA.

essential macromolecules. Focusing his studies on the amino acid tryptophan and the enzyme **tryptophan synthetase**, Monod discovered that if tryptophan is present in sufficient quantity in the growth medium, the enzymes necessary for its synthesis are not produced. It is energetically advantageous for bacteria to repress expression of genes involved in tryptophan synthesis when ample tryptophan is present in the growth medium.

Further investigation showed that a series of enzymes encoded by five contiguous genes on the *E. coli* chromosome are involved in tryptophan synthesis. These genes are part of an operon, and in the presence of tryptophan, all are coordinately repressed, and none of the enzymes is produced. Because of the great similarity between this repression and the induction of enzymes for lactose metabolism, Jacob and Monod proposed a model of gene regulation analogous to the *lac* system (the updated version is shown in Figure 17–12).

To account for repression, they suggested the presence of a *normally inactive repressor* that alone cannot interact with the operator region of the operon. However, the repressor is an allosteric molecule that can bind to tryptophan. When tryptophan is present, the resultant complex of repressor and tryptophan attains a new conformation that binds to the operator, repressing transcription. Thus, when tryptophan, the end product of this anabolic pathway, is present, the system is repressed and enzymes are not made. Since the regulatory complex inhibits transcription of the operon, this repressible system is under negative control. And as tryptophan participates in repression, it is referred to as a **corepressor** in this regulatory scheme.

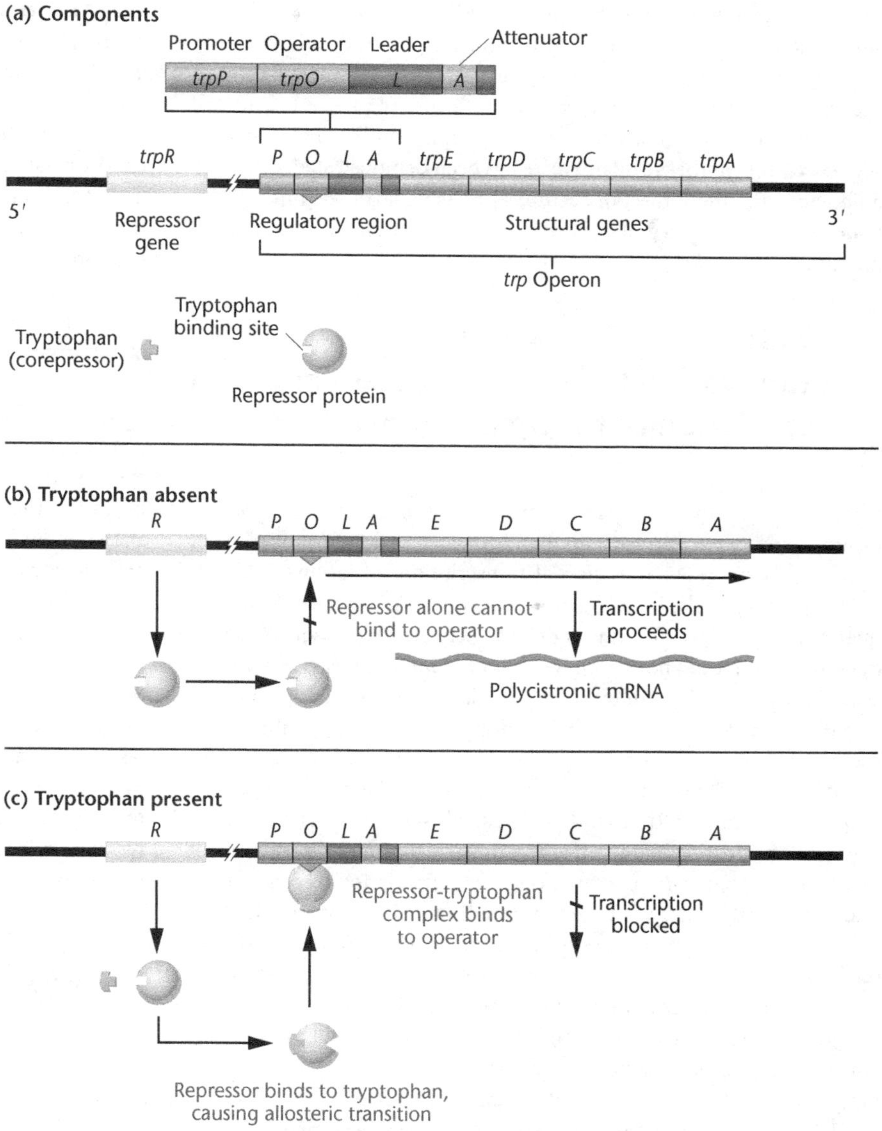


**FIGURE 17–12** A repressible operon. (a) The components involved in regulation of the tryptophan operon. (b) In the absence of tryptophan, an inactive repressor is made that cannot bind to the operator (*O*), thus allowing transcription to proceed. (c) When tryptophan is present, it binds to the repressor, causing an allosteric transition to occur. This complex binds to the operator region, leading to repression of the operon.

## Evidence for the *trp* Operon

Support for the concept of a repressible operon was soon forthcoming, based primarily on the isolation of two distinct categories of constitutive mutations. The first class, *trpR*$^-$, maps at a considerable distance from the structural genes. This locus represents the gene coding for the repressor. Presumably, the mutation inhibits either the repressor's interaction with tryptophan or repressor formation entirely. Whichever the case, no repression ever occurs in cells with the *trpR*$^-$ mutation. As expected, if the *trpR*$^+$ gene encodes a functional repressor molecule, the presence of a copy of this gene will restore repressibility.

The second constitutive mutation is analogous to that of the operator of the lactose operon, because it maps immediately adjacent to the structural genes. Furthermore, the addition of a wild-type operator gene into mutant cells (as an external element) does not restore repression. This is what would be predicted if the mutant operator no longer interacts with the repressor–tryptophan complex.

The entire *trp* operon has now been well defined, as shown in Figure 17–12. The five contiguous structural genes (*trp E, D, C, B,* and *A*) are transcribed as a polycistronic message directing translation of the enzymes that catalyze the biosynthesis of tryptophan. As in the *lac* operon, a promoter region (*trpP*) represents the binding site for RNA polymerase, and an operator region (*trpO*) binds the repressor. In the absence of binding, transcription is initiated within the *trpP–trpO* region and proceeds along a **leader sequence** 162 nucleotides prior to the first structural gene (*trpE*). Within that leader sequence, still another regulatory site has been demonstrated, called an *attenuator*—the subject of Section 17.6. As we shall see, this regulatory unit is an integral part of the control mechanism of this operon.

## 17.6 Attenuation Is a Critical Process in Regulation of the *trp* Operon in *E. coli*

Charles Yanofsky, his coworker Kevin Bertrand, and their colleagues observed that, even when tryptophan is present and the *trp* operon is effectively repressed, initiation of transcription still occurs, leading to synthesis of the initial portion of the mRNA (the 5′-leader sequence). Apparently, the binding of the activated repressor to the operator region does not strongly inhibit the *initiation of transcription* of the operon. This finding suggests that there must be a subsequent mechanism by which tryptophan somehow inhibits transcription of the rest of the operon. Yanofsky discovered that, following initiation of transcription *in the presence of high concentrations of tryptophan*, mRNA synthesis is usually terminated at a point about 140 nucleotides along the transcript. This form of repression, in which the transcription of the operon is greatly reduced rather than prevented entirely, is called **attenuation.*** (In contrast, when tryptophan is absent, or present in very low concentrations, transcription is initiated and *not* subsequently terminated, instead continuing beyond the leader sequence along the DNA encoding the structural genes, starting with the *trpE* gene. As a result, a polycistronic mRNA is produced, and the enzymes essential to the biosynthesis of tryptophan are subsequently translated.)

Identification of the site involved in attenuation was made possible by the isolation of various deletion mutations in the region 115 to 140 nucleotides into the leader sequence. Such mutations abolish attenuation. This site is referred to as the **attenuator**. An explanation of how attenuation occurs and how it is overcome, put forward by Yanofsky and colleagues, is summarized in Figure 17–13. The initial DNA sequence that is transcribed [Figure 17–13(a)] gives rise to an mRNA molecule [Figure 17–13(b)] that has the potential to fold into either of two mutually exclusive stem-loop structures described as "hairpins." In the presence of excess tryptophan, the hairpin that is formed behaves as a **terminator** structure, almost always causing transcription to be terminated prematurely. On the other hand, if tryptophan is scarce, the alternative structure, referred to as the **antiterminator hairpin**, is formed. Transcription in this case proceeds past the antiterminator hairpin region, and the entire mRNA is subsequently produced. These hairpins are illustrated in Figure 17–13(c).

The question is how the absence or scarcity of tryptophan circumvents attenuation. A key point in Yanofsky's model is that the leader transcript must be translated in order for the antiterminator hairpin to form. Yanofsky discovered that the leader transcript contains two triplets (UGG) that encode tryptophan, preceded upstream by the initial AUG sequence that prompts the initiation of translation by ribosomes. When adequate tryptophan is present, charged $tRNA^{trp}$ is present, allowing translation to proceed past the tryptophan-encoding triplets and causing the *terminator hairpin* to form, as illustrated in Figure 17–13(d). If cells are starved of tryptophan, charged $tRNA^{trp}$ is unavailable. As a result, the ribosome "stalls" at the UGG triplets, inducing the formation of the antiterminator hairpin within the leader transcript, as shown in Figure 17–13(d). As a result, attenuation is circumvented and transcription proceeds, leading to expression of the entire set of structural genes.

Attenuation appears to be a mechanism common to other bacterial operons in *E. coli* that regulate the enzymes essential to the biosynthesis of amino acids, including those involved in threonine, histidine, leucine, and phenylalanine metabolism. As with the *trp* operon, attenuators in these operons contain multiple codons calling for the amino acid being regulated. When that amino acid is missing, translation "stalls" and transcription proceeds. For example, the leader sequence in the histidine operon codes for seven histidine residues in a row.

## 17.7 TRAP and AT Proteins Govern Attenuation in *B. subtilis*

Keep in mind that not all organisms, even those of the same type, solve problems such as gene regulation in exactly the same way. Thus, it is not unusual for geneticists to discover that new strategies have arisen during evolution. Often, a new approach is a variation on a well-established theme. Such is the case with the regulation of the *trp* operon in different bacteria.

As we saw previously, *E. coli*, a Gram-negative bacterium, uses charged $tRNA^{trp}$ and a terminator hairpin in the leader sequence of the transcript as machinery for attenuating transcription of its *trp* operon. The Gram-positive bacterium *Bacillus subtilis* also uses attenuation and hairpins to regulate its *trp* operon. In fact, *B. subtilis* relies on attenuation as its sole mechanism of regulation because that bacterium lacks a mechanism that represses transcription entirely in this operon, as is present in *E. coli*.

However, the molecular signals that cause attenuation in *B. subtilis* do not use a process of translation and stalling, as *E. coli* does, to induce the hairpin that terminates transcription. Instead, a

* "Attenuation" is derived from the verb *attenuate*, meaning "to reduce in strength, weaken, or impair."

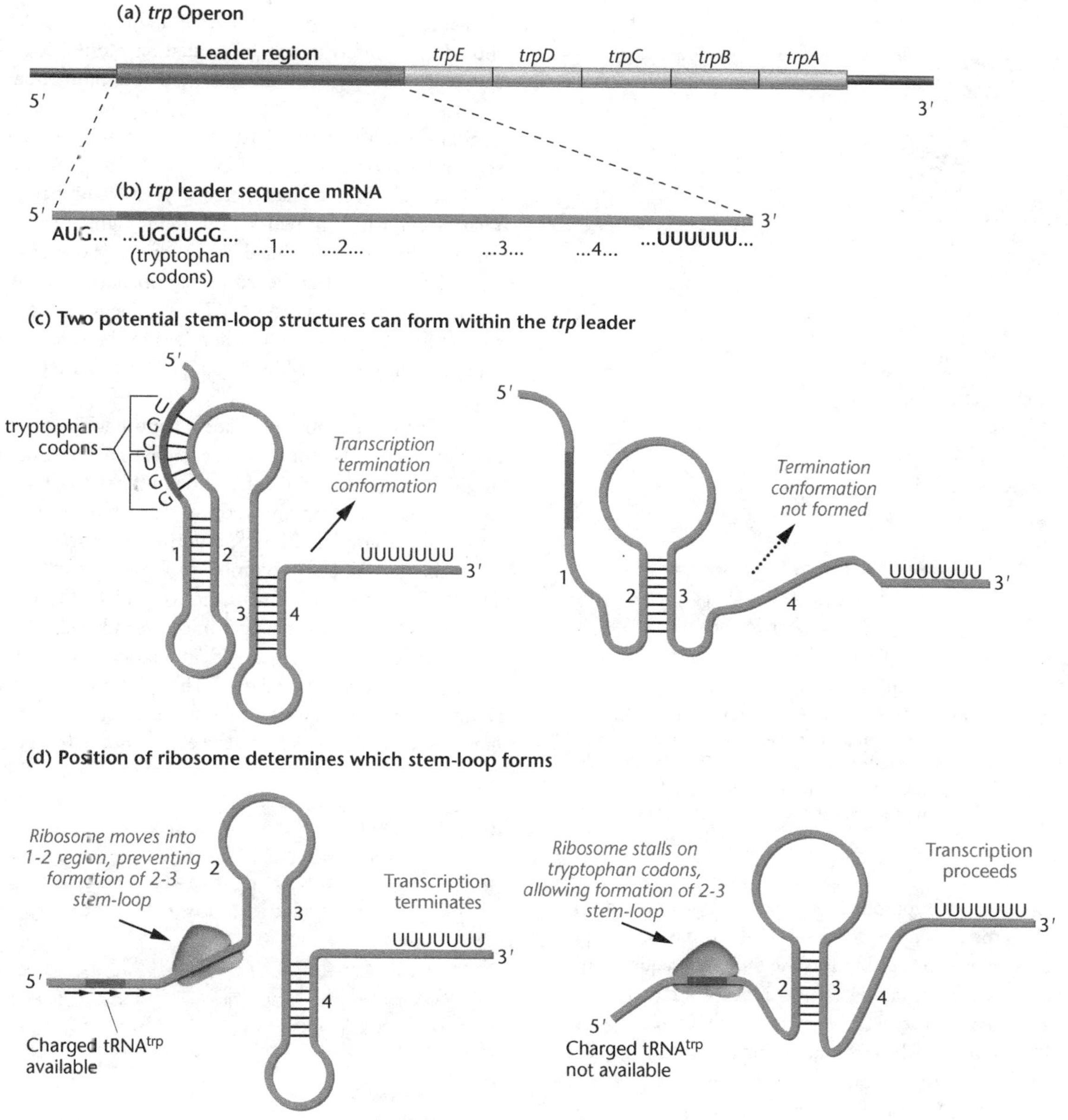


**FIGURE 17–13** Attenuation in the *trp* operon. (a) The *trp* operon, showing the leader region followed by the five *trp* genes. (b) The leader region in the mRNA is enlarged to show the translation start codon (AUG), the two tryptophan codons (UGGUGG), and the string of U residues at the end of the leader. The four regions marked 1, 2, 3, and 4 indicate the four sequence regions that have the potential to form stems by base pairing. (c) The two possible stem-loop structures that can form within the leader sequence mRNA. The first conformation creates a stem-loop involving regions 3 and 4, followed by a string of U residues. This conformation triggers transcription termination by RNA polymerase. The second conformation creates a stem-loop involving regions 2 and 3 that does not allow formation of the terminator signal. (d) During translation, the ribosome either moves past the tryptophan codons or stalls, depending on whether $tRNA^{trp}$ is present. The position of the ribosome determines whether the terminator structure forms or whether transcription can proceed through the terminator region.

specific protein, isolated in the 1990s by Charles Yanofsky and coworkers, either binds or does not bind to the attenuator leader sequence, thereby inducing the alternative terminator or antiterminator configurations, respectively.

Attenuation is accomplished in *B. subtilis* in a unique way. The protein, ***trp* RNA-binding attenuation protein (TRAP)**, binds to tryptophan if it is present in the cell. TRAP consists of 11 subunits, forming a symmetrical quaternary protein structure. Each subunit

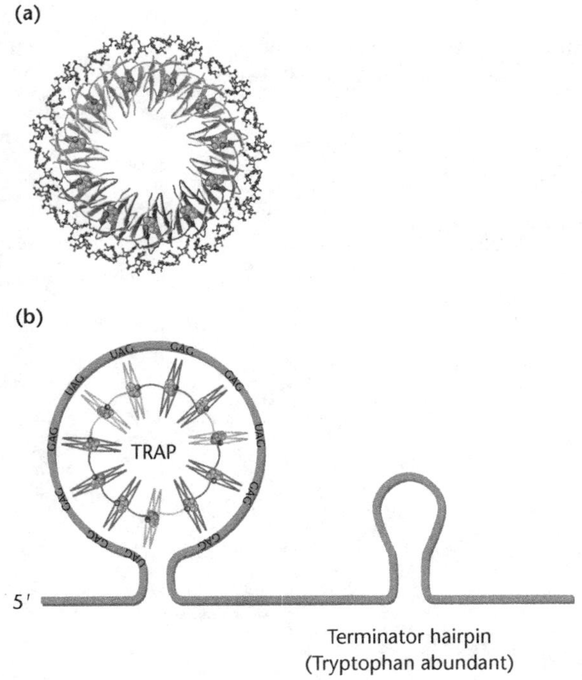


FIGURE 17–14 Model of the *trp* RNA-binding attenuation protein (TRAP). The symmetrical molecule consists of 11 subunits, each capable of binding to one molecule of tryptophan that is embedded within the subunit. (b) The interaction of a tryptophan-bound TRAP molecule with the leader sequence of the *trp* operon of *B. subtilis*. This interaction induces the formation of the terminator hairpin, attenuating expression of the *trp* operon.

can bind one molecule of tryptophan, incorporating it into a deep pocket within the protein [Figure 17–14(a)]. When fully saturated with tryptophan, this protein can bind to the 5′-leader sequence of the RNA transcript, which contains 11 triplet repeats of either GAG or UAG, each separated by several spacer nucleotides. Each triplet is linked to one of the subunits, which contains a binding pocket for the triplet. When bound to the transcript, the TRAP is encircled by a belt of RNA [Figure 17–14(b)], and this conformation prevents the antiterminator hairpin from forming. Instead, the terminator configuration results, leading to premature termination of transcription and, consequently, attenuation of expression of the operon.

The attenuation strategy of *B. subtilis* involves an interesting mechanism of regulation. Two observations had suggested to Yanofsky and his colleagues that the attenuation strategy in *B. subtilis* might be even more complex than had originally appeared. First, regulation of the *trp* operon in *B. subtilis* is extremely sensitive to a wide range of tryptophan concentrations. Such a fine tuning suggests that something more than the simple on–off mechanism attributed to TRAP might be at work. Second, a mutation in the gene encoding tryptophanyl-tRNA synthetase leads to the overexpression of the *trp* operon, even in the presence of excess tryptophan. This enzyme is responsible for charging $tRNA^{trp}$; in its absence, uncharged $tRNA^{trp}$ molecules accumulate, and regulation is interrupted. The interruption of regulation would suggest that uncharged $tRNA^{trp}$ plays some essential role in attenuation, leading Yanofsky and his coworker Angela Valbuzzi to ask how $tRNA^{trp}$ might be involved with TRAP. What they found extends our knowledge of the system of regulation.

They discovered that there exists still another protein, **anti-TRAP (AT)**, which provides a metabolic signal that $tRNA^{trp}$ is uncharged, thereby indicating that tryptophan is very scarce in the cell. Yanofsky and Valbuzzi theorized that uncharged $tRNA^{trp}$ induces a separate operon to express the *AT* gene. The AT protein then associates with TRAP, specifically when it is in the tryptophan-activated state, and inhibits the binding of TRAP to its target leader RNA sequence.

Such a finding not only adds yet another dimension to our understanding of this system of gene regulation, but it also explains the original observation of the mutation in the $tRNA^{trp}$ synthetase gene that leads to the overexpression of the *trp* operon. The mutation prevents charging of $tRNA^{trp}$, even in the presence of tryptophan. As uncharged $tRNA^{trp}$ accumulates, it induces AT, which binds to TRAP. The TRAP is then unable to bind to the leader RNA sequence; as a result, the *trp* operon is overinduced.

The preceding description demonstrates the complexity of regulatory mechanisms in bacteria. That such intricate strategies have resulted from the evolutionary process attests to the critical importance of carefully regulating gene expression in bacteria.

NOW SOLVE THIS

Problem 25 on page 455 concerns the regulation of the *B. subtilis trp* operon, which involves the TRAP protein. You are asked to predict whether or not the structural genes are expressed under various conditions.

■ HINT: *Unlike E. coli, which uses tryptophan as a repressor along with the process of attenuation in regulating the trp operon, B. subtilis relies solely on the process of attenuation and the TRAP protein for regulation.*

## 17.8 The *ara* Operon Is Controlled by a Regulator Protein That Exerts Both Positive and Negative Control

We conclude this chapter with a brief discussion of the **arabinose (*ara*) operon** as analyzed in *E. coli.* This inducible operon is unique because the same regulatory protein is capable of exerting both positive and negative control, and as a result, at some times induces and at others represses gene expression. Figure 17–15(a) shows the operon components, and Figure 17–15(b) and (c) show, respectively, the conditions under which the operon is active or inactive.

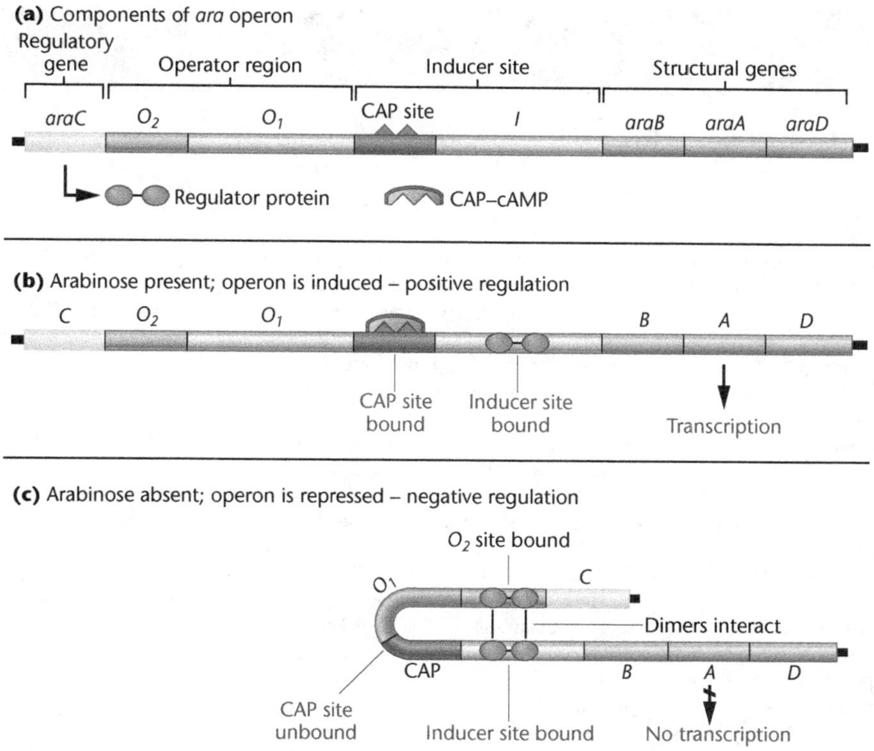


**FIGURE 17–15** Genetic regulation of the *ara* operon. The regulatory protein of the *araC* gene acts as either an inducer (in the presence of arabinose) or a repressor (in the absence of arabinose).

1. The metabolism of the sugar arabinose is governed by the enzymatic products of three structural genes, *ara B*, *A*, and *D* [Figure 17–15(a)]. Their transcription is controlled by the regulatory protein AraC, encoded by the *araC* gene, which interacts with two regulatory regions, *araI* and *ara*$O_2$.* These sites can be bound individually or coordinately by the AraC protein.

2. The *I* region bears that designation because when it is the only region bound by AraC, the system is induced [Figure 17–15(b)]. For this binding to occur, both arabinose and cAMP must be present. Thus, like induction of the *lac* operon, a CAP-binding site is present in the promoter region that modulates catabolite repression in the presence of glucose.

3. In the absence of both arabinose and cAMP, the AraC protein binds coordinately to both the *I* site *and* the $O_2$ site (so named because it was the second operator region to be discovered in this operon). When both *I* and $O_2$ are bound by AraC, a conformational change occurs in the DNA to form a tight loop [Figure 17–15(c)], and the structural genes are repressed.

4. The $O_2$ region is located about 200 nucleotides upstream from the *I* region. Binding at either regulatory region involves a dimer of AraC. When both regions are bound, the dimers interact, producing the loop that causes repression. Presumably, this loop inhibits the access of RNA polymerase to the promoter region. The region of DNA that loops out (the stretch between *I* and $O_2$) is critical to the formation of the repression complex. Genetic alteration here by insertion or deletion of even a few nucleotides is sufficient to interfere with loop formation and, therefore, repression.

Our observations involving the *ara* operon illustrate the degree of complexity that may be seen in the regulation of a group of related genes. As we mentioned at the beginning of this chapter, the development of regulatory mechanisms has provided evolutionary advantages that allow bacterial systems to adjust to a range of natural environments. Without question, these systems are well equipped genetically not only to survive under varying physiological conditions but to do so with great biochemical efficiency.

* An additional operator $O_1$ region is also present, but is not involved in the regulation of the *ara* structural gene.

## GENETICS, TECHNOLOGY, AND SOCIETY

# Quorum Sensing: How Bacteria Talk to One Another

For decades, scientists have regarded bacteria as independent microbial organisms, incapable of cell-to-cell communication. However, recent research has shown that many bacteria can regulate gene expression and coordinate group behavior through a form of communication termed *quorum sensing*. Through this process, bacteria send and receive chemical signals called autoinducers that relay information about population size. When the population size reaches a "quorum," defined in the business world as the minimum number of members of an organization that must be present to conduct business, the autoinducers regulate gene expression in a way that benefits the group as a whole. Quorum sensing has been described in more than 70 species of bacteria, and its uses are manifold. Some marine bacteria use autoinducers to control bioluminescence, while other pathogenic bacteria use similar signals to regulate the expression of toxic virulence factors. After studying the complex mechanisms of quorum sensing, scientists have been forced to reevaluate their understanding of prokaryotic gene regulation and, in addition, have developed some promising practical applications of the phenomenon, including possible alternatives to antibiotic drugs.

The study of quorum sensing began in the 1960s, when researchers noticed that during the day, the squid *Euprymna scolopes* lowers the concentration of the luminescent marine bacterium *Vibrio fischeri* in its light organ, and in response, the bacteria cease to glow. *V. fischeri* has a symbiotic relationship with the squid. While hunting for food at night, the squid uses light emitted by the *V. fischeri* present in its light organ to illuminate the ocean floor, thus perfectly countering the shadows created by moonlight that normally act as a beacon for the squid's predators. This symbiotic relationship provides the bacteria with a protected, nutrient-rich environment (in the squid's light organ).

What turns the bacteria's luminescent (*lux*) genes on in response to high cell density and off in response to low cell density? The chemical "language" that bacteria use for quorum sensing depends largely on two classes of autoinducer molecules: homoserine lactones (HSLs), used by Gram-negative bacteria, and oligopeptides, used by Gram-positive bacteria. In *V. fischeri*, the responsible autoinducer is an HSL. At a critical population size, or quorum, the HSL regulates the *lux* operon by binding directly to transcription factors that regulate gene expression. The oligopeptides used by Gram-positive bacteria also activate transcription factors but do so by initiating phosphorylation cascades.

Because most bacterial species use unique autoinducers with their own distinct chemical structures and properties, quorum sensing was thought to mediate communication only among members of the same species. Then, in 1994, Bonnie Bassler and her colleagues at Princeton University made headlines when they discovered an autoinducer molecule in the marine bacterium *Vibrio harveyi* that was also present in many diverse types of bacteria. This molecule, autoinducer-2 (AI-2), has the potential to mediate "quorum-sensing cross talk" between species and thus serve as a universal language for bacterial communication. Because the accumulation of AI-2 is proportional to cell number, and since the structure of AI-2 may vary slightly between different species, the current hypothesis is that AI-2 can transmit information about both cell density and species composition of a bacterial community. Such a communication system would be highly advantageous in nature, where populations of bacteria commonly contain hundreds of different species. Although presently only the AI-2 structure of *V. harveyi* is known, researchers are studying other bacterial species to determine if slight modifications of this molecule might represent different "words" in the AI-2 bacterial language.

Pathogenic bacteria also use quorum sensing to regulate gene expression. Instead of controlling bioluminescence, these bacteria communicate to coordinate the production of harmful virulence substances or to avoid detection by the immune system. For example, *Vibrio cholerae*, the causative agent of cholera, uses AI-2 and an additional species-specific autoinducer to activate the genes controlling the production of cholera toxin. *Pseudomonas aeruginosa*, the Gram-negative bacterium that often affects cystic fibrosis patients, uses quorum sensing to regulate the production of elastase, a protease that disrupts the respiratory epithelium and interferes with ciliary function. *P. aeruginosa* also uses autoinducers to control the production of biofilms, tough protective shells that resist host defenses and make treatment with antibiotics nearly impossible. Other bacteria determine cell density through quorum sensing to delay the production of toxic substances until the colony is large enough to overpower the host's immune system and establish an infection.

Each year, millions of hospital patients acquire bacterial infections. Although some can be successfully treated with antibiotics, antibiotic-resistant strains of bacteria are evolving faster than scientists can develop alternative treatments. Every year, *Staphylococci* infections alone affect 500,000 patients with implanted medical devices such as catheters and artificial heart valves, resulting in approximately 90,000 deaths each year. Because many bacteria rely on quorum sensing to regulate disease-causing genes, treatments that block quorum sensing may aid in combating infections. In 1999, Bonnie Bassler formed a company called Quorex, with the goal of using quorum-sensing research to develop new antibacterial agents. In June 2004, researchers showed that implanted medical devices coated with a quorum-sensing inhibitor called RIP can reduce the prevalence of *Staphylococcus* infections in rats. Research is also underway to develop drugs that block quorum sensing in *Bacillus anthracis*, the bacterium more commonly known as anthrax, and *Pseudomonas aeruginosa*, the bacterium that may be lethal to those with cystic fibrosis or to others with compromised immune systems. Similar drugs that inhibit mechanisms of bacterial communication are being tested for their ability to inhibit the formation of biofilms, restore the potency of antibiotics, and limit the development of new antibiotic-resistant strains of bacteria. Thus, what began as a fascinating observation in the glowing squid has launched an exciting era of research in bacterial genetics that may one day prove of great clinical significance.

### References

Camara, M., Hardman, A., Williams, P., and Milton, D. 2002. Quorum sensing in *Vibrio cholerae*. *Nature Genetics*. 32: 217–218.

Federle, M. J., and Bassler, B. L. 2003. Interspecies communication in bacteria. *J. Clin. Invest.* 112: 1291–1299.

Suga, H., and Smith, K. M. 2003. Molecular mechanisms of bacterial quorum sensing as a new drug target. *Current Opinion in Chemical Biology*. 7: 586–591.

EXPLORING GENOMICS

# Microarrays and MicrobesOnline

In this chapter we discussed mechanisms of gene expression regulation in prokaryotes, especially transcriptional regulation of bacterial operons. Several of the examples in the chapter demonstrated how bacteria such as *E. coli* respond quickly to changes in environmental conditions to meet the metabolic needs of their cells. For example, you now know that *E. coli* can induce or repress expression of the *lac* operon depending on the availability of lactose as a nutrient.

The Exploring Genomics exercise for Chapter 6 discussed microbial genome projects and some of the reasons microbial genomes are being studied. We will now explore **MicrobesOnline**, a database for gene expression studies of prokaryotic genomes, in order to learn more about the effects of heat shock on gene expression in *E. coli*.

## ■ Exercise I – Gene Microarrays for Analyzing Gene Expression Patterns

**MicrobesOnline** contains genome data for over 400 prokaryotes! Two major purposes of this database are to provide a resource for gene expression data from microbes and to allow scientists to compare gene sequences from different microbes. Much of the gene expression data in MicrobesOnline was derived from gene microarray studies. Microarrays, or gene chips, which will be discussed in detail in Chapter 21 (see Figure 21-20), enable scientists to examine gene expression patterns for thousands of genes simultaneously. Before exploring MicrobesOnline, we will visit the **Microarrays MediaBook** for an excellent overview of how microarrays are made and interpreted, including a tutorial showing microarray data from yeast grown in the presence or absence of oxygen.

1. Access Microarrays MediaBook at http://gcat.davidson.edu/Pirelli/index.htm. After you enter MediaBook, use the "Method," "Exploration," and "Data Interpretation" links, following each step in these tutorials to learn more about how microarrays are made.
2. Answer the following questions:
   a. If you were studying gene expression patterns for bacteria grown in the presence or absence of different antibiotics, would you isolate DNA or RNA from the bacteria to hybridize to the microarray?
   b. What is contained on each "spot" of a microarray?
   c. How can different complementary (cDNA) samples be identified if they are hybridized to the microarray?
   d. How are gene expression values typically reported for microarray data?
   e. When microarray data from an experimental sample are compared to data from a control sample, what does a gene expression ratio of 1 mean?
3. You may also want to use the "Assessment" tool to test your knowledge about microarrays. After selecting an answer, be sure to click the "Submit" button near the bottom of the screen to enter the answer.

## ■ Exercise II – MicrobesOnline: *E. coli* Heat-Shock Genes

Now that you are familiar with microarrays, you will explore MicrobesOnline to evaluate the results of heat shock on gene expression of *E. coli*.

1. Access MicrobesOnline at http://www.microbesonline.org/. Find the link indicating the number of genomes available in the database and review the list of prokaryotic genomes available.
2. Return to the homepage and click the "gene expression data" link to access gene expression data. This part of the database presents microarray data for prokaryotes subjected to a variety of different experimental conditions. Microarrays in this database provide fascinating information on gene expression changes of microbes in response to varying conditions of pH, temperature, nutrients, antibiotics, ultraviolet light, and other experimental parameters. For most of these experiments, microbes were exposed to conditions for different lengths of time. RNA was then isolated from the microbes, and cDNA was synthesized using fluorescent dye-labeled nucleotides and reverse transcriptase. Labeled cDNAs were then hybridized to gene microarrays, and results revealed changes in gene expression patterns.
3. Select "*E. coli* K12" as the organism and "Heat shock" as the experimental condition, then click browse to retrieve available gene expression data.
4. The first results page you will see provides a summary of experimental conditions and control experiments. Links will allow you to access a wealth of information. The most relevant links include:
   - **U**: upregulated genes. The most highly induced genes are shown in dark red. These are genes that showed a statistically significant increase in expression. Z-scores show fold induction as the ratio of experimental to control samples. For example, a Z-score of 3.2 indicates a 3.2-fold induction in expression for that gene in experimental samples compared to control samples.
   - **D**: downregulated genes. The most highly repressed genes are shown in dark gray or gray and are presented at the top of the table. No shading indicates no statistically significant differences in expression.
   - **M**: metabolic pathways. This feature presents maps of metabolic pathways showing induced and repressed genes.
   - **T**: gene functional classifications provided by the Institute for Genomic Research (TIGR). Recall that in Chapter 6 we explored the TIGR Microbial Sequencing Center Web site and discussed the role of TIGR in producing the first complete sequence of a microbial genome (*H. influenzae*).

*Continued on next page*

*Exploring Genomics, continued*

- C: gene functional classifications from the NCBI Clusters of Orthologous Genes (COG) database, which compares protein sequences from complete genomes in different phylogenetic lineages.

5. Use the links to evaluate the results of this experiment and answer the following questions:

   a. What were the experimental conditions for this experiment?

   b. Examine the upregulated gene data showing genes induced by heat shock. What were the top two genes induced by heat shock?

   c. Find the genes *dnaJ* and *GroEL* on the upregulated list. Was the expression of these genes affected by heat shock (and if so, how)?

   d. On the upregulated data page, use the operon list feature. Which operon showed the highest level of induction for all genes in the operon?

6. Each gene on the "Gene List" shows links for **I** (gene information), **A** (results for expression of that gene from all microarray data in the database), and **O** (indication of whether the gene is part of an operon). For the genes you studied in 5b and 5c above, use the I, A, and O links to answer the following questions:

   a. Did any of these genes show changes in expression following any of the other experiments in the database?

   b. Are any of these genes part of an operon?

7. Return to the initial results page for this experiment and use the C link ("clusters of orthologous genes") to look at functional categories of genes affected by heat shock.

   a. Which function categories of genes showed the most induction (upregulation)?

   b. Which two functional categories of genes showed the most repression (downregulation)?

8. Refer to the textbook or do a PubMed search to learn more about the genes studied in steps 5b and 5c. Describe the functions of heat-shock proteins.

9. Explore another experimental condition for a prokaryote of your interest and write a short report summarizing your findings.

## Chapter Summary

1. Genetic regulation prevents the complete genome of an organism from being continuously transcribed in every cell under all conditions. Highly refined mechanisms have evolved that regulate transcription and thus optimize genetic efficiency.
2. Genetic analysis of bacteria has been pursued successfully since the 1940s. The ease of obtaining large quantities of pure cultures of mutant strains of bacteria for experiments has made bacteria organisms of choice in numerous types of genetic studies.
3. The discovery and study of the *lac* operon in *E. coli* pioneered the study of gene regulation in bacteria. Genes involved in the metabolism of lactose are coordinately regulated by a negative control system that responds to the presence or absence of lactose.
4. The catabolite-activating protein (CAP) exerts positive control over *lac* gene expression in *E. coli*. It does so by interacting with RNA polymerase at the *lac* promoter and by responding to levels of cyclic AMP in the bacterial cell.
5. When glucose is present in *E. coli*, cyclic AMP levels are low, CAP binding does not occur, and the structural genes are not expressed. This phenomenon is called catabolite repression. Such a mechanism is efficient energetically, since glucose is preferred, even in the presence of lactose.
6. The *lac* repressor of *E. coli* has been isolated and studied. Crystal structure analysis has shown how it interacts with the DNA of the operon as well as with inducers. These studies have revealed conformational changes in DNA leading to the formation of a repression loop that inhibits binding between RNA polymerase and the promoter region of the operon.
7. The biosynthesis of tryptophan in *E. coli* involves a number of enzymes that are encoded by genes located in an operon. In contrast to the inducible operon controlling lactose metabolism, the *trp* operon is repressible. In the presence of tryptophan, the repressor binds to the regulatory region of the *trp* operon and represses transcription initiation. Like the *lac* operon, the *trp* operon functions under negative control.
8. An additional regulatory step, referred to as attenuation, further regulates the *trp* operon. Attenuation, which occurs when tryptophan is abundant, causes premature termination of transcription within a leader sequence that precedes the structural genes in the mRNA. The mechanism differs in *E. coli* and *B. subtilis*, but both depend on alternative hairpin structures in the leader sequence, that form under different conditions.
9. The *ara* operon is unique in that the regulator protein exerts both positive and negative control over expression of the genes specifying the enzymes that metabolize the sugar arabinose.

# INSIGHTS AND SOLUTIONS

1. A hypothetical operon (*theo*) in *E. coli* contains several structural genes encoding enzymes that are involved sequentially in the biosynthesis of an amino acid. Unlike the *lac* operon, in which the repressor gene is separate from the operon, the gene encoding the regulator molecule is contained within the *theo* operon. When the end product (the amino acid) is present, it combines with the regulator molecule, and this complex binds to the operator, repressing the operon. In the absence of the amino acid, the regulatory molecule fails to bind to the operator, and transcription proceeds.

   Categorize and characterize this operon, then consider the following mutations, as well as the situation in which the wild-type gene is present along with the mutant gene in partially diploid cells (F′):

   (a) Mutation in the operator region.
   (b) Mutation in the promoter region.
   (c) Mutation in the regulator gene.

   In each case, will the operon be active or inactive in transcription, assuming that the mutation affects the regulation of the *theo* operon? Compare each response with the equivalent situation for the *lac* operon.

   **Solution:** The *theo* operon is repressible and under negative control. When there is no amino acid present in the medium (or the environment), the product of the regulatory gene cannot bind to the operator region, and transcription proceeds under the direction of RNA polymerase. The enzymes necessary for the synthesis of the amino acid are produced, as is the regulator molecule. If the amino acid *is* present, either initially or after sufficient synthesis has occurred, the amino acid binds to the regulator, forming a complex that interacts with the operator region, causing repression of transcription of the genes within the operon.

   The *theo* operon is similar to the tryptophan system, except that the regulator gene is within the operon rather than separate from it. Therefore, in the *theo* operon, the regulator gene is itself regulated by the presence or absence of the amino acid.

   a. As in the *lac* operon, a mutation in the *theo* operator gene inhibits binding with the repressor complex, and transcription occurs constitutively. The presence of an F′ plasmid bearing the wild-type allele would have no effect, since it is not adjacent to the structural genes.
   b. A mutation in the *theo* promoter region would no doubt inhibit binding to RNA polymerase and therefore inhibit transcription. This would also happen in the *lac* operon. A wild-type allele present in an F′ plasmid would have no effect.
   c. A mutation in the *theo* regulator gene, as in the *lac* system, may inhibit either its binding to the repressor or its binding to the operator gene. In both cases, transcription will be constitutive, because the *theo* system is repressible. Both cases result in the failure of the regulator to bind to the operator, allowing transcription to proceed. In the *lac* system, failure to bind the corepressor lactose would permanently repress the system. The addition of a wild-type allele would restore repressibility, provided that this gene was transcribed constitutively.

## Problems and Discussion Questions

1. Contrast the need for the enzymes involved in lactose and tryptophan metabolism in bacteria when lactose and tryptophan, respectively, are (a) present and (b) absent.
2. Contrast positive versus negative control of gene expression.
3. Contrast the role of the repressor in an inducible system and in a repressible system.
4. Even though the *lac Z, Y,* and *A* structural genes are transcribed as a single polycistronic mRNA, each gene contains the initiation and termination signals essential for translation. Predict what will happen when a cell growing in the presence of lactose contains a deletion of one nucleotide (a) early in the *Z* gene and (b) early in the *A* gene.
5. For the *lac* genotypes shown in the accompanying table, predict whether the structural genes (*Z*) are constitutive, permanently repressed, or inducible in the presence of lactose.

| Genotype | Constitutive | Repressed | Inducible |
| --- | --- | --- | --- |
| $I^+O^+Z^+$ | | | x |
| $I^-O^+Z^+$ | | | |
| $I^-O^CZ^+$ | | | |
| $I^-O^CZ^+/F'O^+$ | | | |
| $I^+O^CZ^+/F'O^+$ | | | |
| $I^SO^+Z^+$ | | | |
| $I^SO^+Z^+/F'I^+$ | | | |

6. For the genotypes and conditions (lactose present or absent) shown in the accompanying table, predict whether functional enzymes, nonfunctional enzymes, or no enzymes are made.

| Genotype | Condition | Functional Enzyme Made | Nonfunctional Enzyme Made | No Enzyme Made |
|---|---|---|---|---|
| $I^+O^+Z^+$ | No lactose | | | X |
| $I^+O^CZ^+$ | Lactose | | | |
| $I^-O^+Z^-$ | No lactose | | | |
| $I^-O^+Z^-$ | Lactose | | | |
| $I^-O^+Z^+/F'I^+$ | No lactose | | | |
| $I^+O^CZ^+/F'O^+$ | Lactose | | | |
| $I^+O^+Z^-/F'I^+O^+Z^+$ | Lactose | | | |
| $I^-O^+Z^-/F'I^+O^+Z^+$ | No lactose | | | |
| $I^SO^+Z^+/F'O^+$ | No lactose | | | |
| $I^+O^CZ^+/F'O^+Z^+$ | Lactose | | | |

7. The locations of numerous *lacI*$^-$ and *lacI*$^S$ mutations have been determined within the DNA sequence of the *lacI* gene. Among these, *lacI*$^-$ mutations were found to occur in the 5′-upstream region of the gene, while *lacI*$^S$ mutations were found to occur farther downstream in the gene. Are the locations of the two types of mutations within the gene consistent with what is known about the function of the repressor that is the product of the *lacI* gene?
8. Describe the experimental rationale that allowed the *lac* repressor to be isolated.
9. What properties demonstrate the *lac* repressor to be a protein? Describe the evidence that it indeed serves as a repressor within the operon system.
10. Predict the level of genetic activity of the *lac* operon as well as the status of the *lac* repressor and the CAP protein under the cellular conditions listed in the accompanying table.

| | Lactose | Glucose |
|---|---|---|
| (a) | − | − |
| (b) | + | − |
| (c) | − | + |
| (d) | + | + |

11. Predict the effect on the inducibility of the *lac* operon of a mutation that disrupts the function of (a) the *crp* gene, which encodes the CAP protein, and (b) the CAP-binding site within the promoter.
12. Describe the role of attenuation in the regulation of tryptophan biosynthesis.
13. Attenuation of the *trp* operon was viewed as a relatively inefficient way to achieve genetic regulation when it was first discovered in the 1970s. Since then, however, attenuation has been found to be a relatively common regulatory strategy. Assuming that attenuation is a relatively inefficient way to achieve genetic regulation, what might explain its widespread occurrence?
14. Neelaredoxin is a 15-kDa protein that is a gene product common in anaerobic prokaryotes. It has superoxide-scavenging activity, and it is *constitutively expressed.* In addition, its expression is not further *induced* during its exposure to $O_2$ or $H_2O_2$ (Silva, G., et al. 2001. *J. Bacteriol.* 183: 4413–4420). What do the terms *constitutively expressed* and *induced* mean in terms of neelaredoxin synthesis?
15. Milk products such as cheeses and yogurts are dependent on the conversion by various anaerobic bacteria, including several *Lactobacillus* species, of lactose to glucose and galactose, ultimately producing lactic acid. These conversions are dependent on both permease and $\beta$-galactosidase as part of the *lac* operon. After selection for rapid fermentation for the production of yogurt, one *Lactobacillus* subspecies lost its ability to regulate *lac* operon expression (Lapierre, L., et al. 2002. *J. Bacteriol.* 184: 928–935). Would you consider it likely that in this subspecies the *lac* operon is "on" or "off"? What genetic events would likely contribute to the loss of regulation as described above?
16. Assume that the structural genes of the *lac* operon have been fused, through recombinant DNA techniques, to the regulatory apparatus of the *ara* operon. If arabinose is provided in a minimal medium to *E. coli* carrying this gene fusion, would you expect $\beta$-galactosidase to be produced at induced levels? Explain.

**HOW DO WE KNOW?**

17. In this chapter, we focused on the regulation of gene expression in prokaryotes. Along the way, we found many opportunities to consider the methods and reasoning by which much of this information was acquired. From the explanations given in the chapter, what answers would you propose to the following fundamental questions:
    (a) How do we know that bacteria regulate the expression of certain genes in response to the environment?
    (b) What evidence established that lactose serves as the inducer of a gene whose product is related to lactose metabolism?
    (c) What led researchers to conclude that a repressor molecule regulates the *lac* operon?
    (d) What experimental evidence suggested that physical interactions occur between the *lac* repressor and inducer molecules that control the *lac* operon?
    (e) How do we know that the *trp* operon is a repressible control system, in contrast to the *lac* operon, which is an inducible control system?

## Extra-Spicy Problems

18. Bacterial strategies to evade natural or human-imposed antibiotics are varied and include membrane-bound efflux pumps that export antibiotics from the cell. A review of efflux pumps (Grkovic, S., et al. 2002. *Microb. and Mol. Biol. Rev.* 66: 671–701) states that, because energy is required to drive the pumps, activating them in the absence of the antibiotic has a selective disadvantage. The review also states that a given antibiotic may play a role in the regulation of efflux by interacting with either an activator protein or a repressor protein, depending on the system involved. How might such systems be categorized in terms of *negative control* (*inducible* or *repressible*) or *positive control* (*inducible* or *repressible*)?

19. In a theoretical operon, genes *A*, *B*, *C*, and *D* represent the repressor gene, the promoter sequence, the operator gene, and the structural gene, *but not necessarily in the order named.* This operon is concerned with the metabolism of a theoretic molecule (tm). From the data provided in the accompanying table, first decide whether the operon is inducible or repressible. Then assign *A*, *B*, *C*, and *D* to the four parts of the operon. Explain your rationale. (AE = active enzyme; IE = inactive enzyme; NE = no enzyme.)

| Genotype | tm Present | tm Absent |
|---|---|---|
| $A^+B^+C^+D^+$ | AE | NE |
| $A^-B^+C^+D^+$ | AE | AE |
| $A^+B^-C^+D^+$ | NE | NE |
| $A^+B^+C^-D^+$ | IE | NE |
| $A^+B^+C^+D^-$ | AE | AE |
| $A^-B^+C^+D^+/F'A^+B^+C^+D^+$ | AE | AE |
| $A^+B^-C^+D^+/F'A^+B^+C^+D^+$ | AE | NE |
| $A^+B^+C^-D^+/F'A^+B^+C^+D^+$ | AE + IE | NE |
| $A^+B^+C^+D^-/F'A^+B^+C^+D^+$ | AE | NE |

20. A bacterial operon is responsible for the production of the biosynthetic enzymes needed to make the hypothetical amino acid tisophane (tis). The operon is regulated by a separate gene, *R*, deletion of which causes the loss of enzyme synthesis. In the wild-type condition, when tis is present, no enzymes are made; in the absence of tis, the enzymes are made. Mutations in the operator gene ($O^-$) result in repression regardless of the presence of tis. Is the operon under positive or negative control? Propose a model for (a) repression of the genes in the presence of tis in wild-type cells and (b) the mutations.
21. A marine bacterium is isolated and shown to contain an inducible operon whose genetic products metabolize oil when it is encountered in the environment. Investigation demonstrates that the operon is under positive control and that there is a *reg* gene whose product interacts with an operator region (*o*) to regulate the structural genes, designated *sg*.

    In an attempt to understand how the operon functions, a constitutive mutant strain and several partial diploid strains were isolated and tested with the results shown here:

| Host Chromosome | F′ Factor | Phenotype |
|---|---|---|
| wild type | none | inducible |
| wild type | *reg* gene from mutant strain | inducible |
| wild type | operon from mutant strain | constitutive |
| mutant strain | *reg* gene from wild type | constitutive |

Draw all possible conclusions about the mutation as well as the nature of regulation of the operon. Is the constitutive mutation in the *trans*-acting *reg* element or in the *cis*-acting *o* operator element?

22. The SOS repair genes in *E. coli* (discussed in Chapter 16) are negatively regulated by the *lexA* gene product, called the LexA repressor. When a cell sustains extensive damage to its DNA, the LexA repressor is inactivated by the *recA* gene product (RecA), and transcription of the SOS genes is increased dramatically.

    One of the SOS genes is the *uvrA* gene. You are a student studying the function of the *uvrA* gene product in DNA repair. You isolate a mutant strain that shows constitutive expression of the UvrA protein. Naming this mutant strain $uvrA^C$, you construct the following diagram showing the *lexA* and *uvrA* operons:

    (a) Describe two different mutations that would result in a *uvrA* constitutive phenotype. Indicate the actual genotypes involved.
    (b) Outline a series of genetic experiments that would use partial diploid strains to determine which of the two possible mutations you have isolated.
23. A fellow student considers the issues in Problem 22 and argues that there is a more straightforward, nongenetic experiment that could differentiate between the two types of mutations. The experiment requires no fancy genetics and would allow you to easily assay the products of the other SOS genes. Propose such an experiment.
24. Figure 17–13 depicts numerous critical regions of the leader sequence of mRNA that play important roles during the process of attenuation in the *trp* operon. A closer view of the leader sequence, which begins at about position 30 downstream from the 5′-end, is given here (running across both columns of this page):

AUGAAAGCAAUUUUCGUACUGAAAGGUUGGUGGCGCACUUCCUGAAACGGGCAGUGUAUUCACCAUGCGUAAAGCAAUCAGAUACCCAGCCCGCCUAAUGAGCGGGCUUUUUUUU

    Within this molecule are the sequences that cause the formation of the alternative hairpins. It also contains the successive triplets that encode tryptophan, where stalling during translation occurs. Take a large piece of paper (such as manila wrapping paper) and, along with several other students from your genetics class, work through the base sequence to identify the *trp* codons and the parts of the molecule representing the base-pairing regions that form the terminator and antiterminator hairpins shown in Figure 17–13.
25. Consider the regulation of the *trp* operon in *B. subtilis*. For each of the following cases, indicate whether the structural genes in the operon are being expressed or not expressed:
    (a) TRAP present, $tRNA^{trp}$ abundant, tryptophan abundant.
    (b) TRAP present, $tRNA^{trp}$ abundant, tryptophan scarce.
    (c) TRAP present, $tRNA^{trp}$ scarce, tryptophan abundant.
    (d) TRAP absent, $tRNA^{trp}$ abundant, tryptophan abundant.

    In each case, also indicate whether AT is present. Describe the role of TRAP and AT in attenuation.
26. One of the most prevalent sexually transmitted diseases is caused by the bacterium *Chlamydia trachomatis* and leads to blindness if left untreated. Upon infection, metabolically inert cells differentiate, through gene expression, to become metabolically active cells that divide by binary fission. It has been proposed that release from the inert state is dependent on heat-shock proteins that both activate the reproductive cycle and facilitate the binding of chlamydiae to host cells. Researchers made the following observations regarding the heat-shock regulatory system in *Chlamydia trachomatis*: (1) a regulator protein (call it R) binds to a *cis*-acting DNA element (call it D); (2) R and D function as a repressor–operator pair; (3) R functions as a negative regulator of transcription; (4) D is composed of an inverted-repeat sequence; (5) repression by R is dependent on D being supercoiled (Wilson & Tan, 2002).
    (a) Based on this information, present a simple model to explain the heat-dependent regulation of metabolism in *Chlamydia trachomatis*.
    (b) Some bacteria, like *E. coli*, use a heat-shock sigma factor to regulate heat-shock transcription. Are these findings in *Chlamydia* compatible with use of a heat-sensitive sigma factor?

27. A reporter gene—that is, a gene whose activity is relatively easy to detect—is commonly used to study promoter activity under a variety of investigative circumstances. Recombinant DNA technology is used to attach such a reporter gene to the promoter of interest, after which the recombinant molecule is inserted by transformation or transfection into the cell or host organism of choice. Green fluorescent protein (GFP) from the jellyfish *Aequorea victoria* fluoresces green when exposed to blue light. Assume that you wished to study the regulatory apparatus of the *lac* operon and you developed a plasmid (p) that contained all the elements of the *lac* operon except that you replaced the *lac* structural genes with the *GFP* gene (your reporter gene). You used this plasmid to transform *E. coli*. Considering the following genotypes, under which of the following conditions would you expect $\beta$-galactosidase production and/or green colonies when examined under blue light?

| | Medium Condition | | | |
|---|---|---|---|---|
| Genotype | Lactose | Glucose | $\beta$-galactosidase | Green Colonies |
| $I^+O^+Z^+/pI^+O^+GFP$ | − | + | | |
| | + | + | | |
| | + | − | | |
| $I^+O^+Z^-/pI^+O^cGFP$ | − | + | | |
| | + | − | | |
| $I^+O^+Z^+/pI^sO^cGFP$ | − | + | | |
| | + | − | | |
| $I^+O^cZ^+/pI^sO^+GFP$ | − | + | | |
| | + | − | | |

Chromosome territories in a chicken cell nucleus. Each chromosome is stained with a different-colored probe.

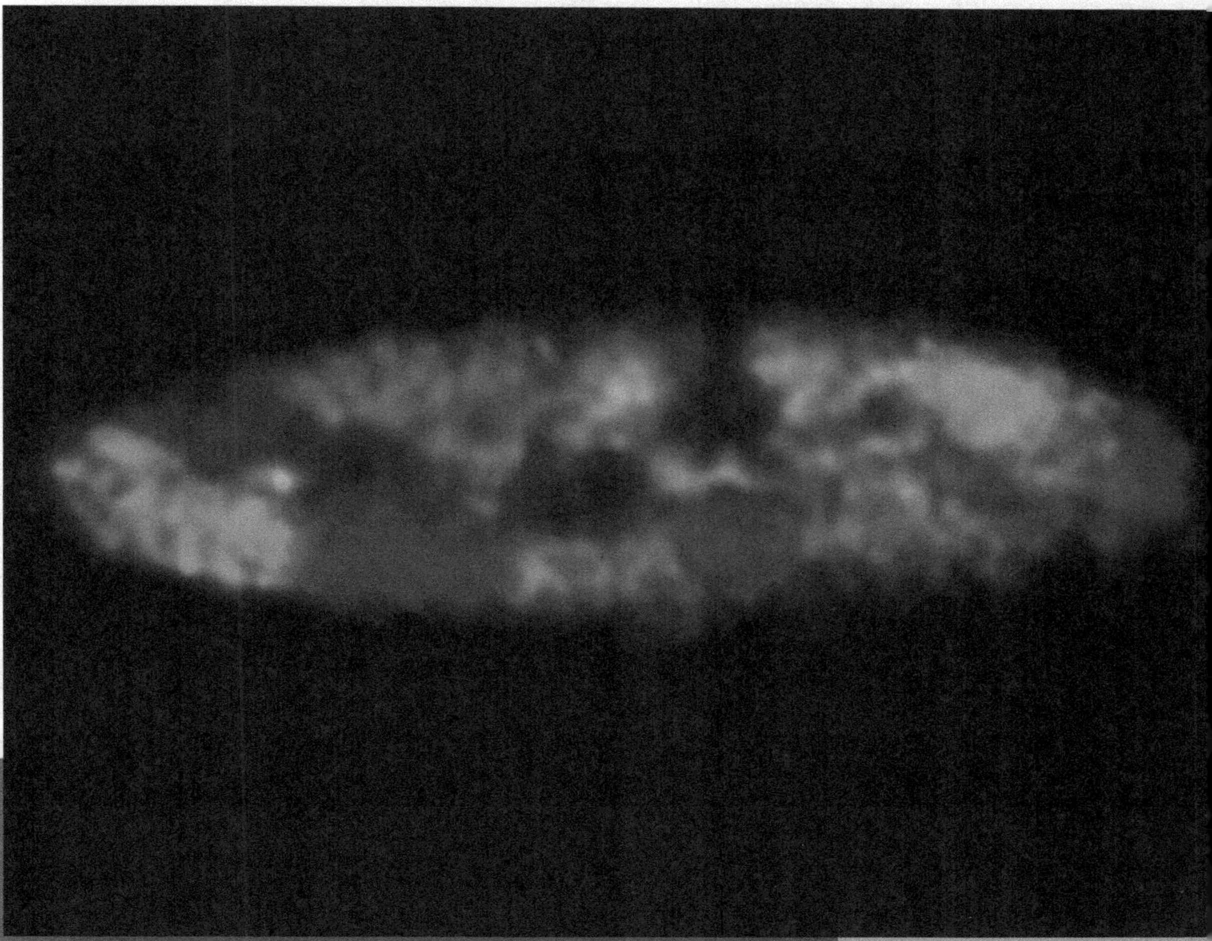

18

# Regulation of Gene Expression in Eukaryotes

## CHAPTER CONCEPTS

- In eukaryotes, expression of genetic information is regulated by mechanisms that exert control over transcription, mRNA stability, translation, and post-translational modifications.
- Eukaryotic gene regulation is more complex than prokaryotic gene regulation.
- The organization of eukaryotic chromatin in the nucleus plays a role in regulating gene expression. Chromatin must be remodeled to provide access to regulatory DNA sequences within it.
- Eukaryotic transcription initiation requires the assembly of transcription regulators at enhancer sites and the assembly of basal transcription complexes at promoter sites.
- Eukaryotic gene expression is also regulated at multiple posttranscriptional steps, including alternative splicing of pre-mRNA, control of mRNA stability, and RNA silencing.
- Gene expression is also regulated at the protein level, by modulation in the translation of mRNA and by the modification of proteins.

REGULATION OF GENE EXPRESSION: EUKARYOTES
WEB TUTORIAL 18.1

In multicellular eukaryotes, differential gene expression is essential. Differences in gene expression from cell to cell and from time to time are at the heart of embryonic development and maintenance of the adult organism. Virtually all cells in a eukaryotic organism contain a complete genome; however, only a subset of genes is expressed in any particular cell type. For example, some white blood cells express genes encoding certain immunoglobulins, allowing these cells to synthesize antibodies that defend the organism from infection and foreign agents. However, skin, kidney, and liver cells do not express immunoglobulins. Pancreatic islet cells, which synthesize and secrete insulin in response to the presence of blood sugars, do not manufacture immunoglobulins; moreover, they do not synthesize insulin when it is not required. Eukaryotic cells, as part of multicellular organisms, do not grow solely in response to the availability of nutrients. Instead, they regulate their growth and division to occur at appropriate places in the body and at appropriate times during development. They do this by expressing genes in a correct temporal and spatial manner. Loss of the gene regulation that controls cell growth and division may lead to cancer or developmental defects.

In this chapter, we will explore the regulation of gene expression in eukaryotes. We will outline some of the general features of gene regulation, including the *cis*-acting DNA elements and *trans*-acting factors that influence the regulation of transcription and how these components interact. In addition, we will consider the many types of posttranscriptional mechanisms that also regulate gene expression in eukaryotes.

**18.1**

## Eukaryotic Gene Regulation Can Occur at Any of the Steps Leading from DNA to Protein Product

In eukaryotes, genes are not simply expressed or not expressed; they are finely tuned to express various levels of gene product at specific times, in specific cell types, and in response to complex changes in the environment. To achieve this degree of fine tuning, eukaryotes take advantage of a wide range of mechanisms for altering the expression of genes. In contrast to prokaryotic gene regulation, which occurs primarily at the level of transcription initiation, regulation of gene expression in eukaryotes can occur at many different levels. These range from the initiation of transcription to the numerous mechanisms that modify mRNAs and their ultimate protein products (Figure 18–1).

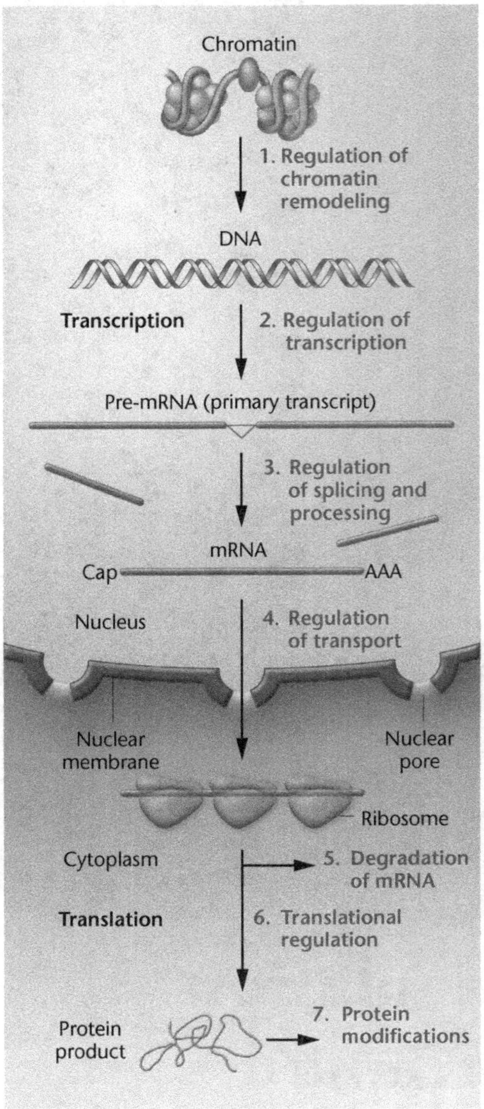


**FIGURE 18–1** Regulation can occur at any stage in the expression of genetic material in eukaryotes. All these forms of regulation affect the degree to which a gene is expressed.

Several features of eukaryotic cells make it possible for them to use more types of gene regulation than are possible in prokaryotic cells:

- Eukaryotic cells contain a much greater amount of DNA than do prokaryotic cells, and this DNA is complexed with histones and other proteins to form highly compact chromatin structures within an enclosed nucleus. Eukaryotic cells modify this structural organization in order to express their genes.
- The mRNAs of most eukaryotic genes are spliced, capped, and polyadenylated prior to transport from the nucleus. Each of

these processes can be regulated in order to influence the numbers and types of mRNAs available for translation.

- Genetic information in eukaryotes is carried on many chromosomes (rather than just one), and these chromosomes are enclosed within a double-membrane-bound nucleus. After transcription, transport of RNAs into the cytoplasm can be regulated in order to modulate the availability of mRNAs for translation.
- Eukaryotic mRNAs can have a wide range of half-lives ($t_{1/2}$). In contrast, the majority of prokaryotic mRNAs decay very rapidly. Rapid turnover of mRNAs allows prokaryotic cells to rapidly respond to environmental changes. In eukaryotes, the complement of mRNAs in each cell type can be more subtly manipulated by altering mRNA decay rates over a larger range.
- In eukaryotes, translation rates can be modulated, and so can the way proteins are processed, modified, and degraded.

In the following sections, we examine some of the major ways in which eukaryotic gene expression is regulated. As most eukaryotic genes are regulated, at least in part, at the transcriptional level, we will emphasize transcriptional control, although we will discuss other levels of control as well. In addition, we will limit our discussion to regulation of genes transcribed by RNA polymerase II. As previously described in Chapter 14, eukaryotic genes use three RNA polymerases for transcription. In contrast, all prokaryotic genes are transcribed by a single RNA polymerase. In eukaryotes, RNA polymerase II transcribes all mRNAs and some small nuclear RNAs, whereas RNA polymerases I and III transcribe ribosomal RNAs, some small nuclear RNAs, and transfer RNAs. The promoter for each type of polymerase has a different nucleotide sequence and binds different transcription factors.

## 18.2 Eukaryotic Gene Expression Is Influenced by Chromosome Organization and Chromatin Modifications

Two structural features of eukaryotic genes distinguish them from the genes of prokaryotes. First, eukaryotic genes are situated on chromosomes that occupy a distinct location within the cell—the nucleus. This sequestering of genetic information in a discrete compartment allows the proteins that directly regulate transcription to be kept apart from those involved with translation and other aspects of cellular metabolism. Second, as described in Chapter 12, eukaryotic DNA is combined with histones and nonhistone proteins to form chromatin. Chromatin's basic structure is characterized by repeating units called nucleosomes that are wound into 30-nm fibers, which in turn form other, even more compact structures. The compactness of these chromatin structures is inhibitory to many processes, including transcription, replication, and DNA repair. In this section, we outline some of the ways in which eukaryotic cells use these structural features of eukaryotic genes to regulate their expression.

### Chromosome Territories and Transcription Factories

During interphase of the cell cycle, chromosomes are unwound and cannot be seen as intact structures by light microscopy. The development of chromosome-painting techniques has revealed that the interphase nucleus is not a bag of tangled chromosome arms, but has a highly organized structure. In the interphase nucleus, each chromosome occupies a discrete domain called a **chromosome territory** and stays separate from other chromosomes (Figure 18–2). Channels between chromosomes contain little or no DNA and are called **interchromosomal domains.**

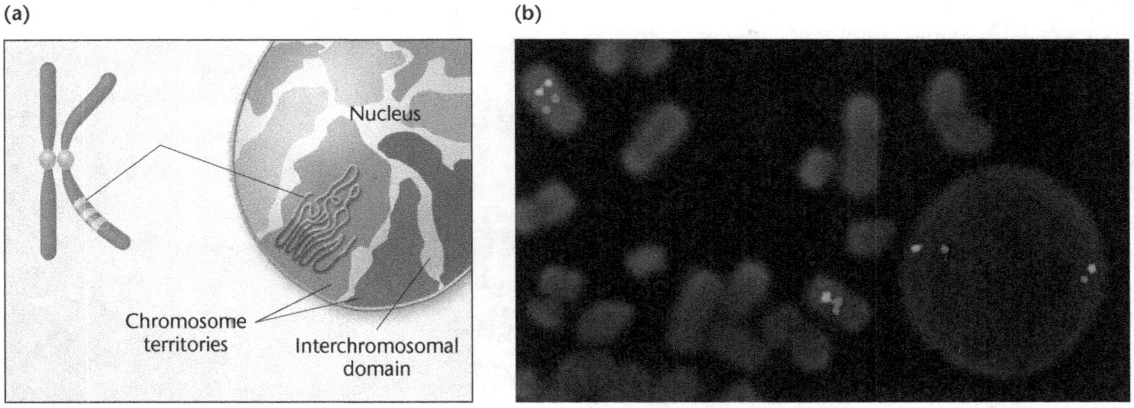


FIGURE 18–2 Chromosome territories. (a) In the nucleus, each chromosome occupies a discrete territory and is separated from other chromosomes by an interchromosomal domain, where mRNA transcription and processing are thought to occur. (b) FISH probes hybridized to human chromosome 7. At the left the probes are hybridized to metaphase chromosomes. In the interphase nucleus (right), the probes reveal the locations occupied by the two copies of chromosome 7. DNA is stained with blue dye.

Chromosome organization appears to be continuously rearranging so that transcriptionally active genes are cycled to the edge of chromosome territories at the border of the interchromosomal domain channels. Although evidence suggests that transcription of many, if not most, genes occurs when they are in direct contact with the interchromosomal compartment, other evidence suggests that transcription can also take place within chromosomal territories. The physiological role of gene relocation within territories is not known. However, it is hypothesized that this organization may bring actively expressed genes into closer association with transcription factors, or with other actively expressed genes, thereby facilitating their coordinated expression.

Another structural feature within the nucleus—the **transcription factory**—may also contribute to regulating gene expression. Transcription factories are specific nuclear sites at which most RNA polymerase II transcription occurs. These sites also contain the majority of active RNA polymerase and other transcription factors. By concentrating transcription proteins and actively transcribed genes in specific locations in the nucleus, the cell may enhance the expression of these genes.

## Chromatin Remodeling

The ability of the cell to alter the association of DNA with other chromatin components is essential to allow regulatory proteins to access DNA. Hence, chromatin modification, referred to as **chromatin remodeling**, is an important step in gene regulation. Chromatin remodeling appears to be a prerequisite for transcription of some eukaryotic genes, although it can occur simultaneously with transcription initiation and elongation of other genes.

Chromatin can be remodeled in two general ways. The first involves changes to nucleosomes, and the second involves modifications to DNA. In this section, we will discuss changes to the nucleosomal component of chromatin. In the next subsection, we present DNA modifications, specifically DNA methylation.

The presence of nucleosomes on DNA and the compaction of chromatin into higher order structures inhibit gene expression. In Chapter 12, we saw how condensation of DNA into heterochromatin regions results in gene silencing. Using molecular biology techniques, scientists have demonstrated that genes in transcriptionally inert regions of the genome are relatively resistant to digestion with the endonuclease DNase I. This resistance is a direct reflection of the gene's association with condensed nucleosomal chromatin. In contrast, when a previously inactive gene begins to be expressed, it becomes sensitive to digestion with DNase I, indicating its presence in a more "open" configuration.

Nucleosomal chromatin can be remodeled in three ways: by altering nucleosome composition, by adding or removing covalent modifications to or from histones, and by repositioning the nucleosome on a gene region. These remodeling functions are performed by specific chromatin remodeling complexes, all of which require ATP hydrolysis. Remodeling complexes are recruited to specific genes that are tagged for transcription by the presence of certain transcription activators or repressors on their promoter regions, or by the presence of modified histones or methylated DNA in the genes' regulatory regions.

Changes in nucleosome composition can affect gene transcription. For example, most nucleosomes contain the normal histone H2A. However, the promoter regions of transcriptionally active and potentially active genes are often flanked by nucleosomes containing variant histones, such as H2A.Z. These variant nucleosomes help keep promoter regions free of repressive nucleosomes, thereby facilitating gene transcription.

A second mechanism of chromatin alteration is histone modification. One such modification is acetylation, a chemical alteration of the histone component of nucleosomes that is catalyzed by **histone acetyltransferase enzymes (HATs)**. When an acetate group is added to specific basic amino acids on the histone tails, the attraction between the basic histone protein and acidic DNA is lessened (Figure 18–3). HATs are recruited to genes by specific transcription factors (such as the "activator" in Figure 18–3). The loosening of histones from DNA facilitates further chromatin

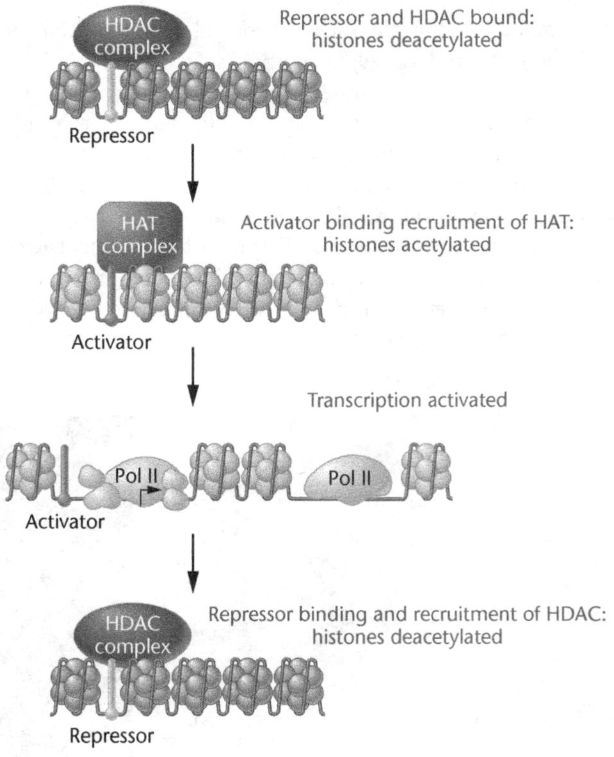


FIGURE 18–3 Proposed model for the actions of HATs and HDACs. Transcription factors (depicted here as repressors and activators) recruit the histone acetylation or deacetylation complexes to the gene region. The enzymes then either add or remove acetyl groups, leading to the opening or closing of chromatin structure.

remodeling catalyzed by ATP-dependent chromatin remodeling complexes, as described below. These modifications make promoter regions available for binding to transcription factors that initiate the chain of events leading to gene transcription, as well as to RNA polymerase. Of course, what can be opened can also be closed. In that case, **histone deacetylases (HDACs)** remove acetate groups from histone tails. Typically, remodeling spreads from enhancer or promoter regions toward the start site of transcription and then into the transcribed regions of a gene. HDACs, like HATs, can be recruited to genes by the presence of certain repressor proteins on regulatory regions (such as the "repressor" shown in Figure 18–3). **Insulator elements**—short DNA sequences that bind specific proteins—can act as barriers to prevent the spread of chromatin remodeling into neighboring genes.

In addition to acetylation, histones can be modified in several other ways, including phosphorylation and methylation. These structural alterations occur at specific amino acid residues in histones. It has been proposed that specific, reversible patterns of covalent histone modifications result in gene activation or gene silencing. In this hypothesis, patterns of dynamic enzyme-mediated modifications constitute a series of signals known as the **histone code**, in which the various chromatin conformations affect gene expression in positive or negative ways.

The third mechanism for chromatin remodeling involves the repositioning of nucleosomes on DNA. Chromatin remodeling complexes that reposition nucleosomes make different regions of the chromosome accessible to transcription proteins, including transcription activators, transcription repressors, and RNA polymerase II. One of the best-studied remodeling complexes is the **SWI/SNF** complex. This is a large multisubunit complex first described in yeast and subsequently found to be widely distributed in other eukaryotes, including humans. Proteins in the SWI/SNF complex were originally identified as transcriptional activators, since their actions lead to increases in gene transcription. Remodelers such as SWI/SNF can act in several different ways (Figure 18–4). They may loosen the attachment between histones and DNA, resulting in the nucleosome sliding along the DNA and exposing regulatory regions. Alternatively, they may loosen the DNA strand from the nucleosome core, or they may cause reorganization of the internal nucleosome components. In all cases, the DNA is left transiently exposed to association with transcription factors and RNA polymerase.

Chromatin remodelers of all types can associate with specific genes and DNA sequences in several ways. They may bind to transcription factors that bind to gene regulatory regions prior to nucleosome repositioning, they may be attracted and stabilized by the presence of acetylated or modified histones, or they may bind to DNA that is methylated.

## DNA Methylation

Another type of change in chromatin that plays a role in gene regulation is the addition or removal of methyl groups to or from bases in DNA. The DNA of most eukaryotic organisms can be modified after DNA replication by the enzyme-mediated addition of methyl groups to bases and sugars. **DNA methylation** most often involves cytosine. In the genome of any given eukaryotic species, approximately 5 percent of the cytosine residues are methylated. However, the extent of methylation can be tissue specific and can vary from less than 2 percent to more than 7 percent of cytosine residues.

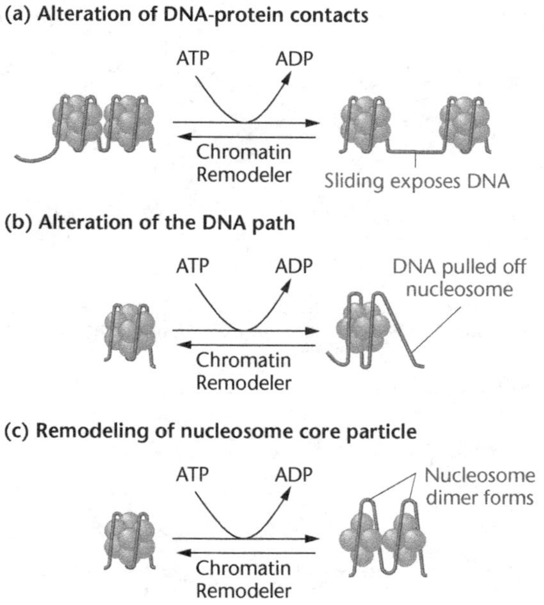


**FIGURE 18–4** Three ways by which chromatin remodelers, such as the SWI/SNF complex, alter the association of nucleosomes with DNA. (a) The DNA-histone contacts may be loosened, allowing the nucleosomes to slide along the DNA, exposing DNA regulatory regions. (b) The path of the DNA around a nucleosome core particle may be altered. (c) Components of the core nucleosome particle may be rearranged, resulting in a modified nucleosome structure.

The ability of base methylation to alter gene expression is known from studies on the *lac* operon in *E. coli*. Methylating DNA in the operator region, even at a single cytosine residue, causes a marked change in the affinity of the repressor for the operator.

Methylation occurs at position 5 of cytosine, causing the methyl group to protrude into the major groove of the DNA helix, where it alters the binding of proteins to the DNA. Methylation occurs most often in the cytosine of CG doublets in DNA, usually in both strands:

$$5' - {}^{m}CpG - 3'$$
$$3' - GpC^{m} - 5'$$

Whether or not DNA is methylated can be determined by restriction-enzyme analysis. The enzyme *HpaII* cleaves at the recognition sequence CCGG; however, if the second cytosine is

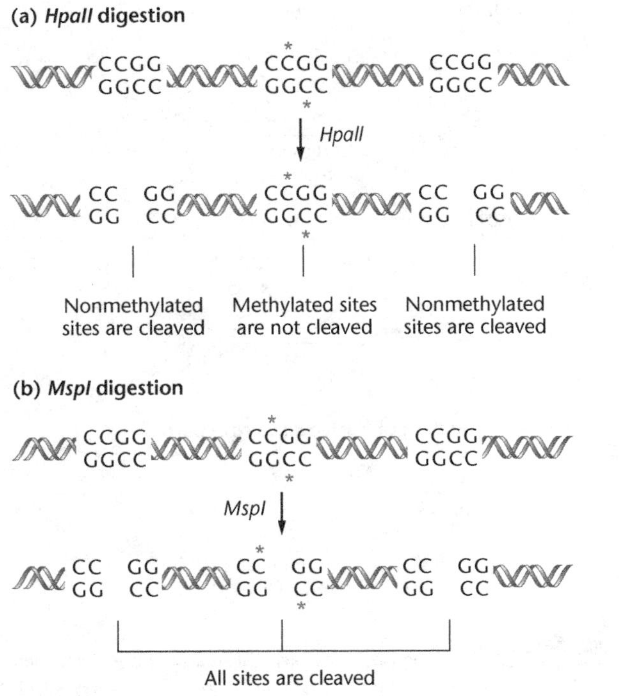


**FIGURE 18–5** The use of restriction enzymes *HpaII* and *MspI* to map methylation sites in DNA. Both of these enzymes cut at CCGG sequences, but *HpaII* will not cut if the second C is methylated (shown with an asterisk). (a) When this DNA is cut with *HpaII*, a long fragment is released, which can be detected with Southern blot analysis. (b) When the DNA is cut with *MspI*, two shorter fragments from the same region are released.

methylated, the enzyme will not cut the DNA. The enzyme *MspI* cuts at the same CCGG sequence, regardless of whether the second cytosine is methylated or unmethylated. If a segment of DNA is unmethylated, both enzymes will produce the same restriction pattern of bands. However, as shown in Figure 18–5, if one site is methylated, digestion with *HpaII* produces a different pattern of fragments from that generated by *MspI* digestion. Using this method, scientists have shown that actively transcribed genes are usually unmethylated or are methylated at a low level.

Evidence of a role for methylation in eukaryotic gene expression is based on a number of observations. First, an inverse relationship exists between the degree of methylation and the degree of expression. That is, low amounts of methylation are associated with high levels of gene expression, and high levels of methylation are associated with low levels of gene expression. Potentially methylatable CpG sequences are not randomly distributed throughout the genome, but are concentrated in CpG-rich regions, called **CpG islands**, located at the 5′ ends of genes, usually in promoter regions. The degree of methylation in CpG islands is inversely related to the propensity of the associated gene to be transcribed. Similarly, large transcriptionally inert regions of the genome are often heavily methylated. In mammalian females, the inactivated X chromosome, which is almost totally transcriptionally inactive, has a higher level of methylation than does the active X chromosome. Within the inactive X chromosome, regions that escape inactivation have much lower levels of methylation than those in adjacent inactive regions. Similarly, the inactive allele of a parentally imprinted gene is often hypermethylated. Transposons and other repetitive sequences in the genome are often methylated, presumably to ensure their silencing.

Second, methylation patterns are tissue specific and, once established, are heritable for all cells of that tissue. It appears that proper patterns of DNA methylation are essential for normal mammalian development. Undifferentiated embryonic cells that are not able to methylate DNA die when they are required to differentiate into specialized cell types. Transgenic mice that are unable to methylate DNA die soon after birth. A loss of normal DNA methylation appears to occur in cancer cells that have lost the ability to regulate their growth and division.

Perhaps the most direct evidence for the role of methylation in gene expression comes from studies using base analogs. The nucleotide **5-azacytidine** can be incorporated into DNA in place of cytidine during DNA replication. This analog cannot be methylated, causing the undermethylation of the sites where it is incorporated. The incorporation of 5-azacytidine into DNA changes the pattern of gene expression and stimulates expression of alleles on inactivated X chromosomes. In addition, the presence of 5-azacytidine in DNA can induce the expression of genes that would normally be silent in certain differentiated cells.

There is growing evidence that abnormal DNA methylation contributes to human diseases. The essay at the end of this chapter discusses the role of methylation in cancer and other disorders.

How might methylation affect gene regulation? Data from *in vitro* studies suggest that methylation can repress transcription by inhibiting the binding of transcription factors to DNA. Methylated DNA may also recruit repressive chromatin remodeling complexes to gene regulatory regions. Another observation is that certain proteins bind to 5-methyl cytosine without regard to the DNA sequence. It is possible that these proteins recruit transcription repressor proteins or repressive chromatin remodeling complexes.

**NOW SOLVE THIS**

Problem 25 on page 483 asks you to interpret the effect of methylation of DNA in different locations on the expression of a gene.

■ HINT: *Remember that the location of various regulatory sequences outside the gene will affect the results.*

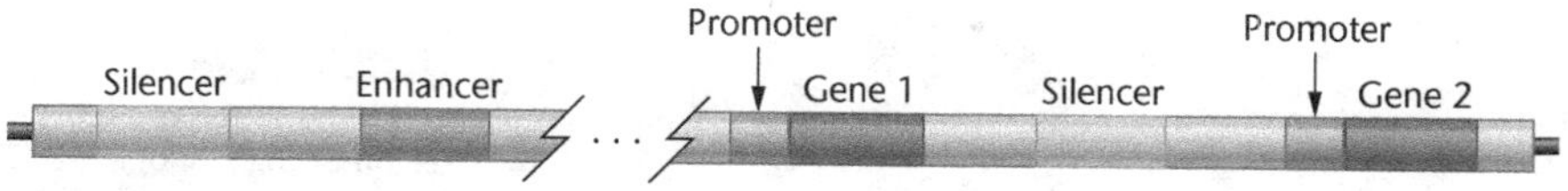


FIGURE 18–6 Transcription of eukaryotic genes is controlled by regulatory elements directly adjacent to the gene (promoters) and by others located at a distance (enhancers and silencers).

18.3

## Eukaryotic Gene Transcription Is Regulated at Specific *Cis*-Acting Sites

Transcription of DNA into an mRNA molecule is a complex, highly regulated process involving several different types of DNA sequences, the interactions of many proteins, chromatin remodeling, and the bending and looping of DNA sequences. We will begin by discussing some of the DNA sequences involved in regulating eukaryotic gene transcription. Eukaryotic genes have several types of regulatory sequences that control transcription, including promoters, silencers, and enhancers (Figure 18–6).

### Promoters

**Promoters** are nucleotide sequences that serve as recognition sites for the transcription machinery. They are necessary in order for transcription to be initiated at a basal level. Promoters are located immediately adjacent to the genes they regulate. These regions are usually several hundred nucleotides in length and specify the site at which transcription begins and the direction of transcription along the DNA.

The promoters of most eukaryotic genes contain one or more elements, including TATA, CAAT, and GC boxes, as well as the transcription start site (Figure 18–7). Some genes utilize elements downstream of the start site of transcription (to about +30) as part of their promoters.

Most, but not all, eukaryotic gene promoters contain TATA boxes. Located about 25 to 30 bases upstream from the transcription start site (a location designated as −25 to −30), the **TATA box** consists of a 7- to 8-bp consensus sequence composed of AT base pairs. The TATA box is often flanked on either side by GC-rich regions. Genetic analysis shows that mutations within TATA sequences reduce transcription (Figure 18–8) and that deletions may alter the initiation point of transcription.

As mentioned earlier, many promoter regions also contain **CAAT boxes.** These elements have the consensus sequence CAAT or CCAAT. The CAAT box frequently appears 70 to 80 bp upstream from the start site. Mutational analysis suggests that CAAT boxes (when present) are critical to the promoter's ability to facilitate transcription. Mutations on either side of this element have no effect on transcription, whereas mutations within the CAAT sequence dramatically lower the rate of transcription (Figure 18–8). The **GC box**, another element we have mentioned as being found in many promoter regions, has the consensus sequence GGGCGG and is

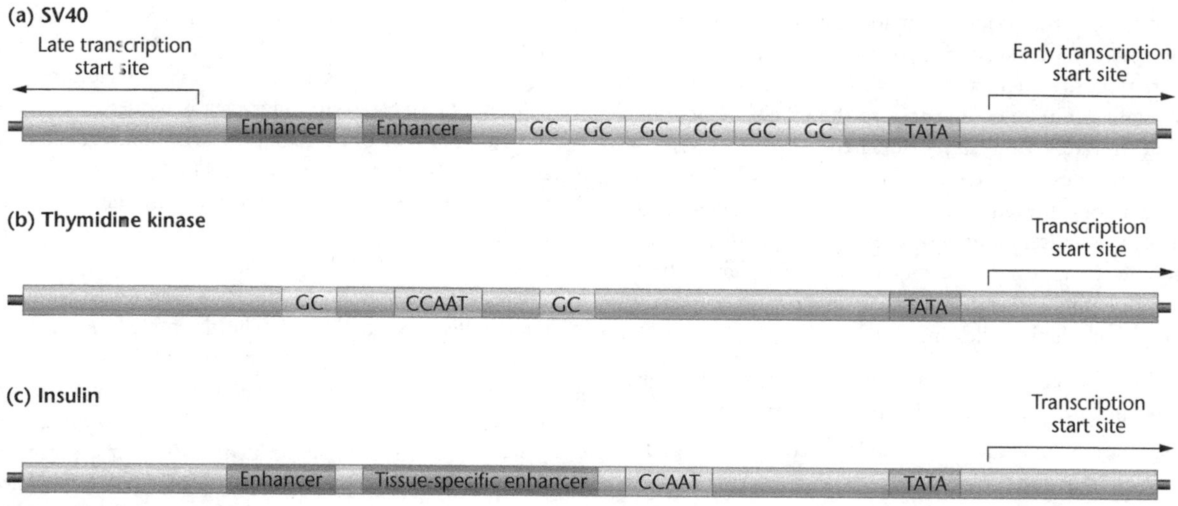


FIGURE 18–7 Organization of transcription regulatory regions in promoters of several genes expressed in eukaryotic cells, illustrating the variable nature, number, and arrangement of controlling elements.

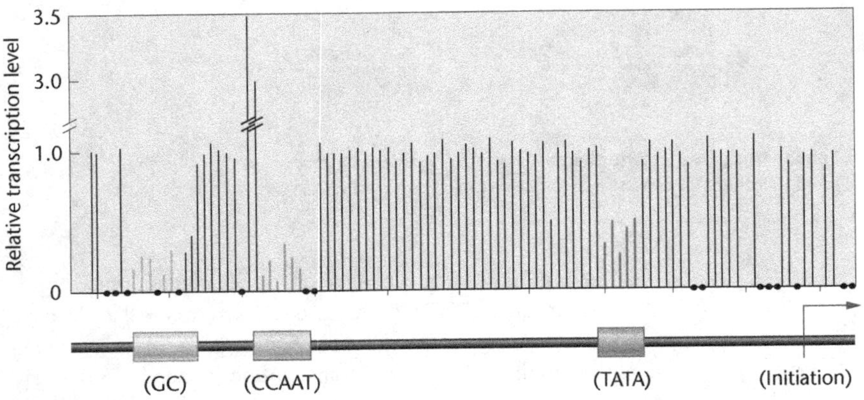


FIGURE 18–8 Summary of the effects on transcription levels of different point mutations in the promoter region of the $\beta$-globin gene. Each line represents the level of transcription produced in a separate experiment by a single-nucleotide mutation (relative to wild-type) at a particular location. Dots represent nucleotides for which no mutation was obtained. Note that mutations within specific elements of the promoter have the greatest effects on the level of transcription.

often located at about position −110. The CAAT and GC elements bind transcription factors and function somewhat like enhancers, which we will cover in the next section.

In summary, each eukaryotic gene contains a number of promoter elements that specify the basal level of transcription from that gene. Different genes may have different subsets of these promoter elements, and they may vary in location and organization from gene to gene.

## Enhancers and Silencers

Transcription of eukaryotic genes is regulated not only by promoters but also by DNA sequences called **enhancers.** Enhancers can be located on either side of a gene, at some distance from the gene, or even within the gene. They are called ***cis* regulators** because they function when adjacent to the structural genes they regulate, as opposed to ***trans* regulators** (such as DNA binding proteins), which can regulate a gene on any chromosome. While promoter sequences are essential for basal-level transcription, enhancers are necessary for achieving the maximum level of transcription. In addition, enhancers are responsible for time- and tissue-specific gene expression.

Enhancers typically interact with multiple regulatory proteins and transcription factors and can increase the efficiency of transcription that initiates from an associated promoter. Within enhancers, binding sites are often found for positive as well as negative gene-regulatory proteins. Thus, there is some degree of analogy between enhancers and operator regions in prokaryotes. However, enhancers appear to be much more complex in both structure and function. Other features that distinguish enhancers from promoters include the following:

1. The position of an enhancer is not fixed; it will function whether it is upstream, downstream, or within the gene it regulates.
2. The orientation of an enhancer can be inverted without significant effect on its action.
3. If an enhancer is experimentally moved adjacent to a gene elsewhere in the genome, or if an unrelated gene is placed near an enhancer, the transcription of the newly adjacent gene is enhanced.

An example of an enhancer located *within* the gene it regulates is the immunoglobulin heavy-chain gene enhancer, which is located in an intron between two coding regions. This enhancer is active only in cells expressing the immunoglobulin genes, indicating that tissue-specific gene expression can be modulated through enhancers. Internal enhancers have been discovered in other eukaryotic genes, including the immunoglobulin light-chain gene. An example of a downstream enhancer is the $\beta$-globin gene enhancer. In chickens, an enhancer located between the $\beta$-globin gene and the $\epsilon$-globin gene works in one direction to control transcription of the $\epsilon$-globin gene during embryonic development and in the opposite direction to regulate expression of the $\beta$-globin gene during adult life.

Enhancers are modular and often contain several different short DNA sequences. For example, the enhancer of the SV40 virus (which is transcribed inside a eukaryotic cell) has a complex structure consisting of two adjacent sequences of approximately 100-bp each, located some 200 bp upstream from a transcriptional start point. One of these is shown in Figure 18–9. Each of the two 100-bp regions contains multiple sequence motifs that contribute to achieving the maximum rate of transcription. If one or the other of these regions is deleted, there is no effect on transcription; but if both are deleted, *in vivo* transcription is greatly reduced.

Another type of *cis*-acting transcription regulatory element, the **silencer**, acts upon eukaryotic genes to repress the level of transcription initiation. Silencers, like enhancers, are short DNA sequence elements that affect the rate of transcription initiated from an associated promoter. They often act in tissue- or temporal-specific ways to control gene expression. An example of a silencer element is found in the human thyrotropin-$\beta$ gene. The thyrotropin-$\beta$ gene encodes one subunit of the thyrotropin hormone, and this gene is expressed only in thyrotrophs (thyrotropin-producing cells) of the pituitary gland. Transcription of this gene is restricted to thyrotrophs due to the actions of a silencer element located −140 bp upstream from the transcription start site. The silencer binds a cellular factor known as Oct-1, which in the context of the thyrotropin-$\beta$ promoter represses transcription in all cell types except thyrotrophs. In thyrotrophs, the action of the silencer is overcome by an enhancer element located over 1.2 kb upstream of the promoter.

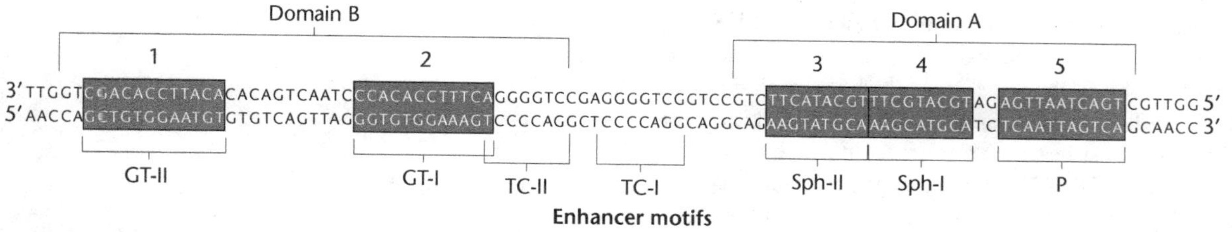


**FIGURE 18–9** DNA sequence of the SV40 enhancer. The blue boxed sequences are those required for maximum enhancer effect. The brackets below the sequence name the various sequence motifs within this region. The two domains of the enhancer (A and B) are indicated.

18.4

## Eukaryotic Transcription Is Regulated by Transcription Factors that Bind to *Cis*-Acting Sites

It is generally accepted that *cis*-acting regulatory sites—including promoters, enhancers, and silencers—influence transcription initiation by acting as binding sites for specific transcription regulatory proteins. These transcription regulatory proteins, known as **transcription factors**, can have diverse and complicated effects on transcription. Some transcription factors are expressed in tissue-specific ways, thereby regulating their target genes for tissue-specific levels of expression. In addition, some transcription factors are expressed in cells at certain times during development or in response to external physiological signals. In some cases, a transcription factor that binds to a *cis*-acting site and regulates a certain gene may be present in a cell and may even bind to its appropriate *cis*-acting site but will only become active when modified structurally (for example, by phosphorylation or by binding to a coactivator such as a hormone). These modifications to transcription factors can also be regulated in tissue- or temporal-specific ways. In addition, different transcription factors may compete for binding to a given DNA sequence or to one of two overlapping sequences. In these cases, transcription factor concentrations and the strength with which each factor binds to the DNA will dictate which factor binds. The same site may also bind different factors in different tissues. Finally, multiple transcription factors that bind to several different enhancers and promoter elements within a gene-regulatory region can interact with each other to fine-tune the levels and timing of transcription initiation.

### The Human Metallothionein IIA Gene: Multiple *Cis*-Acting Elements and Transcription Factors

The **human metallothionein IIA gene (*hMTIIA*)** provides an example of how one gene can be transcriptionally regulated through the interplay of multiple promoter and enhancer elements and the transcription factors that bind to them. The product of the *hMTIIA* gene is a protein that binds to heavy metals such as zinc and cadmium, thereby protecting cells from the toxic effects of high levels of these metals. The gene is expressed at low levels in all cells but is transcriptionally induced to express at high levels when cells are exposed to heavy metals and steroid hormones such as glucocorticoids.

The *cis*-acting regulatory elements controlling transcription of the *hMTIIA* gene include promoter, enhancer, and silencer elements (Figure 18–10). Each *cis*-acting element is a short DNA sequence that has specificity for binding to one or more transcription factors.

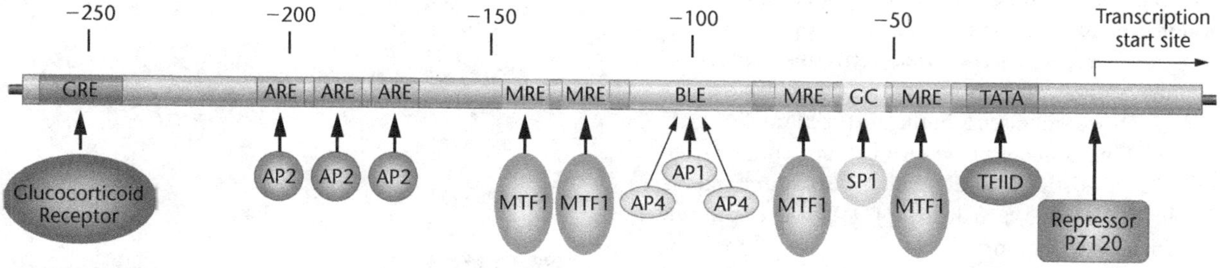


**FIGURE 18–10** The human metallothionein IIA gene promoter and enhancer regions, containing multiple *cis*-acting regulatory sites. The transcription factors controlling both basal and induced levels of MTIIA transcription are indicated below the gene, with arrows showing their binding sites.

The *hMTIIA* gene contains the promoter elements TATA box and start site, which specify the start of transcription. The promoter element, GC, binds the SP1 factor, which is present in most eukaryotic cells and stimulates transcription at low levels in most cells. Basal levels of expression are also regulated by the BLE (basal element) and ARE (AP factor response element) regions. These *cis*-elements bind the activator proteins 1, 2, and 4 (AP1, AP2, and AP4), which are present in various levels in different cell types and can be activated in response to extracellular growth signals. The BLE element contains overlapping binding sites for the AP1 and AP4 factors, providing some degree of selectivity in how these factors stimulate transcription of *hMTIIA* when bound to the BLE in different cell types. High levels of transcription induction are conferred by the MRE (metal response element) and GRE (glucocorticoid response element). The metal-inducible transcription factor (MTF1) binds to the MRE element in response to the presence of heavy metals. The glucocorticoid receptor protein binds to the GRE, but only when the receptor protein is also bound to the glucocorticoid steroid hormone. The glucocorticoid receptor is normally located in the cytoplasm of the cell; however, when glucocorticoid hormone enters the cytoplasm, it binds to the receptor and causes a conformational change that allows the receptor to enter the nucleus, bind to the GRE, and stimulate *hMTIIA* gene transcription. In addition to induction, transcription of the *hMTIIA* gene can be repressed by the actions of the repressor protein PZ120, which binds over the transcription start region.

The presence of multiple regulatory elements and transcription factors that bind to them allows the *hMTIIA* gene to be transcriptionally induced or repressed in response to subtle changes in both extracellular and intracellular conditions.

## Functional Domains of Eukaryotic Transcription Factors

We have described transcription factors as proteins that bind to DNA and activate or repress transcription initiation. To achieve these ends, they have two functional domains (clusters of amino acids that carry out a specific function). One domain, the **DNA-binding domain**, binds to DNA sequences present in the *cis-acting* regulatory site; the other, the ***trans*-activating** or ***trans*-repression domain**, activates or represses transcription through protein–protein interactions. The *trans*-activating and *trans*-repression domains interact with other transcription factors or RNA polymerase, as we will discuss in the next section.

The DNA-binding domains of eukaryotic transcription factors have various characteristic three-dimensional structural motifs. Examples include the helix–turn–helix (HTH), zinc-finger, and basic leucine zipper (bZIP) motifs.

The first DNA-binding domain to be discovered was the **helix–turn–helix (HTH)** motif. This motif, also present in prokaryotic transcription factors, is characterized by a certain geometric conformation rather than a distinctive amino acid sequence (Figure

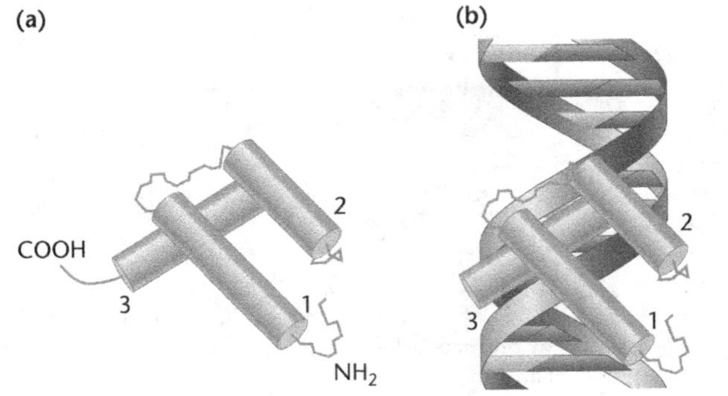


FIGURE 18–11 The helix–turn–helix motif. (a) The three planes of the $\alpha$-helix. (b) How the HTH motif binds in the grooves of the DNA molecule.

18–11). The presence of two adjacent $\alpha$-helices separated by a "turn" of several amino acids (hence the name of the motif) enables the protein to bind to DNA. Amino acid residues outside the HTH motif are also important in regulating the types of DNA sequences to which the HTH motif binds. The HTH motif is an important feature of homeobox-containing transcription factors. The **homeobox** is a stretch of 180 bp that encodes a 60-amino acid sequence known as a **homeodomain**. Transcription factors that contain homeodomains are conserved throughout eukaryotes and are important regulators of animal development.

The **zinc-finger** motif is found in a wide range of transcription factors that regulate gene expression related to cell growth, development, and differentiation. For example, the MTF1 metal-inducible transcription factor discussed in the previous section contains six zinc fingers that comprise the DNA-binding region of this factor. A typical zinc-finger protein contains clusters of two cysteines and two histidines at repeating intervals (Figure 18–12). The consensus

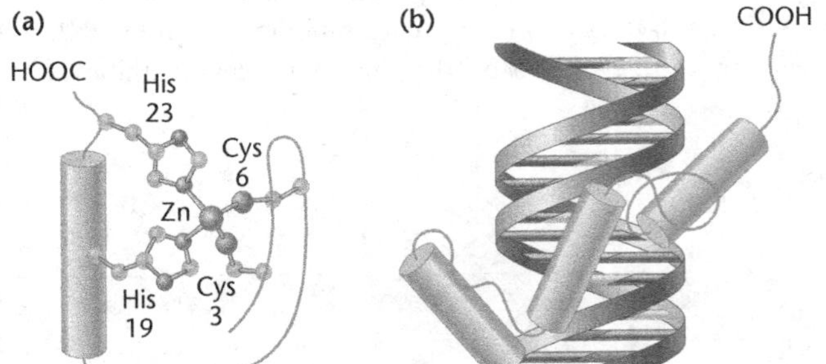


FIGURE 18–12 The zinc-finger motif. (a) The cysteine and histidine residues of the zinc finger bind a $Zn^{2+}$ atom, causing the amino acid chain to loop out into a fingerlike projection that binds in the major groove of the DNA helix (b).

amino acid repeat is $CysN_{2-4}CysN_{12-14}HisN_3His$ (where N is any amino acid). The interspersed cysteine and histidine residues covalently bind zinc atoms, folding the amino acid chain into loops known as zinc fingers. Each finger consists of approximately 23 amino acids (with a loop of 12 to 14 amino acids between the Cys and His residues) and a linker between loops consisting of 7 or 8 amino acids. The amino acids in the loop interact with, and bind to, specific DNA sequences.

A third type of DNA-binding domain is the **basic leucine zipper (bZIP)** motif. This DNA-binding domain contains a region called a leucine zipper that allows protein–protein dimerization. Each of the two molecules that join to form a leucine zipper contains four leucine residues flanked by basic amino acids. When the two molecules dimerize, the leucine residues "zip" together (Figure 18–13). The resulting dimer contains two basic $\alpha$-helical regions adjacent to the zipper that bind to phosphate residues and specific bases in DNA, making the dimer look like a pair of scissors. The transcription factor AP1 is an example of a bZIP transcription factor. AP1 is a heterodimer consisting of two different proteins of the c-fos and c-jun families. The activities of AP1 can be modulated by the mix and match of different family members of c-fos and c-jun that go into the composition of each AP1 molecule.

Transcription factors also contain domains that interact with proteins in the basal transcription complex and control the level of transcription initiation. These *trans*-activating domains, distinct from the DNA-binding domains, can occupy from 30 to 100 amino acids. In addition, many transcription factors also contain domains that bind to chromatin remodeling proteins or to **coactivators**, small molecules such as hormones or metabolites that regulate the transcription factor's activity.

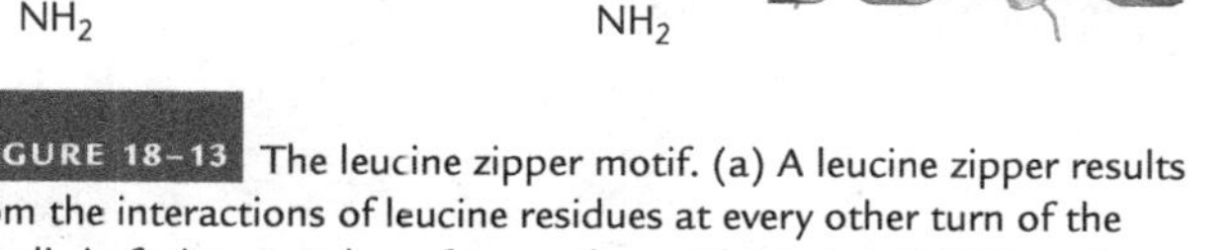


FIGURE 18–13 The leucine zipper motif. (a) A leucine zipper results from the interactions of leucine residues at every other turn of the $\alpha$-helix in facing stretches of two polypeptide chains. (b) When the $\alpha$-helical regions form a leucine zipper, the regions beyond the zipper form a Y-shaped strucure that holds DNA in a scissorlike grip.

18.5

# Activators and Repressors Regulate Transcription by Binding to *Cis*-Acting Sites and Interacting with Other Transcription Factors

We have now discussed the first steps in eukaryotic transcription regulation: first, chromatin must be remodeled and modified in such a way that transcription proteins can bind to their specific *cis*-acting sites; second, transcription factors bind to *cis*-acting sites and bring about a wide range of positive and negative effects on the transcription initiation rate—often in response to extracellular signals or in tissue- or time-specific ways. The next question for us to consider is, how do these *cis*-acting regulatory elements and their DNA-binding factors act to influence transcription initiation? To answer this question, we must first discuss how eukaryotic RNA polymerase II and its basal transcription factors assemble at promoters.

## Formation of the Transcription Initiation Complex

A number of proteins called **basal** or **general transcription factors** are needed to initiate either basal-level or enhanced levels of transcription. These proteins are not part of the RNA polymerase II molecule but assemble at the promoter in a specific order, forming a transcriptional **pre-initiation complex (PIC)** that in turn provides a platform where RNA polymerase is able to recognize and bind to the promoter (Figure 18–14). To initiate formation of the PIC, a complex called **TFIID** binds to the TATA box through one of its subunits, called **TBP** (*TATA Binding Protein*). The TFIID complex is composed of TBP plus approximately 13 proteins called *TAFs* (*TATA Associated Factors*). TFIID then binds additional general factors, such as **TFIIB, TFIIA, TFIIF, TFIIH,** and **TFIIJ**. RNA polymerase II is recruited to the complex along with TFIIF. Once all of the general transcription factors have assembled at a promoter, and RNA polymerase II has made stable contacts with the start site of transcription, the DNA double helix is unwound at the start site and RNA polymerase begins transcription.

## Interactions of the General Transcription Factors with Transcription Activators

There are several ways in which transcription activators alter the rate of transcription initiation. One way is for certain activators to bind to closed chromatin near a gene's promoter. These activators may recruit chromatin remodeling complexes that open regions of the promoter for further interactions with the transcription machinery. A second way involves interactions of the activators with general transcription factors. The first step is the interaction of transcription activators, which are bound to their DNA regulatory elements, with other proteins such as coactivators in a complex known as an **enhanceosome**. The second step is the interaction of

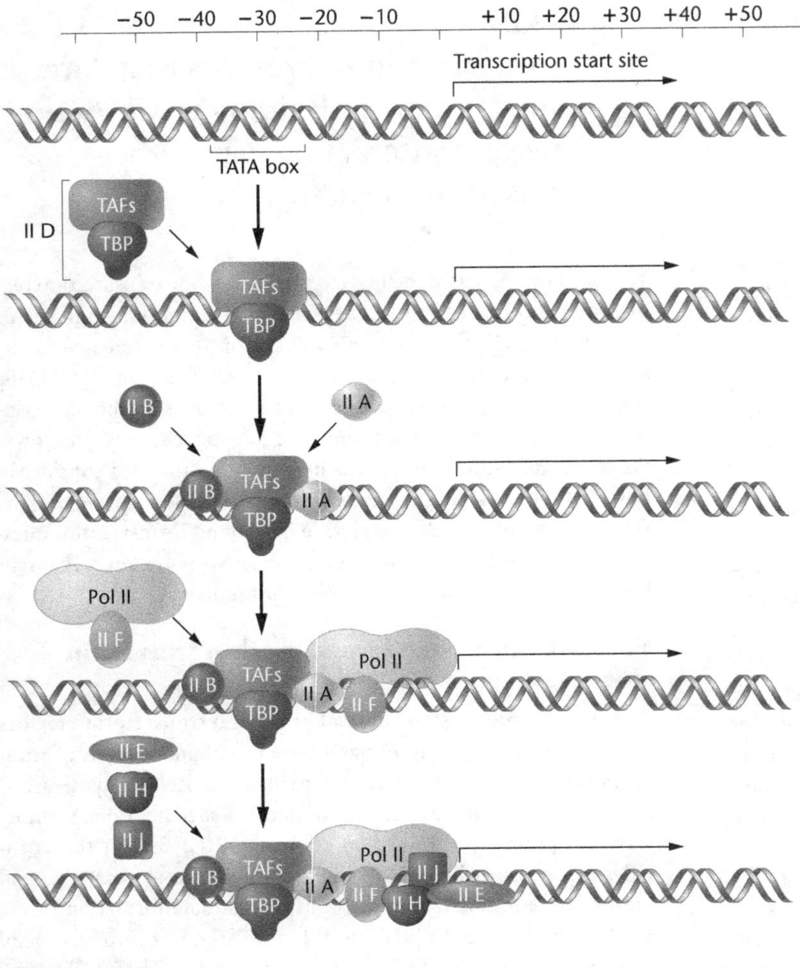


FIGURE 18–14 The assembly of general transcription factors required for the initiation of transcription by RNA polymerase II.

the enhanceosome components with one or more general transcription factors in the PIC (Figure 18–15). In order for proteins bound to the promoter and enhancer regions to interact, the DNA loops out or bends between the promoter and enhancer. By enhancing the rate of PIC assembly, or by influencing its stability, transcription activators stimulate the rate of transcription initiation.

In addition to chromatin remodeling and PIC formation, transcription activators may also increase the rate of DNA unwinding within the gene and accelerate the release of RNA polymerase from the promoter. Conversely, gene transcription can be repressed by the actions of repressor proteins bound at silencer DNA elements. Repressors inhibit the formation of a PIC, block the association of a gene's regulatory elements with activator transcription factors, or stimulate the actions of chromatin remodeling proteins that create repressive chromatin structures.

In summary, the picture of transcription regulation in eukaryotes is complex, but several important generalizations can be drawn. First, alterations in chromatin structure allow (or repress) the binding of transcription factors and RNA polymerase. Second, the interactions of multiple activator or repressor proteins with general transcription factors affect the rate of formation of the PIC and the release of RNA polymerase from the promoter. In addition, different transcription factors may compete for binding at a given DNA sequence or at one of two overlapping sequences. The same DNA sequence may bind different transcription factors in different tissues or at different times or in response to different extracellular signals. Transcription factor concentrations or the efficiency with which each factor binds to the DNA may dictate which factor binds and the

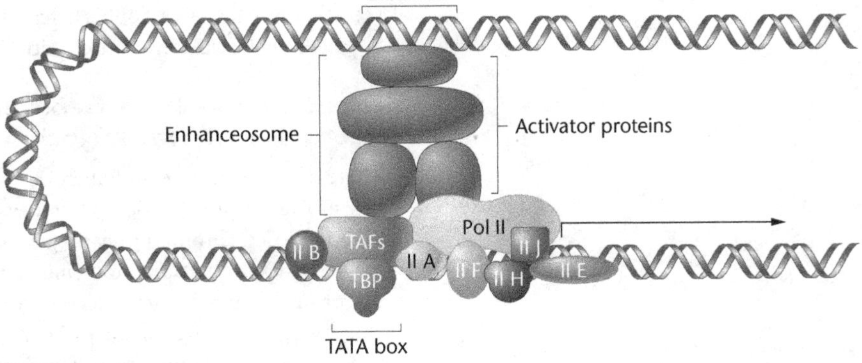


FIGURE 18–15 Formation of a DNA loop allows factors that bind to an enhancer (or silencer) at a distance from a promoter to interact with general transcription factors in the preinitiation complex and to regulate the level of transcription.

effect on transcription levels. Finally, the rate of transcription initiation can be affected simultaneously by multiple enhancers, silencers, and a multiplicity of transcription factors that may bind to these elements in different circumstances.

**NOW SOLVE THIS**

Problem 13 on page 481 asks you to interpret results generated by experiments conducted in two *in vitro* transcription systems (a control system and an experimental system) using a number of DNA fragments containing various deletions.

■ HINT: *Consider the types of promoter and enhancer elements that exist in eukaryotic gene-regulatory regions and their locations. Also, think about what types of proteins bind to these elements and whether the* in vitro *transcription systems would contain these proteins.*

## 18.6 Gene Regulation in a Model Organism: Inducible Transcription of the *GAL* Genes of Yeast

One of the first model systems used to study eukaryotic gene regulation involved the genes in yeast that encode enzymes that break down the sugar galactose. The ***GAL* gene system** comprises four structural genes (*GAL1, GAL10, GAL2,* and *GAL7*) and three regulatory genes (*GAL4, GAL80,* and *GAL3*). The products of the structural genes transport galactose into the cell and metabolize the sugar. The products of the regulatory genes positively and negatively control the transcription of the structural genes. Transcription of the *GAL* structural genes is **inducible**—that is, their transcription is regulated by the presence or absence of the substrate, galactose. In the absence of galactose, the *GAL* genes are not transcribed. If galactose is added to the growth medium, transcription begins immediately, and the mRNA concentration increases a thousandfold. However, transcription is activated only if the concentration of glucose is low. So, like the *lac* and *ara* operons of bacteria, the *GAL* genes are also under a second level of control, catabolite repression. Yeast catabolite repression differs from that in bacteria. This phenomenon in yeast involves an enzyme (a protein kinase), and not cyclic AMP, as in bacteria. Null mutations in the regulatory gene, *GAL4*, prevent activation, indicating that transcription is under **positive control**—that is, the regulator *GAL4* must be present to turn on gene transcription. In this section, we will examine how two of the *GAL* genes, *GAL1* and *GAL10*, are positively regulated at the level of transcription (Figure 18–16).

Transcription of these two genes is controlled by a central control region, called **UAS$_G$** (upstream activating sequence of *GAL* genes), of approximately 170 bp. In yeast, UAS elements are functionally similar to the enhancers found in higher eukaryotes. The chromatin structure of the UAS$_G$ is constitutively open, or **DNase hypersensitive**, meaning that it is free of nucleosomes. Within the UAS$_G$ are four binding sites for the Gal4 protein (**Gal4p**), which is encoded by the *GAL4* gene. These sites are permanently occupied by Gal4p, whether or not the *GAL1* and *GAL10* genes have been transcriptionally activated. Gal4p is, in turn, negatively regulated by the Gal80 protein (**Gal80p**), which is the product of the *GAL80* gene. In the absence of galactose, Gal80p is always bound to Gal4p, covering Gal4p's activation domain, which is shown in Figure 18–16 as a dark patch. Transcriptional activation occurs when galactose interacts with the Gal3 protein (**Gal3p**), encoded by the *GAL3* gene. When bound to galactose, the Gal3p molecule undergoes a conformational change that allows it to interact with the UAS$_G$-bound Gal4p/Gal80p complex. This interaction is thought to disrupt the association of Gal4p with Gal80p, or cause a structural alteration in the complex—either of which exposes the Gal4p activation domain.

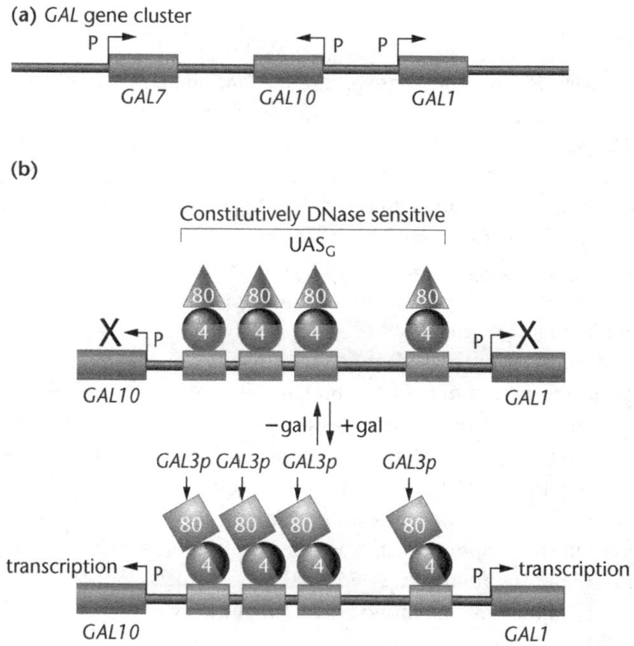


**FIGURE 18–16** Model of *GAL1* and *GAL10* transcriptional activation. (a) The structure of the yeast genome region containing three of the *GAL* genes, including *GAL7, GAL10,* and *GAL1*. Arrows indicate transcription start sites and the direction of transcription. Promoter regions are indicated as "P"s. (b) The *GAL10* and *GAL1* gene UAS$_G$ region. The UAS$_G$ region has four binding sites for Gal4p, shown as purple boxes. Gal4p molecules are shown as blue circles, and Gal80p molecules are shown as red triangles or squares. Induction is indicated by a change in the conformation of Gal80p, or its association with Gal4p, both of which expose the transcription activation domain of Gal4p.

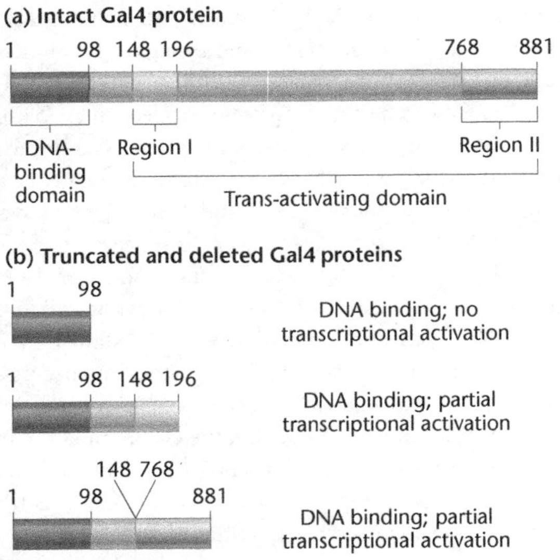


**FIGURE 18–17** Structure and function of the Gal4p activator. (a) Gal4p contains a DNA-binding domain, shown in dark blue, and two transcriptional activation regions, shown in light blue. The Gal80p binding region resides in the amino acid region 851–881. (b) Effects of various deletions on the activity of Gal4p, as described in the text.

Gal4p is a protein consisting of 881 amino acids. It includes a DNA-binding domain that recognizes and binds to sequences in the $UAS_G$ and a *trans*-activating domain that activates transcription [Figure 18–17(a)]. These functional domains were identified by cloning and expressing truncated *GAL4* genes and by assaying the gene products for their ability to bind DNA and to activate transcription. Because Gal4p contains two functional transcription-activating domains, deleting one of these regions allows the other to remain intact and functional.

Mutational analysis identified the DNA-binding domain as a region at the N-terminus of the protein. This region recognizes and binds to the nucleotide sequences in the $UAS_G$. Proteins with deletions from amino acid 98 to the C-terminus (amino acid 881) retain the ability to bind to the $UAS_G$ region, but do not activate transcription. Similar experiments identified two regions of the protein involved in transcriptional activation: region I (amino acids 148–196) and region II (amino acids 768–881). Constructs containing the DNA-binding domain (amino acids 1–96) and region I (amino acids 148–196), or the DNA-binding domain and region II (amino acids 768–881), have reduced transcriptional activity [Figure 18–17(b)].

To explore the nature of Gal4p's activating regions, researchers started with a protein containing the DNA-binding region and the activating region I (amino acids 1–238), and they created point mutations in region I that resulted in increased activation. Most of these mutations caused amino acid substitutions that increased the negative charge of region I. A multiple mutant with four more negative charges than the nonmutant protein causes a ninefold increase in transcriptional activation and has almost 80 percent of the activity of the intact Gal4p. However, negative charges are not the only factor in activation, as some mutants with decreased transcriptional activity have about the same number of negative charges as the parental Gal4p 1–238 molecule.

Data from these and other studies strongly suggest that transcription activation results from contacts between the activating domain of Gal4p and other proteins. How do these protein–protein interactions activate transcription? As we discussed in the previous section, one mechanism of activation is that activator proteins increase the rate of formation of the PIC, or increase its stability, by contacting one or more general transcription factors. An attractive candidate for an activator target is TFIID. The idea that TFIID is a target for activation is supported by the finding that expression of some *GAL4* constructs stimulates transcription in mammalian cell extracts and changes the conformation of DNA-bound TFIID. In addition, purified Gal4p binds to purified TFIID and TFIIB in a test tube, suggesting that two members of the PIC may be recruited by Gal4p. Research shows that Gal4p binds to the chromatin remodeler SWI/SNF, supporting the idea that activation involves chromatin remodeling. During transcription activation, nucleosomes on the promoters of the *GAL* structural genes are displaced, enhancing the ability of RNA polymerase II to bind at the start site of transcription. In addition, Gal4p interacts with molecules that affect RNA polymerase II binding to DNA, as well as affecting its release from the promoter.

Gal4p may activate transcription of the *GAL* genes in several other ways. These include increasing the rate of unwinding of double-stranded DNA within the transcribed region and attracting or stabilizing other factors that bind to the promoter or to the RNA polymerase.

## 18.7 Posttranscriptional Gene Regulation Occurs at All the Steps from RNA Processing to Protein Modification

We have noted that regulation of gene expression occurs at many points along the pathway from DNA to protein. Although transcriptional control is a major type of gene regulation in eukaryotes, **posttranscriptional regulation** also plays a major role. Modification of eukaryotic nuclear RNA transcripts prior to translation includes the removal of noncoding introns, the precise splicing together of the remaining exons, and the addition of a cap at the mRNA's 5′ end and a poly-A tail at its 3′ end. The messenger RNA is then exported to the cytoplasm, where it is translated and degraded. Each of the mRNA processing steps can be regulated to control the quantity of functional mRNA available for synthesis of a protein product. In addition, the rate of translation can be regulated, as well as the stability and activity of protein products. We will examine several mechanisms of posttranscriptional gene regulation that are especially important in eukaryotes—alternative splicing,

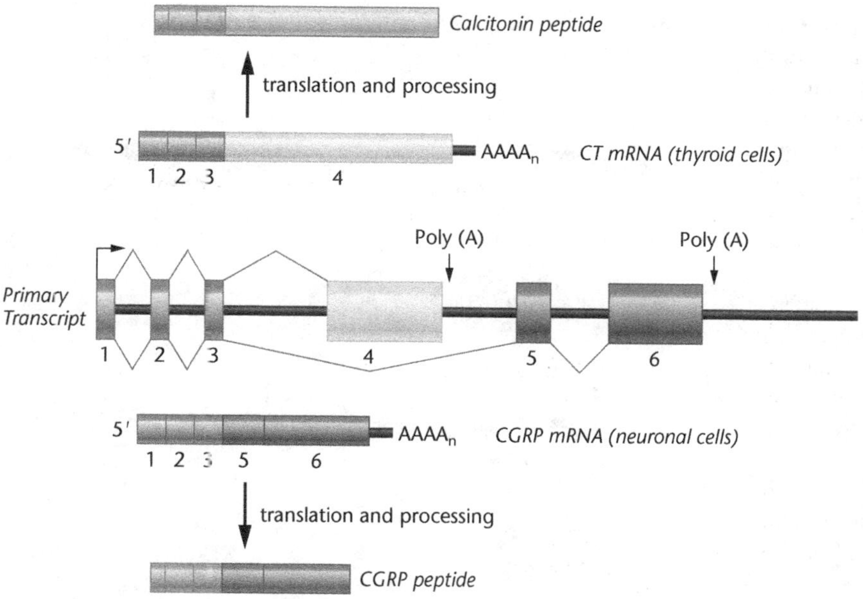


**FIGURE 18–18** Alternative splicing of the *CT/CGRP* gene transcript. The primary transcript, which is shown in the middle of the diagram, contains six exons. The primary transcript can be spliced into two different mRNAs, both containing the first three exons but differing in their final exons. The *CT* mRNA contains exon 4, with polyadenylation occurring at the end of the fourth exon. The *CGRP* mRNA contains exons 5 and 6, and polyadenylation occurs at the end of exon 6. The *CT* mRNA is produced in thyroid cells. After translation, the resulting protein is processed into the calcitonin peptide. In contrast, the *CGRP* mRNA is produced in neuronal cells, and after translation, its protein product is processed into the CGRP peptide.

mRNA stability, translation, and protein stability. In a subsequent section, we will discuss a newly discovered method of posttranscriptional gene regulation—RNA silencing.

## Alternative Splicing of mRNA

**Alternative splicing** can generate different forms of mRNA from identical pre-mRNA molecules, so that expression of one gene can give rise to a number of proteins, with similar or different functions. Changes in splicing patterns can have many different effects on the translated protein. Small changes can alter the protein's enzymatic activity, receptor-binding capacity, or protein localization in the cell. Changes in splicing patterns are therefore important regulatory events that help control aspects of multicellular development, apoptosis, axon-to-axon connection in the nervous system, and many other processes. Mutations that affect regulation of splicing are the basis of several genetic disorders.

Figure 18–18 presents the example of alternative splicing of the pre-mRNA transcribed from the **calcitonin/calcitonin gene-related peptide gene (*CT/CGRP* gene)**. In thyroid cells, the *CT/CGRP* primary transcript is spliced in such a way that the mature mRNA contains the first four exons only. In these cells, the exon 4 polyadenylation signal is used to process the mRNA and add the poly-A tail. This mRNA is translated into the calcitonin peptide, a 32-amino acid peptide hormone that is involved in regulating calcium. In the brain and peripheral nervous system, the *CT/CGRP* primary transcript is spliced to include exons 5 and 6, but not exon 4. In these cells, the exon 6 polyadenylation site is recognized. The *CGRP* mRNA encodes a 37-amino acid peptide with hormonal activities in a wide range of tissues. Through alternative splicing, two peptide hormones with different structures, locations, and functions are synthesized from the same gene.

Alternative splicing increases the number of proteins that can be made from each gene. As a result, the number of proteins that an organism can make—its **proteome**—is not the same as the number of genes in the genome, and protein diversity can exceed gene number by an order of magnitude. Alternative splicing is found in all metazoans but is especially common in vertebrates, including humans. It has been estimated that 30 to 60 percent of the genes in the human genome can undergo alternative splicing. Thus, humans can produce several hundred thousand different proteins (or perhaps more) from the approximately 25,000 genes in the haploid genome.

Given the existence of alternative splicing, how many different polypeptides can be derived from the same pre-mRNA? One answer to that question comes from research on a gene in *Drosophila*. During development, cells of the nervous system must accurately connect with each other. Even in *Drosophila*, with only about 250,000 neurons, this is a formidable task. Neurons have cellular processes called axons that form connections with other nerve cells. The *Drosophila* ***Dscam* gene** encodes a protein that guides axon growth, ensuring that neurons are correctly wired together. In *Dscam* pre-mRNA, exons 4, 6, 9, and 17 each consist of an array of possible alternatives (Figure 18–19). These are spliced into the mature mRNA

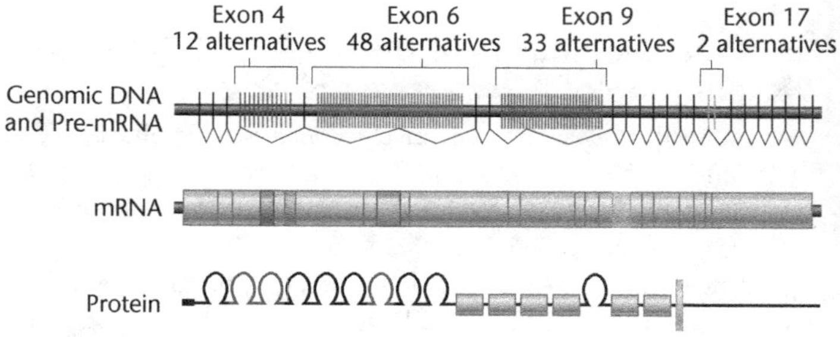


**FIGURE 18–19** Alternative splicing of the *Dscam* gene mRNA. (Top) Organization of the *Dscam* gene in *Drosophila melanogaster* and the transcribed pre-mRNA. The *Dscam* gene encodes a protein that guides axon growth during development. Each mRNA will contain one of the 12 possible exons for exon 4 (red), one of the 48 possible exons for exon 6 (blue), one of the 33 possible exons for exon 9 (green), and one of the 2 possible exons for exon 17 (yellow). Counting all possible combinations of these exons, the *Dscam* gene could encode 38,016 different versions of the DSCAM protein.

in an exclusive fashion, so that each exon is represented by no more than one of its possible alternatives. There are 12 alternatives for exon 4; 48 alternatives for exon 6; 33 alteratives for exon 9; and 2 alternatives for exon 17. The number of possible combinations that could be formed in this way suggests that, theoretically, the *Dscam* gene can produce 38,016 different proteins. Although this is an impressive number of possible different mRNAs and protein isoforms, does the *Drosophila* nervous system require all these alternatives? Recent research suggests that it does.

Scientists have shown that each neuron expresses a different subset of Dscam protein isoforms. In addition, *in vitro* studies show that each Dscam protein isoform can bind to the same Dscam protein isoform but not to others. Even a small change in amino acid sequence reduces or eliminates the binding between two Dscam molecules. *In vivo* studies show that cells expressing the same isoforms of Dscam interact with each other. Therefore, it appears that the diversity of Dscam protein isoforms in neurons provides a kind of molecular identity tag for each neuron, helping to guide it to the correct target, and preventing the tangling of extensions from different neurons.

A more extreme example, ***para***, another gene expressed in the nervous system of *Drosophila*, not only has at least 6 sites of alternative splicing, leading to 48 possible different mRNA variants, but also undergoes another posttranscriptional modification called editing at 11 positions. RNA editing involves base substitutions made after transcription and splicing. With both alternative splicing and editing, the *para* gene can theoretically produce more than 1 million different transcripts.

The *Drosophila* genome contains about 13,000 genes, but the *Dscam* gene alone can produce 2.5 times that many proteins, and while the *para* gene may be an extreme example, it should be obvious that the *Drosophila* proteome is much more complex than its genome. Because alternative splicing is far more common in vertebrates, the combinations of proteins that can be produced from the human genome may be astronomical.

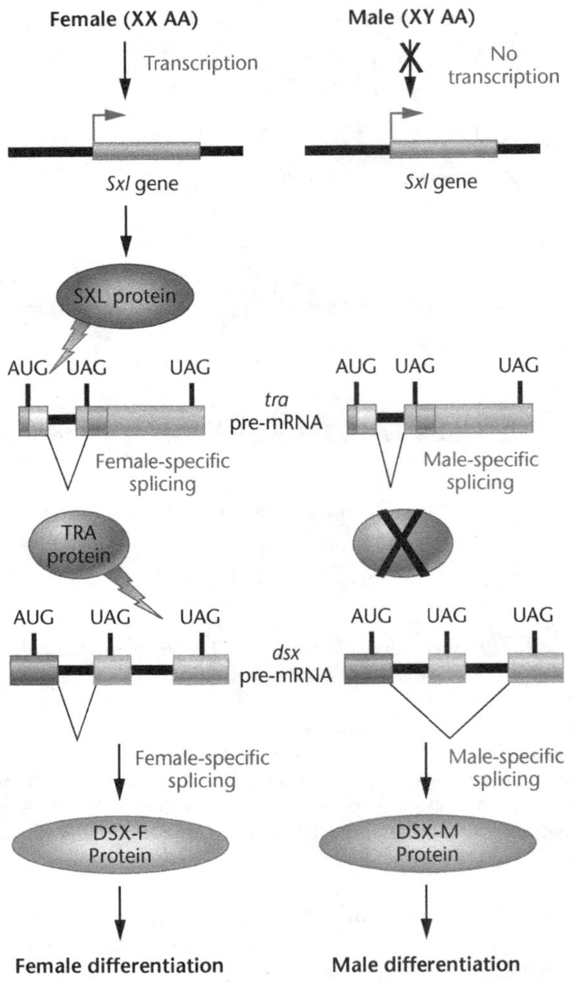


**FIGURE 18–20** Regulation of pre-mRNA splicing that determines male and female sexual development in *Drosophila*. The ratio of X chromosomes to autosomes affects transcription of the *Sxl* gene. The presence of SXL protein begins a cascade of pre-mRNA splicing events that culminate in female-specific gene expression and production of the DSX-F transcription factor. In the absence of SXL protein, a male-specific pattern of pre-mRNA splicing results in male-specific patterns of gene expression induced by the DSX-M protein.

## Sex Determination in *Drosophila*: A Model for Regulation of Alternative Splicing

As outlined in Chapter 7, sex in *Drosophila* is determined by the ratio of X chromosomes to sets of autosomes (X:A). When the ratio is 0.5 (1X:2A), males are produced, even when no Y chromosome is present; when the ratio is 1.0 (2X:2A), females are produced. Intermediate ratios (2X:3A) produce intersexes. Chromosomal ratios are interpreted by a small number of genes that initiate a cascade of splicing events, resulting in the production of male or female somatic cells and the corresponding male or female phenotypes. Three major genes in this pathway are *Sex lethal* (*Sxl*), *transformer* (*tra*), and *doublesex* (*dsx*). We will review some of the key steps in this process.

The regulatory gene at the beginning of this cascade (Figure 18–20) is the gene ***Sex lethal (Sxl)***, which encodes an RNA-binding protein. In females, transcription factors encoded by genes on the X chromosome are thought to activate transcription of the *Sxl* gene. In males, the lower concentration of these transcription factors is not sufficient to activate transcription of *Sxl*. As a result of this differential regulation of transcription, the SXL protein is expressed only in female embryos. The presence (in females) or absence (in males) of SXL protein begins cascades of pre-mRNA splicing events that are specific for females or males. In the presence of SXL, female splicing patterns are expressed, and male splicing patterns are repressed.

One of the targets of SXL protein is the pre-mRNA encoded by the ***transformer*** (***tra***) gene. This pre-mRNA is transcribed in both male and female cells. When SXL protein is present, *tra* pre-mRNAs are spliced to produce mature mRNAs that are translated into functional TRA protein. If SXL is absent, *tra* pre-mRNAs are spliced in such a way that a translation stop codon remains in the mature mRNA. Translation of this mRNA results in a truncated, nonfunctional protein. SXL alters the splicing pattern of *tra* pre-mRNA by binding to the pre-mRNA and altering recognition of splicing signals by the splicing machinery.

The next gene in the cascade, ***doublesex* (*dsx*)**, is a critical control point in the development of sexual phenotype. It produces a functional mRNA and protein in both females and males. However, the pre-mRNA is processed in a sex-specific manner to produce different transcripts and hence different DSX proteins. In females, the functional TRA protein acts as a splicing factor that binds to the *dsx* pre-mRNA and directs splicing in a female-specific pattern. In males, no TRA protein is present, and splicing of the *dsx* pre-mRNA results in a male-specific mRNA and protein. The female DSX protein (DSX-F) and the male protein (DSX-M) are both transcription factors, but act in different ways. DSX-F represses the transcription of genes whose products control male sexual development, whereas DSX-M activates the transcription of genes whose products control male sexual development. In addition, DSX-M represses the transcription of genes that control female sexual development.

In sum, the *Sxl* gene acts as a switch that selects the pathway of sexual development by controlling splicing of the *dsx* transcript. The SXL protein is produced only in embryos with an X:A ratio of 1. Failure to control the splicing of the *dsx* transcript in a female mode results in the default splicing of the transcript in a male mode, leading to the production of a male phenotype.

NOW SOLVE THIS

Problem 17 on page 482 asks you to interpret data showing that a deletion of one extra exon by alternative splicing could rescue protein production by a defective dystrophin gene.

■ HINT: *To answer this question, consider how the deletion or addition of certain numbers of base pairs in DNA can affect reading frames in the resulting mRNA. You might also consider the various outcomes for both the mRNA and the encoded protein that could result from frameshift mutations.*

## Control of mRNA Stability

The **steady-state level** of an mRNA is its amount in the cell as determined by a combination of the rate at which the gene is transcribed and the rate at which the mRNA is degraded. In turn, the steady-state level determines the amount of mRNA that is available for translation. All mRNA molecules are degraded at some point after their synthesis, but the lifetime of an mRNA, defined in terms of its **half-life**, or $t_{1/2}$, can vary widely between different mRNAs and can be regulated in response to the needs of the cell. For example, the abundance of some proteins involved in regulating gene transcription, cell growth, and differentiation is determined more by controlling the rate of degradation of the mRNA for those proteins than by regulating the rate of gene transcription. Some mRNAs are degraded within minutes after their synthesis, whereas others can remain stable for hours, months, or even years (in the case of mRNAs stored in oocytes).

An mRNA may be degraded along three general pathways, and each of these is subject to regulation. First, an mRNA may be targeted for degradation by enzymes that shorten the length of the poly-A tail. In newly synthesized mRNAs, the poly-A tail is about 200 nucleotides long and binds a protein known as the poly-A binding protein. The binding of this protein to the poly-A tail helps to stabilize the mRNA. If the poly-A tail is shortened to less than about 30 nucleotides, the mRNA becomes unstable and acts as a substrate for exonucleases that degrade the RNA in either a 5′ to 3′ or 3′ to 5′ direction. Second, decapping enzymes can remove the 7-methylguanosine cap, which also renders the mRNA unstable. Third, an mRNA may be cleaved internally by an endonuclease, providing unprotected ends at which exonuclease degradation may proceed. Examples of endonucleolytic cleavages are those that occur during **nonsense-mediated decay** and those triggered by RNA interference (which will be discussed in Section 18.8). Nonsense-mediated mRNA decay occurs when translation pauses at premature stop codons. Endonucleases that are present within the ribosome attack the mRNA near the stop codon, leaving unprotected ends that are then degraded by exonucleases.

What are the mechanisms by which a normal mRNA may be targeted for degradation? One way that an mRNA's half-life can be altered is through specific RNA-sequence elements that recruit degrading or stabilizing complexes. One well-studied mRNA stability element is the **adenosine-uracil rich element (ARE)**—a stretch of ribonucleotides that consist of A and U ribonucleotides. These AU-rich elements are usually located in the 3′ untranslated regions of mRNAs that have short, regulated half-lives. These ARE-containing mRNAs encode proteins that are involved in cell growth or transcription control, and need to be rapidly modulated in abundance. In cells that are not growing or require low levels of gene expression, specific complexes bind to the ARE elements of these mRNA molecules, bringing about shortening of the poly-A tail and rapid mRNA degradation. Diseases such as autoimmunities, inflammatory conditions, and cancers appear to be associated with defects in control of mRNA stability through ARE sequences.

Regulation of mRNA stability appears to be closely linked with the process of translation. Several observations demonstrate this link between translation and mRNA stability. First, most mRNA molecules become stable in cells that are treated with translation inhibitors. Second, the presence of premature stop codons in the body of an mRNA, as well as premature translation termination, causes rapid degradation of mRNAs. Third, many of the ribonucleases and mRNA-binding proteins that affect mRNA stability associate with ribosomes.

One of the best studied examples of translational regulation is the synthesis of α- and β-tubulins, the subunit components of eukaryotic microtubules. Treatment of a cell with the drug colchicine leads to a rapid disassembly of its microtubules and an increase in the concentration of free α- and β-subunits. Under these conditions, the synthesis of α- and β-tubulins drops dramatically. However, when cells are treated with vinblastine, a drug that also causes microtubule disassembly, the synthesis of tubulins is increased. Although both drugs cause microtubule disassembly, vinblastine precipitates the subunits out of solution, lowering the concentration of free α- and β-subunits. At low concentrations of free subunits, the synthesis of tubulins is stimulated, whereas at higher concentrations, synthesis is inhibited. This type of translational regulation, where translation is regulated by protein levels within the cell, is known as **autoregulation**.

Research by Don Cleveland and his colleagues has uncovered how the stability of tubulin mRNA is regulated. Fusion of tubulin gene fragments to a cloned thymidine kinase gene shows that the first 13 nucleotides at the 5′ end of the mRNA following the translation start site (encoding the amino terminal region of the protein) cause the hybrid tubulin–thymidine kinase mRNA to be regulated as efficiently as intact tubulin mRNA. These first 13 nucleotides encode the amino acids Met-Arg-Glu-Ile (abbreviated as MREI). Deletions, translocations, or point mutations in this 13-nucleotide segment abolish regulation. Examination of the cytoplasmic distribution of mRNAs demonstrates that only tubulin mRNAs bound to ribosomes are depleted by drug treatment. Copies of tubulin mRNAs not associated with ribosomes are protected from degradation. Also, translation must proceed to at least codon 41 of the mRNA in order for regulation to occur.

A model, shown in Figure 18–21, has been proposed to explain these observations. In this model, regulation occurs after the process of translation has begun. The first four amino acids (MREI) of the tubulin gene product (in this case, β-tubulin) constitute a recognition element to which the regulatory factors bind. The concentration of α- and β-tubulin subunits in the cytoplasm may serve as these regulatory factors. The protein–protein interaction invokes the action of an RNase that may be a ribosomal component or a nonspecific cytoplasmic RNase. Action by the RNase degrades the tubulin mRNA in the act of translation, shutting down tubulin biosynthesis. As an alternative, the binding of the regulatory element may cause the ribosome to stall in its translocation along the mRNA, leaving it exposed to the action of RNase. Translation must proceed to at least codon 41 because the first 30 to 40 amino acids translated occupy a tunnel within the large ribosomal subunit, and only the translation of this number of amino acids will make the MREI recognition sequence accessible for binding.

Such translation-coupled mRNA degradation has been proposed as a regulatory mechanism for other genes as well, including genes encoding histones, some transcription factors, lymphokines, and cytokines.

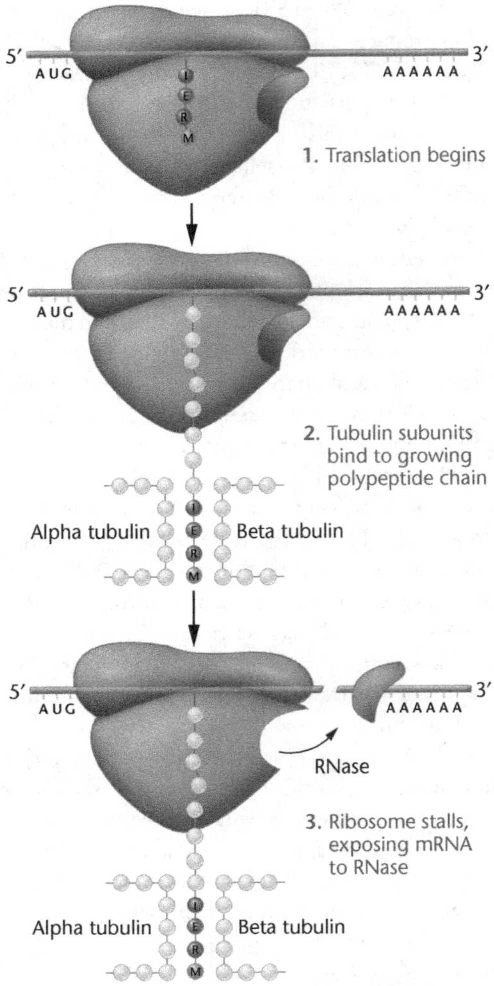


FIGURE 18–21 Model showing how tubulin subunits may control the stability of tubulin mRNA.

## Translational and Post-translational Controls

The ultimate end-point of gene expression is the action or presence of a gene's protein product. In some cases, the translation of an mRNA can be regulated by the extent of the cell's requirement for the protein product. An interesting example of this kind of regulation is the control of ferritin and transferrin receptor mRNA translation.

Soluble iron atoms are necessary for the function of many cellular enzymes; however, an excess of iron is toxic to cells. Within the body, iron is bound to a protein called transferrin. **Transferrin receptor** molecules are present on the surface of cells, and these receptors interact with the transferrin/iron complex and transport the complex into the cytoplasm. Once in the cytoplasm, the iron is released. In order to protect themselves from high levels of

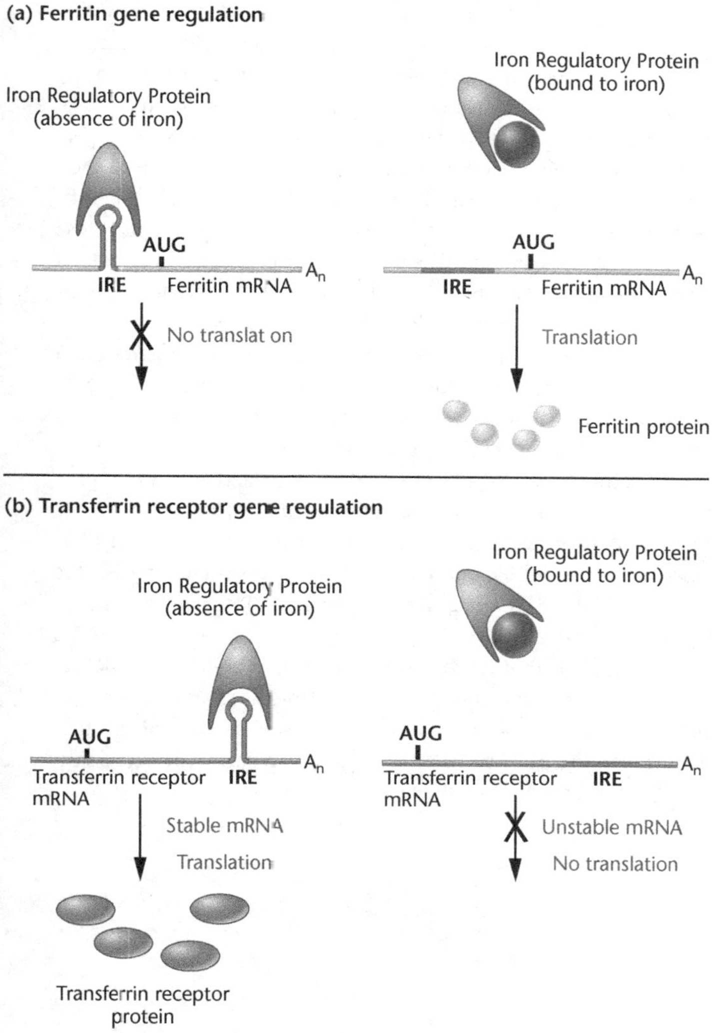


FIGURE 18-22 Posttranscriptional regulation of (a) ferritin and (b) transferrin receptor gene expression. Iron regulatory proteins bind to the IRE stem-loop structure in both ferritin and transferrin receptor mRNAs. In the absence of free iron, the iron regulatory proteins inhibit translation of ferritin mRNAs but stabilize transferrin receptor mRNAs. In the presence of free iron (shown here as dark red circles), the iron regulatory proteins dissociate from the IREs, resulting in increased translation of ferritin and destabilization of transferrin receptor mRNA.

intracellular iron, cells synthesize the protein **ferritin**, which binds to iron atoms, effectively inactivating them within the cytoplasm. Hence, the levels of ferritin protein must be finely regulated in response to iron levels, in order to ensure the correct amount of free iron atoms for cellular metabolism. Similarly, the levels of transferrin receptor must be regulated to provide enough intracellular iron. This dual regulation is achieved by modulating the translatability of both ferritin and transferrin receptor mRNAs.

Within the 5′ untranslated region of the ferritin mRNAs is a 30-nucleotide sequence known as the **iron response element (IRE)**. This element folds into a stem-loop structure that binds a protein known as the **iron regulatory protein** (Figure 18–22). In the absence of excess intracellular iron, the iron regulatory protein binds to the ferritin mRNA IRE. This binding blocks the initiation of translation of ferritin mRNA. In the presence of excess iron, the iron molecules bind to the iron regulatory proteins, causing them to dissociate from the IREs. The ferritin mRNAs are then available for translation.

The same IRE is also present in the 3′ untranslated region of transferrin receptor mRNAs. It also binds iron regulatory proteins in the absence of excess iron. This binding does not affect translation directly, as is the case for ferritin mRNA; instead, the presence of bound iron regulatory proteins increases the stability of the transferrin receptor mRNAs. The increase in stability results in increased mRNA levels, which are then translated into increased levels of transferrin receptor. The presence of more transferrin receptor molecules enhances the transport of iron into the cell. In the presence of excess iron, the iron molecules bind to iron regulatory proteins, dissociating them from transferrin receptor mRNAs, thereby making the mRNAs less stable. The result is that less iron is transported into the cell. The regulation of mRNA translatability provides a rapid means of fine-tuning the levels of these two important proteins.

Protein levels are also controlled by regulation of their stabilities. An example of regulated stability is the **p53 protein.** The p53 protein is essential to protect normal cells from the effects of DNA damage and other stresses. It is a transcription factor that increases the transcription of a number of genes whose products are involved in cell-cycle arrest, DNA repair, and programmed cell death. Under normal conditions, the levels of p53 are extremely low in cells, and the p53 that is present is inactive. When cells suffer DNA damage or metabolic stresses, the levels of p53 protein increase dramatically, and p53 becomes an active transcription factor.

The changes in the levels and activity of p53 are due to a combination of increased protein stability and modifications to the protein. In unstressed cells, p53 is bound by another protein called **Mdm2.** The Mdm2 protein binds to the transcriptional activation domain of the p53 molecule, blocking its ability to induce transcription. In addition, Mdm2 acts as a ubiquitin ligase, adding ubiquitin residues onto the p53 protein. **Ubiquitin** (a "ubiquitous" molecule) is a small protein that tags other proteins for degradation by proteolytic enzymes. The presence of ubiquitin on p53 results in p53 degradation. When cells are stressed, Mdm2 and p53 are modified by phosphorylation and acetylation, resulting in the release of Mdm2 from p53. As a consequence, p53 proteins are stabilized, the levels of p53 increase, and the protein is able to act as a transcription factor. An added level of control is that p53 is a transcription factor that induces the transcription of the *Mdm2* gene. Hence, the presence of active p53 triggers a negative feedback loop that creates more Mdm2 protein, which rapidly returns p53 to its rare and inactive state.

18.8

## RNA Silencing Controls Gene Expression in Several Ways

In the last several years, the discovery that small RNA molecules play an important role in controlling gene expression has given rise to a new field of research. First discovered in plants, short RNA molecules, ~21 nucleotides long, are now known to regulate gene expression in the cytoplasm by repressing translation and triggering the degradation of mRNAs. This form of sequence-specific post-transcriptional regulation is known as **RNA interference (RNAi)**. More recently, short RNAs have been shown to act in the nucleus to alter chromatin structure and bring about repression of transcription. Together, these phenomena are known as RNA-induced **gene silencing**.

RNAi was first discovered during laboratory research, in studies of plant and animal gene expression. For example, Andrew Fire and Craig Mello injected roundworm (*Caenorhabditis elegans*) cells with either single-stranded or double-stranded RNA molecules—both containing sequences complementary to the mRNA of the *unc-22* gene. Although they expected that the single-stranded antisense RNA molecules would suppress *unc-22* gene expression by binding to the endogenous sense mRNA, they were surprised to discover that the injection of double-stranded *unc-22* RNA was 10- to 100-fold more powerful in repressing expression of the *unc-22* mRNA. They studied the phenomenon further and published their results in the journal *Nature* in 1998. They reported that the double-stranded RNA acts to degrade the mRNA if the mRNA is complementary in sequence to one strand of the double-stranded RNA. Only a few molecules of double-stranded RNA are needed to bring about the degradation of large amounts of mRNA. The research of Fire and Mello opened up an entirely new and surprising branch of molecular biology, with far-reaching implications for practical applications. For their insights into RNAi, they were awarded the Nobel Prize in Physiology or Medicine for 2006.

### The Molecular Mechanisms of RNA Silencing

Since Fire and Mello's publication, research by many groups has brought a deeper understanding of how double-stranded RNA silences gene expression. In the cytoplasm, double-stranded RNA molecules undergo a number of processing events and associations with enzymes in order to silence mRNAs. The initial double-stranded RNA can consist of either two linear complementary strands or part of one RNA strand that forms an internal stem-loop structure. The steps in RNAi are outlined in Figure 18–23. First, the double-stranded RNA is recognized by an enzyme complex known as **Dicer**. Dicer has a ribonuclease activity that cleaves long double-stranded RNAs into shorter molecules of about 21 to 23 ribonucleotides long, known as **small interfering RNAs (siRNAs)**. Second, the siRNAs associate with another enzyme complex called the **RNA-induced silencing complex (RISC)**. Within the RISC, the double-stranded siRNA is denatured and the sense strand is degraded. The resulting siRNA/RISC becomes the functional and highly specific agent of RNAi, seeking out mRNA molecules with complementarity to the antisense siRNA contained in the RISC. At this point, RNAi can take two different pathways. If the antisense siRNA in the RISC is perfectly complementary to the mRNA, the RISC will cleave the mRNA. The cleaved mRNA is then degraded by ribonucleases. If the antisense siRNA and the mRNA are not exactly complementary, the siRNA/RISC stays bound to the mRNA, interfering with the ability of ribosomes to translate the mRNA. Hence, RNAi can silence gene expression by affecting either mRNA stability or translation.

Although RNAi and its mode of action were first discovered in laboratory research conditions, it is now becoming clear that RNA-induced gene silencing has profound physiological roles. The mechanisms of RNAi appear to be conserved in all eukaryotes, from plants to humans. Cells use this method of gene regulation to protect themselves from virus infection and the expression of transposons. Many viruses and transposons have double-stranded RNA genomes or use double-stranded RNAs as intermediates in their life cycle. Dicer and RISC act in the cytoplasm to recognize these structures and inactivate them. But perhaps even more significant is the realization that another group of small RNAs—the **microRNAs (miRNAs)**—are major forces in the control of eukaryotic gene expression. These miRNAs are short (21- to 24-nucleotide) RNAs that act through the RISC pathway to negatively regulate the expression of large numbers of cellular genes.

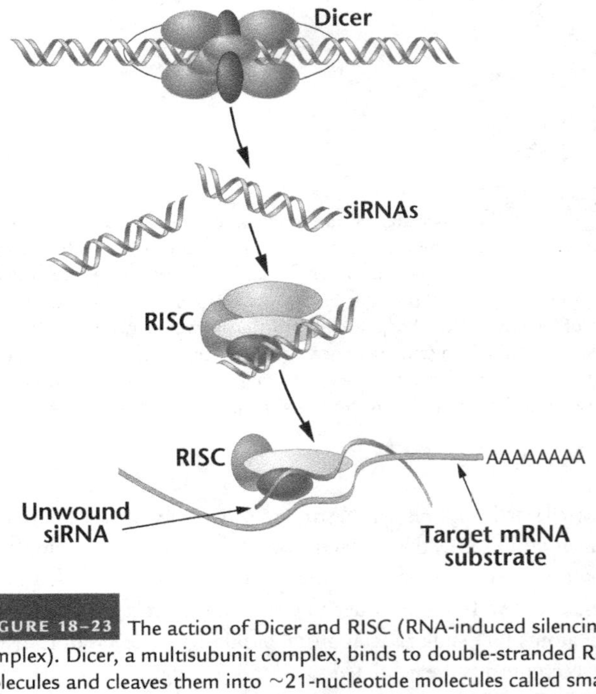


FIGURE 18–23 The action of Dicer and RISC (RNA-induced silencing complex). Dicer, a multisubunit complex, binds to double-stranded RNA molecules and cleaves them into ~21-nucleotide molecules called small interfering RNAs (siRNAs). These bind to a multiprotein RISC complex and are unwound to form single-stranded molecules that target mRNAs with complementary sequences, marking them for degradation by RISC.

Over the last few years, scientists have discovered that significant amounts of eukaryotic genomes are transcribed by RNA polymerase II into RNA products that contain no open reading frames and are not translated into protein products. These RNAs are transcribed either from sequences within the introns of other protein-coding genes or from their own promoters. So far, more than 400 of these noncoding RNA genes have been discovered in the human genome. *Arabidopsis* has more than 130, and *C. elegans* has more than 100. This is probably an underestimate, and scientists speculate that eukaryotic genomes may contain thousands of genes that are transcribed into short noncoding RNAs, which may regulate the expression of more than half of protein-coding genes.

How do these noncoding RNAs work to negatively regulate gene expression? The steps in this process are outlined in Figure 18–24.

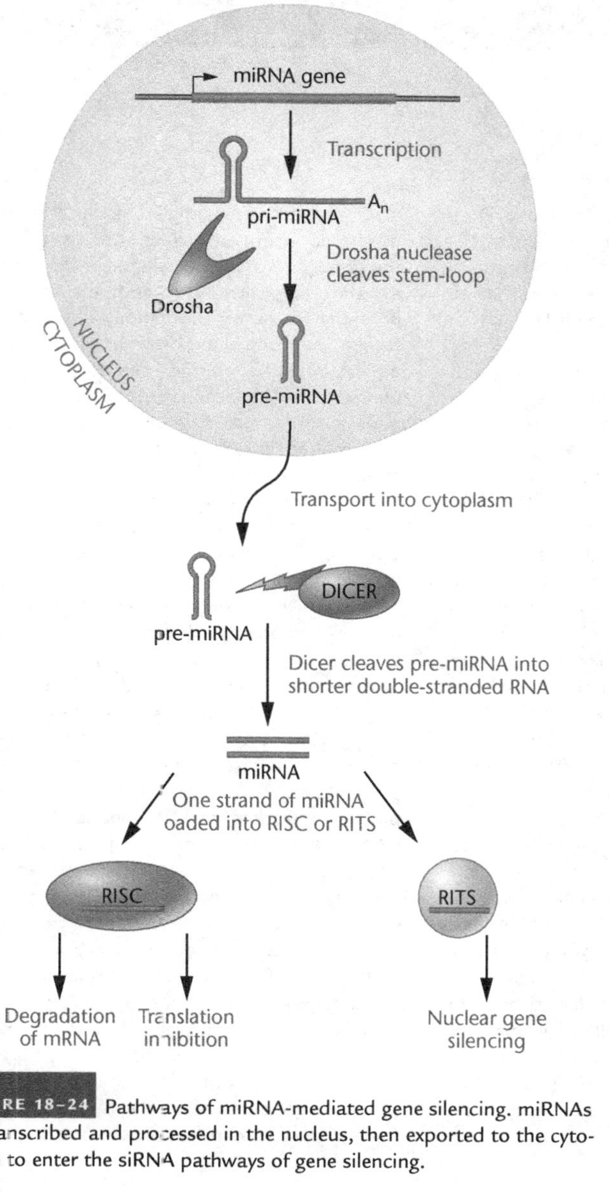


**FIGURE 18-24** Pathways of miRNA-mediated gene silencing. miRNAs are transcribed and processed in the nucleus, then exported to the cytoplasm to enter the siRNA pathways of gene silencing.

After transcription, these noncoding RNAs, called **primary miRNAs (pri-miRNAs)**, remain in the nucleus. They contain regions of about 60–120 ribonucleotides that consist of complementary sequences, which fold into stem-loop structures. These structures are recognized by a protein complex called **Drosha**, which has RNase activity. Drosha cleaves the stem-loop regions of the priRNAs into shorter (less than 70 nucleotides) miRNA precursors, or **pre-miRNAs**. The pre-miRNAs are exported from the nucleus into the cytoplasm. Within the cytoplasm, the pre-miRNA molecules become substrates for the action of Dicer, which cleaves them into smaller (~21 nucleotide) miRNA fragments. Each strand of the miRNA is loaded into a RISC, which then acts as a translational inhibitor or RNA degrader—in the same way as described previously for siRNA/RISC action.

In addition to repressing mRNA translation and triggering mRNA degradation, siRNAs and miRNAs can also repress the transcription of specific genes. They do this by associating with a different complex—the **RITS (RNA-induced initiation of transcriptional silencing)** complex. RITS acts by binding to a gene to be repressed, then recruiting a histone methyltransferase to the site. This enzyme methylates the DNA at CpG dinucleotides and other C residues in promoter regions, triggering the formation of heterochromatin and silencing the gene.

## RNA Silencing in Biotechnology and Therapy

Recently, geneticists have applied RNAi as a powerful research tool. RNAi technology allows investigators to specifically create single-gene defects without having to induce inherited gene mutations. RNAi-mediated gene silencing is relatively specific and inexpensive, and it allows scientists to rapidly analyze gene function. Several dozen scientific supply companies now manufacture synthetic siRNA molecules of specific ribonucleotide sequence for use in research. These molecules can be introduced into cultured cells to knock out specific gene products. The uses of RNAi in research are presented in more detail in Chapter 23.

In addition to its use in laboratory research, RNAi is being developed as a potential pharmaceutical agent and a possible approach for gene therapy. In theory, any disease caused by overexpression of a specific gene, or even normal expression of an abnormal gene product, could be attacked by therapeutic RNAi. Viral infections are an obvious target, and scientists have had promising results using RNAi in tissue cultures to reduce the severity of infection by several types of viruses such as HIV, influenza, and polio. In animal models, siRNA molecules have demonstrated benefits for the treatment of virus infections, eye diseases, nervous system disorders such as Huntington disease and amyotrophic lateral sclerosis (ALS), cancers, and inflammatory bowel disease. In these studies, the synthetic siRNA molecules were applied topically, in aerosols, to the surfaces of mucous membranes, or as preparations introduced into eyes, brain tissues, or the lower intestine. Research into RNAi pharmaceuticals has made rapid progress, and the first clinical trials have completed their Phase I (toxicity) stages. Phase II (efficacy) clinical trials have begun to test the use of siRNAs directed against the product of the vascular endothelial growth factor gene, as a drug to treat age-related macular degeneration. Clinical trials have also been started to test RNAi in the treatment of respiratory syncytial virus infections. Other clinical trials are planned for siRNA treatment of influenza and hepatitis C virus infections.

Scientists are particularly enthusiastic about the potential uses of RNAi in the diagnosis and treatment of cancers. Recent studies show that miRNA genes are tightly regulated during development, and the expression profiles of sets of miRNA genes are characteristic of each tumor type, reflecting the tissue-type origin of the cancer. This observation may lead to more precise methods of diagnosing tumors, predicting their course, and planning treatments. In addition, certain cancers appear to have defects in miRNA gene expression. Treatment of these tumors with synthetic siRNAs may be able to correct these defects and reverse the cancer phenotype. Finally, some cancers are characterized by overexpression or abnormal expression of one or several key proto-oncogenes. If RNAi methods could target these specific gene products, they might help treat cancers that have become resistant to other methods, such as radiation or chemotherapy.

New as it is, the science of RNAi holds powerful promise for molecular medicine. Analysts expect that the first RNAi drugs may be available within the next decade.

GENETICS, TECHNOLOGY, AND SOCIETY

## Gene Regulation and Human Genetic Disorders

When we think of human genetic disorders, we usually think of mutations that alter the coding portion of a gene, or changes in chromosome number or structure that disrupt the delicate balance of gene expression. However, gene expression is a complex process involving a number of sequential steps. Both the DNA encoding genes and the mRNA products of genes undergo epigenetic modifications that alter gene function. These alterations in turn may result in a variety of genetic disorders.

For example, DNA methylation, one of the key examples of epigenetic processing, may lead to gene silencing, and aberrant methylation is often associated with cancer. As a result, assessment of DNA methylation is now a burgeoning field in cancer diagnosis and possible treatment. Recent advancements allow for the noninvasive diagnosis of methylation status using exfoliated cells in urine, sputum, and stool samples. From a urine analysis, aberrant methylation of *laminin-5*, a gene critical to the transport of molecules out of the nucleus, appears to be a useful marker to distinguish invasive from noninvasive bladder cancers. Methylation of the *p16*, *MGMT*, or *RASSF1A* gene in DNA derived from sputum appears to be a sensitive marker for lung cancer. Geneticists have also used stool samples to analyze point mutations in tumor-suppressor genes, and have investigated whether such samples can be used for determining DNA methylation status for these genes as well.

The mechanisms leading to altered DNA methylation in cancer cells are unclear, possibly the result of somatic cell mutations involving the function of the methylation machinery. Widespread demethylation (hypomethylation) of the genome is a phenomenon found across cancer types and appears to be associated with an increase in genetic instability and activation of certain genes involved in tumor growth and metastatic potential. Hypermethylation of DNA at specific chromosomal regions has also been associated with various cancers, seemingly resulting in the silencing of tumor-suppressor genes such as *pRB*, *BRCA1*, and *APC* genes involved in retinoblastoma, breast cancer, and colon cancer, respectively.

Molecular techniques for determining methylation changes in tumor cells may increase the accuracy of cancer diagnosis. For example, the childhood kidney cancer called Wilms tumor occurs when a gene that normally halts cell division in the developing kidney is altered. Normal kidney cells proliferate until the kidney reaches normal size, and then they cease growing. In Wilms tumor, pockets of kidney cells continue dividing as they would otherwise only do in a fetus, and thus they cause kidney tumors. Normal and tumor tissues can be distinguished based on the degree of DNA methylation. DNA in the cells of normal kidney tissue is highly methylated at specific chromosomal locations; however, in tumor cell DNA, methylation at these loci is severely reduced or absent.

Once aberrant methylation patterns are diagnosed, treatments could involve reversing the abnormal DNA methylation using drugs for antitumor therapy. In addition, there is potential for cancer prevention, as tobacco-smoke exposure has been linked to the methylation of a specific gene implicated in non-small-cell lung cancer. In the future, DNA methylation-based diagnostics may potentially change the way physicians diagnose and treat cancer patients.

As we learn more about the process of gene regulation, it is apparent that mutations affecting all steps in the process can result in genetic disorders. For example, one of the first steps in eukaryotic gene expression involves remodeling chromatin to make it accessible for transcription. Remodeling requires the action of the ATP-dependent SWI/SNF protein complex. Mutations of genes encoding the enzymes in the SWI/SNF complex cause diseases with a wide range of phenotypes. These include Williams syndrome, an autosomal dominant disorder causing heart and lung defects as well as growth deficiencies. In this case, the affected gene encodes the *Daxx* cotranscription factor that targets chromatin remodeling at exact promoters and may act by causing aberrant methylation of repetitive DNA sequences.

Histone acetylation and other epigenetic modifications are also important in regulating gene expression. Histone acetyltransferases (HATs) and histone deacetylases (HDACs) cooperate with the SWI/SNF factors to remodel chromatin as a first step in transcription. Mutation in a histone acetyltransferase gene called *CREBBP* (*CREB binding protein*) is responsible for Rubinstein-Taybi syndrome, an autosomal dominant disorder exhibiting growth retardation, facial and hand abnormalities, and mental retardation. In general, inactivation of histone acetylases or deacetylases is associated with developmental disorders with a broad range of complex phenotypes Abnormal activation of HATs or HDACs s also associated with several forms of cancer.

From what we are learning about the molecular mechanisms of human genetic disorders, it is becoming clear that mutations in genes associated with epigenetic processing, chromatin remodeling, and gene expression result in complex, aberrant phenotypes. Work with model organisms is revealing how genetic and epigenetic factors contribute to the phenotypes of complex diseases. Coupled with information from the Human Genome Project, these studies are opening a new and exciting era in human genetics research.

### References

Cho, K. S., Elizondo, L. I., and Boerkoel, C. F. 2004. Advances in chromatin remodeling and human disease. *Curr. Opin. Genet. Develop.* 14: 308–315.

Kisseljova, N. P., and Kisseljov F. L. 2005. DNA demethylation and carcinogenesis. *Biochemistry* (Moscow) 70: 743–752.

Laird, P. L. 2005. Cancer epigenetics. *Hum. Mol. Genet.* 14: R65–R76.

ACTGACNCAC TA TAGGGGCGAATTCGAGCTCGGTACCCGGNGGA CC C AGAG CGACC GCAGGCA GCAAGC GAG A

## EXPLORING GENOMICS

# Tissue-Specific Gene Expression and the ENCODE (ENCyclopedia of DNA Elements) Project

In this chapter, we discussed how eukaryotes can regulate gene expression in many complex ways. Recall that one aspect of gene expression regulation we considered is the way promoter, enhancer, and silencer sequences can govern transcriptional initiation of genes to allow for tissue-specific gene expression. All cells and tissues of an organism possess the same genome, and many genes are expressed in all cell and tissue types. However, muscle cells, blood cells, and all other tissue types express genes that are largely tissue-specific (i.e., they have limited or no expression in other tissue types). A variety of molecular techniques, including several methods you learned about in Chapter 13, enable geneticists to identify tissue-specific gene expression patterns. In this exercise, we use BLAST to learn more about tissue-specific gene expression patterns We then explore a site for the **ENCyclopedia of DNA Elements (ENCODE) Project.**

### Exercise I – Tissue-Specific Gene Expression

In this exercise, we return to the National Center for Biotechnology Information site (NCBI) and use the search tool **BLAST, Basic Local Alignment Search Tool**, which you have used before in other Exploring Genomics exercises.

1. Access BLAST from the NCBI Web site at http://www.ncbi.nlm.nih.gov/BLAST/.

2. Following are GenBank accession numbers for four different genes that show tissue-specific expression patterns. You will perform your searches on these genes.

   NM_021588.1
   NM_00739.1
   AY260853.1
   NM_004917

3. For each gene, carry out a nucleotide BLAST search (refer to Exploring Genomics in Chapter 10 to refresh your memory on BLAST searches) using the accession numbers for your sequence query. Because the accession numbers are for nucleotide sequences, be sure to use the "nucleotide blast" (blastn) program when running your searches. Once you enter "blastn" under the "Choose Search Set" category, you should set the database to "Others (nr etc.)," so that you are not searching an organism-specific database.

4. For the top alignments for each gene, the "Links" column (far right) contains colored boxes labeled U (for UniGene expression data), E (Gene Expression Profiles), and G (Gene Information). Each of these boxes will link you to information about the gene. The UniGene link will show you a UniGene report. For some genes, upon entering UniGene you may need to click a link above the gene name before retrieving a UniGene report. Be sure to explore the "Expression Profile" link under the "Gene Expression" category in each UniGene report. Expression profiles will show a table of gene expression patterns in different tissues.

   Also explore the "GEO profiles" link under the "Gene Expression" category of the UniGene reports, when available. These links will take you to a number of gene expression studies related to each gene of interest. Explore these resources for each gene, and then answer the following questions:

   a. What is the identity of each sequence, based on sequence alignment? How do you know this?

   b. What species was each gene cloned from?

   c. Which tissue(s) are known to express each gene?

   d. Does this gene show regulated expression during different times of development?

   e. Which gene showed the most restricted pattern of expression by being expressed in the fewest tissues?

*Continued on next page*

*Exploring Genomics, continued*

**Exercise II – ENCODE: ENCyclopedia of DNA Elements**

As you learned in this chapter, tissue-specific gene expression patterns are controlled through regulatory elements such as silencers, enhancers, and promoter sequences (refer to Figure 18–6). Recently, the National Human Genome Research Institute (NHGRI) initiated a project called the **ENCyclopedia of DNA Elements (ENCODE)** as a follow-up project to the Human Genome Project.

1. Explore the ENCODE site at http://www.genome.gov/ENCODE/. Read the project "Overview" to learn about the aims for ENCODE. What is the main purpose of this project?
2. Explore the links provided under the "Funded Research Programs," category such as the Functional Analysis Program and the Genetic Variation Program, to discover the goals of these programs.

## Chapter Summary

1. The complete genome is not continuously expressed in all cells under all conditions.
2. In eukaryotes, gene regulation can occur at any of the steps involved in gene expression, from chromatin modification to transcription initiation and posttranscriptional processing.
3. Eukaryotic gene regulation at the chromatin level may involve gene-specific chromatin remodeling, histone modifications, or DNA modifications. These modifications may either allow or inhibit access to promoters and enhancers by transcription factors and RNA polymerase, resulting in increased or decreased levels of transcription initiation.
4. Eukaryotic transcription is regulated at gene-specific promoter and enhancer elements. These *cis*-acting DNA sites may act constitutively or may be active in tissue- or temporal-specific ways.
5. Transcription factors influence transcription rates by binding to *cis*-acting regulatory sites within or adjacent to a gene promoter. They are thought to act by enhancing or repressing the association of basal transcription factors at the core promoter. They may also assist in chromatin remodeling.
6. Posttranscriptional gene regulation includes alternative splicing of nascent RNA, RNA transport, or changes in mRNA stability. Alternative splicing increases the number of gene products encoded by a single gene. Changes in mRNA stability regulate the quantity of mRNA available for translation.
7. Posttranscriptional gene regulation at the levels of translation and protein stability also affect the levels of active gene product.
8. RNA-induced gene silencing is a posttranscriptional mechanism of gene regulation that affects the translatability or stability of mRNA. It acts through the hybridization of small antisense RNAs to specific regions of an mRNA. It also can repress transcription by altering chromatin.
9. RNAi is a powerful and promising tool in research and in the treatment of disease.

## INSIGHTS AND SOLUTIONS

1. Transcription regulatory sites for eukaryotic genes are usually located within a few hundred nucleotides of the transcription start site, but they can be located up to several kilobases away. DNA sequence-specific binding assays have been used to detect and isolate protein factors present at low concentrations in nuclear extracts. In these experiments, short DNA molecules containing DNA-binding sequences are attached to material on a column, and nuclear extracts are passed over the column. The idea is that if proteins that specifically bind to the DNA sequence are present in the nuclear extract, they will bind to the DNA, and they can be recovered from the column after all other nonbinding material has been washed away. Once a DNA-binding protein has been isolated and identified, the problem is to devise a general method for screening cloned libraries for the genes encoding the DNA-binding factor. Determining the amino acid sequence of the protein and constructing synthetic oligonucleotide probes are time consuming and useful for only one factor at a time. Knowing the strong affinity for binding between the protein and its DNA-recognition sequence, how would you screen for genes encoding binding factors?

**Solution:** Several general strategies have been developed using complementary DNA. One of the most promising, which we describe here, was devised by Steve McKnight's laboratory at the Fred Hutchinson Cancer Center. The cDNA isolated from cells expressing the binding factor is cloned into the lambda vector, gt11. Plaques of this library, containing proteins derived from expression of cDNA inserts, are adsorbed onto nitrocellulose filters and probed with double-stranded radioactive DNA corresponding to the binding site. If a fusion protein corresponding to the binding factor is present, it will bind to the DNA probe. After the unbound probe is washed off, the filter is subjected to autoradiography, allowing the plaques corresponding to the DNA-binding proteins to be identified. An added advantage of this particular method is that the bound probe can be washed from the filters, which can then be reused. This ingenious procedure is similar to the colony-hybridization and plaque-hybridization procedures described for library screening in Chapter 13, and it provides a general method for isolating genes encoding DNA-binding factors.

2. The transport of proteins and nucleic acids between the nucleus and cytoplasm provides another mechanism for controlling gene expression. The import and export of these molecules is thought to occur through structures known as nuclear pores, and is accomplished by their association with several classes of proteins, including the importins and exportins. (Macara, I. 2001. *Microbiol. and Mol. Biol. Rev.* 65:570-594.). Suggest a way in which nucleocytoplasmic transport could influence gene expression.

**Solution:** If nucleocytoplasmic transport mechanisms are able to recognize different RNA species and to block, facilitate, or redirect their transport to the cytoplasm, then posttranscriptional regulation could result. For example, if two different cells transcribe similar quantities of a given RNA, differential transport would result in different levels of cytoplasmic mRNA being available for translation. In addition, if transcription activators or repressors are selectively imported or exported from the nucleus, differential gene transcription could result.

## Problems and Discussion Questions

1. Why is gene regulation more complex in a multicellular eukaryote than in a prokaryote? Why is the study of this phenomenon in eukaryotes more difficult?
2. List and define the levels of gene regulation discussed in this chapter.
3. Describe the organization of the interphase nucleus. Include in your presentation a description of chromosome territories, interchromosomal compartments, and transcription factories. Explain how chromosome-painting techniques have helped reveal the organization of the interphase nucleus.
4. A number of experiments have demonstrated that areas of the genome that are relatively inert transcriptionally are resistant to DNase I digestion; however, those areas that are transcriptionally active are DNase I sensitive. Describe why DNase I resistance or sensitivity relates to transcriptional activity.
5. Provide a brief description of two different types of epigenetic modification. Why are epigenetic modifications more elaborate in eukaryotes than in prokaryotes?
6. Present an overview of the manner in which chromatin can be remodeled. In what ways do each of these remodeling processes influence transcription?
7. Distinguish between the *cis*-acting regulatory elements referred to as promoters and enhancers.
8. Is the binding of a transcription factor to its DNA recognition sequence necessary and sufficient for an initiation of transcription at a regulated gene? What else plays a role in this process?
9. Compare the control of gene regulation in eukaryotes and prokaryotes at the level of initiation of transcription. How do the regulatory mechanisms work? What are the similarities and differences in these two types of organisms in terms of the specific components of the regulatory mechanisms? Address how the differences or similarities relate to the biological context of the control of gene expression.
10. Many promoter regions contain CAAT boxes containing consensus sequences CAAT or CCAAT approximately 70 to 80 bases upstream from the transcription start site. How might one determine the influence of CAAT boxes on the transcription rate of a given gene?
11. Contrast and compare the regulation of the *GAL* genes in yeast with the *lac* genes in *E. coli*.
12. Present an overview of RNA silencing achieved through RNA interference (RNAi) and microRNAs (miRNAs). How do the silencing processes begin and what major components participate?
13. You are interested in studying transcription factors and have developed an *in vitro* transcription system that includes a segment of DNA that is transcribed under the control of a eukaryotic promoter. The transcription of this DNA occurs when you add purified RNA polymerase II, TFIID (the TATA binding factor), and the general transcription factors TFIIB, TFIIE, TFIIF, and TFIIH. You perform a series of experiments that compare the efficiency of transcription in the "defined system" with the efficiency of transcription in a crude nuclear extract. You test the two systems with your template DNA and with various deletion templates that you have generated. The following figure shows the results of your study:

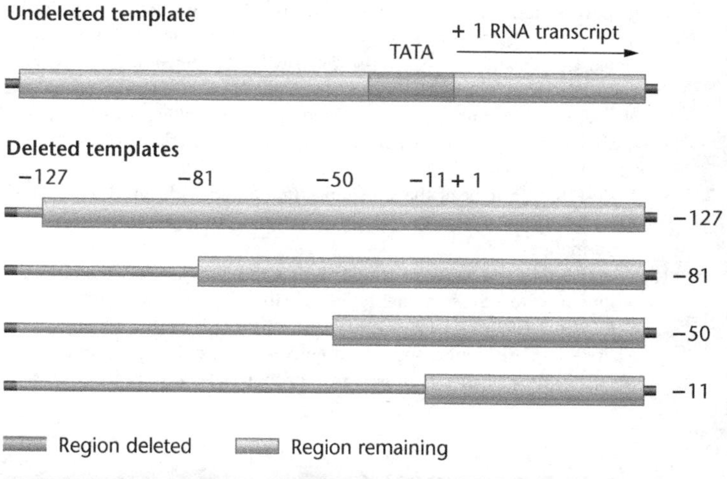


| DNA added | Nuclear extract | Purified system |
|---|---|---|
| undeleted | ++++ | + |
| −127 deletion | ++++ | + |
| −81 deletion | ++++ | + |
| −50 deletion | + | + |
| −11 deletion | o | o |

\+ Low-efficiency transcription
++++ High-efficiency transcription
o No transcription

(a) Why is there no transcription from the −11 deletion template?

(b) How do the results for the nuclear extract and the defined system differ from those for the *undeleted* template? How would you interpret these results?

(c) For the various deletion templates, compare the results obtained from the nuclear extract to those obtained from the purified system.

(d) How would you interpret the results for the deletion templates? Be very specific about what you can conclude from these data.

14. While it is customary to consider transcriptional regulation in eukaryotes as resulting from the positive or negative influence of different factors binding to DNA, a more complex picture is emerging. For instance, researchers have described the action of a transcriptional repressor (Net) that is regulated by nuclear export (Ducret et al., 1999. *Mol. and Cell. Biol.* 19: 7076–7087). Under neutral conditions, Net inhibits transcription of target genes; however, when phosphorylated, Net stimulates transcription of target genes. When stress conditions exist in a cell (for example, from ultraviolet light or heat shock), Net is excluded from the nucleus, and target genes are transcribed. Devise a model (using diagrams) that provides a consistent explanation of these three conditions.

15. DNA supercoiling, which occurs when coiling tension is generated ahead of the replication fork, is relieved by DNA gyrase. Supercoiling may also be involved in transcription regulation. Researchers discovered that transcriptional enhancers operating over a long distance (2500 base pairs) are dependent on DNA supercoiling, while enhancers operating over shorter distances (110 base pairs) are not so dependent (Liu et al., 2001. *Proc. Natl. Acad. Sci. [USA]* 98: 14,883–14,888). Using a diagram, suggest a way in which supercoiling may positively influence enhancer activity over long distances.

16. In some organisms, including mammals, there is an inverse relationship between the presence of 5-methylcytosine ($m^5C$) in CpG sequences and gene activity. In addition, $m^5C$ may be involved in the recruitment of proteins that convert chromatin regions from transcriptionally active to inactive states. Overall, genomic DNA is relatively poor in CpG sequences due to the conversion of $m^5C$ to thymine; however, unmethylated CpG islands are often associated with active genes. Researchers have determined that patterns of DNA methylation in spermatozoa vary with age in rats and suggest that such age-related alterations in DNA methylation may be one mechanism underlying age-related abnormalities in mammals (Oakes et al., 2003. *Proc. Natl. Acad. Sci.* 100: 1775–1780). In light of the above information, provide an explanation that relates paternal age-related alterations in DNA methylation to birth abnormalities.

17. Scientists estimate that approximately 15 percent of disease-causing mutations involve errors in alternative splicing (Philips & Cooper, 2000. *Cell. Mol. Life Sci.* 57: 235–249). However, there is an interesting case where an exon deletion appears to enhance dystrophin production in muscle cells of Duchenne muscular dystrophy (DMD) patients. It turns out that a deletion of exon 45 is the most frequent DMD-causing mutation. But some individuals with Becker muscular dystrophy (BMD), a milder form of muscular dystrophy, have deletions in exons 45 *and* 46 (van Deutekom & van Ommen, 2003. *Nat. Rev. Genetics* 4: 774–783). Recalling that deletions often cause frameshift mutations, provide a possible explanation for why BMD patients, with exon 45 *and* 46 deletions, produce more dystrophin than DMD patients do.

18. In the autoregulation of tubulin synthesis, two models for the mechanism were proposed: (1) the tubulin subunits bind to the mRNA, and (2) the subunits interact with the nascent tubulin polypeptides. To distinguish between these two models, Cleveland and colleagues introduced mutations into the 13-base regulatory element of the $\beta$-tubulin gene. Some of the mutations resulted in amino acid substitution, while others did not. Presented here are the results of a mutagenesis study of the mRNA:

| Wild type | met AUG | arg AGG | glu GAA | lys ATC | Autoregulation + |
|---|---|---|---|---|---|
| Second | —— | UGG | —— | —— | − |
| codon | —— | GGG | —— | —— | − |
| mutations | —— | CGG | —— | —— | + |
| | —— | AGA | —— | —— | + |
| | —— | AGC | —— | —— | − |
| Third | —— | —— | GAC | —— | + |
| codon | —— | —— | AAC | —— | − |
| mutations | —— | —— | UAU | —— | − |
| | —— | —— | UAC | —— | − |

Which of the two models is supported by these results? What experiments would you do to confirm this?

HOW DO WE KNOW?

19. In this chapter, we focused on how eukaryotic genes are regulated at different steps in their expression process, from chromatin modification to control of protein stability. At the same time, we found many opportunities to consider the methods and reasoning by which much of this information was acquired. From the explanations given in the chapter, what answers would you propose to the following fundamental questions:

(a) How do we know that promoter and enhancer sequences control the initiation of transcription in eukaryotes?

(b) How do we know that eukaryotic transcription factors bind to DNA sequences at or near promoter regions?

(c) How do we know that double-stranded RNA molecules can control gene expression?

## Extra-Spicy Problems

20. Because the degree of DNA methylation appears to be a relatively reliable genetic marker for some forms of cancer, researchers have explored the possibility of altering DNA methylation as a form of cancer therapy. Initial studies indicate that while hypomethylation suppresses the formation of some tumors, other tumors thrive. Why would one expect different cancers to respond differently to either hypomethylation or hypermethylation therapies?

21. From chromosomal territories and chromatin remodeling to post-translational modifications, eukaryotic gene expression is clearly multifaceted. If one broadly defines gene regulation, gene amplification might also be classified a form of gene regulation. For instance, in the presence of certain therapeutic drugs, cancer cells often amplify the numbers of resistance genes in their genomes. In insect follicle cells, chorion genes become amplified in number, resulting in the production of large

amounts of eggshell proteins. Insect salivary glands also amplify genes to produce greater amounts of saliva proteins. Would you consider specific gene amplifications, such as those described above, to be forms of genetic regulation?

22. The interphase nucleus appears to be a highly structured organelle with chromosome territories, interchromosomal compartments, and transcription factories. In cultured human cells, researchers have identified approximately 8000 transcription factories per cell, each containing an average of eight tightly associated RNA polymerase II molecules actively transcribing RNA. If each RNA polymerase II molecule is transcribing a different gene, how might such a transcription factory appear? Provide a simple diagram that shows eight different genes being transcribed in a transcription factory and include the promoters, structural genes, and nascent transcripts in your presentation.
23. A particular type of anemia in humans, called $\beta$-thalassemia, results from a severe reduction or absence of a normal $\beta$ chain of hemoglobin. A variety of studies have explored the use of 5-azacytidine for the treatment of such patients. How might administration of 5-azacytidine be an effective treatment for $\beta$-thalassemia? Would you consider adverse side-effects likely? Why?
24. While regulation of gene expression commonly occurs at the transcriptional level, posttranscriptional regulation offers additional opportunities for regulation. The addition of a "5′-cap" and a "3′ poly-A" tail to many nuclear RNA species occurs prior to export of mature mRNAs to the cytoplasm. Such modifications appear to influence mRNA stability. Many assays for gene regulation involve the use of reporter genes such as luciferase from the North American firefly (*Photinus pyralis*). When the luciferase gene is transcribed and translated, the protein product can be easily assayed in a test tube. In the presence of luciferin, ATP, and oxygen, the luceriferase enzyme produces light that can be easily quantified. Assuming that luciferase mRNA can be obtained and differentially modified (5′-capped, poly-A tailed) in a test tube, suggest an assay system that would allow you to determine the influence of 5′-capping and poly-A tail addition on mRNA stability.
25. DNA methylation is commonly associated with a reduction of transcription. The following data come from a study of the impact of the location and extent of DNA methylation on gene activity in human cells. A bacterial gene, luciferase, was cloned next to eukaryotic promoter fragments that were methylated to various degrees, *in vitro*. The chimeric plasmids were then introduced into tissue culture cells, and the luciferase activity was assayed. These data compare the degree of expression of luciferase with differences in the location of DNA methylation (Irvine et al., 2002. *Mol. and Cell. Biol.* 22: 6689–6696). What general conclusions can be drawn from these data?

| DNA Segment | Patch Size of Methylation (kb) | Number of Methylated CpGs | Relative Luciferase Expression |
|---|---|---|---|
| Outside transcription unit (0–7.6 kb away) | 0.0 | 0 | 490X |
| | 2.0 | 100 | 290X |
| | 3.1 | 102 | 250X |
| | 12.1 | 593 | 2X |
| Inside transcription unit | 0.0 | 0 | 490X |
| | 1.9 | 108 | 80X |
| | 2.4 | 134 | 5X |
| | 12.1 | 593 | 2X |

26. Many scientists regard selective nuclear transport of RNAs as a form of genetic regulation in eukaryotes. The discovery of what appears to be an array of nuclear fibrous elements adds to that appeal. In a recent literature review (Pederson, 2000. *Molecular Biology of the Cell* 11: 799–805), the author reported data arising from various labeling experiments designed to determine whether gene regulation in eukaryotes is related to nuclear sorting of pre-mRNAs. The vast majority of the results indicated that nuclear poly-A RNA moves by diffusion. What influence would these results have on models that relate gene regulation to nuclear RNA transport?
27. Noncoding DNA sequences in eukaryotes are often defined as regions that lack promoters, exons, and regulatory sequences. It is often assumed that such noncoding regions lack a function. Nevertheless, it is possible that noncoding DNA may be involved in cell-specific genetic regulation by generating particular chromatin folding patterns that open or conceal certain genetic regions for transcription. What is an inherent weakness in this suggestion, and how does this weakness apply to other suggested models for regulation in eukaryotes?
28. Genome sequencing allows scientists to examine similarities among a variety of organisms and to estimate the frequency of gene homologies. Comparisons of mouse and human genomes indicate that less than one percent of mouse genes have no apparent homolog in the human genome, thereby indicating considerable sequence similarity. But another issue surfaces in terms of the expression of such genomes. Are patterns of alternative splicing also conserved? Recent studies show that approximately 68 percent of alternate splicing patterns in human RNAs are conserved in the mouse (Thanaraj, Clark, and Muilu, 2003. *Nucl. Acids Res.* 31: 2544–2552). Describe how one could determine the conservation of splicing patterns.
29. Mouse quaking is caused by a recessive dysmyelination mutation (*qk*), which appears to involve defects in alternative splicing of an evolutionarily conserved signal transduction protein. Homozygotes (*qk*/*qk*) suffer tremors during exertion. Mature mice may experience seizures and remain motionless for many seconds. Below is a figure that describes the splicing patterns in various genotypes of quaking and normal mice. Portions of brain MAG (myelin-associated glycoprotein) RNA (including exons 11, 12, and 13) from young (14-day) and adult (2-month) mice are represented. The relative concentration of each RNA is indicated in the gel diagram. (Figure and data are modified from Wu, J., et al. 2002. *Proc. Natl. Acad. Sci.* 99: 4233–4238.)

    Given this information, describe the alternative splicing patterns with regards to exons included/excluded as a function of age and genotype.

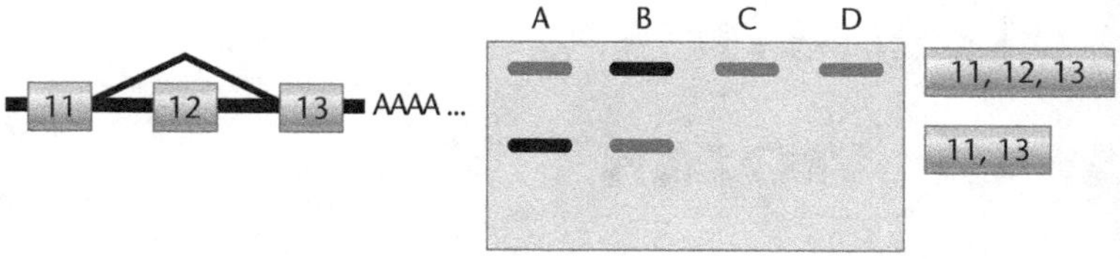


A = +/*qk* young B = *qk*/*qk* young
C = +/*qk* adult D = *qk*/*qk* adult

30. It has been estimated that approximately half of human genes produce alternatively spliced mRNA isoforms. In some cases, incorrectly spliced RNAs lead to human pathologies. Scientists have examined human cancer cells for splice-specific changes and found that many of the changes disrupt tumor-suppressor gene function (Xu and Lee, 2003. *Nucl. Acids Res.* 31: 5635–5643. In general, what would be the effects of splicing changes on these RNAs and the function of tumor-suppressor gene function? How might loss of splicing specificity be associated with cancer?

This unusual four-winged *Drosophila* has developed an extra set of wings as a result of a homeotic mutation.

# 19 Developmental Genetics of Model Organisms

## CHAPTER CONCEPTS

- Gene action during embryonic development is characterized by differential transcription of selected genes and not by retention or loss of genes in the genome.
- All multicellular animals use the same small number of signaling systems and regulatory networks to construct adult body forms from the zygote; thus, animal models are useful for learning about human development.
- Differentiation is controlled by cascades of gene action that stem from the establishment of developmental fate.
- Plants independently evolved developmental mechanisms that parallel those of animals.
- Cell-to-cell signaling programs the developmental fate of adjacent and distant cells.
- Transcriptional networks link together batteries of genes to produce the temporal and spatial patterns of gene expression during development.

In multicellular plants and animals, a fertilized egg undergoes a series of developmental events that ultimately give rise to an adult member of the species. This adult consists of thousands, millions, or even billions of cells organized into a cohesive and coordinated unit that we perceive as a living organism (Figure 19–1). The study of the events through which organisms attain their adult forms is the province of developmental biology, perhaps the most intriguing field of study in biology. The goal of developmental biologists—to attain a comprehensive understanding of developmental processes—requires knowledge of many different biological disciplines, including molecular, cellular, and organismal biology, and a firm grounding in genetics.

At first, developmental biologists focused primarily on describing developmental events, but over the last three decades, their focus has shifted to include sophisticated genetic and molecular analysis of the events that shape embryos into adults. The new techniques have revealed that, in spite of wide diversity in ultimate size and shape, all multicellular organisms share many of the same genes, genetic pathways, and molecular signaling mechanisms in the processes leading from zygote to adult. Genomics, too, has revealed many evolutionary relationships between developmental mechanisms. Thus, scientists can now generalize that at the cellular level, development is marked by three important sets of events: **specification**, in which the first cues conferring cellular identity are established; **determination**, in which a specific developmental fate for a cell becomes fixed; and **differentiation**, in which a cell achieves its final form and function. Genetic analysis has identified specific classes of genes and gene sets that regulate these basic developmental processes in all organisms, and we are beginning to understand how the action and interaction of these genes control development.

In this chapter, the primary emphasis will be on how genetics has been used to study development. The scientific subdiscipline called developmental genetics—at the intersection of genetics and developmental biology—has contributed tremendously to the understanding of development, because the molecular and cellular functions that mediate developmental events and determine the final phenotype of the newly formed organism are ultimately under genetic control.

(a)

(b)

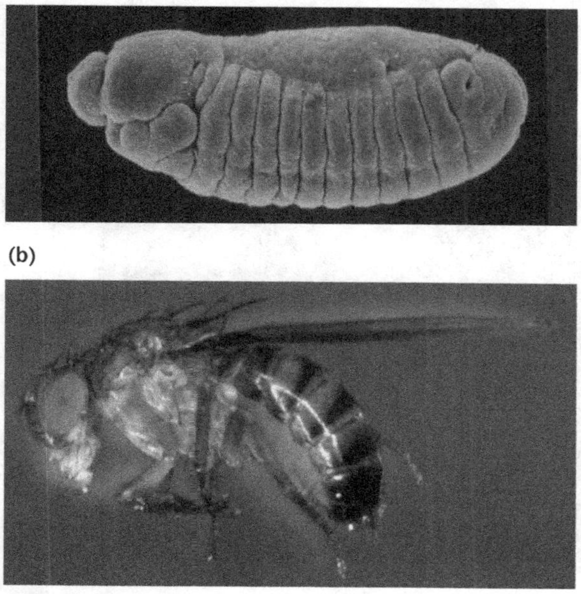

FIGURE 19-1 (a) A *Drosophila* embryo and (b) the adult fly that develops from it.

## 19.1 Developmental Genetics Seeks to Explain How a Differentiated State Develops from Genomic Patterns of Expression

Animal genomes contain tens of thousands of genes, but only a small subset of these control the events that shape the adult body. Developmental geneticists study mutant alleles of these genes to ask important questions about development:

- What genes are expressed?
- When are they expressed?
- In what parts of the developing embryo are they expressed?
- How is the expression of these genes regulated?
- What happens when these genes are defective?

These questions provide a foundation for exploring the molecular basis of developmental processes such as determination, induction, cell–cell communication, and cellular differentiation. The goal is to use genetic analysis to establish a causal relationship between the presence or absence of inducers, receptors, transcriptional events, and cell and tissue interactions, and the observable morphological events that accompany development.

A useful way to define **development** is to say that it is the *attainment of a differentiated state* by all the cells of an organism (except the stem cells). For example, a cell in a blastula-stage embryo (when the embryo is just a ball of uniform-looking cells) is undifferentiated, while a red blood cell synthesizing hemoglobin in the adult body is differentiated. How do cells get from the undifferentiated to the differentiated state? The process involves progressive activation of different gene sets in different cells of the embryo. From a genetics perspective, one way of defining the different cell types that form during development in multicellular organisms is to catalog the genes that are active in each cell type. In other words, development depends on patterns of differential gene expression.

The idea that differentiation is accomplished by activating and inactivating genes at different times and in different cell types is called the **variable gene activity hypothesis**. Its underlying assumptions are, first, that each cell contains an entire genome, and, second, that differential transcription of selected genes controls the development and differentiation of each cell. In multicellular organisms, evolution has conserved the genes involved in development, the patterns of differential transcription, and the ensuing developmental mechanisms. As a result, scientists are able to learn about development in multicellular organisms in general by dissecting these mechanisms in a small number of genetically well-characterized model organisms.

## 19.2 Evolutionary Conservation of Developmental Mechanisms Can Be Studied Using Model Organisms

Genetic analysis of development in a wide range of organisms has demonstrated that there are only a small number of developmental mechanisms and signaling systems used in all multicellular organisms. For example, eye formation in *Drosophila* and eye formation in humans are controlled by the same gene sets working through the same regulatory network. In fact, mouse eye genes inserted into *Drosophila* regulate eye formation in the flies! As another example, most of the differences in size and shape between zebras and zebrafish are controlled by different patterns of expression in a gene set common to both species, not by different genes. Genome sequencing projects have confirmed that genes from a wide range of organisms have been conserved in the course of evolution.

Although all multicellular animals have many developmental mechanisms that are similar, evolution has also generated new and different approaches to the transformation of a zygote into an adult. These mechanisms have been produced through mutation, gene duplication and divergence, the assignment of new functions to old genes, and the recruitment of genes to new developmental pathways. For example, the turtle shell (Figure 19–2), a structure unique among reptiles, was formed by recruiting an existing signaling pathway to a new function. In this new pathway, the fibroblast growth factor 10 (*FGF-10*) gene directs formation of the shell by outgrowth of the dorsal body wall. This gene is part of a signal pathway that controls development of other outgrowths, including limbs and lungs. The emphasis in this chapter, however, will be on the similarities in developmental pathways between species.

### Model Organisms in the Study of Development

Historically, geneticists have used a handful of organisms to study development. These include the yeast *Saccharomyces cerevisiae*, the fruit fly *Drosophila melanogaster*, the nematode *Caenorhabditis elegans*, the zebrafish *Danio rerio*, the mouse *Mus musculus*, and the small flowering plant *Arabidopsis thaliana*. The genetics of these species have been well characterized, and the genomes of each have been sequenced. A large catalog of mutations affecting important steps in development is available for each of these species, and there are well-established procedures for making and maintaining new mutant strains. With the exception of the plants, all these species share gene-regulatory pathways and developmental mechanisms with humans. The use of these organisms has helped identify genes that control disorders of human development and has provided

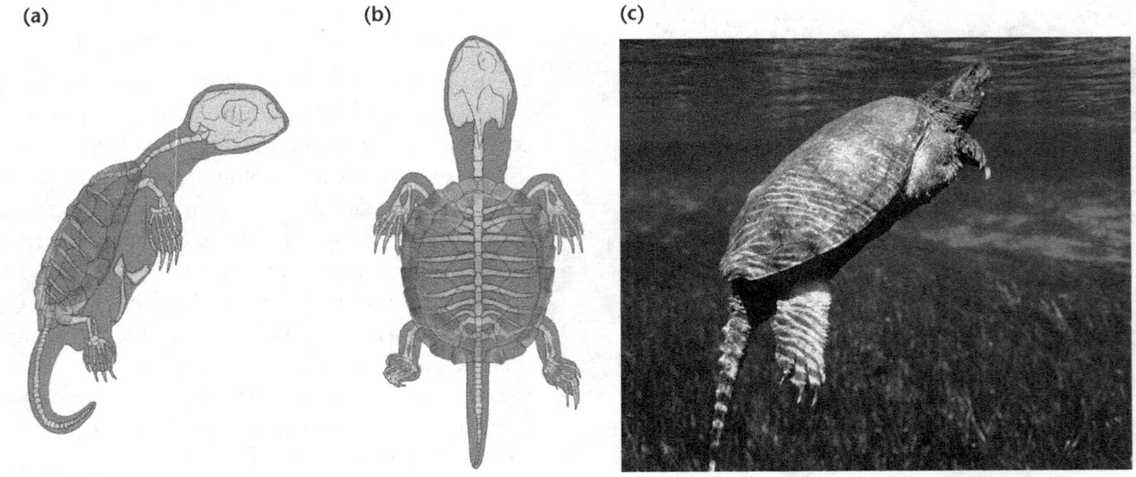


FIGURE 19–2 The shell of a turtle is formed by outgrowth from the dorsal body wall. Its development is controlled by the fibroblast growth factor 10 (FGF10) pathway, which also controls other structures formed by outgrowths, including limbs and lungs.

insights into gene–environment interactions that affect the outcome of key steps in development.

## Analysis of Developmental Mechanisms

In the space of a single chapter, we cannot survey all aspects of development, nor can we explore the genetic analysis of all developmental mechanisms triggered by the fusion of sperm and egg. Instead, we will focus on three general processes in development:

- how the adult body plan of the animal is laid down in the embryo
- the program of gene expression that turns undifferentiated cells into differentiated cells
- the role of cell–cell communication in development

We will use three model systems, *D. melanogaster*, *C. elegans*, and *A. thaliana*, to illustrate these three developmental processes and the related topics. *Drosophila* and *Arabidopsis* will demonstrate the role of differential gene expression in the progressive restriction of developmental options leading to the formation of the body axis. Our discussion will then be expanded to include the selection of pathways that result in differentiated cells in plants and animals, and to consider the role of cell–cell communication in the development of *C. elegans*.

## Basic Concepts in Developmental Genetics

In higher eukaryotes, development begins with the formation of the zygote, a cell generated by the fusion of the sperm and oocyte. The oocyte is a cell with a heterogeneous cytoplasm and a nonuniform distribution of its components, including mRNA and proteins. Thus, following fertilization and early cell divisions, as this nonuniform maternal cytoplasm is distributed into new progeny cells, the nuclei of these cells find themselves in different environments. Evidence suggests that because of these cytoplasmic differences, the cytoplasm exerts cell- or region-specific influences on the genetic material, causing differential transcription at specific points during development. Hence, cytoplasmic distribution, or the inheritance of particular cytoplasmic components by individual embryonic cells, plays a major regulatory role during development. The early gene products subsequently synthesized in the cells reflect these different influences on the zygote's genome, and further alter the cytoplasm of each cell, producing still different cellular environments that, in turn, lead to the activation of yet other different sets of genes, and so on. As the number of cells in the embryo increases, they influence one another by cell–cell interaction. The environmental factors acting on the genetic material within a cell's nucleus, therefore, now include not only the cytoplasm of the individual, differentiating cell but also the signals coming from other cells. Although early in embryogenesis most cells show no evidence of structural or functional specialization, the combination of their internal cytoplasmic components and their position in the developing embryo determine the ultimate structure and functions they will assume. It is as if their fate has been programmed prior to the actual events leading to specialization.

19.3

# Genetic Analysis of Embryonic Development in *Drosophila* Revealed How the Body Axis of Animals Is Specified

How a certain cell turns specific genes on or off at specific stages of development is a central question in developmental biology. At present, there is no simple answer to this question. However, information derived from the study of model organisms provides a starting point for understanding. In particular, the genetic and molecular analysis of embryonic development in *Drosophila* highlights the key role of molecular components distributed in the oocyte cytoplasm during oogenesis.

## Overview of *Drosophila* Development

The life cycle of *Drosophila* is about 10 days long and has several distinct stages: the embryo, three larval stages, the pupal stage, and the adult stage (Figure 19–3). The cycle begins with the fertilized egg, whose cytoplasm is organized into a series of molecular gradients. These gradients, encoded by the maternal genome and created

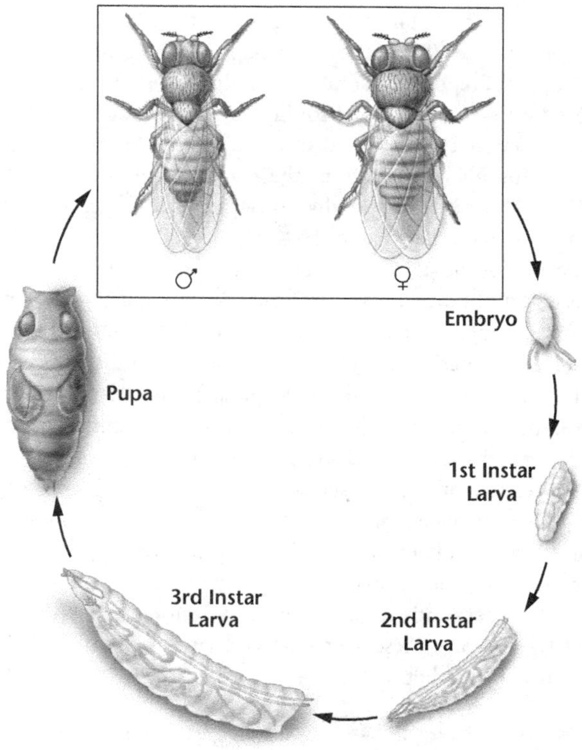


FIGURE 19–3 *Drosophila* life cycle. (Each instar represents one stage in larval development.)

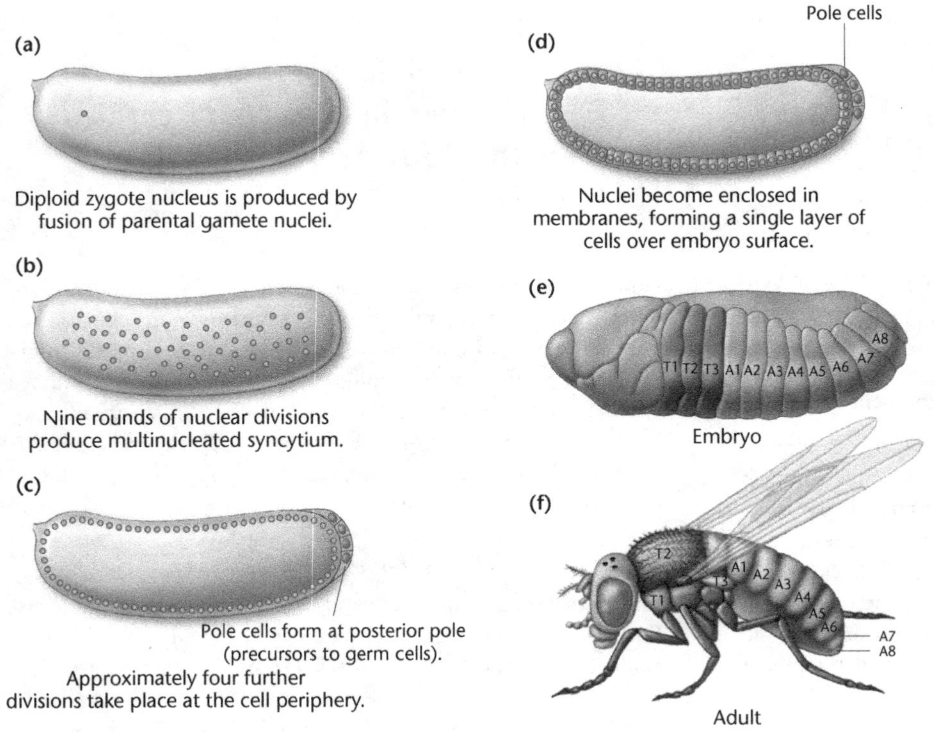


**FIGURE 19–4** Early stages of embryonic development in *Drosophila*. (a) Fertilized egg with zygotic nucleus ($2n$), shortly after fertilization. (b) Nuclear divisions occur about every 10 minutes. Nine rounds of division produce a multinucleate cell, the syncytial blastoderm. (c) At the tenth division, the nuclei migrate to the periphery, or cortex, of the egg, and four additional rounds of nuclear division occur. A small cluster of cells, the pole cells, form at the posterior pole about 2.5 hours after fertilization. These cells will become the germ cells of the adult. (d) About 3 hours after fertilization, the nuclei become enclosed in membranes, forming a single layer of cells—the cellular blastoderm—over the embryo surface. (e) The embryo at about 10 hours after fertilization. At this stage, the segmentation pattern of the body is clearly established. Behind the segments that will form the head, are thoracic segments (T1–T3) and abdominal segments (A1–A8). (f) The adult structures that form from each segment of the embryo.

during oogenesis, play a key role in establishing the developmental fates of nuclei located in specific regions of the embryo.

Immediately after fertilization, the zygote nucleus undergoes a series of DNA replications and nuclear divisions without cytokinesis [Figure 19–4(a) and (b)], forming a syncytial blastoderm (a syncytium is any cell with more than one nucleus). At about the tenth division, the nuclei migrate to the periphery of the egg [Figure 19–4(c)]. After several more divisions, the nuclei become enclosed in plasma membranes [Figure 19–4(d)].

Formation of pole cells (precursors of the germ cells) at the posterior pole of the embryo demonstrates the regulatory role of the localized cytoplasmic components [Figure 19–4(c) and (d)]. Nuclei transplanted experimentally from other regions of the embryo into the posterior cytoplasm (the pole plasm) will form germ cells. Hence, the cytoplasm of the posterior pole must contain maternal components that direct nuclei to form germ cells.

Expression of the transcriptional programs triggered in the non–germ-cell nuclei produces the embryo's anterior–posterior (front to back) and dorsal–ventral (upper to lower) axes of symmetry, leading to the formation of a segmented embryo [Figure 19–4(e)]. Under control of the homeotic selector gene set (discussed in a later section), these segments give rise to the differentiated structures of the adult fly [Figure 19–4(f)].

## Genetic Analysis of Embryogenesis

Two types of genes control embryonic development in *Drosophila*: maternal-effect genes (introduced in Chapter 9) and zygotic genes (see Figure 19–5). Products of maternal-effect genes (mRNA and proteins) are deposited in the developing egg by the "mother" fly. As noted above, many of these products are distributed in a gradient or concentrated in specific regions of the egg cytoplasm. Female flies

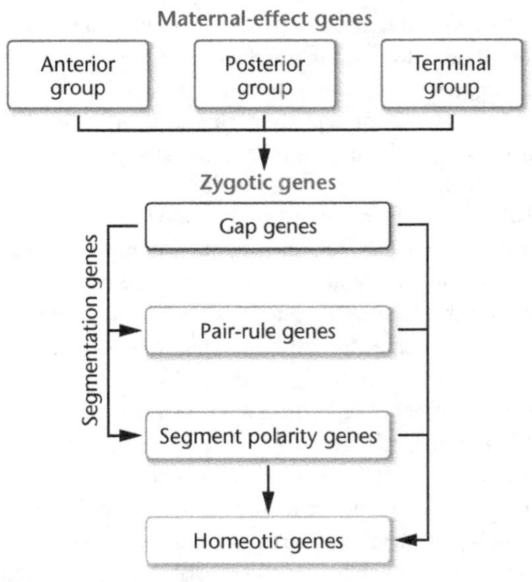


**FIGURE 19–5** The hierarchy of genes involved in establishing the segmented body plan in *Drosophila*. Gene products from the maternal genes regulate the expression of the first three groups of zygotic genes (gap, pair-rule, and segment polarity, collectively called the segmentation genes), which in turn control expression of the homeotic genes.

homozygous for recessive deleterious mutations in maternal-effect genes are sterile: none of their embryos receive wild-type gene products from their mother, so all the embryos develop abnormally. Maternal-effect genes encode transcription factors, receptors, and other proteins that regulate translation. During embryonic development, these gene products activate or repress expression of the zygotic genome in a temporal and spatial sequence.

Zygotic genes are those genes transcribed from the nuclei formed after fertilization and expressed in the developing embryo. Flies with homozygous deleterious mutations in zygotic genes exhibit embryonic lethality. In a cross between two flies heterozygous for a recessive zygotic mutation, one-fourth of the embryos (the homozygotes) therefore fail to develop normally and die. In *Drosophila*, many zygotic genes are transcribed in specific regions of the embryo but not in other regions, as determined by the distribution of maternal-effect proteins.

Much of our knowledge of the genes that regulate *Drosophila* development is based on the work of Christiane Nüsslein-Volhard, Eric Wieschaus, and Ed Lewis, who were awarded the 1995 Nobel Prize for Physiology or Medicine. Ed Lewis initially identified and studied one of these regulatory genes in the 1970s. In the late 1970s, Nüsslein-Volhard and Wieschaus devised a strategy to identify all the genes that control development in *Drosophila*. Their scheme required examining thousands of offspring of randomly mutagenized flies, looking for recessive embryonic lethal mutations with defects in external structures. The parents were thus identified as heterozygous carriers of these mutations, which the researchers grouped into three classes: *gap*, *pair-rule*, and *segment polarity* genes, based on the types of defects observed. In 1980, on the basis of their observations, Nüsslein-Volhard and Wieschaus proposed a model in which embryonic development is initiated by gradients of maternal-effect gene products. The positional information laid down by these molecular gradients is interpreted by two sets of zygotic genes: (1) **segmentation genes** (gap, pair-rule, and segment polarity genes), and (2) **homeotic selector (*Hox*) genes** (Figure 19–5). The segmentation genes divide the embryo into a series of stripes or segments and define the number, size, and polarity of each segment. The homeotic genes specify the fate of each segment and the adult structures formed from the segments.

The model is illustrated in Figure 19–6. Most maternal-effect gene products placed in the egg during oogenesis are activated immediately after fertilization and determine the anterior–posterior axis of the embryo [Figure 19–6(a)]. Many maternal gene products encode transcription factors; these in turn activate transcription of the gap genes, the first of the segmentation genes to be expressed. The action of these genes divides the embryo into a series of regions corresponding to the head, thorax, and abdomen of the adult [Figure 19–6(b)]. Gap proteins are transcription factors that activate so-called pair-rule genes, whose products divide the embryo into smaller regions about two segments wide [Figure 19–6(c)]. The pair-rule genes in turn activate the segment polarity genes, which then divide each segment into anterior and posterior regions [Figure 19–6(d)]. The collective action of the maternal genes and the segmentation genes defines the field of action for the homeotic selector (*Hox*) genes [Figure 19–6(e)].

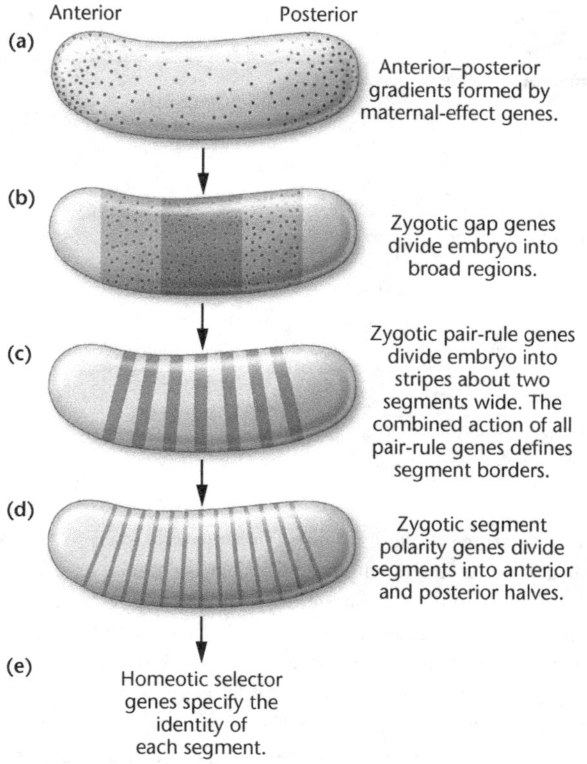


**FIGURE 19–6** Progressive restriction of cell fate during development in *Drosophila*. (a) Gradients of maternal proteins are established along the anterior–posterior axis of the embryo. (b), (c), and (d) Three groups of segmentation genes define progressively more specific body segments. (e) Individual segments are given identity by the homeotic genes.

**NOW SOLVE THIS**

Problem 6 on page 508 involves screening for mutants that affect external structures of the embryo and reconciling the results with those of Wieschaus and Schüpbach.

■ HINT: *In reconciling the results of your study with the conclusions of Wieschaus and Schüpbach, remember the differences between genes and alleles.*

## 19.4 Zygotic Genes Program Segment Formation in *Drosophila*

To summarize, zygotic genes are activated or repressed according to a positional gradient of maternal-effect gene products. The expression of three subsets of segmentation genes divides the embryo into

TABLE 19.1

**Segmentation Genes in *Drosophila***

| Gap Genes | Pair-Rule Genes | Segment Polarity Genes |
|---|---|---|
| *Krüppel* | *hairy* | *engrailed* |
| *knirps* | *even-skipped* | *wingless* |
| *hunchback* | *runt* | *cubitis interruptus*$^{D}$ |
| *giant* | *fushi-tarazu* | *hedgehog* |
| *tailless* | *odd-paired* | *fused* |
| *huckebein* | *odd-skipped* | *armadillo* |
| | *sloppy-paired* | *patched* |
| | | *gooseberry* |
| | | *paired* |
| | | *naked* |
| | | *disheveled* |

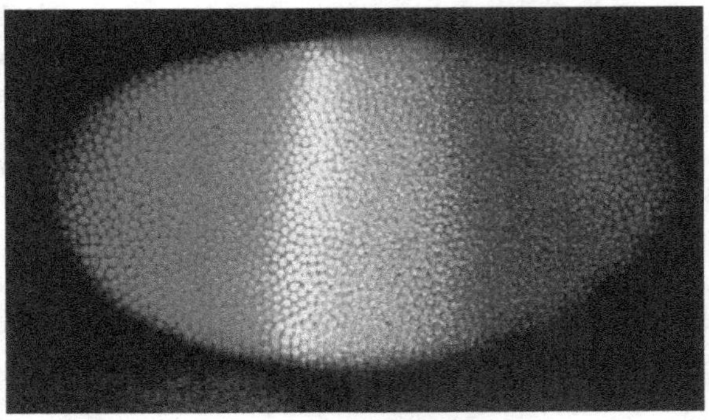

FIGURE 19–7 Expression of gap genes in a *Drosophila* embryo. The hunchback protein is shown in orange, and Krüppel is indicated in green. The yellow stripe is created when cells contain both hunchback and Krüppel proteins. Each dot in the embryo is a nucleus.

a series of segments along its anterior–posterior axis. These segmentation genes are normally transcribed in the developing embryo, and their mutations have embryonic-lethal phenotypes.

Over 20 segmentation gene loci have been identified (Table 19.1). They are classified, as mentioned above, on the basis of their mutant phenotypes: (1) mutations in gap genes delete a group of adjacent segments, (2) mutations in pair-rule genes affect every other segment and eliminate a specific part of each affected segment, and (3) mutations in segment polarity genes cause defects in homologous portions of each segment. Let us now examine each group in greater detail.

## Gap Genes

Transcription of gap genes is activated or inactivated by gene products previously expressed along the anterior–posterior axis and by other genes of the maternal gradient system. When mutated, these genes produce large gaps in the embryo's segmentation pattern: *hunchback* mutants lose head and thorax structures, *Krüppel* mutants lose thoracic and abdominal structures, and *knirps* mutants lose most of the abdominal structures. Transcription of the wild-type gap genes divides the embryo into a series of broad regions that become the head, thorax, and abdomen. Within these regions, different combinations of gene activity eventually specify both the type of segment that forms and the proper order of segments in the body of the larva, pupa, and adult. All gap genes cloned to date encode transcription factors with zinc finger DNA-binding motifs (Chapter 18 introduced this motif). Expression domains of wild-type gap genes in different parts of the embryo (Figure 19–7) correlate roughly with the location of their mutant phenotypes: *hunchback* at the anterior, *Krüppel* in the middle, and *knirps* at the posterior. As mentioned earlier, the transcription factors encoded by gap genes control the expression of pair-rule genes.

## Pair-Rule Genes

Pair-rule genes divide the head–thorax–abdomen regions established by gap genes into sections about two segments wide. Mutations in pair-rule genes eliminate segment-size sections at every other segment. The pair-rule genes are expressed in narrow bands or stripes of nuclei that extend around the circumference of the embryo. As stated earlier, the expression of this gene set first establishes the boundaries of segments, and then it establishes the developmental fate of the cells within each segment by controlling the segment polarity genes. At least eight pair-rule genes act to divide the embryo into a series of stripes. However, the boundaries of these stripes overlap, so that in each area of overlap, the cells express a different combination of pair-rule genes (Figure 19–8). Many pair-rule genes encode transcription factors containing helix–turn–helix homeodomains (another motif described in Chapter 18). The transcription of the pair-rule genes is mediated by the action of gap gene

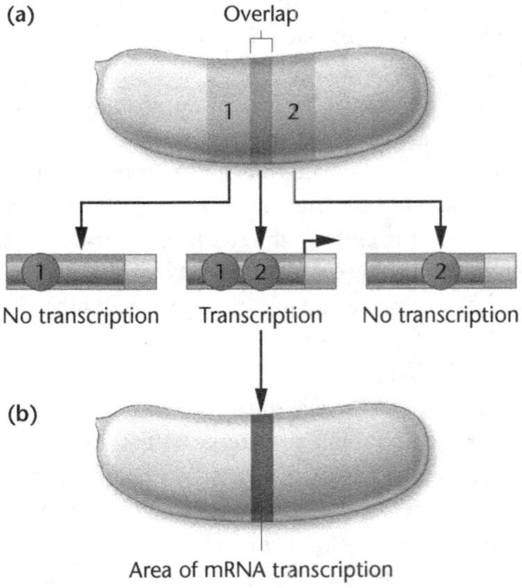


FIGURE 19–8 New patterns of gene expression can be generated by the overlapping of regions containing two different gene products. (a) Transcription factors 1 and 2 are present in overlapping regions of expression. If both transcription factors must bind to the promoter of a target gene to trigger expression, the gene will be active only in cells containing both factors (most likely in the zone of overlap). (b) The expression of the target gene in the restricted region of the embryo.

(a)

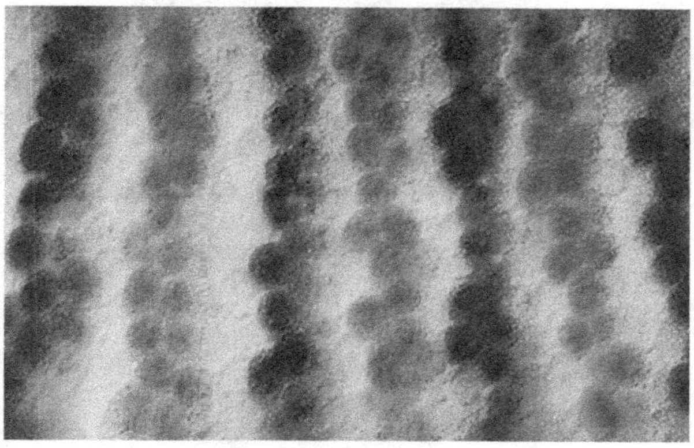

(b)

**FIGURE 19–9** Stripe pattern of pair-rule gene expression in a *Drosophila* embryo. This embryo is stained to show patterns of expression of the genes *even-skipped* and *fushi-tarazu*: (a) low-power view of the embryo; (b) high-power view of the same embryo.

products, but the resolution of the segmentation pattern into highly delineated stripes results from the interaction among the gene products of the pair-rule genes themselves (Figure 19–9).

## Segment Polarity Genes

Expression of segment polarity genes is controlled by transcription factors encoded by pair-rule genes. Within each segment created by pair-rule genes, segment polarity genes become active in a single band of cells that extends around the embryo's circumference (Figure 19–10). This divides the embryo into 14 segments. The products of the segment polarity genes control the cellular identity within each of them and establish the anterior–posterior pattern (the polarity) within each segment.

## Segmentation Genes in Mice and Humans

We have seen that segment formation in *Drosophila* depends on the action of three subsets of segmentation genes. Are these same genes found in humans and other mammals, and do they control aspects of embryonic development in these organisms? To answer these questions, let's examine *runt*, one of the principal pair-rule genes in *Drosophila*. Later in the fly's development, it controls aspects of sex

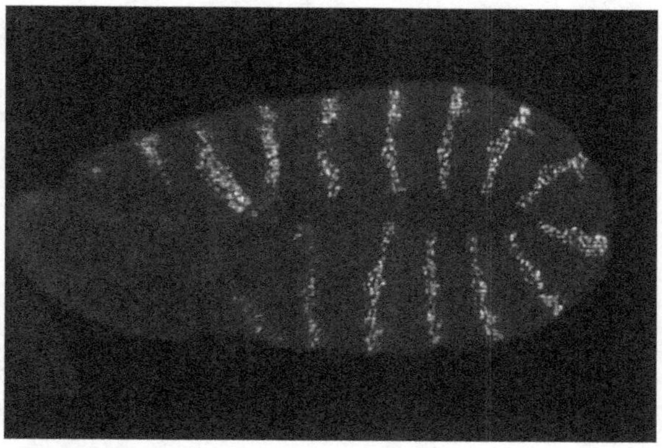

**FIGURE 19–10** The 14 stripes of expression of the segment polarity gene *engrailed* in a *Drosophila* embryo.

determination and nervous system formation. The gene encodes a protein (called runt) that regulates transcription of its target genes. Runt contains a 128-amino acid DNA-binding region called the runt domain that is highly conserved in mouse and human proteins. In fact, *in vitro* experiments show that the *Drosophila* and mouse runt proteins are functionally interchangeable. In the mouse, *runt* is expressed early in development and controls hematopoesis (formation of blood cells), osteogenesis (formation of bone), and formation of the genital system. Although the specific targets are different in flies and mice, expression of *runt* in both organisms specifies the fate of uncommitted cells in the embryo. In humans, mutation in *CBFA*, a homolog of *runt*, causes cleidocranial dysplasia (CCD), an autosomal-dominant trait. Individuals with CCD have a hole in the top of their skull because their fontanel does not close. Their collar bones (clavicles) do not develop, so they are able to fold their shoulders across their chest (Figure 19–11). Mice with one mutant copy of the *runt* homolog have a phenotype

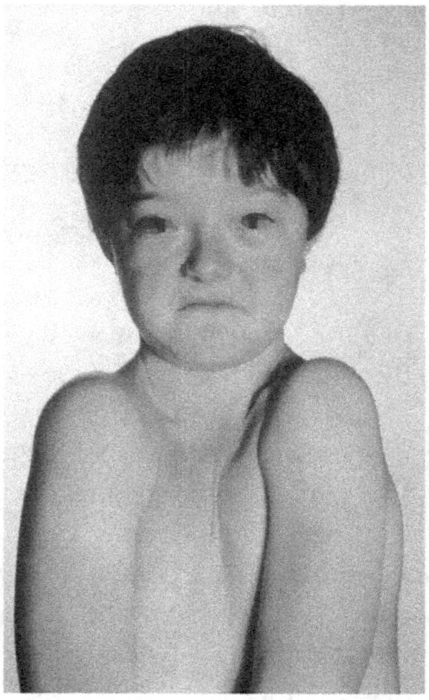

**FIGURE 19–11** A boy affected with cleidocranial dystrophy (CCD). This disorder, inherited as an autosomal dominant trait, is caused by mutation in a human *runt* gene, *CBFA*. Affected heterozygotes have a number of skeletal defects, including a hole in the top of the skull where the infant fontanel fails to close, and collar bones that do not develop or are only small stumps. Because the collar bones do not form, CCD individuals can fold their shoulders across their chests. *Reprinted by permission from Macmillan Publishers Ltd.: Fig 1 on p244 from: British Dental Journal 195: 243–48 2003. Greenwood, M. and Meechan, J.G. General medicine and surgery for dental practitioners. Copyright © Macmillan Magazines Limited.*

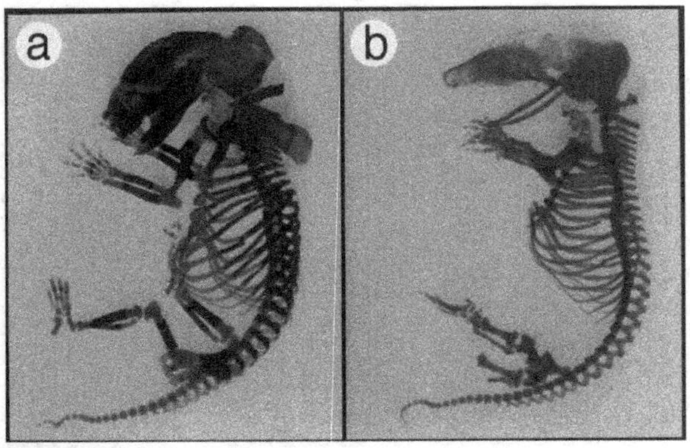


FIGURE 19–12 Bone formation in normal mice and mutants for the *runt* gene *Cbfa1*. (a) Normal mouse embryos at day 17.5 show cartilage (blue) and bone (brown). (b) The skeleton of a 17.5-day homozygous mutant embryo. Only cartilage has formed in the skeleton. There is complete absence of bone formation in the mutant mouse. Expression of a normal copy of the *Cbfa1* gene is essential for specifying the developmental fate of bone-forming osteoblasts.

similar to that seen in humans; mice with two mutant copies of the gene have no bones at all. Their skeletons are made only of cartilage, much like shark skeletons (Figure 19–12), emphasizing the role of the *runt* gene as a developmental switch, in this case, controlling the formation of bone.

## 19.5 Homeotic Selector Genes Specify Parts of the Adult Body

Because segment boundaries are established by action of the segmentation genes, the homeotic genes are activated as targets of the zygotic genes. Expression of homeotic genes determines which adult structures will be formed by each body segment. In *Drosophila*, such structures include the antennae, mouth parts, legs, wings, thorax, and abdomen. Mutants of these genes are called **homeotic mutants**, because an affected segment forms the same structure as that belonging to a neighboring segment. For example, the wild-type allele of *Antennapedia (Antp)* specifies formation of a leg on the second segment of the thorax. Dominant gain-of-function *Antp* mutations cause this gene to be expressed in the head as well, so mutant flies have a leg on their head in place of an antenna (Figure 19–13).

### Homeotic Selector (*Hox*) Genes in *Drosophila*

The *Drosophila* genome contains two clusters of homeotic selector (*Hox*) genes on chromosome 3 (Table 19.2). One cluster, the *Antennapedia (ANT-C)* complex, contains five genes that specify structures in the head and the first two thoracic segments [Figure 19–14(a)]. The second cluster, the *bithorax (BX-C)* complex, contains three genes that specify structures in the posterior portion of the second thoracic segment, the entire third thoracic segment, and abdominal segments [Figure 19–14(b)].

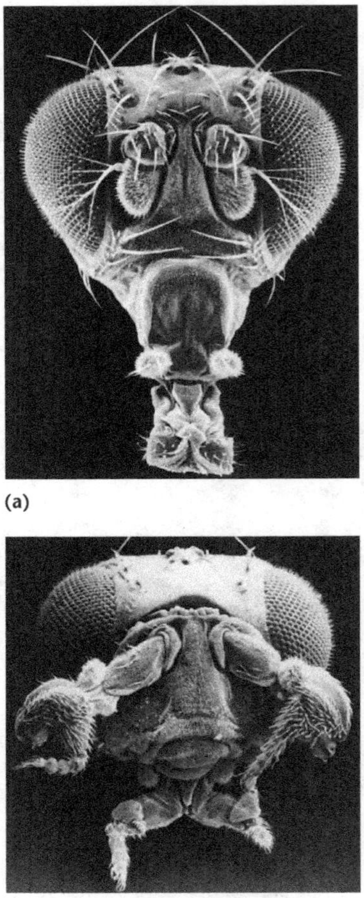

(a)

(b)

FIGURE 19–13 *Antennapedia (Antp)* mutation in *Drosophila*. (a) Head from wild-type *Drosophila*, showing the antenna and other head parts. (b) Head from an *Antp* mutant, showing the replacement of normal antenna structures with legs. This is caused by activation of the *Antp* gene in the head region.

TABLE 19.2

***Hox* Genes of *Drosophila***

| *Antennapedia* Complex | *Bithorax* Complex |
|---|---|
| *labial* | *Ultrabithorax* |
| *Antennapedia* | *abdominal A* |
| *Sex comb reduced* | *Abdominal B* |
| *Deformed* | |
| *proboscipedia* | |

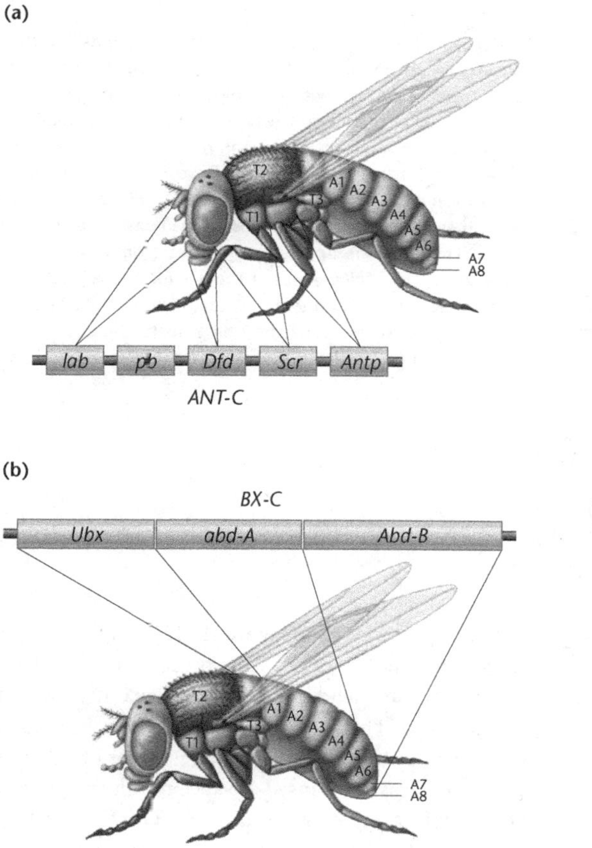


**FIGURE 19–14** Genes of the *Antennapedia* complex and the adult structures they specify. (a) In the *ANT-C* complex, the *labial (lab)* and *Deformed (Dfd)* genes control the formation of head segments. The *Sex comb reduced (Scr)* and *Antennapedia (Antp)* genes specify the identity of the first two thoracic segments. The remaining gene in the complex, *proboscipedia (pb)*, may not act during embryogenesis but may be required to maintain the differentiated state in adults. In *Antp* mutants, the labial palps are transformed into legs. (b) In the *BX-C* complex, *Ultrabithorax (Ubx)* controls formation of structures in the posterior compartment of T2 and structures in T3. The two other genes, *abdominal A (abd-A)* and *Abdominal B (Abd-B)*, specify the segmental identities of the eight abdominal segments (A1–A8).

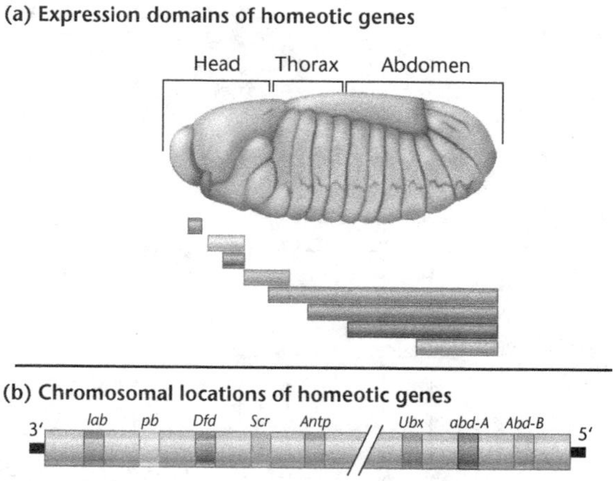


**FIGURE 19–15** The colinear relationship between the spatial pattern of expression and chromosomal locations of homeotic genes in *Drosophila*. (a) A *Drosophila* embryo and the domains of homeotic gene expression in the embryonic epidermis and central nervous system. (b) The chromosomal location of homeotic selector genes. Note that the order of genes on the chromosome correlates with the sequential anterior borders of their expression domains.

The *Hox* genes of *Drosophila* have two properties in common. First, each *Hox* gene encodes a transcription factor that includes a 180-bp DNA-binding domain known as a **homeobox.** (*Hox* is a contraction of homeobox.) The homeobox encodes a sequence of 60 amino acids known as a **homeodomain** (described in Chapter 18). Second, expression of the *Hox* genes is colinear with the structure of the embryo. Genes at the 3′-end of a cluster are expressed at the anterior end of the embryo, those in the middle are expressed in the middle of the embryo, and genes at the 5′-end of a cluster are expressed at the embryo's posterior region (Figure 19–15). Although *Hox* genes were first identified in *Drosophila*, they are found in the genomes of most eukaryotes with segmented body plans, including zebrafish, *Xenopus* (frogs), chickens, mice, and humans (Figure 19–16).

To summarize one last time, genes that control development in *Drosophila* act in a temporally and spatially ordered cascade, beginning with the genes that establish the anterior–posterior (and dorsal–ventral) axis of the egg and early embryo. Gradients of maternal mRNAs and proteins along the anterior–posterior axis activate the gap genes, which subdivide the embryo into broad bands. Gap genes in turn activate the pair-rule genes, which divide the embryo into segments. The final group of segmentation genes, the segment polarity genes, divides each segment into anterior and posterior regions arranged linearly along the anterior–posterior axis. The segments are then given identity by the *Hox* genes. This cascade of gene action (all of which occurs during the first third of embryogenesis) acts on transcription, translation, and cell–cell signaling to impose a progressive restriction of developmental potential on the *Drosophila* embryo's cells.

## *Hox* Genes and Human Genetic Disorders

Although first described in *Drosophila*, *Hox* genes are found in the genomes of all multicellular animals, where they play a fundamental role in shaping the body and its appendages. Humans and most vertebrates have four clusters of *Hox* genes (*HOXA*, *HOXB*, *HOXC*, and *HOXD*), containing a total of 39 genes. The conservation of these clusters, the order of genes within them, and their pattern of expression in vertebrates suggests that these gene sets function in flies and humans to control the pattern of structures along the

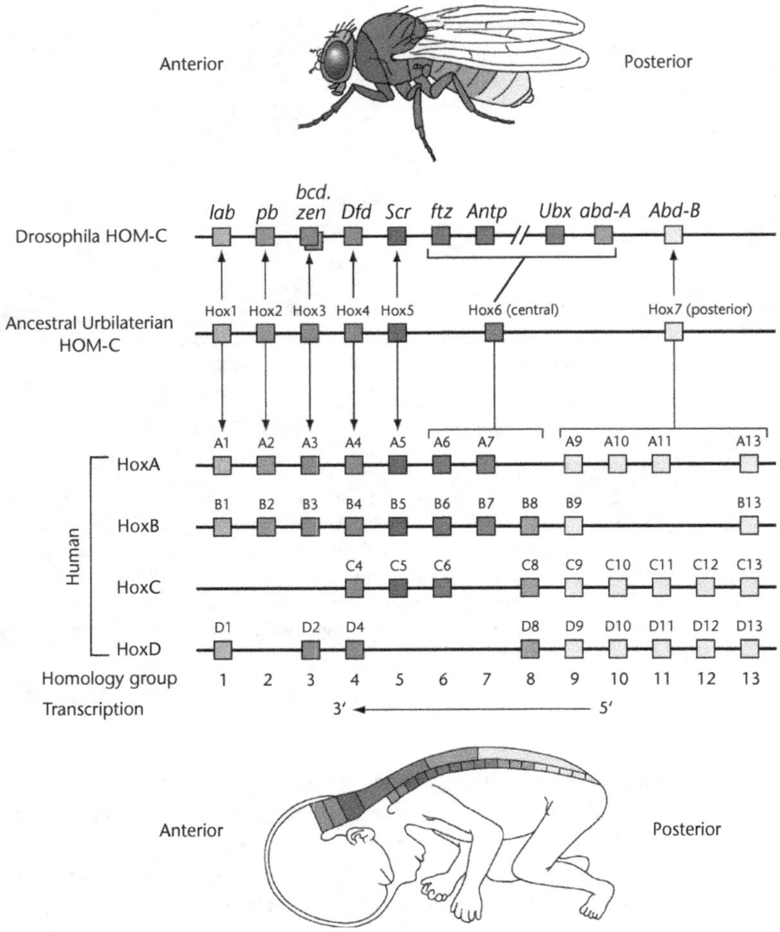


**FIGURE 19-16** Conservation of organization and patterns of expression in *Hox* genes. (Top) The structures formed in adult *Drosophila* are shown, with the colors corresponding to members of the *Hox* cluster that control their formation. (Middle) The reconstructed *Hox* cluster of the common ancestor to all bilateral organisms contains seven genes. (Bottom) The arrangement and expression patterns of the four clusters of *Hox* genes in an early human embryo. Some of the posterior genes are expressed in the limbs. The expression pattern is inferred from that observed in mice. As in *Drosophila*, genes at the 3′-end of the cluster form anterior structures, and genes at the 5′-end of the cluster form posterior structures. Genes homologous to the same ancestral sequence (because of duplications) are indicated by brackets.

anterior–posterior axis (Figure 19–17). Evidence comes from mutational studies in chicks and mice showing that *HOXD* genes near the 5′ end of the cluster play critical roles in limb development. The same role for *HOXD* genes in humans was confirmed by the discovery that a number of inherited limb malformations are caused by specific mutations in *HOXD* genes. For example, mutations in *HOXD13* cause synpolydactyly (SPD), a malformation characterized by extra fingers and toes, and by abnormalities in bones of the hands and feet (Figure 19–18).

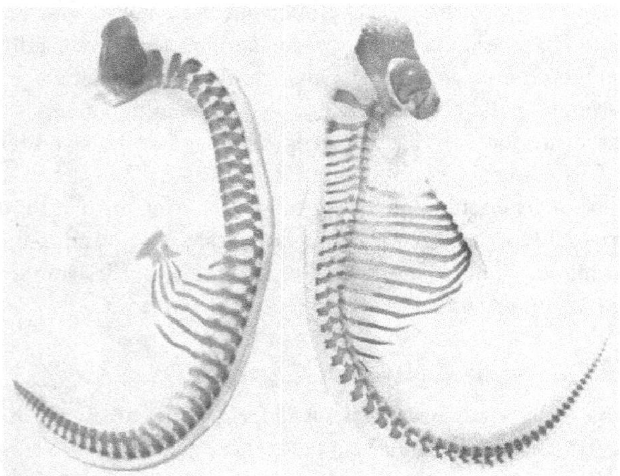

**FIGURE 19-17** Patterns of *Hox* gene expression control the formation of structures along the anterior–posterior axis of bilaterally symmetrical animals in a species-specific manner. In the chick (left) and the mouse (right), expression of the same set of *Hox* genes is differentially programmed in time and space to produce different body forms.

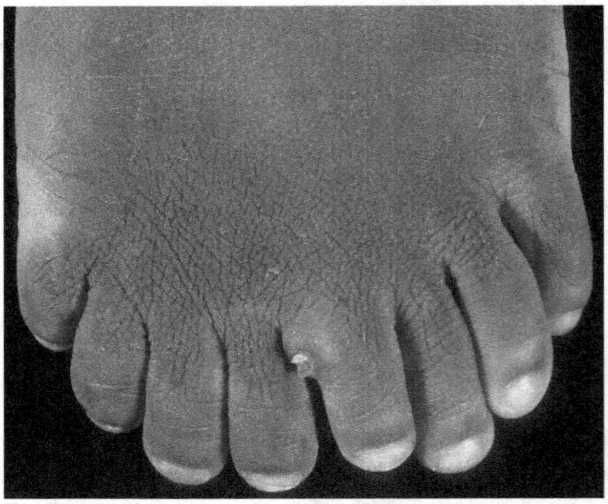

**FIGURE 19-18** Mutations in posterior *Hox* genes (*HOXD13* in this case) in humans result in malformations of the limbs, shown here as extra toes. This condition is known as synpolydactyly. Mutations in *HOXD13* are also associated with abnormalities of the bones in the hands and feet.

## Control of *Hox* Gene Expression

In *Drosophila*, a number of genes controlling *Hox* gene expression have been identified, including *extra sex combs (esc)*, *Polycomb (Pc)*, *supersex combs (sxc)* and *trithorax (trx)*. In the mutant *extra sex combs (esc)*, some of the head and all of the thoracic and abdominal segments develop as posterior segments (Figure 19–19), indicating that this gene normally controls the expression of *BX-C* genes in all body segments. The mutation does not affect either the number or the polarity of the segments. It does affect their developmental fate, indicating that the $esc^+$ gene product stored in the egg by the maternal genome may be needed for correct interpretation of the information gradient in the egg cortex.

The proteins encoded by members of the *Polycomb* family control expression of *Hox* genes by altering chromatin conformation, thus blocking the binding of transcription factors. These proteins assemble at the site of a gene and form multiprotein complexes that modify chromatin and promote gene silencing. In contrast, the *trithorax* proteins reverse the chromatin-based inactivation of genes. These proteins act as transcriptional activators and reverse the blocks put in place by inactivating proteins such as *Polycomb*-encoded products.

At present, little is known about the genetic control of *Hox* genes in other organisms, including humans. Results to date indicate that the regulatory circuits that operate upstream of *Hox* genes are not widely conserved in animal systems. If this is the case, the study of model organisms may not be useful in understanding how *Hox* genes are regulated in humans.

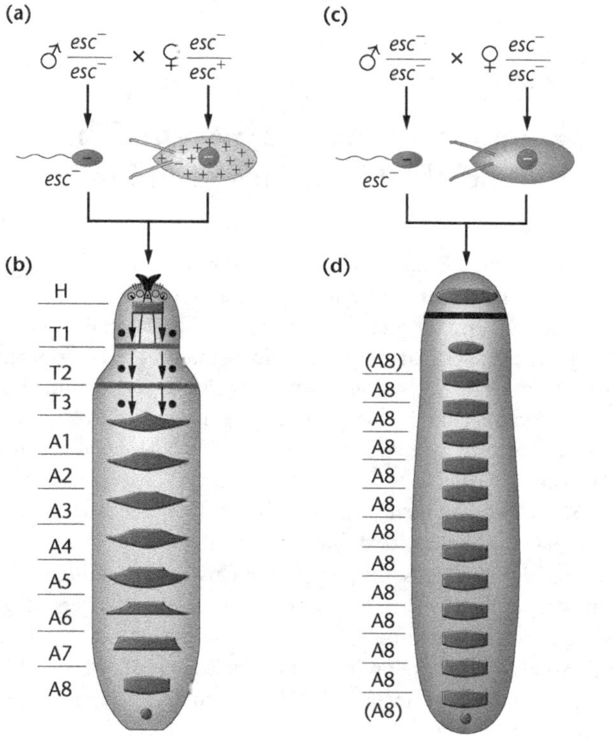


FIGURE 19–19 Action of the *extra sex combs* mutation in *Drosophila*. (a) Heterozygous females form wild-type *esc* gene product and store it in the oocyte. (b) When the $esc^-$ egg formed at meiosis by the heterozygous female is fertilized by an $esc^-$ sperm, the product is a wild-type larva with a normal segmentation pattern (maternal rescue). Borderlines between the head and thorax and between the thorax and abdomen are marked with arrows. (c) Homozygous $esc^-$ females produce defective eggs, which, when fertilized by $esc^-$ sperm, produce a larva (d) in which most of the segments of the head, thorax, and abdomen are transformed into the eighth abdominal segment. Maternal rescue demonstrates that the *esc* gene product is produced by the maternal genome and is stored in the oocyte for use in the embryo. (H, head; T1–T3, thoracic segments; A1–A8, abdominal segments.)

**NOW SOLVE THIS**

Problem 16 on page 508 involves analysis of gene expression in embryos with different genetic backgrounds.

■ HINT: *It is the embryos' genetic background that provides clues to the timing of expression and regulatory patterns of the two genes in question.*

19.6

# Cascades of Gene Action Control Differentiation

In addition to the homeotic genes in the *Hox* gene clusters, there is a large and diverse family of other homeobox-containing genes scattered throughout every eukaryotic genome. In *Drosophila*, one of these, *Distal-less (Dll)*, plays an important role in the development of appendages, including the antennae, mouthparts, legs, and wings. This gene, which maps to chromosome 2 and encodes a transcription factor, is one of the earliest to be expressed in appendage formation. In the antennae, expression of the wild-type *Dll* allele has two roles: It directs cells to form an antenna instead of a leg, and it controls the proximal (closest to the body) to distal (farthest from the body) specification of antennal structures. Mutations in *Dll* produce a wide range of phenotypes, including transformation of the antennae into legs and the formation of shortened legs that are missing distal structures.

The antenna is the ear and olfactory center (nose) of the fly and has several components, including the arista (a bristled structure) and three antennal segments [Figure 19–20(a)]. The arista transfers vibrations from sound waves through the first and second antennal segments to the antennal nerve and then to the brain. The third antennal segment is covered with olfactory receptors. Odors generate signals that are transferred to the first two antennal segments and then via a nerve to the brain.

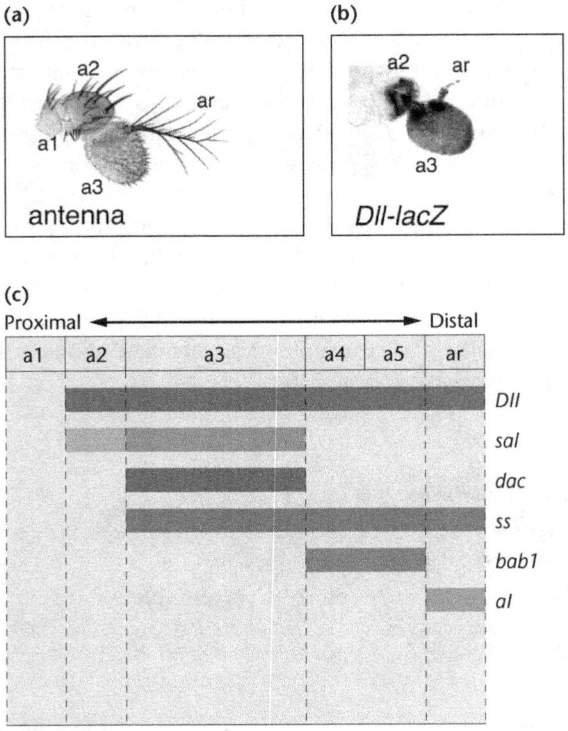


**FIGURE 19–20** Action of the *Dll* gene in *Drosophila* produces (a) the wild-type antenna, divided into the arista (ar) and three antennal segments (a1, a2, a3). (b) *Dll* expression (shown in blue) in a late pupal antenna. Expression at this stage is limited to the arista, part of a2, and a3. (c) Activation of *Dll* target genes (*sal, dac, ss, bab1,* and *al*) in antennal segments during the third larval instar just before the pupal stage. At this stage, *Dll* is expressed in all segments except a1. All of the target genes except *dac* encode transcription factors. Activation of these genes initiates a cascade of gene action in other gene sets.

Expression of *Dll* in the early pupal stage [Figure 19–20(b)] is restricted to the second and third antennal segments and the arista, and is the first step in a cascade of gene expression associated with differentiation. At the larval-pupal transformation, five genes are activated by *Dll,* and each is expressed in a segment-specific pattern [Figure 19–20(c)]. Four of these genes—*spalt (sal), spineless (ss), bric a brac (bab1),* and *aristaless (al)*—encode transcription factors. Each of these genes in turn controls multiple target genes. The fifth gene, *dachshund (dac),* encodes a protein of unknown function that is confined to the nucleus. The genes activated by *Dll* expression initiate a cascade of segment-specific gene expression causing differentiation of the antenna and arista, but the details of how this is accomplished are still being studied.

In humans, there are six homeobox-containing genes (*DLX1–DLX6*) in the *Distal-less* gene family with a sequence closely related to that of the *Drosophila Dll* gene. The *Drosophila Dll* gene is expressed in the head and appendages of the developing fly. The human *DLX3* gene is expressed in the developing head, and mutations in this gene are responsible for an inherited condition called trichodentoosseous syndrome (TDO). This autosomal dominant disorder causes deficiencies in the calcification of bones in the skull and enamel defects in teeth.

The discovery that TDO is caused by a mutation in a homeobox gene began with mutational analysis of *Drosophila* development and illustrates the valuable role that model organisms play in understanding human genetic disorders. Human *Distal-less* genes were discovered by screening a human cDNA library with a cloned probe containing the homeobox sequence of the *Drosophila Dll* gene, and the gene for *DLX3* was mapped to chromosome 17 using fluorescent *in situ* hybridization (FISH) in 1995. TDO was originally described in 1966, but the nature of the gene and its location were unknown. In 1997, geneticists found that TDO was linked to markers on chromosome 17. Further work mapped TDO to the same locus as *DLX3*, and in 1998, analysis of mutations in affected individuals confirmed that TDO is caused by a 4-bp deletion in *DLX3*.

## 19.7 Plants Have Evolved Systems That Parallel the *Hox* Genes of Animals

Plants and animals diverged from a common unicellular ancestor about 1.6 billion years ago, after the origin of eukaryotes and probably before the rise of multicellular organisms. Genome sequencing and genetic analysis of mutants indicate that basic developmental mechanisms have evolved independently in plants and animals. We have already examined the genetic systems that control development and pattern formation in animals, using *Drosophila* as a model organism.

Flower development in *Arabidopsis thaliana* (Figure 19–21), a small plant in the mustard family, has been used to study pattern formation in plants. Adult plants have clusters of undifferentiated cells, called *floral meristems*, which give rise to flowers (Figure 19–22). Each flower consists of four organs—sepals, petals, stamens, and carpels—that develop from concentric rings of cells or whorls, within the meristem [Figure 19–23(a)]. Each organ develops from a different whorl.

### Homeotic Genes in *Arabidopsis*

Three classes of floral homeotic genes control the development of these organs. Class A genes specify sepals, class A and class B genes specify petals, and class B and class C genes control stamen formation. Class C genes alone specify carpels [Figure 19–23]. The genes in each class are listed in Table 19.3. Class A genes are active in whorls 1 and 2 (sepals and petals), class B genes are expressed in whorls 2 and 3 (petals and stamens), and class C genes are expressed in whorls 3 and 4 (stamens and carpels). Each organ's formation depends on the expression pattern of the different genes. Expression of class A genes in whorl 1 causes sepals to form.

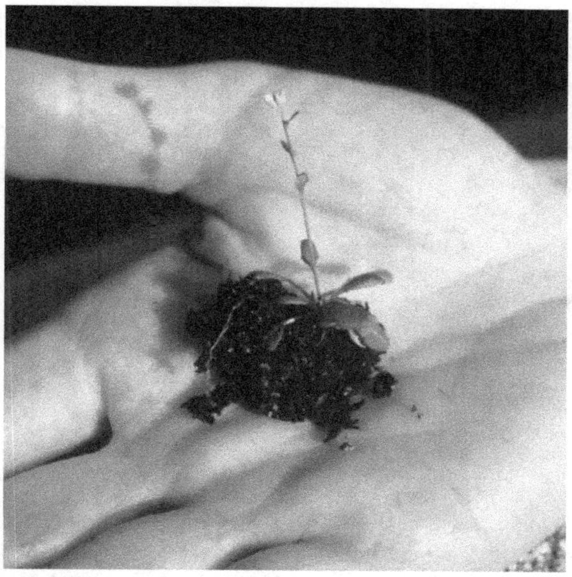

FIGURE 19–21 The flowering plant *Arabidopsis thalania*, used as a model organism in plant genetics.

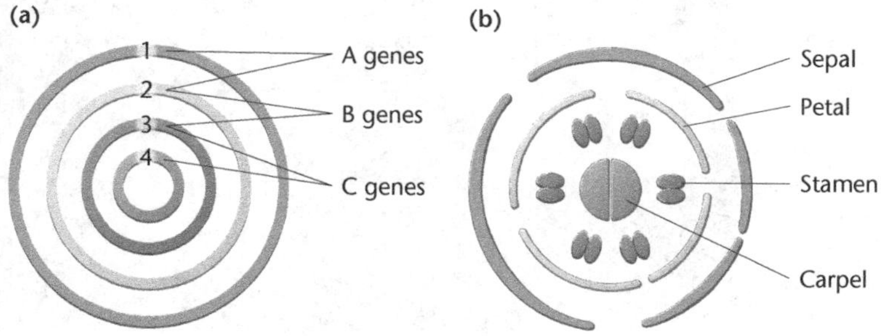


FIGURE 19–23 Cell arrangement in the floral meristem. (a) The four concentric rings, or whorls, labeled 1–4, influenced by genes A, B, and C in the manner shown, give rise to the sepals, petals, stamens, and carpels, respectively. (b) The arrangement of these organs in the mature flower.

Expression of class A *and* class B genes in whorl 2 leads to petal formation. Expression of class B and class C genes in whorl 3 leads to stamen formation. In whorl 4, expression of class C genes causes carpel formation.

As in *Drosophila*, mutations in homeotic genes cause organs to form in abnormal locations. For example, in *AP2* mutants (caused by mutation of a class A gene), the order of outer to inner organs is

TABLE 19.3

**Homeotic Selector Genes in *Arabidopsis***

| | |
|---|---|
| Class A | *APETALA1 (AP1)** |
| | *APETALA2 (AP2)* |
| Class B | *APETALA3 (AP3)* |
| | *PISTILLATA (P1)* |
| Class C | *AGAMOUS (AG)* |

*By convention, wild-type genes in *Arabidopsis* are set in capital letters.

FIGURE 19–22 (a) Parts of the *Arabidopsis* flower. The floral organs are arranged concentrically. The sepals form the outermost ring, surrounding the petals and stamens, with carpels on the inside. (b) View of the flower from above.

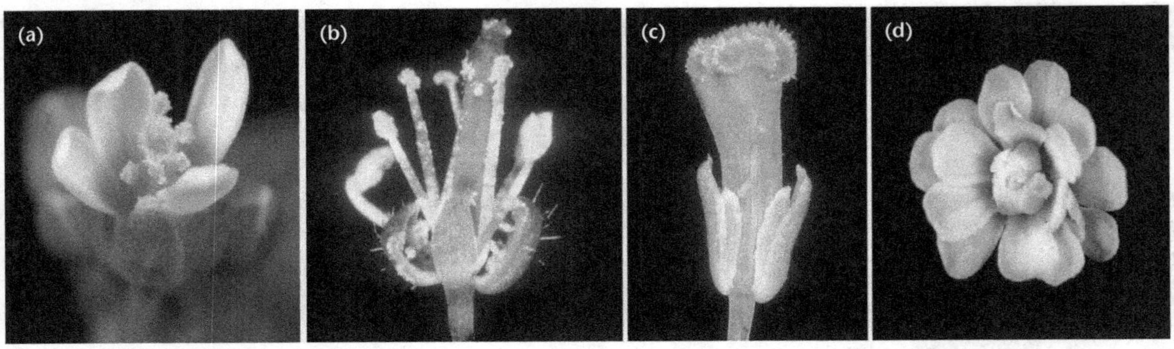


FIGURE 19–24 (a) Wild-type flowers of *Arabidopsis* have (from outside to inside) sepals, petals, stamens, and carpels. (b) Homeotic *APETALA2* mutant flower, with carpels, stamens, stamens, and carpels. (c) *PISTILLATA* mutants have sepals, sepals, carpels, and carpels. (d) *AGAMOUS* mutants have petals and sepals at places where stamens and carpels should form.

carpel, stamen, stamen, and carpel, instead of the normal order of sepal, petal, stamen, and carpel [Figure 19–24(a) and Figure 19–24(b)]. In B loss-of-function mutants, petals become sepals, and stamens are transformed into carpels [Figure 19–24(c)], and the order of organs becomes sepal, sepal, carpel, carpel. Plants carrying a mutation for the class C gene AGAMOUS will have petals in whorl 3 (instead of stamens) and sepals in whorl 4 (instead of carpels), and the order of organs will be sepal, petal, petal, and sepal [Figure 19–24(d)].

## Evolutionary Divergence in Homeotic Genes

From genetic analysis of development in both *Drosophila* and *Arabidopsis*, it is clear that each organism uses a different set of master regulatory genes to establish the body axis and specify the identity of structures along the axis. In *Drosophila*, this task is accomplished in part by the *Hox* genes, which encode a set of transcription factors sharing a homeobox domain. The products of the floral homeotic genes in *Arabidopsis*, however, are members of a different family of transcription factors, called the **MADS-box proteins**. Each member of this family contains a common sequence of 58 amino acids with no similarity in amino acid sequence to the homeobox found in the *Hox* genes. Nevertheless, both gene sets encode transcription factors, both sets are master regulators of development expressed in a pattern of overlapping domains, and both specify identity of structures.

Reflecting the common evolutionary origins of *Drosophila* and *Arabidopsis*, the genomes of both organisms contain members of the homeobox and MADS-box genes, but the genes have been adapted for different uses in the plant and animal kingdoms, indicating that developmental mechanisms evolved independently in each group.

In both plants and animals, the action of transcription factors depends on changes in chromatin structure that make genes available for expression. Mechanisms of transcription initiation are conserved in plants and animals, as is reflected in the homology of genes in *Drosophila* and *Arabidopsis* that maintain patterns of expression initiated by regulatory gene sets. Action of the floral homeotic genes is controlled by a gene called *CURLY LEAF*. This gene shares significant homology with members of the *Drosophila Polycomb* gene family, a group of genes that regulate homeobox genes during development in the fruit fly. Both *CURLY LEAF* and the *Polycomb* genes encode proteins that alter chromatin conformation and shut off gene expression. Thus, although different genes are used to control development, both plants and animals use an evolutionarily conserved mechanism to regulate expression of these gene sets.

# 19.8 Cell–Cell Interactions in Development Are Modeled in *C. elegans*

During development in multicellular organisms, cell–cell interactions influence the transcriptional programs and developmental fate of surrounding cells. Cell–cell interaction is an important factor in the embryonic development of most eukaryotic organisms, including *Drosophila*, as well as vertebrates such as clawed frogs (*Xenopus*), mice, and humans.

## Signaling Pathways in Development

From the earliest stages, animals use a number of signaling pathways to regulate development; after organogenesis begins, other signal pathways are added to those already in use. These newly activated pathways act both independently and in coordinated networks to elicit specific transcriptional responses. Signal networks establish anterior–posterior polarity and body axes, coordinate pattern formation, and direct the differentiation of tissues and organs. The signaling pathways used in embryonic development and some of the developmental processes they control are listed in Table 19.4. After an introduction to the components and interactions of one of these systems—the Notch pathway—we will briefly examine its role in the development of the vulva in the nematode *Caenorhabditis elegans*.

**TABLE 19.4**

**Signaling Systems Used in Early Vertebrate Embryonic Development**

**Wnt Pathway**
Dorsalization of body
Female reproductive development
Dorsal–ventral differences

**TGF-β Pathway**
Mesoderm induction
Left-right asymmetry
Bone development

**Hedgehog Pathway**
Notochord induction
Somitogenesis
Gut/visceral mesoderm

**Receptor Tyrosine Kinase Pathway**
Mesoderm maintenance

**Notch/Delta Pathway**
Blood cell development
Neurogenesis
Retina development

SOURCE: Gerhart, J. 1999. 1998 Warkany lecture: Signaling pathways in development. *Teratology* 60: 226–239.

## The Notch Signaling Pathway

The genes in the **Notch signaling pathway** are named after the *Drosophila* mutants that were used to identify components of this pathway. *Notch* works through direct cell–cell contact to control the developmental fate of the interacting cells. The *Notch* gene (or the equivalent genes in other organisms) encodes a transmembrane signal receptor (Figure 19–25). The signal is another transmembrane protein, encoded by the *Delta* gene (or its equivalents). Because both the signal and receptor are membrane-bound, the Notch signal system works only between adjacent cells. When the Delta protein binds to the Notch receptor, the cytoplasmic tail of the Notch protein cleaves off and moves to the nucleus where it binds to a protein encoded by the *Su(H)* (suppressor of *Hairless*) gene, activating transcription of a gene set that controls a specific developmental pathway directing cell fate (Figure 19–25).

Variations of this pathway control a number of different developmental processes in *Drosophila*, establishing the dorsal–ventral boundary in the wing disc and directing the fate of cells in the nervous system, muscle, and gut. One of the main roles of the Notch signal system is to specify the fate of equivalent cells in a population. In its simplest form, a Notch interaction involves two neighboring cells that are developmentally equivalent. Action of the Notch signaling system may send these cells down different developmental pathways.

In humans, four members of the Notch family (*NOTCH1–NOTCH4*) have been identified. Mutations in these genes and other genes in the Notch pathway are responsible for a number of human developmental disorders, including Alagille syndrome (AGS), lymphoblastic leukemia, and spondylocostal dysostosis (SD).

## Overview of *C. elegans* Development

For a number of reasons, the nematode *C. elegans* is widely used to study the genetic control of development: (1) the genetics of the organism are well known, (2) the genome sequence is available, and (3) adults are formed from a small number of cells through a highly deterministic developmental program that is unchanged from

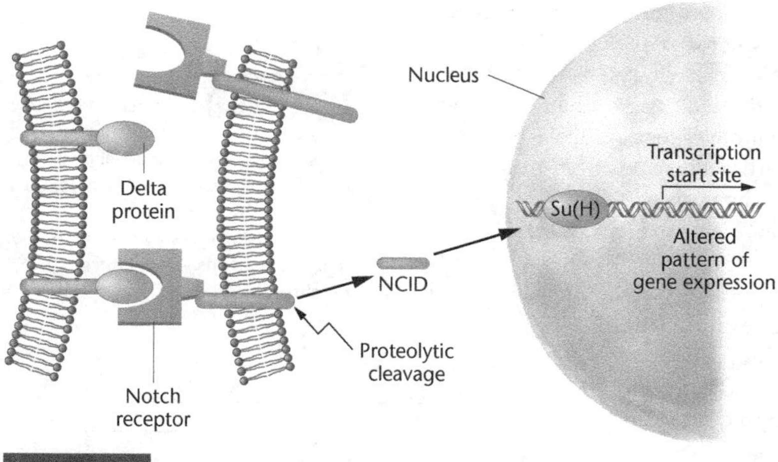


**FIGURE 19–25** Components of the Notch signaling pathway in *Drosophila*. The cell carrying the Delta transmembrane protein is the sending cell; the cell carrying the transmembrane Notch protein receives the signal. Binding of Delta to Notch triggers a proteolysis-mediated activation of transcription. The fragment (NCID, the notch intracellular domain) cleaved from the cytoplasmic side of the Notch protein moves to the nucleus where it combines with the Su(H) protein and activates a program of gene transcription.

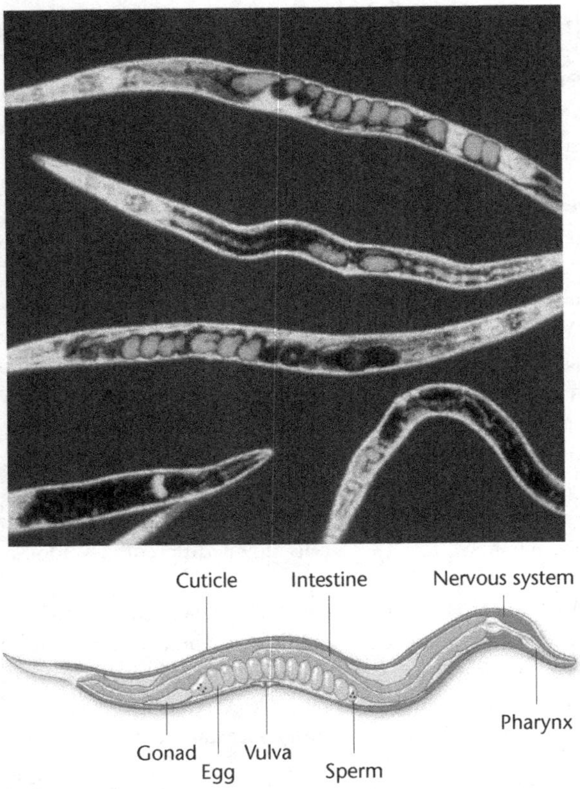


FIGURE 19-26 An adult *Caenorhabditis elegans* hermaphrodite. This nematode, about 1 mm in length, consists of 959 somatic cells and has been used to study many aspects of the genetic control of development.

individual to individual. Adult nematodes are about 1 mm long and mature from a fertilized egg in about two days (Figure 19–26). The life cycle consists of an embryonic stage (about 16 hours), four larval stages (L1 through L4), and the adult stage. Adults are of two sexes: XX self-fertilizing hermaphrodites that can make both eggs and sperm, and XO males. Self-fertilization of mutagen-treated hermaphrodites quickly results in homozygous stocks of mutant strains, and hundreds of mutations have been generated, catalogued, and mapped.

Adult hermaphrodites have 959 somatic cells (and about 2000 germ cells). The exact lineage of each somatic cell, from fertilized egg to adult, has been mapped (Figure 19–27) and is invariant from individual to individual. Knowing the lineage of each cell, researchers can easily follow events resulting from mutations that alter cell fate or from killing specific cells with laser microbeams or ultraviolet irradiation. In *C. elegans* hermaphrodites, the fate of cells in the development of the reproductive system is determined by cell–cell interaction, providing insight into how gene expression and cell–cell interaction work together to specify developmental outcomes.

**NOW SOLVE THIS**

Problem 28 on page 509 involves two genes that control sex determination in *C. elegans*. You are asked to analyze a model of sex determination.

■ HINT: *In solving this problem, remember to consider the action of gene products and the effect of loss-of-function mutations on expression of other genes or the action of other proteins.*

## Genetic Analysis of Vulva Formation

*C. elegans* adult hermaphrodites lay eggs through the vulva, an opening located about midbody (Figure 19–26). The vulva is formed in stages during larval development by means of several rounds of cell–cell interaction. In the first of these, two developmentally equivalent neighboring cells, Z1.ppp and Z4.aaa, interact with each other so that one becomes the gonadal anchor cell and the other becomes a precursor to the uterus. The determination of which cell becomes which occurs during the second larval stage (L2) and is controlled by *lin-12* (one of the Notch receptor genes in *C. elegans*). In recessive *lin-12(0)* mutants (a loss-of-function mutant), both cells become anchor cells. In contrast, the dominant mutation *lin-12(d)* (a gain-of-function mutation) causes both to become uterine precursors. Thus, the expression of the *lin-12* gene seems to cause the selection of the uterine pathway, since in the absence of the LIN-12 (Notch) receptor, both cells become anchor cells.

FIGURE 19-27 A truncated cell lineage chart for *C. elegans*, showing early divisions and the tissues and organs that eventually result. Each vertical line represents a cell division, and horizontal lines connect the two cells produced. For example, the first cell division creates two new cells from the zygote, AB and P1. The cells in this chart are among those present in the first-stage larva, L1. During subsequent larval stages, further cell divisions will produce the 959 somatic cells of the adult hermaphrodite worm.

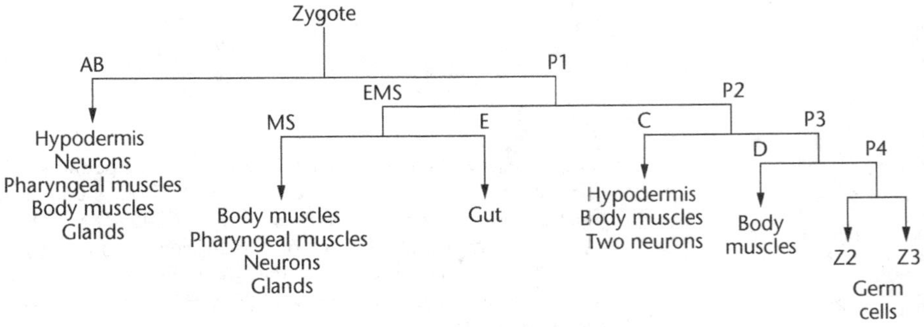

As shown in Figure 19–28, however, the situation is more complex than it first appears. Initially, the two neighboring cells are developmentally equivalent. Each synthesizes low levels of the Delta signal protein (encoded by the *lag-2* gene) *and* the Notch receptor protein. This situation is unstable, and both cells cannot continue simultaneously sending and receiving developmental signals. By chance, the cell secreting more of the signal (LAG-2 ) causes the neighboring cell to increase production of the receptor (LIN-12 protein). The cell producing more receptor protein receives a stronger signal and thus becomes the uterine precursor; the other cell, producing more signal protein, becomes the anchor cell. The critical factor in this first round of cell–cell interaction is the balance between the LAG-2 (Delta) signal gene product and the LIN-12 (Notch) gene product.

A second round of cell–cell communication leads to formation of the vulva. This interaction involves the anchor cell (located in the gonad) and six precursor cells (located in the skin adjacent to the gonad). The precursor cells are named P3.p to P8.p and collectively are called Pn.p cells. The fate of each Pn.p cell is specified by its position relative to the anchor cell. Figure 19–29 shows the developmental pathway described in the following paragraphs.

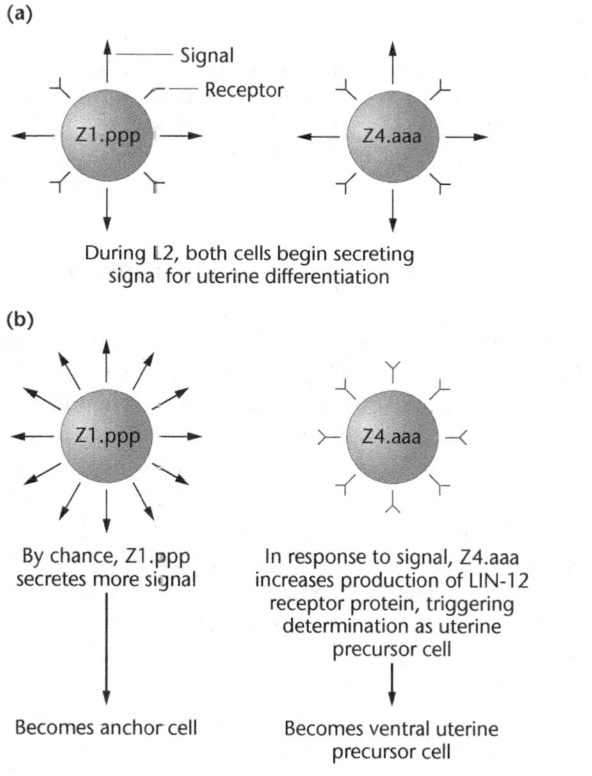


**FIGURE 19–28** Cell–cell interaction in anchor cell determination. (a) During L2, two neighboring cells begin the secretion of chemical signals for the induction of uterine differentiation. (b) By chance, cell Z1.ppp produces more of these signals, causing cell Z4.aaa to increase production of the receptor for signals. The reception of increased signals causes Z4.aaa to become the ventral uterine precursor cell and allows Z1.ppp to become the anchor cell.

In larval stage 3, the *lin-3* gene is activated in the anchor cell. The gene product, LIN-3, is a signal protein related to vertebrate epidermal growth factor (EGF). All six precursor cells express a receptor encoded by *let-23*, a gene homologous to the vertebrate EGF receptor. The binding of LIN-3 to the LET-23 receptor triggers an intracellular cascade of events that determines whether the precursor cells will form the primary vulval precursor cell or secondary vulval cells. In other words, *let-23* establishes the primary and secondary fates of precursor cells. Recessive loss-of-function *let-23* mutations cause all Pn.p cells to act as if they have not received any signal, and as a result, no vulva forms.

Normally, the cell closest to the anchor cell (P6.p) receives the strongest signal initiated by LIN-3 binding to LET-23. The signal activates expression of the *Vulvaless (Vul)* gene (the gene is named for its mutant phenotype) in P6.p, and this cell becomes the primary vulval precursor cell. P6.p then divides three times to produce vulva cells. The two neighboring cells (P5.p and P7.p) receive a lower amount of signal, which initiates a secondary fate. These cells divide asymmetrically to form additional vulva cells.

To reinforce these developmental pathways, a third level of cell–cell interaction occurs. In this interaction, the primary vulval cell (P6.p) sends a signal that activates *lin-12* in the two neighboring cells (P5.p and P7.p). This signal prevents P5.p and P7.p from adopting the division pattern of the primary cell. In other words, cells in which both *Vul* and *lin-12* are active cannot become primary vulva cells. The three remaining precursor cells (P3.p, P4.p, and P8.p) receive no signal from the anchor cell. In these cells the *Multivulva (Muv)* gene is expressed, *Muv* represses *Vul*, and the three cells develop as skin cells.

Thus, three levels of cell–cell interactions are seen in the developmental pathway leading to vulva formation in *C. elegans*. First, two neighboring cells interact to establish the identity of the anchor cell. Second, the anchor cell interacts with three vulval precursor cells to establish the identity of the primary vulval precursor cell (usually P6.p) and two secondary cells (P5.p and P7.p). Third, the primary vulval cell interacts with the secondary cells to suppress their ability to adopt the pathway of the primary cell. In each interaction, a cell produces molecular signals that are received and processed by neighboring cells.

## Notch Signaling Systems in Humans

In humans, there are four Notch genes encoding receptors that play important roles in defining borders during segmentation and in generating cells of the immune system and bone marrow stem cells. Mutations in the Notch signaling pathway are responsible for several inherited disorders, including Alagille syndrome. This autosomal dominant disorder produces developmental abnormalities of the liver, the heart and circulatory system, and the skeleton.

The theme of cell–cell interactions occurring in a spatial and temporal cascade to specify the developmental fates of individual cells is a developmental motif repeated over and over in all living organisms.

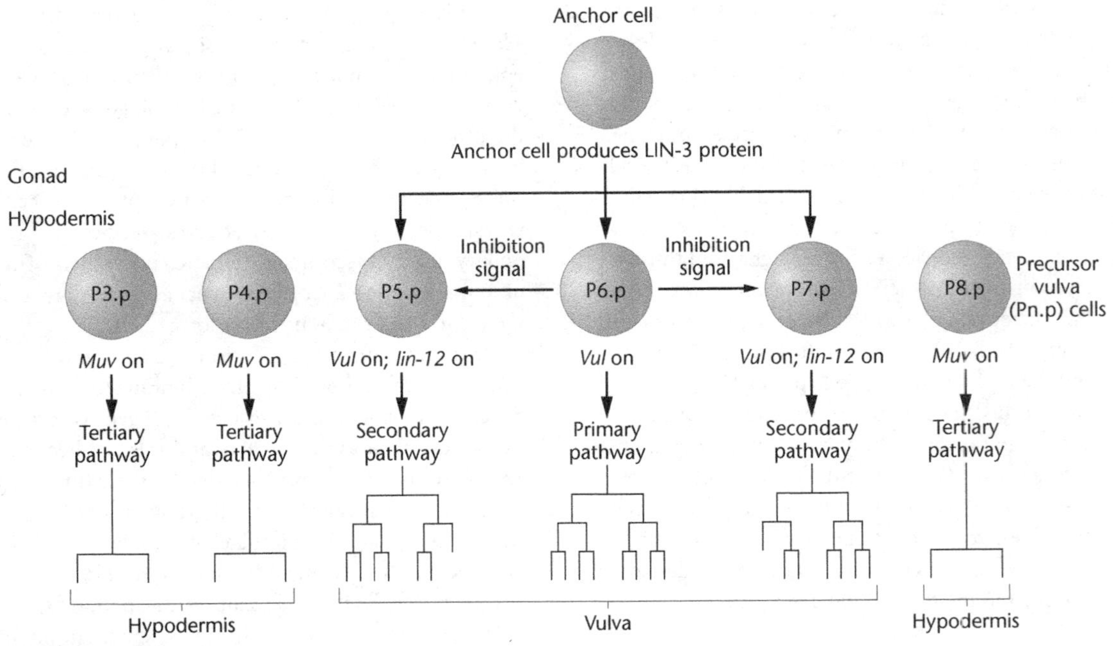


**FIGURE 19–29** Cell lineage determination in *C. elegans* vulva formation. A signal from the anchor cell in the form of LIN-3 protein is received by three precursor vulval cells (Pn.p cells). The cell closest to the anchor cell becomes the primary vulval precursor cell, and adjacent cells become secondary precursor cells. The primary cell produces a signal that activates the *lin-12* gene in secondary cells, preventing them from becoming primary cells. Flanking precursor cells, which receive no signal from the anchor cell, increase the activity of the *Muv* gene and become skin (hypodermis) cells, instead of vulval cells.

## 19.9 Transcriptional Networks Control Gene Expression in Development

As we have seen in examples from *Drosophila* and *C. elegans*, development in higher eukaryotes depends on a closely coordinated pattern of gene expression across both space and time. Understanding development at the level of gene expression will require dissection of the network of transcriptional controls that generates these patterns of expression. Among the key elements coordinating spatiotemporal patterns of gene expression are *cis*-regulatory elements (CREs), also called ***cis*-regulatory modules** (CRMs). These modules can be located upstream from genes, within introns, or downstream from genes. CRMs typically contain binding sites for several transcription factors and usually have multiple binding sites for each of these factors. Binding of transcription factors to sites in CRMs determines whether the gene in question is activated or repressed. If activated, the basal transcription complex is recruited to the core promoter sequence (see Chapter 18 to review eukaryotic transcriptional regulation). Some of the genes regulated by CRMs encode transcription factors, thus forming an interconnected network of regulatory genes and their targets.

### A General Model of a Transcription Network

A general model for a network is presented in Figure 19–30. One of the modules contains binding sites for two different transcription factors [Figure 19–30(a)]. As noted above, CRMs can have sites for more than one transcription factor, and multiple binding sites for each factor. The node in Figure 19–30(b) contains a regulatory gene, B, with six CRMs; some genes contain many more modules. Nodes receive input from other parts of the network and generate output to target genes. Figure 19–30(c) shows a network based on this node. In this example, gene B is the node. A CRM in gene B receives input from transcription factor A (blue). Once activated, this CRM directs the synthesis of a transcription factor (green arrow) that binds to a CRM in gene C. Gene C encodes a transcription factor with two functions. It binds to gene D, generating a developmental response, and it also binds to gene B, repressing further synthesis of the transcription factor encoded by gene B.

### Transcriptional Networks in *Drosophila* Segmentation

We saw earlier that in *Drosophila*, the foundation of the embryo's anterior–posterior axis is established by maternally derived gradients of mRNAs. Figure 19–31(a) reviews the steps by which this

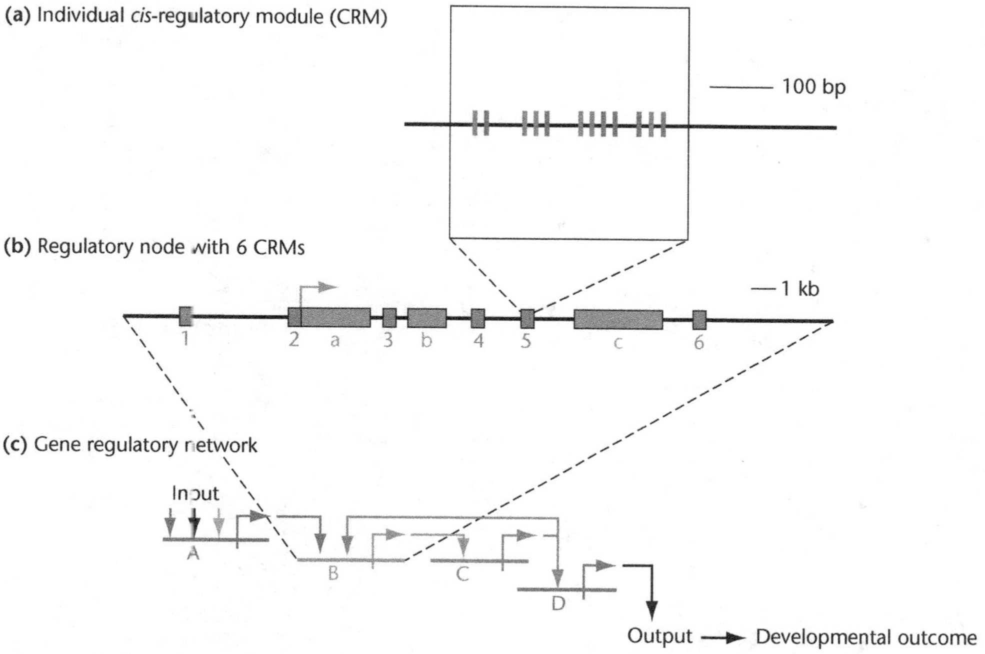


**FIGURE 19–30** The basic components of a transcriptional network. (a) A *cis*-regulatory module (CRM) is a region several hundred nucleotides in length containing binding sites for transcription factors. This example contains multiple copies of two different binding sites (pink and blue). (b) CRMs can be located upstream, downstream, or within introns. The gene shown here, gene B, contains 6 CRMs. The three exons of gene B (a, b, and c) are interspersed with CRMs. (c) The location of gene B in the transcriptional network. Signal input activates gene A, which directs the synthesis of transcription factor A. This transcription factor binds to sites within CRM 5 in gene B, activating gene B, which synthesizes a transcription factor which binds (green arrows) to a CRM in gene C. Gene C directs the synthesis of transcription factor C. Transcription factor C binds to a CRM in gene D, activating that gene, which transcribes a protein that contributes to a developmental event. Transcription factor C also binds to CRM 5 in gene B, as part of a feedback loop, regulating the expression of gene B.

foundation is laid and by which the specification, determination, and differentiation of adult body structures occur.

Mark Schroeder and his colleagues used a combination of genomic sequence information, results from *in vitro* experiments, and computational algorithms to construct a transcriptional network of the *Drosophila* segmentation gene sets, identifying and classifying the CRMs associated with these genes into four categories: maternal/gap gene driven, pair-rule driven, or driven by both types of modules but predominantly maternal/gap or predominantly pair-rule. These researchers then mapped the location of the modules adjacent to and within genes involved in anterior–posterior axis formation.

Their model is shown in Figure 19–31. By examining all the modules associated with a given gene, it is possible to assess the gene's expression pattern and its place in the hierarchy of genes in this network. The model shows that the maternal gradient system, with built-in redundancy, directly controls most or all of the early zygotic patterning events. The maternal transcription factors present in this gradient act as activators of expression within their subnetworks, while gap gene–encoded factors largely act as repressors of expression. Gap and gap-like genes are strongly associated with maternal/gap input modules. Pair-rule and segment polarity genes receive input from pair-rule modules. The homeotic selector genes receive input from both modules.

The model, which correlates strongly with what is known from genetic analysis and the pattern of segmentation gene expression in *Drosophila*, expands the known number of modules by about 50 percent. The computational algorithm is being used to search for more CRMs in the segmentation pathway to provide a complete description of this transcriptional network.

Transcriptional networks have been described for a number of other systems, as well (Table 19.5). These systems come from a wide range of organisms, yet all have underlying similarities in organization and function, indicating that the basic regulatory system is conserved by natural selection and is built on the use of CRMs.

**TABLE 19.5**

**Some Models of Gene Regulatory Networks**

| Organism | Domain Specification |
|---|---|
| Sea urchin (Sp) | Endomesoderm |
| Starfish (Am) | Endoderm |
| Mouse (Mm) | Pancreatic $\beta$-cell |
| Mouse (Mm) | Hematopoetic stem cells; Erythroid lineage |
| Mammals | B-cell specification |
| Mammals | T-cell specification |
| Vertebrates | Heart field specification |
| Frog (Xl) | Mesoderm |
| Ascidian (Ci) | Notochord |
| Ascidian (Ci) | Different territories |
| Fruit fly (Dm) | Dorsal–ventral axis |
| Fruit fly (Dm) | Heart field specification |
| Nematode (Ce) | Left-right ASE taste neuron |
| Nematode (Ce) | Vulva |
| Nematode (Ce) | C-cell lineage |

Abbreviations: Am, *Asterina miniata*; Ce, *Caenorhabditis elegans*; Ci, *Ciona instestinalis*; Dm, *Drosophila melanogaster*; Mm, *Mus musculus*; Sp, *Strongylocentrotus purpuratus*; Xl, *Xenopus laevis*.

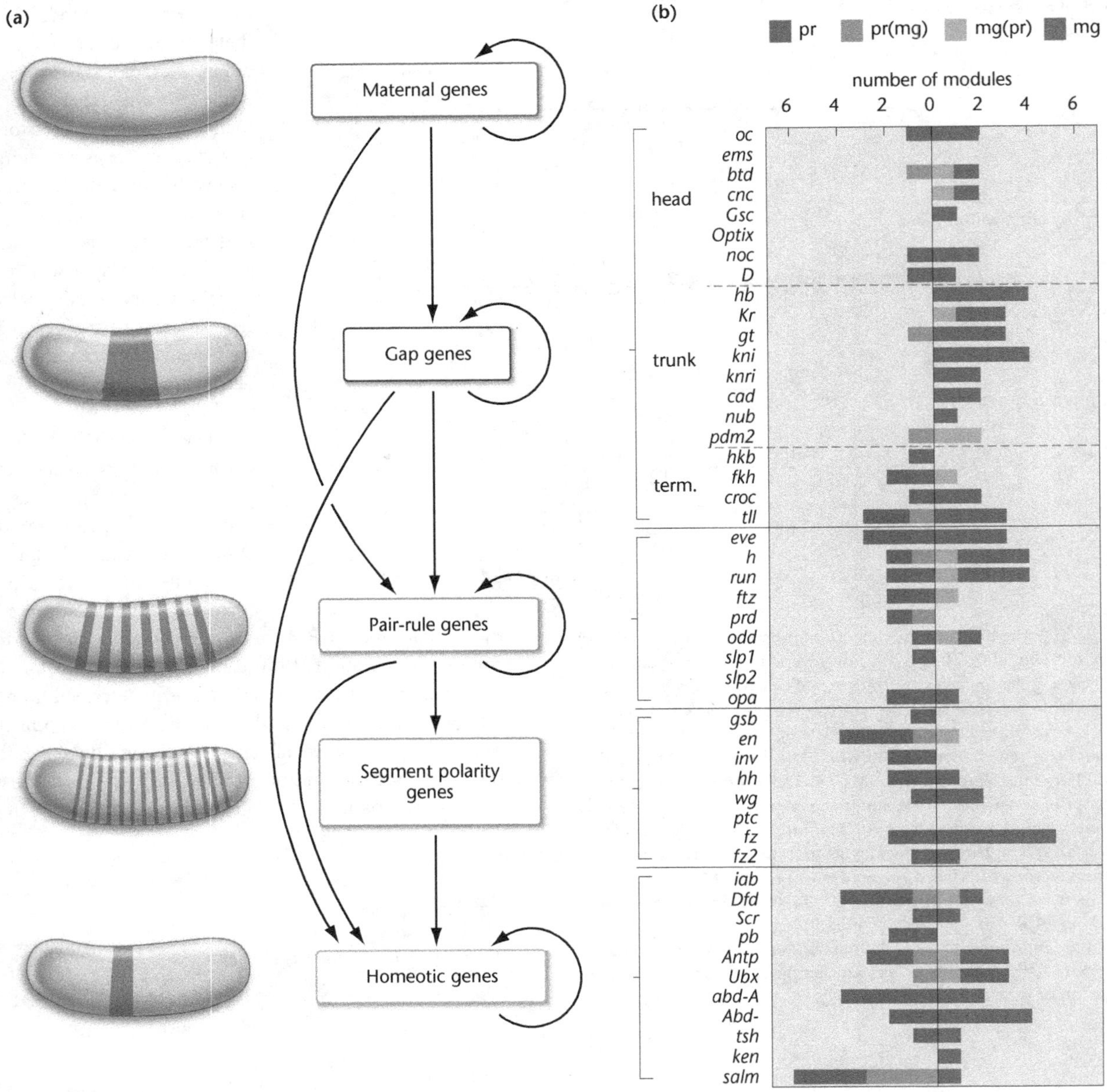


**FIGURE 19-31** The transcriptional network associated with the anterior–posterior axis and segmentation in *Drosophila*. (a) The column at left summarizes the steps in embryo segmentation. Maternal mRNAs form an anterior–posterior gradient in the embryo. This gradient differentially activates the gap genes, and their expression divides the embryo into broad regions of head, thorax, and abdomen. Maternal gradients and the gap genes, in turn, activate the pair-rule genes, which divide the embryo into a series of stripes whose cells are thus assigned to individual segments of the adult body. The pair-rule genes activate the segment polarity genes, which divide each segment into an anterior and a posterior compartment. Homeotic genes receive input from the gap genes, the segment polarity genes, and the pair-rule genes, and specify which structures of the adult body will be formed from each segment. (b) The genes belonging to each class of segmentation genes are listed in the column at the left. (Brackets indicate the class to which the genes belong.) Four different kinds of CRMs have been identified in the segmentation genes: CRMs containing maternal/gap (mg) binding sites (green), CRMs containing pair-rule (pr) binding sites (red), or CRMs containing both types of binding sites but predominantly maternal/gap [mg(pr)] (light green) or predominantly pair-rule [pr(mg)] (light red). The number of each type of module associated with each gene is indicated across the top of the column. Full names of all genes and genetic information about them are available at: http://flybase.bio.indiana.edu/ Information about the actions of each gene in development is available at: http://www.sdbonline.org/fly/aimain/1aahome.htm

## GENETICS, TECHNOLOGY, AND SOCIETY

# Stem Cell Wars

Stem cell research is at the center of a battle being fought by scientists, politicians, advocacy groups, religious leaders, and ethicists. Proponents fighting for the right to carry out stem cell research claim that it has the potential to cure such conditions as diabetes, Parkinson disease, and spinal cord injuries and to improve the quality of life for millions. Critics lobbying for an end to stem cell research warn that it may propel us down a slippery slope toward loss of regard for human life. Although stem cell research is the focus of presidential proclamations, media campaigns, and legislative bans, few people understand it sufficiently to evaluate its pros and cons.

Stem cells are primitive cells that replicate indefinitely and have the unique capacity to differentiate into cells with specialized functions, such as the cells of heart, brain, liver, and muscle tissue. All the cells that make up the approximately 200 distinct types of tissues in our bodies are descended from stem cells. Some types of stem cells are defined as *totipotent*, meaning they have the ability to differentiate into any mature cell type in the body. Other types of stem cells are *pluripotent* and are able to differentiate into any of a smaller number of mature cell types. In contrast, mature, fully differentiated cells do not replicate or undergo transformations into different cell types.

In the last few years, several research teams have isolated and cultured human pluripotent stem cells. These cells remain undifferentiated and grow indefinitely in culture dishes. When treated with growth factors or hormones, these pluripotent stem cells differentiate into cells that have characteristics of neural, bone, kidney, liver, heart, or pancreatic cells.

The fact that pluripotent stem cells grow prolifically in culture and differentiate into more specialized cells has created great excitement. Some foresee a day when stem cells may be a cornucopia from which to harvest unlimited numbers of specialized cells to replace cells in damaged and diseased tissues. Hence, stem cells could be used to treat Parkinson disease, type 1 diabetes, chronic heart disease, kidney and liver failure, Alzheimer disease, Duchenne muscular dystrophy, and spinal cord injuries. Some predict that stem cells will be genetically modified to eliminate transplant rejection or to deliver specific gene products, thereby correcting genetic defects or treating cancers. The excitement about stem cell therapies has been fueled by reports of dramatically successful experiments in animals. For example, mice with spinal cord injuries regained their mobility and bowel and bladder control after they were injected with human stem cells. Both proponents and critics of stem cell research agree that stem cell therapies could be revolutionary. Why, then, should stem cell research be so contentious?

The answer to that question lies in the source of the pluripotent stem cells. To date, all pluripotent stem cell lines have been derived from five-day-old embryonic blastocysts. Blastocysts at this stage consist of about 2110 cells, most of which will develop into placental and supporting tissues for the early embryo. The inner cell mass of the blastocyst consists of about 30 to 40 pluripotent stem cells that develop into all the embryo's tissues. *In vitro* fertilization clinics grow fertilized eggs to the five-day blastocyst stage prior to uterine transfer. Embryonic stem (ES) cell lines can then be created by taking the inner cell mass out of five-day blastocysts and growing those undifferentiated cells in culture dishes. All human ES cell lines have been derived from unused five-day blastocysts that were discarded by *in vitro* fertilization clinics.

The fact that early embryos are destroyed in the process of establishing human ES cell lines disturbs people who believe that preimplantation embryos are persons with rights; however, it does not disturb people who believe that these embryos are too primitive to have the status of a human being. Both sides in the debate put forth lengthy arguments concerning the fundamental question of what constitutes a human being.

Critics of ES cell research argue that we may soon be able to benefit from stem cell therapies without resorting to the use of ES cells. This argument is based on recent reports about the plasticity of adult stem cells. Adult stem cells are undifferentiated cells that are present in certain differentiated tissues such as blood and brain. They divide within the differentiated tissue and differentiate into mature cells that make up the tissue in which they are found. Adult stem cells have been found in bone marrow, the intestines, blood, the retina, the brain, skeletal muscle, the liver, skin, and the pancreas. The best-known adult stem cells are hematopoietic stem cells (HSCs), which are found in bone marrow, peripheral blood, and umbilical cords. HSCs differentiate into mature blood cell types such as red blood cells, lymphocytes, and macrophages and have been used clinically for many years as transplant material to reconstitute the immune systems of patients undergoing treatment for cancer and autoimmune diseases. Recent studies suggest that adult stem cells may have the capacity to differentiate into other cell types.

Although these reports are intriguing, it is still too early to know whether adult stem cells hold the same pluripotent promise as ES cells. Adult stem cells are scarce, are difficult to identify and isolate, and grow poorly, if at all, in culture. However, if these obstacles can be overcome, adult stem cells may provide a more widely accepted alternative to ES cells and calm the raging debate. On the other hand, new philosophical dilemmas could arise. If adult stem cells are found to exhibit the same pluripotency as ES cells, they could have the same potential to create a human embryo—dragging critics and proponents of stem cell research back into the same moral quagmire. At the present time, it is impossible to predict whether either adult or embryonic stem cells will be as miraculous as predicted by scientists and the popular press. But if stem cell research progresses at its current rapid pace, we won't have long to wait.

### ■ References

Green, R. M. 2007. Can we develop ethically universal embryonic stem-cell lines? *Nat. Rev. Genet.* 8: 480–485.

Gruen, L., and Grabel, L. 2006. Concise review: scientific and ethical roadblocks to human embryonic stem cell therapy. *Stem Cells* 24: 2162–2169.

Robertson. J. A. 2001. Human embryonic stem cell research: ethical and legal issues. *Nat. Rev. Gen.* 2: 74–78.

### ■ Websites

National Institutes of Health. 2001. "Stem Cells: Scientific Progress and Future Research Directions." http://stemcells.nih.gov/info/scireport/ (or for the.pdf version: http://stemcells.nih.gov/info/scireport/PDFs/fullrptstem.pdf)

National Institutes of Health. 2004. "Stem Cell Information." http://stemcells.nih.gov/index.asp

National Institutes of Health Stem Cell Information. 2007. http://stemcells.nih.gov/index.asp

EXPLORING GENOMICS

# Gene Collections for Model Organisms

Throughout this book we have emphasized the importance of model organisms for the study of genetics. In this chapter, for example, we discussed aspects of developmental genetics of model organisms. Genome sequencing projects have been completed or are underway for a majority of the model organisms that geneticists have relied on for many years. We have already explored some of these genomes in other Exploring Genomics exercises.

In this exercise we explore genomes of the zebrafish, fruit fly, yeast, African clawed frog, and sea urchin to determine if they share sequence similarities with human genes.

## ■ Exercise I – Model Organisms and Human Genes

1. Visit the National Institutes of Health Model Organisms for Biomedical Research site at http://www.nih.gov/science/models/ and the National Centers for Biotechnology Information Model Organisms Guide at http://www.ncbi.nlm.nih.gov/About/model/index.html. These sites are excellent resources for learning about the genetics of model organisms.

2. Refer to the model organism databases listed in the next column to search for information on the following human genes: *OCT4*, *SOX2*, and *NANOG*. In which model organisms have genes with similar sequences been identified? What can be learned about each gene from these model organisms? Keep in mind that for similar sequences to be found in these databases, the sequence must already have been annotated in the database. These databases change all the time as new genome data are added. If a particular database does not have one of these genes, it does not necessarily mean that a similar gene is not present in the organism. The sequence of that gene may not yet be included in the database.

   *Danio rerio* (Zebrafish) Gene Collection, http://zgc.nci.nih.gov
   *Xenopus laevis* (African clawed frog) Gene Collection, http://xgc.nci.nih.gov/
   *Mus musculus* (Mouse) Genome Resources, http://www.ncbi.nlm.nih.gov/genome/guide/mouse/
   *Saccharomyces pombe* (Yeast) GeneDB, www.genedb.org/genedb/pombe/index.jsp

3. The *Oct4*, *Sox2*, and *Nanog* genes are being studied by many scientists around the world. Why are these genes receiving so much attention? Carry out a PubMed search to learn more about these genes.

## ■ Exercise II – Sea Urchin Genome Revealed

In 2006, scientists announced that a partially completed draft of the genome for the sea urchin, *Strongylocentrotus purpuratus*, had been established. In recent years, sea urchins have become valuable model organisms for developmental biologists, geneticists, and evolutionary biologists, in part because these echinoderms are more closely related to mammals than are other model organisms such as *Drosophila*.

1. One of the best resources for learning about the urchin genome is available through the NCBI at http://www.ncbi.nlm.nih.gov/genome/guide/sea_urchin/index.html. The information found in the sea urchin genome databases is continually being updated as scientists continue to sort through sequence data. Search the sea urchin genome site for the genes from Exercise I and any of the genes discussed in Chapter 19 to determine whether similar genes have been identified in sea urchins.

# Chapter Summary

1. Developmental genetics, which explores the mechanisms by which genetic information controls development and differentiation, is one of the major areas of study in biology. Geneticists are investigating this topic by isolating developmental mutations and identifying the genes involved in developmental processes.
2. Determination is the regulatory process whereby cell fate becomes fixed during early development. Determination precedes the actual differentiation into or specialization of distinctive cell types.
3. During embryogenesis, the activity of specific genes is controlled by the internal environment of the cell, including localized cytoplasmic components. In flies, the regulation of early events is mediated by the maternal cytoplasm, which then influences zygotic gene expression. As development proceeds, both the cell's internal environment and its external environment become further altered by the presence of early gene products and communication with other cells.

4. In *Drosophila*, both genetic and molecular studies have confirmed that the egg contains information specifying the body plan of the larva and adult.
5. Extensive genetic analysis of embryonic development in *Drosophila* has led to the identification of maternal-effect genes that lay down the anterior–posterior axis of the embryo. In addition, these maternal-effect genes activate sets of zygotic segmentation genes, initiating a cascade of gene regulation that ends with the determination of segment identity by the homeotic selector genes. These same gene sets control aspects of embryonic development in all bilateral animals, including humans.
6. Flower formation in *Arabidopsis* is controlled by homeotic genes, but these gene sets are from a different gene family than the homeotic selector genes of *Drosophila* and other animals.
7. In *C. elegans*, the strictly determined lineage of each cell allows developmental biologists to study the cell–cell signaling required for organogenesis.
8. Using experimental data and algorithms, researchers are elucidating the transcriptional networks that control the temporal and spatial patterns of gene expression in developing embryos.

## INSIGHTS AND SOLUTIONS

1. In the slime mold *Dictyostelium*, experimental evidence suggests that cyclic AMP (cAMP) plays a central role in the developmental process leading to spore formation. The genes encoding the cAMP cell-surface receptor have been cloned, and the amino acid sequence of the protein components is known. To form reproductive structures, free-living individual cells aggregate together and then differentiate into one of two cell types, prespore cells or prestalk cells. Aggregating cells secrete waves, or oscillations, of cAMP to foster the aggregation, and then continuously secrete cAMP to activate genes in the aggregated cells at later stages of development. It has been proposed that cAMP controls cell–cell interaction and gene expression. Since every hypothesis ought to be tested using several experimental techniques, what different approaches can you devise to test this one, and what specific experimental systems would you employ?

**Solution:** Two of the most powerful forms of analysis in biology are (1) the use of biochemical analogs (or inhibitors) to block gene transcription or the action of gene products in a predictable way, and (2) the use of mutations to alter the gene and its products. These two approaches can be used to study the role of cAMP in the developmental program of *Dictyostelium*. First, biochemical analogs of cAMP, such as GTP and GDP, can be used to test whether they have any effect on the processes controlled by cAMP. In fact, both GTP and GDP lower the affinity of cell-surface receptors for cAMP, effectively blocking the action of cAMP. To inhibit the synthesis of the cAMP receptor, it is possible to construct a vector containing a DNA sequence that transcribes an antisense RNA (a molecule that has a base sequence complementary to the mRNA). Antisense RNA forms a double-stranded structure with the mRNA, leading to its degradation. If normal cells are transformed with a vector that expresses antisense RNA, no cAMP receptors will be produced. It is possible to predict that such cells will fail to respond to a gradient of cAMP and, consequently, will not migrate to an aggregation center. In fact, that is what happens. Such cells remain dispersed and nonmigratory in the presence of cAMP. Similarly, it is possible to determine whether this response to cAMP is necessary to trigger changes in the transcriptional program by assaying for the expression of developmentally regulated genes in cells expressing this antisense RNA.

Mutational analysis of the components of the cAMP receptor system uses transformation with wild-type genes to restore mutant function. Similarly, because the genes for the receptor proteins have been cloned, it is possible to construct mutants with known alterations in the component proteins and transform them into cells to assess their effects.

2. In the sea urchin, early development up to gastrulation may occur even in the presence of actinomycin D, which inhibits RNA synthesis. However, if actinomycin D is present early in development but is removed at the end of blastula formation, gastrulation does not proceed. In fact, if actinomycin D is present only between the sixth and eleventh hours of development, gastrulation (normally occurring at the fifteenth hour) is arrested. What conclusions can be drawn concerning the role of gene transcription between hours 6 and 15?

**Solution:** A considerable amount of development can take place without transcription of the embryo's genome because maternal mRNAs are present in the fertilized sea urchin egg. The inhibition of gastrulation by prior treatment with actinomycin D suggests that transcripts from the embryo's genome are required to initiate or maintain gastrulation. Apparently, this transcription must take place between the sixth and fifteenth hours of development.

3. If it were possible to introduce one of the homeotic genes from *Drosophila* into an *Arabidopsis* embryo homozygous for a mutation in a homeotic flowering gene, would you expect any of the *Drosophila* genes to negate (rescue) the *Arabidopsis* mutant phenotype? Why or why not?

**Solution:** The *Drosophila* homeotic genes belong to the *Hox* gene family, while *Arabidopsis* homeotic genes belong to the MADS-box protein family. Both gene families are present in *Drosophila* and *Arabidopsis*, but they have evolved different functions in the animal and the plant kingdom. As a result, it is unlikely that a transferred *Drosophila Hox* gene would rescue the phenotype of a MADS-box mutant, but only an actual experiment would confirm this assumption.

## Problems and Discussion Questions

1. Carefully distinguish between the terms *differentiation* and *determination*. During development, which phenomenon occurs first?
2. Nuclei from almost any source may be injected into *Xenopus* oocytes. Studies have shown that these nuclei remain active in transcription and translation. How can such an experimental system be useful in developmental genetic studies?
3. The homunculus doctrine postulated that miniature adult entities are contained within a gamete and merely unfold and grow to give rise to a mature organism. What sorts of isolated evidence presented in this chapter might have led to this doctrine? Why is the epigenetic theory (defined in Chapter 4) held to be correct today?
4. Distinguish between the syncytial blastoderm stage and the cellular blastoderm stage in *Drosophila* embryogenesis.
5. (a) What are maternal-effect genes? (b) When are gene products from these genes synthesized, and where are they located? (c) What aspects of development do maternal-effect genes control? (d) What is the phenotype of maternal-effect mutations?
6. Suppose you conduct a screen for maternal-effect mutations in *Drosophila* affecting external structures of the embryo, and your screen identifies more than 100 mutations that affect external structures. Weischaus and Schüpbach estimated from their screening that there are about 40 maternal-effect genes. How do you reconcile these different results?
7. (a) What are zygotic genes, and when are their gene products made? (b) What is the phenotype associated with zygotic gene mutations? (c) Does the maternal genotype contain zygotic genes?
8. List the main classes of zygotic genes. What is the function of each class of these genes?
9. Experiments have shown that any nuclei placed in the cytoplasm at the posterior pole of the *Drosophila* egg will differentiate into germ cells. If polar cytoplasm is transplanted into the anterior end of the egg just after fertilization, what will happen to nuclei that migrate into this cytoplasm at the anterior pole?
10. How can you determine whether a particular gene is being transcribed in different cell types?
11. You observe that a particular gene is being transcribed during development. How can you tell whether the expression of this gene is under transcriptional or translational control, or under both transcriptional and translational controls?
12. What are *Hox* genes? What properties do they have in common?
13. The homeotic mutation *Antennapedia* causes mutant *Drosophila* to have legs in place of antennae and is a dominant gain-of-function mutation. What are the properties of such mutations? How does the *Antennapedia* gene change antennae into legs?
14. The *Drosophila* homeotic mutation *spineless aristapedia* ($ss^a$) results in the formation of a miniature tarsal structure (normally part of the leg) on the end of the antenna. What insight does $ss^a$ provide about the role of genes during determination?
15. Embryogenesis and oncogenesis (generation of cancer) share a number of features, including cell proliferation, programmed cell death, cell migration and invasion, formation of new blood vessels, and differential gene activity. Embryonic cells are relatively undifferentiated, and cancer cells appear to be undifferentiated or dedifferentiated. Homeotic gene expression directs early development, and mutant expression of these genes leads to loss of the differentiated state or an alternative cell identity. M. T. Lewis (2000. *Breast Can. Res.* 2: 158–169) suggested that breast cancer may be caused by the altered expression of homeotic genes. When he examined eleven such genes in cancers, eight were underexpressed while three were overexpressed compared with controls. Given what you know about homeotic genes, what is the likelihood that they are involved in oncogenesis?
16. In *Drosophila*, both *fushi tarazu (ftz)* and *engrailed* genes encode homeobox transcription factors and are capable of eliciting the expression of other genes. Both genes work at about the same time during development and in the same region to specify cell fate in body segments. To discover if *ftz* regulates the expression of *engrailed*, if *engrailed* regulates *ftz*; or if both are regulated by another gene, you perform a mutant analysis. In $ftz^-$ embryos (*ftz/ftz*), engrailed protein is absent; in $engrailed^-$ embryos (*en/en*), *ftz* expression is normal. What does this tell you about the regulation of these two genes—does the *engrailed* gene regulate *ftz*, or does the *ftz* gene regulate *engrailed*?
17. Early development depends on the temporal and spatial interplay between maternally supplied mRNA and the onset of zygotic gene expression. Maternally encoded mRNAs must be produced, positioned, and degraded (Surdej and Jacobs-Lorena, 1998. *Mol. Cell Biol.* 18: 2892–2900). For example, transcription of the *bicoid* gene that determines anterior–posterior polarity in *Drosophila* is maternal. The mRNA is synthesized in the ovary by nurse cells and then transported to the oocyte, where it localizes to the anterior ends of oocytes. After egg deposition, *bicoid* mRNA is translated, and unstable bicoid protein forms a decreasing concentration gradient from the anterior end of the embryo, where *gap* genes along the anterior half of the embryo are activated. At the start of gastrulation, *bicoid* mRNA has been degraded. Consider two models to explain the degradation of *bicoid* mRNA: (1) degradation may result from signals within the mRNA (intrinsic model), or (2) degradation may result from the mRNA's position within the egg (extrinsic model). Experimentally, how could one distinguish between these two models?
18. In *Arabidopsis*, flower development is controlled by sets of homeotic genes. How many classes of these genes are there, and what structures are formed by their individual and combined expression?
19. Formation of germ cells in *Drosophila* and many other embryos is dependent on their position in the embryo and their exposure to localized cytoplasmic determinants. Nuclei exposed to cytoplasm in the posterior end of *Drosophila* eggs (the pole plasm) form cells that develop into germ cells under the direction of maternally derived components. R. Amikura et al. (2001. *Proc. Nat. Acad. Sci. (USA)* 98: 9133–9138) consistently found mitochondria-type ribosomes outside mitochondria in the germ plasm of *Drosophila* embryos and postulated that they are intimately related to germ-cell specification. If you were studying this phenomenon, what would you want to know about the activity of these ribosomes?
20. One of the most interesting aspects of early development is the remodeling of the cell cycle from cycles of rapid cell divisions, apparently lacking G1 and G2 phases, to slower cell cycles with measurable G1 and G2 phases and checkpoints. During this remodeling, maternal mRNAs that specify cyclins are deadenylated, and zygotic genes are activated to produce cyclins. Audic et al. (2001. *Mol. and Cell. Biol.* 21: 1662–1671) suggest that deadenylation requires transcription of zygotic genes. Prepare a diagram that captures the significant features of these findings.

HOW DO WE KNOW?

21. In this chapter we focused on how differential gene expression guides the processes that lead from the fertilized egg to the adult. At the same time, we found many opportunities to consider the methods and reasoning by which much of this information was acquired. From the explanations given in the chapter, what answers would you propose to the following fundamental questions:
    (a) How do we know how many genes control development in an organism like *Drosophila*?
    (b) What experimental evidence is available to show that molecular gradients in the egg control development?
    (c) How do we know that a genetic program specifying a body part can be changed?
    (d) What genetic evidence shows that chemical signals between cells control developmental events?
    (e) How do we know whether a signaling system in vulva development works only on adjacent cells or uses signals that can affect more distant cells?

## Extra-Spicy Problems

22. In studying gene action during development, it is desirable to be able to position genes in a hierarchy or pathway of action to establish which genes are primary and in what order the genes act. There are several ways of doing this. One is to make double mutants and study the outcome. The gene *fushi-tarazu (ftz)* is expressed in early embryos at the seven-stripe stage. All of the genes involved in forming the anterior–posterior pattern affect the expression of this gene, as do the *gap* genes. However, expression of segment-polarity genes is affected by *ftz*. What is the location of *ftz* in this hierarchy?
23. A number of genes that control expression of *Hox* genes in *Drosophila* have been identified. One of these homozygous mutants is *extra sex combs*, where some of the head and all of the thorax and abdominal segments develop as the last abdominal segment. In other words, all affected segments develop as posterior segments. What does this phenotype tell you about which set of *Hox* genes is controlled by the *extra sex combs* gene?
24. The *apterous* gene in *Drosophila* encodes a protein required for wing patterning and growth. It is also known to function in nerve development, fertility, and viability. When human and mouse genes whose protein products closely resemble *apterous* were used to generate transgenic *Drosophila* (Rincon-Limas et al. 1999. *Proc. Nat. Acad. Sci. [USA]* 96: 2165–2170), the apterous mutant phenotype was *rescued*. In addition, the whole-body expression patterns in the transgenic *Drosophila* were similar to normal *apterous*.
    (a) What is meant by the term *rescued* in this context?
    (b) What do these results indicate about the molecular nature of development?
25. The floral homeotic genes of *Arabidopsis* are MADS-box proteins, while in *Drosophila*, homeotic genes are *Hox* genes, belonging to the homeobox gene family. In both *Arabidopsis* and *Drosophila*, members of the *Polycomb* gene family control expression of these divergent homeotic genes. How do *Polycomb* genes control expression of two very different sets of homeotic genes?
26. Vulval development in *C. elegans* begins when two neighboring cells (Z1.ppp and Z4.aaa) interact with each other by cell–cell signaling involving two components: a membrane-bound signal molecule and a membrane-bound receptor. By chance, one cell produces more signal and causes its neighbor to produce more receptor. The signal-producing cell becomes the anchor cell, and the receptor-producing cell becomes the uterine precursor. This form of cell–cell interaction is called the Notch/Delta signaling system. Although it is a widely used signaling mechanism in metazoans, this pathway works only in adjacent cells. Why is this so, and what are the advantages and disadvantages of such a system?
27. The identification and characterization of genes that control sex determination has been another focus of investigators working with *C. elegans*. As with *Drosophila*, sex in this organism is determined by the ratio of X chromosomes to sets of autosomes. A diploid wild-type male has one X chromosome, and a diploid wild-type hermaphrodite has two X chromosomes. Many different mutations have been identified that affect sex determination. Loss-of-function mutations in a gene called *her-1* cause an XO nematode to develop into a hermaphrodite, but they have no effect on XX development. (That is, XX nematodes are normal hermaphrodites.) In contrast, loss-of-function mutations in a gene called *tra-1* cause an XX nematode to develop into a male. Deduce the roles of these genes in wild-type sex determination from this information.
28. Based on the information in Problem 27 and the analysis of the phenotypes of single- and double-mutant strains, a model for sex determination in *C. elegans* has been generated. This model proposes that the *her-1* gene controls sex determination by establishing the level of activity of the *tra-1* gene, which in turn, controls the expression of genes involved in generating the various sexually dimorphic tissues. Given this information,
    (a) does the *her-1* gene product have a negative or a positive effect on the activity of the *tra-1* gene?
    (b) What would be the phenotype of a *tra-1*, *her-1* double mutant?
29. On the following page is a microarray assessment of gene expression among developmentally significant categories (tissues and organs) in the model organism *Arabidopsis thaliana* (Schmid et al., 2005). Relative gene activities are presented on the ordinate, while several tissue types are presented on the abscissa.
    (a) Are gene expression patterns reasonably compatible with expectations?
    (b) The general developmental program of plants contrasts with that of animals in that plants develop continuously with new organs being added throughout their lifespan. How might such a developmental program help explain the expression of photosynthetic genes in flowers and seeds?

(c) Typically, single or small numbers of genes are studied to determine their impact on development. Here, genome-wide analysis reveals output from general classes containing many genes. How might such a global approach further our understanding of the role of genes in development?

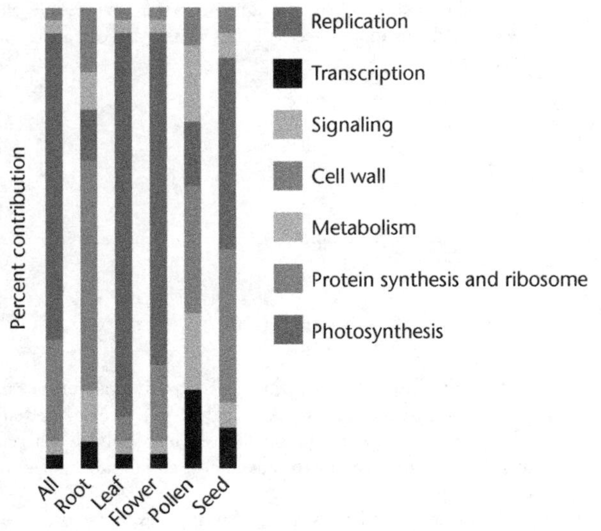


30. Dominguez et al. (2004) suggest that by studying genes that determine growth and tissue specification in the eye of *Drosophila*, much can be learned about human eye development.
   (a) What evidence suggests that genetic eye determinants in *Drosophila* are also found in humans? Include a discussion of orthologous genes in your answer.
   (b) What evidence indicates that the *eyeless* gene is part of a developmental network?
   (c) Are genetic networks likely to specify developmental processes in general? Explain fully and provide an example.

Colored scanning electron micrograph of two prostate cancer cells in the final stages of cell division (cytokinesis). The cells are still joined by strands of cytoplasm.

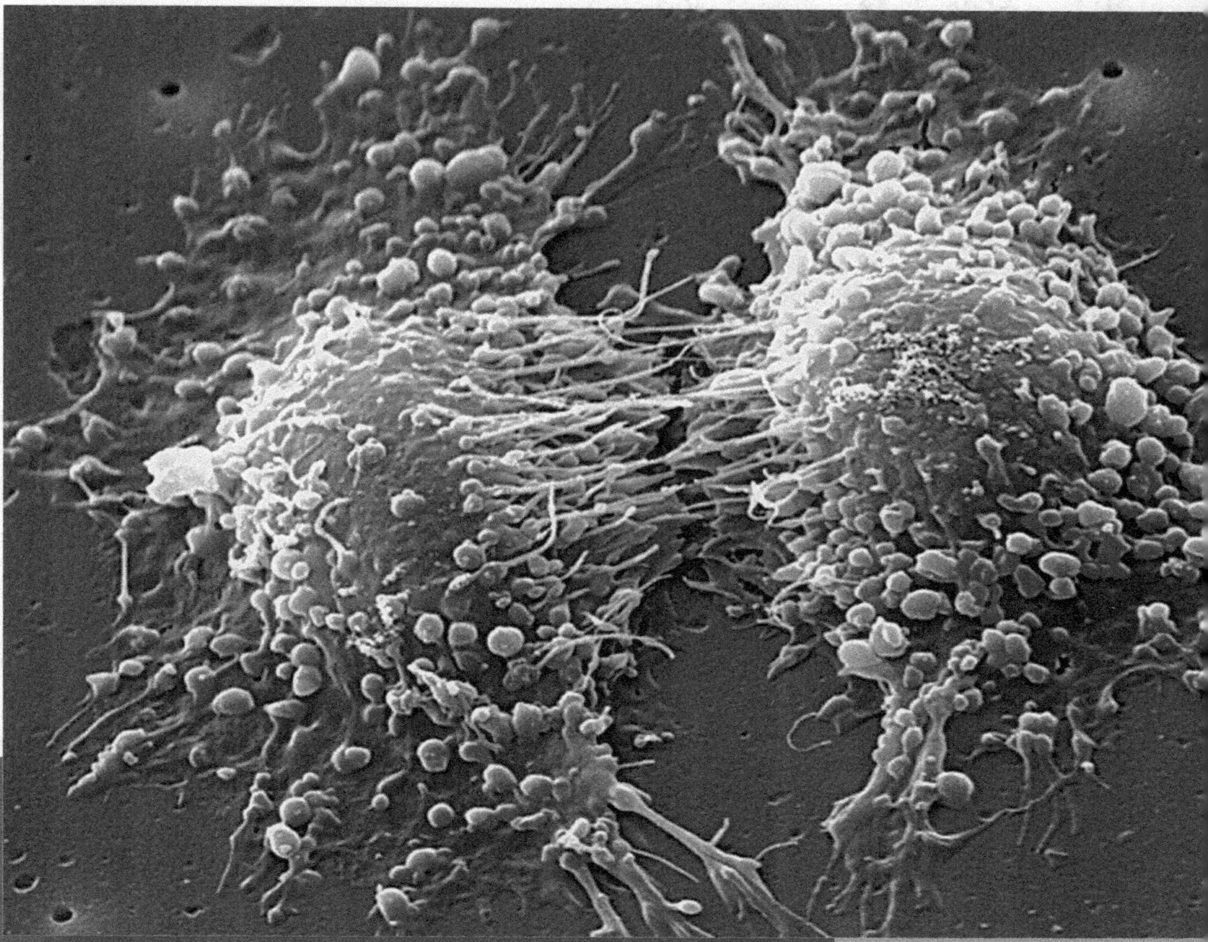

20

# Cancer and Regulation of the Cell Cycle

## CHAPTER CONCEPTS

- Cancer is a group of genetic diseases affecting fundamental aspects of cellular function, including DNA repair, the cell cycle, apoptosis, differentiation, cell migration, and cell–cell contact.
- Most cancer-causing mutations occur spontaneously in somatic cells; only about 1 percent of cancers have a hereditary component.
- Mutations in cancer-related genes lead to abnormal proliferation of cells and loss of control over how cells spread and invade surrounding tissues.
- The development of cancer is a multistep process requiring mutations in genes controlling many aspects of cell proliferation and metastasis.
- Cancer cells show high levels of genomic instability, leading to the accumulation of multiple mutations in cancer-related genes.
- Mutations in proto-oncogenes and tumor-suppressor genes contribute to the development of cancers.
- Oncogenic viruses introduce oncogenes into infected cells and stimulate cell proliferation.
- Environmental agents contribute to cancer by damaging DNA.

Cancer is the second leading cause of death in Western countries, surpassed only by heart disease. It strikes people of all ages, and one out of three people will experience a cancer diagnosis sometime in his or her lifetime. Each year, more than 1 million cases of cancer are diagnosed in the United States and more than 500,000 people die from the disease (Table 20.1).

Over the last 30 years, scientists have discovered that cancer is a genetic disease, characterized by an interplay of mutant forms of oncogenes and tumor-suppressor genes leading to the uncontrolled growth and spread of cancer cells. While some of these mutations may be inherited, as we will see, most mutations that lead to cancer occur in somatic cells that then divide and form tumors. Completion of the Human Genome Project is opening the door to a wealth of new information about the mutations that trigger a cell to become cancerous. This new understanding of cancer genetics is also leading to new gene-specific treatments, some of which are available and others that are now entering clinical trials. Some scientists predict that gene therapies will replace chemotherapies within the next 25 years.

The goal of this chapter is to present an overview of our current understanding of the nature and causes of cancer. As we will see, cancer arises from mutations in genes controlling many basic aspects of cellular function. We will examine the relationship between genes and cancer, and consider how mutations, chromosomal changes, viruses, and environmental agents play roles in the development of cancer.

## 20.1 Cancer Is a Genetic Disease That Arises at the Level of Somatic Cells

Perhaps the most significant development in understanding the causes of cancer has been the realization that cancer is a genetic disease. Genomic alterations that are associated with cancer range from single-nucleotide substitutions to large-scale chromosome rearrangements, amplifications, and deletions (Figure 20–1). However, unlike other genetic diseases, cancer is caused by mutations that occur predominantly in somatic cells. Only about 1 percent of cancers are associated with germ-line mutations that increase a person's susceptibility to certain types of cancer. Another important difference between cancer and other genetic diseases is that cancers rarely arise from a single mutation, but instead result from the accumulation of many mutations—as many as six to twelve. The mutations that lead to cancer affect multiple cellular functions, including repair of DNA damage, cell division, programmed cell death, cellular differentiation, migratory behavior, and cell–cell contact.

### What Is Cancer?

Clinically, cancer is defined as a large number (up to a hundred) of complex diseases that behave differently depending on the cell types from which they originate. Cancers vary in age of onset, growth rate, invasiveness, prognosis, and responsiveness to treatment. However, at the molecular level, all cancers exhibit common characteristics that unite them as a family.

All cancer cells share two fundamental properties: (1) unregulated **cell proliferation,** characterized by abnormal cell growth and division, and (2) **metastasis,** a process that allows cancer cells to spread and invade other parts of the body. In normal cells, cell proliferation and invasion are tightly controlled by genes that are expressed appropriately in time and place. In cancer cells, many of these genes are either mutated or are expressed inappropriately.

It is this combination of uncontrolled cell proliferation and metastatic spread that makes cancer cells dangerous. When a cell simply loses genetic control over its growth and division, it may give rise to a multicellular mass, a **benign tumor.** Such a tumor can often be removed by surgery and may cause no serious harm. However, if

TABLE 20.1

**Probability of Developing Invasive Cancers in the United States**

| Cancer Site | Gender | Age<br>Birth to 39 | 40–59 | 60–69 | 70 and over | Birth–Death |
|---|---|---|---|---|---|---|
| All sites | Male | 1 in 70 | 1 in 12 | 1 in 6 | 1 in 3 | 1 in 2 |
| | Female | 1 in 49 | 1 in 11 | 1 in 9 | 1 in 4 | 1 in 3 |
| Breast | Female | 1 in 210 | 1 in 25 | 1 in 27 | 1 in 15 | 1 in 8 |
| Prostate | Male | <1 in 10,000 | 1 in 39 | 1 in 14 | 1 in 7 | 1 in 6 |
| Lung, bronchus | Male | 1 in 3146 | 1 in 92 | 1 in 38 | 1 in 15 | 1 in 12 |
| | Female | 1 in 2779 | 1 in 117 | 1 in 54 | 1 in 22 | 1 in 16 |
| Colon, rectum | Male | 1 in 1342 | 1 in 107 | 1 in 60 | 1 in 20 | 1 in 17 |
| | Female | 1 in 1469 | 1 in 138 | 1 in 86 | 1 in 22 | 1 in 19 |
| Melanoma of skin | Male | 1 in 775 | 1 in 187 | 1 in 178 | 1 in 76 | 1 in 49 |
| | Female | 1 in 467 | 1 in 237 | 1 in 347 | 1 in 163 | 1 in 73 |
| Non-Hodgkin's lymphoma | Male | 1 in 735 | 1 in 222 | 1 in 176 | 1 in 64 | 1 in 47 |
| | Female | 1 in 1200 | 1 in 313 | 1 in 229 | 1 in 77 | 1 in 55 |

SOURCE: American Cancer Society. *Cancer Facts and Figures 2006*. Atlanta: American Cancer Society, Inc.

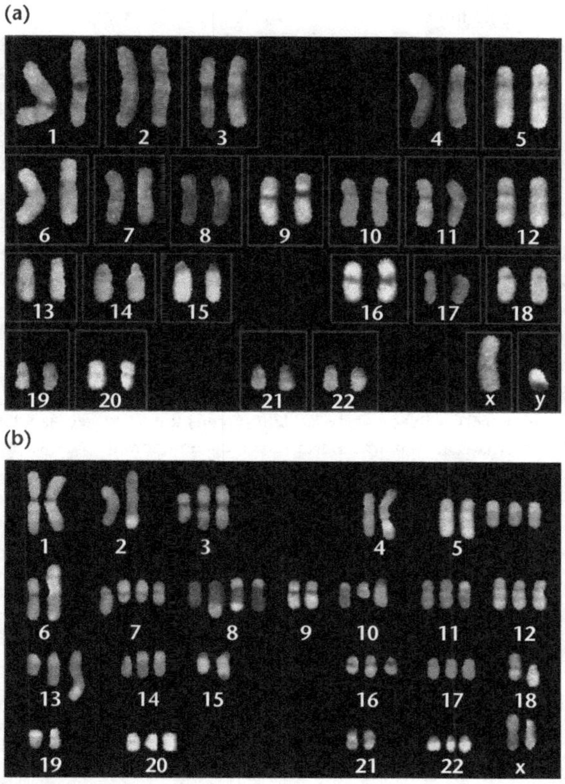


**FIGURE 20-1** (a) Spectral karyotype of a normal cell. (b) Karyotype of a cancer cell showing translocations, deletions, and aneuploidy—characteristic features of cancer cells.

cells in the tumor also acquire the ability to break loose, enter the bloodstream, invade other tissues, and form secondary tumors elsewhere in the body, they become malignant. **Malignant tumors** are difficult to treat and may become life threatening. As we will see later in the chapter, multiple steps and genetic mutations are required to convert a benign tumor into a dangerous malignant tumor.

## The Clonal Origin of Cancer Cells

Although malignant tumors may contain billions of cells, and may invade and grow in numerous parts of the body, all cancer cells in the primary and secondary tumors are clonal. That is, they originated from a common ancestral cell that accumulated numerous specific mutations. This concept is important for understanding the molecular causes of cancer and has implications for its diagnosis.

Numerous data support the concept of cancer clonality. For example, reciprocal chromosomal translocations are characteristic of many cancers, including leukemias and lymphomas (two cancers involving white blood cells). Cancer cells from patients with **Burkitt's lymphoma** show reciprocal translocations between chromosome 8, with translocation breakpoints at or near the *c-myc* gene, and chromosomes 2, 14, or 22, with translocation breakpoints at or near one of the immunoglobulin genes. Each Burkitt's lymphoma patient exhibits unique breakpoints in his or her *c-myc* and immunoglobulin gene DNA sequences; however, all lymphoma cells within that patient contain identical translocation breakpoints. This demonstrates that all cancer cells in each case of Burkitt's lymphoma arise from a single cell, and this cell passes on its genetic aberrations to its progeny.

Another demonstration that cancer cells are clonal is their pattern of X-chromosome inactivation. As explained in Chapter 7, female humans are mosaic in terms of X-chromosome inactivation, with some cells containing an inactivated paternal X chromosome and other cells containing an inactivated maternal X chromosome. X-chromosome inactivation takes place early in development and occurs at random. However, in a given female, all cancer cells within a tumor, whether it is primary or metastatic, contain the same inactivated X chromosome. This supports the idea that all the cancer cells in that patient arose from a common ancestral cell.

## Cancer As a Multistep Process, Requiring Multiple Mutations

Although cancer is a genetic disease initiated by mutations that lead to uncontrolled cell proliferation and metastasis, a single mutation is not sufficient to transform a normal cell into a tumor-forming (tumorigenic), malignant cell. If it were sufficient, then cancer would be far more prevalent than it is. In humans, mutations occur spontaneously at a rate of about $10^{-6}$ per gene, per cell division, mainly due to the intrinsic error rates of DNA replication. As there are approximately $10^{16}$ cell divisions in a human body during a lifetime, a person might suffer up to $10^{10}$ mutations per gene anywhere in the body, during his or her lifetime. However, only about one person in three will suffer from cancer.

The phenomenon of age-related cancer is another indication that cancer develops from the accumulation of several mutagenic events in a single cell. The incidence of most cancers rises exponentially with age. If a single mutation were sufficient to convert a normal cell to a malignant one, then cancer incidence would appear to be independent of age. The age-related incidence of cancer suggests that up to 10 independent mutations, occurring randomly and with a low probability, are necessary before a cell is transformed into a malignant cancer cell. Another indication that cancer is a multistep process is the delay that occurs between exposure to **carcinogens** (cancer-causing agents) and the appearance of the cancer. For example, an incubation period of five to eight years separated exposure of people to the radiation of the atomic explosions at Hiroshima and Nagasaki and the onset of leukemias.

Another indication of the multistep nature of cancer development is the observation that cancers often develop in progressive steps, beginning with mildly aberrant cells and progressing to cells that are increasingly tumorigenic and malignant. This progressive nature of cancer is illustrated by the development of colon cancer, as discussed in Section 20.6.

Each step in **tumorigenesis,** the development of a malignant tumor, appears to be the result of one or more genetic alterations that release the cells progressively from the controls that normally restrain proliferation and malignancy. As we will discover in subsequent sections of this chapter, the genes that undergo mutations leading to cancer (called oncogenes and tumor-suppressor genes) are those that control DNA damage repair, the cell cycle, cell–cell contact, and programmed cell death.

We will now investigate each of these fundamental processes, the genes that control them, and how mutations in these genes may lead to cancer.

20.2

## Cancer Cells Contain Genetic Defects Affecting Genomic Stability, DNA Repair, and Chromatin Modifications

Cancer cells show higher than normal rates of mutation, chromosomal abnormalities, and genomic instability. Many researchers believe that the fundamental defect in cancer cells is a derangement of the cells' normal ability to repair DNA damage. This loss of genomic integrity leads to a general increase in the mutation rate in every gene, including specific genes that control aspects of cell proliferation, programmed cell death, and cell–cell contact. In turn, the accumulation of mutations in genes controlling these processes leads to cancer. The high level of genomic instability seen in cancer cells is known as the **mutator phenotype**. In addition, recent research has revealed that cancer cells contain aberrations in the types and locations of chromatin modifications, particularly DNA methylation patterns.

### Genomic Instability and Defective DNA Repair

Genomic instability in cancer cells is characterized by the presence of gross defects such as translocations, aneuploidy, chromosome loss, DNA amplification, and chromosome deletions (Figures 20–1 and 20–2). Cancer cells that are grown in cultures in the lab also show a great deal of genomic instability—duplicating, losing, and translocating chromosomes or parts of chromosomes. Often cancer cells show specific chromosomal defects that are used to diagnose the type and stage of the cancer. For example, leukemic white blood cells from patients with **chronic myelogenous leukemia (CML)** bear a specific translocation, in which the *C-ABL* gene on chromosome 9 is translocated into the *BCR* gene on chromosome 22. This translocation creates a structure known as the **Philadelphia chromosome** (Figure 20–3). The *BCR-ABL* fusion gene codes for a chimeric BCR-ABL protein. The normal ABL protein is a **protein kinase** that acts within signal transduction pathways, transferring growth factor signals from the external environment to the nucleus. The BCR-ABL protein is an abnormal signal transduction molecule in CML cells, constantly stimulating these cells to proliferate even in the absence of external growth signals.

(a) Double minutes (b) Heterogeneous staining region

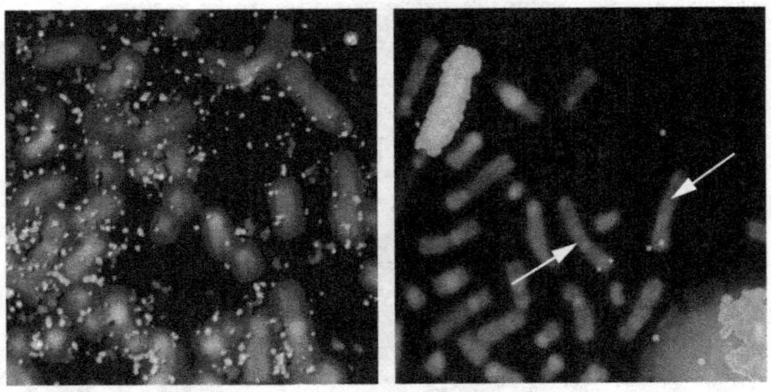

FIGURE 20–2 DNA amplifications in neuroblastoma cells. (a) Two cancer genes (*MYCN* in red and *MDM2* in green) are amplified as small DNA fragments that remain separate from chromosomal DNA within the nucleus. These units of amplified DNA are known as double minute chromosomes. Normal chromosomes are stained blue. (b) Multiple copies of the *MYCN* gene are amplified within one large region called a heterogeneous staining region (green). Single copies of the *MYCN* gene are visible as green dots at the ends of the normal parental chromosomes (white arrows). Normal chromosomes are stained red.

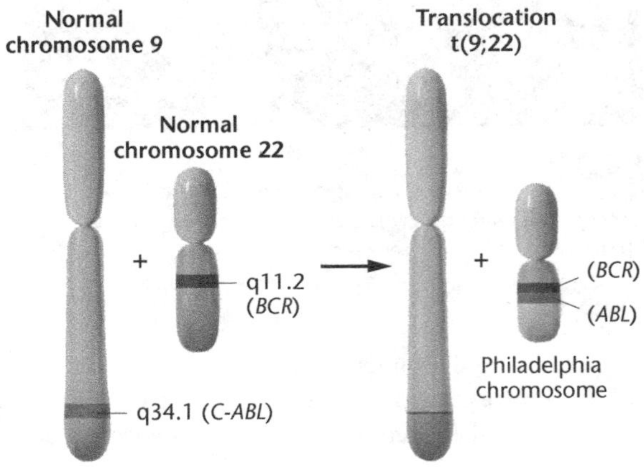


FIGURE 20–3 A reciprocal translocation involving the long arms of chromosomes 9 and 22 results in the formation of a characteristic aberrant chromosome, the Philadelphia chromosome, which is associated with chronic myelogenous leukemia (CML). The t(9;22) translocation results in the fusion of the *C-ABL* proto-oncogene on chromosome 9 with the *BCR* gene on chromosome 22. The fusion protein is a powerful hybrid molecule that allows cells to escape control of the cell cycle, contributing to the development of CML.

In keeping with the concept of the cancer mutator phenotype, a number of inherited cancers are caused by defects in genes that control DNA repair. For example, xeroderma pigmentosum (XP) is a rare hereditary disorder that is characterized by extreme sensitivity to ultraviolet light and other carcinogens. Patients with XP often develop skin cancer. Cells from patients with XP are defective in nucleotide excision repair, with mutations appearing in any one of seven genes whose products are necessary to carry out DNA repair.

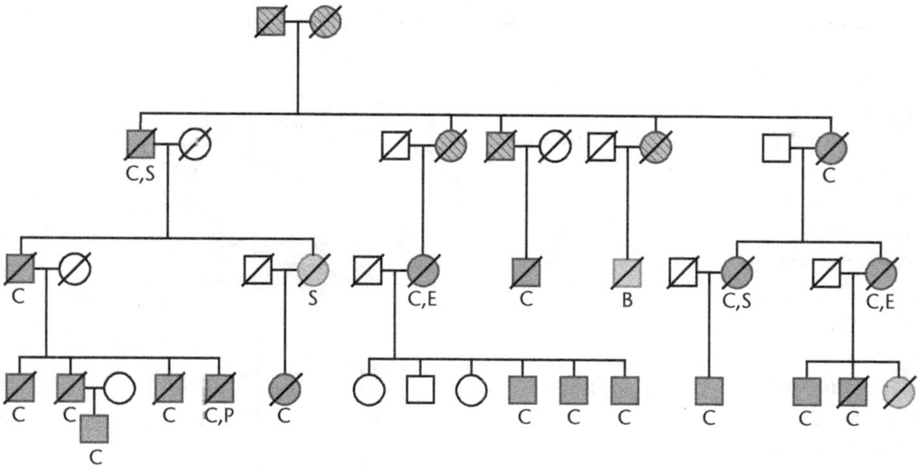


FIGURE 20-4 Pedigree of a family with HNPCC. In familial HNPCC, at least three relatives in two generations have been diagnosed with colon cancer, with one relative diagnosed at less than 50 years of age. Colon cancer, C; stomach cancer, S; endometrial cancer, E; pancreatic cancer, P; bladder/urinary cancer, B. Blue symbols indicate family members with colon cancer; diagonal stripes mean that diagnosis is uncertain; orange symbols indicate other tumors. Symbols with slashes indicate deceased individuals. Reprinted with permission from Aaltonen et al. Clues to the pathogenesis of familial colorectal cancer. *Science* 260: 812–816. Copyright 1993 AAAS.

XP cells are impaired in their ability to repair DNA lesions such as thymine dimers induced by UV light. The relationship between XP and genes controlling nucleotide excision repair is also described in Chapter 16.

Another hereditary cancer, **hereditary nonpolyposis colorectal cancer (HNPCC)**, is also caused by mutations in genes controlling DNA repair. HNPCC is an autosomal dominant syndrome, affecting about one in every 200 to 1000 people (Figure 20–4). Patients affected by HNPCC have an increased risk of developing colon, ovary, uterine, and kidney cancers. Cells from patients with HNPCC show higher than normal mutation rates and genomic instability. At least eight genes are associated with HNPCC, and four of these genes control aspects of DNA mismatch repair. Inactivation of any of these four genes—*MSH2, MSH6, MLH1*, and *MLH3*—causes a rapid accumulation of genome-wide mutations and the subsequent development of colorectal and other cancers.

The observation that hereditary defects in genes controlling nucleotide excision repair and DNA mismatch repair lead to high rates of cancer lends support to the idea that the mutator phenotype is a significant contributor to the development of cancer.

## Chromatin Modifications and Cancer Epigenetics

The field of cancer epigenetics is providing new perspectives on the genetics of cancer. **Epigenetics** is the study of factors that affect gene expression in a heritable way but that do not alter the nucleotide sequence of DNA. DNA methylation and histone modifications such as acetylation and phosphorylation are examples of epigenetic modifications. The genomic patterns and locations of these modifications can be inherited and affect gene expression, as we saw earlier, in Chapter 18. For example, DNA methylation is thought to be responsible for the gene silencing associated with parental imprinting, heterochromatin gene repression, and X-chromosome inactivation.

Cancer cells contain major alterations in DNA methylation. Overall, there is much less DNA methylation in cancer cells than in normal cells. At the same time, the promoters of some genes are hypermethylated in cancer cells. These changes are thought to result in the release of transcription repression over the bulk of genes that would be silent in normal cells—including cancer-causing proto-oncogenes (introduced in Section 20.4), while at the same time repressing transcription of genes that would normally regulate functions such as efficient DNA repair, control of the cell cycle, and cellular differentiation. For example, the DNA repair genes ***MLH1*** and ***BRCA1*** are transcriptionally silenced by hypermethylation in many cancer cells, likely contributing to the mutator phenotype. Virtually every tumor type that has been examined shows hypermethylation on the promoters of genes whose products control growth, DNA repair, and cell invasion. Genome-wide screens of methylation patterns reveal that tumor subtypes bear specific patterns of gene methylation. These methylation profiles are now used to help diagnose tumors and predict their course.

Histone modifications are also disrupted in cancer cells. Genes that encode histone acetylases, deacetylases, methyltransferases, and demethylases are often mutated or aberrantly expressed in cancer cells. The large numbers of epigenetic abnormalities in tumors have prompted some scientists to speculate that there may be more epigenetic defects in cancer cells than there are gene mutations. In addition, because epigenetic modifications are reversible, cancers may lend themselves to epigenetic-based therapies. Although the field of cancer epigenetics is still in its infancy, it has already provided major insights into tumorigenesis as well as new clinical applications.

**NOW SOLVE THIS**

Problem 17 on page 529 asks you to consider how the BCR-ABL hybrid fusion protein, found in leukemic white blood cells of CML patients, could be used as a target for cancer therapy.

■ HINT: *Most cancer therapies, including radiation and chemotherapies, aim to kill cells that are constantly dividing. However, many normal cells in the body also divide and are killed by cancer therapies, leading to side effects. The BCR-ABL fusion protein is found only in CML white blood cells. As you try to answer this problem, you may wish to learn more about the drug Gleevec (see http://www.nci.nih.gov/newscenter/qandagleevec). Also read the Genetics, Technology, and Society essay on page 526, for a discussion of targeted CML therapies.*

20.3

## Cancer Cells Contain Genetic Defects Affecting Cell-Cycle Regulation

One of the fundamental aberrations in all cancer cells is a loss of control over cell proliferation. Cell proliferation is the process of cell growth and division that is essential for all development and tissue repair in multicellular organisms. Although some cells, such as epidermal cells of the skin or blood-forming cells in the bone marrow, continue to grow and divide throughout the organism's lifetime, most cells in adult multicellular organisms remain in a nondividing, quiescent, and differentiated state. **Differentiated cells** are those that are specialized for a specific function, such as photoreceptor cells of the retina or muscle cells of the heart. The most extreme examples of nonproliferating cells are nerve cells, which divide little, if at all, even to replace damaged tissue. In contrast, many differentiated cells, such as those in the liver and kidney, are able to grow and divide when stimulated by extracellular signals and growth factors. In this way, multicellular organisms are able to replace dead and damaged tissue. However, the growth and differentiation of cells must be strictly regulated; otherwise, the integrity of organs and tissues would be compromised by the presence of inappropriate types and quantities of cells. Normal regulation over cell proliferation involves a large number of gene products that control steps in the cell cycle, programmed cell death, and the response of cells to external growth signals. In cancer cells, many of the genes that control these functions are mutated or aberrantly expressed, leading to uncontrolled cell proliferation.

In this section, we will review steps in the cell cycle, some of the genes that control the cell cycle, and how these genes, when mutated, lead to cancer.

### The Cell Cycle and Signal Transduction

The cellular events that occur in sequence from one cell division to the next comprise the **cell cycle** (Figure 20–5). The **interphase** stage of the cell cycle is the interval between mitotic divisions. During this time, the cell grows and replicates its DNA. During **G1**, the cell prepares for DNA synthesis by accumulating the enzymes and molecules required for DNA replication. G1 is followed by **S phase**, during which the cell's chromosomal DNA is replicated. During **G2**, the cell continues to grow and prepare for division. During **M phase**, the duplicated chromosomes condense, sister chromosomes separate to opposite poles, and the cell divides in two.

In early to mid-G1, the cell makes a decision either to enter the next cell cycle or to withdraw from the cell cycle into quiescence. Continuously dividing cells do not exit the cell cycle but proceed through G1, S, G2, and M phases; however, if the cell receives signals to stop growing, it enters the **G0** phase of the cell cycle. During G0, the cell remains metabolically active but does not grow or divide. Most differentiated cells in multicellular organisms can remain in

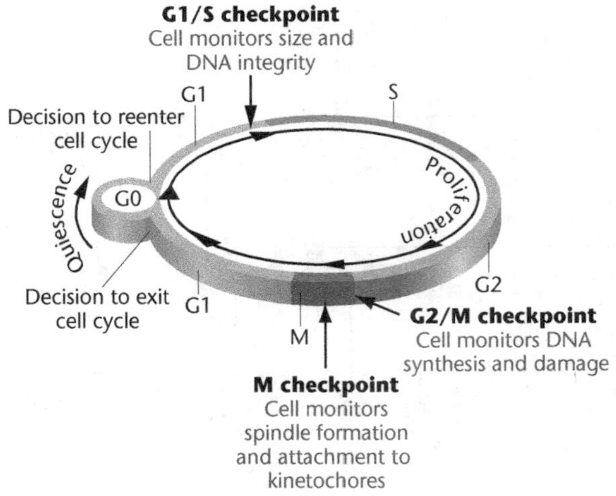


FIGURE 20–5 Checkpoints and proliferation decision points monitor the progress of the cell through the cell cycle.

WEB TUTORIAL 20.1 CELL CYCLE REGULATION

this G0 phase indefinitely. Some, such as neurons, never reenter the cell cycle. In contrast, cancer cells are unable to enter G0, and instead, they continuously cycle. Their *rate* of proliferation is not necessarily any greater than that of normal proliferating cells; however, they are not able to become quiescent at the appropriate time or place.

Cells in G0 can often be stimulated to reenter the cell cycle by external growth signals. These signals are delivered to the cell by molecules such as growth factors and hormones that bind to cell-surface receptors, which then relay the signal from the plasma membrane to the cytoplasm. The process of transmitting growth signals from the external environment to the cell nucleus is known as **signal transduction.** Ultimately, signal transduction initiates a program of gene expression that propels the cell out of G0 back into the cell cycle. Cancer cells often have defects in signal transduction pathways. Sometimes, abnormal signal transduction molecules send continuous growth signals to the nucleus even in the absence of external growth signals. In addition, malignant cells may not respond to external signals from surrounding cells—signals that would normally inhibit cell proliferation within a mature tissue.

### Cell-Cycle Control and Checkpoints

In normal cells, progress through the cell cycle is tightly regulated, and each step must be completed before the next step can begin. There are at least three distinct points in the cell cycle at which the cell monitors external signals and internal equilibrium before proceeding to the next stage. These are the **G1/S**, the **G2/M**, and **M checkpoints** (Figure 20–5). At the G1/S checkpoint, the cell monitors its size and determines whether its DNA has been damaged. If the cell has not achieved an adequate size, or if the DNA has been damaged, further progress through the cell cycle is halted until these conditions are corrected. If cell size and DNA integrity are normal, the G1/S checkpoint is traversed, and the cell proceeds to S phase. The second important checkpoint is the G2/M checkpoint, where

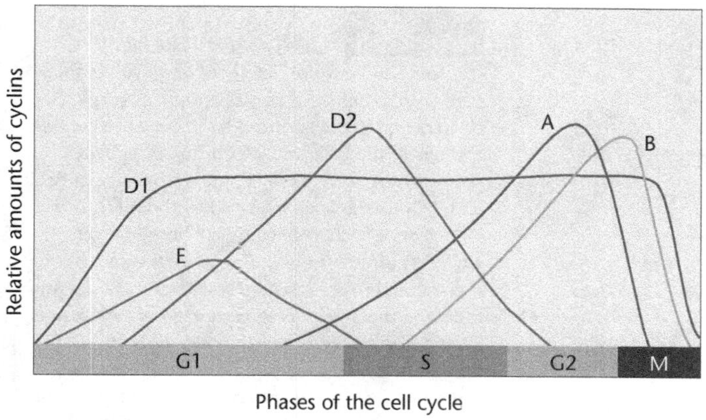


FIGURE 20–6 Relative expression times and amounts of cyclins during the cell cycle. Cyclin D1 accumulates early in G1 and is expressed at a constant level through most of the cycle. Cyclin E accumulates in G1, reaches a peak, and declines by mid-S phase. Cyclin D2 begins accumulating in the last half of G1, reaches a peak just after the beginning of S, and then declines by early G2. Cyclin A appears in late G1, accumulates through S phase, peaks at the G2/M transition, and is rapidly degraded. Cyclin B peaks at the G2/M transition and declines rapidly in M phase.

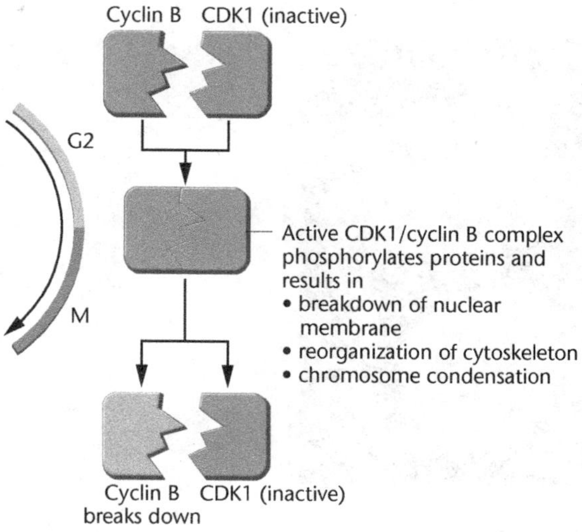


FIGURE 20–7 Transition from G2 to M phase is controlled by CDK1 and cyclin B. These molecules interact to form a complex that adds phosphate groups to cellular components. The phosphorylated components in turn bring about the structural and biochemical changes necessary for mitosis (M phase).

physiological conditions in the cell are monitored prior to mitosis. If DNA replication or repair of any DNA damage has not been completed, the cell cycle arrests until these processes are complete. The third major checkpoint occurs during mitosis and is called the M checkpoint. At this checkpoint, both the successful formation of the spindle-fiber system and the attachment of spindle fibers to the kinetochores associated with the centromeres are monitored. If spindle fibers are not properly formed or attachment is inadequate, mitosis is arrested.

In addition to regulating the cell cycle at checkpoints, the cell controls progress through the cell cycle by means of two classes of proteins: **cyclins** and **cyclin-dependent kinases (CDKs)**. The cell synthesizes and destroys cyclin proteins in a precise pattern during the cell cycle (Figure 20–6). When a cyclin is present, it binds to a specific CDK, triggering activity of the CDK/cyclin complex. The CDK/cyclin complex then selectively phosphorylates and activates other proteins that in turn bring about the changes necessary to advance the cell through the cell cycle. For example, in G1 phase, CDK4/cyclin D complexes activate proteins that stimulate transcription of genes whose products (such as DNA polymerase $\delta$ and DNA ligase) are required for DNA replication during S phase. Another CDK/cyclin complex, CDK1/cyclin B, phosphorylates a number of proteins that bring about the events of early mitosis, such as nuclear membrane breakdown, chromosome condensation, and cytoskeletal reorganization (Figure 20–7). Mitosis can only be completed, however, when cyclin B is degraded and the protein phosphorylations characteristic of M phase are reversed. Although a large number of different protein kinases exist in cells, only a few are involved in cell-cycle regulation.

Both cell-cycle checkpoints and cell-cycle control molecules are genetically regulated. In general, the cell cycle is regulated by an interplay of genes whose products either promote or suppress cell division. Mutation or misexpression of any of the genes controlling the cell cycle can contribute to the development of cancer. For example, if genes that control the G1/S or G2/M checkpoints are mutated, the cell may continue to grow and divide without repairing DNA damage. As these cells continue to divide, they accumulate mutations in genes whose products control cell proliferation or metastasis. Similarly, if genes that control progress through the cell cycle, such as those that encode the cyclins, are expressed at the wrong time or at incorrect levels, the cell may grow and divide continuously and may be unable to exit the cell cycle into G0. The result in both cases is that the cell loses control over proliferation and is on its way to becoming cancerous.

## Control of Apoptosis

As already described, if DNA replication, repair, or chromosome assembly is aberrant, normal cells halt their progress through the cell cycle until the condition is corrected. This reduces the number of mutations and chromosomal abnormalities that accumulate in normal proliferating cells. However, if DNA or chromosomal damage is so severe that repair is impossible, the cell may initiate a second line of defense—a process called **apoptosis**, or **programmed cell death**. Apoptosis is a genetically controlled process whereby the cell commits suicide. Besides its role in preventing cancer, apoptosis is also initiated during normal multicellular development in order to eliminate certain cells that do not contribute to the final adult organism. The steps in apoptosis are the same for damaged cells and for cells

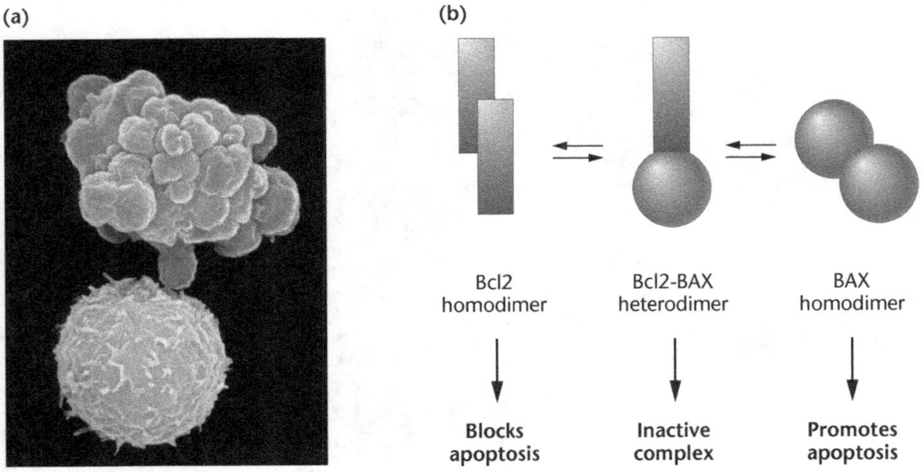


**FIGURE 20–8** (a) A normal white blood cell (bottom) and a white blood cell undergoing apoptosis (top). Apoptotic bodies appear as grape-like clusters on the cell surface. (b) The relative concentrations of the Bcl2 and BAX proteins regulate apoptosis. A normal cell contains a balance of Bcl2 and BAX, which form inactive heterodimers. A relative excess of Bcl2 results in Bcl2 homodimers, which prevent apoptosis. Cancer cells with Bcl2 overexpression are resistant to chemotherapies and radiation therapies. A relative excess of BAX results in BAX homodimers, which induce apoptosis. In normal cells, activated p53 protein induces transcription of the *BAX* gene and inhibits transcription of the *Bcl2* gene, leading to cell death. In many cancer cells, p53 is defective, preventing the apoptotic pathway from removing the cancer cells.

being eliminated during development: nuclear DNA becomes fragmented, internal cellular structures are disrupted, and the cell dissolves into small spherical structures known as apoptotic bodies [Figure 20–8(a)]. In the final step, the apoptotic bodies are engulfed by the immune system's phagocytic cells. A series of proteases called **caspases** are responsible for initiating apoptosis and for digesting intracellular components.

Apoptosis is genetically controlled in that regulation of the levels of specific gene products such as the Bcl2 and BAX proteins shown in Figure 20–8(b) can trigger or prevent apoptosis. By removing damaged cells, programmed cell death reduces the number of mutations that are passed to the next generation, including those in cancer-causing genes. The same genes that control cell-cycle checkpoints can trigger apoptosis. As we will see, these genes are mutated in many cancers. As a result of the mutation or inactivation of these checkpoint genes, the cell is unable to repair its DNA or undergo apoptosis. This inability leads to the accumulation of even more mutations in genes that control growth, division, and metastasis.

## 20.4 Many Cancer-Causing Genes Disrupt Control of the Cell Cycle

As discussed previously, cancer cells are susceptible to accumulating mutations. The question of how many mutations and genetic abnormalities are present in cancer cells has been addressed in recent studies including genomic screens of tumor and normal cells. In one study, researchers determined the DNA sequence of 13,000 genes in breast and colorectal cancer cells, revealing that about 90 genes are mutated in each cancer. Most of these mutated genes probably do not contribute to the cancer phenotype; however, about 11 mutated genes in each tumor likely do contribute, as they are genes affecting aspects of cell growth and invasion. These numbers may be underestimates, because defects originating from chromosomal aberrations and epigenetic changes were not assessed.

Two general categories of cancer-causing genes are mutated or misexpressed in cancer cells—the proto-oncogenes and the tumor-suppressor genes (Table 20.2). **Proto-oncogenes** encode transcription factors that stimulate expression of other genes, signal transduction molecules that stimulate cell division, and cell-cycle regulators that move the cell through the cell cycle. Their products are important for normal cell functions, especially cell growth and division. When normal cells become quiescent and cease division, they repress the expression or activity of most proto-oncogene products. In cancer cells, one or more proto-oncogenes are altered in such a way that their activities cannot be controlled in a normal fashion. This is sometimes due to a mutation in the proto-oncogene resulting in a protein product that acts abnormally. In other cases, proto-oncogenes are overexpressed or cannot be transcriptionally repressed at the correct time. In these cases, the proto-oncogene is continually in an "on" state, which may constantly stimulate the cell to divide. When a proto-oncogene is mutated or aberrantly expressed and contributes to the development of cancer, it is known as an **oncogene**—a cancer-causing gene. Oncogenes are proto-oncogenes that have experienced a gain-of-function alteration. As a result, only one allele of a proto-oncogene needs to be mutated or misexpressed in order to trigger uncontrolled growth. Hence, oncogenes confer a dominant cancer phenotype.

**Tumor-suppressor genes** are genes whose products normally regulate cell-cycle checkpoints and initiate the process of apoptosis. In normal cells, proteins encoded by tumor-suppressor genes halt progress through the cell cycle in response to DNA damage or growth-suppression signals from the extracellular environment. When tumor-suppressor genes are mutated or inactivated, cells are unable to respond normally to cell-cycle checkpoints, or are unable to undergo programmed cell death if DNA damage is extensive.

TABLE 20.2

**Some Proto-oncogenes and Tumor-suppressor Genes**

| Proto-oncogene | Normal Function | Alteration in Cancer | Associated Cancers |
|---|---|---|---|
| *Ha-ras* | Signal transduction molecule; binds GTP/GDP | Point mutations | Colorectal, bladder, many other types |
| *c-erbB* | Transmembrane growth factor receptor | Gene amplification, point mutations | Glioblastomas, breast cancer, cervix |
| *c-myc* | Transcription factor; regulates cell cycle, differentiation, apoptosis | Translocation, amplification, point mutations | Lymphomas, leukemias, lung cancer, many other types |
| *c-kit* | Tyrosine kinase; signal transduction | Mutation | Sarcomas |
| *RARα* | Hormone-dependent transcription factor; differentiation | Chromosomal translocations with PML gene, fusion product | Acute promyelocytic leukemia |
| *E6* | Human papillomavirus encoded oncogene; inactivates p53 | HPV infection | Cervical cancer |
| *Cyclins* | Bind to CDKs, regulate cell cycle | Gene amplification, overexpression | Lung, esophagus, many other types |
| *CDK2, 4* | Cyclin-dependent kinases; regulate cell-cycle phases | Overexpression, mutation | Bladder, breast, many other types |
| **Tumor-suppressor** | **Normal Function** | **Alteration in Cancer** | **Associated Cancers** |
| *p53* | Cell-cycle checkpoints, apoptosis | Mutation, inactivation by viral oncogene products | Brain, lung, colorectal, breast, many other types |
| *RB1* | Cell-cycle checkpoints, binds E2F | Mutation, deletion, inactivation by viral oncogene products | Retinoblastoma, osteosarcoma, many other types |
| *APC* | Cell–cell interaction | Mutation | Colorectal, brain, thyroid |
| *Bcl2* | Apoptosis regulation | Overexpression blocks apoptosis | Lymphomas, leukemias |
| *BRCA2* | DNA repair | Point mutations | Breast, ovarian, prostate cancers |

This leads to a further increase in mutations and to the inability of the cell to leave the cell cycle when it should become quiescent. When both alleles of a tumor-suppressor gene are inactivated, and other changes in the cell keep it growing and dividing, cells may become tumorigenic.

The following are examples of proto-oncogenes and tumor-suppressor genes that contribute to cancer when mutated. More than 300 oncogenes and tumor-suppressor genes are now known, and more will likely be discovered as cancer research continues.

## The *ras* Proto-oncogenes

Some of the most frequently mutated genes in human tumors are those in the ***ras* gene family**. These genes are mutated in more than 30 percent of human tumors. The *ras* gene family encodes signal transduction molecules that are associated with the cell membrane and regulate cell growth and division. Ras proteins normally transmit signals from the cell membrane to the nucleus, stimulating the cell to divide in response to external growth factors (Figure 20–9). Ras proteins alternate between an inactive (switched off) and an active (switched on) state by binding either guanosine diphosphate (GDP) or guanosine triphosphate (GTP). When a cell encounters a growth factor (such as platelet-derived growth factor or epidermal growth factor), growth factor receptors on the cell membrane bind to the growth factor, resulting in autophosphorylation of the cytoplasmic portion of the growth factor receptor. This causes recruitment of proteins known as nucleotide exchange factors to the plasma membrane. These nucleotide exchange factors cause Ras to release GDP and bind GTP, thereby activating Ras. The active, GTP-bound form of Ras then sends its signals through cascades of protein phosphorylations in the cytoplasm. The end-point of these cascades is activation of nuclear transcription factors that stimulate expression of genes whose products drive the cell from quiescence into the cell cycle. Once Ras has sent its signals to the nucleus, it hydrolyzes GTP to GDP and becomes inactive. Mutations that convert

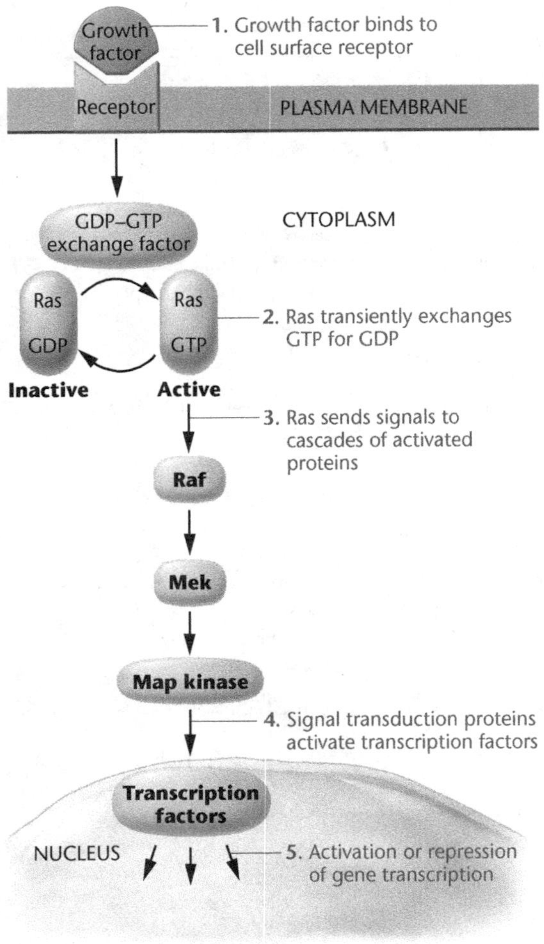


FIGURE 20–9 A signal transduction pathway mediated by Ras.

the proto-oncogene *ras* to an oncogene prevent the Ras protein from hydrolyzing GTP to GDP and hence freeze the Ras protein into its "on" conformation, constantly stimulating the cell to divide.

## The *cyclin D1* and *cyclin E* Proto-oncogenes

As we learned earlier in this chapter, cyclins are synthesized and degraded throughout the cell cycle. They form complexes with CDK molecules, and regulate their activities. These CDK/cyclin complexes are important regulators of each phase of the cell cycle. Several cyclin genes are known to be associated with the development of cancer. For example, the gene encoding **cyclin D1** is amplified in a number of cancers, including those of the breast, bladder, lung, and esophagus. DNA amplification creates multiple copies of the *cyclin D1* gene and hence leads to higher than normal levels of the cyclin D1 protein in these cancer cells. High levels of cyclin D1 protein may contribute to uncontrolled entry into S phase. In other cancers, *cyclin D1* is overexpressed even in the absence of gene amplification. In certain parathyroid tumors and B-cell lymphomas, the *cyclin D1* gene is affected by chromosomal aberrations such as translocations. These alterations to the *cyclin D1* gene may cause it to be expressed abnormally. Similarly, the ***cyclin E*** gene is amplified or overexpressed in some leukemias and cancers of the breast and colon. It is possible that overexpression of these key cell-cycle regulators, or loss of their periodic degradation, keeps cells primed for continuous cell cycling, and they can no longer leave the cell cycle, enter the quiescent G0 phase, or undergo cell differentiation.

## The *p53* Tumor-suppressor Gene

The most frequently mutated gene in human cancers—mutated in more than 50 percent of all cancers—is the ***p53* gene**. This gene encodes a nuclear protein that acts as a transcription factor, repressing or stimulating transcription of more than 50 different genes.

Normally, the p53 protein is continuously synthesized but is rapidly degraded and therefore is present in cells at low levels. In addition, the p53 protein is normally bound to another protein called **Mdm2**, which has several effects on p53. The presence of Mdm2 on the p53 protein tags p53 for degradation and sequesters the transcriptional activation domain of p53. It also prevents the phosphorylations and acetylations that convert the p53 protein from an inactive to an active form. Several types of events bring about rapid increases in the nuclear levels of activated p53 protein. These include chemical damage to DNA, double-stranded breaks in DNA induced by ionizing radiation, and the presence of DNA-repair intermediates generated by exposure of cells to ultraviolet light. In response to these signals, Mdm2 dissociates from p53, making p53 more stable and unmasking its transcription activation domain. Increases in the levels of activated p53 protein also result from increases in protein phosphorylation, acetylation, and other post-translational modifications.

The p53 protein initiates two different responses to DNA damage: either arrest of the cell cycle followed by DNA repair, or apoptosis and cell death if DNA cannot be repaired. Both of these responses are accomplished by p53 acting as a transcription factor that stimulates or represses the expression of genes involved in each response.

In normal cells, p53 can arrest the cell cycle at several phases. To arrest the cell cycle at the G1/S checkpoint, activated p53 protein stimulates transcription of a gene encoding the p21 protein. The p21 protein inhibits the CDK4/cyclin D1 complex, hence preventing the cell from moving from the G1 phase into S phase. Activated p53 protein also regulates expression of genes that retard the progress of DNA replication, thus allowing time for DNA damage to be repaired during S phase. By regulating expression of other genes, activated p53 can block cells at the G2/M checkpoint, if DNA damage occurs during S phase.

Activated p53 can also instruct a damaged cell to commit suicide by apoptosis. It does so by activating the transcription of the *Bax* gene and repressing transcription of the *Bcl2* gene. In normal

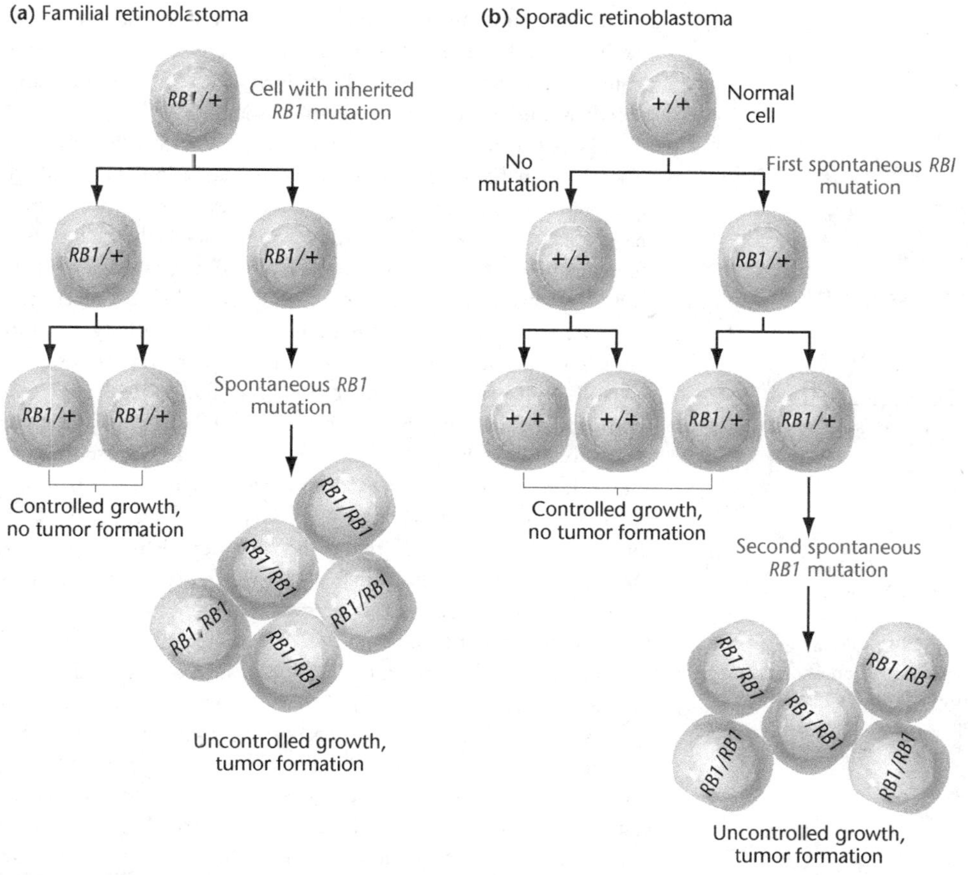


**FIGURE 20–10** (a) In familial retinoblastoma, a mutation in one allele (designated as *RB1*) is inherited and present in all cells. A second mutation in the other allele in any retinal cell contributes to uncontrolled cell growth and tumor formation. (b) In sporadic retinoblastoma, two independent mutations of the wild-type retinoblastoma gene, one in each allele within a single cell, are acquired sequentially, also leading to tumor formation.

cells, the BAX protein is present in a heterodimer with the Bcl2 protein, and the cell remains viable (Figure 20–8). But when the levels of BAX protein increase in response to p53 stimulation of *Bax* gene transcription, BAX homodimers are formed, and these homodimers activate the cellular changes that lead to cellular self-destruction. In cancer cells that lack functional p53, BAX protein levels do not increase in response to cellular damage, and apoptosis may not occur.

Hence, cells lacking functional p53 are unable to arrest at cell-cycle checkpoints or to enter apoptosis in response to DNA damage. As a result, they move unchecked through the cell cycle, regardless of the condition of the cell's DNA. Cells lacking p53 have high mutation rates and accumulate the types of mutations that lead to cancer. Because of the importance of the *p53* gene to genomic integrity, it is often referred to as the "guardian of the genome."

## The *RB1* Tumor-suppressor Gene

The loss or mutation of the ***RB1*** (**retinoblastoma 1**) tumor-suppressor gene contributes to the development of many cancers, including those of the breast, bone, lung, and bladder. The *RB1* gene was originally identified as a result of studies on **retinoblastoma**, an inherited disorder in which tumors develop in the eyes of young children. Retinoblastoma occurs with a frequency of about 1 in 15,000 individuals. In the familial form of the disease, individuals inherit one mutated allele of the *RB1* gene and have an 85 percent chance of developing retinoblastomas as well as an increased chance of developing other cancers. All somatic cells of patients with hereditary retinoblastoma contain one mutated allele of the *RB1* gene. However, it is only when the second normal allele of the *RB1* gene is lost or mutated in certain retinal cells that retinoblastoma develops. In individuals who do not have this hereditary condition, retinoblastoma is extremely rare, as it requires at least two separate somatic mutations in a retinal cell in order to inactivate both copies of the *RB1* gene (Figure 20–10).

The **retinoblastoma protein (pRB)** is a tumor-suppressor protein that controls the G1/S cell-cycle checkpoint. The pRB protein is found in the nuclei of all cell types at all stages of the cell cycle. However, its activity varies throughout the cell cycle, depending on its phosphorylation state. When cells are in the G0 phase of the cell cycle, the pRB protein is nonphosphorylated and binds to transcription factors such as E2F, inactivating them (Figure 20–11). When the cell is stimulated by growth factors, it enters G1 and approaches S phase. Throughout the G1 phase, the pRB protein becomes phosphorylated by the CDK4/cyclin D1 complex. Phosphorylated pRB releases its bound regulatory proteins. When E2F and other regulators are released by pRB, they are free to induce the expression of over 30 genes whose products are required for the transition from G1 into S phase. After cells traverse S, G2, and M phases, pRB reverts to a nonphosphorylated state, binds to regulatory proteins such as E2F, and keeps them sequestered until required for the next cell cycle. In normal quiescent cells, the presence of the pRB protein prevents passage into S phase. In many cancer cells, including retinoblastoma cells, both copies of the *RB1* gene are defective, inactive, or absent, and progression through the cell cycle is not regulated.

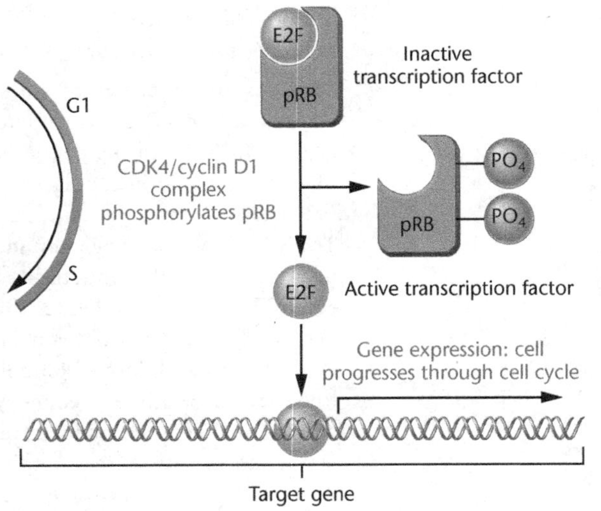


**FIGURE 20-11** In the nucleus during G0 and early G1, pRB interacts with and inactivates transcription factor E2F. As the cell moves from G1 to S phase, a CDK4/cyclin D1 complex forms and adds phosphate groups to pRB. As pRB becomes phosphorylated, E2F is released and becomes transcriptionally active, stimulating the transcription of certain genes whose products allow the cell to pass through S phase. Phosphorylation of pRB is transitory; as CDK/cylin complexes are degraded and the cell moves through mitosis to early G1, pRB phosphorylation declines, allowing pRB to reassociate with E2F.

**NOW SOLVE THIS**

Problem 23 on page 529 asks you to explain how, in the inherited Li-Fraumeni syndrome, mutations in one allele of the *p53* gene can give rise to a wide variety of different cancers.

■ HINT: *To answer this question, you might review the cellular functions regulated by the normal* p53 *tumor-suppressor gene. Consider how each of these functions could be affected if the* p53 *gene product is either absent or defective. To understand how mutations in one allele of a tumor-suppressor gene result in tumors that contain aberrations in both alleles, read about loss of heterozygosity in Section 20.6.*

## 20.5 Cancer Cells Metastasize, Invading Other Tissues

As discussed at the beginning of this chapter, uncontrolled growth alone is insufficient to create a malignant and life-threatening cancer. Cancer cells must also acquire the features of metastasis, which include the ability to disengage from the original tumor site, to enter the blood or lymphatic system, to invade surrounding tissues, and to develop into secondary tumors. In order to leave the site of the primary tumor and invade other tissues, tumor cells must dissociate from other cells and digest components of the **extracellular matrix** and **basal lamina**, which normally surround and separate the body's tissues. The extracellular matrix and basal lamina are composed of proteins and carbohydrates. They form the scaffold for tissue growth and inhibit the migration of cells.

The ability to invade the extracellular matrix is also a property of some normal cell types. For example, implantation of the embryo in the uterine wall during pregnancy requires cell migration across the extracellular matrix. In addition, white blood cells reach sites of infection by penetrating capillary walls. The mechanisms of invasion are probably similar in these normal cells and in cancer cells. The difference is that, in normal cells, the invasive ability is tightly regulated, whereas in tumor cells, this regulation has been lost.

Although less is known about the genes that control metastasis than about those controlling the cell cycle, it is likely that metastasis is controlled by a large number of genes, including those that encode cell-adhesion molecules, cytoskeleton regulators, and proteolytic enzymes. For example, epithelial tumors have a lower than normal level of the **E-cadherin glycoprotein**, which is responsible for cell–cell adhesion in normal tissues. Also, proteolytic enzymes such as **metalloproteinases** are present at higher than normal levels in highly malignant tumors and are not susceptible to the normal controls conferred by regulatory molecules such as **tissue inhibitors of metalloproteinases (TIMPs)**. It has been shown that the level of aggressiveness of a tumor correlates positively with the levels of proteolytic enzymes expressed by the tumor. Hence, inappropriately expressed cell adhesion and proteinase enzymes may assist malignant tumor cells to metastasize by loosening the normal constraints on cell location and creating holes through which the tumor cells can pass into and out of the circulatory system.

Like the tumor-suppressor genes of primary cancers, **metastasis-suppressor genes** control the growth of secondary tumors. Less than a dozen of these metastasis-suppressor genes have been identified so far, but all appear to affect the growth of metastatic tumors and not the primary tumor. The expression of these genes is often reduced by epigenetic mechanisms rather than by mutation. This observation provides hope that researchers can develop antimetastasis therapies that target the epigenetic silencing of metastasis-suppressor genes.

## 20.6 Predisposition to Some Cancers Can Be Inherited

Although the vast majority of human cancers are sporadic, a small fraction (1–2 percent) have a hereditary or familial component. At present, about 50 forms of hereditary cancer are known (Table 20.3).

Most inherited cancer-susceptibility genes, though transmitted in a Mendelian dominant fashion, are not sufficient in themselves to trigger development of a cancer. At least one other somatic mutation in the other copy of the gene must occur in order to drive a cell toward tumorigenesis. In addition, mutations in still other genes are

TABLE 20.3

**Inherited Predispositions to Cancer**

| Tumor Predisposition Syndromes | Chromosome |
|---|---|
| Early-onset familial breast cancer | 17q |
| Familial adenomatous polyposis | 5q |
| Familial melanoma | 9p |
| Gorlin syndrome | 9q |
| Hereditary nonpolyposis colon cancer | 2p |
| Li-Fraumeni syndrome | 17p |
| Multiple endocrine neoplasia, type 1 | 11q |
| Multiple endocrine neoplasia, type 2 | 22q |
| Neurofibromatosis, type 1 | 17q |
| Neurofibromatosis, type 2 | 22q |
| Retinoblastoma | 13q |
| Von Hippel–Lindau syndrome | 3p |
| Wilms tumor | 11p |

usually necessary to fully express the cancer phenotype. As mentioned earlier, inherited mutations in the *RB1* gene predispose individuals to developing various cancers. Although the normal somatic cells of these patients are heterozygous for the *RB1* mutation, cells within their tumors contain mutations in both copies of the gene. The phenomenon whereby the second, wild-type, allele is mutated in a tumor is known as **loss of heterozygosity.** Although loss of heterozygosity is an essential first step in expression of these inherited cancers, further mutations in other proto-oncogenes and tumor-suppressor genes are necessary for the tumor cells to become fully malignant.

The development of hereditary colon cancer illustrates how inherited mutations in one allele of a gene contribute only one step in the multistep pathway leading to malignancy.

About 1 percent of colon cancer cases result from a genetic predisposition to cancer known as **familial adenomatous polyposis (FAP).** In FAP, individuals inherit one mutant copy of the ***APC* (adenomatous polyposis) gene** located on the long arm of chromosome 5. Mutations include deletions, frameshift, and point mutations. The normal function of the *APC* gene product is to act as a tumor-suppressor controlling cell–cell contact and growth inhibition by interacting with the $\beta$-catenin protein. The presence of a heterozygous *APC* mutation causes the epithelial cells of the colon to partially escape cell-cycle control, and the cells divide to form small clusters of cells called **polyps** or adenomas. People who are heterozygous for this condition develop hundreds to thousands of colon and rectal polyps early in life. Although it is not necessary for the second allele of the *APC* gene to be mutated in polyps at this stage, in the majority of cases, the second *APC* allele becomes mutant in a later stage of cancer development. The relative order of mutations in the development of FAP is shown in Figure 20–12.

The second mutation in polyp cells that contain an *APC* gene mutation occurs in the *ras* proto-oncogene. The combined *APC* and *ras* gene mutations bring about the development of intermediate adenomas. Cells within these adenomas have defects in normal cell differentiation. In addition, these cells will grow in culture and are not growth-inhibited by contact with other cells—a process known as **transformation.** The third step toward malignancy requires loss of function of both alleles of the *DCC* (*d*eleted in *c*olon *c*ancer) gene. The *DCC* gene product is thought to be involved with cell adhesion and differentiation. Mutations in both *DCC* alleles result in the formation of late-stage adenomas with a number of finger-like outgrowths (villi). When late adenomas progress to cancerous adenomas, they usually suffer loss of functional *p53* genes. The final steps toward malignancy involve mutations in an unknown number of genes associated with metastasis.

NOW SOLVE THIS

Problem 6 on page 529 asks you to explain why some tobacco smokers and some people with inherited mutations in cancer-related genes never develop cancer.

■ HINT: *When considering the reasons why cancer is not an inevitable consequence of genetic or environmental conditions, you might think about the multiple steps by which cancer develops, the random nature of spontaneous mutations, and the types of genetically controlled functions that are abnormal in cancer cells. It might also be useful to consider how each individual's genetic background may affect mutation rates and differences in DNA repair functions.*

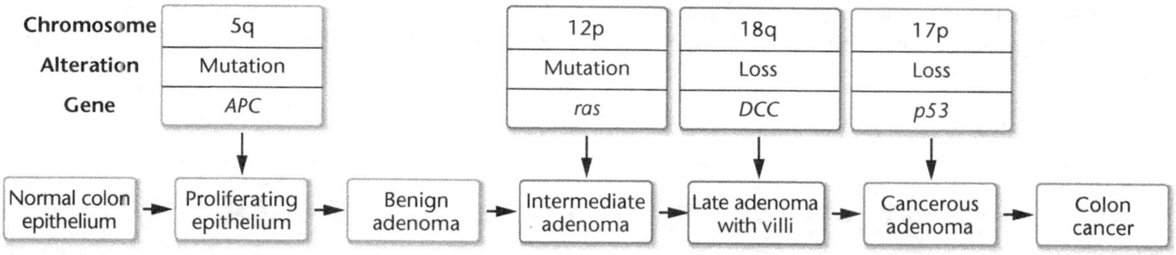


FIGURE 20–12 A model for the multistep development of colon cancer. The first step is the loss or inactivation of one allele of the *APC* gene on chromosome 5. In FAP cases, one mutant *APC* allele is inherited. Subsequent mutations involving genes on chromosomes 12, 17, and 18 in cells of benign adenomas can lead to a malignant transformation that results in colon cancer. Although the mutations on chromosomes 12, 17, and 18 usually occur at a later stage than those involving chromosome 5, the sum of changes is more important than the order in which they occur.

20.7

## Viruses Contribute to Cancer in Both Humans and Animals

Viruses that cause cancer in animals have played a significant role in the search for knowledge about the genetics of human cancer. Most cancer-causing animal viruses are RNA viruses known as **retroviruses.** Because they transform cells into cancer cells, they are known as **acute transforming retroviruses.** The first of these acute transforming retroviruses was discovered in 1910 by Francis Peyton Rous. Rous was studying sarcomas (solid tumors of muscle, bone, or fat) in chickens, and he observed that extracts from these tumors caused the formation of new sarcomas when they were injected into tumor-free chickens. Several decades later, the agent within the extract that caused the sarcomas was identified as a retrovirus and was named the **Rous sarcoma virus (RSV).**

To understand how retroviruses cause cancer in animals, it is necessary to know how these viruses replicate in cells. When a retrovirus infects a cell, its RNA genome is copied into DNA by the **reverse transcriptase** enzyme, which is brought into the cell with the infecting virus. The DNA copy then enters the nucleus of the infected cell, where it integrates at random into the host cell's genome. The integrated DNA copy of the retroviral RNA is called a **provirus.** The proviral DNA contains powerful enhancer and promoter elements in its U5 and U3 sequences at the ends of the provirus (Figure 20–13). The proviral promoter uses the host cell's transcription proteins, directing transcription of the viral genes (*gag, pol,* and *env*). The products of these genes are the proteins and RNA genomes that make up the new retroviral particles. Because the provirus is integrated into the host genome, it is replicated along with the host's DNA during the cell's normal cell cycle. A retrovirus may not kill a cell, but it may continue to use the cell as a factory to replicate more viruses that will then infect surrounding cells.

A retrovirus may cause cancer in two different ways. First, the proviral DNA may integrate by chance near one of the cell's normal proto-oncogenes. The strong promoters and enhancers in the provirus then stimulate high levels or inappropriate timing of transcription of the proto-oncogene, leading to stimulation of host-cell proliferation. Second, a retrovirus may pick up a copy of a host proto-oncogene and integrate it into its genome (Figure 20–13). The new viral oncogene may be mutated during the process of transfer into the virus, or it may be expressed at abnormal levels because it is now under the control of viral promoters. Retroviruses that carry these viral oncogenes can infect and transform normal cells into tumor cells. In the case of RSV, the oncogene that was captured from the chicken genome was the *c-src* gene. Through the study of many acute transforming viruses of animals, scientists have identified dozens of proto-oncogenes.

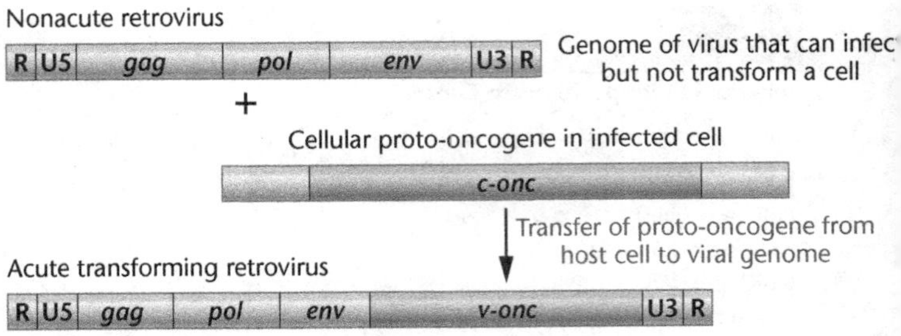


**FIGURE 20–13** The genome of a typical retrovirus is shown at the top of the diagram. The genome contains repeats at the termini (R); U5 and U3 regions that contain promoter and enhancer elements; and three major genes that encode viral structural proteins (*gag* and *env*) and the viral reverse transcriptase (*pol*). RNA transcripts of the entire viral genome become the genomes of the new viral progeny. If the retrovirus acquires all or part of a host-cell proto-oncogene (*c-onc*), this proto-oncogene (now known as a *v-onc*) is expressed along with the viral genes, leading to overexpression or inappropriate expression of *v-onc*. The *v-onc* gene may also acquire mutations that enhance its transforming ability.

So far, no acute transforming retroviruses have been identified in humans. However, RNA and DNA viruses contribute to the development of human cancers in a variety of ways. It is thought that, worldwide, about 15 percent of cancers are associated with viruses, making virus infection the second greatest risk factor for cancer, next to tobacco smoking.

The most significant contributors to virus-induced cancers are the **papillomaviruses (HPV 16 and 18), human T-cell leukemia virus (HTLV-1), hepatitis B virus, human herpesvirus 8,** and **Epstein-Barr virus** (Table 20.4). Like other risk factors for

TABLE 20.4

**Human Viruses Associated with Cancer**

| Virus | Cancer | Oncogenes | Mechanism |
|---|---|---|---|
| *Human papillomavirus 16, 18* | Cervical cancer | *E6, E7* | Inhibit p53 and pRB tumor-suppressors |
| *Hepatitis B virus* | Liver cancer | *HBx* | Signal transduction, stimulates cell cycle |
| *Epstein-Barr virus* | Burkitt's lymphoma, nasopharyngeal | Unknown | Unknown |
| *Human herpesvirus 8* | AIDS-related Kaposi's sarcoma | Several possible | Unknown |
| *Human T-cell leukemia virus* | Adult T-cell leukemia | *pX* | Stimulates cell cycle |

cancer, including hereditary predisposition to certain cancers, virus infection alone is not sufficient to trigger human cancers. Other factors, including DNA damage or the accumulation of mutations in one or more of a cell's oncogenes and tumor-suppressor genes, are required to move a cell down the multistep pathway to cancer.

Because viruses are comprised solely of a nucleic acid genome surrounded by a protein coat, they must utilize the host cell's biosynthetic machinery in order to reproduce themselves. To access the host's DNA-synthesizing enzymes, most viruses require the host cell to be in an actively growing state. Thus, many viruses contain genes encoding products that stimulate the cell cycle. For the host cell, if it survives, this can mean a loss of cell-cycle control and the beginning of tumorigenesis.

## 20.8 Environmental Agents Contribute to Human Cancers

Any substance or event that damages DNA has the potential to be carcinogenic. Unrepaired or inaccurately repaired DNA introduces mutations, which, if they occur in proto-oncogenes or tumor-suppressor genes, can lead to abnormal regulation of the cell cycle or disruption of controls over cell contact and invasion.

Our environment, both natural and human-made, contains abundant carcinogens. These include chemicals, radiation, some viruses, and chronic infections. Perhaps the most significant carcinogen in our environment is tobacco smoke, which contains a number of cancer-causing chemicals. Epidemiologists estimate that about 30 percent of human cancer deaths are associated with cigarette smoking.

Diet is often implicated in the development of cancer. Consumption of red meat and animal fat is associated with some cancers, such as colon, prostate, and breast cancer. The mechanisms by which these substances may contribute to carcinogenesis are not clear but may involve stimulation of cell division through hormones or creation of carcinogenic chemicals during cooking. Alcohol may cause inflammation of the liver and contribute to liver cancer.

Although most people perceive the human-made, industrial environment to be a highly significant contributor to cancer, it may account for only a small percentage of total cancers, and only in special situations. Some of the most mutagenic agents, and hence potentially the most carcinogenic, are natural substances and natural processes. For example, **aflatoxin**, a component of a mold that grows on peanuts and corn, is one of the most carcinogenic chemicals known. Most chemical carcinogens, such as **nitrosamines**, are components of synthetic substances and are found in some preserved meats; however, many are naturally occurring. For example, natural pesticides and antibiotics found in plants may be carcinogenic, and the human body itself creates alkylating agents in the acidic environment of the gut. Nevertheless, these observations do not diminish the serious cancer risks to specific populations who are exposed to human-made carcinogens such as synthetic pesticides or asbestos.

DNA lesions brought about by natural radiation (X rays, ultraviolet light), natural dietary substances, and substances in the external environment contribute the majority of environmentally caused mutations that lead to cancer. In addition, normal metabolism creates oxidative end products that can damage DNA, proteins, and lipids. It is estimated that the human body suffers about 10,000 damaging DNA lesions per day due to the actions of oxygen free radicals. DNA repair enzymes deal successfully with most of this damage; however, some damage may lead to permanent mutations. The process of DNA replication itself is mutagenic. Hence, substances such as growth factors or hormones that stimulate cell division are ultimately mutagenic and perhaps carcinogenic. Chronic inflammation due to infection also stimulates tissue repair and cell division, resulting in DNA lesions accumulating during replication. These mutations may persist, particularly if cell-cycle checkpoints are compromised due to mutations or inactivation of tumor-suppressor genes such as *p53* or *RB1*.

Both ultraviolet (UV) light and ionizing radiation (such as X rays and gamma rays) induce DNA damage. UV in sunlight is well accepted as an inducer of skin cancers. Ionizing radiation has clearly shown itself to be a carcinogen in studies of populations exposed to neutron and gamma radiation from atomic blasts such as those in Hiroshima and Nagasaki. Another significant environmental component, radon gas, may be responsible for up to 50 percent of the ionizing radiation exposure of the U.S. population and could contribute to lung cancers in some populations.

**NOW SOLVE THIS**

Problem 16 on page 529 asks you to provide your own estimate of what percentage of money spent on cancer research should be devoted to research and education aimed at preventing cancer and what percentage should be devoted to research in pursuit of cancer cures.

■ HINT: *In answering this question, think about the relative rates of environmentally induced and spontaneous cancers. Second, consider the proportion of environmentally induced cancers that are due to lifestyle choices. (An interesting source of information on this topic is Ames, B. N. et al. 1995. The causes and prevention of cancer.* Proc. Natl. Acad. Sci. USA 92: 5258–5265.*)*

# GENETICS, TECHNOLOGY, AND SOCIETY

## Cancer in the Cross-Hairs: Taking Aim with Targeted Therapies

Scientists have known about the genetic roots of cancer for more than 30 years. In 1970, the first oncogene was discovered, and in 1976, researchers revealed that oncogenes are defective copies of normal cellular genes. When President Nixon declared a "War on Cancer," there was a flood of optimism that science would soon conquer the disease. Since then, we have spent billions of dollars and invested immense intellectual efforts in order to understand cancer and find effective treatments. So, how have we done?

Some critics say that we have made little progress. Cancer death rates have changed little over the last 50 years. The declines in the death rates of some cancers have been due primarily to lifestyle changes or more effective diagnostic methods. Despite our understanding of the molecular genetic basis of cancer, most cancer treatments are of the "one size fits all" variety, involving surgery, cocktails of chemotherapeutic drugs, and radiation. The goal of most cancer therapies remains an attempt to kill cells that are replicating their DNA, against a background of normal cells that divide infrequently.

Despite this apparently gloomy scenario, a glimmer of hope is emerging. This hope rests on a handful of new, dramatically effective, gene-based drugs, classified as *targeted cancer therapies*.

Targeted cancer therapies are treatments that inhibit the activities of particular oncogene products that are involved in tumor growth. These therapies are effective for subsets of patients with tumor types that express certain oncogene products. Two of these new drugs—Gleevec and Herceptin—illustrate the principles of targeted cancer therapy.

Gleevec (also called imatinib) was developed to treat chronic myelogenous leukemia (CML), a form of blood cancer characterized by excessive growth of certain white blood cells. CML cells contain a reciprocal translocation between chromosomes 9 and 22, with breakpoints within the *C-ABL* and *BCR* genes. This translocation creates a new gene, *BCR-ABL*, whose protein product has constitutive tyrosine kinase activity. The BCR-ABL fusion protein causes CML white blood cells to proliferate without normal cell-cycle controls. Knowing the genetic defect in CML cells, scientists screened many compounds and synthesized others, looking for molecules with tyrosine kinase inhibitor activity. Gleevec was selected, found to inhibit CML-like tumors in mice, and then entered human clinical trials. The results were dramatic. Gleevec brought about strong therapeutic effects in all treated CML patients and caused complete remissions in many patients. Since the initial trials, scientists have discovered that Gleevec is also effective against other forms of cancer, including several other types of leukemia and some gastrointestinal stromal tumors.

Herceptin (also known as trastuzumab) is a monoclonal antibody that binds to the extracellular portion of the HER2 molecule. HER2 is a transmembrane tyrosine kinase involved in transferring growth signals from outside the cell to the nucleus. About 25 percent of invasive breast cancers contain amplifications of the *Her2* gene. When Herceptin binds to HER2 molecules on breast cancer cells, it causes downregulation of HER2 and tumor cell death. Herceptin, when given along with chemotherapy, reduces cancer recurrence by one-half and increases patient survival by one-third, compared with chemotherapy alone.

The effectiveness of Gleevec, Herceptin, and a handful of other targeted cancer therapies has created renewed excitement that knowledge of the molecular roots of cancer will translate into cures for some cancers. However, before we announce a victory in the war against cancer, we must realize that these therapies also pack a bundle of downsides—ranging from drug side effects to economic and political conundrums.

The side effects of targeted therapies are usually mild; however, some side effects—such as a 1 in 200 chance of heart failure for Herceptin or liver failure for Gleevec—may make treatment decisions tricky. Another serious side effect is that virtually all patients develop resistance to these drugs. Often the drug resistance is caused by gene amplifications or secondary mutations that occur in the genes whose products are targeted by these therapies, such as *BCR-ABL*. In response to this problem, scientists are developing new therapeutic agents for tumors that have become resistant.

Perhaps the most difficult negative aspects of targeted cancer therapies have nothing to do with medicine or science—but involve economics and politics. Pharmaceutical companies spend an average of 12 years to develop a new drug and the cost of research can exceed $800 million. In addition, targeted cancer drugs are useful for only a subset of cancer patients whose tumors have specific and known genetic defects. As a result, the cost of therapy can be high. Patients with CML must take Gleevec for the rest of their lives, at an annual cost of about $25,000. Herceptin treatments cost $40,000 a year per patient. Other targeted drugs, often used in combinations, can cost upwards of $100,000 per year. Added to the costs of treatment are the high costs of molecular screening and diagnostics. Sophisticated methods for detecting and monitoring the genetic defects in cancer cells are essential in order to correctly apply targeted therapies and to follow patient progress. In an era of rising health-care costs and an aging population, insurance companies and governments are questioning whether they can continue to fund these therapies.

Layered on top of economic considerations are cultural ones. To develop targeted cancer therapies, drug companies, academics, clinicians, and governments must collaborate and share information. Such collaborative and open structures are still rare in the scientific community. Then, there are issues of inequality. Targeted cancer therapies are not applied evenly across racial, economic, and national boundaries.

Although many problems remain, there is genuine optimism. As new genomic methods reveal more of the genetic defects that contribute to cancer, the potential for even more effective targeted treatments will increase. If we can surmount the tangle of negative issues that surround targeted therapies, there is a real chance that cancer will bend to science's onslaught.

### ■ Reference

Varmus, H., 2006. The New Era in Cancer Research. *Science* 312: 1162–1165.

### ■ Website

National Cancer Institute Factsheet: Targeted Cancer Therapies. http://www.cancer.gov/cancertopics/factsheet/Therapy/targeted

EXPLORING GENOMICS

# The Cancer Genome Anatomy Project (CGAP)

In September 2006, a research group headed by Dr. Victor Velculescu of Johns Hopkins University reported that breast and colon cancers contain about 11 gene mutations that may contribute to the cancer phenotype. The research group analyzed 13,023 of the 21,000 known genes in the human genome, comparing the DNA sequences from normal cells and cancer cells. Most of the mutations that were specific to cancer cells were not previously known to be associated with cancer.

Dr. Velculescu's study is one of the first in the new *Cancer Genome Atlas (TCGA)* project, a $1.5 billion federal project designed to systematically scan the human genome to find genes that are mutated in many different cancers. In this exercise, we will explore several aspects of Dr. Velculescu's research, both by referring to the original research paper and by mining information available in the online database, **The Cancer Genome Anatomy Project (CGAP)**. The purpose of the CGAP is to understand the expression profiles of genes from normal, precancer, and cancer cells. Data within CGAP is made available to all cancer researchers via its Web site, at http://cgap.nci.nih.gov/.

### ■ Exercise I – Genomic Methodologies

1. How did Dr. Velculescu's group detect these cancer-related mutations?

   To answer this question, go to the article "The Consensus Coding Sequences of Human Breast and Colorectal Cancers" in the journal *Science* 314: 268–274 (October 13, 2006) and read the description of their research methods. Access this paper at www.sciencemag.org.

2. Describe how you would use DNA microarrays to perform a mutation screen for genes mutated in cancer cells. (To learn more about DNA microarrays, see the microarrays Mediabook referred to in the Exploring Genomics exercises for Chapter 17, and Chapter 21, Genomics, Bioinformatics and Proteomics)

### ■ Exercise II – Colon Cancer and the *TBX22* Gene

One gene that Dr. Velculescu's research group discovered to be mutated in colon cancers—*TBX22*—was not previously suspected to contribute to this cancer. What is *TBX22*, and how do you think a mutated *TBX22* gene would contribute to the development of colon cancer?

1. To begin your search for the answers, go to CGAP at http://cgap.nci.nih.gov/.
2. Select "Genes" from the links near the top of the page.
3. From the list of "Gene Tools" in the left-hand margin, select "Gene Finder."
4. Select "Homo sapiens" in the "Select organism" box, and type TBX22 in the "Enter a unique identifier" box. Submit the query.
5. Select "Gene Info" in the right-hand column of the table.
6. Explore the many sources of information about *TBX22* from various database links listed on this page.

Prepare a brief written or verbal report on what you learned during your explorations and which sources you used to reach your conclusions about *TBX22*.

## Chapter Summary

1. Cancer is a genetic disease, predominantly of somatic cells. About one percent of cancers are associated with germ-line mutations that increase the susceptibility to certain cancers.
2. Cancer cells show two basic properties: abnormal cell proliferation and a propensity to spread and invade other parts of the body, or metastasize. Genes controlling these aspects of cellular function are either mutated or expressed inappropriately in cancer cells.
3. Cancers are clonal, meaning that all cells within a tumor originate from a single cell that contained a number of mutations.
4. The development of cancer is a multistep process, requiring mutations in several cancer-related genes.
5. Cancer cells show high rates of mutation, chromosomal abnormalities, and genomic instability. This leads to the accumulation of mutations in specific genes that control aspects of cell proliferation, apoptosis, differentiation, DNA repair, cell migration, and cell–cell contact.
6. Abnormal epigenetic modifications account for large numbers of the genetic changes that occur in cancer cells.
7. Cancer cells have defects in the regulation of cell-cycle progression, cell-cycle checkpoints, and signal transduction pathways.
8. Proto-oncogenes are normal genes that promote cell growth and division. When proto-oncogenes are mutated or misexpressed in cancer cells, they are classified as oncogenes.
9. Tumor-suppressor genes normally regulate cell-cycle checkpoints and apoptosis. When tumor-suppressor genes are mutated or inactivated, cells cannot correct DNA damage. This leads to accumulation of mutations that may cause cancer.

10. Inherited mutations in cancer-susceptibility genes are not sufficient to trigger cancer. A second somatic mutation, in the other copy of the gene, is necessary to trigger tumorigenesis. In addition, mutations in other cancer-related genes are necessary for the development of hereditary cancers.
11. RNA and DNA tumor viruses contribute to cancers by stimulating cells to proliferate, introducing new oncogenes, interfering with the cell's normal tumor-suppressor gene products, or stimulating the expression of a cell's proto-oncogenes.
12. Environmental agents such as chemicals, radiation, viruses, and chronic infections contribute to the development of cancer. The most significant environmental factors that affect human cancers are tobacco smoke, diet, and natural radiation.

## INSIGHTS AND SOLUTIONS

1. In disorders such as retinoblastoma, a mutation in one allele of the *RB1* gene can be inherited from the germ line, causing an autosomal dominant predisposition to the development of eye tumors. To develop tumors, a somatic mutation in the second copy of the *RB1* gene is necessary. Given that the first mutation can be inherited, in what ways can a second mutational event occur?

**Solution:** In considering how this second mutation arises, we must look at several types of mutational events, including changes in nucleotide sequence and events that involve whole chromosomes or chromosome parts. Retinoblastoma results when both copies of the *RB1* locus are lost or inactivated. With this in mind, you must first list the phenomena that can result in a mutational loss or the inactivation of a gene.

One way the second *RB1* mutation can occur is by a nucleotide alteration that converts the remaining normal *RB1* allele to a mutant form. This alteration can occur through a nucleotide substitution or through a frameshift mutation caused by the insertion or deletion of nucleotides during replication. A second mechanism involves the loss of the chromosome carrying the normal allele. This event would take place during mitosis, resulting in chromosome 13 monosomy and leaving the mutant copy of the gene as the only *RB1* allele. This mechanism does not necessarily involve loss of the entire chromosome; deletion of the long arm (*RB1* is on 13q) or an interstitial deletion involving the *RB1* locus and some surrounding material would have the same result. Alternatively, a chromosome aberration involving loss of the normal copy of the *RB1* gene might be followed by duplication of the chromosome carrying the mutant allele. Two copies of chromosome 13 would be restored to the cell, but the normal *RB1* allele would not be present. Finally, a recombination event followed by chromosome segregation could produce a homozygous combination of mutant *RB1* alleles.

2. Proto-oncogenes can be converted to oncogenes in a number of different ways. In some cases, the proto-oncogene itself becomes amplified up to hundreds of times in a cancer cell. An example is the *cyclin D1* gene, which is amplified in some cancers. In other cases, the proto-oncogene may be mutated in a limited number of specific ways leading to alterations in the gene product's structure. The *ras* gene is an example of a proto-oncogene that becomes oncogenic after suffering point mutations in specific regions of the gene. Explain why these two proto-oncogenes (*cyclin D1* and *ras*) undergo such different alterations in order to convert them into oncogenes.

**Solution:** The first step to solving this question is to understand the normal functions of these proto-oncogenes and to think about how either amplification or mutation would affect each of these functions.

The cyclin D1 protein regulates progression of the cell cycle from G1 into S phase, by binding to CDK4 and activating this kinase. The cyclin D1/CDK4 complex phosphorylates a number of proteins including pRB, which in turn activates other proteins in a cascade that results in transcription of genes whose products are necessary for DNA replication in S phase. The simplest way to increase the activity of cyclin D1 would be to increase the number of cyclin D1 molecules available for binding to the cell's endogenous CDK4 molecules. This can be accomplished by several mechanisms, including amplification of the *cyclin D1* gene. In contrast, a point mutation in the *cyclin D1* gene would most likely interfere with the ability of the cyclin D1 protein to bind to CDK4. Hence, mutations within the gene would probably repress cell-cycle progression rather than stimulate it.

The *ras* gene product is a signal transduction protein that operates as an on–off switch in response to external stimulation by growth factors. It does so by binding either GTP (producing the "on" state) or GDP (the "off" state). Oncogenic mutations in the *ras* gene occur in specific regions that alter the ability of the Ras protein to exchange GDP for GTP. Oncogenic Ras proteins are locked in the "on" conformation, bound to GTP. In this way, they constantly stimulate the cell to divide. An amplification of the *ras* gene would simply provide more molecules of normal Ras protein, which would still be capable of on–off regulation. Hence, simple amplification of *ras* would be less likely to be oncogenic.

3. Explain why many oncogenic viruses contain genes whose products interact with tumor-suppressor proteins.

**Solution:** In order to answer this question, it is useful to consider how viruses reproduce in cells and what the roles of tumor-suppressors are in normal cells.

Oncogenic viruses enter a cell not for the purpose of causing cancer but for the purpose of viral replication. Viruses are relatively simple entities, consisting of only a nucleic acid genome—either RNA or DNA—and a relatively small number of structural and enzymatic proteins. They depend on their host cells for much of the biosynthetic machinery and structural components necessary to replicate their genomes and assemble new virus particles. For example, many viruses require the host cell's RNA polymerase II enzyme, transcription factors, and ribonucleotide precursors in order to transcribe their viral genes. They also require components of the host cell's DNA replication machinery to replicate viral genomes. Hence, the ideal host cell for a virus infection is one that is in some stage of the cell's proliferation cycle, preferably in late G1 to early S phase.

Most cells in higher eukaryotes such as humans are quiescent (in G0 phase). In order to stimulate the infected cell to enter the cell cycle and become primed for DNA replication, many viruses contain genes that encode growth-stimulating proteins. As we learned in this chapter, tumor-suppressor proteins are those involved in either restraining the cell cycle at checkpoints or in triggering the process of programmed cell death. Both of these functions are contrary to the needs of a typical virus; therefore, many viruses have evolved methods to inactivate tumor-suppressors. One of the ways in which viral proteins can inactivate tumor-suppressor proteins is to bind to them and inhibit their functions. The tumor-suppressors p53 and pRB are common targets of viral regulatory proteins, such as the E6 and E7 proteins of HPV 16 and 18. By inactivating tumor-suppressors, these viruses are able to maintain the cell within the cell cycle. For the host, however, this growth stimulation in the absence of functional cell-cycle checkpoints can lead to increased mutation accumulation and possible tumorigenesis.

## Problems and Discussion Questions

1. As a genetic counselor, you are asked to assess the risk for a couple with a family history of retinoblastoma who are thinking about having children. Both the husband and wife are phenotypically normal, but the husband has a sister with familial retinoblastoma in both eyes. What is the probability that this couple will have a child with retinoblastoma? Are there any tests that you could recommend to help in this assessment?
2. Where are the major regulatory points in the cell cycle?
3. List the functions of kinases and cyclins, and describe how they interact to cause cells to move through the cell cycle.
4. (a) How does pRB function to keep cells at the G1 checkpoint? (b) How do cells get past the G1 checkpoint to move into S phase?
5. What is the difference between saying that cancer is inherited and saying that the predisposition to cancer is inherited?
6. Although tobacco smoking is responsible for a large number of human cancers, not all smokers develop cancer. Similarly, some people who inherit mutations in the tumor-suppressor genes *p53* or *RB1* never develop cancer. Describe some reasons for these observations.
7. What is apoptosis, and under what circumstances do cells undergo this process?
8. Define tumor-suppressor genes. Why is a mutation in a single copy of a tumor-suppressor gene expected to behave as a recessive gene?
9. A genetic variant of the retinoblastoma protein, called PSM-RB (phosphorylation site mutated RB), is not able to be phosphorylated by the action of CDK4/cyclinD1 complex. Explain why PSM-RB is said to have a constitutive growth-suppressing action on the cell cycle.
10. In the Rous sarcoma virus (RSV) genome, the host-cell proto-oncogene is converted into an oncogene. How does this conversion occur?
11. Part of the Ras protein is associated with the plasma membrane, and part extends into the cytoplasm. How does the Ras protein transmit a signal from outside the cell into the cytoplasm? What happens in cases where the *ras* gene is mutated?
12. If a cell suffers damage to its DNA while in S phase, how can this damage be repaired before the cell enters mitosis?
13. Distinguish between oncogenes and proto-oncogenes. In what ways can proto-oncogenes be converted to oncogenes?
14. Of the two classes of genes associated with cancer, tumor-suppressor genes and oncogenes, in which group can mutations be considered gain-of-function mutations? Which group undergoes loss-of-function mutations? Explain.
15. How do translocations such as the Philadelphia chromosome contribute to cancer?
16. Given that cancers can be environmentally induced and that some environmental factors are the result of lifestyle choices such as smoking, sun exposure, and diet, what percentage of the money spent on cancer research do you think should be devoted to research and education aimed at preventing cancer rather than to finding cancer cures?
17. In CML, leukemic blood cells can be distinguished from other cells of the body by the presence of a functional BCR-ABL hybrid protein. Explain how this characteristic provides an opportunity to develop a treatment for CML.
18. How do normal cells protect themselves from accumulating mutations in genes that could lead to cancer? How do cancer cells differ from normal cells in these processes?
19. Describe the difference between an acute transforming virus and a virus that does not cause tumors.
20. Explain how environmental agents such as chemicals and radiation cause cancer.
21. Radiotherapy (treatment with ionizing radiation) is one of the most effective current cancer treatments. It works by damaging DNA and other cellular components. In which ways could radiotherapy control or cure cancer, and why does radiotherapy often have significant side effects?
22. Assume that a young woman in a suspected breast cancer family takes the *BRCA1* and *BRCA2* genetic tests and receives negative results. That is, she does not test positive for the mutant alleles of *BRCA1* or *BRCA2*. Can she consider herself free of risk for breast cancer?
23. People with a genetic condition known as Li–Fraumeni syndrome inherit one mutant copy of the *p53* gene. These people have a high risk of developing a number of different cancers, such as breast cancer, leukemias, bone cancers, adrenocortical tumors, and brain tumors. Explain how mutations in one cancer-related gene can give rise to such a diverse range of tumors.
24. Explain the apparent paradox that both hypermethylation and hypomethylation of DNA are often associated with various cancers.

### HOW DO WE KNOW?

25. In this chapter, we focused on the genetic changes that lead to the development of cancer. At the same time, we found many opportunities to consider the methods and reasoning by which much of this information was acquired. From the explanations given in the chapter, what answers would you propose to the following fundamental questions:
    (a) How do we know that cancers arise from a single cell that contains mutations?
    (b) How do we know that cancer development requires more than one mutation?
    (c) How do we know that cancer cells contain defects in DNA repair?

## Extra-Spicy Problems

26. As part of a cancer research project, you have discovered a gene that is mutated in many metastatic tumors. After determining the DNA sequence of this gene, you compare the sequence with those of other genes in the human genome sequence database. Your gene appears to code for an amino acid sequence that resembles sequences found in some serine proteases. Conjecture how your new gene might contribute to the development of highly invasive cancers.
27. Prostate cancer is a major cause of cancer-related deaths among men. Epigenetic changes that regulate gene expression are involved in both the initiation and progression of such cancers. Following is a table that lists the number of genes known to be hypermethylated in prostate cancer cells (modified from Long-Cheng, L et al., 2005. *J. Natl. Cancer Inst.* 97: 103–115). For each category of genes, speculate on the mechanism(s) by which cancer initiation or progression might be influenced by hypermethylation.

| DNA Hypermethylation of: | Number of Known Genes |
|---|---|
| Hormonal response genes | 5 |
| Cell-cycle control genes | 2 |
| Tumor cell invasion genes | 8 |
| DNA damage repair genes | 2 |
| Signal transduction genes | 4 |

28. Those who inherit a mutant allele of the *RB1* gene are at risk for developing a bone cancer called osteosarcoma. You suspect that in these cases, osteosarcoma requires a mutation in the second *RB1* allele and have cultured some osteosarcoma cells and obtained a cDNA clone of a normal human *RB1* gene. A colleague sends you a research paper revealing that a strain of cancer-prone mice develop malignant tumors when injected with osteosarcoma cells, and you obtain these mice. Using these three resources, what experiments would you perform to determine (a) whether osteosarcoma cells carry two *RB1* mutations, (b) whether osteosarcoma cells produce any pRB protein, and (c) if the addition of a normal *RB1* gene will change the cancer-causing potential of osteosarcoma cells?
29. Table 20.5 summarizes some of the data that have been collected on *BRCA1* mutations in families with a high incidence of both early-onset breast cancer and ovarian cancer.

TABLE 20.5

**Predisposing Mutations in *BRCA1***

| Kindred | Codon | Nucleotide Change | Coding Effect | Frequency in Control Chromosomes |
|---|---|---|---|---|
| 1901 | 24 | −11bp | Frameshift or splice | 0/180 |
| 2082 | 1313 | C→T | Gln→Stop | 0/170 |
| 1910 | 1756 | Extra C | Frameshift | 0/162 |
| 2099 | 1775 | T→G | Met→Arg | 0/120 |
| 2035 | NA* | ? | Loss of transcript | NA* |

SOURCE: 1994. *Science* 266: 66–71.
*NA indicates not applicable, as the regulatory mutation is inferred, and the position has not been identified.

(a) Note the coding effect of the mutation found in kindred group 2082 in Table 20.5. This results from a single base-pair substitution. Draw the normal double-stranded DNA sequence for this codon (with the 5′ and 3′ ends labeled), and show the sequence of events that generated this mutation, assuming that it resulted from an uncorrected mismatch event during DNA replication.
(b) Examine the types of mutations that are listed in Table 20.5 and determine if the *BRCA1* gene is likely to be a tumor-suppressor gene or an oncogene.
(c) Although the mutations in Table 20.5 are clearly deleterious and cause breast cancer in women at very young ages, each of the kindred groups had at least one woman who carried the mutation but lived until age 80 without developing cancer. Name at least two different mechanisms (or variables) that could underlie variation in the expression of a mutant phenotype and propose an explanation for the incomplete penetrance of this mutation. How do these mechanisms or variables relate to this explanation?

30. Table 20.6 shows neutral polymorphisms found in control families (those with no increased frequency of breast and ovarian cancer).

TABLE 20.6

**Neutral Polymorphisms in *BRCA1***

| | | | Frequency in Control Chromosomes* | | | |
|---|---|---|---|---|---|---|
| Name | Codon Location | Base in Codon† | A | C | G | T |
| PM1 | 317 | 2 | 152 | 0 | 10 | 0 |
| PM6 | 878 | 2 | 0 | 55 | 0 | 100 |
| PM7 | 1190 | 2 | 109 | 0 | 53 | 0 |
| PM2 | 1443 | 3 | 0 | 115 | 0 | 58 |
| PM3 | 1619 | 1 | 116 | 0 | 52 | 0 |

*The number of chromosomes with a particular base at the indicated polymorphic site (A, C, G, or T) is shown.
†Position 1, 2, or 3 of the codon.

Examine Table 20.6 and answer the following questions:
(a) What is meant by a neutral polymorphism?
(b) What is the significance of this table in the context of examining a family or population for *BRCA1* mutations that predispose an individual to cancer?
(c) Is the PM2 polymorphism likely to result in a neutral missense mutation or a silent mutation?
(d) Answer part (c) for the PM3 polymorphism.

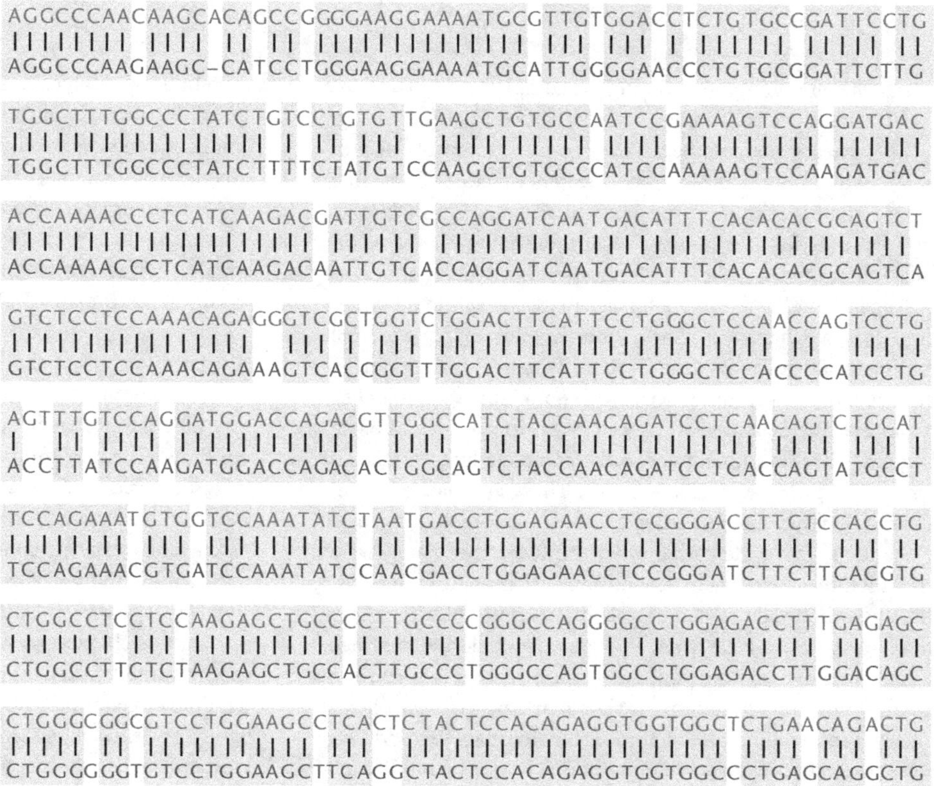


Alignment comparing DNA sequence for the leptin gene from dogs (top) and humans (bottom). Vertical lines and shaded boxes indicate identical bases. *LEP* encodes a hormone that functions to suppress appetite. This type of analysis is a common application of bioinformatics and a good demonstration of comparative genomics.

# 21

# Genomics, Bioinformatics, and Proteomics

## CHAPTER CONCEPTS

- Genomics applies recombinant DNA and DNA sequencing methods, and bioinformatics to sequence, assemble, and analyze genomes.
- Disciplines in genomics encompass several areas of study, including structural and functional genomics, comparative genomics, and metagenomics, and have led to an "omics" revolution in modern biology.
- Bioinformatics is a discipline that merges information technology with biology and mathematics to store, share, compare, and analyze nucleic acid and protein sequence data.
- The Human Genome Project has greatly advanced our understanding of the organization, size, and function of the human genome.
- Genomic analysis of model prokaryotes and eukaryotes has revealed similarities and differences in genome size and organization.
- Metagenomics is the study of genomes from environmental samples and is valuable for identifying microbial genomes.
- Transcriptome analysis provides insight into patterns of gene expression and gene regulatory activity of a genome.
- Proteomics focuses on the protein content of cells and on the structures, functions, and interactions of proteins.
- Systems biology approaches attempt to uncover complex interactions among genes, proteins, and other cellular components.

The term **genome**, meaning the complete set of DNA in a single cell of an organism, was coined in 1920, at a time when geneticists began to turn from the study of individual genes to a focus on the larger picture. To begin to characterize all of the genes in an organism's DNA, geneticists typically followed a two-part approach: (1) identify spontaneous mutations or collect mutants produced by chemical or physical agents, and (2) generate linkage maps using mutant strains as discussed in Chapter 5. These effective but extremely time-consuming strategies were used to identify genes in many of the classic model organisms discussed in this book, such as *Drosophila*, maize, mice, bacteria, and yeast, as well as in viruses, such as bacteriophages. (For an example, refer to the complete linkage maps of the *Drosophila* genome in Figure 5–14). These approaches, developed about 90 years ago, formed the technical backbone of genetic analysis and are still widely used today; however, they have several major limitations. For instance, conventional mutational analysis and linkage would require that at least one mutation for each gene was available before all the genes in a genome could be identified. Obtaining mutants and carrying out linkage studies is very time-consuming, and when mutations are lethal or have no clear phenotype, they can be difficult or impossible to map. In addition, although researchers can generate mutations in animal models in a laboratory, they cannot do the same with humans, so identifying human genes by mutational analysis is largely limited to linkage mapping of inherited or spontaneously acquired mutant genes with clear phenotypes. Another fundamental limitation of these approaches is that, although they can be used for identifying and characterizing gene loci, they do not lead to a determination of DNA sequence. Nor are they particularly useful for studying noncoding areas of the genome such as DNA regulatory sequences.

Beginning in the 1980s, geneticists interested in mapping human genes began using recombinant DNA technology to map DNA sequences to specific chromosomes. Initially, most of these sequences were not actually full-length genes but marker sequences such as restriction fragment length polymorphisms (RFLPs), single nucleotide polymorphisms (SNPs), and other molecular markers. Once assigned to chromosomes, these markers were used in pedigree analysis to establish linkage between the markers and disease phenotypes for genetic disorders. This approach, called **positional cloning**, was used to map, isolate, clone, and sequence the genes for cystic fibrosis, neurofibromatosis, and dozens of other disorders. Positional cloning identified one gene at a time, and yet by the mid-1980s, it had been used to assign more than 3500 genes and markers to human chromosomes. However, at this time it was estimated that there were approximately 100,000 genes in the human genome, and it was readily apparent that mapping using existing methods would be a laborious, time-consuming, and nearly insurmountable task. As you will soon learn, this estimate for gene number turned out to be fairly inaccurate.

In 1977, as recombinant DNA–based techniques were developed, Fred Sanger and colleagues began the field of **genomics**, the study of genomes, by using a newly developed method of DNA sequencing to sequence the 5400-nucleotide genome of the virus $\phi$X174. Other viral genomes were sequenced in short order, but even this technology was slow and labor-intensive, limiting its use to small genomes. During the next two decades, the development of computer-automated DNA sequencing methods made it possible to consider sequencing the larger and more complex genomes of eukaryotes, including the 3.1 billion nucleotides that comprise the human genome.

The development of recombinant DNA technologies coupled with the advent of computer-automated DNA sequencing methods is responsible for accelerating the field of genomics. Genomic technologies have developed so quickly that modern biological research is currently experiencing a genomics revolution. The most recent new subdisciplines of genomics to emerge include *structural* and *functional genomics*, *comparative genomics*, and *metagenomics*.

In this chapter, we will examine basic technologies used in genomics and then discuss examples of genome data and different disciplines of genomics. We will also discuss *transcriptome analysis*, the study of genes expressed in a cell or tissue (the "transcriptome"), and *proteomics*, the study of proteins present in a cell or tissue. The chapter concludes with a brief look at *systems biology*, a new area of contemporary biology that incorporates and integrates genomics, transcriptome analysis, and proteomics data. In Chapter 24, we will continue our discussion of genomics by discussing many modern applications of recombinant DNA and genomic technologies.

## 21.1 Whole-Genome Shotgun Sequencing Is a Widely Used Method for Sequencing and Assembling Entire Genomes

As discussed in Chapter 13, recombinant DNA technology made it possible to generate DNA libraries that could be used to identify, clone, and sequence specific genes of interest. But a primary limitation of library screening and even of most polymerase chain reaction (PCR) approaches is that they typically can identify only relatively small numbers of genes at a time. Genomics allows the sequencing of entire genomes. **Structural genomics** focuses on sequencing genomes and analyzing nucleotide sequences to identify genes and other important sequences such as gene regulatory regions.

Currently, the most widely used strategy for sequencing and assembling an entire genome involves variations of a method called **whole-genome shotgun sequencing**. In simple terms, this

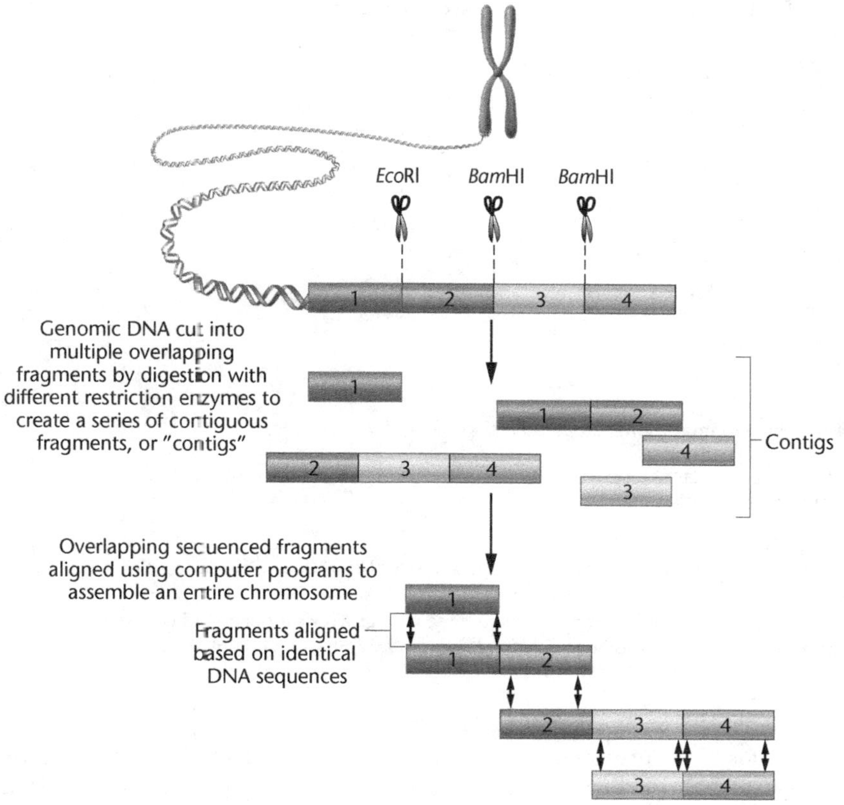


FIGURE 21–1 An overview of whole-genome shotgun sequencing and assembly.

technique is similar to you and a friend taking your respective copies of this genetics textbook and randomly ripping the pages into strips about five to seven inches long. Each chapter represents a chromosome, and all of the letters in the entire book are the "genome." Then you and your friend would go through the painstaking task of comparing the pieces of paper to find places that match, overlapping sentences—areas where there are similar sentences on different pieces of paper. Eventually, in theory, many of the strips containing matching sentences would overlap in ways that you could use to reconstruct the pages and assemble the order of the entire text.

Figure 21–1 shows a basic overview of whole-genome shotgun sequencing. First, an entire chromosome is cut into short, overlapping fragments, either by mechanically shearing the DNA in various ways or, more commonly, by using restriction enzymes to cleave the DNA at different locations. Recall from Chapter 13 that **restriction enzymes** are DNA-digesting enzymes that cut the phosphodiester backbone of DNA at specific sequences.

Different restriction enzymes can be used so that chromosomes are cut at different sites; or sometimes, **partial digests** of DNA using the same restriction enzyme are used. With partial digests, DNA is incubated with restriction enzymes for only a short period of time, so that not every site in a particular sequence is cut to completion by an individual enzyme. Either way, restriction digests of whole chromosomes generate thousands to millions of overlapping DNA fragments. For example, a 6-bp cutter such as *Eco*RI creates about 700,000 fragments when used to digest the human genome! Because these overlapping fragments are adjoining segments that collectively form one continuous DNA molecule within a chromosome, they are called **contiguous fragments,** or "**contigs.**"

Cutting a genome into contigs is not particularly difficult; however, a primary hurdle that initially prevented whole-genome sequencing was the question of how to sequence millions or billions of base pairs in a timely and cost-effective way. The Sanger sequencing method discussed in Chapter 13 was the predominant sequencing technique for a long time; however, its major limitation was that even the best sequencing gels would typically yield only several hundred base pairs in each run. Obviously, it would be very time consuming to manually sequence an entire genome by the Sanger method. The major technological breakthrough that made genomics possible was the development of computer automated sequencers, also discussed in Chapter 13.

## High-Throughput Sequencing

Many **Computer-automated DNA sequencing instruments** utilize dideoxynucleotides (ddNTPs) labeled with fluorescent dyes (refer to Figure 13–28). A single reaction tube is used, and the original manual approach that employed polyacrylamide gels to separate reaction mixtures has been replaced by a single lane of an ultra-thin-diameter polyacrylamide tube gel called a **capillary gel.** As DNA fragments move through the gel, they are scanned with a laser beam. The laser stimulates the fluorescent dye on each DNA fragment, causing each ddNTP to emit different wavelengths of light. The emitted light is collected by a detector, which amplifies and then feeds this information to a computer to process and convert into the DNA sequence.

Certain types of computer-automated sequencers, designed for so-called **high-throughput sequencing,** can process millions of base pairs in a day. These sequencers contain multiple capillary gels that are several feet long. Some run as many as 96 capillary gels at a time, each producing around 900 bases of sequence. Because these sequencers are computer automated, they can work around the clock, generating over 2 million bases of sequence in a day. In the past 10 years, high-throughput sequencing has increased the productivity of DNA-sequencing technology over 500-fold. The total number of bases that could be sequenced in a single reaction was doubling about every 24 months. At the same time, this increase in efficiency brought about a dramatic decrease in cost, from about \$1.00 to less than \$0.001 per base pair.

In the next section, we will discuss the importance of bioinformatics to genomics. One of the earliest bioinformatics applications to be developed for genomic purposes was the use of algorithm-based software programs for creating a DNA-sequence **alignment,**

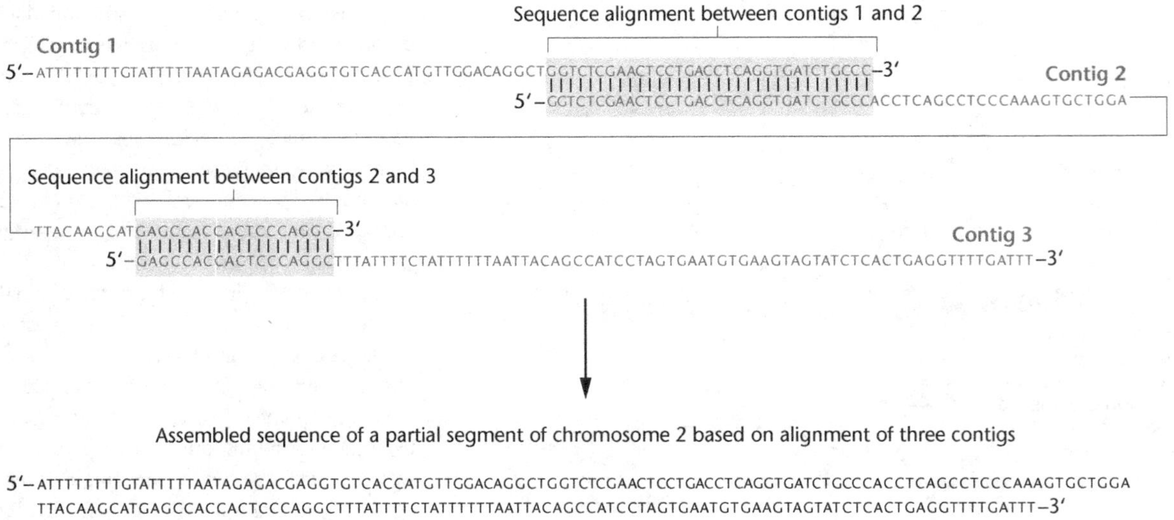


**FIGURE 21–2** DNA-sequence alignment of contigs on human chromosome 2. Single stranded DNA for three different contigs from human chromosome 2 is shown in blue, red, or green. Actual sequence from chromosome 2 is shown, but in reality, contig alignment involves fragments that are several thousand bases in length. Alignment of the three contigs allows a portion of chromosome 2 to be assembled. Alignment of all contigs for a particular chromosome would result in assembly of a completely sequenced chromosome.

in which similar sequences of bases, such as contigs, are lined up for comparison. Alignment identifies overlapping sequences, allowing scientists to reconstruct their order in a chromosome. Figure 21–2 shows an example of contig alignment and assembly for a portion of human chromosome 2. For simplicity, this figure shows relatively short sequences for each contig, which in actuality would be much longer. The figure is also simplified in that, in actual alignments, assembled sequences do not always overlap only at their ends.

The whole-genome shotgun sequencing method was developed by J. Craig Venter and colleagues at The Institute for Genome Research (TIGR). In 1995, TIGR scientists used this approach to sequence the 1.83-million-bp genome of the bacterium *Haemophilus influenzae*. This was the first completed genome sequence from a free-living organism, and it demonstrated "proof of concept" that shotgun sequencing could be used to sequence an entire genome. Even after the genome for *H. influenzae* was sequenced, many scientists were skeptical that a shotgun approach would work on the larger genomes of eukaryotes. But variations of shotgun approaches are now the predominant methods for sequencing genomes, including those of *Drosophila*, dog, several hundred species of bacteria, humans, and many other organisms, as you will read about later in this chapter.

## The Clone-by-Clone Approach

Prior to the widespread use of whole-genome sequencing approaches, genomes were being assembled using a **clone-by-clone** approach, also called **map-based cloning** (Figure 21–3). Initial progress on the Human Genome Project was based on this methodology, in which individual DNA fragments from restriction digests of chromosomes are aligned to create the restriction maps of a chromosome. These restriction fragments are then ligated into vectors such as bacterial artificial chromosomes (BACs) or yeast artificial chromosomes (YACs) to create libraries of contigs. Recall from Chapter 13 that BACs and YACs are good cloning vectors for replicating large fragments of DNA. Often, libraries from individual chromosomes were prepared.

Prior to the development of high-throughput approaches capable of sequencing several thousand bases, DNA fragments in BACs and YACs would often be further digested into smaller, more easily manipulated pieces that were then subcloned into cosmids or plasmids so that they could be sequenced in their entirety (Figure 21–3). After each fragment was sequenced and then analyzed for alignment overlaps, a chromosome could be assembled. The bioinformatics approaches we will discuss in the next section would then be used to identify possible protein-coding genes and assign them a location on the chromosome. For example, Figure 21–3 shows the use of map-based cloning to sequence part of chromosome 11, including part of the human $\beta$-globin gene.

Compared to whole-genome sequencing, the clone-by-clone approach is cumbersome and time-consuming, because of the time required to clone DNA fragments into different vectors, transform bacteria or yeast, select individual clones from the library for sequencing, and then carry out sequence analysis and assembly on relatively short sequences. Essentially, the clone-by-clone approach is the organized sequencing of contigs from a restriction map instead of random sequencing and assembly. As whole-genome sequencing approaches have become the most common method for assembling genomes, map-based cloning approaches are now primarily used to

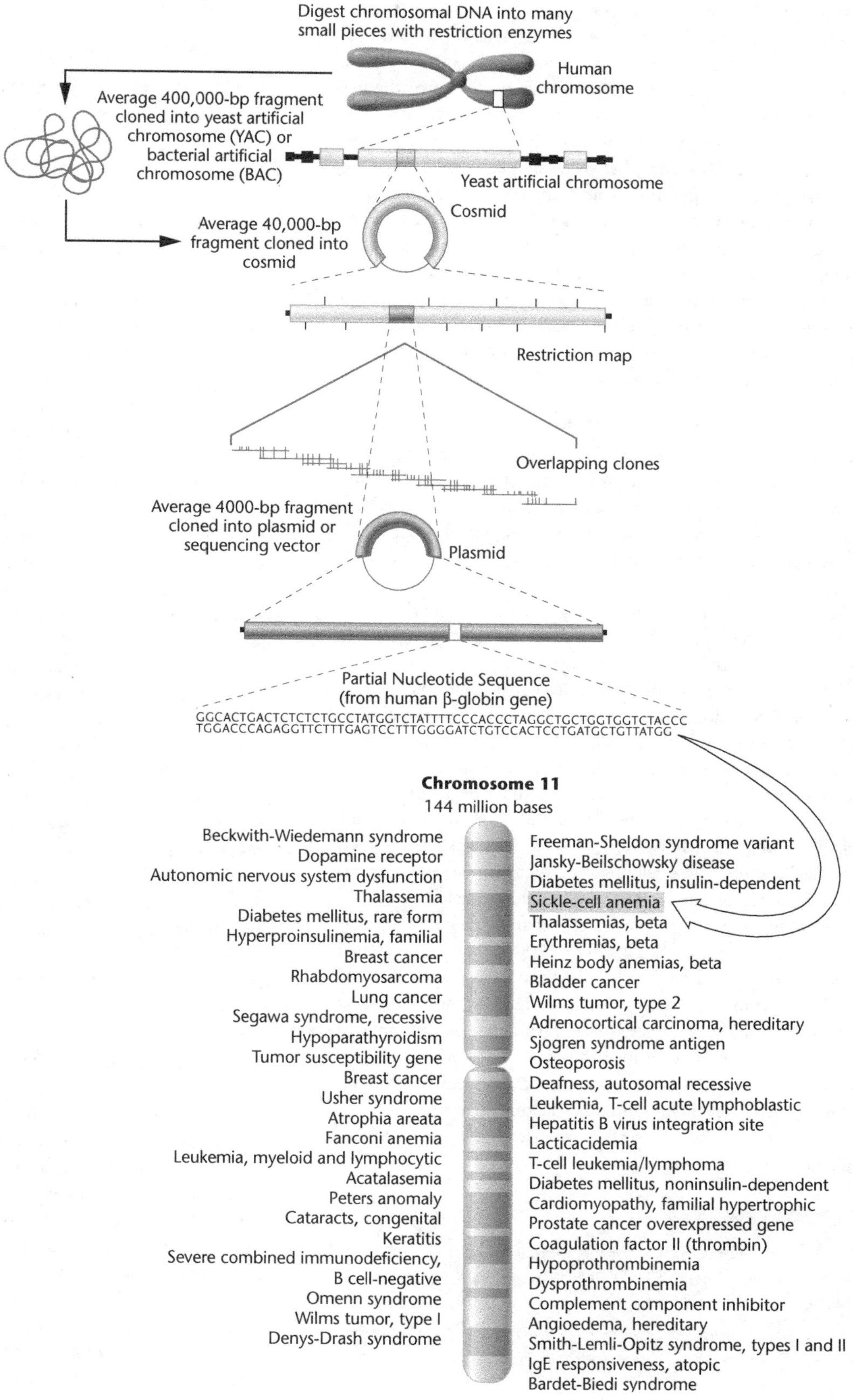


**FIGURE 21-3** A clone-by-clone, or map-based, approach to genome sequencing involves cloning overlapping DNA fragments (contigs) into vectors. Different vectors, such as BACs, YACs, cosmids, and plasmids, are used depending on the size of each DNA fragment being analyzed. Overlapping clones are then sequenced and aligned to assemble an entire chromosome.

resolve the problems often encountered during whole-genome sequencing. For example, highly repetitive sequences in a chromosome can be difficult to align correctly in order to identify overlaps, because with such sequences one cannot know for sure whether portions that are nearly identical are overlapping fragments or belong to different parts of a highly repetitive chromosome. Frequently, there are also gaps between aligned contigs. In the textbook-ripping analogy, after you compare all your pieces of paper, you may be left with some very small ones that contain too few words to be matched with certainty to any others and with some pieces for which you just could not find matches. In these instances, a clone-by-clone approach may enable you to assemble the necessary contigs and complete the chromosome. Thus, whole-genome sequencing and map-based sequencing are commonly combined in the task of sequencing a genome.

### Draft Sequences and Checking for Errors

It is common for a draft sequence of a genome to be announced several years before a final sequence is released. Draft sequences often contain gaps in areas that, for any number of reasons, may have been difficult to analyze. The decision to designate a sequence as "final" is dictated by the amount of error genome scientists are willing to accept as a cutoff.

Chromosome segments are typically sequenced more than once to ensure a high level of accuracy. The assembly of a final genomic sequence from multiple sequencing runs is known as **compiling**. One way to compile and error check is to sequence complementary strands of a DNA molecule separately and then use base-pairing rules to check for errors. In one case, researchers using the shotgun method on the genome of the bacterium *Pseudomonas aeruginosa* sequenced the 6.3 million nucleotides seven times to ensure that the final sequence would be accurate. Yet even with this level of redundancy, the assembler software recognized 1604 regions that required further clarification. These regions were then reanalyzed and resequenced. Finally, relevant parts of the shotgun sequence were compared with the sequences of two widely separated genomic regions obtained by conventional cloning. The 81,843 nucleotides cloned and sequenced by the clone-by-clone method were in perfect agreement with the sequence obtained by the shotgun method. This level of care in checking for accuracy is not unusual; similar precautions are taken in almost every genome project.

Once compiled, a genome is analyzed to identify gene sequences, regulatory elements, and other features that reveal important information. In the next section we discuss the central role of bioinformatics in this process.

**NOW SOLVE THIS**

In Problem 23 on page 572, involving an analysis of error rates in GC-rich genomes, you are asked to suggest explanations for the trend shown in the graph presented.

■ HINT: *Consider the potential challenges of aligning contigs from GC-rich genomes or from any sequences that have an abundance of repeated bases.*

## 21.2 DNA Sequence Analysis Relies on Bioinformatics Applications and Genome Databases

Genomics necessitated the rapid development of **bioinformatics**, the use of computer hardware and software and mathematics applications to organize, share, and analyze data related to gene structure, gene sequence and expression, and protein structure and function. However, even before whole-genome sequencing projects had been initiated, a large amount of sequence information from a range of different organisms was accumulating as a result of gene cloning by recombinant DNA techniques. Scientists around the world needed databases that could be used to store, share, and obtain the maximum amount of information from protein and DNA sequences. Thus, bioinformatics software was already being widely used to compare and analyze DNA sequences and to create private and public databases. Once genomics emerged as a new approach for analyzing DNA, however, bioinformatics became even more important than before. Today, it is a dynamic area of biological research, providing new career opportunities for anyone interested in merging an understanding of biological data with information technology, mathematics, and statistical analysis.

Among the most important applications of bioinformatics are to compare DNA sequences, as in contig alignment, discussed in the previous section; to identify genes in a genomic DNA sequence; to find gene regulatory regions, such as promoters and enhancers; to identify structural sequences, such as telomeric sequences, in chromosomes; to predict the amino acid sequence of a putative polypeptide encoded by a cloned gene sequence; to analyze protein structure and predict protein functions on the basis of identified domains and motifs; and to deduce evolutionary relationships between genes and organisms on the basis of sequence information, an application you will learn more about in Chapter 28.

High-throughput DNA sequencing techniques were developed nearly simultaneously with the expansion of the Internet. As genome data accumulated, many DNA-sequence databases became freely available online. Databases are essential for archiving and sharing data with other researchers and with the public. One of the most important genomic databases, called **GenBank**, is maintained by the National Center for Biotechnology Information (NCBI) in Washington, D.C., and is the largest publicly available database of DNA sequences. GenBank shares and acquires data from databases in Japan and Europe; it contains more than 100 billion bases of sequence data from over 100,000 species; and it doubles in size roughly every 14 months! The Human Genome Nomenclature Committee, supported by the NIH, establishes rules for assigning names and symbols to newly cloned human genes. As sequences are identified and genes are named, each sequence deposited into GenBank is provided with an **accession number** that scientists can use to access and retrieve that sequence for analysis.

The NCBI is an invaluable source of public access databases and bioinformatics tools for analyzing genome data. You have already been introduced to NCBI and GenBank through several Exploring Genomics exercises. In Exploring Genomics for this chapter, you will use NCBI and GenBank to compare and align contigs in order to assemble a chromosome segment.

## Annotation to Identify Gene Sequences

One of the fundamental challenges of genomics is that, although genome projects generate tremendous amounts of DNA sequence information, these data are of little use until they have been analyzed and interpreted. Genome projects accumulate nucleotide sequences, and then scientists have to make sense of those sequences. Thus, after a genome has been sequenced and compiled, scientists are faced with the task of identifying gene regulatory sequences and other sequences of interest in the genome so that gene maps can be developed. This process, called **annotation,** relies heavily on bioinformatics, and a wealth of different software tools are available to carry it out.

One initial approach to annotating a sequence is to compare the newly sequenced genomic DNA to the known sequences already stored in various databases. The NCBI provides access to **BLAST (Basic Local Alignment Search Tool)**, a very popular software application for searching through banks of DNA and protein sequence data. Using BLAST, we can compare a segment of genomic DNA to sequences throughout major databases such as GenBank to identify portions that align with or are the same as existing sequences. Figure 21–4 shows a representative example of a sequence alignment based on a BLAST search. Here a 280-bp chromosome 12 contig from the rat was used to search a mouse database to determine whether there was a sequence in the rat contig that matched a known gene in mice. Notice that the rat contig (the query sequence in the BLAST search) aligned with base pairs 174,612 to 174,891 of mouse chromosome 8. BLAST searches calculate a **similarity score**—also called the **identity** value—determined by the sum of identical matches between aligned sequences divided by the total number of bases aligned. Gaps, indicating missing bases in the two sequences, are usually ignored in calculating similarity scores. The aligned rat and mouse sequences were 93 percent similar and showed no gaps in the alignment. Notice that the BLAST report also provides an "Expect" value, or **E value**, based on the number of matches that would be expected by chance in the aligned sequences. Significant alignments, indicating that DNA sequences are significantly similar, typically have E values less than 1.0.

Because this mouse sequence on chromosome 8 is known to contain an insulin receptor gene (encoding a protein that binds the hormone insulin), it is highly likely that the rat contig sequence also contains an insulin receptor gene. We will return to the topic of similarity in Sections 21.3 and 21.7, where we consider how similarity between gene sequences can be used to infer function and to identify evolutionarily related genes through comparative genomics.

## Hallmark Characteristics of a Gene Sequence Can Be Recognized During Annotation

A major limitation of this approach to annotation is that it only works if similar gene sequences are already in a database. Fortunately,

```
ref | NT_039455.6 | Mm8_39495_36
Mus musculus chromosome 8 genomic contig, strain C57BL/6J
Features in this part of subject sequence: insulin receptor
Score = 418 bits (226), Expect = 2e-114
Identities = 262/280 (93%), Gaps = 0/280 (0%)

Query  1       CAGGCCATCCCGAAAGCGAAGATCCCTTGAAGAGGTGGGCAATGTGACAGCCACTACACC  60
               ||||||||||||||||||||||||||||||||||||||| |||||||||||||| |||| 
Sbjct  174891  CAGGCCATCCCGAAAGCGAAGATCCCTTGAAGAGGTGGGGAATGTGACAGCCACCACACT  174832

Query  61      CACACTTCCAGATTTTCCCAACATCTCCTCCACCATCGCGCCCACAAGCCACGAAGAGCA  120
               ||||||||||||||| |||||| ||||||| ||||| | ||||||||| || || |||||
Sbjct  174831  CACACTTCCAGATTTCCCCAACGTCTCCTCTACCATTGTGCCCACAAGTCAGGAGGAGCA  174772

Query  121     CAGACCATTTGAGAAAGTAGTAAACAAGGAGTCACTTGTCATCTCTGGCCTGAGACACTT  180
               ||| |||||||||||||| || ||||||||||||||||||||||||||||||||||||||
Sbjct  174771  CAGGCCATTTGAGAAAGTGGTGAACAAGGAGTCACTTGTCATCTCTGGCCTGAGACACTT  174712

Query  181     CACTGGGTACCGCATTGAGCTGCAGGCATGCAATCAGGACTCCCCAGAAGAGAGGTGCAG  240
               |||||||||||||||||||||||||||||||||||| || |||||||| |||||||||||
Sbjct  174711  CACTGGGTACCGCATTGAGCTGCAGGCATGCAATCAAGATTCCCCAGATGAGAGGTGCAG  174652

Query  241     CGTGGCTGCCTACGTCAGTGCCCGGACCATGCCTGAAGGT  280
                |||||||||||||||||||||||||||||||||||||||
Sbjct  174651  TGTGGCTGCCTACGTCAGTGCCCGGACCATGCCTGAAGGT  174612
```

**FIGURE 21-4** BLAST results showing a 280-base sequence of a chromosome 12 contig from rats (*Rattus norvegicus*, the "query") aligned with a portion of chromosome 8 from mice (*Mus musculus*, the "subject") that contains a partial sequence for the insulin receptor gene. Vertical lines indicate exact matches. The rat contig sequence was used as a query sequence to search a mouse database in GenBank. Notice that the two sequences show 93 percent identity, strong evidence that this rat contig sequence contains a gene for the insulin receptor.

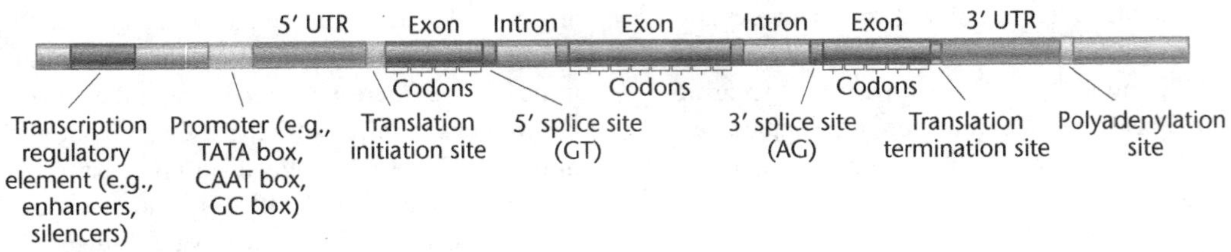


**FIGURE 21-5** Characteristics of a gene that can be used during annotation to identify a gene in an unknown sequence of genomic DNA. Most eukaryotic genes are organized into coding segments (exons) and noncoding segments (introns). When annotating a genome sequence to determine whether it contains a gene, it is necessary to distinguish between introns and exons, gene regulatory sequences, such as promoters and enhancers, untranslated regions (UTRs), and gene termination sequences.

it is not the only way to identify genes. Whether the genome under study is from a eukaryote or a prokaryote, several hallmark characteristics of genes can be searched for using bioinformatics software (Figure 21–5). We discussed many of these characteristics of a "typical" gene in Chapters 14 and 18. They include gene regulatory sequences found upstream of genes, such as promoters, enhancers, and silencers; downstream elements, such as termination sequences; triplet nucleotides that are part of the coding region of the gene; 5′ and 3′ splice sites used to distinguish **introns**, the noncoding regions of a gene, from **exons**, the coding regions of a gene; and polyadenylation sites. Annotation can sometimes be a little bit easier for prokaryotic genes than for eukaryotic genes because there are no introns in prokaryotic genes.

Consider the sequence presented in Figure 21–6(a), which shows a portion of the human genome. From a casual inspection, it is not clear whether this sequence contains any genes and, if so, how many. Analysis of the sequence, however, reveals identifiable features that provide clues to the presence of a protein-coding gene. For instance, control regions at the beginning of genes are marked by identifiable sequences such as promoters and enhancers. Recall from Chapter 18 that TATA box, GC box, and CAAT box sequences are often present in the promoter regions of eukaryotic genes. Recall also that splice sites between exons and introns contain a predictable sequence (most introns begin with GT and end with AG). Splice-site sequences are important for determining intron and exon boundaries. There are also well-defined sequences at the end of the gene, where a polyadenylation sequence signals the addition of a poly(A) tail to the 3′ end of an mRNA transcript (Figure 21–5).

(a)

```
gagccacacc ctagggttgg ccaatctact cccaggagca gggagggcag gagccagggc
tgggcataaa agtcagggca gagccatcta ttgcttacat ttgcttctga cacaactgtg
ttcactagca acctcaaaca gacaccatgg tgcacctgac tcctgaggag aagtctgccg
ttactgccct gtggggcaag gtgaacgtgg atgaagttgg tggtgaggcc ctgggcaggt
tggtatcaag gttacaagac aggtttaagg agaccaatag aaactgggca tgtggagaca
gagaagactc ttgggtttct gataggcact gactctctct gcctattggt ctattttccc
acccttaggc tgctggtggt ctacccttgg acccagaggt tctttgagtc ctttggggat
ctgtccactc ctgatgctgt tatgggcaac cctaaggtga aggctcatgg caagaaagtg
ctcggtgcct ttagtgatgg cctggctcac ctggacaacc tcaagggcac ctttgccaca
ctgagtgagc tgcactgtga caagctgcac gtggatcctg agaacttcag ggtgagtcta
tgggaccctt gatgttttct ttccccttct tttctatggt taagttcatg tcataggaag
gggagaagta acagggtaca gtttagaatg ggaaacagac gaatgattgc atcagtgtgg
aagtctcagg atcgttttag tttcttttat ttgctgttca taacaattgt tttcttttgt
ttaattcttg ctttcttttt ttttcttctc cgcaattttt actattatac ttaatgcctt
aacattgtgt ataacaaaag gaaatatctc tgagatacat taagtaactt aaaaaaaaac
tttacacagt ctgcctagta cattactatt tggaatatat gtgtgcttat ttgcatattc
ataatctccc tactttattt tcttttattt ttaattgata cataatcatt atacatattt
atgggttaaa gtgtaatgtt ttaatatgtg tacacatatt gaccaaatca gggtaatttt
gcatttgtaa ttttaaaaaa tgctttcttc ttttaatata ctttttttgtt tatcttattt
ctaatacttt ccctaatctc tttctttcag ggcaataatg atacaatgta tcatgcctct
ttgcaccatt ctaaagaata acagtgataa tttctgggtt aaggcaatag caatatttct
gcatataaat atttctgcat ataaattgta actgatgtaa gaggtttcat attgctaata
gcagctacaa tccagctacc attctgcttt tattttatgg ttgggataag gctggattat
tctgagtcca agctaggccc ttttgctaat catgttcata cctcttatct tcctcccaca
gctcctgggc aacgtgctgg tctgtgtgct ggcccatcac tttggcaaag aattcacccc
accagtgcag gctgcctatc agaaagtggt ggctggtgtg gctaatgccc tggcccacaa
gtatcactaa gctcgctttc ttgctgtcca atttctatta aaggttcctt tgttccctaa
gtccaactac taaactgggg gatattatga agggccttga gcatctggat tctgcctaat
aaaaaacatt tattttcatt gcaatgatgt atttaaatta tttctgaata ttttactaaa
```

(b)

```
gagccacacc ctagggttgg ccaatctact cccaggagca gggagggcag gagccagggc
tgggcataaa agtcagggca gagccatcta ttgcttacat ttgcttctga cacaactgtg
ttcactagca acctcaaaca gacaccatgg tgcacctgac tcctgaggag aagtctgccg  ⎤
ttactgccct gtggggcaag gtgaacgtgg atgaagttgg tggtgaggcc ctgggcaggt  ⎦ Exon 1
tggtatcaag gttacaagac aggtttaagg agaccaatag aaactgggca tgtggagaca
gagaagactc ttgggtttct gataggcact gactctctct gcctattggt ctattttccc
acccttaggc tgctggtggt ctacccttgg acccagaggt tctttgagtc ctttggggat  ⎤
ctgtccactc ctgatgctgt tatgggcaac cctaaggtga aggctcatgg caagaaagtg  ⎥ Exon 2
ctcggtgcct ttagtgatgg cctggctcac ctggacaacc tcaagggcac ctttgccaca  ⎥
ctgagtgagc tgcactgtga caagctgcac gtggatcctg agaacttcag ggtgagtcta  ⎦
tgggaccctt gatgttttct ttccccttct tttctatggt taagttcatg tcataggaag
gggagaagta acagggtaca gtttagaatg ggaaacagac gaatgattgc atcagtgtgg
aagtctcagg atcgttttag tttcttttat ttgctgttca taacaattgt tttcttttgt
ttaattcttg ctttcttttt ttttcttctc cgcaattttt actattatac ttaatgcctt
aacattgtgt ataacaaaag gaaatatctc tgagatacat taagtaactt aaaaaaaaac
tttacacagt ctgcctagta cattactatt tggaatatat gtgtgcttat ttgcatattc
ataatctccc tactttattt tcttttattt ttaattgata cataatcatt atacatattt
atgggttaaa gtgtaatgtt ttaatatgtg tacacatatt gaccaaatca gggtaatttt
gcatttgtaa ttttaaaaaa tgctttcttc ttttaatata ctttttttgtt tatcttattt
ctaatacttt ccctaatctc tttctttcag ggcaataatg atacaatgta tcatgcctct
ttgcaccatt ctaaagaata acagtgataa tttctgggtt aaggcaatag caatatttct
gcatataaat atttctgcat ataaattgta actgatgtaa gaggtttcat attgctaata
gcagctacaa tccagctacc attctgcttt tattttatgg ttgggataag gctggattat
tctgagtcca agctaggccc ttttgctaat catgttcata cctcttatct tcctcccaca
gctcctgggc aacgtgctgg tctgtgtgct ggcccatcac tttggcaaag aattcacccc  ⎤ Exon 3
accagtgcag gctgcctatc agaaagtggt ggctggtgtg gctaatgccc tggcccacaa  ⎦
gtatcactaa gctcgctttc ttgctgtcca atttctatta aaggttcctt tgttccctaa
gtccaactac taaactgggg gatattatga agggccttga gcatctggat tctgcctaat
aaaaaacatt tattttcatt gcaatgatgt atttaaatta tttctgaata ttttactaaa
```

(c)

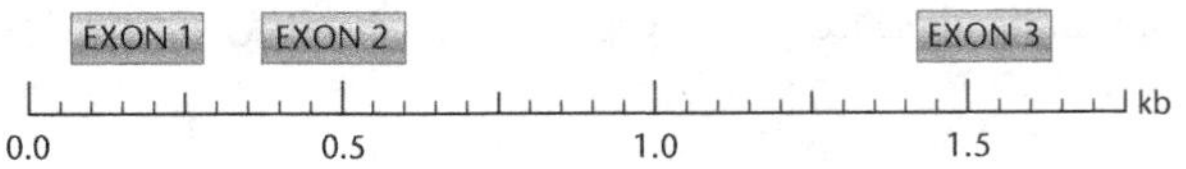


**FIGURE 21-6** Annotation of a DNA sequence containing part of the human $\beta$-globin gene. By convention, the sequence is presented in groups of ten nucleotides, although in reality the sequence is continuous. (a) The location of genes, if any, in this sequence is not readily apparent from a cursory glance. (b) The analyzed sequence, showing the location of an upstream regulatory sequence (green). The red box indicates a start triplet representing a start codon in mRNA. Open reading frames for three exons of the human $\beta$-globin gene are shown in blue. (c) Diagrammatic representation of three exons (Exons 1, 2, and 3) for the human $\beta$-globin gene encoded by the sequence shown in (a).

In addition, protein-coding genes contain one or more **open reading frames (ORFs),** sequences of triplet nucleotides that, after transcription and mRNA splicing, are translated into the amino acid sequence of a protein. ORFs typically begin with an initiation sequence, usually ATG, which transcribes into the AUG start codon of an mRNA molecule, and end with a termination sequence, TAA, TAG, or TGA, which correspond to the stop codons of UAA, UAG, and UGA in mRNA. Genetic information is encoded in groups of three nucleotides (triplets), but it is not always clear whether to begin the analysis of a sequence at the first nucleotide, the second or the third. Typically, the sequence adjacent to a promoter is examined for a start (initiation) triplet; however, ORFs can be used to identify a gene even when a promoter sequence is not apparent. Software programs can then analyze the ORFs three nucleotides at a time. The discovery of an ORF starting with an ATG followed at some distance by a termination sequence is usually a good indication that the coding region of a gene has been identified.

The way genes are organized in eukaryotic genomes (including the human genome) makes direct searching for ORFs more difficult in them than in prokaryotic genomes. First, many eukaryotic genes have introns. As a result, many, if not most, eukaryotic genes are not organized as continuous ORFs; instead, the gene sequences consist of ORFs (exons) interspersed with introns. Second, genes in humans and other eukaryotes are often widely spaced, increasing the chances of finding false ORFs in the regions between gene clusters.

Annotation of the sequence shown in Figure 21–6(a) reveals several identifiable indicators that the sequence contains a protein-coding gene: it includes a promoter sequence, an initation codon, and three exons [Figure 21–6(b)]. The two unshaded regions between the exons represent introns that would be spliced out following transcription when the mRNA is processed [Figure 21–6(b, c)]. Using this sequence as the query in a search of genomic databases would reveal that it is the sequence of a single gene, the human $\beta$-globin gene.

Software designed for ORF analysis of eukaryotic genomes is highly efficient. In addition to the features already mentioned, such software can be used to "translate" ORFs into possible polypeptide sequences as a way to predict the polypeptide encoded by a gene. Shown in Figure 21–7 is a partial sequence for the first exon of the human tubulin alpha 3c gene (*TUBA3C*). Prediction programs scan potential ORFs in both the 5′ to 3′ and 3′ to 5′ direction to predict possible polypeptides from the three possible reading frames in each direction. Figure 21–7 shows the results for the six possible reading frames in the sequence of interest. Amino acids are shown using the single-letter code for each residue. Notice the very different results obtained for each of the six frames. For instance the 5′ to 3′ ORF 1 contains several stop codons interspersed among amino acids but no methionine residues that are evidence of a start codon. Other ORFs would contain too many methionines to produce a functional polypeptide. For this exon of *TUBA3C*, the 5′ to 3′ ORF 2 is correct.

**(a)** *Homo sapiens TUBA3C* (bp 1-300)

1 ggttgaggtcaagtagtagcgttgggctgcggcagcggaggagctcaacatgcgtgagtg
61 tatctctatccacgtggggcaggcaggagtccagatcggcaatgcctgctgggaactgta
121 ctgcctggaacatggaattcagcccgatggtcagatgccaagtgataaaaccattggtgg
181 tggggacgactccttcaacacgttcttcagtgagactggagctggcaagcacgtgcccag
241 agcagtgtttgtggacctggagcccactgtggtcgatgaagtgcgcacaggaacctatag (300)

**(b)** Predicted polypeptides

5'3' Frame 1
G **Stop** G Q V V A L G C G S G G A Q H A **Stop** V Y L Y P R G A G R S P D R Q C L L G T V L P G T W N S A R W S D A K **Stop** **Stop** N H W W W G R L L Q H V L Q **Stop** D W S W Q A R A Q S S V C G P G A H C G R **Stop** S A H R N L **Stop**
5'3' Frame 2
V E V K **Stop** **Stop** R W A A A A E E L N **Met** R E C I S I H V G Q A G V Q I G N A C W E L Y C L E H G I Q P D G Q **Met** P S D K T I G G G D D S F N T F F S E T G A G K H V P R A V F V D L E P T V V D E V R T G T Y
5'3' Frame 3
L R S S S S V G L R Q R R S S T C V S V S L S T W G R Q E S R S A **Met** P A G N C T A W N **Met** E F S P **Met** V R C Q V I K P L V V G T T P S T R S S V R L E L A S T C P E Q C L W T W S P L W S **Met** K C A Q E P I
3'5' Frame 1
L **Stop** V P V R T S S T T V G S R S T N T A L G T C L P A P V S L K N V L K E S S P P P **Met** V L S L G I **Stop** P S G **Stop** I P C S R Q Y S S Q Q A L P I W T P A C P T W I E I H S R **Met** L S S S A A A A Q R Y Y L T S T
3'5' Frame 2
Y R F L C A L H R P Q W A P G P Q T L L W A R A C Q L Q S H **Stop** R T C **Stop** R S R P H H Q W F Y H L A S D H R A E F H V P G S T V P S R H C R S G L L P A P R G **Stop** R Y T H A C **Stop** A P P L P Q P N A T T **Stop** P Q
3'5' Frame 3
I G S C A H F I D H S G L Q V H K H C S G H V L A S S S L T E E R V E G V V P T T N G F I T W H L T I G L N S **Met** F Q A V Q F P A G I A D L D S C L P H V D R D T L T H V E L L R C R S P T L L L D L N

**FIGURE 21–7** Predicted polypeptide sequences translated from potential ORFs in the human *TUBA3C* gene. (a) Nucleotides 1–300 of the first exon in the human *TUBA3C* gene. (b) A translation program predicts six possible polypeptide sequences from this exon. Which predicted sequence is correct?

Prediction programs can also search for **codon bias**, the more frequent use of one or two codons to encode an amino acid that can be specified by a number of different codons. For example, alanine can be encoded by GCA, GCT, GCC, and GCG. If the codons were used randomly, each would be used about 25 percent of the time. Yet in the human genome, GCC is used 41 percent of the time, and GCG only 11 percent of the time. Codon bias is present in exons but should not be present in introns or intergenic spacers.

**NOW SOLVE THIS**

Insights and Solutions problem 1 on page 570 involves searching a genomic sequence for ORFs.

■ **HINT:** *In setting limits for the size of ORFs that may represent genes, consider the average gene length in the organism's genome.*

21.3

## Functional Genomics Attempts to Identify Potential Functions of Genes and Other Elements in a Genome

Reading a genome sequence is a surefire cure for insomnia. What is exciting is not the sequence of the nucleotides but the information that the sequence contains. After a genome has been annotated and ORFs have been identified, the next analytical task is to assign putative functions to all possible genes in the sequence. As the term suggests, **functional genomics** is the study of gene functions, based on the resulting RNAs or possible proteins they encode, and the functions of other components of the genome, such as gene regulatory elements. Functional genomics also considers how genes are expressed and the regulation of gene expression. Aspects of gene-expression analysis in functional genomics will be discussed in Section 21.9.

### Predicting Gene and Protein Functions by Sequence Analysis

Some newly identified genomic sequences may already have had functions assigned to their genes by classic methods such as mutagenesis and linkage mapping, but many other genes that have been sequenced have not yet been correlated with a function. One approach to assigning functions to genes is to use sequence similarity searches, as described in the previous section. Programs such as BLAST are used to search through databases to find alignments between the newly sequenced genome and genes that have already been identified, either in the same or in different species. You were introduced to this approach for predicting gene function in Figure 21–4, when we demonstrated how sequence similarity to the mouse gene was used to identify a gene in a rat contig as the insulin receptor gene. Inferring gene function from similarity searches is based on a relatively simple idea. If a genome sequence shows statistically significant similarity to the sequence of a gene whose function is known, then it is likely that the genome sequence encodes a protein with a similar or related function. Figure 21–8 indicates functional categories that have been assigned to genes in the *Arabidopsis* genome on the basis of similarity searches.

Another major benefit of similarity searches is that they are often able to identify **homologous genes**, genes that are evolutionarily related. After the human genome was sequenced, many ORFs in it were identified as protein-coding genes based on their alignment with related genes of known function in other species. As an example, Figure 21–9 compares portions of the human

(a)

(b)

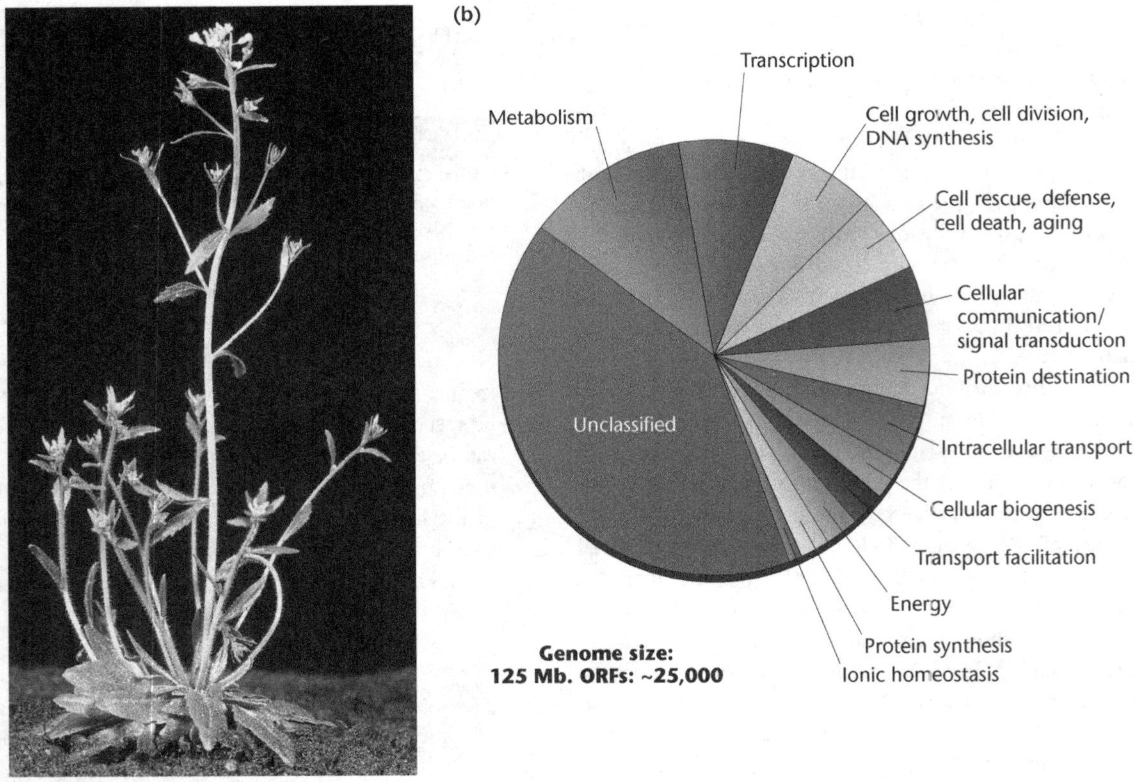


FIGURE 21-8 *Arabidopsis thaliana* (a), a small plant used as a genetic model for studying genome organization and development of flowering plants. (b) Assignment of *Arabidopsis* genes to functional categories based on homology searches.

**Human *LEP* gene**

```
GTCACCAGGATCAATGACATTTCACACACG---TCAGTCTCCTCCAAACAGAAAGTCACC
||||||||||||||||||||||||||||||   || || ||| |||| ||||  ||||| 
GTCACCAGGATCAATGACATTTCACACACGCAGTCGGTATCCGCCAAGCAGAGGGTCACT
```

**Mouse *ob* gene**

```
GGTTTGGACTTCATTCCTGGGCTCCACCCCATCCTGACCTTATCCAAGATGGACCAGACA
|| |||||||||||||||||||| |||||||| ||||  || ||||||||||||||||| 
GGCTTGGACTTCATTCCTGGGCTTCACCCCATTCTGAGTTTGTCCAAGATGGACCAGACT

CTGGCAGTCTACCAACAGATCCTCACCAGTATGCCTTCCAGAAACGTGATCCAAATATCC
||||||||||| |||||| ||||||||||  ||||||||  ||| ||| | || ||| ||
CTGGCAGTCTATCAACAGGTCCTCACCAGCCTGCCTTCCCAAAATGTGCTGCAGATAGCC
```

**FIGURE 21–9** Comparison of the human *LEP* and mouse *ob/Lep* genes. Partial sequences for these homologs are shown with the human *LEP* gene on top and the mouse *ob/Lep* gene sequence below it. Notice from the number of identical nucleotides, indicated by vertical lines, that the nucleotide sequence for these two genes is very similar. Gaps are indicated by horizontal dashes.

leptin gene (*LEP*) with its homolog in mice (*ob/Lep*). These two genes are over 85 percent identical in sequence. The leptin gene was first discovered in mice. The match between the *LEP*-containing DNA sequence in humans and the mouse homolog sequence confirms the identity and leptin-coding function of this gene in human genomic DNA.

As an interesting aside, the leptin gene (also called *ob*, for obesity, in mice) is highly expressed in fat cells (adipocytes). This gene produces the protein hormone leptin, which targets cells in the brain to suppress appetite. Knockout mice lacking a functional *ob* gene grow dramatically overweight. A similar phenotype has been observed in small numbers of humans with particular mutations in *LEP*. Although it is important to note that weight control is not regulated by a single gene, the discovery of leptin has provided significant insight into lipid metabolism and weight disorders in humans. Further studies on leptin will be important for understanding more about the genetics of weight disorders.

If homologous genes in different species are thought to have descended from a gene in a common ancestor, the genes are known as **orthologs**. In Section 21.7 we will consider the globin gene family. Mouse and human α-globin genes are orthologs evolved from a common ancestor. Homologous genes in the same species are called **paralogs**. The α- and β-globin subunits in humans are paralogs resulting from a gene-duplication event. Paralogs often have similar or identical functions.

## Predicting Function from Structural Analysis of Protein Domains and Motifs

When a gene sequence is used to predict a polypeptide sequence, the polypeptide can be analyzed for specific structural domains and motifs. Identification of **protein domains**, such as ion channels, membrane-spanning regions, DNA-binding regions, secretion and export signals, and other structural aspects of a polypeptide that are encoded by a DNA sequence, can in turn be used to predict protein function. Recall from Chapter 18, for example, that the structures of many DNA-binding proteins have characteristic patterns, or **motifs**, such as the helix-turn-helix, leucine zipper, or zinc finger motifs (see Figures 18–11, 18–12, 18–13). These can easily be searched for using bioinformatics software, and their identification in a sequence is a common strategy for inferring possible functions of a protein.

# 21.4 The Human Genome Project Reveals Many Important Aspects of Genome Organization in Humans

Now that you have a general idea of the strategies used for analyzing a genome, let's look at the largest genomics project completed to date. The **Human Genome Project (HGP)** was a coordinated international effort to determine the sequence of the human genome and to identify all the genes it contains. It has produced a plethora of information, much of which is still being analyzed and interpreted. What is clear so far, from all the different kinds of genomes sequenced, is that humans and all other species share a common set of genes essential for cellular function and reproduction, confirming that all living organisms arose from a common ancestor. These evolutionary relationships facilitate the analysis of the human genome and underlie the use of model organisms to study inherited human diseases.

## Origins of the Project

The publicly funded Human Genome Project began in 1990 under the direction of James Watson, the co-discoverer of the double-helix structure of DNA. Eventually the public project was led by Dr. Francis Collins, who had previously led a research team involved in identifying the *CFTR* gene as the cause of cystic fibrosis. In the United States, the Collins-led HGP was coordinated by the Department of Energy and the National Center of Human Genome Research, a division of the National Institutes of Health. It established a 15-year plan with a proposed budget of $3 billion to identify all human genes, originally thought to number between 80,000 and 100,000, to sequence and map them all, and to sequence the approximately 3 billion base pairs thought to comprise the 24 chromosomes (22 autosomes, plus X and Y) in humans. Other primary goals of the HGP included the following:

- To establish functional categories for all human genes
- To analyze genetic variations between humans, including the identification of single-nucleotide polymorphisms (SNPs)

- To map and sequence the genomes of several model organisms used in experimental genetics, including *E. coli*, *S. cerevisiae*, *C. elegans*, *D. melanogaster*, and *M. musculus* (mouse)
- To develop new sequencing technologies, such as high-throughput computer-automated sequencers, in order to facilitate genome analysis
- To disseminate genome information, among both scientists and the general public

Lastly, to deal with the impact that genetic information would have on society, the HGP set up the **ELSI program** (standing for Ethical, Legal, and Social Implications) to consider ethical, legal, and social issues arising from the HGP and to ensure that personal genetic information would be safeguarded and not used in discriminatory ways.

As the HGP grew into an international effort, scientists in 18 countries were involved in the project. Much of the work was carried out by the International Human Genome Sequence Consortium, involving nearly 3000 scientists working at 20 centers in six countries (China, France, Germany, Great Britain, Japan, and the United States).

In 1999, a privately funded human genome project led by J. Craig Venter at Celera Genomics (aptly named from a word meaning "swiftness") was announced. Celera's goal was to use whole-genome shotgun sequencing and computer-automated high-throughput DNA sequencers to sequence the human genome more rapidly than HGP. The public project had proposed using a clone-by-clone approach to sequence the genome. Recall that Venter and colleagues had proven the potential of shotgun sequencing in 1995 when they completed the genome for *H. influenzae*. Celera's announcement set off an intense competition between the two teams, which both aspired to be first with the human genome sequence. This contest eventually led to the HGP finishing ahead of schedule and under budget after scientists from the public project began to use high-throughput sequencers and whole-genome sequencing strategies as well.

## Major Features of the Human Genome

In June 2000, the leaders of the public and private genome projects met at the White House with President Clinton and jointly announced the completion of a draft sequence of the human genome. In February 2001, they each published an analysis covering about 96 percent of the euchromatic region of the genome. The public project sequenced euchromatic portions of the genome 12 times and set a quality control standard of a 0.01 percent error rate for their sequence. Although this error rate may seem very low, it still allows about 600,000 errors in the human genome sequence. Celera sequenced certain areas of the genome more than 35 times when compiling the genome.

The remaining work of completing the sequence by filling in gaps clustered around centromeres, telomeres, and repetitive sequences, correcting misaligned segments, and re-sequencing portions of the genome to ensure accuracy was completed in 2003. The major features of the human genome are summarized in Table 21.1. As you can see in this table, many unexpected observations have provided us with major new insights. In many ways, the HGP has revealed just how little we know about our genome.

Two of the biggest surprises discovered by the HGP were that less than 2 percent of the genome codes for proteins and that there are only around 20,000 protein-coding genes. Recall that the number of genes had originally been estimated to be about 100,000, based in part on a prediction that human cells produce about 100,000 proteins. At least half of the genes show sequence similarity to genes shared by many other organisms, and as you will learn in Section 21.7, a majority of human genes are similar in sequence to genes from closely related species such as chimpanzees. The exact number of human genes is still not certain. One reason is that it is

**TABLE 21.1**

**Major Features of the Human Genome**

- The human genome contains 3.1 billion nucleotides, but protein-coding sequences make up only about 2 percent of the genome.
- The genome sequence is ~99.9 percent similar in individuals of all nationalities. Single-nucleotide polymorphisms (SNPs) and copy number variations (CNVs) comprise the majority of sequence differences from person to person.
- At least 50 percent of the genome is derived from transposable elements, such as LINE and *Alu* sequences, and other repetitive DNA sequences.
- The human genome contains approximately 20,000 protein-coding genes, far fewer than the predicted number of 80,000–100,000 genes.
- Many human genes produce more than one protein through alternative splicing, thus enabling human cells to produce a much larger number of proteins (perhaps as many as 200,000) from only ~20,000 genes.
- More than 50 percent of human genes show a high degree of sequence similarity to genes in other organisms; however, more than 40 percent of the genes identified have no known molecular function.
- Genes are not uniformly distributed on the 24 human chromosomes. Gene-rich clusters are separated by gene-poor "deserts" that account for 20 percent of the genome. These deserts correlate with G bands seen in stained chromosomes. Chromosome 19 has the highest gene density, and chromosome 13 and the Y chromosome have the lowest gene densities.
- Chromosome 1 contains the largest number of genes, and the Y chromosome contains the smallest number.
- Human genes are larger and contain more and larger introns than genes in the genomes of invertebrates, such as *Drosophila*. The largest known human gene encodes dystrophin, a muscle protein. This gene, associated in mutant form with muscular dystrophy, is 2.5 Mb in length (Chapter 14), larger than many bacterial chromosomes. Most of this gene is composed of introns.
- The number of introns in human genes ranges from 0 (in histone genes) to 234 (in the gene for *titin*, which encodes a muscle protein).

unclear whether or not many of the presumed genes produce functional proteins. Genome scientists continue to annotate the genome and hope to answer these questions soon.

The number of genes is much lower than the number of predicted proteins in part because many genes code for multiple proteins through **alternative splicing**. Recall from Chapter 18 that alternative splicing patterns can generate multiple mRNA molecules, and thus multiple proteins, from a single gene, through different combinations of intron-exon splicing arrangements. Estimates suggest that over 50 percent of human genes undergo alternative splicing to produce multiple transcripts and multiple proteins.

Functional categories have been assigned for many human genes, primarily on the basis of (1) functions determined previously (for example, from recombinant DNA cloning of human genes and known mutations involved in human diseases), (2) comparision to known genes and predicted protein sequences from other species, and (3) predictions based on annotation and analysis of protein functional domains and motifs (Figure 21–10). Although functional categories and assignments continue to be revised, the functions of over 40 percent of human genes remain unknown. Determining human gene functions, deciphering complexities of gene-expression regulation and gene interaction, and uncovering the relationships between human genes and phenotypes are among the many challenges for genome scientists.

The HGP has also shown us that in all humans, regardless of racial and ethnic origins, the genomic sequence is approximately 99.9 percent the same. As you will learn elsewhere in the book, most genetic differences between humans result from single-nucleotide polymorphisms (SNPs) and copy number variations (CNVs).

It is now possible to access databases and other sites on the Internet that display maps for all human chromosomes. You will visit a number of these databases in Exploring Genomics exercises. Figure 21–11(a) displays a partial gene map for chromosome 12 that was taken from an NCBI database called Map Viewer. You may already have used Map Viewer for the Exploring Genomics exercises in Chapters 5 and 12. This image shows an ideogram, or cytogenetic map, of chromosome 12. To the right of the ideogram is a column showing the contigs (arranged lying vertically) that were aligned to sequence this chromosome. The Hs UniG column displays a histogram representation of gene density on chromosome 12. Notice that relatively few genes are located near the centromere. Gene symbols, loci, and gene names (by description) are provided for selected genes; in this figure only 20 genes are shown. When accessing these maps on the Internet, one can magnify, or zoom in on, each region of the chromosome, revealing all genes mapped to a particular area.

You can see that most of the genes listed here have been assigned descriptions based on the functions of their products, some of which are transmembrane proteins, some enzymes such as kinases, some receptors, including several involved in olfaction, and so on. Other genes are described in terms of hypothetical products; they are presumed to be genes based on the presence of ORFs, but their function remains unknown [Figure 21–11(a)].

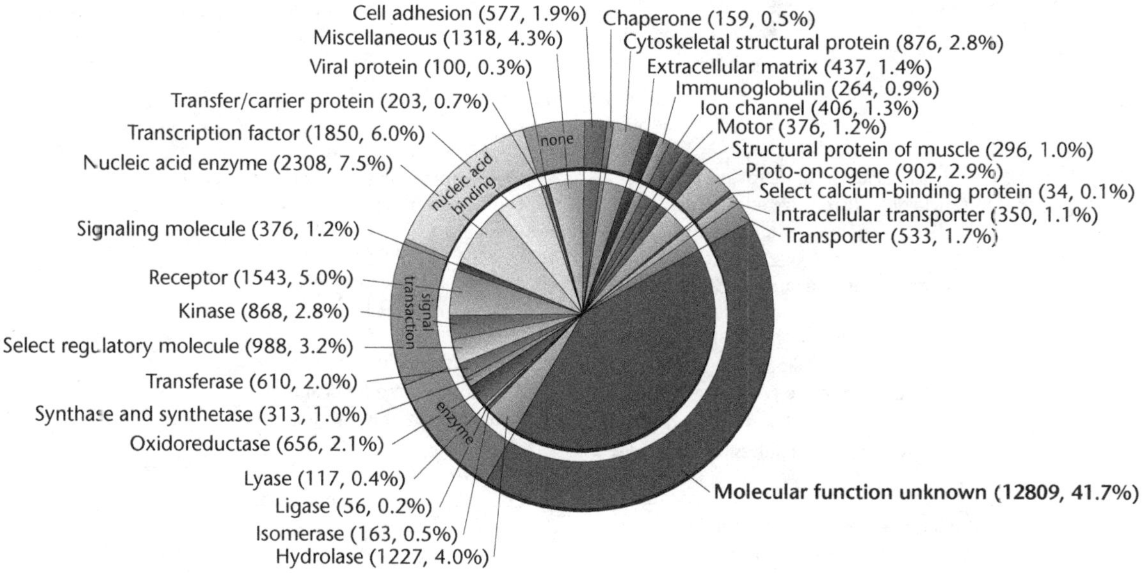


FIGURE 21–10 A preliminary list of the functional categories to which genes in the human genome have been assigned on the basis of similarity to proteins of known function. Among the most common genes are those involved in nucleic acid metabolism (7.5 percent of all genes identified), transcription factors (6.0 percent), receptors (5 percent), hydrolases (4 percent), protein kinases (2.8 percent), and cytoskeletal structural proteins (2.8 percent). A total of 12,809 predicted proteins (41 percent) have unknown functions, indicative of the work that is still needed to fully decipher our genome.

(a)

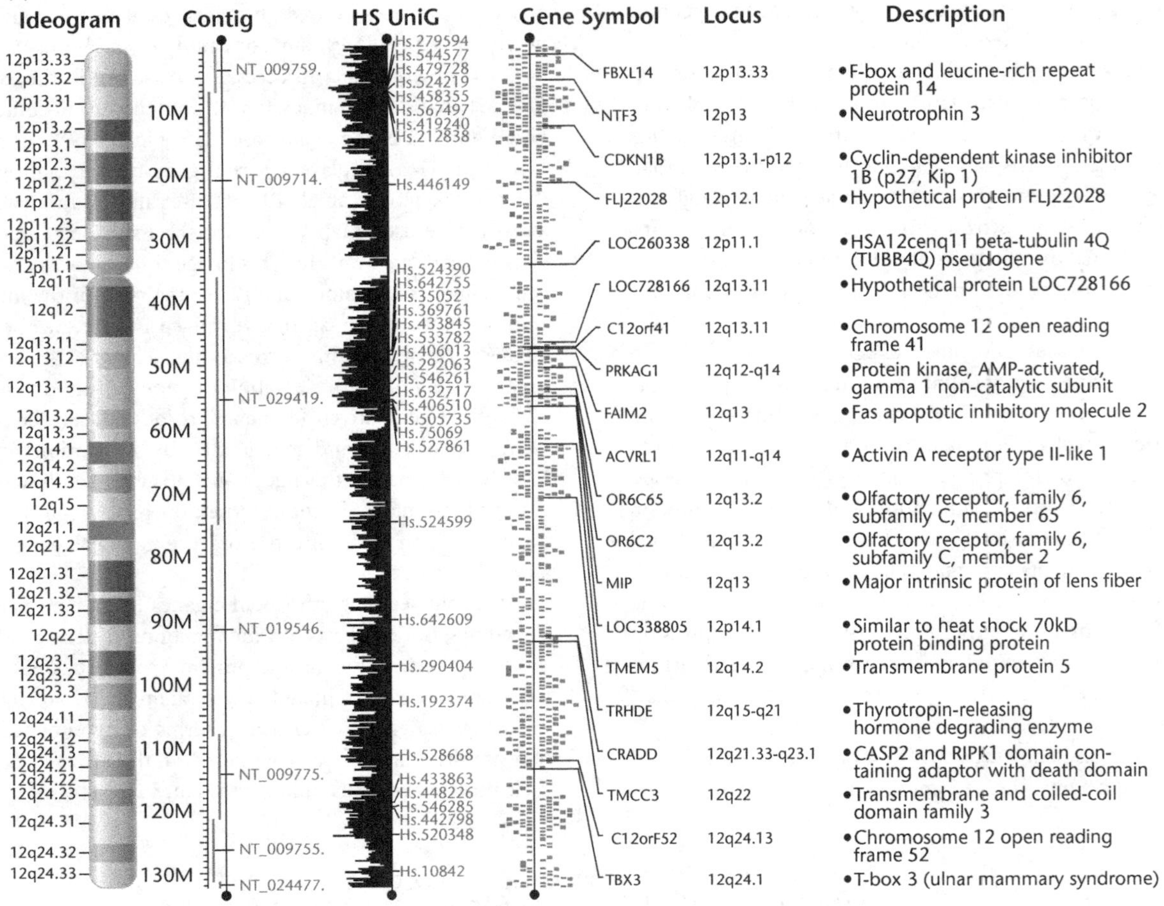


(b)

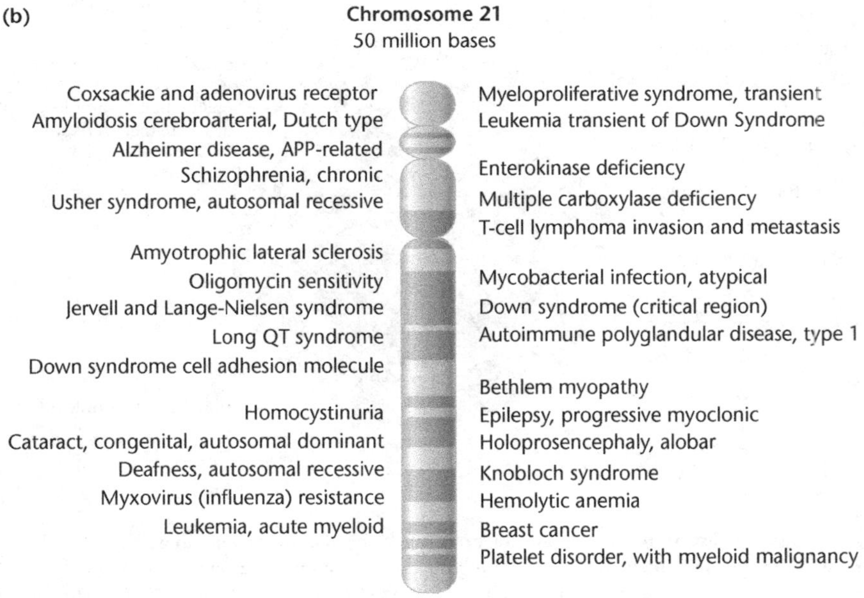


**FIGURE 21–11** (a) A gene map for chromosome 12 from the NCBI database Map Viewer. (b) Partial map of disease genes on human chromosome 21. Maps such as this depict genes thought to be involved in human genetic disease conditions.

Perhaps the most valuable contribution of the HGP will be the identification of disease genes and the development of new treatment strategies as a result. Thus, extensive maps have also been developed for genes implicated in human disease conditions. The disease gene map of chromosome 21 shown in Figure 21–11(b) indicates genes involved in amyotrophic lateral sclerosis (ALS), Alzheimer disease, cataracts, deafness, and several different cancers. In Chapter 24 we will discuss implications of the HGP for the identification of genes involved in human genetic diseases, and for disease diagnosis, detection, and gene therapy applications.

As the Human Genome Project was being completed, a group of about 3 dozen research teams around the world began the **Encyclopedia of DNA Elements (ENCODE) Project**. The main goal of ENCODE is to use both experimental approaches and bioinformatics to identify and analyze functional elements (such as transcriptional start sites, promoters, and enhancers) that regulate expression of human genes. Prior to ENCODE approximately 532 promoters had been identified, but now in excess of 775 promoters have been identified in the human genome, with many other potential promoter sequences being analyzed.

## 21.5 The "Omics" Revolution Has Created a New Era of Biological Research Methods

The Human Genome Project and the development of genomics techniques has been largely responsible for launching a new era of biological research—the era of "omics." It seems that every year, more areas of biological research are being described as having an omics connection. Some examples of "omics" are

- proteomics—the analysis of all the proteins in a cell or tissue
- metabolomics—the analysis of proteins and enzymatic pathways involved in cell metabolism
- glycomics—the analysis of the carbohydrates of a cell or tissue
- toxicogenomics—the analysis of the effects of toxic chemicals on genes, including mutations created by toxins and changes in gene expression caused by toxins
- metagenomics—the analysis of genomes of organisms collected from the environment
- pharmacogenomics—the development of customized medicine based on a person's genetic profile for a particular condition
- transcriptomics—the analysis of all expressed genes in a cell or tissue

We will consider several of these genomics disciplines in other parts of this chapter.

As further evidence of the impact of genomics, a new field of nutritional science called nutritional genomics, or **nutrigenomics**, has emerged. Nutrigenomics focuses on understanding the interactions between diet and genes. We have all had routine medical tests for blood pressure, blood sugar levels, and heart rate. Based on these tests, your physician may recommend that you change your diet and exercise more to lose weight, or that you reduce your intake of sodium to help lower your blood pressure. Now several companies claim to provide nutrigenomics tests that analyze your genomes for genes thought to be associated with different medical conditions or aspects of nutrient metabolism. The companies then provide a customized nutrition report, recommending diet changes for improving your health and preventing illness, based on your genes! It remains to be seen whether this approach as currently practiced is of valid scientific or nutritional value.

In yet another example of how genomics has taken over areas of DNA analysis, a number of labs around the world are involved in analyzing "ancient" DNA. These so called **stone-age genomics** studies are generating fascinating data from miniscule amounts of ancient DNA obtained from bone and other tissues that are tens of thousands of years old. Analysis of DNA from a 2400-year-old Egyptian mummy, mammoths, Pleistocene-age cave bears, and Neanderthals are some of the most prominent examples of stone age genomics. In 2005, researchers from McMaster University in Canada and Pennsylvania State University published about 13 million bp from a 27,000-year-old woolly mammoth. This study revealed a ~98.5 percent sequence identity between mammoths and African elephants. Such work is also a great demonstration of how stable DNA can be under the right conditions, particularly when frozen.

A team of scientists led by Svante Pääbo at the Max Planck Institute for Evolutionary Anthropology in Germany is working to produce a rough draft of the Neanderthal (*Homo neanderthalensis*) genome. In 1997, Pääbo's lab sequenced portions of Neanderthal mitochondrial DNA from a fossil. In late 2006, Pääbo's group along with a number of scientists in the United States reported the first sequence of ~65,000 bp of nuclear DNA isolated from bone of a 38,000-year-old Neanderthal sample from Croatia. Because Neanderthals are close relatives of humans, sequencing the Neanderthal genome is expected to provide an unprecedented opportunity to use comparative genomics (see Section 21.7) to advance our understanding of evolutionary relationships between modern humans and our predecessors.

## 21.6 Prokaryotic and Eukaryotic Genomes Display Common Structural and Functional Features and Important Differences

The genomes of over 400 prokaryotic and eukaryotic organisms—including many model organisms and a number of viruses—have been sequenced. Among these organisms are yeast (*Saccharomyces cerevisiae*), bacteria such as *E. coli*, the nematode roundworm (*Caenorhabditis elegans*), the thale cress plant (*Arabidopsis thaliana*), mice (*Mus musculus*), zebrafish (*Danio rerio*), and of course *Drosophila*. In the past few years, genomes for chimpanzees, dogs, chickens, sea urchins, honey bees, pufferfish, rice, and wheat have all been sequenced.

These studies have demonstrated significant differences in genome organization between prokaryotes and eukaryotes but also many similarities between genomes of nearly all species. In this section we provide a basic overview of genome organization in prokaryotes and eukaryotes and discuss interesting aspects of genomes in selected organisms.

## Unexpected Features of Prokaryotic Genomes

Since most prokaryotes have small genomes amenable to shotgun cloning and sequencing, many genome projects have focused on prokaryotes, and more than 900 additional projects to sequence prokaryotic genomes are now under way. Many of the prokaryotic genomes already sequenced are from organisms that cause human diseases, such as cholera, tuberculosis, and leprosy. On the basis of genome project results, we can make a number of generalizations about the size of bacterial genomes. Traditionally, the bacterial genome has been thought of as relatively small (less than 5 Mb) and contained within a single circular DNA molecule. *E. coli*, used as the prototypical bacterial model organism in genetics, has a genome with these characteristics. However, the flood of genomic information now available has challenged the validity of this viewpoint for bacteria in general (Table 21.2). Although most prokaryotic genomes are small, their sizes vary across a surprisingly wide range. In fact, there is some overlap in size between larger bacterial genomes (30 Mb in *Bacillus megaterium*) and smaller eukaryotic genomes (12.1 Mb in yeast). Gene number in bacterial genomes also demonstrates a wide range, from less than 500 to more than 5000 genes, a tenfold difference. In addition, although many bacteria have a single, circular chromosome, there is substantial variation in chromosome organization and number among bacterial species.

TABLE 21.2

**Genome Size and Gene Number in Selected Prokaryotes**

| | Genome Size (Mb) | Number of Genes |
|---|---|---|
| Archaea | | |
| *Archaeoglobus fulgidis* | 2.17 | 2437 |
| *Methanococcus jannaschii* | 1.66 | 1783 |
| *Thermoplasma acidophilium* | 1.56 | 1509 |
| Eubacteria | | |
| *Escherichia coli* | 4.64 | 4289 |
| *Bacillus subtilis* | 4.21 | 4779 |
| *Haemophilus influenzae* | 1.83 | 1738 |
| *Aquifex aeolicus* | 1.55 | 1749 |
| *Rickettsia prowazekii* | 1.11 | 834 |
| *Mycoplasma pneumoniae* | 0.82 | 680 |
| *Mycoplasma genitalium* | 0.58 | 483 |

TABLE 21.3

**Bacterial Genome Organization**

| Bacterial species | Chromosome Number and Organization | Plasmid Number |
|---|---|---|
| *Agrobacterium tumefaciens* | one linear + one circular | two circular |
| *Bacillus subtilis* | one circular | |
| *Bacillus thuringiensis* | one circular | six |
| *Borrelia burgdorferi* | one linear | >17 circular + linear |
| *Buchnera sp.* | one circular | two circular |
| *Deinococcus radiodurans* | two circular | two circular |
| *Escherichia coli* | one circular | |
| *Rhodobacter sphaeroides* | two circular | five circular |
| *Sinorhizobium meliloti* | one circular | two megaplasmids |
| *Vibrio cholerae* | two circular | |

Although most prokaryotic genomes sequenced to date are single, circular DNA molecules, an increasing number of genomes composed of linear DNA molecules are being identified, including the genome of *Borrelia burgdorferi*, the organism that causes Lyme disease. Perhaps more important, genome data are blurring the distinction between plasmids and bacterial chromosomes, thus redefining our view of the bacterial genome as a single DNA molecule (Table 21.3). **Plasmids** are small, circular DNA molecules containing genes that are not essential to their bacterial host cell (discussed in Chapter 8). Plasmids self-replicate and are distributed to daughter cells along with the host chromosome during cell division. Because plasmids carry nonessential genes and can be transferred from one cell to another, and because the same plasmid is often present in bacteria of different species, plasmid genes are usually not considered to be part of a bacterial genome. However, *B. burgdorferi* carries approximately 17 plasmids, containing a total of at least 430 genes. Some of these genes are essential to the bacterium, including genes for purine biosynthesis and membrane proteins, which in most species are carried on the bacterial chromosome.

Sequencing of the *Vibrio cholerae* genome (this is the organism responsible for cholera) revealed the presence of two circular chromosomes (Figure 21–12). Chromosome 1 (2.96 Mb) contains 2770 ORFs, and chromosome 2 (1.07 Mb) contains 1115 ORFs, some of which encode essential genes, such as ribosomal proteins. Its nucleotide sequence suggests that chromosome 2 is derived from a plasmid captured by an ancestral species, or it could represent a plasmid with many inserted chromosomal genes. Two chromosomes are also present in other species of *Vibrio*.

Other bacteria that have genomes with two or more chromosomes include *Agrobacterium tumefaciens*, *Deinococcus radiodurans*, and *Rhodobacter sphaeroides* (Table 21.3). The finding that some bacterial species have multiple chromosomes raises questions about how replication and segregation of their chromosomes are coordinated during cell division, and about what undiscovered mechanisms of gene regulation may exist in bacteria. Answers to these and

(a)

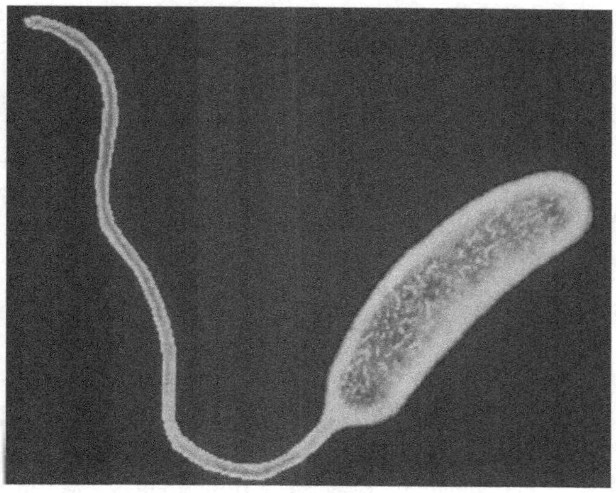

(b)

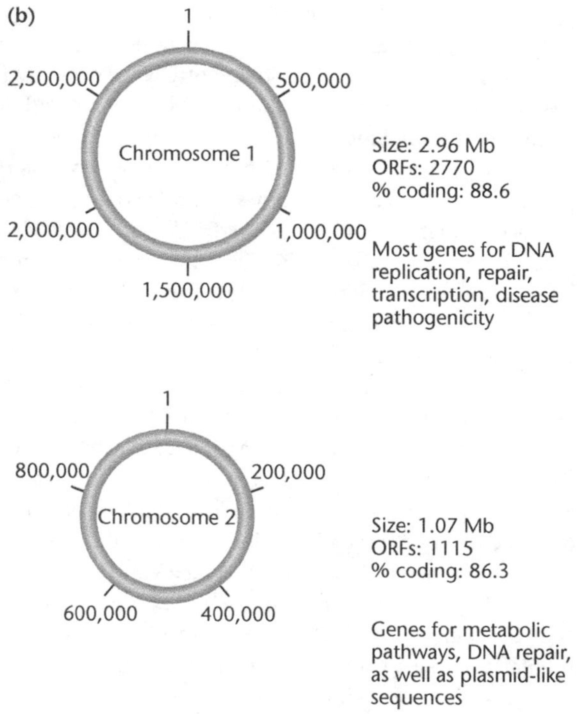


**FIGURE 21-12** (a) *Vibrio cholerae*, the organism that causes cholera. (b) The *Vibrio* genome is contained in two chromosomes. The larger chromosome (chromosome 1) contains most of the genes for essential cellular functions and infectivity. Most of the genes on chromosome 2 (52 percent of 1115) are of unknown function. The imbalance in the chromosomes' gene content and the presence of plasmid-like sequences on chromosome 2 suggest that this chromosome was a megaplasmid captured by an ancestral *Vibrio* species or arose by transfer of chromosomal genes to a plasmid.

other questions raised by genomic discoveries will redefine some ideas about prokaryotic genomes, the nature of plasmids, and the differences between plasmids and chromosomes, and may provide clues about the evolution of multichromosome eukaryotic genomes.

The genome of *E. coli* strain K12 was sequenced using the shotgun method in 1997 and was the second prokaryotic genome to be sequenced. *E. coli* has a genome of 4.1 Mb, and annotation shows that it contains 4289 ORFs organized as a single circular chromosome. ORFs occupy almost 88 percent of the genome, regulatory sequences about 11 percent, and repetitive sequences only about 0.7 percent. Even though the genetics of *E. coli* had been intensively studied for almost 50 years using classic genetic methods, almost 38 percent of the annotated ORFs were for genes with no known functions.

To reach the goal of assigning functions to all the genes in *E. coli*, and to understand how these genes and their products interact in the biology of the organism, researchers are creating (1) knockout mutations for every gene, (2) a clone for every ORF, (3) a gene expression database for patterns of expression under a variety of physiological conditions, and (4) a three-dimensional model of every protein encoded in the genome. The current classifications of *E. coli* genes are shown in Figure 21–13. Genes with no known homology and no known functions now comprise 19 percent of the 4289 ORFs in the genome.

(a)

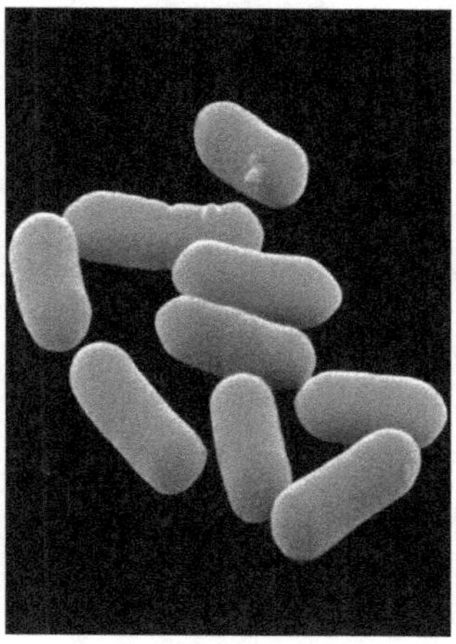

(b)

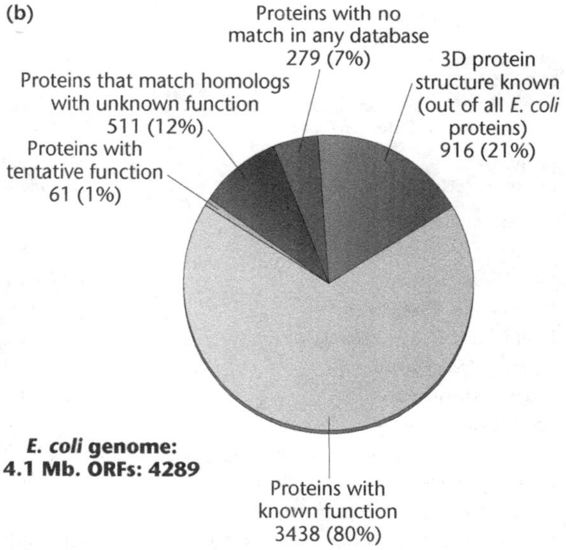


**FIGURE 21-13** *E. coli*, a key model organism for studies of bacterial genetics. (b) Status of gene functional assignments for the *E. coli* genome. The total exceeds 100 percent because of overlap between proteins with known functions and those whose 3D structure is known.

We can make two generalizations about the organization of protein-coding genes in bacteria. First, gene density is very high, averaging about one gene per kilobase of DNA. For example, *E. coli* has a 4.6-Mb genome containing 4289 protein-coding genes. *Mycoplasma genitalium* has a small genome (0.58 Mb), with 483 genes, and its gene density is also close to one gene per 1000 base pairs. This close packing of genes in prokaryotic genomes means that a very high proportion of the DNA (approximately 85 to 90 percent) serves as coding DNA. Typically, only a small amount of a bacterial genome is noncoding DNA, often in the form of regulatory sequences or of transposable elements that can move from one place to another in the genome.

A second generalization we can make is that bacterial genomes contain operons (recall from Chapter 17 that operons contain multiple genes functioning as a transcriptional unit whose protein products are part of a common biochemical pathway). In *E. coli*, 27 percent of all genes are contained in operons (almost 600 operons). In other bacterial genomes, the organization of genes into transcriptional units is challenging our ideas about the nature of operons. For example, in *Aquifex aeolicus*, one polygenic transcription unit contains six genes involved in several different cellular processes with no apparent common relationships: two genes for DNA recombination, one for lipid synthesis, one for nucleic acid synthesis, one for protein synthesis, and one that encodes a protein for cell motility. Other polygenic transcription units in this species also contain genes with widely different functions. This finding, combined with similar results from other genome projects, raises interesting questions about the consensus that operons encode products that control a single metabolic pathway in bacterial cells.

## Organizational Patterns of Eukaryotic Genomes

We now turn our attention to eukaryotic genomes. The human genome has been the most extensively studied eukaryotic genome to date, and we discussed many of its structural features in the previous section when we considered major findings of the Human Genome Project. Here we provide an overview of organizational patterns in eukaryotic genomes and discuss interesting aspects of the genomes for yeast and *Arabidopsis*, two important eukaryotic model organisms. Our discussion on eukaryotic genomes continues in Section 21.7, where we review genomic data for the dog, chimpanzee, sea urchin, and Rhesus monkey.

Although the basic features of eukaryotic genomes are similar in different species, genome size is highly variable (Table 21.4). Genome sizes range from about 10 Mb in fungi to over 100,000 Mb in some flowering plants (a ten-thousandfold range); the number of chromosomes per genome ranges from two into the hundreds (about a hundredfold range), but the number of genes varies much less dramatically than either genome size or chromosome number.

Eukaryotic genomes have several features not found in prokaryotes:

- **Gene density.** In prokaryotes, gene density is close to 1 gene per kilobase. In eukaryotic genomes, there is a wide range of gene density. In yeast, there is about 1 gene/2 kb, in *Drosophila*, about 1 gene/13 kb, and in humans, gene density varies greatly from chromosome to chromosome. Human chromosome 22 has about 1 gene/64 kb, while on chromosome 13 there is 1 gene/155 kb of DNA.

TABLE 21.4

**Comparison of Selected Genomes**

| Organism (Scientific Name) | Approximate Size of Genome (Date Completed) | Number of Genes | Approximate Percentage of Genes Shared with Humans |
|---|---|---|---|
| Bacterium (*Escherichia coli*) | 4.1 Mb (1997) | 4,403 | not determined |
| Chicken (*Gallus gallus*) | 1 Gb (2004) | ~20,000-23,000 | 60% |
| Dog (*Canis familiaris*) | 2.5 Gb (2003) | ~18,400 | 75% |
| Chimpanzee (*Pan troglodytes*) | ~3 Gb (2005) | ~20,000-24,000 | 98% |
| Fruit fly (*Drosophila melanogaster*) | 165 Mb (2000) | ~13,600 | 50% |
| Human (*Homo sapiens*) | ~2.9 Gb (2004) | ~20,000 | 100% |
| Mouse (*Mus musculus*) | ~2.5 Gb (2002) | ~30,000 | 80% |
| Rat (*Rattus norvegicus*) | ~2.75 Gb (2004) | ~22,000 | 80% |
| Rhesus macaque (*Macaca mulatta*) | 2.87 Gb (2007) | ~20,000 | 93% |
| Rice (*Oryza sativa*) | 389 Mb (2005) | ~41,000 | not determined |
| Roundworm (*Caenorhabditis elegans*) | 97 Mb (1998) | 19,099 | 40% |
| Sea urchin (*Strongylocentrotus purpuratus*) | 814 Mb (2006) | ~23,500 | 60% |
| Thale cress (plant) (*Arabidopsis thaliana*) | 140 Mb (2000) | ~27,500 | not determined |
| Yeast (*Saccharomyces cerevisiae*) | 12 Mb (1996) | ~5,700 | 30% |

Adapted from Palladino, M. A. *Understanding the Human Genome Project*, 2nd ed. Benjamin Cummings, 2006.

NOTE: Billion bp (gigabase, Gb).

- **Introns.** Most eukaryotic genes contain introns. There is wide variation among genomes in the number of introns they contain and also wide variation from gene to gene. The entire yeast genome has only 239 introns, whereas just a single gene in the human genome can contain more than 100 introns. Regarding intron size, generally the size in eukaryotes is correlated with genome size. Smaller genomes have smaller average introns, and larger genomes have larger average intron sizes.
- **Repetitive sequences.** The presence of introns and the existence of repetitive sequences are two major reasons for the wide range of genome sizes in eukaryotes. In some plants, such as maize, repetitive sequences are the dominant feature of the genome. The maize genome has about 2500 Mb of DNA, and more than two-thirds of that genome is composed of repetitive DNA. In the human, as discussed previously, about half of the genome is repetitive DNA.

## The Yeast Genome

The yeast (*Saccharomyces cerevisiae*) genome was the first eukaryotic genome sequenced. As outlined in Chapter 13, yeast can be grown and manipulated as easily as some prokaryotes. That feature, along with yeast's highly developed genetic map and a large collection of yeast mutants, made it an excellent candidate for genome sequencing.

The yeast genome was sequenced chromosome by chromosome using a map-based approach. The genome contains 12.1 Mb of DNA, distributed over 16 chromosomes. In the years of analysis following publication of the yeast genome, the estimated number of ORFs has changed several times. Originally thought to have about 6200 genes, yeast is currently thought to have ~5700 genes; the number will probably change again as more evidence is gathered. Meanwhile, the number of characterized genes has risen to over 4000, and an additional 1400 or so yeast genes have been assigned likely functions on the basis of homology. This leaves about several hundred genes to be characterized, some of which may turn out not to be genes at all.

## Plant Genomes

To geneticists, the flowering plant *Arabidopsis thaliana* is the fruit fly of the plant world [Figure 21–8(a)]. *Arabidopsis* is small (several hundred can be grown in an area the size of this page), has a short generation time, and has a relatively small genome. Because the plant kingdom evolved independently from the animal kingdom, the analysis of *Arabidopsis* can provide insight into the ways in which evolutionary and coevolutionary adaptations have shaped genomes in both kingdoms. The *Arabidopsis* genome was sequenced using map-based sequencing. Its 140 Mb genome is distributed in five chromosomes containing an estimated total of 27,500 genes, with a gene density of 1 gene per 5 kb [Figure 21–8(b)]. At least half the genes in the *Arabidopsis* genome are identical to or closely related to genes found in bacteria and humans, but it also contains genes encoding dozens of protein families that may be unique to plants.

Both genome duplications and gene duplications have played a large role in the evolution of the *Arabidopsis* genome. Many of the 25,000 genes are duplicated, and there are fewer than 15,000 different genes. For example, chromosomes 2 and 4 of *Arabidopsis* contain many tandem gene duplications (239 duplications on chromosome 2, involving 539 genes) as well as larger duplications involving four blocks of DNA sequences, spanning 2.5 Mb. There is some interchromosomal gene duplication, too: genes on chromosome 4 are also present on chromosome 5.

Many crop plants, such as maize, rice, and barley, have much larger genomes than *Arabidopsis*, but some have about the same number of genes. In contrast to *Arabidopsis*, genes in these large-genome plants are clustered in stretches of DNA separated by long stretches of intergenic spacer DNA. Collectively, these gene clusters occupy only about 12 to 24 percent of the genome. In maize, the intergenic DNA is composed mainly of transposons (genes that can shift location in the genome; they are discussed in Chapter 22).

Sequencing of the rice genome was completed in 2005. Analysis shows that this cereal crop's genome contains about 389 Mb of DNA, encoding approximately 41,000 genes on 12 chromosomes. About 15 percent of all rice genes are found in duplicated segments of the genome. Close to 90 percent of the genes in *Arabidopsis* are found in rice, but only 70 percent of genes in rice are found in *Arabidopsis*, indicating that in addition to genes found in other flowering plants, cereal crops may also contain unique gene sets. Identification of these genes and their functions in rice and other cereal plants will be critical in helping improve crop yields to feed the Earth's growing population.

Recently, the first genome for a tree, the black cottonwood, a type of poplar, was sequenced. The poplar's 45,555 genes is the highest number found in a genome to date. Scientists anticipate using cottonwood genome data to help the forestry industry make better products, such as biofuels or genetically engineered poplars capable of removing high levels of carbon dioxide from the atmosphere.

## The Minimum Genome for Living Cells

Studying genomes has led to the fundamental question of what is the minimum number of genes necessary to support life. Although we do not yet know the answer to this question for free-living organisms and multicellular eukaryotes, we can use the small genomes of obligate parasites to speculate on the minimum number of genes *required to maintain life*. For example, we can compare sequence information from the bacterial genomes of *Mycoplasma genitalium* and *M. pneumoniae*, two closely related human parasitic pathogens. These organisms are among the simplest self-replicating prokaryotes known and can serve as model systems for understanding the essential functions of a self-replicating cell. *M. genitalium* has a genome of 580 kb, and that of *M. pneumoniae* is 816 kb. Both bacteria cause diseases in a wide range of hosts, including insects, plants, and humans. (In humans, they cause genital and respiratory infections.)

The *M. genitalium* genome, with 483 protein-coding genes, is the smallest bacterial genome sequenced to date. The *M. pneumoniae* genome has those same 483 genes and an additional 194, for a total of 677 protein-coding genes. In contrast, the 1.8 Mb genome of *Haemophilus influenzae* (the first bacterial genome sequenced) has 1783 genes.

The availability of genome sequences allows us to ask whether the 483 genes carried by *M. genitalium* (and shared with *M. pneumoniae*) are close to the minimum gene set needed for life. In other words, can we define life in terms of a number of specific genes? A combination of comparative and experimental methods can be used to answer this question. The comparative approach is based on the premise that genes shared by distantly related organisms are likely to be essential for life. By comparing the gene sets of different organisms, it should be possible to catalog those that are shared and develop a list of genes needed for life.

*M. genitalium* and *H. influenzae* last shared a common ancestor about 1.5 billion years ago; this is another clue that the genes shared between these species may represent the ones essential for life. By comparing the nucleotide sequences of the *M. genitalium* genes with the *H. influenzae* genes, researchers identified 240 genes that are orthologous for these species. In addition to the 240 shared genes, 16 genes with different sequences but identical functions were identified. These represent essential functions performed by nonorthologous genes. Thus, comparative genomics estimates that 256 genes may represent the minimum gene set needed for life.

J. Craig Venter and his colleagues used an experimental approach to determine how many of the 480 *M. genitalium* genes are essential for life. In a method described in Chapter 22, they used transposons to selectively mutate genes in *M. genitalium*. Mutations in essential genes would produce a lethal phenotype, but mutation of nonessential genes would not affect viability. They found that many of the 480 genes were nonessential and that the minimum gene set for *M. genitalium* is about 265 to 300 genes. This is close to the estimate of 256 derived from comparative genome analysis.

Other scientists have used recombinant DNA technology to construct synthetic copies of viral genomes. This approach has demonstrated that a minimal genome can be created and assembled for nonliving entities such as viruses. For example, the genome for polio virus and the 1918 influenza strain responsible for the pandemic flu have been assembled this way. Ongoing projects of this kind are likely to provide interesting perspectives on the minimal genome in the near future.

## 21.7 Comparative Genomics Analyzes and Compares Genomes from Different Organisms

Analysis of the growing number of genome sequences confirms that all living organisms are related and descended from a common ancestor. Similar gene sets are used in all organisms for basic cellular functions, such as DNA replication, transcription, and translation. These genetic relationships are the rationale for the use of model organisms to study inherited human disorders, the effects of the environment on genes, and interactions of genes in complex diseases, such as cardiovascular disease, diabetes, neurodegenerative conditions, and behavioral disorders.

**Comparative genomics** compares the genomes of different organisms in order to answer questions about genetics and other aspects of biology. It is a field with many research and practical applications, including gene discovery and the development of model organisms to study human diseases. It also incorporates the study of gene and genome evolution and the relationship between organisms and their environment. Comparative genomics uses a wide range of techniques and resources, such as the construction and use of nucleotide and protein databases containing nucleic acid and amino acid sequences, fluorescent *in situ* hybridization (FISH), and the creation of gene knockout animals. Comparative genomics can reveal genetic differences and similarities between organisms to provide insight into how those differences contribute to differences in phenotype, life cycle, or other attributes, and to ascertain the evolutionary history of those genetic differences.

In this section, we discuss aspects of comparative genomics in model systems and consider how comparative genomics is contributing to identification of evolutionarily conserved gene families. As mentioned earlier, the Human Genome Project sequenced genomes from a number of model nonhuman organisms too, including *E. coli, Arabidopsis thaliana, Saccharomyces cerevisiae, Drosophila melanogaster,* the nematode roundworm *Caenorhabditis elegans,* and the mouse *Mus musculus.* Complete genome sequences of such organisms have been invaluable for comparative genomics studies of gene function in these organisms and in humans. As shown in Table 21.4, the number of genes humans share with other species is very high, ranging from about 30 percent of the genes in yeast to ~80 percent in mice and ~98 percent in chimpanzees. The human genome even contains around 100 genes that are also present in many bacteria. Comparative genomics has also shown us that many mutated genes involved in human disease are also present in model organisms. For instance, approximately 60 percent of genes mutated in nearly 300 human diseases are also found in *Drosophila.* These include genes involved in prostate, colon, and pancreatic cancers; cardiovascular disease; cystic fibrosis; and several other conditions.

### The Dog as a Model Organism

Recently, the genome for "man's best friend" was completed, and it revealed that we share about 75 percent of our genes with dogs. A rough draft of the dog (*Canis familiaris*) genome was published in 2003, and a more thorough analysis was completed in 2005, providing one of the most useful models with which to study our own genome [Figure 21–14]. Dogs have a genome that is similar in size to the human genome: about 2.5 billion base pairs with an estimated 18,400 genes [Figure 21–14(b)].

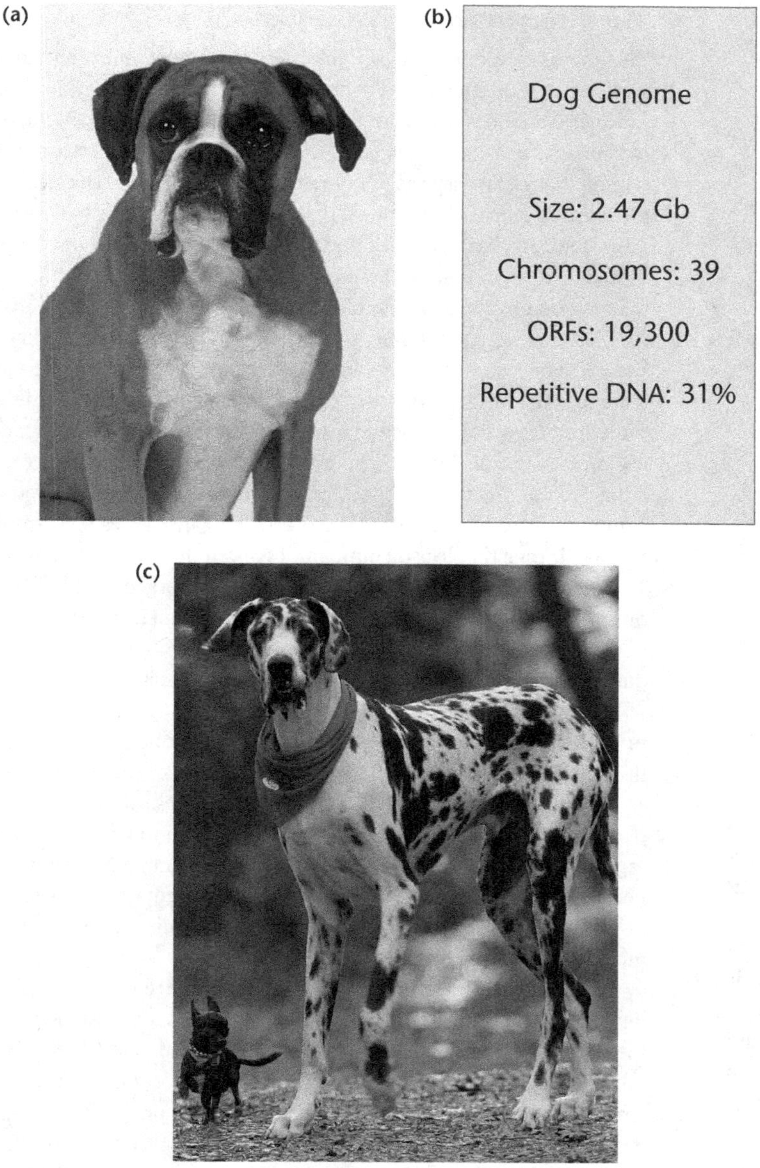


FIGURE 21–14 (a) A boxer named Tasha was one dog chosen for the dog genome project because this breed, like humans, shows relatively little genetic variation. (b) Basic characteristics of the dog genome. (c) A Chihuahua alongside a Great Dane that is over 50 times its mass. The genetic basis for varied phenotypes in dogs, such as body size, is being revealed by analyzing the dog genome.

The dog offers several advantages for studying heritable human diseases. Dogs share many genetic disorders with humans, including over 400 single-gene disorders, sex chromosome aneuploidies, multifactorial diseases (such as epilepsy), behavioral conditions (such as obsessive-compulsive disorder), and genetic predispositions to cancer, blindness, heart disease, and deafness. About 90 conserved blocks of the dog genome can be mapped to human chromosomes by comparative FISH studies. These blocks contain DNA sequences with a high degree of similarity between the dog and the human genomes, reflecting the evolutionary relationship between dogs and our species.

Genomic analysis indicates that the molecular causes of at least 60 percent of inherited diseases in dogs, such as point mutations and deletions, are similar or identical to those found in humans. In addition, at least 50 percent of the genetic diseases in dogs are breed-specific, so that the mutant allele segregates in relatively homogeneous genetic backgrounds. Dog breeds resemble isolated human populations in having a small number of founders and a long period of relative genetic isolation. These properties make individual dog breeds useful as models of human genetic disorders.

In addition, differences in biology and behavior among dog breeds are well documented. Domestic dogs show greater variation in body size than all other living terrestrial vertebrates. The size difference between a Chihuahua and a Great Dane is an excellent example [Figure 21–14(c)]. Mapping DNA sequence differences (polymorphisms) among breeds may help identify genes that contribute to both physiological and behavioral differences. For instance, in 2007, genome-wide analysis of different large and small dog breeds revealed a locus on chromosome 15 where a single-nucleotide polymorphism in the insulin-like growth factor 1 gene (*Igf1*) is common in all small breeds of dogs but virtually absent from large dogs. It is well known that *Igf1* plays important roles in growth-hormone-regulated increases in muscle mass and bone growth during adolescence in humans. This study provides very strong evidence that mutation of *Igf1* is a primary determinant of body size in small dogs.

Dog breeders are now using genetic tests to screen dogs for inherited disease conditions, for coat color in Labrador retrievers and poodles, and for fur length in Mastiffs. Undoubtedly, we can expect many more genetic tests for dogs in the near future, including DNA analysis for size, type of tail, speed, sense of smell, and other traits deemed important by breeders and owners.

## The Chimpanzee Genome

Although the chimpanzee (*Pan troglodytes*) genome was not part of the HGP, its nucleotide sequence was completed in 2004. Overall, the chimp and human sequences differ by less than 2 percent, and 98 percent of the genes are the same. Comparisons between these genomes offer some interesting insights into what makes some primates humans and others chimpanzees.

The speciation events that separated humans and chimpanzees occurred less than 6.3 million years ago (mya). Recent evidence from genomic analysis indicates that these species initially diverged but then exchanged genes again before separating completely. Their separate evolution after this point is exhibited in such differences as are seen between the sequence of chimpanzee chromosome 22 and

TABLE 21.5

**Comparisons Between Human Chromosome 21 and Chimpanzee Chromosome 22**

| | Human 21 | Chimpanzee 22 |
|---|---|---|
| Size (bp) | 33,127,944 | 32,799,845 |
| G + C Content | 40.94 | 41.01 |
| CpG Islands | 950 | 885 |
| SINEs (*Alu* elements) | 15,137 | 15,048 |
| Genes | 284 | 272 |
| Pseudogenes | 98 | 89 |

its human ortholog, chromosome 21 (Table 21.5; chimps have 48 chromosomes and humans have 46, so the numbering is different). These chromosomes have accumulated nucleotide substitutions that total 1.44 percent of the sequence. The most surprising difference is the discovery of 68,000 nucleotide insertions or deletions, collectively called **indels,** in the chimp and human chromosomes, a frequency of 1 indel every 470 bases. Many of these are *Alu* insertions in human chromosome 21. Although the overall difference in the nucleotide sequence is small, there are significant differences in the encoded genes. Only 17 percent of the genes analyzed encode identical proteins in both chromosomes; the other 83 percent encode genes with one or more amino acid differences.

Comparison of these two chromosomes also demonstrates **synteny**, another organizational feature of genomes that has become evident through comparative genomics. In synteny, segments of compared chromosomes, both coding and noncoding stretches, occur in the same conserved order along the chromosome. There are several different classes of synteny; these include conservation of entire chromosomes and conservation of large segments of a chromosome. Other segments of the human and chimpanzee genome also show synteny. Substantial portions of the human and mouse genome display synteny in over 90 percent of the genome.

Differences in the time and place of gene expression also play a major role in differentiating the two primates. Using DNA microarrays (discussed in Section 21.9), researchers compared expression patterns of 202 genes in human and chimp cells from brain and liver. They found more species-specific differences in expression of brain genes than liver genes. To further examine these differences, Svante Pääbo and colleagues compared expression of 10,000 genes in human and chimpanzee brains and found that 10 percent of genes examined differ in expression in one or more regions of the brain. More importantly, these differences are associated with genes in regions of the human genome that have been duplicated subsequent to the divergence of chimps and humans. This finding indicates that genome evolution, speciation, and gene expression are interconnected. Further work on these segmental duplications and the genes they contain may identify genes that help make us human.

## The Rhesus Monkey Genome

The Rhesus macaque monkey (*Macaca mulatta*), another primate, has served as one of the most important model organisms in biomedical research. Macaques have played central roles in our understanding of cardiovascular disease, aging, diabetes, cancer, depression, osteoporosis, and many other aspects of human health. They have been essential for research on AIDS vaccines and for the development of polio vaccines. You may also be familiar with these monkeys from the discovery of Rh factor (named after Rhesus monkeys), the red blood cell protein that determines whether your blood type is Rh – or Rh +.

The macaque's genome is the first monkey genome to have been sequenced. A main reason geneticists are so excited about the completion of this sequencing project is that macaques provide a more distant evolutionary window that is ideally suited for comparing and analyzing human and chimpanzee genomes. As we discussed in the preceding section, humans and chimpanzees shared a common ancestor approximately 6 mya. But macaques split from the ape lineage that led to chimpanzees and humans about 25 mya. The macaque and human genome have thus diverged farther from one another, as evidenced by the ~93 percent sequence identity between humans and macaques compared to the ~98 percent sequence identity shared by humans and chimpanzees.

The macaque genome was published in 2007, and it was no surprise to learn that it consists of 2.87 billion bp (similar to the size of the human genome) contained in 22 chromosomes (20 autosomes, an X, and a Y) with ~20,000 protein-coding genes. Although comparative analyses of this genome are ongoing, a number of interesting features have been revealed so far. As in humans, about 50 percent of the genome consists of repeat elements (transposons, LINEs, SINEs). Gene duplications and gene families are abundant, including cancer gene families found in humans.

A number of interesting surprises have also been observed. For instance, recall from Chapter 4 and elsewhere our discussion about the genetic disorder phenylketonuria (PKU), an autosomal recessive inherited condition in which individuals cannot metabolize the amino acid phenylalanine due to mutation of the phenylalanine hydroxylase (*PAH*) gene. The histidine substitution encoded by a mutation in the *PAH* gene of humans with PKU appears as the wild-type amino acid in the protein from healthy macaques. As another example, the macaque has more than three times the number of **major histocompatibility complex (MHC)** genes present in humans. MHCs encode tissue antigens that are important for immune responses designed to recognize and distinguish foreign cells from body cells. Further analysis of the macaque genome and comparison to the human and chimpanzee genome will be invaluable for geneticists studying genetic variations that played a role in primate evolution.

## The Sea Urchin Genome

In 2006, researchers from the Sea Urchin Genome Sequencing Consortium completed the 814 million bp genome of the sea urchin *Strongylocentrotus purpuratus* [pictured in Figure 21–22(a)]. Sea urchins are shallow-water marine invertebrates that have served as important model organisms, particularly for developmental biologists. One reason is that the sea urchin is a nonchordate deuterostome, and humans, with their spinal cord, are chordate deuterostomes. Fossil records indicate that sea urchins appeared during the Early Cambrian period, around 520 mya.

A combination of whole-genome shotgun sequencing and map-based cloning in BACs was used to complete the genome. Sea urchins have an estimated 23,500 genes, including representative genes for just about all major vertebrate gene families. Sequence alignment and homology searches demonstrate that the sea urchin contains many genes with important functions in humans, yet interestingly, important genes in flies and worms, such as certain cytochrome P-450 genes that play a role in the breakdown of toxic compounds, are missing from sea urchins. The sea urchin genome also has an abundance (~25 to 30 percent) of **pseudogenes,** nonfunctional relatives of protein-coding genes (we meet pseudogenes again in the next subsection). Sea urchins have a smaller average intron size than humans, supporting the general trend revealed by comparative genomics that intron size is correlated with overall genome size.

Another genome trend that urchins share with other eukaryotes is the presence of genes involved in innate immunity, the inborn defense mechanisms that provide broad-spectrum protection against many pathogens. Sea urchins have an extraordinarily rich number of genes providing innate immunity. For example, one very important category of innate immunity genes, the Toll-like receptors (TLRs), produce transmembrane proteins that are essential for pathogen recognition in nearly every cell type of vertebrates. Sea urchins have over 200 *Tlr* genes compared to 11 in humans. The abundance of these and other important innate immunity genes in sea urchins has led to categorizing these genes as the urchin "defensome." This characteristic of sea urchins may help explain how these organisms have adapted so well to the pathogen-loaded environments of sea beds.

Urchins have nearly 1000 genes for sensing light and odor, indicative of great sensory abilities. In this respect, their genome is more typical of vertebrates than invertebrates. A number of orthologs of human genes involved in hearing and balance are present in the sea urchin, as are many human-disease-associated orthologs, including protein kinases, GTPases, transcription factors, TLRs, transporters, and low-density lipoprotein receptors. Approximately 7000 orthologs are shared between sea urchins and humans.

Another interesting aspect of the sea urchin genome project is that it has identified genes previously thought to be vertebrate-specific. One example, the *WntA* gene, important for patterning during embryonic development, as discussed in Chapter 19, was thought to be absent from non-vertebrate deuterostomes. The sea urchin genome also contains genes that are not present in chordates. Further analysis of the urchin genome is expected to make important contributions to our understanding of evolutionary transitions between invertebrates and vertebrates.

## Evolution and Function of Multigene Families

Comparative genomics has also proven to be valuable for identifying members of **multigene families,** groups of genes that share similar but not identical DNA sequences through duplication and descent from a single ancestral gene. Their gene products frequently have similar functions, and the genes are often, but not always, found at a single chromosomal locus. A group of related multigene families is called a **superfamily.** Sequence data from genome projects is providing evidence that multigene families are present in many, if not all, genomes. To demonstrate how analysis of multigene families provides insight into eukaryotic genome evolution and function, we will now examine the globin gene superfamily, whose members encode very similar but not identical polypeptide chains with closely related functions. Other well-characterized gene superfamilies include the histone, tubulin, actin, and immunoglobulin (antibody) gene superfamilies.

Recall that paralogs, which we defined in Section 21.3, are homologous genes present in the same single organism, believed to have evolved by gene duplication. The globin genes that encode the polypeptides in hemoglobin molecules are a paralogous multigene superfamily that arose by duplication and dispersal to occupy different chromosomal sites. One of the best-studied examples of gene family evolution is the **globin gene superfamily** (Figure 21–15). In this

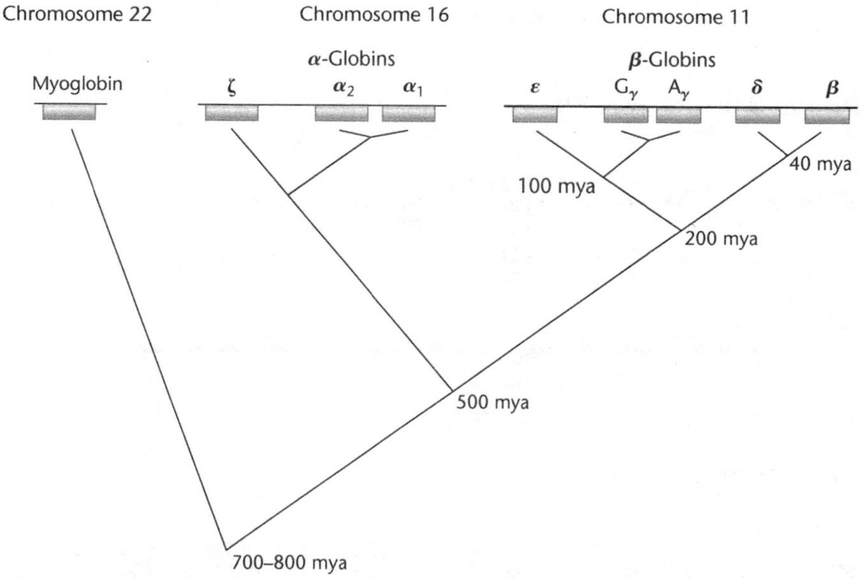


FIGURE 21–15 The evolutionary history of the globin gene superfamily. A duplication event in an ancestral gene gave rise to two lineages about 700 to 800 million years ago (mya). One line led to the myoglobin gene, which is located on chromosome 22 in humans; the other underwent a second duplication event about 500 mya, giving rise to the ancestors of the $\alpha$-globin and $\beta$-globin gene subfamilies. Duplications beginning about 200 mya formed the $\beta$-globin gene subfamilies. In humans, the $\alpha$-globin genes are located on chromosome 16, and the $\beta$-globin genes are on chromosome 11.

$\alpha$-globin V – L S P A D K T N V K A A W G K V G A H A G E Y G A E A L E R M F L S F P T T K T Y F P H F – D L S H
$\beta$-globin V H L T P E E K S A V T A L W G K V – – N V D E V G G E A L G R L L V V Y P W T Q R F F E S F G D L S T

$\alpha$-globin – – – G S A Q V K G H G K K V A D A L T N A V A H V D D M P N A L S A L S D L H A H K L R V D P V N
$\beta$-globin A V M G N P K V K A H G K K V L G A F S D G L A H L D N L K G T F A T L S E L H C D K L H V D P E N

$\alpha$-globin L L S H C L L V T L A A H L P A E F T P A V H A S L D K F L A S V S T V L T S K Y R 141 amino acids
$\beta$-globin L L G N V L V C V L A H H F G K E F T P P V Q A A Y Q K V V A G V A N A L A H K Y H 146 amino acids

**FIGURE 21–16** The amino acid sequences of the $\alpha$- and $\beta$-globin proteins, depicted using the single-letter abbreviations for the amino acids (see Figure 15–16). Shaded areas indicate identical amino acids. The two proteins are slightly different in length. $\alpha$-globin contains 141 amino acids while $\beta$-globin is 146 amino acids long. Gaps in the two sequences, representing areas that do not align, are indicated by horizontal dashes (-).

family, an ancestral gene encoding an oxygen transport protein was duplicated about 800 mya, producing two sister genes, one of which evolved into the modern-day myoglobin gene. **Myoglobin** is an oxygen-carrying protein found in muscle. The other gene underwent further duplication and divergence about 500 mya and formed prototypes of the $\alpha$-globin and $\beta$-globin genes. These genes encode proteins found in **hemoglobin**, the oxygen-carrying molecule in red blood cells. Additional duplications within these genes occurred within the last 200 million years. Events subsequent to each duplication dispersed these gene subfamilies to different chromosomes, and in the human genome, each now resides on a separate chromosome. Similar patterns of evolution are observed in other gene families, including the trypsin–chymotrypsin family of proteases, the homeotic selector genes of animals, and the rhodopsin family of visual pigments.

Adult hemoglobin is a tetramer containing two $\alpha$- and two $\beta$-polypeptides (refer to Figure 15–20 for the structure of hemoglobin). Each polypeptide incorporates a heme group that reversibly binds oxygen. The $\alpha$-globin gene cluster on chromosome 16 and the $\beta$-globin gene cluster on chromosome 11 share nucleotide- and amino acid–sequence similarity (Figure 21–16), but the highest degree of sequence similarity is found within subfamilies.

The **$\alpha$-globin** gene subfamily [Figure 21–17(a)] contains three genes: the $\zeta$ (zeta) gene, expressed only in early embryogenesis, and two copies of the $\alpha$ gene, expressed during the fetal ($\alpha_1$) and adult stages ($\alpha_2$). In addition, the cluster contains two pseudogenes (similar to $\zeta$ and $\alpha_1$), which in this family are designated by the prefix $\psi$ (psi) followed by the symbol of the gene they most resemble. Thus, the designation $\psi\alpha_1$ indicates a pseudogene of the fetal $\alpha_1$ gene.

The organization of the $\alpha$-globin subfamily members and the locations of their introns and exons demonstrate several characteristic features [Figure 21–17(a)]. First, as is common in eukaryotes,

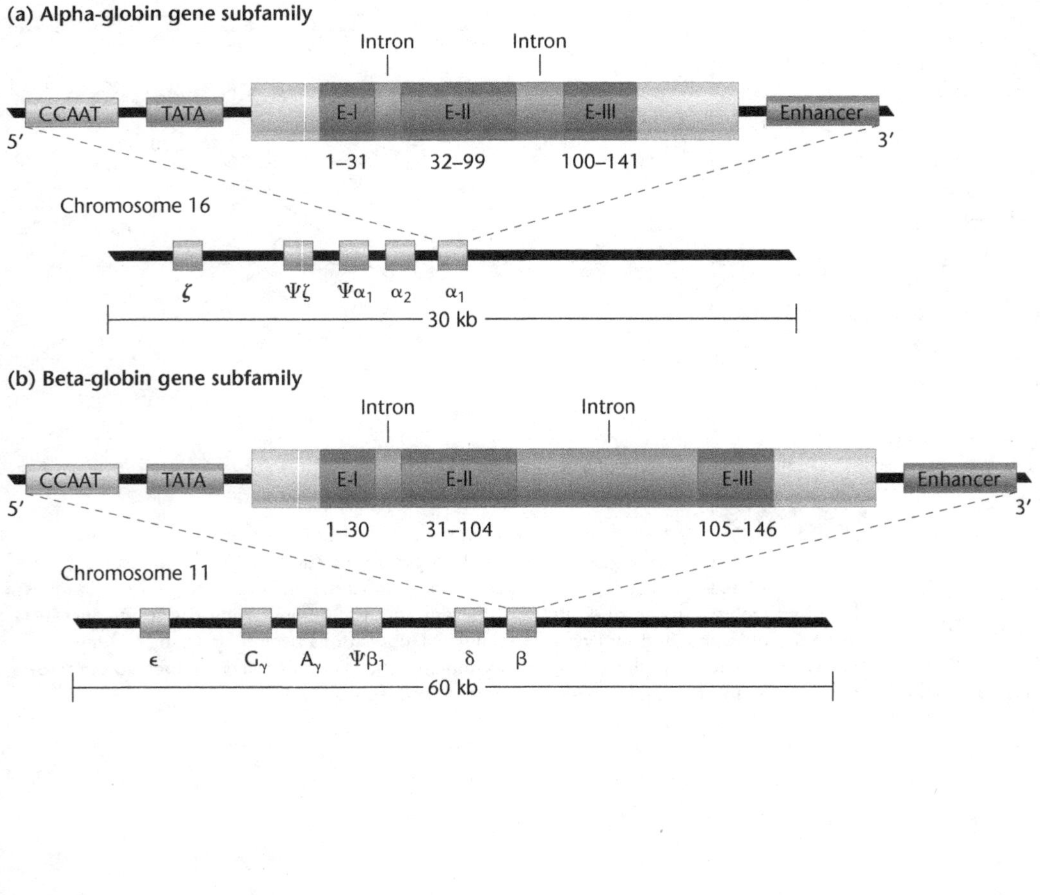


**FIGURE 21–17** Organization of (a) the $\alpha$-globin gene subfamily on chromosome 16 and (b) the $\beta$-globin gene subfamily on chromosome 11. Also shown in each case is the internal organization of the $\alpha_1$ gene and the $\beta$ gene, respectively. Both of the two genes contain three exons (E-I, E-II, E-III) and two introns. The numbers below the exons indicate the location of amino acids in the gene product encoded by each exon.

the DNA encoding the three functional $\alpha$ genes occupies only a small portion of the 30-kb region containing the subfamily. Most of the DNA in this region is intergenic spacer DNA. Second, each functional gene in this subfamily contains two introns at precisely the same positions. Third, the nucleotide sequences within corresponding exons are nearly identical in the $\zeta$ and $\alpha$ genes. Each of these genes encodes a polypeptide chain of 141 amino acids. However, their intron sequences are highly divergent, even though the introns are about the same size. Note that much of the nucleotide sequence of each gene is contained in these noncoding introns.

The human **$\beta$-globin** gene cluster contains five genes spaced over 60 kb of DNA [Figure 21–17(b)]. In this and the $\alpha$-globin gene subfamily, the order of genes on the chromosome parallels their order of expression during development. Three of the five genes are expressed before birth. The $\varepsilon$ (epsilon) gene is expressed only during embryogenesis, while the two nearly identical $\gamma$ genes ($G_\gamma$ and $A_\gamma$) are expressed only during fetal development. The polypeptide products of the two $\gamma$ genes differ only by a single amino acid. The two remaining genes, $\delta$ and $\beta$, are expressed after birth and throughout life. A single pseudogene, $\psi\beta_1$, is present in this subfamily. All five functional genes in this cluster encode proteins with 146 amino acids and have two similar-sized introns at exactly the same positions. The second intron in the $\beta$-globin subfamily is significantly larger than its counterpart in the functional $\alpha$-globin subfamily. These features reflect the evolutionary history of each subfamily and the events such as gene duplication, nucleotide substitution, and chromosome translocations that produced the present-day globin superfamily.

## 21.8 Metagenomics Applies Genomics Techniques to Environmental Samples

**Metagenomics**, also called **environmental genomics**, is the use of whole-genome shotgun approaches to sequence genomes from entire communities of microbes in environmental samples of water, air, and soil. Oceans, glaciers, deserts, and virtually every other environment on Earth are being sampled for metagenomics projects. Human genome pioneer J. Craig Venter left Celera in 2003 to form the J. Craig Venter Institute, and his group has played a central role in developing metagenomics as an emerging area of genomics research.

One of the institute's major initiatives has been a global expedition to sample marine and terrestrial microorganisms from around the world and to sequence their genomes. Called the *Sorcerer II* Global Ocean Sampling (GOS) Expedition, Venter and his researchers traveled the globe by yacht, in a sailing voyage described as a modern day version of Charles Darwin's famous voyage on the *H.M.S. Beagle*. The Discovery Channel has even chronicled this journey.

One of the key benefits of metagenomics is its potential for teaching us more about millions of species of bacteria, of which only a few thousand have been well characterized. Many new viruses, particularly bacteriophages, are also identified through metagenomics studies of water samples. Metagenomics is providing important new information about genetic diversity in microbes that is key to understanding complex interactions between microbial communities and their environment, as well as allowing phylogenetic classification of newly identified microbes. Metagenomics also has great potential for identifying genes with novel functions, some of which have potentially valuable applications in medicine and biotechnology.

The general method used in metagenomics to sequence genomes for all microbes in a given environment involves isolating DNA directly from an environmental sample without requiring cultures of the microbes or viruses. Such an approach is necessary because, often, it is difficult to replicate the complex array of growth conditions needed by the microbes if they are to survive in culture. For the *Sorcerer II* GOS project, samples of water from different layers in the water column were passed through high-density filters of various sizes to capture the microbes. DNA was then isolated from the microbes and subjected to shotgun sequencing and genome assembly. High-throughput sequencers on board the yacht operated nearly around the clock. One of the earliest expeditions by this group sequenced bacterial genomes from the Sargasso Sea off Bermuda. This project yielded over 1.2 million novel DNA sequences from 1800 microbial species, including 148 previously unknown bacterial species, and identified hundreds of photoreceptor genes. Many aquatic microorganisms rely on photoreceptors for capturing light energy to power photosynthesis. Scientists are interested in learning more about photoreceptors to help develop ways in which photosynthesis may be used to produce hydrogen as a fuel source. Medical researchers are also very interested in photoreceptors because in humans and many other species, photoreceptors in the retina of the eye are key proteins that detect light energy and transduce electrical signals that the brain eventually interprets to create visual images.

By early 2007, the GOS database contained approximately 6 billion bp of DNA from more than 400 uncharacterized microbial species! These sequences included 7.7 million previously uncharacterized sequences, encoding more than 6 million different potential proteins. This is almost twice the total number of previously characterized proteins in all other known databases worldwide (such as the Swiss-Prot and Uniprot databases discussed in the Exploring Genomics exercise for Chapter 15). Figure 21–18(a) shows the kingdom assignments for predicted protein sequences in publicly available databases worldwide, such as the NCBI-nonredundant protein database (NCBInr), which accesses GenBank, Ensembl, and other well-known databases. Eukaryotic sequences comprise the majority (63 percent) of predicted proteins in these databases. Reviewing the kingdom assignments of approximately 6 million predicted proteins in the Global Ocean Sampling (GOS) dataset shows that, in contrast, the largest majority (90.8 percent) of sequences in this database are from the bacterial kingdom [Figure 21–18(b)].

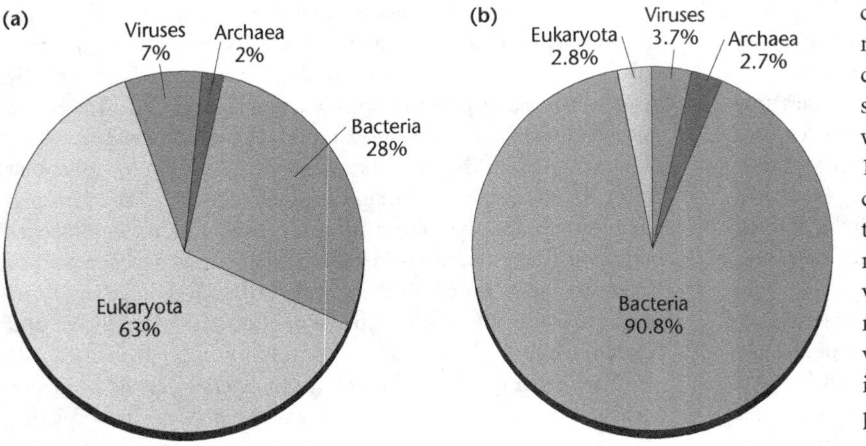


**FIGURE 21–18** (a) Kingdom identifications for predicted proteins in NCBInr, NCBI Prokaryotic Genomes, The Institute for Genomics Research Gene Indices, and Ensembl databases. Notice that the publicly available databases of sequenced genomes and the predicted proteins they encode are dominated by eukaryotic sequences. (b) Kingdom identifications for novel predicted proteins in the Global Ocean Sampling (GOS) database. Bacterial sequences dominate this database, demonstrating the value of metagenomics for revealing new information about microbial genomes and microbial communities.

The GOS Expedition also examined protein families corresponding to the predicted proteins encoded by the genome sequences in the GOS database: 17,067 families were medium (between 20 and 200 proteins) and large-sized (>200 proteins) clusters. A **Venn diagram**, like the image shown in Figure 21–19, is a common way to represent overlapping data in genomics datasets. In this figure, overlapping ovals indicate numbers of protein families belonging to certain sets of categories. Each area of the diagram is labeled with the number and percentage of families out of the 17,067 GOS medium- and large-sized clusters in each category or area of overlap. Of the 17,067 clusters in the GOS database, 3995 did not show significant homology to known protein families in prokaryotes, viruses, or eukaryotes (Figure 21–19). The results summarized in Figures 21–18 and 21–19 demonstrate the value of the GOS Expedition and of metagenomics for identifying novel microbial genes and potential proteins.

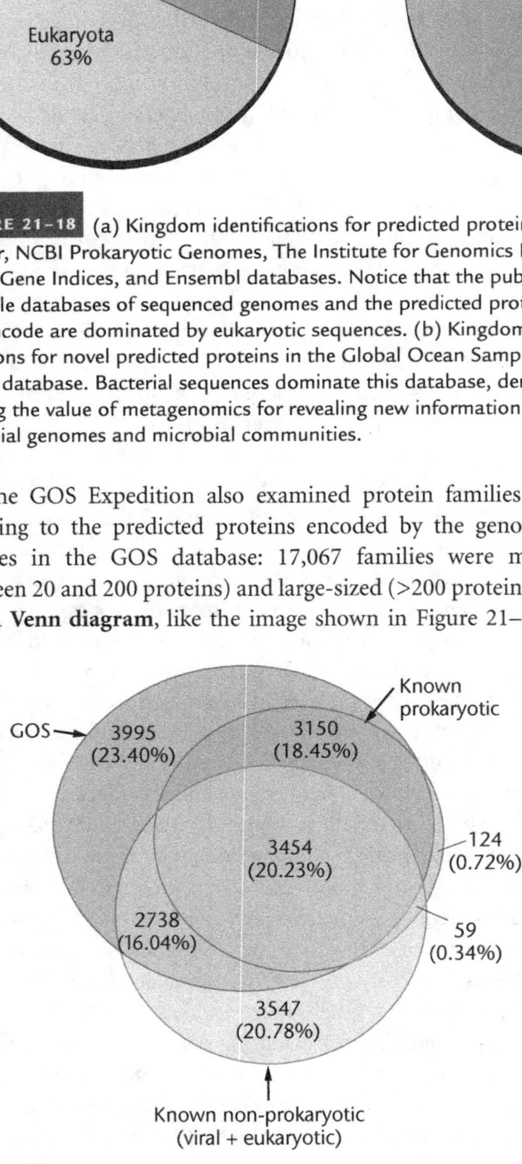


**FIGURE 21–19** Venn diagram representation of 17,067 medium and large clusters of protein families grouped according to three categories: GOS, known prokaryotic sequences, and known nonprokaryotic sequences. Notice that the GOS project identified 3995 medium and large clusters of gene families that are unique to the GOS database and do not show significant homology to known protein families in prokaryotes or nonprokaryotes (viral and eukaryotes).

## 21.9 Transcriptome Analysis Reveals Profiles of Expressed Genes in Cells and Tissues

Sequencing a genome is a major endeavor, and even once any genome has been sequenced and annotated, a formidable challenge still remains: that of understanding genome function by analyzing the genes it contains and the ways the genes expressed by the genome are regulated. **Transcriptome analysis**, also called **transcriptomics** or **global analysis of gene expression,** studies the expression of genes by a genome both qualitatively—by identifying which genes are expressed and which genes are not expressed—and quantitatively—by measuring varying levels of expression for different genes.

As we know, all cells of an organism possess the same genome, but in any cell or tissue type, certain genes will be highly expressed, others expressed at low levels, and some not expressed at all. Transcriptome analysis provides gene-expression profiles that for the same genome may vary from cell to cell or from tissue type to tissue type. Identifying genes expressed by a genome is essential for understanding how the genome functions. Transcriptome analysis provides insights into (1) normal patterns of gene expression that are important for understanding how a cell or tissue type differentiates during development, (2) how gene expression dictates and controls the physiology of differentiated cells, and (3) mechanisms of disease development that result from or cause gene expression changes in cells. In Chapter 24, we will consider why gene-expression analysis is gradually becoming an important diagnostic tool in certain areas of medicine. For example, examining gene-expression profiles in a cancerous tumor can help diagnose tumor type, determine the likelihood of tumor metastasis (spreading), and develop the most effective treatment strategy.

A number of different techniques can be used for transcriptome analysis. PCR-based methods are useful because of their ability to detect genes that are expressed at low levels. **DNA microarray**

**analysis** is widely used because it enables researchers to analyze all of a sample's expressed genes simultaneously.

Most microarrays, also known as **gene chips**, consist of a glass microscope slide onto which single-stranded DNA molecules are attached, or "spotted," using a computer-controlled high-speed robotic arm called an arrayer. Arrayers are fitted with a number of tiny pins. Each pin is immersed in a small amount of solution containing millions of copies of a different single-stranded DNA molecule. For example, many microarrays are made with single-stranded sequences of complementary DNA (cDNA) or expressed sequenced tags (ESTs, described in Chapter 13). The arrayer fixes the DNA onto the slide at specific locations (points, or spots) that are recorded by a computer. A single microarray can have over 20,000 different spots of DNA, each containing a unique sequence for a different gene. Entire genomes are available on microarrays, including the human genome. As you will learn in Chapter 24, researchers are also using microarrays to compare patterns of gene expression in tissues in response to different conditions, to compare gene expression patterns in normal and diseased tissues, and to identify pathogens.

To prepare a microarray for use in transcriptome analysis, scientists typically begin by extracting mRNA from cells or tissues (Figure 21–20). The mRNA is usually then reverse transcribed to synthesize cDNA tagged with fluorescently labeled nucleotides. The mRNA or cDNA can be labeled in a number of ways, but most methods involve the use of fluorescent dyes. Typically, microarray studies often involve comparing gene expression in

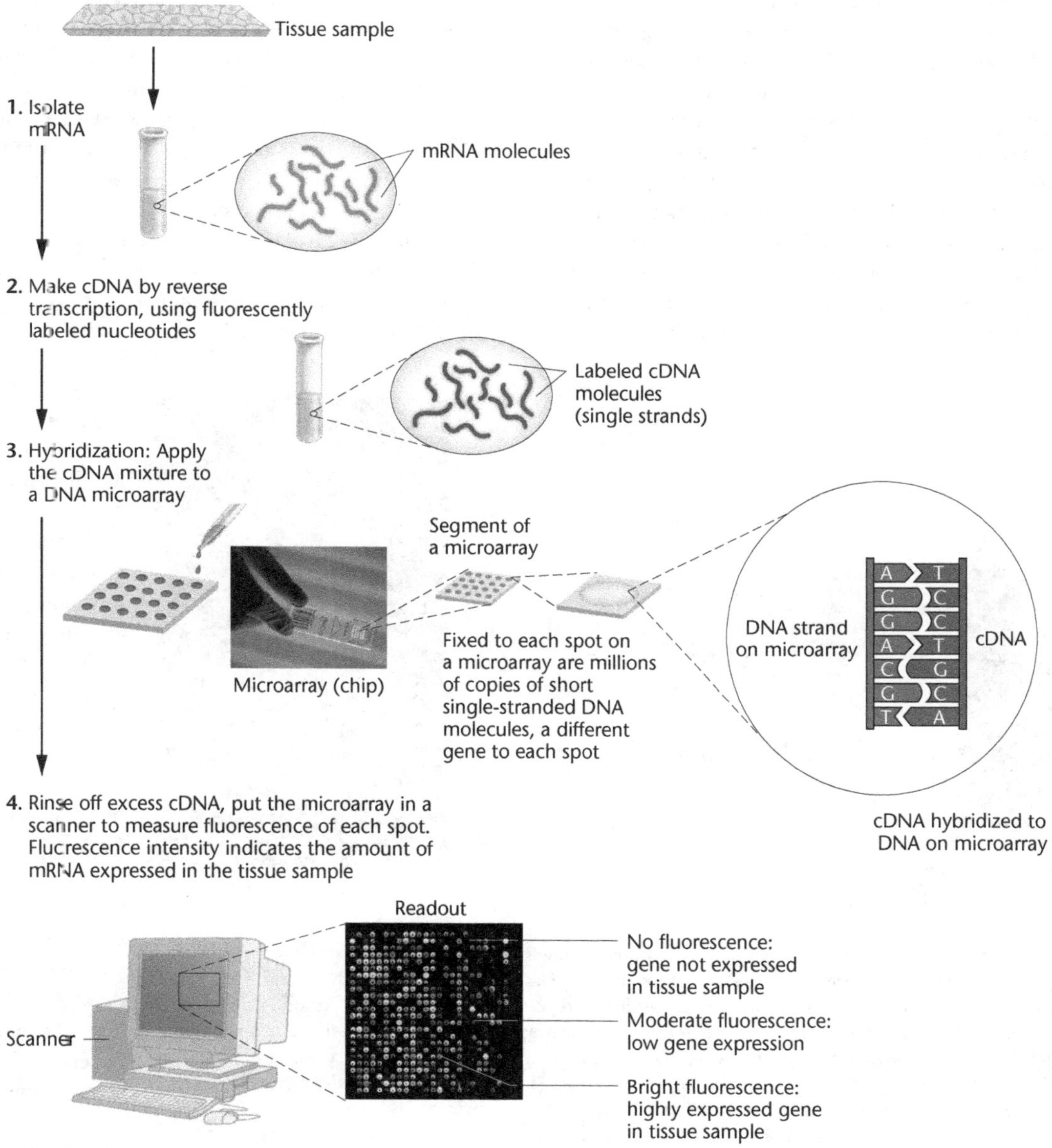


**FIGURE 21–20** Microarray analysis for analyzing gene expression patterns in a tissue.

different cell or tissue samples. cDNA prepared from one tissue is usually labeled with one color dye, red for example, and cDNA from another tissue labeled with a different colored dye, such as green. Labeled cDNAs are then denatured and incubated overnight with the microarray so that they will hybridize to spots on the microarray that contain complementary DNA sequences. Next, the microarray is washed, and then it is scanned by a laser that causes the cDNA hybridized to the microarray to fluoresce. The patterns of fluorescent spots reveal which genes are expressed in the tissue of interest, and the intensity of spot fluorescence indicates the relative level of expression. The brighter the spot, the more the particular mRNA is expressed in that tissue.

Microarrays are dramatically changing the way gene expression patterns are analyzed. As discussed in Chapter 13, northern blot analysis was one of the earliest methods used for analyzing gene expression. Then PCR techniques proved to be rapid and more sensitive approaches. The biggest advantage of microarrays is that they enable thousands of genes to be studied simultaneously. As a result, however, they can generate an overwhelming amount of gene expression data. In addition, even when properly controlled, microarrays often yield variable results. For example, one experiment under certain conditions may not always yield similar patterns of gene expression as another identical experiment. Some of these differences can be due to real differences in gene expression, but others can be due to variability in chip preparation, cDNA synthesis, probe hybridization, or washing conditions, all of which must be carefully controlled to limit such variability. Commercially available microarrays can reduce the variability that can result when individual researchers make their own arrays.

Computerized microarray data analysis programs are essential for organizing gene expression profile data from microarrays. For instance, **cluster algorithm** programs can be used to retrieve spot-intensity data from different locations on a microarray and to group gene expression data from one or multiple microarrays into cluster images incorporating results from many experiments. Cluster analysis groups genes according to whether they show increased (upregulated) or decreased (downregulated) expression under the experimental conditions examined. Figure 21–21(a) shows hierarchical clusters of upregulated and downregulated gene expression patterns for the yeast *Saccharomyces cerevisiae* grown under different experimental culture conditions. Notice that different culture times reveal different patterns of downregulation (yellow oval) and upregulation (white oval). The identities of the genes in these regulated clusters indicate that many of the genes affected by the growth conditions of this experiment are genes involved in cell division.

Figure 21–21(b) reveals an interesting gene expression profile for genes in *Drosophila* in which they display repeating patterns of expression as part of a **circadian rhythm** response. Circadian rhythms are oscillations in biological activity that occur on a regular cycle of time, such as 24 hours. Reductions in brain wave electrical activity that occur when you sleep, followed by the increased brain wave activity that occurs to wake you up, just before your alarm clock sounds in the morning, are examples of circadian responses. In the *Drosophila* microarray, each horizontal row shows expression results for a different gene. Notice how every several hours the expression for most genes is upregulated (red) and then downregulated (green) several hours later in a rhythmic pattern that repeats over the time course of this experiment (6 days). Results such as these are representative of the types of gene expression profiles that can be revealed by microarrays. You will see similar representations of microarray data in Chapter 24.

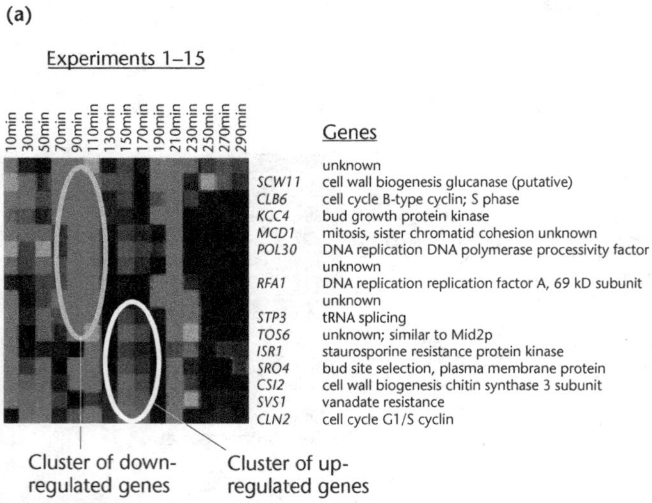


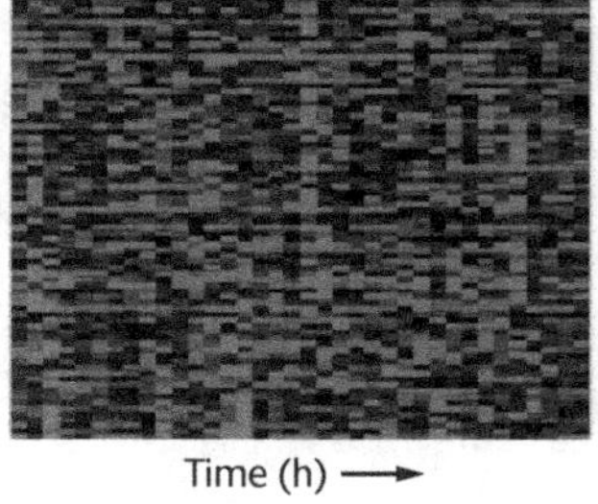


**FIGURE 21–21** (a) Hierarchical clusters in a microarray experiment using RNA samples from *Saccharomyces cerevisiae* grown for varying times in culture (experiments 1–15). Gene identities are labeled to the right of the array image. Red color in the array indicates upregulation in the experimental sample, and green indicates downregulation in the sample, as compared to a control culture. The intensity of the color indicates the magnitude of up- or downregulation. Brighter spots represent higher levels of expression than dimmer or black spots. The yellow oval highlights a cluster of downregulated genes, and the white oval highlights a cluster of upregulated genes. (b) High (red), intermediate (black), and low (green) levels of gene expression for *Drosophila* genes that exhibit a circadian rhythm. RNA samples from *Drosophila* were collected every four hours for six days and then used for microarray analysis. Each row shows all 36 responses of a single gene. Each column shows the gene expression pattern for each time point sampled. Genes were arranged from top to bottom according to the time of peak activity.

(a)

(b)

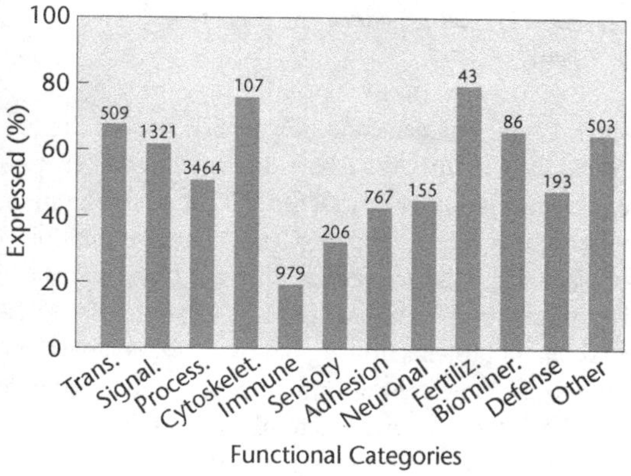


**FIGURE 21-22** (a) The sea urchin *Strongylocentrotus purpuratus*. (b) Transcriptome analysis of genes expressed in the sea urchin embryo. The *y*-axis of the histogram displays the percentage of annotated genes in different functional categories expressed in the embryo. The number at the top of each bar represents the total number of annotated genes in the corresponding functional category. Trans., transcription factors; Signal, signaling genes; Process, basic cellular processes, such as metabolism; Cytoskelet., cytoskeletal genes; Fertiliz., fertilization; and Biominer., biomineralization.

In Section 21.7 we briefly discussed the sea urchin genome and the importance of the sea urchin as a key model organism. Sea urchins have a relatively simple body plan [Figure 21–22(a)]. They contain approximately 1500 cells with only a dozen cell types. Yet sea urchin development progresses through complex changes in gene expression that resemble vertebrate patterns of gene expression changes during development. This is one reason urchins are a valuable model organism for developmental biology. Recently, scientists at NASA's Ames Genome Research Facility in Moffett, California, used microarrays to carry out a transcriptome study on sea urchin gene expression during the first two days of development (to the mid-late gastrula stage, about 48 hours postfertilization).

This work revealed that approximately 52 percent of all genes in the sea urchin are active during this period of development: 11,500 of the sea urchin's 23,500 known genes were expressed in the embryo. The functional categories of genes expressed in the embryo were diverse, including genes for about 70 percent of the nearly 300 transcription factors in the sea urchin genome, along with genes involved in cell signaling, immunity, fertilization, and metabolism [Figure 21–22(b)]. Incredibly, 51,000 RNAs of unknown function were also expressed. Further studies are underway to explain the differences between gene number and transcripts expressed, although it is already known that many sea urchin genes are extensively processed through alternative splicing. Further analysis of the sea urchin genome will undoubtedly reveal interesting aspects of gene function during sea urchin development and advance our understanding of the genetics of embryonic development in both invertebrates and vertebrates.

Now that we have considered genomes and transcriptomes, we turn our attention to the ultimate end products of most genes, the proteins encoded by a genome.

## 21.10 Proteomics Identifies and Analyzes the Protein Composition of Cells

As more genomes have been sequenced and studied, biologists in many different disciplines have focused increasingly on understanding the complex structures and functions of the proteins the genomes encode. This is not surprising given that in most of the genomes sequenced to date, many newly discovered genes and their putative proteins have no known function. Keep in mind, in the ensuing discussion, that although every cell in the body contains an equivalent set of genes, all cells do not express the same genes and proteins. **Proteome** is a term that represents the complete set of proteins encoded by a genome, but it is also often used to mean the entire complement of proteins in a cell. This definition would then include proteins that a cell acquired from another cell type.

**Proteomics**—the identification and characterization of all proteins encoded by the genome of a cell, tissue, or organism—can be used to reconcile differences between the number of genes in a genome and the number of different proteins produced. But equally important, proteomics also provides information about a protein's structure and function; posttranslational modifications; protein–protein, protein-nucleic acid, and protein–metabolite interactions; cellular localization of proteins; and relationships (shared domains, evolutionary history) to other proteins. Proteomics is also of clinical interest, because it allows comparison of proteins in normal and diseased tissues, which can lead to the identification of proteins as biomarkers for disease conditions. We will expand on this aspect of proteomics in Chapter 24.

## Reconciling the Number of Genes and the Number of Proteins Expressed by a Cell or Tissue

Recall from Chapter 15 the one-gene:one polypeptide hypothesis of George Beadle and Edward Tatum. As we have discussed in that chapter and elsewhere, genomics has revealed that the link between gene and gene product is often much more complex. Genes can have multiple transcription start sites that produce several different types of transcripts. Alternative splicing and editing of pre-mRNA molecules can generate dozens of different proteins from a single gene. Remember the current estimate that over 50 percent of human genes produce more than one protein by alternative splicing. As a result, proteomes are substantially larger than genomes. For instance, the ~20,000 genes in the human genome encode ~100,000 proteins, but some estimates suggest that the human proteome may be as large as 150,000–200,000 proteins.

Proteomes undergo dynamic changes that are coordinated in part by regulation of gene-expression patterns—the transcriptome. However, a number of other factors affect the proteome profile of a cell, further complicating the analysis of protein function. For instance, many proteins are modified by co-translational or post-translational events, such as cleavage of signal sequences that target a protein for an organelle pathway, propeptides, or initiator methionine residues; linkage to carbohydrates and lipids; or by the addition of chemical groups through methylation, acetylation, and phosphorylation. Over a hundred different mechanisms of post-translational modification are known. In addition, many proteins work via elaborate protein–protein interactions or as part of a large molecular complex.

Well before a draft sequence of the human genome was available in 2001, scientists were already discussing the possibility of a "Human Proteome Project." One reason such a project never came to pass is that there is no single human proteome: different tissues produce different sets of proteins. But the idea of such a project has led to a **Protein Structure Initiative** by the National Institute of General Medical Sciences (NIGMS), a division of the National Institutes of Health. Initiated in 2000, PSI is a ten-year project designed to analyze the structures of more than 4000 protein families. Proteins with interesting potential therapeutic properties are a top priority for the PSI. There also are a number of other ongoing projects dedicated to identifying proteome profiles that correlate with diseases such as cancer and diabetes.

## Proteomics Technologies: Two-Dimensional Gel Electrophoresis for Separating Proteins

In Chapter 13, we explained how in the early days of DNA cloning and recombinant DNA technology, before genomics, scientists often cloned, sequenced, and studied one or a few genes at a time over a span of several years. Until relatively recently, the same approach typically applied to studying proteins. But now, with proteomics technologies, scientists have the ability to study thousands of proteins simultaneously, generating enormous amounts of data quickly and dramatically changing ways of analyzing the protein content of a cell.

The early history of proteomics dates back to 1975 and the development of **two-dimensional gel electrophoresis (2DGE)** as a technique for separating hundreds to thousands of proteins with high resolution. In this technique, proteins isolated from cells or tissues of interest are loaded onto a polyacrylamide tube gel and first separated by **isoelectric focusing**, which causes proteins to migrate according to their electrical charge in a pH gradient. During isoelectric focusing, proteins migrate until they reach the location in the gel where their net charge is zero compared to the pH of the gel (Figure 21–23). Then in a second migration, perpendicular to the

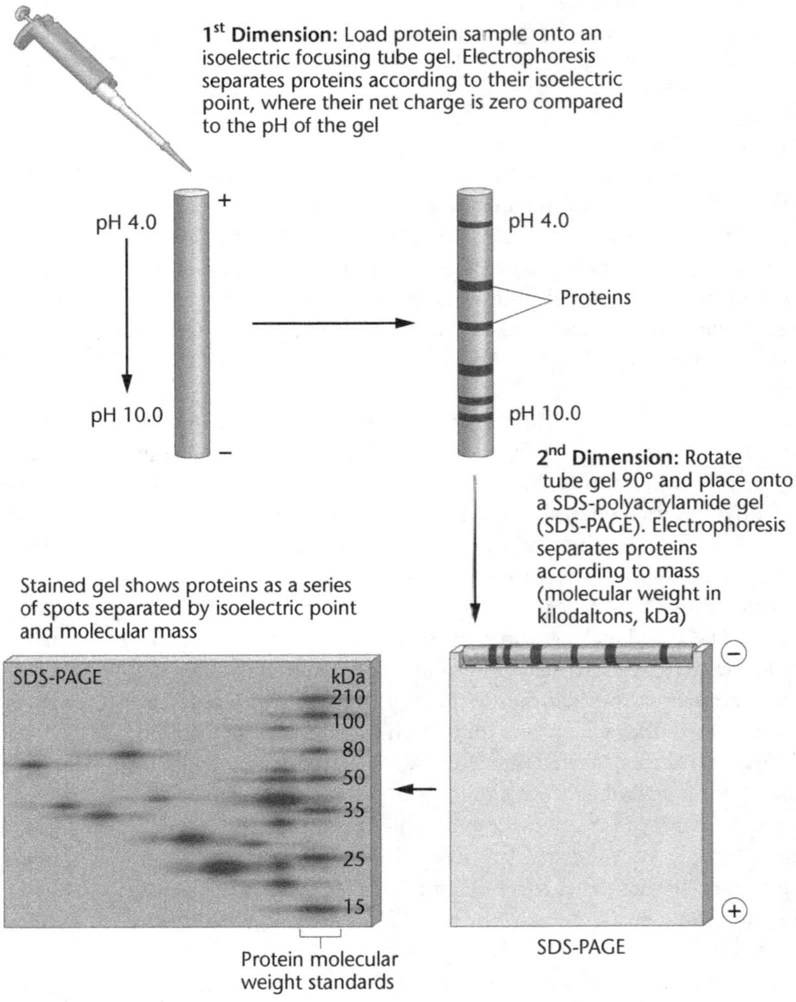


**FIGURE 21–23** Two-dimensional gel electrophoresis (2DGE) is a useful method for separating proteins in a protein extract from cells or tissues that contains a complex mixture of proteins with different biochemical properties.

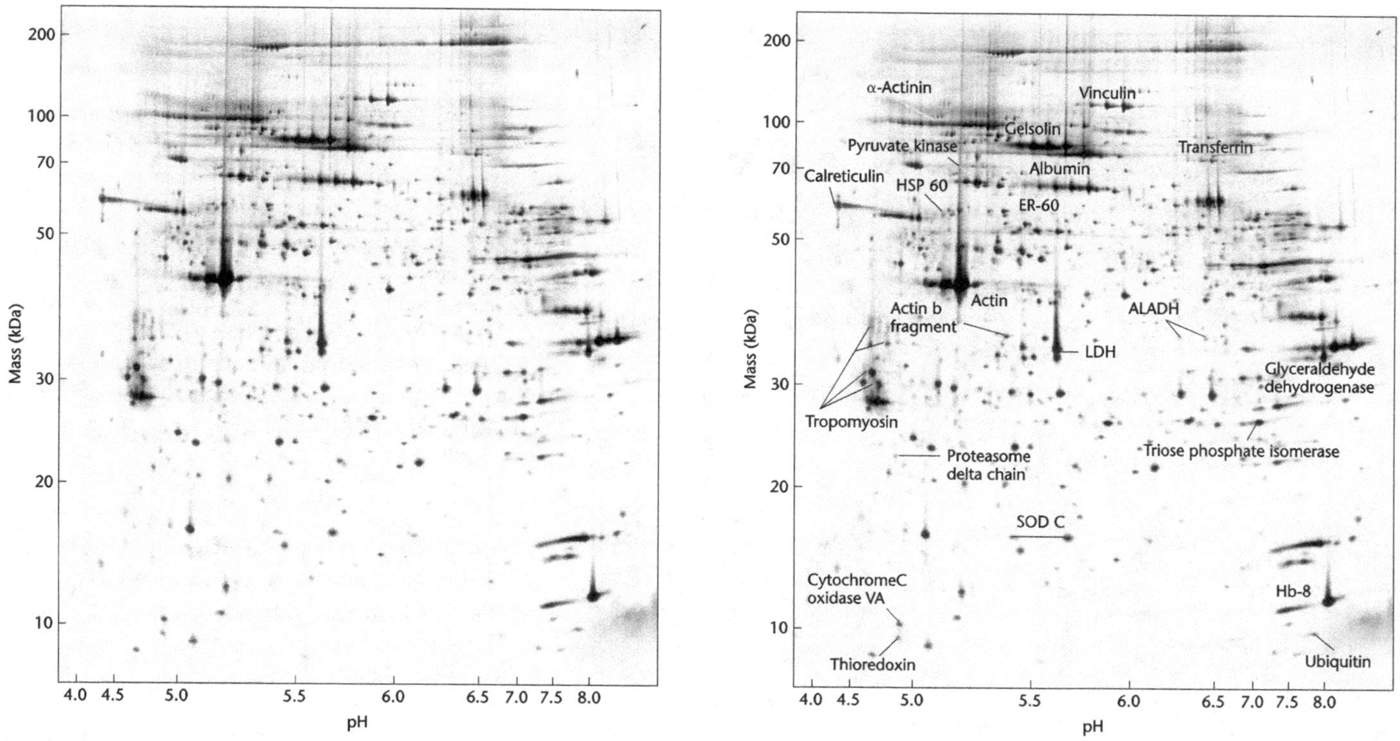


**FIGURE 21–24** Two-dimensional gel separations of human platelet proteins. (a) Unlabeled 2D gel. Each spot represents a different polypeptide separated by molecular weight (*y*-axis) and isoelectric point, pH (*x*-axis). (b) The same gel labeled to identify the locations of known proteins, identified by comparison to a reference gel or by determination of protein sequence using mass spectrometry techniques. Notice that many spots on the gel are unlabeled, indicating proteins of unknown identity.

first, the proteins are separated by their molecular mass using **sodium dodecyl sulfate polyacrylamide gel electrophoresis (SDS-PAGE)**. In this step, the tube gel is rotated 90° and placed on top of an SDS polyacrylamide gel; an electrical current is applied to the gel to separate the proteins by mass.

Proteins in the 2D gel are visualized by staining with Coomassie blue, silver stain, or other dyes that reveal the separated proteins as a series of spots in the gel (Figure 21–24). It is not uncommon for a 2D gel loaded with a complex mixture of proteins to show several thousand spots, as in Figure 21–24(a), which displays the complex mixture of proteins in human platelets (thrombocytes). Figure 21–24(b) is the same 2D gel as in Figure 21–24(a), but it has been labeled with the names of identified proteins. With thousands of different spots on the gel, how are the identities of the proteins ascertained?

In some cases, 2D gel patterns from experimental samples can be compared to gels run with reference standards containing known proteins with well-characterized migration patterns. Many reference gels for different biological samples such as human plasma, are available, and computer software programs can be used to align and compare the spots from different gels. In the early days of 2DGE, proteins were often identified by cutting spots out of a gel and sequencing the amino acids the spots contained. Only relatively small sequences of amino acids can typically be generated this way; rarely can an entire polypeptide be sequenced using this technique. BLAST and similar programs can be used to search protein databases containing amino acid sequences of known proteins. However, because of alternative splicing or post-translational modifications, peptide sequences may not always match easily with the final product, and the identity of the protein may have to be confirmed by another approach. As you will learn in the next section, proteomics has incorporated other techniques to aid in protein identification, and one of these is mass spectrometry.

## Proteomics Technologies: Mass Spectrometry for Protein Identification

As important as 2DGE has been for protein analysis, **mass spectrometry** is a relatively new technique that has been instrumental to the development of proteomics. Mass spectrometry techniques analyze ionized samples in gaseous form and measure the **mass-to-charge (m/z) ratio** of the different ions in a sample. Proteins analyzed by mass spectra generate m/z spectra that can be correlated with an m/z database containing known protein sequences to discover the protein's identity. Some of the most valuable proteomics applications of this technology are to identify an unknown protein or proteins in a complex mix of proteins, to sequence peptides, to identify post-translational modifications of proteins, and to characterize multiprotein complexes.

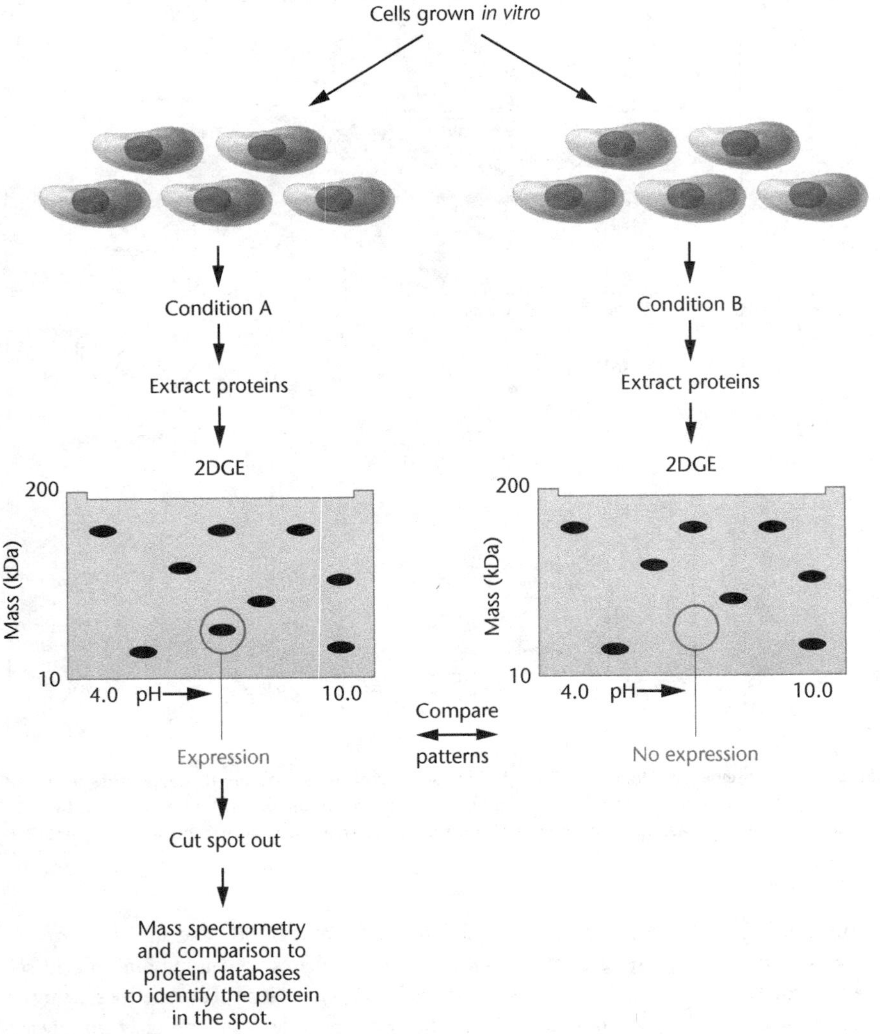


**FIGURE 21–25** In a typical proteomic analysis, cells are exposed to different conditions (such as different growth conditions, drugs, or hormones). Then proteins are extracted from these cells and separated by 2DGE, and the resulting patterns of spots are compared for evidence of differential protein expression. Spots of interest are cut out from the gel, digested into peptide fragments, and analyzed by mass spectrometry to identify the protein they contain.

One commonly used mass spectrometry approach is **matrix-assisted laser desorption ionization (MALDI)**. MALDI is ideally suited for identifying proteins and is widely used for proteomic analysis of tissue samples treated under different conditions. The proteins are first extracted from cells or tissues of interest and separated by 2DGE, after which MALDI (described below) is used to identify the proteins in the different spots. Figure 21–25 shows an example in which two different sets of cells grown in culture are analyzed for protein differences. Just about any source providing a sufficient number of cells can be used: blood, whole tissues, and organs; tumor samples; microbes; and many other substances. Many proteins involved in cancer have been identified by the use of MALDI to compare protein profiles in normal tissue and tumor samples.

Protein spots are cut out of the 2D gel, and proteins are purified out of each gel spot. Computer automated high-throughput instruments are available that can pick all of the spots out of a 2D gel. Isolated proteins are then enzymatically digested (lysed) with a protease (a protein-digesting enzyme) such as trypsin to create a series of peptides. This proteolysis produces a complex mixture of peptides determined by the cleavage sites for the protease in the original protein. Each type of protein produces a characteristic set of peptide fragments, and these are identified by MALDI as follows.

In MALDI, peptides are mixed with a low molecular weight, ultraviolet (UV) light-absorbing acidic matrix material (such as dihydroxybenzoic acid) then applied to a metal plate. A UV laser, often a nitrogen laser at a wavelength of 337 nm, is then fired at the sample. As the matrix absorbs energy from the laser, heat accumulating on the matrix vaporizes and ionizes the peptide fragments. Released ions are then analyzed for mass; MALDI displays the m/z ratio of each ionized peptide as a series of peaks representative of the molecular masses of peptides in the mixture and their relative abundance (Figure 21–26). Because different proteins produce different sets of peptide fragments, MALDI produces a peptide "fingerprint" that is characteristic of the protein being analyzed.

Databases of MALDI-generated m/z spectra for different peptides can be analyzed to look for matches between m/z spectra of unknown samples and those of known proteins. One limitation of this approach is database quality. An unknown protein from a 2D gel can only be identified by MALDI if proteomics databases have a MALDI spectrum for that protein. But as is occurring with genomics databases, proteomics databases with thousands of well-characterized proteins from different organisms are rapidly developing.

MALDI is often coupled with a protein biochemistry technique for mass analysis called **time of flight (TOF)**. TOF moves ionized peptide fragments through an electrical field in a vacuum. Each ion has a speed that varies with its mass. The speed with which each ion crosses the vacuum chamber can be measured, and differences in the ions' kinetic energy can be used to develop a mass-dependent velocity profile—a MALDI-TOF spectrum—that shows each ion's "time of flight." MALDI-TOF spectra can then be compared to databases of spectra for known proteins, as described above for MALDI spectra.

Many other methods involve mass spectrometry. Some incorporate liquid chromatography (LC) to separate proteins by mass and then employ **tandem mass spectrometry (MS/MS)** approaches to generate m/z spectra. Also emerging are new mass spectrometry techniques that do not involve the running of gels. As we mentioned when discussing genomics, high-throughput 2DGE instruments and mass spectrometers can process thousands of samples in a single day. Instruments with faster sample-processing times and increased sensitivity are under development. These instruments may soon make "shotgun proteomics" a viable approach for characterizing entire proteomes.

**Protein microarrays** are also becoming valuable tools for proteomics research. These are designed around the same basic concept as microarrays (gene chips) and are often constructed with antibodies that specifically recognize and bind to different proteins. These microarrays are used, among other applications, for examining protein–protein interactions, for detecting protein markers for disease diagnosis, and for studying in biosensors designed to detect pathogenic microbes and potentially infectious bioweapons.

**NOW SOLVE THIS**

In problem 30 on page 573, a database is searched for matches to a query sequence, and the matches prove to correspond to a certain type of protein domain. You are asked to explain why the function of the query sequence is not related to the function of the known proteins, even though they have the same domain.

■ HINT: *Remember that although protein domains may have related functions, proteins can contain several different domains interacting to determine protein function.*

## Identification of Collagen in *Tyrannosaurus rex* and *Mammut americanum* Fossils

Recently, a team of scientists reported results of mass spectrometry analysis of bone tissue from a *Tyrannosaurus rex* skeleton excavated from the Hell Creek Formation in eastern Montana and estimated to be 68 million years old. As mentioned earlier, DNA has been recovered from fossils, but the general assumption has been that proteins degrade in fossilized materials and cannot be recovered. This study demonstrated that fossilization does not fully destroy all proteins in well preserved fossils under certain conditions. This research also demonstrates the power and sensitivity of mass spectrometry as a proteomics tool.

In this work, medullary tissue was removed from the inside of the left and right femoral bones. Medullary tissue is porous, spongy bone that contains bone marrow cells, blood vessels, and nerves. *T. rex* proteins extracted from the tissue showed cross-reactivity with

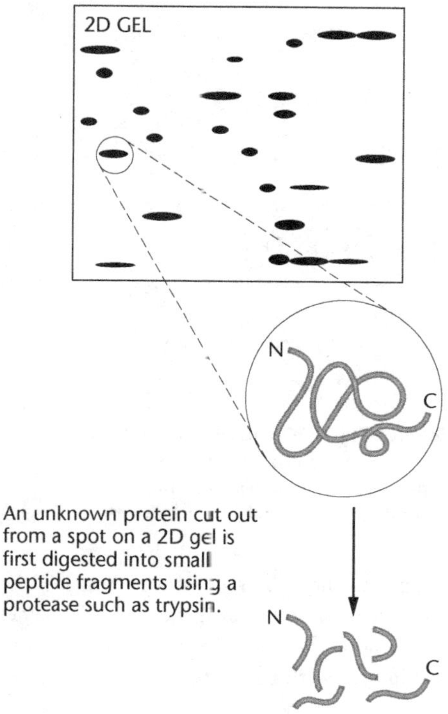


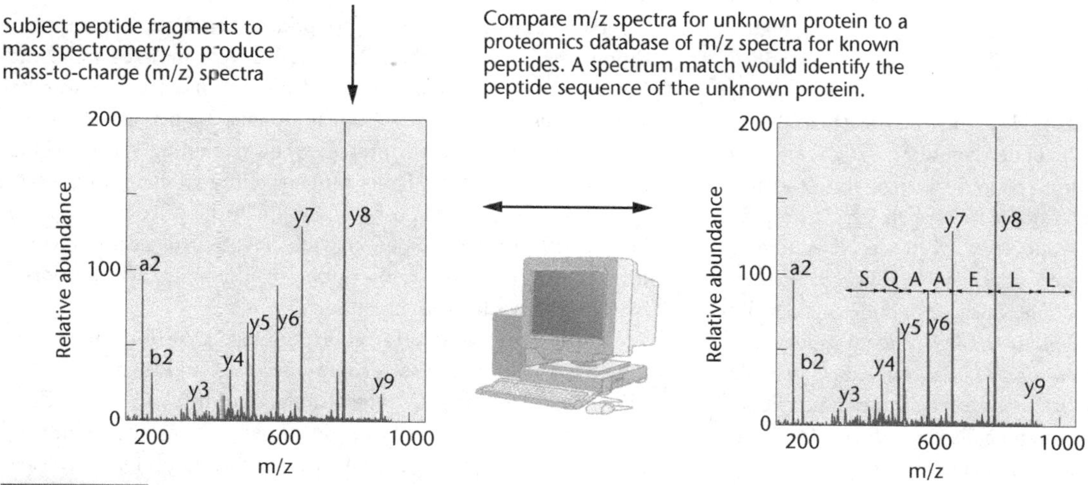


FIGURE 21–26 Mass spectrometry for identifying an unknown protein isolated from a 2D gel. The mass-to-charge spectrum (m/z) (determined, for example, by MALDI) for trypsin-digested peptides from the unknown protein can be compared to a proteomics database for a spectrum match to identify the unknown protein. The peptide in this example was revealed to have the amino acid sequence serine (S)-glutamine (Q)-alanine (A)-alanine (A)-glutamic acid (E)-leucine (L)-leucine (L), shown in single-letter amino acid code.

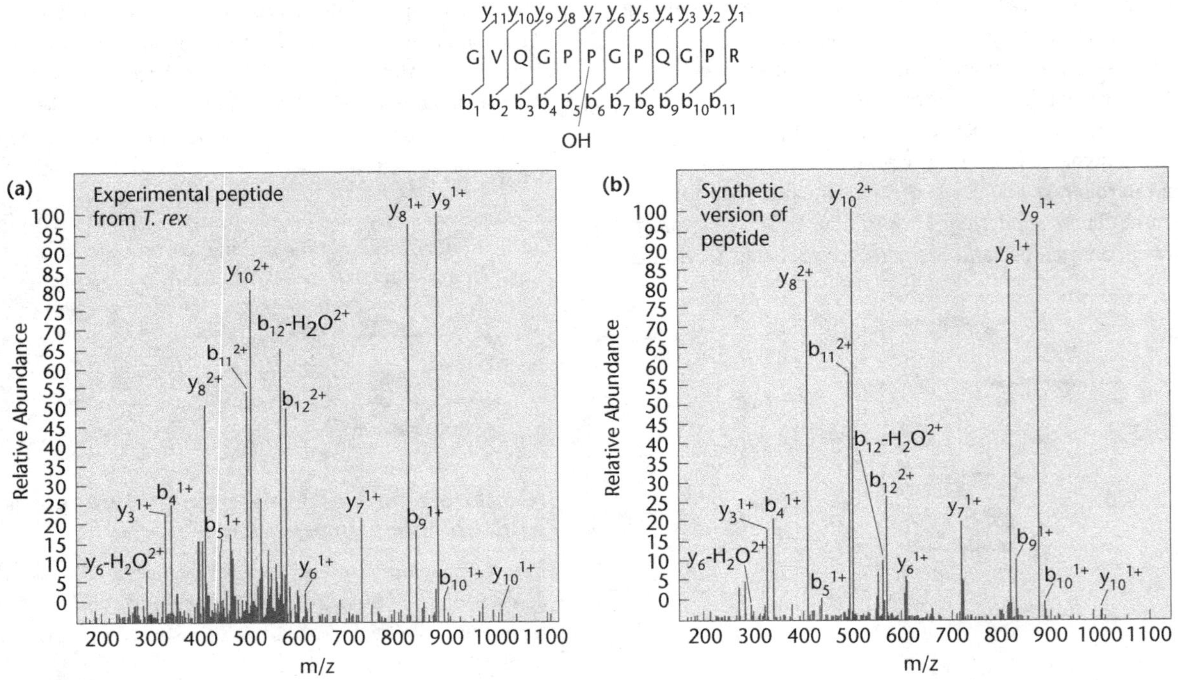


FIGURE 21–27 (a) Mass spectrometry (MS) patterns for a trypsin-digested peptide sequence—GVQPP(OH)GPQGPR—from *T. rex*. The peptide sequence, shown here in single-letter amino acid code, contains a charged hydroxyl group characteristic of collagen. (b) Mass spectrometry of a synthetic version of collagen peptide shows good alignment with the m/z spectra for fragmented ions from the *T. rex* peptide, thus confirming the *T. rex* sequence as collagen and demonstrating the value of MS techniques.

antibodies to chicken collagen and were digested by the collagen-specific protease collagenase. These results suggested that the *T. rex* protein samples contained collagen, a major matrix component of bone, ligaments, tendons, and skin. To definitively identify the presence of collagen, tryptic peptides from the *T. rex* samples were analyzed by liquid chromatography and mass spectrometry (LC/MS; Figure 21–27). The m/z spectra for one of the *T. rex* peptides was identified from a database of m/z spectra as corresponding to collagen. Compare the spectrum for a collagen peptide in Figure 21–27(a) to that of a synthetic version of a collagen peptide [Figure 21–27(b)], and you will notice that the m/z ratios for all major ions align almost identically, confirming that the *T. rex* sequence is collagen. The *T. rex* peptide also contained a hydroxyl group attached to a proline residue. Proline hydroxylation is a characteristic feature of collagen. Furthermore, the amino acid sequence of *T. rex* collagen peptide aligned with an isoform of chicken collagen, demonstrating sequence similarity.

Similar results were obtained for 160,000- to 600,000-year-old mastodon (*Mammut americanum*) peptides that showed matches to collagen from extant species, including collagen isoforms from humans, chimps, dogs, cows, chickens, elephants, and mice.

## Environment-Induced Changes in the *M. genitalium* Proteome

As outlined in Section 21.6, *M. genitalium*, with a genome of 480 genes, is one of the simplest living organisms known. Valerie Wasinger and her colleagues at the University of New South Wales in Australia used proteomics to provide a snapshot of which genes in this genome are expressed under two different growth conditions in *M. genitalium*: exponential growth and the stationary phase following rapid growth.

Using 2DGE, Wasinger's group identified 427 protein spots in exponentially growing cells. Of these, 201 were analyzed and identified by peptide digestion, mass spectrometry, and database searches. The analysis uncovered 158 known proteins (33 percent of the proteome) and 17 unknown proteins. The remaining spots included fragments derived from larger proteins, different forms of the same protein (isoforms), and post-translationally modified products. The identified proteins included enzymes involved in energy metabolism, DNA replication, transcription, translation, and transport of materials across the cell membrane.

During the transition from exponential growth to the stationary phase, cell division in the culture slowed and then stopped. At this time, there was a 42 percent reduction in the number of proteins synthesized. In addition, some new proteins appeared, while other proteins underwent dramatic changes in abundance. These changes are apparently a consequence of nutrient depletion, increased acidity of the growth medium, and other environmental changes (Figure 21–28).

Wasinger's analysis helps establish the minimum number of expressed genes required for maintenance of the living state and

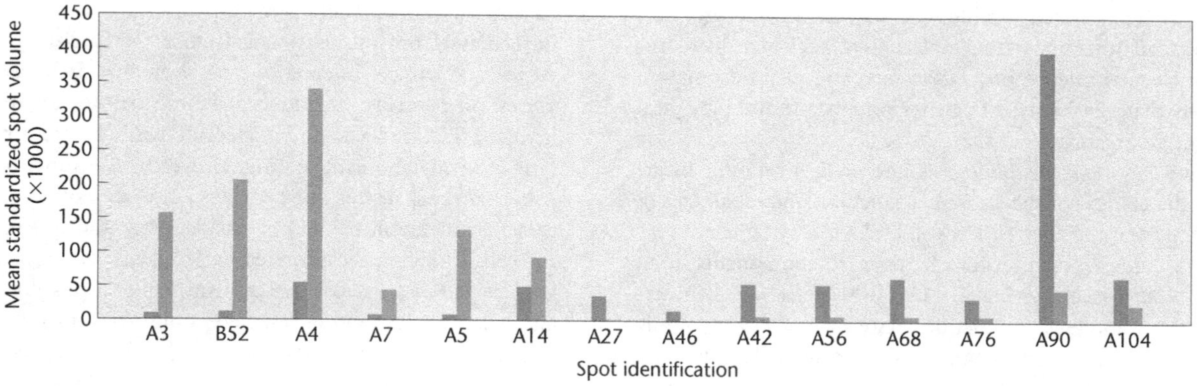


FIGURE 21–28 Changes in spot size for selected proteins during exponential growth (blue bars) and during stationary growth (orange bars) of *M. Genitalium*. These differences represent changes in gene expression and the upregulation or downregulation of individual proteins as growth conditions change.

the changes in gene expression that accompany the transition to the stationary phase. This study also points out one of the limitations of current proteomics technology: only the most abundantly expressed proteins can be detected with 2DGE. In this study, most of the proteome was either unexpressed or undetected because the proteins were present in very low amounts (too low to be detected on gels) or were not solubilized and recovered by the methods used in the study. In spite of these limitations, proteomic analysis provides a wide range of information that cannot be obtained by genome sequencing.

NOW SOLVE THIS

In Problem 34 on page 573, you are asked to suggest criteria that might justify proteomic analysis of clinical samples for disease diagnosis based on protein profiles.

■ HINT: *Consider proteomics data that would support the use of one or more proteins for the diagnosis of a disease condition. For example, what experimental evidence would you want to evaluate to determine whether a particular protein is involved in cancer?*

## 21.11 Systems Biology Is an Integrated Approach to Studying Interactions of All Components of an Organism's Cells

We conclude this chapter by discussing **systems biology**, a newly emerging discipline that incorporates data from genomics, transcriptomics, proteomics, and other areas of biology, as well as engineering applications and problem-solving approaches. Identifying genes and proteins by mutational analysis of genomes has been a very important and successful approach for characterizing genes when mutants showing visible phenotypes are found to be part of similar biochemical pathways. In Chapter 22 we will consider international efforts to create mutant knockout mice as a resource for studying gene function. However, even extensive mutational analysis and screening will not provide a full understanding of complex cellular processes such as signal transduction pathways, metabolic pathways, and regulation of cell division, DNA replication, and gene expression. A more comprehensive, more integrated approach is needed.

As we mentioned earlier in this chapter, until relatively recently, much of what has been learned about gene and protein functions at the cellular, molecular, and biochemical levels has been acquired primarily through decades of work by scientists studying the functions of individual genes or relatively small numbers of genes and proteins. Many researchers have spent entire careers studying one gene or protein. However, just when it seems as if we know all there is to know about even the most well-characterized protein, another study reveals that it possesses novel functions. Such revelations demonstrate the incompleteness of our understanding of the extreme complexity of genes and proteins in a cell. As a simple analogy, you could study the individual components of your cell phone, but until you focused on how the many components interact, you would not truly understand how a cell phone works.

Proteins occasionally function alone, but more typically they work in complex interconnected networks under the regulation and control of other proteins or metabolites. Networks of interacting proteins form the regulatory framework controlling how cells respond to environmental signals, metabolize nutrients, move organelles, divide, and carry out many other processes. As genomics and proteomics have advanced, the discipline of systems biology has emerged as a more holistic approach to studying cell function by analyzing interactions among all of the molecular components of a biological system. Systems biology considers genes, proteins, metabolites, and other interacting molecules of a cell in order to understand molecular interactions and to integrate such information into models that can be used to better understand the biological functions of an organism.

In many ways, systems biology is interpreting genomic information in the context of the structure, function, and regulation of biological pathways. As is well known, biological systems are very complex. By studying relationships between all components in an organism, biologists are trying to build a "systems"-level understanding of how organisms function. Systems biologists typically combine recently acquired genomics and proteomics data with years of more traditional studies of gene and protein structure and function. Much of this data is retrieved from databases such as PubMed, GenBank, and other newly emerging genomics, transcriptomics, and proteomics resources. Systems models are used to diagram interactions within a cell or an entire organism, such as protein–protein interactions, protein–nucleic acid interactions, and protein–metabolite interactions (e.g., enzyme-substrate binding). These models help systems biologists understand the components of interacting pathways and the interrelationships of molecules in an interacting pathway. In recent years, the term **interactome** has arisen to describe the interacting components of a cell.

Systems biologists use several different types of models to diagram protein interaction pathways. One of the most common model types is a **network map**—a sketch showing interacting proteins, genes, and other molecules. These diagrams are essentially the equivalent of an electrical wiring diagram. One disadvantage of network maps is that they are static diagrams that typically lack information about when and where each interaction occurs. Even so, they are a useful foundation for generating computational models that allow the running of simulations to determine how signaling events occur.

Figure 21–29 shows an example of a network map. This map depicts a model of galactose metabolism in yeast, based on many years' worth of data from the scientific literature. Information about individual genes and proteins that function in galactose metabolism, data from mutagenesis studies affecting genes involved in galactose metabolism, and mRNA expression profiles for approximately 6000 genes (the entire yeast transcriptome) were integrated with protein/protein and protein/DNA data in the development of this map. One aspect of the map that should be immediately obvious is that galactose metabolism in yeast involves and affects interactions between broad functional categories of proteins. Included in these categories are proteins involved in amino acid and nucleotide synthesis, RNA processing, fatty acid metabolism, mating, and cell-cycle functions. A network map such as this one provides a snapshot of how changes in galactose metabolism can potentially affect multiple functions of the cell. Knowing the proteins involved in RNA processing that are affected by galactose metabolism makes it possible to predict and better understand how different biological processes in yeast (for instance, cell-cycle activities) would be affected by alterations in galactose metabolism.

Systems biology is becoming increasingly important in the drug discovery and development process, where its approaches

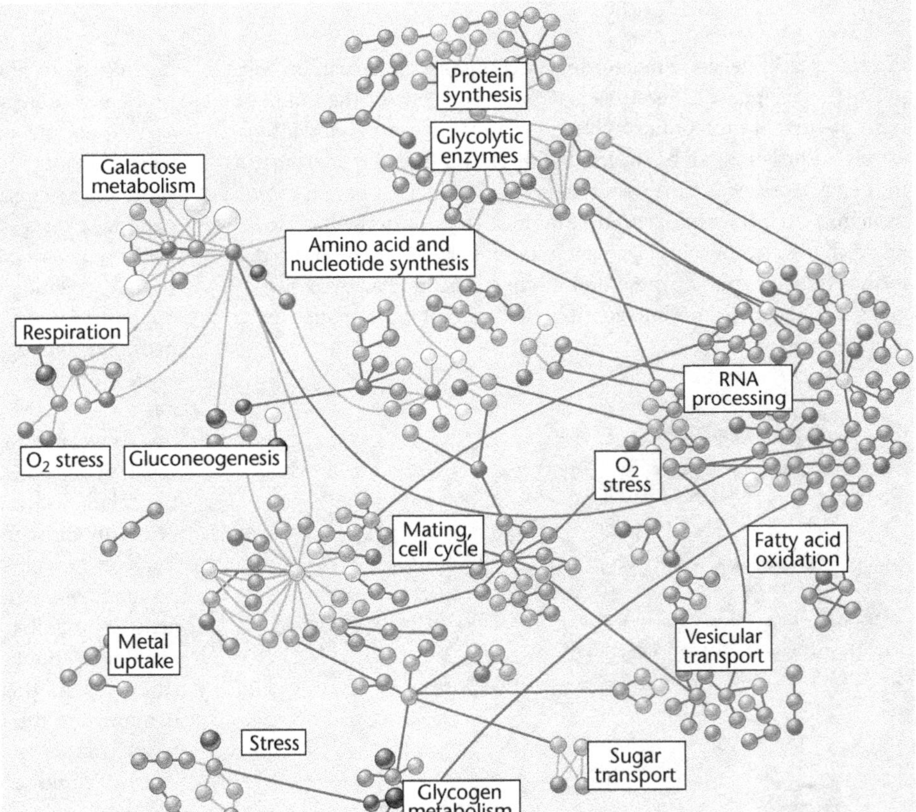


**FIGURE 21–29** A systems biology model of galactose utilization in yeast. The model integrates mRNA expression data for the 6000 yeast genes (shown as circles) under 20 different genetic or environmental conditions, together with protein/protein and protein/DNA interactions. Blue, straight lines show protein/protein interactions; yellow lines are protein/DNA interactions. Gray shading of circles indicates the level of mRNA expression, with darker shades representing abundant expression and white representing low-level expression. The red circle represents yeast with a knockout mutation of the *gal4* gene and protein, a key transcription factor in galactose metabolism. Models such as these enable systems biologists to examine the effects of mutations like *gal4* knockout on genes and proteins that interact directly or indirectly with the mutated gene or its product.

can help scientists and physicians develop a conceptual framework of gene and protein interactions in human disease that can then serve as the rationale for effective drug design. Understanding disease development and progression by defining interaction networks of molecules in normal and diseased tissue will be important for detecting and treating complex diseases such as cancer. Many databases are now being developed to model interactomes for human diseases, including breast and prostate cancer, diabetes, asthma, and cardiovascular disease. Systems biology is also being used to create biofuels and to design genetically modified organisms for cleaning up the environment, among a range of other exciting applications.

## GENETICS, TECHNOLOGY, AND SOCIETY

### Personalized Genome Projects and the Quest for the $1000 Genome

The Human Genome Project, initiated in 1990 as a 15-year project, had a budget of $3 billion. Since its inception, the development of high-throughput sequencing technologies, capable of generating longer sequence reads at higher speeds with greater accuracy, have greatly reduced the cost of DNA sequencing, and expectations for continued cost reductions along with continued technological advances are high. These expectations have led several companies to propose personalized genome sequencing for individual people. In 2006, the X Prize Foundation announced the Archon X Prize for Genomics, an award of $10 million to the first private group that develops technology capable of sequencing 100 human genomes with a high degree of accuracy in 10 days for under $10,000 per genome. Other groups are working on sequencing a personalized genome for a mere $1000! Two programs funded by the National Institutes of Health are challenging scientists to develop sequencing technologies to complete a human genome for $100,000 by 2009 and a $1000 genome by 2014.

Pursuit of the $1000 genome has become an indicator that DNA sequencing may eventually be affordable enough for individuals to consider acquiring a readout of their own genetic blueprint. The genome of James D. Watson, who together with Francis Crick discovered the structure of DNA, was the focus of "Project Jim" by a Connecticut company called 454 Life Sciences, which wanted to sequence the genome of a high-profile person and decided that the co-discoverer of DNA stucture and the first director of the U.S. Human Genome Project should be that person. This company has developed an innovative DNA sequencer that uses a DNA ligase-based procedure technique called **Supported Oligonucleotide Ligation Detection Technology (SOLiD)** that reads two-bases at a time. This technique is a lower-cost format that increase read length and accuracy. The sequencer reads ~1700 bases per second to generate ~40 million bp in a single read. James Watson provided the company with a blood sample in 2005, and by early 2007, it claimed to have sequenced 10 billion bp of overlapping sequences of his DNA in a few weeks. As of mid-2007, the company expected that it would soon be able to provide individual genomes for $100,000.

In April 2007, 454 Life Sciences reported that the work on Watson's genome was about 97 percent complete, after it had been sequenced in triplicate. Not wanting to make any mistakes in their analysis of Dr. Watson's DNA, the group pressed ahead, and by June 2007 announced that sixfold coverage of the genome was complete at a rough cost of just under $1 million. James Watson was then presented with two DVDs containing his genome sequence.

When he first agreed to have his genome sequenced, Watson said that the results would be made available to researchers except for the sequence of his apolipoprotein E gene (*ApoE*). *ApoE* variants can indicate a predisposition for Alzheimer's disease. Watson urged others to have their genomes made available to the public so that researchers could correlate physical and mental traits with genes. Human genome pioneer, J. Craig Venter, whose accomplishments we have discussed in several chapters, had his genome completed by the J. Craig Venter Institute and deposited into GenBank in May 2007. Venter has agreed to make his entire genome available, exposing any genetic imperfections for the whole world to see and study, including potentially revealing information about genetic disorders. Shortly after the Watson sequence was published, geneticists were making plans to compare the Venter and Watson genomes!

George Church of Harvard and his colleagues have started a Personal Genome Project (PGP) and have recruited volunteers to provide DNA for individual genome sequencing on the understanding that the genome data will be made publically available. Church's genome has already been made available online. The concept of a personalized genome project raises the obvious question: would you have your genome sequenced for $10,000, $1000, or even for free? If your genome were sequenced, would you make it publically available? What assurances would you want regarding an acceptable error rate for your sequence? If sequencing costs were minimal, should a personal genome be a routine part of medical diagnostics, the way a blood-typing test is, at the time of birth? The race continues to lower the cost and time required to complete an individual genome, and PGPs aside, new sequencing technologies will be very valuable for the genomics research community.

#### ■ Reference

Church, G. M. 2006. Genomes for All. *Scientific American*, 294(1): 47–54.

EXPLORING GENOMICS

# Contigs, Shotgun Sequencing, and Comparative Genomics

In this chapter, we discussed how whole-genome shotgun sequencing methods can be used to assemble chromosome maps. Recall that in the technique of shotgun cloning, chromosomal DNA is first digested with different restriction enzymes to create a series of overlapping DNA fragments called contiguous sequences, or "contigs." The contigs are then subjected to DNA sequencing, after which bioinformatics-based programs are used to arrange the contigs in their correct order on the basis of short overlapping sequences of nucleotides.

In Exercise I, you will carry out a simulation of contig alignment to help you to understand the underlying logic of this approach to creating sequence maps of a chromosome. For this purpose, you will return to the **National Center for Biotechnology Information BLAST** site that was used for many other Exploring Genomics exercises and apply a DNA alignment program called bl2seq. In Exercise II, you will use bl2seq in a comparative genomics exercise to compare DNA sequences from a gene in humans and dogs.

## Exercise I – Arranging Contigs to Create a Chromosome Map

For a well-produced animated tutorial on contigs and whole-genome sequencing methods, visit http://smcg.cifn.unam.mx/enp-unam/03-EstructuraDelGenoma/animaciones/humanShot.swf. Explore this site before beginning Exercise I.

1. Access BLAST from the NCBI Web site at http://www.ncbi.nlm.nih.gov/BLAST/. Locate and select the "Align two sequences using BLAST (bl2seq)" category at the bottom of the BLAST homepage. The bl2seq feature allows you to compare two DNA sequences at a time to check for sequence similarity alignments.

2. Below are eight contig sequences taken from an actual human chromosome sequence deposited in GenBank. For this exercise we have used short fragments; however, in reality contigs are usually several thousand base pairs long. To complete this exercise, copy and paste two sequences into the Align feature of BLAST and then run an alignment (by clicking on "Align"). (*Hint*: Access these sequences from the Companion Web site so you can copy and paste the sequences easily.) Repeat these steps with other combinations of two sequences to determine which sequences overlap, and then use your findings to create a sequence map that places overlapping contigs in their proper order. A few tips to consider:

- Develop a strategy to be sure that you analyze alignments for all pairs of contigs.
- Only consider alignment overlaps that show 100 percent sequence similarity.

**Sequence A**

CTAATTTTTTTTGTATTTTTAAT
AGAGACGAGGTGTCACCATGTTG
GACAGGCTGGTCTCGAACTCCTGA
CCTCAGGTGATCTGCCCACCTCAG
CCTCCCAAAGTGCTGGGATTACAAG
CATGAGCCACCACTCCCAGGC
TTTATTTTCTATTTTTTAATTACAGC
CATCCTAGTGAATGTGAAGTAG
TATCTCACTGAGGTTTTGATTT
GCATTTTTCTATGACAATGAACA
ATGTTTCATGTGCTTGTTGGCT
GTTTGTATATCCTTTTTGGAGAAAT
ACCAATTCATGTCCTTTGCCCA
TTTTTAAAGTGGATTGCATGTCT
TTTTGTTGTTTAGTTGTAAAGATGT
GGGTTTTTCTTTTGAGACGGAGTC
TCGCTGTCGCCTAGGCTGGAGT
GCAGAGGCATGATCTCGGCTGACT
GCAATCCCCACCTCCTGGCATCAAG
AAGTTCTCCTGCCTCAGCCTTC
CAAGTAGCTGGGTTTACAGATGC

**Sequence B**

CTTTATCTCAGGACAATGAACC
CGCAAGGAGAGGAAGAGCCAGTA
ATTCTATAGAGACTCGGAGGCGCAG
GGGGCACGCTTAGTTAGAGTGGTG
GTGGTATTTTCAGTGTTTTCTGGTTT
TATGATAAACACAAGCATCA
ATGTCTCAAGACTTTCATCTTTATCTT
TTTTTTTTTTTTTTTTTTTTTTCTTGA
GACAGGGTTTCCCTCTG
TCACCCAGGCTGGAGTGCATTGGT
GGTGTGATCTTGGCTTTCTGTAAC
CTCGGGCTTCTGGGCTCAAGCC
GTTCTACTACCTCAGCCTCCCAAAT
AGCTAGAACTACAAGCGTGTGCTGC
CACACCTGGCTAATTTGTTG
TATTTTATTTATTCATTTATTTTTGT
GAAGACAAGGTCTTGCCATGTTGC
CCAGGCTGGTCTCAGACTCC
TGGGCTCAAGCAATCCACCCACC
TTAGTCTCCCAAAGTGCTGGGAT
TACAGGCGTGAGCCACCACACCCA

**Sequence C**

GGAATTTCACTCTTGTTGCCCAAGTT
GGAGTGCAATGGCGCGATCTCAGCT
CACTCCAACCTCCGCCTCC
CAGGTTCAAACGATTCTCCTGC
TTCACTCTCCCCAGTAGCTGGGAT
TACAGGCTGCACCACCACACCTGG
CTAATTTTTTTTGTATTTTTAAT
AGAGACGAGGTGTCACCATGTTG
GACAGGCTGGTCTCGAACTCCTGA
CCTCAGGTGATCTGCCCACCTCAG
CCTCCCAAAGTGCTGGGATTACAAG
CATGAGCCACCACTCCCAGGC

**Sequence D**

GCTTCATCTTTCTCTTCACCGTAAAA
CAGGAAAGTGTGTGGTGACCAGT
ATTTTAAGGGAAAGGCACTTA
CAGAGAATTAAGCATTTGACA
AAATTTATTTACAGATATTTGTCT
GTGGACCACTTCCGCACCAGCTGTG
CATGAGAGGGCTCATTGCTCT
GAATTTGCCTCCTTGTCTGCACCC
AGGACACCGTTTCCCAGATCACGCA
AACGCTGCCTTCTCCCCACACCAG
GGCCCTCAGCATGGGAATGAC
CTTCCAGCGCTGCACGTTTCCAATC
CATGCTCTGTTTTTCAGTTCTGG CT
CACAGAGGACTGCTGGTTGCAAG
CAAACTTGTATCTGGGTCTTCA

**Sequence E**

CCTTAACTGATCTACCTGTCTCTG
CCTCTCAAAGTGCTGGGATTG
CAGGCATAAGCCGCCATGCCCGGCC
CAAAGTTTCTTTATATGTGCTGGATAC
TAGGCCCGTAACAGATATACAATTTGT
AAATATTTTCTCTCATTTTGAAG
ATTTTCTTTTCACTTTCTTGATAAT
GTCCTTTGTGTATTTTTTTGATAATG
TCCTTTCATACACAAAAGTTTT
TAAGTTTGATGAAGTTCAATTTACC
TATTATTTTCTTTTGTTGTTCAT

**Sequence F**

TGTTTGTTTGTTTGTTTTGTTTCATTT
TGTTTTTGAGACAGAGTCTTGTTCT
GTCGCCCAGTGTAGAGTG
CAGTGGCATAATCTCGGCTCACCG
TAACCTCCGCCTCCCGGGTTCAAG
CAACTCTGCCTGCCTCAGCCTC
CCAAGGAGCTGGGATTATAGACG
CCCACCACCATGCCTGGTTAATTT
TTGTAGTTTTTTTTAGTAGAGAT
GGGGTTTTGCCATTTTGGCCAGGCT
GGTCTTGAACTCCTGACCTCAGGT
GATCTGCCCACCCTGGCCTCT
CAAAGTGCTGGGATTACAGGTG
TGAGCTGCCACACTCGGCCACAA
CAAATTTTTGCACCAGTTGCTCACA

**Sequence G**

TCCTTTGATACACAAAAGTTTTTAAG
TTTGATGAAGTTCAATTTACCTAT
TATTTTCTTTTGTTGTTCATTCATTTTG
TGTCCTATGTAGGAATCTATTGCCAAA
TTCAAGGTGATAAAGATTTACCCCTAT
GTTTCCTTCTAAGAGTTTTATTGTTTT
AGCCCTGATATTTAGCTAAACTTAATT
GATTTATTAAGTTTAATTTTCCTATGTG
GTATGAAGTCATTTATCTTCTTTAGTTC
AGGATCCAAGTGAAAGGGGCATCTTC
TATCTGGGACATGCCATTCTCATGACA
GAGGAAAAAGACAAAAAACTGACACA
TACAATGACTTTAAAACTTCACTCA

**Sequence H**

GGGTTTTTCTTTTGAGACGGAGTC
TCGCTGTCGCCTAGGCTGGAGT
GCAGAGGCATGATCTCGGCTGACT
GCAATCCCCACCTCCTGGCATCAAG
AAGTTCTCCTGCCTCAGCCTTC
CAAGTAGCTGGGTTTACAGATGC
CCACCACCATGCCTGGCTGGTTT
TTGTATTTTTAGTAGACACGGGGTTT
TACCATGTTGGCCGGGCTGGT
CTGGAACTCCTAACCTTAAGTGATC
TACCTGTCTCTGCCTCTCAAAGT
GCTGGGATTGCAGGCATAAGCC
GCCATGCCCGGCCCAAAGTTT
CTTTATATGTGCTGGATACTAGGCC
CGTAACAGATATACAATTTGTAAA

3. On the basis of your alignment results, answer the following questions, referring to the sequences by their letter codes (A through H):
   a. What is the correct order of overlapping contigs?
   b. What is the length, measured in number of nucleotides, of each sequence overlap between contigs?
   c. Did you find any contigs that do not overlap with any of the others? Explain.
4. Run a nucleotide-nucleotide BLAST search (BLASTn) on any of the overlapping contigs to determine which chromosome these contigs were taken from, and report your answer.

### ■ Exercise II – Comparative Genomics and the Leptin (*LEP*) Gene

This chapter also discussed the new field of comparative genomics, with its focus on analyzing and comparing DNA sequences from different organisms. Here we use blast2seq to compare sequences from the *LEP* gene in humans and dogs. You may recall learning about this gene (also called *OB* for obesity) in the Exploring Genomics exercise for Chapter 10. *LEP* encodes a peptide hormone produced by fat cells (adipocytes) that plays a role in regulating lipid metabolism.

1. Below are partial sequences for the *LEP* gene in humans and dogs (*Canis familiaris*). Align the sequences using bl2seq, and then answer the following questions:
   a. Are these sequences similar to each other? How many nucleotides in the alignment are exact matches?
   b. What is the percentage of identical nucleotides in these two sequences?

***Homo sapiens***
**(Accession number NM_000230)**

AAGCAAAGCACCAGCTTCTCCAGGCT
CTTTGGGGTCAGCCAGGGCCAGGGG
TCTCCCTGGAGTGCAGTTT
CCAATCCCATAGATGGGTCTGGCTGA
GCTGAACCCATTTTGAGTGACTCGAG
GGTTGGGTTCATCTGAGCAAGAGCTG
GCAAAGGTGGCTCTCCAGTTAGTTCT
CTCGTAACTGGTTTCATTTCTACTGTG
ACTGATGTTACATCACAGTGTTTGCAA
TGGTGTTGCCCTGAGTGGATCTCCAA
GGACCAGGTTATTTTAAAAAGATTTG
TTTTGTCAAGTGTCATATGT

***Canis familiaris***
**(Accession number NM_001003070)**

ATTTCTAGTGACTTGAGGGCTCTCAA
GTTAGTTCTTTGGTAACTGGCTATGTT
TCTACTGTGACGGATGTTAAATTCAGT
GTTTGCAATGGCATTGCCCTGAGCGG
ATCTCCAAGGACCAGGTTATTTCAAAA
AGAAGA

## Chapter Summary

1. Whole-genome sequencing strategies allow the sequencing of entire genomes. Genome sequences are then analyzed to ensure that they are accurate, to identify all encoded genes, to classify known genes into functional categories, and to identify gene regulatory elements. When completed, the sequences are deposited into searchable databases.
2. Bioinformatics technology is essential for the analysis of genomes and proteomes. Bioinformatics applies computer hardware and software together with statistical approaches to analyze biological sequence data.
3. Rapid advances in genomics have led to an "omics" revolution in modern biology, reflected in the number of newly emerging genomics-related

disciplines, including functional genomics, comparative genomics, transcriptomics, metagenomics, and proteomics. Genomics has impacted virtually every area of modern biology, and genome studies are ongoing for a wide range of organisms and viruses. The genomes for many important model organisms have been completed.

4. Bacterial genomes have a very high gene density, averaging one gene per 1000 base pairs of DNA. Typically, as much as 90 percent of a bacterial chromosome encodes genes. Many of the genes are organized into polycistronic transcription units defined as operons. In general, bacterial genes do not contain introns.
5. Eukaryotic genomes have a gene density much lower than that in bacteria. The genes typically are not organized into operons; rather, each is a separate transcription unit. Often, eukaryotic genes are interrupted by introns.
6. Complex multicellular eukaryotes differ from the less complex eukaryote yeast in a number of ways. Complex eukaryotes have more genes and much more DNA than yeast and therefore have gene densities as low as 1 gene per 5 kb or even less than 1 gene per 10–20 kb. A higher proportion of multicellular eukaryotic genes have introns. The number of introns per gene increases, and the size of introns increases as complexity increases from yeast to humans. Some plants, such as *Arabidopsis*, have a gene structure and organization that are indistinguishable from those of animals.
7. The human genome contains 3.1 billion nucleotides, but only about 2 percent of this DNA contains protein-coding genes. The human genome contains an abundance of highly repetitive, noncoding sequences. It is estimated that humans have ~20,000 genes. Many human genes encode more than one polypeptide. Gene density differs greatly on different chromosomes, with gene-rich regions alternating with gene-poor regions. The protein-coding genes of humans have larger and more numerous introns than those of other eukaryotes, such as fruit flies. The genome is approximately 99.9 percent the same between humans of all nationalities.
8. Studies in comparative genomics are valuable for identifying similarities and differences in organization and gene content between the genomes of different organisms. Comparative genomics studies are also important for studying genetic relatedness of species and for identifying gene families.
9. Many eukaryotic genes have undergone duplication followed by sequence divergence, leading to multigene families. The globin gene cluster is a prime example of this phenomenon.
10. Metagenomics involves sequencing genomes from environmental samples, primarily microbial genomes, and it is achieving important insights regarding microbial genetics, microbe diversity and evolutionary relatedness, and complex aspects of microbial communities.
11. Transcriptome analysis produces gene expression profiles for cells and tissues that can be used to study gene expression and regulation. Studying the transcriptome is valuable for understanding gene expression by a genome, for understanding how gene expression affects phenotypes, and for analyzing gene expression regulation in response to changing conditions in both normal and diseased states.
12. Proteomics is used to study the protein content of cells and tissues, to identify unknown proteins, and to provide insight into protein profiles of eukaryotic and prokaryotic cells under different growth conditions.
13. Systems biology approaches are designed to provide an integrated understanding of the interactions between genes, proteins, and other molecules that govern complex biological processes.

## INSIGHTS AND SOLUTIONS

1. One of the main problems in annotation is deciding how long a putative ORF must be before it is accepted as a gene. Shown here are three different ORF scans of the same *E. coli* genome region—the region containing the *lacY* gene. Regions shaded in brown indicate ORFs. The top scan was set to accept ORFs of 50 nucleotides as genes. The middle and bottom scans accepted ORFs of 100 and 300 nucleotides as genes, respectively. How many putative genes are detected in each scan? The longest ORF covers 1254 bp; the next longest, 234 bp; and the shortest, 54 bp. How can we decide the actual number of genes in this region? In this type of ORF scan, is it more likely that the number of genes in the genome will be overestimated or underestimated? Why?

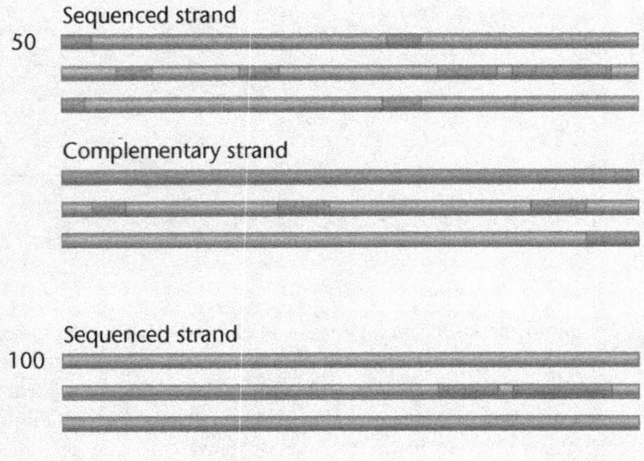


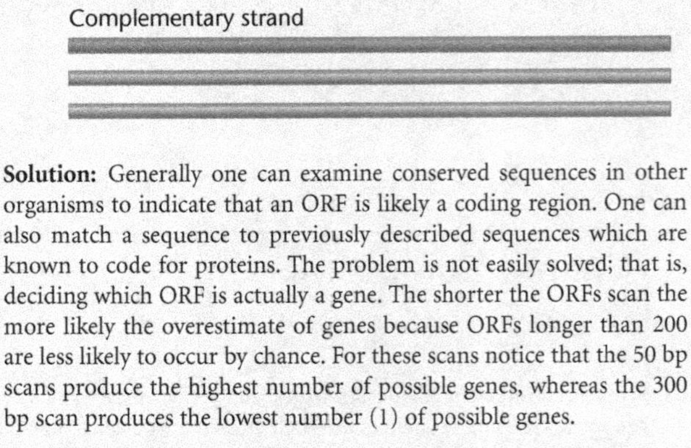


**Solution:** Generally one can examine conserved sequences in other organisms to indicate that an ORF is likely a coding region. One can also match a sequence to previously described sequences which are known to code for proteins. The problem is not easily solved; that is, deciding which ORF is actually a gene. The shorter the ORFs scan the more likely the overestimate of genes because ORFs longer than 200 are less likely to occur by chance. For these scans notice that the 50 bp scans produce the highest number of possible genes, whereas the 300 bp scan produces the lowest number (1) of possible genes.

2. Recent sequencing of the heterochromatic regions (repeat-rich sequences concentrated in centromeric and telomeric areas) of the

*Drosophila* genome indicates that within 20.7 Mb, there are 297 protein-coding genes (Bergman et al. 2002. *genomebiology3 (12)@genomebiology.com/2002/3/12/RESEARCH/0086*). Given that the euchromatic regions of the genome contain 13,379 protein-coding genes in 116.8 Mb, what general conclusion is apparent?

**Solution:** Gene density in euchromatic regions of the *Drosophila* genome is about one gene per 8730 base pairs, while gene density in heterochromatic regions is one gene per 70,000 bases (20.7 Mb/297). Clearly, a given region of heterochromatin is much less likely to contain a gene than the same-sized region in euchromatin.

## Problems and Discussion Questions

1. What is functional genomics? How does it differ from comparative genomics?
2. Compare and contrast whole-genome shotgun sequencing to a map-based cloning approach.
3. What, if any, features do bacterial genomes share with eukaryotic genomes?
4. Plasmids can be transferred between species of bacteria, and most carry nonessential genes. For these and other reasons, plasmid genes have not been included as part of the genomes of bacterial species. The bacterium *B. burgdorferi* contains 17 plasmids carrying 430 genes, some of which are essential for life. Should plasmids carrying essential genes be considered part of an organism's genome? What about other plasmids that do not carry such genes? In other words, how do we define an organism's genome in these cases?
5. What is bioinformatics, and why is this discipline essential for studying genomes? Provide two examples of bioinformatics applications.
6. List and describe three major goals of the Human Genome Project.
7. How do high-throughput techniques facilitate research in genomics and proteomics? Explain.
8. BLAST searches and related applications are essential for analyzing gene and protein sequences. Define BLAST, describe basic features of this bioinformatics tool, and provide an example of information provided by a BLAST search.
9. What are pseudogenes, and how are they produced?
10. Describe the human genome in terms of genome size, the percentage of the genome that codes for proteins, how much is composed of repetitive sequences, and how many genes it contains. Describe two other features of the human genome.
11. Based on a comparison of general features of eukaryotic and prokaryotic genomes, why might we predict that the organization of eukaryotic genetic material is more complex than that of viruses or bacteria?
12. Compare the organization of bacterial genes to that of eukaryotic genes. What are the major differences?
13. The Human Genome Project has demonstrated that in humans of all races and nationalities approximately 99.9 percent of the sequence is the same, yet different individuals can be identified by DNA fingerprinting techniques. What is one primary variation in the human genome that can be used to distinguish different individuals? Briefly explain your answer.
14. Annotation involves identifying genes and gene regulatory sequences in a genome. List and describe characteristics of a genome that are hallmarks for identifying genes in an unknown sequence. What characteristics would you look for in a prokaryotic genome? A eukaryotic genome?
15. It can be said that modern biology is experiencing an "omics" revolution. What does this mean? Explain your answer.
16. Metagenomics studies generate very large amounts of sequence data. Provide examples of genetic insight that can be learned from metagenomics.
17. What are gene microarrays? How are microarrays used?
18. Used in gene annotation, gene-prediction programs allow researchers to identify likely coding regions in DNA sequences. Annotation is complicated when genes are complex, contain multiple initiation sites, or contain numerous exons. For example, one study concluded that even for the most well-studied organisms, such programs can predict correct exon boundaries only about 80 percent of the time (Pavy et al., 1999. *Bioinformatics* 15: 887–899). Given this percentage, what is the likelihood of determining the correct exon boundaries in a gene with five exons?
19. Recent genome-sequencing efforts have provided considerable insight into the molecular workings of living systems. However, it has become increasingly apparent that to fully comprehend the genome, reannotation, manual verification, and more advanced techniques will be needed. Recently, a group of researchers reannotated the *Arabidopsis* genome and found 240 new genes, 92 of which are homologous to known proteins (Haas et al. 2002. *genomebiology3(6)@genomebiology.com/2002/3/6/RESEARCH/0029*). In addition, they identified a new class of exons, called micro-exons, which vary in length from 3 to 25 base pairs. Based on this information, how would you qualify the statement that "to know the sequence of DNA is to know the blueprint of life"?
20. In a recent draft annotation and overview of the human genome sequence, F.A. Wright et al. (*Genome Biol.* 2001: 2(7):Research0025) presented a graph similar to the one shown here. The graph details the approximate number of genes from each chromosome that are expressed only in embryos. Review earlier information in the text on human chromosomal aneuploids and correlate that information with the graph. Does this graph provide insight as to why some aneuploids occur and others do not?

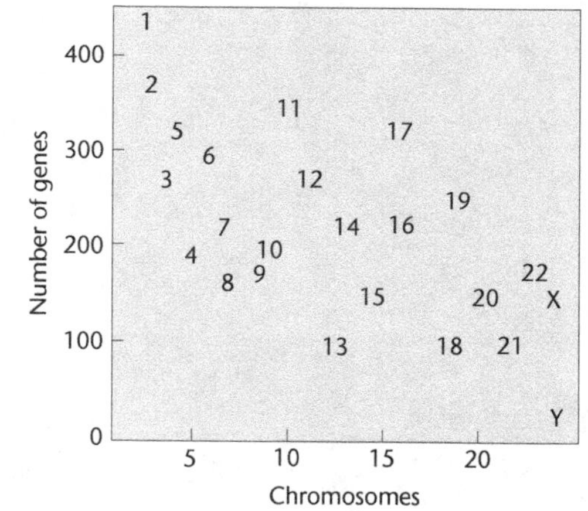


21. The process of annotating a sequenced genome is continual. In March 2000, the first annotated sequence of the *Drosophila* genome was released, which predicted 13,601 protein-coding genes within the euchromatic region of the genome. Shown here are selected data from the next two

annotated versions that were released (modified from Misra et al., 2002. *genomebiology3(12)@genomebiology.com/2002/3/12/RESEARCH/0083*).

(a) Assuming a uniform distribution in Release 3, approximately how many base pairs of DNA lie between protein-coding genes in *Drosophila*?

(b) On average, approximately how many exons are reported per gene in Release 3?

(c) Approximately how many introns are there per gene?

(d) What appears to be the most significant difference between Release 2 and Release 3?

| Criteria | Release 2 | Release 3 |
|---|---|---|
| Total length of euchromatin | 116.2 Mb | 116.8 Mb |
| Total protein-coding genes | 13,474 | 13,379 |
| Protein-coding exons | 50,667 | 54,934 |
| Introns | 48,381 | 48,257 |
| Genes with alternative transcripts | 689 | 2729 |

22. The data given in Problem 21 indicate that the more closely researchers examine genome sequences, the more complex the interpretations of those data will become. S. Misra and colleagues (2002) found that nested and overlapping genes are common in *Drosophila*. They determined that approximately 7.5 percent of all Release 3 genes were nested within the introns of other genes and that the majority of the nested genes are transcribed from the opposite strand of the gene in which they are nested. In addition, they found that about 15 percent of the annotated genes involve the overlap of mRNAs on opposite strands. What impact will this information have on genome annotation, and what clinical significance might it have?

23. To deal with the problems of correctly annotating microbial genomes, Marie Skovgaard and her colleagues (2001. *Trends Genet.* 17: 425–428) compared the annotated number of genes in genomes derived from sequence analysis to the number of known proteins in each organism as reported in a protein database. The results of their study are summarized in the accompanying graph. The errors range from a few percent for *M. genitalium* to almost 100 percent for *A. pernix*. The general trend shown in the graph is that the error rate increases as the GC content of the genome increases. What might account for this? What precautions should be taken in annotating the genomes with high GC content?

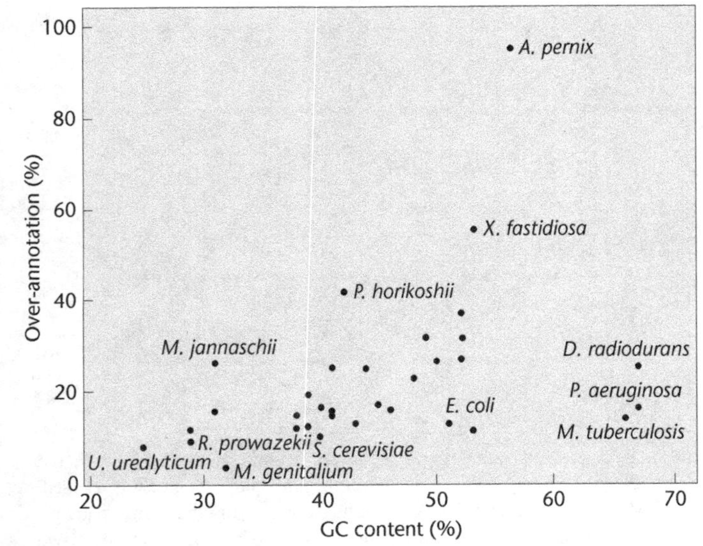


24. Annotation of the human genome sequence reveals a discrepancy between the number of protein-coding genes and the number of predicted proteins actually expressed by the genome. Proteomic analysis indicates that human cells are capable of synthesizing more than 300,000 different proteins. What is the discrepancy, and how can it be reconciled?

25. The discovery that *M. genitalium* has a genome of 0.58 Mb and only 470 protein-coding genes has sparked interest in determining the minimum number of genes needed for a living cell. In the search for organisms with smaller and smaller genomes, a new species of Archaea, *Nanoarchaeum equitans*, was discovered in a high-temperature vent on the ocean floor (Huber et al. 2002. *Nature* 417: 63–67). This prokaryote has one of the smallest cell sizes ever discovered, and its genome is only about 0.5 Mb. However, organisms such as *M. genitalium*, *N. equitans*, and other microbes with very small genomes are either parasites or symbionts. How does this affect the search for a minimum genome? Should the definition of the minimum genome size for a living cell be redefined?

26. In the search for the smallest bacterial genome and the minimum number of genes necessary for life, attention has turned to species of *Buchnera*, which live as intracellular symbionts in aphid intestinal cells. As symbionts, they need not maintain the genes necessary for infection and for evasion of the host's immune system, as do parasites or pathogens, and may have smaller genomes. The genome of one species of *Buchnera*, designated APS, has been sequenced. It has 564 genes in a circular chromosome of 640 kb (Shigenobou et al. 2000. *Nature* 407: 81–86). To determine whether *Buchnera* genome size is conserved across different groups of aphids, R. Gil and colleagues (2002. *Proc. Nat. Acad. Sci. [USA]* 99: 4454–4458) physically mapped the genome sizes of nine *Buchnera* genomes that were isolated from five aphid families. The genomes were sized by digestion with restriction enzymes, and separation of the resulting fragments was done by gel electrophoresis. The data for some *Buchnera* species are given in the following table. Although there are some discrepancies in sizes, the sum of the fragment sizes corresponds to the size of the chromosome that appeared on the gel without restriction digestion. From your analysis of the data, is genome reduction in *Buchnera* still occurring? How do the genome sizes obtained for these species compare with the genome of *M. genitalium*? The APS species of *Buchnera* contains 564 coding genes in a 641-kb genome. How many genes should be present in species CCE? How does this compare to the number of genes in *M. genitalium*? Are there other ways to determine the minimum genome needed for life without searching for other bacterial species with small genomes?

**Size of DNA Fragments Produced by Restriction Enzymes (kb)**

| *Buchnera* species | *ApaI* | *RsrII* | *ApaI* + *RsrII* | Total DNA Length, kb |
|---|---|---|---|---|
| APS | 286, 226, 73, 52, 3.4 | 277, 264, 99 | 240, 104, 99, 73, 52, 45, 24, 3.4 | 640± |
| THS | 545 | 545 | 320, 234 | 544± |
| CCE | 440 | 450 | 405, 46 | 448± |
| CCU | 265, 135, 50, 25 | 475 | 200, 136, 64, 50, 30 | 476± |

**HOW DO WE KNOW?**

27. In this chapter, we focused on the analysis of genomes, transcriptomes, and proteomes and considered important applications and findings from these endeavors. At the same time, we found many opportunities to consider the methods and reasoning by which much of this information was acquired. From the explanations given in the chapter, what answers would you propose to the following fundamental questions:

(a) How do we know which contigs are part of the same chromosome?
(b) How do we know if a genomic DNA sequence contains a protein-coding gene?
(c) What evidence supports the concept that humans share substantial sequence similarities and gene functional similarities with model organisms?
(d) How can proteomics identify differences between the number of protein-coding genes predicted for a genome and the number of proteins expressed by a genome?
(e) What evidence indicates that gene families result from gene duplication events?
(f) How have microarrays demonstrated that, although all cells of an organism have the same genome, some genes are expressed in almost all cells whereas other genes show cell- and tissue-specific expression?

## Extra-Spicy Problems

28. Genomic sequencing has opened doors to numerous studies that help us understand the evolutionary forces shaping the genetic makeup of organisms. Using databases containing the sequences of 25 genomes, scientists (Kreil, D.P. and Ouzounis, C.A., *Nucl. Acids Res.* 29: 1608–1615, 2001) examined the relationship between GC content and global amino acid composition. They found that it is possible to identify thermophilic species on the basis of their amino acid composition alone, which suggests that evolution in a hot environment selects for a certain whole organism amino acid composition. In what way might evolution in extreme environments influence genome and amino acid composition? How might evolution in extreme environments influence the interpretation of genome sequence data?
29. The $\beta$-globin gene family consists of 60 kb of DNA, yet only 5 percent of the DNA encodes gene products. Account for as much of the remaining 95 percent of the DNA as you can.
30. Annotation of a proteome attempts to relate each protein to a function in time and space. Traditionally, protein annotation depended on an amino acid sequence comparison between a query protein and a protein with known function If the two proteins shared a considerable portion of their sequence, the query would be assumed to share the function of the annotated protein. Following is a representation of this "look-the-same" method of protein annotation involving a query sequence and three different human proteins (modified from Rigoutsos et al. 2002. *Nucl. Acids Res.* 30: 3901–3916). Note that the query sequence aligns to common domains within the three other proteins. What argument might you present to suggest that the function of the query is not related to the function of the other three proteins?

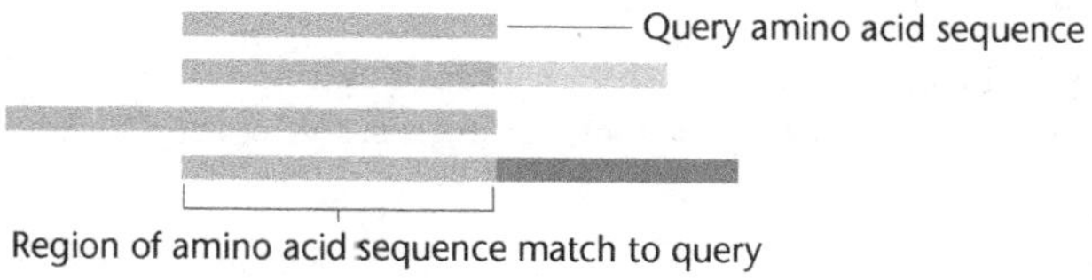


31. In a sequence of 99.4 percent of the euchromatic regions of human chromosome 1, Gregory et al. (Gregory, S.G. et al., *Nature*, 441: 315–321, 2006) have identified 3141 gene structures.
(a) How does one identify a gene within a raw sequence of bases in DNA?
(b) What procedures are often used to verify likely gene assignments?
(c) Given that chromosome 1 contains approximately 8 percent of the human genome, and assuming that there are approximately 20,000–25,000 genes, would you consider chromosome 1 to be "gene rich"?
32. M. Stoll and colleagues have compared candidate loci in humans and rats in search of loci in the human genome that are likely to contribute to the constellation of factors leading to hypertension. Through this research, they identified 26 chromosomal regions that they consider likely to contain hypertension genes. How can comparative genomics aid in the identification of genes responsible for such a complex human disease? The researchers state that comparisons of rat and human candidate loci to those in the mouse may help validate their studies. Why might this be so?
33. Comparisons between human and chimpanzee genomes indicate that a gene that may function as a wild type or normal gene in one primate may function as a disease-causing gene in another (The Chimpanzee Sequence and Analysis Consortium, *Nature*, 437: 69–87, 2005). For instance, the *PPARG* locus (regulator of adipocyte differentiation) is associated with type 2 diabetes in humans but functions as a wild-type gene in chimps. What factors might cause this apparent contradiction? Would you consider such apparent contradictions to be rare or common? What impact might such findings have on the use of comparative genomics to identify and design therapies for disease-causing genes in humans?
34. Because of its accessibility and biological significance, the proteome of human plasma has been intensively studied and used to provide biomarkers for such conditions as myocardial infarction (troponin) and congestive heart failure (B-type natriuretic peptide). Polanski and Anderson (Polanski, M., and Anderson, N. L., *Biomarker Insights*, 2: 1–48, 2006) have compiled a list of 1261 proteins, some occurring in plasma, that appear to be differentially expressed in human cancers. Of these 1261 proteins, only 9 have been recognized by the FDA as tumor-associated proteins. First, what advantage should there be in using plasma as a diagnostic screen for cancer? Second, what criteria should be used to validate that a cancerous state can be assessed through the plasma proteome?

Human immunodeficiency virus (HIV), the causative agent of acquired immunodeficiency syndrome (AIDS), budding from the surface of a human T-lymphocyte.

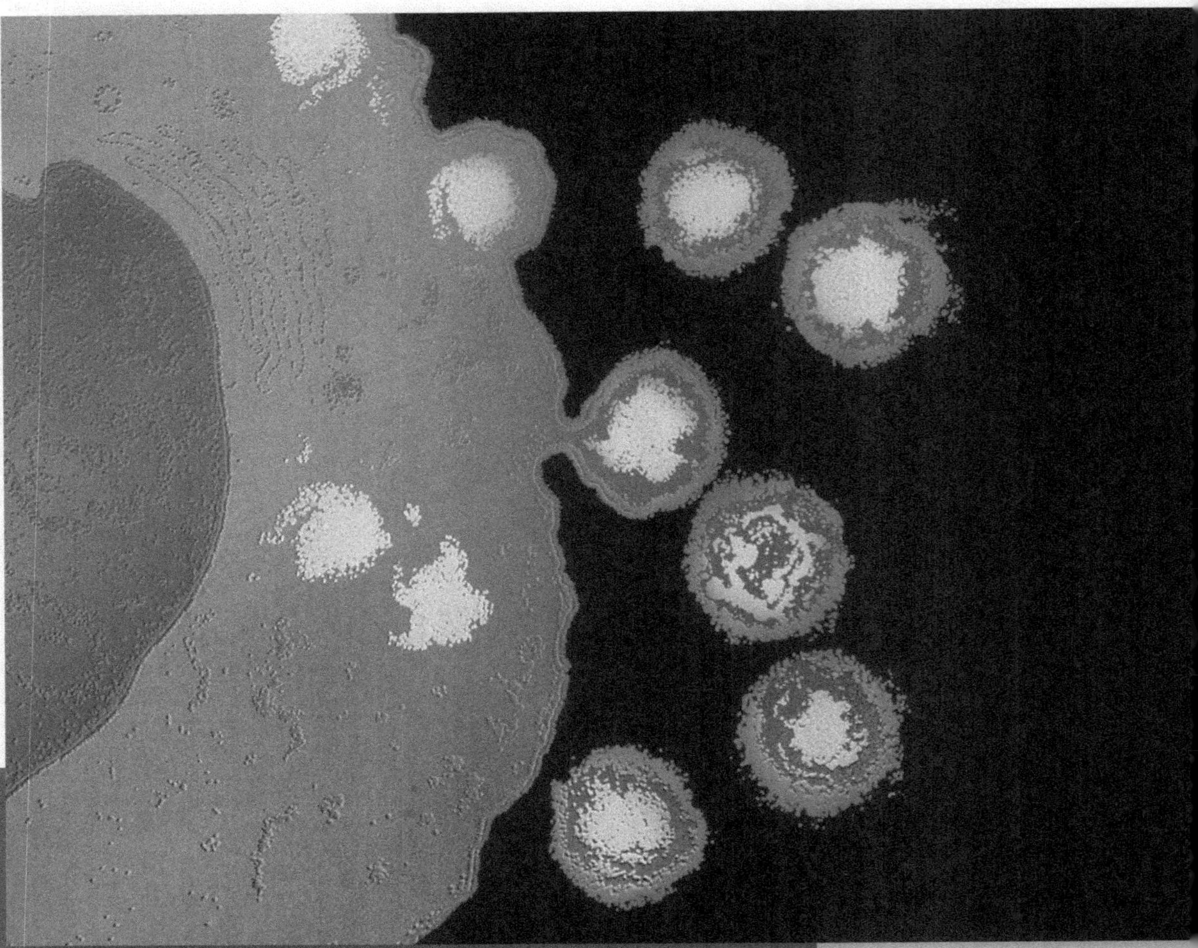

# 22

# Genome Dynamics: Transposons, Immunogenetics, and Eukaryotic Viruses

## CHAPTER CONCEPTS

- Transposable elements make up significant portions of the genomes of all organisms, from bacteria to humans.
- Transposons move within and between genomes, creating mutations, changing patterns of gene expression, and providing new genetic raw material for evolution.
- Immunoglobulin and T-cell receptor genes are created by the rearrangement of genomic segments during the development of B and T lymphocytes.
- Antigenic diversity is the result of imprecise gene rearrangements, hypermutability, and combinations of multiple different gene segments.
- Eukaryotic viruses exhibit high levels of genetic variability and gene exchange with other viruses and host cells.
- Eukaryotic viruses evolve rapidly due to their use of error-prone RNA and DNA polymerases and their ability to capture gene sequences from other genomes.

The genomes of all organisms maintain a high degree of genetic stability. Each species contains a specific number of chromosomes, present as pairs in diploid organisms or as single chromosomes in germ cells or haploid organisms. The stability of genes—in terms of both chromosomal location and DNA sequence—is essential for preserving a species and for the faithful storage, replication, and transmission of genetic information. Genetic stability in general is essential for accurate multicellular development, coordinated cellular functions, and the passage of traits from generation to generation.

In contrast to this indispensable core of genetic stability are dramatic examples of genomic alterations that threaten to disrupt that stability. In earlier chapters, we encountered gene mutations—changes in the base-pair sequences of DNA that arise spontaneously or from environmental assaults. We also learned about chromosomal rearrangements such as deletions, insertions, translocations, and changes in chromosome numbers. If these genetic alterations occur in germ cells, they may be inherited and form the basis of genetic disorders and phenotypic variation. If they occur only in somatic cells, they will not be inherited but may lead to somatic disorders such as cancers or collections of phenotypically aberrant tissue cells. Clearly, these gene mutations, chromosomal rearrangements, and other genome alterations are often detrimental for individual organisms, but we have also come to appreciate that they are the raw material for evolution.

In this chapter, we will describe three other significant genome-altering forces. Two of these—transposons and eukaryotic viruses—reshape genomes by moving in and out of them, introducing new genetic information, altering gene expression, triggering mutations, and creating new genes. The third—immunoglobulin gene rearrangements—are programmed genetic events that are essential for the proper functioning of the immune system. As more genomic data emerge from the Human Genome Project and sequencing projects of other organisms, it is becoming clear that both prokaryotic and eukaryotic genomes are in a constant state of flux. The balance between stability and genomic "creativity" helps to shape the variety of life on Earth.

## 22.1 Transposable Elements Are Present in the Genomes of Both Prokaryotes and Eukaryotes

**Transposable elements**, also known as **transposons** or "jumping genes," are DNA elements that move, or transpose, within the genome and between genomes. Transposable elements were first discovered more than 50 years ago by Barbara McClintock in her studies of maize. However, the idea that some genetic information may not be fixed within the genome was slow to find acceptance. The concept of movable genes was quite alien to the classical understanding of genes as discrete loci that can be mapped to specific chromosomes. It was not until transposable elements were discovered in other organisms, and their molecular basis revealed, that the existence and significance of transposable elements were confirmed.

Transposons are present in the genomes of all organisms, from bacteria to humans. Not only are they ubiquitous, but they also make up large portions of some eukaryotic genomes. For example, recent genomic sequencing has revealed that almost 50 percent of the human genome is derived from transposable elements. Some organisms with unusually large genomes, such as salamanders and barley, contain hundreds of thousands of copies of various types of transposable elements. The function of these elements is unknown, and it is remarkable that eukaryotic genomes tolerate such a load of apparently useless, or "junk," DNA. Data from human genome sequencing suggest that some genes may have evolved from transposons, and that the activity of transposons helps to modify and reshape the genome. In this sense, some transposable elements may confer benefits upon their hosts.

Transposable elements have also proved to be valuable tools in genetic research. As will be described in Chapter 23, geneticists have harnessed transposons as mutagens, as cloning tags, and as vehicles for introducing foreign DNA into model organisms.

In this chapter, we will discuss transposable elements as significant forces for genome change. The movement of transposons from one place in the genome to another has the capacity to disrupt genes and cause mutations, as well as to create source materials for evolution.

### Insertion Sequences

There are two types of transposable elements in bacteria: insertion sequences and bacterial transposons. **Insertion sequences**, also called **IS elements**, were first characterized at the molecular level in the 1970s. They can move from one location to another and, if they insert into a gene or gene regulatory region, may cause mutations.

IS elements were first identified during analyses of mutations in the *gal* operon of *E. coli*. Researchers discovered that certain mutations in this operon were due to the presence of several hundred base pairs of extra DNA inserted into the beginning of the operon. Surprisingly, the segment of mutagenic DNA could spontaneously excise from this location, restoring wild-type function to the *gal* operon.

Subsequent research revealed that several other DNA elements could behave in a similar fashion, inserting into bacterial chromosomes and affecting gene function. These DNA elements are relatively short, not exceeding 2000 bp (2 kb). The first insertion sequence to be characterized in *E. coli*, IS1, is about 800 bp long. Other IS elements, such as IS2, 3, 4, and 5, are about 1250 to 1400 bp in length. IS elements are present in multiple copies in bacterial

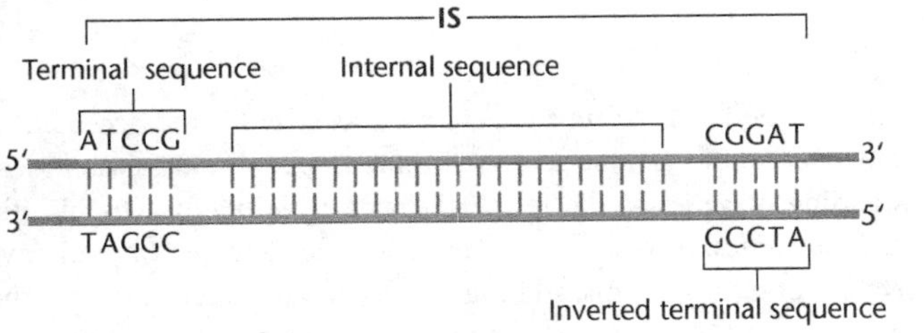


FIGURE 22–1 An insertion sequence (IS), shown in purple. The terminal sequences are perfect inverted repeats of one another.

genomes. For example, the *E. coli* chromosome contains five to eight copies of IS1 and five copies each of IS2 and IS3, as well as copies of IS elements on plasmids such as F factors.

All IS elements contain two features that are essential for their movement. First, they contain a gene that encodes an enzyme called **transposase.** This enzyme is responsible for making staggered cuts in chromosomal DNA, into which, or out of which, the IS element can insert. Second, the ends of IS elements contain inverted terminal repeats (ITRs). ITRs are short segments at opposite ends of a segment of DNA that have the same nucleotide sequence as each other but are oriented in opposite directions (Figure 22–1). Although the ITRs are drawn in Figure 22–1 with only a few nucleotides, many more nucleotides are actually present. For example, IS1 ITRs contain about 20 nucleotide pairs, those of IS2 and IS3 contain about 40 nucleotide pairs, and those of IS4 contain about 18 nucleotide pairs. ITRs are essential for transposition and act as recognition sites for binding the transposase enzyme.

## Bacterial Transposons

Bacterial transposons, known as **Tn elements,** are larger than IS elements and contain protein-coding genes that are unrelated to their transposition properties. Some Tn elements, such as Tn10, consist of a drug-resistance gene flanked by two IS elements present in opposite orientations. The IS elements encode the transposase enzyme that is necessary for transposition of the Tn element. Other types of Tn elements, such as Tn3, have shorter inverted repeat sequences at their ends and encode their transposase enzyme from a transposase gene located in the middle of the Tn element. Like IS elements, Tn elements are mobile in both bacterial chromosomes and in plasmids, and can cause mutations if they insert into genes or gene regulatory regions.

The presence of inverted repeats at the 5′ and 3′ termini of Tn elements has been revealed by electron microscopic analysis (Figure 22–2). When double-stranded DNA from a plasmid containing a Tn element is separated into single strands, the two complementary segments that each single strand contains can anneal to form a heteroduplex within that strand. All other areas of the Tn and plasmid remain single-stranded and form loops on either end of the double-stranded stem.

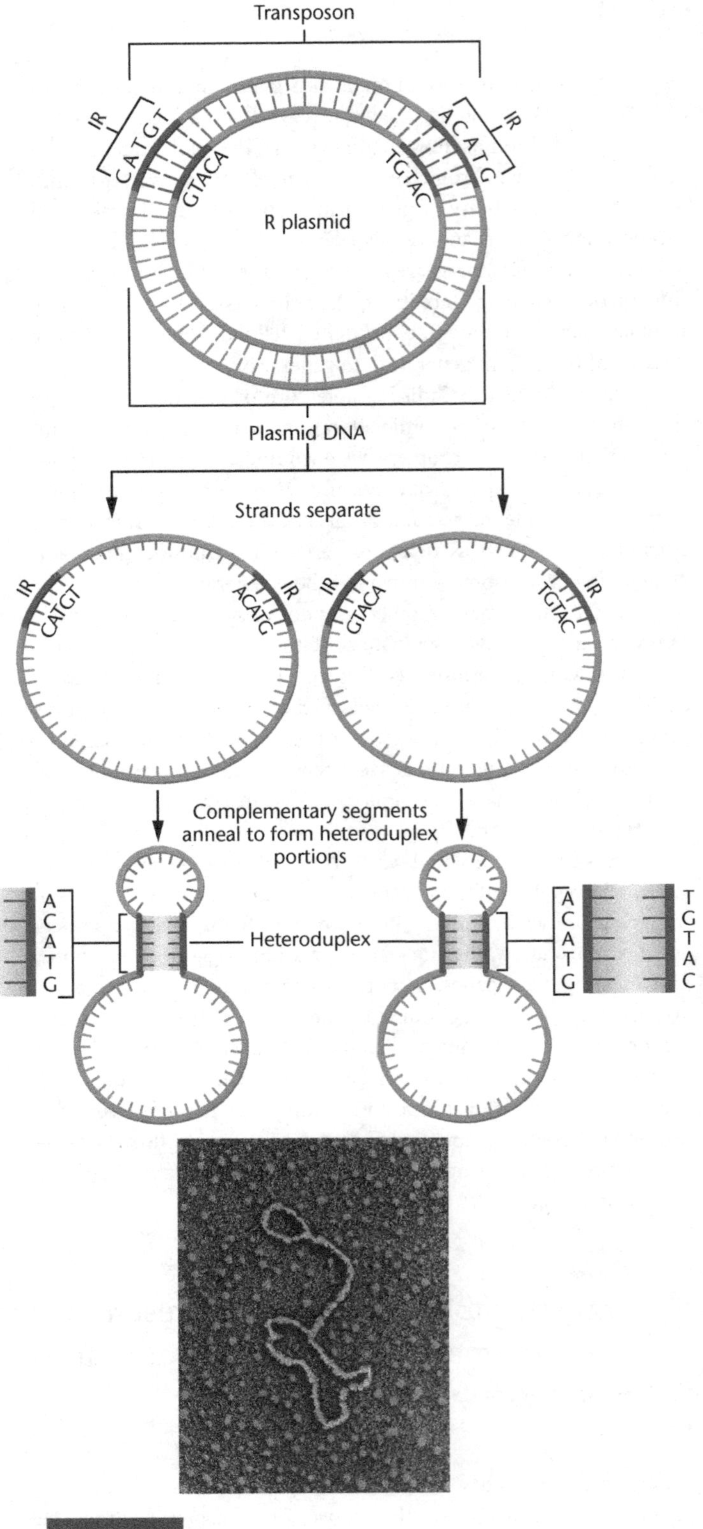


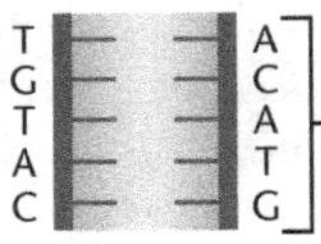


FIGURE 22–2 Heteroduplex formation resulting from the presence of inverted repeat sequences of a transposon inserted into a bacterial plasmid. The electron micrograph illustrates the final heteroduplex.

Tn elements are currently of particular interest because they can introduce multiple drug resistance into bacterial plasmids. These plasmids, called R factors, may contain many Tn elements conferring simultaneous resistance to heavy metals, antibiotics, and other drugs. The Tn elements can move from plasmids onto bacterial chromosomes and can spread multiple drug resistance between different strains of bacteria.

## The *Ac–Ds* System in Maize

About 20 years before the discovery of transposons in bacteria, Barbara McClintock discovered mobile genetic elements in corn plants (maize). She did this by analyzing the genetic behavior of two mutations, ***Dissociation* (*Ds*)** and ***Activator* (*Ac*)**, expressed in either the endosperm or aleurone layers of corn kernels (Figure 22–3). She then correlated her genetic observations with cytological examinations of the maize chromosomes. Initially, McClintock determined that *Ds* was located on chromosome 9. If *Ac* was also present in the genome, *Ds* often induced breakage at a point on the chromosome adjacent to its own location. If chromosome breakage occurred in somatic cells during their development, progeny cells sometimes lost part of the broken chromosome, causing a variety of phenotypic effects.

(a)

(b)

FIGURE 22–3 (a) Barbara McClintock analyzing corn variants in her laboratory. McClintock earned the Nobel Prize in Physiology or Medicine in 1983, at the age of 81, for her pioneering research on mobile genetic elements. (b) Ear of corn, showing a range of pigment variations between and within kernels, due to transposition of the *Ds* element.

Subsequent analysis suggested to McClintock that both *Ds* and *Ac* elements sometimes moved to new chromosomal locations. While *Ds* moved only if *Ac* was also present, *Ac* was capable of autonomous movement. Where *Ds* came to reside determined its genetic effects—that is, it might cause chromosome breakage, or it might inhibit expression of a certain gene. In cells in which *Ds* caused a gene mutation, *Ds* might move again, restoring the gene mutation to wild-type.

Figure 22–4 illustrates the types of movements and effects brought about by *Ds* and *Ac* elements. In McClintock's original observation, pigment synthesis was restored in cells in which the *Ds* element jumped out of chromosome 9. McClintock concluded that

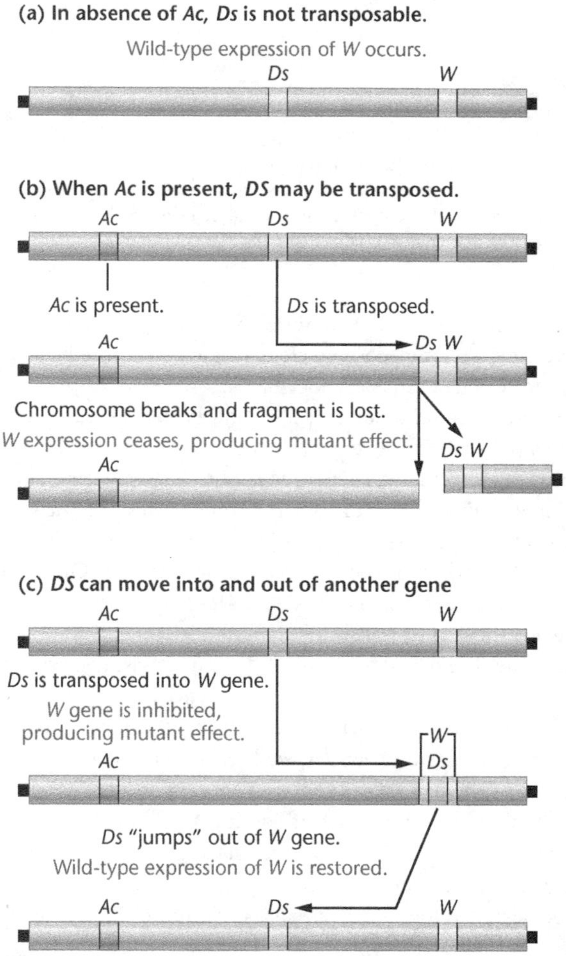


FIGURE 22–4 Effects of *Ac* and *Ds* elements on gene expression. (a) If *Ds* is present in the absence of *Ac*, there is normal expression of a distantly located hypothetical gene *W*. (b) In the presence of *Ac*, *Ds* may transpose to a region adjacent to *W*. *Ds* can induce chromosome breakage, which may lead to loss of a chromosome fragment bearing the *W* gene. (c) In the presence of *Ac*, *Ds* may transpose into the *W* gene, disrupting *W*-gene expression. If *Ds* subsequently transposes out of the *W* gene, *W*-gene expression may return to normal.

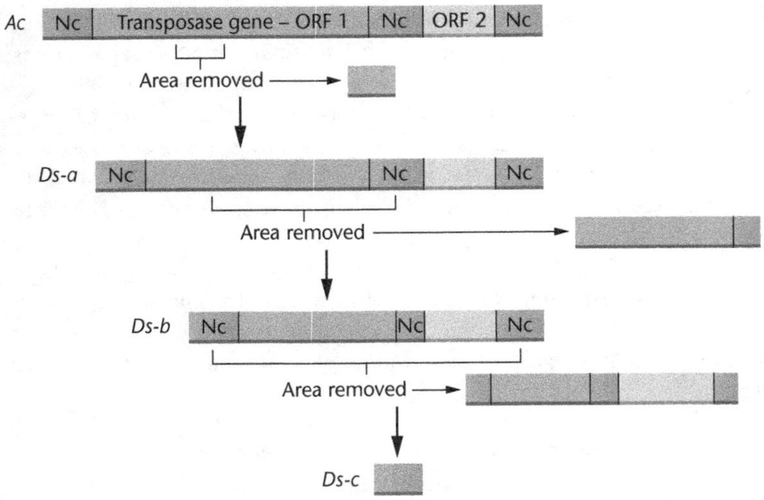


FIGURE 22-5 A comparison of the structures of an *Ac* element and three *Ds* elements. The transposase gene is in the open reading frame designated ORF 1. No function has yet been assigned to ORF 2. (Noncoding regions are designated Nc.) As this illustration shows, *Ds-a* appears to be an *Ac* element that has a small deletion in the gene encoding the transposase enzyme.

the *Ds* and *Ac* genes were **mobile controlling elements**. We now refer to them as transposable elements, a term coined by another great maize geneticist, Alexander Brink.

Several *Ac* and *Ds* elements have now been analyzed, and the relationship between the two elements has been clarified (Figure 22–5). The *Ac* element is 4563 nucleotides long, and its structure is strikingly similar to that of bacterial transposons. The *Ac* sequence contains two 11-base-pair imperfect ITRs, two open reading frames (ORFs), and three noncoding regions. One of the two ORFs encodes the *Ac* transposase enzyme. The first *Ds* element studied (*Ds-a*) is nearly identical to *Ac* except for a 194-bp deletion within the transposase gene. The deletion of part of the transposase gene in the *Ds-a* element explains the *Ds* element's dependence on the *Ac* element for transposition. Several other *Ds* elements have also been sequenced, and each contains an even larger deletion within the transposase gene. In each case, however, the ITRs are retained.

Although the significance of Barbara McClintock's mobile controlling elements was not fully appreciated at the time of her initial observations, molecular analysis has since verified her conclusions. She was awarded the Nobel Prize in Physiology or Medicine in 1983.

## Mobile Genetic Elements in Peas: Mendel Revisited

Recent work on transposable elements in plants leads us back to Gregor Mendel and his observations of the inheritance of round and wrinkled peas. The two phenotypes are controlled by alleles of a single gene, *rugosus,* also referred to as the *wrinkled* gene. It is now known that the wrinkled phenotype is caused by the absence of an enzyme, **starch-branching enzyme I (SBEI)**, that controls the formation of branch points in starch molecules. The lack of starch synthesis in wrinkled peas leads to the accumulation of sucrose and a higher water content and osmotic pressure in the developing seeds. As the seeds mature, those with genotype *rr* lose more water, and thus become wrinkled, while those with genotype *RR* or *Rr* lose less water and remain smooth.

The structural gene for SBEI has been cloned and characterized in both wild-type and mutant genotypes. In the *rr* genotype, the SBEI protein is nonfunctional because the *SBEI* gene is interrupted by a 0.8-kb insertion, resulting in the production of an abnormal RNA transcript. The inserted DNA has 12-bp inverted repeats at each end that are highly homologous to the ITRs of the transposable element *Ac* from maize and to other *Ac*-like elements from snapdragons and parsley.

## *Copia* Elements in *Drosophila*

In 1975, David Hogness and his colleagues David Finnegan, Gerald Rubin, and Michael Young identified a class of DNA elements in *Drosophila melanogaster* that they designated as ***copia***. These elements are transcribed into "copious" amounts of RNA (hence their name). *Copia* elements are present in 10 to 100 copies in the genomes of *Drosophila* cells. Mapping studies show that they are transposable to different chromosomal locations and are dispersed throughout the genome.

Each *copia* element consists of approximately 5000 to 8000 bp of DNA, including a long **direct terminal repeat (DTR)** sequence of 276 bp at each end. Within each DTR is an inverted terminal repeat (ITR) of 17 bp (Figure 22–6). The short ITR sequences are characteristic of *copia* elements. The DTR sequences are found in other transposons in other organisms, but they are not universal.

Insertion of *copia* is dependent on the presence of the ITR sequences and seems to occur preferentially at specific target sites in the genome. The *copia*-like elements demonstrate regulatory effects at the point of their insertion in the chromosome. Certain mutations, including those affecting eye color and segment formation, are due to *copia* insertions within genes. For example, the eye color mutation *white–apricot* ($w^a$), an allele of the *white* (*w*) gene, contains a *copia* element within the gene. Transposition of the copia element out of the $w^a$ allele can restore the allele to wild-type.

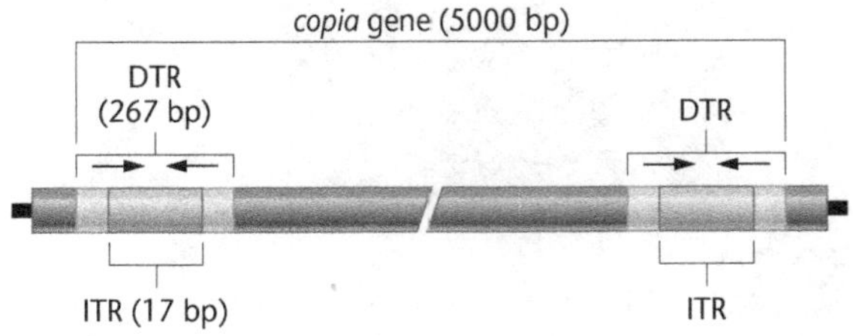


FIGURE 22-6 Structural organization of a *copia* transposable element in *Drosophila melanogaster*, showing the terminal repeats.

*Copia* elements are only one of approximately 50 families of transposable elements in *Drosophila*, each of which is present in as many as 20 to 50 copies in the genome. Together, these families constitute about 12 percent of the *Drosophila* genome and over half of the middle repetitive DNA of this organism. It is estimated that 50 percent of all visible mutations in *Drosophila* are the result of the insertion of transposons into otherwise wild-type genes.

## *P* Element Transposons in *Drosophila*

Perhaps the most significant transposable elements in *Drosophila* are the ***P* elements.** These were discovered in studies of hybrid dysgenesis, a condition characterized by sterility, elevated mutation rates, and chromosome rearrangements in the offspring of crosses between certain strains of fruit flies. Hybrid dysgenesis is caused by high rates of *P*-element transposition in the germ line, in which the transposons insert themselves into or near genes, thereby causing mutations. *P* elements range from 0.5 to 2.9 kb long, with 31-bp ITRs. Full-length *P* elements encode at least two proteins, one of which is the transposase enzyme that is required for transposition, and another is a repressor protein that inhibits transposition. The transposase gene is expressed only in the germ line, accounting for the tissue specificity of *P* element transposition. Strains of flies that contain full-length *P* elements inserted into their genomes are resistant to further transpositions due to the presence of the repressor protein encoded by the *P* elements.

Mutations can arise from several kinds of insertional events. If a *P* element inserts into the coding region of a gene, it can terminate transcription of the gene and destroy normal gene expression. If it inserts into the promoter region of a gene, it can affect the level of expression of the gene. Insertions into introns can affect splicing or cause the premature termination of transcription.

As will be described in Chapter 23, geneticists have harnessed *P* elements as tools for genetic analysis. One of the most useful applications of *P* elements is as vectors to introduce cloned genes into *Drosophila*—a technique known as **germ-line transformation.** *P* elements are also used to generate mutations and to clone mutant genes. In addition, researchers are perfecting methods to target *P* element insertions to precise single-chromosomal sites, which should increase the precision of germ-line transformation in the analysis of gene activity.

## Transposable Elements in Humans

The human genome, like that of other eukaryotes, is riddled with DNA derived from transposons. Recent genomic sequencing data reveal that approximately half of the human genomic DNA is composed of transposable-element DNA.

As we saw in Chapter 12, the major families of human transposons are the **long interspersed elements (LINEs)** and **short interspersed elements (SINEs)**. LINEs consist of DNA elements of about 6 kilobase pairs in length, present in up to 850,000 copies. In all, LINEs account for 21 percent of human genomic DNA. SINEs are about 100 to 500 bp long, with about 1.5 million copies present in human cells. SINEs make up about 13 percent of human genomic DNA. Other families of transposable elements account for a further 11 percent of the human genome. As coding sequences make up only about 5 percent of the human genome, there is about tenfold more transposable-element DNA in the human genome than DNA in functional genes.

Although most human transposons appear to be inactive, the potential mobility and mutagenic effects of transposable elements have far-reaching implications for human genetics, as can be seen in a recent example of a transposon "caught in the act." The case involves a male child with hemophilia. One cause of hemophilia is a defect in blood-clotting factor VIII, the product of an X-linked gene. Haig Kazazian and his colleagues found LINEs inserted at two points within the gene from this child. Researchers were interested in determining whether one of the mother's X chromosomes also contained this specific LINE. If so, the unaffected mother would have been heterozygous and would have passed the LINE-containing chromosome to her son. The surprising finding was that the LINE sequence was *not* present on either of her X chromosomes but *was* detected on chromosome 22 of both parents. This discovery suggests that the mobile element may have transposed from one chromosome to another in the gamete-forming cells of the mother, prior to being transmitted to the son.

LINE insertions into the human *dystrophin* gene have resulted in at least two separate cases of Duchenne muscular dystrophy. In one case, a transposon was inserted into exon 48, and in another case, a transposon was inserted into exon 44; both events led to frameshift mutations and premature termination of translation of the dystrophin protein. There are also reports that LINEs have become inserted into the *APC* and *c-myc* genes, leading to mutations that may have contributed to the development of some colon and breast cancers. In the latter cases, the transposition had occurred within one or a few somatic cells.

SINE insertions are also responsible for a number of human disease cases. In one instance, a type of SINE called an ***Alu* element** integrated into the *BRCA2* gene, inactivating this tumor-suppressor gene and leading to a familial case of breast cancer. Other genes that have been mutated by *Alu* integrations are the *factor IX* gene (leading to hemophilia B), the *ChE* gene (leading to acholinesterasemia), and the *NF1* gene (leading to neurofibromatosis).

# 22.2 Transposons Use Two Different Methods to Move Within Genomes

Transposable elements fall into two groups, based on their mechanisms of transposition. Members of one group, the **DNA transposons**, move from one part of the genome to another by a cut-and-paste mechanism mediated by the enzyme transposase. Members of the second group, the **retrotransposons**, resemble RNA viruses and transpose via an RNA intermediate, using the enzyme reverse transcriptase.

The origins of these different types of transposable elements and the effects they have on the genome are reflected in their mechanisms of transposition.

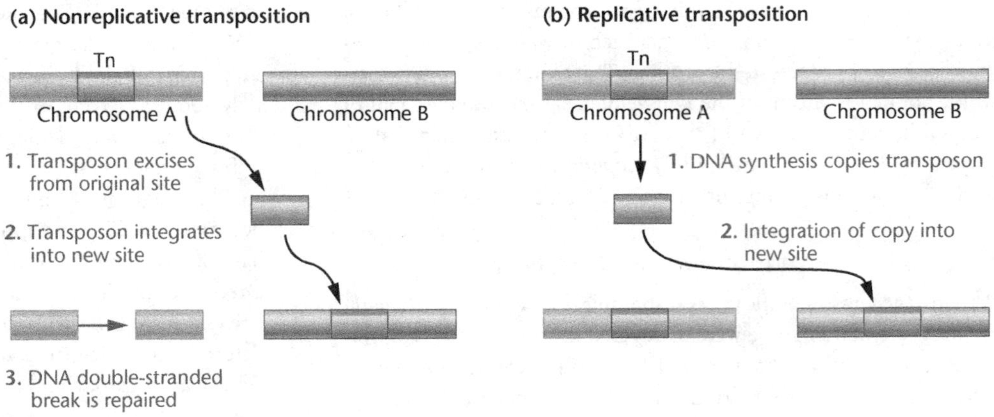


FIGURE 22–8 Nonreplicative and replicative methods of DNA transposition. (a) In nonreplicative transposition, the transposon moves from its original site to a new one, creating a double-stranded break at its original site. (b) In replicative transposition, a copy of the transposon is generated, and the copy becomes integrated into the new site. Several steps and intermediates are involved in replicative transposition and are not shown here.

## DNA Transposons and Transposition

Most prokaryotic transposable elements and many eukaryotic elements are DNA transposons. These elements characteristically contain inverted terminal repeats at their ends. We have already described several examples of DNA transposons, including bacterial Tn elements, *Drosophila P* elements, and the *Ac/Ds* elements in maize. All DNA transposons are flanked by repeated DNA sequences in the host DNA. These repeats are known as **target-site duplications**. The steps leading to creation of these repeated sequences during transposition are shown in Figure 22–7. In the first step, the transposase enzyme makes a staggered cut in the sugar-phosphate backbone of double-stranded DNA at the site of integration of the new transposon. These staggered breaks resemble those created by restriction endonucleases, and leave overhanging single-stranded DNA ends. The transposon inserts itself between the staggered ends of the host DNA, and the ends of the transposon are sealed to the overhanging ends of the host DNA. Finally, the gaps in the host DNA are filled in and sealed by the host cell's DNA polymerase and ligase enzymes. The transposase enzyme is necessary for cutting the target-site DNA and for recognizing the ends of the transposon. Most target sites are chosen at random, although some DNA transposons have preferences for certain regions of the genome.

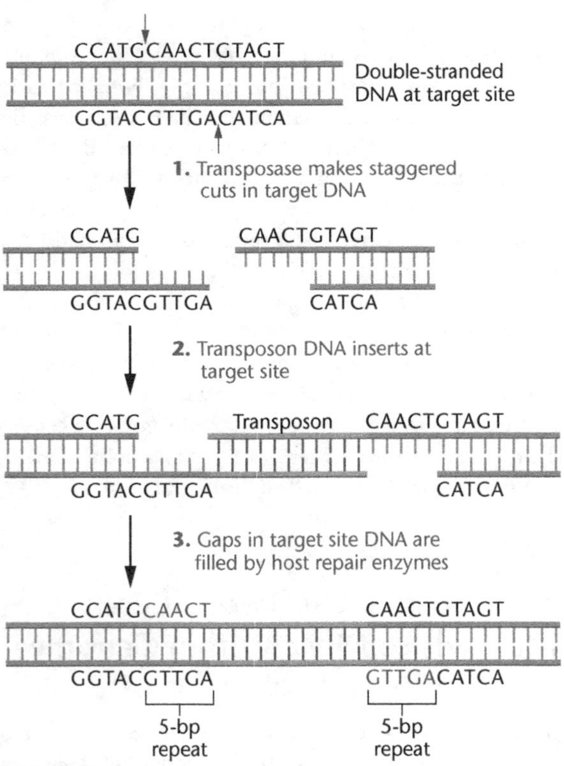


FIGURE 22–7 Creation of target-site duplications in host chromosome DNA during transposition. Transposase makes staggered cuts in the target site DNA. The transposon inserts itself at the target site, after which DNA synthesis and ligation are carried out by host-cell enzymes.

Although the creation of target-site duplications is common to all DNA transposons, transposons can move in two different ways: **nonreplicative transposition** and **replicative transposition** (Figure 22–8). During nonreplicative transposition, the transposon leaves its original site and inserts itself into a new location. The double-stranded cut at the original site of the transposon may be repaired, accurately or inaccurately, using the host cell's double-stranded DNA repair systems. During replicative transposition, the transposon is copied, leaving the original element in its original location. The replicated copy is inserted into the new location, resulting in an increase in the number of transposons in the host's genome.

## Retrotransposons and Transposition

The second group of transposons, the retrotransposons, are structurally distinct from DNA transposons and move by entirely different methods. Examples of retrotransposons that we met earlier are *copia* in *Drosophila* and LINEs and SINEs in humans. All retrotransposons are thought to transpose through an RNA intermediate that is converted back to DNA by reverse transcriptase. **Reverse transcriptase** is a unique polymerase that catalyzes the synthesis of a double-stranded DNA molecule from a single-stranded RNA template, using short RNA molecules as primers.

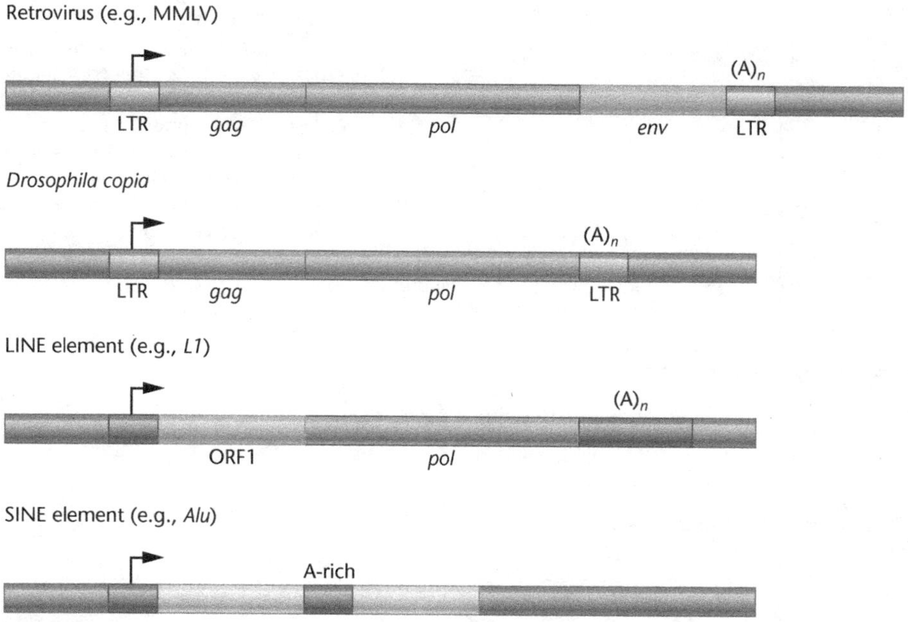


**FIGURE 22–9** Structural features of retroviruses and retrotransposons. Retroviruses, such as the Moloney murine leukemia virus (MMLV), are flanked by long terminal repeats (LTRs) containing transcription promoters (arrow) and polyadenylation signals ($A_n$). Typical viral genes are *gag* (encoding viral particle proteins), *pol* (encoding reverse transcriptase and integrase), and *env* (encoding viral particle proteins that interact with host-cell membranes). The *copia* element contains LTRs and *gag* and *pol* genes, but no *env* gene. Human LINE elements lack LTRs but contain a *pol* gene that encodes reverse transcriptase as well as an ORF that encodes an RNA-binding protein. The promoter, which is marked with an arrow, is recognized by the host cell's RNA polymerase II. Human SINE elements lack LTRs and *pol* genes and are transcribed by RNA polymerase III.

One class of retrotransposons, typified by the *Drosophila copia* element, bears a striking resemblance to the proviral form of eukaryotic retroviruses, which we discuss in more detail in Section 22.6. These elements have **long terminal repeat (LTR)** regions at their ends and contain one or more viral genes that encode reverse transcriptase, integrase, and viral structural proteins (Figure 22–9). Like retroviruses, retrotransposon LTRs contain promoters for RNA polymerase II transcription initiation and polyadenylation signals for processing mRNA. Because of the retrotransposons' structural resemblance to retroviruses, scientists hypothesize that these genomic elements descended from infectious retroviruses that entered the host by infecting germ-line cells. After entering the genome, they amplified themselves by reverse-transcriptase–mediated transposition but did not pass through an infective stage. Once stable in the genome, they were transmitted in a Mendelian manner.

Another class of retrotransposons, to which the human LINE elements belong, bears some similarity to retroviruses. These retrotransposons contain genes encoding reverse transcriptase, but are not flanked by LTRs. LINE elements contain functional RNA polymerase II promoters and polyadenylation signals. They are thought to have arisen by reverse transcription of mRNAs and integration of the double-stranded DNA copies into the genome. SINE elements lack both LTRs and genes encoding reverse transcriptase. In addition, they contain promoters for RNA polymerase III transcription initiation. SINE elements are thought to be derived from genes encoding tRNAs or 7S RNAs. Hence, LINEs and SINEs may be thought of as mobile pseudogenes.

The mechanism by which retrotransposons transpose is summarized in Figure 22–10. The example shown is that of the LTR-containing retrotransposon *copia*. First, the retrotransposon is transcribed into a functional mRNA by the host cell's RNA polymerase. Second, in the cytoplasm, the mRNA may be translated by the host-cell machinery into encoded proteins such as reverse transcriptase. Third, reverse transcriptase creates a double-stranded DNA copy of the mRNA, using short RNA fragments as primers for DNA synthesis. Finally, the double-stranded DNA copy of the mRNA reenters the nucleus, where it integrates into the genome at a new location. This process of transposition leaves the original element intact and creates at least one new copy.

There is considerable evidence that retrotransposons use RNA as an intermediate. In some cases, scientists have detected the intracellular mRNAs encoded by retrotransposons, and some of this RNA is present in virus-like particles (Figure 22–11). Although viral RNA and virus-like particles are present in some cells, they do not create

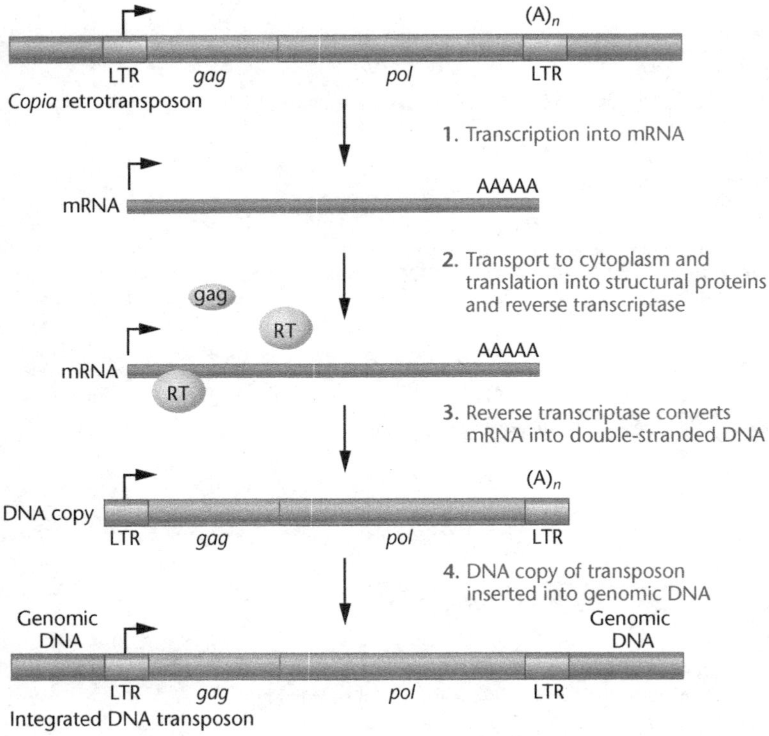


FIGURE 22–10 Transposition by an LTR-containing retrotransposon, *copia*. RNA polymerase II transcribes the retrotransposon into an mRNA that is transported to the cytoplasm and translated. Reverse transcriptase converts the mRNA into a double-stranded DNA copy of the retrotransposon, which is transported to the nucleus and becomes integrated into a new target site in the genome. Reverse transcription involves a number of priming, extension, and processing steps that are not shown here.

infectious retroviruses, due to the presence of deletions and mutations in the retrotransposon genes that encode both viral RNAs and proteins. Given that about 45 percent of the human genome consists of retrotransposon sequences, but only about 3 percent of the genome consists of DNA transposons, their mechanisms of replication must be considered highly efficient. The effects of retrotransposons on the genome are profound and will be discussed in the next section.

**NOW SOLVE THIS**

Problem 3 on page 602 asks you to assess whether the term *jumping genes* is an appropriate one for transposons.

■ HINT: *Consider the differences between DNA transposons and retrotransposons, and the two ways in which DNA transposons can move.*

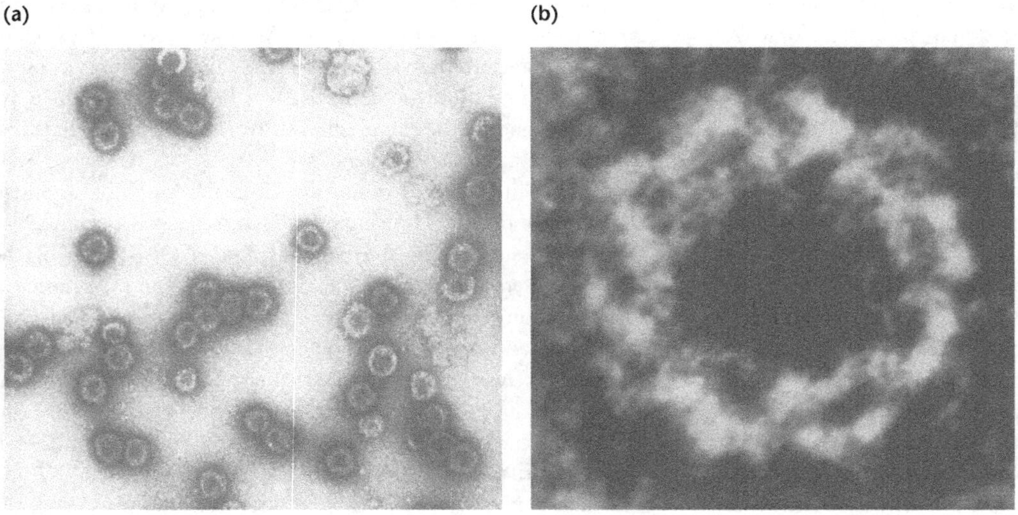


FIGURE 22–11 Electron microscopic images of virus-like particles in yeast cells that contain the retrotransposon *Ty*. (a) Virus-like particles consist of retrotransposon RNA and reverse transcriptase enzymes within a coat of gag-like proteins. They are structurally and functionally analogous to retrovirus particles but are not infectious. (b) One virus-like particle at higher magnification.

22.3

## Transposons Create Mutations and Provide Raw Material for Evolution

One of the most surprising outcomes of the large genome sequencing projects is the realization that a high percentage of plant and animal genomic DNA is noncoding and is derived from transposable elements. Transposons constitute about 15 percent of the *Drosophila* genome, more than 45 percent of the human genome, and up to 90 percent of the genomes of some plants and amphibians. In addition, data from these large-scale genome projects are now shedding light on the importance of these elements for disease causation and evolution.

Transposons have a wide range of effects on gene expression and chromosome structure (Table 22.1). Their presence within or near genes can create mutations that lead to genetic variability or to disease. Transposons can alter gene expression by modulating the levels and cell-type specificity of transcription initiation and by disrupting RNA processing. By stimulating genetic recombination, they trigger duplications, deletions, and translocations. Any of these changes may be detrimental, but some may confer a selective advantage to the organism. In this section, we summarize some of the genetic effects brought about by these moving genetic elements.

### Transposon Silencing

How are plants and animals able to cope with such a large amount of transposon DNA in their genomes? Luckily, most transposons are inactive and unable to transpose due to the accumulation of mutations. Most of the remainder are silenced by host regulatory mechanisms. These mechanisms are usually forms of epigenetic silencing, some of which may also affect the host genome's pattern of gene expression.

Mammalian genomes commonly suppress retrotransposon activity by subjecting transposon DNA to **cytosine methylation**, which leads to chromosome condensation in the methylated region. Elements such as the SINEs are particularly susceptible to cytosine methylation, due to their high GC content. It is thought that silenced transposons may act as starting points for the spread of repressive chromatin structures into nearby genomic DNA. In this way, transposons act through epigenetic mechanisms to affect gene expression. Although most retrotransposons appear to be hypermethylated, some escape this repression and may be active. Patterns of transposon methylation are stable and are passed on to succeeding generations of cells.

### Transposons, Mutations, and Gene Expression

The insertion of transposons into the coding regions of genes often results in mutations, as the presence of large quantities of foreign DNA will most likely disrupt the gene's reading frame or induce premature termination of translation. Insertion of transposons into introns may also cause mutations, as these insertions can bring about termination of transcription within the transposon or aberrant splicing if the transposon disrupts normal splicing signals. Insertions into gene regulatory regions also cause mutations, as transposons may disrupt the gene's normal regulatory sequences or may cause the associated gene to be expressed from the transposon's own promoter and enhancer sequences.

New germ-line transpositions are estimated to occur once in every 50 to 100 human births. Most of these do not cause disease or a change in phenotype; however, it is thought that about 0.2 percent of detectable human mutations are due to retrotransposon insertions.

TABLE 22.1

**Potential Effects of Transposons on the Genome**

| Effect | Cause |
|---|---|
| Gene mutation or inactivation | Insertion into coding region or intron of a gene |
| | Inaccurate excision from the coding region of a gene |
| Change in gene regulation | Insertion into or near the gene's transcription regulatory region |
| | Alteration of splicing by insertion within an intron |
| | Alteration in translatability of mRNA |
| | Epigenetic change in nearby chromatin |
| Addition of new exon | Insertion of transposon into a gene when transposon carries an exon from another gene (exon shuffling) |
| Gene duplication | Replication of a transposon that carries a host gene |
| | Recombination between two regions containing transposons |
| Gene deletion | Recombination between two transposons |
| Chromosome breakage | Excision of a transposon, followed by inadequate DNA repair |
| Chromosome rearrangement (including inversions and translocations) | Recombination between transposons located at two different places in the genome |

Other organisms appear to suffer more damage due to transposition. About 10 percent of new mouse mutations, and 50 percent of *Drosophila* mutations are due to insertions of retrotransposons in or near genes.

One of the most significant ways transposons affect their hosts is by altering transcription. Many types of transposons contain enhancers and promoters for transcription initiation as well as splice sites and polyadenylation signals. When located near a host gene, these transcription regulatory elements can influence how, when, and where the neighboring gene is expressed. For example, the LINE element L1 is present in the transcription control region of the human *apolipoprotein(a)* gene. In this position, it stimulates a ten-fold increase in the level of transcription initiation of this gene. The effects of transposons on gene expression may be more widespread than first thought. Genome sequencing has revealed that about 25 percent of human and mouse genes contain transposons in their promoters or upstream untranslated regions.

In some cases, transcription of a retrotransposon can read through the element's polyadenylation region into genes downstream from the transposon. This read-through may result in abnormal expression of the downstream gene, or in gene duplication if the chimeric transcript is reverse transcribed and inserted back into the genome.

The presence of two or more identical transposable elements in the genome creates the potential for unequal homologous recombination between them, leading to duplications, deletions, inversions, or chromosomal translocations (Figure 22–12). Any of

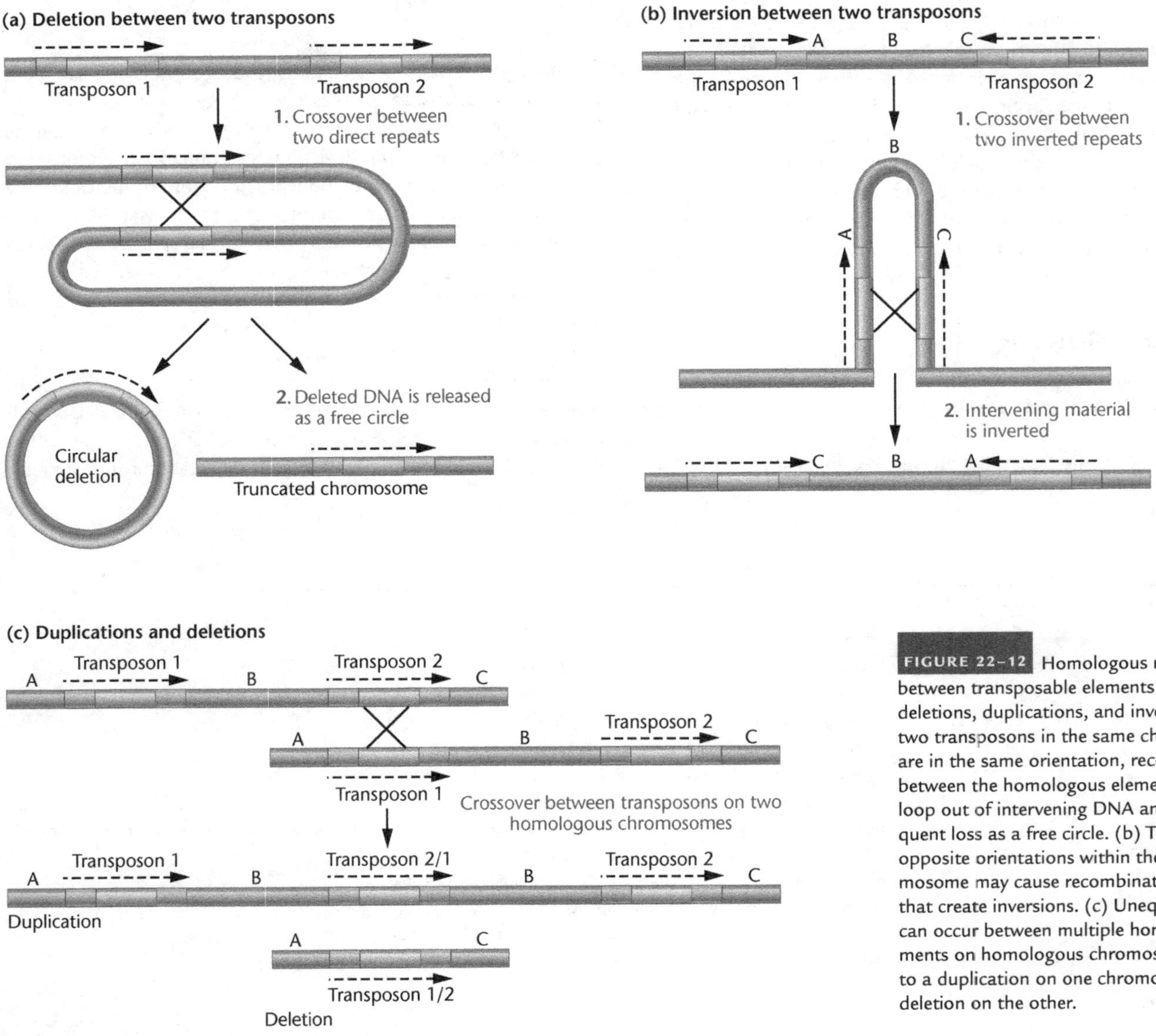


**FIGURE 22–12** Homologous recombination between transposable elements can result in deletions, duplications, and inversions. (a) If two transposons in the same chromosome are in the same orientation, recombination between the homologous elements creates a loop out of intervening DNA and its subsequent loss as a free circle. (b) Transposons in opposite orientations within the same chromosome may cause recombination events that create inversions. (c) Unequal crossovers can occur between multiple homologous elements on homologous chromosomes, leading to a duplication on one chromosome and a deletion on the other.

these chromosome rearrangements may bring about phenotypic changes or disease.

As discussed in Section 22.1, human transposons such as LINE and SINE elements are responsible for a number of genetic diseases, including cases of hemophilia, muscular dystrophy, and cancer. *Alu* elements are associated with over 20 human diseases, and usually insert within coding exons or in regions of the gene that affect RNA splicing. Some *Alu*-caused diseases include cases of Lesch-Nyhan syndrome, thalassemia, and type II diabetes. Transposons in mice also cause mutations. It is estimated that approximately 15 percent of disease-causing mutations in mice are due to transposon insertions.

### Transposons and Evolution

Although transposons were thought to be examples of "**selfish DNA**"—parasitic elements that are tolerated but contribute nothing to their hosts—it is becoming clear that they may cause genetic variability that contributes to evolution. For example, the Tn elements of bacteria carry antibiotic resistance genes between organisms, conferring a survival advantage to the bacteria under certain conditions. Another example of an evolved transposon is provided by *Drosophila* telomeres. LINE-like elements are present at the ends of *Drosophila* chromosomes, and these elements act as telomeres, maintaining the length of *Drosophila* chromosomes over successive cell divisions. Other examples of evolved transposons are the ***RAG1*** and ***RAG2*** genes in humans. These genes encode recombinases that are essential for V(D)J recombination in the immune system (which we discuss in the next section). These two genes appear to have evolved from DNA transposons.

Retrotransposons such as LINE elements also facilitate a phenomenon known as **exon shuffling.** If transcription of a LINE element extends past the LINE polyadenylation region into nearby exons, the resulting composite RNA may be reverse transcribed into a new type of retrotransposon containing a copy of the exon. If the new retrotransposon becomes inserted within another gene, it may introduce this new exon into the gene. If the insertion occurs outside a coding region, the shuffled exon may become the starting point for creating a new gene. Similarly, gene regulatory regions can be duplicated and introduced into new genomic regions, conferring changes in gene expression of nearby genes.

Transposons may also affect the evolution of genomes by altering gene-expression patterns in ways that are subsequently retained by the host. For example, the human *amylase* gene contains an enhancer specific for gene expression in the parotid gland. This tissue-specific enhancer evolved from a retrotransposon LTR. The retrotransposon appears to have become inserted into the *amylase* regulatory region early in primate evolution. Other examples of novel gene expression patterns brought about by retrotransposon insertion include T-cell-specific gene expression of the *CD8* and *zinc-finger 80* genes, and placenta-specific expression of the *leptin* and *CYP19* genes.

## 22.4 Immunoglobulin Genes Undergo Programmed Genome Rearrangements

In the previous sections, we learned that genomes are not static entities but can be invaded and colonized by a wide variety of DNA elements. These noncoding elements may constitute major portions of plant and animal genomic DNA. In the next example of genome dynamics, we will see that some types of developmental gene regulation, for example, regulation of immune-system gene expression, require precise, programmed DNA rearrangements. These rearrangements result in loss of gene segments; however, as the rearrangements occur only in particular types of somatic cells, they are not passed on to progeny.

### The Immune System and Antibody Diversity

The immune system protects organisms against infections and the presence of foreign substances that may enter blood or body tissues. The immune system is able to eliminate foreign or infectious agents by recognizing substances known as **antigens**. Antigens are defined as molecules, usually proteins, that bring about an immune response. To recognize antigen molecules that are present in or on foreign substances, components of the immune system must physically interact with them and bind to them, in a kind of lock-and-key configuration. Antigen recognition is the molecular fit between the antigen and molecules synthesized by cells of the immune system.

Although the immune system has many components, they can all be grouped into two major classifications: **humoral immunity** and **cellular immunity**. The humoral response involves the production of proteins called **immunoglobulins**, or **antibodies**, that bind directly to antigens. Some of these molecules are found on the surface of immune system cells, and some circulate in the body. The binding of antibody to antigen sends signals to other components of the immune system that engulf, lyse, or inactivate the antigen or antigen-bearing foreign entity. Antibodies are synthesized by a type of blood cell known as a **B lymphocyte**, or **B cell**, which undergoes development and maturation in the bone marrow (Figure 22–13).

The cellular response involves the recognition of antigens that are present on the surface of cells. For example, when eukaryotic cells are infected by viruses, viral proteins are processed by the cell and are presented on the cell surface. Immune system cells known as **T lymphocytes**, or **T cells**, directly recognize and bind to these cell-surface antigens, marking infected cells for destruction by other immune system components, such as macrophages. T cells, which develop in the thymus, express molecules known as **T-cell receptors (TCRs)** on their surfaces, and these TCRs interact directly with antigens.

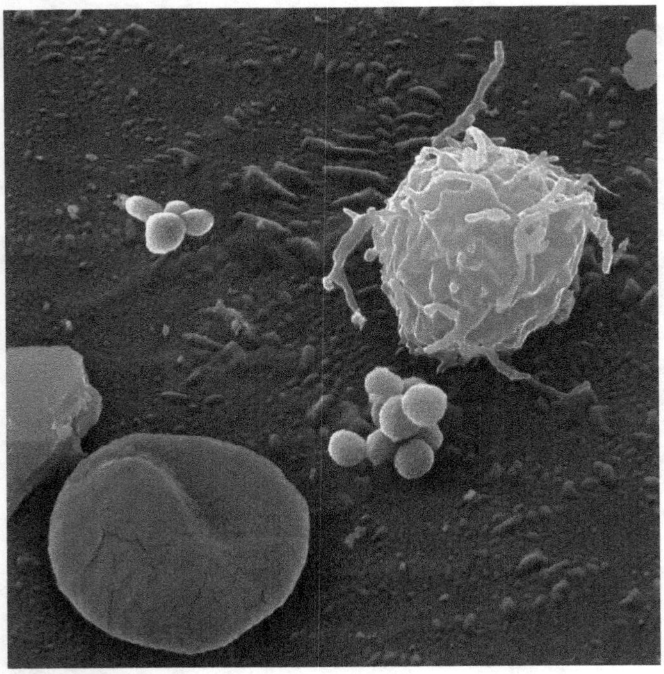

FIGURE 22–13 Colored scanning electron micrograph showing *Staphylococcus aureus* bacteria (in yellow), an erythrocyte (red blood cell), and a lymphocyte (white blood cell). There are two general types of lymphocytes: B cells and T cells. B cells bind to antigens present on pathogens or foreign substances. Encounter with an antigen stimulates the B cell to divide and form a pool of plasma cells that secrete antibodies. T cells recognize and bind to antigens, and then send signals to other components of the immune system which secrete antibodies (B cells) or kill body cells that are infected with a pathogen.

One of the most remarkable properties of the immune system is its ability to recognize a large number of foreign antigens—even antigens that the organism may never have encountered previously or may never encounter again. For example, mammals are capable of producing hundreds of millions of different types of antibodies in response to the presence of a wide variety of foreign substances. The immune system is able to synthesize more different varieties of antibodies and TCRs than there are genes in the genome. In addition, it is known that each B cell or T cell is able to synthesize only one type of antibody or TCR; however, when an antigen is encountered, the immune system mounts a strong response to the antigen, leading to the presence of large numbers of lymphocytes that respond to that one antigen.

If only one B cell or T cell can recognize a particular antibody, how does the immune response develop? An answer is provided by the **clonal selection theory**. This theory explains that, prior to the appearance of an antigen, there are hundreds of millions of immature lymphocytes in the body. Each lymphocyte expresses one type of immunoglobulin or TCR on its surface and is capable of recognizing and binding to one particular antigen. In the absence of an antigen that can interact with the lymphocyte's immunoglobulin, there is no change in the makeup of the pool of lymphocytes. When a foreign antigen interacts with an immunoglobulin molecule on the surface of one of the lymphocytes, a series of signals is sent to the lymphocyte, stimulating it to divide into a pool of clones—genetically identical cells that synthesize the immunoglobulin that recognizes the antigen.

The second question is how can a genome containing only tens of thousands of genes encode hundreds of millions of different antibody molecules? To answer this question, we first need to understand the structure of antibodies and TCRs.

## Immunoglobulin and TCR structure

Immunoglobulin molecules consist of four polypeptide chains, held together by disulfide bonds. Two of these polypeptides are identical **light (L) chains**, and two are identical **heavy (H) chains** (Figure 22–14). Together, they form a Y-shaped immunoglobulin structure. Each of the light and heavy chains contains a **constant region** at the C-terminus of the polypeptide and a **variable region** at the N-terminus. The four variable regions, when combined together into the immunoglobulin molecule, form a unique molecular structure that recognizes one specific antigen. Each B cell synthesizes only one type of immunoglobulin.

There are two types of light chains in humans: the **kappa (κ) chains** that are encoded by genes on chromosome 2, and the **lambda (λ) chains** that are encoded by genes on chromosome 22.

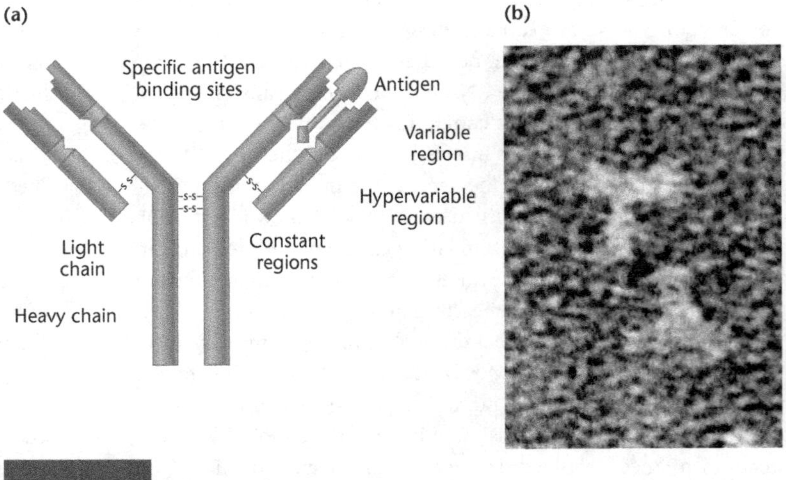


FIGURE 22–14 Structure of immunoglobulin molecules. (a) Immunoglobulin molecules are Y-shaped and consist of four polypeptide chains. Each chain contains a constant region (purple), a variable region (red), and a hypervariable region (blue). The variable and hypervariable regions form an antigen-recognition site that interacts with a specific antigen in a lock-and-key arrangement. The chains are joined by disulfide bonds. (b) This colored scanning electron micrograph reveals the Y-shaped structure of immunoglobulin molecules, which have been negatively stained prior to electron microscopy.

Only one type of heavy chain is encoded by gene segments on chromosome 14. Immunoglobulin molecules that contain either κ or λ light chains can exist in five classes, called **IgM, IgD, IgE, IgG**, and **IgA**. Each of these classes performs different functions. For example, IgM molecules are present on the surface of immature B cells, IgE molecules are involved in allergic responses, and IgG and IgA molecules (also called antibodies) are secreted by mature B cells known as **plasma cells**. Less is known about the roles played by the IgD class. When antigen is present, the B cells that recognize the antigen begin to divide and develop—first expressing their immunoglobulins in an IgM form, and then later, when the B cells have undergone maturation and development, in an IgA, IgG, or IgE form. The variable regions (and hence antigen-recognition sites) of each of these varieties remain the same as the B cell undergoes development, even though the class of immunoglobulin undergoes a switch. As we will see, the type of immunoglobulin class is determined by the constant regions of the heavy chains, which undergo class switching during B-cell development.

TCR molecules are structurally similar to immunoglobulin molecules. They are comprised of two identical light chains and two identical heavy chains. Each contains a constant region and a variable region similar to those of immunoglobulins. Each TCR, like each immunoglobulin, recognizes only one antigen. For the sake of simplicity, we will confine our discussion of immune system gene rearrangements to those associated with creation of the light and heavy chains of immunoglobulins. Similar mechanisms exist for TCR rearrangements.

## The Generation of Antibody Diversity and Class Switching

How does the human immune system generate antibody proteins in hundreds of millions of varieties, when there are only 20,000 to 30,000 genes in the human genome? The answer lies in the fact that the DNA encoding the immunoglobulin genes is not fixed but undergoes multiple programmed changes, including deletions, translocations, and random mutations.

To illustrate this principle of genomic malleability, we will describe the formation of κ light chains and heavy chains in humans. In humans, the mature κ light chain is made up of three regions: V (variable), J (joining), and C (constant). The κ protein is not encoded by a single germ-line gene, but by a gene that is assembled during B-cell development, from multiple gene regions located along chromosome 2. In most somatic cells, as well as in germ cells, these DNA regions are organized as shown in Figure 22–15. There are 70 to 100 different LV regions, about half of which are functional and can encode the V portion of a κ chain. Each LV region is preceded by a transcription promoter. The LV region is located at a distance of about 6 kilobases from the J region, a region that contains five possible J-encoding exons and is located upstream from the single C exon. A strong transcription enhancer is located between the J and C regions.

During development of a mature B cell, one of the LV regions ($L_2V_2$ in Figure 22–15), along with its promoter, is randomly joined by a recombination event to one of the five J regions ($J_3$ in our example). This event deletes all of the intervening DNA of chromosome 2. After this somatic recombination event, the LV promoter is activated by the nearby presence of the enhancer, resulting in transcription of the gene. The resulting RNA is spliced to remove the introns between the L and V regions, and between the J and C regions. After translation of the mature mRNA, the polypeptide is processed by removal of the leader (L) amino acids. The rearranged κ gene is then maintained and passed on to all progeny of the B cell.

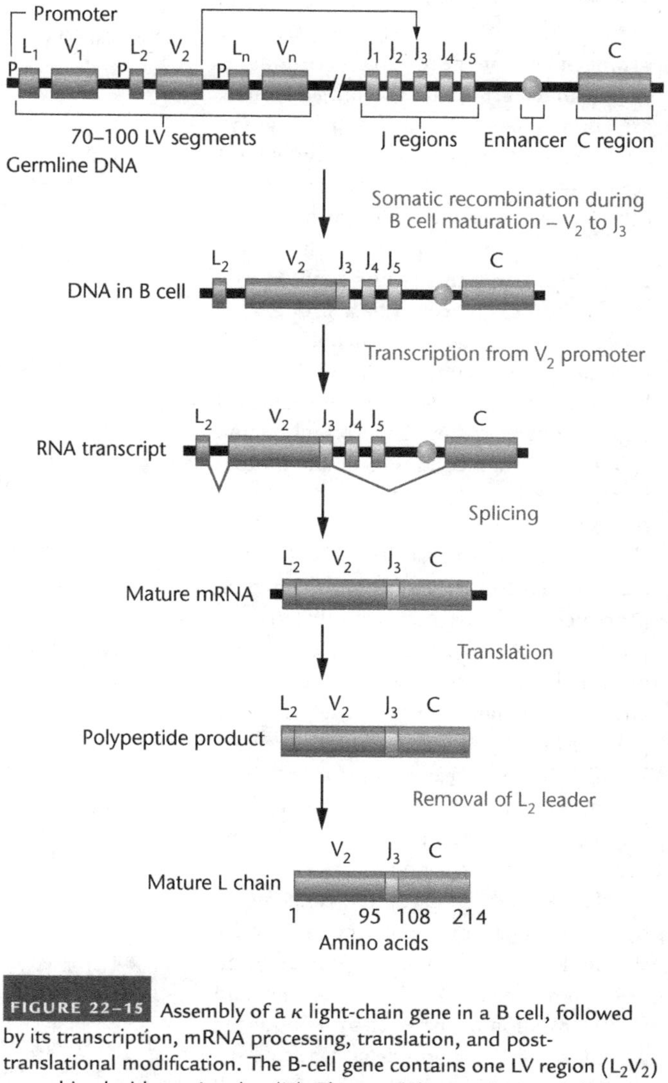


**FIGURE 22–15** Assembly of a κ light-chain gene in a B cell, followed by its transcription, mRNA processing, translation, and post-translational modification. The B-cell gene contains one LV region ($L_2V_2$) recombined with one J region ($J_3$). The remaining J regions and the enhancer-containing intron are removed from the transcript by intron splicing. Following translation, the leader sequence ($L_2$) is cleaved off as the mature polypeptide chain crosses the cell membrane.

Antibody diversity in the $\kappa$ chains results in part from the random recombination of one of 35 to 50 different functional LV regions with any one of five different J regions. Two other mechanisms increase the level of diversity. The recombination event that joins an LV region to a J region is not precise and can occur anywhere within a region containing several base pairs. Because of this imprecision, the recombination between any particular pair of LV and J regions still shows considerable variation. In addition, V regions are susceptible to high rates of random somatic mutation—**hypermutation**—during B-cell development. Hypermutation introduces even more variation into the LVJ region's sequence.

Heavy-chain gene rearrangements are even more complex than those leading to the creation of $\kappa$ genes. The germ-line arrangement of heavy-chain regions is shown in Figure 22–16. In humans, there are approximately 300 different V regions, 20 different D regions, and 6 different J regions—all located far apart from each other on chromosome 14. In addition, there are 9 different C regions, each with a polyadenylation signal at its 3′ end. The C-region cluster is preceded by a strong transcription enhancer. During B-cell development, somatic recombination brings one of the D regions to sit next to a J region (Figure 22–16). This recombination is followed by a second recombination that brings one of the V regions to sit next to the rearranged DJ region. Intervening DNA on chromosome 14 is lost. The resulting H-chain gene is transcribed from the V region promoter, which is now activated by the presence of the C region enhancer. The H-chain RNA is spliced to remove the introns, and cleavage and polyadenylation adds poly(A) tails to the end of the C$\mu$ region. Like $\kappa$ light-chain rearrangements, H-chain rearrangements are further diversified by imprecise recombination and by hypermutability in the V regions.

These first H-chain rearrangements occur in immature B cells that synthesize the IgM form of their immunoglobulin molecules (Figure 22–16). Cleavage and polyadenylation of the H-chain RNA at the 3′ end of the C$\mu$ region determine that the polypeptide translated from that RNA will be of the IgM class. The C regions that lie downstream from the C$\mu$ region are not transcribed as part of the IgM mRNA. However, if the B cell contacts an antigen that is recognized by the IgM immunoglobulin on the B-cell's surface, the gene undergoes further maturation steps that change it from the IgM class to one of the others—most often IgG, a secreted form of the immunoglobulin. Class switching occurs by a further recombination event. In this case, the S (switch) signal at the 5′ end of the C$\mu$ region and the S signal at the 5′ end of the C$\gamma$ region serve as the sites of the next recombination, which deletes the intervening DNA. Transcripts synthesized from the switched immunoglobulin gene are processed and

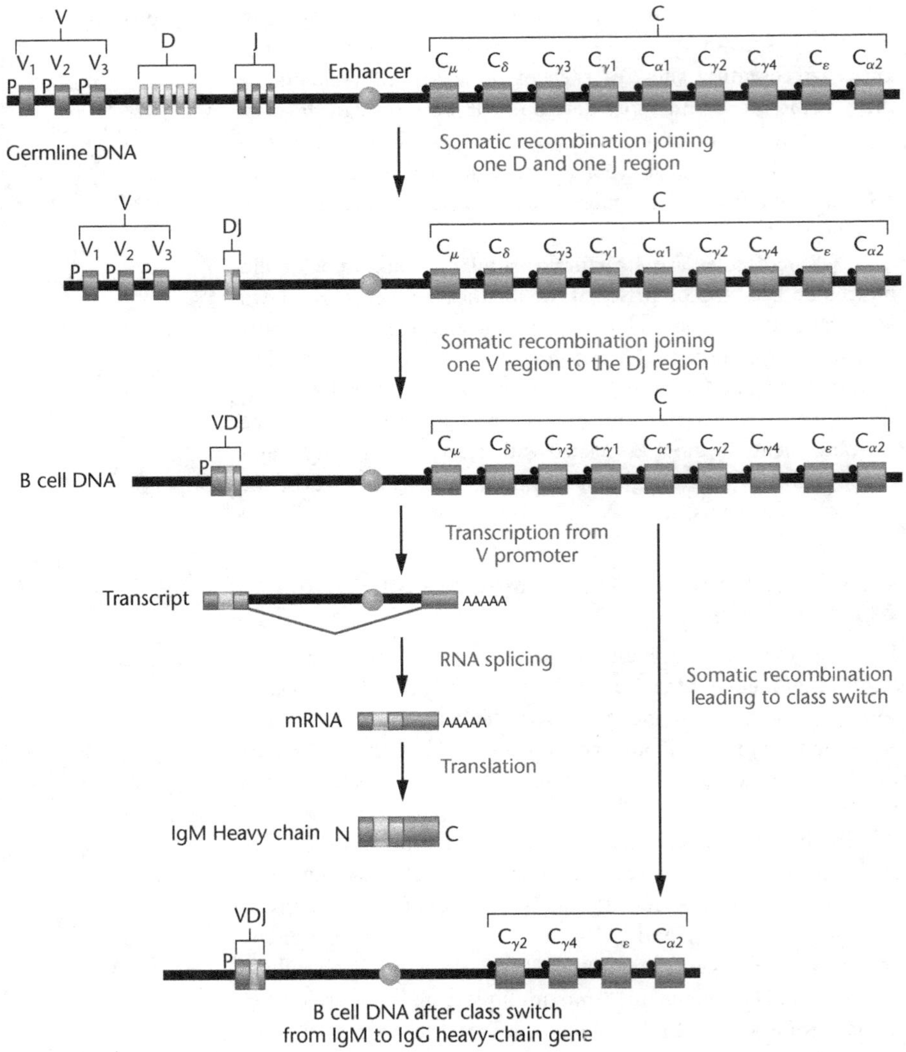


**FIGURE 22–16** Gene rearrangements leading to creation of immunoglobulin heavy-chain genes. Two recombinations juxtapose one D, one J, and one V region (containing the V promoter). Before a class switch, transcription initiates at the V promoter, and transcripts are cleaved and polyadenylated at the 3′ end of the C$\mu$ region. Translation of the mRNA produces an IgM heavy-chain polypeptide. After stimulation by antigen binding, developing B cells recombine the C regions at the switch junctions (shown as black dots), leading to the loss of intervening DNA. In this example, the class of immunoglobulin has switched from IgM to IgG.

polyadenylated at the end of the C$\gamma$ region, leading to translation of the IgG form of the immunoglobulin.

In all instances of immunoglobulin gene rearrangement, somatic recombination is brought about by the **RAG1** and **RAG2** proteins. These proteins create double-stranded breaks in recombination sequences located at the junctions of the V, D, J, and C gene regions. Recombination is completed by DNA-repair enzymes that join the double-stranded DNA ends together.

In summary, the following features combine to produce immense diversity in antibody structures:

- Multiple different V regions
- Multiple different D and J regions
- Imprecise recombination leading to inclusion or deletion of different numbers of nucleotides at recombination junctions
- Hypermutability of the V regions
- Class switching within the C regions of the heavy chains
- Various combinations of L and H chains to form the mature immunoglobulin protein

NOW SOLVE THIS

Problem 12 on page 602 asks you to predict the appearance of chromosomes after an H-chain rearrangement.

■ HINT: *Consider what happens to intervening DNA after V and DJ regions undergo recombination.*

22.5

## Eukaryotic Viruses Shuttle Genes Within and Between Genomes

Viruses are simple biological entities that must invade host cells in order to multiply and are small enough to pass through filters that retain bacteria. Viruses are composed of a small genome (DNA or RNA) enclosed in a protein coat. They can move between cells and commandeer the cell's biosynthetic mechanisms for their own replication. Many viruses cause diseases in their host, usually because their replication leads to the death of the cells they infect. Their apparent simplicity and obligate parasitism have generated controversy about whether they can be classified as living organisms.

In the last decade, large-scale genomic studies have significantly altered our view of viruses and their roles in evolution. Metagenomic analyses reveal that viruses occupy an extraordinary array of environments on Earth and infect every type of cellular organism that has been examined. It is estimated that there are $10^{31}$ virus particles on the planet—ten-fold more than the number of cells. More than 5000 different types of viruses can occupy about 200 liters of seawater, and $10^6$ types of viruses can be found in a kilogram of marine mud. Human intestines contain over 1000 different viral genotypes. Virus concentrations range from 60,000 to several hundred million per milliliter of lake water. Scientists conclude that viruses are the most abundant and pervasive biological entities on Earth.

Comparative genomics is highlighting the remarkable genetic diversity of viruses and is also providing surprising evidence of their capacity to move genes between genomes—a phenomenon known as **horizontal gene transfer.** Not only do viruses exchange genetic information with each other, but they also capture cellular genes, rearrange them, exchange them with other viruses, and reinsert these genes into other host cells. From these discoveries, it is now emerging that viruses are major players in the evolution of life on Earth.

It is well known that bacteriophages move genetic material between host cells, and this gene transfer can provide the host with new traits that may confer a competitive advantage. For example, some toxins that transform bacteria from benign to pathogenic forms are encoded by prophages. Bacteriophages mediate an enormous amount of horizontal gene transfer in prokaryotes and act as agents of bacterial evolution. It is estimated that as many as 10 percent of bacterial genes originated in bacteriophages. It is now emerging that eukaryotic viruses also act as vehicles of horizontal gene transfer and genetic diversity. Comparative genomic analyses of large DNA viruses of plants and animals are revealing that these viral genomes contain large numbers of genes that were exchanged with other eukaryotic viruses, their hosts, and even bacteriophages.

In the next three sections, we will discuss three types of eukaryotic viruses and how these viruses contribute to genome dynamics. We will see that viruses hijack, rearrange, and transfer eukaryotic genes. This genetic plasticity has implications for both viral pathogenicity and eukaryotic gene evolution.

22.6

## Retroviruses Move Genes In and Out of Genomes and Alter Host Gene Expression

**Retroviruses** are viruses that have single-stranded RNA genomes and undergo an obligate proviral stage in their life cycle. Retroviruses copy their RNA genomes into double-stranded DNA forms that become integrated into the host cell's genome, where the viral genome is called a **provirus**. From this genomic location, the provirus is transcribed by the host cell's RNA polymerase into new retroviral RNA genomes. Some examples of retroviruses are the

TABLE 22.2

**Some Retroviruses and Their Features**

| Retrovirus | Natural Host | Features |
|---|---|---|
| Rous sarcoma virus (RSV) | Chicken | Carries *v-src* oncogene, causes tumors |
| Avian leukosis virus (ALV) | Poultry | Causes blood cell cancers |
| Mouse mammary tumor virus (MMTV) | Mouse | Causes tumors |
| Abelson leukemia virus | Mouse | Carries *v-myc* oncogene, causes tumors |
| Moloney murine sarcoma | Mouse | Carries *v-sis* oncogene, causes cancers |
| Mason-Pfizer monkey virus | Monkey | Causes T-cell immunosuppression |
| Human immunodeficiency virus (HIV) | Human | Infects T cells, causes acquired immune deficiency syndrome (AIDS) |
| Human T-lymphotropic virus (HTLV) | Human | Infects T cells, causes a form of leukemia |

human immunodeficiency virus (HIV), which causes AIDS; the Rous sarcoma virus (RSV), which is a cancer-causing virus in chickens; and the mouse mammary tumor virus (MMTV), which causes cancer in mice (Table 22.2).

## The Retroviral Life Cycle

Retrovirus particles consist of two single-stranded RNA genomes (each between 7 and 10 kilobases long) and several molecules of reverse transcriptase and integrase enzymes, all enclosed in a protein coat known as a **capsid** (Figure 22–17). The capsid is surrounded by a **viral envelope** that consists partly of viral glycoproteins and partly of host-cell membrane components. The envelope glycoproteins are responsible for recognizing and binding to cell-surface proteins called **receptors** on the surface of host cells. The molecular fit between the virus glycoprotein and its target cell receptors determines the host-cell specificity of the retrovirus.

A typical proviral genome consists of at least three protein-coding genes—*gag*, *pol*, and *env*—flanked by long terminal repeat (LTR) regions (Figure 22–18). The *gag* gene encodes proteins of the virus particle, the *pol* gene encodes reverse transcriptase and integrase, and the *env* gene encodes envelope glycoproteins. Each LTR region contains a direct repeat sequence (R) and two unique (U) regions. The U3 regions contain viral transcription promoters and enhancers, and the U5 regions contain RNA cleavage and polyadenylation signals.

After the retrovirus attaches to the host cell's receptors, the viral envelope fuses with the host-cell membrane, and the viral

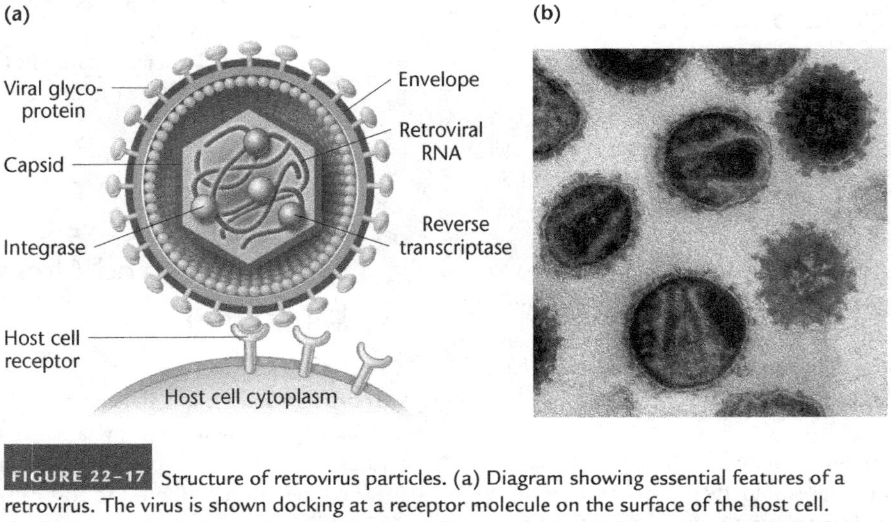


FIGURE 22–17 Structure of retrovirus particles. (a) Diagram showing essential features of a retrovirus. The virus is shown docking at a receptor molecule on the surface of the host cell. (b) Colored transmission electron micrograph of human immunodeficiency viruses (HIV). Three of the viruses are shown in cross section. The envelope is shown in yellow and green. It encloses the capsid and genomic RNA, shown in red.

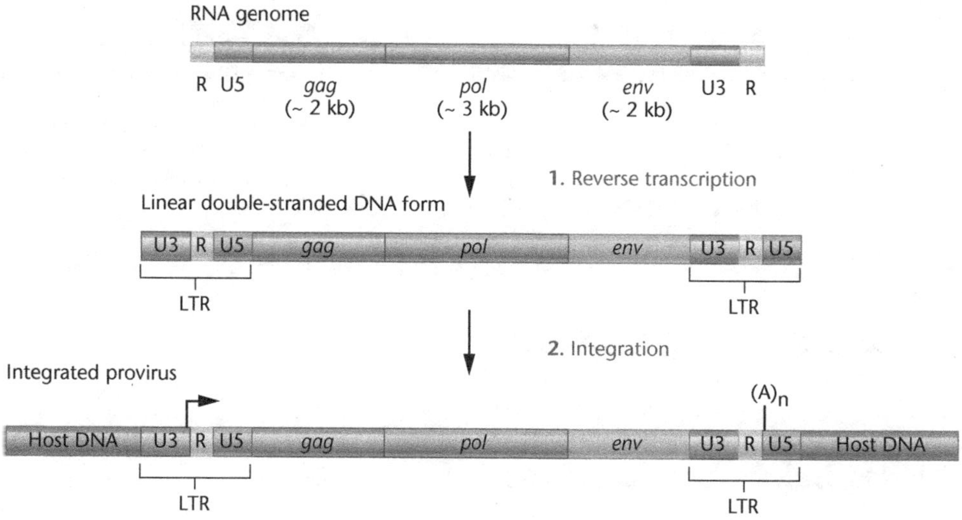


**FIGURE 22–18** Retroviral genome forms. The RNA genome consists of the three protein-coding genes (*gag, pol,* and *env*) flanked by direct repeats (R) that are 10 to 80 nucleotides long, a U5 region that is 80 to 100 nucleotides long, and a U3 region that is between 150 and 1300 nucleotides long. After reverse transcription, the linear DNA form of the virus is flanked by two LTR regions that range from 240 to 1480 base pairs long. The linear DNA integrates into the host genome and generates a short (4–6 bp) direct repeat at the insertion point. The U3 region contains the transcription promoter (shown as an arrow) and enhancer elements. Sequences within the R/U5 region at the 3′ end of the integrated provirus act as cleavage and polyadenylation signals (A) during transcription.

contents enter the cell. Once in the cytoplasm, reverse transcriptase begins its synthesis of a double-stranded DNA copy of the retroviral RNA. The RNA, DNA, and proviral forms of the virus genome are shown in Figure 22–18. The method of double-stranded DNA synthesis by reverse transcriptase, though complicated, is an important feature of the retroviral life cycle and contributes to the ability of retroviruses to pick up host genes and to recombine their genomes with other retroviral genomes. The method of reverse transcriptase DNA synthesis is summarized in Figure 22–19.

Reverse transcriptase, like other DNA polymerases, requires a primer on which to begin synthesis of DNA. In this case, the primer is a molecule of host cell tRNA, which was packaged with the viral genome during the virus's previous infection. The primer anneals to the retroviral RNA near the U5 region. After reverse transcriptase has copied the 5′ end of the viral RNA into a single-stranded stretch of DNA, it degrades part of the viral RNA genome using its RNase activity. At this point, reverse transcriptase performs a feat known as **template switching**. Along with its new strand of DNA, reverse transcriptase disengages from the 5′ end of the genomic RNA and moves to the 3′ end, base pairing with the R region next to U3. From this new location, DNA synthesis can continue to the 5′ end of the RNA. After the tRNA primer is removed, reverse transcriptase partially degrades the genomic RNA, leaving fragments that can act as primers for the synthesis of the second strand of DNA. Synthesis of the second DNA strand continues from the primers to the 5′ end of the template. Reverse transcriptase then performs a second template switch, moving the second DNA strand into a new position at the other end of the template, pairing the U5 regions. After removal of the remaining RNA primers, double-stranded DNA synthesis now continues to completion from the 3′ ends of the partial DNA strands, producing repeats of the U3 and U5 regions, as shown in Figure 22–18.

After the double-stranded DNA form of the virus is synthesized, it enters the nucleus and is integrated into the host genome by means of the retroviral integrase enzyme. The target sites in the host genome are random, and the insertion events are similar to those of retrotransposons, discussed previously in this chapter. The life cycle of the retrovirus is completed when the proviral DNA is transcribed by the host's RNA polymerase enzyme and the RNA is translated into viral proteins. The new RNA genomes are packaged into capsids, and the virus particles bud through the host-cell membrane, acquiring their envelopes. Retrovirus infection usually kills the infected cell; however, some viruses are able to use the cell as a virus factory without directly killing it (Figure 22–20).

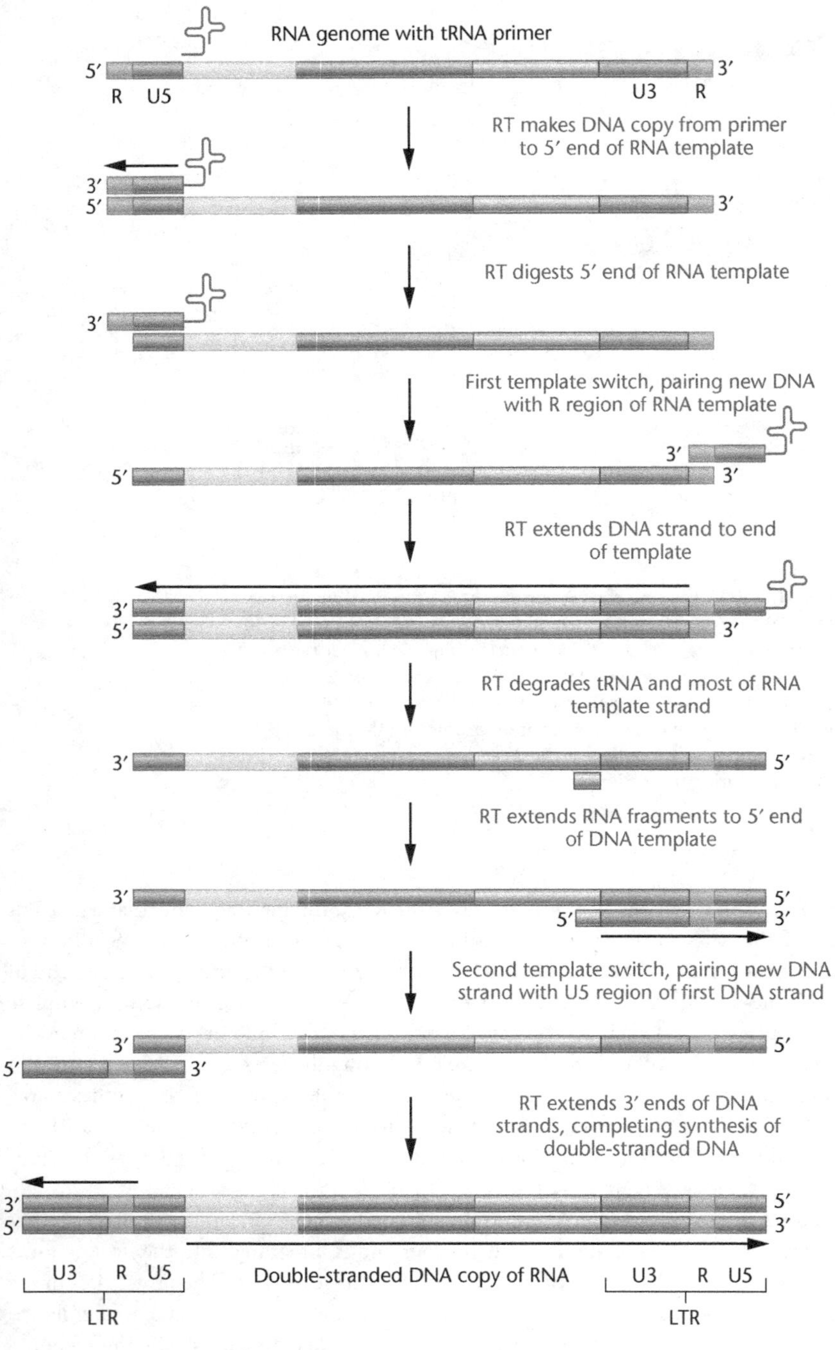


**FIGURE 22–19** Steps in the synthesis of a double-stranded DNA copy of a retroviral RNA genome. DNA synthesis begins with reverse transcriptase (RT) making a single-stranded DNA copy of the RNA genome, by extending synthesis from the 3′ end of a tRNA primer. Two template switches are required in order to generate full-length LTR regions.

## Retroviral Repercussions for Genome Rearrangement

The methods by which retroviruses infect cells and reproduce lead to high levels of genetic change, both in the retroviruses themselves and within their hosts. Retroviruses undergo rapid genetic modification owing to two features of their genome replication. First, reverse transcriptase is a highly error-prone DNA polymerase. Unlike most DNA polymerases, reverse transcriptase does not have a 3′ to 5′ exonuclease activity, and it is not able to proofread replication errors. As a result, each round of replication accumulates large numbers of mutations. Reverse transcriptase generates about one nucleotide mutation for every 30,000 nucleotides synthesized. Second, because of template switching during reverse transcription, retroviruses can recombine with other RNA molecules. If a cell is infected by two or more different retroviruses, a growing strand of DNA may switch not merely to a new part of the same template but to a different viral genome, leading to the creation of a chimeric proviral DNA. This form of recombination may also occur between retroviral and cellular RNAs. In this way, retroviruses acquire new genes or new sequences. If the resulting provirus maintains sufficient sequence similarity to its original form, it may still be expressed and packaged into infective retrovirus particles, which may then infect new host cells. This high level of mutation and recombination is one reason it is difficult to design vaccines that remain effective against retroviruses. It also contributes to the speed with which retroviruses develop resistance to the host's immune system and to drug treatments.

If the new provirus is defective and is not able to express infective virus particles, it may simply remain an endogenous proviral sequence that is maintained in the cell and replicated along with the host's DNA. This, too, has implications for genetic change. If the provirus inserts near a cellular gene, the gene's transcription may be activated by the presence of viral enhancers in the nearby U3 region. This can confer a new expression profile on the cellular gene. This new gene expression may be detrimental to the host cell if, for example, the newly activated host gene encodes a product that contributes to the development of cancer. If the provirus inserts upstream of, or within, a cellular gene, it is possible for transcripts to initiate at the promoter in the U3 region at the 3′ end of the provirus and read through into adjacent cellular gene sequences. This situation may also result in unusual expression patterns for the host gene, or expression of a new chimeric virus-host gene.

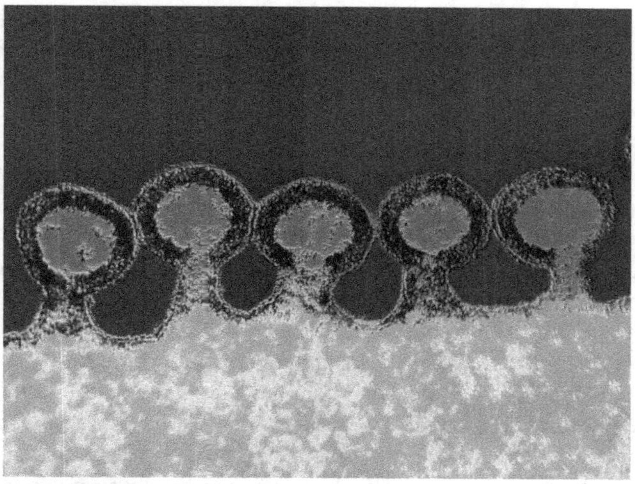

FIGURE 22-20 Colored transmission electron micrograph of retroviruses budding from the surface of a host cell. Human immunodeficiency virus (HIV) particles bud from the surface of a T lymphocyte—the target cell type of HIV—and acquire their envelopes from the host-cell membrane.

One of the most interesting examples of retroviral genetic gymnastics is the creation of **transducing viruses**—viruses that incorporate host genes into their genomes and transfer these genes to new hosts. The most famous examples are the **transforming retroviruses** that contain host oncogenes and cause cancer in their infected hosts. (Transforming retroviruses are also discussed in Chapter 20.) In these retroviruses, part of the virus genome has been replaced with a viral version of a cellular oncogene—called a *v-onc*. The method by which transforming retroviruses are formed is outlined in Figure 22–21. Because essential proviral genes are deleted, the transforming retrovirus is **replication-defective**, that is, unable to replicate; however, in the presence of a normal retrovirus—called a **helper virus**—the essential viral proteins are synthesized, and the replication-defective genome may be packaged into a new virus. The transforming retrovirus genomes may subsequently gain access to a new host and be inserted into the host genome. Expression of the *v-onc* gene from proviral transcription promoters may then lead to the development of cancer in the infected cell.

The ability of retroviruses to transfer genes from one host to another is being exploited for human gene therapy. Scientists insert therapeutic genes into the genomes of retroviruses such as members of the lentivirus family (to which HIV belongs). The resulting defective viruses are packaged in cultured cells, either with the assistance of a helper virus or by expression of cloned genes that produce the essential proteins of the retrovirus. The engineered transducing viruses are harvested and used to infect target cells. If the replication-defective retroviral genome is inserted as a provirus into the target cell, it may express the therapeutic gene. Because it is integrated into the cell genome, the provirus will be inherited by each new generation when the cell divides.

**NOW SOLVE THIS**

Problem 16 on page 603 asks you to predict a proviral DNA sequence and determine whether the provirus is infectious.

■ HINT: *Analyze the DNA sequence and its potential codons.*

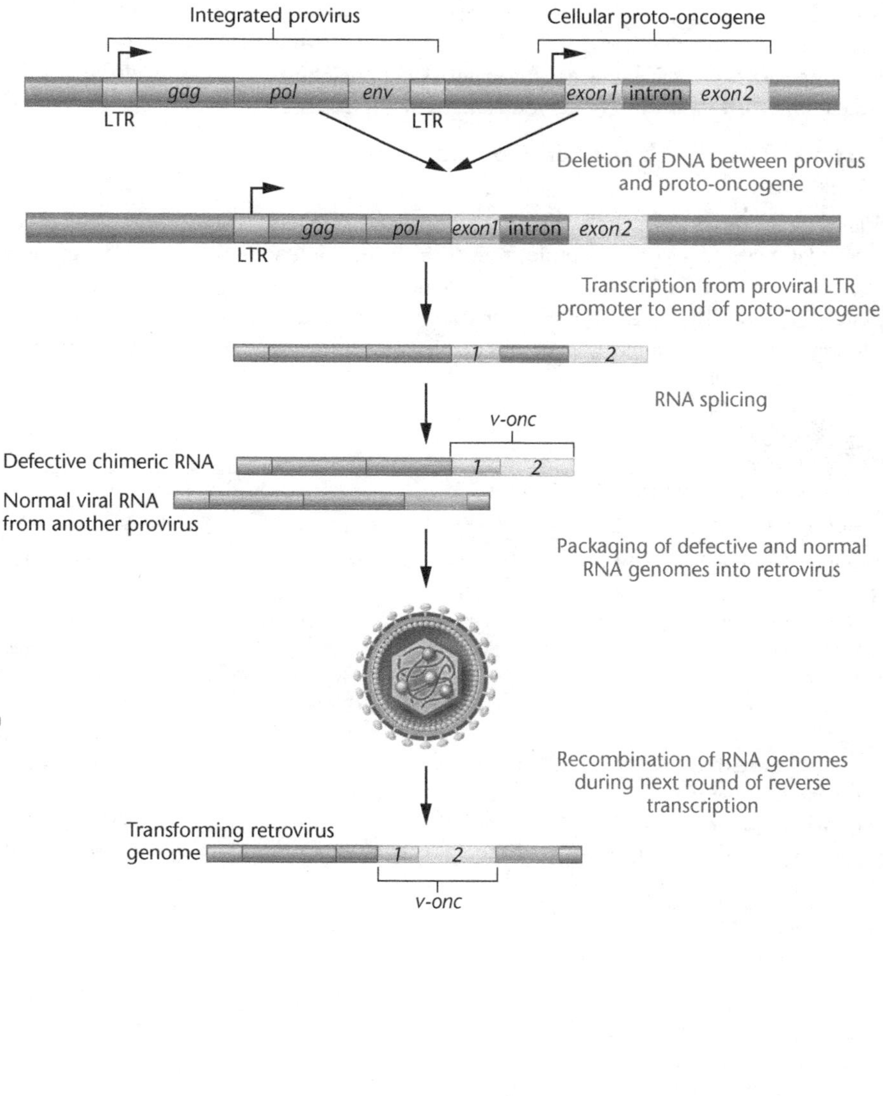


FIGURE 22-21 Formation of a transforming retrovirus. Following the occurrence of a deletion between a provirus and a cellular proto-oncogene, transcription can result in a retroviral RNA that is missing essential genes and has gained cellular gene exons. After splicing, this defective RNA can be packaged into a new virus particle if another viral RNA genome is present in the cell, presumably transcribed from another copy of the provirus. Template switching during the next round of reverse transcription creates the transforming viral genome.

22.7

## Large DNA Viruses Gain Genes by Recombining with Other Host and Viral Genomes

Eukaryotic DNA viruses take many forms and replicate by many different types of mechanisms. Their genomes may consist of single-stranded or double-stranded DNA in either circular or linear forms. Some DNA viruses replicate in the nucleus, and others in the cytoplasm of host cells. Their genomes can range from a few kilobases to 1.2 megabases in length. Some DNA viruses rely extensively on host proteins and enzymes to replicate their genomes and produce infectious progeny; however, others are almost autonomous from their hosts, synthesizing most of the molecules they require from their own genomes. Because viruses have such diverse characteristics, scientists have been puzzled about the origins of viruses and the methods by which they evolved. The availability of genome sequences and advances in the field of comparative genomics are beginning to provide some answers to questions of viral origins.

In this section, we will examine the large double-stranded DNA viruses of eukaryotes, with an emphasis on their genome evolution. For this discussion, we define large DNA viruses as those with genomes over 100 kilobases long. We will see that these viruses have acquired genes by exchanging genetic material with their hosts, with each other, and even with prokaryotes and bacteriophages.

### Gene Transfer between Cellular and Viral Genomes

The human herpesvirus family provides an instructive example of the types of genes that viruses capture from their hosts. These viruses have double-stranded linear DNA genomes about 200 kilobases long, encoding about 100 genes. Of the hundreds of herpesviruses that infect eukaryotes, eight infect human cells (Table 22.3).

A herpesvirus begins infection by entering its target cell and transporting its genome to the cell's nucleus. There, the virus can embark on one of two infection pathways: **lytic infection** or **latent infection.** In lytic infection, viral genes are translated, viral DNA is replicated, and new viruses are assembled and released from the infected cell. Lytic infection kills the host cell. In latent infection, the herpesvirus genome remains relatively silent, expressing only one or a few genes. The latent virus may remain in this silent state throughout the lifetime of the host, remaining sequestered within certain cell types (Table 22.3). These latent genomes may be activated into lytic infections in response to certain external signals (such as exposure to UV light) or if the host becomes immune deficient.

A typical lytic infection has three stages: immediate-early, delayed-early, and late. During the immediate-early stage, the virus expresses only a few viral genes. These genes encode viral gene transcription regulators and molecules that repress the normal molecular functions of the cell, such as transcription. During the delayed-early stage, the virus expresses more of its genes, particularly those involved in viral DNA replication. The virus genome is replicated to high levels in infected cells, rapidly filling the host-cell nucleus with viral DNA (Figure 22–22). At the late stage, the virus expresses the genes whose products form new virus particles, the viral DNA is packaged, and the new virus particles are released from the cell. Throughout the lytic life cycle, herpesviruses transcribe their genes using the host's RNA polymerase II transcription apparatus—including RNA polymerase II, basal transcription factors such as TFIID and TFIIF, and transcription activators such as SP1. At the same time, herpesviruses repress host gene transcription. In contrast, herpesviruses replicate their genomes using a DNA replication apparatus partly derived from the host and partly encoded by virus genes.

Recent comparative genomics studies indicate that over 30 herpesvirus proteins show significant homology to human proteins (Figure 22–23), and the genes encoding these proteins were likely derived from the human genome. This may be an underestimate, as the herpesvirus mutation rate is at least two-fold greater than that

TABLE 22.3

**Human Herpesviruses and Their Features**

| Herpesvirus | Other Names | Features |
|---|---|---|
| Human herpesvirus 1 (HHV-1) | Herpes simplex type 1 (HSV-1) | Infects epithelial cells, latent in neurons; causes cold sores and genital herpes |
| Human herpesvirus 2 (HHV-2) | Herpes simplex type 2 (HSV-2) | Infects epithelial cells; latent in neurons; causes genital herpes |
| Human herpesvirus 3 (HHV-3) | Varicella zoster virus (VZV) | Infects epithelial cells; latent in sensory ganglia of CNS; causes chickenpox and shingles |
| Human herpesvirus 4 (HHV-4) | Epstein-Barr virus (EBV) | Infects B cells and epithelial cells; latent in B cells; causes infectious mononucleosis; associated with Burkitt's lymphoma |
| Human herpesvirus 5 (HHV-5) | Cytomegalovirus (HCMV) | Causes various lesions in immune-compromised individuals |
| Human herpesvirus 6 (HHV-6) | Human B-lymphotrophic virus (HBLV) | Infects T and B cells; causes roseola infantum rash in children |
| Human herpesvirus 7 (HHV-7) | None | Infects lymphocytes; disease association unclear |
| Human herpesvirus 8 (HHV-8) | Kaposi's sarcoma herpesvirus (KSHV) | Infects epithelial cells and B-lymphocytes; latent in B-cells; causes Kaposi's sarcoma, other tumors in immunosuppressed individuals |

**FIGURE 22–22** Confocal microscopic image of cells containing viral replication compartments. Cultured human cells were infected with herpes simplex virus type 1 (HSV-1), incubated for approximately 6 hours, then fixed and treated with stains that recognize cell nuclear DNA (blue) and HSV-1 replicating DNA (red). Within 8 hours of incubation, viral DNA replicates to high levels and fills most of the host-cell nucleus. In this photo, two of the infected cells are in interphase (upper images) and one is in mitosis (lower image).

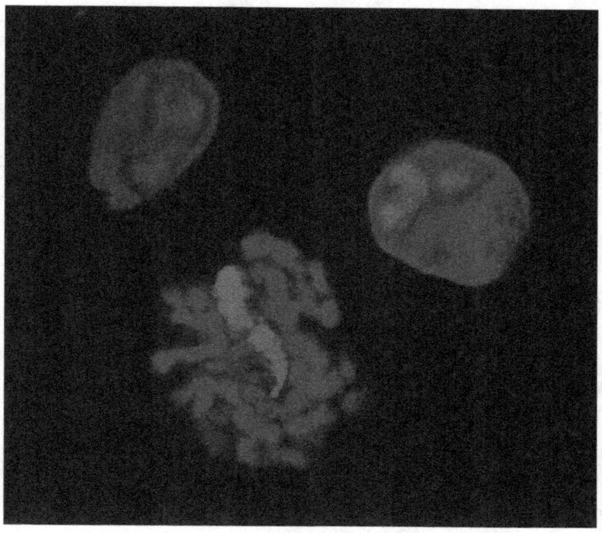

**FIGURE 22–23** Herpesvirus proteins with human homologs. Each column represents homologs in a different virus, from HHV-1 to HHV-8, as shown. The labels on the right are the names of the human proteins and their viral homologs. The color of each block indicates the class of protein: light green, DNA replication protein; dark blue, nucleotide repair and metabolism; light blue, enzymes; purple, gene regulators; yellow, glycoproteins; red, proteins involved in host–virus interactions; black, proteins of unknown function. Diagonal lines mean two gene copies in each viral genome; vertical lines mean three copies; horizontal lines mean 10 copies.
(From: Holzerlandt, R. et al., Identification of New Herpesvirus Gene Homologs in the Human Genome, *Genome Research* **12**: 1739–1748, 2002.)

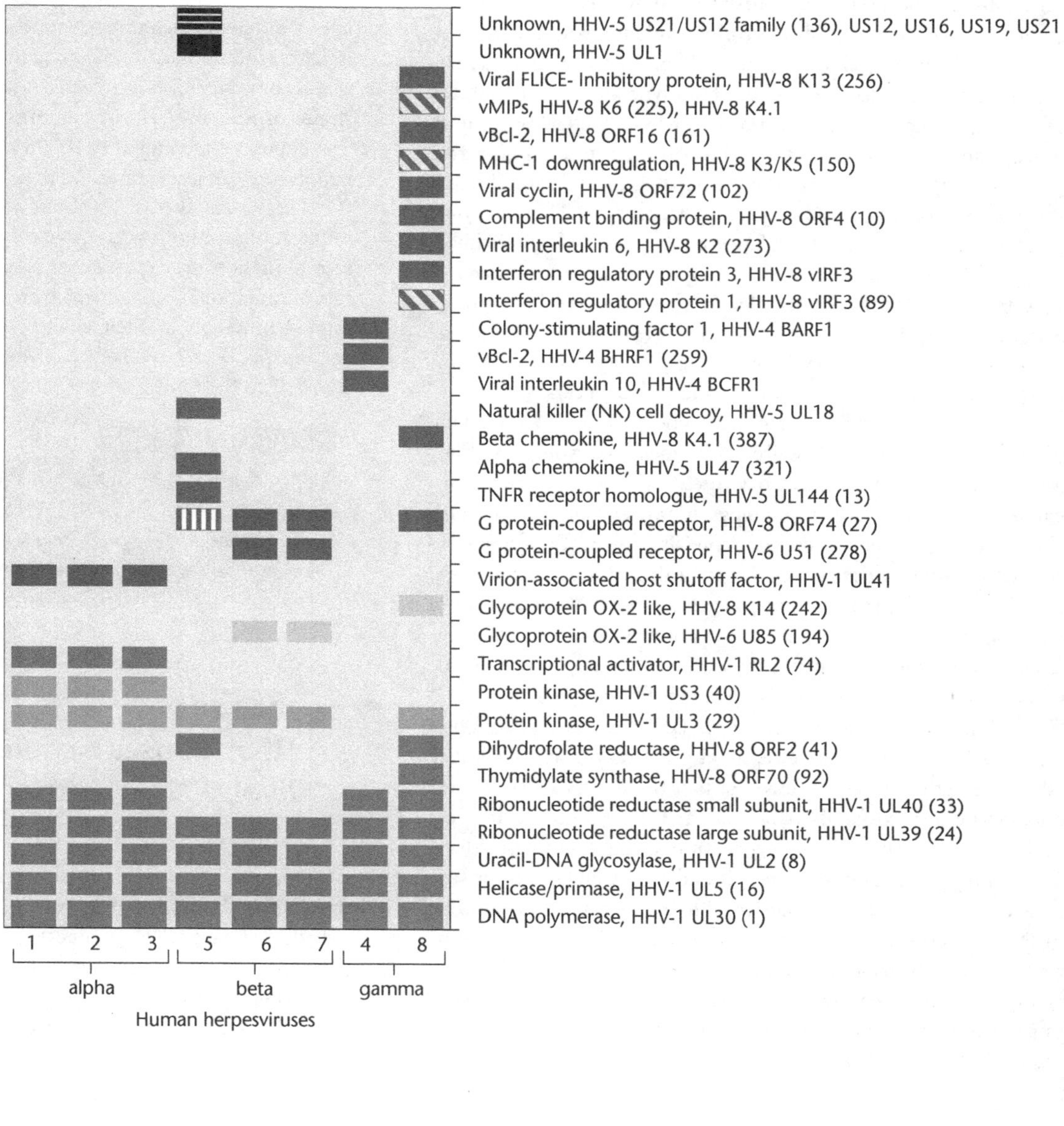

of humans. Over time, some genes captured from the host genome may have diverged significantly. A few of the captured genes are involved in DNA replication, such as those that encode the DNA-dependent DNA polymerase, helicase, primase, uracil-DNA glycosylase, and ribonucleotide reductases. The majority, however, appear to be genes whose products modulate the host's immune system, control the cell cycle, and inhibit programmed cell death, or **apoptosis**. In contrast, most herpesvirus genes—those that encode virus particle structural proteins—are not shared with humans.

An example of how viruses hijack cellular genes for their own advantage is provided by the *cyclin* gene in **human herpesvirus 8 (HHV-8)**. HHV-8, also known as **Kaposi's sarcoma herpes virus (KSHV)**, causes **Kaposi's sarcoma**, a skin tumor that develops in immune-deficient hosts such as patients suffering from AIDS. The virus is also associated with some B-cell lymphomas. HHV-8 infects B lymphocytes and is present as a latent virus in Kaposi's sarcoma tumors. Sequencing of the HHV-8 genome revealed the presence of a gene that encodes a cyclin protein, **v-cyclin**, that is 31 percent identical to the human **cyclin D2** protein. In uninfected cells, the cell's cyclin D2 binds to one of the **cyclin-dependent kinase (CDK)** proteins, stimulating the cell's entry into the S phase of the cell cycle. The v-cyclin protein, which is present in both lytic and latently infected cells, binds to the CDK6 kinase protein. The v-cyclin/CDK6 complex is active and phosphorylates the **retinoblastoma tumor-suppressor protein (pRB)**—a key regulator of the cell division cycle. It is thought that the v-cyclin of HHV-8 may cause its target cells (B lymphocytes) to grow and divide, providing a larger pool of targets for virus infection. In addition, the cyclin may contribute to the development of cancers by stimulating the cell cycle. The presence of cyclin homologs is not unique to HHV-8. At least seven other herpesviruses and one retrovirus encode human cyclin homologs. All of these viruses stimulate cell division in their hosts, and some cause cancers.

Although herpesviruses represent one example of horizontal gene exchange between DNA viruses and their eukaryotic hosts, they are not unique in this regard. Poxviruses, such as the vaccinia virus, contain 80 or more cellular homologs, mostly encoding proteins involved in genome replication and repressing the host's immune response. The mimivirus (a large DNA virus that infects *Acanthamoeba polyphaga*) bears a large number of genes (about 75) derived from bacteria. In addition, almost 200 mimivirus proteins are homologous to eukaryotic proteins. Thirty of these are most closely related to proteins in *Dictyostelium discoideum*—a possible host of this virus.

Although it seems clear that large DNA viruses capture genes from their hosts and that the captured genes may confer a selective advantage to the viruses, it is less clear how this occurs. Direct recombination between viral DNA and nuclear DNA is possible, but it is more likely that gene exchange occurs through an RNA intermediate. The fact that most captured viral genes do not contain introns suggests that these genes were transferred to the viral genome when an mRNA transcribed from the cellular gene was copied into a cDNA by reverse transcriptase activity in the nucleus. This is possible if the cell is infected simultaneously with a retrovirus and a DNA virus. Interestingly, retrovirus sequences have been observed in the genomes of herpesviruses and some other DNA viruses, strengthening the theoretical connection between retroviruses and DNA virus gene transfer. Whatever the mechanisms of viral-cellular gene transfer, they have contributed to the evolution of viral genomes.

### Gene Transfer between Viruses

Horizontal gene transfer and recombination between DNA viruses is another source of viral genetic diversity. Gene transfer of this kind is well known in bacteriophages, but the extent of eukaryotic virus gene exchange is only now being addressed with large-scale genomic sequencing studies. Homologous recombination between herpesvirus genomes has been observed in laboratory conditions, and likely exists in nature as well. This interviral recombination requires that two or more viruses coinfect the same cell. Recent sequencing studies of HSV-1 isolates from individuals in widely separated populations show evidence that recombination has occurred between herpesviruses. Patches of nucleotide sequences in one isolate bear strong similarities to the same regions in another isolate—but these patches are surrounded by highly divergent sequences. In any other two isolates, a different patch of nucleotide sequence may be similar in both isolates, with the surrounding regions being divergent. Studies show that recombination has occurred between the genomes of different strains of Epstein-Barr virus, human cytomegalovirus, and HHV-8. The Kaposi's sarcoma virus HHV-8 also shows evidence of recombination between its genome and that of an unknown related virus.

These new discoveries about genetic exchange in viruses has implications for both virus evolution and pathogenicity. The rate of recombination in herpesviruses is estimated to be similar to their rate of mutation—both of which are about ten-fold greater than the rate of mutation of their eukaryotic hosts. Such mutability and genome plasticity have likely contributed to the success of viruses in adapting to their hosts.

**NOW SOLVE THIS**

Problem 15 on page 602 asks you to explain how DNA viruses capture host genes.

■ **HINT:** *Think about the structure of most captured host genes and the location of the viral DNA during replication.*

**22.8**

## RNA Viruses Acquire Host Genes and Evolve New Forms

RNA viruses are viruses that have either single-stranded or double-stranded RNA genomes. They can be distinguished from retroviruses (which also have RNA genomes) by the fact that they do not integrate their genomes into host-cell DNA as part of their replication cycle. There are two types of single-stranded RNA viruses. Negative-strand RNA viruses have an antisense RNA genome that does not code for protein. The negative-strand RNA must be converted into a complementary positive (protein-coding) strand before the genome can be translated. In contrast, the genomes of positive-strand RNA viruses can be translated immediately after entry into the host cell.

Some examples of negative-strand RNA viruses are the Ebola virus and influenza viruses. Examples of positive-strand RNA viruses are the poliovirus, measles, Norwalk, Hepatitis C, and SARS viruses. Double-stranded RNA viruses include the Colorado tick fever virus and rotaviruses (Table 22.4).

In this discussion of RNA viruses, we will emphasize the positive-strand viruses. These viruses, like the DNA viruses and retroviruses, exhibit considerable genome variability, and exchange genetic information with other viruses and with their hosts.

TABLE 22.4

**Characteristics of Some Human RNA Viruses**

| Virus | Genome Type | Associated Disease |
|---|---|---|
| Poliovirus | Positive-strand RNA | Poliomyelitis |
| Rubella virus | Positive-strand RNA | Measles |
| Norwalk virus | Positive-strand RNA | Gastroenteritis, diarrhea |
| Hepatitis C virus | Positive-strand RNA | Hepatitis C |
| SARS-associated virus | Positive-strand RNA | Severe acute respiratory syndrome |
| Rabies virus | Negative-strand RNA | Rabies in animals and humans |
| Ebola virus | Negative-strand RNA | Ebola |
| Influenza viruses | Negative-strand RNA | Influenza A, B, C |
| Colorado tick fever virus | Double-stranded RNA | Infection of red blood cells, tick fever |
| Rotavirus A | Double-stranded RNA | Gastrointestinal symptoms, diarrhea |

## The Life Cycle of RNA Viruses

In order to synthesize new RNA genomes, RNA viruses must encode **RNA-dependent RNA polymerases (RdRps)**. Molecules of RdRps are enclosed within the virus particles of both negative-strand RNA viruses and double-stranded RNA viruses. After entry into the host cell, these RdRps act immediately to transcribe the viral genomes into positive-strand translatable RNAs. Positive-strand RNA viruses do not enclose RdRps within their virus particles. The positive strands of RNA are translated when the genome enters the host cell, and the newly translated RdRps begin synthesis of the RNA genomes.

RNA replication begins at the 3′ end of the genomic RNA and progresses to the 5′ end of the template. Some RNA viruses use a ribonucleoprotein to prime RNA synthesis, whereas others initiate from structures inherent at the 3′ end of the genomic RNA.

The replication cycle of a generic positive-strand RNA virus is shown in Figure 22–24. Infection begins when the envelope proteins on the viral surface bind to their receptors on the surface of the target cell. The virus enters the cell by endocytosis, and the envelope fuses with host membranes, releasing the capsid and RNA into the cytoplasm. After the capsid is dissolved, the RNA is translated by host-cell ribosomes. The newly synthesized RdRps copy the viral genome into a number of antisense (minus-strand) RNA molecules. The RdRps then use these molecules as templates to synthesize the new viral positive-strand RNAs. These are translated into viral capsid and envelope proteins and become the genomes of the new virus particles. The newly synthesized capsid proteins bind to the new RNA genomes, and the envelope proteins are inserted into the host-cell membranes. As the viral nucleocapsids pass through the host membrane by a budding process, they acquire their new viral envelopes.

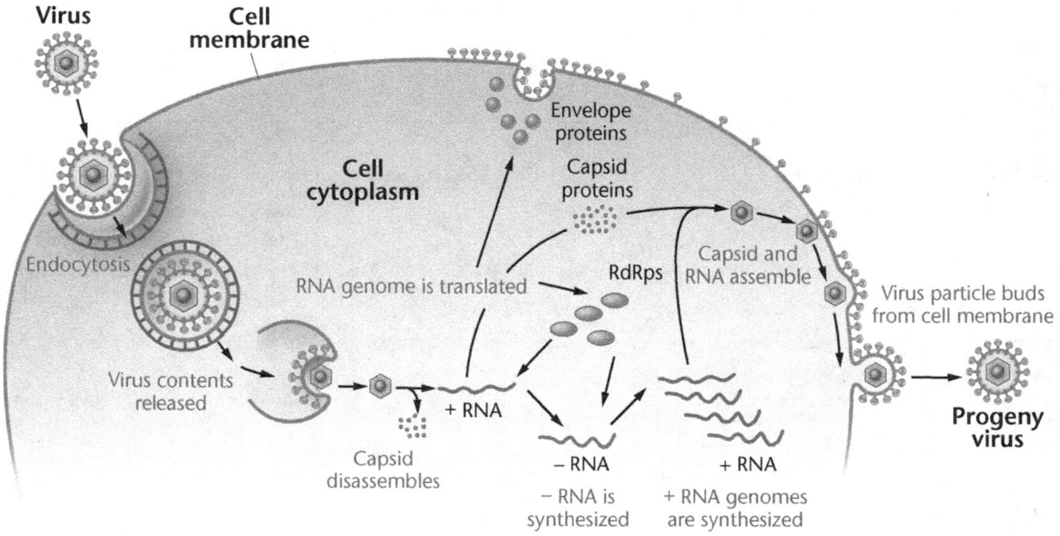


FIGURE 22–24 Stages in the life cycle of a positive-strand RNA virus. The incoming positive-strand RNA (+RNA) is translated into viral proteins including RNA-dependent RNA polymerases (RdRps). RdRps replicate the viral genome using a negative-strand RNA (– RNA) intermediate as a template. The new +RNA strands are packaged into capsids, and the new viral particles bud from the surface of the cell.

## Gene Transfer and Genome Variability in RNA Viruses

Like reverse transcriptase, viral RNA-dependent RNA polymerases have no proofreading capability and hence are error-prone. These enzymes generate about 1 mutation for every 10,000 nucleotides synthesized and at least one mutation per genome in each round of replication. This feature alone generates a great deal of genome variability and rapid evolution. In addition, like retroviruses, RNA viruses can recombine genomes during RNA synthesis. If two or more RNA genomes are present in an infected cell, the RdRp enzyme can switch templates during RNA replication, creating chimeric RNA genomes and hence mosaic viruses. By template switching, both large and small segments of the genome can recombine. Now that the genomes of individual viral isolates from many natural sources can be sequenced, the extent of this genome shuffling among viruses is becoming evident.

An interesting medically relevant example of viral genomic exchange is seen in the virus that caused the recent epidemic of **severe acute respiratory syndrome (SARS)**. SARS first appeared in February 2003 in Guangdong Province, the People's Republic of China. The disease had not been seen previously and was highly contagious. The symptoms began with a cough and fever but progressed rapidly to pneumonia that was often fatal. Within a few months, SARS spread to over 30 countries, infecting more than 8000 people and killing 774. Within a month of the outbreak, scientists had identified the causative agent as a newly emerged type of coronavirus, called the **SARS-associated coronavirus**, or **SARS-CoV**.

**Coronaviruses** are a group of large, enveloped positive-strand RNA viruses that infect a wide variety of animals, causing respiratory and intestinal symptoms (Figure 22–25). They are distributed worldwide. Coronavirus infections in humans usually cause mild cold-like symptoms. The genomes of coronaviruses encode the RdRp enzyme, a hemagglutinin-esterase (in some viruses), a spike protein (which is a glycoprotein found on the viral surface), an envelope protein, a membrane glycoprotein, and the nucleocapsid protein. Several other genes encode some regulatory proteins and proteins of unknown function. There are three groups of coronaviruses, with the characteristics listed in Table 22.5.

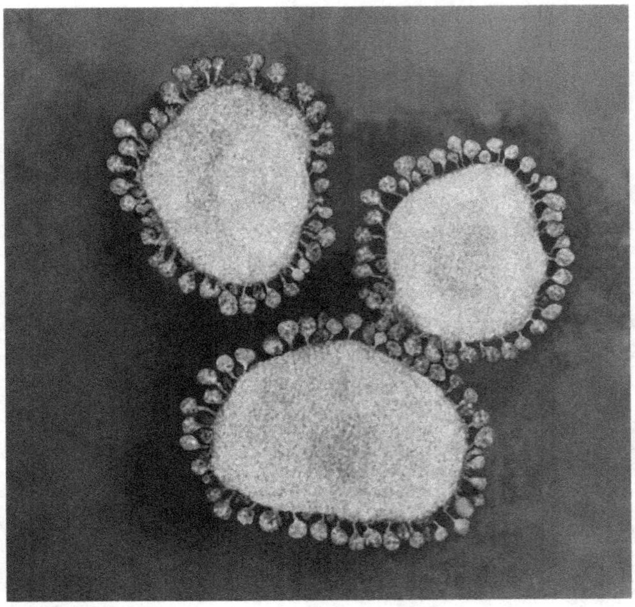

FIGURE 22–25 Colored transmission electron micrograph of three coronavirus particles. Coronaviruses cause diseases such as the common cold, gastroenteritis, and severe acute respiratory syndrome (SARS). These viruses appear to have a crown, or "corona," surrounding them. The proteins of the corona bind to host-cell membrane receptors and help the virus penetrate into the cell cytoplasm.

In an effort to determine the origins of SARS-CoV, scientists gathered the genome sequences of as many coronaviruses as possible—both laboratory strains and primary isolates from humans, domestic animals, and representatives of the wild animals found in the region of China where SARS first appeared. The RNA genome of SARS-CoV is about 29 kilobases long and is capped and

TABLE 22.5

**The Coronaviruses**

| Group | Features | Examples |
|---|---|---|
| Group I | No hemagglutinin protein<br>M protein N-glycosylated<br>S protein not cleaved | Transmissible gastroenteritis virus (TGEV) of pigs<br>Feline infectious peritonitis virus (FIPV)<br>Human coronavirus (HCoV) strain 229E<br>Porcine epidemic diarrhea virus (PEDV) |
| Group II | Have hemagglutinin protein<br>M protein O-glycosylated<br>S protein cleaved | Mouse hepatitis virus (MHV)<br>Bovine coronavirus (BCoV)<br>Human coronavirus (HCoV) strain OC43 |
| Group III | No hemagglutinin protein<br>M protein N-glycosylated<br>S protein cleaved | Infectious bronchitis virus (IBV) of chickens<br>Turkey coronavirus (TCoV) |

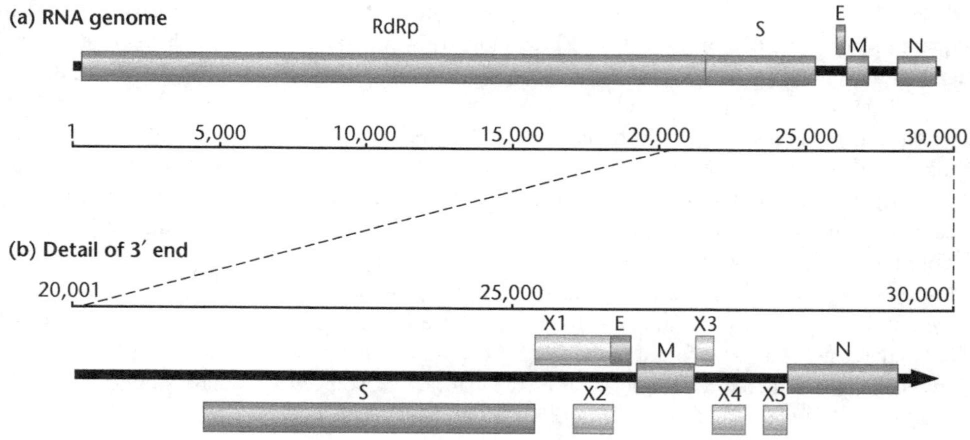


**FIGURE 22-26** Genome organization of SARS-CoV. (a) The 29-kilobase RNA genome of SARS-CoV. The open reading frames encode the RNA-dependent RNA polymerase (RdRp), surface spike protein (S), envelope protein (E), membrane protein (M), and nucleocapsid protein (N). (b) The 3′ end of the genome is magnified to show the open reading frames for the five proteins of unknown function (X1 to X5). (Adapted from: Rota et al., Characterization of a novel coronavirus associated with severe acute respiratory syndrome, *Science* **300**: 1394, 2003.)

polyadenylated. It encodes the characteristic coronavirus products RdRp, spike protein (S), envelope protein (E), membrane protein (M), and nucleocapsid protein (N) (Figure 22–26). In addition, the 3′ end of the viral genome contains five open reading frames that encode proteins of unknown function.

The SARS-CoV does not fit neatly into any of the three coronavirus groups. Its RdRp protein sequence most closely resembles those in Group II, particularly coronaviruses that infect mammals; however, its lack of a hemagglutinin-esterase gene is a characteristic of Groups I and III. Its membrane protein and S-glycoprotein modifications most closely resemble those in Group III. In addition, the organization of genes at the 3′ end of the SARS-CoV genome most closely resembles what is seen in the Group III avian coronaviruses. However, the sequences of the 3′ end genes are unique, with no homologs in the gene databases. At the sequence level, the M and N proteins most closely resemble those of avian coronaviruses. The S protein appears to be a mosaic of sequences from an avian infectious bronchitis virus and a feline infectious peritonitis virus. The recombination event responsible for this feature occurred in the middle of the *S* gene. The S protein is particularly interesting, as it is responsible for attaching the virus to its host cell, contributing to the virus host specificity. One study found evidence of at least seven recombination events in the RdRp and *S* genes, suggesting recombinations of SARS-CoV with six other coronaviruses (Figure 22–27), leading to a mosaic-like structure through these regions. The levels of nucleotide homology in these regions range from 80 to 100 percent. Together, these findings indicate that horizontal gene transfer between coronaviruses, followed by mutation and divergence, has contributed to the genome of SARS-CoV.

Although these nucleotide and amino acid sequence analyses show that recombination events have occurred between coronaviruses, it does not show what the direct origin of the SARS-CoV is and why it infected humans. The answers to these questions are beginning to emerge from sequencing studies of individual wild animals in the vicinity of the first outbreak. The first hypothesis about the origin of SARS-CoV was that the virus jumped to humans from a wild animal present in the live markets of Guangdong Province. The main suspect was the palm civet, which was shown to be infected with a virus very similar to that of SARS-CoV. However, when individual palm civets from the wild were tested, there was no evidence of SARS-CoV infection. If palm civets were the immediate source of the human infection, they likely contracted the virus at the markets and are not natural reservoirs for the virus in the wild. Recently, several species of horseshoe bats have been identified as possible sources of the virus. These bats show high levels of antibodies that recognize SARS-CoV. Scientists collected fecal samples from these bats, amplified RNA using PCR primers that are specific for sequences in the SARS-CoV, and sequenced the PCR products. The

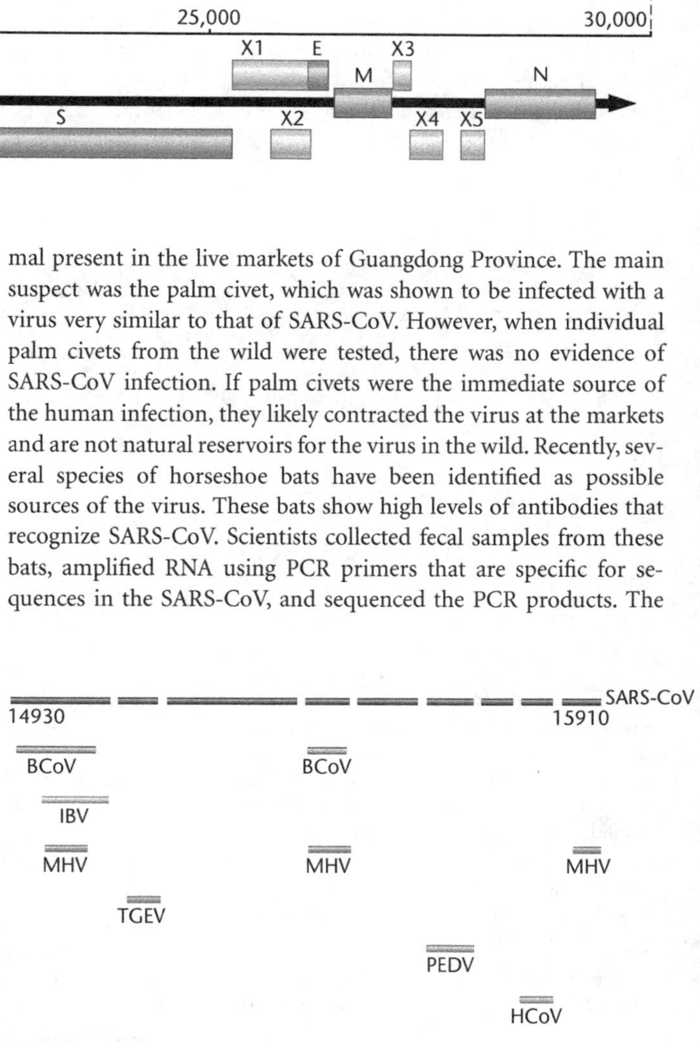


**FIGURE 22-27** Mosaic structure of the RpRd gene region between nucleotides 14930 and 15910 of the SARS-CoV genome. The nucleotide sequences from six coronaviruses were compared with that of SARS-CoV. Regions of greater than 50 nucleotides that showed more than 80 percent sequence identity are shown in the diagram. BCoV = bovine coronavirus, IBV = avian infectious bronchitis virus, MHV = murine hepatitis virus, TGEV = transmissible gastroenteritis virus, PEDV = porcine epidemic diarrhea virus, and HCoV = human coronavirus. (Adapted from Zhang, X.W., et al., Testing the hypothesis of a recombinant origin of the SARS-associated coronavirus. *Archives of Virology* **150**: 1-20, 2005.)

SARS-like virus in bats is almost identical (92 percent nucleotide sequence identity) to that of SARS-CoV. Three short regions in the *S* gene show differences between these two viruses. Interestingly, these same regions are known to be highly variable in comparisons of different isolates of SARS-CoV from humans and palm civets, suggesting that these regions are subject to high mutation rates. These mutation hot spots in the gene that encodes the spike protein likely contribute to rapid changes in host specificity. Another sign of the close relationship between the bat SARS-like virus and SARS-CoV is that both contain the five genes at the 3′ end of the genome that are not found in any other known coronavirus. These five unique coronavirus genes are also found in the SARS-CoV present in palm civets. It appears, therefore, that bats may be the natural reservoir of SARS-CoV. The current favored hypothesis about the origin of SARS-CoV is that it transferred from bats to palm civets, both of which are present in live-animal markets. After transfer to civets, the virus rapidly mutated in the S region of the genome, facilitating a second transfer to humans. A similar rapid accumulation of mutations after a shift to an intermediate host has also been observed for avian influenza A viruses.

These recent studies on the genetics and variability of SARS-CoV suggest that the virus may still be present in wild populations of bats and that transfer to intermediate hosts followed by rapid evolution may give rise to new epidemics in the future. Not only do these high rates of mutation and intergenomic recombination between RNA viruses lead to the emergence of new viruses, but they have implications for the effectiveness of future drug treatments and vaccines.

EXPLORING GENOMICS

# Avian Influenza Information and Databases

In this chapter, we learned that eukaryotic viruses mutate rapidly and exchange gene segments with other viruses and organisms. Recently, the highly pathogenic H5N1 strain of avian influenza virus jumped the species barrier from birds to mammals, including humans. Scientists warn that, through mutation or recombining with other influenza viruses, this virus could acquire the ability to spread between humans and cause a pandemic. In this exercise, we will access updated information on the H5N1 strain of avian influenza.

## ■ Exercise I – The World Health Organization Avian Influenza Web Site

1. Access the **WHO Avian Influenza** Web site at http://www.who.int/csr/disease/avian_influenza/en/index.html.
2. Select the "Fact sheet" link in the right column. Then select the "The disease in humans" link. What features of this strain make it a significant threat to human health?
3. Return to the WHO Avian Influenza Web site homepage. Select the link under "Cumulative Number of Confirmed Human Cases of Avian Influenza A/(H5N1) Reported to WHO." How many human cases have been reported, and what is the death rate? How many countries have been affected so far? Based on the data, does it appear that H5N1 virus infections are under control?

## ■ Exercise II – The NCBI Influenza Virus Resource

**The Influenza Virus Resource** is a principal source of influenza virus DNA and protein sequences, along with tools to analyze these sequences. It allows users to retrieve viral sequences, do multiple alignments, and build phylogenetic trees to analyze sequence relatedness. In this exercise, we will investigate the viral RNA polymerase subunit PB2 and its genetic variability.

1. Access the NCBI Influenza Virus Resource at http://www.ncbi.nlm.nih.gov/genomes/FLU/FLU.html.
2. To select virus sequences for further analysis, first select "Advanced database search."
3. At the top of the page, select "Search for Protein sequence." Under the species column, check "Influenzavirus A." Under the hosts column, check "Humans." Under the countries column, check "All." Under the segments column, check: "PB2." Check "Only these subtypes" and type in "H5N1" in the box.
   a. Click on the "Get sequences" button at the bottom of the page. You will now see a list of each of the viruses that you retrieved, along with some information about each one. (*Hint*: When working with this table, clicking on the heading of each column will sort the table according to that particular feature.) How many sequences are in the database?
   b. Click on the heading at the top of the "year" Column. This will re-sort the data in chronological order. When was the sequence of the first human H5N1 isolate deposited in the database? Where did the virus originate? How many amino acids make up the PB2 protein?
   c. Examine the right-hand columns. What patterns do you see in the ages and

genders of humans infected with the H5N1 strain of Avian Influenza virus?

4. To perform a multiple alignment of these PB2 amino acid sequences, make sure that all of the sequences are checked off in their boxes in the left-hand column of the list. Then click on "Do multiple alignment."

In the alignment, the approximate position of each amino acid is shown at the top of the diagram (1 to 759). The consensus sequence of PB2 is shown below, in blue. Capital letters denote amino acids (see Chapter 15 for amino acid letter designations). A dot indicates that the virus sequence at that position is identical to the consensus sequence. A different amino acid indicates a difference from the consensus sequence. As all of these viral isolates caused a human infection, what can you conclude about the essential regions of the PB2 protein?

5. The presence of a lysine (K) in the sequence AAPPKQ (at approximately position 630) appears to be important for viral pathogenicity in mice. How many of the human PB2 isolates have this sequence? Would you conclude that this sequence is also necessary for H5N1 pathogenicity in humans?

6. Using the same strategy, determine whether the AAPPKQ sequence is important for pathogenicity in any other species besides birds and humans.

## Chapter Summary

1. Transposable elements are found in the genomes of all organisms, including bacteria, plants, and animals. In some genomes, they make up the majority of genetic material.
2. DNA transposons and retrotransposons have different characteristics and move within genomes using different mechanisms.
3. Transposons have a wide range of effects on genomes, including generating mutations, causing chromosome rearrangements, copying exons, and altering gene expression. Any of these changes may contribute to the evolution of their host organisms.
4. The immunoglobulin genes are created by rearrangements of DNA and high rates of mutation within somatic cells of the immune system (B lymphocytes). These rearrangements account for antibody diversity.
5. Eukaryotic viruses act as vehicles for horizontal gene transfer. They capture genes from other viruses and host cells and transfer genetic material to new hosts. This genome shuffling affects the viruses' own evolution and that of their hosts.
6. Retroviruses insert their genomes into those of their hosts. They mutate at high rates and recombine their genomes with other genomes during the reverse transcription phase of their life cycle.
7. Large DNA viruses pick up genes from their hosts and from other DNA viruses, probably through RNA intermediates.
8. RNA viruses have high rates of mutation and recombine their genomes with those of other viruses, creating mosaic viruses. High mutation rates and interviral recombination contributes to their rapid evolution and changes in host specificity.

## INSIGHTS AND SOLUTIONS

1. In the research that led to the initial discovery of IS elements in bacteria, three genes controlling galactose metabolism were affected simultaneously by a single IS insertion, upstream (5′) of those genes. Offer an explanation as to how this might be possible.

**Solution:** In *E. coli*, there are three *gal* genes, known as *galE*, *galT*, and *galK*. They are present in an operon, in the order *galE—galT—galK*, and their products are essential for proper metabolism of galactose. The operon is transcribed from two overlapping promoters upstream of the *galE* gene. Insertions of IS elements anywhere within the *gal* operon will create cells that are *gal* mutants, unable to metabolize galactose. This is because each of the gene products is necessary, and interruption of the reading frame of any one of the genes will affect the metabolism of galactose. In order to affect all three genes simultaneously, the IS element would need to be inserted within the promoter region or the *galE* gene. If the IS element was present upstream of *galE* but downstream of the promoter, transcripts initiating at the *gal* promoter would likely terminate within the IS element, due to the presence of strong transcription terminators within these elements. Similarly, insertions of an IS element within the *galE* gene would cause termination of transcription within the gene, inactivating it as well as the two genes downstream.

2. A hypothetical mammalian species synthesizes immunoglobulin molecules made up of $\lambda$ light chains, $\kappa$ light chains, and H chains, as described in this chapter. The genome of this species contains 150 different V segments and 6 different J segments that contribute to the $\lambda$ light-chain genes. The genome contains 80 different V segments and 10 different J segments that contribute to the $\kappa$ light-chain genes. There are 100 different V segments, 8 different J segments, and 5 different D segments that contribute to the H-chain genes. Assuming that only random somatic recombinations are responsible for constructing the immunoglobulin genes, how many different types of immunoglobulin molecules can be synthesized by this species? Describe why this is likely an underestimate of the numbers of different antibodies that can be produced.

**Solution:** There are $150 \times 6 = 900$ possible combinations of V and J segments that make up the $\lambda$ light-chain genes. There $80 \times 10 = 800$ possible combinations of V and J segments that make up the $\kappa$ light-chain genes. For the H-chain genes, there are $100 \times 8 \times 5 = 4000$ different possible combinations of V, J, and D segments. Given that there are $900 + 800 = 1700$ different possible light chains and 4000 possible heavy chains, there are $1700 \times 4000 = 6{,}800{,}000$ possible immunoglobulin molecules. This estimate is likely low, as imprecise recombination

*Continued on next page*

*Insights and Solutions, continued*

at the V, J, and D junctions, as well as hypermutation, will contribute to immunoglobulin diversity.

3. Retroviruses are used as vectors for gene therapy. A recent clinical trial for gene therapy was halted because several of the patients developed leukemias as a result of the treatment under study. Describe how this might have occurred.

**Solution:** Retroviruses are useful for gene therapy because they insert their proviral DNA into the host-cell genome, causing the provirus to be replicated along with the cell's genome. Usually, the viral genes are removed from the retroviral genome, and the therapeutic gene, along with a strong transcription promoter, is inserted into it. This manipulation is undertaken in the test tube, using recombinant DNA technology. The double-stranded DNA version of the therapeutic retrovirus is then transformed into cells that express the missing viral gene products, and more copies of the engineered therapeutic retroviruses are harvested from these helper cells. When used for gene therapy, the therapeutic retroviruses enter the target cells, convert their genomes to double-stranded DNA forms, and insert the provirus DNA into the target cell genome.

One of the major problems with the use of retroviruses is that the proviral DNA can insert anywhere in the target cell genome. If the insertion occurs within a target cell gene, it may inactivate the gene. If the inactivated gene is a tumor-suppressor gene, and the other allele of that gene is mutant, the cell may become malignant. If the insertion occurs near a target cell gene, the gene may become transcriptionally activated, due to the presence of strong viral enhancers in the provirus. Similarly, transcription initiating in the proviral LTRs may read through into adjacent cellular gene sequences, causing aberrant expression. If the adjacent gene is a proto-oncogene, its overexpression may contribute to the development of cancer.

In the study referred to above, the therapeutic retrovirus became inserted near a gene called *LMO-2,* causing the gene to be expressed at high levels. This gene appears to be involved in the development of childhood cancers. For a report on the study, see *Nature* **420:** 116–118, 2002. Scientists are now developing new retroviral vectors that will be less likely to activate nearby genes.

## Problems and Discussion Questions

1. Compare several transposable elements in bacteria, maize, *Drosophila,* and humans. What properties do they share?
2. *Ty,* a transposable element in yeast, contains an open reading frame (ORF) that encodes the enzyme reverse transcriptase. This enzyme synthesizes DNA from an RNA template. Speculate on the role of the enzyme in the transposition of *Ty* within the yeast genome.
3. Transposons are often referred to as jumping genes, implying that they jump from one position to another within a genome. Think about the mechanisms of transposition and discuss whether the term jumping genes is an accurate description of how transposons move.
4. It is considered likely that some types of transposons arose from retrovirus infections. Describe the events that may have occurred to convert a retrovirus to a transposon.
5. The *Ds* elements in maize appear to be derived from *Ac* elements. What are the differences between the two elements? Explain how the *Ds* element transposes in the maize genome and how its structure affects its transposition.
6. Describe two ways in which a transposon can move genomic sequences that lie outside of the element itself.
7. A *Drosophila* genome contains three copies of a *copia* transposon. Two copies are located near one end of chromosome 1, in opposite orientations to each other and about 7 kilobases apart. The other copy is in the middle of chromosome 3. Draw a set of diagrams showing the types of genome rearrangements that could arise from the presence of these three transposons.
8. A bacterial IS element may contain a large deletion in its transposase gene but still be mobile within the bacterial genome. Explain how this could occur.
9. *P* elements can give rise to hybrid dysgenesis in *Drosophila* by bringing about a variety of genetic alterations, including sterility, elevated mutation, recombination, and rearrangement rates. When the female carries the *P* element and the male does not, no dysgenic offspring are produced. However, in the reciprocal cross (females without *P* elements mated with males containing *P* elements) hybrid dysgenesis results. Explain the differing results of reciprocal crosses described here.
10. Describe the clonal selection theory and how it explains the way the immune system can recognize such a large array of different antigens.
11. How does the inherent imprecision of recombination events during immunoglobulin gene rearrangement contribute to the diversity of the immune response?
12. In the hypothetical extraterrestrial organism, the Quagarre, immunoglobulin genes undergo rearrangement similar to that in humans, with two main differences. One difference is that the heavy-chain V regions are located on chromosome 4 and the remainder of the heavy-chain gene segments are located on chromosome 10. Another difference is that the Quagarre immunoglobulin genes do not rearrange until the creature is 13 years old. By examining a mitotic spread of chromosomes from a Quagarre B cell, how could you estimate the creature's age?
13. How does an IgG molecule differ from an IgM molecule and an IgE molecule? If a B cell switches its class from IgM to IgE, can it switch again, to make an IgG molecule? Explain.
14. Severe combined immunodeficiency syndrome (SCIDS) is a primary immune defect in both the T and B lymphocytes that results in the onset of life-threatening infections in the first few months of life. SCIDS can be caused by mutations in at least nine different genes, two of which are *RAG1* and *RAG2.* Describe the roles of T and B lymphocytes, and the *RAG1* and *RAG2* gene products, in mounting a normal immune response to foreign antigens, and how these may be affected in SCIDS.
15. Explain two ways in which eukaryotic DNA viruses could capture host genes. How might they acquire genes from a retrovirus?

16. A plant retrovirus genome contains the following sequence in the coding region of its *gag* gene:

    5′—AAUAGCUAGCUAAGGCGAUGCGCGAU—3′

    Write out the sequence of the proviral DNA in this region.
    Based on this sequence, would you expect the infected cell to produce infectious new retroviruses?
17. You have just discovered a new virus that infects kidney cells. This virus will infect and reproduce when added to kidney cells in a culture dish. Design two experiments that would tell you whether this new virus was a DNA virus or a retrovirus.
18. What features of retroviruses make them potential vectors for human gene therapy? What might be some drawbacks to using retroviruses for this purpose?
19. The drug zidovudine (AZT) is used to treat patients with acquired immune deficiency syndrome (AIDS). AZT works by inhibiting the actions of reverse transcriptase. It does this by acting as an analog of a normal deoxyribonucleotide, which reverse transcriptase binds and incorporates into the growing DNA chain. Because AZT does not have a 3′ hydroxyl, DNA-chain synthesis stops after AZT is added to the chain. Unfortunately, the AIDS virus rapidly develops resistance to AZT. Explain how this can occur.
20. How does a bacteriophage prophage resemble a retroviral provirus? How do these entities transfer genetic information between host cells?
21. When scientists examined a bone tumor of chickens, they discovered that all cells in the tumor contained several copies of retroviral sequences encoding reverse transcriptase. How might this retrovirus be causing the tumor?
22. The S-glycoprotein of coronaviruses is located on the surface of the virus and interacts with the host cell's receptors, thus playing a role in the virus pathogenicity. What features of the coronavirus replication cycle contribute to the types of changes in the S-glycoprotein gene that result in host shifting?

**HOW DO WE KNOW?**

23. In this chapter, we focused on ways that transposons, immunoglobulin gene rearrangements, and virus infections alter the genomes of organisms. At the same time, we found many opportunities to consider the methods and reasoning through which much of this information was acquired. From the explanations given in the chapter, what answers would you propose to the following fundamental questions:
    (a) How do we know that the insertion of a transposable element in or near a gene can alter an organism's phenotype?
    (b) How do we know that viruses acquire genes and gene segments from their hosts?
    (c) How do we know that human transposons are capable of causing disease?

## Extra-Spicy Problems

24. In a genetics laboratory, it is discovered that a new *Drosophila* eye color mutation (creating a green eye) results from insertion of a *copia* transposon into the coding region of the *rg* eye color gene. Geneticists in the lab selected the mutant and maintained the stock as a homozygous mutant line. After several generations, they noticed that a few flies had red eyes. Explain two ways in which this event could have occurred.
25. Transposase enzymes create staggered cuts in the target site prior to transposon integration. What are the consequences of making such cuts in DNA? Now that the sequence of the human genome is available, how would you use this information to predict the location and numbers of transposons in the human genome?
26. Describe how you would use a *Drosophila P* element to introduce a new gene into a strain of *Drosophila*.
27. In maize, a *Ds* or *Ac* transposon can cause mutations in genes at or near the site of transposon insertion. It is possible for these elements to transpose away from their original site, causing a reversion of the mutant phenotype. In some cases, however, even more severe phenotypes appear, due to events at or near the mutant allele. What might be happening to the transposon or the nearby gene to create more severe mutations?
28. Burkitt's lymphoma is a malignancy affecting B lymphocytes. The B cells of patients with Burkitt's lymphoma grow and divide in an uncontrolled manner. These same cells contain abnormally high levels of the C-MYC protein, a transcription activator that stimulates the expression of a number of genes whose products regulate the cell cycle. Burkitt's lymphoma cells also contain reciprocal chromosome translocations between chromosome 8 and either chromosome 2, 14, or 22. The translocation breakpoints occur near or within the *c-myc* gene on chromosome 8 and the immunoglobulin light or heavy chain genes on chromosomes 2 ($\kappa$ light chain), 14 (H chain), or 22 ($\lambda$ light chain). In one patient, the normal and translocated chromosomes resemble the diagram below. Explain why these Burkitt's lymphoma cells produce high levels of C-MYC protein. Would you expect that these cells would express immunoglobulins? Why do you think that Burkitt's lymphoma B cells contain these kinds of translocations?

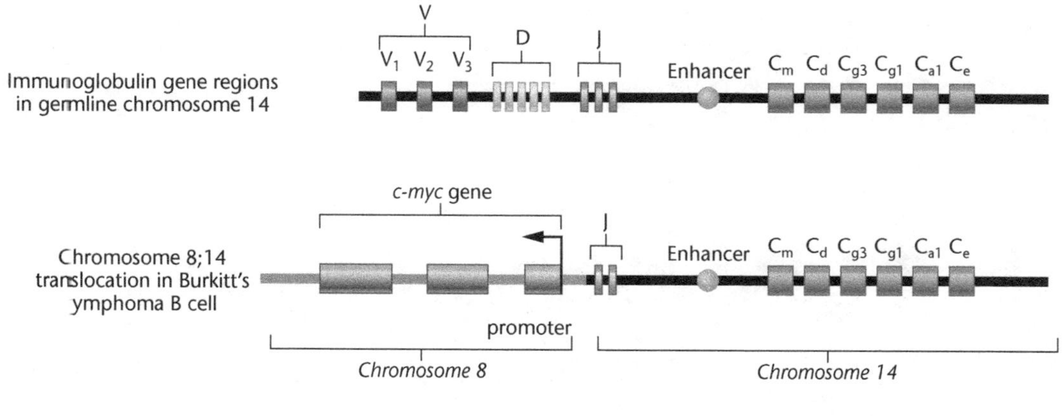

29. You have decided to work in a genomics research lab for the summer. The research team has just completed sequencing the entire genome of a rare species of New Guinea wallaby. You have been given the task of analyzing the genome to locate retroviral provirus genomes. What features would you look for? How would you distinguish proviral sequences from sequences of retrotransposons?
30. The hemagglutinin (HA) glycoprotein of influenza viruses makes up part of the viral envelope and interacts with receptors on the surface of the host cell, determining host-cell specificity. One study identified an avian influenza virus that could infect chicken cells, which are not the normal host cells of this virus (Khatchikian, D. et al. *Nature* **340**: 156–157, 1989). The researchers sequenced the influenza virus genome and discovered that it contained a segment of RNA sequence that was derived from the host 28S ribosomal RNA. This segment was inserted in the middle of the HA gene. Speculate on how the 28S ribosomal RNA sequences entered the viral genome. How might this insertion cause a host shift?

A mouse carrying mutations in its *dystrophin* gene that make it a useful model for research into human muscular dystrophy. The mouse was treated with gene therapy that introduced the normal gene into its muscle cells.

23

# Genomic Analysis—Dissection of Gene Function

## CHAPTER CONCEPTS

- Geneticists use a wide variety of research tools and genetic model organisms to explore the relationships between genotype and phenotype.
- A good model organism for genetic analysis must be easy to grow, have a short generation time, have a small genome, produce abundant progeny, and be readily mutagenized and crossed.
- Forward genetic analysis begins with the isolation of mutants and is followed by defining genetic pathways, cloning the genes, and creating more mutants to identify genetic interactions.
- Reverse genetic analysis begins with a cloned wild-type gene, a DNA sequence in a database, or a purified protein and progresses to site-directed mutagenesis and phenotypic analysis—the opposite of the order followed in forward genetic analysis.
- Functional genomics and high-throughput technologies allow geneticists to dissect the interactions of thousands of gene products simultaneously, supporting the new sciences of genomics, proteomics, and systems biology.

A major goal of genetics is to understand what genes are and how they work. As genes influence every aspect of biological activity, knowledge of which genes are involved in each process, and how the products of these genes control phenotype and function, is central to biology and medicine. Genetic analysis has dramatically extended our understanding of biological processes as simple as biochemical reactions in a single cell or as complex as the developmental steps that lead to the creation of multicellular organisms.

But how do geneticists dissect gene function? How do they discover which genes are involved in which biological processes and how these genes control phenotype?

In a classical genetic analysis, geneticists attempt to answer genetic questions by first collecting a number of individuals that display mutations affecting the phenotype of interest. From there, they determine whether the phenotype is controlled by one or more genes, which genes are responsible for which steps in the pathway leading to the phenotype, and how each gene product controls phenotype at the biochemical level. In short, the classical approach starts with a series of mutants in order to identify, clone, and characterize the function of genes.

Modern genomics and molecular biology have vastly expanded the tools available for genetic analysis. It is now possible to begin a genetic analysis with a cloned gene of unknown function, create specific mutations within the cloned gene, and test the phenotype and function of these mutant genes within model organisms—so-called **reverse genetics**. In addition, tools such as DNA microarrays, automated sequencers, and computerized bioinformatics allow geneticists to explore global patterns of gene expression that control specific biological processes.

In this chapter, we explore some of the methods by which geneticists dissect gene function. We will learn why genetic analysis requires the use of model organisms and how these organisms allow geneticists to answer specific biological questions. We will examine forward and reverse genetic techniques, as well as the new molecular methods based on gene targeting, transgenics, and high-throughput functional genomics.

## 23.1 Geneticists Use Model Organisms to Answer Genetic and Genomic Questions

In order to discover genes and dissect the genetic processes that regulate biological functions, geneticists make extensive use of **model organisms**. Not all organisms are ideal candidates for genetic analysis. Geneticists cannot effectively perform mutagenesis, do rapid controlled matings, or clone genes in most complex, slow-growing life forms such as our own species. In addition, investigators must take into account ethical considerations when undertaking genetic research in humans and many other animals.

A good model organism for genetic analysis must be easy to grow, have a short generation time, produce abundant progeny, and be readily mutagenized and crossed. In addition, the organism must carry out the biological process that is to be studied. One might question, however, whether knowledge gained about gene functions in simple laboratory organisms is relevant to understanding gene function in higher plants and animals.

Fortunately, genetics, molecular biology, and genome sequencing reveal that many of the genes and molecular processes governing biological functions are shared across evolutionary relationships. Many gene sequences are conserved from yeast to higher vertebrates, and these genes often carry out similar functions in a wide range of organisms. For example, the genes that regulate cell-cycle checkpoints in yeast have homologs in humans. The human versions of these genes regulate the cell cycle and act as tumor-suppressor genes. Approximately 200 of the 300 currently known human disease genes have sequence similarities to genes in fruit flies and nematodes. Approximately 100 of these genes are also similar to genes in yeast. Over half of the known human cancer genes have homologs in *Drosophila*. In addition, many early developmental processes, and the genes that control them, are conserved between flies, nematodes, mice, and humans. Due to these genetic similarities, geneticists can study the fundamentals of metabolism, development, and disease in simple laboratory organisms, and then apply this knowledge to more complex eukaryotes.

### Features of Genetic Model Organisms

The eukaryotic organisms used most extensively in genetic research are listed in Table 23.1, along with some of their important features and research uses. For each of these model organisms, researchers have compiled a large store of genetic knowledge acquired over decades of study, including databases of the entire genome sequences and collections of strains bearing specific deletions and mutations, making genetic analysis rapid and efficient. In addition, each organism has features that make it the model of choice for particular kinds of study. Many other important model organisms are also used in genetic research, including prokaryotes, such as *E. coli*, and bacterial viruses, as described in Chapters 6 and 17. Other eukaryotes, such as zebrafish, corn, garden peas, and single-celled protozoans have made significant contributions to genetics.

In this chapter, we will focus on the genetic dissection techniques used in three model organisms: the yeast *Saccharomyces cerevisiae*, the fruit fly *Drosophila melanogaster*, and the mouse *Mus musculus*.

### Yeast as a Genetic Model Organism

The yeast ***Saccharomyces cerevisiae***, also known as budding yeast or brewer's yeast, is one of the most popular model organisms for genetic research. Geneticists often refer to the "awesome power of

**TABLE 23.1**

**Eukaryotic Model Organisms Used in Genetic Research**

| Organism | Common Name | Genes | Genome size | Sequenced |
|---|---|---|---|---|
| ***Saccharomyces cerevisiae*** | **Budding yeast, baker's yeast** | 6600 | 12 Mb | 1996 |
| Key Features—Single-celled organism, rapid generation time, basic nutritional requirements. Intercellular hormonal signaling determines mating type. Haploid and diploid life stages. Grows in liquid culture or as single colonies originating from one cell. Haploid stage of life cycle useful for mutant detection. Easy gene targeting and transgenics. Large storehouse of mutants available. | | | | |
| Research Uses—Cell-cycle regulation, meiosis, development (mating-type switching), gene interactions, recombination, genomics. | | | | |
| Internet information and databases—*Saccharomyces* Genome Database (http://www.yeastgenome.org/) | | | | |
| ***Drosophila melanogaster*** | **Fruit fly** | 13,600 | 180 Mb | 2000 |
| Key Features—Small genome on four chromosomes, rapid generation time, large numbers of progeny, easy to grow. Large, banded polytene chromosomes allow easy detection of chromosome rearrangements using microscopy. Extensive data available on developmental stages and body-segment development. Many mutant and balancer stocks available. *P* element transformation and gene knockouts possible. | | | | |
| Research Uses—Basic genetic processes; development, population, evolution, chromosomal genetics; gene regulation. | | | | |
| Internet information and databases—FlyBase (http://www.flybase.org/) | | | | |
| ***Arabidopsis thaliana*** | **Thale cress** | 25,000 | 125 Mb | 2000 |
| Key Features—Small flowering plant of brassica family, short life cycle, and small genome. Self- or cross-pollination. Large collection of knockout mutation stocks. Ease of seed storage. | | | | |
| Research Uses—Model for genetics, evolution, and development of more complex plants such as corn or wheat. Other studies are in gene regulation, environmental interactions, and plant genomics. | | | | |
| Internet information and databases - The *Arabidopsis* Initiative Resource (http://www.arabidopsis.org) | | | | |
| ***Caenorhabditis elegans*** | **Nematode, roundworm** | 19,000 | 97 Mb | 1998 |
| Key Features—1-mm-long, transparent worm; development can be visualized under the microscope. Contains exact numbers of cells (959 in females and 1031 in males) that originate from specific progenitor cells. Can reproduce sexually or by selfing. Large numbers of progeny. Possible to make transgenics, RNAi knockdowns, and gene knockouts. | | | | |
| Research Uses—Eukaryotic development, apoptosis, cell signaling, aging. | | | | |
| Internet information and databases—WormBase (http://www.wormbase.org) | | | | |
| ***Neurospora crassa*** | **Orange bread mold** | 10,000 | 43 Mb | 2003 |
| Key Features—Haploid and diploid life cycles. Grows on simple defined nutritional medium. Has ordered ascospores, rapid growth; wide range of mutant stocks available. Haploid life cycle, with diploid phase possible with heterokaryons. Can be transformed with plasmid genes. | | | | |
| Research Uses—Studies of meiosis, metabolism, mitochondria-nuclear interactions, cytogenetics, fungal evolution. | | | | |
| Internet information and databases—Trans-NIH *Neurospora* Initiative (http://www.nih.gov/science/models/neurospora) | | | | |
| ***Mus musculus*** | **House mouse** | 30,000 | 2,600 Mb | 2002 |
| Key Features—Mammalian species with genetic similarity to humans (99% of genes have human homologs), mutant stocks available, rapid generation time, many progeny. Transgenics and gene knockouts possible. | | | | |
| Research Uses—Mutation studies, human disease models, development, immunogenetics, carcinogenesis. | | | | |
| Internet information and databases—Mouse Genome Informatics (http://www.informatics.jax.org/) | | | | |

yeast genetics" because of the ease with which genes can be manipulated and characterized in this organism.

Yeast cells undergo both haploid ($1n$) and diploid ($2n$) phases in their life cycle (Figure 23–1). In both phases, cell division occurs by budding, a mitotic process by which a smaller, genetically identical daughter cell buds from the surface of the mother cell, grows, and eventually splits from the mother (Figure 23–2). Haploid yeast cells occur in two mating types, a and $\alpha$. Both a and $\alpha$ cells reproduce mitotically by budding until a chemical signal known as a pheromone stimulates them to mate. The fusion of one a with one $\alpha$ cell is followed by fusion of their nuclei and formation of a diploid yeast cell. The diploid cell either continues to bud or undergoes meiosis and sporulates, depending on the availability of nutrients. The process of meiosis results in genetic reshuffling and formation of four haploid spores.

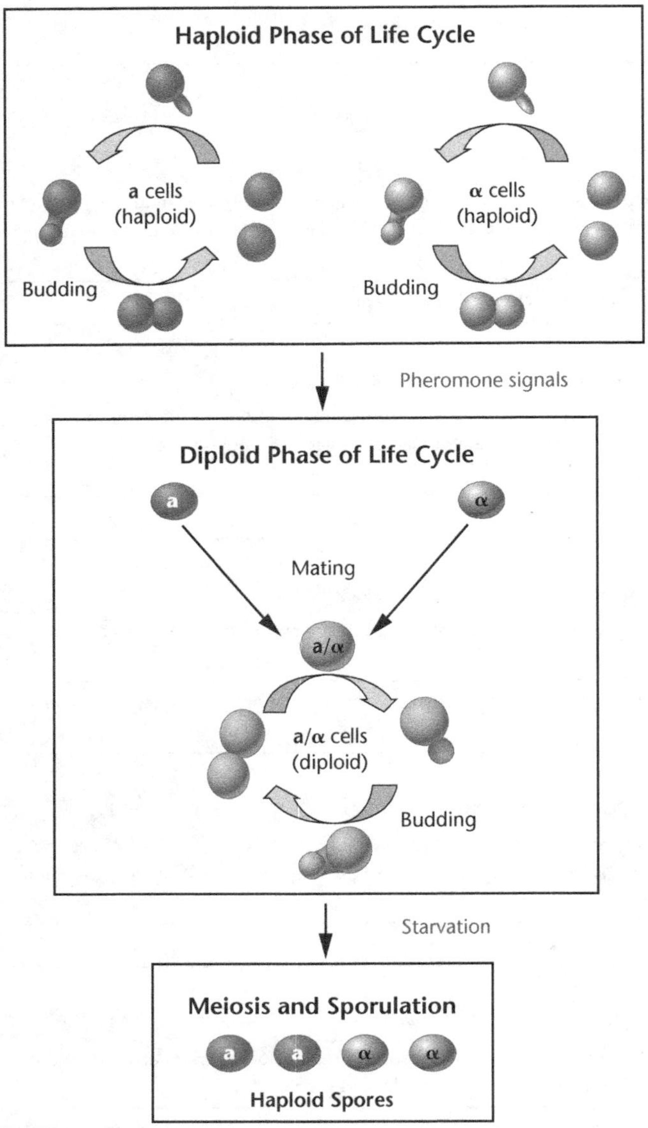


FIGURE 23-1 Life cycle of the budding yeast *S. cerevisiae*. Yeast can grow as either haploid or diploid cells; both kinds divide by budding. Haploid cells, when stimulated by pheromones, fuse to form diploids. In response to starvation, diploid cells undergo meiosis and sporulation, yielding four haploid spores.

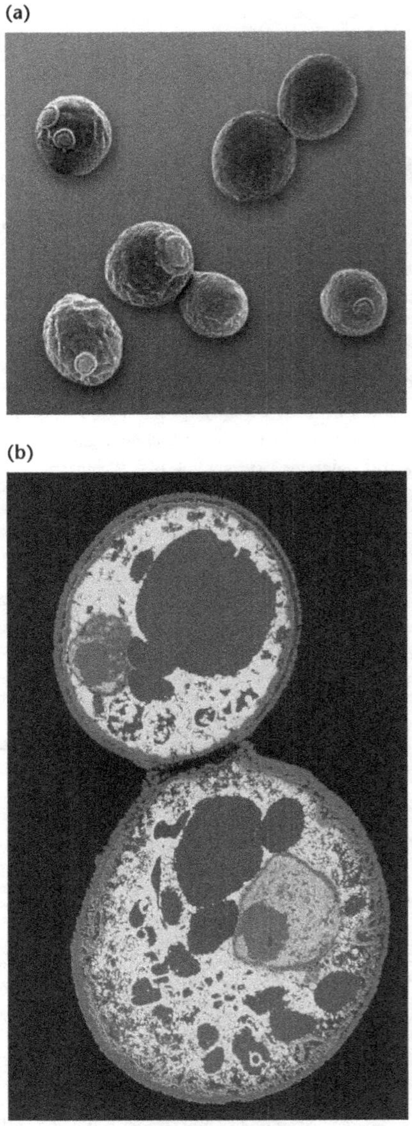


FIGURE 23-2 *Saccharomyces cerevisiae* cells. (a) Yeast cells in various stages of the cell cycle. Scars on a cell's surface indicate sites of previous budding. (b) Photomicrograph of budding yeast cell. Cell wall is shown in red, cytoplasm in blue/yellow, and nucleus in red/green.

The alternating haploid and diploid nature of the yeast life cycle is an advantage during genetic analysis. It is easy to detect recessive mutations in haploid cells, because the recessive allele is not masked by a wild-type allele. The diploid phase makes further studies, such as complementation analysis, possible. Also, recessive lethal mutations can be maintained in a diploid yeast strain that carries the mutation on one chromosome and the wild-type allele on the other chromosome.

Another major advantage of yeast as a genetic model is the availability of DNA sequence information and collections of mutants and deletion strains. *S. cerevisiae* was the first eukaryote to have its genome sequenced, the project being completed in 1996. Bioinformatic analysis of the yeast genome reveals that there are approximately 6600 yeast genes. About 5700 of these genes have a known function or an anticipated function based on sequence similarities to other known genes. Yeast geneticists also have available a library of yeast strains containing deletions in each open reading frame (ORF) of the genome, allowing convenient access to null strains for each yeast gene.

These features, combined with the ease with which yeast can be grown, mutagenized, transformed by plasmids, and genetically manipulated, make yeast a popular model for genetic dissection. Molecular techniques for deleting genes or replacing them with genes that have been mutagenized *in vitro* also lend power to yeast genetic analysis. The major disadvantage of yeast is its low utility for studies of multicellular communication and development. These types of studies require the use of model organisms such as *Drosophila, C. elegans, Arabidopsis*, or mouse.

## *Drosophila* as a Genetic Model Organism

Like yeast, ***Drosophila melanogaster*** is amenable to genetic analysis in part because of the ease with which it can be grown and the size of its genome—about 13,000 genes on four chromosomes (Figure 23–3). *Drosophila* has a short generation time of about 10 days from fertilized egg to adult. Each female fly produces about 3000 offspring in her lifetime. The stages of the *Drosophila* life cycle and some of the genes controlling development are described in Chapter 19.

Perhaps *Drosophila*'s greatest strength for genetic analysis is its easily observed body plan development in embryonic, larval, and adult stages. The fly's outer skeleton presents scientists with abundant phenotypic features, such as eye colors, wing shapes, bristles, and segment organization, that can be easily identified using a light microscope. Changes in these features reflect mutations in genes controlling differentiation and developmental processes.

An important feature in *Drosophila* genetics is the absence of crossing over in males and the moderate degree of crossing over in females. The absence of male crossing over means that it is possible to retain the chromosome linkage relationships of genes that are inherited through the male parent. This feature simplifies several aspects of genetic analysis, including the use of sophisticated genetic screen techniques, as described later.

Unlike yeast, *Drosophila* does not have a haploid phase in its life cycle. Therefore, geneticists have devised ingenious experimental tools to examine recessive mutations and to maintain stocks of mutant organisms bearing recessive lethal mutations. One of the most important of these tools is the **balancer chromosome**. Researchers created these chromosomes by bombarding *Drosophila* with X rays to create multiple overlapping chromosomal inversions, whose presence prevents crossing over between the balancer chromosome and its nonmutated homolog. As a result, a balancer chromosome and its normal homolog retain their differences from one generation to the next, with no recombinant chromosomes passing to the progeny. Balancer chromosomes also bear a dominant marker gene, such as a gene for eye color or wing shape. This allows geneticists to visually identify the presence of a balancer chromosome in individual flies during crosses. In addition, balancer chromosomes contain a recessive lethal gene that prevents homozygous *balancer/balancer* flies from surviving. Therefore, the only offspring that survive matings of heterozygotes with recessive lethal mutations are heterozygotes and wild-type homozygotes. Balancer chromosomes have been created for each of the three autosomal chromosomes, as well as for the X chromosome. The use of a balancer chromosome to recover X-linked recessive lethal mutations is described in the next section.

(a)

(b)

**FIGURE 23–3** Growing and manipulating *Drosophila* cultures. (a) Fruit flies feed and lay their eggs in agar medium poured into milk bottles. Larvae feed within the medium, then crawl up the sides of the jar to pupate. (b) Geneticists observe anaesthetized *Drosophila* under a dissecting microscope and sort flies into groups using a small brush.

*P* element transposons are also powerful genetic tools in *Drosophila*. ***P* elements** are mobile transposable elements that can move into and out of the *Drosophila* genome. *Drosophila* geneticists have harnessed *P* elements as vectors to introduce cloned genes into the *Drosophila* genome. They first insert a cloned gene of interest into the middle of a *P* element that also contains a gene for a visible characteristic such as eye color. Next, they inject the recombinant *P* element DNA into eggs, along with a helper plasmid that encodes the transposase gene (Figure 23–4). The transposase gene is transcribed in the germ cells of the early embryo and is translated into the transposase enzyme that enables the *P* element DNA to insert

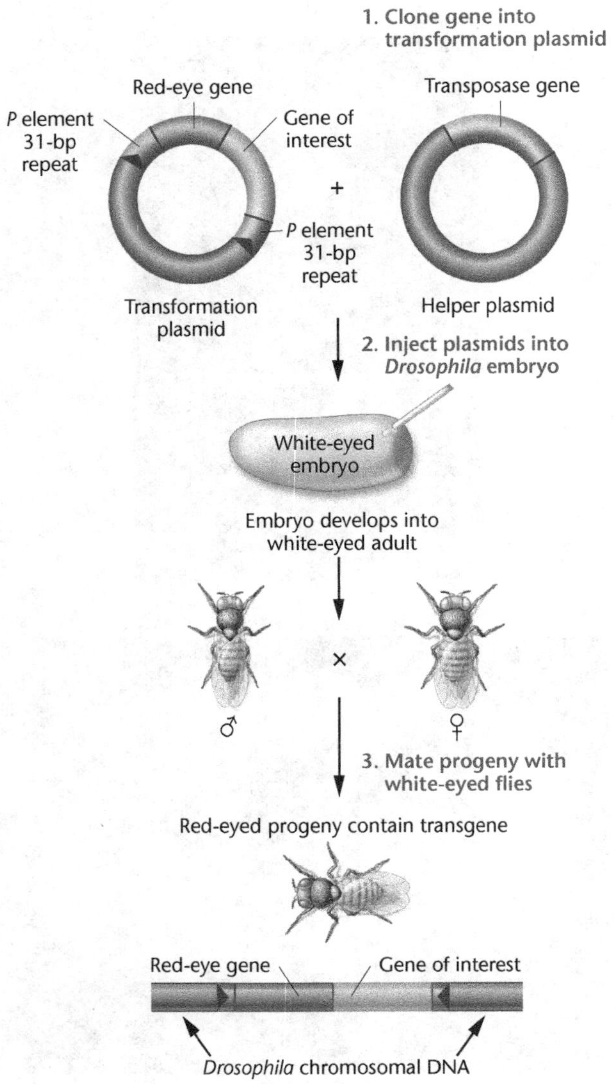


**FIGURE 23-4** *P* element–mediated transformation of *Drosophila*. The gene of interest is cloned into the transformation plasmid, which also contains a dominant gene for eye color (e.g., wild-type red eye). The genes are flanked by *P* element ends. The transformation plasmid and the helper plasmid containing the transposase gene are co-injected into fertilized eggs from a homozygous-recessive mutant white-eye female. The *P* element sequences along with the two transgenes become inserted into the DNA of the embryo's germ cells. Progeny of this fly will express the eye color marker (red) if the transformation plasmid becomes inserted into its germ cells.

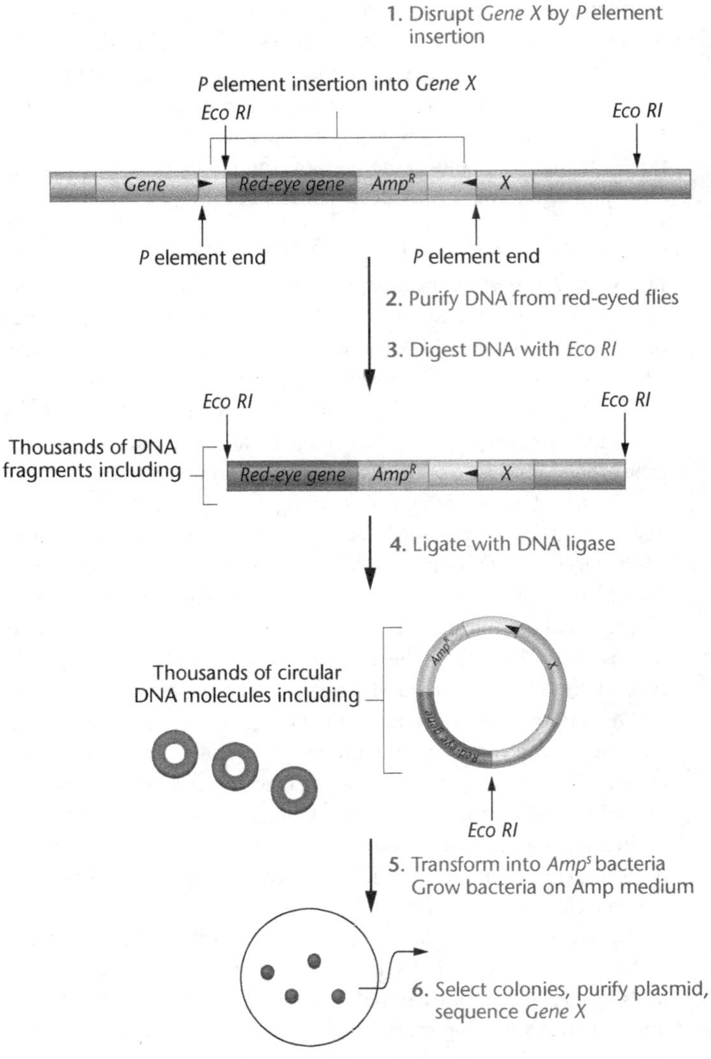


**FIGURE 23-5** *P* element-mediated mutagenesis and cloning. *Gene X* is disrupted by the insertion of a *P* element containing a gene for red eyes, bacterial plasmid sequences, and the ampicillin resistance gene. DNA is purified from red-eyed flies and then digested with a restriction enzyme (e.g., *EcoRI*) that cuts once within the *P* element and at thousands of sites throughout the *Drosophila* genome. Next the digested DNA is ligated into circles and transformed into ampicillin-sensitive bacteria. Surviving bacteria contain the plasmid DNA, which also contains *P* element and *Gene X* sequences.

into the germ-cell genome. Because the helper plasmid does not integrate into the genome or persist during development, further transposition does not occur. *P* element–mediated transformation is one of the most efficient methods for introducing cloned genes into higher eukaryotes.

*Drosophila* geneticists also harness *P* elements as mutagens and as tools to assist gene cloning. To create *P* element insertion mutations, they cross two strains of flies. One strain contains a *P* element that encodes the transposase enzyme but lacks the *P* element ends that are necessary for transposition. The other strain contains a *P* element with intact ends, as well as a gene that encodes a visible marker such as eye color. This second *P* element also contains a gene for antibiotic resistance and origin of replication sequences from bacterial plasmid DNA (Figure 23–5). The germ cells of $F_1$ progeny of a cross between these two strains synthesize the transposase enzyme and carry the mobile *P* element. In the $F_1$ germ cells, the mobile *P* element will excise and insert itself at random throughout the genome. Some of these insertions will be within and around important genes, disrupting their function. Researchers then screen those $F_2$ offspring that show the visible marker (e.g., eye color) for mutant phenotypes that are due to the insertion of *P* elements into new positions in the *Drosophila* genome. When a mutation of interest has been identified, the presence of the *P* element in the mutated gene allows researchers to clone the gene, following the steps shown in Figure 23–5.

## The Mouse as a Genetic Model Organism

Of all the model organisms used in genetic research, the mouse (*Mus musculus*) is perhaps the most relevant model for human disease studies. Mice and humans have similar body plans and undergo similar steps in development. Mice have relatively short generation times of eight to nine weeks, produce eight or more offspring per mating, and can be easily kept in laboratory situations.

Mouse and human genomes are approximately the same size (about 3 billion base pairs of DNA) with similar numbers of chromosomes (20 pairs in mice). Most human genes have homologs in mice. Interestingly, genes that are linked on the same chromosome in humans are often also linked in the mouse. Geneticists use this similarity in gene organization to identify and map genes in one species once the genes have been mapped in the other species.

Genes that have sequence similarity in humans and mice often, but not always, control the same biological processes. As exon sequences are usually well conserved between mouse and human, probes prepared from cloned genes in one species can often be used to detect and clone the corresponding genes in the other species.

Despite the many advantages of using mice experimentally, they are more difficult to grow, cross, and mutagenize than lower eukaryotes such as yeast or *Drosophila*. Large-scale genetic screens to identify genes of interest cannot be readily performed in mice. Nevertheless, specific genes can be inserted, deleted, or subjected to gene targeting in the mouse, as we describe next. This makes the mouse a useful system in which to determine the function and regulation of specific genes.

One of the most important genetic dissection techniques in mice is the creation of transgenic organisms. A mouse with a foreign piece of DNA introduced into its genome is referred to as a **transgenic mouse**. The foreign DNA may be a wild-type gene from another mouse, animal, or plant, or may be a gene that has been mutated *in vitro*.

The method of creating a transgenic mouse is conceptually very simple (Figure 23–6). Researchers isolate newly fertilized eggs from a female mouse and inject purified cloned DNA into the nucleus of the egg. The eggs are then placed in the oviduct of a pseudopregnant female mouse. In 20 to 50 percent of the injected eggs, the transgenic DNA becomes inserted into a chromosome by recombination, due to the actions of naturally occurring DNA repair enzymes.

(a) (b)

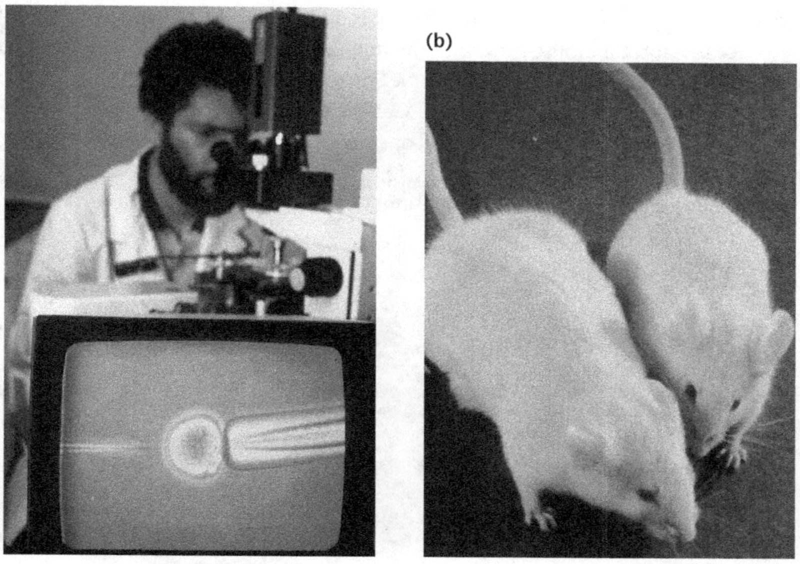

FIGURE 23–6 (a) Scientist microinjecting cloned DNA into a fertilized egg. The injections are performed by manipulating the egg and microinjection needle under a light microscope, seen in the background. The injection procedure is displayed on the screen in the foreground. The egg is held by a suction pipette (seen to the right of the egg). (b) A transgenic mouse with its nontransgenic sibling. The mouse on the left is transgenic for a rat growth hormone gene, cloned downstream from a mouse metallothionein promoter. When the transgenic mouse was fed zinc, the metallothionein promoter induced the transcription of the growth hormone gene, stimulating the growth of the transgenic mouse.

Researchers screen the transgenic mice by obtaining DNA from a sample of tail tissue, purifying the DNA, and performing either a Southern blot or a polymerase chain reaction (PCR) assay, to verify that the transgene is present in the animal's genome. As long as the integrated DNA is present in the germ-line cells, the transgene will be inherited in all of the mouse's offspring. Sibling matings can then generate homozygous transgenic lines.

One of the most common uses of mouse transgenics is to determine the function of human genes that have previously been cloned. For example, as described in Chapter 7, geneticists verified the function of the human *SRY* gene by injecting the mouse *Sry* gene into normal XX (female) mouse eggs. The transgenic XX *Sry* mice developed as males, demonstrating that *SRY* plays an important role in bringing about male development.

A second important genetic dissection technique in mice is the *gene knockout* and its sister technique, *targeted gene replacement*. Together, they are known as *gene targeting*. A knockout mouse is a true-breeding mouse strain lacking the function of a gene because the gene has been replaced with either a null allele or a specific mutagenized allele. Geneticists use knockout mice as models for some human genetic disorders. To create a mouse model of a human genetic disease, researchers clone the mouse gene that is homologous to the gene that causes the human disorder, subject it to site-directed mutagenesis (as described in Section 23.3), and then replace the normal mouse gene with the mutagenized allele. Mouse models of cystic fibrosis and Duchenne muscular dystrophy have been created using this technology. Geneticists also use mouse knockouts to study the genetic control of early development and behavior. We will discuss the technology for creating gene knockouts and gene replacements in mice in Section 23.3.

NOW SOLVE THIS

Problem 2 on page 631 asks you to select an appropriate model organism in which to study each of a number of genetic processes and human diseases, from cell division to cystic fibrosis.

- HINT: *To perform this task, consider the limitations and advantages of using each model organism discussed in this chapter. Then, think about how these features would affect research into each of the research topics.*

23.2

## Geneticists Dissect Gene Function Using Mutations and Forward Genetics

In this section, we explore the methods used in classical genetic analysis, sometimes called **forward genetics**. Forward genetic analysis begins with the isolation of mutants that show differences in phenotype for the process of interest. Mutant isolation is followed by defining gene pathways, cloning the gene, and creating more mutants in order to understand the biological pathway. In this way, mutants define the normal function of the gene.

### Generating Mutants with Radiation, Chemicals, and Transposon Insertion

The first step in forward genetic analysis is to define an experimental question—for example, which genes control the early development of an organ system? The next step is to predict the types of phenotypes that would emerge if the genes involved were mutated. Geneticists then begin to collect mutants displaying those phenotypes.

Geneticists sometimes examine spontaneous, naturally occurring mutations; however, as we saw in Chapter 16, mutations are generally rare in nature. For example, mice experience one mutation per 100,000 genes per generation. Researchers would need to screen millions of fruit flies, yeast cells, or mice in order to detect one relevant mutant.

The process of mutant hunting is greatly accelerated by mutagenizing the model organism before screening for mutations. The goal is to create one mutation at random in the genome of each organism in the population, so that only one gene product is disrupted in each organism, and the remainder of the genome is wild type. When mutagenesis is thorough enough that each gene is mutagenized at least once in the treated population, the mutagenesis is said to be **saturated**.

In genetic analysis, certain mutagens are preferred because of the types of mutations they trigger. For example, ionizing radiation causes chromosome breaks, deletions, translocations, and other major rearrangements. Mutations created by ionizing radiation are likely to be null mutations, which may have severe effects on the phenotype. In contrast, ultraviolet light and certain chemicals such as ethyl methane sulfonate (EMS) and nitrosoguanidine cause single base-pair changes or small deletions and insertions. With these mutagenic agents, a range of mild to severe mutant phenotypes may emerge, depending on where in the gene the lesion occurs. In addition, single base-pair mutations are more likely to result in the creation of conditional mutations, such as temperature-sensitive mutations, which are particularly useful for the study of essential gene functions.

Geneticists also use transposons to create mutations. The random insertion of a transposon, such as a *Drosophila P* element, into the genome can cause major disruptions of gene function. These large insertions usually create null mutations if they transpose into a gene's ORF. They may have a number of different effects on gene expression if they transpose into a gene's regulatory sequences, either in flanking regions around the gene or within the gene's introns.

### Screening for Mutants

After the model organism has been mutagenized, the relevant mutants must be identified. How are mutants detected, grown, and maintained once mutagenesis has occurred?

The most straightforward method that geneticists use to detect mutants is to do a **genetic screen**, or systematic examination. Frequently, a genetic screen involves the visual examination of large numbers of mutagenized organisms. For example, the classic 1974 study by Dr. Leland Hartwell and colleagues that first identified genes controlling the cell cycle involved a visual screen of yeast cells. Their screen required time-lapse photomicroscopic analysis of thousands of yeast colonies growing at two different temperatures. Researchers detected cells with mutations in cell-cycle regulatory genes by examining the ratio of yeast bud size to mother cell size, indicative of growth arrest at specific points in the cell cycle (Figure 23–7). Using this screening technique, Hartwell's group identified 32 yeast genes that controlled progress of cells through the cell cycle. Similarly, the genetic screens used to detect the wide range of *Drosophila* developmental mutations described in Chapter 19 required visual microscopic inspection of tens of thousands of *Drosophila* larvae for the presence of abnormal body-segment patterns.

It is easy to see how dominant mutations can be detected during a genetic screen. As long as the dominant mutation is not lethal, the phenotype will be visible immediately in any organism that bears a dominant mutation in the relevant gene. The mutant organism can then be crossed and heterozygous or homozygous stocks maintained. If the dominant mutation is lethal, it will not usually be detected.

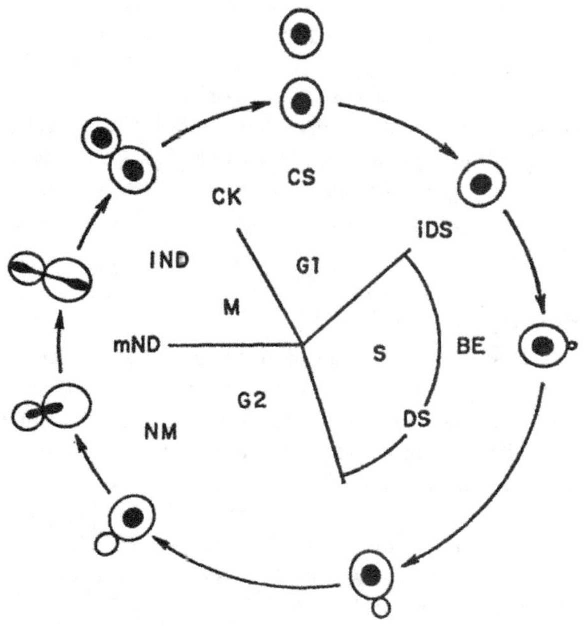


FIGURE 23–7 Yeast cell morphology throughout the cell cycle. Diagram showing stages of the cell cycle (G1, S, G2, and M) and steps in the cell cycle defined by the mutants in Hartwell's study. iDS, initiation of DNA synthesis; BE, bud emergence; DS, DNA synthesis; NM, nuclear migration; mND, medial nuclear division; IND, late nuclear division; CK, cytokinesis; CS, cell separation.

*Adapted from Figure 1 from L. H. Hartwell et al.1974. Genetic control of the cell division cycle in yeast.* Science *183: 46–51. Copyright American Association for the Advancement of Science.*

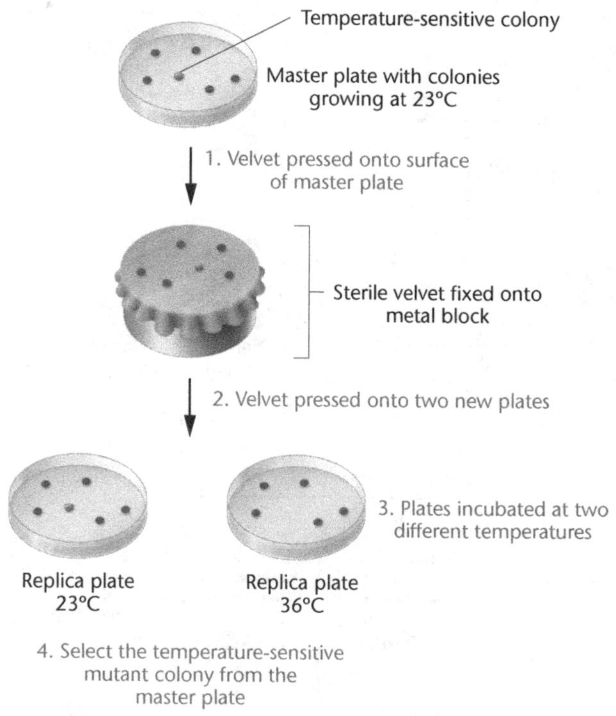


FIGURE 23–8 Replica plating technique to detect conditional temperature-sensitive mutations in yeast.

It is more likely, however, that mutations will be recessive. Haploid organisms, such as yeast, have a significant advantage for the detection of recessive mutations, as the mutant phenotype will be immediately evident in the mutated organism. Detecting recessive mutations in diploid organisms, such as *Drosophila* or mice, requires the mutated organism to be mated and the $F_1$ progeny crossed with each other to reveal the one-quarter $F_2$ that will be homozygous for the recessive allele.

But how do geneticists detect *recessive* mutations that occur in essential genes?

Recessive lethal mutations will not be detectable in haploid organisms, such as yeast. To overcome this limitation, yeast geneticists isolate conditional mutants. A **conditional mutation** is one that allows the mutant gene product to function normally under the permissive condition (usually a normal growth temperature) but causes it to function abnormally under the restrictive condition (usually a higher temperature). Temperature sensitivity occurs because the protein encoded by the mutant gene misfolds and becomes nonfunctional at higher temperatures. The strategy for selecting temperature-sensitive lethal recessive mutations in yeast is simple (Figure 23–8). After mutagenesis, the yeast cells are allowed to grow into colonies on growth medium in a petri dish, at the permissive temperature of 23°C. A piece of sterile velvet is then pressed onto the colonies, and some yeast cells from each colony transfer onto the velvet. The velvet is then pressed onto two new plates containing sterile growth medium. The yeast cells transfer from the velvet onto the replica plates. One replica plate is incubated at the permissive temperature, and one is incubated at the restrictive temperature. Yeast colonies that do not grow at the restrictive temperature may contain temperature-sensitive recessive lethal mutations. Researchers then return to the original plate, growing at the permissive temperature, and select the colony for further analysis.

Diploid organisms that are heterozygous for a recessive lethal allele will not show the mutant phenotype, and those that are homozygous for the lethal allele will die. So how is it possible to detect recessive lethal mutations in a population of diploid organisms?

Geneticists have devised some intricate strategies to detect and recover recessive lethal mutations in diploid organisms. An example of a method to detect recessive lethal mutations in *Drosophila* is the ***ClB* technique.** Hermann Muller devised this technique in the 1920s in order to demonstrate that X rays cause mutations in *Drosophila.* The *ClB* technique detects recessive lethal mutations on the X chromosome (Figure 23–9). In this technique, geneticists treat wild-type males with a mutagen such as radiation and then mate the males to untreated heterozygous *ClB* females. These *ClB* female flies carry a

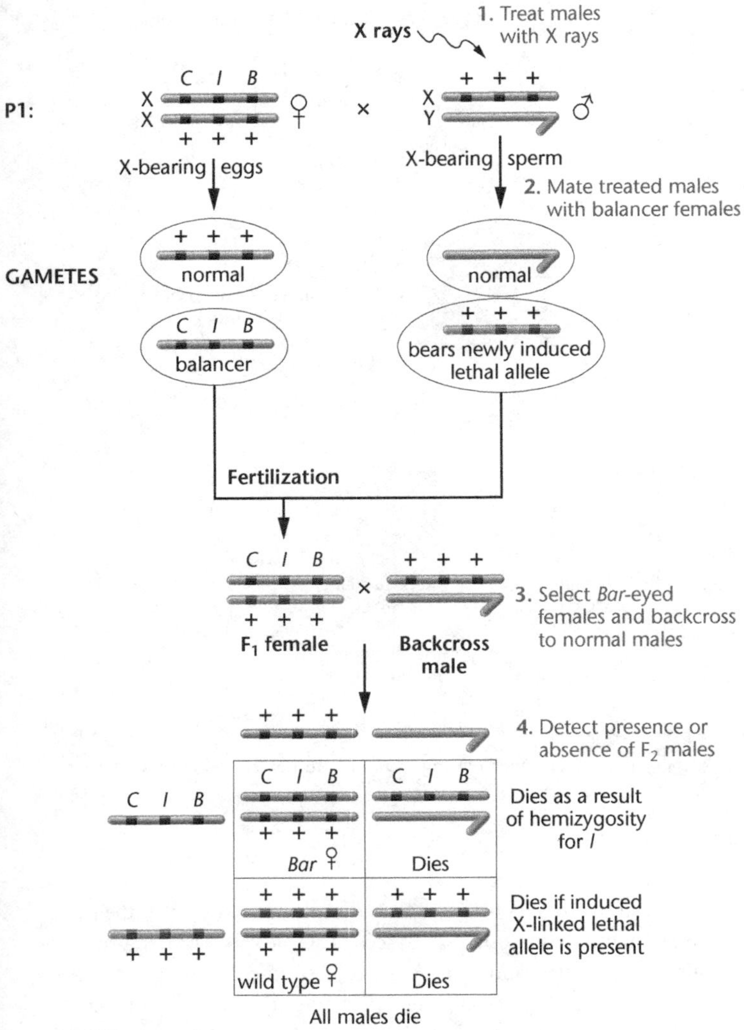


FIGURE 23-9 The *ClB* technique for the detection of induced, X-linked recessive lethal mutations in *Drosophila*.

balancer X chromosome with the three important features mentioned earlier in the chapter. First, the balancer chromosome contains inversions that prevent crossing over (designated by the *C*). Second, it contains a recessive lethal allele (*l*), which eliminates any offspring that are homozygous for the balancer chromosome. Third, the balancer chromosome bears a dominant marker gene, in this case the gene for *Bar* eye (*B*), making it possible to visually inspect offspring for the presence of the *ClB* chromosome.

Next, the researchers select individual $F_1$ females with *Bar* eyes. These flies have one X chromosome from their mother (the *ClB* chromosome) and one X chromosome from their father (which may or may not contain a newly induced recessive lethal mutation). Researchers then backcross each *Bar*-eyed female to a wild-type male. If a recessive lethal mutation is present on the paternal X chromosome, there will be no viable males in the backcross progeny. Half of them will die because they are hemizygous for the *ClB* chromosome (the *l* gene is lethal), and the other half will die because they are hemizygous for the newly induced lethal allele. The absence of males signals the presence of a recessive lethal mutation on the X chromosome. The mutation can be maintained in the population by selecting the wild-type (red-eyed) females and backcrossing them.

*Drosophila* geneticists have devised similar methods using balancer chromosomes to detect recessive lethal mutations in autosomes.

## Selecting for Mutants

In a genetic screen, each organism must be individually examined for the phenotype of interest. Often tens of thousands of individuals must be screened to find the required number of mutants. Though effective, this strategy is obviously slow and labor-intensive. In some cases, a different method, **selection**, can be employed to reduce the amount of time and labor. The goal of selection is to create conditions that remove the irrelevant wild-type or mutant organisms from the population, leaving only the mutants that are sought. Usually, this is accomplished by killing or inhibiting the growth of organisms that do not display the relevant phenotype.

Selection is simplest if the mutant phenotype enhances survival under certain conditions. For example, a mutant gene that confers resistance to a drug can be easily selected by growing the organisms on a medium containing the drug. Also, a mutation that corrects a defect in a metabolic pathway, such as an inability to metabolize galactose, can be selected if the organism is grown in a medium that only contains galactose as a carbon source.

Although some mutations lend themselves to direct selection, most do not because the majority of mutations are loss-of-function mutations. However, geneticists can use selection to study *suppressor mutations* (mutations that restore normal function to an existing loss-of-function mutant). Suppressor mutations help define other genes in multistep pathways leading to the phenotype of interest, as described on page 616.

## Defining the Genes

One of the goals of a forward genetic analysis is to discover all the genes that affect a phenotype and to determine how the genes function. Hence, the next step in a classical genetic dissection is to determine the number of genes that control the phenotype.

As we have seen in previous chapters, a single gene can have several different alleles—weak alleles, null alleles, or variants with different phenotypes. For example, the ABO blood groups described in Chapter 4 represent three different alleles at the same locus that can lead to six different genotypes and four different phenotypes. The $I^A$ and $I^B$ alleles direct the synthesis of the A and B antigens, and the $I^O$ allele is a null, nonfunctional allele. The $I^A$ and $I^B$ alleles are codominant, and the $I^O$ allele is recessive to the other two. It is also possible for a gene to act within a multigene pathway leading to a particular phenotype. A mutation in any one of the genes within a pathway may result in the same phenotype, as each mutant gene product prevents the pathway from functioning. For example, the gene represented by the $I^A$, $I^B$, and $I^O$ alleles acts within the same pathway as the gene represented by the *H* allele. People with a homozygous null mutation in the *H* gene show the same phenotype as those with a homozygous $I^O$ genotype.

How can geneticists determine whether the mutations in the collection of mutant organisms represent alleles of the same gene, or whether they represent mutations in a number of different genes within a pathway?

To make this determination, geneticists may perform either complementation or recombination analysis. **Complementation analysis**, which is described in detail in Chapter 4, allows investigators to determine whether two mutations are in the same gene—that is, whether they are alleles—or whether they represent mutations in separate genes.

Complementation analysis may be used to screen any number of individual mutations that result in the same phenotype. All mutations that are present in any single gene are said to fall into the same **complementation group**, and they will complement mutations in all other groups. If large numbers of mutations affecting the same trait are available and studied using complementation analysis, it is possible to predict the total number of genes involved in determining that trait. In Chapter 16, we discussed how complementation analysis was used to study the genes involved in the inherited human disorder xeroderma pigmentosum. Results of these studies revealed that mutations in any one of seven complementation groups (genes) can lead to the disorder.

Dominant mutations cannot be tested by complementation, but they can be tested by **recombination analysis**. Recombination analysis provides an estimate of genetic linkage by calculating the amount of recombination (crossing over) that occurs between mutant loci when these loci are present in the same chromosome. If two mutations occur in the same gene, they will be tightly linked. If present in different genes, they will most likely be unlinked. Methods used to calculate linkage by crossover and recombination in yeast, *Drosophila*, and other organisms are described in Chapter 5.

## Dissecting Genetic Networks and Pathways

Products of several different genes may act in a pathway leading to a particular phenotype. If complementation or recombination analysis reveals that two or more genes contribute to the same phenotype, they may be acting in one pathway. How, then, do geneticists determine where these genes operate within a single pathway? One way is to examine how two mutant genes interact during epistasis analysis. As we saw in Chapter 4, **epistasis** occurs when the effect of one gene masks or modifies the effect of another gene. If one gene is epistatic to another, the mutation in the gene whose effects are manifest earliest in the pathway will be the one whose phenotype is evident in a double mutant. In order for epistasis analysis to work, each mutant must show a slightly different phenotype.

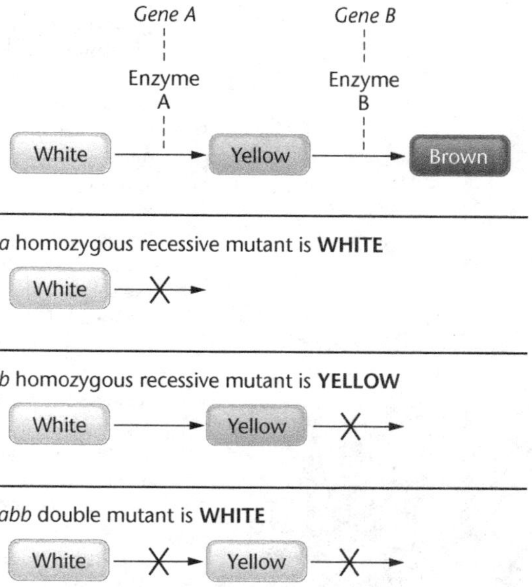


**FIGURE 23–10** Example of epistasis analysis for two genes that control steps in a biochemical pathway that converts a colorless coat pigment to yellow, and then to brown. Wild-type coat color is brown. The double mutant (aabb) displays the same phenotype as the single mutant (aa). This shows that *Gene A* controls a step prior to the step controlled by *Gene B*.

For an epistasis analysis, two homozygous recessive mutants with discernible phenotypes are crossed, and the $F_2$ progeny are examined. The phenotypic ratios of the $F_2$ progeny reveal which of the two genes controls the earlier or later steps in the pathway. Alternatively, a double mutant organism, homozygous for both mutations, can be created and the phenotype of this double mutant determined. If the phenotype of the double mutant is the same as one of the single mutants, that single mutation likely represents the earlier step in the pathway (Figure 23–10). However, one must keep in mind that other types of gene interactions can be epistatic to each other, although they may not operate within the same biochemical pathway. Hence, other types of genetic and biochemical tests need to be completed before a full genetic pathway can be dissected.

In the genetic analysis of the yeast cell cycle discussed previously, Hartwell and colleagues used epistasis analysis to determine the order in which their 32 yeast genes acted in the cell cycle. They performed this analysis by constructing diploid double mutants from pairs of their haploid single mutants. Their analysis showed that a mutant with a defect in a gene controlling the first of six stages in the cell cycle could not complete the remaining five stages. Similarly, mutants with a defect in a gene controlling the second of these stages were arrested at the second stage and could not complete the remaining four stages. A double mutant containing both mutations was arrested at the first stage, confirming that the first mutant contained a mutation in the gene controlling a step prior to

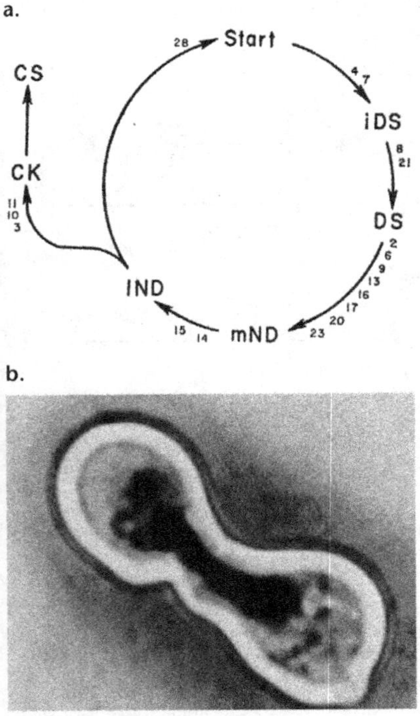


**FIGURE 23–11** Yeast cell-cycle mutants, ordered into steps in the cell cycle. (a) Hartwell's model of cell-cycle steps, showing the steps affected by 18 of his cell division cycle (*cdc*) mutants. Abbreviations are the same as those in Figure 23–7. (b) Microscopic view of a mutant yeast cell, growing at the restrictive temperature. The cell has the characteristic morphology of one arrested at G2/M. *Figure 5 and adaptation of Figure 3 from L. H. Hartwell et al. 1974. Genetic control of the cell division cycle in yeast.* Science *183: 46–51. Copyright American Association for the Advancement of Science.*

the second mutant. Using this strategy, the investigators ordered all 32 cell-cycle mutations into six stages of the yeast cell cycle pathway (Figure 23–11).

## Extending the Analysis: Suppressors and Enhancers

A forward genetic analysis may identify only some of the genes within a pathway. It is possible that the original mutagenesis procedure did not mutagenize each of the genes in the pathway, or that not all genes in the pathway were recovered in the screen. How can geneticists find the other genes in the pathway, if only some have been mutated?

One way to identify other genes in a pathway is to carry out a second mutagenesis screen. The goal of the second screen is to identify genes that dominantly enhance or suppress the first mutant phenotype. To perform the second screen, geneticists subject the first mutant strain to a second round of mutagenesis and then screen for more severe or milder phenotypes.

A **suppressor mutation** is a second mutation that "rescues" the original mutant phenotype, restoring the wild type. For example, a yeast temperature-sensitive mutant that cannot replicate its genome at the restrictive temperature may undergo a second mutation that returns the mutant to wild type by restoring DNA replication. There are a number of mechanisms by which suppressor mutations work (Figure 23–12). If two wild-type proteins interact, but a mutation in one of them causes misfolding and prevents the interaction, a suppressor mutation in the second protein could alter the second protein in such a way that the interaction is restored. Another type of suppressor mutation may occur in a different pathway, enabling that pathway to substitute for or bypass the pathway blocked by the first mutation. A third type of suppressor is the high-copy suppressor. This suppressor rescues the first mutant by supplying high levels of another gene product that may compensate for or stabilize the abnormal mutated gene product. Geneticists use these types of suppressors to define other genes involved in a genetic pathway.

Not all suppressor mutations are useful for identifying other genes in a pathway. Sometimes, the suppressor mutation occurs in the same gene as the first, correcting the defect. This could be a direct reversion of the original mutation, or it could be a second mutation in another region of the same gene. One example is a mutation that reestablishes the correct reading frame. Also, a suppressor may be a nonsense suppressor mutation in a tRNA gene, causing a termination codon in the mutant gene to be recognized as an amino acid codon and thus allowing a functional protein to be translated from the mutant gene.

**Enhancer mutations** are the opposite of suppressor mutations; they increase the intensity of the first mutant phenotype. Geneticists use enhancer mutations in ways similar to their use of suppressor mutations, because both can point the way to other genes in a genetic network.

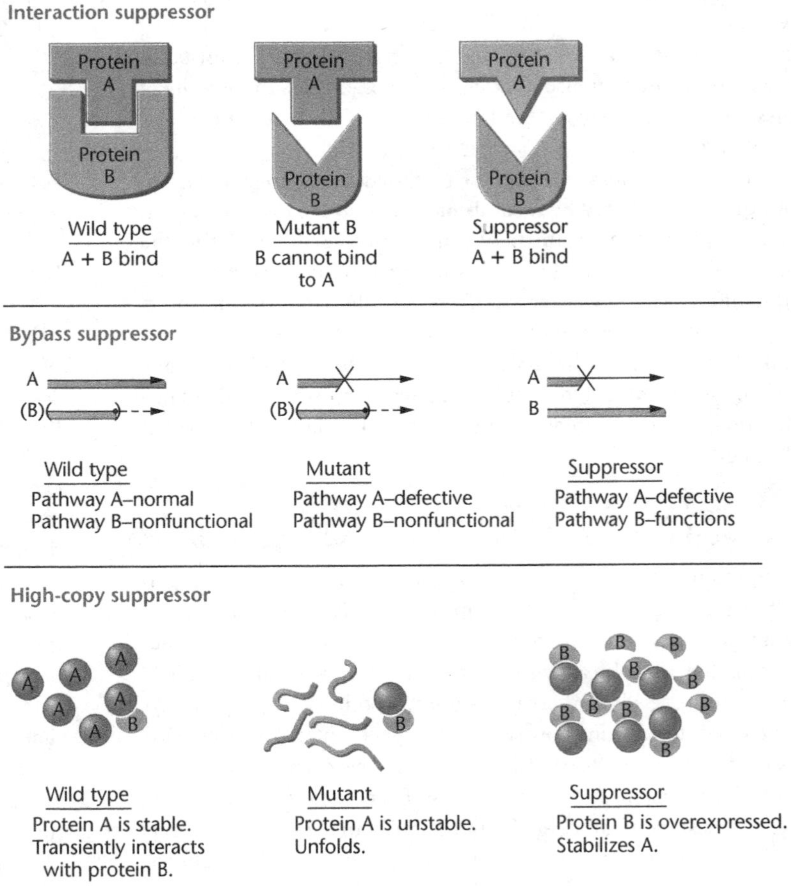


**FIGURE 23–12** Three types of suppressor mutations.

## Extending the Analysis: Cloning the Genes

The next step in a forward genetic analysis is to clone the genes that are relevant to the phenotype of interest. A large number of methods are available for cloning genes identified within a collection of mutant organisms. In a classical genetic approach, the first step in cloning genes from higher organisms is for the researcher to map the position of the gene by linkage analysis. The gene can be mapped to a chromosome and then to a general chromosome region using standard linkage analysis, such as that described in Chapter 5. When the approximate location of the gene is known, the candidate gene can be located by techniques such as positional cloning and chromosome walking.

In **positional cloning**, researchers first identify a chromosomal marker, such as a restriction fragment length polymorphism (RFLP) or a single nucleotide polymorphism (SNP), that is genetically linked to the mutation of interest. After narrowing the region further using more markers, they select DNA clones that contain the closest marker from a DNA library, using the marker as a probe. **Chromosome walking** begins the same way. After these first steps, investigators subclone a fragment of DNA from one end of the DNA clone, and use this subclone to select other clones from a DNA library. Some of these clones will contain DNA that overlaps the first clone, and also lies beyond the end of the first DNA clone. A fragment from one end of the second clone is then subcloned and used to select other clones that in turn overlap and extend the first and second clones. This is repeated until the entire region has been cloned.

After a clone is selected by positional cloning and chromosomal walking, researchers identify potential genes in the region, by sequencing the clone and searching for ORFs and other features indicative of genes. The relevant gene is identified by comparing the DNA sequences from mutant and wild-type individuals. A gene that contains a mutation in a mutant, but not in a wild-type, individual is selected as the candidate gene, and tested by other methods to verify its contribution to the trait in question.

In *Drosophila*, mutations generated by *P* element insertion can be readily cloned, as described previously (Figure 23–5). In addition, the *P* element itself can be used as a tag with which to select the correct DNA from a genomic library created from the mutant organism.

Now that the genomes of many model organisms are fully sequenced and their genomes' ORFs identified, the job of cloning is greatly simplified.

In yeast, it is possible to clone genes by **functional complementation** with a yeast cDNA library. In this method, plasmids encoding all the wild-type genes from yeast are methodically transformed into the mutant strain, until one cDNA clone restores the mutant to wild type. The cDNA clone that restores the mutant to wild type is then purified from the transformed strain of yeast, and the DNA sequence of the cDNA is determined. The complementing gene is then identified by searching the yeast genomic sequence for sequences that match the plasmid sequence. Once the wild-type gene corresponding to the mutated gene has been identified, this cloned wild-type gene can be used to isolate the mutant gene from the mutant organism. Alternatively, geneticists can design PCR primers based on the DNA sequence of the wild-type gene. They then use these primers to amplify all or part of the mutated gene in the mutant organism. The amplified DNA fragments can be directly sequenced to identify the nature of the mutation that led to the mutant phenotype.

In yeast and other model organisms, identifying and cloning the relevant genes may be as simple as obtaining the DNA sequence of a large chromosomal region of the mutant organism and comparing the sequence to the wild-type sequence available in a genomic database. Using this approach, researchers first narrow down the chromosomal location of a mutant gene by performing standard recombination or linkage analyses. These types of mapping studies correlate the locations of known chromosomal markers with the phenotype that is caused by the mutation. Mapping in model organisms uses the large numbers of *simple sequence length polymorphism (SSLP)* and *single-nucleotide polymorphism (SNP)* markers that have been catalogued throughout genomes. Once the location of the mutation is narrowed to a region of about 1 Mb in length, the researchers read the wild-type DNA sequence of that region of the genome, available from a genome sequence database for the model organism used in the study. Annotations for that region's DNA sequence will indicate the presence of ORFs and candidate genes within the region. Next, the researchers synthesize PCR primers whose sequences are complementary to DNA regions flanking each of the candidate genes in the region. DNA is extracted from the mutant organism, and the PCR primers are used to amplify the DNA of each candidate gene in the mutant organism. The amplified DNA is then subjected to DNA sequencing, using automated DNA sequencers. The resulting DNA sequence of each candidate gene is compared to the wild-type sequence of each gene, available in the genome database. This comparison is usually performed using software designed for this purpose. The presence of a mutation in one of the candidate genes in the region suggests that this gene may be responsible for the mutant phenotype. Once the mutant gene has been identified, a clone of the wild-type version can be obtained in one of several ways. The gene can be constructed by PCR amplification of the genomic region containing the gene, and the subsequent fragment can be inserted into a plasmid vector. Alternatively, a PCR primer complementary to sequences within the gene can be used to probe a preexisting cDNA or genomic DNA library, and the relevant clone can be selected from the library.

Once the gene has been identified, its involvement in the phenotype of interest must be confirmed. There are several ways in which the confirmation can be performed. Investigators will often observe the expression patterns of the gene in order to determine whether the tissue location and timing of expression of the mRNA and protein products are consistent with the proposed role of the gene. In addition, independent alleles of the same gene will be examined. Mutations in similar genes from other organisms may help verify the involvement of the gene in the phenotype. The ultimate verification of gene involvement is obtained by the use of gene knockouts and transgenic expression experiments, as described in Section 23.3.

## Extending the Analysis: Gene Product Functions

After the gene is cloned, its DNA sequence determined, and its involvement in the phenotype verified, genetic dissection encounters its next challenge—that of defining the functions of the gene product. This phase of the analysis uses a wide range of bioinformatics, genomic, genetic, and biochemical technologies.

Geneticists may begin the analysis by comparing the gene's DNA sequence to sequences of other genes in the same or different organisms. If a gene of similar sequence exists in another species and the function of that gene is known, it is possible to use this information to deduce a possible function for the unknown gene product. Investigators may analyze the gene's DNA sequence for the presence of amino acid **sequence motifs** in the gene product that may provide clues as to the gene product's function. For example, the gene may contain a sequence that encodes a kinase motif, suggesting that the gene encodes an enzyme that phosphorylates other proteins. Or the gene may contain a sequence that encodes a DNA-binding region found in certain types of transcription factors, suggesting that the unknown gene regulates expression of other genes. Hypotheses such as these can then be tested by further biochemical or genetic tests.

Researchers apply a wide range of **molecular genetic tools** to extend the analysis of candidate genes. These tools include techniques to analyze gene expression in specific tissues, during specific times in development, or in certain regions of the cell during the cell cycle. Investigators can also examine whether other proteins interact with the protein encoded by the gene, using a variety of biochemical and genetic tests. A range of powerful tools that are now used to define gene function are site-directed mutagenesis, gene replacement, gene knockouts, and examination of phenotypes in transgenic organisms. These methods are described in the next sections.

**NOW SOLVE THIS**

Problem 25 on page 632 involves planning a genetic screen for genes that affect a visible phenotype, in this case, the tanning and hardening of adult *Drosophila* cuticle.

■ HINT: *Although the phenotype is a visible one, you must also consider the possibility of lethality at any stage of the fly's life cycle. In Section 23.2, you learned about methods used to recover lethal mutations in* Drosophila. *The last part of the problem deals with a negative outcome of the screen, as no mutants were isolated. To suggest reasons for this outcome, you might consider factors such as the method used for mutagenesis or how the screening was done. Think about the fact that experiments do not always proceed according to plan.*

## 23.3 Geneticists Dissect Gene Function Using Genomics and Reverse Genetics

Forward genetic analysis begins with a collection of mutants and progresses to defining genetic pathways, cloning the wild-type gene, and analyzing the functions of the gene product. Although this approach continues to be a powerful one with which to dissect gene function, modern genomics and molecular biology are challenging forward genetics as the dissection tool of choice. Researchers are debating whether forward genetics will soon become obsolete as a research tool. However, most genomics and molecular-based technologies still require mutational analysis of candidate genes, as it remains the ultimate demonstration of biological function.

**Reverse genetics** is a somewhat broad term describing a number of gene analysis methods that begin with a cloned wild-type gene, a DNA sequence from a database, or purified protein, and progress to site-directed mutagenesis and phenotypic analysis—the opposite of the order used in forward genetics.

Reverse genetics is made possible by the completion of genome sequencing projects from a number of organisms, including yeast, *Drosophila*, and humans. Researchers can now begin with a novel gene sequence in a genomics database and probe the gene's function. They can also analyze genome-wide patterns of gene expression before knowing the function of individual genes being analyzed. In some model organisms, such as yeast and *Drosophila*, investigators are systematically deleting each of the genome's ORFs and studying the resulting phenotypes. These gene-specific and genome-wide approaches make up the field known as **functional genomics**. Functional genomics is also discussed in Chapter 21.

Reverse genetics and functional genomics approaches are useful in studies that aim to identify human disease genes. As humans cannot be mutagenized and mated in controlled situations, it is not possible to use classical forward genetics to identify disease genes. Using reverse genetics techniques, geneticists begin with a disease phenotype, or in some cases a purified mutant protein, and eventually proceed to cloning and studying the responsible genes. Functional genomics adds to this capability by allowing investigators to identify large numbers of genes whose expression levels change in certain disease states, and by facilitating the identification of these genes.

In this section, we explore how geneticists use reverse genetic approaches to dissect gene function. In the next section we will describe a few of the newer methods used in functional genomics.

### Genetic Analysis Beginning with a Purified Protein

One starting point for a reverse genetic analysis is a protein suspected to be involved in a phenotype of interest. If sufficient quantities of the candidate protein can be purified, geneticists can determine its amino acid sequence. This is done by sequentially cleaving the N-terminal amino acids from the protein and identifying them in an automated amino acid sequencer. Investigators then examine the amino acid sequence and use it to deduce the gene's DNA sequence. Because the DNA code is degenerate, several possible DNA sequences could be responsible for coding a particular amino acid sequence. Therefore, researchers synthesize a mixture of different oligonucleotides that correspond to all possible codon combinations (Figure 23–13). They then label the mixture of synthetic DNA oligonucleotides with radioactivity or indicator dyes and use this mixture to probe a genomic or cDNA library. If one or more oligonucleotides hybridize to a clone in the library, the clone is selected for further analysis.

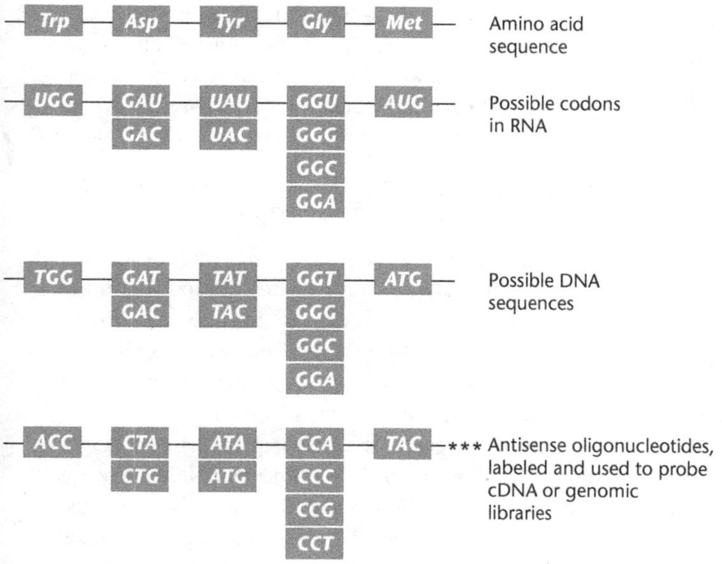


FIGURE 23-13 Strategy used to clone the *Factor VIII* gene from a porcine cDNA library. The purified Factor VIII protein was cleaved into peptides, and the amino acid sequence of each peptide was determined. All possible codon combinations that could code for a given five-amino-acid sequence were calculated, as well as the corresponding DNA sequences. A mixture of antisense oligonucleotides was synthesized for each possible DNA sequence that could encode the five-amino-acid peptide. This mixture was labeled (***) and used to probe a porcine cDNA library in order to select the porcine *Factor VIII* gene.
*Adapted from Gitschier, J. et al. 1984. Characterization of the human factor VIII gene.* Nature *312: 326–30, copyright 1984 Macmillan Publishers Ltd.*

Researchers used this protein-first strategy to clone the gene responsible for **hemophilia A**, a disease causing defects in blood clotting. From pedigree studies and linkage analysis, geneticists discovered that hemophilia A is an X-linked recessive trait, likely controlled by a single mutated gene. A reasonable hypothesis was that the gene responsible for the defect encoded a blood-clotting factor. Biochemical analysis of clotting factors in the blood of normal individuals and hemophiliacs revealed that one protein, **Factor VIII**, was absent in hemophiliacs. Because Factor VIII is present in very low quantities even in normal blood, investigators purified large quantities of the protein from pig blood. They then determined the amino acid sequence of a portion of the Factor VIII protein and predicted the DNA sequence that would encode the protein (Figure 23–13). They synthesized DNA oligonucleotides complementary to this DNA sequence, probed a porcine genomic DNA library, and selected the *Factor VIII* gene from the library. Next they used the porcine genomic clone to probe a human genomic library. By comparing the DNA sequences of the *Factor VIII* gene from normal and hemophiliac individuals, they determined that the *Factor VIII* gene was mutated in people with hemophilia, confirming that the *Factor VIII* gene is responsible for the disease.

An alternative approach to cloning a gene beginning with the purified protein involves the use of antibodies. To generate an antibody to a purified protein, investigators inject the protein into laboratory animals, such as rabbits or mice. These animals will respond by synthesizing antibodies that recognize the injected protein, and the antibodies can be collected from the animals' blood samples. Investigators then use the purified antibodies as probes. They first label the antibodies with radioactivity or fluorescent dyes and hybridize the antibodies to cDNA expression libraries. Expression libraries contain cDNA clones inserted into special vectors containing a promoter that will drive the expression of the cDNA within the bacterial host cells (Figure 23–14). Colonies of bacteria containing cDNA expression clones will produce small quantities of the proteins encoded by the cDNAs. Antibodies that are raised against the purified protein will specifically recognize the protein produced in the bacteria from the cDNA vector, allowing investigators to select the relevant clone. The gene encoding the enzyme **tyrosinase**, which is responsible for **albinism**, was cloned in this way, using tyrosinase-specific antibodies.

## Genetic Analysis Beginning with a Mutant Model Organism

In some cases, it is possible to clone a gene by functional complementation. In the previous section, we learned that geneticists use functional complementation in mutant yeast strains to clone wild-type yeast genes. Functional complementation can also involve transforming a mutant organism with cloned genes from another organism in order to identify the gene controlling the mutant phenotype in the other organism.

An example of functional complementation is the use of mutant yeast strains to clone homologous genes from humans. Many genes and gene products are highly conserved during evolution, both in their sequences and their functions. Hence, it is sometimes possible to functionally complement a yeast mutation with a wild-type human gene. Researchers have used this approach to identify human genes that encode transcription factors, purine and pyrimidine biosynthesis enzymes, and cell-cycle proteins. One of the first successful functional complementations between yeast and human genes was the identification of the human ***CDC2*** gene. The *CDC2* gene encodes a kinase protein that regulates several stages of the cell cycle. In 1987, Paul Nurse and colleagues cloned the human *CDC2* gene by transforming a yeast temperature-sensitive *cdc2* mutant strain with clones from a human cDNA library. One of the human cDNA clones functionally complemented the yeast mutation, returning the yeast strain to wild type. They recovered the complementing plasmid from the yeast strain, sequenced it, and discovered 63 percent sequence identity between the yeast and human Cdc2 proteins. In both organisms, the *CDC2* gene controls key steps in the cell division cycle.

Geneticists also use functional complementation in cultured cells to clone human genes. For example, the Chinese hamster cell line EM9 is defective for DNA repair, making the cells hypersensitive to ionizing radiation. Functional complementation between cloned wild-type human genes and this cell line led to the identification of the human DNA repair gene, ***XRCC1***. The introduction of *XRCC1* into the EM9 cell line corrected the cell line's hypersensitivity to ionizing radiation. In addition, some human tumor-suppressor genes have been

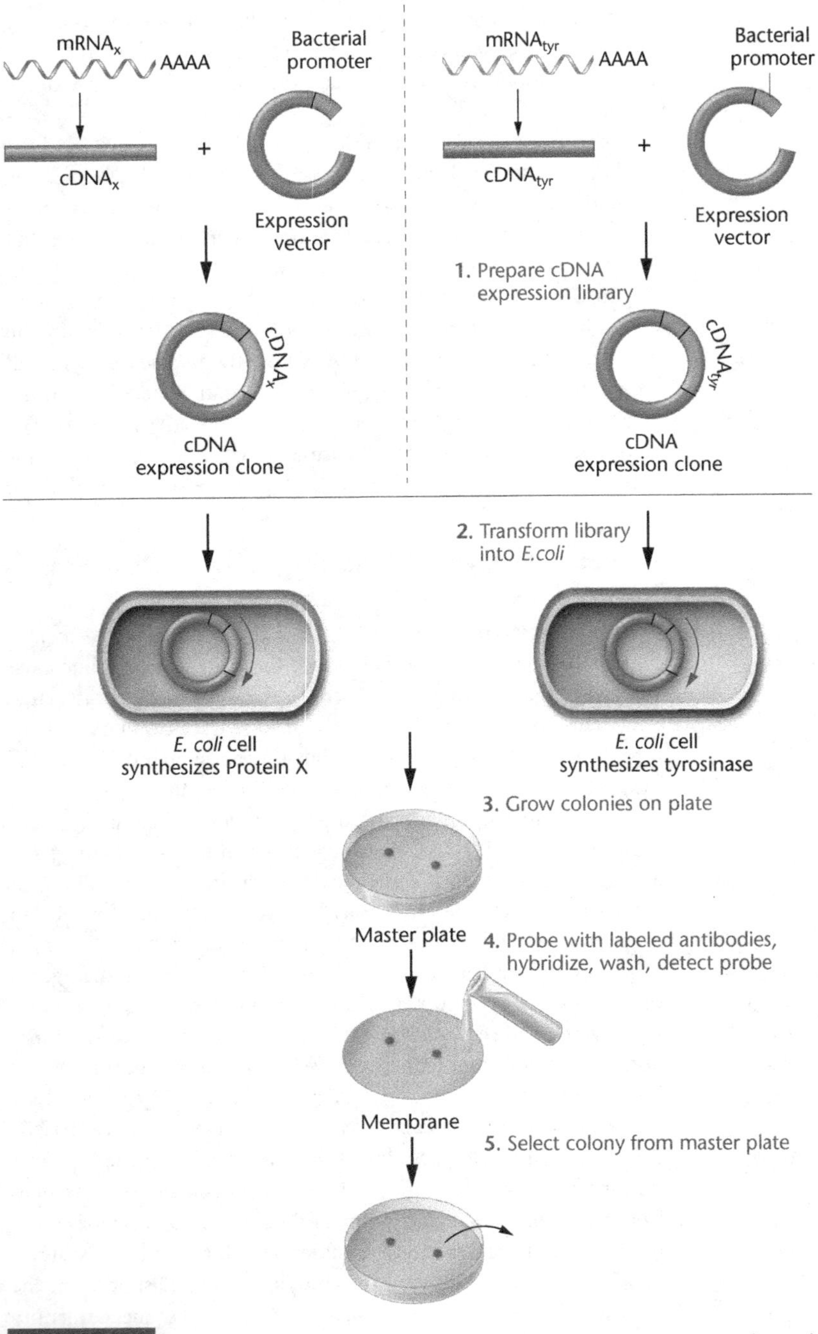


**FIGURE 23-14** Cloning the tyrosinase gene. A cDNA library was prepared by inserting the cDNAs into vectors that contain a bacterial promoter. After transformation into bacterial cells, the cDNA inserts were transcribed from the bacterial promoter, and the mRNAs were translated into proteins. The protein-containing bacteria were grown into colonies on plates. Cells from the colonies were transferred to a membrane, and the membrane was probed with antibodies that recognize the tyrosinase protein. Colonies that bound the antibody were selected from the master plate.

identified and cloned by transforming wild-type human genes into tumor cell lines. Genes that suppress the growth of the tumor cells were selected, sequenced, and further studied to verify their roles as tumor suppressors.

## Genetic Analysis Beginning with the Cloned Gene or DNA Sequence

Once a candidate gene is identified—from a forward mutational analysis, a protein-first approach, a functional complementation assay, or simply by selecting an interesting open reading frame from sequences in a genomic database—it must be subjected to genetic and biochemical analysis to verify its function.

Often the geneticist's first step in characterizing a candidate gene is to compare the gene's DNA sequence to sequences available in DNA databases, such as GenBank. As we also describe in Chapter 21, investigators employ computerized sequence comparison programs to reveal regions of sequence similarity and to set up sequence alignments between the candidate gene and other genes in the database. If a similar gene is present in another organism, anything that is known about its function may provide clues to the function of the candidate gene. Once researchers know the gene's DNA sequence, they translate the DNA sequence into the amino acid sequence of the protein product and compare the amino acid sequence to the sequences of other proteins, also available in public sequence databases, such as SwissProt. Researchers use computer programs such as BLAST to search protein sequences for regions of similarity to known amino acid motifs in other proteins. If the candidate protein contains a motif such as a DNA-binding region, a kinase region, a membrane-spanning motif, or a secretion signal, the presence of the motif may suggest the protein's function. These putative functions can then be tested further by molecular biological or biochemical techniques.

Another approach to discovering the function of a candidate gene is to analyze its patterns of gene expression. Investigators can determine the tissue-specific or temporal-specific patterns of gene expression using a number of techniques. An RNA ***in situ* hybridization** may reveal the presence of the mRNA in one or more tissues in a multicellular organism, or in one or more structures within a single cell (Figure 23–15). To perform an *in situ* hybridization, investigators create a probe by labeling a cDNA clone with fluorescent or colored dyes; then they either hybridize the labeled probe to a thin section of tissue or a fixed preparation of a whole organism. After washing away any unhybridized probe, they observe the section or organism under a light or fluorescent microscope. The presence of stain defines where the mRNA encoded by the gene is expressed and may suggest the gene product's function. For example, if the mRNA is present only in the liver of embryonic mice, the gene may play a

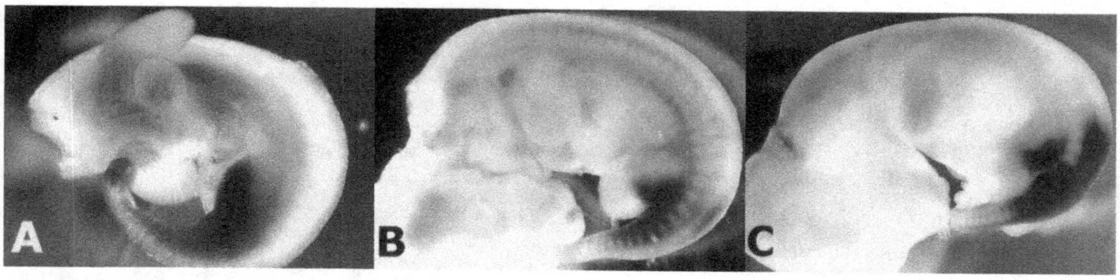


FIGURE 23–15 *In situ* hybridization of whole mouse embryos, showing distribution of the *Hoxc11* mRNA during development. RNA is detectable as dark blue staining, near the posterior end of the embryo. Head of the embryo is at the left; tail at the right. At 10.5 (A), 11.5 (B) and 12.5 (C) days of gestation, *Hoxc11* mRNA is concentrated in hindlimbs, vertebrae, and cells that will later form kidney and reproductive organs. These data suggest that the *Hoxc11* gene product is involved in the early development of these structures.

critical role in early development of the liver. Alternatively, if the mRNA is present in all cells at all stages of development, the gene may encode a ubiquitous housekeeping protein.

Another way to determine the time and place of a gene's expression is to perform a northern blot analysis, which is described in Chapter 13. The technique involves purifying mRNA from the tissue of interest, subjecting the mRNA to electrophoresis, transferring the separated mRNAs onto a filter, and probing the filter with a labeled DNA probe (Figure 23–16). Northern blots can be more labor-intensive than *in situ* hybridization, but are more useful for quantitating the mRNA.

Although the presence of mRNA often reveals the tissue in which a candidate gene is expressed and the temporal pattern of that expression, the expression patterns of the protein may be more revealing, especially if the gene's regulation occurs at posttranscriptional stages. If the expression of the candidate gene occurs at the level of translation or protein stability, the protein and mRNA profiles may not be identical. To verify the time and location of protein expression, investigators assay the protein's expression profiles by **immunofluorescence staining**. Immunofluorescence staining uses an antibody that specifically binds to the protein of interest. For the staining, the antibodies are either directly labeled with a fluorescent tag or are bound to a second antibody that recognizes the first antibody. The second antibody is conjugated to the fluorescent tag. As in *in situ* hybridization, the labeled probe is hybridized to a thin section of tissue, the excess is washed away, and the section is observed using a fluorescence microscope (Figure 23–17).

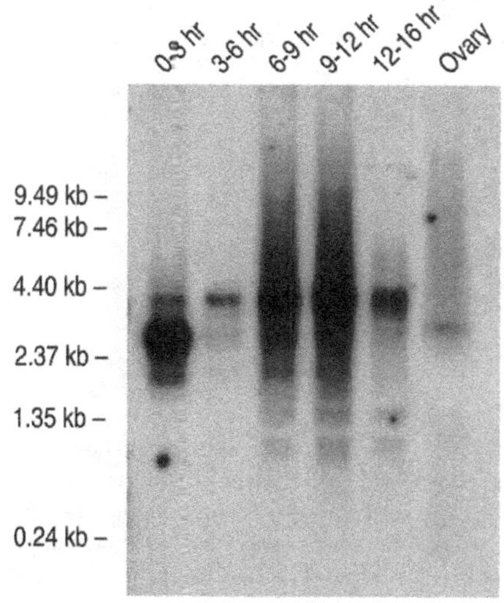


FIGURE 23–16 Northern blot analysis of *dfmr1* gene expression in *Drosophila* ovaries and embryos. A *dfmr1* transcript of approximately 2.8 kb is present in ovaries and 0- to 3-hr-old embryos. The *dfmr1* transcript peaks in abundance between 9 and 12 hr of embryonic development, when it measures 4.0 kb. These data suggest that *dfmr1* gene expression may be regulated at the levels of transcription or transcript processing during embryogenesis. The *dfmr1* gene is a homolog of the human *FMR1* gene. Loss-of-function mutations in *FMR1* result in human fragile X mental retardation.

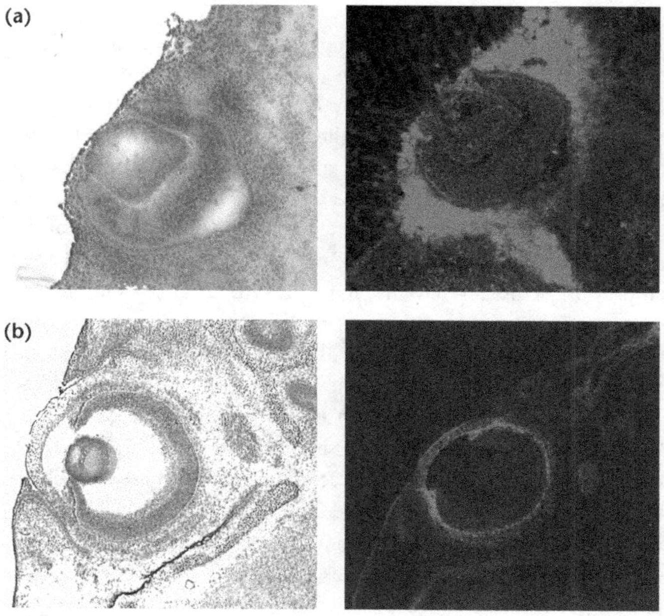


FIGURE 23–17 Immunofluorescence staining of the developing mouse eye, showing the location of the Pitx2 protein. The two right panels show mouse embryonic eyes stained with a DNA-specific stain that denotes the nucleus of each cell. The two left panels are the corresponding immunofluorescence-stained samples, viewed with a fluorescence microscope. Embryonic ages are 11.5 days (a) and 15.5 days (b). The pattern of Pitx2 protein expression is consistent with the Pitx2 gene's proposed role in early eye development.

The search for a candidate gene's function often involves a large number of biochemical and molecular biological assays whose descriptions are beyond the scope of this chapter. For example, if the gene is suspected to encode a protein kinase, investigators may assay the ability of the purified gene product to phosphorylate various substrates *in vitro*. They may examine the interactions of the gene product with other cellular proteins by precipitating specific complexes from living cells and identifying the proteins within the complex. If the gene product is suspected to be a transcription factor or other DNA-binding protein, tools exist that allow researchers to identify the DNA sequences that bind to these factors, and ultimately the genes that are controlled by them. An example of a genome-wide screen involving this type of technology is described in Section 23.4.

## Genetic Analysis Using Gene-Targeting Technologies

The ultimate test of a gene's function comes from the study of *in vivo* phenotypes of mutant organisms. In the reverse genetic approach, this *in vivo* study follows the isolation of the gene. It is made possible by geneticists' ability to manipulate and introduce mutations into genes with great precision. Once researchers make the requisite mutation in the cloned candidate gene, they introduce the mutated gene into a model organism and examine the phenotype. In addition, specific genes can be deleted and moved from one organism to another.

Two powerful techniques that geneticists employ to understand gene function are gene knockouts and gene replacements following site-directed mutagenesis, together known as **gene targeting**.

A targeted **gene knockout** is the deletion or disruption of a specific gene. This can be achieved in most model organisms, including yeast, *Drosophila*, and mice. In yeast, researchers take advantage of the fact that gene replacement is efficient due to high levels of recombination in this organism. One gene-knockout method they employ is demonstrated in Figure 23–18. In this example, researchers insert the *KanMX* gene (which confers resistance to the antibiotic G418) into the middle of the cloned gene of interest, replacing a portion of the gene. Linear pieces of DNA containing the *KanMX* gene flanked by sequences surrounding the gene to be knocked out are then transformed into diploid yeast cells. The DNA fragments undergo homologous recombination with the yeast chromosome, replacing the wild-type gene with the *KanMX* cassette. After the cells undergo sporulation, the spores are grown in the presence of G418 to select for those with the *KanMX* cassette. The resulting G418-resistant haploid cells will be null for the gene of interest.

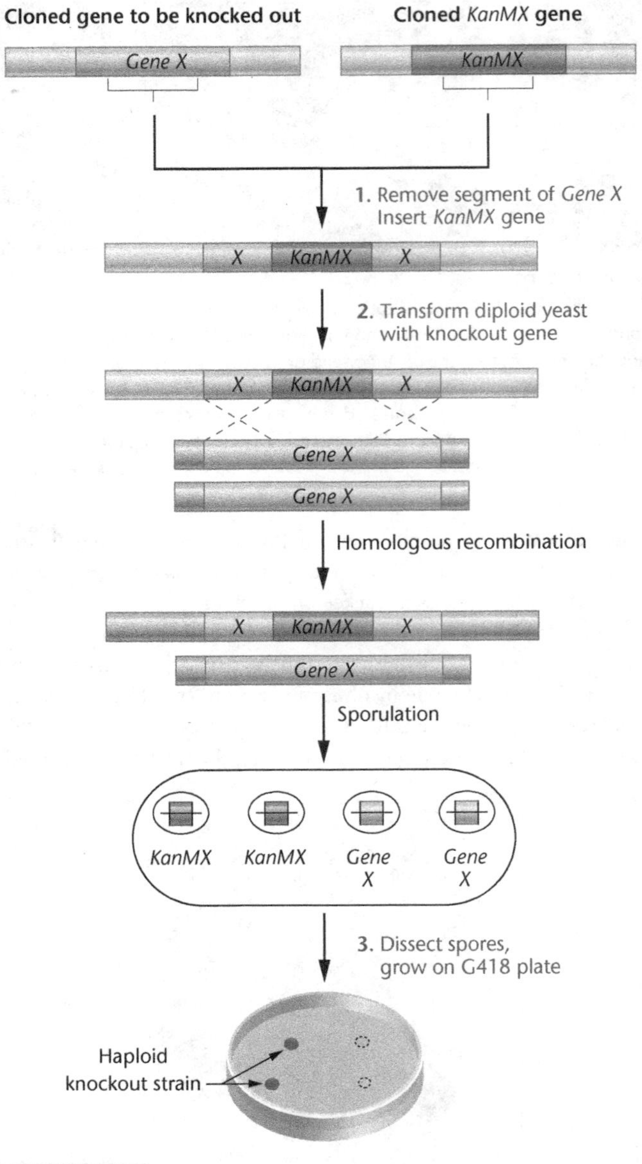


FIGURE 23–18 Targeted gene knockout in yeast. The *KanMX* gene is inserted into a cloned copy of the gene to be disrupted (*Gene X*). The *KanMX*-disrupted *Gene X* is then transformed into diploid yeast cells. Homologous recombination occurs between the cloned knockout gene and one chromosome of the diploid yeast cell, yielding a heterozygous knockout yeast strain. After sporulation, the four haploid spores are dissected from the ascus and grown on a plate containing medium and the antibiotic G418. Any spores that grow contain the *KanMX*-disrupted *Gene X*.

In mice, gene knockouts are more complicated to engineer (Figure 23–19). The first requirement is a culture of mouse **embryonic stem (ES) cells.** These are cells removed from early embryos at the blastocyst stage and grown in tissue culture. Under the right conditions, ES cells will grow and divide as single cells without differentiating. The second requirement is a disrupted version of the gene of interest. Researchers disrupt the cloned gene by splicing in a piece of DNA containing an antibiotic-resistance gene, such as $neo^r$. This fragment then becomes part of a larger recombinant DNA molecule that also contains the herpes simplex virus *tk* gene. The $neo^r$ gene confers resistance to the antibiotic neomycin. The viral *tk* gene encodes an enzyme, thymidine kinase, that selectively phosphorylates nucleoside analogs such as ganciclovir. If a cell

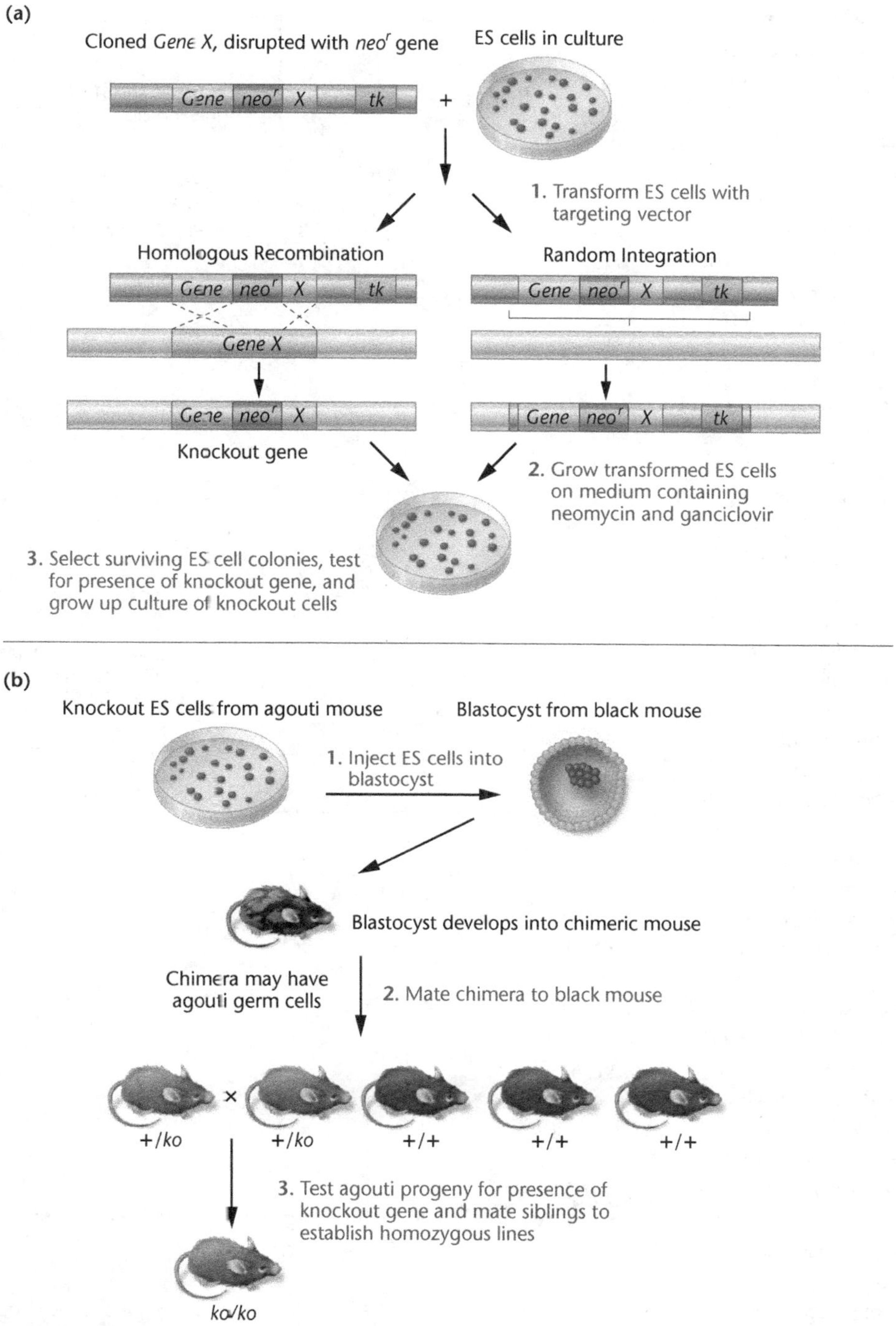


**FIGURE 23-19** Creating a knockout mouse. (a) Inserting the knockout gene into embryonic stem (ES) cells derived from an agouti (brown) mouse. (b) Creating the knockout mouse strain from knockout ES cells.

expressing the *tk* gene is grown in the presence of ganciclovir, the phosphorylated nucleoside analog becomes a toxic inhibitor of cellular DNA replication. The plasmid containing the knocked out gene and a selection gene is known as a **targeting vector**.

When the targeting vector is transformed into the ES cells in culture, two possible recombination events can take place. First, the targeting vector may undergo homologous recombination with the ES cell gene of interest, in the regions of sequence identity that flank the $neo^r$ gene. Second, the targeting vector may undergo random integration into the ES cell genome. To select ES cells that have undergone homologous recombination, geneticists grow the transformed cells in medium containing neomycin and ganciclovir. The only cells that survive the presence of neomycin have incorporated the targeting vector that contains the $neo^r$ gene. The only neomycin-resistant cells that survive in the presence of ganciclovir have undergone homologous recombination with the targeting vector, and hence have not incorporated the *tk* gene. After this drug selection, the surviving ES cells are tested by Southern blot or PCR analysis to verify that the gene of interest in the ES cells has been disrupted with the targeting vector DNA.

Once the ES cells with the targeted disruption are obtained, they are used to create the knockout mouse. Researchers take blastocysts from a mouse that has a coat color different from the mouse that donated the ES cells. For example, the ES cells may have been derived from a mouse with a dominant coat color, such as brown (agouti), whereas the blastocyst may have been donated by a recessive, black mouse. The ES cells are injected into the blastocyst, and these cells become part of the growing embryo, sometimes including the germ line. The resulting mouse will therefore be a **chimera**, an organism containing genetically distinct cells that were derived from different sources. These chimeric mice are easy to identify, as they have patches of brown and black fur. Next, the chimeric mice are mated to black mice. If the knockout gene is present in the germ line of a chimeric mouse, half of its progeny will be heterozygous for agouti coat color as well as the knockout gene. The DNA of the chimeric mouse's progeny is then tested, and progeny containing the knockout gene are mated to establish homozygous knockout mice.

This procedure is used to establish knockout mice as models for human diseases, as well as for testing the function of cloned genes. For example, this approach was used to verify that mutations in the *CFTR* gene are responsible for the disease cystic fibrosis.

**Targeted gene replacement** is an equally valuable tool with which to dissect the function of a cloned gene. Gene replacement usually involves substituting a gene containing specific mutations for the wild-type version of the gene of interest. Geneticists introduce specific mutations into a cloned gene by means of site-directed mutagenesis.

**Site-directed mutagenesis,** first developed by Michael Smith at the University of British Columbia in the 1980s, allows researchers to introduce a specific mutation at a precise site within a cloned gene. By means of this technique, one or more nucleotides in a DNA clone may be deleted, inserted, or changed. Dr. Smith received the Nobel Prize in Chemistry in 1993 for this significant contribution to modern genetic analysis. Several variations of his technique now exist, including those based on the use of PCR. Here, we will discuss the classic method based on his work.

The first step in the classic method of site-directed mutagenesis is to clone the gene of interest, determine its DNA sequence, and examine the DNA sequence to decide on the site and exact nature of the mutation (Figure 23–20). The gene, or a fragment from the gene, is then cloned into the M13 bacteriophage. M13 exists in two forms: the double-stranded DNA form known as the replicative form and a single-stranded DNA form that is extruded from the bacterial cell. Once the gene of interest is cloned into M13, the single-stranded form of the M13 genome is harvested and used *in vitro* as the template on which to create the site-directed mutation.

The second step in site-directed mutagenesis is to design and synthesize an oligonucleotide that is complementary to a region of the gene of interest but contains the desired mutation. The oligonucleotide may be as short as 18 nucleotides. The mutation may be a change in one or more nucleotides, or it may be a small deletion or insertion. The synthetic oligonucleotide is then hybridized to the single-stranded M13 DNA molecule so that it anneals to the regions of complementarity between it (the oligonucleotide) and the cloned gene in M13. The only part of the oligonucleotide that does not anneal is the mutation itself. A purified DNA polymerase is added to the reaction, along with a mixture of nucleotides (dATP, dCTP, dTTP, and dGTP). The DNA polymerase extends the 3′ end of the oligonucleotide to create a double-stranded M13 DNA molecule. DNA ligase is added to seal the newly synthesized DNA strand. The resulting double-stranded M13 molecule is then transformed into *E. coli* bacteria, which proceed to replicate the double-stranded M13 DNA. Because DNA replication is semiconservative, half of the replicated DNA molecules will bear the mutation and half will bear the wild-type version of the cloned gene. The mutated versions of the M13 molecules are then selected using a number of molecular screening procedures.

After the mutated version of the gene is synthesized, it may be removed from the M13 vector and cloned into any number of different vectors and introduced into cultured cells or model organisms to test the mutant gene function. Geneticists can introduce the newly mutated gene into an organism using transgenic technologies. If the organism is null for the gene of interest, the transgene will be the only functional copy of the gene, and its phenotype can be assessed. This same strategy is currently being used to investigate mutations in the human *CFTR* gene, which is responsible for the disease cystic fibrosis. Mutated *CFTR* genes have been introduced into *CFTR* knockout mouse models to determine how each mutation controls the range of symptoms that occur in cystic fibrosis.

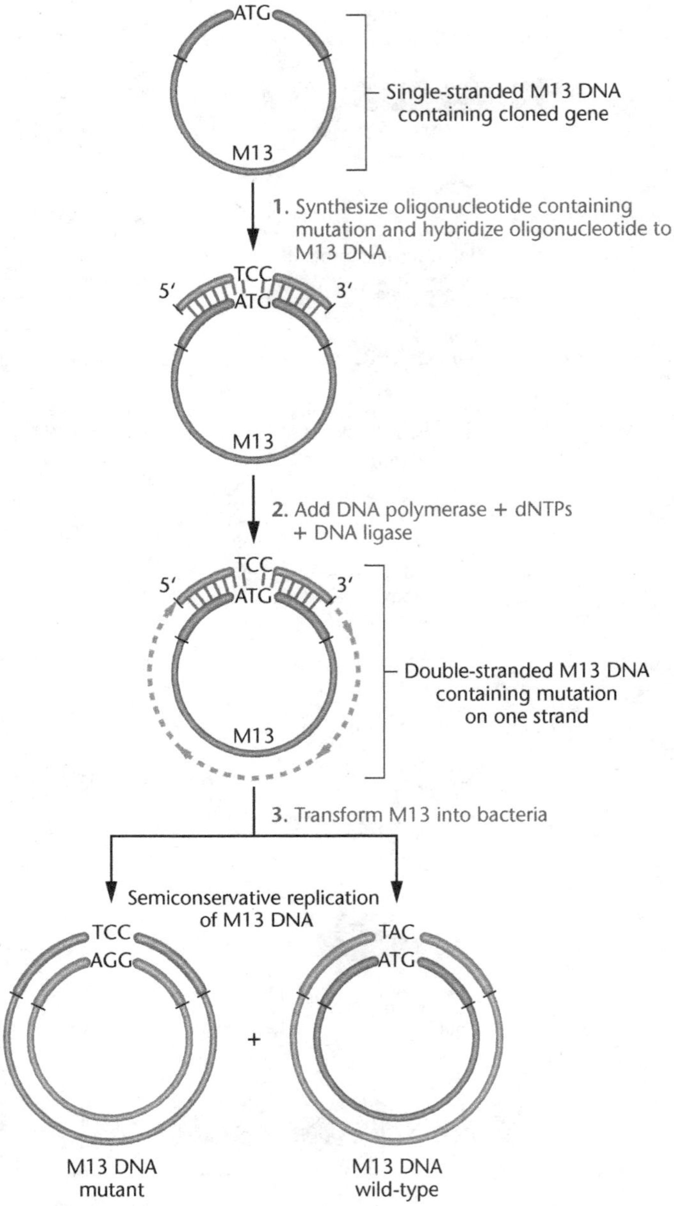


FIGURE 23–20 Oligonucleotide-mediated site-directed mutagenesis.

Sometimes it is desirable to directly replace an endogenous gene in the model organism with a site-directed mutagenized copy. This is done in yeast and mice following procedures similar to the ones used to create gene knockouts in those organisms. Instead of replacing the wild-type gene with a null gene containing an antibiotic resistance gene, the wild-type gene is replaced by the *in vitro* mutagenized copy.

23.4

# Geneticists Dissect Gene Function Using RNAi, Functional Genomic, and Systems Biology Technologies

Both forward and reverse genetic techniques involve the creation, recovery, and analysis of gene mutations. However, in recent years, geneticists have devised alternative approaches to genetic dissection that do not necessarily require the use of gene mutants. These alternative methods make use of nongenetic inactivation of gene functions and various high-throughput technologies that allow simultaneous examination of expression from a large number of genes. In this section, we discuss several of these new technologies, including RNA interference, high-throughput functional genomics methods, and the new science of systems biology.

## RNAi: Genetics without Mutations

Both forward and reverse genetic analyses center on the creation and examination of gene mutants. These approaches, though highly effective, are labor-intensive. Also, they cannot be used in organisms in which gene targeting is not possible. Recently, researchers have devised a new and potentially powerful dissection tool, **RNA interference (RNAi)**. This new technology allows investigators to specifically create single-gene defects without resorting to the creation of heritable mutations. These RNAi-mediated **gene-silencing** technologies are relatively inexpensive and allow rapid analysis of gene function. Researchers hope to use RNAi technology to systematically knock out the function of each gene in model organisms. The technique also may be useful for gene therapy, in which specific disease genes can be silenced.

In nature, cells employ RNAi to defend themselves from viruses or invading transposons. As we learned in Chapter 18, the double-stranded RNA molecules produced by these agents are recognized by enzymes within the cell that degrade double-stranded RNAs. When the RNA is degraded, no protein is translated, and gene expression is blocked. Another way in which RNAi works is to inhibit transcription of viral or transposon genes. In this case, the antisense strands of double-stranded RNA molecules bind to the complementary regions of a gene, leading to recruitment of proteins that modify DNA and inhibit transcription. Cells also use RNAi to regulate expression of their endogenous genes. Short fragments of antisense RNA complementary to a sense mRNA can bind to the mRNA and inhibit translation.

To use RNAi as a research tool, investigators introduce short double-stranded RNA molecules into cells (Figure 23–21). These short RNA molecules, known as **short interfering RNAs (siRNAs)** can be synthesized in the laboratory, or can be transcribed from plasmids introduced into the cell. These molecules trigger the same RNA-degradation pathway that is triggered by viral or transposon double-stranded RNAs. The antisense strand from the double-stranded RNA hybridizes to the complementary mRNA in the cells, targeting the mRNA for degradation.

Alternatively, investigators can transform cells with vectors that express RNA hairpin structures that are partially double-stranded and have an RNAi effect. If the vectors that express the double-stranded RNA molecules become integrated into a cell's genome, all the progeny of that cell will express the RNAi molecules, and one specific gene will be silenced in those cells. Silencing may result from RNA degradation or inhibition of transcription or translation. Another variation on the RNAi technique is to control the expression of the RNAi gene by means of an inducible promoter. For example, if the *Drosophila* heat-shock gene promoter is cloned next to the gene encoding the RNAi, and the RNAi vector is then transformed into *Drosophila* by *P* element-mediated transformation, the RNAi gene in the resulting flies can be expressed by simply raising the temperature. This approach is particularly useful for silencing genes that are essential for viability during development.

RNAi methods have been used successfully to investigate gene functions in organisms that are usually difficult to manipulate genetically. These include mosquitoes, trypanosomes, and mammalian cells. RNAi has also been used to silence the expression of genes that are aberrantly expressed in cancer cells. When these cancer-related genes are silenced, the cancer cells lose their ability to form tumors when introduced into mice. Although RNAi technologies, sometimes referred to as "genetics without mutations," are just beginning to be developed, they hold great promise for both investigative and therapeutic uses.

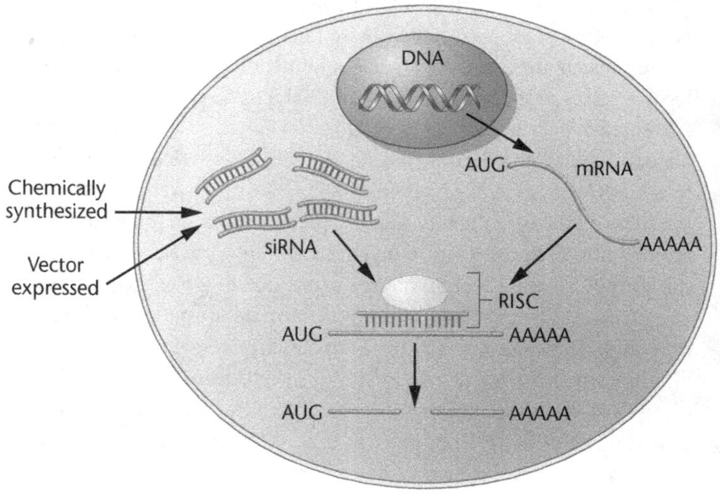


**FIGURE 23–21** RNA interference (RNAi). Short double-stranded RNA molecules called short interfering RNAs (siRNAs) are introduced into a cell. In the cell, they become part of the RNA-induced silencing complex (RISC). The antisense strand of the siRNA binds to the complementary mRNA within the cell, targeting the cellular mRNA for cleavage by enzymes within the RISC. The cleaved mRNA is subsequently degraded and is not translated.

## High-Throughput and Functional Genomics Techniques

The major strength of classical genetic analysis is that investigators can examine the effects of altering one gene at a time against a background of normal function in the rest of the organism's genome. However, since the advent of modern genomics, various **high-throughput technologies** have been developed that allow investigators to probe the genetic interactions of thousands of genes simultaneously. These technologies include the use of **DNA-** and **protein-expression microarrays**, automated methods for the isolation and dissection of large protein complexes and genome-wide searches for protein-DNA interaction sites. These techniques are based on the wealth of data provided by completion of genome sequencing projects, and are important tools in functional genomics. As described in Chapter 21, functional genomics not only examines RNA and protein expression profiles but also assigns putative functions to genomic sequences by comparing DNA sequences between different organisms.

These methods provide a more global picture of gene expression than is revealed by conventional genetic analysis, and may help dissect aspects of genetic pathways and gene interactions that a mutational approach might miss. We will now briefly describe two high-throughput methods used in functional genomic analysis.

**Gene Expression Microarrays** DNA microarrays (DNA chips) are used to examine the expression of thousands of genes simultaneously, by estimating the quantity of mRNA transcribed from each gene in the genome. This technology is sometimes referred to as transcriptomics. As described in Chapter 21, DNA microarrays are simply pieces of glass (chips) onto which DNA samples are applied in an ordered pattern—that is, an array. Often these microarrays are manufactured by automated machines that place microscopic droplets of specific DNA samples in specific positions on the chip. The DNA samples that are applied may be any type of cloned DNA but are often short oligonucleotides that were synthesized *in vitro*. More than 20,000 different DNA samples may be applied to a DNA microarray, representing thousands of different genes. Theoretically, oligonucleotides representing all the genes in an organism's genome can be present on a microarray, allowing analysis of gene expression of the entire genome.

DNA microarrays may be used to compare the patterns of gene expression in two or more different tissues, in tissues at two or more different times during development, or in cells from normal or diseased tissues. The use of DNA microarrays in transcriptomics is discussed in Chapter 21. They are also used in other ways, such as the ChIP-on-chip method that we describe next.

**Genome-Wide Mapping of Protein–DNA Binding Sites: ChIP-on-Chip** Another functional genomic technique is designed to map protein–DNA interactions and is useful for identifying genes that are regulated by DNA-binding transcription factors. The technique is known as **chromatin immunoprecipitation assayed by DNA microarray**, or **ChIP-on-chip**. To perform this technique, researchers

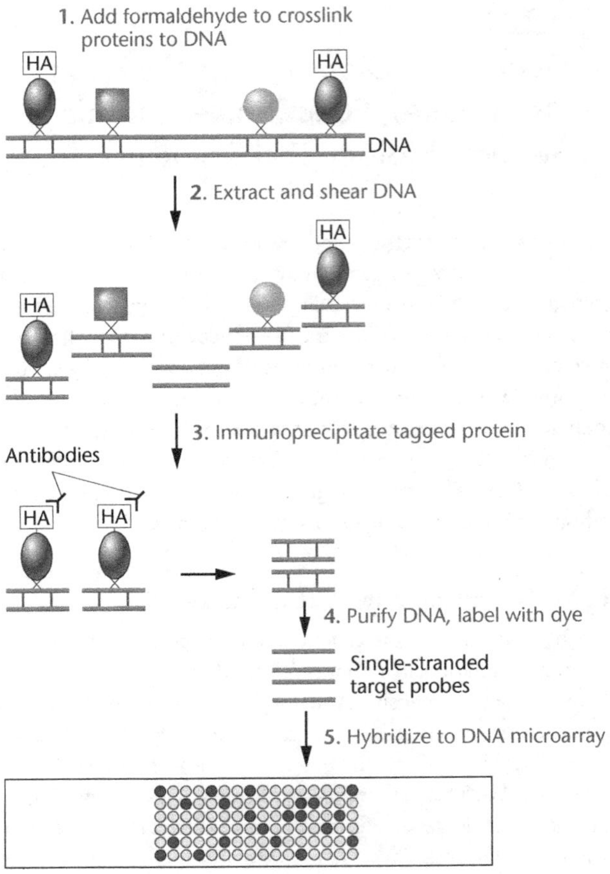


FIGURE 23-22 Genome-wide screen for transcription factor binding sites. The top diagram shows various proteins bound to their DNA-binding sites. One protein is tagged with a protein fragment (HA) that can be recognized by a specific antibody. Formaldehyde is added to tissues or cultured cells to crosslink all DNA-binding proteins to the DNA. Then the DNA is extracted from cells and sheared into small fragments. The antibody that recognizes the HA tag is added to the mixture and attaches to the tagged fragments. Then the antibody, together with its protein-DNA fragment, is pulled out of the mixture (immunoprecipitated). The immunoprecipitated DNA fragments are purified, labeled with a fluorescent dye, and hybridized to a DNA microarray. A positive hybridization signal indicates a DNA sequence that is bound by the HA-tagged protein. *A. Kumar and M. Snyder. 2001. Emerging technologies in yeast genomics.* Nature Reviews Genetics 2(4): *302–312, Fig. 3. Copyright 2001 Macmillan Publishers Ltd.*

treat tissues or single cells such as yeast with formaldehyde (Figure 23–22). Any proteins that are tightly bound to DNA will be crosslinked to the DNA by the formaldehyde. The researchers then extract DNA from the cells and shear it into small fragments. To isolate those DNA fragments that are bound to a specific protein, they add an antibody that recognizes (attaches to) the protein. The antibody is then precipitated from the mixture, along with the protein to which it binds and any DNA fragments that are crosslinked to the protein. The investigators purify the immunoprecipitated DNA fragments, amplify these DNA fragments by PCR, and label the amplified fragments with a fluorescent tag. The labeled DNA fragments are hybridized to a DNA microarray containing synthetic oligonucleotides or cloned DNAs representing the organism's entire genome. Any spot on the microarray that hybridizes to the labeled DNA represents a DNA sequence that bound the protein. Because each spot on the DNA microarray is known, the identity of each positive signal can be determined. In one application of this genome-wide method, more than 200 previously unknown targets of a transcription activator protein that functions at the G1/S cell cycle interface were identified.

## Systems Biology and Gene Networks

Within the last decade, a new science has emerged that makes use of the vast amount of genomics, proteomics, and metabolic information generated from high-throughput, genome-wide studies. This new science, known as **systems biology**, has a different focus from that of classical, reverse, or genomics research. With these other research methods, geneticists concentrate first on identifying individual genes that affect phenotypes and then on discovering the mechanisms by which these genes bring about their effects. Such approaches are often described as reductionist because they study complex phenomena by breaking them down into simpler parts. Systems biology, in contrast, attempts to gather all the information about genes, proteins, and metabolic functions that influence a particular system, and to put these parts together in such a way that the entire system can be examined as a whole. Systems biology views a biological system (such as a mitochondrion, a cell, an organ, or a mouse) as an integrated network of interacting genes, gene-regulatory effects, proteins, post-translational controls, and biochemical processes. It then attempts to analyze these networks in such a way that predictions can be made about how environmental effects or perturbations of internal network components change the functioning of the entire system. We also discuss systems biology in Chapter 21.

Although systems biology is just beginning to develop as a science, it promises to provide major insights into how genes affect phenotype. It also has practical applications, such as providing models to predict the effects of drugs or medical therapies on the workings of cells, tissues, and organs. As the amount of data from high-throughput methods increases, it should be possible for systems biology to create more accurate and complex models of biological systems.

**NOW SOLVE THIS**

Problem 15 on page 631 asks you to think critically about the limitations of one of the new high-throughput functional genomics methods—DNA microarrays (DNA chips)—and how to overcome them.

■ HINT: *When thinking about limitations, you might consider potential technical problems due to the nature of the material on the chips. You might also think about interpreting the data that come from these types of experiments. In addition, think about the number of steps involved and what could go wrong at each step.*

## GENETICS, TECHNOLOGY, AND SOCIETY

### Whose DNA Is It, Anyway?

Few things are more personal than our genetic material. Our genes establish the framework of our physical, mental and even emotional makeups. We pass on our genes to our offspring, thereby ensuring the continuation of our genetic traits into the future. There is something deep within our psyches that holds our genetic material at the center of our beings.

It may be jarring, then, to realize that DNA sequences derived from about 20 percent of our genes are owned by others and are a potential source of commercial profit.

Over the last 30 years, the U.S. Patent and Trademark Office has issued patents for more than 4300 human gene sequences. These patents have been granted to universities, governments, and private corporations. DNA-related patents cover cloned genes, their regulatory sequences, small sequences such as ESTs (expressed sequence tags) and SNPs (single-nucleotide polymorphisms), and various methods and diagnostic tools based on these sequences. Some human genes have multiple patents covering gene variants and fragments, as well as numerous potential applications, including diagnostic tests, or future uses of the sequences as drug-screening probes. Even more remarkably, over three million patent applications have been filed for genes, gene fragments, and other genome-related materials. How many of these will end up as patents, and what the patents might cover, are still unknown.

This ever-increasing "land-grab" over the human genome has raised serious concerns about the nature of patents, the meaning of genetic information, and the balance between commercial gain and public interest.

Proponents of gene patenting point out that patent protection is essential for the development of DNA-based diagnostics and pharmaceuticals. Without exclusive rights, no one would be willing to invest the years of research and millions of dollars required to develop and test new medicines. They also argue that patenting reduces secrecy, as patent holders are required to publicly disclose the

*Continued on next page*

*Genetics, Technology, and Society, continued*

details of their patent, thereby speeding research in related areas.

In contrast, many people are uncomfortable with the idea that human genes and gene sequences can be owned and used for commercial gain. They argue that the human genome is naturally occurring and unique, and cannot be treated like other inventions, such as mousetraps or works of art. Some worry that patent protection will stifle research and health care, as more and more genes, diagnostic tests, and treatments based on DNA sequences are held by commercial interests and the fees for using these patented items are controlled by a few.

Supporters of gene patenting counter by explaining that patents over DNA sequences are not the same things as ownership over the genes from which the sequences are derived. Patents cover only isolated or cloned versions of sequences and not genes as they occur naturally in anyone's body—even though the sequences may match. In addition, patents give exclusive rights to someone to exploit the patented item for only a limited period of time—usually 20 years. During this time, patent holders can license the use of their patent to anyone, often for a fee, meaning that the exclusiveness of the patent is not written in stone.

Is there any evidence that DNA patents have restricted the progress of basic research or the development of new diagnostics and treatments?

Some recent data support the idea that patents have had neutral effects on basic research. A study conducted by the National Academy of Sciences in 2005 revealed that only about 9 percent of university researchers conducted work on patented gene sequences. The remainder either did not know about any relevant patents or did not need to use patented reagents.

In contrast to basic research, the potential conflicts between DNA patents and medical diagnostics may be more problematic. Critics of gene patenting often cite the case of the breast cancer gene *BRCA1* and the patents held by Myriad Genetics Laboratories in Salt Lake City, Utah. In the 1990s, Myriad developed genetic tests for breast cancer, based on specific mutations in the *BRCA1* gene. Their initial patent claimed rights over the normal *BRCA1* sequence, various mutations, *BRCA1* diagnostic tests, the methods used to screen tumor samples, any information derived from the *BRCA1* gene, and all methods to diagnose and treat hereditary breast and ovarian cancer. Myriad has enforced many aspects of this patent, preventing other research centers from offering their own versions of the tests, at lower costs. For example, in Canada and Europe, Myriad required that all samples be sent to their laboratories in Utah for testing. Publicly funded laboratories in Canada had been using their own *BRCA1* tests at lower cost, but stopped the practice after threats of legal action. Although one Canadian province continues to use its own *BRCA1* tests despite possible future law suits, other provinces now send their samples to Myriad Genetics for testing. Some scientists worry that Myriad will build up the world's largest bank of *BRCA1* samples and monopolize the information derived from these samples. In 1994, European countries challenged Myriad's patent claims and have won the right to bypass the *BRCA1* patents. Recently, the American College of Medical Genetics expressed concern that wide-ranging patent claims such as those held by Myriad Genetics may increase the costs of genetic testing, slow the development of quality control, and restrict the development of related gene tests.

Many questions raised by the race to patent the human genome remain unanswered: Should naturally derived materials such as DNA be patentable? Will patents over DNA sequences slow the development of competitive and cost-effective diagnostics and treatments? In the future, will the entire human genome be patented, increasing the costs of basic research? As more DNA sequences become patented, will we be able to reconcile the needs of private industry with those of the public good?

**References**

Stix, G. 2006. Owning the stuff of life. *Scientific American* 294: 76–83.

**Web Site**

Genetics and patenting. Human Genome Program, Ethical, Social and Legal Issues, U.S. Department of Energy, 2006. http://www.ornl.gov/sci/techresources/Human_Genome/elsi/patents.shtml

# EXPLORING GENOMICS

## The Knockout Mouse Project

As we learned in this chapter, the laboratory mouse is one of the most valuable model organisms for research into gene functions and genetic diseases that affect humans. Geneticists can now mutate and remove specific mouse genes for a wide range of studies. In 2005, the U.S. National Institutes of Health established the **Knockout Mouse Project (KOMP)**, which aims to collect and create a library of mouse strains containing a null mutation for every gene of the mouse genome. These mouse strains, along with information on each strain, will be made available to all researchers. KOMP also allows researchers to track the progress of mouse knockout production. In this exercise, we will explore KOMP and use its resources to investigate the functions of an important developmental gene in the mouse.

■ **Exercise I – Exploring the Knockout Mouse Project (KOMP)**

1. Access the KOMP home page at http://www.nih.gov/science/models/mouse/knockout/.
2. Using the information from the first two links on this page, and their sublinks, answer the following questions:
   a. What are knockout mice, and why are they useful?
   b. There are two strategies for creating knockout mice in the KOMP project. Briefly describe these two methods.
3. Visit the KOMP Data Coordination Center at http://www.knockoutmouse.org. Describe the activities and goals of the KOMP Data Coordination Center.
4. At the bottom of the page, select the "KOMP Gene List" link. How many mouse genes are now listed in the KOMP Gene List?

■ **Exercise II – Defining the function of the Mouse *Hoxa1* gene**

In mammals, the *Hox* genes are four families of genes (*Hox A* to *Hox D*) that are involved in establishing body patterns during early development. In this exercise, we will investigate whether *Hoxa1* knockout mice have been created and what is known about the functions of the HOXa1 protein.

1. Begin by accessing the KOMP Web site at http://www.knockoutmouse.org.
2. In the "Search" box, type "Hoxa1" and then click on the "Search" button.
3. The search results show that there are several sources of information about mouse *Hoxa1*. For example, several columns in the right half of the table indicate whether the gene is slated to be knocked out by one of the mouse knockout consortium members, or whether it has already been knocked out in these projects. What is the knockout status of *Hoxa1* as indicated on the search result table?
4. To determine whether other knockouts exist for *Hoxa1*, access the **Mouse Genome Informatics (MGI)** Web site, by selecting the MGI Identifier Number in the "MGI ID" column, or visit http://www.informatics.jax.org/. To find sources of knockouts, scroll down to the "Phenotypes" section and select the "All phenotypic alleles" link. A table appears showing sources of mutant mice and their characteristics.
   a. How many targeted knockouts exist for *Hoxa1*?
   b. The "Allele Symbol" column shows the name of the mouse strain, the type of mutation, and the name of the principal investigator in whose laboratory the mouse strain was created. How many laboratories have generated knockout mice for the *Hoxa1* gene?
5. Return to the MGI Web site "*Hoxa1* Gene Detail" page. Using links and information from this page, answer the following questions:
   a. What are the predominant phenotypes of *Hoxa1* knockout mice?
   b. What are the patterns of gene expression in the developing embryo (both tissue and times of expression of the *Hoxa1* mRNA)?
   c. What do you conclude about the function of the mouse *Hoxa1* gene?
   d. What is the human homolog of the mouse *Hoxa1* gene?
   e. Are any human disorders due to mutations in the human homolog of the *Hoxa1* gene?

# Chapter Summary

1. Geneticists dissect gene function in model organisms that display the biological process to be studied. These model organisms are chosen because they produce abundant progeny in a short period of time, are easily mutagenized, and can be subject to controlled crosses.
2. It is possible to use model organisms such as yeast or *Drosophila* for genetic studies because many of the genes and molecular processes governing biological functions are shared by diverse forms of life.
3. Forward genetic analysis begins with the isolation of mutants that show differences in phenotype for the process of interest. Mutant isolation is followed by defining gene pathways, cloning of the gene, and creating more mutants in order to better understand the biological pathway. In this way, mutants define the normal function of the gene.
4. In forward genetic screens, the goal is to disrupt one gene at a time and to disrupt each of the genes involved in a specific genetic pathway. In order to create enough mutations to disrupt every gene in a pathway, geneticists must subject the model organism to mutagenesis, usually by treating the organisms with chemicals or radiation.
5. To complete a forward genetic analysis, geneticists may mutagenize the first mutant and screen for second mutations that suppress or enhance the original mutant phenotype.
6. Reverse genetic analysis begins with a cloned wild-type gene, purified protein, or DNA sequence and progresses to site-directed mutagenesis and phenotypic analysis—the opposite order from that used in forward genetics.
7. Once a gene is cloned, its function can be studied using a wide array of genetic and biochemical assays. These include computerized homology searches, gene-expression assays, gene-targeting techniques, and site-directed mutagenesis.
8. RNAi techniques allow geneticists to create gene knockouts at the expression level, in the absence of any genetic mutation to the gene to be disrupted.
9. DNA microarrays, high-throughput functional genomics techniques, and systems biology methods provide insight into the interactions of thousands of gene products simultaneously.

## INSIGHTS AND SOLUTIONS

1. You have isolated eight strains of *Drosophila*, each bearing a recessive lethal mutation somewhere on chromosome 3. You have designated these strains as $el^1$, $el^2$, $el^3$, and so on, because the mutations cause missing thoracic segments accompanied by embryonic lethality (*el*). Each strain is maintained by a balancer chromosome. You want to determine whether these mutations are allelic or occur in separate genes on chromosome 3. To do this, you perform complementation tests by crossing all combinations of stocks and looking for 25 percent lethality with segmentation defects. You obtain the following data. (The parental genotypes for each cross are indicated in the top row and left column. Viable progeny are indicated by a "+," and 25 percent embryonic lethality with segmentation defects is indicated by a "–.")

| | $el^1$ | $el^2$ | $el^3$ | $el^4$ | $el^5$ | $el^6$ | $el^7$ | $el^8$ |
|---|---|---|---|---|---|---|---|---|
| $el^1$ | – | – | – | + | + | – | – | + |
| $el^2$ | | – | – | + | + | – | – | + |
| $el^3$ | | | – | + | + | – | – | + |
| $el^4$ | | | | – | – | + | + | + |
| $el^5$ | | | | | – | + | + | + |
| $el^6$ | | | | | | – | – | + |
| $el^7$ | | | | | | | – | + |
| $el^8$ | | | | | | | | – |

How many genes are represented by these mutations? Which complementation group does each mutant fall into?

**Solution:** One way to analyze complementation data is to systematically examine each cross. Obviously, crossing each mutant with itself leads to the lethal segmentation phenotype. Moreover, crossing $el^1$ to $el^2$, $el^3$, $el^6$, or $el^7$ also leads to the lethal segmentation phenotype. Therefore, all of these mutants ($el^1$, $el^2$, $el^3$, $el^6$, and $el^7$) belong to one complementation group. That leaves $el^4$, $el^5$, and $el^8$ to be analyzed. When the $el^4$ and $el^5$ mutants are crossed, they produce the mutant embryonic lethal with segmentation defects phenotype. Therefore, these two mutants ($el^4$ and $el^5$) belong to one complementation group, and this complementation group is distinct from the first one. That leaves $el^8$, which complements all the other mutants and is therefore in a complementation group of its own. Hence, these data represent three complementation groups.

2. In forward genetic dissection, the goal is to conduct a saturation genetic screen. In such a screen, each gene that might affect a phenotype should be mutated at least once in a population of mutagenized organisms. However, some genes may not appear to have been mutagenized, even in a saturation screen. Can you suggest two reasons why a gene might never be revealed in a genetic screen?

**Solution:** The principle behind mutagenesis screens is that mutations in relevant genes will affect a detectable phenotype. If the phenotype cannot be detected, a mutation may be missed. For example, several genes may independently contribute to the same phenotype. This is known as gene redundancy. Mutating one of these genes will not affect the phenotype; therefore, the gene will not be revealed by a forward genetic screen. Another reason that a mutation may be missed is that most or all mutations in the gene lead to dominant lethality. In haploid organisms such as yeast, the mutagenized organisms would die and thus fail to be screened for a phenotype. Similarly, in diploid organisms such as *Drosophila*, any $F_1$ progeny of the mutagenized flies that inherit the dominant lethal mutated gene would die and would not be detected, unless the lethality was accompanied by a phenotype that was observable prior to or in spite of the fly's death.

3. You have isolated a temperature-sensitive mutant strain of *S. cerevisiae* that is null for the *CDC53* gene. This gene functions during the yeast cell cycle, allowing cells to progress from mitosis into the G1 phase of the cycle. At the restrictive temperature, $cdc53^{ts}$ cells arrest in mid-mitosis, with a characteristic phenotype.

(a) What would be the phenotype of haploid $cdc53^{ts}$ cells after 36 hours of growth at the restrictive temperature?

(b) How would you use this mutant strain of yeast to identify genes whose products interact with the CDC53 protein, allowing CDC53 to function properly at the M/G1 transition?

(c) How would you use this mutant strain to clone the *CDC53* gene from human cells?

**Solution:**

(a) To answer this question, you need to know the phenotypes of *S. cerevisiae* cells during the cell cycle (see Figure 23–7). Cells that cannot progress from mitosis into G1 remain joined, although the replicated DNA may migrate into the daughter cell. Therefore, at the restrictive temperature, all the cells in the $cdc53^{ts}$ culture will have this phenotype.

(b) One way to do this is to undertake a suppressor screen. In a suppressor screen, the $cdc53^{ts}$ strain would be subjected to a second round of mutagenesis, and progeny would be screened for a normal cell cycle at the restrictive temperature. The suppressor mutants would then be examined to determine whether the second mutation occurred in the same gene (a reversion) or at other sites (a second-site suppressor). Second-site-suppressor mutations may occur in genes whose products interact with the CDC53 protein. A second approach would be to use an antibody that recognizes CDC53 to immunoprecipitate CDC53 from extracts of wild-type yeast cells. Under the right conditions, any interacting proteins will immunoprecipitate along with the CDC53 protein. These proteins could be purified, their amino acid sequences determined, and the sequences used in a reverse genetic screen to clone their genes.

(c) It may be possible to use the $cdc53^{ts}$ strain in a functional complementation assay. In some cases, a human gene will substitute for the loss of function in a yeast mutant, returning the yeast mutant to wild type. For functional complementation, yeast cells are systematically transformed with clones from a human cDNA library. The transformed $cdc53^{ts}$ cells are then screened for normal cell division at the restrictive temperature. Any cDNA clones that revert the $cdc53^{ts}$ strain to wild type may contain a functionally complementary human clone. The transformed, reverted yeast strain can be grown, the cDNA clone recovered from the yeast strain, and the identity of the cDNA insert determined by DNA sequencing.

## Problems and Discussion Questions

1. What is a suppressor mutation, and how do geneticists use these mutations to characterize gene function?
2. Which model organism—yeast, *Drosophila*, mouse, or human—would you choose to study the following genetic conditions and processes? Explain why.
   (a) Control of kidney development
   (b) Cancer
   (c) Cystic fibrosis
   (d) Purine metabolism
   (e) Immune function
   (f) Cell division
3. You are trying to create a knockout mouse bearing a homozygous null mutation in the proto-oncogene *c-myc*. You begin by transforming ES cells from a normal mouse strain with your cloned DNA, which bears a neomycin-resistance gene inserted into the middle of the second exon of the murine *c-myc* gene. You now have 20 colonies of ES cells that grow in the presence of medium containing neomycin. Describe two methods you could use to verify that these ES cells are *c-myc* knockouts.
4. Describe the method of oligonucleotide-mediated site-directed mutagenesis.
5. Geneticists often employ suppressor mutations to define gene function. Suppressor mutants are generated in the same way as other mutants—by exposing the organism to the effects of a mutagen. In one experiment, *Drosophila* that have a recessive mutation in an eye color gene (pink eye) are exposed to the mutagen ethane methyl sulfonate (EMS). The progeny of the mutagenized flies are screened for eye color. Twenty-three individual flies with wild-type (red) eyes appear in the mutagenesis screen.
   (a) How would you determine whether the suppressor mutations were due to EMS mutagenesis or were spontaneous mutations?
   (b) How would you determine whether the suppressor mutation was a reversion of the first mutation or was due to a second-site mutation?
   (c) How would you locate the site of the suppressor mutation?
   (d) One of the suppressor mutations occurred in the same gene as the original mutation; however, it was not a simple reversion of the original mutation. What other possibilities could explain the suppression?
6. What are *Drosophila* balancer chromosomes, what features do they have, and what do geneticists use them for?
7. How do transposons generate mutations? What sorts of mutations do these mobile elements create?
8. How do geneticists use mouse coat color genes as part of the method to make knockout mice? In the absence of coat color markers, is there another way that the knockout method could work?
9. Yeast with mutations in the *pro-1* gene are unable to synthesize the amino acid proline, which is required for normal growth. The wild-type strain of yeast is capable of synthesizing proline. How would you select suppressor mutants of this yeast strain?
10. In the *ClB* technique for detection of mutations in *Drosophila*,
    (a) What type of mutation may be detected?
    (b) What is the importance of the *l* gene?
    (c) Why is it necessary to have the crossover suppressor (*C*) on the X chromosome?
    (d) What is *C*?
11. Why are geneticists so enthusiastic about yeast as a model organism? What is the advantage of its having a haploid life cycle?
12. In order to study the cell cycle, geneticists generate recessive temperature-sensitive mutants in *S. cerevisiae*. If you were designing an experiment to collect a number of temperature-sensitive recessive mutations in the cell division cycle in yeast, how would you select these mutants? How could you distinguish between *cdc* mutations and other types of temperature-sensitive mutations?
13. As part of your research into how the cell walls of budding yeast are formed, you purify a protein that is synthesized as the cells divide and is incorporated into the new cell wall as the cells grow. Your next task is to clone the gene.
    (a) What strategy would you use to clone the cell wall protein gene?
    (b) Once you have the gene cloned, what would you do to determine whether this protein was important for growth of the yeast cell wall?
14. You have selected an interesting-looking gene from the mouse genome sequence. You think that this gene might be involved in mouse development, as DNA in one part of the gene encodes a classic homeobox domain. You want to find out where and when this gene is expressed. What experiments would you do to determine the temporal- and tissue-specific expression of your gene?
15. Although modern high-throughput genomics methods such as gene-expression microarrays are generally lauded as technologies of the future, many scientists are skeptical about their value. Can you think of two limitations in using these DNA arrays? How might these limitations be overcome?
16. Describe the steps involved in *P* element-mediated transformation in *Drosophila*. What are the differences between this mode of transformation and the methods used to generate transgenic mice?
17. The Berkeley *Drosophila* Genome Project is creating a collection of *Drosophila* strains, each strain bearing a *P*-element insertion in a different gene. The goal is to generate null mutations in each essential gene of the *Drosophila* genome. How will this collection of stocks be useful for genetic analysis? Devise an experiment that would use the Berkeley collection.
18. One complication of mouse transgenics is that the transgene may integrate at random into the coding region, or the regulatory region, of an endogenous mouse gene. What might be the consequences of such random integrations? How might this complicate genetic analysis of the transgene?
19. Is it possible to undertake a forward genetic analysis in *Drosophila* without using a mutagen? If so, explain how this might be done.
20. Describe the difference between selecting for mutants and screening for mutants.
21. Over a period of several years, you have purified a small amount of protein from human blood, a protein that you suspect is involved in blood clotting. Because there is so little protein, you cannot generate antibodies to use as probes of a cDNA expression library. As an alternative, you obtain a bit of the amino acid sequence from the amino terminal end of your protein. The amino acid sequence is:

    Met – Pro – Lys – Glu – Ala – Ile – Gly –

    (a) How would you use this information to clone the gene?
    (b) Once you have cloned the gene, how could you verify that it is the gene that encodes your protein?
22. What is RNAi, and how does it work? Describe two ways that investigators can introduce RNAi molecules into cells.
23. Various strategies are used to generate organisms in which certain genes are removed or disrupted. In such cases, gene output is essentially zero when the knockout is homozygous. Why is a different term, *knockdown*, often used with RNAi-mediated gene-silencing technologies?

HOW DO WE KNOW?

24. In this chapter, we focused on the methods that geneticists use to discover genes and their functions. At the same time, we found many opportunities to consider the methods and reasoning by which much of this information was acquired. From the explanations given in the chapter, what answers would you propose to the following fundamental questions:
    (a) How do we know that organisms such as yeast or *Drosophila* are appropriate models in which to study genes that affect human biological processes?
    (b) How do we know that multiple genes operate in pathways leading to a phenotype?
    (c) How do we know that a large number of genes regulate the progress of cells through the cell cycle in organisms as diverse as yeast and humans?

## Extra-Spicy Problems

25. Your Ph.D. thesis project is to identify all the genes involved in a biosynthetic pathway in *Drosophila*. The pathway leads to the tanning and hardening of the adult cuticle after the fly emerges from the pupa.
    (a) What type of mutagen would you choose and why?
    (b) Would you expect that your mutations would be lethal? Why? How could you tell whether or not mutations in this pathway were lethal?
    (c) How would you screen for the mutants if you didn't know whether or not they would be lethal?
    (d) After you perform numerous mutational screens using thousands of flies, you obtain no mutations that affect the tanning and hardening of the adult cuticle. What possibilities could explain this? What would you do to test your hypotheses?
26. You are interested in the genetic basis of a human disease characterized by neural degeneration and abnormal protein deposition in the brain. A candidate gene has been cloned from humans, and it is possible that this gene is involved in causing the disease. However, it is difficult to analyze this gene in humans. You are working in a research lab that uses *S. cerevisiae* as a model organism for metabolic studies. Would you consider using yeast as a model organism to study this human disease? Explain why you would or would not, and what experiments you would do before making your final decision.
27. As part of a genomics project, you obtain the DNA sequence of a fragment of genomic DNA from a wild-type strain of yeast. On examining the sequence, you see that it contains a long ORF preceded by a promoter and has a polyadenylation sequence at the 3′ end of the ORF—indications that this fragment of DNA likely contains a gene. You are interested in determining what gene it is and what function the gene performs. Your first step is to create a mutant yeast strain that is null for this gene. How could you use your genomic clone of this gene to create a mutant? Provide details of each step you would use in this procedure.
28. While analyzing some new genomic information from the mouse genome project, you notice a DNA sequence that would code for an interesting stretch of amino acids. This amino acid sequence forms a motif known as a helix-loop-helix domain. These domains are found in dozens of proteins in organisms as diverse as flies, *Arabidopsis*, and humans. How would you verify that this DNA sequence was part of a gene and was not simply a random piece of "junk" DNA? Assuming that the DNA sequence was part of a *bona fide* gene, how would you go about investigating the identity and potential function of this mouse gene?
29. A number of mouse models for human cystic fibrosis (CF) exist. Each of these mouse strains is transgenic and bears a different specific *CFTR* gene mutation. The mutations are the same as those seen in the varieties of human CF. These transgenic CF mice are being used to study the range of different phenotypes that characterize CF in humans. They are also used as models to test potential CF drugs. Unfortunately, most transgenic mouse CF strains do not show one of the most characteristic symptoms of human CF, that of lung congestion. Can you think of a reason why mouse CF strains do not display this symptom of human CF?
30. An understanding of interactions among complex genetic networks is often accomplished by complementation tests to determine how many genes are involved and by epistasis analysis to determine how gene products relate to one another. Hartwell and his colleagues used such an approach to unravel genetic regulation of the cell cycle. Following is a description of several genes involved in cell-cycle regulation.

    *wee1*$^+$ = wild-type allele that makes a protein kinase (Wee1) whose function is to phosphorylate tyrosine 15 of the Cdc2/cyclin complex. When the Cdc2/cyclin complex is phosphorylated at tyrosine 15, the complex is inactive.
    *cdc2*$^+$ = wild-type allele that makes a protein kinase (Cdc2) that combines with cyclin. When the Cdc2/cyclin complex is phosphorylated at tyrosine 15, the complex is inactive. When the phosphate on tyrosine 15 is removed by Cdc25, the Cdc2/cyclin complex is active and moves the cell into mitosis from G2.
    *cdc13*$^+$ = wild-type allele that makes a cyclin that complexes with Cdc2 to make the Cdc2/cyclin complex.
    *cdc25*$^+$ = wild-type allele that makes a phosphatase (Cdc25) to remove the phosphate of tyrosine 15 from the Cdc2/cyclin complex, thus activating the complex.

    Based on information contained in Chapter 20 and this chapter, predict the cellular response (progression to mitosis) for the following single- and double-mutant genotypes.

| Mutations | Progression to Mitosis? |
|---|---|
| cdc25 | |
| wee1 | |
| Cdc25 wee1 | |
| Cdc13 wee1 | |
| Cdc2 wee1 | |
| Cdc25 cdc13 | |

Transgenic pigs generated by incorporating a viral vector carrying the jellyfish gene encoding green fluorescent protein into the pig genome.

# 24 Applications and Ethics of Genetic Engineering and Biotechnology

## CHAPTER CONCEPTS

- Recombinant DNA technology, genetic engineering, and biotechnology have revolutionized medicine and agriculture.
- Genetically modified plants and animals can serve as bioreactors to produce therapeutic proteins and other valuable protein products.
- Genetic modifications of plants have resulted in herbicide and pest resistant crops, and crops with improved nutritional value; similarly, transgenic animals are being created to produce therapeutic proteins and to protect animals from disease.
- Applications of recombinant DNA technology and genomics have become essential for diagnosing genetic disorders, determining genotypes, and scanning the human genome to detect diseases.
- Pharmacogenomics and rational drug design have led to customized medicines based primarily on a person's genotype.
- Gene therapy by transfer of cloned copies of functional alleles into target tissues is used to treat genetic disorders.
- DNA fingerprinting can identify specific individuals in a large population and is widely used for forensics and paternity testing.
- Almost all applications of genetic engineering and biotechnology present unresolved ethical dilemmas that involve important moral, social, and legal issues.

Since the dawn of recombinant DNA technology in the 1970s, scientists have harnessed **genetic engineering** not only for biological research, but also for direct applications in medicine, agriculture, and biotechnology. Genetic engineering refers to the alteration of an organism's genome and typically involves the use of recombinant DNA technologies to add genes to a genome, but it can also involve gene removal. The ability to manipulate DNA *in vitro* and to introduce genes into living cells has allowed scientists to generate new varieties of plants, animals, and other organisms with specific gene traits, and to manufacture cheaper and more effective therapeutic products. These new varieties of organisms are called **genetically modified organisms**, or **GMOs**. Industry analysts estimate that genetic engineering will lead to U.S. commercial products worth over $45 billion by 2009. Many of these commercial products will be developed by the biotechnology industry.

**Biotechnology** is the use of living organisms to create a product or a process that helps improve the quality of life for humans or other organisms. As you will soon learn, biotechnology as a modern industry began in earnest when recombinant DNA technology developed. But biotechnology is actually a science dating back to ancient civilization and the use of microbes to make many important products, including beverages such as wine and beer, vinegar, breads, and cheeses. Modern biotechnology relies heavily on recombinant DNA technology, genetic engineering, and genomics applications. Existing products and new developments that occur seemingly every day make the biotechnology industry one of the most rapidly developing branches of the workforce worldwide, encompassing nearly 5000 companies in 54 countries. Over 350 biotechnology products are currently on the market.

The development of the biotechnology industry and the rapid growth in the number of applications for DNA technologies have raised serious concerns about using our power to manipulate genes and to apply gene technologies. Genetic engineering and biotechnology have the potential to provide solutions to major problems globally and to significantly alter how humans deal with the natural world; hence, they raise ethical, social, and economic questions that are unprecedented in human experience. These complex issues cannot be fully explored in the context of an introductory genetics textbook. This chapter will therefore present only a selection of applications that illustrate the power of genetic engineering and biotechnology and the complexity of the dilemmas they engender. We will begin by explaining how genetic engineering has modified agriculturally important plants and animals. Next, we examine the impact of genetic technologies on the diagnosis and treatment of human diseases. In addition, we will briefly describe how genetic engineering has affected the production of pharmaceutical products and how DNA technologies are now used for forensic applications. Finally, we explore some of the social, ethical, and legal implications of genetic engineering and biotechnology.

## 24.1 Genetically Engineered Organisms Synthesize a Wide Range of Biological and Pharmaceutical Products

The most successful and widespread application of recombinant DNA technology has been the production of recombinant proteins as **biopharmaceutical** products—pharmaceutical products produced by biotechnology—particularly, therapeutic proteins to treat diseases. Prior to the recombinant DNA era, biopharmaceutical proteins such as insulin, clotting factors, or growth hormones were purified from tissues such as pancreas, blood, or pituitary glands. Clearly, these sources were in limited supply, and the purification processes were expensive. In addition, products derived from these natural sources could be contaminated by disease agents such as viruses. Now that human genes encoding important therapeutic proteins can be cloned and expressed in a number of nonhuman host cell types, we have more abundant, safer, and less expensive sources of biopharmaceuticals. **Biopharming** is a commonly used term to describe the production of valuable proteins in genetically modified (GM) animals and plants.

In this section, we outline several examples of therapeutic products that are produced by expression of cloned genes in transgenic host cells and organisms. It should not surprise you that cancers, arthritis, diabetes, heart disease, and infectious diseases such as AIDS are among the major diseases that biotechnology companies are targeting for treatment by recombinant therapeutic products. Table 24.1 provides a short list of important recombinant products currently synthesized in transgenic bacteria, plants, yeast, and animals.

### Insulin Production in Bacteria

Several therapeutic proteins have been produced by introducing human genes into bacteria. In most cases, the human gene is cloned into a plasmid, and the recombinant vector is introduced into the bacterial host. Large quantities of the transformed bacteria are grown, and the recombinant human protein is recovered and purified from bacterial extracts.

The first human gene product manufactured by recombinant DNA technology was human insulin, called Humulin, which was licensed for therapeutic use in 1982 by the **U.S. Food and Drug Administration (FDA)**, the government agency responsible for regulating the safety of food and drug products and medical devices. In 1977, scientists at Genentech, the San Francisco biotechnology company cofounded in 1976 by Herbert Boyer (one of the pioneers of using plasmids for recombinant DNA technology) and Robert Swanson isolated and cloned the gene for insulin and

TABLE 24.1

**Some Genetically Engineered Pharmaceutical Products Now Available or under Development**

| Gene Product | Condition Treated | Host Type |
|---|---|---|
| Erythropoitin | Anemia | *E. coli*; cultured mammalian cells |
| Interferons | Multiple sclerosis, cancer | *E. coli*; cultured mammalian cells |
| Tissue plasminogen activator tPA | Heart attack, stroke | Cultured mammalian cells |
| Human growth hormone | Dwarfism | Cultured mammalian cells |
| Monoclonal antibodies against vascular endothelial growth factor (VEGF) | Cancers | Cultured mammalian cells |
| Human clotting factor VIII | Hemophilia A | Transgenic sheep, pigs |
| C1 inhibitor | Hereditary angioedema | Transgenic rabbits |
| Recombinant human antithrombin | Hereditary antithrombin deficiency | Transgenic goats |
| Hepatitis B surface protein vaccine | Hepatitis B infections | Cultured yeast cells, bananas |
| Immunoglobulin IgG1 to HSV-2 | Herpesvirus infections | Transgenic soybeans glycoprotein B |
| Recombinant monoclonal antibodies | Passive immunization against rabies (also used in diagnosing rabies), cancer, rheumatoid arthritis | Transgenic tobacco, soybeans, cultured mammalian cells |
| Norwalk virus capsid protein | Norwalk virus infections | Potato (edible vaccine) |
| *E. coli* heat-labile enterotoxin | *E. coli* infections | Potato (edible vaccine) |

expressed it in bacterial cells. Genentech, short for "genetic engineering technology," is also generally regarded as the first biotechnology company.

Previously, insulin was chemically extracted from the pancreas of cows and pigs obtained from slaughterhouses. **Insulin** is a protein hormone that regulates glucose metabolism. Individuals who cannot produce insulin have diabetes, a disease that, in its more severe form (type I), affects more than 2 million individuals in the United States. Although synthetic human insulin can now be produced by another process, a look at the original genetic engineering method is instructive, as it shows both the promise and the difficulty of applying recombinant DNA technology.

Clusters of cells embedded in the pancreas synthesize a precursor polypeptide known as preproinsulin. As this polypeptide is secreted from the cell, amino acids are cleaved from the end and the middle of the chain. These cleavages produce the mature insulin molecule, which contains two polypeptide chains (the *A* and *B* chains) joined by disulfide bonds. The *A* subunit contains 21 amino acids, and the *B* subunit contains 30.

In the original bacterial bioengineering process, synthetic genes that encode the *A* and *B* subunits were constructed by oligonucleotide synthesis (63 nucleotides for the *A* polypeptide and 90 nucleotides for the *B* polypeptide). Each synthetic oligonucleotide was inserted into a separate vector, adjacent to the *lacZ* gene encoding the bacterial form of the enzyme $\beta$-galactosidase. When transferred to a bacterial host, the *lacZ* gene and the adjacent synthetic oligonucleotide were transcribed and translated as a unit. The product is a **fusion protein**—that is, a hybrid protein consisting of the amino acid sequence for $\beta$-galactosidase attached to the amino acid sequence for one of the insulin subunits (Figure 24–1). The fusion proteins were purified from bacterial extracts and treated with cyanogen bromide, a chemical that cleaves the fusion protein from the $\beta$-galactosidase. When the fusion products were mixed, the two insulin subunits spontaneously united, forming an intact, active insulin molecule. The purified injectable insulin was then packaged for use by diabetics.

Recombinant forms of insulin that are delivered via an air inhaler are now available. Shortly after insulin became available, growth hormone—used to treat children who suffer from a form of dwarfism—was cloned. Soon, recombinant DNA technology made that product readily available too, as well as a wide variety of other medically important proteins that were once difficult to obtain in adequate amounts. Since recombinant insulin ushered in the biotechnology era, well over 200 recombinant products have entered the market worldwide.

## Transgenic Animal Hosts and Pharmaceutical Products

Although bacteria have been widely used to produce therapeutic proteins, there are some disadvantages in using prokaryotic hosts to synthesize eukaryotic proteins. One problem is that bacterial cells often cannot process and modify eukaryotic proteins. As a result, they frequently cannot add the carbohydrates and phosphate groups to proteins that are needed for full biological activity. In addition, eukaryotic proteins produced in prokaryotic cells often do not fold into the proper three-dimensional configuration and are

(a)

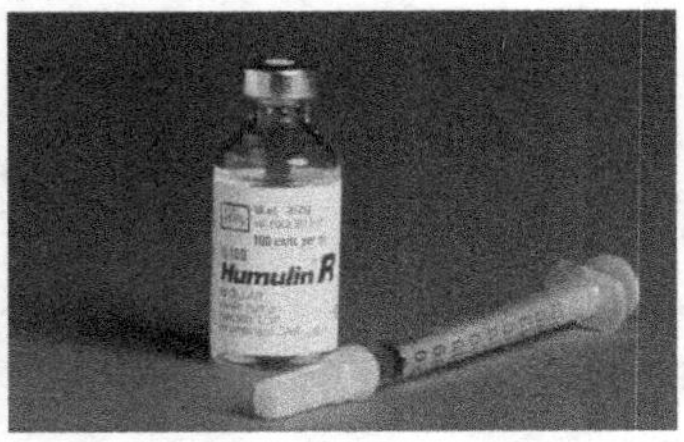


(b)

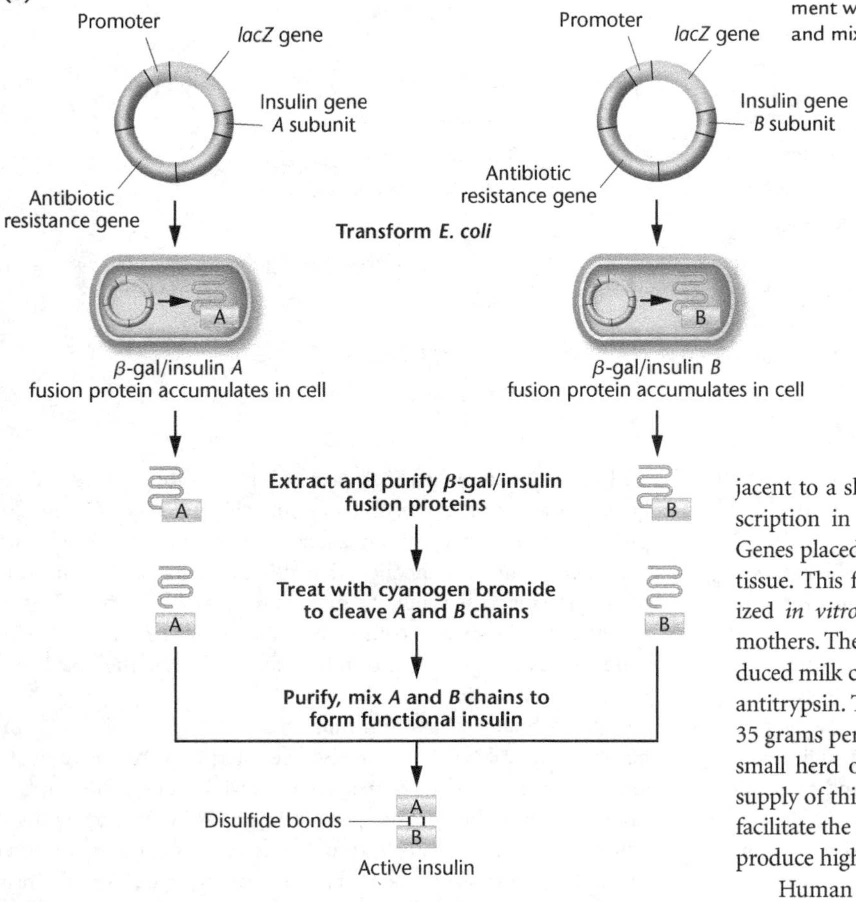


**FIGURE 24–1** (a) Humulin, a recombinant form of human insulin, was the first therapeutic protein produced by recombinant DNA technology to be approved for use in humans. (b) To synthesize recombinant human insulin, synthetic oligonucleotides encoding the insulin *A* and *B* chains were inserted (in separate vectors) at the tail end of a cloned *E. coli lacZ* gene. The recombinant plasmids were transformed into *E. coli* host cells, where the $\beta$-gal/insulin fusion protein was synthesized and accumulated in the cells. Fusion proteins were then extracted from the host cells and purified. Insulin chains were released from $\beta$-galactosidase by treatment with cyanogen bromide. The insulin subunits were purified and mixed to produce a functional insulin molecule.

therefore inactive. To overcome these difficulties and increase yields, many biopharmaceuticals are now produced in eukaryotic hosts. As seen in Table 24.1, eukaryotic hosts may include cultured eukaryotic cells (plant or animal) or transgenic farm animals. A herd of goats or cows serve as very effective **bioreactors** or **biofactories**—living factories—that will continuously make milk containing the desired therapeutic protein that can then be isolated in a non-invasive way.

Yeast are also valuable hosts for expressing recombinant proteins. Even insect cells are valuable for this purpose, through the use of a gene delivery system (virus) called **baculovirus**. Recombinant baculovirus containing a gene of interest are used to infect insect cell lines, which then express the protein at high levels. Baculovirus-insect cell expression is particularly useful for producing human recombinant proteins that are heavily glycosylated. Regardless of the host, therapeutic proteins may then be purified from the host cells—or when transgenic farm animals are used, isolated from animal products such as milk.

An example of a biopharmaceutical product synthesized in transgenic animals is the human protein **$\alpha$1-antitrypsin**. A deficiency of the enzyme $\alpha$1-antitrypsin is associated with the heritable form of emphysema, a progressive and fatal respiratory disorder common among people of European ancestry. To produce $\alpha$1-antitrypsin for use in treating this disease, the human gene was cloned into a vector at a site adjacent to a sheep promoter sequence that specifically activates transcription in milk-producing cells of the sheep mammary glands. Genes placed next to this promoter are expressed only in mammary tissue. This fusion gene was microinjected into sheep zygotes fertilized *in vitro*. The fertilized zygotes were transferred to surrogate mothers. The resulting transgenic sheep developed normally and produced milk containing high concentrations of functional human $\alpha$1-antitrypsin. This human protein is present in concentrations of up to 35 grams per liter of milk and can be easily extracted and purified. A small herd of lactating transgenic sheep can provide an abundant supply of this protein. In fact, the famous sheep, Dolly, was cloned to facilitate the establishment of a flock of sheep that would consistently produce high levels of human proteins.

Human proteins produced in transgenic animals undergo clinical testing as a first step toward ensuring their safety for therapeutic use. A recombinant human enzyme, $\alpha$-glucosidase, produced in rabbit milk, is now in clinical tests for treating children with **Pompe disease.** This progressive and fatal metabolic disorder is caused by a lack of the lysosomal enzyme $\alpha$-glucosidase and is inherited as an autosomal recessive condition. $\alpha$-glucosidase breaks down the complex carbohydrate glycogen into glucose monosaccharides. Individuals with Pompe disease cannot adequately breakdown glycogen causing it to accumulate, particularly in muscle cells. Progressive muscle weakness and other complications occur primarily as a result of glycogen accumulation. In early-onset Pompe disease, infants have poor muscle tone and may never sit or stand. Most die before two years of age from respiratory or cardiac complications. In one of the first tests, the recombinant enzyme was given to affected children weekly, with no significant side effects. All of the children showed normal enzyme activity in the tissues analyzed, and all showed improvements in their symptoms. If large-scale trials are successful, recombinant $\alpha$-glucosidase from transgenic animals will become the preferred method of treatment for this disorder.

In 2006, recombinant human **antithrombin**, an anticlotting protein, became the world's first drug extracted from the milk of farm animals to be approved for use in humans. Scientists at GTC Biotherapeutics of Framingham, Massachusetts, introduced the human antithrombin gene into goats. By placing the gene adjacent to a promoter for beta casein, a common protein in milk, GTC scientists were able to target antithrombin expression in the mammary gland. As a result, antithrombin protein is highly expressed in the milk. In one year, a single goat will produce the equivalent amount of antithrombin that in the past would have been isolated from ~90,000 blood collections.

In a similar example involving a nonbiopharmaceutical application, transgenic "silk-milk" goats have been generated that express spider-silk proteins in their milk. These goats are a rich source of silk proteins used for various commercial applications such as manufacturing bulletproof vests.

## Recombinant DNA Approaches for Vaccine Production and Transgenic Plants with Edible Vaccines

One of the most promising applications of recombinant DNA technology for therapeutic purposes may be the production of vaccines. Vaccines stimulate the immune system to produce antibodies against disease-causing organisms and thereby confer immunity against specific diseases. Traditionally, two types of vaccines have been used: **inactivated vaccines**, which are prepared from killed samples of the infectious virus or bacteria; and **attenuated vaccines**, which are live viruses or bacteria that can no longer reproduce but can cause a mild form of the disease. Inactivated vaccines include the vaccines for rabies and influenza; vaccines for tuberculosis, cholera, and chickenpox are examples of attenuated vaccines.

Genetic engineering is being used to produce a relatively new type of vaccine called a **subunit vaccine**, which consists of one or more surface proteins from the virus or bacterium but not the entire virus or bacterium. This surface protein acts as an antigen that stimulates the immune system to make antibodies that act against the organism from which it was derived. One of the first subunit vaccines was made against the **hepatitis B virus**, which causes liver damage and cancer. The gene that encodes the hepatitis B surface protein was cloned into a yeast expression vector, and the cloned gene was expressed in yeast host cells. The protein was then extracted and purified from the host cells and packaged for use as a vaccine.

In 2005, the FDA approved **Gardasil**, a recombinant subunit vaccine produced by the pharmaceutical company Merck and the first cancer vaccine to receive FDA approval. Gardasil targets four strains of **human papillomavirus** (HPV) that cause ~70 percent of cervical cancers. Approximately 70 percent of sexually active women will be infected by an HPV strain during their lifetime. Gardasil is designed to provide immune protection against HPV prior to infection but is not effective against existing infections. You may have heard of Gardasil through media coverage of the legislation pending in several states that would require all adolescent school children to receive a Gardasil vaccination regardless of whether or not they are sexually active.

Developing countries face serious difficulties in manufacturing, transporting, and storing vaccines. Most vaccines need refrigeration and must be injected under sterile conditions. In many rural areas, refrigeration and sterilization facilities are not available. In addition, in many cultures people are fearful of being injected with needles. To overcome these problems, scientists are attempting to develop vaccines that can be synthesized in edible food plants (Figure 24–2). These vaccines would be inexpensive to produce, would not require refrigeration, and would not have to be given under sterile conditions by trained medical personnel.

Plants offer several other advantages for expressing recombinant proteins. For instance, once a transgenic plant is made, it can easily be grown and replicated in a greenhouse or field, and it will provide a constant source of recombinant protein. In addition, the cost of expressing a recombinant protein in a transgenic plant is typically much lower than making the same protein in bacteria, yeast, or mammalian cells.

No recombinant proteins expressed in transgenic plants have yet been approved for use by the FDA. In one model system, the gene encoding an antigenic subunit of the hepatitis B virus has been introduced into the tobacco plant and expressed in its leaves. For use

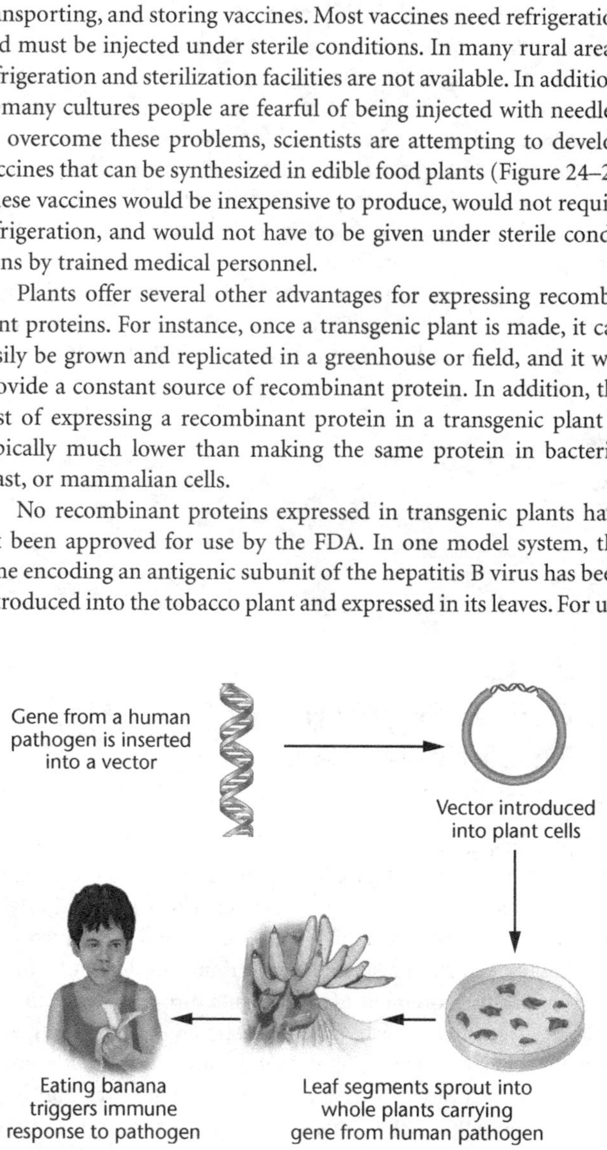


FIGURE 24–2 To make an edible vaccine, a gene from a pathogen (a disease-causing agent, such as a virus or bacterium) is transferred into a vector, which is then introduced into plant cells. In this example, infection of banana plant leaf segments transfer the vector and the pathogen's gene into the nuclei of banana leaf cells. The leaf segments are grown into mature banana trees that express the pathogenic gene. Eating the raw banana produced by these plants triggers an immune response to the protein encoded by the pathogen's gene, conferring immunity to infection by this pathogen.

as a source of vaccine, however, the gene would have to be inserted into food plants such as grains, vegetables, or fruits. Some edible vaccines are now in clinical trials. A vaccine against a bacterium that causes diarrhea has been produced in genetically engineered potatoes and used to successfully vaccinate human volunteers who ate small quantities (50–100 g) of the potatoes. In another vaccine project, transgenic spinach expressing rabies viral antigens was fed to volunteers, who showed significant increases in rabies-specific antibodies. Tests using vaccine-producing bananas and potatoes are currently under way. Bananas are considered to be perhaps the best edible vaccine candidate for a hepatitis B vaccine. Genetically engineered edible plants are being used for trials to vaccinate infants, children, and adults against many infectious diseases. But a number of technical questions about vaccine delivery in plants need to be answered if edible plant vaccines are to become more widely used. For example, how can vaccine dose be carefully controlled when fruits and vegetables grow to different sizes and express different amounts of the vaccine? Will vaccine proteins pass through the digestive tract unaltered so that they maintain their ability to provide immune protection?

NOW SOLVE THIS

Problem 5 on page 665 asks you to reflect on questions concerning the development of an edible vaccine.

■ HINT: *Stimulation of antibody formation by the smallest possible portion of a protein is important to ensure vaccine specificity.*

FIGURE 24–3 Selective breeding is one of the oldest methods of genetic alteration of plants. Shown here is teosinte (*Zea canina*, left), a selectively bred hybrid (center), and modern corn (*Zea mays*).

24.2

## Genetic Engineering of Plants Has Revolutionized Agriculture

For millennia, farmers have manipulated the genetic makeup of plants and animals to enhance food production. Until the advent of genetic engineering 30 years ago, these genetic manipulations were restricted to **selective breeding**—the selection and breeding of naturally occurring or mutagen-induced variants. In the last 50 to 100 years, genetic improvement of crop plants through the traditional methods of artificial selection and genetic crosses has resulted in dramatic increases in productivity and nutritional enhancement. For example, maize yields have increased fourfold over the last 60 years, and more than half of this increase is due to genetic improvement by artificial selection and selective breeding (Figure 24–3). Modern maize has substantially larger ears and kernels than the predecessor crops, including hybrids from which it was bred.

Recombinant DNA technology provides powerful new tools for altering the genetic constitution of agriculturally important organisms. Scientists can now identify, isolate, and clone genes that confer desired traits, then specifically and efficiently introduce these into organisms. As a result, it is possible to quickly introduce insect resistance, herbicide resistance, or nutritional characteristics into farm plants and animals, a primary purpose of **agricultural biotechnology**. In this section we primarily consider genetic manipulations to produce transgenic crop plants of agricultural value. In Section 24.3, we will discuss examples of genetic manipulations of agriculturally important animals.

The gene-based revolution in agriculture began in 1996 with the introduction of herbicide-resistant soybeans. Since then, dozens of transgenic agricultural crops and animals have entered agricultural use. Worldwide, over 200 million acres are planted with genetically engineered crops, particularly herbicide- and pest-resistant soybeans, corn, cotton, and canola; over 50 different transgenic crop varieties are available, including alfalfa, corn, rice, potatoes, tomatoes, tobacco, wheat, and cranberries [Figure 24–4(a)]. These crops are planted by over 8 million farmers in 17 countries. As evident in Figure 24–4(b), both industrialized and developing countries are taking advantage of transgenic crops. Since 1996, there has been a 4000 percent increase in GM crop acreage worldwide.

In 2005, the 10-year anniversary of commercial biotech crops, the one-billionth biotech acre was planted. American farmers planted 111 million acres of GM corn, soybeans, and cotton in 2004, a 17 percent increase from the year before. In the United States 86 percent of the soybeans, 78 percent of the cotton, and 46 percent of the corn are genetically engineered to resist pest or herbicides.

Several of the main reasons for generating transgenic crops include:

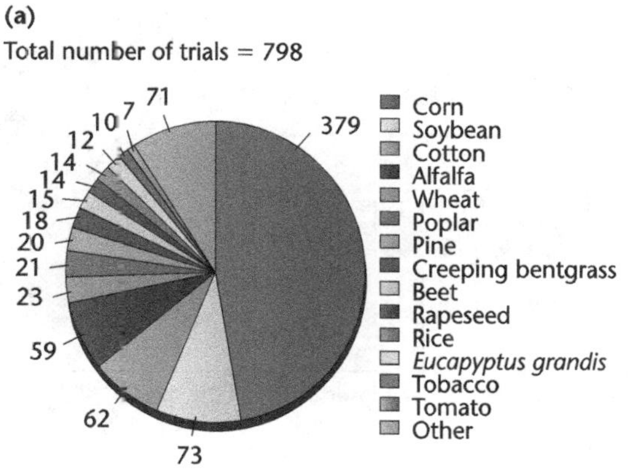


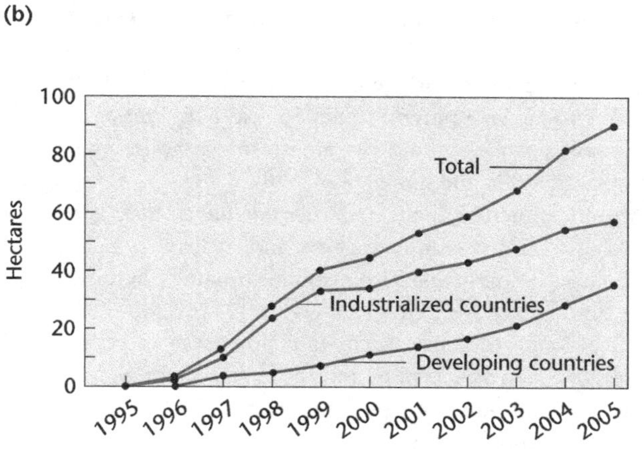


FIGURE 24–4 (a) A recent analysis of nearly 800 transgenic crop trials worldwide shows that GM varieties of corn, soybean, cotton, alfalfa, and wheat are among the most commonly manipulated crops (*Nature Biotechnology*, 23(3), p. 281, March 2005). (b) During the past 10 years, transgenic crops have been rapidly adopted in both industrialized and developing countries (Ernst & Young, Beyond Borders: Global Biotechnology Report 2006, www.ey.com/beyondborders).

- Improving the growth characteristics and yield of agriculturally valuable crops
- Increasing the nutritional value of crops
- Providing crop resistance against insect and viral pests, drought, and herbicides

In addition, many new GM crops that will soon be on the market will be designed for ethanol production and for making biodiesel fuel—that is, for providing sustainable sources of energy.

The first commercially available GM food was called the Flavr Savr tomato. Designed by researchers at Calgene (now a division of Monsanto, Inc.), the Flavr Savr was designed to increase the shelf life of tomatoes by allowing them to stay ripe for several weeks without softening—a common problem for tomatoes. Calgene scientists used antisense RNA technology to inhibit an enzyme called polygalacturonase, which digests pectin in the cell wall of tomatoes. Pectin digestion occurs naturally once tomatoes are picked from the plant, and this process is a main reason why tomatoes soften as they age. This GM approach was generally effective, but the attempts to remedy shipping problems that continued to cause bruising increased the cost of these tomatoes. This and public skepticism about the safety of the first GM food are two of the reasons Flavr Savr was eventually taken off the market.

We will now examine other, more successful examples of genetically engineered plants used in agriculture. Some of the ethical and social concerns associated with these practices will be examined in Section 24.7.

## Transgenic Crops for Herbicide and Pest Resistance

Damage from weed infestation destroys about 10 percent of crops worldwide. In an attempt to combat this problem, farmers often apply herbicides to the soil to kill weeds prior to seeding a field crop. As the most efficient herbicides also kill crop plants, herbicide uses are limited. The creation of herbicide-resistant crops has opened the way to efficient weed control and increased yields of some major agricultural crops. At present, over 75 percent of soybeans and cotton in the United States are resistant to the herbicide **glyphosate.** You may be familiar with glyphosate because it is the active herbicide in Roundup that is commonly sold through home improvement stores for keeping sidewalks and patios weed-free (see Figure 24–5).

(a)

(b)

FIGURE 24–5 (a) Glyphosate is the active chemical in Roundup, a commonly used herbicide. (b) A weed-infested glyphosate-resistant soybean plot before (left) and after Roundup treatment (right).

Glyphosate is effective at very low concentrations, is not toxic to humans, and is rapidly degraded by soil microorganisms. It kills plants by inhibiting the action of a chloroplast enzyme called **EPSP synthase**. This enzyme is important in amino acid biosynthesis in both bacteria and plants. Without the ability to synthesize vital amino acids, plants wither and die.

Recall from Chapter 13 that *Agrobacterium tumefaciens* is a soil microbe that can infect wounded plants and create crown gall tumors. *A. tumefaciens* contains **Ti plasmids**, so named because they contain tumor-inducing genes (see Figure 13–15). Modified versions of Ti plasmids that lack tumor-inducing genes and contain other features, such as antibiotic resistance, have been widely used as vectors for introducing genes into plants. To produce a glyphosate-resistant crop plant, researchers began by isolating and cloning an EPSP synthase gene from a glyphosate-resistant strain of *E. coli*. Next, they cloned the *EPSP* gene into a Ti plasmid between promoter sequences derived from a plant virus and transcription termination sequences derived from a plant gene (Figure 24–6). This recombinant vector was then transformed into *A. tumefaciens*.

The Ti plasmid-carrying bacteria were then used to infect plant cells derived from plant leaves. The clumps of cells (calluses) that formed after infection with *A. tumefaciens* were tested for their ability to grow in the presence of glyphosate. Glyphosate-resistant calluses were grown into transgenic plants and sprayed with glyphosate at concentrations four times higher than that needed to kill wild-type plants. Transgenic plants that expressed EPSP synthase grew and developed, while the control plants withered and died. Figure 24–5(b) demonstrates the effectiveness of glyphosate as a herbicide and the resistance of glyphosate-resistant soybeans.

Similar transgenic techniques have been used to make plants resistant to several other herbicides, to pathogens such as viruses, and also to insect pests. Some of the most well-described and controversial GM crops are the so-called **Bt crops**, designed to be resistant to insects. The bacterium *Bacillus thuringiensis* (Bt) produces a protein that when ingested by insects and larvae will crystallize in the gut, killing pests such as corn-borer larvae that are responsible for millions of dollars of crop damage worldwide. Initially, applications of Bt involved spraying these bacteria on crops. But recombinant DNA technology has enabled scientists to produce Bt transgenic crops with built-in insecticide protection. The *cry* genes that encode the Bt crystalline protein have been effectively introduced into a number of different crops, including corn, cotton, tomatoes, and tobacco.

Bt crops have been hailed as one of the greatest success stories of agricultural biotechnology, but they have also been one of the most controversial. Some studies had suggested a correlation between decreases in Monarch butterfly populations and ingestion of pollen from Bt corn (Monarchs do not feed on the corn itself). More recently, several long-term studies have demonstrated that exposure to Bt crops has no apparent effects on the Monarch; however, the possibility of danger to nontarget insect species must be considered whenever pest-resistant crops are used in the wild. Based on the success of Bt crops, many other transgenic crops are under development, including plants with increased tolerance to viral pests, drought, and salty soils.

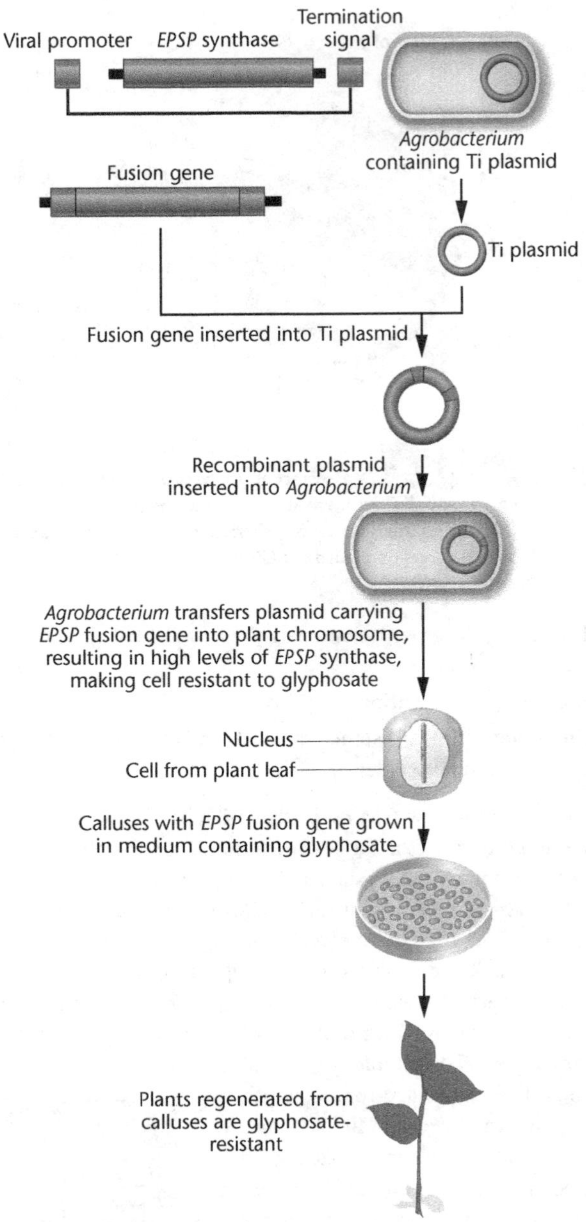


**FIGURE 24–6** To create glyphosate-resistant transgenic plants, the EPSP synthase gene from bacteria is fused to a promoter such as the promoter from the cauliflower mosaic virus. This fusion gene is then ligated into a Ti-plasmid vector, and the recombinant vector is transformed into an *Agrobacterium* host. *Agrobacterium* infection of cultured plant cells transfers the EPSP synthase fusion gene into a plant-cell chromosome. Cells that acquire the gene are able to synthesize large quantities of EPSP synthase, making them resistant to the herbicide glyphosate. Resistant cells are selected by growth in herbicide-containing medium. Plants regenerated from these cells are herbicide-resistant.

NOW SOLVE THIS

Problem 4 on page 665 asks you to predict whether the gene conferring glyphosate resistance could escape from a transgenic crop into weed plants.

■ HINT: *Consider the methods by which plants breed and the types of selective pressures on both wild plants and domestic crops.*

## Nutritional Enhancement of Crop Plants

Gene transfer by recombinant DNA techniques offers a new way to enhance the nutritional value of plants. Many crop plants are deficient in some of the nutrients required in the human diet, and biotechnology is being used to produce crops that meet these dietary requirements. One example is the production of **"golden rice,"** with enhanced levels of β-carotene, a precursor to vitamin A (Figure 24–7). Vitamin A deficiency is prevalent in many areas of Asia and Africa, and more than 500,000 children a year become permanently blind as a result of this deficiency. Rice is a major staple food in these regions but does not contain vitamin A.

To create golden rice, scientists transferred into the rice genome, by recombinant DNA technology, three genes encoding enzymes required for the biosynthetic pathway leading to β-carotenoid synthesis. Two of these genes came from the daffodil and one from a bacterium. Although golden rice is currently available for planting, it produced only moderate levels of β-carotene. New varieties with higher levels of β-carotene production are in development. One of these strains, Golden Rice 2, incorporates the phytoene synthase gene from maize instead of daffodils and produces about 20 times more β-carotene than the original golden rice. Acceptance of genetically engineered varieties of rice and efficient distribution of golden rice remain challenges to its wider use. Nonetheless, encouraged by the effectiveness of Golden Rice 2, researchers are working on developing rice with enhanced iron and protein content.

Many other varieties of nutritionally enhanced food crops have been, and are being, developed. These include plants with augmented levels of key fatty acids, antioxidants, and other vitamins and minerals. These efforts are directed at addressing nutrient deficiencies affecting more than 40 percent of the world's population. Other expected developments include decaffeinated teas and coffees, as well as crops enhanced for traits affecting taste, growth rates, yields, color, storage, ripening, and similar characteristics.

(a)

(b)

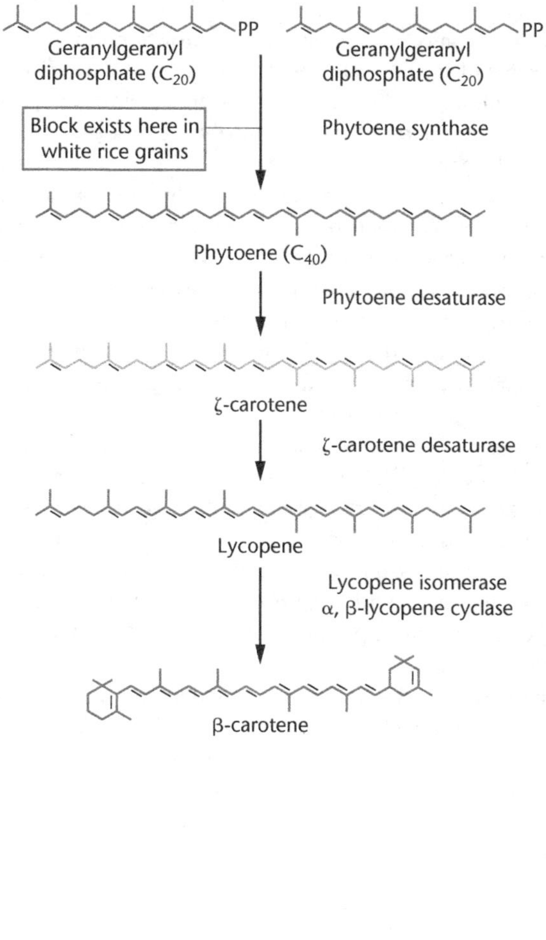


FIGURE 24–7 (a) Golden rice, a strain genetically modified to produce β-carotene, a precursor to vitamin A. Many children in countries where rice is a dietary staple lose their eyesight because of diets deficient in vitamin A. (b) White rice lacks the enzyme phytoene synthase, which is responsible for converting $C_{20}$ into phytoene, a rate-limiting step in the production of β-carotene. Introducing the phytoene synthase gene into rice is one way to overcome this block and produce golden rice enriched in β-carotene.

## 24.3 Transgenic Animals with Genetically Enhanced Characteristics Have the Potential to Serve Important Roles in Agriculture and Biotechnology

Although genetically engineered plants are major players in modern agriculture, transgenic animals are less widespread. Nonetheless, some high-profile examples of genetically engineered farm animals

FIGURE 24–8 Transgenic Atlantic salmon (bottom) overexpressing a growth hormone (GH) gene display rapidly accelerated rates of growth compared to wild strains and nontransgenic domestic strains (top). GH salmon weigh an average of nearly 10 times more than nontransgenic strains.

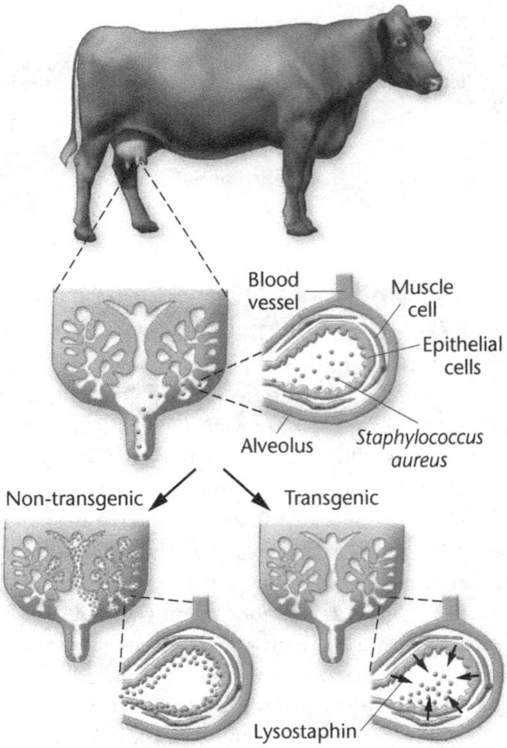


FIGURE 24–9 Transgenic cows for battling mastitis. The mammary glands of nontransgenic cows are highly susceptible to infection by the skin microbe *Staphylococcus aureus*. Transgenic cows express the lysostaphin transgene in milk, where it can kill *S. aureus* before they can multiply in sufficient numbers to cause inflammation and damage mammary tissue.

have aroused public interest and controversy. The first genetically engineered livestock were created in 1985, using pronuclear injection methods. Recall from Chapter 23 that these methods involve the introduction of cloned DNA into a pronucleus of a fertilized egg. The injected DNA integrates at random into the host's genome, and the eggs are then introduced into surrogate mothers. The offspring that contain the integrated DNA are selected and crossed in order to produce homozygous lines. There are now also several other ways besides pronuclear injection to introduce transgenes into an animal.

Oversize mice containing a human growth hormone transgene were some of the first transgenic animals created. Attempts to create farm animals containing transgenic growth hormone genes have not been particularly successful, probably because growth is a complex, multigene trait. One notable exception is the transgenic Atlantic salmon, bearing copies of a Chinook salmon growth hormone gene adjacent to a constitutive promoter. These salmon mature quickly, grow 400 to 600 percent faster than nontransgenic salmon, and appear to have no adverse health effects (Figure 24–8).

As discussed in Section 24.1, currently, the major uses for transgenic farm animals are as bioreactors to produce useful pharmaceutical products, but a number of other interesting transgenic applications are under development. Several of these applications are designed to increase milk production or increase nutritional value of milk. Significant research efforts are also being made to protect farm animals against common pathogens that cause disease and animal loss (including potential bioweapons that could be used in a terrorist attack on food animals) and put the food supply at risk. For instance, controlling mastitis in cattle by creating transgenic cows has shown promise (Figure 24–9). **Mastitis** is an infection of the mammary glands. It is the most costly disease affecting the dairy industry, leading to over $2 billion in losses in the United States. Mastitis can block milk ducts, reducing milk output, and can also contaminate the milk with pathogenic microbes. Infection by the bacterium *Staphylococcus aureus* is the most common cause of mastitis, and most cattle with mastitis typically do not respond well to conventional treatments with antibiotics. As a result, mastitis is a significant cause of herd reduction.

In an attempt to create cattle resistant to mastitis, transgenic cows were generated that possessed the lysostaphin gene from *Staphylococus simulana*. Lysostaphin is an enzyme that specifically cleaves components of the *S. aureus* cell wall. Transgenic cows expressing this protein in milk produce a natural antibiotic that wards off *S. aureus* infections. These transgenic cows do not completely solve the mastitis problem because lysostaphin is not effective against other microbes such as *E. coli* and *S. uberis* that occasionally cause mastitis; moreover, there is also the potential that *S. aureus* may develop resistance to lysostaphin. Nonetheless, scientists are cautiously optimistic that transgenic approaches have a strong future for providing farm animals with a level of protection against major pathogens.

Several groups recently produced cattle that lack the prion protein *PrP* gene. Misfolded configurations of the PrP protein result in **mad cow disease**, or bovine spongioform encephalitis. Early results indicate

FIGURE 24–10 GloFish, marketed as the world's first GM-pet, are a controversial product of genetic engineering.

that these animals are resistant to the development of mad cow disease, but it is not yet known if they are fully protected from developing the disease. Another successful transgenic farm animal is **EnviroPig**, a pig that expresses the gene encoding the enzyme phytase. These pigs are able to break down dietary phosphorus, thereby reducing excretion of phosphorus, which is a major pollutant in pig farms.

Scientists at Yorktown Industries of Austin, Texas, created the **GloFish**, a transgenic strain of zebrafish (*Danio rerio*) containing a red fluorescent protein gene from sea anemones. Marketed as the first GM pet in the United States, GloFish fluoresce bright pink when illuminated by ultraviolet light (Figure 24–10). GM critics describe these fish as an abuse of genetic technology. However, GloFish may not be as frivolous a use of genetic engineering as some believe. Recently, a variation of this transgenic model, incorporating a heavy-metal-inducible promoter adjacent to the red fluorescent protein gene, has shown promise in a bioassay for heavy metal contamination of water. When these transgenic zebrafish are in water contaminated by mercury and other heavy metals, the promoter becomes activated, inducing transcription of the red fluorescent protein gene. In this way, zebrafish fluorescence can be used as a bioassay to measure heavy metal contamination and uptake by living organisms.

## 24.4 Genetic Engineering and Genomics Are Transforming Medical Diagnosis

Geneticists are now applying knowledge about the human genome and the genetic basis of many diseases to a wide range of medical applications. Gene-based technologies have already had a major impact on the diagnosis of disease and will soon revolutionize medical treatments and the development of specific and effective pharmaceuticals. As more information emerges from the Human Genome Project, researchers are identifying increasing numbers of genes involved in both single-gene diseases and complex genetic traits. Methods based on recombinant DNA technologies are speeding up the identification of specific mutations that lead to hereditary disorders and somatic diseases such as cancers.

In this section, we provide a brief overview of how gene-based technologies are being used to diagnose genetic diseases. In the next section, we will describe several exciting technologies that will transform medical treatments over the next few decades.

### Genetic Tests Based on Restriction Enzyme Analysis

Using DNA-based tests, scientists can directly examine a patient's DNA for mutations associated with disease. Gene testing was one of the first successful applications of recombinant DNA technology, and currently more than 900 gene tests are in use. These tests usually detect DNA mutations associated with single-gene disorders that are inherited in a Mendelian fashion. Examples of such genetic tests are those that detect sickle-cell anemia, cystic fibrosis, Huntington disease, hemophilias, and muscular dystrophies. Other genetic tests have been developed for complex disorders such as breast and colon cancers. Gene tests are used to perform prenatal diagnosis of genetic diseases, to identify carriers, to predict the future development of disease in adults, to confirm the diagnosis of a disease detected by other methods, and to identify genetic diseases in embryos created by *in vitro* fertilization.

A classic method of genetic testing is **restriction fragment length polymorphism (RFLP) analysis**. To illustrate this method, we examine the prenatal diagnosis of **sickle-cell anemia**. This disease is an autosomal recessive condition common in people with family origins in areas of West Africa, the Mediterranean basin, and parts of the Middle East and India. It is caused by a single amino acid substitution in the $\beta$-globin protein, as a consequence of a single-nucleotide substitution in the $\beta$-globin gene. The single-nucleotide substitution also eliminates a cutting site in the $\beta$-globin gene for the restriction enzymes *Mst*II and *Cvn*I. As a result, the mutation alters the pattern of restriction fragments seen on Southern blots. These differences in restriction cutting sites are used to prenatally diagnose sickle-cell anemia and to establish the parental genotypes and the genotypes of other family members who may be heterozygous carriers of this condition.

For testing adults by RFLP analysis (and most of the genetic tests we will discuss here), blood samples are generally required, and DNA from white blood cells is typically used or RFLP is carried out on check cells collected by swabbing the inside of the mouth. Some genetic testing can be carried out on gemetes. For prenatal diagnosis, fetal cells are obtained by **amniocentesis** or **chorionic villus**

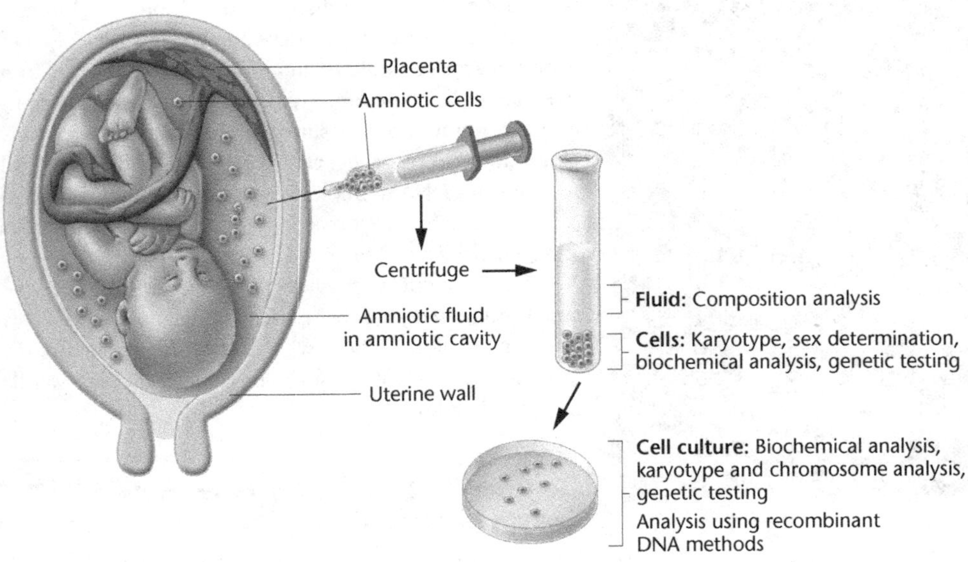


**FIGURE 24–11** For amniocentesis, the position of the fetus is first determined by ultrasound, and then a needle is inserted through the abdominal and uterine walls to recover amniotic fluid and fetal cells for genetic or biochemical analysis.

RESTRICTION FRAGMENT LENGTH POLYMORPHISMS ◀ WEB TUTORIAL 24.1

**sampling.** Figure 24–11 shows the procedure for amniocentesis, in which a small volume of the amniotic fluid surrounding the fetus is removed. The amniotic fluid contains fetal cells that can be used for karyotyping, genetic testing, and other procedures. For chorionic villus sampling, cells from the fetal portion of the placental wall (the chorionic villi) are sampled through a vacuum tube, and analyses can be carried out on this tissue.

DNA is extracted from the samples and digested with *Mst*II. This enzyme cuts three times within a region of the normal $\beta$-globin gene, producing two small DNA fragments. In the mutant sickle-cell allele, the middle *Mst*II site is destroyed by the mutation, and one large restriction fragment is produced by *Mst*II digestion (Figure 24–12). The restriction-enzyme-digested DNA fragments are separated by gel electrophoresis, transferred to a nylon membrane, and visualized by Southern blot hybridization, using a probe from this region. Alternatively, a gene can be amplified by PCR, subjected to RFLP analysis and analyzed by agarose gel electrophoresis without Southern blotting. This approach is much faster because it eliminates the time required for Southern blot hybridization.

Figure 24–12 shows the results of RFLP analysis for sickle-cell anemia in one family. Both parents (I-1 and I-2) are heterozygous carriers of the mutation. *Mst*II digestion of the parents' DNA produces a large band (because of the mutant allele) and two smaller bands (from the normal allele) in each case. The parents' first child (II-1) is homozygous normal because she has only the two smaller bands. The second child (II-2) has sickle-cell anemia; he has only one large band and is homozygous for the mutant allele. The fetus (II-3) has a large band and two small bands and is therefore heterozygous for sickle-cell anemia. He or she will be unaffected but will be a carrier.

Only about 5 to 10 percent of all point mutations can be detected by restriction enzyme analysis because most mutations occur in regions of the genome that do not contain restriction enzyme cutting sites. However, if the gene of interest has been sequenced and the disease-associated mutations are known, geneticists can employ synthetic oligonucleotides to detect these mutations, as described next.

**NOW SOLVE THIS**

Problem 21 on page 666 asks you to use RFLP analysis to determine whether sisters in a family are carriers of hemophilia.

■ HINT: *Differences in the number and location of restriction enzyme sites create RFLPs that can be used to determine genotypes.*

## Genetic Tests Using Allele-Specific Oligonucleotides

Another method of genetic testing involves the use of synthetic DNA probes known as **allele-specific oligonucleotides (ASOs)**. Scientists use these short, single-stranded fragments of DNA to identify alleles that differ by as little as a single nucleotide. In contrast to restriction

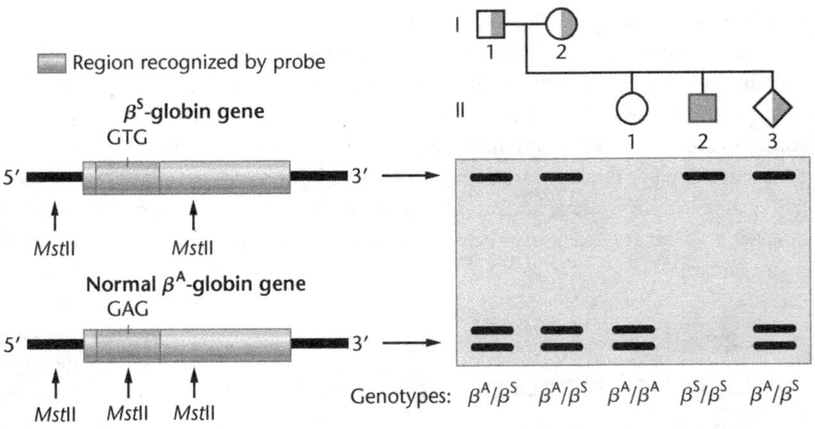


**FIGURE 24–12** RFLP diagnosis of sickle-cell anemia. In the mutant $\beta$-globin allele ($\beta^S$), a point mutation (GAG→GTG) has destroyed a cutting site for the restriction enzyme *Mst*II, resulting in a single large fragment on a Southern blot. In the pedigree, the family has one unaffected homozygous normal daughter (II-1), an affected son (II-2), and an unaffected carrier fetus (II-3). The genotype of each family member can be read directly from the blot and is shown below each lane.

enzyme analysis, which is limited to cases for which a mutation changes a restriction site, ASOs detect single-nucleotide changes of all types, including those that do not affect restriction enzyme cutting sites. As a result, this method offers increased resolution and wider application. Under proper conditions, an ASO will hybridize only with its complementary DNA sequence and not with other sequences, even those that vary by as little as a single nucleotide.

Genetic testing using ASOs and PCR analysis are now available to screen for many disorders, including sickle-cell anemia. In the case of sickle-cell screening, DNA is extracted, and a region of the $\beta$-globin gene is amplified by PCR. A small amount of the amplified DNA is spotted onto strips of a DNA-binding membrane, and each strip is hybridized to an ASO synthesized to resemble the relevant sequence from either a normal or mutant $\beta$-globin gene (Figure 24–13). The ASO is tagged with a molecule that is either radioactive or fluorescent, in order to allow visualization of hybridization of the ASO on the membrane. This rapid, inexpensive, and highly accurate technique is used to diagnose a wide range of genetic disorders caused by point mutations. Although highly effective, point mutations (single-nucleotide polymorphisms) can affect probe binding leading to false positive or false negative results that may not reflect a genetic disorder. Sometimes DNA sequencing is carried out on amplified gene segments to confirm identification of a mutation. Because the Human Genome Project has revealed point mutations involved in many human diseases, the use of ASOs for genetic testing is increasing.

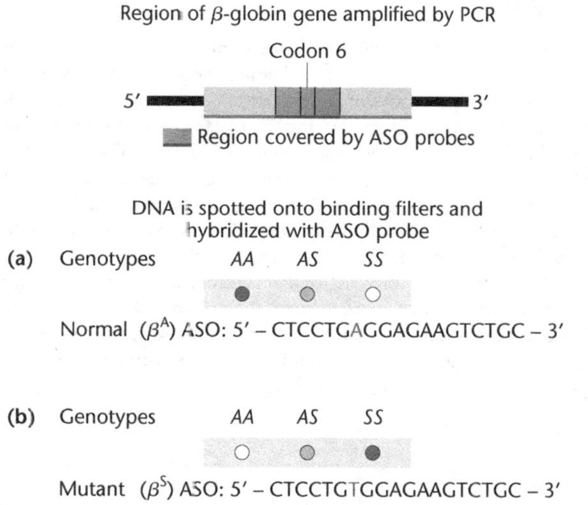


**FIGURE 24–13** Allele-specific oligonucleotide (ASO) testing for the $\beta$-globin gene and sickle-cell anemia. The $\beta$-globin gene is amplified by PCR, using DNA extracted from white blood cells or cells obtained by amniocentesis. The amplified DNA is then denatured and spotted onto strips of DNA-binding membranes. Each strip is hybridized to a specific ASO and visualized on X-ray film after hybridization and exposure. (a) Results observed when the three possible genotypes are hybridized to an ASO from the normal $\beta$-globin allele: *AA*-homozygous individuals have normal hemoglobin that has two copies of the normal $\beta$-globin gene and will show heavy hybridization; *AS*-heterozygous individuals carry one normal $\beta$-globin allele and one mutant allele and will show weaker hybridization; *SS*-homozygous sickle-cell individuals carry no normal copy of the $\beta$-globin gene and will show no hybridization to the ASO probe for the normal $\beta$-globin allele. (b) Results observed when DNA for the three genotypes are hybridized to the probe for the sickle-cell $\beta$-globin allele: no hybridization by the *AA* genotype, weak hybridization by the heterozygote (*AS*), and strong hybridization by the homozygous sickle-cell genotype (*SS*).

Because ASO testing makes use of PCR, small amounts of DNA can be analyzed. As a result, ASO testing is ideal for **preimplantation genetic diagnosis (PGD)**. PGD is the genetic analysis of single cells from embryos created by *in vitro* fertilization (Figure 24–14). When sperm and eggs are mixed to create zygotes, the early-stage embryos are grown in culture. A single cell can be removed from an early-stage embryo using a vacuum pipette to gently aspirate one cell away from the embryo [Figure 24–14(a)]. This could possibly kill the embryo, but if it is done correctly the embryo will often continue to divide normally. DNA from the removed cell is then typically analyzed by FISH (for chromosome analysis) or by ASO testing [Figure 24–14(b)]. The genotypes for each cell can then be used to decide which embryos will be implanted into the uterus. Any alleles that can be detected by ASO testing can be used for PGD. Sickle-cell anemia, cystic fibrosis, and dwarfism are often tested for by PGD, but many other alleles are often also analyzed, depending on the genetic history of the egg and sperm donors.

ASOs can also be used to screen for disorders that involve deletions instead of single-nucleotide mutations. An example is the use of ASOs to diagnose **cystic fibrosis (CF)**. CF is an autosomal recessive disorder associated with a defect in a protein called the **cystic fibrosis transmembrane conductance regulator (CFTR)**, which regulates chloride ion transport across the plasma membrane. A small deletion called *Δ508* is found in 70 percent of all mutant copies of the *CFTR* gene. To detect carriers of the *Δ508* mutation, ASOs are made by PCR from cloned samples of the normal and mutant alleles. DNA extracted from white blood cells of the individuals to be tested is amplified by PCR and then hybridized to each ASO (Figure 24–15). In affected individuals, only the ASO from the mutant allele hybridizes. In heterozygotes, both ASOs hybridize, and in normal homozygotes, only the ASO from the normal allele hybridizes.

CF affects approximately 1 in 2000 individuals of northern European descent. Screening for CF can be used in these populations to detect carriers and to counsel people about their genetic status with respect to CF. However, geneticists have evidence of more than 1000 different mutations for this gene, but not all of them can be screened because they have not all been specifically characterized. Thus, a negative result does not eliminate someone as a heterozygous carrier. Furthermore, it is likely that still more CF mutations remain to be identified. Consequently, CF screening is not widespread but will no doubt become commonplace when tests can detect 98 to 99 percent of all possible mutations.

**NOW SOLVE THIS**

Problem 20 on page 666 asks whether DNA sequences from normal and sickle-cell alleles of the $\beta$-globin gene will bind to a given ASO.

■ HINT: *ASO analysis is done under conditions that allow only identical nucleotide sequences to hybridize to the ASO on the filter.*

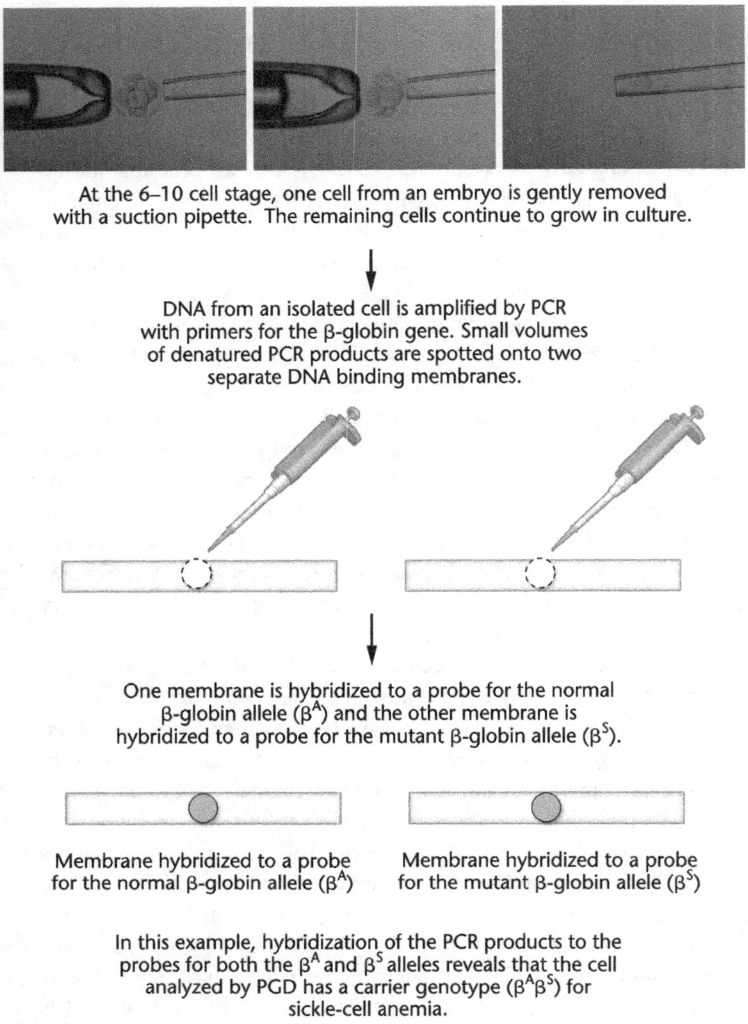


**FIGURE 24–14** A single cell from an early-stage human embryo created by *in vitro* fertilization can be removed and subjected to preimplantation genetic diagnosis (PGD) by ASO testing. DNA from each cell is isolated, amplified by PCR with primers specific for the gene of interest, then subjected to ASO analysis as shown in Figure 24-13. In this example, a region of the *β*-globin gene was amplified and analyzed by ASO testing to determine the sickle-cell genotype for this cell.

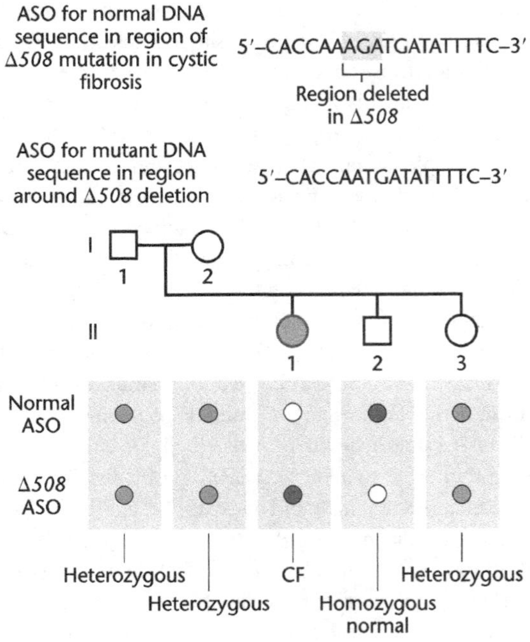


**FIGURE 24–15** Detecting a deletion in the CFTR gene by ASO testing. ASOs for the region spanning the most common mutation in CF (the Δ*508* allele) are prepared from cloned copies of the normal allele and the Δ*508* allele, which contains a small deletion, and spotted onto DNA-binding membranes. In screening, the CF alleles carried by an individual are amplified by PCR, labeled, and hybridized to the ASOs on the membranes. The genotype of each family member can then be read directly from the membranes. DNA from the parents (I-1 and I-2) hybridizes to both ASOs, indicating that they each carry one normal allele and one mutant allele and are therefore heterozygous. DNA from II-1 hybridizes only to the Δ*508* ASO, indicating that this family member is homozygous for the mutation and has cystic fibrosis. DNA from II-2 hybridizes only to the ASO from the normal CF allele, indicating that this individual carries two normal alleles. DNA from II-3 shows two hybridization spots, so the person is heterozygous and a carrier for CF.

## Genetic Testing Using DNA Microarrays and Genome Scans

Both RFLP and ASO analyses are efficient methods of screening for gene mutations; however, they can only detect the presence of one or a few specific mutations whose identity and locations in the gene are known. There is also a need for genetic tests that detect complex mutation patterns or previously unknown mutations in genes associated with genetic diseases and cancers. For example, the gene that is responsible for cystic fibrosis (the *CFTR* gene) contains 27 exons and encompasses 250 kilobases of genomic DNA. As mentioned earlier, of the 1000 mutations that have been counted for the *CFTR* gene, only about half have been characterized—that is, described as specific point mutations, insertions, and deletions—and these are widely distributed throughout the gene. Moreover, additional *CFTR* mutations may yet be discovered. Similarly, over 500 different mutations are known to occur within the tumor suppressor *p53* gene, and any of these mutations may be associated with, or predispose a patient to, a variety of cancers. In order to screen for mutations in these genes, comprehensive, high-throughput methods are required, such as those described in Chapter 21.

Recall from Chapter 21 that one emerging high-throughput screening technique is based on the use of **DNA microarrays.** DNA

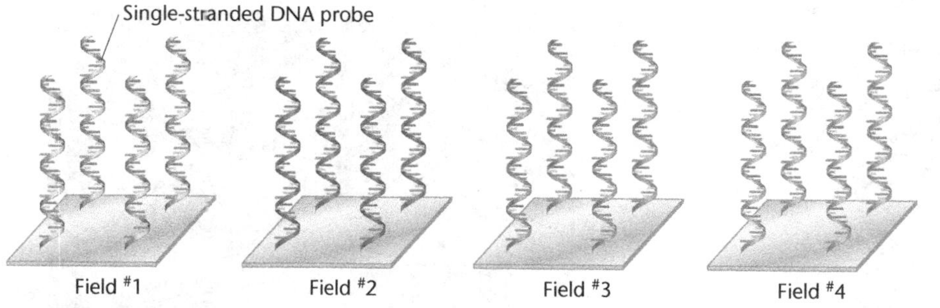


FIGURE 24–16 On a microarray, short, single-stranded DNA molecules (probes) of known sequence are attached to a glass substrate. Each cluster of identical molecules occupies an area known as a field on the microarray. Each field is about half the width of a human hair.

microarrays (also called DNA chips or gene chips) are small, solid supports, usually glass or polished quartz-based, on which known fragments of DNA are deposited in a precise pattern. Each spot on a DNA microarray is called a **field**. The DNA fragments that are deposited in a DNA microarray field—called probes—are single-stranded and may be oligonucleotides synthesized *in vitro* or longer fragments of DNA created from cloning or PCR amplification. There are typically over a million identical molecules of DNA in each field (Figure 24–16). The numbers and types of DNA sequences on a microarray are dictated by the type of analysis that is required. For example, each field on a microarray might contain a DNA sequence derived from each member of a gene family, or sequence variants from one or several genes of interest, or a sequence derived from each gene in an organism's genome. Some microarrays use identical sequences as probes in each field (as shown in Figure 24–26) for a particular gene, other microarrays use many different probes for the same gene. Scientists use DNA microarrays as platforms on which to hybridize a DNA (or RNA) sample to be analyzed.

What makes DNA microarrays so amazing is the immense amount of information that can be simultaneously generated from a single array. DNA microarrays the size of postage stamps (just over 1 cm square) can contain up to 500,000 different fields, each representing a different DNA sequence (Figure 24–16). In Chapter 21, you learned about the use of microarrays for transcriptome analysis. Scientists are now using DNA microarrays in a wide range of applications, including the detection of mutations in genomic DNA and the detection of gene expression patterns in diseased tissues.

Many different microarrays for genome analysis of eukaryotes, prokaryotes, and viruses are now commercially available. For instance, most human genes are available on a human genome microarray (Figure 24–17). Geneticists often use a type of DNA microarray known as a **genotyping microarray** to detect mutations in specific genes. The probes on a genotyping microarray consist of short oligonucleotides, about 20 nucleotides long, synthesized directly on the fields of the microarray. These probes are designed to methodically scan through the gene of interest, one nucleotide at a time, checking for the presence of a mutation at each position in the gene. Each position in the gene is tested by a set of five oligonucleotides (and hence five fields, arranged in a column) that are identical in sequence except for one nucleotide that differs in each of them, being either A, C, G, T, or a deletion (Figure 24–18).

The genomic DNA sample to be tested on such a microarray is cleaved into small fragments, and the DNA regions of interest are amplified by PCR, labeled with a fluorescent dye, and then hybridized to the microarray. DNA molecules with a sequence that exactly matches the sequence present in a microarray field will hybridize to that field. DNA molecules that differ by one or more nucleotides will hybridize less efficiently or not at all, depending on the hybridization conditions. After washing off material that does not hybridize to the microarray, scientists analyze the microarray with a fluorescence scanner to determine which fields hybridized to the test DNA sample and which fields did not hybridize.

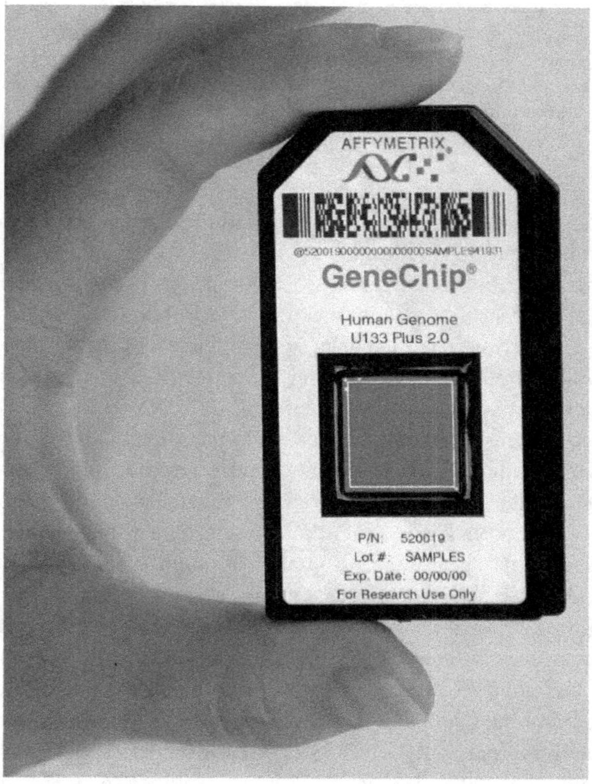


FIGURE 24–17 A commercially available DNA microarray, called a GeneChip, marketed by Affymetrix, Inc. This microarray can be used to analyze expression for approximately 50,000 RNA transcripts. It contains 22 different probes for each transcript and allows scientists to simultaneously assess the expression levels of most of the genes in the human genome.

```
Region of wild-type
p53 gene to be tested:   5' - acttgtcatg gcgactgtcc acctttgtgc - 3'

Set 1  Probe 1   3' - tgaacagtaa cgctgacagg - 5'
       Probe 2   3' - tgaacagtac cgctgacagg - 5'
       Probe 3   3' - tgaacagtat cgctgacagg - 5'
       Probe 4   3' - tgaacagtag cgctgacagg - 5'
       Probe 5   3' - tgaacagta- cgctgacagg - 5'

Set 2  Probe 1   3' - gaacagtac agctgacagg t - 5'
       Probe 2   3' - gaacagtac cgctgacagg t - 5'
       Probe 3   3' - gaacagtac tgctgacagg t - 5'
       Probe 4   3' - gaacagtac ggctgacagg t - 5'
       Probe 5   3' - gaacagtac -gctgacagg t - 5'

Set 3  Probe 1   3' - aacagtac cactgacagg tg - 5'
       Probe 2   3' - aacagtac ccctgacagg tg - 5'
       Probe 3   3' - aacagtac ctctgacagg tg - 5'
       Probe 4   3' - aacagtac cgctgacagg tg - 5'
       Probe 5   3' - aacagtac c-ctgacagg tg - 5'
```

**FIGURE 24-18** Example of the oligonucleotide probe design for use on a genotyping DNA microarray. Each probe would occupy a different field on the microarray. Each set of five probes is aligned under the nucleotide position to be tested (highlighted in blue). Each of the otherwise identical probes in a set contains either an A, C, T, G or a deletion at the test position. One probe in each set is complementary to the wild-type DNA sequence; all other probes are complementary to potential mutations at the same position. Each nucleotide in the gene is tested with a set of five probes, and each probe set is offset from the previous set by one nucleotide. The pattern is repeated throughout the gene until every nucleotide has been tested. These particular probes are designed to test for mutations in the human *p53* gene.

Under stringent hybridization conditions, the PCR-amplified DNA fragment will hybridize most efficiently to its exact complement but less efficiently to the other possible sequence variants. Hence, the strongest hybridization among the five fields will be to the oligonucleotide probe that contains the correct nucleotide at the tested position. The sequence of the test DNA can therefore be determined by reading through the columns and noting the fields to which the hybridization is most intense (Figure 24–19). DNA microarrays have been designed to scan for mutations in many disease-related genes, including the *p53* gene, which is mutated in a majority of human cancers, and the *BRCA1* gene, which, when mutated, predisposes women to breast cancer. Figure 24–19 shows results for a genotypic microarray for the *p53* gene. In this example, hybridization of PCR-amplified DNA fragments from a patient appears as green or blue-green spots on the microarray [Figure 24–19(b)]. Interpreting the hybridization pattern reveals the *p53* gene sequence from the patient ("target").

In addition to testing for mutations in single genes, DNA microarrays can contain probes that detect markers known as **single-nucleotide polymorphisms (SNPs)**. SNPs occur randomly about every 100 to 300 nucleotides throughout the human genome, both inside and outside of genes. Scientists have discovered that certain SNP sequences at a specific locus are shared by certain segments of the population. In addition, certain SNPs cosegregate with genes associated with some disease conditions. By correlating the presence or absence of a particular SNP with a genetic disease, scientists are able to use the SNP as a genetic testing marker. The presence of SNPs as

(a)

(b)

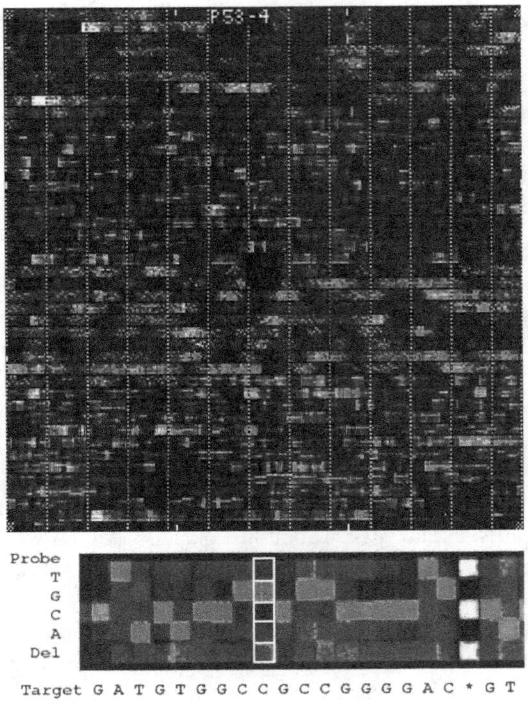


**FIGURE 24-19** A *p53* GeneChip (Affymetrix, Inc.) after hybridization to *p53* PCR products. (a) This DNA microarray contains over 65,000 fields, each about 50 micrometers square. The microarray tests for mutations within the entire 1.2-kb coding sequence of the human *p53* gene. The dotted vertical lines are generated by control fields and allow for alignment of the DNA chip during scanner analysis. (b) An enlarged portion of the *p53* microarray. Each square is a field to which the test DNA has or has not hybridized. Each row represents one of the four nucleotides or a deletion in the test position of the probe. Each column represents the nucleotide position being tested. The column outlined in white shows that the test DNA sample has a C at this position in the *p53* sequence based on its hybridization to the G in this field. The column marked with an asterisk (*) is the alignment control. The sequence of the *p53* gene is shown under the photo.

probes on a DNA microarray allows scientists to simultaneously screen thousands of genes that might be involved in single-gene diseases as well as those involved in disorders exhibiting multifactorial inheritance (a pattern discussed in Chapter 25). This technique, known as **genome scanning**, makes it possible to analyze a person's DNA for dozens or hundreds of disease alleles, including those that might predispose the person to heart attacks, asthma, diabetes, Alzheimer disease, and other genetically defined disease subtypes. Although genome scanning approaches are not widely used yet, perhaps in decades to come, all newborns will undergo genome scanning to determine their lifetime risks for suffering from genetic disorders.

## Genetic Analysis Using Gene Expression Microarrays

When we discussed microarrays in Chapter 21, we focused on their research applications for **transcriptome analysis**, that is, in studying patterns of gene expression. Microarrays are also effective for analyzing gene-expression patterns in genetic diseases because the progression of a tissue from a healthy to a diseased state is almost always accompanied by changes in expression of hundreds to thousands of genes. Microarrays used for this purpose are known as **gene expression microarrays**, and they provide a powerful tool for diagnosing genetic disorders and gene expression changes. Gene expression microarrays differ somewhat from those used for genotyping in that the probes on expression microarrays may be either cDNA fragments or synthetic oligonucleotides that represent the coding regions of genes to be profiled. An expression array may contain probes for only a few specific genes thought to be expressed differently in two cell types or may contain probes representing each gene in the genome. Although microarray techniques provide novel information about gene expression, keep in mind that DNA microarrays do not directly provide us with information about protein levels in a cell or tissue. We often infer what predicted protein levels may be based on mRNA expression patterns but this may not always be accurate.

In a typical expression microarray analysis, mRNA is isolated from two different cell or tissue types—for example, normal cells and cancer cells arising from the same cell type [Figure 24–20(a)]. The mRNA samples contain transcripts from each gene that is expressed in that cell type. Some genes are expressed more efficiently than others; therefore, each type of mRNA is present at a different level. The level of each mRNA can be used to develop a gene expression profile that is characteristic of the cell type. Isolated mRNA molecules are converted into cDNA molecules, using reverse transcriptase. The cDNAs from the normal cells are tagged with fluorescent dye-labeled nucleotides (for example, green), and the cDNAs from the cancer cells are tagged with a different fluorescent dye-labeled nucleotide (for example, red). The labeled cDNAs are mixed together and applied to a DNA microarray. The cDNA molecules bind to complementary single-stranded probes on the microarray but not to other probes. After washing off the nonbinding cDNAs, scientists scan the microarray with a laser, and a computer captures the fluorescent image pattern for analysis. The pattern of hybridization appears as a series of colored dots, with each dot corresponding to one field of the microarray [Figure 24–20(b)].

The colors on the microarray fields [Figure 24–20(c)] provide a sensitive measure of the relative levels of each cDNA in the mixture. In the example shown here, if an mRNA is present only in normal cells, the probe representing the gene encoding that mRNA will appear as a green dot because only "green" cDNAs have hybridized to it. Similarly, if an mRNA is present only in the cancer cells, the microarray probe for that gene will appear as a red dot. If both samples contain the same cDNA, in the same relative amounts, both cDNAs will hybridize to the same field, which will appear yellow [Figure 24–20(b)]. Intermediate colors indicate that the cDNAs are present at different levels in the two samples.

Expression microarray profiling has revealed that certain cancers have distinct patterns of gene expression and that these patterns correlate with factors such as the cancer's stage, clinical course, or response to treatment. In one such experiment, scientists examined

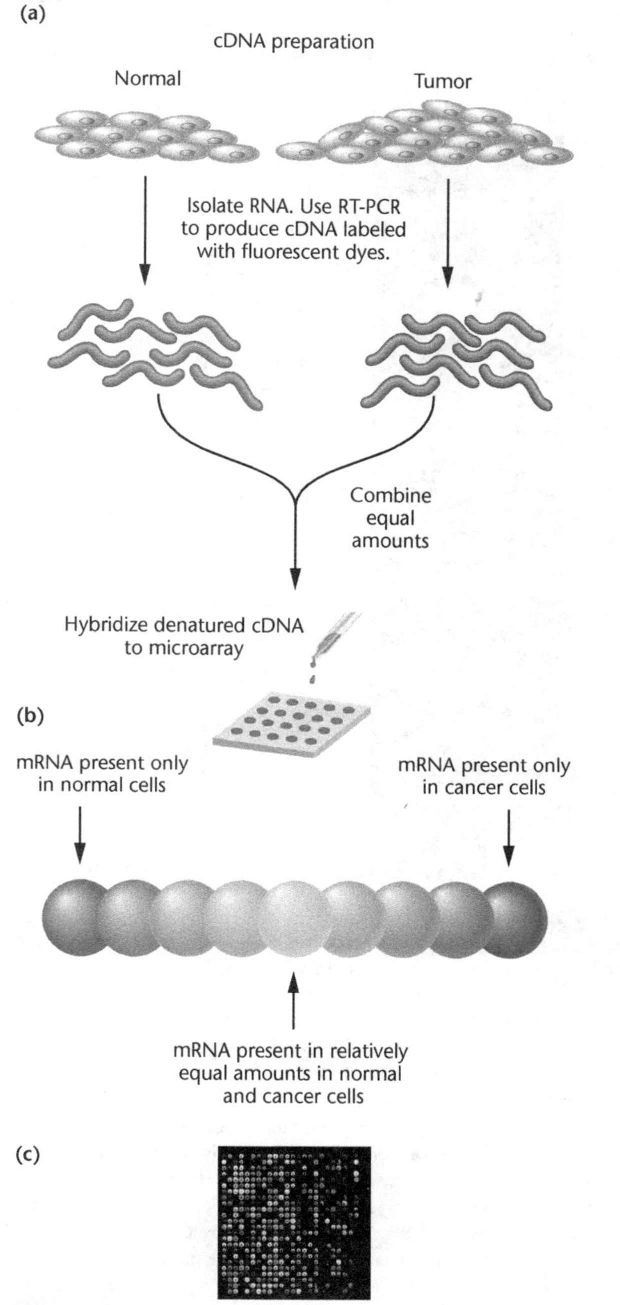


FIGURE 24–20 (a) Microarray procedure for analyzing gene expression in normal and cancer cells. (b) The method shown here is based on a two-channel microarray in which cDNA samples from the two different tissues are competing for binding to the same probe sets. Colors of dots. Colors of dots on an expression microarray represent levels of gene expression. In this example, green dots represent genes expressed only in one cell type (e.g., normal cells), and red dots represent genes expressed only in another cell type (e.g., cancer cells). Intermediate colors represent different levels of expression of the same gene in the two cell types. (c) A small portion of a DNA microarray, showing different levels of hybridization to each field.

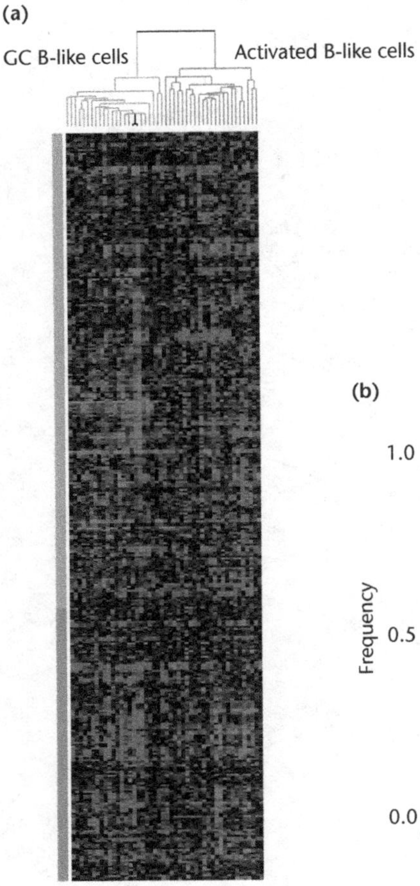


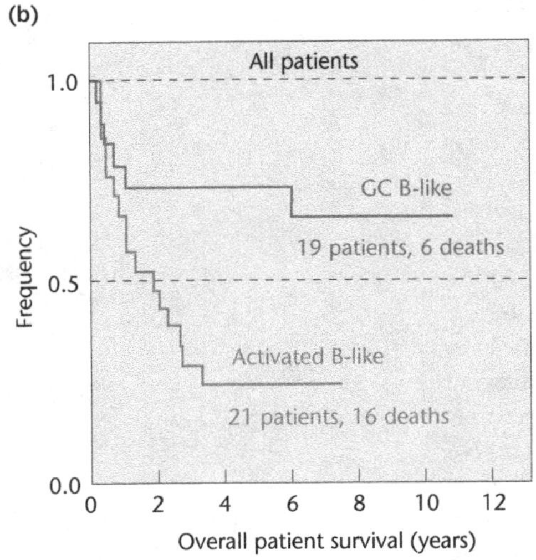


**FIGURE 24–21** (a) Gene expression analysis generated from expression DNA microarrays that analyzed 18,000 genes expressed in normal and cancerous lymphocytes. Each row represents a summary of the gene expression from one particular gene; each column represents data from one cancer patient's sample. The colors represent ratios of relative gene expression compared to normal control cells. Red represents expression greater than the mean level in controls, green represents expression lower than in the controls, and the intensity of the color represents the magnitude of difference from the mean. In this summary analysis, the cancer patients' samples are grouped by how closely their gene expression profiles resemble each other. The cluster of cancer patients' samples marked with orange at the top of the figure are GC B-like DLBCL cells. The blue cluster contains samples from cancer patients within the activated B-like DLBCL group. (b) Gene expression profiling and survival probability. Patients with activated B-like profiles have a much higher rate of death (16 in 21) than those with GC B-like profiles (6 in 19). Data such as these demonstrate the value of microarray analysis for diagnosing disease conditions.

gene expression in both normal white blood cells and in cells from a white blood cell cancer known as **diffuse large B-cell lymphoma (DLBCL)**. About 40 percent of patients with DLBCL respond well to chemotherapy and have long survival times. The other 60 percent respond poorly to therapy and have short survival. The investigators assayed the expression profiles of 18,000 genes and discovered that there were two types of DLBCL, with almost inverse patterns of gene expression (Figure 24–21). One type of DLBCL, called *GC B-like*, had an expression pattern dramatically different from that of a second type, called *activated B-like*. Patients with the activated B-like pattern of gene expression had much lower survival rates than patients with the GC B-like pattern. The researchers concluded that DLBCL is actually two different diseases with different outcomes. Once this type of analysis is introduced into routine clinical use, it may be possible to adjust therapies for each group of cancer patients and to identify new specific treatments based on gene expression profiles. Similar gene expression profiles have been generated for many other cancers, including breast, prostate, ovarian, and colon cancer. Gene expression microarrays are providing tremendous insight into both substantial and subtle variations in genetic diseases.

The Chapter 21 discussion of microarrays mentioned that several companies are now promoting "nutrigenomics" services in which they claim to use genotyping and gene expression microarrays to identify allele polymorphisms and gene expression patterns for genes involved in nutrient metabolism. For example, polymorphisms in genes such as that for apolipoprotein A (*APOA1*), involved in lipid metabolism, and that for *MTHFR* (methylenetetrahydrofolate reductase), involved in metabolism of folic acid, have been implicated in cardiovascular disease. Nutrigenomics companies claim that microarray analysis of a patient's DNA sample for genes such as these and others enables them to judge whether a patient's allele variations or gene expression profiles warrant dietary changes to potentially improve health and reduce the risk of diet-related diseases.

## Application of Microarrays for Gene Expression and Genotype Analysis of Pathogens

Microarrays are also providing infectious disease researchers with powerful new tools for studying pathogens. Genotyping microarrays are being used to identify strains of emergent viruses, such as the virus that causes the highly contagious condition called Severe Acute Respiratory Syndrome (SARS) as well as the H5N1 avian influenza virus, the cause of bird flu, which has killed about two dozen people in Asia, leading to the slaughter of over 80 million chickens and causing concern about possible pandemic outbreaks.

Whole-genome transcriptome analysis of pathogens is being used to inform researchers about genes that are important for pathogen infection and replication (Figure 24–22). In this approach, bacterial, yeast, protists or viral pathogens are used to infect host cells *in vitro*, and then expression microarrays are used to analyze pathogen gene expression profiles. Patterns of gene activity during pathogen infection of host cells and replication are useful for identifying pathogens and understanding mechanisms of infection.

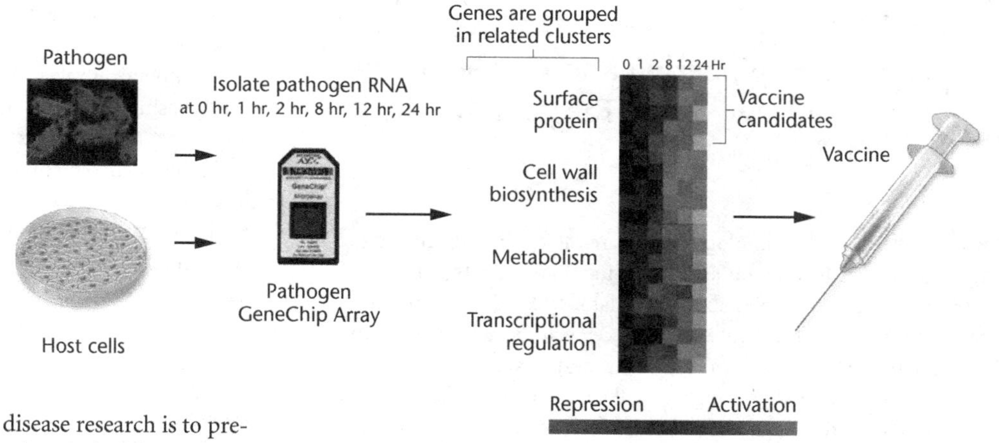


**FIGURE 24–22** Whole-genome expression profiling enables researchers to identify genes that are actively expressed when pathogens infect host cells and replicate. Knowing which genes are critical for pathogen infection and replication can help scientists target proteins as vaccine and drug-treatment candidates for preventing or treating infectious diseases.

But of course a primary goal of infectious disease research is to prevent infection. Gene expression profiling is also a valuable approach for identifying important pathogen genes and the proteins they encode that may prove to be useful targets for subunit vaccine development or for drug treatment strategies to prevent or control infectious disease. This strategy primarily informs researchers about how a pathogen responds to its host.

Similarly, researchers are evaluating host responses to pathogens (Figure 24–23). This type of detection has been accelerated in part by the need to develop pathogen-detection strategies for military and civilian use both for detecting outbreaks of naturally emerging pathogens such as SARS and avian influenza and for potential detection of outbreaks such as anthrax (caused by the bacterium *Bacillus anthracis*) that could be the result of a bioterrorism event. Host-response gene expression profiles are developed by exposing a host to a pathogen and then using expression microarrays to analyze host gene expression patterns.

Figure 24–23 shows the different gene expression profiles for mice following exposure to *Neisseria meningitidis*, the SARS virus, or *E. coli*. In this example, although there are several genes that are upregulated or downregulated by each pathogen, notice how each pathogen strongly induces different prominent clusters of genes that reveal a host gene expression response to the pathogen and provide a signature of pathogen infection. Comparing such host gene expression profiles following exposure to different pathogens provides researchers with a way to quickly diagnose and classify infectious diseases. In the future, scientists expect to develop databases of both pathogen and host response expression profile data that can be used to identify pathogens efficiently.

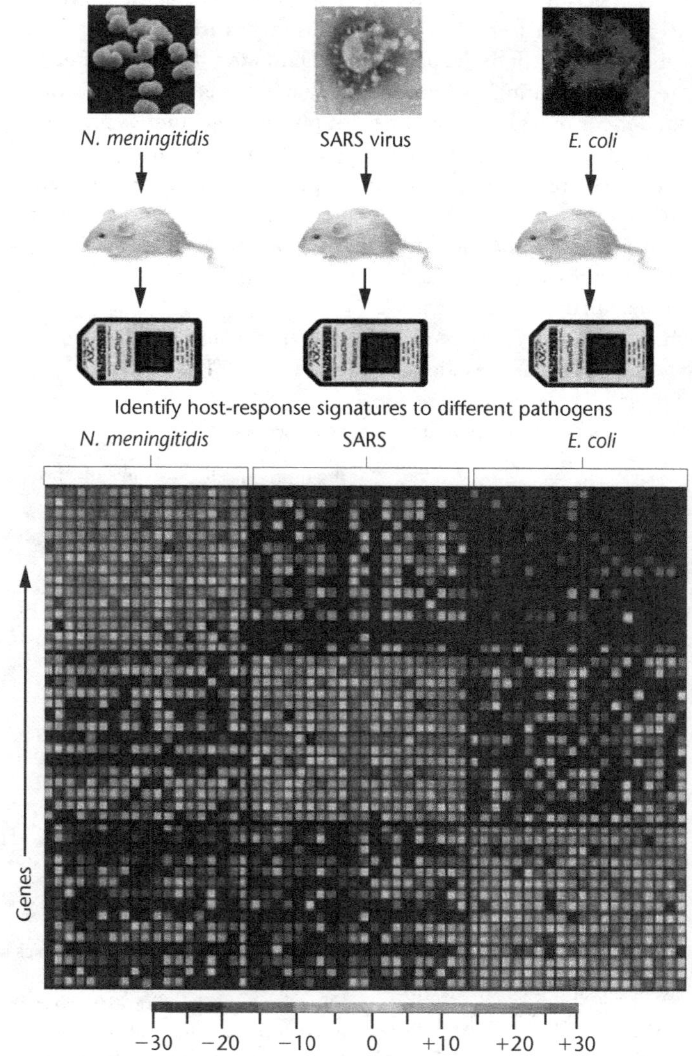


**FIGURE 24–23** Gene expression microarrays can reveal host-response signatures for pathogen identification. In this example, mice were infected with different pathogens: *Neisseria meningitidis*, the virus that causes Severe Acute Respiratory Syndrome (SARS), and *E. coli*. Mouse tissues were then used as the source of mRNA for gene expression microarray analysis. Increased expression compared to uninfected control mice is shown in shades of yellow. Decreased expression compared to uninfected controls is indicated in shades of blue. Notice that each pathogen elicits a somewhat different response in terms of which major clusters of host genes are activated by pathogen infection.

24.5

# Genetic Engineering and Genomics Promise New, More Targeted Medical Therapies

Recombinant DNA technologies are changing medical diagnosis and allowing scientists to manufacture abundant and effective therapeutic proteins. The examples already available today are a strong indication that in the near future, we will see even more transformative medical treatments based on genomics and advanced DNA-based technologies. In this section, we will examine two exciting new methodologies that have the potential to cure genetic diseases and yield specific drug treatments.

## Pharmacogenomics and Rational Drug Design

Every year, more than 2 million Americans experience serious side effects of medications, and more than 100,000 die from adverse drug reactions. In addition, most drugs are effective in only about 60 percent of the population. Until now, the selection of effective medications for each individual has been a random, trial-and-error process. The new field of **pharmacogenomics** promises to lead to more specific, effective, and customized drugs that are designed to complement each person's individual genetic makeup.

Pharmacogenomics began in the 1950s, when scientists discovered that reactions to drugs had a hereditary component. We now know that many genes affect a person's reaction to drugs. These genes encode products such as cell-surface receptors that bind a drug and allow it to enter a cell, as well as enzymes that metabolize drugs. For example, liver enzymes encoded by the cytochrome *P450* gene family affect the metabolism of many modern drugs, including those used to treat cardiovascular and neurological conditions. DNA sequence variations in these genes result in enzymes with different abilities to metabolize and utilize these drugs. Thus, gene variants that encode inactive forms of the cytochrome P450 enzymes are associated with a patient's inability to break down drugs in the body, leading to drug overdoses. A genetic test that recognizes some of these variants is currently being used to screen patients who are recruited into clinical trials for new drugs.

Another example is the reaction of certain people to the thiopurine drugs used to treat childhood leukemias (Figure 24–24). Some individuals have sequence variations in the gene encoding the enzyme thiopurine methyltransferase (TPMT), which breaks down thiopurines. Anticancer drugs such as 6-mercaptopurine (6-MP) that are commonly used to treat leukemia have a thiopurine structure. In individuals with mutations in the *TPMT* gene, thiopurine cancer drugs can build up to toxic levels. As a result, although some patients, such as those who are homozygous for the wild-type *TPMT* gene, respond well to 6-MP treatment, others who are heterozygotes or homozygous for mutations in *TPMT* can have severe or even fatal reactions to 6-MP. At first a genetic test was developed to detect *TPMT* gene variants, but now a simple blood test is enough to enable clinicians to tailor the drug dosage to the individual. As a result of this new technology, toxic effects of 6-MP have decreased, and survival rates for childhood leukemia patients treated with 6-MP have increased—a great example of pharmacogenomics in action.

Several methods are being developed for expanding the uses of pharmacogenomics. One promising method involves the detection of single-nucleotide polymorphisms (SNPs). Earlier in this chapter, we discussed SNPs in the context of detecting mutations through tests based on ASOs, as well as through tests based on genome scans. Perhaps researchers will be able to identify a shared SNP sequence in the DNA of people who also share a heritable reaction to a drug. If the SNP segregates with a part of the genome containing the gene

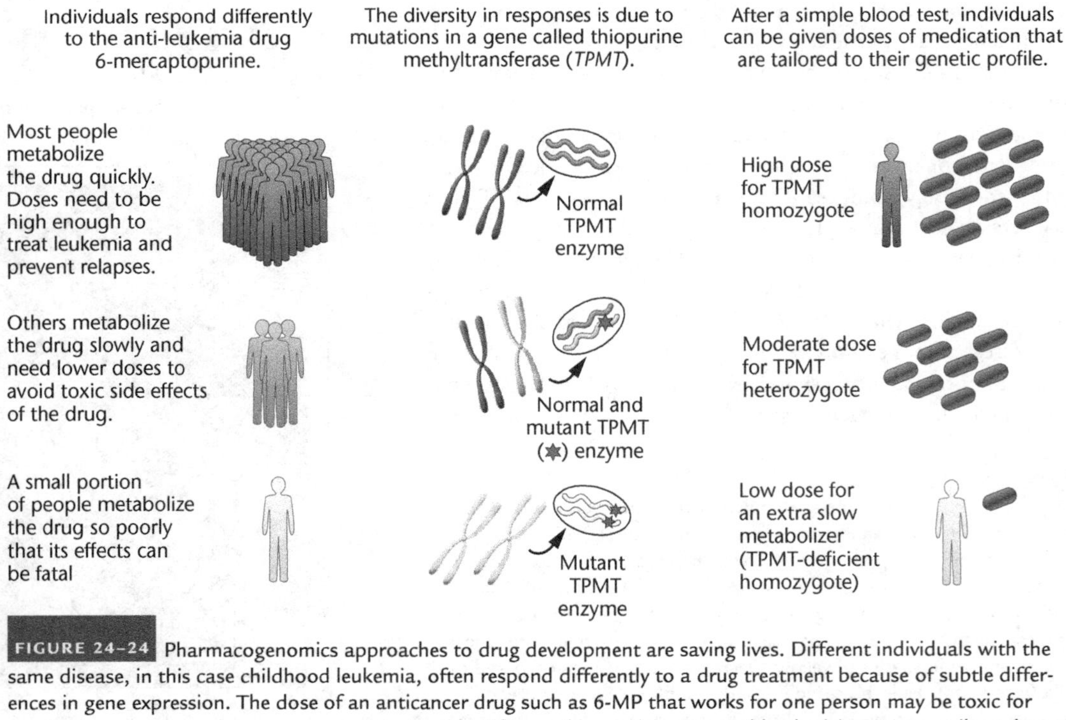


FIGURE 24–24 Pharmacogenomics approaches to drug development are saving lives. Different individuals with the same disease, in this case childhood leukemia, often respond differently to a drug treatment because of subtle differences in gene expression. The dose of an anticancer drug such as 6-MP that works for one person may be toxic for another person. A simple gene or enzyme test to identify genetic variations can enable physicians to prescribe a drug treatment and dosage based on a person's genetic profile.

responsible for the drug reaction, it may be possible to devise gene tests based on the SNP, without even knowing the identity of the gene responsible for the drug reaction. In the future, DNA microarrays may be used to screen a patient's genome for multiple drug reactions. Scientists predict that such gene tests—a type of genome scan—will be available by about 2010.

Knowledge from genetics and molecular biology is also contributing to the development of new drugs targeted at specific disease-associated molecules. Most drug development is currently based on trail-and-error testing of chemicals in lab animals, in the hope of finding a chemical that has a useful effect. In contrast, **rational drug design** involves the synthesis of specific chemical substances that affect specific gene products. An example of a rational drug design product is the new drug imatinib, trade name **Gleevec,** used to treat chronic myelogenous leukemia (CML). Geneticists had discovered that CML cells contain the Philadelphia chromosome, which results from a reciprocal translocation between chromosomes 9 and 22. Gene cloning revealed that the t(9;22) translocation creates a fusion of the *C-ABL* proto-oncogene with the *BCR* gene. This *BCR-ABL* fusion gene encodes a powerful fusion protein that causes cells to escape cell-cycle control. The fusion protein, which acts as a tyrosine kinase, is not present in noncancer cells from CML patients.

To develop Gleevec, chemists used high-throughput screens of chemical libraries to find a molecule that bound to the BCR-ABL enzyme. After chemical modifications to make the inhibitory molecule bind more tightly, tests showed that it specifically inhibited BCR-ABL activity. Clinical trials revealed that Gleevec was effective against CML, with minimal side effects and a higher remission rate than that seen with conventional therapies. Gleevec is now used to treat CML and several other cancers. With scientists discovering more genes and gene products associated with diseases, rational drug design promises to become a powerful technology within the next decade.

## Gene Therapy

For more than two decades, gene products such as insulin have been among the drugs used to treat genetic disorders. Although drug treatments are often effective in controlling symptoms, the ideal outcome of medical treatment is to cure these diseases. In an effort to cure genetic diseases, scientists are actively investigating **gene therapy**—a therapeutic technique that aims to transfer normal genes into a patient's cells. In theory, the normal genes will be transcribed and translated into functional gene products, which, in turn, will bring about a normal phenotype. Delivery of these normal genes and their regulatory sequences would be accomplished by using a vector or gene transfer system.

In many gene therapy trials, scientists often used genetically modified retroviruses as vectors. An example is a vector based on a mouse virus called **Moloney murine leukemia virus (MLV)** (Figure 24–25). The vectors are created by removing a cluster of three genes from the virus and inserting a cloned human gene. After being packaged in a viral protein coat, the recombinant vector is used to infect cells. Once inside a cell, the virus cannot replicate itself because of the missing viral genes. In the cell, the recombinant virus with the inserted human gene moves to the nucleus, integrates into a site on a chromosome, and becomes part of the genome. If the inserted gene is expressed, it produces a normal gene product that may be able to correct the mutation carried by the affected individual. In initial attempts at gene therapy, several heritable disorders, including severe combined immunodeficiency (SCID), familial hypercholesterolemia, and cystic fibrosis were treated. After examining the first attempt to use gene therapy, to treat a young girl with SCID, we will review the trials that are currently underway using a new generation of viral vectors. We will then finish this section by discussing barriers to gene therapy.

Human gene therapy began in 1990 with the treatment of a young girl named Ashanti DeSilva (Figure 24–26), who has a heritable

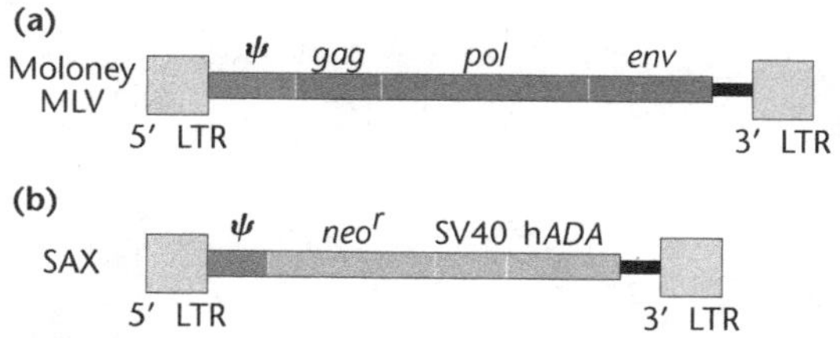


**FIGURE 24–25** (a) The native Moloney MLV genome contains a sequence required for encapsulation ($\psi$), as well as genes that encode viral coat proteins (*gag*), an RNA-dependent DNA polymerase (*pol*), and surface glycoproteins (*env*). At each end, the genome is flanked by long terminal repeat (LTR) sequences that control transcription and integration into the host genome. (b) The SAX vector retains the LTR and $\psi$ sequences and includes a bacterial neomycin resistance ($neo^r$) gene that can be used as a selection marker. As shown, the vector carries a cloned human adenosine deaminase (h*ADA*) gene, which is fused to an SV40 promoter-enhancer. The SAX construct is typical of retroviral vectors that are used in human gene therapy.

**FIGURE 24–26** Ashanti DeSilva, the first person to be treated by gene therapy.

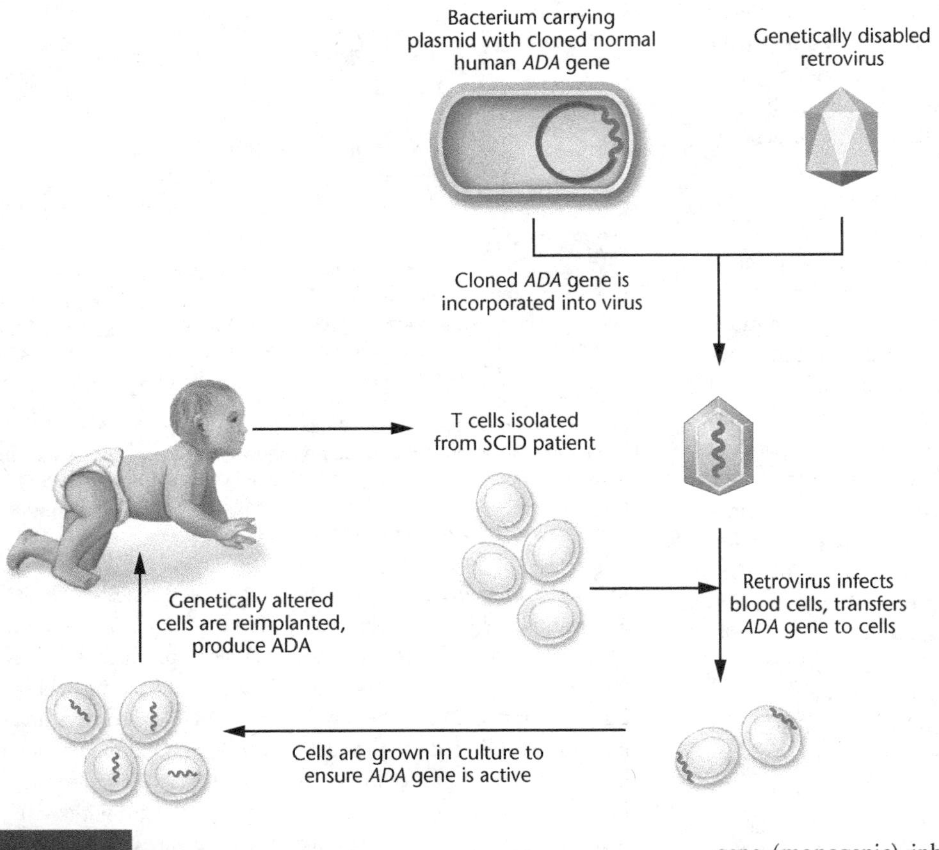


FIGURE 24-27 To treat SCID using gene therapy, a cloned human *ADA* gene is transferred into a viral vector, which is then used to infect white blood cells removed from the patient. The transferred *ADA* gene is incorporated into a chromosome and becomes active. After growth to enhance their numbers, the cells are inserted back into the patient, where they produce ADA, allowing the development of an immune response.

disorder called **severe combined immunodeficiency (SCID)**. Individuals with SCID have no functional immune system and usually die from what would normally be minor infections. Ashanti has an autosomal form of SCID caused by a mutation in the gene encoding the enzyme **adenosine deaminase (ADA)**. Her gene therapy began when clinicians isolated some of her white blood cells, called T cells (Figure 24–27). These cells, which are key components of the immune system, were mixed with a retroviral vector carrying an inserted copy of the normal *ADA* gene. The virus infected many of the T cells, and a normal copy of the *ADA* gene was inserted into the genome of some T cells. After being mixed with the vector, the T cells were grown in the laboratory and analyzed to make sure that the transferred *ADA* gene was expressed. Then a billion or so genetically altered T cells were injected into Ashanti's bloodstream. Some of these T cells migrated to her bone marrow and began dividing and producing ADA. She now has ADA protein expression in 25 to 30 percent of her T cells, which is enough to allow her to lead a normal life.

In later trials, attempts were made to transfer the *ADA* gene into the bone marrow cells that form T cells, but these attempts were mostly unsuccessful. To date, gene therapy has successfully restored the health of about 20 children affected by SCID. Although gene therapy was originally developed as a treatment for single-gene (monogenic) inherited diseases, the technique was quickly adapted for the treatment of acquired diseases such as cancer, neurodegenerative diseases, cardiovascular disease, and infectious diseases, such as HIV. Today, most gene therapy treatments and trials involve these disorders [Figure 24–28(a)].

In addition to retroviral vectors, other viruses and methods are being used to transfer genes into human cells [Figure 24–28(b)]. These methods include the use of other viral vectors, chemically assisted transfer of genes across cell membranes, and fusion of cells with artificial vesicles containing cloned DNA sequences.

Over a 10-year period, from 1990 to 1999, more than 4000 people underwent gene therapy for a variety of genetic disorders. These trials often failed and thus led to a loss of confidence in gene ther-

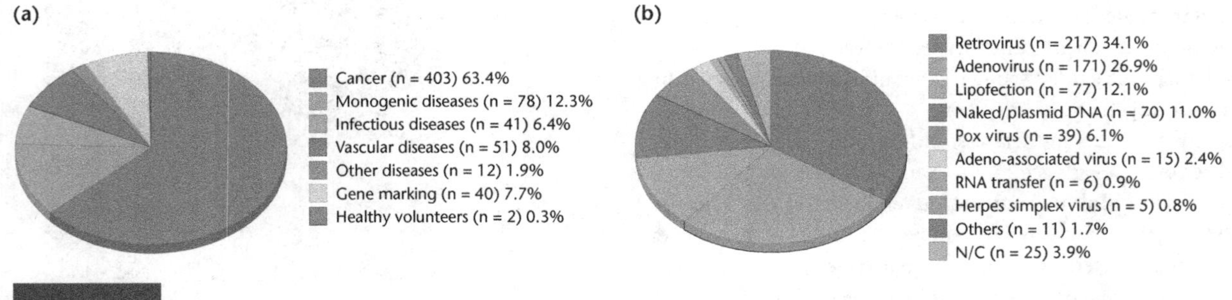


FIGURE 24-28 (a) Pie chart summarizing, by disease, more than 630 gene therapy trials under way worldwide. Most trials involve cancer treatment. (b) Gene therapy trials ranked by the vectors used. Retroviruses are the most widely used vectors, accounting for 34 percent of the total. Newer vectors, such as adeno-associated virus, account for only a small percentage of vectors. N/C = not classified.

apy. Hopes for gene therapy plummeted even further in September 1999 when teenager Jesse Gelsinger died while undergoing gene therapy. His death was triggered by a massive inflammatory response to the vector, a modified **adenovirus**, one of the viruses that cause colds and respiratory infections. The outlook for gene therapy brightened in 2000, when a group of French researchers reported the first large-scale success in gene therapy. Nine children with a fatal X-linked form of SCID developed functional immune systems after being treated with a retroviral vector carrying a normal gene. However, three of these children later developed leukemia, and one died as a result of the treatment. In two of the children, their cancer cells contained the retroviral vector, inserted near or into a gene called *LMO2*. This insertion activated the *LMO2* gene, causing uncontrolled white blood cell proliferation and development of leukemia. The Gelsinger and French X-SCID stories, as well as the future of gene therapy, are discussed further in the Genetics, Technology, and Society essay at the end of this chapter.

Most problems associated with gene therapy have been traced to the vectors used to transfer therapeutic genes into cells. These vectors, including MLV and adenovirus, have several serious drawbacks. First, integration of retroviral genomes (including the human therapeutic gene) into the host cell's genome occurs only if the host cells are replicating their DNA. In the body, only a small number of cells in any tissue are dividing and replicating their DNA. Second, most viral vectors are capable of causing an immune response in the patient, as happened in Jesse Gelsinger's case. Third, insertion of viral genomes into host chromosomes can activate or mutate an essential gene, as in the case of the three French patients. (Unfortunately, it is not yet possible to reliably target insertion of therapeutic genes into specific locations in the genome.) Fourth, retroviruses cannot carry DNA sequences much larger than 8 kb. Many human genes exceed this size. Finally, there is a possibility that a fully infectious virus could be created if the vector were to recombine with another viral genome already present in the host cell.

To overcome these problems, new viral vectors and strategies for transferring genes into cells are being developed in an attempt to improve the action and safety of vectors. Researchers hope that the use of new gene delivery systems will circumvent the problems inherent in earlier vectors, as well as allow regulation of both insertion sites and the levels of gene product produced from the therapeutic genes.

In addition to the vector delivery issues addressed above, a number of other barriers must be overcome if gene therapy is to become a viable approach for reliably treating many genetic disorders. Issues include:

- What is the proper route for gene delivery in different kinds of disorders? For example, what is the best way to treat brain or muscle tissues?
- What percentage of cells in an organ or tissue need to express a therapeutic gene to alleviate the effects of a genetic disorder?
- What amount of a therapeutic gene product must be produced to provide lasting improvement of the condition, and how can sufficient production be ensured? Currently many gene therapy approaches provide only short-lived delivery of the therapeutic gene and its protein.
- Will it be possible to use gene therapy to treat diseases that involve multiple genes? Currently, most gene therapy trials target diseases that are caused by a single gene defect.
- Can expression of therapeutic genes be controlled in a patient?
- Will it be possible to use gene therapy to control the timing of gene expression?

Scientists are also working on gene replacement approaches that involve removing a defective gene from the genome. Encouraging breakthroughs have taken place in this area using model organisms such as mice; however, this technology has not advanced sufficiently for use in humans. Attempts have been made to use antisense oligonucleotides (refer to the GTS in Chapter 14) in order to inhibit translation of mRNAs from defective genes, but this approach to gene therapy has generally not yet proven to be reliable. However, the recent emergence of RNA interference as a powerful gene-silencing tool has reinvigorated gene therapy approaches by gene silencing.

As you learned in Chapters 18 and 23, **RNA interference (RNAi)** is a form of gene expression regulation (see Figures 18–23 and 18–24). In animals short, double-stranded RNA molecules are delivered into cells where the enzyme dicer chops them into 21-nt long pieces called **small interfering RNAs (siRNAs)**. siRNAs then join with an enyzme complex called the **RNA inducing silencing complex (RISC)**, which shuttles the siRNAs to their target mRNA, where they bind by complementary base pairing. The RISC complex can block siRNA-bound mRNAs from being translated into protein or can lead to degradation of siRNA-bound mRNAs so they cannot be translated into protein.

A main challenge to RNAi-based therapeutics so far has been *in vivo* delivery of double-stranded RNA or siRNA. RNAs degrade quickly in the body. It is also hard to get them to penetrate cells and to target the right tissue. Two common delivery approaches are to inject the siRNA directly or to deliver them via a plasmid vector that is taken in by cells and transcribed to make double-stranded RNA that can be cleaved by Dicer into siRNAs.

Several RNAi clinical trials to treat blindness are underway in the United States. One RNAi strategy to treat a form of blindness called macular degeneration targets a gene called *VEGF*. The VEGF protein promotes blood vessel growth. Overexpression of this gene, causing excessive production of blood vessels in the retina, leads to impaired vision and eventually blindness. Many expect that this disease will soon become the first condition to be treated by RNAi therapy. Other disease candidates for treatment by RNAi include several different cancers, diabetes, multiple sclerosis, and arthritis.

**NOW SOLVE THIS**

Problem 16 on page 665 asks you to explain why diseases such as muscular dystrophy would be difficult to cure with gene therapy.

■ HINT: *Consider the types of tissues that are affected in muscular dystrophy patients.*

24.6

# DNA Profiles Help Identify Individuals

DNA FINGERPRINTING
WEB TUTORIAL 24.2

As discussed previously, the presence or absence of restriction sites at specific locations within the human genome can be used as genetic markers. Another type of genetic marker, discovered in the mid-1980s, is based on variations in the length of repetitive DNA sequence clusters. The number of repeats within these clusters varies between individuals and between population groups. These polymorphisms in DNA serve as the basis for **DNA profiling**—also called **DNA fingerprinting** or DNA typing. DNA profiling is used in a wide variety of applications, such as paternity testing, forensics, identification of human or animal remains, archaeology, and conservation biology.

## DNA Profiling Based on DNA Minisatellites (VNTRs)

The first DNA profiling method was developed in the United Kingdom by Sir Alec Jeffreys and was used in 1986 to convict the murderer of two English schoolgirls. This method involves the analysis of large repeated regions of DNA, called **minisatellites**, or **variable numbers of tandem repeats (VNTRs)**.

VNTRs are repeating clusters of about 10 to 100 nucleotides. For example, the VNTR

5′-GACTGCCTGCTAAGAT**GACTGCCTGCTAAGAT**GACT
GCCTGCTAAGAT-3′

is composed of three tandem repeats of the 16-nucleotide sequence GACTGCCTGCTAAGAT. Clusters of such sequences are widely dispersed in the human genome. The number of repeats at each locus ranges from 2 to more than 100, each repeat length representing one allele of that locus. VNTRs are particularly useful for DNA profiling because there are about 30 different possible alleles (repeat lengths) at any VNTR within a population. This creates a large number of possible genotypes. For example, if one examined four different VNTRs within a population, and each VNTR had 20 possible alleles, there would be about 2 billion possible genotypes in this four-locus profile.

To create a VNTR profile, DNA is extracted from the tissue sample and is digested with a restriction enzyme that cleaves on either side of the VNTR repeat region. The digested DNA is separated by gel electrophoresis and subjected to Southern blot analysis (recall our discussion of Southern blotting in Chapter 13). Briefly, separated DNA is transferred from the gel to a membrane, and hybridized with a radioactive probe that recognizes DNA sequences within the VNTR region. After exposing the membrane to X-ray film, the pattern of bands is measured, with larger VNTR repeat alleles remaining near the top of the gel and smaller VNTRs migrating closer to the bottom (Figure 24–29). The pattern of bands is always the same for a given individual, no matter what tissue is used as the source of the DNA. If enough VNTRs are analyzed, each person's DNA profile will be unique (except, of course, for identical twins) because of the huge number of possible VNTRs and alleles. In practice, about five or six VNTR loci are analyzed to create a DNA profile.

A significant limitation of VNTR profiling is that it requires a relatively large sample of DNA (10,000 cells or about 50 $\mu$g of DNA)—more than is usually found at a typical crime scene—and

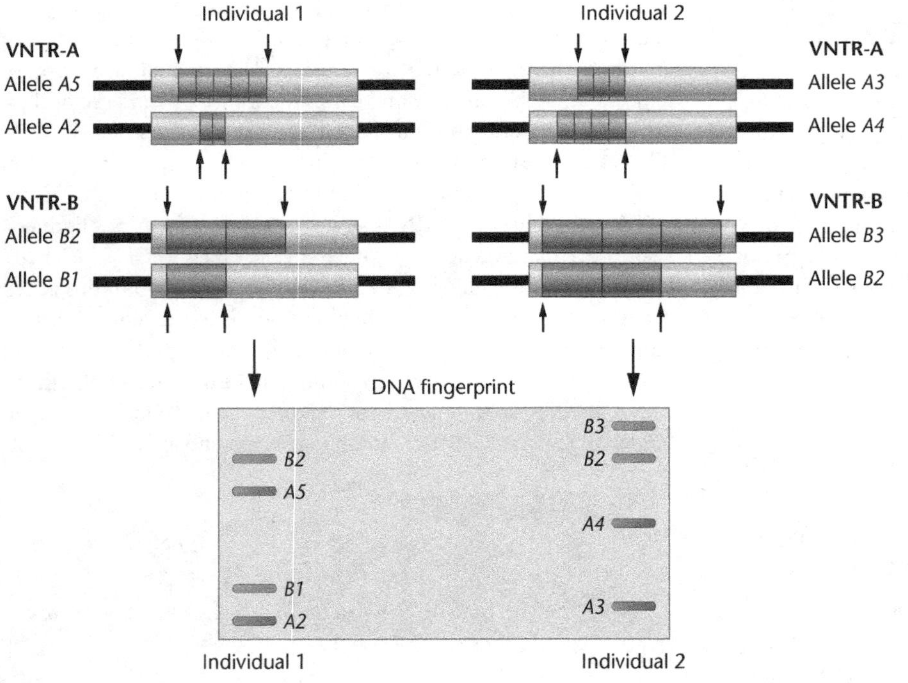


FIGURE 24–29 VNTR alleles at two loci (*A* and *B*) are shown for two different individuals. (Arrows mark restriction enzyme cutting sites flanking the VNTRs.) Restriction enzyme digestion produces a series of fragments that can be detected as bands on a Southern blot (bottom). The number of repeats at each locus is variable, so the overall pattern of bands is distinct for each individual. The DNA fingerprint (profile) shows that these individuals share one allele (*B2*).

the DNA must be relatively intact (nondegraded). As a result, VNTR profiling is used most frequently when large tissue samples are available—such as in paternity testing. Blood samples drawn from the child, mother, and alleged father provide abundant fresh, intact cells for DNA extraction and analysis.

## DNA Profiling Based on DNA Microsatellites

To overcome the problems of DNA sample size and other limitations of VNTR profiling, scientists have developed alternative techniques based on the use of PCR to amplify shorter polymorphic DNA markers. One of these methods examines sequences called **microsatellites,** or **short tandem repeats (STRs)**. STRs are similar to VNTRs, but the repeated motif is shorter—between two and nine base pairs. In addition, only about 7 to 40 repeats are found in each STR locus. Although hundreds of STR loci are found in the human genome, only a subset is used for DNA profiling. The FBI and other law enforcement agencies have selected 13 STR loci to be used as a core set for forensic analysis. DNA profiles based on these loci are stored in a national DNA database called the **Combined DNA Index System (CODIS)**. Profiles generated in criminal cases, as well as for missing persons and from samples collected from crime scenes, are stored in the CODIS.

To create an STR profile, DNA is extracted from the sample and amplified using sets of primers that specifically hybridize to regions of DNA flanking each of the STR loci to be analyzed. Each primer set within the mix is labeled with a different fluorescent dye, making the PCR products amplified from each STR locus a different color. The sizes of the amplified fragments may be measured by standard gel electrophoresis (Figure 24–30). However, recently developed methods for generating STR profiles use faster, more sensitive measures of allele size.

The most common current method is to apply the amplified mixture to the top of a thin capillary tube filled with a gel-like substance and then pass an electric current through the tube. Larger DNA fragments will be retarded in the capillary, and smaller ones will pass through more quickly. This is known as **capillary gel electrophoresis.** At the bottom of the tube, a laser detects when each fluorescent DNA fragment exits the capillary tube and assigns a size to the fragment. All fragments are represented as peaks on a graph (Figure 24–31). This entire process is automated—often within self-contained kits that are used in research labs or in mobile crime units.

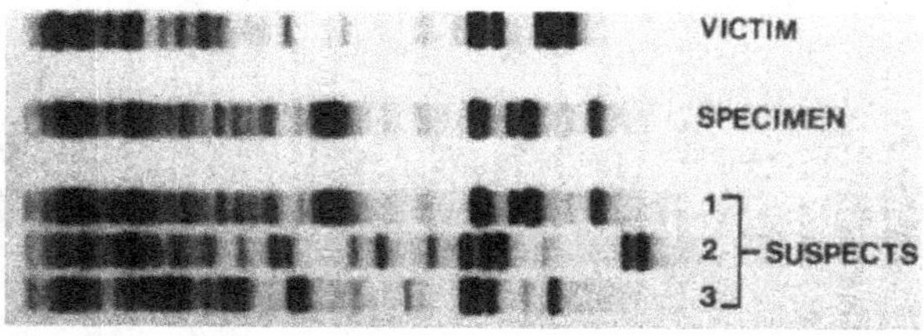


**FIGURE 24–30** STRs were used to create these DNA profiles of the victim, an evidence specimen, and three suspects in a criminal case. (One of the suspects' profiles matches that of the evidence sample.) In this analysis, PCR-amplified DNA was visualized by gel electrophoresis.

(a)

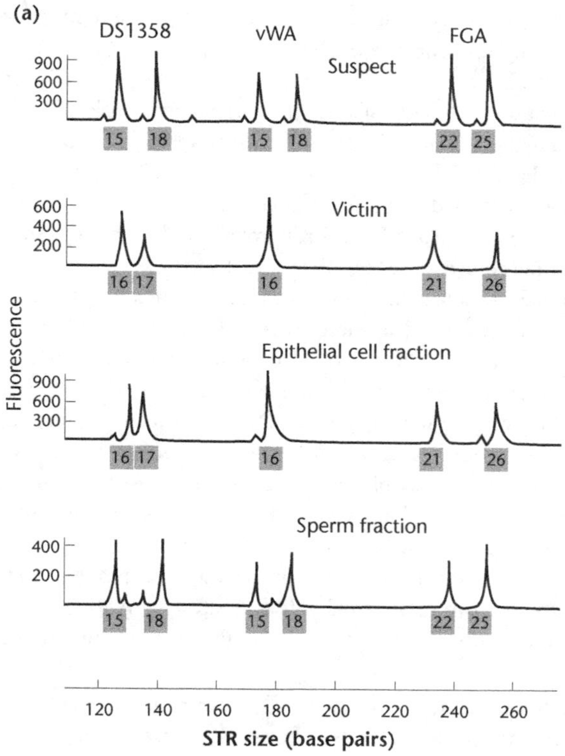


(b)

| | | Probability of identity | | |
|---|---|---|---|---|
| *Locus* | *Number of alleles* | *African American* | *US Caucasian* | *US West Coast Hispanic* |
| DS1358 | 8 | 0.12 | 0.078 | 0.103 |
| vWA | 11 | 0.058 | 0.065 | 0.094 |
| FGA | 14 | 0.035 | 0.036 | 0.030 |
| Combined | — | 0.00021 | 0.00018 | 0.00029 |

**FIGURE 24–31** ABI Prism system electrophoretic readout showing STR profiles of four samples from a rape case. Three STR loci were examined (DS1358, vWA, and FGA) from samples taken from the suspect and victim and from epithelial cells and sperm cells isolated from a vaginal swab taken from the victim. The *x*-axis shows the size of each STR in bases. The *y*-axis represents arbitrary values of fluorescence intensity. Notice that the STR profile of the sperm sample taken from the victim matches the STR profile of the suspect. (b) Population frequencies for STR alleles can be calculated to determine the probability of STR allele combinations in different ethnic groups.

A major advantage of STR profiles is that they can be generated from trace samples (e.g., single hairs or saliva left on a cigarette butt) and even from samples that are old or degraded (e.g., a skull found in a field or an ancient Egyptian mummy). As a result, STRs have replaced VNTRs in most forensic laboratories. In addition, STR typing is less expensive, less labor-intensive, and much faster than VNTR analysis, so it is rapidly replacing VNTR typing in most paternity laboratories as well.

The results of STR analysis are interpreted using statistics, probability, and population genetics [Figure 24–31(b)]. The population frequency of each STR allele in the standard set has been measured in many groups of people throughout the United States and the world. Using this information, scientists can calculate the probability of any combination of these 13 alleles. For example, if an allele of locus 1 is carried by 1 in 333 individuals, and an allele of locus 2 is carried by 1 in 83 individuals, the probability that someone will carry both alleles is equal to the product of their individual frequencies, or nearly 1 in 28,000 ($1/333 \times 1/83$). This overall probability may not be especially convincing, but if an allele of a third locus—carried by 1 in 100 individuals—and an allele of a fourth locus—carried by 1 in 25 individuals—are included in the calculations, the combined frequency becomes almost 1 in 70 million. That is, only about four individuals in the U.S. population will carry this combination of alleles by chance. When a profile containing all 13 CODIS STR loci is generated, the chance of anyone having the same profile is 1 in 10 billion. Since the planet's population is only about 6 billion, it is easy to see why STR analysis is so powerful in its ability to discriminate between samples from different individuals.

STR analysis was one of the primary approaches used to identify victims at two mass casualty disaster sites, the World Trade Center and in countries affected by the South Asian tsunami of 2004.

## Terrorism and Natural Disasters Force Development of New Technologies

**World Trade Center** Shortly after the twin towers of the World Trade Center in New York City were destroyed by the terror attacks of September 11, 2001, forensic scientists came together and rapidly accelerated efforts to use DNA-based techniques to identify the remains of victims. Issues such as the tremendous amount of debris at the site, dangerous working conditions, and heat and microbial decomposition of remains, coupled with the hundreds of thousands of tissue samples (primarily bone fragments) at the site from the nearly 3,000 individuals lost, made it evident that new strategies would need to be employed to quickly prepare and organize DNA profiles and compare them to DNA profiles from relatives. How would scientists establish DNA identity for those who perished in over 1.5 million tons of rubble?

State agencies such as the New York Department of Health and the Medical Examiner's Lab, and federal agencies, including the U.S. Department of Defense, the National Institutes of Health, and the U.S. Department of Justice, immediately responded to help with this task. Within 24 hours after the disaster, the New York City Police Department had established collection points throughout the city, where family members could file missing person's reports and provide cheek cell swabs for DNA isolation. Personal items (combs, toothbrushes) from the missing were also collected for DNA profiling.

Myriad Genetics, Inc., Gene Codes Forensics, DNA-VIEW, Celera Genomics, Bode Technology Group, and Orchid Genescreen were among the companies assisting in this effort. Several companies were involved in developing new software programs to help match samples submitted from family members to DNA profiles obtained from WTC victims. Because tissue samples from the victims primarily consisted of small bone fragments and teeth, which provided fragmented DNA because of degradation due to the intense heat at the site, forensic scientists primarily used STR, mtDNA, and SNP analysis of DNA fragments to develop profiles.

DNA analysis was conducted on over 15,000 tissue samples, although fewer than 1700 of the estimated 2819 people who died at the site were ultimately identified. This tragedy forced the development of new forensic strategies for analyzing and organizing remains recovered from the site. The forensic results provided closure for a number of families that lost loved ones in the attack.

**South Asian Tsunami** The South Asian tsunami of December 2004 was a tragedy that claimed over 225,000 lives and devastated areas of Indonesia, Sri Lanka, and Thailand. The Michigan-based company Gene Codes modified its software system called **Mass Fatality Identification System (M-FISys)** to help with the Thailand Tsunami Victim Identification effort. Because M-FISys was essentially built in response to the 9/11 tragedy, Gene Codes did not have to write entirely new software and were able to customize M-FISys as necessary. In addition to analyzing mitochondrial DNA, M-FISys incorporated male-specific variations in the Y chromosome called Y-STRs to aid in identifying individuals. Within three months approximately 800 victims had been identified.

Recently, Gene Codes established the DNA Shoah Project ("Shoah" is the Hebrew name for the Holocaust), an effort to use M-FISys to establish a genetic database of Naza-era Holocaust survivors with an overall goal of reuniting an estimated 10,000 postwar orphans around the world.

## Forensic Applications of DNA Profiling

Over the last 20 years, DNA profiling has become a powerful tool to help convict the guilty, identify the remains of victims, and exonerate the innocent. Its power emerges from the extreme sensitivity of the technology and its ability to discriminate accurately between crime scene samples. It has rapidly replaced older, more imperfect techniques and is helping to transform the police and justice systems. Since 1992, over 200 convicted persons in the United States have been exonerated of their alleged crimes, based on DNA profile analysis of old crime samples. Many of these convicted individuals had served decades in prison for crimes they did not commit—some incarcerated on death row and at least one wrongly executed prior to having access to DNA profiling evidence. As national and state DNA databases grow, they are becoming important tools in solving so-called **cold cases,** criminal cases where no arrest has been made and no suspects have been identified. By matching DNA from crime scenes with DNA profiles in the database, more than 1600 cold cases were solved between 2000 and 2003.

Although forensic DNA profiling is a powerful technology, it is important to understand what it can and cannot tell us about guilt and innocence. For example, in a criminal case, if a suspect's DNA profile does not match that of the evidence sample, the suspect can be excluded *as the source of that sample.* Whether or not this finding indicates innocence would depend on all other evidence in the case.

For instance, a suspect in a rape case may not have contributed the semen sample evidence but may have been involved in the crime by restraining the victim.

The meaning of a match between an evidence sample and a suspect's DNA profile is also not entirely certain. When all 13 core STR loci are used to create DNA profiles, there is less than one chance in 10 billion that two matching DNA profiles could have come from two different individuals. Although this sounds convincing, in some circumstances such matches might occur. For example, the two profiles could match by chance. Although this is unlikely, it cannot be ruled out. There is a greater chance of a match between related individuals—siblings, parents, or other relatives. It is even possible that the two profiles may have arisen from identical twins who have identical genomes. It is also possible to find a match between a crime scene sample and an innocent person's profile as a result of sample mixup, contamination, or even unintended or deliberate tampering. The very sensitivity of DNA profiling—sufficient to generate a profile from a single hair or from saliva on the back of a postage stamp—makes it highly prone to contamination if sampling and analysis conditions are not stringently controlled.

## 24.7 Genetic Engineering, Genomics, and Biotechnology Create Ethical, Social, and Legal Questions

Geneticists now use recombinant DNA technology to identify genes, diagnose and treat genetic disorders, produce commercial and pharmaceutical products, and solve crimes. However, the applications that arise from genetic engineering raise important ethical, social, and legal issues that must be identified, debated, and resolved. Resolutions often take the form of laws or public policy. Here we present a brief overview of some current ethical debates concerning the uses of gene technologies.

### Concerns about Genetically Modified Organisms and GM Foods

Most genetically modified food products contain an introduced gene encoding a protein that confers a desired trait (for example, herbicide resistance or insect resistance). Much of the concern over genetically modified plants centers on issues of consumer safety and environmental consequences. Are genetically modified plants that contain the new protein safe to eat? In general, if the proteins are not found to be toxic or allergenic and do not have other negative physiological effects, they are not considered to be a significant hazard to health. One such case is that of plants containing EPSP synthase. EPSP synthase is quickly degraded by digestive fluids, is nontoxic to mice at doses thousands of times higher than any potential human exposure, and has no amino acid sequence similarity to known protein toxins or allergens. Standardized methods for evaluating proteins in genetically modified foods are being developed in the United States and Europe. In Europe and Asia, labeling of food containing genetically modified ingredients is mandatory. But in the United States such labeling is not required at the present time, and foods with less than 5 percent of their content from genetically modified organisms (GMOs) can be labeled as GMO-free.

Environmental concerns generally have to do with any risks posed by releasing genetically modified organisms into the environment. Environmental risks include possible gene transfer by cross breeding with wild plants, toxicity, and invasiveness of the modified plant, resulting in loss of natural species (loss of biodiversity). Although laboratory and field studies suggest that cross-pollination and gene transfer can occur between some genetically engineered plants and wild relatives, there is little evidence that this has occurred in nature. If, for example, glyphosate resistance was transferred from cultivated plants such as canola into wild relatives, the herbicide-resistant weeds could make herbicide treatment ineffective. Biotechnology companies have engineered transgenic plants into sterile forms that are unable to transfer their genes into other plants. Built-in sterility was also designed to ensure that farmers could not produce their own seed from genetically modified crops, guaranteeing that biotechnology companies would have exclusive distribution of each year's crop. This, in itself, is an ethical issue, particularly in underdeveloped countries with limited resources to purchase genetically modified seeds.

### Genetic Testing and Ethical Dilemmas

We have considered examples of prenatal diagnosis, heterozygote screening of adults for single recessive disorders, and genome scanning. Although these technologies are valuable additions to medical diagnosis, they also present ethical problems that will not be easy to resolve. For example, what information should people have before deciding to have a genome scan or a genetic test for a single disorder? How can we protect the information revealed by such tests? How can we define and prevent genetic discrimination? As identification of genetic traits becomes more routine in clinical settings, physicians will need to ensure genetic privacy for their patients. There are significant concerns about how genetic information could be used in negative ways by employers, insurance companies, governmental agencies, or the general public. Genetic privacy and prevention of genetic discrimination will be increasingly important in the coming years. Currently, no federal laws regarding genetic privacy and genetic discrimination exist, so individuals must rely on the trustworthiness of the people with whom they are dealing. In 2007, the U.S. House of Representatives passed the **Genetic Information Nondiscrimination Act**. This act prohibits the improper use of genetic information in health insurance and employment. At the time of publication of this book, this act had still not reached the floor of the Senate.

Many of the potential risks and benefits of genetic testing are still unknown. In addition, the results and their consequences are not always clear or certain. For example, we can test for genetic diseases for which there are no effective treatments. Should we test people for these disorders? With present technology, a negative result does not necessarily rule out future development of a disease; nor does a positive result always mean that an individual will get the disease. How can we effectively communicate the results of testing and the actual risks to those being tested? Lawmakers and other groups made up of scientists, health-care professionals, ethicists, and consumers are debating these issues and formulating policy options.

Earlier in this chapter we discussed preimplantation genetic diagnosis (PGD), which provides couples with the ability to screen embryos created by *in vitro* fertilization for genetic disorders. As we learn more about genes involved in human traits, will other, non-disease-related genes be screened for by PGD? Will couples be able to select embryos with certain genes encoding desirable traits for height, weight, intellect, and other physical or mental characteristics? What do you think of using genetic testing to purposely select for an embryo with a genetic disorder? For instance, recently there have been several well-publicized cases of couples seeking to use prenatal diagnosis or PGD to select for embryos with dwarfism and deafness.

## The Ethical Concerns Surrounding Gene Therapy

Gene therapy raises several ethical concerns, and many forms of therapy are sources of intense debate. At present, all gene therapy trials are restricted to using somatic cells as targets for gene transfer. This form of gene therapy is called **somatic gene therapy**; only one individual is affected, and the therapy is done with the permission and informed consent of the patient or family.

Two other forms of gene therapy have not been approved, primarily because of the unresolved ethical issues surrounding them. The first is called **germ-line therapy**, whereby germ cells (the cells that give rise to the gametes—i.e., sperm and eggs) or mature gametes are used as targets for gene transfer. In this approach, the transferred gene is incorporated into all the future cells of the body, including the germ cells. This means that individuals in future generations will also be affected, without their consent. Is this kind of procedure ethical? Do we have the right to make this decision for future generations? Thus far, the concerns have outweighed the potential benefits, and such research is prohibited.

The second unapproved form of gene therapy—which raises an even greater ethical dilemma—is termed **enhancement gene therapy**, whereby people may be "enhanced" for some desired trait. This use of gene therapy is extremely controversial and is strongly opposed by many people. Should genetic technology be used to enhance human potential? For example, should it be permissible to use gene therapy to increase height, enhance athletic ability, or extend intellectual potential? Presently, the consensus is that enhancement therapy, like germ-line therapy, is an unacceptable use of gene therapy. However, there is an ongoing debate, and many issues are still unresolved. For example, the U.S. Food and Drug Administration now permits growth hormone produced by recombinant DNA technology to be used as a growth enhancer, in addition to its medical use for the treatment of growth-associated genetic disorders. Critics charge that the use of a gene product for enhancement will lead to the use of transferred genes for the same purpose. The outcome of these debates may affect not only the fate of individuals but the direction of our society as well.

## The Ethical, Legal, and Social Implications (ELSI) Program

When the Human Genome Project was first discussed, scientists and the general public raised concerns about how genome information would be used and how the interests of both individuals and society can be protected. To address these concerns, the **Ethical, Legal, and Social Implications (ELSI) Program** was established as an adjunct to the Human Genome Project. The ELSI Program considers a range of issues, including the impact of genetic information on individuals, the privacy and confidentiality of genetic information, and implications for medical practice, genetic counseling, and reproductive decision making. Through research grants, workshops, and public forums, ELSI is formulating policy options to address these issues.

ELSI focuses on four areas in its deliberations concerning these various issues: (1) privacy and fairness in the use and interpretation of genetic information, (2) ways to transfer genetic knowledge from the research laboratory to clinical practice, (3) ways to ensure that participants in genetic research know and understand the potential risks and benefits of their participation and give informed consent, and (4) public and professional education. It is hoped that, as the Human Genome Project moves from generating information about the genetic basis of disease to improving treatments, promoting prevention, and developing cures, these and other ethical concerns will have been thoroughly studied and an international consensus developed on appropriate policies and laws.

## DNA and Gene Patents

Intellectual property rights are also being debated as an aspect of the ethical implications of genetic engineering, genomics, and biotechnology. Patents on intellectual property (isolated genes, new gene constructs, recombinant cell types, GMOs) can be potentially lucrative for the patent-holders but may also pose ethical and scientific problems. For example, consider the possibilities for a human gene that has been cloned and then patented by the scientists who did the cloning. The person or company holding the patent could require that anyone attempting to do research with the patented gene pay a licensing fee for its use. Should a diagnostic test or therapy result from the research, more fees and royalties may be demanded and as a result the costs of a genetic test may be too high for many patients to afford. But limiting or preventing the holding of patents for genes or genetic tools could reduce the incentive for pursuing the research that produces such genes and tools, especially for companies that need to profit from their research. Should scientists and companies be allowed to patent DNA sequences from naturally living organisms? And should there be a lower or upper limit to the size of those sequences? For example, should patents be awarded for small pieces of genes, such as expressed sequence tags (ESTs), just because some individual or company wants to claim a stake at having cloned a piece of DNA first, even if no one knows whether the DNA sequence has a use? Can or should investigators be allowed to patent the entire genome of any organism they have sequenced?

Since 1980, the U.S. Patent and Trademark Office has granted patents for more than 20,000 genes or gene sequences, including an estimated 20 percent of human genes. Some scientists are concerned that to award a patent for simply cloning a piece of DNA is awarding a patent for too little work. Given that computers do most of the routine work of genome sequencing, who should get the patent? What about individuals who figure out *what* to do with the gene? What if a gene sequence has a role in a disease for which a genetic therapy may be developed? Many scientists believe that it is more appropriate to patent novel technology and applications that make use of gene sequences than to patent the gene sequences themselves.

## GENETICS, TECHNOLOGY, AND SOCIETY

### Gene Therapy—Two Steps Forward or Two Steps Back?

In September 1999, 18-year-old Jesse Gelsinger received his first dose of gene therapy. Large numbers of adenovirus vectors bearing the *ornithine transcarbamylase* (*OTC*) gene were injected into his hepatic artery. The virus vectors were expected to lodge in his liver, enter the liver cells, and trigger the production of OTC protein. In turn, the OTC protein might correct his genetic defect and perhaps cure him of his liver disease. However, within hours, a massive immune reaction surged through Jesse's body. He developed a high fever, his lungs filled with fluid, multiple organs shut down, and he died four days later of acute respiratory failure. In the aftermath of the tragedy, several government and scientific inquiries were conducted. Investigators learned that clinical trial scientists had not reported other adverse reactions to gene therapy and that some of the scientists were affiliated with private companies that could benefit financially from the trials. They found that serious side effects seen in animal studies were not explained to patients during informed-consent discussions, and that some clinical trials were proceeding too quickly in the face of data suggesting a need for caution. The U.S. Food and Drug Administration (FDA) scrutinized gene therapy trials across the country, halted a number of them, and shut down several gene therapy programs. Other research groups voluntarily suspended their gene therapy studies. Jesse's death had dealt a severe blow to the struggling field of gene therapy—a blow from which it was still reeling when a second tragedy hit.

In April 2000, a French group announced that their gene therapy for X-linked severe combined immunodeficiency (X-SCID) had succeeded. If what they claimed was true, it was the first unequivocal success in the gene therapy field. In this study, the young patients' bone marrow cells were removed, treated with a retrovirus bearing the $\gamma c$ transmembrane protein gene, and transplanted back to the patients. Ten of 11 patients were cured of their immune deficiency and were pronounced able to lead normal lives. Published reports of the study were greeted with enthusiasm by the gene therapy community. But elation turned to despair in 2003, when it became clear that 2 of the 10 children who had been cured of X-SCID had developed leukemia as a direct result of their therapy. The FDA immediately halted 27 similar gene therapy clinical trials and, once again, gene therapy underwent a profound reassessment. In 2005, a third child in the French X-SCID study developed leukemia, likely as a result of gene therapy.

Up until the apparent success of the French X-SCID clinical trials, gene therapy had suffered not only from the scandals and scrutiny that emerged from Jesse Gelsinger's death but also from the skepticism of many scientists and the general public about the feasibility of this much-ballyhooed therapeutic technique.

Since the first clinical trial for gene therapy began in 1990, over 600 gene therapy clinical trials involving over 4000 patients have been initiated. These trials aim to cure cancers, inherited diseases such as hemophilia and cystic fibrosis, and infectious diseases such as AIDS. Despite high expectations of the proponents and intense publicity, therapeutic benefits have been unclear at best and, more frequently, absent.

The most significant positive outcome of gene therapy came from the first clinical trial in 1990. Ashanti DeSilva, who received retroviral-transduced T cells for severe combined immunodeficiency (SCID), now leads a normal life. However, the reasons for Ashanti's success are not entirely clear. She had also been given a new drug treatment that replaced her missing ADA enzyme prior to and after gene therapy. Hence, it is still not known how much of her cure is due to gene therapy and how much is due to drug treatment.

To date, no human gene therapy product has been approved for sale. Critics of gene therapy continue to criticize research groups for undue haste, conflicts of interest, and sloppy clinical trial management, and for promising much but delivering little. In the mid-1990s, a National Institutes of Health review committee concluded that significant problems remain in all basic aspects of gene therapy, including the problems that led to Jesse Gelsinger's death and the X-SCID leukemias—adverse immune reactions to viral vectors and the side effects of retroviral vector integration into the host genome.

The question remains whether gene therapy can ever recover from these setbacks and fulfill its promise as a cure for genetic diseases. At present, hundreds of gene therapy clinical trials are under way in the United States. Most are for cancer and are in phase I trials, which examine safety and dosage but not efficacy. Tighter restrictions on clinical trial protocols are now imposed to correct some of the procedural problems that emerged from the Gelsinger case. In addition, basic science is proceeding with development

*Continued on next page*

*Genetics, Technology, and Society, continued*

of safe, effective vectors, such as vectors that insert sequences into specific regions of the genome (reducing the possibility of cancer or gene mutation due to vector insertion) and vectors bearing receptors on their surfaces that allow them to infect specific cell types only. However, the path ahead is still a long one. There is work to be done to optimize tissue-specific expression of therapeutic genes, to efficiently transduce cells in culture, to predict and control immune reactions to vectors, and to develop better animal models in which to test gene therapies prior to clinical trials.

Many scientists feel that we should continue gene therapy research and clinical trials despite the setbacks. However, those working today have a more sober view of its progress. Clinical trials for any new therapy are potentially dangerous, and often animal studies will not accurately reflect the reaction of individual humans to a new drug or procedure. Inevitably, more adverse reactions to gene therapy will emerge in the clinical trials, even as the methods become more effective. Perhaps we should view gene therapy as we have antibiotics, organ transplants, and manned space travel. There will be setbacks and even tragedies, but step by small step, we will move toward a technology that could—someday—provide cures for many severe genetic diseases.

**■ Reference**

Thomas, C.E., Ehrhardt, A., and Kay, M.A. 2003. Progress and problems with the use of viral vectors for gene therapy. *Nat. Rev. Gen.* 4: 346–358.

EXPLORING GENOMICS

# Genomics Applications to Identify Gene Expression Signatures of Breast Cancer

In Chapter 20 we discussed how multiple mutations in different genes are part of the genetic basis for the development of many cancers. Microarray analysis to identify gene expression profiles, or "signatures," of different types of cancer has become both a valuable research approach and a diagnostic tool for cancer scientists. Understanding the gene expression signatures of different types of cancers has already led to the development of more effective treatment regimes through pharmacogenomics—the use of customized medicines based on a person's genetics.

In Chapter 20 and elsewhere we have discussed the role of *BRCA* genes in breast cancer, and we have also examined how breast cancer, like most cancers, involves mutations or gene expression alterations in multiple genes. In this exercise we will explore two research papers available on the Internet in order to learn more about genomics approaches for studying gene expression profiles in breast cancer.

## ■ Exercise I – Gene Expression Signatures as a Predictor of Survival in Breast Cancer

In 2002, in the *New England Journal of Medicine* (*NEJM*), Marc J. van de Vijver and colleagues published the paper, "A Gene Expression Signature as a Predictor of Survival in Breast Cancer" (*NEJM* 347: 1999–2009). This paper was one of the first to describe gene expression profiles in breast cancer. In this exercise you will access the paper to learn more about the genomics applications involved in this work.

1. Use PubMed http://www.ncbi.nlm.nih.gov/entrez/query.fcgi?db=PubMed to find the paper by an author-journal search, using "van de Vijver and New England Journal of Medicine" as the key words. Be sure you have identified the correct paper in *NEJM*, and use the PubMed link provided for free access.

   *Note:* To access the paper, you may be required to set up a free registration account with *NEJM*.

2. Read the abstract carefully first before reading the introduction and methods. In the results section, focus on Figure 1. Answer the following questions about this research:

   a. What was the primary genomics technique involved in this study? Briefly describe what was done.

   b. How many genes were evaluated in this study? Why were these genes chosen for analysis?

   c. Explain important features of Figure 1. For example, what do the colors in the image represent? What does each row of the *y*-axis represent? Each column on the *x*-axis?

   d. What criteria were used to determine whether patients had a good or poor prognosis for survival?

   e. What is adjuvant therapy?

   f. How do you think scientists and physicians may use gene expression profile data in the future?

## ■ Exercise II – Gene Expression Profiles and p53 Status in Breast Cancer

More recently, Melissa A. Troester and colleagues published a paper entitled "Gene Expression Patterns Associated with p53 Status in Breast Cancer" in the freely accessible journal *BMC Cancer* (6: 276, 2006). Recall from Chapter 20 that the *p53* gene encodes a tumor-suppressor protein and that mutations in *p53* have been implicated in more than 50 percent of human cancers.

1. Use PubMed to access the paper by Troester et al. Read and analyze the paper as you did for Exercise I. In the results section, focus on Figures 1, 3, 4, and 5, then answer the following:
   a. What was the main purpose of this research?
   b. What technique was used to deplete p53 in the experiments shown in Figure 1?
   c. Summarize the results reported in Figure 3 for p53-inhibited cells. Which cell lines showed the most pronounced changes in gene expression following p53-inhibition?
   d. Describe these results. Review the data presented in Figure 4; then answer the following questions:
      i What tissue samples were used for the microarray analysis shown in this figure?
      ii What do the dendrogram colors indicate?
      iii What were the two main functional categories or "clusters" of affected genes identified by microarray analysis?
      iv Which functional category of genes was primarily upregulated in *p53*-mutant breast tumors?
      v Which functional category of genes was primarily upregulated in *p53* wild-type tumors?
   e. Review data presented in Figure 5. What genes are in clusters A, B, C, and D in this figure? What conclusions can be made from these results regarding gene expression profiles and the p53 status of tumors?
2. How might an understanding of *p53* mutation status and gene expression profiles help physicians, scientists, and breast cancer patients in the future?

## Chapter Summary

1. Genetic engineering and biotechnology have led to the production of recombinant therapeutic proteins for treating human diseases. Genetically engineered bacteria, plants, yeasts, cultured cells, and animals are being used to synthesize abundant quantities of useful biological and pharmaceutical protein products such as hormones, antibodies, and vaccines.
2. Genetically modified crop plants have had a great impact on agriculture. Improvements in agriculturally important crop plants as a result of genetic engineering include creating herbicide and insect-resistant crops and enhancing the nutritional value of food crops.
3. Genetic engineering has also been used to enhance qualities of farm animals by creating transgenic animals with improved growth characteristics and resistance to disease.
4. Techniques involving recombinant DNA technology and genomics have resulted in a number of methods for genetic testing to detect mutations associated with genetic diseases. These include genetic tests used for making prenatal diagnosis, completing preimplantation genetic diagnosis, identifying carriers, predicting the course of genetic diseases, and predicting treatment outcomes based on gene profiles.
5. Pharmacogenomics and rational drug design are emerging from our understanding of how genes and genotype affect reactions to drugs and the genetic basis of diseases such as cancer. For certain diseases it is now possible to predict a patient's reaction to drugs prior to taking the medication and to make drugs specifically target a patient's condition based on his or her genotype.
6. Gene therapy, the use of therapeutic genes to cure or treat genetic diseases, continues despite mixed success and recent setbacks. Currently limited to treating a small number of single-gene conditions, many gene therapy clinical trials are underway for a variety of genetic diseases. New approaches in RNA interference and gene removal may also prove to be valuable for gene therapy in the future.
7. DNA fingerprinting is used for forensic applications as well as for paternity testing, identification of human and animal remains, and a wide range of other fields including archaeology, conservation biology, and public health.
8. Applications of genetic engineering, genomics, and biotechnology raise important ethical, social, and legal concerns that must be addressed in the future as these applications continue to become more commonplace.

## INSIGHTS AND SOLUTIONS

1. Probes for DNA fingerprinting can be derived from a single locus or multiple loci. Two multiple-loci probes have been widely employed in both criminal and civil cases and are derived from minisatellite loci on chromosome 1 (1cen-q24) and chromosome 7 (7q31.3). These probes, used because they produce a highly individual fingerprint, have determined paternity in thousands of cases over the last few years. The results of one such DNA fingerprinting of a mother (M), putative father (F), and child (C) are shown in the accompanying figure. The child has 6 maternal bands, 10 paternal bands, and 6 bands shared between the mother and the alleged father. Based on this fingerprint, can you conclude that this man is the father of the child?

**Solution:** All bands present in the child can be assigned as coming from either the mother or the father. In other words, all the bands in

*Continued on next page*

*Insights and Solutions, continued*

the child's DNA fingerprint that are not maternal are present in the father. Since the father and the child share 10 bands and the child has no unassigned bands, paternity can be assigned with confidence. The chance that this man is not the father is on the order of $10^{-13}$.

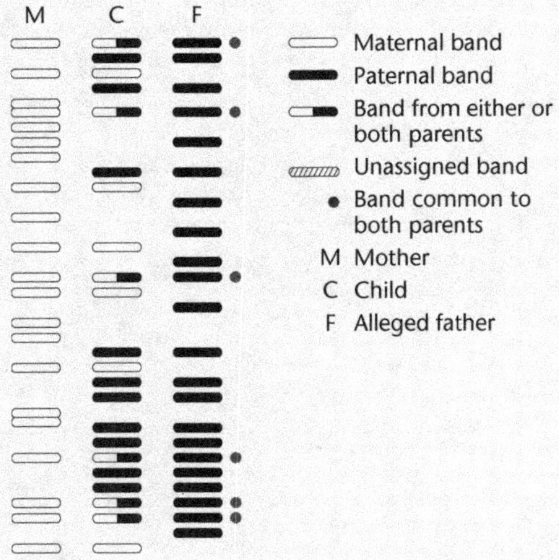


2. The DNA fingerprints of a mother, a child, and the alleged father in a second case are shown in the accompanying figure. The child has 8 maternal bands, 14 paternal bands, 6 bands that are common to both the mother and the alleged father, and 1 band that is not present in either the mother or the alleged father. What are the possible explanations for the presence of the last band? Based on your analysis of the band pattern, which explanation is most likely?

**Solution:** In this case, one band in the child cannot be assigned to either parent. Two possible explanations are that the child is mutant for one band or that the man tested is not the father. To estimate the probability of paternity, the mean number of resolved bands ($n$) is determined, and the mean probability ($x$) that a band in individual $A$ matches that in a second, unrelated individual $B$ is calculated. In this case, because the child and the father share 14 bands in common, the probability that the man tested is not the father is very low (probably $10^{-7}$ or lower). As a result, the most likely explanation is that the child is mutant for a single band. In fact, in 1419 cases of genuine paternity resolved by the minisatellite probes on chromosomes 1 and 7, single unmatching bands in the children were recorded in 399 cases, accounting for 28 percent of all cases.

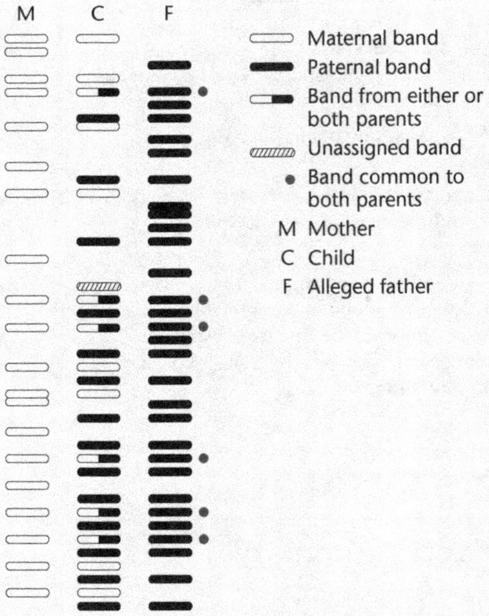


3. Infection by HIV-1 (human immunodeficiency virus) weakens the immune system and results in the symptoms of AIDS (acquired immunodeficiency syndrome). Specifically, HIV infects and kills cells of the immune system that carry a cell-surface receptor known as CD4. An HIV surface protein known as gp120 binds to the CD4 receptor and allows the virus to enter the cell. The gene encoding the CD4 protein has been cloned. How might this clone be used along with recombinant DNA techniques to combat HIV infection?

**Solution:** Researchers hope that clones of the *CD4* gene can be used in the design of systems for the targeted delivery of drugs and toxins to combat the infection. For example, because infection depends on an interaction between the viral gp120 protein and the CD4 protein, the cloned *CD4* gene has been modified to produce a soluble form of the protein (sCD4) that, because of its solubility, would circulate freely in the body. The idea is that HIV might be prevented from infecting cells if the gp120 protein of the virus first encounters and binds to extra molecules of the soluble form of the CD4 protein. Once bound to the extra molecules, the virus would be unable to bind to CD4 proteins on the surface of immune system cells. Studies in cell culture systems indicate that the presence of sCD4 effectively prevents HIV infection of tissue culture cells. However, studies in HIV-positive humans have been somewhat disappointing, mainly because the strains of HIV used in the laboratory are different from those found in infected individuals. In another strategy, the *CD4* gene has been fused with genes encoding bacterial toxins. The resulting fusion protein contains CD4 regions that should bind to gp120 on the surface of HIV-infected cells and toxin regions that should then kill the infected cell. In tissue culture experiments, cells infected with HIV are killed by this fusion protein, whereas uninfected cells survive.

## Problems and Discussion Questions

1. What are some of the reasons why GM crops are controversial? Describe some of the primary concerns that have been raised about GM foods.

2. Should the United States require mandatory labeling of all foods that contain GMOs? Explain your answer.

3. Provide examples of major questions that need to be answered if gene therapy is to become a safe and reliable treatment for genetic diseases.
4. Outline the steps involved in transferring glyphosate resistance to a crop plant. Do you envision that this trait is likely to escape from the crop plant and make weeds glyphosate-resistant? Why or why not?
5. In order to vaccinate people against diseases by having them eat antigens (such as the cholera toxin), the antigen must reach the cells of the small intestine. What are some potential problems of this method?
6. What are the advantages of using STRs instead of VNTRs for DNA profiling?
7. Suppose you develop a screening method for cystic fibrosis that allows you to identify the predominant mutation *Δ508* and the next six most prevalent mutations. What must you consider before using this method to screen a population for this disorder?
8. One of the main safety issues associated with genetically modified crops is the potential for allergenicity caused by introducing an allergen or by changing the level of expression of a host allergen. Based on the observation that common allergenic proteins often contain identical stretches of a few (six or seven) amino acids, researchers developed a method for screening transgenic crops to evaluate potential allergenic properties (Kleter & Peijnenburg, 2002. *BMC Struct. Biol.* 2: 8). How do you think they accomplished this?
9. Why are most recombinant human proteins produced in animal or plant hosts instead of bacterial host cells?
10. There are more than 1000 cloned farm animals in the United States. In the near future, milk from cloned cows and their offspring (born naturally) may be available in supermarkets. These cloned animals have not been transgenically modified, and they are no different than identical twins. Should milk from such animals and their natural-born offspring be labeled as coming from cloned cows or their descendants? Why?
11. One of the major causes of sickness, death, and economic loss in the cattle industry is *Mannheimia haemolytica*, which causes bovine pasteurellosis, or shipping fever. Noninvasive delivery of a vaccine using transgenic plants expressing immunogens would reduce labor costs and trauma to livestock. An early step toward developing an edible vaccine is to determine whether an injected version of an antigen (usually a derivative of the pathogen) is capable of stimulating the development of antibodies in a test organism. The following table assesses the ability of a transgenic portion of a toxin (Lkt) of *M. haemolytica* to stimulate development of specific antibodies in rabbits.
   (a) What general conclusion can you draw from the data?
   (b) With regards to development of a usable edible vaccine, what work remains to be done?

| Immunogen Injected | Antibody Production in Serum |
|---|---|
| Lkt50*—saline extract | + |
| Lkt50—column extract | + |
| Mock injection | − |
| Pre-injection | − |

*Lkt50 is a smaller derivative of Lkt that lacks all hydrophobic regions. + indicates at least 50 percent neutralization of toxicity of Lkt; − indicates no neutralization activity.
*Source:* Modified from Lee et al. 2001. *Infect. and Immunity* 69: 5786–5793.

12. Recombinant adenoviruses have been used in a number of preclinical studies to determine the efficacy of gene therapy for rheumatoid arthritis and osteoarthritis. In the viruses, genes can be delivered by injection to the tissues that need them. Christopher Evans and colleagues (2001. *Arthritis Res.* 3: 142–146) estimated that approximately 20 percent of all human gene therapy trials have used adenoviruses for gene delivery. The death of a patient in 1999 after infusion of adenoviral vectors has caused concern. As you consider the use of viral vectors as therapy-delivery vehicles for human pathologies, what factors seem of paramount concern?
13. Define somatic gene therapy, germ-line therapy, and enhancement gene therapy. Which of these is currently in use?
14. *Transductional targeting* is a preferred route for the delivery of therapeutics for human diseases. It involves the development of tissue-specific interactions between the viral vector and a specific tissue. A genetic approach that has been used involves engineering the capsid of an adeno-associated virus (type 2) vector to target specific human cell types (Ponnazhagan et al., 2002. *J. Virol.* 76: 12,900–12,907). Nongenetic approaches are also possible. Speculate on problems associated with the genetic approach of capsid alteration and problems that might be associated with nongenetic approaches to transductional targeting.
15. The development of safe vectors for human gene therapy has been a goal since 1990. Among the problems associated with viral-based vectors is that many of the viruses (i.e., SV40) have transformation properties thought to be mediated by binding and inactivating gene products such as p53, retinoblastoma protein (pRB), and others. Mark Cooper and colleagues (1997. *Proc. Nat. Acad. Sci. (USA)* 94: 6450–6455) developed SV40-based vectors that are deficient in binding p53, pRB, and other proteins. Why would you specifically want to avoid inactivating p53, pRB, and related proteins?
16. Gene therapy for human genetic disorders involves transferring a copy of the normal human gene into a vector and using the vector to transfer the cloned human gene into target tissues. Presumably, the gene enters the target tissue and becomes active, and the gene product relieves the symptoms.
   (a) Why are disorders such as muscular dystrophy difficult to treat by gene therapy?
   (b) What are the potential problems of using retroviruses as vectors?
   (c) Should gene therapy involve germ-line tissue instead of somatic tissue? What are some of the potential ethical problems associated with the former approach?
17. Sequencing the human genome and the development of microarray technology promises to improve our understanding of normal and abnormal cell behavior. How are microarrays dramatically changing our understanding of complex diseases such as cancer?
18. A couple with European ancestry seeks genetic counseling before having children because of a history of cystic fibrosis (CF) in the husband's family. ASO testing for CF reveals that the husband is heterozygous for the *Δ508* mutation and that the wife is heterozygous for the *R117* mutation. You are the couple's genetic counselor. When consulting with you, they express their conviction that they are not at risk for having an affected child because they each carry different mutations and cannot have a child who is homozygous for either mutation. What would you say to them?
19. Dominant mutations can be categorized according to whether they increase or decrease the overall activity of a gene or gene product. Although a loss-of-function mutation (a mutation that inactivates the gene product) is usually recessive, for some genes, one dose of the normal gene product, encoded by the normal allele, is not sufficient to produce a normal phenotype. In this case, a loss-of-function mutation in the gene will be dominant, and the gene is said to be *haploinsufficient*. A second category of dominant mutation is the gain-of-function mutation, which results in a new activity or increased activity or expression of a gene or gene product. The gene therapy technique currently used in clinical trials involves the "addition" to somatic cells of a normal copy of a gene. In other words, a normal copy of the gene is inserted into the genome of the mutant somatic cell, but the mutated copy of the gene is not removed or replaced. Will this strategy work for either of the two aforementioned types of dominant mutations?

20. The DNA sequence surrounding the site of the sickle-cell mutation in the $\beta$-globin gene, for normal and mutant genes, is as follows.

5′-GACTCCTGAGGAGAAGT-3′
3′-CTGAGGACTCCTCTTCA-5′
Normal DNA

5′-GACTCCTGTGGAGAAGT-3′
3′-CTGAGGACACCTCTTCA-5′
Sickle-cell DNA

Each type of DNA is denatured into single strands and applied to a DNA-binding membrane. The membrane containing the two spots is hybridized to an ASO of the sequence

5′-GACTCCTGAGGAGAAGT-3′

Which spot, if either, will hybridize to this probe?

21. One form of hemophilia, an X-linked disorder of blood clotting, is caused by a mutation in clotting factor VIII. Many single-nucleotide mutations of this gene have been described, making the detection of mutant genes by Southern blots inefficient. There is, however, an RFLP for the enzyme *Hind*III contained in an intron of the factor VIII gene. The two restriction fragments generated from digestion of either wild-type or mutant DNA are shown below.

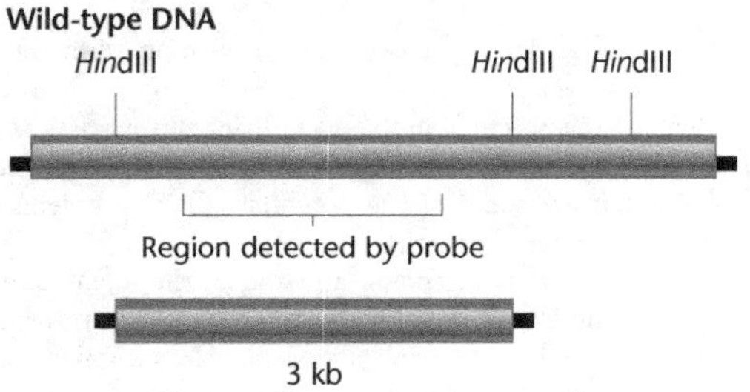


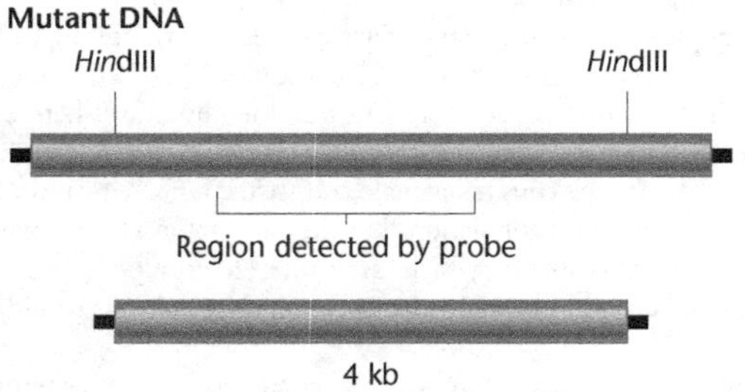


A female whose brother has hemophilia has a 50 percent risk of being a carrier of this disorder. To test her status, DNA is obtained from her white blood cells and those of family members, cut with *Hind*III, and the fragments are probed and visualized by Southern blotting. Using the following results, determine whether either of the females in generation II is a carrier for hemophilia.

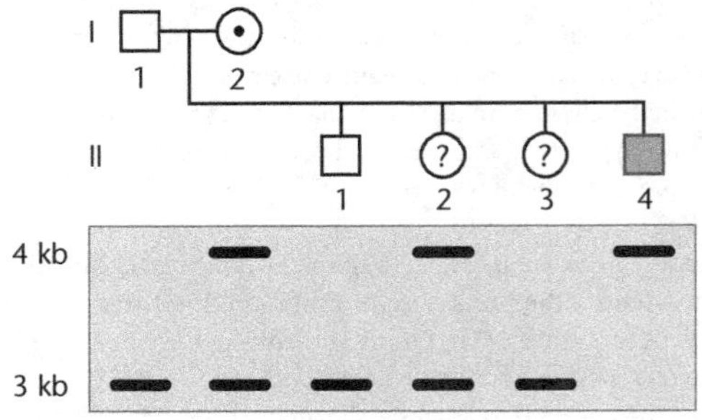


22. The human insulin gene contains introns. Since bacterial cells will not excise introns from mRNA, how can a gene like this be cloned into a bacterial cell that will produce insulin?

23. In mice transfected with the rabbit $\beta$-globin gene, the rabbit gene is active in several tissues, including the spleen, brain, and kidney. In addition, some transfected mice suffer from thalassemia (a form of anemia) caused by an imbalance in the coordinate production of $\alpha$ and $\beta$-globins. Which problems associated with gene therapy are illustrated by these findings?

24. When genome scanning technologies become widespread, medical records will contain the results of such testing. Who should have access to this information? Should employers, potential employers, or insurance companies be allowed to have this information? Would you favor or oppose having the government establish and maintain a central database containing the results of individuals' genome scans?

25. What limits the use of differences in restriction enzyme sites as a way of detecting point mutations in human genes?

**HOW DO WE KNOW?**

26. In this chapter, we focused on a number of interesting applications of genetic engineering and biotechnology. At the same time, we found many opportunities to consider the methods and reasoning by which much of this information was acquired. From the explanations given in the chapter, what answers would you propose to the following fundamental questions:
    (a) How do we determine whether genetically modified plants are safe for human consumption?
    (b) What experimental evidence confirms that we have introduced a useful gene into a transgenic organism and that it performs as we anticipate?
    (c) How can we use DNA analysis to determine that a human fetus has sickle-cell anemia?
    (d) How can DNA microarray analysis be used to identify specific genes that are being expressed in a specific tissue?
    (e) How do we know whether a forensic DNA profile comes from only one individual?

## Extra-Spicy Problems

27. You are asked to assist with a prenatal genetic test for a couple, each of whom is found to be a carrier for a deletion in the $\beta$-globin gene that produces $\beta$-thalassemia when homozygous. The couple already has one child who is unaffected and is not a carrier. The woman is pregnant, and the couple wants to know the status of the fetus. You receive DNA samples obtained from the fetus by amniocentesis and from the rest of the

family by extraction from white blood cells. Using a probe for the deletion, you obtain the following blot. Is the fetus affected? What is its genotype for the $\beta$-globin gene?

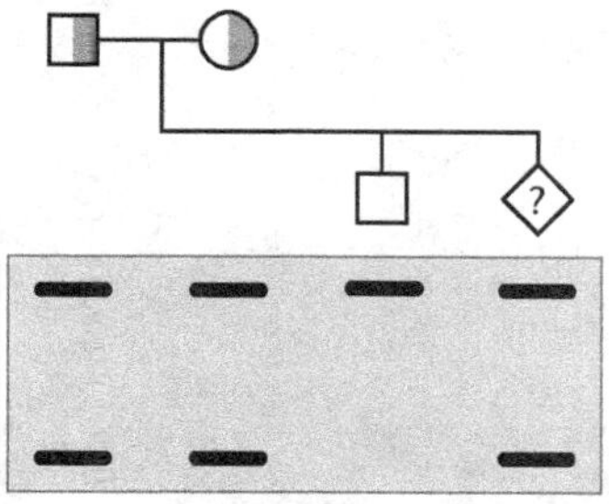

28. Shown here is a pedigree tracking the inheritance of a rare disease.
   (a) Which mode or modes of inheritance are excluded or consistent with this pedigree?

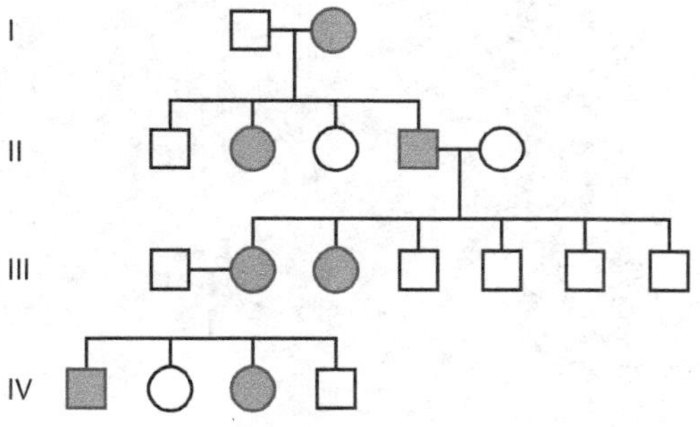


   (b) DNA samples from generations II and III above are obtained and subjected to RFLP linkage analysis (the results are shown in the next column). One RFLP is found on chromosome 10 (identified by probe *A*), and the other is found on chromosome 21 (identified by probe *B*). Assume that additional data were gathered on this family and that they were consistent with the data shown and statistically significant. On which chromosome is the disease gene located?
   (c) Individual III-1 is married to a normal man whose RFLP genotype is *A1A2* and *B1B1*. What kind of prenatal diagnostic test can be done to determine whether the child this couple is expecting will be normal? Describe what you can conclude regarding the result, and indicate the accuracy of the test.
   (d) Individual III-4 marries a woman of genotype *B1B1*. This couple has a child who is *B1B1*. The father assumes that the child is illegitimate, but the mother, who has taken a genetics course, argues that the child could be the result of an event that occurred in the father's germ line. She cites two possibilities. What are they?

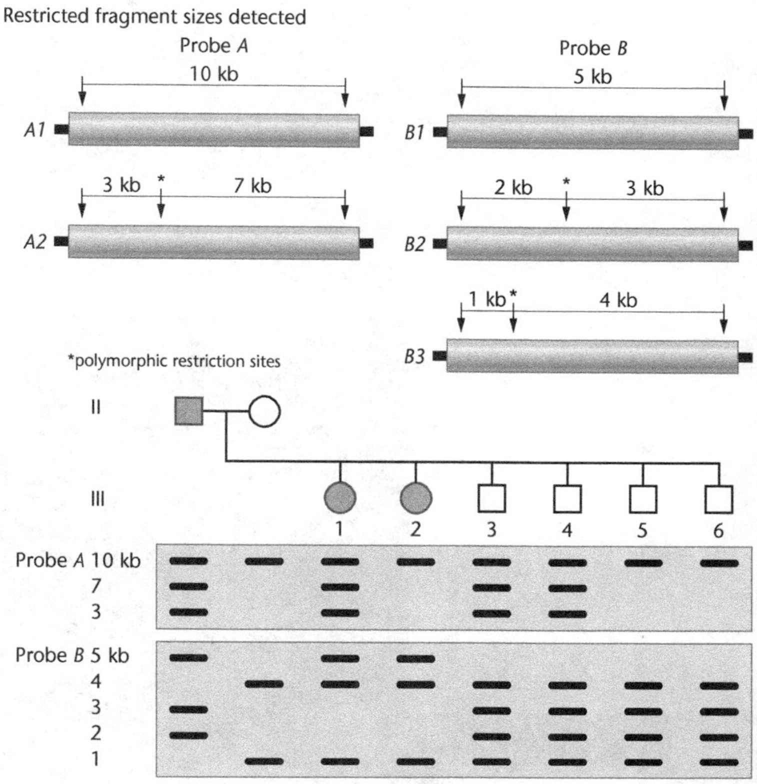


29. In forensic applications, if a suspect's ABO blood type does not match that of a blood sample collected from a crime scene, the suspect can be excluded. While *exclusionary* data are useful, testing using nucleotide repeats is often *inclusionary*. One study estimated an *average* mutation rate for STRs (small tandem repeats) at approximately $7.0 \times 10^{-4}$ mutations per STR per 25 years with considerable variation among STR sites and among populations (African, European, and Polynesian; Zhivotovsky et al. 2004).
   (a) Why can STRs provide inclusionary information, whereas ABO blood types provide only exclusionary information?
   (b) How might a relatively high mutation rate for STR sites offer an advantage over loci with lower mutation rates in forensic applications?
   (c) In computing overall probabilities from STR profiles, how are frequency variations among different STR sites handled?
   (d) How might interpopulational variations of STR loci influence DNA identity testing?
30. Host transgenic mammals are often made by injecting transgenic DNA into a pronucleus, a process that is laborious, technically demanding, and typified by yields of less than 1 percent. Lately, the engineered lentivirus, a retrovirus, has been used to generate transgenic pigs, rats, mice, and cattle with greater than 10 percent efficiency (Whitelaw, 2004). Lentivectors have reverse transcriptase activity, can infect both dividing and nondividing cells, and are replication-defective. A lentivector carrying the reporter gene, which codes a green fluorescent protein, has produced a remarkable set of pigs (refer to the chapter opening figure). Present, in a labeled diagram, a strategy for producing a transgenic mammal carrying a pharmaceutically important gene using a lentivector.

A field of pumpkins, where size is under the influence of additive alleles.

# 25 Quantitative Genetics and Multifactorial Traits

## CHAPTER CONCEPTS

- Quantitative inheritance results in a range of measurable phenotypes for a polygenic trait.
- With some exceptions, polygenic traits tend to demonstrate continuous variation.
- Quantitative traits can be explained in Mendelian terms whereby certain alleles have an additive effect on the traits under study.
- The study of polygenic traits relies on statistical analysis.
- Heritability values estimate the genetic contribution to phenotypic variability under specific environmental conditions.
- Twin studies allow an estimation of heritability in humans.
- Quantitative trait loci (QTLs) can be mapped and identified.

Up to this point, most of our examples of phenotypic variation have been ones that could have been assigned to distinct and separate categories: pea plants were tall or dwarf; squash fruit shape was spherical, disc shaped, or elongated; and fruit fly eye color was red or white (see Chapter 4). Traits such as these, which have a small number of discrete phenotypes, are said to show **discontinuous variation.** Typically, in these traits, a genotype will produce a single identifiable phenotype, although phenomena such as variable penetrance and expressivity, pleiotropy, and epistasis can obscure the relationship between genotype and phenotype, even for simple discontinuous traits.

Now we will look at traits that are more complex, including many that have medical or agricultural importance. The traits we examine in this chapter show much more variation, with a continuous range of phenotypes that cannot be as easily classified into distinct categories. Examples of traits showing **continuous variation** include human height or weight, milk or meat production in cattle, crop yield, and seed protein content. Continuous variation across a range of phenotypes is measured and described in quantitative terms, so this genetic phenomenon is known as **quantitative inheritance**. Because the varying phenotypes result from the input of genes at multiple loci, quantitative traits are sometimes said to be **polygenic** (literally "of many genes").

For traits showing continuous variation, the genotype generated at fertilization establishes the quantitative range within which a particular individual can fall. However, the final phenotype is often also influenced by environmental factors to which that individual is exposed. Human height, for example, is partly genetically determined, but is also affected by environmental factors such as nutrition. Those phenotypes that result from both gene action and environmental influences are sometimes termed **complex**, or **multifactorial, traits.**

In this chapter, we will examine examples of quantitative inheritance and some of the statistical techniques used to study complex traits. We will also consider how geneticists assess the relative importance of genetic versus environmental factors contributing to continuous phenotypic variation, and we will discuss approaches to identifying and mapping genes that influence quantitative traits.

## 25.1 Not All Polygenic Traits Show Continuous Variation

When a continuous trait is examined in a population, individual measurements often lie along a continuum of phenotypes with no clear separation into categories. It is possible for a single-gene trait to show such a continuum of phenotypes within a population, if, for example, there are different degrees of gene penetrance among individuals. Single-gene traits can also be multifactorial, if the interaction of alleles with the environment produces a range of different phenotypes. More often, however, a continuous quantitative trait is the result of polygenic inheritance. Furthermore, polygenic traits are frequently multifactorial, with environmental factors contributing to the range of phenotypes observed.

In addition to continuous quantitative traits, where phenotypic variation can lie at any point within a range of measurement, there are two other classes of polygenic traits. **Meristic** traits are those in which the phenotypes are described by whole numbers. Examples of meristic traits include the number of seeds in a pod or the number of eggs laid by a chicken in a year. These are quantitative traits, but they do not have an infinite range of phenotypes: for example, a pod may contain 2, 4, or 6 seeds, but not 5.75. **Threshold traits** are polygenic (and frequently, environmental factors affect the phenotypes, making them also multifactorial), but they are distinguished from continuous and meristic traits by having a small number of discrete phenotypic classes. Threshold traits are currently of heightened interest to human geneticists because an increasing number of diseases are now thought to show this pattern of polygenic inheritance. One example is **Type II diabetes**, also known as adult-onset diabetes because it typically affects individuals who are middle aged or older. A population can be divided into just two phenotypic classes for this trait—individuals who have Type II diabetes and those who do not—so at first glance, it may appear to more closely resemble a simple monogenic trait. However, no single adult-onset diabetes gene has been identified. Instead, the combination of alleles present at multiple contributing loci gives an individual a greater or lesser likelihood of developing the disease. These varying levels of liability form a continuous range: at one extreme are those at very low risk for Type II diabetes, while at the other end of the distribution are those whose genotypes make it highly likely they will develop the disease (Figure 25–1). As with many threshold traits, environmental factors also play a role in determining the final phenotype, with diet

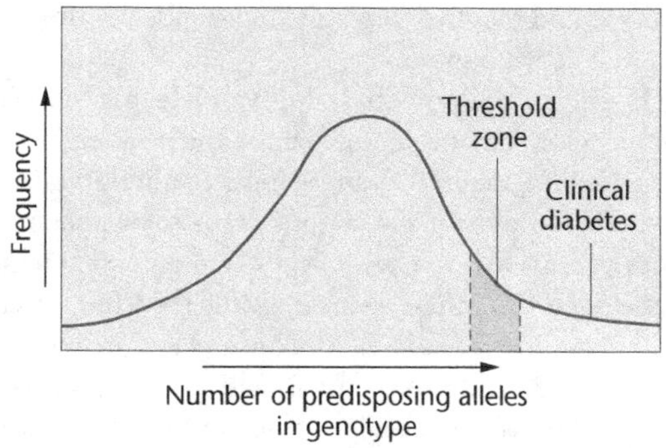


**FIGURE 25–1** A graphic depiction of predisposing alleles characteristic of a threshold trait within a population, illustrated by Type II diabetes.

and lifestyle having significant impact on whether an individual with moderate to high genetic liability will actually develop Type II diabetes.

## 25.2 Quantitative Traits Can Be Explained in Mendelian Terms

The question of whether continuous phenotypic variation could be explained in Mendelian terms caused considerable controversy in the early 1900s. Some scientists argued that, although Mendel's unit factors, or genes, explained patterns of discontinuous segregation with discrete phenotypic classes, they could not also account for the range of phenotypes seen in quantitative patterns of inheritance. However, geneticists William Bateson and Gudny Yule, adhering to a Mendelian explanation, proposed the **multiple-factor** or **multiple-gene hypothesis**, in which many genes, each individually behaving in a Mendelian fashion, contribute to the phenotype in a *cumulative* or *quantitative* way.

### The Multiple-Gene Hypothesis for Quantitative Inheritance

The **multiple-gene hypothesis** was initially based on a key set of experimental results published by Hermann Nilsson-Ehle in 1909. Nilsson-Ehle used grain color in wheat to test the concept that the cumulative effects of alleles at multiple loci produce the range of phenotypes seen in quantitative traits. In one set of experiments, wheat with red grain was crossed to wheat with white grain (Figure 25–2). The $F_1$ generation demonstrated an intermediate pink color, which at first sight suggested incomplete dominance of two alleles at a single locus. However, in the $F_2$ generation, Nilsson-Ehle did not observe the 3:1 segregation typical of a monohybrid cross. Instead, approximately 15/16 of the plants showed some degree of red grain color, while 1/16 of the plants showed white grain color. Careful examination of the $F_2$ revealed that grain with color could be classified into four different shades of red. Because the $F_2$ ratio occurred in sixteenths, it appears that two genes, each with two alleles, control the phenotype and that they segregate independently from one another in a Mendelian fashion.

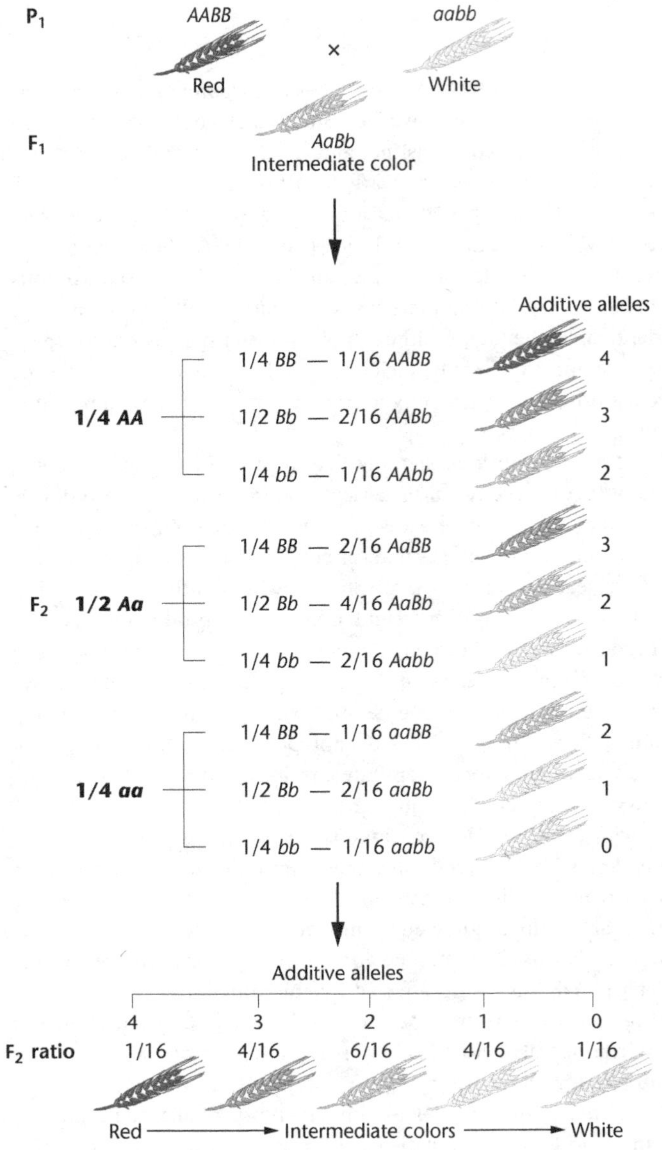


**FIGURE 25–2** How the multiple-factor hypothesis accounts for the 1:4:6:4:1 phenotypic ratio of grain color when all alleles designated by an uppercase letter are additive and contribute an equal amount of pigment to the phenotype.

If each gene has one potential **additive allele** that contributes to the red grain color and one potential **nonadditive allele** that fails to produce any red pigment, we can see how the multiple-factor hypothesis could account for the various grain color phenotypes. In the $P_1$ both parents are homozygous; the red parent contains only additive alleles (*AABB* in Figure 25–2), while the white parent contains only nonadditive alleles (*aabb*). The $F_1$ plants are heterozygous (AaBb), contain two additive (*A* and *B*) and two nonadditive (*a* and *b*) alleles, and express the intermediate pink phenotype. Each of the $F_2$ plants has 4, 3, 2, 1, or 0 additive alleles. $F_2$ plants with no additive alleles are white (*aabb*) like one of the $P_1$ parents, while $F_2$ plants with 4 additive alleles are red (*AABB*) like the other $P_1$ parent. Plants with 3, 2, or 1 additive alleles constitute the other three categories of red color observed in the $F_2$ generation. The greater the number of additive alleles in the genotype, the more intense the red color expressed in the phenotype, as each additive allele present contributes equally to the cumulative amount of pigment produced in the grain.

Nilsson-Ehle's results showed how continuous variation could still be explained in a Mendelian fashion, with additive alleles at multiple loci influencing the phenotype in a quantitative manner, but each individual allele segregating according to Mendelian rules. As we saw in Nilsson-Ehle's initial cross, if two loci, each with two

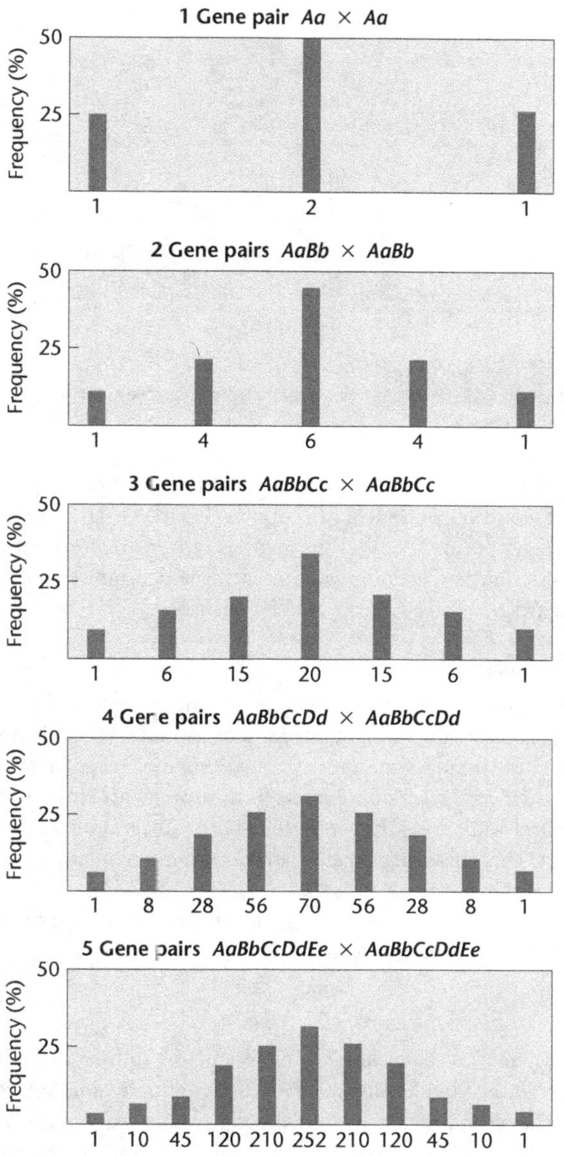


**FIGURE 25–3** The genetic ratios (on the X-axis) resulting from crossing two heterozygotes when polygenic inheritance is in operation with 1–5 gene pairs. The histogram bars indicate the distinct $F_2$ phenotypic classes, ranging from one extreme (left end) to the other extreme (right end). Each phenotype results from a different number of additive alleles.

alleles, were involved, then five $F_2$ phenotypic categories in a 1:4:6:4:1 ratio would be expected. However, there is no reason why three, four, or more loci cannot function in a similar fashion in controlling various quantitative phenotypes. As more quantitative loci become involved, greater and greater numbers of classes appear in the $F_2$ generation in more complex ratios. The number of phenotypes and the expected $F_2$ ratios for crosses involving up to five gene pairs are illustrated in Figure 25–3.

## Additive Alleles: The Basis of Continuous Variation

The multiple-gene hypothesis consists of the following major points:

1. Phenotypic traits showing continuous variation can be quantified by measuring, weighing, counting, and so on.
2. Two or more gene loci, often scattered throughout the genome, account for the hereditary influence on the phenotype in an *additive way*. Because many genes may be involved, inheritance of this type is called *polygenic*.
3. Each gene locus may be occupied by either an *additive* allele, which contributes a constant amount to the phenotype, or a *nonadditive* allele, which does not contribute quantitatively to the phenotype.
4. The contribution to the phenotype of each additive allele, though often small, is approximately equal.
5. Together, the additive alleles contributing to a single quantitative character produce substantial phenotypic variation.

## Calculating the Number of Polygenes

Various formulas have been developed for estimating the number of **polygenes**, the genes contributing to a quantitative trait. For example, if the ratio of $F_2$ individuals resembling *either* of the two extreme $P_1$ phenotypes can be determined, the number of polygenes involved ($n$) may be calculated as follows:

$1/4^n$ = ratio of $F_2$ individuals expressing either extreme phenotype

In the example of the red and white wheat grain color summarized in Figure 25–2, 1/16 of the progeny are either red *or* white like the $P_1$ phenotypes. This ratio can be substituted on the right side of the equation to solve for $n$:

$$\frac{1}{4^n} = \frac{1}{16}$$

$$\frac{1}{4^2} = \frac{1}{16}$$

$$n = 2$$

Table 25.1 lists the ratio and the number of $F_2$ phenotypic classes produced in crosses involving up to five gene pairs.

**TABLE 25.1**

**Determination of the Number of Polygenes ($n$) Involved in a Quantitative Trait**

| $n$ | Individuals Expressing Either Extreme Phenotype | Distinct Phenotypic Classes |
|---|---|---|
| 1 | 1/4 | 3 |
| 2 | 1/16 | 5 |
| 3 | 1/64 | 7 |
| 4 | 1/256 | 9 |
| 5 | 1/1024 | 11 |

For low numbers of polygenes ($n$), it is sometimes easier to use the equation

$(2n + 1) =$ the number of distinct phenotypic categories observed

For example, when there are two polygenes involved ($n = 2$), then $(2n + 1) = 5$ and each phenotype is the result of 4, 3, 2, 1, or 0 additive alleles. If $n = 3$, $2n + 1 = 7$ and each phenotype is the result of 6, 5, 4, 3, 2, 1, or 0 additive alleles. Thus, working backwards with this rule and knowing the number of phenotypes, we can calculate the number of polygenes controlling them.

It should be noted, however, that both these simple methods for estimating the number of polygenes involved in a quantitative trait assume that all the relevant alleles contribute equally and additively, and also that phenotypic expression in the $F_2$ is not affected significantly by environmental factors. As we will see later, for many quantitative traits, these assumptions may not be true.

NOW SOLVE THIS

Problem 3 on page 683 gives $F_1$ and $F_2$ ranges for a quantitative trait and asks you to calculate the number of polygenes involved.

■ HINT: *Remember that unless you know the total number of distinct phenotypes involved, then the ratio (not the number) of parental phenotypes reappearing in the $F_2$ is the key.*

25.3

## The Study of Polygenic Traits Relies on Statistical Analysis

Before considering the approaches that geneticists use to dissect how much of the phenotypic variation observed in a population is due to genotypic differences among individuals and how much is due to environmental factors, we need to consider the basic statistical tools they use for the task. It is not usually feasible to measure expression of a polygenic trait in every individual in a population, so a random subset of individuals is usually selected for measurement to provide a *sample*. It is important to remember that the accuracy of the final results of the measurements depends on whether the sample is truly random and representative of the population from which it was drawn. Suppose, for example, that a student wants to determine the average height of the 100 students in his genetics class, and for his sample he measures the two students sitting next to him, both of whom happen to be centers on the college basketball team. It is unlikely that this sample will provide a good estimate of the average height of the class, for two reasons: first, it is too small; second, it is not a representative subset of the class (unless all 100 students are centers on the basketball team).

If the sample measured for expression of a quantitative trait is sufficiently large and also representative of the population from which it is drawn, we often find that the data form a **normal**

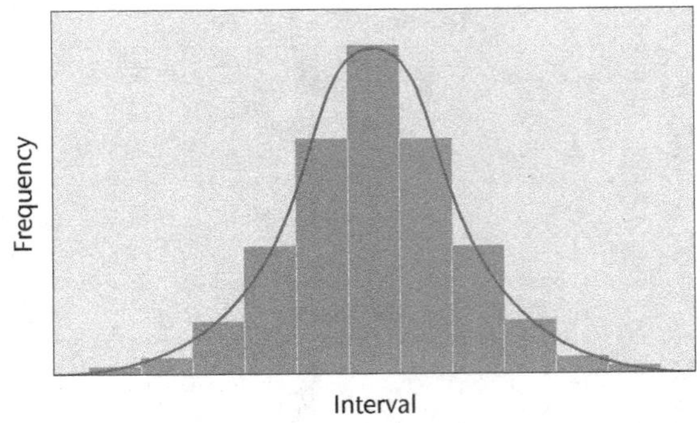


FIGURE 25-4 Normal frequency distribution, characterized by a bell-shaped curve.

**distribution**; that is, they produce a characteristic bell-shaped curve when plotted as a frequency histogram (Figure 25–4). Several statistical concepts are useful in the analysis of traits that exhibit a normal distribution, including the mean, variance, standard deviation, standard error of the mean, and covariance.

### The Mean

The mean provides information about where the central point lies along a range of measurements for a quantitative trait. Figure 25–5 shows the distribution curves for two different sets of phenotypic measurements. Each of these sets of measurements clusters around a central value (as it happens, they both cluster around the same value). This clustering is called a **central tendency**, and the central point is the mean.

Specifically, the **mean** ($\overline{X}$) is the arithmetic average of a set of measurements and is calculated as

$$\overline{X} = \frac{\Sigma X_i}{n}$$

where $\overline{X}$ is the mean, $\Sigma X_i$ represents the sum of all individual values in the sample, and $n$ is the number of individual values.

The mean provides a useful descriptive summary of the sample, but it tells us nothing about the range or spread of the data. As

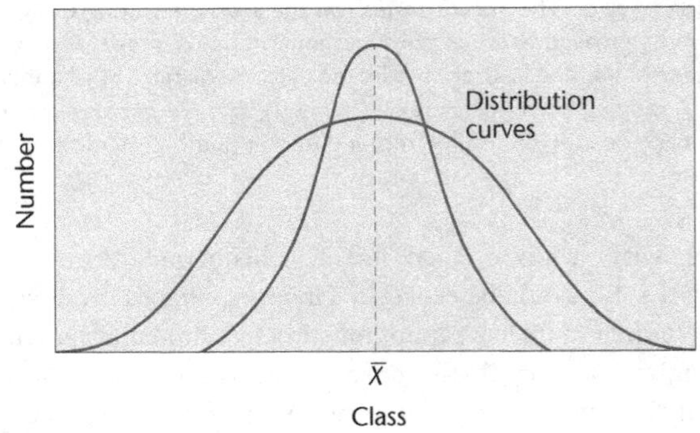


FIGURE 25-5 Two normal frequency distributions with the same mean but different amounts of variation.

illustrated in Figure 25–5, a symmetrical distribution of values in the sample may, in one case, be clustered near the mean. Or a set of measurements may have the same mean but be distributed more widely around it. A second statistic, the variance, provides information about the spread of data around the mean.

## Variance

The **variance** ($s^2$) for a sample is the average squared distance of all measurements from the mean. It is calculated as

$$s^2 = \frac{\Sigma(X_i - \overline{X})^2}{n - 1}$$

where the sum ($\Sigma$) of the squared differences between each measured value ($X_i$) and the mean ($\overline{X}$) is divided by one less than the total sample size $n - 1$.

As Figure 25–5 shows, it is possible for two sets of sample measurements for a quantitative trait to have the same mean but a different distribution of values around it. This range will be reflected in different variances. Estimation of variance can be useful in determining the degree of genetic control of traits when the immediate environment also influences the phenotype.

## Standard Deviation

Because the variance is a squared value, its unit of measurement is also squared ($m^2$, $g^2$, etc.). To express variation around the mean in the original units of measurement, we can use the square root of the variance, a term called the **standard deviation** (*s*):

$$s = \sqrt{s^2}$$

Table 25.2 shows the percentage of individual values within a normal distribution that fall within different multiples of the standard deviation. The values that fall within one standard deviation to either side of the mean represent 68 percent of all values in the sample. More than 95 percent of all values are found within two standard deviations to either side of the mean. This means that the standard deviation *s* can also be interpreted in the form of a probability. For example, a sample measurement picked at random has a 68 percent probability of falling within the range of one standard deviation.

## Standard Error of the Mean

If multiple samples are taken from a population and measured for the same quantitative trait, we might find that their means vary. Theoretically, larger, truly random samples will represent the population more accurately, and their means will be closer to each other.

**TABLE 25.2**

**Sample Inclusion for Various *s* Values**

| Multiples of *s* | Sample Included (%) |
|---|---|
| $\overline{X}\pm1s$ | 68.3 |
| $\overline{X}\pm1.96s$ | 95.0 |
| $\overline{X}\pm2s$ | 95.5 |
| $\overline{X}\pm3s$ | 99.7 |

To measure the accuracy of the sample mean we use the **standard error of the mean** ($S_{\overline{X}}$), calculated as

$$S_{\overline{X}} = \frac{s}{\sqrt{n}}$$

where *s* is the standard deviation and $\sqrt{n}$ is the square root of the sample size. Because the standard error of the mean is computed by dividing *s* by $\sqrt{n}$, it is always a smaller value than the standard deviation.

## Covariance

Often geneticists working with quantitative traits find they have to consider two phenotypic characters simultaneously. For example, a poultry breeder might investigate the correlation between body weight and egg production in hens: Do heavier birds tend to lay more eggs? The **covariance** statistic measures how much variation is common to both quantitative traits. It is calculated by taking the deviations from the mean for each trait (just as we did for estimating variance) for each individual in the sample. This gives a pair of values for each individual. The two values are multiplied together, and the sum of all these individual products is then divided by one fewer than the number in the sample. Thus the covariance $cov_{XY}$ of two sets of trait measurements, *X* and *Y*, is calculated as

$$\text{cov}_{XY} = \frac{\Sigma[(X_i - \overline{X})(Y_i - \overline{Y})]}{n - 1}$$

The covariance can then be standardized as yet another statistic, the **correlation coefficient** (*r*). The calculation of *r* is

$$r = cov_{XY}/S_X S_Y$$

where $S_X$ is the standard deviation of the first set of quantitative measurements *X*, and $S_Y$ is the standard deviation of the second set of quantitative measurements *Y*. Values for the correlation coefficient *r* can range from −1 to +1. Positive *r* values mean that an increase in measurement for one trait tends to be associated with an increase in measurement for the other, while negative *r* values mean that increases in one trait are associated with decreases in the other. Therefore, if heavier hens do tend to lay more eggs, a positive *r* value can be expected. A negative *r* value, on the other hand, suggests that greater egg production is more likely from less heavy birds. One important point to note about correlation coefficients is that even significant *r* values—close to +1 or −1—do not prove that a cause-and-effect relationship exists between two traits. Correlation simply tells us the extent to which variation in one quantitative trait is associated with variation in another, not what causes that variation.

**NOW SOLVE THIS**

Problem 15 on page 685 gives data for two quantitative traits in a flock of sheep. In Part (d), you are asked to determine if the traits are correlated.

■ HINT: *Once the calculation of the correlation coefficient (r) is completed, you must analyze and interpret that value.*

## Analysis of a Quantitative Character

To apply these statistical concepts, let's consider a genetic experiment that crossed two different homozygous varieties of tomato. One of the tomato varieties produces fruit averaging 18 oz in weight, whereas fruit from the other averages 6 oz. The $F_1$ obtained by crossing these two varieties has fruit weights ranging from 10 to 14 oz. The $F_2$ population contains individuals that produce fruit ranging from 6 to 18 oz. The results characterizing both generations are shown in Table 25.3.

The mean value for the fruit weight in the $F_1$ generation can be calculated as

$$\overline{X} = \frac{\Sigma X_i}{n} = \frac{626}{52} = 12.04$$

The mean value for fruit weight in the $F_2$ generation is calculated as

$$\overline{X} = \frac{\Sigma X_i}{n} = \frac{872}{72} = 12.11$$

Although these mean values are similar, the frequency distributions in Table 25.3 show more variation in the $F_2$ generation. The range of variation can be quantified as the sample variance $s^2$, calculated, as we saw above, as the sum of the squared differences between each value and the mean, divided by one less than the total number of observations.

$$s^2 = \frac{\Sigma(X_i - \overline{X})^2}{n - 1}$$

When the above calculation is made, the variance is found to be 1.29 for the $F_1$ generation and 4.27 for the $F_2$ generation. When converted to the standard deviation $\left(s = \sqrt{s^2}\right)$, the values become 1.13 and 2.06, respectively. Therefore, the distribution of tomato weight in the $F_1$ generation can be described as 12.04 ± 1.13, and in the $F_2$ generation it can be described as 12.11 ± 2.06.

Assuming both tomato varieties are homozygous at the loci of interest and that the alleles controlling fruit weight act additively, we can estimate the number of polygenes involved in this trait. Since 1/72 of the $F_2$ offspring have a phenotype that overlaps one of the parental strains (72 total $F_2$ offspring; one weighs 6 oz, one weighs 18 oz; see Table 25.3), the use of the formula $1/4^n = 1/72$ indicates that $n$ is between 3 and 4, providing evidence of the number of genes that control fruit weight in these tomato strains.

## 25.4 Heritability Values Estimate the Genetic Contribution to Phenotypic Variability

The question most often asked by geneticists working with multifactorial traits is how much of the observed phenotypic variation in a population is due to genotypic differences among individuals and how much is due to environment. The term **heritability** is used to describe what proportion of total phenotypic variation in a population is due to genetic factors. For a multifactorial trait in a given population, a high heritability estimate indicates that much of the variation can be attributed to genetic factors, with the environment having less impact on expression of the trait. With a low heritability estimate, environmental factors are likely to have a greater impact on phenotypic variation within the population.

The concept of heritability is frequently misunderstood and misused. It should be emphasized that heritability indicates neither how much of a trait is genetically determined, nor the extent to which an individual's phenotype is due to genotype. In recent years, such misinterpretations of heritability for human quantitative traits have led to controversy, notably in relation to measurements such as "intelligence quotients," or IQs. Variation in heritability estimates for IQ among different "racial" groups tested led to incorrect suggestions that unalterable genetic factors control differences in intelligence levels among humans of different ancestries. Such suggestions misrepresented the meaning of heritability and ignored the contribution of genotype-by-environment interaction variance (see p. 675) to phenotypic variation in a population. Moreover, heritability is not fixed for a trait. For example, a heritability estimate for egg production in a flock of chickens kept in individual cages might be high, indicating that differences in egg output among individual birds are largely due to genetic differences, as they all have very similar environments. For a different flock kept outdoors, heritability for egg production might be much lower, as variation among different birds may also reflect differences in their individual environments. Such differences could include how much food each bird manages to find and whether it competes successfully for a good roosting spot at night. Thus a heritability estimate tells us the proportion of phenotypic variation that can be attributed to genetic variation *within a certain population in a particular environment.* If we measure heritability for the same trait among

TABLE 25.3

**Distribution of $F_1$ and $F_2$ Progeny Derived from a Theoretical Cross Involving Tomatoes**

| | | Weight (oz.) | | | | | | | | | | | | |
|---|---|---|---|---|---|---|---|---|---|---|---|---|---|---|
| | | 6 | 7 | 8 | 9 | 10 | 11 | 12 | 13 | 14 | 15 | 16 | 17 | 18 |
| Number of | $F_1$ | | | | | 4 | 14 | 16 | 12 | 6 | | | | |
| Individuals | $F_2$ | 1 | 1 | 2 | 0 | 9 | 13 | 17 | 14 | 7 | 4 | 3 | 0 | 1 |

different populations in a range of environments, we frequently find that the calculated heritability values have large standard errors. This is an important point to remember when considering heritability estimates for traits in human populations. A mean heritability estimate of 0.65 for human height does not mean that your height is 65 percent due to your genes, but rather that in the populations sampled, on average, *65 percent of the overall variation in height could be explained by genotypic differences among individuals.*

With this subtle, but important distinction in mind, we will now consider how geneticists divide the phenotypic variation observed in a population into genetic and environmental components. As we saw in the previous section, variation can be quantified as a sample variance: taking measurements of the trait in question from a representative sample of the population and determining the extent of the spread of those measurements around the sample mean. This gives us an estimate of the total **phenotypic variance** in the population (**$V_P$**.) Heritability estimates are obtained by using different experimental and statistical techniques to partition $V_P$ into **genotypic variance ($V_G$)** and **environmental variance ($V_E$)** components.

An important factor contributing to overall levels of phenotypic variation is the extent to which individual genotypes affect the phenotype differently depending on the environment. For example, wheat variety A may yield an average of 20 bushels an acre on poor soil, while variety B yields an average of 17 bushels. On good soil, variety A yields 22 bushels, while variety B averages 25 bushels an acre. There are differences in yield between the two genotypically distinct varieties, so variation in wheat yield has a genetic component. Both varieties yield more on good soil, so yield is also affected by environment. However, we also see that the two varieties do not respond to better soil conditions equally: The genotype of wheat variety B achieves a greater increase in yield on good soil than does variety A. Thus, we have differences in the interaction of genotype with environment contributing to variation for yield in populations of wheat plants. This third component of phenotypic variation is **genotype-by-environment interaction variance ($V_{G\times E}$)** (Figure 25–6).

We can now summarize all the components of total phenotypic variance $V_P$ using the following equation:

$$V_P = V_G + V_E + V_{G\times E}$$

In other words, total phenotypic variance can be subdivided into genotypic variance, environmental variance, and genotype-by-environment interaction variance. When obtaining heritability estimates for a multifactorial trait, researchers often assume that the genotype-by-environment interaction variance is small enough that it can be ignored or combined with the environmental variance. However, it is worth remembering that this kind of approximation is another reason heritability values are *estimates* for a given population in a particular context, not a *fixed attribute* for a trait.

Animal and plant breeders use a range of experimental techniques to estimate heritabilities by partitioning measurements of phenotypic variance into genotypic and environmental components. One approach uses inbred strains containing genetically homogeneous individuals with highly homozygous genotypes.

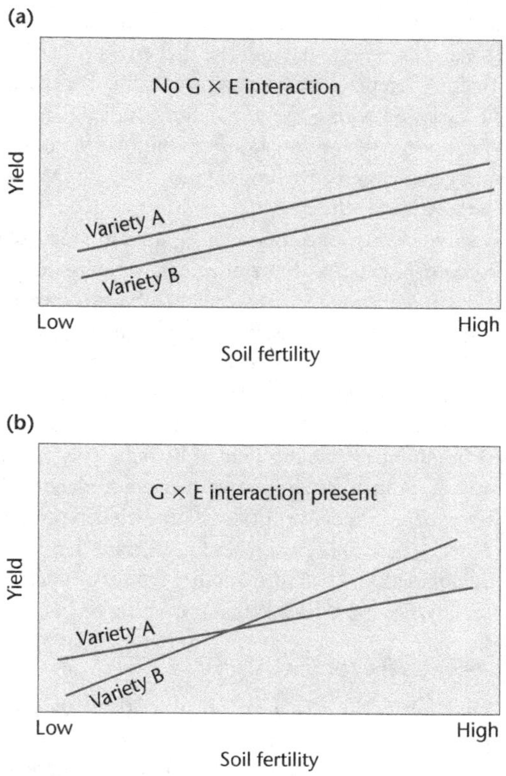


FIGURE 25–6 Differences in yield between two wheat varieties at different soil fertility levels. (a) No genotype-by-environment, or G × E, interaction: The varieties show genetic differences in yield but respond equally to increasing soil fertility. (b) G × E interaction present: Variety A outyields B at low soil fertility, but B yields more than A at high fertility levels.

Experiments are then designed to test the effects of a range of environmental conditions on phenotypic variability. Variation *between* different inbred strains reared in a constant environment is due predominantly to genetic factors. Variation *among* members of the same inbred strain reared under different conditions is more likely to be due to environmental factors. Other approaches involve analysis of variance for a quantitative trait among offspring from different crosses, or comparing expression of a trait among offspring and parents reared in the same environment.

## Broad-Sense Heritability

**Broad-sense heritability** (represented by the term $H^2$) measures the contribution of the genotypic variance to the total phenotypic variance. It is estimated as a proportion:

$$H^2 = \frac{V_G}{V_P}$$

Heritability values for a trait in a population range from 0.0 to 1.0. A value approaching 1.0 indicates that the environmental conditions have little impact on phenotypic variance, which is therefore largely due to genotypic differences among individuals in the population. Low values close to 0.0 indicate that environmental factors, not genotypic differences, are largely responsible for the observed phenotypic variation within the population studied. Few quantitative traits have very high or very low heritability estimates, suggesting that both genetics and environment play a part in the expression of most phenotypes for the trait.

The genotypic variance component $V_G$ used in broad-sense heritability estimates includes all types of genetic variation in the population. It does not distinguish between quantitative trait loci with alleles acting additively as opposed to those with epistatic or dominance effects. Broad-sense heritability estimates also assume that the genotype-by-environment variance component is negligible. While broad-sense heritability estimates for a trait are of general genetic interest, these limitations mean this kind of heritability is not very useful in breeding programs. Animal or plant breeders wishing to develop improved strains of livestock or higher-yielding crop varieties need more precise heritability estimates for the traits they wish to manipulate in a population. Therefore, another type of estimate, narrow-sense heritability, has been devised that is of more practical use.

## Narrow-Sense Heritability

**Narrow-sense heritability ($h^2$)** is the proportion of phenotypic variance due to additive genotypic variance alone. Genotypic variance can be divided into subcomponents representing the different modes of action of alleles at quantitative trait loci. As not all the genes involved in a quantitative trait affect the phenotype in the same way, this partitioning distinguishes among three different kinds of gene action contributing to genotypic variance. **Additive variance, $V_A$,** is the genotypic variance due to the additive action of alleles at quantitative trait loci. **Dominance variance, $V_D$,** is the deviation from the additive components that results when phenotypic expression in heterozygotes is not precisely intermediate between the two homozygotes. **Interactive variance, $V_I$,** is the deviation from the additive components that occurs when two or more loci behave epistatically. The amount of interactive variance is often negligible, and so this component is often excluded from calculations of total genotypic variance.

The partitioning of the total genotypic variance $V_G$ is summarized in the equation

$$V_G = V_A + V_D + V_I$$

and a narrow-sense heritability estimate based only on that portion of the genotypic variance due to additive gene action becomes

$$h^2 = \frac{V_A}{V_P}$$

Omitting $V_I$ and separating $V_P$ into genotypic and environmental variance components, we obtain

$$h^2 = \frac{V_A}{V_E + V_A + V_D}$$

Heritability estimates are used in animal and plant breeding to indicate the potential response of a population to artificial selection for a quantitative trait. Narrow-sense heritability, $h^2$, provides a more accurate prediction of selection response than broad-sense heritability, $H^2$, and therefore $h^2$ is more widely used by breeders.

## Artificial Selection

**Artificial selection** is the process of choosing specific individuals with preferred phenotypes from an initially heterogeneous population for future breeding purposes. Theoretically, if artificial selection based on the same trait preferences is repeated over multiple generations, a population can be developed containing a high frequency of individuals with the desired characteristics. If selection is for a simple trait controlled by just one or two genes subject to little environmental influence, generating the desired population of plants or animals is relatively fast and easy. However, many traits of economic importance in crops and livestock, such as grain yield in plants, weight gain or milk yield in cattle, and speed or stamina in horses, are polygenic and frequently multifactorial. Artificial selection for such traits is slower and more complex. Narrow-sense heritability estimates are valuable to the plant or animal breeder because, as we have just seen, they estimate the proportion of total phenotypic variance for the trait that is due to additive genetic variance. Quantitative trait alleles with additive action are those most easily manipulated by the breeder. Alleles at quantitative trait loci that generate dominance effects or interact epistatically (and therefore contribute to $V_D$ or $V_I$) are less responsive to artificial selection. Thus narrow-sense heritability, $h^2$, can be used to predict the impact of selection. The higher the estimated value for $h^2$ in a population, the greater the change in phenotypic range for the trait that the breeder will see in the next generation after artificial selection.

Partitioning the genetic variance components to calculate $h^2$ and predict response to selection is a complex task requiring careful experimental design and analysis. The simplest approach is to select individuals with superior phenotypes for the desired quantitative trait from a heterogeneous population and breed offspring from those individuals. The mean score for the trait of those offspring ($M2$) can then be compared to that of: (1) the original population's mean score ($M$) and (2) the selected individuals used as parents ($M1$). The relationship between these means and $h^2$ is

$$h^2 = \frac{M2 - M}{M1 - M}$$

This equation can be further simplified by defining $M2 - M$ as the **selection response ($R$)**—the degree of response to mating the selected parents—and M1 − M as the **selection differential ($S$)**—the difference between the mean for the whole population and the mean for the selected population—so $h^2$ reflects the ratio of the response observed to the total response possible. Thus,

$$h^2 = \frac{R}{S}$$

A narrow-sense heritability value obtained in this way by selective breeding and measuring the response in the offspring is referred to as an estimate of **realized heritability**.

As an example of a realized heritability estimate, suppose that we measure the diameter of corn kernels in a population where the mean diameter $M$ is 20 mm. From this population, we select a group with the smallest diameters, for which the mean $M1$ equals 10 mm. The selected plants are interbred, and the mean diameter $M2$ of the progeny kernels is 13 mm. We can calculate the realized heritability $h^2$ to estimate the potential for artificial selection on kernel size:

$$h^2 = \frac{M2 - M}{M1 - M}$$

$$h^2 = \frac{13 - 20}{10 - 20}$$

$$= \frac{-7}{-10}$$

$$= 0.70$$

This value for narrow-sense heritability indicates that the selection potential for kernel size is relatively high.

The longest running artificial selection experiment known is still being conducted at the State Agricultural Laboratory in Illinois. Since 1896, corn has been selected for both high and low oil content. After 76 generations, selection continues to result in increased oil content (Figure 25–7). With each cycle of successful selection, more of the corn plants accumulate a higher percentage of additive alleles involved in oil production. Consequently, the narrow-sense heritability $h^2$ of increased oil content in succeeding generations has declined (see parenthetical values at generations 9, 25, 52, and 76 in Figure 25–7) as artificial selection comes closer and closer to optimizing the genetic potential for oil production. Theoretically, the process will continue until all individuals in the population possess a uniform genotype that includes all the additive alleles responsible for high oil content. At that point, $h^2$ will be reduced to zero, and response to artificial selection will cease. The decrease in response to selection for low oil content shows that heritability for low oil content is approaching this point.

Table 25.4 lists narrow-sense heritability estimates expressed as percentage values for a variety of quantitative traits in different organisms. As you can see, these $h^2$ values vary, but heritability tends to be low for quantitative traits that are essential to an organism's survival. Remember, this does not indicate the absence of a genetic contribution to the observed phenotypes for such traits. Instead, the low $h^2$ values show that natural selection has already largely optimized the genetic component of these traits during evolution. Egg production, litter size, and conception rate are examples of how such physiological limitations on selection have already been reached. Traits that are less critical to survival, such as body weight, tail length, and wing length, have higher heritabilities because more genotypic variation for such traits is still present in the population. Remember also that any single heritability estimate can only provide information about one population in a specific environment. Therefore, narrow-sense heritability is a more valuable predictor of response to selection when estimates are calculated for many populations and environments and show the presence of a clear trend.

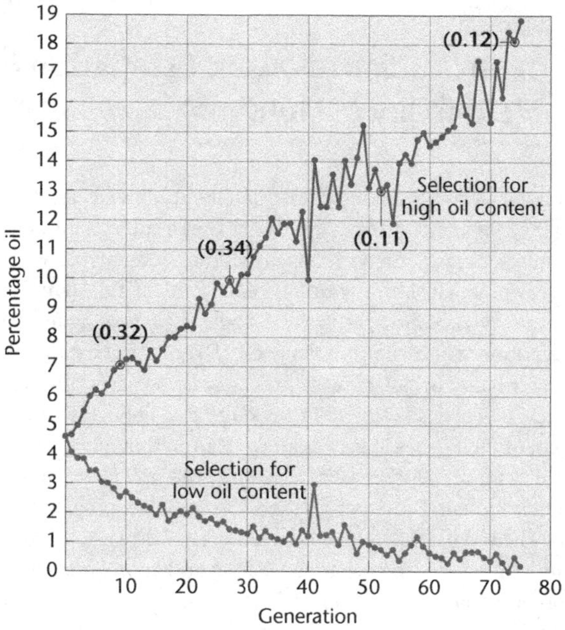


**FIGURE 25–7** Response of corn selected for high and low oil content over 76 generations. The numbers in parentheses at generations 9, 25, 52, and 76 for the "high oil" line indicate the calculation of heritability at these points in the continuing experiment.

**TABLE 25.4**

**Estimates of Heritability for Traits in Different Organisms**

| Trait | Heritability ($h^2$) |
|---|---|
| Mice | |
| Tail length | 60% |
| Body weight | 37 |
| Litter size | 15 |
| Chickens | |
| Body weight | 50 |
| Egg production | 20 |
| Egg hatchability | 15 |
| Cattle | |
| Birth weight | 45 |
| Milk yield | 44 |
| Conception rate | 3 |

**NOW SOLVE THIS**

Problem 12 on page 684 gives data for two quantitative traits in a herd of hogs and asks you to choose the trait that the farmer should select that will respond favorably to selection.

■ **HINT:** *Consider which variance component's proportion is the best indicator of potential response to selection.*

## 25.5 Twin Studies Allow an Estimation of Heritability in Humans

For obvious reasons, traditional heritability studies are not possible in humans. However, human twins can be useful subjects for examining how much variance for a multifactorial trait is genotypic as opposed to environmental. **Monozygotic (MZ)**, or **identical, twins**, derived from the division and splitting of a single egg following fertilization, are genotypically identical. **Dizygotic (DZ)**, or **fraternal, twins**, on the other hand, originate from two separate fertilization events and are only as genetically similar as any other two siblings, with about 50 percent of their genes in common. For a given trait, therefore, phenotypic differences between pairs of identical twins will be equivalent to the environmental variance ($V_E$) (because the genotypic variance is zero). Phenotypic differences between dizygotic twins, however, represent both environmental variance ($V_E$) and approximately half the genotypic variance ($V_G$). Comparison of phenotypic variances for the same trait in monozygotic and dizygotic sets of twins provides an estimate of broad-sense heritability for the trait.

Another approach used in twin studies has been to compare identical twin pairs reared together with pairs that were separated and raised in different settings. For any particular trait, average similarities or differences can be investigated. Twins are said to be **concordant** for a given trait if both express it or neither expresses it. If one expresses the trait and the other does not, the pair is said to be **discordant**. Comparison of concordance values of MZ versus DZ twins reared together illustrates the potential value for heritability assessment. (See the following "Now Solve This" feature.)

Before any conclusions can be drawn, such data must be examined very carefully. If the concordance value approaches 90 to 100 percent in monozygotic twins, we might be inclined to interpret that value as indicating a large genetic contribution to the expression of the trait. In some cases—for example, blood types and eye color—we know that this is indeed true. In the case of contracting measles, however, a high concordance value merely indicates that the trait is almost always induced by a factor in the environment—in this case, a virus.

It is more meaningful to compare the *difference* between the concordance values of monozygotic and dizygotic twins. If these values are significantly higher for monozygotic twins than for dizygotic twins, we suspect a strong genetic component in the determination of the trait. We reach this conclusion because monozygotic twins, with identical genotypes, would be expected to show a greater concordance than genetically related, but not genetically identical, dizygotic twins. In the case of measles, where concordance is high in both types of twins, the environment is assumed to contribute significantly. Such an analysis is useful because phenotypic characteristics that remain similar in different environments are likely to have a strong genetic component.

Interesting as they are, human twin studies contain some unavoidable sources of error. For example, identical twins are often treated more similarly by parents and teachers than are nonidentical twins, especially where the nonidentical siblings are of different sex. This may inflate the environmental variance for nonidentical twins. Another possible error source is genotype-by-environment interaction, which can increase the total phenotypic variance for nonidentical twins compared to identical twins raised in the same environment. Heritability estimates for human traits based on twin studies should therefore be considered approximations and examined very carefully before any conclusions are drawn.

**NOW SOLVE THIS**

Problem 26 on page 686 lists the percentages of monozygotic (MZ) and dizygotic (DZ) twins expressing the same phenotype for different traits and asks you to evaluate the relative importance of genetic and environmental factors.

■ HINT: *Consider how MZ twins differ genetically from DZ twins. If each pair of twins was raised in the same environment, consider how that would affect expression of genetically determined versus environmentally determined phenotypes.*

## 25.6 Quantitative Trait Loci Can Be Mapped

The kind of pedigree analysis we looked at in earlier chapters (see Chapters 3 and 4) is of little use in identifying the multiple genes involved in quantitative traits. Environmental effects, interaction among segregating alleles, and the number of genes contributing to a polygenic phenotype make it difficult to isolate the effect of any one individual gene. However, because many quantitative traits are of economic or medical importance, it is desirable to identify the genes involved and determine their location in the genome. Multiple genes contributing to a quantitative trait are known as **quantitative trait loci (QTLs)**. A single quantitative trait locus is designated a QTL. Identifying and studying these loci help geneticists estimate how many genes are involved in a given quantitative trait and whether they all contribute equally to the trait or whether some genes influence the phenotype more strongly than others. Mapping reveals whether the QTLs associated with a trait are grouped on a single chromosome or scattered throughout the genome.

To find and map QTLs, researchers look for associations between particular DNA sequences within the genome and phenotypes falling within a certain range of the quantitative phenotype. One way to do this is to cross homozygous inbred lines that have different phenotypic extremes for the trait of interest to create an $F_1$

generation whose members will be heterozygous at most of the loci contributing to the trait. Additional crosses, either among $F_1$ individuals or between the $F_1$ and the inbred parent lines, result in $F_2$ generations with a high degree of segregation for different QTL genotypes and associated phenotypes. This segregating $F_2$ is known as the **QTL mapping population**. Researchers measure expression of the trait by individuals in the mapping population and identify different genotypes using DNA markers such as restriction fragment length polymorphisms (RFLPs) and microsatellites. (See Chapter 24 for a more detailed description of marker technology.) Computer-based statistical analysis is then used to examine the correlation between marker genotypes and phenotypic variation for the trait. If a DNA marker is not linked to a QTL, then the phenotypic mean score for the trait will not vary among individuals with different genotypes at that marker locus. However, if a DNA marker is linked to a QTL, then genotypes differing at that marker locus will also differ in their expression of the trait. When this occurs, the marker locus and the QTL are said to *cosegregate*. Consistent cosegregation establishes the presence of a QTL at or near the DNA marker along the chromosome: The marker and QTL are linked. When numerous QTLs for a given trait have been located, a genetic map is created giving the positions of the genes involved on the different chromosomes.

DNA markers are now available for many organisms of agricultural importance, making possible systematic mapping of QTLs. For example, hundreds of RFLP markers have been located in the tomato. QTL analysis is performed by crossing plants with extreme, opposite phenotypes and following the crosses through several generations. Many loci have been identified that are responsible for quantitative traits such as fruit size, shape, soluble solid content, and acidity. These loci are distributed on all 12 chromosomes representing the haploid genome of this plant. Several chromosomes contain loci for more than one of these traits.

Mapping and characterizing QTLs in the tomato have been the focus of a highly successful research effort conducted by Steven Tanksley and his colleagues at Cornell University. Their research shows that approximately 30 QTLs contribute to tomato fruit size and shape. However, these QTLs do not have equal effects: 10 genes scattered among seven chromosomes account for most of the observed phenotypic variation for these traits. One locus in particular, *fw2.2*, on chromosome 2, has been isolated, cloned, and transferred between plants, with interesting results. A difference in alleles at *fw2.2* can change fruit size by up to 30 percent. While the cultivated tomato can weigh up to 1000 grams, fruit from the related wild species thought to be the progenitor of the modern tomato weighs only a few grams. Two distinct alleles of *fw2.2*, identified as a result of RFLP mapping studies, appear to have played a major part in bringing this difference about. One allele is present in all wild small-fruited varieties of tomatoes investigated. The other allele is present in all domesticated large-fruited varieties. When the cloned *fw2.2* allele from small-fruited varieties is transferred to a plant that normally produces large tomatoes, the transformed plant produces fruits that are greatly reduced in weight (Figure 25–8). In the varieties studied by Tanksley's group, the reduction averaged 17 grams, a significant phenotypic change caused by the action of a single QTL. Further analysis of *fw2.2* revealed that the large-fruited allele at this locus codes for a protein acting as a negative repressor of cell division, resulting in increased mitosis in the cortical tissue that forms the fruit. The small-fruited *fw2.2* allele codes for the same protein but has mutations in the promoter region that reduce transcription levels, resulting in lower rates of cell division and much smaller fruit.

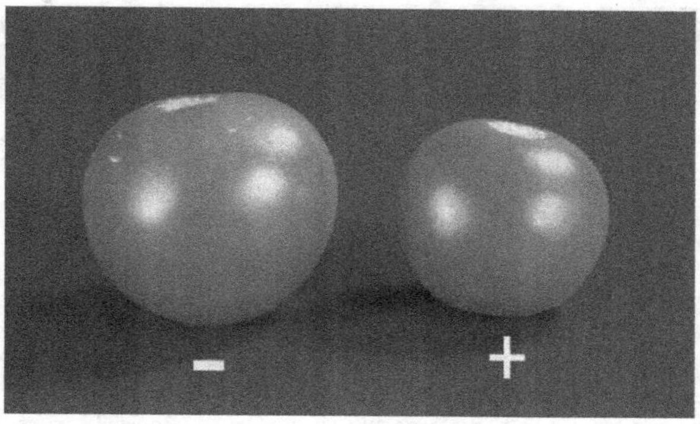


**FIGURE 25–8** Phenotypic effect of the *fw2.2* transgene in the tomato. When the allele causing small fruit is transferred to a plant that normally produces large fruit, the fruit is reduced in size (+). A control fruit (−) that has not been transformed is shown for comparison.

Marker-based mapping can identify chromosomal regions where QTLs are found, but identifying individual quantitative genes usually requires additional techniques. One approach is to develop DNA markers that are tightly linked to a gene of interest and then use positional cloning (see Chapter 24). Clearly, finding and characterizing all the genes involved in a quantitative trait is a long and complex task. At present, *fw2.2* is one of few quantitative genes that have been isolated, cloned, and studied in detail. However, mapping QTLs and defining the function of genes present in these regions in agriculturally important plants will help in the development of strategies for improving crop yields. These studies also pave the way for similar research on animal genes.

GENETICS, TECHNOLOGY, AND SOCIETY

# The Green Revolution Revisited: Genetic Research with Rice

Of the more than 6 billion people now living on Earth, about 750 million do not have enough to eat. And despite efforts to limit population growth, that number is expected to grow by an additional 1 million people each year for the next several decades.

Will we be able to solve this problem? The past gives us some reasons to be optimistic. In the 1950s and 1960s, in the face of looming population increases, plant scientists around the world set about to increase the production of crop plants, including the three most important grains, wheat, rice, and maize. These efforts became known as the Green Revolution. The approach was three-pronged: (1) to increase the use of fertilizers, pesticides, and irrigation water; (2) to bring more land under cultivation; and (3) to develop improved varieties of crop plants by intensive plant breeding. Though highly successful initially, the rate of increase in grain yields has slowed. If food production is to keep pace with the projected increase in the world's population, plant breeders will have to depend more and more on the genetic improvement of crop plants to provide higher yields. But is this possible, or are we approaching the theoretical limits of yield in important crop plants? Recent work with rice suggests that the answer to the second question is a resounding no.

Rice ranks third in worldwide production, just behind wheat and maize. About 2 billion people, fully one-third of Earth's population, depend on rice for their basic nourishment. The majority of the world's rice crop is grown and consumed in Asia, but it is also a dietary staple in Africa and Central America. The Green Revolution for rice began in 1960, with the establishment of the International Rice Research Institute (IRRI), headquartered at Los Baños, Philippines. The goal was to breed rice with improved disease resistance and higher yield. The breeding program was almost too successful: The first high-yield varieties were so top-heavy with grain that they tended to fall over, demonstrating what plant breeders call "lodging." To reduce lodging, IRRI breeders crossed a high-yield line with a dwarf native variety to create semidwarf lines, which were introduced to farmers in 1966. Due in large part to the adoption of the semidwarf lines, the world production of rice doubled in the next 25 years. By the mid-1990s, 75 percent of Asian rice fields were sown with new genetic varieties.

Rice breeders cannot afford to rest on their laurels. Predictions suggest that a 70 percent increase in the annual rice harvest may be necessary to keep pace with anticipated population growth during the next 30 years. Breeders are now looking to wild rice varieties for further crop improvement. Leading the way are Susan McCouch, Steven Tanksley, and their coworkers at Cornell University. To test the hypothesis that wild-rice species carry genes that will improve the yield of cultivated varieties, they crossed cultivated rice (*Oryza sativa*) with a low-yield wild ancestor species (*Oryza rufipogon*) and then successively backcrossed the interspecific hybrid to cultivated rice for three generations. In theory, this would create lines whose genomes were about 95 percent from *O. sativa* and 5 percent from *O. rufipogon*. When testing these backcrossed lines for grain yield, they found that several of them outproduced cultivated rice by as much as 30 percent. These results demonstrated strikingly that even though wild-rice relatives have low yields and appear to be inferior to cultivated rice, they still carry genes that will increase the yield of elite rice varieties. It will now be up to breeders to exploit the wild-rice relatives.

But introducing favorable genes from a wild relative into a cultivated variety by conventional breeding is a long and involved process, often requiring a decade or more of crossing, selection, backcrossing, and more selection. Future improvements in cultivated species must be quicker if crop yields are to keep up with population growth. Fortunately, modern gene-mapping techniques are now leading to the identification of QTLs that control complex traits such as yield and disease resistance. This makes possible a more direct approach to crop improvement, the *advanced backcross QTL method*.

First, a cultivated variety is crossed with a wild relative, just as *O. sativa* was interbred with *O. rufipogon*. The hybrid is then backcrossed to the cultivated variety to generate lines that contain only a small fraction of the "wild" genome. The backcross lines with the best qualities (e.g., highest yield and most disease resistance) are selected, and the "wild QTLs" responsible for the superior performance are identified, using a detailed molecular linkage map. Once the beneficial QTLs are discovered by this method, they may be introduced into other cultivated varieties. In order for this strategy to succeed, it is essential that wild crop relatives be preserved as a storehouse of potentially useful genes. Efforts were begun in the 1970s to protect the existing crop relatives of many plants in their natural habitats and to preserve them in seed banks. As the work of McCouch and Tanksley and others has shown, it is not possible to predict which wild varieties may be needed decades or even centuries from now to contribute their beneficial alleles to cultivated varieties. To prevent the loss of superior genes, the widest possible spectrum of wild species must be preserved, even those that have no obvious favorable characteristics.

Almost 60 years ago, the great Russian plant geneticist N. I. Vavilov suggested that wild relatives of crop plants could be the source of genes to improve agriculture. In the coming century, Vavilov's vision may finally be realized, as genes from long-neglected wild crop relatives, identified by new molecular methods, spark a revitalized Green Revolution.

## ■ References

Evenson, R. E., and Gollin, D. 2003. Assessing the impact of the Green Revolution. *Science* 300: 758.

Goff, S. A., and Salmeron, J. M. 2004. Back to the future of cereal. *Sci. Am.* 291: 42–49.

Monna, L. 2002. Positional cloning of a rice semidwarfing gene, *sd-1*. *DNA Res.* 9: 11–17.

Tanksley, S.D., and McCouch, S.R. 1997. Seed banks and molecular maps: Unlocking genetic potential from the wild. *Science* 277: 1063–1066.

Xiao, J., et al. 1996. Genes from wild rice improve yield. *Nature* 384: 223–224.

# EXPLORING GENOMICS

## ALFRED and Quantitative Trait Loci (QTLs)

In this chapter we discussed how multiple alleles contribute to quantitative inheritance, and we considered mathematical and statistical analysis that can be used to calculate allele frequencies. We also discussed the importance of quantitative trait loci (QTLs), multiple genes that contribute to quantitative trait phenotypes. In this exercise you will explore the **Allele Frequency Database (ALFRED)** to learn more about gene frequencies for human polymorphisms; then you will visit recently developed databases for QTLs.

### ■ Exercise I – ALFRED

**ALFRED** was developed by the National Science Foundation as a resource for learning about allele frequencies in human populations. Prior to working on this exercise, you may want to refer to Chapter 27 for more information on allele frequency calculations.

1. Access ALFRED at http://alfred.med.yale.edu/alfred/. Click on the "Documentation" tab at the top of the screen and use the links listed under it to learn more about ALFRED.

2. Go back to the homepage and run a keyword search in ALFRED for the *ALDH1A1* gene, searching any part of the database for information about this gene.

3. When the keyword search page appears, click on the + symbol in the "select" column to learn more about the gene you searched for. Clicking the + symbol will take you to ALFRED locus information for the gene, including links to information such as:

   Frequency Display Formats: Allele frequency data in different populations. When viewing allele frequency tables, use the link for each population to learn more about that population.

   External Resources: Access to a wealth of information through the Entrez Gene database, OMIM, and other databases, depending on how much is known about the gene.

4. Explore the ALFRED links for the *ALDH1A1* gene; then answer the following questions:

   a. What chromosome is *ALDH1A1* located on?

   b. What is the full name of this gene, and what is the primary function of the ALDH1A1 protein? Describe a phenotype associated with mutations in this gene.

   c. What is the main polymorphism for *ALDH1A1* that is included in ALFRED?

   d. Who are the Mbuti in the Frequency Display Format report for *ALDH1A1*?

   e. In which population is the T polymorphism most frequent? Which population in ALFRED shows the lowest frequency of the G polymorphism? Explain what you found. When you access the allele frequency page as you did in step 3, click the "Aldehyde dehydrogenase 1 family, member A1 link" at the top of the page, and the frequency display format on the next page will provide access to a map feature. Click on the map to see a geographic distribution of the frequency of these polymorphisms in different populations. This feature provides an excellent illustration of the dramatic differences in *ALDH1A1* polymorphisms by geography.

   f. Which population shows the highest frequency of the G polymorphism?

5. Search ALFRED for the *LDLR* gene and use the link under the site column "loci"; then answer the following questions:

   a. What gene is this, and what chromosome is it located on?

   b. What human disease condition is this gene involved in?

   c. A number of different polymorphisms occur in the *LDLR* gene. What is the main polymorphism for *LDLR* found in the ALFRED database that can be detected by restriction fragment length polymorphism (RFLP) analysis?

   d. Which restriction enzyme cutting site is affected by this mutation, and in which exon is this site located?

### ■ Exercise II – Quantitative Trait Loci (QTLs)

Recall that when multiple genes contributing to quantitative traits are clustered on a chromosome, these genes are called *quantitative trait loci (QTLs)*. Genomics projects are rapidly revealing new information about QTLs in complex human behavioral conditions involving multiple genes and in human diseases such as cancer. Geneticists studying livestock animal species to better understand livestock traits are very interested in QTLs, as are plant geneticists, who are studying QTLs such as those that contribute to desirable growth characteristics of agriculturally important crops. Here we explore the **Animal QTL database (Animal QTLdb)** and the **Gramene QTL database**. The Gramene QTL database collects data on QTLs in crop plants.

1. Access the **Animal QTLdb** at http://www.animalgenome.org/QTLdb/.

2. Animal QTLdb is a database of QTLs in livestock animals. Click the "Chicken QTL" link to enter the part of the database pertaining to chickens; then click the "data summary" link to reveal chicken traits inherited by QTLs. What are the top four traits in terms of number of QTLs in the database?

3. Given the popularity of chicken as a food source, it should not surprise you that the genetics of drumstick weight are of interest to scientists working on livestock. Let's learn more about QTLs involved in drumstick weight in chickens. Return to the previous page and click "Search." Use "Option 2: Query by chicken traits" and search for "drumstick weight." Click the drumstick weight link that appears on the next page; then click the "Find all mapped QTLs on this trait" link to reveal a map of

*Continued on next page*

*Exploring Genomics, continued*

QTLs for drumstick weight. Notice how many chromosomes contain QTLs for drumstick weight (labeled DS). Also notice that some QTLs are indicated in blue whereas others are shown in red. What does this mean?

4. Based on these maps, is there any evidence that drumstick weight is a sex-linked quantitative trait? Recall from Chapter 7 that chickens follow a ZZ/ZW mode of sex determination. Explain your answers.

5. Click on any of the DS links to learn more about the drumstick weight QTLs on a particular chromosome. The QTL report that will appear when you click on these links provides information on the experiments used to identify these QTLs, PubMed references, when available, and links to other databases that you can use to learn about each QTL.

6. Explore the Pig QTL and Chicken QTL databases on your own to find QTLs of interest.

7. The Gramene QTL Database http://www.gramene.org/qtl/ contains information on identified QTLs in important agricultural crops such as rice, maize, and oats. Explore this database to see if you can find interesting QTLs affecting crop traits.

## Chapter Summary

1. Quantitative inheritance results in a range of phenotypes for a trait due to the action of multiple genes combined with environmental factors.
2. Many polygenic, or quantitative, traits show continuous variation that cannot be classified into discrete phenotypic classes. Threshold traits, which have a limited number of distinct phenotypic classes, are also polygenic.
3. Numerous statistical methods are used to analyze quantitative traits, including the mean, variance, standard deviation, standard error, covariance, and the correlation coefficient.
4. Heritability is an estimate of the relative contribution of genetic versus environmental factors to the range of phenotypic variation seen in a quantitative trait in a particular population and environment.
5. Twin studies are used to estimate heritabilities for polygenic traits in humans.
6. Multiple genes contributing to a quantitative trait are known as quantitative trait loci, or QTLs. Mapping techniques using DNA markers can help locate and characterize QTLs.

## INSIGHTS AND SOLUTIONS

1. In a certain plant, height varies from 6 to 36 cm. When 6-cm and 36-cm plants were crossed, all $F_1$ plants were 21 cm. In the $F_2$ generation, a continuous range of heights was observed. Most were around 21 cm, and 3 of 200 were as short as the 6-cm $P_1$ parent.

(a) What mode of inheritance does this illustrate, and how many gene pairs are involved?
(b) How much does each additive allele contribute to height?
(c) List all genotypes that give rise to plants that are 31 cm.

**Solution:**

(a) Polygenic inheritance is illustrated when a trait is continuous and when alleles contribute additively to the phenotype. The 3/200 ratio of $F_2$ plants is the key to determining the number of gene pairs. This reduces to a ratio of 1/66.7, very close to 1/64. Using the formula $1/4^n = 1/64$ (where 1/64 is equal to the proportion of $F_2$ phenotypes as extreme as either $P_1$ parent), $n = 3$. Therefore, three gene pairs are involved.
(b) The variation between the two extreme phenotypes is

$$36 - 6 = 30 \text{ cm}$$

Because there are six potential additive alleles (*AABBCC*), each contributes

$$30/6 = 5 \text{ cm}$$

to the base height of 6 cm, which results when no additive alleles (*aabbcc*) are part of the genotype.

(c) All genotypes that include 5 additive alleles will be 31 cm (5 alleles × 5 cm/allele + 6 cm base height = 31 cm). Therefore, *AABBCc*, *AABbCC*, and *AaBBCC* are the genotypes that will result in plants that are 31 cm.

2. In a cross separate from the above-mentioned $F_1$ crosses, a plant of unknown phenotype and genotype was testcrossed, with the following results:

1/4 11 cm

2/4 16 cm

1/4 21 cm

An astute genetics student realized that the unknown plant could be only one phenotype but could be any of three genotypes. What were they?

**Solution:** When testcrossed (with *aabbcc*), the unknown plant must be able to contribute either one, two, or three additive alleles in its gametes in order to yield the three phenotypes in the offspring. Since no 6-cm offspring are observed, the unknown plant never contributes all nonadditive alleles (*abc*). Only plants that are homozygous at one locus and heterozygous at the other two loci will meet these criteria. Therefore, the unknown parent can be any of three genotypes, all of which have a phenotype of 26 cm:

*AABbCc*
*AaBbCC*
*AaBBCc*

For example, in the first genotype (*AABbCc*),

$$AABbCc \times aabbcc$$

yields

1/4 *AaBbCc* 21 cm
1/4 *AaBbcc* 16 cm
1/4 *AabbCc* 16 cm
1/4 *Aabbcc* 11 cm

which is the ratio of phenotypes observed.

3. The results shown in the following table were recorded for ear length in corn:

**Length of Ear in cm**

| | 5 | 6 | 7 | 8 | 9 | 10 | 11 | 12 | 13 | 14 | 15 | 16 | 17 | 18 | 19 | 20 | 21 |
|---|---|---|---|---|---|---|---|---|---|---|---|---|---|---|---|---|---|
| **Parent A** | 4 | 21 | 24 | 8 | | | | | | | | | | | | | |
| **Parent B** | | | | | | | | | 3 | 11 | 12 | 15 | 26 | 15 | 10 | 7 | 2 |
| **$F_1$** | | | | | 1 | 12 | 12 | 14 | 17 | 9 | 4 | | | | | | |

For each of the parental strains and the $F_1$ generation, calculate the mean values for ear length.

**Solution:** The mean values can be calculated as follows:

$$\overline{X} = \frac{\Sigma X_i}{n}$$

$$P_A{:}\ \overline{X} = \frac{\Sigma X_i}{n} = \frac{378}{57} = 6.63$$

$$P_B{:}\ \overline{X} = \frac{\Sigma X_i}{n} = \frac{1697}{101} = 16.80$$

$$F_1{:}\ \overline{X} = \frac{\Sigma X_i}{n} = \frac{836}{69} = 12.11$$

4. For the corn plant described in Exercise 3, compare the mean of the $F_1$ with that of each parental strain. What does this tell you about the type of gene action involved?

**Solution:** The $F_1$ mean (12.11) is almost midway between the parental means of 6.63 and 16.80. This indicates that the genes in question may be additive in effect.

5. The mean and variance of corolla length in two highly inbred strains of *Nicotiana* and their progeny are shown in the following table. One parent ($P_1$) has a short corolla, and the other parent ($P_2$) has a long corolla. Calculate the broad-sense heritability ($H^2$) of corolla length in this plant.

| Strain | Mean (mm) | Variance (mm) |
|---|---|---|
| $P_1$ short | 40.47 | 3.12 |
| $P_2$ long | 93.75 | 3.87 |
| $F_1$ ($P_1 \times P_2$) | 63.90 | 4.74 |
| $F_2$ ($F_1 \times F_1$) | 68.72 | 47.70 |

**Solution:** The formula for estimating heritability is $H^2 = V_G/V_P$, where $V_G$ and $V_P$ are the genetic and phenotypic components of variation, respectively. The main issue in this problem is obtaining some estimate of two components of phenotypic variation: genetic and environmental factors. $V_P$ is the combination of genetic and environmental variance. Because the two parental strains are true breeding, they are assumed to be homozygous, and the variance of 3.12 and 3.87 is considered to be the result of environmental influences. The average of these two values is 3.50. The $F_1$ is also genetically homogeneous and gives us an additional estimate of the impact of environmental factors. By averaging this value along with that of the parents,

$$\frac{4.74 + 3.50}{2} = 4.12$$

we obtain a relatively good idea of environmental impact on the phenotype. The phenotypic variance in the $F_2$ is the sum of the genetic ($V_G$) and environmental ($V_E$) components. We have estimated the environmental input as 4.12, so 47.70 minus 4.12 gives us an estimate of $V_G$ of 43.58. Heritability then becomes 43.58/47.70, or 0.91. This value, when interpreted as a percentage, indicates that about 91 percent of the variation in corolla length is due to genetic influences.

## Problems and Discussion Questions

1. What is the difference between continuous and discontinuous variation? Which of the two is most likely to be the result of polygenic inheritance?
2. Define the following: (a) polygene, (b) additive alleles, (c) correlation, (d) monozygotic and dizygotic twins, (e) heritability, and (f) QTL.
3. A homozygous plant with 20-cm diameter flowers is crossed with a homozygous plant of the same species that has 40-cm diameter flowers. The $F_1$ plants all have flowers 30 cm in diameter. In the $F_2$ generation of 512 plants, 2 plants have flowers 20 cm in diameter, 2 plants have flowers 40 cm in diameter, and the remaining 508 plants have flowers of a range of sizes in between.
   (a) Assuming that all alleles involved act additively, how many genes control flower size in this plant?
   (b) What frequency distribution of flower diameter would you expect to see in the progeny of a backcross between an $F_1$ plant and the large-flowered parent?
   (c) Calculate the mean, variance, and standard deviation for flower diameter in the backcross progeny from (b).
4. A dark-red strain and a white strain of wheat are crossed and produce an intermediate, medium-red $F_1$. When the $F_1$ plants are interbred, an $F_2$ generation is produced in a ratio of 1 dark-red: 4 medium-dark-red: 6 medium-red: 4 light-red: 1 white. Further crosses reveal that the dark-red and white $F_2$ plants are true breeding.
   (a) Based on the ratios in the $F_2$ population, how many genes are involved in the production of color?

(b) How many additive alleles are needed to produce each possible phenotype?
(c) Assign symbols to these alleles and list possible genotypes that give rise to the medium-red and the light-red phenotypes.
(d) Predict the outcome of the $F_1$ and $F_2$ generations in a cross between a true-breeding medium-red plant and a white plant.

5. Height in humans depends on the additive action of genes. Assume that this trait is controlled by the four loci R, S, T, and U and that environmental effects are negligible. Instead of additive versus nonadditive alleles, assume that additive and partially additive alleles exist. Additive alleles contribute two units, and partially additive alleles contribute one unit to height.
(a) Can two individuals of moderate height produce offspring that are much taller or shorter than either parent? If so, how?
(b) If an individual with the minimum height specified by these genes marries an individual of intermediate or moderate height, will any of their children be taller than the tall parent? Why or why not?

6. An inbred strain of plants has a mean height of 24 cm. A second strain of the same species from a different geographical region also has a mean height of 24 cm. When plants from the two strains are crossed together, the $F_1$ plants are the same height as the parent plants. However, the $F_2$ generation shows a wide range of heights; the majority are like the $P_1$ and $F_1$ plants, but approximately 4 of 1000 are only 12 cm high, and about 4 of 1000 are 36 cm high.
(a) What mode of inheritance is occurring here?
(b) How many gene pairs are involved?
(c) How much does each gene contribute to plant height?
(d) Indicate one possible set of genotypes for the original $P_1$ parents and the $F_1$ plants that could account for these results.
(e) Indicate three possible genotypes that could account for $F_2$ plants that are 18 cm high and three that account for $F_2$ plants that are 33 cm high.

7. Erma and Harvey were a compatible barnyard pair, but a curious sight. Harvey's tail was only 6 cm long, while Erma's was 30 cm. Their $F_1$ piglet offspring all grew tails that were 18 cm. When inbred, an $F_2$ generation resulted in many piglets (Erma and Harvey's grandpigs), whose tails ranged in 4-cm intervals from 6 to 30 cm (6, 10, 14, 18, 22, 26, and 30). Most had 18-cm tails, while 1/64 had 6-cm tails and 1/64 had 30-cm tails.
(a) Explain how these tail lengths were inherited by describing the mode of inheritance, indicating how many gene pairs were at work, and designating the genotypes of Harvey, Erma, and their 18-cm-tail offspring.
(b) If one of the 18-cm $F_1$ pigs is mated with one of the 6-cm $F_2$ pigs, what phenotypic ratio would be predicted if many offspring resulted? Diagram the cross.

8. In the following table, average differences of height, weight, and fingerprint ridge count between monozygotic twins (reared together and apart), dizygotic twins, and nontwin siblings are compared:

| Trait | MZ Reared Together | MZ Reared Apart | DZ Reared Together | Sibs Reared Together |
|---|---|---|---|---|
| Height (cm) | 1.7 | 1.8 | 4.4 | 4.5 |
| Weight (kg) | 1.9 | 4.5 | 4.5 | 4.7 |
| Ridge count | 0.7 | 0.6 | 2.4 | 2.7 |

Based on the data in this table, which of these quantitative traits has the highest heritability values?

9. What kind of heritability estimates (broad sense or narrow sense) are obtained from human twin studies?

10. List as many human traits as you can that are likely to be under the control of a polygenic mode of inheritance.

11. Corn plants from a test plot are measured, and the distribution of heights at 10-cm intervals is recorded in the following table:

| Height (cm) | Plants (no.) |
|---|---|
| 100 | 20 |
| 110 | 60 |
| 120 | 90 |
| 130 | 130 |
| 140 | 180 |
| 150 | 120 |
| 160 | 70 |
| 170 | 50 |
| 180 | 40 |

Calculate (a) the mean height, (b) the variance, (c) the standard deviation, and (d) the standard error of the mean. Plot a rough graph of plant height against frequency. Do the values represent a normal distribution? Based on your calculations, how would you assess the variation within this population?

12. The following variances were calculated for two traits in a herd of hogs.

| Trait | $V_P$ | $V_G$ | $V_A$ |
|---|---|---|---|
| Back fat | 30.6 | 12.2 | 8.44 |
| Body length | 52.4 | 26.4 | 11.7 |

(a) Calculate broad-sense ($H^2$) and narrow-sense ($h^2$) heritabilities for each trait in this herd.
(b) Which of the two traits will respond best to selection by a breeder? Why?

13. The mean and variance of plant height of two highly inbred strains ($P_1$ and $P_2$) and their progeny ($F_1$ and $F_2$) are shown here:

| Strain | Mean (cm) | Variance |
|---|---|---|
| $P_1$ | 34.2 | 4.2 |
| $P_2$ | 55.3 | 3.8 |
| $F_1$ | 44.2 | 5.6 |
| $F_2$ | 46.3 | 10.3 |

Calculate the broad-sense heritability ($H^2$) of plant height in this species.

14. A hypothetical study investigated the vitamin A content and the cholesterol content of eggs from a large population of chickens. The variances ($V$) were calculated, as shown here:

| Variance | Trait: Vitamin A | Cholesterol |
|---|---|---|
| $V_P$ | 123.5 | 862.0 |
| $V_E$ | 96.2 | 484.6 |
| $V_A$ | 12.0 | 192.1 |
| $V_D$ | 15.3 | 185.3 |

(a) Calculate the narrow-sense heritability ($h^2$) for both traits.
(b) Which trait, if either, is likely to respond to selection?

15. The following table shows measurements for fiber lengths and fleece weight in a small flock of eight sheep.

| | Sheep Fiber Length (cm) | Fleece Weight (kg) |
|---|---|---|
| 1 | 9.7 | 7.9 |
| 2 | 5.6 | 4.5 |
| 3 | 10.7 | 8.3 |
| 4 | 6.8 | 5.4 |
| 5 | 11.0 | 9.1 |
| 6 | 4.5 | 4.9 |
| 7 | 7.4 | 6.0 |
| 8 | 5.9 | 5.1 |

(a) What are the mean, variance, and standard deviation for each trait in this flock?
(b) What is the covariance of the two traits?
(c) What is the correlation coefficient for fiber length and fleece weight?
(d) Do you think greater fleece weight is caused by an increase in fiber length? Why or why not?

16. In a herd of dairy cows the narrow-sense heritability for milk protein content is 0.76, and for milk butterfat it is 0.82. The correlation coefficient between milk protein content and butterfat is 0.91. If the farmer selects for cows producing more butterfat in their milk, what will be the most likely effect on milk protein content in the next generation?

17. In an assessment of learning in *Drosophila*, flies were trained to avoid certain olfactory cues. In one population, a mean of 8.5 trials was required. A subgroup of this parental population that was trained most quickly (mean = 6.0) was interbred, and their progeny were examined. These flies demonstrated a mean training value of 7.5. Calculate realized heritability for olfactory learning in *Drosophila*.

18. Suppose you want to develop a population of *Drosophila* that would rapidly learn to avoid certain substances the flies could detect by smell. Based on the heritability estimate you obtained in Problem 17, do you think it would be worth doing this by artificial selection? Why or why not?

19. In a population of tomato plants, mean fruit weight is 60 g and ($h^2$) is 0.3. Predict the mean weight of the progeny if tomato plants whose fruit averaged 80 g were selected from the original population and interbred.

20. In a population of 100 inbred, genotypically identical rice plants, variance for grain yield is 4.67. What is the heritability for yield? Would you advise a rice breeder to improve yield in this strain of rice plants by selection?

21. Many traits of economic or medical significance are determined by quantitative trait loci (QTLs) in which many genes, usually scattered throughout the genome, contribute to expression.
(a) What general procedures are used to identify such loci?
(b) What is meant by the term *cosegregate* in the context of QTL mapping? Why are markers such as RFLPs, SNPs, and microsatellites often used in QTL mapping?

**HOW DO WE KNOW?**

22. In this chapter, we focused on a mode of inheritance referred to as quantitative genetics, as well as many of the statistical parameters utilized to study quantitative traits. Along the way, we found opportunities to consider the methods and reasoning by which geneticists acquired much of their understanding of quantitative genetics. From the explanations given in the chapter, what answers would you propose to the following fundamental questions:
(a) How do we know that threshold traits are actually polygenic even though they may have as few as two discrete phenotypic classes?
(b) How can we ascertain the number of polygenes involved in the inheritance of a quantitative trait?
(c) What findings led geneticists to postulate the multiple-factor hypothesis that invoked the idea of additive alleles to explain inheritance patterns?
(d) How do we assess environmental factors to determine if they impact the phenotype of a quantitatively inherited trait?

## Extra-Spicy Problems

23. A mutant strain of *Drosophila* was isolated and shown to be resistant to an experimental insecticide, whereas normal (wild-type) flies were sensitive to the chemical. Following a cross between resistant flies and sensitive flies, isolated populations were derived that had various combinations of chromosomes from the two strains. Each was tested for resistance, as shown in the following histogram. Analyze the data and draw any appropriate conclusion about which chromosome(s) contain a gene responsible for inheritance of resistance to the insecticide.

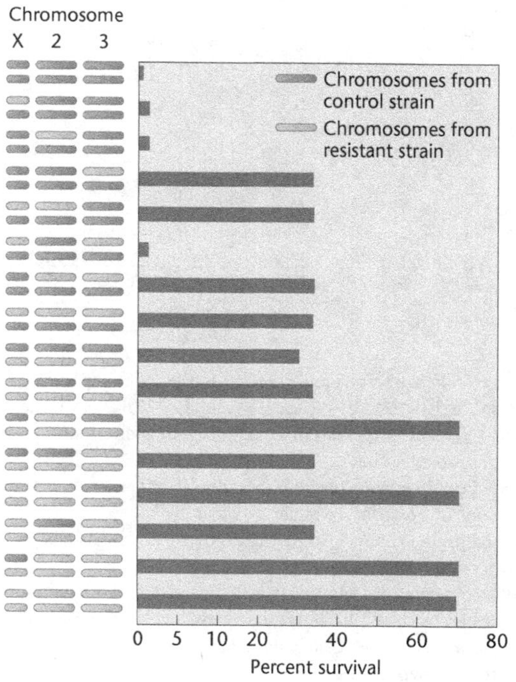

24. In 1988, Horst Wilkens investigated blind cavefish, comparing them with members of a sibling species with normal vision that are found in a lake [Wilkens, H. (1988). *Ecol. Biol.* 23: 271–367]. (We will call them cavefish and lakefish.) Wilkens found that cavefish eyes are about seven times smaller than lakefish eyes. $F_1$ hybrids have eyes of intermediate size. These data, as well as the $F_1 \times F_1$ cross and those from backcrosses ($F_1 \times$ cavefish and $F_1 \times$ lakefish), are depicted as follows:

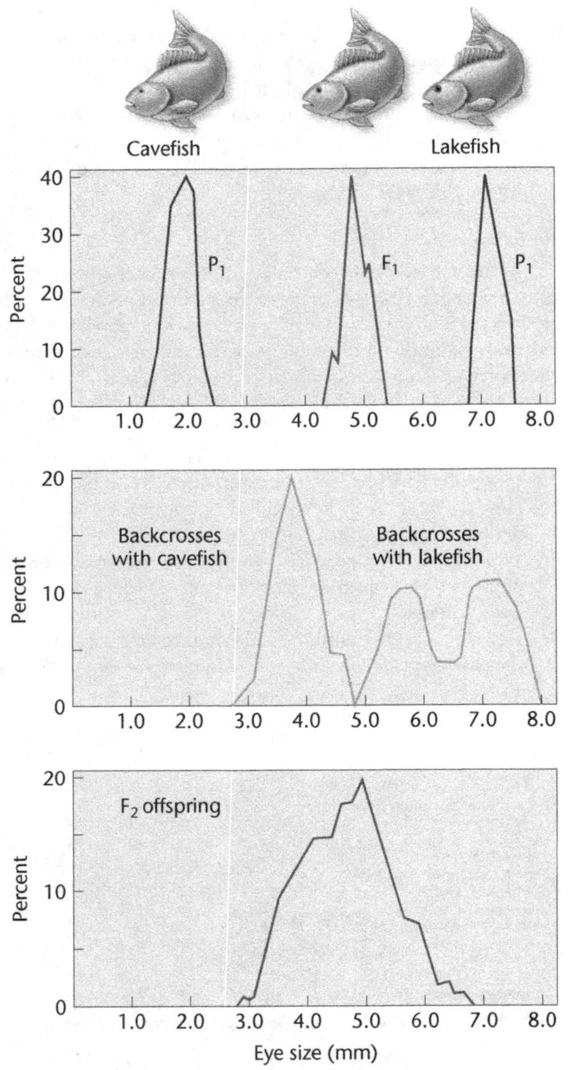


Examine Wilkens's results and respond to the following questions:
(a) Based strictly on the $F_1$ and $F_2$ results of Wilkens's initial crosses, what possible explanation concerning the inheritance of eye size seems most feasible?
(b) Based on the results of the $F_1$ backcross with cavefish, is your explanation supported? Explain.
(c) Based on the results of the $F_1$ backcross with lakefish, is your explanation supported? Explain.
(d) Wilkens examined about 1000 $F_2$ progeny and estimated that 6–7 genes are involved in determining eye size. Is the sample size adequate to justify this conclusion? Propose an experimental protocol to test the hypothesis.
(e) A comparison of the embryonic eye in cavefish and lakefish revealed that both reach approximately 4 mm in diameter. However, lakefish eyes continue to grow, while cavefish eye size is greatly reduced. Speculate on the role of the genes involved in this problem.

25. A 3-inch plant was crossed with a 15-inch plant, and all $F_1$ plants were 9 inches. In the $F_2$ plants exhibited a "normal distribution," with heights of 3, 4, 5, 6, 7, 8, 9, 10, 11, 12, 13, 14, and 15 inches.
(a) What ratio will constitute the "normal distribution" in the $F_2$?
(b) What will be the outcome if the $F_1$ plants are testcrossed with plants that are homozygous for all nonadditive alleles?

26. The following table gives the percentage of twin pairs studied in which both twins expressed the same phenotype for a trait (concordance). Percentages listed are for concordance for each trait in monozygotic (MZ) and dizygotic (DZ) twins. Assuming that both twins in each pair were raised together in the same environment, what do you conclude about the relative importance of genetic versus environmental factors for each trait?

| Trait | MZ% | DZ% |
|---|---|---|
| Blood types | 100 | 66 |
| Eye color | 99 | 28 |
| Mental retardation | 97 | 37 |
| Measles | 95 | 87 |
| Hair color | 89 | 22 |
| Handedness | 79 | 77 |
| Idiopathic epilepsy | 72 | 15 |
| Schizophrenia | 69 | 10 |
| Diabetes | 65 | 18 |
| Identical allergy | 59 | 5 |
| Cleft lip | 42 | 5 |
| Club foot | 32 | 3 |
| Mammary cancer | 6 | 3 |

27. In a cross between a strain of large guinea pigs and a strain of small guinea pigs, the $F_1$ are phenotypically uniform with an average size about intermediate between that of the two parental strains. Among 1014 individuals, 3 are about the same size as the small parental strain and 5 are about the same size as the large parental strain. How many gene pairs are involved in the inheritance of size in these strains of guinea pigs?

28. Type A1B brachydactyly (short middle phalanges) is a genetically determined trait that maps to the short arm of chromosome 5 in humans. If you classify individuals as either having or not having brachydactyly, the trait appears to follow a single-locus, incompletely dominant pattern of inheritance. However, if one examines the fingers and toes of affected individuals, one sees a range of expression from extremely short to only slightly short. What might cause such variation in the expression of brachydactyly?

29. In a series of crosses between two true-breeding strains of peaches, the $F_1$ generation was uniform, producing 30-g peaches. The $F_2$ fruit mass ranges from 38 to 22 g at intervals of 2 g.
(a) Using these data, determine the number of polygenic loci involved in the inheritance of peach mass.
(b) Using gene symbols of your choice, give the genotypes of the parents and the $F_1$.

30. Students in a genetics laboratory began an experiment in an attempt to increase heat tolerance in two strains of *Drosophila melanogaster*. One strain was trapped from the wild six weeks before the experiment was to begin; the other was obtained from a *Drosophila* repository at a university laboratory. In which strain would you expect to see the most rapid and extensive response to heat-tolerance selection, and why?

31. Consider a true-breeding plant, *AABBCC*, crossed with another true-breeding plant, *aabbcc*, whose resulting offspring are *AaBbCc*. If you cross the $F_1$ generation, and independent assortment is operational, the expected fraction of offspring in each phenotypic class is given by the expression $N![M!(N - M!)]$ where $N$ is the total number of alleles (six in this example) and $M$ is the number of uppercase alleles. In a cross of *AaBbCc* × *AaBbCc*, what proportion of the offspring would be expected to contain two uppercase alleles?
32. Canine hip dysplasia is a quantitative trait that continues to affect most large breeds of dogs in spite of approximately 40 years of effort to reduce the impact of this condition. Breeders and veterinarians rely on radiographic and universal registries to facilitate the development of breeding schemes for reducing its incidence. Recent data (Wood and Lakhani. 2003. *Vet. Rec.* 152: 69–72) indicate that there is a "month-of-birth" effect on hip dysplasia in Labrador retrievers and Gordon setters, whereby the frequency and extent of expression of this disorder vary depending on the time of year dogs are born. Speculate on how breeders attempt to "select" out this disorder and what the month-of-birth phenomenon indicates about the expression of polygenic traits?
33. Floral traits in plants often play key roles in diversification, in that slight modifications of those traits, if genetically determined, may quickly lead to reproductive restrictions and evolution. Insight into genetic involvement in flower formation is often acquired through selection experiments that expose realized heritability. Lendvai and Levin (2003) conducted a series of artificial selection experiments on flower size (diameter) in *Phlox drummondii*. Data from their selection experiments are presented below in modified form and content.

| Year | Treatment | Mean (mm) |
|---|---|---|
| 1997 | Control | 30.04 |
| | Selected parents | 34.13 |
| | Offspring | 32.21 |
| 1998 | Control | 28.11 |
| | Selected parents | 31.98 |
| | Offspring | 31.90 |
| 1999 | Control | 29.68 |
| | Selected parents | 31.81 |
| | Offspring | 33.74 |

(a) Considering that differences in control values represent year-to-year differences in greenhouse conditions, calculate (in mm) the average response to selection over the three-year period.
(b) Calculate the realized heritability for each year and the overall realized heritability.
(c) Assuming that the realized heritability in phlox is relatively high, what factors might account for such a high response?
(d) In terms of evolutionary potential, is a population with high heritability likely to be favored compared to one with a low realized heritability?

Genetically engineered *Caenorhabditis elegans* roundworms that have turned blue in response to an environmental stress, such as toxins or heat.

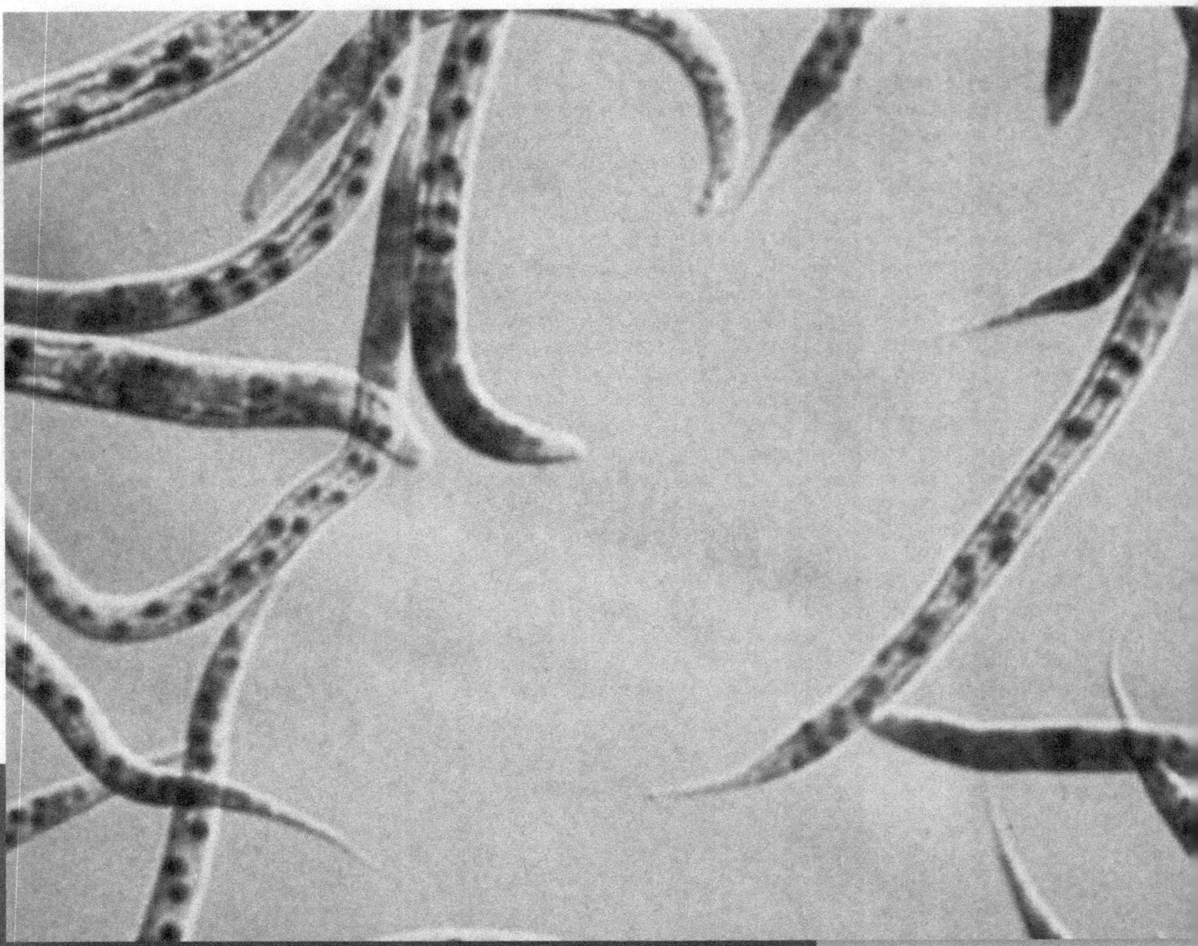

26

# Genetics and Behavior

## CHAPTER CONCEPTS

- Behavior is a complex response to stimuli and is mediated both by the genotype and by the environment.
- Bidirectional selection experiments (also called the behavior-first approach) are used to identify heritable behavioral traits. These traits are further studied by genetic crosses to identify the number and chromosomal location of genes contributing to the trait.
- Mutagenesis experiments (also called the gene-first approach) establish strains carrying single-mutant genes that contribute to a behavioral response. Molecular functional analyses of the normal and mutant alleles provide insight into the underlying mechanism controlling the relationship between the gene and a behavioral response.
- The bidirectional and mutagenesis approaches to studying behavior genetics have been successfully used to study many behavioral responses in *Drosophila*, making it a model organism for the study of nervous system function and human behavioral disorders.
- Many human behavioral disorders are complex responses to environmental stimuli. Mediated by the genotype and polygenically controlled, they may be analyzed by an array of genomic techniques.

**Behavior** is generally defined as a reaction to stimuli, whether internal or external. In broad terms, every action, reaction, and response represents a type of behavior. Animals run, remain still, or counterattack in the presence of a predator; birds build complex and distinctive nests in response to a combination of internal and external signals; fruit flies execute intricate courtship rituals; plants bend toward light; and humans reflexively avoid painful stimuli as well as "behaving" in a variety of ways as guided by their intellect, emotions, and culture.

Even though clear-cut cases of genetic influence on behavior were known in the early 1900s, the study of behavior was of less interest to the early geneticists than to psychologists, who were primarily concerned with learned or conditioned behavior. Although some behavioral traits were recognized as innate or instinctive, behavior that could be modified by prior experience received the most attention. Such traits or patterns of behavior were thought to reflect events in the environmental setting to the exclusion of the organism's genotype. This philosophy served as the basis of the behaviorist school of psychology.

Such thinking provided a somewhat distorted view of the nature of behavioral patterns. After all, any behavior must rely on the expression of the individual's genotype for its execution; and the expression of that genotype takes place within a hierarchy of environmental settings (that is, gene expression depends on interactions within the cell, the tissue, the organ, the organism, and finally the population and surrounding environment). Without genes and their environments, there could be no behavior. Nevertheless, the nature–nurture controversy flourished well into the 1950s. By that time it was clear that while certain behavioral patterns, particularly in less complex animals, seemed to be innate, others were the result of environmental modifications limited by genetic influences. The latter description is particularly applicable to organisms with more complex nervous systems. It is after all, the nervous system that senses the environment, processes the information, and initiates the response that we perceive as behavior. For that reason, much of the study of behavior in genetics and molecular biology focuses on the development, structure, and function of the nervous system.

Since about 1950, research into the genetic components of behavior has intensified, and appreciation for the importance of genetics in understanding behavior has increased. Behavioral genetics originally employed two approaches to study the genetic basis of behavior. Historically, the first was a top-down, or *behavior-first*, approach, in which, first, a behavioral response to a stimulus was identified in an organism, and then, high-scoring responders were interbred and low-scoring responders were interbred to produce strains that bred true for either a high level or a low level of response. After these strains were established, crosses were made to identify and analyze the genetic components of the behavioral response. The second mode of research was a bottom-up, or *gene-first*, approach, in which organisms were subjected to mutagenesis and then screened to identify single-gene mutations associated with abnormal behaviors. Analysis of the molecular mechanism of gene action in these mutant strains provided a direct explanation of the behavior. As we will discuss in subsequent sections of this chapter, each approach has its advantages and shortcomings. In spite of their differences, both approaches share the same goals: to establish the inherited nature of a specific behavior, to identify and enumerate the genes or gene systems involved in the behavior being studied, to map these genes or gene systems to specific chromosomes, and to elucidate the molecular mechanisms by which these genes influence a behavioral response.

These approaches have been supplemented with the use of modern genetic techniques, including the analysis of quantitative trait loci (QTLs), the use of transgenic and knockout models of human behavior, genome-wide screening by microsatellite markers and microarrays, and the cloning and DNA sequencing of selected genes. The net effect of this technology has been to bring the two historical approaches together to provide a unified picture of genetic influences on behavior.

The prevailing view today is that all behavior patterns are influenced by the genotype and by the environment. The genotype provides the physical underpinnings or mental ability (or both) required to execute the behavior and further determines the limitations of environmental influences. Behavioral genetics has grown into an interdisciplinary specialty within genetics as more and more behaviors have been found to be under genetic control.

This chapter examines many examples of how behaviors are influenced by genes. There seems to be no question that the field of behavioral genetics, with its focus on the development, structure, and function of the nervous system, is growing and will be one of the most exciting areas of genetics in the years to come.

## 26.1 Behavioral Differences Between Genetic Strains Can Be Identified

One of the classic methods of behavioral genetics is the screening of existing strains of highly inbred organisms to find behavior differences that are strain-specific. This behavior-first approach takes advantage of the fact that genetically distinct strains of experimental organisms are already available and being used in genetic research, and may harbor previously undescribed strain-specific behaviors in response to environmental stimuli. If such genetic differences exist, they can serve as the starting point for the analysis of the genes that influence the behavior.

## Inbred Mouse Strains: Differences in Alcohol Preference

Inbred strain studies use populations of animals from different strains to study behavior in rigorously controlled environments. For example, classic studies of alcohol preference in mice have compared strain differences in preference for or aversion to ethanol. The behavior phenotype most often studied in these experiments is free-choice alcohol consumption. In one typical study, alcohol consumption rates in four strains of inbred mice were measured over a three-week period. Each strain was presented with seven vessels containing either pure water or alcohol varying in strength from 2.5 to 15.0 percent. The daily consumption was measured. Table 26.1 shows the proportion of alcohol to total liquid consumed on a weekly basis. The examination of these data shows that the C57BL and C3H/2 strains exhibit a preference for alcohol, while BALB/c and A/3 demonstrate an aversion to it. As the mice have been raised over many generations in a constant environment, the differences in alcohol preference are attributed to genotypic differences between the strains. Presumably, these strains vary only in the alleles that they carry at particular loci.

Despite the observed differences in alcohol consumption among mouse strains, the underlying mechanisms remain unclear. It has been suggested that differences in alcohol preference, metabolism, and severity of withdrawal symptoms seen in various mouse strains are related to differences in the major enzymes of alcohol metabolism, particularly alcohol dehydrogenase (ADH) and acetaldehyde dehydrogenase (AHD). These enzymes and their isozymes have been isolated and characterized with respect to biochemical and kinetic properties, and their distribution and activities in different mouse strains have been catalogued. Genetic variants of these enzymes identified in different strains of inbred mice have been used to try to establish an association between a given form of the enzyme and a form of alcohol-related behavior. To date, no correlation between alcohol preference or metabolism and these biochemical markers has been established in mice. Other genetic markers, including markers for neuropharmacological effects of alcohol, may be needed to establish a link between specific genes and alcohol-related behavior in mice before these findings can be extrapolated to humans.

Overall, comparisons of behaviors in inbred strains are useful in establishing that a behavioral phenotype has genetic components, but these experiments are not designed to dissect the phenotype and identify individual genes that contribute to the behavior. To identify genes involved in specific behaviors, researchers are using other techniques, such as transgenic and knockout animals. Transgenic animals are described in another section of this chapter.

## Emotional Behavior Differences in Inbred Mouse Strains

***Open-field tests*** are used to study exploratory and emotional behavior in mice. Open-field testing uses an unfamiliar standard environment much different from the mouse's normal environment to study behaviors, including emotionality or approach/avoidance responses. When mice are placed in a new environment for open-field testing (Figure 26–1), they will normally explore the unfamiliar surroundings, but they are a bit cautious or "nervous" at the same time. The cautious response is evidenced by their elevated rate of defecation and urination. To study these two responses (the exploration and the anxiety) in the laboratory, researchers place the mice in an enclosed, brightly illuminated box; exploration is tracked by recording movements into different areas of the box, and emotion is measured by counting the number of defecations.

**TABLE 26.1**

**Alcohol Consumption in Mice**

| Strain | Week | Proportion of Absolute Alcohol to Total Liquids | Mean (X) |
|---|---|---|---|
| | 1 | 0.085 | |
| C57BL | 2 | 0.093 | 9.4% |
| | 3 | 0.104 | alcohol |
| | 1 | 0.065 | |
| C3H/2 | 2 | 0.066 | 6.9% |
| | 3 | 0.075 | alcohol |
| | 1 | 0.024 | |
| BALB/c | 2 | 0.019 | 2.0% |
| | 3 | 0.018 | alcohol |
| | 1 | 0.021 | |
| A/3 | 2 | 0.016 | 1.7% |
| | 3 | 0.015 | alcohol |

*Source*: Modified from Rogers and McClearn, 1962. Reprinted by permission from *Quarterly Journal of Studies on Alcohol*, Vol. 23, pp. 26–33, 1962. Copyrighted by Journal of Studies on Alcohol, Inc. New Brunswick, NJ 08903.

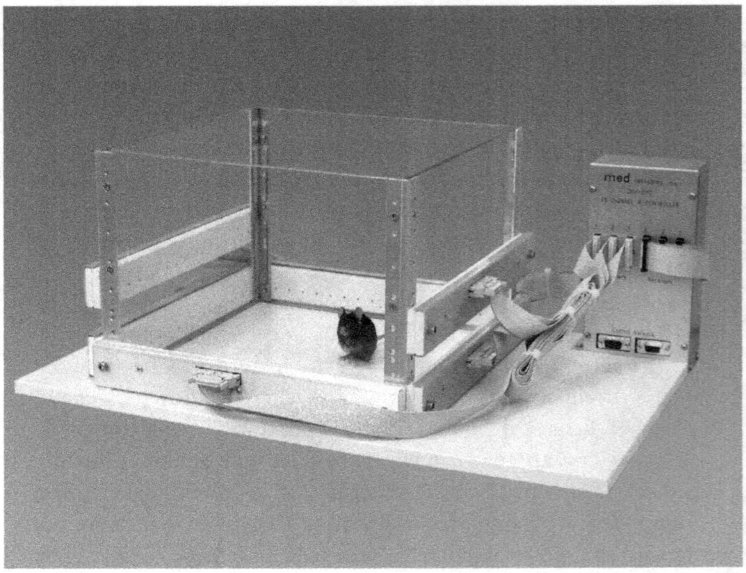

**FIGURE 26–1** A setup to measure open-field behavior in mice.

As in the study of alcohol preference, inbred strains of mice differ significantly in their response to the open-field setting. John C. DeFries and his associates focused their work on two contrasting strains, BALB/cJ and C57BL/6j. The BALB strain is homozygous for a coat color allele, *c*, and is albino, whereas the C57 strain has normal pigmentation (*CC*). BALB mice have low exploratory activity and are highly emotional, whereas C57 mice are active in exploration and relatively unemotional.

To test for genotypic differences, the two strains were crossed, and then the offspring were interbred for several generations. Each generation beyond the $F_1$ contained albino and nonalbino mice, and these were tested for open-field behavior. In all cases, pigmented mice behaved as strain C57, whereas albino mice behaved as BALB. The general conclusion is that the *c* allele behaves pleiotropically, affecting both coat color and behavior.

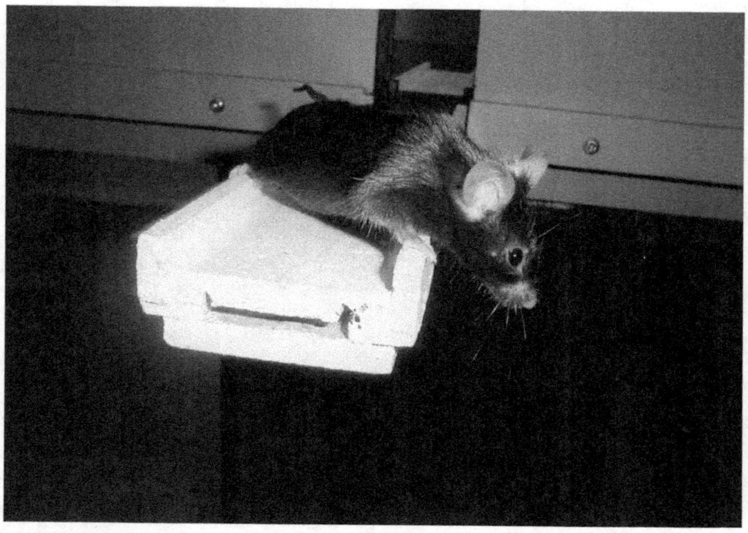

**FIGURE 26–2** A mouse explores a new environment from the open arm of an elevated plus maze, a standard device for measuring fear in mice. This experiment was one of a series designed to identify genes involved in fear and anxiety in mice.

Heritability analysis (see Chapter 25) has been used to assess the input of the *c* gene to these behavioral patterns. The results show that this locus accounts for 12 percent of the variance in open-field activity and 26 percent of the variance in defecation-related emotion. These values indicate that the behaviors are polygenically controlled.

To assess the relationship between albinism and behavior, albino and nonalbino mice were tested for open-field behavior under white light and red light. Red light provides little visual stimulation to mice. The behavioral differences between mice with the two types of coat pigmentation disappeared under red light, indicating that the open-field responses of albino mice are visually mediated. This is not surprising; albino mice are photophobic, because the lack of pigmentation in albinos extends to the iris as well as to the coat.

Many behaviors are genetically complex and polygenically controlled. Genetic crossing between inbred strains can reveal some aspects of genetic involvement in behavior but has not been a successful strategy for identifying specific genes related to polygenic behavioral phenotypes. Environmental variables make important contributions to many such phenotypes, as illustrated by the disappearance of strain-specific behavior in red as opposed to white light. As a result, phenotypes can be inconsistent and more difficult to define. In addition, pleiotropic effects are common in behavior phenotypes and add to the problems of identifying genes that control complex traits.

To narrow the search for genes controlling emotional behavior in mice, quantitative trait loci (QTL) mapping (see Chapter 25 for a discussion of QTLs) has been successfully used. In one experiment, heterozygotes from crosses between C57BL/6J (a nonemotional, low-anxiety strain) and BALB/cJ (an emotional, high-anxiety strain) were intercrossed, and the progeny were selected over multiple generations for extremes of high (H1a) or low (L1a) anxiety. These inbred strains were subjected to several different behavioral tests of anxiety, including the open field, the elevated plus maze (Figure 26–2), and a test called the light–dark box. In this last-named test, mice are placed in an environment in which there is a dark, enclosed box, from which they can emerge into a brightly lit area. Fearful animals prefer to remain in the box and do not explore the lighted areas of the environment. The phenotypically most extreme progeny recovered from these anxiety tests were analyzed for 84 microsatellite markers and four different measures of emotionality. A QTL on chromosome 1 was identified and localized to a small region of the chromosome covering a small region spanning 66 Mb. Over 1600 mice in replicated populations consistently gave the same results. Similar experiments studying microsatellite markers on the other mouse chromosomes have shown that other QTLs control other subsets of behavior (Figure 26–3).

To advance beyond the mapping of a QTL region to identifying specific genes involved in emotional behavior requires the use of genetics and genomics. Genes in the chromosome 1 QTL region are

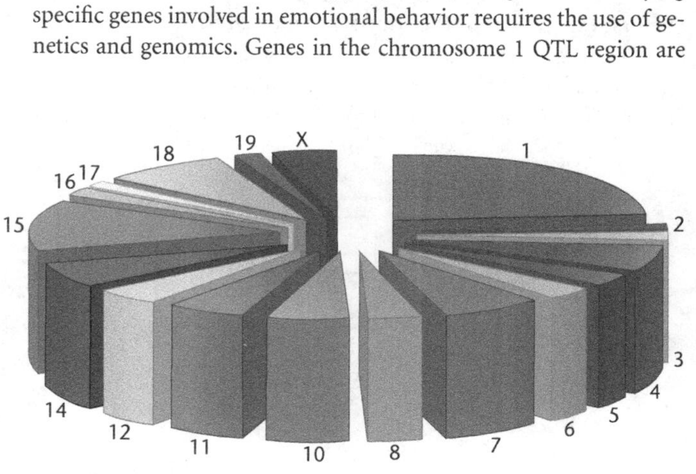


**FIGURE 26–3** Loci for genes controlling emotional behavior in the mouse are distributed on 17 of 19 autosomes. Significant numbers of genes controlling this behavior are located on chromosomes 1, 15, and 18.

being tested in genetic complementation assays to detect interactions among loci. Once these genes are identified, their mechanisms of action and relevance to human behavior remain to be assessed.

**NOW SOLVE THIS**

Problem 3 on page 707 asks for a comparison of inbred strain analysis and QTL analysis in identifying the molecular mechanisms involved in alcohol preference in mice.

■ HINT: *Begin by comparing the strengths and weaknesses of each approach.*

## 26.2 Artificial Selection Can Establish Genetic Strains with Behavioral Differences

The behavior-first approach to studying behavioral genetics selects organisms exhibiting a specific behavior from among a genetically heterogeneous population. If genetic strains can be established that uniformly express this behavior, and if the trait can be transferred by genetic crosses to another strain that initially does not exhibit the behavior under study, genetic involvement in the behavior is confirmed. Often, two groups are selected: one showing high levels and one showing low levels of the behavior under investigation. This bidirectional selection creates two lines with progressively greater differences in behavior. The behavior-first approach maximizes differences in genes that control the trait of interest, and can minimize differences in all other genes (discussed next). The study of maze learning in rats and geotaxis in *Drosophila* are two examples we will use to illustrate this approach.

### Maze Learning in Rats

The first experiment of this kind was reported by E. C. Tolman in 1924. He began with 82 white rats of heterozygous ancestry and measured their ability to "learn" to obtain food at the end of a multiple-T maze (Figure 26–4) by recording the number of errors and

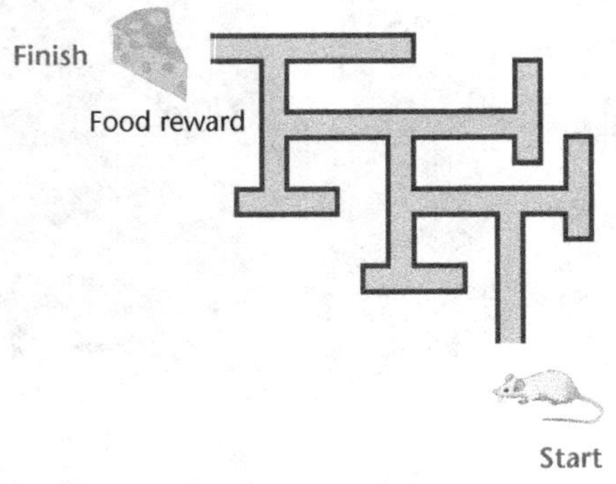


FIGURE 26–4 A multiple T-maze used in learning studies with rats.

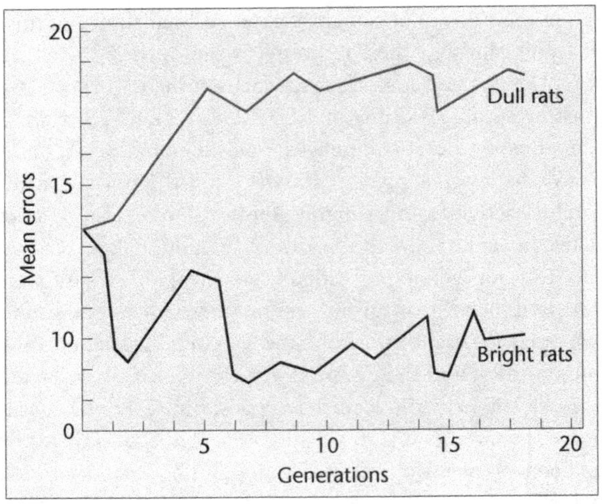


FIGURE 26–5 Selection for the ability and inability of rats to learn to negotiate a maze.

trials. When first exposed to the maze, a rat explores all alleys and eventually arrives at the end, where it is rewarded with food. In succeeding trials, as the rat learns the correct route, it makes fewer and fewer mistakes. Eventually, a hungry rat may proceed directly to the food with no wrong turns.

From the initial 82 rats, nine pairs of the "brightest" and "dullest" rats were selected and mated to produce two lines. In each generation, selection was continued. Even in the first generation, Tolman demonstrated that he could select and breed rats whose offspring performed more efficiently in the maze. Subsequently, Tolman's approach was adopted by others, notably R. C. Tryon, who, in 1942, published results of 18 generations of selection.

As shown in Figure 26–5, two lines with significant differences in maze-learning ability were established. There was some variation around the mean, but by the eighth generation, there was no overlap between lines. That is, by the eighth generation, the dullest of the bright rats were superior to the brightest of the dull rats.

These two lines were also used to study other behavior traits, with varying results. Bright rats were found to be better in solving hunger-motivation problems but inferior in escape-from-water tests. Bright rats were also found to be more emotional in open-field experiments. These results show that selection of genetic strains superior in certain traits is possible, but care must be taken not to generalize such studies to overall intelligence, which is composed of many learning parameters.

Selective breeding has established that selected lines can be used to demonstrate that a behavior has a genetic basis. Selective breeding also can set the stage for further work to identify chromosomes that carry genes involved in the behavior (as described for *Drosophila* in the next section).

Finally, selective breeding can identify genes controlling behavioral traits. However, for this approach to be effective, the population studied should be large, because inbreeding and the subsequent generation of large-scale genetic differences between the selected lines will result if the population is too small.

**NOW SOLVE THIS**

Problem 2 on page 707 asks how to determine if a behavioral trait is genetic or nongenetic. If it is genetic, you are asked to determine the mechanism of gene expression if it is an X-linked trait.

■ HINT: *Hemizygous males express all X-linked recessive traits.*

## Artificial Selection for Geotaxis in *Drosophila*

A **taxis** is a movement response toward or away from an external stimulus. The response may be positive (moving toward the stimulus) or negative (away from it). Many kinds of stimuli elicit this form of behavior, including chemicals (chemotaxis), gravity (geotaxis), and light (phototaxis).

To investigate geotaxis in *Drosophila*, Jerry Hirsch and his colleagues designed a mass-screening device that tests about 200 flies per trial, as shown in Figure 26–6. In this test, flies are lured into a vertical maze that repeatedly requires them to choose between moving up or down. Flies that turn up at each intersection will exit at the top of the maze; those that always turn down will exit at the bottom; and those making both "up" and "down" decisions will exit somewhere in between. Flies can be selected both for positive and for negative geotaxis, establishing the existence of a genetic influence on this behavioral response.

As shown in Figure 26–7, mean scores of flies in this maze vary from about +4 to −6, indicating that selection is stronger for negative geotaxis than for positive geotaxis. The two lines have now undergone selection for almost 40 years, encompassing over 1000 generations and the testing of more than 80,000 flies. Throughout the experiment, clear-cut, but fluctuating, differences have been observed. These results indicate that this complex behavior in *Drosophila* is genetically controlled and is a polygenic trait.

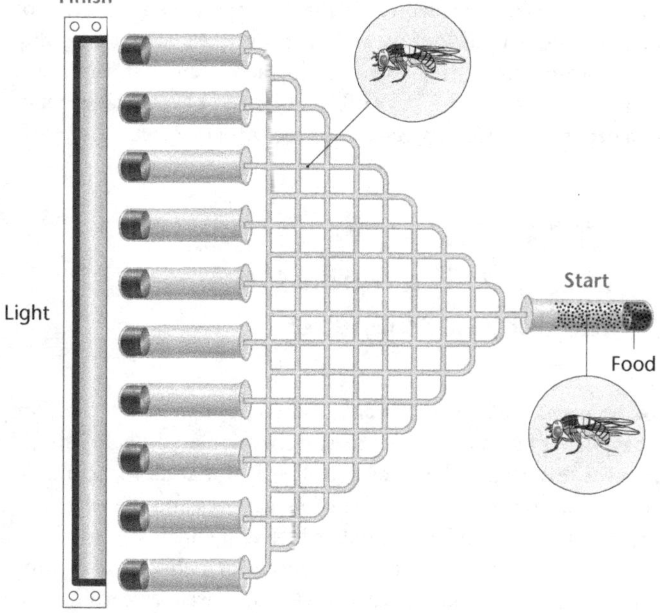


FIGURE 26–6 Schematic drawing of a maze used to study geotaxis in *Drosophila*.

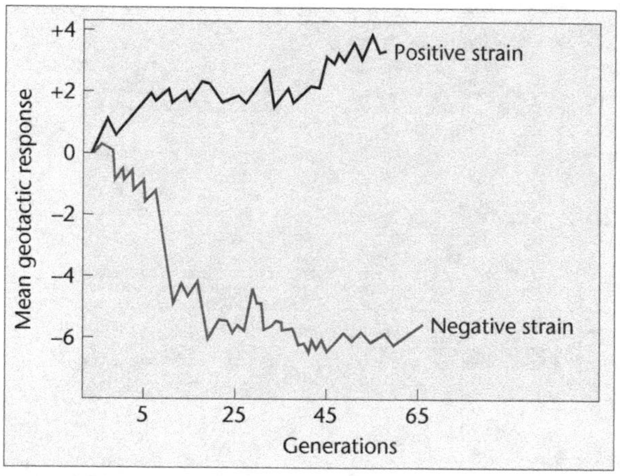


FIGURE 26–7 Selection for positive and negative geotaxis in *Drosophila* over many generations.

Using the lines selected for positive and negative geotaxis, Hirsch and his colleagues analyzed the relative contribution of loci on different chromosomes to geotaxis. Loci on chromosomes 2 and 3 and on the X were identified in an ingenious way. Crosses produced flies that were either heterozygous or homozygous for a given chromosome. Figure 26–8 shows how this is accomplished. From a selected line, a male is crossed to a "tester" female, whose chromosomes are each marked with a dominant mutation (these chromosomes, called balancer chromosomes, are described in detail in Chapter 23). Each marked chromosome carries an inversion to suppress the recovery of any crossover products. An $F_1$ female heterozygous for each chromosome is backcrossed to a male from the original line. The resulting female offspring contain all combinations of the chromosomes. The dominant mutations make it possible to recognize which chromosomes from the selected lines are present in homozygous or heterozygous configurations.

Flies with genotypes O and X (Figure 26–8) are homozygous and heterozygous, respectively, for the X chromosome. Similarly, flies with genotypes 0 and 2 are homozygous and heterozygous, respectively, for chromosome 2, and flies with genotypes 0 and 3 are homozygous and heterozygous, respectively, for chromosome 3. By testing these different flies in the maze, it is possible to assess the behavioral influence of loci on any given chromosome.

For negatively geotaxic flies (the ones that go up in the maze), genes on the second chromosome make the largest contribution to the phenotype, followed by loci on chromosome 3 and on the X chromosome (a gradient of chromosomes 2 > 3 > X). For the positive line (flies that go down in the maze), the reverse arrangement (a gradient

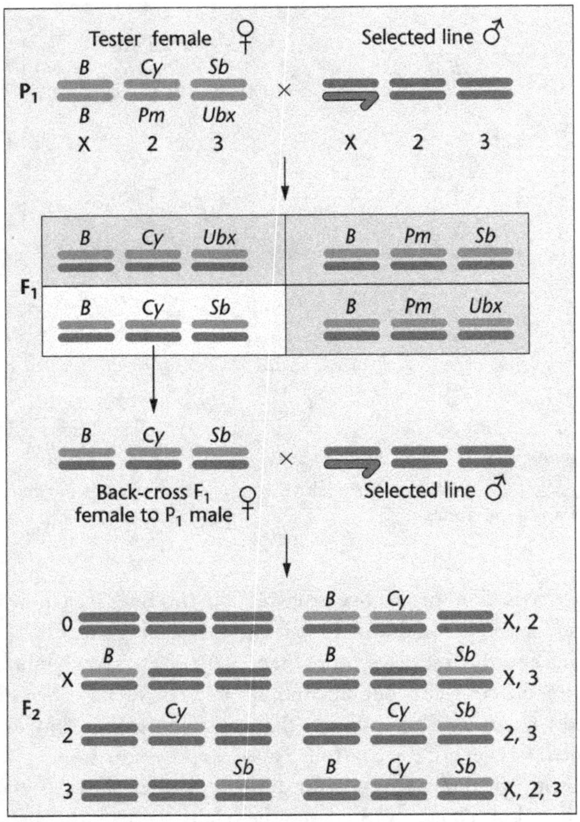


**FIGURE 26-8** In this mating scheme in *Drosophila*, the effect of genes located on specific chromosomes that contribute to geotaxis can be assessed. The progeny produced by backcrossing the female contain all combinations of chromosomes. Examining the presence or absence of dominant marker phenotypes in these flies makes it possible to determine which chromosomes from the selected strain are present. Subsequent testing for geotaxis is then performed (*B* = *Bar eyes*; *Cy* = *Curly wing*; *Sb* = *Stubble bristles*; *Ubx* = *ultrabithorax*). The designations alongside each genotype in the $F_2$ generation (e.g., X, 2, and 3) indicate which chromosomes are heterozygous.

of chromosomes X > 3 > 2) is seen. The overall results indicate that geotaxis is under polygenic control and that the loci controlling this trait are distributed on all three major chromosomes of *Drosophila*.

Further genetic testing in which each chromosome from a selected line has been isolated in homozygous form in an unselected background has been used to estimate the number of genes controlling the geotactic response in *Drosophila*. Once the chromosome has been placed in a genetic background that has not been selected for geotactic behavior, it is possible to estimate the number of genes that control the phenotype. This work indicates that a small number of genes, perhaps two to four loci, are responsible for geotaxis in *Drosophila*.

As ingenious as this work is, it is also arduous, illustrating one of the limitations of behavioral selection. Although selection for geotaxis showed that this behavior is genetically controlled and that loci on all major chromosomes are involved, it cannot identify specific genes involved in this behavior. This is partly because the behavior is polygenically controlled and influenced by several to many genes, each of which may make only a small contribution to the phenotype.

The development of new technology, including microarrays (described in Chapters 21 and 23), offers a new approach to identifying genes that influence complex polygenic behavioral traits. Microarrays are one example of the high-throughput experiments that are revolutionizing genetics and molecular biology. Instead of measuring changes in expression one gene at a time, microarrays make it possible to measure changes in dozens, hundreds, or even thousands of genes in a single experiment. Genes used in microarray analysis can be selected on the basis of their expression patterns in specific tissues, in specific cells, or in specific structures.

To identify genes involved in geotaxis, researchers used two steps. First, they used cDNA microarrays containing about one-third of the genes in the *Drosophila* genome to analyze mRNA levels from the strains with the highest (*Hi5*) and lowest (*Lo*) geotaxis behavior. A small number of genes showed reproducible differences in expression and were selected for further analysis. To narrow the list of candidate genes further, the researchers selected mutants of these genes that led to neurological defects. Ten mutant lines, including several control lines, were selected for the second step in this gene-identification strategy: analysis of geotaxis behavior.

The geotaxis behavior and mRNA levels of three genes in the mutant flies corresponded to levels of mRNA expression and behavior in *Hi5* and *Lo* flies. These genes each made small, incremental contributions to geotaxis, as would be expected for a polygenically controlled trait. The mechanism by which these three genes, *cryptochrome* (*cry*), *Pendulin* (*Pen*), and *Pigment-dispersing-factor* (*Pdf*), are involved in geotaxis is unknown. However, all three genes are known to be involved in the development and function of the brain and nervous system, and their role in geotaxis is currently being investigated. The use of this two-step strategy is significant because it illustrates the successful integration of classical genetics with genomics to identify genes involved in a polygenic trait.

## 26.3 *Drosophila* Is a Model Organism for Behavior Genetics

We discussed geotaxis in *Drosophila* as an example of selection for behavior, but much more extensive information on the behavior genetics of this organism is available. This is not at all surprising in view of our extensive knowledge of its genetics and the ease with which *Drosophila* can be manipulated experimentally.

*Drosophila* has many complex forms of behavior, including courtship, feeding, and circadian rhythms, as well as learning and memory. We will use two behaviors, courtship and learning, to illustrate how *Drosophila* serves as a model organism for studying the molecular aspects of behavior.

## Genetic Control of Courtship

As early as 1915, Alfred Sturtevant observed that, in addition to its effects on body pigmentation, the X-linked recessive gene *yellow* affects mating preference in *Drosophila* females. He found that both wild-type and *yellow* females, when given the choice of wild-type or *yellow* males, prefer to mate with wild types. Both wild-type and *yellow* males prefer to mate with *yellow* females. These conclusions were based on measurements of mating success in all combinations of *yellow* and wild-type (gray-bodied) males and females.

In 1956, Margaret Bastock extended these observations by investigating which, if any, component of courtship behavior was affected by the *yellow* mutant gene. Courtship in *Drosophila* is a complex ritual. The male first shows orientation. He follows the female, often circling her, and then orients himself at a right angle to her body and taps her on the abdomen. Once he has her attention, the male begins wing display. He raises the wing closest to the female and vibrates this wing rapidly for several seconds (the "singing" shown in Figure 26–9). He then moves behind her and makes contact with her genitalia. If she signals acceptance by remaining in place, he mounts her, and copulation occurs (Figure 26–9). Bastock compared the courtship ritual in wild-type and *yellow* males. The *yellow* males prolong orientation, but spend much less time in the vibrating and genital contact phases. Her observations indicate that the *yellow* mutation alters the pattern of male courtship, making these males less successful in mating.

From these beginnings, *Drosophila* geneticists identified, cloned, and sequenced many genes that affect courtship. Each of the steps in the courtship ritual involves different parts of the nervous system. One gene, *fruitless* (*fru*), affects all aspects of male courtship but has no effect on female behavior. The gene has a complex organization (Figure 26–10) and encodes a transcription factor. *fru* can be transcribed from any of four promoter sites (P1–P4), but only transcripts from the P1 promoter may be spliced in a male-specific manner, and proteins translated from these mRNAs control male courtship. (Transcripts from that promoter may also be spliced in a female-specific manner.) Male-specific *fru* transcripts are expressed during pupal development in about 20 cells in the central nervous system, including both the brain and the ventral nerve cord. Presumably, *fru* expression results in male-specific neural connections that process sensory information during courtship.

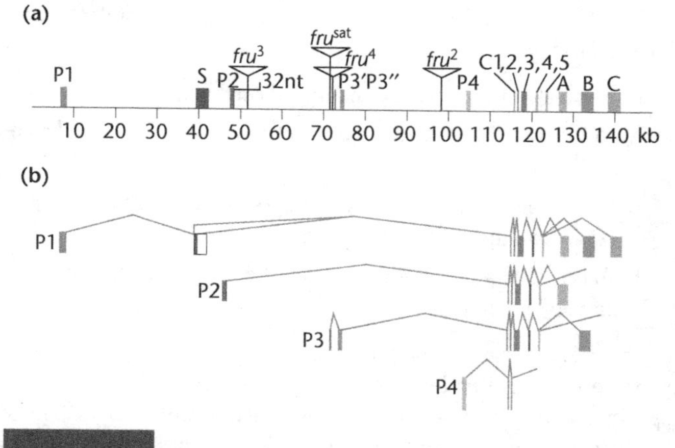


FIGURE 26–10 The patterns of transcription and splicing for the *fruitless* (*fru*) gene of *Drosophila*. (a) The locations of the promoters (P1–P4) and exons of the *fru* gene are shown along the gene. (b) Transcripts from the P2, P3, and P4 promoters are identical in the two sexes. Genetic analysis of mutant alleles of *fru* (the inverted triangles in part a represent deletion alleles of this gene) indicate that some transcripts from P2–P4 are found throughout the central nervous system and other tissues, and encode proteins essential for viability. Transcripts from P1 are spliced in a sex-specific manner. Female-specific transcripts from the P1 promoter introduce a small number of amino acids on the N-terminal side of the domain represented by exons C1–C5. Male-specific transcripts from P1 add over 100 amino acids to this region.

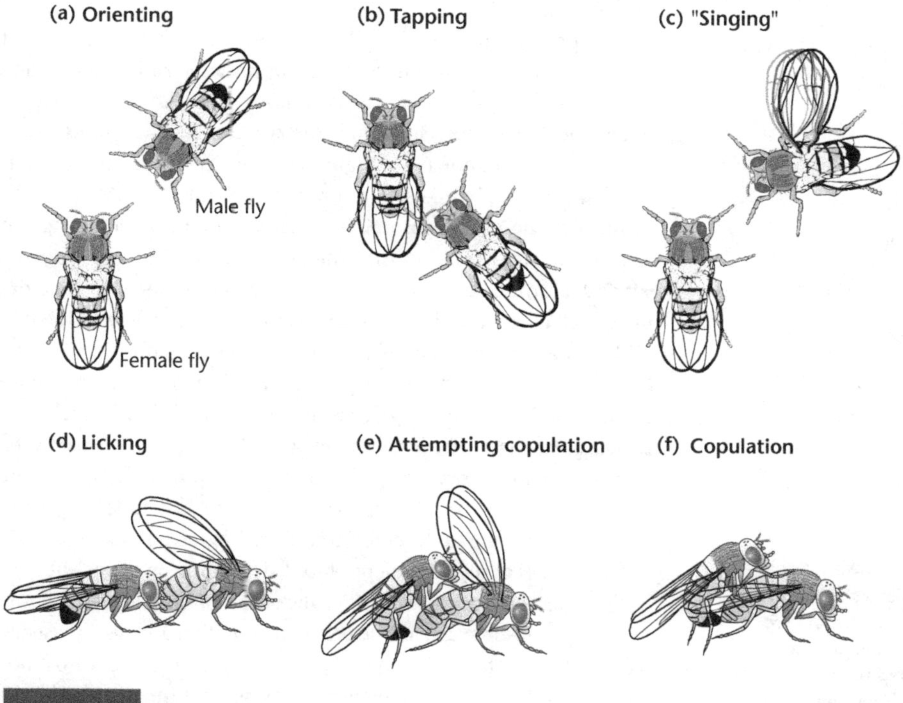


FIGURE 26–9 The stages of male behavior during courtship and mating in *Drosophila*.

## Dissecting Behavior with Genetic Mosaics

In the late 1960s, Seymour Benzer and his colleagues pioneered the use of the gene-first method to study behavior in *Drosophila*. This approach begins by generating mutant alleles with a behavioral phenotype and using them to "dissect" a complex biological phenomenon into its simpler components. His approach had several goals:

1. To isolate mutations that disrupt normal behavior.
2. To identify the mutant genes by chromosome localization and mapping.
3. To determine the structural component through which gene expression influences the behavioral response.
4. To establish a causal link between the mutant allele and its behavioral phenotype.

All four steps can be illustrated using the example of phototaxis, one of the first behaviors studied by Benzer. Normal flies are positively phototactic—that is, they move toward a light source. Mutations were induced by feeding male flies sugar water containing ethylmethanesulfonate (EMS, a potent mutagen). These males were then mated to attached-X virgin females. These flies carry two physically attached X chromosomes and a Y chromosome. As shown in Figure 26–11, the $F_1$ male offspring of such females receive their X chromosome from their fathers. If recessive mutations were induced in the males exposed to EMS, those mutations will be phenotypically expressed in the $F_1$ males. The $F_1$ males were tested for response to light (phototaxis), and flies with abnormal behavior were isolated. Benzer found runner mutants, which move quickly to and from light; negative phototactic mutants, which move away from light; and nonphototactic mutants, which show no preference for light or darkness.

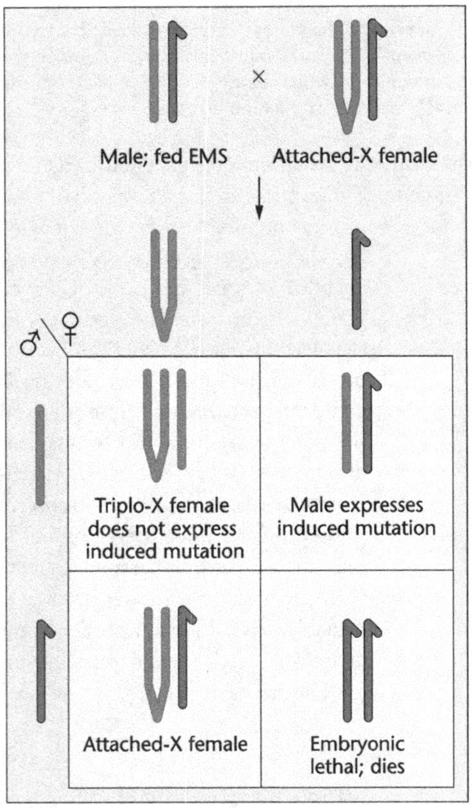


FIGURE 26–11 A genetic cross in *Drosophila* that facilitates the recovery of X-linked-induced mutations. The female parent contains two X chromosomes that are attached, in addition to a Y chromosome. In a cross between this female and a normal male that has been fed the mutagen ethylmethanesulfonate, all surviving males receive their X chromosomes from their father and express all mutations induced on that chromosome.

The genetic basis of these behavioral changes was confirmed by mating mutant $F_1$ males to attached-X virgin females. The male progeny of this cross also showed the abnormal phototactic responses, indicating that these behavioral alterations are the result of X-linked recessive mutations.

Nonphototactic mutants walk normally in the dark but show no phototactic response to light. Benzer and Yoshiki Hotta tested electrical activity at the surface of mutant eyes in response to a flash of light. The pattern of electrical activity was recorded as an electroretinogram. Various types of abnormal responses were detected in different nonphototactic mutant strains. None of the mutants had a normal pattern of electrical activity. When these mutations were mapped, they were not all traced to alleles of the same gene; instead, they were shown to occupy several loci on the X chromosome. This analysis indicates that several gene products encoded by genes on the X chromosome contribute to the phototactic response.

Where, within the fly, must gene expression occur to produce a normal electroretinogram? In an ingenious strategy for answering this question, Benzer turned to the use of mosaics. In **mosaic flies**, some tissues are mutant and others are wild type. If analysis can determine which body part must be mutant in order to yield the abnormal behavior, the primary focus of the genetic alteration can be determined.

To produce mosaic flies, Benzer used a *Drosophila* strain that carries one of its X chromosomes in an unstable ring shape. When the "ring-X" is present in a zygote undergoing mitosis, it is frequently lost by nondisjunction. If the zygote is female and its two X chromosomes are one normal and one ring-X, loss of the ring-X at the first mitotic division will result in one of the two daughter cells having a single X chromosome (normal X) and the other having two X chromosomes (one normal and one ring-X). The cell with the single X chromosome goes on to produce male tissue (XO) and expresses all alleles on the remaining X, whereas the cell with the two X chromosomes produces female tissue and does not express heterozygous recessive X-linked genes. This is illustrated in Figure 26–12. The loss of the ring-X results in mosaic flies with male and female parts that differ in the expression of X-linked recessive traits.

When and where the ring-X is lost in development determines the pattern of mosaicism. The loss usually occurs early in development, before cells migrate to the embryo surface. As shown in Figure 26–13, different types of mosaics are produced, depending on the orientation of the mitotic spindle when loss of the ring-X takes place. If the normal X chromosome carries the behavior mutation and also a visually recognizable mutant allele (*yellow*, for example), the pattern of mosaicism will be easy to distinguish. By examining the distribution of the visible mutation in flies with a combination of normal and mutant

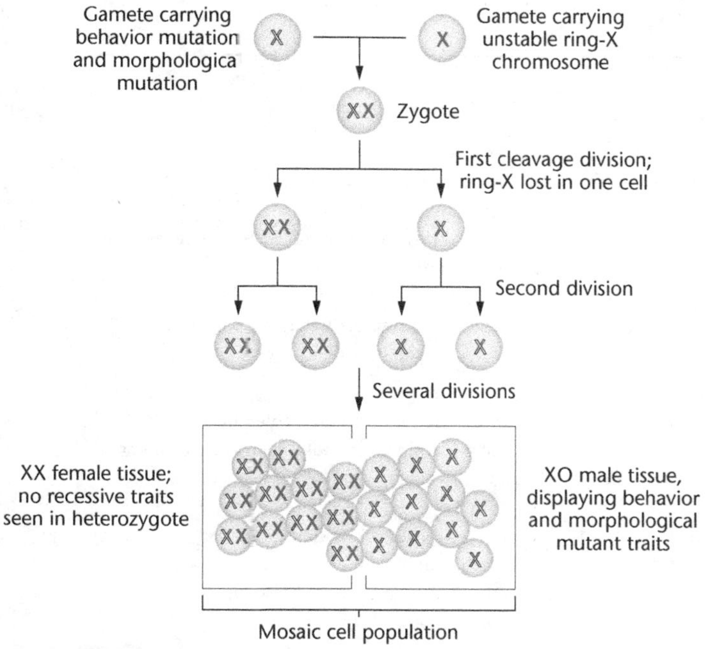


**FIGURE 26-12** Production of a mosaic *Drosophila* as a result of fertilization by a gamete carrying an unstable ring-X chromosome (shown in color). If this chromosome is lost in one of the two cells following the first mitotic division, the body of the fly will consist of one side that is male (XO) and one side that is female (XX). The male side will display recessive mutant traits contained on the X chromosome.

structures—such as a fly with a mutant head on a wild-type body, a normal head on a mutant body, or one normal and one mutant eye on a normal or mutant body—researchers can identify the parts of the body in which the behavioral mutation may be expressed.

When nonphototactic mosaics were studied, the focus of the genetic defect was found to be in the eye itself. In mosaics, where every part of the fly except the eye was normal, abnormal behavior was still detected. When one eye was mutant and the other normal, the fly had an unusual behavior. Instead of crawling straight up toward light as the normal fly does, the mosaic fly with one mutant eye crawls upward toward light in a spiral pattern. In the dark, this fly moves in a straight line.

Using a combination of genetics, physiology, and biochemistry, Benzer and other workers in the field have identified and analyzed a large number of genes affecting behavior in *Drosophila*. As shown in Table 26.2, mutations that affect many different aspects of behavior have been isolated. In keeping with the long-standing tradition for the naming of mutants in *Drosophila* genetics, some of these mutations have received very descriptive and often humorous names.

Many mutations have been analyzed with the mosaic technique to localize the focus of gene expression. It was easy to predict that the focus of the nonphototactic mutation would be in the eye, but other mutations were not so predictable. For example, researchers did not know whether the mutation for a wings-up phenotype would be discovered to be a defect in the wings or in muscle-attachment sites in the thorax, or some neuromuscular defect or abnormality of muscle formation. The analysis of mosaics for this

**FIGURE 26-13** The effect of spindle orientation on the production of mosaic flies such as the ones created in Figure 26-12.

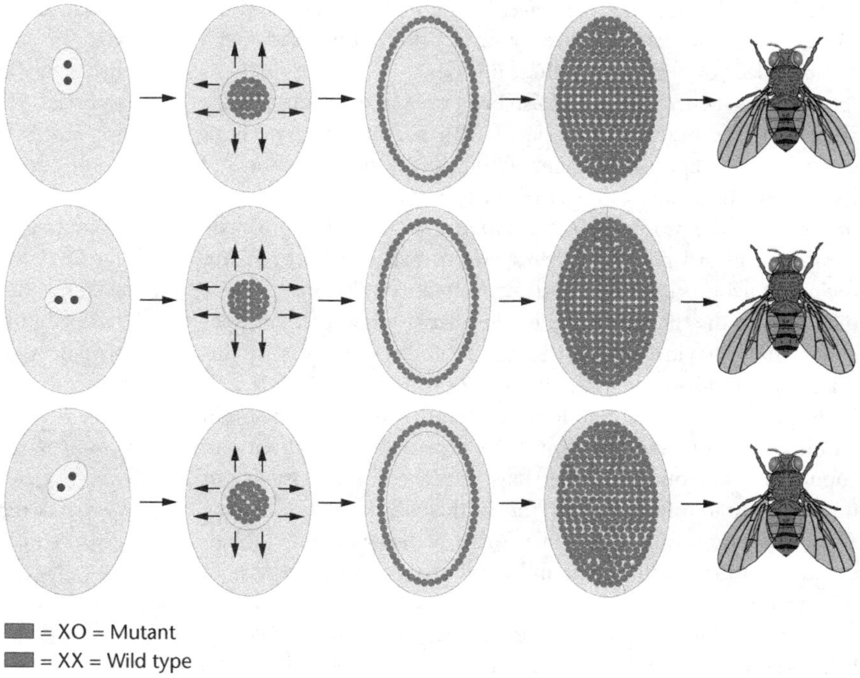

TABLE 26.2

**Some *Drosophila* Mutations That Affect Courtship and Learning**

| Behavior Genes | Molecular Function | Behavior Phenotype, Function |
|---|---|---|
| **Courtship** | | |
| *fruitless (fru)* | Transcription factor | All aspects of male courtship |
| *doublesex (dsx)* | Transcription factor | Song defect |
| *dissatisfaction (dsf)* | Steroid hormone receptor | Poor sex discrimination, reduced female receptiveness |
| *courtless (crl)* | Ubiquitin-conjugating enzyme | Failure to court |
| *slowpoke (slo)* | Calcium-activated potassium channel | Song defect |
| *cacophony (cac); nightblind A (nbA)* | Voltage-sensitive calcium channel | Song defect, optomotor behavior, photophobia |
| *dissonance (diss); no on-or-off transient (nonA)* | RNA binding | Song defect, optomotor behavior |
| **Learning and memory** | | |
| *dunce (dnc)* | cAMP-specific phosphodiesterase | Locomotor rhythms, ethanol tolerance |
| *rutabaga (rut)* | Adenylate cyclase | Courtship learning, ethanol tolerance, grooming |
| *amnesiac (amn); cheapdate (chpd)* | Neuropeptide | Ethanol tolerance |
| *latheo (lat)* | DNA-replication factor | Larval feeding |
| *Shaker (Sh)* | Voltage-sensitive potassium channel | Courtship suppression, gustation defect, ether sensitivity |
| *G protein s* α 60 A (*G-s* α60A) | Heterotrimeric G protein | Visual behavior, cocaine sensitivity |
| *DCO; cAMP-dependent protein kinase I (Pka-C1)* | Protein Ser/Thr kinase | Locomotor rhythms, ethanol tolerance |
| *cAMP-response-element-binding protein B at 17A (CrebB17A); dCREB* | Transcription factor | Locomotor rhythms |
| *Calcium/calmodulin-dependent protein kinase II (CaMKII)* | Protein Ser/Thr kinase | Courtship suppression |
| *Neurofibromatosis 1 (Nf1)* | Ras GTPase activator | Embryonic, nervous system defects |

*Source*: Sokolowski, M. 2001. *Drosophila*: Genetics meets behaviour. *Nature Reviews Genetics* 2: 879–890, Table 1, p. 882

mutation established that the defect is in the flight muscles of the thorax. Cytological studies have confirmed this finding, showing a complete lack of myofibrils in these muscles.

The mosaic technique has also been used to determine which regions of the brain are associated with sex-specific aspects of courtship and mating behavior. Jeffrey Hall and his associates have shown that mosaics with male cells in the most dorsal region of the brain, the protocerebrum, exhibit the initial stages of male courtship toward females. Later stages of male sexual behavior, including wing vibrations and attempted copulations, require male cells in the thoracic ganglion. Similar studies of female-specific sexual behavior have shown that the ability of a mosaic to induce courtship by a male depends on the presence of female cells in the posterior thorax or abdominal region. A region of the brain within the protocerebrum must be female for receptivity to copulation. Anatomical studies have confirmed the existence of fine structural differences in the brains of male and female *Drosophila*, indicating that some forms of behavior may be dependent on the development and maturation of specific parts of the nervous system.

The gene-first approach has several advantages over the behavior-first approach. Using a bottom-up approach allows the impact of single genes on a behavioral response to be quantified. By rigorous application of this method, all genes contributing to the behavior can eventually be identified, mapped, cloned, and sequenced. The disadvantage is that it requires the use of an organism for which researchers have developed an extensive and sophisticated set of genetic tools, especially marker chromosomes and collections of strains with deleted and duplicated chromosome regions, as well as mutants with cytological markers for mosaic analysis. The fruit fly, *Drosophila*, is the organism that best fits this requirement, and fortunately, it is an excellent model system for studying not only behavior but its underlying basis in the structure and function of the nervous system.

NOW SOLVE THIS

Problem 4 on page 707 involves analysis of behavior in a mosaic fly constructed using an unstable ring-X chromosome and distinguishing between the origins of male- and female-specific behavior in the mosaics.

■ HINT: *The markers on the cuticle reflect the genotype of the underlying structures, including brain regions.*

## Functional Analysis of the Nervous System

The analysis of behavior mutants in *Drosophila* has also led to an understanding of functional mechanisms in the nervous system.

Figure 26–14 depicts a simplified nerve cell (neuron), in which nerve impulses are generated at one end (on a dendrite) and move along the cell to the end of the axon, which transmits the impulse to adjacent neurons. During this process, the impulse is propagated by the movement of sodium and potassium ions across the neuron's plasma membrane. The movement of these electrically charged ions can be monitored by measuring changes in the electrical potential of the membrane. To screen for genes that control the generation and transmission of nerve impulses, Barry Ganetzky and his colleagues screened behavioral mutants of *Drosophila* to identify those with electrophysiological abnormalities in the generation and propagation of nerve impulses. Two general classes of such mutants have been isolated: those with defects in the movement of sodium and those with defects in potassium transport (Table 26.3).

One mutant identified in the screening procedure is a temperature-sensitive allele called *paralytic.* Flies homozygous for this mutant allele become paralyzed when exposed to temperatures at or above 29°C, but they recover rapidly when the temperature is lowered to 25°C. Mosaic studies revealed that both the brain and thoracic ganglia are the focus of this abnormal behavior. Electrophysiological studies showed that mutant flies have defective sodium transport associated with the conduction of nerve impulses. Subsequently, Ganetzky and his colleagues mapped, isolated, and cloned the *paralytic* gene. This locus encodes a protein called the sodium channel, which controls the movement of sodium across the membrane of nerve cells in many different organisms, from flies to humans.

A second mutant, called *Shaker,* originally isolated over 40 years ago as a behavioral mutant, encodes a potassium channel gene that has also been cloned and characterized. Because the mechanism of nerve impulse conduction has been highly conserved during animal evolution, the cloned *Drosophila* genes were used as probes to isolate the equivalent human ion channel genes. The cloned human genes are being used to provide new insights into the molecular basis of neuronal activity. One of the human ion channel genes first identified in *Drosophila* is defective in a heritable form of cardiac arrhythmia. The identification and cloning of the human gene now makes it possible to screen for family members at risk for this potentially fatal condition.

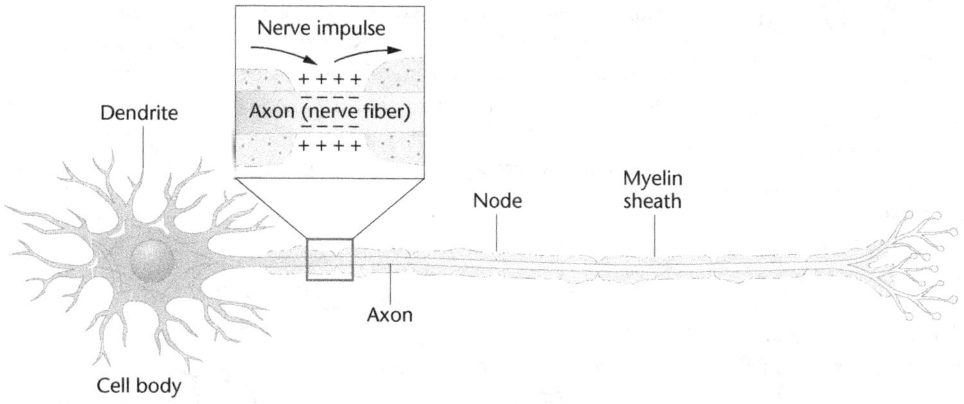


FIGURE 26–14 A nerve cell (neuron) has extensions called dendrites that carry impulses toward the cell body and also has one or more axons that carry impulses away from the cell body. The axon is encased in a sheath of myelin, except at nodes spaced at intervals along the axon. Nerve impulses are accelerated by jumping from node to node. Electrodes placed on either side of the neuron's plasma membrane can record the electric potential across the membrane. When the neuron is at rest, there is more sodium outside the cell and more potassium inside the cell. When a nerve impulse is generated, sodium moves into the cell, and potassium moves out, altering the electric potential. As the impulse moves away, ions are pumped across the membrane to restore the original electric potential.

TABLE 26.3

**Behavioral Mutants of *Drosophila* in Which Nerve Impulse Transmission Is Affected**

| Mutation | Map Location | Ion Channel Affected | Phenotype |
|---|---|---|---|
| *$nap^{ts}$* | 2-56.2 | Sodium | Adults and larvae paralyzed at 37.5°C; reversible at 25°C |
| *$para^{ts}$* | 1-53.9 | Sodium | Adults paralyzed at 29°C, larvae at 37°C; reversible at 25°C |
| *Tip-E* | 3-13.5 | Sodium | Adults and larvae paralyzed at 39–40°C; reversible at 25°C |
| *$sei^{ts}$* | 2-10.6 | Sodium | Adults paralyzed at 38°C, larvae unaffected; adults recover at 25°C |
| *Sh* | 1-57.7 | Potassium | Aberrant leg shaking in adults exposed to ether |
| *eag* | 1-50.0 | Potassium | Aberrant leg shaking in adults exposed to ether |
| *Hk* | 1-30.0 | Potassium | Ether-induced leg shaking |
| *sio* | 3-85.0 | Potassium | At 22°C, adults are weak fliers; at 38°C, adults are weak, uncoordinated |

## *Drosophila* Can Learn and Remember

Researchers gain a great advantage when they can study the genetics of a complex behavior, such as learning, in a model organism that is as convenient for them to manipulate as *Drosophila*. However, this is only possible if the model organism can perform the complex behavior the researchers are interested in studying. Thus, before using *Drosophila* to study learning, researchers needed to know whether *Drosophila* could learn.

Benzer's lab was one of the first to isolate genes that control learning and memory in *Drosophila*. To do so, they used an olfactory-based shock-avoidance learning system in which flies are presented with a pair of odors, one of which is associated with an electric shock. Flies quickly learn to avoid the odor associated with the shock. For several reasons, this response is thought to be learned. First, performance is associated with the pairing of a stimulus or response with reinforcement. Second, the response is reversible: flies can be trained to select an odor they previously avoided. Third, flies exhibit short-term memory for the training they have received.

The demonstration that *Drosophila* can learn opened the way to selecting mutants that are defective in learning and memory. To accomplish this part of the study, males from an inbred wild-type strain were mutagenized and mated to females from the same strain. Their progeny were recovered and mated to produce stocks that carried a mutagenized X chromosome. Mutations that affected learning were selected by testing for responses in the olfactory shock apparatus. A number of learning-deficient mutants, including *dunce*, *turnip*, *rutabaga*, and *cabbage*, were isolated in this way (Table 26.2). In addition, a memory-deficient mutant, *amnesia*, which learns normally, but forgets four times faster than normal, was recovered. Each of these mutations represents a single-gene defect that affects a specific form of behavior. Because of the method used to recover them, all the mutants found so far are X-linked genes. Presumably, similar genes that control behavior are also located on the autosomes.

Since, in many cases, mutation results in the alteration or abolition of a single protein, the nervous system mutants described here can provide a link between behavior and molecular biology. One group of *Drosophila* learning mutations, for example, involves defects in a cellular signal transduction system that proved to play a role in learning.

In cells of the nervous system, and in many other cell types, cyclic adenosine monophosphate (cAMP) is produced from adenosine 5′ triphosphate (ATP) in the cytoplasm in response to signals received at the cell's surface. Cyclic AMP (cAMP) activates protein kinases, which, in turn, phosphorylate proteins and initiate a cascade of metabolic effects that control gene expression.

Behavioral mutants of *Drosophila* were among the first to show the link between cAMP and learning (Figure 26–15). The *rutabaga* (*rut*) locus encodes adenylyl cyclase, the enzyme that synthesizes cAMP from ATP. In the mutant allele of *rutabaga*, a missense mutation destroys the catalytic activity of adenylyl cyclase and is associated with learning deficiency in homozygous flies. The *dunce* (*dnc*) locus encodes the structural gene for the enzyme cAMP phosphodiesterase, which degrades adenylyl cyclase. The *turnip* (*tur*) mutation occurs in a gene encoding a G protein, a class of molecules that bind guanosine 5′-triphosphate (GTP) and, in turn, activate adenylyl cyclase.

Cells in the *Drosophila* nervous system, called mushroom body neurons, integrate and process sensory input from olfactory stimuli and from electric shocks. When both stimuli arrive at a mushroom body simultaneously, adenylyl cyclase is stimulated and activates a G-protein-linked receptor, which leads to elevated cAMP levels. The increase in cAMP leads to changes in the excitability of the mushroom body neuron. Elevated levels of cAMP upregulate expression of a protein kinase, which enhances short-term and long-term memory. Short-term memory is mediated by changes in potassium channels. Long-term memory is associated with changes in mushroom body neuron gene expression controlled by a transcriptional activator, CREB (cAMP response element-binding protein). CREB binds to elements upstream of cAMP-inducible genes and changes the pattern of gene expression, encoding memory.

The biochemical pathways that control learning and memory in *Drosophila* are similar to those in other organisms, including

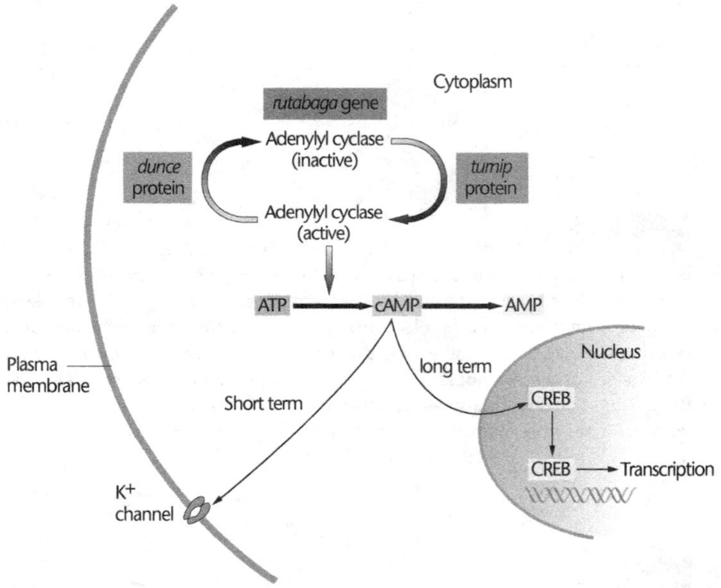


**FIGURE 26–15** The metabolic pathway for cyclic AMP (cAMP) synthesis and degradation. In this pathway, ATP is converted into cAMP by the enzyme adenylyl cyclase. cAMP has dual activities in signal transduction. It affects short-term memory through the potassium channel in the plasma membrane, and it affects long-term memory by its impact on CREB, a transcription factor. cAMP is broken down into AMP. In *Drosophila*, the *rutabaga* gene encodes adenylyl cyclase. When produced, the enzyme is inactive; it is converted to its active form by a protein encoded by the *turnip* gene. The enzyme is inactivated by a protein encoded by the *dunce* gene. Mutations in any of these genes affect learning and memory in *Drosophila*.

mice. The mammalian homologs of the *Drosophila* genes are active in parts of the mouse brain involved in learning and memory. New genetic screening methods in both mice and flies are being used to identify new genes and other biochemical pathways involved in learning.

## 26.4 Human Behavior Has Genetic Components

The genetic control of behavior has proven more difficult to characterize in humans than in other organisms. Not only are humans unavailable as experimental subjects in genetic investigations, but the types of responses considered to be the most interesting forms of human behavior, including aspects of *intelligence*, *language*, *personality*, and *emotion*, are difficult to study. Two problems arise in studying such behaviors. First, all behaviors are difficult to define objectively and to measure quantitatively. Second, they are affected by environmental factors. In each case, the environment is extremely important in shaping, limiting, or facilitating the final phenotype for each trait.

Historically, the study of human behavior genetics has been hampered by other factors as well. Many early studies of human behavior were performed by psychologists without adequate input from geneticists. Moreover, traits involving intelligence, personality, and emotion have great social and political significance. Consequently, research findings concerning these traits are likely to be distorted by sensationalism when reported to the public. In fact, because the study of these traits often comes close to infringing upon individual liberties, such as the right to privacy, the studies themselves, much less their conclusions, very frequently stir up controversy.

In lamenting the gulf between psychology and genetics in the study of human behavior, C. C. Darlington wrote in 1963, "Human behavior has thus become a happy hunting ground for literary amateurs. And the reason is that psychology and genetics, whose business it is to explain behavior, have failed to face the task together." Since 1963, some progress has been made in bridging this gap, but the genetics of human behavior remains controversial. Like other areas of behavioral genetics, the study of human behavior depends on the analysis of single genes, as well as on the investigation of complex, polygenic traits with environmental components.

### Single Genes and Behavior: Huntington Disease

Some single-gene human genetic disorders that affect the development, structure, or function of the brain and nervous system result in a behavioral abnormality. One well-studied example is **Huntington disease (HD)**. Inherited as an autosomal dominant disorder that affects about 1 in 10,000 people, HD has its principal impact on the brain. Symptoms usually appear in the fifth decade of life as a gradual loss of motor function and coordination. As structural degeneration of the brain progresses, personality changes occur. The symptoms are due to brain-cell death taking place in specific brain regions, including the cerebral cortex and the striatum. Most victims die within 10 to 15 years after the onset of the disease.

Because HD usually appears in a person after he or she has already started a family, all the children of the affected person must live with the knowledge that they face a 50 percent probability of developing the disorder (affected individuals are usually heterozygotes). The *HD* gene, located on the short arm of chromosome 4, was one of the first genes to be mapped by using restriction fragment length polymorphisms (RFLPs). The gene encodes a large protein (350 kDa) called huntingtin (Htt) that is essential for the survival of certain neurons in adult brains. Mutant HD alleles have an expanded number of cytosine-adenine-guanine (CAG) trinucleotide repeats in exon 1 (see Chapter 16). Normal individuals have from 7 to about 34 repeats, encoding the insertion of multiple copies of the amino acid glutamine near the amino terminus of the huntingtin protein. In mutant alleles, expansion of CAG repeats leads to expansion of glutamine residues in the encoded protein. Individuals carrying more than 40 CAG repeats have polyglutamine tract expansions that cause Htt to become toxic and kill nerve cells in the brain. Genetic anticipation is seen in HD, and the number of CAG repeats is inversely related to the age of onset.

HD is a member of a class of inherited neurodegenerative disorders characterized by expansion of CAG repeats within exons, resulting in the expansion of polyglutamine tracts in the encoded proteins. These polyglutamine diseases are all inherited as dominant traits with symptoms first appearing in adulthood, and they are all characterized by the intracellular accumulation of polyglutamine proteins and cell death. These similarities suggest that expansion of the polyglutamine tract in the encoded protein is important in the pathogenesis of these diseases. To study the molecular events associated with the formation of the cellular aggregates in HD, transgenic animal models of the disease have been developed.

### A Transgenic Mouse Model of Huntington Disease

Several animal models of HD have been developed to study the normal function of huntingtin, to duplicate the disease process and study its effects at the molecular level, and to develop drugs that slow or stop the degeneration and death of brain cells.

To study the relationship between CAG repeat length and disease progression, Danilo Tagle and his colleagues constructed transgenic mice carrying a full-length mutated copy of a human *HD* gene with 16, 48, or 89 copies of the CAG repeat. The vector used for transferring the gene into the mice had the *HD* gene inserted at a site adjacent to a promoter and enhancer to achieve high levels of expression.

Mice carrying the transgene were monitored from birth to death to determine the age of onset and progression of abnormal behavioral phenotypes. Animals carrying 48- or 89-repeat *HD* genes showed behavioral abnormalities as early as 8 weeks, and by 20 weeks, these mice showed both behavioral and motor-coordination abnormalities compared with the control animals.

At various ages, brain sections of wild-type mice and of the transgenic animals carrying mutant alleles with 16, 48, and 89 copies of the CAG repeat were examined for changes in brain structure. Degenerating neurons and cell loss were evident in mice carrying 48 and 89 repeats, but no changes were seen in brains of wild-type mice or of those carrying a 16-repeat transgene (Figure 26–16).

The behavioral changes and the degeneration in specific brain regions in the transgenic mice parallel the progression of HD in humans. These mice are now used to study early changes in brain structure that occur before the onset of locomotor changes, and in the development of experimental treatments to slow or reverse cell loss. Having a mouse model for the disease allows researchers to administer treatment at specific times in disease progression and to evaluate the outcome of treatments in the presymptomatic stages of HD.

## Mechanisms of Huntington Disease

Although the *HD* gene was mapped, isolated, and characterized almost 15 years ago, the mechanism by which mutant Htt causes HD is still unknown. Htt is expressed in many cell types but causes cell death only in striatal cells in the brain. Several mechanisms have been proposed, including abnormal mitochondrial metabolism, increased activity of proteases associated with cell death (apoptosis), misfolding of Htt, and the formation of nuclear inclusions that contain Htt fragments. Recently, attention has focused on the role of mutant Htt in the disruption of transcriptional regulation.

FIGURE 26–16 Relative levels of neuronal loss in HD transgenic mice. Cell counts show a significant reduction of certain neurons (small-medium neurons) in the corpus striatum of the brains of HD48 mutants (middle column) and HD89 mutants (right column) compared to wild-type (left column) mice. Cell loss in this brain tissue is also found in humans with HD, making these transgenic mice valuable models to study the course of this disease.

In affected individuals, onset of HD is associated with decreases in mRNA levels for genes encoding certain neurotransmitter receptors. In a transgenic HD mouse model, researchers used DNA arrays to screen nearly 6000 mRNAs from striatal cells. They found lower levels of mRNA from a small set of genes involved in signaling pathways that are critical to striatal cell function. Similar results were found in another transgenic HD mouse model and are consistent with findings from HD patients, implicating aberrant gene expression as an underlying cause of HD.

Transcriptional regulation involves the action of transcription factors that alter chromatin structure by histone modification (see Chapter 18 for a detailed discussion). In general, acetylation of histones increases transcriptional activity, and deacetylation lowers or turns off transcription. The mutant form of Htt interacts with and binds to several transcription factors, and may alter patterns of gene expression by making these proteins unavailable. Some proteins that interact with Htt have histone-modifying activity and are found in Htt protein aggregates in brains of transgenic HD mice and patients with HD. In addition, mutant versions of amino terminal Htt fragments have been shown to bind to and dissociate proteins from transcriptional complexes on gene promoters.

Using transgenic HD mice, several groups are studying the use of drugs that inhibit histone deacetylation. Sodium butyrate and phenylbutyrate both increased survival and reduced cell death in the transgenic mice. Drugs that inhibit histone deacetylation are now being used in human clinical trials on HD patients, and other drugs are being screened in mouse models for therapeutic effects.

## Multifactorial Behavioral Traits: Schizophrenia

Although RFLP linkage studies have been successful in mapping and cloning a small number of single genes with behavioral phenotypes (such as *HD*), other behavioral disorders are polygenic and have strong environmental components, making their genetics more difficult to dissect. Schizophrenia is a complex brain disorder affecting about 1 percent of the population and is the example we will focus on in this section.

**Schizophrenia** is a mental disorder characterized by avoidance of social contact and by bizarre and sometimes delusional behavior. Those affected by the disease are unable to lead normal lives and are periodically disabled by the condition. It is clearly a familial disorder, with relatives of schizophrenics having a much higher incidence of the condition than the general population (Figure 26–17). Furthermore, the closer the relationship to an affected individual, the greater is a person's probability of developing the disorder.

Concordance for schizophrenia in monozygotic and dizygotic twins has been the subject of many studies. In almost every investigation, concordance has been higher in monozygotic twins (~50%) than in dizygotic twins (~17%). Although these results suggest that a genetic component exists, they do not reveal the precise genetic basis of schizophrenia. Other studies implicate the importance of shared environmental risk factors. The current view is that schizo-

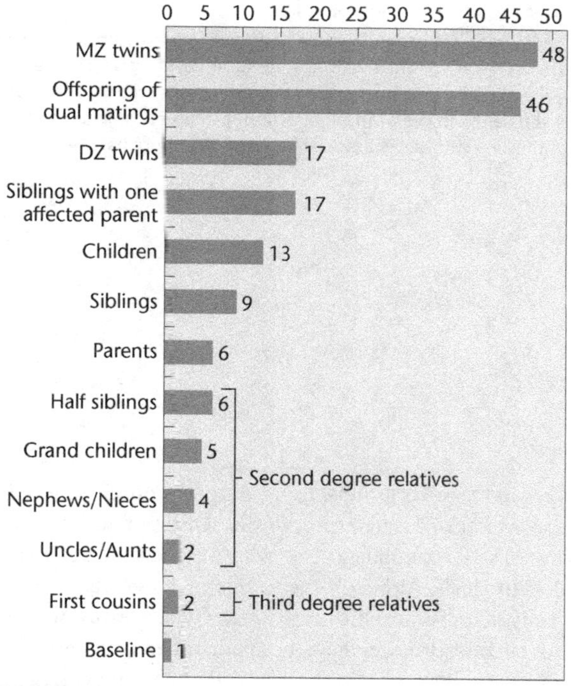


**FIGURE 26–17** Relative lifetime risk of schizophrenia for nonrelated individuals and for relatives of a schizophrenic proband.

phrenia is genetically mediated but not genetically determined. The genetic contributions to schizophrenia have been difficult to identify. Simple monohybrid inheritance, dihybrid inheritance, and multiple gene control have been proposed for schizophrenia. It does not seem likely that only one or two loci are involved, nor is it likely that the disorder is strictly quantitative, as in polygenic inheritance. Several attempts to identify genes involved in schizophrenia have failed, and new approaches are being used. We will describe one of these new approaches: the use of DNA microarrays to carry out genome-wide scans to identify genes whose expression is altered in schizophrenia.

In one important study, researchers isolated RNA from the prefrontal cortex (a region of the brain known to be affected in this disease) of normal and schizophrenic individuals obtained at autopsy. The isolated RNA was used to prepare cDNA for hybridization to the microarrays. The microarrays carried over 6000 probes for human genes. Results from the study are shown in Figure 26–18. The genes tested have been organized into clusters with similar biological functions. Some clusters have lower levels of expression in schizophrenia, whereas others have increased levels of expression. The analysis focused on 89 genes that were previously found to have altered levels of expression in schizophrenia. A cluster of genes involved in myelination had lower levels of expression in schizophrenia, while all others had increased levels of expression. This suggests that schizophrenia is associated with functional disruption in oligodendrocytes, the cells that form the myelin sheath that wraps

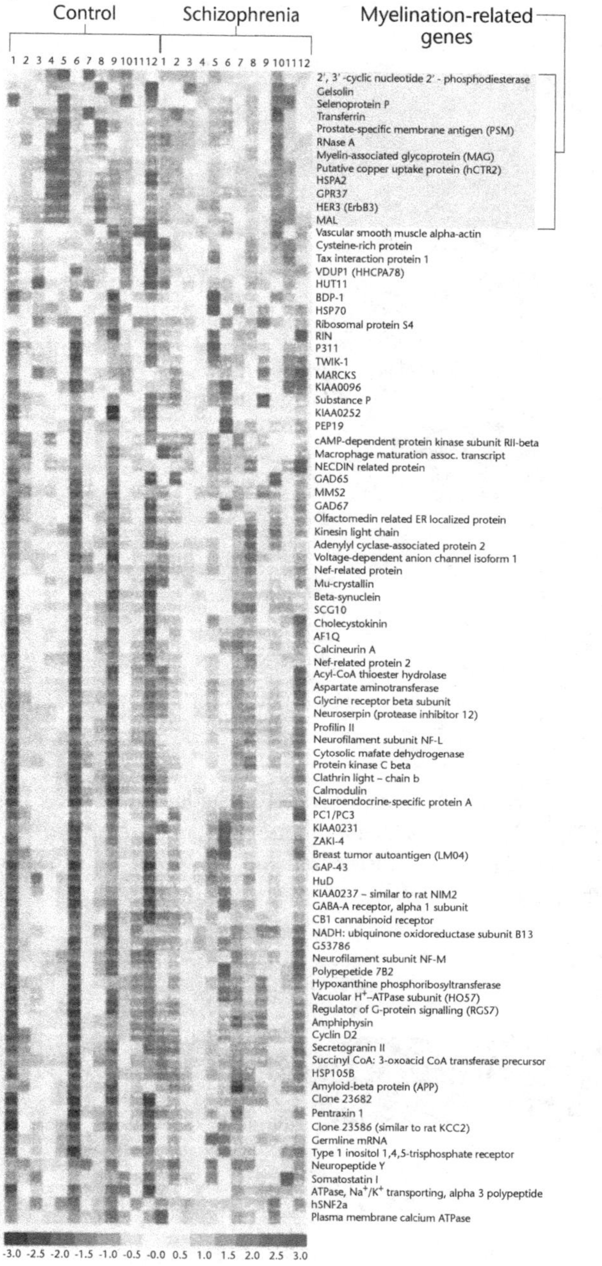


**FIGURE 26–18** Relative levels of expression for 89 genes that are differentially expressed in individuals with schizophrenia. Each column shows the gene-expression level (blue is low, red is high) in one individual, and each row shows the levels of expression for a given gene in a number of different individuals. Although a small number of genes show increases in expression, there is an overall significant reduction in expression of myelin-related and cytoskeletal genes in schizophrenic individuals compared to normal control individuals, leading to the conclusion that there may be a disruption in oligodendrocytes (the cells that make myelin) in schizophrenia.

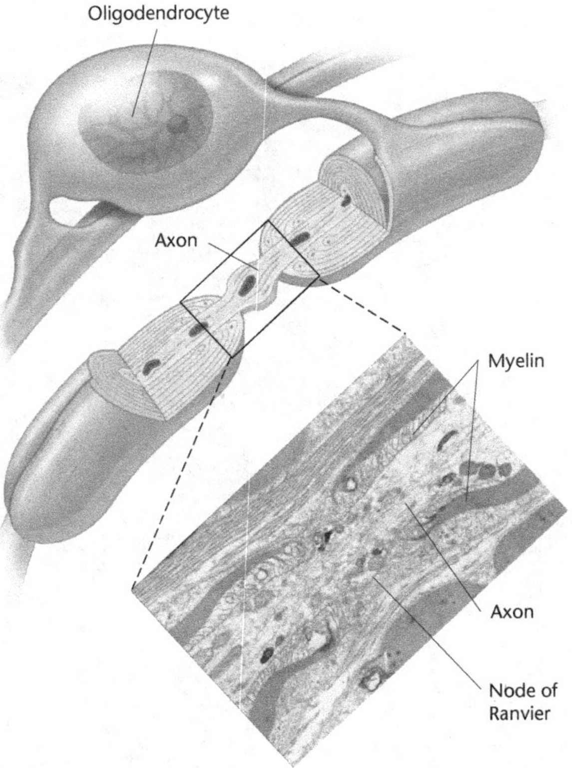


**FIGURE 26–19** Oligodendrocytes and the myelin sheath. Oligodendrocytes (left) are cells in the nervous system that synthesize and deposit the myelin sheaths that surround neurons. A cross section of a neuron (right) shows the layers of myelin. Genomic analysis indicates that abnormal myelin synthesis is associated with schizophrenia.

around nerve cells and ensures rapid conduction of nerve impulses (Figure 26–19). Myelination in the prefrontal cortex occurs during late adolescence and early adulthood, which is usually the time that the behavioral symptoms of schizophrenia first appear.

Various research groups, using over two dozen genome-wide microarray scans of gene expression in schizophrenia, have identified several sets of candidate genes whose expression is altered in affected individuals. Although the results of these studies have not been consistent, six myelin-related genes whose expression is specific to oligodendrocytes have been identified. In schizophrenia, expression of all six of these genes is reduced. The emerging evidence indicates that myelination abnormalities are an important part of the disease process in schizophrenia, but that environmental components also play a major role.

Strains of knockout mice, each missing a myelin-related gene, are being developed to connect changes in gene expression to specific behavioral phenotypes and to design therapies targeted at specific gene clusters.

## GENETICS, TECHNOLOGY, AND SOCIETY

### Genetics of Sexual Orientation

"Sex is an emotion in motion," remarked Mae West. Are we prepared to believe that genes may influence choices so personal and emotional as our sexual attraction to others? In the early 1990s, scientists who study sexual orientation found some support for genetic influences on this behavior, when they determined that identical twins are much more likely to share sexual orientation than are fraternal twins. It may be that some aspects of male sexual orientation are transmitted with the X chromosome or through a form of maternal inheritance. Researchers have posited this theory based on their finding that certain homosexual men have a larger-than-random number of homosexual relatives on the mother's side of their families. Although in 1993 some researchers announced linkage of a particular region of the X chromosome to male homosexuality, subsequent studies failed to replicate this provocative result. Thus, no specific human genes that influence sexual orientation have yet been identified.

Moreover, though the twin statistics implicate genes as a factor in sexual orientation, they simultaneously make clear that environment plays a strong role in this complex human behavior—nearly half of the gay identical twins do *not* share the homosexual orientation with their twin brother. Perhaps a glimpse into the sexual behaviors of fruit flies could provide a simpler, and less politically and emotionally charged example of how genetic pathways influence such intricate behaviors as courtship and mating.

*Drosophila melanogaster* males court females with an elaborate series of steps: dancing, singing, and licking females before attempting to mate. A gene called *fruitless* controls aspects of the courting and mating behavior. The *fruitless* gene's mRNA is spliced sex-specifically, so that males produce a different set of fruitless proteins ($Fru^M$) than do females ($Fru^F$). In both sexes of flies, the Fru proteins are expressed in cells of many sensory organs involved in courtship behavior, including the olfactory receptors and other neurons. Fru seems to be especially important in the regions where pheromones are detected.

Studies of *fruitless* mutants confirm the role of *fruitless* in mating. Males with null mutations in the gene do not court females, but instead court and attempt copulation with other males. Males who express the normal "female" spliced *fruitless* mRNA instead of the "male" version also prefer to mate with males rather than females. Furthermore, female flies that are made to express the "male" spliced *fruitless* mRNA actually reject males and court females themselves! In all three of these cases, if placed in large same-sex groups, these mutant flies will form "courtship chains," with each fly attempting to catch and mate with the fly in front of it. Thus, the splicing pattern of the *fruitless* mRNA into a male or female pattern determines the sexual behavior of the flies, regardless of their reproductive organs.

Human courtship is usually much more involved than a simple song and dance. Nonetheless we share some fundamentals of mating with flies. For one, the natural mechanism by which we pass our genetic contributions to the next generation is through heterosexual mating. If an allele promotes homosexual pairing, how is it passed on? Presumably, homosexuality genes would not be passed on to subsequent generations. However, the frequency of humans (and other animals) who show homosexual behavior is greater than can be explained without some selective advantages associated with a hypothetical "homosexual" allele or alleles.

Sergey Gavilets and William Rice recently modeled answers to this riddle, using the theoretical framework of population genetics. They evaluated two suggested explanations for the preservation and spread of homosexual alleles in populations: overdominance and sexually antagonistic selection.

In overdominance, or "heterozygote advantage," an individual homozygous for a particular allele is less reproductively fit than a wild-type homozygote, yet the *heterozygote* is actually more reproductively fit than either type of homozygote. A well-known human example of overdominance is found in sickle-cell anemia, in which a heterozygous carrier of the sickle-cell allele has a reduced chance of infection with malaria and thus possesses a survival advantage in regions where malaria is common. In the case of an allele that increases homosexual behavior, possible reproductive advantages for heterozygous bearers of the allele might be enhanced attractiveness to members of the opposite sex, boosting the odds of mating, or more viable gametes that improve the chances of a successful reproduction resulting from the mating. When Gavilets and Rice modeled an overdominance scenario involving a theoretical "gay" allele, they showed that the allele could easily be maintained in a population by overdominance, and that maintenance would be most likely if the gene were on an autosome, rather than the X chromosome.

With sexually antagonistic selection, an allele has two different, opposite effects for individuals of the two sexes. For example, a "gay" allele in a male might reduce his reproductive fitness, but the same allele expressed in a female might increase hers. Gavilets and Rice's model of sexually antagonistic selection demonstrated that it is another very plausible mechanism for preserving and spreading a homosexual-behavior-promoting allele. Interestingly, the converse chromosomal placement is favored in this scenario: Sexually antagonistic selection is much more likely to maintain an allele that increases homosexual behavior if its gene is on the X chromosome, rather than an autosome.

Although the identity of any sexual orientation-influencing gene still eludes human geneticists, these theoretical models stimulate further research. They tentatively address the apparent conundrum of how homosexual-behavior-promoting alleles might be maintained, and they bring forth specific predictions relating these mechanisms to the mode of chromosomal transmission. As we tread through the thorny territory of explaining human sexual orientation, we may be sure that tantalizing discoveries are yet to come.

### References


Kyriacou, C.P. 2005. Sex in fruitflies is *fruitless*. *Nature*. 436: 334–335.

Savolainen, V., and Lehmann, L. 2007. Genetics and bisexuality. *Nature*. 445:158–159.

Wickelgren, I. 1999. Discovery of "gay gene" questioned. *Science*. 284: 571.

## EXPLORING GENOMICS

# HomoloGene: Searching for Behavioral Genes

This chapter discussed some of the complexities of understanding behavior—in humans and other species—as a product of both genetics and environment. In this exercise, we will explore the NCBI database **HomoloGene** to search for genes implicated in behavioral conditions. HomoloGene is a database of homologs from several eukaryotic species and it contains genes that have been implicated in behavioral conditions both in animal models and in humans. Some of these are mutant genes, which, on the basis of single-nucleotide polymorphisms or linkage mapping, are thought to contribute to a behavioral phenotype. Keep in mind that correlation of a gene with a particular behavioral condition does not necessarily

*Continued on next page*

*Exploring Genomics, continued*

necessarily mean that the gene is the cause of the condition. Nor does it mean that the gene is the only one involved in the behavioral condition or that the condition is entirely due to genetics. Nonetheless, exploring HomoloGene is an interesting way to search for putative behavioral genes and learn more about them.

**Exercise I – HomoloGene**

1. Access **HomoloGene** from the ENTREZ site at http://www.ncbi.nlm.nih.gov/gquery/gquery.fcgi.
2. Search HomoloGene using the following list of terms related to behavioral conditions in humans. What human genes are listed in the top two categories of gene homologs that result from the HomoloGene search for each condition? Name these genes and identify their chromosomal loci. Notice the list of other species thought to share homologs of each gene.
   a. alcoholism
   b. depression
   c. nicotine addiction
   d. seasonal affective disorder
3. For each of the top two categories of homologs, click on the category title to open a report on that category of genes. Then use the links for individual genes to learn more about the functions of those genes. In addition to other good resources, the "Additional Links" category at the bottom of each gene report page will provide you with access to the OnLine Mendelian Inheritance in Man (OMIM) site that we have used for other exercises.
4. Search HomoloGene for any behavioral conditions you are interested in to see if the database contains genes that may be implicated in those conditions.

**Exercise II – Behavioral Gene Expression Patterns**

1. For each of the genes identified in Step 2 above, determine which tissues express the gene and speculate about why the gene may be involved in the behavioral condition in question. In the HomoloGene report for each gene, under the "Additional Links" heading, use the UniGene link or links to determine which tissues express this gene. The "Expression Profile" link under the "Gene Expression" heading of a UniGene report is one place to find this information, based on UniGene Expressed Sequence Tag (EST) data.
2. For example, in what tissues or organs is alcohol dehydrogenase (*ADH1B*) most highly expressed in humans? Based on its expression pattern, do you think this gene is responsible for the *behavior* that causes humans to drink, or is it more likely to contribute to a tolerance for alcohol in an alcoholic or heavy drinker?
3. What did you find when you looked at the UniGene Expression Profile for the *CYP2A6* gene involved in nicotine addiction?

## Chapter Summary

1. Behavioral genetics has emerged as an important specialty within genetics because both the genotype and environment have been found to have an impact in determining an organism's behavioral response.
2. Historically, two approaches have been used in studying the genetic control of behavior: (1) the artificial bidirectional selection of behavior in an organism to develop strains with high and low responses and (2) mutagenesis to identify and isolate single genes that have an effect on a behavioral response.
3. Studies done on alcohol preference and open-field behavior in mice illustrate behaviors strongly influenced by the genotype.
4. Studies of maze learning in rats and geotaxis in *Drosophila* have successfully established both bright and dull lines of rats and positively and negatively geotaxic *Drosophila*. These results have led researchers to identify certain genes on specific chromosomes as being associated with these behaviors, and to calculate their numbers and relative contributions.
5. Using *Drosophila* as a model organism, researchers have been able to isolate mutations that cause deviations from normal behavior and have then established the role of the corresponding wild-type allele in the normal response. The numerous examples include genes responsible for courtship, phototaxis, and the functional architecture of the nervous system.
6. Many aspects of human behavior are difficult to study, since the individual's environment makes an important contribution to trait development. In humans, although multifactorial traits such as schizophrenia have genetic components, studies have shown that expression of these traits may be modified by the environment.

## INSIGHTS AND SOLUTIONS

1. Bipolar disorder is a mental illness associated with pervasive and wide mood swings. It is estimated that 1 in 100 individuals has or will suffer from this disorder at least once in his or her lifetime. Genetic studies indicate that bipolar disorder is familial and that mutations in several genes are involved.

   In an attempt to identify genes associated with bipolar disorder (this condition was formerly called manic depression), researchers began by using conventional genetic analysis. In 1987, two separate studies using pedigree analysis, RFLP analysis, and a variety of other genetic markers reported a linkage between manic depression and, in one report, markers on the X chromosome and, in the other report, markers on the short arm of chromosome 11. At the time, these reports were hailed as landmark discoveries. It was thought that genes controlling this behavioral disorder could be identified, mapped, and sequenced in much the same way as genes for cystic fibrosis, neurofibromatosis, and other disorders. It was hoped that this discovery would open the way to the development of therapeutic strategies based on knowledge of the nature of the gene product and its mechanism of action. Soon after, however, further work on the same populations (reported in 1989 and 1990) demonstrated that the original results were invalid after individuals in the study who did not carry the markers subsequently developed bipolar disorder. As a result, it became clear that no linkage to markers on the X chromosome or to markers on chromosome 11 could be established. These findings do not exclude the role of major genes on the X chromosome and chromosome 11 in bipolar disorder, but they do exclude linkage to the markers used in the original reports.

This setback not only caused embarrassment and confusion, but also forced a reexamination of the validity of the newly developed recombinant DNA-based methods used in mapping other human genes and of the analysis of data from linkage studies involving complex behavioral traits. Several factors have been proposed to explain the flawed conclusions reported in the original studies. What do you suppose some of these factors to be?

**Solution:** While some of the criticisms were directed at the choice of markers, most of the factors at work in this situation appear to be related to the phenotype of manic depression. At least three confounding factors have been identified. One factor relates to the diagnosis of bipolar disorder itself. The phenotype is complex and not as easily quantified as height or weight. In addition, mood swings are a universal part of everyday life, and it is not always easy to distinguish transient mood alterations and the role of environmental factors from disorders having a biological or genetic basis (or both). A second confounding factor is that, in the populations studied, there may in fact be more than one major X-linked gene and more than one major autosomal gene that trigger bipolar disorder. A third factor is age of onset. There is a positive correlation between age and the appearance of the symptoms of bipolar disorder. Therefore, at the time of a pedigree study, younger individuals who will be affected later in life may not show any signs of bipolar disorder. Taken together, these factors point up the difficulty in researching the genetic basis of complex behavioral traits. Individually or in combinations, the factors described might skew the results enough so that the guidelines for proof that are adequate for other traits are not stringent enough for behavioral traits with complex phenotypes and complex underlying causes.

## Problems and Discussion Questions

1. Contrast the two principal approaches used in studying behavioral genetics. What are the advantages of each method with respect to the type of information gained?
2. Assume that you discovered a fruit fly that walked with a limp and was continually off balance as it moved. Describe how you would determine whether this behavior was due to an injury (induced in the environment) or to an inherited trait. Assuming that it was inherited, what are the various ways a gene might lead to this trait? Describe how you would determine the mechanism of gene expression experimentally if the gene were X-linked.
3. Certain inbred strains of mice (i.e., C57BL and C3H/2) appear to exhibit a preference for alcohol in free-choice consumption experiments, and a variety of techniques have been applied to assess underlying genetic predispositions. How might the analysis of such inbred strains provide insight into molecular mechanisms that drive such a behavior? Tarantino et al. (1998) applied QTL analysis to alcohol preference in mice using microsatellite markers. They found three significant QTLs on chromosomes 1, 4, and 9 and three suggestive QTLs on chromosomes 2, 3, and 10. How might QTL analysis lead to an understanding of alcohol preference in mice? Summarize the strengths and weaknesses of each approach.
4. Mosaic analysis using an unstable ring-X chromosome has been successfully applied to determine the primary focus of a number of X-linked behavioral genes. Konopka et al. (1983) used gynandromorphic analysis to examine the X-chromosome gene *short-period* (*per$^s$*) circadian clock mutation and determined the focus to be located in the brain. Interestingly, some of the flies with male/female heads, as determined by cuticular markers, had circadian activity patterns with both male and female components. What does this suggest about the functional parameters of the *per$^s$* gene?
5. In humans, the chemical phenylthiocarbamide (PTC) is either tasted or not. When the offspring of various combinations of taster and nontaster parents are examined, the following data are obtained:

| Parents | Offspring |
|---|---|
| Both tasters | All tasters |
| Both tasters | 1/2 tasters<br>1/2 nontasters |
| Both tasters | 3/4 tasters<br>1/4 nontasters |
| One taster<br>One nontaster | All tasters |
| One taster<br>One nontaster | 1/2 tasters<br>1/2 nontasters |
| Both nontasters | All nontasters |

Based on these data, how is PTC tasting behavior inherited?

6. Jerry Hirsch and colleagues have used the behavior-first approach to study the genes responsible for geotaxis in *Drosophila*. Their elegant genetic analysis has shown that this behavior is under genetic control, and they estimate the number of genes responsible for geotactic behavior. Although the approach has been successful, what are the limitations of the behavior-first approach as used here? What steps did Hirsch and his colleagues take to overcome those limitations? Has this approach been successful?
7. Various approaches have been applied to study the genetics of problem and pathological gambling (PG), and within-family vulnerability has been well documented. However, family studies, while showing clusters within blood relatives, cannot separate genetic from environmental influences. Eisen (2001) applied "twin studies" using 3359 twin pairs from the Vietnam-era Twin Registry and found that a substantial portion of the variance associated with PG can be attributed to inherited factors. How might twin studies be used to distinguish environmental from genetic factors in complex behavioral traits such as PG?

8. *Caenorhabditis elegans* has become a valuable model organism for the study of development and genetics for a variety of reasons. The developmental fate of each cell (1031 in males and 959 in hermaphrodites) has been mapped. *C. elegans* has only 302 neurons whose pattern of connectivity is known, and it displays a variety of interesting behaviors including chemotaxis, thermotaxis, and mating behavior, to name a few. In addition, isogenic lines have been established. What advantage would the use of *C. elegans* have over other model organisms in the study of animal behavior? What are likely disadvantages?
9. Using the behavioral phenotypes of a series of 18 mutants falling into five phenotypic classes, Thomas (1990) described the genetic program that controls the cyclic defecation motor program in *Caenorhabditis elegans*. The first step in each cycle is contraction of the posterior body muscles, followed by contraction of the anterior body muscles. Finally, anal muscles open and expel the intestinal contents. Below is a list of selected mutants (simplified) that cause defects in defecation.

| Gene | Phenotype |
|---|---|
| *exp* | failure to open the anus and expel intestinal contents |
| *unc* | failure to contract anterior body muscles |
| *pbo* | failure to contract posterior body muscles |
| *aex* | failure to open the anus and expel intestinal contents and failure to contract anterior body muscles |
| *cha* | alteration of cycle periodicity, all other components being completely functional |

Present a simple schematic that illustrates genetic involvement in *C. elegans* defecation. Account for the action of the *aex* gene and speculate on the involvement and placement of the *cha* gene.

10. Discuss why the study of human behavior genetics has lagged behind that of other organisms.
11. J. P. Scott and J. L. Fuller studied 50 traits in five pure breeds of dogs. Almost all the traits varied significantly in the five breeds, but very few bred true in crosses. What can you conclude with respect to the genetic control of these behavioral traits?

### HOW DO WE KNOW?

12. In this chapter we focused on how genes that control the development, structure, and function of the nervous system and interactions with environmental factors produce behavior. At the same time, we found many opportunities to consider the methods and reasoning by which much of this information was acquired. From the explanations given in the chapter, what answers would you propose to the following fundamental questions:
    (a) How do studies of alcohol preference in strains of mice indicate that this behavior has a genetic basis?
    (b) What experimental evidence is available to show that maze behavior in rats is a selectable trait?
    (c) How does the work on geotaxis in *Drosophila* unify the behavior-first and the gene-first approaches to the study of behavior?
    (d) What genetic evidence shows that certain behaviors in *Drosophila* are controlled by specific brain regions?
    (e) How do we know that defects in the structure of the nervous system are associated with schizophrenia?

## Extra-Spicy Problems

13. Although not discussed in this chapter, *C. elegans* is a model system whose life cycle makes it an excellent choice for the genetic dissection of many biological processes. *C. elegans* has two natural sexes: hermaphrodite and male. The hermaphrodite is essentially a female that can generate sperm as well as oocytes, so reproduction can occur by hermaphrodite self-fertilization or hermaphrodite–male mating. In the context of studying mutations in the nervous system, what is the advantage of hermaphrodite self-fertilization with respect to the identification of recessive mutations and the propagation of mutant strains?
14. Hypothetical data concerning the genetic effect of *Drosophila* chromosomes on geotaxis are shown here:

| | Chromosome | | |
|---|---|---|---|
| | *X* | *2* | *3* |
| Positive geotaxis | +0.2 | +0.1 | +3.0 |
| Unselected | +0.1 | −0.2 | +1.0 |
| Negative geotaxis | −0.1 | −2.6 | +0.1 |

What conclusions can you draw?

15. In July 2006, a population of flies, *Drosophila melanogaster*, rode the space shuttle *Discovery* to the International Space Station (ISS) where a number of graviperception experiments and observations are to be conducted over a nine-generation period. Eventually, frozen specimens will be collected by astronauts and returned to Earth. Researchers expect to correlate behavioral and physiological responses to microgravity with changes in gene activity by analyzing RNA and protein profiles. The title of the project is "*Drosophila* Behavior and Gene Expression in Microgravity." If you were in a position to conduct three experiments on the behavioral aspects of these flies, what would they be? How would you go about assaying changes in gene expression in response to microgravity? Given that humans share over half of the genome and proteins of *Drosophila*, how would you justify the expense of such a project in terms of improving human health?
16. Of a variety of investigational approaches that have been applied to the study of schizophrenia, two separate distinct approaches, twin studies and microarray analysis, have provided strong support for a genetic component.
    (a) Provide a summary of the contribution that each has played in understanding schizophrenia.
    (b) Speculate on the value that microarray analysis using samples from monozygotic and dizygotic twins might have for enhancing research on schizophrenia.
17. An interesting and controversial finding by Wedekind and colleagues (1995) suggested that a sense of smell, flavored with various HLA (Human Leukocyte Antigen) haplotypes, plays a role in mate selection in humans. Women preferred T-shirts from men with HLA haplotypes

unlike their own. The HLA haplotypes are components of the Major Histocompatibility Complex (MHC) and are known to have significant immunological functions. In addition to humans, mice and fish also condition mate preference on MHC constitution. The three-spined stickleback (*Gasterosteus aculeatus*) has recently been studied to determine whether females use an MHC odor-based system to select males as mates. Examine the following data (modified from Milinski et al. 2005) and provide a summary conclusion. Why might organisms evolve a mate selection scheme for assessing and optimizing MCH diversity?

| | Male with optimal MHC | Male with non-optimal MHC |
|---|---|---|
| Gravid female time spent (seconds) | 350 | 250 |
| Number spawning females | 11 | 3 |

Odor-based choices made by gravid females for males with optimal MHC alleles and males with non-optimal MHC alleles. Gravid females chose different amounts of time (seconds) for exposure to males with optimal MHC alleles versus those with non-optimal MHC alleles. The number of females spawning when exposed to different males with different MHC complements is also presented.

These lady-bird beetles, from the Chiricahua Mountains in Arizona, show considerable phenotypic variation in the number, size, and distribution of spots.

27

# Population Genetics

## CHAPTER CONCEPTS

- Allele frequencies in population gene pools can vary over time.
- The relationship between allele frequencies and genotype frequencies in an ideal population is described by the Hardy–Weinberg law.
- Natural selection is a major force driving allele frequency change and population evolution.
- Migration and drift can also cause changes in allele frequency.
- Nonrandom mating changes population genotype frequency but not allele frequency.
- Mutation creates new alleles in a population gene pool.

At the time Alfred Russel Wallace and Charles Darwin first identified natural selection as the mechanism of adaptive evolution in the mid-nineteenth century, there was no accurate model of the mechanisms responsible for variation and inheritance. Gregor Mendel published his work on the inheritance of traits in 1866, but it received little notice at the time. The rediscovery of Mendel's work in 1900 began a 30-year effort to reconcile Mendel's concept of genes and alleles with the theory of evolution. In a key insight, biologists realized that the frequency of phenotypic traits in a population is linked to the relative abundance of the alleles influencing those traits. This insight laid the foundation for **population genetics**—the study of genetic variation in populations and how it changes over time. Population geneticists investigate patterns of **genetic variation** within and among groups of interbreeding individuals. Because changes in genetic structure form the basis for evolution of a population, population genetics has become an important subdiscipline of evolutionary biology.

Early in the twentieth century a number of workers, including G. Udny Yule, William Castle, Godfrey Hardy, and Wilhelm Weinberg, formulated the basic principles of population genetics. Theoretical geneticists Sewall Wright, Ronald Fisher, and J.B.S. Haldane developed mathematical models to describe the genetic structure of a population. More recently, field researchers have been able to test these models using biochemical and molecular techniques that measure variation directly at the protein and DNA levels. These experiments examine allele frequencies and the forces that act on them, such as selection, mutation, migration, and genetic drift. In this chapter, we will consider these general aspects of population genetics and discuss the relationship of population genetics to evolution.

## 27.1 Allele Frequencies in Population Gene Pools Vary in Space and Time

A **population** is a group of individuals who share a common set of genes, live in the same geographic area, and actually or potentially interbreed. All the alleles shared by these individuals constitute the **gene pool** for the population. Often when we examine a single genetic locus in a population, we find that combinations of the alleles at this locus result in individuals with different genotypes. In studying a population, geneticists face three important tasks: computing the frequencies of various alleles in the gene pool, the frequencies of different genotypes in the population, and the changes in frequency that occur from one generation to the next. Population geneticists use these calculations to address questions such as, How much genetic variation is present in a population? Are genotypes randomly distributed in time and space, or is there a pattern in their distribution? What processes affect the composition of a population's gene pool? Do these processes produce genetic divergence among populations?

Populations are dynamic; they expand and contract through changes in birth and death rates, migration, or contact with other populations. Often, some individuals within a population will produce more offspring than others, contributing a disproportionate amount of their alleles to the next generation. Thus the dynamic nature of a population can, over time, lead to changes in the population's gene pool.

## 27.2 The Hardy–Weinberg Law Describes the Relationship between Allele Frequencies and Genotype Frequencies in an Ideal Population

The theoretical relationship between the relative proportions of alleles in the gene pool and the frequencies of different genotypes in the population was elegantly described in the early 1900s in a mathematical model developed independently by the British mathematician Godfrey H. Hardy and the German physician Wilhelm Weinberg. This model, called the **Hardy–Weinberg law**, describes what happens to alleles and genotypes in an "ideal" population, meaning one that is infinitely large, is not subject to any evolutionary forces such as mutation, migration, or selection, and in which mates are selected randomly. Under these conditions the Hardy–Weinberg model makes two predictions:

1. The frequencies of the alleles in the gene pool do not change over time.
2. If a locus may be occupied by either of two alleles, $A$ or $a$, then after one generation of random mating, the frequencies of the resulting genotypes $AA$, $Aa$, and $aa$ in the population can be represented by

$$p^2 + 2pq + q^2 = 1$$

where $p$ = frequency of allele $A$ and $q$ = frequency of allele $a$.

A population that meets the Hardy–Weinberg criteria and in which the frequencies $p$ and $q$ of two alleles at a locus do result in the predicted genotypic frequencies is said to be in Hardy–Weinberg equilibrium. As we will see later in this chapter, it is rare for a real population to conform totally to the Hardy–Weinberg model and for all allele and genotype frequencies to remain unchanged for generation after generation.

The Hardy–Weinberg model uses Mendelian principles of segregation and simple probability to explain the relationship between allele and genotype frequencies in a population. To demonstrate this relationship, let's consider the hypothetical example of a single locus with two alleles, *A* and *a*, in a population where the frequency of *A* is 0.7 and the frequency of *a* is 0.3. This means that 70 percent of the copies of the gene in the population are present as the dominant allele *A* and 30 percent are present as the recessive allele *a*. Note that $0.7 + 0.3 = 1$, indicating that all (100 percent) of the alleles for gene *A* present in the gene pool are accounted for. We assume that individuals mate randomly, following Hardy–Weinberg requirements, so for any one zygote the probability that the female gamete will contain *A* is 0.7, and the probability that the male gamete will contain *A* is also 0.7. The probability that *both* gametes will contain *A* is $0.7 \times 0.7 = 0.49$. Thus we predict that genotype *AA* will occur 49 percent of the time. The probability that a zygote will be formed from a female gamete carrying *A* and a male gamete carrying *a* is $0.7 \times 0.3 = 0.21$, and the probability of a female gamete carrying *a* being fertilized by a male gamete carrying *A* is $0.3 \times 0.7 = 0.21$. The frequency of genotype *Aa* is therefore $0.21 + 0.21 = 0.42 = 42$ percent. Finally, the probability that a zygote will be formed from two gametes carrying *a* is $0.3 \times 0.3 = 0.09$, so the frequency of genotype *aa* is 9 percent. As a check on our calculations, note that $0.49 + 0.42 + 0.09 = 1.0$, confirming that we have accounted for all of the zygotes. These calculations are summarized in Figure 27–1.

POPULATION GENETICS WEB TUTORIAL 27.1

We started with the frequency of a particular allele in a specific gene pool and calculated the probability that certain zygote genotypes would be produced from this pool. When the zygotes develop into adults and reproduce, what will be the frequency distribution of alleles in the new gene pool? Under the Hardy–Weinberg law, we assume that all genotypes have equal rates of survival and reproduction. This means that in the next generation, all genotypes contribute equally to the new gene pool. The *AA* individuals constitute 49 percent of the population, and we can predict that the gametes they produce will constitute 49 percent of the gene pool. These gametes all carry allele *A*. Similarly, *Aa* individuals constitute 42 percent of the population, so we predict that their gametes will make up 42 percent of the new gene pool. Half (0.5) of these gametes will carry allele *A*. Thus, the frequency of allele *A* in the gene pool is $0.49 + (0.5)0.42 = 0.7$. The other half of the gametes produced by *Aa* individuals will carry allele *a*. The *aa* individuals constitute 9 percent of the population, so their gametes will constitute 9 percent of the new gene pool. These gametes all carry allele *a*. Thus, we can predict that the allele *a* in the new gene pool is $(0.5)0.42 + 0.09 = 0.3$. As a check on our calculation, note that $0.7 + 0.3 = 1.0$, accounting for all of the gametes in the gene pool of the new generation. We have arrived back at the beginning, with a gene pool in which the frequency of allele *A* is 0.7 and the frequency of allele *a* is 0.3.

To write the general equation for the Hardy–Weinberg law as it applies to two alleles (let's keep calling them *A* and *a*), we use variables instead of numerical values for the allele frequencies. The frequency of one allele (say, *A*) is represented by the variable *p*, and the frequency of the other allele (*a*) is represented by the variable *q*, such that $p + q = 1$. If we randomly draw male and female gametes from the gene pool and pair them to make a zygote, the probability that both will carry allele *A* is $p \times p$. Thus, the frequency of genotype *AA* among the zygotes is equal to $p^2$. The probability that the female gamete carries *A* and the male gamete carries *a* is $p \times q$, and the probability that the female gamete carries *a* and the male gamete carries *A* is $q \times p$. Thus, the frequency of genotype *Aa* among the zygotes is $2pq$. Finally, the probability that both gametes carry *a* is $q \times q$, making the frequency of genotype *aa* among the zygotes $q^2$. Therefore, the distribution of genotypes among the zygotes is

$$p^2 + 2pq + q^2 = 1$$

This logic is summarized in Figure 27–2.

| Eggs \ Sperm | fr(*A*) = 0.7 | fr(*a*) = 0.3 |
|---|---|---|
| fr(*A*) = 0.7 | fr(*AA*) = 0.7 × 0.7 = 0.49 | fr(*Aa*) = 0.7 × 0.3 = 0.21 |
| fr(*a*) = 0.3 | fr(*aA*) = 0.3 × 0.7 = 0.21 | fr(*aa*) = 0.3 × 0.3 = 0.09 |

**FIGURE 27–1** Calculating genotype frequencies from allele frequencies. Gametes represent withdrawals from the gene pool to form the genotypes of the next generation. In this population, the frequency of the *A* allele, or fr(*A*), is 0.7, and the frequency of the *a* allele, or fr(*a*), is 0.3. The frequencies of the genotypes in the next generation are calculated as 0.49 for *AA*, 0.42 for *Aa*, and 0.09 for *aa*. Under the Hardy–Weinberg law, the frequencies of *A* and *a* remain constant from generation to generation.

| Eggs \ Sperm | fr(*A*) = *p* | fr(*a*) = *q* |
|---|---|---|
| fr(*A*) = *p* | fr(*AA*) = $p^2$ | fr(*Aa*) = *pq* |
| fr(*a*) = *q* | fr(*aA*) = *qp* | fr(*aa*) = $q^2$ |

**FIGURE 27–2** The general description of allele and genotype frequencies under Hardy–Weinberg assumptions. The frequency of allele *A* is *p*, and the frequency of allele *a* is *q*. Random mating produces the three genotypes *AA*, *Aa*, and *aa* in the frequencies $p^2$, $2pq$, and $q^2$, respectively.

These calculations demonstrate the two main predictions of the Hardy–Weinberg law: allele frequencies in our hypothetical population do not change from one generation to the next, and genotype frequencies after one generation of random mating can be predicted from the allele frequencies. In other words, this population does not change or evolve with respect to the locus we have examined. Remember, however, the assumptions about the ideal population described by the Hardy–Weinberg model:

1. Individuals of all genotypes have equal rates of survival and equal reproductive success—that is, there is no selection.
2. No new alleles are created and no old alleles are lost by mutation.
3. Individuals do not migrate into or out of the population.
4. The population is infinitely large, which in practical terms means it is large enough that sampling errors and other random effects are negligible.
5. Individuals in the population mate randomly.

These assumptions are what make the Hardy–Weinberg law so useful in population genetic research. By specifying the conditions under which the population cannot evolve, the Hardy–Weinberg model identifies the real-world forces that cause allele frequencies to change. In other words, by holding certain conditions constant, Hardy–Weinberg isolates the forces of evolution and allows them to be quantified. Application of the Hardy–Weinberg model can also reveal "neutral genes" in a population gene pool—genes that are not being operated on by the forces of evolution.

The Hardy–Weinberg law has three additional important consequences. First, it shows that dominant traits do not necessarily increase from one generation to the next. Second, it demonstrates that genetic variation can be maintained in a population, since, once established in an ideal population, allele frequencies remain unchanged. Third, if we invoke Hardy–Weinberg assumptions, then knowing the frequency of just one genotype for a locus enables us to calculate the frequencies of all other genotypes at that locus. This is particularly useful in human genetics because we can calculate the frequency of heterozygous carriers for recessive genetic disorders even when all we know is the frequency of affected individuals. (See Section 27.4.)

**NOW SOLVE THIS**

Problem 1 on page 734 asks you to calculate frequencies of a dominant and a recessive allele in a population where you know the phenotype frequencies.

■ **HINT:** *Determine which allele frequency (*p *for the dominant allele or* q *for the recessive allele) you must estimate first when homozygous dominant and heterozygous genotypes have the same phenotype. Then calculate the frequency of the other allele before using the Hardy–Weinberg equation to calculate genotype frequencies.*

27.3

## The Hardy–Weinberg Law Can Be Applied to Human Populations

To show how allele frequencies are measured in a real population and to demonstrate the kind of questions asked by population genetics researchers, we will take an example from research into genetic factors that influence an individual's susceptibility to infection by HIV-1, the virus responsible for most AIDS cases worldwide. Of particular interest to AIDS researchers are the small number of individuals who practice high-risk behaviors (such as having unprotected sex with HIV-positive partners) and yet remain uninfected. In 1996, Rong Liu and colleagues discovered that two such exposed but uninfected individuals were homozygous for a mutant allele of a gene called *CC-CKR5*.

### Calculating an Allele's Frequency

The *CC-CKR5* gene, located on chromosome 3, encodes a protein called the C–C chemokine receptor-5, often abbreviated CCR5. Chemokines are signaling molecules associated with the immune system. When chemokines bind at the CCR5 receptor proteins on the surface of white blood cells, the cells respond by moving into inflamed tissues to fight infection.

The CCR5 protein is also a receptor for strains of HIV-1. A protein called Env (short for envelope protein) on the surface of HIV-1 binds to a glycoprotein called the CD4 protein on the surface of a prospective host cell. Binding to CD4 causes Env to change shape and form a second binding site. This second site binds to CCR5, which, in turn, initiates the fusion of the viral protein coat with the host-cell membrane. The merging of the viral envelope with the cell membrane transports the HIV viral core into the host cell's cytoplasm.

The mutant allele of the *CCR5* gene contains a 32-bp deletion in one of its coding regions. As a consequence, the protein encoded by the mutant allele is incomplete and nonfunctional. The abnormal receptor protein never makes it to the cell membrane, and without it, HIV-1 cannot enter the cell. The gene's normal allele is called *CCR51* (we will designate this allele *1*), and the allele with the 32-bp deletion is called *CCR51-Δ32* (we will designate this allele *Δ32*). The two uninfected individuals described by Liu both had the genotype *Δ32/Δ32*. As a result, they had no CCR5 receptors on the surface of their cells and were resistant to infection by strains of HIV-1 that require CCR5 as a co-receptor. Curiously, researchers have not discovered any adverse effects associated with the *Δ32/Δ32* genotype.

Unfortunately, homozygosity for the *Δ32* mutation does not provide complete resistance to infection. At least three *Δ32/Δ32* individuals infected with HIV-1 have been found. Heterozygotes with genotype *1/Δ32* are susceptible to HIV-1 infection, but evidence suggests that their infections progress more slowly to AIDS.

TABLE 27.1

**CCR5 Genotypes and Phenotypes**

| Genotype | Phenotype |
|---|---|
| *1/1* | Susceptible to sexually transmitted strains of HIV-1 |
| *1/Δ32* | Susceptible but may progress to AIDS slowly |
| *Δ32/Δ32* | Resistant to most sexually transmitted strains of HIV-1 |

Table 27.1 summarizes the genotypes possible at the *CCR5* locus and the phenotypes associated with each.

The fact that the *CCR5-Δ32* allele exists and provides some protection against AIDS generates two important questions: Which human populations harbor the *Δ32* allele, and how common is it? To address these questions, several teams of researchers surveyed a large number of people from a variety of populations. Genotypes were determined by direct analysis of DNA (Figure 27–3) and then used to calculate the allele frequencies. For example, in one sample population of 100 French individuals from a survey in Brittany, 79 of the individuals have genotype *1/1*, 20 have genotype *1/Δ32*, and 1 has genotype *Δ32/Δ32*. In this sample, 158 *1* alleles are carried by the *1/1* individuals, and 20 *1* alleles are carried by the *1/Δ32* individuals, for a total of 178 *1* alleles in the sample. The frequency of the *CCR51* allele in the sample population is thus 178/200 = 0.89 = 89 percent. The same sample contains 20 *CCR5-Δ32* alleles carried by *1/Δ32* individuals plus 2 *CCR5-Δ32* alleles carried by the *Δ32/Δ32* individual, for a total of 22. The frequency of the *CCR5-Δ32* allele is thus 22/200 = 0.11 = 11 percent. Notice that the frequencies of both alleles added together (0.89 + 0.11) equal 1, or $p + q = 1$, which confirms that we have accounted for all the alleles for this gene in the entire gene pool. Table 27.2 shows two methods for computing the frequencies of the *1* and *Δ32* alleles in the Brittany sample.

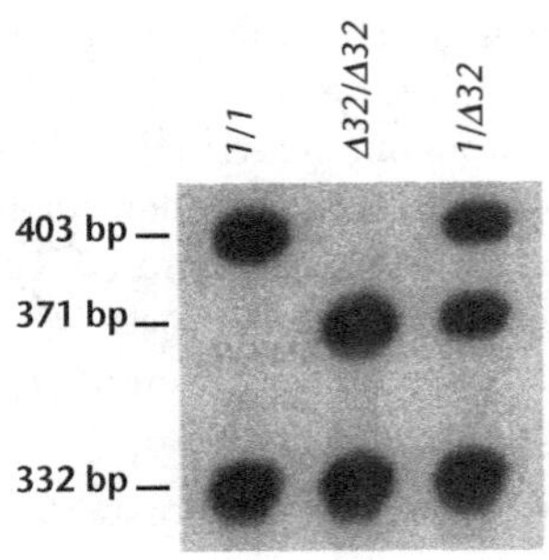


FIGURE 27–3 Allelic variation in the *CCR5* gene. Michel Samson and colleagues used PCR to amplify a part of the *CCR5* gene containing the site of the 32-bp deletion, cut the resulting DNA with a restriction enzyme, and ran the fragments on an electrophoresis gel. Each lane reveals the genotype of a single individual. The *1* allele produces a 332-bp fragment and a 403-bp fragment; the *Δ32* allele produces a 332-bp fragment and a 371-bp fragment. Heterozygotes produce three bands.

Figure 27–4 shows the frequency of the *CCR5-Δ32* allele in each of the 18 European populations surveyed. The studies show that populations in northern Europe around the Baltic Sea have the highest frequencies of the *Δ32* allele. There is a sharp gradient in allele frequency from north to south, and populations in southern Europe have very low frequencies. In populations without European ancestry, the *Δ32* allele was found to be essentially absent. The clearly recognizable pattern in the global distribution of the *Δ32* allele presents an evolutionary puzzle that we will return to later in the chapter.

Can we expect the *CCR5-Δ32* allele to increase in human populations in which it is currently rare? From what we now know about the Hardy–Weinberg law, we can say that if (1) individuals of all genotypes have equal rates of survival and reproduction, (2) there is no mutation, (3) there is no migration into or out of the population that alters allele frequencies, (4) the population is extremely large, and (5) the individuals in the population choose their mates randomly with respect to the locus of interest, then the frequency of the *Δ32* allele in human populations will not change. However, as we shall see in the last half of this chapter, when the assumptions of the Hardy–Weinberg law are not met—because of natural selection, mutation, migration, or genetic drift—the allele frequencies in a population may change from one generation to the next. Nonrandom mating does not, by itself, alter *allele* fre-

TABLE 27.2

**Methods of Determining Allele Frequencies from Data on Genotypes**

**A. Counting Alleles**

| | Genotype | | | |
|---|---|---|---|---|
| | *1/1* | *1/Δ32* | *Δ32/Δ32* | Total |
| Number of individuals | 79 | 20 | 1 | 100 |
| Number of 1 alleles | 158 | 20 | 0 | 178 |
| Number of *Δ32* alleles | 0 | 20 | 2 | 22 |
| Total number of alleles | 158 | 40 | 2 | 200 |

Frequency of *CCR51* in sample: 178/200 = 0.89 = 89%

Frequency of *CCR5-Δ32* in sample: 22/200 = 0.11 = 11%

**B. From Genotype Frequencies**

| | Genotype | | | |
|---|---|---|---|---|
| | *1/1* | *1/Δ32* | *Δ32/Δ32* | Total |
| Number of individuals | 79 | 20 | 1 | 100 |
| Genotype frequency | 79/100 = 0.79 | 20/100 = 0.20 | 1/100 = 0.01 | 1.00 |

Frequency of *CCR51* in sample: 0.79 + (0.5)0.20 = 0.89

Frequency of *CCR5-Δ32* in sample: (0.5)0.20 + 0.01 = 0.11

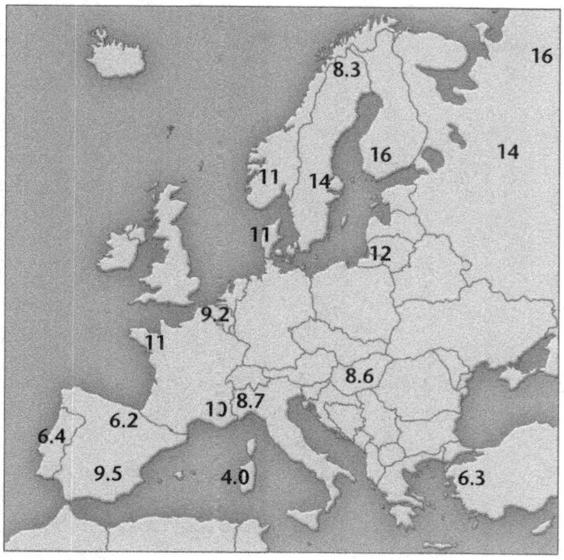


**FIGURE 27-4** The frequency (percentage) of the *CCR5-Δ32* allele in 18 European populations.

quencies, but by altering *genotype* frequencies, it indirectly affects the course of evolution. We return to the *CCR5-Δ32* allele later in the chapter to see how a population genetics perspective has generated fruitful hypotheses for further research.

## Testing for Hardy–Weinberg Equilibrium

As mentioned in Section 27.2, the Hardy–Weinberg law tells geneticists where to look to find the causes of evolution in populations. One way to establish whether one or more of the Hardy–Weinberg assumptions do not hold in a given population (and thus to establish the reason for any change in the population's genetic structure) is by determining whether the population's genotypes are in equilibrium. To do so, we first determine the frequencies of the genotypes, either directly from the phenotypes (if heterozygotes are recognizable) or by analyzing proteins or DNA sequences. We then calculate the allele frequencies from the genotype frequencies, as demonstrated earlier. Finally, we use the allele frequencies in the parental generation to predict the next generation's genotype frequencies. If the genotype frequencies do not fit the $p^2 + 2pq + q^2 = 1$ relationship predicted by the Hardy–Weinberg law, then one or more of the assumptions are invalid for the population in question.

We will use the *CCR5* genotypes of a population in Britain to demonstrate the Hardy–Weinberg law. The population consists of 283 individuals, 223 of whom have genotype *1/1*, 57 genotype *1/Δ32*, and 3 genotype *Δ32/Δ32*. These numbers represent genotype frequencies of 223/283 = 0.788, 57/283 = 0.201, and 3/283 = 0.011, respectively. From the genotype frequencies, we compute the *CCR51* allele frequency as 0.89 (89%) and the frequency of the *CCR5-Δ32* allele as 0.11 (11%). We can apply these allele frequencies to the Hardy–Weinberg law to determine whether this population is in equilibrium. Using $p$ to represent the frequency of *1* and $q$ to represent the frequency of *Δ32*, we find that the allele frequencies predict the genotype frequencies as follows:

Expected frequency of genotype *1/1*:

$$1/1 = p^2 = (0.89)^2 = 0.792$$

Expected frequency of genotype *1/Δ32*:

$$1/\Delta 32 = 2pq = 2(0.89)(0.11) = 0.196$$

Expected frequency of genotype *Δ32/Δ32*:

$$\Delta 32/\Delta 32 = q^2 = (0.11)^2 = 0.012$$

These expected frequencies are nearly identical to the observed frequencies. Our test of this population has thus failed to provide evidence that Hardy–Weinberg assumptions are being violated. That conclusion is confirmed by a $\chi^2$ analysis, as described in Chapter 3. The $\chi^2$ value in this case is tiny: 0.00023. To reject the null hypothesis at even the most generous accepted level, $p = 0.05$, the $\chi^2$ value would have to be 3.84. (In a test for Hardy–Weinberg equilibrium, the degrees of freedom are given by $k - 1 - m$, where $k$ is the number of genotypes and $m$ is the number of independent allele frequencies estimated from the data. Here, $k = 3$ and $m = 1$, because calculating only one allele frequency allows us to determine the other by subtraction. Thus, we have $3 - 1 - 1 = 1$ degree of freedom.)

On the other hand, if the Hardy–Weinberg test had demonstrated that the population is not in equilibrium, it would indicate that one or more assumptions were not being met. A recently studied example of a real human population that is not in Hardy–Weinberg equilibrium is a small, isolated population of Native Americans in the American Southwest. Researchers determined the genotype of each individual in this population at two loci affecting the major histocompatibility complex (MHC). The two genes, *HLA-A* and *HLA-B*, encode proteins involved in the immune system's discrimination between self and nonself. The immune systems of individuals heterozygous at MHC loci appear to recognize a greater diversity of foreign invaders and thus may be better able to fight disease. The results of the survey showed significantly more individuals who are heterozygous at both loci and significantly fewer homozygous individuals than would be expected under the Hardy–Weinberg law. Violation of either, or both, of two Hardy–Weinberg assumptions could explain this excess of heterozygotes. First, fetuses, children, and adults who are heterozygous for *HLA-A* and *HLA-B* may have higher rates of survival than homozygous individuals. Second, rather than choosing their mates randomly, individuals in this population may somehow prefer mates whose MHC genotypes differ from their own.

**NOW SOLVE THIS**

Problem 6 on page 734 asks you to determine whether two populations are in Hardy–Weinberg equilibrium, based on their genotype frequencies.

■ HINT: *Start by determining the allele frequencies from the data provided. Then use the allele frequencies to predict the genotype frequencies using the Hardy–Weinberg equation.*

27.4

## The Hardy–Weinberg Law Can Be Used to Study Multiple Alleles, X-Linked Traits, and Heterozygote Frequencies

Having explained the reasoning behind the Hardy–Weinberg law, we can now discuss how the law is used for estimating allele frequencies within populations. We will examine calculations that involve multiple alleles, X-linked traits, and heterozygote frequencies.

### Calculating Frequencies for Multiple Alleles in Hardy–Weinberg Populations

We commonly find several alleles for a single locus in a population. The locus that determines the ABO blood group in humans (discussed in Chapter 4) is an example. This locus, *I* (isoagglutinin), has three alleles $I^A$, $I^B$, and $I^O$, yielding six possible genotypic combinations ($I^AI^A$, $I^BI^B$, $I^OI^O$, $I^AI^B$, $I^AI^O$, $I^BI^O$). You may recall that $I^A$ and $I^B$ are codominant alleles and that both are dominant to $I^O$. The result is that homozygous $I^AI^A$ and heterozygous $I^AI^O$ individuals are phenotypically identical, as are $I^BI^B$ and $I^BI^O$ individuals, so only four phenotypic combinations are distinguishable.

By adding a third variable to the Hardy–Weinberg equation, we can calculate both the genotype frequencies and allele frequencies for a situation involving three alleles. Let $p$, $q$, and $r$ represent the frequencies of alleles $I^A$, $I^B$, and $I^O$, respectively. Note that because there are three alleles

$$p + q + r = 1$$

Under Hardy–Weinberg assumptions, the frequencies of the genotypes are given by

$$(p + q + r)^2 = p^2 + q^2 + r^2 + 2pq + 2pr + 2qr = 1$$

If we know the frequencies of blood phenotypes for a population, we can then estimate the frequencies for the three alleles of the ABO system. For example, in one population sampled, the following blood-type frequencies are observed: A = 0.53, B = 0.13, O = 0.26, and AB = 0.08. Because the $I^O$ allele is recessive, the population's frequency of type O blood equals the proportion of the recessive genotype $r^2$. Thus,

$$r^2 = 0.26$$
$$r = \sqrt{0.26}$$
$$r = 0.51$$

Using $r$, we can estimate the allele frequencies for the $I^A$ and $I^B$ alleles. The $I^A$ allele is present in two genotypes, $I^AI^A$ and $I^AI^O$. The frequency of the $I^AI^A$ genotype is represented by $p^2$ and the $I^AI^O$ genotype by $2pr$. Therefore, the combined frequency of type A blood and type O blood is given by

$$p^2 + 2pr + r^2 = 0.53 + 0.26$$

If we factor the left side of the equation and take the sum of the terms on the right, we get

$$(p + r)^2 = 0.79$$
$$p + r = \sqrt{0.79} = 0.89$$
$$p = 0.89 - r$$
$$p = 0.89 - 0.51 = 0.38$$

Having estimated $p$ and $r$, the frequencies of allele $I^A$ and allele $I^O$, we can now estimate the frequency of the $I^B$ allele:

$$(p + q + r) = 1$$
$$q = 1 - p - r$$
$$= 1 - 0.38 - 0.51$$
$$= 0.11$$

The phenotypic frequencies and genotypic frequencies for this population are summarized in Table 27.3.

### Calculating Frequencies for X-linked Traits

The Hardy–Weinberg law can be used to calculate allele and genotype frequencies for X-linked traits, as long as we remember that in an XY sex-determination system, the homogametic (XX) sex will have two copies of an X-linked allele, whereas the heterogametic sex (XY) only has one copy. Thus, for mammals (including humans) where the female is XX and the male is XY, the frequency of the X-linked allele in the gene pool and the frequency of males expressing the X-linked trait will be the same. This is because each male only has one X chromosome, and the probability of any individual male receiving an X chromosome with the allele in question must be equal to the frequency of the allele. The probability of any individual female having the allele in question on both X chromosomes will be $q^2$, where $q$ is the frequency of the allele.

To illustrate this for a recessive X-linked trait, let's consider the example of red-green color blindness, which affects 8 percent of

TABLE 27.3

**Calculating Genotype Frequencies for Multiple Alleles Where the Frequency of Allele $I^A$ = 0.38, Allele $I^B$ = 0.11, and Allele $I^O$ = 0.51**

| Genotype | Genotype Frequency | Phenotype | Phenotype Frequency |
|---|---|---|---|
| $I^AI^A$ | $p^2 = (0.38)^2 = 0.14$ | A | 0.53 |
| $I^AI^O$ | $2pr = 2(0.38)(0.51) = 0.39$ | | |
| $I^BI^B$ | $q^2 = (0.11)^2 = 0.01$ | B | 0.12 |
| $I^BI^O$ | $2qr = 2(0.11)(0.51) = 0.11$ | | |
| $I^AI^B$ | $2pq = 2(0.38)(0.11) = 0.084$ | AB | 0.08 |
| $I^OI^O$ | $r^2 = (0.51)^2 = 0.26$ | O | 0.26 |

human males. The frequency of the color blindness allele is therefore 0.08; in other words, 8 percent of X chromosomes carry it. The other 92 percent of X chromosomes carry the dominant allele for normal red-green color vision. If we define $p$ as the frequency of the normal allele and $q$ as the frequency of the color-blindness allele, then $p = 0.92$ and $q = 0.08$. The frequency of color-blind females (with two affected X chromosomes) is $q^2 = (0.08)^2 = 0.0064$, and the frequency of carrier females (having one normal and one affected X chromosome) is $2pq = 2(0.08)(0.92) = 0.147$. In other words, 14.7 percent of females carry the allele for red-green color blindness and can pass it to their children, although they themselves have normal color vision.

An important consequence of the difference in allele frequency for X-linked genes between male and female gametes is that for a rare recessive allele, the trait will be expressed at a much higher frequency among XY individuals than among those who are XX. So, for example, diseases such as hemophilia and Duchenne muscular dystrophy (DMD) in humans, both of which are caused by recessive mutations on the X chromosome, are much more common in boys, who need only inherit a single copy of the mutated allele to suffer from the disease. Girls who inherit two affected X chromosomes will also have the disease; but with a rare allele, the probability of this occurrence is small.

**NOW SOLVE THIS**

Problem 18 on page 735 asks you to calculate the number of heterozygous carriers for an X-linked trait in a human population.

■ HINT: *First determine the genotype of carriers. Then calculate what proportion of the population can be a carrier.*

## Calculating Heterozygote Frequency

Another application of the Hardy–Weinberg law is to estimate the frequency of heterozygotes in a population. The frequency of a recessive trait can usually be determined by counting homozygous recessive individuals in a sample of the population. With this information, we can calculate the allele and Hardy–Weinberg equilibrium genotype frequencies for the entire population.

Cystic fibrosis, an autosomal recessive trait, has an incidence of about $1/2500 = 0.0004$ in people of northern European ancestry. Individuals with cystic fibrosis are easily distinguished from the population at large by such symptoms as salty sweat, excess amounts of thick mucus in the lungs, and susceptibility to bacterial infections. Because this is a recessive trait, individuals with cystic fibrosis must be homozygous. Their frequency in a population is represented by $q^2$ (provided that mating has been random in the previous generation). The frequency of the recessive allele therefore is

$$q = \sqrt{q^2} = \sqrt{0.0004} = 0.02$$

Since $p + q = 1$, the frequency of $p$ is

$$p = 1 - q = 1 - 0.02 = 0.98$$

In the Hardy–Weinberg equation, the frequency of heterozygotes is $2pq$, so

$$\begin{aligned} 2pq &= 2(0.98)(0.02) \\ &= 0.04 \text{ or } 4 \text{ percent, or } 1/25 \end{aligned}$$

Thus, heterozygotes for cystic fibrosis are rather common in the population (4 percent), even though the incidence of homozygous recessives is only 1/2500, or 0.0004 percent. Calculations such as these are only estimates, because the population may not meet all Hardy–Weinberg expectations.

In general, for a locus with a dominant and a recessive allele, the frequencies of all three genotypes (homozygous dominant, homozygous recessive, and heterozygous) can be estimated once the frequency of either allele is known and Hardy–Weinberg assumptions are invoked. The relationship between genotype and allele frequency is shown in Figure 27–5. It is important to note that heterozygotes increase rapidly in a population as the values of $p$ and $q$ move from 0 or 1.0 toward 0.5. This observation confirms our conclusion that when a recessive trait such as cystic fibrosis is rare, the majority of those carrying the allele are heterozygotes. In populations in which the frequencies of $p$ and $q$ are between 0.33 and 0.67, heterozygotes occur at higher frequency than either homozygote.

**NOW SOLVE THIS**

Problem 17 on page 735 asks you to calculate the frequency of heterozygous carriers of the recessive trait albinism in a human population.

■ HINT: *Solve the parts of this problem in the order presented, first determining how to calculate the frequency of the albinism allele in this population. Then calculate the frequency of the normal allele, frequency of heterozygotes, and how often matings between heterozygotes occur.*

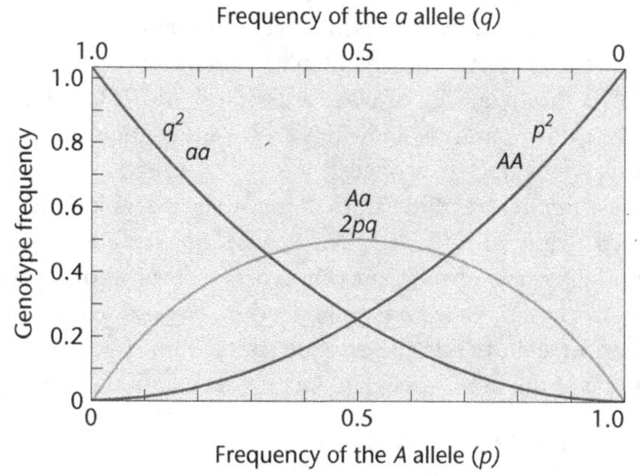


**FIGURE 27–5** The relationship between genotype and allele frequencies derived from the Hardy–Weinberg equation.

27.5

## Natural Selection Is a Major Force Driving Allele Frequency Change

We have noted that the Hardy–Weinberg law allows us to estimate allele and genotype frequencies in ideal populations, that is, in populations described by the assumptions of equal viability and fertility, absence of mutation, absence of migration, sufficient size, and random mating. Obviously, it is almost impossible to find natural populations in which all these assumptions hold for all loci. In nature, populations are dynamic, and changes in size and gene pool are common. However, an understanding of the Hardy–Weinberg law allows us to investigate populations that vary from the ideal. In this and the following sections, we discuss factors that prevent populations from reaching Hardy–Weinberg equilibrium, or that drive populations toward a different equilibrium, and the relative contributions of these factors to evolutionary change.

### Natural Selection

The first assumption of the Hardy–Weinberg law is that individuals of all genotypes have equal survival rates and equal reproductive success. If this assumption does not hold, allele frequencies may change from one generation to the next. To see why, let's imagine a population of 100 individuals in which allele *A* has a frequency of 0.5 and allele *a* has a frequency of 0.5. Assuming the previous generation mated randomly, we find that the genotype frequencies in the present generation are $(0.5)^2 = 0.25$ for *AA*, $2(0.5)(0.5) = 0.5$ for *Aa*, and $(0.5)^2 = 0.25$ for *aa*. Because our population contains 100 individuals, we have 25 *AA* individuals, 50 *Aa* individuals, and 25 *aa* individuals. Now suppose that individuals with different genotypes have different rates of survival: all 25 *AA* individuals survive to reproduce; 90 percent, or 45, of the *Aa* individuals survive to reproduce; and 80 percent, or 20, of the *aa* individuals survive to reproduce. When the survivors reproduce, each contributes two gametes to the new gene pool, giving us $2(25) + 2(45) + 2(20) = 180$ gametes. What are the frequencies of the two alleles in the surviving population? We have 50 *A* gametes from *AA* individuals, plus 45 *A* gametes from *Aa* individuals, so the frequency of allele *A* is $(50 + 45)/180 = 0.53$. We have 45 *a* gametes from *Aa* individuals, plus 40 *a* gametes from *aa* individuals, so the frequency of allele *a* is $(45 + 40)/180 = 0.47$.

These frequencies differ from those we started with. Allele *A* has increased, while allele *a* has declined. A nonrandom difference among individuals of different genotypes in survival or reproduction rate or both is called **natural selection**. Natural selection is the principal force that shifts allele frequencies within large populations and is one of the most important factors in evolutionary change.

### Fitness and Selection

Selection occurs whenever individuals with a particular genotype enjoy an advantage in survival or reproduction over other genotypes. However, selection may vary from less than 1 to 100 percent (the latter indicating a lethal allele). In the previous hypothetical example, selection was strong. Weak selection might involve just a fraction of a percent difference in the survival rates of different genotypes. Advantages in survival and reproduction ultimately translate into increased genetic contribution to future generations. An individual organism's genetic contribution to future generations is called its **fitness**. Genotypes associated with high rates of reproductive success are said to have high fitness, whereas genotypes associated with low reproductive success are said to have low fitness.

Hardy–Weinberg analysis also allows us to examine fitness. By convention, population geneticists use the letter $w$ to represent fitness. Thus, $w_{AA}$ represents the relative fitness of genotype *AA*, $w_{Aa}$ the relative fitness of genotype *Aa*, and $w_{aa}$ the relative fitness of genotype *aa*. Assigning the values $w_{AA} = 1$, $w_{Aa} = 0.9$, and $w_{aa} = 0.8$ would mean, for example, that all *AA* individuals survive, 90 percent of the *Aa* individuals survive, and 80 percent of the *aa* individuals survive, as in the previous hypothetical case.

Let's consider selection against deleterious alleles. Fitness values $w_{AA} = 1$, $w_{Aa} = 1$, and $w_{aa} = 0$ describe a situation in which *a* is a homozygous lethal allele. As homozygous recessive individuals die without leaving offspring, the frequency of allele *a* will decline. The decline in the frequency of allele *a* is described by the equation

$$q_g = \frac{q_0}{1 + gq_0}$$

where $q_g$ is the frequency of allele *a* in generation $g$, $q_0$ is the starting frequency of *a* (i.e., the frequency of *a* in generation zero), and $g$ is the number of generations that have passed.

Figure 27–6 shows what happens to a lethal recessive allele with an initial frequency of 0.5. At first, because of the high percentage of *aa* genotypes, the frequency of allele *a* declines rapidly. The frequency of *a* is halved in only two generations. By the sixth generation, the frequency is halved again. By now, however, the majority of *a* alleles are carried by heterozygotes. Because *a* is recessive, these heterozygotes are not selected against. Consequently, as more time passes, the frequency of allele *a* declines ever more slowly. As long as heterozygotes continue to mate, it is difficult for selection to completely eliminate a recessive allele from a population.

Of course, a deleterious allele need not be recessive; many other scenarios are possible. A deleterious allele may be codominant, so that heterozygotes have intermediate fitness. Or an allele may be deleterious in the homozygous state but beneficial in the heterozygous state, like the allele for sickle-cell anemia. Figure 27–7 shows examples in which a deleterious allele is codominant, but in the different examples, the intensity of selection varies from strong to weak. In each case, the frequency of the deleterious allele, *a*, starts at 0.99 and declines over time. However, the rate of decline depends heavily on the strength of selection. When only 90 percent of the heterozygotes and 80 percent of the *aa* homozygotes survive (red curve), the frequency of allele *a* drops from 0.99 to less than 0.01 in 85 generations. However, when 99.8 percent of the heterozygotes and 99.6 percent of the *aa* homozygotes survive (blue curve), it takes 1000 generations for the frequency of allele *a* to drop from

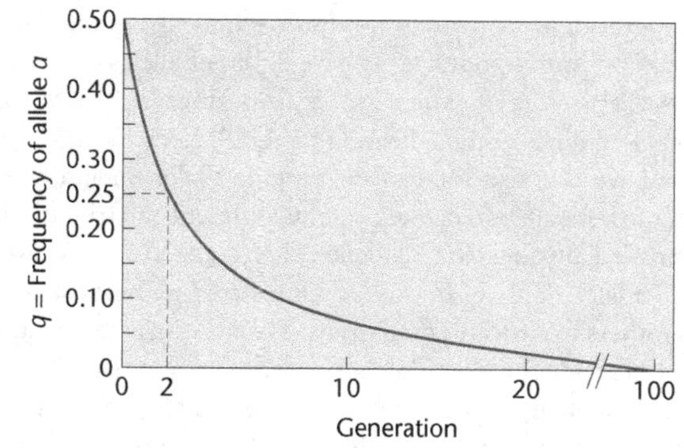


| Generation | $p$ | $q$ | $p^2$ | $2pq$ | $q^2$ |
|---|---|---|---|---|---|
| 0 | 0.50 | 0.50 | 0.25 | 0.50 | 0.25 |
| 1 | 0.67 | 0.33 | 0.44 | 0.44 | 0.12 |
| 2 | 0.75 | 0.25 | 0.56 | 0.38 | 0.06 |
| 3 | 0.80 | 0.20 | 0.64 | 0.32 | 0.04 |
| 4 | 0.83 | 0.17 | 0.69 | 0.28 | 0.03 |
| 5 | 0.86 | 0.14 | 0.73 | 0.25 | 0.02 |
| 6 | 0.88 | 0.12 | 0.77 | 0.21 | 0.01 |
| 10 | 0.91 | 0.09 | 0.84 | 0.15 | 0.01 |
| 20 | 0.95 | 0.05 | 0.91 | 0.09 | < 0.01 |
| 40 | 0.98 | 0.02 | 0.95 | 0.05 | < 0.01 |
| 70 | 0.99 | 0.01 | 0.98 | 0.02 | < 0.01 |
| 100 | 0.99 | 0.01 | 0.98 | 0.02 | < 0.01 |

FIGURE 27–6 Change in the frequency of a lethal recessive allele, *a*. The frequency of *a* is halved in two generations and halved again by the sixth generation. Subsequent reductions occur slowly because the majority of *a* alleles are carried by heterozygotes.

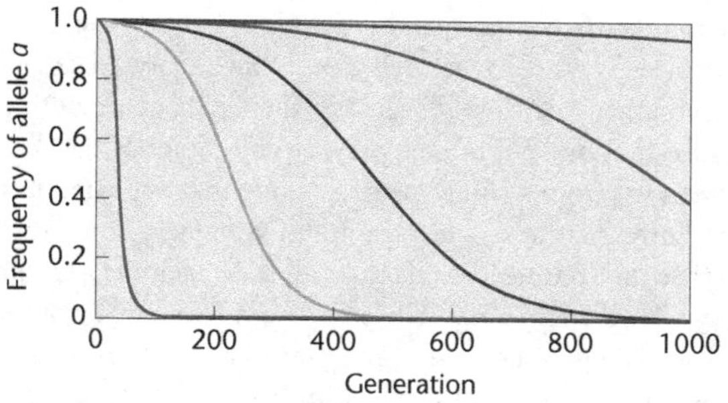


| Selection Against Allele *a* | | | | | |
|---|---|---|---|---|---|
| | Strong ← | | | | → Weak |
| | — | — | — | — | — |
| $w_{AA}$ | 1.0 | 1.0 | 1.0 | 1.0 | 1.0 |
| $w_{Aa}$ | 0.90 | 0.98 | 0.99 | 0.995 | 0.998 |
| $w_{aa}$ | 0.80 | 0.96 | 0.98 | 0.99 | 0.996 |

FIGURE 27–7 The effect of selection on allele frequency. The rate at which a deleterious allele is removed from a population depends heavily on the strength of selection.

0.99 to 0.93. Two important conclusions can be drawn from this graph. First, given thousands of generations, even weak selection can cause substantial changes in allele frequencies; because evolution generally occurs over a large number of generations, selection is a powerful force of evolution. Second, for selection to produce rapid changes in allele frequencies—changes measurable within a human life span—the differences in fitness among genotypes must be large, or generation time must be short.

These observations concerning the effects of selection on allele frequencies allow us to make some inferences about the *CCR5*-$\Delta 32$ allele that we discussed earlier. Because individuals with genotype $\Delta 32/\Delta 32$ are resistant to most sexually transmitted strains of HIV-1, while individuals with genotypes *1/1* and $1/\Delta 32$ are susceptible, we might expect AIDS to act as a selective force causing the frequency of the $\Delta 32$ allele to increase over time. Indeed, it probably will, but the increase in frequency is likely to be slow in human terms.

We can do a rough calculation as follows: Imagine a population in which the current frequency of the $\Delta 32$ allele is 0.10. Under Hardy–Weinberg assumptions, the genotype frequencies in this population are 0.81 for *1/1*, 0.18 for $1/\Delta 32$, and 0.01 for $\Delta 32/\Delta 32$. Imagine also that 1 percent of the and individuals in this population will contract HIV and die of AIDS.

Based on our assumptions, we can assign fitness levels to the genotypes of $w_{1/1}$ = 0.99; $w_{1/\Delta 32}$ = 0.99; and $w_{\Delta 32/\Delta 32}$ = 1.0. Given the assigned fitness levels, we can predict that the frequency of the *CCR5*-$\Delta 32$ allele in the next generation will be 0.100091. In fact, it will take about 100 generations (about 2000 years) for the frequency of the $\Delta 32$ allele to reach just 0.11 (Figure 27–8). In other words, the frequency of the $\Delta 32$ allele will probably not change much over the next few generations in most populations that currently harbor it.

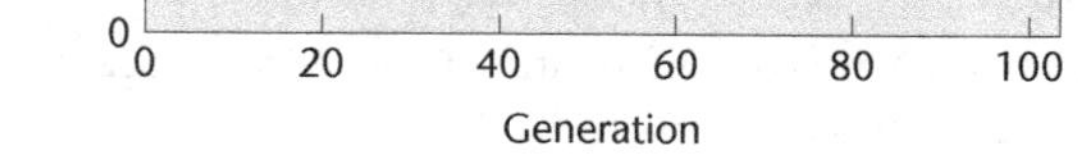


FIGURE 27–8 The rate at which the frequency of the *CCR5*-$\Delta 32$ allele changes in hypothetical populations with different initial frequencies and different fitnesses.

A population genetics perspective sheds light on the *CCR5*-Δ*32* story in other ways. Two research groups have analyzed genetic variation at marker loci closely linked to the *CCR5* gene. Both groups concluded that most, if not all, present-day copies of the Δ*32* allele are descended from a single ancestral copy that appeared in northeastern Europe at most a few thousand years ago. One group estimates that the common ancestor of all Δ*32* alleles existed just 700 years ago. How could a new allele rise from a frequency of virtually zero to as high as 20 percent in roughly 30 generations?

It would seem that, at some point in the past, there was strong selection in favor of the Δ*32* allele, most likely in the form of an infectious disease. The agent of selection cannot have been HIV-1 because HIV-1 moved from chimpanzees to humans too recently. The question is, which disease might have favored strong selection for the Δ*32* allele? One theory holds that selection began about 700 years ago and that the agent of selection was bubonic plague. During the Black Death of 1346–1352, between a quarter and a third of all Europeans died from plague. Bubonic plague is caused by the bacterium *Yersinia pestis*. This bacterium manufactures a protein that kills certain kinds of white blood cells. J. Claiborne Stephens and colleagues hypothesize that the process through which the bacterial protein kills white cells involves the *CCR5* gene product and that some mechanism related to this product makes individuals who are homozygous for the Δ*32* allele more likely to survive plague epidemics.

A second proposal for the selective agent for the Δ*32* allele comes from a study of the *CCR5* locus in Jewish and European populations. These researchers suggested that the Δ*32* allele may have arisen as early as the eighth century and that smallpox may have been the selective agent responsible for the rapid increase in its frequency. Variola, the virus responsible for smallpox, also uses the CCR5 receptor for infecting cells, and with a fatality rate of 25 percent, waves of smallpox would be a powerful selective agent. Work by other researchers gave support to the hypothesis that smallpox was the selective agent.

More recently, work using DNA extracted from Bronze Age skeletons (~ 3000 years old) has revealed two important findings: (1) the *CCR5*-Δ*32* allele was present in ancient populations, and (2) the frequency of this deletion allele in the ancient population sampled was similar to frequencies found in modern German populations.

These findings indicate that the Δ*32* allele is far older than originally thought and may even have arisen about 5000 years ago, long before epidemics of plague swept through human populations and perhaps before smallpox began its ravages. This does not mean that the Δ*32* allele did not confer resistance to one or more infectious diseases over the past several thousand years, but it points out the difficulty of identifying the selective agent or agents at such a distance.

If ancient or recent epidemics of an infectious disease are responsible for the high frequency of the *CCR5*-Δ*32* allele in European populations, then the virtual absence of the allele in non-European populations is at first somewhat puzzling. Perhaps plague has been more common in Europe than elsewhere. Alternatively, perhaps other populations have different alleles of the *CCR5* gene that also confer protection against disease. Teams of researchers looking for other alleles of the *CCR5* gene in various populations have found a total of 20 mutant alleles, including Δ*32*. Sixteen of these alleles encode proteins different in structure from that encoded by the *CCR51* allele. Some, like Δ*32*, are loss-of-function alleles. Some of the alleles appear confined to Asian populations, others to African populations. Their frequencies are as high as 3 to 4 percent. Together, these discoveries are consistent with the hypothesis that alteration or loss of the CCR5 protein protects against an as-yet-unidentified infectious disease or diseases.

**NOW SOLVE THIS**

Problem 8 on page 734 describes a set of alleles and their frequencies and asks you to calculate the effect of different fitness levels on the allele frequencies in the next generation.

■ HINT: *First calculate the genotype frequencies in the next generation. Once these are known, calculate allele frequencies.*

## Selection in Natural Populations

Geneticists have done a great deal of research on the effects of natural selection on allele frequencies in both laboratory and natural populations. Among the most detailed studies of natural populations are those concerning insects exposed to pesticides. Christine Chevillon and colleagues, for example, studied the effect of the insecticide chlorpyrifos on allele frequencies in populations of house mosquitoes. Chlorpyrifos kills mosquitoes by interfering with the function of the enzyme acetylcholinesterase (ACE), which under normal circumstances breaks down the neurotransmitter acetylcholine. An allele of the gene coding for ACE called $Ace^R$ encodes a slightly altered version of ACE that is immune to interference by chlorpyrifos.

Chevillon measured the frequency of the $Ace^R$ allele in nine populations. The first four populations lived in locations where chlorpyrifos had been used to control mosquitoes for 22 years; the last five lived in locations where chlorpyrifos had never been used. Chevillon predicted that the frequency of $Ace^R$ would be higher in the exposed populations. The researchers also predicted that the frequencies of alleles for enzymes unrelated to the physiological effects of chlorpyrifos would show no such pattern. Among the control enzymes studied was aspartate amino transferase 1. The results appear in Figure 27–9. As the researchers predicted, the frequency of the $Ace^R$ allele was significantly higher in the exposed populations. Also as predicted, the frequencies of the most common alleles of the control enzyme gene showed no such trends. The explanation is that during the 22 years of exposure to the pesticide, mosquitoes had higher rates of survival if they carried the $Ace^R$ allele. In other words, the $Ace^R$ allele was favored by selection.

When mosquito populations are exposed to chlorpyrifos, the rate of change in allele frequencies at the ACE locus will depend on

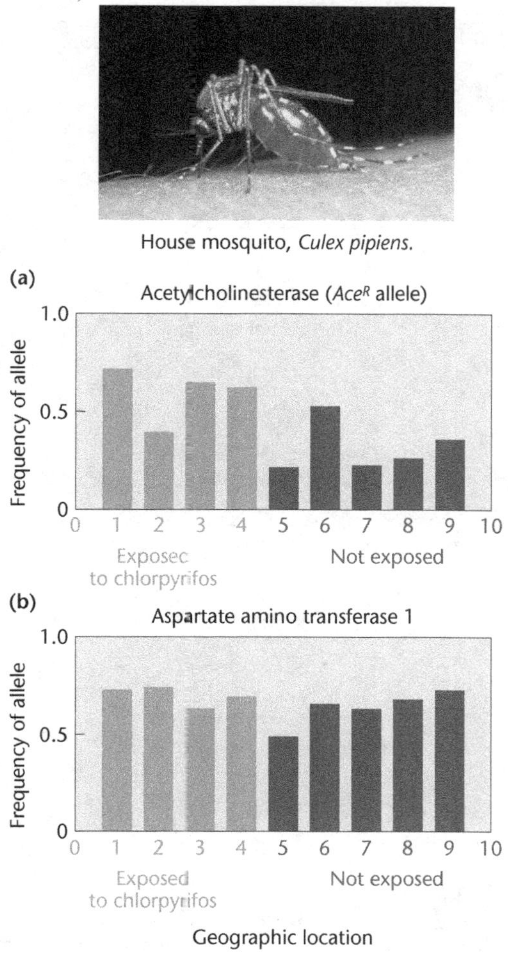


**FIGURE 27-9** The effect of selection on allele frequencies in natural populations. (a) The frequency of the $Ace^R$ allele, which confers resistance to the insecticide chlorpyrifos, is higher in house mosquito populations exposed to chlorpyrifos. (b) The frequency of an allele for an enzyme unrelated to chlorpyrifos metabolism (aspartate amino transferase 1) shows no such pattern.

how strongly mosquitoes that do not have the $Ace^R$ allele are selected against. The strength of selection is introduced into allele frequency calculations as a selection coefficient. Thus, if selection is operating against an allele in the population with an initial frequency $q$, the new frequency $q'$ after one generation can be calculated using the following equation:

$$q' = \frac{q(1 - sq)}{1 - sq^2}$$

where $s$ is the selection coefficient that describes the survival disadvantage of individuals homozygous for the pesticide-sensitive allele. Thus, if the relative fitness of mosquitoes that do not have the $Ace^R$ allele and are sensitive to chlorpyrifos is 0.75 and the initial frequency of the pesticide-sensitive *Ace* allele is 0.95, then after one generation of exposure to the insecticide, the new frequency $q'$ of the pesticide-sensitive allele will be:

$$q' = \frac{0.95(1 - 0.25 \times 0.95)}{1 - 0.25 \times 0.95^2} = 0.935$$

This shows that even with relatively strong selection in its favor, a new allele (in this case, the $Ace^R$ allele) present in the population at a low frequency has a small impact on the frequency of the normal allele, and will require multiple generations to produce a significant change in the gene pool.

**NOW SOLVE THIS**

Problem 23 on page 735 asks you to calculate the change in frequency of a resistance allele in an insect population feeding on a transgenic corn crop.

■ HINT: *Start by answering these questions: Which allele is being selected against? What is the relative fitness of individuals homozygous for that allele? What is the selection coefficient? Then calculate the change in allele frequency.*

## Natural Selection and Quantitative Traits

As we saw in Chapter 25, many phenotypic traits are not controlled by alleles at a single locus but instead are quantitative: the phenotype is the result of the combined influence of the individual's genotype at many different loci and the environment. Selection acting on these quantitative traits can be classified as (1) directional, (2) stabilizing, or (3) disruptive.

In **directional selection,** the genotypes conferring one or the other phenotypic extremes are selected, resulting in a change in the population mean over time. This form of selection is widely practiced in plant and animal breeding. If the desired trait is polygenic, the most extreme phenotype that the genotype can express will appear in the population only after prolonged selection.

In nature, directional selection can occur when one of the phenotypic extremes becomes selected for or against, usually as a result of changes in the environment. A carefully documented example comes from research by Peter and Rosemary Grant and their colleagues, who studied the medium ground finches (*Geospiza fortis*) of Daphne Major Island in the Galapagos Islands. The beak size of these birds varies enormously. In 1976, for example, some birds in the population had beaks less than 7 mm deep, whereas others had beaks more than 12 mm deep. Beak size is heritable, which means that large-beaked parents tend to have large-beaked offspring, and small-beaked parents tend to have small-beaked offspring. In 1977, a severe drought on Daphne killed some 80 percent of the finches. Big-beaked birds survived at higher rates than small-beaked birds because when food became scarce, the big-beaked birds were able to eat a greater variety of seeds. When the drought ended and the survivors paired off and bred, the offspring inherited their parents' big beaks

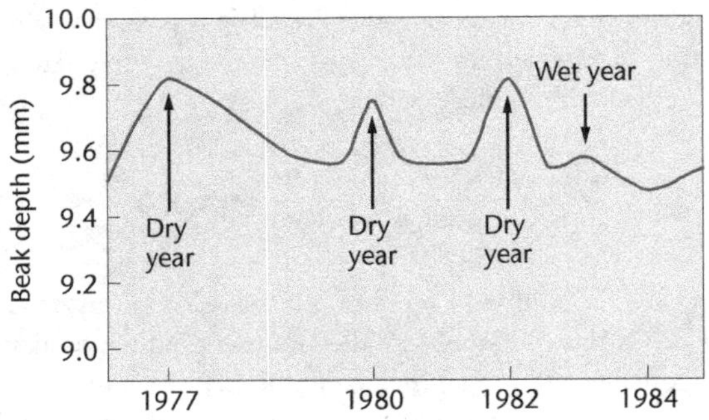


FIGURE 27–10 Beak size in finches during dry years increases because of strong selection. Between droughts, selection for large beak size is not as strong, and birds with smaller beak sizes survive and reproduce, increasing the number of birds with smaller beaks.

(Figure 27–10). Between 1976 and 1978, the beak depth of the average finch in the Daphne Major population increased by just under 0.5 mm, shifting the average beak size toward one phenotypic extreme, emphasizing the role of directional selection on phenotypes.

**Stabilizing selection**, in contrast, tends to favor intermediate phenotypes, with phenotypes at both extremes being selected against. Over time, this form of selection will reduce the population variance, but without causing a significant shift in the mean. One of the clearest demonstrations of stabilizing selection is provided by the data of Mary Karn and Sheldon Penrose on human birth weight and survival for 13,730 children born over an 11-year period. Figure 27–11 shows the distribution of birth weight among the children and the percentage of those infants who died by 4 weeks of age. Infant mortality increases on either side of the optimal birth weight of 7.5 pounds (the increase is quite dramatic at the low end). At the genetic level, stabilizing selection acts to keep a population well adapted to its environment. In this situation, individuals closer to the average for a given trait will have higher fitness.

**Disruptive selection** is selection against intermediates and for both phenotypic extremes. It can be viewed as the opposite of stabilizing selection because the intermediate types are selected against.

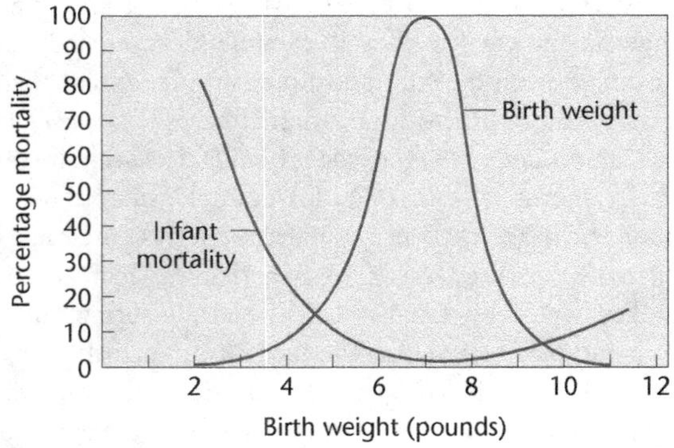


FIGURE 27–11 Relationship between birth weight and mortality in humans.

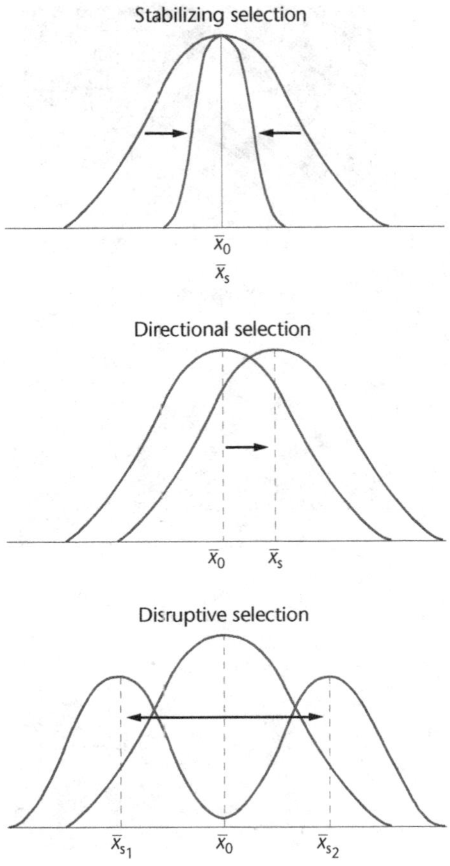


FIGURE 27–12 The impact of directional, stabilizing, and disruptive selection. In each case, the mean of an original population (green) and the mean ($\bar{x}$) of the population following selection (red) is shown.

This form of selection will result in a population with an increasingly bimodal distribution for the trait, as we can see in Figure 27–12. In one set of experiments, John Thoday applied disruptive selection to a population of *Drosophila* on the basis of bristle number. In every generation, he allowed only the flies with high or low bristle numbers to breed. After several generations, most of the flies could be easily placed in a low- or high-bristle category (Figure 27–13). In natural populations, such a situation might exist for a population in a heterogeneous environment.

## 27.6 Mutation Creates New Alleles in a Gene Pool

Within a population, the gene pool is reshuffled each generation to produce new genotypes in the offspring. Because the number of possible genotypic combinations is so large, population members alive at any given time represent only a fraction of all possible geno-

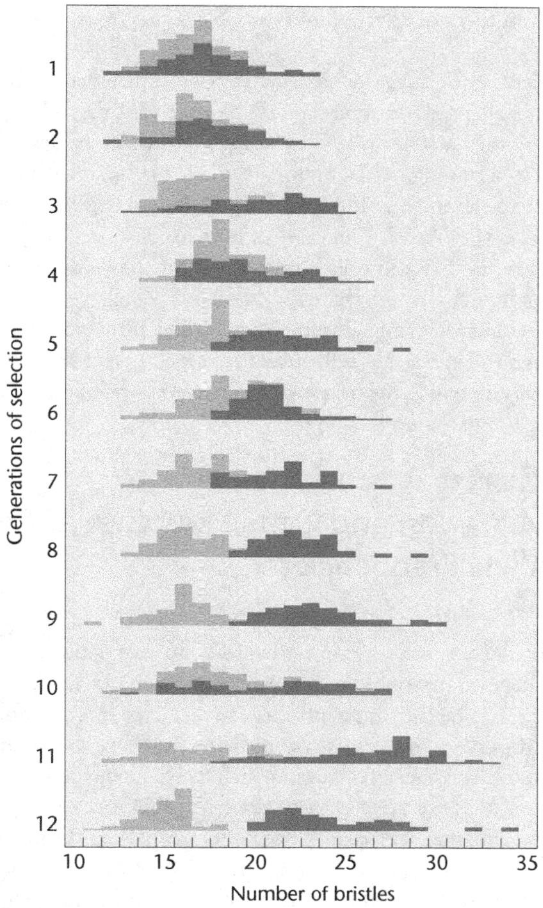


**FIGURE 27–13** The effect of disruptive selection on bristle number in *Drosophila*. When individuals with the highest and lowest bristle number were selected as parents for each succeeding generation, the population showed a nonoverlapping divergence in only 12 generations.

types. The enormous genetic reserve that exists in the gene pool allows Mendelian assortment and recombination to produce new genotypic combinations perpetually. But assortment and recombination do not produce new alleles. Mutation alone acts to create new alleles. It is important to keep in mind that mutational events occur at random—that is, without regard for any possible benefit or disadvantage to the organism. In this section, we consider whether mutation is, by itself, a significant factor in causing allele frequencies to change.

To determine whether mutation is a significant force in changing allele frequencies, we must first measure the rate at which mutations are produced. As most mutations are recessive, it is difficult to observe mutation rates directly in diploid organisms. Indirect methods that use probability and statistics or large-scale screening programs are employed. For certain dominant mutations, however, a direct method of measurement can be used. To ensure accuracy, several conditions must be met:

1. The allele being evaluated must produce a distinctive phenotype that can be distinguished from similar phenotypes produced by recessive alleles.
2. The trait must be fully expressed, or completely penetrant, so that mutant individuals can be identified.
3. An identical phenotype must never be produced by nongenetic agents such as drugs or chemicals.

Mutation rates can be stated as the number of new mutant alleles of a given gene per given number of gametes. Suppose that a given gene undergoes mutation from a recessive allele to a dominant allele at a rate that produces 2 births that exhibit the mutant phenotype in every 100,000 births. The parents of these two offspring are phenotypically normal. Because the zygotes that produced the 100,000 births each carry two copies of the gene, we have actually surveyed 200,000 copies of the gene (carried in 200,000 gametes). If we assume that the affected births are each heterozygous, we have uncovered two mutant alleles out of 200,000. Thus, the mutation rate for this gene is 2/200,000 or 1/100,000, which in scientific notation is written as $1 \times 10^{-5}$.

In humans, a dominant form of dwarfism known as **achondroplasia** fulfills the requirements for measuring mutation rates. Individuals with this skeletal disorder have an enlarged skull but short arms and legs and can be diagnosed by X-ray examination at birth. In a survey of almost 250,000 births, the mutation rate ($\mu$) for achondroplasia has been calculated as

$$\mu = 1.4 \times 10^{-5} \pm 0.5 \times 10^{-5}$$

Knowing a gene's rate of mutation, we can estimate the extent to which mutation alone can cause the allele frequencies to change from one generation to the next. We represent the normal allele as *d* and the allele for achondroplasia as *D*.

Imagine a population of 500,000 individuals in which everyone has genotype *dd*. The initial frequency of *d* is 1.0, and the initial frequency of *D* is 0. If each individual contributes two gametes to the gene pool, the gene pool will contain 1,000,000 gametes, all carrying allele *d*. While the gametes are in the gene pool, 1.4 of every 100,000 *d* alleles mutates into a *D* allele. The frequency of allele *d* is now $(1{,}000{,}000 - 14)/1{,}000{,}000 = 0.999986$, and the frequency of *D* is $14/1{,}000{,}000 = 0.000014$. Given these numbers, it will clearly be a long time before mutation, by itself, causes any appreciable change in the allele frequencies in this population.

More generally, if we have two alleles, *A* with frequency *p* and *a* with frequency *q*, and if $\mu$ represents the rate of mutations converting *A* into *a*, then the frequencies of the alleles in the next generation are given by

$$p_{g+1} = p_g - \mu p_g \quad \text{and} \quad q_{g+1} = q_g + \mu p_g$$

where $p_{g+1}$ and $q_{g+1}$ represent the allele frequencies in the next generation, and $p_g$ and $q_g$ represent the allele frequencies in the present generation.

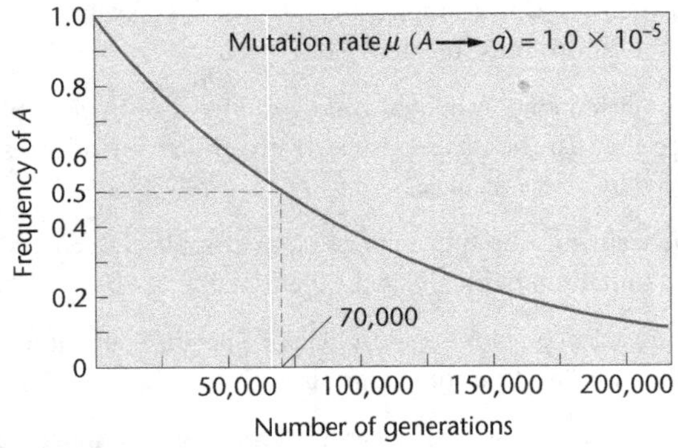


FIGURE 27–14 Replacement rate of an allele by mutation alone, assuming an average mutation rate of $1.0 \times 10^{-5}$.

Figure 27–14 shows the replacement rate (change over time) in allele *A* for a population in which the initial frequency of *A* is 1.0 and the rate of mutation ($\mu$) converting *A* into *a* is $1.0 \times 10^{-5}$. At this mutation rate, it will take about 70,000 generations to reduce the frequency of *A* to 0.5. Even if the rate of mutation increases through exposure to higher levels of radioactivity or chemical mutagens, the impact of mutation alone on allele frequencies will be extremely weak. In short, the ultimate source of genetic variability, mutation, provides the raw material for evolution but *by itself* plays a relatively insignificant role in changing allele frequencies. Instead, the fate of alleles created by mutation is determined to a greater extent by natural selection (discussed previously) and genetic drift (discussed later).

An evolutionary perspective on mutation can lead to medical discoveries. The autosomal recessive disease cystic fibrosis, which we discussed previously in relation to calculating heterozygote frequencies, is caused when a person has two alleles containing any of various loss-of-function mutations in the gene for a cell-surface protein called the cystic fibrosis transmembrane conductance regulator (CFTR). The frequency of the mutant alleles causing cystic fibrosis is about 2 percent in European populations; yet until recently, most individuals with two of these mutant alleles died before reproducing, meaning that selection against homozygous recessive individuals was rather strong. This situation presents a puzzle: In the face of selection against them, what has maintained the mutant alleles at an overall frequency of 2 percent? Two possible explanations have been proposed.

The **mutation-selection balance** hypothesis posits that mutation is constantly creating new alleles to replace the ones being eliminated by selection. However, for this scenario to work, the rate of mutations creating new alleles would have to be rather high, on the order of $5 \times 10^{-4}$, to counteract the effect of selection. Many evolutionary geneticists instead prefer an alternative explanation, the **heterozygote superiority hypothesis**. According to the heterozygote superiority hypothesis, selection against homozygous mutant individuals is counterbalanced by selection in favor of heterozygotes. Among proponents of this theory, the most popular candidate for an agent of selection in favor of heterozygotes is resistance to a disease.

Recent work suggests that cystic fibrosis heterozygotes may have an enhanced resistance to typhoid fever. Typhoid fever is caused by the bacterium *Salmonella typhi*, which infiltrates cells of the intestinal lining. In laboratory studies, mouse intestinal cells that were heterozygous for *CFTR-Δ508,* the analog of the most common cystic fibrosis mutation in humans, acquired 86 percent fewer bacteria than did cells homozygous for the wild-type allele. Whether humans heterozygous for *CFTR-Δ508* also enjoy resistance to typhoid fever remains to be established. If they do, then cystic fibrosis will join sickle-cell anemia as an example of heterozygote superiority, in which the fitness of heterozygotes is superior to that of either homozygous genotype.

## 27.7 Migration and Gene Flow Can Alter Allele Frequencies

Occasionally, a species becomes divided into populations that are separated geographically. Various evolutionary forces, including selection, can establish different allele frequencies in such populations. **Migration** occurs when individuals move between the populations. Imagine a species in which a given locus has two alleles, *A* and *a*. There are two populations of this species, one on a mainland and one on an island. The frequency of *A* on the mainland is represented by $p_m$, and the frequency of *A* on the island is $p_i$. If there is migration from the mainland to the island, the frequency of *A* in the next generation on the island ($p_{i'}$) is given by

$$p_{i'} = (1 - m)p_i + mp_m$$

where $m$ represents migrants from the mainland to the island.

As an example of how migration might affect the frequency of *A* in the next generation on the island ($p_{i'}$), assume that $p_i = 0.4$ and $p_m = 0.6$ and that 10 percent of the parents of the next generation are migrants from the mainland ($m = 0.1$). In the next generation, the frequency of allele *A* on the island will therefore be

$$\begin{aligned} p_{i'} &= [(1 - 0.1) \times 0.4] + (0.1 \times 0.6) \\ &= 0.36 + 0.06 \\ &= 0.42 \end{aligned}$$

In this case, migration from the mainland has changed the frequency of *A* on the island from 0.40 to 0.42 in a single generation.

These calculations reveal that the change in allele frequency attributable to migration is proportional to the differences in allele frequency between the donor and recipient populations and to the rate of migration. If either $m$ is large or $p_m$ is very different from $p_i$, then a rather large change in the frequency of *A* can occur in a single generation. If migration is the only force acting to change the allele frequency on the island, then equilibrium will be attained only when $p_i = p_m$. These guidelines can often be used to estimate

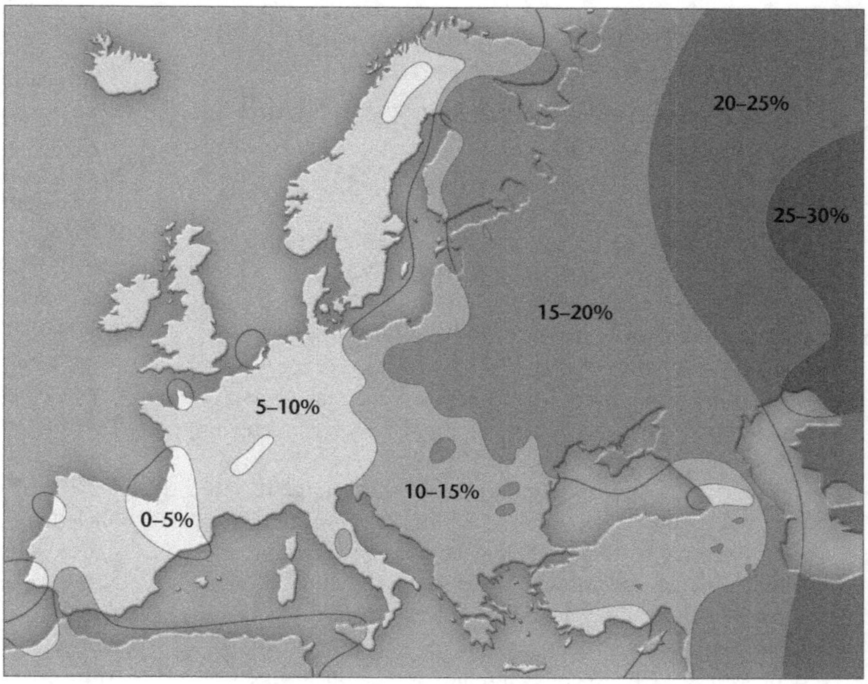


**FIGURE 27–15** Migration as a force in evolution. The $I^B$ allele of the *ABO* locus is present in a gradient from east to west. This allele shows the highest frequency in central Asia, and the lowest is in northeastern Spain. The gradient parallels the waves of Mongol migration into Europe following the fall of the Roman Empire and is a genetic relic of human history.

migration in cases where it is difficult to quantify. As $m$ can have a wide range of values, the effect of migration can substantially alter allele frequencies in populations, as shown for the $I^B$ allele of the ABO blood group in Figure 27–15.

Migration can also be regarded as the flow of genes between populations that were once, but are no longer, geographically isolated. Esteban Parra and colleagues measured allele frequencies for several different DNA-sequence polymorphisms in African-American and European-American populations and in African and European populations representative of the ancestral populations from which the two American populations are descended. One locus they studied, a restriction-site polymorphism called *FY-NULL*, has two alleles, *FY-NULL*1* and *FY-NULL*2*. Figure 27–16 shows the frequency of *FY-NULL*1* in each population. The frequency of this allele is 0 in the three African populations and 1.0 in the three European populations, but lies between these extremes in all African-American and European-American populations. The simplest explanation for these data is that genes have mixed between American populations with predominantly African ancestry and American populations with predominantly European ancestry. Based on *FY-NULL* and several other loci, researchers estimate that African-American populations derive between 11.6 and 22.5 percent of their ancestry from Europeans and that European-American populations derive between 0.5 and 1.2 percent of their ancestry from Africans.

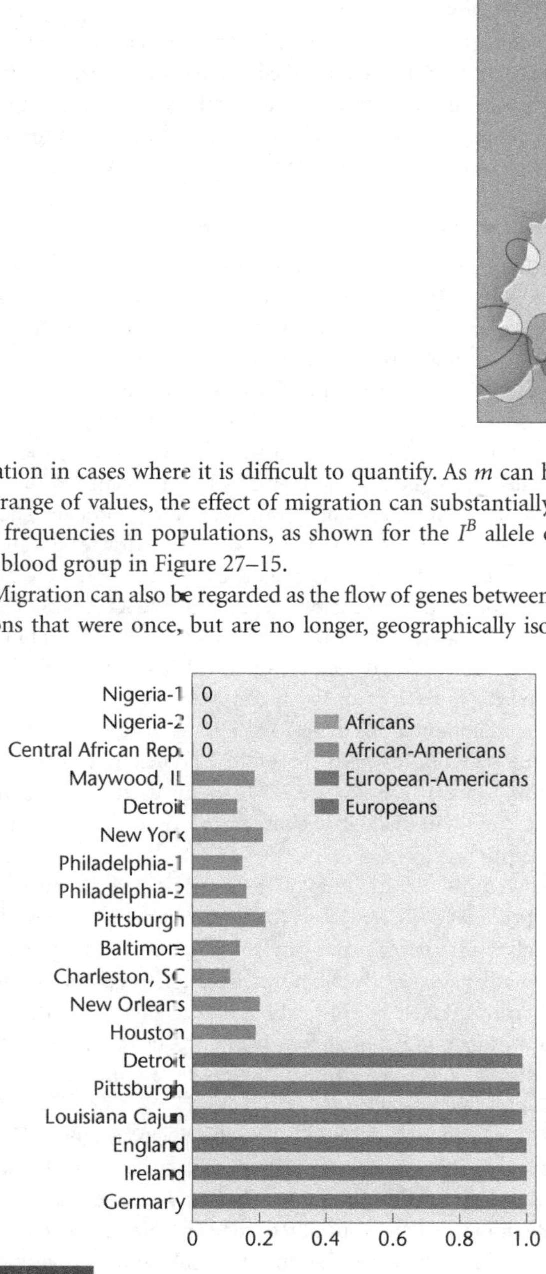


**FIGURE 27–16** Frequency histogram of the *FY-NULL*1* allele, a restriction-site polymorphism allele, in several human populations. These and other data demonstrate that African-American and European-American populations are of mixed ancestry.

**NOW SOLVE THIS**

Problem 25 on page 735 asks you to calculate the change in frequency of a mutant allele in a population of bighorn sheep that has been augmented by artificially introducing individuals from another population.

■ HINT: *If the introduced sheep came from a population that lacked the mutation, would you expect the allele frequency in the augmented population to increase or decrease? Then use the Hardy–Weinberg equation to calculate the frequency of the mutant allele in the next generation.*

27.8

## Genetic Drift Causes Random Changes in Allele Frequency in Small Populations

In small populations, significant random fluctuations in allele frequencies are possible through chance deviation. The degree of fluctuation increases as the population size decreases. This fluctuation due to chance is known as **genetic drift**. In addition to small population size, genetic drift can also arise through the **founder effect**, the genetic consequences seen when a population originates from a small number of individuals. Although the population may later increase to a large size, the genes carried by all members are derived from those of the founders (assuming no mutation, migration, or selection, and the presence of random mating). Drift can also be induced by a **genetic bottleneck**. Bottlenecks develop when a large population undergoes a drastic but temporary reduction in numbers. Even though the population recovers, its genetic diversity has been greatly reduced.

Genetic drift occurs when the number of reproducing individuals in a population is too small to ensure that all the alleles in the gene pool will be passed on to the next generation in their existing frequencies. Imagine a sexually reproducing diploid population consisting of 25 males and 25 females. Even if *all* the individuals in the population pair up into 25 couples to mate (an unlikely scenario in real life) and then generate 25 offspring, the number of gametes in the gene pool of the next generation will be just 50. It is unlikely that such a small number of gametes will accurately reflect the genetic structure of the parent population (which contained twice that number of gametes). Consequently, by chance alone, individual alleles may be over- or underrepresented in the gamete pool, resulting in random changes in frequency from one generation to the next. For any one locus with two alleles, *A* and *a*, genetic drift may result in one of the alleles eventually disappearing while the other becomes fixed, meaning it becomes the only version of that gene present in the gene pool of the population. The probability that an allele will become fixed through drift is the same as its initial frequency. So, if $p(A) = 0.8$, the probability that *A* will become fixed is 0.8, or 80 percent; the probability that *A* will be lost through drift is $(1 - 0.8) = 0.2$, or 20 percent.

To study genetic drift in laboratory populations of *Drosophila melanogaster*, Warwick Kerr and Sewall Wright set up over 100 separately breeding lines, with four males and four females as the parents for each line. Within each line, the frequency of the sex-linked bristle mutant *forked* (*f*) and its wild-type allele ($f^+$) was 0.5. In each generation, four males and four females were chosen at random to parent the next generation. After 16 generations, the complete loss of one allele and the fixation of the other had occurred in 70 lines—29 of which ended up with only the *forked* allele and 41 of which ended up with the wild-type allele becoming fixed. The remaining lines still contained both alleles or had gone extinct. If fixation occurred randomly, then each allele would have the same chance of becoming fixed. In fact, the experimental results do not differ statistically from the ratio of 35:35, which would be expected to be produced by random fixation. These results demonstrate that alleles can spread through a population and eliminate other alleles by chance alone.

How are small populations created in nature? A disaster such as an epidemic might occur, leaving a small number of survivors to breed. Or a small group might emigrate from the larger population and become founders in a new environment, such as a volcanically created island. Among endangered plants and animals, habitat loss often results in the creation of small isolated populations in which loss of genetic diversity through drift can be a threat to the species' long-term survival. (This is discussed in more detail in Chapter 29.)

### Founder Effects in Human Populations

Allele frequencies in certain human populations demonstrate the role of genetic drift as an evolutionary force in natural populations. Native Americans living in the southwestern United States, for example, have a high frequency of oculocutaneous albinism (OCA). In the Navajo, who live primarily in northeast Arizona, albinism occurs with a frequency of 1 in 1500–2000, compared with 1 in 36,000 in whites and 1 in 10,000 in African-Americans. There are four different forms of OCA (OCA1–4), and each is associated with varying degrees of melanin deficiency in the skin, eyes, and hair. The OCA phenotype in the Navajo is similar to the phenotypes produced by OCA2 and OCA4. OCA2 is caused by mutations in the *P* gene, which encodes a plasma membrane protein; OCA4 is caused by mutations in a gene called *MAPT*. To investigate whether albinism in the Navajo was OCA2 or OCA4, Murray Brilliant and his colleagues screened for mutations in the *P* and *MAPT* genes. In their study, they found no mutations in the *MAPT* gene. Instead, all Navajo with albinism were homozygous for a 122.5-kb deletion in the *P* gene, spanning exons 10–20 (Figure 27–17). This deletion allele was not present in 34 individuals belonging to other Native American populations.

After molecular analysis of the breakpoints, the investigators developed a set of PCR primers to identify homozygous affected individuals and heterozygous carriers (Figure 27–18). Using these primers, they surveyed 134 normally pigmented Navajo and 42 members of the Apache tribe, who are closely related to the Navajo. Based on this sample, the heterozygote frequency in the Navajo is estimated to be 4.5 percent. No carriers were found in the Apache population that was studied.

The fact that the 122.5-kb deletion allele causing OCA2 was found only in the Navajo population and not in members of other Native American tribes in the southwestern United States suggests that the mutant allele is specific to the Navajo and may have arisen in a single individual who was one of a small number of founders of the Navajo population. Such founder mutations originate on a single chromosome and have a characteristic set of closely linked flanking markers. Over time, mutation and recombination modify or replace this set of markers, or both. If the rate of recombination and

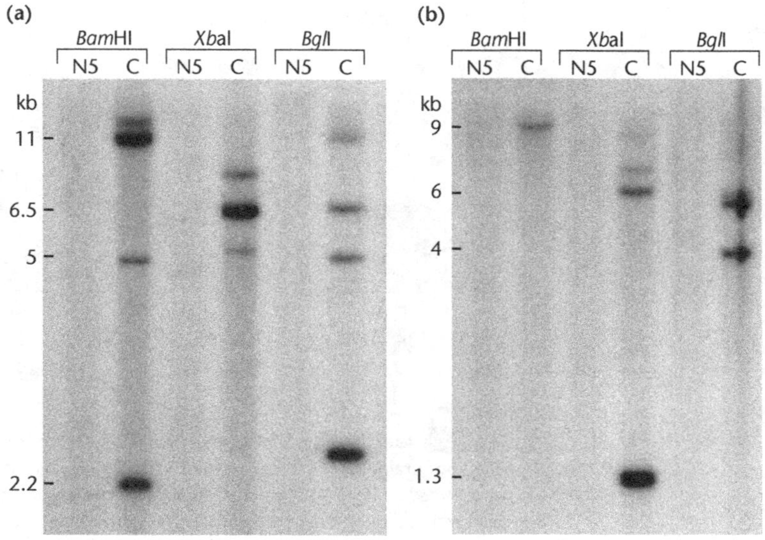


FIGURE 27-17 Genomic DNA digests from a Navajo affected with albinism (N5) and a normally pigmented individual (C). (a) Hybridization with a probe covering exons 11–15 of the *P* gene; there are no hybridizing fragments detected in N5. (b) Hybridization with a probe covering exons 15–20 of the *P* gene; again, there are no hybridizing fragments detected in N5. These results confirm the presence of a deletion in affected individuals. *(a) Courtesy of Murray Brilliant, A 122.5 kilobase deletion of P gene underlies the high prevalence of oculocutaneous albinism type 2 in the Navajo population: From: American Journal Human Genetics 72: 62–72, fig. 1 p. 65. Published by University of Chicago Press. (b) Courtesy of Murray Brilliant, A 122.5 kilobase deletion of P gene underlies the high prevalence of oculocutaneous albinism type 2 in the Navajo population: From: American Journal Human Genetics 72: 62–72, fig. 2 p. 66. Published by University of Chicago Press.*

the rate of mutation in this region of the genome are known, the origin of the founder mutation can be dated. Using these parameters and studying flanking alleles in deletion and nondeletion chromosomes, Brilliant and his colleagues estimated the age of the mutation to be between 400 and 11,000 years. To narrow this range, they searched for clues in tribal history. Navajo oral tradition indicates that the Navajo and Apache became separate populations between 600 and 1000 years ago. Because the deletion is not found in the Apaches, it probably arose in the Navajo population after the tribes split. On this basis, the deletion is estimated to be 400 to 1000 years old and is believed to have arisen as a founder mutation.

## Allele Loss during a Bottleneck

Although populations of many species have undergone well-documented bottlenecks, and as a result now exhibit low genetic variation in the surviving population, the amount of genetic variation present in the pre-bottleneck population—and therefore the amount of genetic variation lost in the bottleneck—often cannot be measured. Fortunately for population geneticists interested in the effects of bottlenecks, the loss of genetic variation following a bottleneck in populations of one species, the greater prairie chicken, has been measured in several studies, using museum specimens and samples collected and stored over time. Across the Midwestern United States, prairie chickens were once abundant. The species numbered millions of birds in the 1860s. Loss of habitat and other human activity caused a drastic reduction in these numbers. In central Wisconsin, for example, the prairie chicken population had been reduced to about 2500 by 1950, and over the next ten years, this number fell to about 1500 birds, where it has remained, with minor fluctuations.

In one study designed to determine whether the population bottleneck caused loss of genetic variation, Renee Bellinger and colleagues used PCR to amplify alleles at six microsatellite loci (see Chapter 24 for a discussion of microsatellites). Samples were obtained from specimens collected in 1951 and from specimens collected from 1996 to 1999 in the same geographic region. This collection strategy minimized geographic differences in allele frequency.

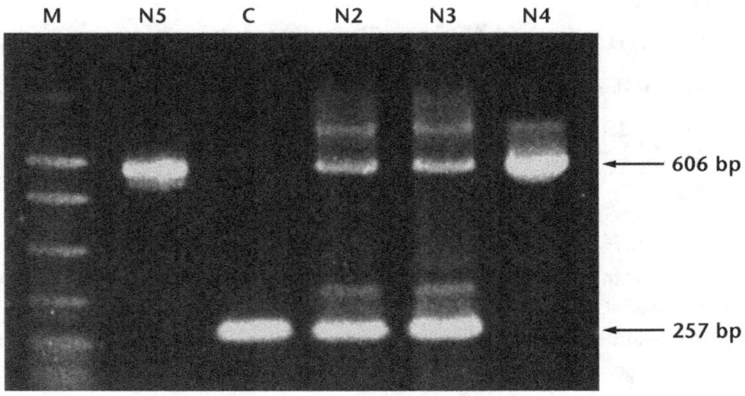


FIGURE 27-18 PCR screens of Navajo affected with albinism (N4 and N5) and the parents of N4 (N2 and N3). Affected individuals (N4 and N5), heterozygous carriers (N2 and N3), and a homozygous normal individual (C) each give a distinctive band pattern, allowing detection of heterozygous carriers in the population. Molecular size markers (M) are in the first lane. *Courtesy of Murray Brilliant, A 122.5 kilobase deletion of P gene underlies the high prevalence of oculocutaneous albinism type 2 in the Navajo population: From: American Journal Human Genetics 72: 62–72, fig. 3 p. 67. Published by University of Chicago Press.*

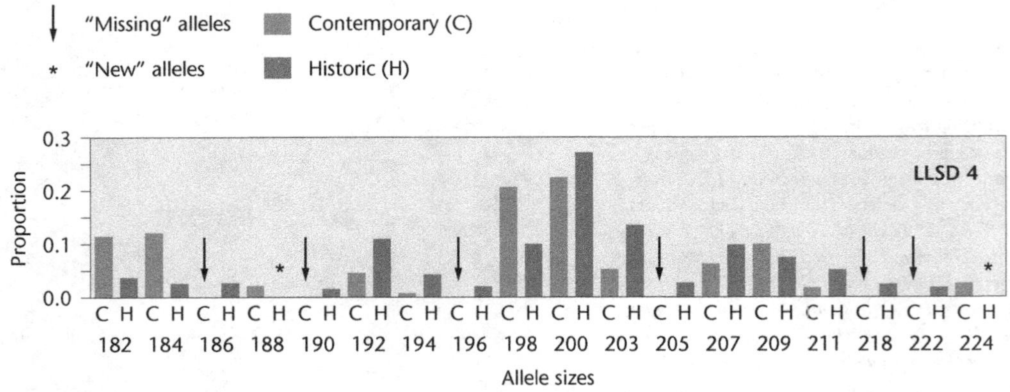


FIGURE 27–19 Microsatellite alleles (variations in the size of the microsatellite repeated DNA) at the *LLSD4* locus. DNA samples of prairie chickens from central Wisconsin collected in 1951 [designated as the historic (H) population] and collected in 1996–1999 [designated as the contemporary (C) population]. The *LLSD4* locus contains 18 alleles, each with a different microsatellite length shown in base pairs along the bottom of the figure. Alleles lost between 1951 and 1996–1999 are indicated by arrows. Alleles present in the modern population but not in the 1951 population are indicated by asterisks. In addition to allele loss, shifts in allele frequency are found at all loci. Source: Bellinger, M. R., et al. 2003. Loss of genetic variation in Greater prairie chickens following a population bottleneck in Wisconsin, U.S.A. *Conservation Biology 17* (3): 717–724, their figure 2, p. 721.

The observed changes in genetic variation at one of these loci are presented in Figure 27–19. The results show that there are significantly fewer alleles at this locus in the modern population than in the 1951 population. Across all six loci, 16 of 55 alleles were lost, representing a 29 percent decline in genetic variability. In addition to allele loss, the modern population also experienced a significant decline in heterozygosity. (The presence of new alleles in the modern population that are not seen in the 1951 population may be due to the small number of individuals in the older population sample, but mutation and migration cannot be ruled out as explanations.)

The decline in genetic variation quantified in this population sample is the result of genetic drift in the form of a bottleneck. Unless the historic trend of allele loss in these small, fragmented populations of prairie chickens is reversed by management or intervention, it is likely that these populations will continue to lose alleles and, as their fitness declines—for reasons discussed in the next section and in Chapters 28 and 29—eventually become extinct.

## 27.9 Nonrandom Mating Changes Genotype Frequency but Not Allele Frequency

We have explored how violations of the first four assumptions of the Hardy–Weinberg law, in the form of selection, mutation, migration, and genetic drift, can cause allele frequencies to change. The fifth assumption is that the members of a population mate at random; in other words, any one genotype has an equal probability of mating with any other genotype in the population. Nonrandom mating can change the frequencies of genotypes in a population. Subsequently, selection for or against certain genotypes has the potential to affect the overall frequencies of the alleles they contain, but it is important to note that nonrandom mating *does not itself directly change allele frequencies.*

Nonrandom mating can take one of several forms. In **positive assortive mating,** similar genotypes are more likely to mate than dissimilar ones. This often occurs in humans: a number of studies have indicated that many people are more attracted to individuals who resemble them physically (and are therefore more likely to be genetically similar as well). **Negative assortive mating** occurs when dissimilar genotypes are more likely to mate; some plant species have inbuilt pollen/stigma recognition systems that prevent fertilization between individuals with the same alleles at key loci. However, the form of nonrandom mating most commonly found to affect genotype frequencies in population genetics is inbreeding.

### Coefficient of Inbreeding

**Inbreeding** occurs when mating individuals are more closely related than any two individuals drawn from the population at random; loosely defined, inbreeding is mating among relatives. For a given allele, inbreeding increases the proportion of homozygotes in the population. A completely inbred population will theoretically consist only of homozygous genotypes. To demonstrate this phenomenon, let us consider the most extreme form of inbreeding, **self-fertilization,** which though rare in animals is widespread in plant species. Figure 27–20 shows the results of four generations of self-fertilization, starting with a single individual heterozygous for one pair of alleles. By the fourth generation, only about 6 percent of the individuals are still heterozygous, and 94 percent of the population is homozygous. Note, however, that the frequencies of alleles *A* and *a* remain unchanged at 50 percent.

$P_1$ — Self-fertilization *Aa*

| | *AA* | *Aa* | *aa* |
|---|---|---|---|
| $F_1$ | 0.250 | 0.500 | 0.250 |
| $F_2$ | 0.375 | 0.250 | 0.375 |
| $F_3$ | 0.437 | 0.125 | 0.437 |
| $F_4$ | 0.468 | 0.063 | 0.468 |
| $F_n$ | $\frac{1-\frac{1}{2^n}}{2}$ | $\frac{1}{2^n}$ | $\frac{1-\frac{1}{2^n}}{2}$ |

**FIGURE 27–20** Reduction in heterozygote frequency brought about by self-fertilization. The frequencies of the genotypes after *n* generations can be calculated according to the formulas in the bottom row.

Not all inbreeding in populations occurs through self-fertilization, and different degrees of inbreeding exist. To describe the intensity of inbreeding in a population, geneticist Sewall Wright devised the **coefficient of inbreeding (*F*)**. *F* quantifies the probability that the two alleles of a given gene in an individual are identical *because they are descended from the same single copy of the allele in an ancestor*. If $F = 1$, all individuals in the population are homozygous, and both alleles in every individual are derived from the same ancestral copy. If $F = 0$, no individual has two alleles derived from a common ancestral copy.

One simple method of estimating *F* for a population is based on the inverse relationship between inbreeding and the frequency of heterozygotes: As the level of inbreeding increases, the proportion of heterozygotes declines. Therefore, *F* can be calculated as

$$F = \frac{H_e - H_o}{H_e}$$

where $H_e$ is the expected heterozygosity based on the Hardy–Weinberg law and $H_o$ the observed heterozygosity in a population. Note that in a completely random mating population the expected and observed levels of heterozygosity will be equal and $F = 0$.

A different method can be used to estimate *F* for an individual. Figure 27–21 shows a pedigree of a first-cousin marriage. The fourth-generation female (shaded pink) is the daughter of first cousins (yellow). Suppose her great-grandmother (green) was a carrier of a recessive lethal allele, *a*. What is the probability that the fourth-generation female will inherit two copies of her great-grandmother's lethal allele? For this to happen, (1) the great-grandmother has to pass a copy of the allele to her son, (2) her son has to pass it to his daughter, and (3) his daughter has to pass it to her daughter (the pink female). Also, (4) the great-grandmother has to pass a copy of the allele to her daughter, (5) her daughter has to pass it to her son, and (6) her son has to pass it to his daughter (the pink female). Each of the six necessary events has an individual probability of 1/2, and they *all* have to happen, so the probability that the pink female will inherit two copies of her great-grandmother's lethal allele is $(1/2)^6 = 1/64$. To calculate an overall value of *F* for the pink female as a child of a first-cousin marriage, remember that she could also inherit two copies of any of the other three alleles present in her great-grandparents. Because any of four possibilities would give the pink female two alleles identical by descent from an ancestral copy,

$$F = 4 \times (1/64) = 1/16.$$

**NOW SOLVE THIS**

Problem 22 on page 735 asks you to calculate the inbreeding coefficient of a population based on DNA marker data.

■ HINT: *First, work out the frequencies of the DNA marker alleles. Then, if this population is in Hardy–Weinberg equilibrium, how many heterozygotes would you expect to see?*

## Outcomes of Inbreeding

Inbreeding results in the production of individuals homozygous for recessive alleles that were previously concealed in heterozygotes. Because many recessive alleles are deleterious when homozygous, one consequence of inbreeding is an increased chance that an individual will be homozygous for a recessive deleterious allele. Inbred populations often have a lowered mean fitness. **Inbreeding depression** is a measure of the loss of fitness caused by inbreeding. In

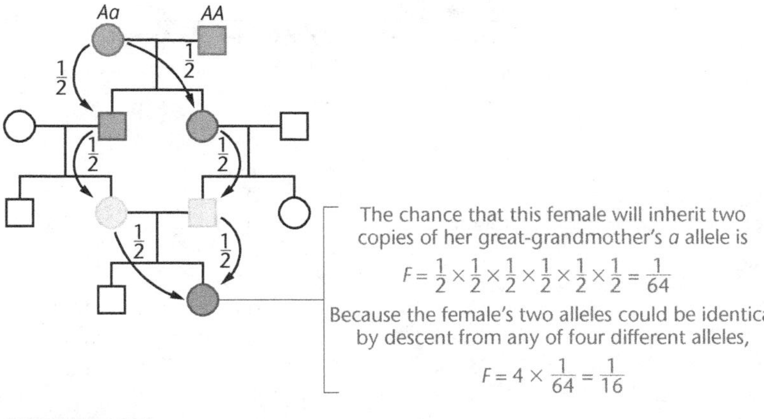


**FIGURE 27–21** Calculating the coefficient of inbreeding *(F)* for the offspring of a first-cousin marriage.

domesticated plants and animals, inbreeding and selection have been used for thousands of years, and these organisms already have a high degree of homozygosity at many loci. Further inbreeding will usually produce only a small loss of fitness. However, inbreeding among individuals from large, randomly mating populations can produce high levels of inbreeding depression. This effect can be seen by examining the mortality rates in offspring of inbred animals in zoo populations (Table 27.4). The potential impact of inbreeding on populations of threatened and endangered species is discussed further in Chapter 29.

In humans, inbreeding increases the risk of spontaneous abortions, neonatal deaths, congenital deformities, and recessive genetic disorders. Although less common than in the past, inbreeding occurs in many regions of the world where social customs favor marriage between first cousins. Alan Bittles and James Neel analyzed data from numerous studies on different cultures. They found that the rate of child mortality (i.e., death in the first several years of life) varies dramatically from culture to culture. No matter what the baseline mortality rate for children of unrelated parents, however, children of first cousins virtually always have a higher death rate—typically by about 4.5 percentage points.

Inbreeding is not always harmful. Indeed, inbreeding has long been recognized as a useful tool for breeders of domesticated plants and animals. When an inbreeding program is initiated, homozygosity increases, causing some breeding stocks to become fixed for favorable alleles and others to become fixed for unfavorable alleles. If the more viable and vigorous plants or animals are selected for further breeding, the proportion of individuals carrying desirable traits can be increased.

If members of two separately inbred lines are mated, hybrid offspring are often more vigorous in desirable traits than either of the parental lines. This phenomenon is called **hybrid vigor**. When such an approach was used in breeding programs established for maize, crop yields increased tremendously. However, hybrid vigor is highest in the $F_1$ generation and typically declines in subsequent generations due to additional allelic segregation and recombination. Consequently, the $F_1$ hybrids must be regenerated for each planting by crossing the original inbred parental lines.

Hybrid vigor has been explained in two ways. The first theory, the **dominance hypothesis**, is based on the immediate reversal of inbreeding depression, which inevitably must occur in outcrossing. Consider a cross between two strains of maize with the following genotypes:

$$\begin{array}{ccccc} \text{Strain A} & & \text{Strain B} & & F_1 \\ aaBBCCddee & \times & AAbbccDDEE & \longrightarrow & AaBbCcDdEe \end{array}$$

The $F_1$ hybrids are heterozygotes at all loci shown. Any deleterious recessive alleles present in the homozygous form in the parents are masked by the more favorable dominant alleles in the hybrids. Such masking is thought to cause hybrid vigor.

The second theory, **overdominance**, holds that in many cases the heterozygote is superior to either homozygote perhaps because two forms of a gene product may be present in the heterozygote, thereby providing a form of biochemical diversity. Thus, the cumulative effect of heterozygosity at many loci accounts for the hybrid vigor. Most likely, hybrid vigor results from a combination of the phenomena cited in these hypotheses.

We have seen that under the Hardy–Weinberg law nonrandom mating can drive genotype frequencies in a population away from their expected values. This can indirectly affect the course of evolution. As in stocks purposely inbred by animal and plant breeders, inbreeding in a natural population may increase the frequency of homozygotes for a deleterious recessive allele. With domestic stocks, however, this increase will in turn increase the efficiency with which selective breeding can remove the deleterious allele from the population. Once the deleterious genes are removed, inbreeding no longer causes problems.

**TABLE 27.4**

**Mortality in Offspring of Inbred Zoo Animals**

| Species | *n* | | Noninbred | Inbred | Inbreeding Coefficient |
|---|---|---|---|---|---|
| Zebra | 32 | Lived: | 20 | 3 | 0.250 |
| | | Died: | 7 | 2 | |
| Eld's deer | 24 | Lived: | 13 | 0 | 0.250 |
| | | Died: | 4 | 7 | |
| Giraffe | 19 | Lived: | 11 | 2 | 0.250 |
| | | Died: | 3 | 3 | |
| Oryx | 42 | Lived: | 35 | 0 | 0.250 |
| | | Died: | 2 | 5 | |
| Dorcas gazelle | 92 | Lived: | 36 | 17 | 0.259 |
| | | Died: | 14 | 25 | |

# GENETICS, TECHNOLOGY, AND SOCIETY

## Tracking Our Genetic Footprints out of Africa

Where did we humans come from? Are we one big family with only minor differences from one person to the next, or are we separated into distinct races, each having profound and ancient roots? For millennia, our efforts to answer these questions invoked legends, mythologies, and the creation stories of our many religions. Over the last century, paleoanthropologists have applied a variety of sophisticated scientific tools to explore our origins and human kinships. The evolving story of our beginnings, based on modern genetics, is as fascinating and controversial as any creation myth.

Based on the physical traits and distribution of hominid fossils, most paleoanthropologists agree that a large-brained, tool-using hominid they call *Homo erectus* appeared in east Africa about 2 million years ago. This species used simple stone tools, hunted but did not fish, did not build houses or fireplaces, and did not follow ritual burial practices. About 1.7 million years ago, *H. erectus* spread into Eurasia and south Asia. Most scientists also agree that *H. erectus* likely developed into several hominid types, including Neanderthals (in Europe) and Peking man or Java man (in Asia). These hominids were anatomically robust, with large, heavy skeletons and skulls. Neanderthals and other *H. erectus* groups disappeared 50,000 to 30,000 years ago—around the same time that anatomically modern humans (*H. sapiens*) appeared all over the world.

It is at this point in our history—when ancient hominids gave way to anatomically modern humans—that controversy arises.

At present, there are two dominant hypotheses to explain human origins: the multiregional hypothesis and the out-of-Africa hypothesis. The multiregional hypothesis is based primarily on archaeological and fossil evidence. It proposes that *H. sapiens* developed gradually and simultaneously all over the world from existing *H. erectus* groups, including Neanderthals. Interbreeding between these groups eventually made *H. sapiens* a genetically homogeneous species. Natural selection over 1.5 million years then created the regional variants (races) that we see today. In the multiregional view, our genetic makeup should include contributions from Neanderthals and other *H. erectus* groups. In contrast, the out-of-Africa hypothesis, based primarily on genetic analyses of modern human populations, contends that *H. sapiens* evolved from the descendants of *H. erectus* in sub-Saharan Africa about 200,000 to 400,000 years ago. A small band of *H. sapiens* (fewer than 10,000) then left Africa, expanded, and migrated into Europe and Asia around 100,000 years ago. By about 60,000 years ago, populations of *H. sapiens* reached Australia and later migrated into North America. In the out-of-Africa model, *H. sapiens* replaced all the preexisting *H. erectus* types, without interbreeding. In this way, *H. sapiens* became the only species in the genus by about 30,000 years ago.

Although the out-of-Africa hypothesis is still debated, most genetic evidence appears to support it. Humans all over the globe are remarkably similar genetically. DNA sequences from any two people chosen at random are 99.9 percent identical. More genetic identity exists between two persons chosen at random from a human population than between two chimpanzees chosen at random from a chimpanzee population. Interestingly, about 90 percent of the genetic differences that do exist occur between individuals rather than between populations. This unusually high degree of genetic relatedness in all humans around the world supports the idea that our species arose recently from a small founding group of humans.

Studies of mitochondrial DNA sequences from current human populations reveal that the highest levels of genetic variation occur within African populations. Africans show twice the mitochondrial DNA-sequence diversity of non-Africans. This implies that the earliest branches of *H. sapiens* diverged in Africa and had a longer time to accumulate mitochondrial DNA mutations, which are thought to accumulate at a constant rate over time.

DNA sequences from mitochondrial, Y-chromosome, and chromosome-21 markers support the idea that human roots are in east Africa and that the migration out of Africa occurred through Ethiopia, along the coast of the Arabian Peninsula, and outward to Eurasia and southeast Asia. Recent data based on DNA-sequence diversity in nuclear microsatellites further support the notion that humans migrated out of Africa and dispersed throughout the world from a small founding population. Sub-Saharan African populations show the highest levels of microsatellite heterozygosity, followed by those in the Middle East, Europe, East Asia, Oceania, and the Americas—in that order. Native American populations show about 15 percent less microsatellite heterozygosity than that seen in Africans.

By comparing DNA-sequence differences between populations around the world and by extrapolating back to a time when all sequences would have been the same, paleoanthropologists propose that modern *H. sapiens* developed from a small group in Africa between 200,000 and 400,000 years ago. The time of the out-of-Africa migration is calculated to be 50,000 to 100,000 years ago.

The recent sequencing of Neanderthal mitochondrial DNA shows that it is so different from ours that Neanderthals were likely a separate species and that Neanderthals and *H. sapiens* diverged about 600,000 years ago. Hence, it appears unlikely that Neanderthals and perhaps other *H. erectus* groups such as Peking man contributed significantly to the *H. sapiens* gene pool.

If all people on Earth are so similar genetically, how did we come to have such a range of physical differences, which some describe as racial differences? Many geneticists believe that the genetic changes responsible for these characteristics, such as skin color and facial features, could have accumulated over short periods of time, especially if they were adaptive to particular climatic and geographic conditions.

As with any explanation of human origins, the out-of-Africa hypothesis is actively debated and may undergo mutation—or even extinction—over time. As methods to sequence DNA from ancient fossils improve, it may be possible to fill the gaps in the genetic pathway leading out of Africa and to resolve those age-old questions about our origins.

### ■ References


Cavalli-Sforza, L.L., and Feldman M.W. 2003. The application of molecular genetic approaches to the study of human evolution. *Nat. Gen. (Suppl.)* 33: 266–275.

Wade, N. 2006. *Before the Dawn: Recovering the Lost History of our Ancestors.* New York: Penguin Press.

Wells, S. 2003. *The journey of man: A genetic odyssey.* Princeton, NJ: Princeton University Press.


### ■ Web Sites


Johanson, D. 2001. *Origins of modern humans: Multiregional or out of Africa?* http://www.actionbioscience.orglevolutionljohanson.html

EXPLORING GENOMICS

# Single-Nucleotide Polymorphisms (SNPs) and the Y Chromosome Haplotype Reference Database (YHRD)

In this chapter we examined important principles of population genetics, and we considered examples of allele frequency variations in human populations. Among the variations studied by population geneticists are *single-nucleotide polymorphisms (SNPs)*, which are single base changes, and *short tandem repeats (STRs)*, which are short sequences of tandem base repeats that are typically 2 to 9 bp in length. Both SNPs and STRs can be inherited as *haplotypes*, clusters or blocks of sequence variations that are closely linked on a chromosome and typically inherited as a unit.

In Exercise I you will explore a SNP database to learn more about SNPs in the gene for cystic fibrosis. In Exercise II you will use the **Y Chromosome Haplotype Reference Database (YHRD)** to learn about population frequencies for STRs on the Y chromosome.

## ■ Exercise I – Single-Nucleotide Polymorphisms (SNPs) and the International HapMap Project

The **SNP Research Facility** Web site produced by Washington University (St. Louis) School of Medicine is an excellent resource for learning about SNPs. The scientists at this facility are part of the **International HapMap Project,** a partnership of scientists from Canada, China, Japan, Nigeria, the United Kingdom, and the United States. Visit http://www.hapmap.org/abouthapmap.html for an excellent overview of SNPs and haplotype variations in humans.

A major goal of the HapMap project is to identify patterns of SNP sequence variations in humans, particularly those SNPs involved in disease and individual response to medicines. HapMap scientists expect that SNP discovery will lead to the development of better customized medicines (pharmacogenomics), and this is a major reason many private companies have invested heavily in the HapMap Project. To date, over 7 million SNPs from different populations around the world have been identified.

1. Access the SNP Research Facility homepage at http://snp.wustl.edu/index.html.
2. Click on the "SNPS" tab at the top of the page to learn more about SNPs and the International Hapmap Project.
3. Back on the homepage, use the "SNPseek Database" link to access a search engine for this database. Search the Main Menu for *CFTR* to retrieve SNP data for the cystic fibrosis transmembrane conductance regulator gene. The *CFTR* gene, whose allele frequencies we discussed earlier in this chapter, encodes a cell-surface protein that functions as a chloride channel. It has been particularly challenging to develop drug and gene therapy treatments for cystic fibrosis because of the number of SNPs that occur in the *CFTR* gene.
4. In the table showing "Known Genes Matching *CFTR*," click the link for NM_000492. The next screen will provide a summary of SNP details, including a table of "Annotated Features" that shows whether a SNP occurs in an intron or an exon or other regions of the gene.
   a. What chromosome contains the *CFTR* gene?
   b. How many SNPs for this gene have been identified so far?
5. In the Gene Info box, click the link to the right of "Variation," indicating the number of SNPs in the database. On the next page, leaving all of the parameters at their default values to maximize information retrieved from the database, click "Search." You will now see a comprehensive table providing information on the position of the SNP, a reference sequence (RS) number for the SNP, the nucleotide variation at this position, the gene region affected, and whether this SNP is conserved in other species. The "Minor Allele Freqs" columns provide allele frequency data from different populations, if such data is in the database (see the bottom of the page for definitions of abbreviations). Let's use these features to learn more about a *CFTR* SNP.
6. Find the SNP with the RS number rs1042077. Then answer the following questions:
   a. Is this SNP located in an intron or an exon of the *CFTR* gene?
   b. What nucleotides are involved in this SNP?
   c. Is this SNP conserved in other species?
   d. What population in the database shows the highest allele frequency for this SNP?
7. Click on the RS number and then use the "SNP Info" box to link to the NCBI Single-Nucleotide Polymorphism site (dbSNP) and the International HapMap Project site. Both links will open directly to pages on the rs1042077 SNP. Explore these sites to learn more about this *CFTR* SNP. For example, use the HapMap site to see a chromosome map showing the locus for this SNP. Click the color pie charts to reveal homozygous and heterozygous allele frequencies for this SNP. Both of these sites are excellent resources for learning about SNPs.
8. Return to the SNPseek page and search for SNPs in a gene of interest to you.

## ■ Exercise II – The Y Chromosome Haplotype Reference Database (YHRD)

The Genetics, Technology, and Society essay for this chapter mentions that Y chromosome sequences have been used to track human migration patterns around the world. The **Y Chromosome Haplotype Reference Database (YHRD)** is a great program for examining haplotype frequencies for Y chromosome loci, particularly STRs on the Y chromosome. Y chromosome STR haplotypes have been very useful for tracing paternal lineages in genealogical and kinship testing and for identifying male DNA in crime scene samples.

1. Visit the YHRD site at http://www.yhrd.org/index.html. Explore the links at this site to learn more about the purpose of

YHRD and features of the Y chromosome that can be used for genealogy and forensic applications.

2. Use the "Haplotype" characteristics link to learn more about short terminal repeats (STRs) on the Y chromosome and the position of specific Y-STR loci; then answer the following questions:

   a. What are the minimal haplotypes (minHt)?

   b. Is the locus for DYS390 located on the p or q arm of the Y?

3. Use the "Population analyses" link to learn more about the frequency of minHts in different populations. The "Per population feature" will show you a drop-down menu of different choices. For example, what is the most frequent minHt for people in Verona, Italy? View other populations you are interested in.

4. Use the "Search database" link "GeoSearch" feature to search for a particular haplotype frequency of your choice.

   a. What does it mean if a particular population has a "10" for a particular STR locus such as DYS390?

   b. Explore the "GeoSearch" feature by entering a haplotype combination of your choice. What did you find? Can you find a haplotype combination that does not match any of the haplotypes in the YHRD? Various answers are possible.

5. Use the "Compare populations" feature of "Population analyses" to compare minHt frequencies for people from "USA[European American]" with frequencies for people from "Birmingham, UK." Do the same to compare "USA[European American]" with "Virginia, USA[Hispanic American]" and to compare "USA[European American]" with "West Africa." What do these results tell us about the ancestry of these populations? Are these data consistent with what you have learned about migration and gene flow? What do these results tell us about the relatedness of these populations based on the Y-minHts evaluated?

6. Explore the site to learn more about Y-STRs in other populations that interest you.

## Chapter Summary

1. Populations evolve as a result of changes in allele frequencies at a number of loci over a period of time. Population genetics studies the factors driving change in allele frequencies and the amount and distribution of genetic variation in populations.
2. The Hardy–Weinberg law provides a simple mathematical model describing the relationship between allele frequency and genotype frequency in a population. This allows prediction of allele or genotype frequencies at a given locus in a population under a set of simple assumptions. If there is no selection, mutation, or migration; if the population is large; and if individuals mate at random, then allele frequencies will not change from one generation to the next.
3. The Hardy–Weinberg formula can be used to investigate whether or not a population is in equilibrium at a given locus and to estimate the frequency of heterozygotes in a population from the frequency of homozygous recessives.
4. By specifying the conditions under which allele frequencies will not change, the Hardy–Weinberg law identifies the forces that can drive evolution in a population. Selection, mutation, migration, and genetic drift can cause change in allele frequencies.
5. Nonrandom mating alters genotype frequency but not allele frequency in a population. Inbreeding, or mating between relatives, is the form of nonrandom mating with the most significant impact: it increases the frequency of homozygotes in a population and decreases the frequency of heterozygotes.
6. Natural selection is the most powerful of the forces affecting allele frequency. The rate of change under natural selection depends on initial allele frequencies, selection intensity, and the relative fitness of different genotypes.
7. Mutation and migration introduce new alleles into a population, but usually have only small effects on allele frequencies. Persistence of new alleles in a population depends on the fitness they confer and the action of selection.
8. Genetic drift produces random change in allele frequencies and can have a major impact in small populations.

## INSIGHTS AND SOLUTIONS

1. Tay–Sachs disease, inherited as an autosomal recessive condition, is caused by loss-of-function mutations in a gene on chromosome 15 that encodes a lysosomal enzyme. Among Ashkenazi Jews of central European ancestry, about 1 in 3600 children is born with the disease. What fraction of the individuals in this population are carriers?

**Solution:** If we let $p$ represent the frequency of the wild-type enzyme allele and $q$ the total frequency of recessive loss-of-function alleles, and if we assume that the population is in Hardy–Weinberg equilibrium, then the frequencies of the genotypes are given by $p^2$ for homozygous normal, $2pq$ for carriers, and $q^2$ for individuals with Tay–Sachs. The frequency of Tay–Sachs alleles is thus

$$q = \sqrt{q^2} = \sqrt{\frac{1}{3600}} = 0.017$$

Since $p + q = 1$, we have

$$p = 1 - q = 1 - 0.017 = 0.983$$

Therefore, we can estimate that the frequency of carriers is

$$2pq = 2(0.983)(0.017) = 0.033 \text{ or } 1 \text{ in } 30$$

*Continued on next page*

*Insights and Solutions, continued*

2. *Eugenics* is the term employed for the selective breeding of humans to bring about improvements in populations. As a eugenic measure, it has been suggested that individuals suffering from serious genetic disorders should be prevented (sometimes by force of law) from reproducing (by sterilization, if necessary) in order to reduce the frequency of the disorders in future generations. Suppose that a harmful recessive trait were present in the population at a frequency of 1 in 40,000 and that affected individuals did not reproduce. In 10 generations, or about 250 years, what would be the frequency of the condition? Would the eugenic measures be effective in this case?

**Solution:** Let $q$ represent the frequency of the recessive allele responsible for the disorder. Because the disorder is recessive, we can estimate that

$$q = \sqrt{\frac{1}{40{,}000}} = 0.005$$

If all affected individuals are prevented from reproducing, then in an evolutionary sense, the disorder is lethal: affected individuals have zero fitness. This means that we can predict the frequency of the recessive allele 10 generations in the future by using the equation

$$q_g = q_0/(1 + gq_0)$$

Here, $q_0 = 0.005$, and $g = 10$, so we have

$$q_{10} = \frac{(0.005)}{[1 + (10 \times 0.005)]} = 0.0048$$

If $q_{10} = 0.0048$, then the frequency of homozygous recessive individuals will be roughly

$$(q_{10})^2 = (0.0048)^2 = 0.000023 = \frac{1}{43{,}500}$$

The frequency of the genetic disorder has been reduced from 1 in 40,000 to 1 in 43,500 in 10 generations, indicating that this eugenic measure has limited effectiveness.

## Problems and Discussion Questions

1. The ability to taste the compound PTC is controlled by a dominant allele $T$. Individuals homozygous for the recessive allele $t$ are unable to taste PTC. In a genetics class of 125 students, 88 can taste PTC and 37 cannot. Calculate the frequency of the $T$ and $t$ alleles in this population and the frequency of the genotypes.
2. Calculate the frequencies of the *AA*, *Aa*, and *aa* genotypes after one generation if the initial population consists of 0.2 *AA*, 0.6 *Aa*, and 0.2 *aa* genotypes and meets the requirements of the Hardy–Weinberg relationship. What genotype frequencies will occur after a second generation?
3. Each of the following values is the frequency in a certain population of a different rare disorder caused by an autosomal recessive mutation. For each disorder, calculate the percentage of heterozygous carriers in the population:
   (a) 0.0064
   (b) 0.000081
   (c) 0.09
   (d) 0.01
   (e) 0.10
4. What must be assumed in order to validate the answers in Problem 3?
5. In a population where only the total number of individuals with the dominant phenotype is known, how can you calculate the percentage of carriers and homozygous recessives?
6. Determine whether the following two sets of data represent populations that are in Hardy–Weinberg equilibrium (use $\chi^2$ analysis if necessary):
   (a) *CCR5* genotypes: *1/1*, 60 percent; *1/Δ32*, 35.1 percent; *Δ32/Δ32*, 4.9 percent
   (b) Sickle-cell hemoglobin: *AA*, 75.6 percent; *AS*, 24.2 percent; *SS*, 0.2 percent
7. If 4 percent of a population in equilibrium expresses a recessive trait, what is the probability that the offspring of two individuals who do not express the trait will express it?
8. Consider a population in which the frequency of allele $A$ is $p = 0.7$ and the frequency of allele $a$ is $q = 0.3$, and where the alleles are codominant. For each of the following sets of fitness levels, what will be the allele frequencies in the population after one generation?
   (a) $w_{AA} = 1$, $w_{Aa} = 0.9$, and $w_{aa} = 0.8$
   (b) $w_{AA} = 1$, $w_{Aa} = 0.95$, $w_{aa} = 0.9$
   (c) $w_{AA} = 1$, $w_{Aa} = 0.99$, $w_{aa} = 0.98$
   (d) $w_{AA} = 0.8$, $w_{Aa} = 1$, $w_{aa} = 0.8$
9. If the initial allele frequencies are $p = 0.5$ and $q = 0.5$ and allele $a$ is a lethal recessive, what will be the frequencies after 1, 5, 10, 25, 100, and 1000 generations?
10. Under what circumstances might a lethal dominant allele persist in a population?
11. Determine the frequency of allele $A$ in an island population after one generation of migration from the mainland under the following conditions:
    (a) $p_i = 0.6$; $p_m = 0.1$; $m = 0.2$
    (b) $p_i = 0.2$; $p_m = 0.7$; $m = 0.3$
    (c) $p_i = 0.1$; $p_m = 0.2$; $m = 0.1$
12. Assume that a recessive autosomal disorder occurs in 1 of 10,000 individuals (0.0001) in the general population and that in this population about 2 percent (0.02) of the individuals are carriers for the disorder. Estimate the probability of this disorder occurring in the offspring of a marriage between first cousins. Compare this probability to that for the population at large.
13. What is the basis of inbreeding depression?
14. Describe how inbreeding can be used in the domestication of plants and animals. Discuss the theories underlying these techniques.
15. Evaluate the following statement: Inbreeding increases the frequency of recessive alleles in a population.

16. In a breeding program to improve crop plants, which of the following mating systems should be employed to produce a homozygous line in the shortest possible time?
 (a) self-fertilization
 (b) brother–sister matings
 (c) first-cousin matings
 (d) random matings
 Illustrate your choice with pedigree diagrams.
17. If the albino phenotype occurs in 1/10,000 individuals in a population at equilibrium, and albinism is caused by an autosomal recessive allele *a*, calculate the frequency of
 (a) the recessive mutant allele
 (b) the normal dominant allele
 (c) heterozygotes in the population
 (d) matings between heterozygotes
18. In a human population of 4000 there are two males diagnosed with hemophilia (an X-linked disorder). Assuming this population is 50 percent male and 50 percent female, how many female carriers of hemophilia will there be?
19. One of the first Mendelian traits identified in humans was a dominant condition known as *brachydactyly*. The dominant allele causes an abnormal shortening of the fingers or toes (or both). At the time, some thought that the trait would spread until 75 percent of the population would be affected (because the phenotypic ratio of dominant to recessive is 3:1). Show that this reasoning is incorrect.
20. Achondroplasia is a dominant trait that causes a characteristic form of dwarfism. In a survey of 50,000 births, five infants with achondroplasia were identified. Three of the affected infants had affected parents, while two had parents who were not affected. Calculate the mutation rate for achondroplasia and express the rate as the number of mutant genes in a given number of gametes.
21. A prospective groom has a sister with cystic fibrosis (CF), an autosomal recessive disease, though he himself does not have it, nor do their parents. The brother plans to marry a woman who has no history of CF in her family. What is the probability that they will produce a CF child? They are both Caucasian, and the overall frequency of CF in the Caucasian population is 1/2500—that is, 1 affected child per 2500. (Assume the population meets the Hardy–Weinberg assumptions.)
22. A botanist studying waterlilies in an isolated pond observed three leaf shapes in the population: round, arrowhead, and scalloped. Marker analysis of DNA from 125 individuals showed the round-leafed plants to be homozygous for allele *r1*, while the plants with arrowhead leaves were homozygous for a different allele at the same locus, *r2*. Plants with scalloped leaves showed DNA profiles with both the *r1* and *r2* markers. Frequency of the *r1* marker was estimated at 0.81. If the botanist counted 20 plants with scalloped leaves in the pond, what is the inbreeding coefficient *F* for this population?
23. A farmer plants transgenic Bt corn that is genetically modified to produce its own insecticide. Of the corn borer larvae feeding on these Bt corn plants, only 10 percent survive unless they have at least one copy of the dominant resistance allele *B* that confers resistance to the Bt insecticide. When the farmer first plants Bt corn, the frequency of the *B* resistance allele in the corn borer population is 0.02. What will be the frequency of the resistance allele after one generation of corn borers fed on Bt corn?
24. In an isolated population of 50 desert bighorn sheep, a mutant recessive allele *c* has been found to cause curled coats in both males and females. The normal dominant allele *C* produces straight coats. A biologist studying these sheep counts four with curled coats and takes blood samples for DNA marker analysis, which reveals that 17 of the straight-coated sheep are carriers of the *c* allele. What is the inbreeding coefficient *F* for this population?
25. To increase genetic diversity in the bighorn sheep population described in Problem 24, 10 sheep are introduced from a population in a different region where the *c* mutation is absent. Assuming that random mating occurs between the original and the introduced sheep, and that the *c* allele is selectively neutral, what will be the frequency of *c* in the next generation?

**HOW DO WE KNOW?**

26. In this chapter we focused on how population geneticists study changes in allele frequency in population gene pools over time, the nature and amount of genetic diversity in populations, and patterns of distribution of different genotypes. At the same time, we found many opportunities to consider the methods and reasoning by which much of their understanding was acquired. From the explanations given in the chapter, what answers would you propose to the following fundamental questions:
 (a) How do geneticists detect the presence of different alleles in a population? How are different genotypes identified?
 (b) Are differences in morphological phenotype (the physical appearance of an individual) the only way that genetically distinct individuals in a population can be identified?

## Extra-Spicy Problems

27. A form of dwarfism known as Ellis–van Creveld syndrome was first discovered in the late 1930s, when Richard Ellis and Simon van Creveld shared a train compartment on the way to a pediatrics meeting. In the course of conversation, they discovered that they each had a patient with this syndrome. They published a description of the syndrome in 1940. Affected individuals have a short-limbed form of dwarfism and often have defects of the lips and teeth, as well as polydactyly (extra fingers). The largest pedigree for the condition was reported in an Old Order Amish population in eastern Pennsylvania by Victor McKusick and his colleagues (1964). In that community, about 5 per 1000 births are affected, and in the population of 8000, the observed frequency is 2 per 1000. All affected individuals have unaffected parents, and all affected cases can trace their ancestry to Samuel King and his wife, who arrived in the area in 1774. It is known that neither King nor his wife was affected with the disorder. There are no cases of the disorder in other Amish communities, such as those in Ohio or Indiana.
 (a) From the information provided, derive the most likely mode of inheritance of this disorder. Using the Hardy–Weinberg law, calculate the frequency of the mutant allele in the population and the frequency of heterozygotes, assuming Hardy–Weinberg conditions.
 (b) What is the most likely explanation for the high frequency of the disorder in the Pennsylvania Amish community and its absence in other Amish communities?

28. The following graph shows the variation in frequency of a particular allele ($A$) over time in two relatively small, independent populations exposed to very similar environmental conditions. A student has analyzed the graph and concluded that the best explanation for these data is that selection for allele $A$ is occurring in Population 1. Is the student's conclusion correct? Explain.

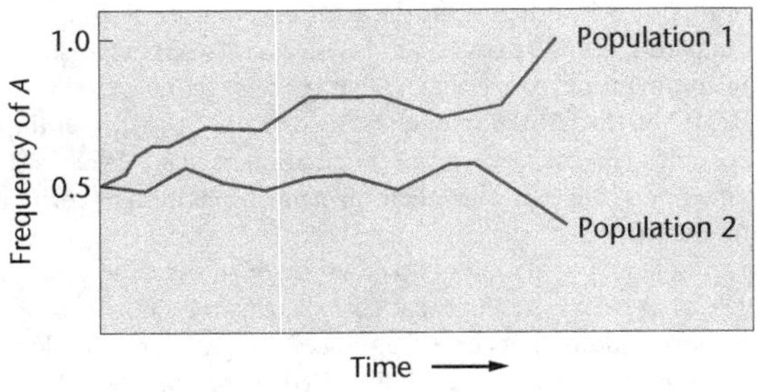


29. The original source of new alleles, upon which selection operates, is mutation, a variety of random events without regard to selectional value to the organism. Although many model organisms have been used to study mutational events in populations, some investigators have developed abiotic, molecular models. Soll (2006) examined one such model to study the relationship between both deleterious and advantageous mutations and population size in a ligase molecule composed of RNA (ribozyme). Soll found that the smaller the population of molecules, the more likely it was that not only deleterious mutations but also advantageous mutations would disappear. Why would population size influence the survival of both types of mutations (deleterious and advantageous) in populations?

30. A number of comparisons of nucleotide sequences among hominids and rodents indicates that inbreeding may have occurred more in hominid than in rodent ancestry. When a population bottleneck of approximately 10,000 individuals occurred, Knight (2005) and Bakewell (2007) suggest that this may have left early humans with a greater chance for genetic disease. Why would a population bottleneck influence the frequency of genetic disease?

31. In an attempt to determine the underlying molecular basis of hybrid vigor Hedgecock et al. (2007) examined shell length and other parameters in Pacific oysters (*Crassostrea gigas*). The accompanying table presents the shell length and growth rate (at 5 days of age) in two inbred ($F = 0.375$) and two hybrid oyster lines.
    (a) In this context what is meant by the equation $F = 0.375$?
    (b) Identify lines that are likely to be inbred and those that are likely to be hybrid.
    (c) Suggest a molecular basis for the differences between the two groups (inbred versus hybrid).

| Line | Shell Length ($\mu$m on day 5) |
|---|---|
| 33 | 102 |
| 35 | 105 |
| 53 | 109 |
| 55 | 99 |

Light and dark forms of the peppered moth *Biston betularia* on light tree bark. These moths are often cited as examples of rapid evolutionary change resulting from selective pressures during the industrial revolution in Great Britain.

28

# Evolutionary Genetics

## CHAPTER CONCEPTS

- Speciation can occur by transformation or by splitting of gene pools.
- Most populations and species harbor considerable genetic variation.
- The genetic structures of populations change across space and time.
- The definition of species is a great challenge for evolutionary biology.
- A reduction in gene flow between populations, accompanied by divergent selection or genetic drift, can lead to speciation.
- Genetic differences between populations or species are used to reconstruct evolutionary history.

As we embark on a discussion of genetics and evolution, consider two very different scenarios:

1. In late 1986, a Florida dentist tested positive for HIV. Several months later, he was diagnosed with AIDS. He continued to practice general dentistry for two more years, until one of his patients, a woman with no known risk factors, discovered that she, too, was infected with HIV. When the dentist publicly urged his other patients to have themselves tested, several more were found to be HIV positive. Did this dentist transmit HIV to his patients, or did the patients become infected by some other means?
2. Fossil evidence indicates that Neanderthals (*Homo neanderthalensis*) lived in Europe and western Asia from 300,000 to 30,000 years ago. For some of that time, Neanderthals coexisted with anatomically modern humans (*Homo sapiens*). Because they lived in the same places at the same time, did Neanderthals and modern humans interbreed, so that descendants of the Neanderthals are alive today, or did the Neanderthals die off, so that their lineage is now extinct? How were scientists able to establish our relationship to an extinct species?

These scenarios pose distinctly different kinds of questions. However, in one way the questions are closely related: they all can be addressed by using genetic data to reconstruct evolution. The acquisition and analysis of such data is the focus of the current chapter. We will use the methods of genetic analysis and the reconstruction of evolutionary history to address these two sets of questions at the end of the chapter.

Evolution results from two processes: the transformation of lineages and the splitting of lineages. Together, they produce a diversity of populations and species. In Chapter 27, we described the evolution of populations in terms of changes in allele frequencies, and we outlined the forces that can cause allele frequencies to change. In this chapter, we examine the population genetics processes of **microevolution**—defined as evolutionary change within species—and briefly consider how these processes can be extended to **macroevolution**—defined as evolutionary events leading to the emergence of new species and other taxonomic groups.

Mutation, migration, selection, and drift, individually and collectively, alter allele frequencies and bring about evolutionary divergence that eventually may result in **speciation**, the formation of new species. Speciation is facilitated by environmental diversity. If a population is spread over a geographic range encompassing a number of ecologically distinct subenvironments with different selection pressures, the populations occupying these areas may gradually adapt and become genetically distinct from one another. Genetically differentiated populations may remain in existence, become extinct, reunite with each other, or continue to diverge until they become reproductively isolated and form a new species. Genetic changes within populations can modify a species over time, transform it into another species, or cause it to split into two or more species.

A **species** can be defined as a group of actually or potentially interbreeding organisms that is reproductively isolated in nature from all other such groups. In sexually reproducing organisms, speciation transforms the gene pool of the parental species or divides a single gene pool into two or more separate gene pools. Changes in morphology or physiology, and adaptation to an ecological niche may also occur but are not necessary components of the speciation event. Speciation can take place gradually or within a few generations.

Because population and species divergence are accompanied by genetic differentiation, we can use patterns of genetic differences to reconstruct evolutionary history. After exploring the genetic structure of populations, the characteristics of divergence across space and time, and the process of speciation, we will discuss how genetic data can be used to answer questions that have an evolutionary context, including those in the two scenarios at the beginning of the chapter.

## 28.1 Speciation Can Occur by Transformation or by Splitting Gene Pools

Figure 28–1 shows a **phylogenetic tree** (a diagram of evolutionary change over time) that describes the evolutionary history of several hypothetical lizard species. The passage of time is plotted horizontally as changes in the phenotype of the lizards occur and the lineages diverge or separate.

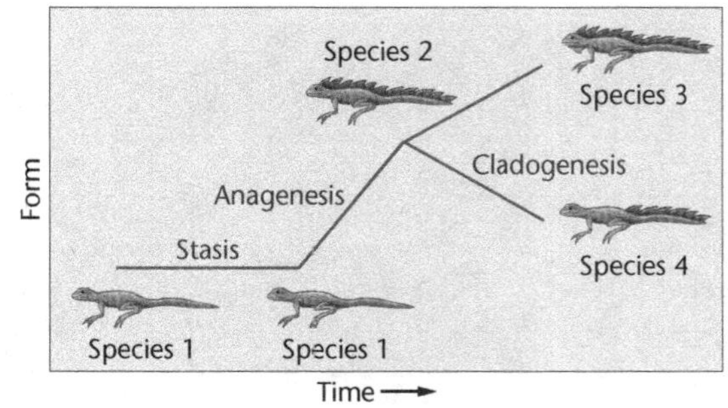


FIGURE 28–1 In phyletic evolution, or anagenesis, one species is transformed over time into another species. At all times during this transformation, only one species exists. In cladogenesis, one species splits into two or more species.

The history of these lizards begins with species 1. For some time, that species experiences evolutionary **stasis**; that is, it does not change. Species 1 then steadily transforms into species 2. This form of speciation is called **phyletic evolution**, or **anagenesis**. During anagenesis, there is never more than one species present, but the precise time at which species 1 becomes species 2 is difficult to identify. After species 2 forms, Figure 28–1 shows it undergoing **cladogenesis**, another form of speciation that gives rise to two distinct and independent daughter species. Further transformation of these daughter species produces the two species that we see today, species 3 and species 4. The distinctness of the extant species is in many cases experimentally verifiable. That is, they may be tested to determine their reproductive compatibility (to see if they are one species) or incompatibility (two species).

In his 1859 book, *On the Origin of Species*, Charles Darwin amassed evidence that all species derive from a single common ancestor by transformation and speciation:

> "All living things have much in common, in their chemical composition, their germinal vesicles, their cellular structure, and their laws of growth and reproduction.... Therefore I should infer... that probably all the organic beings which have ever lived on this earth have descended from some one primordial form."

Everything biologists have since learned supports Darwin's conclusion that only one tree of life exists. To understand evolution, we must understand the mechanisms that transform one species into another and that split one species into two or more. Chapter 27 described the mechanisms responsible for the transformation of species at the level of populations (mechanisms of transformation at the molecular level were discussed in Chapters 16 and 22). Chief among the mechanisms transforming populations is **natural selection**, discovered independently by Darwin and by Alfred Russel Wallace. The Wallace–Darwin concept of natural selection can be summarized as follows:

1. Individuals of a species exhibit variations in phenotype—for example, differences in size, agility, coloration, defenses against enemies, ability to obtain food, courtship behaviors, flowering times, and so on.
2. Many of these variations, even small and seemingly insignificant ones, are heritable and passed on to offspring.
3. Organisms tend to reproduce in an exponential fashion. More offspring are produced than can survive. This causes members of a species to engage in a struggle for survival, competing with other members of the community for scarce resources. Offspring also must avoid predators, and in sexually reproducing species, adults must compete for mates.
4. In the struggle for survival, individuals with particular phenotypes will be more successful than others, allowing the former to survive and reproduce at higher rates.

As a consequence of natural selection, species change. The phenotypes that confer improved ability to survive and reproduce become more common, and the phenotypes that confer poor prospects for survival and reproduction may eventually disappear. Under certain conditions, populations that at one time could interbreed may lose that capability, thus segregating their adaptations into particular niches. If selection continues, it may result in the appearance of new species.

Although Wallace and Darwin proposed that natural selection explains how evolution occurs, they could not explain either the origin of the variations that provide the raw material for evolution or the mechanisms by which such variations are passed from parents to offspring. In the twentieth century, as biologists applied the principles of Mendelian genetics to populations, the source of variation (mutation) and the mechanism of inheritance (segregation of alleles) were both explained. We now view evolution as a consequence of changes in genetic material through mutation and changes in allele frequencies in populations over time. This union of population genetics with the theory of natural selection generated a new view of the evolutionary process, called *neo-Darwinism*.

In this chapter, we consider the mechanisms responsible for speciation. As with natural selection, our understanding of speciation is built on key insights about genetic variation.

## 28.2 Most Populations and Species Harbor Considerable Genetic Variation

Your initial thought may be that members of a well-adapted population must be highly homozygous because you assume that the most favorable allele at each locus has become fixed. Certainly, an examination of most populations of plants and animals reveals many phenotypic similarities among individuals. However, a large body of evidence indicates that, in reality, most populations contain a high degree of heterozygosity. This built-in genetic diversity is often concealed, so to speak, because it is not necessarily apparent phenotypically; hence, detecting it is not a simple task. Nevertheless, with the use of the techniques discussed next, the diversity within a population can be revealed.

### Artificial Selection

One way to determine whether genetic variation affects a phenotypic character is to impose artificial selection on the character. A phenotype that is not affected by genetic variation will not respond to selection; if genetic variation does have an effect, the phenotype will change over a few generations. A dramatic example of this test is the domestic dog. The broad array of sizes, shapes, colors, and behaviors seen in different breeds of dogs all arose from the effects of selection on the genetic variation present in wild wolves, from which all domestic dogs are descended. Genetic and archaeological evidence indicates that the domestication of dogs took place at least 15,000 years ago, and possibly much earlier. On a shorter time scale, laboratory selection experiments on the fruit fly *Drosophila melanogaster* have caused significant changes over a few generations in almost every phenotype imaginable, including size, shape, developmental rate, fecundity, and behavior.

## Variations in Amino Acid Sequence

Gel electrophoresis separates protein molecules on the basis of differences in size and electrical charge. If a nucleotide variation in a structural gene results in the substitution of a charged amino acid (such as glutamic acid) for an uncharged amino acid (such as glycine), the net electrical charge on the protein will be altered. This difference in charge can be detected as a difference in the rate at which proteins migrate through an electrical field. In the mid-1960s, John Hubby and Richard Lewontin used gel electrophoresis to measure protein variation in natural populations of *Drosophila*, and Harry Harris used the same technique to measure protein variations in human populations. In subsequent years, researchers have used the technique to study genetic variation in a wide range of organisms (Table 28.1), although subsequently developed DNA techniques have in many cases replaced the use of protein electrophoresis.

The electrophoretically distinct forms of an enzyme produced by different alleles are called **allozymes**. As shown in the table, a surprisingly large percentage of loci examined from diverse species produce distinct allozymes. For the populations listed, approximately 30 loci per species were examined, and about 30 percent of the loci were polymorphic, with an average of 10 percent allozyme heterozygosity per diploid genome. These estimates clearly support the notion that organismal genotypes harbor vast amounts of DNA-based variability.

These values apply only to genetic variation detectable by altered protein migration in an electric field. Electrophoresis probably detects only about 30 percent of the actual variation caused by amino acid substitutions because many substitutions do not change the net electric charge on the molecule. Richard Lewontin estimated that about two-thirds of all loci in a population are polymorphic. In any individual within the population, about one-third of the loci exhibit genetic variation in the form of heterozygosity. The significance of genetic variation detected by electrophoresis is controversial. Some argue that allozymes are functionally equivalent and therefore do not play any role in adaptive evolution. We address this argument later in this section.

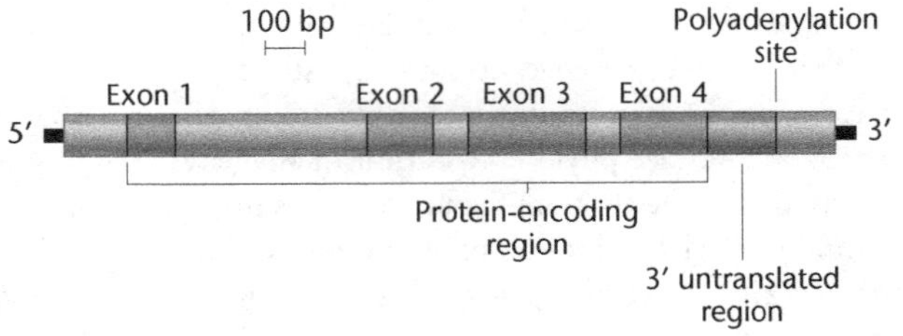


FIGURE 28–2 Organization of the *Adh* locus of *Drosophila melanogaster*.

## Variations in Nucleotide Sequence

The most direct way to estimate genetic variation is to compare the nucleotide sequences of genes carried by individuals in a population. With the development of techniques for cloning and sequencing DNA, and with the proliferation of genome projects, nucleotide sequence variations have been catalogued for an increasing number of genes and genomes. In an early study, Alec Jeffreys used restriction enzymes to detect polymorphisms in 60 unrelated individuals. His aim was to estimate the total number of DNA sequence variants in humans. His results showed that within intergenic spacers of the $\beta$-globin cluster, 1 in 100 base pairs showed polymorphic variation. If that region is representative of the genome, this study indicates that at least $3 \times 10^7$ nucleotide variants per genome are possible.

In another study, Martin Kreitman examined the *alcohol dehydrogenase* locus (*Adh*) in *Drosophila melanogaster* (Figure 28–2). This locus encodes two allozymic variants: the *Adh-f* and the *Adh-s* alleles. These differ by only a single amino acid (thr versus lys at codon 192). To determine whether the amount of genetic variation detectable at the protein level (one amino acid difference) corresponds to the variation at the nucleotide level, Kreitman cloned and sequenced *Adh* loci from five natural populations of *Drosophila*.

The 11 cloned loci contained a total of 43 nucleotide variations in the consensus *Adh* sequence of 2721 base pairs. These variations are distributed throughout the gene: 14 in the exon-coding regions, 18 in the introns, and 11 in the untranslated and flanking regions. Of the 14 variations in coding regions, only one leads to an amino acid replacement—the one in codon 192, producing the two observed electrophoretic variants. The other 13 coding-region nucleotide substitutions do not lead to amino acid replacements. Are the differences in the number of allozyme and nucleotide variants the result

TABLE 28.1

**Allozyme Heterozygosity at the Protein Level**

| Species | Number of Populations Studied | Number of Loci Examined | Polymorphic Loci* per Population (%) | Heterozygotes per Locus (%) |
|---|---|---|---|---|
| *Homo sapiens* (humans) | 1 | 71 | 28 | 6.7 |
| *Mus musculus* (mouse) | 4 | 41 | 29 | 9.1 |
| *Drosophila pseudoobscura* (fruit fly) | 10 | 24 | 43 | 12.8 |
| *Limulus polyphemus* (horseshoe crab) | 4 | 25 | 25 | 6.1 |

*A polymorphic locus is one for which a population harbors more than one allele.
*Source:* Adapted from Lewontin, 1974, p. 117.

of natural selection? If so, of what significance is this fact? Let's examine these questions.

Among the most intensively studied loci to date is the locus encoding the cystic fibrosis transmembrane conductance regulator (CFTR). Recessive loss-of-function mutations in the *CFTR* locus cause **cystic fibrosis**, a disease with symptoms including salty skin and the production of an excess of thick mucus in the lungs, which leads to susceptibility to bacterial infections. Geneticists have closely examined the *CFTR* locus in many populations and have found more than 1500 different mutations that can cause the disease. Among these mutations are missense mutations, amino acid deletions, nonsense mutations, frameshifts, and splice defects.

Figure 28–3 shows a map of the 27 exons in the *CFTR* locus, with most exons identified by function. The histogram above the map shows the locations of some of the disease-causing mutations and the number of copies of each that have been found. One mutation, a 3-bp deletion in exon 10 called $\Delta F508$ accounts for 67 percent of all mutant cystic fibrosis alleles, but several other mutations were found in at least 100 of the chromosomes surveyed. In populations of European ancestry, between 1 in 44 and 1 in 20 individuals are heterozygous carriers of mutant alleles. Note that Figure 28–3 includes only the sequence variants that alter the function of the CFTR protein. There are undoubtedly many more *CFTR* alleles with silent sequence variants that do not change the structure of the protein and do not affect its function.

Studies of other organisms, including the rat, the mouse, and the mustard plant *Arabidopsis thaliana* have produced similar estimates of nucleotide diversity in various genes. These studies indicate that there is an enormous reservoir of genetic variability within most populations and that, at the level of DNA, most, and perhaps all, genes exhibit diversity from individual to individual.

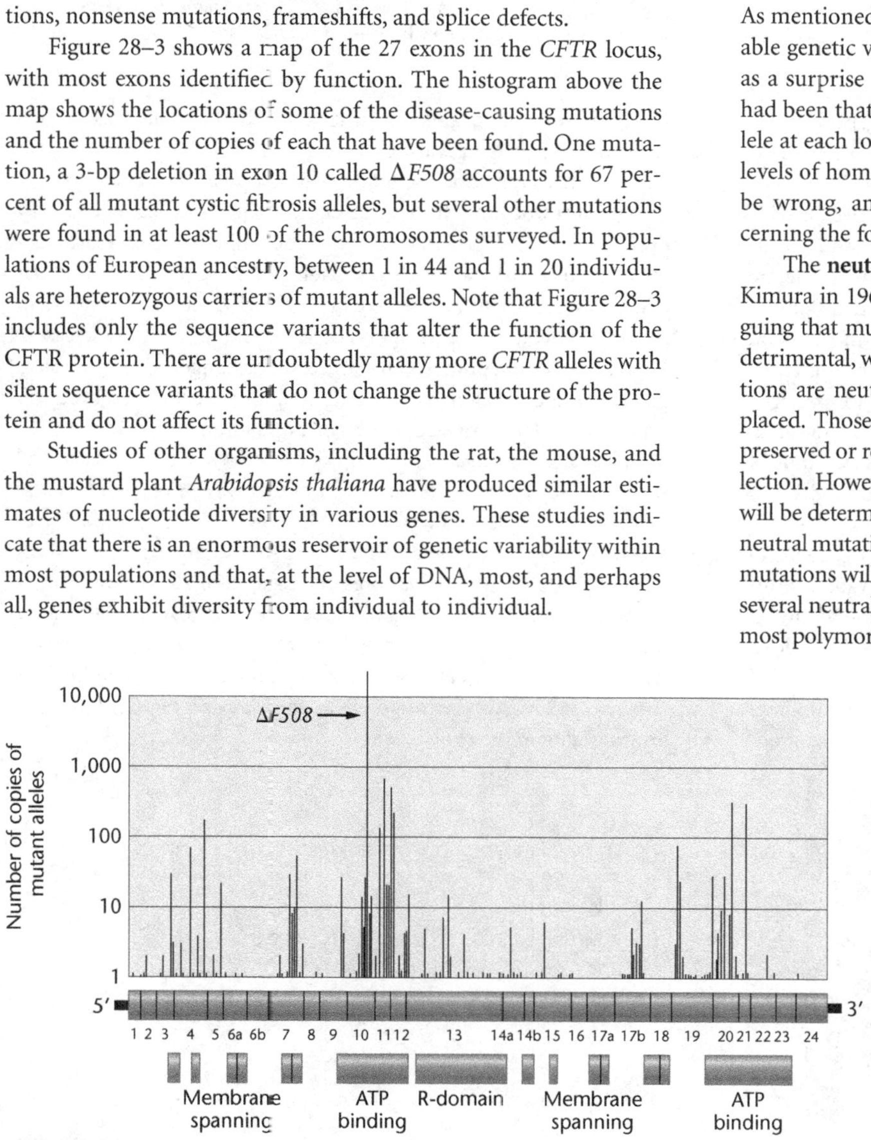


**FIGURE 28–3** The locations of disease-causing mutations in the cystic fibrosis gene. The histogram shows the number of copies of each mutation geneticists have found. (The vertical axis is on a logarithmic scale.) The genetic map below the histogram shows the locations and relative sizes of the 27 exons of the *CFTR* locus. The boxes at the bottom indicate the functions of different domains of the CFTR protein.

**NOW SOLVE THIS**

Problem 11 on page 759 asks what types of nucleotide substitutions will not be detected by electrophoretic studies of proteins.

■ HINT: *Keep in mind that allozymes that migrate at different speeds in the gel matrix do so because of their variable charge. Then take into account the effect of charges that do not alter migration speeds.*

## Explaining the High Level of Genetic Variation in Populations

As mentioned earlier, the finding that populations harbor considerable genetic variation at the amino acid and nucleotide levels came as a surprise to many evolutionary biologists. The early consensus had been that selection would favor a single optimal (wild-type) allele at each locus and that, as a result, populations would have high levels of homozygosity. This expectation was shown conclusively to be wrong, and considerable research and argument ensued concerning the forces that maintain genetic variation.

The **neutral theory** of molecular evolution, proposed by Motoo Kimura in 1968, explains these high levels of genetic variation by arguing that mutations leading to amino acid substitutions are usually detrimental, with a very small fraction that are favorable. Some mutations are neutral, or functionally equivalent to the allele that is replaced. Those polymorphisms that are favorable or detrimental are preserved or removed from the population, respectively, by natural selection. However, the frequency of the neutral alleles in a population will be determined by mutation rates and random genetic drift. Some neutral mutations will drift to fixation in the population; other neutral mutations will be lost. At any given time, the population may contain several neutral alleles at any particular locus. The diversity of alleles at most polymorphic loci does not, however, reflect the action of natural selection, but instead is a function of population size (larger populations have more variation) and the fraction of mutations that are neutral.

The alternative explanation for the surprisingly high variation is natural selection. There are several examples in which enzyme or protein polymorphisms are maintained by adaptation to certain environmental conditions. The well-known advantage of sickle-cell anemia heterozygotes when infected by malarial parasites is such an example, and another was discussed in Chapter 27, when we considered evidence that polymorphism at the *CFTR* locus may reflect the superior fitness of heterozygotes in areas where typhoid fever is common.

Fitness differences of a fraction of a percent would be sufficient to maintain a polymorphism, but at that level their presence would be difficult to measure. Current data are therefore insufficient to determine what fraction of molecular genetic variation is neutral and what fraction is subject to selection. The neutral theory nonetheless serves a crucial function: by pointing out that some genetic variation is expected simply as a result of mutation and drift, the neutral theory provides a working hypothesis for studies of molecular evolution. In other words, biologists must find positive evidence that selection is acting on allele frequencies at a particular locus before they can reject the simpler assumption that only mutation and drift are at work.

## 28.3 The Genetic Structure of Populations Changes across Space and Time

As population geneticists discovered that most populations harbor considerable genetic diversity, they also found that the genetic structure of populations varies across space and time. To illustrate, we consider studies on *Drosophila pseudoobscura* conducted by Theodosius Dobzhansky and his colleagues. This species is found over a wide range of environmental habitats, including the western and southwestern United States. Although the flies throughout this range are morphologically similar, Dobzhansky's team discovered that populations from different locations vary in the arrangement of genes on chromosome 3. They found several different inversions in this chromosome that can be detected as loop formations in larval polytene chromosomes. Each inversion sequence is named after the locale in which it was first discovered (e.g., AR = Arrowhead, British Columbia; and CH = Chiricahua Mountains, Arizona). The inversion sequences were compared with one standard sequence, arbitrarily designated ST.

Figure 28–4 compares the frequencies of three arrangements of chromosome 3 at different elevations in the Sierra Nevada Mountains in California. The ST arrangement is most common at low elevations; at 8000 feet, AR is the most common and ST least common. In the populations studied, the frequency of the CH arrangement gradually increases with elevation, a phenomenon that is probably the result of natural selection and that parallels the gradual environmental changes occurring at ascending elevations, such as decreasing air temperature.

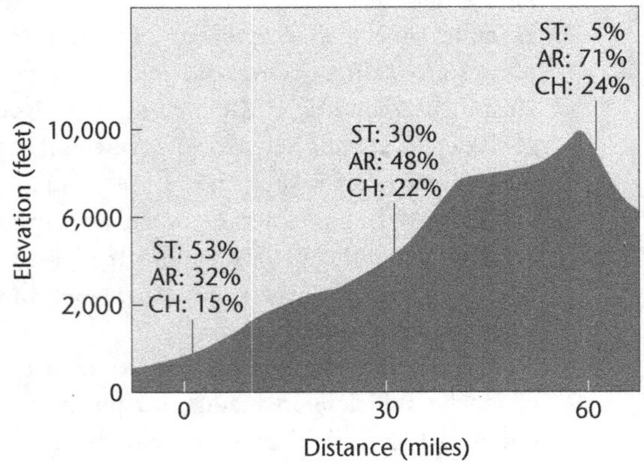


FIGURE 28–4 Inversions in chromosome 3 of *D. pseudoobscura* at different elevations in the Sierra Nevada range near Yosemite National Park.

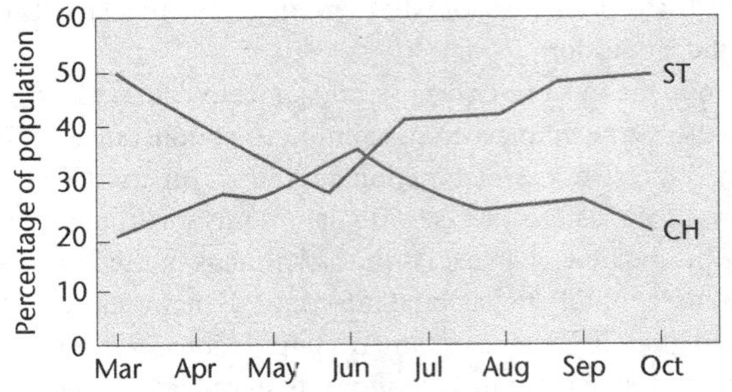


FIGURE 28–5 Changes in the ST and CH arrangements in *D. pseudoobscura* through the year, in each location studied.

Dobzhansky's team also found that if populations are collected at a single site throughout the year, inversion frequencies change cyclically over time as well. That is, cyclic variations in chromosome arrangements occur as the seasons change, as shown in Figure 28–5. This variation was consistently observed over a period of several years. In spring, the frequency of ST always declines, and that of CH increases.

To test the hypothesis that this cyclic change is a response to natural selection, Dobzhansky and his group devised a laboratory experiment. They constructed large population cages from which samples of *D. pseudoobscura* could be removed periodically and studied. They began with a population that was 88 percent CH and 12 percent ST. The flies were maintained at 25°C and sampled over a 1-year period. As shown in Figure 28–6, the frequency of ST increased gradually until it was present at a level of 70 percent. At that point, an equilibrium between ST and CH was reached. When the same experiment was performed at 16°C, no change in inversion frequency occurred. The researchers concluded that the equilibrium reached at 25°C was in response to the elevated temperature, the only variable in the experiment.

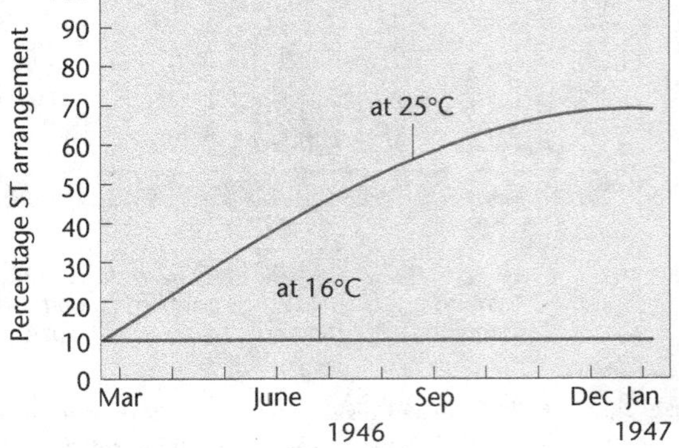


FIGURE 28–6 Increase in the ST arrangement of *D. pseudoobscura* in population cages under laboratory conditions.

The results of the study indicate that, for the population as a whole, a balance in the frequency of the two inversions and the gene arrangements they contain is superior to either inversion by itself. The equilibrium reached in the experiment presumably represents the highest mean fitness in the population under controlled laboratory conditions. This interpretation of the experiment suggests that natural selection is the driving force maintaining the diversity in chromosome 3 inversions.

In a more extensive study, Dobzhansky and his colleagues sampled *D. pseudoobscura* populations over a broad geographic range. They found 22 different chromosome arrangements in populations from 12 locations. In Figure 28–7, the frequencies of five of these inversions are shown according to geographic location. Most of the populations differed in the relative frequencies of these inversions. Collectively, Dobzhansky's data show that the genetic structure of *D. pseudoobscura* populations changes from place to place and from one time to another. At least some of this variation in population genetic structure is the result of natural selection.

Another example of how the genetic structure of a species varies among populations is provided by the work of Dennis A. Powers and Patricia Schulte on the mummichog (*Fundulus heteroclitus*), a small fish (5 to 10 cm long) that lives in inlets, bays, and estuaries along the Atlantic coast of North America from Florida to Newfoundland. These workers measured allele frequencies at the locus encoding the enzyme lactate dehydrogenase-B (LDH-B), which is made in the liver, heart, and red skeletal muscle. LDH-B converts lactate to pyruvate and is thus pivotal in both the manufacture of glucose and aerobic metabolism. Two allozymes of LDH-B are seen on gels; they differ at two amino acid positions. The alleles encoding the allozymes are designated *Ldh-B*$^a$ and *Ldh-B*$^b$.

Frequencies of the *Ldh-B* alleles vary dramatically among mummichog populations [Figure 28–8(a)]. In northern populations, where the mean water temperature is about 6°C, *Ldh-B*$^b$ predominates. In southern populations, where the mean water temperature is about 21°C, *Ldh-B*$^a$ predominates. Between the geographic extremes, allele frequencies are intermediate.

To determine whether the geographic variation in *Ldh-B* allele frequencies is due to natural selection, Powers and Schulte studied the biochemical properties of the LDH-B allozymes. They found that the enzyme encoded by *Ldh-B*$^b$ has higher catalytic efficiency than *Ldh-B*$^a$ at low temperatures, whereas the gene product of *Ldh-B*$^a$ is more efficient than *Ldh-B*$^b$ at high temperatures [Figure 28–8(b)]. A mixture of the two forms has intermediate efficiency at all temperatures.

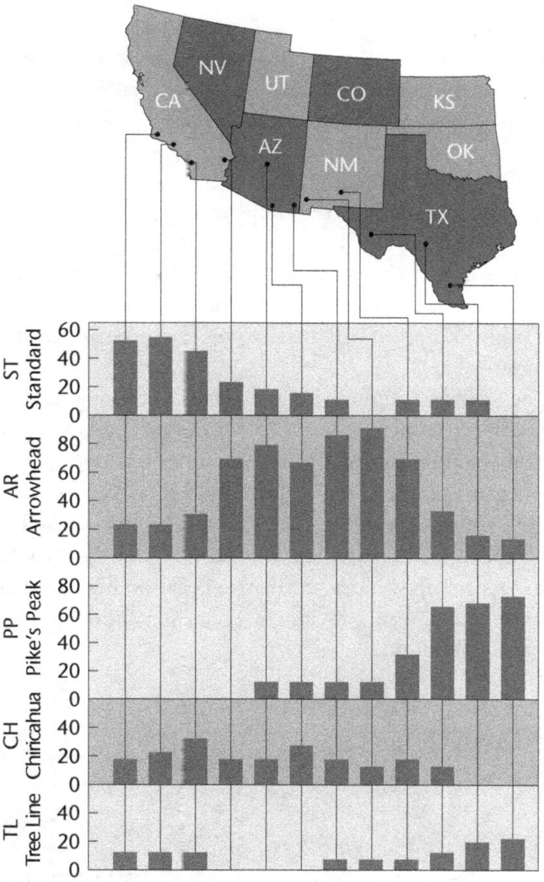


**FIGURE 28–7** Relative frequencies (percentages) of five chromosomal inversions in *D. pseudoobscura* in different geographic regions.

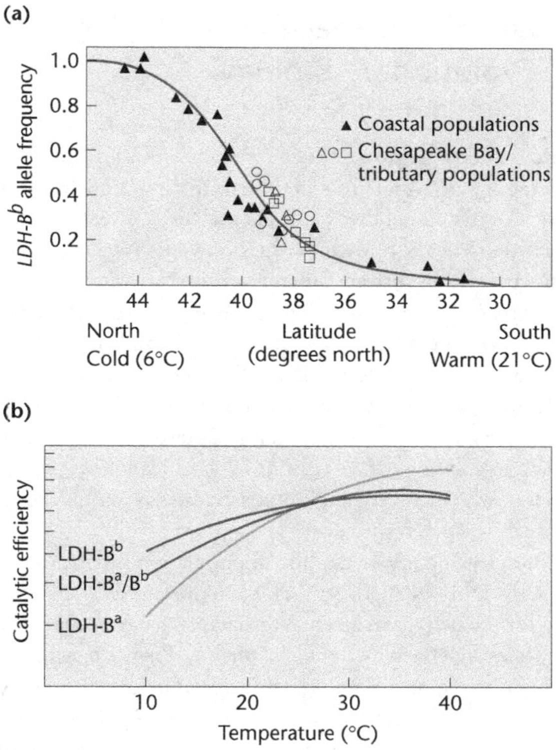


**FIGURE 28–8** Variation in genetic structure among mummichog populations. (a) Frequencies of the $B^b$ allele in populations along the Atlantic coast of North America. (b) Catalytic efficiency of LDH-B allozymes as a function of temperature.

In addition to the functional differences between the LDH-B allozymes, the *Ldh-B* alleles have nucleotide sequence differences in their regulatory regions. One such difference results in the *Ldh-B*$^b$ allele's rate of transcription being more than twice that of the *Ldh-B*$^a$ allele. As a result, northern fish have higher concentrations of the LDH-B enzyme in their cells.

Differences in the catalytic efficiency and transciption rate of the two *Ldh-B* alleles are consistent with the hypothesis that mummichog populations are adapted to the temperatures at which they live. The fish are ectotherms: their body temperature is determined by their environment. In general, low body temperatures slow an ectotherm's metabolic rate. The higher transcription rate of the *Ldh-B*$^b$ allele and the superior low-temperature catalytic efficiency of its gene product appear to help northern fish compensate for the tendency of their cold environment to reduce their metabolic rate. The superior high-temperature catalytic efficiency and lower transcription rate associated with the *Ldh-B*$^a$ allele appear to allow southern fish to economize on the resources devoted to glucose production and aerobic metabolism. On the basis of this and other evidence, Powers and Schulte suggest that differences among mummichog populations in *Ldh-B* allele frequencies are the result of natural selection.

## 28.4 Defining a Species Is a Challenge for Evolutionary Biology

Before we consider how genetic divergence can lead to speciation, let's examine how biologists in various disciplines define a species. Charles Darwin and other naturalists of the nineteenth century relied on *morphology* to define a species. Closely related species are often separated by distinct differences in morphological characters. This approach is useful with living species as well as those from the fossil record, but it also poses several problems. Even within recognized species there are often large differences in visible characters; and, conversely, there are many examples of morphologically very similar species that nonetheless cannot interbreed and therefore are distinct. In addition, where conflict in classification arises, it is often not clear which morphological characteristics should be used to classify the disputed species.

Our growing understanding of population genetics and evolution in the mid-twentieth century led to the *biological species concept* (used for the definition given at the chapter's beginning). A biological species is defined as *a group of interbreeding or potentially interbreeding populations reproductively isolated in nature from all other such groups*. According to this definition, species have distinct gene pools; when one gene pool divides into two or more separate gene pools, speciation has occurred. Changes in morphology or physiology, and adaptation to an ecological niche may also occur, but are not considered necessary components of speciation, which can take place gradually or within a few generations.

Zoologists in particular favor the biological species concept, since it is easily applied to animals, most of which have distinct sexes and exchange genes through sexual reproduction. However, this interpretation also has problems. Many species—for example, most prokaryotes—do not reproduce sexually, and so individuals within the species do not interbreed. Yet clearly, these individuals are not all different species. Also, about 20 percent of plants and 10 percent of birds and butterflies can interbreed and yet are recognized as distinct species. The presence in genomes of genes resulting from horizontal gene transfer also suggests that hybridization across species is not uncommon, and it argues against defining species solely by reproductive isolation. In addition, it is impossible to use the biological species concept for extinct species, since we are not able to determine the presence of reproductive isolation in fossils. Even for extant species it may be impossible to test for reproductive isolation in a cluster of closely related putative species.

With the establishment of databases that contain a wealth of molecular and genomic sequence information, our ability to compare the genomes of putative species has been enhanced. This has resulted in a third way of defining species, based on *phylogeny*. According to this definition, a species consists of those organisms that are monophyletic (belong to the same taxonomic group) and share one or more uniquely derived characteristics. This approach has its problems as well, because we cannot be certain that we have determined the correct relationships among all possible relatives and thus correctly identified the monophyletic groups, called **clades**.

Finally, some biologists use an *ecological species concept*, in which species are defined as groups of individuals that live in the same type of environment and have shared ecological requirements, filling the same niche. The main problem here is that there are examples in which populations that seem to be the same species (i.e., they readily interbreed with one another) have very distinct ecological requirements.

The scientific community is engaged in an ongoing dialogue about how to define a species. This is one of the most important and challenging questions in evolutionary biology, but for now, it has no clear answer. The biological species concept is most often used by population geneticists studying evolution and will be used in the following discussions about the process of species formation. Its emphasis on reproductive isolation not only helps define species but also provides a focus for thinking about the process of speciation.

**NOW SOLVE THIS**

In Problem 4 on page 759 you are asked to justify the classification of phenotypically similar species as separate species rather than as the same species with some phenotypic variation.

■ HINT: *The species concept that you rely upon will influence the answer to this question. Then, using one or more species concepts, determine whether they are different species.*

28.5

# Reduced Gene Flow, Selection, and Genetic Drift Can Lead to Speciation

We have learned that most populations harbor considerable genetic variation and that different populations within a species may have different alleles or allele frequencies at a variety of loci. The genetic divergence of these populations can be caused by natural selection, genetic drift, or both. In Chapter 27, we saw that the migration of individuals between populations, together with the gene flow that accompanies that migration, tends to homogenize allele frequencies among populations. In other words, migration counteracts the tendency of populations to diverge.

When gene flow between populations is reduced or absent, the populations may diverge to the point that members of one population are no longer able to interbreed successfully with members of the other. When populations reach the point where they are reproductively isolated from one another, they have become different species, according to the biological species concept.

The biological barriers that prevent or reduce interbreeding between populations are called **reproductive isolating mechanisms** and are classified in Table 28.2. These mechanisms may be ecological, seasonal or temporal, behavioral, mechanical, or physiological.

**Prezygotic isolating mechanisms** prevent individuals from mating in the first place. Individuals from different populations may not find each other at the right time, may not recognize each other as suitable mates, or may try to mate, but find that they are unable to do so.

**Postzygotic isolating mechanisms** create reproductive isolation even when the members of two populations are willing and able to mate with each other. For example, genetic divergence may have reached the stage where the viability or fertility of hybrids is reduced. Hybrid zygotes may be formed, but all or most may be inviable. Alternatively, the hybrids may be viable, but may be sterile or suffer from reduced fertility. Yet again, the hybrids themselves may be fertile, but their progeny may have lowered viability or fertility. In these situations, hybrids will not reproduce and are genetic dead ends. These postzygotic mechanisms act at or beyond the level of the zygote and are generated by genetic divergence.

In some models of speciation, postzygotic isolation evolves first and is followed by prezygotic isolation. Postzygotic isolating mechanisms waste gametes and zygotes and lower the reproductive fitness of hybrid survivors. Selection will therefore favor the spread of alleles that reduce the formation of hybrids, leading to the development of prezygotic isolating mechanisms, which in turn prevent interbreeding and the formation of hybrid zygotes and offspring. In animal evolution, the most effective prezygotic mechanism is behavioral isolation, involving courtship behavior.

Genes that control development of reproductive organs may play a vital role in the evolution of major new lineages, even at macroevolutionary levels beyond species. Recent research on *Arabidopsis thaliana* and other taxa indicates that evolutionary changes in MADS-box homeotic gene regulation and function have been a driving force in the evolution of flowers. This regulation is a molecular key to reproductive isolation and divergence, since successful pollination and therefore mating within a plant species depends on floral anatomy. Research on *A. thaliana* in the laboratories of Vivian Irish and her colleagues has focused on *APETALA3* (*AP3*) and *PISTILLATA* (*PI*), floral homeotic genes that encode MADS-domain–containing transcription factors that determine stamen and petal identities during floral development (Chapter 19). They have shown that these genes are differentially expressed in the petals and sepals of flowering plants with differences in flower structure. This finding suggests that changes in the expression of developmental

TABLE 28.2

**Reproductive Isolating Mechanisms**

| |
|---|
| **Prezygotic Mechanisms**<br>**Prevent fertilization and zygote formation** |
| 1. **Geographic or ecological**: The populations live in the same regions but occupy different habitats. |
| 2. **Seasonal or temporal**: The populations live in the same regions but are sexually mature at different times. |
| 3. **Behavioral** (only in animals): The populations are isolated by different and incompatible behavior before mating. |
| 4. **Mechanical**: Cross-fertilization is prevented or restricted by differences in reproductive structures (genitalia in animals, flowers in plants). |
| 5. **Physiological**: Gametes fail to survive in alien reproductive tracts. |
| **Postzygotic Mechanisms**<br>**Fertilization takes place and hybrid zygotes are formed, but these are nonviable or give rise to weak or sterile hybrids.** |
| 1. **Hybrid nonviability or weakness** |
| 2. **Developmental hybrid sterility**: Hybrids are sterile because gonads develop abnormally or meiosis breaks down before completion. |
| 3. **Segregational hybrid sterility**: Hybrids are sterile because of abnormal segregation into gametes of whole chromosomes, chromosome segments, or combinations of genes. |
| 4. **$F_2$ breakdown**: $F_1$ hybrids are normal, vigorous, and fertile, but the $F_2$ contains many weak or sterile individuals. |

*Source:* From G. Ledyard Stebbins, *Processes of organic evolution*, 3rd ed., copyright 1977, p. 143. Reprinted by permission of Prentice Hall, Upper Saddle River, NJ.

genes may change floral morphology and be a force in evolution. These researchers have also investigated evolutionary changes in the function of these genes. They grew transgenic *A. thaliana* plants carrying an *AP3* mutant gene with a segment replaced with one from an evolutionarily older plant lineage that does not produce petals in its flowers. The resulting transgenic plants produced stamens, but not petals, suggesting that the function of the gene and its encoded protein changed when the *A. thaliana* lineage diverged from the ancestral lineage. Such so-called evo-devo collaborations between evolutionary biologists and developmental geneticists are leading to exciting new discoveries about the pivotal role of developmental genes in major evolutionary events such as the origin of new lineages and species.

## Examples of Speciation

In this section, we consider two examples of speciation, one from a laboratory study and the other from a field study. Diane Dodd and her colleagues studied the evolution of digestive physiology in *Drosophila pseudoobscura*. They collected flies from a wild population and established several separate laboratory populations. Some of the laboratory populations were raised on starch-based medium and others on maltose-based medium. Both food sources were stressful for the flies.

Dodd's team wanted to know whether the starch-adapted populations and the maltose-adapted populations, which diverged under strong selection and in the absence of gene flow, had become different species. Roughly a year after the populations were established, a series of mating trials were performed. For each trial, 48 flies were placed in a bottle: 12 males and 12 females from a starch-adapted population and 12 males and 12 females from a maltose-adapted population. The researchers then noted which flies mated.

Dodd predicted that if the two populations, having adapted to different media, had speciated, then the flies would prefer to mate with members of their own population. If the populations had not speciated, then the flies would mate at random. The results appear in Table 28.3. Roughly 600 of the 900 matings observed were between males and females from the same population. In other words, the differently adapted fly populations showed partial premating isolation. Dodd and her colleagues concluded that the populations had begun to speciate but had not yet completed the process.

TABLE 28.3

**Number of Matings between Laboratory Populations of *Drosophila pseudoobscura***

| | Female | |
|---|---|---|
| Male | Starch-adapted | Maltose-adapted |
| **Starch-adapted** | 290 | 153 |
| **Maltose-adapted** | 149 | 312 |

*Source:* Compiled from Dodd, D.M.B. 1989. Reproductive isolation as a consequence of adaptive divergence in *Drosophila pseudoobscura*. *Evolution* 43: 1308–1311.

The Isthmus of Panama formed roughly 3 million years ago, creating a land bridge connecting North and South America and simultaneously separating the Caribbean Sea from the Pacific Ocean. Nancy Knowlton and colleagues took advantage of a natural experiment that the formation of the Isthmus of Panama had performed on several species of snapping shrimps (Figure 28–9). After identifying seven Caribbean species of snapping shrimp, researchers matched a member of each one with a member from a Pacific species to form a pair. The members of each matched pair were more similar to each other in structure and appearance than either was to any other species in its own ocean. Analysis of allozyme allele frequencies and mitochondrial DNA sequences confirmed that the members of each pair were one another's closest genetic relatives.

Prior to the formation of the isthmus, the ancestors of each pair were a single species. When the isthmus closed, each of the seven ancestral species was divided into two separate populations, one in the Caribbean and the other in the Pacific. The researchers interpreted the results of the mating experiments as follows.

Meeting in a dish in Knowlton's lab for the first time in 3 million years, would Caribbean and Pacific members of a species pair recognize each other as suitable mates? Knowlton placed males and females of a species pair together and noted their behavior toward each other. She then calculated the relative inclination of Caribbean–Pacific couples to mate versus that of Caribbean–Caribbean or Pacific–Pacific couples. Three of the seven transoceanic species pairings refused to mate altogether. Of the other four species pairings, transoceanic couples were 33, 45, 67, and 86 percent as likely to mate with each other as were same-ocean pairs. Of the same-ocean couples that mated, 60 percent produced viable clutches of eggs. Of the transoceanic couples that mated, only 1 percent produced viable clutches. We can conclude from these results that 3 million years of separation has resulted in complete or nearly complete speciation, involving strong pre- and postzygotic isolating mechanisms for all seven species pairs.

FIGURE 28–9 A snapping shrimp (genus *Alpheus*).

## The Minimum Genetic Divergence for Speciation

How much genetic separation is required between two populations before they become different species? We shall consider two examples, one from an insect and the other from a plant, which demonstrate that in some cases the answer is "not very much."

(a)

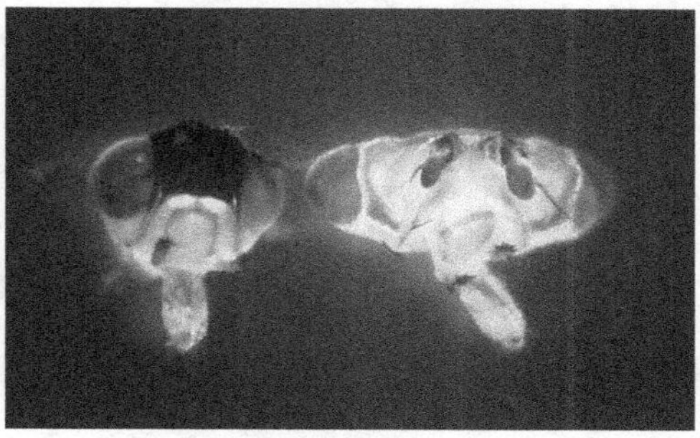

(b)

FIGURE 28-11 (a) Differences in pigmentation patterns in *D. sylvestris* (left) and *D. heteroneura* (right). (b) Head morphology in *D. sylvestris* (left) and *D. heteroneura* (right).

Researchers estimate that *Drosophila heteroneura* and *D. silvestris*, two species found only on the island of Hawaii, diverged from a common ancestral species only about 300,000 years ago. The two species are thought to be descended from *D. planitibia*, which originated on the older island of Maui (Figure 28–10). The two species are clearly separated from each other by different and incompatible courtship and mating behaviors (a prezygotic isolating mechanism), by morphology, and by body and wing pigmentation (Figure 28–11).

In spite of the morphological and behavioral divergence between these species, researchers have had a hard time demonstrating significant protein polymorphisms or differences in chromosomal inversion patterns. DNA hybridization studies carried out on the two species indicate that the sequence diversity between them is only about 0.55 percent (Figure 28–12). This finding suggests that nucleotide sequence diversity may precede the development of protein or chromosomal polymorphisms.

Genetic evidence suggests that the differences between *D. heteroneura* and *D. silvestris* are controlled by a relatively small number of genes. For example, as few as 15 to 19 major loci may be responsible for the morphological differences between the species, demonstrating that it is possible for speciation to involve only a small number of genes.

Studies using two closely related species of the monkey flower, a plant that grows in the Rocky Mountains and areas west, confirm that species can be separated by only a few genetic differences. One species, *Mimulus cardinalis*, is fertilized by hummingbirds and does not interbreed with *Mimulus lewisii*, which is fertilized by bumblebees. H. D. Bradshaw and his colleagues studied genetic differences related to reproduction in the two species—namely, flower shape, size, and color and nectar production. For each trait, a difference in a single gene provided at least 25 percent of the variation observed among laboratory-created hybrids. *M. cardinalis* makes 80 times

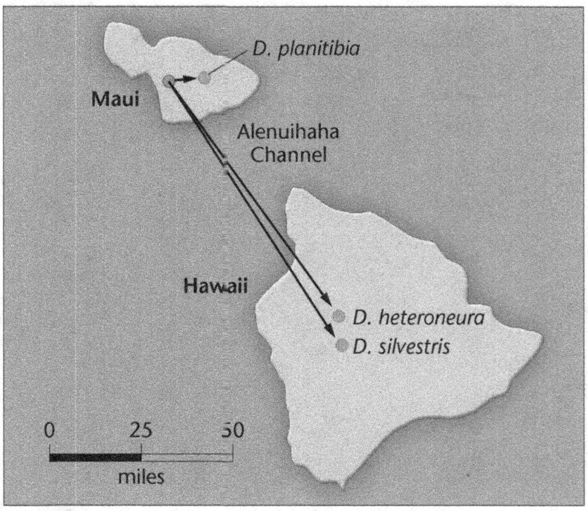


FIGURE 28-10 Proposed pathway for Hawaii's colonization by members of the *D. planitibia* species. The purple circle represents a population ancestral to the three present-day species.

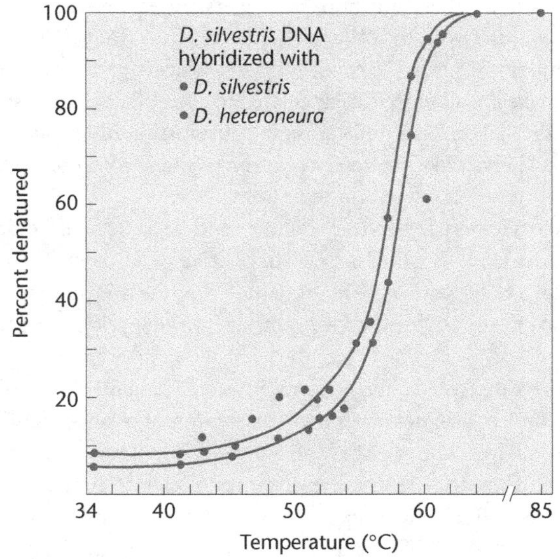


FIGURE 28-12 Nucleotide sequence diversity in species descended from *Drosophila planitibia*. The shift to the left by the heterologous hybrids indicates the degree of nucleotide sequence divergence.

FIGURE 28–13 Flowers of two closely related species of monkey flowers, (a) *Mimulus cardinalis* and (b) *Mimulus lewisii.*

more nectar than does *M. lewisii*, and a single gene is responsible for at least half the difference. A single gene also controls a large part of the difference in flower color between the two species (Figure 28–13). In this case, as in the Hawaiian *Drosophila*, species differences can be traced to a relatively small number of genes.

## The Rate of Speciation

How much time is required for speciation? In many cases, speciation takes place slowly over a long period, but in other cases, speciation can be surprisingly rapid.

The Rift Valley lakes of east Africa support hundreds of species of cichlid fish. Lake Victoria (Figure 28–14), for example, has more than 400 species. Cichlids have some morphological diversity, but, more dramatically, they are highly specialized for different ecological niches (Figure 28–15). Some eat algae floating on the water's surface, whereas others are bottom feeders, insect feeders, mollusk eaters, or predators on other fish species. Lake Tanganyika has a similar array of species. Genetic analyses indicate that the species in a given lake are all more closely related to each other than to species from other lakes. The implication is that most or all the species in, say, Lake Tanganyika are descended from a single common ancestor and that they evolved within their home lake.

Lake Victoria is between 250,000 and 750,000 years old, and there is evidence that the lake may have dried out almost completely less than 14,000 years ago. Is it possible that the 400 species in the lake today evolved from a common ancestral species in less than 14,000 years?

In a study of cichlid origins in Lake Tanganyika, Norihiro Okada and colleagues examined insertions of a novel family of DNA sequences called short interspersed elements (SINEs) (discussed in Chapter 12) into the genomes of cichlid species in Lake Tanganyika. SINEs are a type of retrotransposon, and the random integration of a SINE at a locus is most likely an irreversible event. The presence of SINE at the same locus in the genome of all species examined is strong evidence that all of those species descend from a common ancestor. Using a SINE called AFC, Okada's team screened 33 species of cichlids belonging to four groupings of species (called species tribes). In each tribe, the SINE was present at all the sites tested, indicating that the species in each tribe are descended from a single ancestral species (Figure 28–16). If research of the same kind on the Lake Victoria species produces similar results, it would mean that the 400 cichlid species in the lake most likely evolved in less than 14,000 years. If

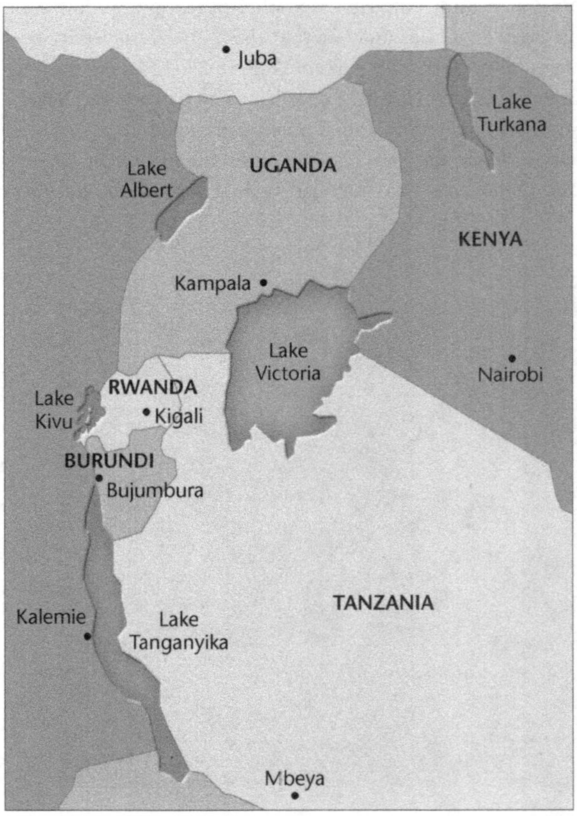


FIGURE 28–14 Lake Victoria in the Rift Valley of east Africa is home to more than 400 species of cichlids.

**FIGURE 28–15** Cichlids occupy a diverse array of niches, and each species is specialized for a distinct food source.

confirmed, this finding would represent the fastest evolutionary radiation ever documented in vertebrates.

Even faster speciation is possible through the mechanism of polyploidy. The formation of animal species by polyploidy is rare, but polyploidy is an important factor in plant evolution. It is estimated that one-half of all flowering plants have evolved by this mechanism. One form of polyploidy is allopolyploidy (see Chapter 8), produced by doubling the chromosome number in a hybrid formed by crossing two species.

If two species of related plants have the genetic constitution SS and TT (where S and T represent the haploid set of chromosomes in each species), the hybrid would have the chromosome constitution ST. Normally, such a plant would be sterile because few or no homologous chromosome pairs would be present at meiosis. However, if the hybrid underwent a spontaneous doubling of its chromosome number, a tetraploid SSTT plant would be produced. This chromosomal aberration might occur during mitosis in somatic tissue, giving rise to a partially tetraploid plant that produces some tetraploid flowers. Alternatively, aberrant meiotic events may produce ST gametes that, when fertilized, would yield SSTT zygotes. The SSTT plants would be fertile because they would possess homologous chromosomes producing viable ST gametes. This new, true-breeding tetraploid would combine characteristics derived from both parental species and would be reproductively isolated from those species because hybrids would be triploids and consequently sterile.

The tobacco plant *Nicotiana tabacum* ($2n = 48$) is the result of a doubling of the chromosome number in the hybrid between *N. otophora* ($2n = 24$) and *N. silvestris* ($2n = 24$). The origin of *N. tabacum* is an example of virtually instantaneous speciation.

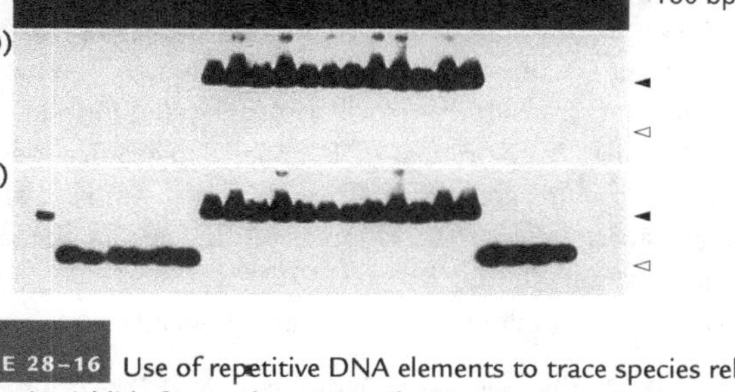


**FIGURE 28–16** Use of repetitive DNA elements to trace species relationships in cichlids from Lake Tanganyika. (a) Photograph of an agarose gel showing PCR fragments generated from primers that flank the AFC family of SINES. Large fragments containing this family are present in all samples of genomic DNA from members of the Lamprologini tribe of cichlids (lanes 9–20). DNA from the species in lane 2 has a similar but shorter fragment, which may represent another repetitive sequence. DNA from other species (lanes 3–8 and 21–24) produce short, non-SINE-containing fragments. (b) A Southern blot of the gel from (a) probed with DNA from the AFC family of SINES. AFC is present in DNA from all species in the Lamprologini tribe (lanes 9–20), but not in DNA from other species (lanes 2–8 and 21–24). (c) A second Southern blot of the gel from (a) probed with the genomic sequence at which the AFC SINE inserts All species examined (lanes 2–24) contain the insertion sequence. The larger fragments in Lamprologini DNA (lanes 9–20) correspond to those in the previous blot, showing that the SINE is inserted at the same site in all tribal species. In sum, these data show that the AFC SINE is present only in members of this tribe, is present in all member species, and is inserted at the same site in all cases. These results are interpreted as showing a common origin for the species of the tribe in question.

**NOW SOLVE THIS**

In Problem 27 on page 760 you are asked to interpret whether neutral sequence polymorphisms found among the cichlid species in African lakes help to explain their evolution.

■ **HINT:** *If a sequence is neutral, then it should not be subject to any form of selection through time, and it should only be subject to random mutation and genetic drift. Then examine whether these polymorphisms can be used to determine ancestry.*

28.6

## Genetic Differences Can Be Used to Reconstruct Evolutionary History

Early in this chapter, we noted that evolutionary history consists of the creation of new species both by transformation and by splitting. Our examples have demonstrated that speciation is associated with changes in the genetic structure of populations and with genetic divergence. Therefore, we should be able to use genetic differences among present-day species to reconstruct their evolutionary histories.

In an important early example of phylogeny reconstruction, W. M. Fitch and E. Margoliash assembled data on the amino acid sequence for cytochrome c in a variety of organisms. **Cytochrome c** is a respiratory molecule found in the mitochondria of eukaryotes, and its amino acid sequence has evolved very slowly. For example, its amino acid sequence in humans and chimpanzees is identical, and humans and rhesus monkeys show only one amino acid difference. This similarity is remarkable considering that the fossil record indicates that the lines leading to humans and monkeys diverged from a common ancestor approximately 20 million years ago.

Column (a) of Table 28.4 shows the number of amino acid differences between cytochrome c in humans and in various other species. The table is broadly consistent with our intuitions about how closely related we are to these other species. For example, we are more closely related to other mammals than we are to insects, and we are more closely related to insects than we are to yeast. Similarly, our cytochrome c differs in 10 amino acids from that of dogs, in 24 amino acids from that of moths, and in 38 amino acids from that of yeast.

More than one nucleotide change, however, may be required to change a given amino acid. When the nucleotide changes necessary for all amino acid differences observed in a protein are totaled, the **minimal mutational distance** between the genes of any two species is established. Column (b) in Table 28.4 shows such an analysis of the genes encoding cytochrome c. As expected, these values are larger than the corresponding number of amino acids separating humans from the other nine organisms listed.

TABLE 28.4

**Amino Acid Differences and Minimal Mutational Distances between Cytochrome c in Humans and Other Organisms**

| Organism | (a) Amino Acid Differences | (b) Minimal Mutational Distance |
|---|---|---|
| Human | 0 | 0 |
| Chimpanzee | 0 | 0 |
| Rhesus monkey | 1 | 1 |
| Rabbit | 9 | 12 |
| Pig | 10 | 13 |
| Dog | 10 | 13 |
| Horse | 12 | 17 |
| Penguin | 11 | 18 |
| Moth | 24 | 36 |
| Yeast | 38 | 56 |

*Source:* From W.M. Fitch and E. Margoliash, Construction of phylogenetic trees, *Science* 155:279–284, January 20, 1967. Copyright 1967 by the American Association for the Advancement of Science.

Fitch used data on the minimal mutational distances between the cytochrome c genes of 19 organisms to reconstruct their evolutionary history. The result is an estimate of the phylogenetic tree that unites the species studied (Figure 28–17). The black dots on the tips of the branches represent existing species, whose inferred common ancestors are linked to them by green lines and represented by red dots. The ancestral species evolved and diverged to produce the modern species. The common ancestors are connected to still earlier common ancestors, culminating in a single common ancestor for all the species on the tree, represented by the red dot on the extreme left.

### Constructing Evolutionary Trees from Genetic Data

Many methods are available for estimating phylogenies on the basis of genetic differences. Rather than survey them all, we will use one method, called the *u*nweighted *p*air *g*roup *m*ethod using *a*rithmetic averages, or UPGMA, as an example. This is not the most powerful method for estimating phylogenies from genetic data, but in spite of its lengthy name, it is intuitively straightforward. Furthermore, UPGMA works reasonably well under many circumstances.

The starting point for UPGMA analysis is a table of genetic distances among a group of species [Figure 28–18(a)]. To demonstrate, we will use data from studies by Charles Sibley and Jon Alquist, who used differences in the results of DNA:DNA hybridization experiments to compute genetic distances between humans and four species of apes: the common chimpanzee, the gorilla, the siamang, and the common gibbon.

The basic procedure consists of the following steps:

1. Search for the smallest genetic distance between any pair of species. In Figure 28–18(a), this distance is 1.628, the distance between human and chimpanzee. Once this distance is identified, the species pair it connects is placed on neighboring branches of an evolutionary tree [Figure 28–18(b)]. The length of each branch is half the genetic distance between the species (1.628/2), so the branches connecting human and chimpanzee to their common ancestor are each about 0.81 unit long.

2. Next, calculate the genetic distances between this species pair and all other species [Figure 28–18(c)]. The genetic distance between the human–chimpanzee (Hu–Ch) cluster and the other species is the average of the distances between each member of the cluster and the other species.* For example,

* We say "cluster" even though what we really have is a pair; the reason is that we can (and will) form larger groups, and it is convenient to call them by the same, more inclusive name—ergo *cluster*.

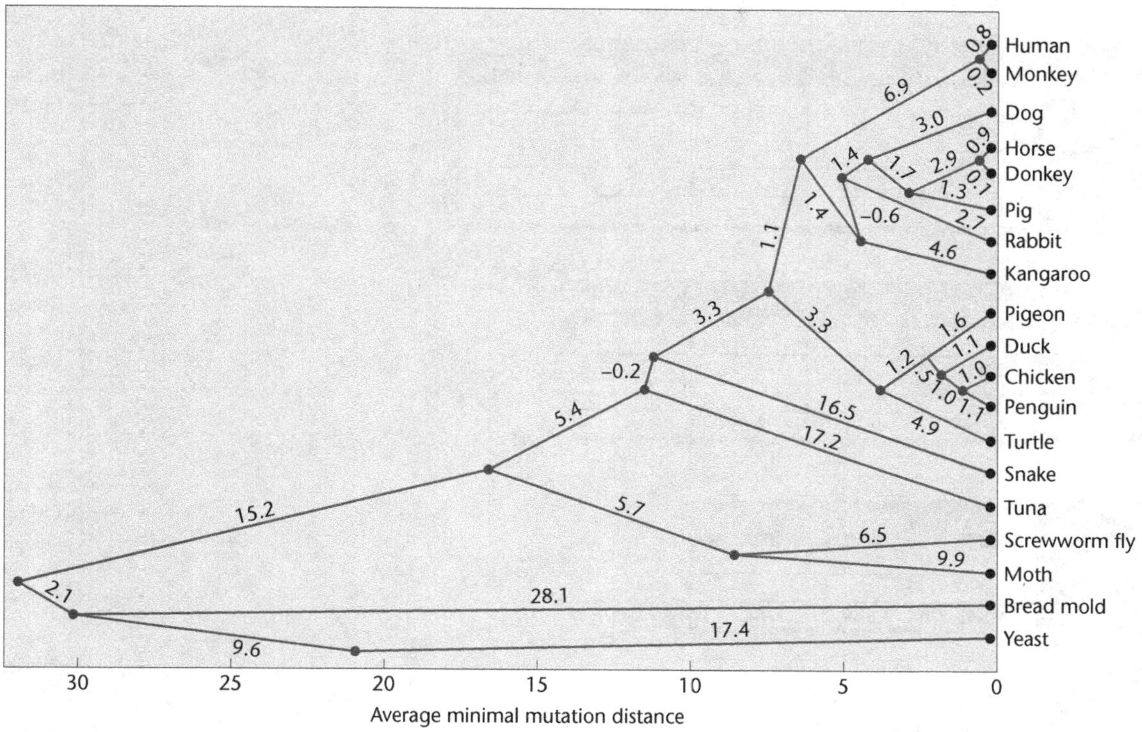


**FIGURE 28-17** Phylogenetic tree constructed by comparing homologies in cytochrome c amino acid sequences.
*Reprinted with permission from Fitch, W.M., and Margoliash E. 1967. Construction of phylogenetic trees.* Science *279: 279–284. Figure 2. Copyright 1967 AAAS.*

the genetic distance between the human–chimp cluster and the gorilla is the average of the human–gorilla distance and the chimp–gorilla distance, or

$$(2.267 + 2.21)/2 = 2.2385$$

3. Repeat steps 1 and 2 (as demonstrated below), each time using the newly computed table, until all the species have been added to the tree.

The smallest distance in the recalculated table is 1.95, the distance between siamang and gibbon [Figure 28–18(c)]. To build the next section of the tree, siamang and gibbon are placed on neighboring branches, with lengths equal to half of 1.95, or 0.98 [Figure 28–18(d)].

We now recalculate the table to display the two clusters we have isolated plus the gorilla [Figure 28–18(e)]. The genetic distance between the human–chimp cluster and the siamang–gibbon cluster is the average of four distances: human–siamang, human–gibbon, chimp–siamang, and chimp–gibbon.

In the new table, the smallest genetic distance is 2.239, the distance between the human–chimp cluster and the gorilla. We therefore add a gorilla branch to the tree and connect it to the common ancestor of the human–chimp cluster [Figure 28–18(f)]. The branches are drawn so that the distance between the tips of any two branches in the human–chimp–gorilla cluster is 2.239.

To recalculate the table for the final time [Figure 28–18(g)], we compute the genetic distance between the human–chimp–gorilla cluster and the siamang–gibbon cluster as the average of six genetic distances: human–siamang, human–gibbon, chimp–siamang, chimp–gibbon, gorilla–siamang, and gorilla–gibbon. This distance is 4.778, which allows us to complete our evolutionary tree [Figure 28–18(h)]. The tree indicates that humans and chimpanzees are one another's closest relatives. That is not to say that humans evolved from chimpanzees; rather, humans and chimpanzees share a more recent common ancestor than either share with other species on the tree.

The chief shortcoming of UPGMA is that it provides no means of determining how well the resulting trees fit the data, compared with other possible trees. For example, how much better does the tree we just created fit our data than a tree that shows chimpanzees and gorillas as closest relatives?

Evolutionary geneticists have developed several techniques for searching the set of all possible trees connecting a group of species and

(a)

| | Human | Chimp | Gorilla | Siamang | Gibbon |
|---|---|---|---|---|---|
| Human | – | | | | |
| Chimp | **1.628** | – | | | |
| Gorilla | 2.267 | 2.21 | – | | |
| Siamang | 4.7 | 5.133 | 4.543 | – | |
| Gibbon | 4.779 | 4.76 | 4.753 | 1.95 | – |

(c)

| | Hu-Ch | Gorilla | Siamang | Gibbon |
|---|---|---|---|---|
| Hu-Ch | – | | | |
| Gorilla | 2.2385 | – | | |
| Siamang | 4.9165 | 4.543 | – | |
| Gibbon | 4.7695 | 4.753 | **1.95** | – |

(e)

| | Hu-Ch | Gorilla | Si-Gi |
|---|---|---|---|
| Hu-Ch | – | | |
| Gorilla | **2.239** | – | |
| Si-Gi | 4.843 | 4.648 | – |

(g)

| | Hu-Ch-Go | Si-Gi |
|---|---|---|
| Hu-Ch-Go | – | |
| Si-Gi | **4.778** | – |

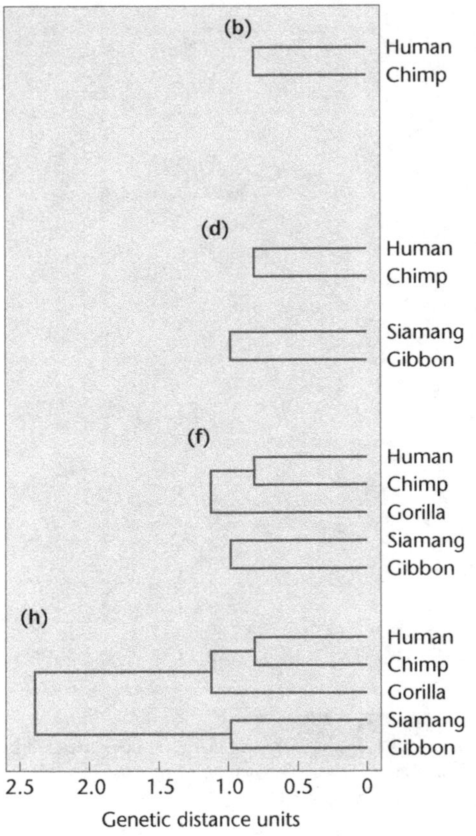


FIGURE 28–18 Phylogeny reconstruction by UPGMA.

calculating the relative veracity of each. One set of techniques, called **parsimony** methods, compares trees according to the minimum number of evolutionary changes each requires and then selects the simplest possible tree. Often, however, there are two or more equally parsimonious trees, so these methods often must be combined with other criteria to develop the most accurate phylogeny. Another set of analytical tools, called **maximum-likelihood** methods, starts with a model of the evolutionary process, calculates how likely it is that evolution will produce each possible tree under the model, and selects the most likely tree. It is important to remember that a phylogenetic tree produced from data on genetic distances does not necessarily describe the way species actually evolved, but instead is a reasonable estimate of their evolutionary histories given the methods used. Phylogenies are considered more reliable if multiple methods are used.

## Molecular Clocks

In many cases, we would like to estimate not only which members of a set of species are most closely related, but also when their common ancestors lived. Sometimes we can do so, thanks to **molecular clocks**—amino acid sequences or nucleotide sequences in which evolutionary changes accumulate at a constant rate over time.

Research by Walter M. Fitch and colleagues on the influenza A virus shows how molecular clocks are used. Fitch and his associates sequenced part of the hemagglutinin gene from flu viruses that had been isolated at different times over a 20-year period. They then calculated the numbers of nucleotide differences between the various virus samples and constructed an evolutionary tree [Figure 28–19(b)]. Most of the strains that had been collected have gone extinct, leaving no descendants among the more recently isolated viruses. Fitch's group then plotted the number of nucleotide substitutions between the first virus and each subsequent virus against the year in which the virus was isolated [Figure 28–19(a)]. The points all fall very close to a straight line, indicating that nucleotide substitutions in this gene have accumulated at a steady rate. Nucleotide substitutions in the hemagglutinin gene thus serve as a molecular clock. Molecular clocks are used to compare the sequences of new flu viruses as they appear each year and to estimate the time that has passed since each diverged from a common ancestor.

Molecular clocks must be carefully calibrated and used with caution. For example, Fitch's data indicate that strains of influenza A that jump from birds to humans have evolved much more rapidly

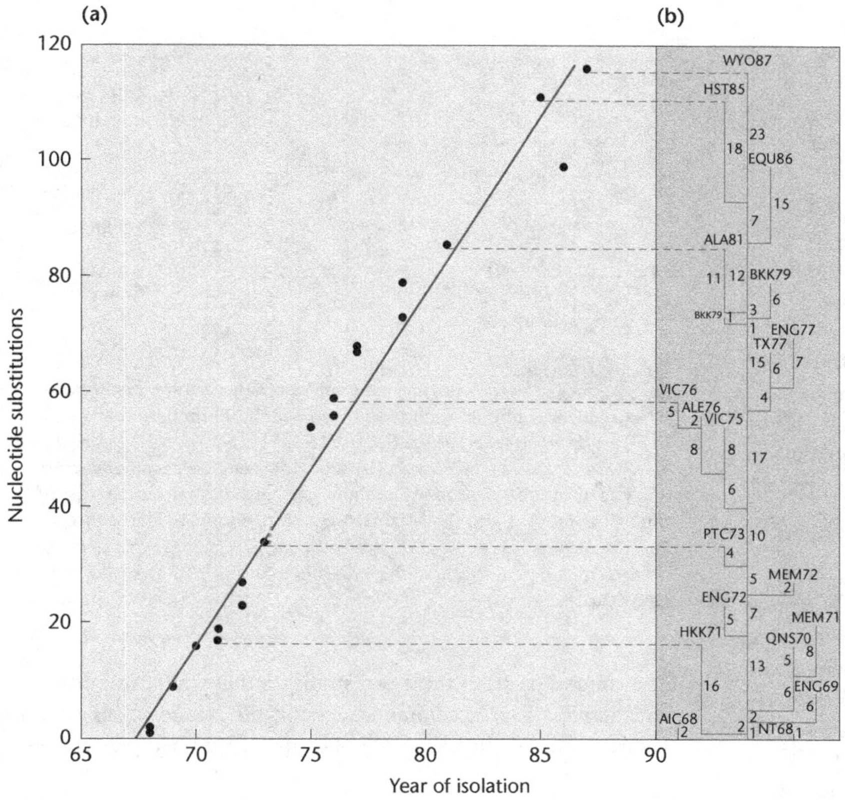


FIGURE 28-19 Molecular clock in the influenza A hemagglutinin gene. (a) Number of nucleotide differences between the first isolate and each subsequent isolate as a function of year of isolation. (b) Estimate of the phylogeny of the isolates.

than strains that have remained in birds. Hence, a molecular clock calibrated from human strains of the virus would be highly misleading if applied to bird strains.

## 28.7 Reconstructing Evolutionary History Allows Us to Answer Many Questions

Evolutionary analysis addresses a wide range of questions, some of which pertain directly to evolution, whereas others are concerned with contemporary issues and even criminal activity. To conclude this chapter, we show how evolutionary genetics can provide answers to the questions posed in the scenarios set forth in the chapter's beginning.

### Transmission of HIV

Recall the case of a Florida dentist who tested positive for HIV, and later developed AIDS. Two years later, one of his patients, a young woman with no known risk factors, became infected with HIV. Several other patients were found to be HIV positive. The question is, did this dentist transmit HIV to his patients, or did the patients become infected by some other means?

At first glance, this situation appears to be a long way from a discussion of evolution; however, the movement of a virus from one individual to another is similar to the founding of a new island population by a small number of migrants. Gene flow between the ancestral population and the new population is nonexistent, and the populations are free to diverge, as a result of genetic drift, adaptation to different environments, or both.

If the dentist passed his HIV infection to his patients, then the viral strains isolated from these patients should be more closely related to each other and to the dentist's strain than to strains from any other individuals living in the same area. Following this line of reasoning, Chin-Yih Ou and colleagues sequenced portions of the gene for the HIV envelope protein from viruses collected from the dentist, 10 of his patients, and several other HIV-infected individuals from the area who served as local controls.

An evolutionary tree for the HIV strains produced from these sequence data is shown in Figure 28–20.

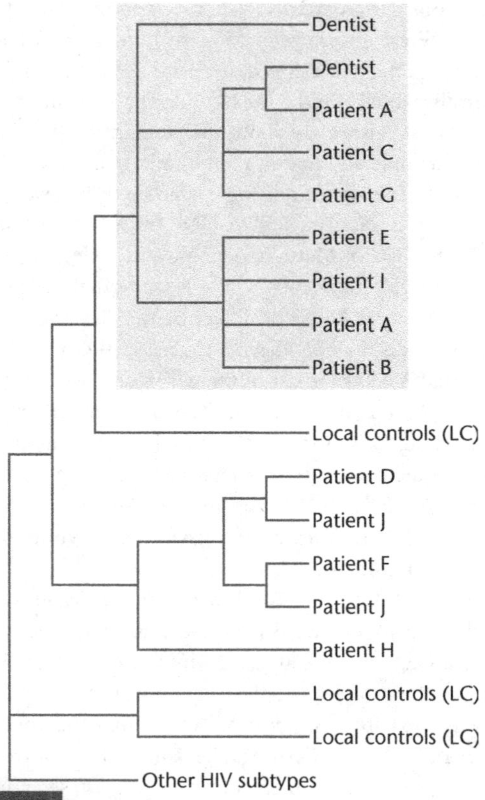


FIGURE 28-20 A phylogenetic tree of HIV strains taken from a dentist, his patients, and several local controls (LC). The group of viral strains in the shaded portion are all derived from a common ancestral strain.

The viruses isolated from the participants in the study are at the tips of the branches; branch points join these to the inferred common ancestors. The HIV samples from patients A, B, C, E, G, and I all share a more recent common ancestor with each other and with the dentist's strains than with any of the HIV samples from other local individuals. The evolutionary relationships among these HIV strains indicate that the dentist did, indeed, transmit his infection to those patients. In contrast, the viruses taken from patients D, F, H, and J are all more closely related to strains from local controls (LC) than they are to the dentist's strains. These patients got their infection from someone other than the dentist. Patient J, in fact, seems to have acquired his infection from two different sources. The Centers for Disease Control, which tracks HIV cases, has not been able to determine how the dentist-to-patient transmission took place. There have been no other documented cases of provider-to-patient transmission in a health-care setting.

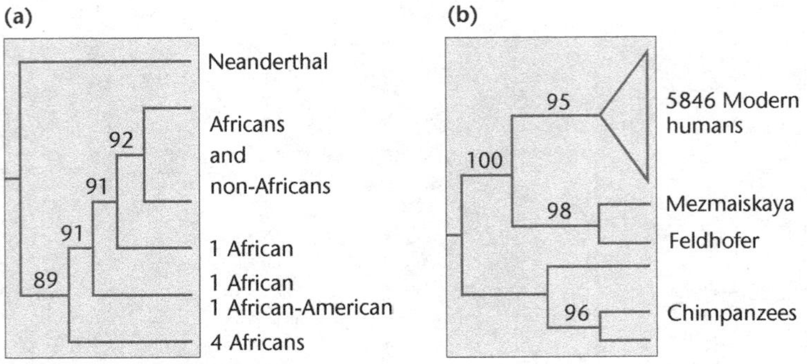


FIGURE 28–21 (a) A phylogenetic tree estimated from mitochondrial DNA sequences of one Neanderthal and over 2000 modern humans. (a) *From Krings, M. et al., 1997.* Cell *90: 19–30. Figure 7A, p.26, copyright 1997 with permission from Elsevier*; (b) A phylogenetic tree estimated from analysis of over 5000 modern humans, the Neanderthal samples from the Feldhofer cave and the Mezmaiskaya cave, and from chimpanzees. *From Ovchinnikov, I.V. et al., 2000. Molecular analysis of Neanderthal DNA from the northern Caucasus.* Nature *404: 490–493 copyright 2000 Macmillan Publishers Ltd.*

## Neanderthals and Modern Humans

Paleontological evidence indicates that the Neanderthals, *Homo neanderthalensis*, lived in Europe and western Asia from some 300,000 to 30,000 years ago. For at least 30,000 years, Neanderthals coexisted with anatomically modern humans (*H. sapiens*) in several areas. Genetic analysis using DNA sequencing answered several questions about Neanderthals and modern humans: (1) Were Neanderthals direct ancestors of modern humans? (2) Did Neanderthals and *H. sapiens* interbreed, so that descendants of the Neanderthals are alive today? Or did the Neanderthals die off and become extinct? (3) What can we say about the similarities and differences between our genome and that of the Neanderthals?

These and other questions were answered by several groups who were able to extract and analyze DNA from Neanderthal bones. In 1997, Svante Pääbo, Matthias Krings, and colleagues extracted mitochondrial DNA fragments from a Neanderthal skeleton found in Feldhofer cave near Düsseldorf, Germany. After placing the Neanderthal sequences on a phylogenetic tree together with sequences from more than 2000 modern humans [Figure 28–21(a)], the researchers concluded that Neanderthals are a distant relative of modern humans. Using a molecular clock calibrated with chimpanzee and human sequences, the researchers calculated that the last common ancestor of Neanderthals and modern humans lived roughly 600,000 years ago, four times as long ago as the last common ancestor of all modern humans.

In a second study, Igor Ovchinnikov and colleagues analyzed mitochondrial DNA extracted from Neanderthal bones found in Mezmaiskaya cave in the Caucasus Mountains east of the Black Sea. Although the two Neanderthal sequences (i.e., from Feldhofer and Mezmaiskaya) are from locations more than 1000 miles apart, they vary by only about 3.5 percent. This indicates that they derive from a single gene pool. Furthermore, the variation seen in the Neanderthal sequences is comparable to that observed among modern humans. Phylogenetic analysis places the two Neanderthals in a group that is distinct from modern humans [Figure 28–21(b)]. The conclusion from these two studies is that while Neanderthals and humans have a common ancestor, the Neanderthals were a separate hominid line and did not contribute mitochondrial genes to *H. sapiens*.

## Neanderthal Genomics

Although results from a number of studies on Neanderthals using mitochondrial DNA support the conclusion that Neanderthals and modern humans are only distantly related and there is little or no Neanderthal contribution to our genome, these studies and their conclusions have some important limitations. First, only a few hundred base pairs of mitochondrial DNA were analyzed, and sequencing of more mitochondrial DNA might show interbreeding between the two species. Second, because mitochondria are maternally inherited, studies of genomic sequences from cell nuclei are needed to complete the picture and clarify what relationship our species might have with Neanderthals.

Recently, two research groups have successfully recovered, sequenced, and analyzed genomic sequences from Neanderthal remains. The two teams worked in parallel, first ensuring that the Neanderthal specimens were not contaminated with DNA from modern humans. Through use of different methods developed independently by each team, about 65,000 base pairs of Neanderthal sequence were obtained by one team, and more than 1,000,000 nucleotides of the Neanderthal genome were sequenced by the other team. Sequence analysis indicates that Neanderthals and modern humans have genomes that are more than 99.5 percent identical. The Neanderthal genomic DNA analyzed to date represents sequences present on all human chromosomes (Figure 28–22), indicating that the DNA is broadly representative of the Neanderthal genome.

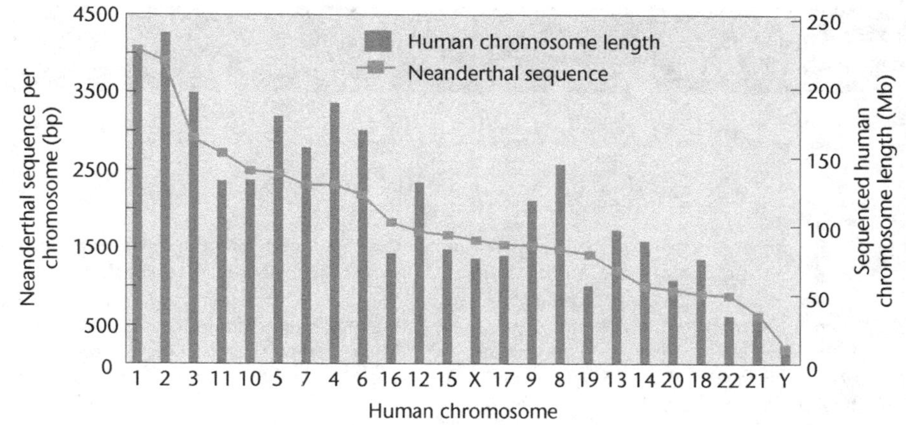


**FIGURE 28–22** Ranking of human chromosomes (across the bottom) by the amount of Neanderthal DNA sequences that align with sequences from that chromosome. The amount of Neanderthal DNA recovered and sequenced is generally proportional to the amount of DNA present in human chromosomes, as shown by the order of human chromosomes, indicating that the Neanderthal DNA represents sequences from across the genome. *From Noonan, J.P. et al. 2006. Sequencing and analysis of Neanderthal genomic DNA.* Science *314: 1113–1118.*

These sequence data have been analyzed and compared to the sequences of modern humans and chimpanzees to determine when Neanderthals and humans last shared a common ancestor, assuming that chimpanzees and humans last shared a common ancestor about 6,500,000 years ago. The different methods used by the two teams produced slightly divergent dates for the events in question. One team combined their genomic data with the fossil data to show that Neanderthals and modern humans last shared a common ancestor about 700,000 years ago, and that the split between Neanderthals and modern humans occurred about 370,000 years ago (Figure 28–23). The other team concluded that the human and Neanderthal lines split about 500,000 years ago, a number supported by the analysis of mitochondrial DNA. These studies confirmed the earlier studies and reinforce the conclusions about the relationship between our species and the Neanderthal species. First, Neanderthals are not direct ancestors of our species. Second, Neanderthals and members of our species may have interbred, but from the results available, it appears that Neanderthals did not make major contributions to our genome. As a species, Neanderthals are extinct, but some of their genes may survive as part of our genome. Third, from what little we know about the Neanderthal genome, we share most genes and other sequences with them.

More exciting answers to the question about the similarities and differences between our genome and that of Neanderthals could be derived by sequencing the entire Neanderthal genome. Using several new methods, a number of research groups are now working to isolate, sequence, and analyze the Neanderthal genome, with the goal of having a draft sequence completed by 2008. Analysis of the genomes of chimpanzees, Neanderthals, and modern humans would allow us to identify key differences that define our species and, in the process, revolutionize the field of human evolution.

**FIGURE 28–23** Estimated times of divergence of human and Neanderthal genomic sequences and, subsequently, of their populations, relative to landmark events in both human and Neanderthal evolution. These estimates are based on sequencing about 65,000 base pairs of Neanderthal DNA (y.a. = years ago). *From Noonan, J.P. et al. 2006. Sequencing and analysis of Neanderthal genomic DNA.* Science *314: 1113–1118.*

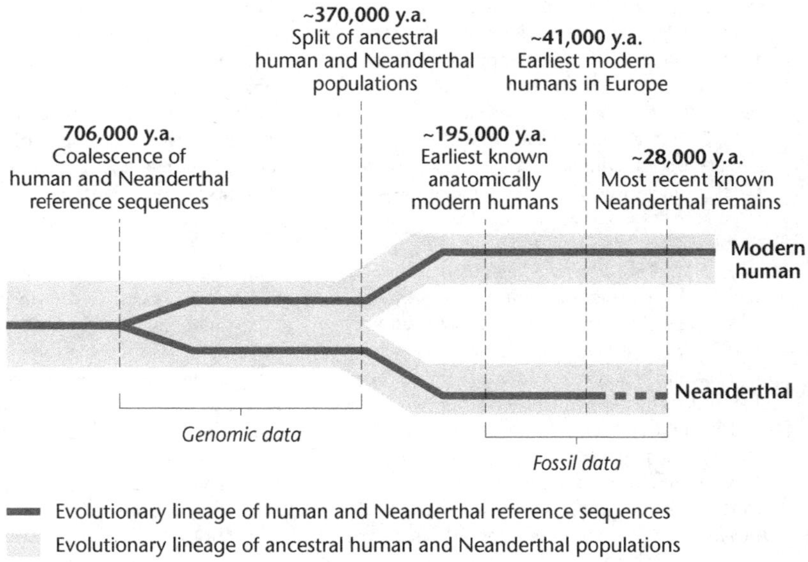

GENETICS, TECHNOLOGY, AND SOCIETY

# What Can We Learn from the Failure of the Eugenics Movement?

The eugenics movement had its origins in the ideas of the English scientist Francis Galton, who became convinced from his study of the appearance of geniuses within families (including his own) that intelligence is inherited. Galton concluded in his 1869 book *Hereditary Genius* that it would be "quite practicable to produce a highly gifted race of men by judicious marriages during several consecutive generations." The term *eugenics*, coined by Galton in 1883, refers to the improvement of the human species by such selective mating. Once Mendel's principles were rediscovered in 1900, the eugenics movement flourished.

The eugenicists believed that a wide range of human attributes were inherited as Mendelian traits, including many aspects of behavior, intelligence, and moral character. Their overriding concern was that the presumed genetically "feebleminded" and immoral in the population were reproducing faster than those who might be considered genetically superior, and that this differential birthrate would result in the progressive deterioration of the intellectual capacity and moral fiber of the human race. Several remedies were proposed. Positive eugenics called for the encouragement of especially "fit" parents to have more children. More central to the goals of the eugenicists, however, was the negative eugenics approach aimed at discouraging—or, if possible, preventing—the reproduction of the "genetically inferior."

In the United States, the eugenics movement enjoyed wide popular support for a time and had a significant impact on public policy. Partially at the urging of prominent eugenicists, 30 states passed laws compelling the sterilization of criminals, epileptics, and inmates in mental institutions; most states enacted laws invalidating marriages between the "feebleminded" and others considered eugenically unfit. The crowning legislative achievement of the eugenics movement, however, was the passage of the Immigration Restriction Act of 1924, which severely limited the entry of immigrants from eastern and southern Europe due to their perceived mental inferiority.

Throughout the first two decades of the century, most geneticists passively accepted the views of eugenicists, but by the 1930s critics recognized that the goals of the eugenics movement were determined more by racism, class prejudice, and anti-immigrant sentiment than by sound genetics. Increasingly, prominent geneticists began to speak out against the eugenics movement, among them William Castle, Thomas Hunt Morgan, and Hermann Muller. When the horrific extremes to which the Nazis took eugenics became known, a strong reaction developed that all but ended the eugenics movement.

In a way, it is ironic that the eugenics movement arose at the same time basic Mendelian principles were being developed. Although the movement misapplied those principles, those who understood them properly eventually used them to undermine the theoretical foundation of eugenics. Today, any student who has completed an introductory course in genetics should be able to identify several fundamental mistakes the eugenicists made.

1. They assumed that complex human traits such as intelligence and personality were strictly inherited, completely disregarding any environmental contribution to the phenotype. Their reasoning was that because certain traits ran in families, they must be genetically determined.
2. They assumed that these complex traits were determined by single genes with dominant and recessive alleles. This belief persisted despite research showing that multiple genes contribute to many phenotypes.
3. They assumed that a single ideal genotype would be desirable for humans. Presumably, such a genotype would be highly homozygous in order to be sustained. This precept runs counter to current evidence suggesting that a high level of homozygosity is often deleterious, supporting the superiority of the heterozygote.
4. They assumed that the frequency of recessively inherited defects in the population could be significantly lowered by preventing homozygotes from reproducing. In fact, for recessive traits that are relatively rare, most of the recessive alleles in the population are carried by asymptomatic heterozygotes who are spared from such selection. Negative eugenic practices, no matter how harsh, are relatively ineffective at eliminating such traits (see Insights and Solutions, Chapter 27).
5. They thought that those deemed genetically unfit in the population might outreproduce those thought to be genetically fit. This is the exact reverse of the Darwinian concept of fitness, which equates reproductive success with fitness. (Galton should have understood this, being Darwin's first cousin!)

More than seven decades have passed since the eugenics movement was at its height. We now have a much more sophisticated understanding of genetics, as well as a greater awareness of its potential misuses. But the application of current genetic technologies makes possible a "new eugenics" of a scope and power that Francis Galton could not have imagined. In particular, prenatal genetic screening and *in vitro* fertilization enable the selection of children according to their genotype, a power that will dramatically increase as more and more genes are associated with inherited diseases and perhaps even behaviors.

As we move into this new genetic age, we must not forget the misunderstandings disseminated by the early eugenicists. We must remember that the phenotype is a complex interaction between the genotype and the environment, and so we must not fall into the error of treating a person as only a collection of genes. We must keep in mind that many genes may contribute to a particular phenotype, whether a disease or a behavior, and that the alleles of these genes may interact in unpredictable ways. We must not fall prey to the assumption that an ideal genotype exists. The success of all populations in nature is believed to be enhanced by genetic diversity. Most of all, we must not use genetic information to advance ideological goals. We may find that there is a fine line between the legitimate uses of genetic technologies, such as having healthy children, and eugenic practices that may infringe on human rights. It will be up to us to decide exactly where the line falls.

## ■ References

Allen, G.E., Jacoby, R., and Glauberman, N. (eds.). 1995. Eugenics and American social history, 1880–1950. *Genome* 31: 885–889.

Hartl, D.L. 1988. *A primer of population genetics*, 2nd ed. Sunderland, MA: Sinauer.

Kevles, D.J. 1985. *In the name of eugenics: Genetics and the uses of human heredity*. Berkeley: University of California Press.

# EXPLORING GENOMICS

## ClustalW and Phylogenetic Analysis

One of the topics discussed in this chapter is how genetic differences between species can be used to reconstruct evolutionary history through the development of phylogenetic trees. A number of different programs for phylogeny analysis are available on the Internet, and as genome data accumulates, it is becoming increasingly common for geneticists to use molecular data for such analysis. In this exercise, you will use NCBI to obtain amino acid sequence data for a protein from different species, and then you will build a small phylogenetic tree based on a conserved amino acid sequence in that protein.

### ■ Exercise I – Phylogenetic Analysis of Histone H4

The polypeptide you will use for phylogenetic analysis is histone H4. (Based on what you know about histones, would you expect this gene to show substantial genetic variation in different species? This exercise will tell you if you're correct.) First, you will retrieve polypeptide sequence data for histone H4 from different species and run a multiple sequence alignment to compare the sequences. Then you will build a small phylogenetic tree using a program called **ClustalW**, which is hosted by the **European Molecular Biology Laboratory (EMBL)** site, a major bioinformatics resource.

1. Begin this exercise by accessing the NCBI site at http://www.ncbi.nih.gov. From the homepage or from Entrez, use the accession numbers provided in the following list to retrieve histone H4 protein sequence reports from five different species. Each time you retrieve a report, check the box next to the accession number for the sequence, and then use the "Send to" drop-down tab near the top right of the screen to send the report to your clipboard. *Hint:* You can retrieve all sequences at the same time by inserting a space between each accession number that you enter into Entrez.

   XP_001275769
   CAM12247
   XP_425458
   NP_563797
   AAT78470

2. Once you have saved reports for all species in the clipboard, click on the "Clipboard" tab and you should see the five reports listed. Which species will be used for this analysis? Click on the link for each report to reveal this information if you are not familiar with the scientific and common names for each species.

3. Return to the clipboard. Check the box next to each report, and under the "Display" drop-down menu, select "FASTA." This will save sequence data in a format we will need for our analysis. You should now see a page containing the five sequences.

4. Copy and paste each sequence into a Microsoft Word file or a similar text format that you can use to make a few simple edits. Leave one line space between each sequence. For each sequence, start directly after the > symbol and delete any text (such as the accession number and protein name) until you reach the scientific name. Each sequence should now have a > symbol followed by the scientific name of the organism, then the amino acid sequence.

5. Access the ClustalW program at http://www.ebi.ac.uk/clustalw/#. This program has a number of complex features; however, you only need to be familiar with a few basic features to run a multiple sequence alignment and generate a phylogeny. The default features for ClustalW are set up for a multiple-sequence alignment. Paste the five sequences into the text box, then click "Run." Results will appear in a separate window.

6. The Scores Table provides a summary of sequence alignment results. Use the "Sort by" feature to sort results by alignment score and you will see a table indicating the most similar sequences on top and the most divergent sequences on the bottom, with a percent score for similarity. The sequence alignment is shown below the scores table. Click "Show Colors" to highlight amino acid residues by their properties. Answer the following questions:

   a. How many amino acids were analyzed for each species?

   b. What does the * symbol identify in the alignment?

   c. Which sequences were most similar? Which were most divergent? How do you know this? Did you expect these results?

7. At the bottom of the page, a cladogram will appear. A *cladogram* shows an estimate of phylogeny based on predicted common ancestry. The branches in a cladogram are assumed to be of equal length, and it shows only the branching order of nodes. Click "Show as Phylogram Tree" to see a simple *phylogram* (phylogenetic tree). This program calculates a quantitative score for each pair of aligned sequences and then uses those scores to develop the tree. Recall that a phylogenetic tree is an estimate of degrees of phylogenetic development and that the length of each branch is proportional to the amount of inferred evolutionary change based on the sequences analyzed. Click on the "Show Distances" button to see calculated genetic distances.

   It is important to remember that there are many different ways to carry out phylogeny reconstruction and that this simple exercise relies on a number of assumptions that may not be true. Nevertheless, it does demonstrate how sequence information can be used to infer relatedness. Examine the cladogram and phylogram trees and then answer the following questions:

   a. What does each tree tell us? What can be estimated from these trees?

   b. What does the position of *Aspergillus* in the phylogram suggest?

*Continued on next page*

*Exploring Genomics, continued*

**Exercise II - Exploring ClustalW**

Now that you have learned how to use some basic features of ClustalW, can you think of another highly conserved protein that may be used for phylogenetic analysis? Return to the NCBI site and see if you can retrieve the amino acid sequence for this protein from at least four different species. Repeat the procedure you carried out for Exercise I to learn more about the relatedness of the species based on the protein you selected for the analysis. Good luck!

## Chapter Summary

1. Today's species are the products of an evolutionary history consisting of transforming, splitting, and diverging lineages. Alfred Russel Wallace and Charles Darwin formulated the theory of natural selection, which provides a major mechanism for the transformation of lineages. The genetic basis of evolution and the role of natural selection in changing allele frequencies were discovered in the twentieth century.
2. When population geneticists began studying the genetic structure of populations, they found that most populations harbor considerable genetic diversity, which becomes apparent in electrophoretic studies of proteins and at the level of DNA sequences. Whether the genetic diversity of populations is maintained primarily by mutation plus genetic drift or by natural selection is a matter of some debate.
3. The geographic ranges of most species exhibit a degree of environmental diversity. As a result of both adaptation to different environments and genetic drift, different populations within a species may have different alleles or allele frequencies at many loci.
4. Gene flow among populations tends to homogenize their genetic composition. When gene flow is reduced, genetic drift and adaptation to different environments can cause populations to diverge. Eventually, populations may become so different that the individuals in one population either will not or cannot mate with the individuals in the other. At this point, the divergent populations have become different species.
5. Because speciation is associated with genetic divergence, we can use the genetic differences among species to infer their evolutionary history. By comparing amino acid or nucleotide sequences, we can determine the genetic distances between species. We can then use the genetic distances to reconstruct evolutionary trees. The simplest methods for reconstructing phylogenetic trees are based on the assumption that the least divergent species are one another's closest relatives.
6. The reconstruction of evolutionary trees is a key technique for answering a diversity of interesting questions. Examples include tracing the route of transmission of a disease in an epidemic, deciphering recent events in human evolution, and distinguishing our genetic heritage from that of the Neanderthals and other extinct human species.

## INSIGHTS AND SOLUTIONS

1. Influenza accompanied by pneumonia is one of the top ten causes of death in the United States. Each year, a different strain of the influenza virus sweeps across the world. Viral genes mutate rapidly, but most mutant strains die out, leaving one infective strain. In an attempt to predict the evolution of the influenza A strain to facilitate the preparation of an effective vaccine, scientists are studying the pattern of mutations in the HA1 domain of the hemagglutinin gene of the H3 influenza subtype. Studies of nucleotide substitutions in the 329 codons of HA1 have identified 18 codons in which mutations leading to amino acid replacement have been selectively advantageous, leading to successful infective strains. Construction of phylogenetic trees shows that flu strains do not produce a multiply-branched tree, but because one main strain survives, the tree resembles a staircase. These trees show that strains with more mutations in these 18 codons were founders of the annual influenza strains in 9 of 11 influenza seasons surveyed. Can these observations be used to predict the course of influenza virus evolution in future years, allowing the preparation of a strain-specific flu vaccine before the annual epidemic begins?

   **Solution:** The strategy for predicting the evolution of the H3 strain of influenza A would depend on the early isolation of multiple mutant strains, the construction of phylogenetic trees, and the monitoring of all strains for mutations in these 18 codons. One of the strains that accumulates the most mutations in the 18 monitored codons in the HA1 domain is most likely to be the strain responsible for the next influenza epidemic. Based on the phylogenetic relationship to the previous strain, and the number of mutations acquired in the monitored codons, it may be possible to identify the strain before an epidemic begins and to use that virus to develop a strain-specific vaccine that will reduce infections and related deaths.

2. A single plant twice the size of others in the same population suddenly appears. Normally, plants of that species reproduce by self-fertilization and by cross-fertilization. Is this new giant plant simply a variant, or could it be a new species? How would you determine which it is?

   **Solution:** One of the most common mechanisms of speciation in higher plants is polyploidy, the multiplication of entire sets of chromosomes. The result of polyploidy is usually a larger plant with larger flowers and seeds. There are two ways of testing the new variant to determine whether it is a new species. First, the giant plant should be crossed with a normal-sized plant to see whether the giant plant produces viable, fertile offspring. If it does not, then the two different types of plants would appear to be reproductively isolated. Second, the giant plant should be cytogenetically screened to discover its chromosome complement. If it has twice the number of its normal-sized neighbors, it is a tetraploid that may have arisen spontaneously. If the chromosome number differs by a factor of two and the new plant is reproductively isolated from its normal-sized neighbors, it is a new species.

## Problems and Discussion Questions

1. What is the neo-Darwinian definition of evolution?
2. Define speciation. How does this term differ from evolution?
3. Wallace and Darwin identified natural selection as a force acting on phenotypic variation. What two processes related to phenotypic variation were they unable to explain?
4. Two closely related species may have very few phenotypic differences. How can we justify classifying them as two different species instead of one species exhibiting a small range of phenotypic variation?
5. Natural selection can lead to speciation. Is this the only way species can form?
6. The maintenance of allozymic diversity in natural populations remains a controversial topic. One study examined polymorphisms among 111 above-ground and 132 subterranean mammalian species and found that subterranean mammals are less polymorphic than those living above ground (Nevo, 2001. *Proc. Natl. Acad. Sci. [USA]* 98: 6233–6240). Provide a possible explanation.
7. Discuss the rationale behind the statement that inversions in chromosome 3 of *D. pseudoobscura* represent genetic variation.
8. Dobzhansky's studies on populations of *D. pseudoobscura* showed changes in the frequency of the ST and CH chromosome arrangements throughout the year. Why hasn't one of these arrangements been eliminated over a long period of time?
9. Describe how populations with substantial genetic differences can form within a species. What is the role of natural selection?
10. Researchers conducted a genetic study of the toxin transport protein (PA) of *Bacillus anthracis*, the bacterium that causes anthrax in humans (Price et al., 1999. *J. Bacteriol.* 181: 2358–2362). In 26 strains, they identified five point mutations—two missense and three synonyms (silent mutations that do not change an encoded amino acid)—among different isolates of the 2294-nucleotide gene. Necropsy samples from an anthrax outbreak in 1979 revealed a novel missense mutation and five unique nucleotide changes in samples collected from ten victims. The authors concluded that these data indicate little or no horizontal transfer between different *B. anthracis* strains.
    (a) Which types of nucleotide changes (missense or synonyms) cause amino acid changes?
    (b) What is meant by horizontal transfer?
    (c) On what basis did the authors conclude that evidence of horizontal transfer is absent from their data?
11. What types of nucleotide substitutions will not be detected by electrophoretic studies of a gene's protein product?
12. A recent study examining the mutation rates of 5669 mammalian genes (17,208 sequences) indicates that, contrary to popular belief, mutation rates among lineages with vastly different generation lengths and physiological attributes are remarkably constant (Kumar, 2002. *Proc. Natl. Acad. Sci. [USA]* 99: 803–808). The average rate is estimated at $1.22 \times 10^{-9}$ per bp per year. What is the significance of this finding in terms of mammalian evolution?
13. Discuss the arguments supporting the neutral theory of molecular evolution. What counterarguments do the selectionists propose? Of what value is the debate concerning the neutral theory?
14. What genetic changes take place during speciation?
15. The mummichog shows *Ldh* allele frequency differences from cold northern waters to warm southern waters.
    (a) Is there a correlation between allele type, allele frequency, and water temperature?
    (b) Why are differences in catalytic efficiency *and* differences in transcription rates important in this adaptive strategy?
16. List the barriers that prevent interbreeding, and give an example of each.
17. What are the two groups of reproductive isolating mechanisms? Which of these is regarded as more efficient, and why?
18. In the tobacco plant *N. tabacum*, a polyploid formed as a hybrid between *N. otophora* and *N. silvestris*, would you expect to find higher levels of heterozygosity in the polyploid species than in the diploid parental species? Is this an advantage or disadvantage?
19. Why are many species of flowering plants polyploid, but only a few animal species are polyploid?
20. Some critics have warned that the use of gene therapy to correct genetic disorders will affect the course of human evolution. Evaluate this criticism in light of what you know about population genetics and evolution, distinguishing between somatic gene therapy and germ-line gene therapy.

**HOW DO WE KNOW?**

21. In this chapter we focused on how changes in gene frequency are related to the process of species formation. At the same time, we found many opportunities to consider the methods and reasoning by which much of this information was acquired. From the explanations given in the chapter, what answers would you propose to the following fundamental questions:
    (a) How do we determine how much genetic variation exists in a population?
    (b) What evidence has been obtained from laboratory studies to indicate that natural selection is the cause of genetic differences among natural populations of a species?
    (c) How can the minimum genetic divergence between two species be determined?
    (d) How can we determine the last common ancestor shared by two divergent species?

## Extra-Spicy Problems

22. Shown below are two homologous lengths of the $\alpha$ and $\beta$ chains of human hemoglobin. Consult the genetic code dictionary (Figure 14–7) and determine how many amino acid substitutions may have occurred as a result of a single-nucleotide substitution. For any that cannot occur as the result of a single change, determine the minimal mutational distance.

| | | | | | | | | | |
|---|---|---|---|---|---|---|---|---|---|
| $\alpha$: | Ala | Val | Ala | His | Val | Asp | Asp | Met | Pro |
| $\beta$: | Gly | Leu | Ala | His | Leu | Asp | Asn | Leu | Lys |

23. Determine the minimal mutational distances between these amino acid sequences of cytochrome c from various organisms. Compare the distance between humans and each organism.

| | | | | | | |
|---|---|---|---|---|---|---|
| Human: | Lys | Glu | Glu | Arg | Ala | Asp |
| Horse: | Lys | Thr | Glu | Arg | Glu | Asp |
| Pig: | Lys | Gly | Glu | Arg | Glu | Asp |
| Dog: | Thr | Gly | Glu | Arg | Glu | Asp |
| Chicken: | Lys | Ser | Glu | Arg | Val | Asp |
| Bullfrog: | Lys | Gly | Glu | Arg | Glu | Asp |
| Fungus: | Ala | Lys | Asp | Arg | Asn | Asp |

24. The following data represent short DNA sequences from five species: human, chimpanzee, gorilla, orangutan, and baboon. The sequences are each 50 bp long. They represent a short piece of a gene for testis-specific protein Y, located on the Y chromosome. The complete human sequence is given; the sequences for the other four species are shown only where they differ from the human sequence. Calculate the genetic difference between each pair of species. (For example, chimps and gorillas differ from each other in 2 out of 50 bases, or 4 percent; thus, the genetic difference between chimps and gorillas is 0.04.) Then use UPGMA to reconstruct the phylogeny for these five species. Is your phylogeny consistent with those shown in Figure 28–18?

```
Human:      AGAGGTTTTTCAGTGAATGAAGCTATTTTTAAGGGAGTGTGATTGCTGCC
Chimpanzee:                  C
Gorilla:             T       C                     C
Orangutan:   T     G T      CC            C        C
Baboon:      CC G G  T       C    G     G C    C   CG
```

25. The genetic difference between two *Drosophila* species, *D. heteroneura* and *D. sylvestris*, as measured by nucleotide diversity, is about 1.8 percent. The difference between chimpanzees (*P. troglodytes*) and humans (*H. sapiens*) is about the same, yet *H. sapiens* and (*P. troglodytes*) are classified in different genera. In your opinion, is this valid? Explain why.

26. The use of nucleotide sequence data to measure genetic variability is complicated by the fact that the genes of higher eukaryotes are complex in organization and contain 5′ and 3′ flanking regions as well as introns. Researchers compared the nucleotide sequence of two cloned alleles of the *γ-globin* gene from a single individual and found a variation of 1 percent. The differences include 13 substitutions of one nucleotide for another and 3 short DNA segments that have been inserted in one allele or deleted in the other. None of the changes takes place in the gene's exons (coding regions). Why do you think these changes occur outside exon-coding regions, and should it change our concept of genetic variation?

27. In a study of cichlid fish inhabiting Lake Victoria in Africa, researchers examined presumably neutral sequence polymorphisms thought to be located in noncoding genomic loci in 12 lake species and their putative riverine ancestors (Nagl et al. 1998. *Proc. Natl. Acad. Sci. [USA]* 95: 14,238–14,243). At all loci, the same polymorphism was found in nearly all of the tested species from Lake Victoria, both lacustrine and riverine. Different polymorphisms at these loci were found in cichlids at other African lakes.
    (a) Why would you suspect neutral sequences to be located in noncoding genomic regions?
    (b) What conclusions can be drawn from these polymorphism data in terms of cichlid ancestry in these lakes?

28. Given that there are approximately 400 cichlid species in Lake Victoria and that it dried up almost completely about 14,000 years ago, what evidence indicates that extremely rapid evolutionary adaptation rather than extensive immigration occurred?

29. Comparisons of Neanderthal mitochondrial DNA with that of modern humans indicate that the Neanderthals are not related to modern humans and did not contribute to our mitochondrial heritage. However, because Neanderthals and modern humans are separated by at least 25,000 years, this does not rule out that some forms of interbreeding may have caused the modern European gene pool to be derived from both Neanderthals and early humans (called Cro-Magnons). To resolve this question, researchers analyzed mitochondrial DNA sequences from 25,000-year-old Cro-Magnon remains and compared them to four Neanderthal specimens and a large data set derived from modern humans (Caramelli et al., 2003. *Proc. Natl. Acad. Sci. [USA]* 100: 6593–6597). The results are shown in the graph below.
    (a) What can you conclude about the relationship between Cro-Magnons and modern Europeans? What about the relationship between Cro-Magnons and Neanderthals?
    (b) From these data, does it seem likely that Neanderthals made any contributions to the Cro-Magnon gene pool or the modern European gene pool?

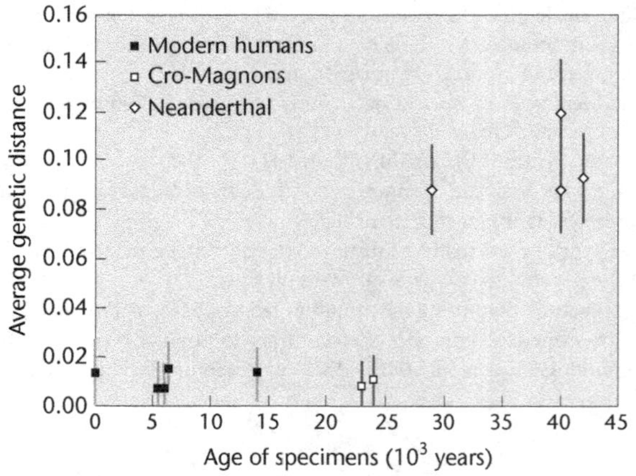


30. Recent reconstructions of evolutionary history are often dependent on assigning divergence in terms of changes in amino acid or nucleotide sequences. For example, a comparison of cytochrome c shows 10 amino acid differences between humans and dogs, 24 differences between humans and moths, and 38 differences between humans and yeast. Such data provide no information as to the absolute times of divergence for humans, dogs, moths, and yeast. How might one calibrate the molecular clock to an absolute time clock? What problems might one encounter in such a calibration?

31. Polyploidy in plants appears to be a significant factor in evolution. However, recent studies (reviewed by Adams and Wendel, 2005) explain that after polyploidization, differential gene silencing and loss reshapes and subfunctionalizes the genome. In the next column are inferred polyploidy events (**x**) in selected lineages of familiar plants (modified from Blanc and Wolfe, 2004).
   (a) Would you expect to see a similar pattern of polyploidization in animal lineages?
   (b) In what way would the merging of two genomes (allopolyploidy) support rapid evolution?
   (c) Polyploidization in maize occurred approximately 11 million years ago, yet approximately half of all the duplicated genes have been lost. How might polyploidization followed by gene loss contribute to speciation?
   (d) Apparently, gene loss is not random. In one study, genes involved in transcription and signal transduction have been retained more often than those involved in DNA repair. In addition, genes for organellar proteins seem to be preferentially lost. How might one explain these findings?
   (e) Present a drawing that depicts the union of two different genomes in an initial allopolyploidization event followed by selective gene loss that returns the genome to a genetically diploid amalgam.

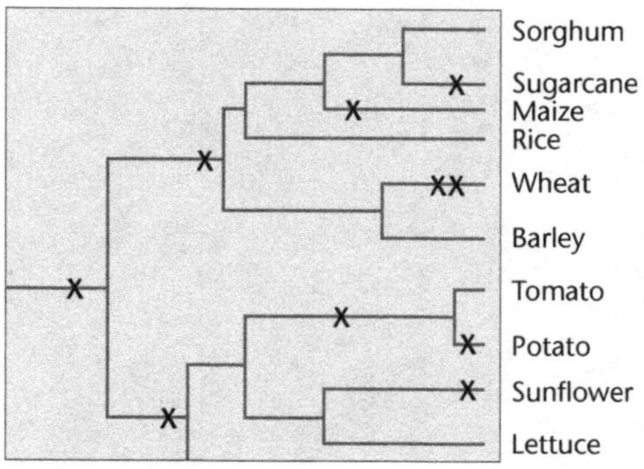

Seeds of rare crop varieties, cryogenically preserved at the U.S. Department of Agriculture National Seed Storage Laboratory.

# 29

# Conservation Genetics

## CHAPTER CONCEPTS

- Species require genetic diversity for long-term survival and adaptation.
- Small, isolated populations are particularly vulnerable to genetic effects.
- Endangered populations that recover in size may not redevelop genetic diversity.
- Conservation and breeding efforts focus on retaining genetic diversity for long-term species survival.

As the twenty-first century progresses, the diversity of life on Earth is under increasing pressure from the direct and indirect effects of explosive human population growth. Approximately 10 million *Homo sapiens* lived on the planet 10,000 years ago. This number grew to 100 million 2000 years ago and to 2.5 billion by 1950. Within the span of a single lifetime, the world's human population more than doubled to 5.5 billion in 1993 and is projected to reach as high as 19 billion by 2100 (Figure 29–1).

The effect of accelerating human population growth on other species has been dramatic. Data from the 2006 World Conservation Union (IUCN) Red List of Threatened Species show that, globally, 23 percent of all mammals, 12 percent of birds, 51 percent of reptiles, 31 percent of amphibians, and 40 percent of fish species are threatened. Plants are not faring much better: the 2006 IUCN Red List also includes over 8300 vascular plant species worldwide. Not only wild species are at risk. Genetic diversity in domesticated plants and animals is also being lost as many traditional crop varieties and livestock breeds disappear. The Food and Agriculture Organization (FAO) estimates that since 1900, 75 percent of the genetic diversity in agricultural crops has been lost. Out of approximately 5000 different breeds of domesticated farm animals worldwide, one-third are at risk.

Why should we be concerned about losing **biodiversity**—that is, the biological variation represented by these different plants and animals? After all, the fossil record shows that many different plants and animals have become extinct in the course of evolution, even before humans existed, and other species have taken their places. Biologists are concerned, however, at the accelerated rate of species extinctions we are witnessing today, all of which can be ascribed directly or indirectly to human impact. Deliberate hunting or harvesting of plants and animals by humans, and habitat destruction through human development activities have greatly reduced the

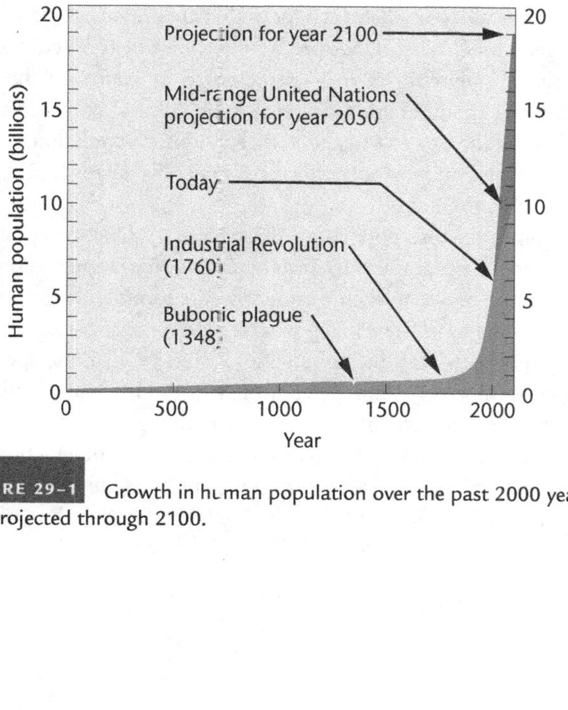


FIGURE 29–1 Growth in human population over the past 2000 years and projected through 2100.

FIGURE 29–2 The brown-tree snake (*Boiga irregularis*).

populations of many species. An additional problem in many parts of the world is the deliberate or accidental introduction by humans of invasive nonnative plants or animals that prey on or compete with native species, further jeopardizing their survival. For example, the brown tree snake, *Boiga irregularis* (Figure 29–2), was accidentally transported by ship in the late 1940s to the island of Guam. An aggressive predator, the brown tree snake has so far caused the extinction of 12 bird and 4 lizard species that were previously native to Guam. This introduced species continues to threaten native biodiversity on the island.

One of the greatest threats to biodiversity, however, may be climate change due to global warming. Scientists are increasingly concerned that this phenomenon will lead to the extinction of many species, especially those adapted to live in colder environments. For example, in 2006 the IUCN reclassified the 22,000 remaining polar bears (Figure 29–3) as vulnerable due to fears that their habitat will be lost from global warming.

FIGURE 29–3 The polar bear (*Ursus maritimus*).

As a result of species extinction, biologists fear that **ecosystems**—the complex webs of interdependent but diverse plants and animals found together in the same environment—may collapse if key sustaining species are lost. This portends possible consequences for our own long-term survival. Other scientists have pointed out that we lose unknown economic potential of unexploited plants and animals if we allow them to become extinct. Rare species sometimes turn out to be immensely valuable. For example, an obscure bacterium, *Thermus aquaticus*, was first discovered in a thermal pool in Yellowstone National Park and is known to exist in only a few hot springs throughout the world. DNA polymerase isolated from this microorganism functions at very high temperatures and is critical to the polymerase chain reaction (PCR), a process for *in vitro* replication of DNA that we have alluded to throughout this book. This research technique is essential to the multibillion dollar global biotechnology industry. Quite apart from the practical benefits of biodiversity, scientists and nonscientists have made the case that human life will be diminished both spiritually and aesthetically if we do not strive to maintain the fascinating and often beautiful variety of living organisms that share the planet with us.

Conservation biologists work to understand and maintain biodiversity, studying the factors that lead to species decline and the ways species can be preserved. The new field of **conservation genetics** has emerged in the last 20 years as scientists have begun to recognize that genetics will be an important tool in maintaining and restoring population viability. The applications of genetics to conservation biology are multifaceted; in this chapter, we will explore just a few of them. Underlying the increasingly important role of genetics in conservation biology is the recognition that biodiversity depends on genetic diversity and that maintaining biodiversity in the long term is unlikely if genetic diversity is lost.

## 29.1 Genetic Diversity Is the Goal of Conservation Genetics

While *biodiversity* refers to the variation represented by all existing species of plants and animals on this planet at any given time, **genetic diversity** is not as easy to define or to study. Genetic diversity can be considered on two levels: *interspecific diversity* and *intraspecific diversity*.

Diversity between species, or **interspecific diversity**, is the diversity reflected in the number of different plant and animal species present in an ecosystem. Some ecosystems have a very high level of interspecific diversity, such as a tropical rainforest in which hundreds of different plant and animal species may be found within a few square meters [Figure 29–4(a)]. Other ecosystems, especially where plants and animals must adapt to a harsh environment, may have much lower numbers of species [Figure 29–4(b)]. Species inventories—lists of the different plant and animal species found in a particular environment—are useful for identifying diversity hot spots: geographic areas with especially high interspecific diversity, where conservation efforts can be focused. Conservation biologists working at the ecosystem level are interested in preventing species from being lost and in restoring species that were once part of the system but are no longer present. The reestablishment of the gray wolf (*Canis lupus*) in Yellowstone National Park and the release of captive-bred California condors (*Gymnogyps californianus*) into the mountain ranges they previously occupied in the southwestern United States are examples of current attempts by conservation biologists to restore missing species to their ecosystems.

FIGURE 29–4 (a) A tropical rain forest is an ecosystem with high interspecific diversity. (b) A coastal marsh in North Carolina exemplifies an ecosystem with low interspecific diversity.

**Intraspecific diversity**—the diversity within a species—itself has two components: *intrapopulation diversity* is the genetic variation occurring between individuals within a single population of a given species; and *interpopulation diversity* is the variation occurring between different populations of the same species. Genetic variation within populations can be measured as the frequency of individuals in the population that are heterozygous at a given locus or as the number of different alleles at a locus that are present in the population gene pool. When DNA-profiling techniques are used, the percentage of polymorphic loci—represented by bands on the DNA profile that are different in different individuals—can be calculated to indicate the extent of genetic diversity in a population. In outbreeding species, most intraspecific genetic diversity is found at the intrapopulation level.

Significant interpopulation diversity can occur if populations are separated geographically and there is no migration or exchange of gametes between them. Similarly, predominantly inbreeding species (such as self-fertilizing plants) tend to have greater levels of interpopulation than intrapopulation diversity. In these species, a limited number of genotypes dominate an individual population, but there is greater differentiation between populations. Understanding the breeding system and the distribution of genetic variation in an endangered species is important to its conservation, not only to ensure continued production of offspring but also to determine the best strategy for maintaining intraspecific diversity. For example, would it be more effective to preserve a few large populations or many small, distinct ones? This information can then be used to guide conservation or restoration efforts.

## Loss of Genetic Diversity

Loss of genetic diversity in nondomesticated species is usually associated with a reduction in population size due in part to excessive hunting or harvesting. For example, biologists blame commercial overfishing for the collapse in the early 1990s of the deep-sea cod populations off the Newfoundland coast. The cod fishery in that area was once one of the most productive in the world, but despite being protected from fishing since 1992, these fish populations have not fully recovered. Habitat loss is also a major cause of population decline. As the global human population increases, more land is developed for housing and transport systems or is put into agricultural production, reducing or eliminating areas that were once home to wild plants and animals. The shrinking of available habitat reduces populations of wild species and often also isolates them from each other, as individual populations become trapped in pockets of undeveloped land surrounded by areas taken over for agriculture, urban development, or other human uses. This process is known as **population fragmentation**. When populations are no longer in contact with each other, **gene flow**, the gradual exchange of alleles between two populations, brought about through migration or gamete exchange between them, ceases, and an important mechanism for maintaining genetic variation is lost.

In domesticated species, loss of genetic diversity is not usually the result of habitat loss or collapsing population numbers; there is little risk that cows or corn as species will become extinct any time soon. Reduction in diversity within domesticated species can instead be traced to changes in agricultural practice and consumer demand. Modern farming techniques have greatly increased production levels, but they have also led to greater genetic uniformity. As farmers switch to new crop varieties or improved livestock strains on a large scale, they abandon cultivation of many older local types, which may then disappear if efforts are not made to preserve them.

For example, in 1900, more than 100 different kinds of potatoes were available in the United States. Today, three quarters of commercial potato production in this country depends on just nine varieties. A single type, Russet Burbank, makes up 45 percent of the total acreage planted, with many farmers also growing related potato varieties such as Ranger Russet and Russet Norkotah. Modern varieties such as these are better adapted for modern agricultural production, but older types often contain useful genes that can still play a vital role in survival functions, such as resistance to disease, cold, or drought.

## Identifying Genetic Diversity

For many years, population geneticists based estimates of intraspecific diversity on phenotypic differences between individuals, such as different colors of seeds or flowers or variation in markings

FIGURE 29–5 Phenotypic variation in seed color and markings in the common bean (*Phaseolus vulgaris*) reveals high levels of intraspecific diversity.

(Figure 29–5). As techniques were developed to analyze genes and gene products at the molecular level, identifying genetic diversity became more precise. For example, a simple molecular method that has been widely used is allozyme analysis. **Allozymes** are multiple versions of a single enzyme occurring within a single species. Allozymes catalyze the same reactions, but their polypeptide chains differ because of slight variations in the allelic DNA sequence at the locus that codes them. The resulting proteins therefore differ in size or net charge and can be separated by electrophoresis. The presence of allozymes in a population can be used as an indicator of the lower limit of genetic variation because the different versions of the molecule signify that different alleles are present at a locus.

DNA analysis is a more modern and direct molecular approach for detecting and quantifying genetic differences between individuals. This technique has become an important tool for assessing intra- and interpopulation genetic variation. Nuclear, mitochondrial, and chloroplast DNA can all be analyzed to determine levels of genetic diversity. We have already described applications of DNA analysis using either restriction enzymes or PCR in Chapter 24. Conservation biologists have used similar techniques to examine levels and distribution of genetic variation both within and between populations and to subsequently guide conservation efforts.

One technique that has been widely used to analyze genetic diversity in plant and animal populations takes advantage of the presence of **STRs (short tandem repeats)**, also known as **microsatellites.** STRs consist of short sequences (two to nine bases) repeated a variable number of times and are found throughout the genome. The number of times a repeat is present at a given STR locus often varies between individuals. Although the STR regions themselves do not contain expressed genes, conservation biologists can use microsatellite variation among individuals within populations as an indirect estimate of the amount of overall genetic variation present.

For example, researchers from the University of Arizona recently used microsatellite analysis to compare genetic variation in the three remaining populations of the Mauna Loa silversword, an endangered plant found only in Hawaii. Biologists working to conserve this plant needed to know whether they should try to increase genetic diversity by redistributing seed among the populations. However, STR analysis not only revealed unexpected amounts of diversity still present within populations but also showed that the three populations were genetically quite different from one another. The seed redistribution plan was abandoned in favor of preserving the unique identity of each population.

In another recent study utilizing DNA markers, *restriction fragment length polymorphisms (RFLPs)* (see Chapter 24) were analyzed in the only remaining population of a critically endangered plant species, *Limonium cavanillesii*, growing on the Mediterranean coast of Spain. Researchers found very low levels of diversity but did detect genetically distinct individual plants within the population from which seeds could be collected.

In still another recent RFLP study, DNA marker analysis of the southwestern willow flycatcher, an endangered migratory bird species that nests in the southwestern United States, found significant levels of genetic diversity within populations at different breeding sites but little interpopulation diversity. The scientists studying the flycatcher therefore concluded that migration of individual birds between breeding populations is important in maintaining diversity in the species and that conservation efforts should be directed toward preserving nesting sites that allow such migration to continue.

Another application of DNA markers in conservation involves *PCR-based DNA fingerprinting*. The use of DNA-profiling techniques in forensic science was discussed in Chapter 24. Similar techniques can be used to uncover illegal trade in endangered plant and animal species that are protected by law. For example, since 1986, an international moratorium on commercial whaling has been in place, with only limited hunting of a few species permitted. In 1994, scientists based in New Zealand and Hawaii used PCR to amplify mitochondrial DNA (mtDNA) sequences extracted from meat on sale in markets in Japan, where whale meat is prized as a delicacy. The DNA fingerprints revealed several meat samples from humpback and fin whales, which are protected species. Two meat samples were not from whales at all but were dolphin meat! This information assisted Japanese customs officials in tracking down sources of illegal whale meat imports.

Unlawful trade in animal products also occurs in the United States. In a recent study by scientists at the University of Florida, amplification of mtDNA sequences from turtle meat sold in Louisiana and Florida revealed that one quarter of the samples were actually alligator, not turtle. Most of the rest were from small freshwater turtles, indicating that populations of freshwater turtle species were declining through overharvesting and thus were in need of protection. Further refinements to DNA extraction and profiling techniques allowed accurate identification of illegally harvested sea turtle eggs cooked and served in restaurants, and led to prosecution of the poachers.

New methods for analyzing DNA marker data also permit estimation of past changes in populations, based on the rate at which mtDNA is known to accumulate spontaneous point mutations. In a controversial recent study, scientists at Stanford University led by Dr. Stephen Palumbi measured sequence diversity at key points in mtDNA from humpback whales and compared their findings with the amount of diversity expected for a given population size. The Stanford researchers found much more mtDNA sequence diversity than expected and concluded that before being hunted almost to extinction in the nineteenth and twentieth centuries, the global humpback whale population must have numbered over one million, ten times higher than historical estimates based on records from whaling ships. As current international agreements could permit hunting of whales to resume when populations have recovered to 54 percent of what is thought to be their original carrying capacity, Palumbi's findings are important for the future management of this species.

## 29.2 Population Size Has a Major Impact on Species Survival

Some species have never been numerous, especially those that are adapted to survive in unusual habitats. Biologists refer to such species as *naturally rare*. *Newly rare* species, on the other hand, are those whose numbers are in decline because of pressures such as habitat loss. Populations of such species may not only be small in number but also fragmented and isolated from other populations. Both decreased population size and increased isolation have important genetic consequences, leading to an increased risk of loss of diversity.

How small must a population be before it is considered endangered? It varies somewhat with species, but small populations can quickly become vulnerable to genetic phenomena that increase the risk of extinction. In general, a population of fewer than 100 individuals is considered extremely sensitive to these problems, which include genetic drift, inbreeding, and reduction in gene flow. The effects of such problems on species survival are substantial. Studies of bighorn sheep, for example, have shown that populations of fewer than 50 are highly likely to become extinct within 50 years. Projections based on computer models show that for all species, populations of fewer than 10,000 are likely to be limited in adaptive genetic variation, and that at least 100,000 individuals must be present if a population is to show long-term sustainability.

Determining the number of individuals a population must contain in order to have long-term sustainability is complicated by the fact that not all members of a population are equally likely to produce offspring: some will be infertile, too young, or too old. The *effective population size* ($N_e$) is defined as the number of individuals in a population having an equal probability of contributing gametes to the next generation. $N_e$ is almost always smaller than the *absolute population size* ($N$). The effective population size can be calculated in different ways, depending on the factors that are preventing individuals in a population from all contributing equally to the next generation. In a sexually reproducing population that contains different numbers of males and females, for example, the effective population size is calculated as

$$N_e = \frac{4(N_m N_f)}{N_m + N_f}$$

where $N_m$ is the number of males and $N_f$ the number of females in the population. Hence a population of 100 males and 100 females would have an effective size of $4(100 \times 100)/(100 + 100) = 200$. In contrast, if there were 180 males and only 20 females, the effective population size would be $4(180 \times 20)/(180 + 20) = 72$.

Effective population size is also influenced by fluctuations in absolute population size from one generation to the next. Here, the effective population size is the harmonic mean of the numbers in each generation, so

$$N_e = 1 \Big/ \frac{1}{t}\left(\frac{1}{N_1} + \frac{1}{N_2} \cdots + \frac{1}{N_t}\right)$$

where $t$ is the total number of generations being considered. For example, if a population went through a temporary reduction in size in generation 2, so that $N_1 = 100$, $N_2 = 10$, and $N_3 = 100$, then

$$N_e = 1 \Big/ \frac{1}{3}\left(\frac{1}{100} + \frac{1}{10} + \frac{1}{100}\right) = \frac{1}{0.04} = 25$$

In this case, although the actual mean number of individuals in the population over three generations was 70, the effective population size during that time was only 25. A severe temporary reduction in size such as this is known as a **population bottleneck.** Bottlenecks occur when a population or species is reduced to a few reproducing individuals whose offspring then increase in numbers over subsequent generations to reestablish the population. Although the number of individuals may be restored to healthier levels, genetic diversity in the newly expanded population is often severely reduced because gametes from the handful of surviving individuals functioning as parents only represent a subset of the original gene pool.

**NOW SOLVE THIS**

Question 13 on page 778 asks you to calculate effective population size $N_e$ for an endangered gorilla population.

■ HINT: *Consider what unusual feature of the social structure of the gorilla population will affect $N_e$.*

Captive-breeding programs, in which a few surviving individuals from an endangered species are removed from the wild to rebuild the population in a protected environment, inevitably create population bottlenecks. Bottlenecks also occur naturally when a small number of individuals from one population migrate to establish a new population elsewhere. When a new population derived from a small subset of individuals has significantly less genetic diversity than the original population, it exhibits the **founder effect** (Chapter 27). Reduced levels of genetic diversity due to a founder effect can persist for many generations, as shown by studies of two species of Antarctic fur seal, *Arctocephalus gazella* and *Arctocephalus tropicalis*. Seal hunters in the eighteenth and nineteenth centuries severely reduced the populations of these species, eliminating them from parts of their natural range in the Southern Ocean.* Although the number of Antarctic fur seals has now rebounded and the two species have recolonized much of their original habitat, mtDNA fingerprinting reveals that a founder effect can still be detected in *A. gazella*, with reduced genetic variation in current populations descended from a handful of surviving individuals.

**FIGURE 29–6** The cheetah (*Acinonyx jubatus*).

The cheetah (*Acinonyx jubatus*), shown in Figure 29–6, is another well-studied example of a species with reduced genetic variation resulting from at least one severe population bottleneck that occurred in its recent history. Allozyme studies of South African cheetah populations have shown levels of genetic variation that are less than 10 percent of those found in other mammals. The abnormal spermatozoa and poor reproductive rates commonly observed in cheetahs are thought to be linked to the lack of genetic diversity, although when and how the population bottleneck occurred in this species is still unclear.

**NOW SOLVE THIS**

Question 26 on page 779 asks you to compare allozyme data for cheetah populations with similar data from other cat species.

■ HINT: *Consider what genetic factors potentially reduce the levels of genetic variation reflected in the data. Which species will be most affected by these factors?*

---

* In 2000, this fifth world ocean was delimited by oceanographers, incorporating southern portions of the Atlantic, Indian, and Pacific oceans.

29.3

## Genetic Effects Are More Pronounced in Small, Isolated Populations

Small isolated populations, such as those found in threatened and endangered species or produced by population fragmentation, are especially vulnerable to genetic drift, inbreeding, and reduction in gene flow. These phenomena act on the gene pool in different ways, but ultimately have similar effects in that they all can further reduce genetic diversity and long-term species viability.

### Genetic Drift

If the number of breeding individuals in a population is reduced in size, fewer gametes will form the next generation. The alleles carried by these gametes may not be a representative sample of all those present in the population; purely by chance, some alleles may be underrepresented or not present at all, which will cause changes in allele frequency over time, resulting in **genetic drift**. This phenomenon has already been described in Chapter 27. A serious result of genetic drift in populations with a small effective population size is the loss of genetic variation. Genetic drift is a random process, so either deleterious or advantageous alleles can become fixed within a small population. In other words, one allele becomes the only version of that gene present in the gene pool of the population. This means a useful allele can be lost even if it has the potential to increase fitness or long-term adaptability. The probability that an allele will be fixed through drift is the same as its initial frequency. Thus, if a locus has two alleles, $A$ and $a$, and if the initial frequency of $A$ is 0.8, then the probability of $A$ becoming fixed is 0.8, or 80 percent, and the probability of $A$ being lost through drift is 0.2 or 20 percent. Figure 29–7 is a graph of the effect of genetic drift on a theoretical population.

### Inbreeding

In small populations, the chance of **inbreeding**, meaning mating between closely related individuals, is greater. As described in Chapter 27, inbreeding increases the proportion of homozygotes in a population, thus expanding the possibility that an individual may be homozygous for a deleterious allele. We have already seen in Chapter 27 how the extent of inbreeding in a population can be quantified as the **inbreeding coefficient (*F*)**, which measures the probability that two alleles of a given gene are derived from a common ancestral allele. Remember that the inbreeding coefficient is inversely related to the frequency of heterozygotes in the population and can be calculated as

$$F = \frac{2pq - H}{2pq}$$

where $2pq$ is the expected frequency of heterozygotes based on the Hardy–Weinberg law and $H$ is the actual frequency of heterozygotes in the population.

In a declining population that has become small enough for drift to occur, heterozygosity ($H$) will decrease with each generation. The smaller the effective population size, the more rapid the decrease in $H$ and the resulting increase in $F$, as can be seen from the equation

$$H_t/H_0 = \left(1 - \frac{1}{2N_e}\right)^t$$

where $H_0$ is the initial frequency of heterozygotes, $H_t$ is the frequency of heterozygotes after $t$ generations, and $N_e$ is the effective population size. Figure 29–8 compares rates of increase in $F$ for different effective population sizes.

**NOW SOLVE THIS**

Question 1 on page 777 asks you to calculate effective population sizes, a heterozygote frequency, and an inbreeding coefficient for a jackal population that crashed and then recovered.

■ HINT: *Calculate $N_e$ for each generation first, and remember that heterozygote frequency and inbreeding coefficient are inversely related.*

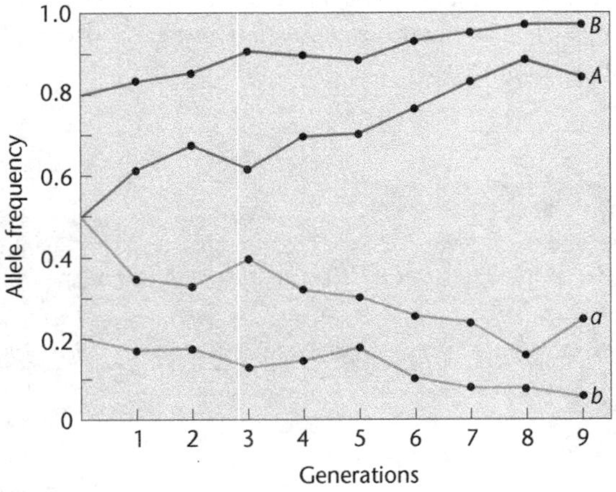


**FIGURE 29–7** Change in frequencies over 10 generations for two sets of alleles, $A/a$ and $B/b$, in a theoretical population subject to genetic drift.

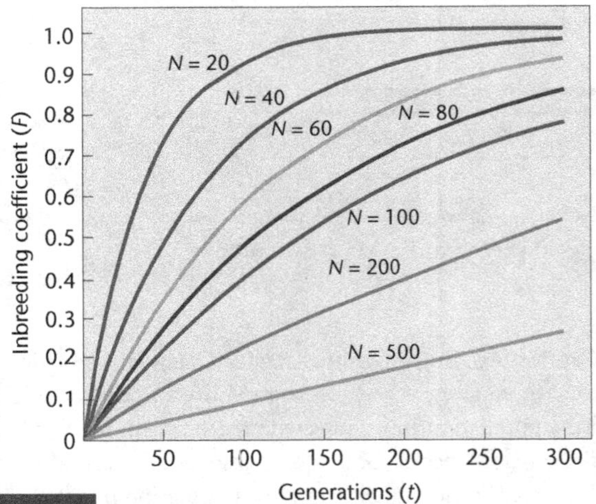


**FIGURE 29–8** Increase in inbreeding coefficient ($F$) in theoretical populations as the population size ($N$) decreases.

What effect does inbreeding have on the long-term survival of a species? If numbers of individuals in a species remain high, inbreeding in a single population may not immediately reduce the amount of genetic variation present in the overall gene pool. Self-pollinating plants, for example, often show high levels of homozygosity and relatively little genetic variation within single populations. However, they tend to have considerable variation between *different populations*, each of which has adapted to slightly different local environmental conditions. On the other hand, in most outbreeding species (those where gene flow occurs freely between populations), including all mammals, inbreeding is associated with reduced fitness and lower survival rates among offspring. This **inbreeding depression** can result from increased homozygosity for deleterious alleles. The number of deleterious alleles present in the gene pool of a population is called the **genetic load** (or genetic burden).

In some species, inbreeding accompanied by selection against less fit individuals homozygous for deleterious alleles has resulted in the elimination of these alleles from the gene pool, a process known as *purging the genetic load*. Species that have successfully purged their genetic load do not show continued reduction in fitness, even after many generations of inbreeding. This is true of numerous domesticated species, especially self-pollinating plants such as wheat. However, computer simulation experiments have shown that it may take 50 generations or more to complete the purging process, during which time inbreeding depression will still occur.

Alternatively, inbreeding depression can result from heterozygous individuals having a higher level of fitness than either of the corresponding homozygotes. In this case, the long-term survival of the population requires that inbreeding be avoided and that the levels of all alleles in the gene pool be maintained. Both of these requirements can be difficult to satisfy in a species that has already suffered a significant reduction in population size.

The effects of inbreeding depression and loss of genetic variation in a small, isolated population have been documented in the case of the Isle Royale gray wolves (Figure 29–9). Around 1950, a pair of gray wolves apparently crossed an ice bridge from the Canadian mainland to Isle Royale in Lake Superior. The island had no other wolves and had an abundance of moose, which became the wolves' main food source. By 1980, the Isle Royale wolf population had increased to over 50 individuals. Over the next decade, however, wolf numbers declined to fewer than a dozen with no new litters being born, despite plentiful food and no apparent sign of disease. The genetic variation of the remaining wolves was examined by mtDNA analysis and nuclear DNA fingerprinting. It was found that the Isle Royale wolves had levels of homozygosity that were twice as high as wolves in an adjacent mainland population. Furthermore, all the wolves possessed the same mtDNA genotype, consistent with descent from the same female. Hence, the degree of relatedness between individual Isle Royale wolves was equivalent to that of full siblings. This finding suggests that the wolves' reproductive failure was due to inbreeding depression, a phenomenon that has also been seen in captive-wolf populations.

FIGURE 29–9 The Isle Royale gray wolf (*Canis lupus*).

## Reduction in Gene Flow

Gene flow, described earlier as the gradual exchange of alleles between populations, is brought about by the dispersal of gametes or the migration of individuals. It is an important mechanism for introducing new alleles into a gene pool and increasing genetic variation. Migration is the main route for gene flow in animals; we examined the effect on population allele frequencies of migration and gene flow in Chapter 27. In plants, gene flow occurs not through movement of individuals but as a result of cross-pollination between different populations and through seed dispersal. Isolation and fragmentation of populations in rare and declining species significantly reduces gene flow and the potential for maintaining genetic diversity. As we have already discussed, habitat loss is a major threat to species survival. It is not unusual for a threatened or endangered species to be restricted to small separate pockets of the remaining habitat. This isolates and fragments the surviving populations so that movement of individuals can no longer occur between them, thus preventing gene flow.

The term **metapopulation** describes a population consisting of spatially separated subpopulations with limited gene flow, especially if local extinctions and replacements of some of the subpopulations occur over time. One well-studied metapopulation is that of the endangered red-cockaded woodpecker (*Picoides borealis*) (Figure 29–10), which was once common in pinewoods throughout the southeastern United States. Habitat loss because of logging has reduced the remaining populations of this bird to small scattered sites isolated from each other, with virtually no migration between them. Studies using allozymes and DNA profiling show that the smallest surviving woodpecker populations (where $N = < 100$) have suffered the greatest loss of genetic diversity and are at most risk of inbreeding depression, compared with larger populations where $N = > 100$. Management of the red-cockaded woodpecker now includes efforts to increase the genetic diversity of the smallest populations by introducing birds from different larger populations to artificially re-create the gene flow by migration that would have occurred in the original unfragmented distribution of the species.

FIGURE 29–10 The red-cockaded woodpecker (*Picoides borealis*).

Another species in which the effects of gene-flow reduction have been studied is the North American brown bear, *Ursus arctos*. A team of Canadian and U.S. researchers measured heterozygosity in different brown bear populations using amplified microsatellite markers to create DNA profiles for individual bears. The researchers found that levels of heterozygosity in the brown bear population living in and around Yellowstone Park were only two-thirds as high as those in brown bear populations from Canada and mainland Alaska. They concluded that this was due to the isolation and reduced migration of the Yellowstone bears. The habitat of the Canadian and Alaskan bear populations was much less fragmented, allowing migration of individuals and consequent gene flow between populations. The researchers also examined an island population of brown bears on the Kodiak archipelago off the Alaskan coast. Here, heterozygosity was even lower—less than one-half that of the mainland populations—providing further evidence for the effect of small population size, restricted migration, and restricted gene flow on the reduction of genetic diversity.

## 29.4 Genetic Erosion Threatens Species' Survival

The loss of previously existing genetic diversity from a population or species is referred to as **genetic erosion**. Why does it matter if a population loses genetic diversity, especially if the numbers of individuals remain high? Genetic erosion has two important effects on a population. First, it can result in the loss of potentially useful alleles from the gene pool, thus reducing the ability of the population to adapt to changing environmental conditions and increasing its risk of extinction. Several decades before the dawn of modern genetics, Charles Darwin recognized the importance of diversity to long-term species survival and evolutionary success. In *On the Origin of Species* (1859), Darwin wrote:

> "The more diversified the descendants of any one species . . . by so much will they be better enabled to seize on many widely diversified places in the polity of nature, and so enabled to increase in numbers."

While the importance of allelic diversity for population survival under changing environmental conditions can be presumed for an endangered species, it has actually been demonstrated in several weed species whose gene pools contain alleles conferring resistance to chemical herbicide. If the weeds are not sprayed, the plants with these resistance alleles enjoy no particular selective advantage, and the resistance allele often persists at a relatively low frequency. However, if herbicide is applied, the individuals with the resistance allele are much more likely to survive and generate a new resistant population, while populations whose gene pools lack the resistance allele will be eliminated.

The second important effect of genetic erosion is a reduction in levels of heterozygosity. At the population level, reduced heterozygosity will be seen as an increase in the number of individuals homozygous at a given locus. At the individual level, a decrease in the number of heterozygous loci within the genotype of a particular plant or animal will occur. As we have seen, loss of heterozygosity is a common consequence of reduced population size. Alleles may be lost through genetic drift or because individuals carrying them die without reproducing. Smaller populations also increase the likelihood of inbreeding, which inevitably increases homozygosity.

Obviously, once an allele is lost from a gene pool, the potential for heterozygosity is greatly reduced. If there were originally only two alleles at a given locus, loss of one means that heterozygosity at the locus is completely eliminated and the other allele is now fixed. The level of homozygosity that can be tolerated varies with species. Studies of populations showing higher-than-normal levels of homozygosity have documented a range of deleterious effects, including reduced sperm viability and reproductive abnormalities in African lions, increased offspring mortality in elephant seals, and reduced nesting success in woodpeckers.

As yet, no evidence has emerged from field studies that conclusively links the extinction of a wild population to genetic erosion. However, laboratory studies using *Drosophila melanogaster* to model evolutionary events show that loss of genetic variation does reduce the ability of a population to adapt to changing environmental conditions. Fruit flies, with their small size and rapid generation time of only 10 to 14 days, are a useful model organism for studying evolutionary events, especially because large populations can be readily developed and maintained through multiple generations.

In one set of experiments carried out by scientists at Macquarie University in Sydney, Australia, *Drosophila* populations were reduced to a single pair for up to three generations to simulate a

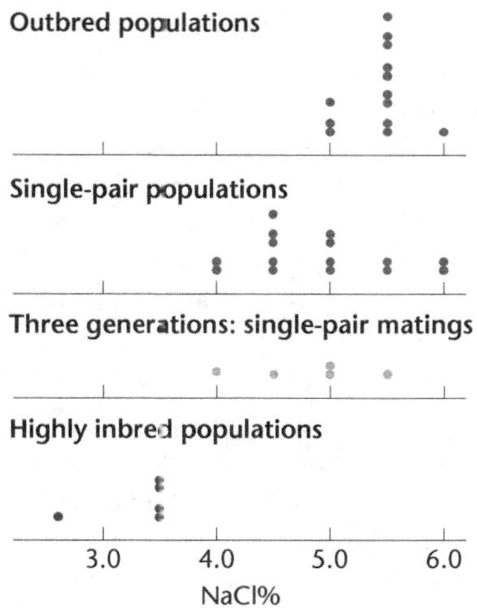


FIGURE 29–11 Effects of bottlenecks in various populations on evolutionary potential in *Drosophila*, as shown by distributions of NaCl concentrations at extinction.

population bottleneck and were then allowed to increase in number. The capacity of the bottlenecked populations to tolerate increasing levels of sodium chloride was compared with that of normal outbred populations. Researchers found that bottlenecked populations with low levels of heterozygosity were less tolerant of high salt concentrations (Figure 29–11).

Other experiments comparing inbred and outbred *Drosophila* populations exposed to environmental stresses such as high temperatures and ethanol presence have shown that populations with even a low level of inbreeding have a much greater probability of extinction at lower levels of environmental stress than outbred populations. Experimental results such as these indicate that genetic erosion does indeed reduce the long-term viability of a population by reducing its capacity to adapt to changing environmental conditions.

## 29.5 Conservation of Genetic Diversity Is Essential to Species Survival

Scientists working to maintain biological diversity face several dilemmas. Should they focus on preserving individual populations, or should they take a broader approach by trying to conserve not just one species but all the interdependent plants and animals in an ecosystem? How can genetic diversity be maintained in a species whose numbers are declining? Can genetic diversity lost from a population be restored? Early conservation efforts often focused only on the population size of an endangered species. Biologists now recognize that a complex interplay of factors must be considered during conservation efforts, including the habitat and role of a species within an ecosystem, as well as the importance of genetic variation for long-term survival.

### *Ex Situ* Conservation: Captive Breeding

***Ex situ*** (Latin for off-site) **conservation** involves removing plants or animals from their original habitat to an artificially maintained location such as a zoo or botanic garden. This living collection can then form the basis of a captive-breeding program.

These programs, where a few surviving individuals from an endangered species are removed from the wild to breed, raise their offspring, and rebuild the population in a protected environment, have been instrumental in bringing a number of species back from the brink of extinction. However, such programs can have undesirable genetic consequences that potentially jeopardize the long-term survival of the species even after population numbers have been restored. A captive-breeding program is rarely initiated until very few individuals are left in the wild, when the original genetic diversity of the species is already depleted. The breeding program is frequently based on a small number of captured individuals, so genetic diversity is further reduced by the founder effect. Since captive-breeding facilities often accommodate only a small number of individuals in the breeding group, inbreeding is difficult to avoid, especially for animal species that form harems (breeding groups dominated by a single male), so that $N_e$ is considerably smaller than $N$. Finally, unintended selection for genotypes more suited to captive-breeding conditions over time can reduce the overall capacity of the recovered population to adapt and survive in the wild.

How can this type of program be managed to minimize these genetic effects? As we have already seen, loss of genetic diversity measured as heterozygosity over $t$ generations can be expressed as

$$H_t/H_0 = \left(1 - \frac{1}{2N_e}\right)^t$$

This equation indicates that the loss of genetic diversity $H_t/H_0$ will be greater with a smaller effective population size $N_e$ and a larger number of generations $t$. We can expand this equation for a captive-breeding population as follows:

$$H_t/H_0 = [1 - (1/2N_{fo})]\{1 - 1/[2N(N_e/N)]\}^{t-1}$$

where $N_{fo}$ is the effective size of the founding population, $N$ is the mean population size, and $N_e$ is the mean effective population size of the group over $t$ generations. Examination of this equation shows that maximum genetic diversity in a captive-breeding group will be maintained by (1) using the largest possible number of founding individuals to maximize $N_{fo}$, (2) maximizing $N_e/N$ so that as many individuals as possible produce offspring each generation, and (3) minimizing the number of generations in captivity to reduce $t$. The importance of maximizing $N_{fo}$ to increase allelic diversity in the founding population can also be seen in Figure 29–12. Successful programs of this sort for endangered species are managed with these three goals in mind. In addition, keeping good pedigree records and exchanging individuals between different breeding programs when possible will reduce inbreeding.

**NOW SOLVE THIS**

Question 14 on page 778 asks you to estimate the potential for genetic erosion in a captive population of rare red pandas.

■ HINT: *If $N_e/N$ is maintained at 0.42 and actual population size at 50, you can calculate $N_e$. Then use the equation given in the discussion above.*

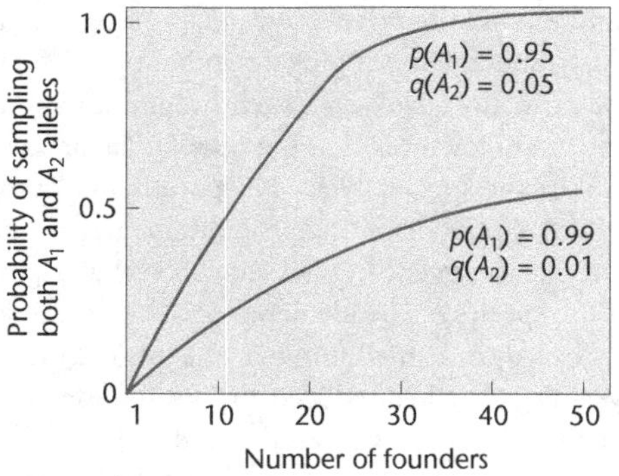


FIGURE 29–12 Effect of captive-population founder number on the probability of maintaining both $A_1$ and $A_2$ alleles at a locus.

## Rescue of the Black-Footed Ferret through Captive Breeding

The black-footed ferret, *Mustela nigripes* (Figure 29–13), is an excellent example of a species that has been successfully subjected to captive breeding. This ferret was once widespread throughout the plains of the western United States, but by the 1970s it was considered to be extinct. Decades of trapping and poisoning of both the ferret and its main prey, the prairie dog, had decimated their populations. However, in 1981 a small surviving colony of black-footed ferrets was discovered on a ranch near Meeteetse, Wyoming. Conservation biologists first tried to conserve this ferret population *in situ*, but in 1985 the colony was infected with canine distemper, which nearly wiped it out. Of the 18 ferrets that were saved and transferred to a captive facility, only 8 were considered to be sufficiently unrelated to found a new population. The breeding program has produced more than 3000 ferrets, and the species is now being reintroduced to parts of its original range. The goal is to establish populations of 1500 ferrets in sustainable wild colonies by 2010. By early 2006, more than 400 captive-bred ferrets had been released, and the first ferret kits born in the wild were reported from a reintroduction site in Colorado.

FIGURE 29–13 The black-footed ferret (*Mustela nigripes*).

Although the recovery of the black-footed ferret appears to be a success, the small founder group of eight individuals caused a severe bottleneck for this species. Additional loss of genetic diversity occurred in the captive-breeding program because of drift, limited reproduction from several of the founder females, and a high rate of breeding by one founder male. In short, the ferrets still face the risks of inbreeding and further genetic erosion that are characteristic of this type of program.

Genetic management strategies, such as using DNA markers to identify the most genetically varied individuals, maintaining careful pedigree records to avoid mating closely related animals, and developing techniques for artificial insemination and sperm cryopreservation have helped to conserve the genetic diversity that remained in the species. Conservation geneticists working on the recovery project estimate that all existing black-footed ferrets share about 12 percent of their genome. This is roughly the equivalent of being full cousins. The effects of this degree of genetic similarity in the expanding ferret population are unclear. Occasional abnormalities such as webbed feet and kinked or short tails have been observed in the captive population, but without a noninbred population available for comparison, it is unclear whether this is evidence that inbreeding is increasing the homozygosity for deleterious alleles.

Scientists disagree as to whether the long-term future of the black-footed ferret is jeopardized by the severe bottleneck and subsequent loss of genetic diversity the species has experienced. Some geneticists suggest that because the one surviving Meeteetse population was so isolated, it may have already become sufficiently inbred to purge any deleterious alleles. Other researchers point out that studies of black-footed ferret DNA extracted from museum specimens show that the ferrets alive today have lost significant genetic diversity compared with earlier prebottleneck populations, and they suggest that loss of fitness because of inbreeding will inevitably be seen over time.

## *Ex Situ* Conservation and Gene Banks

Another form of *ex situ* conservation is provided by establishing **gene banks**. In contrast to housing entire animals or plants, these collections instead provide long-term storage and preservation for reproductive components, such as sperm, ova, and frozen embryos in the case of animals, and seeds, pollen, and cultured tissue in the case of plants. Many more individual genotypes can be preserved for longer periods in a gene bank than in a living collection. Cryopreserved gametes or seeds can be used to reconstitute lost or endangered animals or plants after many years in storage. Because they are expensive to construct and maintain, most gene banks are used to conserve the genetic legacy of domesticated species having economic value.

Gene banks have been established in many countries to help preserve genetic material of agricultural importance, such as traditional crop varieties that are no longer grown or old livestock breeds that are becoming rare. One of the most important *ex situ* collections in the United States is the National Center for Genetic Resources Preservation, a Department of Agriculture (USDA) facility in Fort Collins, Colorado, which maintains more than 300,000 different crop varieties and related wild species. Some of the accessions are stored as seeds and others as cryogenically preserved tissue from which whole plants can be regenerated. (See this chapter's opening photograph.) Animal genetic resources, including frozen semen and embryos from endangered livestock breeds, are also preserved at this facility.

*Ex situ* conservation using gene banks, though often vital, has several disadvantages compared to other methods of conservation. A major problem with gene banks is that even large collections cannot contain all the genetic variation that is present in a species. Conservation geneticists attempt to address this problem by identifying a **core collection** for a species. The core collection is a subset of individual genotypes, carefully chosen to contain as much as possible of the species' genetic variation; preserving the core collection takes priority over randomly collecting and preserving large numbers of genotypes. Another disadvantage of *ex situ* conservation is that the artificial conditions under which a species is preserved in a living collection or gene bank often create their own selection pressures. When seeds of a rare plant species are maintained in cold storage, for example, selection may occur for those genotypes better adapted to withstand the low temperatures, and genotypes may be lost that would actually have greater fitness in the plant's natural environment. Yet another problem posed by *ex situ* conservation is that while the greatest biological diversity in both domesticated and nondomesticated species is frequently found in underdeveloped countries, most *ex situ* collections are situated in developed countries that have the resources to establish and maintain them. This leads to conflict over who owns and has access to the potentially valuable genetic resources maintained in such collections.

### *In Situ* Conservation

***In situ*** (Latin for on-site) **conservation** attempts to preserve the population size and biological diversity of a species while it remains in its original habitat. The use of species inventories to identify diversity hot spots is an important tool for determining the best places to establish parks and reserves where plants and animals can be protected from hunting or collecting and where their habitat can be preserved.

For domesticated species, there is increasing interest in "on-farm" preservation, in which additional resources and financial incentives encourage farmers to maintain traditional crop varieties and livestock breeds. Nondomesticated species with economic potential have also been targeted for *in situ* preservation. In 1998, the USDA established its first *in situ* conservation sites for a wild plant in order to protect populations of the native rock grape (*Vitis rupestris*) in several eastern states. The rock grape is prized by winegrowers not for its fruit but for its roots; grape vines grafted onto wild rock grape rootstock are resistant to phylloxera, a serious pest of wine grapes.

The advantage of *in situ* conservation for the rock grape, as with other species, is that larger populations with greater genetic diversity can be maintained. Another advantage is that species conserved *in situ* continue to live and reproduce in the environments to which they are adapted, which reduces the likelihood that novel selection pressures will produce undesirable changes in allele frequency. However, the continuing increase in global human population makes setting aside suitable areas for *in situ* conservation an ever greater challenge. For example, even in a large preserve such as Yellowstone National Park, the migration and gene flow of species such as the North American brown bear may eventually be eliminated, with a consequent loss of genetic diversity. This problem is even more acute in smaller, more fragmented areas of protected habitat.

### Population Augmentation

What genetic considerations should accompany efforts to restore populations or species that are in decline? As we have already seen, populations that experience bottlenecks continue to suffer from low levels of genetic diversity, even after their numbers have recovered. We have also seen that inbreeding and drift contribute to genetic erosion in small populations, and fragmentation interrupts migration and gene flow, further reducing diversity. Captive-breeding programs designed to restore a critically endangered species beginning with only a few surviving individuals risk genetic erosion in the renewed population from founder effect and inbreeding, as in the case of the black-footed ferret.

An alternative strategy used by conservation biologists is **population augmentation**—boosting the numbers of a declining population by transplanting and releasing individuals of the same species captured or collected from more numerous populations elsewhere. Attempts to reestablish gene flow in severely fragmented populations of the endangered red-cockaded woodpecker by augmenting the smallest populations were described earlier. Other population augmentation projects in the United States have involved bighorn sheep and grizzly bears in the Rocky Mountains. This method has also been employed with the Florida panther, *Felis coryi* (Figure 29–14), using an isolated population of less than 50 animals confined to the area around the Big Cypress Swamp and Everglades National Park in south Florida. As we will discuss in the Genetics, Technology, and Society essay at the end of this chapter, DNA-profiling patterns show high levels of inbreeding in the Florida panther, with reduced fitness because of severe reproductive abnormalities and increased susceptibility to parasite infections. In this effort, seven unrelated animals from a captive population of South and North American panthers were released into the Everglades in the 1960s and allowed to interbreed with the Florida

FIGURE 29–14 The Florida panther (*Felis coryi*).

population, thus producing more genetically diverse family groups. The case of the Florida panther shows how augmentation can increase population numbers, as well as genetic diversity, if the transplanted individuals are unrelated to those in the population to which they are introduced. However, the Florida panther also demonstrates a potential problem with population augmentation, namely, **genetic swamping.** This occurs when the gene pool of the original population is overwhelmed by different genotypes from the transplanted individuals and loses its identity. Further augmentation of the Florida panther population remains controversial, as some biologists argue that the unique features that allow the Florida panther to be classified as a separate subspecies will be lost if breeding takes place with other panthers.

Another difficulty that can be caused by population augmentation is **outbreeding depression**, where reduced fitness occurs in the progeny of matings between genetically diverse individuals. Outbreeding depression occurring in the $F_1$ generation is thought to be due to the offspring being less well adapted to local environmental conditions than either parent. This phenomenon has been documented in some plant species in which seeds of the same species but from a different location were used to revegetate a damaged area by cross-pollination with the remaining local plants. Outbreeding depression that occurs in the $F_2$ and later generations is due to the disruption of **coadapted gene complexes**—groups of alleles that have evolved to work together to produce the best level of fitness in an individual. This type of outbreeding depression has been documented in $F_2$ hybrid offspring from matings between fish from different salmon populations in Alaska. These studies suggest that restoring the most beneficial type and amount of genetic diversity in a population is more complicated than previously thought, reinforcing the argument that the best long-term strategy for species survival is to prevent the loss of diversity in the first place.

## GENETICS, TECHNOLOGY, AND SOCIETY

### Gene Pools and Endangered Species: The Plight of the Florida Panther

In the last 400 years, more than 700 of Earth's animal and plant species have become extinct. In the United States alone, at least 30 species have suffered extinction in the last decade. In addition, hundreds of genetically distinct plant and animal species are now endangered. This dramatic loss of biological diversity is a direct consequence of human activity. As humans increase in number and spread over Earth's surface, we harness more and more of Earth's resources for our use. As a consequence of these activities, we have cleared the forests, polluted the water, and permanently altered Earth's natural balance.

Although the destruction continues, we are beginning to understand the implications of our actions and to make efforts to save some of the more endangered life forms. However, despite our best efforts to save threatened species, we often intercede too late—after population numbers have suffered severe declines and after important ecosystems have been permanently compromised. As a result, much of the genetic diversity that existed in these species is lost, and the populations must be rebuilt from reduced gene pools. This genetic uniformity can reduce fitness, as well as expose genetic diseases. Reduced fitness may further deplete the numbers of the threatened plant or animal, and the population may spiral downward to extinction.

The story of the Florida panther provides a dramatic example of an animal brought to the verge of extinction and the challenges that must be overcome to restore it to a healthy place in the ecosystem. The Florida panther is one of 30 subspecies of cougar and one of the most endangered mammals in the world. These panthers once roamed the southeastern corner of the United States, from South Carolina and Arkansas to the southern tip of Florida. People who settled in the panthers' habitats considered the panthers to be a threat to livestock and humans. So the panthers were killed by hunting, poisoning, highway collisions, and loss of the habitat that supports their prey—primarily wild deer and hogs. By 1967, Florida panthers were listed as endangered, and only about 30 of the animals were left. Population estimates predicted that the Florida panther would be extinct by the year 2055.

Because of about 20 generations of geographical isolation and inbreeding, Florida panthers had the lowest levels of genetic heterozygosity of any subspecies of cougar. The loss of genetic diversity manifested itself in the appearance of some severe genetic defects. For example, almost 80 percent of panther males born after 1989 in the Big Cypress region showed a rare (in other animals), heritable (autosomal dominant or X-linked recessive) condition known as cryptorchidism—failure of one or both testicles to descend. This defect is associated with low testosterone levels and reduced sperm count. Life-threatening congenital heart defects also appeared in the Florida panther population, possibly because of an autosomal dominant gene defect. In addition, some immune deficiencies emerged, making the animals more susceptible to diseases and further contributing to the population's decline. Other less serious genetic features arose, such as a kink in the tail and a whorl of fur on the back.

Over the last two decades, a faint glimmer of hope has appeared for the Florida panther. Federal and state agencies, as well as private individuals, have implemented a Florida Panther Recovery Program. The program's goal is to exceed 500 breeding animals by the year 2010. If it is successful, the panther species would have a 95 percent probability of survival, while retaining up to 90 percent of its genetic diversity. The plan includes a captive-breeding program, strict protection, enlargement and improvement of the panther's habitat, and education of the public and private landowners. Wildlife underpasses have been constructed along highways in panther territory, and these have significantly reduced panther highway fatalities (which account for about half of panther deaths). By 2002, Florida panther numbers had increased to about 100 animals.

To retard the detrimental effects of inbreeding, eight wild female Texas panthers (a related subspecies from western Texas) were released into Florida panther territory in 1995. Three of the females died prior to breeding; however, the remaining Texas females gave birth to litters of healthy kittens. It is estimated that 40 to 70 of the 100 panthers in south Florida are now hybrids of Texas panthers and Florida panthers. None of the hybrids appears to have the kinked tail or other traits characteristic of the inbred Florida panthers. The hybrid cats are more diverse genetically than purebred Florida panthers, with up to 20 to 30 percent of their genetic material being contributed by the Texas panthers. This is of concern to some biologists, who worry that the Texas cats may genetically swamp the distinct Florida panther population.

Paradoxically, the success of the restoration program has become a problem. Now that the population has reached about 100 animals, the Florida panther has almost exceeded the capacity of its existing habitat. Biologists estimate that the population must reach at least 250 individuals in order to be self-sustaining. To reach this number, the panthers will need to expand their territory, again putting them in direct competition with human expansion in Florida. Biologists are now evaluating potential new territories for the growing population of Florida panthers—including parts of Louisiana, Arkansas, and South Carolina.

Despite the recent success of the restoration program, the survival of the Florida panther is far from certain. The panther's comeback will require years of monitoring and frequent intervention. In addition, people must be willing to share their land with wild creatures that are dangerous and do not directly further human interests. Public support for the return of the Florida panther has been strong, however, so there may be hope for this unique, impressive animal.

### References

Maehr, D.S., and Lacy, R.C. 2002. Avoiding the lurking pitfalls in Florida panther recovery. *Wildlife Soc. Bull.* 30(3): 971–978.

Mansfield, K.G., and Land, E.D. 2002. Cryptorchidism in Florida panthers: Prevalence, features and influence of genetic restoration. *J. of Wildlife Diseases* 38(4): 693–698.

### Web Sites

Derr, M. 2002. Florida panther's great leap hits a wall [online]. *The New York Times*, October 15, 2002. http://www.nytimes.com/2002/10/15/science/life/15PANT.html

Florida Panther Net [online]. http://www.panther.state.fl.us/

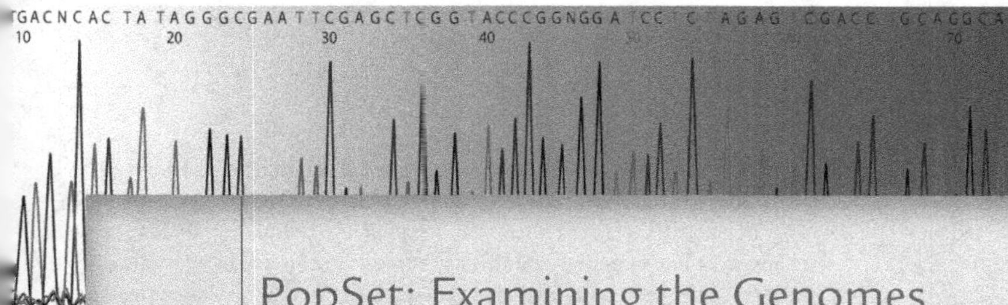

## EXPLORING GENOMICS

# PopSet: Examining the Genomes of Endangered Species

In this chapter, we discussed some of the ways genetic diversity and population size impact long-term species survival. As you learned, conservation geneticists can study genetic variation in a population to examine degrees of relatedness and track the loss of genetic diversity in a declining population. Until fairly recently, conservation geneticists studying genetic diversity typically examined relatively few loci. However, genomics has provided scientists with the ability to use genome-wide DNA fingerprinting techniques and whole-genome shotgun sequencing for genetic diversity analysis. In this chapter, you

*Continued on next page*

*Exploring Genomics, continued*

also learned how techniques such as DNA fingerprinting can be used to uncover illegal harvesting of endangered species. In this exercise we will explore the NCBI site **PopSet** to learn more about DNA sequences from different species that are threatened, endangered, or extinct.

### ■ Exercise I – PopSet

**PopSet** is a DNA-sequence database designed for analyzing the evolutionary relatedness of a population, primarily on the basis of sequence alignment. It contains genome data for many threatened and endangered species.

1. Access PopSet at http://www.ncbi.nlm.nih.gov/entrez/query.fcgi?db=PopSet.
2. Search PopSet using the complete names of the following organisms: *Mammuthus primigenius, Microbatrachella capensis, Zapus hudsonius preblei,* and *Pseudobarbus quathlambae.* The first page you will see after initiating your search provides a list of reports in the database. Click the link for the first report, and then, on the page that appears next, use the BLAST feature to generate an alignment of the search sequence with other sequences in PopSet (i.e., click the "Generate Alignment" button).
3. From the alignment page, explore the links represented by the blue-colored accession number located to the left of each aligned sequence. Provide the common name, the DNA sequence analyzed, the geographic location and threatened or endangered status, and the names of other organisms that show sequence similarity for each of the four organisms in this exercise: *Mammuthus primigenius, Microbatrachella capensis, Zapus hudsonius preblei,* and *Pseudobarbus quathlambae.*

To find this information you will need to navigate through PopSet into other databases.

### ■ Exercise II – Mitochondrial DNA Sequences and the Iberian Wolf

One of the examples presented in this chapter was the effect of inbreeding depression and loss of genetic variation in a population of Isle Royale gray wolves (*Canis lupus*). The Iberian wolf, *Canis lupus signatus*, is an endangered species in Portugal with a rapidly declining population. About 80 years ago these wolves were found throughout Portugal; now their population consists of approximately 200 individuals. A captive breeding program is in place to help conserve *C. lupus signatus.*

1. For a number of reasons, relatively few studies have examined genetic diversity in *C. lupus signatus*. Search PopSet and find the *C. lupus signatus* (lobo Ibérico) dataset provided by the scientist D. Parra and colleagues with the PopSet ID number [125742365].
2. Click on "Parra" to open this dataset. DNA sequences from four wolves were compared in this study. What DNA sequence was analyzed?
3. Generate an alignment for the sequences. Notice that aligned sequences are numbered 1 through 4. Placing the pointer over each colored link on the left side of the alignments reveals that the four wolves in this study were designated ClupMIT1, ClupMIT2, ClupMIT3, and ClupMIT4. Answer the following questions:
   a. What was the length of the DNA sequence analyzed?
   b. Look carefully at the aligned sequences. Recall from Chapter 27 that *single-nucleotide polymorphisms (SNPs)* are single-base changes in the genome that represent a significant cause of genetic variation in a population. Which SNP distinguishes the wolf ClupMIT1 from the other wolves in this study? What nucleotide change was noted in this SNP?
   c. Conservation geneticists studying this species are interested in determining which females in the remaining population of wolves have provided viable offspring in recent years. How do you think studies such as this one may help scientists in their efforts to conserve these wolves?
4. Search the Internet using the phrase "Iberian Wolf Recovery Centre" to learn more about *C. lupus signatus* conservation efforts.
5. Search PopSet for any of the species discussed in this chapter to see what sequence information may be available in the database.

## Chapter Summary

1. Biodiversity is lost as increasing numbers of plants and animal species are threatened with extinction. Conservation genetics applies principles of population genetics to the preservation and restoration of threatened species. A major concern of conservation geneticists is the maintenance of genetic diversity.
2. Genetic diversity includes interspecific diversity, which is reflected by the number of different species present in an ecosystem, and intraspecific diversity, which is reflected by genetic variation within a population or between different populations of the same species. Genetic diversity can be measured by examining different phenotypes in the population or, at the molecular level, by using allozyme analysis or DNA-profiling techniques.
3. Major declines in a species' population numbers reduce genetic diversity and contribute to the risk of extinction as a result of genetic drift, inbreeding, or loss of gene flow. Populations that suffer severe reductions in effective size and then recover have passed through a population bottleneck and often show reduced genetic diversity.
4. Loss of genetic diversity reduces the capacity of a population to adapt to changing environmental conditions, because useful alleles may disappear from the gene pool. Reduced genetic diversity also results in greater levels of homozygosity in a population, often leading to the manifestation of deleterious alleles and inbreeding depression.
5. Conservation of genetic diversity depends on *ex situ* methods, such as living collections, captive-breeding programs, and gene banks, as well as *in situ* approaches, such as the establishment of parks and preserves.
6. Population augmentation, in which individuals are transplanted into a declining population from a more numerous population of the same species located elsewhere, can be used to increase numbers and genetic diversity. However, the risk of outbreeding depression and genetic swamping accompanies this process.

# INSIGHTS AND SOLUTIONS

1. Is a rare species found as several fragmented subpopulations more vulnerable to extinction than an equally rare species found as one larger population? What factors should be considered when managing fragmented populations of a rare species?

**Solution:** A rare species in which the remaining individuals are divided among smaller, isolated subpopulations can appear to be less vulnerable. If one subpopulation becomes extinct through local causes, such as disease or habitat loss, then the remaining subpopulations may still survive. However, genetic drift will cause smaller populations to experience more rapidly increasing homozygosity over time compared with larger populations. Even with random mating, the change in heterozygosity from one generation to the next because of drift can be calculated as

$$H_1 = H_0(1 - 1/2N)$$

where $H_0$ is the frequency of heterozygotes in the present generation, $H_1$ is the frequency of heterozygotes in the next generation, and $N$ is the number of individuals in the population. Thus, in a small population of 50 individuals with an initial heterozygote frequency of 0.5, heterozygosity will decline to $0.5(1 - 1/100) = 0.495$, a loss of 0.5 percent in just one generation. In a larger population of 500 individuals and the same initial heterozygote frequency, heterozygosity after one generation will be $0.5(1 - 1/1000) = 0.4995$, a loss of only 0.05 percent.

Therefore, smaller populations are likely to show the effects of homozygosity for deleterious alleles sooner than larger populations, even with random mating. If populations are fragmented so that movement of individuals or gametes between them is prevented, management options could include transplanting individuals from one subpopulation to another to enable gene flow to occur. Establishment of "wildlife corridors" of undisturbed habitat that connect fragmented populations could be considered. In captive populations, exchange of breeding adults (or their gametes through shipment of preserved semen or pollen) can be undertaken. Promotion of increased population numbers, however, is vital to prevent further genetic erosion through drift.

## Problems and Discussion Questions

1. A wildlife biologist studied four generations of a population of rare Ethiopian jackals. When the study began, there were 47 jackals in the population, and analysis of microsatellite loci from these animals showed a heterozygote frequency of 0.55. In the second generation, an outbreak of distemper occurred in the population, and only 17 animals survived to adulthood. These jackals produced 20 surviving offspring, which in turn gave rise to 35 progeny in the fourth generation.
   (a) What was the effective population size for the four generations of this study?
   (b) Based on its effective population size, what is the heterozygote frequency of the jackal population in generation 4?
   (c) What is the inbreeding coefficient in generation 4, assuming an inbreeding coefficient of $F = 0$ at the beginning of the study, no change in microsatellite allele frequencies in the gene pool, and random mating in all generations?
2. Chondrodystrophy, a lethal form of dwarfism, has recently been reported in captive populations of the California condor and has killed embryos in 5 out of 169 fertile eggs. Chondrodystrophy in condors appears to be caused by an autosomal recessive allele with an estimated frequency of 0.09 in the gene pool of this species.
   (a) How do you think California condor populations should be managed in the future to minimize the effect of this lethal allele?
   (b) What are the advantages and disadvantages of attempting to eliminate it from the gene pool?
3. A geneticist is studying three loci, each with one dominant and one recessive allele, in a small population of rare plants. She estimates the frequencies of the alleles at each of these loci as $A = 0.75$, $a = 0.25$; $B = 0.80$, $b = 0.20$; $C = 0.95$, $c = 0.05$. What is the probability that all the recessive alleles will be lost from the population through genetic drift?
4. How are genetic drift and inbreeding similar in their effects on a population? How are they different?
5. You are the manager of a game park in Africa with a native herd of just 16 black rhinos, an endangered species worldwide. Describe how you would manage this herd to establish a viable population of black rhinos in the park. What genetic factors would you take into consideration in your management plan?
6. Compare the causes and effects of inbreeding depression and outbreeding depression.
7. Cloning, using the techniques similar to those pioneered by the Scottish scientists who produced Dolly the sheep, has been proposed as a way to increase the numbers of some highly endangered mammalian species. Discuss the advantages and disadvantages of using such an approach to aid long-term species survival.
8. In a population of wild poppies found in a remote region of the mountains of eastern Mexico, almost all of the members have pale yellow flowers, but breeding experiments show that pale yellow is recessive to deep orange. Using the tools of the conservation geneticists described in this chapter, how could you experimentally determine whether the prevalence of the recessive phenotype among the eastern Mexican poppy population is due to natural selection or simply due to the effects of genetic drift and/or inbreeding?
9. Contrast *ex situ* conservation techniques with *in situ* conservation techniques.
10. Describe how the captive-breeding *ex situ* conservation approach is applied to a severely endangered species.
11. Explain why a low level of genetic diversity in a species is a detriment to the survival of that species.
12. Contrast allozyme analysis with RFLP analysis as measures of genetic diversity.

13. A population of endangered lowland gorillas is studied by conservation biologists in the wild. The biologists count 15 gorillas but observe that the population consists of two harems, each dominated by a different single male. One harem contains eight females and the other has five. What is the effective population size, $N_e$, of the gorilla population?
14. Twenty endangered red pandas are taken into captivity to found a breeding group.
    (a) What is the probability that at least one of the captured pandas has the genotype $B_1/B_2$ if $p(B_1) = 0.99$?
    (b) Careful management keeps the $N_e/N$ ratio for the red panda breeding population at 0.42. If the overall size of the captive population is maintained at 50 individuals, what proportion of the heterozygosity present in the founding population will still be present after five generations in captivity?
15. Use your analysis of the red panda in Problem 14 to make suggestions for managing the captive-breeding population of red pandas to maintain as much genetic diversity as possible.
16. *Antechinus agilis*, a small mouse-like marsupial found in Australia, mates for a 10–15 day period in August during which females ovulate and after which males die of stress. Individual females rely on stored sperm from multiple male sources to fertilize their eggs. Using DNA profiling, researchers determined that females are capable of releasing a mix of sperm for fertilization—regardless of the time of male access to the female or her time of ovulation—that maintains high genetic heterogeneity in the face of male absenteeism (Shimmin et al., 2000). Does the reproductive strategy of *Antechinus* provide any clues as to the lengths organisms will go to maintain genetic variation?

17. As revealed by microsatellite analysis, habitat loss and fragmentation have led to a significant loss of heterozygosity among nine populations of Florida black bear (*Ursus americanus floridanus*). Researchers determined that although each population was in Hardy–Weinberg equilibrium, mean heterozygosity was lower in smaller, less interconnected populations, and each population was genetically distinct (Dixon et al., 2007). No relationship was found between the degree of genetic differences and the distance from neighboring populations.
    (a) Is their finding of a lower mean heterozygosity in smaller, less interconnected populations expected?
    (b) Given that the Florida black bear inhabits relatively populated areas, what would explain the lack of a relationship between genetic differences and interpopulational distances?
    (c) Considering that low heterozygosity often contributes to reduced species survival, what steps might be taken to enhance the survival of the Florida black bear?
18. Once nearly extinct, black-footed ferrets, which prey on prairie dogs, have increased in numbers from 18 in 1985 to approximately 700 presently occupying a recovery area of Buffalo Gap National Grassland (O'Neill, 2007). However, because prairie dogs eat grass that feeds cattle, some ranchers want to reduce prairie dog populations in that area, where approximately 30,000 of their cattle graze. A plan for balancing the welfare of the ranchers, the prairie dogs, and the black-footed ferret is under consideration by the U.S. Forest Service.
    (a) What impact would a reduction of prairie dog populations have on black-footed ferret populations?
    (b) Conservationists argue that exposing the black-footed ferret population to a second bottleneck would irreversibly harm ferret survival. Some might argue that since the ferret population has in the past recovered from one bottleneck, it should be able to recover from a second. Do you agree?
    (c) Given that the black-footed ferret is one of America's most endangered mammals, what kind of compromise would you negotiate among the farmer, the prairie dog, and the ferret?

**HOW DO WE KNOW?**

19. In this chapter, we focused on conservation genetics, emphasizing how geneticists assess genetic diversity and work to maintain species survival. Along the way, we found many opportunities to consider the methods and reasoning by which several of these practices were developed. From the explanations given in the chapter, what answers would you propose to the following fundamental questions:
    (a) How do we know the extent of genetic diversity in a species?
    (b) How do we know that diminished genetic diversity is detrimental to species survival?
    (c) How do we determine the most effective approach for countering decreased population size?
    (d) How do we attempt to "conserve" existing genetic diversity?

## Extra-Spicy Problems

20. Przewalski's horse (*Equus przewalskii*), thought by some biologists to represent the ancestral species from which modern horses were domesticated, is classified as a separate species from the domestic horse, although members of the two species can mate and produce fertile offspring. Przewalski's horse was hunted to extinction in its native habitat in the Asian steppes by the 1920s. The few hundred Przewalski's horses alive today are descended from 12 surviving individuals that had been taken into captivity. The founder breeding group also included a domestic mare. Currently, there is interest in reintroducing Przewalski's horse to its original range.
    (a) What genetic factors associated with the animals available for reintroduction do you think should be considered before a decision is reached?
    (b) What advice would you give to the conservation biologists managing the reintroduction project?

21. DNA profiles based on different kinds of molecular markers are increasingly used to measure levels of genetic diversity in populations of threatened and endangered species. Do you think these marker-based estimates of genetic diversity are reliable indicators of a population's potential for survival and adaptation in a natural environment? Why, or why not?
22. Seed banks provide managed protection for the conservation of species of economic and noneconomic importance. One study quantified considerable fitness decay in seeds maintained in long-term storage (Schoen et al., 1998. *Proc. Natl. Acad. Sci. [USA]* 95: 394–399). When germination falls below 65–85 percent, regeneration (planting and seed collection) of a finite sample of the stored seed type is recommended. What genetic consequences might you expect to accompany the conservation practice of long-term seed storage and regeneration?
23. According to one writer, "One of the first questions a resource manager asks about threatened and endangered species is: How bad is it?" (Holmes, 2001. *Proc. Natl. Acad. Sci. [USA]* 98: 5072–5077). In other words, the manager seeks a population viability analysis that includes estimates of extinction risk. What factors would you consider significant in providing an estimate of extinction risk for a species?
24. Microsatellite loci are short (2-to-5 base pair) tandem repeats, abundantly and somewhat randomly distributed in all eukaryotic chromosomes. They show high mutability such that new alleles are produced more frequently than in traditional protein-coding genes. It is likely that most microsatellites are selectively neutral and highly heterozygous in natural populations. Considering that cheetahs underwent a population bottleneck approximately 12,000 years ago, North American pumas 10,000 years ago, and Gir Forest lions 1,000 years ago, in which of these species would you expect to see the highest degree of microsatellite polymorphism? the lowest?
25. Using the behavior and evolution of microsatellites described in Problem 24, create a graph that relates microsatellite polymorphism (variance) to the number of years since a species' bottleneck. Select variances from 1.0 to 9.0, where increasing variance represents increasing polymorphism, and set the range of years since the bottleneck from 10,000 to 50,000.
26. Allozymes are electrophoretically distinct forms of a particular protein, while microsatellites and minisatellites are repetitive DNA sequences that have been found in all eukaryotes studied to date. Such repetitive sequences are rarely associated with coding sequences of DNA. The data shown here represent the percent heterozygosity of allozymes and microsatellite and minisatellite DNAs in nuclear genomes of four species of felines (modified from Driscoll et al., 2002. *Genome Res.* 12: 414–423).
    (a) Which species appears to contain the greatest genetic variability? Why might this species be so variable?
    (b) Which species appears to have the least genetic variability?
    (c) Why are allozymes less variable than minisatellite or microsatellite DNAs?

| | % Heterozygosity | | | |
|---|---|---|---|---|
| Marker | Cheetah | Lion | Puma | Domesticated Cat |
| Allozyme | 1.4 | 0.0 | 1.8 | 8.2 |
| Minisatellite | 43.3 | 2.9 | 10.3 | 44.9 |
| Microsatellite | 46.7 | 7.9 | 14.7 | 68.1 |

27. If we assume that one of the species in Problem 26 underwent a recent population crisis in which the number of effective breeders reached critically low levels, which species do you think it would be?
28. For many conservation efforts, scientists lack sufficient data to make definitive conservation decisions. Yet the allocation of habitats for conservation cannot be delayed until such data are available. In these cases, conservation efforts are often directed toward three classes of species: flagships (high-profile species), umbrellas (species requiring large areas for habitat), and biodiversity indicators (species representing diverse, especially productive habitats) (Andelman and Fagan, 2000. *Proc. Natl. Acad. Sci. [USA]* 97: 5954–5959). In terms of protecting threatened species, what advantages and disadvantages might accompany investing scarce talent and resources in each of these classes?

# Appendix A

## GLOSSARY

**abortive transduction** An event in which transducing DNA fails to be incorporated into the recipient chromosome. See *transduction.*
**accession number** An identifying number or code assigned to a nucleotide or amino acid sequence for entry and cataloging in a database.
**acentric chromosome** Chromosome or chromosome fragment with no centromere.
**acridine dyes** A class of organic compounds that bind to DNA and intercalate into the double-stranded structure, producing local disruptions of base pairing. These disruptions result in nucleotide additions or deletions in the next round of replication.
**acrocentric chromosome** Chromosome with the centromere located very close to one end. Human chromosomes 13, 14, 15, 21, and 22 are acrocentric.
**active site** The substrate-binding site of an enzyme; in other proteins, the portion whose structural integrity is required for function.
**adaptation** A heritable component of the phenotype that confers an advantage in survival and reproductive success. The process by which organisms adapt to current environmental conditions.
**additive genes** See *polygenic inheritance.*
**additive variance** Genetic variance attributed to the substitution of one allele for another at a given locus. This variance can be used to predict the rate of response to phenotypic selection in quantitative traits.
**A-DNA** An alternative form of right-handed, double-helical DNA. Its helix is more tightly coiled than the more common B-DNA, with 11 base pairs per full turn. In the A form, the bases in the helix are displaced laterally and tilted in relation to the longitudinal axis. It is not yet clear whether this form has biological significance. See *B-DNA.*
**albinism** A condition caused by the lack of melanin production in the iris, hair, and skin. In humans, it is most often inherited as an autosomal recessive trait.
**alkaptonuria** An autosomal recessive condition in humans caused by lack of the enzyme homogentisic acid oxidase. Urine of homozygous individuals turns dark upon standing because of oxidation of excreted homogentisic acid. The cartilage of homozygous adults blackens from deposition of a pigment derived from homogentisic acid. Affected individuals often develop arthritic conditions.
**allele** One of the possible mutational forms of a gene, often distinguished from other alleles by phenotypic effects.
**allele-specific oligonucleotide (ASO)** Synthetic nucleotides, usually 15–20 bp in length, that under carefully controlled conditions will hybridize only to a perfectly matching complementary sequence.
**allelic exclusion** In a plasma cell heterozygous for an immunoglobulin gene, the selective action of only one allele.
**allelism test** See *complementation test.*
**allolactose** A lactose derivative that acts as the inducer for the *lac* operon.
**allopatric speciation** Process of speciation associated with geographic isolation.
**allopolyploid** Polyploid condition formed by the union of two or more distinct chromosome sets with a subsequent doubling of chromosome number.
**allosteric effect** Conformational change in the active site of a protein brought about by interaction with an effector molecule.
**allotetraploid** An allopolyploid containing two genomes derived from different species.
**allozyme** An allelic form of a protein that can be distinguished from other forms by electrophoresis.
**alternative splicing** Generation of different protein molecules from the same pre-mRNA by incorporation of a different set and order of exons into the mRNA product.
***Alu* sequence** A DNA sequence of approximately 300 bp found interspersed within the genomes of primates that is cleaved by the restriction enzyme *Alu*I. In humans, 300,000–600,000 copies are dispersed throughout the genome and constitute some 3–6 percent of the genome. See *short interspersed elements.*
**amber codon** The codon UAG, which does not code for an amino acid but for chain termination.
**Ames test** A bacterial assay developed by Bruce Ames to detect mutagenic compounds; it assesses reversion to histidine independence in the bacterium *Salmonella typhimurium.*
**amino acids** Aminocarboxylic acids comprising the subunits that are covalently linked to form proteins.
**aminoacyl tRNA** Covalently linked combination of an amino acid and a tRNA molecule. Also referred to as a charged tRNA.
**amniocentesis** A procedure in which fluid and fetal cells are withdrawn from the amniotic layer surrounding the fetus; used to test for fetal defects.
**amphidiploid** Same as *allotetraploid.*
**anabolism** The metabolic synthesis of complex molecules from less complex precursors.
**analog** A chemical compound that differs structurally from a similar compound and whose chemical behavior is the same. Used experimentally to provide improved detection during analysis. See *base analog.*
**anaphase** Stage of mitosis or meiosis in which chromosomes begin moving to opposite poles of the cell.
**anaphase I** The stage in the first meiotic division during which members of homologous pairs of chromosomes separate from one another.
**aneuploidy** A condition in which the chromosome number is not an exact multiple of the haploid set.
**annotation** Analysis of genomic nucleotide sequence data to identify the protein-coding genes, the nonprotein-coding genes, and the regulatory sequences and function(s) of each gene.
**antibody** Protein (immunoglobulin) produced in response to an antigenic stimulus with the capacity to bind specifically to an antigen.
**anticipation** See *genetic anticipation.*
**anticodon** In a tRNA molecule, the nucleotide triplet that binds to its complementary codon triplet in an mRNA molecule.
**antigen** A molecule, often a cell-surface protein, that is capable of eliciting the formation of antibodies.
**antiparallel** A term describing molecules in parallel alignment but running in opposite directions. Most commonly used to describe the opposite orientations of the two strands of a DNA molecule.
**antisense RNA** An RNA molecule (synthesized *in vivo* or synthetic) with a ribonucleotide sequence that is complementary to part of an mRNA molecule.
**apoptosis** A genetically controlled program of cell death, activated as part of normal development or as a result of cell damage.

**artificial selection** See *selection.*
**ascospore** A meiotic spore produced in certain fungi.
**ascus** In fungi, the sac enclosing the four or eight ascospores.
**asexual reproduction** Production of offspring in the absence of any sexual process.
**assortative mating** Nonrandom mating between males and females of a species. Positive assortative mating selects mates with the same genotype; negative selects mates with opposite genotypes.
**ATP** Adenosine triphosphate, a nucleotide that is the main energy source in cells.
**attached-X chromosome** Two conjoined X chromosomes that share a single centromere and thus migrate together during cell division.
**attenuator** A nucleotide sequence between the promoter and the structural gene of some bacterial operons that regulates the transit of RNA polymerase, reducing transcription of the related structural gene.
**autogamy** A process of self-fertilization resulting in homozygosis.
**autoimmune disease** The production of antibodies that results from an immune response to one's own molecules, cells, or tissues. Such a response results from the inability of the immune system to distinguish self from nonself. Diseases such as arthritis, scleroderma, systemic lupus erythematosus, and juvenile-onset diabetes are examples of autoimmune diseases.
**autonomously replicating sequences (ARS)** Origins of replication, about 100 nucleotides in length, found in yeast chromosomes. ARS elements are also present in organelle DNA.
**autopolyploidy** Polyploid condition resulting from the duplication of one diploid set of chromosomes.
**autoradiography** Production of a photographic image by radioactive decay. Used to localize radioactively labeled compounds within cells and tissues or to identify radioactive probes in various blotting techniques. See *Southern blotting.*
**autosomes** Chromosomes other than the sex chromosomes. In humans, there are 22 pairs of autosomes.
**autotetraploid** An autopolyploid condition composed of four copies of the same genome.
**auxotroph** A mutant microorganism or cell line that requires a nutritional substance for growth that can be synthesized, and is not required by the wild-type strain.
**backcross** A cross between an $F_1$ heterozygote and one of the $P_1$ parents (or an organism with a genotype identical to one of the parents).
**bacteriophage** A virus that infects bacteria, using it as the host for reproduction (also, *phage*).
**balanced lethals** Recessive, nonallelic lethal genes, each carried on different homologous chromosomes. When organisms carrying balanced lethal genes are interbred, only organisms with genotypes identical to the parents (heterozygotes) survive.
**balanced polymorphism** Genetic polymorphism maintained in a population by natural selection.
**Barr body** Densely staining DNA-positive mass seen in the somatic nuclei of mammalian females. Discovered by Murray Barr, this body represents an inactivated X chromosome.
**base analog** A purine or pyrimidine base that differs structurally from one normally used in biological systems whose chemical behavior is the same. Used experimentally to provide improved detection during analysis, for example, 5-bromouracil, which "looks like" thymidine, substitutes for it, and can be detected because of its increased mass. See *analog.*
**base pair** See *nucleotide pair.*
**base substitution** A single base change in a DNA molecule that produces a mutation. There are two types of substitutions: *transitions*, in which a purine is substituted for a purine, or a pyrimidine for a pyrimidine; and *transversions*, in which a purine is substituted for a pyrimidine or vice versa.
**B-DNA** The conformation of DNA which is most often found in cells and serves as the basis of the Watson-Crick double-helical model. There are 10 base pairs per full turn of its right-handed helix, with the nucleotides stacked 0.34 nm apart. The helix has a diameter of 2.0 nm.
**$\beta$-galactosidase** A bacterial enzyme, encoded by the *lacZ* gene, that converts lactose into galactose and glucose.
**bidirectional replication** A mechanism of DNA replication in which two replication forks move in opposite directions from a common origin.
**biodiversity** The genetic diversity present in populations and species of plants and animals.
**bioinformatics** The design and application of software and computational methods for the storage, analysis, and management of biological information such as nucleotide or amino acid sequences.
**biometry** The application of statistics and statistical methods to biological problems.
**biotechnology** Commercial and/or industrial processes that utilize biological organisms or products.
**bivalents** Synapsed homologous chromosomes in the first prophase of meiosis.
**BLAST (Basic Local Alignment Search Tool)** Any of a family of search engines designed to compare or query nucleotide or amino acid sequences with sequences in databases. BLAST also calculates the statistical significance of the matches.
**Bombay phenotype** A rare variant of the ABO antigen system in which affected individuals do not have A or B antigens and thus appear to have blood type O, even though their genotype may carry unexpressed alleles for the A and/or B antigens.
**bottleneck** See *population bottleneck.*
**bovine spongiform encephalopathy (BSE)** A fatal, degenerative brain disease of cattle (transmissible to humans and other animals) caused by prion infection. Also known as mad cow disease.
**BrdU (5-bromodeoxyuridine)** A mutagenically active analog of thymidine in which the methyl group at the 5′ position in thymine is replaced by bromine; also abbreviated BUdR.
**broad heritability** That proportion of total phenotypic variance in a population that can be attributed to genotypic variance.
**buoyant density** A property of particles (and molecules) that depends on their actual density, as determined by partial specific volume and degree of hydration. It provides the basis for density-gradient separation of molecules or particles.
**CAAT box** A highly conserved DNA sequence found in the untranslated promoter region of eukaryotic genes. This sequence is recognized by transcription factors.
**CAP** Catabolite activator protein; a protein that binds cAMP and regulates the activation of inducible operons.
**carcinogen** A physical or chemical agent that causes cancer.
**carrier** An individual heterozygous for a recessive trait.
**catabolism** A metabolic reaction in which complex molecules are broken down into simpler forms, often accompanied by the release of energy.
**catabolite activator protein** See *CAP.*
**catabolite repression** The selective inactivation of an operon by a metabolic product of the enzymes encoded by the operon.
***cdc* mutation** A class of cell division cycle mutations in yeasts that affect the timing and progression through the cell cycle.
**cDNA (complementary DNA)** DNA synthesized from an RNA template by the enzyme reverse transcriptase.
**cDNA library** A collection of cloned cDNA sequences.
**cell cycle** The sequence of growth phases of an individual cell; divided into G1 (gap 1), S (DNA synthesis), G2 (gap 2), and M (mitosis). A cell may temporarily or permanently be withdrawn from the cell cycle, in which case it is said to enter the G0 stage.

**cell-free extract** A preparation of the soluble fraction of cells, made by lysing cells and removing the particulate matter, such as nuclei, membranes, and organelles. Often used to carry out the synthesis of proteins by the addition of specific, exogenous mRNA molecules.
**CEN** The DNA region of centromeres critical to their function. In yeasts, fragments of chromosomal DNA, about 120 bp in length, that when inserted into plasmids confer the ability to segregate during mitosis.
**centimorgan (cM)** A unit of distance between genes on chromosomes representing 1 percent crossing over between two genes. Equivalent to 1 map unit (mu).
**central dogma** The concept that genetic information flow progresses from DNA to RNA to proteins. Although exceptions are known, this idea is central to an understanding of gene function.
**centric fusion** See *Robertsonian translocation.*
**centriole** A cytoplasmic organelle composed of nine groups of microtubules, generally arranged in triplets. Centrioles function in the generation of cilia and flagella and serve as foci for the spindles in cell division.
**centromere** The specialized heterochromatic chromosomal region at which sister chromatids remain attached after replication, and the site to which spindle fibers attach to the chromosome during cell division. Location of the centromere determines the shape of the chromosome during the anaphase portion of cell division. Also known as the primary constriction.
**centrosome** Region of the cytoplasm containing a pair of centrioles.
**chaperone** A protein that regulates the folding of a polypeptide into a functional three-dimensional shape.
**character** An observable phenotypic attribute of an organism.
**charon phages** A group of genetically modified lambda phages designed to be used as vectors (carriers) for cloning foreign DNA. Named after the ferryman in Greek mythology who carried the souls of the dead across the River Styx.
**chemotaxis** Movement of a cell or organism in response to a chemical gradient.
**chiasma (pl., chiasmata)** The crossed strands of nonsister chromatids seen in diplotene of the first meiotic division. Regarded as the cytological evidence for exchange of chromosomal material, or crossing over.
**ChIP-on-chip** A technique that combines chromatin immunoprecipitation (ChIP) with microarrays (chips) to identify and localize DNA sites that bind DNA-binding proteins of interest. These sites may be enhancers, promoters, transcription initiation sites, or other functional DNA regions.
**chi-square ($\chi^2$) analysis** Statistical test to determine whether or not an observed set of data is equivalent to a theoretical expectation.
**chloroplast** A self-replicating cytoplasmic organelle containing chlorophyll. The site of photosynthesis.
**chorionic villus sampling (CVS)** A technique of prenatal diagnosis in which chorionic fetal cells are retrieved intravaginally and used to detect cytogenetic and biochemical defects in the embryo.
**chromatid** One of the longitudinal subunits of a replicated chromosome.
**chromatin** The complex of DNA, RNA, histones, and nonhistone proteins that make up uncoiled chromosomes, characteristic of the eukaryotic interphase nucleus.
**chromatography** Technique for the separation of a mixture of solubilized molecules by their differential migration over a substrate.
**chromocenter** In polytene chromosomes, an aggregation of centromeres and heterochromatic elements where the chromosomes appear to be attached together.
**chromomere** A coiled, beadlike region of a chromosome, most easily visualized during cell division. The aligned chromomeres of polytene chromosomes are responsible for their distinctive banding pattern.
**chromosomal aberration** Any duplication, deletion, or rearrangement of the otherwise diploid chromosomal content of an organism.
**chromosomal mutation** The process resulting in the duplication, deletion, or rearrangements of the diploid chromosomal content of an organism. See *chromosomal aberration.*
**chromosomal polymorphism** Possession of alternative structures or arrangements of a chromosome among members of a population.
**chromosome** In prokaryotes, a DNA molecule containing the organism's genome; in eukaryotes, a DNA molecule complexed with RNA and proteins to form a threadlike structure containing genetic information arranged in a linear sequence and visible during mitosis and meiosis.
**chromosome banding** Technique for the differential staining of mitotic or meiotic chromosomes to produce a characteristic banding pattern; or selective staining of certain chromosomal regions such as centromeres, the nucleolus organizer regions, and GC- or AT-rich regions. Not to be confused with the banding pattern present in polytene chromosomes, which is produced by the alignment of chromomeres.
**chromosome map** A diagram showing the location of genes on chromosomes.
**chromosome puff** A localized uncoiling and swelling in a polytene chromosome, usually regarded as a sign of active transcription.
**chromosome theory of inheritance** The idea put forward independently by Walter Sutton and Theodore Boveri that chromosomes are the carriers of genes and the basis for the Mendelian mechanisms of segregation and independent assortment.
**chromosome walking** A method for analyzing long stretches of DNA. The end of a cloned segment of DNA is subcloned and used as a probe to identify other clones that overlap the first clone.
***cis*-acting element** A DNA sequence that regulates the expression of a gene located on the same chromosome. This contrasts with a trans-acting element where regulation is under the control of a sequence on the homologous chromosome. See *trans-acting element.*
***cis* configuration** The arrangement of two genes (or two mutant sites within a gene) on the same homolog, such as

$$\frac{a^1 \quad a^2}{+ \quad +}$$

This contrasts with a *trans* arrangement, where the mutant alleles are located on opposite homologs. See *trans configuration.*
***cis–trans* test** A genetic test to determine whether two mutations are located within the same cistron (or gene).
**cistron** That portion of a DNA molecule, often referring to a gene, coding for a single-polypeptide chain; defined by a genetic test as a region within which two mutations cannot complement each other.
**cline** A gradient of genotype or phenotype distributed over a geographic range.
**clonal selection theory** Proposed explanation in immunology that antibody diversity precedes exposure to the antigen and that the antigen functions to select the cells containing its specific antibody to undergo proliferation.
**clone** Identical molecules, cells, or organisms derived from a single ancestor by asexual or parasexual methods; for example, a DNA segment that has been inserted into a plasmid or chromosome of a phage or a bacterium and replicated to produce many copies, or an organism of identical genetic composition to that used in its production.
**codominance** Condition in which the phenotypic effects of a gene's alleles are fully and simultaneously expressed in the heterozygote.
**codon** A triplet of nucleotides that specifies a particular amino acid or a start or stop signal in the genetic code. Sixty-one codons specify the amino acids used in proteins, and three codons, called stop codons, signal termination of growth of the polypeptide chain.

**coefficient of coincidence** A ratio of the observed number of double crossovers divided by the expected number of such crossovers.
**coefficient of inbreeding** The probability that two alleles present in a zygote are descended from a common ancestor.
**coefficient of selection (*s*)** A measurement of the reproductive disadvantage of a given genotype in a population. For example, if for genotype *aa* only 99 of 100 individuals reproduce, then the selection coefficient is 0.01.
**colchicine** An alkaloid compound that inhibits spindle formation during cell division. In the preparation of karyotypes, used for collecting a large population of cells inhibited at the metaphase stage of mitosis.
**colinearity** The linear relationship between the nucleotide sequence in a gene (or the RNA transcribed from it) and the order of amino acids in the polypeptide chain specified by the gene.
**competence** In bacteria, the transient state or condition during which the cell can bind and internalize exogenous DNA molecules, making transformation possible.
**complementarity** Chemical affinity between nitrogenous bases of nucleic acid strands as a result of hydrogen bonding. Responsible for the base-pairing between the strands of the DNA double helix and between DNA and RNA strands during genetic expression in cells and during the use of molecular hybridization techniques.
**complementation test** A genetic test to determine whether two mutations occur within the same gene (or cistron). If two mutations are introduced into a cell simultaneously and produce a wild-type phenotype (i.e., they complement each other), they are often nonallelic. If a mutant phenotype is produced, the mutations are noncomplementing and are often allelic.
**complete linkage** A condition in which two genes are located so close to each other that no recombination occurs between them.
**complexity (X)** The total number of nucleotides or nucleotide pairs in a population of nucleic acid molecules as determined by reassociation kinetics.
**complex locus** A gene within which a set of functionally related pseudoalleles can be identified by recombinational analysis (e.g., the *bithorax* locus in *Drosophila*).
**complex trait** A trait whose phenotype is determined by the interaction of multiple genes and environmental factors.
**concatemer** A chain or linear series of subunits linked together. The process of forming a concatemer is called concatenation (e.g., multiple units of a phage genome produced during replication).
**concordance** Pairs or groups of individuals identical in their phenotype. In twin studies, a condition in which both twins exhibit or fail to exhibit a trait under investigation.
**conditional mutation** A mutation that is expressed only under a certain condition, that is, a wild-type phenotype is expressed under certain (permissive) conditions and a mutant phenotype under other (restrictive) conditions.
**conjugation** Temporary fusion of two single-celled organisms for the sexual transfer of genetic material.
**consanguineous** Related by a common ancestor within the previous few generations.
**consensus sequence** The sequence of nucleotides in DNA or amino acids in proteins most often present in a particular gene or protein under study in a group of organisms.
**conservation genetics** The branch of genetics concerned with the preservation and maintenance of genetic diversity and existing species of plants and animals in their natural environments.
**contig** A continuous DNA sequence reconstructed from overlapping DNA sequences derived by cloning or sequence analysis.
**continuous variation** Phenotype variation in which quantitative traits range from one phenotypic extreme to another in an overlapping or continuous fashion.
**cosmid** A vector designed to allow cloning of large segments of foreign DNA. Cosmids are composed of the *cos* sites of phage $\lambda$ inserted into a plasmid. In cloning, the recombinant DNA molecules are packaged into phage protein coats, and after infection of bacterial cells, the recombinant molecule replicates and can be maintained as a plasmid.
**coupling conformation** Same as *cis configuration.*
**covalent bond** A nonionic chemical bond, formed by the sharing of electrons.
**CpG island** A short region of regulatory DNA found upstream of genes that contain unmethylated stretches of sequence with a high frequency of C and G nucleotides.
**Creutzfeldt–Jakob disease (CJD)** A progressive degenerative and fatal disease of the brain and nervous system caused by mutations that lead to aberrant forms of the encoded protein (prions). CJD is inherited as an autosomal dominant trait.
**cri-du-chat syndrome** A clinical syndrome produced by a deletion of a portion of the short arm of chromosome 5 in humans. Afflicted infants have a distinctive cry that sounds like a cat.
**crossing over** The exchange of chromosomal material (parts of chromosomal arms) between homologous chromosomes by breakage and reunion. The exchange of material between nonsister chromatids during meiosis is the basis of genetic recombination.
**cross-reacting material (CRM)** A nonfunctional enzyme produced by a mutant gene, yet recognized by antibodies made against the normal enzyme.
**C-terminal amino acid** In a peptide chain, the terminal amino acid that carries a free carboxyl group.
**C terminus** In a polypeptide, the end that carries a free carboxyl group of the last amino acid. By convention, the structural formula of polypeptides is written with the C terminus at the right.
***C* value** The haploid amount of DNA present in a genome.
***C* value paradox** The apparent paradox that there is no relationship between the size of the genome and the evolutionary complexity of species. For example, the *C* value (haploid genome size) of amphibians varies by a factor of 100.
**cyclic adenosine monophosphate (cAMP)** An important regulatory molecule in both prokaryotic and eukaryotic organisms.
**cyclins** In eukaryotic cells, a class of proteins that are synthesized and degraded in synchrony with the cell cycle and regulate passage through stages of the cycle.
**cytogenetics** A branch of biology in which the techniques of both cytology and genetics are used in genetic investigations.
**cytokinesis** The division or separation of the cytoplasm during mitosis or meiosis.
**cytoplasmic inheritance** Non-Mendelian form of inheritance in which genetic information is transmitted through the cytoplasm rather than the nucleus, usually by self-replicating cytoplasmic organelles such as mitochondria and chloroplasts.
**cytoskeleton** An internal array of microtubules, microfilaments, and intermediate filaments that confers shape and the ability to move to a eukaryotic cell.
**dalton (Da)** A unit of mass equal to that of the hydrogen atom, which is $1.67 \times 10^{-24}$ gram. A unit used in designating molecular weights.
**Darwinian fitness** Same as *fitness*.
**deficiency** A chromosomal mutation, also referred to as a deletion, involving the loss of chromosomal material.
**degenerate** Applied to the genetic code, refers to the fact that a given amino acid may be represented by more than one codon. For example, some amino acids (leucine) have six codons, while others (isoleucine) have three.

**deletion** See *deficiency.*
**deme** A local interbreeding population.
**denatured DNA** DNA molecules that have separated into single strands.
***de novo*** Newly arising; synthesized from less complex precursors rather than being produced by modification of an existing molecule.
**density gradient centrifugation** A method of separating macromolecular mixtures by the use of centrifugal force and solutions of varying density. In buoyant density gradient centrifugation using cesium chloride, the centrifugal field produces a density gradient in a cesium solution through which a mixture of macromolecules such as DNA will migrate until each component reaches a point equal to its own density. Sometimes referred to as equilibrium sedimentation centrifugation.
**deoxyribonuclease (DNAse)** A class of enzyme that breaks down DNA into oligonucleotide fragments by introducing single-stranded or double-stranded breaks into the double helix.
**deoxyribonucleic acid (DNA)** A macromolecule usually consisting of polymers of nucleotides comprising antiparallel chains in which the sugar residues are deoxyribose and which are held together by hydrogen bonds. The primary carrier of genetic information.
**deoxyribose** The five-carbon sugar associated with the deoxyribonucleotides in DNA.
**dermatoglyphics** The study of the surface ridges of the skin, especially of the hands and feet.
**determination** Establishment of a specific pattern of future gene activity and developmental fate for a given cell, usually prior to any manifestation of the cell's future phenotype.
**diakinesis** The final stage of meiotic prophase I, in which the chromosomes become tightly coiled and compacted and move toward the periphery of the nucleus.
**dicentric chromosome** A chromosome having two centromeres, which is pulled in opposite directions during anaphase of cell division.
**dicer** An enzyme (a ribonuclease) that cleaves double-stranded RNA (dsRNA) and pre-micro RNA (miRNA) to form small interfering RNA (siRNA) molecules about 20–25 nucleotides long that serve as guide molecules for the degradation of mRNA molecules with sequences complementary to the siRNA.
**dideoxynucleotide** A nucleotide containing a deoxyribose sugar lacking a 3′ hydroxyl group. It stops further chain elongation when incorporated into a growing polynucleotide and is used in the Sanger method of DNA sequencing.
**differentiation** The complex process of change by which cells and tissues attain their adult structure and functional capacity.
**dihybrid cross** A genetic cross involving two characters in which the parents possess different forms of each character (e.g., yellow, round × green, wrinkled peas).
**diploid ($2n$)** A condition in which each chromosome exists in pairs; having two of each chromosome.
**diplotene** The stage of meiotic prophase I immediately following pachytene. In diplotene, the sister chromatids begin to separate, and chiasmata become visible. These cross-like overlaps move toward the ends of the chromatids (terminalization).
**directed mutagenesis** See *gene targeting.*
**directional selection** A selective force that changes the frequency of an allele in a given direction, either toward fixation or toward elimination.
**discontinuous replication of DNA** The synthesis of DNA in discontinuous fragments on the lagging strand during replication. The fragments, known as Okazaki fragments, are subsequently joined by DNA ligase to form a continuous strand.
**discontinuous variation** Pattern of variation for a trait whose phenotypes fall into two or more distinct classes.
**discordance** In twin studies, a situation where one twin expresses a trait but the other does not.
**disjunction** The separation of chromosomes during the anaphase stage of cell division.
**disruptive selection** Simultaneous selection for phenotypic extremes in a population, usually resulting in the production of two phenotypically discontinuous strains.
**dizygotic twins** Twins produced from separate fertilization events; two ova fertilized independently. Also known as fraternal twins.
**DNA** See *deoxyribonucleic acid.*
**DNA fingerprinting** A molecular method for identifying an individual member of a population or species. A unique pattern of DNA fragments is obtained by restriction enzyme digestion followed by Southern blot hybridization using minisatellite probes. See also *DNA profiling, STR sequences.*
**DNA footprinting** A technique for identifying a DNA sequence that binds to a particular protein, based on the idea that the phosphodiester bonds in the region covered by the protein are protected from digestion by deoxyribonucleases.
**DNA gyrase** One of a class of enzymes known as topoisomerases that converts closed circular DNA to a negatively supercoiled form prior to replication, transcription, or recombination. The enzyme acts during DNA replication to reduce molecular tension caused by supercoiling.
**DNA helicase** An enzyme that participates in DNA replication by unwinding the double helix near the replication fork.
**DNA ligase** An enzyme that forms a covalent bond between the 5′ end of one polynucleotide chain and the 3′ end of another polynucleotide chain. It is also called polynucleotide-joining enzyme.
**DNA microarray** An ordered arrangement of DNA sequences or oligonucleotides on a substrate (often glass). Microarrays are used in quantitative assays of DNA-DNA or DNA-RNA binding to measure profiles of gene expression (for example, during development or to compare the differences in gene expression between normal and cancer cells).
**DNA polymerase** An enzyme that catalyzes the synthesis of DNA from deoxyribonucleotides utilizing a template DNA molecule.
**DNA profiling** A method for identification of individuals that uses variations in the length of short tandem repeating DNA sequences (STRs) that are widely distributed in the genome.
**DNase** Deoxyribonucleosidase, an enzyme that degrades or breaks down DNA into fragments or constitutive nucleotides. See also *endonuclease, exonuclease, restriction endonuclease.*
**dominance** The expression of a trait in the heterozygous condition.
**dosage compensation** A genetic mechanism that equalizes the levels of expression of genes at loci on the X chromosome. In mammals, this is accomplished by random inactivation of one X chromosome, leading to Barr body formation.
**double crossover** Two separate events of chromosome breakage and exchange occurring within the same tetrad.
**double helix** The model for DNA structure proposed by James Watson and Francis Crick, in which two antiparallel hydrogen-bonded polynucleotide chains are wound into a right-handed helical configuration 2 nm in diameter, with 10 base pairs per full turn.
**Duchenne muscular dystrophy** An X-linked recessive genetic disorder caused by a mutation in the gene for dystrophin, a protein found in muscle cells. Affected males show a progressive weakness and wasting of muscle tissue. Death ensues by about age 20 caused by respiratory infection or cardiac failure.
**duplication** A chromosomal aberration in which a segment of the chromosome is repeated.
**dyad** The products of tetrad separation or disjunction at meiotic prophase I. Each dyad consists of two sister chromatids joined at the centromere.
**dystrophin** A protein that attaches to the inside of the muscle cell plasma membrane and stabilizes the membrane during muscle

contraction. Mutations in the gene encoding dystrophin cause Duchenne and Becker muscular dystrophy. See *Duchenne muscular dystrophy.*
**effective population size** In a population, the number of individuals with an equal probability of contributing gametes to the next generation.
**effector molecule** Small, biologically active molecule that regulates the activity of a protein by binding to a specific receptor site on the protein.
**electrophoresis** A technique that separates a mixture of molecules by their differential migration through a stationary medium (such as a gel) under the influence of an electrical field.
**ELSI (Ethical, Legal, Social Implications)** A program established by the National Human Genome Research Institute in 1990 as part of the Human Genome Project in order to sponsor research on the ethical, legal, and social implications of genomic research and its impact on individuals and social institutions.
**embryonic stem cells** Cells derived from the inner cell mass of early blastocyst mammalian embryos. These cells are pluripotent, meaning they can differentiate into any of the embryonic or adult cell types characteristic of the organism.
**endocytosis** The uptake by a cell of fluids, macromolecules, or particles by pinocytosis, phagocytosis, or receptor-mediated endocytosis.
**endomitosis** An increase of DNA content in multiples of the haploid amount occurring in the absence of nuclear or cytoplasmic division. Polytene chromosomes are formed by endomitosis.
**endonuclease** An enzyme that hydrolyzes internal phosphodiester bonds in a single- or double-stranded polynucleotide chain.
**endoplasmic reticulum (ER)** A membranous organelle system in the cytoplasm of eukaryotic cells. In rough ER, the outer surface of the membranes is ribosome-studded; in smooth ER, it is not.
**endopolyploidy** The increase in chromosome sets within somatic nuclei that results from endomitotic replication.
**endosymbiont theory** The proposal that self-replicating cellular organelles such as mitochondria and chloroplasts were originally free-living organisms that entered into a symbiotic relationship with nucleated cells.
**enhancer** A DNA sequence that enhances transcription, enhancing the expression of nearby structural genes. Enhancers can act over a distance of thousands of base pairs and can be located upstream, downstream, or internal to the gene they affect, differentiating them from promoters.
**environment** The complex of geographic, climatic, and biotic factors within which an organism lives.
**enzyme** A protein or complex of proteins that catalyzes a specific biochemical reaction by lowering the energy of activation that is normally required to initiate the reaction.
**epigenesis** The idea that an organism or organ arises through the sequential appearance and development of new structures, in contrast to preformationism, which holds that development is the result of assembly of structures already present in the egg.
**epigenetics** The study of modifications in an organism's gene function or phenotypic expression that are not attributable to alterations in the nucleotide sequence (gene mutation) of the organism's DNA.
**episome** In bacterial cells, a circular genetic element that can replicate independently of the bacterial chromosome or integrate and replicate as part of the chromosome.
**epistasis** Nonreciprocal interaction between nonallelic genes such that one gene influences or interferes with the expression of another gene, leading to a specific phenotype.
**epitope** That portion of a macromolecule or cell that acts to elicit an antibody response; an antigenic determinant. A complex molecule or cell can contain several such sites.
**equational division** A division stage where the number of centromeres is not reduced by half but where each chromosome is split into longitudinal halves that are distributed into two daughter nuclei. Chromosome division in mitosis and the second meiotic division are examples of equational divisions. See also *reductional division.*
**equatorial plate** See *metaphase plate.*
**euchromatin** Chromatin or chromosomal regions that are lightly staining and are relatively uncoiled during the interphase portion of the cell cycle. Euchromatic regions contain most of the structural genes.
**eugenics** A movement advocating the improvement of the human species by selective breeding. Positive eugenics refers to the promotion of breeding between people thought to possess favorable genes, and negative eugenics refers to the discouragement of breeding among those thought to have undesirable traits.
**eukaryotes** Organisms having true nuclei and membranous organelles and whose cells demonstrate mitosis and meiosis.
**euphenics** Medical or genetic intervention to reduce the impact of defective genotypes.
**euploid** Polyploid with a chromosome number that is an exact multiple of a basic chromosome set.
**evolution** Descent with modification. The emergence of new kinds of plants and animals from preexisting types.
**excision repair** Removal of damaged DNA segments followed by repair. Excision can include the removal of individual bases (base repair) or of a stretch of damaged nucleotides (nucleotide repair).
**exon** The DNA segments of a gene that contain the sequences that, through transcription and translation, are eventually represented in the final polypeptide product.
**exonuclease** An enzyme that breaks down nucleic acid molecules by breaking the phosphodiester bonds at the 3′- or 5′-terminal nucleotides.
**expressed sequence tag (EST)** All or part of the nucleotide sequence of a cDNA clone. ESTs are used as markers in the construction of genetic maps.
**expression vector** Plasmids or phages carrying promoter regions designed to cause expression of inserted DNA sequences.
**expressivity** The degree or range in which a phenotype for a given trait is expressed.
**extranuclear inheritance** Transmission of traits by genetic information contained in cytoplasmic organelles such as mitochondria and chloroplasts.
**$F^-$ cell** A bacterial cell that does not contain a fertility factor and that acts as a recipient in bacterial conjugation.
**$F^+$ cell** A bacterial cell that contains a fertility factor and that acts as a donor in bacterial conjugation.
**F factor** An episomal plasmid in bacterial cells that confers the ability to act as a donor in conjugation (also, *fertility factor*).
**F′ factor** A fertility factor that contains a portion of the bacterial chromosome.
**$F_1$ generation** First filial generation; the progeny resulting from the first cross in a series.
**$F_2$ generation** Second filial generation; the progeny resulting from a cross of the $F_1$ generation.
**F pilus** On bacterial cells possessing an F factor, a filament-like projection that plays a role in conjugation.
**familial trait** A trait transmitted through and expressed by members of a family. Usually used to describe a trait that runs in families, but whose precise mode of inheritance is not clear.
**fate map** A diagram of an embryo showing the location of cells whose developmental fate is known.
**fertility factor** See *F factor.*
**filial generations** See *$F_1$, $F_2$ generations.*
**fingerprint** The unique pattern of ridges and whorls on the tip of a human finger. Also, the pattern obtained by enzymatically cleaving a protein or nucleic acid and subjecting the digest to two-dimensional chromatography or electrophoresis. See also *DNA fingerprinting.*
**FISH** See *fluorescence* in situ *hybridization.*
**fitness** A measure of the relative survival and reproductive success of a given individual or genotype.

**fixation** In population genetics, a condition in which all members of a population are homozygous for a given allele.
**fluctuation test** A statistical test developed by Salvadore Luria and Max Delbrück that demonstrated that bacterial mutations arise spontaneously, in contrast to being induced by selective agents.
**fluorescence *in situ* hybridization (FISH)** A method of *in situ hybridization* that utilizes probes labeled with a fluorescent tag, causing the site of hybridization to fluoresce when viewed using ultraviolet light.
**flush–crash cycle** A period of rapid population growth followed by a drastic reduction in population size.
**f-met** See *formylmethionine*.
**folded-fiber model** A model of eukaryotic chromosome organization in which each sister chromatid consists of a single chromatin fiber composed of double-stranded DNA and proteins wound like a tightly coiled skein of yarn.
**footprinting** See *DNA footprinting*.
**formylmethionine (f-met)** A molecule derived from the amino acid methionine by attachment of a formyl group to its terminal amino group. This is the first monomer used in the synthesis of all bacterial polypeptides. Also known as *N*-formyl methionine.
**forward genetics** The classical approach used to identify a gene controlling a phenotypic trait in the absence of knowledge of the gene's location in the genome or its DNA sequence. Accomplished by isolating mutant alleles and mapping the gene's location, most traditionally using recombination analysis. Once mapped, the gene may be cloned and further studied at the molecular level. An approach contrasted with *reverse genetics*.
**founder effect** A form of genetic drift. The establishment of a population by a small number of individuals whose genotypes carry only a fraction of the different kinds of alleles in the parental population.
**fragile site** A heritable gap, or nonstaining region, of a chromosome that can be induced to generate chromosome breaks.
**fragile X syndrome** A human genetic disorder caused by the expansion of a CGG trinucleotide repeat and a fragile site at Xq27.3 within the *FMR-1* gene. Fragile X syndrome is the most common form of mental retardation. Affected males have distinctive facial features and are mentally retarded. Carrier females lack physical symptoms but as a group have higher rates of mental retardation than normal individuals.
**frameshift mutation** A mutational event leading to the insertion of one or more base pairs in a gene, shifting the codon reading frame in all codons that follow the mutational site.
**fraternal twins** Same as *dizygotic twins*.
**G1 checkpoint** A point in the G1 phase of the cell cycle when a cell becomes committed to initiating DNA synthesis and continuing the cycle or withdraws into the G0 resting stage.
**G0** A nondividing but metabolically active state that cells may enter from the G1 phase of the cell cycle.
**gain of function mutation** A mutation that produces a phenotype different from that of the normal allele and from any loss of function alleles.
**gamete** A specialized reproductive cell with a haploid number of chromosomes.
**gap genes** Genes expressed in contiguous domains along the anterior–posterior axis of the *Drosophila* embryo that regulate the process of segmentation in each domain.
**GC box** In eukaryotes, a region in a promoter containing a 5′-GGGCGG-3′ sequence, which is a binding site for transcriptional regulatory proteins.
**gene** The fundamental physical unit of heredity, whose existence can be confirmed by allelic variants and which occupies a specific chromosomal locus. A DNA sequence coding for a single polypeptide.
**gene amplification** The process by which gene sequences are selected and differentially replicated either extrachromosomally or intrachromosomally.
**gene chip** See *DNA microarray*.
**gene conversion** The process of nonreciprocal recombination by which one allele in a heterozygote is converted into the corresponding allele.
**gene duplication** An event leading to the production of a tandem repeat of a gene sequence during replication.
**gene family** A number of closely related genes derived from a common ancestral gene by duplication and divergence over evolutionary time.
**gene flow** The gradual exchange of genes between two populations; brought about by the dispersal of gametes or the migration of individuals.
**gene frequency** The percentage of alleles of a given type in a population.
**gene interaction** Production of novel phenotypes by the interaction of alleles of different genes.
**gene knockout** The introduction of a *null mutation* into a gene that is subsequently introduced into an organism using transgenic techniques, whereby the organism loses the function of the gene. Often used in mice. See also *gene targeting*.
**gene mutation** See *point mutation*.
**gene pool** The total of all alleles possessed by all reproductive members of a population.
**gene targeting** A transgenic technique used to create and introduce into an organism a specifically altered gene. In mice, gene targeting often involves the induction of a specific mutation in a cloned gene that is then introduced into the genome of a gamete involved in fertilization. The organism produced is bred to produce adults homozygous for the mutation, for example, the creation of a *gene knockout*.
**generalized transduction** The process of transduction whereby recombination of any gene in the bacterial genome is recombined in a process mediated by a bacteriophage.
**genetically modified organism (GMO)** A plant or animal whose genome carries a gene transferred from another species by recombinant DNA technology and expressed to produce a gene product.
**genetic anticipation** The phenomenon in which the severity of symptoms in genetic disorders increases from generation to generation and the age of onset decreases from generation to generation. It is caused by the expansion of trinucleotide repeats within or near a gene and was first observed in myotonic dystrophy.
**genetic background** The collective genome of an organism, as it impacts on the expression of a gene under investigation.
**genetic code** The deoxynucleotide triplets that encode the 20 amino acids or specify termination of translation.
**genetic counseling** Analysis of risk for genetic defects in a family and the presentation of options available to avoid or ameliorate possible risks.
**genetic drift** Random variation in allele frequency from generation to generation, most often observed in small populations.
**genetic engineering** The technique of altering the genetic constitution of cells or individuals by the selective removal, insertion, or modification of individual genes or gene sets.
**genetic equilibrium** A condition relevant to population genetic studies in which allele frequencies are neither increasing nor decreasing.
**genetic erosion** The loss of genetic diversity from a population or a species.
**genetic fine structure analysis** Intragenic recombinational analysis that provides intragenic mapping information at the level of individual nucleotides.
**genetic load** Average number of recessive lethal genes carried in the heterozygous condition by an individual in a population.
**genetic polymorphism** The stable coexistence of two or more distinct genotypes for a given trait in a population. When the frequencies of two alleles for such a trait are in equilibrium, the condition is called balanced polymorphism.
**genetics** The branch of biology concerned with study of inherited variation. More specifically, the study of the origin, transmission, and expression of genetic information.

**genome** The set of hereditary information encoded in the DNA of an organism, including both the protein-coding and non-protein-coding sequences.
**genomic imprinting** The process by which the expression of a gene depends on whether it has been inherited from a male or a female parent. Also referred to as parental imprinting.
**genomic library** A collection of clones that contains all the DNA sequences of an organism's genome.
**genomics** A subdiscipline of the field of genetics generated by the union of classical and molecular biology with the goal of sequencing and understanding genes, gene interaction, genetic elements, and the structure of genomes.
**genotype** The allelic or genetic constitution of an organism; often, the allelic composition of one or a limited number of genes under investigation.
**germ line** An embryonic cell lineage that forms the reproductive cells (eggs and sperm).
**germ plasm** Hereditary material transmitted from generation to generation.
**GMO** See *genetically modified organism (GMO).*
**Goldberg–Hogness box** A short nucleotide sequence 20–30 bp upstream from the initiation site of eukaryotic genes to which RNA polymerase II binds. The consensus sequence is TATAAAA. Also known as a TATA box.
**gynandromorphy** An individual composed of cells with both male and female genotypes.
**gyrase** See *DNA gyrase.*
**haploid (*n*)** A cell or organism having one member of each pair of homologous chromosomes. Also, the gametic chromosome number.
**haplotype** A set of alleles from closely linked loci carried by an individual inherited as a unit.
**HapMap Project** An international effort by geneticists to identify haplotypes (closely linked genetic markers on a single chromosome) shared by certain individuals as a way of facilitating efforts to identify, map, and isolate genes associated with disease or disease susceptibility.
**Hardy–Weinberg law** The principle that genotype frequencies will remain in equilibrium in an infinitely large, randomly mating population in the absence of mutation, migration, and selection.
**heat shock** A transient genetic response following exposure of cells or organisms to elevated temperatures. The response includes activation of a small number of loci, inactivation of some previously active loci, and selective translation of heat shock mRNA.
**helicase** See *DNA helicase.*
**helix–turn–helix (HDH) motif** In DNA-binding proteins, the structure of a region in which a turn of four amino acids holds two $\alpha$ helices at right angles to each other.
**hemizygous** Having a gene present in a single dose in an otherwise diploid cell. Usually applied to genes on the X chromosome in heterogametic males.
**hemoglobin (Hb)** An iron-containing, oxygen-carrying multimeric protein occurring chiefly in the red blood cells of vertebrates.
**hemophilia** An X-linked trait in humans that is associated with defective blood-clotting mechanisms.
**heredity** Transmission of traits from one generation to another.
**heritability** A relative measure of the degree to which observed phenotypic differences for a trait are genetic.
**heterochromatin** The heavily staining, late-replicating regions of chromosomes that are prematurely condensed in interphase. Thought to be devoid of structural genes.
**heteroduplex** A double-stranded nucleic acid molecule in which each polynucleotide chain has a different origin. It may be produced as an intermediate in a recombinational event or by the *in vitro* reannealing of single-stranded, complementary molecules.
**heterogametic sex** The sex that produces gametes containing unlike sex chromosomes. In mammals, the male is the heterogametic sex.
**heterogeneous nuclear RNA (hnRNA)** The collection of RNA transcripts in the nucleus, consisting of precursors to and processing intermediates for rRNA, mRNA, and tRNA. Also includes RNA transcripts that will not be transported to the cytoplasm, such as snRNA.
**heterokaryon** A somatic cell containing nuclei from two different sources.
**heterozygote** An individual with different alleles at one or more loci. Such individuals will produce unlike gametes and therefore will not breed true.
**Hf** Strains of bacteria exhibiting a high frequency of recombination. These strains have a chromosomally integrated F factor that is able to mobilize and transfer part of the chromosome to a recipient $F^-$ cell.
**histocompatibility antigens** See *HLA.*
**histone code** Various chemical modifications applied to histone tails (the free ends of histone molecules, projecting from nucleosomes). These modifications influence DNA–histone interactions and promote or repress transcription.
**histones** Positively charged proteins complexed with DNA in the nucleus. They are rich in the basic amino acids arginine and lysine, and function in coiling DNA to form nucleosomes.
**HLA** Cell-surface proteins produced by histocompatibility loci and involved in the acceptance or rejection of tissue and organ grafts and transplants.
**hnRNA** See *heterogeneous nuclear RNA.*
**Holliday structure** In DNA recombination, an intermediate seen in transmission electron microscope images as an X-shaped structure showing four single-stranded DNA regions.
**homeobox** A sequence of about 180 nucleotides that encodes a sequence of 60 amino acids called a *homeodomain,* which is part of a DNA-binding protein that acts as a transcription factor.
**homeotic mutation** A mutation that causes a tissue normally determined to form a specific organ or body part to alter its differentiation and form another structure.
**homogametic sex** The sex that produces gametes that do not differ with respect to sex chromosome content; in mammals, the female is homogametic.
**homologous chromosomes** Chromosomes that synapse or pair during meiosis and that are identical with respect to their genetic loci and centromere placement.
**homozygote** An individual with identical alleles for a gene or genes of interest. These individuals will produce identical gametes (with respect to the gene or genes in question) and will therefore breed true.
**horizontal gene transfer** The nonreproductive transfer of genetic information from an organism to another, across species and higher taxa (even domains). This mode is contrasted with vertical gene transfer, which is the transfer of genetic information from parent to offspring. In some species of bacteria and archaea, up to 5 percent of the genome may have originally been acquired through horizontal gene transfer.
**H substance** The carbohydrate group present on the surface of red blood cells to which the A and/or B antigen may be added. When unmodified, it results in blood type O.
**human immunodeficiency virus (HIV)** An RNA-containing human retrovirus associated with the onset and progression of AIDS.
**hybrid** An individual produced by crossing parents from two different genetic compositions or strains.
**hybrid vigor** The general superiority of a hybrid over a purebred for the traits health and resilience.

**hydrogen bond** A weak electrostatic attraction between a hydrogen atom covalently bonded to an oxygen or nitrogen atom and an atom that contains an unshared electron pair.
**hypervariable regions** The regions of antibody molecules that attach to antigens. These regions have a high degree of diversity in amino acid content.
**identical twins** See *monozygotic twins.*
**Ig** See *immunoglobulin.*
**immunoglobulin (Ig)** The class of serum proteins having the properties of antibodies.
**imprinting** See *genomic imprinting.*
**inborn error of metabolism** A genetically controlled biochemical disorder; usually an enzyme defect that produces a clinical syndrome.
**inbreeding** Mating between closely related organisms.
**inbreeding depression** A decrease in viability, vigor, or growth in progeny after several generations of inbreeding.
**incomplete dominance** Expressing a heterozygous phenotype that is distinct from the phenotype of either homozygous parent. Also called *partial dominance.*
**independent assortment** The independent behavior of each pair of homologous chromosomes during their segregation in meiosis I. The random distribution of maternal and paternal homologs into gametes.
**inducer** An effector molecule that activates transcription.
**inducible enzyme system** An enzyme system under the control of an inducer, a regulatory molecule that acts to block a repressor and allow transcription.
**initiation codon** The nucleotide triplet AUG that in an mRNA molecule codes for incorporation of the amino acid methionine as the first amino acid in a polypeptide chain.
**insertion sequence** See *IS element.*
***in situ* hybridization** A cytological technique for pinpointing the chromosomal location of DNA sequences complementary to a given nucleic acid or polynucleotide.
**interference (*I*)** A measure of the degree to which one crossover affects the incidence of another crossover in an adjacent region of the same chromatid. Negative interference increases the chance of another crossover; positive interference reduces the probability of a second crossover event.
**interphase** In the cell cycle, the interval between divisions.
**intervening sequence** See *intron.*
**intron** Any segment of DNA that lies between coding regions in a gene. Introns are transcribed but are spliced out of the RNA product and are not represented in the polypeptide encoded by the gene.
**inversion** A chromosomal aberration in which a chromosomal segment has been reversed.
**inversion loop** The chromosomal configuration resulting from the synapsis of homologous chromosomes, one of which carries an inversion.
***in vitro*** Literally, *in glass*; outside the living organism; occurring in an artificial environment.
***in vivo*** Literally, *in the living*; occurring within the living body of an organism.
**IS element** A mobile DNA segment that is transposable to any of a number of sites in the genome.
**isoagglutinogen** An antigenic factor or substance present on the surface of cells that is capable of inducing the formation of an antibody (e.g., the A and B antigens on the surface of human red blood cells).
**isochromosome** An aberrant chromosome with two identical arms and homologous loci.
**isolating mechanism** Any barrier to the exchange of genes between different populations of a group of organisms. In general, isolation can be classified as spatial, environmental, or reproductive.
**isotopes** Alternative forms of atoms with identical chemical properties that have the same atomic number but differ in the number of neutrons (and thus their mass) contained in the nucleus.
**isozyme** Any of two or more distinct forms of an enzyme that have identical or nearly identical chemical properties but differ in some property such as net electrical charge, pH optima, number and type of subunits, or substrate concentration.
**$\kappa$ particles** DNA-containing cytoplasmic particles found in certain strains of *Paramecium aurelia* capable of releasing a toxin, paramecin, that kills other sensitive strains.
**karyokinesis** The process of nuclear division.
**karyotype** The chromosome complement of a cell or an individual. Often used to refer to the arrangement of metaphase chromosomes in a sequence according to length and centromere position.
**kilobase (kb)** A unit of length consisting of 1000 nucleotides.
**kinetochore** A fibrous structure with a size of about 400 nm, located within the centromere. It appears to be the site of microtubule attachment during division.
**Klinefelter syndrome** A genetic disorder in human males caused by the presence of one or more extra X chromosomes. Klinefelter males are usually XXY instead of XY. This syndrome includes enlarged breasts, small testes, sterility, and mild mental retardation.
**knockout mice** Mice created by a process in which a normal gene is cloned, inactivated by the insertion of a marker (such as an antibiotic resistance gene), and transferred to embryonic stem cells, where the altered gene will replace the normal gene (in some cells). These cells are injected into a blastomere embryo, producing a mouse that is then bred to yield mice homozygous for the mutated gene. See also *gene targeting.*
**Kozak sequence** A short nucleotide sequence adjacent to the initiation codon that is recognized as the translational start site in eukaryotic mRNA.
***lac* repressor protein** A protein that binds to the operator in the *lac* operon and blocks transcription.
**lagging strand** During DNA replication, the strand synthesized in a discontinuous fashion, in the direction opposite of the replication fork. See also *Okazaki fragment.*
**lampbrush chromosomes** Meiotic chromosomes characterized by extended lateral loops that reach maximum extension during diplotene. Although most intensively studied in amphibians, these structures occur in meiotic cells of organisms ranging from insects to humans.
**lariat structure** A structure formed by an intron by means of a 5′ to 3′ bond during processing and removal of that intron from an mRNA molecule.
**leader sequence** That portion of an mRNA molecule from the 5′ end to the initiating codon, often containing regulatory or ribosome binding sites.
**leading strand** During DNA replication, the strand synthesized continuously in the direction of the replication fork.
**leptotene** The initial stage of meiotic prophase I, during which the chromosomes become visible and are often arranged with one or both ends gathered at one spot on the inner nuclear membrane (the so-called bouquet configuration).
**lethal gene** A gene whose expression results in death of the organism at some stage of its life cycle.
**leucine zipper** In DNA-binding proteins, a structural motif characterized by a stretch in which every seventh amino acid residue is leucine, with adjacent regions containing positively charged amino acids. Leucine zippers on two polypeptides may interact to form a dimer that binds to DNA.
**linkage** The condition in which genes have their loci present on the same chromosome, causing them to be inherited as a unit, provided that they are not separated by crossing over during meiosis.

**linking number** The number of times that two strands of a closed, circular DNA duplex cross over each other.
**locus (pl., loci)** The site or place on a chromosome where a particular gene is located.
**lod score (LOD)** A statistical method used to determine whether two loci are linked or unlinked. A lod (log of the odds) score of 4 indicates that linkage is 10,000 times more likely than nonlinkage. By convention, lod scores of 3–4 are signs of linkage.
**long interspersed elements (LINES)** Long, repetitive sequences found interspersed in the genomes of higher organisms.
**long terminal repeat (LTR)** A sequence of several hundred base pairs found at both ends of a retroviral DNA.
**loss of function mutation** Mutations that produce alleles with reduced or no function.
**Lyon hypothesis** The proposal describing the random inactivation of the maternal or paternal X chromosome in somatic cells of mammalian females early in development. All daughter cells will have the same X chromosome inactivated as in the cell they descended from, producing a mosaic pattern of expression of X chromosome genes.
**lysis** The disintegration of a cell brought about by the rupture of its membrane.
**lysogenic bacterium** A bacterial cell carrying the DNA of a temperate bacteriophage integrated into its chromosome.
**lysogeny** The process by which the DNA of an infecting phage becomes repressed and integrated into the chromosome of the bacterial cell it infects.
**lytic phase** The condition in which a bacteriophage invades, reproduces, and lyses the bacterial cell. Lysogenic bacteria may be induced to enter the lytic phase.
**major histocompatibility (MHC) loci** Loci encoding antigenic determinants responsible for tissue specificity. In humans, the HLA complex; and in mice, the H2 complex. See also *HLA*.
**mapping functions** A mathematical formula that relates map distances to recombination frequencies.
**map unit** A measure of the genetic distance between two genes, corresponding to a recombination frequency of 1 percent. See also *centimorgan*.
**maternal effect** Phenotypic effects in offspring attributable to genetic information transmitted through the oocyte derived from the maternal genome.
**maternal influence** Same as *maternal effect*.
**maternal inheritance** The transmission of traits strictly through the maternal parent, usually due to DNA found in the cytoplasmic organelles, the mitochondria, or chloroplasts.
**mean** The arithmetic average.
**median** In a set of data points, the one below and above which there are equal numbers of other data points.
**meiosis** The process of cell division in gametogenesis or sporogenesis during which the diploid number of chromosomes is reduced to the haploid number.
**melting profile ($T_m$)** The temperature at which a population of double-stranded nucleic acid molecules is half-dissociated into single strands.
**merozygote** A partially diploid bacterial cell containing, in addition to its own chromosome, a chromosome fragment introduced into the cell by transformation, transduction, or conjugation.
**messenger RNA (mRNA)** An RNA molecule transcribed from DNA and translated into the amino acid sequence of a polypeptide.
**metabolism** The collective chemical processes by which living cells are created and maintained. In particular, the chemical basis of generating and utilizing energy in living organisms.
**metacentric chromosome** A chromosome that has a centrally located centromere and therefore chromosome arms of equal lengths.
**metafemale** In *Drosophila*, a poorly developed female of low viability with a ratio of X chromosomes to sets of autosomes exceeding 1.0. Previously called a *superfemale*.
**metagenomics** The study of DNA recovered from organisms obtained from the environment as opposed to laboratory cultures. Often used for estimating the diversity of organisms in an environmental sample.
**metamale** In *Drosophila*, a poorly developed male of low viability with a ratio of X chromosomes to sets of autosomes below 0.5. Previously called a *supermale*.
**metaphase** The stage of cell division in which condensed chromosomes lie in a central plane between the two poles of the cell and during which the chromosomes become attached to the spindle fibers.
**metaphase plate** The plane in which mitotic or meiotic chromosomes collect at the equator of the cell during metaphase.
**methylation** Enzymatic transfer of methyl groups from S-adenosylmethionine to biological molecules, including phospholipids, proteins, RNA, and DNA. Methylation of DNA is associated with the regulation of gene expression and with epigenetic phenomena such as imprinting.
**MHC** See *major histocompatibility loci*.
**microRNA** Single-stranded RNA molecules approximately 20–23 nucleotides in length that regulate gene expression by participating in the degradation of mRNA.
**microsatellite** A short, highly polymorphic DNA sequence of 1–4 base pairs, widely distributed in the genome, that are used as molecular markers in a variety of methods. Also called *simple sequence repeats (SSRs)*.
**migration coefficient** A measure of the proportion of migrant genes entering the population per generation.
**minimal medium** A medium containing only the essential nutrients needed to support the growth and reproduction of wild-type strains of an organism. Usually comprised of inorganic components that include a carbon and nitrogen source.
**minisatellite** Series of short tandem repeat sequences (STRs) 10–100 nucleotides in length that occur frequently throughout the genome of eukaryotes. Because the number of repeats at each locus is variable, the loci are known as variable number tandem repeats (VNTRs). Used in DNA fingerprinting. See also *VNTRs* and *STR sequences*.
**mismatch repair** A form of excision repair of DNA in which the repair mechanism is able to distinguish between the strand with the error and the strand that is correct.
**missense mutation** A mutation that alters a codon to that of another amino acid and thus results in an alteration in the translation product.
**mitochondrial DNA (mtDNA)** Double-stranded, self-replicating circular DNA found in mitochondria that encodes mitochondrial ribosomal RNAs, transfer RNAs, and proteins used in oxidative respiratory functions of the organelle.
**mitochondrion** The so-called power house of the cell—a self-reproducing, DNA-containing, cytoplasmic organelle in eukaryotes involved in the generation of the high-energy compound ATP.
**mitogen** A substance that stimulates mitosis in nondividing cells.
**mitosis** A form of cellular reproduction producing two progeny cells, identical genetically to the progenitor cell, that is, the production of two cells from one, each with the same chromosome complement as the parent cell.
**mode** In a set of data, the value occurring with the greatest frequency.
**model genetic organism** An experimental organism conducive to efficiently conducted research whose genetics is intensively studied on the premise that the findings can be applied to other organisms; for example, the fruit fly (*Drosophila melanogaster*) and the mouse (*Mus musculus*) are model organisms used to study the causes and development of human genetic diseases.

**molecular clock** In evolutionary studies, a method that counts the number of differences in DNA or protein sequences in order to measure the time elapsed since two species diverged from a common ancestor.
**monohybrid cross** A genetic cross involving only one character (e.g., *AA* × *aa*).
**monosomic** An aneuploid condition in which one member of a chromosome pair is missing; having a chromosome number of $2n - 1$.
**monozygotic twins** Twins produced from a single fertilization event; the first division of the zygote produces two cells, each of which develops into an embryo. Also known as *identical twins.*
**mRNA** See *messenger RNA.*
**mtDNA** See *mitochondrial DNA.*
**multigene family** A gene set descended from a common ancestor exhibiting duplication and subsequent divergence. The globin genes are an example of a multigene family.
**multiple alleles** In a population of organisms, three or more alleles of the same gene.
**multiple-factor inheritance** Same as *polygenic inheritance.*
**mu (*μ*) bacteriophage** A phage group in which the genetic material behaves like an insertion sequence (i.e., capable of insertion, excision, transposition, inactivation of host genes, and induction of chromosomal rearrangements).
**mutagen** Any agent that causes an increase in the spontaneous rate of mutation.
**mutant** A cell or organism carrying an altered or mutant gene.
**mutation** The process that produces an alteration in DNA or chromosome structure; in genes, the source of new alleles.
**mutation rate** The frequency with which mutations take place at a given locus or in a population.
**muton** In fine structure analysis of the gene, the smallest unit of mutation, corresponding to a single base change.
**natural selection** Differential reproduction among members of a species owing to variable fitness conferred by genotypic differences.
**neutral mutation** A mutation with no immediate adaptive significance or phenotypic effect.
**nonautonomous transposon** A transposable element that lacks a functional transposase gene.
**noncrossover gamete** A gamete whose chromosomes have undergone no genetic recombination.
**nondisjunction** A cell division error in which homologous chromosomes (in meiosis) or the sister chromatids (in mitosis) fail to separate and migrate to opposite poles; responsible for defects such as monosomy and trisomy.
**nonsense codons** The nucleotide triplets (UGA, UAG, and UAA) in an mRNA molecule that signals the termination of translation.
**nonsense mutation** A mutation that changes codon encoding an amino acid into a termination codon, leading to premature termination during translation of mRNA.
**NOR** See *nucleolar organizer region.*
**normal distribution** A probability function that approximates the distribution of random variables. The normal curve, also known as a Gaussian or bell-shaped curve, is the graphic display of a normal distribution.
**northern blot** An analytic technique in which RNA molecules are separated by electrophoresis and transferred by capillary action to a nylon or nitrocellulose membrane. Specific RNA molecules can then be identified by hybridization to a labeled nucleic acid probe.
**N-terminal amino acid** In a peptide chain, the terminal amino acid that carries a free amino group.
**N terminus** The free amino group of the first amino acid in a polypeptide. By convention, the structural formula of polypeptides is written with the N terminus at the left.
**nuclease** An enzyme that breaks bonds in nucleic acid molecules.
**nucleoid** The DNA-containing region within the cytoplasm in prokaryotic cells.
**nucleolar organizer region (NOR)** A chromosomal region containing the genes for rRNA; most often found in physical association with the *nucleolus.*
**nucleolus** The nuclear site of ribosome biosynthesis and assembly; usually associated with or formed in association with the DNA comprising the *nucleolar organizer region.*
**nucleoside** In nucleic acid chemical nomenclature, a purine or pyrimidine base covalently linked to a ribose or deoxyribose sugar molecule.
**nucleosome** In eukaryotes, a nuclear complex consisting of four pairs of histone molecules wrapped by two turns of a DNA molecule. The major structure associated with the organization of chromatin in the nucleus.
**nucleotide** In nucleic acid chemical nomenclature, a nucleoside covalently linked to one or more phosphate groups. Nucleotides containing a single phosphate linked to the 5′ carbon of the ribose or deoxyribose are the building blocks of nucleic acids.
**nucleotide pair** The base-paired nucleotides (A with T or G with C) on opposite strands of a DNA molecule that are hydrogen-bonded to each other.
**nucleus** The membrane-bound cytoplasmic organelle of eukaryotic cells that contains the chromosomes and nucleolus.
**null allele** A mutant allele that produces no functional gene product. Usually inherited as a recessive trait.
**null hypothesis ($H_0$)** Used in statistical tests, the hypothesis that there is no real difference between the observed and expected datasets. Statistical methods such as chi-square analysis are used to test the probability associated with this hypothesis.
**nullisomic** Term describing an individual with a chromosomal aberration in which both members of a chromosome pair are missing.
**Okazaki fragment** The small, discontinuous strands of DNA produced on the lagging strand during DNA synthesis.
**oligonucleotide** A linear sequence of about 10–20 nucleotides connected by 5′-3′ phosphodiester bonds.
**oncogene** A gene whose activity promotes uncontrolled proliferation in eukaryotic cells. Usually a mutant gene derived from a *proto-oncogene.*
**Online Mendelian Inheritance in Man (OMIM)** A database listing all known genetic disorders and disorders with genetic components. It also contains a listing of all known human genes and links genes with genetic disorders.
**open reading frame (ORF)** A nucleotide sequence organized as triplets that encodes the amino acid sequence of a polypeptide, including an initiation codon and a termination codon.
**operator region** In bacterial DNA, a region that interacts with a specific repressor protein to regulate the expression of an adjacent gene or gene set.
**operon** A genetic unit consisting of one or more structural genes encoding polypeptides, and an adjacent operator gene that regulates the transcriptional activity of the structural gene or genes.
**origin of replication (ori)** Sites where DNA replication begins along the length of a chromosome.
**outbreeding depression** Reduction in fitness in the offspring produced by mating genetically diverse parents. It is thought to result from a lowered adaptation to local environmental conditions.
**overdominance** The phenomenon in which heterozygotes have a phenotype that is more extreme than either homozygous genotype.
**overlapping code** A hypothetical genetic code in which any given triplet is shared by more than one adjacent codon.
**pachytene** The stage in meiotic prophase I when the synapsed homologous chromosomes split longitudinally (except at the centromere), producing a group of four chromatids called a tetrad.

**pair-rule genes** Genes expressed as stripes around the blastoderm embryo during development of the *Drosophila* embryo.
**palindrome** In genetics, a sequence of DNA base pairs that reads the same backward or forward. Since strands run antiparallel to one another in DNA, the base sequences on the two strands read the same backward and forward. For example:

5′-GAATTC-3′
3′-CTTAAG-5′

Palindromic sequences are noteworthy as recognition sites along DNA, often providing the substrates for restriction endonucleases.
**paracentric inversion** A chromosomal inversion that does not include the region containing the centromere.
**parental gamete** Same as *noncrossover gamete.*
**parthenogenesis** Development of an egg without fertilization.
**partial diploids** See *merozygote.*
**partial dominance** See *incomplete dominance.*
**patroclinous inheritance** A form of genetic transmission in which the offspring have the phenotype of the father.
**pedigree** In human genetics, a diagram showing the ancestral relationships and transmission of genetic traits over several generations in a family.
**P element** In *Drosophila,* a transposable DNA element responsible for hybrid dysgenesis.
**penetrance** The frequency, expressed as a percentage, with which individuals of a given genotype manifest at least some degree of a specific mutant phenotype associated with a trait.
**peptide bond** The covalent bond between the amino group of one amino acid and the carboxyl group of another amino acid, creating a dipeptide.
**pericentric inversion** A chromosomal inversion that involves both arms of the chromosome and thus the centromere.
**phage** See *bacteriophage.*
**pharmacogenomics** The study of how genetic variation influences the action of pharmaceutical drugs in individuals.
**phenotype** The overt appearance of a genetically controlled trait.
**phenylketonuria (PKU)** A hereditary condition in humans that is associated with the inability to metabolize the amino acid phenylalanine due to the loss of activity of the enzyme phenylalanine hydroxylase.
**Philadelphia chromosome** The product of a reciprocal translocation in humans that contains the short arm of chromosome 9, carrying the *C-ABL* oncogene, and the long arm of chromosome 22, carrying the *BCR* gene.
**phosphodiester bond** In nucleic acids, the system of covalent bonds by which a phosphate group links adjacent nucleotides, extending from the 5′ carbon of one pentose sugar (ribose or deoxyribose) to the 3′ carbon of another pentose sugar in the neighboring nucleotide. Phosphodiester bonds create the backbone of nucleic acid molecules.
**photoreactivation enzyme (PRE)** An exonuclease that catalyzes the light-activated excision of ultraviolet-induced thymine dimers from DNA.
**photoreactivation repair** Light-induced repair of damage caused by exposure to ultraviolet light. Associated with an intracellular enzyme system.
**phyletic evolution** The gradual transformation of one species into another over time; so-called vertical evolution.
**pilus** A filamentlike projection from the surface of a bacterial cell. Often associated with cells possessing F factors.
**plaque** On an otherwise opaque bacterial lawn, a clear area caused by the growth and reproduction of a single bacteriophage.
**plasmid** An extrachromosomal, circular DNA molecule that replicates independently of the host chromosome.
**pleiotropy** Condition in which a single mutation causes multiple phenotypic effects.
**ploidy** A term referring to the basic chromosome set or to multiples of that set.
**pluripotent** See *totipotent.*
**point mutation** A mutation that can be mapped to a single locus. At the molecular level, a mutation that results in the substitution of one nucleotide for another. Also called a *gene mutation.*
**polar body** Produced in females at either the first or second meiotic division of gametogenesis, a discarded cell that contains one of the nuclei of the division process, but almost no cytoplasm as a result of an unequal cytokinesis.
**polycistronic mRNA** A messenger RNA molecule that encodes the amino acid sequence of two or more polypeptide chains in adjacent structural genes.
**polygenic inheritance** The transmission of a phenotypic trait whose expression depends on the additive effect of a number of genes.
**polylinker** A segment of DNA that has been engineered to contain multiple sites for restriction enzyme digestion. Polylinkers are usually found in engineered vectors such as plasmids.
**polymerase chain reaction (PCR)** A method for amplifying DNA segments that depends on repeated cycles of denaturation, primers, and DNA polymerase–directed DNA synthesis.
**polymerases** Enzymes that catalyze the formation of DNA and RNA from deoxynucleotides and ribonucleotides, respectively.
**polymorphism** The existence of two or more discontinuous, segregating phenotypes in a population.
**polynucleotide** A linear sequence of 20 or more nucleotides, joined by 5′-3′ phosphodiester bonds. See also *oligonucleotide.*
**polypeptide** A molecule made up of amino acids joined by covalent peptide bonds. This term is used to denote the amino acid chain before it assumes its functional three-dimensional configuration.
**polyploid** A cell or individual having more than two haploid sets of chromosomes.
**polyribosome** See *polysome.*
**polysome** A structure composed of two or more ribosomes associated with mRNA and engaged in translation. Also called a *polyribosome.*
**polytene chromosome** Literally, a many stranded chromosome, one that has undergone numerous rounds of DNA replication without separation of the replicated strands, which remain in exact parallel register. The result is a giant chromosome with aligned chromomeres displaying a characteristic banding pattern, most often studied in *Drosophila* larval salivary gland cells.
**population** A local group of individuals belonging to the same species that are actually or potentially interbreeding.
**population bottleneck** A drastic reduction in population size and consequent loss of genetic diversity, followed by an increase in population size. The rebuilt population has an altered gene pool with reduced variability caused by genetic drift.
**positional cloning** The identification and subsequent cloning of a gene in the absence of knowledge of its polypeptide product or function. Based upon the determination of its precise chromosomal location, the process begins with knowledge of an approximate location of the gene within the genome that is then refined, often by examining the cosegregation of mutant phenotypes with known genetic map locations or DNA markers (e.g., *RFLP polymorphisms*).
**position effect** Change in expression of a gene associated with a change in the gene's location within the genome.
**post-translational modification** The processing or modification of the translated polypeptide chain by enzymatic cleavage, or the addition of phosphate groups, carbohydrate chains, or lipids.
**posttranscriptional modification** Changes made to pre-mRNA molecules during conversion to mature mRNA. These include the addition of a methylated cap at the 5′ end and a poly-A tail at the 3′ end, excision of introns, and exon splicing.
**postzygotic isolation mechanism** A factor that prevents or reduces inbreeding by acting after fertilization to produce nonviable, sterile hybrids or hybrids of lowered fitness.

**preadaptive mutation** A mutational event that later becomes of adaptive significance.
**preimplantation genetic diagnosis (PGD)** The removal and genetic analysis of unfertilized oocytes, polar bodies, or single cells from an early embryo (3–5 days old) following *in vitro* fertilization.
**prezygotic isolation mechanism** A factor that reduces inbreeding by preventing courtship, mating, or fertilization.
**Pribnow box** In prokaryotic genes, a 6-bp sequence to which the sigma ($\sigma$) subunit of RNA polymerase binds, upstream from the beginning of transcription. The consensus sequence for this box is TATAAT.
**primary protein structure** The sequence of amino acids in a polypeptide chain.
**primary sex ratio** Ratio of males to females at fertilization, often expressed in decimal form (e.g., 1.06).
**primer** In nucleic acids, a short length of RNA or single-stranded DNA required for the initiating synthesis directed by polymerases.
**prion** An infectious pathogenic agent devoid of nucleic acid and composed of a protein, PrP, with a molecular weight of 27,000–30,000 Da. Prions are known to cause scrapie, a degenerative neurological disease in sheep; bovine spongiform encephalopathy (BSE, or mad cow disease) in cattle; and similar diseases in humans, including kuru and Creutzfeldt–Jakob disease.
**probability** Ratio of the frequency of a given outcome to the frequency of all possible outcomes.
**proband** An individual who is the focus of, or who called attention to, the genetic study leading to the production of a pedigree tracking the inheritance of a genetically determined trait of interest. Formerly known as a *propositus.*
**probe** A macromolecule such as DNA or RNA that has been labeled and can be detected by an assay such as autoradiography or fluorescence microscopy. Probes are used to identify target molecules, genes, or gene products.
**product law** In statistics, the law holding that the probability of two independent events occurring simultaneously is equal to the product of their independent probabilities.
**progeny** The offspring produced from a mating.
**prokaryotes** Organisms lacking nuclear membranes and true chromosomes. Bacteria and blue–green algae are examples of prokaryotic organisms.
**promoter** An upstream region of a gene serving a regulatory function and to which RNA polymerase binds prior to the initiation of transcription.
**proofreading** A molecular mechanism for scanning and correcting errors in replication, transcription, or translation.
**prophage** A bacteriophage genome integrated into a bacterial chromosome that is replicated along with the bacterial chromosome. Bacterial cells carrying prophages are said to be *lysogenic* and to be capable of entering the *lytic cycle*, whereby the phage is reproduced.
**propositus (female, proposita)** See *proband.*
**protein** A molecule composed of one or more polypeptides, each composed of amino acids covalently linked together. Proteins demonstrate *primary, secondary, tertiary,* and often, *quaternary structure.*
**proteome** The entire set of proteins expressed by a cell, tissue, or organism at a given time.
**proteomics** The study of the expressed proteins present in a cell at a given time.
**proto-oncogene** A gene that functions to initiate, facilitate, or maintain cell growth and division. Proto-oncogenes can be converted to *oncogenes* by mutation.
**protoplast** A bacterial or plant cell with the cell wall removed. Sometimes called a *spheroplast.*
**prototroph** A strain (usually of a microorganism) that is capable of growth on a defined, minimal medium. Wild-type strains are usually regarded as prototrophs and contrasted with *auxotrophs.*
**pseudoalleles** Genes that behave as alleles to one another by complementation but can be separated from one another by recombination.
**pseudoautosomal region** A region present on the human Y chromosome that is also represented on the X chromosome. Genes found in this region of the Y chromosome have a pattern of inheritance that is indistinguishable from genes on autosomes.
**pseudodominance** The expression of a recessive allele on one homolog owing to the deletion of the dominant allele on the other homolog.
**pseudogene** A nonfunctional gene with sequence homology to a known structural gene present elsewhere in the genome. It differs from the functional version by insertions or deletions and by the presence of flanking direct-repeat sequences of 10–20 nucleotides.
**puff** Same as *chromosome puff.*
**punctuated equilibrium** A pattern in the fossil record of long periods of species stability, punctuated with brief periods of species divergence.
**quantitative inheritance** Same as *polygenic inheritance.*
**quantitative trait loci (QTLs)** Two or more genes that act on a single polygenic trait.
**quantum speciation** Formation of a new species within a single or a few generations by a combination of selection and drift.
**quaternary protein structure** Types and modes of interaction between two or more polypeptide chains within a protein molecule.
**race** A genotypically or geographically distinct subgroup within a species.
**rad** A unit of absorbed dose of radiation with an energy equal to 100 ergs per gram of irradiated tissue.
**radioactive isotope** An unstable isotope with an altered number of neutrons that emits ionizing radiation during decay as it is transformed to a stable atomic configuration. See also *isotope.*
**random amplified polymorphic DNA (RAPD)** A PCR method that uses random primers about 10 nucleotides in length to amplify unknown DNA sequences.
**random mating** Mating between individuals without regard to genotype.
**reading frame** A linear sequence of codons in a nucleic acid.
**reannealing** Formation of double-stranded DNA molecules from denatured single strands.
**recessive** An allele whose potential genetic expression is overridden in the heterozygous condition by a dominant allele.
**reciprocal cross** A pair of crosses in which the genotype of the female in one is present as the genotype of the male in the other, and vice versa.
**reciprocal translocation** A chromosomal aberration in which nonhomologous chromosomes exchange parts.
**recombinant DNA** A DNA molecule formed by joining two heterologous molecules. A term also applied to the technology associated with the use of DNA molecules produced by *in vitro* ligation of DNA from two different organisms.
**recombinant gamete** A gamete containing a new combination of genes produced by crossing over during meiosis.
**recombination** The process that leads to the formation of new gene combinations on chromosomes.
**recon** A term utilized in fine structure analysis studies to denote the smallest intragenic units between which recombination can occur.
**reductional division** The chromosome division that halves the number of centromeres and thus reduces the chromosome number by half. The first division of meiosis is a reductional division. See also *equational division.*
**redundant genes** Gene sequences present in more than one copy per haploid genome (e.g., ribosomal genes).

**regulatory site** A DNA sequence that functions in the control of expression of other genes, usually involving an interaction with another molecule.
**rem** Radiation equivalent in humans; the dosage of radiation that will cause the same biological effect as one roentgen of X rays.
**renaturation** The process by which a denatured protein or nucleic acid returns to its normal three-dimensional structure.
**repetitive DNA sequence** A DNA sequence present in many copies in the haploid genome.
**replicating form (RF)** Double-stranded nucleic acid molecules present as an intermediate during the reproduction of certain RNA-containing viruses.
**replication** The process whereby DNA is duplicated.
**replication fork** The Y-shaped region of a chromosome associated with the site of replication.
**replicon** The unit of DNA replication, originating with DNA sequences necessary for the initiation of DNA replication. In bacteria, the entire chromosome is a replicon.
**replisome** The complex of proteins, including DNA polymerase, that assembles at the bacterial replication fork to synthesize DNA.
**repressible enzyme system** An enzyme or group of enzymes whose synthesis is regulated by the intracellular concentration of certain metabolites.
**repressor** A protein that binds to a regulatory sequence adjacent to a gene and blocks transcription of the gene.
**reproductive isolation** Absence of interbreeding between populations, subspecies, or species. Reproductive isolation can be brought about by extrinsic factors, such as behavior, or intrinsic barriers, such as hybrid inviability.
**resistance transfer factor (RTF)** A component of R plasmids that confers the ability to transfer the R plasmid between cells by conjugation.
**restriction endonuclease** A bacterial nuclease that recognizes specific nucleotide sequences in a DNA molecule, often a *palindrome*, and cleaves or nicks the DNA at those sites. Provides defense for invading DNA segments in bacteria and is useful in the construction of *recombinant DNA molecules*.
**restriction fragment length polymorphism (RFLP)** Variation in the length of DNA fragments generated by restriction endonucleases. These variations are caused by mutations that create or abolish cutting sites for restriction enzymes. RFLPs are inherited in a codominant fashion and are extremely useful as genetic markers.
**restrictive transduction** See *specialized transduction*.
**retrotransposon** Major components of many eukaryotic genomes, mobile genetic elements that can be copied by means of an RNA intermediate and inserted at a distant locus.
**retrovirus** A type of virus that uses RNA as its genetic material and employs the enzyme reverse transcriptase during its life cycle.
**reverse genetics** An experimental approach used to discover the function of a gene, after the gene has been identified, cloned, and sequenced. The cloned gene may be knocked out (e.g., by *gene targeting*) or have its expression altered (e.g. by *RNA interference* or *transgenic overexpression*) and the resulting phenotype studied. An approach contrasted with *forward genetics*.
**reverse transcriptase** A polymerase that uses RNA as a template to transcribe a single-stranded DNA molecule as a product.
**reversion** A mutation that restores the wild-type phenotype.
**R factor (R plasmid)** A bacterial plasmid that carries antibiotic resistance genes. Most R plasmids have two components: an r-determinant, which carries the antibiotic resistance genes, and the resistance transfer factor (RTF).
**RFLP** See *restriction fragment length polymorphism*.
**Rh factor** An antigenic system first described in the rhesus monkey. Recessive $r/r$ individuals produce no Rh antigens and are Rh negative, while $R/R$ and $R/r$ individuals have Rh antigens on the surface of their red blood cells and are classified as Rh positive. The genetic basis of the immunological incompatibility disease called erythroblastosis fetalis (hemolytic disease of the newborn).
**ribonucleic acid (RNA)** A nucleic acid similar to DNA but characterized by the pentose sugar ribose, the pyrimidine uracil, and the single-stranded nature of the polynucleotide chain. Several forms are recognized, including ribosomal RNA, messenger RNA, transfer RNA, and a variety of small regulatory RNA molecules.
**ribose** The five-carbon sugar associated with ribonucleosides and ribonucleotides associated with RNA.
**ribosomal RNA (rRNA)** The RNA molecules that are the structural components of the ribosomal subunits. In prokaryotes, these are the 16*S*, 23*S*, and 5*S* molecules; in eukaryotes, they are the 18*S*, 28*S*, and 5*S* molecules. See also *Svedberg coefficient (S)*.
**ribosome** A ribonucleoprotein organelle consisting of two subunits, each containing RNA and protein molecules. Ribosomes are the site of translation of mRNA codons into the amino acid sequence of a polypeptide chain.
**RNA** See *ribonucleic acid*.
**RNA editing** Alteration of the nucleotide sequence of an mRNA molecule after transcription and before translation. There are two main types of editing: substitution editing, which changes individual nucleotides, and insertion/deletion editing, in which individual nucleotides are added or deleted.
**RNA polymerase** An enzyme that catalyzes the formation of an RNA polynucleotide strand using the base sequence of a DNA molecule as a template.
**RNase** A class of enzymes that hydrolyzes RNA.
**Robertsonian translocation** A form of chromosomal aberration in which breaks occur in the short arms of two acrocentric chromosomes and the long arms of these chromosomes fuse at the centromere. Also called *centric fusion*.
**roentgen (R)** A unit of measure of radiation emission, corresponding to the amount that generates $2.083 \times 10^9$ ion pairs in one cubic centimeter of air at 0°C and an atmospheric pressure of 760 mm of mercury.
**rolling circle model** A model of DNA replication in which the growing point, or replication fork, rolls around a circular template strand; in each pass around the circle, the newly synthesized strand displaces the strand from the previous replication, producing a series of contiguous copies of the template strand.
**rRNA** See *ribosomal RNA*.
**RTF** See *resistance transfer factor*.
**$S_1$ nuclease** A deoxyribonuclease that cuts and degrades single-stranded molecules of DNA.
**satellite DNA** DNA that forms a minor band when genomic DNA is centrifuged in a cesium salt gradient. This DNA usually consists of short sequences repeated many times in the genome.
**SCE** See *sister chromatid exchange*.
**secondary protein structure** The $\alpha$-helical or $\beta$-pleated-sheet formations in a polypeptide, dependent on hydrogen bonding between certain amino acids.
**secondary sex ratio** The ratio of males to females at birth, usually expressed in decimal form (e.g., 1.06).
**secretor** An individual who has soluble forms of the blood group antigens A and/or B present in saliva and other body fluids. This condition is caused by a dominant autosomal gene independent of the *ABO* locus (*I* locus).
**sedimentation coefficient (*S*)** See *Svedberg coefficient*.
**segment polarity genes** Genes that regulate the spatial pattern of differentiation within each segment of the developing *Drosophila* embryo.

**segregation** The separation of maternal and paternal homologs of each homologous chromosome pair into gametes during meiosis.
**selection** The changes that occur in the frequency of alleles and genotypes in populations as a result of differential reproduction.
**selection coefficient (*s*)** A quantitative measure of the relative fitness of one genotype compared with another. Same as *coefficient of selection.*
**selfing** In plant genetics, the fertilization of a plant's ovules by pollen produced by the same plant. Reproduction by self-fertilization.
**semiconservative replication** A mode of DNA replication in which a double-stranded molecule replicates in such a way that the daughter molecules are each composed of one parental (old) and one newly synthesized strand.
**semisterility** A condition in which a percentage of all zygotes are inviable.
**sex chromatin body** See *Barr body.*
**sex chromosome** A chromosome, such as the X or Y in humans, which is involved in sex determination.
**sexduction** Transmission of chromosomal genes from a donor bacterium to a recipient cell by means of the F factor.
**sex-influenced inheritance** Phenotypic expression that is conditioned by the sex of the individual. A heterozygote may express one phenotype in one sex and the alternate phenotype in the other sex (e.g., pattern baldness in humans).
**sex-limited inheritance** A trait that is expressed in only one sex even though the trait may not be X-linked.
**sex ratio** See *primary sex ratio* and *secondary sex ratio.*
**sexual reproduction** Reproduction through the fusion of gametes, which are the haploid products of meiosis.
**Shine–Dalgarno sequence** The nucleotides AGGAGG that serve as a ribosome-binding site in the leader sequence of prokaryotic genes. The 16*S* RNA of the small ribosomal subunit contains a complementary sequence to which the mRNA binds.
**short interspersed elements (SINES)** Repetitive sequences found in the genomes of higher organisms, such as the 300-bp *Alu* sequence.
**short tandem repeats** See *STR sequences.*
**shotgun experiment** The cloning of random fragments of genomic DNA into a vehicle such as a plasmid or phage, usually to produce a library from which clones of specific interest can be selected.
**sibling species** Species that are morphologically almost identical but that are reproductively isolated from one another.
**sickle-cell anemia** A genetic disease in humans caused by an autosomal recessive gene, often fatal in the homozygous condition if untreated. Caused by a mutation leading to an alteration in the amino acid sequence of the $\beta$ chain of globin.
**sickle-cell trait** The phenotype exhibited by individuals heterozygous for the sickle-cell gene.
**sigma($\sigma$) factor** In RNA polymerase, a polypeptide subunit that recognizes the DNA binding site for the initiation of transcription.
**single-stranded binding proteins (SSBs)** In DNA replication, proteins that bind to and stabilize the single-stranded regions of DNA that result from the action of unwinding proteins.
**sister chromatid exchange (SCE)** A crossing over event that can occur in meiotic and mitotic cells involving the reciprocal exchange of chromosomal material between sister chromatids joined by a common centromere. Such exchanges can be detected cytologically after BrdU is incorporated into the replicating chromosomes.
**site-directed mutagenesis** A process that uses a synthetic oligonucleotide containing a mutant base or sequence as a primer for inducing a mutation at a specific site in a cloned gene.
**small nuclear RNA (snRNA)** Abundant species of small RNA molecules ranging in size from 90 to 400 nucleotides that in association with proteins form RNP particles known as snRNPs or *snurps.* Located in the nucleoplasm, snRNAs have been implicated in the processing of pre-mRNA and may have a range of cleavage and ligation functions.
**snurps** See *small nuclear RNA.*
**solenoid structure** A feature of eukaryotic chromatin conformation that is generated by the supercoiling of nucleosomes.
**somatic cell genetics** The use of cultured somatic cells to investigate genetic phenomena by means of parasexual techniques involving the fusion of cells from different organisms.
**somatic cells** All cells other than the germ cells or gametes in an organism.
**somatic mutation** A nonheritable mutation occurring in a somatic cell.
**somatic pairing** The pairing of homologous chromosomes in somatic cells.
**SOS response** The induction of enzymes for repairing damaged DNA in *Escherichia coli.* The response involves activation of an enzyme that cleaves a repressor, activating a series of genes involved in DNA repair.
**Southern blotting** Developed by Edwin Southern, a technique in which DNA fragments produced by restriction enzyme digestion are separated by electrophoresis and transferred by capillary action to a nylon or nitrocellulose membrane. Specific DNA fragments can be identified by hybridization to a complementary radioactively labeled nucleic acid probe using the technique of *autoradiography.*
**spacer DNA** DNA sequences found between genes. Usually, the genes are repetitive DNA segments.
**specialized transduction** Genetic transfer of only specific host genes by transducing phages.
**speciation** The process by which new species of plants and animals arise.
**species** A group of actually or potentially interbreeding individuals that is reproductively isolated from other such groups.
**spheroplast** See *protoplast.*
**spindle fibers** Cytoplasmic fibrils formed during cell division that are involved with the separation of chromatids at the anaphase stage of mitosis and meiosis as well as their movement toward opposite poles in the cell.
**spliceosome** The nuclear macromolecule complex within which splicing reactions occur to remove introns from pre-mRNAs.
**spontaneous mutation** A mutation that is not induced by a mutagenic agent.
**spore** Produced by some bacteria, plants, and invertebrates, a unicellular body or cell encased in a protective coat. It is capable of surviving in unfavorable environmental conditions and gives rise to a new individual upon germination. In plants, spores are the haploid products of meiosis.
***SRY*** The sex-determining region of the Y chromosome, found near the chromosome's pseudoautosomal boundary. Accumulated evidence indicates that this gene's product is the testis-determining factor (TDF).
**stabilizing selection** Preferential reproduction of those individuals having genotypes close to the mean for the population. A selective elimination of genotypes at both extremes.
**standard deviation (s)** A quantitative measure of the amount of variation in a sample of measurements from a population.
**standard error** A quantitative measure of the amount of variation in a sample of measurements from a population.
**strain** A group with common ancestry that has physiological or morphological characteristics of interest for genetic study or domestication.
**STR sequences** Short tandem repeats 2–9 base pairs long that are found within minisatellites. These sequences are used to prepare DNA profiles in forensics, paternity identification, and other applications.
**structural gene** A gene that encodes the amino acid sequence of a polypeptide chain.
**sublethal gene** A mutation causing lowered viability, with death before maturity in less than 50 percent of the individuals carrying the gene.

**submetacentric chromosome** A chromosome with the centromere placed so that one arm of the chromosome is slightly longer than the other.
**subspecies** A morphologically or geographically distinct interbreeding population of a species.
**sum law** The law that holds that the probability of one of two mutually exclusive events occurring is the sum of their individual probabilities.
**supercoiled DNA** A DNA configuration in which the helix is coiled upon itself. Supercoils can exist in stable forms only when the ends of the DNA are not free, as in a covalently closed circular DNA molecule.
**superfemale** See *metafemale.*
**supermale** See *metamale.*
**suppressor mutation** A mutation that acts to completely or partially restore the function lost by a mutation at another site.
**Svedberg coefficient unit (*S*)** A unit of measure for the rate at which particles (molecules) sediment in a centrifugal field. This rate is a function of several physicochemical properties, including size and shape. A rate of $1 \times 10^{-13}$ sec is defined as one Svedberg coefficient unit.
**symbiont** An organism coexisting in a mutually beneficial relationship with another organism.
**sympatric speciation** Speciation occurring in populations that inhabit, at least in part, the same geographic range.
**synapsis** The pairing of homologous chromosomes at meiosis.
**synaptonemal complex (SC)** An organelle consisting of a tripartite nucleoprotein ribbon that forms between the paired homologous chromosomes in the pachytene stage of the first meiotic division.
**syndrome** A group of characteristics or symptoms associated with a disease or abnormality. An affected individual may express a number of these characteristics but not necessarily all of them.
**synkaryon** The fusion of two gametic or somatic nuclei. Also, in somatic cell genetics, the product of nuclear fusion.
**syntenic test** In somatic cell genetics, a method for determining whether two genes are on the same chromosome.
**systems biology** A field that identifies and analyzes gene and protein networks to gain an understanding of intracellular regulation of metabolism, intra- and intercellular communication, and complex interactions within, between, and among cells.
**TATA box** See *Goldberg–Hogness box.*
**tautomeric shift** A reversible isomerization in a molecule, brought about by a shift in the location of a hydrogen atom. In nucleic acids, tautomeric shifts in the bases of nucleotides can cause changes in other bases at replication and are a source of mutations.
**TDF (testis-determining factor)** The product of the *SRY* gene on the Y chromosome; it controls the developmental switch point for the development of the indifferent gonad into a testis.
**telocentric chromosome** A chromosome in which the centromere is located at its very end.
**telomerase** The enzyme that adds short, tandemly repeated DNA sequences to the ends of eukaryotic chromosomes.
**telomere** The heterochromatic terminal region of a chromosome.
**telophase** The stage of cell division in which the daughter chromosomes have reached the opposite poles of the cell and reverse the stages characteristic of prophase, re-forming the nuclear envelopes and uncoiling the chromosomes. Telophase ends with the cytokinesis, which divides the cytoplasm and splits cell into two.
**telophase I** The stage in the first meiotic division when duplicated chromosomes reach the poles of the dividing cell.
**temperate phage** A bacteriophage that can become a prophage, integrating into the chromosome of the host bacterial cell and making the latter lysogenic.
**temperature-sensitive mutation** A conditional mutation that produces a mutant phenotype at one temperature range and a wild-type phenotype at another.
**template** The single-stranded DNA or RNA molecule that specifies the complementary nucleotide sequence of a strand synthesized by DNA or RNA polymerase.
**terminalization** The movement of chiasmata toward the ends of chromosomes during the diplotene stage of the first meiotic division.
**tertiary protein structure** The three-dimensional conformation of a polypeptide chain in space, brought about by the polypeptide's *primary structure*. The tertiary structure achieves a state of maximum thermodynamic stability.
**testcross** A cross between an individual whose genotype at one or more loci may be unknown and an individual who is homozygous recessive for the gene or genes in question.
**tetrad** The four chromatids that make up paired homologs in the prophase of the first meiotic division. In eukaryotes with a predominant haploid stage (some algae and fungi), tetrad denotes the four haploid cells produced by a single meiotic division.
**tetrad analysis** A method that analyzes gene linkage and recombination in organisms with a predominant haploid phase in their life cycle. See *tetrad.*
**tetranucleotide hypothesis** An early theory of DNA structure proposing that the molecule was composed of repeating units, each consisting of the four nucleotides represented by adenine, thymine, cytosine, and guanine.
**theta ($\theta$) structure** An intermediate configuration in the bidirectional replication of circular DNA molecules. At about midway through the cycle of replication, the intermediate resembles the Greek letter theta.
**thymine dimer** In a polynucleotide strand, a lesion consisting of two adjacent thymine bases that become joined by a covalent bond. Usually caused by exposure to ultraviolet light, this lesion inhibits DNA replication.
$T_m$ See *melting profile.*
**topoisomerase** A class of enzymes that convert DNA from one topological form to another. During replication, a topoisomerase, *DNA gyrase*, facilitates DNA replication by reducing molecular tension caused by supercoiling upstream from the *replication fork.*
**totipotent** The capacity of a cell or embryo part to differentiate into all types characteristic of an adult. This capacity is usually progressively restricted during development. Used interchangeably with *pluripotent.*
**trait** Any detectable phenotypic variation of a particular inherited character.
***trans*-acting element** A gene product (usually a diffusible protein or an RNA molecule) that acts to regulate the expression of a target gene.
*trans* **configuration** An arrangement in which two mutant sites are on opposite homologs, such as

$$\frac{a^1 \quad +}{+ \quad a^2}$$

in contrast to a *cis arrangement*, in which the sites are located on the same homolog.
**transcription** Transfer of genetic information from DNA by the synthesis of a complementary RNA molecule using a DNA template.
**transcriptome** The set of mRNA molecules present in a cell at any given time.
**transdetermination** Change in developmental fate of a cell or group of cells.
**transduction** Virally mediated bacterial recombination. Also used to describe the transfer of eukaryotic genes mediated by a retrovirus.
**transfer RNA (tRNA)** A small ribonucleic acid molecule playing an essential role in *translation*. tRNAs contain: (1) a three-base segment (anticodon) that recognizes a codon in mRNA; (2) a binding site for the specific amino acid corresponding to the anticodon; and (3) recognition sites for interaction with ribosomes and with the enzyme that links the tRNA to its specific amino acid.

**transformation** Heritable change in a cell or an organism brought about by exogenous DNA. Known to occur naturally and also used in *recombinant DNA* studies.
**transgenic organism** An organism whose genome has been modified by the introduction of external DNA sequences into the germ line.
**transition** A mutational event in which one purine is replaced by another or one pyrimidine is replaced by another.
**translation** The derivation of the amino acid sequence of a polypeptide from the base sequence of an mRNA molecule in association with a ribosome and dependent on *tRNAs*.
**translocation** A chromosomal mutation associated with the reciprocal or nonreciprocal transfer of a chromosomal segment from one chromosome to another. Also denotes the movement of mRNA through the ribosome during translation.
**transmission genetics** The field of genetics concerned with heredity and the mechanisms by which genes are transferred from parent to offspring.
**transposable element** A DNA segment that translocates to other sites in the genome, essentially independent of sequence homology. Usually, such elements are flanked at each end by short inverted repeats of 20–40 base pairs. Insertion into a structural gene can produce a mutant phenotype. Insertion and excision of transposable elements depend on two enzymes, transposase and resolvase. Such elements have been identified in both prokaryotes and eukaryotes.
**transversion** A mutational event in which a purine is replaced by a pyrimidine or a pyrimidine is replaced by a purine.
**trinucleotide repeat** A tandemly repeated cluster of three nucleotides (such as CTG) within or near a gene. Certain diseases (myotonic dystrophy, Huntington disease) are caused by expansion in copy number of such repeats.
**triploidy** The condition in which a cell or organism possesses three haploid sets of chromosomes
**trisomy** The condition in which a cell or organism possesses two copies of each chromosome except for one, which is present in three copies (designated $2n + 1$).
**tRNA** See *transfer RNA*.
**tumor-suppressor gene** A gene that encodes a product that normally functions to suppress cell division. Mutations in tumor-suppressor genes result in the activation of cell division and tumor formation.
**Turner syndrome** A genetic condition in human females caused by a 45,X genotype. Such individuals are phenotypically female but are sterile because of undeveloped ovaries.
**unequal crossing over** A crossover between two improperly aligned homologs, producing one homolog with three copies of a region and the other with one copy of that region.
**unique DNA** DNA sequences that are present only once per genome.
**universal code** A genetic code used by all life forms. Some exceptions are found in mitochondria, ciliates, and mycoplasmas.
**unwinding proteins** Nuclear proteins that act during DNA replication to destabilize and unwind the DNA helix ahead of the replicating fork.
**variable number tandem repeats (VNTRs)** Short, repeated DNA sequences (of 2–20 nucleotides) present as tandem repeats between two restriction enzyme sites. Variation in the number of repeats creates DNA fragments of differing lengths following restriction enzyme digestion. Used in early versions of *DNA fingerprinting*.
**variable region** Portion of an immunoglobulin molecule that exhibits many amino acid sequence differences between antibodies of differing specificities.
**variance ($s^2$)** A statistical measure of the variation of values from a central value, calculated as the square of the standard deviation.
**variegation** Patches of differing phenotypes, such as color, in a tissue.
**vector** In recombinant DNA, an agent such as a phage or plasmid into which a foreign DNA segment will be inserted and utilized to transform host cells.
**viability** The ability of individuals in a given phenotypic class to survive as compared to another class (usually wild type).
**virulent phage** A bacteriophage that infects and lyses the host bacterial cell.
**VNTRs** See *variable number tandem repeats*.
**western blot** An analytical technique in which proteins are separated by gel electrophoresis and transferred by capillary action to a nylon membrane or nitrocellulose sheet. A specific protein can be identified through hybridization to a labeled antibody.
**wild type** The most commonly observed phenotype or genotype, designated as the norm or standard.
**wobble hypothesis** An idea proposed by Francis Crick, stating that the third base in an anticodon can align in several ways to allow it to recognize more than one base in the codons of mRNA.
**W, Z chromosomes** The sex chromosomes in species where the female is the heterogametic sex (WZ).
**X chromosome** The sex chromosome present in species where females are the homogametic sex (XX).
**X inactivation** In mammalian females, the random cessation of transcriptional activity of either the maternally or paternally derived X chromosome. This event, which occurs early in development, is a mechanism of dosage compensation, and all progeny cells inactivate the same X chromosome. See also *Barr body*, *Lyon hypothesis*, *XIST*.
***XIST*** A locus in the X-chromosome inactivation center that controls inactivation of the X chromosome in mammalian females.
**X-linkage** The pattern of inheritance resulting from genes located on the X chromosome.
**X-ray crystallography** A technique for determining the three-dimensional structure of molecules by analyzing X-ray diffraction patterns produced by crystals of the molecule under study.
**YAC** A cloning vector in the form of a yeast artificial chromosome, constructed using chromosomal elements from yeast, including telomeres (from a ciliate), centromeres, origin of replication, and marker genes. YACs are used to clone long stretches of eukaryotic DNA.
**Y chromosome** The sex chromosome in species where the male is heterogametic (XY).
**Y-linkage** Mode of inheritance shown by genes located on the Y chromosome.
**Z-DNA** An alternative "zig-zag" structure of DNA in which the two antiparallel polynucleotide chains form a left-handed double helix. Implicated in regulation of gene expression.
**zein** Principal storage protein of maize endosperm, consisting of two major proteins with molecular weights of 19,000 and 21,000 Da.
**zinc finger** A class of DNA-binding domains seen in proteins. They have a characteristic pattern of cysteine and histidine residues that complex with zinc ions, throwing intermediate amino acid residues into a series of loops or fingers.
**zygote** The diploid cell produced by the fusion of haploid gametic nuclei.
**zygotene** A stage of meiotic prophase I in which the homologous chromosomes synapse and pair along their entire length, forming bivalents. The synaptonemal complex forms at this stage.

# Appendix B

## ANSWERS TO SELECTED PROBLEMS

### Chapter 1

2. Based on the parallels between Mendel's model of heredity and the behavior of chromosomes, the chromosome theory of inheritance emerged. It states that inherited traits are controlled by genes residing on chromosomes that are transmitted by gametes.
4. A gene variant is called an allele. There can be many such variants in a population, but for a diploid organism, only two such alleles can exist in any given individual.
6. *Genes*, linear sequences of nucleotides, usually exert their influence by producing proteins through the process of transcription and translation. Genes are the functional units of heredity. They associate, sometimes with proteins, to form *chromosomes*.
8. The central dogma of molecular genetics refers to the relationships among DNA, RNA, and protein. The processes of *transcription* and *translation* are integral to understanding these relationships.
10. Restriction enzymes (endonucleases) cut double-stranded DNA at particular base sequences. When a vector is cleaved with the same enzyme, complementary ends are created such that ends, regardless of their origin, can be combined and ligated to form intact double-stranded structures. Such recombinant forms are often useful for industrial, research, and/or pharmaceutical efforts.
12. Supporters of organismic patenting argue that it is needed to encourage innovation and allow the costs of discovery to be recovered. Capital investors assume that there is a likely chance that their investments will yield positive returns. Others argue that natural substances should not be privately owned and that once owned by a small number of companies, free enterprise will be stifled. Individuals and companies needing vital, but patented, products may have limited access.
14. Model organisms are not only useful, but necessary, for understanding genes that influence human diseases. Given that genetic/molecular systems are highly conserved across broad phylogenetic lines, what is learned in one organism is usually applied to all organisms. Most model organisms have peculiarities, such as ease of growth, genetic understanding, or abundant offspring, that make them straightforward and especially informative in genetic studies.
16. Safeguards should probably include tests for allergenicity, environmental impact, and likelihood of cross-pollination. In addition, concern would increase if such a crop contained antibiotic-resistant genetic markers and genes conferring toxicity to pests. While not required in the United States, in the interest of the consumer, one might consider labeling such products as genetically modified. On a broader scale, one might reduce vulnerability by using multiple suppliers (if available) and help minimize the domination of the world food supply by a few companies.

### Chapter 2

2. Chromosomes that are homologous share many properties including:

   Overall length
   Position of the centromere
   Banding patterns
   Type and location of genes
   Autoradiographic pattern

   *Diploidy* is a term often used in conjunction with the symbol $2n$. It means that both members of a homologous pair of chromosomes are present. The change from a diploid ($2n$) to haploid ($n$) occurs during *reduction division* when tetrads become dyads during meiosis I.
6. Centromere placement may be medial (metacentric), off center (submetacentric), near one end (acrocentric), or at one end (telocentric). Notice the different anaphase shapes of chromosomes as they move to the poles: metacentric (a), submetacentric (b), acrocentric (c), telocentric (d).

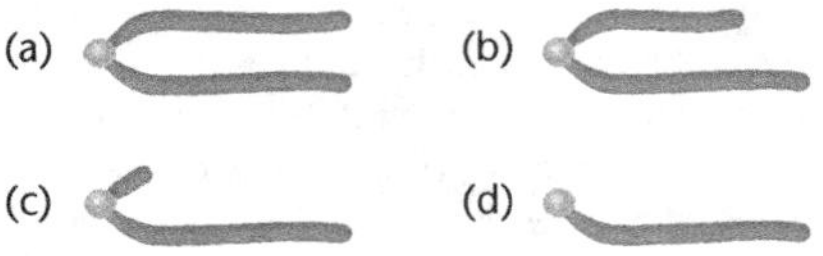


8. Carefully read the section on mitosis and cell division in the text. Major divisions of the cell cycle include Interphase and Mitosis. Interphase is composed of four phases: G1, G0, S, and G2. During the S phase, chromosomal DNA doubles. Karyokinesis involves nuclear division while cytokinesis involves division of the cytoplasm.
10. Compared with mitosis which maintains a chromosomal constancy, meiosis provides for a reduction in chromosome number and an opportunity for exchange of genetic material between homologous chromosomes. In mitosis there is no change in chromosome number or kind in the two daughter cells, whereas in meiosis numerous potentially different haploid ($n$) cells are produced.
12. Sister chromatids are genetically identical, except where mutations may have occurred during DNA replication. Nonsister chromatids are genetically similar if on homologous chromosomes or genetically dissimilar if on nonhomologous chromosomes. If crossing over occurs, then chromatids attached to the same centromere will no longer be identical.
14. (a) If there are 16 chromosomes, there should be 8 tetrads.
    (b) Also note that, after meiosis I and in the second meiotic prophase, there are as many dyads as there are pairs of chromosomes. There will be 8 dyads.
    (c) Because the monads migrate to opposite poles during meiosis II (from the separation of dyads), there should be 8 monads migrating to *each* pole.
16. Through independent assortment of chromosomes at anaphase I of meiosis, daughter cells (secondary spermatocytes and secondary oocytes) may contain different sets of maternally and paternally derived chromosomes. Crossing over, which happens at a much higher frequency in meiotic cells as compared to mitotic cells, allows maternally and paternally derived chromosomes to exchange segments, thereby increasing the likelihood that daughter cells are variable.
18. If there are eight combinations possible for part (c) in the previous problem, there will be 16 combinations with the addition of another chromosome pair.
20. One-half of each tetrad will have a maternal homolog: $(1/2)^{10}$.
22. In angiosperms, meiosis results in the formation of microspores (male) and megaspores (female), which give rise to the haploid male

and female gametophyte stage. Micro- and megagametophytes produce the pollen and the ovules, respectively. Following fertilization, the sporophyte is formed.

**24.** The folded-fiber model is based on each chromatid consisting of a single fiber wound like a skein of yarn. Each fiber consists of DNA and protein. A coiling process occurs during the transition of interphase chromatin to more condensed chromosomes during prophase of mitosis or meiosis. Such condensation leads to a 5000-fold contraction in the length of the DNA within each chromatid.

**26.** (a) The lengths and centromere placements of nearly all such chromosomes can be matched into pairs. (b) DNA synthesis can be detected by the incorporation of labeled precursors into DNA, and DNA content in a G2 nucleus is twice that of a G1 nucleus. (c) If the fibers comprising the mitotic chromosomes are loosened, they reveal fibers like those of interphase chromatin. Electron microscopic observations indicate that mitotic chromosomes are in varying states of extensively folded structures derived from chromatin. Various labeling procedures would also be effective.

**28.** Side-by-side alignment of $A^m$, $A^p$, $B^m$, $B^p$, $C^m$, and $C^p$ will occur in various arrangements at metaphase I. Eight possible combinations of products will occur at the completion of anaphase: $A^m$, $B^p$, $C^m$, for example (each with sister chromatids). In other words, after meiosis I, the two product cells would be as follows: $A^m$ or $A^p$, $B^m$ or $B^p$, $C^m$ or $C^p$.

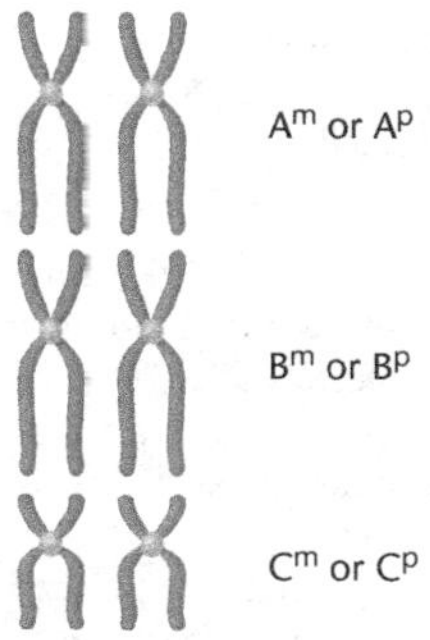


**30.** Eight (2 × 2 × 2) combinations are possible at metaphase I. During anaphase II, the centromeres will split, and chromatids will go to opposite poles.

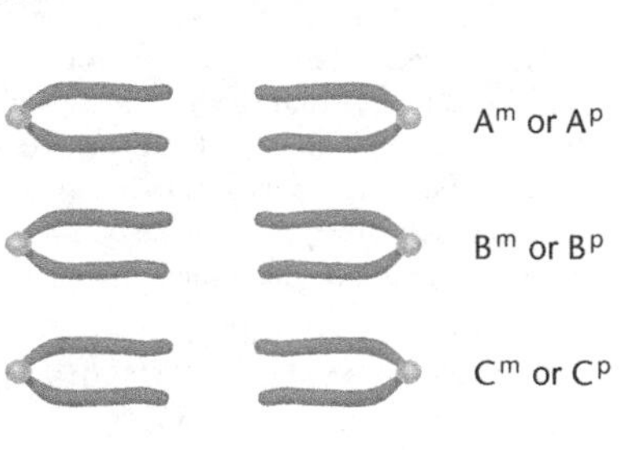


**32.** Taking this question exactly as it is described—nondisjunction of the *C* chromosome at meiosis I and dyad separation at meiosis II—you will end up, after fertilization, with the following combinations under the conditions described in Problem 31.

zygote 1: two copies of chromosome *A*
two copies of chromosome *B*
three copies of chromosome *C*

zygote 2: two copies of chromosome *A*
two copies of chromosome *B*
one copy of chromosome *C*

**34.** (a) If there were two dyads of chromosome 21 in the first polar body, the secondary oocyte would completely lack chromosome 21. The resulting zygote would have one copy of chromosome 21 (from the father) and two copies of all the other chromosomes. (b) If the polar body lacked chromosome 21, the secondary oocyte would have two dyads and the resulting zygote would have three number 21 chromsomes, two coming from the mother and one coming from the father. (c) The secondary oocyte would have a dyad and a monad from chromosome 21. Depending on how the monad partitioned at meiosis II, you would have either a normal chromosome 21 complement (the zygote that did not receive the monad) or a chromosome 21 trisomy in which the zygote received two number 21 chromosomes from the mother and one from the father.

## Chapter 3

**2.** Since albinism is inherited as a recessive trait, genotypes *AA* and *Aa* should produce the normal phenotype, while *aa* will give albinism.
(a) Both parents must be heterozygous (*Aa*).
(b) The female must be *aa*. Since all the children are normal, one would consider the male to be *AA* instead of *Aa*. However, the male could be *Aa*. Under that circumstance, the likelihood of having six children, all normal, is 1/64.
(c) The female must be *aa*. The fact that half of the children are normal and half are albino indicates a typical testcross in which the *Aa* male is mated to the *aa* female.
(d)

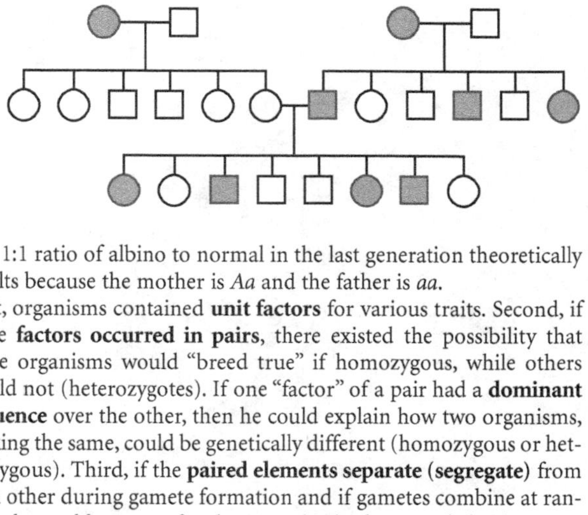

The 1:1 ratio of albino to normal in the last generation theoretically results because the mother is *Aa* and the father is *aa*.

**4.** First, organisms contained **unit factors** for various traits. Second, if these **factors occurred in pairs**, there existed the possibility that some organisms would "breed true" if homozygous, while others would not (heterozygotes). If one "factor" of a pair had a **dominant influence** over the other, then he could explain how two organisms, looking the same, could be genetically different (homozygous or heterozygous). Third, if the **paired elements separate (segregate)** from each other during gamete formation and if gametes combine at random, he could account for the 3:1 ratios in the monohybrid crosses. The fourth postulate, independent assortment, cannot be demonstrated by a monohybrid cross because two gene pairs must be involved to do so.

**6.** $P$ = checkered; $p$ = plain. Checkered is tentatively assigned the dominant function because in a casual examination of the data, especially cross (b), we see that checkered types are more likely to be produced than plain types.

Cross (a): *PP* × *PP* or *PP* × *Pp*
Cross (b): *PP* × *pp*
Cross (c): Because all the offspring from this cross are plain, there is no doubt that the genotype of both parents is *pp*.

Genotypes of all individuals:

| | $F_1$ Progeny | |
|---|---|---|
| $P_1$ Cross | Checkered | Plain |
| (a) *PP* × *PP* | *PP* | |
| (b) *PP* × *pp* | *Pp* | |
| (c) *pp* × *pp* | | *pp* |
| (d) *PP* × *pp* | *Pp* | |
| (e) *Pp* × *pp* | *Pp* | *pp* |
| (f) *Pp* × *Pp* | *PP*, *Pp* | *pp* |
| (g) *PP* × *Pp* | *PP*, *Pp* | |

**8.** *WWgg* = 1/16

**10.** In Problem 9, (d) fits this description.

**12.** Mendel's four postulates are related to the diagram below.

1. Factors occur in pairs. Notice *A* and *a*.
2. Some genes have dominant and recessive alleles. Notice *A* and *a*.
3. Alleles segregate from each other during gamete formation. When homologous chromosomes separate from each other at anaphase I, alleles will go to opposite poles of the meiotic apparatus.
4. One gene pair separates independently from other gene pairs. Different gene pairs on the same homologous pair of chromosomes (if far apart) or on nonhomologous chromosomes will separate independently from each other during meiosis.

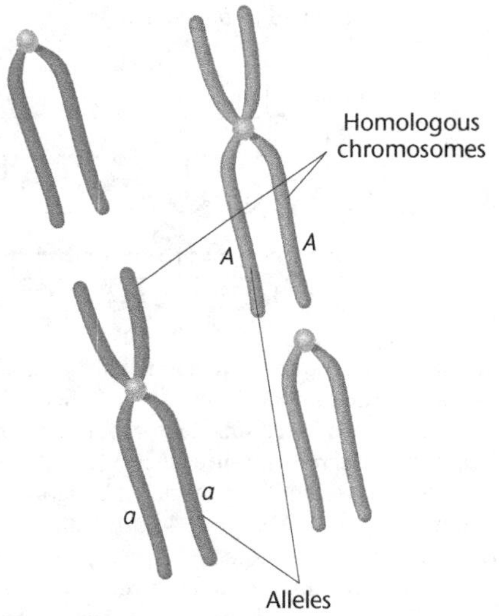


**14.** Homozygosity refers to a condition where both genes of a pair are the same (i.e., *AA* or *GG* or *hh*), whereas heterozygosity refers to the condition where members of a gene pair are different (i.e., *Aa* or *Gg* or *Bb*).

**16.** The general formula for determining the number of kinds of gametes produced by an organism is $2^n$ where n = number of *heterozygous* gene pairs.

(a) 4: *AB, Ab, aB, ab*
(b) 2: *AB, aB*
(c) 8: *ABC, ABc, AbC, Abc, aBC, aBc, abC, abc*
(d) 2: *ABc, aBc*
(e) 4: *ABc, Abc, aBc, abc*
(f) $2^5 = 32$

**18.** *G* = yellow seeds, *g* = green seeds.

| Phenotypes | Genotypes |
|---|---|
| $P_1$: Yellow × green | *GG* × *gg* |
| $F_1$: all yellow | *Gg* |
| $F_2$: 6022 yellow | 1/4 *GG*; 2/4 *Gg* |
| 2001 green | 1/4 *gg* |

*Of the yellow* $F_2$ *offspring*, notice that 1/3 of them are *GG* and 2/3 are *Gg*. If you selfed the 1/3 *GG* types, then all the offspring (the 166) would breed true, whereas the others (353 which are *Gg*) should produce offspring in a 3:1 ratio when selfed.

*GG* × *GG* = all *GG*

*Gg* × *Gg* = 1/4 *GG*; 2/4 *Gg*; 1/4 *gg*

**20.** Symbols:

| Seed shape | Seed color |
|---|---|
| *W* = round | *G* = yellow |
| *w* = wrinkled | *g* = green |

$P_1$: *WWgg* × *wwGG*

$F_1$: *WwGg* cross to *wwgg* (which is a typical testcross)

The offspring will occur in a typical 1:1:1:1 as

1/4 *WwGg* (round, yellow)
1/4 *Wwgg* (round, green)
1/4 *wwGg* (wrinkled, yellow)
1/4 *wwgg* (wrinkled, green)

**22.** (a) $\chi^2 = .064$ probability ($p$) value between 0.9 and 0.5. We would therefore say that there is a "good fit" between the observed and expected values.

(b) $\chi^2 = 0.39$ The $p$ value in the table for 1 degree of freedom is still between 0.9 and 0.5. The deviation in each case can be attributed to chance.

**24.** For the test of a 3:1 ratio, the $\chi^2$ value is 33.3 with an associated $p$ value of less than 0.01 for 1 degree of freedom. For the test of a 1:1 ratio, the $\chi^2$ value is 25.0 again with an associated $p$ value of less than 0.01 for 1 degree of freedom. Based on these probability values, both null hypotheses should be rejected.

**26.** If a gene is dominant, it will not skip generations, nor will it be passed to offspring unless the parents have the gene. On the other hand, genes that are recessive can skip generations and exist in a carrier state in parents. Notice that II-4 and II-5 produce a female child (III-4) with the affected phenotype. On these criteria alone, the gene must be viewed as being recessive. *Note*: If a gene is recessive and X-linked (to be discussed later), the pattern will often be from affected male to carrier female to affected male.

I-1 (*Aa*), I-2 (*aa*), I-3 (*Aa*), I-4(*Aa*)
II-1 (*aa*), II-2 (*Aa*), II-3 (*aa*), II-4 (*Aa*), II-5 (*Aa*),
II-6 (*aa*), II-7 (*AA* or *Aa*), II-8 (*AA* or *Aa*)
III-1 (*AA* or *Aa*), III-2 (*AA* or *Aa*), III-3 (*AA* or *Aa*),
III-4 (*aa*), III-5 (probably *AA*), III-6 (*aa*)
IV-1 through IV-7 all *Aa*.

**28.** 1/8

**30.** 3/4

**32.** 4/9

**34.** 1/9

**36.** $P = [8!\,(3/4)^6(1/4)^2]/6!2!$

**38.** (a) First consider that each parent is homozygous (true-breeding in the question) and since in the $F_1$ only round, axial, violet, and full phenotypes were expressed, each must be dominant.
(b) Round, axial, violet, and full would be the most frequent phenotypes:

$$3/4 \times 3/4 \times 3/4 \times 3/4$$

(c) Wrinkled, terminal, white, and constricted would be the least frequent phenotypes:

$$1/4 \times 1/4 \times 1/4 \times 1/4$$

(d) $3/4 \times 1/4 \times 3/4 \times 1/4$
plus $1/4 \times 3/4 \times 1/4 \times 3/4 = 18/256$
(e) There would be 16 different phenotypes in the testcross offspring just as there are 16 different phenotypes in the $F_2$ generation.

**40.** (a) Notice in the first cross that a 3:1 ratio exists for the spiny to smooth phenotypes. Thus, we would predict that the *spiny* allele is dominant to *smooth*. We would also predict that the *purple* allele is dominant to *white*. (b) One could cross a homozygous purple, spiny plant to a homozygous white, smooth plant. The purple, spiny $F_1$ would support the hypothesis that *purple* is dominant to *white* and *spiny* is dominant to *smooth*. In the $F_2$, a 9:3:3:1 ratio would not only support the above hypothesis, but it would also indicate the independent inheritance and expression of the two traits.

**42.** $P = [5!\,(3/4)^3(1/4)^2]/3!2!$
$= 135/512$

**44.** (a) For Set I the $\chi^2 = 2.15$, with $p$ being between 0.2 and 0.05, so one would accept the null hypothesis of no significant difference between the expected and observed values.
For Set II, the $\chi^2 = 21.43$ and $p < 0.001$. One would reject the null hypothesis and assume a significant difference between the observed and expected values.
(b) In most cases, more confidence is gained as the sample size increases; however, depending on the organism or experiment, there are practical limits on sample size.

**46.** The probability that their first child will be a male with dentinogenesis imperfecta would be 1/2 (passage of the allele) × 1/2 (probability of child being male) = 1/4. The probability that three of their six children will have the disease would be 5/16 or about .31 (31%).

## Chapter 4

**2.** *Incomplete dominance* can be viewed more as a quantitative phenomenon where the heterozygote is intermediate (approximately) between the limits set by the homozygotes. *Codominance* can be viewed in a more qualitative manner where both of the alleles in the heterozygote are expressed.

**4.** Cross 1:

short tail × normal long tail ⇒

approximately 1/2 short, 1/2 long

This tells you that one type is heterozygous, the other homozygous.

Cross 2:

short tail × short tail ⇒

6 short tail, 3 long tail (2/3 short, 1/3 long)

At this point one would consider that the 2/3 *short* are heterozygotes, and *long* is the homozygous class. Also, short is dominant to long. Since these ratios were repeated and verified, one can conclude that a 2:1 ratio is not a statistical artifact and that the following genotypic model would hold. Because long is the "normal" does not mean that it is dominant.
Symbolism:

$S$ = short, $s$ = long

Cross 1:

$Ss \times ss$ ⇒

1/2 $Ss$ (short), 1/2 $ss$ (long)

Cross 2:

$Ss \times Ss$ ⇒

1/4 $SS$ (lethal), 2/4 $Ss$ (short), 1/4 $ss$ (long)

**6.**

$$I^AI^O \times I^BI^O$$

| | $I^B$ | $I^O$ |
|---|---|---|
| $I^A$ | $I^AI^B$(AB) | $I^AI^O$(A) |
| $I^O$ | $I^BI^O$(B) | $I^OI^O$(O) |

The ratio would be

1(A):1(B):1(AB):1(O)

**8.** Symbolism:

$Se$ = secretor  $se$ = nonsecretor

$I^AI^B\ Sese \times I^OI^O\ Sese$

(a)

| | $I^O\ Se$ | $I^O\ se$ |
|---|---|---|
| $I^A\ Se$ | A, secretor | A, secretor |
| $I^A\ se$ | A secretor | A, nonsecretor |
| $I^B\ Se$ | B, secretor | B, secretor |
| $I^B\ se$ | B, secretor | B, nonsecretor |

Overall ratio: 3/8 A, secretor
1/8 A, nonsecretor
3/8 B, secretor
1/8 B, nonsecretor

(b) 1/4 of all individuals will have blood type O.

**10.** (a) Phenotypes:

Himalayan × Himalayan ⇒ albino
Genotypes: $c^hc^a$  $c^hc^a$  $c^ac^a$

The Himalayan parents must both be heterozygous to produce an albino offspring.
Phenotypes:

full color × albino ⇒ chinchilla
Genotypes: $Cc^{ch}$  $c^ac^a$  $c^{ch}c^a$

Therefore, the cross of albino with chinchilla would be as follows:

$c^ac^a \times c^{ch}c^a$ ⇒

1/2 chinchilla; 1/2 albino

(b) Phenotypes:

albino × chinchilla $\Rightarrow$ albino

Genotypes: $c^ac^a$ $c^{ch}c^a$ $c^ac^a$

Phenotypes:

full color × albino $\Rightarrow$ full color

Genotypes: $C\_$ $c^ac^a$ $Cc^a$

It is impossible to determine the complete genotype of the full color parent, but the full color offspring must be as indicated, $Cc^a$.
Therefore, the cross of the albino with full color would be as follows:

$c^ac^a \times Cc^a \Rightarrow$

1/2 full color ; 1/2 albino

(c) Phenotypes:

chinchilla × albino $\Rightarrow$ Himalayan

Genotypes: $c^{ch}c^h$ $c^ac^a$ $c^hc^a$

The chinchilla parent must be heterozygous for Himalayan because of the Himalayan offspring.
Phenotypes:

full color × albino $\Rightarrow$ Himalayan

Genotypes: $Cc^h$ $c^ac^a$ $c^hc^a$

Therefore, a cross between the two Himalayan types would produce the following offspring:

$c^hc^a \times c^hc^a$

$\Downarrow$

3/4 Himalayan; 1/4 albino

**12.** 18/64

**14.** (a) $C^{ch}C^{ch}$ = chestnut

$C^cC^c$ = cremello

$C^{ch}C^c$ = palomino

(b) The $F_1$ resulting from matings between cremello and chestnut horses would be expected to be all palomino. The $F_2$ would be expected to fall in a 1:2:1 ratio as in the third cross in part (a) above.

**16.** (a) In a cross of

$AACC \times aacc$

the offspring are all *AaCc* (agouti) because the *C* allele allows pigment to be deposited in the hair, and when it is it will be agouti. $F_2$ offspring would have the following "simplified" genotypes with the corresponding phenotypes:

$A\_C\_$ = 9/16 (agouti)

$A\_cc$ = 3/16

(colorless because *cc* is epistatic to *A*)

$aaC\_$ = 3/16 (black)

$aacc$ = 1/16

(colorless because *cc* is epistatic to *aa*)

The two colorless classes are phenotypically indistinguishable; therefore the final ratio is 9:3:4.

(b) Results of crosses of female agouti

($A\_C\_$) × *aacc* (males)

are given in three groups:

1. To produce an even number of agouti and colorless offspring, the female parent must have been *AACc* so that half of the offspring are able to deposit pigment because of *C*, and when they do, they are all agouti (having received only *A* from the female parent).
2. To produce an even number of agouti and black offspring, the mother must have been *Aa*, and so that no colorless offspring were produced, the female must have been *CC*. Her genotype must have been *AaCC*.
3. Notice that half of the offspring are colorless: therefore the female must have been *Cc*. Half of the pigmented offspring are black and half are agouti. Therefore the female must have been *Aa*. Overall, the *AaCc* genotype seems appropriate.

**18.** (a) *AaBbCc* $\Rightarrow$ gray (*C* allows pigment)

(b) $A\_B\_Cc$ $\Rightarrow$ gray (*C* allows pigment)

(c) 16/32 albino;
9/32 gray;
3/32 yellow;
3/32 black;
1/32 cream

(d) 9/16 (gray);
3/16 (black);
4/16 (albino)

(e) 3/8 (gray);
1/8 (yellow);
4/8 (albino)

**20.** (a) Cross A:

$P_1$: *AABB* × *aaBB*

$F_1$: *AaBB*

$F_2$: 3/4 $A\_BB$: 1/4 *aaBB*

Cross B:

$P_1$: *AABB* × *AAbb*

$F_1$: *AABb*

$F_2$: 3/4 $AAB\_$: 1/4 *AAbb*

Cross C:

$P_1$: *aaBB* × *AAbb*

$F_1$: *AaBb*

$F_2$: 9/16 $A\_B\_$: 3/16 $A\_bb$:

3/16 $aaB\_$: 1/16 *aabb*

(b) The genotype of the unknown $P_1$ individual would be *AAbb* (brown), while the $F_1$ would be *AaBb* (green).

**22.** (a) Assign the phenotypes as given; then see if patterns emerge.

$A\_B\_$ = 9/16 (yellow)

$A\_bb$ = 3/16 (blue)

$aaB\_$ = 3/16 (red)

$aabb$ = 1/16 (mauve)

See that each type can exist as a full homozygote. If plants with blue flowers (homozygotes) are crossed to red-flowered homozygotes, the $F_1$ plants will have yellow flowers. If yellow-flowered plants are crossed with mauve-flowered plants, the $F_1$ plants will be yellow and the $F_2$ will occur in a 9:3:3:1 ratio. All of the observations fit the model as proposed.

(b) If one crosses a true-breeding red plant (*aaBB*) with a mauve plant (*aabb*), the $F_1$ should be red (*aaBb*). The $F_2$ would be as follows:

$aaBb \times aaBb \Rightarrow$

3/4 *aaB_* (*red*): 1/4 *aabb* (mauve)

**24.** (a) 1/4
(b) 1/2
(c) 1/4
(d) zero

**26.** Symbolism: Normal wing margins = $sd^+$; scalloped = $sd$

(a) $P_1$: $X^{sd}X^{sd} \times X^+/Y \Rightarrow$

$F_1$: 1/2 $X^+X^{sd}$ (female, normal)
1/2 $X^{sd}/Y$ (male, scalloped)

$F_2$: 1/4 $X^+X^{sd}$ (female, normal)
1/4 $X^{sd}X^{sd}$ (female, scalloped)
1/4 $X^+/Y$ (male, normal)
1/4 $X^{sd}/Y$ (male, scalloped)

(b) $P_1$: $X^+/X^+ \times X^{sd}/Y \Rightarrow$

$F_1$: 1/2 $X^+X^{sd}$ (female, normal)
1/2 $X^+/Y$ (male, normal)

$F_2$: 1/4 $X^+X^+$ (female, normal)
1/4 $X^+X^{sd}$ (female, normal)
1/4 $X^+/Y$ (male, normal)
1/4 $X^{sd}/Y$ (male, scalloped)

If the *scalloped* gene were not X-linked, then all of the $F_1$ offspring would be wild (phenotypically) and a 3:1 ratio of normal to scalloped would occur in the $F_2$.

**28.** $P_1$: $X^+X^+$; *su-v*/*su-v* × $X^v/Y$; *su-v*$^+$/*su-v*$^+$

$F_1$: 1/2 $X^+X^v$; *su-v*$^+$/*su-v* (female, normal)
1/2 $X^+/Y$; *su-v*$^+$/*su-v* (male, normal)

$F_2$: 2/4 females, $X^+/\_$ → 3/4 *su-v*$^+$/_ ; 1/4 *su-v*/*su-v*
1/4 males, $X^+/Y$ → 3/4 *su-v*$^+$/_ ; 1/4 *su-v*/*su-v*
1/4 males, $X^v/Y$ → 3/4 *su-v*$^+$/_ ; 1/4 *su-v*/*su-v*

8/16 wild-type females; (none of the females are homozygous for the *vermilion* gene)

5/16 wild-type males; (4/16 because they have no *vermilion* gene and 1/16 because the X-linked, hemizygous *vermilion* gene is suppressed by *su-v*/*su-v*)

3/16 vermilion males; (no suppression of the *vermilion* gene)

**30.** (a) *w*/*w*; *se*$^+$/*se*$^+$ × *w*$^+$/Y; *se*/*se*

$F_1$: ⇓

*w*$^+$/*w*; *se*$^+$/*se* = wild females
*w*/Y; *se*$^+$/*se* = white-eyed males

$F_2$: 3/16 males wild
4/16 males white
1/16 males sepia
3/16 females wild
4/16 females white
1/16 females sepia

(b) *w*$^+$/*w*$^+$; *se*/*se* × *w*/Y; *se*$^+$/*se*$^+$

⇓

$F_1$: *w*$^+$/*w*; *se*$^+$/*se* = wild females
*w*$^+$/Y; *se*$^+$/*se* = wild males

$F_2$: 3/16 males wild
4/16 males white
1/16 males sepia
6/16 females wild
2/16 females sepia

**32.** Let *a* represent the mutant gene and *A* represent its normal allele.
(a) This pedigree is consistent with an X-linked recessive trait because the male would contribute an X chromosome carrying the *a* mutation to the *aa* daughter. The mother would have to be heterozygous *Aa*.
(b) This pedigree is consistent with an X-linked recessive trait because the mother could be *Aa* and transmit her *a* allele to her one son (*a*/Y) and her *A* allele to her other son.
(c) This pedigree is not consistent with an X-linked mode of inheritance because the *aa* mother has an *A*/Y son.

**34.** $F_1$: all hen-feathering
$F_2$: All of the offspring would be hen-feathered except for 1/8 males, which are cock-feathered.

**36.** Phenotypic expression is dependent on the genome of the organism, the immediate molecular and cellular environment of the genome, and numerous interactions between a genome, the organism, and the environment.

**38.** *Anticipation* occurs when a heritable disorder exhibits a progressively earlier age of onset and an increased severity in successive generations. *Imprinting* occurs when phenotypic expression is influenced by the parental origin of the chromosome carrying a particular gene.

**40.** Symbolism:

*A_B_* = black
*A_bb* = golden
*aabb* = golden
*aaB_* = brown

The combination of *bb* is epistatic to the *A* locus.
(a) *AAB_* × *aaBB* (other configurations are possible, but each must give all offspring with *A* and *B* dominant alleles)
(b) *AaB_* × *aaBB* (other configurations are possible, but no types can be produced)
(c) *AABb* × *aaBb*
(d) *AABB* × *aabb*
(e) *AaBb* × *Aabb*
(f) *AaBb* × *aabb*
(g) *aaBb* × *aaBb*
(h) *AaBb* × *AaBb*

Those genotypes that will breed true will be as follows:

black = *AABB*

golden = all genotypes that are *bb*

brown = *aaBB*

**42.** The homozygous dominant type is lethal. Polled is caused by an independently assorting dominant allele, while horned is caused by the recessive allele to polled.

**44.** Because of the relatively high frequency of occurrence of precocious puberty in the pedigree, one might consider a dominant gene to be involved. Indeed, there is no skipping of generations typical of recessive traits. Notice, however, that there is an apparent skipping of generations in giving rise to the IV-5 son. This is because females are not capable of expressing the gene. Given the degree of outcrossing, that the gene is probably quite rare and therefore heterozygotes are uncommon, and that the frequency of transmission is high, it is likely that this form of male precious puberty is caused by an autosomal dominant, sex-limited gene.

**46.** The reduced ratio is 12 white, 3 orange, and 1 brown, and in a dihybrid cross ($AaBb \times AaBb$) the following would occur:

| | |
|---|---|
| 12 white | *A_B_* or *aaB_* |
| 3 orange | *A_bb* |
| 1 brown | *aabb* |

**48.** Since proto-oncogenes stimulate a cell to progress through a cell cycle, loss of function of such genes should inhibit such cellular progress. In this case, the gene would likely function as a recessive. However, if overproduction of a proto-oncogene should occur, then the cell would be stimulated to undergo more rapid cycles (perhaps), which would lead to an expressed phenotype such as a tumor or cancer. Under this condition (gain of function), the gene would behave as a dominant. If the regulatory region of a proto-oncogene is mutated such that loss of control occurs and the proto-oncogene is overexpressed, then it would "gain function" and be dominant. If the proto-oncogene product is defective, there would be a loss of function and it would more than likely behave as a recessive.

## Chapter 5

**2.** First, in order for chromosomes to engage in crossing over, they must be in proximity. It is likely that the side-by-side pairing that occurs during synapsis is the earliest time during the cell cycle that chromosomes achieve that necessary proximity. Second, chiasmata are visible during prophase I of meiosis, and it is likely that these structures are intimately associated with the genetic event of crossing over.

**4.** Because crossing over occurs at the four-strand stage of the cell cycle (that is, after S phase), each single crossover involves only two of the four chromatids.

**6.** Interference is often explained by a physical rigidity of chromatids such that they are unlikely to make sufficiently sharp bends to allow crossovers to be close together.

**8.** *dp*-- *cl*----------------*ap*
3 *mu.* 39 *mu.*

**10.** The most frequent phenotypes in the offspring, the parentals, are colored, green (88) and colorless, yellow (92). This indicates that the heterozygous parent in the testcross is coupled

$$RY/ry \times ry/ry$$

with the two dominant genes on one chromosome and the two recessives on the homolog. Seeing that there are 20 crossover progeny among the 200, or 20/200, the map distance would be 10 map units ($20/200 \times 100$ to convert to percentages) between the *R* and *Y* loci.

**12.** The most frequent classes are *PZ* and *pz*. These classes represent the parental (noncrossover) groups, which indicates that the original parental arrangement in the test cross was

$$PZ/pz \times pz/pz$$

Adding the crossover percentages together ($6.9 + 7.1$) gives 14 percent, which would be the map distance between the two genes.

**14.**

| | *female* A: | *female* B: | *Frequency:* |
|---|---|---|---|
| NCO | 3, 4 | 7, 8 | first |
| SCO | 1, 2 | 3, 4 | second |
| SCO | 7, 8 | 5, 6 | third |
| DCO | 5, 6 | 1, 2 | fourth |

The single-crossover classes, which represent crossovers between the genes that are closer together (*d-b*), would occur less frequently than the classes of crossovers between more distant genes (*b-c*).

**16.** (a) *y w* + / + + *ct* × *y w* + /Y

(b) *y*------*w*-----------------*ct*
0.0 1.5 20.0

(c) There were

$$.185 \times .015 \times 1000 = 2.775$$

double crossovers expected.

(d) Because the cross to the $F_1$ males included the normal (wild-type) gene for *cut wings*, it would not be possible to unequivocally determine the genotypes from the $F_2$ phenotypes for all classes.

**18.** Provide the genotypes of the parents in the original cross and the reciprocal. Use a semicolon to indicate that two different chromosome pairs are involved.

$P_1$:

females: + / + ; *p e*/*p e*

×

males: *dp*/*dp*; + + / + +

$F_1$:

females: + /*dp*; + + /*p e*

×

males: *dp*/*dp*; *p e*/*p e*

0.20 wild type
0.05 ebony
0.05 pink
0.20 pink, ebony
0.20 dumpy
0.05 dumpy, ebony
0.05 dumpy, pink
0.20 dumpy, pink, ebony

For the reciprocal cross:

.25 wild type
.25 pink, ebony
.25 dumpy
.25 dumpy, pink, ebony

The results would change because of no crossing over in males.

**20.** (a,b) $+ \ b \ c/a \ + \ +$

$$a - b = \frac{32 + 38 + 0 + 0}{1000} \times 100$$

$$= 7 \text{ map units}$$

$$b - c = \frac{11 + 9 + 0 + 0}{1000} \times 100$$

$$= 2 \text{ map units}$$

(c) The progeny phenotypes that are missing are $+ \ + \ c$ and $a \ b \ +$, which, of 1000 offspring, 1.4 (.07 × .02 × 1000) would be expected. Perhaps by chance or some other unknown selective factor, they were not observed.

**22.** Because sister chromatids are genetically identical (with the exception of rare new mutations), crossing over between sisters provides no increase in genetic variability. Individual genetic variability could be generated by somatic crossing over because certain patches on the individual would be genetically different from other regions. This variability would be of only minor consequence in all likelihood. Somatic crossing over would have no influence on the offspring produced.

**24.** (a) There would be $2^n = 8$ genotypic and phenotypic classes, and they would occur in a 1:1:1:1:1:1:1:1 ratio.
(b) There would be two classes, and they would occur in a 1:1 ratio.
(c) There are 20 map units between the *A* and *B* loci, and locus *C* assorts independently from both *A* and *B* loci.

**26.** Assign the following symbols for example:

$R$ = Red  $r$ = yellow

$O$ = Oval  $o$ = long

Progeny A: *Ro/rO* × *rroo* = 10 map units

Progeny B: *RO/ro* × *rroo* = 10 map units

**28.** The percentage of second-division segregation is 20/100 or 20%. Dividing by 2 (because only two of the four chromatids are involved in any single-crossover event) gives 10 map units.

**30.** For Cross 1:

$$\frac{36 + 14}{100} = 50 \text{ map units}$$

Because there are 50 map units between genes *a* and *b*, they are not linked.

For Cross 2:

$$\frac{3 + 9}{100} = 12 \text{ map units}$$

Because genes *a* and *b* are not linked, they could be on nonhomologous chromosomes or far apart (50 map units or more) on the same chromosome. Because genes *c* and *b* are linked and therefore on the same chromosome, it is also possible that genes *a* and *c* are on different chromosome pairs. Under that condition, the NP and P (parental ditypes) would be equal; however, there is a possibility that the following arrangement occurs and that genes *a* and *c* are linked.

<50   12
*a*--------------*c*--------*b*
>50

**32.** (a)

| *Tetrad in Problem* | *Class* |
|---|---|
| 1 | NP |
| 2 | T |
| 3 | P |
| 4 | NP |
| 5 | T |
| 6 | P |
| 7 | T |

(b) If P > NP, then the genes are linked. If P = NP they are independently assorting. In the problem given here, P = 44 and NP = 2. Therefore the genes are linked.
(c) For the centromere to *c* distance: = 7.2 map units
For the centromere to *d* distance: = 15.9 map units
(d) 20 map units
(e) C--------*c*--------*d*
7.9   16.4
The discrepancy between the two mapping systems is caused by the manner in which first- and second-division segregation products are scored. For instance, in tetrad arrangement #1, there are actually two crossovers between the *d* gene and the centromere, but it is still scored as a first division segregation. In tetrad arrangement #4, three crossovers occur between the *d* gene and the centromere, but they are scored as one. If one draws out all the crossovers needed to produce the tetrad arrangements in this problem, it would become clear that many crossovers between the *d* gene and the centromere go undetected in the scoring of the arrangements of the *d* gene itself. This will cause one to underestimate the distance and give the discrepancy noted.
One could account for these additional crossover classes to make the map more accurate.
(f) Tetrad class #6

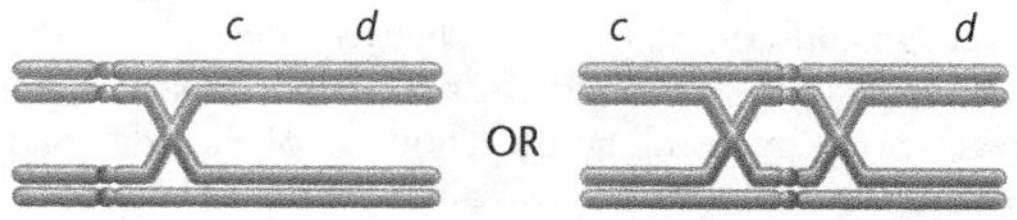


**34.** (a) Morgan and his students, especially Alfred Sturtevant, correlated chiasma frequency with the distance between linked genes. The farther apart two genes, the higher the chiasma and therefore, crossover frequency. The most important hint was that recombination frequency between gene *a* and *c* could equal to recombination frequency between *a* and *b* plus recombination frequency between *b* and *c*. (b) The discovery of linkage, genes segregating together during gamete formation, indicated a physical association among genes. (c) Two experimental lines, one using maize (Creighton and McClintock) and the other using *Drosophila* (Stern), showed that each time a crossover occurred, an actual physical exchange of chromosomes also occurred. Each experiment demonstrated a switch in chromosomal markers when genetic markers exchanged. (d) Even when sister chromatid exchanges do not produce new allelic combinations, they can be demonstrated using molecular markers such as bromodeoxyuridine. (e) Linkage analysis in humans was historically accomplished by Lod score analysis, somatic cell hybridization (synteny testing), and pedigree analysis. Modern methods combine these historical approaches with database analyses often employing a variety of physical markers (microsatellites, minisatellites, RFLPs, and SNPs).

**36.** The map distances would be computed as follows:

$$a - b = \frac{20 + 20 + 2 + 2}{400} \times 100$$
$$= 11 \text{ map units}$$
$$b - c = \frac{10 + 10 + 2 + 2}{400} \times 100$$
$$= 6 \text{ map units}$$

**38. (a)** Any gene on the same chromosome will be completely linked to any other gene on the same chromosome. Since you can get *pink* by itself, *short* cannot be completely linked to it. This leaves linkage to *black* on the second chromosome, the fourth chromosome, or the X chromosome. Since the distribution of phenotypes in males and females is essentially the same, the gene cannot be X-linked. In addition, the $F_1$ males were wild and if the *short* gene is on the X, the $F_1$ males will be short.

It is also reasonable to state that the gene cannot be on the fourth chromosome because there would be eight phenotypic classes (independent assortment of three genes) instead of the four observed. Through these insights, one could conclude that the *short* gene is on chromosome 2 with the *black* gene.

Another way to approach this problem is to make three chromosomal configurations possible in the $F_1$ male. By producing gametes from this male, the answer becomes obvious.

| *Case A* | *Case B* | *Case C* |
|---|---|---|
| $\frac{p\ b}{+\,+}\ \frac{sh}{+}$ | $\frac{p\ sh}{+\,+}\ \frac{b}{+}$ | $\frac{b\ sh}{+\,+}\ \frac{p}{+}$ |

Develop the gametes from Case C and cross them out to the completely recessive triple mutant. You will get the results in the table.

(b) The parental cross is now the following:

Females: $\frac{b\ sh}{+\,+}\ \frac{p}{+}$ × Males: $\frac{b\ sh}{b\ sh}\ \frac{p}{p}$

The new gametes resulting from crossing over in the female would be $b$ + and + $sh$. Since the gene $p$ is assorting independently, it is not important in this discussion. Because 15 percent of the offspring now contain these recombinant chromatids, the map distance between the two genes must be 15.

**40.** Once the pedigree is established as requested in part (a), one can address part (b) by detailing the genotypes of each daughter and her husband. To symbolize the two alleles at the EMWX locus, a " + " superscript is used for the normal allele and a "−" superscript is used for the mutant allele.

Daughter 1:

$$EMWX^+ Xg^+/EMWX^- Xg^- \times Xg^+/Y$$

Daughter 2:

$$EMWX^+ Xg^+/EMWX^- Xg^- \times Xg^-/Y$$

Daughter 3:

$$EMWX^+ Xg^+/EMWX^- Xg^- \times Xg^-/Y$$

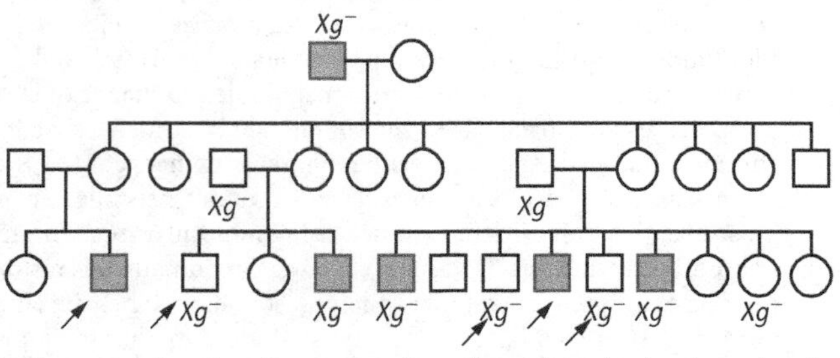


By examining the chromosomal configurations in the three daughters along with their husbands, one can determine, at least for the male offspring, whether a crossover was required to produce the given phenotypes. Crossover offspring are noted by an arrow.

## Chapter 6

**2.** (a) The requirement for physical contact between bacterial cells during conjugation was established by placing a filter in a U-tube such that the medium can be exchanged but the bacteria cannot come in contact.

(b) By treating cells with streptomycin, an antibiotic, it was shown that recombination would not occur if one of the two bacterial strains was inactivated. However, if the other was similarly treated, recombination would occur.

(c) An $F^+$ bacterium contains a circular, double-stranded, structurally independent, DNA molecule that can direct recombination.

**4.** Mapping the chromosome in an Hfr × $F^-$ cross takes advantage of the oriented transfer of the bacterial chromosome through the conjugation tube. For each F type, the point of insertion and the direction of transfer are fixed. Therefore breaking the conjugation tube at different times produces partial diploids with corresponding portions of the donor chromosome being transferred. The length of the chromosome being transferred is contingent on the duration of conjugation. Thus mapping of genes is based on time.

**6.** The $F^+$ element can enter the host bacterial chromosome, and upon returning to its independent state, it may pick up a piece of a bacterial chromosome. When combined with a bacterium with a complete chromosome, a partial diploid, or merozygote, is formed.

**8.** In the first dataset, the transformation of each locus, $a^+$ or $b^+$, occurs at a frequency of .031 and .012, respectively. To determine if there is linkage, one would determine whether the frequency of double transformants $a^+b^+$ is greater than that expected by a multiplication of the two independent events. Multiplying .031 × .012 gives .00037, or approximately 0.04%. From this information, one would consider no linkage between these two loci. Notice that this frequency is approximately the same as the frequency in the second experiment, where the loci are transformed independently.

**10.** In their experiment a filter was placed between the two auxotrophic strains, which would not allow contact. F-mediated conjugation requires contact, and without that contact, such conjugation cannot occur. The treatment with DNase showed that the filterable agent was not naked DNA.

**12.** In *generalized transduction* virtually any genetic element from a host strain may be included in the phage coat and thereby be transduced. In *specialized transduction* only those genetic elements of the host that are closely linked to the insertion point of the phage can be transduced. Specialized transduction involves the process of lysogeny.

**14.** Viral recombination occurs when there is a sufficiently high number of infecting viruses so that there is a high likelihood that more than one type of phage will infect a given bacterium. Under this condition phage chromosomes can recombine by crossing over.

**16.** Starting with a single bacteriophage, one lytic cycle produces 200 progeny phage; three more lytic cycles would produce $(200)^4$ or 1,600,000,000 phage.

**18.** For Group A, *d* and *f* are in the same complementation group (gene), while *e* is in a different one. Therefore

$$e \times f = +$$

For Group B, all three mutations are in the same gene, hence

$$b \times i = -$$

In Group C, *j* and *k* are in different complementation groups as are *j* and *l*. It would be impossible to determine whether *l* and *k* are in the same or different complementation group if the *r*II region had more than two cistrons. However, because only two complementation regions exist, and both are not in the same one as *j*, *k* and *l* must both be in the other.

**20.** $5 \times 10^{-4}$
**22.** T C H R O M B A K
**24.** (a)

| *Combination* | *Complementation* |
|---|---|
| 1, 2 | − |
| 1, 3 | + |
| 2, 4 | + |
| 4, 5 | − |

**26.** Because the frequency of double transformants is quite high (compare the $try^+tyr^+$ transformants in A and B experiments), one may conclude that the genes are quite closely linked together. Part B in the experiment gives one the frequencies of transformations of the individual genes and the frequency of transformants receiving two pieces of DNA (2 in the data table). One must know these numbers in order to estimate the actual number of $try^+tyr^+$ cotransformations.

**28.** Because 1/10 ml of a $10^{-6}$ dilution is used to add to the bacterial suspension, the concentration of the original phage suspension would be:

$$10 \times 10^6 \times 17 = 1.7 \times 10^8$$

plaque forming units per ml or pfu/ml.

**30.** Since *g* cotransforms with *f* it is likely to be in the *c b f* "linkage group" and would be expected to cotransform with each. One would not expect transformation with *a*, *d*, or *e*.

**32.** (a) No, all functional groups do not impact similarly on conjugative transfer of R27. Regions 1, 2, and 4 appear to be least influenced by mutation because transfer is at 100 percent.
(b) Regions 3, 5, 6, 8, 9, 10, 12, 13, and 14 appear to have the most impact on conjugation because when mutant, conjugation is abolished.
(c) Regions 7 and 11, when mutant, only partially abolish conjugation; therefore they probably have less impact on conjugation than those listed in part (b).
(d) Notice that regions 1, 2, 4, 7, and 11, those that appear to have little, if any, impact on conjugation, are functionally related as indicated by their shading.

## Chapter 7

**4.** Sexual differentiation is the response of cells, tissues, and organs to signals provided by the genetic mechanisms of sex determination. In other words, genes are present which signal developmental pathways whereby the sexes are generated. Sexual differentiation is the complex set of responses to those genetic signals.

**6.** In *Drosophila* it is the balance between the number of X chromosomes and the number of haploid sets of autosomes which determines sex. In humans a small region on the Y chromosome determines maleness.

**8.** Sexual differentiation of the genital ridge does not occur until around the seventh week of development, even though the eventual fate of the tissue is determined. In the absence of the Y chromosome, no male development occurs and the cortex of the genital ridge forms ovarian tissue.

**10.** In *primary* nondisjunction, half of the gametes contain two X chromosomes, while the complementary gametes contain no X chromosomes. Fertilization, by a Y-bearing sperm cell, of those female gametes with two X chromosomes would produce the XXY Klinefelter syndrome. Fertilization of the "no-X" female gamete with a normal X-bearing sperm will produce the Turner syndrome.

**12.** No. Since the Y chromosome cannot be detected in these crosses, there is no way to distinguish the two modes of sex determination.

**14.** Because attached-X chromosomes have a mother-to-daughter inheritance and the father's X is transferred to the son, one would see daughters with the white-eye phenotype and sons with the miniature wing phenotype.

**16.** Because synapsis of chromosomes in meiotic tissue is often accompanied by crossing over, it would be detrimental to sex-determining mechanisms to have sex-determining loci on the Y chromosome transferred, through crossing over, to the X chromosome.

**18.**

| | |
|---|---|
| Klinefelter syndrome (XXY) | = 1 |
| Turner syndrome (XO) | = 0 |
| 47, XYY | = 0 |
| 47, XXX | = 2 |
| 48, XXXX | = 3 |

**20.** Unless other markers, cytological or molecular, are available, one cannot test the Lyon hypothesis with homozygous X-linked genes. The test requires identification of allelic alternatives to see differences in X-chromosome activity.

**22.** Dosage compensation and the formation of Barr bodies occurs only when there are two or more X chromosomes. Males normally have only one X chromosome; therefore such mosaicism cannot occur. Females normally have two X chromosomes. There are cases of male calico cats that are XXY.

**24.** In mammals, the scheme of sex determination is dependent on the presence of a piece of the Y chromosome. If present, a male is produced. In *Bonellia viridis*, the female proboscis produces some substance that triggers a morphological, physiological, and behavioral developmental pattern that produces males. To elucidate the mechanism, one could attempt to isolate and characterize the active substance by testing different chemical fractions of the proboscis. Second, mutant analysis usually provides critical approaches into developmental processes. Depending on characteristics of the organism, one could attempt to isolate mutants that lead to changes in male or female development. Third, by using micro-tissue transplantations, one could attempt to determine which anatomical "centers" of the embryo respond to the chemical cues of the female.

**26.** One could account for the significant departures from a 1:1 ratio of males to females by suggesting that at anaphase I of meiosis, the Y chromosome more often goes to the pole that produces the more viable sperm cells. One could also speculate that the Y-bearing sperm has a higher likelihood of surviving in the female reproductive tract, or that the egg surface is more receptive to Y-bearing sperm. At this time the mechanism is unclear.

**28.** Because of the homology between the *red* and *green* genes, there exists the possibility for an irregular synapsis (see the figure below), which, following crossing over, would give a chromosome with only one (*green*) of the duplicated genes. When this X chromosome combines with the normal Y chromosome, the son's phenotype can be explained.

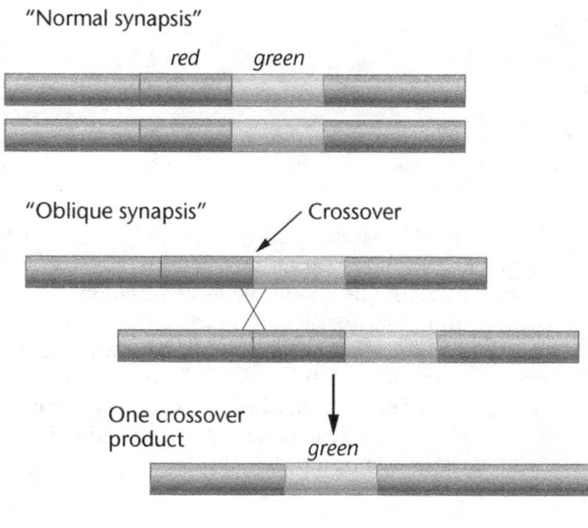

**30.** The presence of the Y chromosome provides a factor (or factors) that leads to the initial specification of maleness. Subsequent expression of secondary sex characteristics must be dependent on the interaction of the normal X-linked *Tfm* allele with testosterone. Without such interaction, differentiation takes the female path.

**32.** Since all haploids are male and half of the eggs are unfertilized, 50 percent of the offspring would be male at the start; adding the $X_a/X_a$ types gives 25 percent more male; the remainder $X_a/X_b$ would be female. Overall, 75 percent of the offspring would be male, while 25 percent would be female.

**34.** Different cells manage X-chromosome inactivation in different ways. The absence of orange patches is due to the fact that in gonadal tissue, while oogonia have a single active X chromosome, the inactive X chromosome is reactivated at, or more likely, shortly before, entry into meiotic prophase (Kratzer and Chapman 1981). Thus, X-chromosome inactivation does not remain in certain ovarian cells as in somatic tissue. However, since the timing of reactivation is variable in different cell lines, there is some uncertainty as to which cell is in which state of inactivation.

**36.** If one assumes that the somatic ovarian cell engaged in X-chromosome inactivation, the ovarian somatic cell that Rainbow donated to create CC contained an activated black gene and an inactivated orange gene (from X inactivation). This would mean that as CC developed, her cells did not change that inactivation pattern. Therefore, unlike Rainbow, CC developed without any cells that specified orange coat color. The result is CC's black and white tiger-tabby coat.

**38.** (a) The figures below depict only those chromosomes relevant to the *Mm* and *Dd* genotypes. With the *MmDd* genotype, both dyads separate intact to the secondary spermatocytes and are potentially capable of fertilizing the egg. One configuration of the MmDd arrangement occurs below (crossing over is not considered in this answer).

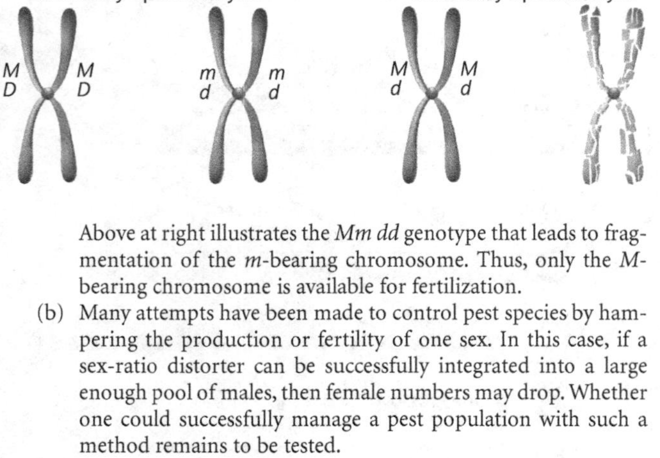


Above at right illustrates the *Mm dd* genotype that leads to fragmentation of the *m*-bearing chromosome. Thus, only the *M*-bearing chromosome is available for fertilization.

(b) Many attempts have been made to control pest species by hampering the production or fertility of one sex. In this case, if a sex-ratio distorter can be successfully integrated into a large enough pool of males, then female numbers may drop. Whether one could successfully manage a pest population with such a method remains to be tested.

# Chapter 8

**4.** Because an allotetraploid has a possibility of producing bivalents at meiosis I, it would be considered the most fertile of the three. Having an even number of chromosomes to match up at the metaphase I plate, autotetraploids would be considered to be more fertile than autotriploids.

**6.** American cultivated cotton has 26 pairs of chromosomes; 13 large, 13 small. Old world cotton has 13 pairs of large chromosomes, and American wild cotton has 13 pairs of small chromosomes. It is likely that an interspecific hybridization occurred, followed by chromosome doubling. These events probably produced a fertile amphidiploid (allotetraploid). Experiments have been conducted to reconstruct the origin of American cultivated cotton.

**8.** While there is the appearance that crossing over is suppressed in inversion "heterozygotes," the phenomenon extends from the fact that the crossover chromatids end up being abnormal in genetic content. As such, they fail to produce viable (or competitive) gametes or lead to zygotic or embryonic death.

**10.** The mutant *Notch* in *Drosophila* produces flies with abnormal wings. It is a sex-linked dominant gene (a deletion) which also behaves as a recessive lethal. A deficiency (compensation) loop indicates that bands 3C2 through 3C11 are involved. Loci near *Notch* display pseudodominance. The *Bar* gene on the other hand results from a duplication of a sex-linked region (16A) and results in abnormal eye shape.

Females: $N^+/N$ × males: $B/Y$

| | | |
|---|---|---|
| 1/4 | $N^+/B$ | females; Bar |
| 1/4 | $N/B$ | females; Notch, Bar |
| 1/4 | $N^+/Y$ | males; wild |
| 1/4 | $N/Y$ | lethal |

The final phenotypic ratio would be 1:1:1 for the phenotypes shown above.

**12.** By having the genes in an inversion, crossover chromatids are not recovered and therefore are not passed on to future generations. Translocations offer an opportunity for new gene combinations by associations of genes from nonhomologous chromosomes. Under certain conditions, such new combinations may provide selective advantage, and meiotic conditions have evolved so that segregation of translocated chromosomes yields a relatively uniform set of gametes.

**14.** The primrose, *Primula kewensis*, with its 36 chromosomes, is likely to have formed from the hybridization and subsequent chromosome doubling of a cross between the two other species, each with 18 chromosomes.

**16.** The rare double crossovers in the boundaries of a paracentric or pericentric inversion produce only minor departures from the standard chromosomal arrangement as long as the crossovers involve the same two chromatids. With two-strand double crossovers, the second crossover negates the first. However, three-strand and four-strand double crossovers have consequences that lead to anaphase bridges as well as a high degree of genetically unbalanced gametes.

**18.** In the trisomic, segregation will be 2 × 1 as illustrated below:

$P_1$: $b/b/b \times b^+/b^+$

gametes: (bb)(b) ($b^+$)

$F_1$: $b^+/b^1/b^2$ × $b^+/b$

(normal bristles) (normal bristles)

Notice that there are several segregation patterns created by the trivalent at anaphase I.
gametes:

| | |
|---|---|
| $b^+$ | $b^+$ |
| $b^1b^2$ | $b$ |
| $b^+b^1$ | |
| $b^+b^2$ | |
| $b^2$ | |
| $b^1$ | |

$F_2$:

$b^+b^+$ = normal bristles
$b^+b^1b^2$ = normal bristles
$b^+b^+b^1$ = normal bristles
$b^+b^+b^2$ = normal bristles
$b^+b^2$ = normal bristles
$b^+b^1$ = normal bristles
$b^+b$ = normal bristles
$bb^1b^2$ = bent bristles
$b^+bb^1$ = normal bristles
$b^+bb^2$ = normal bristles
$bb^2$ = bent bristles
$bb^1$ = bent bristles

**20.** Given some of the information in the above problem, the expression would be as follows:

$$(35/36W{:}1/36w)(35/36A{:}1/36a) \rightarrow$$

$(35/36)^2$ $W$---$A$---

$35/(36)^2$ $W$---$aaaa$

$35/(36)^2$ $wwwwA$---

$1/(36)^2$ $wwwwaaaa$

**22.** (a) Before the advent of polymorphic markers, maternal involvement in trisomy 21 was strongly suspected because of the striking influence of maternal age on incidence. (b) Karyotype analysis of spontaneously aborted fetuses has shown that a significant percentage of abortuses are trisomic and every chromosome can be involved. (c) A variety of studies, many tracing to early work with specialized (polytene) chromosomes in *Drosophila* and aneuploidy in other organisms demonstrated that as chromosome structures or numbers are altered, phenotypic consequences are likely. (d) By examining the polytene chromosomes of *Drosophila*, Bridges and Muller determined that the Bar-eye phenotype was caused by a chromosomal duplication of the 16A region on the X chromosome. In addition, unequal crossing over reverted and enhanced the phenotype. (e) Ohno suggested that duplications provide a "reservoir" for new genes by buffering mutations, thus allowing a duplicated gene to evolve with little or no consequence. This model is supported by findings that many genes have a substantial amount of their DNA sequence in common. Gene families also support the model of gene origin by duplication.

**24.** (a) Reciprocal translocation
(b)

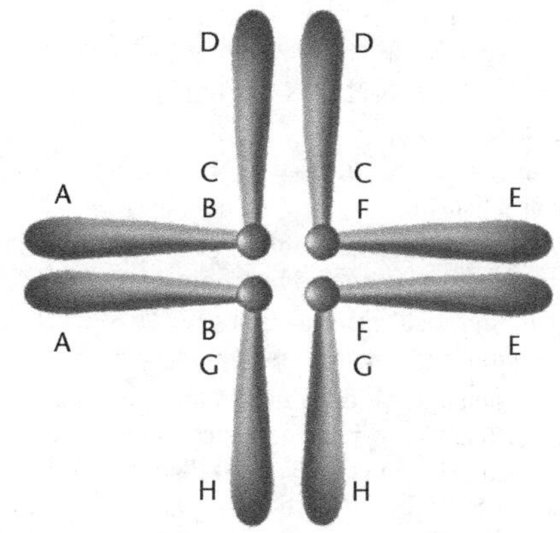


(c) Notice that all chromosomal segments are present and there is no apparent loss of chromosomal material. However, if the breakpoints for the translocation occur within genes, then an abnormal phenotype may be the result. In addition, a gene's function is sometimes influenced by its position (its neighbors, in other words). If such "position effects" occur, a different phenotype may result.

**26.** (a) The father must have contributed the abnormal X-linked gene.
(b) Since the son is XXY and heterozygous for anhidrotic dysplasia, he must have received both the defective gene and the Y chromosome from his father. Thus nondisjunction must have occurred during meiosis I.
(c) This son's mosaic phenotype is caused by X-chromosome inactivation, a form of dosage compensation in mammals.

**28.** Below is a description of breakage/reunion events that illustrate a translocation in relatively small, similarly sized, chromosomes 19 (metacentric) and 20 (metacentric/submetacentric). The case described here is shown occurring before S phase duplication. The same phenomenon is shown in the text as occurring after S phase. Since the likelihood of such a translocation is fairly small in a general population, inbreeding played a significant role in allowing the translocation to "meet itself."

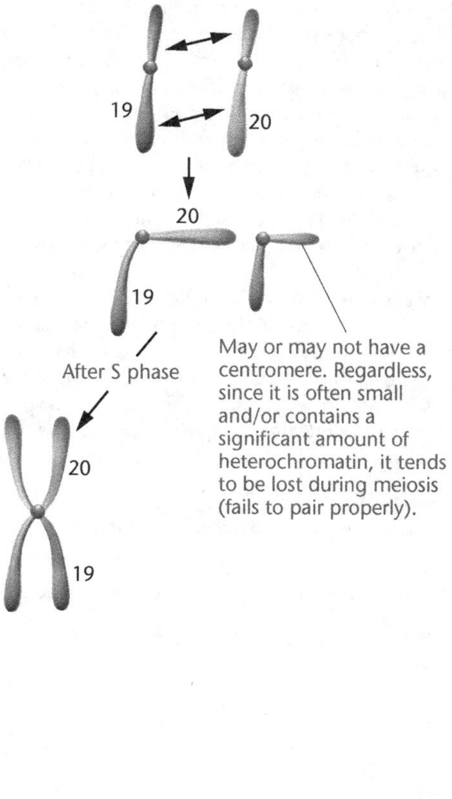

**30.** This female will produce meiotic products of the following types:

*normal:* 18 + 21
*translocated:* 18/21
*translocated plus* 21: 18/21 + 21
*deficient:* 18 only

Fertilization with a normal 18 + 21 sperm cell will produce the following offspring:

*normal:* 46 chromosomes
*translocation carrier:* 45 chromosomes 18/21 + 18 + 21
*trisomy 21:* 46 chromosomes 18/21 + 21 + 21
*monosomic:* 45 chromosomes 18 + 18 + 21, lethal

**32.** (a, b) Polyploidization provides an opportunity for novel combinations of relatively diverse genomes. However, because meiosis is often complicated by polyploidy, diploidization is often an evolutionary advantage. Since diploidization is apparently a gradual and relatively haphazard process, some genomic regions may become diploid while others may not. Such differential diploidization may lead to advantageous genomic combinations and serve as genetic reservoirs that may be tested by selection. (c) Generally, diploidization is accomplished by deletions directly as well as chromosomal rearrangements, such as inversions and translocations, that lead to deletions.

## Chapter 9

**2.** The $mt^+$ strain (resistant for the nuclear and chloroplast genes) contributes the "cytoplasmic" component of streptomycin resistance, which would negate any contribution from the $mt^-$ strain. Therefore, all the offspring will have the streptomycin resistance phenotype. In the reciprocal cross, with the $mt^+$ strain being streptomycin sensitive, all the offspring will be sensitive.

**4.** (a) neutral
(b) segregational (nuclear mutations)
(c) suppressive

**6.** The inheritance patterns for the two, *segregational* and *neutral*, are quite different. The segregational mode is dependent on nuclear genes, while that of the neutral type is dependent on cytoplasmic influences, namely, mitochondria. If the two are crossed as stated in the problem, then one would expect, in the diploid zygote, the *segregational* allele to be "covered" by normal alleles from the neutral strain. On the other hand, as the nuclear genes are again "exposed" in the haploid state of the ascospores, one would expect a 1:1 ratio of normals to petites. The petite phenotype is caused by the nuclear, *segregational* gene.

**8.** (a) It is likely that mitochondria and chloroplasts evolved from bacteria in a symbiotic relationship, therefore it is not surprising that certain antibiotics that influence bacteria will also influence all mitochondria and chloroplasts.
(b) Clearly, the $mt^+$ strain is the donor of the *cp*DNA since the inheritance of resistance or sensitivity is dependent on the status of the $mt^+$ gene.

**10.** The fact that all of the offspring ($F_1$) showed a dextral coiling pattern indicates that one of the parents (maternal parent) contains the *D* allele. Taking these offspring and seeing that their progeny (call these $F_2$) occur in a 1:1 ratio indicates that half of the offspring ($F_1$) are *dd*. In order to have these results, one of the original parents must have been *Dd*, while the other must have been *dd*.

Parents: *Dd* × *dd*

Offspring (F1): 1/2 *Dd*, 1/2 *dd*
(all dextral because of the maternal genotype)

Progeny ($F_2$):
All those from *Dd* parents will be dextral, while all those from *dd* parents will be sinistral.

**12.** Since there is no evidence for segregation patterns typical of chromosomal genes and Mendelian traits, some form of extranuclear inheritance seems possible. If the *lethargic* gene is dominant, a maternal effect may be involved. In that case, some of the $F_2$ progeny would be hyperactive because maternal effects are only temporary, affecting only the immediate progeny. If the lethargic condition is caused by some infective agent, then perhaps injection experiments could be used. If caused by a mitochondrial defect, then the condition would persist in all offspring of lethargic mothers, through more than one generation.

**14.** Since an initial mutation does not involve all copies of mtDNA within a mitochondrion, the original state is heteroplasmic and the mutated mtDNA is rare. If the new mutation confers no selective advantage to the host cell, the frequency of the mutation is likely to diminish. However, if the mutation confers a selective advantage to the cell, it is likely to gain in frequency. Depending on a number of factors, including chance as well as the extent of the selective advantage, a mtDNA mutation may become prominent and eventually establish homoplasmy.

**16.** Mitochondrial defects often involve processes of oxidative phosphorylation and/or other essential mitochondrial functions that are dependent not only on the mitochondrial genome, but also on the nuclear genome. When a nuclear genome is transferred, it may contain mutant genes that negatively influence mitochondrial function that were compensated for in the original donor cells, but not in the recipient. In addition, enucleated eggs invariably contain both normal and defective mitochondria. When a nuclear genome is transferred to an enucleated egg, mitochondrial defects may arise from a defective nuclear genome, the heteroplasmic condition of the egg, or a combination of the two. A disease occurs when the mitochondrial mutational load exceeds a tissue-specific threshold that is generally low in highly metabolic tissues such as brain, heart, and muscle.

**18.** (a) In general, organelle heredity is detected when the maternal parent has more influence over the phenotype of the offspring. In addition, if X-linkage can be eliminated, different results from reciprocal crosses also may support inheritance by extranuclear elements. (b) While *segregational petites* exhibited Mendelian inheritance, both *neutral* and *suppressive petites* followed non-Mendelian patterns that were consistent with the involvement of an extranuclear agent. (c) Electron micrographs show that DNA of mitochondria and chloroplasts looks like that seen in bacteria. In addition, the molecular components, notably ribosomal RNAs, are more like those in prokaryotes than eukaryotes. (d) When eggs from a *Dd* (but sinistral because of its parent) snail are self-fertilized, all the offspring, even those that are *dd*, are coiled dextrally. Thus, the phenotype of the offspring is determined not by the mother's phenotype or its own genotype, but by the mother's genotype. The genotype of the sperm is not influential in determining the direction of shell coiling in offspring.

**20.** (a) The presence of $bcd^-/bcd^-$males can be explained by the maternal effect: mothers were $bcd^+/bcd^-$.
(b) The cross

female $bcd^+/bcd^-$ × male $bcd^-/bcd^-$

will produce an $F_1$ with normal embryogenesis because of the maternal effect. In the $F_2$, any cross having $bcd^+/bcd^-$ mothers will have phenotypically normal embryos. Offspring of any cross involving homozygous $bcd^-/bcd^-$ mothers will have problems with embryogenesis.

**22.** (a) A locus, *Segregation Distortion* (*SD*), is present on the wild-type chromosome. Some aspect of *SD* causes a shift in the segregation ratio by allowing sperm to carry the *SD* chromosome at the expense of the homolog.
(b) One could use this *SD* chromosome in a variety of crosses and

determine that the abnormal segregation is based on a particular chromosomal element. One could even map the *SD* locus on the second chromosome (as has been done). (c) Segregation distortion describes a condition in which typical Mendelian segregation is distorted from the 50:50.

**24.** Deficiencies that remove histone genes contribute to increased survival of progeny of *abo/abo* mothers. Since deficiencies in histone genes reduce the severity of the maternal effect, it is likely that an overabundance of histones is involved. The observation that the addition of heterochromatin also reduces the severity of the maternal effect may be due to a sequestering of the histone overload by the heterochromatin. One might therefore speculate that the *abo* gene is a regulator of histone production. The normal allele specifies a negative regulator of histone genes, while the mutant *abo* gene fails to exert such negative control.

## Chapter 10

**2.** Proteins are composed of as many as 20 different subunits (amino acids), thereby providing ample structural and functional variation for the multiple tasks that must be accomplished by the genetic material. The tetranucleotide hypothesis (structure) provided insufficient variability to account for the diverse roles of the genetic material.

**4.** Specific degradative enzymes, proteases, RNase, and DNase were used to selectively eliminate components of the extract, and, if transformation is concomitantly eliminated, then the eliminated fraction is the transforming principle. DNase eliminates DNA and transformation; therefore it must be the transforming principle.

**6.** Actually phosphorus is found in approximately equal amounts in DNA and RNA. Therefore labeling with $^{32}P$ would "tag" both RNA and DNA. However, the T2 phage, in its mature state, contains very little if any RNA; therefore DNA would be interpreted as being the genetic material in T2 phage.

**8.** By comparing DNA content in various cell types (sperm and somatic cells) and observing that the *action* and *absorption* spectra of ultraviolet light were correlated, DNA was considered to be the genetic material. This suggestion was supported by the fact that DNA was shown to be the genetic material in bacteria and some phage. Direct evidence for DNA being the genetic material comes from a variety of observations, including gene transfer, which has been facilitated by recombinant DNA techniques.

**10.** Linkages among the three components require the removal of water ($H_2O$).

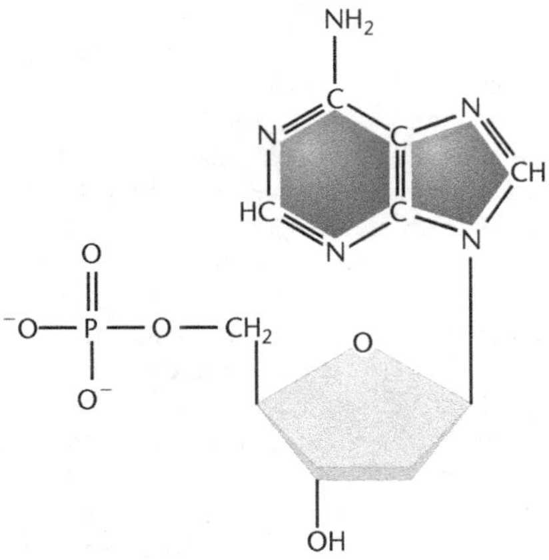


**12.** Guanine: 2-amino-6-oxypurine
Cytosine: 2-oxy-4-aminopyrimidine
Thymine: 2,4-dioxy-5-methylpyrimidine
Uracil: 2,4-dioxypyrimidine

**16.** Because in double-stranded DNA, A = T and G = C (within limits of experimental error), the data presented would have indicated a lack of pairing of these bases in favor of a single-stranded structure or some other nonhydrogen-bonded structure.

Alternatively, from the data it would appear that A = C and T = G, which would negate the chance for typical hydrogen bonding since opposite charge relationships do not exist. Therefore, it is quite unlikely that a tight helical structure would form at all. In conclusion, Watson and Crick might have concluded that hydrogen bonding is not a significant factor in maintaining a double-stranded structure.

**18.** Three main differences between RNA and DNA are the following:

1. Uracil in RNA replaces thymine in DNA.
2. Ribose in RNA replaces deoxyribose in DNA.
3. RNA often occurs as both single- and partially double-stranded forms, whereas DNA most often occurs in a double-stranded form.

**20.** The nitrogenous bases of nucleic acids (nucleosides, nucleotides, and single- and double-stranded polynucleotides) absorb UV light maximally at wavelengths 254 to 260 nm. Using this phenomenon, one can often determine the presence and concentration of nucleic acids in a mixture. Since proteins absorb UV light maximally at 280 nm, this is a relatively simple way of dealing with mixtures of biologically important molecules.

UV absorption is greater in single-stranded molecules (hyperchromic shift) as compared to double-stranded structures. Therefore, by applying denaturing conditions, one can easily determine a nucleic acid is in the single- or double-stranded form. In addition, A—T rich DNA denatures more readily than G—C rich DNA. Therefore one can estimate base content by denaturation kinetics.

**22.** Guanine and cytosine are held together by three hydrogen bonds, whereas adenine and thymine are held together by two. Because G—C base pairs are more compact, they are more dense than A—T pairs. The percentage of G—C pairs in DNA is thus proportional to the buoyant density of the molecule.

**24.** For curve A in the problem, there is evidence for a rapidly renaturing species (repetitive) and a slowly renaturing species (unique). The fraction that reassociates faster than the *E. coli* DNA is highly repetitive, and the last fraction (with the highest $C_ot_{1/2}$ value) contains primarily unique sequences. Fraction B contains mostly unique, relatively complex DNA.

**26.** Because G—C base pairs are formed with three hydrogen bonds while A—T base pairs by two such bonds, it takes more energy (higher temperature) to separate G—C pairs.

**28.** In one sentence of Watson and Crick's paper in *Nature*, they state, "It has not escaped our notice that the specific pairing we have postulated immediately suggests a possible copying mechanism for the genetic material." The model itself indicates that unwinding of the helix and separation of the double-stranded structure into two single strands immediately expose the specific hydrogen bonds through which new bases are brought into place.

**30.** MS-2 = 200 base pairs

*E. coli* = $2 \times 10^6$ base pairs

**32.** Since cytosine pairs with guanine and uracil pairs with adenine, the result will be a base substitution of G:C to A:T after rounds of replication.

**34.** Fluorescence *in situ* hybridization employs fluorescently labeled DNA that hybridizes to metaphase chromosomes and interphase nuclei. Because of the relatively high likelihood of aneuploidy for chromosomes 13, 18, 21, X, and Y, they are routine candidates for analysis.

**36.** (i) The X-ray diffraction studies would indicate a helical structure, for it is on the basis of such data that a helical pattern is suggested. The fact that it is irregular may indicate different diameters (base pairings), additional strands in the helix, kinking, or bending.

(ii) The hyperchromic shift would indicate considerable hydrogen bonding, possibly caused by base pairing.
(iii) Such data may suggest irregular base pairing in which purines bind purines (all the bases presented are purines) thus giving the atypical dimensions.
(iv) Because of the presence of ribose, the molecule may show more flexibility, kinking, and/or folding.

While there are several situations possible for this model, the phosphates are still likely to be far apart (on the outside) because of their strong like charges. Hydrogen bonding probably exists on the inside of the molecule, and there is probably considerable flexibility, kinking, and/or bending.

**38.** Heat application will yield a hyperchromic shift if the DNA is double-stranded. One could also get a rough estimation of the GC content from the kinetics of denaturation and the degree of sequence complexity from comparative renaturation studies. Determination of base content by hydrolysis and chromatography could be used for comparative purposes and could also provide evidence as to the strandedness of the DNA. Antibodies for Z-DNA could be used to determine the degree of left-handed structures, if present. Sequencing the DNA from both viruses would indicate sequence homology. In addition, through various electronic searches readily available on the Internet (Web site: *blast@ncbi.nlm.nih.gov*, for example) one could determine whether similar sequences exist in other viruses or in other organisms.

**40.** In general, superhelical/supercoiled DNA (form I) migrates the fastest, followed by linear DNA (form III). The slowest to migrate is usually the loose circle (form II).

## Chapter 11

**2.** By labeling the pool of nitrogenous bases of the DNA of *E. coli* with a heavy isotope $^{15}N$, it would be possible to "follow" the "old" DNA.

**4.** (a) Under a conservative scheme, all of the newly labeled DNA will go to one sister chromatid, while the other sister chromatid will remain unlabeled.
(b) Under a dispersive scheme all of the newly labeled DNA will be interspersed with unlabeled DNA.

**6.** The *in vitro* replication requires a DNA template, a divalent cation ($Mg^{++}$), and all four of the deoxyribonucleoside triphosphates: dATP, dCTP, dTTP, and dGTP.

**8.** Two general analytical approaches showed that the products of DNA polymerase I were probably copies of the template DNA. Because *base composition* can be similar without reflecting sequence similarity, the least stringent test was the comparison of base composition. By comparing *nearest neighbor frequencies*, Kornberg determined that there is a very high likelihood that the product was of the same base sequence as the template.

**10.** The *in vitro* rate of DNA synthesis using DNA polymerase I is slow, being more effective at replicating single-stranded DNA than double-stranded DNA. In addition, it is capable of degrading as well as synthesizing DNA. Such degradation suggested that it functioned as a repair enzyme. In addition, DeLucia and Cairns discovered a strain of *E. coli* (*pol*A1), which still replicated its DNA but was deficient in DNA polymerase I activity.

**12.** Biologically active DNA implies that the DNA is capable of supporting typical metabolic activities of the cell or organism and is capable of faithful reproduction.

**14.** DNA polymerase I and DNA ligase are used to synthesize and label an RF duplex. DNase is used to nick one of the two strands. The heavy ($^{32}P$/BU-containing) DNA is isolated by denaturation and centrifugation, and DNA polymerase and DNA ligase are used to make a synthetic complementary strand. When isolated, this synthetic strand is capable of transfecting *E. coli* protoplasts, from which new *ϕX*174 phages are produced.

**18.** Given a stretch of double-stranded DNA, one could initiate synthesis at a given point and either replicate strands in one direction only (unidirectional) or in both directions (bidirectional). Synthesis of complementary strands occurs in a *continuous* 5′ > 3′ mode on the leading strand in the direction of the replication fork, and in a *discontinuous* 5′ > 3′ mode on the lagging strand opposite the direction of the replication fork.

**20.** *Okazaki fragments* are relatively short (1000 to 2000 bases in prokaryotes) DNA fragments that are synthesized in a discontinuous fashion on the lagging strand during DNA replication. Such fragments appear to be necessary because template DNA is not available for 5′ > 3′ synthesis until some degree of continuous DNA synthesis occurs on the leading strand in the direction of the replication fork. The isolation of such fragments provides support for the scheme of replication. DNA *ligase* is required to form phosphodiester linkages in gaps that are generated when DNA polymerase I removes RNA primer and meets newly synthesized DNA ahead of it. The discontinuous DNA strands are ligated together into a single continuous strand. *Primer* RNA is formed by RNA primase to serve as an initiation point for the production of DNA strands on a DNA template. None of the DNA polymerases is capable of initiating synthesis without a free 3′ hydroxyl group. The primer RNA provides that group and thus can be used by DNA polymerase III.

**22.** Because there is a much greater amount of DNA to be replicated and DNA replication is slower, there are multiple initiation sites for replication in eukaryotes (and increased DNA polymerase per cell) in contrast to the single replication origin in prokaryotes. Replication occurs at different sites during different intervals of the S phase. The proposed functions of four DNA polymerases are described in the text.

**24.** (a) In *E. coli*, 100 kb are added to each growing chain per minute. Therefore the chain should be about 4,000,000 bp.
(b) Given $(4 \times 10^6 \text{ bp}) \times 0.34 \text{ nm/bp} = 1.36 \times 10^6 \text{ nm}$ or 1.3 mm

**26.** *Gene conversion* is likely to be a consequence of genetic recombination in which nonreciprocal recombination yields products in which it appears that one allele is "converted" to another. Gene conversion is now considered a result of heteroduplex formation, which is accompanied by mismatched bases. When these mismatches are corrected, the "conversion" occurs.

**28.** Telomerase activity is present in germ-line tissue to maintain telomere length from one generation to the next. In other words, telomeres cannot shorten indefinitely without eventually eroding genetic information.

**30.** Since synthesis is bidirectional, one can multiply the rate of synthesis by two to come up with a figure of 18,000 bases replicated per five minutes (30 bases/second $\times$ 300 seconds). Dividing $1.6 \times 10^8$ by $1.8 \times 10^4$ gives $0.88 \times 10^4$, or about 8800 replication sites.

**32.** (a) DNA polymerase would catalyze a bond between the 5′ end of the last nucleotide added and the 3′ end of the incoming nucleotide. In this reaction, the energy would be provided by the cleavage of the gamma- and beta-phosphates of the last nucleotide added to the chain rather than of the incoming nucleotide.
(b) If DNA polymerase removed a base, it would not be able to add any more bases to the chain because the penultimate base would have a monophosphate rather than a triphosphate and there would be no source of energy for the polymerization reaction.

**34.** (a) 5′ACCUAAGU
(b) U

**36.** (a) DNA, since one of the nitrogenous bases is T; also, notice the lack of an OH group at the 2′ carbon. (b) 3′ (c) Since spleen diesterase cuts between the 5′ carbon and the phosphate, the original 5′ phosphate is transferred to the 3′ carbon of the 5′ neighbor. Therefore deoxyadenosine would obtain the phosphate at its 3′ position.

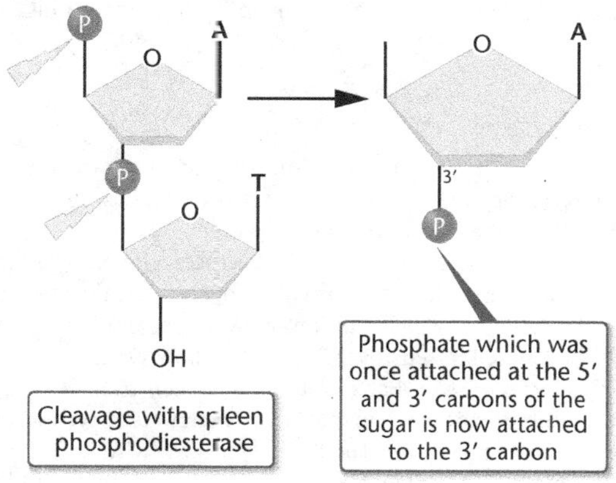


**38.** *Strain A is gyr$^{ts}$. Strain B is gyr$^{ts,\ r}$. Strain C is wild type. Strain D is gyr$^{r}$.*

**40.**

| *Initial Labeled Base* | *Labeled Base after Spleen Labeled Phosphodiesterase Digestion* | |
|---|---|---|
| | *ANTIPARALLEL* | *PARALLEL* |
| G | A,T | C,T |
| C | G,A,G | G,A |
| T | C,T,G | C,T,A,G |
| A | T,C,A,T | T,C,A,G |

One can determine which model occurs in nature by comparing the pattern in which the labeled phosphate is shifted following spleen phosphodiesterase digestion. Focus your attention on the antiparallel model and notice that the frequency of which "C" (for example) is the 5′ neighbor of "G" is not necessarily the same as the frequency of which "G" is the 5′ neighbor of "C." However, in the parallel model (b), the frequency of which "C" is the 5′ neighbor of "G" is the same as the frequency of which "G" is the 5′ neighbor of "C." By examining such "digestion frequencies," it can be determined that DNA exists in the opposite polarity.

## Chapter 12

**2.** By having a circular chromosome, no free ends present the problem of linear chromosomes, namely, complete replication of terminal sequences.

**4.** Since eukaryotic chromosomes are "multirepliconic" in that there are multiple replication forks along their lengths, one would expect to see multiple clusters of radioactivity.

**6.** Long interspersed elements (LINEs) are repetitive transposable DNA sequences in humans. The most prominent family, designated **L1**, is about 6.4 kb each and is represented about 100,000 times. LINEs are often referred to as retrotransposons because their mechanism of transposition resembles that used by retroviruses.

**8.** Because of the diverse cell types of multicellular eukaryotes, a variety of gene products is required, which may be related to the increase in DNA content per cell. In addition, the advantage of diploidy automatically increases DNA content per cell. However, seeing the question in another way, it is likely that a much higher *percentage* of the genome of a prokaryote is actually involved in phenotype production than in a eukaryote.

Eukaryotes have evolved the capacity to obtain and maintain what appears to be large amounts of "extra," perhaps "junk," DNA. Prokaryotes on the other hand, with their relatively short life cycle, are extremely efficient in their accumulation and use of their genome.

Given the larger amount of DNA per cell and the requirement that the DNA be partitioned in an orderly fashion to daughter cells during cell division, certain mechanisms and structures (mitosis, nucleosomes, centromeres, etc.) have evolved for packaging and distributing the DNA. In addition, the genome is divided into separate entities (chromosomes) to perhaps facilitate the partitioning process in mitosis and meiosis.

**10.** Nucleosomes are octomeric structures of two molecules of each histone (H2A, H2B, H3, and H4) except H1. Between the nucleosomes and complexed with linker DNA is histone H1. A 146-base pair sequence of DNA wraps around the nucleosome.

**12.** *Heterochromatin* is chromosomal material that stains deeply and remains condensed when other parts of chromosomes, euchromatin, are otherwise pale and decondensed. Heterochromatic regions replicate late in S phase and are relatively inactive in a genetic sense because there are few genes present, or if they are present, they are repressed. Telomeres and the areas adjacent to centromeres are composed of heterochromatin.

**14.** Volume of DNA: $3.14 \times 10\ \text{Å} \times 10\ \text{Å} \times (50 \times 10^4\ \text{Å}) = 1.57 \times 10^8\ \text{Å}^3$

Volume of capsid: $4/3\ (3.14 \times 400\ \text{Å} \times 400\ \text{Å} \times 400\ \text{Å}) = 2.67 \times 10^8\ \text{Å}^3$

Because the capsid head has a greater volume than the volume of DNA, the DNA will fit into the capsid.

**16.** about 36.3 percent

**18.** (a) In the absence of a product, the identification of pseudogenes is mainly dependent on extensive mining of DNA sequence databases to identify sections of DNA that resemble functional genes. Mutations within pseudogenes also make them difficult to identify. (b) Because gene spacing, chromatin folding, and various silencer/enhancer domains of chromosomes clearly interact in normal genomic function, pseudogenes may play important roles by sponsoring or inhibiting such interactions. Recent evidence indicates that what can appear to be an inactive pseudogene may actually be transcribed in one setting (tissue or organism) and may function in regulation.

**22.** Chromosomes are not randomly distributed within nuclei. Homologous chromosomes tend to distribute themselves opposite each other and in an antiparallel manner, meaning that their positions are in reverse order on opposite sides of the nucleus. Assuming that such patterns are maintained throughout the entire cell cycle, it is possible that chromosomal positions may influence gene function and/or chromosomal behavior during mitosis and/or meiosis. If gene function is influenced not only by gene position in a chromosome, but also by gene position in a nucleus, then an alternative explanation for position effect exists.

**24.** Nucleosomes follow a *dispersive* pattern, with each daughter chromatid containing a mixture of new and original nucleosomes. One could test the distribution of nucleosomes by conducting an autoradiographic experiment similar to Taylor-Woods-Hughes, but instead of labeling the DNA with $^3$H-thymidine, one would label some or all the histones H2A, H2B, H3, and H4 in nucleosomes.

**26.** Bacteriophage lambda is composed of a double-stranded, linear DNA molecule of about 48,000 base pairs. It is capable of forming a closed, double-stranded circular molecule because of a 12-base pair, single-stranded, complementary "overhanging" sequence at the 5′ end of each single strand.

**28.** The general frequency and pattern of various trinucleotide repeat motifs are similar in all taxonomic groups. Within-gene trinucleotide repeats are the most frequent repeat motif in all taxonomic groups followed by hexanucleotide repeats. One explanation might be that various microsatellite types (mono, di, tri, etc.) are generated at different rates in different genomic regions (within and between genes). A second possibility is that selection acts differentially depending on the type and location of a repeat. The correlation between the high frequency of tri- and hexanucleotide repeats within genes and a triplet code specifying particular amino acids within genes may not be coincidental.

**30.** If microsatellites in general are flanked by a conserved sequence, those conserved sequences may be involved in the generation and/or maintenance of the microsatellite. Alternatively, the microsatellite may generate the nonmicrosatellite region. Any hypothesis presented is in need of additional investigation before definitive statements can be made.

**32.** Generally, the higher the AT content of a DNA strand, the lower the temperature of melting or denaturation. Since one can monitor the degree of strand separation with absorption of ultraviolet light (optical density), one can get an estimate of the AT/GC content in a given stretch of DNA. For relatively short strands of DNA, the temperature of melting is also related to strand length. Other factors being equal, the shorter the strand, the lower the $T_m$.

**34.** Since both genes mentioned in the problem are located near the end of chromosome 16, it is possible that erosion of the end of the chromosome is related to each disease. Examination of the gene by *in situ* hybridization and molecular cloning indicates that thalassemia involves a terminal deletion in the distal portion of 16p.

## Chapter 13

**2.** Eukaryotic mRNAs typically have a 3′ polyA tail. The poly-dT segment provides a double-stranded section that serves to prime the production of the complementary strand.

**4.** It is believed that the protein interacts with the major groove of the DNA helix. This information comes from the structure of the proteins, which have been sufficiently well studied to suggest that the DNA major groove and "fingers" or extensions of the protein form the basis of interaction.

**6.** This segment contains the palindromic sequence of GGATCC, which is recognized by the restriction enzyme *Bam*HI. The double-stranded sequence is the following:

CCTAGG
GGATCC

**8.** Plasmids were the first to be used as cloning vectors, and they are still routinely used to clone relatively small fragments of DNA. Because of their small size, they are relatively easy to separate from the host bacterial chromosome, and they have relatively few restriction sites. They can be engineered fairly easily (i.e., polylinkers and reporter genes added). For cloning larger pieces of DNA such as entire eukaryotic genes, cosmids are often used. For instance, when modifications are made in the bacterial virus lambda (λ), relatively large inserts of about 20 kb can be cloned. This is an important advantage when one needs to clone a large gene or generate a genomic library from a eukaryote. In addition, some cosmids will only accept inserts of a limited size, which means that small, less meaningful perhaps, fragments will not be cloned unnecessarily. Both plasmids and cosmids suffer from the limitation that they can only use bacteria as hosts.

YACs (yeast artificial chromosomes) contain telomeres, an origin of replication, and a centromere and are extensively used to clone DNA in yeast. With selectable markers (TRP1 and URA3) and a cluster of restriction sites, DNA inserts ranging from 100 kb to 1000 kb can be cloned and inserted into yeast. Since yeast, being eukaryotes, undergo many of the typical RNA and protein processing steps of other, more complex eukaryotes, the advantages are numerous when working with eukaryotic genes.

**10.** This problem can be solved by the following expressions:

| | |
|---|---|
| *Not*I | $4^8$ |
| *Hin*fI | $4 \times 4 \times 1 \times 4 \times 4$ |
| *Xho*II | $2 \times 4 \times 4 \times 4 \times 4 \times 2$ |

A "1" is used in the *Hin*fI portion because any of the four bases can be inserted for the "N," whereas only two bases can be used for "Pu" and "Py" in the *Xho*II portion.

**12.** (a) Because the *Drosophila* DNA has been cloned into the *Pst*I site in the ampicillin resistance gene of the plasmid, the gene will be mutated and any bacterium with the recombinant plasmid will be ampicillin sensitive. The tetracycline resistance gene remains active, however. Bacteria that have been transformed with the recombinant plasmid will be resistant to tetracycline, and therefore tetracycline should be added to the medium.

(b) Colonies that grow on a tetracycline medium should be tested for growth on an ampicillin medium either by replica plating or some similar controlled transfer method. Those bacteria that do not grow on the ampicillin medium probably contain the *Drosophila* DNA insert.

(c) Resistance to both antibiotics by a transformed bacterium could be explained in several ways. First, if cleavage with the *Pst*I was incomplete, then no change in biological properties of the uncut plasmids would be expected. Second, it is possible that the cut ends of the plasmid were ligated together in the original form with no insert.

**14.** Given that there is only one site for the action of *Hin*dIII, the following will occur. Cuts will be made such that a four-base single-stranded set of sticky ends will be produced. For the antibiotic resistance to be present, the ligation will reform the plasmid into its original form. However, two of the plasmids can join to form a dimer as indicated in the following diagram.

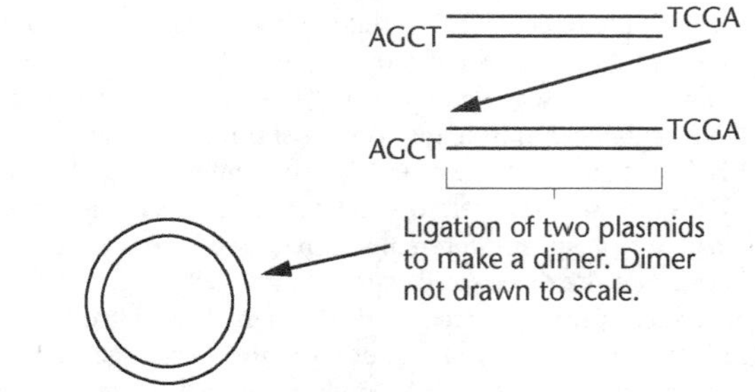


**16.** Because of complementary base pairing, the 3′ end of the DNA strand often loops back onto itself, thereby providing a primer for DNA polymerase I.

**18.** A filter is used to bind the DNA from the colonies containing recombinant plasmids. A labeled probe is constructed for the protein sequence of EF-1a. Since it is highly conserved, it should show considerable complementation to the human EF-1a cDNA. It is used to detect, through hybridization, the DNA of interest. Cells with the desired clone are then picked from the original plate and the plasmid is isolated from the cells.

**20.** The problem can be best solved by drawing out the strands, then placing the restriction sites in the appropriate positions as follows:

enzyme I   __350_|__950______
enzyme II   200|______1100______

To determine the orientation of the restriction sites to each other, examine the results of the double digested DNA and note that there is a 150 bp fragment meaning that enzyme II cuts within the 350 bp fragment of enzyme I. Therefore the final map is as follows:

II I
200 _|__|______950__________
150

**22.** Option (b) fits the expectation because the thick band in the offspring probably represents the bands at approximately the same position in both parents. The likelihood of such a match is expected to be low in the general population.

**24.** One must know at least a portion of the amino acid sequence of the protein product or its nucleic acid sequence in order for the procedure to be applied. Some problems can occur through degeneracy in the genetic code (not allowing construction of an appropriate probe), the possible existence of pseudogenes in the library (hybridizations with inappropriate related fragments in the library), and variability of DNA sequences in the library due to introns (causing poor or background hybridization). To overcome some of these problems, one can construct a variety of relatively small probes of different types that take into account the degeneracy in the code. By varying the conditions of hybridization (salt and temperature), one can reduce undesired hybridizations.

**26.** (a) Heating to 90–95°C denatures the double-stranded DNA so that it dissociates into single strands. It usually takes about five minutes, depending on the length and GC content of the DNA. (b) Lowering the temperature to 50–70°C allows the primers to bind to the denatured DNA. (c) Bringing the temperature to 70–75°C allows the heat stable DNA polymerase an opportunity to extend the primers by adding nucleotides to the 3′ ends of each growing strand. Each PCR is designed with specific temperatures (not ranges) based on the characteristics of the DNAs (template and primers).

**28.** Since there is no 3′-hydroxyl group, chain elongation cannot take place, and resulting fragments are formed which can be separated by electrophoresis. Where the ddNTP was incorporated, the length of each strand and therefore the position of the particular ddNTP is established and used to eventually provide the base sequence of the DNA.

**32.** $T_m(°C) = 81.5 + 0.41(\%GC) - (675/N) = 81.5 + 0.41(33.3) - (675/21) =$ about 63°C. Subtracting 5°C gives us a good starting point of about 58°C for PCR with this primer. Notice that as the % of GC and length increase, the $T_m$(°C) increases. GC pairs contain three hydrogen bonds rather than two as between AT pairs.

**34.** (a)

| $Na^+$ | *% Formamide(F)* | $T_m$ |
|---|---|---|
| 0.825 | 20 | 84.16 |
| 0.825 | 40 | 69.76 |
| 0.165 | 20 | 72.56 |
| 0.165 | 40 | 58.16 |

(b) Sodium, a monovalent cation, interacts with the negative phosphates that make up the nucleic acid backbone. The repulsive forces between negative phosphates are reduced and the double helix becomes more stable, therefore requiring a higher temperature for melting. Formamide competes for hydrogen bond locations of the bases and lessens the attractions that hold each double helix together. As the competition for hydrogen bonding increases with increased formamide, the lower the melting temperature.

**36.** (a) By amplifying Y-STRs by PCR and separating the amplified products by electrophoresis, one can genotypically type an individual as one does with a standard fingerprint. While not being 100 percent absolute, linking an individual with the time and place of a significant event has multiple forensic applications. Eliminating individuals is also extremely useful. (b) The nonrecombining region of the Y is maintained strictly in the male population because it does not recombine with other chromosomes. (c) Because different ethnic groups show different levels of Y-STR polymorphism, different final probabilities occur as products of individual probabilities. Since these probabilities are used to match individuals in forensics, ethnic variations must be taken under consideration. (d) If an individual's genotype matches that found in DNA at a crime scene, depending on the product of frequencies of the haplotypes, one might be able to say that the individual was at the crime scene. However, contamination, inappropriate genotyping, and laboratory expertise may give both false positive or negative results. With identical twins, both will have identical DNA profiles.

## Chapter 14

**2.** Given a sextuplet code, restoration of the reading frames would only occur with the addition or loss of 6 nucleotides.

**4.** Because of a triplet code, a trinucleotide sequence will, once initiated, remain in the same reading frame and produce the same code all along the sequence regardless of the initiation site. If a tetranucleotide is used, there are four different initiation sites for reading a triplet code. Therefore there will be four different sequences.

**6.** From the repeating polymer ACACA . . . one can say that threonine is either CAC or ACA. From the polymer CAACAA . . . with ACACA . . . , ACA is the only codon in common. Therefore, threonine would have the codon ACA.

**8.** The basis of the technique is that if a trinucleotide contains bases (a codon) that are complementary to the anticodon of a charged tRNA, a relatively large complex is formed, which contains the ribosome, the tRNA, and the trinucleotide. This complex is trapped in the filter, whereas the components by themselves are not trapped. If the amino acid on a charged, trapped tRNA is radioactive, the filter becomes radioactive.

**10.** Apply the most conservative pathway of change.

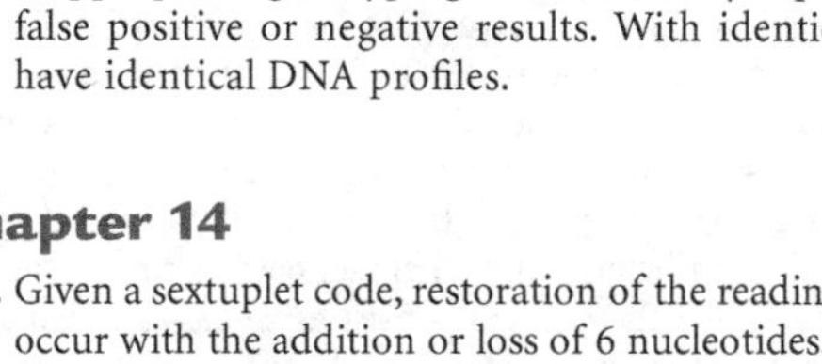


**12.** Because Poly U is complementary to Poly A, double-stranded structures will be formed. In order for an RNA to serve as a messenger RNA, it must be single stranded, thereby exposing the bases for interaction with ribosomal subunits and tRNAs.

**14.** (a)

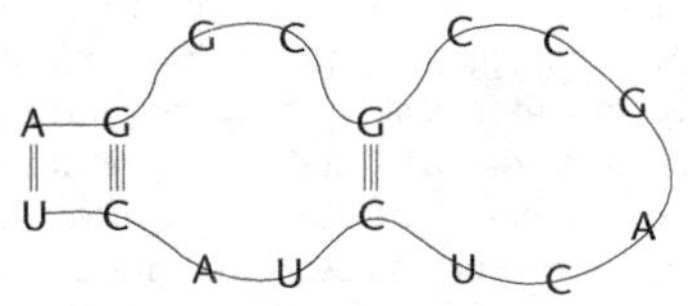

(b) TCCGCGGCTGAGATGA (use complementary bases, substituting T for U)
(c) GCU
(d) Assuming that the AGG. . . is the 5′ end of the mRNA, the sequence would be

*arg-arg-arg-leu-tyr*

**16.** (a) met-his-thr-tyr-glu-thr-leu-gly

met-arg-pro-leu-asp (or glu)

(b) In the shorter of the two reading sequences (the one using the internal AUG triplet), a UGA triplet was introduced at the second codon. While not in the reading frames of the longer polypeptide (using the first AUG codon), the UGA triplet eliminates the product starting at the second initiation codon.

**18.** The central dogma of molecular genetics and to some extent, all of biology, states that DNA produces, through transcription, RNA, which is "decoded" (during translation) to produce proteins.

**20.** RNA polymerase from *E. coli* is a complex, large (almost 500,000 daltons) molecule composed of subunits $\alpha, \beta, \beta', \sigma$) in the proportion $\alpha 2, \beta, \beta', \sigma$ for the holoenzyme. The $\beta$ subunit provides catalytic function while the sigma ($\sigma$) subunit is involved in recognition of specific promoters. The core enzyme is the protein without the sigma.

**22.** Although some folding (from complementary base pairing) may occur with mRNA molecules, they generally exist as single-stranded structures that are quite labile. Eukaryotic mRNAs are generally processed such that the 5′ end is "capped" and the 3′ end has a considerable string of adenine bases. It is thought that these features protect the mRNAs from degradation. Such stability of eukaryotic mRNAs probably evolved with the differentiation of nuclear and cytoplasmic functions. Because prokaryotic cells exist in a more unstable environment (nutritionally and physically, for example) than many cells of multicellular organisms, rapid genetic response to environmental change is likely to be adaptive. To accomplish such rapid responses, a labile gene product (mRNA) is advantageous. A pancreatic cell, which is developmentally stable and exists in a relatively stable environment, could produce more insulin on stable mRNAs for a given transcriptional rate.

**24.** The immediate product of transcription of an RNA destined to become an mRNA involves modification of the 5′ end to which a 7-methylguanosine cap is added. In addition, a stretch of as many as 250 adenylic acid residues is added to the 3′ end after removal of a AAUAAA sequence. The vast majority of eukaryotic genes also contain intervening sequences that are removed, often in a variety of combinations, during the maturation process. In some organisms, RNA editing occurs in one of two ways; substitution editing where nucleotides are altered and insertion/deletion editing that changes the total number of bases.

| *Processing location* | *Example* |
|---|---|
| 5′-end | addition of 7-mG |
| 3′-end | poly-A addition |
| internal | removal of internal sequences |
| | RNA editing |
| | substitution |
| | insertion/deletion |

**28.**

| | |
|---|---|
| Proline: | $C_3$, and one of the $C_2A$ triplets |
| Histidine: | one of the $C_2A$ triplets |
| Threonine: | one $C_2A$ triplet and one $A_2C$ triplet |
| Glutamine: | one of the $A_2C$ triplets |
| Asparagine: | one of the $A_2C$ triplets |
| Lysine: | $A_3$ |

**30.** (a, b) Use the code table to determine the number of triplets that code each amino acid, then construct a graph and plot such as this one:

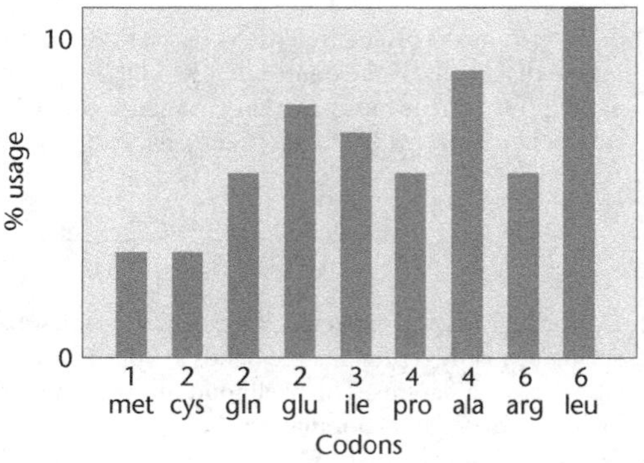


(c) There appears to be a weak correlation between the relative frequency of amino acid usage and the number of triplets for each.
(d) To continue to investigate this issue, one might examine additional amino acids in a similar manner. In addition, different phylogenetic groups use code synonyms differently. It may be possible to find situations in which the relationships are more extreme. One might also examine more proteins to determine whether such a weak correlation is stronger with different proteins.

**32.** (a) guccc aaccaugcccaccgaucuuccgccugcuucugaagAUGCGGGCC CAG
(b) 5′ gtc cca acc **atg** ccc acc gat ctt ccg cct gct tct gaa gAT GCG GGC CCA G
(c) 5′ gtcccaaccatgcccaccgatcttccgcctgcttctgaag **ATG** CGG GCC CAG
The two initiator codons are not in phase.
(d) 5′ gtc cca acc **atg** ccc acc gat ctt ccg cct gct tct gaa gAT GCG GGC CCA G
**met pro thr asp leu pro pro ala ser glu asp ala gly pro**
5′ gtcccaaccatgcccaccgatcttccgcctgcttctgaag **ATG** CGG GCC CAG
**met arg ala gln**
The amino acid sequences in the region of overlap are not the same.
(e) One might argue for the conservation of DNA by having the same region code for a multiple of products, and that might be the case in viruses and prokaryotes where genomic efficiency is more of an issue. However, eukaryotes appear to be much less likely to evolve strategies that conserve DNA sequences per se. However, if functionally and/or structurally related products can be conveniently regulated by such an arrangement, then perhaps an evolutionary advantage exists. The most obvious disadvantage is that if a mutation occurs in the common region, two gene products are altered instead of one.

**34.** (a)

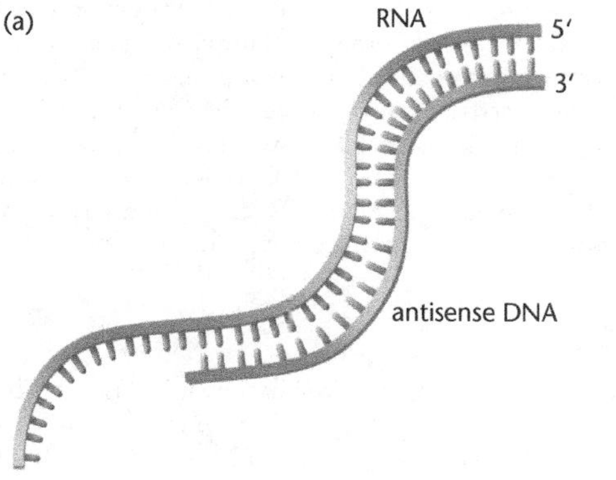


(b)

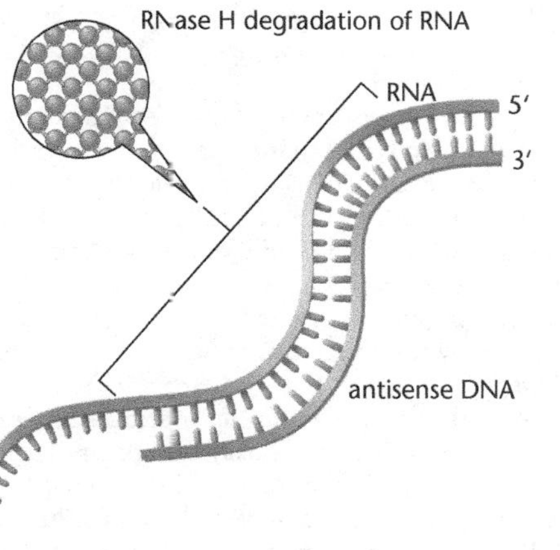


(c) Since the antisense strand is relatively short, under physiological conditions, nonspecific binding of the antisense strand to other RNAs in the cell may cause the degradation of nontargeted RNAs.

(d) Since the exact behavior of antisense DNA in a cell is not completely known, the answer to this question is elusive. However, given the complexity of intracellular events, it is likely that reduction of expression of a repressor gene group may induce other genes by eliminating such repressors.

(e) Among the major drawbacks to antisense therapy are the lack of specificity described here and the relative unknown behavior of DNA oligodeoxynucleotides in living cells.

## Chapter 15

**4.** The sequence of base triplets in mRNA constitutes the sequence of codons. A three-base portion of the tRNA constitutes the anticodon.

**6.** Dividing 20 by 0.34 gives the number of nucleotides (about 59) occupied by a ribosome. Dividing 59 by 3 gives the approximate number of triplet codes, approximately 20.

**8.** The four sites in tRNA that provide for specific recognition are the following: attachment of the specific amino acid, interaction with the aminoacyl tRNA synthetase, interaction with the ribosome, and interaction with the codon (anticodon).

**10.** Both phenylalanine and tyrosine can be obtained from the diet. Even though individuals with PKU cannot convert phenylalanine to tyrosine, it is obtained from the diet.

**12.** When an expectant mother returns to consumption of phenylalanine in her diet, she subjects her baby to higher than normal levels of phenylalanine throughout its development. Since increased phenylalanine is toxic, many (approximately 90 percent) newborns are severely and irreversibly retarded at birth. Expectant mothers (who are genetically phenylketonurics) should return to a low phenylalanine intake during pregnancy.

**14.**

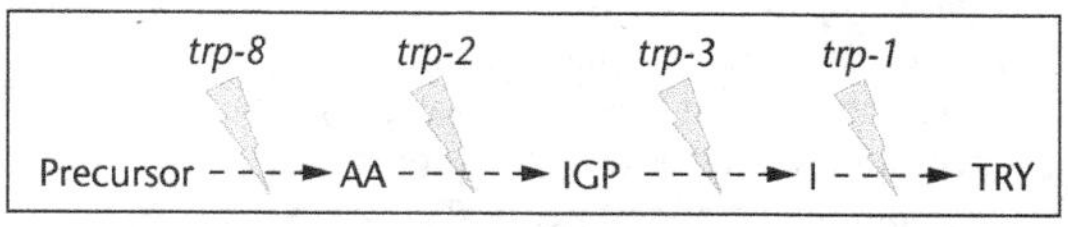


**16.** The fact that enzymes are a subclass of the general term *protein*, a *one-gene:one-protein* statement might seem to be more appropriate. However, some proteins are made up of subunits, each different type of subunit (polypeptide chain) being under the control of a different gene. Under this circumstance, the *one-gene:one-polypeptide* might be more reasonable, however, many additional complexities exist, and a simple statement regarding the relationship of a stretch of DNA to its physical product is difficult to justify.

**18.** The following types of normal hemoglobin:

| *Hemoglobin* | *Polypeptide chains* |
|---|---|
| HbA | $2\alpha 2\beta$ (alpha, beta) |
| $HbA_2$ | $2\alpha 2\delta$ (alpha, delta) |
| HbF | $2\alpha 2\gamma$ (alpha, gamma) |
| Gower 1 | $2\zeta 2\varepsilon$ (zeta, epsilon) |

The *alpha* and *beta* chains contain 141 and 146 amino acids, respectively. The *zeta* chain is similar to the *alpha* chain, while the other chains are like the *beta* chain.

**20.** In the late 1940s, Pauling demonstrated a difference in the electrophoretic mobility of HbA and HbS (sickle-cell hemoglobin) and concluded that the difference had a chemical basis. Ingram determined that the chemical change occurs in the primary structure of the globin portion of the molecule using the fingerprinting technique. He found a change in the sixth amino acid in the $\beta$ chain.

**22.** One would expect individuals with HbC to suffer some altered hemoglobin function and perhaps be resistant to malaria as well. In fact, HbC homozygotes suffer mild hemolytic anemia (a benign hemoglobinopathy). The HbC gene is distributed particularly in malarial-infested areas, suggesting that some resistance to malaria is conferred. Recent studies indicate that HbC may be protective against severe forms of malaria, but not to more uncomplicated forms.

**24.** Having the precise intragenic location of mutations as well as the ability to isolate the products, especially mutant products, allows scientists to compare the locations of lesions within genes. Mutations occurring near the 5′ end of a gene will produce proteins with defects near the N-terminus. In this problem, the lesions cause chain termination; therefore, the nearer the mutations are to the 5′ end of the mRNA, the shorter will be the polypeptide product and thus demonstrate the colinear relationship of genes and proteins.

**26.** Yanofsky's work on the *trp* A locus in bacteria involved the mapping of mutations and the finding that a relationship exists between the position of the mutation in a gene and the amino acid change in a protein. Work with the MS2 by Fiers showed, by sequencing of the coat protein (129 amino acids) and the gene (387 nucleotides), a linear relationship as predicted by the code word dictionary. Because Fiers showed a direct relationship between codons, amino acids and punctuation (initiation and termination), one would consider it direct evidence for colinearity.

**28.** There are probably as many different types of proteins as there are different types of structures and functions in living systems. Your text lists the following:

*Oxygen transport:* hemoglobin, myoglobin
*Structural*: collagen, keratin, histones
*Contractile*: actin, myosin
*Immune system:* immunoglobins
*Cross-membrane transport*: a variety of proteins in and around membranes, such as receptor proteins
*Regulatory:* hormones, perhaps histones
*Catalytic:* enzymes

**30.** An exon often encodes a functional domain within a protein rather than a random region. Evidence supporting exon shuffling comes from discoveries of DNA sequences related to functional domains of proteins that appear to have been recruited and shuffled during evolution. In addition, the partial architectures of genes with respect to exons appear to be conserved but also modified as if altered by shuffling. The "intron-early" theory suggests that introns appeared early in evolution but were selected against. This model is supported by DNA sequence data whereby similar gene architecture is shared by distantly related species. The intron-late theory suggests that introns arose late in evolution with the origin of eukaryotes. The fact that introns are almost totally absent in prokaryotes and infrequent in yeast supports this view.

**32.** One can conclude that the amino acid is not involved in recognition of the codon.

**34.** Because cross **(a, b)** is essentially a monohybrid cross, there would be no difference in the results if crossing over occurred (or did not occur) between the *a* and *b* loci.

**36.** (a) outcomes from the last cross with its 9:4:3 ratio suggest two gene pairs.
(b) orange = *Y_R_*
yellow = *yyrr*, *yy*R_
red = *Y_rr*
(c) white → yellow —*y*-> red —*r*-> orange (pathway V)

**38.**

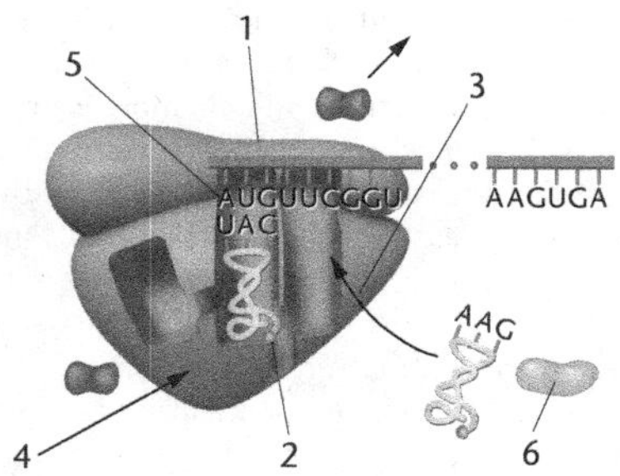


**40.** (a) Since protein synthesis is dependent on the passage of mRNA from DNA to ribosomes, any circumstance that compromises this flow will cause a reduction in protein synthesis. The more specific the binding of the antisense oligonucleotide to the target mRNA, the more specific the influence on protein synthesis. The ideal situation would be where a particular species of antisense oligonucleotide impacted on one and only one protein population.
(b) Clearly, a length of around 15–16 nucleotides is most effective in causing RNA degradation.
(c) A number of factors, including length of the oligonucleotide, are probably involved *in vivo*. Stability of the oligonucleotide is likely dependent on its base composition and length. The oligonucleotide must be small enough to diffuse effectively throughout the cell in order to "locate" the targeted mRNA, and it must not assume a folded conformation that blocks opportunities for base pairing. Since the oligonucleotide is so much smaller than the target mRNA, it is also likely that the actual location of binding to the target is important in mRNA degradation. One of the main problems of antisense therapy is the introduction of the oligonucleotide into the interior of target cells.

## Chapter 16

**2.** If a single somatic cell of a multicellular organism is mutated, it is highly unlikely that the organism will be sufficiently altered to respond to a screen because none of the other cells in that organism will have the same mutation. That's not to say that somatic mutations cannot influence the organism. Cancer cells generally originate from a single altered cell and can have a profound influence on the fate of an organism.

**4.** Each gene and its product function in an environment that has also evolved, or coevolved. A coordinated output of each gene product is required for life. Deviations from the norm, caused by mutation, are likely to be disruptive because of the complex and interactive environment in which each gene product must function. However, on occasion a beneficial variation occurs.

**6.** A *conditional* mutation is one that produces a wild-type phenotype under one environmental condition and a mutant phenotype under a different condition.

**8.** All three of the agents are mutagenic because they cause base substitutions. Deaminating agents oxidatively deaminate bases such that cytosine is converted to uracil and adenine is converted to hypoxanthine. Uracil pairs with adenine and hypoxanthine pairs with cytosine. Alkylating agents donate an alkyl group to the amino or keto groups of nucleotides, thus altering base-pairing affinities. 6-ethyl guanine acts like adenine, thus pairing with thymine. Base analogs such as 5-bromouracil and 2-amino purine are incorporated as thymine and adenine, respectively, yet they base-pair with guanine and cytosine, respectively.

**10.** X rays are of higher energy and shorter wavelength than UV light. They have greater penetrating ability and can create more disruption of DNA.

**12.** *Photoreactivation* can lead to repair of UV-induced damage. An enzyme, photoreactivation enzyme, will absorb a photon of light to cleave thymine dimers. *Excision repair* involves the products of several genes, DNA polymerase I, and DNA ligase to clip out the UV-induced dimer, fill in, and join the phosphodiester backbone in the resulting gap. The excision repair process can be activated by damage which distorts the DNA helix. *Recombinational repair* is a system that responds to DNA that has escaped other repair mechanisms at the time of replication. If a gap is created on one of the newly synthesized strands, a "rescue operation or SOS response" allows the gap to be filled. Many different gene products are involved in this repair process: *rec*A, *lex*A. In SOS repair, the proofreading by DNA polymerase III is suppressed and this therefore is called an "error-prone system."

**14.** Each involves amplification of trinucleotide repeats. As the degree of amplification increases, so does the degree of expression of the abnormal phenotype. (See the text for a detailed description of the role of trinucleotide repeats in a variety of human diseases.) Genetic anticipation is the occurrence of an earlier age of onset of a genetic disease in successive generations.

**16.** *Xeroderma pigmentosum* is a form of human skin cancer caused by perhaps several rare autosomal genes that interfere with the repair of damaged DNA. Studies with heterokaryons provided evidence for complementation, indicating that as many as seven different genes may be involved. The photoreactivation repair enzyme appears to be involved.

**18.** It is possible that through the reduction of certain environmental agents that cause mutations, mutation rates might be reduced. On the other hand, certain industrial and medical activities actually concentrate mutagens (radioactive agents and hazardous chemicals). Unless human populations are protected from such agents, mutation rates might actually increase. If one asks about the accumulation of mutations (not rates) in human populations as a result of improved living conditions and medical care, it is likely that as the environment becomes less harsh (through improvements), more mutations will be tolerated as selection pressure decreases. However, as individuals live longer and have children at a later age, some studies indicate that older males accumulate more gametic mutations.

**20.** Both major forms of muscular dystrophy include muscular wasting of differing severity and age of onset. Both forms are caused by mutations in the *dystrophin* gene, which is very large, composed of 97 exons and 2.6 Mb in length. Given the size of this gene and the number of exons/introns, many opportunities exist for mutational upset.

**22.** An unexpected mutant gene may enter a pedigree in several ways. If a gene is incompletely penetrant, it may be present in a population and only express itself under certain conditions. It is unlikely that the gene for hemophilia behaved in this manner. If a gene's expression is suppressed by another mutation in an individual, it is possible that offspring may inherit a given gene and not inherit its suppressor. Such offspring would have hemophilia. Since all genetic variations must arise at some point, it is possible that the mutation in the Queen Victoria family was new, arising in the father. Lastly, it is possible that the mother was heterozygous and by chance, no other individuals in her family were unlucky enough to receive the mutant gene.

**24.** Unscheduled DNA synthesis represents DNA repair. One can determine complementation groupings by placing each heterokaryon giving a "−" into one group and those giving a "+" into a separate group. For instance, *XP1* and *XP2* are placed into the same group because they do not complement each other. However, *XP1* and *XP5* do complement ("+"); therefore they are in different groups. Completing such pairings allows one to determine the following groupings:

| | | |
|---|---|---|
| *XP1* | *XP4* | *XP5* |
| *XP2* | | *XP6* |
| *XP3* | | *XP7* |

The groupings (complementation groups) indicate that at least three "genes" form products necessary for unscheduled DNA synthesis. All of the cell lines that are in the same complementation group are defective in the same product.

**26.** The cystic fibrosis gene produces a complex membrane transport protein that contains several major domains: a highly conserved ATP binding domain, two hydrophobic domains, and a large cytoplasmic domain, which probably serves in a regulatory capacity. The protein is like many ATP-dependent transport systems, some of which have been well studied. When a mutation causes clinical symptoms, fluid secretion is decreased and dehydrated mucus accumulates in the lungs and air passages. Mutations that radically alter the structure of the protein (frameshift, splicing, nonsense, deletions, duplications, etc.) would probably have more influence on protein function than those that cause relatively minor amino acid substitutions, although this generalization does not always hold true. A protein with multiple functional domains would be expected to react to mutational insult in a variety of ways.

**28.** Nonsense mutation in coding regions: shortened product somewhat less than 375 amino acids depending on where the chain termination occurred. Insertion in Exon 1, causing frameshift: a variety of amino acid substitutions and possible chain termination downstream. Insertion in Exon 7, causing frameshift: a variety of amino acid substitutions and possible chain termination downstream involving less of the protein than the insertion in Exon 1. Missense mutation: an amino acid substitution. Deletion in Exon 2, causing frameshift: depending on the size of the deletion, a few to many amino acids may be missing. The frameshift would cause additional amino acid changes and possible termination. Deletion in Exon 2, in frame: amino acids missing from Exon 2 without additional changes in the protein. Large deletion covering Exons 2 and 3: significant loss of amino acids toward the N-terminal side of the protein.

**30.** (a, b, c) Although a positive Ames test does not prove carcinogenicity, it is highly likely to be the case. The control plate indicates that some mutations occur as normal background in the buffer. In Test #1, the relatively high concentration of compound Z is so mutagenic, that all bacteria were killed. In Test #3, while the extract partially reduced the mutagenicity of compound Z, it is still mutagenic. Test #2 shows some clearing (killing) of bacteria around the disk because of considerable mutagenicity. As the more dilute solution of compound Z diffuses out to the periphery of the plate, some $his^+$ bacteria survive. (d) There is no clear space around the disk in Test #3 because the extract reduced the mutagenicity of compound Z to a considerable degree. The similarity of the control plate with Test #3 indicates that the extract is quite effective in reducing the mutagenicity of compound Z. (e) The Ames test is useful as a preliminary screen for testing the mutagenicity of compounds; however, within an organism, compounds are metabolized differentially. In addition, different organisms metabolize compounds in different ways. (f) Because the extract was apparently capable of altering the mutagenicity of compound Z, one might fractionate the extract into its components to determine the nature and concentration of the component(s) involved in reducing the mutagenicity of compound Z.

## Chapter 17

**2.** Under *negative* control, the regulatory molecule interferes with transcription while in *positive* control, the regulatory molecule stimulates transcription.

**4.** (a) Due to the deletion of a base early in the *lac Z* gene, there will be "frameshift" of all the reading frames downstream from the deletion. It is likely that either premature chain termination of translation will occur (from the introduction of a nonsense triplet in a reading frame) or the normal chain termination will be ignored. Regardless, a mutant condition for the *Z* gene will be likely. If such a cell is placed on a lactose medium, it will be incapable of growth because $\beta$-galactosidase is not available.

(b) If the deletion occurs early in the *A* gene, one might expect impaired function of the *A* gene product, but it will not influence the use of lactose as a carbon source.

**6.** $I^+O^+Z^+$ = Because of the function of the active repressor from the $I^+$ gene, and no lactose to influence its function, there will be **No Enzyme Made**.

$I^+O^cZ^+$ = There will be a **Functional Enzyme Made** because the constitutive operator is in *cis* with a *Z* gene. The lactose in the medium will have no influence because of the constitutive operator. The repressor cannot bind to the mutant operator.

$I^-O^+Z^-$ = There will be a **Nonfunctional Enzyme Made** because with $I^-$ the system is constitutive but the *Z* gene is mutant. The absence of lactose in the medium will have no influence because of the nonfunctional repressor. The mutant repressor cannot bind to the operator.

$I^-O^+Z^-$ = There will be a **Nonfunctional Enzyme Made** because with $I^-$ the system is constitutive but the *Z* gene is mutant. The lactose in the medium will have no influence because of the nonfunctional repressor. The mutant repressor cannot bind to the operator.

$I^-O^+Z^+/F'I^+$ = There will be **No Enzyme Made** because in the absence of lactose, the repressor product of the $I^+$ gene will bind to the operator and inhibit transcription.

$I^+O^cZ^+/F'O^+$ = Because there is a constitutive operator in *cis* with a normal *Z* gene, there will be **Functional Enzyme Made.** The lactose in the medium will have no influence because of the mutant operator.

$I^+O^+Z^-/F'I^+O^+Z^+$ = Because there is lactose in the medium, the repressor protein will not bind to the operator and transcription will occur. The presence of a normal *Z* gene allows a **Functional and Nonfunctional Enzyme to Be Made.** The repressor protein is diffusable, working in *trans.*

$I^-O^+Z^-/F'I^+O^+Z^+$ = Because there is no lactose in the medium, the repressor protein (from $I^+$) will repress the operators and there will be **No Enzyme Made.**

$I^sO^+Z^+/F'O^+$ = With the product of $I^s$ there is binding of the repressor to the operator and therefore **No Enzyme Made.** The lack of lactose in the medium is of no consequence because the mutant repressor is insensitive to lactose.

$I^+O^cZ^+/F'O^+Z^+$ = The arrangement of the constitutive operator ($O^c$) with the *Z* gene will cause a **Functional Enzyme to Be Made.**

8. A single *E. coli* cell contains very few molecules of the *lac* repressor. However, the *lac* $I^q$ mutation causes a 10× increase in repressor protein production, thus facilitating its isolation. With the use of dialysis against a radioactive gratuitous inducer (IPTG), Gilbert and Muller-Hill were able to identify the repressor protein in certain extracts of *lac* $I^q$ cells. The material that bound the labeled IPTG was purified and shown to be heat labile and to have other characteristics of protein. Extracts of *lac* $I^-$ cells did not bind the labeled IPTG.
10. (a) With no lactose and no glucose, the operon is off because the *lac* repressor is bound to the operator. Although CAP is bound to its binding site, it will not override the action of the repressor.
(b) With lactose added to the medium, the *lac* repressor is inactivated and the operon is transcribing the structural genes. With no glucose, the CAP is bound to its binding site, thus enhancing transcription.
(c) With no lactose present in the medium, the *lac* repressor is bound to the operator region, and since glucose inhibits adenyl cyclase, the CAP protein will not interact with its binding site. The operon is therefore "off."
(d) With lactose present, the *lac* repressor is inactivated. However, since glucose is also present, CAP will not interact with its binding site. Under this condition, transcription is severely diminished and the operon can be considered to be "off."
12. Attenuation functions to reduce the synthesis of tryptophan when it is in full supply. It does so by reducing transcription of the *tryptophan* operon. The same phenomenon is observed when tryptophan activates the repressor to shut off transcription of the *tryptophan* operon.
14. Neelaredoxin appears to be a protein that defends anaerobic and perhaps aerobic organisms from oxidative stress brought on by the metabolism of oxygen. The generation of oxygen free radicals (creates the oxidative stress) is dependent on several molecular species, including $O_2$ and $H_2O_2$. Apparently, relatively high levels of neelaredoxin are produced at all times (*constitutively expressed*), even when potential inducers of gene expression are not added to the system. Additional neelaredoxin gene expression is not responsive (*induced*) as a result of $O_2$ and $H_2O_2$ treatment.
16. When arabinose is present in the medium, the structural genes for the *arabinose* operon are transcribed. If the structural genes for the *lac* operon replaced the structural genes for the *ara* operon, then in the presence of arabinose, the *lac* structural genes would be transcribed and $\beta$-galactosidase would be produced at induced levels.
18. Since a substance supplied in the medium (the antibiotic) causes the synthesis of the efflux pump components, two situations seem appropriate. Under a *negative control* system, the antibiotic would interrupt the repressor to bring about induction (this would be an inducible system). Under *positive control,* the antibiotic would activate an activator (this again would be an inducible system).
20. Because the deletion of the regulatory gene causes a loss of synthesis of the enzymes, the regulatory gene product can be viewed as one exerting *positive control.*

When *tis* is present, no enzymes are made, therefore, *tis* must inactivate the positive regulatory protein. When *tis* is absent, the regulatory protein is free to exert its positive influence on transcription. Mutations in the operator negate the positive action of the regulator. The model below illustrates these points.

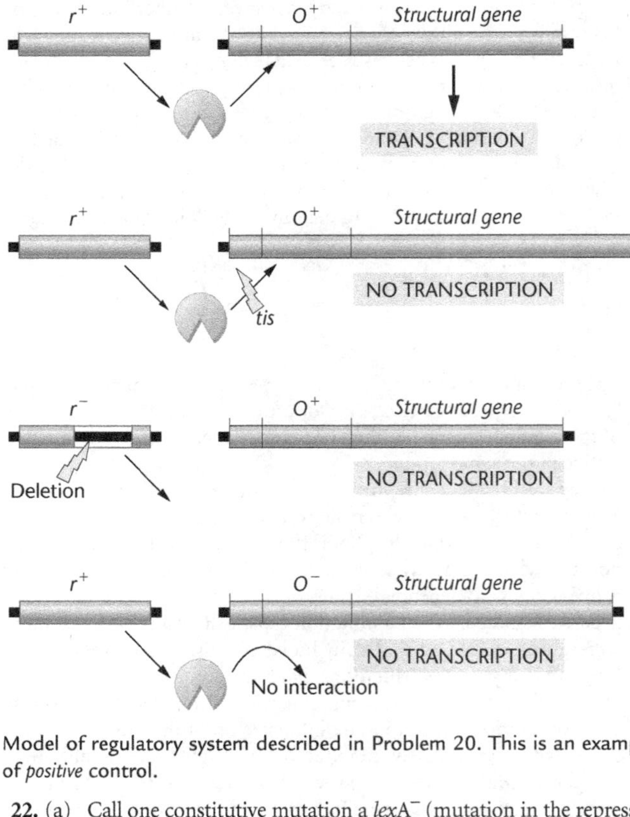


Model of regulatory system described in Problem 20. This is an example of *positive* control.

22. (a) Call one constitutive mutation a *lexA*$^-$ (mutation in the repressor gene product) and the other $O^{uvrA^-}$ (mutation in the operator).

(b) One can make partial diploid strains using F′. $O^{uvrA^-}$ will be dominant to $O^{uvrA^+}$ and *lex*A$^-$ will be recessive to *lex*A$^+$. $O^{uvrA^-}$ will act in *cis*.

24. To get started, find the CACUUCC sequence. It pairs, with one mismatch, with a second region. *Hint*: The third region is composed of seven bases and starts with an AG.

26. (a) The simplest model for the action of R and D in *Chlamydia* would be to have the repressor element (R) become ineffective in binding the *cis*-acting element (D) when heat-shocked. This could happen in two ways. Either the supercoiled DNA alters its conformation and becomes ineffective at binding R or the D-binding efficiency of the R protein is altered by heat. In either case, the genes for infectivity are transcribed in the presence of the heat shock.

(b) The most straightforward comparison between the heat-shock R and D system in *Chlamydia* and the heat-shock sigma factor in *E. coli* would be where the R-D system is inactivated in *Chlamydia* and a sigma factor is activated by heat in *E. coli*.

## Chapter 18

4. When DNA is transcriptionally active, it is in a less condense state and as such, is more open to DNase digestion.

6. In general, chromatin is remodeled when there are significant changes in chromatin organization. Such remodeling involves changes in DNA methylation and interaction of DNA with histones in nucleosomes. Nucleosome remodeling complexes alter nucleosome structure and position by a number of processes including histone modification.

8. Transcription factors are proteins that are *necessary* for the initiation of transcription. However, they are not *sufficient* for the initiation of transcription. To be activated, RNA polymerase II requires a number of transcription factors. Transcription factors contain at least two functional domains: one binds to the DNA sequences of promoters and/or enhancers, while the other interacts with RNA polymerase or other transcription factors. Some transcription factors bind to other transcription factors without themselves binding to DNA.

10. In general, one determines the influence of various regulatory elements by removing necessary elements or adding extra elements. Assay systems determine the relative levels of gene expression after such action.

12. RNA interference begins with a double-stranded RNA being processed by a protein called Dicer, which, in combination with RISC, generates short interfering RNA (siRNA). Unwinding of siRNA produces an antisense strand that combines with a protein to cleave mRNA complementary sequences. Short RNAs called microRNAs pair with the 3′-untranslated regions of mRNAs and block their translation.

14.

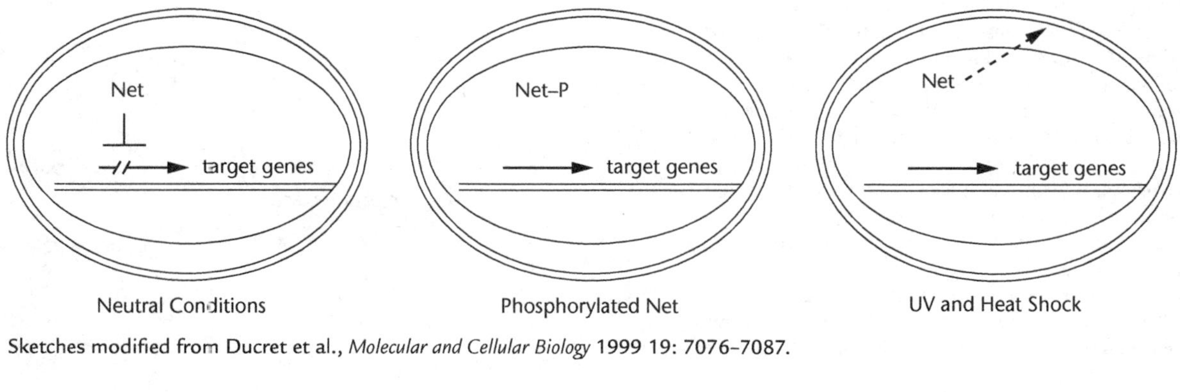


Sketches modified from Ducret et al., *Molecular and Cellular Biology* 1999 19: 7076–7087.

16. Given that DNA methylation plays a role in gene expression in mammals, any change in DNA methylation, plus or minus, can potentially have a negative impact on progeny development. In addition, since m$^5$C >>> thymine, transitions are likely to cause mutations in coding regions of DNA, when methylation patterns change, new sites for mutation arise. Should mutations occur at a higher rate in previously unmethylated sites (genes), embryonic development is likely to be affected.

18. The work of Cleveland and colleagues allowed selective changes to be made in the *met-arg-glu-lys* sequence. Only the engineered mRNA sequences that caused an amino acid substitution negated the autoregulation, indicating that it is the sequence of the amino acids, not the mRNA, that is critical in the process of autoregulation. Notice that code degeneracy allows for changes in mRNA sequence without changes in the amino acid sequence. Therefore, the model that depicts binding of factors to the nascent polypeptide chain is supported. A variety of experiments could be used to substantiate such a model. One might stabilize the proposed MREI-protein complex with "crosslinkers," treat with RNase to digest mRNA and to break up polysomes, and then isolate individual ribosomes. One may use some specific antibody or other method to determine whether tubulin subunits contaminate the ribosome population.

20. Since multiple routes may lead to cancer, one would expect complex regulatory systems to be involved. More specifically, while in some cases, downregulation of a gene, such as an oncogene, may be a reasonable cancer therapy, downregulation of a tumor-suppressor gene would be undesirable in therapy.

22. Below is a sketch of several RNA polymerase molecules (filled circles) in what might be a transcription factory. This diagram shows eight RNA pol II molecules being transcribed. Nascent transcripts are shown in red. For simplicity, only one promoter (blue) and one structural gene (yellow) are shown.

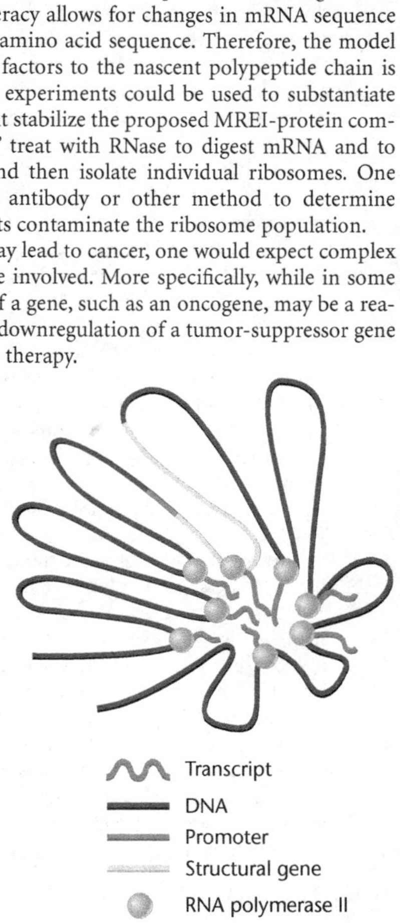

**24.** Since mRNA stability is directly linked to the likelihood of translation, one could use a number of different constructs as shown below and test for luciferase activity in the assay system described in the problem.

| *5′-cap* | *3′-tail* | *mRNA (no cap, no tail)* |
|---|---|---|
| − | − | + |
| + | − | + |
| − | + | + |
| + | − | − |
| − | + | − |

**26.** If mRNA transport is achieved by diffusion, then regulation can be achieved by the positioning of different concentrations of ribosomes. The closer a ribosome pool is to the source of a particular mRNA source, the higher the likelihood that a given mRNA would be translated. In addition, it may be that certain territories in cells are less hostile to traveling mRNAs, which would also favor the translation of certain mRNAs. If nuclear territories purge their RNAs at different locations around the nucleus, then even if diffusion is the mode of mRNA transport, different cellular domains would receive different types of mRNAs. If such domains have different translation efficiencies, then genetic regulation is achieved.

**28.** Criteria for determining the conservation of alternative splicing patterns would likely include the following:
1. similar mRNA length
2. conservation of splice junctions
3. positioning of homologous introns
4. size of homologous introns
5. exon nucleotide sequence homology
6. predicted amino acid physicochemical similarity
7. same 5′-3′ orientation
8. conserved use of alternative stop codons in frameshift splicing events
9. conserved use of alternative frames of translation

**30.** When splice specificity is lost, one might observe several classes of altered RNAs: (1) a variety of nonspecific variants producing RNA pools with many lengths and combinations of exons and introns, (2) incomplete splicing where introns and exons are erroneously included or excluded in the mRNA product, and (3) a variety of nonsense products, which result in premature RNA decay or truncated protein products. It is presently unknown as to whether cancer-specific splices initiate or result from tumorigenesis. Given the complexity of cancer induction and the maintenance of the transformed cellular state, gene products that are significant in regulating the cell cycle may be influenced by alternative splicing and thus contribute to cancer.

## Chapter 19

**2.** The fact that nuclei from almost any source remain transcriptionally and translationally active substantiates the fact that the genetic code and the ancillary processes of transcription and translation are compatible throughout the animal and plant kingdoms. Because the egg represents an isolated, "closed" system that can be mechanically, environmentally, and, to some extent, biochemically manipulated, various conditions may be developed that allow one to study facets of gene regulation.

**4.** The syncytial blasterm is formed as nuclei migrate to the egg's outer margin or cortex, where additional divisions take place. Plasma membranes organize around each of the nuclei at the cortex, thus creating the cellular blastoderm.

**6.** It is possible that your screen was more inclusive; that is, it identified more subtle alterations than the screen of Wieschaus and Schupbach. In addition, your screen may have included some zygotic effect mutations, which were dependent on the action of maternal-effect genes. You may have identified several different mutations in some of the same genes.

**8.** The three main classes of zygotic genes are (a) *gap* genes, which specify adjacent segments, (2) *pair-rule* genes, which specify every other segment and a part of each segment, and (3) *segment polarity* genes, which specify homologous parts of each segment.

**10.** If protein products of a given gene are present in different cell types, it can be assumed that the responsible gene is being transcribed. If one is able to actually observe, microscopically, gene activity, as is the case in some specialized chromosomes (polytene chromosomes), gene activity can be inferred by the presence of localized chromosomal puffs. A more direct and common practice to assess transcription of particular genes is to use labeled probes. If a labeled probe can be obtained that contains base sequences complementary to the transcribed RNA, then such probes will hybridize to that RNA if present in different tissues. This technique is called *in situ* hybridization and is a powerful tool in the study of gene activity during development.

**12.** *Hox* genes in the *Drosophila* genome can be found in two clusters on chromosome 3: Antp-C and BX-C. They have two properties in common: encoding of transcription factors and colinear gene expression. A *homeotic* gene alters the identity of a segment or field within a segment. Not all homeotic genes are *Hox* genes.

**14.** Many of the appendages of the head, including the mouth parts and the antennae, are evolutionary derivatives of ancestral leg structures. In *spineless aristapedia,* the distal portion of the antenna is replaced by its ancestral counterpart, the distal portion of the leg (tarsal segments). Because the replacement of the arista (end of the antenna) can occur by a mutation in a single gene, one would consider that one "selector" gene distinguishes aristal from tarsal structures.

**16.** Because the engrailed product is absent in *ftz/ftz* embryos and *ftz* expression is normal in *en/en* embryos, one can conclude that the *ftz* gene product regulates *en* either directly or indirectly. Because the *ftz* gene is expressed normally in *en/en* embryos, the product of the *engrailed* gene does not regulate expression of *ftz.*

**18.** Three classes of flower *homeotic* genes are known that are activated in an overlapping pattern to specify various floral organs. Class *A* genes give rise to sepals. Expression of *A* and *B* class genes specify petals, *B* and *C* genes control stamen formation, and expression of *C* genes gives rise to carpels.

**20.**

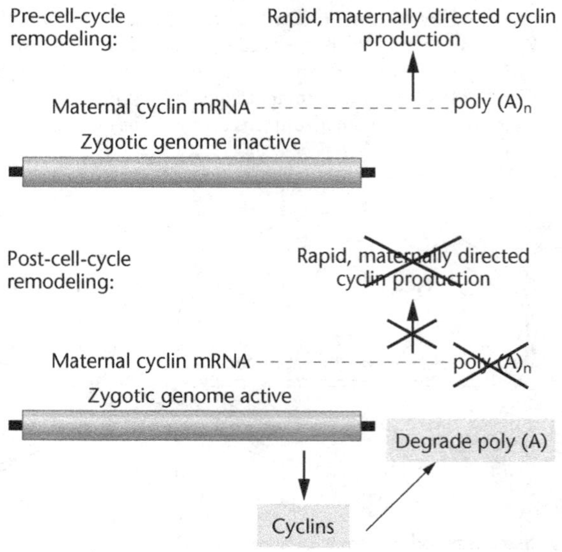

**22.** The typical developmental sequence for axis and segment formation in *Drosophila* proceeds from the gap genes to the pair-rule genes to the segment polarity genes. The fact that *fushi-tarazu (ftz)* is affected by early (anterior-posterior determining genes) and gap genes indicates that *ftz* functions after those genes. That segment polarity genes are influenced by *ftz* indicates that *ftz* functions earlier, thus placing *ftz* before the segment-polarity genes or in the pair-rule group of genes.

**24.** (a) The term *rescued* is often used when the introduction of genes from an outside source (within or among species) restores the wild-type phenotype from a mutant organism.

(b) Results such as these—and there are many like them—indicate the extreme conservation of protein structure and function across phylogenetically distant organisms. Such results attest to the conservation from a distant common ancestor of fundamental molecular species during development. Failure to adhere to a common developmental theme is rewarded by death.

**26.** Because signal-receptor interactions depend on membrane-bound structures, the pathway can only work with adjacent cells. The advantage of such a system is that only cells in a certain location will be influenced—those in contact. A disadvantage would occur if large groups of cells were to be induced in a particular developmental pathway or if cells not in contact needed to be induced.

**28.** If the *her-1*$^+$ product acts as a negative regulator, then when the gene is mutant, suppression over *tra-1*$^+$ is lost and hermaphroditism would be the result. This hypothesis fits the information provided. The double mutant should be male because even though there is no suppression from *her-1*$^-$, there is no *tra-1*$^+$ product to support hermaphrodite development.

**30.** (a) A number of studies indicate that genes in *Drosophila* have evolutionary counterparts (orthologs) in other organisms, including humans. A number of similar genes influence eye development in both insects and vertebrates.

(b) Genes that produce eyes are part of a complex network of at least seven genes that constitute the master regulators of eye development. Each gene functions in coordination with others in a conserved network that is used by broad evolutionary groups. Such genes, descended from common ancestral genes that have the same function in different species, are called orthologs.

(c) Since development is dependent on the coordinated output of numerous genes, genetic networks are probably the rule rather than the exception. The fact that a single genetic change (in the case of the mouse homolog of the fly *eyeless* gene) can trigger the formation of ectopic eyes in *Drosophila* shows that entire networks are evolutionarily conserved.

## Chapter 20

**2.** The major regulatory points of the cell cycle include the following:
1. late G1 (G1/S)
2. the border between G2 and mitosis (G2/M)
3. in mitosis (M)

**4.** The retinoblastoma gene (*RB1*), located on chromosome 13, encodes a protein designated pRB. Cells progress through the G1/S transition when pRB is phosphorylated and CDK4 binds to cyclin D. In the absence of phosphorylation of pRB, it binds to members of the E2F family of transcription factors, which control the expression of genes required to move the cell from G1 to S. When E2F and other regulators are released by pRB, they are free to induce the expression of over 30 genes whose products are required for the transition from G1 into S phase. After cells traverse S, G2, and M phases, pRB reverts to a nonphosphorylated state, binds to regulatory proteins such as E2F, and keeps them sequestered until required for the next cell cycle.

**6.** Cancer is a complex alteration in normal cell-cycle controls. Even if a major "cancer-causing" gene is transmitted, other genes, often new mutations, are usually necessary in order to drive a cell toward tumor formation. Full expression of the cancer phenotype is likely to be the result of an interplay among a variety of genes and therefore show variable penetrance and expressivity.

**8.** A tumor-suppressor gene is a gene that normally functions to suppress cell division. Since tumors and cancers represent a significant threat to survival and therefore Darwinian fitness, strong evolutionary forces would favor a variety of coevolved and perhaps complex conditions in which mutations in these suppressor genes would be recessive. Looking at it in another way, if a tumor-suppressor gene makes a product that regulates the cell cycle favorably, cellular conditions have evolved in such a way that sufficient quantities of this gene product are made from just one gene (of the two present in each diploid individual) to provide normal function.

**10.** Proto-oncogenes are converted to oncogenes in a number of ways: point mutations in which a mutant gene acts as a positive "switch" in the cell cycle, translocations whereby a hybrid gene might be formed, and overexpression whereby a gene might acquire a new promoter and/or enhancer. In the case of RSV, an oncogene (*c-src*) was captured from the chicken genome.

**12.** Various kinases can be activated by breaks in DNA. One kinase called ATM, and/or a kinase called Chk2, phosphorylate BRCA1 and p53. The activated p53 arrests replication during the S phase to facilitate DNA repair. The activated BRCA1 protein, in conjunction with BRCA2, mRAD51, and other nuclear proteins, is involved in repairing the DNA.

**14.** Proto-oncogenes are those that normally function to promote or maintain cell division. In the mutant state (oncogenes), they induce or maintain uncontrolled cell division; that is, there is a gain-of-function. Generally, this gain-of-function takes the form of increased or abnormally continuous gene output. On the other hand, loss-of-function is generally attributed to tumor-suppressor genes, which function to halt passage through the cell cycle. When such genes are mutant, they have lost their capacity to halt the cell cycle.

**16.** Unfortunately, it is common to spend enormous amounts of money on dealing with diseases after they occur rather than concentrating on disease prevention. Too often pressure from special interest groups or lack of political stimulus retards advances in education and prevention. Obviously, it is less expensive, in terms of both human suffering and money, to seek preventive measures for as many diseases as possible. However, having gained some understanding of the mechanisms of disease, in this case cancer, it must also be stated that, no matter what preventive measures are taken, it will be impossible to completely eliminate disease from the human population. It is extremely important, however, that we increase efforts to educate and protect the human population from as many hazardous environmental agents as possible.

**18.** Normal cells are often capable of withstanding mutational assault because they have checkpoints and DNA repair mechanisms in place. When such mechanisms fail, cancer may be a result. Through mutation, such protective mechanisms are compromised in cancer cells, and as a result they show higher than normal rates of mutation, chromosomal abnormalities, and genomic instability.

**20.** Certain environmental agents such as chemicals and X rays cause mutations. Since genes control the cell cycle, mutations in cell cycle control genes, or those that impact on cell-cycle control can lead to cancer.

**22.** No, she will still have the general population risk of about 10 percent. In addition, it is possible that genetic tests will not detect all breast cancer mutations.

**24.** Various levels of methylation (hypermethylation and hypomethylation) influence gene activity. Changes in gene activity are often associated with cancer. Changes in gene methylation, therefore, can cause cancer.

**26.** Proteases, in general, and serine proteases, specifically, are considered tumor-promoting agents because they degrade proteins, especially those in the extracellular matrix. When such proteolysis occurs, cellular invasion and metastasis is encouraged. Consistent with this observation are numerous observations that metastatic tumor cells are associated with higher than normal amounts of protease expression. Inhibitors of serine proteases are often tested for their anticancer efficacy.

**28.** (a) A genomic library could be constructed of both osteosarcoma cell DNA and noncancerous cells from the same organism. You could then screen the library using labeled probes from the clones carrying the *RB1* gene available to you as stated in the problem. At this point, some indications might emerge because if there is a significant alteration in mutant *RB1* genes, probes may not successfully hybridize to any clones in the cancerous cell DNA library. Assuming that control hybridization occurs in the noncancerous cells, lack of hybridization in the library derived from the osteosarcoma cell line might indicate deletions. However, assuming that hybridization does allow one to identify clones containing putative *RB1* alleles, subcloning into appropriate vectors would allow sequencing to reveal sequence changes in the *RB1* alleles when compared with nonmutant genes. A second approach combines an immunoassay described in part (b) of this problem. Assuming that one can successfully make antibodies to the normal *RB1* gene product (pRB), lack of cross-reactivity of the pRB antibodies to proteins from the cancerous cell line would indicate that both *RB1* alleles are mutant.

(b) As indicated in the last portion of part (a) above, one can make antibodies to pRB from the noncancerous cells and test these antibodies for reactivity against proteins from the cancerous cell lines. A pRB-antibody reaction would indicate that the pRB protein is made.

(c) To determine whether addition of a normal *RB1* gene will change the cancer-causing potential of osteosarcoma cells, one could transfer the cloned normal *RB1* gene into the cells by transformation or transfection (often by electroporation or ultrasound). Transformed cells would then be introduced into the cancer-prone mice to determine whether their cancer-causing potential had been altered.

**30.** (a, b) Even though there are changes in the *BRCA1* gene, they do not always have physiological consequences. Such neutral polymorphisms make screening difficult in that one cannot always be certain that a mutation will cause problems for the patient.

(c) The polymorphism in *PM2* is probably a silent mutation because the third base of the codon is involved.

(d) The polymorphism in *PM3* is probably a neutral missense mutation because the first base is involved.

## Chapter 21

**2.** Whole-genome shotgun sequencing involves randomly cutting the genome into numerous smaller segments. Overlapping sequences are used to identify segments that were once contiguous, eventually producing the entire sequence. Users of this approach often encounter difficulty with repetitive regions of the genome. Map-based sequencing relies on known landmarks (genes, nucleotide polymorphisms, etc.) to orient the alignment of cloned fragments that have been sequenced. Compared to whole-genome sequencing, the map-based approach is somewhat cumbersome and time consuming. Whole-genome sequencing has become the most common method for assembling genome, with map-based cloning being used to resolve the problems often encountered during whole-genome sequencing.

**4.** The question of how to define an organism's genome is complicated by a variety of symbiotic relationships that are known to exist in virtually all organisms. Plasmids are capable of carrying both essential and nonessential genes of the host. To complicate the matter, all cells likely contain nonessential genes. An organism's genome will probably come to encompass all genetic elements that can be shown to be stable cellular inhabitants.

**6.** The main goals of the Human Genome Project are to establish, categorize, and analyze functions for human genes.

**8.** One initial approach to annotating a sequence is to compare the newly sequenced genomic DNA to the known sequences already stored in various databases. The National Center for Biotechnology Information (NCBI) provides access to BLAST (Basic Local Alignment Search Tool) software that directs searches through databanks of DNA and protein sequences. A segment of DNA can be compared to sequences in major databases such as GenBank to identify matches that align in whole or in part. One might seek similarities of a sequence on chromosome 11 in a mouse and find that or similar sequences in a number of taxa. BLAST will compute a similarity score or identity value to indicate the degree of similarity of sequences.

**10.** The human genome is composed of over 3 billion nucleotides in which about 2 percent code for genes. Genes are unevenly distributed over chromosomes with clusters of gene-rich separated by gene-poor ones (deserts). Human genes tend to be larger and contain more and larger introns than in invertebrates such as *Drosophila.* It is estimated that at least half of the genes generate products by alternative splicing. Hundreds of genes have been transferred from bacteria into vertebrates. Duplicated regions are common, which may facilitate chromosomal rearrangement. The human genome appears to contain approximately 20,000 protein-coding genes, though uncertainty as to the total number remains.

**12.** Bacterial genes are densely packed in the chromosome. The protein-coding genes are mostly organized in polycistronic transcription units without introns. Eukaryotic genes are less densely packed in chromosomes, and protein-coding genes are mostly organized as single-transcription units with introns.

**14.** One usually begins to annotate a sequence by comparing it, often using BLAST, to the known sequences already stored in various databases. Similarity to other annotated sequences often provides insight as to a sequence's function. Hallmarks to annotation include the identification of gene-regulatory sequences found upstream of genes (promoters, enhancers, and silencers), downstream elements (termination sequences), and triplet nucleotides that are part of the coding region of the gene. In addition, 5′ and 3′ splice sites that are used to distinguish exons from introns, as well as polyadenylation sites, are also used in annotation. Similar hallmarks are used to annotate prokaryotic genes. Because prokaryotic genes do not contain introns, however, their annotation is sometimes less complicated.

**16.** Metagenomics is a relatively new discipline that examines the genomes from entire communities of microorganisms in environmental samples of water, air, and soil. Virtually every environment on Earth is being sampled in metagenomics projects. A major initiative is a global expedition called the *Sorcerer II* Global Ocean Sampling (GOS), in which researchers travel the globe by yacht and sample flora and fauna. Metagenomics is teaching us more about millions of species of bacteria and viruses, of which only a few thousand have been well characterized.

**18.** $0.8^5 = 33\%$

**20.** Aneuploidy in humans occurs for the sex chromosomes (X and Y) and three of the autosomes (13, 18, and 21). Other aneuploids are apparently not compatible with survival. Extra or missing X chromosomes appear to be tolerated because of dosage compensation, while Y chromosome aneuploids are most likely compatible with survival because of general paucity of Y-linked genes. Notice that the number of genes on chromosomes 13, 18, and 21 are the lowest for the autosomes. It is probably not coincidental that chromosomes with the fewest genes and no known mechanism for dosage compensation are the only ones that survive as human aneuploids.

**22.** Accurate annotation of genomes will not be a simple, straightforward process. Limitations on interpretation of sequence data will mean that new, uncharted levels of genomic, transcriptomic, and proteomic complications will be discovered. Since computer programs can only identify what is already predicted, manual verification and examination will be needed to bridge the gap between what we think we know and what has actually evolved over a few billion years. All the information will be highly significant in terms of understanding how organisms go about daily living and how we manage our relationships among them. For instance, one clinical application of genome sequence information is based on the development of antisense DNA to nullify the function of harmful RNAs and proteins within a cell. Opposite-strand RNA transcription overlap generates the possibility of natural antisense interactions for gene regulation *in vivo* and may provide insight into the development of antisense therapies presently being developed.

**24.** Increased protein production from approximately 20,000 genes is probably related to alternative splicing and various post-translational processing schemes. In addition, a particular DNA segment may be read in a variety of ways and in two directions.

**26.** Assuming that the APS strain is the ancestral strain, the remaining strains appear to have smaller genome sizes indicating genome reduction. The smallest *Buchnera* genome is approximately 448 kb compared with the genome size of *M. genitalium* of about 600 kb, with about 480 protein-coding genes. The APS genome codes for about 564 genes in its 641 kb genome. A gene is coded every 1136 bp (641,000/564) for the APS strain and every 1250 bp (600,000/480) for *M. genitalium.* Given these data, the CCE species should code for approximately

$$(448{,}000/1193) = 375.5$$

genes. (*Note*: 1193 was obtained as the average gene spacing of the two bacterial species mentioned above.) Using these calculations, we find that the CCE strain would contain fewer genes than *M. genitalium.* Other possible approaches to determine minimum genome size to sustain life include computational studies whereby one might estimate the number of essential chemical reactions that are needed for life. Another would be to take an organism with a small number of genes and then systematically mutate genes to see if elimination of genes caused reduced survival. By eliminating individual and groups of genes by mutation, the minimum number might be obtainable.

**28.** In general, one would expect certain factors (such as heat or salt) to favor evolution to increase protein stability: distribution of ionic interactions on the surface, density of hydrophobic residues and interactions, and number of hydrogen and disulfide bonds. By examining the codon table, a high GC ratio would favor amino acids Ala, Gly, Pro, Arg, and Trp and minimize the use of Ile, Phe, Lys, Asn, and Tyr. How codon bias influences actual protein stability is not yet understood. Most genomic sequences change by relatively gradual responses to mild selection over long periods of time. They strongly resemble patterns of common descent; that is, they are conserved. While the same can be said for organisms adapted to extreme environments, extraordinary physiological demands may dictate unexpected sequence bias.

**30.** Since structural and chemical factors determine the function of a protein, several proteins will likely share a considerable amino acid sequence identity but will not be functionally identical. Because the *in vivo* function of such a protein is determined by secondary and tertiary structures, as well as local surface chemistries in active or functional sites, the nonidentical sequences may have considerable influence on function. Note that the query matches to different site positions within the target proteins. A number of other factors suggesting different functions include: associations with other molecules (cytoplasmic, membrane, or extracellular), chemical nature and position of binding domains, post-translational modification, and signal sequences.

**32.** Any time a DNA sequence is conserved in other species, it is likely that that sequence has an influence on similar phenotypes. The higher the number of species that conserve the sequence, the higher the likelihood of determining its function. Coupled with mutation analysis and physical mapping, comparative genomics provides a powerful method for linking DNA sequences with complex human diseases.

**34.** Because blood is relatively easy to obtain in a pure state, its components can be analyzed without fear of tissue-site contamination. Blood is intimately exposed to virtually all cells of the body and may therefore carry chemical markers to certain abnormal cells it represents, theoretically, an ideal probe into the human body. However, when blood is removed from the body, its proteome changes and those changes are dependent on a number of environmental factors including age and sex. Thus, what might be a valid diagnostic for under one condition might not be so under others.

In addition, the serum proteome is subject to change depending on the patient's genetic, physiologic, and environmental state. Validation of a plasma proteome for a particular cancer would be strengthened by demonstrating that the stage of development of the cancer correlates with a commensurate change in the proteome. In addition, if large-scale samples can be archieved, they could be compared for particular biomarkers over time. Significant changes in cancer biomarkers correlated with the onset of cancer within an individual would also provide validation. The types of changes in the proteome should be reproducible and, at least until complexities are clarified, should involve tumorigenic proteins.

## Chapter 22

**2.** It is likely that the reverse transcriptase, in making DNA, provides a DNA segment capable of integrating into the yeast chromosome as other types of DNA are known to do.

**4.** Some transposons contain genes (reverse transcriptase, integrase, structural genes) and structural elements (terminal repeats, polyadenylation signals, etc.) common to proviral forms of retroviruses. They may produce viral particles that fail to generate a virus. If, after entering the genome as retroviruses do, they lose the ability to continue through the infective state but retain their ability to transpose, a transposon has evolved.

**6.** One way in which a transposon can move genes within the genome is by read-through transcription, producing a chimeric transcript containing a downstream gene(s). When this chimeric transcript is reverse transcribed and inserted into the genome, it could include the downstream gene(s). When two or more identical genetic elements exist in a genome, an opportunity exists for unequal homologous recombination that can lead to a variety of chromosomal aberrations: duplications, deletions, inversions, and/or translocations.

**8.** In the presence of a "helper element," a defective element may be able to transpose. Within a single bacterium, a similar transposase gene works in *trans* because a cytoplasmic transposase enzyme is produced. Therefore, the intact element may compensate for the loss of the deletion-defective transposase.

**10.** During the maturation of millions of lymphocytes, each produces one type of immunoglobulin or T-cell receptor capable of recognizing and binding to one particular antigen. When a foreign antigen interacts with a "matching" lymphocyte, a chemical signal stimulates proliferation of that lymphocyte. A pool of such stimulated lymphocytes represents a selected clone of cells.

**12.** Assuming that you can identify chromosomes 4 and 10, it should be possible to determine their intimate association (translocated chromosomes) and subsequent recombination microscopically. Since formation of a complete heavy chain is dependent on recombination, if such chromosomes are intimately associated, the creature is likely to be at least 13 years of age.

**14.** B cells synthesize antibodies, while T cells directly recognize and mark infected cells for destruction. Immunoglobulin gene rearrangement brought about by somatic recombination is dependent on the *RAG1* and *RAG2* genes. They produce two proteins that create double-stranded breaks at the junctions of the V, D, J, and C regions. Without such recombination, immunoglobulin diversity is not possible in either the T or B lymphocytes.

**16.** The sequence of the proviral DNA would be as follows:

3′-TTATCGATCGATTCCGCTACGCGCTA
5′-AATAGCTAGCTAAGGCGATGCGCGAT

Notice that regardless of the reading frame, one encounters a termination codon (underlined). Since the *gag* gene produces a protein necessary for viral maturation and infection, it is highly unlikely that the sequence would produce new infectious retroviruses.

5′-AAUAGCUAGCUAAGGCGAUGCGCGAU

**18.** Retroviruses are relatively small RNA viruses that generate double-stranded DNA that is integrated into the host's genome. With proper genetic engineering, a modified, inserted virus can become an endogenous provirus and, under ideal conditions, produce desired gene products at the appropriate time and place. Unfortunately, regulating the output of introduced genes is difficult. In addition, the host may mount an immune response to the vector, and because the site(s) of integration is uncertain at present, insertional mutagenesis of host genes may occur. The provirus may also contain enhancers that cause aberrant expression of nearby genes.

**20.** Both a bacterial prophage and retroviral provirus are integrated as double-stranded DNAs that replicate with the host chromosomes. By different processes, each can produce progeny viruses, and each can acquire and transfer host genes. Bacteriophage engage in transduction whereby the generation of a defective phage provides for the acquisition of host DNA. For a retrovirus, occurrence of a deletion between a provirus and a cellular gene can result in a retroviral RNA that is missing some of its own genes and has gained cellular DNA. This defective RNA can be packaged into a new virus particle if another viral RNA genome is present in the cell transcribed from another copy of the provirus. Template switching during the next round of reverse transcription creates a viral genome that contains host DNA.

**22.** Host shifting can occur when the S protein changes its specificity. Both its high mutation rate and recombination frequency generate a mosaic of sequences. The high mutation rate is caused by error-prone RNA polymerases and lack of proofreading that copy the RNA genome. In addition RNA genomes can recombine during RNA synthesis by strand switching.

**24.** The original insertion of a *copia* transposon within the normal allele of the *rg* eye color gene can directly disrupt function by altering coding and/or splicing, or by causing termination of transcription or translation. Although suppression of the *rg* mutation by a nonallelic gene may explain the origin of a few red-eyed flies in the mutant cell line, a more likely explanation is the excision of the *copia* element. Because there are sizable terminal repeats at each end of *copia* elements, loop formation and synapsis of these elements followed by crossing over can lead to excision. If the excision is precise, the original wild-type gene may be restored. In addition, once the mutant line was established, there was an opportunity for unequal synapsis of homologous chromosomes to occur at terminal repeats. If followed by crossing over, a chromosome somewhat free of the *copia* element can be generated.

**26.** To introduce a gene into a strain of *Drosophila,* the gene of interest is cloned into the middle of a *P* element that also contains a gene that produces a visible marker. The *P* elements are injected into embryos along with a helper plasmid that carries the transposase gene. If the *P* element inserts into the germ-line genome of the embryo, the resulting progeny can be identified by the visible marker and are likely to carry the gene of interest (transgene).

**28.** The type of Burkitt's lymphoma described in the question is caused by the translocation of elements from chromosome 8 to within the H chain immunoglobulin gene of chromosome 14. Formation of the translocated chromosome misregulates the expression of *c-myc* by combining enhancer elements in chromosome 14 with c-*myc,* leading to high levels of a C-MYC transcription factor. In addition, some negative regulatory elements within c-*myc* are often removed or altered as a result of the translocation, thereby contributing to increased c-*myc* activity. This C-MYC transcription factor plays a central role in the control of a diverse set of cellular processes, including cell-cycle progression and programmed cell death (apoptosis). Because chromosome 14 contains the only location for the production of the heavy immunoglobulin chain in humans, and because the translocation disrupts the region, usually in the joining (J) sites, production of heavy chains and therefore normal immunoglobulins from a t(8:14) chromosome is unlikely. However, because human cells are diploid, H chains can be synthesized from a nontranslocated chromosome 14 within a cell. Studies reveal that in the majority of cases, H-chain synthesis is drastically reduced in cells containing a t(8:14) chromosome. The translocation probably occurs during attempted V(D)J recombination under the influence of RAG1 and RAG2; however, sequencing studies have shown no homologies between c-myc and V(D)J. Therefore, translocations are probably common in the normal process of immunoglobulin chain recombination. Burkitt's lymphoma is but one example of the activation of an oncogene by an interrupted immunoglobulin gene.

**30.** Replication of an RNA virus such as the influenza virus is dependent on the viral RNA-dependent RNA polymerases that, like reverse transcriptases, can sponsor recombination between cellular RNAs. Since 28s ribosomal RNA is present in a cell, template switching could occur during RNA replication and enable the incorporation of a 28s-specifying segment in the viral genomic RNA. Altering the conformation of the HA glycoprotein alters the interaction of the virus with host receptors. In this way host shifting can occur by shifting HA glycoproteins.

## Chapter 23

**2.** (a) *Control of kidney development* could be studied most directly in humans and/or mice since both are similar physiologically and anatomically. Since mice, rather than humans, can be genetically manipulated, the study of the genetics of kidney development would probably be more productive. While *Drosophila* has excretory organs (Malpighian tubules), they are not closely related to the mammalian kidney. Yeast do not have excretory organs, but both yeast and *Drosophila* engage in membrane transport processes; the genetics of such processes will be fundamental and most likely apply to all organisms.

(b) Cancer can be studied in virtually all model organisms because it involves fundamentals of cell-cycle control. However, much of what we know about cell-cycle control derives from work in yeast, and, given that cell cycle systems are highly conserved processes, information obtained from any organism is often universally applicable. Yeast are well understood genetically, and they can be more easily manipulated (genetically and environmentally) than the other organisms listed. Mammalian cell lines are often used to discover regulatory elements homologous to those seen in yeast.

(c) Cystic fibrosis is caused by mutations in the cystic fibrosis transmembrane conductance regulator gene. This gene normally funnels chloride ions out of a cell, leading to a saltier cellular external environment, which in turn draws water out of the cell by osmosis. The human CFTR gene has been cloned and expressed in yeast, so a viable way to look at the gene, its regulation, and product is available in a yeast model. In addition, *Drosophila* cell lines have been developed to study vertebrate expression systems for anion channel proteins. Using these transgenic systems, one could expand knowledge of the transport systems in general. Seemingly, the most appropriate system to study human cystic fibrosis from a genetic standpoint would be the mouse. However, while strains have been generated with numerous mutations in the CFTR gene, mice do not develop the most serious symptom of human cystic fibrosis, namely, chronic lung infections by *Pseudomonas aeruginosa*. Recently, this problem has been overcome by the development of a mouse strain that is hypersusceptible to this bacterium (Coleman et al. 2003) and holds promise as a model organism for CF. Depending on the experimental approach, either human cell lines, transgenic *Drosophila* or yeast, or new mouse models may be most appropriate.

(d) Molecular aspects of purine metabolism are highly conserved throughout the animal and plant kingdom. To study the genetics of purine metabolism, it might be easiest to take advantage of the ease of mutant screens in yeast. In addition, certain strains of *Drosophila* respond differently to purine nutritional supplements. Therefore, one might use *Drosophila* to screen or select strains with altered purine metabolism.

(e) Genetic analyses of vertebrate immune function could be studied most directly in mouse models where genetic manipulation is possible. However, a relatively large number of forkhead-box (FOX) transcription factors have been found and characterized in *Drosophila melanogaster.* These transcription factors are members of a broader FOX family and have crucial roles in various aspects of immune regulation ranging from lymphocyte survival to thymic development. It would therefore be possible to study the immune system in a variety of organisms, taking advantage of the peculiarities of each.

(f) Like the study of cancer, the genetics of cell division might be most directly studied using yeast. The vast genetic understanding coupled with relative ease of genetic manipulation has already greatly enhanced our understanding of cell division processes. Because cell-cycle mutations hamper the proliferation of the organism, conditional mutants are often successfully employed. Cultured mouse and human cells are also useful for studying specific aspects of cell division. However, their genetic manipulation is more difficult than yeast cells. In most cases, knockout/rescue strategies are useful.

**4.** The basic procedure in oligonucleotide site-directed mutagenesis is to synthesize an appropriate oligonucleotide that is homologous to the gene or portion of gene of interest. The oligonucleotide contains a desired mutation that may involve one or more nucleotides. The oligonucleotide is hybridized to M13 DNA, which contains a cloned gene (or gene fragment) of interest and anneals to a homologous region. DNA polymerase is added along with other *in vitro* DNA polymerizing components to yield a double-stranded M13 molecule, which is then transformed into *E. coli* where replication occurs. Semiconservative replication provides mutant molecules that are then screened or selected using a variety of standard techniques.

**6.** Balancer chromosomes contain multiple overlapping inversions that greatly reduce the recovery of crossover chromatids. By using such chromosomes, the mutagenized chromosome remains intact. Such chromosomes are most useful when they contain dominant marker genes such as wing or eye shape or eye color. Balancer chromosomes usually contain a recessive lethal gene, so that the balancer chromosome will not exist in the homozygous state.

**8.** Coat color markers are useful in generating knockout mice because when embryonic stem cells are injected into blastocysts, the resulting chimeras will have patches of different fur colors, making them easy to recognize. Theoretically, any other surface marker (hair or hairless, etc.) would be usable as long as expression and penetrance of the dominant marker are high. Any test, molecular, biochemical, or visible, that allows one to determine the presence of a gene in an organism would be usable. Clearly visible tests are much more efficient, however. One could imagine using a gene that provided some needed substance for survival that circulates through body fluids. If the blastocyst is deficient for this vital substance, only the chimeras would survive.

**10.** (a) While autosomal and dominant mutations may be detected with the *ClB* technique, it is specifically designed to detect recessive mutations, especially but not exclusively lethals, that are X-linked.

(b) The recessive, X-linked *l* gene is used to exclude hemizygous *ClB/Y* flies (males) in the F2 generation. If no males appear in the F2, then one (or more) X-linked recessive lethal was introduced on the mutagenized male.

(c) The multiply-inverted X chromosome (C) suppresses the recovery of crossing-over products on the X, thereby leaving the mutagenized X chromosome intact.

(d) *C* is a balancer chromosome that contains multiple overlapping inversions to greatly reduce the recovery of crossover products (chromatids).

**12.** Temperature-sensitive mutations allow one to collect mutations that impact on the vital functions of cells. To select such mutations, one conducts a mutagenesis experiment in which the yeast cells are grown at 23°C for example. In general, the mutagen selected is one that causes minor changes in DNA, such as base analogs, ultraviolet light, or nitrosoguanidine. The resulting colonies are then replica plated at the permissive temperature (23°C, for example), and one sample is incubated at the restrictive temperature (36°C, for example). Colonies that do not grow at the restrictive temperature are likely to contain temperature-sensitive lethal mutations. Returning to the original plate (permissive temperature) allows one to select cells for further analysis. Since yeast cells produce buds, the size of which characterizes the stage of the cell cycle, it is possible to isolate cell division cycle (*cdc*) defects by the morphology of the cells. In addition, if a gene is controlling a specific point in the cell cycle, a uniform morphology is likely.

**14.** Tissue-specific and/or temporal-specific patterns of gene expression are often studied by *in situ* hybridization that may reveal the pattern of mRNA in an organism. A labeled cDNA is used as a probe to hydrogen bond with the RNA. One may also use the northern blot technique, which involves purification of mRNA from a tissue(s) of interest. Lastly, immunostaining may be used to generate a protein's expression profile.

**16.** P-elements containing a visible marker in the host fly are modified to contain a gene of interest. The recombinant P-element is then injected into *Drosophila* embryos, along with a helper plasmid that contains the transposase gene, which is transcribed and translated in the embryo. This enables the P-element to insert into the embryo's DNA. Recombinant adult flies can be identified by expression of the visible marker. Offspring from such a fly may contain the gene of interest. To generate a transgenic mouse, DNA is injected into the haploid nucleus of the embryo, which is then placed into the oviduct of a pseudopregnant female mouse. Resulting offspring are then screened for evidence of transgenesis. The difference between the two techniques resides in the type of DNA used and the cell in which the DNA is injected.

**18.** If transgenesis occurs in a vital gene, the organism may not survive, and that could seriously complicate a study. In addition, genes are often influenced by position effects whereby their function is altered by their neighbors or broad chromosomal location (in heterochromatin, for example). If such position effects occur, accurate interpretation of gene action will be compromised.

**20.** Selecting for mutants occurs when one creates conditions that remove irrelevant organisms, leaving only the mutants of interest. Screening for mutants is often more labor intensive and involves visual examination of organisms.

**22.** RNAi (RNA interference) provides a new technology that allows researchers to create single-gene defects without creating heritable mutations. Short double-stranded RNA molecules are introduced into cells that trigger RNA-degradation pathways. To introduce RNAi into cells, researchers may transform cells with vectors that express RNA hairpin structures that may integrate into a cell's genome. Alternatively, an inducible promoter may be used to drive the expression of the RNAi gene. Often, heat-shock gene promoters are effective.

**26.** Yeast may process the candidate gene in a form that is similar to that seen in humans. Because abnormal accumulation of a protein is involved, yeast may accumulate the same protein if transformed with the candidate gene.

**28.** The first step in gaining insight into the functionality of a DNA sequence is to annotate that sequence. Given that an interesting motif is possibly produced, suggests, but does not prove, that the sequence is not "junk" DNA. Annotating the sequence involves the use of a variety of annotation tools that scan the sequence in search of open reading frames (ORFs), which begin with an initiation sequence and end with a termination sequence. In addition, other landmarks are typically present in genes: intron/exon topology, intron/exon junction sequences, codon bias, upstream regulatory sequences, a 3′ polyadenylation signal, and, in some cases, CpG islands. Once a sequence is identified as a likely gene, comparisons with other known, well-studied genes may provide clues as to function. From that DNA sequence, the amino acid sequences can be predicted that can be compared with amino acid sequences in databases such as SwissProt or BLAST. From such searches, one may discover homologies to other proteins with known function. It is possible that additional significant motifs may be found to provide clues to function. One could eventually analyze tissue-specific and temporal-specific patterns of gene expression using any number of techniques such as *in situ* hybridization, northern blots, immunostaining, or microarray technology.

**30.**

| ***Mutations*** | ***Progress to mitosis?*** |
|---|---|
| *cdc25* | No, block not removed |
| *wee1* | Yes, no block installed |
| *cdc25 wee1* | Yes, no block installed |
| *cdc13 wee1* | No, no cyclin |
| *cdc2 wee1* | No, no cyclin-dependent kinase |
| *cdc25 cdc13* | No, no cyclin |

## Chapter 24

**2.** Because of the recent rise in food sensitivities (allergies and adverse reactions), the public should probably have access to the contents of all foods. GMOs have the potential for possessing suites of gene products that might be atypical for a given food, and, unless consumers know of that possibility, harm could result. Some argue that consumers have a right to know about GMOs as a matter of principle, regardless of potential health risks.

**4.** Glyphosate (a herbicide) inhibits EPSP, a chloroplast enzyme involved in the synthesis of the amino acids phenylalanine, tyrosine, and tryptophan. To generate glyphosate resistance in crop plants, a fusion gene was created that introduced a viral promoter to control the EPSP synthetase gene. The fusion product was placed into the Ti vector and transferred to *A. tumifaciens,* which was used to infect crop cells. Calluses were selected on the basis of their resistance to glyphosate. Resistant calluses were later developed into transgenic plants. There is a remote possibility that such an "accident" can occur as suggested in the question. However, in retracing the steps to generate the resistant plant in the first place, it seems more likely that the trait will not "escape" from the plant; rather, the engineered *A. tumifaciens* may escape, infect, and transfer glyphosate resistance to pest species.

**6.** Short tandem repeats are very similar to VNTRs, but the repeat motif is much shorter (2–9 base pairs). STRs have been used to generate a marker panel for DNA profiling. STR typing is less expensive, less labor intensive, and quicker to perform than traditional DNA typing.

**8.** Kleter and Peijnenburg used the BLAST tool from the http://www.ncbi.nlm.nih.gov/BLAST Web site to conduct a series of alignment comparisons of transgenic sequences with sequences of known allergenic proteins. Of 33 transgenic proteins screened for the identities of at least 6 contiguous amino acids found in allergenic proteins, 22 gave positive results.

**10.** From a purely scientific viewpoint, there will be no added danger to consuming cow's milk from cloned animals. However, some individuals may have an aversion to organismic cloning, and supporting such activities through consumption of products of cloned organisms may be viewed negatively on moral grounds. It is likely that public pressure will compel the labeling of "cloned products" on the grounds that consumers should be able to make an informed choice as to the origin of the products they consume.

**12.** As with all therapies, the cure must be less hazardous than the disease. In the case of viral-mediated gene therapy, the antigenicity of the virus must not interfere with the delivery system; such antigenicity can cause inflammation or more severe immunologic responses. Combating the host immune response may involve the use of immunosuppressive drugs or modification of the vector. The duration of desired gene expression at the diseased site is an issue. Short-period expression may require repeated exposure to the vehicle, which may present undesired responses. For some diseases, local gene therapy through inhalation or injection may produce fewer side effects than systemic exposure. Adenoviruses appear to be particularly useful for gene therapy because they can infect nondividing cells and they can accept relatively large amounts of additional DNA (30 kb or more).

**14.** A major problem associated with engineering the capsid to specifically engage target cells is that the capsid itself is now altered. A reconfigured capsid, either by size or shape, may no longer serve its packaging and infecting roles properly. One may end up with a very specific viral-target interaction, but the virus may be incapable of replicating efficiently. A nongenetic approach is to use bispecific molecules to conjugate vectors with target cells. Such approaches often employ desired electrostatic bridges or monoclonal antibody conjugates. While such approaches often work in the test tube, their application *in vivo* is often limited due to instability.

**16.** (a, b) One of the main problems with gene therapy involves the effective delivery of the desired virus to the target tissue. Several of the problems associated with the use of retroviral vectors are the following: (1) Integration into the host must be cell specific so as not to damage nontarget cells. (2) Retroviral integration into host-cell genomes occurs only if the host cell is replicating. (3) Insertion of the viral genome might influence nontarget, but essential, genes. (4) Retroviral genomes have a low cloning capacity and cannot carry large inserted sequences as are many human genes. (5) There is a possibility that recombination with host viruses will produce an infectious virus that may do harm.

(c) The question posed here plays on the practical versus the ethical. It would certainly be more efficient (though perhaps more difficult technically) to engineer germ tissue, for once it is done in a family, the disease would be eliminated. However, germ-plasm therapy presents considerable ethical problems. It recalls previous attempts of the eugenics movements of past decades, which involved the use of selective breeding to purify the human stock. Some present-day biologists have said publicly that germ-line gene therapy will *not* be conducted.

**18.** Since both mutations occur in the CF gene, children who possess both alleles will suffer from CF. With both parents heterozygous, each child born will have a 25 percent chance of developing CF.

**20.** It will hybridize by base complementation to the normal DNA sequence.

**22.** One method is to use the amino acid sequence of the protein to produce the gene synthetically. Alternatively, since the introns are spliced out of the hnRNA in the production of mRNA, if mRNA can be obtained, it can be used to make DNA (cDNA) through the use of reverse transcriptase.

**24.** At this point, there is considerable reluctance to allow the open sharing of genetic information among institutions. In general, the establishment of governmental databases containing our most intimate information is viewed with skepticism. It is likely that considerable time and discussion will elapse before such databases are established.

**28.** (a) Y-linked excluded, X-linked recessive excluded, autosomal recessive is possible, but unlikely, because the gene is stated as being rare; X-linked dominant is possible if heterozygous, autosomal dominant is possible.

(b) Chromosome 21 with the B1 marker probably contains the mutation.

(c) The disease gene is segregating with some certainty with the B1 RFLP marker in the family. Since the mother also has the B3 marker, the offspring could be tested. If the child carries the B3 marker, then he or she does not carry the B1 marker, which has been segregating with the defective gene. However, this prediction is not completely accurate because a crossover in the mother could put the undesirable gene with the B3 marker.

(d) The possibilities would include a crossover between the restriction sites in the father giving a B1 chromosome, or a mutation eliminating either the B2 or B3 restriction site.

**30.** Numerous methods are available to generate transgenic mammals using retroviral vectors. One method involves injection of an engineered retroviral RNA into embryonic (pronuclear) cells directly. Another scheme uses cultured embryonic stem cells (ES) and is presented in the following mouse example.

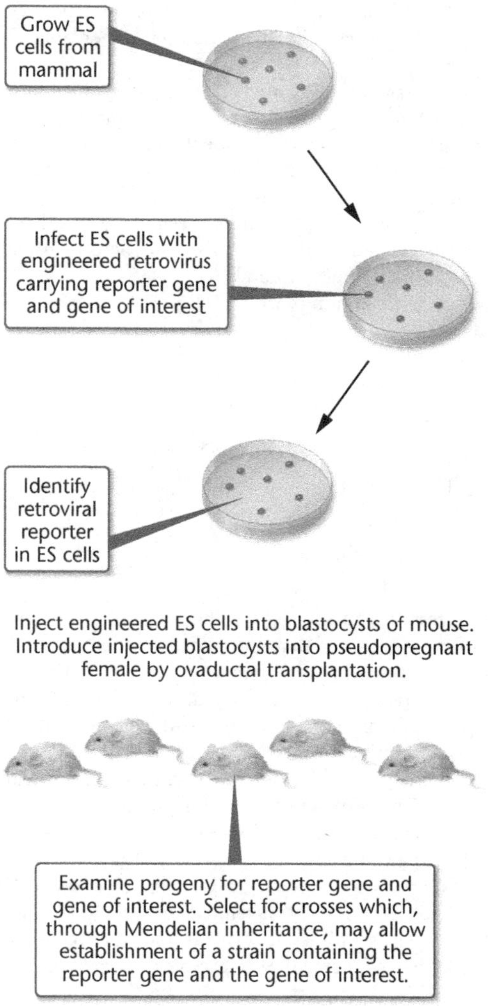

## Chapter 25

**2.** (a) *Polygenes* are those genes involved in determining continuously varying or multiple-factor traits.

(b) *Additive alleles* are those alleles that account for the hereditary influence on the phenotype in an additive way.

(c) *Correlation* is a statistic that varies from −1 to +1 and describes the extent to which variation in one trait is associated with variation in another. It does not imply that a cause-and-effect relationship exists between two traits.

(d) *Monozygotic twins* are derived from a single fertilized egg and are thus genetically identical to each other. They provide a method for determining the influence of genetics and environment on certain traits. *Dizygotic twins* arise from two eggs fertilized by two sperm cells. They have the same genetic relationship as siblings.

The role of genetics and the role of the environment can be studied by comparing the expression of traits in monozygotic and dizygotic twins. The higher concordance value for monozygotic twins as compared to the value for dizygotic twins indicates a significant genetic component for a given trait.

(e) *Heritability* is a measure of the degree to which the phenotypic variation of a given trait is due to genetic factors. A high heritability indicates that genetic factors are major contributors to phenotypic variation, whereas environmental factors have little impact.

(f) *QTL* stands for Quantitative Trait Loci, which are situations in which multiple genes contribute to a quantitative trait.

**4.** If you add the numbers given for the ratio, you obtain the value of 16, which is indicative of a dihybrid cross. The distribution is that of a dihybrid cross with additive effects.

(a) Because a dihybrid result has been identified, there are two loci involved in the production of color. There are two alleles at each locus for a total of four alleles.

(b, c) 6/16 = medium red = *AAbb*
*4AaBb*
*aaBB*

4/16 = light red = *2aaBb*
*2Aabb*

(d) $F_1$ = all light red
$F_2$ = 1/4 medium red
2/4 light red
1/4 white

**6.** As you read this question, notice that the strains are inbred, and therefore homozygous, and that approximately 1/250 represent the shortest and tallest groups in the $F_2$ generation. See $1/4^n$ formula in the text.

(a, b) Referring to the text, see that where four gene pairs act additively, the proportion of one of the extreme phenotypes to the total number of offspring is 1/256 (add the numbers in each phenotypic class). The same may be said for the other extreme type. The extreme types in this problem are the 12 cm and 36 cm plants. From this observation one would suggest that four gene pairs are involved.

(c) If there are four gene pairs, there are nine $(2n + 1)$ phenotypic categories and eight increments between these categories. Since there is a difference of 24 cm between the extremes, 24 cm/8 = 3 cm for each increment (each of the additive alleles).

(d) A typical $F_1$ cross, which produces a "typical" $F_2$ distribution would be where all gene pairs are heterozygous (*AaBbCcDd*), independently assorting, and additive. Many possible sets of parents would give an $F_1$ of this type.

The limitation is that each parent has genotypes that give a height of 24 cm as stated in the problem. Because the parents are inbred, they are expected to be fully homozygous. An example:

*AABBccdd* × *aabbCCDD*

(e) Since the *aabbccdd* genotype gives a height of 12 cm and each upper-case allele adds 3 cm to the height, there are many possibilities for an 18 cm plant:

*AAbbccdd,*
*AaBbccdd,*
*aaBbCcdd,* etc.

Any plant with seven upper-case letters will be 33 cm tall:

*AABBCCDd,*
*AABBCcDD,*
*AABbCCDD,* for examples.

**8.** For height, notice that average differences between MZ twins reared together (1.7 cm) and those MZ twins reared apart (1.8 cm) are similar (meaning little environmental influence) and considerably less than differences of DZ twins (4.4 cm) or sibs (4.5) reared together. These data indicate that genetics plays a major role in determining height. However, for weight, notice that MZ twins reared together have a much smaller (1.9 kg) difference than MZ twins reared apart, indicating that the environment has a considerable impact on weight. By comparing the weight differences of MZ twins reared apart with DZ twins and sibs reared together, one can conclude that the environment has almost as great an influence on weight as genetics. For ridge count, the differences between MZ twins reared together and those reared apart are small. For the data in the table, it would appear that ridge count and height have the highest heritability values.

**10.** Many traits, especially those that we view as quantitative, are likely to be determined by a polygenic mode, with possible environmental influences. The following are some common examples: height, general body structure, skin color, and perhaps most common behavioral traits including intelligence.

**12.** (a) Using the following equations, $H^2$ and $h^2$ can be calculated as follows.

*For back fat:*
Broad-sense heritability $= H^2 = 12.2/30.6 = 0.398$

Narrow-sense heritability $= h^2 = 8.44/30.6 = 0.276$

*For body length:*
Broad-sense heritability $= H^2 = 26.4/52.4 = 0.504$

Narrow-sense heritability $= h^2 = 11.7/52.4 = 0.223$

(b) In animal and plant breeding, a measure of potential response to selection based on additive variance and dominance variance is termed narrow heritability ($h^2$). A relatively high narrow heritability is a prediction of what impact selection may have in altering an initial randomly breeding population. Therefore, of the two traits, selection for back fat would produce more response.

**14.** (a) For Vitamin A

$$h_{A^2} = V_A/V_P = V_A/(V_E + V_A + V_D) = 0.097$$

For Cholesterol $h_{A^2} = 0.223$

(b) Cholesterol content should be influenced to a greater extent by selection.

**16.** Given that both narrow-sense heritability values are relatively high, it is likely that a farmer would be able to alter both milk protein content and butterfat by selection. The value of 0.91 for the correlation coefficient between protein content and butterfat suggests that if one selects for butterfat, protein content will increase. However, correlation coefficients describe the extent to which variation in one quantitative trait is associated with variation in another and does not reveal the underlying causes of such variation. Assuming that these dairy cows had been selected for high butterfat in the past and increased protein content followed that selection (for butterfat), it is likely that selection for butterfat would continue to correlate with increased protein content. However, there may well be a point where physiological circumstances change and selection for high butterfat may be at the expense of protein content.

**18.** Given the realized heritability value of 0.4, it is unlikely that selection experiments would cause a rapid and/or significant response to selection. A minor response might result from intense selection.

**20.** Since the rice plants are genetically identical, $V_G$ is zero and $H^2 = V_G/V_P =$ zero. Broad-sense heritability is a measure to which the phenotypic variance is due to genetic factors. In this case, with genetically identical plants, $H^2$ is zero, and the variance observed in grain yield is due to the environment. Selection would not be effective in this strain of rice.

**24.** (a) The most direct explanation would involve two gene pairs, with each additive gene contributing about 1.2 mm to the phenotype.

(b) The fit to this backcross supports the original hypothesis.

(c) These data do not support the simple hypothesis provided in part (a).

(d, e) With these data, one can see no distinct phenotypic classes suggesting that the environment may play a role in eye development or that there are more genes involved.

**26.** *Monozygotic twins* are derived from a single fertilized egg and are thus genetically identical to each other. They provide a method for determining the influence of genetics and environment on certain traits. *Dizygotic twins* arise from two eggs being fertilized by two sperm cells. They have the same genetic relationship as siblings. The role of both genetics and the environment can be studied by comparing the expression of traits in monozygotic and dizygotic twins. The higher concordance value for monozygotic twins as compared with the value for dizygotic twins indicates a significant genetic component for a given trait. Notice that for traits including blood type, eye color, and mental retardation, a fairly significant difference exists between MZ and DZ groups.

However, for measles, the difference is not as significant, indicating a greater role of the environment. Hair color has a significant genetic component, as do idiopathic epilepsy, schizophrenia, diabetes, allergies, cleft lip, and club foot. The genetic component of mammary cancer is present but minimal according to these data.

**28.** As with many traits that are caused by numerous loci acting additively, some genes have more influence on expression than others. In addition, environmental factors may play a role in the expression of some polygenic traits. In the case of brachydactyly, numerous modifier genes in the genome can influence brachydactyly expression. Examination of OMIM (*Online Mendelian Inheritance of Man*) through http://www.ncbi.nlm.nih.gov/ will illustrate this point.

**30.** It is likely that the flies maintained in the *Drosophila* repository are more highly inbred and less heterozygous than those recently obtained from the wild. Response to selection is dependent on genetic variation. The greater the genetic variation in a species, the more likely and dramatic the response to selection. Therefore, one would expect a greater response to selection in the wild population.

**32.** Breeders attempt to "select" out this disorder by first maintaining complete and detailed breeding records of afflicted strains. Second, they avoid breeding dogs whose close relatives are afflicted. The molecular-developmental mechanism that causes the "month of birth" effect in canine hip dysplasia is unknown. However, with many, perhaps all, quantitative traits, it is clear that there is a significant environmental influence on both the penetrance and/or expression of the phenotype. With many genes acting in various ways to influence a phenotype, opportunities exist for varied molecular and developmental intraorganismic microenvironments. Stated another way, the longer and more complex the molecular distance from the genome to the phenotype, the greater the likelihood for environmental factors to be involved in expression.

## Chapter 26

**2.** One of the easiest ways to determine whether a genetic basis exists for a given abnormality is to cross the abnormal fly to a normal fly. If the trait is determined by a dominant gene, that trait should appear in the offspring, probably half of them if the gene was in the heterozygous state. If the gene is recessive and homozygous, then one may not see expression in the offspring of the first cross, however, if one crosses the $F_1$, the trait might appear in approximately 1/4 of the offspring. If the gene is caused by more than one gene locus, additional analysis would be needed. Modifications of these patterns would be expected if the mode of inheritance is X-linked or shows other modifications of typical Mendelian ratios. One might hypothesize that the trait influences the nervous, cuticular, or muscular system. Mapping the primary focus of the gene could be accomplished using the unstable ring-X-chromosome. Given that the gene is X-linked, one would use classical recombination methods to place recessive markers (*singed bristles* and/or *yellow body*) on the X chromosome with the gene causing the limp. This would help one identify the male/female boundary. One would then cross homozygous females (for the trait and markers) to ring-X-males, or the reciprocal, and then examine the offspring for gynandromorphs (and marker mosaics). If one obtained a pool of gynandromorphs, one could then assess the phenotype (limp or normal) with respect to exposure of the recessive gene in the male tissue. Correlating such would allow one to provide an educated guess as to the primary focus of the gene causing the limp.

**4.** First, when one examines cuticular markers in a gynandromorphic analysis, one is not always certain that underlying tissues and organs follow the surface mosaic pattern. The finding that flies with male/female cuticular markers had circadian activity patterns with both male and female components indicates that the brain can be genetically mosaic if cuticular markers are mosaic. The observation also suggests that the male and female oscillators function autonomously.

**6.** One limitation of this approach is the intensity of effort required to establish genetically uniform strains reflecting a particular behavior. In addition, since geotaxis is a complex trait and many genes may have multiple functions, selection experiments offer little indication of the number and location of selected genes. New technologies, including microarrays, have been applied to this problem that enable the analysis of the expression patterns of many genes simultanteously. Strains having high and low responses to geotaxis selection showed reproducible differences in gene expression and revealed some of the molecular and physiological characteristics of geotaxis behavior.

**8.** Because of the rigidity of development in *Caenorhabditis* and the extensive knowledge of cellular fates and connectivity, it represents an excellent experimental organism for the study of development. In addition, because of their fixed fate, cells can be altered, physically and/or genetically, and resulting development and behavior can be studied. However, the behavioral repertoire of *C. elegans* is somewhat narrow. In addition, *C. elegans* has more protein-coding genes than *Drosophila,* another organism whose behavioral genetics has been highly studied.

**10.** A number of issues limit the study of human behavior including the following:

1. With relatively small numbers of offspring per mating, standard genetic methods are difficult.
2. Records on family illnesses, especially behavioral illnesses, are difficult to obtain.
3. The long generation time makes longitudinal studies difficult.
4. The scientist cannot direct matings.
5. There are limits to the experimental treatments that can be applied.
6. Traits that are interesting to study are often complex and difficult to quantify.
7. Culture and family background may strongly influence behavior.

**14.** Examine the data and notice that the magnitude of change attributed to the X chromosome is relatively strong when compared to the unselected line. Notice that chromosome 2 contributes relatively strongly to negative geotaxis, while chromosome 3 contributes fairly strongly to positive geotaxis. Because this is a whole chromosome comparison, it is possible that there are strong positive geotaxis alleles and strong negative geotaxis alleles on the same chromosome that cancel each other.

**16.** Schizophrenia is a complex familial brain disorder, with relatives of schizophrenics having a higher incidence than the general population. The closer the relationship to the schizophrenic, the greater is the probability of the disorder occurring. Concordance for schizophrenia is higher in monozygotic twins (~ 50 percent) than dizygotic twins (~ 17 percent) and suggests that a genetic component exists along with shared environmental risk factors. Recently, microarrays have been used to identify genes whose expression is altered in schizophrenia. Researchers have isolated RNA from the brains of normal and schizophrenic individuals and used the RNA to prepare cDNA for hybridization to the microarrays. The microarrays carried over 6000 probes for the human genome. Some of the identified genes have low levels of expression in schizophrenia, while others express at high levels. One cluster of genes involved in myelination has lower levels of expression in schizophrenics, while some other clusters have increased levels of expression, suggesting that schizophrenia is associated with functional disruption in oligodendrocytes. A variety of genome-wide microarray scans of gene expression in schizophrenia have identified candidate genes whose expression are altered in schizophrenics. Knockout mice missing a myelin-related gene are being studied in order to relate changes in gene expression with specific behavioral phenotype. From such studies, it is hoped that therapies can be targeted at specific genes involved in schizophrenia.

## Chapter 27

**2.**

$$p = \text{frequency of A} = 0.2 + 0.3 = 0.5$$

$$q = 1 - p = 0.5$$

$$\text{Frequency of } AA = p^2 = (0.5)^2 = 0.25 \text{ or } 25\%$$

$$\text{Frequency of } Aa = 2pq = 2(0.5)(0.5) = 0.5 \text{ or } 50\%$$

$$\text{Frequency of } aa = q^2 = (0.5)^2 = 0.25 \text{ or } 25\%$$

**4.** In order for the Hardy-Weinberg equations to apply, the population must be in equilibrium.

**6.** (a) The equilibrium values will be as follows:

$$\text{Frequency of } l/l = p^2 = (0.7755)^2 = 0.6014 \text{ or } 60.14\%$$

$$\text{Frequency of } l/\Delta 32 = 2pq = 2(0.7755)(0.2245) = 0.3482 \text{ or } 34.82\%$$

$$\text{Frequency of } \Delta 32/\Delta 32 = q^2 = (0.2245)^2 = 0.0504 \text{ or } 5.04\%$$

Comparing these equilibrium values with the observed values strongly suggests that the observed values are drawn from a population in equilibrium.

(b) The equilibrium values will be as follows:

$$\text{Frequency of } AA = p^2 = (0.877)^2 = 0.7691 \text{ or } 76.91\%$$

$$\text{Frequency of } AS = 2pq = 2(0.877)(0.123) = 0.2157 \text{ or } 21.57\%$$

$$\text{Frequency of } SS = q^2 = (0.123)^2 = 0.0151 \text{ or } 1.51\%$$

Comparing these equilibrium values with the observed values suggests that the observed values may be drawn from a population that is not in equilibrium. Notice that there are more heterozygotes than predicted, and fewer *SS* types.

To test for a Hardy-Weinberg equilibrium, apply the chi-square test as follows.

$$\chi^2 = \frac{\Sigma(o - e)^2}{e} = 1.47$$

In calculating degrees of freedom in a test of gene frequencies, the "free variables" are reduced by an additional degree of freedom because one estimated a parameter ($p$ or $q$) used in determining the expected values. Therefore, there is one degree of freedom, even though there are three classes. Checking the $\chi^2$ table with one degree of freedom gives a value of 3.84 at the 0.05 probability level. Since the $\chi^2$ value calculated here is smaller, the null hypothesis (the observed values fluctuate from the equilibrium values by chance and chance alone) should not be rejected. Thus the frequencies of *AA, AS, SS* sampled a population that is in equilibrium.

**8.** The following formula calculates the frequency of an allele in the next generation for any selection scenario, given the frequencies of $a$ and $A$ in this generation and the fitness of all three genotypes.

$$q_{g+1} = [w_{Aa}p_g q_g + w_{aa}q_g{}^2]/[w_{AA}p_g{}^2 + w_{Aa}2p_g q_g + w_{aa}q_g{}^2]$$

where $q_{g+1}$ is the frequency of the $a$ allele in the next generation, $q_g$ is the frequency of the $a$ allele in this generation, $p_g$ is the frequency of the $A$ allele in this generation, and each $w$ represents the fitness of their respective genotypes.

(a) $q_{g+1} = [0.9(0.7)(0.3) + 0.8(0.3)^2/$

$[1(0.7)^2 + 0.9(2)(0.7)(0.3) + 0.8(0.3)^2]$

$q_{g+1} = 0.278 \qquad p_{g+1} = 0.722$

(b) $q_{g+1} = 0.289 \qquad p_{g+1} = 0.711$
(c) $q_{g+1} = 0.298 \qquad p_{g+1} = 0.702$
(d) $q_{g+1} = 0.319 \qquad p_{g+1} = 0.681$

**10.** Since a dominant lethal gene is highly selected against, it is unlikely that it will exist at too high a frequency, if at all. However, if the gene shows incomplete penetrance or late age of onset (after reproductive age), it may remain in a population.

**12.** What one must do is predict the probability of one of the grandparents being heterozygous in this problem. Given the frequency of the disorder in the population as 1 in 10,000 individuals (0.0001), then $q^2 = 0.0001$, and $q = 0.01$. The frequency of heterozygosity is $2pq$ or approximately 0.02 as also stated in the problem. The probability for one of the grandparents to be heterozygous would therefore be $0.02 + 0.02$ or 0.04 or 1/25. If one of the grandparents is a carrier, then the probability of the offspring from a first-cousin mating being homozygous for the recessive gene is 1/16. Multiplying the two probabilities together gives $1/16 \times 1/25 = 1/400$.

Following the same analysis for the second-cousin mating gives $1/64 \times 1/25 = 1/1600$. Notice that the population at large has a frequency of homozygotes of 1/10,000, therefore one can easily see how inbreeding increases the likelihood of homozygosity.

**14.** Because heterozygosity tends to mask expression of recessive genes, which may be desirable in a domesticated animal or plant, inbreeding schemes are often used to render strains homozygous so that such recessive genes can be expressed. In addition, assume that a particularly desirable trait occurs in a domesticated plant or animal. The best way to increase the frequency of individuals with that trait is by self-fertilization (not often possible) or by matings to blood relatives (inbreeding). In theory, one increases the likelihood of a gene "meeting itself" by various inbreeding schemes. There are disadvantages to increasing the degree of homozygosity by inbreeding. *Inbreeding depression* is a reduction in fitness often associated with an increase in homozygosity.

**16.** The quickest way to generate a homozygous line of an organism is to *self-fertilize* that organism. Because this is not always possible, brother-sister matings are often used.

**18.** If there were two males with hemophilia, $q = 2/2000$ or 1/1000 and $p = 1998/2000$. The frequency of heterozygous females would be $2pq$ or $2(1998/2000 \times 2/2000) = 0.001998$. The number of heterozygous females would be 3.996 or approximately 4.

**20.** Because three of the affected infants had affected parents, only two "new" genes, from mutation, enter into the problem.

The gene is dominant; therefore, each new case of achondroplasia arose from a single new mutation. There are 50,000 births; therefore, 100,000 gametes (genes) are involved. The frequency of mutation is therefore given as follows: 2/100,000 or $2 \times 10^{-5}$.

**22.** Since $r1 = 0.81$ and $r2 = 0.19$, the expected frequency of heterozygotes would be $2pq \times 125$ or $2(0.81 \times 0.19) \times 125 = 38.475$. Given the following equation and substituting the values:

$$F = (H_e - H_o)/H_e$$

$$F = (38.475 - 20)/38.475 = 0.48$$

**24.** The following distribution of genotypes occurs among the 50 desert bighorn sheep in which the normal dominant $C$ allele produces straight coats.

$CC$ = 29 = straight coats

$Cc$ = 17 = straight coats

$cc$ = 4 = curled coats

Computing, $p = 0.75$ and $q = 0.25$ and $2pq = 0.375$ for the expected frequency of heterozygotes. Since 17/50 or (0.34) are observed as heterozygotes, the following equation applies:

$$F = (H_e - H_o)/H_e$$

$$F = (0.375 - 0.34)/0.375 = 0.093$$

This problem could also be solved using the actual numbers of sheep in each category where there would be 18.75 heterozygotes expected $(2pq)(50)$:

$$F = (18.75 - 17)/18.75 = 0.093$$

**28.** Given small populations and very similar environmental conditions, it is more likely that "sampling error" or genetic drift is operating. Under such conditions (small population sizes), large fluctuations in gene frequency are likely, regardless of selection pressures. Since the same gene is behaving differently under similar environmental conditions, selection is an unlikely explanation.

**30.** When a population bottleneck occurs and the number of effective breeders is reduced in a population, two phenomena usually follow. First, because the population is small, wide fluctuations in genotypic frequencies occur, thereby revealing deleterious alleles by chance. Second, inbreeding often occurs in small populations, thereby increasing the chance for homozygosity. With increased homozygosity comes an increased likelihood that recessive alleles will be expressed. Since many disease-producing genes are recessive, an increase in genetic diseases is a likely aftermath to a population bottleneck.

## Chapter 28

**2.** A species is a group of interbreeding or potentially interbreeding populations that is reproductively isolated from all other such groups. Speciation is the process that leads to the formation of species. Evolution is the change in a population over time. Speciation is one of many results of evolution.

**4.** Organisms may appear to be similar but may be reproductively isolated for a sufficient period to justify their species identity. If significant genetic differences occur, they can be considered separate species.

**6.** The subterranean niche is less broad and less dynamic than above ground. The underground microhabitat consists of a narrower range of climatic changes; thus genetic polymorphism is not selected for. This conclusion has been supported by additional studies indicating that genetic diversity is positively correlated with niche-width.

**8.** Results from laboratory studies indicated that there was a selective advantage in having the two inversions present rather than either one. Thus, natural selection favored the maintenance of both inversions over the loss of either.

**10.** (a) Missense mutations cause amino acid changes.
(b) Horizontal transfer refers to the process of passing genetic information from one organism to another without producing offspring. In bacteria, plasmid transfer is an example of horizontal transfer.
(c) The fact that none of the isolates shared identical nucleotide changes indicates that there is little genetic exchange among different strains. Each alteration is unique, most likely originating in an ancestral strain and maintained in descendants of that strain only.

**12.** The approximate similarity of mutation rates among genes and lineages should provide more credible estimates of divergence times of species and allow for broader interpretations of sequence comparisons. It also provides for increased understanding of the mutational processes that govern evolution among mammalian genomes. For instance, if the rate of mutation is fairly constant among lineages or cells that have a more rapid turnover, it indicates that replication-related errors do not make a significant contribution to mutation rates.

**14.** In general, speciation involves the gradual accumulation of genetic changes to a point where reproductive isolation occurs. Depending on environmental or geographic conditions, genetic changes may occur slowly or rapidly. They can involve point or chromosomal changes.

**16.** Reproductive isolating mechanisms are grouped into prezygotic and postzygotic and include those listed here. See the text for specific examples and illustrations.

- geographic or ecological
- seasonal or temporal
- behavioral
- mechanical
- physiological
- hybrid inviability or weakness
- developmental hybrid sterility
- segregational hybrid sterility
- $F_2$ breakdown

**18.** Polyploid plants, which result from hybridization of two species, would be expected to be more heterozygous than the diploid parental species because two distinct genomes are combined. Generally, genetic variation is an advantage unless a significant degree of that variation is outside acceptable physiological tolerance.

**20.** Somatic gene therapy, like any therapy, allows some individuals to live more normal lives than those not receiving therapy. As such, the ability of such individuals to contribute to the gene pool increases the likelihood that less fit genes will enter and be maintained in the gene pool. This is a normal consequence of therapy, genetic or not, and in the face of disease control and prevention, societies have generally accepted this consequence. Germ-line therapy could, if successful, lead to limited, isolated, and infrequent removal of a gene from a gene lineage. However, given the present state of the science, its impact on the course of human evolution will be diluted and negated by a host of other factors that afflict humankind.

**22.** All of the amino acid substitutions

(Ala - Gly, Val - Leu, Asp - Asn, Met - Leu)

require only one nucleotide change. The last change from

Pro (CC-) - Lys (AAA,G)

requires two changes (the minimal mutational distance).

**24.** Construct a chart similar to the one below, which indicates the number of base changes between each pair:

| | H | C | G | O |
|---|---|---|---|---|
| H | – | – | – | – |
| C | 1 | – | – | – |
| G | 3 | 2 | – | – |
| O | 7 | 6 | 4 | – |
| B | 12 | 11 | 9 | 10 |

Following the instructions given in the text, develop the relationships in the following manner:

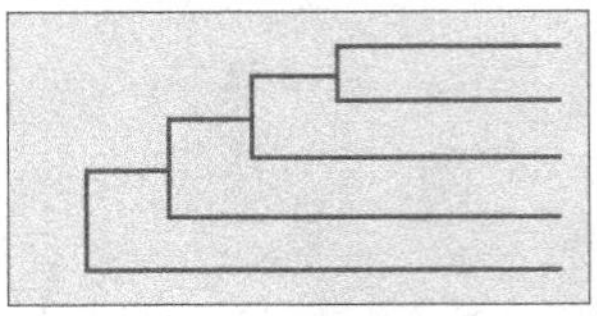


**26.** Many sections of DNA in a eukaryotic genome are not reflected in a protein product. Indeed, many sections of DNA are not even transcribed and/or have no apparent physiological role. Such regions are more likely to tolerate nucleotide changes compared with those regions with a necessary physiological impact. Introns, on one hand, show sequence variation, which is not reflected in a protein product. Exons, on the other hand, code for products that are usually involved in production of a phenotype and, as such, are subject to selection.

**28.** A number of studies using SINES, repetitive DNA, and neutral polymorphisms (see above problem) indicate that most, if not all, cichlid species in Lake Victoria evolved from a single ancestral species. If that is the case, then this finding would represent the most rapid evolutionary radiation ever documented for vertebrates.

**30.** In general, there are two methods for calibrating molecular data, amino acid and nucleotide substitutions, to absolute times of divergence. First, molecular data are compared with the existing fossil record. Second, major paleontological events such as the Bryophyta/Tracheophyta split during the Ordovician (443–490 Mya) or the Actinopterygii/Mammalia split during the Devonian (354–417 Mya), can provide some clues for calibration. Both methods are subject to error due to uncertainty of both the fossil record and the times of major paleontological events. In addition, mutation rates vary among taxa.

## Chapter 29

**2.** The frequency of the lethal gene in the captive population ($q^2 = 5/169$ and $q = 0.172$) is approximately double that in the gene pool as a whole ($q = 0.09$). Applying the formula

$$q_n = q_o/(1 + nq_o)$$

one can estimate that it would take 10 generations to reduce the lethal gene's frequency to 0.063 in the captive population with no intervention (random mating assumed). Since condors produce very few eggs per year, a more proactive approach seems justified.

(a, b) First, if detailed records are kept of the breeding partners of the captive birds, then knowledge of heterozygotes should be available. Breeding programs could be established to restrict matings between those carrying the lethal gene. Such "kinship management" is often used in captive populations. If kinship records are not available, it is often possible to establish kinship using genetic markers such as DNA microsatellite polymorphisms. Using such markers, one can often identify mating partners and link them to their offspring.

By coupling knowledge of mating partners with the likelihood of producing a lethal genetic combination, selective matings can often be used to minimize the influence of a deleterious gene. In addition, such markers can be used to establish matings that optimize genetic mixing, thus reducing inbreeding depression.

**4.** Both genetic drift and inbreeding tend to drive populations toward homozygosity. Genetic drift is more common when the effective breeding size of the population is low. When this condition prevails, inbreeding is also much more likely. They are different in that inbreeding can occur when certain population structures or behaviors favor matings between relatives, regardless of the effective size of the population. Inbreeding tends to increase the frequency of both homozygous classes at the expense of the heterozygotes. Genetic drift can lead to fixation of one allele or the other, thus producing a single homozygous class.

**6.** Inbreeding depression, over time, reduces the level of heterozygosity, usually a selectively advantageous quality of a species. When homozygosity increases (through loss of heterozygosity), deleterious alleles are likely to become more of a load on a population. Outbreeding depression occurs when there is a reduction in fitness of progeny from genetically diverse individuals. It is usually attributed to offspring being less well-adapted to the local environmental conditions of the parents. Even though forced outbreeding may be necessary to save a threatened species, where population numbers are low, it significantly and permanently changes the genetic makeup of the species.

**8.** Often, molecular assays of overall heterozygosity can indicate the degree of inbreeding and/or genetic drift. As inbreeding (and genetic drift for that matter) occurs, the degree of heterozygosity decreases. An allele whose frequency is dictated by inbreeding and/or genetic drift will not be uniquely influenced. That is, other alleles would be characterized by decreased heterozygosity as well. So, if the genome in general has a relatively high degree of heterozygosity, the gene is probably influenced by selection rather than by inbreeding and/or genetic drift.

**10.** Generally, threatened species are captured and bred in an artificial environment until sufficient population numbers are achieved to ensure species survival. Next, genetic management strategies are applied to breed individuals in such a way as to increase genetic heterozygosity as much as possible. If plants are involved, seed banks are often used to maintain and facilitate long-term survival.

**12.** Allozymes are variants of a given allele often detected by electrophoresis. Such variation may or may not impact on the fitness of an individual. The greater the allozyme variation, the more genetically heterogeneous the individual. It is generally agreed that such genetic diversity is essential for long-term survival. All other factors being equal, allozyme variation is more likely to reflect physiological variation than RFLP (restriction fragment length polymorphism) variation because RFLP regions are not necessarily found in protein-coding regions of the genome. RFLP allows one to detect very small amounts of genetic diversity in a population and is unlikely to encounter an organism that is not in some way variable in terms of RFLP with respect to other organisms (within and among species).

**14.** (a) The probability of being a heterozygote is $2pq = 2(0.99)(0.01) = 0.0198$. Multiplying this value by 20 gives the probability of being heterozygous:

$$0.0198 \times 20 = 0.396$$

(b) To determine $N_e$, use the expression:

$$N_e/N = 0.42$$
$$N_e = 0.42 \times 50 = 21$$
$$H_t = (1 - 1/2N_e)^t H_o$$
$$= (1 - 1/42)^5 \times 0.0198$$
$$= 0.01755$$
$$H_t/H_o = 0.01755/0.0198 = 0.886$$

Therefore there is a loss of approximately 11.4 percent heterozygosity after five generations.

**16.** Data from *Antechinus* provide insight as to the significance of genetic diversity to the survival of a species. Because such mechanisms (i.e., sperm mixing) are in place, there must be considerable evolutionary rewards. In this case, maintaining diversity must offset the cost (if any) of evolving such a mechanism.

**18.** (a) Since prairie dogs are the main food source for black-footed ferrets, a reduction in prairie dogs would likely stress black-footed ferrets. Unless alternate food sources are available and utilized by the ferrets, their numbers would decline.

(b) Because a population survives a population bottleneck does not mean that the population is in a healthy state. Usually, bottlenecks reduce genetic variability upon which survival and adaptation depend. A second bottleneck, though perhaps not having immediate ramifications, would likely have a negative impact on the long-term survival of the species. One would expect additional reductions in genetic diversity.

(c) Because extinction of an organism is irreversible, one might consider the fate of the ferret as the highest priority. If the prairie dog population is reduced very slowly, the ferret population may succeed in finding alternative food sources, but this is doubtful. Since the ferret population is the most fragile of the three (ferret, prairie dog, and cattle) and represents one of America's most endangered mammals, this case will test the strength of laws designed to protect such species. In some situations, compromise to the point of mutual agreement is not possible.

**20.** First, it will be necessary to determine whether the native habitat in the Asian steppes of the 1920s is suitable to any introduction. If the original range is supportive of reintroduction, care must be taken to introduce horses with maximum genetic diversity possible. To do so, you might monitor RFLP (restriction fragment length polymorphisms) patterns or other indicators of genetic diversity. Since the founder breeding group included a domestic mare, it may be desirable to select for reintroduction those that are least like the domestic mare genetically. It might be desirable to release reasonably sized breeding groups in separate locations within the range to enhance eventual genetic diversity.

**22.** From a physiological standpoint, cryogenic preservation in liquid nitrogen can allow 100 years or more of seed storage for some species. However, such elaborate storage can only be offered to a small fraction of the world's seeds. Thus, seeds of most species undergo storage loss, which decreases genetic diversity. Seeds of tropical plants are somewhat intolerant to cold storage and must be regenerated frequently, a practice that is prone to a loss of genetic diversity arising from genetic drift. Only a finite number of seeds can be used in each regeneration procedure, and the restriction of sample size (often fewer than 100 plants) reduces genetic diversity. To somewhat counteract this problem, plants are grown under optimum conditions to reduce selection. Another problem with preserved seeds is the accumulation of deleterious mutations as a result of both seed storage and regeneration. Some studies indicate increased frequencies of chromosomal and mtDNA lesions, chlorophyll deficiency mutations, and decreased DNA polymerase activity associated with long-term seed storage.

**24.** The longest bottleneck-to-present interval occurred with cheetahs, and one would expect cheetahs to show the highest degree of microsatellite polymorphism. The shortest bottleneck-to-present interval occurred with the Gir Forest lion, so it would be expected to have the least polymorphism. Data from Driscoll et al. (2002 *Genome Research* 12: 414–423) include the following estimates of microsatellite polymorphism in the three feline groups mentioned above: cheetahs (84.1 percent), pumas (42.9 percent), and Gir Forest lions (19.3 percent).

**26.** (a) The species with the greatest genetic variability, as estimated by these markers, is the domesticated cat. Domesticated cats share an immense and variable gene pool. Their staggering numbers and outbreeding behaviors allow them to maintain a high degree of genetic variability.

(b) The lion has the least genetic variability.

(c) Since allozymes code for proteins and proteins often provide a significant function in an organism, selection is stronger and mutations are less tolerated. Selection would be expected to be harsher on DNA segments that are related to function. In addition, by their very nature, microsatellites and minisatellites are more mutable.

**28.** While flagship species (often large mammals) may make it possible to gather considerable public support and funding, they may reduce support for species that may have a greater impact on a community of species. Primary producers (plants) are a necessary component of a diverse and supportive habitat. If one focuses on a flagship species within an area, it is possible that other areas will suffer more dramatically because foundational species are lost. Using umbrella species to protect a large geographic area in hopes of protecting other species in that area is a reasonable approach. However, the size of an area is not necessarily a primary factor in determining species success. Diversity and productivity of a habitat are major contributors to species success. Since land is at a premium, it may be wiser in the long run to select umbrella species in diverse and productive habitats rather than on the basis of land size. By selecting sets of species that show considerable biodiversity, one increases the likelihood of protecting a sufficiently rich habitat to support many species. Such habitats are often of considerable economic value, thereby making their availability limited.

# Appendix C

## SELECTED READINGS


### Chapter 1 Introduction to Genetics

Barnum, S. R. 2005. *Biotechnology,* 2nd ed. Belmont, CA: Brooks-Cole.

Bilen, J., and Bonini, N. M. 2005. *Drosophila* as a model for human neurodegenerative disease. *Annu. Rev. Genet.* 39: 153–171.

Campbell, A. M., and Heyer, L. J. (2007). *Discovering genomics, proteomics, and bioinformatics,* 2nd ed. San Francisco, CA: Benjamin Cummings.

Dale, P. J., Clarke, B., and Fontes, E. M. G. 2002. Potential for the environmental impact of transgenic crops. *Nature Biotech.* 20: 567–574.

Fortini, M., and Bonini, N. M. 2000. Modeling human neurodegenerative diseases in *Drosophila. Trends Genet.* 16: 161–167.

Lonberg, N. 2005. Human antibodies from transgenic animals. *Nature Biotech.* 23: 1117–1125.

Lurquin, P. 2002. *High tech harvest.* Boulder, CO: Westview Press.

Pearson, H. 2006. What is a gene? *Nature* 441: 399–401.

Potter, C. J., Turenchalk, G. S., and Xu, T. 2000. *Drosophila* in cancer research: An expanding role. *Trends Genet.* 16: 33–39.

Pray, C. E., Huang, J., Hu, R., and Rozelle, S. 2002. Five years of Bt cotton in China—The benefits continue. *The Plant Journal* 31: 423–430.

Primrose, S. B., and Twyman, R. M. 2004. *Genomics: Applications in human biology.* Oxford: Blackwell Publishing.

Weinberg, R. A. 1985. The molecules of life. *Sci. Am.* (Oct.) 253: 48–57.

Wisniewski, J-P., Frange, N., Massonneau, A., and Dumas, C. 2002. Between myth and reality Genetically modified maize, an example of a sizeable scientific controversy. *Biochimie* 84: 1095–1103.

### Chapter 2 Mitosis and Meiosis

Alberts, B., et al. 2007. *Molecular biology of the cell,* 5th ed. New York: Garland Publ.

Brachet, J., and Mirsky, A. E. 1961. *The cell: Meiosis and mitosis,* Vol. 3. Orlando, FL: Academic Press.

DuPraw, E. J. 1970. *DNA and chromosomes.* New York: Holt, Rinehart & Winston.

Glotzer, M. 2005. The molecular requirements for cytokinesis. *Science* 307: 1735–1739.

Glover, D. M., Gonzalez, C., and Raff, J. W. 1993. The centrosome. *Sci. Am.* (June) 268: 62–68.

Golomb, H. M., and Bahr, G. F. 1971. Scanning electron microscopic observations of surface structures of isolated human chromosomes. *Science* 171: 1024–1026.

Hartwell, L. H., and Karstan, M. B. 1994. Cell cycle control and cancer. *Science* 266: 1821–1828.

Hartwell, L. H., and Weinert, T. A. 1989. Checkpoint controls that ensure the order of cell cycle events. *Science* 246: 629–634.

Mazia, D. 1961. How cells divide. *Sci. Am.* (Jan.) 205: 101–120.

———. 1974. The cell cycle. *Sci. Am.* (Jan.) 235: 54–64.

McIntosh, J. R., and McDonald, K. L. 1989. The mitotic spindle. *Sci. Am.* (Oct.) 261: 48–56.

*Nature Milestones.* 2001. *Cell Division.* Nature Publishing Group, London.

Westergaard, M., and von Wettstein, D. 1972. The synaptinemal complex. *Annu. Rev. Genet.* 6: 71–110.

### Chapter 3 Mendelian Genetics

Bennett, R. L., et al. 1995. Recommendations for standardized human pedigree nomenclature. *Am. J. Hum. Genet.* 56: 745–752.

Carlson, E. A. 1987. *The gene: A critical history,* 2nd ed. Philadelphia: Saunders.

Dunn, L. C. 1965. *A short history of genetics.* New York: McGraw-Hill.

Henig, R. M. 2001. *The monk in the garden: The lost and found genius of Gregor Mendel, the father of genetics.* New York: Houghton-Mifflin.

Klein, J. 2000. Johann Mendel's field of dreams. *Genetics* 156: 1–6.

Miller, J. A. 1984. Mendel's peas: A matter of genius or of guile? *Sci. News* 125: 108–109.

Olby, R. C. 1985. *Origins of Mendelism,* 2nd ed. London: Constable.

Orel, V. 1996. *Gregor Mendel: The first geneticist.* Oxford: Oxford University Press.

Peters, J., ed. 1959. *Classic papers in genetics.* Englewood Cliffs, NJ: Prentice-Hall.

Sokal, R. R., and Rohlf, F. J. 1995. *Biometry,* 3rd ed. New York: W. H. Freeman.

Stern, C., and Sherwood, E. 1966. *The origins of genetics: A Mendel source book.* San Francisco: W. H. Freeman.

Stubbe, H. 1972. *History of genetics: From prehistoric times to the rediscovery of Mendel's laws.* Cambridge, MA: MIT Press.

Sturtevant, A. H. 1965. *A history of genetics.* New York: Harper & Row.

Tschermak-Seysenegg, E. 1951. The rediscovery of Mendel's work. *J. Hered.* 42: 163–172.

Welling, F. 1991. Historical study: Johann Gregor Mendel 1822–1884. *Am. J. Med. Genet.* 40: 1–25.

### Chapter 4 Extensions of Mendelian Genetics

Bartolomei, M. S., and Tilghman, S. M. 1997. Genomic imprinting in mammals. *Annu. Rev. Genet.* 31: 493–525.

Brink, R. A., ed. 1967. *Heritage from Mendel.* Madison: University of Wisconsin Press.

Bultman, S. J., Michaud, E. J., and Woychik, R. P. 1992. Molecular characterization of the mouse *agouti* locus. *Cell* 71: 1195–1204.

Carlson, E. A. 1987. *The gene: A critical history,* 2nd ed. Philadelphia: Saunders.

Cattanach, B. M., and Jones, J. 1994. Genetic imprinting in the mouse: Implications for gene regulation. *J. Inherit. Metab. Dis.* 17: 403–420.

Dunn, L. C. 1966. *A short history of genetics.* New York: McGraw-Hill.

Feil, R., and Khosla, S. 1999. Genomic imprinting in mammals: An interplay between chromatin and DNA methylation. *Trends in Genet.* 15: 431.

Foster, M. 1965. Mammalian pigment genetics. *Adv. Genet.* 13: 311–339.

Grant, V. 1975. *Genetics of flowering plants.* New York: Columbia University Press.

Harper, P. S., et al. 1992. Anticipation in myotonic dystrophy: New light on an old problem. *Am. J. Hum. Genet.* 51: 10–16.

Howeler, C. J., et al. 1989. Anticipation in myotonic dystrophy: Fact or fiction? *Brain* 112: 779–797.

Morgan, T. H. 1910. Sex-limited inheritance in *Drosophila*. *Science* 32: 120–122.
Peters, J. A., ed. 1959. *Classic papers in genetics*. Englewood Cliffs, NJ: Prentice-Hall.
Phillips, P. C. 1998. The language of gene interaction. *Genetics* 149: 1167–1171.
Race, R. R., and Sanger, R. 1975. *Blood groups in man*, 6th ed. Oxford: Blackwell.
Sapienza, C. 1990. Parental imprinting of genes. *Sci. Am.* (Oct.) 363: 52–60.
Siracusa, L. D. 1994. The *agouti* gene: Turned on to yellow. *Cell* 10: 423–428.
Waters, D. J., and Wildasin, K. 2006. Cancer clues from pet dogs. *Sci. Am.* (Dec.) 295: 96–101.
Yoshida, A. 1982. Biochemical genetics of the human blood group ABO system. *Am. J. Hum. Genet.* 34: 1–14.

## Chapter 5 Chromosome Mapping in Eukaryotes

Allen, G. E. 1978. *Thomas Hunt Morgan: The man and his science*. Princeton, NJ: Princeton University Press.
Chaganti, R., Schonberg, S., and German, J. 1974. A manyfold increase in sister chromatid exchange in Bloom syndrome lymphocytes. *Proc. Natl. Acad. Sci.* 71: 4508–4512.
Creighton, H. S., and McClintock, B. 1931. A correlation of cytological and genetical crossing over in Zea mays. *Proc. Natl. Acad. Sci.* 17: 492–497.
Douglas, L., and Novitski, E. 1977. What chance did Mendel's experiments give him of noticing linkage? *Heredity* 38: 253–257.
Ellis, N. A., et al. 1995. The Bloom's syndrome gene product is homologous to RecQ helicases. *Cell* 83: 655–666.
Ephrussi, B., and Weiss, M. C. 1969. Hybrid somatic cells. *Sci. Am.* (Apr.) 220: 26–35.
Latt, S. A. 1981. Sister chromatid exchange formation. *Annu. Rev. Genet.* 15: 11–56.
Lindsley, D. L., and Grell, E. H. 1972. *Genetic variations of Drosophila melanogaster*. Washington, DC: Carnegie Institute of Washington.
Morgan, T. H. 1911. An attempt to analyze the constitution of the chromosomes on the basis of sex-linked inheritance in *Drosophila*. *J. Exp. Zool.* 11: 365–414.
Morton, N. E. 1955. Sequential test for the detection of linkage. *Am. J. Hum. Genet.* 7: 277–318.
———. 1995. LODs—Past and present. *Genetics* 140: 7–12.
Neuffer, M. G., Jones, L., and Zober, M. 1968. *The mutants of maize*. Madison, WI: Crop Sci. Soc. of America.
Perkins, D. 1962. Crossing over and interference in a multiply marked chromosome arm of *Neurospora*. *Genetics* 47: 1253–1274.
Ruddle, F. H., and Kucherlapati, R. S. 1974. Hybrid cells and human genes. *Sci. Am.* (July) 231: 36–49.
Stahl, F. W. 1979. *Genetic recombination*. New York: W. H. Freeman.
Stern, C. 1936. Somatic crossing over and segregation in *Drosophila melanogaster*. *Genetics* 21: 625–631.
Sturtevant, A. H. 1913. The linear arrangement of six sex-linked factors in *Drosophila*, as shown by their mode of association. *J. Exp. Zool.* 14: 43–59.
———. 1965. *A history of genetics*. New York: Harper & Row.
Voeller, B. R., ed. 1968. *The chromosome theory of inheritance: Classical papers in development and heredity*. New York: Appleton-Century-Croft.
Wellcome Trust Case Control Consortium, 2007. Genome association study of 14,000 cases of seven common diseases and 3,000 shared controls. *Nature* 447: 661–676.
Wolff, S., ed. 1982. *Sister chromatid exchange*. New York: Wiley-Interscience.

## Chapter 6 Genetic Analysis and Mapping in Bacteria and Bacteriophages

Adelberg, E. A. 1960. *Papers on bacterial genetics*. Boston: Little, Brown.
Benzer, S. 1962. The fine structure of the gene. *Sci. Am.* (Jan.) 206: 70–86.
Birge, E. A. 1988. *Bacterial and bacteriophage genetics—An introduction*. New York: Springer-Verlag.
Brock, T. 1990. *The emergence of bacterial genetics*. Cold Spring Harbor, NY: Cold Spring Harbor Press.
Bukhari, A. I., Shapiro, J. A., and Adhya, S. L., eds. 1977. *DNA insertion elements, plasmids, and episomes*. Cold Spring Harbor, NY: Cold Spring Harbor Press.
Cairns, J., Stent, G. S., and Watson, J. D., eds. 1966. *Phage and the origins of molecular biology*. Cold Spring Harbor, NY: Cold Spring Harbor Press.
Campbell, A. M. 1976. How viruses insert their DNA into the DNA of the host cell. *Sci. Am.* (Dec.) 235: 102–113.
Hayes, W. 1968. *The genetics of bacteria and their viruses*, 2nd ed. New York: Wiley.
Hershey, A. D., and Rotman, R. 1949. Genetic recombination between host range and plaque-type mutants of bacteriophage in single cells. *Genetics* 34: 44–71.
Hotchkiss, R. D., and Marmur, J. 1954. Double marker transformations as evidence of linked factors in deoxyribonucleate transforming agents. *Proc. Natl. Acad. Sci. (USA)* 40: 55–60.
Jacob, F., and Wollman, E. L. 1961. Viruses and genes. *Sci. Am.* (June) 204: 92–106.
Kohiyama, M., et al. 2003. Bacterial sex: Playing voyeurs 50 years later. *Science* 301: 802–803.
Kruse, H., and Sorum, H. 1994. Transfer of multiple drug resistance plasmids between bacteria of diverse origins in natural microenvironments. *Appl. Environ. Microbiol.* 60: 4015–4021.
Lederberg, J. 1986. Forty years of genetic recombination in bacteria: A fortieth anniversary reminiscence. *Nature* 324: 627–628.
Luria, S. E., and Delbruck, M. 1943. Mutations of bacteria from virus sensitivity to virus resistance. *Genetics* 28: 491–511.
Lwoff, A. 1953. Lysogeny. *Bacteriol. Rev.* 17: 269–337.
Miller, J. H. 1992. *A short course in bacterial genetics*. Cold Spring Harbor, NY: Cold Spring Harbor Press.
Miller, R. V. 1998. Bacterial gene swapping in nature. *Sci. Am.* (Jan.) 278: 66–71.
Morse, M. L., Lederberg, E. M., and Lederberg, J. 1956. Transduction in *Escherichia coli* K12. *Genetics* 41: 141–156.
Novick, R. P. 1980. Plasmids. *Sci. Am.* (Dec.) 243: 102–127.
Smith-Keary, P. F. 1989. *Molecular genetics of* Escherichia coli. New York: Guilford Press.
Stahl, F. W. 1987. Genetic recombination. *Sci. Am.* (Nov.) 256: 91–101.
Stent, G. S. 1966. *Papers on bacterial viruses*, 2nd ed. Boston: Little, Brown.
Wollman, E. L., Jacob, F., and Hayes, W. 1956. Conjugation and genetic recombination in *Escherichia coli* K12. *Cold Spring Harb. Symp. Quant. Biol.* 21: 141–162.
Zinder, N. D. 1958. Transduction in bacteria. *Sci. Am.* (Nov.) 199: 38–46.

## Chapter 7 Sex Determination and Sex Chromosomes

Amory, J. K., et al. 2000. Klinefelter's syndrome. *Lancet* 356: 333–335.
Court-Brown, W. M. 1968. Males with an XYY sex chromosome complement. *J. Med. Genet.* 5: 341–359.

Davidson, R., Nitowski, H., and Childs, B. 1963. Demonstration of two populations of cells in human females heterozygous for glucose-6-phosphate dehydrogenase variants. *Proc. Natl. Acad. Sci. (USA)* 50: 481–485.

Erickson, J. D. 1976. The secondary sex ratio of the United States, 1969–71: Association with race, parental ages, birth order, paternal education and legitimacy. *Ann. Hum. Genet.* (London) 40: 205–212.

Hodgkin, J. 1990. Sex determination compared in *Drosophila* and *Caenorhabditis*. *Nature* 344: 721–728.

Hook, E. B. 1973. Behavioral implications of the humans XYY genotype. *Science* 179: 139–150.

Irish, E. E. 1996. Regulation of sex determination in maize. *BioEssays* 18: 363–369.

Jacobs, P. A., et al. 1974. A cytogenetic survey of 11,680 newborn infants. *Ann. Hum. Genet.* 37: 359–376.

Jegalian, K., and Lahn, B. T. 2001. Why the Y is so weird. *Sci. Am.* (Feb.) 284: 56–61.

Koopman, P., et al. 1991. Male development of chromosomally female mice transgenic for *Sry*. *Nature* 351: 117–121.

Lahn, B. T., and Page, D. C. 1997. Functional coherence of the human Y chromosome. *Science* 278: 675–680.

Lucchesi, J. 1983. The relationship between gene dosage, gene expression, and sex in *Drosophila*. *Dev. Genet.* 3: 275–282.

Lyon, M. F. 1972. X-chromosome inactivation and developmental patterns in mammals. *Biol. Rev.* 47: 1–35.

McMillen, M. M. 1979. Differential mortality by sex in fetal and neonatal deaths. *Science* 204: 89–91.

Ng, K., Pullirsch, D., Leeb, M., and Wutz, A. 2007. Xist and the order of silencing. *EMBO Reports* 8: 34–39.

Penny, G. D., et al. 1996. Requirement for Xist in X chromosome inactivation. *Nature* 379: 131–137.

Pieau, C. 1996. Temperature variation and sex determination in reptiles. *BioEssays* 18: 19–26.

Straub, T., and Becker, P. B. 2007. Dosage compensation: the beginning and end of a generalization. *Nat. Rev. Genet.* 8: 47–57.

Westergaard, M. 1958. The mechanism of sex determination in dioecious flowering plants. *Adv. Genet.* 9: 217–281.

Whitelaw, E. 2006. Unravelling the X in sex. *Dev. Cell* 11: 759–762.

Witkin, H. A., et al. 1996. Criminality in XYY and XXY men. *Science* 193: 547–555.

Xu, N., Tsai, C-L., and Lee, J. T. 2006. Transient homologous chromosome pairing marks the onset of X inactivation. *Science* 311: 1149–1152.

## Chapter 8 Chromosome Mutations: Variation in Chromosome Number and Arrangement

Antonarakis, S. E. 1998. Ten years of genomics, chromosome 21, and Down syndrome. *Genomics* 51: 1–16.

Ashley-Koch, A. E., et al. 1997. Examination of factors associated with instability of the *FMR1* CGG repeat. *Am. J. Hum. Genet.* 63: 776–785.

Beasley, J. O. 1942. Meiotic chromosome behavior in species, species hybrids, haploids, and induced polyploids of *Gossypium*. *Genetics* 27: 25–54.

Blakeslee, A. F. 1934. New jimson weeds from old chromosomes. *J. Hered.* 25: 80–108.

Boue, A. 1985. Cytogenetics of pregnancy wastage. *Adv. Hum. Genet.* 14: 1–58.

Carr, D. H. 1971. Genetic basis of abortion. *Ann. Rev. Genet.* 5: 65–80.

Croce, C. M. 1996. The *FHIT* gene at 3p14.2 is abnormal in lung cancer. *Cell* 85: 17–26.

DeArce, M. A., and Kearns, A. 1984. The fragile X syndrome: The patients and their chromosomes. *J. Med. Genet.* 21: 84–91.

Feldman, M., and Sears, E. R. 1981. The wild gene resources of wheat. *Sci. Am.* (Jan.) 244: 102–112.

Galitski, T., et al. 1999. Ploidy regulation of gene expression. *Science* 285: 251–254.

Gersh, M., et al. 1995. Evidence for a distinct region causing a catlike cry in patients with 5p deletions. *Am. J. Hum. Genet.* 56: 1404–1410.

Hassold, T. J., et al. 1980. Effect of maternal age on autosomal trisomies. *Ann. Hum. Genet.* (London) 44: 29–36.

Hassold, T. J., and Hunt, P. 2001. To err (meiotically) is human: The genesis of human aneuploidy. *Nat. Rev. Gen.* 2: 280–291.

Hassold, T., and Jacobs, P. A. 1984. Trisomy in man. *Annu. Rev. Genet.* 18: 69–98.

Hecht, F. 1988. Enigmatic fragile sites on human chromosomes. *Trends Genet.* 4: 121–122.

Hulse, J. H., and Spurgeon, D. 1974. *Triticale*. *Sci. Am.* (Aug.) 231: 72–81.

Kaiser, P. 1984. Pericentric inversions: Problems and significance for clinical genetics. *Hum. Genet.* 68: 1–47.

Lewis, E. B. 1950. The phenomenon of position effect. *Adv. Genet.* 3: 73–115.

Lewis, W. H., ed. 1980. *Polyploidy: Biological relevance*. New York: Plenum Press.

Lynch, M., and Conery, J. S. 2000. The evolutionary fate and consequences of duplicate genes. *Science* 290: 1151–1154.

Madan, K. 1995. Paracentric inversions: A review. *Hum. Genet.* 96: 503–515.

Nelson, D. L., and Gibbs, R. A. 2004. The critical region in trisomy 21. *Science* 306: 619–621.

Ohno, S. 1970. *Evolution by gene duplication*. New York: Springer-Verlag.

Oostra, B. A., and Verkerk, A. J. 1992. The fragile X syndrome: Isolation of the *FMR-1* gene and characterization of the fragile X mutation. *Chromosoma* 101: 381–387.

Patterson, D. 1987. The causes of Down syndrome. *Sci. Am.* (Aug.) 257: 52–61.

Patterson, D., and Costa, A. 2005. Down syndrome and genetics—A case of linked histories. *Nature Reviews Genetics* 6: 137–145.

Shepard, J., et al. 1983. Genetic transfer in plants through interspecific protoplast fusion. *Science* 21: 683–688.

Shepard, J. F. 1982. The regeneration of potato plants from protoplasts. *Sci. Am.* (May) 246: 154–166.

Stranger, B. E., et al., 2007. Relative impact of nucleotide and copy number variation on gene expression phenotypes. *Science* 315: 848–854.

Taylor, A. I. 1968. Autosomal trisomy syndromes: A detailed study of 27 cases of Edwards syndrome and 27 cases of Patau syndrome. *J. Med. Genet.* 5: 227–252.

Tjio, J. H., and Levan, A. 1956. The chromosome number of man. *Hereditas* 42: 1–6.

Wilkins, L. E., Brown, J. X., and Wolf, B. 1980. Psychomotor development in 65 home-reared children with cri-du-chat syndrome. *J. Pediatr.* 97: 401–405.

## Chapter 9 Extranuclear Inheritance

Adams, K. L., et al. 2000. Repeated, recent and diverse transfers of a mitochondrial gene to the nucleus in flowering plants. *Nature* 408: 354–357.

Choman, A. 1998. The myoclonic epilepsy and ragged-red fiber mutation provides new insights into human mitochondrial function and genetics. *Am. J. Hum. Genet.* 62: 745–751.

Freeman, G., and Lundelius, J. W. 1982. The developmental genetics of dextrality and sinistrality in the gastropod *Lymnaea peregra*. *Wilhelm Roux Arch.* 191: 69–83.

Gray, M. W., Burger, G., and Lang, B. 1999. Mitochondrial evolution. *Science* 283: 1476–1481.

Green, B. R., and Burton, H. 1970. Acetabularia chloroplast DNA: Electron microscopic visualization. *Science* 168: 981–982.

Grivell, L. A. 1983. Mitochondrial DNA. *Sci. Am.* (Mar.) 248: 78–89.

Hiendleder, S. 2007, Mitochondrial DNA inheritance after SCNT. *Adv. Exp. Med Biol.* 591: 103–116.

Larson, N. G., and Clayton, D. A. 1995. Molecular genetic aspects of human mitochondrial disorders. *Ann. Rev. Genet.* 29: 151–178.

Margulis, L. 1970. *Origin of eukaryotic cells.* New Haven, CT: Yale University Press.

Mitchell, M. B., and Mitchell, H. K. 1952. A case of maternal inheritance in *Neurospora crassa. Proc. Natl. Acad. Sci. (USA)* 38: 442–449.

Nüsslein-Volhard. C. 1996. Gradients that organize embryo development. *Sci. Am.* (Aug.) 275: 54–61.

Preer, J. R. 1971. Extrachromosomal inheritance: Hereditary symbionts, mitochondria, chloroplasts. *Annu. Rev. Genet.* 5: 361–406.

Sager, R. 1965. Genes outside the chromosomes. *Sci. Am.* (Jan.) 212: 70–79.

_______. 1985. Chloroplast genetics. *BioEssays* 3: 180–184.

Schwartz, R. M., and Dayhoff, M. O. 1978. Origins of prokaryotes, eukaryotes, mitochondria and chloroplasts. *Science* 199: 395–403.

Sonneborn, T. M. 1959. Kappa and related particles in *Paramecium. Adv. Virus Res.* 6: 229–256.

Sturtevant, A. H. 1923. Inheritance of the direction of coiling in *Limnaea. Science* 58: 269–270.

Taylor, R. W., and Turnbull, D. M., 2005. Mitochondrial DNA mutations in human disease. *Nature Reviews Genetics* 6: 389–402.

Wallace, D. C. 1997. Mitochondrial DNA in aging and disease. *Sci. Am.* (Aug.) 277: 40–59.

_______. 1999. Mitochondrial diseases in man and mouse. *Science* 283: 1482–1488.

## Chapter 10 DNA Structure and Analysis

Adleman, L. M. 1998. Computing with DNA. *Sci. Am.* (Aug.) 279: 54–61.

Alloway, J. L. 1933. Further observations on the use of pneumococcus extracts in effecting transformation of type *in vitro. J. Exp. Med.* 57: 265–278.

Avery, O. T., MacLeod, C. M., and McCarty, M. 1944. Studies on the chemical nature of the substance inducing transformation of pneumococcal types: Induction of transformation by a desoxyribonucleic acid fraction isolated from pneumococcus type III. *J. Exp. Med.* 79: 137–158. (Reprinted in Taylor, J. H. 1965. *Selected papers in molecular genetics.* Orlando, FL: Academic Press.)

Britten, R. J., and Kohne, D. E. 1968. Repeated sequences in DNA. *Science* 161: 529–540.

Chargaff, E. 1950. Chemical specificity of nucleic acids and mechanism for their enzymatic degradation. *Experientia* 6: 201–209.

Darnell, J. E. 1985. RNA. *Sci. Am.* (Oct.) 253: 68–87.

Dawson, M. H. 1930. The transformation of pneumococcal types: I. The interconvertibility of type-specific *S. pneumococci. J. Exp. Med.* 51: 123–147.

Dickerson, R. E., et al. 1982. The anatomy of A-, B-, and Z-DNA. *Science* 216: 475–485.

Dubos, R. J. 1976. *The professor, the Institute and DNA: Oswald T. Avery, his life and scientific achievements.* New York: Rockefeller University Press.

Felsenfeld, G. 1985. DNA. *Sci. Am.* (Oct.) 253: 58–78.

Franklin, R. E., and Gosling, R. G. 1953. Molecular configuration in sodium thymonucleate. *Nature* 171: 740–741.

Griffith, F. 1928. The significance of pneumococcal types. *J. Hyg.* 27: 113–159.

Guthrie, G. D., and Sinsheimer, R. L. 1960. Infection of protoplasts of *Escherichia coli* by subviral particles. *J. Mol. Biol.* 2: 297–305.

Hershey, A. D., and Chase, M. 1952. Independent functions of viral protein and nucleic acid in growth of bacteriophage. *J. Gen. Phys.* 36: 39–56. (Reprinted in Taylor, J. H. 1965. *Selected papers in molecular genetics.* Orlando, FL: Academic Press.)

Levene, P. A., and Simms, H. S. 1926. Nucleic acid structure as determined by electrometric titration data. *J. Biol. Chem.* 70: 327–341.

McCarty, M. 1985. *The transforming principle: Discovering that genes are made of DNA.* New York: W. W. Norton.

Olby, R. 1974. *The path to the double helix.* Seattle: University of Washington Press.

Pauling, L., and Corey, R. B. 1953. A proposed structure for the nucleic acids. *Proc. Natl. Acad. Sci. (USA)* 39: 84–97.

Rich, A., Nordheim, A., and Wang, A. H.-J. 1984. The chemistry and biology of left-handed Z-DNA. *Annu. Rev. Biochem.* 53: 791–846.

Spizizen, J. 1957. Infection of protoplasts by disrupted T2 viruses. *Proc. Natl. Acad. Sci. (USA)* 43: 694–701.

Varmus, H. 1988. Retroviruses. *Science* 240: 1427–1435.

Watson, J. D. 1968. *The double helix.* New York: Atheneum.

Watson, J. D., and Crick, F. C. 1953a. Molecular structure of nucleic acids: A structure for deoxyribose nucleic acids. *Nature* 171: 737–738.

———. 1953b. Genetic implications of the structure of deoxyribose nucleic acid. *Nature* 171: 964.

Wilkins, M. H. F., Stokes, A. R., and Wilson, H. R. 1953. Molecular structure of desoxypentose nucleic acids. *Nature* 171: 738–740.

## Chapter 11 DNA Replication and Recombination

Blackburn, E. H. 1991. Structure and function of telomeres. *Nature* 350: 569–572.

DeLucia, P., and Cairns, J. 1969. Isolation of an *E. coli* strain with a mutation affecting DNA polymerase. *Nature* 224: 1164–1166.

Gilbert, D. M. 2001. Making sense of eukaryotic DNA replication origins. *Science* 294: 96–100.

Greider, C. W. 1996. Telomeres, telomerase, and cancer. *Sci. Am.* (Feb.) 274: 92–97.

_______. 1998. Telomerase activity, cell proliferation, and cancer. *Proc. Natl. Acad. Sci. (USA)* 95: 90–92.

Holliday, R. 1964. A mechanism for gene conversion in fungi. *Genet. Res.* 5: 282–304.

Holmes, F. L. 2001. *Meselson, Stahl, and replication of DNA: A history of the "most beautiful experiment in biology."* New Haven, CT: Yale University Press.

Kim, J., Kaminker, P., and Campisi, J. 2002. Telomeres, aging and cancer: In search of a happy ending. *Oncogene* 21: 503–511.

Kornberg, A. 1960. Biological synthesis of DNA. *Science* 131: 1503–1508.

Kornberg, A., and Baker, T. 1992. *DNA replication,* 2nd ed. New York: W. H. Freeman.

Meselson, M., and Stahl, F. W. 1958. The replication of DNA in *Escherichia coli. Proc. Natl. Acad. Sci. (USA)* 44: 671–682.

Mitchell, M. B. 1955. Aberrant recombination of pyridoxine mutants of *Neurospora. Proc. Natl. Acad. Sci. (USA)* 41: 215–220.

Okazaki, T., et al. 1979. Structure and metabolism of the RNA primer in the discontinuous replication of prokaryotic DNA. *Cold Spring Harbor Symp. Quant. Biol.* 43: 203–222.

Radman, M., and Wagner, R. 1988. The high fidelity of DNA duplication. *Sci. Am.* (Aug.) 259: 40–46.

———. 1987. Genetic recombination. *Sci. Am.* (Feb.) 256: 90–101.

Taylor, J. H., Woods, P. S., and Hughes, W. C. 1957. The organization and duplication of chromosomes revealed by autoradiographic studies using tritium-labeled thymidine. *Proc. Natl. Acad. Sci. (USA)* 48: 122–128.

Wang, J. C. 1987. Recent studies of DNA topoisomerases. *Biochim. Biophys. Acta* 909: 1–9.
Whitehouse, H. L. K. 1982. *Genetic recombination: Understanding the mechanisms.* New York: Wiley.

## Chapter 12 DNA Organization in Chromosomes

Angelier, N., et al. 1984. Scanning electron microscopy of amphibian lampbrush chromosomes. *Chromosoma* 89: 243–253.
Beerman, W., and Clever, U. 1964. Chromosome puffs. *Sci. Am.* (Apr.) 210: 50–58.
Carbon, J. 1984. Yeast centromeres: Structure and function. *Cell* 37: 352–353.
Chen, T. R., and Ruddle, F. H. 1971. Karyotype analysis utilizing differential stained constitutive heterochromatin of human and murine chromosomes. *Chromosoma* 34: 51–72.
DuPraw, E. J. 1970. *DNA and chromosomes.* New York: Holt, Rinehart & Winston.
Gall, J. G. 1981. Chromosome structure and the *C*-value paradox. *J. Cell Biol.* 91: 3s–14s.
Hewish, D. R., and Burgoyne, L. 1973. Chromatin substructure. The digestion of chromatin DNA at regularly spaced sites by a nuclear deoxyribonuclease. *Biochem. Biophys. Res. Comm.* 52: 504–510.
Korenberg, J. R., and Rykowski, M. C. 1988. Human genome organization: *Alu*, LINES, and the molecular organization of metaphase chromosome bands. *Cell* 53: 391–400.
Kornberg, R. D. 1975. Chromatin structure: A repeating unit of histones and DNA. *Science* 184: 868–871.
Kornberg, R. D., and Klug, A. 1981. The nucleosome. *Sci. Am.* (Feb.) 244: 52–64.
Luger, K., et al. 1997. Crystal structure of the nucleosome core particle at 2.8 resolution. *Nature* 389: 251–256.
Moyzis, R. K. 1991. The human telomere. *Sci. Am.* (Aug.) 265: 48–55.
Olins, A. L., and Olins, D. E. 1974. Spheroid chromatin units (*n* bodies). *Science* 183: 330–332.
———. 1978. Nucleosomes: The structural quantum in chromosomes. *Am. Sci.* 66: 704–711.
Singer, M. F. 1982. SINES and LINES: Highly repeated short and long interspersed sequences in mammalian genomes. *Cell* 28: 433–434.

## Chapter 13 Recombinant DNA Technology and Gene Cloning

Brownlee, C. 2005. Danna and Nathans: Restriction enzymes and the boon to modern molecular biology. *Proc. Nat. Acad. Sci.* 103: 5909.
Burger, S. L., and Kimmel, A. R. 1987. *Guide to molecular cloning techniques. Methods in enzymology,* Vol. 152. San Diego: Academic Press.
Cohen, S. N. 1975. The manipulation of genes. *Sci. Am.* (Jan.) 233: 24–33.
Flotte, T. R. 2005. Adeno-associated virus-based gene therapy for inherited disorders. *Pediatr. Res.* 58: 1143–1147.
McKusick, V. A. 1988. The new genetics and clinical medicine. *Hosp. Pract.* 23: 177–191.
Metzker, M. L. 2005. Emerging technologies in DNA sequencing. *Genome Res.* 15: 1767–1775.
Micklos, D. A., et al. 2002. *DNA science: A first course.* Cold Spring Harbor, NY: Cold Spring Harbor Press.
Mullis, K. B. 1990. The unusual origin of the polymerase chain reaction. *Sci. Am.* (Apr.) 262: 56–65.
Pingoud, A., et al. 2005. Type II restriction endonucleases: structure and mechanism. *Cell Mol. Life Sci.* 62: 685–707.
Sambrook, J., and Russell, D. 2001. *Molecular cloning: A laboratory manual,* 3rd ed. Cold Spring Harbor, NY: Cold Spring Harbor Press.
Sanger, F., et al. 1977. DNA sequencing with chain-terminating inhibitors. *Proc. Natl. Acad. Sci. (USA)* 74: 5463–5467.
Southern, E. 1975. Detection of specific sequences among DNA fragments separated by gel electrophoresis. *J. Mol. Biol.* 98: 503–507.

## Chapter 14 The Genetic Code and Transcription

Barrell, B. G., Air, G., and Hutchinson, C. 1976. Overlapping genes in bacteriophage $\phi$X174 *Nature* 264: 34–40.
Barrell, B. G., Banker, A. T., and Drouin, J. 1979. A different genetic code in human mitochondria. *Nature* 282: 189–194.
Bass, B. L., ed. 2000. *RNA editing.* Oxford: Oxford University Press.
Brenner, S., Jacob, F., and Meselson, M. 1961. An unstable intermediate carrying information from genes to ribosomes for protein synthesis. *Nature* 190: 575–580.
Brenner, S., Stretton, A. O. W., and Kaplan, D. 1965. Genetic code: The nonsense triplets for chain termination and their suppression. *Nature* 206: 994–998.
Cattaneo, R. 1991. Different types of messenger RNA editing. *Annu. Rev. Genet.* 25: 71–88.
Cech, T. R. 1986. RNA as an enzyme. *Sci. Am.* (Nov.) 255(5): 64–75.
_______. 1987. The chemistry of self-splicing RNA and RNA enzymes. *Science* 236: 1532–1539.
Chambon, P. 1981. Split genes. *Sci. Am.* (May) 244: 60–71.
Cramer, P., et al. 2000. Architecture of RNA polymerase II and implications for the transcription mechanism. *Science* 288: 640–649.
Crick, F. H. C. 1962. The genetic code. *Sci. Am.* (Oct.) 207: 66–77.
_______. 1966a. The genetic code: III. *Sci. Am.* (Oct.) 215: 55–63.
_______. 1966b. Codon–anticodon pairing: The wobble hypothesis. *J. Mol. Biol.* 19: 548–555.
Crick, F. H. C., Barnett, L., Brenner, S., and Watts-Tobin, R. J. 1961. General nature of the genetic code for proteins. *Nature* 192: 1227–1232.
Darnell, J. E. 1983. The processing of RNA. *Sci. Am.* (Oct.) 249: 90–100.
Dickerson, R. E. 1983. The DNA helix and how it is read. *Sci. Am.* (Dec.) 249: 94–111.
Dugaiczk, A., et al. 1978. The natural ovalbumin gene contains seven intervening sequences. *Nature* 274: 328–333.
Fiers, W., et al. 1976. Complete nucleotide sequence of bacteriophage MS2 RNA: Primary and secondary structure of the replicase gene. *Nature* 260: 500–507.
Gamow, G. 1954. Possible relation between DNA and protein structures. *Nature* 173: 318.
Hall, B. D., and Spiegelman, S. 1961. Sequence complementarity of T2-DNA and T2-specific RNA. *Proc. Natl. Acad. Sci. (USA)* 47: 137–146.
Hamkalo, B. 1985. Visualizing transcription in chromosomes. *Trends Genet.* 1: 255–260.
Khorana, H. G. 1967. Polynucleotide synthesis and the genetic code. *Harvey Lectures* 62: 79–105.
Miller, O. L., Hamkalo, B., and Thomas, C. 1970. Visualization of bacterial genes in action. *Science* 169: 392–395.
Nirenberg, M. W. 1963. The genetic code: II. *Sci. Am.* (Mar.) 190: 80–94.
O'Malley, B., et al. 1979. A comparison of the sequence organization of the chicken ovalbumin and ovomucoid genes. In *Eucaryotic gene regulation,* R. Axel et al., eds., pp. 281–299. Orlando, FL: Academic Press.
Reed, R., and Maniatis, T. 1985. Intron sequences involved in lariat formation during pre-mRNA splicing. *Cell* 41: 95–105.
Ridley, M., 2006. *Francis Crick: Discoverer of the genetic code.* New York: HarperCollins.

Sharp, P. A. 1994. Nobel lecture: Split genes and RNA splicing. *Cell* 77: 805–815.

Steitz, J. A. 1988. Snurps. *Sci. Am.* (June) 258(6): 56–63.

Volkin, E., and Astrachan, L. 1956. Phosphorus incorporation in *E. coli* ribonucleic acids after infection with bacteriophage T2. *Virology* 2: 149–161.

Watson, J. D. 1963. Involvement of RNA in the synthesis of proteins. *Science* 140: 17–26.

Woychik, N. A., and Jampsey, M. 2002. The RNA polymerase II machinery: Structure illuminates function. *Cell* 108: 453–464.

## Chapter 15 Translation and Proteins

Anfinsen, C. B. 1973. Principles that govern the folding of protein chains. *Science* 181: 223–230.

Bartholome, K. 1979. Genetics and biochemistry of phenylketonuria—Present state. *Hum. Genet.* 51: 241–245.

Beadle, G. W., and Tatum, E. L. 1941. Genetic control of biochemical reactions in *Neurospora. Proc. Natl. Acad. Sci. (USA)* 27: 499–506.

Beet, E. A. 1949. The genetics of the sickle-cell trait in a Bantu tribe. *Ann. Eugenics* 14: 279–284.

Brenner, S. 1955. Tryptophan biosynthesis in *Salmonella typhimurium. Proc. Natl. Acad. Sci. (USA)* 41: 862–863.

Doolittle, R. F. 1985. Proteins. *Sci. Am.* (Oct.) 253: 88–99.

Frank, J. 1998. How the ribosome works. *Amer. Scient.* 86: 428–439.

Garrod, A. E. 1902. The incidence of alkaptonuria: A study in chemical individuality. *Lancet* 2: 1616–1620.

_______. 1909. *Inborn errors of metabolism.* London: Oxford University Press. (Reprinted 1963, Oxford University Press, London.)

Garrod, S. C. 1989. Family influences on A. E. Garrod's thinking. *J. Inher. Metab. Dis.* 12: 2–8.

Ingram, V. M. 1957. Gene mutations in human hemoglobin: The chemical difference between normal and sickle-cell hemoglobin. *Nature* 180: 326–328.

Koshland, D. E. 1973. Protein shape and control. *Sci. Am.* (Oct.) 229: 52–64.

Lake, J. A. 1981. The ribosome. *Sci. Am.* (Aug.) 245: 84–97.

Maniatis, T., et al. 1980. The molecular genetics of human hemoglobins. *Annu. Rev. Genet.* 14: 145–178.

Neel, J. V. 1949. The inheritance of sickle-cell anemia. *Science* 110: 64–66.

Nirenberg, M. W., and Leder, P. 1964. RNA codewords and protein synthesis. *Science* 145: 1399–1407.

Nomura, M. 1984. The control of ribosome synthesis. *Sci. Am.* (Jan.) 250: 102–114.

Pauling, L., Itano, H. A., Singer, S. J., and Wells, I. C. 1949. Sickle-cell anemia: A molecular disease. *Science* 110: 543–548.

Ramakrishnan, V. 2002. Ribosome structure and the mechanism of translation. *Cell* 108: 557–572.

Rich, A., and Houkim, S. 1978. The three-dimensional structure of transfer RNA. *Sci. Am.* (Jan.) 238: 52–62.

Rich, A., Warner, J. R., and Goodman, H. M. 1963. The structure and function of polyribosomes. *Cold Spring Harbor Symp. Quant. Biol.* 28: 269–285.

Richards, F. M. 1991. The protein-folding problem. *Sci. Am.* (Jan.) 264: 54–63.

Rould, M. A., et al. 1989. Structure of *E. coli* glutaminyl-tRNA synthetase complexed with $tRNA^{gln}$ and ATP at 2.8 resolution. *Science* 246: 1135–1142.

Srb, A. M., and Horowitz, N. H. 1944. The ornithine cycle in *Neurospora* and its genetic control. *J. Biol. Chem.* 154: 129–139.

Warner, J., and Rich, A. 1964. The number of soluble RNA molecules on reticulocyte polyribosomes. *Proc. Natl. Acad. Sci. (USA)* 51: 1134–1141.

Wimberly, B. T., et al. 2000. Structure of the 30*S* ribosomal subunit. *Nature* 407: 327–333.

Yusupov, M. M., et al. 2001. Crystal structure of the ribosome at 5.5A resolution. *Science* 292: 883–896.

## Chapter 16 Gene Mutation and DNA Repair

Ames, B. N., McCann, J., and Yamasaki, E. 1975. Method for detecting carcinogens and mutagens with the *Salmonella*/mammalian microsome mutagenicity test. *Mut. Res.* 31: 347–364.

Bates, G., and Leharch, H. 1994. Trinucleotide repeat expansions and human genetic disease. *BioEssays* 16: 277–284.

Cairns, J., Overbaugh, J., and Miller, S. 1988. The origin of mutants. *Nature* 335: 142–145.

Cleaver, J. E. 1990. Do we know the cause of xeroderma pigmentosum? *Carcinogenesis* 11: 875–882.

Comfort, N. C. 2001. *The tangled field: Barbara McClintock's search for the patterns of genetic control.* Cambridge, MA: Harvard University Press.

Friedberg, E. C., Walker, G. C., and Siede, W. 1995. *DNA repair and mutagenesis.* Washington, DC: ASM Press.

Jiricny, J. 1998. Eukaryotic mismatch repair: An update. *Mutation Research* 409: 107–121.

Landers, E. S., et al., 2001. Initial sequencing and analysis of the human genome. *Nature* 409: 860–921.

Miki, Y. 1998. Retrotransposal integration of mobile genetic elements in human disease. *J. Human Genet.* 43: 77–84.

O'Hare, K. 1985. The mechanism and control of *P* element transposition in *Drosophila. Trends Genet.* 1: 250–254.

Radman, M., and Wagner, R. 1988. The high fidelity of DNA duplication. *Sci. Am.* (Aug.) 259: 40–46.

## Chapter 17 Regulation of Gene Expression in Prokaryotes

Antson, A. A., et al. 1999. Structure of the trp RNA-binding attenuation protein, TRAP, bound to RNA. *Nature* 401: 235–242.

Beckwith, J. R., and Zipser, D., eds. 1970. *The lactose operon.* Cold Spring Harbor, NY: Cold Spring Harbor Laboratory Press.

Bertrand, K., et al. 1975. New features of the regulation of the tryptophan operon. *Science* 189: 22–26.

Gilbert, W., and MŸller-Hill, B. 1966. Isolation of the *lac* repressor. *Proc. Natl. Acad. Sci. (USA)* 56: 1891–1898.

_______. 1967. The *lac* operator is DNA. *Proc. Natl. Acad. Sci. (USA)* 58: 2415–2421.

Jacob, F., and Monod, J. 1961. Genetic regulatory mechanisms in the synthesis of proteins. *J. Mol. Biol.* 3: 318–356.

Lewis, M., et al. 1996. Crystal structure of the lactose operon repressor and its complexes with DNA and inducer. *Science* 271: 1247–1254.

Stroynowski, I., and Yanofsky, C. 1982. Transcript secondary structures regulate transcription termination at the attenuator of *S. marcescens* tryptophan operon. *Nature* 298: 34–38.

Valbuzzi, A., and Yanofsky, C. 2001. Inhibition of the *B. subtilis* regulatory protein TRAP by the TRAP-inhibitory protein, AT. *Science* 293: 2057–2061.

Yanofsky, C. 1981. Attenuation in the control of expression of bacterial operons. *Nature* 289: 751–758.

## Chapter 18 Regulation of Gene Expression in Eukaryotes

Becker, P. B., and Hörz, W. 2002. ATP-dependent nucleosome remodeling. *Annu. Rev. Biochem.* 71: 247–273.

Beckwith, J. R., and Zipser, D., eds. 1970. *The lactose operon.* Cold Spring Harbor, NY: Cold Spring Harbor Laboratory Press.

Bertrand, K., et al. 1975. New features of the regulation of the tryptophan operon. *Science* 189: 22–26.
Black, D. L. 2000. Protein diversity from alternative splicing: a challenge for bioinformatics and postgenomic biology. *Cell* 103: 367–370.
Black, D. L. 2003. Mechanisms of alternative pre-mRNA splicing. *Annu. Rev. Biochem.* 72: 291–336.
Butler, J. E. F., and Kadonaga, J. T. 2002. The RNA polymerase II core promoter: A key component in the regulation of gene expression. *Genes Dev.* 16: 2583–2592.
Dillon, N., and Sabbattini, P. 2000. Functional gene expression domains: Defining the functional unit of eukaryotic gene regulation. *Bioessays* 22: 657–665.
Gilbert, W., and Müller-Hill, B. 1966. Isolation of the *lac* repressor. *Proc. Natl. Acad. Sci. (USA)* 56: 1891–1898.
_______. 1967. The *lac* operator is DNA. *Proc. Natl. Acad. Sci. (USA)* 58: 2415–2421.
Jackson, D. A. 2003. The anatomy of transcription sites. *Curr. Opin. Cell Biol.* 15: 311–317.
Jacob, F., and Monod, J. 1961. Genetic regulatory mechanisms in the synthesis of proteins. *J. Mol. Biol.* 3: 318–356.
Lee, T. I., and Young, R. A. 2000. Transcription of eukaryotic protein-coding genes. *Ann. Rev. Genet.* 34: 77–137.
Meister, G., and Tuschl, T. 2004. Mechanisms of gene silencing by double-standard RNA. *Nature* 431: 343–349.
Mello, C. C., and Conte, D., Jr. 2004. Revealing the world of RNA interference. *Nature* 431: 338–342.
Turner, B. M. 2000. Histone acetylation and an epigenetic code. *Bioessays* 22: 836–845.
Yanofsky, C. 1981. Attenuation in the control of expression of bacterial operons. *Nature* 289: 751–758.
Yanofsky, C., and Kolter, R. 1982. Attenuation in amino acid biosynthetic operons. *Annu. Rev. Genet.* 16: 113–134.

## Chapter 19 Developmental Genetics of Model Organisms

De-Leon, S. B-T., and Davidson, E. H. 2007. Gene regulation: Gene control networks in development. *Annu. Rev. Biophys. Biomol. Struct.* 36: 191–212.
Gilbert, S. F., et al. 2001. Morphogenesis of the turtle shell: development of a novel structure in tetrapod evolution. *Evol Dev.* 3: 47–58.
Goodman, F. 2002. Limb malformations and the human *HOX* genes. *Am. J. Med. Genet.* 112: 256–265.
Gridley, T. 2003. Notch signaling and inherited human diseases. *Hum. Mol. Genet.* 12: R9–R13.
Inoue, T., et al. 2005. Gene regulatory special feature: Transcriptional network underlying *Caenorhabditis elegans* vulval development. *Proc. Nat. Acad. Sci.* 102: 4972–4977.
Krizek, B. A., and Fletcher, J. C. 2006. Molecular mechanisms of flower development: An armchair guide. *Nat. Rev. Genet.* 6: 688–698.
Lall, S., and Patel, N. H. 2001. Conservation and divergence in molecular mechanisms of axis formation. *Annu. Rev. Genet.* 35: 407–437.
Moens, C. B., and Selleri, L. 2006. Hox cofactors in vertebrate development. *Dev. Biol.* 291: 193–206.
Nüsslein-Volhard, C., and Weischaus, E. 1980. Mutations affecting segment number and polarity in *Drosophila*. *Nature* 287: 795–801.
Ochoa-Espinosa, A., and Small, S. 2005. Developmental mechanisms and *cis*-regulatory codes. *Curr. Opin. Genet. Develop.* 16: 165–170.
Reyes, J. C. 2006. Chromatin modifiers that control plant development. *Curr. Opin. Plant Biol.* 9: 21–27.
Rudel, D., and Sommer, R. J. 2003. The evolution of developmental mechanisms. *Dev. Biol.* 264: 15–37.
Stathopoulos, A., and Levine, M. 2005. Genomic regulatory networks and animal development. *Dev. Cell* 9: 449–462.
Sundaram, M. V. 2005. The love-hate relationship between Ras and Notch. *Genes Dev.* 19: 1825–1839.
Tickle, C. 2006. Making digit patterns in the vertebrate limb. *Nat. Rev. Mol. Cell Biol.* 7: 45–53.
Verakasa, A., Del Campo, M., and McGinnis, W. 2000. Developmental patterning genes and their conserved functions: from model organisms to humans. *Mol. Genet. Metabol.* 69: 85–100.
Wang, M., and Sternberg, P. W. 2001. Pattern formation during *C. elegans* vulval induction. *Curr. Top. Dev. Biol.* 51: 189–220.

## Chapter 20 Cancer and Regulation of the Cell Cycle

Ames, B. N., Gold, L. S., and Willett, W. C. 1995. The causes and prevention of cancer. *Proc. Natl. Acad. Sci. (USA)* 92: 5258–5265.
Bernards, R., and Weinberg, R. A. 2002. A progression puzzle. *Nature* 418: 823.
Brown, M. A. 1997. Tumor suppressor genes and human cancer. *Adv. Genet.* 36: 45–135.
Compagni, A., and Christofori, G. 2000. Recent advances in research on multistage tumorigenesis. *Brit. J. Cancer* 83: 1–5.
Cornelis, J. F., et al. 1998. Metastasis. *Am. Scient.* 86: 130–141.
Esteller, M. 2007. Cancer epigenomics: DNA methylomes and histone-modification maps. *Nature Reviews Cancer* 8: 286–297.
Futreal, P. A., et al. 2004. A census of human cancer genes. *Nature Reviews Cancer* 4: 177–183.
Hartwell, L. H., and Kastan, M. B. 1994. Cell cycle control and cancer. *Science* 266: 1821–1827.
Lengauer, C., Kinzler, K. W., and Vogelstein, B. 1997. Genetic instability in colorectal cancer. *Nature* 386: 623–627.
Nurse, P. 1997. Checkpoint pathways come of age. *Cell* 91: 865–867.
Raff, M. 1998. Cell suicide for beginners. *Nature* 396: 119–122.
Sherr, C. J. 1996. Cancer cell cycles. *Science* 274: 1672–1677.
Vogelstein, B., and Kinzler, K.W. 1993. The multistep nature of cancer. *Trends in Genetics* 9: 138–141.
Weinberg, R. A. 1995. The retinoblastoma protein and cell cycle control. *Cell* 81: 323–330.
———. 1996. How cancer arises. *Sci. Am.* (Sept.) 275: 62–70.

## Chapter 21 Genomics, Bioinformatics, and Proteomics

Andersen, J., et al. 2002. Directed proteomic analysis of the human nucleolus. *Curr. Biol.* 12: 1–11.
Asara, J. M., et al. 2007. Protein sequences from Mastodon and *Tyrannosaurus Rex* revealed by mass spectrometry. *Science* 316: 280–285.
Butler, J. M. 2006. Genetics and genomics of core short tandem repeat loci used in human identity testing. *J. Forensic Sci.* 51: 253–265.
Chandonia, J. M. 2006. The impact of structural genomics: expectations and outcomes. *Science* 311: 347–351.
Church, G. M. 2006. Genomes for all. *Sci. Am.* 47–54.
Domon, B., and Aebersold, R. 2006. Mass spectrometry and protein analysis. *Science* 312: 212–217.
Fleming, K., et al. 2006. The proteome: structure, function and evolution. *Philos. Trans. R. Soc. Lond. B Biol. Sci.* 361: 441–451.
Hood, L., Heath, J. R., Phelps, M. E., and Lin, B. 2004. Systems biology and new technologies enable predictive and preventative medicine. *Science* 306: 640–643.
Hoskins, R. A., et al., 2007. Sequence finishing and mapping of *Drosophila melanogaster* heterochromatin. *Science* 316: 1625–1628.

International Human Genome Sequencing Consortium 2004. Finishing the euchromatic sequence of the human genome. *Nature* 431: 931–945.

International Human Genome Sequencing Consortium. 2001. Initial sequencing and analysis of the human genome. *Nature* 409: 860–921.

Joyce, A. R., and Palsson, B. O. 2006. The model organism as a system integrating "omics" data sets. *Nat. Rev. Mol. Cell Biol.* 7: 901–909.

Nei, M., and Rooney, A. P. 2005. Concerted and birth-and-death evolution of multigene families. *Annu. Rev. Genet.* 39: 197–218.

Noonan, J. P., et al. 2006. Sequencing and analysis of Neanderthal DNA. *Science* 314: 1113–1118.

Ochman, H., and Davilos, L. M. 2006. The nature and dynamics of bacterial genomes. *Science* 311: 1730–1733.

Pennisi, E. 2006. The dawn of Stone Age genomics. *Science* 314: 1068–1071.

Rhesus Macque Genome Sequencing and Analysis Consortium. 2007. Evolutionary and biomedical insights from the Rhesus Macaque genome. *Science* 316: 222–234.

Schweitzer, M. H., et al., 2007. Analyses of soft tissue from *Tyrannosaurus rex* suggest the presence of protein. *Science* 316: 277–280.

Sea Urchin Genome Sequencing Consortium. 2006. The genome of the sea urchin *Stronglyocentrotus purpuratus*. *Science* 314: 941–956.

Switnoski, M., Szczerbal, I., and Nowacka, J. 2004. The dog genome map and its use in mammalian comparative genomics. *J. Appl. Physiol.* 45: 195–214.

Thieman, W. J. and Palladino, M. A. 2009. *Introduction to biotechnology*, 2nd ed. San Francisco, CA: Benjamin Cummings.

Venter, J. C., et al. 2001. The sequence of the human genome. *Science* 291: 1304–1351.

Venter, J. C., et al. 2004. Environmental genome shotgun sequencing of the Sargasso Sea. *Science* 304: 6–74.

Yooseph, S., Sutton, G., Rusch, D. B., Halpern, A. L., Williamson, S .J., et al. 2007. *The Sorcerer II* global ocean sampling expedition: Expanding the universe of protein families. *PLoS Biology* 5(3): 0432–0466.

## Chapter 22 Genome Dynamics: Transposons, Immunogenetics, and Eukaryotic Viruses

Alt, F. W., et al. 1987. Development of the primary antibody repertoire. *Science* 238: 1079–1087.

Becker, Y. 2000. Evolution of viruses by acquisition of cellular RNA or DNA nucleotide sequences and genes: an introduction. *Virus Genes* 21: 7–12.

Comfort, N. C. 2001. *The tangled field: Barbara McClintock's search for the patterns of genetic control.* Cambridge, MA: Harvard University Press.

deParseval, N., and Heidmann, T. 2005. Human endogenous retroviruses: from infectious elements to human genes. *Cytogenet. Genome Res.* 110: 318–332.

Drucker, R., and Whitelaw, E. 2004. Retrotransposon-derived elements in the mammalian genome: A potential source of disease. *J. Inherit. Metab. Dis.* 27: 319–330.

Hamilton, G. 2006. The gene weavers. *Nature* 441: 68–685.

Honjo, T., et al. 2002. Molecular mechanism of class switch recombination: linkage with somatic hypermutation. *Ann. Rev. Immunol.* 20: 165–196.

Jung, D., et al. 2006. Mechanism and control of V(D)J recombination at the immunoglobulin heavy chain locus. *Ann. Rev. Immunol.* 24: 541–570.

Lakshminarayan, M. I., et al. 2006. Evolutionary genomics of nucleo-cytoplasmic large DNA viruses. *Virus Research* 117: 156–184.

Medstrand, P., et al. 2005. Impact of transposable elements on the evolution of mammalian gene regulation. *Cytogenet. Genome Res.* 110: 34–352.

Miki, Y. 1998. Retrotransposal integration of mobile genetic elements in human disease. *J. Human Genet.* 43: 77–84.

O'Hare, K. 1985. The mechanism and control of *P* element transposition in *Drosophila. Trends Genet.* 1: 250–254.

Ochman, H., Lawrence, J. G., and Groisman, E. A. 2000. Lateral gene transfer and the nature of bacterial innovation. *Nature* 405: 299–305.

Raftery, M., Müller, A., and Schönrich, G. 2000. Herpesvirus homologues of cellular genes. *Virus Genes* 21: 65–75.

Tonegawa, S. 1983. Somatic generation of antibody diversity. *Nature* 302: 575–581.

Zhang, X. W., Yap, Y. L., and Danchin, A. 2005. Testing the hypothesis of a recombinant origin of the SARS-associated coronavirus. *Arch. Virol.* 150: 1–20.

## Chapter 23 Genome Analysis—Dissection of Gene Function

Capecchi, M. R. 1989. Altering the genome by homologous recombination. *Science* 244: 1288–1292.

Dykxhoorn, D. M., et al. 2003. Killing the messenger: Short RNAs that silence gene expression. *Nature Reviews Molecular Cell Biology* 4: 457–467.

Forsburg, S. L. 2001. The art and design of genetic screens: yeast. *Nature Reviews Genetics* 2: 659–668.

Fraser, A. 2004. RNA interference: human genes hit the big screen. *Nature* 428: 375–378.

Gitschier, J., et al. 1984. Characterization of the human *factor VIII* gene. *Nature* 312: 326–330.

Hartwell, L. H., et al. 1974. Genetic control of the cell division cycle in yeast. *Science* 183: 46–51.

Hartwell, L. H., et al. 1970. Genetic control of the cell-division cycle in yeast, I. Detection of mutants. *Proc. Natl. Acad. Sci.* 66: 352–359.

Kile, B. T., and Hilton, D. J. 2005. The art and design of genetic screens: mouse. *Nature Reviews Genetics* 6: 557–567.

Kwon, B. S, et al. 1987. Isolation and sequence of a cDNA clone for human tyrosinase that maps at the mouse *c-albino* locus. *Proc. Natl. Acad. Sci. (USA)* 84: 7473–7477.

Leder, A., et al. 1986. Consequences of widespread deregulation of the *c-myc* gene in transgenic mice: multiple neoplasms and normal development. *Cell* 45: 485–495.

Oliver, S. G. 1997. From gene to screen with yeast. *Curr. Opin. Genetics and Development* 7: 405–409.

Steinmetz, L. M., and Davis, R. W. 2004. Maximizing the potential of functional genomics. *Nature Reviews Genetics* 5: 190–201.

St. Johnston, D. 2002. The art and design of genetic screens: *Drosophila melanogaster. Nature Reviews Genetics* 3: 176–188.

## Chapter 24 Applications and Ethics of Genetic Engineering and Biotechnology

Cavanna-Calvo, M., et al. 2000. Gene therapy of severe combined immunodeficiency (SCID)-XI disease. *Science* 288: 669–672.

Chrispeels, M. J., and Sadava, D. E. 2003. *Plants, Genes, and Crop Biotechnology*, 2nd ed. Sudbury, MA: Jones and Barlett Publishers.

Dale, P. J., Clarke, B., and Fontes, E. M. G. 2002. Potential for the environmental impact of transgenic crops. *Nat. Biotechnol.* 20: 567–574.

Dykxhoorn, D. M., and Liberman, J. 2006. Knocking down disease with siRNAs. *Cell* 126: 231–235.

Engler, O. B., et al. 2001. Peptide vaccines against hepatitis B virus: From animal model to human studies. *Mol. Immunol.* 38: 457–465.

Friend, S. H., and Stoughton, R. B. 2002. The magic of microarrays. *Sci. Am.* (Feb.) 286: 44–50.

Hunter, C. V., Tiley, L. S., and Sang, H. M. 2005. Developments in transgenic technology: applications for medicine. *Trends Mol. Med.* 11: 293–300.

Knoppers, B. M., Bordet, S., and Isasi, R. M. 2006. Preimplantation genetic diagnosis: an overview of socio-ethical and legal considerations. *Annu. Rev. Genom. Hum. Genet.* 7: 201–221.

O'Connor, T. P., and Crystal, R. G. 2006. Genetic medicines: treatment strategies for hereditary disorders. *Nature Reviews Genetics* 7: 261–276.

Pray, C. E., Huang, J., and Rozelle, S. 2002. Five years of Bt cotton in China: The benefits continue. *The Plant J.* 31: 423–430.

Schillberg, S., Fischer, T., and Emans, N. 2003. Molecular farming of recombinant antibodies in plants. *Cell Mol. Life Sci.* 60: 433–445.

Shastry, B. S. 2006. Pharmacogenetics and the concept of individualized medicine. *Pharmacogenetics Journal* 6: 16–21.

Stix, G. 2006. Owning the stuff of life. *Sci. Am.* 294(1): 76–83.

Wang, D. G., et al. 1998. Large-scale identification, mapping, and genotyping of single-nucleotide polymorphisms in the human genome. *Science* 280: 1077–1082.

Whitelaw, C. B. A. 2004. Transgenic livestock made easy. *Trends in Biotechnology* 22(4): 257–259.

Varsha, M. S. 2006. DNA fingerprinting in the criminal justice system: An overview. *DNA Cell Biol.* 25: 181–188.

## Chapter 25 Quantitative Genetics and Multifactorial Traits

Browman, K. W. 2001. Review of statistical methods of QTL mapping in experimental crosses. *Lab Animal* 30: 44–52.

Crow, J. F. 1993. Francis Galton: Count and measure, measure and count. *Genetics* 135: 1.

Falconer, D. S., and Mackay, F. C. 1996. *Introduction to quantitative genetics*, 4th ed. Essex, England: Longman.

Farber, S. 1980. *Identical twins reared apart.* New York: Basic Books.

Feldman, M. W., and Lewontin, R. C. 1975. The heritability hangup. *Science* 190: 1163–1166.

Frary, A., et al. 2000. *fw2.2*: A quantitative trait locus key to the evolution of tomato fruit size. *Science* 289: 85–88.

Haley, C. 1991. Use of DNA fingerprints for the detection of major genes for quantitative traits in domestic species. *Anim. Genet.* 22: 259–277.

———. 1996. Livestock QTLs: Bringing home the bacon. *Trends Genet.* 11: 488–490.

Lander, E., and Botstein, D. 1989. Mapping Mendelian factors underlying quantitative traits using RFLP linkage maps. *Genetics* 121: 185–199.

Lander, E., and Schork, N. 1994. Genetic dissection of complex traits. *Science* 265: 2037–2048.

Lynch, M., and Walsh, B. 1998. *Genetics and analysis of quantitative traits.* Sunderland, MA: Sinauer Associates.

Macy, T. F. C. 2001. Quantitative trait loci in *Drosophila. Nature Reviews Genetics* 2: 11–19.

Newman, H. H., Freeman, F. N., and Holzinger, K. T. 1937. *Twins: A study of heredity and environment.* Chicago: University of Chicago Press.

Paterson, A., Deverna, J., Lanini, B., and Tanksley, S. 1990. Fine mapping of quantitative traits loci using selected overlapping recombinant chromosomes in an interspecific cross of tomato. *Genetics* 124: 735–742.

Tanksley, S. D. 2004. The genetic, developmental and molecular bases of fruit size and shape variation in tomato. *Plant Cell* 16: S181–S189.

Zar, J. H. 1999. *Biostatistical analysis*, 4th ed. Upper Saddle River, NJ: Prentice Hall.

## Chapter 26 Genetics and Behavior

Benzer, S. 1973. Genetic dissection of behavior. *Sci. Am.* (Dec.) 229: 24–37.

Buck, K. J. 1995. Strategies for mapping and identifying quantitative trait loci specifying behavioral responses to alcohol. *Alcohol Clin. Exp. Res.* 19: 795–801.

Craddock, N., and Forty, L. 2006. Genetics of affective (mood) disorders. *Eur. J. Hum. Genet.* 14: 660–668.

Davis, R. L. 2005. Olfactory memory formation in *Drosophila*: From molecular to systems neuroscience. *Annu. Rev. Neurosci.* 28: 275–302.

de Belle, J. S. 2002. Unifying the genetics of behavior. *Nat. Genet.* 31: 1–2.

Ganetzky, B. 1994. Cysteine strings, calcium channels and synaptic transmission. *BioEssays* 16: 461–463.

Grahame, N. J. 2000. Selected lines and inbred strains: Tools in the hunt for the genes involved in alcohol. *Alcohol Res. Health* 24: 159–163.

Hakak, Y., et al., 2001. Genome-wide expression analysis reveals dysregulation of myelination-related genes in chronic schizophrenia. *Proc. Nat. Acad. Sci. (USA)* 98: 4746–4751.

Hotta, Y., and Benzer, S. 1972. Mapping behavior of *Drosophila* mosaics. *Nature* 240: 527–535.

Mackay, T. F. C. and Anholt, R. H. 2006. Of flies and man: *Drosophila* as a model for human complex traits. *Annu. Rev. Hum. Genet.* 7: 39–367.

Melo, J. A., Shendure, J., Pociask, K., and Silver, L. M. 1996. Identification of sex-specific quantitative trait loci controlling alcohol preference in C57BL/6 mice. *Nature Genet.* 13: 147–153.

Reddy, P. H., et al. 1999. Transgenic mice expressing mutated full-length *HD* cDNA: a paradigm for locomotor changes and selective neuronal loss in Huntington's disease. *Phil. Trans. R. Soc. Lond.* B: 354: 1035–1045.

Ricker, J. P., and Hirsch, J. 1988. Genetic changes occurring over 500 generations in lines of *Drosophila melanogaster* selected divergently for geotaxis. *Behav. Genet.* 18: 13–24.

Riley, B., and Kendler, K. S. 2006. Molecular genetic studies of schizophrenia. *Eur. J. Hum. Genet.* 14: 669–680.

Sequeira, A., and Turecki, G. 2006. Genome-wide gene expression in mood disorders. *OMICS* 10: 444–454.

Siegel, R. W., Hall, J. C., Gailey, D. A., and Kyriacou, C. P. 1984. Genetic elements of courtship in *Drosophila*: Mosaics and learning mutants. *Behav. Genet.* 14: 383–410.

Stoltenberg, S. F., Hirsch, J., and Berlocher, S. H. 1995. Analyzing correlations of three types in selected lines of *Drosophila melanogaster* that have evolved stable extreme geotactic performance. *J. Comp. Psychol.* 105: 85–94.

Toma, D. P., White, K. P., Hirsch, J., and Greenspan, R. J. 2002. Identification of genes involved in *Drosophila melanogaster* geotaxis, a complex behavioral trait. *Nat. Genet.* 31: 349–353.

Turri, M. G., et al. 2001. QTL analysis identifies multiple behavioral dimensions in ethological tests of anxiety in laboratory mice. *Curr. Biol.* 11: 725–734.

Tully, T. 1996. Discovery of genes involved with learning and memory: An experimental synthesis of Hirschian and Benzerian perspectives. *Proc. Nat. Acad. Sci. (USA)* 93: 13460–13467.

Willis-Owen, S. A. G., and Flint, J. 2006. The genetic basis of emotional behavior in mice. *Eur. J. Hum. Genet.* 14: 721–728.

## Chapter 27 Population Genetics

Ansari-Lari, M. A., et al. 1997. The extent of genetic variation in the *CCR5* gene. *Nature Genetics* 16: 221–222.

Ballou, J., and Ralls, K. 1982. Inbreeding and juvenile mortality in small populations of ungulates: A detailed analysis. *Biol. Conserv.* 24: 239–272.

Bellinger, M. R., Johnson, J. A., Toepfer, J., and Dunn, P. 2003. Loss of genetic variation in greater prairie chickens following a population bottleneck in Wisconsin, U.S.A. *Conserv. Biol.* 17: 717–724.

Bittles, A. H., and Neel, J. V. 1994. The costs of human inbreeding and their implications for variations at the DNA level. *Nature Genet.* 8: 117–121.

Carrington, M., Kissner, T., et al. 1997. Novel alleles of the chemokine-receptor gene *CCR5*. *Am. J. Hum. Genet.* 61: 1261–1267.

Chevillon, C., et al. 1995. Population structure and dynamics of selected genes in the mosquito *Culex pipiens*. *Evolution* 49: 997–1007.

Dudley, J. W. 1977. 76 generations of selection for oil and protein percentage in maize. In *Proc. Intern. Conf. on Quant. Genet.*, E. Pollack et al., eds., pp. 459–473. Ames: Iowa State University Press.

Freire-Maia, N. 1990. Five landmarks in inbreeding studies. *Am. J. Med. Genet.* 35: 118–120.

Hummell, S., Schmidt, D., Kremeyer, B., Herrmann, B., and Oppermann, M. 2005. Detection of the *CCR5-Δ32* HIV resistance gene in Bronze age skeletons. *Genes and Immun.* 6: 371–374.

Karn, M. N., and Penrose, L. S. 1951. Birth weight and gestation time in relation to maternal age, parity and infant survival. *Ann. Eugen.* 16: 147–164.

Kerr, W. E., and Wright, S. 1954. Experimental studies of the distribution of gene frequencies in very small populations of *Drosophila melanogaster*. *Evolution* 8: 172–177.

Laikre, L., Ryman, N., and Thompson, E. A. 1993. Hereditary blindness in a captive wolf (*Canis lupus*) population: Frequency reduction of a deleterious allele in relation to gene conservation. *Conservation Biol.* 7: 592–601.

Leibert, F., et al. 1998. The *DCCR5* mutation conferring protection against HIV-1 in Caucasian populations has a single and recent origin in northeastern Europe. *Hum. Molec. Genet.* 7: 399–406.

Markow, T., et al. 1993. HLA polymorphism in the Havasupai: Evidence for balancing selection. *Am. J. Hum. Genet.* 53: 943–952.

Parra, E. J., et al. 1998. Estimating African American admixture proportions by use of population-specific alleles. *Am. J. Hum. Genet.* 63: 1839–1851.

Pier, G. B., et al. 1998. *Salmonella typhi* uses CFTR to enter intestinal epithelial cells. *Nature* 393: 79–82.

Woodworth, C. M., Leng, E. R., and Jugenheimer, R. W. 1952. Fifty generations of selection for protein and oil in corn. *Agron. J.* 44: 60–66.

Yi, Z., Garrison, N., Cohen-Barak, O., Karafet, T. A., King, R. A., Erickson, R. P., Hammer, M. F., and Brilliant, M. H. 2003. A 122.5 kilobase deletion of the *P* gene underlies the high prevalence of oculocutaneous albinism type 2 in the Navajo population. *Am. J. Hum. Genet.* 72: 62–72.

## Chapter 28 Evolutionary Genetics

Anderson, W., et al. 1975. Genetics of natural populations: XLII. Three decades of genetic change in *Drosophila pseudoobscura*. *Evolution* 29: 24–36.

Ayala, F. J., 1984. Molecular polymorphism: How much is there, and why is there so much? *Dev. Genet.* 4: 379–391.

Bradshaw, H. D., Jr., Otto, K. G., and Frewen, B. E. 1998. Quantitative trait loci affecting differences in floral morphology between two species of monkey flower (*Mimulus*). *Genetics* 149: 367–382.

Dobzhansky, T. 1947. Adaptive changes induced by natural selection in wild populations of *Drosophila*. *Evolution* 1: 1–16.

Green, R. E., et al. 2006. Analysis of one million base pairs of Neanderthal DNA. *Nature* 444: 330–336.

Ke, Y., et al. 2001. African origin of modern humans in East Asia: A tale of 12,000 Y chromosomes. *Science* 292: 1151–1153.

Kimura, M. 1989. The neutral theory of molecular evolution and the world view of the neutralists. *Genome* 31: 24–31.

King, M. C., and Wilson, A. C. 1975. Evolution at two levels: Molecular similarities and biological differences between humans and chimpanzees. *Science* 188: 107–116.

Knowlton, N., et al. 1993. Divergence in proteins, mitochondrial DNA, and reproductive compatibility across the Isthmus of Panama. *Science* 260: 1629–1632.

Kreitman, M. 1983. Nucleotide polymorphism at the alcohol dehydrogenase locus of *Drosophila melanogaster*. *Nature* 304: 412–417.

Lamb, R. S., and Irish, V. F. 2003. Functional divergence within the *APETALA3/PISTILLATA* floral homeotic gene lineages. *Proc. Natl. Acad. Sci.* 100: 6558–6563.

Lewontin, R. C., and Hubby, J. L. 1966. A molecular approach to the study of genic heterozygosity in natural populations: II. Amount of variation and degree of heterozygosity in natural populations of *Drosophila pseudoobscura*. *Genetics* 54: 595–609.

Noonan, J. P., et al. 2006. Sequencing and analysis of Neanderthal genomic DNA. *Science* 314: 111–1118.

Ou, C.-Y., et al. 1992. Molecular epidemiology of HIV transmission in a dental practice. *Science* 256: 1165–1171.

Relethford, J. H. 2001. Ancient DNA and the origin of modern humans. *Proc. Nat. Acad. Sci.* 98: 390–391.

Stiassny, M. L. J., and Meyer, A. 1999. Cichlids of the Rift Lakes. *Sci. Am.* (Feb.) 280: 64–69.

Storz, J. F. 2005. Using genome scans of DNA polymorphism to infer adaptive population divergence. *Mol. Ecol.* 14: 671–688.

Takahashi, K., 1998. A novel family of short interspersed repetitive elements (SINES) from cichlids: The pattern of insertion of SINES at orthologous loci support the proposed monophyly of four major groups of cichlid fishes in Lake Tanganyika. *Mol. Biol. Evol.* 15: 391–407.

## Chapter 29 Conservation Genetics

Baker, C. S., and Palumbi, S. R. 1994. Which whales are hunted? A molecular genetic approach to monitoring whaling. *Science* 265: 1538–1539.

Daniels, S. J., and Walters, J. R. 2000. Inbreeding depression and its effects on natal dispersal in red-cockaded woodpeckers. *Condor* 102: 482–491.

Dobson, A., and Lyles, A. 2000. Black-footed ferret recovery. *Science* 288: 985.

Frankham, R. 1995. Conservation genetics. *Ann. Rev. Genet.* 29: 305–327.

Friar, E. A., et al. 2001. Population structure in the endangered Mauna Loa silversword, *Argyroxiphium kauense*, and its bearing on reintroduction. *Molecular Ecology* 10: 1657–1663.

Gharrett, A. J., and Smoker, W. W. 1991. Two generations of hybrids between even-year and odd-year pink salmon (*Oncorhynchus gorbuscha*): A test for outbreeding depression? *Canadian J. Fish. Aquat. Sci.* 48: 426–438.

Hedrick, P. W. 2001. Conservation genetics: Where are we now? *Trends Ecol. Evol.* 16: 629–636.

Lacy, R. C. 1997. Importance of genetic variation to the viability of mammalian populations. *J. Mammalogy* 78: 320–335.

Moore, M. K., et al. 2003. Use of restriction fragment length polymorphisms to identify sea turtle eggs and cooked meats to species. *Conserv. Genet.* 4: 95–103.

Paetkau, D., et al. 1998. Variation in genetic diversity across the range of North American brown bears. *Conserv. Biol.* 12: 418–429.

Ralls, K., et al. 2000. Genetic management of chondrodystrophy in California condors. *Animal Conserv.* 3: 145–153.

Roman, J., and Bowen, B. W. 2000. The mock turtle syndrome: Genetic identification of turtle meat purchased in the southeastern United States of America. *Animal Conserv.* 3: 61–65.

Roman, J., and Palumbi, S. R. 2003. Whales before whaling in the North Atlantic. *Science* 301: 508–510.

Wayne, R. K., et al. 1991. Conservation genetics of the endangered Isle Royale gray wolf. *Conserv. Biol.* 5: 41–51.

Wilson, E. O., ed. 1988. *Biodiversity*. Washington, DC: National Academy of Sciences.

Wynen, L. P., et al. 2000. Postsealing genetic variation and population structure of two species of fur seal. *Mol. Ecol.* 9: 299–314.

# Credits

## Front Matter

**p. ix** Dr. Andrew S. Bajer, University of Oregon; **p. xi** Mark J. Barrett/Alamy; **p. xii** Biozentrum, University of Basel/Science Photo Library/Photo Researchers, Inc.; **p. xiv** Dr. Gopal Murti/Science Photo Library/Photo Researchers, Inc.; **p. xv** K.G. Murti/Visuals Unlimited; **p. xvii** Prof. Oscar L. Miller/Science Photo Library/Photo Researchers, Inc.; **p. xviii** James King-Holmes/Photo Researchers, Inc.; **p. xxi** Edward B. Lewis, California Institute of Technology; **p. xxii** From The Roslin Institute—roslin.design@bbsrc.ac.uk http://www.roslin.ac.uk/imagelibrary/index.php Trends in Biotechnology Volume 22, Issue 4, April 2004, Pages 157–159 doi:10.1016/j.tibtech.2004.02.005; **p. xxiii** Edward S. Ross, California Academy of Sciences; **p. xxiv** Sarah Ward/Department of Soil and Crop Services/Colorado State University.

## Chapter 1: Introduction to Genetics

**p. 1, CO1** Omikron/Science Source/Photo Researchers, Inc.; **p. 3, 1–1** Malcolm Gutter/Visuals Unlimited; **p. 3, 1–2** Biophoto Associates/Science Source/Photo Researchers, Inc.; **p. 3, 1–3** Sovereign/Phototake NYC; **p. 4, 1–4** ISM/Phototake NYC; **p. 4, 1–6 (top and bottom)** Carolina Biological Supply Company/Phototake NYC; **p. 5, 1–7** Biozentrum, University of Basel/Science Photo Library/Photo Researchers, Inc.; **p. 7, 1–10** Manuel C. Peitsch/Corbis/Bettmann; **p. 7, 1–11** MRC-LMB, MRC/Photo Researchers, Inc.; **p. 8, 1–13** Oliver Meckes & Nicole Ottawa/Photo Researchers, Inc.; **p. 9, 1–15** Photo courtesy of Roslin Institute; **p. 10, 1–16** Science Museum/Science & Society Picture Library; **p. 11, 1–18** Alfred Pasieka/SPL/Photo Researchers, Inc.; **p. 11, 1–19** Luis de la Maza, Ph.D., M.D./Phototake NYC; **p. 12, 1–20(a)** John Paul Endress/Pearson Learning Photo Studio; **p. 12, 1–20(b)** Dr. David M. Phillips/Visuals Unlimited; **p. 13, 1–21(a)** Dr. Jeremy Burgess/Photo Researchers, Inc.; **p. 13, 1–21(b)** © Dr. David Phillips/Visuals Unlimited; **p. 13, 1–22(a)** Sinclair Stammers/Photo Researchers, Inc.; **p. 13, 1–22(b)** Wally Eberhart/Visuals Unlimited; **p. 13, 1–22(c)** © Mark Smith/Photo Researchers.

## Chapter 2: Mitosis and Meiosis

**p. 18, CO2** Dr. Andrew S. Bajer, University of Oregon; **p. 21, 2–2** CNRI/Science Photo Library/Photo Researchers, Inc.; **p. 22, 2–4** Dr. David Ward, Yale University; **pp. 26–27, 2–7(a-g)** Dr. Andrew S. Bajer, University of Oregon; **p. 35, 2–13(a)** Biophoto Associates/Photo Researchers, Inc.; **p. 35, 2–13(b)** Andrew Syred/SPL/Photo Researchers, Inc.; **p. 35, 2–13(c)** Biophoto Associates/Science Source/Photo Researchers, Inc.

## Chapter 3: Mendelian Genetics

**p. 42, CO3** Archiv/Photo Researchers, Inc.; **p. 48 (left and right)** Mike Dunton/VictorySeeds.com; **p. 66 (left)** Joyce Photographics/Photo Researchers, Inc.; **p. 66 (right)** R.J. Erwin/Photo Researchers, Inc.

## Chapter 4: Extensions of Mendelian Genetics

**p. 70, CO4** Mark J. Barrett/Alamy; **p. 73, 4–1(a)** John D. Cunningham/Visuals Unlimited; **p. 77, 4–4(b)** Photo courtesy of Stanton K. Short (The Jackson Laboratory, Bar Harbor, ME); **p. 77, 4–4(c)** Fred Habegger/Grant Heilman Photography, Inc.; **p. 82, 4–9** Irene Vandermolen/Animals Animals/Earth Scenes; **p. 86, 4–12 (left and right)** Carolina Biological Supply Company/Phototake NYC; **p. 88, 4–14(a)** Mary Teresa Giancoli/Mary Teresa Giancoli; **p. 89, 4–15** Hans Reinhard/Bruce Coleman Inc.; **p. 90, 4–16** Debra P. Hershkowitz/Bruce Coleman Inc.; **p. 90, 4–17(a, c)** Tanya Wolff, Washington University School of Medicine; **p. 90, 4–17(b)** Joel C. Eissenberg, Ph.D., Dept. of Biochemistry and Molecular Biology, St. Louis University Medical Center; **p. 91, 4–18(a)** Tanya Wolff, Washington University School of Medicine; **p. 91, 4–18(b)** Dr. Steven Henikoff, Howard Hughes Medical Institute, Fred Hutchinson Cancer Research Center, Seattle, Washington; **p. 92, 4–19(a)** Jane Burton/Bruce Coleman Inc.; **p. 92, 4–19(b)** Steve Klug/Dr. William S. Klug; **p. 97,** (Walnut and Rose) Dr. Ralph Somes; **p. 97,** (Pea and Single) J James Bitgood/University of Wisconsin, Animal Sciences Dept.; **p. 100,** (Chestnut) Corbis/Bettmann; **p. 100,** (Palomino) John Daniels/Ardea London Limited; **p. 100,** (Cremello) Dusty L. Perin; **p. 102,** Daniel Sambraus/Photo Researchers, Inc.; **p. 103 (top),** Dale C. Spartas/Dale C. Spartas/DCS Photo, Inc.; **p. 103 (bottom),** (Kerry cow) Nigel J.H. Smith/Animals Animals/Earth Scenes; **p. 103 (bottom),** (Dexter bull) Francois Gohier/Photo Researchers, Inc.; **p. 104,** Hans Reinhard/Bruce Coleman Inc.

## Chapter 5: Chromosome Mapping in Eukaryotes

**p. 105, CO5** B. John Cabisco/Visuals Unlimited; **p. 112, 5–7** ©2002 Clare A. Hasenkampf/Biological Photo Service; **p. 127, 5–19** Dr. Sheldon Wolff & Judy Bodycote/Laboratory of Radiobiology and Environmental Health, University of California, San Francisco; **p. 128, 5–20** Biophoto Assoc./Photo Researchers, Inc.; **p. 128, 5–21(a)** John D. Cunningham/Visuals Unlimited; **p. 128, 5–21(b)** Science Source/Photo Researchers, Inc.; **p. 129, 5–22** James W. Richardson/Visuals Unlimited.

## Chapter 6: Genetic Analysis and Mapping in Bacteria and Bacteriophages

**p. 143, CO6** Dr. L. Caro/Science Photo Library/Photo Researchers, Inc.; **p. 145, 6–2** Michael G. Gabridge/Visuals Unlimited; **p. 147, 6–5** Dennis Kunkel/Phototake NYC; **p. 153, 6–12(a)** K.G. Murti/Visuals Unlimited; **p. 155, 6–14(a)** M. Wurtz/Biozentrum, University of Basel/Science Photo Library/Photo Researchers, Inc.; **p. 157, 6–16(b)** Bruce Iverson/Bruce Iverson, Photomicrography; **p. 172, figure and Table for problem #32** Lawley, T. D. et al. 2002. Functional and mutational analysis of conjugative transfer region 1 (Tra1) from the incHI1 plasmid R27, *Journal of Bacteriology 184*: 2173-2180, fig. 1 and table 3. Copyright © 2002 American Society for Microbiology.

## Chapter 7: Sex Determination and Sex Chromosomes

**p. 173, CO7** Wellcome Trust Medical Photographic Library; **p. 175, 7–1(a)** Biophoto Assoc./Photo Researchers, Inc.; **p. 176, 7–3(a)** Bill Beatty/Visuals Unlimited; **p. 177, 7–4(a)** Dr. Maria Gallegos, University of California, San Francisco; **p. 179, 7–6(a-b)** Courtesy of the Greenwood Genetic Center, Greenwood, SC; **p. 179, 7–7(a-b)** Catherine G. Palmer, Indiana University; **p. 184, 7–10 (top and bottom)** Stuart Kenter Associates; **p. 185, 7–12(a)** W. Layer/Okapia/Photo Researchers, Inc.; **p. 185, 7–12(b)** Marc Henrie © Dorling Kindersley; **p. 187, 7–14** Adapted from L. Carrel. 2006. Molecular Biology: Enhanced: "X"-rated chromosomal rendezvous. *Science* 311: 1107–1109, fig. on p. 1108; **p. 189, 7–16** From "Dosage compensation: the beginning and end of generalization" Nature Reviews/Genetics Vol 8, Jan 2007 p.47, fig 6, Tobias Straub, Peter B. Becker. Nature Journals reprinted by permission from Macmillan Publishers, Ltd.; **p. 196,** Associated Press/Courtesy of Texas A & M University.

## Chapter 8: Chromosome Mutations: Variation in Chromosome Number and Arrangement

**p. 198, CO8** Evelin Schrock, Stan du Manoir and Tom Reid, NIH; **p. 201, 8–2** TH Foto-Werbung/Phototake NYC; **p. 202, 8–4 (left)** Courtesy of the Greenwood Genetic Center, Greenwood, SC; **p. 202, 8–4 (right)** William McCoy/Rainbow; **p. 203, 8–6** David D. Weaver, M.D., Indiana University; **p. 204, 8–7** David D. Weaver, M.D., Indiana University; **p. 207, 8–11** Ken Wagner/Phototake NYC; **p. 208, 8–12(b)** Pfizer, Inc./Phototake NYC; **p. 210, 8–15 (left)** Courtesy of University of Washington Medical Center Pathology; **p. 210, 8–15 (right)** Ray Clarke, Cri du chat Syndrome Support Group, UK; **p. 212, 8–18 (top to bottom)** Mary Lilly/Carnegie Institution of Washington; **p. 219, 8–25** Dr. Jorge Yunis. From Yunis and Chandler, 1979; **p. 219, 8–26** Science VU/Visuals Unlimited.

## Chapter 9: Extranuclear Inheritance

**p. 227, CO9** Tania Midgley/Corbis/Bettmann; **p. 229, 9–1** Pam Collins/Photo Researchers, Inc.; **p. 229, 9–2** RMF/Visuals Unlimited; **p. 230, 9–3** John D. Cunningham/Visuals Unlimited; **p. 230, 9–4 (left and right)** Dr. Ronald A. Butow, Department of Molecular Biology and Oncology, University of Texas Southwestern Medical Center; **p. 232, 9–6(a)** Dr. Richard D. Kolodnar/Dana-Farber Cancer Institute; **p. 232, 9–6(b)** Adapted from E. Passarge. 2006. *Color Atlas of Genetics*, 3rd ed., pg.133. Thieme Medical Publishers, Thieme, NY; **p. 232, 9–7(a)** Don W. Fawcett/Kahri/Dawid/Science Source/Photo Researchers, Inc.; **p. 232, 9–7(b)** Adapted from E. Passarge. 2006. *Color Atlas of Genetics*, 3rd ed., pg.135. Thieme Medical Publishers, Thieme, NY; **p. 235, 9–9(a-b)** Dr. Alan Pestronk, Dept. of Neurology, Washington University School of Medicine, St. Louis; **p. 236, 9–10** John D. Cunningham/Visuals Unlimited; **p. 238, 9–13** Robert & Linda Mitchell Photography; **p. 244, Table for problem #24** Berloco et al. 2001. The maternal effect gene, *abnormal oocyte* (abo), of *Drosophila melangaster* encodes a specific negative regulator of histones. Proc. Natl. Acad. Sci. 98:12126-31, table 1.

## Chapter 10: DNA Structure and Analysis

**p. 245, CO10** Richard Megna/Fundamental Photographs, NYC; **p. 248, 10–3** Bruce Iverson/Bruce Iverson, Photomicrography; **p. 251, 10–5** Oliver Meckes/Max-Planck-Institut-Tubingen/Photo Researchers, Inc.; **p. 254, 10–8 (top)** Runk/Schoenberger/Grant Heilman Photography, Inc.; **p. 254, 10–8 (bottom)** Biology Media/Photo Researchers, Inc.; **p. 259, 10–13** Photo by M.H.F. Wilkins. Courtesy of King's College London, England; **p. 263, 10–17** Ken Eward/Science Source/Photo Researchers, Inc.; **p. 268, 10–22** Ventana Medical Systems Inc.; **p. 270, 10–27** Dr. William S. Klug.

## Chapter 11: DNA Replication and Recombination

**p. 278, CO11** Dr. Gopal Murti/Science Photo Library/Photo Researchers, Inc.; **p. 282, 11–5 (top)** Walter H. Hodge/Peter Arnold, Inc.; **p. 282, 11–5(a-b)** Figure from "Molecular Genetics," Pt. 1 pp. 74–75, J.H. Taylor (ed). Copyright ©1963 and renewed 1991, reproduced with permission from Elsevier Science Ltd.; **p. 283, 11–6** Reprinted from CELL, Vol. 25, 1981, pp 659, Sundin and Varshavsky, (1 figure), with permission from Elsevier Science. Courtesy of A. Varshavsky; **p. 290, 11–14** Reproduced by permission from H.J. Kreigstein and D.S. Hogness, Proceedings of the National Academy of Sciences 71:136 (1974), p. 137, Fig. 2; **p. 291, 11–15** Dr. Harold Weintraub, Howard Hughes Medical Institute, Fred Hutchinson Cancer Center/"Essential Molecular Biology" 2e, Freifelder & Malachinski, Jones & Bartlett, Fig. 7–24, pp. 141.

## Chapter 12: DNA Organization in Chromosomes

**p. 302, CO12** Science VU/BMRL/Visuals Unlimited; **p. 303, 12–1(a)** Dr. M. Wurtz/Biozentrum, University of Basel/Science Photo Library/Photo Researchers, Inc.; **p. 304, 12–2** Science Source/Photo Researchers, Inc.; **p. 304, 12–3** Dr. Gopal Murti/Science Photo Library/Photo Researchers, Inc.; **p. 306, 12–5** Image courtesy of Brian Harmon and John Sedat, University of California, San Francisco; **p. 307, 12–6** Science Source/Photo Researchers, Inc.; **p. 307, 12–7(b)** Omikron/Photo Researchers, Inc.; **p. 312, 12–12** Douglas Chapman/University of Washington Department of Pathology; **p. 320, figure for problem #22** Reprinted with permission from R. Nagele et al. 1995. Precise spatial positioning of chromosomes during prometaphase: Evidence for chromosomal order. *Science* 270: 1831-35, Fig. 1. Copyright © 1995 AAAS.

## Chapter 13: Recombinant DNA Technology and Gene Cloning

**p. 322, CO13** Michael Gabridge/Visuals Unlimited; **p. 324, 13–1** Riken Biomolecules Gallery/Kyushu Institute of Technology (KIT); **p. 325, 13–4** K.G. Murti/Visuals Unlimited; **p. 326, 13–6** Michael Gabridge/Visuals Unlimited; **p. 329, 13–11** Dr. Michael Blaber, Florida State University, Tallahassee, FL 32306. pET figure from http://www.bio.davidson.edu/courses/Molbio/MolStudents/spring2003/Causey/pET.html.; **p. 331, 13–15** Jon Gordon/Phototake NYC; **p. 335, 13–18** Bio-Rad Laboratories Diagnostics Group; **p. 338, 13–21** National Institutes of Health/Custom Medical Stock Photo, Inc.; **p. 341, 13–24(a-b)** Laurie Ann Achenbach, Southern Illinois University at Carbondale; **p. 342, 13–27** Dr. Suzanne McCutcheon.

## Chapter 14: The Genetic Code and Transcription

**p. 352, CO14** Prof. Oscar L. Miller/Science Photo Library/Photo Researchers, Inc.; **p. 369, 14–11** Bert W. O'Malley, M.D., Baylor College of Medicine.

## Chapter 15: Translation and Proteins

**p. 389, 15–9(a)** "The Structure and Function of Polyribosomes." Alexander Rich, Jonathan R. Warner and Howard M. Goodman, 1963. Reproduced by permission of the Cold Spring Harbor Laboratory Press. Cold Spring Harbor Symp. Quant. Biol. 28 (1963) fig. 4C (top), p. 273 ©1964; **p. 389, 15–9(b)** E.V. Kiseleva/E.V. Kiseleva; **p. 395, 15–13(a)** Dennis Kunkel/Phototake NYC; **p. 395, 15–13(b)** Francis Leroy/Biocosmos/Science Photo Library/Photo Researchers, Inc.; **p. 399, 15–19** Kenneth Eward/BioGrafx/Science Source/Photo Researchers, Inc.

## Chapter 16: Gene Mutation and DNA Repair

**p. 410, CO16** Francis Leroy/Biocosmos/Science Photo Library/Photo Researchers, Inc.; **p. 425, 16–15 (left and right)** W. Clark Lambert, M.D., Ph. D./University of Medicine & Dentistry of New Jersey; **p. 427,** Mary Evans Picture Library/Photo Researchers, Inc.

## Chapter 17: Regulation of Gene Expression in Prokaryotes

**p. 435, CO17** Nature vol 401 Sept 16, 1999 (Cover). Article: "Structure of the trp RNA-binding attenuation protein, TRAP, bound to RNA" by Alfred A. Antson, Eleanor J. Dodson, Guy Dodson, Richard B. Greaves, Xiao-ping Chen & Paul Gollnick; **p. 444, 17–11(a)** Science Lewis, et al/Johnson Research Foundation; **p. 444, 17–11(b)** Science Lewis, et al 271 pg. 1247–1254/Johnson Research Foundation; **p. 444, 17–11(c)** Science Lewis, et al/Johnson Research Foundation.

## Chapter 18: Regulation of Gene Expression in Eukaryotes

**p. 457, CO18** T. Cremer/Dr. I. Solovei/Biozentrum (LMU); **p. 459, 18–2(b)** From: Qumsiyeh, Mazin B. 1999 "Structure and function of the nucleus: anatomy and physiology of chromatin" Cellular Molecular Life Science Vol. 55, pp. 1129–1140, Fig. 1C, pg. 1132; **p. 471, 18–19** Reprinted from *Cell*, Vol. 103, D. L. Black, Protein diversity from alternative splicing: A challenge for bioinformatics and post-genome biology, 367–370, Figure 1, p. 368, Copyright 2000, with permission from Elsevier; **p. 476, 18–23** From G. J. Hannon, "RNA Interface." Reprinted by permission from *Nature* Vol. 481, p. 244–251, Figure 2, p. 246. Copyright © 2002 Macmillan Publishers Ltd. and the author.

## Chapter 19: Developmental Genetics of Model Organisms

**p. 484, CO19** Edward B. Lewis, California Institute of Technology; **p. 485, 19–1(a)** F.R. Turner/Visuals Unlimited; **p. 485, 19–1(b)** R. Calentine/Visuals Unlimited; **p. 486, 19–2(a-b)** From S. F. Gilbert et al. 2001. Morphogenesis of the turtle shell: The development of a novel structure in tetrapod evolution. *Evolution and Development* 3: 47-58, Figure 3, parts K & L; **p. 486, 19–2(c)** Bill Curtsinger/Getty–National Geographic Society; **p. 490, 19–7** Jim Langeland, Stephen Paddock, and Sean Carroll, University of Wisconsin at Madison; **p. 491, 19–9(a)** Peter A. Lawrence and P. Johnston, "Development", 105, 761–767 (1989); **p. 491, 19–9(b)** Peter A. Lawrence "The Making of a Fly", Blackwells Scientific, 1992; **p. 491, 19–10** Jim Langeland, Stephen Paddock, and Sean Carroll, University of Wisconsin at Madison; **p. 492, 19–12(a-b)** Reproduced by permission from Fig. 2 on p. 766 in Cell 89: 765–71, May 30, 1997. Copyright © by Elsevier Science Ltd. Image courtesy of Mike J. Owen; **p. 492, 19–13(a-b)** Reproduced by permission from T. Kaufmann, et al., Advanced Genetics 27:309–362, 1990. Image courtesy of F. Rudolf Turner, Indiana University; **p. 494, 19–17** Reproduced with permission from [Figure 1a on page 332 from "Kmita, M. and Duboule, D. 2003. Organizing Axes in Time and Space; 25 Years of Colinear Tinkering. Science 2003 301: 331–333".]. Copyright American Association for the Advancement of Science; **p. 494, 19–18** P. Barber/Custom Medical Stock Photo, Inc.; **p. 497, 19–21** Elliot M. Meyerowitz, California Institute of Technology, Division of Biology; **p. 497, 19–22(a-b)** Max-Planck-Institut fur Entwicklungsbiologie; **p. 498, 19–24(a-d)** Dr. Jose Luis Riechmann, Division of Biology, California Institute of Technology. From Science 2002: 295, pp. 1482–85; **p. 500, 19–26** James King-Holmes/Photo Researchers, Inc; **p. 503, 19–30** From S. B-T. de-Leon and E. H. Davidson. 2007. Gene regulation: Gene control network in development. *Ann. Rev. Biophysics, Biomolecular Structure*, Vol. 36: 191–212, Figure 1, pg. 193. Originally published in E. H. Davidson, 2006. *The Regulatory Genome: Gene Regulatory Networks in Development and Evolution*. San Diego, CA: Academic Press; **p. 503, Table 19.5** From S. B-T. de-Leon and E. H. Davidson. 2007. Gene regulation: Gene control network in development. *Ann. Rev. Biophysics, Biomolecular Structure*, Vol. 36: 191–212, Table 1 p. 194; **p. 504, 19–31** Copyright © 2004, M. D. Schroeder et al. Transcription control in the segmentation gene network of Drosophila. PLoS Biology 2: 1396–1410, Figure 6, pg. 1405; **p. 510, figure for problem #29** Adapted from M. Schmid et al. 2005. A gene expression map of *Arabidopsis thaliana* development. *Nature Genetics* 37: 501–506, Fig. 4, p. 504.

## Chapter 20: Cancer and Regulation of the Cell Cycle

**p. 511, CO20** SPL/Photo Researchers, Inc.; **p. 513, 20–1(a)** Courtesy of Hesed M. Padilla-Nash, Antonio Fargiano, and Thomas Ried. Affiliation is Section of Cancer Genomics, Genetics Branch, Center for Research, National Cancer Institute, National Institutes of Health, Bethesda, MD 20892; **p. 513, 20–1(b)** Courtesy of Hesed M. Padilla-Nash and Thomas Ried. Affiliation is Section of Cancer Genomics, Genetics Branch, Center for Research, National Cancer Institute, National Institutes of Health, Bethesda, MD 20892; **p. 514, 20–2(a-b)** Courtesy of Professor Manfred Schwab, DKFZ, Heidelberg, Germany; **p. 518, 20–8(a)** Dr. Gopal Murti/Photo Researchers, Inc; **p. 530, Table 20.5** Y. Miki et al. 1994. A strong candidate for the breast and ovarian cancer susceptibility gene BRCA1. *Science* 266: 66–71.

## Chapter 21: Genomics, Bioinformatics, and Proteomics

**p. 531, CO21** BLAST search of GenBank database: http://www.ncbi.nlm.nih.gov/BLAST/; **p. 537, 21–4(a)** BLAST search of GenBank database: http://www.ncbi.nlm.nih.gov/BLAST/; **p. 547, 21–12(a)** Moredun Animal Health Ltd./Photo Researchers, Inc.; **p. 540, 21–8(a)** Holt Studios International/Photo Researchers, Inc.; **p. 547, 21–12(a)** Moredun Animal Health Ltd./Photo Researchers, Inc.; **p. 547, 21–13(a)** David McCarthy/SPL/Photo Researchers, Inc.; **p. 551, 21–14(a)** DLILLC/Corbis; **p. 551, 21–14(c)** Science Magazine and Deanne Fitzmaurice/Corbis; **p. 556, 21–18** and **21–19** From S. Yooseph et al. 2007. The Socercer II Global ocean sampling expedition: Expanding the universe of protein families. PLoS Biology, 5 (3): 0432–0466. Fig. 1, p. 0436 and Fig. 3, p. 0438, respectively; **p. 557, 21–20 (top)** Volker Steger/SPL/Photo Researcher, Inc.; **p. 557, 21–20 (bottom)** Cancer Genetics Branch/National Human Genome Research Institute/NIH; **p. 558, 21–21(a)** from Sherlock, G., et al. (2001) The Stanford Microarray Database. Nucleic Acids Research, 29(1):152–155. Fig 2; **p. 558, 21–21(b)** Reproduced by permission from Claridge-Chang A, Wijnen H, Naef F, Boothroyd C, Rajewsky N, Young MW. Circadian regulation of gene expression systems in the Drosophila head. Neuron. 2001 Nov 20; 32(4):657–71. Copyright by Elsevier Science Ltd. Image courtesy of Adam Calridge-Chang/The Rockefeller University; **p. 559, 21–22(a)** Jeff Rotman/Nature Picture Library; **p. 559, 21–22(b)** From M. P. Samanta et al. 2006. The transcriptome of the sea urchin empbryo. Science 314: 960–962, Fig. 2. p. 961; **p. 561, 21–24(a-b)** Swiss Institute of Bioinformatics http://www.proteome.org.au/General-Enquiries/default.aspx apafinfo@ proteome.org.au http://au.expasy.org/contact.html. **p. 564, 21–27** From J. M. Asara et al. 2007. Protein sequences from Mastodon and *Tyrannosaurus* Rex revealed by mass spectrometry. Science 316: 280-285, Fig. 3, p. 283; **p. 565, 21–28** From V. Wasinger et al. 2000. The proteome of *Mycoplasma Genitaliaum. Eur. J. Biochem 367*: 1571–82, Fig 5, p. 1575. Reprinted with permission from the author; **p. 566, 21–29** From L. Hood et al. 2004. Systems biology and new technologies enable predictive and preventative medicine. *Science* 306: 640–643, Fig. 1, p. 641.

## Chapter 22: Genome Dynamics: Transposons, Immunogenetics, and Eukaryotic Viruses

**p. 574, CO22** Chris Bjornberg/Photo Researchers, Inc.; **p. 576, 22–2** Stanley Cohen/Science Photo Library/Photo Researchers, Inc.; **p. 577, 22–3(a)** Courtesy of the Barbara McClintock Papers, American Philosophical Society; **p. 577, 22–3(b)** MaizeGDB/Courtesy M.G. Neuffer; **p. 582, 22–11(a)** Reprinted by permission from Nature "Electron Microscopy of negatively stained p1–381VLPs from EMBO Journal 11(3):1155–1164, 1992. Copyright ©Macmillan Magazines Limited. Photo by Alan J. Kingsman; **p. 582, 22–11(b)** Reprinted by permission from Nature. Image Enhancement of p1–381 VLPs. EMBO Journal 11(3):1155–1164, 1992. Copyright ©Macmillan Magazines Limited. Photo by Alan J. Kingsman; **p. 586, 22–13** Juergen Berger/Photo Researchers, Inc.; **p. 586, 22–14(b)** Dr. Klaus Boller/Photo Researchers, Inc.; **p. 590, 22–17(b)** Eye of Science/Photo Researchers, Inc.; **p. 593, 22–20** NIBSC/Photo Researchers, Inc.; **p. 595, 22–22** Charlotte A. Spencer; **p. 595, 22–23** From R. Holzerlandt et al. 2002. Identification of New Herpesvirus Gene homologs in the human genome, *Genome Research* 12: 1739–1748. Cold Spring Laboratory Press; **p. 598, 22–25** Dr. Linda Stannard, UCT/Photo Researchers, Inc. **p. 599, 22–26** Adapted from P. A. Rota et al. 2003. Characterization of a novel coronavirus associated with severe acute respiratory syndrome. *Science*, 300:1394–1399, Fig. 1. Copyright 2003 AAAS; **p. 599, 22–27** Adapted from X. W. Zhang et al. 2005. Testing the hypothesis of a recombinant origin of the SARS-associated coronavirus. *Archives of Virology* 150: 1–20, Fig. 5.

## Chapter 23: Genome Analysis—Dissection of Gene Function

**p. 605, CO23** Jean Claude Revy - ISM/Phototake NYC; **p. 608, 23–2(a)** Dr. Stanley Flegler/Visuals Unlimited; **p. 608, 23–2(b)** A.B. Dowsett/Science Photo Library/Photo Researchers, Inc.; **p. 609, 23–3(a)** The Royal Alberta Museum; **p. 609, 23–3(b)** Dr. Kimberly A. Hughes and Dr. Kevin A. Dixon/Department of Animal Biology/University of Illinois at Urbana-Champaign; **p. 611, 23–6(a)** Hank Morgan; **p. 611, 23–6(b)** Dr.R.L. Brinster/Peter Arnold, Inc.; **p. 616, 23–11(a)** Adapted from Figure 3 from "Hartwell, L. et al. 1974. Genetic control of the cell division cycle in yeast. Science 183:46–51". Copyright American Association for the Advancement of Science; **p. 616, 23–11(b)** Figure 5 from "Hartwell, L. et al. 1974. Genetic control of the cell division cycle in yeast. Science 183:46–51." Copyright American Association for the Advancement of Science; **p. 621, 23–15(a-c)** Dr. Mario Capecchi—Figure 3a in: Hostikka SL, Capecchi MR, "The mouse Hoxc11 gene: genomic structure and expression pattern." Mech Dev. 1998 Jan; 70(1–2):133–45; **p. 621, 23–16** From Molecular & Cellular Biology (20:8536–47). Image courtesy of Dr. Gideon Dreyfuss; **p. 621, 23–17(a-b)** Reproduced with permission from Figure 2 from: Hjalt TA, Semina EV, Amendt BA, Murray JC (2000) The Pitx2 protein in mouse development. Developmental Dynamics 218:195–200). Copyright © John Wiley & Sons, Inc.

## Chapter 24: Applications and Ethics of Genetic Engineering and Biotechnology

**p. 633, CO24** From The Roslin Institute—roslin.design@bbsrc.ac.uk http://www.roslin.ac.uk/imagelibrary/index.php Trends in Biotechnology Volume 22, Issue 4, April 2004, Pages 157–159 doi:10.1016/j.tibtech.2004.02.005; **p. 636, 24–1(a)** SIU BioMed/Custom Medical Stock Photo, Inc.; **p. 637, 24–2** Reprinted with permission from Biodesign Institute, Arizona State University, Box 5001 Tempe, AZ 85287-5001; **p. 638, 24–3** From Doebley, J. Plant Cell, 2005 Nov; 17(11): 2859–72. Courtesy of John Doebley/University of Wisconsin; **p. 639, 24–4(a)** From S. Lawrence. 2005. Agbio keeps on growing. *Nature Biotechnology*, 23: 281; **p. 639, 24–4(b)** Adapted from Ernst & Young, Beyond Borders: Global Biotechnology Report 2006. Ernst & Young © 2006. http://www.ey.com/beyondborders; **p. 639, 24–5(a)** Michael A. Palladino; **p. 639, 24–5(b)** Pfizer/Pharmacia Corporation; **p. 641, 24–7(a)** Reprinted with permission from "New Genes Boost Rice Nutrients" by I. Potrykus and P. Beyer, Science, August 13, 1999, Vol. 285, pp 994; **p. 641, 24–7(b)** From M. A. Grusak, 2005. Golden Rice gets a boost from maize. *Nature Biotechnology* 23: 429; **p. 642, 24–8** Choy Hew, National University of Singapore, and Garth Fletcher, Memorial University of Newfoundland, St. John's, Newfoundland; **p. 642, 24–9** From P. Rainard. 2005. Tackling mastitis in dairy cows. *Nature Biotechnology* 23(4): 430–432; **p. 643, 24–10** www.glofish.com; **p. 646, 24–14(a-c)** Pascal Goetgheluck/Photo Researchers, Inc.; **p. 647, 24–17** GeneChip® Human Genome U133 Plus 2.0 Array. Courtesy of Affymetrix, Inc.; **p. 648, 24–19(a)** Fig. 1b and c from Clinical Chemistry 46:10 2000, Fredrik P. Wikman et al. Evaluation of the performance of a p53 sequencing microarray chip using 140 previously sequenced bladder tumor samples; **p. 648, 24–19(b)** From Fig. 1d from Clinical Chemistry 46:10 2000, Fredrik P. Wikman et al. Evaluation of the performance of a p53 sequencing microarray chip using 140 previously sequenced bladder tumor samples; **p. 649, 24–20(c)** Cancer Genetics Branch/National Human Genome Research Institute/NIH; **p. 650, 24–21(a)** Reprinted with permission from Ash Alizadeh, Nature Magazine 2000: 403, pgs. 503–511, figure 4 left panel. Copyright 2000 Macmillian Magazines Limited; **p. 651, 24–22** Whole genome expression profiling. From Lorence, M. Application of High-Density Microarrays for Expression and Genotype Analysis in Infectious Disease, American Biotechnology Laboratory, p. 10–12, January 2006. Figure, p.11 Matthew Lorence, Ph.D., M.B.A.; **p. 651, 24-23** Whole genome expression profiling. From Lorence, M. Application of High-Density Microarrays for Expression and Genotype Analysis in Infectious Disease, American Biotechnology Laboratory, p. 10–12, January 2006. Figure, p.10 Matthew Lorence, Ph.D., M.B.A.; **p. 653, 24–26** Van De Silva; **p. 657, 24–30** Courtesy of Orchid Cellmark, Inc.

## Chapter 25: Quantitative Genetics and Multifactorial Traits

**p. 668, CO25** Ed Reschke/Peter Arnold, Inc.; **p. 679, 25–8** Courtesy of Dr. Steven D. Tanksley, Cornell University; **p. 687,** Nature's Images/Photo Researchers, Inc.

## Chapter 26: Genetics and Behavior

**p. 688, CO26** Dr. Eve. G. Stringham and Dr. E. Peter M. Candido, Journal of Experimental Zoology 266: 227–233 (1993). Publication of Wiley-Liss Inc. University of British Columbia; **p. 690, 26–1** Photo provided by MED Associates, Inc., St. Albans, VT 05478, www.med-associates.com; **p. 691, 26–2** Dr. Jonathan Flint; **p. 691, 26–3** Adapted from S. A. G. Willis-Owen and J. Flint. 2006. The genetic basis of emotional behavior in mice. *European Journal of Human Genetics* 14: 721–728, plot 1, pg. 723. Copyright © 2006 Nature; **p. 692, 26–4** Adapted from Fig. 1, p. 881 in M. B. Sokolowski. 2001. *Drosophila*: Genetics meets behavior. *Nature Reviews Genetics* 2: 879-890. Copyright © 2001 by Macmillan Magazines Ltd. Reprinted by permission from Nature Reviews Genetics; **p. 695, 26–10** Adapted from B. S. Baker et al. 2001. Are complex behaviors specified by dedicated regulatory genes? Reasoning from *Drosophila*. Cell 105: 13–24, fig. 3, p. 20. Copyright © 2001 with permission from Elsevier Science; **p. 698, Table 26.2** Adapted from Table 1, p. 882 in M. B. Sokolowski. 2001. *Drosophila*: Genetics meets behavior. *Nature Reviews Genetics* 2: 879-890. Copyright © 2001 by Macmillan Magazines Ltd. Reprinted by permission from Nature Reviews Genetics; **p. 703, 26–17** Adapted from B. Riley and K. S. Kendler. 2006. Molecular genetics studies of schizophrenia. *European Journal of Human Genetics* 14: 669-680, fig. 1, p. 670; **p. 704, 26–19** ISM/CMEABG/UCBL1/Phototake NYC.

## Chapter 27: Population Genetics

**p. 710, CO27** Edward S. Ross, California Academy of Sciences; **p. 714, 27–3** Michel Samson, Frederick Libert, et al., "Resistance to HIV-1 infection in Caucasian individuals bearing mutant alleles of the CCR-5 chemokine receptor gene." Reprinted with permission from Nature [vol. 382, 22 August 1996, p. 725, Fig. 3]. Copyright 1996 Macmillan Magazines Limited; **p. 715, 27–4** Adapted from F. Leibert et al. 1998. The ¢ccr5 mutation conferring protection against HIV-1 in Caucasian populations has a single and recent origin in Northeastern Europe. *Human Molecular Genetics* 7: 399–406, by permission of Oxford University Press; **p. 721, 27–9 (top)** Hans Pfletschinger/Getty Images Inc—Science Faction; **p. 728, 27–19** Adapted from M. R. Bellinger et al. 2003. Loss of genetic variation in Greater Prairie chickens following a population bottleneck in Wisconsin, USA. *Conservation Biology* 17: 717–724, fig. 2, p. 721.

## Chapter 28: Evolutionary Genetics

**p. 737, CO28** Breck P. Kent/Animals Animals/Earth Scenes; **p. 741, 28–3** Reprinted from *Trends in Genetics* 8: 392–398, L. Tsui, The spectrum of cystic fibrosis mutations, fig. 1. Copyright © 1992, with permission from Elsevier; **p. 743, 28–8(a)** From D. A. Powers et al. 1991. Genetic mechanisms for adapting to a changing environment, *Annual Review of Genetics* 25: 629660, fig. 3, pp. 639. Reprinted with permission form the *Annual Review of Genetics*, Vol. 25. Copyright © 1991 by Annual Reviews www.annualreviews.org; **p. 743, 28–8(b)** Adapted with permission from A. R. Place and D. A. Powers. 1979. Genetic variation and relative catalytic efficiencies: LDH-B allozymes of *Fundulus heteroclitus*. *Proc. Natl. Acad. Sci. USA*, 76: 2354–2358, fig. 4, p. 2357; **p. 746, 28–9** Carl C. Hansen/Nancy Knowlton/Smithsonian Institution Photo Services; **p. 747, 28–11(a-b)** Kenneth Kaneshiro/University of Hawaii/CCRT; **p. 748, 28–13(a-b)** Courtesy of Toby Bradshaw and Doug W. Schemske, University of Washington. Photo by Jordan Rehm; **p. 749, 28–15(a-b)** Dr. Paul V. Loiselle; **p. 749, 28–16** Fig. 5 in Molecular Biology Evolution 15 (4) pages 391–407, Kazuhiko Takahasi/Norihiro Okada. The Society for Molecular Biology and Evolution. **p. 752, 28–18** From C. G. Sibley and J. E. Alhquist. 1987. DNA hybridization evidence of hominoid phylogeny" results from an expanded data set. *Journal of Molecular Evolution*, 26: 99–121 Table 1, p. 101 with permission from Springer-Verlag GmbH; **p. 752, 28–19** From W. M. Fitch et al, 1991. Positive Darwinian evolution in human influenza A viruses. *Proc. Natl. Acad. Sci. USA* 88: 4270–73; **p. 753, 28–20** Reprinted from *Current Biology* Vol. 7, No. 3, D. M. Hillis, Phylogenetic analysis, pp. R129-R131, figure 1. Copyright © 1998, with permission from Elsevier; **p. 755, 28–22** and **28–23** Reprinted from J. P. Noonan et al. 2006. Sequencing and analysis of Neanderthal genomic DNA *Science* 314: 1113–1118, Figure 3a, p. 1115 and Figure 6, p. 1117; **p. 760, figure for problem #29** Reprinted from D. Caramelli et al. 2003. Evidence for a genetic discontinuity between Neanderthals and 24,000-year-old anatomically modern Europeans. *Proc. Natl. Acad. Sci. USA* 100: 6593–6597, Figure 2, p. 6596.

## Chapter 29: Conservation Genetics

**p. 762, CO29** Scott Bauer/USDA/ARS/Agricultural Research Service; **p. 763, 29–2** John Mitchell/Photo Researchers, Inc.; **p. 763, 29–3** Daniel J Cox/Getty Images Inc.—Stone Allstock; **p. 764, 29–4(a)** David Austen/David R. Austen Photography Group; **p. 764, 29–4(b)** Sarah Ward/Department of Soil and Crop Services/Colorado State University; **p. 765, 29–5** Sarah Ward/Department of Soil and Crop Services/Colorado State University; **p. 767, 29–6** Johnny Johnson/DRK Photo; **p. 769, 29–9** Art Wolfe/Getty Images Inc.—Stone Allstock; **p. 770, 29–10** Tim Thompson/Corbis/Bettmann; **p. 772, 29–13** Jim Brandenburg/Minden Pictures; **p. 774, 29–14** Lynn Stone/Animals Animals/Earth Scenes; **p. 778,** NHPA/Photo Researchers, Inc.

# Index

## C

## H

# I

# J



# K



# L



# M

## N

## O



## P

# Q